A Guidebook for

GATE

Instrumentation Engineering

Theory

Objective Questions with Detailed Answers

Other CBS Titles on GATE

A Guidebook for

GATE

Instrumentation Engineering

Theory

Objective Questions with Detailed Answers

Satish K Karna ME
Ex-Educational Consultant
Karna Institute of Engineering
Chandigarh

Ravi Verma
Guest Faculty

CBS Publishers & Distributors Pvt Ltd

New Delhi • Bengaluru • Chennai • Kochi • Mumbai • Pune
Hyderabad • Kolkata • Nagpur • Patna • Vijayawada

A Guidebook for GATE
Instrumentation Engineering

Theory
Objective Questions with Detailed Answers

ISBN: 978-81-239-2369-7

First Edition: 2014

Published by Satish Kumar Jain for
CBS Publishers & Distributors Pvt Ltd
4819/XI Prahlad Street, 24 Ansari Road, Daryaganj, New Delhi 110 002, India.
Ph: 23289259, 23266861, 23266867 Website: www.cbspd.com
Fax: 011-23243014 e-mail: delhi@cbspd.com; cbspubs@airtelmail.in.
Corporate Office: 204 FIE, Industrial Area, Patparganj, Delhi 110 092
Ph: 4934 4934 Fax: 4934 4935 e-mail: publishing@cbspd.com; publicity@cbspd.com

Branches

- **Bengaluru:** Seema House 2975, 17th Cross, K.R. Road, Banasankari 2nd Stage, Bengaluru 560 070, Karnataka
  Ph: +91-80-26771678/79 Fax: +91-80-26771680 e-mail: bangalore@cbspd.com
- **Chennai:** 20, West Park Road, Shenoy Nagar, Chennai 600 030, Tamil Nadu
  Ph: +91-44-26260666, 26208620 Fax: +91-44-42032115 e-mail: chennai@cbspd.com
- **Kochi:** 36/14 Kalluvilakam, Lissie Hospital Road, Kochi 682 018, Kerala
  Ph: +91-484-4059061-65 Fax: +91-484-4059065 e-mail: kochi@cbspd.com
- **Mumbai:** 83-C, Dr E Moses Road, Worli, Mumbai-400018, Maharashtra
  Ph: +91-9833017933 e-mail: mumbai@cbspd.com
- **Pune:** Bhuruk Prestige, Sr. No. 52/12/2+1+3/2 Narhe, Haveli (Near Katraj-Dehu Road Bypass), Pune 411 041, Maharashtra
  Ph: +91-20-64704058, 64704059, 32392277 Fax: +91-20-24300160 e-mail: pune@cbspd.com

Representatives

- **Hyderabad** 0-9885175004
- **Kolkata** 0-9831437309, 0-9051152362
- **Nagpur** 0-9021734563
- **Patna** 0-9334159340
- **Vijayawada** 0-9000660880

Printed at India Binding House, Noida (UP)

A Guidebook for
GATE

Instrumentation Engineering

Preface

Keeping in view the latest requirements according to the new syllabus of GATE, necessary amendments have been made in this edition. The section on General Ability has been included according to the latest pattern for the examination. The present edition contains **9** Chapters of **Instrumentation Engineering**. It has a compulsory chapter on Engineering Mathematics. All these chapters have been updated as per the revised guidelines of the examining authority of GATE.

Each chapter follows theory/study material in brief according to the latest syllabus of GATE with a fundamental approach and necessary multiple choice type questions, with detailed solutions.

The number of MCQs (one-mark and two-marks, linked/common data questions) has been greatly increased which is a unique feature of this edition. In view of the fact that previous GATE examination papers had objective type questions only, answer sheets for all objective type questions have also been included in the interest of students.

Keeping in view the material available in the market, we have extensively added the new pattern of questions for the students. We hope this effort will help in instilling confidence among the students.

Extensive efforts have been made in revising the content according to the latest syllabus, incorporating all the topics and implementing various suggestions given by experts from different colleges/universities and students so that book is comprehensive, elegant, authoritative and error-free.

It is pleasure in bringing out this edition of the book for the GATE examination. Seeing the trend of the previous year's papers and suggestions received from various readers have been incorporated. Linked type questions–answers with detailed solution at the end of every section to help the readers gain confidence. The author and his team members have taken great pains in compiling and editing the book material with least number of mistakes. The publisher has also worked with dedication to bring out the book at a very reasonable price.

Suggestions from readers are welcome: Appropriate and suitable suggestions of any kind from readers will be duly acknowledged

I do hope that the GATE aspirants will find the book extremely useful. Although every effort has been made by me and my team to minimize the errors in the book. Any inadvertent error when pointed out by any person will be duly appreciated and rectified in the subsequent edition.

Satish K Karna
satish_karna@ymail.com

Acknowledgements

This book would be incomplete without expressing our gratitude to those wonderful people who provided real inspiraration, motivation and encouragement in taking up this challenging project. A special gratitute to:

Dr. S. Chatterjee, Head, Department of Electrical Engineering, TTTI, Chandigarh, and Dr.B.S. Pabala, Assistant Professor, TTTI, Chandigarh.

I am grateful to these people for their support in revising the contents according to latest syllabus and solving the problems and developing the book:

Mrs Manmeet Kaur ME (Electrical, Power System)
Mr Pravin Kumar ME (Instrumentation and Control)
Miss Suman ME (Electrical, Power System)
Mr Divesh Kumar ME (Electrical, Power System)
Mrs Ritula Thakaur ME (Electrical, Power System)
Mr Anand Parkash ME (Electronics and Communication)
Mr Hitendar Singh Pawar ME (Microelectronics)
Mr Jitender Virmani ME (Control and Instrumentation)
Mr Anupam Anand MSc (Statistics) and Mech. (Computer Sc.)
Mr Shailendra Tiwari MTech (Computer Science)
Mr Surendra Singh MTech (Computer Science)
Mr Deepak Singla MTech (Computer Science)
Mr Asish Saxsena BTech (Electronics and Communication)
Mr M.S Rao BTech (Electronics and Communication)

Our students:

Miss Kritika Singh BE (Electronics and Communication)
Mr Amandeep BE (Electronics and Communication)
Miss Baljinder Kaur BE (Electronics and Communication)
Miss Rupinder Kaur BE (Electronics and Communication)
Miss Suman BE (Electronics and Communication)
Mrs Monika BE (Electronics and Communication)
Mrs Rina BE (Electronics and Instrumentation)
Miss Kawalpreet Kaur BE (Electronics and Communication)
Miss Kritika BE (Electronics and Communication)
Mr Chirag BE (Electronics and Communication)
Mr Rajneesh Kumar BE (Electrical)
Mr Prateek Srivastava BE (Electronics and Intrumentation)
Miss Purva Kaul BE (Electronics and Communication)
Rohil Khanna BE (Electronics and Intrumentation)
Mr Ravi Verma BE (Electronics and intimidation)
Mr Sudhir Kumar BE (Electronics and Communication).
Mr Gautam Khera BE (Electrical)
Mr Inderpreet Singh BE (Electrical)

I owe a debt of gratitude to Mayank and Nilophar for their patience and understanding. I am thankful to Mr Mukesh Kumar Koundal, Miss Preety Rani, Mr Kamal Kumar and Ram Raja for efficiently word processing the manuscript.

Satish K Karna

A Guidebook for GATE

Instrumentation Engineering

Contents

Syllabus

ENGINEERING MATHEMATICS

Linear Algebra: Matrix Algebra, Systems of linear equations, Eigen values and eigen vectors.

Calculus: Mean value theorems, Theorems of integral calculus, Evaluation of definite and improper integrals, Partial Derivatives, Maxima and minima, Multiple integrals, Fourier series. Vector identities, Directional derivatives, Line, Surface and Volume integrals, Stokes, Gauss and Green's theorems.

Differential Equations: First order equation (linear and nonlinear), Higher order linear differential equations with constant coefficients, Method of variation of parameters, Cauchy's and Euler's equations, Initial and boundary value problems, Partial Differential Equations and variable separable method.

Complex variables: Analytic functions, Cauchy's integral theorem and integral formula, Taylor's and Laurent' series, Residue theorem, solution integrals.

Probability and Statistics: Sampling theorems, Conditional probability, Mean, median, mode and standard deviation, Random variables, Discrete and continuous distributions, Poisson, Normal and Binomial distribution, Correlation and regression analysis.

Numerical Methods: Solutions of non-linear algebraic equations, single and multi-step methods for differential equations.

Transform Theory: Fourier transform, Laplace transform, Z-transform.

INSTRUMENTATION ENGINEERING

Basics of Circuits and Measurement Systems: Kirchoff's laws, mesh and nodal Analysis. Circuit theorems. One-port and two-port Network Functions. Static and dynamic characteristics of Measurement Systems. Error and uncertainty analysis. Statistical analysis of data and curve fitting.

Transducers, Mechanical Measurement and Industrial Instrumentation: Resistive, Capacitive, Inductive and piezoelectric transducers and their signal conditioning. Measurement of displacement, velocity and acceleration (translational and rotational), force, torque, vibration and shock. Measurement of pressure, flow, temperature and liquid level. Measurement of pH, conductivity, viscosity and humidity.

Analog Electronics: Characteristics of diode, BJT, JFET and MOSFET. Diode circuits. Transistors at low and high frequencies, Amplifiers, single and multi-stage. Feedback amplifiers. Operational amplifiers, characteristics and circuit configurations. Instrumentation amplifier. Precision rectifier. V-to-I and I-to-V converter. Op-Amp based active filters. Oscillators and signal generators.

Digital Electronics: Combinational logic circuits, minimization of Boolean functions. IC families, TTL, MOS and CMOS. Arithmetic circuits. Comparators, Schmitt trigger, timers and mono-stable multi-vibrator. Sequential circuits, flip-flops, counters, shift registers. Multiplexer, S/H circuit. Analog-to-Digital and Digital-to-Analog converters. Basics of number system. Microprocessor applications, memory and input-output interfacing. Microcontrollers.

Signals, Systems and Communications: Periodic and aperiodic signals. Impulse response, transfer function and frequency response of first- and second order systems. Convolution, correlation and characteristics of linear time invariant systems. Discrete time system, impulse and frequency response. Pulse transfer function. IIR and FIR filters. Amplitude and frequency modulation and demodulation. Sampling theorem, pulse code modulation. Frequency and time division multiplexing. Amplitude shift keying, frequency shift keying and pulse shift keying for digital modulation.

Electrical and Electronic Measurements: Bridges and potentiometers, measurement of R,L and C. Measurements of voltage, current, power, power factor and energy. A.C & D.C current probes. Extension of instrument ranges. Q-meter and waveform analyzer. Digital voltmeter and multi-meter. Time, phase and frequency measurements. Cathode ray oscilloscope. Serial and parallel communication. Shielding and grounding.

Control Systems and Process Control: Feedback principles. Signal flow graphs. Transient Response, steady-state-errors. Routh and Nyquist criteria. Bode plot, root loci. Time delay systems. Phase and gain margin. State space representation of systems. Mechanical, hydraulic and pneumatic system components. Synchro pair, servo and step motors. On-off, cascade, P, P-I, P-I-D, feed forward and derivative controller, Fuzzy controllers.

Analytical, Optical and Biomedical Instrumentation: Mass spectrometry. UV, visible and IR spectrometry. X-ray and nuclear radiation measurements. Optical sources and detectors, LED, laser, Photo-diode, photo-resistor and their characteristics. Interferometers, applications in metrology. Basics of fiber optics. Biomedical instruments, EEG, ECG and EMG. Clinical measurements. Ultrasonic transducers and Ultrasonography. Principles of Computer Assisted Tomography.

A Guidebook for

GATE

Instrumentation Engineering

Theory

Objective Questions with

Detailed Answers

1 Engineering Mathematics

Linear Algebra: Matrix Algebra, Systems of linear equations, Eigen values and eigen vectors. **Calculus:** Mean value theorems, Theorems of integral calculus, Evaluation of definite and improper integrals, Partial Derivatives, Maxima and minima, Multiple integrals, Fourier series. Vector identities, Directional derivatives, Line, Surface and Volume integrals, Stokes, Gauss and Green's theorems. **Differential equations:** First order equation (linear and nonlinear), Higher order linear differential equations with constant coefficients, Method of variation of parameters, Cauchy's and Euler's equations, Initial and boundary value problems, Partial Differential Equations and variable separable method. **Complex variables:** Analytic functions, Cauchy's integral theorem and integral formula, Taylor's and Laurent' series, Residue theorem, solution integrals. **Probability and Statistics:** Sampling theorems, Conditional probability, Mean, median, mode and standard deviation, Random variables, Discrete and continuous distributions, Poisson, Normal and Binomial distribution, Correlation and regression analysis. **Numerical Methods:** Solutions of non-linear algebraic equations, single and multi-step methods for differential equations. **Transform Theory:** Fourier transform, Laplace transform, Z-transform.

1.1 LINER ALAGEBRA

1.1.1 MATRICES

A Matrix is an arrangements of m n elements along m rows and n columns in the form of a rectangle, It is represented by A, B, C, X, Y, Z ...

Example:

$$A = \begin{pmatrix} a_{11} & a_{12} & \rightarrow & a_{in} \\ a_{21} & a_{22} & \rightarrow & a_{2n} \\ a_{31} & a_{32} & \rightarrow & a_{3n} \\ a_{m1} & a_{m2} & \rightarrow & a_{mn} \end{pmatrix}_{m \times n}$$

m = Number of rows
n = number of columns
Where, m × n is called order of A

Rectangular matrix: A matrix of order m × n is called rectangular matrix if m ≠ n.

Example:

$$\begin{pmatrix} 1 & 2 & 3 \\ 4 & 0 & 1 \end{pmatrix}_{2 \times 3}$$

$$\begin{pmatrix} 1 & 0 \\ 0 & 2 \\ 1 & 0 \end{pmatrix}_{2 \times 3} \quad \begin{pmatrix} 1 & 2 & 5 & 7 \\ 2 & 4 & 6 & 8 \end{pmatrix}_{2 \times 4}$$

are rectangular matrices because m ≠ n.

Square matrix: A matrix of order m × n is called square matrix if m = n.

Example:

$$\begin{pmatrix} 1 & 2 \\ 3 & 4 \end{pmatrix}_{2 \times 3} \quad \begin{pmatrix} 1 & 1 & 2 \\ 0 & 1 & 2 \\ 1 & 0 & 3 \end{pmatrix}_{3 \times 3}$$

are square matrix (m = n)

Row matrices: If a matrix Contain only one row then it is called row matrix.

$[a_{11}\ a_{12}\ a_{13} ... a_{1n}]_{1 \times n,}\ [1\ 2\ 3\ 4]_{1 \times 4}$
$[a\quad b]_{1 \times 2}$ are row matrices

Column matrices: If a matrix contain only one column then it is called column matrix.

Example:

$$\begin{pmatrix} 1 \\ 2 \\ 3 \\ 4 \end{pmatrix}_{4 \times 1} \quad \begin{pmatrix} 1 \\ 2 \\ 3 \end{pmatrix}_{3 \times 1} \quad \begin{pmatrix} a_{11} \\ a_{21} \\ a_{31} \end{pmatrix}_{3 \times 1}$$

are column matrices

Zero matrix is null matrix: A matrix is called zero or null matrix if all its elements are zero. It is denoted by 0.

Example:

$$[0];\ [0,0] \begin{pmatrix} 0 \\ 0 \end{pmatrix} \begin{pmatrix} 0 & 0 \\ 0 & 0 \end{pmatrix} \begin{pmatrix} 0 & 0 & 0 \\ 0 & 0 & 0 \end{pmatrix}$$

Diagonal matrix: In a square matrix if all elements except diagonal are zero then it is called diagonal matrix.

Example:

$$\begin{pmatrix}1&0&0\\0&2&0\\0&0&1\end{pmatrix}_{3\times 3}\begin{pmatrix}0&0&0\\0&1&0\\0&0&1\end{pmatrix}_{3\times 3}\begin{pmatrix}0&0\\0&0\end{pmatrix}_{2\times 2}$$

Scalar matrix: A diagonal matrix whose all diagonal element are same is called a scalar matrix.

Example:

$$\begin{pmatrix}1&0&0\\0&1&0\\0&0&1\end{pmatrix}\begin{pmatrix}2&0&0\\0&2&0\\0&0&2\end{pmatrix}\begin{pmatrix}a&0&0\\0&a&0\\0&0&a\end{pmatrix}\begin{pmatrix}7&0\\0&7\end{pmatrix}$$

Unit matrix: A diagonal matrix is called unit matrix if all the diagonal entries are unity.

Example:

$$\begin{pmatrix}1&0\\0&1\end{pmatrix}\begin{pmatrix}1&0&0\\0&1&0\\0&0&1\end{pmatrix}\begin{pmatrix}1&0&0&0\\0&1&0&0\\0&0&1&0\\0&0&0&1\end{pmatrix}$$

Permutation matrix: Permutation Matrix is a (0, 1) matrix that has exactly one entry 1 in each row and each column and 0's elsewhere.

Permutation Matrix representation is given by:

$$\begin{bmatrix}1&0&0\\0&1&0\\0&0&1\end{bmatrix}$$

Hermitian matrix: A Hermitian matrix (or self-aqdjoint matrix) is a square matrix with complex entries which is equal to its own cónjugate transpose, i.e. the element in row and column is equal to complex conjugate of the element in the row and column, for all indicas i and j.

$a_{ji} = \bar{a}_{ji}$, e.g: $\begin{bmatrix}4 & s+i\\ s-i & 1\end{bmatrix}$ is a Hermitian matrix.

1. The entries on the main diagonal of any Hermitan matrix are necessarily real.
2. A matrix that has only real entries is Hermitan iff it is a symmetric matrix.

Skew hermitian matrix: A square matrix a is said to be skew - Hermitian if its conjugate transpose is also negative, i.e. $\overline{a_{ij}} = -a_{ij}\ \forall i, j$, e.g: $\begin{bmatrix}i & 2+i\\ -2+i & 3i\end{bmatrix}$ is a skew hermitian matrix.

1. All entries on the main diagonal of a skew-Hermitian matrix have to ve pure imaginary.
2. If A isSkew-Hermitian, then is Hermitian.
3. If A is Skew-Hermitian then A raised to an odd power is Skew-hermitian.

Orthogonal matrices

Definition: A matrix A is said to be orhtogonal of $AA' = I_n$.

Symmetric and skew symmetric matrices

A square Matrix is called symmetric if

$A^t = A$

Where A^t = transpose of A

Example:

$$A = \begin{pmatrix}a&b&c\\b&l&d\\c&d&m\end{pmatrix} = A$$

$$A^t = \begin{pmatrix}a&b&c\\b&l&d\\c&d&m\end{pmatrix} = A$$

Clearly $A^t = A$

∴ A is symmetric matrix

A square matrix is called skew symmetric if

$\mathbf{A^t = -A}$ where A^t = transpose of A

Cleary $A^t = A$

A is symmetric e matrix

A square matrix is called skew symmetric if

$A^t = -A$

Example:

$$A = \begin{pmatrix}0&-h&-g\\h&0&f\\-g&f&0\end{pmatrix}$$

$$\text{Here } A^t = \begin{pmatrix}0&h&g\\-h&0&-f\\g&f&0\end{pmatrix}$$

Clearly then A is skew symmetric

$A^t = -A$

Triangular Matrix

Triangular Matrix are of two types

a) Upper triangular matrix .

b) Lower triangular matrix

$\mathbf{A^t = -A}$

a) ***Upper matrix:*** A square matrix is called upper Matrix if all the non diagonal entries below the diagonal are zero.

Example: $\begin{pmatrix} 2 & 0 & 0 \\ 0 & 1 & 1 \\ 0 & 0 & 0 \end{pmatrix}$

It is upper triangular matrix be cause all elements below the diagonal are zero.

b) ***Lower triangular matrix:*** A square matrix is called lower triangular matrix if all the non diagonal elements above the diagonal are zero.

Example:

$$B = \begin{pmatrix} 2 & 0 & 0 \\ 1 & 0 & 0 \\ -2 & 2 & -1 \end{pmatrix}$$

Here all elements above the diagonal are zero.
A is called lower triangular matrix .

Equal matrix: Two matrices are said to be equal if they are of the same order and each corresponding element of A are equal to B.

Example: $\begin{pmatrix} a & b \\ c & d \end{pmatrix}_{2 \times 2} \begin{pmatrix} 1 & 2 \\ 3 & 4 \end{pmatrix}_{2 \times 2}$

if a = 1, b = 2, c = 3, d = 4

A joint of a square matrix A is the transpose of cofactors matrix A.

Note: $(AB)^1 = B^1 A^1$

Inverse of matrix: Inverse of a square matrix. A is written as A^{-1} and obtained by using the formula

$A^{-1} = \frac{1}{|A|}$. Adj A, where $|A| \neq 0$.

Echelon form of a matrix: If A be the given matrix of order m x n than it can be reduced to echelon form by row transformations if we can make all the entries equal to zero . Then R (A) = Number of non zero rows.

Definition: A determinant is a function depending on n that associates a scalar det (A) to every $(n \times n)$ square matrices A deter,omamt of nth order is consist of n rows and n columns and is of the form:

$$\begin{vmatrix} a_{11}a_{12}\ldots a_{1n} \\ a_{21}a_{22}\ldots a_{2n} \\ \ldots\ldots\ldots\ldots\ldots \\ a_{n1}a_{n2}\ldots a_{nn} \end{vmatrix}$$

This determinant is the elimination of variables $X_1, X_2, \ldots X_n$ from the equations.

$a_{11}x_1 + a_{12}x_2 + \ldots + a_{1n}x_n = 0$

$a_{21}x_1 + a_{22}x_2 + ..a_{2n}x_n = 0$

$a_{n1}x_1 + a_{n2}x_2 + \ldots + a_{nn}x_n, = 0$

Consistency and so/ⁿ of algebraic equations

In argument matrix A* we reduce echelon for if

a) Rank of A* = Rank of A then equation are consistent.

b) If Rank of A* ≠ Rank of A then the system is inconsistent. (i.e. no solution).

In case of homogenous system of equations A x = 0

a) If rank = Number of unknowns then it has zero solution.

b) If rank < Number of unknowns then it has infinitely many solution.

Definition of rank: Let A be a matrix of order m x n : than r is the rank of the matrix if all (r + 1) th order minors are zero an at least one minor of order r is non zero.

Note:

1. Rank of a matrix is written as ρ.
2. Rank of a Null matrix is zero.
3. Rank of a square matrix which is non singular if order n is n.
4. Rank of matrix A of order $m \times n \leq$ min of {m.n}.

Definition of Normal Form

A matrix is said to be in Normal form if it can be expressed.

$$\begin{pmatrix} In-r & 0 \\ 0 & 0 \end{pmatrix}$$

Where In – r is Identity matrix of order (n – r)

Hermitian matrix: A matrix is called her Hermitian if $fA^* = A$ or $A^\theta = A$. Where A* means (A^1).

Note:

Matrix a must be square matrix. Each diagonal entry of skew Hermitian matrix is either purely imaginary or zero.

Unitary Matrix

A matrix is called unitary if A* A = I

Similar Matrices

A matrix B is said to be similar to another matrix a if ∃. An in veritable matrix P such that AP = PB or $A = PBP^{-1}$ or $B = p^{-1}$ AB. Note. A, B, P are square matrix of same order over a field F.

Note:

A and B have same eigen values. Characteristic equation Eigen values and Eigen vectors. Cayley hamilton the order:

Definition

If A square matrix of order n, then its characteristic equation can be written by:

$|A - \lambda I = 0$ is called the Characteristic of A, i.e. $|A - I| = 0$

Where, λ is any scalar where I is Identity matrix of order *n*.

Some Properties of Matrix

1. Matrix addition is commutative, i.e. $A + B = B + A$
2. Matrix multiplication is not commutative, i.e. $AB \neq BA$ generally. But if AB = BA then, A and B are square matrix of the same order.
3. Matrix multination is associative, i.e. A(BC) = (AB) C.
4. Matrix multiplication is distributive w.r.t. matrix addition, i.e. A (B + C) = AB + AC, (B + C) A = BA + CA (provided the requirements of orders are fulfilled).

1.1.2 SYSTEMS OF LINEAR EQUATIONS

The system of linear equation can be written in the form:

$$\left.\begin{array}{l} a_{11}X_1 + a_{12}X_2 + \ldots + a_{1n}X_n = b_1 \\ a_{21}X_1 + a_{22}X_2 + \ldots + a_{2n}X_n = b_2 \end{array}\right\} \rightarrow A$$

$$\vdots$$

$$a_{s1}X_1 + a_{s2}X_2 + \ldots + A_{sn}X_n = b_n$$

or, $AX = b \rightarrow$ Matrix form:

$$\text{Where } [A]_{s\times n} = \begin{pmatrix} a_{11} & a_{12} & \cdots & a_{1n} \\ a_{21} & a_{22} & \cdots & a_{2n} \\ \cdots & \cdots & \cdots & \cdots \\ a_{s1} & a_{s2} & \cdots & a_{sn} \end{pmatrix}.$$

$$x = \begin{pmatrix} x_1 \\ x_2 \\ \vdots \\ x_n \end{pmatrix}, \quad b = \begin{pmatrix} b_1 \\ b_2 \\ \vdots \\ b_n \end{pmatrix}.$$

- If a system of linear equations doesn't have any solution, it is been called inconsistent.
- If a system of linear equations has solutions it is termed consistent.

Consistent system [Determinate if it has unique solution / Indeterminate if there is more than one solution

Solution method of system of linear equations:

Cramer's Rule

Let equation (A) be the system of linear, equation, (Full form) $\rightarrow$ AX = B.

Let det $A \neq 0 \Rightarrow A^{-1}$ exists.

Therefore the system has unique solution $X = A^{-1}B$

Gauss Elimination Method

Let, we know that arbitrary system of liner equation are given:

$$\left.\begin{array}{l} a_{11}X_1 + a_{12}X_2 + \ldots + a_{1n}X_n = b_1 \\ a_{21}X_1 + a_{22}X_2 + \ldots + a_{2n}X_n = b_2 \end{array}\right\} \quad (1)$$

$$\vdots$$

$$a_{n1}X_1 + a_{n2}X_2 + \ldots + A_{nn}X_n = b_n$$

Let the coefficient $a_{11} \neq 0$

Now transform equation (1) by eliminating the unknown X_1 from all equations except the first. To do this, multiply members of the first equation by the number $\frac{a_{21}}{a_{11}}$ and subtract form the corresponding members of the equation. Then multiply all members of the first equation b $\frac{a_{31}}{a_{11}}$ and subtract them form the corresponding members of the third equation, and so on. Continuing in this way we finally get the following equivalent system of equations at step:

$$a'_{11}X_1 + a'_{12}X_2 + a'_{13}X_3 + \ldots + a'_{1n}X_n = b_1$$

$$a^2_{22}X_2 + a^2_{23}X_3 + a^2_{2n}X_n = b_2$$

$$a^3_{33}X_3 + \ldots + a^3_{3n}X_n = b_3$$

$$\vdots$$

$$a^n_{nn}X_n = b_n$$

Existence and number of solutions of the non-homogeneous system, $[A]_{M\times n}$ *Matrix*

Case 1:

$m = n$ [Number of equation = number of variables]

The system is consistent if rank A = rank $\overline{A}$ $\overline{A}$:the augmented matrix.

1. Rank A = rank $\overline{A}$ = n (the number of variables) *A* in non singular and $\exists A^{-1}$.

   The system has unique solution, $X = A^{-1}B$
2. Rank *A* rank $\overline{A} < n$

The system has infinitely many solutions.

Case 2:

$m > n$ [number of equation < number of varibles

1. The system is consistent if rank A = rank $\overline{A} < m$.
2. If the system is consistent then rank A = rank $\overline{A} < n$.

The system possesses infinitely many solutions:

Case 3:

$m > n$ [number of equation > number of variables]

1. Rank A = rank $\bar{A} = n$

The system possesses only one solution, i.e. zero solution.

2. Rank A = rank $\bar{A} < n$

The system possesses infinitely many solution.

Homogeneous System Of Equations

$AX = 0$

$A : m \times n$ matrix

1. If $m < n$ [number of equations < number of variables] The system admits of a non-zero solution:
2. The homogenous system of equation containing n equations in n unknown has a non-zero solution if rand $A < n$.

   In this case the system possess infinite solutions

1.1.3 EIGEN VALUES AND EIGEN VECTORS

The roots of the Characteristic $|(A - \lambda I)| = 0$ are called Characteristic. roots or latent or eigen values of A.

Note: Eigen vector Corresponding to one eigen value is not unique because if x is the eigen vector so is K, X, K is an arbitrary constant.

Trace A = Sum of all eigen values of A.

Where trace A = Sum of diagonal entries of A.

Cayley-hamilton theorem: Every square matrix satisfies its own Characteristic.

Proof: Let A be a square matrix of order n and its Characteristic equation.

$|A - I| = (-1)^n \lambda^n + a_1 \lambda^{n-1} + a_2 \lambda^{n-2} \ldots + a_n = 0$

Then to prove that A also satisfies

Characteristic this equation

i.e. $(-1)^n A^n + a_1 A^{n-1} + a_2 A^{n-2} + \ldots a_n I = 0$ Where λ is eigen value of a now $(A-\lambda I)$ is of order n then ad joint $(A-\lambda I)$ is of order n–1.

We can write Ad joint $(A-\lambda I) = P_1 \lambda^{n-1} + P_2{}^{n-2} + \ldots p_{n-1} \lambda + p_n$

Where $P_1, P_2 \ldots P_n$ are square matrices.

We k now A. adjA = | A | I

$(A-\lambda I)$ Adj $(A-\lambda I) = |A - \lambda I| I$

$(A - \lambda I)(P_1 \lambda^{n-1} + P_2 \lambda n^{-2} \ldots P_{n-1} + Pn) = [(-1)^n \lambda^n + a_1 \lambda_{n-1} a_2 \lambda n_{-2} l \ldots a_{n-1} + a_n]$

Properties of Eigen Value

1. Any square matrix A and its transport A' have the same eigen values. we have.
2. The eight value of a triangular matrix are just the diagonal element of the matrix.
3. The eight values of an idempotent matrix are either zero either zero or unity.
4. The sum of the eigen values of a matrix is the sum of the element of the principal diagonal.
5. The Product of the eigen values of matrix A is equal to its determinant. Putting $\lambda = 0$ in (iii), we get the result.
6. If is an eigen value of a matrix A, then $1/\lambda$ is the eigen value of A^{-1}.
7. If λ is an eigen value of an orthogonal matrix, than $1/\lambda$ is also its eigen value.
8. If $\lambda_1, \lambda_2 \ldots \lambda_n$ are the eigen values of matrix A, then A^m has the eigen values $\lambda_1^m, \lambda_2^m, \ldots \lambda_n^m$,

1.2 CALCULUS

1.2.1 MEAN VALUE THEOREMS

Rolle's Theorem

If a function f(x) is continuous in closed interval [a, b], derivable in the open interval (a, b) and f(a) = f(b), then exists at east one real C in (a, b) such that f(c) = 0.

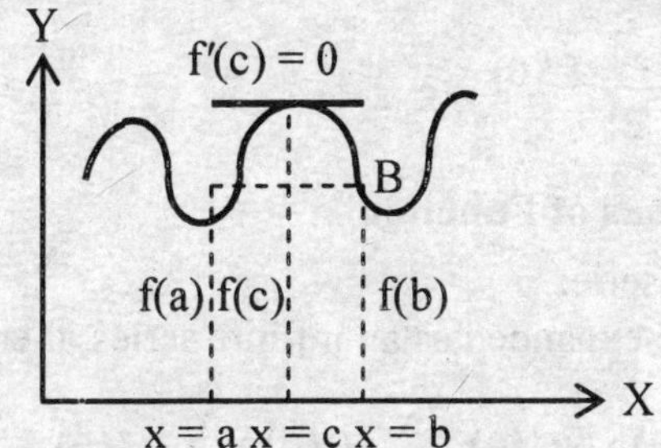

1. f (x) is continuous
2. f′ (x) exist
3. If f(a) = f(b), then there exist a point C ∈ [a b] such that f′(c) = 0.

Lagrange's Mean Value Theorem

If f(x) be a function such that

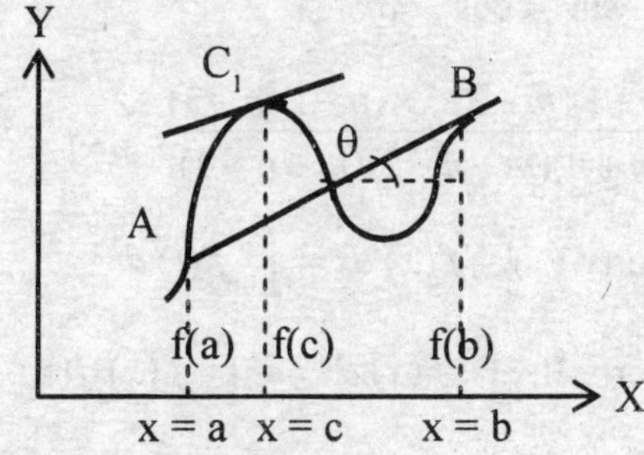

1. f(x) is continuous in (a, b)
2. f′(x) exists in (a, b)

Then there exist at least one C, such that

$$f'(c) = \frac{f(b) - f(a)}{b - a}$$

Cauchy's Mean Value Theorem

If f(x) and g(x) be two functions such that

1. f(x) and g(x) both are continuous in [a b]
2. f′(x) and g′(x) both exist in (a, b)
3. $g'(x) \neq 0 \ \forall \ x \in (a\ b)$

Then there exist latest one point C in (a b) such that

$$\frac{f'(C)}{g'(C)} = \frac{f(b) - f(a)}{g(b) - g(a)}$$

Taylor's Theorem

a) If (x) be a continuous function such that f′(x), f″x), f‴x)... f^{n-1} (x) all are continuous in [a a + h]

b) f″ (x) exist in (a, a + h), where h = b – a.

Then $f(a + h) = f(a) + hf'(a) + \frac{h^2}{2!} f''(a) + \ldots \frac{h^{n-1}}{n-1!} f'^{n-1}(a) + \frac{h^n}{n!} f''(a)$

If we put a = 0, h = x, then the series becomes $f(x) = f(0) + x f'(0) + \frac{x^2}{2!} f''(0) + \ldots$

4. Expansions of Function

Maclaurin' series

$f(x)$ can be expanded as an infinite series, then

$$f(x) = f(0) + xf'(0) + \frac{x^2}{2!} f'(0) + \frac{x^3}{3!} f''(0) + \ldots\infty$$

1.2.2 THEOREM OF INTEGRAL CALCULUS

Integral Calculus

1. Useful Formula to Integration and Result
2. $\int_0^{\pi/2} \sin^n x\,dx = \int_0^{\pi/2} \cos^n x dx$
3. $\int_0^{\pi/2} \sin^m x \cos^n x\,dx$

   $= \frac{(m-1)(m-3)\ldots\times(n-1)(n-3)}{(m+n)(m+n-2)(m+n-4)}$
4. Property I. $\int_a^b f(x)\,dx = \int_a^b f(t)\,dt$
5. Property II. $\int_a^b f(x)\,dx = -\int_b^a f(x)\,d(x)$
6. Property III. $\int_a^b f(x)\,dx = \int_a^c f(x)\,dx + \int_c^b f(x)\,dx.$
7. Property IV. $\int_0^a f(x)\,dx = \int_0^a f(a-x)\,dx$
8. $V. \int_{-a}^a f(x)\,dx = 2\int_0^a f(x)\,dx$ if f (x) is an even function.
9. $VI. \int_0^{2a} f(x)\,dx = 2\int_0^a f(x)\,dx$
10. $\int_0^{\pi/2} \log \sin x\,dx = \int_0^{\pi/2} \log \cos x\,dx = -\frac{\pi}{2}\log 2$
11. $\int_0^{\pi} \log \sin x\,dx = -\pi \log 2$

1.2.3 EVALUATION OF DEFINITE AND IMPROPER INTEGRALS

Improper Integral

When a limit is infinite or integral is infinite in a $a \le x \le b$, in case of $\int_a^b f(x)dx$, then comes the notion of improper integral.

Types of Improper Integral

The interval increasers without limit

1. Let $f(x)$ be bounded and integrable in $a \le x \le B$, $B > a$. Then $\int_a^\infty f(x)dx$ is said to converge or to exist $\lim_{B\to\infty} \int_a^B f(x)dx$ exits finitely and $\int_a^\infty f(x)dx$

   $= \lim_{B\to\infty} \int_a^B f(x)dx$
2. Let $f(x)$ be bounded and integrable in $A \le x \le b$, for every $A < b$. Then is $\int_{-\infty}^b f(x)dx$ said to converge or to exist in $\lim_{A\to\infty} \int_A^b f(x)dx$ exits finitely and $\int_{-\infty}^b f(x)dx$

   $= \lim_{A\to-\infty} \int_A^b f(x)dx$.
3. Let $f(x)$ be bounded and integrable in $A \le x \le a$, for every $A < a$ and in $a \le x \le B$ for every $B > a$

   Then $\int_{-\infty}^\infty f(x)dx$ is said to be convergent if $\lim_{A\to-\infty} \int_A^a f(x)dx$ and $\lim_{B\to\infty} \int_a^B f(x)dx$ exits finitely and

   we write, $\int_{-\infty}^\infty f(x)dx = \int_{-\infty}^a f(x)dx + \int_a^\infty f(x)dx$

   $= \lim_{A\to-\infty} \int_A^a f(x)dx + \lim_{A\to-\infty} \int_a^b f(x)dx$

1.2.4 PARTIAL DERIVATIVES

The partial differential coefficient of z = f(x, y) with respect to x is defined as:

$$\underset{\delta \to 0}{\text{Lt}} \frac{f(x + \delta x, y) - f(x, y)}{\delta x}$$

Provided this limit exists and is denoted by:

$$\frac{\partial f}{\partial y} \text{ or } \frac{\partial z}{\partial y} \text{ or } f_x \text{ or } z_x$$

Similarly, the partial derivatives of f(x, y) with respect to y is defined as

$$\underset{\delta x \to 0}{\text{Lt}} \frac{f(x, y + \delta y) - f(x, y)}{\delta x}$$

Provided this limit exists and is denoted by:

$$\frac{\partial f}{\partial x} \text{ or } \frac{\partial z}{\partial y} \text{ or } f_y \text{ or } z_y$$

are called first order partial derivatives of z.
The partial derivation of f(x,y) with respect to y is the ordinary derivative of f(x,y) when x is regarded as constant.

Partial Derivatives of Higher order

Since f_x and f_y are also function of x and y and so we can further differentiate f_x and f_y partially with respect to x or y. The further partial derivatives of f_x or f_y are called partial derivative of higher order of u.

- The partial derivatives of [∂u/∂x] with respect to x and y are $\frac{\partial u}{\partial x}$ or $\frac{\partial^2 u}{\partial x\, \partial x}$ respectively.
- The partial derivatives

$$\frac{\partial^2 u}{\partial^2 x} \text{ or } \frac{\partial^2 u}{\partial y\, \partial x} \quad \frac{\partial^2 u}{\partial y \partial x}, \quad \frac{\partial^2 u}{\partial y^2}$$

are also denoted by f_{xx}, f_{yx}, f_{xy}, f_{yy}, respectively.

Total Differentiation

We can find the total differentiation of only composite function. If u is function of x, y and x, y are function of t then u is called *composite function of t*.
Let u = f(x, y) and x = g(t) y = h(t)

Then the total derivatives of u with respect to t is given by:

$$\frac{du}{dt} = \frac{\partial u}{\partial x} \cdot \frac{dx}{dt} + \frac{\partial u}{\partial y} \cdot \frac{dx}{dt} \cdot \frac{dy}{dt}$$

Also,

$$\frac{\partial u}{\partial t} = \frac{\partial^2 u}{\partial x} \cdot \frac{\partial x}{\partial t} + \frac{\partial u}{\partial y} \cdot \frac{\partial y}{\partial t}$$

Differential of Implicit functions

If f(x, y) is a function of x and y, then

$$\frac{\partial f}{dx} = \frac{\partial f}{\partial x} \cdot \frac{dx}{dx} + \frac{\partial f}{\partial y} \cdot \frac{dy}{dx}$$

$$= \frac{\partial f}{\partial x} + \frac{\partial f}{\partial y} \cdot \frac{dy}{dx}$$

Now, if f (x, y) = 0, the [df/dx] = 0

$$\therefore \quad 0 = \frac{\partial f}{\partial x} + \frac{\partial f}{\partial y} \frac{dy}{dx}$$

$$\therefore \quad \frac{\partial y}{\partial x} = -\frac{[\partial f/\partial x]}{[\partial f/\partial y]} = -1 \frac{f_x}{f_y}$$

Second derivative of an implicit function is

$$\frac{d^2 y}{dx^2} = \frac{f_{xx} f_y^2 - 2 f_x f_y f_{xy} + f_{yy} f_x^2}{f_y^3}$$

Definitions: Linear differential equations are the in which the dependent variable and its derivatives occur only in the first degree and are in not multiplied together. Thus the general linear differential equation of the nth order is of the form.

$$\frac{d^n y}{dx^n} + P_1 \frac{d^{n-1} y}{dx^{n-1}} + P_2 \frac{d^{n-2} y}{dx^{n-2}} + \ldots + P_n y = X, \text{ b}$$

where $p_1, p_2 \ldots\ldots, p_n$ and X are functions of x only.

Linear differential equations with constant efficient are of the form.

$$\frac{d^n y}{dx^n} + k_1 \frac{d^{n-1} y}{dx^{n-1}} + k2 \frac{d^{n-2} y}{dx^{n-2}} + \ldots + k_n y = X$$

where $k_1, k_2, \ldots, k_n$ are constants. Such equations are most important in the study of electro-mechanical vibrations and other engineering problems.

Theorem: If y_1, y_2 are only two solution of the equation

$$\frac{d^n y}{dx^n} + k_1 \frac{d^{n-1} y}{dx^{n-1}} + k_2 \frac{d^{n-2} y}{dx^{n-2}} + \ldots + k_n y = 0 \quad \ldots(1)$$

then $c_1 y_1 + c_2 y_2 (= u)$ is also only its solution.
Since and x are solution of (1).

$$\therefore \frac{d^n y^1}{dx^n} + k_1 \frac{d^{n-1} y1}{dx^{n-1}} + k_2 \frac{d^{n-2} y1}{dx^{n-2}} + \ldots + k_n y_1 = 0 \text{ and}$$

$$\frac{d^n y_2}{dx^n} + k_1 \frac{d^{n-1} y_2}{dx^{n-1}} + k_2 \frac{d^{n-2} y_2}{dx^{n-2}} + \ldots + k_n y_{2_1} = 0$$

If c_1, c_2 be two arbitrary constants, then

$$\frac{d^n(c_1y_1+c_2y_2)}{dx^n}+k_1\frac{d^{n-1}(c_1y_1+c_2y_2)}{dx^{n-1}}+...+k_n(c_1y_1+c_2y_2)$$

$$=\begin{bmatrix} c_1\left(\dfrac{d^n y_1}{dx^n}+k_1\dfrac{d^{n-1}y1}{dx^{n-1}}+...+k_ny_1\right) \\ +c_2\left(\dfrac{d^n y_2}{dx^n}+k_1\dfrac{d^{n-1}y2}{dx^{n-1}}+...+k_ny_2\right)\end{bmatrix}$$

$$=c_1(0)+c_2(0)=0$$

$$\frac{d^nu}{dx^n}+k_1\frac{d^{n-1}u}{dx^{n-1}}+...+k_ny=0$$

This prove the theorem.

(2) since the general solution of a differential equation of the nth order contains n arbitrary constants, it follows, from above, that if $y_1, y_2, y_3,..., y_n$, are n indepenty solutions of (1), then $c_1y_1+c_2y_2+...+c_ny_n(=u)$ is its complete solution.

(3) If $y=v$ be any particular solution of

$$\frac{d^ny}{dx^n}+k_1\frac{d^{n-1}y}{dx^{n-1}}+...+k_ny=x \quad ...(5)$$

then

$$\frac{d^nv}{dx^n}+k_1\frac{d^{n-1}v}{dx^{n-1}}+...+k_nv=x \quad ...(6)$$

Adding (4) and (6), we have

$$\frac{d^n(u+v)}{dx^n}+k_1\frac{d^{n-1}(u+v)}{dx^{n-1}}+...+k_n(u+v)=x$$

This shows that is the complete solutions of (5).

The part u is called the complementary function (CF) and the part v is called the particular integral (PL) of (5).

The complete solution (CS) of (5) is $y=CF+PI$.

Thus in order to solve the equation (5), we have to first find the CF, i.e... the complete solutions of (1), and then the PI, i.e. a particular solution of (5).

Operator D: Denoting $\frac{d}{dx}, \frac{d^2}{dx^2}, \frac{d^3}{dx^3}$, etc by D, D^2, D^3, etc. so that, etc. the equation (5) above can be written in the symbolic form.

$$(D^n+k_1D^{n-1}+....+k_n)y=x,$$

i.e. $\quad f(D)y=x,$

where $f(D)=D^n+k_1D^{n-1}+...+k_n$, *i.e.* a polynomial in D.

Thus the symbol D stands for the operation of differentiation and can be treated much the same as an algebraic quantity, i.e. can be factories by ordinary rules of algebra and the factors may be taken in any order. For instance.

$$\frac{d^2y}{dx^2}+2\frac{dy}{dx}-3y=(D^2+2D-3)y=(D+3)(D-1)y \text{ or}$$

$(D-1)(D+3)y.$

Rules for Finding the Complementary Function

To solve equation $\frac{d^ny}{dx^n}+k_1\frac{d^{n-1}y}{dx^{n-1}}+k_2\frac{d^{n-2}y}{dx^{n-2}}+...+k_ny=0$ ...(1)

where k's are constants.

The equation (1) in symbolic form is

$$(D^n+k_1D^{n-1}+k_2D^{n-2}+...+k_n)y=0 \quad ...(2)$$

Its symbolic co-efficient equated to zero, i.e. $D^n+k_1D^{n-1}+k_2D^{n-2}+...+k_n=0$ is called the auxiliary equation (AE). Let $m_1, m_2,....., m_n$ be its roots.

Case 1: If all the roots be real and different, then (2) is equivalent to $(D-m_1)(D-m_2)...(D-m_n)y=0$ ...(3)

Now (3) will be satisfied by the solution of $(D-m_n)y=0$, i.e. by $\frac{dy}{dx}-m_ny=0$,

This is a Leibnitz linear and $L.F.=e^{-m_nx}$

Its solution is $ye^{-m_nx}=c_n$, $i.e.\ y=c_ne^{m_nx}$

Similarly, since the factors in (3) can be taken in any order, it will be satisfied by the solutions of $(D-m_1)y=0$, $(D-m_2)y=0$, etc., i.e. by, etc.

Thus the complete solutions of (1) is

$$y=c_1e^{m_1x},\ y=c_2e^{m_2x}+...+c_ne^{m_nx} \quad ...(4)$$

Case 2: If two roots are equal then (4) become

$$y=(C_1+C_2)e^{m_1x}+C_3e^{m_3x}+...+c_ne^{m_nx}$$

$$y=C\,e^{m_1x}+c_3e^{m_3x}+...+c_ne^{m_nx}$$

It has only $n-1$ arbitrary constants and is, therefore, not in the complex solution in case, we process as follows:

The part of the complete solution corresponding to the repeated root is the complete solutions of $(D-m_1)(D-m_1)y=0.$

Putting $(D-m_1)y=z$, it becomes $(D-m)z=0$ or $\frac{dz}{dx}-m_1z=0$

This is a Leibnitiz's linear in z and Its Its solution

$I.F. = e^{-m_1x}$. is $ze^{-m_1x} = c_1$ or $z = c_1e^{m_1x}$

Thus $(D - m_1)y = z = c_1e^{m_1x}$

or

$$\frac{dy}{dx} - m_1 y = c_1 e^{m_1x} \quad (5)$$

Its IF being e^{-m1x}, the solutions of (5) is

$ye^{-m_1x} = \int c_1e^{m_1x}dx + c_2 = c_1x + c_2$ or $y = (c_1x + c_2)e^{m_1x}$

Thus the complete solution of (1) is $y = (c_1x + c_2)e^{m_1x}$ $+ c_3e^{m_3x} + ... + c_ne^{m_nx}$

If, however, the AE has three equal roots $(i.e. m_1 = m_2 = m_3)$, then the complete solution is $y = (c_1x^2 + c_2x + c_3)e^{m_1x} + c_4e^{m_4x} + ... + c_ne^{m_nx}$

Case 3: If one pair of roots be imaginary, i.e. $m_1 = a + i\beta, m_2 = a - i\beta$, then the complete solution is

$y = c_1e^{(a+i\beta)x} + c_2e^{(a-i\beta)x} + c_3e^{m_3x} + ... + c_ne^{m_nx}$

$= e^{ax}(c_1e^{i\beta x} + c_2e^{-i\beta x}) + c_3e^{m_3x} + ... + c_ne^{m_nx}$

$e^{ax}[c_1(\cos\beta x + i\sin\beta x) + c_2(\cos\beta x - i\sin\beta x)] +$

$c_3e^{m_3x} + ... + c_ne^{m_nx}$

$= e^{ax}(C_1\cos\beta x + C_2\sin\beta x) + c_3e^{m_3x} + ... + c_ne^{m_nx}$

$C_1 = c_1 + c_2 = i(c_1 - c_2)$.

Case 4: If two pairs of imaginary roots be equal i.e. $m_1 = m_2 = a + i\beta,\ m_3 = m_4 = a - i\beta$, then by case II, the complete solution is

$y = e^{ax}[(c_1x + c_2)\cos\beta x + (c_3x + c_4)\sin\beta x] + ... + c_ne^{m_nx}$

Inverse Operator: $\frac{i}{f(D)}X$ is that function of x, not containing arbitrary constant which when operated upon by $f(D)$ gives X, i.e. $f(D)\left\{\frac{1}{f(D)}X\right\} = X$

Thus $\frac{1}{f(D)}X$ satisfies the equation $f(D)y = X$ and is, therefore, its particular integral. Obviously, $f(D)$ and $\frac{1}{f(D)}$ are inverse operators.

2. $\frac{1}{D}X = \int X dx$

Let $\frac{1}{D}X = y$

Operating by D, $D\frac{1}{D}X = Dy$, i.e. $X = \frac{dy}{dx}$

Integrating both sides w.r.t x, $y = \int X\,dx$, no constant being added as (1) does not contain any constant.

Thus $\frac{1}{D}X = \int X dx$.

3. $\frac{1}{D-a}X = e^{ax}\int Xe^{-ax}dx$.

Let $\frac{1}{D-a}X = y$

Operating by, $(D-a)$, $D - a$

$\frac{1}{D-a}X = (D-a)y$.

or $X = \frac{dy}{dx} - ay$, i.e. $\frac{dy}{dx} - ay = X$ which is a leibnitiz's linear equation.

IF being e^{-ax}, its solution is $ye^{-ax} = \int Xe^{-ax}dx$, no constant being added as (ii) doesn't any contain any constant.

Thus $\frac{1}{D-a}X = y = e^{ax}\int Xe^{-ax}dx$.

Rules for Finding the Particular Integral

Consider the equation

$$\frac{d^ny}{dx^n} + k1\frac{d^{n-1}y}{dx^{n-1}} + k_2\frac{d^{n-2}y}{dx^{n-2}} + ... + k_ny = X$$

Which symbolic form is

$$(D^n + k_1D^{n-1} + k_2D^{n-2} + ... + k_n)y = X.$$

$$P.I. = \frac{1}{D^n + k_1D^{n-1} + k_2D^{n-2} + ... + k_n}X$$

Case 1: When $X = e^{ax}$

Since $\frac{1}{f(D)}e^{ax} = \frac{1}{f(a)}e^{ax}$ provided $f(a) \neq 0$

If the above rule fails sand we proceed further.

Case 2: When $X = \sin(ax + b)$ or $\cos(ax + b)$.

Since D sin $(ax + b) = a\cos(ax + b)$ $D^2\sin(ax + b) = a^2$ sin $(ax + b)$

$D^3\sin(ax + b) = -a^3\cos(ax + b)$

$D^4\sin(ax + b) = a^4\sin(ax + b)$

i.e. $D^2\sin(ax + b) = (-a^2)\sin(ax + b)$

$(D^2)^2\sin(ax + b = (-a^2)^2\sin(ax + b)$

In general $(D^2)^r\sin(ax + b) = (-a^2)^r\sin(ax + b)$.

$\therefore\ f(D^2)\sin(ax+b) = f(-a^2)\sin(ax+b).$

Operating on both sides $1/f(D^2)$

$$\frac{1}{f(D^2)} f(D^2)\sin(ax+b) = \frac{1}{f(D^2)} f(-a^2)\sin(ax+b)$$

Or $\sin(ax+b) = f(-a^2)\frac{1}{f(D^2)}\sin(ax+b)$

Dividing by

$$f(-a^2)\frac{1}{f(D^2)}\sin(ax+b) = \frac{1}{f(-a^2)}\sin(ax+b)$$

provided $f(-a^2) \neq 0$ ...(4)

If $f(-a^2)=0$, the above rule fails and we proceed further.

Since $\cos(ax+b) + i\sin(ax+b) = e^{i(ax+b)}$

$$\frac{1}{f(D^2)}\sin(ax+b) = I.P.\text{ of } \frac{1}{f(D^2)} e^{i(ax+b)}$$

$$= I.P.\text{ of } x\frac{1}{f'(D^2)} e^{i(ax+b)}$$

$$\frac{1}{f(D^2)}\sin(ax+b) = x\frac{1}{f'(-a^2)}\sin(ax+b),$$

provided a $f'(-a^2) \neq 0$, and so on.

Similarly, $\frac{1}{f(D^2)}\cos(ax+b) = \frac{1}{f(-a^2)}\cos(ax+b),$

provided $f(-a^2) \neq 0,$

If $f(-a^2)=0$, $\frac{1}{f(D^2)}\cos(ax+b) = X\frac{1}{f(-a^2)}$

$\cos(ax+b)$, provided $f'(-a^2) \neq 0$.

If $f'(-a^2)=0$, $\frac{1}{f(D^2)}\cos(ax+b)$

$= x^2\frac{1}{f''(-a^2)}\cos(ax+b)$ provided $f''(-a^2) \neq 0$ and so on.

Case 3: When $X = x^m$

Here $P.I. = \frac{1}{f(D)}x^m = [f(D)]^{-1}x^m.$

Expand $[f(D)]^{-1}$ in ascending powers of D as far as the term D^m in and operate on x^m term. Since the $(m+1)$th and higher derivatives of x^m are zero, we need not consider terms beyond.

Case 4: When $X = e^{ax}V$, being a function of x. If u is a function x, then

$$D(e^{ax}u) = ae^{ax}u + ae^{ax}u = e^{ax}u(D+a)u$$

$$D^2(e^{ax}u) = a^{ax}D^2u + 2ae^{ax}Du + a^2e^{ax}u$$

$$= e^{ax}(D+a)^{2u}$$

$$D^n(e^{ax}u) = e^{ax}(D+a)^n u$$

$$f(D)(e^{ax}u) = e^{ax}(D+a)^n u$$

and Operating both sides by 1/f (D).

$$\frac{1}{f(D)} f(D)(e^{ax}u) = \frac{1}{f(D)}[e^{ax}f(D+a)u]$$

$$e^{ax}u = \frac{1}{f(D)}[e^{ax}f(D+a)u]$$

Now put $f(D+a)u = V$, i.e. $u = \frac{1}{f(D+a)}V$, so that

$$e^{ax}\frac{1}{f(D+a)}V = \frac{1}{f(D)}(e^{ax}V)$$

i.e. $\frac{1}{f(D)}(e^{ax}V) = e^{ax}\frac{1}{f(D+a)}V.$

Case 5: When x is any other function of x.

Here $P.I. = \frac{1}{f(D)}X.$

If $f(D) = (D-m_1)(D-m_2)...(D-m_n)$, resolving into partial fractions,

$$\frac{1}{f(D)} = \frac{A_1}{D-m_1} + \frac{A_2}{D-m_2} + ... + \frac{A_n}{D-m_n}$$

$$P.I. = \left[\frac{A_1}{D-m_1} + \frac{A_2}{D-m_2} + ... + \frac{An}{D-m_n}\right]X$$

$$= A_1\frac{1}{D-m_1}X + A_2\frac{1}{D-m_2}X + ... + A_n\frac{1}{D-m_n}X$$

$$A_1\, e^{m_1x}\int Xe^{-m_1x}dx + A_2\, e^{m_2x}\int Xe^{-m_2x}\,dx + ... +$$

$$A_n\, e^{m_nx}\int Xe^{-m_nx}dx$$

Working Procedure to Solve the Equation

$\frac{d^ny}{dx^n} + k_1\frac{d^{n-1}y}{dx^{n-1}} + ... + k_{n-1}\frac{dy}{dx} + k_ny = x$ of which of the symbolic form is $(D^n + k_1D^{n-1} + ... + k_{n-1}D + k_n)y = X.$

Step 1: To find the complementary function

(a) Write the AE and solve it for D.

(b) Write the CF as follows:

Roots of AE and CF

1. If ($m_1, m_2, m_3 \ldots$ real and different roots) then $c_1e^{m_1x} + c_2e^{m_2x} + c_3e^{m_3x} + \ldots$
2. If (m_1, m_1, m_3,... two real and equal roots) then $(c_1 + c_2x)e^{m_1x} + c_3e^{m_3x} + \ldots$
3. If (m_1, m_1, m_1, m_4,... three real and equal roots) then $(c_1 + c_2x + c_3x^2)\, e^{m_1x} + c_4e^{m_4x} + \ldots$
4. $a + i\beta$, $a - i\beta$, $m_3 \ldots$ a pair of imaginary roots then $e^{ax}(c_1 \cos \beta x + c_2 \sin \beta x) + c_3e^{m_3x} + \ldots$
5. $a \pm i\beta$, $a \pm i\beta$, $m_5 \ldots$ 2 pairs of equal imaginary roots then $e^{ax}[(c_1 + c_2x)\cos\beta x + (c_3 + c_4x)\sin\beta x] + c_5e^{m_5x} + \ldots$

Step 2: To fond the particular integral

Form symbolic form

$$P.I. = \frac{1}{D^n + k_1D^{n-1} + \ldots + k^{n-1}D + k_n} \cdot X = \frac{1}{f(D)}$$

or $\dfrac{1}{\phi(D^2)}x$

1. When $X = e^{ax}$

$$P.I. = \frac{1}{f'(D)}e^{ax}, \quad \text{put } D = a$$

$$= x\frac{1}{f(D)}e^{ax}, \quad \text{put } D = a$$

$$x^2\frac{1}{f''(D)}e^{ax}, \quad \text{put } D = a$$

and so on.

where $f'(D) = diff.coeff.of\ f(D)\ w.r.t.\ D$

$f''(D) = diff.coeff.of\ f'(D)\ w.r.t.\ D$

2. When or $X = \sin(ax + b)$ or $\cos(ax + b)$. put

$$P.I. = \frac{1}{\phi(D^2)}\sin(ax + b) \text{ or } [\cos(ax + b)],$$

put $\quad D^2 = -a^2$

$$= x\frac{1}{\phi'(D^2)}\sin(ax + b)\ or\ [\cos(ax + b)], \text{put } D^2 = -a^2$$

where $= x^2\dfrac{1}{\phi''(D^2)}\sin(ax + b) or [\cos(ax + b)]$, and so on.

$\phi'(D^2) = diff.coeff.of\ \phi(D^2)\ w.r.t.\ D,$

$\phi''(D^2) = diff.coeff.of\ \phi(D^2)\ w.r.t.\ D$, etc.

3. When $X = x^m$ being a positive integer.

$$P.I. = \frac{1}{f(D)}x^m = [f(D)]^{-1}x^m$$

To evaluate it, expand $[f(D)]^{-1}$ in ascending powers of D by Binomial theorem as far D^m as and operate on x^m term by term.

4. When $X = x^m$ where V is a function of x.

$$P.I. = \frac{1}{f(D)}e^{ax}V = e^{ax}\frac{1}{f(D + a)}V$$

and then evaluate $\dfrac{1}{f(D + a)}V$. as in (i), (ii) and (iii).

5. When X is any function of x.

$$P.I. = \frac{1}{f(D)}X.$$

Resolve $\dfrac{1}{f(D)}$ into partial fractions and operate each partial fraction on X remembering that

$$\frac{1}{D - a}X = e^{ax}\int Xe^{-ax}\,dx.$$

Step 3: To find the complete solution

Then the C.S. is $y = C.F. + P.I.$

1.2.5 MAXIMA AND MINIMA

Maxima and Minima of Function of Two Variables

A function f(x, y) is said to have a maximum or minimum at x = a , y = b , according as f(a + h, b + k) is less or greater that f (a, b) for all positive or negative small values of h and k

In order words, if Δ = f(a + h, b + k) - f(a, b) is of the same sign for all small value of h, k and if this sign is negative, then f(a,b) is a maximum. If this sign is positive, then f(a, b) is a minimum.

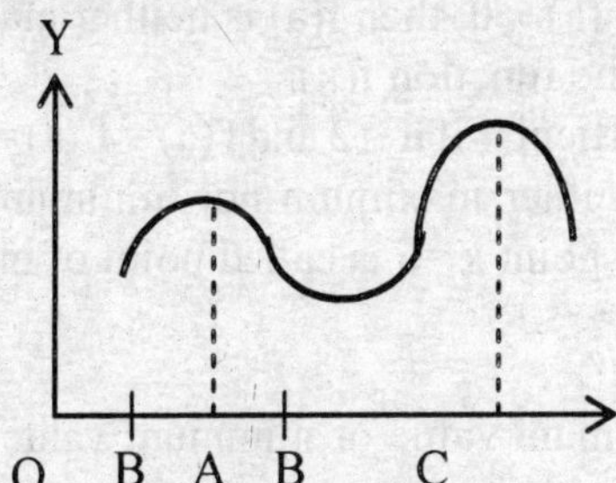

A maximum or minimum value of a function is called its extreme value. The figure shows a graph of the function f(x) and OA = a, i.e. f(x) has a maximum value

for x = a because of f(a) has a value more than the values of f(x) for every value of x between B and B'. f(x) is said to be a maximum at x = a, even thought the value of f(x) at x = a is less than the value of f(x) at x = a should be greater than all other values of f(x) in some small neighborhood. Thus a maximum value of f(x) is not necessarily the greatest value of f(x). In fact, a curve might have several maxima (and minima).

Working rules for finding absolute maxima and minima values.

If f is a differentiable function in [a, b] except, at finitely many points, then to find absolute maximum and absolute minimum values, adopt the following procedure:

1. Evaluate f (x) at the point , where f'(x) = 0
2. Evaluate f(a) and f (b)

Then the maximum of these value is the absolute maxima and minimum of these values is called absolute minima

Working rules to find local maximum and local minima.

For Maxima

The function f(x) has a maximum value f(a) if f'(a) = 0 f (x) changes sign from positive to negative as x passes through a from left to right.

- In general, for any even number n,
  $f'(a) = f''(a) = f(a) \ldots f^{n-1}(a) = 0$ and $f^n(a) < 0$, then
f(a) is a maximum value of f(x)
- In particular, if $^n = 2$, then f'(a) = 0 and f''(a) < 0, then the function is maximum at x = a.

Condition for minima

- The function f(x) has a minimum value f(a) if f'(a) = 0 and f'(x) changes sign from negative to positive as x passes through x = 1 , from left to right.
- In general, for any even number n.
  $f'(a) = f''(a) = f'''(a) = \ldots f^{n-1}(a) = 0$, and $f^n(a) > 0$, then f(a) is minimum value of f(x).
- In particular, if n = 2, then f (a) = 0 and f''(a) < 0
  Then the function is minimum at x = a.

Note: If n is a even number such that $f'(a) = f'(a) = \ldots f^n(a) = 0$, $f^{n+1}(a) \neq 0$, then f(a) is neither maximum nor minimum of the function f(x).

In small Particular if n = 2 and f'(a) = f''(a) = 0. f''(a)≠0, then f(a) is neither maximum nor minimum values of f(x). Then the point x =a is called point of inflexion.

Extreme Value

Either a maximum value or minimum value f(a) of the function f(x) is said to be extreme value.

Stationary Value

If f'(a) = 0, the f(a) is said to be stationary value

Note:

Every extreme value is stationary but every stationary value of the function need not be extreme value.

Euler's therorem on Homogeneous Functions

If u is a function of x, y of degree n, then $x\dfrac{\partial u}{\partial x} + y\dfrac{\partial u}{\partial y} = nu$

$$x^2\frac{\partial^2 u}{\partial x^2} + 2xy\frac{\partial^2 u}{\partial x \partial y} + y^2\frac{\partial^2 u}{\partial y^2} = n(n-1)u$$

1.2.6 MULTIPLE INTEGRALS

Double Integrals

The definite integral $\int_a^b f(x)$ is defined as the limit of the sum $f(x_1)\,\delta x_1 + f(x_2)\,\delta x_2 + \ldots + f(x_n)\,\delta x_n$

where $n \to \infty$ and each of the lengths tends to zero. A double integral is its counterpart in two dimensions.

Consider a function $f(x, y)$ of the independent variables x, y defined at each point in the finite region R of the *xy*-plane. Divide R into n elementary areas $\delta A_1, \delta A_2 \ldots \delta A_n$. Let (x_r, y_r) be any point within the rth elementary area δA_r. Consider the sum

$$f(x_1, y_1)\,\delta A_1 + f(x_2, y_2)\,\delta A_2 + \ldots + f(x_n, y_n)\,\delta A_n$$

i.e. $\sum_{r=1}^{n} f(x_r, y_r)\,\delta A_r$

The limit of this sum, if it exits as the number of sub-divisions increases indefinitely and area of each sub-division decreases to zero, is defined as the double integral $f(x, y)$ of over the region R and is written $\iint_R f(x, y)\,dA$.

Thus $\iint_R f(x, y)\,dA = \underset{\substack{n \to \infty \\ \delta A \to 0}}{LT} \sum_{r=1}^{n} f(x_r, y_r)\,\delta A_r$

The utility of double integrals would be limited if it were required to take limit of sums to evaluate them. However, there is another method of evaluating double integrals by successive single integrations.

For purposes of evaluation (1) is expressed as the repeated integral $\int_{x_1}^{x_2}\int_{y_1}^{y_2} f(x, y)\,dx\,dy$. Its value is found as follows.

1. When y_1, y_2 are functions of x and x_1, x_2, are constants, $f(x, y)$ is first integrates w.r.t y keeping x fixed between limits and then the resulting expression is integrates w.r.t x within the limit x_1, x_2, i.e.

$$I_1 = \int_{x_1}^{x_2} \left(\int_{y_1}^{y_2} f(x, y)\, dy \right) dx$$

Where integration is carried from the inner to the outer rectangle.

Figure illustrates this process. Here *AB* and *CD* are the two curves whose equation are $y_1 = f_1(x)$ and $y_2 = f_2(x)$. *PQ* is a vertical strip of with *dx*.

Then the inner rectangle integral means that the integration is along one edge of the strip *PQ* from *P* to *Q* (*x* remaining constant), while the outer rectangle integral corresponds to the sliding of the edge from *AC* to *BD*.

Thus the whole region of integration is the area *ABDC*.

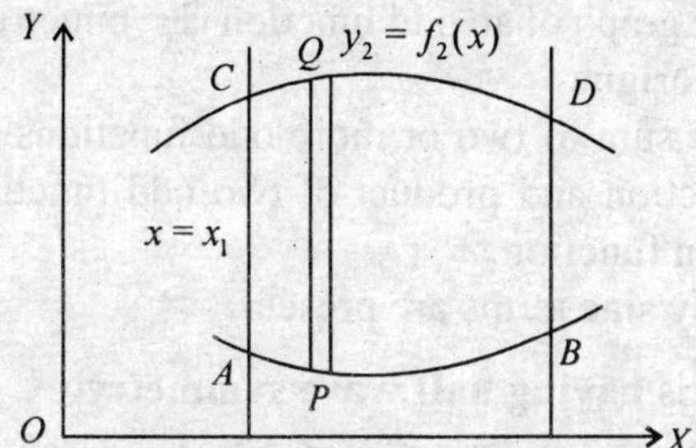

2. When x_1, x_2 are functions of y and y_1, y_2 are constants is first integrates w.r.t. *y* keeping *x* fixed within the limit x_1, x_2 and the resulting expression is integrated w.r.t y between the limits x_1, x_2,

   i.e. $I_2 = \int_{y_1}^{y_2} \left(\int_{x_1}^{x_2} f(x, y) dx \right) dy$

Here *AB* and *CD* are the curves $x_1 = f_1(y)$. *PQ* and $x_2 = f_2(y)$ is a horizontal strip of width *dy*.

Then inner rectangle indicates that the integration is along one edge of the strip from *P* to *Q* while the outer rectangle corresponds t the sliding of this edge from *AC* to *BC*.

Thus the whole region of integration is the area *ABDC*.

3. When both pairs of limits are constants, the region of integration is the rectangle *ABDC*.

In I_1, we integrate along the horizontal strip *PQ* and then slide it from *AC* to *BD*.

Here obviously $I_1 = I_2$.

Thus for constant limits, it hardly matters whether we first integrate w.r.t. x and then w.r.t. y or vice versa.

Example

Evaluate $\int_0^5 \int_0^{x^2} x(x^2 + y^2) dx\, dy$

Solution:

$$I = \int_0^5 dx \int_0^{x^2} (x^3 + xy^2) dy$$

$$= \int_0^5 \left[x^3 y + x \cdot \frac{y^3}{3} \right]_0^{x^2} dx$$

$$= \int_0^5 \left[x^3 \cdot x^2 + x \cdot \frac{x^6}{3} \right] dx = \int_0^5 \left(x^5 + \frac{x^7}{3} \right) dx$$

$$= \left| \frac{x^6}{6} + \frac{x^8}{24} \right|_0^5 = \left[\frac{1}{6} + \frac{5^2}{24} \right] \cdot 5^6$$

$$= 18880.2 \text{ nearly.}$$

Triple Integrals

Consider a function $f(x,y,z)$ defined at every point of the 3-dimensional finite region *V*. Divide *V* into *n* elementary volumes $\delta V_1, \delta V_2, \ldots, \delta V_n$. Let (x_r, y_r, z_r) be any point within the *r*th sub-divisions δV_r. Consider the sum $\sum_{r=1}^{\infty} f(x_r, y_r, z_r)\, \delta V_r$

The limit of this sum, if it exists, as $n \to \infty$ and $\delta V_r \to 0$ is called the triple integral of $f(x,y,z)$ over the region V and is denoted by

$$\iiint f(x, y, z)\, dV$$

For purpose of evaluation, it is can also be expressed as the repeated integral.

$$\int_{x_1}^{x_2} \int_{y_1}^{y_2} \int_{z_1}^{z_2} f(x, y, z)\, dx\, dy\, dz$$

If x_1, x_2 are constants y_1, y_2 are either constants or function of x and z_1, z_2 are either constants or functions of *x* and *y*, then this integral is evaluated as follows:

First $f(x,y,z)$ is integrated w.r.t. *z* between the limits and keeping *x* and *y* fixed. The resulting expression is integrated w.r.t. *y* between the limits y_1 and y_2 keeping *x* constant. The result obtained is finally integrated w.r.t. from x_1 to x_2 Thus, $I = \int_{x_1}^{x_2} \int_{y_1(x)}^{y_2(x)} \int_{z_1(x,y)}^{z_2(x,y)} f(x, y, z)\, dz dy dx$

Where the integration is carried out form the innermost rectangle to the outermost rectangle.

1.2.7 FOURIER SERIES

Any periodic signal can be represent by some infinite number of sinusoids of Harmonic frequencies where the composite set of function consisting of a set cos n o t and Sin n o t for (n = 0, 1, ...2) form a complete ***orthogonal set***.

Mutually orthogonal function: Two real time functions are said to be "mutually orthogonal" over an interval between t1 and t2 if the integral of their product over this interval is zero; i.e. $f(t)$ is orthogonal to h(t), if

$$\int_{t_1}^{t_2} f(t)\ h(t)\ dt = 0$$

Therefore, a set of time function $\{g_1(t), g_2(t), \ldots\}$ forms an orthogonal set over the interval between t_1 and t_2 if

$$\int_{t_1}^{t_2} g_j(t)\ g_k(t)\ dt = 0,\ (j \neq k).$$

Orthogonality in complex function: If $f_1(t)$ and $f_2(t)$ are complex functions of the real variable t, then the two functions are orthogonal over the interval between t_1 and t_2, if $\int_{t_1}^{t_2} f_1(t)\ f_2^*(t)\ dt = \int_{t_1}^{t_2} f_1^*(t)\ f_2(t)\ dt = 0$

where $f_1^*(t)$ and $f_1^*(t)$ are complex conjugates of $f_1(t)$ and $f_2(t)$, respectively. The complex functions $e^{jn\omega_o t}$ (n = 0, ± 1, ± 2, ...) form a set of orthogonal functions. Fourier series of a periodic function $f(t)$.

A periodic signal is defined by the relation.

$$f(t) = f(t \pm T)$$

where, T = Time period and $\omega_o = 2\pi/T$

A periodic signal $f(t)$ of time period T can be represented by presenting the sum of infinite number of sinusoids of harmonic frequencies ω_o, $2\omega_o$, $3\omega_o$ ...... called the fourier series.

1. $f(t)$ must satisfy the dirichlet's condition, i.e.
   a) f (t) has only a finite number of maxima and minima in any one period.
   b) f (t) is discontinuous and has a finite number of discontinuities in any one period.
   c) $f(t)$ discontinuities are bonded, i.e. it has a finite average values over the period T.

Waveform Symmetry.

A. If $f(t)$ is an even function, i.e.

$f(t) = f(-t)$. For example cos t, t^2, t sin t.

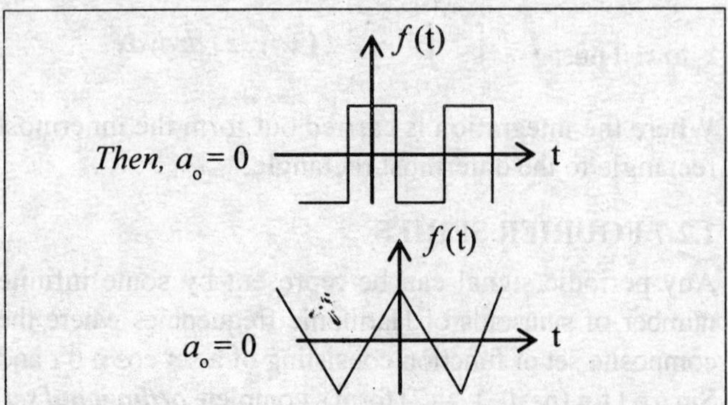

$$a_n = \frac{4}{T}\int_0^{T/2} f(t)\cos(n\ \omega_o t)\ dt$$

a) $b_n = 0$. The graph of an even function will be symmetrical with respect the to y-axis and only cosine terms are present (dc term optional).

b) The sum or product of two or more even function is an even function.

B. If $f(t)$ is an odd function

$-f(t) = f(-t)$; Then,

$a_o = 0$; $a_n = 0$

$$b_n = \frac{4}{T}\int_0^{T/2} f(t)\sin(n\ \omega_o t)\ dt$$

a) The graph of an odd function is symmetrical about the origin.

b) The sum of two or more odd functions is an odd function and product of two odd functions is an even function.

c) Only sine terms are present.

C. If $f(t)$ is having half wave symmetry

$-f(t \pm T/2) = f(t)$: Then $a_o = 0$

$$a_n = \frac{4}{T}\int_0^{T/2} f(t)\cos(n\ \omega_o t)\ dt$$

$$b_n = \frac{4}{T}\int_0^{T/2} f(t)\sin(n\ \omega_o t)\ dt$$

a) Only odd harmonic terms are present.

b) If $f(t \pm T/2) = f(t)$, then only even harmonic are present.

D. If $f(t)$ is having quarter wave symmetry, i.e.

(a) Half wave and odd symmetry

$a^o = 0$ all n

$a_n = 0$ all n

$$b_n = \frac{8}{T}\int_0^{T/4} f(t)\sin(n\ \omega_o t)\ dt \quad \text{n odd}$$

$b_n = 0$ n even

(b) Half wave and even symmetry

a_o = Find out.

$a_n = 0$ n even

$$a_n = \frac{8}{T}\int_0^{T/4} f(t)\cos(n\ \omega_o t)\ dt \quad \text{n odd}$$

$b_n = 0$ all n

Exponential Fourier series:

$$f(t) = \sum_{n=-\infty}^{\infty} F_n\ e^{jn\omega_o t}$$

where $F_n = \frac{1}{T}\int_0^T f(t)e^{-jn\omega_0 t}dt$

$$F_n = |F_n|\ e^{-j\theta n} \quad \theta_n = \tan^{-1}\ \frac{I_m[F_n]}{R_e[F_n]}$$

|Fn| gives magnitude (discrete) spectrum, i.e. always even function of n, θ_n gives phase (discrete) spectrum, i.e. always an odd function of n.

1.2.8 VECTOR IDENTITIES

Vector: A quality which is completely specified by its magnitude and direction is called a vector.

Vector is represented by a directed line segment. Thus $\overline{PQ}$ represents a vector where magnitude is the length PQ and direction is form P (starting point) to Q (end point)

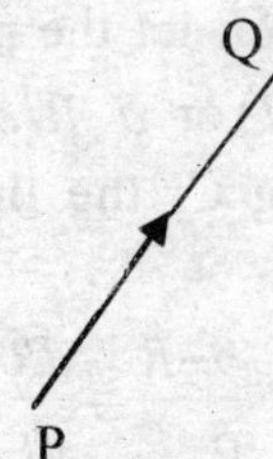

Unit Vector

A vector of unit magnitude is called unit vector. Unit vector corresponding to vector A is written as $\hat{A}$

Null Vector

A vector of zero magnitude, which can have no direction associated with it is called zero (or null) vector and is denoted by o-a thick zero.

Product of two Vectov

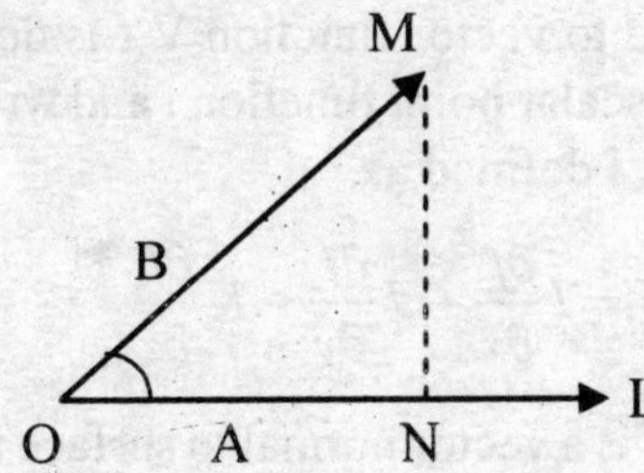

1. Scalar or Dot Product

The Scalar or dot product of two vector A and B is defined as scalar ab cosθ, where θ is angle between A and B.

$$A.B = ab\cos\theta$$

Properties

1. Two vector to be perpendicular is that their scalar product should be zero.

A.B = ab cos 90° $[\theta = 90°]$

A.B = 0

2. For mutually perpendicular unit vector I, J, K, are have relation.

I.J = JK = KI = 0

$I^2 = J^2 = K^2 = 1$

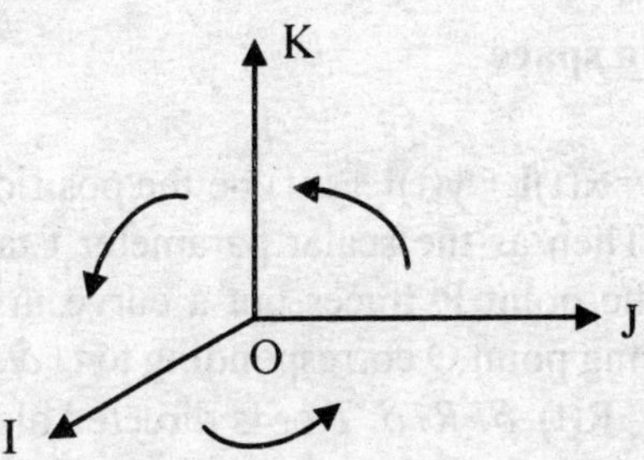

(II) Vector or cross product

Vector or cross product of two vector A and B defined as a Vector, ab sinθ, where θ is angle between A and B.

Mathematically

$$A \times B = ab\sin\theta N$$

N is unit vector normal to plane A and B.

Properties

a) Two non zero vector to be parallel is that their vector product should be zero.

$\theta = 0^0$ or 180°

A × B = Sin 0^0 or Sin 180°

A × B = 0

b) For the orthonormal vector I, J, K, we have the relation:

I × I = J × J = K × K = 0

I × J = K   J × I = – K

J × k = I   K × J = – I

K × I = J   I × K = – J

Differentiation of Vector

If a vector R varies continuously as a scalar variable t changes, then R is said to be function of t and is written as R = f(t).

A in scalar calculus, we define derivative of vector function

$$R = f(t) \text{ as } \underset{\delta t \to 0}{Lt} = F\ \frac{(t+\delta t) - F(t)}{\delta t} \text{ and it write as}$$

$$\frac{dr}{dt}\ or\ \frac{df}{dt}\ or\ F'(t)$$

a) If F (t) has constant magnitude, then $F.\frac{df}{dt} = 0$

For F (t) . F (t) = $[F\ (t)]^2$ = constant

$F.\frac{dF}{dt}=0$ i.e. $\frac{dF}{dt}\perp F$

b) If F(t) has constant or fixed direction,

Then $F \times \frac{dF}{dt} = 0$

Curves in space

Tangent

Let R(t) = x(t)I + y(t)I + z(t) be the position vector of a point P. Then as the scalar parameter t takes different values, the point P. traces out a curve in space. If the neighboring point Q corresponding to t+δt, then δR = R (t + δt) – R(t) $\delta\ R/\delta\ T$ or is directed along the chord PQ. As $\delta\ t\to 0$, $Lt\ \delta\ R/\delta\ T$ becomes the tangent (vector) to curve at P whenever it exists and is not zero, thus the vector R' '= (dr/dt) us a tangent to space curve R = F(t).

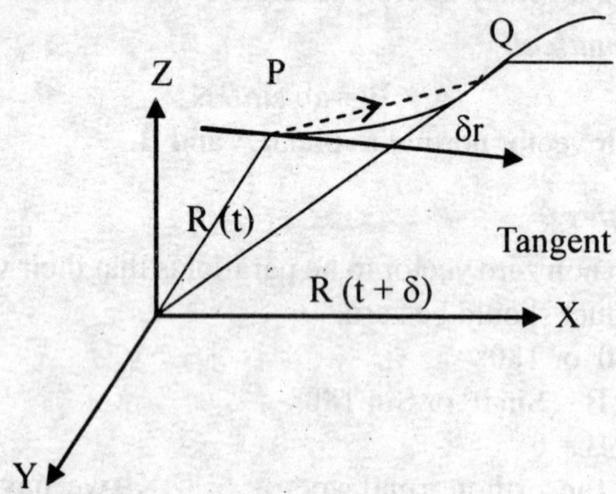

If we take S as the parameter in place of t then magnitude of tangent vector, = 1. Thus denoting the unit tangent vector by T, he have

$$T=\frac{dR}{ds}$$

Principal of Normal – since T is a unit vector, we have

$$\frac{dr}{ds}\,.\,T=0$$

[dT/ds] is perpendicular to T , in which case T is constant vector wrt. the arc length s and so has fixed direction, i.e. curve is straight line.

If we denote a unit normal vector to curve a + p by N then [dt/ds] is known as principal normal to the space curve at P.

Binomial

A third unit vector B defined by B = $T \times N$, is called the binomial at P. Since T and N are unit vector, B is also a unit vector perpendicular to both T and N and then normal to plane at P.

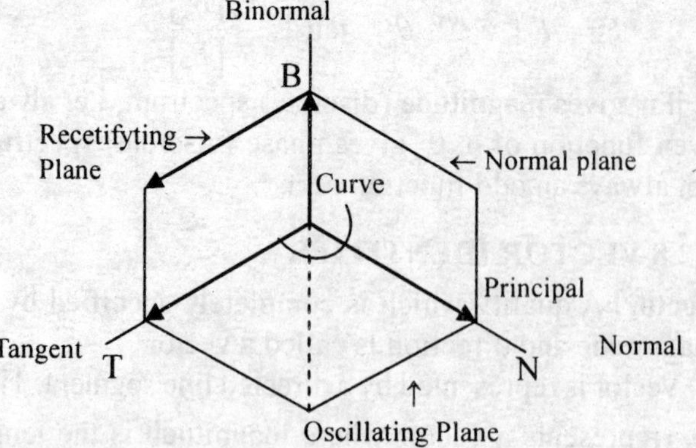

Velocity And Acceleration

Let the position of particle P at time t on a path (curve) C be R(t). At time t + δt, let the particle be at Q, then δR = $R(t+\delta\ t)-R(t)$ or $\delta\ R/\delta\ T$ is directed along PQ. *As* $Q\to P$ along C, the line PQ becomes the tangent at P to C.

$$\lim_{\delta T\to 0}\frac{\delta\ R}{\delta\ T}=\frac{dR}{dt}=V$$

The derivative of the velocity vector V(t) is called the acceleration (vector) A (t), which gives as

$$A(t)=\frac{dV}{dt}=\frac{d^2R}{dt^2}$$

Vector Calculus

Consider a vector operator $\nabla = I\frac{\partial}{\partial x}+J\frac{\partial}{\partial y}+K\frac{\partial}{\partial z}$

Gradient

Del (∇) applied to vector function ∇ f is defined as the gradient of the scalar point function f and written as grad f. thus gradient f defined as,

$$\text{Grad f} = \nabla f = I\frac{\partial f}{\partial x}+J\frac{\partial f}{\partial y}+K\frac{\partial f}{\partial z}$$

Note: Grad f is a vector normal to surface f = constant, and has a magnitude equal to the rate of charge of along this normal.

1.2.9 DIRECTIONAL DERIVATIVE

If δr denotes the length pp' and N' is a unit vector in the direction pp' then the limiting value of [δf/δr] as δr → 0 (i.e. ∂f/∂r) is known as directional derivative of f at p along the direction PP'

Since δr = δn | cos α = δn | N.N'

$$\therefore \frac{\partial f}{\partial r} = \underset{\delta r \to 0}{Lt} \left[N.N' \frac{\delta f}{\delta r} \right] = N'. \frac{\partial f}{\partial n} \quad N = N'\nabla f$$

Thus the directional derivative of f in the direction of N' is the resolved of f in the direction N'.

Since ∇f. N' = | ∇ f | cos α ≤ | ∇f |

It is follows that ∇f gives the maximum rate of charge of f.

Divergence

The divergence of a continuously differentiable vector point function F is denoted by divergence F and is define by the equation.

$$div\ F = \nabla.F = I\ .\frac{\partial F}{\partial x} + J\cdot\frac{\partial F}{\partial y} + K.\frac{\partial F}{\partial z}$$

if F = ƒI + ϕJ + ψK

then

$$div\ F = \nabla.F = \left(I\ .\frac{\partial F}{\partial x} + J\frac{\partial F}{\partial y} + \quad K\ .\ \frac{\partial F}{\partial z} \right) (fI + \phi\ J + \Psi K)$$

$$div\ F = \left(I\ .\frac{\partial f}{\partial x} + J\frac{\partial\ \phi}{\partial y} + K\ .\ \frac{\partial \Psi}{\partial z} \right)$$

Curl

The curl of a continuously differentiable vector point function F is defined by the equation.

$$\text{Curl},\ F = \nabla\times F = I\times\frac{\partial F}{\partial x} + J\times\frac{\partial F}{\partial y} + K\times\frac{\partial F}{\partial x}$$

If F = ƒI + ϕJ + ψK

$$F = \nabla\ \text{x}\ F = \left(I\ .\frac{\partial}{\partial x} + J\ \frac{\partial}{\partial y} + \quad K\ .\frac{\partial}{\partial z} \right) \text{x}\ (fI + \phi\ J + \Psi K)$$

$$= \begin{vmatrix} I & J & K \\ \frac{\partial}{\partial x} & \frac{\partial}{\partial y} & \frac{\partial}{\partial z} \\ f & \phi & \Psi \end{vmatrix}$$

$$= I\left(\frac{\partial \Psi}{\partial y} - \frac{\partial \phi}{\partial z} \right) + J\left(\frac{\partial \Psi}{\partial y} - \frac{\partial \phi}{\partial z} \right) + K\left(\frac{\partial \phi}{\partial x} - \frac{\partial f}{\partial y} \right)$$

DEL applied twice to point function

1. Div grad f

$$f = \nabla.\ \nabla f = \nabla^2 f = \frac{\partial^2 f}{\partial x^2} + \frac{\partial^2 f}{\partial y^2} + \frac{\partial^2 f}{\partial z^2}$$

2. Curl grad f = $\nabla\times\nabla f = 0$

3. Div curl F = $\nabla.\ \nabla\times F = 0$

4. Curl F = grad div F- $\nabla^2 f$

i.e. $\nabla\times(\nabla\times F) = \nabla(\nabla.\ F) - \nabla^2 F$

5. Grad div F = curl l F + $\nabla^2 F$

i.e. $\nabla(\nabla.F) = \nabla\times\ (\nabla\times F) + \nabla^2 F$

Del (∇) applied to product of point functions

1. $\nabla\ (fg) = f\ \nabla g + g\ \nabla f$
2. $\nabla.(fG) = \nabla f.G + f\ \nabla.G$
3. $\nabla(F.G) = (F.\nabla)G + (G.\nabla)F$
   $+ F \times (\nabla \times G) + G \times (\nabla \times F)$
4. $\nabla.(F\times G) = G.(\nabla\times F) - F.(\nabla\times F)$
5. $\nabla\times(F\times G) = F(\nabla.G) - G(\nabla.F) + G.\ \nabla F - F.\nabla G$

1.2.10 LINE INTEGRAL

Consider a continuous vector function F (R) which is defined at each point of curve C in space. Divide C into n parts at the points $A = P_0, P_1, \ldots, P_1, - P_i - 1,\ P_i \ldots, P_n = B$ Let their position vectors be $R_0, R_1, \ldots, R_i - 1, R_i, \ldots R_n$. Let U_i be the position vector of any point on the are $P_{i-1}P_i$.

Now consider the sum

$$S = \sum_{i=0}^{n} F(U_i).\delta R,\ \text{where}\ \delta R_i = R_i - R_{i-1}$$

The *limit of this sum as n → ∞ in such a way* that | δRi | → 0, provided it exists, is called the tangential line integral of **F (R)** along C and is symbolically written as.

$$\int_C F(R).dR\ \ or\ \int_C F.\ \frac{dR}{dt}dt\ .$$

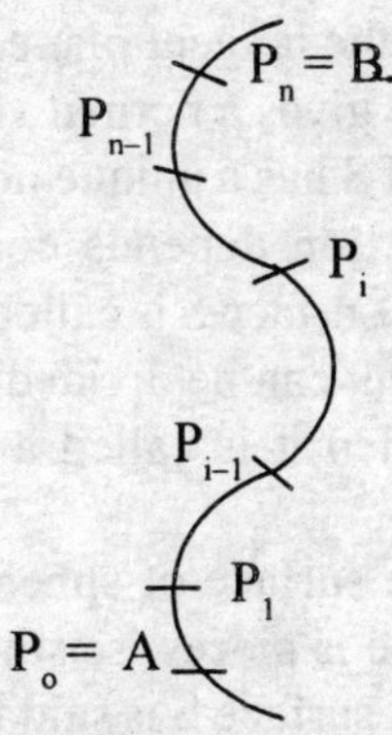

When the path of integration is a closed curve, this fact is denoted by using ∮ in place of ∫.

If F(R) = I f(x, y, z) + J ϕ (x, y, z) + Kψ (x, y, z) and dR = Idx + Jdy + Kdz.

then $\int_C$ F(R). dR = $\int_C$ (ƒdx + ϕdy + ψdz).

Two other types of line integrals are $\int_C F \times dR$ and $\int_C f\,dR$ which are both vectors.

Circulation: If F represents the velocity of a fluid particle then the line integral $\int_C F.\,dR$ is called the circulation F around the curve. When the circulation of F around every closed curve in a region E vanishes, F is said to be irrigational in E.

Work: If F represents the force acting on a particle moving along an arc AB then the work done during the small displacement δR. = F. δR.

∴ The total work done by F during the displacement from A to B is given by the line integral $\int_A^B F.dR$.

1.2.11 SURFACES INTEGRAL

As seen in chapter 4, a surface S may be represented by F (x, y, z) = 0.

The parametric representation of S is of the form R (u, v) = x (u, v)I + y (u, v) J + z (u, v) K and the continuous functions u = ϕ (t) and v = ψ (t) of a real parameter represent a curve C on this surface S.

For example, the parametric representation of the circular cylinder $x^2 + y^2 = a^2$, $-1 \le z \le 1$, (radius a and height 2) is R (u, v) = a cos u I + a sin u J + vK.

Where the parameters u and v vary in the rectangle $0 \le u \le 2\pi$ and $-1 \le v \le 1$. Also y = t, v = bt represent a circular helix on this circular cylinder . The equation of the circular helix is R = a cos t I + a sin t J + bt K.

Differentiating R = R (u, v), w.r.t. t, we get.

$$\frac{dR}{dt} = \frac{\partial R}{\partial u}\cdot\frac{du}{dt} + \frac{\partial R}{\partial v}\cdot\frac{dv}{dt}$$

The vectors (∂R / ∂u) and (∂R /∂v) are tangential to S at P and determine the tangent plane of S at P.N = (∂R/ ∂u) x (∂R /∂v) (≠ 0) gives a normal vector N of S at P.

Definition (1): If S has a unique normal at each of its points whose direction depends continuously on the points of S. then the surface S is called a smooth surface. If S is not smooth but can be divided into finitely many smooth portions, then it is called a *piecewise smooth surface*.

For instance, the surface of sphere is smooth while the surface of a cube is piecewise smooth.

Definition (2): A surface S is said to be orient able or two sided if the positive normal direction at any point P of S can be continued in a unique and continuous way to the entire surface. If the positive direction of the normal is reversed as we move around a curve on S passing through P, then the surface generate is non - orient able.

Surface Integral

Consider a continuous function F (R)and a surface S. Divide S into a finite number of sub- surfaces . Let the surface element surrounding any point P(R) be δS which can be regarded as a vector; its magnitude being the area and its direction that of the outward normal to the element.

Consider the sum ΣF(R). δS, where the summation extends over all the sub-surfaces. The limit of this sum as the number of sub-surfaces tends to infinity and the areas of each sub-surface tends to zero, is called the normal surface integral of F(R) over S and is denoted by $\int_S F.dS$ or $\int_S F.\,N ds$ where N is a unit outward normal at P to S.

Other types of surface integrals are $\int_S F \times dS$ and $\int_S fds$ which are both vectors.

Notation: Only one integral sign in when there is one differential (say dR or dS) and two (or three) signs when there are two (or three) differentials.

Flux Across a Surface

If F represents the velocity of a fluid particle then the total outward flux of F across a closed surface S is the surface integral $\int_S F.dS$.

When the flux of F across every surface S in a region E vanishes, F is said to be a solenoid vector point function in E.

It may be noted that F could equally well be taken as any other physical quantity, e.g. gravitational force, electric forces and magnetic force.

1.2.12 VOLUME INTEGRAL

Consider a continuous vector F(R) and surface S enclosing the region E. Divide E into finite number of sub-regions $E_1, E_2, \ldots E_n$. Let δvi be the volume of the sub-region Ei enclosing any point whose position vector is R_i.

Consider the sum $V = \sum_{i=1}^{n} F(R_i)\,\delta v_i$

The limit of this sum as n → ∞ in such a way that δvi → o, is called the volume integral of F(R) over E and is symbolically written as $\int_E Fdv$.

If F(R) = f (x, y, z) I + ϕ (x, y, z) J + ψ (x,y,z) K so that δv = δxδyδz, then

$$\int_E Fdv = I \iint \int_E fdxdydz + J \iint \int_E \phi dxdydz + K\,\psi dxdydz$$

1.2.13 STOKES THEOREM

$$\oint_C \cdot \overline{A}\cdot\overline{dl} = \iint_s (\nabla \times \overline{A})\,ds$$

The close line integral of any vector $\overline{A}$ integral over any close curve C is always equal to the surface integral of the curl of vector A integrated over the surface area S which is in closed by the closed curve C

1.2.14 GREEN THEOREM

Let ϕ (x, y) . ψ (x, y), ϕy and ψx be continuous in a region E of the xy plane bound by a closed curve C, then

$$\int_C \phi \; dx + \psi dy = \iint_E \left(\frac{\partial \psi}{\partial x} - \frac{\partial \phi}{\partial y} \right) dx \; dy \quad ...(1)$$

1.2.15 GAUSS'S DIVERGENCE THEOREM

Relation between Surface and Volume Integrals

Definition

This theorem states that a the volume integrals of a vector field F taken over a any volume V is equal to the surface integral of F taken over the closed surface surrounding the volume V.

Mathematically, for any vector-field F

$$F \int_S F \cdot N \; ds = \int_E \text{div } F dv$$

Where N is the unit external vector

If F(R) = f (x, y, z)I + ϕ (x, y, z) J + ψ (x, y, z) k

Then,

$$\int \int (F \; dy \; dz + \phi \; dzdx + \psi \; dx \; dy)$$

$$\int \int \int_E \left(\frac{\partial f}{\partial x} + \frac{\partial \phi}{\partial y} + \frac{\partial \psi}{\partial z} \right) dx \; dy \; dx \quad ...(1)$$

1.3 DIFFERENTIAL EQUATIONS

Equations in which unknown function or the vector function appears under sign of derivative are called differential equations.

Following are examples:

$$\frac{dx}{dt} = kx$$

$$m\frac{d^2r}{dt^2} = F\left(t, r, \frac{dr}{dt}\right)$$

$$\frac{d^2u}{dx^2} + \frac{d^2y}{dy^2} + \frac{d^2u}{dz^2} = 4\pi \rho (x, y, z)$$

If the function which is unknown like x in (1), r in (2) or u in (3) is a function of one variable the differential equation is ordinary. In (1) x is function of t only. Similarly in (2) r is function of t only. Hence (1) and (2) are ordinary differential equations. (3) is a partial differential equation in which u is function of x, y and z.

The *order* of highest derivative is the *order* of the differential equation. Thus (1) is first order while (2) and (3) are second order differential equations.

1. 3.1 FIRST ORDER EQUATION (LINEAR AND NON LINEAR)

A first order linear differential equation is linear in the unknown function and its derivative, e.g.

$$\frac{dy}{dx} + p(x)y = f(x)$$

p(x) and f(x) are continuous function of x if f(x) = 0 in

1. then it is $\frac{dy}{dx} + p(x)y = 0$
2. Is homogeneous linear equation
3. Can be separated in variables lime

$$\frac{dy}{y} = - p(x) \; dx$$

The two sides then can integrated separately.

It is possible to reduce a given be equation in variable separable form by changing variables. This will be demonstrated in solved examples.

An equation of the form

$$M(x, y) \; dx + N(x, y) \; dy = 0$$

Is called *exact differential equation* if left hand side is *total differential* of some function u(x, y) where y is a function of x. y (x) is the solution of the equation . In this case

$$du = M \; dx + N \; dy = 0$$

$$\text{But, } du = \frac{\partial y}{\partial x} dx + \frac{\partial u}{\partial y} dy$$

$$\therefore \quad \frac{\partial u}{\partial x} = M(x, y) \text{ and } \frac{\partial u}{\partial y} = N(x, y)$$

Principle of Supper position whereby two solutions of an equations may be combined linearly to obtain yet another solution is applicable to linear homogeneous equations.

A general form of second PDE is

$$A\frac{\partial^2 u}{\partial x^2} + B\frac{\partial^2 u}{\partial x \, \partial y} + C\frac{\partial^2 u}{\partial y^2}$$

$$+D\frac{\partial u}{\partial x}+E\frac{\partial u}{\partial y}+Fu=G$$

Where A, B, C, D, E, F and G are functions of x and y.

If G is identically zero, the above equations is *homogenous*.

The equation is elliptic if $B^2 - 4AC < 0$

Hyperbolic if $B^2 - 4AC > 0$

Parabolic if $B2 - 4AC = 0$

For obtaining solution of a linear homogenous equations of second order one can remember the following:

The equation $\frac{d^2y}{dx^2}+a\frac{dy}{dx}+by=0$

Write corresponding algebraic equation $D^2 + aD + b = 0$

The roots of the algebraic equations may be called m1 and m2 and they assume three distinct values depending upon the constants a and b. Based upon these values solution of above equation can be written.

1. If m_1 and m_2 are real and distinct, i.e. $a^2 - 4b > 0$ the solution is $y = C_1 e^{m1x} + C_2 e^{m2x}$.
2. If m_1 and m_2 are real and equal, i.e. $a^2 - 4b = 0$ the solution is $y = C_1 e^{m1x} + C_2 xe^{m2x}$.
3. If m_1 and m_2 are complex, i.e. $m_1 = p + q_{i,}$ $m_2 = p - q_{i,}$ i.e. when $a_2 - 4b < 0$ the solution is $y = e^{px}(C_1 qx + C_2 \sin qx)$.
   where $p = -(a/2)$, $q = \sqrt{[(b - a^2)/4]}$

Differential equations of ordinary and partial type arise in several physical phenomena. The examples are vibration, fluid mechanics, heat flow, stability of structure, motion of particles and rigid bodies. Several of them were dealt with in earlier sections The method of their solutions will be illustrated in solved examples.

Some important partial differential equations are described below:

One dimensional heat equation $\frac{\partial u}{\partial t}=c^2\frac{\partial^2 u}{\partial x^2}$

u is temperature along length (i.e. x-axis). Temperature is u(t, x), c^2 is a constant which is directly proportional to thermal conductivity (k), inversely proportional to specific heat (σ) and density (ρ).

A three dimensional form is

$$\frac{\partial u}{\partial t}=c^2\nabla^2 u \text{ where } \nabla^2=\frac{\partial^2}{\partial x^2}+\frac{\partial^2}{\partial y^2}+\frac{\partial^2}{\partial z^2}$$

One dimensional wave equations or equation to vibration to string $\frac{\partial^2 u}{\partial t^2}=c^2\frac{\partial^2 u}{\partial x^2}$

u is the displacement from mean position and function of time (t) and distance (x) from origin

Two dimensional Laplace equation

$$\frac{\partial^2 u}{\partial x^2}=c^2\frac{\partial^2 u}{\partial y^2}=0$$

Two dimensional Poisson's equation

$$\frac{\partial u^2}{\partial x^2}+\frac{\partial u^2}{\partial x^2}=f(x, y)$$

Three dimensional Laplace equation

1.3.2 HIGHER ORDER LINEAR DIFFERENTIAL EQUATIONS WITH CONSTANT COEFFICIENT

Linear Equations

A differential equation is said to be linear if the dependent variable and its differential coefficients occur only in the first degree and not multiplied together.

Thus, the standard form of a linear equation of the first order, commonly known as Leibnitz's linear equation's* is $\frac{dy}{dx}+Py=Q$

Where *P, Q* are the function of *x*,

To solve the equation, multiply both side by $e^{\int Pdx}$ so that we get

$$\frac{dy}{dx}\cdot e^{\int Pdx}+y\left(e^{\int Pdx}P\right)=Qe^{\int Pdx}$$

i.e. $\frac{d}{dx}\left(ye^{\int Pdx}\right)=Qe^{\int Pdx}$

Integrating both side, we get $ye^{\int Pdx}=\int Qe^{\int Pdx}dx+c$ as the required solution.

Bernoulli's Equation

The equation $\frac{dy}{dx}+Py=Qy^n$ ...(1)

Where *P, Q* are functions of *x*, is reducible to the Leibniz's linear equation and is usually called the Bernoulli's equations.

To solve (1), divide both side by y^n, so that

$$y^{-n}\frac{dy}{dx}+Py^{1-n}=Q$$

Put $y^{1-n}=z$ so that $(1-n)y^{-n}\frac{dy}{dx}=\frac{dz}{dx}$.

(2) becomes $\frac{1}{1-n}\frac{dz}{dx}+Pz=Q$

or $\frac{dz}{dx}+P(1-n)z=Q(1-n)$

Which is Leibniz's linear in z and can be solved easily.

1.3.3 METHOD OF VARIATION OF PARAMETERS

Two other Methods of Finding (PI)

1. Method of Variation Parameters

This method is quite and applies to equation of the form

$$y''+py'+qy=X \quad ...(1)$$

where p, q, and X are functions of x. It gives

$$P.I.=-y_1\int\frac{y_2x}{w}dx+y_2\int\frac{y_1X}{W}dx \quad ...(2)$$

where and are the solutions of

$$y''+py'+qy=0 \quad ...(3)$$

and $W=\begin{vmatrix} y_1 & y_2 \\ y_1' & y_2' \end{vmatrix}$ is called the woolskin* of y_1, y_2,

proof. Let the CF of (1) be $y=c_1y_1+c_2y_2$

Replacing c_1, c_2 (regarded as parameters) by unknown function u (x) and v (x), Let the PI be

$$y=uy_1+vy_2 \quad ...(4)$$

Differentiating (4) w.r.t. x, we get

$$y'=uy'_1+vy'_2+u'y_1+v'y_2=uy'_1+vy'_2 \quad ...(5)$$

on assuming that $u'y_1+v'y_2=0$ ...(6)

Differentiate (4) and substitute in (1). Then noting that y_1 and y_2, satisfy (3), we obtain $u'y'_1+v'y'_2=X$ ...(7)

Solving (6) and (7), we get

$$u'=-\frac{y_2x}{W},\quad v'=\frac{y_1X}{W}$$

where $W=y_1y_2-y_2y_1$

Integrating $u=-\int\frac{y_2X}{W}dx$, Substituting these in (4), we get (2).

2. Method of Undetermined Coefficients

1. To find the PI of $f(D)y=x$, we assume a trial solution containing unknown constants which are determined by substitution in the given equation. The trial solution to be assumed in each case, depends on the form of X. Thus when (i) $X=2e^{3x}$, trial solution $=ae^{3x}$
2. $X=3\sin 2x$, trial solution $=a_1\sin 2x+a_2\cos 2x$
3. $X=2x^3$, trial solution $a_1x^3+a_2x^2+a_3x+a_4$

However when $X=\tan x$ or sec x, this method fails, since the number of terms obtained by differentiating X = tan x or sec x is infinite.

The above method hold so long as no term in the trial solution appears in the CF. If any terms of the trial solution appears in the CF, We multiply this trial solution by the lowest positive integral power of x which is large enough so that none of the terms which are then present, appear in the CF.

1.3.4 CAUCHY'S AND EULER'S EQUATIONS

A linear differential equation of the form

$$x^n\frac{d^ny}{dx^n}+a_1\frac{d^{n-1}}{dx^{n-1}}+\ldots a_nY=X \quad ...(1)$$

$$x^nD^n+a_1x^{n-1}D^{n-1}+\ldots a_nY=X \quad ...(2)$$

Where $a_1, a_2,\ldots a_n$ are constants and X is a function of x or constant is called **Homogeneous Linear Differential Equation**. Also known as **Cauchy-Euler Equations.**

To solve the equations of type (1) and (2) change of variables is generally employed.

Substituting $x=e^z$, i.e. $z=\log_e x$ we can transform (1) and (2) into linear equation with constant coefficients.

Using $x=e^z$

$$\frac{dy}{dx}=\frac{dy}{dz}\frac{dx}{dz}=\frac{1}{x}\frac{dy}{dz}$$

$$\frac{xdy}{dx}=\frac{dy}{dz}\text{ or } xD=D_1\left(D_1=\frac{d}{dz}\right) \quad ...(3)$$

$$\frac{d^2y}{dx^2}=\frac{d}{dx}\left(\frac{1}{x}\frac{dy}{dz}\right)$$

$$=\frac{-1}{x^2}\frac{dy}{dz}+\frac{1}{x}\frac{d}{dx}\left(\frac{d}{dz}\right)$$

$$=\frac{-1}{x^2}\frac{dy}{dz}+\frac{1}{x}\frac{d^2y}{dz^2}\cdot\frac{1}{x}$$

$$=\frac{-1}{x^2}\frac{dy}{dz}+\frac{1}{x^2}\frac{d^2y}{dz^2}$$

$$=\frac{1}{x^2}\left(D_1^2-D_1\right)y$$

$$\frac{x^2d^2y}{dx^2}=\left(D_1^2-D_1\right)y=D_1\left(D_1-1\right)y$$

$$\frac{x^nd^ny}{dx^n}=D_1\left(D_1-1\right)\left(D_2-2\right)\ldots\left(D_1-n+1\right)y$$

Substituting the above values in the original we get a linear differential equation that can be solved for y and z. finally z can be re-substituted by x using the substitution $z=\log_e x$.

Equations Reducible to Homogeneous Form

A differential equation of the type

Where a, b, a_1, a_n are constant and X is a function of x. can be reduced to Cauchy Euler's equation by substituting $(a+bx)=v$.

We get on substitution for various derivative, the following equation:

$$\frac{v^n d^n y}{dv^n}+a_1 v^{n-1}\frac{d^{n-1}}{dv^{n-1}}+\ldots\frac{a_n}{b_n}y=\frac{V}{b^n}$$

Where V is x expressed in terms of v.

1.3.5 INITIAL AND BOUNDARY VALUE PROBLEMS

When solving concrete problem, we shall most often be concerned with particular solutions specified by the so called initial conditions which are set in order to isolate the solutions we are interested in.

An initial condition for a first order equation is usually introduced by setting a pair a values and of the independent variable x and of the unknown function y which corresponding to each other. This condition is written as $y|x=x_o=y_o$ ...(1)

A differential equation along wit the initial condition (1) is called an initial value problem.

The initial conditions for a second order equation are usually of the form

$$y|_{x=x_o}=y_o, y'|_{x=x_o}=y'_o$$

Where y_o, a y'_o nd are some given numbers.

1.3.6 PARTIAL DIFFERENTIAL EQUATIONS AND VARIABLE SEPARABLE METHOD

In our discussion related to partial differential equations we will take x an y as the independent variable and z as the dependent variable such that $z=f(x,y)$. We will use the following notation in our discussions.

$$\frac{\partial z}{\partial x}=p,\frac{\partial z}{\partial y}=q,\frac{\partial^2 z}{\partial x^2}=r,\frac{\partial^2 z}{\partial x\partial y}=s,\frac{\partial^2 z}{\partial y^2}=t$$

The partial differential equation can be formed either by the elimination of arbitrary constant or by the elimination of arbitrary functions from a relation involving three or more variables.

Linear Equation of First Order

They are also called Lagrange's linear equation form for these equations is

$$Pp+Qq=R \quad ...(1)$$

Where *P, Q* and *R* are function of *x, y* and *z*. When *P, Q* and *R* independent of *z*, the equation is known as linear equation otherwise it is called quasi-linear equations.

Working rule to solve $Pp+Qq=R$.

1. Form the subsidiary equations $\frac{dx}{P}=\frac{dy}{Q}=\frac{dz}{R}$
2. Solve these simultaneous equation giving $u=a$ and $v=b$ as its solutions.
3. Write the complete solution as $\phi(u,v)=0$ or $u=f(v)$

Exact Differential Equations

1. Def.: A differential equation of the form $M(x,y)\,dx+N(x,y)\,dy=0$ is said to be exact of lits left hand member is the exact differential of some function $u(x,y)$ *i.e.* $du=Mdx+Ndy=0$. Its solution, therefore, is $u(x,y)=c$.

2. Theorem: The necessary and sufficient condition for the differential equation $Mdx+Ndy=0$ to be exact is

$$\frac{\partial M}{\partial y}=\frac{\partial N}{\partial x}$$

Condition is necessary:

The equation $Mdx+Ndy=0$ will be exact, if

$$Mdx+Ndy=du \quad ...(1)$$

where *u* is some function of *x* and *y*

But $$du=\frac{\partial u}{\partial x}dx+\frac{\partial u}{\partial y}dy \quad ...(2)$$

Equating coefficient of *dx* and *dy* in equation (1) and (2)

we get $M=\frac{\partial u}{\partial x}$ and $N=\frac{\partial u}{\partial y}$

$\frac{\partial M}{\partial y}=\frac{\partial N}{\partial x}$ which is the necessary condition for exactness.

Condition is sufficient: i.e. if $\frac{\partial M}{\partial y}=\frac{\partial N}{\partial x}$ then $Mdx+Ndy=0$ is exact.

then solution of is $Mdx+Ndy=0$

$$\int Mdx+\int(\text{terms of N not containing } x)\,dy=c$$

Provided $\frac{\partial M}{\partial y}=\frac{\partial N}{\partial x}$

Equations Reducible to Exact Equations

Sometimes a differential equation which is not exact can be made so on multiplication by a suitable factor called an integrating factor. The rules for finding integrating factors of the equating $Mdx+Ndy=0$ are as follows:

(1) IF found by inspection: In a number of cases, the integrating factor can be found after regrouping the terms of the equation and recognizing each group as being a part of an exact differential. In this connection the following integral combinations prove quite useful:

$$xdy + ydx = d(xy)$$

$$\frac{xdy - ydx}{x^2} = d\left(\frac{y}{x}\right)$$

$$\frac{xdy - ydx}{xy} = d\left[\log\left(\frac{y}{x}\right)\right]$$

$$\frac{xdy - ydx}{y^2} = -d\left(\frac{y}{x}\right)$$

$$\frac{xdy - ydx}{x^2 + y^2} = d\left(\tan^{-1}\frac{y}{x}\right)$$

$$\frac{xdy - ydx}{x^2 - y^2} = d\left(\frac{1}{2}\tan\frac{x+y}{x-y}\right)$$

(2) IF of a homogeneous equation: If $Mdx + Ndy = 0$ be a homogeneous equations in x and y, then $\frac{1}{Mx + Ny}$ is an integrating factor $Mx + Ny \neq 0$

(3) IF for an equation of the type:

$f_1(xy)\,ydx + f_2(xy)\,xdy = 0$

If the equation be of this form, then $\frac{1}{Mx - Ny}$ is an integrating factor ($Mx - Ny \neq 0$).

(4) In the equation: $Mdx + Ndy = 0$

(a) if $\dfrac{\frac{\partial M}{\partial y} - \frac{\partial N}{\partial x}}{N}$ be a function of x only $= f(x)$, then $e^{\int f(x)dx}$ is an integrating factor.

(b) if $\dfrac{\frac{\partial N}{\partial x} - \frac{\partial M}{\partial y}}{M}$ be a function of y only $= F(y)$ say then $e^{\int F(y)dy}$ is an integrating factor.

(5) For the equation of the type

$x^a x^b(my\,dx + nx\,dy) + x^{a'} x^{b'}(m'y\,dx + n'x\,dy) = 0$ an integrating factor is $x^h y^k$

where $\frac{a+h+1}{m} = \frac{b+k+1}{n}$

$$\frac{a'+h+1}{m'} = \frac{b'+k+1}{n'}$$

1.4 COMPLEX VARIABLES

1.4.1 ANALYTIC FUNCTIONS

A function f(z) which is single-valued and possesses a unique derivative with respect to z at all points of regions R, is called an analytic or regular function of z in that region.

A point at which an analytic function cease to possess a derivative is called a singular point of the function. Thus if u and v are real single-valued functions of x and y such that $[\partial u / \partial x]$, $[\partial u / \partial y]$, $[\partial v / \partial x]$ $[\partial v / \partial y]$ are continuous thought out a region R, then the Cauchy-Riemann equation.

$$\frac{\partial u}{\partial x} = \frac{\partial v}{\partial y} \text{ and } \frac{\partial u}{\partial y} = -\frac{\partial v}{\partial x}$$

are both necessary are sufficient conditions for the function f(z) = u + iv to be analytic in R.

The real and imaginary parts of an analytic function are called conjugate functions. The relation between two conjugate functions is given by C-R equation (1).

Harmonics Function

If $f(z) = u + iv$ be a analytic function in some region of the z-plane, then the Cauchy-Riemann equations are satisfied.

i.e. $$\frac{\partial u}{\partial x} = \frac{\partial v}{\partial y} \quad ...(1)$$

and $$\frac{\partial u}{\partial x} = -\frac{\partial v}{\partial y} \quad ...(2)$$

Differentiating (1) with respect of x and (2) with respect to y, we obtain

$$\frac{\partial^2 u}{\partial x^2} = \frac{\partial^2 v}{\partial x \partial y} \quad ...(3)$$

and $$\frac{\partial^2 u}{\partial x^2} = -\frac{\partial^2 v}{\partial y dx} \quad ...(4)$$

Adding (3) and (4) and assuming that $\frac{\partial^2 v}{\partial x \partial y} = \frac{\partial^2 v}{\partial y \partial x}$, we get $\frac{\partial^2 u}{\partial x^2} + \frac{\partial^2 u}{\partial y^2} = 0$

Similarly, by differentiating (1) with respect to y and (2) with respect to x and subtracting we obtain

$$\frac{\partial^2 v}{\partial x^2} + \frac{\partial^2 v}{\partial y^2} = 0$$

Thus both the functions u and v satisfy the Laplace's equation in two variables. For this reason, they are known as harmonic functions and their theory is called potential theory.

Orthogonal System

Consider the two families of curves

$$u(x, y) = c_1$$

and $\quad v(x, y) = c_2$

Differentiating (7) we get $\dfrac{\partial u}{\partial x} + \dfrac{\partial u}{\partial y} \cdot \dfrac{dy}{dx} = 0$

Or $\quad \dfrac{dy}{dx} = -\dfrac{\partial u/\partial x}{\partial u/\partial y}$

$$= \frac{\partial v/dy}{\partial v/dx} = m_1 \text{ (say)}$$

Similarly (8) gives $\dfrac{dy}{dx} = -\dfrac{\partial v/\partial x}{\partial v/\partial y} = m_2 \text{ (say)}$

$m_1 m_2 = -1$, i.e. (7) and (8) form orthogonal system.

Hence every analytic function $f(z) = u + iv$ defines two families of carves $u(x, y) = c_1$ and $v(x, y) = c_2$ which form an orthogonal system.

1.4.2 CAUCHY'S THEOREM

If f (z) is an analytic function and f'(z) is continuous at each point each within or on a closed curve C, then $\int_C f(z)\, dz = 0$

Writing f(z) = u(x, y) + iv(x, y) and noting that dz = dx + idy

$\int_C f(z)\, dz = C\,(udx - vdy) + i \int_C (vdx + udy)$ (1) Since f'(z) is continuous, therefore,

$$\frac{\partial u}{\partial x} = \frac{\partial v}{\partial y} \text{ and } \frac{\partial u}{\partial y} = -\frac{\partial v}{\partial x}$$

are also continuous in the region D enclosed by C. Hence the Green's theorem can be applied to (1), giving

$$\int_C f(z)\, dz = \begin{pmatrix} -\iint_D \left(\dfrac{\partial v}{\partial x} + \dfrac{\partial u}{\partial y}\right) dx\, dy \\ +i \iint_D \left(\dfrac{\partial u}{\partial x} - \dfrac{\partial y}{\partial y}\right) dx\, dy \end{pmatrix}$$

Now f(z) being analytic, u and v necessarily satisfy the Cauchy-Riemann equation and thus the integrands of the two double integrals in (2) vanish identically

Hence, $\int_C f(z)\, dz = 0$

1.4.3 CAUCHY'S INTEGRAL FORMULA

If f(z) is analytic within and on a closed curve and if a is any point with C, then

$$f(a) = \frac{1}{2\pi i} \int_C \frac{f(z)\, dz}{z - a}$$

Consider the function [f(z) / {z - a }] which is analytic at all points within C except at z = a. With the point a as center and radius r, draw a small circle C_1 lying entirely within C.

Now [f(z) / {z - a }] being analytic in the region enclosed by C and C_1, we have by Cauchy's theorem,

$$\int_C \frac{f(z)}{z - a} dz = \int_{C1} \frac{f(z)}{z - a} dz$$

For any point on C_1,

$z - a = re^{i\theta}$ and $dz = ire^{i\theta}$ d

= f (a + rei) . ire i = i f(a + rei) dθ

C1 rei C1

In the limiting form, as the circle C_1 shrinks to the point a, i.e. as $r \to 0$, the integral (1) will approach to

$$i \int_{C1} f(a)\, d\theta = if(a) \int_0^{2\pi} d\theta = 2\pi\, if(a).$$

$$\text{Thus } \int_C \frac{f(z)}{z - a} dz = \frac{2\pi\, if(a)}{z - a}$$

$$\text{i.e. } f(a) = \frac{1}{2\pi i} \int_C \frac{f(z)}{z - a} dz \qquad \text{...(2)}$$

Which is the desired Cauchy's integral formula.

Cor. Differentiation both sides of (2) w.r.t a,

$$f'(a) = \frac{1}{2\pi i} \int_C \frac{\partial}{\partial a}\left(\frac{f(z)}{z - a}\right) dz = \frac{1}{2\pi i} \int_C \frac{f(z)}{(z - a)^2} dz \qquad \text{...(3)}$$

Similarly,

$$f''(a) = \frac{2!}{2\pi i} \int_C \frac{f(z)}{(z - a)^3} dz \qquad \text{...(4)}$$

and in general,

$$f^n(a) = \frac{2!}{2\pi i} \int_C \frac{f(z)}{(z - a)^{n+1}} dz. \qquad \text{...(5)}$$

Thus it follows from the results (2) to (5) that if a function f(z) is known to be analytic on the simple closed curve C then the values of the functions and all its derivative can be found at any point of C. Incident, we have established a remarkable fact that an *analytic function possesses derivative of all orders and these are themselves all analytic*.

1.4.4 TAYLOY'S SERIES

If f(z) is analytic inside a circle C with center at a, then for z inside C,

$$f(z)= f(a) + f'(a)(z-a) + \frac{f''(a)}{2!}(z-a)2$$

$$+ \ldots \frac{f^n(a)}{n!}(z-a)^n + \ldots \qquad \ldots(1)$$

1.4.5 LAURENT'S SERIES

If f(z) is analytic in the ring-shaped region R bounded by two concentric circles C and C_1 of radii r and r_1 $(r > r_1)$ and with centre at a, then for all z in R.

$$f(z) = a_0 + a_1(z-a) + a_2(z-a)^2 + \ldots$$
$$+ a_{-1}(z-a)^{-1} + a_{-2}(z-a)^{-2} + \ldots$$

Where,

$$a_n = \frac{1}{2\pi i}\int_r \frac{f(t)}{(t-a)^{n+1}}dt$$

being any curve is R, encircling C_1

1.4.6 RESIDUE: THEOREM AND SOLUTION INTEGRAL

Residue: The co-efficient $(z-a)^{-1}$ of in the expansion of $f(z)$ around an isolated singularity is called the residue of $f(z)$ at that point. Thus in the Laurent's series expansion of $f(z)$ around, i.e..

$$f(z) = \begin{bmatrix} a_0 + a_1(z-a) + a_2(z-a)^2 + \ldots + a_{-1}(z-a)^{-1} \\ +a_{-2}(z-a)^{-2} + \ldots \end{bmatrix},$$

the residue $f(z)$ of at $z = a$ is. a_{-1}.

$$Res\ f(a) = \frac{1}{2\pi i}\int_C f(z)dz$$

i.e. $\int_C f(z)dz = 2\pi i Res\ f(a)$

Residue Theorem: If *f*(z) is analytic in a closed curve C except at a finite number of singular points within. C, then $\int_C f(z)dz = 2\pi i \times$ (Sum of the residues at the singular points within C).

Let us surround each of the singular points $a_1, a_2, \ldots, a_n$ by a small circle such that it encloses no other singular point. Then these circles $C_1, C_2, \ldots, C_n$ together with C, form a multiply connected region in which $f(z)$ is analytic.

Applying Cauchy's theorem, we have

$$\int_C f(z)dz = \int_{C_1} f(z)dz + \int_{C_2} f(z)dz + \ldots + \int_{C_n} f(z)dz$$

$= 2\pi i\,[Res\ f(a_1) + Res\ f(a_2) + \ldots + Res\ f(a_n)]$ which is the desired result.

Calculation of Residues

1. If *f*(*z*) has a simple pole at *z* = *a*, then
2. If *f*(*z*) has a pole of order *n* at *z* = *a*, then

$$Res\ f(a) = \underset{z\to a}{Lt}\left[(z-a)f(z)\right].$$

1.4.7 EVALUATION OF REAL DEFINITE INTEGRAL

An integral of the type $\int_0^{2\pi} f(\sin\theta, \cos\theta)\,d\theta$ where the integral in a rational function of $\sin\theta$, $\cos\theta$ can be evaluated by writing $e^{\sqrt{\theta}} = z$

Then

$$\sin\theta = \frac{1}{2j}\left(z - \frac{1}{z}\right)$$

$$\cos\theta = \frac{1}{2}\left(z + \frac{1}{z}\right)$$

This integral and takes the form $\int_c f(z)dz$ where *f*(*z*) in a rational function of *z* and *c* is a unit circle.

1.5 PROBABILITY AND STATISTICS

Sampling: We know that a small section selected from the population is called a sample and the process of drawing a sample is called sampling. It is essential that the sample must be a random selection so that each member of the population has the same chance of being includes in the sample. Thus the fundamental assumption underlying theory of sampling is Random sampling

A special case of random sampling in which each event has the same probability p of success and the chance of success of different events are independent whether previous trials have been made or not, is known as simple sampling.

The statistical constants of the population such as mean (μ), standard deviation (σ), etc. are called the parameters. Similarly, constants for the sample drawn from the given population, i.e. mean ($\overline{x}$), standard deviation (S), etc. are called the statistic. The population parameters are in general, not known and their estimates given by the corresponding sample statistic are used. We use the Greek letters to denoted the population parameters and Roman letters for sample statistic.

Objectives of sampling: Sampling aims at gathering the maximum information about the population with the minimum effort, cost and time. The object of sampling studies is to obtain the best possible value of the parameters under specific conditions. Sampling determines the reliability of these estimates. The logic of the sampling theory is the logic of induction in which we pass from a particular (sample) to general (population). Such a generalization from sample to population is called Statistical inference.

Sampling distribution: Consider all possible of size n which can be drawn from a given population at random . For each sample, we can compute the mean . The mean of the samples will not be identical. If we group these different means according of their frequencies, the frequency distribution so formed is known as sampling distribution of the mean. Similarly we can have sampling distribution of the standard divination, etc.

When drawing each sample, we put back the previous sample so that the parent population remains the same. This is called sampling with replacement and all subsequent formulae will pertain to sampling with replacements.

1.5.2 MEAN AND MODE

- Mode and median are also measures of central tendency of observations in addition to arithmetic mean. While median is the middle value of all observations , the mode is that observation which occurs most frequently.
  It can be shown that arithmetic mean is the best measure of central tendency
- The dispersion of a sample is measured as variance,

$$S^2 . S^2 = \frac{\sum_{i}^{n} (X_i - \overline{X}_i)^2}{n-1}$$

$$\text{or, } S^2 = \frac{n \sum_{i}^{n} X_i^2 - \left(\sum_{i}^{n} X_i\right)^2}{n(n-1)}$$

(n – 1) is the degree of freedom.

Probability Function/Frequency Function/Probability of Event

Suppose that a random experiment is repeated n times. If an event A occurs nA times, then its probability p(A) is defined as

$$p(A) = p(x_k) = \lim_{n\to\infty} \frac{n_A}{n}$$

$$= p(A) = \frac{N_A}{N}$$

= Probability of outcome A.

nA = times occur A

n = total time experiment

Note:

1. That this limit may not exist.
2. That this definition presumes that all outcomes are equally likely to occur.
   p(A + B) = p(A) + p(B) – p(AB)
   p(AB) is joint probability.

   Mutually exclusive event P(AB) = 0

   p(A + B) = p(A) + p(B)

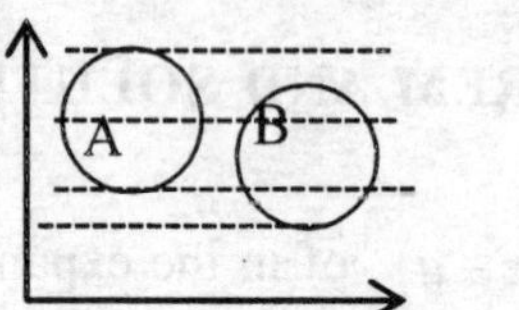

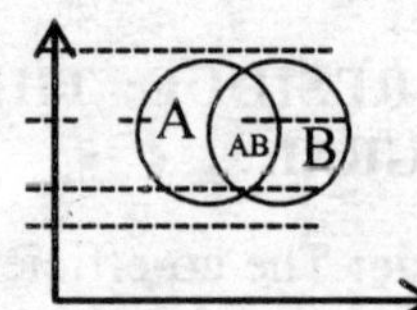

1.5.3 STANDARD DEVIATION

The most important and the most power full measure of dispersion is the standard deviation (SD) dented by σ. It is computed as the square root of the mean of the square of the difference of the variant values from their mean Standard deviation (SD)

$$\sigma = \sqrt{\frac{\Sigma f_i \left(X_i - \overline{X}\right)^2}{N}}$$

Where N is the total frequency $= \Sigma f_i$. The deviations are measured from any other value say A. Instead of $\overline{X}$, it is called the root mean square deviation. The square of the standard deviation is known as the variance.

1.5.4 CONDITIONAL PROBABILITY

p(A B) = p(A) . p(B / A)

p(AB) = p(B) . p(A/B)

$$p(A/B) = \frac{p(AB)}{p(B)}$$

$$p(A/B) = \frac{p(A) \cdot p(B/A)}{p(B)} \ldots \text{Bay's relation}$$

Where, (B/A) is probability of happening B, when A has already happened.

Statistically independent event

$$p(AB) = p(A) \cdot p(B/A)$$
$$= p(A) \cdot p(B)$$

i.e. if event A and B are not dependent on each other, the conditional probability simply equal to unconditional probability, i.e.

$$p(B/A) = p(B)$$
$$\& \; p(A/B) = p(A)$$

Repeated Trials

We know that the probability of getting a head or a tail on tossing a coin is $\frac{1}{2}$. If the coin is tossed thrice, the probability of getting one head and two tails can be combined as $H-T-T$, $T-H-T$, $T-T-H$. The Probability of each one of these being $\frac{1}{2}\times\frac{1}{2}\times\frac{1}{2}$, i.e. $\left(\frac{1}{2}\right)^3$ their total probability shall be $3\left(\frac{1}{2}\right)^3$.

Similarly if a trial is repeated *n* times and if *p* is the probability of a success and *q* that of a failure, then the probability of *r* successes and $n-r$ failures is given by $p^r q^{n-r}$ but these are successes and failures can occur in any of the ${}^nC_r$ ways in each of which the probability is same.

Thus the probability of r successes is ${}^nC_r p^r q^{n-r}$.

Cor. The probabilities of at least r successes in n trials = the sum of the probabilities of $r, r+1, \ldots, n$

$$= {}^nC_r p^r q^{n-r} + {}^nC_{r+1} p^{r+1} q^{n-r-1} + \ldots + {}^nC_n p^n$$

Cumulative distributive function (CDF)
Probability distributive function (PDF)

The distributive function yields the probability that the random variable will be less than or equal to some specified value x_j thus.

$$P(X \le x_i) = \sum_{K=1}^{i} p(x_k)$$

$$= p(x_1) + p(x_2) + p(x_3) + \ldots p(x_j)$$

Random variable for discrete signal

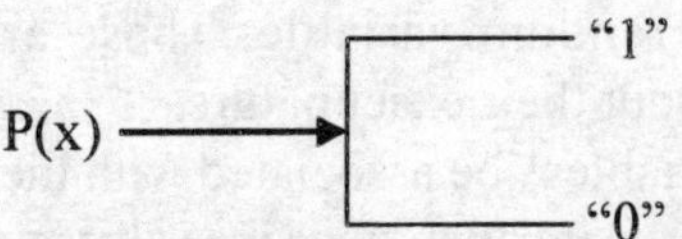

Statistical Average/Moments

Expectation of random variable/(mean)

Consider N independent observation of a discrete random variable. If the event $X = x_i$ occurs N_i time, the sum of the observation value is

$$N_1 x_1 + N x_2 \cdots N_k x_k = \frac{1}{N} \sum_{i=1}^{k} N_i x_i$$

Upon dividing the sum N and letting $N \to \infty$ the relative frequency Ni/N becomes $p(X = x_i) = px(x_i)$ Hence.

$$m = \bar{x} = E(X) = \sum_{i=1}^{k} X_k \, p(x_k)$$

This is dc value of random variable.

$$\text{Where, } p(x) = \frac{N_{i0}}{N}$$

$$\bar{x} = E(X) = p_1 x_1 + p_2 x_2 + p_3 + \ldots$$

Second Moment or Mean Square Value

Consider the random variable X which assume the possible value $x_1, x_2, x_3 \ldots x_m$. In a sequence of n experiment, let event. x1 occur n1 time, x_2 occur n_2 times, etc. The arithmetic average or mean of x_k is

$$X_{avg} = \frac{n_1 x_1 + n_2 x_2 + \cdots + x_m n_m}{n}$$

$$= \frac{1}{n} \sum_{k=1}^{m} x_k n_k$$

If $n \to \infty$ then $n_1/n \to p(x_1)$

$$= p_1 x_1 + p_2 x_2 + \ldots p_m x_m$$

$$= \sum_{k=1}^{m} p_k x_k$$

$$X^2_{avg} = \frac{n_1 x_1^2 + n_2 x_2^2 + \cdots}{N}$$

1.5.5 RANDOM VARIABLE

Many variable of interest in a system are not expressible on explicit function of time, all such variable are

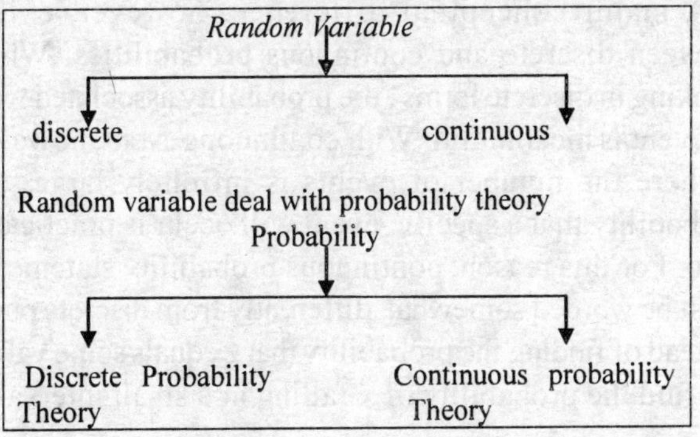

classified as random variable. These are implying uncertainty about their exact nature.

If a real variable *X* be associated with the outcome of a random experiment, then since the values which X takes depend on chance, it is called a random variable or a stochastic variable or simply a variate. For instance, if a random experiment *E* consists of tossing a pair of dice, the sum *X* of the two numbers which turn up have the value 2, 3, 4,... 12 depending on chance. Then *X* is the random variable. It is function whose values are real number and depend on chance.

If in a random experiment , the event corresponding to a number *a* occurs, then the corresponding random variable *X* is said to assume he value *a* and the probability of the vent is denoted by $P(X=a)$. Similarly the probability of the event *X* assuming any value in the interval $a>X<b$ is denoted by $a>X<b$. The probability of the event $X\le c$ is written as $P(X\le c)$.

If a random variable takes a finite set of values, it is called a discrete variate. On the other hand, if it assumes an infinite number of uncountable value, it is called a continuous variate.

1.5.6 DISCRETE PROBABILITY DISTRIBUTION

Suppose a discrete variate *X* is the outcome experiment. If the probability that X takes the values x_i is p_i, then

$$P(X=x_1)=pi \text{ or } i=1,2\ldots$$

Where

$(i)\ p(x_i)\ge 0$ for all values of *i*,

2. $\Sigma p(x_i)=1$

The set of value x_i with their probabilities p_i constitute a discrete probability distribution of the discrete variable *X*.

1.5.7 CONTINUOUS PROBABILITY DISTRIBUTION

When a variate *X* takes every value in an interval, it gives rise to continuous distribution of *X*. The distributions defined by the variates like heights or weights are continuous distributions.

A major conceptual difference, however, exists between discrete and continuous probabilities. When thinking in discrete terms , the probability associated with an event is meaningful. With continuous event, however , where the number of events is infinitely large, the probability that a specific event will occur is practically zero. For this reason, continuous probability statements must be worded somewhat, differently from discrete ones. Instead of finding the probability that z equals some value, we find the probability of x falling in a small interval.

Thus the probability distribution of a continuous variate x is defined by a function f(x) such that the probability of the variate x falling in the small interval $[x-(1/2)\,dx]$ to $[x+(1/2)\,dx]$ is f(x) dx. Symbolically it can be expressed as $P[x-(1/2)\,dx\le x\le x+(1/2)\,dx]=f(x)\,dx$. Then f(x) is called the probability density function and the continuous curve = f(x) is called the probability curve.

The range of the variable may be finite or infinite. But even when the range is finite, it is convenient of consider it as infinite by supposing the density function to be zero outside the given range. Thus if $f(x)=\phi(x)$ be the density function denoted for the variate x in the interval (a, b), then it can be written as

$$\begin{aligned} f(x) &= 0, & x<a \\ &= \phi(x) & a\le x\le b \\ &= 0, & x>b \end{aligned}$$

The density function f(x) is always positive and $\int_{-\infty}^{\infty} f(x)\,dx = 1$ (i.e. the total area under the probability curve and the x-axis is unity which corresponds to the requirements that the total probability of happening of an event is unity).

1.5.8 POISSON DISTRIBUTION

It is a distribution related to the probabilities of events which are extremely rare, but which have a large number of independent opportunities for occurrence.

This distribution can be derived as limiting case of the binomial distribution by making n very large an p very small keeping np fixed (= m, say)

The probability of r successes in a binomial-distribution is

$$P(r) = {}^nC_r p^r q^{n-r}$$

$$= \frac{n(n-1)(n-2)\ldots(n-r+1)\,p^r q^{n-r}}{r!}$$

$$= \frac{np(np-p)(np-2p)\ldots(np-r-1p)(1-p)^{n-r}}{r!}$$

As $n\to\infty$, $p\to 0$ (np = m), we have

$$p(r) = \frac{m^r}{r!} \operatorname*{Lt}_{n\to\infty} \frac{(1-m/n)^n}{(1-m/n)^r} = \frac{m^r e^{-m}}{r!}$$

So that the probabilities of 0, 1, 2…, r,… successes in poisson distribution are give by

$$e^{-m},\ me^{-m} = \frac{e^{-m}}{2!},\ldots \frac{m^r e^{-m}}{r!},\ldots$$

The sum of these probabilities is unity as it should be.

1.5.9 NORMAL (OR GAUSSIAN) DISTRIBUTION:

A random variable X is called normal (or Gaussian) if its pdf is of the form.

$$f(x) = \frac{1}{\sqrt{2\pi}\sigma} \exp - \frac{(x-m)^2}{2\sigma^2}$$

The distribution function of X is

$$F(x) = \frac{1}{\sqrt{(2\pi\sigma)}} \int_{-\infty}^{x} \exp - \frac{(x-m)^2}{2\sigma^2} dx$$

$[(x - m)^2/\sigma]$

$$= \frac{1}{\sqrt{(2\pi)}} \int_{-\infty} \exp - \frac{\lambda^2}{2} d\lambda$$

This integral cannot be evaluated in a closed form and must evaluated numerically. It is convenient to use the function Q(z) , defined as

$$Q(z) = \frac{1}{\sqrt{(2\pi)}} \int_{-\infty}^{\infty} \exp - \frac{\lambda^2}{2} d\lambda$$

Then above equation can be written as

$$F(x) = 1 - Q \frac{[x - m]}{\sigma}$$

The function Q(z) is shown as the complementary error function, or simply the Q function.

The pdf and the distribution function of X are plotted in respectively The mean and variance of X are

$$m_x = E[X] = m$$
$$\sigma_x^2 = E[(X - m_x)^2] = \sigma^2$$

We shall use the notation $N(m; \sigma^2)$ to denote that X is normal (or Gaussian) with mean *m* and variance σ^2. In particular, X = N (0; 1), that is, X with zero mean and unit variance, is defined as a normalized Gaussian random variable.

$$f(x) = \frac{1}{\sqrt{(2\pi\sigma)}} \exp - \frac{(x-m)^2}{2\sigma^2}$$

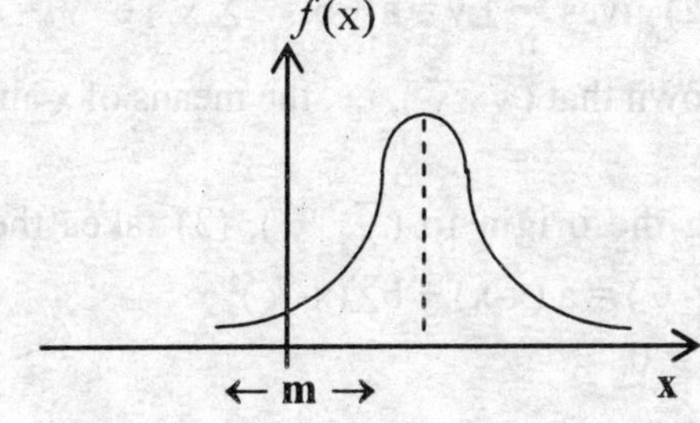

If mean = 0 i.e. m = 0

$$f(x) = \frac{1}{\sqrt{(2\pi\sigma)}} \exp - \frac{(x)^2}{2\sigma^2}$$

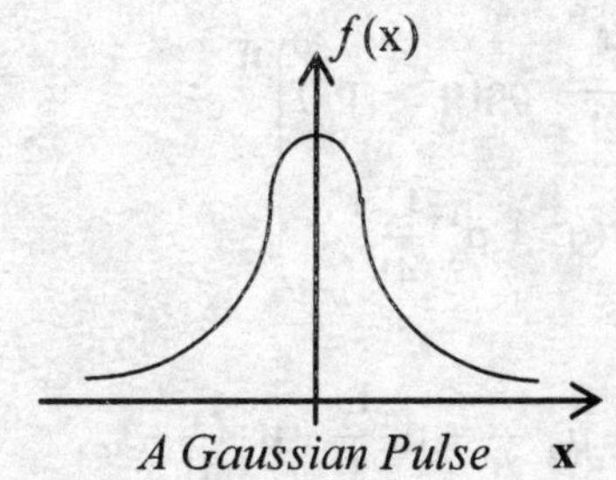

A Gaussian Pulse

Gaussian Process

$$f(y) = \frac{1}{\sqrt{(2\pi\sigma_y^2)}} \exp \frac{[-(y - m_y)^2]}{2\sigma_y^2}$$

Where, $m_y \rightarrow$ mean

$\sigma_y^2 \rightarrow$ variance

Property 1: If a Gaussian process x(t) is applied to a stable linear filter, then the random process y(t) developed. The output of the filter is also gaussian.

Property 2: Consider the set of random variables or samples $x(t_1)$, $x(t_2)$, $x(t_3)$... obtained by observing a random process x (t) at time t_1, t_2,... If the process is Gaussian then this set of random variable is to...

1.5.10 BINOMIAL DISTRIBUTION

It is concerned with trials of a repetitive nature in which only the occurrence or non-occurrence, success or failure, acceptance or rejection, yes or no of a particular event is of interest

If we perform a series of independent trials such that for each trial p is the probability of success and q that of a failure, then the probability of r successes in a series of n trials is gives by ${}^nC_r\, p^r q^{n-r}$, where r takes any integral value from 0 to n. The probabilities of 0, 1, 2,…r,…,n successes are therefore , given by q^n, ${}^nC_1\, p\, q^{n-1}$, ${}^nC_2\, p^2q^{n-2}$, … ${}^nC_r\, p^rq^{n-r}$, …,pn.

The probability of the number of successes so obtained is called the binomial distribution for the simple reason that the probability are the successive terms in the expansion of the binominal $(q + p)^n$.

∴ The sum of the probabilities

$$= q^n + {}^nC_1\, p\, q^{n-1} + {}^nC_2\, p^2q^{n-2} + \ldots + p^n = (q + p)^n = 1$$

Constants of the binomial distribution.

The moment generating function about the origin is

$$M_0(t) = E(e^{tx}) = {}^nC_1\, p^xq^{n-x}\, e^{tx}$$
$$= \Sigma^n\, C_x\, (pe^t)\, x\, q^{n-x}, = (q + pe^t)^n$$

Differentiating with respect to t and putting t and putting = 0, we get the mean given by

$$M_m(t) = e^{-npt}\, (q + pe^t)^n = (qe^{-pt} + pe^{qt})^n$$

$$= \begin{pmatrix} 1+pq\dfrac{t^2}{2!}+pq(q^2-p^2) \\ \dfrac{t^3}{3!}+pq\,(q^3+p^3)\dfrac{t^4}{4!}+.. \end{pmatrix}^n$$

Or

$$1+\mu_1 t+\mu_2\frac{t^2}{2!}+\mu_3\frac{t_3}{3!}+\mu_4\frac{t_4}{4!}+...$$

$$= 1+npq\,\frac{t^2}{2!}+npq\,(q-p)\,\frac{t^3}{3!}$$

$$+\,npq\,(1+3(n-2)\,pq)\frac{t^4}{4!}+....$$

Equating the coefficients of like power of t on either side, we have

$\mu_2 = npq$, $\mu_3 = npq\,(q-p)$,

$\mu_4 = npq\,[1+3(n-2)pq]$

Also

$$\beta_1 = \frac{\mu_3{}^2}{\mu_2} = \frac{(q-p)^2}{npq} = \frac{(1-2p)^2}{npq}$$

$$\beta_2 = \frac{\mu_4}{\mu_2{}^2} = 3+\frac{1-6pq}{npq}$$

Thus mean = np, standard deviation = $\sqrt{(npq)}$

Skewness = $[(1-2p)/\sqrt{(npq)}]$, kurtosis = β_2

Obs. The skewess is positive for p(1/2) and negative p>(1/2). When p = (1/2). When p = (1/2), the skewness is zero, i.e. the probability curve of binomial distribution will be symmetrical (bell-shaped).

As n the number of trials increase indefinitely, $1 \to 0$ and $\beta_2 \to 3$.

Binomial Frequency Distribution

If n independent trials constitute one experiment an this experiment be repeated N times, then the frequency of r successes is $N^n\,C_r\,p^r\,q^{n-r}$. The possible number of successes together with these expected frequencies constitute the binomial frequency distribution.

Applications of Binomial Distribution

This distribution is applied to problems concerning:

1. Number of defectives in a sample from production line.
2. Estimation of reliability of system.
3. Number of rounds fired from a gun hitting a target.
4. Radar detection.

1.5.11 CORRELATION: AND REGRESSION ANALYSIS

So far we have confined our attention to the analysis of observations on a single variable. There are, however, many phenomena where the changes in one variables are related to the changes in the other variable. For instance, the yield of a crop varies with the amount of rainfall, the price of a commodity increase with the reduction in its supply and so on. Such a simultaneous variation, i.e. when the changes in one variable are associated or followed by changes in the other, is called correlation. Such a data connecting two variables is called bivariate population.

If an increase (or decrease) in the value corresponds to an increase (or decrease) in the other, the correlations said to be positive. If the increase (or decrease) in one correlation to the decrease (or increase) in the other, the correlation is said to be negative. If there is no relationship indicated between the variables, they are said to be independent or uncorrelated.

If frequency happens that the dots of the scatter diagram generally, tend to cluster along a well defined direction which suggests a linear relationship between the variables x and y . Such a line of best - fit for the given distribution of dots is called the line of regression (1). In fact there are two such lines, one giving the best possible mean values of y for each specified value of x and the other giving the best possible mean values for x for given values of y. The former is known as the line for regression of y on x and the latter as the line of regression of x on y.

Consider first the line of regression of y on x. Let the straight line satisfying the general trend of n dots in a scatter diagram be

$$y = a + bx \qquad ...(1)$$

We have to determine the constants a and b so that (1) gives for each value of x, the best estimate for the average value of y in accordance with the principle of least squares, therefore, the normal equations for a and be are

$$\Sigma y = na + b\Sigma x \qquad ...(2)$$

and

$$\Sigma xy = a\Sigma x + b\Sigma x^2 \qquad ...(3)$$

equation (2) gives $\frac{1}{n}\Sigma y = a + b,\ \frac{1}{n}\Sigma\, x$, i.e. $\bar{y} = a + b\bar{x}$

This shown that $(\bar{x}, \bar{y})$, i.e. the means of x and y, lie on (1).

Shifting the origin to $(\bar{x}, \bar{y})$, (3) takes the form

$\Sigma(x-\bar{x})(y-\bar{y}) = a\,(x-x) + b\Sigma(x-\bar{x})^2$

But $(x-\bar{x}) = 0$

$$b = \frac{\Sigma\,(x-x)\,(y-\bar{y})}{\Sigma\,(x-x)^2} = \frac{\Sigma XY}{\Sigma X^2} = \frac{\Sigma XY}{n\sigma_x{}^2} = r\,\frac{\sigma_y}{\sigma_x}$$

$$\left(\therefore\ r = \frac{\Sigma XY}{n\sigma_x\sigma_y}\right)$$

Thus the line of best fit becomes

$$(y - \bar{y}) = r\frac{\sigma_y}{\sigma_x}(x - \bar{x}) \quad ...(4)$$

Which is the equation of the line of regression of y on x. Its slope is called the regression co-efficient of y on x Interchanging x and y, we find that the line of regression of x on y is

$$x\ \bar{x} = r\frac{\sigma_x}{\sigma_y}(y - \bar{y}) \quad ...(5)$$

Thus the regression co-efficient of y on x

$$= r\sigma_y / \sigma_z \quad ...(6)$$

and the regression co-efficient of x on y

$$= r\sigma_x / \sigma_y \quad ...(7)$$

Cor. The correlation co-efficient is the geometric mean between the two regression co-efficient.

For $= \frac{\sigma_y}{\sigma_x} \times \frac{\sigma_y}{\sigma_y} = r^2$

Rank Correlation

A group of n individuals may be arranged in order of merit respect to some characteristic. The same group would given different order for different characteristics. Considering the orders corresponding to two characteristic A and B, the correlation between these n pairs of ranks is called the rank correlation in the characteristics A and B for that group of individuals.

Let x_i, y_i be the ranks of the ith individuals in A and B respectively Assuming that no two individuals are bracketed equal in either case, each of the variables taking the values 1, 2, 3,…, n, we have

$$\bar{x} = \bar{y} = \frac{1 + 2 + 3 + ...+n}{n} = \frac{n(n+1)}{2n} = \frac{n+1}{2}$$

If X, Y be the deviation of x, y from their mean then

$$\Sigma X_i^2 = \Sigma(x^i - \bar{x})^2 = \Sigma x_i^2 + n(\bar{x})^2 - 2\,\bar{x}\,\Sigma x_i$$

$$= \Sigma n^2 + \frac{n(n+1)^2}{4} - 2\,\frac{n+1}{2}.\,\Sigma n$$

$$= \frac{n(n+1)(2n+1)}{4} + \frac{n(n+1)2}{4} - \frac{n(n+1)^2}{2}$$

$$= \frac{1}{12}(n^3 - n)$$

Similarly $\Sigma Y_i^2 \frac{1}{12} \cdot (n^3 - n)$

Now let $d_i = x_i - y_i$

so that $d_i = (x_i - x) - (y_i - \bar{y}) = X_i - Y_i$

$$\Sigma d_i^2 = \Sigma X_i^2\ 2\Sigma X_i Y_i$$

$$\Sigma X_i Y_i = \frac{1}{2}(\Sigma X_i^2 + \Sigma Y_i^2 - \Sigma d_i^2)$$

$$= \frac{1}{12}(n^3 - n) - \frac{1}{2}\Sigma d_i^2$$

Hence the correlation coefficient between these variables is

$$r = \frac{\Sigma X_i Y_i}{\sqrt{(\Sigma X_i^2 Y_i^2)}} = \frac{[1/12](n^3 - n) - [1/2]\,\Sigma d_i^2}{[1/12](n^3 - n)}$$

$$= 1 - \frac{6\Sigma d_i^2}{n^3 - n}$$

This is called the rank correlation coefficient and the denoted by

1.6 NUMERICAL METHOD

1.6.1 SOLUTION OF ALGEBRAIC AND LINEAR EQUATIONS

To find the roots of an equation $f(x) = 0$, we start with a known approximate solution and apply any to the following methods:

1. Bisection method: This method consists in location the root of the equation $f(x) = 0$ between ***a*** and ***b***. If $f(x)$ is continuous between ***a*** and ***b***, and $f(a)$ and $f(b)$ are of opposite signs then there is a root between ***a*** and ***b***. For definiteness, let $f(a)$ be negative and be positive. Then the first approximation to the root is $x_1 = \frac{1}{2}(a+b)$.

If $f(x_1) = 0$, then x_1 is a root of $f(x) = 0$. Otherwise , the root lies between a and x_1 or x_1 b according as $f(x_1)$ is positive or negative. Then we bisect the interval as before and continue the process until the root is found to desired accuracy.

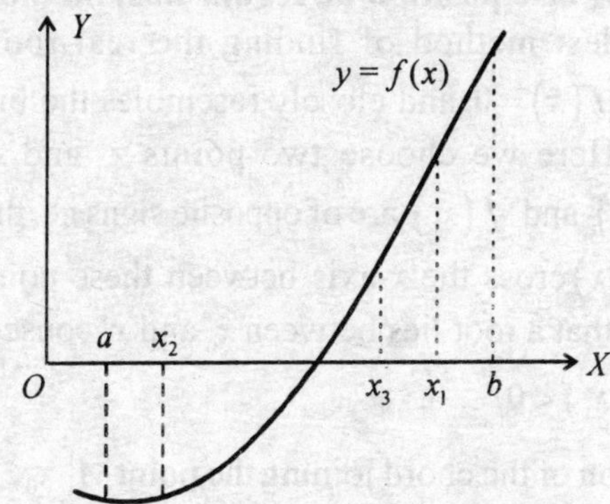

In the figure $f(x_1)$ is + ve, so that root lies between a and x_1. Then the second approximation to the root is $x_2 = \frac{1}{2}(a + x_1)$. If $f(x_2)$ is–ve, the root lies between x_1 and x_2. Then the third approximation to the root is $x_3 = \frac{1}{2}(x_1 + x_2)$ and so on.

1.6.2(A) SOLUTION OF NON-LINER EQUATIONS

Consider the equations

$$f(x, y) = 0, \; g(x, y) = 0 \qquad ...(1)$$

If an Initial approximation to a solution has been found by grasphical method or otherwise, then a better approximation (x_1, y_1) can be obtained as follows:

Let $x_1 = x_0 + h,\; y_1 = y_0 + k$, so that

$$f(x_0 + h,\; y_0 + k) = 0,$$
$$g(x_0 + h,\; y_0 + k) = 0 \qquad ...(2)$$

Expanding each of the function on (2) by Taylor's series to forst degree terms, we get approximately.

$$\left.\begin{aligned} f_0 + h\frac{\partial f}{\partial x_0} + k\frac{\partial f}{\partial y0} = 0 \\ g_0 + h\frac{\partial g}{\partial x_0} + k\frac{\partial g}{\partial y_0} = 0 \end{aligned}\right\} \qquad ...(3)$$

Where $f_0 = f(x_0, y_0),\quad \frac{\partial f}{\partial x_0} = \left(\frac{\partial f}{\partial x}\right)_{x_0 y_0}$ etc

Solving the equation (3) for h and k, we get a new approximation to the root as

$$x_1 = x_0 + h$$
$$y_1 = y_0 + k$$

This process is repeated till we get the values to the desired accuracy.

1.6.2(B) SINGLE STEP METHOD

Method of fase position or regula-falsi method: This is the oldest method of finding the real root of an equation $f(x) = 0$ and closely resembles the bisection method. Here we choose two points x_0 and x_1 such that $f(x_0)$ and $f(x_1)$ are of opposite signs i.e. the graph of $x = f(x)$ cross the x-axis between these point. This indicates that a root lies between x_0 and x_1 consequently $f(x_0)f(x_1) < 0$.

Equation of the chord joining the point $A[x_0, f(x_0)]$ and $B[x_1, f(x_1)]$ is $y - f(x_0) = \frac{f(x_1) - f(x_0)}{x_1 - x_0}(x - x_0)$

The method consist in replacing the curve AB by means of the chord AB and taking the point of intersection of the chord with the x-axis as an approximation to the root $(y = 0)$. So the abscissa of the point where the chord cuts the x-axis is given by

$$x_2 = x_0 - \frac{x_1 - x_0}{f(x_1) - f(x_0)} f(x_0)$$

Which is an approximation to the root.

If now $f(x_0)$ and $f(x_2)$ are of opposite signs, then the lies between x_0 and x_2. So replacing x_1 by x_2 in (2), we obtain the next approximation x_3. (The root could as well lie between x_1 and x_2 and would obtain x_3 accordingly). This procedure is repeated till the root is found to desired accuracy. The iteration process based on (1) is known as the method of false position.

Network-Raphson Method

Let be an approximate root of the equation $f(x) = 0$. If $x_1 = x_0 + h$ be the exact root, then $f(x_1) = 0$.

$\therefore$ Expanding $f(x_0 + h)$ Taylor's series

$$f(x_0) + hf'(x_0) + \frac{h^2}{2!}f''(x_0) + ... = 0$$

Since h is small, neglecting h^2 and higher powers of h, we get

$$f(x_0) + hf'(x_0) = 0 \text{ or } h = -\frac{f(x_0)}{f'(x_0)}$$

$\therefore$ A closer approximation to the root is given by

$$x_1 = x_0 - \frac{f(x_0)}{f'(x_0)}.$$

Similarly, starting with x_1 a still better approximation x_2 is given by

$$x_2 = x_1 - \frac{f(x_1)}{f'(x_1)} \quad \text{In general,} \quad x_n + 1 = x_n - \frac{f(x_n)}{f'(x_n)}$$

Which is known as the Newton's-Raphson formula or Newton's iteration formula.

1.6.3 MULTI STEP METHOD

The process of evaluating a definite integral form a set of tabulated values of the integrand $f(x)$ is called numerical integration. This process when applied to a function of a single variable. Is known as quadrature.

Newton- Cote's Quadrate Formula

Let $I = \int_a^b f(x)\,dx$

Where $f(x)$ takes the values $y_0, y_1, y_2 ... y_n$ for $x = x_0, x_1, x_2, ... x_n$

Let us divide the interval (a, b) into n sub-intervals of width h so that

$x_0 = a, x_1 = x_0 + h,$ so that $x_0 = a, x_1 = x_0 + h,$

$x_2 = x_0 + 2h, ..., x_n = x_0 + nh = b.$

Then

$$I = \int_{x_0}^{x_0+nh} f(x)\,dx$$

$$= h\int_0^n f(x_0 + rh)dr = h\int_0^n$$

$$\left[y_0 + r\Delta y_0 + \frac{r(r-1)}{2!}\Delta^2 y_0 + \frac{r(r-1)(r-2)}{3!}\Delta^3 y_0 + ... \right] dr$$

Integrating term by term, we obtain

$$\int_{x_0}^{x_0+nh} f(x)\,dx = nh$$

$$\left[y_0 + \frac{n}{2}\Delta y_0 + \frac{n(2n-3)}{12}\Delta^2 y + \frac{n(n-2)^2}{24}\Delta^3 y + ... \right]$$ which

is known as the Newton-Cote's quadrature formula, Being general formula, we deduce form this the following quadrate rules by taking $n = 1, 2, 3, ...$

Trapezoidal Rule

Putting $n = 1$ in (1) and taking the curve though, (x_0, y_0) and (x_1, y_1) as a straight line, i.e. a polynomial of first order so that differences of order higher than first becomes zero, we get

$$\int_{x_0}^{x_0+h} f(x)\,dx = h\left(y_0 + \frac{1}{2}\Delta y_0 \right) = \frac{h}{2}(y_0 + y_1)$$ Similarly

$$\int_{x_0+h}^{x_0+2h} f(x)\,dx = h\left(y_1 + \frac{1}{2}\Delta y_1 \right) = \frac{h}{2}(y_1 + y_2)$$

$$\int_{x_0+(n-1)h}^{x_0+nh} f(x)\,dx = \frac{h}{2}(y_{n-1} + y_n)$$

Adding these n integrals, we obtain

$$\int_{x0}^{x_0+nh} f(x)\,dx = \frac{h}{2}[(y_0 + y_n) + 2(y_1 + y_3 + ... + y_{n-1})]$$

This is know as the trapezoidal rule.

Simpsom's One-Third Rule

Putting $n = 2$ in (1) above and taking the curve thought (x_0, y_0), (x_1, y_1) and (x_2, y_2) as a parabola, i.e. a polynomial of second order so that differences of orders higher than second vanish, we get.

$$\int_{x_0}^{x_0+2h} f(x)\,dx = 2h(y_0 + \Delta y_0 + \frac{1}{6}\Delta^2 y_0) = \frac{h}{3}(y_0 + 4y_1 + y_2)$$ Similarly\

$$\int_{x_0+2h}^{x_0+4h} f(x)\,dx = \frac{h}{3}(y_2 + 4y_3 + y4)$$

$$\int_{x_0+(n-1)h}^{x_0+nh} f(x)\,dx = \frac{h}{3}(y_{n-2} + 4y_{n-1} + y_n)$$, n being even

Adding all these integrals, we have when n is even

$$\int_{x_0}^{x_0+nh} f(x)\,dx = \frac{h}{3}\begin{bmatrix} [(y_0 + y_n) + 4(y_1 + y_3 + ... + y_{n-1}) \\ +2(y_2 + y_4 + ... + y_{n-2})] \end{bmatrix} \quad (3)$$

This known as the simpson's one-third rule or simply simpson's rule and is most commonly used.

Obs. While applying (3), the given interval must be divided into even number of equal sub-intervals since we find the area of two strips at a time.

Simpson's Three-Eight Rule

Putting $n = 3$ in (1) above and taking the curve through. $(x_i, y_i): i = 0, 1, 2, 3$ as a polynomial of third order so that differences above the third order vanish, we get

$$\int_{x_0}^{x_0+3h} f(x)dx = 3h\left(y_0 + \frac{3}{2}\Delta y_0 + \frac{3}{4}\Delta^2 y_0 + \frac{1}{8}\Delta^3 y_0 \right) = \frac{3h}{8}(y_0 + 3y_1 + 3y_2 + y_3)$$

Similarly

$$\int_{x_0+3h}^{x_0+6h} f(x)dx = \frac{3h}{8}(y_3 + 3y_4 + 3y_5 + y_6)$$

Adding all such expression form to , where n is a multiple of 3, we obtain

$$\int_{x_0}^{x_0+nh} f(x)\,dx = \frac{3h}{8}\begin{bmatrix} (y_0 + y_n) + 3(y_1 + y_2 + y_4 + y_5 + ... + y_{n-1}) \\ +2(y_3 + y_6 + y_{n-3}) \end{bmatrix} \quad (4)$$

Which is known as simpsom's one-third rule.

Runge-Kutta Method

Working rule for finding the increment k of y corresponding to an increment h of x by Runge-Kutta method form.

$\frac{dy}{dx} = f(x, y),\ y(x_0) = y_0$ is an follows:

Calculate successively

$$k_1 = hf(x_0, y_0)$$

$$k_2 = hf(x_0 + \frac{1}{2}h, y_0 + \frac{1}{2}k_1)$$

$$k_3 = hf(x_0 + \frac{1}{2}h, y + \frac{1}{2}k_2)$$

$$k_4 = hf(x_0 + h, y_0 + k_3)$$

Finally compute

$$k = \frac{1}{6}(k_1 + 2k_2 + 2k_3 + k_4)$$

Which gives the required approximate value

$y_1 = y_0 + k$

1.7 TRANSFORM THEORY

1.7.1 FOURIER TRANSFORM

Fourier Transform $f(x)$ of $F(s)$ is and is defined by

$$F(s) = \int_{-\infty}^{\infty} f(x)^{+isx} dx$$

$$f(x) = \frac{1}{2\pi}\int_{-\infty}^{\infty} F(s)^{-isx} ds$$

Fourier Sine and Cosine transform

$$F_s(s) = \int_0^{\infty} f(x)\sin(sx)dx$$

$$f(x) = \frac{2}{\pi}\int_0^{\infty} F_s(s)\sin(sx)ds$$

Is known as the Fourier Sine transform of $f(x)$ and inversion in $[0, \infty]$

Similarly

$$F_C(s) = \int_0^{\infty} f(x)\cos(sx)\,dx \text{ and}$$

$$f(x) = \frac{2}{x}\int_0^{\infty} F_C(s)\cos(sx)\,ds$$

Is known as Fourier Cosine Transform of $f(x)$

Finite fourier sine and cosine transform

The Finite Fourier transform of $f(x)$ in $0 < x < c$

$$F_s(x) = \int_0^C f(x)\sin\left(\frac{n\pi x}{C}\right)dx,$$

$$f(x) = \frac{2}{C}\sum_{n=1}^{\infty} F_s(n)\sin\left(\frac{n\pi x}{C}\right)$$

Finite Fourier cosine transform of $f(x)$ in [0, C]

$$F_c(x) = \int_0^C f(x)\cos\left(\frac{n\pi x}{C}\right)dx$$

Properties of Fourier Transform

Property	Function	F.T
Linearity	$af(x) + bg(x)$	$aF(s) + bG(s)$
Change of scale	$f(ax)$	$\frac{1}{a}F\left(\frac{s}{a}\right)$
Shifting	$f(x-a)$	$e^{-sa}F(s)$
Modulation	$f(x)\cos ax$	$\frac{1}{2}F[s+a] + F(s-a)$

1.7.2 LAPLACE TRANSFORM

The Laplace transformation is a power full method for solving linear differential equations usually encountered in electrical circuits.

The laplace transformation reduces the problem of solving differential equations to an algebraic problem.

While Performing Laplace Transform we Perform Two Steps

1. Multiply function f(t) by e^{-st} = f(t). e^{-st}
2. Integrate from limit 0 to ∞

The direct Laplace transformation of a function f(t) is define as.

$$F(s) = LT\ f(t) = \int_0^{\infty} f(t)\ e^{-st}\ dt$$

where, s = Complex frequency = σ + jω

where, σ = Real

ω = Imaginary

For fourier transform, σ → 0

LT f(t) = F(s) ; LT^{-1} F(s) = f(t)

f(t) ↔ F(s)

f(t) and F(s)are Laplace Transform pair

Inverse Laplace Transform, i.e.

$$LT^{-1}[F(s)] = f(t) = \frac{1}{2\pi j}\int_{\sigma-j\infty}^{\sigma+j\infty} F(s)\,e^{st}\,ds$$

Condition for Laplace Transform to Exists

$$\int_0^{\infty}\left|f(t)\ e^{-st}\right|\ dt < \infty$$

$$\Rightarrow \int_0^{\infty}\left|f(t)\ e^{-(\sigma+j\omega)t}\right|\ dt < \infty$$

Now, $= |e^{-(+j\omega)t}| = |e^{-\sigma t}|\,|e^{-j\omega t}|$

$= e^{-\sigma t}\,|\cos t - \omega j \sin \omega t|$

$= e^{-\sigma t}\,(\cos^2 t + \sin^2 \omega t|^{1/2} = e^{-\sigma t}$

Then, $\int_0^{\infty}|f(t)|\ |e^{-st}|$ must be finite

Important laplace transform are given

1. f(t) = δ(t) ← delta function, i.e. unit Impulse function.

$$F(s) = \int_0^\infty f(t)\ e^{-st}\ dt = \int_0^\infty \delta(t)\ e^{-st}\ dt$$

2. f(t) = u(t) ← unit step function
   = 0; t < 0
   = 1; t > 1

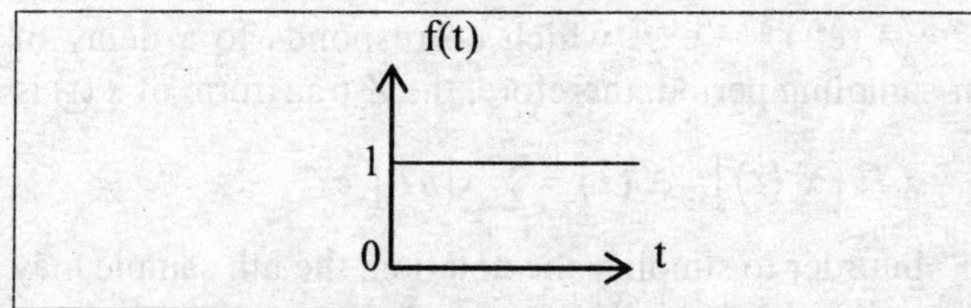

$$F(s) = \int_0^\infty f(t)\ e^{-st}\ dt = \frac{1}{s}$$

3. Unit doublet function $f(t) = \frac{d}{dt}\ \delta(t)$

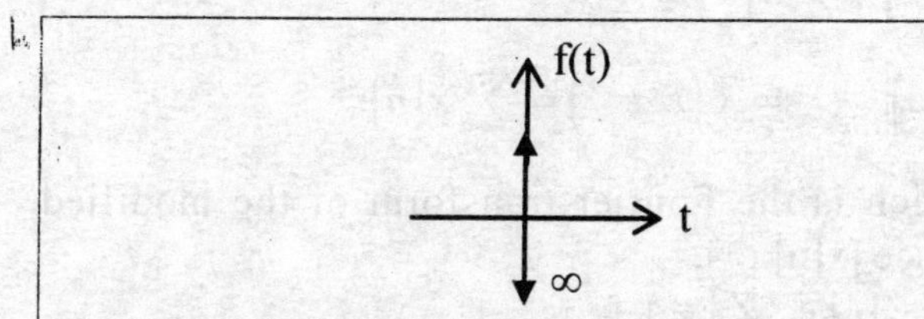

4. Unit triplet function parabola

$$LT\ \left(\frac{d}{dt}\ \delta(t)\right) = s$$

5. f(t) = t^n

$$\therefore\ F(s) = \frac{n\ !}{s^{n+1}} \leftrightarrow t^n$$

6. Unit parabola $\frac{t^2}{2} \to \frac{1}{s^3}$

7. f(t) = e^{-at} ← exponential function

$$F(s) = \int_0^\infty a^{-at}\ e^{-st}\ dt = \int_0^\infty e^{-(s+a)a}\ dt = \frac{1}{s+a}$$

8. f(t) = sin *at* ← Sinusoidal function
9. f(t) = cos *at* ← Co-sinusoidal function

$$LT\ [\sin\ a\ t] = \frac{a}{s^2 + a^2}$$

10. f(t) = sinh a t ← hyperbolic function

$$LT[\cos\ a\ t] = \frac{s}{s^2 + a^2}$$

11. f(t) = cosh a t

$$LT[\cosh a\,t] = \frac{s}{s^2 - a^2}$$

12. f(t) = e^{-bt}. sin *at*

$$LT\left[e^{-bt}\ \sin at\right] \leftrightarrow \frac{a}{(s+b)^2 + a^2}$$

13. f(t) = e^{-bt}. cos *at*

$$LT\left[e^{-bt}\ \cos at\right] \leftrightarrow \frac{s+b}{(s+b)^2 + a^2}$$

14. f(t) = e^{-bt} sin *h at*

$$LT\left[e^{-bt}\ \sin h\ at\right] \leftrightarrow \frac{a}{(s+b)^2 - a^2}$$

15. $f(t) = e^{-bt} \cdot \cos h\ at \leftrightarrow \frac{s+b}{(s+b)^2 - a^2}$

16. f(t) = t e^{-at}

$$LT\left[\ e^{-at}\right] \leftrightarrow \frac{1}{(s+a)^2}$$

Laplace transform of derivative

$$LT\left(\frac{d\,f(t)}{dt}\right) = \int_0^\infty \frac{d}{dt}\ f(t)\ e^{-st}\ dt = sF(s) - f(0)$$

where (0) represent initial value.

Laplace of second derivative

$$LT\left(\frac{d^2\,f(t)}{dt}\right)$$

$$= s^2\ F(s) - s\ f(0) - f'(0)$$

Laplace transform of an integral of time

$$LT\ \int f^n(t) = \frac{F(s)}{s^n} + \frac{f^{-1}(0)}{s^n} + \frac{f^{-2}(0)}{s^{n-1}} + \ldots$$

Initial Value Theorem

$$f(0^+) = \lim_{t \to 0}\ f(t) = \lim_{s \to \infty}\ s\ F(s)$$

Final Value Theorem

$$f(\infty) = \lim_{t \to \infty}\ f(t) = \lim_{s \to o}\ [s\ F(s)]$$

1.7.3 Z-TRANSFORM

Z-transform provides the solution for linear constant coefficient differential equations relating discrete time system in the frequency domain.

$$H(s) = \frac{K(s - Z_1)\ (s - Z_2)\ (s - Z_3)\ldots}{(s - P_1)\ (s - P_2)\ (s - P_3)\ldots}$$

Where K is constant

S = jω and 0 ≤ ω ≤ ∞

P_1, P_2, P_3 → are poles

Z_1, Z_2, Z_3 → are zero

$$H(z) = \frac{K(z-z_1)(z-z_2)(z-z_3)}{(z-p_1)(z-p_2)(z-p_3)}$$

$Z = e^{j\omega}$ $0 < \omega \le \pi$

$\omega \rightarrow$ angular frequency, i.e.

$\rightarrow$ rad/second

The analysis of any sampled signal or sampled data system in the frequency domain is extremely difficult using s-plane representation because of signal or system equation will contain infinite long polynomials due to the characteristic infinite number of poles and zeros.

Fortunately, this problem can be over come by using the Z-transform which reduce the poles and zero to finite number in Z-plane.

The transformation of any point $s = \sigma + j\omega$ in the s-plane to the Z-plane by corresponding point $Z[r\angle\theta]$ in the s-plane is carried out by the relationship.

$Z = e^{ST}$ where T is the sampling Period.

$\sigma = 0,$	$\omega_s = [2\pi/T]$			
$j\omega$ 1	$(\omega_s/8)$	$(\omega_s/4)$	$(3\omega_s/8)$	$(\omega_s/2)$
$z(1/\omega T)=$ $1\angle 0$	$1\angle 45^\circ$	$1\angle 90^\circ$	$1\angle 135^\circ$	$1\angle 180^\circ$
$(5\omega_s/8)$	$(3\omega_s/4)$	$(7\omega_s/8)$ ω_s		
$1\angle 225^\circ$	$1\angle 270^\circ$	$1\angle 135$	$1\angle 30$	

Under this mapping, the imaginary axis $\sigma = 0$ maps on to the unit circle $|Z| = 1$ in the Z-plane. Also the LHS of s-plane i.e. $\sigma < 0$ corresponds to the interior of the unit circle $|Z| = 1$ in the s-plane.

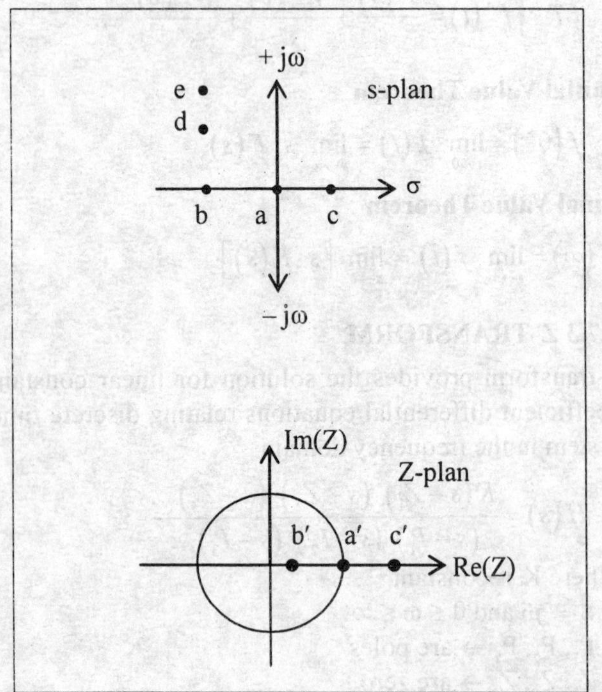

When $\sigma = 0$, $Z = e^{sT} = e^{j\omega T} = 1 \angle \pm j\omega T$.

Which gives the value in polar co-ordinate.

$$LT\left[x(t)\right] = \int_0^\infty x(t)\, e^{-sT}\, dt$$

$$LT\left[x_s(t)\right] = X(s) = \sum_{n=0}^{\infty} x[nT]\, e^{-nsT}$$

But we have $Z = e^{sT}$ for Z-transform and hence $Z^{-n} = (e^{sT})^{-n} = e^{-nsT}$ which corresponds to a delay of n-sampling period, therefore, the Z-transform of $x_s(t)$ is

$$LT\left[x_s(t)\right] = X(z) = \sum_{n=0}^{\infty} x[nT]\, e^{-n}$$

In order to simplify the notation, the nth sample may be written as x[n] and hence the above equation becomes.

$$X[z] = \sum_{n=0}^{\infty} x[n]\, e^{-n}$$

Region of convergence (ROC)

$$X[z] = \sum_{n=0}^{\infty} x[n]\, z^{-n}$$

$$X[z]\Big|_{z=r\,e^{j\omega}} = X\left(r\, e^{j\omega}\right) = \sum_{n=-\infty}^{\infty} x[n]\, r^{-n}\, e^{-j\omega_n}$$

Which is the Fourier transform of the modified sequence $[x[n]\, r^{-n}]$.

If r = 1, i.e. $|Z| = 1$

X[z] reduces to its Fourier transform

The series of the above equation converges if $x[n]\, r^{-n}$ is absolutely sum able, i.e.

$$= \sum_{n=0}^{\infty} x[n]\, \left|r^{-n}\right| < \infty$$

If the output signal magnitude of the digital signal system x[n] is to be finite, the magnitude of z-transform X[z] must be finite. The set of Z-Value in the Z-Plane for which the magnitude of X[z] is finite is called the region of convergence (ROC).

Therefore, the condition for X[z] to be finite is $|z| > 1$. In other words, the ROC of X[z] is the area outside the unit circle in the z-plane.

The ROC of a rational Z-transform is bounded by the location of its poles.

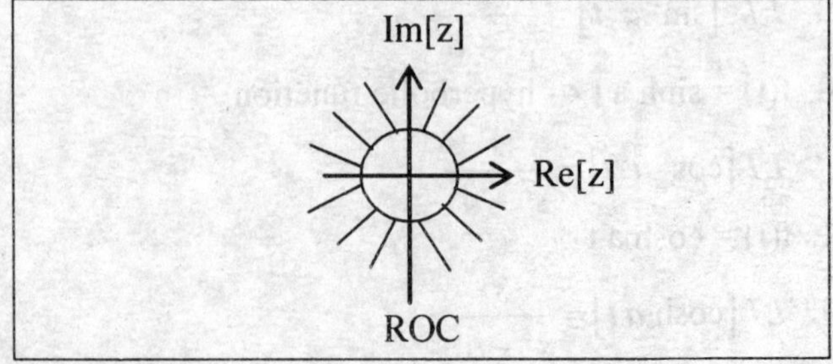

Example: X[z] = [z/z – 1]

Pole, i.e. z – 1 > 0

| z | > 1; ROC | z | > |

And extending all the way to ∞

Important properties of the ROC of Z-transform

1. X[z] converges uniformly if and only if the ROC of Z-transform X[z] of the sequence includes the unit circle.

$$= \sum_{n=-\infty}^{\infty} \left|x[n]\, r^{-n}\right| < \infty.$$

2. The ROC does not contain any pole.
3. When x[n] is of finite duration, the ROC is the entire z-plane except possibly z = 0 and/or z = ∞.
4. If x[n] is right sided sequence, the ROC will not include infinity.
5. If x[n] is left sided sequence, the ROC will not include z = 0. However if x[n] = 0 for all n > 0, then ROC will include z = 0.
6. If x[n] is two sided, and if the circle | z | = ro is in the ROC, then ROC will consist of ring in the z-plane that include the circle | z | = r_o, that is the ROC includes the intersection of ROC of the component.
7. If X[z] is rational, then the ROC extends to infinity, i.e. ROC is bounded by poles.
8. If x[n] is causal, then the ROC include z = ∞ except z = 0.
9. If x[n] is anti casual, the ROC include z = 0 except z = ∞.

To determine the ROC for the series expressed by the equation.

$$ZT\ (x[n]) = X(z) = \sum_{n=-\infty}^{\infty} x[n]\ z^{-n}$$

$$= \sum_{n=-\infty}^{-1} x[n]\ z^{-n} + \sum_{n=0}^{\infty} x[n]\ z^{-n}$$

$$= \sum_{n=1}^{\infty} x[-n]\ z^{+n} + \sum_{n=0}^{\infty} x[n]\ z^{-n}$$

The first series, a non-causal sequence, converges for $|z| < r_1$ and the second series a causal sequence converges for $|z| > r_2$ resulting in an annular region of convergence.

Properties of Z-transform

(a) Linearity: If it is homogenous and satisfy super Position:

If x[n] = $A_1 x_1[n] + A_2 x_2[n]$

X[z] = $A_1 X_1[z] + A_2 X_2[z]$

With ROC = R_1 and R_2

R = $R_1 R_2$

Time Shifting

if x[n] → X[z] with ROC = R

Then x[n – k] → X[z] z – k

With ROC = R except addition or deleted of the origin infinity

If for k > 0 → origin deleted

Scaling

If x[n] → X[z] with ROC = R

A^n x[n] → X[Z/A] with ROC = [| A | R].

Time Reversal

x[n] → X[z] with ROC = R

x[-n] → X[1/z] with ROC = [1/R]

Time Expansion

x[n] → X[z] ; ROC = R

$$x_k[n] = \begin{cases} x[n/k] & ;\quad \text{if } n \text{ is multiple of } k \\ 0 & ;\quad \text{if } n \text{ is not multiple of } k \end{cases}$$

$x_k[n] \rightarrow X[z^k]$ with ROC = $R^{(1/k)}$.

Differentiation in Z-Domain

x[n] → X[z]

$$n\ x[n] \rightarrow -\frac{z\ d\ X[z]}{dz}$$

$$\rightarrow z^{-1} \frac{d\ X[z]}{d\ z^{-1}}$$

Convolution

$x_1[n] \rightarrow X_1[z]$; ROC = R_1

$x_2[n] \rightarrow X_2[z]$; ROC = R_2

$$x[n] = x_1[n] * x_2[n] = \sum_{k=-\infty}^{n} x_1(k)\ x_2[n-k]$$

= $X_1[z]. X_2[z]$; ROC = $R_1 \cap R_2$

Correlation

$x_1[n]\ X_1[z]$; ROC = R_1

$x_2[n]\ X_2[z]$; ROC = R_2

$$r_{x1}\ x_2[p] = \sum_{n=-\infty}^{\infty} x_1(n)\ x_2[n-p]$$

$$Rx_1x^2[z] = X_1[z]\ X_2[z^{-1}];\ \text{ROC} = R_1 \cap R_2$$

Auto Correlation

$x_1[n] = x_2[n]$

$X[z] = X_1[z]\, X_1(z^{-1})$

Initial value theorem

If x[n] = 0, for n < 0 i.e. x[n] is casual

The initial value is given by

$$x[0]=\lim_{n\to 0} x[n]=\lim_{|z|\to\infty} X[z]$$

Final Value Theorem

If X[z] = ZT(x [n]) and the poles of X[z] are all inside the unit circle

$$x(\infty)=\lim_{n\to\infty} x[n]=\lim_{|z|\to 1} \left(1-z^{-1}\right) X[z]$$

One sided Z-transform

Time Delay

x[n] → X[z]

$$x[n-k]\to z^{-k}\left(X[z]+\sum_{n=1}^{k} x[-n]\, z^{n}\right) k>0$$

Time Advance

x[n] → X[z]

$$x[n+k]\to z^{-k}\left(X[z]-\sum_{n=0}^{k-1} x[n]\, z^{-n}\right)$$

Evaluation of Inverse Z-transform

$$x[n]=\frac{1}{2\pi j}\oint_C X[z]\, z^{n-1}\, dz\ldots$$

The above equation represent a contour integral over a closed path C that encloses the origin and lies within the ROC of X[z].

MCQ Set–1

1. X is a uniformly distributed random variable that takes values between 0 and 1. The value of $E\{X^3\}$ will be.
   (a) 0 (b) 1/8
   (c) 1/4 (d) 1/2
2. The characteristic equation of a (3×3) matrix P is defined as $\alpha(\lambda)=|\lambda I-P|=\lambda^3+\lambda^2+2\lambda+1=0$. If I denote indentity matrix, then the inverse of matrix P will be.
   (a) (P^2+P+2I) (b) (P^2+P+I)
   (c) $-(P^2+P+I)$ (d) $-(P^2+P+2I)$
3. If the rank of a f(5×6) matrix Q is 4, then which one of the following statements is correct?
   (a) Q will have four linearly independent rows and four linearly independent columns.
   (b) Q will have four linearly independent rows and five linearly independent columns.
   (c) QQ^T will be invertible
   (d) Q^TQ will be invertible
4. Consider function $f(x)=\left(x^2-4\right)^2$ where x is a real number. Then the function has.
   (a) Only one minimum
   (b) Only two minima
   (c) Three minima
   (d) Three maxima
5. Eqation $e^x-1=0$ is required to be solved using Newton's method with an initial gaues $x_0=-1$. Then, after one step of Newton's method estimate x_1 of the solution will be given by.
   (a) 0.71828 (b) 0.36784
   (c) 0.20587 (d) 0.00000
6. A is $m\times n$ full rank matrix with $m>n$ and I is an identity matrix. $A^+=\left(A^TA\right)^{-1}A^T$. Then which one of the following statements is FALSE?
   (a) $AA^+A=A$ (b) $\left(AA^+\right)^2=AA^+$
   (c) $A^+A=1$ (d) $AA^+A=A^+$
7. A differential equation $\frac{dx}{dt}=e^{-2t}u(t)$ has to be solved using trapezoidal rule of integration with a step size $h=0.01\,s$. Function $u(t)$ indicates a unit step function. If $x(0-)=0$, then value of x at $t=0.01\,s$ will be given by
   (a) 0.00099 (b) 0.00495
   (c) 0.0099 (d) 0.0198
8. Let P be a 2×2 real orthogonal matrix and $\vec{X}$ is a real vector $[x_1,x_2]^T$ with length $\|\vec{X}\|=\left(x_1^2+x_2^2\right)^{1/2}$. Then, which one of the following statements is correct?
   (a) $\|P\vec{X}\|\le\|\vec{X}\|$ where at least one vector satisfies $\|P\vec{X}\|<\|\vec{X}\|$
   (b) $\|P\vec{X}\|=\|\vec{X}\|$ for all vectors $\vec{X}$
   (c) $\|P\vec{X}\|\ge\|\vec{X}\|$ where at least one vector satisfies $\|P\vec{X}\|>\|\vec{X}\|$
   (d) NO relationship can be established between $\|\vec{X}\|$ and $\|P\vec{X}\|$
9. The trace and determinant of a 2×2 matrix are known to be –2 and –35 respectively. Its eigenval ues are

(a) −30 and −5 (b) −37 and −1
(c) −7 and 5 (d) 17.5 and −2

10. $f(x,y)$ is a continous function defined over $(x,y)\in[0\ \ 1]\times[0\ \ 1]$. Given the two constraints, $x>y^2$ and $y>x^2$, the volume under $f(x\,y)$ is

(a) $\int_{y=0}^{y=1}\int_{x=y^2}^{x=\sqrt{y}} f(x,y)\,dxdy$

(b) $\int_{y=x^2}^{y=1}\int_{x=y^2}^{x=1} f(x,y)\,dxdy$

(c) $\int_{y=0}^{y=1}\int_{x=0}^{x=1} f(x,y)\,dxdy$

(d) $\int_{y=0}^{y=\sqrt{x}}\int_{x=0}^{x=\sqrt{y}} f(x,y)\,dxdy$

11. Assume for simplicity that N people, all born in April (amonth of 30 days), are collected in a room. Consider the event of at least two people in the room being born on the same date of the month, event if in different years, e.g. 1980 and 1985. What is the smallest N so that the probability of this event exceesds 0.5?
(a) 20 (b) 7
(c) 15 (d) 16

12. A cubic polynomial with real coefficients
(a) Can possibly have no extrema and no zero crossings
(b) May have up to three extrema and upto 2 zero crossings
(c) Cannot have more than two extrema and more than three zero crossings
(d) Will always have an equal number of extrema and zero crossings.

13. Let $x^2-117=0$. The iterative steps for the solution using Network Raphson's method is given by

(a) $x_{k+1}=\frac{1}{2}\left(x_k+\frac{117}{x_k}\right)$

(b) $x_{k+1}=x_k-\frac{117}{x_k}$

(c) $x_{k+1}=x_k-\frac{x_k}{117}$

(d) $x_{k+1}=x_k-\frac{1}{2}(x_k+\frac{117}{x_k})$

14. All the four entries of the of 2×2 matrix P $=\begin{bmatrix}p_{11} & p_{12}\\ p_{21} & p_{22}\end{bmatrix}$ are nonzero, and one of its eigenvalues is zero. Which of the following statements is true?
(a) $p_{11}p_{22}-p_{12}p_{21}=1$
(b) $p_{11}p_{22}-p_{12}p_{21}=-1$
(c) $p_{11}p_{22}-p_{12}p_{21}=0$
(d) $p_{11}p_{22}+p_{12}p_{21}=0$

15. The system of liner equation
$4x+2y=7$
$2x+y=6$
has
(a) A unique solution
(b) No solution
(c) An infinite number of solutions
(d) Exactly two distinct solutions

16. The equation sin (z) =10 has
(a) No real or complex solution
(b) Exactly two distinct complex solutions
(c) A unique solution
(d) An infinite number of complex solutions

17. For real values of x, the minimum value of function $f(x)=\exp(x)+\exp(-x)$ is
(a) 2 (b) 2
(c) 0.5 (d) 0

18. Which of the following functions would have only odd powers of x in its Taylor series expansion about the point x = 0?
(a) sin (x^3) (b) sin (x^2)
(c) cos (x^3) (d) cos (x^2)

19. Which of the following is a solution to the differential equation $\frac{dx(t)}{dt}+3x(t)=0$?
(a) $x(t)=3e^{-t}$ (b) $x(t)=2e^{-3t}$
(c) $x(t)=-\frac{3}{2}t^2$ (d) $x(t)=3t^2$

20. The recursion relation to solve $x=e^{-x}$ using Newton-Raphson method is
(a) $x_{n+1}=e^{-x_n}$
(b) $x_{n+1}=x_n-e^{-x_n}$
(c) $x_{n+1}=(1+x_n)\frac{e^{-x_n}}{1+e^{-x_n}}$
(d) $x_{n+1}=\frac{x_n^2-e^{-x_n}(1+x_n)-1}{x_n-e^{-x_n}}$

21. The residue of the function $f(z) = \dfrac{1}{(z+2)^2(z-2)^2}$ at z = 2 is

(a) $-\frac{1}{32}$ (b) $-\frac{1}{16}$

(c) $\frac{1}{16}$ (d) $\frac{1}{32}$

22. Consider the matrix $P = \begin{bmatrix} 0 & 1 \\ -2 & -3 \end{bmatrix}$. The value of e^P is

(a) $\begin{bmatrix} 2e^{-2}-3e^{-1} & e^{-1}-e^{-2} \\ 2e^{-2}-2e^{-1} & 5e^{-2}-e^{-1} \end{bmatrix}$

(b) $\begin{bmatrix} e^{-1}+e^{-2} & 2e^{-2}-e^{-1} \\ 2e^{-1}-4e^{-2} & 3e^{-1}+2e^{-2} \end{bmatrix}$

(c) $\begin{bmatrix} 5e^{-2}-e^{-1} & 3e^{-2}-e^{-2} \\ 2e^{-2}-6e^{-1} & 4e^{-2}+e^{-1} \end{bmatrix}$

(d) $\begin{bmatrix} 2e^{-1}-e^{-2} & e^{-1}-e^{-2} \\ -2e^{-1}+2e^{-2} & -e^{-1}+2e^{-2} \end{bmatrix}$

23. In the Taylor series expansion of $\exp(x)+\sin(x)$ about the point $x=\pi$, the coefficient of $(x-\pi)^2$ is

(a) exp (π) (b) 0.5 exp(π)

(c) exp(π) + 1 (d) exp(π)-1

24. The value of the integral of the function $g(x,y) = 4x^3 + 10y^4$ along the straight line segment from the point (0.0) to the point (1, 2) in the x–y plane is

(a) 33 (b) 35

(c) 40 (d) 56

25. Consider point P and Q in the x-y plane, with $p = (1.0)$. The line integral $2\int_P^Q (xdx + ydx)$ along the samicircle with the line segment P Q as its diameter

(a) is–1 (b) is 0

(c) is 1

(d) depends on the direction (clickwise or anti-clockwise) of the semicircle

26. The value of the quantity P, where $P = \int_0^1 xe^x dx$, is equal to

(a) 0 (b) 1

(c) e (d) $1/e$

27. Divergence of the three-dimensional radial vector field $\vec{r}$ is

(a) 3 (b) $1/r$

(c) $\hat{i}+\hat{j}+\hat{k}$ (d) $3(\hat{i}+\hat{j}+\hat{k})$

28. The period of the signal $x(t) = 8\sin\left(0.8\pi t + \frac{\pi}{4}\right)$ is

(a) $0.4\pi\, s$ (b) $0.8\pi\, s$

(c) $1.25\, s$ (d) $2.5\, s$

29. The system represented by the input-output relationship $y(t) = \int_{-\infty}^{5t} x(\tau)d\tau,\ t>0$ is

(a) Linear and causal

(b Linear but not causal

(c) Causal but not linear

(d) Neither linear nor causal

30. A fair coin is tossed independently four times. The probability of the event "the number of times heads show up is more than the number of times tails show up" is

(a) $\frac{1}{16}$ (b) $\frac{1}{8}$

(c) $\frac{1}{4}$ (d) $\frac{5}{16}$

31. If $\vec{A} = xy\,\hat{a}_x + x^2\hat{a}_y$ then $\oint_c \vec{A}\cdot\vec{dl}$ over the path shown in the figure is

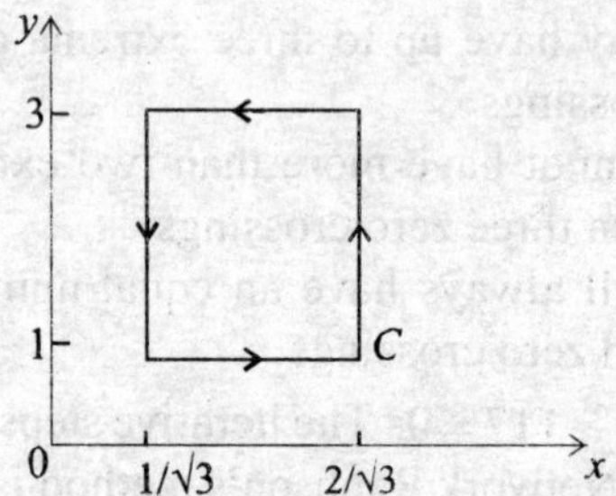

(a) 0 (b) $\frac{2}{\sqrt{3}}$

(c) 1 (d) $2\sqrt{3}$

32. The residues of a complex function $X(z) = \dfrac{1-2z}{z(z-1)(z-2)}$ at its poles are

(a) $\frac{1}{2}, -\frac{1}{2}$ and 1 (b) $\frac{1}{2}, \frac{1}{2}$ and -1

(c) $\frac{1}{2}$, 1 and $-\frac{3}{2}$ (d) $\frac{1}{2}$, -1 and $\frac{3}{2}$

33. Consider a differential equation $\frac{dy(x)}{dx} - y(x) = x$ with the initial condition $y(0) = 0$. Using Euler' s first order method with a step size of 0.1, the value of $y(0.3)$ is

(a) 0.01 (b) 0.031
(c) 0.0631 (d) 0.1

34. Given $f(t) = LT^{-1}\left[\frac{3s+1}{s^3+4s^2+(K-3)s}\right]$.

If $\lim_{t\to\infty} f(t) = 1$, then the value of K is

(a) 1 (b) 2
(c) 3 (d) 4

35. At $t = 0$, the function $f(t) = \frac{\sin t}{t}$ has

(a) a minimum
(b) a discontinuity
(c) a point of inflection
(d) a maximum

36. A box contains 4 white balls and 3 red balls. In succession, two balls are randomly selected and removed from the box.Given that the first removed ball is white, the probability that the second removed ball is red is

(a) 1/3 (b) 3/7
(c) 1/2 (d) 4/7

37. An eigenvector of $P = \begin{pmatrix} 1 & 1 & 0 \\ 0 & 2 & 2 \\ 0 & 0 & 3 \end{pmatrix}$ is

(a) $[-111]^T$ (b) $[121]^T$
(c) $[1-1\ 2]^T$ (d) $[21-1]^T$

38. For the differential equation $\frac{d^2x}{dt^2} + 6\frac{dx}{dt} + 8x = 0$ with initial conditions $x(0) = 1$ and $\left.\frac{dx}{dt}\right|_{t=0} = 0$, the solution is

(a) $x(t) = 2e^{-6t} - e^{-2t}$
(b) $x(t) = 2e^{-2t} - e^{-4t}$
(c) $x(t) = -e^{-6t} + 2e^{-4t}$
(d) $x(t) = e^{-2t} + 2e^{-4t}$

39. For the set of equations

$$x_1 + 2x_2 + x_3 + 4x_4 = 2$$
$$3x_1 + 6x_2 + 3x_3 + 12x_4 = 6$$

the following statement is true:

(a) Only the trivial solutions $x_1 = x_2 = x_3 = x_4 = 0$ exists
(b) There are no solutions.
(c) A unique nontrivial solution exists.
(d) Multiple non-trivial solutions exist.

40. $x(t)$ is a positive rectangular pulse from $t = -1$ to $t = +1$ with unit height as shown in the figure. The value of $\int_{-\infty}^{\infty} |X(\omega)|^2 d\omega$ {where $X(\omega)$ is the Fourier transform of $x(t)$} is

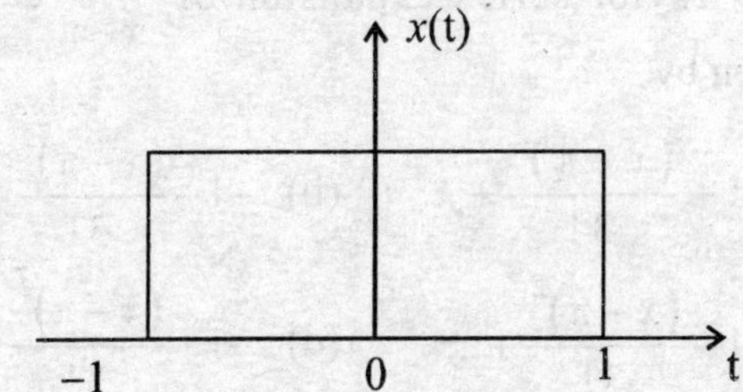

(a) 2 (b) 2π
(c) 4 (d) 4π

41. The order of the differential equation $\frac{d^2y}{dt^2} + \left(\frac{dy}{dt}\right)^3 + y^4 = e^{-t}$ is

(a) 1 (b) 2
(c) 3 (d) 4

42. The Fourier series of a real periodic function has only.

P. cosine terms if it is even
Q. sine terms if it is even
R. cosine terms if it is odd
S. sine terms if it odd

Which of the above statements are correct?

(a) P and S (b) P and R
(c) Q and S (d) Q and R

43. A function is given by $f(t) = \sin^2 t + \cos 2t$. Which of the following is true?

(a) f has frequency components at 0 and $1/2\pi$ *Hz*
(b) f has frequency components at 0 and $1/\pi$ *Hz*
(c) f has frequency components at $1/2\pi$ and $1/\pi$ *Hz*

44. The ROC of Z-transform of the discrete time sequence $x(n) = \left(\frac{1}{3}\right)^n u(n) - \left(\frac{1}{2}\right)^n u(-n-1)$ is

(a) $\frac{1}{3} < |z| < \frac{1}{2}$ (b) $|z| > \frac{1}{2}$
(c) $|z| < \frac{1}{3}$ (d) $2 < |z| < 3$

45. Consider two independent random variables X and Y with identical distributions. The variables X and

Y take values 0, 1 and 2 with probabilities $\frac{1}{2}, \frac{1}{4}$ and $\frac{1}{4}$ respectively. What is the conditional probability $P(X+Y=2|X-Y=0)$?

(a) 0 (b) $\frac{1}{16}$

(c) $\frac{1}{6}$ (d) 1

46. The Taylor series expansion of $\frac{\sin x}{x-\pi}$ at $x=\pi$ is given by

(a) $1+\frac{(x-\pi)^2}{3!}+...$ (b) $-1-\frac{(x-\pi)^2}{3!}+...$

(c) $1-\frac{(x-\pi)^2}{3!}+...$ (d) $-1+\frac{(x-\pi)^2}{3!}+...$

47. Given that $F(s)$ is the one-sided Laplace transform of $f(t)$, the Laplace transform of $\int_0^t f(\tau)\,d\tau$ is

(a) $sF(s)-f(0)$ (b) $\frac{1}{s}F(s)$

(c) $\int_0^s F(\tau)\,d\tau$ (d) $\frac{1}{s}[F(s)-f(0)]$

48. Match each differential equation in Group I to its family of solution curves from Group II.

Group I	Group II
P. $\frac{dy}{dx}=\frac{y}{x}$	1. Circles
Q. $\frac{dy}{dx}=-\frac{y}{x}$	2. Straight lines
R. $\frac{dy}{dx}=\frac{x}{y}$	3. Hyperbolas
S. $\frac{dy}{dx}=-\frac{x}{y}$	

(a) $P-2, Q-3, R-3, S-1$

(b) $P-1, Q-3, R-2, S-1$

(c) $P-2, Q-1, R-3, S-3$

(d) $P-3, Q-2, R-1, S-2$

49. The eigen values of the following matrix are

$$\begin{bmatrix} -1 & 3 & 5 \\ -3 & -1 & 6 \\ 0 & 0 & 3 \end{bmatrix}$$

(a) $3, 3+5j, 6-j$ (b) $-6+5j, 3+j, 3-j$

(c) $3+j, 3-j, 5+j$ (d) $3, -1+3j, -1-3j$

50. A discrete random variable X takes values from 1 to 5 with probabilities as shown in the table.

k	1	2	3	4	5
P(X = k)	0.1	0.2	0.4	0.2	0.1

A student calculates the mean of X as 3.5 and her teacher calculates the variance of X as 1.5. Which of the following statements is true?

(a) Both the student and the teacher are right

(b) Both the student and the teacher are wrong

(c) The student is wrong but the teacher is right

(d) The student is right but the teacher is wrong

51. The eigenvalues of a skew-symmetric metric are

(a) always zero

(b) always pure imaginary

(c) either zero or pure imaginary

(d) always real

52. The trigonometric fourier series for the waveform $f(t)$ shown below contains

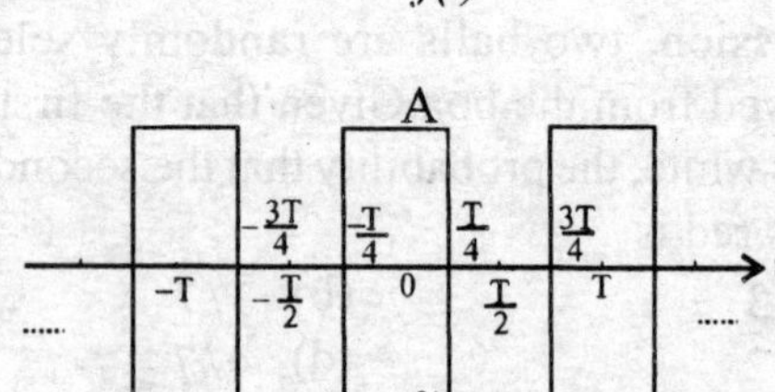

(a) Only cosine terms and zero value for the dc component

(b) Only cosine terms and a positive value for the dc component

(c) Only cosine terms and a negative value for the dc component

(d) Only sine terms and a negative value for the dc component

53. A function $n(x)$ satisfies the differential equation $\frac{d^2n(x)}{dx^2}-\frac{n(x)}{L^2}=0$ where L is a constant. The boundary conditions are: $n(0)=K$ and $n(\infty)=0$. The solution to this equation is

(a) $n(x)=K\exp(x/L)$

(b) $n(x)=K\exp(-x/\sqrt{L})$

(c) $n(x)=K^2\exp(-x/\sqrt{L})$

(d) $n(x)=K\exp(-x/L)$

54. Consider the z-transform $X(z) = 5z^2 + 4z^{-1} + 3$; $0 < |z| < \infty$. The inverse z-transform $x[n]$ is

(a) $5\,\delta[n+2] + 3\,\delta[n] + 4\,\delta[n-1]$

(b) $5\,\delta[n-2] + 3\,\delta[n] + 4\,\delta[n+1]$

(c) $5\,u[n+2] + 3\,u[n] + 4\,u[n-1]$

(d) $5\,u[n-2] + 3\,u[n] + 4\,u[n+1]$

55. If $e^y = x^{\frac{1}{x}}$, then y has a

(a) maximum at $x = e$ (b) minimum at $x = e$

(c) maximum at $x = e^{-1}$ (d) minimum at $x = e^{-1}$

Answers Set-1

1. Ans. (c)

Solution:

Give $f(x) = \begin{matrix} 1 & 0 < x < 1 \\ 0 & \text{otherwise} \end{matrix}$

$E(X) = \int x dx$

$E(X)^3 = \int x^3 dx = \left.\frac{x^4}{4}\right|_0^1 = \frac{1}{4}$

2. Ans. (d)

Solution: The characteric equation

$\lambda^3 + \lambda^2 + 2\lambda + I = 0$, must satisfy the P-matrix

$\Rightarrow P^3 + P^2 + 2P + 1 = 0$

$\Rightarrow P\left(P^2 + P + 2I + P^{-1}\right) = 0$

$P \neq 0$ *and* $P^2 + P + 2I + P^{-1} = 0$

$P^{-1} = -\left(P^2 + P + 2I\right)$

3. Ans. (a)

Solution: Rank is 4, so Q will have four linearly indenpendent columns and four linearly independent rows.

4. Ans. (d)

Solution:

$f(x) = \left(x^2 - 4\right)^2$

$f'(x) = 2\left(x^2 - 4\right).2x$

$f'(x) = 0$

$\Rightarrow 4x\left(x^2 - 4\right) = 0,$

$\Rightarrow x = 0, 2, -2$

$f''(x) = 12x^2 - 16$

$x = 0, \Rightarrow f'' = -16 \rightarrow \max$

$x = -2 \Rightarrow f'' = -64 \rightarrow \max$

$f''' = 24x$

$x = 2 \Rightarrow f''' = 24 \times 2 = 48 \rightarrow \max$

5. Ans. (b)

Solution: $f(x_n) = e^{x_n} - 1$

$x_{n+1} = x_n - \frac{f(x_n)}{f'(x_n)}$

$x_1 = x_0 - \frac{f(x_0)}{f'(x_0)}$

$f'(x_n) = e^{x_n}$

$x_1 = -1 - \frac{e^{x_o} - 1}{e^{x_o}} = -1 - \left(1 - \frac{1}{e^{x_o}}\right) = 0.367$

6. Ans. (d)

Solution:

$A^+ = \left(A^T A\right)^{-1} A^T = A^{-1}\left(A^T\right)^{-1}\left(A^T\right) = A^{-1}$

By checking options

$AA^+A = AA^{-1}A = A$

$\therefore AA^+A \neq A^+$

7. Ans.(c)

Solution: $x = \frac{h}{2}[y_0 + 2y_2]$

8. Ans.(b)

Solution:

Let an orthogonal matrix, $P = \begin{bmatrix} \cos\theta & -\sin\theta \\ \sin\theta & \cos\theta \end{bmatrix}$

Property of ordhogonal matrix P is PPT = I

Given $\bar{x} = [x_1 \quad x_2]^T$

$= \begin{bmatrix} x_1 \\ x_2 \end{bmatrix}$

so, now $P\,\bar{x} = \begin{bmatrix} \cos\theta & -\sin\theta \\ \sin\theta & \cos\theta \end{bmatrix}\begin{bmatrix} x_1 \\ x_2 \end{bmatrix}$

$P\,\bar{x} = \begin{bmatrix} x_1\cos\theta - x_2\sin\theta \\ x_1\cos\theta + x_2\cos\theta \end{bmatrix}$

$\|P\,\bar{x}\| = \sqrt{\begin{matrix}(x_1\cos\theta - x_2\sin\theta)^2 \\ +(x_1\sin\theta + x_2\cos\theta)^2\end{matrix}}$

$$\|P\,\bar{x}\| = \sqrt{x_1^2 + x_2^2}$$

$$\|P\,\bar{x}\| = \|\,\bar{x}\,\| \text{ for all vector } \bar{x}$$

9. Ans. (c)

Solution:

Second method

Letd λ_1 and λ_2 are eigen given values of matrix

$$\lambda_1 + \lambda_2 = \text{Trace}$$

$$\lambda_1 \times \lambda_2 = \text{Determinant}$$

$$\lambda_1 + \lambda_2 = -2 \quad \text{...(1)}$$

$$\lambda_1 \times \lambda_2 = -35 \quad \text{...(2)}$$

$$\lambda_2 = -2 - \lambda_1$$

Put this value in (2) equation

$$\lambda_1 \times (-2 - \lambda_1) = -35$$

$$-2\lambda_1 - \lambda_1^2 = -35$$

$$\lambda_1^2 + 2\lambda_1 - 35 = 0$$

$$\lambda_1^2 + 7\lambda_1 - 5\lambda_1 - 35 = 0$$

$$\lambda_1(\lambda_1 + 7) - 5(\lambda_1 + 7) = 0$$

$$(\lambda_1 - 5)(\lambda_1 + 7) = 0$$

$$\lambda_1 = 5 \text{ or } -7$$

Put value $\lambda_1 = 5$ in equation 1

$$5 + \lambda_2 = -2$$

$$\lambda_2 = -7$$

$\therefore$ Eigen values are $5, -7$

10. Ans. (a)

Solution:

11. Ans. (b)

Solution:

12. Ans. (c)

Solution:

13. Ans. (a)

Solution:

14. Ans. (c)

Solution: Use $[\lambda I - P] = 0$

$$\begin{bmatrix} \lambda & 0 \\ 0 & \lambda \end{bmatrix} - \begin{pmatrix} P_{11} & P_{12} \\ P_{21} & P_{22} \end{pmatrix} = 0$$

$$\begin{bmatrix} \lambda - P_{11} & -P_{12} \\ -P_{21} & \lambda - P_{22} \end{bmatrix} = 0$$

$$\Rightarrow (\lambda - P_{11})(\lambda - P_{22}) - P_{21}P_{12} = 0$$

Putting $\lambda = 0$, we get $P_{11}P_{22} - P_{21} = 0$

15. Ans. (b)

Solution: Given $4x + 2y = 7$ ...(1)

$$2x + y = 6 \quad \text{...(2)}$$

Equation (1) Can be written as

$$2x + y = 3.5 \quad \text{...(3)}$$

Now LHS of equation (2) and (3) are same but RHS is different which is not possible. Hence above equations have no solution.

16. Ans. (a)

Solution: Since sine value lies between –1 and + 1. Therefore it has no real or complex solution

17. Ans. (a)

Solution: Given function

$$f(x) = e^x + e^{-x}$$

$$f'(x) = e^x - e^{-x}$$

For maximama or minimum

$$f'(x) = 0 \Rightarrow e^x - e^{-x} = 0$$

$$f''(x) = e^x + e^{-x}$$

$$f''(0) \Rightarrow \text{Positive}$$

Thus minimum of

$$f(x) = e^0 + e^0 = 2$$

18. Ans. (a)

Solution: $\sin x = x - \dfrac{x^3}{3!} + \dfrac{x^5}{5!} + \cdots$

$$\cos(x) = 1 - \frac{x^2}{2!} + \frac{x^2}{4!} + \cdots$$

Thus $\sin(x^3)$ will have odd powers

$\sin(x^2)$ will have even powers

$\cos(x^3)$ will have even powers

$\cos(x^2)$ will have even powers

19. Ans. (b)

Solution: $\dfrac{dx(t)}{dt} + 3x(t) = 0$

$$\frac{dx(t)}{dt} = -3x(t)$$

$$\frac{dx(t)}{x(t)} = -3dt$$

or $$\int \frac{dx(t)}{x(t)} = \int 3dt$$

$$\log x(t) = 3t$$

$$x(t) = k.e^{-3t}$$

20. Ans. (c)

Solution: $f(x_n) = x_n e^{-x_n}$

$$f'(x_n) = 1 + e^{-x_n}$$

$$x_{n+1} = x_n - \frac{x_n - e^{-x_n}}{1 + e^{-x_n}}$$

$$= \frac{x_n + x_n e^{-x_n} - x_n + e^{-x_n}}{1 + e^{-x_n}} = \frac{(x_n + 1)e^{-x_n}}{1 + e^{-x_n}}$$

21. Ans. (a)

Solution: Residue at z = 2

$$= \lim_{z\to 2} \frac{1}{1!} \frac{d}{d^2}\left[(z-2)^2 f(z)\right]$$

$$\lim_{z\to 2} \frac{1}{1!} \frac{d}{dz}(z-2)^2 \frac{1}{(z+2)^2 (z-2)^2}$$

$$= \lim_{z\to 2} \frac{-Z}{(z+2)^3} \Rightarrow \text{Residue} = \frac{-1}{32}$$

22. Ans. (d)

Solution: $P = \begin{bmatrix} 0 & 1 \\ -2 & -3 \end{bmatrix}$

$$e^P = LT^{-1}\,[sI - P]^{-1}$$

$$\left[\begin{bmatrix} s & 0 \\ 0 & s \end{bmatrix} - \begin{bmatrix} 0 & 1 \\ -2 & -3 \end{bmatrix}\right]^{-1} = \begin{bmatrix} s & -1 \\ 2 & s+3 \end{bmatrix}^{-1}$$

$$= LT^{-1}\left[\frac{1}{(s+1)(s+2)}\begin{bmatrix} s+3 & 1 \\ -2 & s \end{bmatrix}\right]$$

$$= \begin{bmatrix} \frac{s+3}{(s+1)(s+2)} & \frac{1}{(s+1)(s+2)} \\ \frac{-2}{(s+1)(s+2)} & \frac{s}{(s+1)(s+2)} \end{bmatrix}$$

$$e^P = LT^{-1}\begin{bmatrix} \frac{2}{(s+1)} - \frac{1}{(s+2)} & \frac{1}{(s+1)} - \frac{1}{(s+2)} \\ \frac{-2}{(s+1)} + \frac{2}{(s+2)} & \frac{2}{(s+2)} - \frac{1}{(s+1)} \end{bmatrix}$$

$$= \begin{bmatrix} 2e^{-1} - e^{-2} & e^{-1} - e^{-2} \\ -2e^{-1} + 2e^{-2} & 2e^{-2} - e^{-1} \end{bmatrix}$$

23. Ans. (b)

Solution: $f(x) = e^x + \sin x$

Taylor's series,

$$f(x) = f(a) + (x-a)f'(a) + \frac{(x-a)^2}{2!} f''(a)$$

Where a = p

$$f(x) = f(\pi) + (x-\pi)f'(\pi) + \frac{(x-\pi)^2}{2!} f''(\pi)$$

Cofficient of $(x-\pi)^2 = \frac{1}{2!} f''(x)$

$$f'(x) = e^x + \cos x$$

$$f''(x) = e^x - \sin x$$

$$f''(x)\Big|_{x=\pi} = e^\pi$$

Thus coefficient of $(x-\pi)^2 = 0.5$

24. Ans. (a)

Solution:

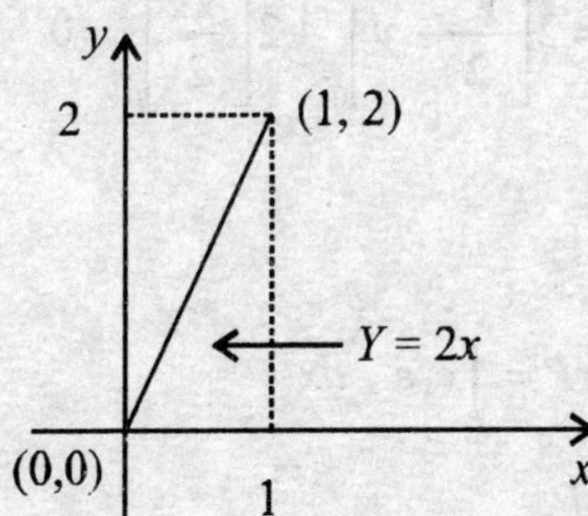

The straight line can be expressed as

$$y = 2x$$

Then $g(x, y) = 4x^3 + 10y^4 = 4x^3 + 10(2x)^4$

$$\therefore \quad I = \int_0^1 \left[4x^3 + 10(2x)^4\right] dx$$

$$= \int_0^1 (4x^3 + 160\ x^4) dx$$

$$= \left[\frac{4x^4}{4} + \frac{160}{5}x^5\right]_0^1 = 33$$

25. Ans. (b)

Solution: $Y = mx + c$

$C = 1$

$m = -1$

$Y = -1 \cdot x + c$

or $Y + x = 1$

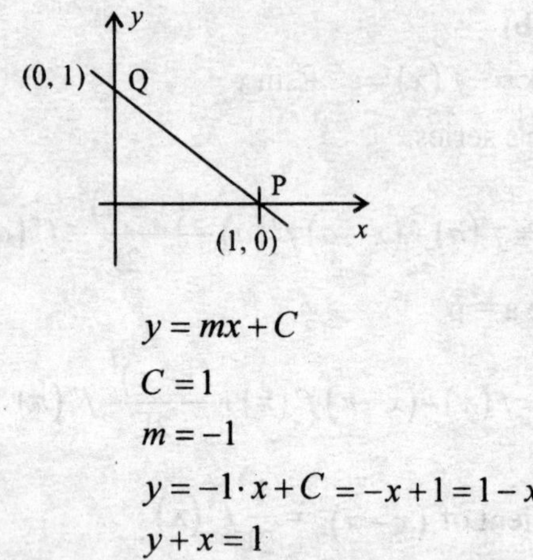

$$y = mx + C$$
$$C = 1$$
$$m = -1$$
$$\therefore \quad y = -1 \cdot x + C = -x + 1 = 1 - x$$
$$\text{or,} \quad y + x = 1$$

The straight line equation is $x + y = 1$

$$I = \int_P^Q (x\,dx + y\,dy)$$

Then, $I = 2\int_0^1 (1-y)\cdot(-dy) + 2\int_0^1 y\ dy$

$$= 2\left[\frac{y^2}{2} - y\right]_0^1 + 2\left[\frac{y^2}{2}\right]_0^1 = 0$$

26. Ans. (b)

Solution:

As given $P = \int_0^1 x\ e^x\ dx$

Let, $x = I$

$e^x = II$

So using product rule of integration.

$$\int I \cdot II\ dx = I \int II\ dx - \int \left[\left(\frac{dI}{dx}\right)\int II\ dx\right] dx$$

$$\int_0^1 xe^x\ dx = xe^x\Big|_0^1 - \int_0^1 e^x\ dx$$

$$= [e - 0] - e^x\Big|_0^1$$

$$= e - [e - 0] = e - e + 1$$

$$= 1$$

27. Ans. (a)

Solution:

As we know that divergence of any veetol $\vec{r}$ is

$$\nabla \cdot \vec{r}$$

$\vec{r}$ is given three dimensional

so suppose veeto $\vec{r}$ as

$\vec{r} = x\tilde{i} + y\hat{j} + z\hat{k}$

Where

$\tilde{i}, \hat{j}, \hat{k}$ are unit vector in the direction of (x, y, z) respectively, so

$$\nabla \cdot \vec{r} = \left(\frac{d}{dx}\tilde{i} + \frac{d}{\partial y}\hat{j} + \frac{d}{\partial z}\hat{k}\right) \cdot \left(x\tilde{i} + y\hat{j} + z\hat{k}\right)$$

$$= \frac{d}{dx}x + \frac{d}{dy}y + \frac{d}{dz}z = 1 + 1 + 1 = 3$$

28. Ans. (d)

Solution:

$$x(t) = \sin(2\pi + \omega t)$$

$$x(t) = 8\sin\left(0.8\,\pi t + \frac{\pi}{4}\right)$$

$$\omega t = 0.8\,\pi t$$

$$\omega = \frac{2\,\pi}{T}$$

$$0.8\,\pi = \frac{2\,\pi}{T}$$

$$T = \frac{2\,\pi}{0.8\,\pi} = 2.5\,s$$

29. Ans. (b)

Solution:

Given input and output

$$y(t) = \int_{-\infty}^{5t} x(\tau)d\tau,\ t > 0$$

For first input $x_1(\tau)$

$$y_1(t) = \int_{-\infty}^{5t} x_1(\tau)\,d\tau$$

For second input $x_2(\tau)$

$$y_2(t) = \int_{-\infty}^{5t} x_2(\tau)d\tau$$

For input $x_1(\tau) + x_2(\tau)$

$$y_3(t) = \int_{-\infty}^{5t} \left[x_1(\tau) + x_2(\tau)\right]d\tau$$

$$y_3(t) = y_1(\tau) + y_2(\tau)$$

So system is linear.

Causal: Since system depends on the value for $\tau < 0$, so that the system is non-causal.

30. Ans. (d)

Solution: 4 heads & 0 tail or 3 head & 1 tail

$$P={}^4C_0\left(\frac{1}{2}\right)^2\left(\frac{1}{2}\right)^0 + {}^4C_3\left(\frac{1}{2}\right)^3\left(\frac{1}{2}\right)^1 = \frac{5}{16}$$

Second Method

In this question we can use the binomial probability distribution in which P be the probability of success and q that failure

then $P(X=r) = {}^nC_r P^r q^{n-r}$

we required either 3 Head or all four head in outcome for sauces

Probaliestoy of $P(H) = P(T) = \frac{1}{2}$

Required probability

$$= {}^4C_4\left(\frac{1}{2}\right)^4 \left(\frac{1}{2}\right)^0 + {}^4C_3\left(\frac{1}{2}\right)^3 \left(\frac{1}{2}\right)^1$$

$$= \frac{1}{16} + \frac{4}{16} = \frac{5}{16}$$

31. Ans. (c)

Solution:

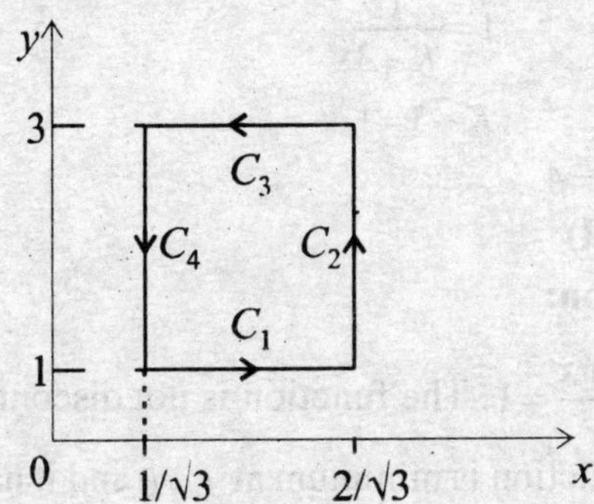

As $\overline{A} = xy\,\hat{a}_x + x^2\hat{a}_y$

$$\oint_C \vec{A}.\vec{dl}$$

$$= \oint_C \left(xy\hat{a}_x + x^2\hat{a}_y\right).\left(\hat{a}_x dx + \hat{a}_y dy\right)$$

$$= \oint_C xy\,dx + x^2 dy.$$

Now along this closed loop total integral will be.

$$= \left[\begin{array}{l} \int_{C_1} xydx + x^2dy + \int_{C_2} xy\,dx + x^2dy \\ + \int_{C_3} xy\,dx + x^2dx + \int_{C_4} xy\,dx + x^2dy \end{array}\right]$$

Along C_1 : $y = 1$ and $x = \frac{1}{\sqrt{3}}$ to $2/\sqrt{3}$

$dy = 0$

$$\int xydx + x^2dy = \int_{\frac{1}{\sqrt{3}}}^{2/\sqrt{3}} x\,dx + 0$$

$$= \frac{1}{2}x^2\Big|_{1/\sqrt{3}}^{2/\sqrt{3}} = \frac{1}{2}\left(\frac{4}{3} - \frac{1}{3}\right) = \frac{1}{2} \qquad ...(1)$$

Along $C_2 : x = 2/\sqrt{3}$ and $y = 1$ to 3

$dx = 0$

$$\text{so, } \int_{C_2} xydx + x^2dy = \int_1^3 0 + \left(\frac{2}{\sqrt{3}}\right)^2 dy$$

$$= \frac{4}{3}[y]_1^3 = \frac{4}{3}[3-1] = \frac{8}{3} \qquad ...(2)$$

Along $C_3 : y = 3$ and $x = \frac{2}{\sqrt{3}}$ to $\frac{1}{\sqrt{3}}$

$dy = 0$

$$\int_{C_3} xydx + x^2dy = \int_{2/\sqrt{3}}^{1/\sqrt{3}} 3x\,dx \qquad (3)$$

$$= \frac{3}{2}x^2\Big|_{2/\sqrt{3}}^{\frac{1}{\sqrt{3}}} = \frac{3}{2}\left[\frac{1}{3} - \frac{4}{3}\right] = \frac{-3}{2}$$

Along $C_4 : x = \frac{1}{\sqrt{3}}$ and $y = 3$ to 1

$$\int_{C_4} xydx + x^2dy = \int_3^1 \left(\frac{1}{\sqrt{3}}\right)^3 dy$$

$$= \frac{1}{3}[y]_3^1 = \frac{1}{3}[1-3] = \frac{-2}{3} \qquad ...(4)$$

Put all the. value from equation 1, 2, 3 and 4

we get $\oint_C A.dl = \frac{1}{2} + \frac{8}{3} - \frac{3}{2} - \frac{2}{3}$

$$= \frac{3+16-9-4}{6}$$

$$= \frac{6}{6}$$

$$\oint_C A.dl = 1$$

32. Ans. (c)

Solution:

Residue at pole $= \left.\frac{X(z)}{(\text{Pole})}\right|_{z=\text{pole}}$

i.e. $z\, x\, X(z)\big|_{z=0} = \left.\frac{1-2z}{(z-1)(z-2)}\right|_{z=0}$

$$= \frac{1}{-1\times -2} = \frac{1}{2}$$

$$(z-1)X(z)\big|_{z=1} = \left.\frac{1-2z}{z(z-2)}\right|_{z=1} = \frac{1-2\times(1)}{1(1-2)}$$

$$= \frac{1-2}{1\times -1} = \frac{-1}{-1} = 1$$

$$(z-2)X(z)\big|_{z=2} = \left.\frac{1-2z}{z(z-1)}\right|_{z=2} = \frac{1-2\times(+2)}{+2(+2-1)}$$

$$= \frac{1-4}{2\times 1} = \frac{-3}{2}$$

33. Ans. (b)

Solution:

Given that $\frac{dy(x)}{dx} - y(x) = x$

$y(0) = 0$ and $h = 0.1$

$x_1 = 0$

$y_1 = 0$

$$\therefore \left(\frac{dy}{dx}\right) = 0 - 0 = 0$$

$$y_2 = y_1 + h\left(\frac{dy}{dx}\right)_1$$

$$= 0 + 0.1\times 0 = 0$$

$x_2 = x_1 + h = 0.1$

$y_2 = 0$

$$= x_2 + y_2(x)$$

$$\therefore \left(\frac{dy}{dx}\right)_2 = x_2 + y_2(x)$$

$$= 0.1 + 0 = 0.1$$

$$y_3 = y_2 + h\left(\frac{dy}{dx}\right)_2 = 0 + 0.1\times 0.1 = 0.01$$

$x_3 = x_2 + h = 0.2$

$y_3 = 0.01$

$$\therefore \left(\frac{dy}{dx}\right)_3 = x_3 + y_3(x)$$

$$= 0.2 + 0.01$$
$$= 0.21$$

$$\therefore y_4 = y_3 + h\left(\frac{dy}{dy}\right)$$

$$= 0.01 + 0.1\times 0.21$$
$$= 0.01 + 0.021$$
$$= 0.031$$

$x_4 = x_3 + h = 0.3$

$y_4 = 0.31$

$\therefore y(0.3) = 0.031$

34. Ans. (d)

Solution:

Apply find value the one $\lim_{t\to\infty} f(t) = \lim_{s\to 0} s\cdot F(s)$

$$1 = \left.\frac{s(3s+1)}{s^3+4s^2+(k-3)s}\right|_{s\to 0} = \left.\frac{3s+1}{s^2+4s+(k-3)}\right|_{s\to 0}$$

or $1 = \frac{1}{K-3}$

or $K-3 = 1$

$K = 4$

35. Ans. (d)

Solution:

$\underset{X\to 0}{Lt} \frac{\sin x}{x} = 1$. The function is not discontinuous

$\therefore$ The function is minimum at $t = 0$ and it has no point of re-flexion, at $t = 0$ and it is also not maximum

36. Ans. (d)

Solution:

Given

Number of white balls $= 4$

Number of red balls $= 3$

Total Number of balls $= 4 + 3 = 7$

out of 7 balls, 2 can be drawn in ${}^7C_2$ ways

Exhaustine number of cases

$${}^7C_2 = \frac{7!}{2!5!} = \frac{7\times 6}{2} = 21$$

Our of 4 white balls, 1 can be crown

$${}^4C_2 = \frac{4!}{3!} = 4$$

Out of 3 red balls, 1 can be drawn ${}^3C_1 = 3$.

So total favorable number of cases $= 3 \times 4 = 12$

So Probability $= \frac{12}{21} = \frac{4}{7}$

37. Ans. (b)

Solution:

given matrix

$$P = \begin{pmatrix} 1 & 1 & 0 \\ 0 & 2 & 2 \\ 0 & 0 & 3 \end{pmatrix}$$

$$|P - \lambda I| = 0$$

$$\begin{vmatrix} 1-\lambda & 1 & 0 \\ 0 & 2-\lambda & 2 \\ 0 & 0 & 3-\lambda \end{vmatrix}$$

$$(3-\lambda)\left[(1-\lambda)(2-\lambda)-0\right] = 0$$

$\lambda = 3,\ 1, 2 \rightarrow$ eigen values

Now eigen vector for $\lambda = 3$

$$|A - 3I|X = 0$$

or $\begin{vmatrix} -2 & 1 & 0 \\ 0 & -1 & 2 \\ 0 & 0 & 0 \end{vmatrix} \begin{vmatrix} x_1 \\ x_2 \\ x_3 \end{vmatrix} = 0$

or $\quad -2x_1 + x_2 + 0x_3 \qquad ...(1)$

$\quad 0x_1 - x_2 + 2x_3 = 0 \qquad ...(2)$

Using equations (1) and (2)

$$\frac{x_1}{2-0} = \frac{x_2}{0+4} = \frac{x_3}{2-0}$$

$$= \frac{x_1}{2} = \frac{x_2}{4} = \frac{x_3}{2}$$

or $\frac{x_1}{1} = \frac{x_2}{2} = \frac{x_3}{1}$

So eigne vectors are $X = \begin{vmatrix} 1 \\ 2 \\ 1 \end{vmatrix}$

or $\begin{bmatrix} 1 & 2 & 1 \end{bmatrix}^T$

38. Ans. (b)

Solution:

Given differential equation is

$$\frac{d^2x}{dt^2} + \frac{6dx}{dt} + 8x = 0 \qquad ...(1)$$

This differential equation is homogenous so it will only have complementey function. No particular ineegral will be true

Put $\frac{d}{dE} = D$

from equation (1) $\left(D^2x + 6Dx + 8x\right) = 0$

$\left(D^2 + 6D + 8\right)x = 0$

Auxullay equation $D^2 + 6D + 8 = 0$

$$D^2 + 4D + 2D + 8 = 0$$

$$D(D+4) + 2(D+4) = 0$$

$$(D+2)(D+4) = 0$$

$$D = -2, -4$$

As roots are real and distinct so solution will be

$$x(t) = A_1e^{-2t} + A_2e^{-4t} \qquad ...(2)$$

Now $x(0) = A_1e^0 + A_2e^0 \qquad A_1 + A_2 = 1$

Equation (2) of (3) satisfies the option (b) only

39. Ans. (d)

Solution:

Since there are only two constraints and there are four variables there will be multiple non trivial solutions.

40. Ans. (d)

Solution:

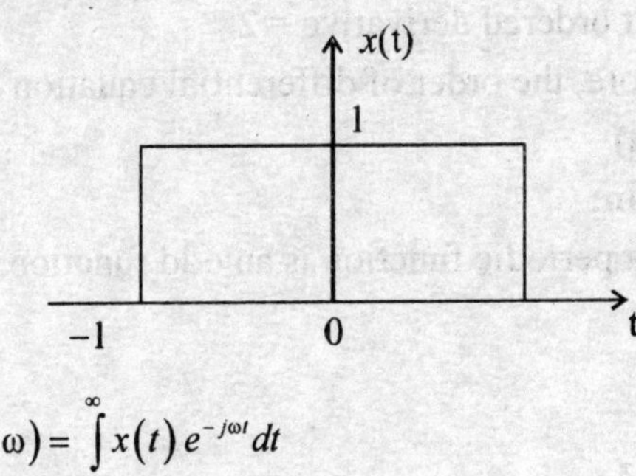

$$X(\omega) = \int_{-\infty}^{\infty} x(t)\, e^{-j\omega t}\, dt$$

$$= \int_{-1}^{+1} 1 \cdot e^{-j\omega t}\, dt = 2\left(\frac{\sin\omega}{\omega}\right)$$

find $\int_{-\infty}^{\infty} |X(\omega)|^2 d\omega$

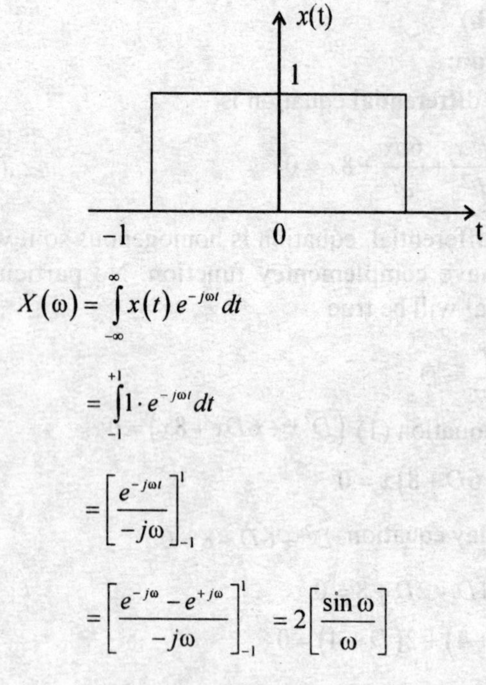

$$X(\omega) = \int_{-\infty}^{\infty} x(t)\, e^{-j\omega t}\, dt$$

$$= \int_{-1}^{+1} 1 \cdot e^{-j\omega t}\, dt$$

$$= \left[\frac{e^{-j\omega t}}{-j\omega}\right]_{-1}^{1}$$

$$= \left[\frac{e^{-j\omega} - e^{+j\omega}}{-j\omega}\right]_{-1}^{1} = 2\left[\frac{\sin\omega}{\omega}\right]$$

$$\int_{-\infty}^{\infty} |X(\omega)|^2 d\omega = \int_{-\infty}^{\infty} 4\left(\frac{\sin\omega}{\omega}\right)^2 d\omega$$

$$= 4\int_{-\infty}^{\infty}\left(\frac{\sin\omega}{\omega}\right)^2 d\omega$$

$$= 4\pi$$

41. Ans. (b)

Solution:

The order of a differential equation is the order of the highest ordered derivative present in the equation. The given equation is.

$$\frac{d^2y}{dt^2} + \left(\frac{dy}{dx}\right)^3 + y^4 = e^{-t}$$

Highest ordered derivative = 2

Therefore, the order of differential equation is 2 .

42. Ans. (a)

Solution:

When a periodic function is an odd function, then

$$a_0 = 0$$

$$a_n = 0$$

$$b_n = \frac{2}{\pi}\int_{o}^{\pi} f(x) \sin n\, x\, dx$$

when a periodic function is an even function, then

$$a_o = \frac{2}{\pi}\int_{0}^{\pi} f(x)\, dx$$

$$a_n = \frac{2}{\pi}\int_{0}^{\pi} f(x) \cos n\, x dx$$

$$b_n = o$$

Thus, if function is odd , then its fourier expansion contains only sine terms and if function is even, then its fourier expansion contains only cosine terms.

43. Ans. (b)

Solution: Given that

$$f(t) = \sin^2 t + \cos 2t$$

$$= \frac{1-\cos 2t}{2} + \cos 2t$$

$$= \frac{1-\cos 2t + 2\cos 2t}{2}$$

$$= \frac{1+\cos 2t}{2}$$

$$f(t) = \frac{1}{2} + \frac{1}{2}\cos 2t$$

$$f(t) = f_1(t) + f_2(t)$$

where $f_1(t) = \frac{1}{2}$

$$\omega_1 = 0$$

$$f_1 = 0 \text{ Hz}$$

And, $f_2(t) = \frac{1}{2}\cos 2t$

$$\omega_2 = 2$$

$$2\pi f_2 = 2$$

$$f_2 = \frac{1}{\pi}\text{ Hz}$$

Therefore, it has 0 and $\frac{1}{\pi}$ Hz frequency component.

44. Ans. (a)

Solution: Given that

$$x(n) = \left(\frac{1}{3}\right)^n u(n) - \left(\frac{1}{2}\right)^n u(-n-1)$$

$$X(z) = \frac{1}{1-\frac{1}{3}z^{-1}} + \frac{1}{1-\frac{1}{2}z^{-1}}$$

$$X(z) = X_1(z) + X_2(z)$$

Where $X_1(z) = \dfrac{1}{1-\frac{1}{3}z^{-1}}$

ROC $\Rightarrow |z| > \dfrac{1}{3}$

And $X_2(z) = \dfrac{1}{1-\frac{1}{2}z^{-1}}$

ROC $\Rightarrow |z| < \dfrac{1}{2}$

$\therefore$ $X(z)$ has ROC

$\dfrac{1}{3} < |z| < \dfrac{1}{2}$

45. Ans. (c)

Solution:

$P\left(\dfrac{x-y=2}{x-y=0}\right) = ?$

Probability of

$$\begin{array}{c|c|c} X=0 & Y=0 & (1/2) \\ X=1 & Y=1 & (1/4) \\ X=2 & Y=2 & (1/4) \end{array}$$

Let $P(X+Y=2) = P(A)$

$P(X-Y=0) = P(B)$

and we are suppose to find $P(A/B)$

So, $P(X+Y=2) = P(A)$

$$= P(X=0)P(Y=2) + P(X=1)P(Y=1) + P(X=2)P(Y=0)$$

Because X and Y are independent variable

So, $P(A) = \dfrac{1}{2}\cdot\dfrac{1}{4} + \dfrac{1}{4}\cdot\dfrac{1}{4} + \dfrac{1}{2}\cdot\dfrac{1}{4} = \dfrac{5}{16}$

$P(X-Y=0) = P(B)$

$$= P(X=0)P(Y=0) + P(X=1)P(Y=1) + P(X=2)P(Y=2)$$

$$= \frac{1}{2}\cdot\frac{1}{2} + \frac{1}{4}\cdot\frac{1}{4} + \frac{1}{4}\cdot\frac{1}{4} = \frac{6}{16}$$

Now $P\left(\dfrac{A}{B}\right) = \dfrac{P(A\cap B)}{P(B)}$

$$P(A\cap B) = P(X+Y=2)\cap(X-Y=0)$$

$$= P(X=1)P(Y=1)$$

$$= \frac{1}{4}\cdot\frac{1}{4} = \frac{1}{16}$$

As only one case is possible when both $X \neq Y = 1$

Hence $P\left(\dfrac{A}{B}\right) = \dfrac{(1/16)}{(6/16)} = \dfrac{1}{6}$

$P\left(\dfrac{X+Y=2}{X-Y=0}\right) = \dfrac{1}{6}$

46. Ans. (d)

Solution:

$$\sin x = x - \frac{x^3}{3!} + \frac{x^5}{5!}$$

$$\sin(x-\pi) = (x-\pi) - \frac{(x-\pi)^3}{3!} + \frac{(x-\pi)^5}{5!}$$

$$\Rightarrow -\frac{\sin x}{(x-\pi)} = 1 - \frac{(x-\pi)^2}{3!} + \frac{(x-\pi)^4}{5!}$$

$$\Rightarrow \frac{\sin x}{(x-\pi)} = -1 + \frac{(x-\pi)^2}{3!} + \ldots$$

47. Ans. (b)

Solution:

The laplace transform of $\int_o^t f(\tau)\,dz$

$= \dfrac{1}{s}[F(s) - f(0)]$

Since it is one sided, take initial conditions equal to zero

$$\int_o^t f(\tau)\,d\tau = \frac{1}{s}F(s)$$

48. Ans. (a)

Solution:

Using $\dfrac{dy}{dx} = \dfrac{y}{x}$

$\dfrac{1}{y}\,dy = \dfrac{1}{x}\,dx$

Integrate both side

$\log y = \log x + \log c$

$\log y - \log x = \log c$

$\log \frac{y}{x} = \log c$

$\frac{y}{x} = c$

Nor $y = cx$

Which is equation of straight line.

Using $\frac{dy}{dx} = \frac{-y}{x}$

$\frac{1}{y} dy = -\frac{1}{x} dx$

$\log y = -\log x + \log c$

$\log x + \log y = \log c$

$xy = c$

Using $\frac{dy}{dx} = \frac{x}{y}$ $y\,dy = x\,dx$

Integrating on both side, we get

$\frac{y^2}{2} = \frac{x^2}{2} + c$

$\frac{y^2}{2} - \frac{x^3}{2} = c$

$y^2 - x^2 = 2c$

$y^2 - x^2 = \text{constant}$

Which is an equation of hyperbola

Using $\frac{dy}{dx} = \frac{-x}{y}$

$y\,dy = -x\,dx$

Integrating on both side, we get

$\frac{y^2}{2} = \frac{-x^2}{2} + c$

$x^2 + y^2 = 2c$

$x^2 + y^2 = \text{constant}$

Which is an equation of circle .

49. Ans. (d)

Solution:

Given that, $A = \begin{bmatrix} -1 & 3 & 5 \\ -3 & -1 & 6 \\ 0 & 0 & 3 \end{bmatrix}$

The characteristic equation of A is

$|A - \lambda I| = o$

Where I is an identity matrix

$\lambda I = \lambda \begin{bmatrix} 1 & 0 & 0 \\ 0 & 1 & 0 \\ 0 & 0 & 1 \end{bmatrix}$

$= \begin{bmatrix} \lambda & 0 & 0 \\ 0 & \lambda & 0 \\ 0 & 0 & \lambda \end{bmatrix}$

$|A - \lambda I| = \begin{bmatrix} -1 & 3 & 5 \\ -3 & -1 & 6 \\ 0 & 0 & 3 \end{bmatrix} - \begin{bmatrix} \lambda & 0 & 0 \\ 0 & \lambda & 0 \\ 0 & 0 & \lambda \end{bmatrix}$

$= \begin{bmatrix} -1-\lambda & 3 & 5 \\ -3 & -1-\lambda & 6 \\ 0 & 0 & 3-\lambda \end{bmatrix}$

$|A - \lambda I| = \lambda^3 - \lambda^2 + 4\lambda - 30$

From $|A - \lambda I| = 0$

$\lambda^3 - \lambda^2 + 4\lambda - 30 = 0$

$(\lambda - 3)(\lambda^2 + 2\lambda + 10) = 0$

$(\lambda - 3)(\lambda^2 + 2\lambda + 10) = 0$

$(\lambda - 3)(\lambda + 1 - 3j)(\lambda + 1 + 3j) = 0$

$\lambda = 3,\ -1 + 3j,\ -1 - 3j$

The eigen value are 3, $-1 + 3j$, $-1 - 3j$

$= \frac{-150 \times 3K}{3K} v_i$

$= -150\left(A \cos 20t + \beta \sin 10^6 t\right)$

Since coupling capacitor is large, so low frquency effect is to be considered that filter out the low.

50. Ans. (b)

Solution:

k(x)	p(f)	kp	$x = X - \overline{X}$	x^2	px^2
1	0.1	0.1	− 2	4	0.4
2	0.2	0.4	− 1	1	0.2
3	0.4	1.2	0	0	0
4	0.2	0.8	1	1	0.2
5	0.1	0.5	2	4	0.4

$\Sigma p(f) = N = 1$

$\Sigma kp = 3$

$\Sigma px^2 = 1.2$

$$\text{Mean} = \frac{\sum k\,p}{N} = \frac{3}{1} = 3$$

$$\text{Strandard deviation } \sigma = \sqrt{\frac{\sum P\,x^2}{N}}$$

$$\text{Varience } \sigma^2 = \frac{\sum p\,x^2}{N} = \frac{1.2}{1}$$

Variance = 1.2

51. Ans. (a)

Solution:

Let,

$$\text{Matrix given A} \Rightarrow \begin{bmatrix} 2 & 4 \\ 5 & 8 \end{bmatrix} \text{ Then}$$

$$A' \rightarrow \begin{bmatrix} 2 & 5 \\ 4 & 8 \end{bmatrix}$$

$$\text{skew-symmetric matrix} \rightarrow \frac{1}{2}\left[A - A^1\right]$$

$$\Rightarrow \frac{1}{2}\begin{bmatrix} 0 & -1 \\ 1 & 0 \end{bmatrix} \Rightarrow \begin{bmatrix} 0 & \frac{-1}{2} \\ \frac{1}{2} & 0 \end{bmatrix}$$

Since maginitude of diagonal element is same only sign is changed, so sign value must be either zero or pure imaginary to satisfy this relation.

52. Ans. (c)

Solution:

Since given waveform is even, thus it will have only cosine terms. Function $f(t)$ has more negative value then positive, so it will have negative DC Component.

Second Method

Trigonometric Fourier series is expressed as

$$f(t) \Rightarrow \frac{a_o}{2} + \left[\sum_{n=1}^{\infty}(a_n \cos n\omega_o t) + b_n \sin n\,\omega_o t\right]$$

$$\omega_0 \rightarrow \frac{2\pi}{T}$$

$a_0 \rightarrow$ Reprcnts the d.c. value

i.e. the area of the waveform for one period. T above & below t axis

$$\text{Area from } -\frac{T}{4} \text{ to } \frac{T}{4}$$

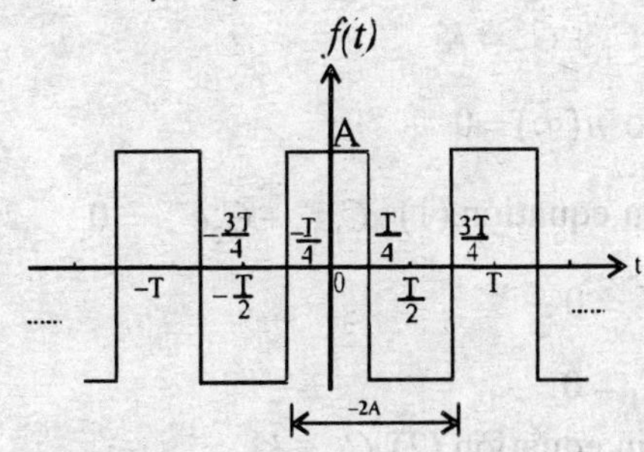

$$\Rightarrow \frac{T}{2} \times A \rightarrow \frac{AT}{2}$$

$$\text{Area from } \frac{T}{4} \text{ to } \frac{3T}{4}$$

$$\Rightarrow \frac{T}{2} \times -2A = -AT$$

$\Rightarrow$ Net area is in negative

$\therefore$ Negative value of d.c. component more over the waveform has even symmetry.

$\therefore$ Only a_n coefficient are present, i.e. cosine terms are present.

53. Ans. (d)

Solution:

Given differential equation.

$$\frac{d^2 n(x)}{dx^2} - \frac{n(x)}{L^2} = 0$$

$$\text{Put } \frac{\partial^2}{dx} = D \text{ or } \frac{\partial^2}{\partial x^2} = D^2$$

$$\text{Then } D^2\,n(x) - \frac{1}{L^2}n(x) = 0$$

$$\left(D^2 - \frac{1}{L^2}\right)n(x) = 0$$

$$\text{A.E } D^2 - \frac{1}{L^2} = 0$$

$$D^2 = \frac{1}{L^2}$$

$$D = \pm\frac{1}{L}$$

So complementary function

$CF = n(x) = C_1 e^{\frac{1}{L}x} + C_2 e^{\frac{-1}{L}x}$ ...(1)

given data $n(0) = K$

from equation (1) $C_1 e^0 + C_2 e^0 = K$

or $C_1 + C_2 = K$ ...(2)

also $n(\infty) = 0$

from equation (1) $C_1 e^{\infty} + C_2 e^{-\infty} = 0$

$C_1 e^{\infty} = 0$

$C_1 = 0$ ...(3)

Put in equation (2) $C_2 = K$ ...(4)

using equation (3) and (4) in equation (1)

$$n(x) = 0 + Ke^{\frac{-1}{L}x}$$

$$= Ke^{\frac{-1}{L}x}$$

54. Ans. (a)

Solution:

$$X(z) = 5z^2 + 4z^{-1} + 3$$

$$= 5z^2 + 3 + 3z^{-1}$$

$$X(n) = \left\{ \begin{matrix} 5, 0, 3, 4 \\ \uparrow \end{matrix} \right\}$$

55. Ans. (a)

Solution:

As given $e^y = x^{\frac{1}{x}}$

To find the maxima and minima find it differentiation and equate to zero

i.e. $e^y = x^{\frac{1}{x}}$

take log both side

$$\log e^y = \log x^{\frac{1}{x}}$$

$$y = \frac{1}{x} \log x$$

Now $\frac{dy}{dx} = \frac{1}{x^2} - \frac{1}{x^2} \log x$

put $\frac{dy}{dt} = 0$ for maximum or minima

$$\frac{1}{x^2} - \frac{1}{x^2} \log x = 0$$

$\frac{1}{x^2} = 0$ or $1 - \log x = 0$

$\log x = 1$

or maxima at $x = e$

One Mark Questions

1. The Laplace transform of $i(t)$ is given by $I(s) = \frac{2}{s(1+s)}$. As $t \to \infty$, the value of $i(t)$ tends to
   (a) 0 (b) 1
   (c) 2 (d) ∞
2. The Fourier transform of a conjugate symmetric function is always
   (a) imaginary
   (b) conjugate anti-symmetric
   (c) real
   (d) conjugate symmetric
3. The following differential equation has
   $$3\left(\frac{d^2y}{dt^2}\right) + 4\left(\frac{dy}{dt}\right)^3 + y^2 + 2 = X$$
   (a) Degree = 2, order = 1
   (b) Degree = 1, order = 2
   (c) Degree = 4, order = 3
   (d) Degree = 2, order = 3
4. A fair dice is rolled twice. The probability that an odd number will follow an even number is
   (a) [1/2] (b) [1/6]
   (c) [1/3] (d) [1/4]
5. A solution of the following differential equation is given by
   $$\frac{d^2y}{dx^2} - 5\frac{dy}{dx} + 6y = 0$$
   (a) $y = e^{2x} + e^{-3x}$ (b) $y = e^{2x} + e^{3x}$
   (c) $y = e^{-2x} + e^{3x}$ (d) $y = e^{-2x} + e^{-3x}$
6. The rank of the matrix
   $$\begin{pmatrix} 1 & 1 & 1 \\ 1 & -1 & 0 \\ 1 & 1 & 1 \end{pmatrix}$$ is
   (a) 0 (b) 1
   (c) 2 (d) 3
7. $\nabla \times \nabla \times P$, Where P is a vector, is equal to
   (a) $P \times \nabla \times P - \nabla^3 P$ (b) $\nabla^2 P + \nabla(\nabla . P)$
   (c) $\nabla^2 P + \nabla \times P$ (d) $\nabla(\nabla . P) - \nabla^2 P$
8. $\iint (\nabla \times P) . ds$, where P is a vector, is equal to

(a) $\oint P \cdot dl$ (b) $\oint \nabla \times \nabla \times P \cdot dl$
(c) $\oint \nabla \times P \cdot dl$ (d) $\iiint \nabla \cdot P dv$

9. A probability density function is of the form $P(x) = Ke^{-a|x|}, x \in (-\infty, \infty)$
The value of K is.
(a) 0.5 (b) 1
(c) 0.5 α (d) α

10. A solution for the differential equation
$$x'(t) + 2x(t) = \delta(t)$$
With initial condition x (0 –) = 0 is
(a) $e^{-2t} u(t)$ (b) $e^{2t} u(t)$
(c) $e^{-1t} u(t)$ (d) $e^{t} u(t)$

11. The following plot shows a function y which varies linearly with x. The value of the integral $I = \int_1^2 y \, dx$ is

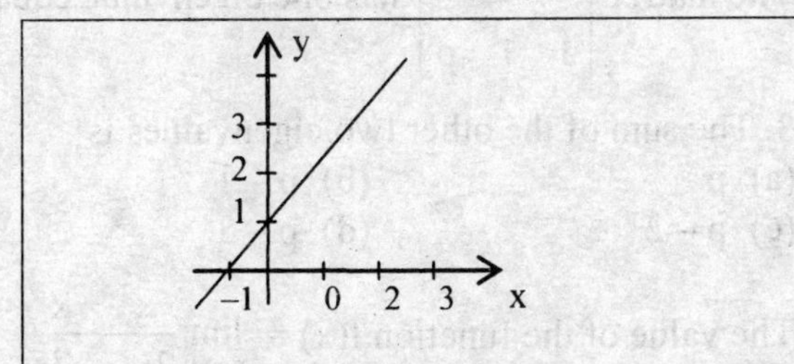

(a) 1.0 (b) 2.4
(c) 4.0 (d) 5.0

12. For $|x| << 1$, coth (x) can be approximated as
(a) x (b) x^2
(c) $1/x^2$ (d) $1/x^2$

13. $\lim_{\theta \to 0} \sin \frac{(\theta/2)}{\theta}$ is
(a) 0.5 (b) 1
(c) 2 (d) Not defined

14. Which one of the following function is strictly bounded?
(a) $1/x^2$ (b) e^x
(c) x^2 (d) e^{-x^2}

15. For the function e^{-x}, the linear approximation around $x = 2$ is
(a) $(3-x)e^{-2}$
(b) $1-x$
(c) $\left[3+2\sqrt{2}-\left(1+\sqrt{2}\right)x\right] e^{-2}$
(d) e^{-2}

16. In the matrix equation Px = q, which of the following is a necessary condition for the existence of at least one solution for the unknown vector x:
(a) Augmented matrix [Pq] must have the same rank as matrix P
(b) Vector q must have only non-zero elements
(c) Matrix P must be singular
(d) Matrix P must be square

17. If P and Q are two random events, then the following is true.
(a) Independence of P and Q implies that probability (P∩Q) = 0
(b) Probability (P∪Q) ≥ Probability (P) + Probability (Q)
(c) If P and Q are mutually exclusive, then they must be independent
(d) Probability (P∩Q) ≤ Probability (P)

18. If $S = \int_1^\infty x^{-3} dx$, then S has the value
(a) –1/3 (b) 1/4
(c) 1/2 (d) 1

19. The solution of the first order differential equation
$x'(t) - 3x(t), x(0) = x_0$ is
(a) $x(t) = x_0 e^{-3t}$ (b)
(c) $x(t) = x_0 e^{-1/3}$ (d) $x(t) = x_0 e^{-1}$

20. $x = \begin{bmatrix} x_1 & x_2 & \dots & x_n \end{bmatrix}^T$ is an n-tuple nonzero vector. The n × n matrix $V = xx^T$
(a) Has rank zero (b) Has rank 1
(c) Is orthogonal (d) Has rank n

21. Consider the matrices $X_{(4\times3)}$, $Y_{(4\times3)}$ and $P_{(2\times3)}$. The order of $[P(X^TY)^{-1}P^T]^T$ will be
(a) (2 × 2) (b) (3 × 3)
(c) (4 × 3) (d) (3 × 4)

22. Real matrices $[A]_{3\times1}$, $[B]_{3\times3}$, $[C]_{3\times5}$, $[D]_{5\times3}$, $[E]_{5\times5}$ and $[F]_{5\times1}$ are given. Matrices [B] and [E] are symmetric. Following statements are made with respect to these matrices.
1. Matrix product $[F]^T [C]^T [B] [C] [F]$ is a scalar.
2. Matrix product $[D]^T [F] [D]$ is always symmetric.

With reference to above statements, which of the following applies?
(a) Statement 1 is true but 2 is false
(b) Statement 1 is false but 2 is true
(c) Both the statements are true
(d) Both the statements are false

23. A square matrix B is skew-symmetric if
(a) $B^T = -B$ (b) $B^T = B$
(c) $B^{-1} = B$ (d) $B^{-1} = B^T$

24. The product of matrices $(PQ)^{-1} P$ is
(a) P^{-1} (b) Q^{-1}
(c) $P^{-1} Q^{-1} P$ (d) $PQ P^{-1}$

25. For a matrix $[M] = \begin{bmatrix} \frac{3}{5} & \frac{4}{5} \\ x & \frac{3}{5} \end{bmatrix}$, the transpose of the matrix is equal to the inverse of the matrix, $[M]^T = [M]^{-1}$. The value of x is given by

(a) $-\frac{4}{5}$ (b) $-\frac{3}{5}$
(b) $\frac{3}{5}$ (d) $\frac{4}{5}$

26. Let A, B, C, D be n × n matrices, each with non-zero determinant, if ABCD = I, then B^{-1} is
(a) $D^{-1} C^{-1} A^{-1}$
(b) CDA
(c) ADC
(d) does not necessarily exist

27. Given matrix $[A] = \begin{bmatrix} 4 & 2 & 1 & 3 \\ 6 & 3 & 4 & 7 \\ 2 & 1 & 0 & 1 \end{bmatrix}$, the rank of the matrix is
(a) 4 (b) 3
(c) 2 (d) 1

28. Solution for the system defined by the set of equations
$4y + 3z = 8$;
$2x - z = 2$ and
$3x + 2y = 5$ is
(a) x = 0; y = 1; z = 4/3 (b) x = 0; y = 1/2; z = 2
(c) x = 1; y = 1/2; z = 2 (d) non-existent

29. A is a 3 × 4 and matrix and A x = b is inconsistent system of equation. The highest possible rank of A is
(a) 1 (b) 2
(3) 3 (d) 4

30. For the matrix $\begin{bmatrix} 4 & 1 \\ 1 & 4 \end{bmatrix}$ the eigen value are
(a) 3 and – 3 (b) – 3 and – 5
(b) 3 and 5 (d) 5 and 0

31. The minimum and the maximum eigen values of the matrix $\begin{bmatrix} 1 & 1 & 3 \\ 1 & 5 & 1 \\ 3 & 1 & 1 \end{bmatrix}$ are –2 and 6, respectively.

What is the other eigen value?
(a) 5 (b) 3
(c) 1 (d) – 1

32. The sum of the eigen values of the matrix given below is $\begin{bmatrix} 1 & 2 & 3 \\ 1 & 5 & 1 \\ 3 & 1 & 1 \end{bmatrix}$
(a) 5 (b) 7
(c) 9 (d) 18

33. If a square matrix A is real and symmetric, then the eigenvalues.
(a) are always real
(b) are always real and positive
(c) are always real and non-negative
(d) occur in complex conjugate pairs

34. The matrix $\begin{bmatrix} 1 & 2 & 4 \\ 3 & 0 & 6 \\ 1 & 1 & p \end{bmatrix}$ has one eigenvalue equal to 3. The sum of the other two eigenvalues is
(a) p (b) p – 1
(c) p – 2 (d) p – 3

35. The value of the function $f(x) = \lim_{x \to 0} \frac{x^3 + x^2}{2x^3 - 7x^2}$ is

(a) 0 (b) $-\frac{1}{7}$
(b) $\frac{1}{7}$ (d) ∞

36. The $\lim_{x \to 0} \frac{\sin\left[\frac{2}{3}x\right]}{x}$ is
(a) 2/3 (b) 1
(c) 3/2 (d) ∞

37. $\underset{x \to 0}{\text{Lt}} \frac{\sin^2 x}{x}$ is equal to
(a) 0 (b) ∞
(c) 1 (d) – 1

38. The value of $\lim_{x \to 8} \frac{x^{1/3} - 2}{(x - 8)}$

(a) $\frac{1}{16}$ (b) $\frac{1}{12}$

(c) $\frac{1}{8}$ (d) $\frac{1}{4}$

39. If x = a (θ + sin θ) and y = a (1 – cos θ), then dy/dx will be equal to

(a) $\sin\left(\frac{\theta}{2}\right)$ (b) $\cos\left(\frac{\theta}{2}\right)$

(c) $\tan\left(\frac{\theta}{2}\right)$ (d) $\cot\left(\frac{\theta}{2}\right)$

40. The minimum value of functin $y = x^2$ in the interval [1, 5] is

(a) 0 (b) 1
(c) 25 (d) undefined

41. In the Taylor series expansiion of e^x about $x = 2$, the coefficient of $(x - 2)^4$ is

(a) 1/4! (b) $2^4/4!$
(c) $e^2/4!$ (d) $e^4/4!$

42. The value of the integral $\int_{-\infty}^{\infty} \frac{dx}{1-x^2} dx$ is

(a) $-\pi$ (b) $-\pi/2$
(b) $\pi/2$ (d) π

43. The parabolic arc $y = \sqrt{x}$, $1 \le x \le 2$ is revolved around the x-axix. The volume of the solid of the revolution is

(a) $\pi/4$ (b) $\pi/2$
(c) $3\pi/4$ (d) $3\pi/2$

44. Changing the order of the integration in the double integral $I = \int_0^8 \int_{x/4}^2 f(x, y)\, dydx$ leads to

(a) 4y (b) $16y^2$
(c) x (d) 8

45. For a scalar functin $f(x, y, z) = x^2 + 3y^2 + 2z^2$, the gradient at the point P(1, 2, –1) is

(a) $2\vec{i} + 6\vec{j} + 4\vec{k}$ (b) $2\vec{i} + 12\vec{j} - 4\vec{k}$

(c) $2\vec{i} + 12\vec{j} + 4\vec{k}$ (d) $\sqrt{56}$

46. The divergence of the vector field $3xz\hat{i} + 2zy\hat{j} - yz^2\hat{k}$ at a point (1, 1, 1) is equal to

(a) 7 (b) 4
(c) 3 (d) 0

47. Stokes theorem connects

(a) a line integral and a surface integral
(b) a surface integral and a volume integral
(c) a line integral and a volume integral
(d) gradient of a function and its surface integral

48. The degree of the differential equation $\frac{d^2x}{dt^2} + 2x^3 = 0$ is

(a) 0 (b) 1
(c) 2 (d) 3

49. The order and degree of the differential equation $\frac{d^3y}{dx^3} + 4\sqrt{\left(\frac{dy}{dx}\right)^3} + y^2 = 0$ are respectively.

(a) 3 and 2 (b) 2 and 3
(c) 3 and 3 (d) 3 and 1

50. The Blasius equation, $\frac{d^3f}{d\eta^3} + \frac{f}{2}\frac{d^2f}{d\eta^2} = 0$ is a

(a) second order nonlinear ordinary differential equation
(b) third order nonlinear ordinary differential equation
(c) third order linear ordinary differential equation
(d) mixed order nonlinear ordinary differential equation

51. The solution for the differential equation $\frac{dy}{dx} = x^2y$ with the condition that y = 1 at x = 0 is

(a) $y = e^{\frac{1}{2x}}$ (b) $\ln(y) = \frac{x^3}{3} + 4$

(c) $\ln(y) = \frac{x^2}{2}$ (d) $y = e^{\frac{x^3}{3}}$

52. The solution of the differential equation $\frac{dy}{dt} + 2xy = e^{-x^2}$ with y (0) = 1 is

(a) $(1 + x)e^{+x^2}$ (b) $(1 + x)e^{-x^2}$

(c) $(1 - x)e^{+x^2}$ (d) $(1 - x)e^{-x^2}$

53. The general solution of $\frac{d^2y}{dx^2} + y = 0$ is

(a) y = p cos x + Q sin x
(b) y = P cos x
(c) y = P sin x
(d) $y = P \sin^2 x$

54. Given that $\ddot{x} + 3x = 0$, and $x(0) = 1$, $\dot{x}(0) = 0$, what is x (1)?

(a) – 0.99 (b) – 0.16
(c) 0.16 (d) 0.99

55. The modulus of the complex number $\left(\frac{3+4i}{1-2i}\right)$ is

(a) 5 (b) $\sqrt{5}$

(c) $1/\sqrt{5}$ (d) 1/5

56. The analytic function $f(z)=\dfrac{z-1}{z^2+1}$ has singularities at

(a) 1 and – 1 (b) 1 and i
(c) 1 and – i (d) i and – i

57. Which one of the following statements is NOT true?

(a) The measure of skewness is dependent upon the amount of dispersion
(b) In a symmetric distribution, the values of mean, mode and median are the same
(c) In a positively skewed distribution: mean > median > mode
(d) In a negatively skewed distribution: mode > mean > median

58. A box contains 10 screws, 3 of which are defective. Two screws are drawn at random with replacement. The probability that none of the two screws is defective will be

(a) 100% (b) 50%
(c) 49% (d) None of these

59. A lot has 10% defective items. Ten item are chosen randomly from this lot. The probability that exactly 2 of the chosen items are defective is

(a) 0.0036 (b) 0.1937
(c) 0.2234 (d) 0.3874

60. A coin is tossed 4 times. What is the probability of getting heads exactly 3 times?

(a) 1/4 (b) 3/8
(c) 1/2 (d) 3/4

61. If three coins are tossed simultaneously, the probability of getting at least one head is

(a) 1/8 (b) 3/8
(c) 1/2 (d) 7/8

62. A box contains 20 defective items and 80 non-defective items. If two items are selected at random without replacement, what will be the probability that both items are defective?

(a) $\dfrac{1}{5}$ (b) $\dfrac{1}{25}$

(c) $\dfrac{20}{99}$ (d) $\dfrac{19}{495}$

63. The accuracy of Simpson's rule quadrature for a step size h is

(a) $O(h^2)$ (b) $O(h^3)$
(c) $O(h^4)$ (d) $O(h^5)$

64. The differential equation $(dy/dx) = 0.25\, y^2$ is to be solved using the backward (implicit) Euler's method with the boundary condition y = 1 at x = 0 and with a step size of 1. What would be the value of y at x = 1?

(a) 1.33 (b) 1.67
(c) 2.00 (d) 2.33

65. The inverse Laplace transform of $\dfrac{1}{(s^2+s)}$ is

(a) $1 + e^t$ (b) $1 - e^t$
(c) $1 - e^{-t}$ (d) $1 + e^{-t}$

Answers
Engineering Mathematics

1. Ans. (c)

Solution:

$$I(s)=\frac{2}{s(1+s)}, \text{ By final value theorem}$$

$$\lim_{t\to\infty} i(t)=\lim_{s\to 0} sI(s)=2$$

2. Ans. (c)

Solution:

Let $z(t) = x(t) + j\,y(t)$ be conjugate symmetric

$z(-t) = z(t)$ which implies

$x(t) \rightarrow$ even function and

$y(t) \rightarrow$ odd function

$$Z(\omega)=\int_{-\infty}^{\infty} z(t)\,e^{-j\omega t} = \int_{-\infty}^{\infty}(x+jy)\,e^{-j\omega t}dt$$

$$=\left[\begin{array}{l}\displaystyle\int_{-\infty}^{\infty}\big(x(t)\cos\omega t+y(t)\sin\omega t\big)dt\\ \displaystyle+j\int_{-\infty}^{\infty}\big(x(t)\sin\omega t-y(t)\cos\omega t\big)dt\end{array}\right]$$

$$=\int_{-\infty}^{\infty}A(t)\,dt+j\int_{-\infty}^{\infty}B(t)\,dt$$

Now $A(t)$ is real and even and

$B(t)$ is real and odd

$$\int_{-\infty}^{\infty}B(t)\,dt=0$$

$\therefore Z(\omega)$ is always real

3. Ans. (b)

Solution:

$$3\left(\frac{d^2y}{dt^2}\right)+4\left(\frac{dy}{dt}\right)^3+y^2+2=x$$

Order is the highest derivative term present in the given equation and degree is the power of highest derivative term in the given equation

Hence order = 2 and

degree = 1

4. Ans. (d)

Solution: n (E) = (1, 2) (3, 4) (5, 6)

(1, 4) (3, 2) (5, 2)
(1, 6) (3, 6) (5, 4)

n(s) = 6 × 6 = 36

P(E) = [n(E)/n(s)] = [9/36] = [1/4]

5. Ans. (a)

Solution:

For the given equation, auxiliary equation is

$m^2 - 5m + 6 = 0$

$(m - 3)(m - 2) = 0$

$m = 3, 2$

$y = e^{2x} + e^{3x}$

6. Ans. (c)

Solution: Because determinant of given matrix of 3 × 3 is 0 and there exist a 2 × 2 matrix whose determinant is non-zero.

7. Ans. (d)

Solution: $\nabla \times \nabla \times P = \nabla(\nabla . P) - \nabla^2 P$

It is vector identity.

8. Ans. (a)

Solution: By stroke's theorem

$\oint P . dl = \iint (\nabla \times P). ds$

where P is a vector

9. Ans. (c)

Solution: For a pdf the total probability is given by

$$\int_{-\infty}^{\infty} p(x) = 1$$

$$\Rightarrow \int_{-\infty}^{\infty} ke^{-\alpha|x|} dx \Rightarrow \frac{-k}{\alpha}(e^{-\infty} - e^{0})$$

$$\Rightarrow \frac{-k}{\alpha}(e^{0} - e^{-\infty}) \Rightarrow \frac{k}{\alpha} + \frac{k}{\alpha} \Rightarrow \frac{2K}{\alpha} = 1$$

$$\Rightarrow k = \frac{1}{2}\alpha = 0.5\,\alpha$$

10. Ans. (a)

Solution: $\frac{dx}{dt} + 2x = \delta(t)$

Taking laplace transform both side

$s X(s) - x(0) + 2X(s) = 1$

assume $x(0) = 0$

$X(s) = \frac{1}{s+2}$

Taking inverse Laplace $x(t) = e^{-2t}\,u(t)$

11. Ans. (b)

Solution:

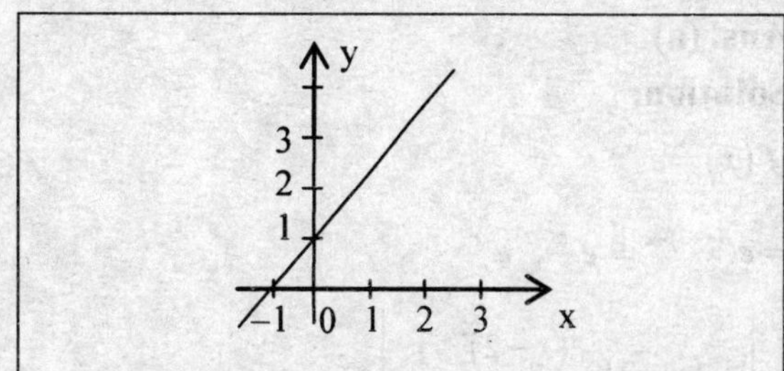

$$\int_1^2 y\,dx = ?$$

⇒ Equation of line ⇒ $y = mx + C$

Slope, $m = \frac{y_2 - y_1}{x_2 - x_1} = \frac{1-0}{0-(-1)} = 1$

Intercept on y at x = 0 is 1, i.e. C = 1

$\therefore y = x + 1$

$$\int_1^2 y\,dx = \int_1^2 (x+1)\,dx = \left|\frac{x^2}{2} + x\right|_1^2 = \left(\frac{4}{2}+2\right) - \left(\frac{1}{2}+1\right)$$

$$= 4 - \frac{3}{2} = 2.5$$

12. Ans. (c)

Solution: $\sinh x = x$ for $x \ll 1$;

$\cosh x = 1$ for $x \ll 1$;

Thus $\coth x = 1/x$

13. Ans. (a)

Solution: $\underset{\theta\to 0}{Lt} \frac{\sin\theta/2}{\theta}$

Apply L' Hospital Rule

$$\underset{\theta\to 0}{Lt} \frac{\frac{d}{d\theta}\sin\frac{\theta}{2}}{\frac{d}{d\theta}\theta} = \underset{\theta\to 0}{Lt} \frac{1}{2}\cos\frac{\theta}{2} = 0.5 = \frac{1}{2}$$

$[\because \cos 0 = 1]$

14. Ans. (d)

Solution: Strictly bounded function is e^{-x^2} In given options only 2 function are bounded, i.e. $\frac{1}{x^2}$ and e^{-x^2}

at $x = 0$; $\frac{1}{x^2} = \infty$; $e^{-x^2} = 1$

at $x = 10$; $\frac{1}{x^2} = .01$

$e^{-x^2} = 3.7 \times 10^{-44}$

Therefore e^{-x^2} is bounded

15. Ans. (a)

Solution:

$f(x) = e^{-x}$

$= e^{-(x-2)-2} = e^{-(x-2)}e^{-2}$

$= \left|1-(x-2)+\frac{(x-2)^2}{2!}+\ldots\right|e^{-2}$

$= |1-(x-2)|e^{-2}$

$= |(3-x)|e^{-2}$

16. Ans. (a)

Solution:

The system is consistent if and only if coefficient matrix and the argumented matrix are of the same rank, otherwise the system is in consistent.

17. Ans. (d)

Solution:

For two random events conditional probability is given by probability $(P \cap Q)$ = probability (P) probability (Q).

Probability $(Q) = \frac{\text{probability}(P \cap Q)}{\text{probability}(P)} \leq 1$

So probability $(P \cap Q) \leq$ probability (P)

18. Ans. (c)

Solution:

Given integral is

$s = \int_1^{\infty} x^{-3}dx$

$= \left[\frac{x^{-2}}{-2}\right]_1^{\infty} = \left[\frac{\infty^{-2}-1^{-2}}{-2}\right]$

$= \frac{\left[\frac{1}{\infty}-\frac{1}{1}\right]}{-2} = \frac{0-1}{-2} = \frac{1}{2}$

19. Ans. (a)

Solution:

Given $x'(t) = -3x(t)$ or

$x'(t) + 3x(t) = 0$

A.E. $D + 3 = 0$

Thus solution is

$x(t) = C_1e^{-3t}$

From $x(0) = x_0$ we get

$C_1 = x_0$

Thus

20. Ans. (d)

Solution: Given $X = [x_1 \ x_2 \ \ldots \ x_n]^T$

Then $X = \begin{bmatrix} x_1 \\ x_2 \\ \vdots \\ x_n \end{bmatrix}$

$V = XX^T = \begin{bmatrix} x_1 \\ x_2 \\ \vdots \\ x_n \end{bmatrix} [x_1 \ x_2 \ \ldots \ x_n]$

$= \begin{bmatrix} x_1^2 & x_1x_2 & \ldots & x_1x_n \\ x_1x_2 & x_2^2 & \ldots & \\ \vdots & \vdots & x_{n-1}^2 & x_{n-1}x_n \\ x_1x_n & \vdots & x_{n-1}x_n & x_n^2 \end{bmatrix}_{n\times n}$

For example let n = 3

$V = \begin{bmatrix} x_1^2 & x_1x_2 & x_1x_3 \\ x_1x_2 & x_2^2 & x_2x_3 \\ x_1x_3 & x_2x_3 & x_3^2 \end{bmatrix}$

Rank = 1 become all (3 × 3) and (2 × 2) minors are zero therefore $V_{n\times n}$ has Rank = 1.

21. Ans. (a)

Solution:

From the given order of matrices

$X^T \to 3 \times 4$

$Y \to 4 \times 3$

$X^TY \to 3 \times 3$

$(X^TY)^{-1} \to 3 \times 3$

$P \to 2 \times 3$

$P^T \to 3 \times 2$

$P(X^TY)^{-1}P^T \to (2 \times 3)\,(3 \times 3)\,(3 \times 2) \to 2 \times 2$

$(P(X^TY)^{-1}P^T)^T \to 2 \times 2$

22. Ans. (d)

Solution:

Since the matrix product when it exists is always another matrix and not a scalar. Hence Statement 1 is false, 2: $D^T F D$ is always symmetric.

Now D F D does not exist since $D_{3\times5}$, $F_{5\times1}$ and $D_{5\times3}$ are not compatible for multiplication since, $D_{3\times5}$ $F_{5\times1}$ = $X_{3\times1}$ and $X_{3\times1}$ $D_{5\times3}$ does not exist.

Hence, statement 2 is false.

23. Ans. (a)

Solution:

A square matrix B is defined as skew-symmetric if and only if $B^T = -B$.

24. Ans. (b)

Solution:

$$(PQ)^{-1} P = (Q^{-1} P^{-1}) P = (Q^{-1})(P^{-1} P) = Q^{-1}$$

25. Ans. (a)

Solution:

$$A^{-1} = \frac{\text{adj } A}{|A|}$$

$A^T = A^{-1}$, then A is orthogonal matrix.

Therefore $A.A^{-1} = A^{-1} A = I$ and $A^T A = A.A^T = I$

Since M is orthognal matrix.

$M^T M = I$

$$\begin{bmatrix} \frac{3}{5} & x \\ \frac{4}{5} & \frac{3}{5} \end{bmatrix} \begin{bmatrix} \frac{3}{5} & \frac{4}{5} \\ x & \frac{3}{5} \end{bmatrix} = \begin{bmatrix} 1 & 0 \\ 0 & 1 \end{bmatrix}$$

or
$$\begin{bmatrix} \left(\frac{3}{5}\right)^2 + x^2 & \left(\frac{3}{5}\cdot\frac{4}{5}\right) + \frac{3}{5}x \\ \left(\frac{4}{5}\cdot\frac{3}{5}\right) + \frac{3}{5}\cdot x & \left(\frac{4}{5}\right)^2 + \left(\frac{3}{5}\right)2 \end{bmatrix} = \begin{bmatrix} 1 & 0 \\ 0 & 1 \end{bmatrix}$$

or Compare both sides a_{12}

$$a_{12} = \left(\frac{3}{5}\right)\left(\frac{4}{5}\right) + \left(\frac{3}{5}\right) x = 0$$

or
$$\frac{3}{5}x = -\frac{3}{5}\cdot\frac{4}{5}$$

or
$$x = -\frac{4}{5}$$

26. Ans. (b)

Solution:

Given

A, B, C, D is n × n matrix.

and $ABCD = I$

or $ABCDD^{-1} C^{-1} = D^{-1} C^{-1}$

or $AB = D–1 C–1$

or $A–1 AB = A–1 D–1 C–1$

or $B = A–1 D–1 C–1$

$= (C^{-1})^{-1}. (D^{-1})^{-1}. (A^{-1})^{-1}$

$= CDA$

27. Ans. (c)

Solution:

Since maximum possible rank is 3 Consider first 3 × 3 minors.

$$\begin{vmatrix} 4 & 2 & 1 \\ 6 & 3 & 4 \\ 2 & 1 & 0 \end{vmatrix} = 0$$

$$\begin{vmatrix} 2 & 1 & 3 \\ 3 & 4 & 7 \\ 1 & 0 & 1 \end{vmatrix} = 0$$

$$\begin{vmatrix} 4 & 1 & 3 \\ 6 & 4 & 7 \\ 2 & 0 & 1 \end{vmatrix} = 0$$

$$\begin{vmatrix} 4 & 2 & 3 \\ 6 & 3 & 7 \\ 2 & 1 & 1 \end{vmatrix} = 0$$

Since all 3 × 3 minors are zero, now consider 2 × 2 minors.

$$\begin{vmatrix} 4 & 2 \\ 6 & 3 \end{vmatrix} = 0$$

$$\begin{vmatrix} 2 & 1 \\ 3 & 4 \end{vmatrix} = 8 - 3 = 5 \neq 0$$

rank = 2

28. Ans. (d)

Solution:

On augmented matrix apply gausss elimination method

$$\left[\begin{array}{ccc|c} 0 & 4 & 3 & 8 \\ 2 & 0 & -1 & 2 \\ 3 & 2 & 0 & 5 \end{array}\right] \xrightarrow{R_1 \leftrightarrow R_2} \left[\begin{array}{ccc|c} 2 & 0 & -1 & 2 \\ 0 & 4 & 3 & 8 \\ 3 & 2 & 0 & 5 \end{array}\right]$$

$$\left[\begin{array}{ccc|c} 2 & 0 & -1 & 2 \\ 0 & 4 & 3 & 8 \\ 3 & 2 & 0 & 5 \end{array}\right] \xrightarrow{R_3 - \frac{3}{2}R_1} \left[\begin{array}{ccc|c} 2 & 0 & -1 & 2 \\ 0 & 4 & 3 & 8 \\ 0 & 2 & 3/2 & 2 \end{array}\right]$$

$$\xrightarrow{R_3 - \frac{2}{4}R_2} \left[\begin{array}{ccc|c} 2 & 0 & -1 & 8 \\ 0 & 4 & 3 & 8 \\ 0 & 0 & 0 & -2 \end{array}\right]$$

For last row 0 = – 2 which is inconsistent. Also notice that r(A) = 2, while r(A) | B) = 3, (r(A) ≠ r(A | B) means inconsistent).

29. Ans. (c)

Solution:

$$r(A_{m\times n}) \le \min(m, n)$$

Highest possible rank = Least value of 3 and 4.

30. Ans. (c)

Solution:

Given $A = \begin{bmatrix} 4 & 1 \\ 1 & 4 \end{bmatrix}$

$$A - \lambda I = 0$$

Therefore $\begin{bmatrix} 4-\lambda & 1 \\ 1 & 4-\lambda \end{bmatrix} = 0$

$$(4-\lambda)^2 - 1 = 0$$

$$(4-\lambda)^2 - (1)^2 = 0$$

or, $(4-\lambda+1)(4-\lambda-1) = 0$

or, $(5-\lambda)(3-\lambda) = 0$

Therefore $\lambda = 3, \lambda = 5$

31. Ans. (b)

Solution:

Sum of eigen values of given matrix = sum of diagonal element of given matrix

$$\Sigma\lambda_1 = \text{Trace (A)}$$

$$\lambda_1 = 1 + 5 + 1 = 7$$

Given $\lambda_1 = -2, \lambda_2 = 6$

Therefore $-2 + 6 + \lambda_3 = 7$

$$\lambda_3 = 3$$

32. Ans. (b)

Solution:

Sum of eigen values of given matrix = sum of diagonal element of given matrix = 1 + 5 + 1 = 7.

33. Ans. (a)

Solution:

The eigen values of any real matrix is always real.

34. Ans. (c)

Solution:

Sum of the eigen values of matrix = sum of diagonal elements of given matrix

Therefore $1 + 0 + p = 3 + \lambda_2 + \lambda_3$

or $p + 1 = p + 1 - 3$

$= p - 2$

35. Ans. (b)

Solution:

Given

$$f(x) = \lim_{x\to 0}\left[\frac{x^3 + x^2}{2x^3 - 7x^2}\right]$$

Has $\frac{0}{0}$ form,

Apply L' Hospitals rule.

$$f(x) = \lim_{x\to 0}\left[\frac{3x^2 + 2x}{6x^2 - 14x}\right]$$

$$= \lim_{x\to 0}\left[\frac{6x + 2}{12x - 14}\right]$$

$$= \left[\frac{6 \times 0 + 2}{12 \times 0 - 14}\right] = -\frac{1}{7}$$

36. Ans. (a)

Solution:

Modify the seven limit as

$$\lim_{x\to 0}\frac{\sin\left[\frac{2}{3}x\right]}{x} = \lim_{\frac{2}{3}x\to 0}\frac{\frac{2}{3}\sin\left[\frac{2}{3}x\right]}{\frac{3}{2}x}$$

Let $\theta = \frac{2}{3}x$

$$\lim_{\frac{2}{3}x\to 0}\frac{\frac{2}{3}\sin\left[\frac{2}{3}x\right]}{\frac{3}{2}x} = \frac{2}{3}\lim_{\theta\to 0}\frac{\sin\theta}{\theta}$$

$$= \frac{2}{3} \times 1$$

$$= \frac{2}{3}$$

37. Ans. (a)

Solution:

$$\underset{x\to 0}{\text{Lt}} \frac{\sin^2 x}{x} = \underset{x\to 0}{\text{Lt}} \left(\frac{\sin x}{x^2}\right)^2 x$$

$$= 1 \times 0 = 0$$

38. Ans. (b)

Solution:

Let $h = (x - 8)$

or $x = 8 + h$

therefore $\underset{h\to 0}{\text{Lt}} \frac{(8+h)^{1/3} - 2}{h}$

Since this becomes $\left(\frac{0}{0}\right)$ form

Apply L' hospital rule

$$\lim_{h\to 0} \frac{\frac{1}{3}(8+h)^{\left(\frac{1}{3}-1\right)} - 2}{1} = \frac{1}{3}(8)^{-\frac{2}{3}}$$

$$= \frac{1}{12}$$

39. Ans. (c)

Solution:

Given, $x = a(\theta + \sin\theta)$,

$y = a(1 - \cos\theta)$

$$\frac{dx}{d\theta} = a(1 + \cos\theta),$$

$$\frac{dy}{d\theta} = a \sin\theta$$

$$\frac{dy}{dx} = \frac{a \sin\theta}{a(1 + \cos\theta)}$$

$$= \frac{2a \sin\left(\frac{\theta}{2}\right) . \cos\left(\frac{\theta}{2}\right)}{a \times 2\cos^2\left(\frac{\theta}{2}\right)}$$

$$= \tan \theta/2$$

40. Ans. (b)

Solution:

Given, $y = x^2$

or $\frac{dy}{dx} = 2x$

$= 0$ at $x = 0$

or $\left.\frac{d^2y}{dx^2}\right]_{x=0} = 2$

which is + ve value that indicate a local minima at $x = 0$

but it is not a in candidate for minima or maxima range

At the end point $x = 1$

$y = 1$

at second end point $x = 5$

$y = 25$

Hence, minimum value of function in the interval [1, 5] is 1.

41. Ans. (c)

Solution:

f(x) in the neighbourhood of a is,

$$f(x) = \sum_{n=0}^{\infty} b_n (x-a)^n$$

$$b_n = \frac{f^n(a)}{n!}$$

$f^4(x) = e^x$;

$f^4(2) = e^2$

therefore Coefficient of

$$(x-2)^4 = b_4 = \frac{f^4(2)}{4!} = \frac{e^2}{4!}$$

42. Ans. (d)

Solution:

Given, $\int_{-\infty}^{\infty} \frac{dx}{1-x^2} = \left[\tan^{-1} x\right]_{-\infty}^{\infty}$

$= \tan^{-1}(\infty) - \tan^{-1}(\infty)$

$= \frac{\pi}{2} - \left[\frac{-\pi}{2}\right]$

$= \pi$

43. Ans. (d)

Solution:

The volume of a solid generatedby revlolution about the x-axis, ofthe area bounded by curve y = f(x), the x-axis and the ordinates x = 4, y = b is

$$\text{Volume} = \int_a^b \pi y^2 dx$$

Here, a = 1, b = 2 and $y = \sqrt{x} \Rightarrow y^2 = x$

Therefore $\text{Volume} = \int_1^2 \pi \cdot x \cdot dx$

$$= \pi \cdot \left[\frac{x^2}{2}\right]_1^2$$

$$= \frac{\pi}{2}\left[x^2\right]_1^2 = \frac{\pi}{2}\left[2^2 - 1^2\right]$$

$$= \frac{3}{2}\pi$$

44. Ans. (a)

Solution:

When $I = \int_0^8 \int_{x/4}^2 f(x \cdot y)\, dy dx$

i.e.

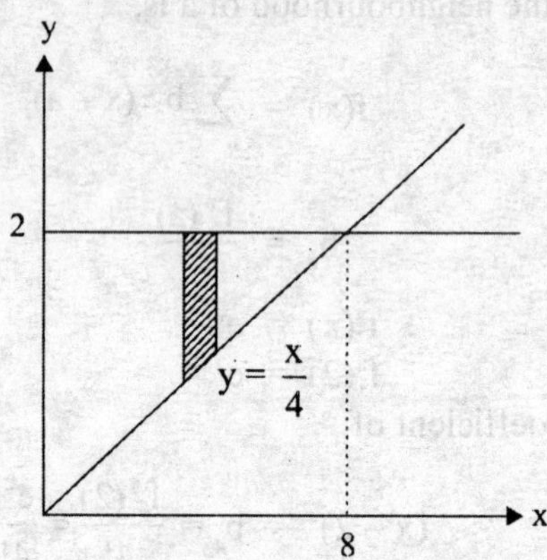

Now,

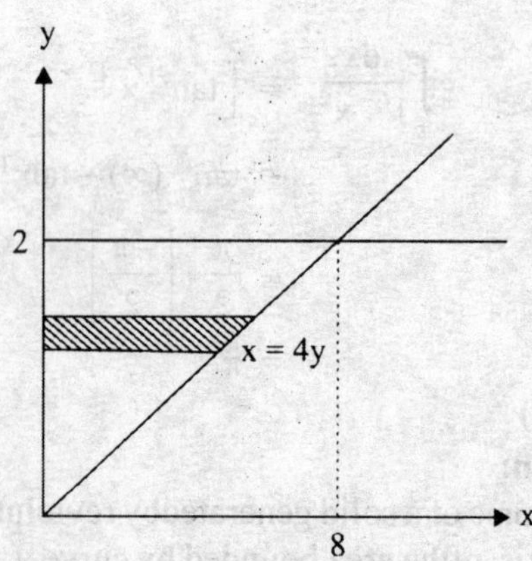

$$I = \int_0^2 \int_0^{4y} f(x, y)\, dy\, dx$$

therefore $q = 4y$

45. Ans. (b)

Solution:

Given, $f = x^2 + 3y^2 + 2z^2$

$$\Delta f = \text{grad } f = i\frac{\partial f}{\partial x} + j\frac{\partial f}{\partial y} + k\frac{\partial f}{\partial z}$$

$$= i\,2x + j6y + k4z$$

The gradient at P(1, 2, –1) is

$$= i\,(2 \times 1) + j\,(6 \times 2) + k\,(4 \times -1)$$

$$= 2i + 12j - 4k$$

46. Ans. (c)

Solution:

Given, Vector field,

$$\hat{f} = 3xz\hat{i} + 2zy\hat{j} - yz2\hat{k}$$

$$= v_x i + v_y j + v_z k$$

Divergence of vector field

$$\text{Div (f)} = \Delta.f$$

$$= \frac{\partial v_x}{\partial x} + \frac{\partial v_y}{\partial y} + \frac{\partial v_z}{\partial z}$$

$$= \frac{\partial}{\partial x}[3xz] + \frac{\partial}{\partial y}[2xy] + \frac{\partial}{\partial z}[-2yz^2]$$

$$= 3z + 2x - 2zy$$

$$\text{Div (f)}\,|_{(1,1,1)} = 3(1) + 2(1) - 2(1)(1) = 3$$

47. Ans. (a)

Solution:

A line integral and a surface integral is related by stoke's theorem, i.e. $\oint A \cdot \overline{dl} = \iint_S (\nabla \times A)\, ds$

48. Ans. (b)

Solution:

Degree of a differential equation is the power of its highest order derivative after the differential equation is made free of radeicals and fractions if any, in derivative power.

Here the degree is 1, whih is power of $\frac{d^2x}{dt^2}$.

49. Ans. (a)

Solution:

Degree of a differential equation is the power of its highest order derivative after the differential equation is made free of radeicals and fractions if any, in derivative power.

$$\frac{d^3y}{dx^3} + 4\sqrt{\left(\frac{dy}{dx}\right)^3 + y^2} = 0$$

Remove radicals

$$\left(\frac{d^3y}{dx^3}\right)^2 = 16\left[\left(\frac{dy}{dx}\right)^3 + y^2\right]$$

The order is 3 since highest differential is $\frac{d^3y}{dx^3}$

The degree is 2 since power of highest differential is 2.

50. Ans. (b)

Solution:

Given equation

$\frac{d^3f}{dn^3} + \frac{f}{2}\frac{d^2f}{dn^2} = 0$ it is third order $\left(\frac{d^3f}{dn^3}\right)$

It is non linear, the product $f \times \frac{d^2f}{dn^2}$

51. Ans. (d)

Solution:

Given

$$\frac{dy}{dx} = x^2y$$

This is solve by variable separable form

$$\frac{dy}{y} = x^2dx$$

$$\int\frac{dy}{y} = \int x^2dx$$

or $\log_e y = \frac{x^3}{3} + C_1$

or $y = e^{\frac{x^3}{3}+C_1}$

$= e^{C1} \times e^{\frac{x^3}{3}}$

at $x = 0, y = 1$

$1 = Ce^{\frac{0}{3}}$

or $C = 1$

Therefore $y = e^{\frac{x^3}{3}}$

52. Ans. (b)

Solution:

Given

$$\frac{dv}{dt} + 2xy = e^{-x2}$$

This is a first order linear differential equation

Itegrating factor I.F. $= e^{\int 2xdx} = e^{x^2}$

Solution is $y(\text{I.F.}) = \int Q(\text{IF})dx + C$

$ye^{x^2} = \int e^{-x^2} e^{x^2}\, dx + C$

$ye^{x^2} = x + C$

at $x = 0, y = 1$ (given)

therefore $1e^0 = 0 + C$

or $C = 1$

So, $ye^{-x^2} = x + 1$

or $y = e^{x^2}(x + 1)$

53. Ans. (a)

Solution:

Given

$$\frac{d^2y}{dx^2} + y = 0$$

$$D^2 + 1 = 0$$

$$D = \pm i$$

Therefore, General solution is

$$y = e^{0x}[C_1 \cos(1\,x) + C_2 \sin(1\,x)]$$

$$= C_1 \cos x + C_2 \sin x$$

$$= P \cos x + Q \sin x$$

where P and Q are constants.

54. Ans. (d)

Solution:

Given

$$\ddot{x} + 3x = 0$$

A.E. is

$$D^2 + 3 = 0$$

i.e. $D = \pm\sqrt{3}\, i$

Therefore $x = A\cos\sqrt{3}t + B\sin\sqrt{3}t$

at $t = 0, x = 1$

or $A = 1$

Now, $\dot{x} = \sqrt{3}\,(B\cos\sqrt{3}t - A\sin\sqrt{3}t)$

At $t = 0, \dot{x}\, 0$

or $B = 0$

So, $x = \cos\sqrt{3}t$

$$x(1) = \cos\sqrt{3}t = 0.99$$

55. Ans. (b)

Solution:

Let, $Z = \frac{3 + 4i}{1 - 2i}$

$$= \frac{(3+4i)(1+2i)}{(1-2i)(1+2i)}$$

$$= \frac{-5+10i}{5} = 1+2i$$

$$|Z| = \sqrt{(-1)^2+(2)^2} = \sqrt{5}$$

56. Ans. (d)

Solution:

$$f(z) = \frac{z-1}{z^2+1} = \frac{z-1}{z^2-i^2} = \frac{z-1}{(z-i)(z+i)}$$

Therefore, The gingularities arc at z = i and –i

57. Ans. (d)

Solution:

In a negatively skewed distribution, mode > median > mean.

58. Ans. (c)

Solution:

Since population is finite, sampling is done with replacement and so probability does not change from trial to trial. This problem can is to be solved by binormal distribution, as

Here, $n = 2$

$x = 0$ (no defective)

$$p = p(\text{defective}) = \frac{3}{10}$$

So, $$p(x=0) = \left(1-\frac{3}{10}\right)^2$$

$= 0.49 = 49\%$

Therefore, p (none of 2 screws is defective)

$$= \frac{3C_0 \times 7C_2}{10C_2} \times 100\%$$

$$= \frac{7}{15} \times 100\%$$

$= 46.6 \simeq 47\%$

59. Ans. (b)

Solution:

Since population is infinite. This problem can be done using binomial distrubution.

Probability of defective item,

$p = 0.1$

Probability of getting head exactly 3 times is

$$p(X=3) = {}^4C_3\,(0.5)^3\,(0.5)^1 = 1/4$$

61. Ans. (d)

Solution:

Since this problem involves coins. Binomial distribution is used.

Here, $n = 3$

$p = p(H) = 1/2$

Now, $p(x \geq 1) = 1 - p(x = 0)$

$$= 1 - 3C_0\left(\frac{1}{2}\right)^0\left(1-\frac{1}{2}\right)^3$$

$$= 1 - \frac{1}{8} = \frac{7}{8}$$

62. Ans. (d)

Solution:

Use hypergeometric distribution

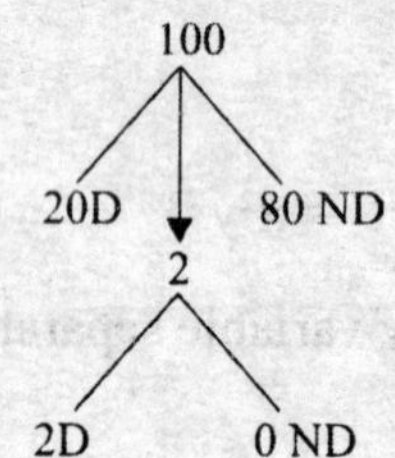

$$p(X=2) = \frac{20C_2 \times 80C_0}{100C_2} = \frac{19}{495}$$

63. Ans. (d)

Solution:

Here quadrature mean fifth order, Hance Accuracy $= O(h^5)$.

64. Ans. (c)

Solution:

Given

$$\frac{dy}{dx} = 0.25y^2 \qquad (y = 1 \text{ at } x = 0)$$

$h = 1$

Iterative equation for backward (implicit) Euler's methods is

$$y_{k+1} = y_k + h\, f(x_{k+1}, y_{k+1})$$

$$y_{k+1} = y_k + h \times 0.25\, y^2_{k+1}$$

or $0.25h\, y^2_{k+1} - y_{k+1} + y_k = 0$

put $k = 0$ in above equation

$$2.25\, y_1^2 - y_1 + y_0 = 0$$

or $$y_1 = \frac{1 \pm \sqrt{1-1}}{2 \times 0.25} = 2$$

or $y_1 = 2$

65. Ans. (c)

Solution:

$$LT^{-1}\left(\frac{1}{s^2+s}\right) = ?$$

$$\frac{1}{s^2+s} = \frac{1}{s(s^2+1)} = \frac{1}{s} - \frac{1}{s+1}$$

$$LT^{-1}\left(\frac{1}{s^2+s}\right) = LT^{-1}\left(\frac{1}{s}\right) - L^{-1}\left(\frac{1}{s+1}\right)$$

$$= 1 - e^{-t}$$

Two Marks Questions

1. For the matrix $p = \begin{bmatrix} 3 & -2 & 2 \\ 0 & -2 & 1 \\ 0 & 0 & 1 \end{bmatrix}$, one of the eigen values is equal to-2. Which of the following is an eigen vector?

(a) $\begin{bmatrix} 3 \\ -2 \\ 1 \end{bmatrix}$ (b) $\begin{bmatrix} -3 \\ 2 \\ -1 \end{bmatrix}$

(c) $\begin{bmatrix} 1 \\ -2 \\ 3 \end{bmatrix}$ (d) $\begin{bmatrix} 2 \\ 5 \\ 0 \end{bmatrix}$

2. If $R = \begin{bmatrix} 1 & 0 & -1 \\ 2 & 1 & -1 \\ 2 & 3 & 2 \end{bmatrix}$, then top row of R^{-1} is

(a) [5 6 4] (b) [5 –3 1]
(c) [2 0 –1] (d) [2 –2 1/2]

3. A fair coin is tossed three times in succession. If the first toss produces a head, then the probability of getting exactly two hands in three tosses is

(a) $\frac{1}{8}$ (b) $\frac{1}{2}$

(c) $\frac{3}{8}$ (d) $\frac{3}{4}$

4. For the function $f(x) = x^2e^{-x}$ the maximum occurs when x is equal to

(a) 2 (b) 1
(c) 0 (d) –1

5. For the scalar field $u = \frac{x^2}{2} + \frac{y^2}{3}$, magnitude of the gradient at the point (1, 3) is

(a) $\sqrt{13/9}$ (b) $\sqrt{9/2}$

(c) $\sqrt{5}$ (d) $\frac{9}{2}$

6. For the equation $\ddot{x}(t) + 3\dot{x}(t) + 2x(t) = 5$ the solution x (*t*) approaches which of the following values as $t \to \infty$?

a) 0 (b) $\frac{5}{2}$

c) 5 (d) 10

7. The Laplace transform of a function f(t) if F (s) $= \frac{5s^2+23s+6}{s\left(s^2+2s+2\right)}$. As $t \to \infty$, approaches

(a) 3 (b) 5

(c) $\frac{17}{2}$ (d) ∞

8. The expression $V = \int_0^H \pi R^2 [(1-h/H)]^2$ dh for the volume of a cone is equal to

(a) $\int_0^R \pi R^2(1-h/H)\, dr$

(b) $\int_0^R \pi R^2(1-h/H)^2\, dh$

(c) $\int_0^H 2\pi r H(1-r/R)\, dh$

(b) $\int_0^R 2\pi r H\left(1-\frac{r}{R}\right)^2 dr$

9. A surface S (x, y) = 2x + 5y – 3 is integrated once over a path consisting of the points that satisfy $(x+1)^2 + (y-1)^2 = \sqrt{2}$. The integral evaluates to

(a) $17\sqrt{2}$ (b) $17/\sqrt{2}$

(c) $\sqrt{2}/17$ (d) 0

10. Two fair dice are rolled and the sum r of the numbers turned up is considered

(a) Pr (r > 6) = $\frac{1}{6}$

(b) Pr (r/3 is an integer) = $\frac{5}{6}$

(c) Pr (r = 81 r/4 is an integer) = $\frac{5}{9}$

(d) Pr (r = 61 r/5 is an integer) = $\frac{1}{18}$

11. $p=[-10\ 1\ 3]^T$, $q=[-2\ -5\ 9]^T$, $r=[2\ -7\ 12]^t$ are three vectors an orthogonal set of vectors having a span that contains p, q, r is

(a) $[-6-3\ \ 6]^T$, $[4-2\ \ 3]^T$

(b) $[-4\ 2\ \ 4]^T$, $[4\ \ 7-11]^T$, $[8\ \ 2-3]^T$

(c) $[6\ \ 7-1]^T$, $[-3\ \ 12-2]^T$, $[3\ \ 9-4]^T$

(d) $[4\ \ 3\ \ 11]^T$, $[1\ \ 31\ \ 3]^T$, $[5\ \ 3\ \ 4]^T$

12. $p=[-10\ 1\ 3]^T$, $q=[-2\ -5\ 9]^T$, $r=[2\ -7\ 12]^t$ are three vectors.
The following vector is linearly dependent upon the solution to the previous problem.

(a) $[8\ \ 9\ \ 3]^T$ (b) $[-2\ \ -17\ \ 30]^T$

(c) $[4\ \ 4\ \ 5]^T$ (d) $[13\ \ 2\ \ -3]^T$

13. The differential equation $\frac{dx}{dt}=\frac{1-x}{\tau}$ is discretised using Euler's numerical integration method with a time step $\Delta T > 0$. What is the maximum permissible value of ΔT to ensure stability of the solution of the corresponding discrete time equation?

(a) 1 (b) τ/2

(c) τ (d) τ

14. The integral $\frac{1}{2\pi}\int_0^{2\pi} \sin\,(t-\tau)\,\cos\,\tau\ d\tau$ equal.

(a) sin *t* cos *t* (b) 0

(c) $\left(\frac{1}{2}\right)\cos t$ (d) $\left(\frac{1}{2}\right)\sin t$

15. A loaded dice has following probability distribution of occurrences.

Dice Value	1	2	3	4	6	7
Probability	1/4	1/8	1/8	1/8	1/8	1/4

If three identical dice as the above are thrown, the probability of occurrence of values 1, 5 and 6 on the three dice is

(a) Same as that of occurrence of 3, 4, 5
(b) Same as that of occurrence of 1, 2, 5
(c) 1/128
(d) 5/8

16. Let x and y be two vectors in a 3 dimensional space and <*x*, *y*> denote their dot product. Then the determinant.

(a) Is zero when *x* and *y* are linearly independent
(b) Is positive when *x* and *y* are linearly independent
(c) Is non-zero for all non-zero *x* and *y*
(d) Is zero only when either *x* or *y* is zero

17. The linear operator *L* (*x*) is defined by the cross product $L(x)=b\,X_x$ where $b=[0\ 1\ 0]^T$ and $x=[\dot{x}_1\ \ x_2\ \ x_3]^T$ are three dimensional vectors. The 3 × 3 matrix M of this operation satisfies

$$L(x)=M\begin{bmatrix}x_1\\x_2\\x_3\end{bmatrix}$$

Then the eigen values of M are

(a) 0, + 1, –1 (b) 1 –1, 1
(c) i, –i, 1 (d) i, –i, 0

18. Cayley–Hamilton theorem states that a square matrix satisfies its own characteristic question. Consider a matrix

$$A=\begin{bmatrix}-3 & 2\\-1 & 0\end{bmatrix}$$

A satisfies the relation

(a) $A+3I+2A^{-1}=0$

(b) $A^2+2A+2I=0$

(c) $(A+I)\,(A+2I)=0$

(d) $e^{(A)}=0$

19. Cayley–Hamilton theorem states that a square matrix satisfies its own characteristic question. Consider a matrix

$$A=\begin{bmatrix}-3 & 2\\-1 & 0\end{bmatrix}$$

A^9 equals

(a) $511A+510\ I$ (b) $309\ A+104\ I$

(c) $154A+155\ I$ (d) $e^{(9A)}$

20. In what range should Re(s) remain so that the Laplace transform of the function $e^{(a+2)t+5}$ exits?

(a) Re (s) > a + 2 (b) Re (s) > a + 7

(c) Re (s) < 2 (d) Re (s) > a + 5

21. Given the matrix $\begin{pmatrix} -4 & 2 \\ 4 & 3 \end{pmatrix}$, the eigenvector is

(a) $\begin{pmatrix} 3 \\ 2 \end{pmatrix}$ (b) $\begin{pmatrix} 4 \\ 3 \end{pmatrix}$

(c) $\begin{pmatrix} 2 \\ -1 \end{pmatrix}$ (d) $\begin{pmatrix} -1 \\ 2 \end{pmatrix}$

22. Let, $A = \begin{pmatrix} 2 & -0.1 \\ 0 & 3 \end{pmatrix}$ and $A^{-1} = \begin{pmatrix} [1/2] & a \\ 0 & b \end{pmatrix}$ Then $(a+b) =$

(a) [7/20] (b) [3/20]

(c) [19/60] (d) [11/60]

23. The value of the integral

$$I = [1/\sqrt{(2\pi)}] \int_0^{\infty} \exp[-(x)^2/8]\, dx \text{ is}$$

(a) 1 (b) π

(c) 2 (d) 2π

24. The derivative of the symmetric function drawn in given figure will look like

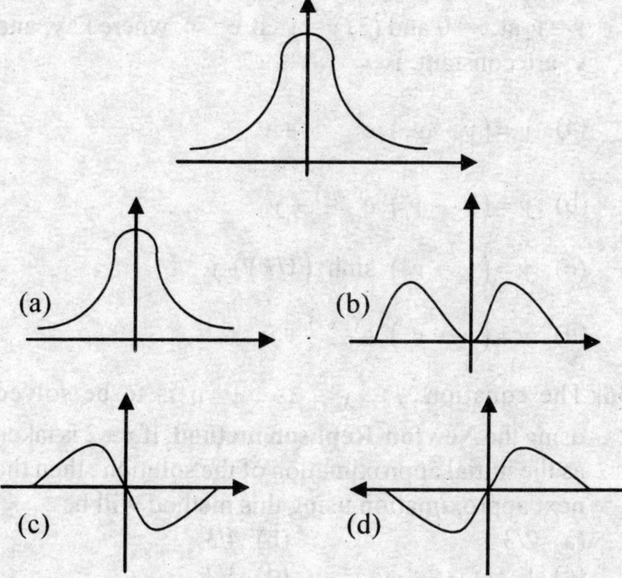

25. Match the following and choose the correct combination:

Group (1)

A. Newton-Raphson method
B. Ruge - kutta method
C. Simpson's Rule
D. Gauss elimination

Group (2)

1. Solving nonlinear equations
2. Solving linear simultaneous equations
3. Solving ordinary differential equations
4. Numerical integration
1) Interpolation
2) Calculation of Eigen values

	A	B	C	D
(a)	6	1	5	3
(b)	1	6	4	3
(c)	1	3	4	2
(d)	5	3	4	1

26. Given an orthogonal matrix

$$A = \begin{pmatrix} 1 & 1 & 1 & 1 \\ 1 & 1 & -1 & -1 \\ 1 & -1 & 0 & 0 \\ 0 & 0 & 1 & -1 \end{pmatrix}$$

(a) $\begin{pmatrix} [1/4] & 0 & 0 & 0 \\ 0 & [1/4] & 0 & 0 \\ 0 & 0 & [1/2] & 0 \\ 0 & 0 & 0 & [1/2] \end{pmatrix}$

(b) $\begin{pmatrix} [1/2] & 0 & 0 & 0 \\ 0 & [1/2] & 0 & 0 \\ 0 & 0 & [1/2] & 0 \\ 0 & 0 & 0 & [1/2] \end{pmatrix}$

(c) $\begin{pmatrix} 1 & 0 & 0 & 0 \\ 0 & 1 & 0 & 0 \\ 0 & 0 & 1 & 0 \\ 0 & 0 & 0 & 1 \end{pmatrix}$

(d) $\begin{pmatrix} [1/4] & 0 & 0 & 0 \\ 0 & [1/4] & 0 & 0 \\ 0 & 0 & [1/4] & 0 \\ 0 & 0 & 0 & [1/4] \end{pmatrix}$

27. The eigen values and the corresponding eigen vectors of a 2 × 2 matrix are given by

Eigen value	Eigen vector
$\lambda 1 = 8$	$v_1 = \begin{pmatrix} 1 \\ 1 \end{pmatrix}$
$\lambda_2 = 4$	$v_2 = \begin{pmatrix} 1 \\ -1 \end{pmatrix}$

The matrix is

(a) $\begin{pmatrix} 6 & 2 \\ 2 & 6 \end{pmatrix}$ (b) $\begin{pmatrix} 4 & 6 \\ 6 & 4 \end{pmatrix}$

(c) $\begin{pmatrix} 2 & 4 \\ 4 & 2 \end{pmatrix}$ (d) $\begin{pmatrix} 4 & 8 \\ 8 & 4 \end{pmatrix}$

28. For the function of a complex variable W = In Z (where, W = u + jv and Z = x + jy).
The u = constant lines get mapped in Z – plane as
(a) Set of radial straight lines
(b) Set of concentric circles
(c) Set of confocal hyperbolas
(d) Set confocal ellipses

29. The value of the contour integral

$\oint_{|z-j|=2} [1/(z^2+4)]\, dz$ in positive sense is

(a) $[j\pi/2]$ (b) $[-\pi/2]$
(c) $[-j\pi/2]$ (d) $[\pi/2]$

30. The integral $\int_0^{\pi} \sin^3\theta \, d\theta$ d is given by

(a) [1/2] (b) [2/3]
(c) [4/3] (d) [8/3]

31. Three companies X, Y and Z supply computers to a university. The percentage of computers supplied by them and the probability of those being defective are tabulated below.

Company	% of computer supplied	Probability of being defective
X	60%	0.01
Y	30%	0.02
Z	10%	0.03

Given that a computer is defective, the probability it was supplied by Y is.
(a) 0.1 (b) 0.2
(c) 0.3 (d) 0.4

32. For the matrix $\begin{pmatrix} 4 & 2 \\ 2 & 4 \end{pmatrix}$ the eigen value corresponding to the eigenvector $\begin{pmatrix} 101 \\ 101 \end{pmatrix}$ is

(a) 2 (b) 4
(c) 6 (d) 8

33. For the differential equation $[d^2y/dx^2] + k^2y = 0$ the boundary conditions are
i. y = 0 for x = 0 and
ii. y = 0 for x = a
The form of non-zero solutions of y (where m varies over all integers) are

(a) $y = \sum_m A_m \sin(m\pi x/a)$

(b) $y = \sum_m A_m \cos(m\pi x/a)$

(c) $y = \sum_m A_m x(m\pi/a)$

(d) $y = \sum_m A_m \exp(-m\pi/a)$

34. As x is increased from – ∞ to ∞, the function. $f(x) = [e^x/(1+e^x)]$
(a) Monotonically increases
(b) Monotonically decreases
(c) Increases to a maximum value and then decreases
(d) Decreases to a minimum value and then increases.

35. The solution of the differential equation $k^2 \frac{d^2y}{dx^2} = y - y_2$, under the boundary conditions (1) $y = y_1$ at $x = 0$ and (2) $y = y_2$ at $x = \infty$, where k, y_1 and y_2 are constant, is

(a) $y = (y_1 - y_2)\, e^{(-x/k^2)} + y_2$

(b) $y = (y_2 - y_1)\, e^{(-x/k)} + y_1$

(c) $y = (y_1 - y_2)\, \sinh(x/k) + y_1$

(d) $y = (y_1 - y_2)\, e^{(-x/k)} + y_2$

36. The equation $x^3 - x^2 + 4x - 4 = 0$ is to be solved using the **Newton-Rephson method**. If $x = 2$ is taken as the initial approximation of the solution , then the next approximation using this method will be
(a) 2/3 (b) 4/3
(c) 1 (d) 3/2

37. Three function $f_1(t)$, $f_2(t)$ and $f_3(t)$, which are zero out side the interval [0, T], are shown in the

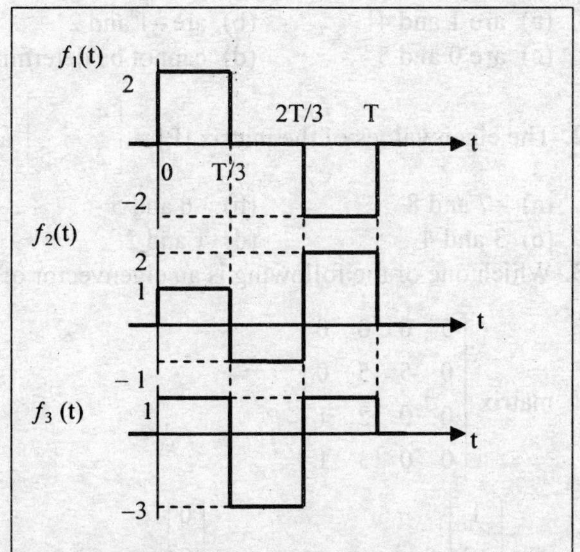

figure. Which of the following of the following statements is correct?

(a) $f_1(t)$ and $f_2(t)$ are orthogonal

(b) $f_1(t)$ and $f_3(t)$, are orthogonal

(c) $f_2(t)$ and, are orthogonal

(d) $f_1(t)$ and $f_2(t)$ are orthonormal

38. If the semi-circular contour D of radius 2 is as shown in the figure, then the value of the integral $\oint_D \frac{1}{(s^2-1)} ds$ is

(a) jπ (b) − jπ
(c) (d) π

39. It is given that $X_1, X_2, \ldots X_M$ are M non-zero, orthogonal vectors. The dimension of the vector space spanned by the 2M vectors $X_1, X_2, \ldots X_M$, $-X_1 - X_2, \ldots - X_M$ is

(a) 2M
(b) M + 1
(c) M
(d) Dependent on the choice of $X_1, X_2, \ldots X_M$

40. Consider the function $f(x) = x^2 - x - 2$. The maximum of $f(x)$ in the closed interval [−4, 4] is

(a) 18 (b) 10
(c) −2.25 (d) Indeterminate

41. An examinations consists of two papers, Paper 1 and Paper 2. The probability of failing in Paper 1 is 0.3 and that is Paper 2 is 0.2. Given that student has failed in Paper 2, the probability failing in Paper 1 is 0.6. The probability of student failing in both the paper is

(a) 0.5 (b) 0.18
(c) 0.12 (d) 0.06

42. Multiplication of matrices E and F is G. Matrices E and G are

$$E = \begin{bmatrix} \cos\theta & -\sin\theta & 0 \\ \sin\theta & \cos\theta & 0 \\ 0 & 0 & 1 \end{bmatrix} \text{ and } G = \begin{bmatrix} 1 & 0 & 0 \\ 0 & 1 & 0 \\ 0 & 0 & 1 \end{bmatrix}$$

What is the matrix F?

(a) $\begin{bmatrix} \cos\theta & -\sin\theta & 0 \\ \sin\theta & \cos\theta & 0 \\ 0 & 0 & 1 \end{bmatrix}$ (b) $\begin{bmatrix} \cos\theta & \cos\theta & 0 \\ -\cos\theta & \sin\theta & 0 \\ 0 & 0 & 1 \end{bmatrix}$

(c) $\begin{bmatrix} \cos\theta & \sin\theta & 0 \\ -\sin\theta & \cos\theta & 0 \\ 0 & 0 & 1 \end{bmatrix}$ (d) $\begin{bmatrix} \sin\theta & -\cos\theta & 0 \\ \cos\theta & \sin\theta & 0 \\ 0 & 0 & 1 \end{bmatrix}$

43. Match **List-I** with **List-II** and select the correct answer using the codes given below the lists:

List -I	**List-II**
A. Singular matrix	1. Determinant is not defined
B. Non-square matrix	2. Determinant is always one
C. Real symmetric	3. Determinant is zero
D. Orthogonal matrix	4. Eigenvalues are always real
	5. Eigenvalues are not defined

Codes:

	A	B	C	D
(a)	3	1	4	2
(b)	2	3	4	1
(c)	3	2	5	4
(d)	3	4	2	1

44. The inverse of the 2 × 2 matrix $\begin{bmatrix} 1 & 2 \\ 5 & 7 \end{bmatrix}$ is

(a) $\frac{1}{3}\begin{bmatrix} -7 & 2 \\ -5 & 1 \end{bmatrix}$ (b) $\frac{1}{3}\begin{bmatrix} -7 & 2 \\ 5 & 1 \end{bmatrix}$

(c) $\frac{1}{3}\begin{bmatrix} 7 & -2 \\ -5 & 1 \end{bmatrix}$ (d) $\frac{1}{3}\begin{bmatrix} -7 & -2 \\ -5 & -1 \end{bmatrix}$

45. The inverse of the matrix $\begin{bmatrix} 3+2i & -i \\ i & 3-2i \end{bmatrix}$ is

(a) $\frac{1}{12}\begin{bmatrix} 3+2i & -i \\ i & 3-2i \end{bmatrix}$ (b) $\frac{1}{12}\begin{bmatrix} 3-2i & -i \\ i & 3+2i \end{bmatrix}$

(c) $\frac{1}{14}\begin{bmatrix} 3+2i & -i \\ i & 3-2i \end{bmatrix}$ (d) $\frac{1}{14}\begin{bmatrix} 3-2i & -i \\ i & 3+2i \end{bmatrix}$

46. For which value of x will the matrix given below become singular?

$$\begin{bmatrix} 8 & x & 20 \\ 4 & 0 & 2 \\ 12 & 6 & 0 \end{bmatrix}$$

(a) 4 (b) 6
(c) 8 (d) 12

47. For what values of α and β, the following simultaneous equations have an infinite number of solutions?
$x + y + z = 5;$
$x + 3y + 3z = 9;$
$x + 2y + \alpha z = \beta$
(a) 2, 7 (b) 3, 8
(c) 8, 3 (d) 7, 2

48. The following simultaneous equations
$x + y + z = 3$
$x + 2y + 3z = 4$
$x + 4y + kz = 6$
will not have a uique solution for k equal to
(a) 0 (b) 5
(c) 6 (d) 7

49. Consider the system of simultaneous equations
$x + 2y + z = 6$
$2x + y + 2z = 6$
$x + y + z = 5$
This system has
(a) unique solution
(b) infinite number of solutions
(c) no solution
(d) exactly two solutions

50. For what value of a, if any, will the following system of equations in x, y and z have a solution?
$2x + 3y = 4$
$x + y + z = 4$
$x + 2y - z = a$
(a) Any real number (b) 0
(c) 1 (d) There is no such value

51. The eigen values of the miatrix $\begin{bmatrix} 4 & -2 \\ -2 & 1 \end{bmatrix}$
(a) are 1 and 4 (b) are –1 and 2
(c) are 0 and 5 (d) cannot be determined

52. The eigen values of the matrix $[P] = \begin{bmatrix} 4 & 5 \\ 2 & -5 \end{bmatrix}$ are
(a) –7 and 8 (b) –6 and 5
(c) 3 and 4 (d) 1 and 2

53. Which one of the following is an eigenvector of the matrix $\begin{bmatrix} 5 & 0 & 0 & 0 \\ 0 & 5 & 5 & 0 \\ 0 & 0 & 2 & 1 \\ 0 & 0 & 3 & 1 \end{bmatrix}$?

(a) $\begin{bmatrix} 1 \\ -2 \\ 0 \\ 0 \end{bmatrix}$ (b) $\begin{bmatrix} 0 \\ 0 \\ 1 \\ 0 \end{bmatrix}$

(c) $\begin{bmatrix} 1 \\ 0 \\ 0 \\ -2 \end{bmatrix}$ (d) $\begin{bmatrix} 1 \\ -1 \\ 2 \\ 1 \end{bmatrix}$

54. The number of lineraly independent eigenvectors of $\begin{bmatrix} 2 & 1 \\ 0 & 2 \end{bmatrix}$ is
(a) 0 (b) 1
(c) 2 (d) infinite

55. The eigenvectors of the matrix $\begin{bmatrix} 1 & 2 \\ 0 & 2 \end{bmatrix}$ are written in the form $\begin{bmatrix} 1 \\ a \end{bmatrix}$ and $\begin{bmatrix} 1 \\ b \end{bmatrix}$. What is a + b?
(a) 0 (b) 1/2
(c) 1 (d) 2

56. One of the eigen vectors of the matrix $A = \begin{bmatrix} 2 & 2 \\ 1 & 3 \end{bmatrix}$ is

(a) $\begin{bmatrix} 2 \\ -1 \end{bmatrix}$ (b) $\begin{bmatrix} 2 \\ 1 \end{bmatrix}$

(c) $\begin{bmatrix} 4 \\ 1 \end{bmatrix}$ (d) $\begin{bmatrix} 1 \\ -1 \end{bmatrix}$

57. Consider the system of equations $A_{(n\times n)} X_{(n\times l)} = \lambda_{(n\times l)}$ where, λ is scalar. Let (λ_i, x_i) be an eigen-pair of an eigen value and its corresponding eigen vector for real matrix A. Let I be a (n × n) unit matrix. Which one of the following statement is NOT correct?

(a) For a homogeneous n × n system of linear equations, $(A - \lambda I)\,x = 0$ having a nontrivial solution, the rank of $(A - \lambda I)$ is less than n

(b) For matrix A^m, m being a positive integer, (λ_i^m, x_i^m) will be the eigen-pair for all i

(c) If $A^T = A^{-1}$, then $|\lambda_i| = 1$ for all i

(d) If $A^T = A$, then λ_i is real for all i

58. For a given matrix

$$A = \begin{bmatrix} 2 & -2 & 3 \\ -2 & -1 & 6 \\ 1 & 2 & 0 \end{bmatrix}$$

One of the eigenvalues is 3. The other two eigenvalues are

(a) 2, –5 (b) 3, –5
(c) 2, 5 (d) 3, 5

59. Eigenvalues of a matrix $S = \begin{bmatrix} 3 & 2 \\ 2 & 3 \end{bmatrix}$ are 5 and 1.

What are the eigenvalues of the matrix. $S_2 = SS$?

(a) 1 and 25 (b) 6 and 4
(c) 5 and 1 (d) 2 and 10

60. If $f(x) = \dfrac{2x^2 - 7x + 3}{5x^2 - 12x - 9}$, then $\lim_{x \to 3} f(x)$ will be

(a) –1/3 (b) 5/18
(c) 0 (d) 2/5

61. $\displaystyle \operatorname{Lt}_{x \to 0} \frac{e^x - \left(1 + x + \frac{x^2}{2}\right)}{x^3} =$

(a) 0 (b) 1/6
(c) 1/3 (d) 1

62. A rail engine accelerates from its stationary position for 8 secods and travels a distance of 280m. According to the Mean Value Theorem, the speedometer at a certain time during acceleration must read exactly.

(a) 0 (b) 8 kmph
(c) 75 kmph (d) 126 kmph

63. Given a function $f(x, y) = 4x^2 - 6y^2 - 8x - 4y + 8$. The optimal value of f(x, y).

(a) is a minimum equal to 10/3
(b) is a maximum equal to 10/3
(c) is a minimum equal to 8/3
(d) is a maximum equal to 8/3

64. The distance between the origin and the point nearest to it on the surface $z^2 = 1 + xy$ is

(a) 1 (b) $\dfrac{\sqrt{3}}{2}$
(c) $\sqrt{3}$ (d) 2

65. The function $f(x) = 2x^3 - 3x^2 - 36x + 2$ has its maxima at

(a) x = – 2 only (b) x = 0 only
(c) x = –3 only (d) both x = – 2 and x = –3

66. The right circular cone of largest volume that can be enclosed by a sphere of 1 m radius has a height of

(a) 1/3 m (b) 2/3 m
(c) $\dfrac{2\sqrt{2}}{3}$ m (d) 4/3 m

67. Let $f = y^x$. What is $\dfrac{\partial^2 f}{\partial x \partial y}$ at x = 2, y = 1?

(a) 0 (b) ln 2
(c) 1 (d) $\dfrac{1}{\ln 2}$

68. Assuming $i = \sqrt{-1}$ and t is a real number, $\int_0^{\pi/3} e^{it}\,dt$ is

(a) $\dfrac{\sqrt{3}}{2} + i\,\dfrac{1}{2}$ (b) $\dfrac{\sqrt{3}}{2} - i\,\dfrac{1}{2}$
(c) $\dfrac{1}{2} + i\left(\dfrac{\sqrt{3}}{2}\right)$ (d) $\dfrac{1}{2} + i\left(1 - \dfrac{\sqrt{3}}{2}\right)$

69. Evaluate $\int_0^{\infty} \dfrac{\sin t}{t}\,dt$

(a) π (b) $\dfrac{\pi}{2}$
(c) $\dfrac{\pi}{4}$ (d) $\dfrac{\pi}{8}$

70. Which of the following integrals is unbounded?

(a) $\int_0^{\pi/4} \tan x\,dx$ (b) $\int_0^{\infty} \dfrac{1}{x^2 + 1}\,dx$
(c) $\int_0^{\infty} x e^{-x}\,dx$ (d) $\int_0^{1} \dfrac{1}{1 - x}\,dx$

71. What is the area common to the circles r = a and r = 2a cos θ?

(a) $0.524\ a^2$ (b) $0.614\ a^2$
(c) $1.047\ a^2$ (d) $1.228\ a^2$

72. The length of the curve $y = \frac{2}{3}x^{2/3}$ between x = 0 and x = 1 is
(a) 0.27 (b) 0.67
(c) 1 (d) 1.22

73. A path AB in the form of one quarter of a circle of unit radius is shown in the figure. Integration of $(x + y)^2$ on path AB traversed in a counter-clockwise sense is

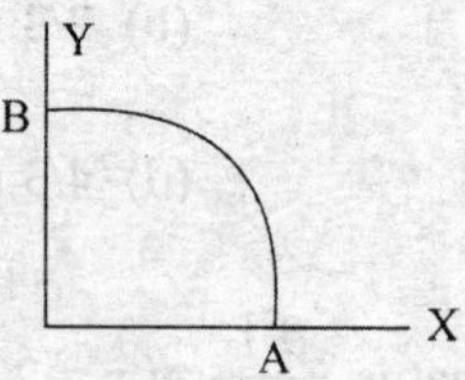

(a) $\frac{\pi}{2}$ (b) $\frac{\pi}{2}+1$
(c) $\frac{\pi}{2}$ (d) 1

74. The area enclosed between the curves $y^2 = 4x$ and $x^2 = 4y$ is
(a) $\frac{16}{3}$ (b) 8
(c) $\frac{32}{3}$ (d) 16

75. The value of $\int_0^3\int_0^x (6 - x - y)\, dx\, dy$ is
(a) 13.5 (b) 27.0
(c) 40.5 (d) 54.0

76. The volume of an object expressed in spherical co-ordinates is given by $V = \int_0^{2\pi}\int_0^{\pi/3}\int_0^1 r^2 \sin\phi\, dr\, d\phi\, d\theta$

The value of the intergal is
(a) $\frac{\pi}{3}$ (b) $\frac{\pi}{6}$
(c) $\frac{2\pi}{6}$ (d) $\frac{\pi}{4}$

77. By a change of variable x (u, v) = uv, y (u, v) = v/u is double integral, the integrand f(x, y) changes to f(uv, v/u) ϕ (u, v). Then, ϕ (u, v) is
(a) 2 u/v (b) 2 uv
(c) v^2 (d) 1

78. Consider the shaded triangular region P shown in the figure. What is $\iint_P xy\,dx\,dy$?

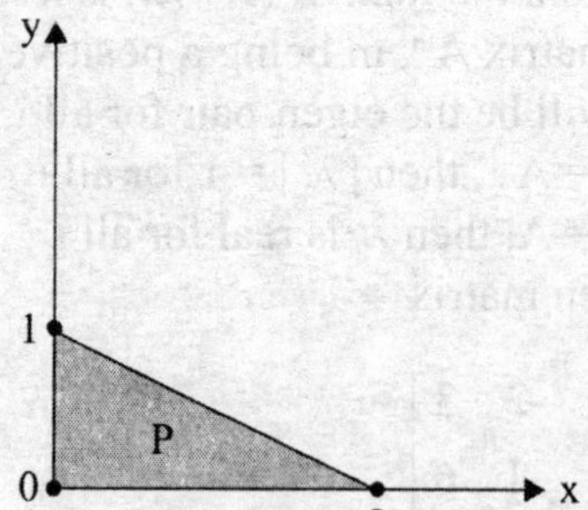

(a) $\frac{1}{6}$ (b) $\frac{2}{9}$
(c) $\frac{7}{16}$ (d) 1

79. The area of a triangle formed by the tips of vectors $\bar{a}$, $\bar{b}$ and $\bar{c}$ is
(a) $\frac{1}{2}(\bar{a}-\bar{b}).(\bar{a}-\bar{c})$ (b) $\frac{1}{2}\left|(\bar{a}-\bar{b}).(\bar{a}-\bar{c})\right|$
(c) $\frac{1}{2}|a \times b \times c|$ (d) $\frac{1}{2}(a \times b).\bar{c}$

80. Let x and y be two vectors in 3 dimensional space and <x, y> denoe their dot product. Then the determinant det $\begin{bmatrix} <x, x> & <x, y> \\ <y, x> & <y, y> \end{bmatrix}$.
(a) is zero when x and y are linearly independent
(b) is positive when x and y are linearly independent
(c) is non-zero for all non-zero x and y
(d) is zero only when either x or y is zero

81. The directional derivative of $f(x, y, z) = 2x^2 + 3y^2 + z^2$ at the point P (2, 1, 3) in the direction of the vector a = i −2 k is
(a) −2.785 (b) −2.145
(c) −1.789 (d) 1.000

82. For a scalar function $f(x, y, z) = x^2 + 3y^2 + 2z^2$, the directional derivative at the point P (1, 2, −1) in the direction of a vector $\vec{i} - \vec{j} + 2\vec{k}$ is
(a) −18 (b) $-3\sqrt{6}$
(c) $3\sqrt{6}$ (d) 18

83. Equation of the line normal to function $f(x) = (x-8)^{2/3} + 1$ at P (0, 5) is

(a) $y = 3x - 5$ (b) $y = 3x + 5$
(c) $3y = x + 15$ (d) $3y = x - 15$

84. Velocity vector of a flow field is given as $\vec{V} = 2xy\hat{i} - x^2z\hat{j}$. The vorticity vector at (1, 1, 1) is
(a) $4\hat{i} - \hat{j}$ (b) $4\hat{i} - \hat{k}$
(c) $\hat{i} - 4\hat{j}$ (d) $\hat{i} - 4\hat{k}$

85. The line integral $\int \vec{V}.d\vec{r}$ of the vector $\vec{V}.(\vec{r}) = 2xyz\hat{i} + x^2z\hat{j} + x^2y\hat{k}$ from the origin to the point P(1, 1, 1).
(a) is 1
(b) is –1
(c) is zero
(d) cannot be determined without specifying the path

86. Value of the integral $\oint_c (xydy - y^2dx)$, where, c is the square cut from the first quadrant by the lines x = 1 and y = 1 will be.
(a) $\frac{1}{2}$ (b) 1
(c) $\frac{3}{2}$ (d) $\frac{5}{3}$

87. Biotransformation of an organic compound having concentration (x) can be modeled using an ordinary differential equation $\frac{dx}{dt} + kx^2 = 0$, where k is the reaction rate constant. If x = a at t = 0, the solution of the equation is
(a) $x = ae^{-kt}$ (b) $\frac{1}{x} = \frac{1}{a} + kt$
(c) $a = a(1 - e^{-kt})$ (d) $x = a + kt$

88. A spherical naphthalene ball exposed to the atmosphere loses volume at a rate proportional to its instantaneous surface area due to evaporation. If the initial diameter of the ball is 2 cm and the diameter reduces to 1 cm after 3 months, the ball completely evaporates in
(a) 6 months (b) 9 months
(c) 12 months (d) infinite time

89. Solution of $\frac{dy}{dx} = -\frac{x}{y}$ at x = 1 and y = $\sqrt{3}$ is
(a) $x - y^2 = -2$ (b) $x + y^2 = 4$
(c) $x^2 - y^2 = -2$ (d) $x^2 + y^2 = 4$

90. Solution of the differential equation $3y\frac{dy}{dx} + 2x = 0$ represents a family of
(a) ellipses (b) circles
(c) parabolas (d) hyperbolas

91. The solution of the differential equation $\frac{dy}{dx} + y^2 = 0$ is
(a) $y = \frac{1}{x + c}$
(b) $y = \frac{-x^3}{3} + c$
(c) ce^x
(d) unsolvable as equation is non-linear

92. The solution of dy/dx = y^2 with initial value y (0) = 1 bounded in the interval
(a) $-\infty \le x \le \infty$ (b) $-\infty \le x \le 1$
(c) $x < 1, x > 1$ (d) $-2 \le x \le 2$

93. Transformation to linear form by substituting $v = y^{1-n}$ of the equation $\frac{dy}{dt} + p(t)y = q(t)y^n$; n > 0 will be
(a) $\frac{dv}{dt} + (1 - n)pv = (1 - n)q$
(b) $\frac{dv}{dt} + (1 - n)pv = (1 + n)q$
(c) $\frac{dv}{dt} + (1 + n)pv = (1 - n)q$
(d) $\frac{dv}{dt} + (1 + n)pv = (1 + n)q$

94. The solution of $x\frac{dy}{dx} + y = x^4$ with the condition $y(1) = \frac{6}{5}$ is
(a) $y = \frac{x^4}{5} + \frac{1}{x}$ (b) $y = \frac{4x^4}{5} + \frac{4}{5x}$
(c) $y = \frac{x^4}{5} + 1$ (d) $y = \frac{x^5}{5} + 1$

95. The solution of $\frac{d^2y}{dx^2} + 2\frac{dy}{dx} + 17y = 0$; y(0) = 1,

$\frac{dy}{dx}\left(\frac{\pi}{4}\right) = 0$ in the range $0 < x < \frac{\pi}{4}$ is given by

(a) $e^{-x}\left(\cos 4x + \frac{1}{4}\sin 4x\right)$

(d) $e^{x}\left(\cos 4x - \frac{1}{4}\sin 4x\right)$

(c) $e^{-4x}\left(\cos 4x - \frac{1}{4}\sin x\right)$

(d) $e^{-4x}\left(\cos 4x - \frac{1}{4}\sin 4x\right)$

96. It is given that $y'' + 2y' + y = 0, y(0) = 0, y(1) = 0$. What is y (0.5)?

(a) 0 (b) 0.37
(c) 0.62 (d) 1.13

97. For $\frac{d^2y}{dx^2} + 4\frac{dy}{dx} + 3y = 3e^{2x}$, the particular integral is

(a) $\frac{1}{15}e^{2x}$ (b) $\frac{1}{5}e^{2x}$
(c) $3e^{2x}$ (d) $C_1e^{-x} + C_2e^{-3x}$

98. An analytic function of a complex variable $z = x + iy$ is expressed as $f(z) = u(x, y) + i\,v(x, y)$ where $i = \sqrt{-1}$. If $u = xy$, the expression for v should be

(a) $\frac{(x+y)^2}{2} + k$ (b) $\frac{x^2 - y^2}{2} + k$
(d) $\frac{y^2 - y^2}{2} + k$ (d) $\frac{(x-y)^2}{2} + k$

99. Using Cauchy's integral theorem, the value of the integral (integration being taken in counterclockwise direction) $\oint_C \frac{z^3 - 6}{3z - i}dz$ is

(a) $\frac{2\pi}{81} - 4\pi i$ (b) $\frac{\pi}{8} - 6\pi i$
(c) $\frac{4\pi}{81} - 6\pi i$ (d) 1

100. The value of the integral $\int_C \frac{\cos 2\pi z}{(2z-1)(z-3)}dz$ (wherre C is a closed curve given by $|z| = 1$) is

(a) $-\pi i$ (b) $\frac{\pi}{5}$
(c) $\frac{2\pi i}{5}$ (d) πi

101. Consider likely applicability of Cauchy's Integral Theorem to evaluate the following integral counter clockwise around the unit circle c.

$$I = \oint_c \sec z dz,$$

z being a complex variable. The value of I will be

(a) $I = 0$: singularities set $= \phi$
(b) $I = 0$: singularities set =

$$\left\{\pm\frac{2n+1}{2}\pi;\ n = 0, 1, 2...\right\}$$

(c) $I = \pi/2$: singularities set $= \{\pm n\pi;\ n = 0, 1, 2...\}$
(d) None of above

102. The integral $\oint f(z)\,dz$ evaluated around the unit circle on the complex plane for $f(z) = \frac{\cos z}{z}$ is

(a) $2\pi i$ (b) $4\pi i$
(c) $-2\pi i$ (d) 0

103. The value of the contour integral $\oint_{|z-i|=2} \frac{1}{z^2+4}dz$ in positive sense is

(a) $i\pi/2$ (d) $-\pi/2$
(c) $-i\pi/2$ (d) $\pi/2$

104. A single die is thrown twice. What is the probability that the sum is neither 8 nor 9?

(a) 1/3 (b) 5/36
(c) 1/4 (d) 3/4

105. A hydraulic structure has four gates which operate independently. The probability of failure of each gate is 0.2. Given that gate 1 has failed, the probability that both gates 2 and 3 will fail is

(a) 0.240 (b) 0.200
(c) 0.040 (d) 0.008

106. A person on a trip has a choice between private car and public transport. The probability of using a private car is 0.45. While using the public transport, further choices available are bus and metro, out of which the probability of commuting by a bus is 0.55. In such a situation, the probability (rounded up to two decimals) of using a car, bus and metro, respectively would be

(a) 0.45, 0.30 and 0.25 (b) 0.45, 0.25 and 0.30
(c) 0.45, 0.55 and 0.00 (d) 0.45, 0.35 and 0.20

107. A box contains 2 washers, 3 nuts and 4 bolts. Items are drawn from the box at random one at a time without replacement. The probability of drawing 2 washers first followed by 3 nuts and subsequently the 5 bolts is

(a) 2/315 (b) 1/630
(c) 1/1260 (d) 1/250

108. If the standard deviation of the spot speed of vehicles in a highway is 8.8 kmph and the mean speed of the vehicles is 33 kmph, the coefficient of variation in speed is

(a) 0.1517 (b) 0.1867
(c) 0.2666 (d) 0.3646

109. There are 25 calculators in a box. Two of them are defective. Suppose 5 calculators are randomly picked for inspection (i.e., each has the same chance of being selected), what is the probability that only one of the defective calculators will be included in the inspection?

(a) $\frac{1}{2}$ (b) $\frac{1}{3}$
(c) $\frac{1}{4}$ (d) $\frac{1}{5}$

110. A box contains 5 black and 5 red balls. Two balls are randomly picked one after another from the box, without replacement. The probability for both balls being red is

(a) $\frac{1}{90}$ (b) $\frac{1}{2}$
(c) $\frac{19}{90}$ (d) $\frac{2}{9}$

111. From a pack of regular playing cards, two cards are drawn at random. What is the probability that both cards will be Kings, if first card in NOT replaced?

(a) $\frac{1}{26}$ (b) $\frac{1}{52}$
(c) $\frac{1}{169}$ (d) $\frac{1}{122}$

112. If probability density function of a random variable X is

$$f(x) = x^2 \text{ for } -1 \le x \le 1, \text{ and}$$
$$= 0 \text{ for any other value of x}$$

then, the percentage probability $P\left(-\frac{1}{3} \le x \le \frac{1}{3}\right)$ is

(a) 0.247 (b) 2.47
(c) 24.7 (d) 247

113. Consider the continuous random variable with probability density function

$$f(t) = 1 + 1 \text{ for } -1 \le t \le 0$$
$$= 1 - t \text{ for } 0 \le t \le 1$$

The standard deviation of the random variable is

(a) $\frac{1}{\sqrt{3}}$ (b) $\frac{1}{\sqrt{6}}$
(c) $\frac{1}{3}$ (d) $\frac{1}{6}$

114. Let X and Y be two independent random variaables. Which one of the relations between expectation (E), variance (Var) and covariance (Cov) given below is FALSE?

(a) E (XY) = E (X) E (Y)
(b) Cov (X, Y) = 0
(c) Var (X + Y) = Var (X) + Var (Y)
(d) $E(X^2 Y^2) = (E(X))^2 (E(Y))^2$

115. The standard devition of a uniformly distributed random variable between 0 and 1 is

(a) $\frac{1}{\sqrt{12}}$ (b) $\frac{1}{\sqrt{3}}$
(c) $\frac{5}{\sqrt{12}}$ (d) $\frac{7}{\sqrt{12}}$

116. A class of first year B. Tech. students is composed of four batches A, B, C and D, each conssisting of 30 students. It is found that the sessional marks of students in Engineering Drawing in batch C have a mean of 6.6 and standard deviation of 2.3. It is decided by the course instructor to normalize the marks of the students of all batches to have the same mean and standard deviation as that of the entire class. Due to this, the marks of a student in batch C are changed from 8.5 to

(a) 6.0 (b) 7.0
(c) 8.0 (d) 9.0

117. In the solution of the following set of linear equations by Gauss elimination using partial pivoting 5x + y + 2z = 34; 4y – 3z = 12; and 10x – 2y + z = – 4; the pivots for elimination of x and y are

(a) 10 and 4 (b) 10 and 2
(c) 5 and 4 (d) 5 and – 4

118. Given that one root of the equation $x^3 - 10x^2 + 31x - 30 = 0$ is 5 the other two roots are

(a) 2 and 3 (b) 2 and 4
(c) 3 and 4 (d) −2 and −3

119. Starting from $x_0 = 1$, one step of Newton-Raphson method in solving the equation $x^3 + 3x - 7 = 0$ gives the next value (x_1) as

(a) $x_1 = 0.5$ (b) $x_1 = 1.406$
(c) $x_1 = 1.5$ (d) $x_1 = 2$

120. The table below gives values of a function F(x) obtained for values of x at intervals of 0.25.

x	0	0.25	0.5	0.75	1.0
F(x)	1	0.9412	0.8	0.64	0.50

The value of the integral of the function between the limits 0 to 1 using Simpson's rule is

(a) 0.7854 (b) 2.3562
(c) 3.1416 (d) 7.5000

121. A 2nd degree polynomial, f(x) has values of 1, 4 and 15 at x = 0, 1 and 2, respectively. The integral $\int_0^2 f(x)\,dx$ is to be estimated by applying the trapezoidal rule to this data. What is the error (defined as "true value-approximate value") in the estimate?

(a) $-\frac{4}{3}$ (b) $-\frac{2}{3}$

(c) 0 (d) $\frac{2}{3}$

122. A calculator has accuracy up to 8 digits after decimal place. The value of $\int_0^{2\pi} \sin x\,dx$ when evaluated using this calculator by trapezoidal method with 8 equal intervals, to 5 significant digits is

(a) 0.00000 (b) 1.0000
(c) 0.00500 (d) 0.00025

123. Torque exerted on a flywheel over a cycle is listed in the table. Flywheel energy (in J per unit cycle) using Simpson's rule is

Angle (degree)	0	60	120	180	240	300	360
Torque (N m)	0	1066	−323	0	323	−355	0

(a) 542 (b) 993
(c) 1444 (d) 1986

124. If F(s) is the Laplace transform of function f (t), then Laplace transform of $\int_0^1 f(\tau)\,d\tau$ is

(a) $\frac{1}{s}F(s)$ (b) $\frac{1}{s}F(s) - f(0)$

(c) $sF(s) - f(0)$ (d) $\int F(s)\,ds$

125. Evaluate $\int_0^\infty \frac{\sin t}{t}\,dt$

(a) π (b) $\frac{\pi}{2}$

(c) $\frac{\pi}{4}$ (d) $\frac{\pi}{8}$

126. The Laplace transform of a function f(t) is $\frac{1}{s^2(s+1)}$. The function f(t) is

(a) $t - 1 + e^{-t}$ (b) $t + 1 + e^{-t}$
(c) $-1 + e^{-t}$ (d) $2t + e^t$

127. A delayed unit step function is defined as

$$u(t-a) = \begin{cases} 0, & \text{for } t < a \\ 1, & \text{for } t \geq a \end{cases}.$$

Its Laplace transform is

(a) a.e–as (b) $\frac{e^{-as}}{s}$

(c) $\frac{e^{as}}{s}$ (d) $\frac{e^{as}}{a}$

Statement for linked answer question number 128 and 129.

128. The complete solution of the ordinary differential equation $\frac{d^2y}{dx^2} + p\frac{d^2y}{dx} + qy = 0$ is $y = C_1 e^{-x} + C_2 e^{-3x}$ Then, p and q are

(a) p = 3, q = 3 (b) p = 3, q = 4
(c) p = 4, q = 3 (d) p = 4, q = 4

129. Which of the following is a solution of the differential equation $\frac{d^2y}{dx^2} + p\frac{dy}{dx} + (q+1) = 0$?

(a) e^{-3x} (b) xe^{-x}
(c) xe^{-2x} (d) x^2e^{-2x}

Answers
Engineering Mathematics

1. **Ans. (d)**
Solution:

$$P = \begin{bmatrix} 3 & -2 & 2 \\ 0 & -2 & 1 \\ 0 & 0 & 1 \end{bmatrix}$$

$$|P - vI| = \begin{bmatrix} 3-v & -2 & 2 \\ 0 & -2v & 1 \\ 0 & 0 & 1-v \end{bmatrix}$$

$$= (3-v)\ [(-2-v)\ (1-v)] - 0;\ (p - vI) = 0$$

$$= (3-v)\ (-2+2v-v+v^2) = 1;\ \text{v} = 3$$

$v^2 + v - 2 = 0$; $v^2 + 2v - v - 2 = 0$

$v(v+2) + (v+1)$; v + 2; v – 1

$v = -2$

$v = 1$

v 1, 2, 3

∴ eigen values of P are 1, 2, 3 for eigen vector corresponding to v = 1.

$$[A - vI]\ X = \begin{bmatrix} 3-v & -2 & 2 \\ 0 & -2-v & 1 \\ 0 & 0 & 1-v \end{bmatrix} \begin{bmatrix} x \\ y \\ z \end{bmatrix} = 0$$

Put v = 1

$$= \begin{bmatrix} 3-1 & -2 & 2 \\ 0 & -2-1 & 1 \\ 0 & 0 & 0 \end{bmatrix} \begin{bmatrix} x \\ y \\ z \end{bmatrix} = 0$$

or $$\begin{bmatrix} 2 & -2 & 2 \\ 0 & -3 & 1 \\ 0 & 0 & 0 \end{bmatrix} \begin{bmatrix} x \\ y \\ z \end{bmatrix} = 0$$

$2x - 2y + 3v = 0$ ...(1)

$-3y + z = 0$ ...(2)

Form equation (1) and (2)

– 3y = – z

$\frac{y}{1} = \frac{z}{3}$ ...(3)

Put,

z = 3y equation (1) from (3) (2) and (4)

$2x - 2y + 6y = 0 \qquad \frac{x}{-2} = \frac{y}{1} = \frac{z}{3}$

$2x + 4y = 0$ eigen vector (– 2, 1, 3)

$2x = 4y$

$\frac{x}{-2} = \frac{y}{1}$ ...(4)

Case II

When 1 = –2

$$\begin{bmatrix} 3+2 & -2 & 2 \\ 0 & -2+2 & 1 \\ 0 & 0 & 1+2 \end{bmatrix} \begin{bmatrix} x \\ y \\ z \end{bmatrix} = 0$$

$$\begin{bmatrix} 5 & -2 & 2 \\ 0 & 0 & 1 \\ 0 & 0 & 3 \end{bmatrix} \begin{bmatrix} x \\ y \\ z \end{bmatrix} = 0$$

5x – 2y = 0

5x = 2y

$\frac{x}{2} = \frac{y}{5} = \frac{z}{0}$

$5x - 2y + 2z = 0$ ...(1)

Z = 0 ...(2)

3z = 0 ...(3)

∴ Eigen vector

(2, 5, 0)

Method-2

Let $x = \begin{pmatrix} x_1 \\ x_2 \\ x_3 \end{pmatrix}$ be an eigen vector corresponding to – 2

$Ax = \lambda x$ gives $3x_1 - 2x_2 + 2x_3 = -2x_1$

$-2x_2 + x_3 = -2x_2$

$x_3 = -2x_3$

∴ $x_3 = 0$

Let $x_2 = t$ then $x_1 = \frac{2}{5}t$

∴ An eigen vector $\begin{bmatrix} 2 \\ 5 \\ 0 \end{bmatrix}$ corresponding to eigen value (–2)

2. **Ans. (b)**

Solution:

If $R = \begin{bmatrix} 1 & 0 & -1 \\ 2 & 1 & -1 \\ 2 & 3 & 2 \end{bmatrix}$

Co-factors are as follows

$= 1(-1+3)+0(1+(-1))(6-2)\ 2+0-4=-2$

$C_{11} = +(2+3) = +5 \quad C_{12} = -(4+2) = -6 \quad C_{13} = +(6-2) = 4$

$C_{21} = -(0+3) = -3 \quad C_{22} = +(2+2) = +4 \quad C_{23} = -(3-0) = -3$

$C_{31} = +(0+13) \quad C_{32} = -(-1+2) = -1 \quad C_{33} = +(1-0) = 1$

$$C = \begin{bmatrix} 5 & -6 & 4 \\ -3 & 4 & -3 \\ 1 & -1 & 1 \end{bmatrix}$$

$$\therefore A+ = \frac{1}{|A|} A\ dj\ A$$

$$C' = \begin{bmatrix} 5 & -3 & 1 \\ -6 & 4 & -1 \\ 4 & -3 & 1 \end{bmatrix}$$

$$= -\frac{1}{2}\begin{bmatrix} 5 & -3 & 1 \\ -6 & 4 & -1 \\ 4 & -3 & +1 \end{bmatrix} = \begin{bmatrix} -5/2 & 3/2 & -1/2 \\ 3 & -2 & 1/2 \\ -2 & 3/22 & -1/2 \end{bmatrix}$$

$-5/2 \quad 3/2 \quad -1/2$

or $5 \quad -3 \quad 1$

3. **Ans. (d)**

Solution:

Given first toss produces a head

$\therefore$ Probability of tails is 2nd and 3rd toss

$$= \frac{1}{2} \times \frac{1}{2} = \frac{1}{4} \text{ y}$$

$\therefore$ Probability of exactly 2 heads in 3 tosses

$$= 1 - \frac{1}{4} = \frac{3}{4}$$

4. **Ans. (b)**

Solution:

Given function $f(x) = x^2\ e^{-x}$

For maxima, $\dfrac{d\ f(x)}{dx} = 0$ for $x = 0,\ 2$

$$\left.\frac{d^2\ f(x)}{dx^2}\right|_{x=2} < 0$$

$\therefore$ Maxima occurs at x = 1

5. **Ans. (c)**

Solution:

$$\nabla.\text{u} = \text{Grad } u = \hat{i}\frac{\partial u}{\partial x} + \hat{j}\frac{\partial u}{\partial y} = x\hat{i} + \frac{2}{3}y\hat{j}$$

$$\text{At (1, 3), Grad } u = \sqrt{1+\left(\frac{2}{3}\times 3\right)^2} = \sqrt{5}$$

6. **Ans. (b)**

Solution:

$$(s^2 + 3s + 2)\ X(s) = \frac{5}{s}$$

$$\therefore X(s) = \frac{5/s}{s^2+3s+2}$$

Apply final value theorem.

$$x(\infty) = \underset{t\to\infty}{Lim}\, x(t) = \underset{s\to 0}{Lim}\, sX(s) = \frac{5}{2}$$

7. **Ans. (a)**

Solution: Apply find value theorem

$$x(\infty) = \underset{t\to\infty}{Lim}\, f(t) = \underset{s\to 0}{Lim}\, sF(s)$$

$$= \underset{s\to 0}{Lim}\frac{s(5s^2+23s+6)}{(s^2+2s+2)s} = 3$$

8. **Ans. (d)**

Solution: Option is correct but solution is to be verified.

$$V = \int_0^H \pi R^2 \left(1-\frac{h}{H}\right)^2 = \frac{\pi R^2 H}{3}$$

Option (a) and Option (c) → tends to 0 since there is no r in the integrating function

Option (b)

$$= \int_0^{\pi} \pi R^2\left(1+\frac{h}{H^2}-\frac{2h}{3}\right)dh$$

$$= \pi R^2\left(h\frac{h^2}{H^2}-\frac{h}{H}\right)_0^R$$

$$\pi R^2\left(R\frac{R^2}{3R^2}-\frac{R^2}{R}\right) = \frac{\pi R^2}{3}$$

Option (d)

$$= \int_0^R 2\pi R^2\left(1-\frac{r}{R}\right) dr = \frac{\pi R^2 H}{3}$$

9. Ans. (d)

Solution:

Path equation given as

$(x+1)^2 + (y-1)^2 = \sqrt{2}$

But $X = x+1,\ Y = y-1$

So it can be written as ...(1)

Which is the equation of circle and path of circle is closed and is integrated once over this path.

So $\oint s(x,y) = 0$

10. Ans. (c)

Solution:

By throwing dice twice $6 \times 6 = 36$ possibilities will occur. Out of these sample space consist of sum 4, 8 and 12 because $\frac{r}{4}$ is an integer. This can occur in following way :

$$\begin{bmatrix}(1,3)(2,2)(2,6)(3,1)\\(3,5)(4,4)(5,3)(6,2)\,and\,(6,6)\end{bmatrix}$$

Sample Space = 9

Favourable space is coming out of 8 =5

Probability of coming out 8

11. Ans. (d)

Solution:

12. Ans. (b)

$[-2\ -17\ \ 30]T$ is linearly dependent as it can be derived from the solution of Q (15).

13. Ans. (b)

14. Ans. (d)

Solution: $\frac{1}{2\pi}\int_0^{2\pi} \sin(t-\tau)\cos\tau\, d\tau$

$$= \frac{1}{2\pi}\int\left(\sin t\cos^2\tau - \sin\tau\cos\tau\cos t\right)d\tau$$

$$= \frac{1}{2\pi}\int\frac{1}{2}\sin t(\cos 2\tau + 1)\,d\tau - \frac{1}{2\pi}\int\frac{1}{2}\sin 2\tau\cos t\,d\tau$$

$$= \frac{1}{2\pi}\frac{\sin t}{2}\int(\cos 2\tau + 1)\,d\tau - \frac{1}{2\pi}\frac{\cos t}{2}\int\sin 2\tau\,d\tau$$

$$= \begin{bmatrix}\frac{1}{4\pi}\sin t\left|\frac{\sin 2\tau}{2}\right|_0^{2\pi} + \frac{\sin t}{4\pi}\cdot t\Big|_0^{2\pi}\\ -\frac{1}{4\pi}\cos t\left|-\frac{\cos 2\tau}{2}\right|_0^{2\pi}\end{bmatrix}$$

$$= \frac{1}{4\pi}\sin t\,(0) - \frac{1}{4\pi}\cos t\,(0) + \frac{2\pi}{4}\sin t$$

$$= \frac{1}{2}\sin t$$

15. Ans. (c)

Solution: The probability of occurrence of value 1, 5 and 6 on three dice = Product of probability of individual occurrence = $\frac{1}{4}\times\frac{1}{8}\times\frac{1}{4} = \frac{1}{128}$.

16. Ans. (d)

Solution:

$$\det\begin{bmatrix}x\cdot x\cdot x\cdot y\\ y\cdot x\cdot y\cdot y\end{bmatrix} = (x\cdot x)(y\cdot y) - (x\cdot y)(y\cdot x)$$

= 0 only when x or y is zero

17. Ans. (c)

18. Ans. (a)

Solution: Given $A = \begin{bmatrix}-3 & 2\\ -1 & 0\end{bmatrix}$

Then characteristic equation is

$|A - \lambda I| = 0$

or, $\begin{vmatrix}-3-\lambda & 2\\ -1 & -\lambda\end{vmatrix} = 0$

or, $(-3-\lambda)\ (-\lambda)\ +2 = 0$

or, $3\lambda + \lambda^2 + 2 = 0$

or, $\lambda^2 + 3\lambda + 2 = 0$ ...(1)

or, $(\lambda+1)\ (\lambda+2) = 0$

or, $\lambda = -1, -2$ [Eigen values]

Put $\lambda = A$ in equation (1)

$A^2 + 3A + 2 = 0$

Multiplying both sides by A^{-1}

$A^{-1}(A^2) + 3A^{-1}(A) + 2A^{-1} = 0$

or, $(A^{-1}A)A + 3(A^{-1}A) + 2A^{-1} = 0$

or, $IA + 3I + 2A^{-1} = 0$ $[\because AA^{-1} = I]$

$[\because IA = A]$

or, $A + 3I + 2A^{-1} = 0$

19. Ans. (d)

Solution: Refer the solution of Q: 78 and apply here, as

$A = -3I - 2A^{-1}$

$A^2 = -3A - 2I$

$A^3 = -3A^2 - 2A$

$= -3(-3A - 2I) - 2A$

$= 9A + 6I - 2A$

$= 7A + 6I$

$A^4 = 7A^2 + 6A$

$= 7(-3 - 2I) + 6A$

$= -21A - 14I + 6A$

$= -15A - 14I$

$A^5 = -15A^2 - 14A$

$= -15(-3A - 2I) - 14A$

$= 45A + 30I - 14A$

$= 31A + 30I$

$A^6 = 31A^2 + 30I$

$= 31(-3A - 2I) + 30A$

$= -93A + 62I + 30A$

$= -63A - 62I$

$A^7 = -63A^2 - 62A$

$= -63(-3A - 2I) - 62A$

$= 189A + 126I - 62A$

$= 127A + 126I$

$A^8 = -127A^2 - 126A$

$= 127(-3A - 2I) + 126A$

$= -381A - 254I + 126A$

$= -255A - 254I$

$A^9 = -255A^2 - 254A$

$= -225(-3A - 2I) - 254A$

$= 756A + 510I + 254A$

$= 511A + 510I$

Note: In examination need not to solve in detail, they can observe the pattern at A^3 and A^4 stage and get the correct answer.

20. Ans. (a)

Solution: $e^{(a+2t)+5} = e^5 \cdot e^{(a+2)t}$

Laplace transform exists if Rs (s) > a + 2

Condition for LT to exist

$$\int_{-\infty}^{\infty} | f(t)e^{-st} | dt < \infty$$

21. Ans. (c)

Solution:

$$A = \begin{pmatrix} -4 & 2 \\ 4 & 3 \end{pmatrix}$$

$\therefore$ $(A - \lambda I) = 0$

$\Rightarrow$ $(4 - \lambda)(3 - \lambda) - 2 \times 4 = 0$

$\Rightarrow$ $-12 + \lambda^2 - 3\lambda + 4\lambda - 2 \times 4 = 0$

$\Rightarrow$ $\lambda^2 + -20 = 0$

$= -5, 4$

Putting $= -5 \begin{pmatrix} 1 & 2 \\ 4 & 8 \end{pmatrix} \begin{pmatrix} x_1 \\ x_2 \end{pmatrix} = 0$

$\Rightarrow$ $x_1 + 2x_2 \times 2 = 0$

$\Rightarrow$ $x_1 = -2x_2$

$\Rightarrow$ $\frac{x_1}{2} = \frac{x_2}{-1}$

Hence $\begin{pmatrix} 2 \\ -1 \end{pmatrix}$ is eigen vector

22. Ans. (a)

Solution: $A = \begin{pmatrix} 2 & \frac{-1}{10} \\ 0 & 3 \end{pmatrix}$ $A^{-1} = \frac{1}{6}\begin{pmatrix} 3 & \frac{-1}{10} \\ 0 & 2 \end{pmatrix}$

$$= \begin{pmatrix} 0.5 & \frac{1}{60} \\ 0 & \frac{1}{3} \end{pmatrix}$$

a = [1/60], b = [1/3]

$$a + b \; \frac{1}{60} + \frac{20}{60} = \frac{21}{60} = \frac{7}{20}$$

Second Method

$$A = \begin{bmatrix} 2 & -0.1 \\ 0 & 3 \end{bmatrix} \text{ and } A^{-1} = \begin{bmatrix} 1/2 & a \\ 0 & b \end{bmatrix}$$

As we know $AA^{-1} = I$

$$\text{So} \begin{bmatrix} 2 & -0.1 \\ 0 & 3 \end{bmatrix} \begin{bmatrix} 1/2 & a \\ 0 & b \end{bmatrix} = \begin{bmatrix} 1 & 0 \\ 0 & 1 \end{bmatrix}$$

$$= \begin{bmatrix} 1 & 2a - 0.1b \\ 0 & 3b \end{bmatrix} = \begin{bmatrix} 1 & 0 \\ 0 & 1 \end{bmatrix}$$

Comparing both side $3b = 1,\; b = \frac{1}{3}$

also $2a - 0.1\,b = 0$

$$2a - \frac{1}{30} = 0$$

$$a = \frac{1}{60}$$

$$a + b = \frac{1}{3} + \frac{1}{60} = \frac{7}{20}$$

23. Ans. (a)

Solution: $\int_0^\infty e^{[-x^2/8]}\, dx = \int_0^\infty e^{[-bx^2]}\, dt$

b = [1/8],

Let, $bx^2 = t$; $2\,bx\,dx = dt$

$dx = [dt/(2\sqrt{bt})]$

$$\frac{1}{2\sqrt{b}} \int_0^\infty e^{-1}\, t^{(-1/2)}\, dt = \frac{\pi^{(1/2)}}{2\sqrt{b}}$$

π (1/2) Gamma function

$$I = \frac{1}{\sqrt{(2\pi)}} \times \frac{\sqrt{\pi}\sqrt{8}}{2} = \frac{(2\sqrt{2}\;\sqrt{\pi})}{(2\sqrt{2}\;\sqrt{\pi})} = 1$$

I = 1

Another method

Given integral, $I = I = \frac{1}{\sqrt{2\pi}} \int_0^\infty \exp - x^2\, dx$

Let, $\frac{x}{2\sqrt{2}} = Z,$

then, $\frac{dx}{2\sqrt{2}} = dz \Rightarrow dx = 2\sqrt{2}dZ$

$$\therefore \quad I = \frac{2\sqrt{2}}{\sqrt{2\pi}} = \int_0^\infty \exp(-Z^2)\, dZ$$

$$= \frac{2\sqrt{2}}{\sqrt{\pi}} = \int_0^\infty \exp(-Z^2)\, dZ$$

$$= \frac{2\sqrt{2}}{\sqrt{\pi}} = \int_0^\infty \exp(-Z^2)\, dZ$$

$= \text{er}\,f(\infty) = 1$

24. Ans. (c)

Solution:

The symmetric function equation is $\exp - t^2$ (Simple Gaussian equation)

$Y(t) = [d/dt] \exp[-t^2] = -2t \exp[-t^2]$

When t is + Ve, y(t) is negative

When t is – Ve, y(t) is positive.

Another Method

$\frac{dy}{dx} > 0$ for $x < 0$

$\frac{dy}{dx} < 0$ for $x > 0$

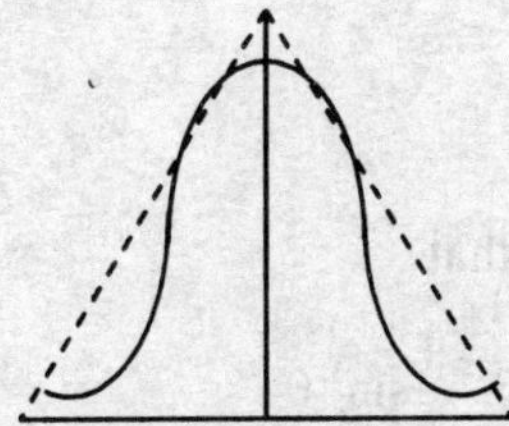

25. Ans (c)

Solution: The correct match is given below.

Newton-Raphson method	Solving nonlinear equations
Ruge - kutta method	Solving ordinary differential equations
Simpson's Rule	Numerical integration
Gauss elimination	Solving linear simultaneous equations

26. Ans (c)

Solution:

The given matrix is orthogonal.

Property of orthogonal matrix.

$AA^T = I^{-1}$

Take inverse Both side

$[AA^T]^{-1} = I^{-1}$

$[AA^T]^{-1} = ...\ [I^{-1} = I]$

27. Ans. (a)

Solution: The sum of eigen values of a matrix is the sum of diagonal elements $\lambda_1 + \lambda_2 = 12$ only option (a) has sum of diagonal elements of 12.

28. Ans. (b)

Solution: $W = \log_e z$

$\Rightarrow u + jv = \log(x + jy)$

$= [1/2] \log(x^2 + y^2) + j \tan^{-1} [y/x]$

u = constant

$\Rightarrow [1/2] \log(x^2 + y^2) = C$

$\Rightarrow x^2 + y^2 = C$

29. Ans. (d)

Solution: $|Z - j| = 2$ is circle with center (0, 1) and radius 2.

$$\frac{1}{z^2 + 4} = \frac{1}{(z + 2j)(z - 2j)}$$

z = (0, 2) lies inside circle.

So, by Cauchy integral formula,

$$\text{Integral} = \frac{2\pi j}{(2j + 2j)(z - 2j)} = \frac{\pi}{2}$$

30. Ans. (c)

Solution:

$$\int_0^{\pi} \sin^3\theta \, d\theta \quad ...(1)$$

As we know that

$$\sin 3\theta = \sin 3\theta - \frac{1}{4}\sin^3\theta$$

$$\text{So } \sin^3\theta = \frac{3}{4}\sin\theta - 4\sin 3\theta$$

$$\int_0^{\pi} \sin^3 d\theta = \int_0^{\pi} \left(\frac{3}{4}\sin\theta - \frac{1}{4}\sin 3\theta\right) d\theta$$

$$= \frac{3}{4}[-\cos\theta]_0^{\pi} - \frac{1}{4\times 3}[-\cos 3\theta]_0^{\pi}$$

$$= \frac{3}{4}[1+1] - \frac{1}{12}[1+1]$$

$$= \frac{3}{2} - \frac{1}{6}$$

$$= \frac{9-1}{6}$$

$$= \frac{4}{3}$$

31. Ans. (d)

Solution:

Let, $E_1 \rightarrow$ Event that X supplied computer.

$E_2 \rightarrow$ Event that Y supplied computer.

$E_3 \rightarrow$ Event that Z supplied computer.

$A \rightarrow$ Even that computer is detective.

We have to find $P\left(\frac{E_i}{A}\right)$.

Now according to Bay's theorem of probability.

$$P\left(\frac{E_i}{A}\right) = \frac{P(E_i)\,P\left(\frac{E_i}{A}\right)}{\sum_{i=-1}^{n} P(E_i)\,P\left(\frac{E_i}{A}\right)}$$

Probability that defective part is supplied by y is:

$$\Rightarrow \frac{0.3\times 0.02}{0.6\times 0.01 + 0.3\times 0.02 + 0.1\times 0.03}$$

$$\Rightarrow \frac{6}{6+6+3} = \frac{6}{15} = \frac{2}{5} = 0.4$$

32. Ans. (c)

Solution: Eigen values of given matrix = 2, 6

For $\lambda = 6$ eigen vector $= \begin{pmatrix} 1 \\ 1 \end{pmatrix}$ or $\begin{pmatrix} 1 & 0 & 1 \\ 1 & 0 & 1 \end{pmatrix}$

33. Ans. (a)

Solution:

The given differential equation is:

$$\frac{\partial^2 y}{\partial x^2} + K^2 y = 0$$

$$(D^2 + K^2)y = 0$$

$$D = \pm\, iK$$

So $y = C_1 \cos Kx + C_2 \sin Kx$ ...(1)

1st boundary condition

$y = 0$ for $x = 0$

From equation (1) $0 = C_1 + 0$

or $C_1 = 0$ ...(2)

2nd boundary condition

$y = 0$ for $x = a$

From equation (1)

$y = C_1 \cos Ka + C_2 \sin Ka$

using equation (2)

$0 = 0 + C_2 \sin Ka$

or $C_2 = 0$

So over all solution becomes zero for the second boundary condition $x = a$

When we put $x = a$ in all the option only option (a) get varnish so correct option is (a).

34. Ans. (a)

Solution: F $(-\infty) = 0$

$F(0) = [1/2]$

$F(\infty) = 1$

Which represent $f(x) = [e^x/(1 + e^x)]$ is monotonically increases.

35. Ans. (d)

Solution:

The given differential is

$$K^2 \frac{\partial^2 y}{dx^2} = y - y_2$$

$$\text{or } \frac{\partial^2 y}{dx^2} = \frac{y}{K^2} - \frac{y_2}{K^2}$$

$$= \frac{d^2 y}{dx^2} - \frac{y}{K^2} = \frac{-y}{K^2}$$

$$= \left(D^2 - \frac{1}{K^2}\right) y = \frac{-y_2}{K^2}$$

A.E equation

$$D^2 - \frac{1}{K^2} = 0$$

$$D = \pm \frac{1}{K}$$

$$\text{C.F} = C_1 e^{+\frac{1}{K}x} + C_2 e^{-\frac{1}{K}x}$$

$$\text{P.I} = \frac{1}{\left(D^2 - \frac{1}{K^2}\right)} \cdot \frac{-y^2 e^{0x}}{K^2}$$

$$= \frac{-y_2}{K^2} \cdot \frac{1}{D^2 - \frac{1}{K^2}} \cdot e^{0x}$$

$$= \frac{-y_2}{K^2} \times \frac{1}{\frac{-1}{K^2}}$$

$\text{P.I} = y_2$

Completer solution = $CF + P.I$

$$y = C_1 e^{\frac{1}{K}x} + C_2 e^{\frac{-1}{K}x} + y_2 \quad ...(1)$$

Boundary condition

$y_1 = y_1$ at $x = 0$

From equation (1)

$$y_1 = C_1 + C_2 + y_2 \quad ...(2)$$

Also $y = y_2$ at $x = \infty$

$y_2 = C_1 e^{\infty} + 0 + y_2$

$$\text{or } \frac{y_2}{e^{\infty}} = C_1 + \frac{y}{e^{\infty}}$$

$$\text{or } C_1 = 0 \quad ...(3)$$

Put in equation (2)

$y_1 = C_2 + y_2$

$$C_2 = y_1 - y_2 \quad ...(4)$$

Use (3) and (4) in equation (1)

$$y_2 = (y_1 - y_2)\, e^{\frac{-1}{K}x} + y^2$$

36. Ans. (b)

Solution: $f(x) = x^3 - x^2 + 4x + 4$

37. Ans. (b)

Solution:

If $\int f_1(t) \cdot f_2(t)\, dt = 0$

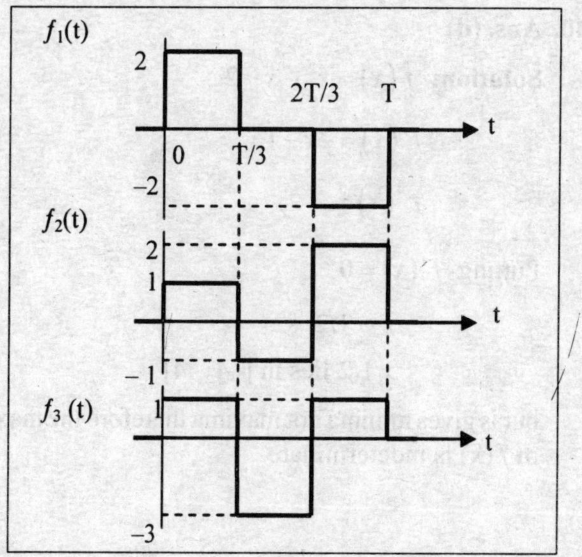

Then $f_1(t)$ and $f_2(t)$ are orthogonal.

In present case

$f_1(t)$ and $f_3(t)$ satisfies this condition.

$0 - T/3 \rightarrow f_1(t) \cdot f_3(t) = 2$

$T/3 - 2T/3 \rightarrow f_1(t) \cdot f_3(t) = 0$

$2T/3 - T \rightarrow f_1(t) \cdot f_3(t) = -2$

Thus is 0 – T $\int f_1(t) \cdot f_3(t)\, dt = 0$

38. Ans. (a)

Solution:

$$\oint_D \frac{1}{(S^2-1)} dS$$

$$\oint_0 \frac{1}{(S-1)(S-1)} dS$$

It has two poles as $S = -1$ and $+1$ only pole $S = +1$ in the given contour then according to Cauchy's integral theorem.

$$\oint_D \frac{1}{(S-1)(S-1)} dS = 2\pi j \frac{1}{(S+1)}\bigg|_{S=1} + 0$$

$$= \frac{2\pi j}{2} = \pi j$$

39. Ans. (c)

Solution: $x_1, -x_1$ for one axis:

$x_2, -x_2$ for another axis:

⋮ ⋮

$x_m, -x_m$ for m[th] axis.

40. Ans. (d)

Solution: $f(x) = x^2 - x - 2$

$$f'(x) = 2x - 1$$

$$f''(x) = 2$$

Putting $f'(x) = 0$

$x = 1/2$

x 1/2 lies in [–4 4]

but is gives minima not maxima therefore the maxima of $f(x)$ is indeterminate.

41. Ans. (c)

Solution: Probability of failing in Paper 1

$$= \bar{P}_1 = 0.3 = P(A)$$

Probability of failing in Paper 2

$$= \bar{P}_2 = 0.2 = P(B)$$

Probability of failing in Paper 1

When failed in 2 = $P\left(\frac{\bar{P}_1}{\bar{P}_2}\right) = 0.6 = P(A/B)$

$$P(A \cap B) = P(A/B) \cdot P(B)$$

$$P(\bar{P}_1 \cap \bar{P}_2) = P\left(\frac{\bar{P}_1}{\bar{P}_2}\right) \times P(\bar{P}_2) = 0.6 \times 0.2 = 0.12$$

42. Ans. (c)

Solution:

According to the given problem

E × F = G

$$\therefore \begin{bmatrix} \cos\theta & \sin\theta & 0 \\ -\sin\theta & \cos\theta & 0 \\ 0 & 0 & 1 \end{bmatrix} \times F = \begin{bmatrix} 1 & 0 & 0 \\ 0 & 1 & 0 \\ 0 & 0 & 1 \end{bmatrix}$$

Hence we see that product of (E × F) is unit matrix so F will be the inverse of E.

$$F = E^{-1} = \frac{\text{Adj(E)}}{|E|} = \begin{bmatrix} \cos\theta & \sin\theta & 0 \\ -\sin\theta & \cos\theta & 0 \\ 0 & 0 & 1 \end{bmatrix}$$

43. Ans. (a)

Solution:

The correct match is given below:

Singular matrix → Determinant is zero

Non-square matrix → Determinant is not defined

Real symmetric → Eigen values are always real

Orthogonal matrix → Determinant is always one

44. Ans. (b)

Solution:

The inverse of $\begin{bmatrix} a & b \\ c & d \end{bmatrix}$ is

$$\begin{bmatrix} a & b \\ c & d \end{bmatrix}^{-1} = \frac{1}{(ad-bc)}\begin{bmatrix} d & -b \\ -c & a \end{bmatrix}$$

$$\therefore \quad \begin{bmatrix} 1 & 2 \\ 5 & 7 \end{bmatrix}^{-1} = \frac{1}{(7-10)}\begin{bmatrix} 7 & -2 \\ -5 & 1 \end{bmatrix} = \frac{1}{3}\begin{bmatrix} -7 & 2 \\ 5 & -1 \end{bmatrix}$$

45. Ans. (b)

Solution:

$$\begin{bmatrix} a & b \\ c & d \end{bmatrix}^{-1} = \frac{1}{(ad-bc)}\begin{bmatrix} d & -b \\ -c & a \end{bmatrix}$$

$$\therefore \begin{bmatrix} 3+2i & i \\ -i & 3-2i \end{bmatrix}^{-1} = \frac{1}{[(3+2i)(3-2i)+i^2]}\begin{bmatrix} 3-2i & -i \\ i & 3+2i \end{bmatrix} = \frac{1}{12}\begin{bmatrix} 3-2i & -i \\ i & 3+2i \end{bmatrix}$$

46. Ans. (a)

Solution:

For singularity matrix $= \begin{vmatrix} 8 & x & 20 \\ 4 & 0 & 2 \\ 12 & 6 & 0 \end{vmatrix} = 0$

or $8(0-12) - x(0 - 2 \times 12) = 0$

Therefore $x = 4$

47. Ans. (a)

Solution:

The augmented matrix for this system is as below:

$$\begin{bmatrix} 1 & 1 & 1 \\ 1 & 3 & 3 \\ 1 & 2 & \alpha \end{bmatrix}\begin{bmatrix} 5 \\ 9 \\ \beta \end{bmatrix}$$

Apply Gauss-elimination method

$$\left[\begin{array}{ccc|c} 1 & 1 & 1 & 5 \\ 1 & 3 & 3 & 9 \\ 1 & 2 & \alpha & \beta \end{array}\right] \xrightarrow[R_3 - R_1]{R_2 - R_1} \left[\begin{array}{ccc|c} 1 & 1 & 1 & 5 \\ 0 & 2 & 2 & 4 \\ 0 & 1 & \alpha-1 & \beta-5 \end{array}\right]$$

$$\xrightarrow{R_3 - \frac{1}{2}R_2} \left[\begin{array}{ccc|c} 1 & 1 & 1 & 5 \\ 0 & 2 & 2 & 4 \\ 0 & 1 & \alpha-1 & \beta-7 \end{array}\right]$$

For infinite solution last row must be zero

i.e. $\alpha - 2$ and $\beta - 7 = 0$

or $\alpha = 2$ and $\beta = 7$

48. Ans. (d)

Solution:

The augmented matrix for given equations is

$$\left[\begin{array}{ccc|c} 1 & 1 & 1 & 3 \\ 1 & 2 & 3 & 4 \\ 1 & 4 & k & 6 \end{array}\right].$$

Apply Gauss elimination method.

$$\left[\begin{array}{ccc|c} 1 & 1 & 1 & 3 \\ 1 & 2 & 3 & 4 \\ 1 & 4 & k & 6 \end{array}\right] \xrightarrow[R_3 - R_1]{R_2 - R_1} \left[\begin{array}{ccc|c} 1 & 1 & 1 & 3 \\ 0 & 1 & 2 & 1 \\ 0 & 3 & k-1 & 3 \end{array}\right]$$

$$\xrightarrow{R_3 - 3R_3} \left[\begin{array}{ccc|c} 1 & 1 & 1 & 3 \\ 0 & 1 & 2 & 1 \\ 0 & 0 & k-7 & 0 \end{array}\right]$$

if $k \neq 7$

rank (A) = rank (A | B) = 3

$\Rightarrow$ unique solution

If $k = 7$, rank (A) = rank (A | B) = 2

which is less than number or variables

$\Rightarrow$ unique solution is not possible and only infinite solution is possible.

49. Ans. (c)

Solution:

Given equations can be written as

$$\begin{bmatrix} 1 & 2 & 1 \\ 2 & 1 & 2 \\ 1 & 1 & 1 \end{bmatrix}\begin{bmatrix} x \\ y \\ z \end{bmatrix} = \begin{bmatrix} 6 \\ 6 \\ 5 \end{bmatrix}$$

Augmented matrix is $\left[\begin{array}{ccc|c} 1 & 2 & 1 & 6 \\ 2 & 1 & 2 & 6 \\ 1 & 1 & 1 & 5 \end{array}\right]$

By gauss elimination method

$$\left[\begin{array}{ccc|c} 1 & 2 & 1 & 6 \\ 2 & 1 & 2 & 6 \\ 1 & 1 & 1 & 5 \end{array}\right] \xrightarrow[R_3 - R_1]{R_2 - 2R_1} \left[\begin{array}{ccc|c} 1 & 2 & 1 & 6 \\ 0 & -3 & 0 & -6 \\ 0 & -1 & 0 & -1 \end{array}\right]$$

$$\xrightarrow{R_3 - \frac{1}{3}R_2} \left[\begin{array}{ccc|c} 1 & 2 & 1 & 6 \\ 0 & -3 & 0 & -6 \\ 0 & 0 & 0 & 1 \end{array}\right]$$

rank r(A) = 2

rank r(A | B) = 3

Since the rank of coefficient matrix is 2 and rank of argument matrix is 3, which is not equal. Hence system has no solution, i.e. system is inconsitent.

50. Ans. (b)

Solution:

Augmented matrix is $\left[\begin{array}{ccc|c} 2 & 3 & 0 & 4 \\ 1 & 1 & 1 & 4 \\ 1 & 2 & -1 & a \end{array}\right]$.

Apply guess-elimination method,

$$\left[\begin{array}{ccc|c} 2 & 3 & 0 & 4 \\ 1 & 1 & 1 & 4 \\ 1 & 2 & -1 & a \end{array}\right] \xrightarrow[R_3 - \frac{1}{2}R_1]{R2 - \frac{1}{2}R_1} \left[\begin{array}{ccc|c} 2 & 3 & 0 & 4 \\ 0 & -1/2 & 1 & 2 \\ 0 & 1/2 & -1 & a-2 \end{array}\right]$$

$$\xrightarrow{R_3 + R_2} \left[\begin{array}{ccc|c} 2 & 3 & 0 & 4 \\ 0 & -1/2 & 1 & 2 \\ 0 & 0 & 0 & a \end{array}\right]$$

If $a \neq 0$, rank (A) = 2 and rank (A | B) = 3, hence equations have no solutions.

If a = 0, rank (A) = rank (A | B) = 2, then equations infinite solution.

51. Ans. (c)

Solution:

Characteristic equation is $|A - \lambda I| = 0$

$$\therefore \quad \begin{vmatrix} 4-\lambda & -2 \\ -2 & 1-\lambda \end{vmatrix} = 0$$

$$(4 - \lambda) \times (1 - \lambda) - [(-2) \times (-2)] = 0$$

$$\lambda^2 - 5\lambda = 0$$

or $\quad \lambda(\lambda - 5) = 0$

Hence, $\quad \lambda = 0$, and 5 are the eigen values.

52. Ans. (b)

Solution:

$$A = \begin{bmatrix} 4 & 5 \\ 2 & -5 \end{bmatrix}$$

Characteristic equation is $|A - \lambda I| = 0$

$$\begin{vmatrix} 4-\lambda & -5 \\ -2 & -5-\lambda \end{vmatrix} = 0$$

or $\quad (4 - \lambda) \times (-5 - \lambda) - 2 \times 5 = 0$

or $\quad \lambda^2 - \lambda - 30 = 0$

$$\lambda = 5, -6$$

These are the eigen values.

53. Ans. (a)

Solution:

Charactreristi equation is $|A - \lambda I| = 0$

$$= \begin{bmatrix} 5-\lambda & 0 & 0 & 0 \\ 0 & 5-\lambda & 5 & 0 \\ 0 & 0 & 2-\lambda & 1 \\ 0 & 0 & 3 & 1-\lambda \end{bmatrix} = 0$$

$$= (5 - \lambda)(5 - \lambda)[(2 - \lambda)(1 - \lambda) - 3] = 0$$

$$= (5 - \lambda)(5 - \lambda)(\lambda^2 - 3\lambda - 1) = 0$$

$$\lambda = 5, 5, \frac{3 \pm \sqrt{13}}{2}$$

Use these eigen value to get eign vector put $\lambda = 5$ in $[A - \lambda I]\ \hat{X} = 0$

$$\begin{bmatrix} 5-\lambda & 0 & 0 & 0 \\ 0 & 5-5 & 5 & 0 \\ 0 & 0 & 2-5 & 1 \\ 0 & 0 & 3 & 1-5 \end{bmatrix} \begin{bmatrix} x_1 \\ x_2 \\ x_3 \\ x_4 \end{bmatrix} = \begin{bmatrix} 0 \\ 0 \\ 0 \\ 0 \end{bmatrix}$$

or $$\begin{bmatrix} 0 & 0 & 0 & 0 \\ 0 & 0 & 5 & 0 \\ 0 & 0 & -3 & 1 \\ 0 & 0 & 3 & -4 \end{bmatrix} \begin{bmatrix} x_1 \\ x_2 \\ x_3 \\ x_4 \end{bmatrix} = \begin{bmatrix} 0 \\ 0 \\ 0 \\ 0 \end{bmatrix}$$

$$5x_3 = 0$$

$$-3x_3 + x_4 = 0$$

$$3x_3 - 4x_4 = 0$$

On solving $x_3 = 0$, x_1 and x_2 may be anything. The eigen vector corresponding to $\lambda = 5$, may be written as

$$\hat{X}_1 = \begin{bmatrix} x_1 \\ x_2 \\ x_3 \\ x_4 \end{bmatrix} = \begin{bmatrix} k_1 \\ k_2 \\ 0 \\ 0 \end{bmatrix}$$

where k_1, k_2 may be any real number. Since correct choice (a) as the matrix in this form with both x_3 and $x_4 = 0$,.

54. Ans. (b)

Solution:

$$A = \begin{bmatrix} 2 & 1 \\ 0 & 2 \end{bmatrix}$$

$$[A - \lambda I] = 0$$

$$\begin{bmatrix} 2-2 & 1 \\ 0 & 2-2 \end{bmatrix} = 0$$

or $(2-\lambda)^2 = 0$

or $\lambda = 2$

Use this eign value to set eigen vector.

put $\lambda = 2$, in $[A - \lambda I]\ \hat{X} = 0$

$$\begin{bmatrix} 2-\lambda & 1 \\ 0 & 2-\lambda \end{bmatrix}\begin{bmatrix} x_1 \\ x_2 \end{bmatrix} = \begin{bmatrix} 0 \\ 0 \end{bmatrix}$$

$$\begin{bmatrix} 0 & 1 \\ 0 & 0 \end{bmatrix}\begin{bmatrix} x_1 \\ x_2 \end{bmatrix} = \begin{bmatrix} 0 \\ 0 \end{bmatrix}$$

$$x_2 = 0$$

$$0 = 0$$

The solution is therefore $x_2 = 0$, and $x_1 =$ anything

$$\begin{bmatrix} x_1 \\ x_2 \end{bmatrix} = \begin{bmatrix} k \\ 0 \end{bmatrix}$$

$\therefore$ There is only one linearly independent eigen vector for this problem which may be written as $\begin{bmatrix} k \\ 0 \end{bmatrix}$.

55. Ans. (b)

Solution: Characteristic equation is

$|A - \lambda I| = 0$

$$\begin{vmatrix} (1-\lambda) & 2 \\ 0 & (2-\lambda) \end{vmatrix} = 0$$

or $(1-\lambda)(2-\lambda) = 0$

therefore $\lambda = 1, 2$

putting the value of $\lambda = 1$

or $\begin{bmatrix} 0 & 2 \\ 0 & 1 \end{bmatrix}\begin{bmatrix} 1 \\ a \end{bmatrix} = 0$

or $a = 0$

putting the value of $\lambda = 2$

$$\begin{bmatrix} -1 & 2 \\ 0 & 0 \end{bmatrix}\begin{bmatrix} 1 \\ b \end{bmatrix} = 0$$

or $-1 + 2b = 0$

and $0 = 0$

or $b = \frac{1}{2}$

therefore $a + b = \frac{1}{2}$

56. Ans. (a)

Solution:

$$A = \begin{bmatrix} 2 & 2 \\ 1 & 3 \end{bmatrix}$$

Characteristic equation of A is $|A - \lambda I| = 0$

$$\begin{bmatrix} 2-\lambda & 2 \\ 1 & 3-\lambda \end{bmatrix} = 0$$

$$(2-\lambda)(3-\lambda) - 2 = 0$$

$$\lambda^2 - 5\lambda = 0$$

$$\lambda = 1, 4$$

The eigen vector problem is $[A - \lambda I]\ \hat{X} = 0$

$$\begin{bmatrix} 2-\lambda & 2 \\ 1 & 3-\lambda \end{bmatrix}\begin{bmatrix} x_1 \\ x_2 \end{bmatrix} = \begin{bmatrix} 0 \\ 0 \end{bmatrix}$$

Putting $\lambda = 1$

$$\begin{bmatrix} 1 & 2 \\ 1 & 2 \end{bmatrix}\begin{bmatrix} x_1 \\ x_2 \end{bmatrix} = \begin{bmatrix} 0 \\ 0 \end{bmatrix}$$

$$x_1 + 2x_2 = 0 \quad (1)$$

$$x_1 + 2x_2 = 0 \quad (2)$$

For solution if $x_2 = k$, Then $2y = -2$

$$\hat{X} = \begin{bmatrix} -2k \\ k \end{bmatrix} \text{i.e. } x_1 : x_2 = -2:1$$

Therefore solution is

$$= \begin{bmatrix} 2y \\ x^2 \end{bmatrix} = \begin{bmatrix} -2 \\ 1 \end{bmatrix}$$

57. Ans. (b)

Solution:

λ_i^m may be the correcsponding eigen values of A^m, x_i^m may not be corresponding eigen vectors.

58. Ans. (b)

Solution:

$$\Sigma\lambda_i = \text{Trace}(A)$$

$$\lambda_1 + \lambda_2 + \lambda_3 = \text{Trace}(A) = 2 + (-1) + 0 = 1$$

given, $\lambda_1 = 3$

therefore $3 + \lambda_2 + \lambda_3 = 1$

or $\lambda_2 + \lambda_3 = -2$

Here choice (b) only satisfies this condition.

59. Ans. (a)

Solution:

If $\lambda_1, \lambda_2, \lambda_3 \ldots \lambda_n$ are the eigen values of A. Then the eigen values of

A^m are $\lambda_1^m, \lambda_2^m, \lambda_3^m, \ldots$

$\therefore$ S^2 matrix has eigen values 1^2 and 5^2 i.e. 1 and 25

60. Ans. (b)

Solution:

$$\underset{x \to 3}{\text{Lt}}\, f(x) = \underset{x \to 3}{\text{Lt}} \left(\frac{2x^2 - 7x + 3}{5x^2 - 12x - 9} \right)$$

This is of the form of $\left(\frac{0}{0}\right)$

Applying L, Hospitals's rule

$$\underset{x \to 3}{\text{Lt}} \left(\frac{4x - 7}{10x - 12} \right) = \frac{5}{18}$$

61. Ans. (b)

Solution: The given limit is of the form of $\left(\frac{0}{0}\right)$

Applying L' Hospital rule

$$\underset{x \to 0}{\text{Lt}} \frac{e^x - \left(1 + x + \frac{x^2}{2}\right)}{x^3}$$

$$\underset{x \to 0}{\text{Lt}} \frac{e^x - 1}{6x}$$

$$\underset{x \to 0}{\text{Lt}} \frac{e^x}{6} = \frac{1}{6}$$

62. Ans. (d)

Solution:

Since the position of rail engine x(t) is continuous and differentiable function, according to lagranges mean value theorem.

t where $0 \le t \le 8$ such that

$$x'(t) = v(t) = \frac{x(8) - x(0)}{8 - 0} = \frac{(280 - 0)}{(8 - 0)} \text{ m/sec}$$

$$= \frac{280}{8} \text{ m/sec}$$

$$= \frac{280}{8} \times \frac{3600}{1000} \text{ kmph}$$

$$= 126 \text{ kmph}$$

63. Ans. (a)

Solution:

The sixth function is

$$f(x, y) = 4x^2 + 6y^2 - 8x - 4y + 8$$

$$\frac{\partial f}{\partial x} = 8x - 8$$

$$\frac{\partial f}{\partial y} = 12y - 4y$$

Put, $\frac{\partial f}{\partial x} = 0$ and $\frac{\partial f}{\partial y} = 0$

$$8x - 8 = 0 \text{ and } 12y - 4y = 0$$

$\therefore$ $x = 1$ and $y = \frac{1}{3}$

$\left(1, \frac{1}{3}\right)$ is the stationary point.

$$r = \left[\frac{\partial^2 f}{\partial x^2}\right]_{\left(1,\frac{1}{3}\right)} = 8$$

$$s = \left[\frac{\partial^2 f}{\partial x dy}\right]_{\left(1,\frac{1}{3}\right)} = 0$$

$$t = \left[\frac{\partial^2 f}{\partial y^2}\right]_{\left(1,\frac{1}{3}\right)} = 12$$

Since, $rt = 8 \times 12 = 48$

$s^2 = 0$

Since, $rt > s^2$,

Hence either a maxima or minima occure at $\left(1, \frac{1}{3}\right)$.

also since, $r = \left[\frac{\partial f}{\partial x^2}\right]_{\left(1,\frac{1}{3}\right)} = 8 > 0$, the point $\left(1, \frac{1}{3}\right)$ is a point of minima.

The minimum value is

$$f\left(1, \frac{1}{3}\right) = 4 \times 1^2 + 6 \times \frac{1}{3^2} - 8 \times 1\ 4 \times \frac{1}{3} + 8 = \frac{10}{3}$$

64. Ans. (a)

Solution:

Let the point (x, y, z) be on surface $z^2 = 1 + xy$

Distance from origin = l

$$= \sqrt{(x-0)^2 + (y-0)^2 + (z-0)^2}$$

$$= \sqrt{x^2 + y^2 + z^2}$$

$$l = \sqrt{x^2 + y^2 + 1 + xy}$$

[since $z^2 = 1 + xy$ is given]

This distance is shortest when l is minimum. Hence find minima of $x^2 + y^2 + 1 + xy$

Let, $u = x^2 + y^2 + 1 + xy$

$$\frac{\partial u}{\partial x} = 2x + y$$

$$\frac{\partial u}{\partial y} = 2y + z$$

$$\frac{\partial u}{\partial x} = 0 \text{ and } \frac{\partial u}{\partial y} = 0$$

or $2x + y = 0$ and $2y + x = 0$

Solving simultaneously, we get

$x = 0$ and $y = 0$

so (0, 0) is the only stationary point.

$$r = \frac{\partial^2 u}{\partial x^2} = 2$$

$$s = \frac{\partial^2 u}{\partial x \partial y^2} = 1$$

$$t = \frac{\partial^2 u}{\partial y^2} = 2$$

Since $rt = 2 \times = 4 > S^2 = 1$

We have case 1, i.e. either a maximum or minimum exists at (0, 0).

Now, since $r = 2 > 0$, so it is a minima at (0, 0).

Now at $x = 0, y = 0, z = \sqrt{1+xy} = \sqrt{1+0} = 1$

The distance $l = \sqrt{0^2 + 0^2 + 1^2} = 1$

65. Ans. (a)

Solution:

Putting

$$f'(x) = 6x^2 - 6x - 36 = 0$$

or $x^2 - x - 6 = 0$

or $x = 3$ or -2

$$f''(x) = 12x - 6$$

$$f''(3) = 30 > 0 \text{ (minima)}$$

$$f''(-2) = -30 < 0 \text{ (minima)}$$

Hence maxima is at $x = -2$ only.

66. Ans. (c)

67. Ans. (c)

Solution:

$$f = y^x$$

Treating x as constant, we get

$$\frac{\partial f}{\partial y} = xy^{x-1}$$

Now we treat y as a constant and get,

$$\frac{\partial^2 f}{\partial x \partial y} = \frac{\partial}{\partial x}\left(y^{x-1}\right) = y^{x-1} + x\,y^{x-1}\ \ln y$$

at $x = 2$ and $y = 1$

$$= \frac{\partial^2 f}{\partial x \partial y} = 1^{(2-1)}(1 + 2.\,n1) = 1$$

68. Ans. (a)

Solution:

$$I = \int_0^{\pi/3} e^{it}\, dt = \left[\frac{e^{it}}{i}\right]_0^{\pi/3}$$

$$= \left[\frac{\cos t + i \sin t}{i}\right]_0^{\pi/3} = \frac{1}{i}\left[\frac{1}{2} + i\frac{\sqrt{3}}{2} - 1\right]$$

$$= \left[-\frac{1}{2i} + \frac{\sqrt{3}}{2}\right] = \left[\frac{\sqrt{3}}{2} + i\,\frac{1}{2}\right]$$

69. Ans. (b)

Solution:

$$\text{Let} \quad I(\alpha) = \int_0^{\infty} e^{-\alpha x} \frac{\sin x}{x} dx \qquad (1)$$

$$\frac{\partial I}{\partial \alpha} = \int_0^{\infty} (-x)\, e^{-\alpha x} \frac{\sin x}{x} dx = -\int_0^{\infty} e^{-\alpha x} \sin x\, dx$$

Then integrating by parts we get,

$$\frac{\partial I}{\partial \alpha} = -\left[\frac{e^{-\alpha x}}{1+\alpha^2}(-\alpha \sin x - 1.\cos x)\right]_0^{\infty} = -\frac{1}{1+\alpha^2}$$

therefore $\dfrac{\partial I}{\partial \alpha} = -\dfrac{1}{1+\alpha^2}$

$$\partial I = -\frac{\partial \alpha}{1+\alpha^2}$$

Integrating, $I = -\tan^{-1}\alpha + C$

$$I(\infty) = -\frac{\pi}{2} + C \text{ but from (1), } |(\infty) = 0$$

therefore $-\frac{\pi}{2} + C = 0$

or $C = \frac{\pi}{2}$

therefore $I(\alpha) = -\tan^{-1}\alpha + \frac{\pi}{2}$

or $I(0) = \frac{\pi}{2} - \tan^{-1} 0$

therefore $I(0) = \frac{\pi}{2}$

But from equation (1),

$$I(0) = \int_0^\infty \frac{\sin}{x} dx$$

therefore $\int_0^\infty \frac{\sin}{x} dx = \frac{\pi}{2}$

70. Ans. (d)

Solution:

$$\int_0^{\pi 4} \tan x \, dx = \log\sqrt{2}$$

$$\int_0^\infty \frac{dx}{x^2+1} = \frac{\pi}{2}$$

$$\int_0^\infty x e^{-x} dx$$

Integrating by part, taking $u = x$ and $dv = e^{-x}$ dx

$$\int_0^\infty x e^{-x} dx = x(-e^{-x}) - \int -e^{-x} dx$$

$$= -x\, ex^{-x} - e^{-x}$$

$$= -e^{-x}(x+1)$$

Now $= [-e - x(x+1)]_0^\infty = 1$

$$\int_0^1 \frac{1}{1-x} dx = \ln 0 - \ln 1 = -\infty - 0 = -\infty$$

71. Ans. (d)

Solution:

Area common to circles, $r = a$ and $r = 2a\cos\theta$ is $1.228a^2$.

72. Ans. (d)

Solution:

$$y = \frac{2}{3}x^{2/3}$$

$$\frac{dy}{dx} = x^{1/2}$$

length of the curve is given by

$$\int_0^1 \sqrt{1+\left(\frac{dy}{dx}\right)^2} dx = \int_0^1 \sqrt{1+x}\, dx$$

$$= \left[\frac{2}{3}(1+x)^{3/2}\right]_{x=0}^{x=1} = 1.22$$

73. Ans. (b)

Solution:

Path AB: $x^2 + y^2 = 1$

$x = \cos\theta$

$y = \sin\theta$

Along path ABθ varies from $[0 \text{ to } \pi/2]$

$$\int_{\text{Path AB}} (x+y)^2 (r d\theta) = \int_0^{\pi/2} (\cos\theta + \sin\theta)^2 1.d\theta$$

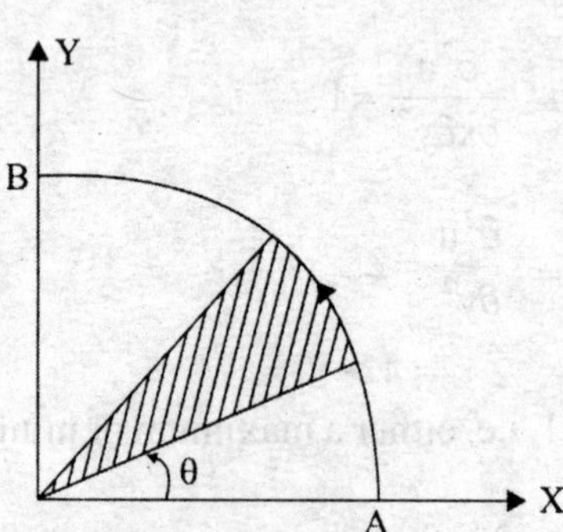

$$= \int_0^{\pi/2} (\sin^2\theta + \cos^2\theta + 2\sin\theta\cos\theta) d\theta$$

$$= \int_0^{\pi/2} (1 + \sin 2\theta)\, d\theta$$

$$= \theta + \left.\frac{(-\cos 2\theta)}{2}\right|_0^{\pi/2}$$

$$= \frac{\pi}{2} - \frac{1}{2}\left[\cos 2\frac{\pi}{2} - \cos 0\right]$$

$$= \frac{\pi}{2} - \frac{1}{2}[-1-1] = \frac{\pi}{2} + 1$$

74. Ans. (a)

Solution:

Curve 1: $y^2 = 4x$

Curve 2: $x^2 = 4y$

Intersection points of curve 1 and 2

$$y^2 = 4x = 4\sqrt{4} = 8\sqrt{y}$$

$$y^4 = 8 \times 8y$$

or $$y(y^3 - 64) = 0$$

$$y = 4 \text{ and } y = 0$$

then $$x = 4 \qquad x = 0$$

Therefore intersection point are A(4, 4) and 0 (0, 0)

The area enclosed between curves 1 and 2 are given by

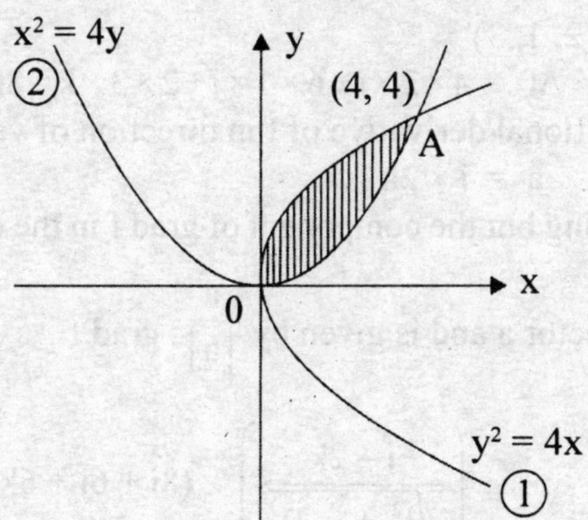

$$\text{Area} = \int_{x_1}^{x_2} y_1\, dx - \int_{x_1}^{x_2} y_2\, dx$$

$$= -\int_0^4 \frac{x^2}{4}\, dx$$

$$= 2\frac{x^{3/2}}{3/2}\Big|_0^4 - \frac{x^3}{3 \times 4}\Big|_0^4$$

$$= \frac{4}{3}(4)^{3/2} - \frac{(4)^3}{3 \times 4} = \frac{16}{3}$$

75. Ans. (a)

Solution:

$$\int_0^3\int_0^x (6 - x - y)\, dxdy$$

$$= \int_0^3 \left[(6-x)y - \frac{y^2}{2}\right]_0^x dx$$

$$= \int_0^3 \left[(6 - x)x - \frac{x^2}{2}\right] dx = 13.5$$

76. Ans. (a)

Solution:

$$V = \int_0^{2\pi}\int_0^{\pi/3}\int_0^1 r^2 \sin\phi \cdot dr \cdot d\phi \cdot d\theta$$

$$\int_0^{2\pi}\int_0^{\pi/3}\left[\frac{r^3}{3}\right]_0^1 \sin\phi d\phi\, d\theta$$

$$= \frac{1}{3}\int_0^{2\pi}[-\cos\phi]_0^{\pi/3} d\theta$$

$$= \frac{1}{3} \times \frac{1}{2} \times \int_0^{2\pi} d\theta$$

$$= \frac{1}{3} \times \frac{1}{2} \times 2\pi = \frac{\pi}{3}$$

77. Ans. (a)

Solution:

$$\frac{\partial x}{\partial u} = v \qquad \frac{\partial x}{\partial v} = u$$

and $$\frac{\partial y}{\partial u} = -\frac{v}{u^2} \qquad \frac{\partial y}{\partial v} = \frac{1}{u}$$

and

$$\phi(u, v) = \begin{vmatrix} \frac{\partial x}{\partial u} & \frac{\partial x}{\partial v} \\ \frac{\partial y}{\partial u} & \frac{\partial y}{\partial v} \end{vmatrix} = \begin{vmatrix} v & u \\ -\frac{v}{u^2} & \frac{1}{u} \end{vmatrix} = \frac{v}{u} + \frac{v}{u} = \frac{2v}{u}$$

78. Ans. (a)

Solution:

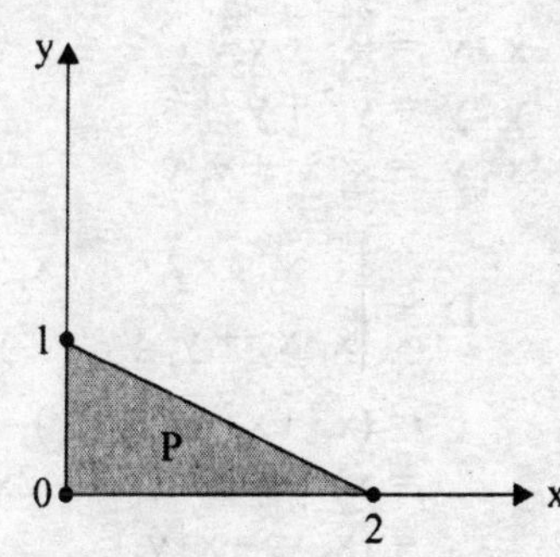

The equation of the straight line with x-intercept = 2 and y-intercept = 1 is

$$\frac{x}{2} + \frac{y}{1} = 1$$

or $$y = 1 - \frac{x}{2}$$

or $\quad x = 2 - 2y$

or $\quad \int_0^1 \int_0^{(2-2y)} (xy\,dx)\,dy = \int_0^1 \left[\left(\frac{yx^2}{2} \right) \right]_0^{2-2y} dy$

$= \int_0^1 \frac{y}{2}(2-2y)^2\,dy = \int_0^1 2y(1-y)^2\,dy = \frac{1}{6}$

79. Ans. (b)

Solution:

Form C, draw CN ⊥ AB. From right-angled ΔCAN.

$$\sin A = \frac{|CN|}{|AC|} \Rightarrow |CN| = |AC| \sin A.$$

$$\text{Area of } \Delta ABC = \frac{1}{2}|AB| \times |CN|$$

$$= \frac{1}{2}|AB|.|CN| \sin A = \frac{1}{2}\left|\overrightarrow{AB} \times \overrightarrow{AC}\right|$$

From above figure, $\overrightarrow{AB} = \vec{b} - \vec{a}$ and $\overrightarrow{AC} = \vec{c} - \vec{a}$.

$$\text{Area of } \Delta ABC = \frac{1}{2}\left|(\vec{b} - \vec{a}) \times (\vec{c} - \vec{a})\right|$$

$$= \frac{1}{2}\left|(\bar{a} - \bar{b})\ (\bar{a} - c)\right|$$

80. Ans. (b)

Solution:

Let $\quad D = \begin{vmatrix} x.x & x.y \\ y.x & y.y \end{vmatrix}$

Let
$$x = x_1 i + x_2 j$$
$$y = y_1 i + y_2 j$$
$$x . y = x_1^2 + x_2^2$$
$$y . y = y_1^2 + y_2^2$$
$$x . y = x_1 x_2 + y_1 y_2$$

Therefore $\quad D = \begin{vmatrix} x_1^2 + x_2^2 & x_1 x_2 + y_1 y_2 \\ x_1 x_2 + y_1 y_2 & y_1^2 + y_2^2 \end{vmatrix}$

$$= (x_1^2 + x_2^2)(y_1^2 + y_2^2) - (x_1 x_2 + y_1 y_2)^2$$
$$= x_2^2 y_1^2 + x_1^2 y_2^2 - 2x_1 y_1 x_2 y_2$$
$$= (x_2 y_1 - x_1 y_2)^2$$

Now, $\quad D = 0$

$$x_2 y_1 - x_1 y_2 = 0$$

Or $\quad \frac{x_1}{x_2} = \frac{y_1}{y_2}$

Or Vector $x_1\ i + x_2\ j$ and $y_1\ i + y_2\ j$ are linearly dependent.

Therefore $\quad$ Linear dependence $\Rightarrow D = 0$

So, $\quad D \neq 0$, i.e. is negetie or positive.

[notice that since $D = (x_2\ y_1 - x_1\ y_2)^2$, it cannot be negative].

So, $\quad$ Liear independence $\Rightarrow$ D is positive

81. Ans. (c)

Solution:

Given

$$f = 2x^2 + 3y^2 + z^2$$

$$\Delta f = i\frac{\partial f}{\partial x} + j\frac{\partial f}{\partial y} + k\frac{\partial f}{\partial z}$$
$$= 4xi + 6yj + 2zk$$

at P (2, 1, 3)

$$\Delta f = 4 \times 2 \times i + 6 \times 1 \times j + 2 \times 3 \times k = 8i + 6j + 6k$$

directional derivative of f in direction of vector

$$a = i \times 2k \text{ is}$$

nothing but the component of grad f in the direction of vector a and is given by $\frac{a}{|a|}$. grad f

$$= \left(\frac{i - 2k}{\sqrt{1^2 + (-2)^2}} \right) \cdot (8i + 6j + 6k)$$

$$= \frac{1}{\sqrt{5}}(1.8 + 0.6 + (-2)6)$$

$$= \frac{-4}{\sqrt{5}} = -1.789$$

82. Ans. (b)

Solution:

$$\Delta f = i\frac{\partial f}{\partial x} + j\frac{\partial f}{\partial y} + k\frac{\partial f}{\partial z}$$

here $\quad f = x^2 + 3y^2 + 2z^2$

therefore $\quad \Delta f = i(2x) + j(6y) + k(4 \times -1)$

$$= 2i + 12j - 4k$$

The directional derivative in direction of vector a = i – j + 2k is given by

$$\frac{a}{|a|} \cdot \text{grad } f = \frac{i + j + 2k}{\sqrt{1^2 + (-1)^2 + 2^2}} . (2i + 12j - 4k)$$

$$= \frac{1}{\sqrt{6}}(1.2 + (-1).\ 12 + 2\ (-4))$$

$$= -\frac{18}{\sqrt{6}} = -\sqrt{6}$$

83. Ans. (b)

Solution:

Given $f(x) = (x-8)^{2/3} + 1$

$$f'(x) = \frac{2}{3}(x-8)^{-1/3}$$

Slope of tangent at point (0, 5)

$$m = \frac{2}{3}(0-8)^{-1/3} = -\frac{1}{3}$$

Slope of normal at point (0, 5)

$$m_1 = \frac{-1}{m} = 3$$

$$y - 5 = 3(x-0)$$

or $y = 3x + 5$

84. Ans. (d)

Solution:

Velocity vector $\vec{V} = 2xy\hat{i} - x^2z\hat{j}$

The vorticity vector = curl (velocity vector)

= curl ($\vec{V}$)

$= \Delta \times \vec{V}$

$$= \begin{vmatrix} \hat{i} & \hat{i} & \hat{k} \\ \frac{\partial}{\partial x} & \frac{\partial}{\partial y} & \frac{\partial}{\partial z} \\ 2xy & -x^2z & 0 \end{vmatrix}$$

$$= \left[\frac{\partial}{\partial y}(0) - \frac{\partial}{\partial z}(-x^2z)\right] i - \left[\frac{\partial}{\partial x}(0) - \frac{\partial}{\partial z}(2xy)\right]\hat{j} + \left[\frac{\partial}{\partial x}(-x^2z) - \frac{\partial}{\partial y}(2xy)\right]\hat{k}$$

$$= x^2\,\hat{i} + [-2xz - 2x]\,\hat{k}$$

at (1, 1, 1),

vorticity vector = $\hat{i} - 4\hat{k}$

85. Ans. (a)

Solution:

$$f_x = 2xyz,\ f_y = x^2z,\ f_z = x^2y$$

By, using line integration

f = Potential function of $V = x^2yz$

Therefore, Line integral of the vector function from point A (0, 0, 0) to the point B (1, 1, 1) is

$= f(B) - f(A)$

$= (x^2yz)_{1,1,1} - (x^2yz)_{(0,0,0)}$

$= 1 - 0 = 1$

86. Ans. (c)

Solution:

By Green's Theorem

$$\oint_C \phi dx + \psi dy = \iint_R \left(\frac{\partial \psi}{\partial x} - \frac{\partial \phi}{\partial y}\right) dxdy$$

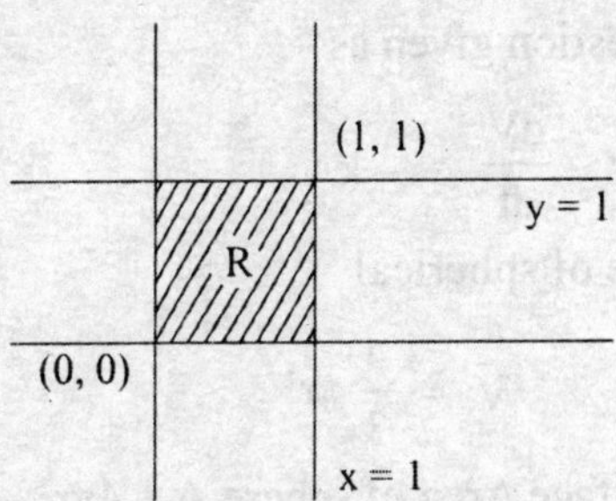

Here

$$I = \oint_C (xydy - y^2dx = \oint_C (-y^2)\,dx + (xy)\,dy$$

Therefore $\phi = -y^2,\ \psi = xy$

$$\frac{\partial \psi}{\partial x} = y,\ \frac{\partial \phi}{\partial y} = -2y$$

Using Green's theorem,

$$I = \int_{y=0}^{1}\int_{x=0}^{1} [y - (2y)]\,dx\,dy = \int_{y=0}^{1}\int_{x=0}^{1} 3y\,dx\,dy$$

$$= \int_{y=0}^{1} [3xy]_{x=0}^{1}\,dy = \int_{y=0}^{1} 3ydy = \frac{3}{2}$$

87. Ans. (b)

Solution:

$$\frac{dx}{dt} = -kx^2$$

or $\dfrac{dx}{x^2} = -kdt$

Integrate both sides, $\int \dfrac{dx}{x^2} = -\int kdt$

$$-\frac{1}{x} = -kt + C$$

or $\dfrac{1}{x} = kt + C'$

at $t = 0,\ x = a$

or $\dfrac{1}{a} = k \times 0 + C'$

or $\quad C' = \frac{1}{a}$

therefore $\quad \frac{1}{x} = kt + \frac{1}{a}$

88. Ans. (a)

Solution:

From question given as

$$\frac{dV}{dt} = -kA \quad (1)$$

If volume of spherical

$$V = \frac{4}{3}\pi r^3$$

Outer surface Area of sphere $A = 4\pi r^2$

or $\quad \frac{dV}{dt} = \frac{4}{3}\pi \times 3r^2 \frac{dr}{dt} = 4\pi r^2 \frac{dr}{dt}$

Substituting these in (1),

$$4\pi r^2 \frac{dr}{dt} = -k\,(4\pi r^2)\frac{dr}{dt}$$

$$\frac{dr}{dt} = -k$$

or $\quad dr = -k dt$

integrating both side

$$r = -kt + C$$

at $t = 0$, $r = 1$

or $\quad 1 = -k \times 0 + C$

or $\quad C = 1$

therefore $\quad r = -kt + 1 \quad (2)$

Now at $t = 3$ months

$$r = 0.5 \text{ cm}$$

or $\quad k = \frac{0.5}{3}$

Now substituting this value of r in equation (2)

$$r = -\frac{0.5}{3}t + 1$$

putting $\quad r = 0$ (ball completely evaporates)

in above and solving for t gives $0 = -\frac{0.5}{3}t + 1$

or $\quad t = 6$ months

89. Ans. (d)

Solution:

$$\frac{dy}{dx} = \frac{-x}{y}$$

or $\quad y\,dy = -x\,dx$

or $\quad \int y\,dy = \int -x\,dx$

$$\frac{y^2}{2} = -\frac{x^2}{2} + C$$

at $x = 1$, $y = \sqrt{3}$

therefore $\quad \frac{(\sqrt{3})^2}{2} = \frac{-1^2}{2} + C$

or $\quad C = 2$

$\therefore$ Solution is $\quad \frac{y^2}{2} = \frac{x^2}{2} + 2$

or $\quad x^2 + y^2 = 4$

90. Ans. (a)

Solution:

Given differential equation.

$$3y\frac{dy}{dx} + 2x = 0$$

or $\quad \frac{dy}{dx} = \frac{-2}{3y}$

or $\quad 3y dy = -2x dx$

or $\quad \int 3y dy = \int -2x dx$

or $\quad \frac{3}{2}y^2 = -2 \times \frac{x^2}{2} + C$

or $\quad 3y^2 + 2x^2 = C$

or $\quad \frac{x^2}{\left(\frac{1}{2}\right)} + \frac{y^2}{\left(\frac{1}{3}C\right)} = C$

which is the equation of a family of ellipes.

91. Ans. (a)

Solution:

Given diffenential equation

$$\frac{dy}{dx} + y^2 = 0$$

or $\quad -\frac{dy}{y^2} = dx$

or $\quad -\int \frac{dy}{y^2} = \int dx$

$$\frac{1}{y} = x + C$$

$$y = \frac{1}{x + C}$$

92. Ans. (c)

Solution:

Given $\frac{dy}{dx} = y^2$

or $\int \frac{dy}{y^2} = \int dx$

or $-\frac{1}{y} = x + C$

Therefore $y = -\frac{1}{x+C}$

When $x = 0$

$y = 1$

Therefore $C = -1$

Therefore $y = -\frac{1}{x-1}$

y is bounded when $x - 1 \neq 0$

i.e. $x \neq 1$

i.e. $x < 1$ or $x > 1$

93. Ans. (a)

Solution:

Given equation is $\frac{dy}{dt} + p(t)\,y = q(t)\,y^n$; $n > 0$ (1)

put $v = y^{1-n}$

$$\frac{dv}{dt} = (1-n)\,y^{-n}\frac{dy}{dt}$$

$$\frac{dy}{dt} = \frac{1}{(1-n)y^{-n}}\frac{dv}{dt}$$

Substituting in, equation (1)

$$\frac{1}{(1-n)y^{-n}}\frac{dv}{dt} + p(t)y = q(t)y^n$$

Multiplying by $(1-n)y^{-n}$,

$$\frac{dv}{dt} + p(t)\,(1-n)y^{1-n} = q(t)\,(1-n)$$

Since $y^{1-n} = v$,

Therefore $\frac{dv}{dt} + (1-n)\,pv = (1-n)q$

(which is linear with v as dependent variable and t at independent variable).

94. Ans. (a)

Solution:

Given equation is

$$x\frac{dy}{dx} + y = x^4$$

or $\frac{dy}{dx} + \left(\frac{y}{x}\right) = x^3$ (1)

Standard form of linear equation is

$$\frac{dy}{dx} + PY = Q \quad (2)$$

where P and Q function of x only and solution is given by

$$y(\text{I.F.}) = \int Q.\,(\text{I.F.})\,dx + C$$

where, (I.F.) $= e^{\int P dx}$

Here in equation (1)

$$P = \frac{1}{x} \text{ and } Q = x^3$$

$$\text{I.F.} = e^{\int \frac{1}{x}dx} = e^{\ln x} = x$$

Solution $y(x) = \int x^3\,x\,dx + C$

$$yx = \frac{x^5}{5} + C$$

given condition $y(1) = \frac{6}{5}$

means at $x = 1;\ y = \frac{6}{5}$

or $\frac{6}{5} \times 1 = \frac{1}{5} + C$

or $C = \frac{6}{5} - \frac{1}{5} = 1$

Therefore $y = \frac{x^4}{5} + \frac{1}{x}$

or $y = \frac{x^4}{5} + \frac{1}{x}$

95. Ans. (a)

Solution:

Given equation is

$$\frac{d^2y}{dx^2} + 2\frac{dy}{dx} + 17 = 0$$

$$y(0) = 1$$

$$\frac{dy}{dx}\left(\frac{\pi}{4}\right) = 0$$

This is a linear differential equation

$$D^2 + 2D + 17 = 0$$
$$D = -1 \pm 4i$$
$$Y = C_1e^{(-1+4i)} + C_2e^{(-1-4i)x}$$
$$= e^{-x}C_1e^{4xi} + C_{2e}{}^{e-4xi}$$
$$= e^{-x}[C_1(\cos 4x + i \sin = 4x)] + C_2[\cos(-4x) + i \sin(-4x)]$$
$$= e^{-x}[(C_1+C_2)\cos 4x + (C_1-C_2) i \sin 4x]$$

Let $C_1 + C_2 = C_3$ and $(C_1-C_2) i = C_4$

$$y = e^{-x}(C_3 \cos 4x + C_4 \sin 4x$$

since $y(0) = 1$

or $1 = e^{-0}(C_3 \cos 0 + C_4 \sin 0)$

or $C_3 = 1$

$$\frac{dy}{dx} = e^{-x}(-4C_3 \sin 4x + 4C_4 \cos 4x)$$
$$-e^{-x}[C_3 \cos 4x + C_4 \sin 4x]$$
$$= e^{-x}[(-4C_3 - C_4) \sin 4x + (4C_4 - C_3) \cos 4x]$$

$$\frac{dy}{dx} \text{ at } x = \frac{\pi}{4} \text{ is } 0$$

therefore

$$-(-4C_4 - C_3)e^{-\pi/4} = 0$$
$$4C_4 = C_3$$
$$C_4 = \frac{C_3}{4} = \frac{1}{4}$$

therefore $C_3 = 1$ and $C_4 = \frac{1}{4}$

$$y = e^{-x}(\cos 4x + \frac{1}{4}\sin 4x)$$

96. Ans. (a)

Solution:

Given equation is

$$y'' + 2y' + y = 0$$
$$(D^2 + 2D + 1)y = 0$$

or $D^2 + 2D + 1 = 0$

or $(D + 1)^2 = 0$

or $D = -1, -1$

Therefore $y = (C_1 + C_2x)e^{-x}$

$y(0) = 0 \Rightarrow 0 = (C1 + C2(0)e^{-0}$

or $C_1 = 0$

$y(1) = 0 \quad 0 = (C_1 + C_2)e^{-1}$

or $C_1 + C_2 = 0$

or $C_2 = 0$

Therefore $y = (0 + 0x)e^{-x} = 0$ is the solution

Therefore $y(0.5) = 0$

97. Ans. (b)

Solution:

Given equtions is

$$\frac{d^2y}{dx^2} + 4\frac{dy}{dx} + 3y = 3e^{2x}$$

or $(D^2 + 4D + 3)y = 3e^{2x}$

Particular integral $\text{P.I.} = \frac{1}{D^2 + 4D + 3}3e^{2x}$

Now since, $\frac{1}{f(D)}e^{ax} = \frac{1}{f(a)}e^{ax}$

$$\text{P.I.} = 3\frac{e^{2x}}{(2)2 + 4(2) + 3} = \frac{3e^{2x}}{15} = \frac{e^{2x}}{5}$$

98. Ans. (c)

Solution:

Given equation is

$$f(z) = u + iv$$

Therefore, It must satisfy the Causchy-Reimann equation

$u_x = v_y$ (1)

and $v_x = -u_y$ (2)

Since, given as $u = xy$

Substit u_x and u_y in equation (1) and (2)

$v_y = y$ (3)

and $v_x = -x$ (4)

Integrating (3) and (4) get v as follow:

$$v_y = y$$

or $\frac{\partial v}{\partial y} = y$

or $\int \partial v = \int y\, \partial y$

or $v = \frac{y^2}{2} + f(x)$ (5)

from (5), $v_x = f'(x)$ (6)

Since from (4), $v_x = -x$

Substituting this is (6),

$$f'(x) = -x$$

or $\frac{df}{dx} = -x$

or $\int df = \int -x\, dx$

or $$f = \frac{-x^2}{2} + k$$

Substitute this is (5),

$$v = \frac{y^2}{2} - \frac{x^2}{2} + k$$

or $$v = \frac{y^2 - x^2}{2} + k$$

99. Ans. (a)

Solution:

Cauchy's integral formula is

$$f(a) = \frac{1}{2\pi i} \oint_C \frac{f(z)}{z-a} dz$$

$$= \oint_C \frac{f(2)}{2-a} dz = 2\pi i\, f(a)$$

i.e. $$\oint_C \frac{z^3 - 6}{3z - i} dz = \frac{1}{3} \oint_C \frac{z^3 - 6}{\left(z - \frac{i}{3}\right)} dz$$

Apply Cauchy's integral theorem, using $f(z) = z^3 - 6$.

$$= \frac{1}{3}\left(2\pi i\, f\left(\frac{i}{3}\right)\right) = \frac{1}{3}\left(2\pi i\left[\left(\frac{i}{3}\right)^3 - 6\right]\right)$$

$$= \frac{1}{3}\left[2\pi i\left(\frac{i^3}{27} - 6\right)\right] = \frac{2\pi}{81} i^4 - 4\pi i = \frac{2\pi}{81} - 4\pi i$$

100. Ans. (c)

Solution:

Given integral.

$$I = \int_C \frac{\cos(2\pi z)}{(2z-1)(z-3)} dz$$

$$= \frac{1}{2} \int_C \frac{\left[\frac{\cos(2\pi z)}{(z-3)}\right]}{\left[z - \frac{1}{2}\right]}$$

Since, $z = 1/2$ is a point within $|z| = 1$ (the closed curve C) Apply Cauchy's integral theorem

Using $$f(z) = \frac{\cos(2\pi z)}{(z-3)}$$

$$I = \frac{1}{2} f\left(\frac{1}{2}\right)$$

Therefore $$I = \frac{1}{2} \frac{\cos\left(2\pi \frac{1}{2}\right)}{\left(\frac{1}{2} - 3\right)}$$

$$= \frac{2\pi i}{5}$$

101. Ans. (a)

Solution:

$$\int \sec z dz = \int \frac{1}{\cos z} dz$$

The poles are at $z_0 = (n + 1/2)\pi = ...-3\pi/2, -\pi/2, \pi/2, +3\pi/2...$

None of these poles lie inside the unit circle $|z| = 1$

Hence, sum of residues at poles = 0

Therefore, singularities set= ϕ and

$I = 2\pi i$ [sum of residues of f(z) at the poles]

$= 2\pi i \times 0 = 0$

102. Ans. (a)

Solution:

Given function

$$f(z) = \frac{\cos z}{z}$$

has simple pole at $z = 0$ and $z = 0$ is inside the unit circle on complex plane

Therefore,Residue of f(z) at $z = 0$

$$\underset{z\to 0}{\text{Lt}}\, f(z)\cdot z\cdot = \underset{z\to 0}{\text{Lt}} \cos z = 1$$

$$\int_C f(z)\, dz = 2\pi i \text{ (Residue at } z = 0)$$

$$= 2\,\pi\, i.\, 1 = 2\pi i$$

103. Ans. (d)

Solution:

$$\frac{1}{z^2 + 4} = \frac{1}{(z+2i)(z-2i)}$$

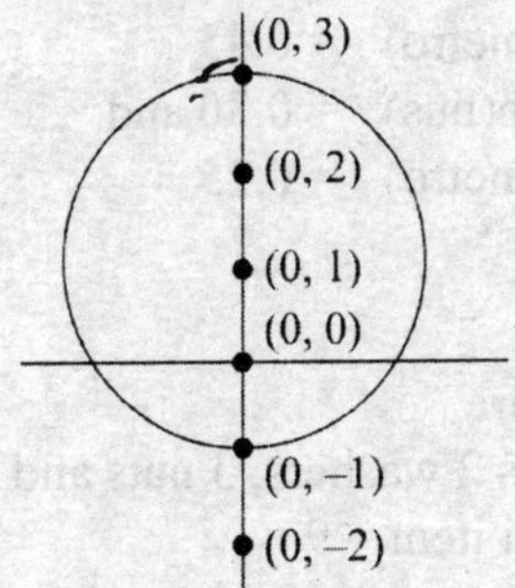

Pole (0, 2) lies inside the circle $|z - i| = 2$ while pole (0, –2) is out side the circle $|2 - i| = 2$ as can be seen from figure below:

$$\int_C f(z)\,dz = 2\pi i \text{ [Residue at those poles which are inside]}$$
$$= 2\pi i \text{ Res } f(2i)$$
$$= 2\pi i \frac{f}{(2i + 2i)} = \frac{\pi}{2}$$

104. Ans. (d)

Solution:

For die sample space = $(6)^2 = 36$

Total ways in which sum is either 8 or 9 is (2, 6), (3, 5), (3, 6) (4, 4) (4, 5) (5, 3), (5, 4) (6, 2) (6, 3) = 9 ways

Therefore, Probability of coming 8 or 9 = $\frac{9}{36} = \frac{1}{4}$

So probability of not coming 8 or 9 = $1 - \frac{1}{4} = \frac{3}{4}$

105. Ans. (c)

Solution:

As given all three gates are independent

p (gate 2 and gate 3 fail when gate 1 already failed)

= p (gate 2 and gate 3 fail)

[As gate 2 and 3 fail independently]

= p (gate 2) × p (gate 3)

= 0.2 × 0.2 = 0.04

106. Ans. (a)

Solution:

Given data are,

p(private car) = 0.45

p(bus public transport) = 0.55

A person has a choice between private car and public transport

p(public transport) = 1 – p(private car) = 1 – 0.45 = 0.55

p(bus) = p(bus ∩ public transport) × p(public transport) = 0.55 × 0.55 = 0.3025 ≃ 0.30

p(metro) = 0.45

Therefore p(bus) = 0.30 and

p(metro) = 0.25

107. Ans. (c)

Solution:

Given data are,

Box contains 2 washers, 3 nuts and 4 bolts

Total no. of i item = 9

p(2 washers, then 3 nuts, then 4 bolts)

$$= \left(\frac{2}{9} \times \frac{1}{8}\right) \times \left(\frac{3}{7} \times \frac{2}{6} \times \frac{1}{5}\right) \times \left(\frac{4}{4} \times \frac{3}{3} \times \frac{2}{2} \times \frac{1}{1}\right) = \frac{1}{1260}$$

108. Ans. (c)

Solution:

The coefficient of variation in speed

$$= \frac{\sigma}{\mu} = \frac{8.8}{33} = 0.2666$$

109. Ans. (b)

Solution:

Population is finite, hypergeometric distribution can be applied as.

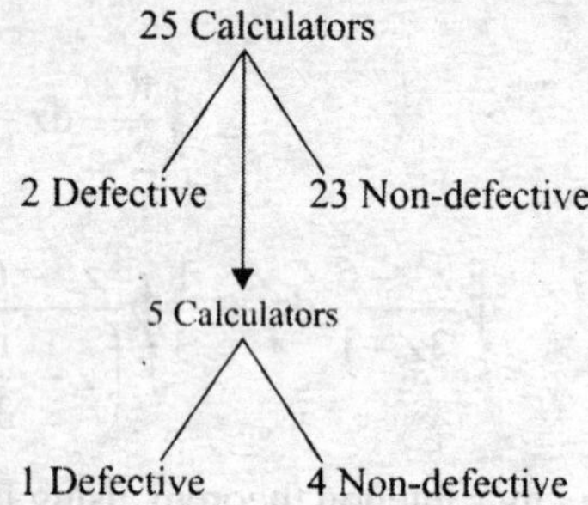

$$p\text{ (1 defective in 5 calculators)} = \frac{2C_1 \times 23C_4}{25C_5} = \frac{1}{3}$$

110. Ans. (d)

Solution:

Probability of drawing two red balls

= p (first is red) × p (second is red)

$$= \frac{5}{10} \times \frac{4}{9} = \frac{2}{9}$$

111. Ans. (d)

Solution:

Hypergeometric distribution can be applied as:

$$p\,(X = 2) = \frac{4C_2 \times 48C_0}{52C_2} = \frac{1}{221}$$

52

4K 48NK

2

2K 0NK

112. Ans. (b)

Solution:

$$P\left(-\frac{1}{3} \le x \le \frac{1}{3}\right) = \int_{-\frac{1}{3}}^{\frac{1}{3}} f(x)\,dx = \int_{-\frac{1}{3}}^{\frac{1}{3}} x^2\,dx$$

$$= \left[\frac{x^3}{3}\right]_{-\frac{1}{3}}^{\frac{1}{3}} = \frac{2}{81}$$

The probability in percentage,

$$= \frac{2}{81} \times 100 = 2.469\% = 2.47\%$$

113. Ans. (b)

Solution:

Mean $\quad E(t) = \int_{-\infty}^{\infty} t \cdot f(t) \cdot dt = \int_{-1}^{1} t \cdot f(t) \cdot dt$

$$= \int_{-1}^{0} t(1+t)\,dt + \int_{0}^{1} t(1-t)\,dt$$

$$= \left[\frac{t^2}{2}+\frac{t^3}{3}\right]_{-1}^{0} + \left[\frac{t^2}{2}-\frac{t^3}{3}\right]_{0}^{1}$$

$$= -\left[\frac{1}{2}-\frac{1}{3}\right] + \left[\frac{1}{2}-\frac{1}{3}\right]$$

$$= -\frac{1}{6}+\frac{1}{6} = 0$$

Variance $(\sigma^2) = E(t^2) - [E(t)]^2$

$$= \int_{-\infty}^{\infty} t^2 f(t)\,dt - [E(t)]^2 = \int_{-\infty}^{\infty} t^2 f(t)\,dt - (0)^2$$

$$= \int_{-\infty}^{\infty} t^2 f(t)\,dt = \int_{-1}^{0} t^2 (1+t)\,dt + \int_{0}^{1} t^2 (t-1)\,dt$$

$$= \int_{-1}^{0} (t^2+t^3)\cdot dt + \int_{0}^{1} t^2(-t^3)\,dt$$

$$= \left[\frac{t^3}{3}+\frac{t^4}{4}\right]_{-1}^{0} + \left[\frac{t^3}{3}-\frac{t^4}{4}\right]_{0}^{1} = -\left[\frac{-1}{3}+\frac{1}{4}\right] + \left[\frac{1}{3}-\frac{1}{4}\right]$$

$$= \frac{1}{12}+\frac{1}{12} = \frac{1}{6}$$

Standard deviation $(\sigma) = \sqrt{\text{variance}} = \frac{1}{\sqrt{6}}$

114. Ans. (d)

Solution:

$$E(X^2 Y^2) = (E(X^2)\,E(Y^2)$$

But since x is not indipendent of x,

$$E(x^2) \neq [E(x)]^2$$

Therefore, $\quad E(X^2 Y^2) \neq [(E(X)]^2\,[(E(Y)]^2$

115. Ans. (a)

Solution:

$$\sigma = \sqrt{\frac{(\beta-\alpha)^2}{12}} = \sqrt{\frac{(1-0)^2}{12}} = \frac{1}{\sqrt{12}}$$

116. Ans. (d)

Solution:

Let μ_c and σ_c are mean and standard deviation of the students of batch C respectively, μ and σ mean and standarrd deviation of entire class of first year students respectively.

Given, $\quad \mu_c = 6.6$

$\sigma_c = 2.3$

$\mu = 5.5$

$\sigma = 4.2$

In order to normalise batch C to entire class, the normalised score (Z scores) must be equated.

since $\quad Z = \frac{x-\mu}{\sigma} = \frac{x-5.5}{4.2} \quad (1)$

$$Z_c = \frac{x_c-\mu_c}{\sigma_c} = \frac{8.8-6.6}{2.3} \quad (2)$$

Equate equation (1) and (2) and solve it

$$\frac{8.5-6.6}{2.3} = \frac{x-5.5}{4.2}$$

Or $\quad x = 8.969 \simeq 9.0$

117. Ans. (a)

Solution:

The given equations are

$$5x + y + 2z = 34$$
$$0x + 4y - 3z = 12$$
$$10x - 2y + z = -4$$

The augmented maxtrix is $\left[\begin{array}{ccc|c} 5 & 1 & 2 & 34 \\ 0 & 4 & -3 & 12 \\ 10 & -2 & 1 & -4 \end{array}\right]$

Since in the first column maximum element in absolute value is 10, exchange row 1 with row 3.

$$\left[\begin{array}{ccc|c} 5 & 1 & 2 & 34 \\ 0 & 4 & -3 & 12 \\ 10 & -2 & 1 & -4 \end{array}\right] \xrightarrow{R(1,3)} \left[\begin{array}{ccc|c} 10 & -2 & 1 & -4 \\ 0 & 4 & -3 & 12 \\ 5 & 1 & 2 & 34 \end{array}\right]$$

So the pivot for eliminating x is $a_{11} = 10$.

Now to eliminate y, compare the elements in second column at and below the diagonal element.

Since, $a_{22} = 4$ is already larger in absolute value compares to $a_{32} = 1$

$\therefore$ The pivot element for eliminating y is $a_{22} = 4$

118. Ans. (a)

Solution:

Since 5 is a root, f(x) is divisible by x – 5.

$$x-5\overline{)\,x^2 - 10x^2 + 31x - 30\,}(x^2 - 5x + 6$$

$$\begin{array}{l} x^2 - 5x \\ \hline -5x^2 + 31x - 30 \\ -5x^2 + 25x \\ \hline 6x - 30 \\ 6x - 30 \\ \hline 0 \end{array}$$

Therefore $(x - 5)(x^2 - 5x + 6) = 0$

Roots of $x^2 - 5x + 6$ are 2 and 3.

119. Ans. (c)

Solution:

Newton–Rephson formula is

$$x_1 = x_0 - \frac{f(x_0)}{f'(x_0)} \qquad (1)$$

Given $\quad f(x) = x^3 + 3x - 7$

$$f'(x) = 3x^2 + 3$$

Put $\quad x_0 = 1$

$$f(x_0) = f(1) = (1)^3 + 3 \times (1) - 7 = -3$$

$$f'(x_0) = f'(1) = 3 \times (1)^2 + 3 = 6$$

Subsite x_0, $f(x_0)$ and $f'(x_0)$ values into (1)

$$x_1 = 1 - \left(\frac{-3}{6}\right) \times 1 = 1.5$$

120. Ans. (a)

Solution:

Simpson's formula is

$$I = \frac{h}{3}\,[f_0 + 4(f_1 + f_3 + f_5 + ... + y_{n-1}) + 2(f_2 + f_4 + f_6 ... + y_{n-2})] + y_n$$

$$I = \frac{1}{3} h\,(f_0 + 4f_1 + 2f_2 + 4f_3 + f_4)$$

$$= \frac{1}{3} \times 0.25\,(1 + 4 \times 0.9412 + 2 \times 0.8 + 4 \times 0.64 + 0.5)$$

$$= 0.7854$$

121. Ans. (a)

Solution:

Given

$f(x) = 1, 4\ 15$ at $x = 0, 1$ and 2 respectively

Trapezoidal rul

$$\int_{x_0}^{2_y} y\,dx = \frac{h}{2}\,[y_0 + 2(y_1 + y_2 + ... y_{n-1}) + y_n]$$

$$\int_0^2 f(x)\,dx = \frac{h}{2}\,(f_1 + 2f_2 + f_3)$$

(3 point Trapezoidal Rule)

here $\quad h = 1$

therefore $\displaystyle\int_0^2 f(x)\,dx = \frac{1}{2}\,(1 + 2 \times 4 + 15) = 12$

Since f(x) is second degree polynomial, let

$$f(x) = a_0 + a_1x + a_2x^2$$

$$f(0) = 1$$

or $\quad a_0 + 0 + 0 = 1$

or $\quad a_0 = 1$

$$f(1) = 4$$

or $\quad a_0 + a_1 + a_2 = 4$

or $\quad 1 + a_1 + a_2 = 4$

or $\quad a_1 + a_2 = 3 \qquad (1)$

$$f(2) = 15$$

or $\quad a_0 + 2a_1 + 4a_2 = 15$

or $\quad 1 + 2a_1 + 4a_2 = 15$

or $\quad 2a_1 + 4a_2 = 14 \qquad (2)$

Solving (1) and (2) $a_1 = -1$ and $a_2 = 4$

$$f(x) = 1 - x + 4x^2$$

Now, exact value of $\displaystyle\int_0^2 f(x)\,dx = \int_0^2 (1 - x + 4x^2)\,dx$

$$= \left[x - \frac{x^2}{2} + \frac{4x^3}{3}\right]_0^2 = \frac{32}{3}$$

Error = Exact – Approximate vaule $= \dfrac{32}{3} - 12 = -\dfrac{4}{3}$

122. Ans. (a)

Solution:

$$h = \frac{2\pi - 0}{8} = \frac{\pi}{4}$$

$$y_0 = \sin(0) = 0$$

i	x	y = sin x
0	0	0
1	$\frac{\pi}{4}$	0.70710
2	$\frac{\pi}{2}$	1
3	$3\frac{\pi}{4}$	0.70710
4	π	0
5	$5\frac{\pi}{4}$	– 0.70710
6	$6\frac{\pi}{4}$	–1
7	$7\frac{\pi}{4}$	–0.70710
8	2π	0

$$y_1 = \sin\left(\frac{\pi}{4}\right) = 0.70710$$

$$y_2 = \sin\left(\frac{\pi}{2}\right) = 1$$

$$y_3 = \sin\left(\frac{3\pi}{2}\right) = 0.70710$$

$$y_4 = \sin(\pi) = 0$$

$$y_5 = \sin\left(\frac{5\pi}{4}\right) = -0.70710$$

$$y_6 = \sin\left(\frac{6\pi}{4}\right) = -1$$

$$y_7 = \sin\left(\frac{7\pi}{4}\right) = -0.70710$$

$$y_8 = \sin\left(\frac{8\pi}{4}\right) = -0$$

By Trapezoidal rule

$$\int_{x_0}^{x_0+nh} f(x)\cdot dx = \frac{h}{2}[y_0 + y_n + 2(y_1 + y_2 + ... + y_{n-1})]$$

$$\int_0^{2\pi} \sin x \cdot dx = \frac{\pi}{8} \times [(0+0) + 2(0.70710 + 1 + 0.70710$$

$$+ 0 - 0.70710 - 0.70710)]$$

$$= 0.00000$$

123. Ans. (b)

Solution:

$$\text{Flywheel energy} = \int_0^{2\pi} T(\theta)\, d\theta$$

where T(θ) is torque exerted.

The integral by using simpson's rule is

$$I = \frac{h}{3}(f_0 + 4f_1 + 2f_2 + 4f_3 + 2f_4 + 4f_5 + f_6)$$

$$h = 60 \text{ degrees} = \frac{\pi}{3} \text{ radious}$$

$$I = \frac{1}{3} \times \frac{\pi}{3} \times [0 + 4 \times 1066 + 2(-323) + 4(0)$$

$$+ 2(323) + 4(-355) + 0]$$

$$= 993$$

124. Ans. (a)

Solution:

$$LT\left[\int_0^t\int_0^t ... \int_0^t f(t)\, dt^n\right] = \frac{1}{s^n}F(s)$$

Given n = 1

$$LT\left[\int_0^t f(\tau)\, d\tau\right] = \frac{1}{s}F(s)$$

125. Ans. (b)

Solution:

$$LT(\sin mt) = \frac{m}{(s^2 + m^2)} = F(s),$$

$$LT\left(\frac{\sin mt}{t}\right) = \int_s^\infty F(s)\, ds = \int_s^\infty \frac{mds}{s_2 + m_2} = \left|\tan^{-1}\frac{s}{m}\right|_s^\infty$$

or by Definition,

$$\int_0^\infty e^{-st}\frac{\sin mt}{t}\, dt = \frac{\pi}{2}\tan^{-1}\frac{s}{m}$$

Now $\underset{s\to 0}{\text{Lt}} \tan^{-1}\left(\frac{s}{m}\right) = 0$ if m > 0 or π if m < 0.

Thus taking limits as s → 0, we get

$$\int_0^\infty \frac{\sin mt}{t}\, dt = \frac{\pi}{2} \text{ if } m > 0 \text{ or } -\frac{\pi}{2} \text{ if } m < 0.$$

Here = 1 which is > 0 therefore the answer is $\frac{\pi}{2}$.

126. Ans. (a)

Solution:

$$f(t) = LT^{-1}\left[\frac{1}{s^2(s+1)}\right]$$

$$\frac{1}{s^2(s+1)} = \frac{A}{s} + \frac{B}{s^2} + \frac{C}{s+1}$$

$$\frac{1}{s^2(s+1)} = \frac{As(s+1) + B(s+1) + C(s^2)}{s^2(s+1)}$$

Matching coefficient of s^2, s and constant in numerator,

$$A + C = 0 \quad (1)$$

$$A + B = 0 \quad (2)$$

$$B = 1 \quad (3)$$

On solving A = –1, B = 1, C = 1

So, $$f(t) = LT^{-1}\left[\frac{-1}{s} + \frac{1}{s^2} + \frac{1}{s+1}\right]$$

$$= -1 + t + e^{-1}$$

$$= t - 1 + e^{-t}$$

127. Ans. (d)

Solution:

$$L[u(t-a)] = \int_0^\infty e^{-st}\, u(t-a)\, dt$$

$$= \int_0^a e^{-st}\cdot 0\cdot dt + \int_0^\infty e^{-st}\cdot 1\cdot dt$$

$$= 0 + \int_a^\infty e^{-st} dt$$

$$= \left[\frac{e^{-st}}{-s}\right]_a^\infty = \frac{e^{-as}}{s}$$

Statement for linked answer question number 128 and 129.

128. Ans. (c)

Solution:

Given equation is

$$\frac{d^2y}{dx^2} + p\frac{dy}{dx} + qy = 0$$

$$\therefore \quad (D^2 + pD + q)y = 0$$

$$y = C_1 e^{-x} + C_2 e^{-3x}$$

So the roots of $D^2 + pD + q = 0$ are $\alpha = -1$ and $\beta = -3$

Sum of roots $= -p = -1 - 3 \Rightarrow p = 4$

Product of roots $= q = (-1)(-3) \Rightarrow q = 3$

129. Ans. (c)

Solution:

Given equation is

$$\frac{d^2y}{dx^2} + p\frac{dy}{dx} + (q+1) = 0$$

Or $[D_2 + PD + (q + 1)]y = 0$

Put $p = 4$ and $q = 3$

Therefore $(D_2 + 4D + 4)y = 0$

$$D^2 + 4D + 4 = 0$$

Therefore $D = 2, -2$

$\Rightarrow y = (C_1x + C_2)\, e^{-2x}$

out of choices given, $y = x\, e^{-2x}$

is the only answer in the required form (i.e. $(c_1x + c_2)\, e^{-2x}$ putting $C_1 = 1$ and $C_2 = 0$

2

Basic of Circuit and Measurement System

Basics of circuits and measurement systems: Kirchoff's laws, mesh and nodal Analysis. Circuit theorems. One-port and two-port Network Functions. Static and dynamic characteristics of Measurement Systems. Error and uncertainty analysis. Statistical analysis of data and curve fitting.

2.1 BASICS OF CIRCUITS AND MEASUREMENT SYSTEMS

Active and Passive Element

Several kind of electric energy sources are known as active element. Examples are voltage and current source.

Voltage Source

Ideal voltage sources.

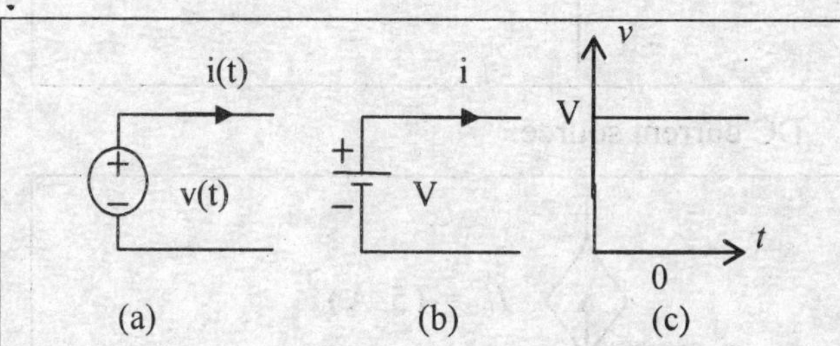

Figure: (a–c) ideal independent voltage source and v-t characteristics For ideal voltage source internal resistance = 0

Practical Voltage Source

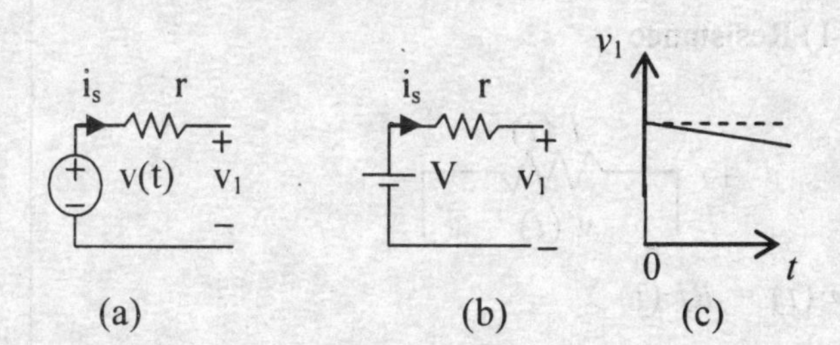

Figure: Practical voltage source and v-t characteristic

Independent Voltage Source

Voltage generated does not vary with time and any circuit element.

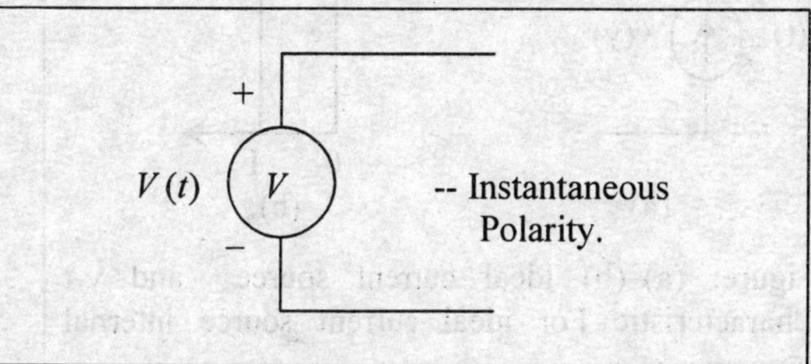

Dependent Voltage Source

Voltage varies with time and depends upon other element voltage or current passing through it.

Voltage Dependent Voltage Source

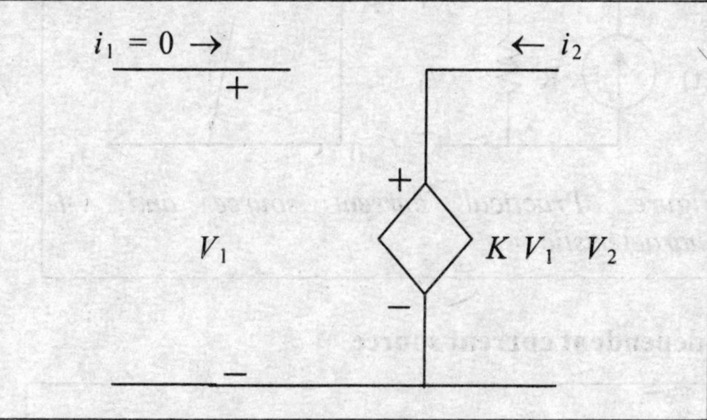

Current Dependent Voltage Source

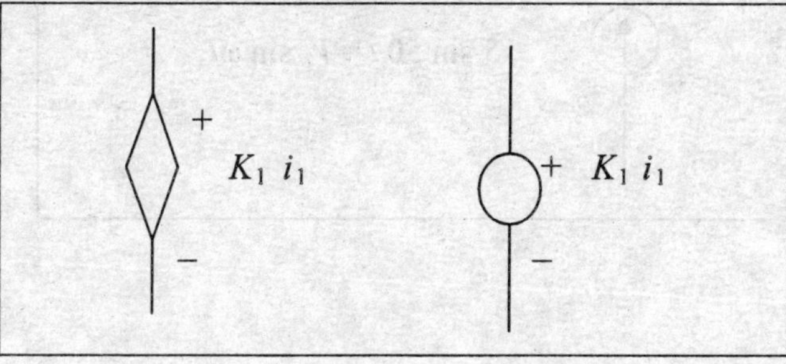

DC Voltage Source

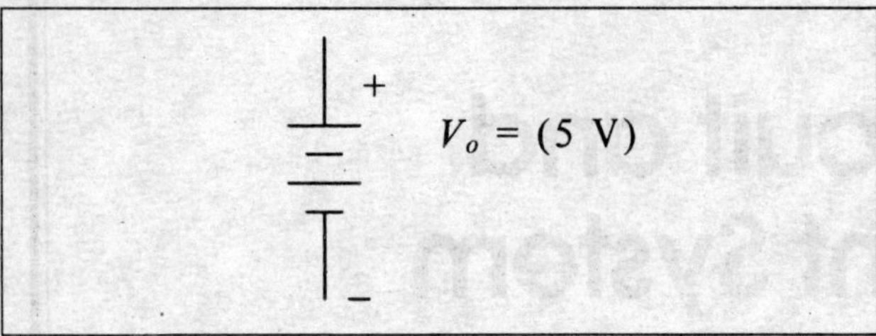

Current source: Examples are transistor, vacuum tube and photoelectric cell.

Current source

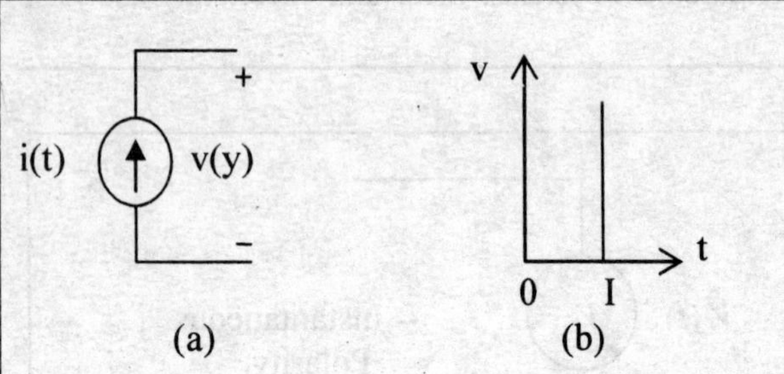

Figure: (a)–(b) Ideal current source and v-t Characteristic For ideal current source internal resistance = ∞

Practical current source

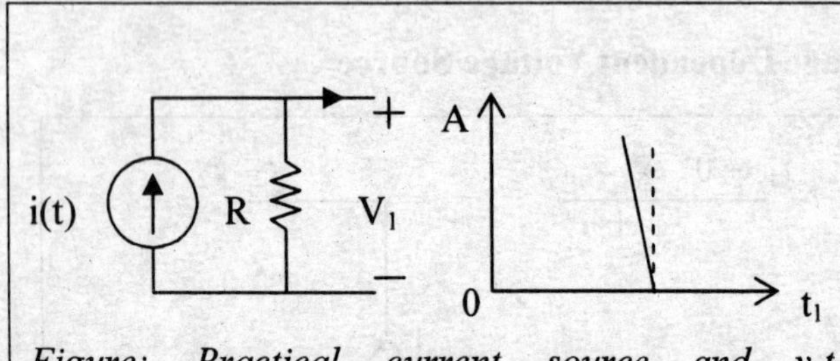

Figure: Practical current source and v-t characteristic

Independent current source

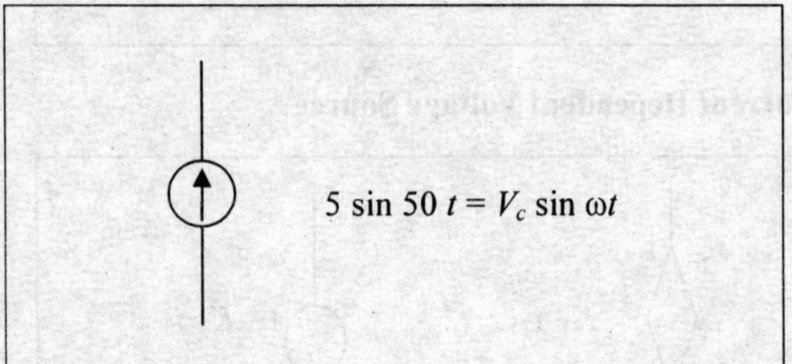

Dependent current source

Current dependent current source.

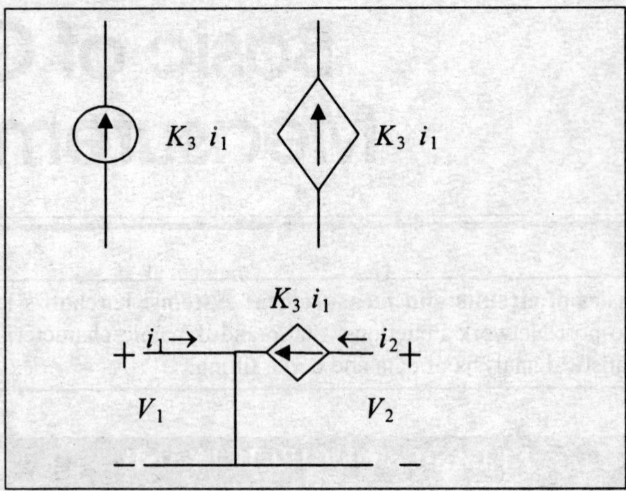

Voltage dependent current source.

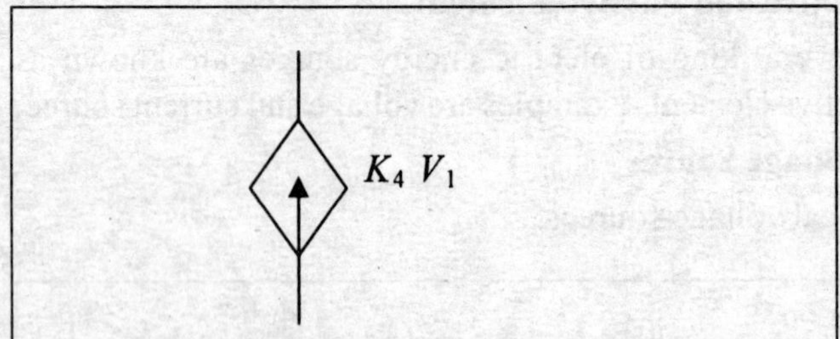

DC current source.

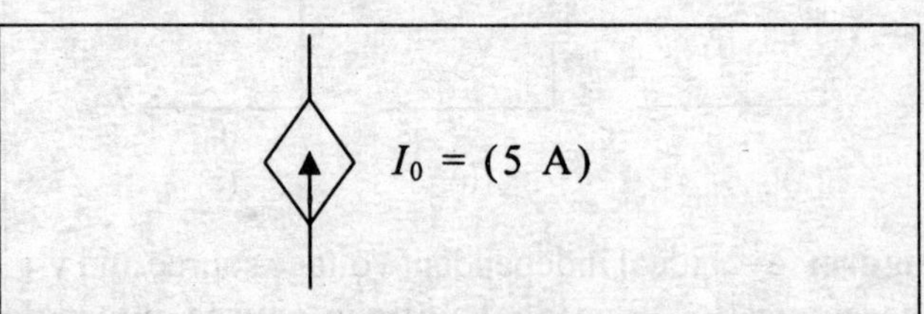

Passive Element

Element not having any energy source are called passive element. Examples are *R*, *L*, and *C*.

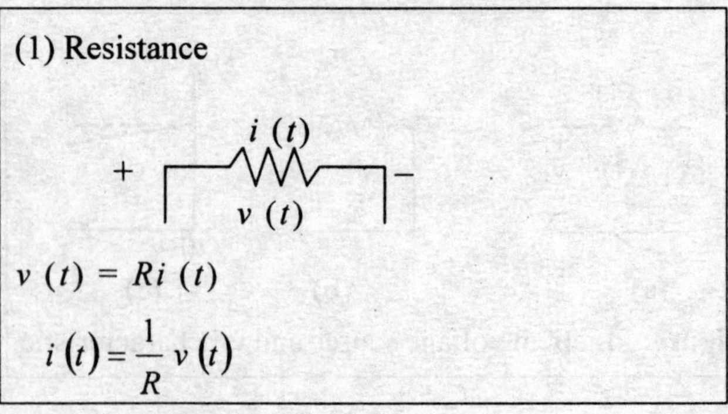

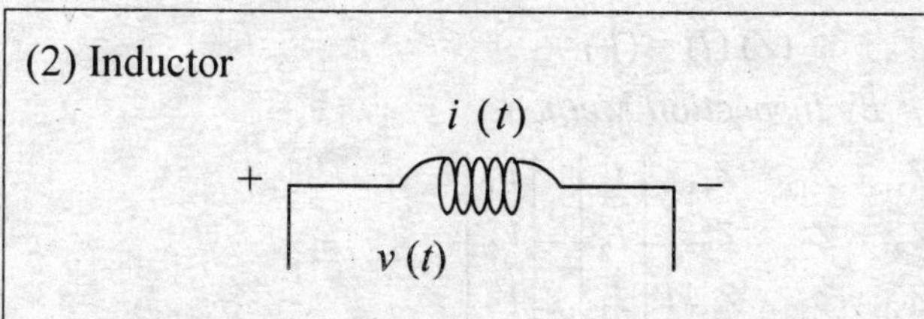

$$v_L(t) = L\frac{di(t)}{dt};\quad i(t) = \frac{1}{L}\int v(t)\,dt$$

(3) Capacitor

$\rightarrow$ $i\ (t)$

+ −

$v\ (t)$

$$i_C(t) = C\frac{dv_C(t)}{dt};\quad v_c(t) = \frac{1}{C}\int i(t)\,dt$$

Unilateral and Bilateral element

A unilateral system has different law relating current and voltage for the two possible directions of currents.

Examples: are vacuum diode, silicon diode and crystal detectors.

In bilateral element the same relationship between current and voltage exists for current flowing in either direction example are *R*, *L*, and *C*.

Network Theory: *Basic law:*

(i) Ohm's Law

- $V \propto I$ or $V = RI$
- $R = \frac{\rho l}{A}$

Potential divider rule

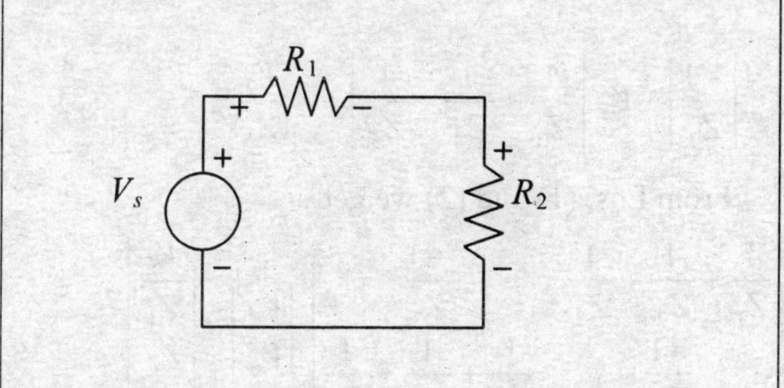

$$\text{Voltage across } R_1 \text{ is} = \frac{V_s \cdot R_1}{R_1 + R_2}$$

$$\text{Voltage across } R_2 \text{ is} = \frac{V_s \cdot R_2}{R_1 + R_2}$$

Current division rule

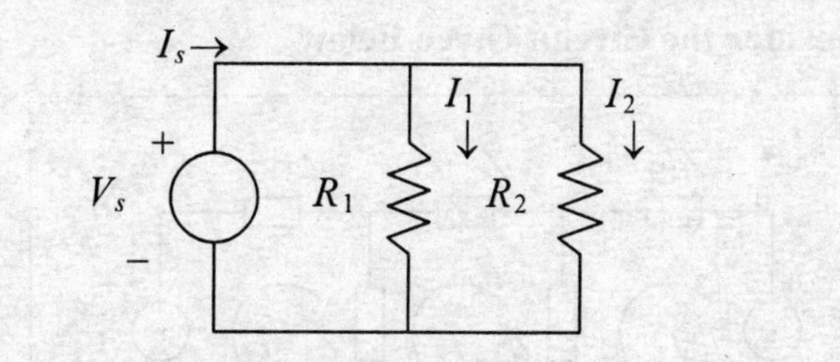

$$\text{Current through } R_1 = \frac{I_s R_2}{R_1 + R_2}$$

$$\text{Current through } R_2 = \frac{I_s R_1}{R_1 + R_2}$$

Element in Series

Resistor $R_{eq} = R_1 + R_2$

Inductor $L_{eq} = L_1 + L_2$

Capacitor $C_{eq} = \frac{1}{[1/C_1] + [1/C_2]}$

Element in Parallel

Resistor $R_{eq} = \frac{1}{[1/R_1] + [1/R_2]}$

Inductor $\frac{1}{L_{eq}} = \frac{1}{L_1} + \frac{1}{L_2}$

Capacitor $C_{eq} = C_1 + C_2$

2.2 KIRCHHOFF'S LAW

Current law state that the algebraic sum of all branch currents entering or leaving a node is zero at all instant of time.

Charge, which enters a node, must leave that node because it cannot be stored there.

$$\sum_{n=1}^{N} i_n = 0$$

Voltage law state that the algebraic sum of all branch voltages around a closed loop path is zero at all instant of time. It is a consequence of the law of conservation of energy.

$$\sum_{n=1}^{N} V_L = 0$$

2.3 Mesh and Nodal Analysis

Consider the Circuit Given Below

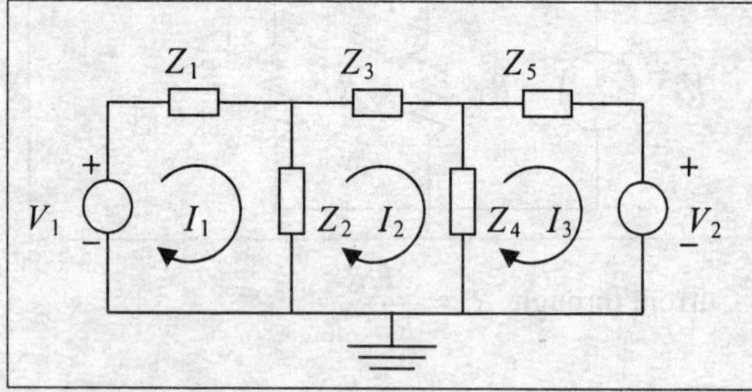

Number of node = $n = 3$

(Reference voltage)

Number of branches = $b = 5$

(reference current)

No. of KVL equation = $b - n + 1$

$= 5 - 3 + 1 = 3$

No. of KCL equation = $n - 1 = 3 - 1 = 2$.

For application of KVL

Convention:

Voltage drop *(+ ve to – ve)* is *+ ve*

Voltage increase *(– ve to + ve) – ve*

I_1, I_2, I_3 are mesh current.

Mesh 1: $-V_1 + I_1 Z_1 + (I_1 - I_2) Z_2 = 0$

$$I_1 (Z_1 + Z_2) + I_2 (-Z_2) + I_3 (0) = V_1 \quad (1)$$

Mesh 2: $(I_2 - I_1) Z_2 + I_2 Z_3 + (I_2 - I_3) Z_4 = 0$

$$1 (-Z_2) + I_2 (Z_2 + Z_3 + Z_4) + I_3 (-Z_4) = 0 \quad (2)$$

Mesh 3: $(I_3 - I_2) Z_4 + I_3 Z_5 + V_2 = 0$

$$I_1 (0) + I_2 (-Z_4) + I_3 (Z_4 + Z_5) = -V_2 \quad (3)$$

From Equations (1), (2), and (3) we get

$$\begin{vmatrix} Z_1 + Z_2 & -Z_2 & 0 \\ -Z_2 & Z_2 + Z_3 + Z_4 & -Z_4 \\ 0 & -Z_4 & Z_4 + Z_5 \end{vmatrix} \begin{vmatrix} I_1 \\ I_2 \\ I_3 \end{vmatrix} = \begin{vmatrix} V_1 \\ 0 \\ -V_2 \end{vmatrix}$$

$$(Z)(I) = (V)$$

By Inspection Method

$$\begin{vmatrix} Z_{11} & Z_{12} & Z_{13} \\ Z_{21} & Z_{22} & Z_{23} \\ Z_{31} & Z_{32} & Z_{33} \end{vmatrix} \begin{vmatrix} I_1 \\ I_2 \\ I_3 \end{vmatrix} = \begin{vmatrix} V_1 \\ V_2 \\ V_3 \end{vmatrix}$$

$$\begin{vmatrix} Z_1 + Z_2 & -Z_2 & 0 \\ -Z_2 & Z_2 + Z_3 + Z_4 & -Z_4 \\ 0 & -Z_4 & Z_4 + Z_5 \end{vmatrix} \begin{vmatrix} I_1 \\ I_2 \\ I_3 \end{vmatrix} = \begin{vmatrix} +V_1 \\ 0 \\ -V_2 \end{vmatrix}$$

Note: If the source potential as current shows then it will be positive otherwise negative

For Application of KCL

Convention: Out going current from node is positive and incoming current to the node is negative.

V_A, V_B, V_C are the node voltage

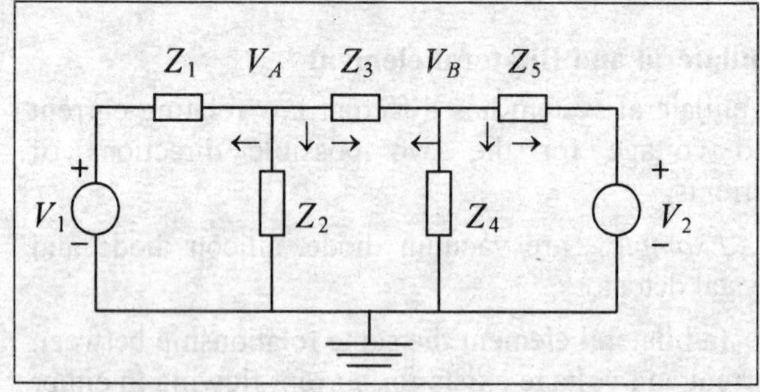

At note A:

$$\frac{V_A - V_1}{Z_1} + \frac{V_A - V_B}{Z_3} + \frac{V_A - 0}{Z_2} = 0$$

$$V_A \left| \frac{1}{Z_1} + \frac{1}{Z_2} + \frac{1}{Z_3} \right| + V_B \left| \frac{-1}{Z_3} \right| = \frac{V_1}{Z_3} \quad (1)$$

At note B:

$$\frac{V_B - V_A}{Z_3} + \frac{V_B - V_2}{Z_5} + \frac{V_B - 0}{Z_4} = 0$$

$$V_A \left| \frac{-1}{Z_3} \right| + V_B \left| \frac{1}{Z_3} + \frac{1}{Z_4} + \frac{1}{Z_5} \right| + = \frac{V_2}{Z_5} \quad (2)$$

From Eqs. (1) and (2) we get

$$\begin{vmatrix} \frac{1}{Z_1} + \frac{1}{Z_2} + \frac{1}{Z_3} & \frac{-1}{Z_3} \\ \frac{-1}{Z_3} & \frac{1}{Z_3} + \frac{1}{Z_4} + \frac{1}{Z_5} \end{vmatrix} \begin{vmatrix} V_A \\ V_B \end{vmatrix} = \begin{vmatrix} \frac{V_1}{Z_1} \\ \frac{V_2}{Z_5} \end{vmatrix}$$

$(Y)\ (V) = (I)$

By Inspection Method

$$\begin{vmatrix} Y_{11} & Y_{12} \\ Y_{21} & Y_{22} \end{vmatrix} \begin{vmatrix} V_1 \\ V_2 \end{vmatrix} = \begin{vmatrix} I_1 \\ I_2 \end{vmatrix}$$

$$\begin{vmatrix} \frac{1}{Z_1}+\frac{1}{Z_2}+\frac{1}{Z_3} & \frac{-1}{Z_3} \\ \frac{-1}{Z_3} & \frac{1}{Z_3}+\frac{1}{Z_4}+\frac{1}{Z_5} \end{vmatrix} \begin{vmatrix} V_A \\ V_B \end{vmatrix} = \begin{vmatrix} \frac{V_1}{Z_1} \\ \frac{V_2}{Z_5} \end{vmatrix}$$

Note: Current sign in matrix is same as the source voltage potential corresponding to the node.

Coupled Elements and Circuits

Self Inductance

When a current is changing in a circuit the magnetic flux linking the same circuit changes and an emf is induced in the circuit. The induced emf is proportional to the rate of change of the current i.e.

$$V_L = \propto \frac{di}{dt}$$

where constant of proportionality –L is called self-inductance of the circuit, If Unit → Henry

If the coil has N terms, the induced emf is given by $V_L = N\frac{d\phi}{dt}$

where Ndϕ is the flux linkage of the circuit $L\frac{di}{dt} = N\frac{d\phi}{dt}$

Result which as $L = N\frac{d\phi}{dt}$

Mutual Inductance

Consider a time varying current i_1 in coil 1. The changing current i_1 establishes a magnetic flux ϕ_1 part of this links only coil 1 and is called self linkage flux ϕ_{11}, The remaining flux ϕ_{12} links also coil 2 as shown.

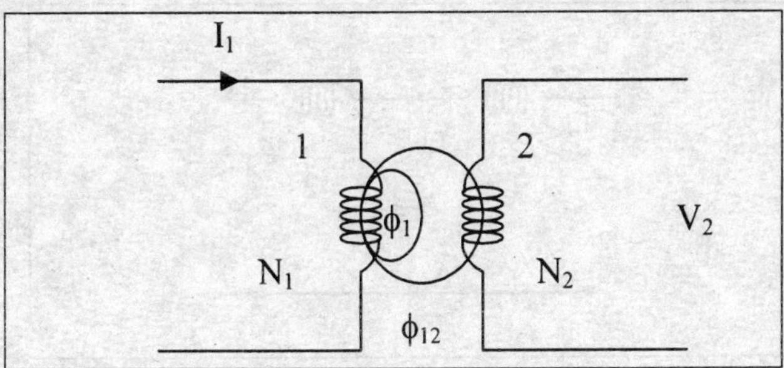

Induced voltage in coil 2 is given by Faraday's Law

$$V_L = N_2\frac{d\phi_{12}}{dt}$$

ϕ_{12} is related to the current i_1 .V_2 is proportional to the rate of change of i_1 or

$$V_2 = M\frac{di_1}{dt}$$

where the constant of proportionality M is called mutual inductance between the two coils. The unit of mutual inductance is Hennery

$$\therefore V_2 = M\frac{di_1}{dt} = N_2\frac{d\phi_{12}}{dt}$$

and $M = N_2\frac{d\phi_{12}}{di_1}$

If the coils are linked with air as the medium, the flux and current are linearly related and the mutual inductance is

$$M = \frac{N_2\phi_{12}}{i_1}$$

Analogous results i.e. $M = \frac{N_1 d\phi_{21}}{di_1}$ and $M = \frac{N_1\ \phi_{21}}{i_2}$ are obtained.

If a time –verying current i_2 is introduced in coil 2.

Coupling Coefficients

Fraction of total flux which links the coils is called the coefficient of coupling K.

$$K = \frac{\phi_{12}}{\phi_1} = \frac{\phi_{21}}{\phi_2}$$

$$M^2 = \left(\frac{N_2\phi_{12}}{i_1}\right)\left(\frac{N_1\phi_{21}}{i_2}\right)$$

$$= \left(\frac{N_1K\phi_1}{i_1}\right)\left(\frac{N_1K\phi_2}{i_2}\right) = K^2\left(\frac{N_1\phi_1}{i_1}\right)\left(\frac{N_2\phi_2}{i_2}\right)$$

Substituting $L_1 = \frac{N_1\phi_1}{i_1}, L_2 = \frac{N_2\phi_2}{i_2}$

$$M = K\sqrt{L_1, L_2}$$

where, $0 \le K \le 1$ and

$\begin{cases} K = 0, \text{ there is no coupling between the coils.} \\ K = 1, \text{ there is perfect coupling} \end{cases}$

Couple circuit

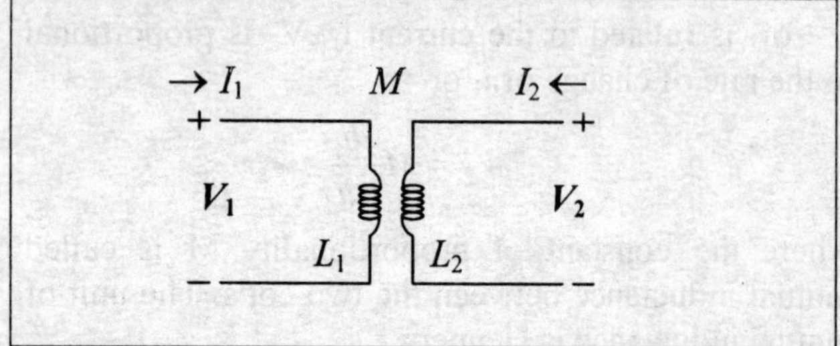

Main Point:

1. Here we consider two inductors placed physically closed to each other.
2. When there is an ac current flow in one inductor the magnetic field will be produced due to this flow of current. Due to it there will be rate of change of magnetic flux in the second coil.
3. Due to faraday's law of electro magnetic induction this rate of change of magnetic flux will produce some induced emf across the second inductor.
4. Thus there is some finite effect in the second inductor due to the flow of the current in the first inductor. This effect is called the **mutual coupling or mutual inductance** between the two inductors.

Dot Convention

M-Mutual inductance and its unit is Henry

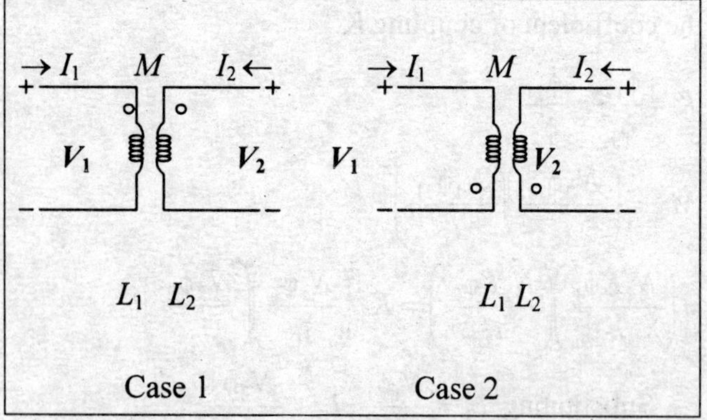

Case 1 and 2: Effect of mutual inductance is positive when I_1 and I_2 are both entering and both leaving out of the **dot.**

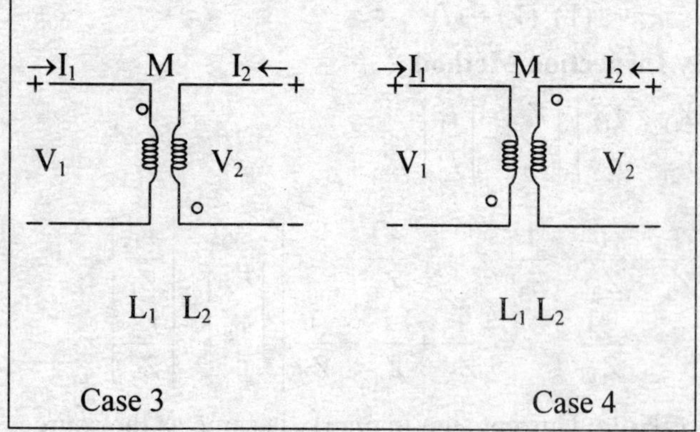

Case 3 and 4: Effect of mutual inductance is negative when one current is entering into the dot and the second current is leaving out of the **dot.**

KVL Equation

Case 1 and 2

$$V_1 = L_1\frac{dI_1}{dt} + M\frac{dI_2}{dt} \;;\quad V_2 = L_2\frac{dI_2}{dt} + M\frac{dI_1}{dt}$$

If the current are sinusoidal, then

$$V_1 = j\omega\, L_1 I_1 + j\omega\, M I_2 \;;\; V_2 = j\omega\, M I_1 + j\omega\, L_2 I_2$$

$$\begin{vmatrix} V_1 \\ V_2 \end{vmatrix} = \begin{vmatrix} j\omega L_1 & j\omega M \\ j\omega\, M & j\omega L_2 \end{vmatrix} \begin{vmatrix} I_1 \\ I_2 \end{vmatrix}$$

Case 3 and 4

$$V_1 = L_1\frac{dI_1}{dt} - M\frac{dI_2}{dt} \;;\quad V_2 = L_2\frac{dI_2}{dt} - M\frac{dI_1}{dt}$$

If I_1 and I_2 are sinusoidal, then

$$V_1 = j\omega\, L_1 I_1 - j\omega\, M I_2$$
$$V_2 = -j\omega\, M I_1 + j\omega\, L_2 I_2$$

$$\begin{vmatrix} V_1 \\ V_2 \end{vmatrix} = \begin{vmatrix} j\omega L_1 & -j\omega M \\ -j\omega\, M & j\omega L_2 \end{vmatrix} \begin{vmatrix} I_1 \\ I_2 \end{vmatrix}$$

where more than two inductors coupled to garter a new pair dots is needed for each pair of inductors coupled several kinds of dots are used.

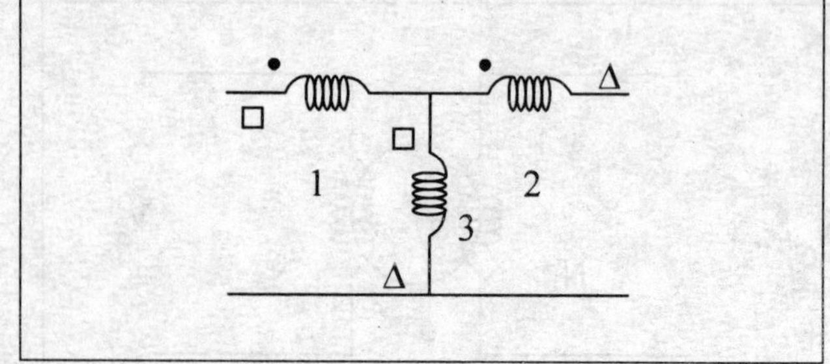

• → coupling between coils 1and 2;

Δ → Represent the coupling between coils 2 and 3;

□ → Represent the coupling between coils 1 and 3.

Inductance of Coupled Coils

(a) series–aiding connection:

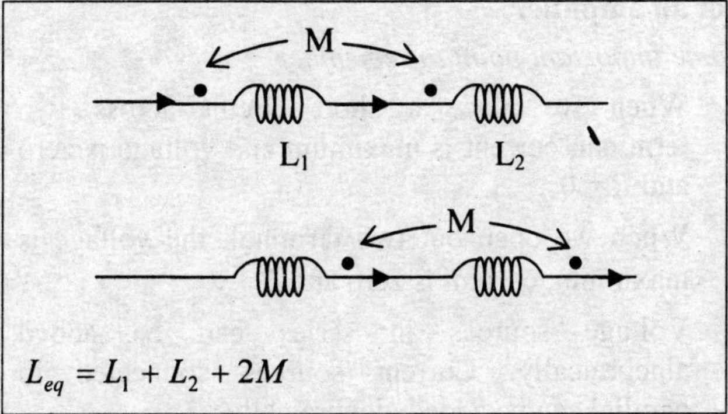

$$L_{eq} = L_1 + L_2 + 2M$$

(b) Series–opposition connection:

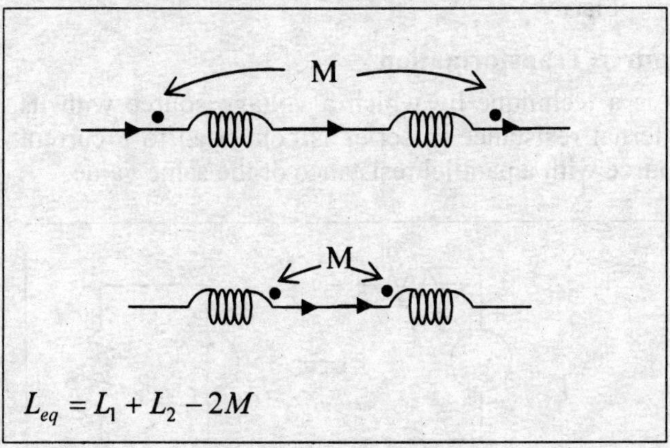

$$L_{eq} = L_1 + L_2 - 2M$$

$$L_{eq} = (10+2-2)+(10+2+2)+(10+2-2)=34$$

(c) Parallel–additive coupled:

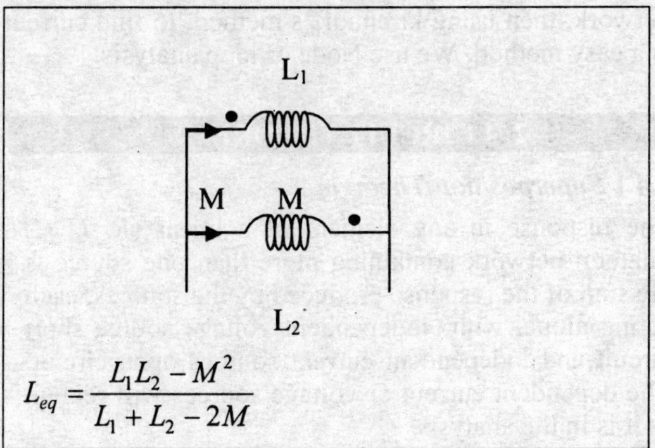

$$L_{eq} = \frac{L_1 L_2 - M^2}{L_1 + L_2 - 2M}$$

Example:

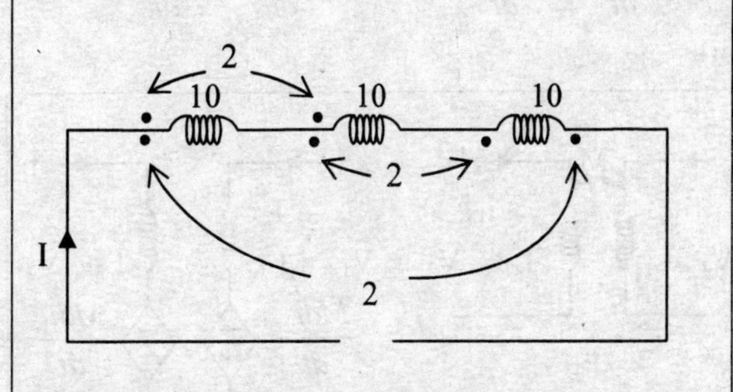

(d) Parallel differentially coupled

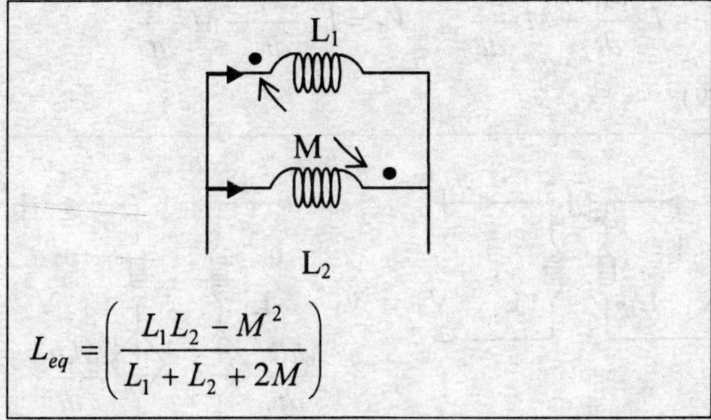

$$L_{eq} = \left(\frac{L_1 L_2 - M^2}{L_1 + L_2 + 2M}\right)$$

Magnetically coupled circuits and their equivalent circuits:

(i)

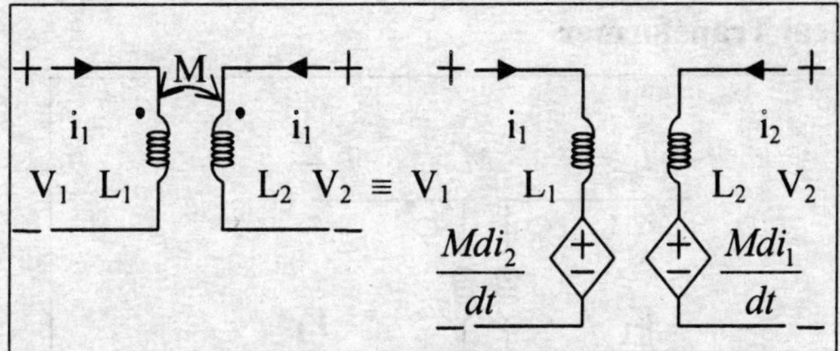

$$V_1 = L_1\frac{di_1}{dt} + M\frac{di_2}{dt} \quad ; \quad V_2 = L_2\frac{di_2}{dt} + M\frac{di_1}{dt}$$

(ii)

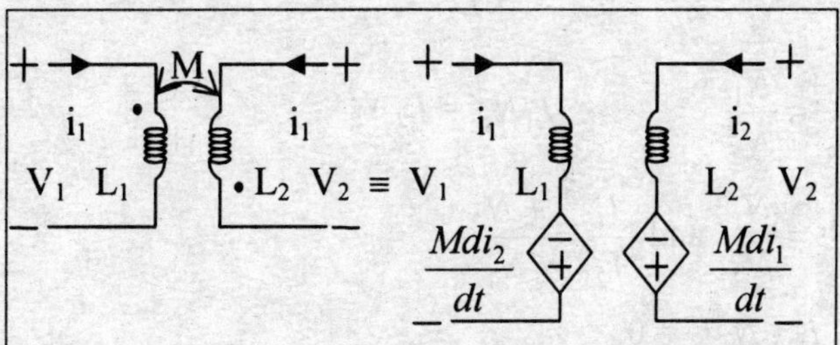

$$V_1 = L_1\frac{di_1}{dt} - M\frac{di_2}{dt};\quad V_2 = L_2\frac{di_2}{dt} - M\frac{di_1}{dt}$$

(iii)

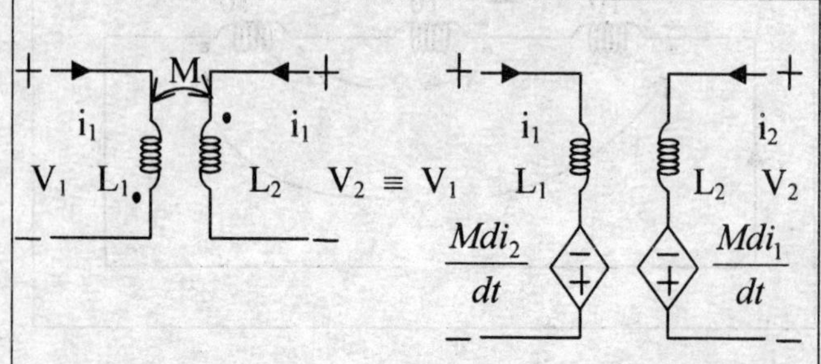

$$V_1 = L_1\frac{di_1}{dt} - M\frac{di_2}{dt};\quad V_2 = L_2\frac{di_2}{dt} - M\frac{di_1}{dt}$$

(iv)

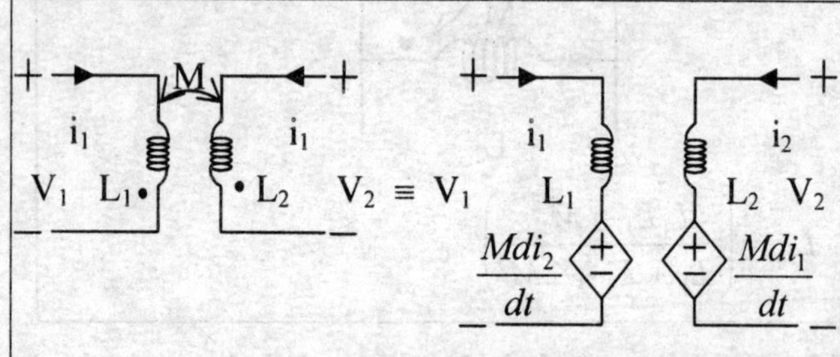

$$V_1 = L_1\frac{di_1}{dt} + M\frac{di_2}{dt};\quad V_2 = L_2\frac{di_2}{dt} + M\frac{di_1}{dt}$$

Ideal Transformer

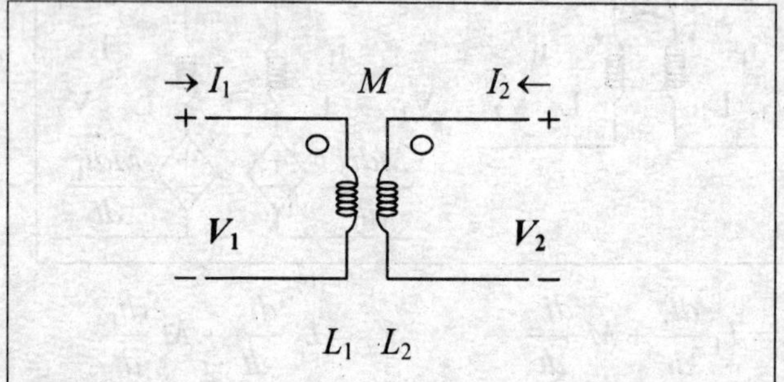

$$L_1 = N_1^2 \quad ; \quad L_2 = N_2^2$$

$$\frac{L_1}{L_2} = \frac{N_1^2}{N_2^2} \quad ; \quad L_1 N_1^2 = L_2 N_2^2$$

$$\frac{V_1}{V_2} = \frac{N_1}{N_2} \text{ and } \frac{I_1}{I_2} = \frac{-N_2}{N_1}$$

$$N_1 I_1 + N_2 I_2 = 0$$

Total Number Ampere Turn is Zero

$N_1 = 10;\ N_2 = 50$

$$\frac{V_2}{V_1} = \frac{N_2}{N_1} = 5$$

Since power is constant, thus transformer is not an amplifier.

Some important point to remember

- When we place a short circuit across two terminal, current is maximum and voltage is zero and $R = 0$.
- When we open out two terminal, the voltage is maximum, current is zero and $R = \infty$.
- Voltage sources in series can be added algebraically. Current sources connected in parallel can be added algebraically.
- Algebraically adding means with due regard to polarity.

Source Transformation

It is a technique by which a voltage source with its internal resistance in series is converted to a current source with a parallel resistance of the same value.

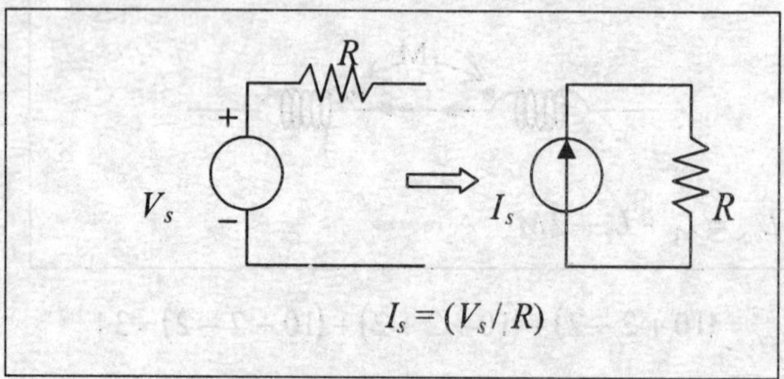

Node and mesh analysis: If we have a complex network, then using kirchhoff's method, to find current is a easy method. We use Node or loop analysis.

2. 4 CIRCUIT THEOREMS

2.4.1 *Superposition Theorem*

The response in any element of a linear (R, L, C) bilateral network containing more than one source is the sum of the response produced by the sources, each acting alone, with independent voltage source short circuit and independent current sources open circuit. The dependent current or voltage sources will remain as it is in the analysis.

The super position does not apply to power since it is not a linear function. Power is a square of current or voltage.

- Only applicable to linear circuit.
- It states that if response of a circuit with input (x_1) is y_1 and with input (x_2) is y_2, then output of the circuit when both inputs is applied simultaneously is given by sum of two output.
- For solving circuit, for applying super-position theorem, the effect of one source is taken at a time by suppressing the other source.
- Current source is opened and voltage source is shorted for suppressing the sources.
- Superposition cannot be applied directly to the computation of power as power is square of current or voltage.
- Dependent sources will remain unchanged in the analysis.

Example: Find the current I in the given circuit diagram.

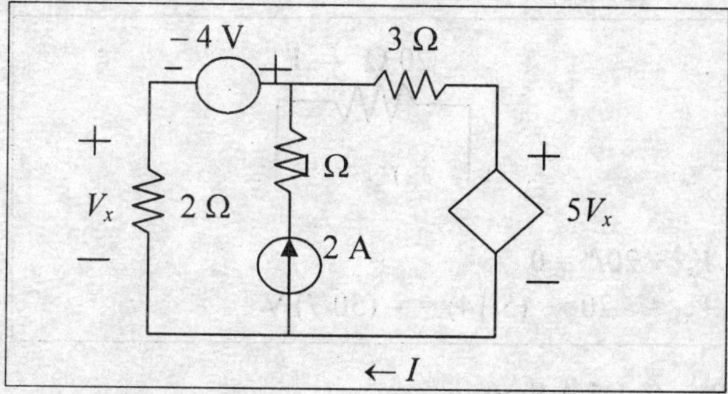

Let,

I_1 = current due to (–4 V) source with current source open circuit.

I_2 = current due to (2 A) source with voltage source short-circuit.

$I = I_1 + I_2$

I_1 (due to –4 V)

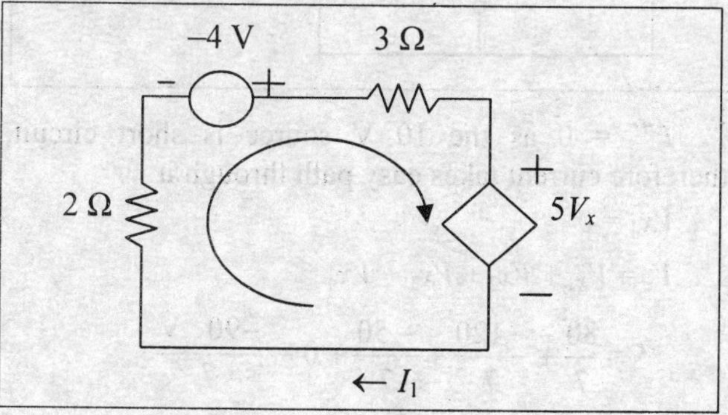

KVL in above loop,

$2I_1 - (-4) + 3I_1 + 5V_x = 0$

$5I_1 + 5V_x = -4$ (1)

To find the V_x use.

KVL in circuit.

$$-V_x - 2I_1 = 0$$

$$V_x = -2I_1 \quad (2)$$

From Eqs. (1) and (2)

$5I_1 + 5(-2I_1) = -4$

$-5I_1 = -4$; $I_1 = (4/5)$

$I_1 = 0.8.$

I_2 (due to 2 A)

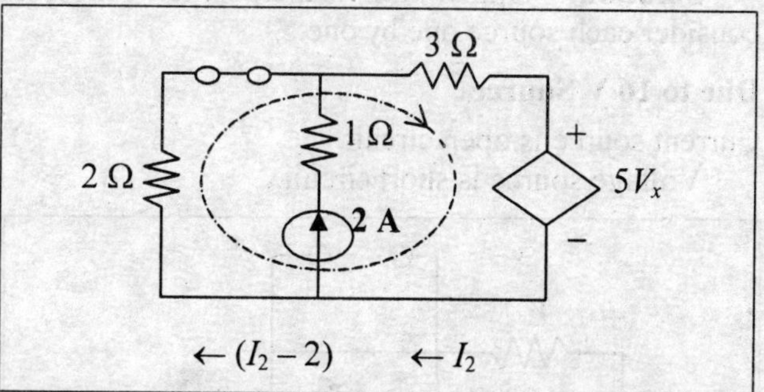

To apply the KVL select a loop which does not contain current source?

To find V_x in second case

KVL in outer loop,

$$2(I_2 - 2) + 3I_2 + 5V_x = 0$$

$$5I_2 + 5V_x = +4 \quad (3)$$

KVL in below circuit.

$$-V_x - 2(I_2 - 2) = 0$$

$$V_x = 2I_2 + 4 \quad (4)$$

From Eqs. (3) and (4)

$$5I_2 + 5(-2I_2 + 4) = 4$$

$$-5I_2 = 4 - 20; \quad I_2 = (16/5)\text{ A}$$

$$I = I_1 + I_2 = (4/5) + (16/5) = (20/5) = 4\text{ A}$$

Example: Find the V_x in the given circuit.

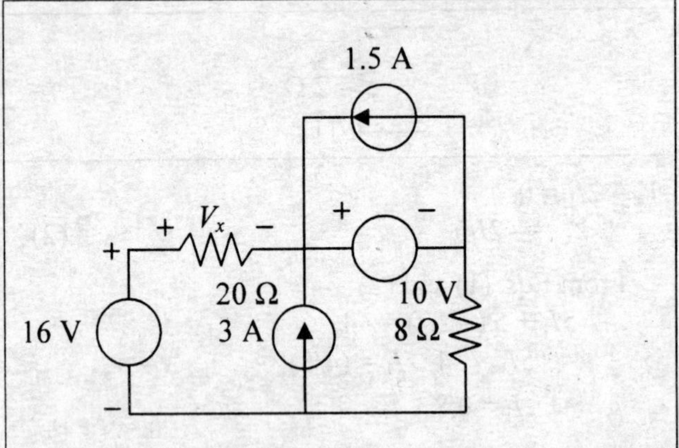

Solution: Apply the superposition theorem consider each source one by one.

Due to 16 V Source

Current source is open circuit.

Voltage source is short circuit.

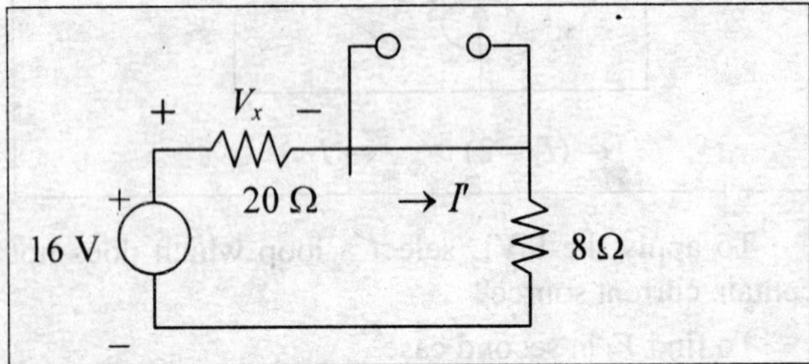

$I' = (16/28)$ A;

$V_{x1} = I' \times 20 = (4/7) \times 20 = 80/7$ volt.

Due to 3 A Source

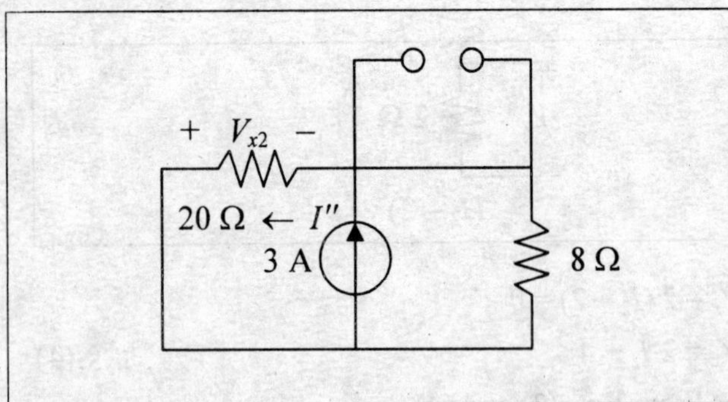

$$I'' = \frac{8 \times 3}{8 + 20} = \frac{8 \times 3}{28} = \frac{6A}{7}$$

20 Ω ← I″

+ V_{x2}

$$+V_{x2} + 20\,I'' = 0$$

$$V_{x2} = -20\,I'' = -20 \times (6/7) = -(120/7)\text{ V}$$

Due to 10 V Source

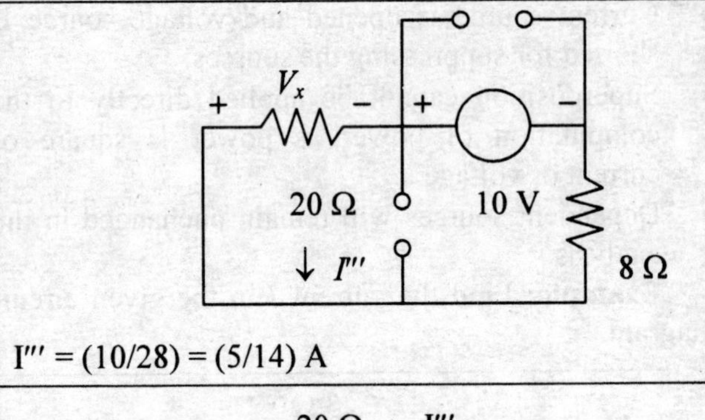

I‴ = (10/28) = (5/14) A

20 Ω ← I‴

+ V_{x3} −

$$V_{x3} + 20I''' = 0$$

$$V_{x3} = -20 \times (5/14) = -(50/7)\text{ V}$$

Due to 1.5 A Source

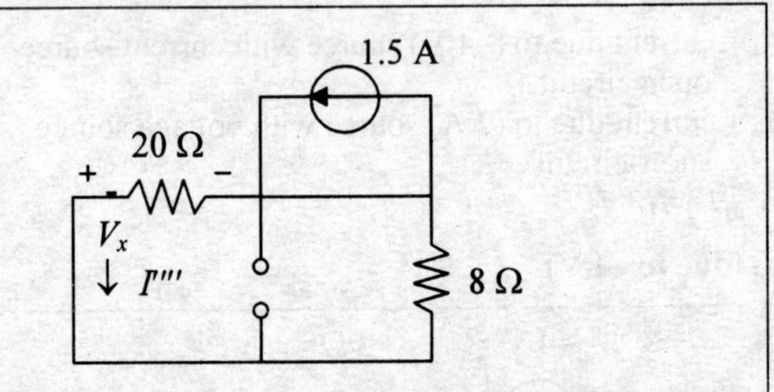

I'''' = 0 as the 10 V source is short circuit, therefore current takes easy path through it.

$$Vx_4 = 0$$

$$V_x = Vx_1 + Vx_2 + Vx_3 + Vx_4$$

$$= \frac{80}{7} + \frac{-120}{7} + \frac{-50}{7} + 0 = \frac{-90\text{ V}}{7}.$$

2.4.2 Thevenin's Theorem:

A linear active network which contain one or more than one voltage or current source can be replace by a single voltage source and a series impedance as shown in figure.

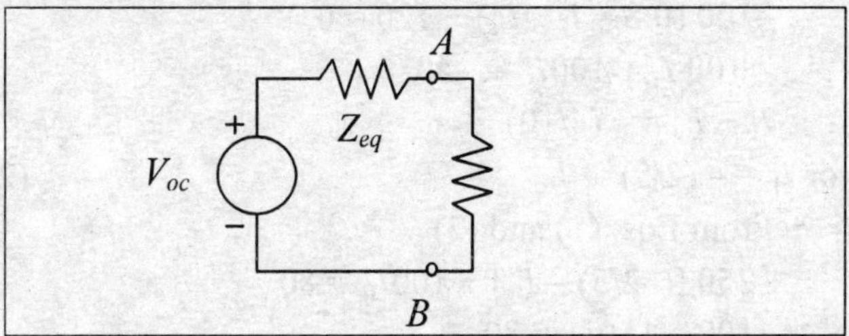

where, V_{oc} is open circuit voltage across the terminal A and B.

Z_{eq} is equivalent impedance of the network between the terminal A and B, when all independent voltage sources are short circuited, all independent current sources are open circuited and all dependent sources are kept as it is.

R_L is that element which was removed to find V_{oc} and through which current and voltage is to be measured.

- This theorem is applied in finding the current, voltage and power in a particular element of Network.
- First isolate the element in which current and voltage is to be found.
- From disconnected pair of terminals, reduce the circuit to a single equivalent resistance Z_{eq} after suppressing sources.
- Find open circuit voltage across that particular terminal.
- We have V_{oc} and series internal resistance Z_{eq}. O
- Equivalent circuit can be drawn finally.

2.4.3 Norton's Theorem

A linear active network which contain one or more than one voltage or current sources can be replace by a single current source and a parallel impedance as shown in figure.

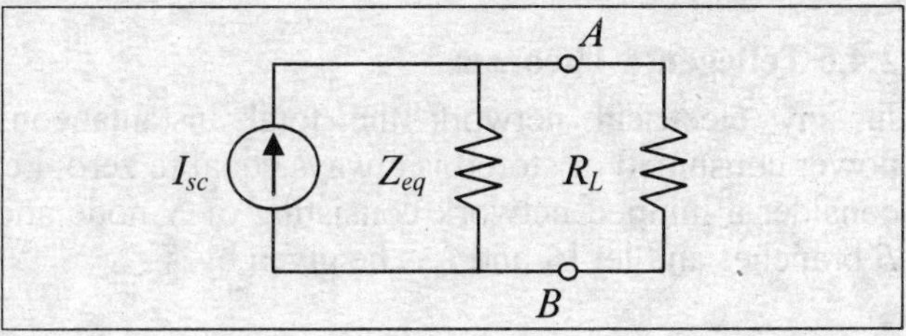

where I_{sc} is the current between terminal when A and B are short circuited.

Z_{eq} is as Thevenin's equivalent model.

- It is dual of Thevenin's theorem.
- After isolating the element, the terminal is shorted and the short-circuit current I_{sc} is calculated.

Example: For the given below circuit, find the Thevenin's and Norton's equivalent circuit.

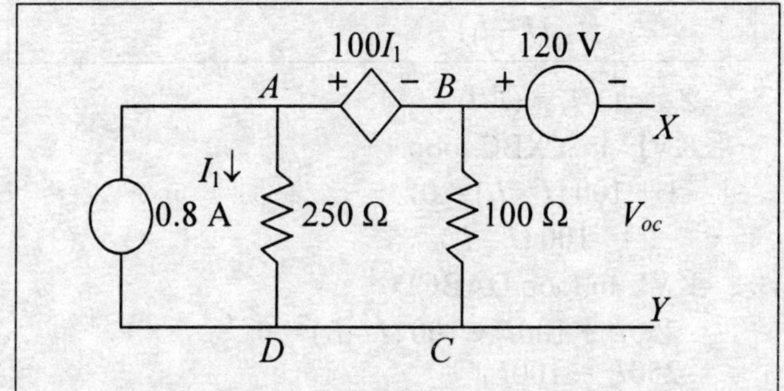

Solution: For Thevenin's

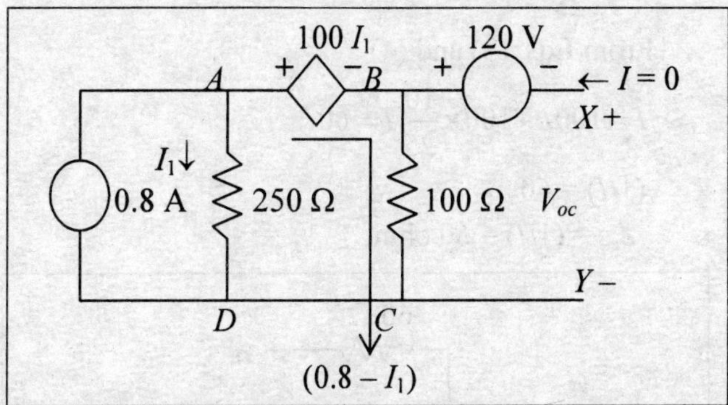

Apply KVL in CYXB loop

$$-V_{oc} - 120 + 100\,(0.8 - I_1) = 0$$

$$V_{oc} = -120 + 100\,(0.8 - I_1)$$

$$= -40 - 100\,I_1 \qquad (1)$$

KVL around the loop ABCD

$$-250I_1 + 100I_1 + 100\,(0.8 - I_1) = 0$$

$$250I_1 = 80$$

$$I_1 = 8/25 \qquad (2)$$

From Eqs. (1) and (2)

$$V_{oc} = -40 - 100 \times (8/25) = -72 \text{ V}$$

Note: When ever there is a dependent source in the circuit, to find the equivalent impedance for the circuit, short circuit voltage source and open circuit current source and apply the 1 volt source at the output terminal and fined the current I. The inverse of I gives the Thevenin's equivalent impedance.

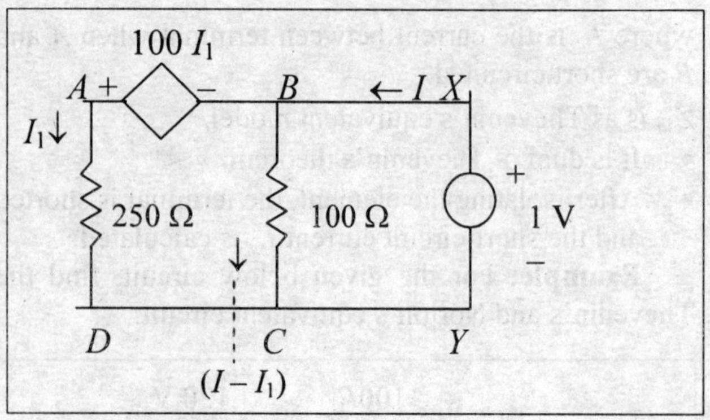

$Z_{eq} = (1/I) = (V/I)$

KVL in YXBC loop.

$-1 + 100\,(I - I_1) = 0$

$1 = 100\,(I - I_1)$ (3)

KVL in loop DABCD

$-250I_1 + 100I_1 + 100\,(I - I_1) = 0$

$-250I_1 = 100I$

$$I_1 = \frac{10}{25} I \quad (4)$$

From Eqs. (3) and (4)

$$I = 100I - 100 \times \frac{10}{25} I = 60I$$

$(1/I) = 60$

$Z_{eq} = (1/I) = 60$ ohm

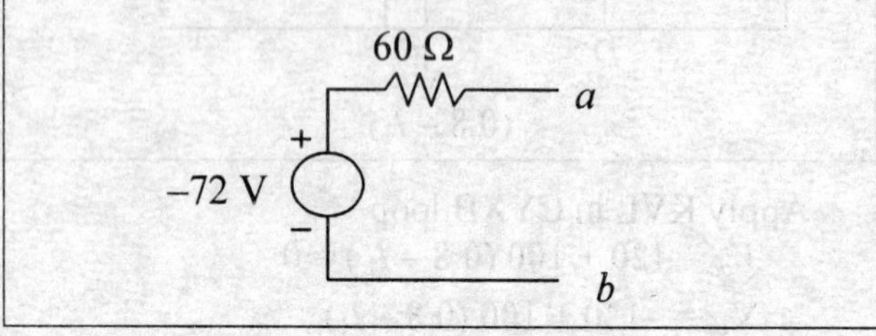

Analysis for Norton's equivalent circuit

$I_{sc} = ?$ $Z_{eq} = 60\ \Omega$, i.e. same as Thevenin's

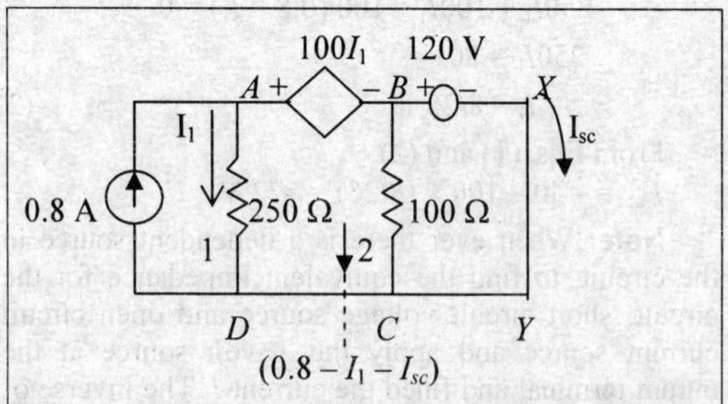

KVL in loop DABC

$-250I_1 + 100I_1 + 100\,(0.8 - I_1 - I_{sc}) = 0$

$-250I_1 + 80 - 100I_{sc} = 0$

$250I_1 + 100I_{sc} = +\,80$ (1)

KVL in loop CBXY

$-100\,(0.8 - I_1 - I_{sc}) + 120 = 0$

$100\,I_1 + 100I_{sc} = -40$

$I_1 + I_{sc} = -\,(4/10)$

or $I_1 = -\,(2/5) - I_{sc}$ (2)

From Eqs. (1) and (2)

$250\,[(-2/5) - I_{sc}] + 100I_{sc} = 80$

$-100 - 150I_{sc} = 80$

$150I_{sc} = -180$

$I_{sc} = (-180/150) = (-6/5) = -1.2$ A

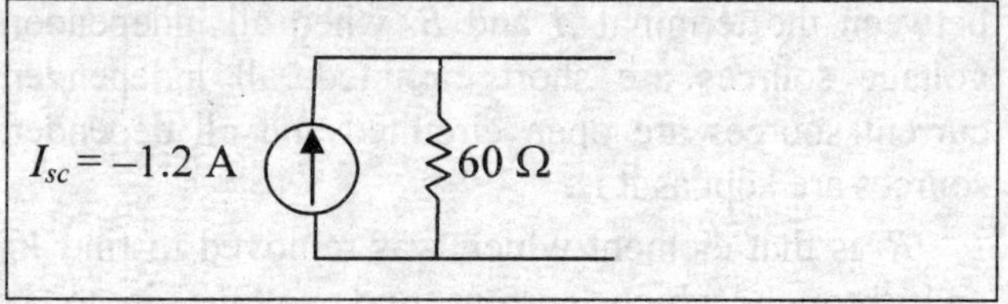

2.4.4 Maximum Power Transfer Theorem

A source with an internal impedance of $R_s + jX_s$, deliver maximum power to a load impedance Z_L.

When, $Z_L = Z_s^*$

When Z_s^* = represents complex conjugate

If $Z_s = R_s + j\,X_s$, then for maximum power transfer.

$$Z_L = Z_s^* = R_s - j\,X_s$$

$R_L = R_s$ and $X_L = -\,X_s$.

2.4.5 Reciprocity Theorem

In a linear bilateral single source network the ratio of excitation (current or voltage) to response (current or voltage) is constant when the position of excitation and response are interchanged. The basis of the theorem is the symmetry of Z and Y matrices.

This theorem is not applicable to the circuit in which dependent source are present.

2.4.6 Tellegen's Theorem

In any electrical network the total instantaneous power consumed or stored is always equal to zero, i.e. consider a lumped network consisting of N node and B branches and let $\boldsymbol{V_b}$ and $\boldsymbol{i_b}$ is be given by.

$V_{bk} = (V_{b1,}, V_{b2} \ldots\ldots\ldots\ldots V_{bB})$ and

$i_{bk} = (i_{b1}, i_{b2} \ldots\ldots\ldots\ldots i_{bB})$

where V_{bk} and i_{bk} denote the voltage and current through the branch K.

Then, $\sum_{k=1}^{B} V_{bk}\, i_{bk} = 0$

Alternatively in any electrical network the total instantaneous power given by the active source must be equal to the total power dissipated or stored in the passive element.

Physical interpretation of Tellegen's theorem is the conservation of power.

2.4.7 Millman's Theorem

This theorem enable a number of voltage (or current) sources to be combined into a single voltage (or current) source, i.e.

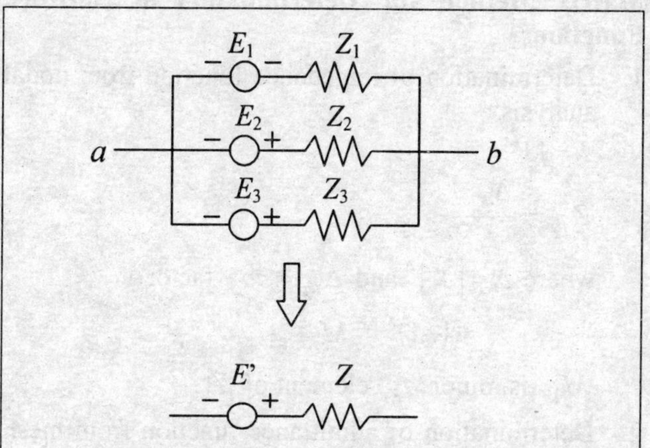

where, $E' = \dfrac{E_1Y_1 + E_2Y_2 + E_3Y_3}{Y_1 + Y_2 + Y_3}$

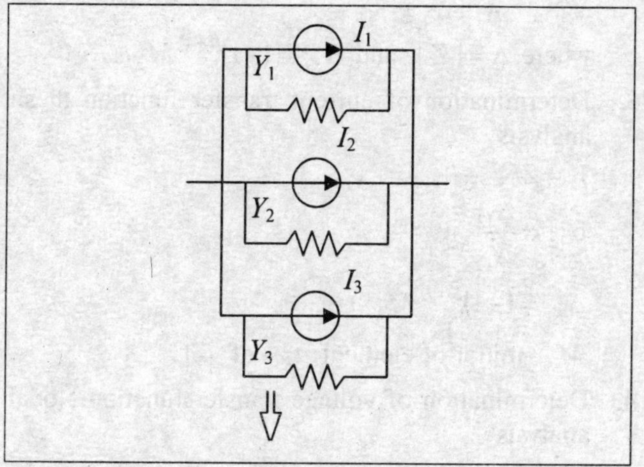

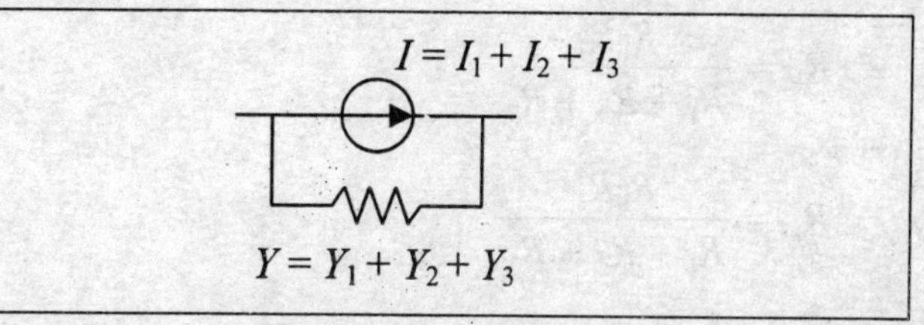

2.4.8 Star-To-Delta Conversion

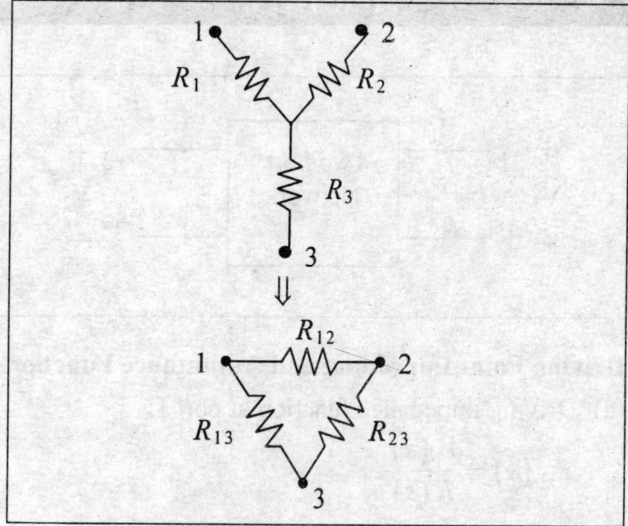

$$R_{12} = \frac{R_1R_2 + R_2R_3 + R_3R_1}{R_3}$$

$$R_{13} = \frac{R_1R_2 + R_2R_3 + R_3R_1}{R_2}$$

$$R_{23} = \frac{R_1R_2 + R_2R_3 + R_3R_1}{R_1}.$$

2.4.9 Delta-to-star Conversion

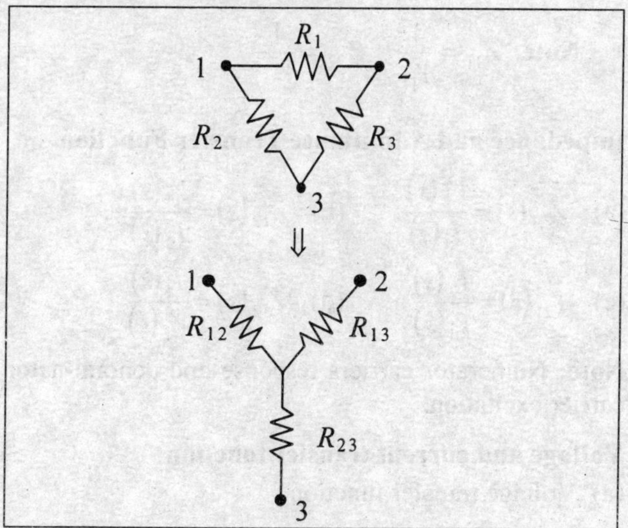

$$R_{12} = \frac{R_1 R_2}{R_1 + R_2 + R_3}$$

$$R_{13} = \frac{R_1 R_3}{R_1 + R_2 + R_3}$$

2.5 ONE PORT AND TWO PORT NETWORK FUNCTIONS

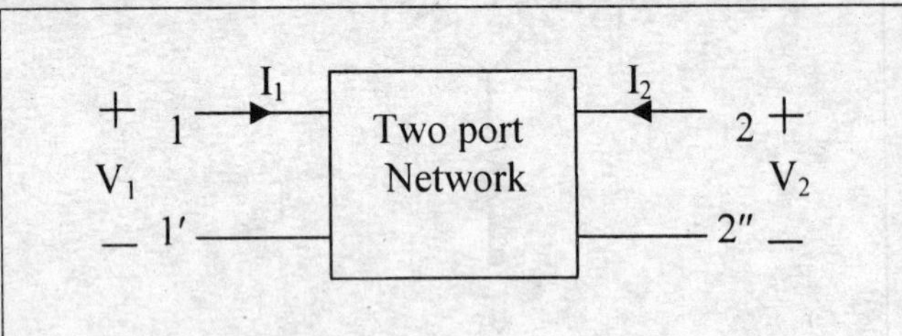

Driving Point Impedance and Admittance Function

(a) Driving impedance function at port 1:

$$Z_{11}(s) = \frac{V_1(s)}{I_1(s)}$$

(b) Driving impedance function port 2:

$$Z_{22}(s) = \frac{V_2(s)}{I_2(s)}$$

(c) Driving admittance function port 1: $Y_{11}(s) = \frac{I_1(s)}{V_1(s)}$

(d) Driving admittance function at port 2:

$$Y_{22}(s) = \frac{I_2(s)}{V_2(s)}$$

Note: $Z_{11} = \frac{1}{Y_{11}}$; $Z_{22} = \frac{1}{Y_{22}}$

Impedance and Admittance Transfer Function

(a) $Z_{21}(s) = \frac{V_2(s)}{I_1(s)}$; (b) $Z_{12}(s) = \frac{V_1(s)}{I_2(s)}$

(c) $Y_{21}(s) = \frac{I_2(s)}{V_1(s)}$; (d) $Z_{12}(s) = \frac{I_1(s)}{V_2(s)}$

Note: Numerator carriers response and denomi-nator carrier excitation.

Voltage and current transfer function

(a) Voltage transfer function

1. $G_{21}(s) = \frac{V_2(s)}{V_1(s)}$; 2. $G_{12}(s) = \frac{V_1(s)}{V_2(s)}$

(b) Current transfer function

1. $\alpha_{21} = \frac{I_2(s)}{I_1(s)}$; 2. $\alpha_{12} = \frac{I_1(s)}{I_2(s)}$

Consider a Network Function

$$H(s) = \frac{a_0 s^n + a_1 s^{n-1} + \ldots + a_n}{b_0 s^m + b_1 s^{m-1} + \ldots + b_m}$$

$$= \frac{(a_0 - b_0)[(s - Z_1)(s - Z_2)(s - Z_3)\cdots(s - Z_N)]}{[(s - P_1)(s - P_2)\cdots(s - P_m)]}$$

$$= \frac{P(s)}{Q(s)}.$$

Matrix Method for Determination of Network Functions

1. Determination of impedance function from nodal analysis

   $I = YV$

   $$Z_{jK} = \frac{\Delta_{Kj}}{\Delta}$$

   where $\Delta = |Y|$ and $\Delta_{Kj} = co -$ factor $= (-1)^{K+j} M_{Kj}$

   M_{Kj} is minor a_{Kj} element of $|Y|$.

2. Determination of admittance function from mesh analysis

   $V = ZI$

   $$Y_{jK} = \frac{\Delta_{Kj}}{\Delta}$$

   where $\Delta = |Z|$ and $\Delta_{Kj} = (-1)^{K+j} M_{Kj}$.

3. Determination of current transfer function mesh analysis

   $V = ZI$

   $$\alpha_{12} = \frac{\Delta_{12}}{\Delta_{11}}$$

   $\Delta_{Kj} = (-1)^{K+j} M_{Kj}$

   M_{Kj}-minor of element a_{Kj} of $|Z|$.

4. Determination of voltage transfer function Nodal analysis

$I = YV$; $G_{12} = \dfrac{\Delta_{12}}{\Delta_{11}}$

$\Delta_{Kj} = (-1)^{K+j}\, M_{Kj}$

M_{kj} Minor of element of a_{kj} of $|Y|$.

2.5.1 Properties of driving point impedance and admittance functions or necessary conditions for driving point impedance and admittance function.

1. All coefficient of $P(s)$ and $Q(s)$ must be positive and real.
2. Complex and imaginary poles and zeros must be conjugate

   i.e. $F(s) = \dfrac{s+1}{(s+3)(s+1-j)}$

   is not a driving point impedance function.
3. (a) Real part of all poles and zeros must not be positive.

   (b) If poles and zeros are imaginary they must be simple (non-repeated).
4. There should not be any missing power of 's' between highest and lowest power in numerator and denominator both except when either all even or odd power are missing.
5. The degree of $P(s)$ and $Q(s)$ can at the most differ by 1.
6. Lowest power of s in numerator and denominator can differ at most '1'.

2.5.2 Network Transform Functions

- A transfer function is the ratio of Laplace transform of the output $Y(s)$ to Laplace transform of the input $X(s)$.
- By setting the denominator of the transfer function to zero, and obtaining the roots by solving the equation, we can know the system's stability by considering by setting $s = 0$, i.e. $G(0)$ give DC gain where $G(s)$ is transfer function.

Procedure for Deriving Transfer Functions

Following assumption is made in deriving the transfer function:

- It is assumed that no loading of one system on other. If the system has more than one non loading element, the transfer function of each element can be determined independently and the overall transfer function can be obtain by multiplying the individual transfer function.
- If the system consisting of elements, which load each other, the overall transfer function should be derived by the basic analysis.

Necessary Conditions for Transfer Function

1. The coefficients in the polynomials $P(s)$ and $Q(s)$ of $H(s) = P(s)/Q(s)$ must be real and those for $Q(s)$ must be positive.
2. Poles and zeros which are complex and imaginary must be conjugate.
3. The real part of the poles must be either negative or zero, if poles are imaginary, then they must be simple (non-repeated). But zeros can have positive real part also.
4. There should not be any missing power of s between lowest and highest power of s in denominator except all odd all even power are missing. Numerator can have i.e. $(P(s))$ can have missing terms of s and some of the coefficients of may be negative.
5. The degree of $P(s)$ may be as small as zero independent of the degree of $Q(s)$.
6. (a) For G_{12} and α_{12}, the maximum degree of $P(s)$ is the degree of $Q(s)$.

   (b) For Z_{12} and Y_{12} the maximum degree of $P(s)$ is the degree of $Q(s)$ pulse one

2.5.3 Impedance Matrix of Gyrator

- It is a device that gyrates the current of one port into a voltage at the other and vice versa.

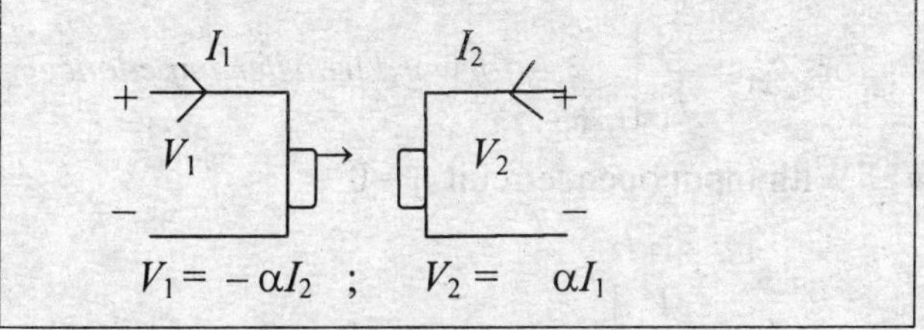

$V_1 = -\alpha I_2$; $V_2 = \alpha I_1$

As impedance transforming element.

V_1 V_2

R_o

$V_1 + R_o I_2 = 0$

$V_2 = R_o I_1$

- Matrix form $\begin{vmatrix} 0 & -\alpha \\ \alpha & 0 \end{vmatrix} \begin{vmatrix} i_1 \\ i_2 \end{vmatrix} = \begin{vmatrix} V_1 \\ V_2 \end{vmatrix}$.

- If the gyrator is terminated in an impedance Z_L, the driving point impedance

$$Z_d = \frac{Z_{11}Z_L + \Delta_z}{Z_{22} + Z_L};\quad Z_{in} = (\alpha^2/Z_L).$$

- If Z_L is capacitor, then $Z_L = (1/sC)$ $Z_{in} = sC\alpha^2$ give inductor of value= $\alpha^2 C$.
- If Z_L is an inductor $Z_L = sL$ $Z_{in} = (\alpha^2/s_L)$ gives capacitor of value = (L/α^2).
- Used to simulate inductor from capacitor.

2.5.4 *Z* or Impedance Parameters

In this, the dependent variables are V_1 and V_2 and independent variable I_1, and I_2.

Defining equations are:

$$V_1 = Z_{11} I_1 + Z_{12} I_2$$
$$V_2 = Z_{21} I_1 + Z_{22} I_2.$$

In Impedance Matrix Form

$$\begin{vmatrix} V_1 \\ V_2 \end{vmatrix} = \begin{vmatrix} Z_{11} & Z_{12} \\ Z_{21} & Z_{22} \end{vmatrix}\begin{vmatrix} I_1 \\ I_2 \end{vmatrix}$$

- To obtain the *Z*-parameters, we open circuit the output and input port alternately.
- With output open circuit $I_2 = 0$

$$V_1 = Z_{11}I_1$$
$$\text{or } Z_{11} = \left.\frac{V_1}{I_1}\right|_{I_2=0} = \textit{input impedance}$$
$$V_2 = Z_{21}I_1$$
$$\text{or } Z_{21} = \left.\frac{V_2}{I_1}\right|_{I_2=0} = \textit{forward transfer impedance}$$

- With input open circuit $I_1 = 0$

$$V_1 = Z_{12} I_2$$
$$Z_{12} = \left.\frac{V_1}{I_2}\right|_{I_1=0} = \textit{reverse transfer impedance}$$
$$V_2 = Z_{22} I_2$$
$$Z_{22} = \left.\frac{V_2}{I_2}\right|_{I_1=0} = \textit{output impedance.}$$

***Z*-parameter for a *T*-network**

$$Z = \begin{vmatrix} Z_A + Z_C & Z_C \\ Z_C & Z_B + Z_C \end{vmatrix}.$$

Property of *Z*-parameter

- $\mathbf{Z_{11}}$, $\mathbf{Z_{22}}$ are called **driving point input and output impedance function.**
- Z_{12} and Z_{21} are called **transfer function.**
- Network for which $\mathbf{Z_{11}=Z_{22}}$ called as symmetrical.
- Network for which $\mathbf{Z_{12}=Z_{21}}$ are known as **reciprocal network.**
- Reciprocal network need not be symmetrical.

2.5.5 *Y* or Admittance Parameter

- Also known as admittance parameter.
- Independent variables are voltage V_1, V_2 and dependent variable current I_1, I_2.
- $I_1 = Y_{11}V_1 + Y_{12} V_2$
  $I_2 = Y_{21} V_1 + Y_{22} V_2$.
- Admittance in matrix form

$$Y = \begin{vmatrix} Y_{11} & Y_{12} \\ Y_{21} & Y_{22} \end{vmatrix}$$

- For reciprocal $\mathbf{Y_{12} = Y_{21}}$.

Symmetric

A two port is symmetric if port can be interchanged, without any effect on the performance.

$$Z_{11} = Z_{22}$$
$$Y_{11} = Y_{22}.$$

2.5.6 Hybrid (h) Parameters

Current (I_1) and voltage (V_2) are the independent variable and current (I_2) and voltage (V_1) are dependent variable.

$$V_1 = h_{11} I_1 + h_{12} V_2$$
$$I_2 = h_{21} I_1 + h_{22}V_2.$$

2.5.7 Inverse Hybrid Parameter (*g*-parameter)

$$I_1 = g_{11} V_1 + g_{12} I_2$$
$$V_2 = g_{21}V_1 + g_{22} I_2.$$

2.5.8 Transmission Parameters (ABCD parameters)

- It relates the voltage and current at input port to voltage and current at output port.
- $V_1 = AV_2 - BI_2$
  $I_1 = CV_2 - DI_2$.

- For symmetrical Network $A = D$.
- For reciprocal $\begin{vmatrix} A & B \\ C & D \end{vmatrix} = 1$

= i.e. $AD - BC = 1$.

2.5.9 Cascading

Two Port Network Connected in Cascade

Overall *ABCD* parameter of two cascaded two port network is multiplication of their *ABCD* parameters.

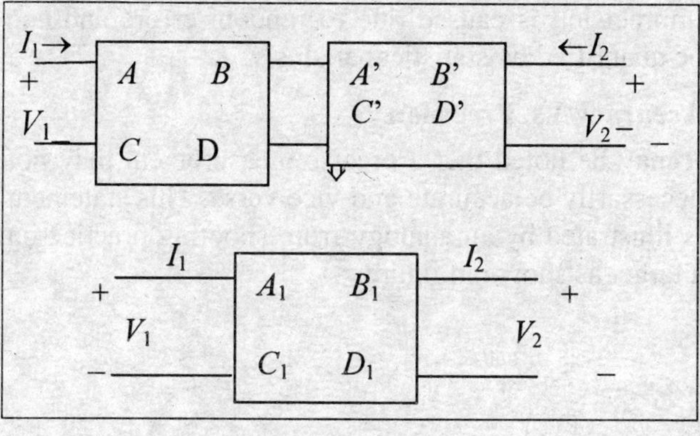

$$\begin{vmatrix} A_1 & B_1 \\ C_1 & D_1 \end{vmatrix} = \begin{vmatrix} A & B \\ C & D \end{vmatrix} \times \begin{vmatrix} A' & B' \\ C' & D' \end{vmatrix}.$$

Two Port Network in Parallel

Overall matrix for two, port network connected in parallel is the sum of the individual network *y*-parameters $(y) = (y_A) + (y_B)$

where

(y) = overall *y*-parameter of combine network.

(y_A) = *y*-parameter of network *A*.

(y_B) = *y*-parameter of network *B*.

Two Port Network Connected in Series

The overall matrix for two-port network connected in series is equal to the sum of the individual network *Z*-parameter matrix.

$$(z) = (z_A) + (z_B).$$

Some Special Results

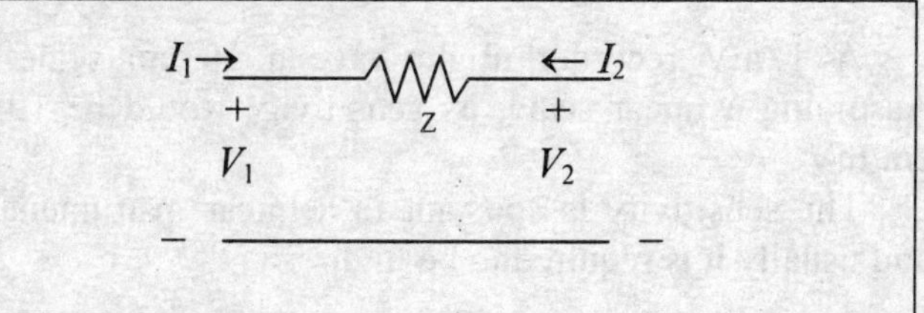

Z-parameter of this network does not exist

$$y = \begin{pmatrix} 1/z & -1/z \\ -1/z & 1/z \end{pmatrix}, ABCD = \begin{pmatrix} 1 & Z \\ 0 & 1 \end{pmatrix}$$

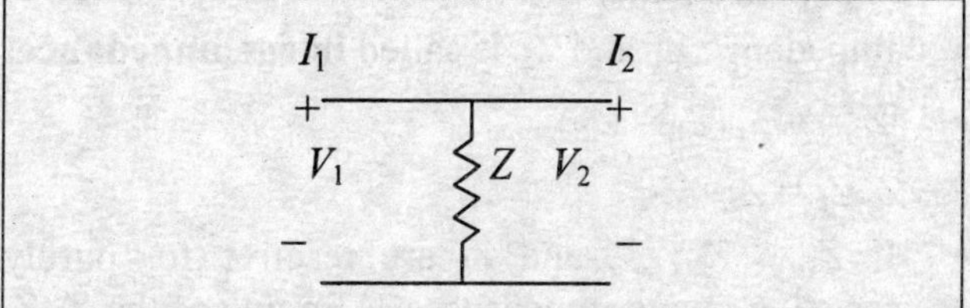

This network cannot have *y*-parameter.

$$Z = \begin{pmatrix} Z & Z \\ Z & Z \end{pmatrix};\ ABCD = \begin{pmatrix} 1 & Z \\ 0 & 1 \end{pmatrix}.$$

Obtain *T*-equivalent of the π-network

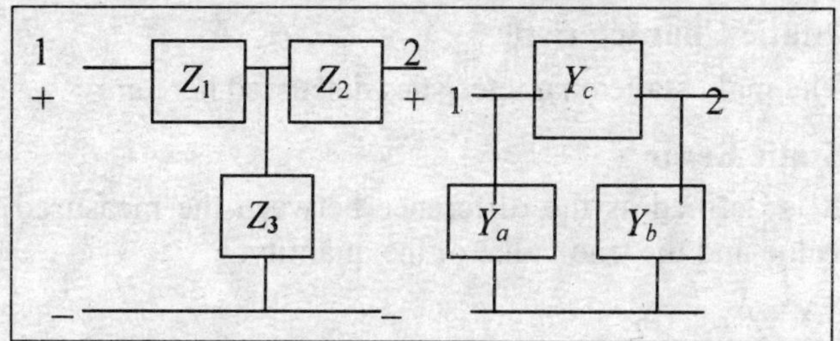

$$Z_{11} = \frac{Z_a(Z_b + Z_c)}{Z_a + Z_b + Z_c}, Z_{12} = Z_{21} = \frac{Z_a Z_b}{Z_a + Z_b + Z_c}$$

$$Z_{22} = \frac{Z_a(Z_b + Z_c)}{Z_a + Z_b + Z_c}$$

$$Z_3 = \frac{Z_a Z_b}{Z_a + Z_b + Z_c}$$

$$Z_{21} = \frac{Z_a Z_c}{Z_a + Z_b + Z_c}$$

$$Z_2 = \frac{Z_a Z_c}{Z_a + Z_b + Z_c}.$$

Impedance Matching

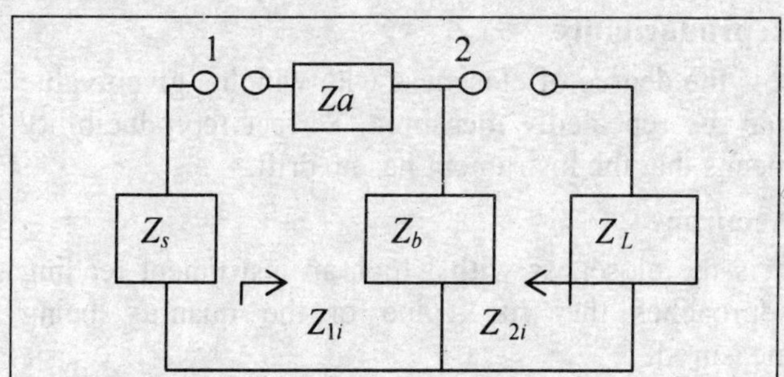

- *L*-section Z_a and Z_b is inserted between impedances Z_s (source) and Z_{1i} (load) so that the source sees an impedance Z_{1i} and seen an impedance Z_{2i}. Such an arrangement is called *impedance matching.*
- Impedance Z_{1i} and Z_{2i} is called **image impedance.**
- $Z_a^2 = Z_{1i}(Z_{1i} - Z_{2i})$

  $Z_a.Z_b = Z_{1i} - Z_{2i}$.
- If $Z_{1i} < Z_{2i}$; Z_a and Z_b are reactive for purely resistive image impedances. Also one of the Z, Z_b is inductive and the other capacitive.

2.6 STATIC AND DYNAMIC CHARACTERISTICS OF MEASUREMENT SYSTEMS

Static Characteristic

The main static characteristics discussed here are:

Static Error

It is defined as the difference between the measured value and the true value of the quantity.

$\delta A = A_m - A_t$

where, δA = error

A_m = measured value of quantity

A_t = true value of quantity

δA is also known as absolute static error of quantity A.

Relative static error ε_r is given by

$$\varepsilon_r = \frac{\text{Absolute error}}{\text{True value}} = \frac{\delta A}{A_t}$$

Static Correction

It is the difference between the true value and the measured value of the quantity.

$$\delta C = A_t - A_m$$

where δC = static correction = $-\delta A$

Reproducibility

It is the degree of closeness with which a given value may be repeatedly measured. Perfect reproducibility means that the instrument has no drift.

Accuracy

It is the closeness with which an instrument reading approaches the true value of the quantity being measured.

It can be specified in term of inaccuracy or limit of error.

Precision

It refers to repeatability or to how closely the individual measured values of a quantity agree with each other when the measurements are carried out under identical conditions at a short interval of time. It can be defined as the ability of the instrument to reproduce a group of measurements of the same measured quantity under the same conditions. Imprecision is caused due to random errors and can be quantified by statistical analysis.

Accuracy Vs. Precision

It may be noted that a precise measurement may not necessarily be accurate and vice versa. This statement is illustrated by an analogy from shooting practice on a target as shown in figure.

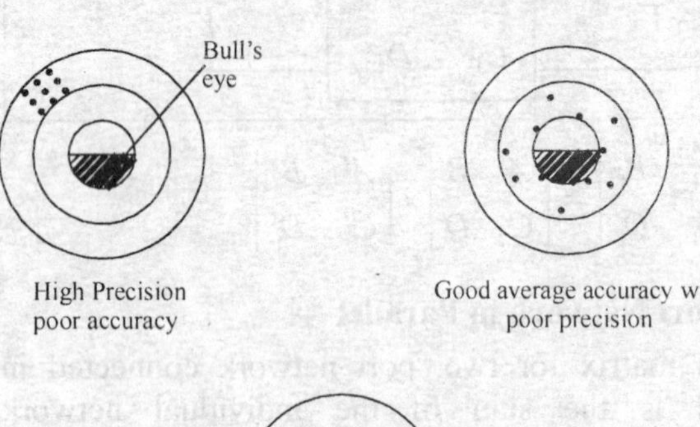

Figure: Accuracy and precision in a typical target shooting

Sensitivity

It is defined as the ratio of the magnitude of output signal to the magnitude of input signal.

$$\text{Sensitivity} = \frac{\text{Change of output signal}}{\text{Change in input signal}}$$

A 1 mV recorder might have a 10 cm scale. Assuming a linear scale, its sensitivity would be 10 cm/mV.

The sensitivity is constant in a linear instrument and usually it is required to be high.

Resolution

This is the smallest change in input or measured value which can be detected by the instrument. The least count of an instrument can be taken as the resolution of an instrument.

Threshold

It is defined as the minimum value of input below which of output can be detected.

Drift

It is defined as the variation or change in output for a given over a period or time. This is occurs slowly. This can be caused by component's instability or temperature changes. Wear, erosion, polarization saturation, and effects like aging may also cause drift.

Hysteresls

It is defined as the difference in the output for a given input when this value is approached from the opposite direction.

Stated simply, it means that loading and unloading curves do not coincide.

When input is increasing [curve *oa*], for input θ_i, ouput θ_0, when input is decreasing [curve *ao*] for the same input θ_i output is θ'_0.

This can be caused by frication, backlash, elastic deformation, magnetic and thermal effects.

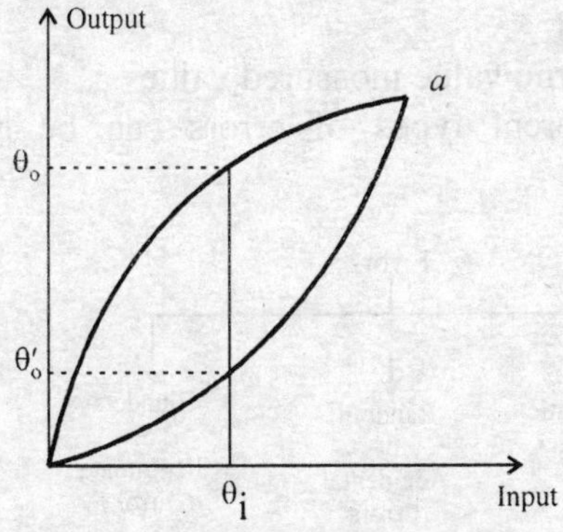

Figure: Hysteresis

Dead Time

It is defined, as the time required by a measurement system to begin to respond to a change in the measured.

Dead Zone

It is defined as the largest change of input quantity for which there is no output of the instrument. For example the input applied to the instrument may not be sufficient to over come the friction so no output.

Loading Effect

The ideal situation in a measurement system is that when an element used for any purpose is introduced into the system the original signal should remain undistorted. However, under practical conditions, it has been found that introduction of any element in a system results distortion of original signal.

The incapability of the system to faithfully measure, record or control the input signal in undistorted form is called loading effect.

So, when an instrument, which is to be connected across the system for measurement, must have high input impedance so that it can't load the circuit.

2.7 ERROR AND UNCERTAINTY ANALYSIS

In most instruments the accuracy is guarantee to be within a certain percentage of full scale reading. Components are guaranteed to be within a certain percentage of the rated value. The limits of these deviations from the specified value are defined as Limiting Errors.

The magnitude of a quantity having a nominal value A_s and a maximum error or limiting error of $\pm$ δA must have a magnitude A_a between the limits $A_s - i\delta A$ and $A_a + \delta A$ or

Actual value of quantity

$$A_a = A_s \pm \delta A$$

where δA = limiting error, A_s = nominal value.

Relative Limiting Error

The relative (fractional) error is defined as the ratio of the error to the specified (nominal) magnitude of a quantity. Therefore,

$$\text{Relative limiting error } \varepsilon_r = \frac{\delta A}{A_s} = \frac{\varepsilon_a}{A_s}$$

or

$$\varepsilon_0 = \delta A = \varepsilon_r A_s$$

$$\text{Relative limiting error, } \varepsilon_r = \frac{A_a \quad A_s}{A_s}$$

$$= \frac{\text{actual value} - \text{nominal value}}{\text{norminal value}}$$

Combination of Quantizes With Limiting Error

When two or more quantities each having a limiting error, are combined, it is advantageous to be able to compute the limiting error of the combination. The limiting error can be easily found by considering the relative increment of the function if the final result is in the form of an algebraic equations.

Sum of two quantities

Let y be the final result which is the sum of measured quantities u and v.

$\therefore \quad y = u + v$

The relative increment of the function is given by

$$\frac{dy}{y} = \frac{d(u+v)}{y} = \frac{du}{y} + \frac{dv}{y}.$$

Expressing the result in terms of relative increment of the component quantities

$$\frac{dy}{y} = \frac{u}{y}\cdot\frac{du}{u} + \frac{v}{y}\cdot\frac{dv}{v}.$$

If the errors in the component quantities are represented by ± δu and ± δv then corresponding *limiting error* δy in y is given by

$$\frac{dy}{y} = \left(\frac{u}{y}\cdot\frac{du}{u} + \frac{v}{y}\cdot\frac{dv}{v}\right).$$

Difference of two quantities

Let $y = u - v$

$$\therefore \quad \frac{dy}{y} = \frac{du}{y} - \frac{dv}{y}.$$

Expressing the result in terms of relative increments of component quantities

$$\frac{dy}{y} = \frac{u}{y}\frac{du}{u} - \frac{v}{y}\frac{dv}{v}.$$

If the error in u, v are ± δy and ±δv respectively, the signs may be interpreted to given the worst possible discrepancy, i.e. when the error in u is + δu, the error in v is –δv and vice versa, then the corresponding relative limiting error δy in y is given by

$$\frac{\delta y}{y} = \pm\left(\frac{u}{y}\cdot\frac{\delta u}{u} + \frac{v}{y}\cdot\frac{\delta v}{v}\right).$$

3. *Product of two quantities*: $X = x_1 . x_2$

$$\frac{\delta X}{X} = \pm\left(\frac{\delta x_1}{x_1} + \frac{\delta x_2}{x_2}\right)$$

Product or Quotient of More Than Two Quantities

Let y = uvw or y = [u/vw] or y = [1/uvw]

We have relative limiting error for y

$$\frac{\delta y}{y} = \pm\left(\frac{\delta u}{u} + \frac{\delta v}{v} + \frac{\delta w}{w}\right).$$

Relative limiting error of product of terms is equal to the sum of relative limiting errors of terms.

Quotient: $X = [x_1 / x_2]$.

$$\frac{\partial X}{X} = \pm\left(\frac{\delta x_1}{x_1} \mp \frac{\delta x_2}{x_2}\right)$$

Composite factor:

Let, $y = u^n \cdot v^{th}$

$\log_e y = n \log_e u + m \log_e v$

or $\frac{1}{y} = \frac{n}{u}\cdot\frac{du}{dy} + \frac{m}{v}\frac{dv}{dy}$

or $\frac{dy}{y} = n\frac{du}{u} + m\frac{dv}{v}$

Relative limiting error of y is

$$\frac{\delta y}{y} = \pm\left(n\frac{\delta u}{u} + m\frac{\delta v}{v}\right)$$

Type of Errors

Errors can define as the difference between the measured value and the true value. True is the actual magnitude of an input signal to a measuring instrument. Measured or estimated value indicated by the instrument.

Error = True value measured value

The different types of errors can be broadly classified as

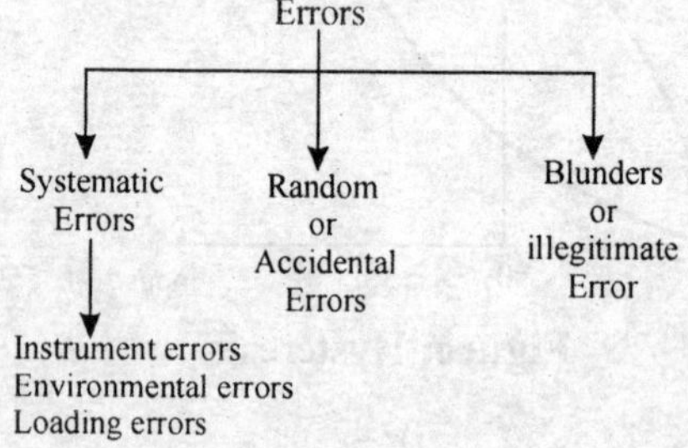

1. Systematic Errors

These types of errors tend to have the same magnitude and sign. These errors are also called fixed or Blass errors. A theoretical estimate of these types of errors may be possible or they can be corrected by proper calibration. They are caused by the followings,

- **Instrument Errors:** These can be built-in errors resulting due to incorrect design and constructions of the instrument e.g. error in the graduated scale, defective gears, inequality of arms etc. These can be corrected by, (1) proper selection of the instrument (2) by proper calibration or applying suitable correction.
- **Environment Errors:** The condition external to the instrument like change in temperature, pressure, humidity, magnetic field etc. can cause such errors. For example, error in the measurement of length caused by the expansion or contraction of the measuring scale due to temperature changes.
- **Loading Errors:** The very act of measurement alters the quantity being measured. Some energy from the input source is required for measurement. Therefore the indicated value and measured value will always be different. Loading can be mechanical, thermal or electrical. Measuring voltmeter may draw some current which may lower the source voltage. A temperature measuring device may take so much of heat so as to lower the source temperature. A flow measuring device may disturb the flow or after the flow pattern or conditions.

The above errors sometimes can be eliminated or minimized. A suitable correction can be applied by a proper calibration of the instrument.

In the case of electrical system consider a transducer connected directly to an indicating device.

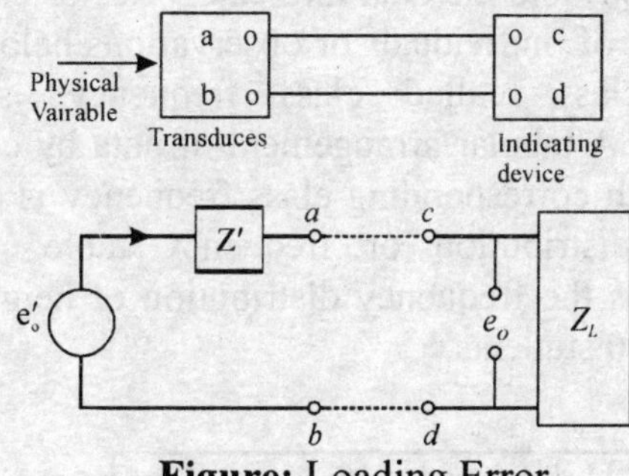

Figure: Loading Error

Let

e'_0 = output voltage of transducer without load

e_0 = Output voltage with load

Z' = Source impedance

Z_L = Load impedance of indicating device

i = current

$$e'_0 = iZ' + i\,Z_L$$

$$e'_0 = i(Z' + Z_L)$$

$$e_0 = i\,Z_L$$

$$\frac{e_0}{e'_0} = \frac{Z_L}{Z' + Z_L}$$

or $$\frac{e_0}{e'_0} = \frac{1}{1 + \frac{Z'}{Z_L}}$$

If the ratio $\frac{Z'}{Z_L}$ is small, output voltage under load will approximately be the same as no load voltage. Measured value and error are related of true value = measured value error = Measure of value + Bial + uncertainty.

Random Errors

The errors which vary in magnitude and can be positive or negative are termed as random error. These errors usually follow a certain statistical distribution but not always. The presence of these errors causes a lack of consistency in the measured value when the same input is given to the instrument. These can be caused due to certain defects in the elements of the measuring system such as backlash, friction and large dimensional tolerance in mating parts. One way to deteal and correct such errors is to measure quantity, first while increasing and decreasing its magnitude. Certain uncontrolled disturbance like vibration or temperature may also cause such errors.

Static Performance Parameters Accuracy
Blunders or Illegitimate Errors

These may invalidate the measurved value and may be due to human errovsin readings or using the instrument, and tauilty instrument or tallty adjustment. It refers to how closely the measured value agrees with the true value. It is the maximum amount by which the measured value differs from the true value. Inaccuracy (also called bias) depends on systematic error involved in the measurement.

The accuracy of an instrument can be specified as given below:

1. Percentage of true value

$$= \frac{\text{Measured value} - \text{True value}}{\text{True value}}$$

2. Percentage of full scale deflection (f.s.d)

$$= \frac{\text{Measured value} - \text{True value}}{\text{Maximum scale value}}$$

2.8 STATISTICAL ANALYSIS OF DATA

Statistics

Plane add diagram some of the imp terms frequency used are declined below.

Data

Elemental bits of information by experimental means.

Population and Sample

While collecting data concerning the characteristic of a group of individuals or objects, such as height and weight of students in a school or number of defective automobile part produced in a factory, it is often impossible to observe the entire group, called population. The practical approach is to randomly select and examine a small part of the group called sample. If a sample is representative of a population, the characteristics of the population can be inferred from the analysis of the sample.

Mean

The arithmetic mean $\overline{x}$ of sample is the sum of the individual data points x_i divided by the number of point n.

$$\overline{x} = \frac{\Sigma x_i}{n}$$

The mean of entire population is denoted by μ

Median

It is the midpoint of a group of data, as many fall to oneside of median as to the otherside. If the number of measurements is odd, the median is the middle value. If the number is even, then the median is the arithmetic mean of two middle values.

Mode

It is the value that occurs most frequently.

Deviation

It is the difference between a single value and mean value.

Average or Mean Deviation

It is an average of the absolute value of deviations from the mean and is defined as

$$\text{W}\frac{\Sigma|x_i - \overline{x}|}{n}$$

Variance

It is found by summing the squared deviation of the data values about the mean and then dividing the total by $(n-1)$, if data set is for a sample and by n if data set is for a population

$$\text{Sample variance } s^2 = \frac{\Sigma(x_i - \overline{x})^2}{n-1}$$

$$\text{Population variance } \sigma^2 = \frac{\Sigma(x_i - \mu)^2}{n}$$

Standard deviation

It is the square root of the sample or population variance

$$s = \sqrt{\frac{\Sigma(x_i - \overline{x})^2}{n-1}} \text{ Sample standard deviation}$$

$$\sigma = \sqrt{\frac{\Sigma(x_i - \mu)^2}{n}} \text{ Population standard deviation}$$

Frequency Distribution

When a large mass of data is presented it is often useful to distribute the data into categories or classes. The number of individuals or observations belonging to each class, called class frequency is the determined. A tabular arrangement of data by classes together with corresponding class frequency is called frequency distribution or frequency table. Table below shows the frequency distribution of heights in inches of 100 students.

Height	No. Of student
60-62	5
63-65	18
66-68	42
69-71	27
72-74	8
	Total 100

Current Practices on evaluation and expression of uncertainty in measurements- ISO guidelines

ISO standard 7025 requires, amongst other things, a statements of compliance to the uncertainty of measurements, even for testing laboratories. The conventional method used for evaluation of uncertainty of measurement is based on treating the uncertainty as having two components systematic and random. Systematic component is one which varies in a predicable manner when conditions change. Random component is one which varies in an unpredictable manner during a number of identical measurements.

In the modern approach the experimental method is the basis of determining uncertainty in measurements . the sources which contribute to uncertainties depend on following factors:

Related to or Affecting the standard	S
Related to or Affecting the work piece	W
Related to or Affecting the Measuring Instrument	I
Related to or Affecting Person	P
Related to Enviroment	E

It can seen from these factors that the status of random and systematic components and some of them may interchange as conditioned by the experiment.

Also, it can be appreciated that a measurement process may involve one of more than one instrument. Further, instruments is only one of the factors affecting the uncertainty of measurement.

Modern approach is to group uncertainty in two categories as follow:

Type 'A' Evaluation

This component of uncertainty is determined by using statistical tools from a series of experimental data obtained by repeated measurements.

Type 'B' Evaluation

This component of uncertainty deals with the analysis of a series of data obtained from sources other than from repeat measurements. Data belonging to this category can be obtained from experience/general knowledge of the instrument, previous measurement data, data provided in the calibration and other certificates and manufacturer's specification etc.

Method of Estimation of Measurement uncertainty as per ISO guideline consists of following steps:

- A careful study of measurement process and identifying all source of uncertainties.
- Grouping these under type 'A' or type 'B' evaluation.
- Evaluating the corresponding standard Uncertainties $(U_1, U_2, ...)$.
- Calculating the combined standard uncertainty (U_c) by the square root of the sum-of-the –quare method. $U_c = \sqrt{U_1^2 + U_2^2 + ... + U_n^2}$.
- Calculating the Expanded uncertainty 'U' by multiplying Combined standard uncertainty by a coverage factor K, general value is $K = 2$, which is equal to a confidence level of 95%

$$U = KU_c = 2U_c$$

If 'y' be the measured value of the parameter and U the expanded uncertainty, then the estimated true value Y is expressed as,

$$Y = y \pm U$$

Arithmetic Mean

The most probable value of measured variable (variate) is the arithmetic mean of the number of reading taken. The best approximation is made when the number of reading s of the same quantity are very large. Theoretically, an infinite number of readings would given the best result, although in practice, only a finite number of measurements can be made.

The arithmetic mean is given by:

$$\overline{X} = \frac{x_1 + x_2 + x_3 + x_4 \cdots\ldots x_n}{n} = \frac{\sum x}{n}$$

where, $\overline{X}$ = arithmetic mean

$x_1, x_2 \ldots x_n$ = Readings or variates or samples

n = number of readings

Dispersion

The property which denotes the extent to which the values are dispersed about the central value, is termed as dispersion.

2.9 CURVE FITTING

Determination of Parameters in Linear Relationship

When the scatter diagram happens to be linear, then the equation representing it is a straight line.

$$y = a_0 + a_1 x$$

Where,

a_0 = intercept on y-axis

a_1 = slope of the line

To determine a_0 and a_1 following methods are employed. (Fig. 2.6)

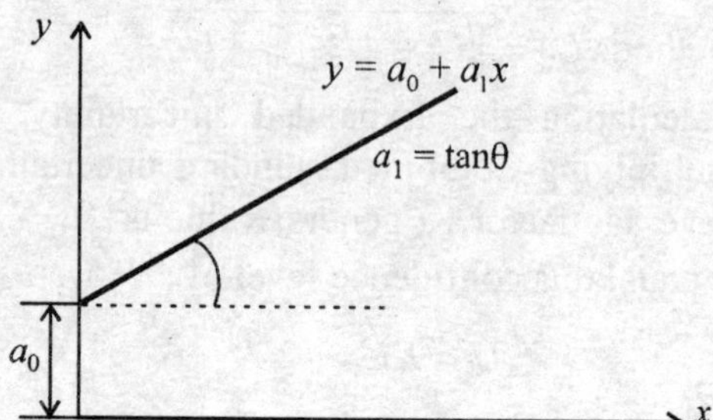

Graphical Method

The given set of data points are plotted on a graph and a straight line is drawn such that data point lie uniformly about the line. The constraints in the equation can then be determined by choosing two point on the line. This method although quite convenient but has following shortcomings.

- The method depends on the judgment of an individual.
- If data is plotted using different scales, then the resulting line drawn may differ.
- The method does not give any idea of how good the straight line is

Method of Sequential Differences

Consider 'n' pair of observations $(x_1, y_1,), (x_2, y_2,) \ldots (x_n, y_n)$ whose scatter diagram is of linear type.

To determine the slope of line represented by 'n' points, the slopes of each pair of adjacent point are determined giving $(n-1)$ values of slopes. The best estimate of slope is given by the average of these $(n-1)$ values.

Slopes for various pair of points are.

$$a_1 = \frac{(y_2 - y_1)}{(x_2 - x_1)}$$

$$a_2 = \frac{(y_3 - y_2)}{(x_3 - x_2)}$$

$$a_{n-1} = \frac{(y_n - y_{(n-1)})}{(x_n - x_{(n-1)})}$$

Mean of slopes.

$$\bar{a} = \frac{\sum_{i=l}^{n-1} a_i}{(n-1)}$$

To determine the intercept a_0, it is assumed that the centroid $(\overline{X}, \overline{Y})$ of the data lies on the straight line,

$$\overline{Y} = \bar{a}\,\overline{X} + a_0$$

Or

$$a_0 = \overline{Y} - \bar{a}\,\overline{X}$$

Where,

$$\overline{X} = \sum_{i-l}^{n} \frac{X_i}{n}$$

$$\overline{Y} = \sum_{i-l}^{n} \frac{Y_i}{n}$$

The limitations of this method are

- If the interval of x_is is equal then slope 'a' is function of first and the last points of the data.
- The method is not very accurate if the range of a_i is large.

MCQ Set–1

1. The conformity (closeness, nearness) of the output (value indicated by an instrument) to the true value of the measurand is called.
   (a) Accuracy (b) Sensitivity
   (c) Precision (d) Constantancy
2. The **least change of the measured** variable which can be detected at the output of measuring system is.
   (a) sensitivity (b) discrimination
   (c) least count (d) tolerance
3. The resolution of a system refers to
   (a) Difference between the true value of the input.
   (b) Retardation of the responce
   (c) Smallest change in the measurand that can be measured.
   (d) Range of enaccuracy which can be tolerated.

4. The **minimum change in the measured** variable which produces an effective response of the instrument is known as
(a) deviation (b) drift
(c) resolution (d) dead zone

5. The ratio of the output change for a given measuring system is referred to as
(a) sensitivity (b) linearity
(c) stability (d) fidelity

6. State which one of the following describes the **linearilty** of instrument.
(a) largest change in the measured variable which produces no instrument response.
(b) closeness of the calibration curve to a specified straight line.
(c) closeness of the indicated value to the true value of the measurand.
(d) range of a inccuracy which can be tolerated.

7. A flow meter having a range of 0 to $200m^3/s$ has a guranteed accuracy of $\pm0.75\%$ above 25 percent of scale reading. If the instrument indicates $160\ m^3/s$. then the static error will be.
(a) $\pm1.5\ m^3/s$ (b) $\pm1.2\ m^3/s$
(c) $\pm0.75\ m^3/s$ (d) $\pm0.3\ m^3/s$

8. Which one of the following constitutes the **most important specification** of an instrument ?
(a) resolution (b) reproducibility
(c) range (d) sensitivity

9. The **minimum value of input** which is necessary to cause a detectable change from zero output is called
(a) least count (b) resolution
(c) threhold (d) drift

10. **Precision** is the
(a) closeness of the instrument output to true value of the measured quantity
(b) ratio of different between measured value and the true value of the measured.
(c) smallest increamsnt in measurand that can be detected with certainty by the instrument
(d) degree of repeatability of serval independent measurements of the desired input at same reference conditions.

11. Which of the following is a **false** statement ?
(a) The specification of the percentage of full scale deflection is more accurate than the percentage of true value.
(b) Hystersis refers to the error in the output for a given value of input when this value is approached from ascending order and then descending order.
(c) Sensitivity defined as the ratio of the magitude of response to the mangitudeof quanitity being measured, has a wide range of units depeding upon the instrument or measurement system being investigaged.
(d) Linearity of an instrument represents the closenss of calibration curve to a specified straight line.

12. Dynamic quantites
(a) vary rapidly with time
(b) remain constant over a period of time
(c) are displaced from zero position
(d) All of the above

13. All of the following statements are correct, except
(a) the sensitivity has a wide range of units depending upon the instrument used.
(b) the magnitude of output for a given output is indepedent of the direction of the change in input.
(c) the amplitude ratio is close to unity at very low input frequencies both for the first and second order instruments
(d) the value of time constant for dynamic measurement should be as small as possible

14. The largest change in the measured variable which produces no instrument response is called.
(a) dead band (b) dynamic error
(c) fidelity (d) threshold

15. The **dead time** of an instrument refers to
(a) large change of input quantity for which there is no output
(b) the time encountered when the instrument has to wait for some reactions to take place.
(c) the time before the instrment begins to response after the quantity has altered
(d) retardation or delay in the response of an instrument to a change in the input signal

16. A step-in put signal
(a) starts from zero and rises uniformly with time
(b) changes rapidly with time
(c) starts from zero, rises to a particular value and returns back to zero.
(d) starts from zero and rises to a value in zero time.

17. Identify the first order instrument
(a) mechnical level
(b) temperature measurement by mercury - in - glass thermometer
(c) linear electrical potentiometer
(d) piezo - electrical pick ups

18. Which of the follwing is a **second order instrument**?
(a) spring mass system for force measurment.
(b) temperature measurement by thermo couples
(c) amplifiers
(d) build up of air pressure in bellows

19. The **speed of response** of a first order system is judged by.
(a) time constant (b) transient response
(c) steady state value (d) rise time

20. A **second order system** would be critically damped when.
(a) damping ratio is less than 1
(b) damping ratio is equal to 1
(c) damping ratio is greater than unity
(d) dampling ratio tands to infinity

21. When the driving frequency approaches the system natural frequency, there is a rapid increase in the output amplitude of an undamped second order system.
This condition is known as.
(a) vibration (b) instability
(c) resonance (d) oscillation

22. The time taken by an - damped second order system to reach peak output in response to step input is known as
(a) settling time (b) responce time
(c) rise time (d) dead time

23. A damping ratio of 0.7 desirable for the measuring instruments because.
(a) it gives minimum possible settling time
(b) it gives the instrument the highest sensitivity.
(c) no resonance occurs at this value
(d) no overshoot occurs when step input signal is applied

24. A reliable instrument.
(a) gives reproducible results within specified limits
(b) has good frequency response
(c) has errors of same magnitude and sign under the same working conditions.
(d) gives linear characteristics within the entire range

25. **Systematic errors** in a bourdom tube pressure gauge may be caused by :
(a) friction in the pins and gears of the amplifying mechanism.
(b) incorrect zero setting of the pointer
(c) variation of atmospheric pressure
(d) incorrect readings of the scale due to parallax.

26. Errors which may be variable both in magnitude and nature (hpositive or negative) are calssfied as
(a) hysteresis error (b) random error
(c) systematic error (d) interaction error

27. The charactically of an observer to read either systematically higher or systematically lower than other observers is referred to as
(a) insensitivity
(b) lack of precision
(c) lack of descrimination
(d) lack of discrimination

28. Which of the following errors can arise as a result of mistakes in reading parallax, improper instrument location nd inadequate lighting
(a) construction error (b) transmission error
(c) observation error (d) translation error

29. Gearing, backlash, friction between moving parts, and scale inaccuracies are generally known as.
(a) instrument errors (b) interference errors
(c) calibration error (d) interaction error

30. The measurment of surface temperature in exposed situations where atmospheric variation can interfere with temperature measurment gives rise to
(a) tranmission error (b) interaction error
(c) operation error (d) interference error

31. What represents the departure of the observed reading from the arithmetic mean of the group of readings.
(a) dispersion (b) deviation
(c) variance (d) midian

32. The maximum probability has a large value for small value of
(a) standard deviation (b) precision index
(c) uncertainty (d) average deviation

33. What refers to the number of changes that a partiduclar reading will occur when the error limit is specified"
(a) probability (b) probable error
(c) odds (d) confidence interval

34. Which is the most critical measurement in the following relations used for the computation result R from a set of varibles.

$$R = \frac{\pi}{6} \frac{x_1^4 x_2^{3/2} x_3}{\sqrt{x_4}}$$

(a) x_1 (b) x_2
(c) x_3 (d) x_4

35. Indicated results are most accurate when the pointer in the analog read out systems lies.
(a) in the first quarter of scale
(b) in the second quarter of scale
(c) around the mid - third of scale
(d) around the full capacity of scale

36. Statistical analysis of measured data accounts for
1. hysteresis in elestic members
2. changes in ambient temperature
3. measuring tape is too long or short
4. mechanical virbration and friction

select the correct answer using the codes given below.
(a) 1, 2 and 3 (b) 1 and 2
(c) 1 and 4 (d) 2 and 4

37. All of the following statements are correct, except.
(a) zero error of a micrometer is systematic error
(b) loading error is caused by the act of measurement on the physical system being tested
(c) randow errors are accidental, small and independent
(d) backlash refers to drift in the instrument output caused by hysteresis due to aging

38. In measurment system, which of the following static characteristics are desirable?
(a) Reproducibility (b) Sensitivity
(c) Accuracy (d) All of these

39. In measurment systems, which of the following are undesirable static characteristics?

(a) Sensitivity and accuracy
(b) Drift, static error and dead zone
(c) Reproducibility and non - linearity
(d) Drift, static error, dead zone and non - linearity

40. An ammeter reads $10.7\ A$ and the true value of current is 10.54 A. Determine the error and correction for this instrument

(a) $0.8, -0.8$ (b) $0.08, -0.08$
(c) $0.16, -0.16$ (d) None of these

41. A meter reads 127.50V and the ture value of the voltage is 127.43V. Determine the static error and static correction of instrument.

(a) $0.7V, -0.7V$ (b) $0.07V, -0.07V$
(c) $0.035, -0.035$ (d) None of these

42. A voltmeter is calibrated 100V to 250V. The accuracy is specified as within ± 0.5 percent of the voltmeter span. What is maximum static error?

(a) 0.75 (b) 1.25
(c) 1.75 (d) 0.5

43. A voltmeter has a uniform scale with 100 divisions, the full scale reading is 200 V and (1/10) of a scale division can be estimates with a fair degree of certainty. Determine the resolution of the instrument in voltmeter.

(a) 0.1 (b) 0.3
(c) 0.2 (d) 20

44. A Wheatstone bridge requires a change of $10\ \Omega$ in the unknown arm of the bridge to produce a change in deflection of 5mm of the galvanometer. Calculate the sensitivity.

(a) $2mm/\Omega$ (b) $10mm/\Omega$
(c) $0.5mm/\Omega$ (d) $5mm/\Omega$

45. Uncertaintly distribution is used for
(a) Analysis of multi - sample data
(b) Analysis of single sample data
(c) Analysis of both single and multi sample data
(d) None of the above

46. The value of inductance of an inductor is specified as $5mH \pm 1\%$ by the manufacturer. Find the limits between which the value of the inductance is guaranteed.

(a) 4.95 to 5.05 mH (b) 5mH to 5.05mH
(c) 4.99 to 5.01mH (d) None of these

47. An 0-10A has a guaranteed accuracy of 1 percent of full scale deflection. The limiting error while reading 2.5 is

(a) 1 % (b) 2 %
(c) 4 % (d) None of these

48. Two resistance R_1 and R_2 are connected in series.

R_1 and R_2 have the following ratings

$R_1 = 1000\Omega \pm 5\%$, $R_2 = 200\Omega \pm 5\%$

Then calulate percentage of limiting error.

(a) $\pm 15\%$ (b) $+5\%$
(c) -5% (d) $\pm 5\%$

49. A 0 to 100A ammeter has an error of $\pm 2\%$ full scale deflection. What would be the range of reading if true current reading is 50 A?

(a) $49.8 - 50.2\ A$ (b) 40 to 60A
(c) 48 to 52A (d) None of these

50. A set of readings has wide range. So it has
(a) High accuracy (b) High precision
(c) Low precision (d) Low accuracy

51. Two resistance $100\Omega \pm 5$ and $250\Omega \pm 10$ are connected in series. If the deviations are standard deviations, the resuitant can be expressed as

(a) 350 ± 15 (b) 350 ± 5
(c) 350 ± 14.7 (d) 350 ± 11.18

52. A circuit was tuned for resonance by eight different students and the value of resonance frequency in KHz were reconded as 532, 548, 543, 535, 546, 531, 543 and 536. Calculate the arithmetic mean

(a) 6.5 kHz (b) 546 kHz
(c) 539.25 kHz (d) None of these

53. A wattmeter has full scale range of 2500 W. It has an error $\pm 1\%$ of trun value. What would be range of rading if true power is 1250 W?

(a) 1225 W-1275 W (b) 1245 W-1255W
(c) 1200W-1300W (d) 1237.5 W-1262

54. A Voltmeter is having sensitivity $1000\Omega/V$. It is connected across an unknown resistor in series with a milli - ammeter. The voltmeter reads 40V on its 150V scale an ammeter reads 800mA. Calculated the % of error due to loading effect of voltmeter.

(a) 0.034 % (b) 0.34 %
(c) -0.034% (d) -0.34%

55. Power in a d.c. circuit is measured by measuring the voltage across and the current through the circuit. The voltage and current measurments are made to an accuracy of $\pm 2\%$ and $\pm 3\%$ respectively. The error are limiting errors .The error in measurment of power is

(a) $\pm 2\%$ (b) $\pm 3\%$
(c) $\pm 6\%$ (d) $\pm 5\%$

56. The power in a circuit is measured by measuring a current through a resistor. The current is measured with an accurancy of $\pm 1.5\%$ and the tolerance band of resistor $\pm 0.5\%$. The errors are limiting errors. The accuracy with which power is measured is.

(a) $\pm 1.125\%$ (b) $\pm 3.5\%$
(c) $\pm 2\%$ (d) $\pm 2.5\%$

57. A thermometer possesing a thermal time constant of 0.5 min is intoduced in a bath where the temperature is increasing at a constant rate of $5^{\circ}C/min$. The steady state error in the thermometer reading is.

(a) $10^{\circ}C$ (b) $2.5^{\circ}C$
(c) $0.1^{\circ}C$ (d) $0.4^{\circ}C$

58. A first order process has a static gain of 1, a time constant of 1 sec and transport delay of 0.1sec. Its transfer function will be.

(a) $\frac{e}{1+0.1s}$ (b) $\frac{e^{-10s}}{1+s}$

(c) $\frac{e^{-0.1s}}{1+s}$ (d) $\frac{e^{-0.1s}}{1+0.1s}$

59. A quantity whose magnitude has a definte repeating cycle is called a
(a) Transient
(b) Steady state periodic
(c) Steady state aperiodic
(d) Transient state periodic

60. A voltage has a true value of 1.50. An analog indicating instrument with a scale range of 0-2.50V, shows a voltage of 1.46V. Calculate error as a fraction of true value and the full scale deflection.

(a) $-2.67\%, -1.60\%$ (b) $-2.67\% - 2.67\%$

(c) $-0.04\%, 0.04\%$ (d) None of the above

61. A $10,000\Omega$ variable resistance has a linearity of 0.1% and the movement of contact arm is $320°$. Determine the resistance devlation in ohm. If this instrument is to be used as a potentiometer with a linear scale of 0 to 1.6v, then determine the maximum voltage error.

(a) $0.1\Omega, 1.6mv$ (b) $10\Omega, 1.6mv$

(c) $10\Omega, 16mv$ (d) None of the above

62. A $160 \pm 10pF$ capacitor, an inductor of $160 \pm 10\%$ mH and a resistor of $1200 \pm 10\Omega$ are connected in series. If all three components are $+10\%$ compute the expected resonance frequency of the combinations and the percantage error compared to normal arrangement.

(a) 1 *MHz*, 10% (b) 1 *MHz*, –10%

(c) 1 *MHz*, ±10% (d) 0.9 *MHz*, - 10%

63. We have a parallel circuit having two branches. the current in one branch is $I_1 = 100 \pm 2A$ and in other is $I_2 = 200 \pm 5A$. Determine value of total current $I = I_1 + I_2$.
i. Considering the errors in I_1 and I_2 as limiting errors
ii. Considering the errors as standard deviations
(a) 1. $300 \pm 7A$ and $300 \pm 3A$
(b) 1. $300 \pm 7A$ and $300 \pm 3.58A$
(c) 1. $300 \pm 3.58A$ and $300 \pm 7A$
(d) None of the above

64. A certain resistor has a voltage drop of 110.2 V and a current of 5.3 A. The uncetainties in the measurements are $\pm 0.2V$ and $\pm 0.6V$ respectively. Calculate the power dissipated in the resister and uncertainty in power

(a) 584, ± 0.26% (b) 584 W, ± 0.14%

(c) 5.84 W ±1.15% (d) None of the above

65. Dynamic response consists of
(a) Two parts, one steady and other transient state response
(b) Only trasnsient state
(c) Only steady state response
(d) Steady state and trasnsient frequency response

66. The following are the desirable dynamic characteristic of a measurement system
(a) Fast response, fidelity, measuring lag and dynanic error
(b) Fast response and measuring lag
(c) Fidelity and measuirng lag
(d) Fast response and fidelity

67. The output of a system is given by the folloiwng equation
$y = ax + b$
Where x is input and a, b are constants
The system exhibits
(a) Linear behaviour
(b) Non-linear behaviour as it does not satisfy the property of super position
(c) Non-linear behaviour as does not satisfy the property of homogeneity
(d) Non-linear behaviour because it does not satisfy properties of both superposition and homogeneity

68. An RC series circuit has a resistance of $2M\Omega$ and a capacitance of $1\mu F$. It is subjected to a step input of $10V$. The dynamic error (refference to voltage across capacitor) at time = 4 sec of application of input is

(a) 1.35 V (b) 0.5 V

(c) 3.68 V (d) 0 V

69. A unit step is applied at t = 0 to a first order system without time delay. The response has a value of 1.264 units at t = 10 mins and 2 units at steady state. The transfer function of the system is

(a) $2\left[1-e^{-t/600}\right]$ (b) $1.264\left[1-e^{-t/600}\right]$

(c) $0.632\left[1-e^{-t/600}\right]$ (d) None of these

70. The relative limiting error can be termed as

(a) norminal error (b) fractional error

(c) guarantee error (d) none of these

71. Let x be the final result which is defined by $x = x_1 - x_2 + x_3$. What will be the relative limiting error.

(a) $\pm\left(\frac{x_1}{x}, \frac{\delta x_1}{x_1} + \frac{x_2}{x}.\frac{\delta x_2}{x_2} + \frac{x_3}{x}.\frac{\delta x_3}{x_3}\right)$

(b) $\pm\left(\frac{x_1}{x}, \frac{\delta x_1}{x_1} - \frac{x_2}{x}.\frac{\delta x_2}{x_2} + \frac{x_3}{x}.\frac{\delta x_3}{x_3}\right)$

(c) $\pm\frac{1}{x}\left(\frac{\delta x_1}{x_1} + \frac{\delta x_2}{x_2} + \frac{\delta x_3}{x_3}\right)$

(d) $\pm\frac{1}{x}\left(\frac{\delta x_1}{x_1} - \frac{\delta x_2}{x_2} + \frac{\delta x_3}{x_3}\right)$

72. For, $x = \frac{x_1}{x_2 x_3}$. Determine the relative limiting error.

(a) $\frac{\delta x}{x} = \pm\left(\frac{\delta x_1}{x_1} - \frac{\delta x_2}{x_2} - \frac{\delta x_3}{x_3}\right)$

(b) $\frac{\delta x}{x} = \pm\left(\frac{x_1}{x}\frac{\delta x_1}{x_1} + \frac{x_2}{x}\frac{\delta x_2}{x_2} + \frac{x_3}{x}\frac{\delta x_3}{x_3}\right)$

(c) $\frac{\delta x}{x} = \pm\left(\frac{\delta x_1}{x_1} + \frac{\delta x_2}{x_2} - \frac{\delta x_3}{x_3}\right)$

(d) $\frac{\delta x}{x} = \pm\left(\frac{x_1}{x}\frac{\delta x_1}{x_1} - \frac{x_2}{x}\frac{\delta x_2}{x_2} - \frac{x_3}{x}\frac{\delta x_3}{x_3}\right)$

73. One user reads $25.8°C$ and record $28.5°C$ instead. This type of error is called .

(a) Systematic errors (b) Random error
(c) Gross error (d) None of these

74. If σ = standard deviation, n = Number of reading Standard devlation of mean (σ_n) is given by

(a) $\sigma_m = \frac{\sigma}{\sqrt{n}}$ (b) $\sigma_m = \frac{\sigma}{n}$

(c) $\sigma_m = \frac{\sigma}{\sqrt{2n}}$ (d) $\sigma_m = \frac{\sigma}{n^2}$

75. Relation between standard deviation of standard deviation (σ_o) and standard deviation of mean (σ_m).

(a) $\sigma_o = \frac{\sigma_m}{2}$ (b) $\sigma_o = \frac{\sigma_m}{\sqrt{2}}$

(c) $\sigma_o = \frac{\sigma_m}{2} \times \sqrt{3}$ (d) $\sigma_o = \frac{1}{2}\sigma_m$

76. Relation between probability of occurrence and odds is:

(a) Probability of occurrence $= \frac{odds}{odds+1}$

(b) Probability of occurrence $= \frac{1+odds}{2+odds}$

(c) Probability of occurrence $= (0dds)^2$

(d) Probability of occurrence $= \frac{odds^2+1}{odds^2+2}$

77. The probability that the value of randomly selected observation will lie in confidence interval is called :

(a) Probability of occurrence
(b) Confidence probability
(c) Confidence level
(d) None of these

78. If $y =$ number of readings at any deviation x.
$x =$ magnitude of deviation from mean.
$n =$ number of reading

$$y = \frac{n}{\sqrt{\pi}}\exp(-h^2x^2)$$

Then $h = ?$

(a) Gaussian probrbility error
(b) Precision index
(c) Variance
(d) None of these

79. A $0-150\,V$ voltmeter has a guranteed accuracy of 1 percent of full scale reading. The voltage measured by this instrument is $75\,V$. Calculate the limiting error in percent.

(a) 3% (b) 2%
(c) 4% (d) 10%

80. Three resistors have following ratings.

$R_1 = 37\Omega + 5\%$

$R_2 = 50\Omega \pm 5\%$

$R_3 = 75\Omega + 5\%$

Determine the percent limiting error of these resistance connected in series.

(a) ±15% (b) ±10%

(c) ±5% (d) $\pm\frac{5}{3}\%$

81. A circuit was turned for resonce by eight different students and the value of resonant frequency in KHz were recorded as 532, 548, 543, 535, 546, 531, 543 and 536. Calculate the variance.

(a) $42.77\,(KHz)^2$ (b) $6.54\,(KHz)^2$

(c) $5.75\,(KHz)^2$ (d) $36\,(KHz)^2$

82. The resistance of an unknown resistor is determined by wheatstone bridge. The solution for the unknown resistance is stated as : $R_4 = \frac{R_1 R_2}{R_3}$

Where $R_1 = 500\Omega \pm 1\%$, $R_2 = 615\Omega \pm 1\%$,

$R_3 = 100\Omega \pm 0.5\%$,

Determine the nominal value of unknown resistor.

(a) $\pm 76.88\Omega$ (b) $\pm 3075\Omega$

(c) 1000Ω (d) None of these

83. Power factor of the circuit can be determined by $\cos\phi = \frac{P}{vI}$. Relative error in power, current and voltages are respectively ±0.5%, ±1%, . Calculate the realtive error for power factor?

(a) $\pm 2.5\%$, (b) $\pm 1.5\%$,
(c) $\mp 1.5\%$, (d) None of these

84. For a Gaussian distribution, the probable error is r. This mean that :
(a) Area under the curve between $\pm$r limits is 0.5.
(b) Half of the observer values lie between $\pm$ r limits.
(c) The changes that an additional observation will lie between $\pm$ r limits are 50 percent.
(d) All of these.

85. A set of readings has a wide range and therefore it has.
(a) low percision (b) high percision
(c) low accuracy (d) high accuracy

86. Uncertainty distrubution is used for
(a) Analysis of multisample data
(b) Analysis of single sample data
(c) Analysis of both multi and single sample data
(d) None of these

87. For a muiltiple sample data, the standard deviation of mean is given as 70.5, what will be its standard deviation
(a) 10 (b) 40
(c) 50 (d) None of these

88. A resistor of 15 $k\Omega$ with a tolerance of 10% is connected in parallel with a 10 $k\Omega$ resistor of 15% tolerance. What is the tolerance limit for the parallel network?
(a) 13% (b) 10%
(c) 12.5% (d) 6.67%

89. A voltmeter with an accuracy of $\pm 3\%$ of full scale is used on $200V$ scale to measure $150V$.
(i) What is the percentage error ? (ii) What range will the actual voltage fall within if the instrument reads $180V$?
(a) $2\%, 177-183V$ (b) $4\%, 174-186V$
(c) $3\%, 181.2V$ (d) $7.8\%, 180V$

90. A voltmeter with an accuracy of $\pm 4\%$ is used on 300v scale of measure 280v.
What is the possible percentage error?
(a) 3.84% (b) 2.81%
(c) 5.2% (d) 4.29%

91. A voltmeter with an accuracy of $\pm 4\%$ is used on 300v scale of measure 280v.
From the above question, find the range where the actual voltage may fail if the instrument reads 230v.
(a) 218 V (b) $242V$
(c) $218-242V$ (d) $268-292V$

92. A certain resistor has a voltage drop of 110.2 V and current of 5.3 A. The uncertanties in the measurment are 0.2V and 0.06V respectively. Calculate the power dissipated in the resistor and the uncertainty in power.
(a) 584 W and 1.15% (b) 584 W and 1.10%
(c) 584 W and 1% (d) 500 W and 1%

93. An $0-10A$ ammeter has a guarantteed **accuracy** of 1 percent of full scale defection. The **limiting error** while reading 2.5 A is:
(a) 1% (b) 2%
(c) 4% (d) none of the above

94. A $0-300\,V$ voltmeter has an error of $\pm 2\%$ of full scale deflection. What would be the range of readings if true voltage is 30 V?
(a) 24 V to 36 V (b) 29.4 V to 30.6 V
(c) $20\,V$ to $40\,V$ (d) None of these

95. A wattmeter has a full scale range of 2500 W. It has an error $\pm 1\%$ of true value. What would be the range of reading if true power is 1250 W ?
(a) $1225\,W-1275\,W$
(b) $1245\,W-1255\,W$
(c) $1200\,W-1300\,W$
(d) $1237.5\,W-1262.5\,W$

96. Power in a d.c. ciruit is measured by measuring the voltage across and current through the circuit. The voltage and current measurements are made to an accuracy of $\pm 2\%$ and $\pm 3\%$ repectively. The error are limiting errors. The error in measurement of power is:
(a) $\pm 2\%$ (b) $\pm 3\%$
(c) $\pm 6\%$ (d) $\pm 5\%$

97. The power in a circuit is measured by measuring a current through a resistor. The current is measured with an accuracy of $\pm 1.5\%$ and the tolerance band of the resistor $\pm 0.5\%$. The errors are limiting or Guarantee eross. The **accuracy** with which power is measured is:
(a) $\pm 1.125\%$ (b) $\pm 3.5\%$
(c) $\pm 2\%$ (d) $\pm 2.5\%$

98. In a permanent magnet moving coll ammeter the deflection of the pointer is proportional to product of flux density of mangetic field produced by the permanent magnet becomes 95% of the original, the meter gives erraneous reading resulting into error. This error can be classified as:
(a) Gross error (b) Systematic error
(c) Random error (d) None of the above

99. The voltage of a circuit is measured by a voltmeter having as input impedance comparable with the output impedance of the circuit thereby causing **error** in voltage measurement. This error may be called.
(a) Gross error
(b) Random error
(c) Error caused by misiuse of instrument
(d) Error caused by loading effect

100. The mean deviation $\overline{D}$ in terms of deviations from the mean value of n readings is:

(a) $\frac{\sum |d|}{n}$ (b) $\frac{\sum d}{n}$

(c) $\frac{\sqrt{\sum d^2}}{n}$ (d) $\sqrt{\frac{\sum d^2}{n}}$

101. A set of readings has a wide range and therfore it has:
(a) low precision (b) high precision
(c) low accuracy (d) high accuracy

102. The Gaussian distribution can be mathematically expressed as:

(a) $y = \frac{h}{\sqrt{\pi}} \exp(h^2 x^2)$

(b) $y = -\frac{h}{\sqrt{\pi}} \exp(h^2 x^2)$

(c) $y = \frac{h}{\pi} \exp(-h^2 x^2)$

(d) $y = \frac{h}{\sqrt{\pi}} \exp(-h^2 x^2)$

103. For a Gaussian distribution, the probable error is r. This meas that:
(a) area under the curve between $\pm r$ limits is 0.5
(b) half of the obeserved values lie between $\pm r$ limits.
(c) the chances that an additional observation will lie between $\pm r$ limits are 50 percent.
(d) all of the above.

104. Two resistance $100\,\Omega \pm 5\,\Omega$ and $150\,\Omega \pm 15\,\Omega$ are connected in series. If the **deviations** are standard deviations, the resultant resistnace can be expressed as:
(a) $250\,\Omega \pm 20\,\Omega$ (b) $250\,\Omega \pm 10\,\Omega$
(c) $250\,\Omega \pm 15.8\,\Omega$ (d) $250\,\Omega \pm 10.6\,\Omega$

105. If the result of a measurement is expressed as $\overline{X} \pm 3\sigma$ where
$\overline{X} =$ **mean value** and
$\sigma =$ **standard deviation**, it mean that:
(a) apporximately 99 percent of the readings lie between $\pm 3\sigma$ limit.
(b) 26 readings out of 1000 will lie outside $\pm 3\sigma$ limit.
(c) the odds for any reading to lie within $\pm 3\sigma$ limit are 256 to 1
(d) All of the above.

106. A batch of resistors have a **mean value** of $100.00\,\Omega$ and a standard deviation $\sigma = 0.2\,\Omega$. The probabllity corresponding to 2σ is 0.9546. The value of odds that randomly selected resistor will lie within $100.00 \pm 0.40\,\Omega$ is :
(a) 1 to 1 (b) 2.15 to 1
(c) 21 to 1 (d) 256 to 1

107. According to Chauvenet's criterion, a reading out of a set of n readings should be rejected if the probability of obtaining the parrticular **deviations** from mean is:
(a) *less than* $1/2n$ (b) *greater than* $1/2n$
(c) less than 1/n (d) less than $\frac{1}{\sqrt{2}}n$

108. If the confidence level is 0.95, then the value tying outside confidence interval are:
(a) 1 in 5 (b) 1 in 20
(c) 1 in 100 (d) 1 in 1000

109. Uncertaintuy distribution is used for:
(a) analysis of multi - sample data
(b) analysis of single sample data
(c) analysis of both single and multi sample data.
(d) none of the above.

110. A quantity whose manitude has a **definite repeating time cycle** is called a:
(a) transient
(b) steady state peridic
(c) steady state aperiodic
(d) transient state periodic

111. Dynamic response consists of:
(a) two parts, one steady state and the other transient state response
(b) only transient state responce
(c) only steady state response
(d) steady state and transient frequency response.

112. The following are the desirable **dynamic characteristics** of a measurement system:
(a) fast response, fidelity, measuring lag and dynamic error.
(b) fast responce and measuring lag.
(c) fidelity and measuring lag.
(d) fast response and fidelity.

113. The output of a system is given by the following equation,
$y = ax + h$
Where x is the input and a and b are constants.
The system exhibits:
(a) linear behaviour
(b) non - linear wbehaciout as it does not satisfy the property of superposition.
(c) non - linear behaviour as it does not sarisgy the property of homogencity
(d) non - linear behaviour because it does not satisy properties of both supersition and homogencity.

114. The **dynamic response** of a system is given by the following differential equation.

$$a_3\frac{d^3c(t)}{dt^3}+a_2\frac{d^2c(t)}{dt^2}+a_1\frac{dc(t)}{dt}+a_0c(t)$$

$$=b_2\frac{d^2r(t)}{dt^2}+b_1\frac{dr(t)}{dt}+b_0r(t)$$

Where c(t) and r(t) are respectively the output and input and a_3, a_2, a_1, a_0, b_2, b_1, and b_0, are constants. It is a:

(a) non - linear system
(b) linear system
(c) linear time variant system
(d) linear time invariant system.

115. In **force - current analogy stiffness constant** K is analogous to:

(a) inductance
(b) reciprocal of inductance
(c) capacitance
(d) reciprocal of capacitance.

116. In **force - voltage analogy viscous friction** constant B is analogous to:

(a) resistance (b) reciprocal of resistance
(c) inductance (d) capacitance

117. In **thermal system** if M is mass of liquid stored, Q is the liquid inflow rate and s is the specific heat, the thermal resistance is given by:

(a) Ms (b) Qs
(c) $1/Ms$ (d) $1/Qs$

118. In **liquid level system**, the units of capacitance of a tank are:

(a) m^2 (b) m^3
(c) m^2/s (d) $m^{-2}s$

119. The **transfer function** of a system is defined as :

(a) ratio of Laplace tansfrom of input to laplace transfrom of out put with intial conditional not equal to zero.
(b) ratio of Laplace transform of output to laplace transform of input with intial conditions not equal to zero.
(c) ratio of Laplace transfrom of input to laplace transfrom of output with all intial conditions equal to zero.
(d) ratio of laplace transform of output to laplace transform of input with all inital conditions equal to zero.

120. A first order system is subjected to a step input. Its **time constant** can be defined as:

(a) time for the output to rise to 63.2% of its final steady value in the case of rising output.
(b) time for the output to drop to 36.8% of its intial value in the case of a decaying output.
(c) time of output to rise and decrease to reach its final steady value if the initial rate of change of output is maintained.
(d) all of the above.

121. A first order system has a time constant of 20s. It is subjected to a step input. The setting time of the out put is assumed to be the time it reaches 95% of its final steady state value. The **setting time** of the system is:

(a) 100 s (b) 60 s
(c) 20 s (d) infinity

122. An RC series circuit has a resistance of $2\ M\Omega$ and a capacitance of $1\ \mu F$. It is subjecteed to a step input of 10 V. The **dynamic error** at time $t = 4$ s of application of input is:

(a) 1.35 V (b) 0.5 V
(c) 3.68 V (d) 0. V

123. The volume of bulb of the liquid filled is $1000\ mm^3$. Its time constant is 300 s. It is desired to reduce the **time constant** 100. What is the new volume of the bulb, if all other conditions remain the same except the dimensions of the bulb:

(a) 333.3 mm^3 (b) 37 mm^3
(c) 10 mm^3 (d) 5773 mm^3

124. The volume of bulb of a the theromometer liquid filled thermometer is $1000\ mm^3$. Its time constant and sensitivity are respect 300 s and $3.6\ mm/^\circ C$. If the volume of the bulb is reduced to $500\ mm^2$, the sensitivity becomes:

(a) $7.2\ mm/^\circ C$ (b) $10.8\ mm/^\circ C$
(c) $1.8\ mm/^\circ C$ (d) $1.2\ mm/^\circ C$

125. A ballon is fitted with a first order temperature measuring device with a time constant of 20 s. It is also fitted with a zero order instrument for altitude measurement. It radious back to earth both temperature and altitiude signals. The decrease in temperture is 0.02 x °C where x is the altitude in metre. The balloon travels vartically at a speed of 5m/s. The ballon radious back altitude signal as 3000 m. The decrease in temperature radioed back to earth is:

(a) $60°c$ (b) $58°c$
(c) $62°c$ (d) None of the above

126. A 2 gm mass is suspended form a simple spring. The deflection caused is 5 mm. The **natural frequency** of the system is:

(a) 7 Hz (b) 22 Hz
(c) 2.2 Hz (d) None of the above

127. In a second order systems, the deamping factor is 0.65 and the natural frequency is 4 Hz. If the stiffness of springs is halved, the new values of **damping factor** and natural frequency are:

(a) 0.92, 2.83 Hz (b) 0.46, 5.66 Hz
(c) 0.92, 5.66 Hz (d) 0.46, 2.83 Hz

128. In a second order system, the steady state error is 0.02. If both viscous frictions constant and spring constant are doubled, the **steady error** will be:

(a) 0.01 (b) 0.04
(c) 0.0025 (d) 0.02

129. In second order suystem the frequency of damped oscillations is 18 rad/s. The value of damping factor is 0.866. The **natural frequency** of oscillartions is:
(a) 15.6 rad/s (b) 19.3 rad/s
(c) 36 rad/s (d) 9 rad/s

130. In a second order system, the peak overshoot is 100%. The value of **damping factor** is:
(a) 1 (b) 0.707
(c) 0 (d) 0.5

131. For a second order system, the **setting time** for $\pm 2\%$ band is:
(a) $\frac{3}{\xi\omega_n}$ (b) $\frac{3}{\xi\omega_0}$
(c) $2\xi\omega_n$ (d) $\frac{4}{\xi\omega_n}$

132. In a second order under damped system, if the damping factor is d the the value of peak overshoot in per unit terms is:

(a) $M_P = \exp\left(\frac{\pi\xi}{\sqrt{1-\xi^2}}\right)$

(b) $M_P = -\exp\left(\frac{\pi\xi}{\sqrt{1-\xi^2}}\right)$

(c) $M_P = \exp\left(\frac{-\pi\xi}{\sqrt{1-\xi^2}}\right)$

(d) $M_P = \left(\frac{\pi\xi}{\sqrt{1-\xi^2}}\right)$

133. A first order thermometer has a time constant of 50 s. It is subjected to a sinusoidal input cycling at 0.002 Hz. The **time lag** of the instrument is:
(a) 50 s (b) 44.6 s
(c) 500 s (d) 0.01 s

134. A first order instrument should indicate an output which is within $\pm 10\%$ of the true value of the output when subjected to a sinusoidal input. The product $\omega\tau$, where ω = angular frequency of input and τ time constant of Instrument should be:
(a) less than 0.5 (b) greater than 0.5
(c) less than 10 (d) greatter than 10

135. A second order system when subjected to a unit step input has a peak overshoot of 10%. The same system when subjected to a sinusiodal input of unit amlitude will have a **resonant peak** of nearly:
(a) 103 % (b) 110 %
(c) 190 % (d) none of the above.

136. A second order underdamped system has a damping factor of 0.8. it is subjected to a sinusoidal input of unit amplitude. It has **resonant peak** of:
(a) 108 % (b) 92 %
(c) 20 % (d) it has no resonant peak

137. Bandwidth a frequency domain concept is indicative of
(a) rise time in time domain
(b) setting time in time domain
(c) steady state error in time domain
(d) all of the above

138. As undamped second order instrument has a natural frequency of 500 Hz. It is subjected to a sinusoidal input of unit amplitude at 500 Hz. The **amplitude of output** is:
(a) unity (b) 0.5
(c) 2.0 (d) infinity

139. The transfer function of a system is:

$$G(s) = \frac{100e^{-st}}{s(s+10)}$$

The system
(a) is a linear system
(b) is a non-linear system
(c) has a transportation lag
(d) none of the above

Answers Set-1

1. (a)	2. (b)	3. (c)	4. (c)
5. (a)	6. (b)	7. (b)	8.(c)
9 .(c)	10. (d)	11. (a)	12. (a)
13. (b)	14. (a)	15. (c)	16. (d)
17 .(b)	18. (a)	19. (a)	20. (b)
21. (c)	22. (c)	23 .(c)	24. (a)
25.(b)	26. (b)	27. (d)	28. (c)
29. (a)	30. (d)	31. (b)	32. (b)
33. (c)	34. (a)	37. (c)	36. (a)
37. (d)	38. (d)	39. (d)	40. (c)
41. (b)	42. (a)	43. (c)	44. (c)
45. (b)	46. (a)	47. (c)	48. (d)
49. (c)	50. (c)	51. (d)	52. (c)
53. (d)	54. (c)	55. (d)	56. (b)
57. (b)	58. (c)	59. (b)	60. (a)
61. (b)	62. (d)	63. (b)	64. (c)
65. (a)	66. (d)	67. (c)	68. (a)
69. (a)	70. (b)	71. (a)	72. (c)
73. (c)	74. (b)	75. (a)	76. (b)
77. (a)	78. (c)	79. (b)	80. (b)
81.(c)	82. (a)	83. (b)	84. (a)
85. (d)	86. (a)	87. (b)	88. (c)

89. (a)	90. (b)	91. (d)	92. (c)
93. (c)	94. (a)	95. (d)	96. (d)
97. (b)	98. (b)	99. (d)	100. (a)
101. (a)	102. (d)	103. (d)	104. (c)
105. (a)	106. (c)	107. (a)	108. (b)
109 (b)	110. (b)	111. (a)	112. (d)
113. (c)	114. (d)	115. (b)	116. (a)
117. (d)	118. (a)	119. (d)	120. (d)
121. (b)	122. (a)	123. (b)	124 (c)
125. (b)	126. (a)	127. (a)	128. (d)
129. (c)	130. (c)	131. (d)	132. (c)
133. (b)	134. (a)	135. (a)	136. (d)
137. (b)	138. (d)	139. (c)	

MCQ Set–2

1. The precision of an instrument indicates its ability to reproduce a certain reading with a given

(a) Drift (b) Resolution
(c) Shift (d) Consistency.

2. The percentage limiting error, in the case of an instrument of 8.3 V with a 0 to 150 voltmeter having a guaranteed accuracy of 1% full-scale reading is

(a) 1.810% (b) 0.181%
(c) 18.10% (d) 0.0018%.

3. Four voltmeters have the following specification. If **accuracy** is the main consideration, which one of these voltmeter should one select?

(a) 100 V; 2 mA (b) 100 V; 100 ohm/volt
(c) 100 V; 1 mA (d) 10,000 V; 10 mA.
(c) 100 Ù (d) 10000 Ù.

4. A 0 to 200 V dc moving coil voltmeter has a guaranteed accuracy of 0.75% of full-scale reading. The voltage measured by instrument is 100 V. The limiting error is

(a) 3% (b) 2%
(c) 1.5% (d) 0.75%.

5. Two resistances R_1 and R_2 are connected as shown in given figure. If R_1 = 528 ± 5Ù and R_2 = 325 ± 3Ù then the total resistor R_T will be

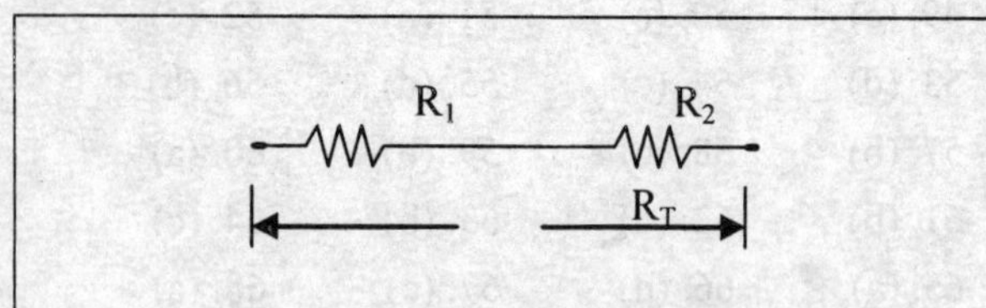

(a) 853 ± 2Ù (b) 853 ± 5Ù
(c) 853 ± 3Ù (d) 853 ± 8Ù.

6. An ammeter of range 0-25 A has a guaranteed accuracy of 1% of full-scale reading. The current measured by the ammeter is 5 A. The limiting error in the reading is

(a) 2% (b) 2.5%
(c) 4% (d) 5%.

7. A 0–150 V voltmeter has an accuracy of 1% of full-scale reading. The voltage measured by this instrument is 75 V. The limiting error is

(a) 1% (b) 2%
(c) 2.5% (d) 3%.

8. Which one of the following is the best definition of accuracy?

(a) It is the measure of consistency or reproducibility of measurements
(b) It is the ratio of the change in output signal to the change in input signal
(c) It is the smallest change in measurable input.
(d) It is the closeness with which an instrument reading approaches the true value of the quantity measured.

9. According to Guassian statistical analysis, if the confidence level is 0.80, then the values lying outside the confidence interval are:

(a) 1 in 5 (b) 1 in 10
(c) 1 in 20 (d) 8 in 10.

10. The measured value of a capacitor is 100 mF. The true value of the capacitor is 110 mF. The percentage relative error is:

(a) 9.99% (b) 9.09%
(c) 10.0% (d) 4 .76%.

11. The voltage across an impedance is measured by a voltmeter having input impedance comparable with the impendence comparable with the impedance causing an error in the reading.
What is this error called?

(a) Random error (b) Gross error
(c) Systematic error (d) Loading effect error.

12. In the circuit given in the figure, the limiting error in the power dissipation '' in the resistor R is

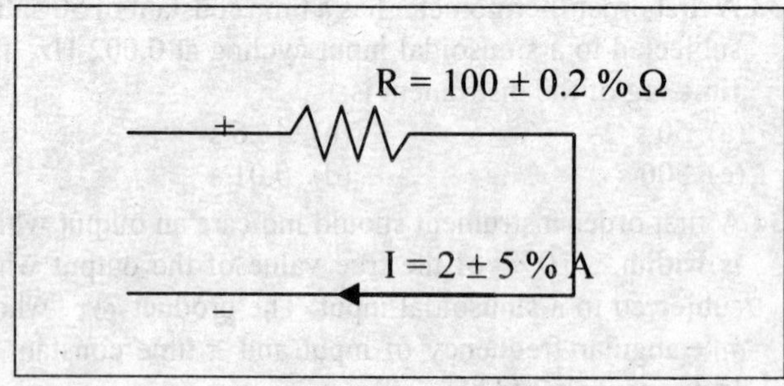

(a) 1.2 % (b) 5.2 %
(c) 10.2 % (d) 25.2 %

13. The maximum percentage **quantization error** for a 12-bit analog to digital converter is

(a) ± 0.00076 % (b) ± 0.012207 %
(c) ± 3.125 % (d) ± 4.17 %

14. Which one of the following statements is **not correct**?

(a) It is not possible to have precise measurements which are not accurate
(b) Correctness in measurements requires both accuracy and precision
(c) Reproducibility and consistency are expressions that best describe precision in measurements
(d) An instrument with 2% accuracy is better than another with 5% accuracy

15. A zero to 300 V voltmeter has an error of ±2% of the full –scale deflection. If the true voltage is 30 V, then the range of readings on this voltmeter would be
(a) 20 V to 40 V (b) 24 V to 36V
(c) 29.4 V to 30.6 V (d) 29.94V to 30.06V

16. A component manufacturer produces certain resistors which have a resistance value anywhere between 1.14 kΩ and 1.26 kΩ and classifies them as 1.2 kΩ resistors. If the resistance values are specified at 25° C and resistance have temperature coefficient of + 500 PPM/°C. Then the maximum resistance that one of these components might have at 75° C is
(a) 1.1111 kΩ (b) 1.2915 kΩ
(c) 1.5959 kΩ (d) 1.7575 kΩ

17. Match List I with List II and select the correct answer using the codes given below the Lists:

List I	List II
A. Cesium	1. Standard resistance
B. Managanin	2. Atomic clock
C .Standard voltage	3. Instrument interface
D. IEEE 488	4. Josephson junction

Codes:

	A	B	C	D
(a)	2	1	3	4
(b)	2	1	4	3
(c)	1	2	3	4
(d)	1	2	4	3

18. The limiting errors of measurement of power consumed by and the current passing through a resistance are $\pm 1.5\%$ and $\pm 1\%$ respectively . The limiting error of measurement of resistance will then be
(a) $\pm 0.5\%$ (b) $\pm 1.0\%$
(c) $\pm 2.5\%$ (d) $\pm 3.5\%$

19. The current 'I' through a resistance R is measured with the following uncertainties $I = 4A \pm 0.5\%$ $R = 100\ \Omega \pm 0.2\%$. If power is computed from these two measured quantities, the uncertainty in the power computed will be
(a) $\pm 0.01\%$ (b) $\pm 0.29\%$
(c) $\pm 0.7\%$ (d) $\pm 1.2\%$

20. Match List I with List II and select the correct answer using the codes given below the lists:

List I	List II
A. Sensitivity	1. Closenees to the true value
B. Resolution	2. A measure of reproducibility of the instrument
C. Accuracy	3. Ration of response to the change in the input signal
D. Precision	4. Smallest change in input to which the instrument can respond

Codes:

	A	B	C	D
(a)	3	4	1	2
(b)	3	4	2	1
(c)	4	3	2	1
(d)	4	3	1	2

21. In PMMC instrument, the central spring stiffness and the strength of the magnet decrease by 0.04 % and 0.02 % respectively due to a rise in temperature by 1°C. With a rise in temperature of 10 °C, the instrument reading will
(a) Increase by 0.2 %
(b) Decrease by 0.2 %
(c) Increase by 0.6 %
(d) Decrease by 0.6 %

22. A symmetrical square wave voltage is read on an average response electronic voltmeter whose scale is calibrated in terms of rms values of a sinusoidal wave. The error in the reading is
(a) – 3.9 % (b) + 3.9 %
(c) – 11 % (d) +11 %

23. A 150 V moving iron voltmeter of accuracy class 1-0 reads 75 V when used in a circuit under standard conditions. The maximum possible percentage error in the reading is
(a) 0.5 (b) 1.0
(c) 2.0 (d) 4.0

24. A resistance of 105 ohms is specified using **significant** figures as indicated below:
1. 105 ohms
2. 105.0 ohms
3. 0.000 105 Mega ohms .

Among these
(a) 1 represents greater precision than 2 and 3
(b) 2 represents greater precision but 1 and 3 represent same precision
(c) 2 and 3 represent greater precision than 1
(d) 1, 2 and 3 represent same precision

Answers Set-2

1. Ans. (d)

Solution: Precision of instrument is a measure of the reproducibility of measurements, i.e. gives a fixed values of quantity, precision is a measure of the degree of agreement with in a group of measurement. Precision is used in measurement to describe the consistency or reproducibility of results.

2. Ans. (c)

Solution: Maximum possible error at any reading

$$=\frac{150}{100}\times = 1.5 \text{ V}$$

$\therefore$ Percentage error at a reading of $8.3\,V = \frac{1.5}{8.3}\times 100$

$$= 18.1\%.$$

3. Ans. (d)

Solution: The resistance of these voltmeter are

$$R_1 = \frac{100 \text{ volt}}{2\times 10^{-3} \text{ amp}} = 50\times 10^3\ \Omega$$

$$R_2 = 100\times 100\ \Omega = 10^4\ \Omega$$

$$R_3 = \frac{100 \text{ volt}}{1\times 10^{-3} \text{ amp}} = 100\times 10^3\,\Omega$$

$$R_4 = \frac{10{,}000}{10\times 10^{-3}} = 100\times 10^3\,\Omega$$

R_4 is largest of all. Hence (d) is the correct choice.

4. Ans. (c)

Solution: Limiting error

$$= \frac{\text{Accuary}\times \text{Range}}{\text{Measured value}}$$

$$\text{Error} = \frac{0.75\times 200}{100} = 1.5\%.$$

Second Method

Relative limitary error is ratio of the error to nominal (Speed field) magnitude of quandary

$$\%\,\varepsilon_r = \frac{\delta A}{A_s} = \frac{(75/100)\times 200}{100} = 1.5\%.$$

5. Ans. (d)

Solution: $R_T = (R_1 + R_2) \pm (\delta R_1 + \delta R_2)$

$R_T = (528 + 325) \pm (5 + 3)$

$= (853 \pm 8)$ ohm.

6. Ans. (d)

Solution: On full scale i.e. 25A

$$\text{deflection error} = \frac{25\times 1}{100} = 0.25$$

$$\text{On 5A deflection error} = \frac{0.25\times 100}{5} = 5\%.$$

7. Ans. (b)

Solution: Magnitude of limiting error,

$$\partial A = \varepsilon_r \cdot A_s = 0.01\times 150 = 1.5 \text{ V}$$

Relative error,

$$\varepsilon_r = \frac{1.5}{75} = 0.02$$

Voltage between the limits

$$75(1\pm 0.02) = 75\pm 1.5 \text{ V}$$

% limiting error

$$= \frac{1.5}{75}\times 100 = 2\%.$$

8. Ans. (d)

9. Ans. (a)

Solution: If the confidence level is 0.800, the confidence interval is $\overline{X} \pm 1.282\sigma$.

Thus, for it values lying outside confidence interval is 1 in 5.

10. Ans. (b)

Solution:

$$\%\text{ relative error} = \frac{\text{True value} - \text{Measured value}}{\text{True value}}$$

$$= \frac{110-100}{110}\times 100$$

$$= \frac{10}{110}\times 100 = 9.09\%.$$

11. Ans. (d)

12. Ans. (c)

Solution: $P = I^2 R$

$$\therefore \quad \frac{dP}{P} = 2\frac{dI}{I} + \frac{dR}{R} = 2\times 5 + 0.2$$

$$= 10 + 0.2 = 10.2$$

13. Ans. (b)

Solution: % of quantization error

$$= \pm\frac{(1/2)\,LSB}{2^n - 1} = \frac{0.5}{4095} = \pm 0.01221\,\%$$

14. Ans. (b)

15. Ans. (b)

Solution: 2 % of 300 $V = \frac{2}{100} \times 300 = 6\,V$

Then % error $= \frac{6}{30} \times 100 = 20\,\%$

Hence range $= 30 \pm 20\,\% = 30 \pm 6$

$= 24\,V$ to $36\,V$

16. Ans. (b)

Solution: $Rt_2 = R\left[1 + \alpha\left(t_2 - t_1\right)\right]$

$Rt_{2\max} = 1.26\left[1 + 500 \times 10^{-6}\left(75 - 25\right)\right] k\Omega$

$R_{\max}$ at $75^o = 1.26\left(1 + 500 \times 10^{-6} \times 50\right) k\Omega$

$= 1.2915\ k\Omega$

17. Ans. (b)

Solution: The correct match is given below

Cesium	Atomic clock
Managanin	Standard resistance
Standard voltage	Joseph son junction
IEEE 488	Instrument interface

18. Ans. (d)

Solution: Power $P = I^2 R$

Then $\frac{dP}{P} = 2\frac{dI}{I} + \frac{dR}{R}$

But $\frac{dR}{R} = \frac{dP}{P} - 2\frac{dI}{I}$

Note: In case of error are ± value i.e. ± dP and ±dI, the sign may be interpreted to give worst possible discrepancy

$\frac{dR}{R} = \frac{dP}{P} + \frac{2dI}{I}$

$= \pm\left(1.5 + 2 \times 1\right)$

$= \pm 3.5\,\%$

Another Method

If $W = U^a / V^b$, then

Limiting error (in 'W') = a ×limiting error in 'U' + 'b' × limiting error in 'V'.

$R = \frac{P}{I^2}$

∴ Limiting error in the measurement of resistance R

= limiting error in the measurement of power + 2 × limiting error in the measurement of current

$= \pm\left(1.5 + 2 \times 1\right) = \pm 3.5\,\%$

19. Ans. (d)

Solution: $P = I^2 R$

$\frac{dP}{p} = 2\frac{dI}{I} + \frac{dR}{R}$

$= \pm 2 \times 0.5 \pm 0.2$

$= \pm\left(1 + 0.2\right) = \pm 1.2\%$

20. Ans. (a)

Solution: The correct match is given below

Sensitivity	Ration of response to the change in the input signal
Resolution	Smallest change in input to which the instrument can respond
Accuracy	Closenees to the true value
Precision	A measure of reproducibility of the instrument

21. Ans. (a)

22. Ans. (d)

23. Ans. (c)

Solution: The error specified is always with respect to full scale reading and this is the maximum error that may be present at any deflection.

Therefore, Percentage error $= \frac{1.0}{100} \times 150 \times \frac{1}{75} \times 100$

$= 20\,\%$

24. Ans. (b)

MCQ Set–3

1. The circuit shown in figure is supplied by a dc source. The switch (S) in the circuit is closed at t = 0 and remains closed for a long time. The steady – state output Voltage (V_o) is.

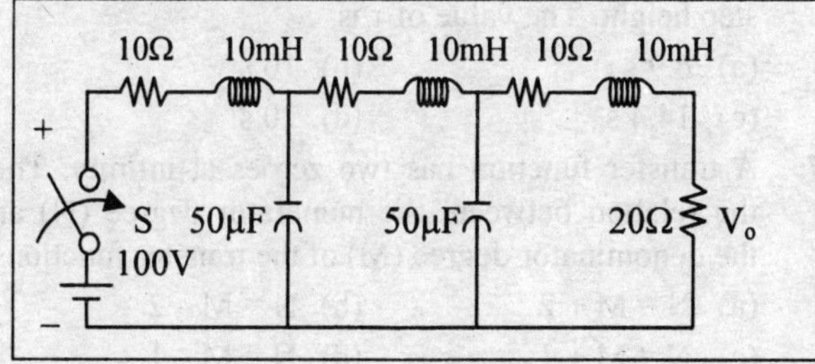

(a) 40 V (b) 100 V

(c) 25 V (d) 20 Sin ωt V

2. The power delivered by a signal source with 100 V_{rms} sinusoidal voltage and 50 Ω source resistance to a load consisting of a diode in series with 50 Ω resistor is.

(a) 25 W (b) 50 W
(c) 141 W (d) 70.7 W

3. For the current in branch AB shown in figure, the voltage V_{in} volt is

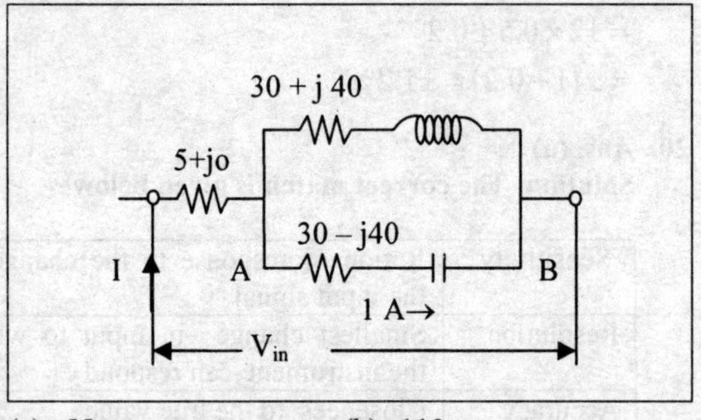

(a) 55 (b) 110
(c) 56 (d) 90

4. In figure the stored energy of the capacitor is.

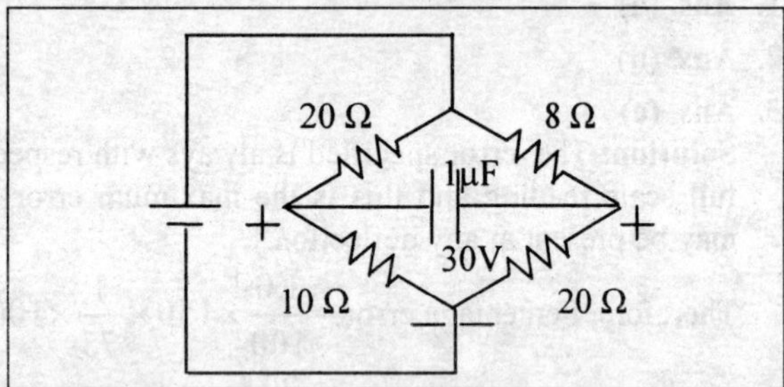

(a) 8 μJ (b) 30μJ
(c) 450 μJ (d) 900 μJ

5. For the waveform V (t) = 2 + cos [ωt + (π / 6)] the ratio $[V_{rms} / V_{average}]$ is
(a) [3 / (2√2)] (b) √(3 / 2)
(c) π (d) π / 2

6. A system with transfer function [1/(sτ + 1)] subjected to a step input takes 10 seconds to reach 50% of the step height. The value of τ is.
(a) 6.9 s (b) 10 s
(c) 14.4 s (d) 20 s

7. A transfer function has two zeroes at infinity. Then the relation between, the numerator degree (N) and the denominator degree (M) of the transfer functions.
(a) N = M + 2 (b) N = M – 2
(c) N = M + 1 (d) N = M – 1

8. The system G (s) = $[0.8/ s(s^2 + s - 2)]$ is subjected to a step input. The system output y(t) as t → ∞ is
(a) 0.8 (b) 0.4
(c) –0.4 (d) unbounded

9. The output voltage of the circuit in figure for t > 0 is

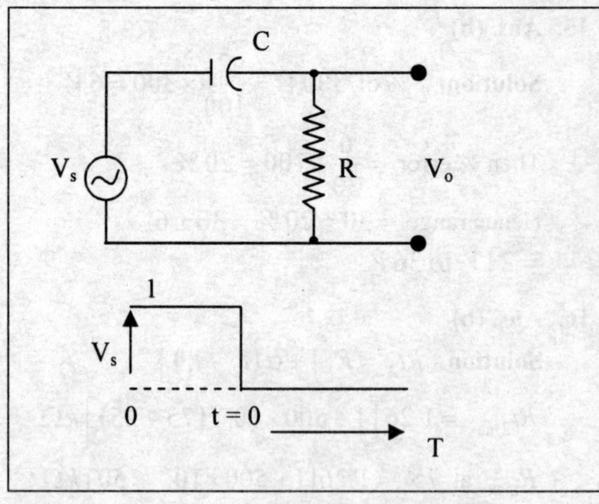

(a) $e^{-t/RC}$ (b) $-e^{-t/RC}$
(c) $1 - e^{-t/RC}$ (d) $e^{-t/RC} - 1$

10. A voltmeter connected across the 10 kΩ resistor in the figure reads 5 V. The voltmeter is rated at 1000 ohms/ volt and has a full scale reading of 10 V. The supply voltage V_s in volt is

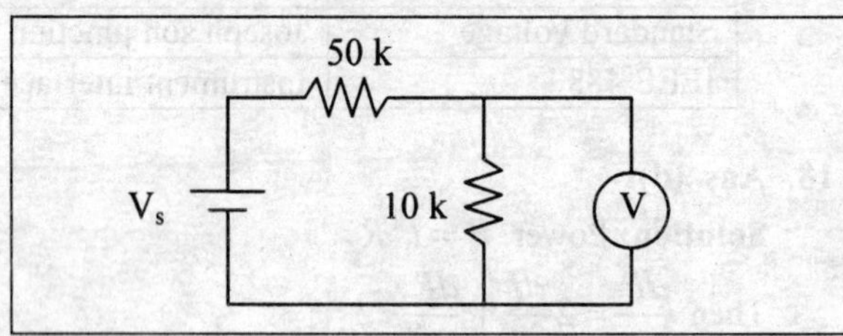

(a) 30 (b) 50
(c) 55 d) 80

11. The transfer function of a passive circuit has is poles and zeroes on
(a) Left and right halves respectively of the s-plane
(b) Right and left halves respectively of the s-plane
(c) Right half of the s-plane
(d) left half of the s-plane

12. For the circuit shown in figure. $V_R = 20\,V$ when $R = 10\,\Omega$ and $V_R = 30\,V$ when $R = 20\,\Omega$. For $R = 80\,\Omega$, V_R will read as

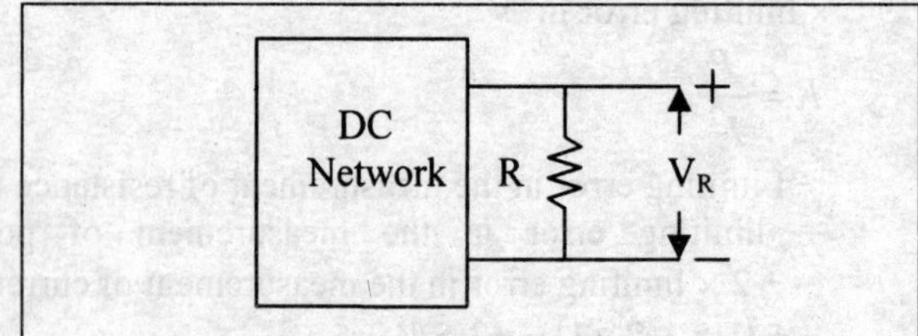

(a) 48 V (b) 60 V
(c) 120 V (d) 160 V

13. For the circuit shown in figure, R is adjusted to have maximum power transfer to it. The maximum power transferred is.

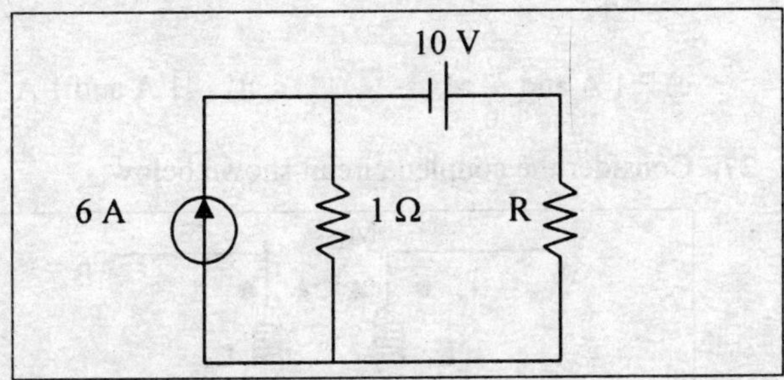

(a) 16 W (b) 32 W
(c) 64 W (d) 100 W

14. The value of the impedance $Z_d(s)$ between 1-1 of the circuit shown in figure as $s \to \infty$ is

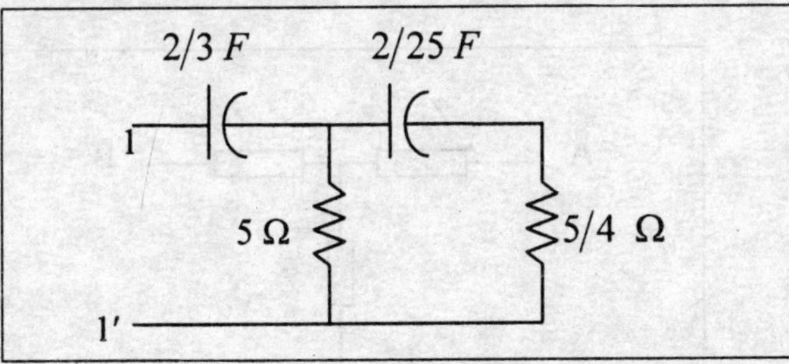

(a) 0 (b) 1
(c) 5/4 (d) 5

15. Superposition theorem does not hold for
a) R-L-C networks
b) Networks containing diodes and R-C elements
c) Linear op-amp circuit
d) None of the above

16. The final value of a function y(t) whose Laplace transform $Y(s)=\frac{4}{s^2+2s+2}$ is

(a) 4 (b) 2
(c) 1 (d) 0

17. In the circuit shown in figure current through the 5 Ω resistor is.

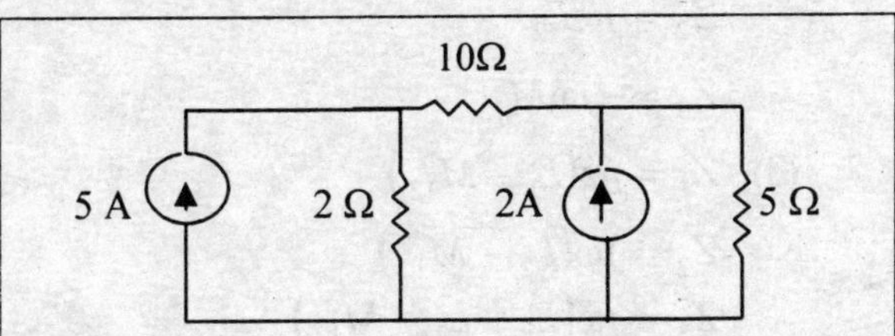

(a) 0 A (b) 2 A
(c) 3 A (d) 7 A

18. Two identical 2 – port networks with y- parameters $y_{11} = -y_{12} = -y_{21} = y_{22} = 1$ S are **connected in scade.** The overall y-parameters will satisfy the condition

(a) $y_{11} = 1$ S (b) $y_{12} = (-1/2)$ S
(c) $y_{21} = -2$ S (d) $y_{22} = -1$ S

19. For an input
V(t) = 5 – 2 sin (100 πt) – cos (200 πt),
The output of a full – wave rectifier average is.
(a) 2 (b) 4
(c) 5 (d) 8

20. The output resistance across the terminals 1 and 2 of the DC bridge in figure is

(a) 12.5 kΩ (b) 24.5 kΩ
(c) 25.0 kΩ (d) 100 kΩ

21. The output resistance across the terminals 1 and 2 of the DC bridge in figure is

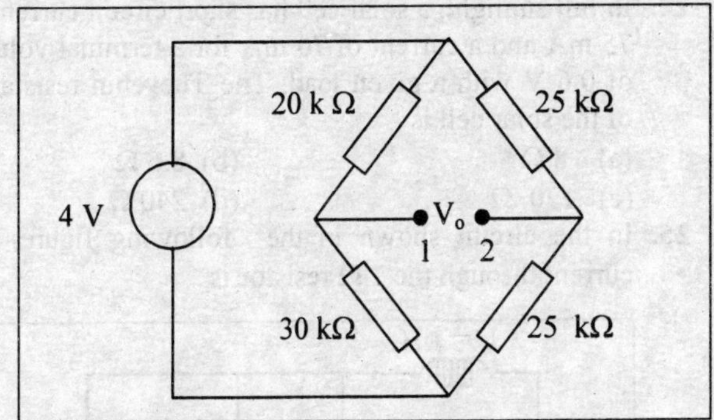

(a) 12.5 kΩ (b) 24.5 kΩ
(c) 25.0 kΩ (d) 100 kΩ

22. In the circuit shown in the figure, the switch S is closed at time t = 0. The supply current at $t = 0^+$ and the capacitor voltage at $t \to \infty$ are, respectively

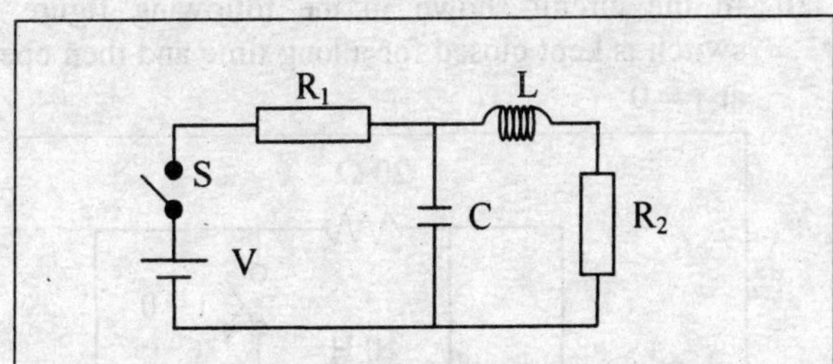

(a) 0 and V

(b) $\frac{V}{R_1}$ and $\frac{VR_2}{R_1+R_2}$

(c) $\frac{V}{R_1 + R_2}$ and $\frac{VR_2}{R_1 + R_2}$

(d) $\frac{V}{R_1}$ and V

23. If the value of the resistance R in the following figure is increased by 50%, then voltage gain of the amplifier shown in the figure will change by

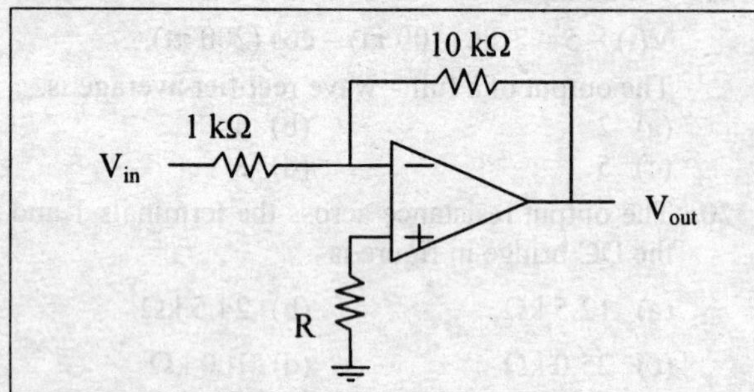

(a) 50% (b) 5%
(c) – 50% (d) negligible amount

24. In full sunlight, a solar cell has short circuit current of 75 mA and a current of 70 mA for a terminal voltage of 0.6 V with a given load. The Thevenin resistance of the solar cell is

(a) 8 Ω (b) 8.6 Ω
(c) 120 Ω (d) 240 Ω

25. In the circuit shown in the following figure, the current through the 1 Ω resistor is

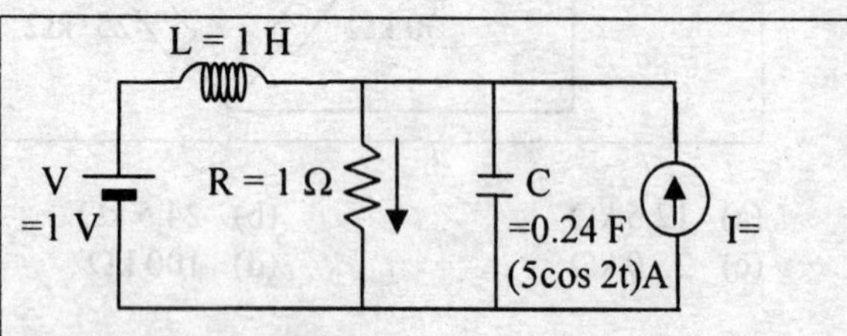

(a) $(1 + 5\cos 2t)\,A$ (b) $(5 + \cos 2t)\,A$
(c) $(1 - 5\cos 2t)\,A$ (d) $6\,A$

26. In the circuit shown in the following figure, the switch is kept closed for a long time and then opened at $t = 0$.

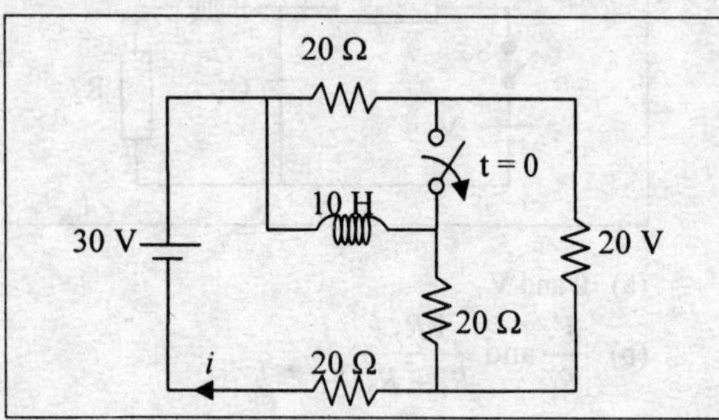

The value of the current i just before opening the switch $(t = 0^-)$ and just a after opening the switch $(t = 0^+)$ are, respectively

a) $\frac{3}{4}A$ and 1 A b) $\frac{3}{4}A$ and $\frac{5}{2}A$

c) 1 A and $\frac{7}{6}A$ d) 1 A and 1 A

27. Consider the coupled circuit shown below:

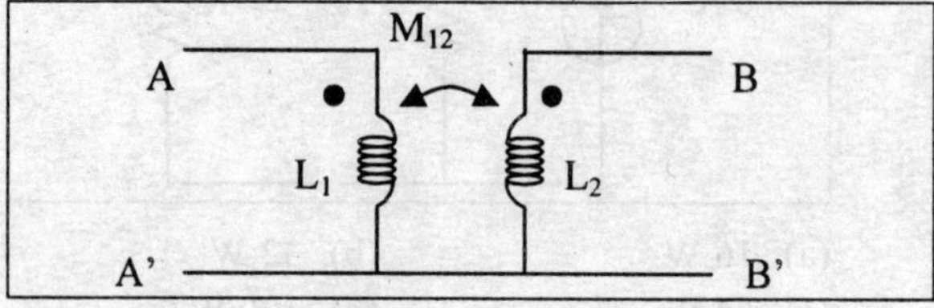

At angular frequency ω, this circuit can be represented.
By the equivalent T-network shown below.

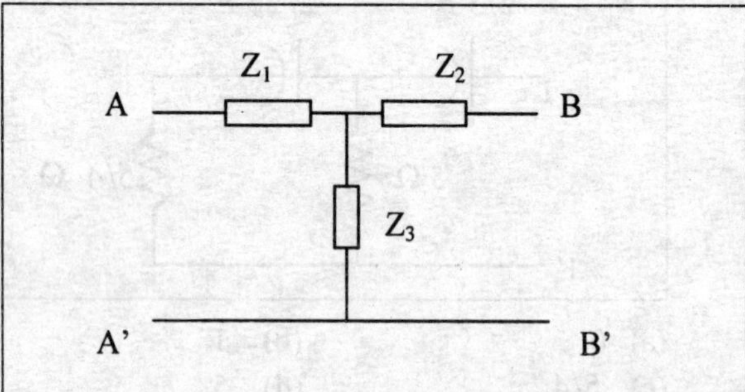

Indicate the correct set of expressions for the impedances of the T-network

(a) $Z_1 = j\omega(L_1 - M_{12})$
$Z_2 = j\omega(L_2 - M_{12})$
$Z_3 = j\omega M_{12}$

(b) $Z_1 = j\omega(L_1 + M_{12})$
$Z_2 = j\omega(L_2 + M_{12})$
$Z_3 = -j\omega M_{12}$

(c) $Z_1 = j\omega L_1$
$Z_2 = j\omega L_2$
$Z_3 = -j\omega M_{12}$

(d) $Z_1 = j\omega(L_1 - M_{12})$
$Z_2 = j\omega(L_2 - M_{12})$
$Z_3 = j\omega(L_1 + L_2 + M_{12})$

Answers Set-3

1. **Ans. (A)**

Solution:

Since at steady state there is no DC voltage drop across the inductor and no current flows through the capacitor, so the circuit reduces to

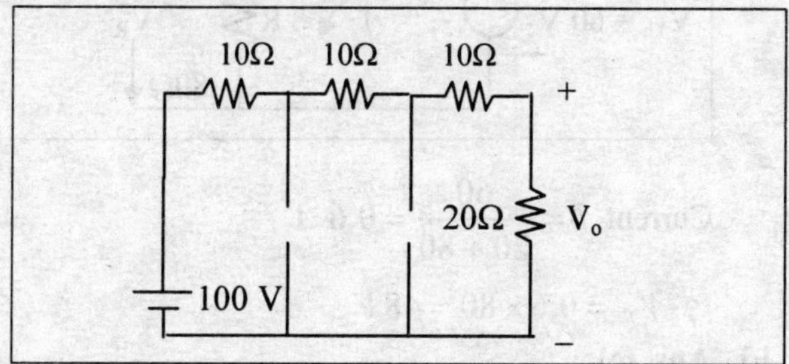

$$V_0 = \frac{20}{50} \times 100 = 40\,V$$

2. **Ans. (b)**

Sooution:

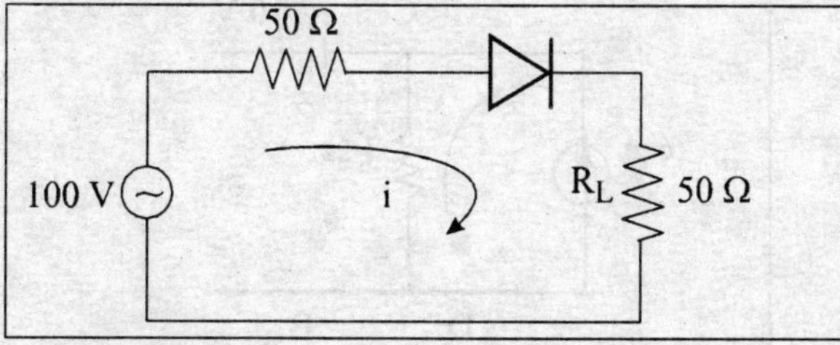

$$I = \frac{100}{50+50} = 1\,A$$

Power to load $R_L = I^2 R_L$

$$= I^2 \times 50 = 50\,Watt$$

3. **Ans. (c)**

Solution:

$$1_A = \left(\frac{30+j40}{30+j40+30-j40}\right) I$$

$$V_{in} = \frac{60}{30+j40} \times 5 + (30-j40) \times 1$$

$$= \frac{300}{30+j40} + (30-j40)$$

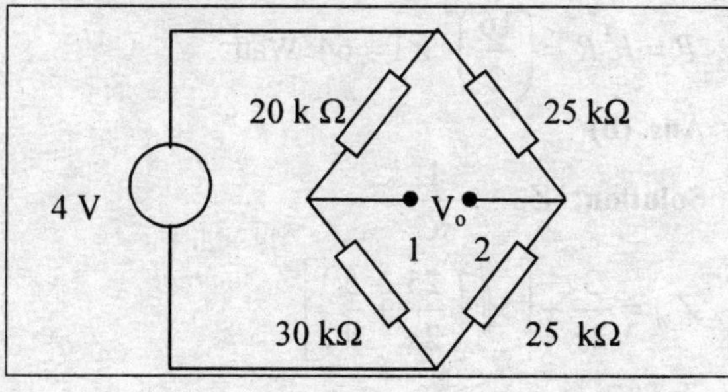

$$= \frac{300+900+1600}{30+j40} + (30-j40)$$

$$= \frac{2800}{50} = 56\,V$$

4. **Ans. (c)**

Solution:

Energy stored in a capacitor $= \frac{1}{2} CV^2$

$$= \frac{1}{2} \times 1 \times 10^{-6} \times 30^2 \text{ t}$$

$$= 450 \times 10^{-6}\,J$$

5. **Ans. (a)**

Solution: $V_{rms} = \sqrt{2^2 + \left(\frac{1}{\sqrt{2}}\right)^2} = \sqrt{\frac{9}{2}} = \frac{3}{\sqrt{2}}$

$$V_{average} = 2$$

$$\therefore \quad \frac{V_{rms}}{V_{average}} = \frac{3}{2\sqrt{2}}$$

6. **Ans. (c)**

Solution:

$$H(s) = \frac{1}{1+\tau}$$

$$V_i(s) = \frac{1}{s}$$

$$V_o(s) = \frac{1}{s(1+\tau)} = \frac{1/\tau}{s\left(1+\frac{1}{\tau}\right)}$$

$$V_o(t) = 1 - e^{-t/\tau}$$

$$0.5 = 1 - e^{-10/\tau}$$

$$e^{-10/\tau} = 0.5$$

$$-\frac{10}{\tau} = \log e^{0.5} = -0.69$$

$$\tau = \frac{10}{0.69} = 14.4 \text{ sec}$$

7. **Ans. (b)**

Solution:

All poles terminate in zeros. If the zero are less than poles, they terminate at infinity.

8. **Ans. (c)**

Solution:

$$\underset{s \to 0}{Lt}\ s\,G(s) = \frac{s\,0.8}{s(s^2+s-2)} = \frac{0.8}{-2} = -0.4$$

9. Ans. (a)

Solution: $TF = \dfrac{s}{s+\dfrac{1}{CR}} \times \dfrac{1}{s} = \dfrac{1}{s+\dfrac{1}{CR}} = e^{-t/CR}$

10. Ans. (d)

Solution:

Meter resistance $5 \times 1000 = 5000\ \Omega$

$$\frac{5\ K \times 10\ K}{15\ K} = \frac{50}{15} K$$

$$\frac{V_S \times \dfrac{50}{15} K}{50\ K + \dfrac{50}{15} K} = 5$$

$$V_S = \frac{5 \times 80}{50} = 80\ V$$

11. Ans. (d)

Solution:

12. Ans. (a)

Solution: Let the thevenins's equivalent network represent the DC network

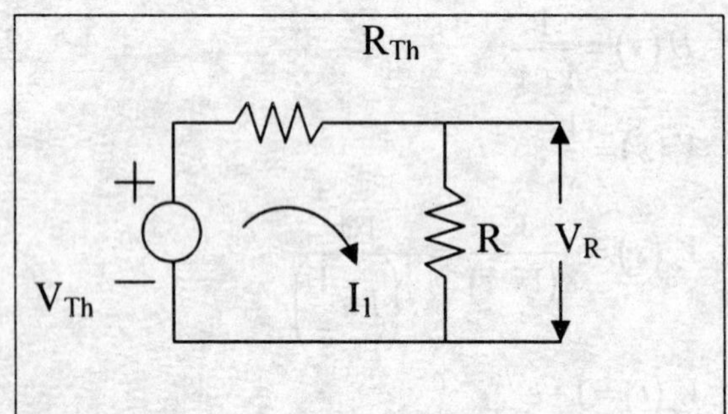

$I_1 = \dfrac{V_{R1}}{R} = \dfrac{20}{10} = 2A$

$V_{Th} = 2R_{Th} + 20$ (1)

Similarly, $I_2 = \dfrac{V_{R2}}{R} = \dfrac{30}{20} = 1.5\ A$

$V_{Th} = 1.5\ R_{Th} + 30$ (2)

From (1) and (2)

$2R_{Th} + 20 = 1.5R_{Th} + 30$

$0.5R_{Th} = 10$

i.e. $R_{Th} = 20\ \Omega$

From (1)

$V_{Th} = 2\ R_{Th} + 30$

$V_{Th} = 2 \times 20 + 20 = 60\ V$

Now, when R = 80 Ω as given then

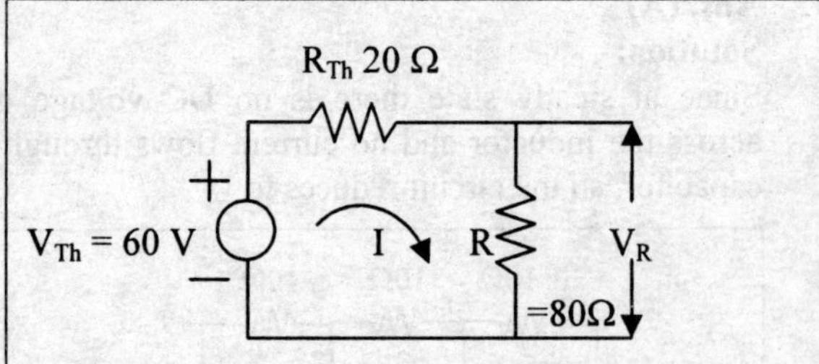

Current $I = \dfrac{60}{20+80} = 0.6\ A$

$\therefore\ V_R = 0.6 \times 80 = 48\ V$

13. Ans. (c)

Solution:

Remove the R and solve for thevenin's equivalent circuit

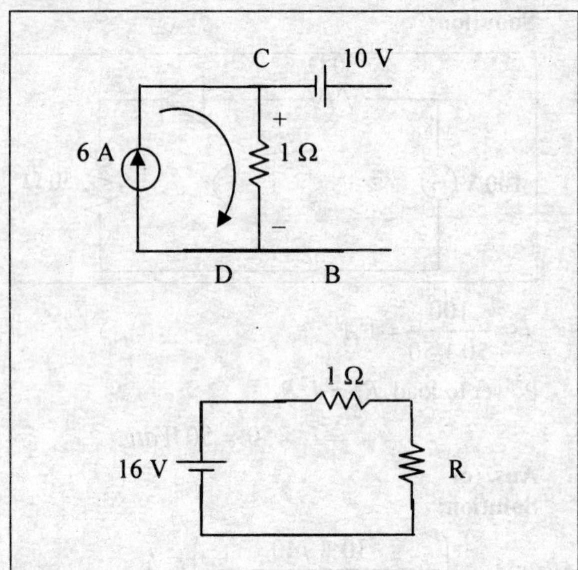

$R_s = 1\ \Omega$; $V_{CD} = 6\ V$

$V_{AB} = V_{OC} = 10 + 6 = 16\ V$

For maximum power transfer

$Z_L = Z_s$

$\therefore\ R = 1\ \Omega$

$P = I^2 R = \left(\dfrac{16}{2}\right)^2 \times 1 = 64$ Watt

14. Ans. (b)

Solution: $Z_c \rightarrow \dfrac{1}{sC}$

$$Z_{in} = \frac{3}{2s} + \left[5 \parallel \left(\frac{25}{2s} + \frac{5}{4}\right)\right]$$

$$Z_{in} = \frac{\left(\frac{25}{2s}+\frac{5}{4}\right)5}{\left(\frac{5}{4}\times\frac{25}{2s}\right)+5} + \frac{3}{2s}$$

$$\underset{s\to\infty}{Lt}\ Z_{in} = \frac{\left(0+\frac{5}{4}\right)5}{\frac{5}{4}+5} + 0 = \frac{\frac{25}{4}}{\frac{25}{4}} = 1$$

Second Method

As $t \to 0$

$C \to$ short circuit

Hence $Z_{in} = 5 \| \frac{5}{4}$

$= 1\,\Omega$

15. **Ans. (b)**
Solution: It is used for only linear network
16. **Ans. (d)**
17. **Ans. (b)**
Solution:

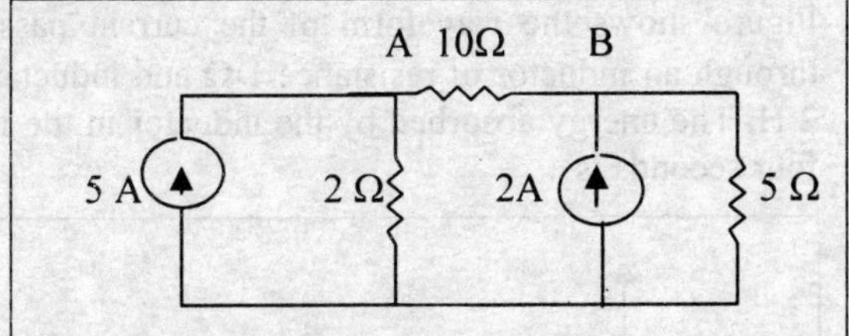

$$-5 + \frac{V_A}{2} + \frac{V_A - V_B}{10} = 0$$

$$-50 + 5V + V_A - V_B = 0$$

$$\Rightarrow 6V_A - V_B = 50 \qquad (1)$$

$$-2 + \frac{V_B - V_A}{10} + \frac{V_B}{5} = 0$$

$$-20 + V_B - V_A + 2V_B = 0$$

$$\Rightarrow -V_A + 3V_B = 20 \qquad (2)$$

From equation (1) and (2)

$$6V_A - V_B = 50$$

$$-6V_A + 18V_B = 120$$

$$-17V_B = 170$$

$$V_B = \frac{-170}{17} = 10$$

$$I_{5\Omega} = \frac{10}{5} = 2\ A$$

18. **Ans. (a)**
19. **Ans. (c)**
20. **Ans. (b)**
Solution:
Shorting the voltage source, we have

$$R_{12} = [20 \| 30] + [25 \| 25]\,k\Omega$$

$$= 24.5\,k\Omega$$

21. **Ans. (b)**
Solution:
Shorting the voltage source, we have

$$R_{12} = [20 \| 30] + [25 \| 25]\,k\Omega\,\text{t}$$

$$= 24.5\,k\Omega$$

22. **Ans. (b)**
Solution: Given circuit is

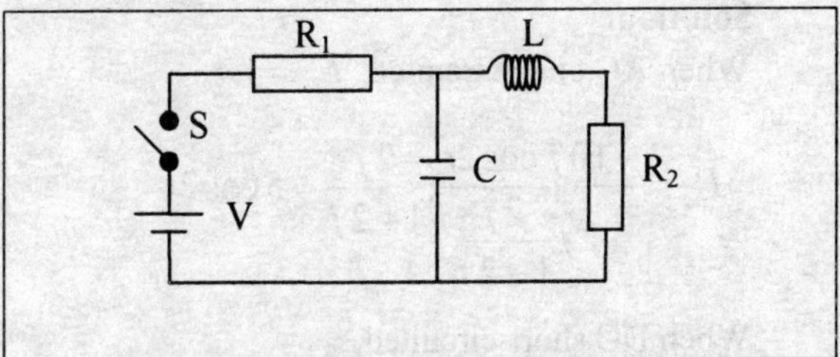

At $t \to 0^+$, inductor will act as open circuited and capacitor as short circuited

So supply current at $t = 0^+$ is $\frac{V}{R_1}$

At $t \to 0^+$, inductor will act as short circuited and capacitor as short circuited

$$V_c = \frac{V\,R_2}{R_1 + R_2}$$

23. **Ans. (d)**
Solution: KVL in input to output loop

$$\frac{0 - v_0}{10\,k\Omega} = \frac{v_{in} - 0}{1\,k\Omega}$$

$$\Rightarrow \qquad \frac{V_o}{V_{in}} \cong -10$$

∵ Gain is independent of resistance R

∴ Gain of amplifier changes negligible amount due to change in R.

24. Ans. (a)

Solution:

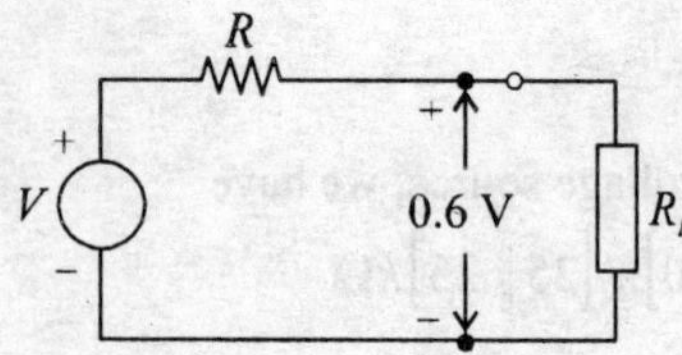

Here,

$$\frac{V}{R_{SC}} = \frac{R}{R+R_L} = 70\ mA$$

$$V - \frac{VR}{R+R_L} = 0.6$$

$$\frac{VR_L}{R+R_L} = 0.6$$

$$70\ mA\ R_L = 0.6$$

$$R_L = 8.6\ \Omega$$

25. Ans (a)

Solution:

When AC open circuited, $I_{dc} = 1A$

$$I_{ac} = \frac{-10j\cos 2t}{-2j\dfrac{2j}{1+2j}} \cdot \frac{2j}{1+2j} = 5\cos 2t$$

When DC short-circuited,

Total current through 1 Ω resistor,

$I = I_{dc} + I_{ac}$

$= (1 + 5\cos 2t)\,A$

26. Ans (d)

Solution:

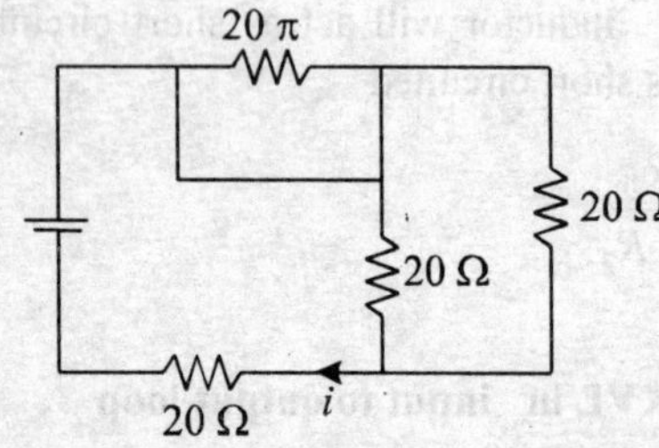

$t = 0-$

$$i = \frac{30}{(20/20) + 20}$$

The steady state condition can't change instaneously, therefore $i(t = 0-)$ will be equal $i(t = 0+)$

27. Ans (a)

Solution:

From givne compled circuit

$$V_1 = j\omega L_1 I_1 + j\omega M_{12} I_2 \quad (1)$$

$$V_2 = j\omega L_2 I_2 + j\omega M_{12} I_1 \quad (2)$$

Frow given network

$$V_1 = I_1 Z_1 + Z_3(I_1 + I_2) = (Z_1 + Z_3)I_1 + Z_3 I_2 \quad (3)$$

$$V_2 = I_2 Z_2 + (I_1 + I_2)Z_3 = (Z_2 + Z_3)I_1 + Z_3 I_1 \quad (4)$$

Comparing, equation (1) with (3) and (2) with (4) we get

$$Z_1 + Z_3 = j\omega L_1$$

$$Z_2 + Z_3 = j\omega L_2$$

$$Z_3 = j\omega M_{12}$$

$$Z_1 = j\omega(L_1 - M_{12})$$

$$Z_2 = j\omega(L_2 - M_{12})$$

$$Z_3 = j\omega M_{12}$$

One Mark Questions

1. Figure shows the waveform of the current passing through an inductor of resistance 1 Ω and inductance 2 H. The energy absorbed by the inductor in the first four seconds is

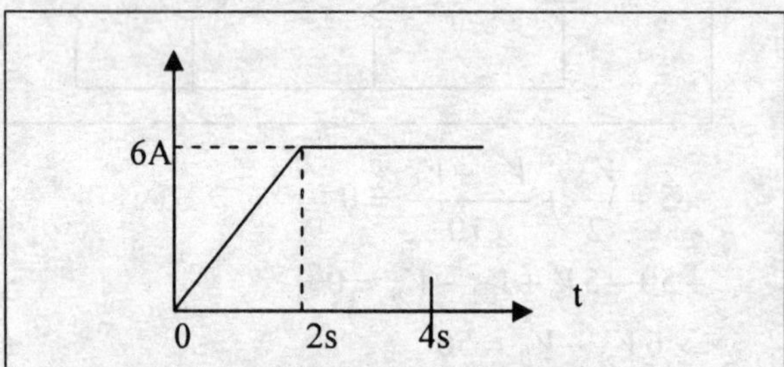

(a) 144 J (b) 98 J
(c) 132 J (d) 168 J

2. A segment of a circuit is shown in Figure. $V_R = 5V$, $V_C = 4 \sin 2t$. The voltage V_L is given by

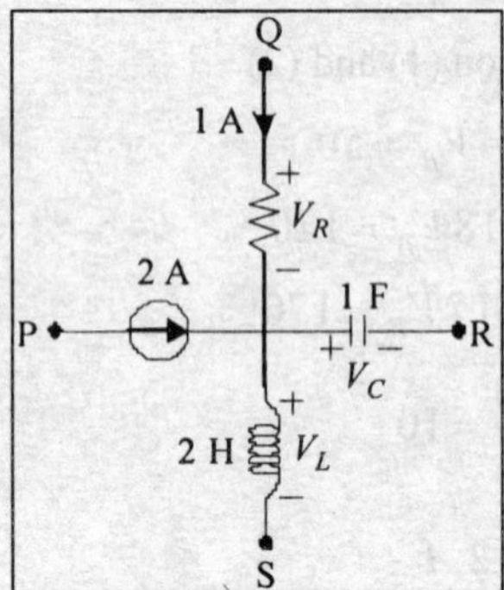

(a) $3 - 8\cos 2T$ (b) $32 \sin 2t$
(c) $16 \sin 2t$ (d) $16 \cos 2t$

3. In the Figure $Z_1 = 10 \angle -60°$, $Z_2 = 10 \angle 60°$, $Z_3 = 50 \angle 53.13°$. Thevenin impedance seen form X – Y is

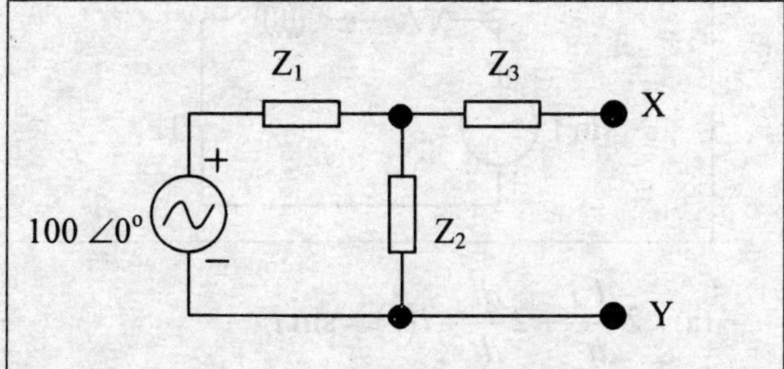

(a) $56.67 \angle 45°$ (b) $60\angle 30°$
(c) $70\angle 30°$ (d) $34.4\angle 65°$

4. The value of Z in figure which is most appropriate to cause parallel resonance at 500 Hz is

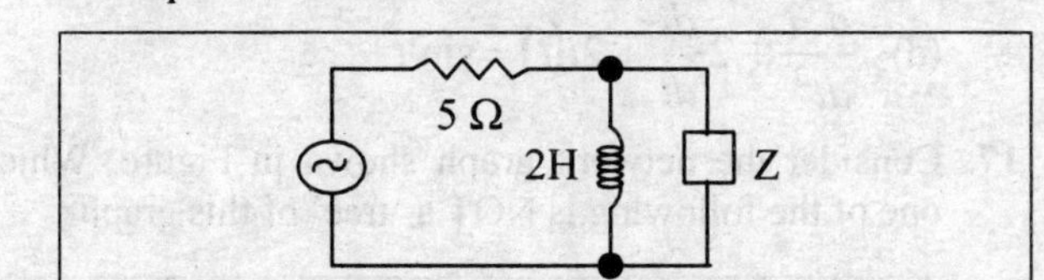

(a) 125.00 mH (b) 304.20 μ F
(c) 2.0 μF (d) 0.05 μ F

5. In the figure given below the value of R is

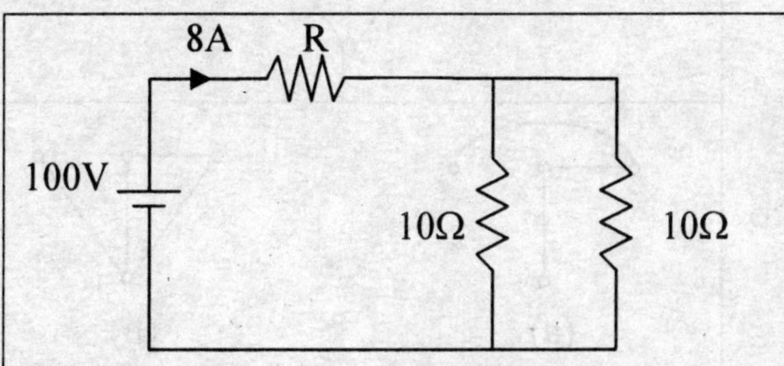

(a) 2.5 Ω (b) 5.0 Ω
(c) 7.5 Ω (d) 10.0 Ω

6. The RMS value of the voltage u(t) = 3+4cos (3t) is
(a) $\sqrt{17}$ V (b) 5 V
(c) 7 V (d) $\left(3+2\sqrt{2}\right)V$

7. For the two port network shown in the figure the Z-matrix is given by

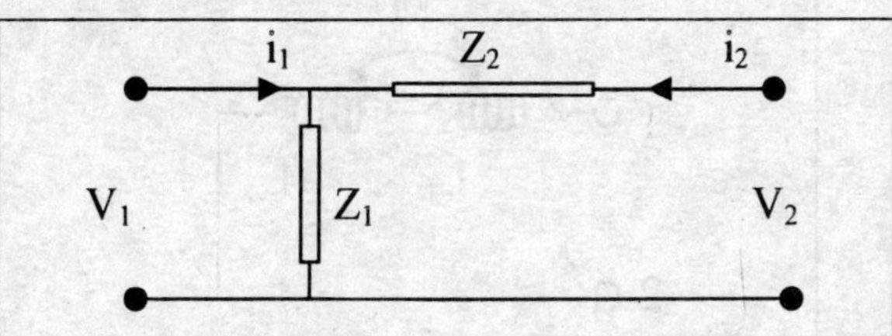

(a) $\begin{bmatrix} Z_1 & Z_1+Z_2 \\ Z_1+Z_2 & Z_2 \end{bmatrix}$ (b) $\begin{bmatrix} Z_1 & Z_1 \\ Z_1+Z_2 & Z_2 \end{bmatrix}$

(c) $\begin{bmatrix} Z_1 & Z_2 \\ Z_2 & Z_1+Z_2 \end{bmatrix}$ (d) $\begin{bmatrix} Z_1 & Z_1 \\ Z_1 & Z_1+Z_2 \end{bmatrix}$

8. In the figure given for, the initial capacitor voltage is zero. The switch is closed at t=0. The final steady-state voltage across the capacitor is

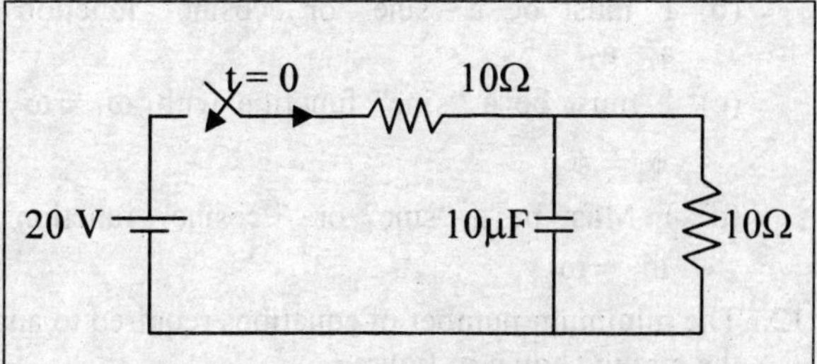

(a) 20 V (b) 10 V
(c) 5 V (d) 0 V

9. In the figure the current source is $1\angle 0\ A$, $R = 1\,\Omega$, the impedances are $Z_C = -j\Omega$, and $Z_1 = 2j\,\Omega$. The thevenin equivalent looking into the circuit across X-Y is

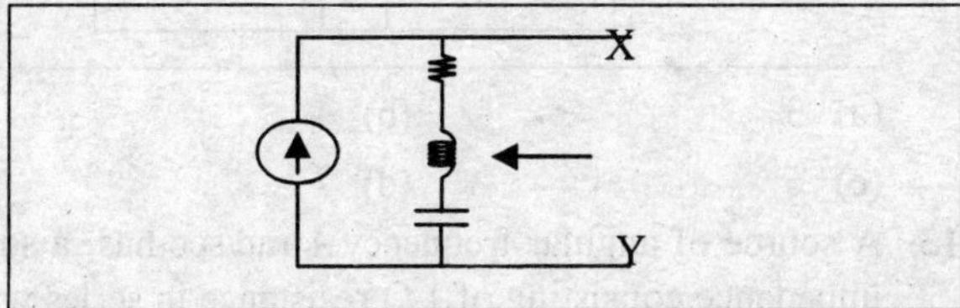

(a) $\sqrt{2}\angle 0\ V, (1+2j)\,\Omega$

(b) $2\angle 45^o\ V, (1-2j)\,\Omega$

(c) $2\angle 45^o\ V, (1+j)\,\Omega$

(d) $\sqrt{2}\angle 45^o\ V, (1+j)\,\Omega$

10. The three limbed non ideal core shown in the figure has three windings with nominal inductances L each when measured individually with a single phase AC source. The inductance of the winding as connected will be

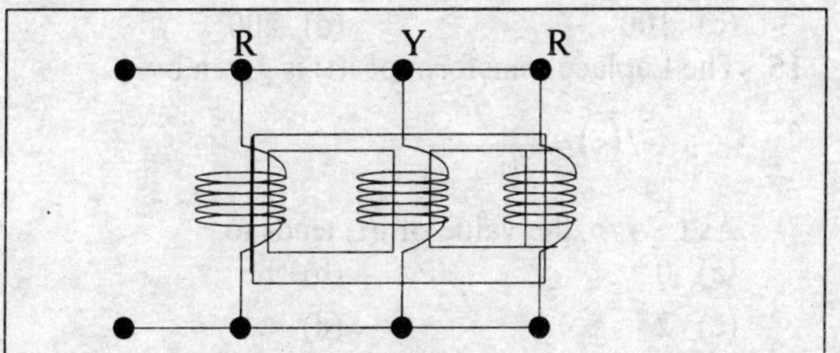

(a) very low (b) L/3
(c) 3L (d) very high

11. Let a signal $a_1 \sin(\omega_1 t + \phi_1)$ be applied to a stable **linear time-invariant** system. Let the corresponding steady state output be represented as $a_2 F(\omega_2 t + \phi_2)$. Then which of the following statements is true?

(a) F is not necessarily a "sine" or "cosine" function but must be periodic with $\omega_1 = \omega_2$
(b) F must be a "sine" or "cosine" function with $a_1 = a_2$
(c) F must be a "sine" function with $\omega_1 = \omega_2$ and $\phi_1 = \phi_2$
(d) F Must be a "sine" or "cosine" function with $\omega_1 = \omega_2$

12. The minimum number of equations required to analyze the circuit shown in figure.

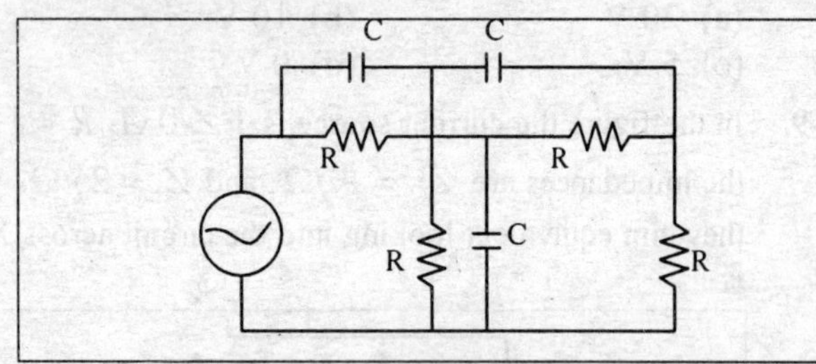

(a) 3 (b) 4
(c) 6 (d) 7

13. A source of angular frequency 1 rad/sec has a source impedance consisting of 1 Ω resistance in series with 1 H inductance. The load that will obtain the maximum power transfer is

(a) 1 Ω resistance
(b) 1Ω resistance in parallel with 1 H inductance
(c) 1Ω resistance in series with 1 F capacitor
(d) 1Ω resistance in parallel with 1 F capacitor

14. A series RLC circuit has resonance frequency of 1 kHz and a quality factor Q = 100. If each of R, L and C is doubled form its original value, the new Q of the circuit is

(a) 25 (b) 50
(c) 100 (d) 200

15. The Laplace transform of i(t) is given by

$$I(s) = \frac{2}{s(1+s)}$$

As t → ∞, the value of i(t) tends to

(a) 0 (b) 1
(c) 2 (d) ∞

16. The differential equation for he current i(t) in the circuit of figure is

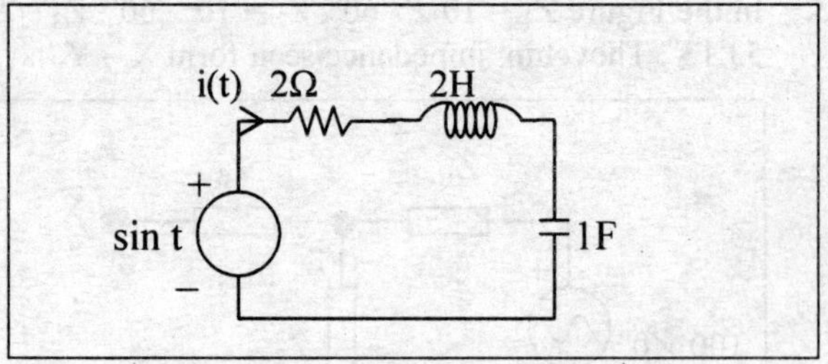

(a) $2\frac{d^2i}{dt^2} + 2\frac{di}{dt} + i(t) = \sin t$
(b) $2\frac{d^2i}{dt^2} + 2\frac{di}{dt} + i(t) = -\cos t$
(c) $2\frac{d^2i}{dt^2} + 2\frac{di}{dt} + i(t) = \cos t$
(d) $\frac{d^2i}{dt^2} + 2\frac{di}{dt} + 2i(t) = \sin t$

17. Consider the network graph shown in Figure. Which one of the following is NOT a 'tree' of this graph?

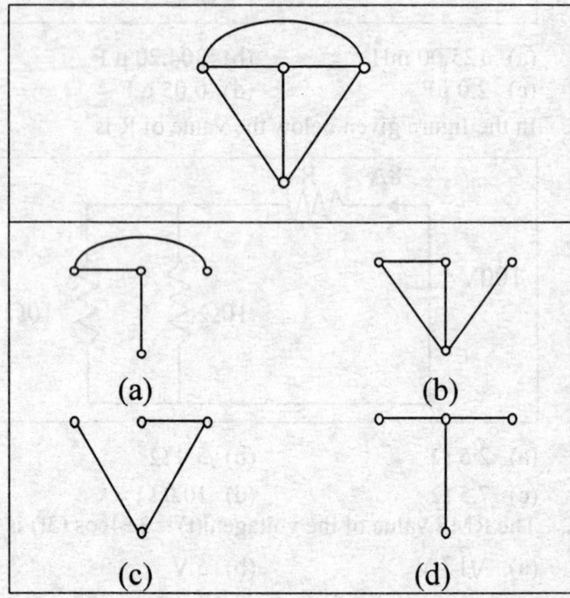

18. The equivalent inductance measured between the terminals 1 and 2 for the circuit shown in figure is

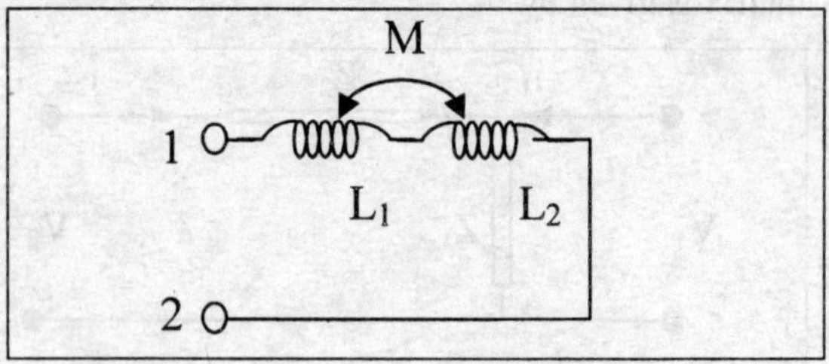

(a) $L_1 + L_2 + M$ (b) $L_1 + L_2 - M$
(c) $L_1 + L_2 + 2M$ (d) $L_1 + L_2 - 2M$

19. The circuit shown in figure with R = (1/3) Ω, L = (1/4) H, C = 3F has input voltage v(t) = sin 2t. The resulting current i(t) is

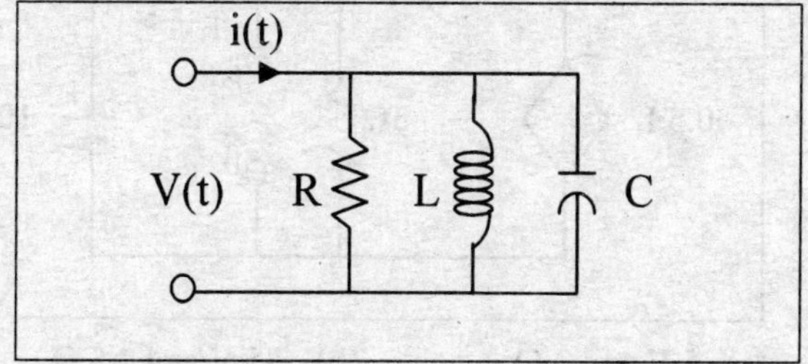

(a) 5 sin (2t + 53.1°)
(b) 5 sin (2t - 53.1°)
(c) 25 sin (2t + 53.1°)
(d) 25 sin (2t - 53.1°)

20. For the circuit shown in figure (a) the time constant RC = 1ms. The input voltage is

$V_i(t) = \sqrt{2}\ \sin 10^3\ t.$

The output voltage $v_o(t)$ is equal to

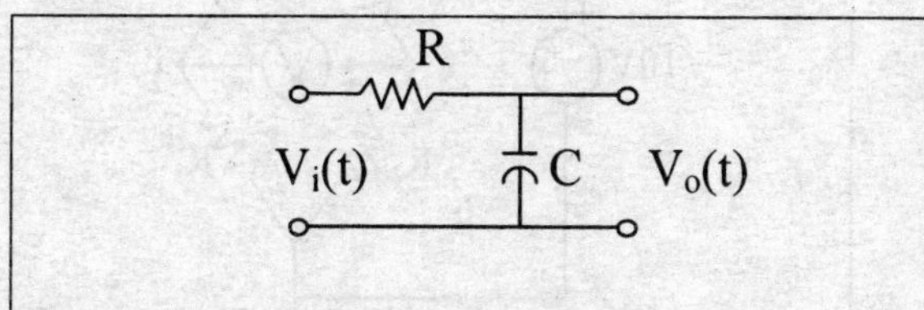

(a) $\sin(10^3t - 45^o)$ (b) $\sin(10^3t + 45^o)$
(c) $\sin(10^3 - 53^o)$ (d) $\sin(10^3t - 53^o)$

21. For the R-L circuit shown in Figure the input voltage $v_i(t) = u(t)$. The current i(t) is

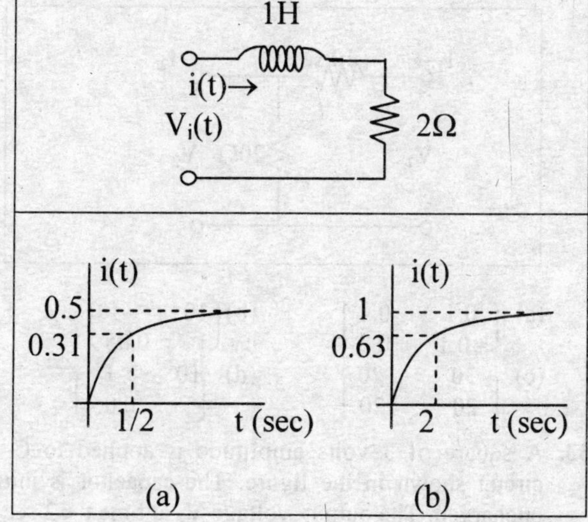

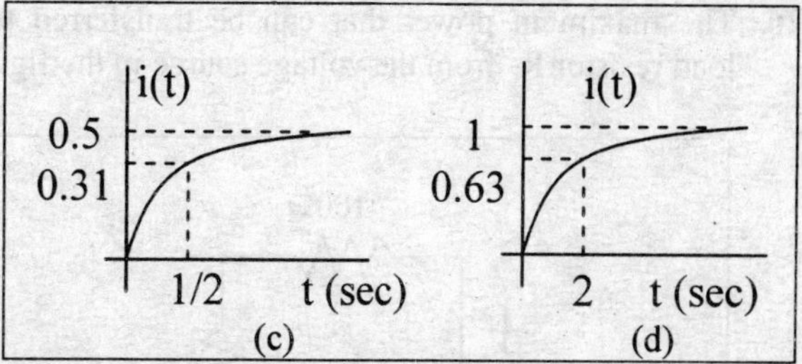

22. The Z-transform of the following real exponential sequence:

$x(n\ T) = a^n\ ;\quad nT > 0$
$x(n\ T) = 0\ ;\quad nT < 0, \eta > 0$

is given by

(a) $1 - z^{-1}\,4\,;\,|\,z\,| > a$
(b) $\dfrac{1}{1 - a\,z^{-1}}\ ;\ |\,z\,| > a$
(c) 1 for all z
(d) $1 - az^{-1}\ ;\ |\,z\,| > \eta$

23. The condition on R, L and C such that the step response y(t) in the figure has no oscillations, is

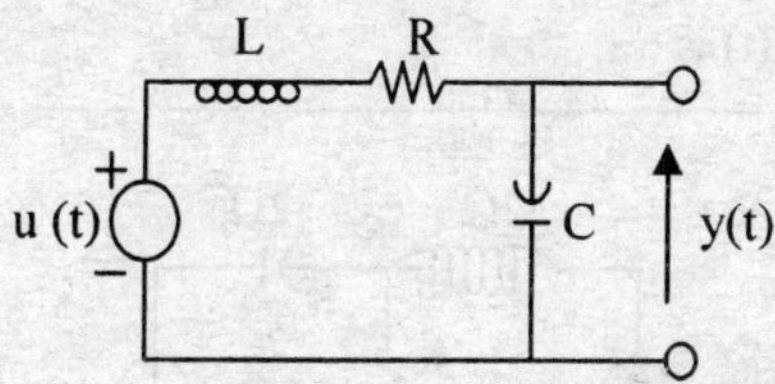

(a) $R \geq [1/2]\sqrt{(L/C)}$ (b) $R \geq \sqrt{(L/C)}$
(c) $R \geq 2\sqrt{(L/C)}$ (d) $R = [1/\sqrt{(LC)}]$

24. The ABCD parameters of an ideal n: 1 transformer shown in the figure are $\begin{pmatrix} n & 0 \\ 0 & X \end{pmatrix}$,

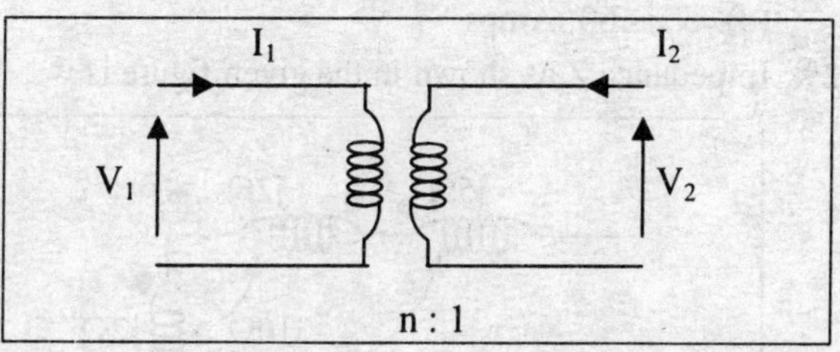

The value of X will be

(a) n (b) [1 / n]
(c) n^2 (d) $[1/n^2]$

25. In a series RLC circuit, R = 2kΩ, L = 1 H, and C = [1 / 400] μF. The resonant frequency is

(a) 2×10^4 Hz (b) $[1/\pi] \times 10^4$ Hz
(c) 10^4 Hz (d) $2\pi \times 10^4$ Hz

26. The maximum power that can be transferred to the load resistor R_L from the voltage source in the figure is

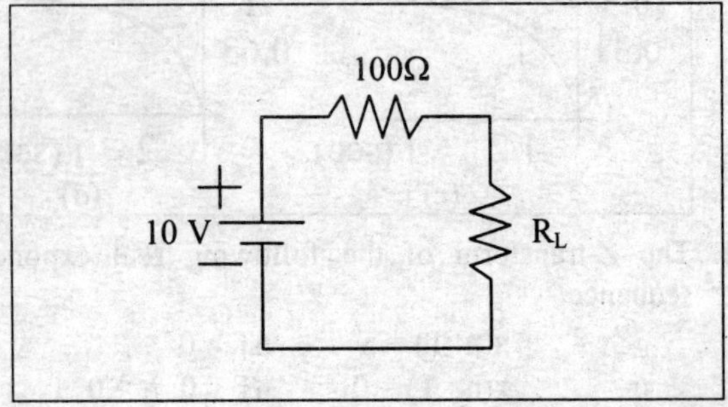

(a) 1 W (b) 10 W
(c) 0.25 W (d) 0.5 W

27. The first and the last critical frequency of an RC-driving point impedance function must respectively be

(a) A zero and a pole
(b) A zero and a zero
(c) A pole and a pole
(d) A pole and a zero

28. For the circuit shown in the figure, the instantaneous current $i_1(t)$ is

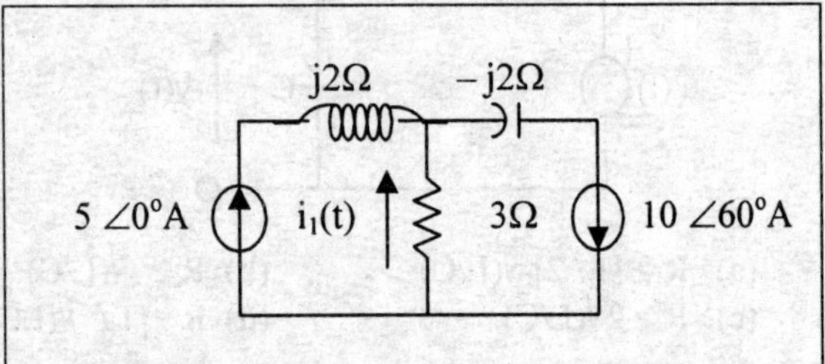

(a) $[(10\sqrt{3})/2]\angle 90^\circ$ Amps
(b) $[(10\sqrt{3})/2]\angle -90^\circ$ Amps
(c) $5\angle 60^\circ$ Amps
(d) $5\angle -60^\circ$ Amps

29. Impedance Z as shown in the given figure is

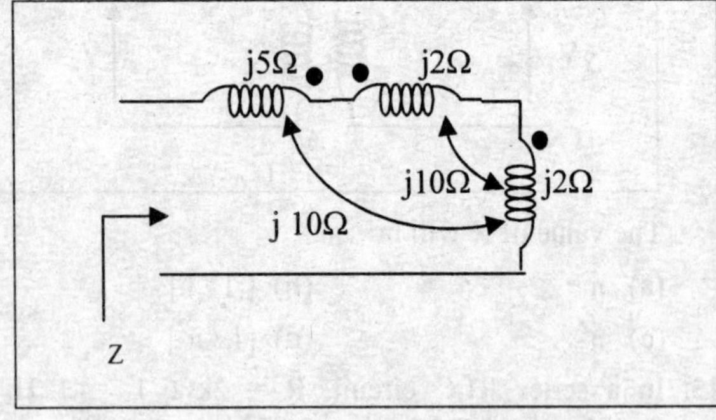

(a) j29Ω (b) j9 Ω
(c) j19Ω (d) j39Ω

30. For the circuit shown in the figure, Thevenin's voltage and Thevenin's equivalent resistance at terminals a – b is

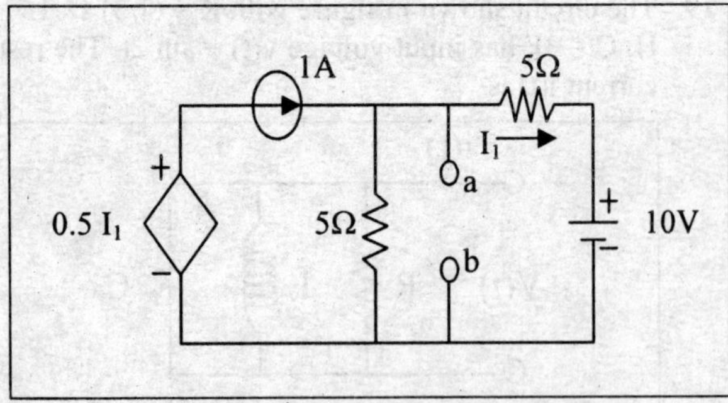

(a) 5 V and 2 Ω (b) 7.5 V and 2.5 Ω
(c) 4 V and 2 Ω (d) 3 V and 2.5 Ω.

31. IF $R_1 = R_2 = R_4 = R$ and $R_3 = 1.1R$ in the bridge circuit shown in the figure, then the reading in the ideal voltmeter connected between a and b is

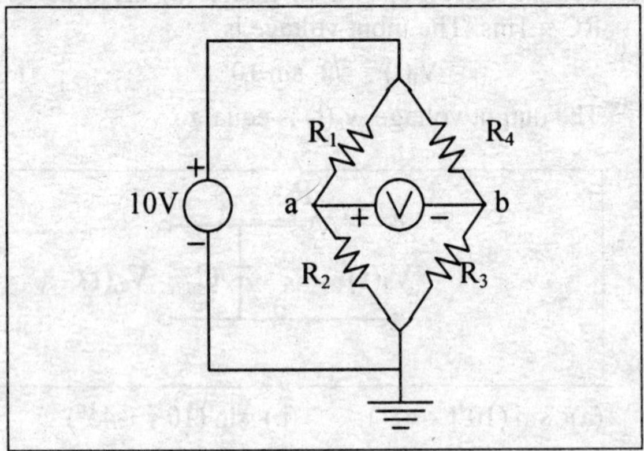

(a) 0.238 V (b) 0.138 V
(c) –0.238 V (d) 1 V

32. The h parameters of the circuit shown in the figure are

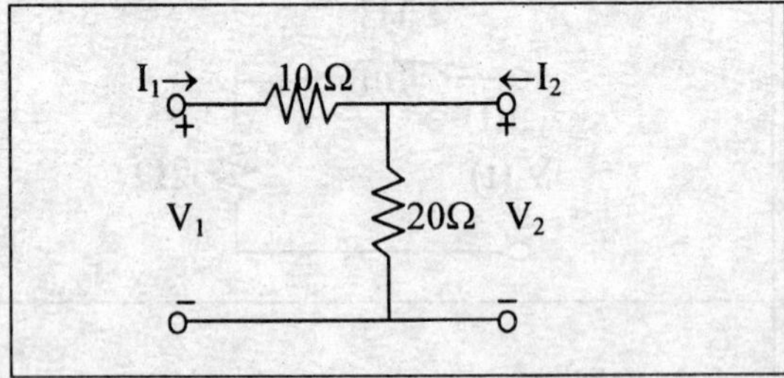

(a) $\begin{bmatrix} 0.1 & 0.1 \\ -0.1 & 0.3 \end{bmatrix}$ (b) $\begin{bmatrix} 10 & -1 \\ 1 & 0.05 \end{bmatrix}$

(c) $\begin{bmatrix} 30 & 20 \\ 20 & 20 \end{bmatrix}$ (d) $\begin{bmatrix} 10 & 1 \\ -1 & 0.05 \end{bmatrix}$

33. A square of 3 volts amplitude is applied to C – R circuit shown in the figure. The capacitor is initially uncharged. The output voltage V_2 at time t = 2 see is

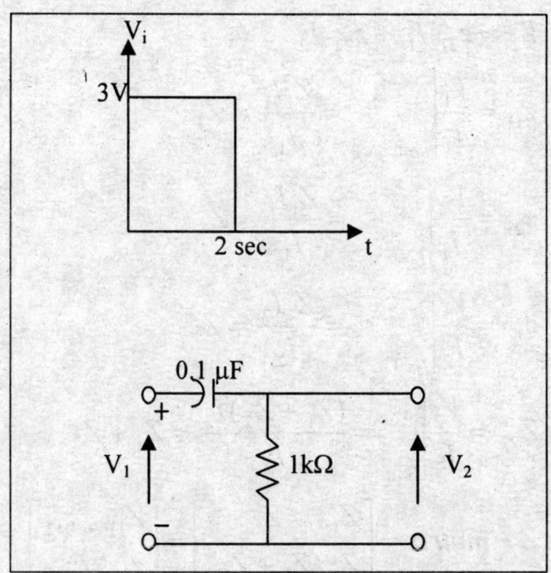

(a) 3 V (b) - 3 V
(c) 4 V (d) - 4 V

34. In the system shown below, x(t) = (sin t) u(t) . In steady state, the response y (t) will be.

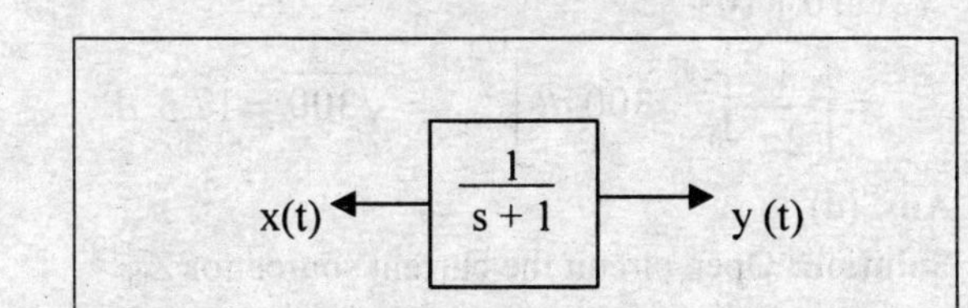

(a) $(1/\sqrt{2}) \sin [t - (\pi/4)]$
(b) $(1/\sqrt{2}) \sin [t + (\pi/4)]$
(c) $(1/\sqrt{2}) e^{-1} \sin t$
(d) $\sin t - \cos t$

35. An independent voltage source in series with an impedance $Z_s = R_s + jX_s$ delivers a maximum average power to a load impedance Z_L when

(a) $Z_L = R_s + jX_s$ (b) $Z_L = R_s$
(c) $Z_L = jZ_s$ (d) $Z_L = R_s - jX_s$

36. The RC circuit shown in the figure is

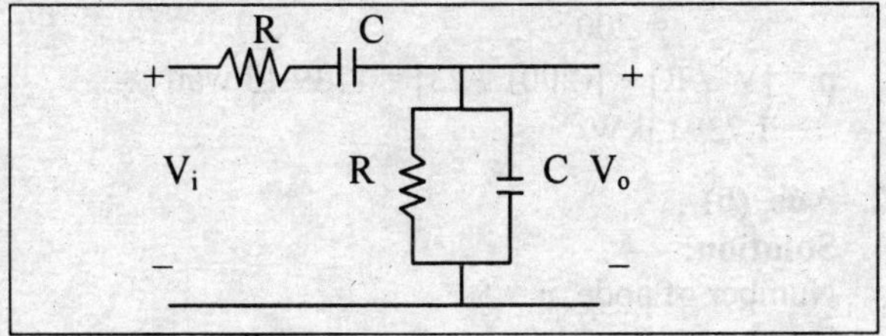

(a) A low-pas filter (b) A high-pass filter
(c) A band-pass filter (d) A band-reject filter

Answers
Basic of Circuit and Measurement System

1. Ans. (a)

Solution: For 0 –2 sec, current. 1 changes with i.e.

$$\frac{di}{dt} = \frac{6}{2} (amp)/s$$

$$V_L = L\frac{di}{dt} = 2 \times \frac{6}{2} = 6 \text{ V},$$

$W = v\,i\,t = 6 \times 6 \times 2 = 72$ J

For 2 – 4 sec; for steady current, V = I × R = 6 × 1 = 6 V

W = 6 × 6 ×2 = 72 J

Total energy absorbed = 72 J + 72J = 144 J

2. Ans. (b)

Solution: Current through capacitor

$$i_c = C\frac{dv_c}{dt} = 1 \times \frac{d}{dt}(4 \sin 2t) = 8 \cos 2t$$

KCL at node

$$-2 + 1 - 8 \cos 2t + i_L = 0$$

$$i_L = 1 + 8 \cos 2t$$

$$v_L = L\frac{di_L}{dt} = 2\frac{d}{dt}(1 + 8 \cos 2t) = -32 \sin 2t$$

3. Ans. (a)

Solution: $Z_1 = 10 \angle -60^\circ = 10 \cos(-60) + j \sin(-60)]$

$Z_2 = 10 \angle 60^\circ = [10 \cos 60 + \sin 60]$;

$Z_3 = 50 \angle 53.13^\circ = 50 [\cos 53.13 + j \sin 53.13]$

$$Z_1 \| Z_2 = \frac{(5 - j8.66)(5 + j8.66)}{5 - j8.66 + 5 + j8.66} = \frac{25 + 75}{10} = 10$$

$Z_{Th} = [Z_1 \| Z_2] + Z_3 = 10 + 30 + j\,40 = 40 + j\,40$

$$= \sqrt{40^2 + 40^2} \angle 45^\circ = 56.6 \angle 45^\circ$$

4. Ans. (d)

Solution: Resistance does not play any role for resonance frequency. The given circuit is parallel LC circuit,

Admittance of given circuit,

$$Y = Y_L + Y_Z = \frac{1}{J\omega L} + \frac{1}{Z}$$

At resonance, j term must be equal to zero i.e.

$$\frac{1}{j\omega L}+\frac{1}{Z}=0$$

or $\frac{-1}{j\omega L}=\frac{1}{Z}$ (1)

or $\frac{j}{\omega L}=\frac{1}{Z}$

or $Z=\frac{\omega L}{j}$ i.e. Capacitive

or $\frac{1}{j\omega L}=\frac{\omega L}{j}$

$$C=\frac{1}{\omega^2 L}=0.05\,\mu F.$$

5. Ans. (c)

Solution. $10\,\Omega \,||\, 10\,\Omega = 5\,\Omega$

KVL in loop, $-100+(R+5)\,8=0$

or $-100+8R+40=0$

or $R=\frac{60}{8}=7\cdot5$

6. Ans. (a)

Solution. Given that u (t) = 3 + 4 cos 3t

Which mean, $\omega = 3$; $\omega = 2\pi f = \frac{2\pi}{T}$

$$T=\frac{2\pi}{\omega}=\frac{2\pi}{3}$$

$$U_{RMS}=\sqrt{\frac{1}{T}\int_O^T[u(t)]^2}$$

$$=\sqrt{\frac{3}{2\pi}\int_o^{2\pi/3}(3+4\cos 3t)^2}$$

$$=\sqrt{17}$$

7. Ans. (d)

Solution:

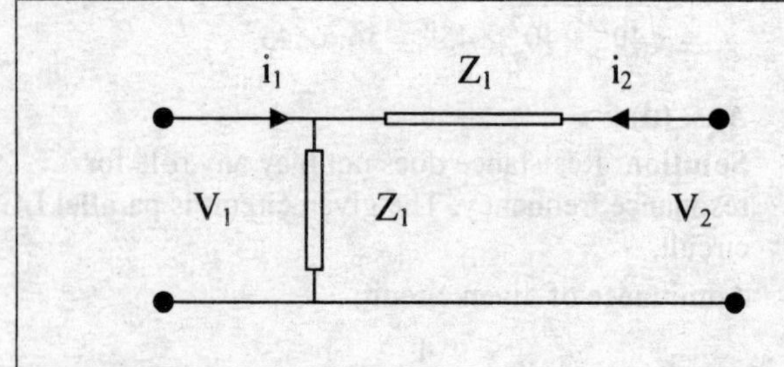

Z – Parameter represented as

$$V_1 = z_{11}\, I_1 + z_{12}\, I_2$$

$$V_2 = z_{21}\, I_1 + z_{22}\, I_2$$

$$z_{11}=\left.\frac{V_1}{I_1}\right|_{I_2=0} = \frac{Z_1 I_1}{I_1}=Z_1$$

$$z_{21}=\left.\frac{V_2}{I_1}\right|_{I_2=0} = \frac{Z_1 I_1}{I_1}=Z_1$$

$$z_{12}=\left.\frac{V_1}{I_2}\right|_{I_1=0} = \frac{Z_1 I_2}{I_2}=Z_1$$

$$Z_{22}=\left.\frac{V_2}{I_2}\right|_{I_1=0} = \frac{(Z_1+Z_2)I_2}{I_2}=Z_1+Z_2$$

$$Z-matrix=\begin{bmatrix} Z_1 & Z_1 \\ Z_1 & Z_1+Z_2 \end{bmatrix}=\begin{bmatrix} Z_{11} & z_{12} \\ z_{21} & z_{22} \end{bmatrix}$$

8. Ans. (b)

Solution: At t = 0, the capacitor voltage $V_c = 0$

At steady state, capacitor is open circuit ie at $t \to \infty$,

$$V_c=\frac{10}{10+10}\times 20 = 10V$$

$$=\left[\frac{1}{2\pi}\int_o^{2\pi}300.dt\right]^{1/2} = \sqrt{300}=17.3\ A.$$

9. Ans. (d)

Solution: Open circuit the current source for Z_{th}

$Z_{th} = 1 - j + j2 = (1 + j)\,\Omega = \sqrt{2}\,\angle 45^\circ$

$V_{th} = \sqrt{2}\,\angle 45^\circ \times 1\angle 0^\circ = \sqrt{2}\,\angle 45^\circ$ V

10. Ans. (d)

11. Ans. (b)

Solution: Energy = Power × time

$$\text{Power}=\frac{(\text{Voltage})^2}{\text{Resistance}}$$

$E = [(V^2/R)]\,t$

$2300 = [(230)^2/R] \times 1$

$\Rightarrow R = 23\,\Omega$

rms value = $\frac{200^2 \times (T/2) + (-200)^2 \times (T/2)}{T}$

= 200

$p = [V^2/R] = [(200)^2/23] = 1739.13$ watt

= 1.73913kW

12. Ans. (b)

Solution:

Number of node n = 5

Number of branches b = 8

Number of KCL equation $= (n - 1) = (5 - 1) = 4$
Number of KVL equation $= b - n + 1 = 8 - 5 + 1 = 4$

13. Ans. (c)

Solution : Source impedance $R_s = R + j\omega L = 1 + j$
For max power transfer $R_L = R_s^* = 1 - j$

The above equation represent 1 Ω resistance with 1 F capacitor i.e.

$1\Omega \quad 1\,F$; $\omega = 1$

$[R + (1/j\omega C)] = [1 + (1/j)] = 1 - j$

i.e., for maximum power transfer the load impedance should be complex conjugate of source impedance

14. Ans. (b)

Solution: Resonance frequency $\omega_o = 2\pi f_o = [1/\sqrt{(LC)}]$

Quality factor $Q = [1/(\omega_o RC)] = [\omega_o L/R]$
If R, L, C are doubled then $\omega'_o = [\omega_o/2]$

$Q' = [1/\omega_o' R'C'] = [\omega_o' L'/R']$

$$= \frac{2}{\omega_o \times 2R \times 2C} = \frac{1}{2\omega_o RC} = \frac{Q_o}{2}$$

$= [100/2] = \mathbf{50}$

15. Ans. (c)

Solution: $i(\infty) = \underset{t\to\infty}{Lt}\, i(t) = \underset{s\to 0}{Lt}\left[sI(s)\right]$

$$= \underset{s\to 0}{Lt} \cdot s \times \frac{2}{s(s+1)} = \underset{s\to 0}{Lt} \cdot \frac{2}{1+s} = 2$$

16. Ans. (c)

Solution: Using KVL equation the loop
$V(t) = R\,i(t) + L\,[di(t)/dt] + [1/c]\int i(t)\,dt$
$\sin t = 2\,i(t) + 2\,[di(t)\,dt] + \int i(t)\,dt$

Differentiate the above equation

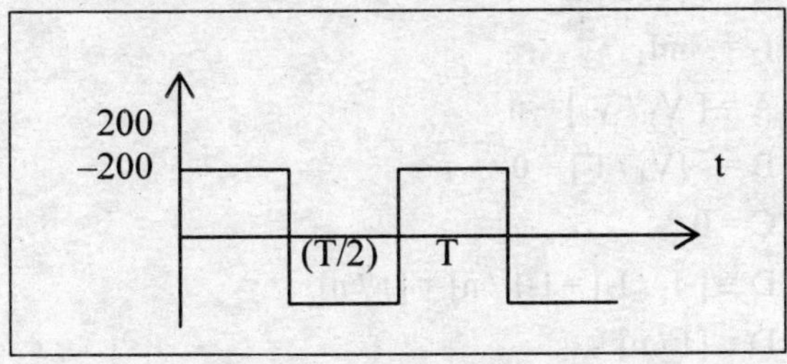

$\cos t = 2\,[di(t)/dt] + 2\,[d^2 i(t)\,dt^2] + i(t)$

17. Ans. (b)

Solution: In a tree there is a no loop but (b) contain close loop.

18. Ans. (d)

Solution:

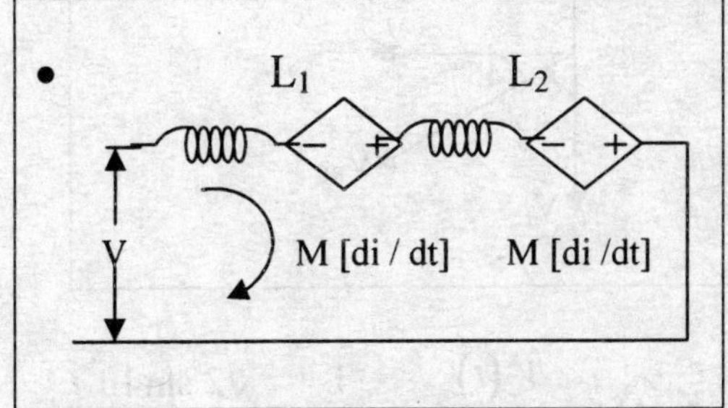

$V = L_1 \frac{di}{dt} - M\frac{di}{dt} + (L_2 - M)\frac{di}{dt}$

$= (L_1 + L_2 - 2M)\,[di/dt] = L\,[di/dt]$

$L = L_1 + L_2 - 2M$

Second Method

$L_{eq} = L_1 + L_2 \pm 2M$

NOTE: + for positively complex and – for negative complex circuit

Hence compiling is negative as dots are on opposite ends.

19. Ans. (a)

Solution:

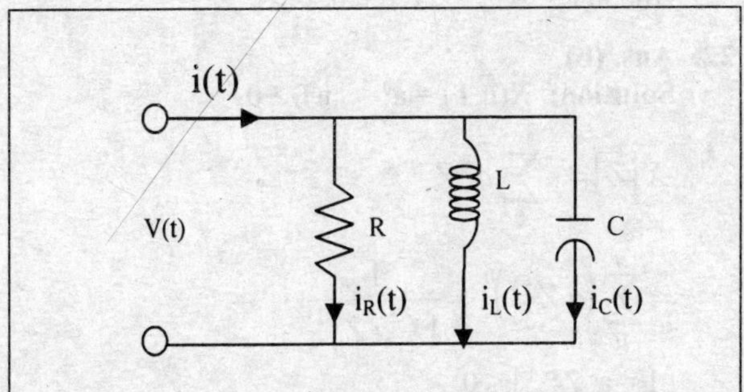

$\omega = 2$

$i(t) = i_R(t) + i_L(t) + i_c(t)$

$$i(t) = \frac{v(t)}{R} + \frac{v(t)}{j\omega L} + v(t) \times (j\omega C)$$

$\Rightarrow \quad i(t) = \sin 2t\,[3 - 2j + 6j] = (3 + 4j)\,\sin 2t$

$= \sqrt{(3^2 + 4^2)}\,\sin 2t \;\angle \tan^{-1}[4/3]$

$i(t) = 5\sin(2t + 53.1^\circ)$

20. Ans. (a)

Solution:

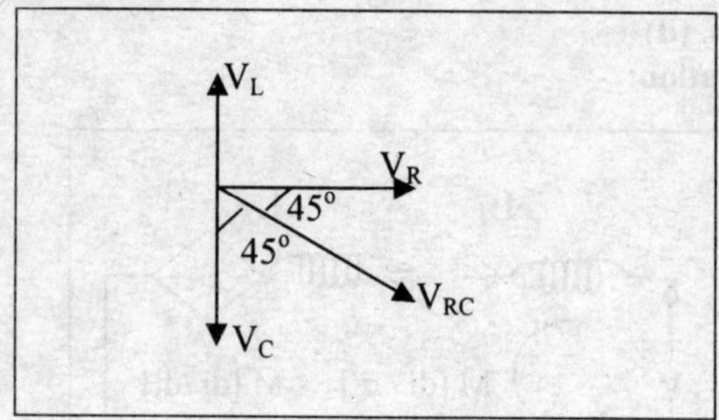

$$V_o(t) = \frac{V_i(t)}{R + \frac{1}{j\omega C}} \times \frac{1}{j\omega C} = \frac{\sqrt{2}\sin 10^3 t}{j+1}$$

$= \sin(10^3 t - 45^o)$

21. Ans. (c)

Solution: $V_i(t) = R\, i(t) + L\,[di(t)/dt]$

With Laplace transformation, we have

$V_i(s) = [R + Ls]\, I(s)$

Or, $$I(s) = \frac{V_i(s)}{R + Ls} = \frac{1}{s(2+s)}$$

[Since, $V_i(s) = [1/s]$]

In time-domain, $i(t) = [u(t)/2]\,[1 - e^{-2t}]$

At $t = [1/2]$, $i(t) = 0.31$

And at $t = \infty$, $i(t) = 0.5$

22. Ans. (b)

Solution: $X(nT) = a^n$; $nT > 0$

$$X[Z] = -\sum_{0}^{\infty} a^n Z^{-n}$$

$$= \sum_{0}^{n} \left(a Z^{-1}\right)^n = \frac{1}{1 - a Z^{-1}}$$

$| 1 - a\, Z^{-1} | < 0$

$| a Z^{-1} | < 1$

$| [a/Z] | < 1$; $Z > a$

23. Ans. (c)

Solution: $u(t) = Ri + [L\, di/dt] + [1/C] \int i\, dt$

By differentiating , 0 =

$0 = [R\,di/dt] + [L\, d^2 i/(dt)^2] + [1/C]\, i$

$$\frac{d^2 i}{dt^2} + \left(\frac{R}{L}\right)\frac{di}{dt} + \left(\frac{1}{LC}\right) i = 0$$

Roots are,

$$\frac{-R}{2L} \pm \sqrt{\left(\frac{R}{2L}\right)^2 - \frac{1}{LC}}$$

For no oscillations, $[R/2L]^2 \geq [1/\{LC\}]$

$R \geq 2\sqrt{(L/C)}$

Second Method

Transfer function is

$$\frac{Y(s)}{U(s)} = \frac{\frac{1}{sC}}{R + sL + \frac{1}{sC}}$$

$$= \frac{1}{s^2 LC + sCR + 1}$$

$$= \frac{\frac{1}{LC}}{s^2 + \frac{R}{L}s + \frac{1}{LC}}$$

Comparing with standard 2nd order system $s^2 + 2\delta_n s + \omega_n^2 = 0$ we have

Here $$2\delta\omega_n = \frac{R}{L}$$

And $$\omega_n = \frac{1}{\sqrt{LC}}$$

Thus $$\delta = \frac{R}{2L}\sqrt{LC} = \frac{R}{2}\sqrt{\frac{C}{L}}$$

For no oscillation $\delta \geq 1$

Thus $$\frac{R}{2}\sqrt{\frac{C}{L}} \geq 1$$

or $$R \geq 2\sqrt{\frac{C}{L}}$$

24. Ans. (b)

Solution: $n_1 = n$; $n_2 = 1$

For and ideal transformer.

$V_1 = n_1 V_2$; $V_2 = n_2 V_1$

$I_1 = -[1/n]\, I_2$

$I_2 = -nI_1$

$A = [V_1/V_2] = n$

$B = -[V_1/I_2] = 0$

$C = 0$

$D = [-I_1/I_2] = [-1/n] = [1/n]$

$D = [1/n]$

25. Ans. (b)

Solution: $f = [1/2\pi]\sqrt{(1/LC)}$

$= [1/2\pi] \,.\, 1\sqrt{(1 \times [1/400])} \times 10^{-6}$

$= [20/2\pi] = [10^4/\pi]$

26. Ans. (c)

Solution: $[V^2 / 4 R_L] = [102 / (4 \times 100)]$

$= [1/4] = 0.25$ W

27. Ans. (d)

Solution:

The first and the last critical frequency of an RC-driving point impedance function must respectively be a pole and a zero

28. Ans. (a)

Solution:

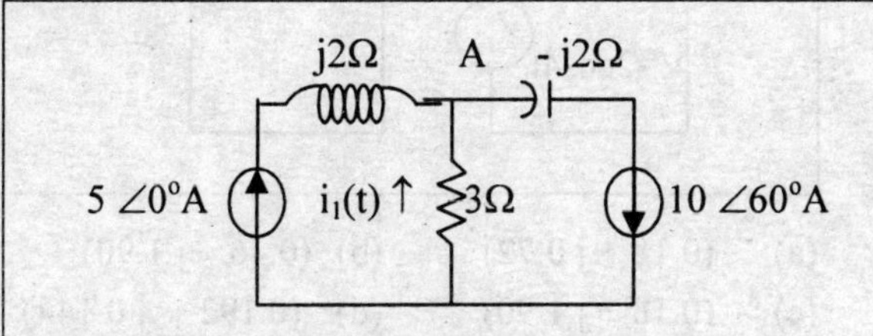

By KCL at Node A

$-5\angle 0^\circ - I_1 + 10\angle 60^\circ = 0$

$I_1 = 10\angle 60^\circ - 5\angle 0^\circ$

$= 10 [\text{Cos } 60 + j \text{ Sin } 60] - 5$

$I_1 = 10 [1/2] + 10 [j\sqrt{3}/2] - 5$

$I_1 = \frac{10\sqrt{3}}{2} \angle 90^\circ$ amp

29. Ans. (b)

Solution: $Z = j5\ \Omega + j2\Omega + j2\Omega + 2\ (j10) - 2\ (j10)$

$= \mathbf{j\ 9\Omega}$

30. Ans. (b)

Solution:

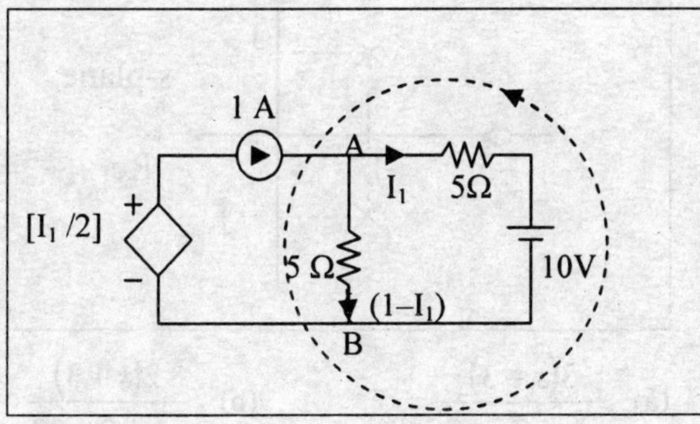

By KVL, in loop

$10 = -5I_1 + 5(1 - I_1) = 5 - 10 I_1$

$I_1 = -[1/2]$

$V_{ab} = 5(1 - I_1) = 5[1 + (1/2)]$

$\mathbf{V_{ab} = 7.5}$

31. Ans (c)

Solution:

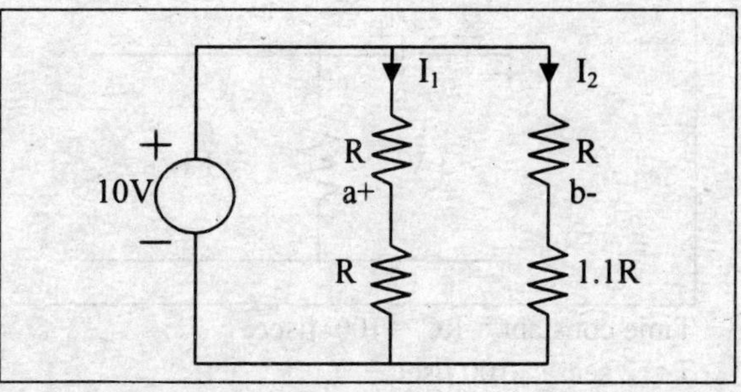

$I_1 = [10/2R] = [5/R]$

$I_2 = [10/2.1R]$

$= 4.76 / R$

$V_{ab} = -R I_1 + I_2 R$

$= R [I_2 - I_1]$

$= [R/R] (4.76 - 5)$

$\mathbf{V_{ab} = -0.238}$

32. Ans. (d)

Solution:

h – parameter standard equation

$V_1 = h_{11} I_1 + h_{12} V_2$ (1)

$I_2 = h_{21} I_1 + h_{22} V_2$ (2)

$V_1 - 10I_1 - 20(I_1 + I_2) = 0$; $V_2 - 20(I_1 + I_2) = 0$

$\Rightarrow V_1 - 30I_1 - 20I_2 = 0$; $V_2 - 20 I_1 - 20I_2) = 0$

$V_1 = 30I_1 + 20I_2$ (3)

$V_2 = 20 I_1 + 20I_2$ (4)

$\Rightarrow 20I_2 = V_1 - 30I_1$; $\Rightarrow 20 I_1 = V_2 - 20I_1$

$V_1 - V_2 = 10I_1 \Rightarrow V_1 = V_2 + 10I_2$

$V_1 = 10I_1 + V_2$ (5)

Now, $V_1 = h_{11} I_1 + h_{12} V_2$ (1)

Comparing equation (1) and (5) we get

$h_{11} = 10$ & $h_{12} = 1$

From equation (3) and (5)

$20I_2 = V_1 - 30I_1 = V_2 + 10I_1 - 30I_1$

$= V_2 - 20I_1$

$I_2 = [1/20] V_2 - I_1$ (6)

$I_2 = h_{21} I_1 + h_{22} V_2$ (2)

Comparing equation (6) and (2) we get

$h_{21} = -1$ and $h_{22} = 0.05$

Hence, h parameter are $\begin{pmatrix} 10 & 1 \\ -1 & 0.05 \end{pmatrix}$

33. Ans (b)

Solution:

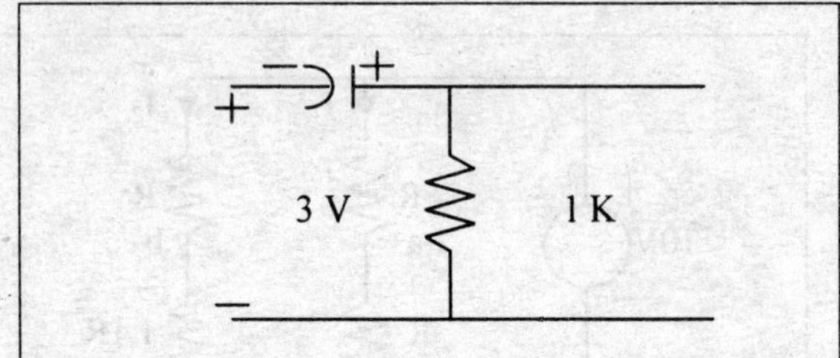

Time constant = RC = 100 μsec.

T = 2 sec >> 100 μsec.

So, C is fully charged (opposite polarity)

$V_{out} = -3$

34. Ans. (a)

Solution: System [1 / (s + 1)] gives change in output as

$\frac{1}{\sqrt{(1+1)}} \angle -\tan^{-1}\frac{1}{1}$

$\frac{1}{\sqrt{2}} \angle -45^\circ$

Therefore output = [1 / √2] Sin [t – (π / 4)]

35. Ans. (d)

Solution: $Z_s = R_s + jX_s$

For maximum power transfer the load impedance should be equal to complex conjugate of source impedance i.e.

$Z_L = Z_s^* = R_s - jX_s$

36. Ans. (c)

Solution:

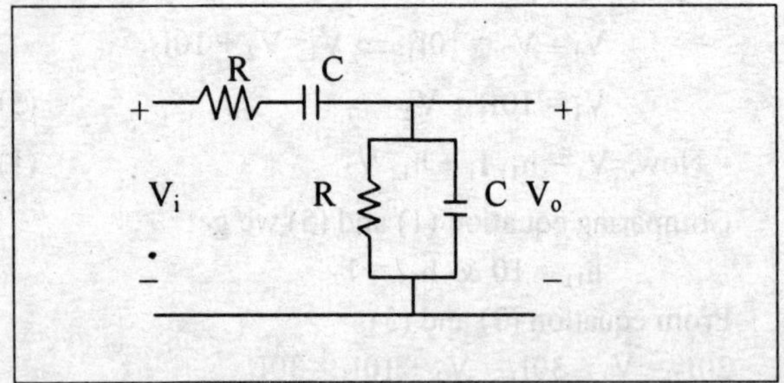

⇒ Band-pass filter

Two Marks Questions

1. Twelve 1Ω resistance are used as edges to form a cube. The resistance between two diagonally opposite corners of the cube is:

(a) (5/6) Ω (b) 1 Ω
(c) (6/5) Ω (d) (3/2) Ω

2. The current flowing through the resistance R in the circuit in Figure has the form P cos 4t, where P is

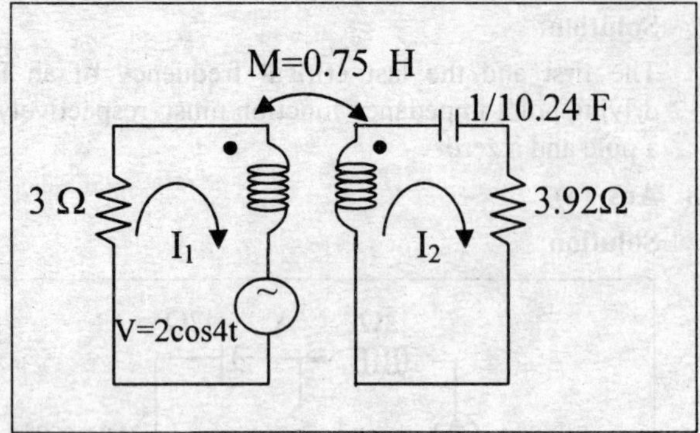

(a) (0.18 + j 0.72) (b) (0.46 + j 1.90)
(c) – (0.18 + j 1.90) (d) (0.192 + j 0.144)

3. An input voltage v(t) = 10√2 cos (t + 10°) + 10 √5 cos (2t + 10°) V is applied to a series combination of resistance R = 1Ω and an inductance L = 1H. The resulting steady-state current i(t) in ampere is

(a) $10 \cos(t+55^\circ) + 10 \cos(2t + 10^\circ + \tan^{-1} 2)$
(b) $10 \cos(t+55^\circ) + 10 [\sqrt{3/2}] \cos (2t + 55^\circ)$
(c) $10 \cos(t - 35^\circ) + 10 \cos(2t + 10^\circ - \tan^{-1} 2)$
(d) $10 \cos(t - 35^\circ) + 10 [\sqrt{3/2}] \cos (2t + 35^\circ)$

4. The driving-point impedance Z(s) of a network has the pole-zero locations as shown in Figure. If Z(0) = 3, then Z(s) is

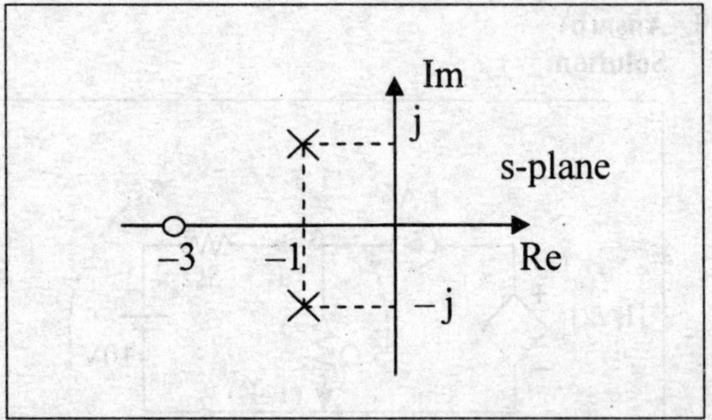

(a) $\frac{3(s+3)}{s^2+2s+3}$ (b) $\frac{2(s+3)}{s^2+2s+2}$
(c) $\frac{3(s+3)}{s^2-2s-3}$ (d) $\frac{2(s+3)}{s^2-2s-2}$

5. The impedance parameters Z_{11} and Z_{12} of the two-port network in Figure are

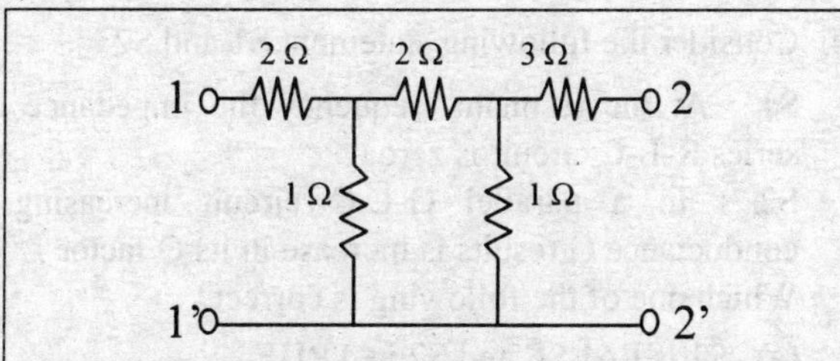

(a) $Z_{11} = 2.75\ \Omega$ and $Z_{12} = 0.25\ \Omega$
(b) $Z_{11} = 3\ \Omega$ and $Z_{12} = 0.5\ \Omega$
(c) $Z_{11} = 3\ \Omega$ and $Z_{12} = 0.25\ \Omega$
(d) $Z_{11} = 2.25\ \Omega$ and $Z_{12} = 0.5\ \Omega$

6. The circuit for Q 11 and Q. 12 is given. Assume that the switch S is in position 1 for a long time and thrown to position 2 at $t = 0$.

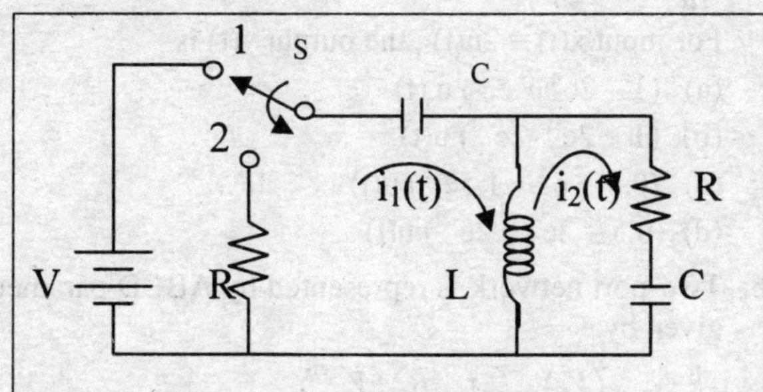

At, t = 0$^+$, the current i_1 is
(a) (-V /2R)
(b) (-V / R)
(c) (-V/4R)
(d) zero

7. The circuit for Q 11 and Q. 12 is given. Assume that the switch S is in position 1 for a long time and thrown to position 2 at $t = 0$.

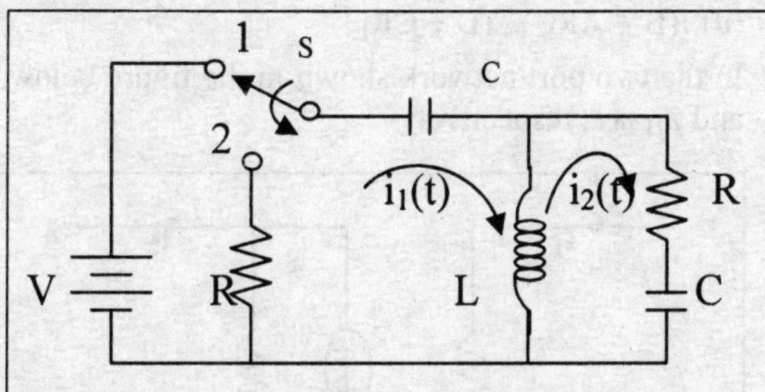

$I_1(s)$ and $I_2(s)$ are the Laplace transforms of $i_1(t)$ and $i_2(t)$ respectively. The equations for the loop currents $I_1(s)$ and $I_2(S)$ for the circuit shown in figure (Q.82) after the switch is brought from position 1 to position 2 at = 0, are

(a) $\begin{pmatrix} R + Ls + \frac{1}{Cs} & -Ls \\ -Ls & R + \frac{1}{Cs} \end{pmatrix} \begin{pmatrix} I_1(s) \\ I_2(s) \end{pmatrix} = \begin{pmatrix} V/s \\ 0 \end{pmatrix}$

(b) $\begin{pmatrix} R + Ls + \frac{1}{Cs} & -Ls \\ -Ls & R + \frac{1}{Cs} \end{pmatrix} \begin{pmatrix} I_1(s) \\ I_2(s) \end{pmatrix} = \begin{pmatrix} -V/s \\ 0 \end{pmatrix}$

(c) $\begin{pmatrix} R + Ls + \frac{1}{Cs} & -Ls \\ -Ls & R + Ls + \frac{1}{Cs} \end{pmatrix} \begin{pmatrix} I_1(s) \\ I_2(s) \end{pmatrix} = \begin{pmatrix} V/s \\ 0 \end{pmatrix}$

(d) $\begin{pmatrix} R + Ls + \frac{1}{Cs} & -Ls \\ -Ls & R + Ls + \frac{1}{Cs} \end{pmatrix} \begin{pmatrix} I_1(s) \\ I_2(s) \end{pmatrix} = \begin{pmatrix} -V/s \\ 0 \end{pmatrix}$

8. The system under consideration is are RC Low-Pass filter (RC-LPF) with $R = 1.0\ \text{k}\Omega$ and $C = 0.01\ \mu F$. Let $H(f)$ denote the frequency response of the RC-LPF. Let f_1 be the highest frequency such that $0 \le |f| \le f_1, \frac{|Hf_1|}{H(0)} \ge 0.95$. Then f_1 (in Hz) is
(a) 327.8
(b) 163.9
(c) 52.2
(d) 104.4

9. The system under consideration is are RC Low-Pass filter (RC-LPF) with $R = 1.0\ \text{k}\Omega$ and $C = 0.01\ \mu F$. Let t_g (f) be the **group delay** function of the given RC-LPF and f_2 = 100 Hz. Then t_g (f_2) in ms, is
(a) 0.717
(b) 7.17
(c) 71.7
(d) 4.505

GATE – 2004 Two Marks Questions

10. For the lattice circuit shown in figure Z_a = j2Ω and Z_b = 2Ω. The values of the open circuit impedance parameters $Z = \begin{pmatrix} Z_{11} & Z_{12} \\ Z_{21} & Z_{22} \end{pmatrix}$ are

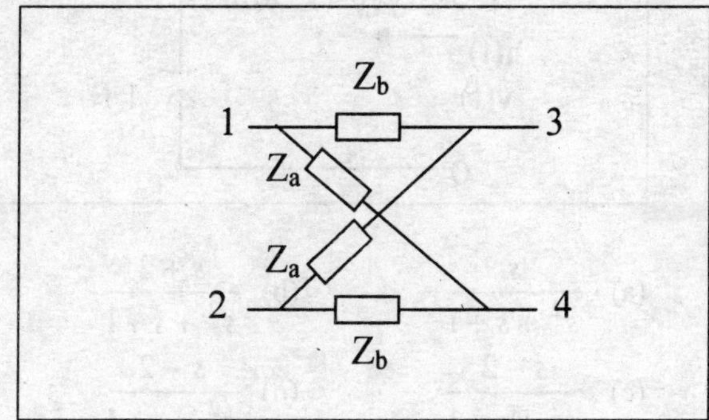

(a) $\begin{pmatrix} 1-j & 1+j \\ 1+j & 1+j \end{pmatrix}$ (b) $\begin{pmatrix} 1-j & 1+j \\ -1+j & 1-j \end{pmatrix}$

(c) $\begin{pmatrix} 1+j & 1+j \\ 1-j & 1-j \end{pmatrix}$ (d) $\begin{pmatrix} 1+j & -1+j \\ -1+j & 1+j \end{pmatrix}$

11. The transfer function $H(s)=\dfrac{V_o(s)}{V_i(s)}$ of an R-L-C circuit is given by $H(s)=\dfrac{10^6}{s^2+20s+10^6}$

The Quality factor (Q-factor) of this circuit is

(a) 25 (b) 50
(c) 100 (d) 5000

12. For the circuit shown in Figure the initial conditions are zero. Its transfer function $H(s)=\dfrac{V_O(s)}{Vi(s)}$ is

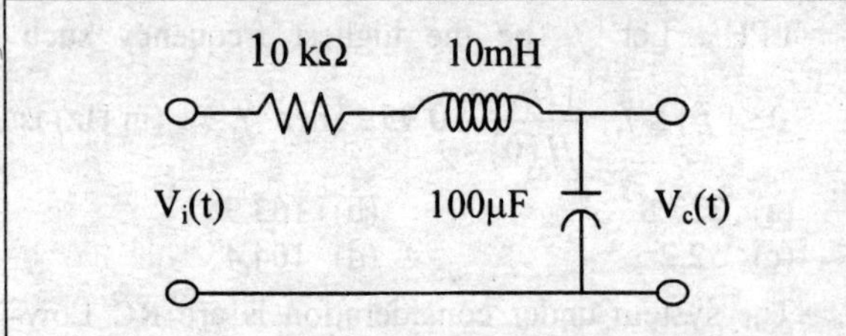

(a) $\dfrac{1}{s^2+10^6 s+10^6}$ (b) $\dfrac{10^6}{s^2+10^3 s+10^6}$

(c) $\dfrac{10^3}{s^2+10^3 s+10^6}$ (d) $\dfrac{10^6}{s^2+10^6 s+10^6}$

13. The circuit shown in Figure has initial current $i_L(0^-) = 1$ A through the inductor and an initial voltage $v_c(0^-) = -1$ V across the capacitor. For input v(t) = u(t), the Laplace transform of the current i(t) for $t \geq 0$ is

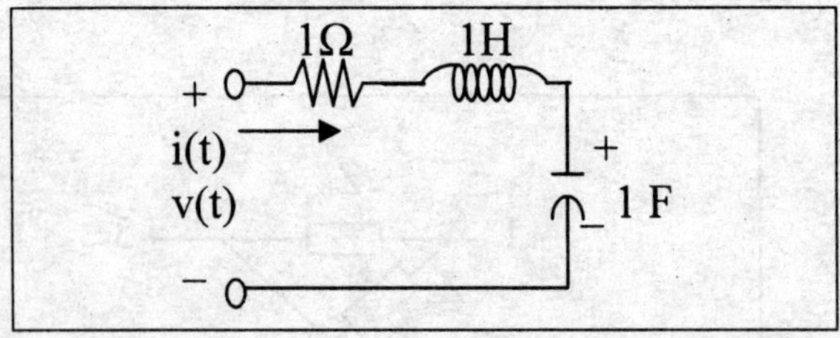

(a) $\dfrac{s}{s^2+s+1}$ (b) $\dfrac{s+2}{s^2+s+1}$

(c) $\dfrac{s-2}{s^2+s+1}$ (d) $\dfrac{s-2}{s^2+s+1}$

14. Consider the following statement S1 and S2

S1 : At the resonant frequency the impedance of a series R-L-C circuit is zero.

S2 : In a parallel G-L-C circuit increasing the conductance G results is increase in its Q factor .

Which one of the following is correct?

(a) S1 is FALSE and S2 in TRUE
(b) Both S1 and S2 are TRUE
(c) S1 is TURE and S2 in FALSE
(d) Both S1 and S2 are FALSE

15. A system described by the following differential equation

$$\frac{d^2y}{dt^2}+3\frac{dy}{dt}+2y=x(t)$$ is initially at rest.

For input x(t) = 2u(t) , the output y(t) is

(a) $(1-2e^{-t}+e^{-2t})$ u (t)
(b) $(1+2e^{-t}-2e^{-2t})$ u(t)
(c) $(0.5+e^{-t}+1.5e^{-2t})$ u(t)
(d) $(0.5-2e^{-t}+2e^{-2t})$ u(t)

16. Two-port network is represented by ABCD parameters given by.

$$\begin{pmatrix} V_1 \\ I_1 \end{pmatrix}=\begin{pmatrix} A & B \\ C & D \end{pmatrix}\begin{pmatrix} V_2 \\ -I_2 \end{pmatrix}$$

If port-2 is terminated by R_L, the input impedance seen at port-1 is given by

(a) $(A + BR_L) / (C + DR_L)$
(b) $(AR_L + C) / (BR_L + D)$
(c) $(DR_L + A) / (BR_L + C)$
(d) $(B + AR_L) / (D + CR_L)$

17. In the two port network shown in the figure below, z_{12} and z_{21} are, respectively

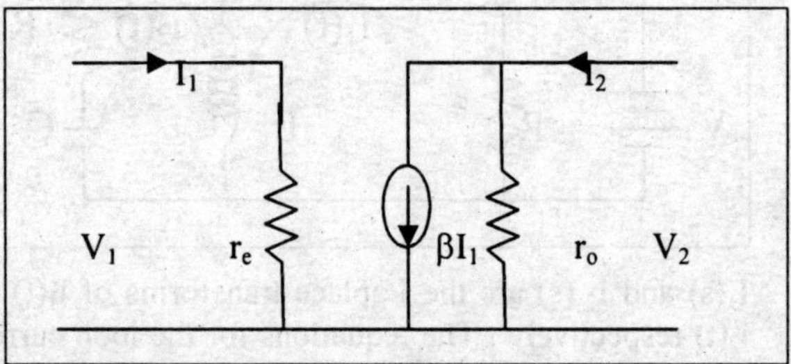

(a) r_e and βr_o (b) 0 and $-\beta r_o$
(c) 0 and βr_o (d) r_e and $-\beta r_o$

18. The first and the last critical frequencies (singularities) of a driving point impedance function of a passive network having two kinds of elements, are a pole and a zero respectively. The above property will be satisfied by.

(a) RL network only
(b) RC network only
(c) LC network only
(d) RC as well as RL networks

19. A 2 mH inductor with some initial current can be represented as shown below, where s is the Laplace Transform variable. The value of initial current is

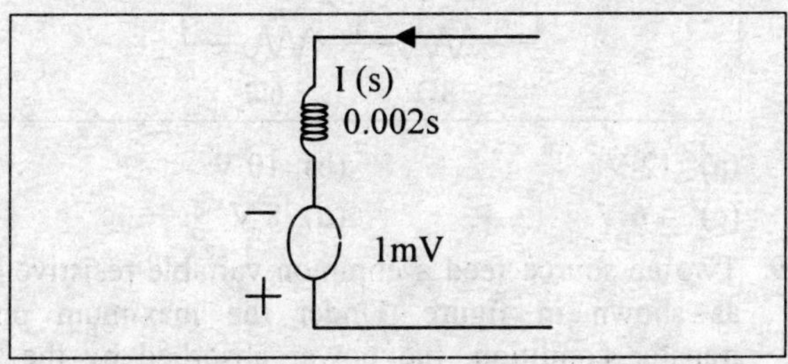

(a) 0.5 A (b) 2.0 A
(c) 1.0 A (d) 0.0 A

20. In the figure shown below, assume that all the capacitors are initially uncharged. If $v_1(t) = 10\ u(t)$ Volts, $V_o(t)$ is given by.

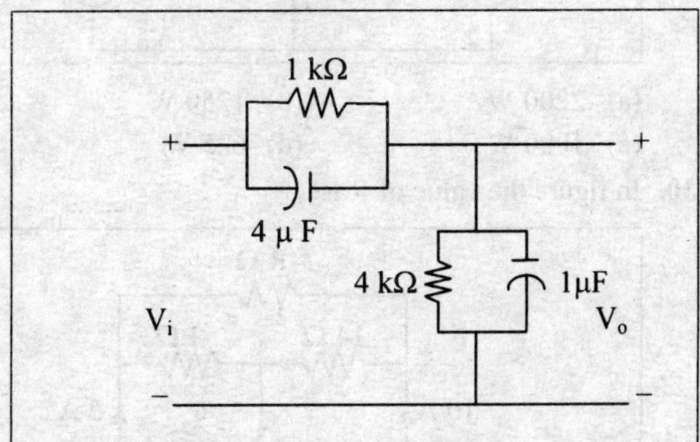

(a) $8e^{-0.004t}$ Volts
(b) $8(1-e^{+0.004t})$ Volts
(c) 8u (t) Volts
(d) 8 Volts

21. A negative resistance R_{neg} is connected to a passive network N having driving point impedance $Z_1(s)$ as shown below. For $Z_2(s)$ to be positive real.

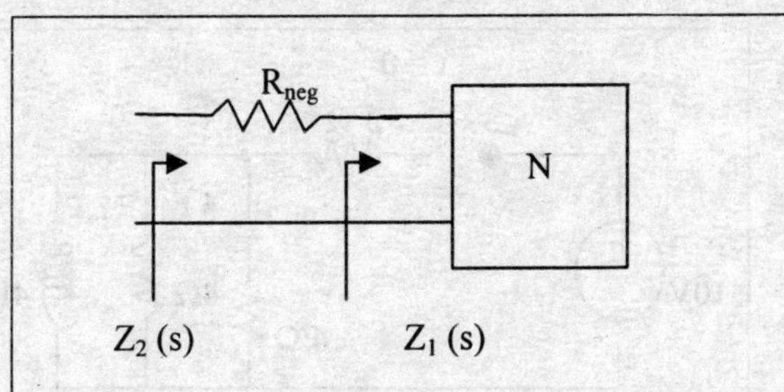

(a) $|R_{neg}| \le \text{Re}\, Z_1(j\omega),\ \forall\omega$
(b) $|R_{neg}| \le |Z_1(j\omega)|,\ \forall\omega$
(c) $|R_{neg}| \le \text{Im}\, Z_1(j\omega),\ \forall\omega$
(d) $|R_{neg}| \le \angle Z_1(j\omega),\ \forall\omega$

22. Two series resonant filter are as shown in the figure. Let the 3-dB bandwidth of Filter 1 be B_1 and that of Filter 2 be B_2. The value of B_1/B_2 is

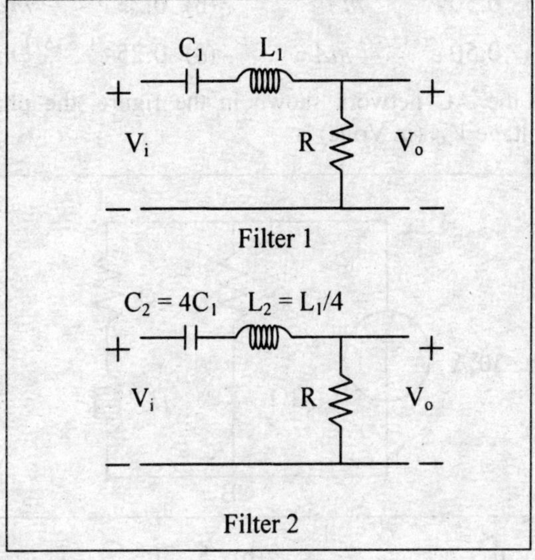

(a) 4 (b) 1
(c) 1/2 (d) 1/4

23. For the circuit shown in the figure, the Thevenin voltage and resistance looking into X – Y are

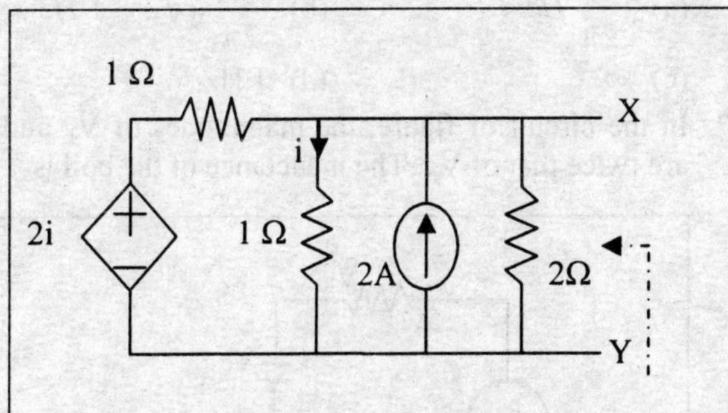

(a) $\frac{4}{3}V, 2\,\Omega$ (b) $4\,V, \frac{2}{3}\,\Omega$
(c) $\frac{4}{3}V, \frac{2}{3}\Omega$ (d) $4\,V, 2\,\Omega$

24. In the circuit shown, V_C is 0 volts at t = 0 sec. For t > 0, the capacitor current $i_c(t)$, where t is in seconds, is given by

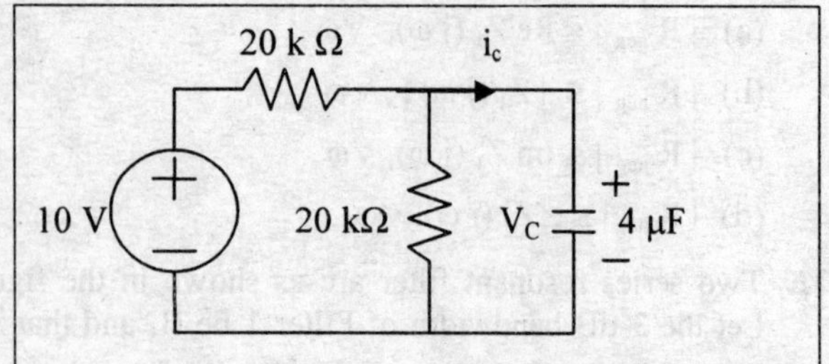

(a) $0.50\,e^{(-25\,t)}\,mA$ (b) $0.25\,e^{(-25\,t)}\,mA$

(c) $0.50\,e^{(-12.5\,t)}\,mA$ (d) $0.25\,e^{(-6.25\,t)}\,mA$

25. In the AC network shown in the figure, the phasor voltage V_{AB} (in Volts) is

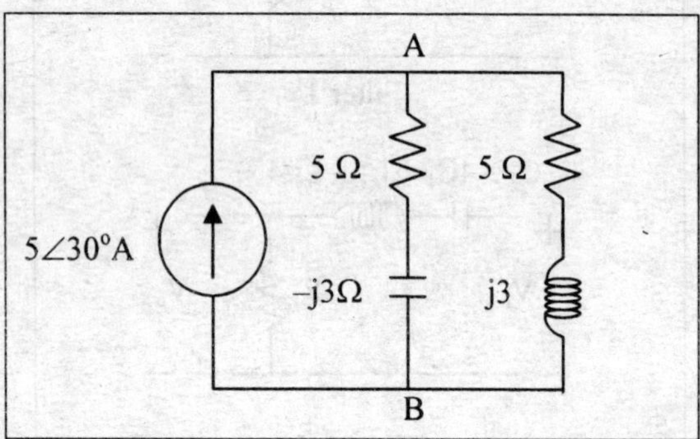

(a) 0 (b) $5\angle 30^o$

(c) $12.5\angle 30^o$ (d) $17\angle 30^o$

26. The 3-dB bandwidth of the low-pass signal $e^{-t}u(t)$, where u(t) is the unit step function , is given by

(a) $\frac{1}{2\pi}\ Hz$ (b) $\frac{1}{2\pi}\sqrt{\sqrt{2}-1}\ Hz$

(c) ∞ (d) 1 Hz

27. In the circuit of figure, the magnitudes of V_L and V_C are twice that of V_R. The inductance of the coil is

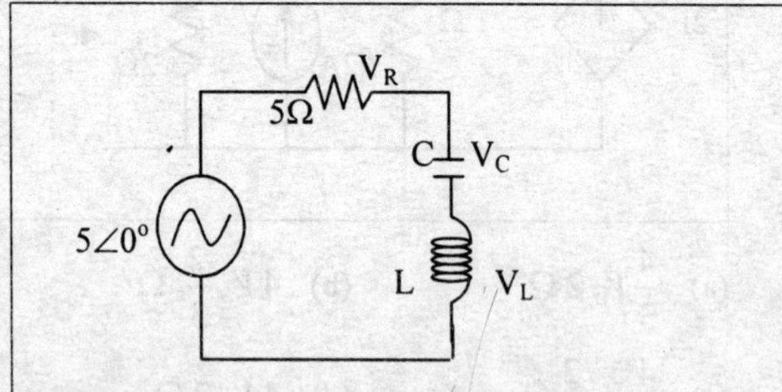

(a) 2.14 mH (b) 5.30 H

(c) 3.18 mH (d) 1.32 H

28. In Figure, the potential difference between points P and Q is

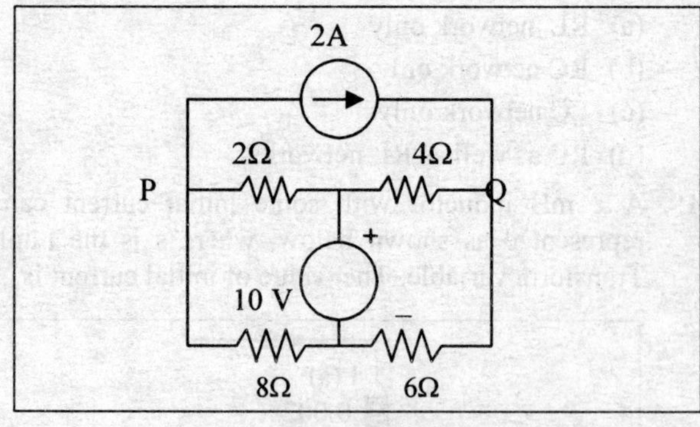

(a) 12 V (b) 10 V

(c) – 6 V (d) 8 V

29. Two ac source feed a common variable resistive load as shown in figure. Under the maximum power transfer condition, the power absorbed by the load resistance R_L is

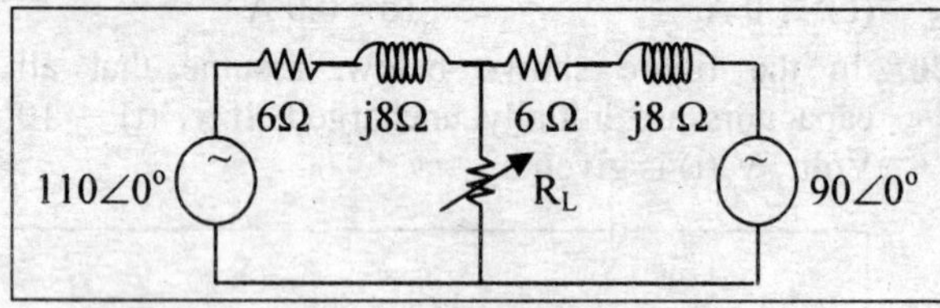

(a) 2200 W (b) 1250 W

(c) 1000 W (d) 625 W

30. In figure the value of R is

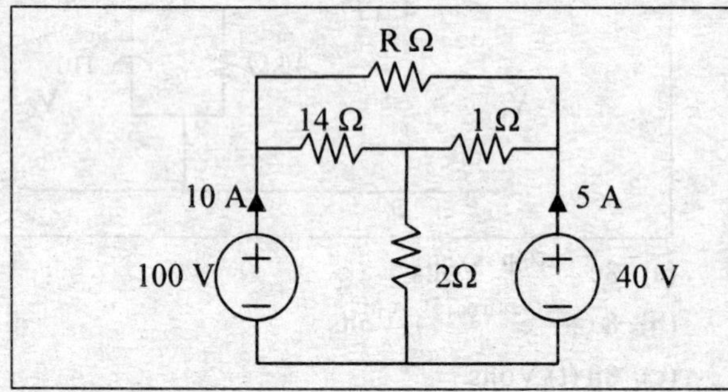

(a) 10 Ω (b) 18 Ω

(c) 24 Ω (d) 12 Ω

31. In the circuit shown in Figure, the switch S is closed at time t = 0. The voltage across the inductance at = 0^+, is

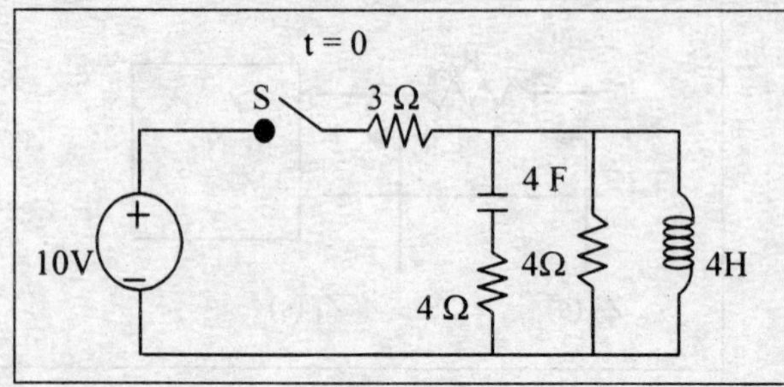

(a) 2 V (b) 4 V
(c) – 6 V (d) 8 V

32. The h parameters for a two-port network are defined by $\begin{bmatrix} E_1 \\ I_2 \end{bmatrix} = \begin{bmatrix} h_{11} & h_{12} \\ h_{21} & h_{22} \end{bmatrix} \begin{bmatrix} I_1 \\ E_2 \end{bmatrix}$

For the two-port network shown in figure the value of h_{12} is given by

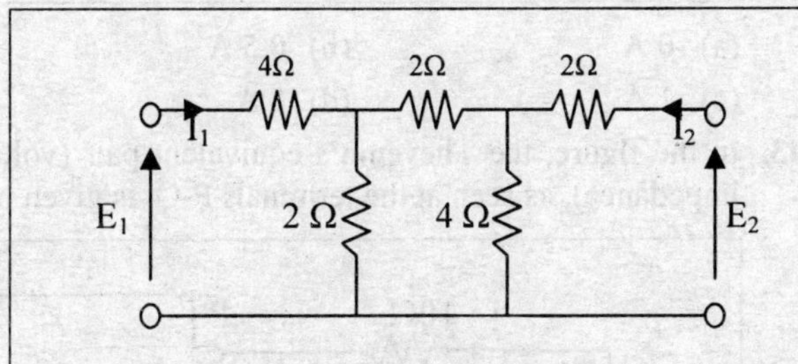

(a) 0.125 (b) 0.167
(c) 0.625 (d) 0.25

33. In figure the value of the source voltage is

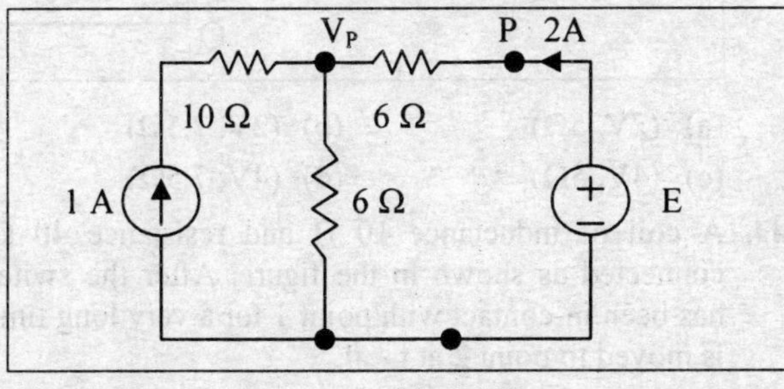

(a) 12 V (b) 24 V
(c) 30 V (d) 44 V

34. In figure R_a, R_b and R_c are 20 Ω, 10 Ω, 10 Ω respectively. The resistance R_1, R_2 and R_3 in Ω of an equivalent star-connection are

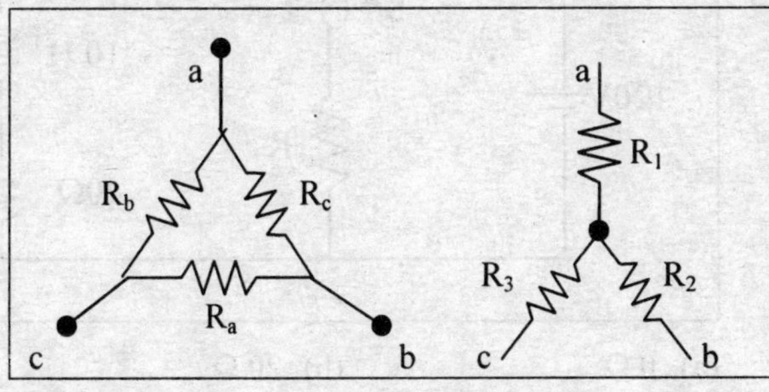

(a) 2.5, 5, 5 (b) 5, 2.5, 5
(c) 5, 5, 2.5 (d) 2.5, 5, 2.5

35. In figure the admittance values of the elements in Siemens are $Y_R = 0.5 + j0$, $Y_L = 0 - j1.5$, $Y_C = 0 + j0.3$ respectively. The value of I as a phasor when the voltage E across the elements is $10\angle 0^\circ$ is

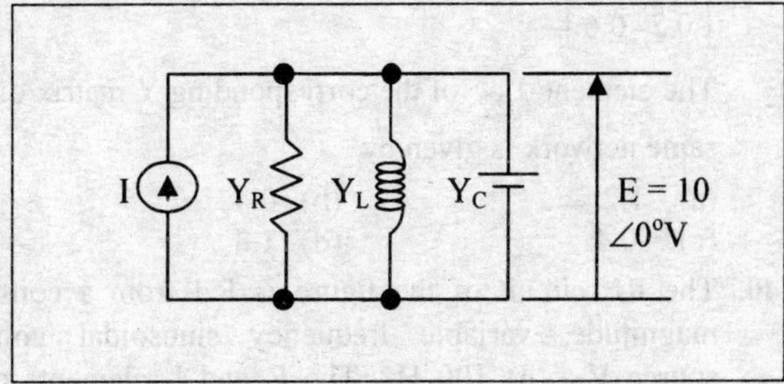

(a) 1.5 + j0.5 (b) 5 – j18
(c) 0.5 + j1.8 (d) 5 – j12

36. In figure the value of resistance R in Ω is

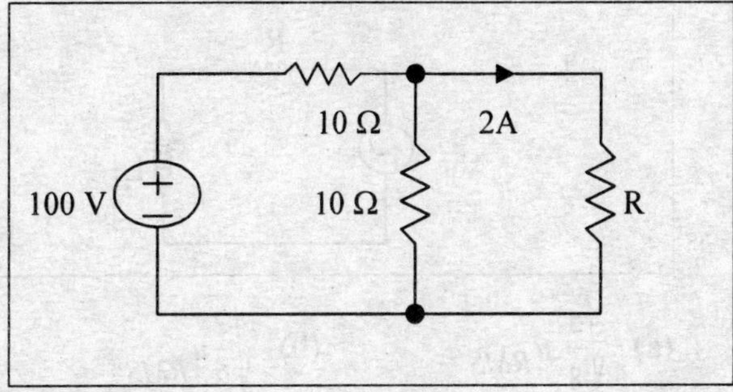

(a) 10 (b) 20
(c) 30 (d) 40

37. In figure the capacitor initially has a charge of 10 Coulomb. The current in the circuit one second after the switch S is closed will be

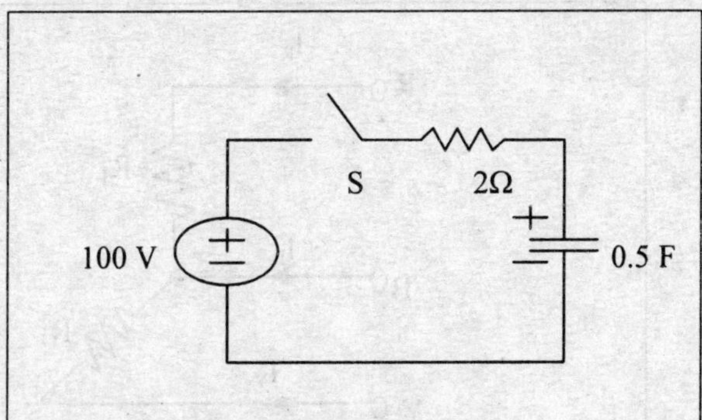

(a) 14.7 A (b) 18.5 A
(c) 40.0 A (d) 50.0 A

38. The rms value of the resultant current in a wire which carries a dc current of 10 A and a sinusoidal alternating current of peak value 20 A is

(a) 14.1A (b) 17.3 A
(c) 22.4 A (d) 30.0 A

39. The Z matrix of a 2-port network as given by $\begin{bmatrix} 0.9 & 0.2 \\ 0.2 & 0.6 \end{bmatrix}$

The element Y_{22} of the corresponding Y matrix of the same network is given by

(a) 1.2 (b) 0.4
(c) –0.4 (d) 1.8

40. The RL circuit of the figure is fed from a constant magnitude, variable frequency sinusoidal voltage source V_{IN}. At 100 Hz. The R and L elements each have a voltage drop V_{RMS}. If the frequency of the source is changed to 50 Hz, then new voltage drop across R is

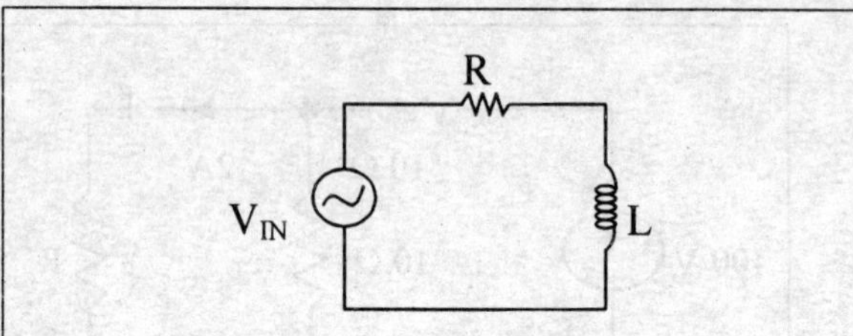

(a) $\sqrt{\frac{5}{8}} u_{RMS}$ (b) $\sqrt{\frac{2}{3}} u_{RMS}$
(c) $\sqrt{\frac{8}{5}} u_{RMS}$ (d) $\sqrt{\frac{3}{2}} u_{RMS}$

41. For the three-phase circuit shown in the figure the ratio of the currents $I_R:I_Y:I_B$ is given by

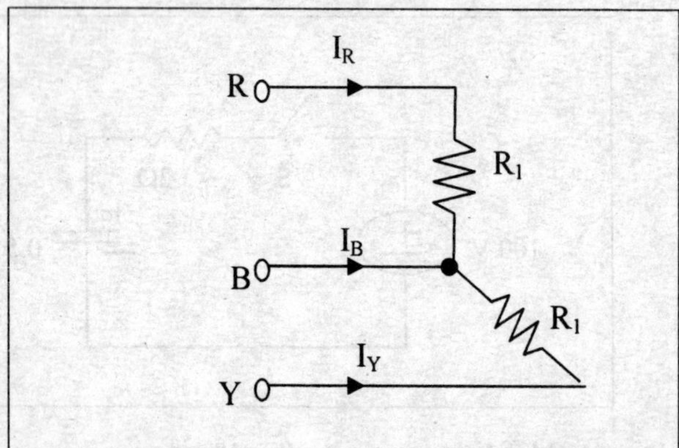

(a) $1:1:\sqrt{3}$ (b) $1:1:2$
(c) $1:1:0$ (d) $1:1:\sqrt{3/2}$

42. The circuit shown in the figure is in steady state, when the switch is closed at t=0. Assuming that the inductance is ideal, the current through the inductor at $t = 0^+$ equals

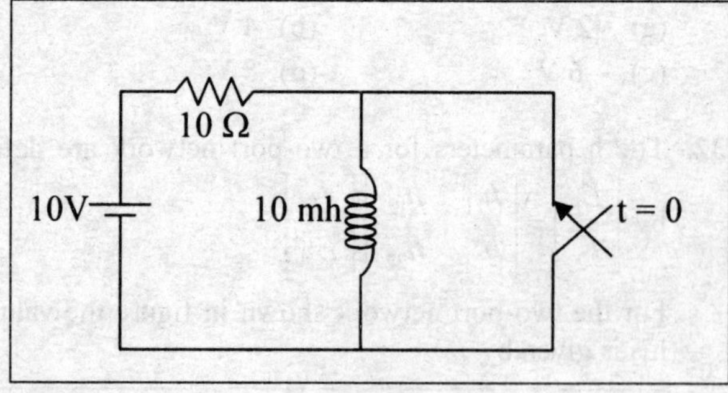

(a) 0 A (b) 0.5 A
(c) 1 A (d) 5 A

43. In the figure, the Thevenin's equivalent pair (voltage, impedance), as seen at the terminals P-Q, is given by

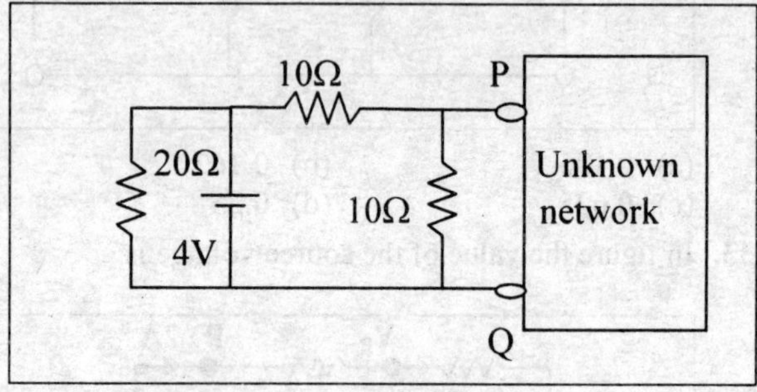

(a) (2V, 5Ω) (b) (2V, 7.5Ω)
(c) (4V, 5Ω) (d) (4V, 7.5Ω)

44. A coil of inductance 10 H and resistance 40 Ω is connected as shown in the figure. After the switch S has been in-contact with point 1 for a very long time, it is moved to point 2 at t = 0.

If, at $t = 0^+$, the voltage across the coil is 120 V, the value of resistance R is

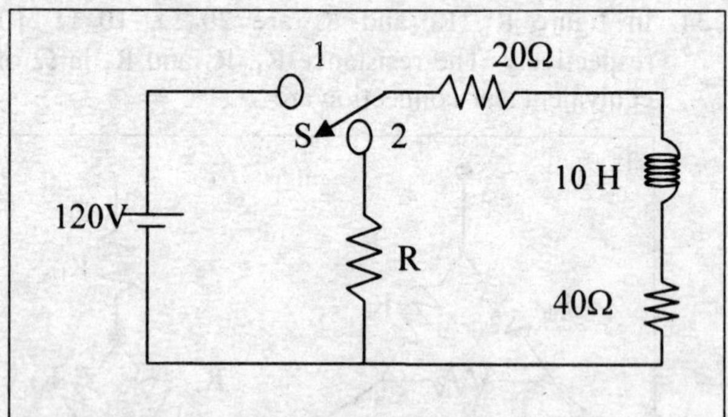

(a) 0 Ω (b) 20 Ω
(c) 40 Ω (d) 60 Ω

45. A coil of inductance 10 H and resistance 40 Ω is connected as shown in the figure. After the switch S has been in-contact with point 1 for a very long time, it is moved to point 2 at t = 0.

For the value of obtained in (a), the time taken for 95% of the stored energy to be dissipated is close to

(a) 0.10 sec (b) 0.15 sec

(c) 0.50 sec (d) 1.0 sec

46. The parameters of the circuit shown in the figure are $R_1 = 1\ M\ \Omega$, $R_0 = 10\Omega$, $A = 10^6$ V/V. If $V_i = 1\mu\ V$ the output voltage, input impedance and output impedance respectively are

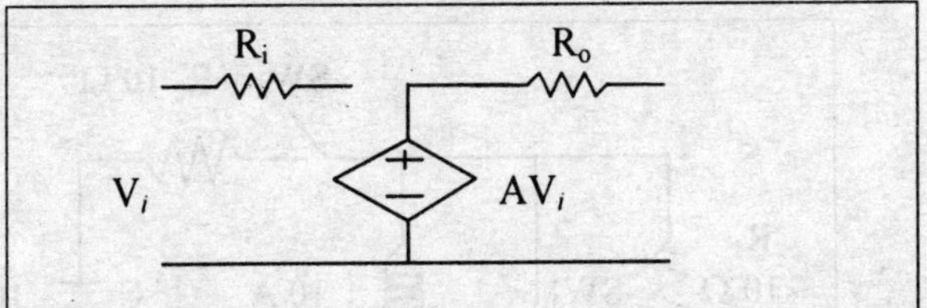

(a) $1\ V, \infty, 10\ \Omega$ (b) $1\ V, 0, 10\ \Omega$

(c) $1\ V, 0, \infty$ (d) $10\ V, \infty, 10\ \Omega$

47. In the circuit shown in the figure, the current source I=1 A, the voltage source V = 5V, $R_1 = R_2 = R_3 = 1\ \Omega$, $L_1 = L_2 = L_3 = 1H$, $C_1 = C_2 = 1F$. The currents (in A) through R_3 and the voltage source V respectively will be

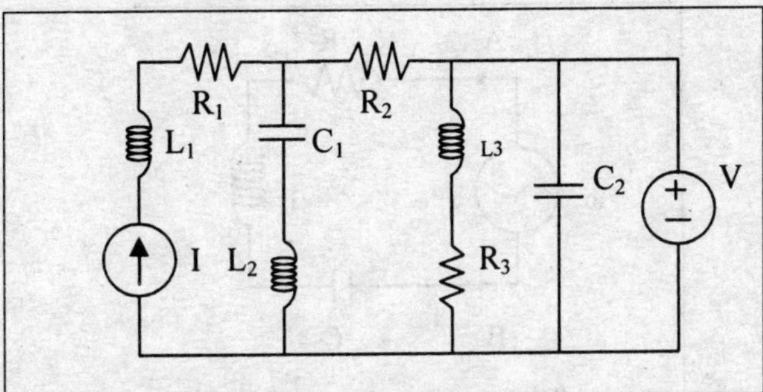

(a) 1, 4 (b) 5, 1
(c) 5, 2 (d) 5, 4

48. The parameter type and the matrix representation of the relevant two part parameter that describe the circuit shown are

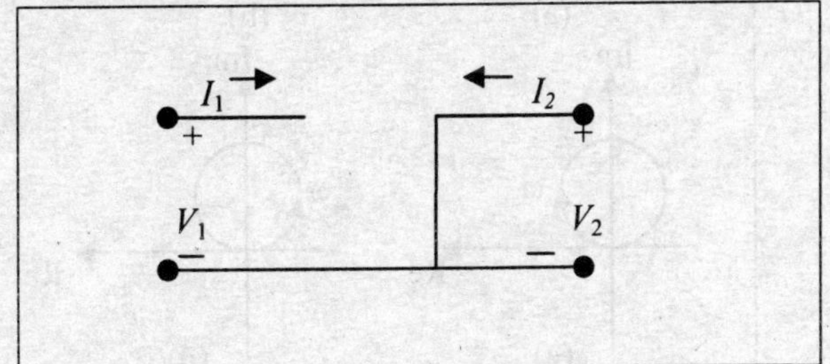

(a) z parameters, $\begin{bmatrix} 0 & 0 \\ 0 & 0 \end{bmatrix}$ (b) h parameters, $\begin{bmatrix} 1 & 0 \\ 0 & 1 \end{bmatrix}$

(c) h parameters, $\begin{bmatrix} 0 & 0 \\ 0 & 0 \end{bmatrix}$ (d) z parameters, $\begin{bmatrix} 1 & 0 \\ 0 & 1 \end{bmatrix}$

49. The circuit shown in the figure is energized by a sinusoidal voltage source V_1 at a frequency which causes resonance with a current of I.

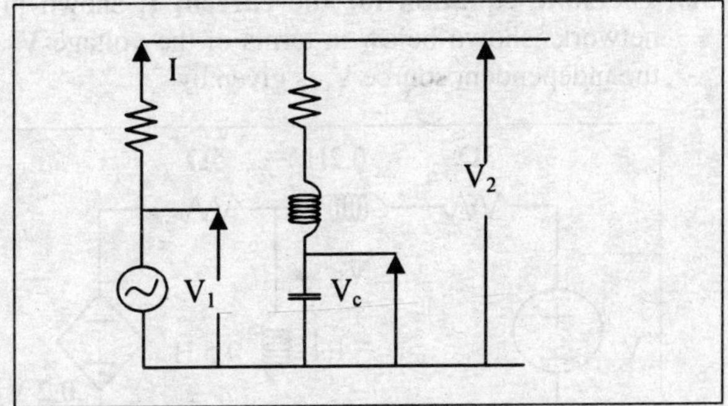

The phasor diagram which is applicable to this circuit is

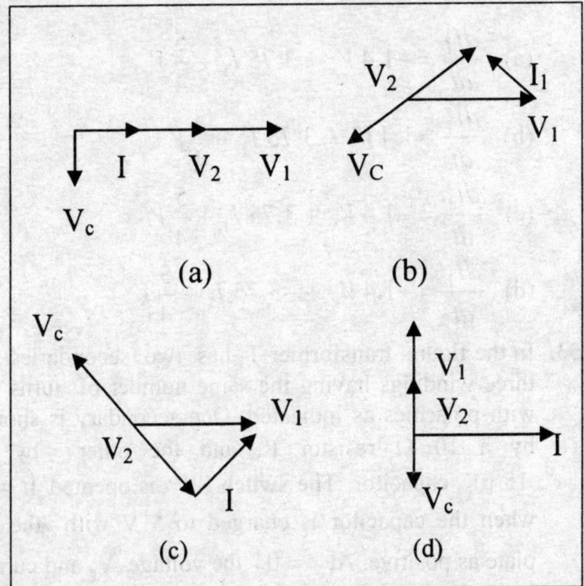

50. An ideal capacitor is charged to a voltage V_0 and connected at t=0 across an ideal inductor L. (The circuit now consists of a capacitor and inductor alone).

If we let $\omega_0 = \frac{1}{\sqrt{LC}}$, the voltage across the capacitor at time t>0 is given by

(a) V_0 (b) $V_0 \cos(\omega_0 t)$

(c) $V_0 \sin(\omega_0 t)$ (d) $V_0 e^{-\omega_0 t} \cos(\omega_o t)$

51. An energy meter connected to an immersion heater (resistive) operating on an AC 230 V, 50 Hz, AC single phase source reads 2.3 units (kWh) in 1 hour. The heater is removed from the supply and now connected to a 400 V peak to peak square wave source of 150 Hz. The power in kW dissipated by the heater will be

(a) 3.478 (b) 1.739
(c) 1.540 (d) 0.870

52. The **state equation for the current I_1** shown in the network shown below in terms of the voltage V_x and the independent source V, is given by

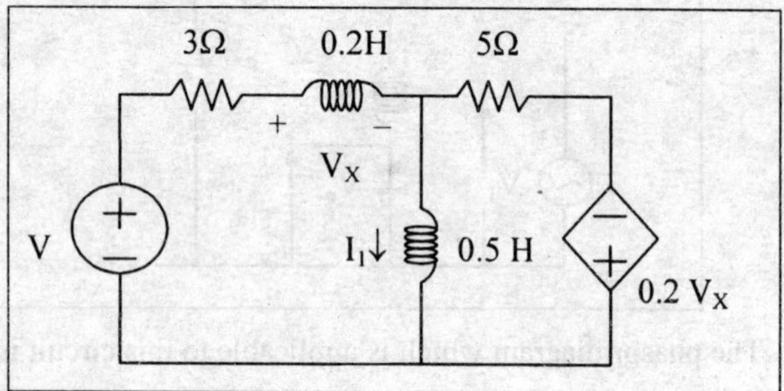

(a) $\frac{dI_1}{dt} = -1.4\,V_x - 3.75\,I_1 + \frac{5}{4}V$

(b) $\frac{dI_1}{dt} = 1.4\,V_x - 3.75\,I_1 - \frac{5}{4}V$

(c) $\frac{dI_1}{dt} = -1.4\,V_x + 3.75\,I_1 + \frac{5}{4}V$

(d) $\frac{dI_1}{dt} = -1.4\,V_x + -3.75\,I_1 - \frac{5}{4}V$

53. In the figure, transformer T_1 has two secondaries, all three windings having the same number of turns and with polarities as indicated. One secondary is shorted by a 10 Ω resistor R, and the other by a 15 μF capacitor. The switch SW is opented $(t = 0)$ when the capacitor is charged to 5 V with the left plate as positive. At $t = 0^+$ the voltage V_p and current I_R are

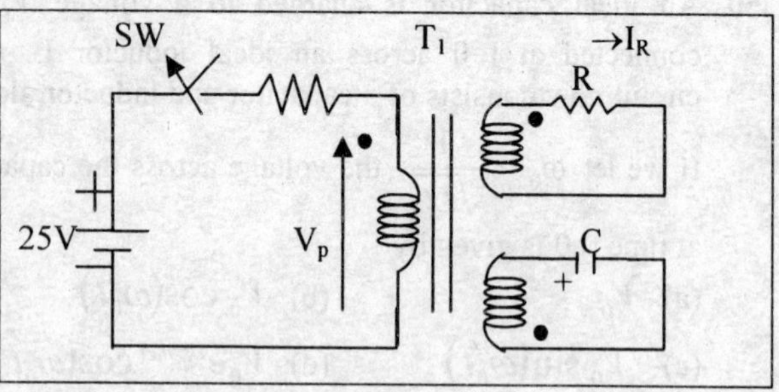

(a) –25 V, 0.0 A
(b) Very large voltage, very large current
(c) 5.0 V, 0.5 A
(d) –5.0 V, –0.5 A

54. In the circuit shown in figure switch SWI is initially CLOSED and SW_2 is OPEN. The inductor L carries a current 10 A and the capacitor is charged to 10 V with polarities as indicated. SW2.is initially CLOSED at $t = 0^-$ and SW1 is OPENTED is $t = 0$. The current through C and the voltage across L at $t = 0^+$ is

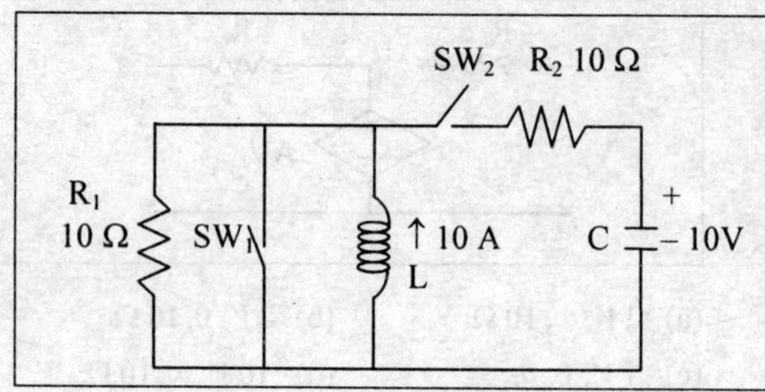

(a) 55 A, 4.5 V (b) 5.5 A, 45 V
(c) 45 A, 5.5 V (d) 4.5 A, 55 V

55. The R-L-C series circuit shown is supplied from a variable frequency voltage source. The admittance-locus of the R-L-C network at terminals AB for increasing frequency ω is

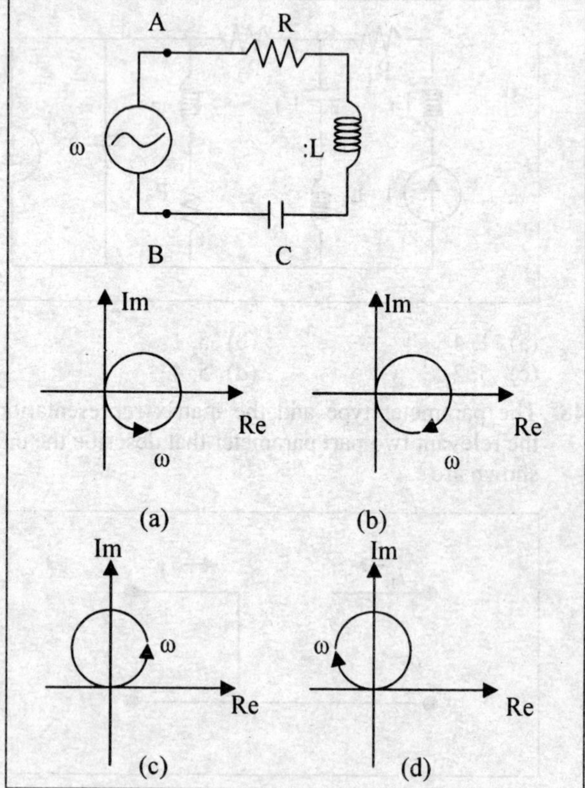

56. In the figure given below all phasors are with reference to the potential at point "O". The locus of voltage phasor V_{YX} as R is varied from zero to infinity is shown by

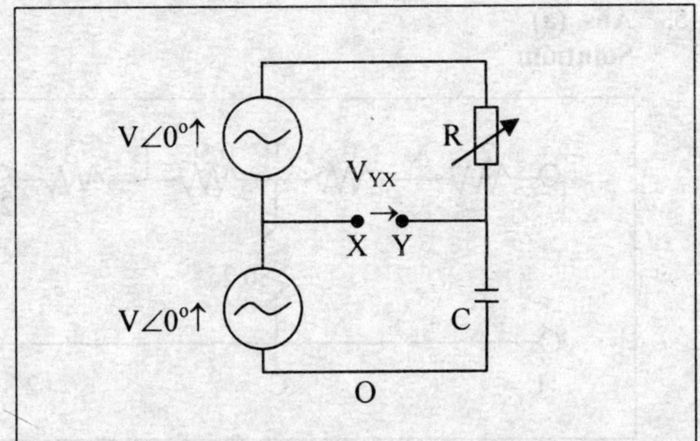

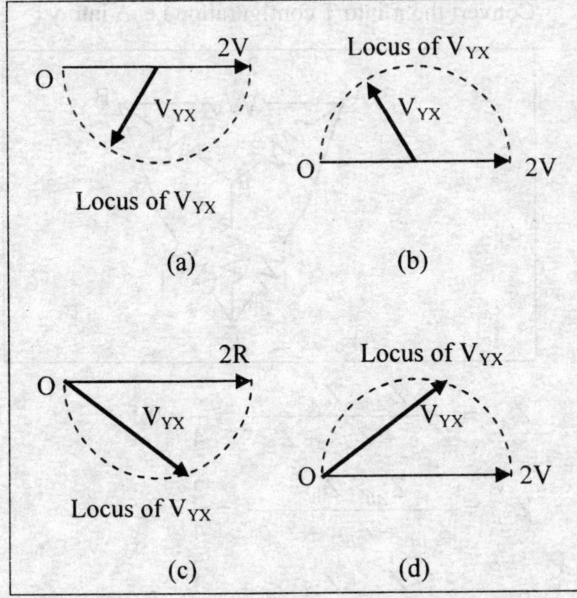

57. A 3 V dc supply with an internal resistance of 2 Ω supplies a passive non-linear resistance characterized by the relation $V_{NL} = I_{NL}^2$. The power dissipated in the non-linear resistance is

(a) 1.0 W (b) 1.5 W
(c) 2.5 W (d) 3.0 W

58. The matrix A give below is the **node incidence** matrix of a network. The columns correspond to branches of the network while the rows correspond to nodes. Let $V = [v_1\ v_2 \ldots v_6]^T$ denote the vector of branch voltage while $I = [i_1\ i_2 \ldots i_6]^T$ that of branch currents. The vector $E = [e_1\ e_2\ e_3\ e_4]^T$ denotes the vectors of node voltages relative of a common ground.

$$A = \begin{bmatrix} 1 & 1 & 1 & 0 & 0 & 0 \\ 0 & -1 & 0 & -1 & 1 & 0 \\ -1 & 0 & 0 & 0 & -1 & -1 \\ 0 & 0 & -1 & 1 & 0 & 1 \end{bmatrix}$$

Which of the following statements is true ?

(a) The equations

$v_1 - v_2 + v_3 = 0,\ v_3 + v_4 - v_5 = 0$

are KVL equations for the network for some loops

(b) The equations

$v_1 - v_3 - v_6 = 0,\ v_4 + v_5 - v_6 = 0$

are KVL equations for the network for some loops

(c) K = AV

(d) AV = 0 are KVL equations for the network

Answers

Basic of Circuit and Measurement System

1. Ans. (a)

Solution:

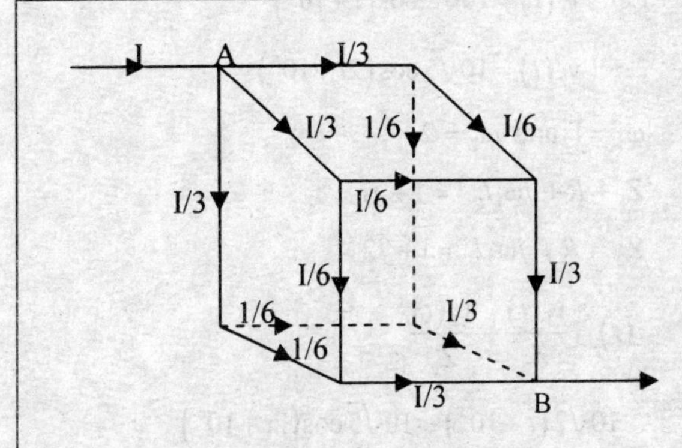

R = 1 $R_{AB} = [V_{AB} / I]$

$$V_{AB} = \left(\frac{1}{3} + \frac{1}{6} + \frac{1}{3}\right) R$$

$$= \frac{2I + 1I + 2I}{6} \cdot R = \frac{5I}{6} \cdot R = [5/6] \cdot 1$$

$V_{AB} = [5I/6]$

$R_{AB} = [V_{AB} / I] = $ **[5 / 6]Ω**

2. Ans. (d)

Solution: Here the value of inductance is not given we can ignore it

$I_1 = \frac{2 \cos 4t}{3} = 0.67 \cos 4t$ i.e. ω = 4

$I_2 = \frac{-j\omega M}{R - [j / \omega C]} = \frac{-j\, 4 \times 0.75\, I_1}{3.92 - 2.56\, j}$

$= \frac{-j3 \times 0.67 \cos 4t}{3.92 - 2.56j} \times \frac{3.92 + 2.56j}{3.92 + 2.5j}$

= (0.23 – 0.35j) cos 4 t

3. Ans. (c)

Solution: $I(t) = i_1(t) + i_2(t)$

$= \frac{10\sqrt{2}\cos(t + 10^o)}{R + j\omega_1 L} + \frac{10\sqrt{5}\cos(2t + 10^o)}{R + j\omega_2 L}$

Given

$R = 1\Omega$ and $L = 1$ H

$\omega_1 = 1$ and $\omega_2 = 2$

$I(t) = \frac{10\sqrt{2}\cos(t + 10^o)}{1 + j} + \frac{10\sqrt{5}\cos(2t + 10^o)}{1 + 2j}$

$= \frac{10\sqrt{2}\cos(t + 10^o)}{\sqrt{2}\angle\tan^{-1}1} + \frac{10\sqrt{5}\cos(2t + 10^o}{\sqrt{5}\angle\tan^{-1}2}$

$= 10\cos(t + 10 - 45) + 10\cos[2t + 10 - \tan^{-1}2)$

$\mathbf{= 10\cos(t - 35) + 10\cos(2t + 10 - \tan^{-1}2)}$

Second Method

$v(t) = 10\sqrt{2}\cos(t + 10^\circ) + 10\sqrt{5}\cos(2t + 10^\circ)$

Let, $v_1(t) = 10\sqrt{2}\cos(t + 10^\circ)$

$v_2(t) = 10\sqrt{5}\cos(2t + 10^\circ)$

$\omega_1 = 1$ and $\omega_2 = 2$

$Z_1 + R + j\omega_1 L = 1 + j1$

$Z_2 = R + j\omega_2 L = 1 + j2$

$i(t) = \frac{v_1(t)}{Z_1} + \frac{v_2(t)}{Z_2}$

$= \frac{10\sqrt{2}(t + 10^\circ)}{1 + j} + \frac{10\sqrt{5}\cos(2t + 10^\circ)}{1 + j2}$

$= \frac{10\sqrt{2}\cos(t + 10^\circ)}{\sqrt{1^2 + 1^2}\angle\tan^{-1}1} + \frac{10\sqrt{5}\cos(2t + 10^\circ)}{\sqrt{1^2 + 2^2}\angle\tan^{-1}2}$

$= \frac{10\sqrt{2}\cos(t + 10^\circ)}{\sqrt{2}\angle 45^\circ} + \frac{10\sqrt{5}\cos(2t + 10^\circ)}{\sqrt{5}\angle 2}$

$i(t) = 10\cos(t - 35^\circ) + 10\cos(2t + 10 - \tan^{-1}2)$

4. Ans. (b)

Solution: $Z(s) = \frac{K(s + 3)}{(s + 1 + j)(s + 1 - j)}$

$= \frac{k(s + 3)}{(s + 1)^2 + 1}$

$Z(o) = 3 = [3K/2]$; $K = 2$

$Z(s) = \frac{2(s + 3)}{(s + 1)^2 + 1} = \frac{2(s + 3)}{s^2 + 2s + 2}$

5. Ans. (a)

Solution:

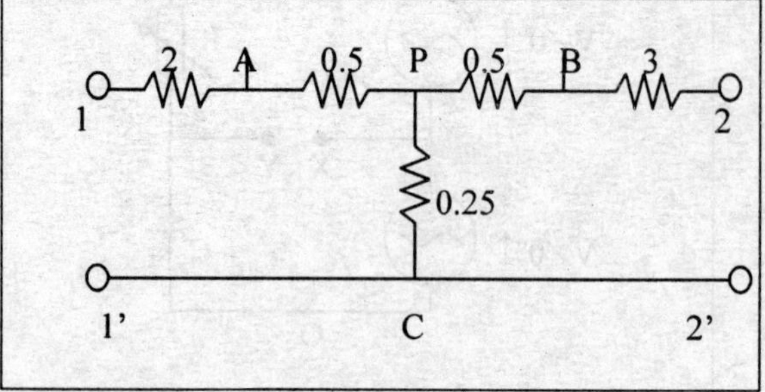

Convert the π into T configuration i.e. Δ into y

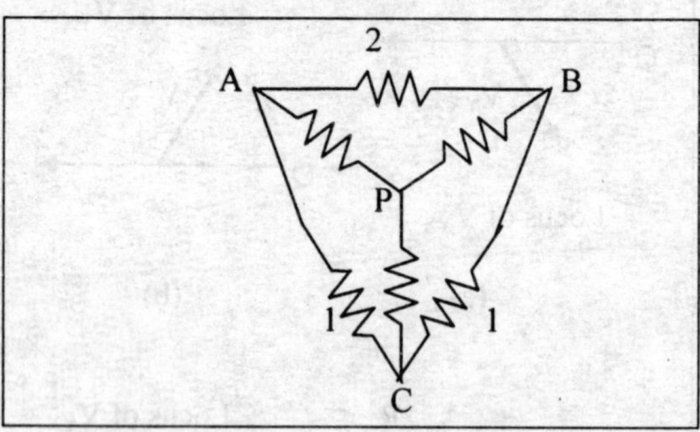

$Z_{AP} = \frac{Z_{AB} \cdot Z_{AC}}{Z_{AB} + Z_{AC} + Z_{BC}} = \frac{2}{4} = 0.5$

$Z_{PB} = \frac{Z_{AB} \cdot Z_{BC}}{Z_{AB} + Z_{BC} + Z_{AC}} = \frac{2}{4} = 0.5$

$Z_{PC} = \frac{Z_{AC} \cdot Z_{BC}}{Z_{AB} + Z_{BC} + Z_{AC}} = \frac{1}{4} = 0.25$

$V_1 = 2.75\, I_1 + 0.25\, I_2$

$V_1 = Z_{11}\, I_1 + Z_{12}\, I_2$

$Z_{11} = 2.75$ & $Z_{12} = 0.25\ \Omega$

6. Ans. (d)

Solution: At, $t = 0^-$ i.e. final, condition it mean capacitor is open circuit and L → Short circuit

At, $t = 0^+$, i.e. initial condition it mean the voltage across capacitors is – V i.e. $V_C(0^-) = V$

At $t = 0^+$ i.e. initial condition

C→ Short and L → Open circuit

Capacitor in series with R is shorted, inductor open

Since loop is open circuit $e_i = 0$

7. Ans. : (d)

Solution: Transformed circuit representation is as shown below

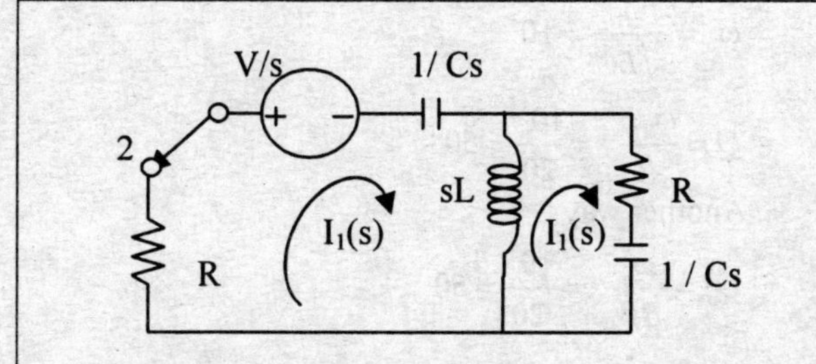

Using KVL in both the loops, we get

$I_1(s) \cdot [1/Cs + sL + R] - I_2 \cdot sL + [V/s] = 0$

And $I_2(s)\,[R + (1/Cs) + Ls] - I_1\, Ls = 0$

Writing in matrix form,

$$\begin{pmatrix} R+sL+\dfrac{1}{Cs} & -Ls \\ -Ls & R+Ls+\dfrac{1}{Cs} \end{pmatrix}\begin{pmatrix} I_1(s) \\ I_2(s) \end{pmatrix} = \begin{pmatrix} -V/s \\ 0 \end{pmatrix}$$

8. Ans. (c)

Solution:

$$H(f) = \frac{1}{j2\pi R\, f\, C + 1}, \quad H(0) = 1$$

Now, as $\dfrac{|H(f_1)|}{H(0)} \geq 0.95$

Then $\dfrac{1}{\sqrt{4\pi^2 R^2 f_1^2 C^2 + 1}} \geq 0.95$

or $4\pi^2 R^2 f_1^2 C^2 \leq 0.108$

or $(f_1)_{\max} = \dfrac{\sqrt{0.018}}{2\pi\, RC} = \dfrac{\sqrt{0.018}}{2 \times 3.14 \times 10^{-3}} = 52.2\ Hz$

9. Ans. (a)

Solution:

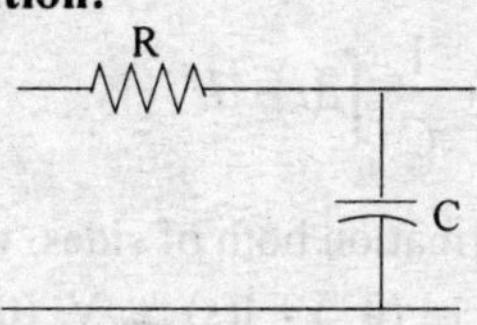

$$H(\omega) = \frac{1}{1 + j\omega RC}$$

$$\theta(\omega) = -\tan^{-1} \omega RC$$

Group delay $\tau_g = \dfrac{-d\theta(\omega)}{-d\omega} = \dfrac{RC}{1 + \omega^2 R^2 C^2}$

$$= \frac{10^{-3}}{1 + 4\pi^2 \times 10^4 \times 10^{-6}}$$

$$= 0.717\ \text{ms}$$

10. Ans. (d)

Solution:

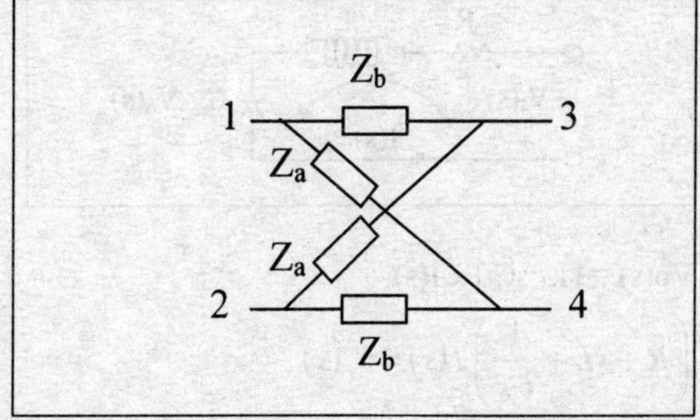

$V_1 = I_1 Z_{11} + I_2 Z_{12}$

$V_2 = I_1 Z_{21} + I_2 Z_{22}$

$$Z_{11} = \left.\frac{V_1}{I_1}\right|_{I_2=0} = 0$$

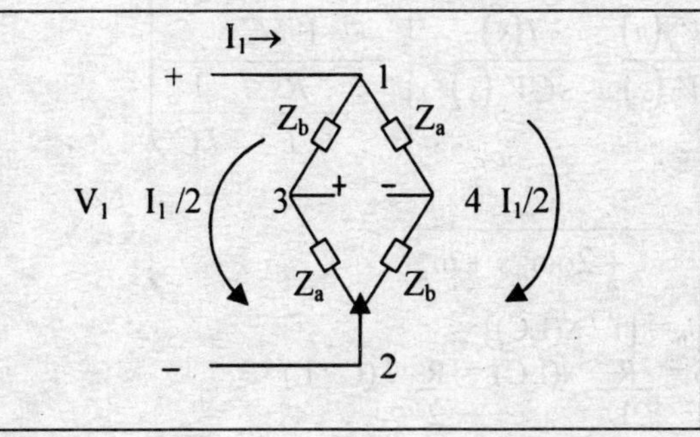

KVL

$V_1 = [I_1/2]\ \{Z_a + Z_b\}$

$0 = -V_2 - [I_1/2]\, Z_b + [I_1/2]\, Z_a$

$V_2 = [I_1/2]\ \{Z_a - Z_b\}$

Where, $Z_1 = Z_b = 2$

$Z_2 = Z_a = 2j$

$Z_{11} = [V_1/I_1] = [1/2]\ \{Z_a + Z_b\} = j + 1$

$Z_{21} = [V_2/I_1] = [1/2]\ \{Z_a - Z_b\} = j - 1 = -1 + j$

If series impedance in a circuit is equal then network is symmetrical. i.e. $Z_{11} = Z_{22}$

If cross arm impedance is equal then network is reciprocal i.e. $Z_{12} = Z_{21}$

$$\begin{pmatrix} Z_{11} & Z_{12} \\ Z_{21} & Z_{22} \end{pmatrix} = \begin{pmatrix} 1+j & -1+j \\ -1+j & 1+j \end{pmatrix}$$

11. Ans. (b)
Solution:

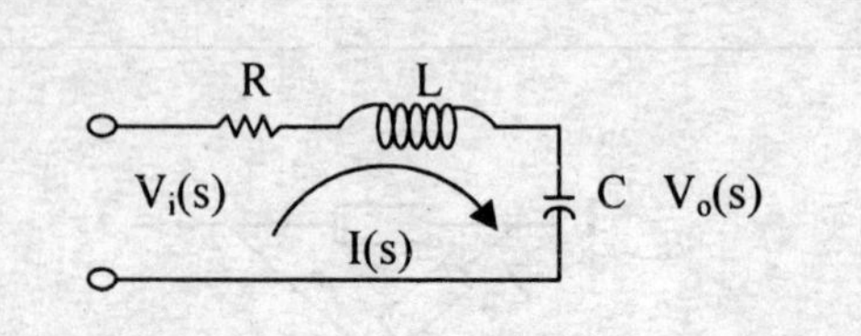

Vo(s) = [1 / sC] x I(s)

$$\left(R + sL + \frac{1}{Cs}\right) I(s) = V_i(s)$$

$$\frac{I(s)}{V_i(s)} = \left(\frac{s}{L\left(s^2 + \frac{Rs}{L} + \frac{1}{LC}\right)} \right)$$

$$\frac{V_o(s)}{V_i(s)} = \frac{I(s)}{sCV_i(s)} = \left(\frac{1/LC}{s^2 + \frac{Rs}{L} + \frac{1}{LC}} \right)$$

$$= \frac{\omega_n^2}{s^2 + 2\delta\omega_n s + \omega_n^2}$$

$\omega_n = [1 / \sqrt{(LC)}]$

$\delta = \frac{R}{2L} \sqrt{(LC)} = \frac{R}{2} \sqrt{(C / L)}$

Given transfer function = $\frac{10^6}{s^2 + 20s + 10^6} = \frac{V_o(s)}{V_i(s)}$

$\omega = 10^3 = [R / Q] [\sqrt{(C / L)}]$

$Q = [\omega L / R] = [1 / R\omega C]$

$Q^2 = [L / R^2 C]$

$Q = \frac{1}{R} [\sqrt{(L / C)}] = \frac{1}{2\delta} = \frac{1}{2 \times 10^{-2}}$

= 50

Second Method

$$H(s) = \frac{10^6}{s^2 + 20\,s + 10^6} \cong \frac{\omega_n^2}{s^2 + 2\delta\omega_n s + \omega_n^2}$$

For series RLC circuit

$$H(s) = \frac{1/LC}{s^2 + \frac{R}{L}s + \frac{1}{LC}}$$

Comparing above equations

$$\frac{1}{LC} = 10^6 \text{ and } \frac{R}{L} = 20$$

$$\omega_o = \frac{1}{\sqrt{LC}} = 10^3$$

$$Q = \frac{\omega_o L}{R} = \frac{10^3}{20} = 50$$

Another way

$$Q_o = \frac{\omega_o}{BW} = \frac{10^3}{20} = 50$$

Hence Bandwidth can be derived as $\frac{R}{L}$ rad/sec

12. Ans. (d)
Solution:

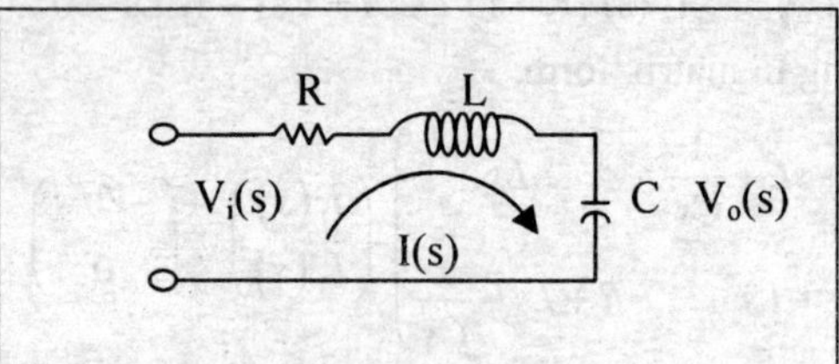

R = 10 kΩ, L = 10 mH , C = 100 μF

$V_i(s) = R_I(s) + Ls \cdot I(s) + [1 / Cs]\ I(s)$

And $V_c(s) = [1 / Cs]\ I(s)$

Then, $\frac{V_c(s)}{V_i(s)} = \frac{1}{LCs^2 + RCs + 1}$

Or, $\frac{V_c(s)}{V_i(s)} = \frac{10^6}{s^2 + 10^6 s + 10^6}$

13. Ans. (b)
Solution: V(t) = u (t)

Using KVL around the loop, we have

$$V(t) = Ri(t) + \frac{Ld\,i(t)}{dt} + \frac{1}{C}\int_0^\infty i(t) \cdot dt$$

Taking Laplace transformation both of sides, we have

$V(s) = RI(s) + Ls\,I(s) - LI(0+) + \frac{I(s)}{sC} + \frac{V_c(0+)}{s}$

$\frac{1}{s} = I(s) + sI(s) - 1 + \frac{I(s)}{s} - \frac{1}{s}$

$\frac{2}{s} + 1 = \frac{I(s)}{s}[s^2 + s + 1]$

$I(s) = \frac{s+2}{s^2 + s + 1}$

14. Ans. (d)

Solution: In series

The impedance, Z = R + j [ωL – (1 / ωC]

At resonance, ωL – [1 / ωL] = 0

Thus Z = R [purely resistive] ; [Q = (ωL / R)]

In parallel R – L – C circuit, Q – Factor varies inversely proportional to G.

$$Q=\frac{R}{\omega L}=\frac{1}{\omega LG}$$

15. Ans. (a)

Solution:

Given , $\frac{d^2y}{dt^2}+3\frac{dy}{dt}+2y=x(t)$

Taking Laplace transformation both sides, we have

$[s^2 + 3s + 2]$ Y (s) = X (s) = [2 /s]

or, $Y(s)=\frac{2}{s(s+1)(s+2)}=\frac{1}{s}-\frac{2}{s+1}+\frac{1}{s+2}$

Inverse Laplace transformation gives,

$Y(t) = \mathbf{(1 - 2e^{-t} + e^{-2t})\, u(t)}$

16. Ans. (d)

Solution: $V_1 = AV_2 - BI_2$

$I_1 = CV_2 - DI_2$

As port 2 is terminated with RL.

Put $V_2 = -I_2R_L$

$\frac{V_1}{I_1} = \frac{AR_L + B}{CR_L + D}$

17. Ans. (b)

Solution:

$V_1 = I_1 r_e$

$V_2 = -r_o (I_2 - \beta I_1)$

$V_2 = -\beta I_1 r_o + I_2 r_o$

Compare with below equation

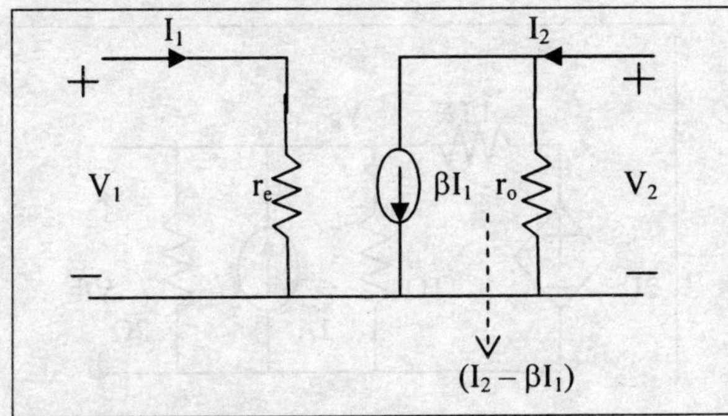

$V_1 = Z_{11}I_1 + Z_{12} I_2$

$V_2 = Z_{21}I_1 + Z_{22} I_2$

We get $Z_{12} = 0$ and $Z_{21} = -\beta r_o$

18. Ans. (b)

Solution:

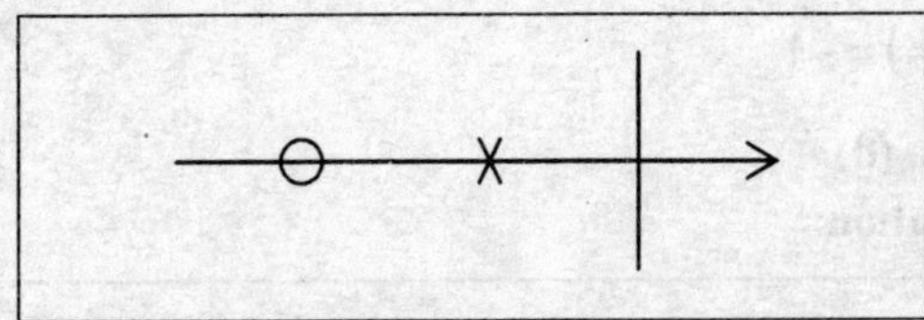

RL – driving point admittance or

RC – driving point impedance

19. Ans. (a)

Solution: Initial current = i (0^-)

$V_L(s) = sLI(s) + L\, i\,(0^-)$

Voltage source is represent by L i (0^-) = 1mV

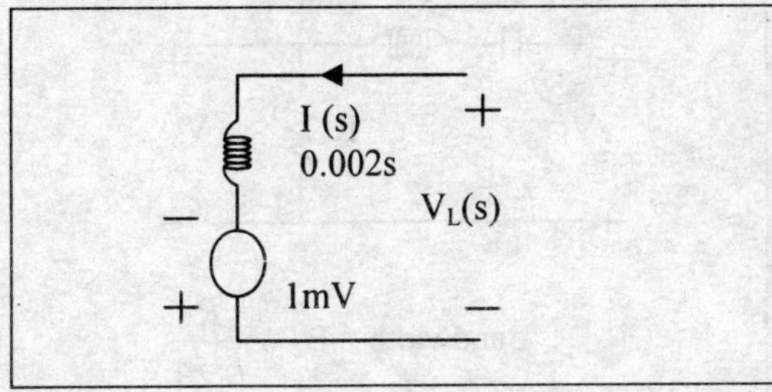

L x i (0^-) = 1 mV

⇒ 2 x 10^{-3} x i(0^-) = 1 x 10^{-3}

⇒ i (0^-) = [1 / 2] = 0.5 A

20. Ans. (b)

Solution:

Under steady state condition

C → open circuit

Final $= V_o(\infty)=\frac{4}{5}\times 10 = 8\,V$

$\tau = R_{eq}\, C_{eq} = (1\,K \parallel 4\,K)\ (4\,\mu F \parallel 1\,\mu F)$

$=\frac{4}{5}k\Omega\times 5\,\mu F = 4\,ms$

$V_o(0)=$ Initial value $= 0\,V$

Apply the formula

$V_o(t)=V_o(\infty)-[V_o(\infty)-V_o(0)]e^{-t/\tau}$

$= 8-[8-0]e^{-t/0.004} = 8[1-e^{-t/0.004}]$ volt

21. Ans. (a)

Solution: $Y(s)=\frac{1}{s(s-1)}$

$$y(t) =: LT^{-1}\left(\frac{-1}{s}+\frac{1}{s-1}\right)=\left(-1+e^{-t}\right)u(t)$$

$$y(\infty)=-1$$

22. Ans. (d)

Solution:

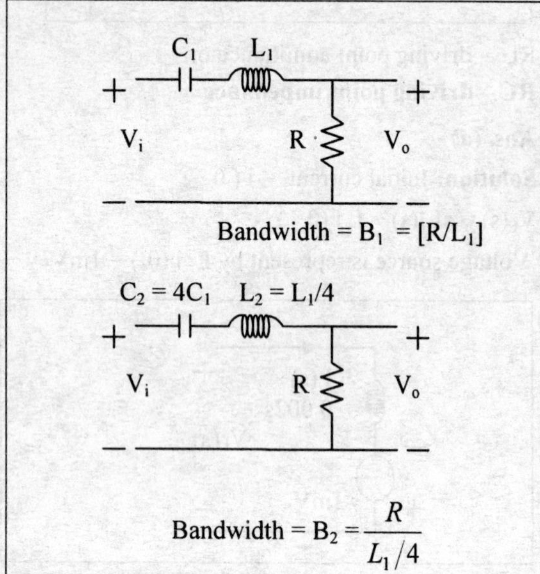

$$\frac{B_1}{B_2}=?$$

For series RLC, Bandwidth = R/L

$$\therefore\ B_1=\frac{R}{L_1}\ ;\ B_2=\frac{R}{L_1/4}$$

$$\therefore\ \frac{B_1}{B_2}=\frac{1}{4}$$

23. Ans. (d)

Solution:

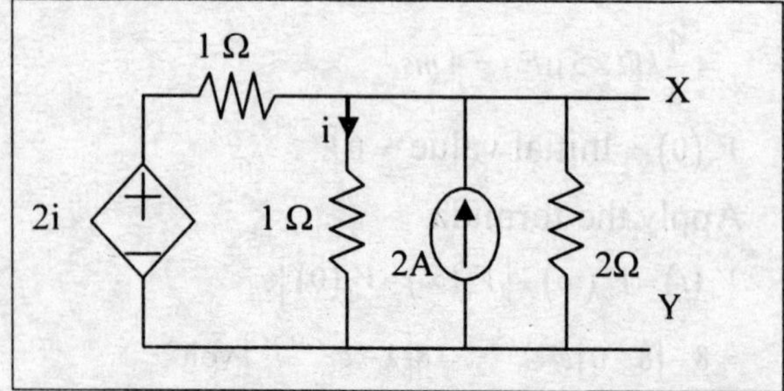

For Thevenin's impedance R_{th} remove the independent source and dependence source remain as it is. Apply 1 volt source to the X, Y terminal and finding R_{th} = [1 /I]

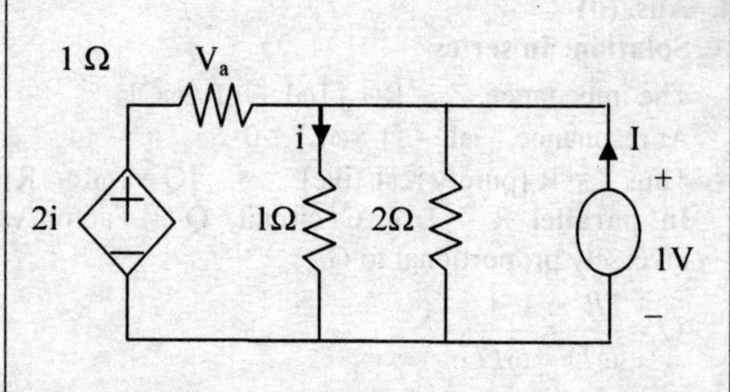

KCL at Node

$$\frac{2i-V_a}{1}=i+\frac{V_a}{2}-I$$

$$V_a=1V,\quad i=\frac{1V}{1\,\Omega}=1\text{A, then}$$

$$\therefore\ 1=1+\frac{1}{2}-I\quad ;\quad I=\frac{1}{2}$$

$$\therefore\ R_{th}=\frac{1}{I}=2\ \Omega$$

For Thevenin's voltage V_{th}

$$\frac{2i-V_a}{1}+2=i+\frac{V_a}{2}$$

$$\text{or, } 2i-i+2=V_a+\frac{V_a}{2}\quad \text{or,}\quad i+2=3\frac{V_a}{2}$$

Also $V_a=i\times 1=i$

$$\therefore\ i+2=3\frac{i}{2}\Rightarrow 2=\frac{i}{2}\Rightarrow i=4$$

$$\text{or } 2=\frac{i}{2}$$

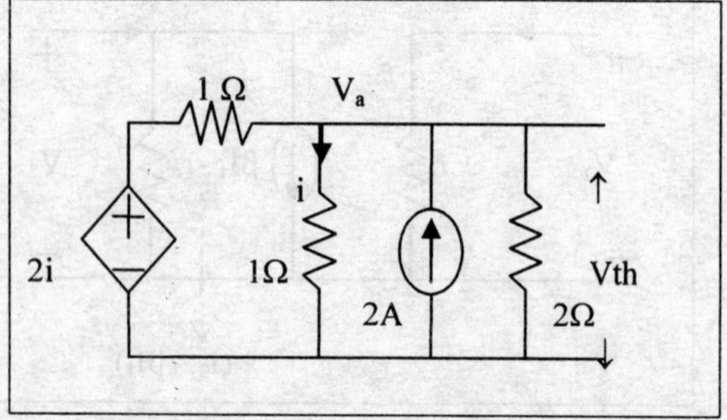

or $\quad i=4$

$$\therefore\quad V_{th}=V_a=i=4V$$

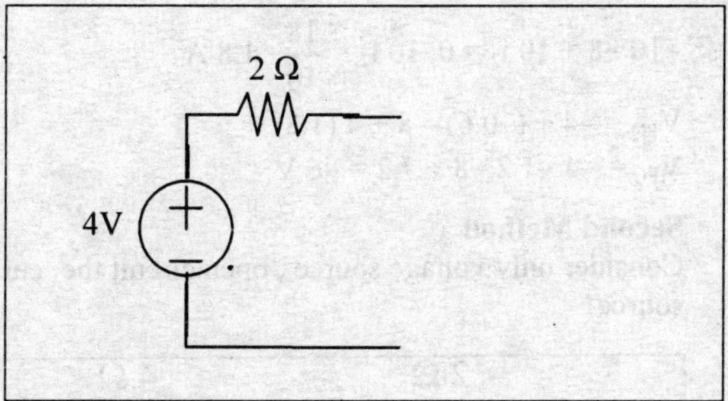

24. Ans. (a)

Solution:

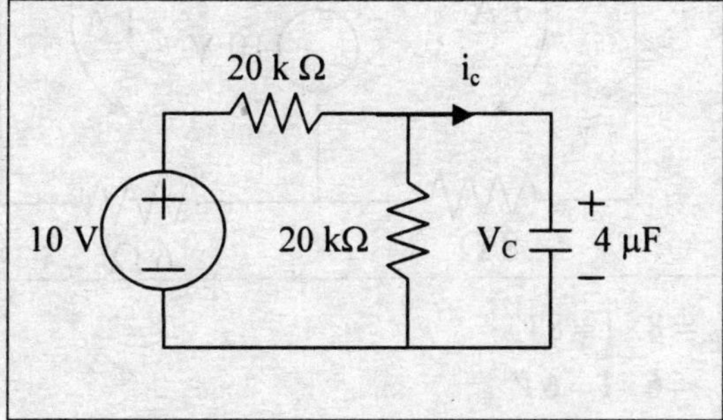

$$\frac{10}{S} = 20,000\,I(S) + \frac{20,000\,I(S)}{20,000 \times 4 \times 10^{-6} S + 1}$$

$$\frac{10}{S} = 20,000\,I(S)\left[1 + \frac{100/8}{S + \frac{100}{8}}\right]$$

$$I(S) = \frac{10}{S} \times 20,000 \times \left[\frac{S + \frac{100}{8}}{S + \frac{200}{8}}\right]$$

$$I_C(S) = \frac{20,000\,I(S)}{20,000 + \frac{1}{4 \times 10^{-6} S}}$$

$$= \left[\frac{\frac{8S}{100}}{\frac{8S}{100} + 1}\right] \times \frac{10}{S} \times \frac{1}{20,000}\left[\frac{S + \frac{100}{8}}{S + \frac{200}{8}}\right]$$

$$= \left[\frac{(S)}{S + \frac{100}{S}}\right] \times \frac{10}{(S)} \times \frac{1}{20,000}\left[\frac{S + \frac{100}{8}}{S + 25}\right]$$

$$\therefore\ i_c .0005\,e^{-25t}$$

$$i_c = 0.5\,e^{-25t} \ \ \textbf{mA}$$

Second Method

$$i_c(t) = i_{c(\text{initial})}\, e^{-\frac{t}{RC}} + i_{c(\text{initial})}\left(1 - e^{-\frac{t}{RC}}\right)$$

$$i_c(t) = i_{c(\text{final})} + \left(\text{i}_{\text{c initial}} - \text{i}_{\text{c final}}\right) e^{-\frac{t}{RC}}$$

$$R' = 20 \parallel 20 \quad = 10 \text{ k ohm's}$$

$$i_{c\ initial} = 0 \ ;\ i_{c\ final} = \frac{10}{20} = 0.5$$

Thus $i_c(t) = 0.5\,e^{-25t}$

25. Ans. (d)

Solution:

$$R_{AB} = \frac{(5-3j)(5+3j)}{(5-3j)+(5+3j)} = \frac{25+9}{10} = 3.4\,\Omega$$

$$V_{AB} = R_{AB} \times I \ = 3.4 \times 5\angle 30 \ = 17\angle 30$$

26. Ans. (a)

Solution:

$$f(t) = e^{-t} u(t)$$

$$F(s) = \frac{1}{s+1}$$

For frequency domain, Put, $s = j\omega$

$$F(j\omega) = \frac{1}{(1+j\omega)} \ = \frac{1}{\sqrt{1+\omega^2}}\angle \tan^{-1}\omega$$

For 2 dB bandwidth, $\frac{1}{\sqrt{1+\omega^2}} = \frac{1}{\sqrt{2}}$

$$\therefore \quad \omega = 1$$

$$\omega = 2\pi f_c$$

For $f_c = \frac{1}{2\pi}$

Second Method

$$f(t) e =^{-t} u(t)$$

$$F(s) = \frac{1}{s+1}$$

$$F(j\omega) = \frac{1}{1+j\omega}$$

At 2-dB bandwidth

$$|F(j\omega)| = \frac{1}{\sqrt{2}}$$

$$\frac{1}{\sqrt{1+\omega^2}} = \frac{1}{\sqrt{2}}$$

$$1+\omega^2 = 2$$

$$\omega = 1$$

$$2\pi f_c = 1$$

$$f_c = \frac{1}{2\pi}\ Hz$$

27. Ans. (c)

Solution: Given that $V_L = V_C = 2V_R$,

Therefore $V_L = V_C$

Or $X_L = X_C$, represent a resonance condition

and $\therefore$ current, $I = \frac{5}{R} = \frac{5}{5} = 1A$

Now, V_L (Voltage across L) $= 2V_R = 10$

or $I\omega L = 10$

or

$$L = \frac{10}{\omega} = \frac{10}{2\pi f} = \frac{10}{2\pi 50} = 31.8\ mH$$

28. Ans. (c)

Solution: The current source is converted into voltage source Apply KVL in loop.

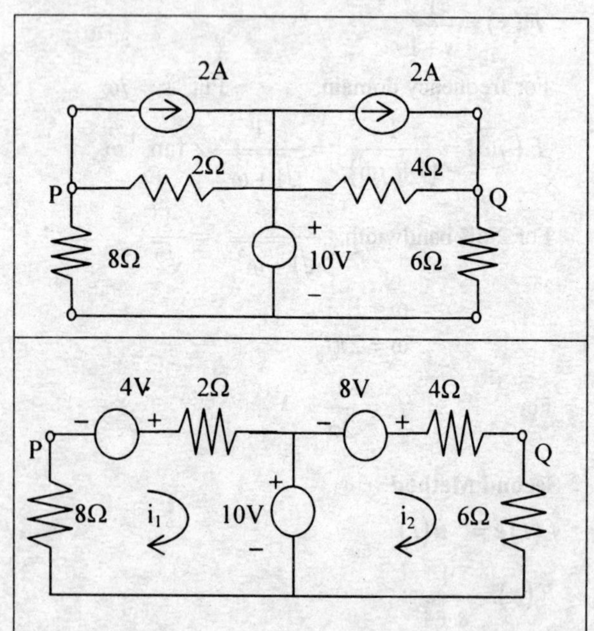

$-4 + 10 + 10\, i_1 = 0,\ i_1 = -\frac{6}{10} = -0.6$ A

$-10 - 8 + 10\, i_2 = 0,\ 10\, i_2 = \frac{18}{10} = 1.8$ A

$V_{PQ} = -4 + (-0.6) - 8 + 4\,(1.8)$

$V_{PQ} = -4 - 1.2 - 8 + 7.2 = -6$ V

Second Method

Consider only voltage source , open circuit the current source

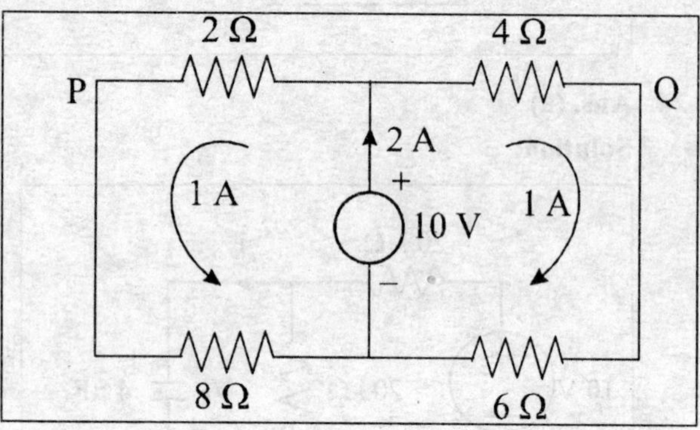

$$V_P = 8 \times 1 = 8\ V$$

$$V_Q = 6 \times 1 = 6\ V$$

$$V_{PQ1} = 8 - 6 = 2\ V$$

Consider only current source, Short circuit the voltage source

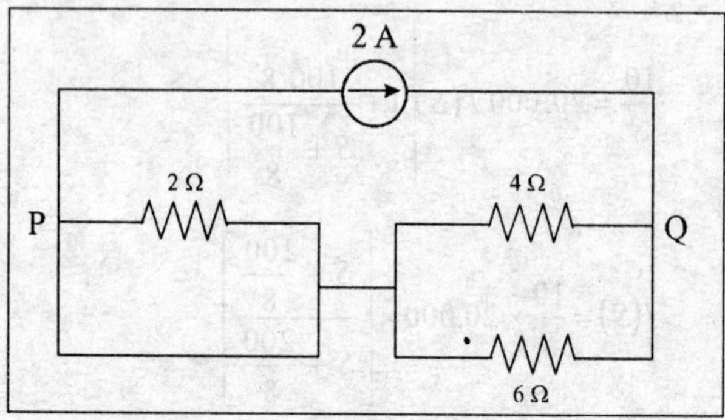

$$V_{PQ2} = -2[2.4 + 1.6] = -8\ V$$

$$V_{PQ} = -8 + 2 = -6\ V$$

29. Ans. (d)

Solution:

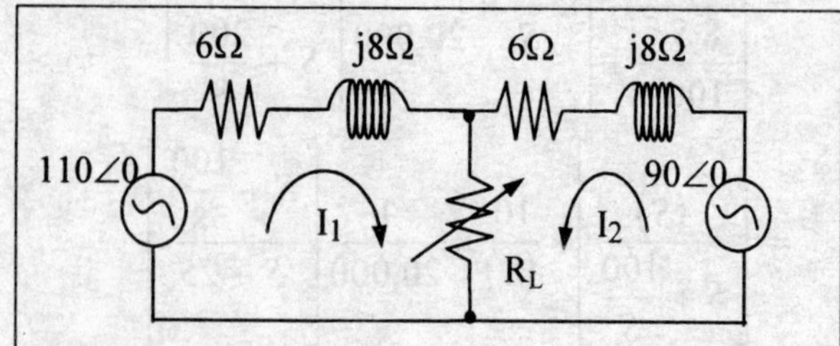

Thevenin resistance looking from R_L after shorting sources is

$$(6+j8) \| (6+j8) = 3+j4$$

$$R_L = \sqrt{3^2+4^2} = 5\Omega$$

Now, $110 = (6+j8+5)I_1 + 5I_2$ (1)

and $90 = (6+j8+5)I_2 + 5I_1$ (2)

Solving these two equations, we get

$$I_1 = 5.5 - 2.75j$$

$$I_2 = 4.5 - 2.2j$$

$\Rightarrow$ Total current in $R_L = I_1 + I_2$

$= (10 - 4.55j)$ A

$= 11.15$ A

Power absorbed by $R_L = I^2R$

$= (11.15)^2 \times 5 = 625$ W

30. Ans. (d)

Solution:

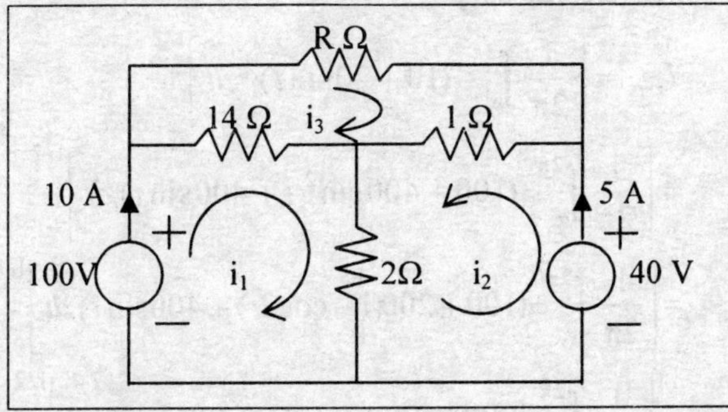

Apply the KVL in all three loop

$100 = 16\, i_1 + 2\, i_2 - 14\, i_3$ (1)

$0 = -14\, i_1 + i_2 + (15+R)\, i_3$ (2)

$i_1 = 10$ A and $i_2 = 5$ A

From equation (1) and (2)

$$100 = 160 + 10 - 14\, i_3$$

$$i_3 = \frac{70}{14} = 5$$

$$0 = -140 + 5 + (15+R)\,5$$

$$-(15+R)\,5 = -135,\ R = 27 - 15 = 12\ \Omega$$

31. Ans. (b)

Solution: When switch is ON at t = 0 capacitor is short circuit and inductor is open circuit.

$\therefore 4 \| 4 = 2\ \Omega$,

circuit total resistance, 3 + 2 = 5Ω

source current $I = \frac{10}{5} = 2A$

current through 4Ω resistance $= \frac{2}{2} = 1A$. Voltage across 4Ω resistance = 4 × 1 = 4V as inductor is parallel with resistance.

32. Ans. (d)

Solution:

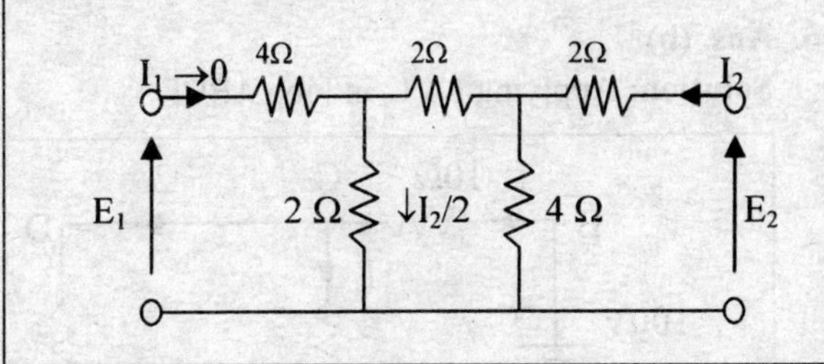

$$h_{12} = \left.\frac{E_1}{E_2}\right|_{I_1=0},\ E_2 = 2I_2 + \frac{I_2}{2}\times 4 = 4I_2$$

$$E_1 = \frac{I_2}{2}\times 2 = I_2$$

$$\therefore \frac{E_1}{E_2} = \frac{I_2}{4I_2} = 0.25$$

33. Ans. (c)

Solution:

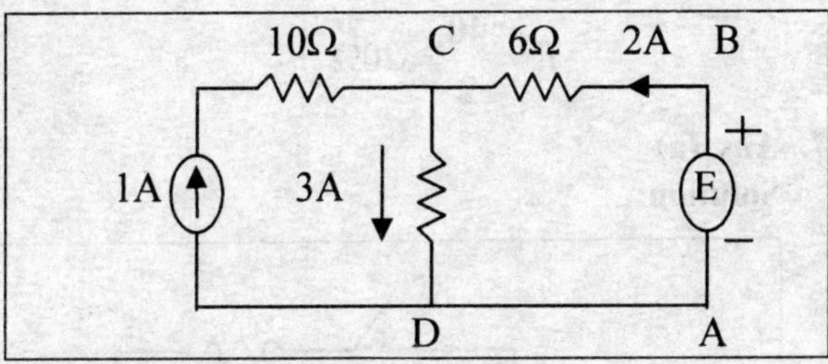

Apply KVL in loop ABCD

$-E + 6\times 2 + 6\times 3 = 0$

$E = 2\times 6 + 3\times 6$

$= 12 + 18 = 30$V

34. Ans. (a)

Solution: Star-delta conversion is carried out by applying the formula w.r.t. diagram.

$$R_1 = \frac{R_b R_c}{R_a + R_b + R_c} = \frac{100}{40} = 2.5\Omega$$

$$R_2 = \frac{R_a R_c}{R_a + R_b + R_c} = \frac{200}{40} = 5\Omega$$

$$R_b = \frac{R_a R_b}{R_a + R_b + R_c} = \frac{200}{40} = 5\Omega$$

35. Ans. (a)

Solution: From the diagram given in the question paper,

$I_R = Y_R . E = 0.5\angle 0° * 10\angle 0° = 5\angle 0°$ A

$I_L = Y_L . E = 1.5\angle -90° * 10\angle 0° = 15\angle -90°$A

$I_C = Y^C . E = 0.3\angle 90° * 10\angle 0° = 3\angle 90°$ A

$\bar{I} = \bar{I}_R + \bar{I}_L + \bar{I}_C = 5 - j15 + j3 = (5 - j12).$

36. Ans. (b)

Solution: Applying KVL, in loop ABCF

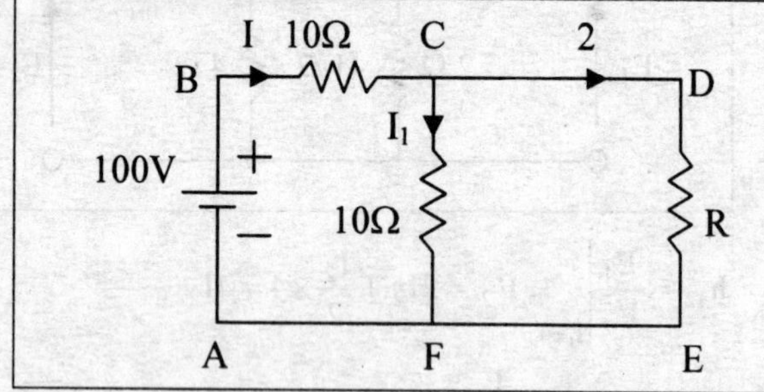

$-100 + 10\,I + 10(I - 2) = 0$

$\Rightarrow\ 20\,I = 100 + 20 = 120$

$\Rightarrow\ I = 6A.$

$I_1 = I - 2 = 6 - 2 = 4A$

Hence voltage drop across resistance R,

$V_R = I_1 . 10 = 4\times 10 = 40V = 2R$

$\therefore\ R = \frac{40}{2} = 20\Omega$

37. Ans. (a)

Solution:

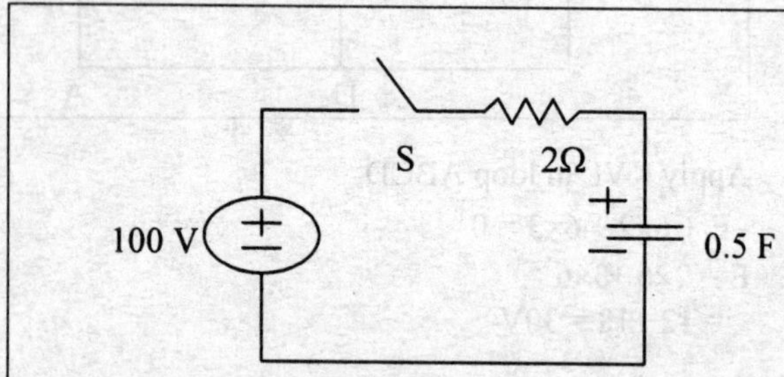

$V = IR + \frac{1}{C}\int I dt$

Or $R\cdot\frac{dq}{dt} + \frac{1}{C}q = 100$

Or $(100 - q) = RC\frac{dq}{dt}$

$$\int_{q_o}^{q}\frac{dq}{(100\,C - q} = \frac{1}{RC}\int_o^t dt$$

$$\Rightarrow\ \ell n[100C - q]_{q_o}^{q} = -\frac{t}{RC}$$

$$\Rightarrow\ \ell n\left[\frac{100C - q}{100C - q_o}\right]^{-1} = \frac{t}{RC}$$

$$\Rightarrow\ \ell n\left[\frac{100C - q}{100C - q_o}\right] = \frac{-t}{RC}$$

$$\Rightarrow\ 100C - q = (100C - q_o)e^{-t/RC}$$

$$\Rightarrow\ -\frac{dq}{dt} = (100C - q_o)e^{-t/RC}\left(t\frac{1}{RC}\right)$$

$$i = \frac{(100C - q_o)e^{-t/RC}}{RC} = 40e^{-1} = 14.7$$

38. Ans. (b)

Solution:

$i = A \sin \omega t\ ;\ \omega = 2\pi/T$

$i(t) = 10 + 20 \sin t$

$$I_{rms} = \left[\frac{1}{2\pi}\int_o^{2\pi}(10 + 20\sin t)^2 .dt\right]^{1/2}$$

$$= \left[\frac{1}{2\pi}\int_o^{2\pi}(100 + 400\sin^2 t + 400\sin t).dt\right]^{1/2}$$

$$= \left[\frac{1}{2\pi}\int_o^{2\pi}(100 + 200(1 - \cos 2t) + 400\sin t).dt\right]^{1/2}$$

$$= \left[\frac{1}{2\pi}\int_o^{2\pi}(300 - 200\cos 2t + 400\sin t).dt\right]^{1/2}$$

39. Ans. (d)

Solution: $Z = \begin{bmatrix} 0.9 & 0.2 \\ 0.2 & 0.6 \end{bmatrix} = \begin{bmatrix} z_{11} & z_{12} \\ z_{21} & z_{22} \end{bmatrix}$

$|Z| = 0.9\times 0.6 - 0.04 = 0.5$

$$Y = \frac{1}{0.5}\begin{bmatrix} 0.6 & -0.2 \\ -0.2 & 0.9 \end{bmatrix} = 2\begin{bmatrix} 0.6 & -0.2 \\ -0.2 & 0.9 \end{bmatrix}$$

$$= \begin{bmatrix} 1.2 & -0.4 \\ -0.4 & 1.8 \end{bmatrix} = \begin{bmatrix} y_{11} & y_{12} \\ y_{21} & y_{22} \end{bmatrix} = \frac{1}{1Z1}\begin{bmatrix} z_{22} & z_{21} \\ z_{12} & z_{12} \end{bmatrix}$$

hence $y_{22} = 1.8$.

40. Ans. (c)

Solution:

At f = 100 HZ: $V_R = V_L$; $R = X_L = \omega L$

$V_{rms} = \frac{V_{in}}{(\sqrt{2})}$; $V_{in} = (\sqrt{2})V_{rms.}$

$$At\ f = 50\ Hz,\ X_L = \frac{\omega L}{2};\ Z = \sqrt{R^2 + \frac{\omega^2 L^2}{4}}$$

$$I_{in} = \frac{V_{in}}{\sqrt{R^2 + \frac{\omega^2 L^2}{4}}} = \frac{V_{in}}{\sqrt{\omega^2 L^2 + \frac{\omega^2 L^2}{4}}} = \frac{2V_{in}}{\omega L\sqrt{5}}$$

$$V_{new} = \frac{2V_{in}}{\omega L\sqrt{5}} \times R = \frac{2V_{in}}{\omega L\sqrt{5}} \cdot \omega L = \frac{2V_{in}}{\sqrt{5}}$$

$$V_{new} = \frac{2.\sqrt{2}\ V_{rms}}{\sqrt{5}} = \sqrt{\frac{8}{5}}\ V_{rms.}$$

41. Ans. (a)

Solution:

$$I_R : I_y : I_B = 1:1:\sqrt{3}$$

∵ I_B is phasor sum of I_R and I_y at angle of 60°

$$I_B = \sqrt{I_R^2 + I_y^2 + 2I_R I_y \cos 60^o} = \sqrt{3}\ I_R$$

42. Ans. (c)

Solution:

At, t = 0^- the circuit is in steady state inductor behave as short circuit, the current through the inductor $i(0^-)$ =10/10 = 1A. The current through the inductor does not change instantaneously. Therefore $i(o^-) = i(o^+) = 1$ A.

43. Ans. (a)

Solution:

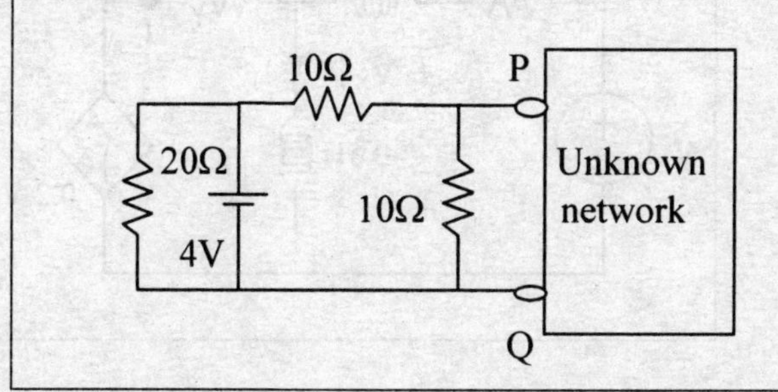

For Z_{eq}, short circuit the voltage source and open circuit the current source, in present case 20Ω is also short circuit, therefore net impedance,

Z_{th} = 10 || 10 = 5 Ω.

For V_{oc}, in Parallel circuit

Current through 4V source

$$I = \frac{4}{20 \parallel (10+10)} = \frac{4}{10} = 0.4$$

I_{PQ} = 0.2 A

$V_{oc} = V_{PQ}$ = 0.2× 10 = 2V.

44. Ans. (a)

Solution:

At, t = 0^-, inductor is short circuit

$$i\ (0^-) = \frac{120}{60} = 2A$$

$i\ (0^-) = i\ (0^+)$ = 2A

$$ie\ 2 = \frac{120}{R + 40 + 20}$$

R = 0 Ω

45. Ans. (c)

46. Ans. (a)

Solution:

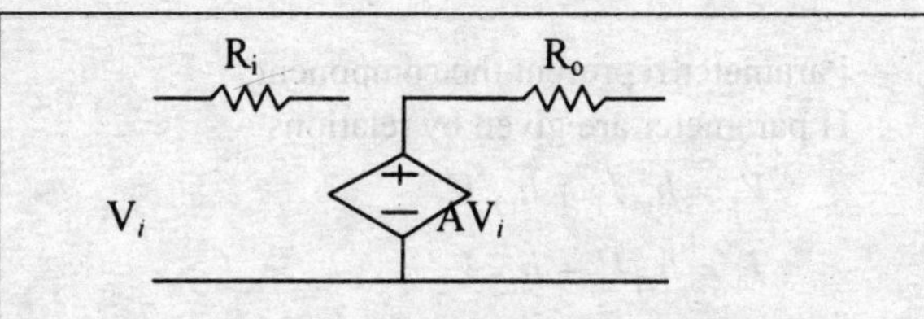

$R_{in} = \infty$

R_{out} = 10Ω

$V_{out} = AV_{in}$ = 1 V

47. Ans. (d)

Solution: By applying superposition theorem

1. Assume the current source = 0

   Since, voltage source is dc, all the inductors will behave like a short circuit while capacitors will behave like a open circuit, therefore after simplifying the circuit, we have

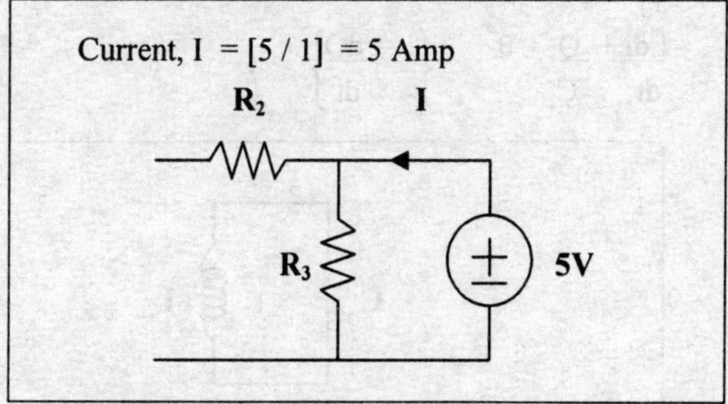

2. Assuming the voltage source to be zero R_3 is shorted

   ∴ No current will flow throw it and I_{ab} = 1 Amp.

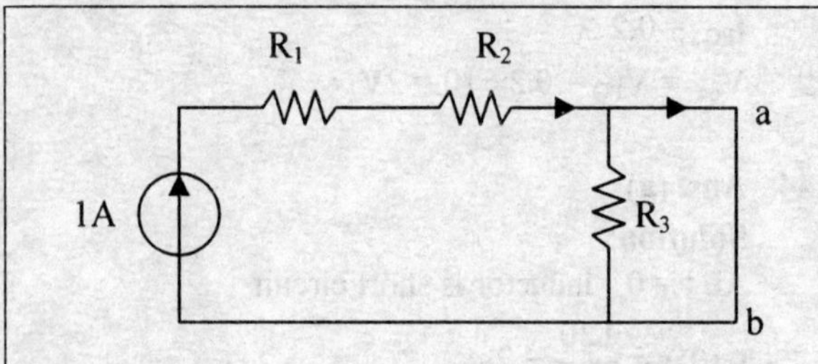

48. Ans. (c)

Solution:

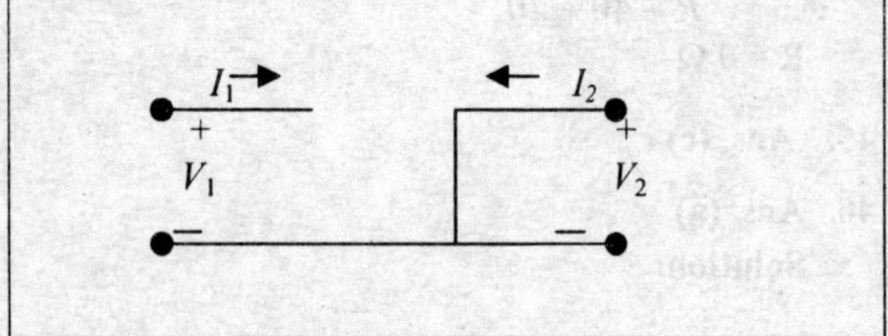

Parameter represent the component

H parameter are given by relations

$$V_1 = h_{11} I_1 + h_{12} V_2$$

$$I_1 = h_{11} I_1' + h_{22} V_2$$

here all the parameter are '0' so

$$[h] = \begin{bmatrix} 0 & 0 \\ 0 & 0 \end{bmatrix}$$

49. Ans. (a)

Solution: In Resonance circuit voltage V_1 and current I are in phase, The voltage across the capacitor lag the current by 90°

Hence answer is (a)

50. Ans. (b)

Solution: By KVL,

$V_L + V_C = 0$

$\frac{Ldi}{dt} + \frac{Q}{C} = 0 \quad ; \left\{ i = \frac{dQ}{dt} \right\}$

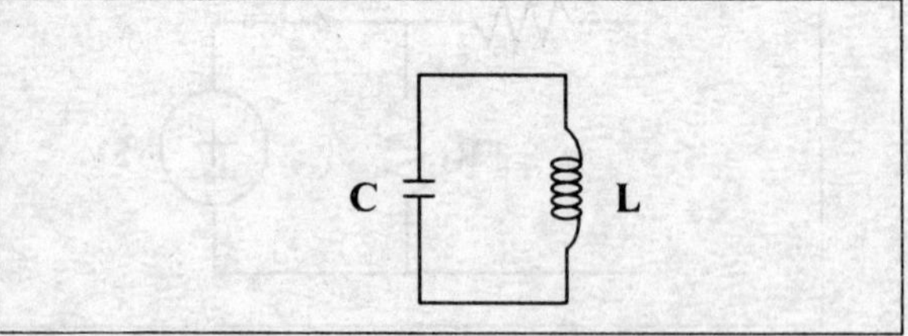

$\frac{Ld^2Q}{dt^2} + \frac{Q}{C} = 0$

$Q(t) = Q_o \cos(\omega t + \phi)$

$Q(t) = Q_o \cos(\omega_0 t)$

$V(t) = \frac{Q(t)}{C} = \frac{Q_o}{C} \cos \omega_o t = V_o \cos \omega(\omega_o t)$

$V(t) = V_o \cos(\omega_o t)$

In other way,

It can be explained as if both capacitor and inductor an ideal then Electrical energy stored in the capacitor changes its form by storing energy in inductor in the form of magnetic energy. The change of the energy is continuous. The energy is in the from of voltage oscillate and voltage across capacitor $V_c = V_o \cos(\omega_o t)$

51. Ans. (b)

Solution: Energy = Power × time

$$\text{Power} = \frac{(\text{Voltage})^2}{\text{Resistance}}$$

$E = [(V^2 / R)]\, t$

$2300 = [(230)^2 / R] \times 1$

$\Rightarrow R = 23\ \Omega$

$$\text{rms value} = \frac{200^2 \times (T/2) + (-200)^2 \times (T/2)}{T}$$

$= 200$

$p = [V^2 / R] = [(200)^2 / 23] = 1739.13$ watt

$= 1.73913$kW

52. Ans. (a)

Solution:

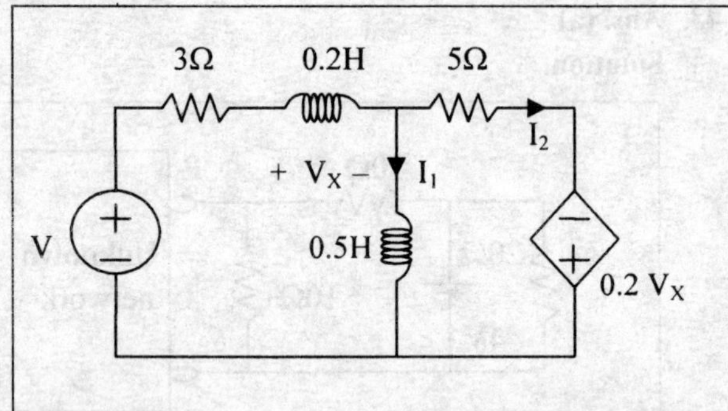

$$-V + 3(I_1 + I_2) + V_X + 0.5\frac{dI_1}{dt} = 0 \qquad (1)$$

$$-0.5\frac{dI_1}{dt} + 5I_2 - 2.V_X = 0 \qquad (2)$$

Solving, equation (1) and (2) by illuminating I_2

$$\frac{dI_1}{dt} = \frac{5}{4}V - \frac{15}{4}I_1 - 1.4\,V_x$$

53. Ans. (d)

54. Ans. (d)

Solution:

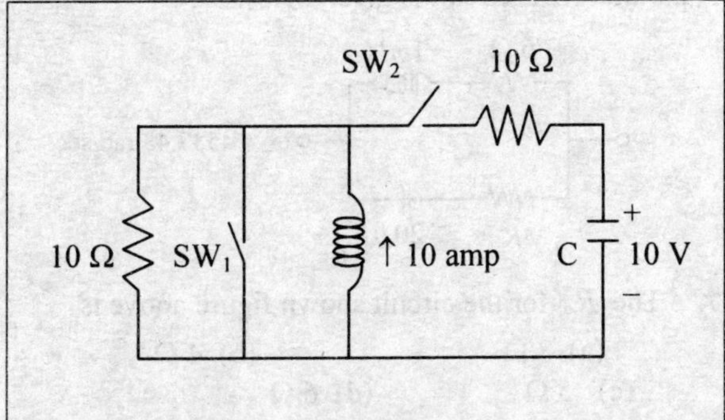

At t = 0⁻, SW_2 is closed and SW_1 is opened

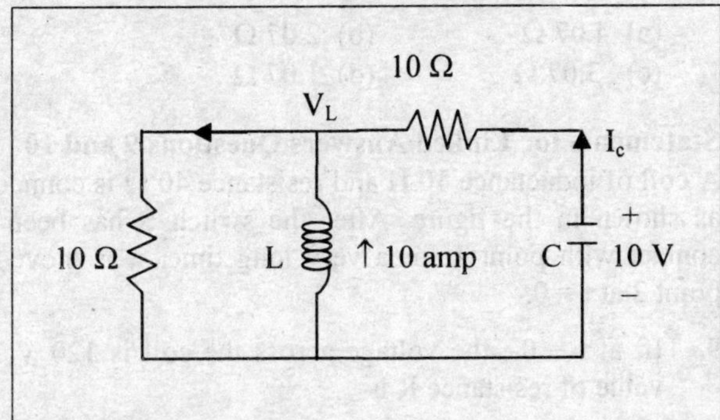

Voltage across $L = V_L$

Apply KCL at Node

$$\frac{V_L}{10} = 10 + \frac{10 - V_L}{10}$$

$$\frac{V_L}{5} = 11\ V;\ V_L = 55\,V$$

Voltage across C and current through the L does not change instantaneously

KVL in load loop

Therefore, at $t = 0^+$

$$-10 + 10 \times I_c + V_L$$

$$V_c = 10 \times I_c + V_L$$

$$10 = 10\,I_c + 55$$

$$I_c = -4.5$$

Another Way

Current through C will be

$$\frac{V_L - 10}{10} = \frac{55 - 10}{10} = 4.5\,A$$

55. Ans. (a)

56. Ans. (c)

Solution:

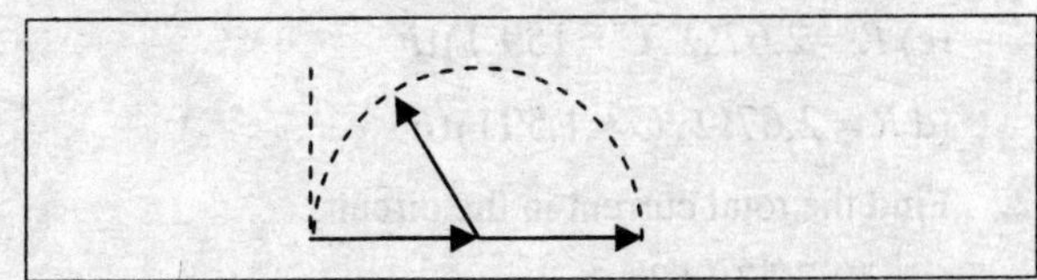

$$V_{YX} = V_X - V_Y\ \text{V}$$

$$= V - 0 \text{ for } R = \propto\ = V - 2V \quad R = 0$$

57. Ans. (a)

Solution: Total voltage drop

$$V = IR + I^2$$

$$3 = 2I + I^2$$

$$I^2 + 2I - 3 = 0$$

Neglect the negative value of I

∴ I = 1 amp

$$\therefore\ V_{NL} = 1V$$

$$\left(\because\ V_{NL} = I^2\right)$$

Power = 1

58. Ans. (c)

Solution:

$$\rightarrow V_1 \quad V_2 \quad V_3 \quad V_4 \quad V_5 \quad V_6$$

$$\downarrow$$

$$\begin{matrix} e_1 \\ e_2 \\ e_3 \\ e_4 \end{matrix} \begin{bmatrix} 1 & 1 & 1 & 0 & 0 & 0 \\ 0 & -1 & 0 & -1 & 1 & 0 \\ -1 & 0 & 0 & 0 & -1 & -1 \\ 0 & 0 & -1 & 1 & 0 & 1 \end{bmatrix}$$

$$E = [e_1\ e_2\ e_3\ e_4]^T = AV$$

Linked Answers Questions 1 and 2

For the gien RLC series circuit

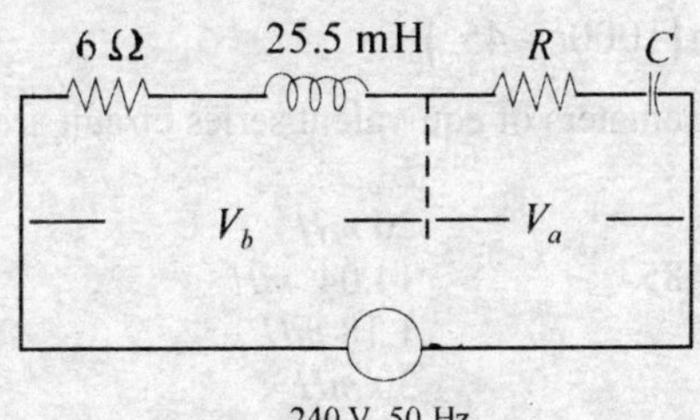

240 V, 50 Hz

1. Find the value of R and C so that $V_b = 3V_a$ and phase angle between V_b and V_a is $90°$.

(a) $R = 2.67\Omega,\ C = 1591\ \mu F$
(b) $R = 26.7\Omega, C = 159.1\ \mu F$
(c) $R = 2.67\Omega, C = 159.1\ \mu F$
(d $R = 2.67\Omega, C = 1.591\ \mu F$

2. Find the total current in the circuit.
(a) $22.77\angle 30.63^\circ$
(b) $22.77\angle 34.68^\circ$
(c) $22.77\angle -34.68^\circ$
(d) $32.13\angle 132.5^\circ$

Statements for Linked Answers Questions 3 and 4

The V–I relation for the network shown in given as $V = 4I - 9$.

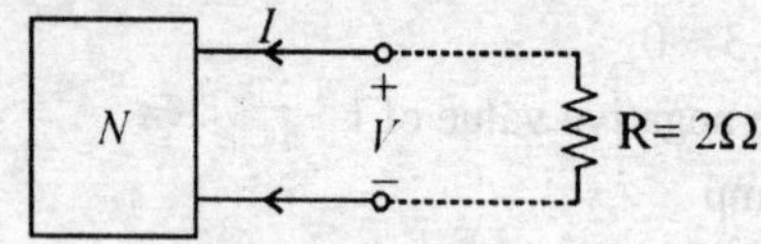

The equivalent circuit to network N is shown as

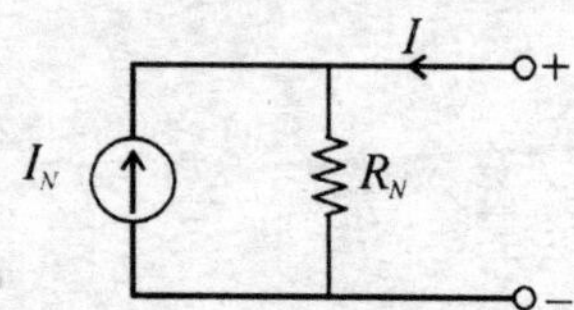

3. The voalues of I_N and R_N are
(a) 2.25 A, 4 Ω (b) −2.25 A, 4Ω
(c) 0, 4 Ω (d) None of these

4. The now a resistance of $R = 2\ \Omega$ is connected across it the current through it is
(a) −4.5 A (b) −1.5 A
(c) 1.5 A (d) 4.5 A

Statements for Linked Answers Questions 5 and 6

A resistance of 5 Ω and inductance of 0.02 H are connected in parallel and fed from a source having $v = 100\sin\left(1000t + 45^\circ\right)$

5. The parameters of equivalent series circuit are

	R_{eq}	L_{eq}
(a)	5	20 mH
(b)	14.85	14.04 mH
(c)	4.7	1.18 mH
(d)	2	30 mH

6. Find the power dissipated in the above circuit
(a) 285.3 W (b) 1000.07 W
(c) 1807.8 W (d) None of the given

Statements for Linked Answers Questions 7 and 8

A parallel RLC circuit is given below

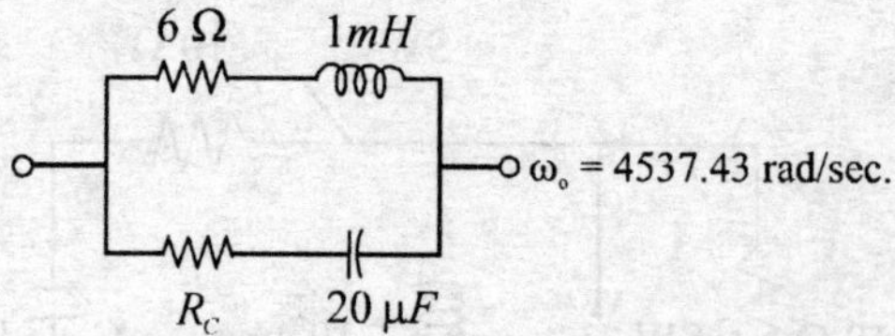

7. The R_C for the circuit shown figure above is
(a) 3 Ω (b) 4 Ω
(c) 5 Ω (d) 6 Ω

8. What is the positive maximum change in the value of R_C for which resonance can occur?
(a) 4.07 Ω (b) 2.07 Ω
(c) 3.07 Ω (d) 1.07 Ω

Statements for Linked Answers Questions 9 and 10

A coil of inductance 10 H and resistance 40 Ω is connected as shown in the figure. After the switch S has been in-contact with point 1 for a very long time, it is moved to point 2 at t = 0.

9. If, at t = 0^+, the voltage across the coil is 120 V, the value of resistance R is

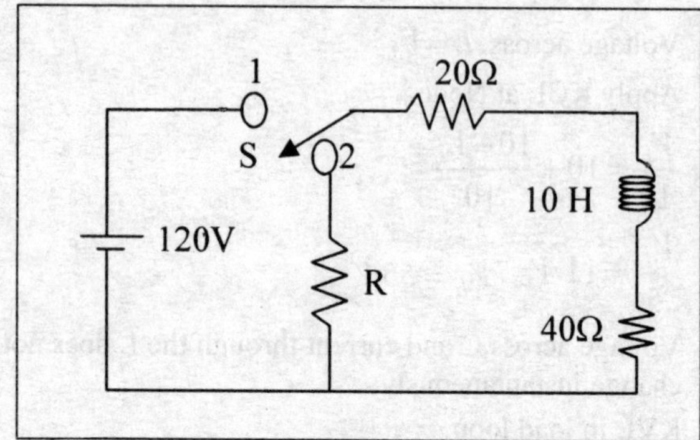

(a) 0 Ω (b) 20 Ω
(c) 40 Ω (d) 60 Ω

10. For the value of obtained in (a), the time taken for 95% of the stored energy to be dissipated is close to
a) 0.10 sec b) 0.15 sec
c) 0.50 sec d) 1.0 sec

Common data Q11 and Q12

The circuit for Q 11 and Q. 12 is given. Assume that the switch S is in position 1 for a long time and thrown to position 2 at $t = 0$.

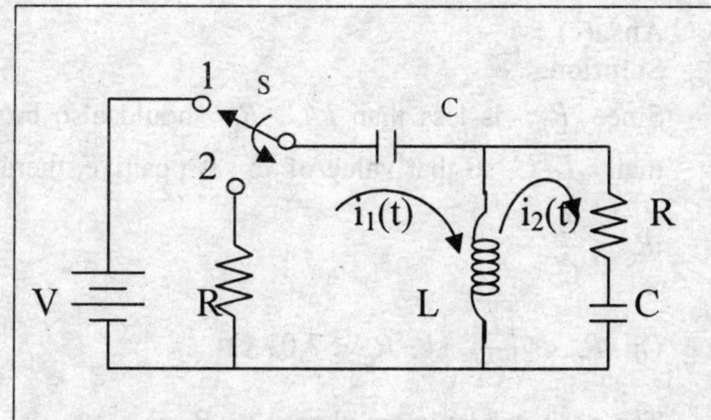

11. At, $t = 0^+$, the current i_1 is:

(a) (–V /2R) (b) (–V / R)
(c) (-V/4R) (d) zero

12. $I_1(s)$ and $I_2(s)$ are the Laplace transforms of $i_1(t)$ and $i_2(t)$ respectively. The equations for the loop currents $I_1(s)$ and $I_2(S)$ for the circuit shown in figure (Q.82) after the switch is brought from position 1 to position 2 at = 0, are

(a) $\begin{pmatrix} R+Ls+\frac{1}{Cs} & -Ls \\ -Ls & R+\frac{1}{Cs} \end{pmatrix}\begin{pmatrix} I_1(s) \\ I_2(s) \end{pmatrix}=\begin{pmatrix} V/s \\ 0 \end{pmatrix}$

(b) $\begin{pmatrix} R+Ls+\frac{1}{Cs} & -Ls \\ -Ls & R+\frac{1}{Cs} \end{pmatrix}\begin{pmatrix} I_1(s) \\ I_2(s) \end{pmatrix}=\begin{pmatrix} -V/s \\ 0 \end{pmatrix}$

(c) $\begin{pmatrix} R+Ls+\frac{1}{Cs} & -Ls \\ -Ls & R+Ls+\frac{1}{Cs} \end{pmatrix}\begin{pmatrix} I_1(s) \\ I_2(s) \end{pmatrix}=\begin{pmatrix} V/s \\ 0 \end{pmatrix}$

(d) $\begin{pmatrix} R+Ls+\frac{1}{Cs} & -Ls \\ -Ls & R+Ls+\frac{1}{Cs} \end{pmatrix}\begin{pmatrix} I_1(s) \\ I_2(s) \end{pmatrix}=\begin{pmatrix} -V/s \\ 0 \end{pmatrix}$

Common Data Questions 13 and 14

The system under consideration is are RC Low-Pass filter (RC-LPF) with $R = 1.0\ \text{k}\Omega$ and $C = 0.01\ \mu F$.

13. Let $H(f)$ denote the frequency response of the RC-LPF. Let f_1 be the highest frequency such that $0 \le |f| \le f_1, \frac{|Hf_1|}{H(0)} \ge 0.95$. Then f_1 (in Hz) is

(a) 327.8 (b) 163.9
(c) 52.2 (d) 104.4

14. Let $t_g(f)$ be the **group delay** function of the given RC-LPF and $f_2 = 100$ Hz. Then $t_g(f_2)$ in ms, is

(a) 0.717 (b) 7.17
(c) 71.7 (d) 4.505

Answers

Basic of Circuit and Measurement System

1. **Ans. (a)**
Solution:
From the given circuits and its data

$X_L = 2\pi \times 50 \times 25.5 \times 10^{-3} = 8\Omega$

$Z_b = 6 + j8 = 10\angle 53.33^\circ \Omega$

$I = |I|\angle\theta$

$V_b = Z_b \cdot I = |10I|\angle\theta + 53.13^\circ$

$V_a = (|I|\angle\theta)(R - jX_C)$

$= |I|[R^2 + X_c^2]^{\frac{1}{2}}\angle\left(\theta - \tan^{-1}\frac{X_C}{R}\right)$

Since $V_b = 3V_a$

$|10I| = 3|I|(R^2 + X_C^2)^{\frac{1}{2}}$

Or $R^2 + X_C^2 = \frac{100}{9}$ and

$\theta + 53.13^\circ - 90^\circ$

$= \theta - \tan^{-1}\frac{X_C}{R}$

Or $\frac{X_C}{R} = 0.75$

Therefore, $R = 2.67\Omega$

$X_C = 2\Omega = \frac{1}{2\pi 50C}$

Therefore, $C = 1591 \times 10^{-6}\ F$

2. **Ans. (c)**
Solutin
Total impedance $Z = (6 + j8) + (2.67 - j2)$

$= 8.67 + j6$

$= 10.54\angle 34.68^\circ\ \Omega$

$I = \frac{V}{Z} = 22.77\angle -34.68^\circ\ A$

3. **Ans. (c)**
Solution:
Given equivalent circuit is Norton's circuit $V = 4I - 9$

Put $V = 0$ $I = I_N = \frac{9}{4}A$

Put $I = 0$, $V = V_{th} = -9\,V$

$$R_{th} = \frac{|-9|}{\frac{9}{4}} = 4\,\Omega$$

4. **Ans. (b)**
Solution:

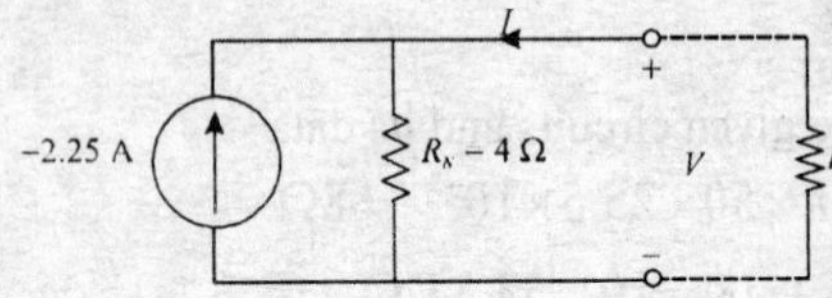

$$I_{R_L} = \frac{4 \times -2.25}{4+2} = \frac{-9}{6} = -1.5A$$

5. **Ans. (c)**
Solution:

$$X_L = \omega L = 1000 \times 0.02 = 20\,\Omega$$

$$Z = \frac{R(jX_L)}{R + jX_L} = \frac{100\angle 90}{50 + j20} = \frac{100\angle 90}{20.62\angle 75.9}$$

$$= 4.85\angle 14.04^\circ\,\Omega = 4.7 + j1.18^\circ\,\Omega$$

$$\therefore\ R_{eq} = 4.7\,\Omega$$

$$L_{eq} = \frac{1.18}{1000} = 1.18 \times 10^{-3}\,H$$

$$= 1.18\,mH$$

6. **Ans. (b)**
Solution:
We know that $P = VI\cos\theta$

$$I = \frac{100}{4.85}\sin\left(1000t + 45^\circ - 1404^\circ\right)$$

$$= 20.62\sin\left(1000t + 30.96^\circ\right)$$

$$P = \frac{100}{\sqrt{2}} \times \frac{20.62}{\sqrt{2}} \times \cos(14.04)$$

$$= 1000.07\,\Omega$$

7. **Ans. (b)**
Solution:
For the parallel RLC circuit the resonance frequency is given by the formula

$$\omega_o = \frac{1}{\sqrt{LC}}\left[\frac{R_L^2 - \frac{L}{C}}{R_C^2 - \frac{L}{C}}\right]^{0.5}$$

When we put the value of R_L, L and R_C, C then get.
$R_C = 4\,\Omega$

8. **Ans. (c)**
Solution:
Since R_L^2 is less than L/C, R_C^2 should also be less than L/C so that value of ω_o is positive, therefore

$$R_C^2 < \frac{L}{C}$$

Or $R_C < \sqrt{\frac{L}{C}}$ i.e. $R_C < 7.07\,\Omega$

The positive maximum change in R_C is
$7.07 - 4 = 3.07\,\Omega$.

9. **Ans. (a)**
Solution:
At, $t = 0^-$, inductor is short circuit

$$i\,(0^-) = \frac{120}{60} = 2A$$

$i\,(0^-) = i\,(0^+) = 2A$

$$\text{ie } 2 = \frac{120}{R + 40 + 20}$$

$R = 0\,\Omega$

10. **Ans. (c)**
Solution:
Let stored energy and dissipated energy are E_1 and E_2 respectively.
Then, Current

$$\frac{i_2^2}{i_1^2} = \frac{E_2}{E_1} = 0.95$$

$$i_2 = \sqrt{0.95}\,i_1 = 0.97 i_1$$

Current at any time t, when the switch is in position 2 is given by

$$i(t) = i_1 e^{-\frac{R}{L}t} = 2e^{-\frac{60}{10}t} = 2e^{-6t}$$

After 95% of energy dissipated current remaining in the circuit is

$$i = 2 - 2 \times 0.97$$
$$= 0.05\text{ A}$$

There fore $0.05 = 2e^{-6t}$

$t \approx 0.50$ sec

11. **Ans. (d)**
Solution: At, t = 0^- i.e. final, condition it mean capacitor is open circuit and L $\rightarrow$ Short circuit
At, t = 0^+, i.e. initial condition it mean the voltage across capacitors is – V i.e. $V_C(0^-) = V$

At $t = 0^+$ i.e. initial condition

C→ Short and L → Open circuit
Capacitor in series with R is shorted, inductor open
Since loop is open circuit $e_i = 0$

12. Ans. (d)
Solution: Transformed circuit representation is as shown below:

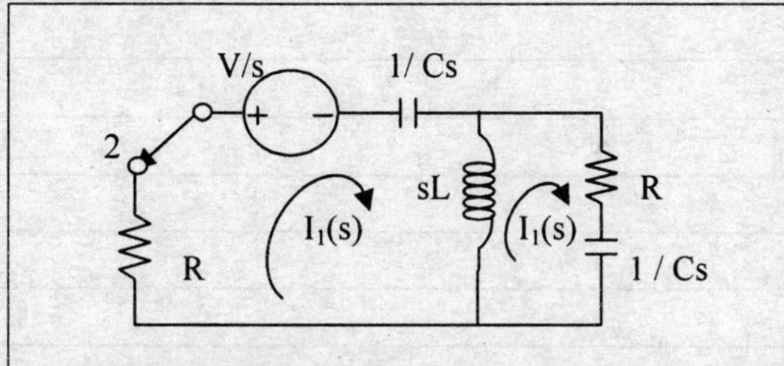

Using KVL in both the loops, we get

$I_1(s) . [1/ Cs + sL + R] – I_2 . sL + [V / s] = 0$

And $I_2 (s) [R + (1 / Cs) + Ls] – I_1 Ls = 0$

Writing in matrix form,

$$\begin{pmatrix} R+sL+\dfrac{1}{Cs} & -Ls \\ -Ls & R+Ls+\dfrac{1}{Cs} \end{pmatrix}\begin{pmatrix} I_1(s) \\ I_2(s) \end{pmatrix}=\begin{pmatrix} -V/s \\ 0 \end{pmatrix}$$

13. Ans. (c)
Solution:

$$H(f)=\frac{1}{j2\pi R\,f\,C+1},\quad H(0)=1$$

Now, as $\dfrac{|H(f_1)|}{H(0)} \geq 0.95$

Then $\dfrac{1}{\sqrt{4\pi^2 R^2 f_1^2 C^2+1}} \geq 0.95$

or $4\pi^2 R^2 f_1^2 C^2 \leq 0.108$

or $(f_1)_{max} = \dfrac{\sqrt{0.018}}{2\pi RC} = \dfrac{\sqrt{0.018}}{2\times 3.14\times 10^{-3}} = 52.2\ Hz$

14. Ans. (a)
Solution:

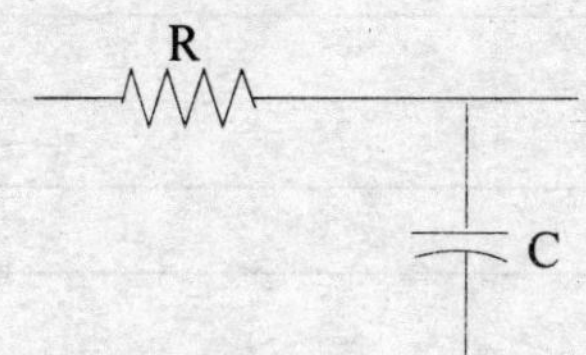

$$H(\omega)=\frac{1}{1+j\omega RC}$$

$$\theta(\omega)=-\tan^{-1}\omega RC$$

Group delay $\tau_g = \dfrac{-d\theta(\omega)}{-d\omega} = \dfrac{RC}{1+\omega^2R^2C^2}$

$$=\frac{10^{-3}}{1+4\pi^2\times 10^4\times 10^{-6}}$$

$$=0.717\ \text{ms}$$

Reader's Notes

3 Transducers, Mechanical Measurement and Industrial Instrumentation

Transducers: Resistive, Capacitive, Inductive and piezoelectric transducers and their signal conditioning. Mechanical Measurement: Measurement of displacement, velocity and acceleration (translational and rotational), force, torque, vibration and shock. **Industrial Instrumentation:** Measurement of pressure, flow, temperature and liquid level. Measurement of pH, conductivity, viscosity and humidity.

3.1 TRANSDUCERS

Transducers and their Applications

A general arrangement of most measuring systems consist of three stages:

1. An input device to sense the measured, i.e. sensor detector to give proportional output.
2. A signal processor or conditioner. This may consist of amplification, filtering unwanted signals and making it suitable for the recording device.
3. An output for read out.

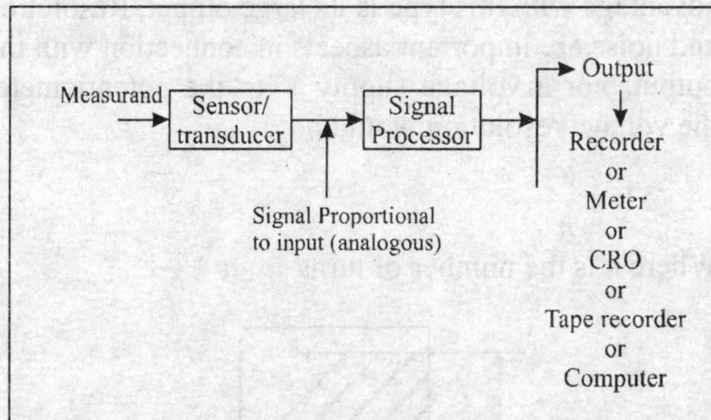

Here we are concerned with the first stage of conversion of a particular physical quantity equivalent electrical signals. This type of device is known as **transducer**.

There are two type of transducers (1) passive or modulating type and (2) self generating.

In the passive type, the non-electrical quantity changes resistance, inductance or capacitance (*R, L* or *C*) in which an external supply is applied.

In generating type, the non-electrical quantity generates a voltage directly by piezoelectric effect and thermo emf.

The choice of a transducer depends a number of factors such as the physical quantity that is to be transformed, ambient and environmental conditions.

3.1.1 RESISTIVE TRANSDUCERS

Resistive elements are required to measure several non-electrical quantities. Resistive transducers are playing vital role to measure different non electrical quantities. Resistive transducers may be used as a primary or a secondary transducer.

The major factors that make resistance elements popular and powerful as transducers are given below:

1. Versatility in regard to their size and resistance value.
2. High speed of response.
3. Availability of large variety of electrical circuitry which yield high resolution in measurements.
4. Suitability for use in circuits containing dc and ac supplied.

The dc resistance R of a conductor of length L and uniform area of construction A and with specific receptivity of its material is given by $R = \frac{\rho L}{A}$.

The resistively of the material varies with the temperature and composition of the material. The resistivity of carbon and some other semiconductor materials is affected by pressure. Resistivity variations of hygroscopied materials such as lithium chloride, due to absorption of moisture are used as hygrometers. Suitably designed electrode systems are used to measure the density of liquids and pressure of gaseous atmosphere ranging from 10^{-1} to 10^{-12} *mm* of Hg.

Resistance Thermometers

Resistance thermometers are primary electrical transducers enabling measurement of temperature changes in terms of resistance change. The resistivity of metals increase with temperature, while of semiconductors and insulators generally decreases. Materials used for restance thermometers have temperature co-efficient of resistivity much larger than the co-efficient of thermal expansion.

The temperature co-efficient of resistance α is given by

$$\alpha = \frac{1}{\Delta T}\frac{\Delta\rho}{\rho_0} = \frac{1}{\Delta T}\frac{\Delta R}{R_0}$$

where,

ΔT = changes in temperature, ^{o}C

$\Delta\rho/\rho_0$ = fractional change in resistivity

$\Delta R/R_0$ = fractional change in resistor

ρ_0, R_0 = resistivity and resistance

respectively at $0°C$.

The Resistance R, at any other temperature $T°C$ is given by:

$$R_T = R_0(1+\alpha\Delta T)$$

Or $R_T = R_0(1+\alpha T)$

Each metal or metallic alloy obeys the above relationship over a range of temperature and each and every one resistance transducer is generally limited to measure temperature with in this range.

For higher temperature non linearity come in to effect. And in such cases the relationship is modified as:

$$R_T = R_0(1 + a_1\, R + a_2T^2 .....a_nT^n)$$

Where, $a_1, a_2....a_n$ are constants applicable for each metal.

Platinum, nickel, copper and tungsten are the commonly used metals for the wire wound elements.

Thermistor

Resistance thermometer elements made from a class of materials known as semiconductor compounds are called thermistors. These compounds are sintered mixtures of sulphides, sillcates, oxides and selenides of marital such as nickel, manganese, cobalt, copper, zinc, iron, aluminum and uranium. Thermistors are thermally sensitive resistors and depending on their composition, the resistivity in the range of 10^{-1} to $10^{9-}\Omega cm$. Resistance values of thermistor at $20°C$ may lie in the range of 100Ω and $1M\,\Omega$.

Most of the semiconductor materials used for thermometry possess high resistivity and high negative temperature coefficient of resistance. There are some compounds mixture having positive co-efficient. The resistance temperature characteristic is far from linear and is governed by the relation.

$$R_T = R_0 \exp\left[\beta\left(\frac{1}{T}-\frac{1}{T_0}\right)\right]$$

Where R_T and R_0 are resistance values of the thermistor at absolute temperatures TK and T_0K respectively and β a constant of the material expressed in degrees Kelvin. The typical value of β at $25°C$ is 4000, for commonly used materials for temperature measurements. The temperature co-efficient of resistance obtained from previous equation is

$$\alpha_T = \frac{\left(\frac{dR_T}{dT}\right)}{R_T} = \frac{-\beta}{T^2}$$

Resistive Potentiometers

A resistive potentiometer is a kind of variable resistance transducer. A typical scheme of a precision wire wound potentiometer is shown below. A major advantage with this type is its large output. Resolution and noise are important aspects in connection with the output. For a voltage supply V to the potentiometer the voltage resolution would be

$$\Delta V = \frac{V}{n}$$

Where n is the number of turns

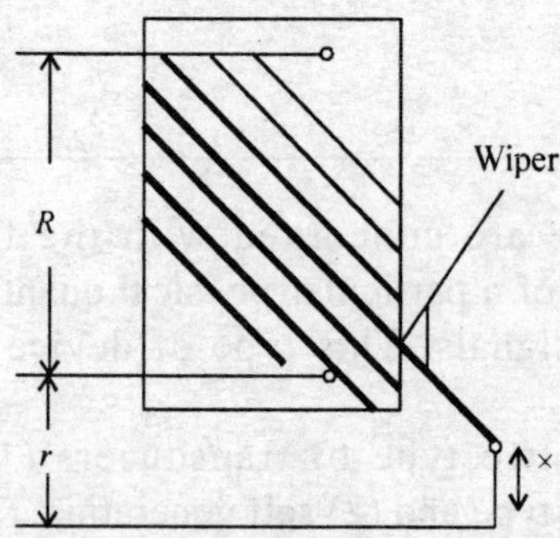

Figure: Wire-wound potentiometer

Strain Gauge

Strain Gauge consists of a fine resistance wire which is pasted on the member subjected to strain. The resistance of the gauge change with strain and this principle is used to measure stress, force, pressure, torque etc.

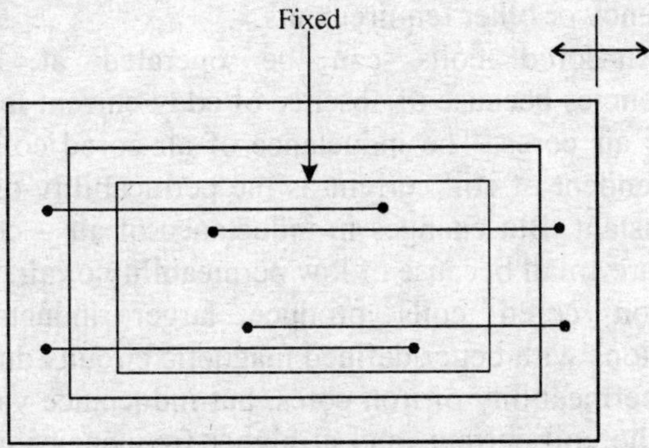

Figure: Unbounded Gauge

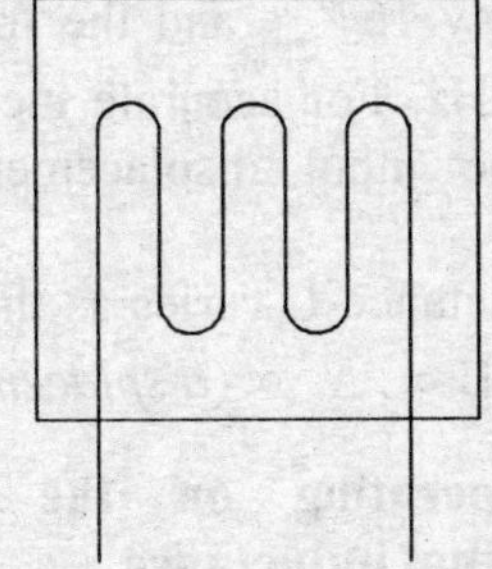

Figure: Bonded Gauge

Deposited thin metal film gauges and sputter-deposited thin metal film gauges are termed as thin film gauges that are recently developed. The sensitivity of the strain gauge is expressed as gauge factor which is given by $\frac{\Delta R/R}{\Delta l/l}$. Where R is the resistance value of the strain gauge, ΔR is the change in resistance due to a strain of $\Delta l/l$. Some of the materials used for the construction of metallic strain gauge are Advance(GF = 0.21), Isoelastic (GF = 3.6), Nickel (GF = -12), Nichrome $V(GF = 2.5)$ and Manganise (GF=0.47). Semiconductor gauges have a large value of gauge factor. The typical values of gauge factor are given below:

Silicon (p-type) → 175 ;
Silicon (n-type) → -133
Germanium (p-type) →102;
Germanium (n-type) → 157

when compared with metallic strain gauges semiconductor gauges have sensitivity but also have higher non-linearly and temperature instability

Temperature Compensation

The two effects of temperature variation on strain gauges are:

1. Change in resistance due to temperature coefficient of resistance of gauge material, and
2. Change in resistance due to difference in temperature coefficient linear expansion of gauge and underlying materials.

Temperature compensation is achieved by using a dummy gauge which is similar to the active gauge in all aspects to another member of same material but none subjected to strain. The dummy gauge is connected in the adjacent arm of the Wheatstone bridge so that he temperature effect in the two gauges get cancelled.

Another method of temperature compensation is to use two active strain gauges in application where strain of opposite polarity could be measured.

When a cantilever is subjected to strain as shown in the figure below, the gauge SG 1.

Under goes tension and gauge SG2 undergoes compression. As the two gauges are connected in the adjacent arms of the bridge, the sensitivity of the bridge circuit gets doubled and temperature error is also compensated.

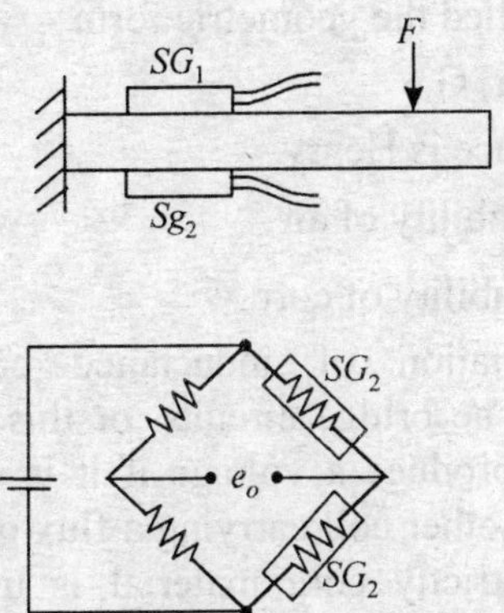

Figure: Two active gauge for temperature compensation

3.1.2 INDUCTIVE TRANSDUCERS

The inductive transducer utilizes the simple principle that a physical quantity, to be measured, like motion, can be made to vary the inductance of coil, maintaining the relation between the two. The inductance of a coil is given by:

Inductiven Transducers

These are analog passive transducers. These transducers operate, generally, upon one of the following three principles.

- Variation of self inductance of the coil.
- Variation of mutual inductance of the coil.
- Production of eddy currents.

Transducers operating on the Principal of Variation of Self inductance.

The self inductance of a coil is given by the expression. $L = \frac{N^2}{l/\mu A} = N^2 \frac{\mu A}{l} = N^2 \mu G$

$$L = \frac{N^2 A \mu_0 \mu_r}{l}$$

Where

N → the number of turns of the coil.

l → the mean length of the magnetic path.

A → the area of the x-section of the magnetic path and

μ → the permeability of the magnetic materials.

A/l → is called the geometric form factor, G.

L → inductance is Henry

μ_0 → permeability of air

μ_r → permeability of core

The variation of inductance can often be measured by ac bridge circuits, or this variation can be made to produce a voltage if it is mangnetically coupled to another coil carrying a flux or a voltage. If a magneto strictly core material is used, force or pressure can change the permeability which can be measured as change in the inductance of a coil around the core.

These transducers are usually used for measurement of displacement. The displacement under measurement causes change in self inductance of the coil L by ΔL by varying any of the three variables viz, N μ or G.

(a) Self inductance of a coil is changed by changing the number of turns on the coil. Such transducers can be used for measurement of linear or angular displacement.

Air or iron- cored coil can be used, in these transducers, depending mainly on the working frequency or other requirements.

Air-cored coils can be operated at high frequencies because of absence of eddy current losses in the air cores. The inductance of air-cored coils is independent of coil current as the permeability of air is constant. But changes in inductance of air – cored coils are small because of low permeability of air.

Iron cored coils produce larger inductance variations with better defined magnetic circuits due to high permeability of iron cores, but inductance varies with the coil current and at higher frequencies eddy current losses become high and affect the coil impedance. For these reasons the supply voltage is usually kept below $15 V_{rms}$ and the frequencies does not exceed 20 KHz. For accurate measurements the frequency of the input displacement should not exceed 2 KHz.

The self inductance L varies as the square of the displacement i.e. $L \propto N^2 \propto (displacement)^2$

Transducers operating on the principle of Variation of Mutual inductance.

Such transducers operate on the fact that mutual inductance between the coils depends upon the self inductances of the coils and coefficient of coupling between them as mutual inductance between two coils is given by the expression $M = K\sqrt{L_1 L_2}$.

In such transducers two to more coils are used and the mutual inductance between the coils is varied as per measured. In such a transducer, both the coils are in series so inductance of both the coils as an algebraic sum of self inductance between the coils. The self inductance of such an arrangement varies form $L_1 + L_2 - 2M$ to $L_1 + L_2 + 2M$.

With displacement, moving irom piece causes change in air gap of the magnetic circuit, so change in magnetic flux linking with the coils resulting change in mutual inductance between the coils. This change in mutual inductance is sensed and indicated by the bridge circuit.

An arrangement for measuring angular displacement consists of two air – cored coils – one moving and another fixed one. The movable coil is made to rotate by the input angular displacement θ. The self inductance of each coil is constant but the mutual inductance between the two coils varies on account of variation of angle θ. The coefficient of coupling is proportional to $\cos\theta$. The mutual inductance is maximum when the axis of two coils coincide. The inductive between terminals A and D varies from $L_1 + L_2 - 2M$ to $L_1 + L_2 + 2M$ as obvious.

Transducers operating on The principle of production of Eddy Currents.

These transducers operate on the principle that when a conducting plate is kept near a coil carrying alternating current, eddy currents are induced in the conducting plate, producing its ouw magnetic field in opposition to the main field created by the coil. Thus eddy currents induced in the conducting plate reduce the net flux linking with the coil and so the inductance of the coil is reduced. The nearer is the plate to the coil, the higher are the induced eddy currents and so higher is the reduction in the inductance of the coil. Thus the inductance of the coil changes with the movement of the plate.

An inductive transducer based on this principle is a coil carrying an alternating current and a cylindrical sleeve of some proper conducting material moving parallel and co-axially to the coil. When the sleeve covers less length of coil, low eddy currents are induced in it so more is the inductance of coil and when sleeve covers more length of the coil. More eddy currents are induced in it and so lesser is the inductance of the coil. This changed in inductance, the measure of displacement, is sensed and indicated by an ac bridge circuit.

Sensitivity of this type of inductive transducer is usually low.

Inductive transducers may be broadly classified in to two types.

1. Self-generating types and
2. Non -self- generating types

Self-generating type

In this type a voltage generator is the transducer because of the relative motion of the conductor and magnetic field.

e.g.: Electrodynamic, Electromagnetic, eddy current type transducer etc.

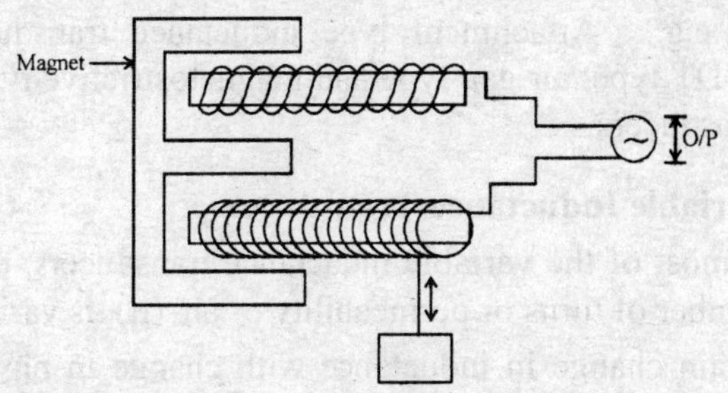

Figure: Self generating type transducer

Examples of **Self generating types**

1. Electro dynamic type transducer for linear motion

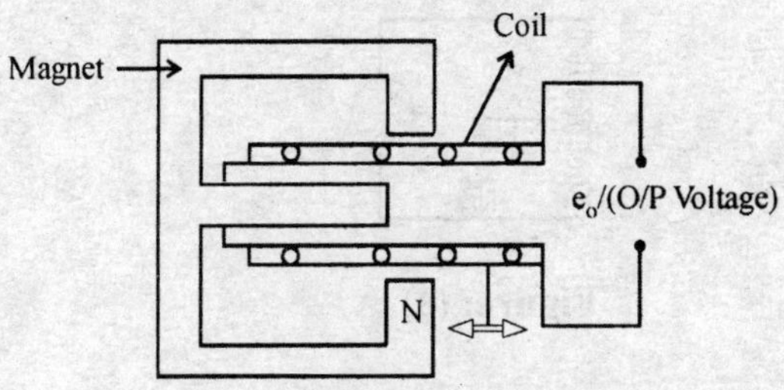

2. Electro dynamic type transducer or rotary motion

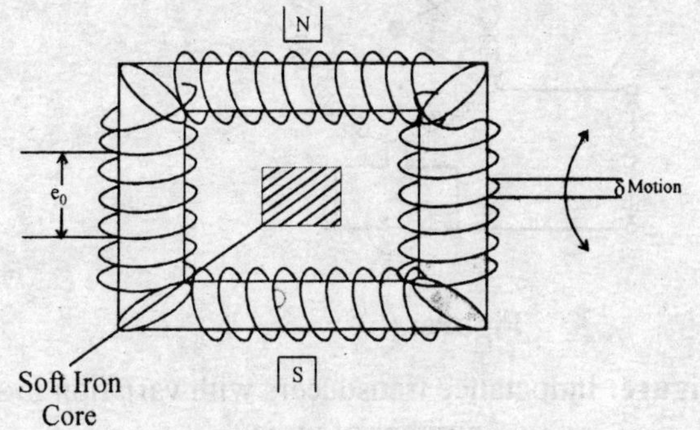

3. Eddy current type transducer

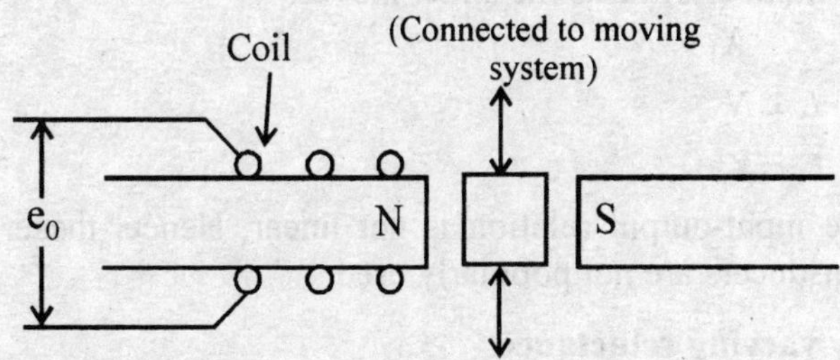

Non-self generating Type

In this type an external source is needed to energies a coil, the inductance of which would change due to the motion of the object.

e.g. : Attachment type inductance transducers, LVDT type, air gap type and magnetostrictive type of transducers.

Variable Inductance Transducer

In most of the variable inductance transducers, either number of turns or permeability of air (μ) is varied to obtain change in inductance with change in physical parameters.

By varying N

Angular or linear displacement can be measured by constructing an inductance transducer similar to that of resistance potentiometer.

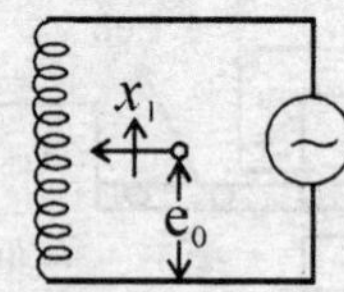

Figure: (a)

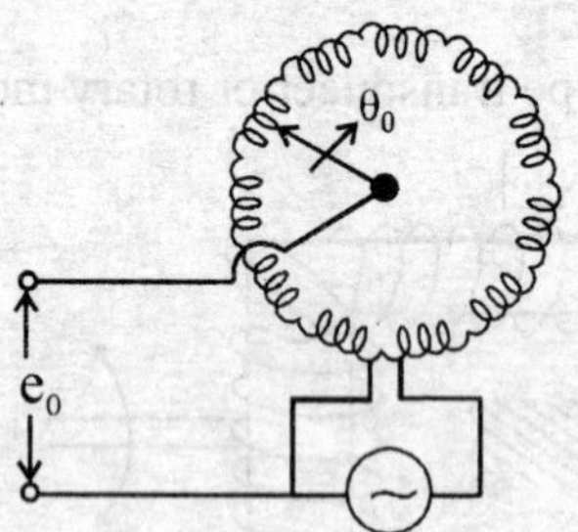

Figure: (b)

Figure: Inductance transducers with variation in number of turns

The linear or angular displacement is proportional to the number of turns the slider moves.

$$X_1 \propto N$$

As $L \propto N^2$,

$$L \propto X_1^2,$$

The input-output relation is not linear, Hence, these transducers are not popularly used.

By varying reluctance

The reluctance of the magnetic path is given by

$$R = \frac{l}{A\mu_0\mu}$$

Where,

R $\rightarrow$ reluctance

$l \rightarrow$ length of magnetic path

A $\rightarrow$ area of cross – section of the core

$\mu_0 \rightarrow$ permeability of air

$\mu_r \rightarrow$ permeability of core

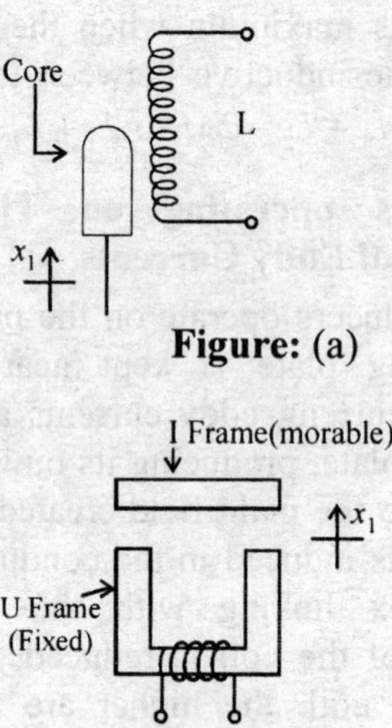

Figure: (a)

Figure: (b)

Figure: Variable Reluctance Transducers

Many transducers have been constructed based on variable reluctance principle. The two ways of constructing a variable reluctance transducer is shown in the above figure.

3.1.3 CAPACITIVE TRANSDUCERS

The capacitance between two parallel plates is given by $C = \frac{\varepsilon_0, \varepsilon_r A}{x}$.

Where,

C $\rightarrow$capacitance in farad.

$\varepsilon_0 \rightarrow$ dielectric constant of air $(8.854 \times 10^{-12} F/m)$

$\varepsilon_r \rightarrow$ dielectric constant of medium between the plates.(relative permittivity)

A $\rightarrow$ the area of cross-section of plates

$x \rightarrow$ the distance between plates.

Since the capacitance of a parallel – plate capacitor is inversely proportional to the spacing'd' between the two plates, any variation on 'd' causes a corresponding variation in the capacitance. This principle is applied in the capacitive transducer for the measurement of displacement.

1. Capacitive transducers based on variation in air gap.

As the capacitance is inversely proportional to 'x', variation in x results in nonlinear relation between x and C

$$\frac{dC}{dx} \propto -\frac{1}{x^2}$$

Sensitivity can be increased by decreasing 'x'

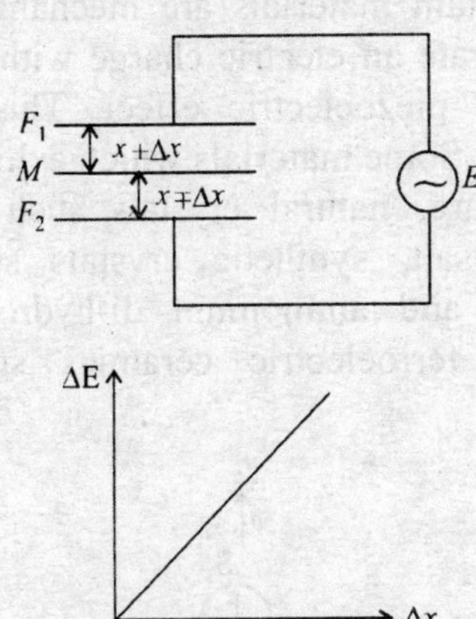

Figure: Differential Capacitor with variation in airgap

The two plates F_1 and F_2 are fixed and the displacement is given to the movable plate M. The capacitor is excited by a.c. voltage of magnitude E.

Let,

C_1 be the capacitance between F_1 and M

C_2 be the capacitance between F_2 and M.

E_1 be the voltage across C_1.

E_2 be the voltage across C_2

$$C_1 = \frac{\varepsilon_0 . \varepsilon_r \ A}{x \pm \Delta x} \quad ; \quad C_2 = \frac{\varepsilon_0 . \varepsilon_r \ A}{x \mp \Delta x}$$

$$E_1 = \frac{C_2}{C_1 + C_2} . E \quad ; \quad E_2 = \frac{C_1}{C_1 + C_2} . E$$

2. Capacitive pressure sensor with thin stretched diaphragm

When a thin stretched diaphragm is used in a capacitive pressure transducer, the change in capacitance is given by.

$$\Delta C = \frac{\pi . \varepsilon_0 . a^4}{8 x^2 S} . \rho$$

Or, $\dfrac{\Delta C}{C} = \dfrac{a^2}{8 \times S} . \rho$

Where

$a \rightarrow$ the radius of diaphragm

S $\rightarrow$ the tension in the

$\rho \rightarrow$ fluid pressure

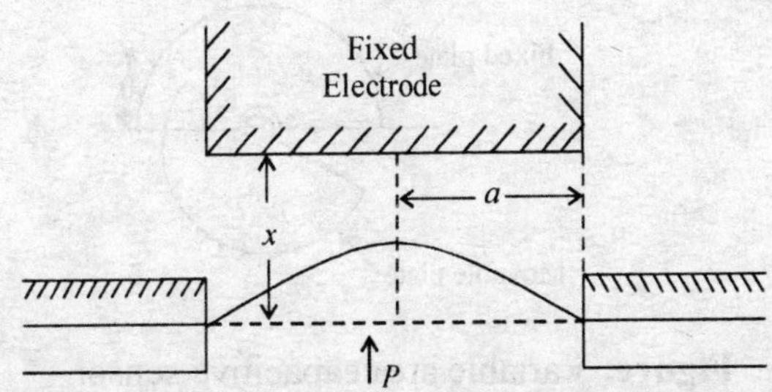

Figure: Fluid pressure capacitive pressure sensor

3. Capacitive Pressure Sensor with Clamped Diaphragm

While the thin stretched diaphragm has negligible stiffness in bending, a Clamped diaphragm has appreciable stiffness in bending. The clamped diaphragm has no initial radial stress and only bending forces are considered. The change in capacitance of a capacitive transducer with clamped diaphragm is given by,

$$\frac{\Delta C}{C} = \frac{(1 - y^2) . a^4 . \rho}{16 x \ E . t^3}$$

Where

y $\rightarrow$ poisson's ratio

a $\rightarrow$ the radius of diagram,

E $\rightarrow$ Young's modulus

t $\rightarrow$ the thickness of diaphragm.

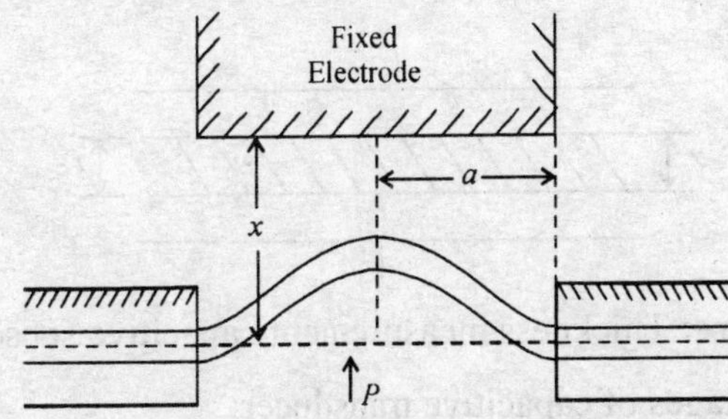

Figure: Capacitive Pressure Sensor with clamped diaphragm

4. Capacitive sensor based on variation in overlapping area.

When the input quantity varies, the overlapping area of the parallel plate capacitor change and a linear relation is obtained between change in capacitance and change in area.

Angular displacement can be measured by using two overlapping plates.

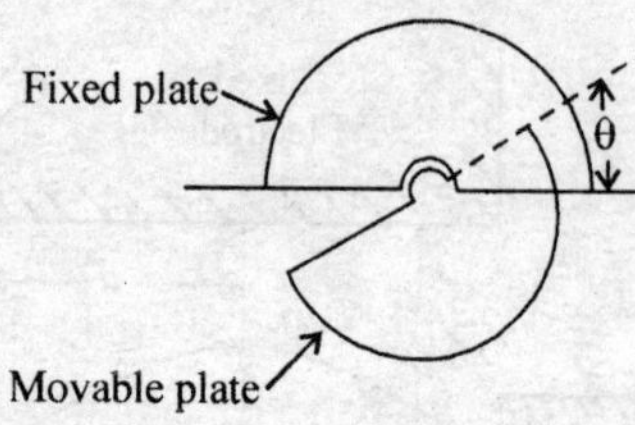

Figure: Variable area capacitive sensor

5. Capacitive sensor based on variation in dielectric constant

When a solid material is placed in between the two parallel plates of a capacitor two measurements are possible:

- Dielectric constant of the solid material can be measured, when thickness is constant.
- Thickness of the solid material can be measured, when dielectric constant of the material is constant.

The capacitance variation for second case is given by

$$\frac{\Delta C}{C} = \frac{\Delta t}{t} . N_4 . \frac{1}{1 - N_4 . \frac{\Delta t}{t}}$$

Where

$$N_4 = \frac{\varepsilon_r - 1}{1 + \varepsilon_r \left(\frac{x - t}{t} \right)}$$

Figure: Thickness measurement capacitive sensor

Advantages of capacitive transducer:

1. They require very little force to operate then and hence are very useful in small systems.
2. They have a good frequency response and can measure both static and dynamic changes

Disadvantages of capacitive transducer:

1. The metallic parts of the capacitor must be insulated form each other.
2. Their performance is severely affected by dirt and other contaminants because they change the dielectric constant.
3. They are sensitive to temperature variations and there are possibilities of erratic or distorted signals due to long lead length.

3.1.4 PIEZOELECTRIC TRANSDUCERS

When certain materials are mechanically deformed, they generate an electric charge within them. This is known as piezoelectric effect. This effect is also reversible. Some materials which exhibit piezoelectric property are, natural crystals such as quartz and Rochelle salt, synthetic crystals such as lithium sulbhite and ammonium di-hydrogen phosphate, polarized ferroelectric ceramics such as barium titanate.

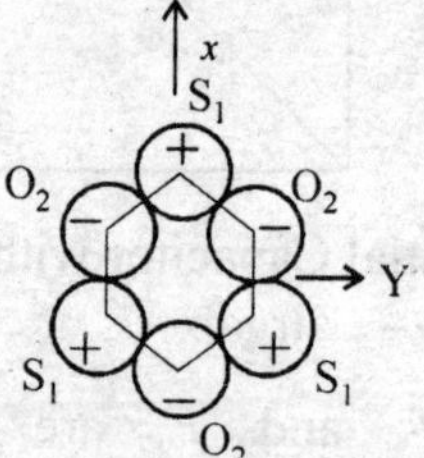

Figure: Quartz crystal model

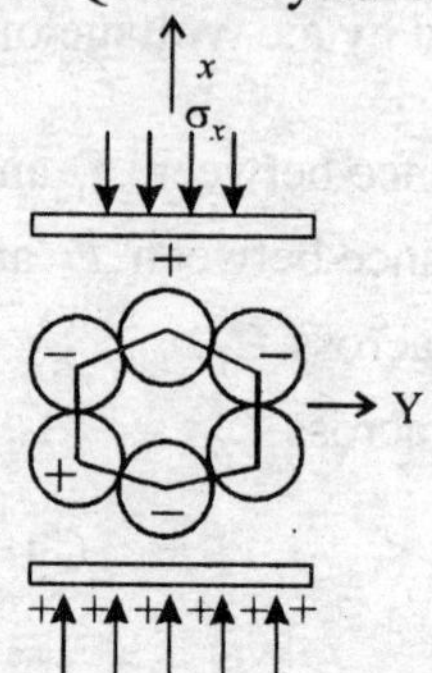

Figure: Crystal under longitudinal effect

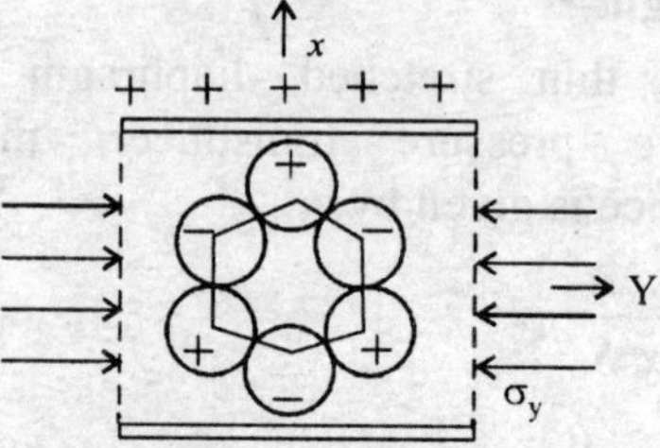

Figure: Crystal under Transverse effect

All piezoelectric crystals are electrical insulators. Two metal electrodes are laced on each side of the material to bring out the charge. Hence piezoelectric element is equivalent to a charge generator and capacitor. The sensitivity of a piezoelectric material is expressed in two ways,

Charge sensitivity,

$$d = \frac{\text{Charge generated}}{\text{Force applied}}$$

Voltage sensitivity,

$$g = \frac{\textit{Electric field generated}}{\textit{Stress applied}}$$

Both sensitivity are related and is given by,

$d = \varepsilon \times g$

Where

$\varepsilon \rightarrow$the dielectric constant of the material

The sensitivity constants are direction sensitive i.e. values of the constants change with direction of applied force and direction of output taken.

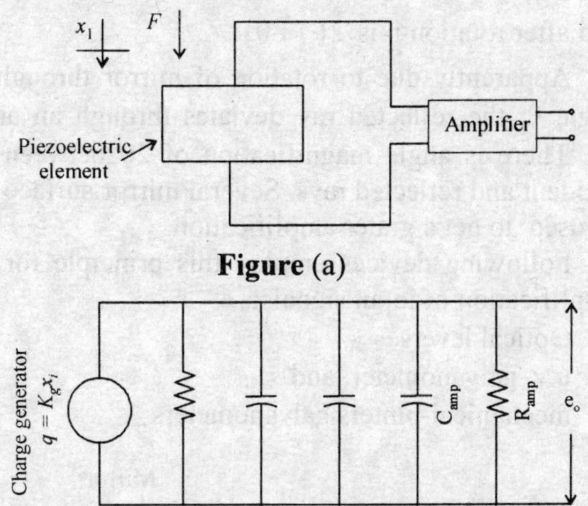

Figure: Piezoelectric Transducer (a) Circuit connection (b) equivalent circuit

3.1.5 SIGNAL CONDITIONING

The output from the detector-transducer stage of a measurement system needs certain modification before it becomes compatible with data presentation stage. The necessary modification is carried out in the intermediate stage, more commonly referred to as the *signal conditioning stage*. The state signal conditioning equipment may be required to perform the following function on the transduced signal:

1. Amplification: Enhance the signal level which is often in the low level range. The amplification system must bring the level of transducer signal upto a value sufficient enough to make it useful for conversion, processing indicating and recording

2. Modification or Modulation: Change the form of signal. The signal may be smoothed, linearised (correction for non-linearity in the preceding conversion process), filtered (elimination of any kind of unwanted or spurious signal) or converted into digital form.

3. Impedance matching: Arrange the input and output impedance of the matching device with a view to prevent loading of the transducer and to maintain a high signal level at the recorder.

4. Data processing: Mathematical operation (addition, subtraction, differentiation and integration etc.) before indication or recording of data.

5. Date transmission: Transmit the signal from one location to another without changing its information content.

The entire task of signal conditioning is by no means simple; it requires conditioning elements are designed to be insensitive to all extraneous inputs. The accuracy, range and dynamic response are all designed to be compatible with the detector transducer.

Amplification

The signal from the detector transducer is normally very weak, and it needs amplification to a certain level where it can be detected for display or record. Amplification is also needed to transmit the signal over some distance. The device used to increase or augment the weak signal is referred to as amplifier; it may operate on mechanical (levels, gears etc.), optical pneumatic and hydraulic, or electrical and electronic principles.

Input → G → Output
θ_i θ_o

Mechanical amplifiers

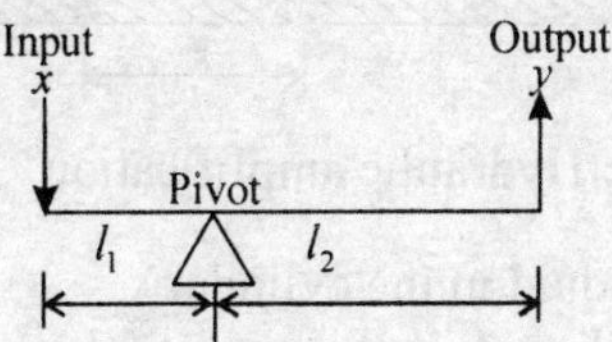

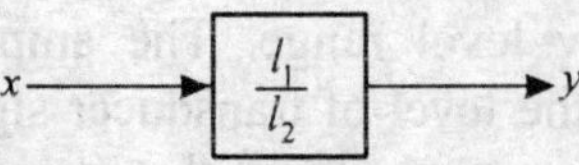

Figure: Simple lever signal amplifier

Figure shows the arrangement and the associated block diagram for a simple level supported on a pivot. For an input displacement x, the simple lever causes the output end to displace by an amount y ; x and y are related to each other by the expression.

$$\frac{y}{x} = \frac{l_2}{l_1}$$

Where, $l_1 \rightarrow$ distance of input end from the pivot,

$l_2 \rightarrow$ distance of output end from the pivot.

$\frac{l_2}{l_1} \rightarrow$ determines the displacement gain

Amplification of the lever can be varied depending on the relative distances of input ends from the pivot. With a simple lever, the input and out displacement are of opposite phase, i.e., if the input x goes down the output y moves up.

Fluid amplifier

1. Hydraulic amplifier

When a small displacement x is applied to a piston operating inside a cylinder containing some liquid, there occurs a large displacement y of the liquid in the output tube which has a small diameter d. From volume balance of the liquid.

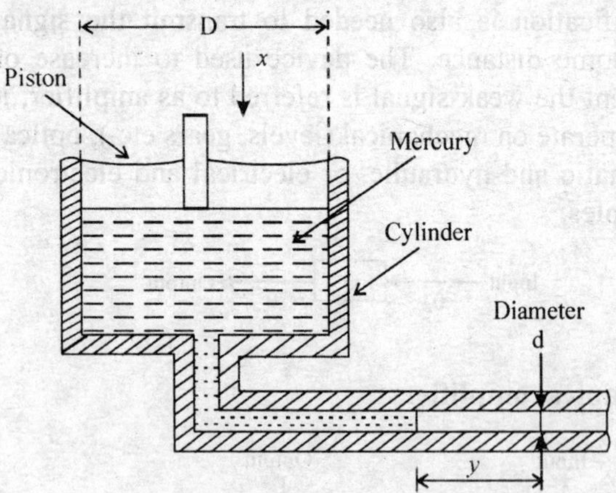

Figure: Hydraulic amplification

Displacement of liquid in the cylinder
= displacement of liquid in the output tube

$$\frac{\pi}{4} D^2 x = \frac{\pi}{4} d^2 y$$

Therefore the gain or amplification is

$$\text{Gain } G = \frac{y}{x} = \left(\frac{D}{d}\right)^2$$

And the output is $y = \left(\frac{D}{d}\right)^2 x$

This principle is employed in the mercury-in-glass thermometer and in the single-column manometers.

2. Pneumatic amplifier

Optical amplifiers

A ray of light strikes a mirror with an angle of incidence i and gets reflected with angle of reflection equal to the angle of incidence. When the mirror rotates through an angle θ, the angle of incidence changes to $(i+\theta)$ and the angle of reflection also takes the same value $(i+\theta)$. Before mirror rotation, the angle between the incident and reflected ray is $2i$ and after rotation it is $2(i+\theta)$.

Apparently due to rotation of mirror through an angle θ, the reflected ray deviates through an angle 2θ. There is angle magnification of 2θ between the incident and reflected rays. Several mirror surface can be used to get a grater amplification.

Following devices employ this principle for the amplification of input signals:

- Optical levers
- u.v. galvanometer, and
- mechanical-pinters galvanometers

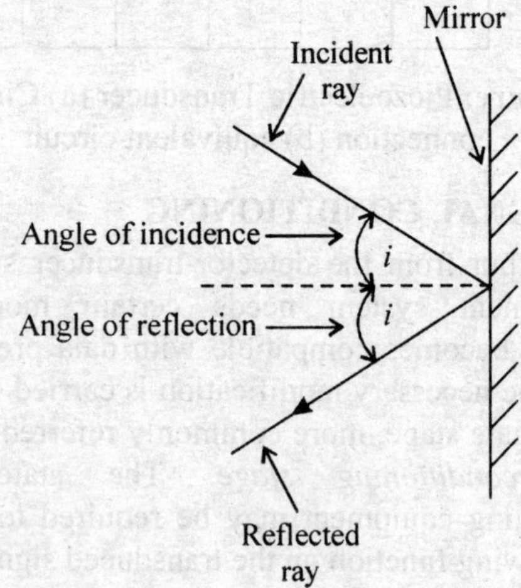

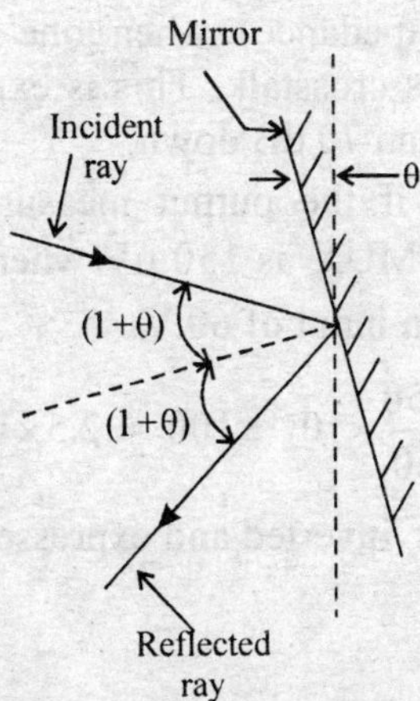

Figure: Schematics of optical amplifications

The optical amplifier operate without the application of any measurable force to displaced element.

Electrical and Electronics amplifier

For detail refer Analog circuit

Modulation Technique

1. Amplitude Modulation
2. Frequency Modulation

Signal Filtering

For correct measuring of physical quantities, it is necessary that the input be reproduced faithfully. That necessitates the elimination of the kind of unwanted or spurious signals which may creep into the measurement system either at the transduction stage or at the signal conditioning stage. The task is accomplished with the help of filters which are designed to be selective, i.e. they pass wanted signal and reject or considerably attenuate (decrease) the unwanted signal. The filters can be introduced into the measurement system at one of its three stages (input, intermediate and output stage); the choice depends upon the application.

Signal Transmission

The terms 'data transmission' and 'telemetry' refer to the process by which the measureand (physical quantity being measured) is transferred to a remote location for the purpose of being processed, recorded and displayed

Telemetering means remote monitoring of functional parameter to maintain the health of the system.

Telemetry is resorted to on account of

(a) Hazards may be to the operating personnel as well as the indicating instruments. Such places are nuclear reactors due to radiation, chemical processor on account of explosive, toxic and pyrometric nature, high voltages such as generators and transmitters of high power.

(b) Inaccessibility-The physical location of the parameters may be inaccessible as for instance satellites and human bodies while taking ECG etc.

Transfer of data from the point of location to the point of observation and recording is done by cable pairs or by wireless or by z combination of both. This is know as telemetry.

There are two type of data acquisition systems, analog and digital

Analog system have wide bandwidths but lower accuracy.

Digital system are slowly varying and hence narrow bandwidth but have high accuracy and low per channel cost.

Analog Data Acquisition System

A block diagram of ADAS is given. There are two types. Either data may be collected from n source and multiplexed into one channel by A/D multiplex or the collected data from a n source are passed through a A/D converter and fed into one channel by digital multiplexer.

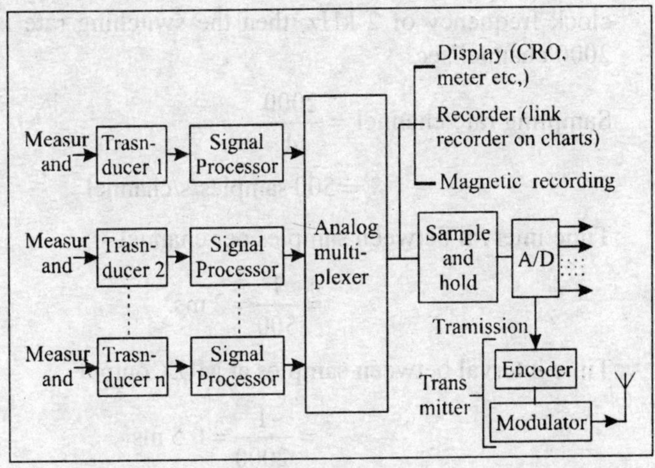

Transducer: Convert electrical and non-electrical quantities into electrical signals.

Signal processing: Signals are of very low strength and they are amplifier by low noise amplifiers and passed through filters to improve S/N ratio.

A/D multiplexer: The input to the multiplexer is through a FET switch which is operated by a pulse. The pulse may cyclically switch on the respective switches at fixed intervals. A/D converter output goes to a A/D converter via S/H circuit is necessary before an A/D converter as the input signal not change during the process of conversion.

Digital Data Acquisition system

The block diagram is as follows

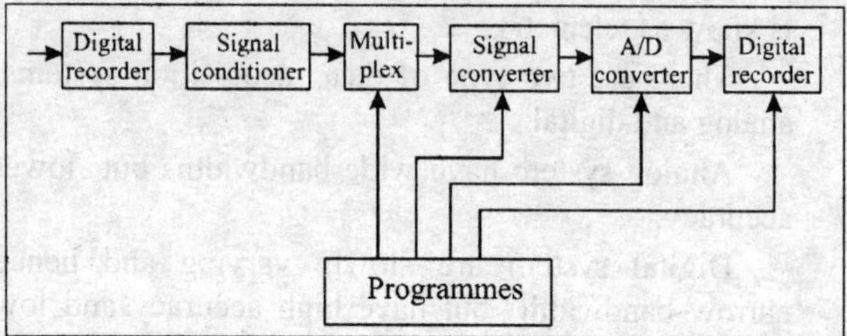

Multiplexer: A multiplexer receives inputs from a number of sources but sends only one output. They are classified as 4×1 or 8×1 MUX

For a digital MUX D1, D2, D3, D4, there will be 3 select lines and one data line.

Switching rate of MUX is the rate at which the addressing input a driven. If 4×1 MUX is addressed by a 2 bit counter which is driven by a clock frequency of 2 kHz, then the switching rate is 2000 channel/sec.

$$\text{Sampling rate/channel} = \frac{2000}{4}$$

$$= 500 \text{ samples/s/channel}$$

Time interval between samples per channel

$$= \frac{1}{500} = 2 \text{ ms.}$$

Time interval between samples at MUX output

$$= \frac{1}{2000} = 0.5 \text{ ms}$$

Crosstalk: The presence of a signal at the output of an off channel (a channel whose input is terminated by its input impedance) when one or the other channels are on is crosstalk. This is expressed in dB should be more than 70 dB down.

For instance if the output measured in an off channel of a 8×1 MUX is 150 μV when all the other 7 channels have an input of 60 V.

$$\text{The crosstalk} = \frac{150}{60}\times10^{-6}\times100 = 2.5\times10^{-4}\%$$

This percentage is inverted and expressed in dB.

$$\frac{1}{2.5\times10^{-4}} = 400$$

$$\text{dB} = 20 \log\ 4000 = 72.04 \text{ dB}$$

$$\text{Crosstalk} = -72.04 \text{ dB}$$

Filtering: Active and passive filters may be used. But active filters such as op amp filter are preferred as both amplification and filtering can be got in one circuit.

3.2 Mechanical Measurement

3.2.1 MEASUREMENT OF DISPLACEMENT,

Application of force to a medium causes displacement as for example in a piston or dash pot assembly. Force is through variable and displacement across. In the this case there are two transducers, one the primary transducer for converting the force into displacement and the second, the displacement causing the variation of an electrical element or voltage/current.

The primary transducer are diaphragm, flat or conjugated, bellows, tube, straight tube, mass cantilever and pivot torque.

The electrical signal is caused by the displacement due to variations in resistance, capacitance, inductance, differential transformer, ionization, oscillation, piezoelectricity, photo electricity.

Resistance Transducers

As per the fundamental equation for resistance

$$R = \rho\frac{L}{A} \tag{1}$$

Where

R = Resistance of the wire, in Ω

ρ = specific resistance of the wire conductor in $\Omega-m$

L = length of the wire, in m
A = cross-sectional area of the wire, in Sq-m

$$= \frac{\pi d^2}{4} \quad (2)$$

d = diameter of the wire

Gauge Factor: The gauge factor of the strain gauge is defined as the unit change in resistance per unit change in length.

$$\therefore G = \frac{\Delta R / R}{\Delta L / L}$$

$$\text{or } G = \frac{\Delta R / R}{\sigma} \quad (3)$$

Where,

μ = Poisson's ratio
ΔR = change in resistance
R = original resistance
ΔL = change in length
L = original length
G = Gauge Factor
σ = Strain in the lateral direction

Again we know, $\Delta R = R\sigma(1+2\mu)$

Now, putting the value of ΔR in the expression for gauge factor i.e. equation (3)

$$G = \frac{R\sigma(1+2\mu)}{R\sigma},$$

$$\text{or } G = 1 + 2\mu \quad (4)$$

Poisson's Ratio for most metals lies in the range 0.25 to 0.35 and accordingly the gauge factor in the range 1.5 to 1.7

Variation of any one of ρ, L or A causes a change in resistance.

Change of L may be caused in a potentiometer through which a steady current passes. Change of resistance creates a voltage change which can be read on a meter calibrated in length (displacement). Potentiometers are called pots and it can be either linear or circular.

Resistance Strain Gage

A more important device in the measurement of displacement is the Strain Gage in which all the 3 variables ρ, L, A are simultaneously affected by the input strain.

There are 3 types of resistance strain gages:

1. Metallic resistance strain gage made of metallic wires such as constantan (Cu-Ni alloy) Nichrome *V* or platinum alloy.
2. Foil strain gage consist of a thin, 8 to 15 μm nitro-cellulose impregnated paper on which is photo etched metal ally filaments as resistance material. **For higher temperature, an epoxy backing is used instead of paper**. The active length of the gage is along the transverse axis. The gage should be mounted with its transverse axis in the same direction as the direction of application of force or strain. Thus elongation of the gage by force (tension) increases by resistance and compressing the gage reduces the length and consequently the resistance.

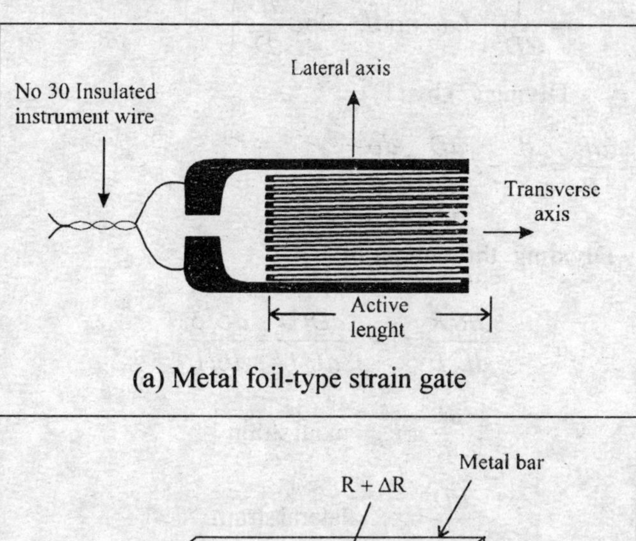

(a) Metal foil-type strain gate

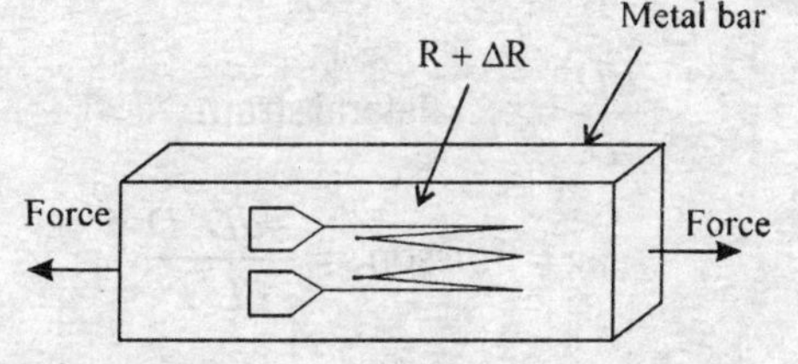

(b) Tension lengthens bar and gage to increase gage resistance by ΔR

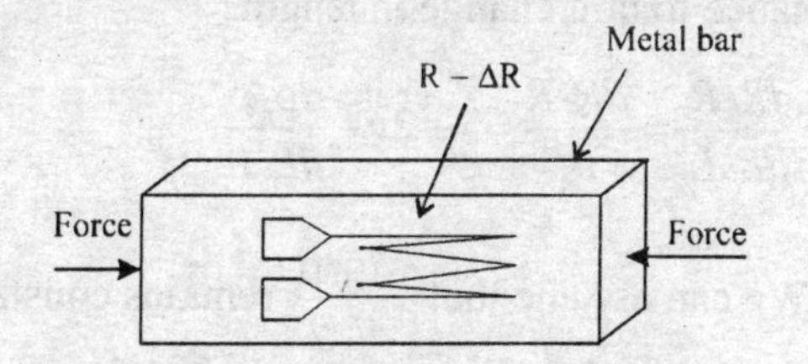

(c) Compression shortens bar and gage to reduce gage resistance by ΔR

Figure: Using a strain gage to measure the change in length of a structure

3. The third type is the **semiconductor gage**. It depends on the piezoresistive properties of silicon and germanium. They have **high sensitivities with gage factors from 50 to 200**. Their chief defects are fluctuations due to temperature and non linear output. The *p* type gages increase resistance with applied tensile strain while *n* type gage resistance decreases.

Assuming that the strain causes a change in each of the 3 variable and differentiating for a small change in resistance.

$$dR = \frac{CD^2[Ld\rho + \rho dL] - 2C\rho DLdD}{CD^2}$$

$$= \frac{1}{CD^2}\left(Ld_2 + \rho dL - 2\rho L\frac{dD}{D}\right)$$

Diving (2) by (1)

$$\frac{dR}{R} = \frac{dL}{L} - 2\frac{dD}{D} + \frac{d\rho}{p}$$

Dividing throughout by $\frac{dL}{L}$

$$\frac{dR/R}{dL/L} = 1 - 2\frac{dD/D}{dL/L} + \frac{d\rho/\rho}{dL/L}$$

$$\frac{dL}{L} = \varepsilon_a = \text{axial strain}$$

$$\frac{dD}{D} = \varepsilon_L = \text{lateral strain}$$

$$v = \text{Poisson's} = \frac{-dD/D}{dL/L}$$

$\frac{dR/R}{dL/L}$ is defined as gage factor *K* i.e. unit change in resistance to unit change in length.

$$K = \frac{dR/R}{dL/L} = \frac{dR/R}{\varepsilon_a} = 1 + 2v + \frac{d\rho/\rho}{dL/L}$$

We can assume that $\frac{d\rho/\rho}{dL/L}$ remains constant.

Poisson's ratio v for most metals lies in the range of 0.25 to 0.35 so that gage factor is of the order of 1.5 to 1.7.

We desire a high sensitivity so that even minute changes in length cause corresponding change in resistance. Hence a large gage factor *K* means a large means a large resistance change which can be easily measured. *K* values of constantan and Nichrome *V* are 2 whereas platinum-iridium.

Equation (8) can be simplified as

$$\frac{dR}{R} = K\frac{dL}{L} = K\varepsilon_a$$

or $$\varepsilon_a = \frac{1}{K}\frac{dR}{R}$$

Hooke's law states that stress, force/unit area, is proportional to strain, elongation/unit length

$S = \varepsilon_a E$

S – Stress, kg.cm^2,

E – Young's modulus, Kg/cm^2,

ε_a – strain, (dL/L)

$$\varepsilon_a = \frac{S}{E}$$

Equation can be written as

$$\frac{S}{E} = \frac{1}{K}\frac{dR}{R}$$

or $$\frac{dR}{R} = \frac{K}{E}\frac{F}{A}$$

Manufacturers specify the value of *K* and *P*.

As the resistance change is of the order of milliohms, this very small will cause a very small change in voltage. If a current of 4 mA flows through the strain gage and the change in resistance 1 mΩ, the voltage change is 4 μV. For this low voltage to be sensed, we need an amplifier. In this respect a bridge circuit offers many advantages.

There are basically two types of strain gauges.

Bonded Strain Gauge: In bonded type of strain gauges, a grid of fine wire is connected to a thin paper sheet or very thin bakelite sheet and covered with a protective sheet of paper to tin bakelite. The paper sheet is bonded with an adhesive material to the structure under study.

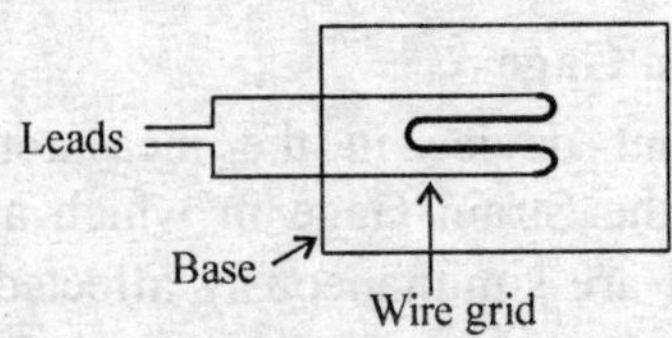

Figure: Bonded Stain Gauge

The strain gauge is useful only for measuring very small displacement

Unbounded Strain Gauge: It consists of a stationary frame and an armature that is supported in the center of the frame.

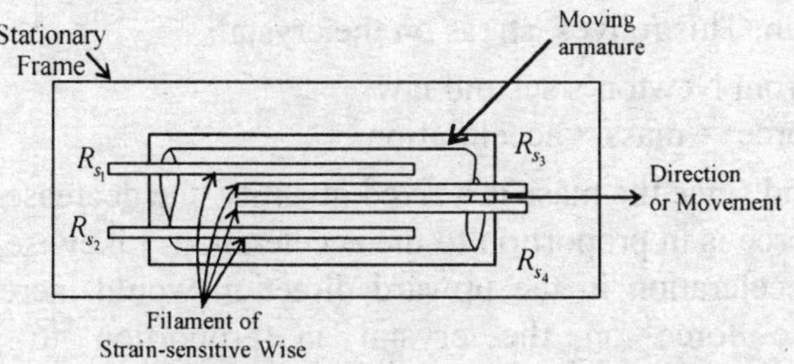

Figure: Unbounded strain Gauge

The unbonded strain gauge transducer can be constructed in a variety of configuration depending on the required use. They are mainly used in force and pressure transducers and accelerometers.

Unbonded strain Gauge: The unbounded strain gauge consists of a stationary frame and an armature that is supported in the centre of the frame. The armature can move only in one direction and its travel in that direction is limited by four filaments of strain – sensitive wire wound between rigid insulators that are mounted on the frame and on the armature. The filaments are of equal length and arranged.

Linear variable differential Transducer (LVDT): The linear variable differential transducer is the most widely used inductive transducers to translate linear motion into electrical.

A differential transducer consists of a primary winding and two secondary windings. The windings are arranged concentrically and next to each other. They are wound over a hollow bobbin which is usually of a non-magnetic and insulating material.

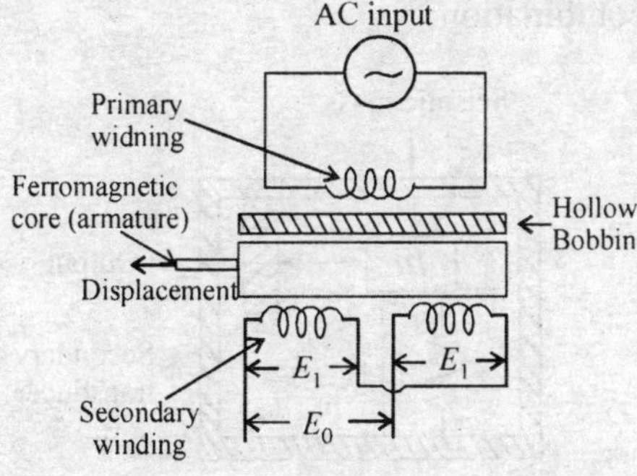

Figure: Schematic diagram of LVDT

A ferromagnetic core (armature) in the shape of a rod or cylinder is attached to the transducer sensing shaft. The core sliders freely within the hollow portion of the bobbin. An ac excitation is applied across the primary winding and the movable core varies the coupling between it and the two secondary coil is equal. As the core away from the centre position, the coupling to one secondary becomes more and hence its output voltage increases, while the coupling and the output voltage of the other decreases. In the simplex winding configuration, the *two secondary windings are connected in series opposition*. Any physical displacement of the core causes the voltage of one secondary winding to increase while simultaneously, reducing the voltage in the other secondary winding. The difference of the two voltage appears across the output terminals of the transducer and gives a measure of the physical position of the core and hence the displacement.

When the core is in the neutral or zero position, voltages induced in the secondary windings are equal and opposite and the net output is negligible. As the core in moved in one direction from the null position, the differential voltage, i.e. the different of the two secondary voltages, will increase while maintaining an in phase relationship with the voltage from the input source. In the other direction from null position, the differential voltage will again increase, but will be 180° out of phase with the voltage from the source. By comparing the magnitude and phase of the output (differential) voltage with the input source, the amount and direction of movement of the core and hence of displacement may be determined.

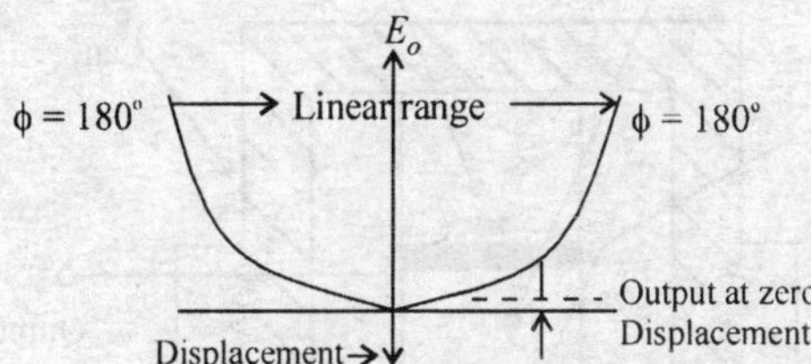

Figure: variation of output voltage with core position in LVDT

Advantages of LVDT:

1. They output voltage of these transducers is practically linear for displacements up to 5 mm
2. They have infinite resolution
3. These transducers posses a high sensitivity

4. They are simple, light in weight and easy to align and maintain
5. These transducers can usually tolerate a high degree of shock and vibration without any adverse effect
6. These transducer have low hysterisis and hence repeatability is excellent under all conditions

Disadvantages of LVDT

Relatively large displacements are required for appreciable differential output

1. They are sensitive to stray magnetic field but shielding is possible
2. They are inherently low in power output.
3. Temperature affects the performance of the transducer. Temperature also causes phase shifting effect which may be minimized by using a capacitor across one of the secondary windings.

3.2.2 MEASUREMENT OF VELOCITY/ ACCELERATION (TRANSLATIONAL AND ROTATIONAL)

Measurement of Acceleration

There are two types of accelerometers generally used for measurement of acceleration:

1. Piezo-electric type
2. Seismic type

Piezo-electric Accelerometer

The unit is perhaps the simplest and most commonly used transducer employed for measuring acceleration.

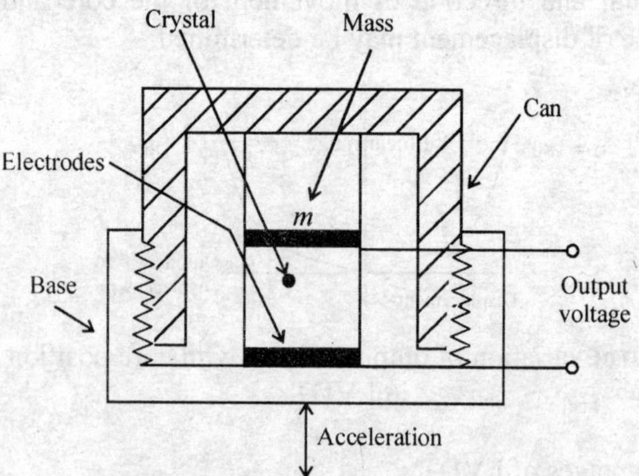

Figure: Piezo-electric accelerometer

The sensor consists of a piezo-electric crystal sandwitched between two electrodes and has a mass placed on it. The unit is fastened to the base whose acceleration characteristics are to be obtained. The can threaded to the base acts as a spring and squeezes the mass against the crystal. Mass exerts a force on the crystal and a certain output voltage is generated. If the base is now accelerated downward, inertial reaction force on the base acts upward the top of the can. This relives stress on the crystal.

From Newton's second law
force = mass × acceleration

and since the mass is a fixed quantity, the decreases in force is in proportion to the acceleration. Likewise, an acceleration in the upward direction would increase the force on the crystal in proportion to the acceleration. The resulting change in the output voltage is recorded and correlated to the acceleration imposed on the base.

Advantages and Limitations

- Rugged and inexpensive devices
- High output impedance
- High frequency response from 10 *Hz* to 50 *kHz*
- High sensitivity varies from 10 to 1000 mv/g where $g = 9.807$ *m/s*2
- Capability to measure acceleration from a fraction of g to thousand of *g*
- Somowhat sensitive to change in temperature
- Subject to hysteresis errors.

Principle of Seismic Instrument

A schematic diagram of seismic instrument consists of a mass connected through parallel spring and viscous damper to the housing frame. The housing is connected to the vibrating source whose characteristic are to be measured. The relative motion between the mass and the housing can be used for either displacement, or acceleration measurement, depending upon are selection of mass, spring and damper combination.

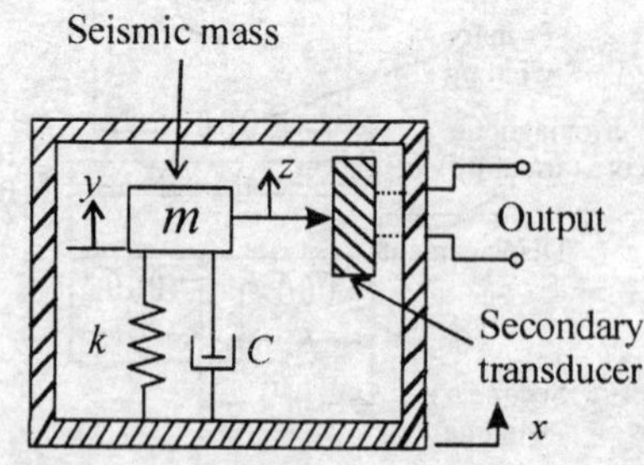

Figure: Seismic Insturment

Generally, a large mass and soft spring (designed for low undamped natural frequency) is used for displacement measurement, while a relatively small mass and stiff spring (designed for high undamped natural frequency) is used for acceleration measurements.

Theory of Seismic instrument

A seismic instrument is a second order system. In a seismic instrument, excitation (or input) is applied to the support or base of the system and not the mass. Also, the instrument output is the relative displacement of the mass with respect to the support or base.

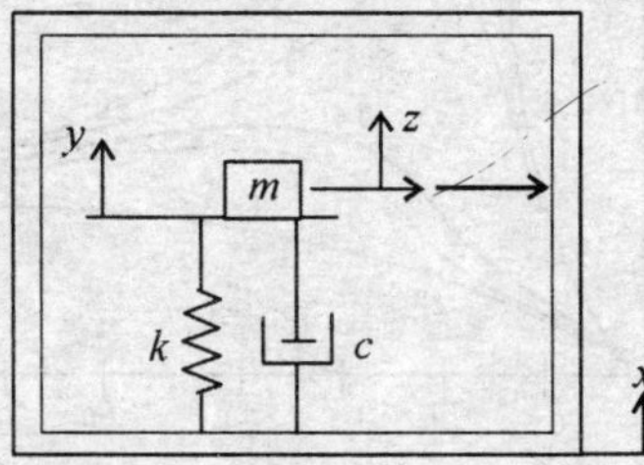

Figure: Seismic Instrument

m = mass of seismic element
k = stiffness of spring
δ = damping coefficient
x = displacement given to the frame
y = absolute displacement of seismic mass
z = relative displacement of seismic mass with respect to the frame

$$z = y - x \qquad (1)$$

ω_n = undamped natural frequency of the system

Let be the harmonic motion given to the frame

$$x = x_o \cos\omega t \qquad (2)$$

Where,

x_o = displacement amplitude
ω = frequency of harmonic or excitation

Equation of motion of the mass

$$my'' = -(y' - x')c - k(y - x) \qquad (3)$$

Where x', y' denote the first differential with respect to time

Using, equation (1) $(y - x) = z$

Above equation can be written as,

$$mz'' + cz' + kz = -mx''$$

$$x = x_o \cos\omega t$$

$$x'' = -\omega^2 x_1 \cos\omega t$$

$$m z'' + c z + k z = m\omega^2 x_o \cos\omega t$$

$$m\frac{d^2z}{dt^2} + c\frac{dz}{dt} + kz = m\omega^2 x_o \cos\omega t \qquad (4)$$

The steady state solution is given by,

$$z = z_o \cos(\omega t - \phi) \qquad (5)$$

where z_o = relative displacement amplitude of seismic mass w.r.t. frame

$$z_o = \frac{(\omega/\omega_n)^2 x_o}{\sqrt{\left(1 - \left(\frac{\omega}{\omega_n}\right)^2\right)^2 + \left(2\delta\frac{\omega}{\omega_n}\right)^2}}$$

$$\frac{z_o}{x_o} = \frac{r^2}{\sqrt{(1 - r^2)^2 + (2\delta r)^2}} \qquad (6)$$

$r = \frac{\omega}{\omega_n}$ is the frequency ratio

$$\omega_n = \sqrt{\frac{k}{m}}$$

$$\frac{\text{Amplitude of relative motion of the seismic mass (output)}}{\text{Amplitude of impressed motion of the frame (input)}} = \frac{z_o}{x_o}$$

$$= \frac{r^2}{\sqrt{(1 - r^2)^2 + (2\delta r)^2}}$$

$$\text{Phase lag } \phi = \tan^{-1}\left[\frac{2\delta\left(\frac{\omega}{\omega_n}\right)}{1 - \left(\frac{\omega}{\omega_n}\right)^2}\right]$$

$$\phi = \tan^{-1}\left(\frac{2\delta r}{1 - r^2}\right)$$

The relative displacement between seismic mass and the frame (z_o) can be used to measure *displacement* (x_o) of the frame.

Seismic Instrument: Acceleration Response

Impressed harmonic motion is,

$x = x_o \cos\omega t$

Acceleration of impressed motion

$x'' = -\omega^2 x_o \cos\omega t$

Amplitude of acceleration

$$A_o = \omega^2 x_o \quad (7)$$

Expression, of equation (6)

$$\frac{z_o}{x_o} = \frac{(\omega/\omega_n)^2}{\sqrt{\left(1-\left(\frac{\omega}{\omega_n}\right)^2\right)^2 + 2\delta\left(\frac{\omega}{\omega_n}\right)^2}}$$

Can be written as

$$\frac{z_o}{\omega^2 x_o} = \frac{1}{\omega_n^2\sqrt{\left(1-\left(\frac{\omega}{\omega_n}\right)^2\right)^2 + 2\delta\left(\frac{\omega}{\omega_n}\right)^2}}$$

$$\frac{z_o}{A_o} = \frac{1}{\omega_n^2}\frac{1}{\sqrt{(1-r^2)^2+(2\delta r)^2}}$$

Where $r = \frac{\omega}{\omega_n}$

The relative displacement z_o between the seismic mass and the frame, therefore, can be used as measured of the amplitude of acceleration of the frame. The term ω_n^2 is fixed for a instrument. The term under the root sign depends on the frequency of excitation ω, damping remaining constant for a instrument.

Displacement Measurement

Ideally, the amplitude ratio should be unity and phase shift be zero or constant for all excitation frequencies ω. The amplitude and phase response is seen to be depending upon the damping (δ) present and the frequency of excitation (ω) relative to ω_n. Inspection of the curve shows that

1. For a damping ratio in the region of 0.6 to 0.7 amplitude ratio becomes near unity $\left(\frac{z_o}{x_o} \approx 1\right)$
2. For frequency ratio (ω/ω_n) between value 1.5 to 2.0.
3. For higher frequency ratio damping is not important. The instrument is therefore, used for displacement measurement at excitation frequency which are greater than its natural frequency. *Hence for displacement measurement it is necessary that the natural frequency of the instrument should be low so that frequency range may include low values.*

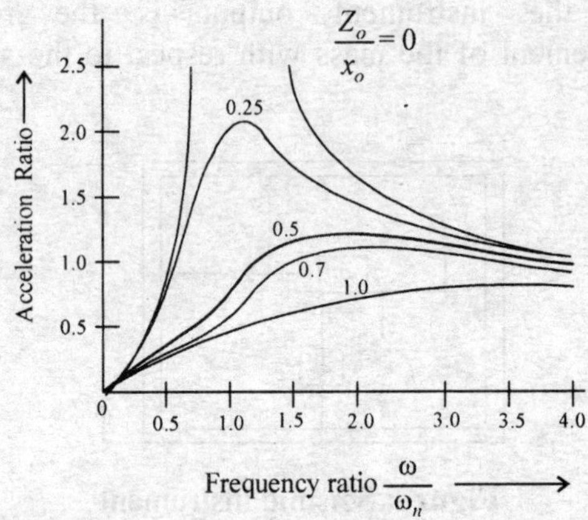

Figure: displacement Response of seismic Instruments

A zero phase lag is not practical. For zero damping the seismic mass and the frame are in phase below resonance $(\omega < \omega_n)$. Above resonance $(\omega > \omega_n)$ they are completely out of phase. A lower damping ratio is favourable for 180o phase shift in the region of higher (ω/ω_n) ratio. A compromise between amplitude response and phase response is made by choosing a damping ration near 0.6 to 0.7

Below listed transducers are displacement transducers.

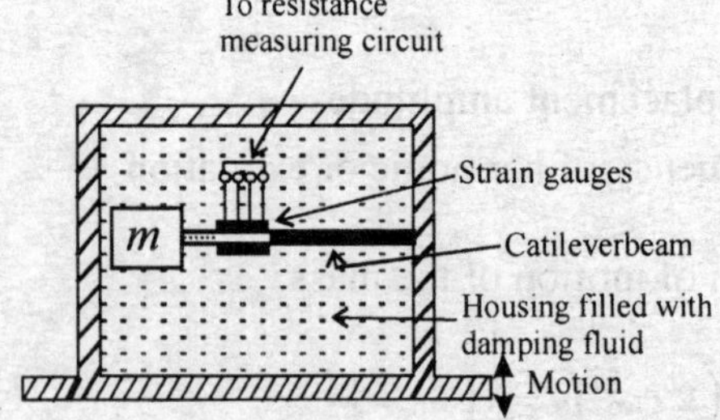

Figure: Seismic Instrument using Bonded Strain Gauges

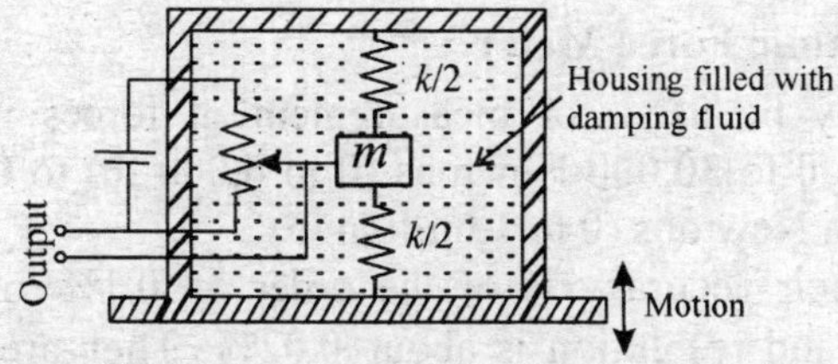

Figure: Seismic Instrument using Resistance Potentiometer

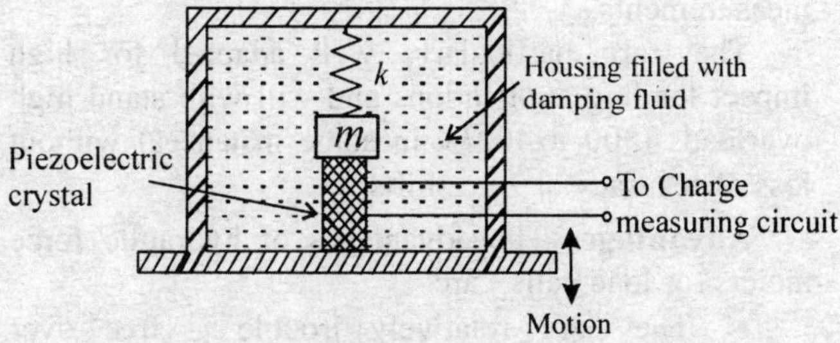

Figure: Seismic Instrument using Piezoelectric Transducer

Acceleration Measurement

Inspection of the curves figure indicates that for getting a constant amplitude ratio for acceleration measurement, excitation frequency 'ω' should be kept below 0.4 of the natural frequency $(\omega < 0.4\omega_n)$ and the damping ratio around 0.7. Hence to have a wide useful range of this type of instrument it must have **high natural frequency.**

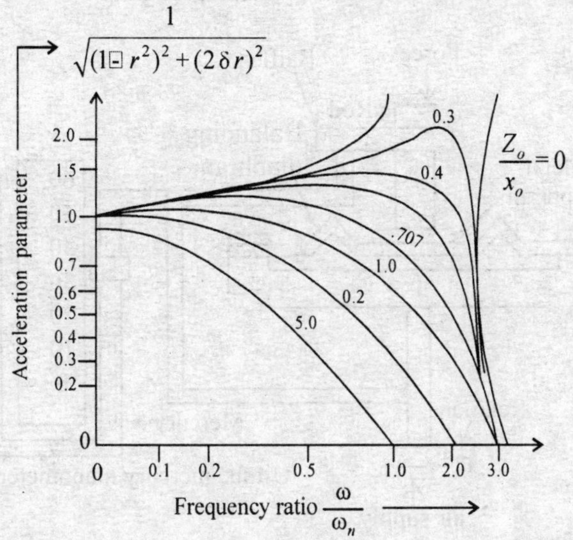

Figure: Acceleration response of seismic Instrument

Seismic Accelertometer

Let the input frequency ω be much smaller that the natural frequency ω_n, then

$$(x_2 - x_1) = \frac{\omega^2 A}{\omega_n^2} = \frac{1}{\omega_n^2} \times \text{maximum acceleration}$$

And this relation remains valid for $\frac{\omega}{\omega_n} \leq 0.4$.

This if the pick up is to be used for acceleration measurement, ω_n should be large, i.e. the system should be have a stiff spring and small mass. That would enable to operate the system over a wide range of frequencies and still keep the response linear.

Velocity Measurement

The same relationship as derived for the displacement are applicable to velocity measurement. But the secondary transducer is a velocity sensitive device

If the secondary transducer used a is a "velocity pickup" type the output becomes a function of velocity.

These transducer incorporate a permanent magnet and a coil. The output voltage induced due to the relative motion between the two is proportional to the rate of change of magnetic flux, hence to the rate of change of displacement or velocity.

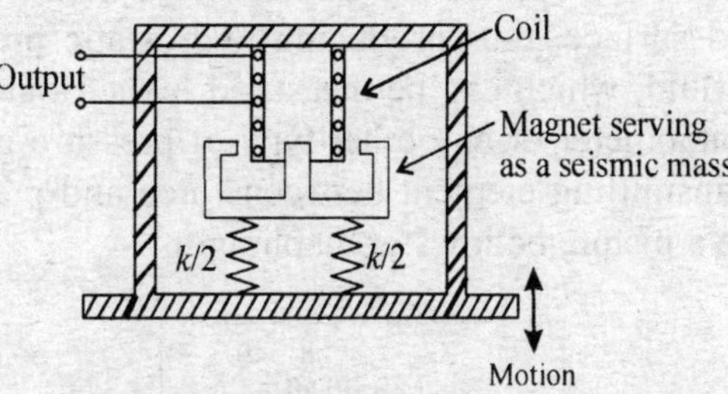

Figure: Seismic Instrument with Velocity Transducer

These secondary transducers are of two types:

1. In this type, seismic mass is in the form of a spring restrained coil, moving in a magnetic field supplied by a permanent magnet. As seismic mass moves relative to a magnetic field, the rate of change flux induces voltage in the pick up terminals. Damping provided is of electrical type
2. Another type uses a permanent magnet as seismic mass the pick up coil for measuring induced voltage is attached to the instrument housing., Damping is provided by the silicon fluid surrounding the seismic mass.

Both the above instruments are velocity sensitive as it is the rate of change of lines of magnetic flux cut, which determine the output from the secondary transducer.

3.2.4 MEASUREMENT OF FORCE

Force may be defined as cause that produces resistance or obstruction to any moving body, or changes the motion of a body, or tends to produces these effects.

According to network's second law of motion, force, F is defined as proportional to acceleration a, that is, $F \propto a$

Force is given as:

$F = Ma$

Where

F = force in Newton

M = mass in *kg*

a = acceleration in m/s^2

Force in addition to an effect along its line of action (push or pull) may also cause a turning effect is called *moment, torque*, or couple which can cause bending in a beam or torsion in a shaft. Measurement of force can be done by following methods:

Hydraulic Force Meter

It operates on the principle of a force counter balance. It involves the application of force to a definite area of fluid surface, thus producing hydrostatic pressure in the fluid, which can be measured by a Bourdon – tube manometer or any other type of pressure gauge. The transmitting element between force and pressure may be a piston, bellows or diaphragm.

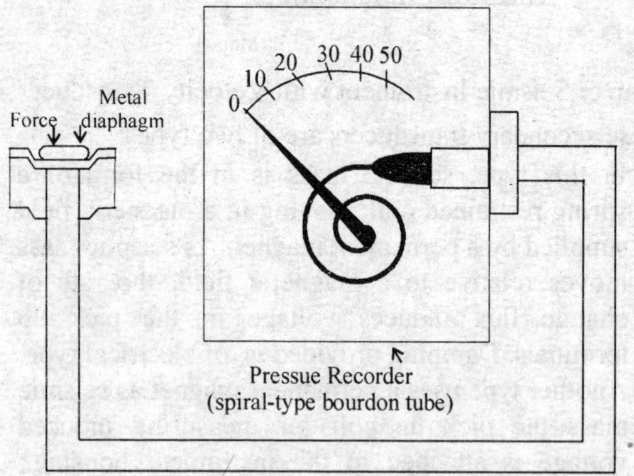

Figure: Hydraulic force Meter

Hydraulic Force Meter

It: may be used for measurement of forces in the ranges 0 to 30,000 Newtons (0 to 6,000 Ib) to 0 to 5 million Newtons (0 to 1 million ib).

Their accuracy is of the order of 0.1% of full scale, and resolution is about 0.02%. They are self-contained and require no outside power for operation.

They are inherently explosion-proof and are available for both tension and compression force measurements.

They are particularly well adapted for high impact loading applications and will with stand high overloads (300 to 400% in some instances) without loss of accuracy or zero shifts.

Advantages The advantages of hydraulic force meters (or load cells) are:

- They are relatively trouble – free over extended periods of use.
- They require no outside force of power for indication.
- They respond quickly to load change.
- They are relatively insensitive to temperature changes.
- They are well – suited for used in hazardous

Pneumatic Force Meter

It operates on the force balance principle. In this type of meter, the variable downward force to be measured is balanced by the upward force of air pressure against the effective area of the diaphragm.

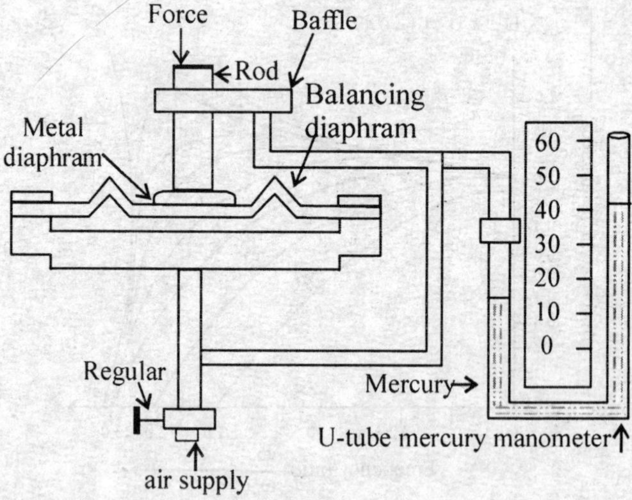

Figure: Pneumatic Force Meter

Pneumatic Force Meter

These types of instruments are available in ranges of 0 to 7 Ibs (0 to35 Newton) to 0 to 2.450 Ibs (0 to 12250 Newton) with a standard 3 to 15 psig output signal. By using non – standard signals, they are manufactured with a maximum load rating of 9,800 Ibs (49,000 Newton) with a 51 – psig output signal. These instruments have an accuracy of about 0.5% of full – scale.

Advantages Following are the advantage of pneumatic force meters (or load cells):

- They are suitable for use in hazardous or explosive areas.
- They require no special transmitting system.
- They cannot cause product contamination due to leakage or hydraulic fluid.
- They are relatively free from temperature – related errors.

Electric Force Transducer

Force measurement is also done by electric means in which the force is first converted into a displacement at an elastic element and then the displacement is measured.

Electric Force Transducers

Proving Ring

With this type of instrument, forces in the range of 0.0045 to 45,000 kg (0.01 to 100,000 Ib) can be measured with a rated output range of 5 to 200 mV per volt of excitation. Its linearity range is from 0.1 to 0.5%, and repeatability is 0.05%.

Advantages : proving rings:

- These instruments furnish relatively high output singal levels.
- They give high accuracies.
- They cover a broad range of measuring capacities.

Analytical Balance force meter

This is an **equal arm beam balance**. This operates on the principle of moment comparison. The moment, produced by the unknown weight, is compared with that produced by a known weight.

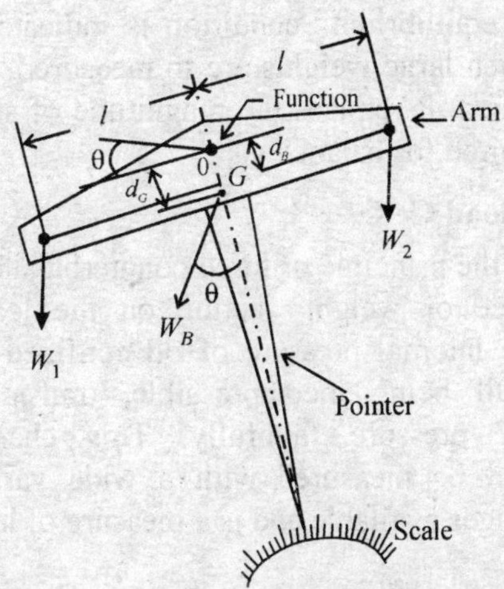

Figure: Schematic of Analytical Balance

In figure, 'O', is the fulcrum or the knife edge, W_B , the weight of the arm and G, its centre of gravity and θ is the angular unbalance. It can be shown that the sensitivity is independent of the loading of the balance provided d_B is zero. It can be improved by decreasing both d_G and W_B and increasing l. Correction in results is to be applied to account for buoyancy forces acting on the weights. Precision balances of accuracy of about 1 part in 10^6 are available

Principle of weight Beam Type Devices

In these devices gravitational force acting on unknown mass is balanced by gravitational force acting on known standard masses.

Standard masses that are used to balance the beam consist of a fixed standard mass and a sliding standard mass. The sliding mass, which can be slided along a notched graduated bar, provides fine adjustment such that a null balance is achieved as shown by the pointer

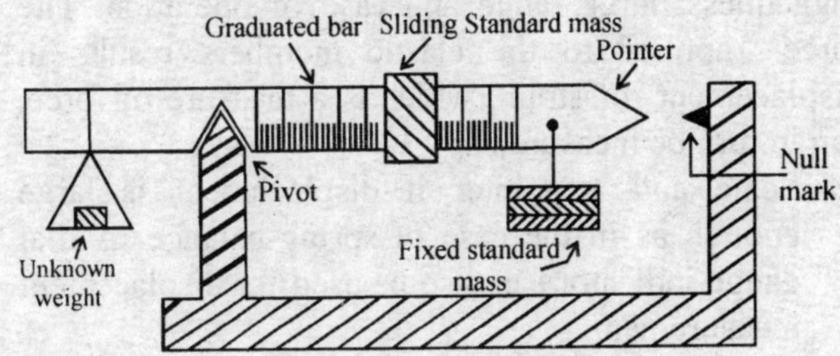

Figure: Schematic of weight beam device

Balance or equilibrium condition is indicated by a pointer. When large weights are to measured, a lever system is used. It reduces the magnitude of standard masses required for balancing.

Hydraulic load Cell

It works on the principle of force counterbalance. The applied force or weight acting on the load cell changes the internal pressure of oil confined within the cell. Oil being incompressible, transmits the increase of pressure faithfully. This change in pressure can be measured with a wide variety of pressure gauges available and is a measure of load.

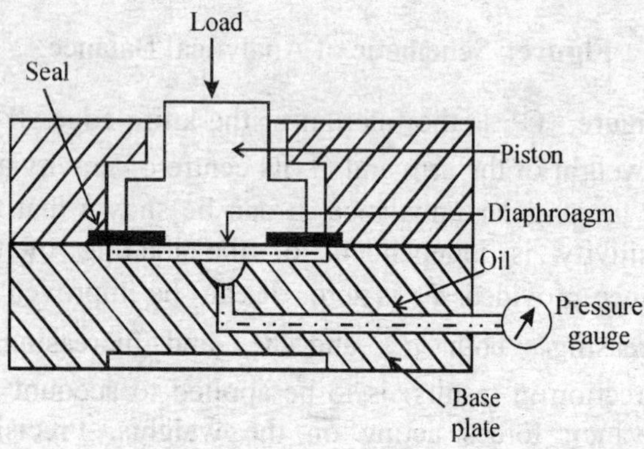

Figure: Hydraulic load cell

Figure shows a hydraulic cell with a floating position and diaphragm type seal. The oil is confined within the diaphragm chamber and the base plate. The load is applied to the top of the piston.

Capacity of cell can be as high as 100 kN and accuracy is ± ½ %. But temperature, if changes during measurement, can cause error. *Dynamic response of the cell is good.*

Elastic Force Transducers

Elastic elements are frequently for both static and dynamic measurement of force because of their ruggedness, large range and ease of operation. The force applied to the elastic members results in displacement or strain, which is a measure of force, can in turn be measured by:

1. Scale and a pointer if displacement is large enough as in the case of spring-balance. A dial gauge indicator can also be used for displacement measurement.
2. An electrical secondary transducer can be used for the purpose. Strain gauge are commonly used for measuring strain.

Various Elastic transducers are:

1. *Spring:* A simple spring is a force-displacement transducer, often used for weighing.

Force and displacement are related as:

$F = k\,x$

Where F = the applied force
x = displacement
k = stiffness or spring constant (force per unit displacement)

2. *Axially Loaded Member:* It is a bar or a rod axially loaded either in tension or compression

Displacement of the free end $x = \dfrac{FL}{AE}$

Stiffness $k = \dfrac{AE}{L}$

strain in axial direction $\varepsilon_1 = \dfrac{F}{AE}$

Strain in lateral direction $\varepsilon_2 = -\nu\varepsilon$

Where F = force
A = area of cross-section
L = Length
E = Young's modulus of the material
ν = Poisson's ratio

Strain Gauge Load Cell

It is an electromechanical transducer which translates changes in force into change in voltage. The change in voltage can be calibrated directly in terms of the force (or load) applied to the cell.

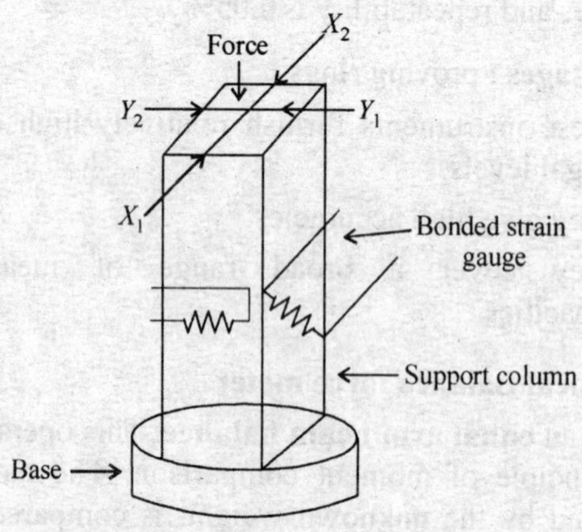

Figure: Strain Gauge Load Cell

Advantage:

1. Strain gauge load cells are small and campact in size.
2. They are well suited for measurements where an electrical output signal is desired.
3. They respond rapidly to load variations. But they have to be protected from angular or non-axial loads as they have no way to discriminate between bending and axial loads.

Pressductor / Magnetostrictive Load Cells

The working of this is based on the magnetoelastic principle in which the permeability of the ferromagnetic material is changed due to applied mechanical stress.

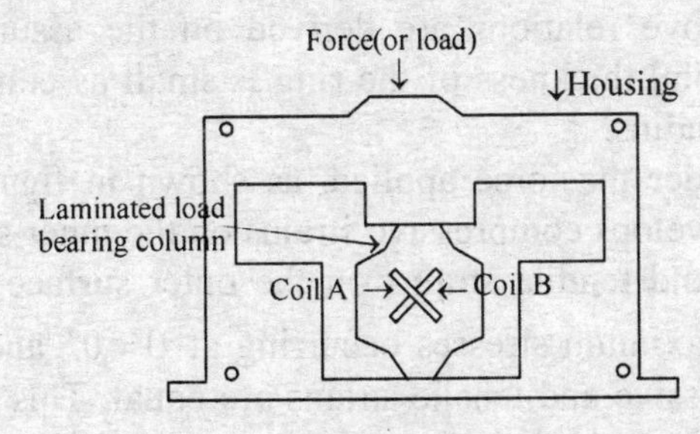

Figure: (a)

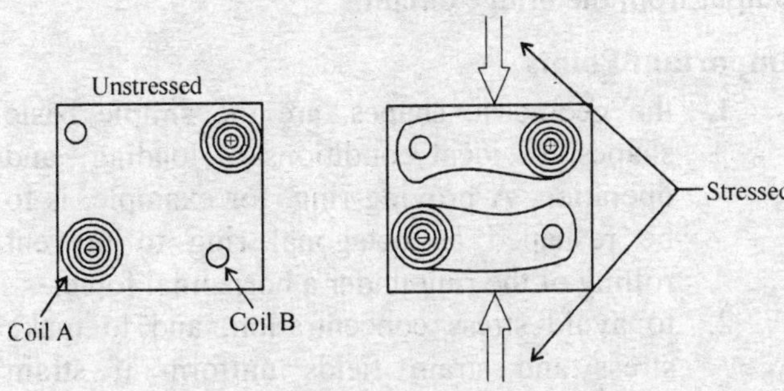

Figure (b)

Figure: The Pressductor load cell

Pressductor load Cells

The pressductor has greater applicability in steel plants for the measurement of roll – force in rolling mills, strip – tension in strip mills, and in weighing cranes in steel – melting shops.

Advantages

- It is an extremely robust transducer.
- It produces relatively high output signal levels.
- Its overload ratings are as high as 15 times the rated loads.

Disadvantages Following are the disadvantage of a pressductor load cell:

- The excessive stress and ageing may cause permanent changes.
- Zero drift and sensitivity changes due to temperature variation.
- It has hysteresis errors errors.

3. Cantilever Beam: In this a force F is applied at a distance L from the fixed end. The force causes a bending moment and the deflection of the free end. Strains ε_1 and ε_2 induced are equal in magnitude but opposite in sign.

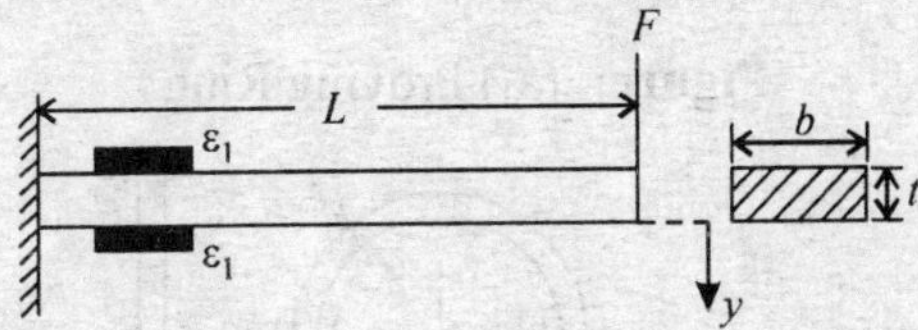

Deflection of the free end $y = \frac{1}{3}\frac{FL^3}{EI} = \frac{4FL^3}{E\,b\,t^3}$

Stiffness $k = \frac{3EI}{L^3} = \frac{E\,b\,t^3}{4\,L^3}$

Where,

$I = \frac{b\,t^3}{12}$ → the moment of inertia of the section about the centroidal axis

Strain, $\varepsilon_1 = \frac{6\,FL}{E\,b\,t^2}$ (tension)

$\varepsilon_2 = \frac{6\,FL}{E\,b\,t^2}$ (compression)

The opposite sign of strains ε_1 and ε_2 (in above case) is of advantage while using strain gauges employing a bridge circuit.

4.Ring (Proving Ring): It is used for measurement of load with a range of 1.5 *kN* to 1.5 *MN*. This device has been the standard for calibrating tensile testing machines. Figure shows that when the ring is subjected to diametrical compression by a force F it deflects by amount y. Deflection can be measured by a dial gauge or micrometer with a vibrating reed.

It utilizes the principle of a linear – variable differential transformer which senses the displacement caused by the force, resulting in a proportional voltage.

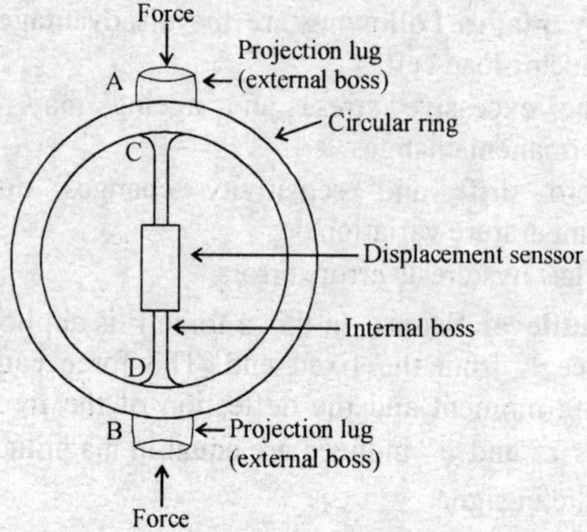

Figure: (a_1) Proving Ring

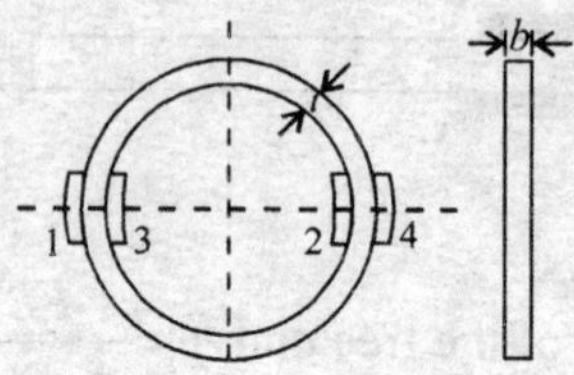

Figure: (a_2) Proving Ring

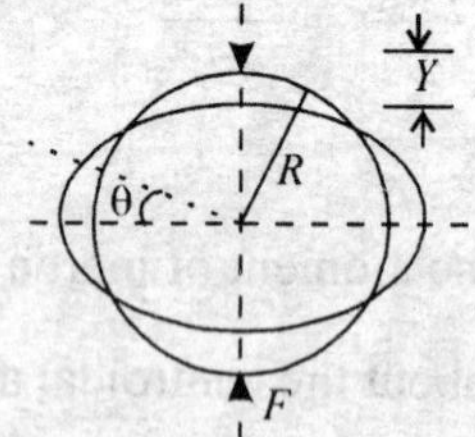

Figure (b): Proving Ring loaded diametrically in compression

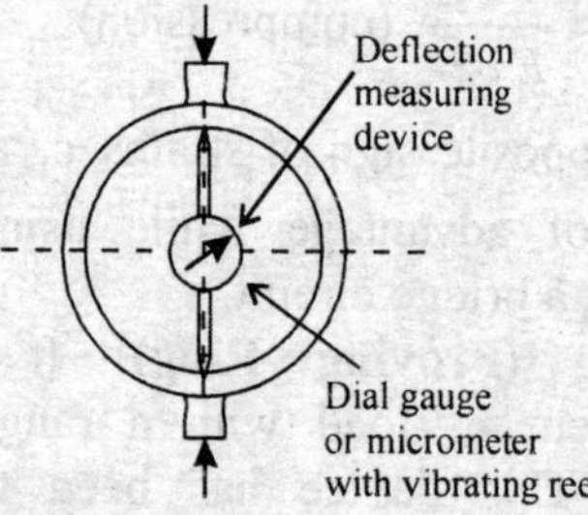

Figure (c): Proving Ring with Deflection Measuring Device

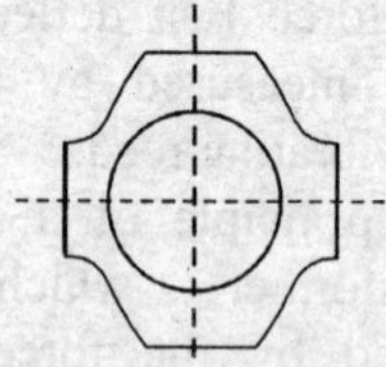

Figure (d): Octagonal Type Ring

Bending moment $M = \frac{FR}{2}\left(\cos\theta - \frac{2}{\pi}\right)$

$= 0.182\ FR\,(M \text{ at } \theta = 0)$

Deflection $y = \frac{1}{16}\left(\frac{\pi}{2} - \frac{4}{\pi}\right)\frac{FD^3}{EI}$

$y = \frac{1.79\ FD^3}{Ebt^3}$

Where,
F = load
D = diameter of ring
E = Young's modulus
I = moment of the inertia of the section about centroidal axis of bending.

Above relations are derived on the assumption that radial thickness of the ring is small as compared to the radius.

Under the force applied, as shown in figure (a), ring develops compressive strains on the inner surface (2, 3) and tendile strains on the outer surface (1, 4) with maximum stresses occurring at $\theta = 0^o$ and both compressive and tensile strains are equal. This fact is used when employing strain gauges to get a higher output from the bridge circuit.

Important Points

1. the geometric shapes, are the simple basic shapes for ideal conditions of loading and operation. A proving ring, for example, is to be reshaped as octagonal ring to prevent rolling of the ring under a horizontal force.
2. to avoid stress concentration, and to make stress and strain fields uniform if strain gauges are to be fixed.
3. Basic consideration for design are that the stresses should remain below elastic limit. Other considerations are residual stresses, sensitivity and response time.

3.2.5 MEASUREMENT OF TORQUE

Torque is defined as the force which tends to change the linear motion or rotation of a body. It may also be defined as the turning of twisting moment of a force about an axis.

Measurement of torque in rotating devices and machine is needed to determine the power required to operate a machine or power developed by a machine. Torque and power are related as:

$$P = \omega T = 2\pi N T$$

Where P is the power (watt)
T is the torque (N-m)
ω is the angular speed (radian/s)
N is the angular speed (revolution per second)

Torque is given as

$$T = FD.$$

Where
T = Torque
F = Force
D = Perpendicular distance from the axis of rotation of the line of action of the force.

1. In-line rotating torque sensor: It is used for the torque measurement of rotating equipments. It consists of a metal shaft with bonded strain gauges electrically connected in the form of wheat stone bridge.

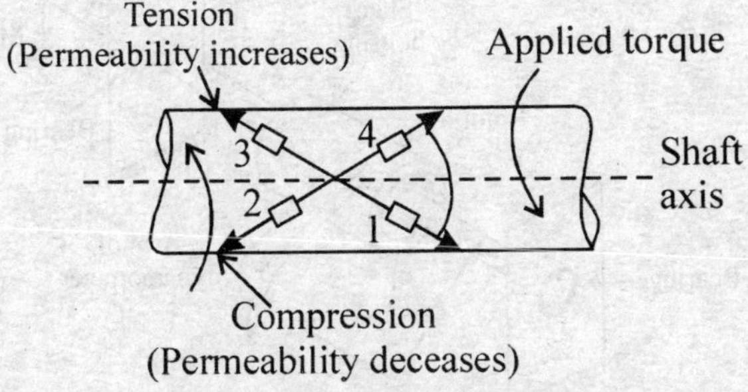

Figure: Tension-compression stress on surface of a circular shaft

The strain gauges 1 and 3 must be diametrically opposite as must strain gauges 2 and 4. In one direction at a 45° angle to the axis, pure tensile stress exists where as 45° in the other direction pure compressive stress is extent. The rotor shaft is elastic and will reflect minutely under the imposed stress. The out put of the wheat stone bridge is in proportion to torsion and hence to the applied torque on the shaft. Bridge power and output voltage may be connected to the sensor through slipping and brushes. But this type of pick-off is limited to rotational speed in the order of $30m/\sec$ at the brush surface. Also, with this type of system, the measurement is affected due to the constant resistance, contact friction and heating effect. A sensor with non-contacting power supply and signal pick-off is in the figure.

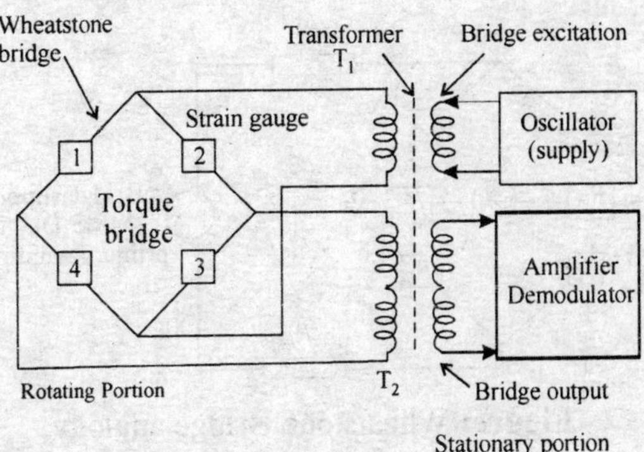

Figure: Wheatstone Bridge

The bridge power and output signals are transmitted between the rotating and stationary members through transformers T_1 and T_2. The bridge power is constant amplitude , high-frequency sing wave and the output is a sine wave of the power frequency whose amplitude is a function of torque. The power supply system consists of an oscillator to generate the carrier frequency. The output circuit consists of a demodulator to produce a dc signal proportional to the output sine wave.

2. In Line Stationary Torque Sensor

It utilizes the *Weidmann magnetostriction* effect for the measurement of torque. The **magnetostriction effect** causes changes in the permeability of the materials subjected to tensile or compressive stresses. The permeability increases under tensile stress and decreases under compressive stress.

It consists of two primary and two secondary windings mounted near the rotating shaft

The secondary coils S_1 and S_2 are coupled by the magnetic lines of flux through the shaft to the primary windings.

When there is no loading on the shaft, the permeability of the shaft is uniform and equal but opposite voltage are induced in secondary coils S_1 and S_2. When the shaft is under torsion, permeability and the number magnetic lines of flux increases in the direction of tension and decreases in the direction of compression. The voltages included in the secondary coils S_1 and S_2 do not cancel each other and their algebraic sum is in proportion to the toque.

The primary and the secondary windings are arranged magnetically to produce a circuit analogous to the wheatstone bridge

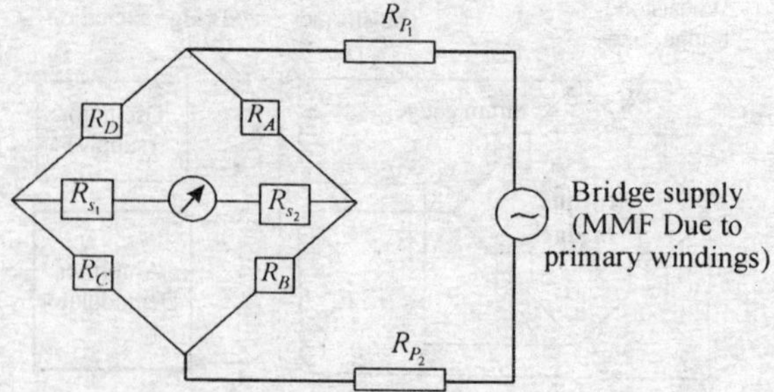

Figure: Wheatstone Bridge analogy

The bridge supply being equivalent to the primary windings P_1, P_2, R_{P_1}, R_{P_2}, R_{s1}, and R_{s2} are the reluctances of the corresponding air gaps between the primary and secondary windings and shaft. They are analogous to electrical resistance. The reluctance of the metal shafts is represented by R_A, R_B, R_c and R_D. When there is no loading on the shaft, the bridge remains in balance condition and the output voltage is zero.

3. Proximity Torque Sensor

Using proximity torque sensor, the rotating shaft torque is determined by measuring the angular displacement between the two sections of a rotating shaft.

It consists of two identical toothed wheels fixed on a shaft, a certain distance a part. Two proximity sensors are placed one at each wheel. When the shaft rotates due to the applied torque, the tooth of the wheel passes the proximity sensors due to which alternating voltage are produced, whose phase differences is proportional to the applied torque.

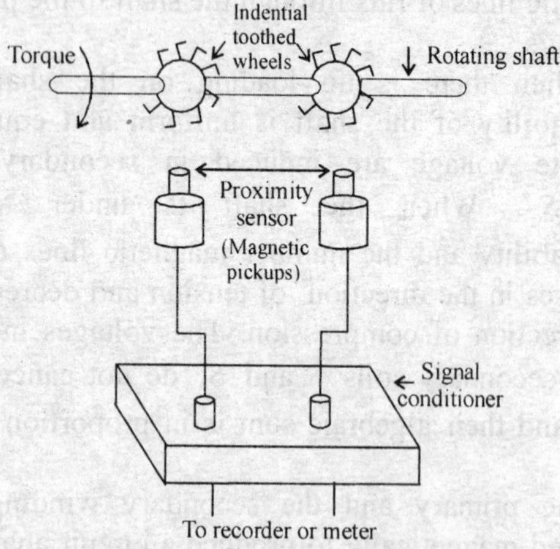

Figure: Proximity Torque Sensor

Craddling

The concept of craddling is widely used in dynamometers and is being explained before describing various types of dynamometers. These dynamometer consist of a rotor and a stator (or housing). The stator is not fixed but can turn relative to the rotor. The rotor is connected by a shaft to the machine to be tested. The rotor and stator are coupled electromagnetically or hydraulically. Te torque is measured as the reaction torque acting on the stator. An arm is attached on the stator and the force acting on the arm is determined. The torque is then give by the product of this force and the arm of the dynamometer because the restraining torque equals the reaction torque. The rotor and stator can be an electric motor (source of power) or a hydraulic dynamometer (a sink).

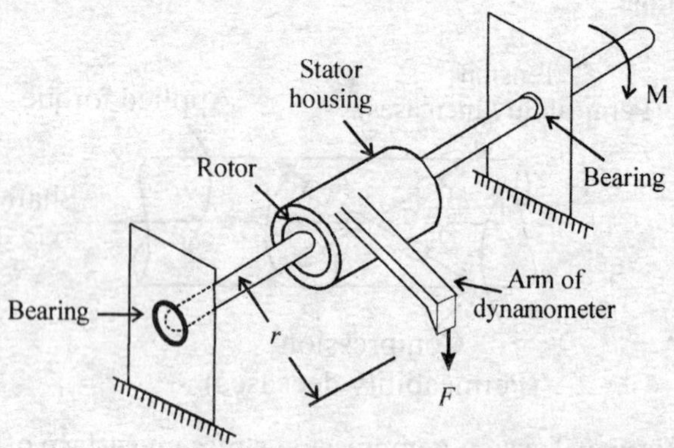

Figure: Schematic of Craddled Dynamometer

Torque measuring devices are commonly known as dynamometers and are classified as,

Absorption dynamometer

In these, the mechanical energy is absorbed and dissipated (generally as heat) as torque of the machine is measured. These type of dynamometers cannot drive test machine. These are particularly useful for measuring power or torque developed by engines and motors. Absorption dynamometer can be of the following types:

A. Mechanical/ Prony Brake Dynamometer

In these dynamometers, frictional resistance is created between a rotating flywheel attached to the test machine and a rope or band or brake shoes. The heat, thus, generated is dissipated by circulating cooling water. One of the most familiar and simple type is known as prony brake dynamometer.

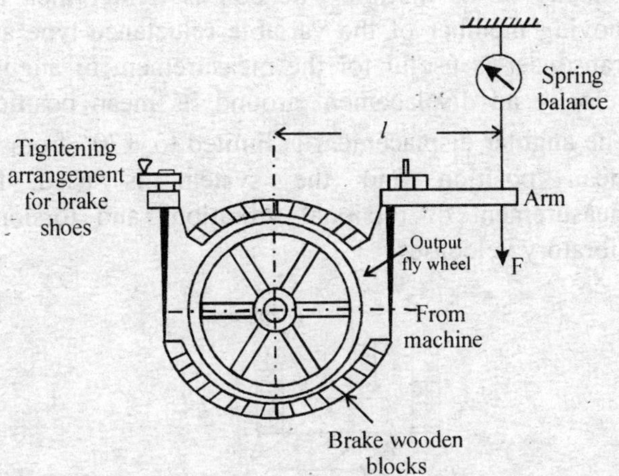

Figure: Schematic of a Prony brake

Includes some arrangement to vary the frictional resistance, e.g., tightening of rope or wooden brakes. The force 'F' can be measured by some force measuring device. In dynamometers involving small forces, an arrangement involving dead weights and spring balance can be used.

Torque = $= T = F \times l$

Power $= P = \dfrac{2\pi NT}{60}$

Where N is speed in R.P.M

$$P = \frac{2\pi\, N F\, l}{60}$$

B. Hydraulic Dynamometer

This is a craddle type dynamometer in which stator and rotor are hydraulically coupled. The stator is in two halves and placed on either side of the rotor. Semi-elliptical groves in the rotor match with that in the casing through which a flow of water is maintained. Water flows through helical paths creating vertices and eddy currents resulting in resulting in dissipation of energy. The tendency of the stator to rotate, which is freely pivoted on bearings (carddling), is opposed by weight applied to the moment arm on the stator. The breaking effect can be varied by controlling the amount of water flowing by some mechanism between rotor and stator. There is no need to provided a separate cooling arrangement in this type of dynamometer because water itself acts as coolant.

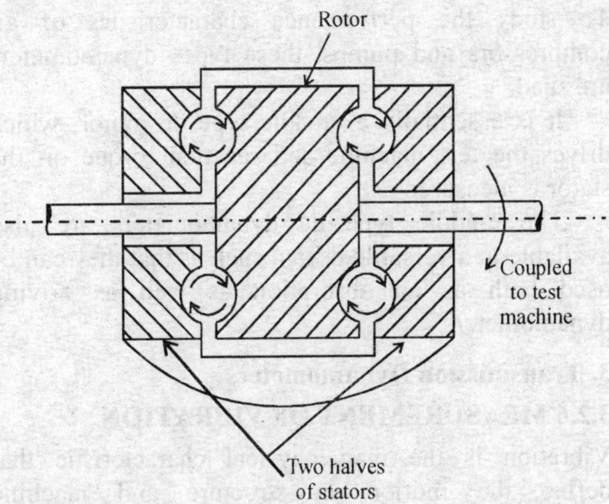

Figure: Schematic of a Hydraulic Absorption Dynamometer

C. Eddy Current Dynamometer

It is a craddle type electrical dynamometer and of absorption type. It consist of a metal disc, which is rotated (by the test machine) in a magnetic field.

The magnetic field is produced by the field coils, housed in the dynamometer (stator) and excited by an external source. Stator is mounted in trunnion bearings (craddling). As the disc rotates in the magnetic field, eddy currents are generated and the reaction with the magnetic field tends to rotate the dynamometer housing (stator) in the trunnion bearing. The torque is measured with the help of moment arm, attached to the stator. The load can be measured with the help of moment arm, attached to the stator. The load can be varied by varying the filed current and the power absorbed is converted into heat and is removed by circulating cooling water.

D. Electrical Generator Type

It is craddle type dynamometer and is essentially a generator driven by the test machine. The mechanical energy is converted into the electrical energy by the electrical generator. The electrical energy is then dissipated through a variable resistance grid, which also serves to vary the load. Craddling in trunnion bearings permits the determination of reaction torque on the stator explained earlier.

2. Driving Dynamometers

Driving dynameters measure the torque as well as provide the energy to operate the device to be tested.

To study the performance characteristics of air compressors and pumps, these types dynamometers are sued.

It is essentially a craddle type dc motor, which drives the test machine and reaction torque on the stator is measured

The craddle type dc dynamometers are also available in a versatile design such as that they can be used both as an absorption as well as driving dynamometers.

3. Transmission Dynamometers

3.2.6 MEASUREMENT OF VIBRATION

Vibration is the man physical characteristic that define the motion in structure and machine components. Acceleration, velocity and displacement as well as force and stresses in vibrating bodies comprise bulk of vibration. Vibration measurements are made for three major reasons:

1. To obtain the response of a body or structure each as response of an aircraft wing to various load conditions.
2. To define the vibratory convironment surrounding a vibratory source.
3. To monitor and control the system. Vibration monitoring is carried out on improatnt machines like power station turbines and generators to gives an early warning of impending conditions which may develop and lead to complete failure and destruction of the equipment.

Vibration, in the general sense, occurs as periodic, as random or us transient motion, the latter more normally being referred to as shock when the transient is amplitude and brief in duration.

To control the vibration (the enemy of man and machine) first it should be measured and analysed to dingnose the course. Vibrato on-analysis gives judgment about internal condition of the machine. Various type of portable and continuous monitoring system are used in industries today. Often these monitors are provided with alarms to alert the operator and in extreme condition to shutdown the machine.

Electro – dynamic Vibration Transducer

Electro- dynamic vibration transducers are based on the principle of electromagnetic induction. The permanent magnet moving coil system is one such transducer in which a relative displacement of the moving coil with respect to the magnet induces emf in the coil and the emf is proportional to the angular velocity of its motion. The coil is lighter then the moving member of the variable reluctance type and transducer is useful for the measurement of angular velocity ad displacement around its mean position. The angular displacement is limited to $\pm 70^o$ from its mean position and the system is used for measurement of torsional vibrations and torsional vibratory velocities.

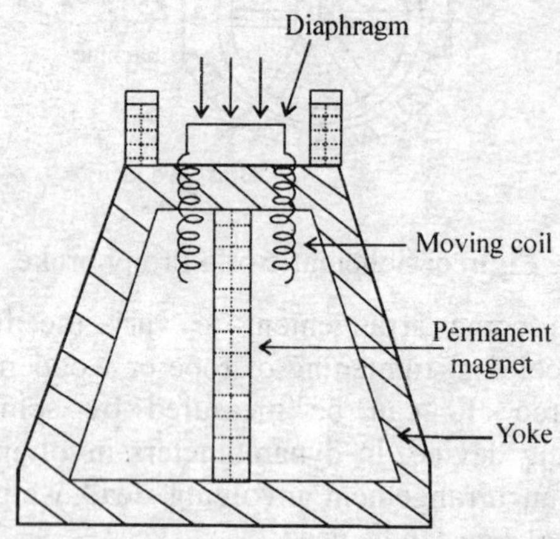

Figure: Electrodynamics Vibration Transducer.

The system is equivalent to Hibbert's magnetic standard in which the light cylindrical hollow coil of n turns cuts the flux set-up in the annular gap of the magnet assembly. The soft iron yoke preservers the stability of the system and the transducer is used for the measurement of linear vibratory velocities from which the dynamic displacements of the coil are obtained around its mean position.

3.2.7 MEASUREMENT SHOCK.

Shock is sudden short impulse of applied force that generator very large acceleration. 100000 g can arise and is not recurrent. It can be regarded as a once-only occurrence of a vibration waveform although sometime it is used to describe a short burst of oscillation.

Understanding the behavior of a given vibration sensor requires Fourier analysis of its response to a turned waveshape. The mathematic becomes more complex. Theoretical study does lead to the generalization that as the waveform becomes more like a single pulse of high magnitude and very short duration the frequency band of the sensor must be widened if the delivered output is to be a satisfactory

replies of the actual input Vibration Parameter. Fidelity increase as the period of the natural frequency of the sensor becomes shorter than the pulse length.
The very large force exerted on the transducer requires a design that recognizes the need to withstand large transient force without altering mechanical strains in the sensor system.
Well designed shock sensor can accurately single half-sine wave pulse as short 5 μs. Some amount of ringing in the output is usually tolerated in order to provided measurement of very short duration shocks.

1.4 Bell type pressure gauge

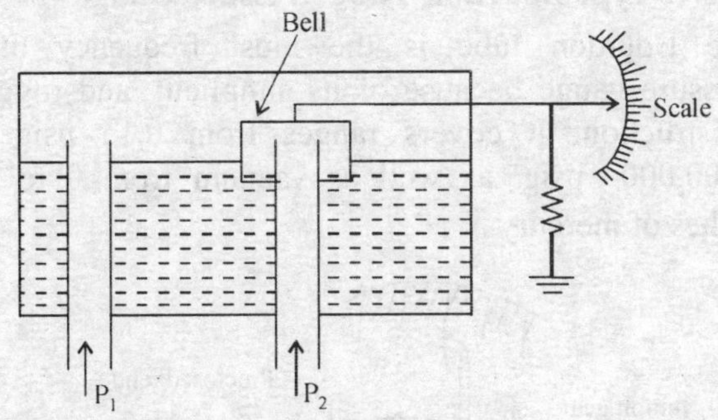

Where,
F – Spring constant
H – Change in spring length which is directly proportional to the amount of bell movement
A – Cross – sectional area of Bell
Pressure difference causes movement of the bell which is taken out by a pressure tight link-lever mechanism to the pressure indicating mechanism. The motion of the bell is opposed by a restricting spring. The bell type pressure gauge have accuracy 1%. The inside area and outside area of the bell are same. The bell material should be thin but the diameter should be large.

3.3 INDUSTRIAL INSTRUMENTATION

3.3.1 MEASUREMENT OF PRESSURE,

Pressure is defined as the amount of force applied to a surface or distributed over it and is measured as force per unit area.

Unit Of Pressure

(a) High Pressure
1 Newton per square meter $(1N/m^2)$= One Pascal 1 Pa)
1 atmospheric pressure
$(1\,\text{atm}) = 14.696\,\text{psi} = 101.325\,\text{kPa}$

(b) Low Pressure
1 mill bar = 100 dyne/cm^2 $= 14.5\times10^{-3}$ psi
1 micron $= 106$ m Hg 19.34×10^{-6} psi
1 torr $= 1$ mm Hg $= 1000$ microns
$= 1934\times10^{-3}$ psi

DIFFERENT TYPES OF PRESSURE

When pressure is measured , it is usually desired to read it in terms of either gauge pressure, absolute pressure or vacuum (or differential) pressure. Different types of pressures are discussed below.

(a) Gauge Pressure

A gauge that indicates zero at atmospheric pressure measures the difference between actual and atmospheric pressure ,This difference is called "gauge pressure", It is abbreviated as psig (pounds per square inch gauge.)

(b) Absolute Pressure

Absolute pressure is actual total pressure (including atmospheric pressure) acting on a surface. It is abbreviated as Pisa (pounds per square inch absolute).

(c) Vacuum or Differential Pressure

Gauges that indicate gauge pressure may be designed to indicate pressures below zero. Such a gauge is called a "vacuum gauge" Gauges that indicate absolute pressure cannot indicate pressures below zero , because zero is a perfect vacuum.
In a differential pressure measurement , the gauge pressure is the difference between the absolute pressure of the fluid and the atmospheric pressure

Static Pressure and Velocity Pressure

When the fluid is in equilibrium , the pressure at a particular point is identical in all directions and independent of orientation. This is called "static pressure" Velocity pressure is the difference between the total pressure and static pressure:
Velocity pressure = Total pressure – Static pressure

METHODS OF PRESSURE MEASUREMENT

Most pressure instruments measure a difference between two pressures one usually being that of the atmosphere.

The different methods of pressure measurement are listed below.

1. Manometers

The manometer is the simplest measuring instrument used for gauge pressure (low-range pressure) measurement

The action of all manometers depends on the effect of pressure exerted by a fluid at a depth.

The Barometer

Since manometers inherently measure the pressure difference between the two ends of the liquid column, if one end is at zero absolute pressure, then the difference in height of the liquid from the zero reference indicates the absolute pressure, This is the principle of the barometer.

A barometer is a well-type absolute pressure gauge whose pressure range is from zero absolute to atmospheric pressure, Its readings are generally in millimeters of mercury (mm Hg). With a barometer, high vacuums are not measured. The pressure in the evacuated portion of the barometer is not really absolute zero but rather the vapour pressure of the filling fluid, mercury, at ambient temperature.

Fluids For Manometers

The most common fluids for manometers are water, red oil and mercury, Water is the best fluid to use if air is the fluid to be measured and the glass length is within reasonable limits To minimize the effects of freezing and freezing and evaporation, kerosene or antifreeze may be used.

Mercury is used when the measured fluid maxes with the manometer fluid or a very long tube length is necessary , while red oil is used as the indicating fluid when it is required to read manometer form a distant place, it is easier to read and it does not combine chemically with some fluids in dry system

Advantages of Manometers

1. Simple and time proven.
2. High accuracy and sensitivity.
3. The availability of a wide range of filling fluids of varying specific gravities.
4. Reasonable cost.
5. Their suitability for low pressure and low differential pressure applications.

Disadvantages of Manometers

1. Large and bulky
2. Need for leveling
3. Lack of portability
4. Measured fluid must be compatible with monometer fluid
5. No over-range protection
6. Condensation may present problems.

2. Elastic Pressure Transducers

This type of pressure transducers use elastic primary sensing elements such as the Bourdon tube, bellows and diaphragm.

The C-Type Bourdon Tube Pressure Gauge

The Bourdon tube is the most frequency used pressure gauge because of its simplicity and rugged construction. It covers ranges from 0.15 psig to 0100,000 psig, as well as vacuum from 0 to 30 inches of mercury.

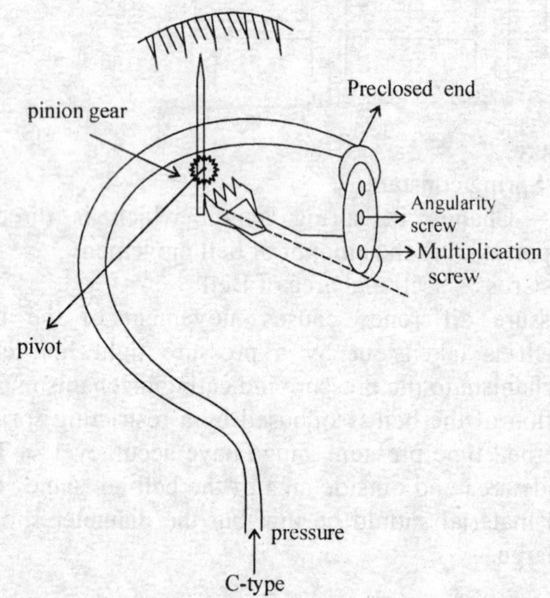

Figure: C-Type Bourdon Tube Pressure Gauge

Diaphragm Pressure Transducers

Diaphragms are widely used for pressure (gauge pressure) and drift measurements particularly in very low ranges. They can detect a pressure differential even in the range of 0 to 4 mm.

The diaphragms can be in the form of flat corrugated or dished plates and the choice depends on the strength and amount of defection desired. In high precision instruments the diaphragms are generally used in a pair, back to back, to form an elastic capsule.

Corrugated Diaphragm

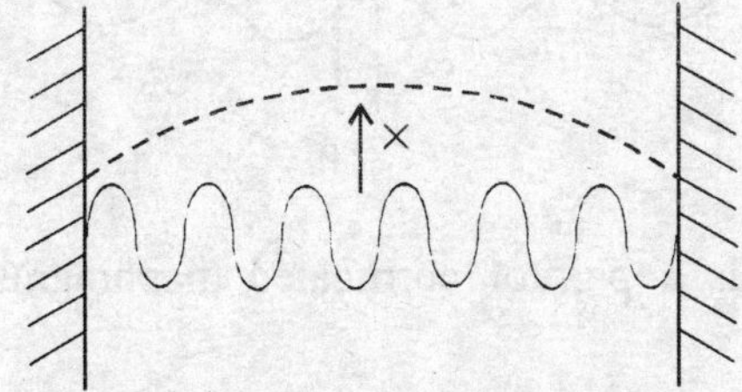

Figure: Corrugated Diaphragm

The corrugation allow increased linear deflection for the same applied pressure. They are having *larger diameter then the flat type diaphragm.* The larger size and the increased deflection reduce the dynamic response of the system. So it is *used only for static pressure measurement.*

With suitable modification, diaphragm elements can be made to cause changes in electrical circuits, thus converting pressure movements to electrical signals which can be transmitted to an indicating or a recording system. Two types of diaphragms are generally used:

1. Metallic diaphragm gauge
2. Stack diaphragm gauge.

(a) Metallic Diaphragm Gauge

It consists of a thin flexible diaphragm made of materials such as brass or bronze. A pointer is attached to the diaphragm. The force of pressure against the effective area of the diaphragm causes a deflection of the diaphragm. In some cases the deflection of the diaphragm is opposed by the spring qualities of the diaphragm itself and in other cases a spring is added to limit the deflection of the diaphragm. The motion of the diaphragm operates an indicting or a recording type of instrument A metallic diaphragm gauge.

This type of gauge is capable of working in any position and is portable, and therefore well adapted for use or for installation in moving equipments such as aircrafts.

(a) Stack Diaphragm Gauge

It is more difficult to measure pressure below the atmospheric pressure because the changes the changes are small. The full range from atmospheric pressure to a pressure to a perfect vacuum is only 14.7 psi $\left(1kg/cm^2\right)$, Pressure in this range can be measured with a slight modification of the diaphragm. A diaphragm with a large area produces a large change in force form a small change in pressure. Similarly, distances the diaphragm slack rather than tight allows it move a large distances in response to a small pressure change. A slack diaphragm can be made for rubber or other flexible materials.

A stack diaphragm gauge with a weak spring and large area can be used over pressure ranges as low as $0.01 - 0{,}40$ mm Hg (torr), It is possible to achieve accuracies of $1 - 2\%$.

Bellows

The bellows-type gauges are used for the measurement of absolute pressures (normal as well as low pressure), It somewhat more sensitive than Bourdon tube gauges .It is generally used for the range down to 155.1 mm Hg (3 psi). It may be used for even lower pressures up to 40mm Hg by making the bellows large enough. The bellows are made of an alloy which is ductile, has strength and retinas its properties over long use, i.e., has very little hysteresis effect. They are used in two forms, In one arrangement, pressure is applied to one side of he bellows and the resulting deflection is counterbalanced by a spring. This arrangement indicates the gauge pressure.

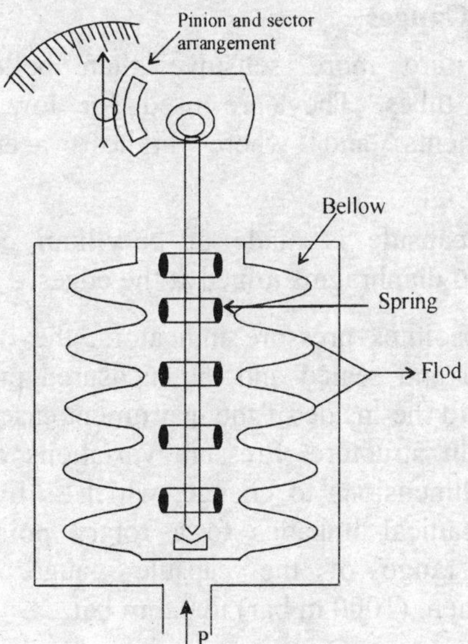

Figure: Bellow

Bellows element is a thin walled cylindrical structure with a number of folds for its cylindrical surface. These folds are also called convolution. The materials used for bellow element are stainless **steel, phosphor-bronze, beryllium copper, plastic materials and rubber**. The diameter of the bellows ranges from 7mm to 150 mm. The wall thickness ranges from 0.08 to 0.3 mm. Pressure range of bellow is 125 mm, water gauge to 1.05 MN/m^2. The pressure is applied to one side of the bellows by suitable linkages. The bellows displacement is magnified and the gauge pressure is indicated by a pointer and a scale.

In another differential arrangement, the differential pressure is also indicated. In this device, one pressure is applied to the inside of one sealed bellow while the other pressure is applied to the inside of another.

Measurement of Vacuum

Vacuum pressures are those which are below, atmospheric. With modern vacuum pressure systems, it is possible to obtain pressures from 1000 m bar (approximately 1 atmosphere) down to 10m bar. There is no single transducer available which covers this full range, Down to 1 m bar, it is possible to use some of the techniques, e.g. manometers and diaphragm type transducers. Some of the methods of vacuum measurement are discussed below.

Capsule Gauges

Capsules are more sensitive than bellows and Bourdon tubes. They are used for low pressure measurements and where highest accuracy is required.

The capsule is made of beryllium copper. It corrugated diaphragms joined at the edges.

In absolutes pressure indicators, the capsule is evacuated and sealed and the measured pressure is admitted to the inside of the instrument case which is of air-tight structure. Pressure variations cause the capsule dimensions to change, which is transmitted via mechanical linkages to a rotary pointer. The pressure range of the capsule gauge, is from 1 atmosphere (1000 m bar) to 0.5 m bar.

Capsule Type

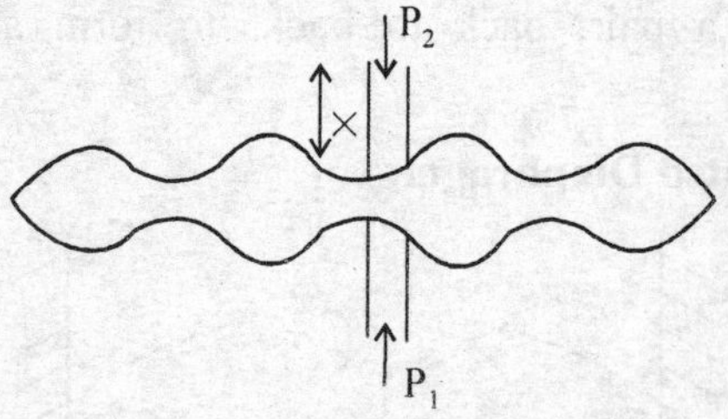

Capsule is a pair of corrugated diaphragms joined together.

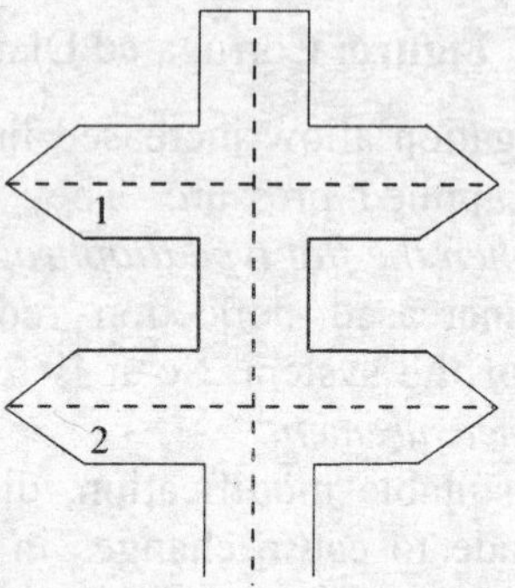

In this case the deflection of the whole capsule is equal to sum of the deflection of individual capsule. The material used for making capsule is **bronze, Stainless steel, ferrous-nickel alloy.**

McLeod Gauge

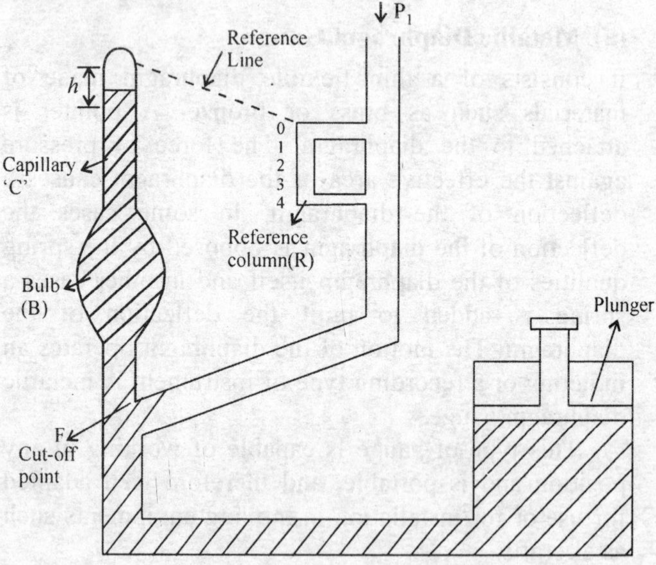

Figure: McLeod Gauge

It is based on the principle of compressing sample of known volume of low pressure gas to a known volume and high pressure which can be read buy a simple manometer.

The McLeod gauge is used for measuring very low pressures down to one hundred thousandth of an inch of mercury (or 10 m bar down to 10^{-s} m bar) The McLeod gauge amplifies pressure by compressing a gas into a small volume. The pressure of the compressed gas is than measured with a mercury manometer.

Construction and Working

The gas enters the gauge through B and fills the tubes down to the level of the mercury reservoir .The pressure is equal throughout the tubes and the bulb mercury is pumped up from the reservoir G as the mercury rises above the cut off , it traps gas inside the bulb The mercury is then pumped higher in the gauge until all the gas in the bulb is compressed into the closed capillary tube The operator allows the mercury to rise until it reference line on the closed capillary tube .The mercury rises faster in the open capillary tube A.

The compression of gas in the closed capillary tube makes the pressure of the trapped gas higher than the measured pressure, This pressure difference causes a difference mercury levels in the two tubes. The difference in height is used to calculate the pressure.

The volume of the bulb can be made quite large and the zero reference line on the closed capillary tube can be placed near the top of the tube. Thus, a large volume of gas can be compressed in to very small volume. This compression multiples the pressure by time. The pressure can be calculated by using the following equations. $P = KH\ H_0(1 - KH)$

Where P = measured pressure

K= a constant, determined by the geometry of the gauge

H = difference in heights of the two mercury columns

H_0 = height of the top of the closed Capillary tube above the line marked one the tube

The McLeod gauge is a very accurate pressure measuring device and often serves as standard for calibrating other low pressure gauges, it can be designed to measure pressures as low as 0.05 microns (00005 torr).

Thermal-Conductivity Gauge

Thermal –conductively gauges measure pressure by measuring the changes in the ability of a gas to conduct heat, The ability of a material to carry heat by conduction change when the pressure changes, until the pressure drops below about one torr. As the pressure continues to drop, the conductivity of the gas decreases and the gas loses its ability to conduct heat. Thus ,at low pressure , the conductivity of a gas has a direct relationship to its pressure the relationship between changes in conductivity and changes in pressure work over a pressure range form about 10^{-4} torr up to about 10^{-2} torr. It is used for the absolute pressure (very low pressure) measurement.

Basic Operating Principles

A pressure gauge based on changes in thermal conductivity is made by enclosing a wire filament in a chamber connected to the pressure source. When voltage is applied to the filament , electricity flows, making it hot . The rising temperature increases the resistance of e filament . The filament then reaches an 'equilibrium temperature', the temperature at which heat is produced in the filament as fast as it is removed. Heat is removed by both radiation and conduction. Convection is so slight that it can be ignored . The voltage applied the filament is held constant and any changes in pressure cusses a change in conductivity of the gas surrounding the filament, The change in conductivity change the equilibrium temperature of the filament which in turn causes the changes the change in the resistance, Therefore the changes in resistance is used to indicates the pressure change.

An increase in conductivity (due to an increase in pressure) increases the flow of heat away from the filament decreasing the temperature of the filament , A decrease in conductivity (due to decrease in pressure) increases the filament temperature

There are two types of thermal conductivity gauges generally used to measure pressure, which are discussed below:

(a) Pirani Gauge

It consists of two wire filament only filament serves as a reference and is sealed in an evacuated glass, while the other.

Filament is kept in a container connected to the source of pressure these two filament are connected in a bridge circuit

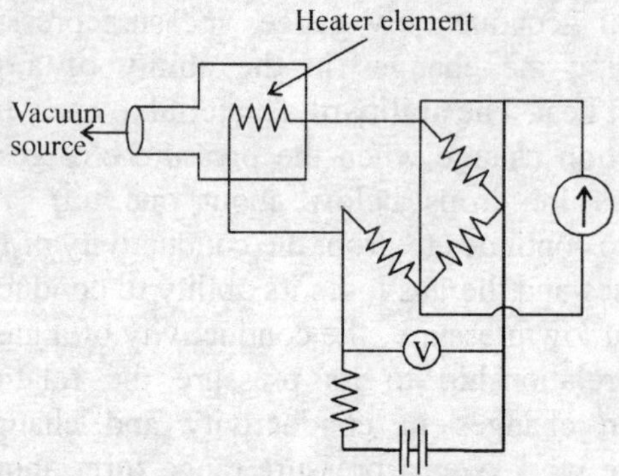

Figure: Pirani Gauge

The operation of pirani gauge depended on variation of the **thermal conductivity** of a gas with pressure. At low pressure there is a relationship between pressure and the thermal conductivity of the gas, the heat conductivity decreases with decrease of pressure. Pressure range of the pirani gauge is 10^{-5} to 1 torr.

(b) Thermocouple Gauge

The thermocouple gauge works on the same basic principle as the Pirani Gauge The only difference is that the filament temperature is measured by a thermocouple gauge, it requires careful calibration but once calibrated it is as accreted as the Pirani Gauge, A curve between thermocouple current (μa) and pressure (torr). Thermocouple gauges are used for the pressure range from about 10^{-4} to 1 torr.

Thermocouple Gauge

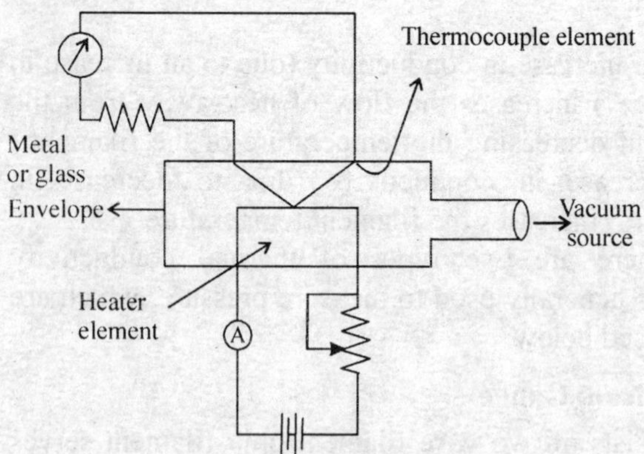

Figure: Thermocouple gauge

The thermocouple gauge consists of a heater element 0.025 *mm* of tungsteim wire and thermocouples welded to it are enclosed in a metal or glass envelope. This envelope communicates with a vacuum source whose pressure is to be measured. The element is supplied with a constant electrical energy and its temperature which is a function of the heat loss and hence thermal conductivity or the pressure of the surrounding gas is instrument can be directly calibrated to read the pressure of the gas. The range this instrument is 10^{-4} Torr to 1 Torr of pressure.

Ionization Gauges

A ionized on gauge measures the density of a gas. The operating principle of the ionization gauge follows Boyle's law, I e at constant temperature, the ratio of pressure of two gages is equal to the ratio of their two densities, i.e.

$$\frac{P}{P_1} = \frac{\rho}{\rho_1}$$

Where P = measured pressure
P_1 = initial pressure
ρ = measured density
ρ_1= initial density

This instrument consists of a chamber in which some of the gas mole-clues are changed to positively charged ions These ions are attracted towards a negatively charged plate and deliver their charge , which 'creates an electric current The current increase in proportion to density of the gas inside the chamber. Thus the pressure is measured proportional to the relative density.

An ionization gauge using alpha radiation is known as an "Alphatron Gauge".

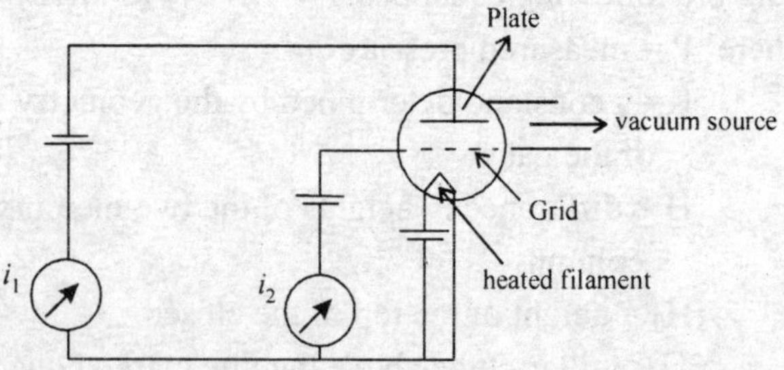

Ionization gauge consists of a heated filament, a positively charged grid and a negatively charged plate enclosed in a glass envelope connected to the pressure source. The grid is maintained at a positive potential of $100-350v$. While the plate is maintained at a negative potential of 2-50v with respect to heated filament. The pressure range is from 10^{-3} torr to 10^{-11} torr. It has got a fast response to pressure change.

3. Force Balance Pressure Gauge

Dead-weight piston gauge is used for the measurement of higher steady pressures, and for checking the elastic diaphragm or Bourdon type of gauge. In the practical form , it is often used as a standard of pressure of measurement. In this type of Instrument the force produced on a piston of known area is measured directly by the weight it will support. A dead-weight piston gauge.

Construction and Working

It consists of a very accurately machined, bored and finished piston which is inserted into a close fitting cylinder The cross sectional areas of both the piston and the cylinder are known At the top of the piston is platform on which the standard weight of known accuracy can be placed An oil reservoir with a cheek valve at its bottom is also provided The oil form the reservist can be sucked by a displacement pump on its upward stroke.

For calibration purposes first a known (calculated) weight is placed on the platform and the fluid pressure is applied on the other end of the piston until enough force is developed to left the piston weight combination and the piston flats freely within the cylinder between limit stops. The error in a dead-weight tester is less than 0.1%.

In order to reduce the friction between the piston and the cylinder, the piston is generally rotated while a reading is being taken.

No oil must enter the gauge when a testing oxygen pressure gauges, because the oil vapor together with the oxygen will produce an explosive mixture. For pressures up to 300 psig oxygen gauges may be tested with water or air.

4. Electrical Pressure Transducers

"A Transducer is a device which converts one form of energy into another from of energy". However in the field of electrical instrumentation "a transducer is defined as a device which converts a physical quantity, a physical condition, or mechanical output into an physical quantity, a physical condition or mechanical output into an electrical signal.

An electrical pressure transducer consists of three elements:

1. Pressure sensing element such as a bellow, a diaphragm or a Bourdon tube.
2. Primary conversion element e.g. resistance or a voltage.
3. Secondary conversion element.

Strain Gauge Pressure Transducer

Strain gauge is a passive type resistance pressure transducer whose electrical resistance changes when it is stretched or compressed. It can be attached to a pressure sensing diaphragm.

Basic Principle

The strain gauge is a fine wire which changes its resistance when mechanically strained, due to physical effects. A strain gauge may be attached to the diaphragm so that when the diaphragm flexes due to the process pressure applied on it, the strain gauge stretches or compresses. This deformation of the strain gauge causes the variation in its length and cross-sectional area due to which its resistance also changes.

The resistance change of a strain gauge is usually converted into voltage by connecting one, two, or four similar gauge, as of a Wheatstone bridge (known as strain gauge bridge), and applying excitation to the bridge. The bridge output voltage is then a measure of the pressure sensed by the strain gauges.

Construction and Working

A bridge circuit with four strain gauges, Reg 1, Rsg 2. Rsg 3 and Rsg 4. Two strain gauges , Rsg 1 and Rsg 4, are mounted so that increasing pressure increases, their resistance Strain gauges Rsg 2 and Rsg 3 are mounted so that increasing pressure decreases their resistance . A change in temperature affects all the four strain gauges in the same way, resulting in no change in the pressure indication.

At balance, when there is no pressure, no current flows through the galvanometer G, and hence there will be no deflection in the galvanometer As soon as the pressure is applied, the strain gauge stretches or compresses accordingly and the bridge circuit is unbalanced due to the change in resistance of the strain gauges. Thus, a current flows in the galvanometer, which is measured by the deflection of the galvanometer. These change affect the output of the bridge circuit, which indicates a change in measured pressure, Now, this change in output voltage may be calibrated for the pressure change.

Potentionetric Pressure Tranducers

In this type of pressure transducer, there is a potentiometer (basically a variable resistance) which is made by windings resistance wire around an insulated cylinder. A movable electrical contact, called a wiper, slides along the cylinder, touching the wire at one point on each turn. The position of the wiper determines how much wire , and therefore, how much resistance is between the end of the wire and the wiper. A mechanical linkage form the pressure sensing element (such as bellows, a diaphragm, and the Bourdon tube) controls the position of the wiper on the potentiometer. Some potentiometer are made curved so that the wiper can plvot in a circular motion rather than moving along a straight line the position of the wiper determines the resistance of the potentiometer, which in turn determines the pressure.

A potentiometer pressure transducer in which the sensing element is a Bourdon tube. An increase in pressure makes the Bourdon tube straighten out partially. The motion causes the linkage to move on the wiper across the winding on the potentiometer , As the wiper moves, it increases the resistance between terminals A and C, which is equivalent to the pressure sensed by the Bourdon tube.

Capacitive Pressure Transducers

The principle of operation of capacitive pressure transducers in based upon the familiar capacitance equation of the parallel plate capacitor, i.e.

$$C = \frac{\varepsilon_0 \varepsilon_r A}{d} \text{ farad}$$

where, C = The capacitance of a capacitor in farad

A = Area of each plate in m^2

D = Distance between the two plates in m

$\varepsilon_0 = \mathcal{E}.854 \times 10^{-12}$ farad/m^2

ε_r = Dielectric constant (relative Permittivity).

Thus, capacitance of a capacitor varies when,

1. The area A of the plate is changed
2. The distance d between the two plates is changed
3. The dielectric constant ε_r is changed.

Now, from the above equation, the capacitance of a parallel plate capacitor is inversely proportional to the spacing between the plates

$$\left(i.e\, C \propto \frac{1}{\propto} \right)$$

This is made use of in the capacitive pressure transducer for detecting pressure changes. illustrates the basic principle of a capacitive pressure transducer.

It consists of a fixed plate and a movable plate which is free to move as the pressure applied changes. According to the change in pressure the movable plate also changes its position, due to which the distance d is changed With increases in pressure, the distance d becomes less, due to which the capacitance C is increased (as C $\propto$ 1/d) With decrease in pressure, the distance d increases and thus capacitance C is decreased. This change in capacitance can be calibrated to measure the change in pressure.

In place of movable plate a diaphragm may be used, which expands and contracts due to change in pressure. The diaphragm plate acts as a movable plate of a capacitor. A fixed plate is placed near the diaphragm. These plates form a parallel plate capacitor which is connected as one of the arms of a bridge, Any change in pressure causes a change in distance between the diaphragm and fixed plate, which unbalances the bridge. The voltage output of the bridge corresponds to the pressure applied to the diaphragm plate.

5. Reluctive Pressure Transducers

Reluctance in a magnetic circuit is equivalent to resistance in an electrical circuit. Whenever the spacing (or coupling) between two magnetic divices (or coils) changes, the reluctance between them also changes. Thus a pressure sensor can be used to change the spacing between two coils by moving one part of the magnetic circuit. This motion changes the reluctance between the coils, which in turn changes the reluctance between the coils, which in turn changes the voltage induced by one coil in the other. The change in the induced voltage can then be interpreted as a change in pressure.

(a) Linear Variable Differential Transformer (LVDT)

It is the most widely used inductive transducer to translate liner motion into an electrical signal. Figure

5.24 shows an LVDT for the measurement of pressure. (also see displacement measurement).

Servo Pressure Transducers

A servo pressure transducer is also called a "force balance pressure transducer". It produces an electrical signal proportional to the pressure transducer is used in many industrial differential pressure cells. Figurer 5.25 shows the construction of a servo pressure transducer. Working Principle An increase in pressure P_1 over P_2 (Fig. 5.25) flexes the diaphragm and moves the short end of the force beam. The force beam pivots, and the long end moves a magnetic material in the reductive detector. The signal from the reductive detector is converted from a.c. power to d.c. power and sent to an amplifier. The amplifier responds by activating an inductive motor that moves the force bean back towards its original position , Very little flexing ever occurs in the entire range of the instrument, As a result, the diaphragm lasts a long time.

Servo pressure transducers are available in a multitude of pressure ranges. The devise are generally used for measurement of pressure below 500 psi.

They do not respond to high frequency pressure oscillations. Other servo pressure instruments use capacitive detectors, and some use a Bourdon tube as the sensing element.

Piezoelectric Pressure Transducers

These devices utilize the piezoelectric characteristics of certain crystalline and ceramic materials (such as quartz) to generate an electrical signal.

Basis Principle

Such transducers depend upon the principle that "when pressure is applied on piezoelectric crystals (such as quartz) an electrical charge is generated" There are about 40 crystalline materials that, when subjected to a squeeze" generate an electric seine red powder, crystal of quartz, tourmaline, and Rochelle salts. Figure 5.26 shows a piezoelectric pressure transducer.

Construction and working. It consists of a diaphragm by which pressure is transmitted to the piezoelectric crystal Y_1. This crystal generates an electrical signal which is amplified by a charge amplifier A second.

Piezoelectric crystal Y_2 is included to compensate for any acceleration of the device during use. This compensation pressure on the rapid acceleration of the transducer creates additional pressure on the piezoelectric crystal, Vibration is a major source of high, rapidly changing acceleration.

Signals form the compensating crystal, are amplified by a second charge amplifier. A different amplifier is used which subtracts pressure alone all effects of acceleration are removed.

Piezoelectric pressure transducers may be used to measure pressures over ranges up to 0-50,000 psi. However, piezoelectric cannot measure steady pressure.

6. Pressure Switches

A pressure switch turns an electric circuit 'On or 'OFF' at a preset pressure. This pressure is called the set point of the switch. A pressure switch is used in some form of control e. g .to operate a solenoid value a given pressure or start up a pump.

The pressure switch is usually a micros witch or a mercury switch, A Bourdon tube, a diaphragm or a bellows can be used to actuate the switch. Figure 5.27 shows the simplest form of a pressure switch used to actuate a relay.

The pressure is fed to the inside of a bellows which carries a contact plate B. When the pressure reaches a sufficient (or preset) value, the contact plate touches contact points C_1 and C_2 thus closing an electrical circuit to an alarm or motor control gear. The flexibility of the bellows ensures that the plate makes adequate contact area clean.

The pressure switch can be modified so as to make a low pressure contact in addition to a high pressure contact .It is adjustable. The contacts in a pressure switch may be normally closed when the pressure is below the set point For example , the contacts in a normally open switch remains open until the pressure rises above the set point. Then the sensing element makes the contacts snap to the closed position The contacts open again when the pressure falls below the set point. The contacts modified so as to make a low pressure contacts in a normally closed switch remain closed until the pressure.

Then the contacts snap open and remain open until the pressure drops below the set point again, Most switches contain two sets of contacts, one normally open and the other normally closed.

A pressure switch has a "dead band" i.e the pressure must fall below the set points before the switch resets to normal position. The amount of dead band is the difference in between the set points and the reset point.

3.3.2 MEASUREMENT OF FLOW

Inferential Flow Measurements

In the inferential type of flow measuring methods, the flow rate is inferred from a characteristic effect of a related phenomenon. A classic example of an inferential measurement is a human body running fever, resulting from a serious sickness or disability, in which the rise in body temperature (fever) is not the cause of sickness but a characteristic phenomenon resulting from the ailment.

Variable Head or Differential Flowmeters

It on the principle that a restriction (or obstruction) in the line (or pipe) of a flowing fluid, introduced by the orifice plate or venturi tube or elbow, produces a differential pressure across the restriction element which is proportional to the flow rate. The proportionality is not a linear one but has a square root of the differential pressure. This relationship was derived form Bernoull's theorem which states that, in a flowing stream, the sum of the pressure head, the velocity head and removed in the direction of flow the first point plus the loss due to friction between the two points. Velocity head is defined as the vertical distance through which a liquid would fall to attain a given velocity Pressure head is the vertical distance which a column of the flowing liquid would rise in an open end tube as result of the static pressure The basic equations are given as,

$$V = K\sqrt{\frac{2gh}{\rho}}$$

$$Q = KA\sqrt{\frac{2gh}{\rho}}$$

$$W = KA\sqrt{\frac{2gh}{\rho}}$$

where V = Velocity of flowing fluid

Q = Volume flow rate

W = Mass flow rate

A = Cross sectional area of pipe through which fluid is flowing

h = Differential head (Pressure) across the restriction element.

g = Acceleration due to gravity

ρ = Density of the flowing liquid

$$K = \frac{C}{\sqrt{1-\beta^4}} = \text{a constant}$$

where C = Discharge coefficient

β = Diameter ratio

$$\beta = \frac{d\,(\text{diameter of restriction element})}{D\,(\text{inside diameter of the pipe})}.$$

Reynolds Number

Reynolds number is a very important reference numbers in the accurate determination of flow. It is used to determine the point at which the flow goes form the viscous to the turbulent stage As the flow changes from viscous to turbulent and vice versa, there is a very little change in the value of the flow coefficient, but there is accuracy are desired when head meters are employed, then corrections will have to be made for the Reynolds numbers. It was discovered by Osborne Reynolds in the late 1800s and was named in his honour. Reynolds number is given by the equation,

$$R_D = \frac{VD\rho}{\mu}$$

where R_D = Reynolds number

V = Average velocity

D = Inside pipe diameter

ρ = Density of flowing fluid.

Reynolds number is a dimensionless unit when consistent units are used in its determination.

Parts of a Differential flow meter

A differential flow meter basically consists of two parts:

1. Primary elements
2. Secondary elements.

The parts of the meter used to restrict the fluid flow in the pipe line in order produce a differential pressure are known as primary elements.

Orifice Plates

Orifice plates, are the simplest and cheapest from of primary elements and are used more frequently than all other types An orifice plate is inserted in the line and he differential pressure across it is measured. There are four types of orifice plates which are listed below:

1. Concentric orifice plate
2. Eccentric orifice plate
3. Segmental orifice plate
4. Quadrant edge orifice plate.

Ventured Tubes

A venturi tube is used where permanent pressure loss is of prime importance, and where, maximum accuracy is desired in the measurement of high viscous fluids, It consists a straight inlet section of the same diameter as the pipe and in which the high pressure tap is located, a converging conical inlet section in which the cross section of the crease of velocity head and decrease of pressure head ,a cylindrical throat which provides for the low pressure tap location of the decreased pressure in an area where flow velocity is neither increases nor decreasing and a diverging recovery cone in which velocity decreases and the decreased velocity head is recovered as pressure head.

Flow Nozzles

The flow nozzles are used for flow measurement at high fluid velocities and are more rugged and more.

Rugged and more resistant to erosion than the sharp edged orifice plate.

Variable Area Flowmetrs

In the variable area flow meters, the size of the restriction is adjusted by an amount necessary to keep the pressure differential constant when the flow rate changes and the amount of adjustment required proportional to the flow rate.

Rotameters

The rotameter is the most extensively used form of the variable area flowmeter It consists of a vertical tapered tube with a float which is free to move up or down within the tube.

Turbine Flometers

The turbine flowmeter is used for the measurement of liquid, gas and very low rates. It works on basic principle of turbine. It consists of a multi-bladed rotor (turbine wheel) which is mounted at right angles to the axis of the flowing liquid.

Thermal Flowmeters

Thermal flowmeters are very popular for the measure flow rate in terms of mass, which is a very desirable feature, specially on gas service. There are two of types of thermal flowmenters:

1. Head transfer flowmeters
2. Hot wire flowmeters.

Heat Transfer Flow meters

This type of flowmeter measures the rise is temperature of he fluid after a known amount of heat has been added to it, Its theory is based upon the specific heat equation which is given as,

$$Q = WC_p(T_2 - T_1)$$

or $$W = \frac{Q}{C_p(T_2 - T_1)}$$

where Q = Heat transfer

W = Mass flow rate of fluid

C_P = Specific heat of fluid

T_1 = Initial temperature of the fluid after heat has been transferred

T_2 = Final temperature after heating the fluid.

Ultrasonic Flowmeters

In ultrasonic flowmeters, the measurement of flow rater is determined by the variation in parameters of ultrasonic oscillations, there are two types of ultrasonic currently in use:

1. Time difference type
2. Doppler flowmeters.

3.3.3 MEASUREMENT OF TEMPERATURE

$$\frac{°C}{1000} = \frac{°F - 32}{180} \qquad 1$$

$$°K = °C + 273.15 \qquad 2$$

$$°C = \frac{5}{9}(°F - 32) \qquad 3$$

The rankine scale is abbreviated as °*R*. On the rankine scale, the ice-point is $491.7°R$ and the steam point is $671.7°R$.

Temperature in Fahrenheit (°*F*) can be converted into rankine (°*R*) by using the equation:

$$°R = °F + 459.69 \qquad 4$$

Both the rankine and Kelvin temperature scales are called *absolute scales* because they use absolute zero as one of their reference points.

Methods of Temperature Measurement

The temperature measuring instruments are classified according to the nature of the change produced in the testing body by the change of temperature. They may be classified as follows:

1. Expansion Thermometers
2. Filled System Thermometers
3. Electrical Temperature Instruments
4. Pyrometers.

1. Expansion Thermometers

Expansion thermometers are classified according to the nature of substances which expands.

Expansion of solids
Bimetallic thermometers

Bimetallic thermometers

The commercial solid expansion type is a bimetallic thermometers. Two strips A and B of dissimilar metals are bonded firmly together at temperature T_o The strips have thickness t_1 and t_2 and coefficients of expansions are α_1 and α_2. One sides of the bonded member is held rigidly. If now the temperature changes from that of the bonding temperature to 'T' because of unequal expansions in the two strips will form a curve of radius 'r'.

$$r = \frac{t[3(1+m)^2 + (1+m)(m^2 + 1/mn)]}{6(\alpha_2 - \alpha_1)(T - T_0)(1+m^2)}$$

n→ ratio of young's modulus of low to hith

expansion materials $\left(n = \frac{E_1}{E_2}\right)$

m → ratio of the thickness of low $\left(m = \frac{t_1}{t_2}\right)$ to

high expansion materials.

$\alpha_1 \rightarrow$ higher coefficient of expansion

$T \rightarrow$ Operating temperature

$T_o \rightarrow$ initial bonding temperature

The deflection of the free tip is thus directly proportional to temperature.

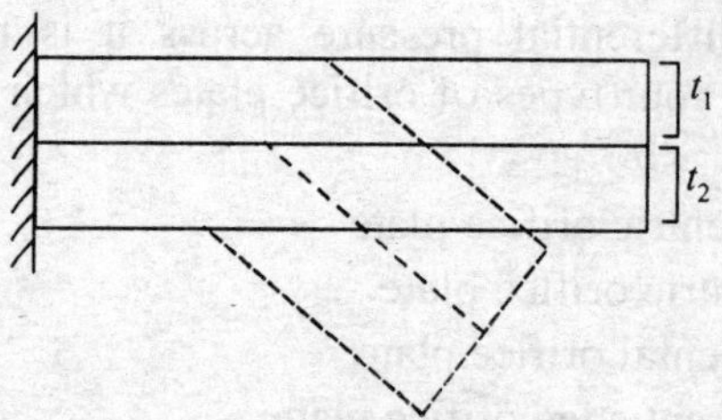

Bimetallic Strip Thermometers

The movement of the bimetallic strip is utilized to deflect a pointer over a calibrated scale. The deflection of the tip is small if the strip is short, and is large if the tip is long. A longer strip can be contained in relatively small space of the strip is wound in a spiral or helix form. Temperature range is -103 to $1004\,^oF$.

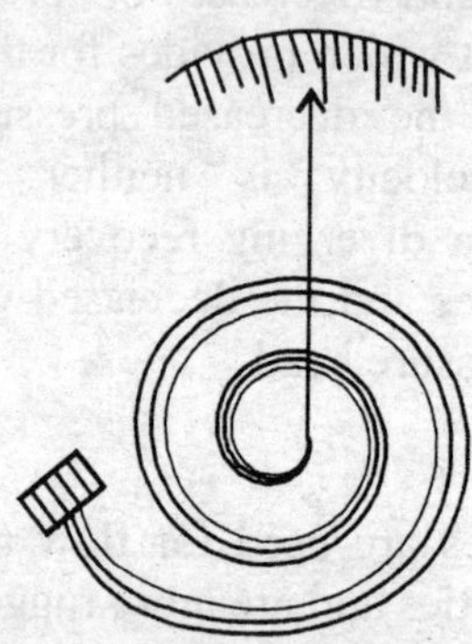

Figure: Bimetallic Strip thermometers

1. Expansion of liquids
   (a) Liquid-in-glass thermometer
   (b) Liquids-in-metal thermometer.

Liquid in Metal Thermometer

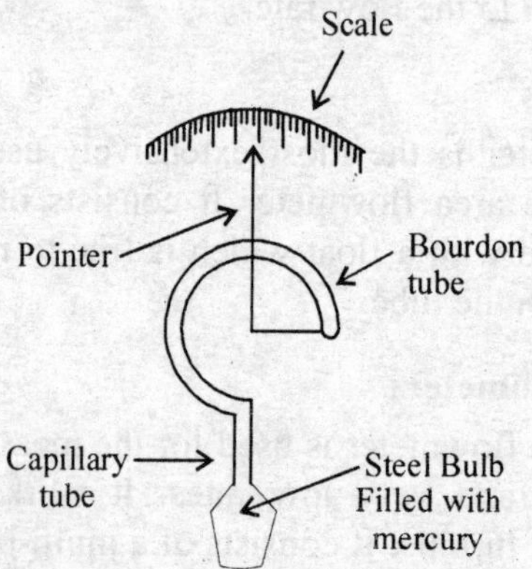

Figure: Liquid in metal thermometer

The liquid in metal thermometer consists of **mercury** has been used as liquid and the metal is steel. The mercury in steel thermometer works exactly in the same principle as the liquid in glass thermometer. A bourdon tubes used to measure the change in its volume. The bourdon tube, the bulb and the capillary tube in to the bourdon tube. As the temperature continues to rise increasing amount of mercury will be driven into the bourdon tube, causing it to bend. One end of the bourdon tube is fixed, while the motion of the other end is communicated to the pointer which moves on a calibrated temperature scale.

Expansion of gases

(a) Gas the thermometers.

2. Filled System Theromometers

The basic components of a filled system thermometer is consists of a Bourdon tube, a capillary tube, and a thermometer bulb all interconnected as shown. The entire system is sealed after filling with.

An appropriate liquid (known as filling-liquid) at a pressure at the normal ambient temperature. The common liquids that are used are mercury, ethyl alcohol ethylene and toluene. When in use, the thermometer bulb is installed inside the substance to be measured. This causes the filling liquid inside the bulb to heat or cool until its temperature matches the temperature of the measured substance This change in temperature causes the filling liquid to expand or contract and thus the Bourdon tube moves. With increases in temperature (heating) the liquid expands and this expansion forces the Bourdon tube to uncoil With decreases in temperature (cooling) the liquid contracts and it forces the Bourdon tube to coil more tightly The movement of the Bourdon tube may be used to drive a pointer for indicating temperature or to drive the pen on a strip chart.

Basically, there are four types of filled system thermometers:

1. Gas filled thermometers
2. Liquid filled thermometers
3. Mercury–filled thermometers
4. Vapour pressure thermometers

Temperature Ranges of Liquids used in Vapour-pressure System

Liquid	Critical temp.	Boiling point	Typical range available
Argon	–122°C	–185.7°C	Used to measure very low temperatures down to –253°C
Methyle chloride	143°C	–23.7°C	30 to 130° F (0° to 50°C)
Sulphur dioxide	157°C	–10°C	30 to 120°C
Ethyl-alcohol	243°C	78.5°C	200 to 350°F
Toluene	321°C	110.5°C	150 to 250°C
Ethyl-chloride	187°C	12.2°C	30 to 100°C
Bulane (n)	154°C	–0.6°C	60 to 130°F (20 to 80°C)
Methy bromide	—	4.6°C	80 to 180°F
Di-ethyle ether	194°C	34.5°C	130 to 300°F (60 to 160°C)
Water	375°C	100°C	120 to 220 °C

3. Electrical Temperature Instruments

There are three types of electrical temperature instruments which are generally used in industries:

1. Resistance thermometer
2. Thermocouple
3. Thermistor.

Resistance Thermometer

The resistance of certain metals changes with temperature change Resistance thermometer utilizes this characteristics, with the increases of temperature, the electrical resistance of certain metals increases in direct proportion to the rise of temperature. Therefore, if the electrical resistance of a wire of known and calibrated material is measured, the temperature of the wire can be determined Platinum, copper and nickel are generally used in resistance thermometers. In this type of thermometer, a temperature sensitive resistance element is fabricated in a suitable form to insert in medium whose temperature is to be measured, and is connected by leads to a wheat stone bridge.

Metals Used for Resistance Thermometers

Material	Temp. coeff. per °C (%)	Range °C		Melting Point °C
		Min.	Max.	
Platinum	0.39	–260	1100	1773
Copper	0.39	0	180	1083
Nickel	0.62	–220	300	1455
Tungsten	0.45	–200	1000	3370

Thermocouples

The working principle of a thermocouple depends on the thermo-electric effect. If two dissimilar metals are joined to gather so as to form a closed circuit, there will be two junctions where they meet each other. If one of these junctions is heated, then, a current flows in the circuit which can be detected by a galvanometer. The amount of the current produced depends on the difference in temperature between the two junctions and on the characteristics of the two metals. This was first observed by Seebeck in 1821 and is known as Seebeck effect.

Instruments which record the variations in current flow are calibrated in terms of temperature and are known as Thermocouple Pyrometers.

Thermistors

Thermistors are semiconductors, made from a specific mixture of pure oxides of nickel, manganese, copper, cobalt, iron, magnesium titanium, and other metals sintered at temperatures above 982°C. Their distinguishing characteristics are a **high temperature coefficient, usually negative,** although it can be positive as well and the fact that their resistance is a function of absolute temperature.

4. Pyro Meters

pyrometers is a technique for measuring temperature without physical contact. It depends upon the relationship between the temperature of a hot body and the electromagnetic radiation emitted by the body When a body is heated, it emits thermal energy known as heat radiation. A black matt surface (or a black body) is a very good absorber of heat radiations and, also a very good emitter of such radiations when heated, Pyrometry us a technique for determining a body's temperature by measuring its electromagnetic radiation.

1. Radiation pyrometers
2. Optical pyrometers.

Radiation Pyrometer

Operation of radiation pyrometer is based upon the measurement of radiant energy emitted by the hot body. The radiation pyrometer consists of a lens to focus the radiated energy from the body whose temperature is to be measured. The radiations are focused to the detector of receiving element. This receiving element may have a variety of forms such as resistance thermometers or a thermocouple or a thermopile.

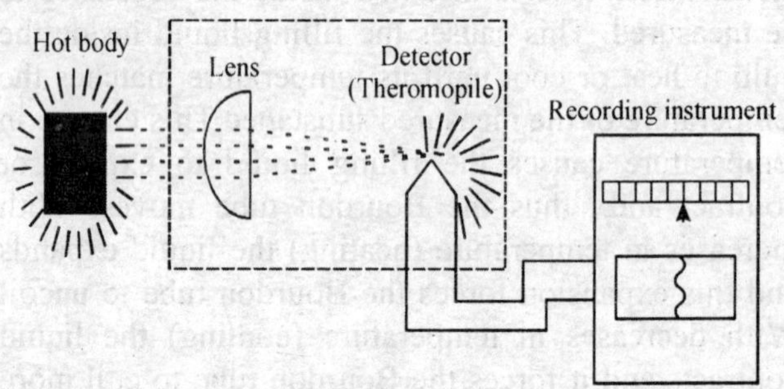

Figure: Radiation Pyrometer

A **thermopile** consists of several thermocouples, connected in series. A temperature indicator is attached with the receiving element to indicate the temperature. When the total energy radiated by a hot body whose temperature is to be measured enters the pyrometer and is focused by the lens on to the detector. The e.m.f. developed by tube is opened at one end to receive radiation from the object. Temperature range $\rightarrow 700^{\circ}C - 2000^{\circ}C$.

Optical Pyrometers

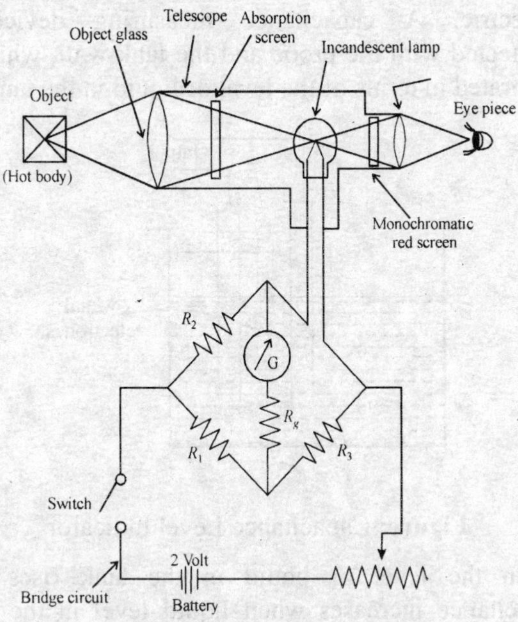

Figure: Optical Pyrometers

A metallic surface/object is usually dark and dull colour at room temperature. When the surface is heated it emits radiation of different wavelengths, this radiations are however not visible at low, temperature. As the temperature is progressively increased beyond 540°C, the surface become dark red then orange and finally white in colour. The principle of temperature measurement by color or brightness comparison is utilized in optical pyrometer design to measure temperature accurately in the range 700°C to 3000°C and are very useful for checking and calibrating radiation pyrometers. But they are not suitable for recording or controlling temperatures.

These pyrometers compare the brightness of radiation emitted by a body at a given wavelength with that of a black body calibrated lamp whose temperature is known. Radiation from the target surface are focused by a objected lens upon the filament of a incandescent lamp. The eyepiece is also adjusted until filament of a incandescent lamp. The eyepiece is also adjusted until is in sharp focus and under this condition the filament is seen supper imposed on the image of the target surface. A red filter is placed between the eyepiece and filament and it allows only a narrow band of wavelength to pass through it. Matching of brightness of the lamp filament with that of target surface is achieved by adjusting current through the standard lamp by changing the value of circuit resistance. The milliammeter reading may then be calibrated in terms of target temperature. When the filament is indistinguishable in terms of brightness, from the image of the target surface then is reading at the same intensity as the eyepiece. When the filament is coldwater than the target surface it appears as a dark wire against the light colour background. Filament brightness is then increased by causing more current to pass through the filament. A filament hotter than the object would appear brighter than the target surface. The current through the filament is then reduced to provide correct imaging of filament and the object.

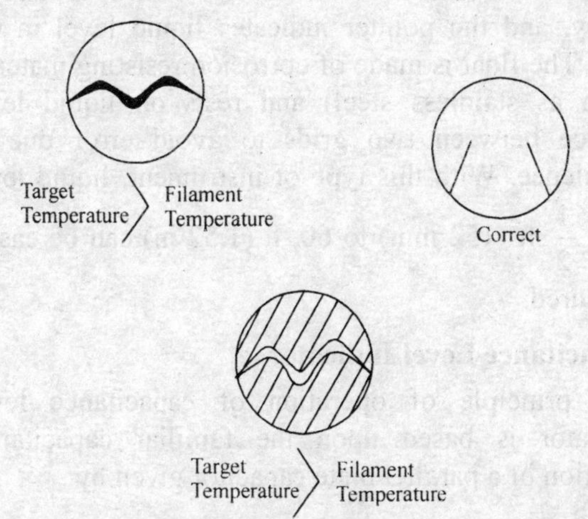

3.3.4 MEASUREMENT OF LIQUID LEVEL

Direct Methods

This is the simples method of measuring liquid level In this methods, the level of liquid is measured directly by means of the following level indicators:

1. Hook type Level Indicator
2. Sight Glass
3. float Type

Float Type Level Indicator

Float operated level indicator is used to measure liquid levels in a tank in which, a float rests on the surface of liquid and follows the changing level of liquid The movement of the float is transmitted to a pointer through a suitable mechanism which indicates the level on a calibrated scale Various type of floats are used such as hollow metal spheres, cylindrical shaped floats and disc shaped floats.

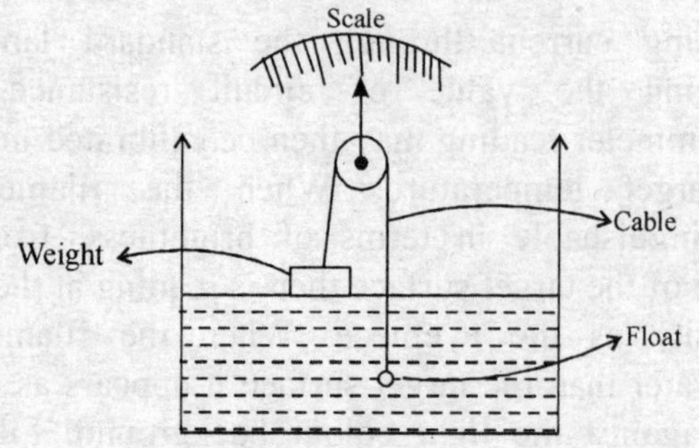

Figure: Float Type Level Indicator

Figure shows a simplest from of float operated mechanism for the continuous liquid level measurement. is the movement of the float is transmitted to the pointer by a stainless steel or phosphor-bronze flexible cable wound around a pulley, and the pointer indicates liquid level in the tank. The float is made of corrosion resisting material (such as stainless steel) and rests on liquid level surface between two grids to avoid error due to turbulence. With this type of instrument, liquid level form $\frac{1}{2}$ ft (152 mm) to 60, ft (1.52 m) can be easily measured.

Capacitance Level Indicator

The principle of operation of capacitance level indicator is based upon the familiar capacitance equation of a parallel plate capacitor given by:

$$C = K\frac{A}{D}$$

where, C = Capacitance, in farad

K = Dielectric constant

A = Area of plate, in m^2

D = Distance between two plates, in m

Therefore, it is seen from the above equation that if A and D are constant, then the capacitance of a capacitor is directly proportional to the dielectric constant and this principle is utilized in the capacitance level indicator.

Construction And Working

A capacitance type liquid level indicator consists of an insulated capacitance probe (which is a metal electrode) firmly fixed near and parallel to the metal wall of the tank If liquid in the tank is non inductive , the capacitance probe and the tank wall form the plates of parallel plate capacitor and liquid in between them acts as the dielectric If liquid is conductive the capacitance probe and liquid form the plated of the capacitor and the insulation of the probe acts as the dielectric. A capacitance measuring device is connected with the probe and the tank wall, which is calibrated in terms of the level of liquid in the tank.

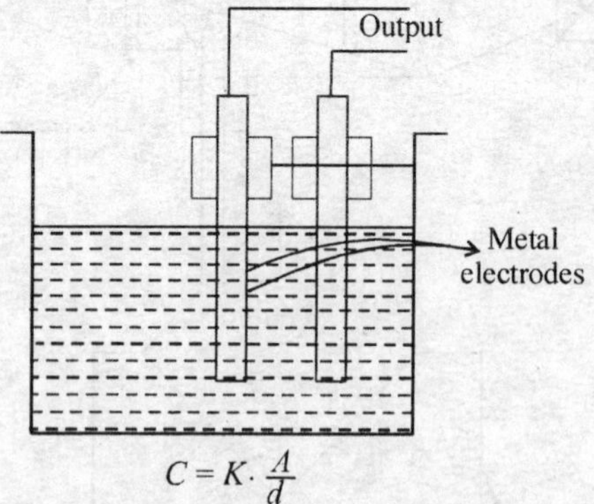

Figure: Capacitance Level Indicator

When the level of liquid in the tank rises, the capacitance increases when liquid level in the tank decreases, the capacitance also decreases this increases and decreases in the capacitance is measured and is displayed on the indicator calibrated in terms of liquid level.

RADIATION LEVEL DETECTOR

Radiation level detectors are used where other electrical methods would not service. Also, the most common reason for using a radiation level detector is that it does not need to come in contact with the liquid being measured.

Construction And Detector

A radiation level detector it consists rays source holder on one side of the tank and a gamma detector on the other side of the tank The gamma rays form the source are directed toward the detector in a thin band of radiation .When the gamma rays penetrated the thick wall of the tank its energy level afterwards is greatly reduced, the radiation received at

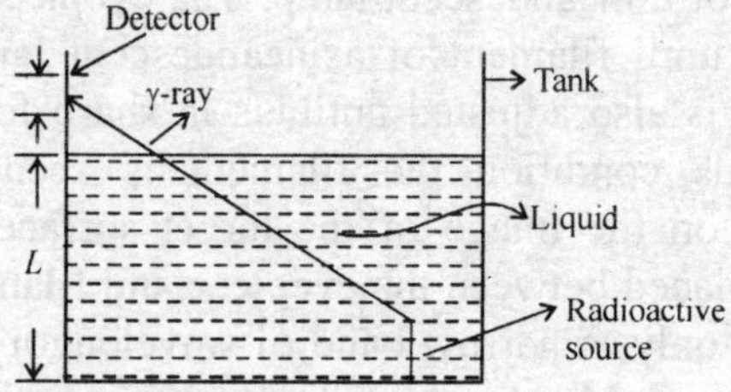

Figure: Radiation Level Detector

The gamma detector is inversely proportional to the thickness of the tank walls and the medium between the radiation source and the detector, that is the thicker the medium between source and detector , the less radiation received by he detector and vice versa, When the tank is empty the gamma rays passes only through the two tank walls and the air or about in the empty tank when liquid enters the tank and its level rises, the radiation beam passes through a path in the liquid ,as well as the tank walls, The liquid in the tank reduces the radiation received by the detector. The liquid in the tank reduces the radiation received by the detector The amount of radiation received is inversely proportional to the amount of liquid between the radiation source and the detector The difference in the amount of radiation received by the detector corresponds to the liquid level in the level in the tank , Thus when liquid level rises the amount of radiation received is reduced and vice versa. The radiation loss received by the tank walls is constant whether the tank is full or empty.

DISPLACER LEVEL DETECTORS

The displacer – type level detectors work on the Archimedes'. Principle which states that a body, wholly or partially immersed in fluid, is buoyed up by a force which is equal to the weight of the fluid displaced. If the cross – sectional area of the displacer and the density of the liquid are constant, then a unit change in level will result in a reproducible unit change in the displacer weight. By detecting the apparent weight of the immersed displacer, the level of the fluid is measured.

PRESSURE GAUGE METHOD

This is the simplest method used for liquid level measurement in an open tank. Hydrostatic pressure of any liquid in an open tank is given by the equation:

$$P = \rho \times h \times s_g \qquad (1)$$

or

$$h = \frac{P}{\rho \times S_g} \qquad (2)$$

where, P = pressure in psi or N/m^2

ρ = the water density

S_g = specific gravity

h = height of liquid (often called head) in inch or metre.

Hydrostatic pressure of a liquid in a close tank or vessel is given by the equation:

$$P = (\rho \times h \times S_g) + \text{ external pressure on liquid} \qquad (3)$$

Therefore, the pressure measured at the bottom of tank containing a liquid of known density and specific gravity is directly proportional to the height or level of liquid in the tank. This principle is utilized in pressure gauge instrument for liquid level measurement.

AIR BELLOWS

Air bellows are used for liquid level measurement where an indicator cannot be conveniently located at the specified datum line.

AIR PURGE SYSTEM

Air purge (also known as bubbler tube) system is one of the most popular hydrostatic pressure type of liquid level measuring system which is suitable for any liquid.

LIQUID PURGE SYSTEM

When an air purge system is unsuitable, because air bubbling through liquid may interfere with its crystallization, a liquid purge system is used. The construction and working of liquid purge system is the same as an air purge system, the only difference is that in place of air, water or light mineral oil is used as the purge liquid. The nature of the purging liquid must be such that the introduction of small quantites of it into the plant will not affect the product on process.

RADIATION LEVEL DETECTOR

Radiation level detectors are used where other electrical methods would not survive. Also, the most common reason for using a radiation level detector is that it does not need to come in contact with the liquid being measured.

3.3.5 MEASUREMENT OF PH

A number of industrial processes require that the acidity or alkalinity of a solution be measured and controlled especially those involving food, drink and bacteriological processes.

The pH of a solution is a measure of the hydrogen ion ($H+$) and quantifies acidity or alkalinity.

The pH of a solution is defined as the negative logarithm of hydrogen ions ($H+$) concentration in g/litre.

$$\text{pH} = -\log_{10}(H^+)$$

or $$\text{pH} = \log_{10}\left(\frac{1}{H^+}\right)$$

The value of pH can range from 0.0 to 14.0.

A 0.0 value corresponds to extreme acidity and 14.0 to extreme alkalinity.

Pure water or a neutral solution has a pH of 7.0. Hence, as the solution becomes more acidic its pH value deceases below 7.0, and as the solution become more alkaline its pH value increases above 7.

Principle: It is based on hydrogen ions (H^+) which react with the surface of the glass, electrode and a potential developed in the glass electrode and a potential developed in the glass electrode with respect to the solution.

Nernst equation: It is the relation between potential developed and H^+ ions concentration.

$$E_g = E_g^0 + \frac{2.303RT}{F}\log_{10}[H^+]$$

Where,

$E_q \rightarrow$ sum of reference potential and the potential of the liquid junction

$E_g^0 \rightarrow$ potential of electrode at $[H^+] = 1$

$2.303 \rightarrow$ logarithmic conversion factor

$R \rightarrow$ molar gas constant 1.986 cal/deg/mole

$T \rightarrow$ temperature in $^\circ K$

$F \rightarrow$ faraday constant = 96487 coulomb-mol^{-1}K

$[H^+] \rightarrow H^+$ ion concentration

Again pH is given by $pH = -\log_{10}[H^+]$

The measurement of different of potential of two measuring cell is done. The two measuring cell are

- Measuring electrode
- Reference electrode

These two electrodes are put in the solution whose pH to be measured.

Reference electrode

Different type of reference electrodes are commercially available for different pH ranges. Industrially the calomel electrode is more often used. It consists of an inner tube containing 'calomel' i.e. Hg and HgCl in which a Pt wire is immersed. Then inner tube also consists of glass wool placed at the bottom where a small role is made for bridging. The inner tube is placed in an outer glass envelope containing saturated solution of KCL. The outer tube is provided with a filling neck and a bridging hole for contact with the measured solution. The hole is covered with a ground glass peeve and bridging occurs through the liquid film in the glass joint. There is very gradual escape of KCL into the measured liquid through this bridging hole. The strength KCL varies from 3.5M to 0.1 M

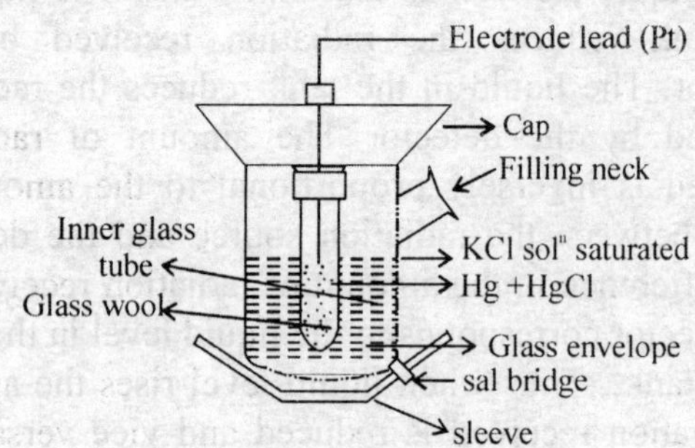

Figure: Reference electrode

Measuring Electrode

The measuring electrode is the glass electrode. A special glass wall separates a buffer solution contained in the bulb from the measuring solution. The electrode wire is dipped in the buffer. Between this buffer whose pH is known and constant and the unknown solution, an emf developed across the glass wall depending on the H^+ ion concentration of the measured solution. The value ranges from -4.3 mV at 10 pH to -65 mV at 2pH.

The special type of glass is made up of soda lime. It consists of

$SiO_2 \rightarrow 72.2\%$

$CaO \rightarrow 6.4\%$

$Na_2O \rightarrow 21.4\%$

It is very suitable because of its low melting point, high hygroscopicity and relatively low electrical receptivity.

This glass is always kept soaking in water because when we use from dry condition it will give unsteady potential. The errors for glass electrodes have been found to be due to high alkali metal ion concentration at high pH, smaller aqueous activation and high temperature. This occurs particularly when pH is greater then 10, and error due to this is too large to be neglected. The error occurs because same type of cation may take part in exchange of ions between the unknown solution and the glass electrode. So in place of Na, K is used to neglect the error. Actually glasses have been manufactured which can give quite low errors both in acid and alkali solutions. **Lithium based all purpose glasses have practically zero** error over the entire pH range. The temperature effect of lithium glass in also less.

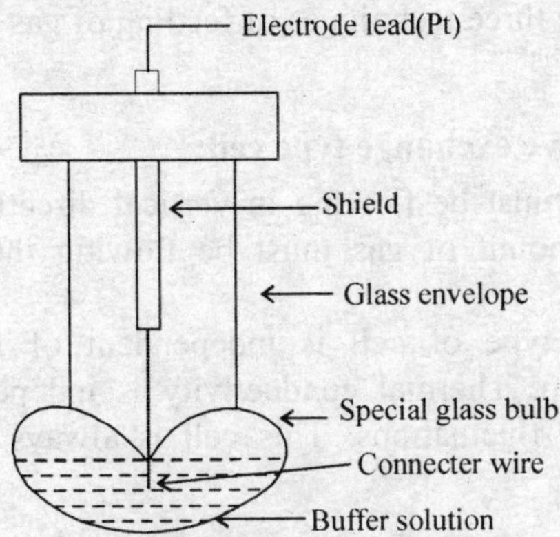

Figure: Measuring electrode

So by measuring the different in potential of both the electrodes we can determine the pH of an unknown solution. Now days a combination of glass and reference electrodes is available. Total diameter of the electrode assembly is about 1.5 cm and length including cap is about 14 cm.

pH-Meter

Measurement of pH requires a special hydrogen ion detector associated circuity. The most common method of pH determination is by a special glass electrode.

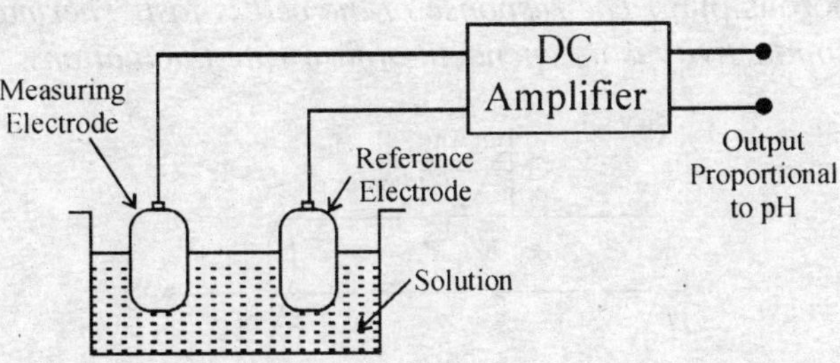

Figure: Schematic of Glass-Electrode pH-meter

The arrangement consists of two electrodes which are to be immersed in the solution whose pH is to determined and measuring the voltage developed across the two electrodes. One the electrodes called measuring electrode is a **glass electrode** which is permeable only to hydrogen ions. The very thin bottom acts as a membrane. The bulb contains chemical solution. The other electrode called **reference electrode** is also glass bulb which contains solution of calomel.

Both electrodes are mounted to form an assembly that is dipped into the solution. The difference of potential between two electrodes, is a function of the pH of the solution, can be measured by a voltmeter calibrated in the units of pH value. Application of the voltage signal is preferred. The pH sensitive glass electrode produces a voltage change of about 9 mV for 1.0 pH solution at $25\,^{\circ}C$.

The glass electrode method of pH measurement has some problem associated with it.

1. Measuring probe has a very high resistance of the order of $10^{8}\,\Omega$ and very low output. The probe is to be properly electrically screened to prevent spurious signal pick-up and a DC amplifier of very high input impedance is required.
2. Deterioration in accuracy may occur as the glass membrane becomes coated with substance that may be present in the measured solution.
3. The response time is high with a time constant of 60-120 seconds. The probe therefore, is to be kept for long time is solution before reading is taken.

3.3.6 MEASUREMENT OF CONDUCTIVITY

This method is used for gas analysis. Thermal conductivity of a gas mixture depends on thermal conductivity of individual gases and their concentration. Thermal conductivity varies from gas to gas.

Unit of thermal conductivity are BTU/hr/ $ft^2/{}^0F$ /in. in FPS system and

Cal / sec / $cm^2/{}^0C$ /cm in CGS system.

The expression for thermal conductivity is given by

$$q = \lambda . A . \frac{dT}{dX}$$

$q \rightarrow$ heat loss

$\lambda \rightarrow$ thermal conductivity

$A \rightarrow$ area of heating surface

$\frac{dT}{dX} \rightarrow$ radial temperature gradient

The measurement of thermal conductivity involves the measurement of heat loss due to conduction. This is done in hot wire cell.

Principle components are

- Measuring cell
- Regulated power supply
- Wheatstone bridge

Measuring cell consists of a relatively large mass of metal in the form of a block.

Heating element → elongated straight wire of tungsten or platinum and glass bead thermistor

Sensing element → two pairs of wheat stone bridge.

One pair → in sample stream

Wheat stone bridge is powered by regulated power supply with current flowing of about 100 to 300 mA.

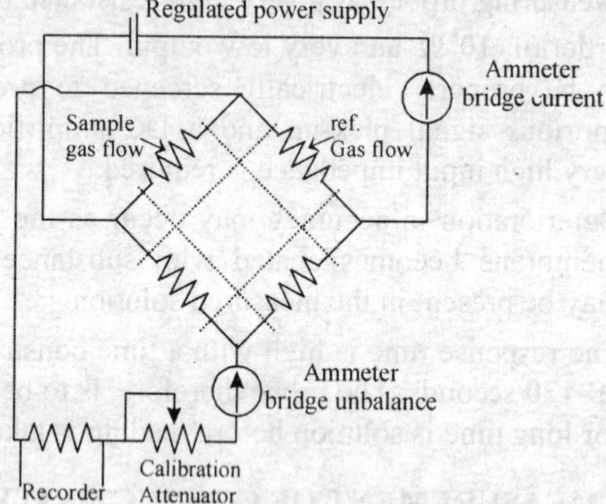

Figure: Typical wheatstone bridge measuring circuit for thermal conductivity analyzer.

Operation: Flow rate of sample gas should be in the range of 5 - 200cc/min. Sample gas passes through the measuring cell and precisely through the hot wire element. This filament element is heated up by the regulated power supply in the Wheatstone bridge. Temperature of this filament is maintained higher than the wall temperature of the measuring cell. After some time an equilibrium is reached such that.

Heat input = Thermal loss

Heat energy is conducted away from the filament, sample gas and the walls of the measuring cell.

Quantity of heat conducted away is a function of the thermal conductivity if the sample gas.

If a sample of less thermal conductivity is tested less heat is conducted away and so the resistive element heat up and resistance increases which will make the bridge unbalanced.

Degree of unbalance is measure of thermal conductivity and gas composition.

Degree of unbalance is calibrated in terms of gas composition.

Thermistors have negative temperature coefficient, so the ammeter is connected reversely. The reference filament is used for better stability (if temperature and barometric pressure varies).

There are three techniques of feeding of gas through a cell.

Convective exchange type cell:

The gas must be flowing in vertical direction and a minor amount of gas must be flowing the heating element.

This type of cell is independent of flow rate fluctuation. Thermal conductivity is independent of flow rate fluctuations. This cell is always mounted vertically.

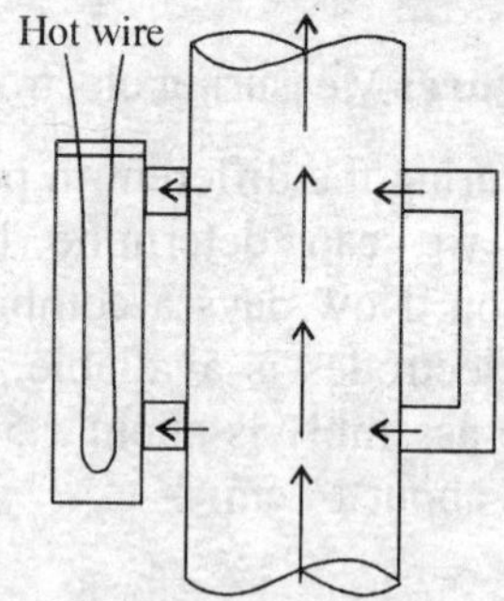

Figure: Convective exchange type cell

Diffusion exchange type cell:

The heating element is separated from the main system by a porous plug. Mixing takes place through porous plug. *The response of the cell is fast. Thermal conductivity is independent of flow rate fluctuations.*

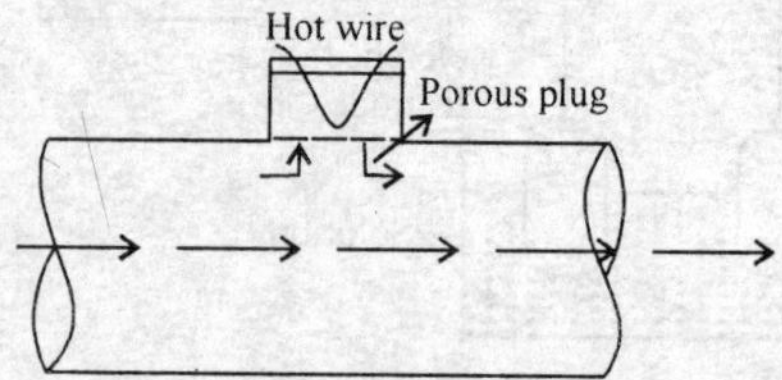

Figure: Diffusion exchange type cell

Direct online type cell:

if flow rate varies, measurement also varies. *Response is fastest among the three.* Hot wire is present in the main line.

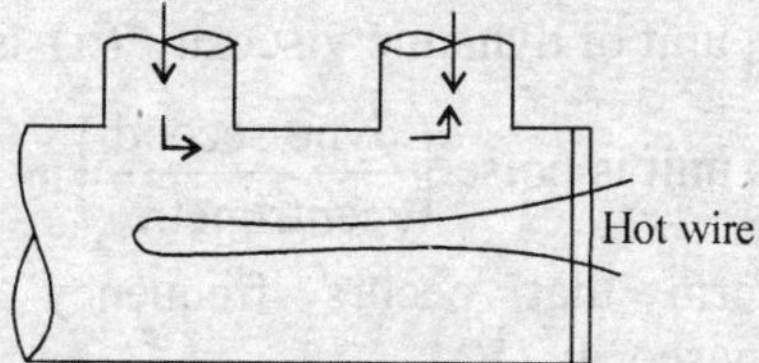

Figure: Direct online type cell

Measurement of Thermal-Conductivity Using Gas analyser

When the chemical purity of a gas is to be measured, the thermal conductivity gas analyzer is used. The operation of this type of analyzer is based on the vast difference in heat conductivity among gases.

Note that a balanced bridge circuit is formed with two sensing resistor acting as balanced legs of the bridge. One sensing resistor is surrounded by the gas to be analysed, whereas the other resistor is surrounded by a reference gas (such as oxygen, pure atmosphere etc). Where practical, both gases are maintained at the same pressure, water content, and so forth.

The bridge is first balanced by exposing both resistors to the same gas. Current flows through both sensing resistors, as well as through resistors P and Q. Resistor Q is adjusted for a "balance" or "zero set" condition on the meter. Then the resistors are exposed the reference sample and gas sample.

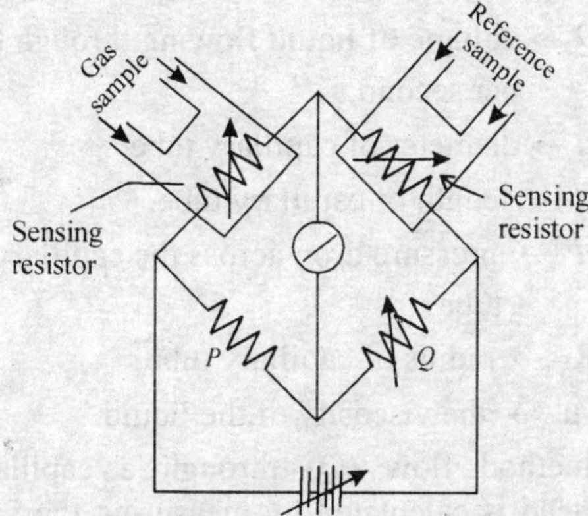

Figure: Basic thermal-conductivity gas analyzer circuit

If the gas sample contains elements having a different thermal conductivity that the reference sample the bridge will be unbalanced. In some cases, the meter reads out in terms of the thermal-conductivity, whereas in other analyzers, the indication is on a go-on go or good-bad basis.

Measurement of thermal Conductivity Using Thermistor

For this application two thermistors are connected in the adjacent arma of a wheastone bridge as shown in figure. The bridge supply is high enough to raise the thermistors above temperature, typically to about $150oC$. One thermistor is mounted in a static area to provide temperature compensation while the other is placed in the medium to be measured.

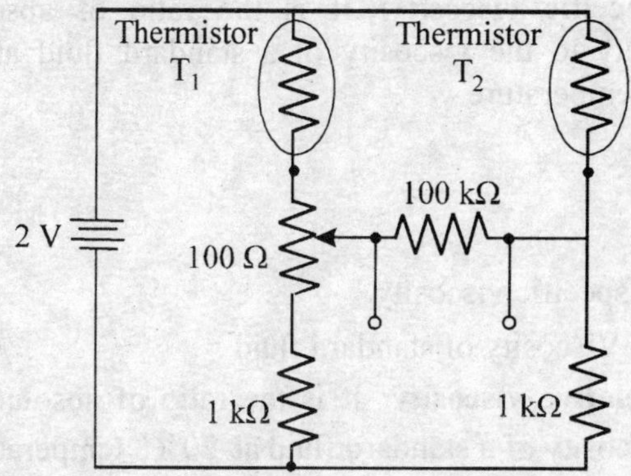

Figure: Measurement of thermal conductivity with thermistors.

3.3.7 MEASUREMENT OF VISCOSITY

Viscosity is a property of a fluid which affect its behavior. If a shearing stress is applied to a fluid it will deform.

Absolute/dynamic viscosity (μ): It is the constant of proportionality between shear stress and shear rate.

$$\textbf{A.V} = \frac{\text{Shear stress}}{\text{Shear rate}} = \frac{\text{shear stress}}{\text{Velocity gradient}}$$

Unit : 1. $gm.cm^{-1}$ $\sec^{-1}$ in C.G.S. units i.e. poise

2. $lb.ft^{-1}$ $\sec^{-1}$ in F.P.S units

$$\text{Pascal-sec} = \frac{Newton - \text{sec}}{m^2}$$

$$= 10 \; poise \; in \; S.I.units$$

Apparent Viscosity: It is applicable for non – Newtonian fluid is defined by resistance offered by the fluid.

Consistency: It is the resistance offered by the fluid to the deformation.

Fluidity: It is reciprocal of viscosity and its unit

$$= \frac{1}{poise} \; in \; C.G.S \; units$$

Kinematic viscosity (v): It is the ratio of the absolute viscosity to the density and its unit is cm^2, sec^{-1} = stokes in C.G.S units

$$V = \frac{\mu}{\rho}$$

Where,

$\mu \rightarrow$ absolute viscosity

$\rho \rightarrow$ density

Specific viscosity: It is the ratio of absolute viscosity to the viscosity of a standard fluid at the same temperature.

$$\mu_s = \frac{\mu}{\mu_{st}}$$

Where,

$\mu_s \rightarrow$ specific viscosity

$\mu_{st} \rightarrow$ Viscosity of standard fluid

Relative viscosity: It is the ratio of absolute to the viscosity of a standard fluid at $20^o C$ temperature

$$\mu_R = \frac{\mu}{\mu_{st}|_{20^o C}}.$$

Viscosity of a fluid is the property which resists shearing and is a comparative measure of resistance to flow within a fluid. For Newtonian fluids, the shearing stress experienced by two layers is directly proportional to the velocity gradient between them. Newton's law of viscosity can be expressed as,

Shearing stress $\propto$ velocity gradient

$$\tau = \text{constant}\ \frac{dv}{dy} = \mu\frac{dv}{dy}, \quad \tau = \frac{F}{A}$$

μ is termed as dynamic /absolute viscosity

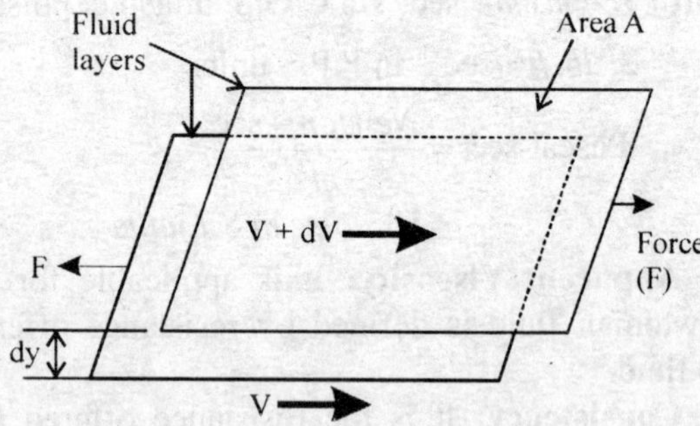

Figure illustrates an infinitely thin film of fluid of thickness dy. The two faces are having velocity of V and $V + dV$, caused by a shearing force F acting on area A.

S.I. unit of dynamic viscosity (μ) is Ns/m^2

cgs unit is poise $\left(\frac{\text{dyne-second}}{\text{centrimetre}}\right)$

Another term that occurs frequency is called kinematics viscosity (V)

$$V = \frac{\mu}{\rho}$$ Where, ρ is mass density

S. I. unit is m^2/s

Cgs unit is cm^2/s

Measurement of viscosity is important in lubrication and wear studies. Also, in the atomisation of fuel oils such as in fuel injection and for paint spraying, viscosity is an important factor.

Viscosity measuring instruments can be classified according to their principle of operation as,

- Falling sphere
- Falling piston
- Rotating cylinder
- Flow through a capillary tube
- Rota meter type device

Capillary Tube Viscometer

Flow through capillary is based on the Hagen-Poiseuille equation for laminar flow in a tube.

$$Q = \frac{\pi d^4 p}{128 \mu l}$$

Where, $Q \rightarrow$ volume of liquid flowing through tube per second.

$d \rightarrow$ diameter of capillary tube.

$l \rightarrow$ length of capillary tube.

$P \rightarrow$ pressure drop across the capillary tube.

R $\rightarrow$ radius of capillary tube

$\mu \rightarrow$ the viscosity of the liquid.

In this method flow (Q) through a capillary at constant head is calculated by measuring the volume of the fluid collected in a given time,

$$Q = \frac{\text{Volume calculated}}{\text{time taken}}$$

$p = \rho g h$, as head is kept constant, viscosity is calculated as,

$$\mu = \frac{\pi d^4 \rho g h}{128 Q l}$$

or $\quad , \mu = \dfrac{\pi P(R)^4}{8Ql}$

It is the ratio of partial pressure of water vapor in mixture to the saturation pressure of pure water at same temperature of mixture. Also, it can be defined as,

In the above expression, diameter d appears as raised to power four, therefore requires an accurate measurement. Further, it needs maintaining a constant head 'h'

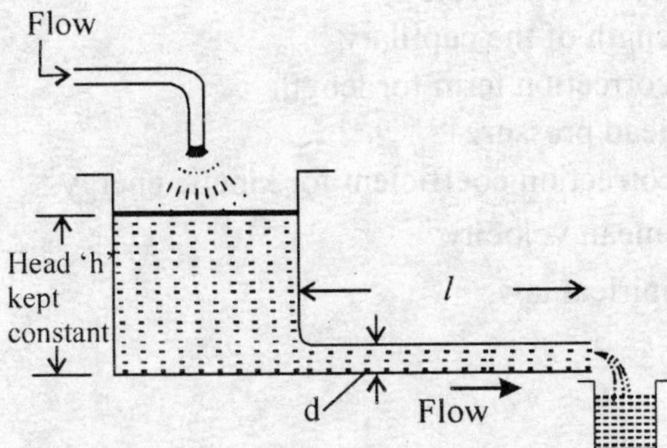

Figure: Capillary Tube Viscometer

Advantages

- Capillary-tube viscometer is a simple and convenient instrument for measuring kinematic viscosity accurately in the range of 0.2 to 12,000 centistokes.
- With controlled temperature, it is capable of very accurate measurement.
- It is relatively inexpensive.

Applications

- Kinematic viscosity measurements.
- Measurement of low – viscosity Newtonian fluids.
- Intrinsic viscosity determination.
- Molecular weight measurements by relating it to intrinsic viscosity
- Study of molecular shapes of natural and synthetic polymers.

Rota meter Type Viscometer

It is a continuous viscosity measuring device. It is based on the principle that the position of the float, in a rota meter tube for a constant flow rate, is a function of the viscosity of the flowing liquid.

Fig. shows a rots meter installed in a bypass line through which a constant flow rate is maintained by a pump regulator. The Rota meter has two floats, one designed to be sensitive ton viscosity and the other insensitive to viscosity but sensitive to flow rate.

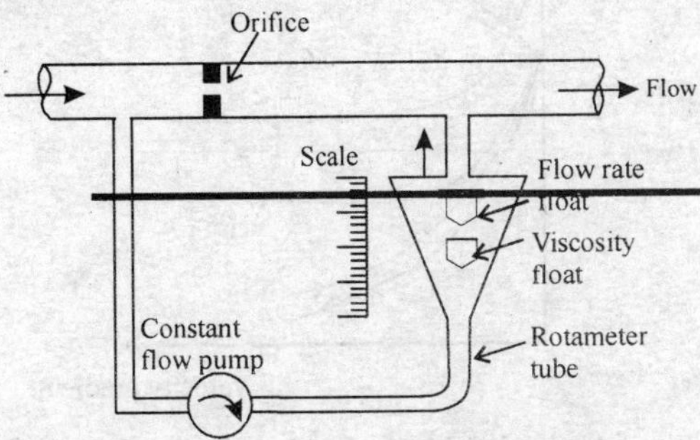

Figure: Rota meter Type Viscometer

The flow rate sensitive float is set at some index mark. The position of the viscosity sensitive float will then directly indicate the viscosity of the flowing.

Advantages Rotational Viscometers

- The cone – and - plate viscometer is the most versatile of all types and is an excellent rhemeter. It has the ability to record automatically for non-Newtonian fluids which exhibit shear dependent and time dependent behavior over a wide range of shear stress and rates.
- Rotational viscometers have ability to perform continous measurement under varying conditions on the same sample or at a given set of boundary conditions for an extended time period.

Different fluid Material

1. Newtonian fluid → water, most oils, salt solution etc.
2. Plastic → chewing gum, tar, most plastics etc.
3. Pseudo plastic → paper pulp, prints, fruit juice etc.
4. Thixotropic → silica gel, most paints, fruit juice etc.
5. Rheopectic → gypsum in water
6. Dilatent → starch, peanut butter

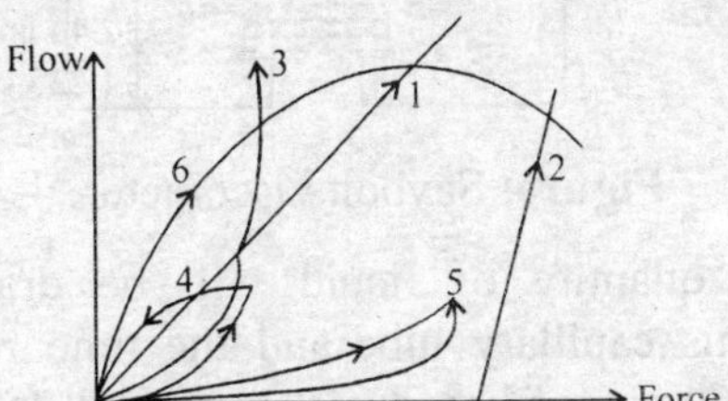

Figure: Flow-velocity relationship

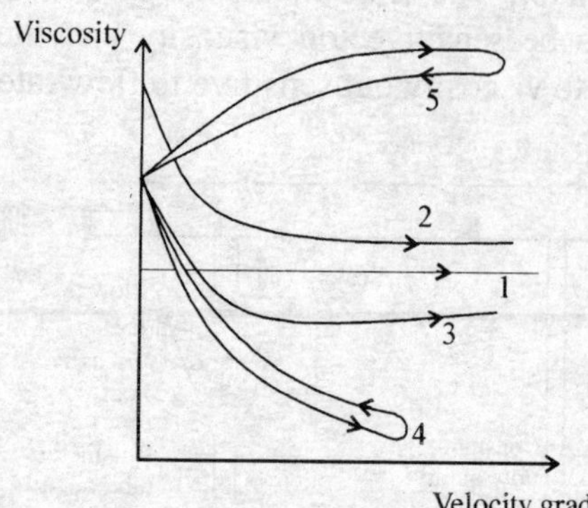

Figure: Velocity-Velocity gradient relationship

For a known volume, pressure different and flow rate we can get the value of 'μ' of a certain fluid.

1. *Viscosity meter* used in oil refinery:

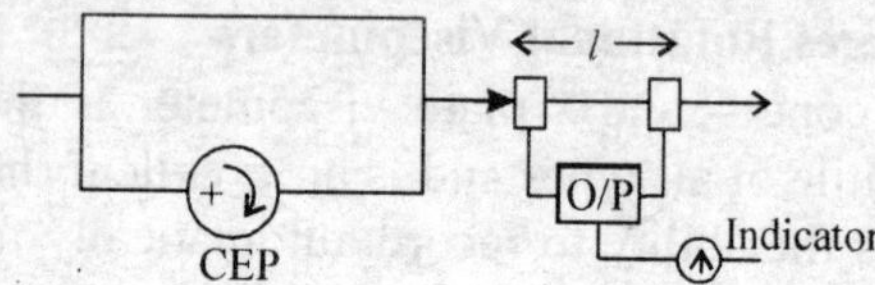

Figure: Viscosity meter used in oil refinery

$P \rightarrow$ known

$Q \rightarrow$ known

Then $\mu = \dfrac{\pi P.R^4}{8.Ql}$

Range → 0 to 3000 Cent Poise

2. Saybolt viscometer: It is used to measure kinematic viscosity. Unit is $\dfrac{cm^2}{\sec} = stokes$.

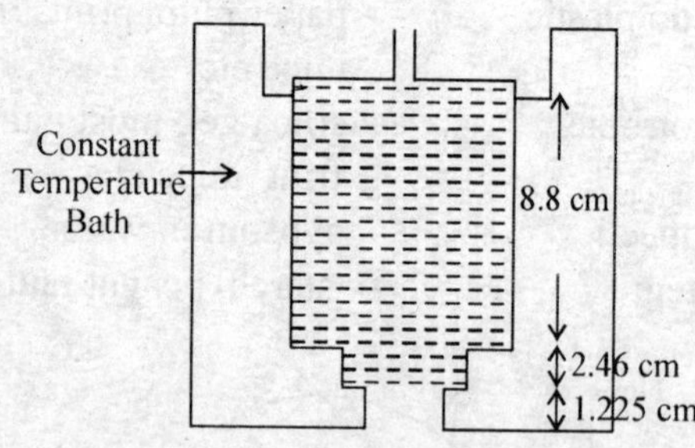

Figure: Saybolt viscometer

Measured quantity of liquid will be drained out through this capillary tube and the time require to drain out 60 c.c. of liquid will be varied, If viscosity is varied.

Time (expressed in seconds) is known as Sayboll's number.

Kinematic viscosity

$$v = \frac{\mu}{\rho} = \frac{\pi g R^4 t}{8Q(l+\lambda)}\left(h - \frac{k.v_m^2}{g}\right)$$

Where

$g \rightarrow$ gravitational constant

R → radius of the tube

t → time required to drain out 60 cc of liquid

Q → volume flow rate

$l \rightarrow$ length of the capillary

$\lambda \rightarrow$ correction term for length

H → head pressure

k → correction coefficient for kinetic energy.

$v_n \rightarrow$ mean velocity

An empirical law,

$$V = At - \frac{B}{t}$$

where $\dfrac{B}{t}$ is a compensation term

For short length an error may occur for this the compensation is required.

When, $t < 100$, $v = 0.226t - \dfrac{195}{t}$ centistokes (CS)

$$t > 100,\ v = 0.220t - \frac{135}{t} CS$$

Range → 0 to 10^3 CP (Viscosity of the liquid flowing)

Continuous viscosity meters

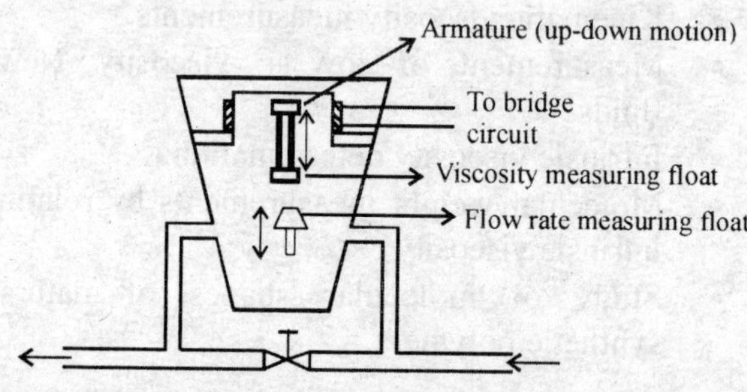

Figure: Continuous viscosity meters

Two floats:

- Lower float → immune to viscosity of the fluid. This acts as flow setting float. To maintain a constant flow rate.
- Upper float → acting the viscous drag force

3.3.8 MEASUREMENT OF HUMIDITY

The quantity of water vapor in the atmosphere or humidity/ moisture is important in a number of processes. Humidity or moisture affects the behavior of many commercial materials such as paper, textiles, leather, wood products, soap powder etc. Humidity measurement and control is required during many industrial processes including air-conditioning and heating for human comfort. Growth of bacteria in pharmaceutical industry, drying process of printer's link and paints and food preservation are some of the examples.

Atmosphere air is considered to be a mixture of dry air and water vapor. Some important terms pertaining to humidity and its measurement are defined below:

1. Relative Humidity

It is ratio of partial pressure of water vapour in maxture to the saturation pressure of pure water at same temperature of mixture. Also, it can be defined as,

$$\text{Relative Humidity, } \phi = \frac{\text{mass of water vapour in a given volume of air at temperature T}}{\text{mass of water vapour when the volume of air is saturated at temperature T}}$$

For dry air, $RH = \phi = 0.0\%$

For saturated air, $RH = \phi = 100\%$

2. Humidity Ratio

It is defined as the mass of water vapor per unit mass of dry air in a mixture of air and water vapor.

3. Absolute Humidity

It is the mass of water vapor in a unit volume of air and is expressed as gm/m^3.

4. Dry Bulb Temperature

It is the temperature of air – water vapor mixture as recorded by a thermometer with a dry bulb.

5. Dew Point

When the temperature of mixture of air and water vapor is reduced by continuous cooling at constant pressure, the water vapor in the mixture starts condensing at a particular temperature, which is called as the dew point temperature.

6. Psychometrics

The study of the properties of air-water vapor mixture is called psychometrics.

For humidity calculations psychometric chart is required. The relationship between various humidity variables such as, dry bulb and wet bulb temperature, dew point temperature, relative humidity, specific humidity etc. at specified atmospheric pressure are given in the form of a chart.

Hygrometers

Electrolytic typed the electrolytic-hygrometer cell analyses for moisture in process streams which are free from foreign material and corrosive gases. The cell consists of a thin film of phosphorus hold by the platinum electrodes. A dc potential is applied across the electrodes for dissociation of the water into hydrogen and oxygen. The electrical current produced by the dissociation is directly proportional to the concentration of water in the sample. Output from the cell can be used to monitor moisture concentration in various process streams, or it can be used to control during equipment in the process in order to maintain moisture concentration at a predetermined value. The hygrometer is a simple yet accurate analyzer for measuring the dew point of gases as low as minus $125° F$ which amounts to 1 part of water in million parts of air.

The controls located on the front panel include a sensitive indicating motor with alarm-adjustment and alarm-rest button, range selector switch, flow control values and flow meters for both the bypass and the analyzer-sample stream.

The non explosion proof hygrometer is an integral unit containing flow controls, analyzer cell, amplifier, and metering equipment. During analysis, a sample of the process gas is continuously passed through the flow-control system of the instrument, where the sample flow is precisely regulated to ensure a high degree of accuracy. Where higher rates of sample flow are necessary to obtain the desired speed of response, a portion of the sample maybe diverted through the bypass system. This reduces time lag in the system without overloading the instrument with sample gas.

From the flow-control system, the sample enters the analysis cell, where moisture is absorbed by a thin film of phosphorus held between platinum electrodes.

A dc potential applied across the electrodes dissociates the absorbed water into hydrogen and oxygen. The electro current produced by this distortion is directly proportional to the concentration of water in the sample. This current becomes the linear output signal which drives the indicating alarm meter. The signal can also be employed to drive any potentiometer recording instrument. For conversion from parts per million H_2O to dew point, special charts are available.

Ranges over which the electrolytic hygrometer is capable of measuring the moisture content of gases make it a valuable instrument for measuring the amount of water in parts per million contained in such process stream as recycle gases , instrument air exit from gas dryers, and welding and hent-treating gases as well as product purity, process efficiency and corrosion control.

The electrolytic hygrometer is very reliable in its operation if used a gas for which it is deuced but by no mean does its principle of operation permit measurement of moisture in all types of gases or ammonia and hydrogen fluoride being two gases in which it does not permit moisture measurement. Therefore this type of instruments should be supplied with an analysis of the gas in which the instruments is to be used so that they can make recommendation before the installation is made.

Wet and Dry bulb

The pressure of the aqueous vapour present in the atmosphere and hence the dew point may be found by the wet and dry bulb thermometer method. Two thermocouple or thermo bulbs are arranged on a stand. One mounted bare is ordina V temperature measuring element that shows or records the temperature of the.

The bulb of the other thermometer or thermocouple has a of clean lamp wick around it which is led to a small basin of water as in figure 108, this keeps the bulb moistened.

If there be much aqueous vapour present in the atmosphere there will be little bit evaporation form the moisture near the bulb of the thermometer or near the junction of the thermocouple and therefore there will be a small cooling effect. The difference in reading of the thermometers or thermocouples will then depends on the hygroscopic state of the atmosphere. Both thermometers or thermocouples read the respective temperature which can be recorded by different pens or indicated by separate pointer. The aqueous vapour pressure in found by use of specially constructed tables. This method does not give very accurate results.

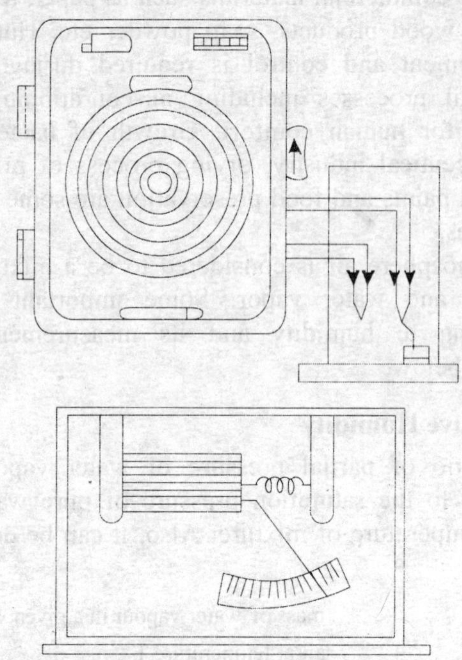

Hair Hygrometer

It has been found that the human hair alters its dimensions in accordance with the moisture of the atmosphere surrounding it. This principle can be made use of if one end of number of such elements are tied to a fixed member and the other to a movable one balanced by a spring and transmitting the movement to a pointer through links. This is also found to be not so accurate.

HUMIDITY MEASURMENT

Humidity can be defined as the amount of water retained by a gas.

Absolute Humidity

$$= \frac{\text{Weight of water vapour in unit}}{\text{Weight of dry gas}}$$

$$H = \frac{\omega_v}{\omega_g}$$

Specific humidity is the ratio of weight of water vapour and weight of gas mixture. Relative humidity is the ratio of amount of moisture in a saturated mixture.

Dew point is the saturation temperature of gas – water vapour mixture at the corresponding vapour pressure.

Where,

$$W_v = \frac{P_v.V}{R_v.T}$$

$$W_g = \frac{P_g.V}{R_g.T}$$

Where,

W → weight of the constituent

P → absolute partial pressure of constituent

R → universal gas constant

T → absolute temperature

V → volume

v → vapour

G → gas

Absolute humidity, $H = \frac{W_v}{W_g} = \frac{R_g}{R_v}.\frac{\rho_v}{P_g}$

Relative humidity, $H_R = \frac{\rho_v}{\rho_{sat}} = \frac{\rho_H}{\rho_{sat}}$

Since, $\rho_v = \rho_H.R_v.T$

1. Psychrometer

It consists c f two glass bulb thermometer

Dry bulb → exposed to atmosphere

Wet bulb → wrapped by a wick, soaked in water

2. Hair Hygrometer

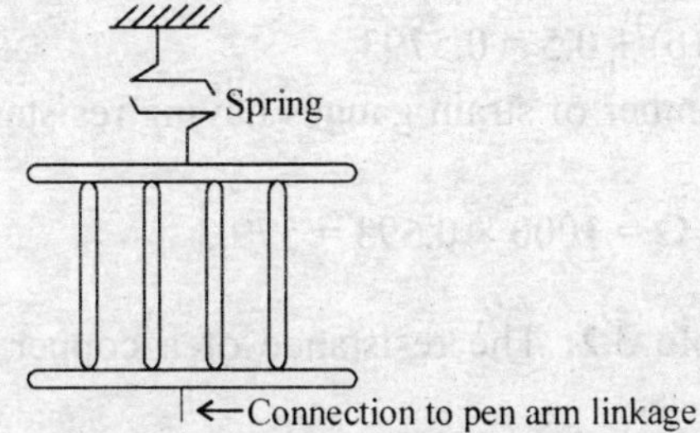

Principle: Change of length of some organic or synthetic fiber when it is exposed to moist atmosphere

Material: Wood, hair, animal membrane, paper etc.

Amount of moisture absorbed by these elements depend on temperature and partial vapour pressure of water of the atmosphere.

Range: 25-95% of relative humidity in a temperature range of 0° to 160° F.

3. Electrical Type Humidity Transducer

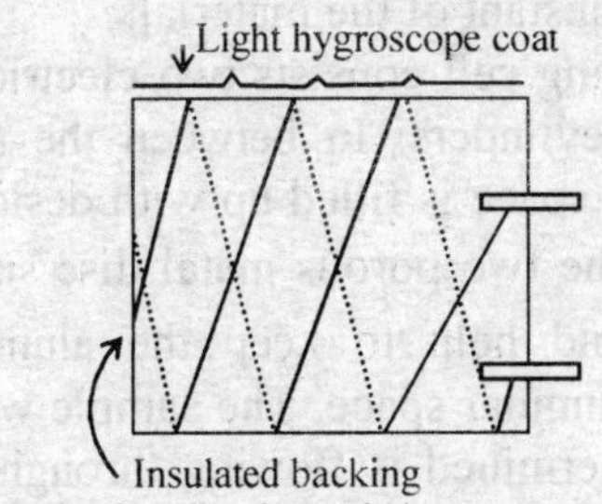

Sketch of a wire electrode hygrometer

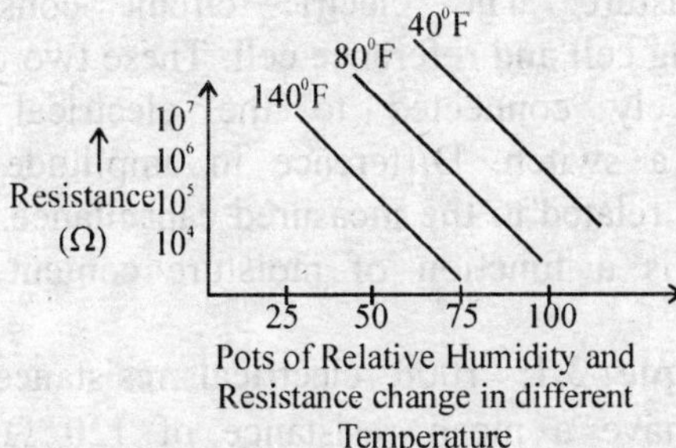

Pots of Relative Humidity and Resistance change in different Temperature

In this two electrodes are used. These two electrodes maintain a certain gap in a specific position. There two electrodes are coated by light hygroscopic materials. They are also backed by insulations.

Materials → $LiCl_2$, $ZnCl_2$, H_3PO_4, $CaCl_2$

The hygroscopic materials dissociates when they absorb moisture and the dissociation depends on the amount of water present. Thus resistance changes of the insulated backing.

Range: Change in resistance → $10\,M\Omega$ to $10\,k\Omega$.

4. Capacitance Hygrometer

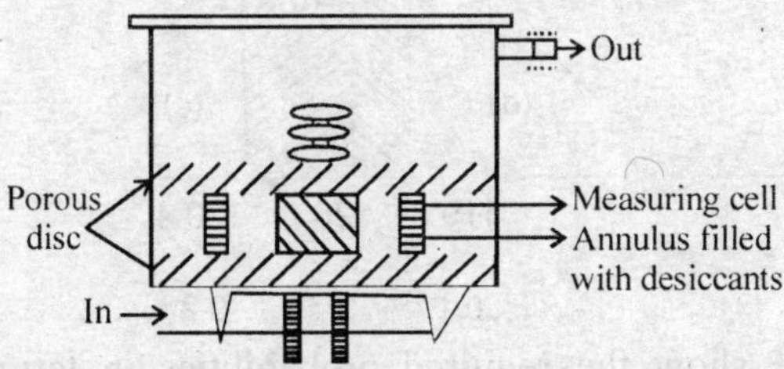

Principle: Change in capacitance if the material having dielectric constant changes

Dielectric constant of air → 1,

mica → 6,

water → 80.

Value of capacitance depends on

1. Plate area
2. Distance between the plates

If the above two are constant then it depends on dielectric constant of the material.

Measuring cell consists two electrically insulated concentric cylinders. In between the two cylinders, the annulus space is filled up with desiccant Alumina (Al_2O_3). The two porous metal disc support the two cylinders and help to keep the alumina desiccant within the annular space. The sample whose moisture is to be determined is flowing through the inlet and the porous plates and the alumina desiccant absorbe the moisture. The electric circuit consists of measuring cell and reference cell. These two cells are alternatively connected to the electrical circuit through a switch. Difference in amplitude in the output is related to the measured capacitance. (which in turn is a function of moisture content of the sample).

Example 3.1: 1000 electrical resistance strain gauges have a mean resistance of 120 Ω and a standard deviation of 2. Calculate,

(i) The number of strain gauges with resistance bwteeen 119.8 Ω and 120.4 Ω.

(ii) The number of strain gauges with resistance large thatn 119.8 Ω.

(iii) The number of strain gauges with resistance less than 120.4 Ω.

Solution:

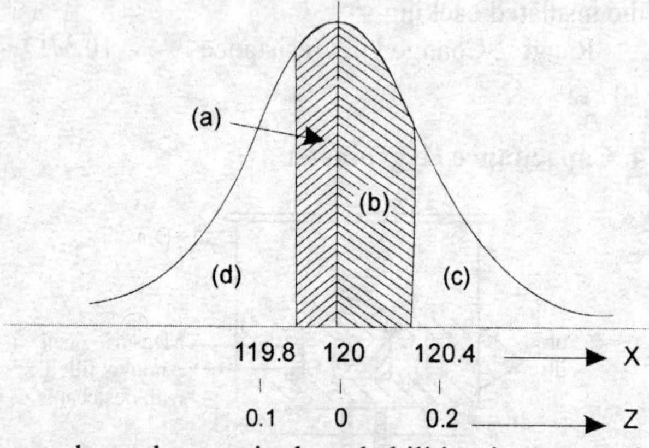

Figure show the required probabilities in terms of areas (a), (b), (c) and (d) and is calculated using standard normal table.

(i) The probability of resistance lying between 119.8 Ω and 120.4 Ω is expressed as,

$p(119.8 \le x \le 120.4)$

$= p\left(\frac{119.8-120}{2} < z \le \frac{120.4-120}{2}\right)$

$= p(-0.1 \le z \le 0.2)$

where, $z = \frac{x - \bar{x}}{\sigma}$

For z = 0.1, p = 0.0398

= [area (a) + area (b)]

= 0.0398 + 0.0793

= 0.1191

Number of strain gauges with resistance 119.8 Ω and 120.4 Ω = 1000 × 0.1191

= 119.1

(ii) The probability of resistance larger than 119.8 Ω is expressed as,

$p(x > 119.8) = p\left(z > \frac{119.8-120}{2}\right)$

= p (z > – 0.1)

= [area (a) + {area (b) + area (c) }]

= area (a) + 0.5

= 0.00398 + 0.5

= 0.5398

The number of strain gauges having resistance larger than

119.8 Ω = 1000 × 0.5398 = 539.8

(iii) The probability of resistance less than 120.4 Ω is expressed as,

$p(x < 120.4) = p\left(z < \frac{120.4-120}{2}\right)$

= p (z < 0.2)

= [area (b) + {area (a) + area (d)}]

= area (b) + 0.5 = 0.5793

The number of strain gauges having resistance less than

120.4 Ω = 1000 × 0.593 = 579.3

Example 3.2: The resistance of a copper wire is expressed as,

$R = R_0 [1 + \alpha (T - 20)]$

where, $R_0 = 5\Omega \pm 0.2\%$, is the resistance at 20°C

α = 0.004 per °C ± 1%, is the temperature coefficient of resistance

T = 30 ± 1°C, is the temperature of the wire.

Calculate the resistance of the wire and its uncertainty.

Solution: The nominal resistance is,

R = 5 [1 + (0.004) (30 – 20)]

R = 5.20Ω

The uncertainty is calculated using the equation is expressed as,

$U_R = \pm$

$$\sqrt{\left(\frac{\partial R}{\partial R}U_{RO}\right)^2+\left(\frac{\partial R}{\partial R}U_\alpha\right)^2+\left(\frac{\partial R}{\partial R}U_T\right)^2}$$

The various terms are calculated as,

$U_{R0} = 5 \times 0.2/100 = 0.01\Omega$

$$\frac{\partial R}{\partial R_0} = 1 + \alpha (T - 20) = 1 + (0.004)(30 - 20)$$

$= 1.04$

$$\frac{\partial R}{\partial \alpha} = R_0 (T - 20) = (5)(30 - 20) = 50$$

$$U_\alpha = 0.004 \times \frac{1}{100} = 4 \times 10^{-5} \text{ per}^\circ\text{C}$$

$$\frac{\partial R}{\partial T} = R_0 \alpha = (5)(0.004) = 0.02$$

$U_T = 1°C$

Substituting the above values, the uncertainty of resistance is

$U_R = \pm$

$$\sqrt{\left[(1.04\times0.01)^2+(50\times4\times10^{-5})^2+(0.02\times1)^2\right]}$$

$U_R = \pm \sqrt{0.000108+0.000004+0.004}$

$UR = \pm 0.02263$

The resistance of the wire can be written as,

$R = (5.2 \pm 0.02263)\ \Omega$

Neglecting the term associated with uncertainty is 'α' which is very small as compared to other uncertainties,

$U_R = \pm 0.02254$

Which show that if the larger uncertainty predominates the result, very little is gained by reducing *small* uncertainties.

Example 3.3: Resistance arrangement of 50Ω is required. For this purpose, two resistances of $(100 \pm 0.12)\ \Omega$ and $(25 \pm 0.02)\ \Omega$ are available. Which should be used (i) series arrangement of 25 Ω resistance or (ii) a parallel arrangement with 100 Ω resistance? Calculate the uncertainty for each arrangement.

Solution :

$R_1 = (100 \pm 0.12)\Omega$

$R_2 = (25 \pm 0.02)\Omega$

(i) Series arrangement of $(25 + 0.02)\Omega$

Resultant resistance

$R = R_1 + R_2$

$R = 25 + 25 = 50\Omega$

$$U_R = \sqrt{\left(\frac{\partial R}{\partial R_1}U_{R1}\right)^2+\left(\frac{\partial R}{\partial R_2}UR_2\right)^2} \qquad (1)$$

$$\frac{\partial R}{\partial R_1} = 1 \qquad U_{R1} = 0.02$$

$$\frac{\partial R}{\partial R_2} = 1 \qquad U_{R2} = 0.2$$

Subtituting the above values, uncertainty,

$U_R = \sqrt{(1\times0.02)^2+(1\times0.02)^2}$

$= \pm 0.0283$

Resultant resistance

$R = (50 \pm 0.08283)\Omega$

(ii) Parallel Resistance of $(100 \pm 0.12)\Omega$

Resultant resistance R is given by

$$\frac{1}{R} = \frac{1}{R_1} + \frac{1}{R_2}$$

or $$R = \frac{R_1R_2}{R_1+R_2} = \frac{100\times100}{(100+100)} = 50\Omega$$

$$\frac{\partial R}{\partial R_1} = \frac{R_2}{(R_1+R_2)} - \frac{R_1R_2}{(R_1+R_2)^2}$$

$$= \frac{100}{(100+100)} - \frac{100\times100}{(100+100)^2} = 0.25$$

$$\frac{\partial R}{\partial R_1} = 0.25 \qquad U_{R1} = 0.12$$

Similarly $$\frac{\partial R}{\partial R_2} = \frac{R_1}{(R_1+R_2)} - \frac{R_1R_2}{(R_1+R_2)^2}$$

$$= \frac{100}{(100+100)} - \frac{100\times100}{(100+100)^2} = 0.25$$

$$\frac{\partial R}{\partial R_2} = 0.25, \qquad U_{R2} = 0.12$$

Uncertainty = ±

$\sqrt{(0.25\times.012)^2+(0.25\times0.12)^2}$

$= 0.0424$

Resultant resistant

$R = (50 \pm 0.0424)\Omega.$

Example 3.4: The coefficient of discharge C_d of an obstructin flow meter is given by

$$C_d = \frac{m}{t\rho A\sqrt{2gh}}$$

Assuming following data, calculate the value of C_d and its uncertainty.

m = mass of water collected = (400 ± 0.2) kg

t = time = (550 ± 2)s

ρ = density = $(1000 \pm 0.1\%)$ kg/m^3

A = Area = $\frac{\pi}{4} d^2$, $d = (13 \pm 0.02)$ mm

$g = (9.81 \pm 0.1\%)$ m/s^2

h = pressure head = (3.5 ± 0.005) m

The uncertainty is to be calculated considering

(i) the uncertainties are absolute value

(ii) the uncertainties are within $\pm 3\,\sigma$ statistical limits.

Solution : Nominal value of C_d

$$C_d = \frac{m}{t\,\rho\,A\sqrt{2gh}} \quad (1)$$

Substituting values, (dimensions of all the variables are to be consistent)

$$A = \frac{\pi}{4}d^2 = \frac{\pi}{4}\left(\frac{13}{1000}\right)^2 = 1.33 \times 10^{-4}\ \text{m}^2$$

$$C_d = \frac{400}{(550)(1000)(1.33\times10^{-4})\sqrt{2\times9.81\times3.5}}$$

= 0.66

Example 3.5: An electrical resistance strain gauge has a gauge factor of 2.0 and resistance of 120 ohms. Find the change in resistance of the gauge caused by a strain of 1 μm/m.

Solution : F = 2.0, R = 120Ω, $\varepsilon = 1 \times 10^{-6}$

$$\text{Gauges factor } F = \frac{\Delta R/R}{\varepsilon}$$

$$2 = \frac{\Delta R}{120\times1\times10^{-6}}, \text{ or } \quad \Delta R = 2.4 \times 10^{-4}\ \Omega$$

= 0.00024 Ω

Percentage change in resistance

$$= \frac{0.00024}{120} \times 100 = 0.0002\%$$

To measure changes in resistance as small as of this magnitude, required sensitive instrumentation or circuits.

Example 3.6: An electrical resistance strain gauge of resistance 120Ω and gauge factor 2.0 is bonded to a specimen of steel. Calculate the resistance change of the gauge, (a) due to a tensile stress of 50 MN/m^2 in the specimen, (b) an increase of 30°C in temperature.

For the gauge

Temperature coefficient of resistance

$= 20 \times 10^{-6}$/°C

Thermal coefficient of linear expansion

$= 16 \times 10^{-6}$/°C

For the steel

Modulus of elasticity E = 200 GN/m^2

Thermal coefficient of linear expansion = 12×10^{-6}/°C

Solution :

(a) Due to stress, σ = 50 MN/m^2 = 50×10^6 N/M^2

$$\text{Strain induced, } \varepsilon = \frac{\sigma}{E} = \frac{50\times10^6}{200\times10^9}$$

$= 2.5 \times 10^{-4}$ μs

$$F = \frac{\Delta R/R}{\varepsilon}$$

$\Delta R = F\,\varepsilon\,R = 2.0 \times 2.5 \times 10^{-4} \times 120 = +\,0.06\Omega$

(b) Due to temperature: It consists of two components,

(i) Due to temperature coefficient of resistance,

$(\Delta R)_{(i)} = R\,\gamma\,\Delta T = 120 \times 20 \times 10^{-6} \times 30$

= (+) 0.072Ω

(ii) Due to differential expansion of gauge material and steel specimen

$(\Delta R)_{(ii)} = FR\,(\alpha_s - \alpha_g)\,\Delta T = 2 \times 120\,(12 - 16)\,10^{-6} \times 30$

$= (-)\ 0.028\ \Omega\ (\alpha_g > \alpha_s)$

Note here the relative magnitude of resistance changes. Positive ΔR is an increase and negative one is decrease in resistance. Further, the two components of resistance change due to temperature have opposite signs.

Example 3.7: An electrical resistance strain gauge of resistance 100Ω and gauge factor 2.0 is connected in a Wheatstone bridge circuit. What value of shunt resistance is required to be connecte to indicate a strain of (i) 100 μs (0.0001) and (ii) 1000 μs (0.001)

Solution:

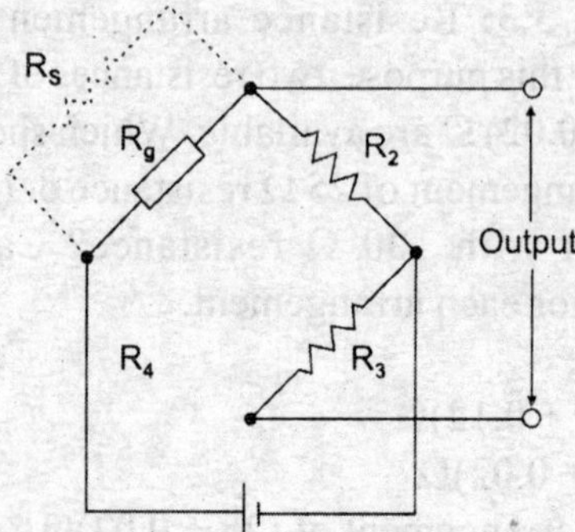

Figure shows a WB circuit shunted by a resistance R_s, strain induced is,

$$\varepsilon = \frac{R_g}{F(R_g + R_s)}$$

$F = 2.0, \quad R_g = 100\Omega$

For, $\varepsilon = 100\ \mu s = 100 \times 10^{-6}$

$$100 \times 10^{-6} = \frac{100}{2(100 + R_s)}$$

$$100 + R_s = \frac{100}{2} \times \frac{1}{100 \times 10^{-6}} = 50 \times 10^4$$

$R_s = 500000 - 100 = 499900\Omega$

$R_s = 499.9\ k\Omega$

For, $\varepsilon = 1000\ \mu s = 1000 \times 10^{-6}$

$$1000 \times 10^{-6} = \frac{100}{2(100 + R_c)}$$

$100 + R_s = 50 \times 10^3 = 50000$

$R_s = 49900\Omega \quad$ or $49.9\ k\Omega$

Example 3.8: Illustrates shunt Resistance method for calibrating a resistance strain gauge system. The method consists of determining the system's response (indiciated strain) to the introduction of a known small resistance change at the gauge and calculating the equivalent applied strain there form.

Solution:

A large resistance (shunt resistance) connected across the gauge serves to change the resistance in the arm by a very small amount and can be calculated. The applied strain can be calculated if the gauge factor is known. The strain in this method is not applied by mechanical loading, but by changing resistance of the gauge electrically. Mathematical relation is derived below.

When the gauge of resistance R_g is shunted by a resistance R_s in any one arm, the resistance of the combination in the arm is,

$$R = \frac{1}{\frac{1}{R_s} + \frac{1}{R_g}} = \frac{R_g R_s}{R_g + R_s}$$

Change in resistance,

$$\Delta R = R_g - \frac{R_g R_s}{(R_g + R_s)} = \frac{-R_g^2}{(R_g + R_s)}$$

$$\Delta R/R = \frac{-R_g^2}{(R_g + R_s) R_g}$$

$$= \frac{-R_g}{(R_g + R_s)} \quad \left(\text{-ve sign mean decrease in resistance}\right)$$

$$F = \frac{\Delta R / R}{\varepsilon}$$

Equivalent applied strain,

$$\varepsilon = \frac{R_g}{F(R_g + R_s)}$$

Example 3.9: An electrical resistance strain gauge of gauge factor F = 2.0 and resistance $R_g = 120\Omega$ is placed in an Wheatstone bridge in which all resistances are equal to 120Ω. The excitation voltage is 4.0 V. Calculate the unbalanced current for a strain of 1 μm/m indicated by a galvanometer of resistance 100Ω.

Also determine the voltage indicated by a high-impedance device for a 1 μm/m strain.

Solution :

$F = 2.0$

$R_1 = R_2 = R_3 = R_4 = 120\Omega$

$V = 4.0\ V$

$\varepsilon = 1\ \mu m/m$

$R_G = 100\Omega$ (resistance of galvanometer)

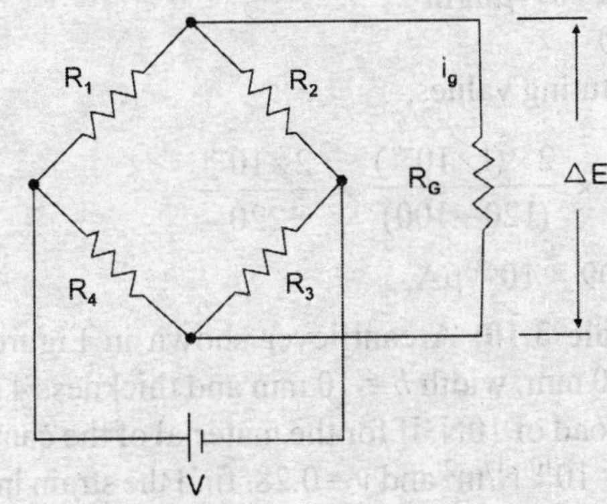

Output voltage (when R_G is very large or open circuit voltage)

$$\Delta E = \frac{V R_1 R_2}{(R_1 + R_2)^2} \times \left(\frac{\Delta R_1}{R_1} - \frac{\Delta R_2}{R_2} + \frac{\Delta R_3}{R_3} - \frac{\Delta R_4}{R_4} \right)$$

$$\Delta E = V \frac{(R_1)^2}{(2R_1)^2} \left[\frac{\Delta R_1}{R_1} \right] = \frac{V}{4} \frac{\Delta R_1}{R_1}$$

$$F = \frac{\Delta R_1 / R_1}{\varepsilon}$$

$$\frac{\Delta R_1}{R_1} = F\varepsilon$$

$$\Delta E = \frac{V}{4} = F\varepsilon$$

Substituting value,

$$\Delta E = \frac{4}{4} \times 2 \times 1 \times 10^{-6}\ V = 2\mu V$$

Unbalanced Current, considering galvanometer resistance

$$i_g = \frac{V}{4(R_1+R_G)} \times \left(\frac{\Delta R_1}{R_1} - \frac{\Delta R_2}{R_2} + \frac{\Delta R_3}{R_3} - \frac{\Delta R_4}{R_4}\right)$$

when only one arm has a strain gauge.

$$i_g = \frac{V}{4(R_1+R_G)} \frac{\Delta R_1}{R_1}$$

$$F = \frac{\Delta R_1 / R_1}{\varepsilon_1}$$

$$\frac{\Delta R_1}{R_1} = F\varepsilon$$

$$i_g = \frac{V}{4} \frac{1}{(R_1+R_G)} = F\varepsilon$$

$\varepsilon = 1 \times 10^{-6}$ μm/m

F = 2.0

Substituting values,

$$i_g = \frac{4}{4} \times \frac{2 \times (1 \times 10^{-6})}{(120+100)} = \frac{2 \times 10^{-6}}{220}$$

$i_g = 9.09 \times 10^{-3}$ μA.

Example 3.10: A cantilever shown in Figure is of length 100 mm. width $b = 10$ mm and thickness 4 mm. It carries a load of 10N. If for the material of the cantilever $E = 22.5 \times 10^{10}$ N/m² and $v = 0.28$. find the strain induced in the strain gauges fixed in the longitudinal (R_1) and transverse directions (R_2) due to load.

Solution:

W = 10N, $l = 100 \times 10^{-3}$ m

$b = 10 \times 10^{-3}$ m, $t = 4 \times 10^{-3}$ m

$E = 22.5 \times 10^{10}$ N/m², $v = 0.28$

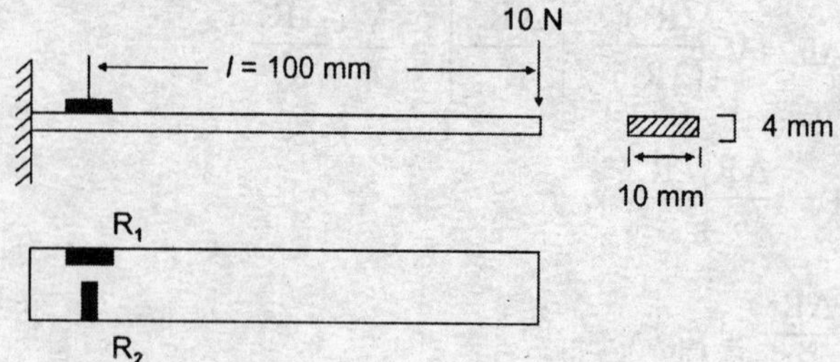

Bending moment

$M = Wl = 10 \times 100 \times 10^{-3} = 1$ Nm

$$\frac{M}{I} = \frac{\sigma}{y}$$

$$I = \frac{1}{12} b t^3 = \frac{1}{12} (10 \times 10^{-3})$$

$(4 \times 10^{-3})\ I = 53.3 \times 10^{-12}$ m⁴

Stress σ_1 on the surface of the cantilever in the longitudinal direction (at $y = \frac{t}{2} = 2 \times 10^{-3}$m)

$$\text{Stress } \sigma_1 = \frac{M}{I} \frac{t}{2} = \frac{1 \times 2 \times 10^{-3}}{53.3 \times 10^{-12}}$$

$= 0.038 \times 10^9$ N/m²

$$\text{Srain } \varepsilon_1 = \frac{\sigma_1}{E} = \frac{0.038 \times 10^9}{22.5 \times 10^{10}}$$

$= 168 \times 10^{-6}$ (Tension)

Strain ε_2 in the transverse direction $= v\varepsilon_1$

$= 168 \times 10^{-6} \times 0.28$

$= 47.3 \times 10^{-6}$ (compression).

Example 3.11: A U-tube mercury manometer is used to measure a differential pressure of 50 k *Pa* of oil density 800 kg/m². What difference in the mercury column in two arm will be observed?

Solution: $p_1 > P_2$

$p_1 - p_2 = 50$ k Pa

= 50000 N/m²

density of oil $\rho_0 = 800$ kg/m³

density of mercury $= \rho_m = 13600$ kg/m³

$p_1 - p_2 = \rho_m g\, h - \rho_0\, g\, h$

$50000 = gh\,(\rho_m - \rho_0)$

$50000 = 9.81 \times h\,(13600 - 800)$

$$h = \frac{50000}{9.81 \times 12600} = 0.40 \text{ m} = 40 \text{ cm}$$

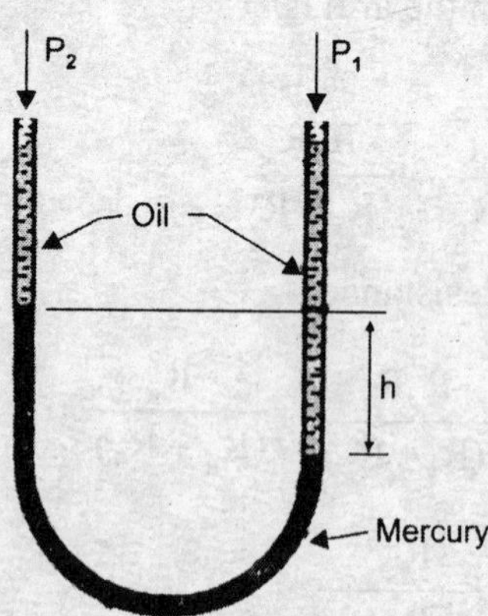

If the same manometer were to measure a differential pressure of 50 k *Pa* of air, the second term in the formula is to be neglected as density of air compared to that of mercury is negligibiliy small. Corresponding deflection.

$$h = \frac{50000}{9.81 \times 13600} = 0.375\text{m} = 37.4 \text{ cm}$$

Example 3.12: A diaphram pressure gauge is constructed of steel of diameter 5 cm. It is designed to measure a maxmum pressur of 1.5 M *Pa*. Find the thickness of the diaphram to limit its maximum delfection to one-third its thickness. Also calculate the natural frequency of the diaphram. Assume following data:

$E = 200$ GN/m², $v = 0.3$

$\sigma \setminus 7.9 \times 10^3$ kg/m³

Solution:

Max, deflection is when $r = 0$

$$y_{max} = \frac{3}{16} \frac{p}{Et^3} R^4 (1 - v_2)$$

$p = 1.5$ M $Pa = 1.5 \times 10^6$ N/m²

$E = 200$ GN/m² $= 200 \times 10^9$ N/m²

$R = 2.5$ cm $= 2.5 \times 10^{-2}$m

$$y_{max} = \frac{t}{3}$$

$$\frac{t}{3} = \frac{3}{16} \times \frac{1.5 \times 10^6 (2.5 \times 10^{-2})^4 [1 - (0.3)^2]}{200 \times 10^9 \, t^3}$$

$$t^4 = 3 \times \frac{3}{16} \times \frac{1.5 \times 10^6 \times 39.06 \times 10^{-8} \times 0.91}{200 \times 10^9}$$

$t = 1.1$ mm $= 1.1 \times 10^{-3}$ m.

$$\omega_n = \frac{10.21}{R^2} \sqrt{\frac{Et^2}{12(1 - v^2)\rho}}$$

$$= \frac{10.21}{(2.5 \times 10^{-2})^2} \sqrt{\frac{200 \times 10^9 \times (1.1 \times 10^{-3})^2}{12(1 - 0.3^2) \times 7.9 \times 10^3}}$$

$\omega_n = 27350$ rad/s

$\omega_n = 4353$ Hz.

Example 3.13: A McLeod gauge has bulb and capillary volume of (till cut off point) 100cm² and a capillary of 1 mm diameter. Determine the pressure indicated by a reading of $h = 2.5$ cm of the gauge.

Solution: Pressure indicated in terms of mercury column.

$$\frac{p_i}{\rho g} = \frac{Ah^2}{V - Ah}$$

$h = 2.5$ cm $= 26$ mm, $d = 1$ mm

$$A = \frac{\pi}{4} d^2 = \frac{\pi}{4} (I)^2 = 0.705 \text{ mm}^2$$

$V = 100$ cm³ $= 10^5$ mm³

Pressure in mm of mercury

$$\frac{p_i}{\rho g} = \frac{0.785 \times (25)^2}{1 \times 10^5 - 0.785 \times 25} = \frac{490.625}{10^5 - 19.625}$$

$$= 490.6 \times 10^{-5}$$

In the above equation, since the unit used is millimetre, the pressure head calculated shall be in mm of Hg. Also, as V >>> A*h*, ($10^5 > 19.265$) so using the approximation,

$$\frac{p_i}{\rho g} \approx \frac{Ah^2}{V}$$

will not introduce much of an error.

Pressure $= 490.6 \times 10^{-5}$ mm of mercury

$= 0.0049$ mm of mercury

$= 0.0049$ torr

Example 3.14: Find the sound pressure level corresponding to sound pressures of (a) N/m² and (b) 10 N/m²

$$\text{(a) SPL} = 20 \log \frac{1}{0.00002}$$

$= 93.98$ dB

$$\text{(b) SPL} = 20 \log \frac{10}{0.00002}$$

$= 113.99$ dB

Example 3.15: Calculate the maximum velocity of air immediately upstrea of a thermocouple if the error due to this in measuring the temperature of the moving air does not exceed 3.5°C. The specific heat at constant pressure for air is $C_p = 1005$ J/kg K.

Solution:

Air at a velocity V and temperature T is brought to rest locally at the theromocouple. The kinetic energy of air is converted into internal energy, resulting in a local rise in the temperature at the tip of the thermocouple, called "stagnation temperature T_s" using the expression,

$$\text{Solution : } R = R_0 \, e \left[\beta \frac{1}{T} - \frac{1}{T_0} \right]$$

R_0 9000Ω, $T_\infty = (0 + 273) = 273$ K

$R = 500Ω$, $T = (100 + 273) = 373$ K

$$500 = 9000e \, \beta \left[\frac{1}{373} - \frac{1}{273} \right]$$

Taking logarithm of both sides

$$\text{In}\,(0.0555) = \beta\left[\frac{1}{373} - \frac{1}{273}\right]$$

$$-2.89 = \beta\,(.00268 - .00366)$$

$\beta = 2949$ K

If R_{50} is the resistance at T = 50°C = 323 K

$$R_{50} = 9000\, e\ \beta\left[\frac{1}{323} - \frac{1}{273}\right]$$

$$\text{In}\,R_{50} = \text{in}\,9000 + \beta\left[\frac{1}{323} - \frac{1}{273}\right]$$

$R_{50} = 4605\Omega$.

Example 3.16: A bimetal strip 55 mm long is made of strips of nickel-chrome alloy and Invar bonded toether at 30°C. Each material has a thickness of 1.5 mm. Find the radius of curvature of the strip subjected to 110°C. Assume following data:

For Invar → $\alpha_1 = 1.5 \times 10^{-6}$°C^{-1}

$E_1 = 1.5 \times 10^{11}$ N/m^2

For Nickel-chrome alloy → $\alpha_2 = 12.5 \times 10^{-6}$°C^{-1},

$E_2 = 2.3 \times 10^{11}$ N/m^2

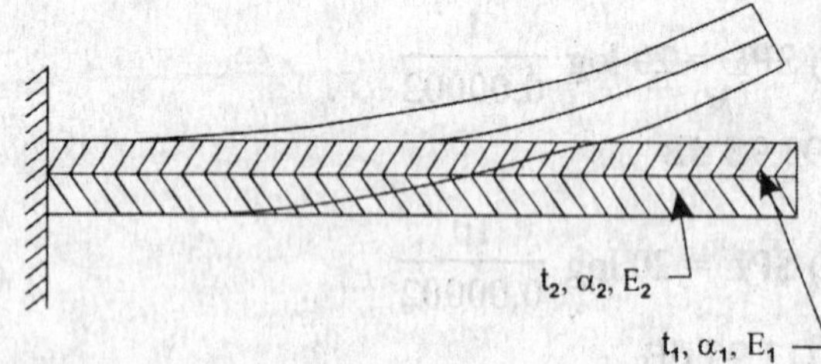

Solution: The radius of curvature of a bimetal strip due to temperature change is given by.

$$r = \frac{t\left[3(1+m)^2 + (1+mn)\left[m^2 + 1/mn\right]\right]}{6(\alpha_2 - \alpha_1)(T - T_0)(1+m)^2}$$

where, t = combined thickness of the strip $= t_1 + t_2$

$m = \dfrac{t_2}{t_1}$ = ratio of thickness of low to high expansion materials

$n = \dfrac{E_2}{E_1}$ = ratio of modulii of elasticity of low to high expansion materials

α_1 = lower coefficient of thermal expansion

α_2 = higher coefficient of thermal expansion

T = operating temperature

T_0 = initial bonding temperature

$t = 1.5 + 1.5 = 3 \times 10^{-3}$ m

$m = 1.5/1.5 = 1.0$

T = 110°C

$$n = \frac{1.5\times10^{11}}{2.3\times10^{11}} = 0.652$$

$T_0 = 30$°C

$$r = \frac{3\times10^{-3}\left[3(1+1)^2 + (1+1\times0.652)\left(1^2 + \frac{1}{1\times0.625}\right)\right]}{6(12.5-1.5)10^{-6}(110-30)(1+1)^2}$$

$$r = \frac{1\times[3\times4 + 1.652\times2.6]10^3}{2(11\times80\times4)}$$

$r = 2.3$ m.

Example 3.17: A platinum resistance thermometer has a resistance os 100Ω at 0°C and 139.5Ω at 100°C. Its resistance wat found to be 295.5Ω when measuring the temperature of a hot gas. Find the temperature of gas. Assume the temperature coefficient of platinum = 0.0039°C^{-1}.

Solution: The temperature resistance relation for platinum is,

$$R_2 = R_1 + R_0\,\alpha\,(T_2 - T_1)$$

$$T_2 = T_1 + \frac{(R_2 - R_1)}{R_0\,\alpha}$$

$$T_2 = 100 + \frac{(295-5-139.5)}{100\times0.0039}$$

$T_2 = 100 + 400 = 500$°C.

Example 3.18: A pitot-static tube is placed in front of a submarine moving horizontally in the sea under water. A U-tube mercury manometer shows a deflection of 15 cm between stagnation pressure and static pressure. Find the speed of the submarine if the density of sea water is 1026 kg/m^3.

Density of mercury, $\rho_m = 13.6 \times 1000$ kg/m^3

Density of sea water, $\rho_m = 1026$ kg/m^3

Manometer deflection = 15 cm of mercury

Pressure difference in terms of sea water column = h_m

$$\left(\frac{\rho_m}{\rho_m} - 1\right)$$

$$h_w = 15\left(\frac{13.6\times1000}{1026} - 1\right) = 183.8\text{ cm}$$

Velocity of fluid relative to submarine

$$V = \sqrt{2\,g\,h_w} = \sqrt{2\times9.81\times1.838} = 6.0\text{ m/s}$$

Example 3.19: A pitot static tube is fitted at the nose of an airplane. A U-tube water manometer connected to it shows a deflection of 10 cm of water. Find the air velocity assuming density of air 1.20 kg/m³.

Solution:

Pressure difference h_w = 10 cm of water

Pressure difference in terms of air column

$$h_a = h_w\left(\frac{\rho_w}{\rho_a} - 1\right)$$

$$\approx h_w \frac{\rho_w}{\rho_a}$$

$$= \frac{10}{100}\left(\frac{1000}{1.2}\right)$$

Flow velocity = $\sqrt{2\,gh_a} = \sqrt{2\times 9.81\times 83.3}$

= 40.43 m/s

In the above solution air has been treated as incompressible.

Example 3.20: A 30 cm × 15 cm venturimeter is used to measure the flow of water in a pipeline. A U-tube mercury manometer connected at the entrance and throat shows a deflection of 25 cm. Find the rate of flow of water. Take the coefficient of discharge of venturimeter C_d = 0.97.

Solution:

using, $Q = C_d \dfrac{A_1 A_2}{\sqrt{A_1^2 - A_2^2}} \sqrt{2g}\sqrt{h}$

where h is differential pressure in terms of water column and is,

$$h = h_m\left(\frac{\rho_m}{\rho_w} - 1\right)$$

where, ρ_w = density of mercury

ρ_w = density of water

$$= \frac{25}{100}\left(\frac{13.6}{1} - 1\right) m = 3.15 \text{ m}$$

h = 3.15 m

$$A_1 = \frac{\pi}{4}(0.3)^2 = 0.0707 \text{ m}^2$$

$$A_2 = \frac{\pi}{4}(0.15)^2 = 0.0176 \text{ m}^2$$

$$Q = 0.97\times\frac{0.0707\times 0.017}{\sqrt{(0.0707)^2 - (0.0176)^2}}\sqrt{2\times 9.81\times 3.15}$$

Q = 0.0246 m³/s

As the value of D_d was given, there is no need to fin flow coefficient.

Flow coefficients are tabulated for a meter size and varying rane of Reynold's numbers.

Example 3.21: A pitot tue is inserted in a pipe of 30 cm diameter carrying water. The difference between stgnation and static pressure head is found to be 2.36 m of water at the centre of the pipe. Calculate the rate of flow of water if the mean velocity of low is 0.85 time the central velocity. Assume C_v = 0.8.

Velocity at the centre = $C_v\sqrt{2gh}$

$= 0.98\sqrt{2\times 9.81\times 2.36}$

= 6.668 m/s

Mean velocity = 5.6678 m/s

Area of flow $= \dfrac{\pi}{4}d^2$

$= \dfrac{\pi}{4}(0.3)^2 = 0.7606m^2$

Rate of low of water = Mean velocity × Area

= 5.6678 × 0.7068

Rate of flow = 0.4006 m³/s

Example 3.22: A seismic vibrometer sensing displacement hs undamped natural frequency of 20 Hz and damping ratio of 0.7. Determine the amplitude ratio an phase angle between input and output if input is sinusoidal displacement of 30 Hz.

Solution: ξ = 0.7 ω = 2π × 20 rad/s

ω = 2π × 30 rad/s

$$r = \frac{\omega}{\omega_n} = \frac{30}{20} = 1.5$$

$$\text{Amplitude ratio} = \frac{z}{x} = \frac{r^2}{\sqrt{(1-r^2)+(2\xi r)^2}}$$

$$= \frac{(1.5)^2}{\sqrt{(1-(1.5)^2+(2\times 0.7\times 1.5)^2}}$$

$$\frac{z}{x} = 0.92$$

At this frequency (30Hz) output error of –8% occurs

$$\phi = \tan^{-1}\frac{2\xi r}{1-r^2} = \tan^{-1}\frac{-20.7\times 1.5}{1-(1.5)^2} = \tan^{-1}(-1.68)$$

$$\phi = 120.8°$$

Example 3.23: The motion of a vibrating machine is given by equation,

$$x = 0.10 \sin 60\,\pi t + 0.05 \sin 120\pi t$$

Determine the vibration record that would be obtained by a seismic istrument of damping ration $\xi = 0.65$ and natural frequency 1500 Hz.

Solution: The two components of excitation can be treated separately and the result then combined by using the principle of superposition.

Input equation,

$x = 0.10 \sin 60\pi t + 0.05 \sin 120\pi t$

can be expressend as,

$x = (X_0)_1 \sin \omega_1 t + (X_0)_2 \sin \omega_2 t$

where,

$(X_0)_1 = 0.1,\ (X_0)_2 = 0.05,\ \omega_n = 2\pi \times 1500$

$\omega_1 = 60\pi,\ \omega_2 = 120\pi,\ \xi = 0.65$

and output as,

$$z = (z_0)_1 \sin(\omega_1 t - \phi_1) + (z_0)_2 \sin\left[\omega^2 t - \phi_2\right] \text{ using,}$$

$$\frac{z_0}{x_2} = \frac{r^2}{\sqrt{(1-r^2)+(2\xi r)^2}} \quad \text{where, } r = \frac{\omega}{\omega_n}$$

$$\phi = \tan^{-1}\left(\frac{2\xi r}{1-r^2}\right)$$

and solving above equations for,

$$r_1 = \frac{60}{2\pi(1500)} = \frac{1}{50}$$

$$r_2 = \frac{120}{2\pi(1500)} = \frac{1}{25}$$

$$(z_0)_1 = \frac{0.10}{2500},\ \phi_1 = \tan^{-1} 0.0261 = 0.026 \text{ rad}$$

$$(z_0)_2 = \frac{0.20}{2500},\ \phi_2 = \tan^{-1} 0.05028 = 0.05 \text{ rad}$$

Resultant output,

$$z = \frac{1}{2500}[0.10 \sin(60\pi t - 0.026) + 0.20 \sin(120\pi t - 0.05)]$$

It can be seen that in the case of a complex waveform, the various components of the input are treated differently in terms of mplitude and phase response.

Example 3.24: A piezoelectri accelerometer has a natural frequency of 10,000 Hz an a damping ratio of 0.2. Find the voltage output for an input acceleration amplitude of 500 g (g being acceleration due to gravity) at a frequency of 2000 Hz. Given the voltage sensitiviry constant of the crystal is 250 V/mm.

Solution: $A_0 = 500g = 500 \times 9.81 = 4905 \text{ m/s}^2$

$\xi = 0.2$

$\omega_n = 10000 \text{ Hz} = 2\pi \times 10000 = 6.28 \times 10^4 \text{ rad/s}$

$\omega = 2000 \text{ Hz} = 2\pi \times 2000 = 1.256 \times 10^4 \text{ rad/s}$

$$r = \frac{\omega}{\omega_n} = 0.2$$

$$\frac{Z_0}{A_0} = \frac{1}{\omega_n^2\sqrt{(1-r^2)2+(2\xi r)^2}}$$

$$Z_0 = \frac{A_0}{\omega_n^2\sqrt{(1-r^2)2+(2\xi r)^2}}$$

$$Z_0 = \frac{4905}{(6.28\times 10^4)^2} \times \frac{1}{\sqrt{(1-(0.2)^2)^2 + (2\times 0.2\times 0.2)^2}}$$

$Z_0 = 1.295 \times 10^{-6}$m

(Amplitude of output displacement)

$= 1.295 \times 10^{-3}$m

Sensitivity of piezoelcetic crystal = 250 V/mm.

Example 3.25: A thermostal for on-off control of room temperature produces a cycle with 30 minute perod and 3°C amplitude. If a thermometer with a time lag of 4 minutes is placed beside it, what cycle should it indicate? Assume the thermostal to be a first order instrument.

Solution: Frequency $\omega = 2\pi f = \frac{2\pi}{t} = \frac{2\pi}{30}$

$= 0.209$ rad/min

Time lag $= \frac{1}{\omega} \times$ phase lag in radians

$= \frac{1}{\omega}(\tan^{-1}\omega\tau) = 4$ minutes (given)

The output has a frequency response given by

$$= \sin\omega\left(t - \frac{1}{\omega}\tan^{-1}\omega\tau\right)$$

$= \sin 0.209\,(30 - 4) = \sin 5.434$

If ω_c is changed angular frequency, then $\sin \omega_c t = \tan 5.434$.

$$\therefore \quad \omega_c = \frac{5.434}{t} = \frac{5.434}{30} = 0.1811 \text{ rad/min}$$

$$f_c = \frac{\omega_c}{2\pi} = \frac{0.1811}{2\pi} = 0.0288 \text{ cpm}$$

Therefore the thermostat has a cycle of $\frac{1}{0.0288} \approx 34.72$ minute period.

Example 3.26: A thermocouple system with a time constant of 8 seconds has been used to measure the temperature of a furnace whose temperature fluctuates sinusiodally between 400°C and 450°C with a periodic time of 60 seconds. Determine the maximum and minimum values that will be indicated by the thermocouple. Calculate the phase angle and the corresponding time lag between the temperature signals and the thermocouple output signals.

Solution: The temperature would oscillate abut the mean value of 425°C with an amplitude of 25°C.

Frequency of oscillation,

$$\omega = 2\pi f = \frac{2\pi}{t} = \frac{2\pi}{60} = 0.104 \text{ rad/s}$$

$$\omega\tau = 0.104 \times 8 = 0.832$$

After the transient effects die away, the amplitude ratio is

$$\frac{\theta_0}{\theta_i} = \frac{1}{\sqrt{1+(\omega\tau)^2}} = \frac{1}{\sqrt{1+(0.832)^2}} = 0.768$$

$\theta_0 = 0.768 \times 25 = 192$°C

Therefore:

Maximum temperature = 425 + 19.2 = 444.2°C

Minimum temperature = 425 − 19.2 = 408.8°C

Phase sift $\phi = \tan^{-1}(\omega\tau) = \tan^{-1}(0.832) = 39.76$°C = 0.6935 radians.

$$\text{Time lag} = \frac{1}{\omega} \times \text{phase shift in adians}$$

$$= \frac{0.6935}{0.104} = 6.669 \text{ seconds}$$

Example 3.27: First order instrument having a time constant of 0.25 seconds has been subjected to a sinusoidal input prescribed by the relation.

$$\theta_i = 0.25 \sin 20t$$

Develop an expression for the corresponding output.

Solution: After the transient effects die away, the dimensionless amplotude ratio is

$$\frac{\theta_0}{\theta_i} = \frac{1}{\sqrt{1+(\omega\tau)^2}} = \frac{1}{\sqrt{1+(20\times 0.25)^2}} = 0.196$$

$\therefore$ Magnitude of output
= 0.196 × magnitude of input
= 0.196 × 0.25 = 0.049

Phase shift $\phi = \tan^{-1}(\omega\tau)$
$= \tan^{-1}(20 \times 0.25) = 78.69°$

Therefore the required output,

$$\theta_0 \text{ M} = 0.049 \sin (20t - 78.69)$$

Example 3.28: Explain why most of the second-order systems are designed with a damping ratio in between 60–70% of critical.

The following data pertains to torque sensing transucer connected to the shaft of an electric motor which drives a load:

Input torque = sinusoidal with frequency varying at 2.15 radian/s

Maximum permissible error = +10%.

Moment of inertia of load J = 015 kg m^2

Torsional constant of the shiaft K = 1.25 Nm/rad.

Make calculatins for the natural frequency, the damping ratio, the damped natural frequency and the time constant of the system.

Solution: The natural frequency

$$\omega_n = \sqrt{\frac{K}{J}} = \sqrt{\frac{1.25}{0.125}} = 289 \text{ rad/s}$$

$$\text{Normalised frequency, } r = \frac{\omega}{\omega_n} = \frac{2.15}{2.89} = 0.743$$

Corresponding to maximum permissible error of +10%, the qmplitude ratio is 1.1 Further the amplitude ratio is prescribed by the relation.

$$M = \frac{1}{\sqrt{(1-r^2)+(2\zeta r)^2}}$$

$$1.1 = \frac{1}{\sqrt{(1-0.743^2)^2+(2\zeta\times 0.743)^2}}$$

$$= \frac{1}{\sqrt{0.20+2.21\zeta^2}}$$

$$0.20 + 2.21\zeta^2 = (1/1.1)^2 = 0.826$$

$$\therefore \text{ Mamping ratio } \zeta = \sqrt{\frac{0.826-0.20}{2.21}} = 0.53$$

Damped natural frequency,

$$\omega_d = \omega_n\sqrt{(1-\zeta^2)} = 2.89\sqrt{1-(0.53)^2}$$
$$= 2.45 \text{ rad/s}$$

Time constant $\tau = \dfrac{1}{\omega_n} = \dfrac{1}{2.89} = 0.346\text{s}$

Example 3.29: During the act of force measurement by a pressure transducer of the diaphragm type, the system is stated to have a natural frequency of 1000 cps and a damping ratio of 60%. Determine the frequency range over which the amplitude ratio corresponding to sinusoidal input deviates by a maximum amount of 10%. Consider the pressure transducer to behave as a second-order system.

Solution: Corresponding to a maximum deviation of 10%, the amplitude ratio is either 0.9 or 1.1. Further, the amplitude ratio is prescribed by the relation.

$$M = \frac{1}{\sqrt{(1-r^2)+(2\zeta r)^2}}$$

where r = normalised requency $= \dfrac{\omega}{\omega_n}$

(*i*) When M = 1.1

$$1.1 = \frac{1}{\sqrt{(1-r^2)+(2\times 2.6\times r)^2}}$$

or $(1-r^2)^2 + (2\times 0.6\times r)^2 = (1/1.1)^2 = 0.826$

This results into the following quadratic equation,

$$r^4 - 0.56r^2 + 0.174 = 0$$

The solution the imaginary roots and this implies that with $\zeta = 0.6$ the output would never be 1.1 times the input for any freqency.

(*ii*) When $M = 0.9$

$$0.9 = \frac{1}{\sqrt{(1-r^2)^2 + (2\times 0.6\times r)^2}}$$

This would give,

$$r^4 - 0.56r^2 - 2.34 = 0$$

$$r^2 = \frac{0.56 \pm \sqrt{(0.56)^2 + 4\times 0.234}}{2} = 0.84$$

$$r = \sqrt{0.84} = 0.916$$

$$\therefore\ \omega = \omega_n \times r = 1000\times 0.916 = 916 \text{ cps}$$

Apparently, the deviation remains within 10% of output for the frequency range 0–916 cps.

Exaple 3.30: A voltameter is used to measure known voltage of 75 volts. Forty percent of the readings are within 0.8 volt of true value. Estimate the standard deviation for the meter and the probability of an error of 1.2 volts.

Solution: Considering a normal distribution curve, 20% of the readings have a positive error and 20% have negative error.

From probability tables, it is found that corresponding to a probability of $P(z) = 0.2$,

$$z = 0.5025$$

$$\frac{x-\bar{x}}{\sigma} = 0.5025$$

∴ Standard deviation

$$\sigma = \frac{x-\bar{x}}{0.5025} = \frac{0.8}{0.5025} = 1.592$$

(*ii*) Now, $z = \dfrac{x-\bar{x}}{\sigma} = \dfrac{12}{1.592} = 0.753$

Corresponding to z = 0.753, the probability as read from the probability tables is $P(z) = 0.2743$

∴ Probability of an error of ± 1.2 volt

= 2 × 0.2743 = 0.5486

Thus about 55% of the redings are within 1.2 volt of true value.

Example 3.31: A tachometer has been used to measure the speed of a hydraulic turbine model that is being run at 1000 rpm. For a smaple of 20 readings made at this speed, how many of the readings you would expect to fall between 990 and 1010 rpm. Presume that the tachometer gives a normal distribution set of deviation with precision index 0.04 at 100 rpm.

Solution: Divation $\delta = x - \bar{x} = \pm 10$ rpm

Standard deviation

$$\sigma = \frac{1}{\sqrt{2}h} = \frac{1}{\sqrt{2}\times 0.04} = 17.68$$

$$\therefore\quad z = \frac{x-\bar{x}}{\sigma} = \frac{10}{17.98} = 0.5056$$

Corresponding to z = 0.5656, the probability tables is $P(z) = 0.214$

∴ Probability of an error of ± 10 rpm

= 2 × 0.214 = 0.428

Thus we expect 42.8% or 0.428 × 20 ≃ 9 readings to lie between 990 and 1010 rpm.

Exaple 3.32: The resistance of a certain wire has been determined by measuring the current flow and power fed into the circuit. Calculate the maximum possible error in resistance when the absolute (liminting) errors in the power and current are 1.8% and 1.2% respectively. The expression relating power, current and resistance is $P = I^2 R$.

Solution: From the given relation, the resistance R of wire is, $R = \frac{P}{I^2}$

Then, the relative error in the computed resistance is

$$\frac{dR}{R} = \pm\left(\frac{dP}{P} + 2\frac{dI}{I}\right)$$

Inserting the appropriate values,

$$\frac{dR}{R} = \pm(1.8 + 2\times 1.2) = \pm 4.2\%$$

Example 3.33: A wheatstone bridge circuit compreses four resistances; one resistance is connected in each limb of the bridge. Under null mode operaion, the solution of the unknwon resistance is:

$$R_4 = \frac{R_2 + R_3}{R_1}$$

The known resistances and the limiting error associated weith them are:

$R_1 = (100 \pm 0.75\%)$ ohm

$R_2 = (1000 \pm 0.5\%)$ ohm

$R_3 = (850 \pm 0.6\%)$ ohm

Determine the magnitude of unknown resistance R_4 and its limiting error. Also workout the guaranteed value of the resistance.

Solution: The relative error in the unknown resistance is

$$\frac{dR_4}{R_4} = \pm\left(\frac{dR_1}{R_1} + \frac{dR_2}{R_2} + \frac{dR_3}{R_3}\right)$$

$$= \pm(0{:}75 + 0.5 + 0.6) = \pm 1.85\%$$

Nominal value of resistance

$$R_4 = \frac{R_2 \times R_3}{R_1} = \frac{1000\times 850}{100} = 8500 \text{ ohm}$$

Limiting error of resistance

$$R_4 = \frac{R_2 \times R_3}{R_1} = \frac{1000\times 850}{100} = 8500 \text{ ohm}$$

$\therefore$ Guaranteed value of resistance $R_4 = (8500 \pm 175.5)$ ohm

Example 3.34: The power output from a rotating shaft, as measured by a dynamometer, is workout from the relation.

$$P = K\frac{FLN}{t} \text{ kW}$$

where F is the force at the end of torque arm (newton), L is length of the torque arm (cm), N is the number of revolutions in time t seconds, and K is a numerical constant of the value 6.15×10^{-4}.

With limiting (absolute) errors, the test data is

$F = (4.65 \pm 0.05)$ newton; $L = (38.5 + 0.15)$ cm

$N = (1160 \pm 2)$ rpm; $t = (60 \pm 0.5)$ secons

Make calculations for the power and the limiting error associated with it.

Solution: The relative error in the estimated power is

$$\frac{dP}{P} + \left(\frac{dF}{F} + \frac{dL}{L} + \frac{dN}{N} + \frac{dt}{t}\right)$$

$$= \pm\left(\frac{0.05}{4.65} + \frac{0.15}{38.5} + \frac{2}{1160} + \frac{0.5}{60}\right) = \pm 0.0247$$

Nominal power $P = K\frac{FLN}{t}$

$$= 6.15\times 10^{-4} \times \frac{4.65\times 38.5\times 1160}{60} = 2.132$$

Limiting error in power = 2.132 × 0.0247 = 0.05267

$\therefore$ Power $P = (2.132 \pm 0.05267)$ kW

Exaple 3.35: The coefficient of friction in a belt drive system can e calculated by using the relation

$$\frac{T_1}{T_2} = e^{\mu\theta}$$

In a particular experimental setup, the tensions T_1 and T_2 and the constant angle θ have been measured to an absolute accuracy of 1%, 1.5% and 1.25% respectively. Workout the maximum error in the computed value of coefficient of friction when the ratio $T_1/T_2 = 2$ and 20. Comment upons the results you obtain.

Solution: From the given expression, the coefficient of friction μ is:

$$\mu = \frac{1}{\theta}\log_e \frac{T_1}{T_2}$$

The relative error in the calculaed coefficient of friction is

$$\frac{d\mu}{\mu} = \frac{d\theta}{\theta} + \frac{1}{\log_e \frac{T_1}{T_2}} \frac{dT_1}{T_1} + \frac{1}{\log_e \frac{T_1}{T_2}} \frac{dT_2}{T_2}$$

(*i*) Inserting the appropriate values for the case when $T_1/T_2 = 2$.

$$\frac{d\mu}{\mu} = 1.25 + \frac{1}{0.693} \times 1 + \frac{1}{0.693} \times 1.5$$

$$= 1.25 + 1.44 + 2.16 = 4.85$$

∴ Maximum possible error in μ is ± 2.08%

(*ii*) Inserting the appropriate values for the case when $T_1/T_2 = 20$

$$\frac{d\mu}{\mu} = 1.25 + \frac{1}{2.995} \times 1 + \frac{1}{2.995} \times 1.5$$

$$= 1.25 + 0.3 + 0.50 + = 2.08$$

∴ Maximum possible in μ is ± 2.08%

The following observation may be made from the calculations made above:

(1) The error in the calculated value is greater than the individual errors.

(2) The higher belt tension ratio gives greater accuracy in the calculated value of the coefficient of frection.

Example 3.36: The discharge over a triangular weir varies as $H^{2.5}$ where H is the head above the weir. If uncertainty in the head reading is 2% of the absolute value, what is the percentage error in discharge measurement? No error results in the measurement of other quantities.

Solution: The expression relating discharge Q to to heat H above the weir is

$$Q = K H^{2.5}$$

where K is a constant.

The error specified pertains to statistical bonunds, therefore the error or uncertainty in discharge measurement is

$$e_Q = \sqrt{\left(\frac{\partial Q}{\partial H} e_H\right)^2}$$

$$\frac{\partial Q}{\partial H} = 2.5\, K\, H^{1.5}$$

e_H = 2% of absolute value = 0.02 H

$$\therefore\ e_Q = \left[(2.5\, K\, H^{1.5} \times 0.02\, H)^2\right]^{1/2} = 0.05\, K\, H^{2.5}$$

The percentage error in discharge measurement becomes,

$$e_Q \frac{0.05\, K\, H^{2.5}}{K\, H^{2.5}} \times 100 = 5\%$$

Example 3.37: A resistance arrangement of 50 ohms is desired. Two resistors of (100 ± 0.1) ohms and two resistors (25 ± 0.02) hms are available. Which should be used a series arrangement with 25 ohm resistors or a parallel arrangement with the 100 ohm resistors?

Solution: When the two resistors of (25 ± 0.02) ohms are connected in series, the resultant resistance is:

$$R = R_1 + R_2$$

$$\therefore\ \frac{\partial R}{\partial R_1} = 1 \quad \text{and} \quad \frac{\partial R}{\partial R_2} = 1$$

Hence uncertainty in the total resistance is

$$e_R = \pm\left[\left(\frac{\partial R}{\partial R_1} e_{R1}\right)^2 + \left(\frac{\partial R}{\partial R_2} e_{R2}\right)^2\right]^{1/2}$$

$$= \pm\left[(1 \times 0.02)^2 + (1 \times 0.02)^2\right]^{1/2}$$

$$= \pm\, 0.02825 \text{ ohm}$$

The total resistance $R = 25 + 25 = 50$ ohm can be expressed as

$$R = (50 + 0.025828) \text{ ohm}$$

(*ii*) When the two resistors of (100 ± 0.01) ohm are connected in parallel, the resultant resistance is

$$R = \frac{R_1 R_2}{R_1 + R_2} = \frac{100 \times 100}{100 + 100} = 50 \text{ ohms}$$

The derivatives are:

$$\frac{\partial R}{\partial R_1} = R_2 (R_1 + R_2)^{-1} - R_1 R_2 (R_1 + R_2)^{-2}$$

$$= \frac{R_2}{R_1 + R_2} - \frac{R_1 R_2}{(R_1 + R_2)}$$

$$= \frac{100}{100 + 100} - \frac{100 \times 100}{(100 + 100)^2} = 0.25$$

Similarly,

$$\frac{\partial R}{\partial R_2} = \frac{R_1}{R_1 + R_2} - \frac{R_1 R_2}{(R_1 + R_2)^2}$$

$$= \frac{100}{100 + 100} - \frac{100 \times 1000}{(100 + 100)^2} = 0.25$$

Hence uncertainty in total resistance is

$$e_R = \pm\left[\left(\frac{\partial R}{\partial R_1} e_{R1}\right)^2 + \left(\frac{\partial R}{\partial R_2} e_{R2}\right)^2\right]^{1/2}$$

$$= \pm\left[(0.25\times 0.1)^2 (0.25\times 0.1)^2\right]^{1/2}$$

$= \pm 0.03535$ ohm

The total resistance can be written as

$R = (50 \pm 0.03535)$ ohm

Undoubtedly the series arrangement with 25 ohm resistors should be preferred because it has a low value of uncertaintly.

Example 3.38: The resistance of a certain size of copper wire is expressed by the relation.

$$R = R_0\left[1 + \alpha\,(T - 30)\right]$$

where $R_0 = (5 \text{ ohm} \pm 0.2\%)$ is the resistance at 30°C; $\alpha = (0.004 \text{ per °C} \pm 1\%)$ is the temperature coefficient of resistance; and $T = (40 \pm 1°\text{C})$ is the temperature of wire. Calculate the resistance of the wire and uncertainty associated with it.

Solution: The nominal resistance of the wire is

$$R = 2\left[1 + 0.004\,(40 - 30)\right] = 5.20 \text{ ohms}$$

The uncertainty in the resistance is given by

$$e_R = \pm\left[\left(\frac{\partial R}{\partial R_0} e_{RO}\right)^2 + \left(\frac{\partial R}{\partial \alpha} e_\alpha\right)^2 + \left(\frac{\partial R}{\partial T} e_T\right)^2\right]^{1/2}$$

The various terms are:

$$\frac{\partial R}{\partial R_0} = 1 + \alpha\,(T - 30) = 1 + 0.004\,(40 - 30) = 1.04$$

$$e_{RO} = 5 \times \frac{0.2}{100} = 0.01 \text{ ohm}$$

$$\frac{\partial R}{\partial \alpha} = R_0\,(T - 30) = 5\,(40 - 30) = 50$$

$$e_a = 0.004 \times \frac{1}{100} = 4 \times 10^{-5} \text{ per °C}$$

and $$\frac{\partial R}{\partial T} = R_0 \alpha = 5 \times 0.004 = 0.02$$

Substiting the values, we obtain the uncertainty in resistance as

$$e_R = \pm\left[(1.04 \times 0.01)^2 + (50 \times 4 \times 10^{-5})^2 + (0.02 \times 1)^2\right]^{1/2}$$

$= \pm 0.02256$ ohm

The resistance can be written as

$R = (5.20 \pm 0.02256)$ ohm

The percentage uncertainly works out as;

$$= \pm \frac{0.02256}{5.20} \times 100 = 0.434\%$$

Example 3.39: The 3 cm region between the two plates of a parallel plate capacitor is filled by two dielectric layers:

(*i*) 1 cm thick with dielectric contstant 5, and

(*ii*) 2 cm thick with dielectric constant 10.

What would be the relative permittivity (dielectric constant) of a material which gives the same capacitance if it completely fills the region between the plates?

Solution: For a multiple parallel plate capacitor with space between adjacent plates filled with different materials having dielectric constants $\varepsilon_{r1}, \varepsilon_{r2}, \varepsilon_{r3}$...... and having respective distances t_1, t_2, t_3 ......, we have:

$$C = \frac{\varepsilon_0 A}{\dfrac{t_1}{\varepsilon_{r1}} + \dfrac{t_2}{\varepsilon_{r2}} + \dfrac{t_3}{\varepsilon_{r3}}}$$

Substituting the values from the given data,

$$C = \frac{\varepsilon_0 A}{\dfrac{0.01}{5} + \dfrac{0.02}{10}} = \frac{\varepsilon_0 A}{0.004}$$

Let ϵ_r be the dielectric constant of the single medium which completely fills the gap of 3 cm between the two parallel plates of area A and gives the same capacitance. Then

$$C = \varepsilon_0 \varepsilon_r \frac{A}{t} = \varepsilon_0 \varepsilon_r \frac{A}{0.03}$$

Equating the capacitance values given by expressions (*i*) and (*ii*).

$$\frac{\varepsilon_0 A}{0.004} = \varepsilon_0 \varepsilon_r \frac{A}{0.03}$$

$$\therefore\ \varepsilon_r = \frac{0.03}{0.004} = 7.5$$

Example 3.40: A capacitive transducer using two quartz diaphragms of area 800 mm^2 and separated by a distance of 4 mm has a capacitance of 350 μF. When a pressure of 1 MN/m^2 is applied to one of the diaphragms, a deflection of 0.75 mm is produced. Workout the change in the capacitance of the system.

Solution: Before application of pressure:

$$C_1 = \varepsilon_0 \varepsilon_r \frac{A}{t_1}$$

After application of pressure: $C_2 = \varepsilon_0 \varepsilon_r \dfrac{A}{t_2}$

Given: $C_2 = 350\mu\text{F}$; $t_1 = 4$ mm and $t_2 = 4 - 0.75 = 3.25$ mm

$$\therefore\ C_2 = 350 \times \frac{4}{3.25} = 430.77\ \mu\text{F}$$

$\therefore$ Change in capacitance $\Delta C = C_2 - C_1$

$$= 430.77 - 350 = 80.77\ \mu\text{F}$$

Example 3.41: A parallel plate air spaced capacitor has an effective palte area of 6.5 × 10^{-4} m^2 and the distance between the plates is 1mm. Worout the displacement sensitivity of the device if air has a relative permittivity of unity.

Solution: For a parrallel plate capacitorl,

$$C = \varepsilon_0 \varepsilon_r \frac{A}{t}$$

Upon differentiation, we get the sensitivity as

$$\frac{dC}{dt} = -\varepsilon_0 \varepsilon_r \frac{A}{t^2}$$

The minus sign indicates a reduction in the capacitance value for increasing t; the the distance between the plates. Apparently, the sensitiviry is high for small values of t.

$$\text{Sensitivity } \frac{dC}{dt} = \frac{8.854 \times 10^{-12} \times 1.00 \times 6.5 \times 10^{-4}}{(0.001)^2}$$

$$= 57.55 \times 10^{-10}\ \text{F/m}$$

Example 3.42: A capacitive transducer consists of five plates arranged as shown in Figure. Each plate measures 25 mm × 25 mm and the distance between the adjacent plates is 0.25 mm. This arrangement is used to measure the displacement x by observing the resulting change in capacitance. Make calculations for the sensitivity of the transducer if air is the medium separating the plates.

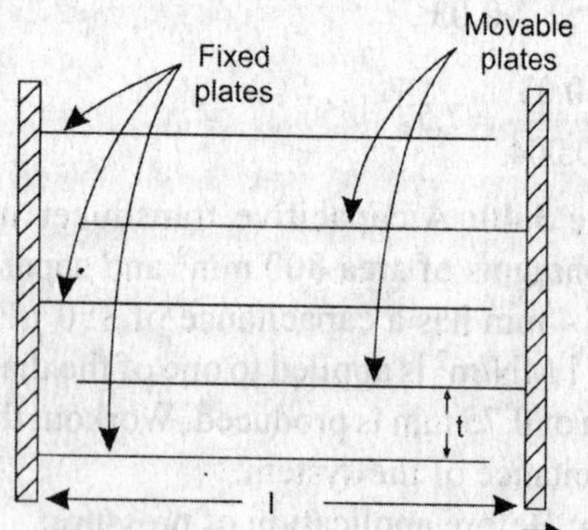

Solution: For a multiple capacitor, the capacitance C is given by

$$C = \varepsilon_0 \varepsilon_r \frac{A}{t}(N-1) \text{ farads}$$

where A = overlapping or effective area between plates

T = distance between plates (m)

N = number of capacitor plates

ϵ_0 = permittivity of free space = 8.95 10^{-12} F/m

ϵ_r = relative permittivity or dielectric constant of the medium between the places. For air, $\varepsilon_r = 1$.

With the given dimension l, width of each plate w and displacement x given to the movable plate on the right side.

$$C = \varepsilon_0 \varepsilon_r \frac{(l-x)w}{t}(N-1)$$

and sensitivity of the transducer is

$$\frac{\partial C}{\partial x} = -\varepsilon_0\ \varepsilon_r \frac{w}{t}(N-1)$$

$$= -8.854 \times 10^{-12} \times 1 \times \frac{25 \times 10^{-3}}{0.25 \times 10^{-3}} \times (5 \times 1)$$

$$= -3541.6 \times 10^{-12}\ \text{F/m}$$

Disregarding the sign, the sensitivity of the transducer is 3541.6 pF/m.

Example 3.43: The diaphragms of a capacitive transducer has a capacitance of 300 pF and is connected in an oscillator circuit having a frequency of 100 kHz. When a pressure of 500 kN/m^2 is applied, an average deflection of 0.28 mm is observed in the diaphragm of the transducer. Determine the change in frequency of the oscillator after the application of pressure to the transducer.

Solution: Let subscripts 1 and 2 refer to the conditions before and after the application pressure. Then

$t_1 = 4$ mm, $t_2 = 4 - 0.28 = 0.28 = 3.72$mm, $C_1 = 300$ pF and $fr_1 = 100$ kHz

For an oscillator circuit having inductance L and capacitance C, the resonant frequency is prescbribed by the relation.

$$f = \frac{1}{2\pi\sqrt{LC}}$$

For the given setup, inductance is constant and the capacitance is variable. Accordingly

$$f_1 = \frac{1}{2\pi\sqrt{LC_1}} \text{ and } f_2 = \frac{1}{2\pi\sqrt{LC_2}}$$

$$\text{Hence } \frac{fr_2}{fr_1} = \sqrt{\frac{C_1}{C_2}}$$

Since $C_1 = \varepsilon_0\ \varepsilon_r \dfrac{A}{t_1}$ and $C_2 = \varepsilon_0\ \varepsilon_r \dfrac{A}{t_2}$, we get

$$\frac{f_2}{f_1} = \sqrt{\frac{\varepsilon_0\ \varepsilon_r A/t_1}{\varepsilon_0\ \varepsilon_r A/t_2}} = \sqrt{\frac{t_2}{t_1}}$$

or $$f_2 = f_1\sqrt{\frac{t_2}{t_1}} = 100\sqrt{\frac{3.72}{4}}$$

∴ Change in frquency of oscillator = 100 – 96.44 = 3.56 kHz

Example 3.44: A quartz crystal having a thickness of 2mm and a voltage sensitiviry of 0.05 Volt-m/newton is subjected to a pressure of 15 × 10^5 N/m^2. Calculate the voltage developed bny the piezo-electric pick up.

Solution: The output voltage for a piezo-electric pick up is given by:

$$V_0 = g\,t\,P$$

$$= 0.05 \times 0.002 \times (15 \times 10^5) = 150\,\text{V}$$

Example 3.45: A quartz pressure tranducer has sensitiviry of 80×10^{-12} coulomb/bar. Workout the capacitance of the device.

Solution: Charge Q = sensitivity K × pressure force F = 80 × 10–12 × 3.5=280 × 10^{-12} coulomb

Output volatage $V_0 = \dfrac{\text{charge } Q}{\text{capacitance } C}$

$$\therefore \text{Capacitance } C = \frac{280 \times 10^{-12}}{1.2} = 10^{-12} \text{ farads}$$

Example 3.46: A force of 8 newton is impressed upon a piezo-electric crystal having dimensions of 6mm × 6mm × 1.5mm thick. For the crystal materials,

Charge sensitiviry = 140 × 10^{-12} C/N
Primitivity = 12 × 10^{-9} N/m, and
Modulus of elasticity = 11.5 × 10^6 N/m^2
Make calculation for the voltage gererated, capacitance, and the strain introduced in the crystal.

Solution: Plate area A = 6 × 6 = 36 mm^2 = 36 × 10^{-6} m^2

∴ Stress or pressure

$$P = \frac{8}{36 \times 10^{-6}} = 0.222 \times 10^{-6}\,\text{N/m}^2$$

$$\text{Strain } \frac{\delta t}{t} = \frac{\text{stress}}{\text{modulus of elasticity}}$$

$$= \frac{0.222 \times 10-6}{11.5 \times 10^{-6}} = 0.0193$$

Voltage sensitivity

$$g = \frac{K}{\varepsilon_0 \varepsilon_r} = \frac{140 \times 10^{-12}}{12 \times 10^{-9} \times 1} = 11.67 \times 10^{-3} \text{ Vm/N}$$

Output voltage $V_0 = g\,t\,P$
= 11.67 × 10^{-3} × (1.5 × 10^{-3}) ×0.222 × 10^6
Charge Q = sensitivity K × pressure force F
= 140 ×10^{-12} × 8 = 1120 × 10^{12} coulomb

$$\text{Capacitance } C = \frac{\text{output voltage } V_0}{\text{charge } Q}$$

$$= \frac{1120 \times 10^{-12}}{3.886} = 288 \times 10^{-12} \text{ farads}$$

Example 3.47: In a commercial null-type Wheatstone bridge circuit as indicated in figure, the variable resistor is accurate to within ± 0.02% and the accuracy of bridge ratio (R_3/R_4) = (1/1000) and the variable resistor is set at 3426.5 Ω. Calculate the value and error in the measurements or resistant R_1.

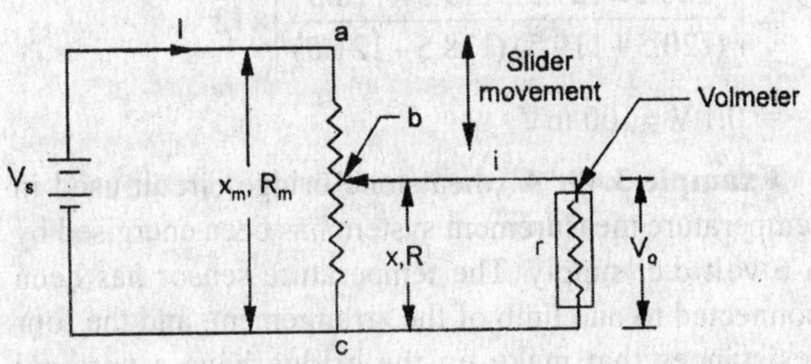

Solution: At balance

$$R_1 = R_2\left(\frac{R_3}{R_4}\right) = 3426.5 \times \frac{1}{1000} = 3.4265\Omega$$

$$\text{Also } R_1\left(\frac{R_3}{R_4}\right)V_2 = K\,V_2$$

Taking logs: log R_1 = log K + log V_2
Upon differentiation,

$$\frac{dR_1}{R_1} = \frac{dK}{K} + \frac{dR_2}{R_2} = \pm\left(\frac{0.025}{100} + \frac{0.02}{100}\right) = \pm 0.00045$$

Thus the possible error in measurement of resistance R_1 is

= ± 3.4265 × 0.00045 = ± 0.001542

Hence $R_1 = (3.4265 \pm 0.001542)\ \Omega$

Example 3.48: Explain the principle of Wheatstone bridge circuit. How this bridge can be used in the null (balanced) and deflection (unbalanced) mode?

A resistance bridge has the configuration as shown in figure in which $R_1 = 120.5\Omega$ $R_2 = 119.5\Omega$ and $R_3 = 118\Omega$ Calculate the resistance R_4 if the bridge is to have resistance balance.

What open-circuit voltage would be indicated on a meter if R_4 has a value of 121.5Ω. The bridge is energised with a 12V d.c. source and the meter has infinite internal impedance.

Solution: For balanced (null) mode of operation,

$$\frac{R_1}{R_2} = \frac{R_3}{R_4} \quad ; \quad \frac{120.5}{119.5} = \frac{118.5}{R_4}$$

$$\therefore \ R_4 = \frac{119.5 \times 118.5}{120.5} = 117.52\Omega$$

The open circuit output voltage is given by the relation,

$$V_0 = \left[\frac{R_1 R_4 - R_2 R_3}{(R_1 + R_2)(R_2 + R_4)}\right] V_s$$

$$= \left[\frac{120.5 \times 121.5 - 119.5 \times 118.5}{(120.5 + 119.5)(118.5 + 121.5)}\right] \times 12$$

$= 0.1\text{ V} = 100\text{ mV}$

Example 3.49: A wheatstone bridge circuit used in temperature measurement system has been energised by a 6 volt d.c. supply. The temperature sensor has been connected to one limb of the arrangement, and the four resistances that make up the bridge have a nominal resistance of $500\,\Omega$. What output voltage would be indicted from the bridge if the temperature change causes ± 5% change in the resistance of the temperature sensor. The measuring instrument has an infinite internal resistance.

Solution: For the single arm active bridge and with the stipulation of infinite internal resistance ofthe measuring instrument, the output volatage is given by:

$$V_0 = \left[\frac{dR}{4R + 2dR}\right] V_s = \left[\frac{dR / R}{4 + 2dR / R}\right] V_s$$

$$= \left[\frac{\pm 0.05}{4 + 2(\pm 0.05)}\right] 6 = 0.0731\text{ V and} - 0.069\text{ V}$$

Example 3.50: A wheatstone bridge circuit (Figure) used in strain measurement system comprises four strain gauges of equal magnitude, one connected into each limb of the bridge. The nominal resistance of each gauge is 120Ω and the bridge supply voltage is 10V. When stained, the resistance changes by 0.50Ω. Obtain a relation between the output voltage V_0 and fractional change in resistance x, calculate the output voltage. The meter has infinite internal impedance.

Solution: Refer Figure.

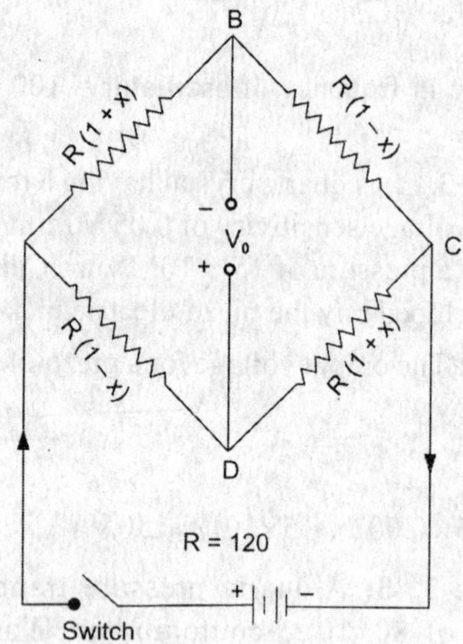

Since no current is drawn by the meter due to its infinite internal impedance, the same current I_1 flows through resistors AB and BC, current I_2 flows through resistors AD and CD. The values of current are:

$$I_1 = \frac{V_s}{R_{ab} + R_{bc}} \text{ and } I_2 = \frac{V_s}{R_{ad} + R_{dc}}$$

The output voltage is

V_{bd} = voltage across terminal B and D

$$= I_1 R_{ab} - I_2 R_{ad}$$

$$= \left(\frac{R(1+x)}{R(1+x)+(1-x)} - \frac{R(1-x)}{R(1-x)+(1+x)}\right) V_s$$

Inserting the appropriate values,

$$V_0 = \left[\frac{R(1+x)}{R(1+x)+R(1-x)} - \frac{R(1-x)}{R(1-x)+R(1+x)}\right] V_s$$

$$= \left(\frac{1+x}{2} - \frac{1-x}{2}\right) V_s = x V_s$$

Given: $x = \dfrac{dR}{R} = \dfrac{0.5}{120} = 4.167 \times 10^{-3}$

$\therefore$ Output voltage $= (4.167 \times 10^{-3}) \times 10$

$= 0.04167\text{ V}$

Since the bridge is fully active (all the limbs aresensitive), the output voltage could also be obtained from the relation:

$$V_0 = 4\left(\frac{dR}{4R+2dR}\right)V_s$$

$$= 4\left(\frac{0.5}{4\times120+2\times0.5}\right)10 = 0.04158 \text{ V}$$

Example 3.51: One end of a U-tube manometer is connected to a piezometer opening in a air duct and the other end is left open to atmosphere. The manometer has mercury as the balancing liquid. If the mercury level in the manometer on the duct side stads 0.2m higher than that on the open end, calculate the absolute pressure of air inside the duct. Take atmospheric pressure $p_a = 10\times10^4$ N/m^2.

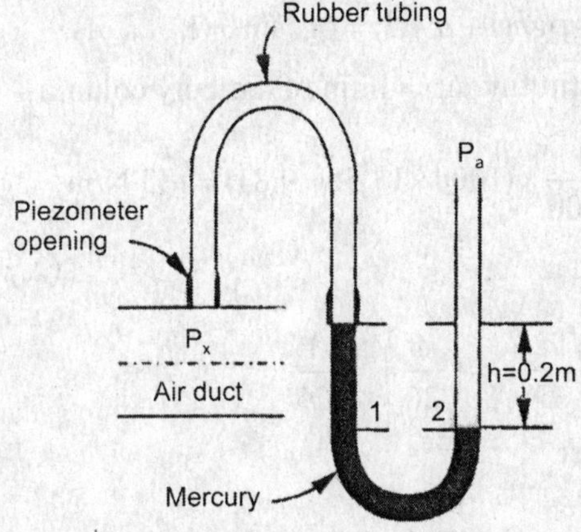

Solution: Refer figure. Let p_x be the gauge pressure in the air duct.

Neglecting the pressure exerted by the air column above mercury in the left limb of the manometer, the governing manometric equation for the points 1 and 2 is:

$$p_x = w_m h = 0 \; ; \; p_x = -w_m h$$

$$p_x = -(9810\times13.6)\times0.2$$

$$= -26683.2 \text{ N/m}^2 \text{ (gauge)}$$

$$= 26683.2 + 10\times10^4 = 73316.8 \text{ N/m}^2 \text{ (absolute)}$$

Example 3.52: The right limb of a U-tube manometer constaining mercury is open to the atmosphere while the left limb is connected to a pipe through which flows a fluid of specific gravity 0.85. The center of the pipe lies 15cm below the level of mercury in the right limb. If the difference of mercury level in the two limb is 25cm, determine the pressure of fluid in the pipe.

Solution: Refer figure. Let p_x be gauge pressure of the fluid in the pipeline. Consider pressure balance in the horizontal plane 0–0; the pressures in the left and right limbs at this plane are equal. That is:

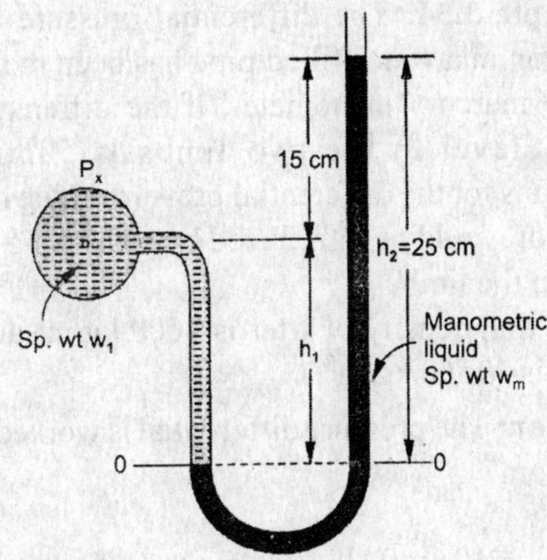

$$p_x + w_1 h_1 = w_m h_2$$

or $p_x = w_m h_2 = w_1 h_1$

$$= (9810\times13.6)\times0.25 - (9810\times0.85)$$

$$\times(0.25-0.15)$$

$$= 33354 - 833.85 \simeq 32520 \text{ N/m}^2$$

Example 3.53: The right limb of a simple U-tube manometer containing mercury is open to atmosphere and the left limb is connected to a pipe through which flows a fluid of specific gravity 0.8. Make calculations for the vacuum pressure in the pipe if the difference of mercury level in the limbs is 30cm and the level of fluid in the limb is 10cm below the centre of pipe.

Solution: Refer figure. Let p_x be gauge pressure of fluid in the pipeline. Consider pressure balance in the horizontal plane 0–0; the pressure in the left and right limb at this level are equal. That is:

$$p_x + w_1 h_1 + w_m h_2 = 0$$

$$px = -(w_1 h_1 + w_m h_2)$$

$$= -(9810\times0.8\times0.1 + 9810\times13.6\times0.3)$$

$$+9810\times13.6\times0.3)$$

$$= -(784.8 + 40024.8) = -40809.6 \text{ N/m}^2$$

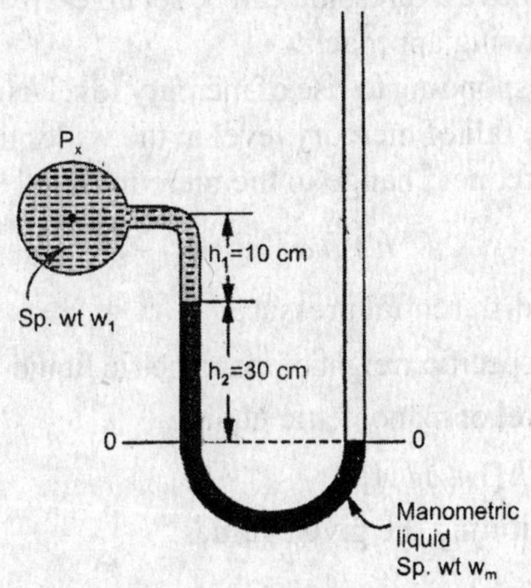

Example 3.54: The differential pressure across an obstruction in a water-filled pipe has been measured by a U-tube mercury manometer. If the difference in the mercury level in the two limbs is 50mm, make calculations for the differential pressure being measured. What error would be introduced by neglecting the water density in the limb?

Given that density of wter is 1000 kg/m^3 and that of mercury is 13560 kg/m^3.

Solution: The pressure differential is worked out from the relation.

$$p_x - p_y = (\rho_m - \rho_w) g\, h$$

$$= (13560 - 1000)$$

$$\times 9.81 \times \frac{50}{1000}$$

$$= 6160.98 \text{ N/m}^2$$

(*b*) Neglecting the density of water,

$$p_x - p_y = \rho_m\, g\, h$$

$$\text{Percentage error} = \frac{\rho_w\, g\, h}{(\rho_m - \rho_w)\, g\, h} \times 100 = \frac{\rho_w}{\rho_m - \rho_w}$$

$$= \frac{1000}{13560 - 1000} \times 100 = 7.96\%$$

Example 3.55: A single-column manometer is using mercury of specific gravity 13.6 as the manometric liquid. To what height will be mercury rise in the narrow limb if a differential pressure of 75 kN/m^2 is applied. The wider and the narrow limb diameters are 160 mm and 6 mm respectively.

Solution: For a single column manometer, the differential pressure is given by:

$$\Delta p = w_m h\,[1 + a / A]$$

The above expression can be set up easily by adopting the following approach:

Corresponding to rise of mercury level *h* in the narrow limb, the fall of mercury level in the wider limb is *h a/A*. Therefore, net change in the mercury level becomes

$$h + h\,a / A = h[1 + a / A]$$

Now differential pressure

Δp = specific weight of manometric liquid × net change in the level of manometric liquid

$= w_m\ h\,[1 + a / A]$

Substituting the given values:

$$75 \times 10^2 = (9810 \times 13.6)\, h\,[1 + (6/160)^2]$$

$$[\because a / A = (d / D)^2]$$

Manometric height h = 0.562 m = 56.2 cm

Example 3.56: A well type manometer using mercury (specific gravity 13.56) as manometric liquid has well and tube diameters as 20mm and 4mm respectively. This manometer has a scale graduated accurately in mm to measure the pressure differenc, i.e. 1 mm scale division indicates 1mm head of pressure change. Work out the angle at which the tube should be inclined to the vertical to accomplish this task.

Solution: For a single column manometer, the differential pressure is given by:

$$\Delta p = \rho g h (1 + a / A) = \rho g l \sin\theta\, (1 + a / A)$$

Substitutin: Δp = 1mm of mercury column

$$= \frac{1}{1000} \times (1000 \times 13.56 \times 9.81) = 133 \text{ N/m}^2$$

$$(\because p = \rho h g)$$

$$\frac{\alpha}{A} = \left(\frac{d}{D}\right)^2 = \left(\frac{4}{20}\right)^2 = \frac{1}{25}$$

we get:

$$133 = (1000 \times 13.56 \times 9.81) \times \frac{1}{1000} \times \sin\theta$$

$$\left[(1 + d / D)^2\right]$$

$$\sin\theta = \frac{133 \times 1000}{1000 \times 13.56 \times 9.81} \times \frac{25}{26} = 0.9614$$

$$\theta = \sin^{-1}(0.9614) = 74°$$

Hence the angle at which the tube is inclined to vertical = 90° − 74° = 16°.

Example 3.57: A new control weight is to be made for a ring balance meter. Given that the average radius of the ring is 200 mm, tube diameter is 100 mm and the distance of the weight from the centre is 275 mm, determine the value of control weight required in order that the maximum deflection of the ring is 30° when a differential pressure of 200 mm water gauge is applied to the ring.

Solution: For a ring balance meter, the equilibrium equation is:

$$\Delta P = \frac{W\, l \sin\theta}{La} \quad \text{with usual notations.}$$

substitutin,

$\Delta P =$ 200 mm, water gauge

$$= 1000 \times 9.81 \times \frac{200}{1000} = 1962 \text{ N/m}^2$$

$$(\because P = \rho gh)$$

$= 1962 \times 10^{-6}$ N/mm

$L = 275$ mm, $r = 200$ mm

$$\sin\theta = \sin 30 = 0.5 \text{ and } a = \frac{\pi}{4}(100)^2 = 7850\text{mm}^2$$

we get,

$$1962 \times 10\text{–}6 = \frac{W \times 275 \times 0.5}{7850 \times 200}$$

$$\text{or} \quad W = \frac{1962 \times 10^{-6} \times 7850 \times 200}{275 \times 0.5} = 22.4 \text{ N}$$

Example 3.58: With walled inverted bell-type differential pressure gauge instrument, the bell diameter is 150mm and it moves vertically 20 mm when a full load pressure of 250 mm water guge is applied. Find the spring rate of the opposing spring.

Solution: For a thin walled inverted bell-type differential gauge, the equilibrium equation is

$$\Delta P = \frac{Fh}{a}$$

where F is the spring constant, h is the bell movement and a is cross sectional area of the bell. Substituting: ΔP = 250 mm water gauge.

$$= 1000 \times 9.81 \times \frac{250}{1000} = 2452.5 \text{ N/m}^2$$

$$(\because P = \rho gh)$$

$$= 2452.5 \times 10^{-6} \text{ N/mm}^2$$

$$\text{h} = 20 \text{ mm and } a = \frac{\pi}{4}(150)^2 17662.5 \text{ mm}^2$$

$$\text{wet get : } 2452.6 \times 10^{-6} = \frac{F \times 20}{17662.5};$$

$$F = 2.166 \text{ N/mm}$$

Example 3.59: Below figure represents the results of a test on 0–55 MN/m² bourdon gauge wherein the gauge reading has been plotted against the true pressure. Point out the error indicated by the gauge and suggest some means to rectify those errors.

Solution: The following observation can be made from the results depicted in the graph:

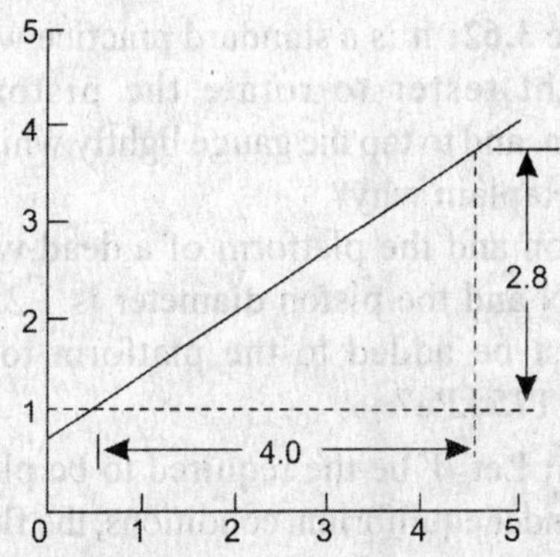

(*i*) The straight line graph indicates that movements of the gauge pointer are directly propotional to pressure changes.

(*ii*) The graph does not pass through the origin andthat indicates a zero error.

$$(iii) \text{ Slope of the graph} = \frac{2.8}{4} = 0.7$$

Since the slope is not unity, the gauge has an.

Example 3.60: It is required to design a diaphragm pressure gauge to measure a maximum pressure of 1.5 MPa gauge. What be the diaphragm thickness if the maximum deflection is limited to one-third of thickness?

The diaphragm is to be constructed of spring steel 5 cm in diameter; modulu of elasticity E = 200GPa and poission's ratio μ = 0.3.

Solution: Using the relation:

$$y_{\max} = \frac{t}{3} = \frac{3\Delta P}{16Et^3} r^4 (1 - \mu^2)$$

$$\therefore \quad t^4 = \frac{9\Delta P r^4 (1 - \mu^2)}{16E}$$

$$= \frac{9 \times 1.5 \times 10^6 \times .25^4 [1 - (0.3)^2]}{16 \times 200 \times 10^9}$$

$$= 149.96 \times 10^{-6} \text{cm}^4; t = 0.1106 \text{ cm}$$

Example 3.61: A Mcleod gauge of $V_0 = 200$ cm³ and capillary cross-sectional area $a = 0.1$ cm² indicates 1 cm of mercury. Express the pressure in microns.

$$\textbf{Solution:} \quad p_0 = \frac{ah^2}{V_0} = \frac{0.1 \times (1 \times 1)}{200}$$

$$= 0.5 \times 10^{-3} \text{ cm of Hg}$$

$$= 5 \times 10^{-3} \text{ mm of Hg}$$

$$= 5 \times 10^{-3} \text{ torr} = 5 \text{ microns}$$

Example 3.62: It is a standard practice when using a dead-weight tester to rotate the piston-platform combination, and to tap the gauge lightly while obtaining readings. Exaplain why?

The piston and the platform of a dead-weight tester weight 20 N and the piston diameter is 1.25 cm. What weight must be added to the platform to produce a pressure of 175 kPa?

Solution: Let W be the required to be placed on the platfrom. Under equilibrium conditions, the fluid pressure is given by the combination of wieght, piston and platform assembly divided by the effective piston-cylinder area. That is:

$$\text{Pressure} = \frac{\text{total weight}}{\text{effective area}}$$

or $$175 \times 10^3 = \frac{W + 20}{\frac{\pi}{4}\left(1.25/100\right)^2}$$

$$W + 20 = 175 \times 10^3 \times \frac{\pi}{4}\left(1.25/100\right)^2 = 21.464$$

$$\therefore \quad W = 21.464 - 20 = 1.464 \text{ N}$$

Example 3.63: A 30 cm × 15 cm venturimeter is provided in a vertical pipe line carrying oil of specific gravity 0.9; the flow being upwards. The difference in elevations of the throat section and entrance section of the venturimeter is 30 cm. The differential U-tube mercury manometer shows a gauge deflection of 25 cm. Calculate (*i*) the discharge of the oil, and (*ii*) the pressure difference between the entrance and throat section. Take the discharge coefficient as 0.98 and the specific gravity of mercury as 13.6.

Solution: The actual discharge through a venturimeter is given by the relation:

$$Q = C_d\, M\, A_2\, E \sqrt{2\left\{\frac{p_1 - p_2}{\rho} + g(z_1 - z_2)\right\}}$$

Presuming thermal expansion factor $E = 1$ and recalling that velocity of approach factor.

$$M = \frac{1}{\sqrt{1 - (A_2/A_1)^2}} = \frac{A_1}{\sqrt{A_1^2 - A_2^2}}$$

we may write:

$$Q = C_d \frac{A_1 A_2}{\sqrt{A_1^2 - A_2^2}} \sqrt{2g\left\{\frac{p_1 - p_2}{\rho g} + (z_1 - z_2)\right\}}$$

Now, $A_1 = \frac{\pi}{4}(0.3)^2 = 0.07065 \text{ m}^2$

and $A_2 = \frac{\pi}{4}(0.15)^2 = 0.01766$

$$\frac{p_1 - p_2}{\rho g} + (z_1 - z_2) = h\left(\frac{\rho_m}{\rho} - 1\right)$$

$$= 0.25\left(\frac{13.6}{0.9} - 1\right) = 3.528 \text{m of oil}$$

Substituting the given data in the discharge equation,

$$Q = 0.98 \times \frac{0.07065 \times 0.01766}{\sqrt{0.07065^2 - 0.01766^2}} \times \sqrt{2 \times 9.81 3.528}$$

$$= 0.1487 \text{ m}^3/\text{s}$$

(*ii*) $$\frac{p_1 - p_2}{\rho g} + (z_1 - z_2) = 3.528$$

$$\therefore \quad \frac{p_1 - p_2}{\rho g} = 3.528 - (z_1 - z_2)$$

$$= 3.528 - (-0.3) = 2.828 \text{ m of oil}$$

Hence the pressure difference between the entrance and throat section,

$$p_1 - p_2 = \rho g \times 3.528$$
$$= (1000 \times 0.9) \times 9.81 \times 3.828$$
$$= 33797.4 \text{ N/m}^2$$

Example 3.64: Air flows through a duct and its velocity is being measured by a pitot-static tube placed properly along the duct axis. The leads from the tubes have been attached to a differential manomerter which shows a difference of hed of 0 cm of water. Workout the air velocity presuming that the coefficient of velocity for the pitot-tube is 0.98. You may take density of air 1.22 kg/m³.

Solution: Manometric head h_m = 10 cm of water

Equivalent head of air column,

$$h = h_m\left(\frac{\rho_m}{\rho} - 1\right)$$

$$= \frac{10}{100}\left(\frac{1000}{1.22} - 1\right) = 91.87 \text{ m of air}$$

∴ Flow velocity, $V = C_v\sqrt{2gh}$

$$= 0.9\sqrt{2 \times 9.81 \times 81.87} = 39.2 \text{ m/s}$$

Example 3.65: A pitot-static probe with a thermometer attached is used to measure the velocity of air in a duct and a differential pressure of 10 cm of water is recorded. The thermocuple shows an air temperature of 300 K in the duct, and a separate static tapping shows that the air pressure in the ductis 10 kN/m^2. The local barometer reads 760 mm of mercury. Calculate the aire velocity if the coeffiecient for the pitot-static probe is 0.98 Take the gas constant R as 287 J/kgK for air and the relative density of mercury as 13.6.

Solution: The atmosphereic pressure is,

$$p_{at} = \rho gh\,(13.6\times10^3)\times 9.81\times\frac{760}{1000}$$

$$= 101.39\times10^3\ \text{N/m}^2 = 101.39\ \text{kN/m}^2$$

$\therefore$ Static pressure in the duct = 101.39 + 10 = 111.39 kN/m^2

From the characteristic gas equation, $pV = mRT$.

$$\therefore \text{ Density of air, } \rho = \frac{m}{V} = \frac{p}{RT} = \frac{111.39\times10^3}{287\times 300}$$

$$= 1.29\ \text{kg/m}^3$$

Manometric head $h_m = 10$ cm of water

Equivalent head of air column,

$$h = h_m\frac{\rho_m}{\rho} = \frac{10}{100}\times\frac{1000}{1.29} = 77.52 \text{ m of air}$$

Flow velocity $V = C_v\sqrt{2gh}$

$$= 0.98\sqrt{2\times9.81\times77.52}$$

$$= 38.22 \text{ m/s}$$

Example 3.66: How signal conditioning is achieved in the liquid-in-glass thermometers, and the filled-in-system thermometers?

A bourdon tube temperature indicator has a range of 0°C to 250°C and produces a corresponding rotation of 0 to 270°. Worlout the sensitivity of the bourdon tube in radians per degree celcius ifthe mechanical levers and gears have an amplification of 30.

Solution: In the liquid-in-glass thermometers, signal conditioning is achieved by the capillary tube which converts the liquied expansion into liquid expansion into liquid movement against a scale.

In the filled-in-system thermometers, the fluid expansion ca nbe transmitted upto 35m via capillary tube to the bourdon tube. Thw small deflection produced at the free end of the boufdon tube is magnified and converted into a pointer motion through a mechanism of gears and links. The pointer moves across a scale calibrated in temperature units.

(*b*) Mechanical ampification

$$= \frac{\text{movement of pointer}}{\text{angular displacement of the bourdon tube}}$$

$$30 = \frac{270}{\text{angular displacement of the bourdon tube}}$$

$\therefore$ Angulardisplacement of bourdon tube

$$= \frac{270}{30} = 9° \text{ or } 0.157 \text{ radian}$$

Sensitivity of bourdon tube

$$= \frac{\text{angular displacement}}{\text{temperature}}$$

$$= \frac{0.157}{250} = 6.28\times10^{-4}\ \text{rad/°C}$$

Example 3.67: In a vapour type filled-system thermometer (filled with pure water), the bulb was located 4.5 m above the indicator. If the actual temperature at the bulb is 200°C, estimate the effect of relative elevation of the bulb and the indicator. From steam tables, thefollowing data has been recorded:

pressure, bar	15	16	17
staturation, °C temperature	197.4	200.0	203.4

Solution: Under the given conditions, the capillary would be full of water; the pressure due to water column equals.

$$= \rho gh = 1000\times9.81\times4.5 = 44145\ \text{N/m}^2$$

$$= 0.441 \text{ bar} \qquad (\because 1 \text{ bar} = 10^5\ \text{N/m}^2)$$

Vapour pressure at saturation temperature of 200°C is 16 bar.

$\therefore$ Pressure at the indicator

= 16 + 0.441 = 16.441 bar

Corresponding to this pressure, the saturating temperature is;

= 200 + 0.441 (203.4 – 200) = 201.5°C

Thus the thermometer would read approximately 1.5°C high because of the head effect.

Example 3.68: Give the general relationship suitable for resistance of metals over a wide temperature range.

Comment upon the constants appearing in the equation.

A platinum resistance thermometer has a resistance of 150Ω at 100°C. When exposed to high temperature gas, the resistance to 300Ω. Determine the temperature of hot gas. Itmay be presumed that at 0°C, the resistance is 100Ω and that the temperature coefficient of resistance forplatium is $0.0039°C^{-1}$.

Solution: Over a limited temperature range around 0°C;

$R = R_0(1+\alpha\theta)$; $R_1 = R_0(1+\alpha\theta_1)$

and $R_2 = R_0(1+\alpha\theta_2)$

Rearrangement gives: $\theta_2 = \theta_1 + \dfrac{R_2 - R_1}{\alpha R_0}$

Inserting the appropriate values,

$$\theta_2 = 100 + \frac{300-150}{0.0039\times 100} = 484.61\ °C$$

The temperature of the gas is therefore 484.61°C.

Example 3.69: A thermistor is characterisedby the relation:

$R_1 = R_0\, ae^{b/T}$

where R_0 is resistance at ice point, R_1 isthe resistance at the absolute temperature TK, and a and b are constants. Determine the temperature indicated by such a thermistor if $R_1 = 1050\Omega$, $R_0 = 4000\Omega$, $a = 300\times10^{-6}$ and $b = 2850$.

Solution:

$R_1 = R_0\, ae^{b/T}$

$$\frac{b}{T} = \log_e R_1 - \log_e (R_0 \times a)$$

$$T = \frac{b}{\log_e R_1 - \log_e (R_0\times a)}$$

$$= \frac{2850}{\log_e 1050 - \log_e (4000\times 300\times 10^{-6})}$$

$$= \frac{2850}{6.956 - 0.1823} = 420\ \text{K}$$

Example 3.70: Calculate the maximum velocity of air immediately upstream of a temperature sensing device if the error due to impact in measuring the temperature of moving air is not to exceed 3.5°C. Assume a recovery factor of 0.95.

Solution: During impact the kinetic energy of the moving air stream is transformed into internal energy, resulting in a local rise in temperature at the measuring probe. The resulting stagnation temperature T, and the acutal air temperature T_d are related by the expression.

$$T_g - Ta = K\frac{V_1^2}{2c_p} \quad \text{where } K \text{ is the recovery factor}$$

Substituting the values, we obtain.

$$3.5 = \frac{0.95V^2}{2\times 1005};\ V = 85.84\ \text{m/s}$$

Example 3.71: During a experimetn to measure the temperature of hot gas flowing through a large pipe with small thermocouple located centrally in the pipe, the following data were recorded under steady state conditions.

Temperature of tube wall = 340°C; indicated temperature thermocouple = 450°C; emissivity of thermocouple = 0.85; convective film coefficient between gas and thermocouple = 46.5 W/m² K⁴.

Determine the true gas temperature. Take radiation constant for black body = 5.67×10^{-8} W/m².

Solution: A part of the convective heat flow from the gas to the thermocouple would be lost by radiation to tube wall. As such, temperature indicated by the thermocouple would be less than the acutal gas temperature.

Tube wall temperature $T_w = 350 + 273 = 623$ K

Temperature indicated by thermocouple

$$T_c = 450 + 273 = 723\ \text{K}$$

Let T_g be the true gas temperature in °K

Convective heat flow from gas to thermocouple

$$= hA(T_g - T_c) = 46.5A(T_g - 723) \quad (1)$$

Heat radiated by thermocouple to tube wall

$$A\sigma_0 \in (T_0^4 - T_w^4)$$

$$= 5.67\times10^{-8} A\times 0.85[(723)^4 - (623)^4] \quad (2)$$

For steady state condition, equate expressions (1) and (2) and get

$$T_g - 723 = \frac{5.67\times10^{-6}\times0.85}{46.5}\left[(723)^4-(623)^3\right]$$
$$=127$$

$\therefore$ True gas temperature $T_g = 723 + 127 = 850$ K

Note: The heatradiated by the gas to the thermocouple is very small as compared with convective heat flow and so has been neglected.

Example 3.72: A 100Ω strain gauge is bonded to a low carbon steel bar which has been subjected to a tensile load. The bar has a preload uniform cross-sectional area of 0.5×10^{-4} m^2 and Young's modulus for low carbon steel is 200 GN/m^2. If a load of 50 kN produces a change of 1Ω in the gauge resistance, determine the gauge factor for the strain gauge.

Solution: Within elastic limits

$$\text{Young's modulus} = \frac{\text{stress}}{\text{strain}}$$

$$200\times10^9 = \frac{50\times10^3/(0.5\times10^{-4})}{dl/l}$$

$$\therefore \quad \frac{dl}{l} = \frac{50\times10^3}{0.5\times10^{-4}\times200\times10^9} = 0.005$$

The gauge factor of a rdsistance strain gauge is defined as the fractional change in gauge resistance divided by the strain applied to it. That is

$$F = \frac{dR/R}{dl/l} = \frac{1/100}{0.005} = 2$$

Example 3.73 A metallic electric resistance strain gauge of 120Ω resistance with gauge factor of 2 is cemented onto an alumnium test surface. If the yield stess is 200 × 10^6 N/m^2 and Young's sodulus of elasticity is 700 × 10^9 N/m^2, determine the change in resistance of the gauge that would be causedby loading the material to yield point.

Solution: The stain at trhe yield point would be:

$$\text{Strain}=\frac{\text{yield point stress}}{\text{Young's modulus}}$$

$$\frac{dl}{l} = \frac{200\times10^6}{700\times10^6}0.286\times10^{-2}$$

This is the maximum possible strain that can be applied to the test-specimen.

$$\text{Now, gauge factor } F = \frac{dR/R}{dl/l}$$

$\therefore$ Change in the strain gauge resistance,

$$dR = F\frac{dl}{l}R$$
$$= 2 \times (0.286 \times10^{-2}) \times 120 = 0.686\Omega$$

Evidently the change in resistance is very small; only about 0.57% over the whole possible region of strain. Such small changes must be accurately determined by employing a wheatstone bridge circuit incorporating very sensitive galvanometer.

Example 3.74: A strain gauge potentiometer circuit uses a strain gauge of 120Ω resistance and a gauge factor 2. When working under maximum sensitivity conditions, the current flow through the system is limited to 20mA. Calculate the supply voltage and the change in output voltage when a stain of 500 microstrain is mpressed upon the gauge.

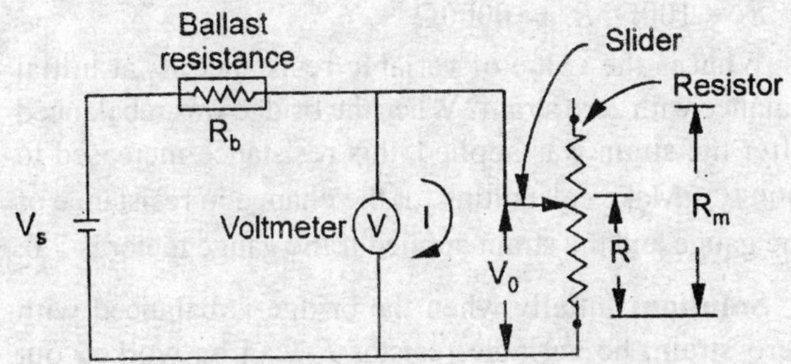

Solution: Refer figure. The current following through the system is given by:

$$i = \frac{V}{R+R_b}$$

For maximum sensitivity $R = R_b$ and therefore

$$i = \frac{V_s}{2R}$$

and su;;ly voltage

$$V_s = 2iR = 2\times(20\times10^{-3})\times120 = 4.8\text{V}$$

When strained, the change in output voltage is:

$$dV_0 = \frac{V_s}{4}F\,\varepsilon$$

$$= \frac{4.8}{4}\times2\times(500\times10^{-6}) = 1.2\times10^{-3}$$

V = 1.2 mV

Example 3.75: In a strain-gauge Wheatstone bridge, figure the four limbs were:

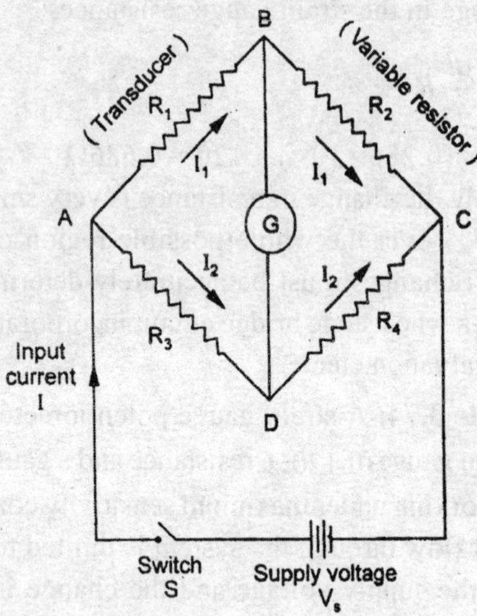

R_1 = =99.75 (strain gauge resistance);

R_2 = variable

$R_3 = 100\Omega$; $R_4 = 10000\Omega$

What is the value of variable resistance R_2 at initial balance with zero strain? When the bridge was rebalanced after the strain was applied, this resistance increased to 9997Ω. Make calcualtins for the change in resistance of the gauge and the strain applied if the gauge factor is 2.0.

Solution: Intially when the bridge is balanced with zero strain, he variable resistor R_2 can be worked out from the equality:

$$\frac{R_1}{R_2} = \frac{R_3}{R_4}$$

$$R_2 = R_1 \times \frac{R_3}{R_4} = 99.75 \times \frac{10000}{100} = 9975\Omega$$

When strain is applied, let dR_2 be the change in variable resistance R_2 to rebalance the bridge. Then

$$dR_2 = 9997 - 9975 = 22\Omega$$

Change in resistance dR_1 of the strain gauge R_1 is then given by:

$$dR_1 = dR_2 \times \frac{R_3}{R_4} = 22 \times \frac{100}{10000} = 0.22\Omega$$

By definition: Gauge factor $F = \dfrac{dR_1 / R_1}{dl / l}$

$$\therefore \text{ Strain } \frac{dl}{l} = \frac{1}{F}\frac{dR_1}{R_1} = \frac{1}{2.0} \times \frac{0.22}{99.75} = 1.103 \times 10^{-3}$$

It would be a tensile strain since $R2$ and hence R_1 have increased.

Example 3.76: A 120 ohm resistance strain gauge having a gauge factor pf 2.0 has been mounted on a steel bar and it froms one arm of a symmetrical Wheatstone bridge circuit. When thesteel bar is subjected to a tensile load, the galvanometer indicates and output voltage equivalent to 5mV. Wokout the mechanical strain if the gauge specifications suplied by the manufactures indicate and operating current of 15 mA.

Solution: Voltage across one resistor

$$= IR = (15 \times 10^{-3}) \times 120 = 1.8 \text{ V}$$

Since there are two resistors in each parallel path, the total exciting volatage will be twice the voltage across each resistor.

Exitation (supply) voltage = 1.8 × 2 = 3.6 V

For the quarter bridge (*i.e.*, one active strain gauge), the output voltage is given by

$$V_0 = \frac{V_s}{4}\frac{dR}{R} = \frac{V_s}{4}F\frac{dl}{l}$$

$$\therefore \text{ Strain } \frac{dl}{l} = \frac{4V_0}{V_s F}$$

$$= \frac{4 \times (5 \times 10^{-3})}{3.6 \times 2} = 2.78 \times 10^{-3}$$

Example 3.77: An electrical resistance strain gauge of resistance 100Ω and gauge factor 2.0 is bounded to steel haveing an elastic-limit stress of 400 MN/m^2 and modulus of elasticity 200 GN/m^2. Determine resistance change of the gauge due to (*i*) change of stress equal to 10% of elastic range (*ii*) an increase of 25°C in temperature. The gauge is made of advance alloy for which temperature coefficient of resistance.

$\alpha_g = 20 \times 10^{-6}$ per °C

temperatue coefficient of linear expansion $\lambda_g = 16 \times 10^{-6}$ per °C

For steel, the temperature coefficient to linear expansion is $\lambda_s = 12 \times 10^{-6}$ per °C

Solution: Change of stress

$$\alpha = \frac{10}{100} \times (400) = 40 \text{ MN/m}^2$$

$$\text{Change of strain } \varepsilon = \frac{\sigma}{E} = \frac{40 \times 10^6}{200 \times 10^9} = 2 \times 10^{-4}$$

$\therefore dR_1 = F \, \varepsilon \, R$

$$= 2 \times (2 \times 10^{-4}) \times 100 = 4.04\Omega$$

(*ii*) Change in resistance due to temperature coefficient of resistance of the gauge is

$dR_2 = R\alpha_g\, dt$

$= 10 \times (20 \times 10^{-6}) \times 25 = 0.05\Omega$

(*iii*) Change in resistance due to differential expansion of the gauge metal and the steel test-specimen is

$dR_3 = FR\,(\lambda_s - \lambda_g)\, dt$

$= 2 \times 100\,(12 - 16) \times 10^{-6} \times 25 = -0.02\Omega$

It is to be noted that changes of resistance due to measured strains and due to temperature changes are of the same order. Evidently, the strain measurements under varying temperature conditions whould be hopelessly inaccurate unless means for temperature compensation are provieded.

Example 3.78: A load of 26500 N was applied to a tension member and strains of 1600 and 450 microstrain were measured in the axial and transverse directions respectively. Presuming that the tension member has an area of 125 mm², make calculatins for modulus of elsticity and possions's ratio for the material of the tension member.

Solution: Axial stress

$$\sigma_x = \frac{26500}{125 \times 10^{-6}} = 212 \times 10^6 \text{ N/m}^2$$

Transverse stress $\sigma_y = 0$

$$\text{Axial strain } \varepsilon_x = \frac{1}{E}(\sigma_x - \mu\sigma_y) = \frac{\sigma_x}{E}$$

$$\therefore \text{ Modulus of elasticity } E = \frac{\sigma_x}{\varepsilon_x}$$

$$= \frac{212 \times 10^6}{1600 \times 10^{-6}} = 132.5 \times 10^6 \text{ N/m}^2$$

$= 132.5 \text{ GN/m}^2$

Strain in the transverse direction,

$$\varepsilon_y = \frac{1}{E}(\sigma_y - \mu\sigma_x) = -\frac{\mu\sigma_x}{E}$$

$$-450 \times 10^{-6} = -\mu \times \frac{212 \times 10^6}{132.5 \times 10^9}$$

Upon solution, we get

Poission's ratio $\mu = 0.281$.

Example 3.79: An osciloscope displays a sine-wave and the distance between the first and the fourth peaks is found to be 5.4 cm. If the time-base setting is 20×10^{-3} s/cm., make calculations for the periodic time and requency of the sine wave.

Solution: We have complete cycles between the first the fourth peak.

∴ Distance for one cycle

$$= \frac{5.4}{3} = 1.8 \text{cm}$$

Periodic time $T = 20 \times 10^{-3} \times 1.8 = 36 \times 10^{-3}$ s

Example 3.80: A rotating disc has five equispaced radial lines marked on it. When a stroboscope is directed at the disc, a true pattern is observed at the highest flash frequency equal to 3000 flashes per minute. What sill be the other flash frequencies w hich produce (*i*) a 5-line pattern (*ii*) a 10-line pattern?

Solution: (*i*) A 5-line patten would be formed at 3000 flashes per minute which is the highest flas frequency. This corresponds to the speed to the speed at which the disc is rotating. An identical pattern would also result when the rotational speed is any whole multiple of flashing frequency thus: $f_r = nf_f$

where $n = 1, 2, 3, 4, 5$

Therefore, the other flash frequencies which give a 5 line pattern are:

$$\frac{3000}{2} = 1500;\ \frac{3000}{3} = 1000;\ \frac{3000}{4} = 750 \text{ and}$$

$$\frac{3000}{5} = 600 \text{ flashes per minute.}$$

(*ii*) A 10-line pattern would be obtained at flash frequencies whose magnitude is twice of those which give the 5-line pattern. These are:

2000 and 1200 flashes per minute. Here the flash frequencies of 3000 and 1500 flashes per minute have not been included because at these flash frequencies a 5-line pattern result.

Example 3.81: (*i*) A strobscope projects 6000 flashes per minute on a disk mounted on the shaft of a machine. Find the speed of the machine if the disk appears stationary and has a singl image of 10 points.

(*ii*) What would be the two possible shaft speeds if the 10 points appear to be revolving once in 12 seconds? How can you determine which is correct?

Solution: (*i*) When the points pattern appears stationary, he shaft speed is given by:

$$\text{Speed} = \frac{\text{number of flashes per minute}}{\text{number of point patterns}}$$

$$= \frac{6000}{10} = 600 \text{ rpm}$$

(*ii*) Since the points appear to be revoloving once in 12 seconds or 5 times in one minute, we have:

Slip = 5 rpm

Possible shaft speed = 600 ± 5 = 605, 595 rpm

The shaft will be rotating at 605 rpm if the points patten appear to move forward in the direction of rotation.

Exmaple 3.82: The speed of a tubocharger ws measured by a stroboscope and for that a radial mark was made on the rotating shaft. The synchronism ws attained for the highest rate of flashing and subsequently the flashing rate ws reduced and a single image observed at reduced flash rates. Calculate the speed of the turbocharger if synchronisation is achieved forstoboscope settings of 3600, 1800, 1200, 900 and 720 rpm.

Solution: The actual speed is given by

$$f_r = \frac{f_1\, f_n\,(n-1)}{f_n - f_1}$$

where f_n = highest flashing frequency = 3600 rpm

f_1 = lowest flashing frequency = 720 rpm

n = number of flashing frequencies = 5

$$\therefore\ f_r = \frac{720\times 3600\times(5-1)}{3600-720}$$

$$= \frac{720\times 3600\times 4}{2880} = 3600 \text{ rpm}$$

Example 3.83: Determine the following vibration amplitudes in decibels of a certain mechanism.

(a) measured vibration velocity amplitude is 3.125 m/s and the reference level is 10^{-3} m/s.

(b) measured vibration acceleration amplitude is 1.5g and the reference level is 10^{-5} m/s^2.

Solution: The vibration level expressed in decibels is given by

$$\text{vibration level} = 120\ \log_{10}\frac{A_1}{A_0}\text{dB}$$

where A_1 is the measured amplitude and A_0 isthe reference amplitude.

∴ velocity vibration level

$$= 20\ \log_{10}\frac{3.125}{10^{-3}} = 69.897 \text{ dB}$$

acceleratin vibration level

$$= 20\ \log_{10}\frac{1.5\times 9.81}{10^{-5}} = 123.355 \text{ dB}$$

Example 3.84: A body is stated to have vibrational acceleration amplitude of 16g at 8 Hz. Make calculations for the following:

(a) the displacement and velocity amplitude,

(b) the velocity vibrational level.

Solution: $\omega = 2\pi f = 2\pi\times 850.24$ rad/s

acceleratopm amplitude

= 16g = 16 × 9.81 = 156.96 m/s^2

$$\text{velocity amplitude} = \frac{\text{acceleration amplitude}}{\omega}$$

$$= \frac{156.96}{50.24} = 3.124 \text{ m/s}$$

$$\text{displacement amplitude} = \frac{\text{vilocity amplitude}}{\omega}$$

$$= \frac{3.124}{50.24} = 0.0622 \text{ m}$$

(b) velocity vibration level = $20\log_{10}\frac{A_1}{A_0}$

$$= 20\log_{10}\frac{3.124}{10^{-3}} = 69.89 \text{ dB}$$

The above calculation is with respect to velocity reference level of 10^{-3} m/s.

Example 3.85: A shaft transmits a maximum power of 50 kW when runnings at constant speed of 1500 rpm. Measurements of torque are made by a pair of strain gauges which are bounded onto a specially machined portion of the shaft. Each gauge has a nominal resistance of $R = 120\Omega$, gauge factor $F = 2.0$ and are connected electrically to the two arms of a half-activated wheatstone bridge circuit which is energised with an exitation voltage of 6 volts. If the gauges have a maximum strain of 0.0015, calculate the shaft diameter. The modulus of elasticity of the shaft material is 200 GN/m^2.

Also work-out the output voltage and the sensitivity of trhe measuring system.

Solution: Power $P = 2\pi\, NT$

$$\therefore\ \text{Torque, } T = \frac{P}{2\pi N} = \frac{50\times 1000}{2\pi\,(1500/60)} = 318.47 \text{ Nm}$$

Maximum strain $\varepsilon = 0.0015$

Maximum stress $f_s = E \in$

$= 0.00015\times\ (200\times 10^9) = 300\times 10^6$

Invoking the relation,

$$\frac{T}{I_p} = \frac{f_s}{r}$$

$$\frac{318.47}{\frac{\pi}{32}d^4} = \frac{300\times10^6}{\frac{d}{2}}$$

$$\therefore \text{ Shaft diameter } d = \left(\frac{16}{\pi}\times\frac{318.47}{300\times10^6}\right)^{\frac{1}{3}}$$

$$= 0.0182 \text{ m}$$

(*b*) Fractional change in resistance,

$$\frac{dR}{R} = F\in = 2\times0.0015 = 3\times10^{-3}$$

The output voltage of an equal arm bridge using two active elements (half-activated bridge) is

$$dV_0 = 2\left(\frac{dR}{R}\frac{V_s}{4}\right) = 2\times\left(3\times10^{-3}\times\frac{6}{4}\right)$$

$$= 9\times10^{-3}\text{V} = 9 \text{ mV}$$

Overall sensitivity of the bridge,

$$= \frac{9}{50} = 0.8 \text{ mV/kW}$$

MCQ Set–1

1. The **transducer** element in a measurment system server to.
   (a) Codify the input signal
   (b) Indicate the measurment
   (c) amplify and transmit the measurment
   (d) Convert the input signal from one physical form into another
2. The intermediate stage of an instrument is used to
   (a) Sense the primary signal
   (b) Condition and transmit the signal after it has been transduced.
   (c) Provide a record or indication of the measurement
   (d) Give a signal that the desired value has been reached.
3. The output stage of a generalised measurement system may comprise a.
   (a) detector - transducer unit
   (b) transmission element to telemeter the data for remote reading and recording.
   (c) indicating or recording unit
   (d) any one of the above depending upon the nature and application of measurment.
4. Which of the following **does not fit into** the functions of an instrument?
   (a) acquistion of information by sensing and perception.
   (b) Processing of the acquired information.
   (c) Presentation of measured data to a human observer.
   (d) Changing of the signal to be measured
5. Which of the following is a **wrong statement**?
   (a) Measurement systems seldom provide exact in formation of the measured quantity.
   (b) Any measurment is expressed by a numerical value alone
   (c) The information collected from the instrumentation system (output) is known as measurment.
   (d) Analog instruments cost less and are easy to maintain and repair.
6. Indicate the **secondary systems** which involve only one traslation.
   (a) Use of beam balance to measure mass
   (b) Matching of two lenghts when determining the length of an object with a rular.
   (c) Pressure measurement by a piezometer
   (d) Measurment of temperature by a themocouple.
7. The measurment of the speed of a rotating shaft by means of an electrical tachometer is a
   (a) direct measurement
   (b) non - contact type measurement
   (c) secondary measurment
   (d) tertiary measurment
8. The term **registering** as applied to a measuring system means.
   (a) conversion of signal from one form into another.
   (b) performing of certain manipulations before the results are indicated or recorded.
   (c) indication of the value of some quanitity by number or by some other symbol of discrete incre ments.
   (d) getting a continuous record of the quanitity being measured.
9. In the digital instruments the physical variables are represented by digits which are discrete and vary in steps. Which of the following devices gives a digital output?
   (a) spring balance
   (b) thermocouple or thermopile
   (c) bourdon tube pressure gauge
   (d) odometer of an automobile
10. The most aproporiate clause pertaining to the development of digital instruments is
    (a) fast reading
    (b) no observation error
    (c) better accuracy
    (d) possibility of output being fed to a memeory circuit
11. The **proximity or non - contact instrument** is one which measures the desired input even - throrght it may not be in physical contact with the system being investigated. Which of the following devices constitutes a proximituy instrument?
    (a) strain gauge load cell
    (b) stroboscope
    (c) clinical thermometer
    (d) dial gauge
12. Tick the odd one out

(a) mercury - in - glass thermometer
(b) bimetallic strip thermometer
(c) optical radiation pyrometer
(d) resistance thermometer

13. In a **self - generated instrument**, the energy requirements are met entirely by the input signal it self. Which of the followiing is a self - generating type instrument?
(a) piezo - electric pickup
(b) thermistor
(c) potentiometric device
(d) linear variable differential transformer

14. Indiate the **externally - powered** type of instrument transducer.
(a) spring balance
(b) centrifugal tachometer
(c) thermo - couple or thermopile
(d) photo emissive cell

15. A **null type instrument** generates an equivalent opposing effect to nullify the physical effect casued by the quantity being measured. The equivalent null causing effect then provides a measure of the quantity. A mongst the following, an instrument with null output is
(a) bourdon gauge
(b) mercury manometer
(c) platform weighing scale
(d) milivoltmeter used to measure the emf generated in a thermo - couple circuit

16. A high grade set of slip gauges preserved in a factory and not put into general use would by a standard of the type.
(a) international (b) primary
(c) secondary (d) working

17. Choose the **false** statement.
(a) any measurement is expressed by a numerical value along with the appropriate unit.
(b) the conversion of force into displacement by means of a spring is a typical example of primary mode to measurment.
(c) the tempeature emeasurement by a mercury - in - glass thermometer is automatic.
(d) the anlog instruments cost less and are easy to meaintain and repair.

18. All of the following statements are correct, **except**.
(a) most of the variable to be measured originate in analoge form.
(b) digital signals are noise resistant during trasnmission.
(c) a mercury - in - glass thermometer has in - built amplifying element.
(d) deflection type instrument are more accurate than the corresponding null type instruments

19. Pointout the device that refer to **self - generating** trasnducer.
(a) resistive (b) capacitive
(c) piezoelectric (d) thermometer

20. The **active transducer** which can be used for linear or angular velocity measurments depends upon:
(a) generation of force by allowing current to flow through the conductor.
(b) variation in mutual inductance of the coils
(c) movement of conductor through a magnetic field
(d) variation in capacitance of a capacitor

21. The LVDT is an inductive trasnducer which function due to:
(a) change in the air gap
(b) change in the amount of core material
(c) mutual inductance
(d) variation in the position of the core

22. Specify the photo - electric device which converts the light information to resistance information :
(a) photo - emissive cell
(b) photo - conductive cell
(c) photo - voltatic cell
(d) All of the above

23. When certain natural or artifcial crystals are deformed, an electric charge is generated. This characteristics is referred to as:
(a) thermo - electric effect
(b) capacitive effect
(c) electro - magnetic effect
(d) piezo - electric effect

24. Specify the variable in a capacitive transducer that does not necessitate a phycisal contact between the trasnducer and the measurand :
(a) effective or overlapping area of plates
(b) distance between plates
(c) dielectric constant of the insulator
(d) All of the above

25. Specify the transducer which is generlly used for dynamic rather that for static measurements.
(a) capacitive (b) resistive
(c) piezo - electricq (d) inductive transducer

26. A potentiometer produce large variation in resistance by:
(a) moving a conductor through a magnetic field
(b) moving a slider across a resistor
(c) streching a metal wire
(d) thermally expanding a conductor

27. The resolution of a wire - wound voltage - dividing displacemet transducer can be improved by:
(a) increasing the applied voltage
(b) reducing the diameter of resistance wire
(c) reducing the applied voltage
(d) increasing the coil diameter

28. Small resistance changes occuring in temperature or strain measuring system are usually measured with:
(a) voltage-sensitive input circuit
(b) current sensitive input circuit
(c) ballast circuit
(d) wheat stone bridge circuit

29. A piezo - ele ctric transducer has all the following advantage except

(a) small size and hgih output
(b) neligible phase shift
(c) good frequency response
(d) capbility to measure both static and dynamic quantities

30. All of the following statements with reference to LVDT, are correct expect one. Indentify that statement. LVDT
(a) works on the principle of mutual induction
(b) is a self - generating type of transducer
(c) cannot be used for the measurement of static caribles.
(d) stands for linear varible differential transformer

31. One atmospheric pressure is not equivalent to
(a) 1.013 $kg\ l\ cm$ (b) $10^3\ N/m^2$
(c) 760 mm of Hg (d) 10.33 m of water column

32. A barometer measures
(a) absolute pressure (b) gauge pressure
(c) vacuum pressure (d) atmospheric pressure

33. Identify the pressure sensing element commonly used in **ameroid barometer**
(a) flat diaphragm (b) bellows
(c) capsule (d) sprial tube

34. A piezometer tube is used to measure the
(a) undisturbed fluid pressure
(b) gauge pressure in a static mass of fluid
(c) pressure difference between two fluids
(d) dynamic pressure of a moving stream

35. A **piezometer tube** is not used for pressure measurements in a pipe- line when.
(a) the fluid is highly viscous
(b) the pressure difference is large
(c) the flow velocity is small
(d) the flowing fluid is a gas

36. The **stagnation (total) pressure** at a point is measured by
(a) static tube (b) piezometer tube
(c) impact tube (d) pitot - static probe

37. A **pitot - static** tube measures
(a) undisturbed fluid pressure
(b) dynamic pressure of a moving stream
(c) pressure differennce between two fluids
(d) difference between the dynamic and static pressure

38. Small pressure changes is low velocity flows can be measured accrately by
(a) U - tube manometer
(b) single column manometer
(c) compound U - tube manometer
(d) draft gauge

39. A U - tube differential manometer is
(a) always used upright
(b) used upright when pressure difference is large
(c) used inverted when pressure difference is small
(d) used inverted when pressure difference is large

40. **A well - type manometer** is used in perference to a simple U - tube manometer to obtain
(a) better accuracy (b) better precision
(c) high sensitivity (d) high resolution

41. Which of the following type of manometer is likely to have the highest sensitivity and accuracy?
(a) U - tube manometer
(b) inverted U - tube manometer
(c) single column manometers
(d) U - tube double reservoir manometer

42. Mercury is used in barometers because it
(a) does not strick to the walls of the tubing
(b) is shining and can be read only
(c) has practically zero vapour pressure
(d) has high dinsity

43. The range of pressure for a particular bourdon tube is determined by the
(a) graduations on the scale
(b) geometry of the mechanism
(c) strength of the return spring
(d) material and thickness of the tubing

44. The elstic element used for converting pressure to deformation in a conventional bourdon tube gauge is of
(a) circular cross-section
(b) elliptical cross - section
(c) square cross - section
(d) rectangular cross - section.

45. Which of the following type of bourdon tube shape has a small tip travel and necessitates amplification?
(a) C - type (b) spiral
(c) helical (d) twisted tube

46. In a bourdon tube pressure gauge, incorrect readings may be encountered due to
(a) hyswteresis (b) fatigue
(c) creep (d) drift

47. The angular error in a bourdon tube refer to
(a) pointer becomes loose on the spindle
(b) gauge tends to give a progressively higher or lower reading
(c) gauge reads correctly both at maximum and minimum
(d) gauge reads correctly onluy at the mid - point scale

48. Most common material chosen for the fabrication of bellow pressure gauge is
(a) leather (b) plastic
(c) brass and bronze (d) animal membrane

49. Which of the following cannot be used for the measurement of negative pressures ?
(a) piezometer tube (b) pirani gauge
(c) bourdon gauge (d) U - tube manometer

50. All of the following devices are used for measuring vacumm, except
(a) kenometer
(b) hydrometer

(c) pirani gauge
(d) thermal conductivity gauge

51. Measurment of low pressure are made by sensing changes in the resistance of a heated filament exposed to that pressure. Which of the following gauges utilze this principle?
(a) ionsation gauge
(b) pirani - gauge
(c) thermocouple vacuum gauge
(d) Mclead gauge

52. Signal conditioning is carried out by the capilary tubes which convert gas pressure into a mercury height. This statement pretains to
(a) Mclead pressure gauge
(b) diaphragm pressure transducer
(c) pirani gauge
(d) thermocouple vacuum gauge

53. Which condition is essential to prevail in the operation of Mclead gauge : an instrument used for vacuum measurement?
(a) isobaric (b) isochoric
(c) isothermal (d) adiabtic

54. High vacuum pressure is most commonlu expressed in
(a) cm of water (b) pascal
(c) torr (d) microm

55. Suggest the instrument for measuring ultra high vacuum; 10^9 torr and beyond
(a) U - tube double reservoir manometer
(b) Mclead gauge
(c) Pirani gauge
(d) Hot filament ionisation gauge

56. A dead weight tester is used for
(a) producing high pressures
(b) accurate measurements of load
(c) testing the magnitude of given weight
(d) calibrating pressure measuring instruments

57. A pitot - static tube measures :
(a) static pressure (b) dynamic pressure
(c) total pressure (d) differencial pressure

58. The local velocity at a point in channel or duct (through which fluid is flowing) is measured by:
(a) current meter (b) flow nozzle
(c) vane anemometer (d) pitot - static probe

59. Of the varios methods of measuring discharge through a pipe line, the one with the least loss of energy and direct readings is by mean of:
(a) transversing a pitot static tube
(b) orifice meter
(c) notch
(d) venturimeter

60. A horizontal pipe line conveys constant rate of flow which is measured by a venturimeter located in it. When the pipe is inclined upward in the direction of flow, the deflection on a differential manometer will:
(a) increase
(b) decrease
(c) remain same
(d) provide erroneous result

61. Which of the following instruments is a **rate meter**?
(a) venturimeter (b) hot wire anemometer
(c) nutating disk meter (d) current meter

62. The **hot wire anemometer** is used to measure:
(a) pressure in gases
(b) liquid discharges
(c) gas vbelocities
(d) wind vellocities at airports

63. The expansion factor Y depends upon:
(a) $\gamma, p_2 / p_1$ and A_2 / A_1
(b) γ, R and A_2 / A_1
(c) $R, p_2 / p_1$ and A_2 / A_1
(d) γ, R and p_2 / p_1

64. The fluid flow between the electrodes of an electromagnetic flow meter generates an emf which is a function of:
(a) dynamic pressure
(b) discharge
(c) flow velocity
(d) All of the above

65. A quantitative estimate of density changes can be made by:
(a) interferometer (b) shadograph
(c) schlieren system (d) any of the above

66. In a venturimeter, perssure at the throat is
(a) less than in the enetrance pipe
(b) greater than that in the entrance pipe
(c) equal to that in the rentrance pipe
(d) independent of the rate of flow

67. The range of coefficient discharge for a **venturimeter** is
(a) 0.6 - 0.7 (b) 0.74 - 0.8
(c) 0.8 - 0.9 (d) 0.95 - 0.97

68. For a verturimeter, the coefficient of discharge
(a) does not depend no the Reynolds number
(b) increase with increase in Reynolds number
(c) decrease with increase in Reynolds number
(d) increase upto a certain value of Reynolds

69. A venturimeter is preferred to an orifice palte because
(a) it is cheaper and eassy to install
(b) its coefficient of discharge has a constant value
(c) energy or head loss is less
(d) accurate measurements of discharges can be made

70. Postive displacement flow meter are
(a) the quantity flow meters
(b) the constant area - variable head meters
(c) the constant head - varible area meters
(d) the varible area - varible head meters

71. Which of the following instrumens is not a **rate meter**?
(a) weirs and flumes (b) orifice plate
(c) nutating disk (d) rotameter

72. Cylinder and piston type flow meters are
(a) variable area flow meters
(b) varible head flow meters
(c) vortex flow meters
(d) positive displacement flow meters

73. Which amongst the follwing is a constant head varible area flow meter?
(a) Venturimeter (b) weir
(c) vane anemometer (d) rotameter

74. Which of the following is not suitable for measuring the flow of a non - conducing fluid ?
(a) orifice plate
(b) nutating disk flow meter
(c) electro - magnetic flow meter
(d) rotameter

75. Which amongst the following gives the flow rate as well as totalised flow?
(a) flow nozzle
(b) nutating disk flow meter
(c) electro - magnetic flow meter
(d) weirs and flumes

76. The fluid density does not affect the working of
(a) orifice plate
(b) rotameter
(c) electro - magnetic flow meter
(d) pitot - static probe traverse

77. The tube of a rotameter
(a) tapers in the upward direction
(b) tapers in the downward direction
(c) tapers in the horizontal direction
(d) is of uniform cross - section

78. The hot wire anemometer used for measuring gas velocities is a variable
(a) resistance transducer
(b) inductance transducer
(c) capacitance transducer
(d) frequecy transducer

79. Which of the following flow meters gives good results when the fluid being metered is exceptionally clean ?
(a) hot wire anomometer
(b) turbine flow meter
(c) ultrasonic flow meter

80. Indicate the correct statement
(a) The cup tupe anemometers are used at high speeds and the vane type at low speeds
(b) In the order of diminishing losses we prefer nozzle, thin orifice plate and enturimeter for flow measurment.
(c) The totalisation of flow is possible is turbine flow meters
(d) Hot wire anemometery uses only the constant resistance type of Wheat stone bridge circuit

81. Choose the wrong statement
(a) The domestic water uses the positive displacement method for flow measurement
(b) For the same flow rate of fluids of different densities a rotameter will give different readings
(c) The venturimeter orifice and nozzle cannot be used for metring flow in the reversed direction
(d) Weirs are used for measuring water flow through open channels and rivers

82. Mark the odd one
(a) orifice plate
(b) hot wire anemometer
(c) electromagnetic - flow meter
(d) nutating disk flow meter

83. Consider the following statements in the context of **flow maters:**
(a) Hot wire anemometer is preferred to measure highly fluctuating velocities.
(b) In an electromagnetic flow meter the induced voltage is proportional to flow rate
(c) The obstruction type flow meters have a linear realtion ship between the pressure difference and the rate of fluid flow
(d) Rotameter is a variable area flow meter and it should always be installed in a vertical pipeline

Select the correct answer using the codes given blow:
(a) 1 and 2 (b) 1, 2 and 4
(c) 2 and 4 (d) 1, 2, 3 and 4

84. The lowest temperature limit for mercuru - in - glass thermometers is.
(a) $-200°C$ (b) $-80°C$
(c) $-35°C$ (d) $-50°C$

85. Which of the following substacnes can be used as thermometric substance for temperature measurements below $-40°C$.
1. alcohol 2. pentane
3. mercury 4. coresote

Select the correct answer using the codes given below
(a) 1 and 2 (b) 2 and 3
(c) 3 and 4 (d) 1, 2 and 3

86. Recording is not possible with
(a) liquid - in - glass thermometers
(b) thermocouples
(c) filled in system thermometers
(d) resistance thermometers

87. The bending of bi - metallic strips during rise in temperature is due to difference in their.
(a) coefficient of linear expansion
(b) elastic proprieties
(c) thermal conductivities
(d) none of these

88. Thermocouples are generally used for temperature measurements upto

(a) $500°C$ (b) $1000°C$
(c) $1500°C$ (d) $2000°C$

89. Which of the following thermocouple can measure temperature in a comparatively high range ?
(a) irom - constant on (b) copper - constant on
(c) chormel - alumel (d) tungsten - molybdenum

90. The electrical output from a thermocouple circuit is detected by using
(a) wheatstone bridge
(b) current sensitive device
(c) voltage - balancing circuit
(d) voltage sensitive device

91. These units are employed where it is not possible to have physical contact with the system whose temperature is to be measured.
(a) thermocouples (b) resistance thermometers
(c) radiation pyrometers (d) thermostors

92. A parallel combination of thermocouples is used for the measurement of
(a) small temperature difference between the two junctions
(b) large temperature difference between the two junctions
(c) average temperature of a number of points
(d) None of the above

93. Which of the following metals is to be preferred in a resistance - thermometer employed for measuring temperture in the range $-200°C$ to $1000°C$?
(a) Platinum (b) Copper
(c) Nickel (d) Tungsten

94. Which of the following temperature measuring device depends for its action upon a change in the electrical resistance of a metallic element.
(a) bolometer (b) thermistor
(c) bi - metallic strip (d) thermoplie

95. Thermistors are:
(a) bi - metallic strips used in thermometers
(b) semi - conductor which have negative coefficient of resistance
(c) metals having small postivie coefficient of resistance
(d) transducers used for measuring the thermal properties of materials

96. Self - compensating capillary tube is used in the filled - ni - system thermometer to eliminiate or minimise.
(a) elevation effect (b) immersion effect
(c) barometric effect (d) temperature effect

97. Which arrangement has the capability to measure the temperature of an object which may be either stationary or moving?
(a) thermocouples
(b) thermistors
(c) total radiation pyrometer
(d) filled - in - system thermometers

98. Indicate the externally - powered type of instrument transducer
(a) spring balance
(b) centrifugal tachometer
(c) thermo - couple or thermopile
(d) photo emissive cell

99. Which of the following thermometers is most suitable for the measurement of surface temperature?
(a) liquid - in - glass thermometer
(b) thermo - couples
(c) thermistors
(d) resistance thermometer

100. Indicate fthe appropriate thermometer to the used for measuring the temperature of a the furnace of a boiler
(a) bi- metallic thermometer
(b) resistance thermometer
(c) thermo - couples arranged in parallel
(d) radiation pyrometer

101. It is required to measure the temperature of a red hot moving material like molten steel. Which of the following devices is most suitable ?
(a) filled - in - system thermometer
(b) themistor
(c) thermo - couples arranged in series
(d) radiation pyrometer

102. Spot the odd one
(a) mercury - in - glass thermometer
(b) bi - metallic thermometer
(c) resistance thermometer
(d) optical purometers

103. Which arrangement has the manual null - balance oepration.
(a) optical pyrometer
(b) resisitance thermometer
(c) liquid - in - glass thermometer

104. Which metal/non - metal has the highest temperature range?
(a) semi - conductors (b) nickel
(c) copper (d) platinum

105. All of the following statements are correct, except
(a) Thermocouple combination in series is used when average temperture is to be measured.
(b) Law of intermediate metal in thermocouple allow them to use extension wires of meterials
(c) Thermistors are semicounductors which have negaitive coefficient of resistance
(d) Bi - metallic strips made of two different materials bend during a rise in temperature on account of difference in their coefficient of linear expansion.

106. Consider the following statements pertaising to the act of temperature measurement:
1. Bi - metallic thermometer is unsuitable d
2. Optical pyrometer is the most suitable thermometer for the measurement of temperature of the wire of an incandescent bulb.

3. Response and senitivity of a thermocouple is improved by using thick wires

select the correct answer using the codes given below:

(a) 1 and 2 (b) 2 and 3
(c) 1 and 3 (d) 1, 2 and 3

107. Mechanical strain gauges can measure:

(a) Static strains only
(b) Dynamic strains only
(c) Static and quasi - static strains
(d) Static and dynamci strains

108. The gauge factor of a resistance gauge depends upon:

(a) the gauge material
(b) the configuration of the gauge wire
(c) the mechanical loading
(d) all the above factors

109. A strain gauge should have high value of gauge factor:

(a) to reduce hysteresis effects
(b) to increase sensitivity
(c) to reduce/eliminate the effect of variation in ambient temperature
(d) to give a linear relation between applied strain and resistance change

110. Which of the following gauges can be detached from the test specimen and used again:

(a) semi - conductor gauges
(b) metal foil gauge
(c) bonded metal wire gauges
(d) unbounded metal wire gauges

111. In a semi - conductor strain gauge, the change in the value of resistance occurs primarily due to change in

(a) only length
(b) only cross - sectional area
(c) length as well as cross - sectional area
(d) resistivity of the strain gauge material

112. The use of semi - conductor strain gauge is limited due to their

(a) high frequency response
(b) high value of gauge factor
(c) small size and high cost auxiliary equipment
(d) poor linearity and sensitivity to temperature changes

113. Which of the following materials is used as the carrier with strain gauge at room temperature?

(a) rubber (b) impregnated paper
(c) bakelite (d) epoxy resisn

114. The use of a dummy strain gauge in a Wheatstone bridge circuit helps to

(a) increase the sensitivity of the measuring instrument
(b) measure lateral strain
(c) eliminate the errors due to variation in ambient temperature
(d) improve the stability of the system

115. Dummy strain gauges

(a) are substandard strain gauges which are connected in limbes which lie opposite to those active gauges
(b) undergo resistance chage due to temperature as well as mechanical strain
(c) are bonded to separate unstrained component and suffer resistance change due to temperature only
(d) poor linearity and sensitivity to temperature changes

116. The bridge constant (sensitivity) to half - bridge arrangement where two strain gauges subjected to equal and opposite strains is:

(a) unity (b) doubled
(c) quandrupled (d) $(1+\mu)$ times

117. While measuring pure torsion, the strains is gauge are bonded to the shaft:

(a) in the direction of the applied torque
(b) longitudinally along to axis of the shaft
(c) transversely to the shaft axis
(d) at $45°$ to the axis of rotation

118. With zero shaft between the two sine waves, the Lissajou figure is:

(a) a straight line with a slope of $45°$ to the right
(b) a circle
(c) a straight line with a slope of $45°$ to the left
(d) None of the above

119. The average speed measurements are given by:

(a) centrifugal tachometer
(b) drag cup tachometer
(c) revolution counter and a timer
(d) Stroboscope

120. Tachometer are used to measure:

(a) displacement (b) angular velocity
(c) vibration and (d) time

121. The speed of a sealed compressor units can be measured by a

(a) strobscope
(b) vibrating reed tachometer
(c) capacitive pick - up tachometer
(d) tachoscope

122. The stroboscope of speed measurement has the advantage that:

(a) the method is simple and straight forward
(b) the multiples of angular speed can be measured
(c) a stationary image can be observed
(d) there is no need of any physical cantact between the instrument and the rotating shaft.

123. Identify the instrument that indicates the rotational speed of a shaft in revolution per minute

(a) tachometer (b) tachosocope
(c) odometer (d) hand speed indicator

124. A disc mounted on a rotating shft has one mark and a stroboscope is used for measurement of shaft speed. If the shaft is turning N revolutions per second, three stationary marks will be seen if the flash frequency f is

(a) $f = n$ (b) $f = 3n$

(c) $f = \frac{n}{3}$ (d) $f = \frac{2n}{3}$

125. Consider the following statements:

1. A phenomenon that repeats periodically with a constant time period can be used for the measurement of time
2. The frequency of a dynamic input signal can be determined by Lissajou diagram on an oscilloscope
3. Photo - electric type speed measurement is a non-contact type analog device
4. Seismic type accelectrimeters are designed with low natural frequency

Select the correct answer using the codes given below

(a) 1 and 2 (b) 1, 2 and 3

(c) 1, 2 and 4 (d) 2 and 3

Answers Set-1

1. (d)	2. (b)	3. (c)	4. (d)
5. (b)	6. (c)	7. (d)	8. (c)
9. (d)	10. (d)	11. (b)	12. (c)
13. (b)	14. (d)	15. (c)	16. (c)
17. (b)	18. (d)	19. (c)	20. (c)
21. (d)	22. (b)	23. (d)	24. (c)
25. (c)	26. (b)	27. (b)	28. (d)
29. (d)	30. (b)	31. (b)	32. (d)
33. (c)	34. (b)	35. (d)	36. (c)
37. (b)	38. (d)	39. (c)	40. (c)
41. (d)	42. (d)	43. (d)	44. (b)
45. (a)	46. (c)	47. (c)	48. (c)
49. (a)	50. (b)	51. (b)	52. (a)
53. (c)	54. (c)	55. (d)	56. (d)
57. (b)	58.(d)	59. (d)	60. (c)
61. (a)	62. (b)	63. (a)	64. (c)
65. (a)	66. (a)	67. (d)	68. (d)
69. (c)	70. (a)	71. (c)	72. (d)
73. (d)	74. (c)	75. (b)	76. (c)
77. (b)	78. (c)	79. (a)	80. (c)
81. (c)	82. (c)	83. (b)	84. (c)
85. (a)	86. (a)	87. (a)	88. (c)
89(d)	90 (c)	91. (c)	92 (c)
93. (a)	94. (a)	95. (b)	96. (d)
97. (c)	98. (a)	99. (b)	100. (d)
101. (d)	102. (d)	103. (a)	104. (d)
105. (a)	106. (a)	107. (c)	108. (d)
109. (b)	110. (d)	111. (d)	112. (d)
113. (b)	114. (c)	115. (c)	116. (b)
117. (d)	118. (d)	119. (c)	120. (b)
121. (b)	122. (d)	123. (a)	124. (b)
125. (c)			

MCQ Set–2

1. The measurement of a quantity is an act of comparison of an unknown quantity with

(a) another quantity.

(b) a known quantity whose sccuracy may be known or may not be knowm

(c) a Pedefined acceptable standard which is accurately knowm.

(d) none of the above.

2. In present day measurement systems.

(a) direct methods are commonly used.

(b) use of direct methods is limited but indirect methods are commonly used.

(c) both direct and indirect methods are commonly used.

(d) all of the above.

3. Purely **mechanical instruments** cannot be used for dunamic measurements because they have:

(a) high inertia (b) large time constant

(c) higher response time (d) all of the above

4. The usage of **electronic Instrument** is becoming more extensive because they have.

(a) a high sensitiveity and reliability.

(b) a fast response and compatibility with digital computers.

(c) the capability to respond to signals from remote places.

(d) all of the above.

5. A **null type of instrument** as compared to a deflection type instrument has

(a) a higher accuracy (b) a lower sensitivity

(c) a faster reponse (d) all of the above

6. In an **intelligent information system** used for the measurement of a physical quantity.

(a) the digital computer must perform all or nearly all the signal and information processing.

(b) the observer must perform all or nearly all the signal and information processing.

(c) the digital computer is not required.

(d) none of the above.

7. In a **generalized measurement system**, the function of the signal manipulating element is to

(a) change the quantity under measurement to an analogous signal.

(b) change the magnitude of the input signal while retaining its identity.

(c) to perform linear operations like addition and multiplication.

(d) none of above.

8. A modifying input to a measurement system can be defined as an input which change the input-output relationship for

(a) desired inputs only.

(b) desired as well as interfering inputs.

(c) interfering inputs only.
(d) none of the above.

9. In a measurement system, the open loop transfer function is 10 and changes by 10 percent. If a closed loop system is adopted where in the feedback system has a transfer function of 100. What will be the change in output in the case of open - loop and in closed loop systems.
(a) 10 percent, 10×10^{-3} percent.
(b) 10 percent 1×10^{-3} percent
(c) 100 percent 10×10^{-3} percent
(d) 100 percent 1×10^{-3} percent

10. In a **permanent magent moving** coil type of ammeter, the current is,

$$I = \frac{K\theta}{NBA}$$

where
θ = deflection,
K = spring constant,
N = number of turns of moving coil,
A = area of moving coil and
B = flux density of field produced permanent magnet.
Temperature is an interfering input to the system. It is known that strength of magnets reduces by 0.02 percent per $^{\circ}C$. Which method should be adopted to apply corrections for the interfering input ?
(a) Method of inherent insensitivity.
(b) Method of high gain feed back.
(c) Method of calculated output corrections.
(d) Method of opposing inputs.

11. In measurement systems, which of the following static **characteristice are disirabel**:
(a) Accuracy (b) Sensitivity
(c) Reproducibility (d) All of the above

12. In measurement systems, which of the following are undesirable static characteristics:
(a) Sensitivity and accuracy
(b) Drift, statoc error anjd dead zone
(c) Reproducibility and non - linearity
(d) Drift, static error, dead zone and non - linearity.

13. A **pressure measurement instrument** is callbrated between 10 bar and 250 bar. The scale span of the instrment is
(a) 10 bar, (b) 250 bar,
(c) 240 bar, (d) 260 bar,

14. The output of an instrument under particular enviromental conditions is given by the relationship $q_0 = a + b\, q_i =$ input and a and b are constants $b \neq 1$. The theoretical input - output relationship is repesented by a stralght passing through zero. This is a case of.
(a) zero drift only
(b) sensitivity drift only
(c) zero and sensitivity drift together
(d) none of the above

15. A reading is recorded as $23.90\, C^{o}$. The reading has significant figures of
(a) three (b) five
(c) four (d) none of the above

16. The scale of a $0-500\,V$ voltmeter is divided into ten large divisions representimg $50\,V$ each and each large division is further subdivided into 10 small divisions, each representing $5\,V$. It is used for measurment of output voltage of a **potpentiometer** which can be varied from 0 to 500 V. It is observed that when the sliding contact is moved from its zero position, there is no perceptible movement of pointer of the voltmeter till then sliding contant reaches a position where the output voltage should be $5\,V$. Therefore, it can be concluded that.
(a) The threshold of the voltmeter is $5\,V$.
(b) The resolution of the voltmeter is $5\,V$.
(c) The sensitivity of the voltmeter is $5\,V$.
(d) None of the above.

17. In the centre zero **analog ammeter** having a range of $-10\,A$ to $+10\,A$, there is a detectable change of the pointer from its zero position on either side of the scale only if the current reaches a value of 1A. The ammeter has a
(a) resolution of 1 A (b) dead zone of 1A
(c) dead zone of 2 A (d) sensitivity of 1A

18. A **pressure gauge** is calibrated from $0-50\text{ kN/m}^2$. It has a uniform scale with 100 scale divisions. One fifth of a scale division can be read with certainty. The gaguge has a:
(a) resolution of $0.1\,kN/m^2$
(b) threshold of $0.1\,kN/m^2$
(c) dead zone of $0.2\,kN/m^2$
(d) a resolution of $0.5\,kN/m^2$

19. A d.c. circuit can be represented by an internal voltage source of 50 V with an output resistance of $100\,k\Omega$. In order to achieve 99 percent **accuracy** for voltage measurement across its terminals, the voltage measuring device should have a resistance of
(a) atleast $10\,M\Omega$ (b) $100\,k\Omega$
(c) atleast $10\,\Omega$ (d) none of the above.

20. A d.c. circuit can be represented by a voltage source of 10 V. in series with as output resistance of $1\,k\Omega$. An ammeter of $50\,\Omega$ resistance is connected to the source terminals for measurement of current. The accuracy of measurement is nearly.

(a) –4.8 percent (b) +4.8 percent
(c) 99 percent (d) 95.2 percent

21. In a.c. circuits, the connection of measuring instruments cause **loading errors** which may effect.
(a) only the magnitude of quantity being measured.
(b) only the phase of the quantity being measured
(c) both the magnitude and phase of the quantity being measured
(d) magnitude, phase and also the waveform of the quantity being measured.

22. The input resistance of a **cathode ray oscilloscope** is of the order of.
(a) tens of ohm (b) megohm
(c) kilohm (d) fraction of an ohm

23. A resistor, at room temperature of $290^\circ K$ has a noise voltage of $2\,\mu V$ for a 50 kHz bandwidth. The **temperature** at which the noise voltage is 20% of its value at room temperture is,
(a) $58^\circ K$ (b) $11.6^\circ K$
(c) $241.7^\circ K$ (d) None of these

24. An amplifier has an input signal voltage of $25\,\mu V$ and a notise $2.5\,\mu V$. The signal to noise ratio is
(a) 10 (b) 100
(c) 10 db (d) none of the above

25. Johonson noise is:
(a) caused by thermal agitation of free electrons carrying current thereby modulating the current.
(b) noise carried into a circuit through conductors
(c) noise of an electromagnetic origin that is radiated into a circuit.
(d) shot noise which results from randoim emission of electrons across a PN junction

26. In a measurement system if a q_1 is an across variable and its associated throuhg variable is q_2 (power passed) the static impedance of the system is :
(a) $q_1 \times q_2$ (b) q_1 / q_2
(c) q_2 / q_1 (d) q_1^2 / q_2

27. In a measurement system 'force' is an across variable. Its associates.
(a) translational displacement
(b) translational
(c) translational velocity
(d) rotational displacement

28. The **power can be controlled** in a circuit by:
(a) controlling the voltage applied to load.
(b) controlling the current level in the circuit.
(c) using pulse modulation
(d) all of the above

29. The value of **static** stiffness and **compliance** in a measurement system determine the amount of:
(a) power drain from a system.
(b) current drain from a system.
(c) energy drain from a system.
(d) none of the above.

30. A **wheatstone bridge** is balanced with all the four resistance equal to $1\,k\Omega$ each. The bridge supply voltage is 100 V. The value of one of the resistance is changed to $1010\,\Omega$. The output voltage is measured with a voltage measuring device of infinite resistance. The **bridge sensitivity** is:
(a) $10\,V/\Omega$ (b) $25\,mV/\Omega$
(c) $2.5\,mV/\Omega$ (d) None of the above.

31. A co - herent system of units is to be evolved which has N kinds of quantities to evaluate and M independent physical quantities expressing relationships between them, we can independently choose the sizes of **units** of:
(a) N + M quantities (b) $N - M$ quantities
(c) N quantities (d) M quantities

32. The **units** whose sizes cannot be choosen independently are called:
(a) Derived units
(b) Fundamental units
(c) Absolute units
(d) Auxlliary fundamental units

33. The dimensions of inductance in CGS e.s. system are:
(a) $[\varepsilon L^{-1} T^{-2}]$ (b) $[\varepsilon^{-1} ML^{-1} T^{2}]$
(c) $[\varepsilon^{-1} LT^{-2}]$ (d) $[\varepsilon^{-1} L^{-1}T^{2}]$

34. The dimensions of resistance CGS e.m. system are :
(a) $[\mu^{-1}LT^{-1}]$ (b) $[\mu^{-1}MLT^{-1}]$
(c) $[\mu LT^{-1}]$ (d) $[\mu LT^{-2}]$

35. The base units in SI system are
(a) metre, kilogramme, second
(b) metre, kilogramme, second, ampere
(c) metre, kilogramme, second, ampere, kelvin , can dela, mole
(d) metre, kilogramme, second, ampere, kelvin, can dela.

36. The **symbol** nm stands for :
(a) newton metre (b) nano metre
(c) nano milli (d) number of moles

37. In measuring the absolute value of ampere using **Rayleigh's balance,** the value of gravitational constant must be accurately known because :
(a) the expression for the force is in newton and weights used in counterbalancing the force are in kg.
(b) value of g is different at different places
(c) corrections have to be made for the weight of moving coil.
(d) none of the above.

38. In **Lorenz method for measurement of ohm**, two measurement are made one with current in one direction and the other with current in the opposite direction. This is done to:
(a) eliminate the effect of thermal emfs.
(b) eliminate the effect of earth's magnetic field
(c) eliminate the effects of both thermal emfs and earth's magnetic field.
(d) eliminate the effects of temperture on coil and disc.

39. The **international standard of lenght** is defined in terms of:
(a) length of earth's meridian passing through paris
(b) distance between two lines engraved on a platinum-iridium bar.
(c) wavelength in vacum, of rediation of krypion - 86 atom in its two specified transitions
(d) none of the above.

40. The most stable **primary** atomic **standand for frequency** is:
(a) Caesium beam standard
(b) Hydrogen maser standard
(c) quarts standard
(d) Rubidium vapout standard.

41. Caeslum bearm standard is normally used primary atomic **standard for frequency** because:
(a) Caesium beam standard has an accuracy 3 parts in 10^{12} has short term stability, low drift and small size.
(b) Hydrogen maser standard, though it the most stable frequency source but has a large size.
(c) Quartz crystal operating frequency changes with changes with change in load capacitance and also teh drift in frequency is asymptotic with time.
(d) all of the above.

42. The disadvantages of atomic standards are:
(a) their accuracy is lower than that of standards based upon astronmical measurements
(b) the time taken for making measurements is long.
(c) The results obtained with atomic standard are not uniform.
(d) all of the above.

43. The maximum current that can be drawn from a saturated Weston cell without damaging it, is.
(a) 100 mA momentarily (b) 100 μA continuously
(c) 100 μA *momentarily* (d) 1 mA momenterily

44. The internal resistance of a saturated Weston cell is of the order of:
(a) a few ohm (b) a few kilo ohm
(c) a few hundred of ohm (d) a fractions ohm

45. The decrease in the value of emf of an unsaturated Weston cell is:
(a) about 1 μV per year
(b) about 1 mV per year
(c) about 5–10 μV per year
(d) about 30 -50 μV per year

46. Zener sources when used as voltage standards:
(a) are reverse biased
(b) can be designed to conver a wide range of volt ages
(c) are immune to the effects of short circuits
(d) all of the above.

47. The output voltage josephon effect standard:
(a) can be changed in a continous manner.
(b) can be changed only in discrete steps.
(c) can be changed continously as well as in discrete steps
(d) cannot be changed.

48. Candela can be defined as luminous intensity in the perpendicular, of a full radiator surface of:
(a) 1/2 59000 square meter
(b) 1/600 square centimetre
(c) 1/5,89,000 square metre
(d) 1/600,000 square metre

49. The material of wires used for making resistance standards is usually:
(a) Manganin (b) Nichrome
(c) Copper (d) Phosphor

50. In order to reduce **skin effect** in resistance standards when they are used no high ferquency a.c.:
(a) the diameter of the resistance wire should be small.
(b) adjacent turns of the resistance coil should carry currents in the opposite direction
(c) they should be provided with four terminals
(d) all of the above

51. A **PMMC** meter has an internal resistance 200 Ω and the current required for its full scale deflection is 50 μA. The meter is capable of measuring, on its own, maximum voltage of:
(a) 5 mV (b) 10 mV
(c) 5 μV (d) 10 μV

52. A **PMMC** meter has an internal resistance of 500 Ω and the current required for its full scale deflection is 100 μA. The power consumed by the meter is:
(a) 5 μW (b) 5 mW
(c) 5 W (d) 0.5 μW

53. An average reading **VTVM** uses one diode connected in series with an external resistance. A high value of series resistance is used so that the instrument should have:
(a) a high input impedance
(b) a linear v - i characteristics
(c) low power consumption
(d) all of the above

54. In a compensated shunt type peak reading **VTVM** a bucking battery with a series adjustable resistance is used for:
(a) providing power for the circuit
(b) zero setting
(c) the purpose of amlifying low level signals
(d) all of the above

55. In a differential amplifler type electric voltmeter using two FETs, a resistance has to be used to reduce the meter current to zero even when no voltage is applied to the circuit. This is necessitated on account of:
(a) mismatch between characteristics of the FET's
(b) difference between tolerance values of resistors used in the circuit even though they are marked nomainally equal.
(c) variations in the operating voltage of the circuit.
(d) both (a) and (b).

56. The restances, R_D, connected in the bridge circuit of a difference amplifler type of Electronic Voltmeter are each $10\ k\Omega$. The value of R_D is much smaller than the value a.c. drain resistance, r_d, of FETs. The transconductance of FETs is 0.005 mho. The value of current through the meter when an input voltage of 1 V is applied to the gate of left FET with the gate of right hand FET grounded, and the meter resistance almost regligible is:
(a) 5 mA (b) 2.5 mA
(c) 10 mA (d) 1 mA

57. In a source follower type electronic voltmeter, the transconductance of FETs is 0.005 mho and the resistance of the meter is $100\ \Omega$. The approximate value of current through the meter, when the input voltage is 1V, is:
(a) 2.5 mA (b) 5 mA
(c) 2 mA (d) 1 mA

58. The power consumption of a d.c. voltmeter using a direct coupled amplifier when measuring a voltage 0.5 V is the order of:
(a) a few watt (b) a few milliwatt
(c) a few microwatt (d) a few nanowatt

59. A chopper stabllized a.c. amplifier may use:
(a) an electro-mechanical chopper
(b) bipolar trnsistors as choppers
(c) MOS field effect transistors as choppers
(d) All if the light activated devices as choppers

50. Electronic voltmeter which use rectifiers employ negative feedback. This is done.
(a) to increase the overall gain
(b) to improve stability
(c) to overcome nonlinearity of diodes
(d) none of the above

61. A voltage having symmetrical square waveshape is applied to an average voltmeter with scale calibrated in terms of rms value of a sinusoidal wave has an error of:
(a) −3.9% (b) +3.9%
(c) −11% (d) +11%

62. A true reading voltmeter uses two thermocouples in order.
(a) to increase sensitivity
(b) that the second themocouple cancels out the no-linear effects of the first themocouple
(c) to prevent drift in the d.c. amplifier
(d) all of the above

63. In an **electronic ohmmeter**, an OP-AMP is used as a:
(a) Summer (b) Multiplier
(c) Buffer amplifier (d) Integrator

64. A voltmeter with a broad band width has:
(a) low noise level and high sensitivity
(b) high noise level and high sensitivity
(c) high noise level and low sensitivity
(d) low noise level and low sensitiviy

65. In circuits of **RF voltmeters**
(a) vacuum tube diodes are used
(b) conventional PN junction doides are used
(c) Schotiky barrier or point contact type diodes are used
(d) all of the above.

66. The source of emission of electrons in a CRT is:
(a) PN function diode
(b) a barium and strntium oxide coated cathode
(c) accelerating anodes
(d) post - acceletating anodes

67. In a CRT the focusing anode is located
(a) between pre-accelerating and accelerating anodes
(b) after accelerating anode.
(c) before pre-accelerating anode
(d) none of the above

68. The deflection of an electron beam on a CRT screens is 10 mm. Suppose the pre-accelerating anode voltage is halved and the potential between deflecting plates is doubled, the deflection of the electron beam will be:
(a) 80 mm (b) 40 mm
(c) 20 mm (d) 10 mm

69. If the distance of screen from a CRT to cener of deflection plates is 15 cm. The length of deflection plates is 2 cm, the distance between plates is 1 cm and the accelerating voltage is 500 V, the deflection sensitivity is
(a) 33.2 V/cm (b) 0.03 cm/V
(c) 66.4 V/cm (d) 0.015 cm/V

70. Post acceleration is needed in a CRO if the frequency of the signal is:
(a) less than 1 MHz (b) more than 1 MHz
(c) more than 10 MHz (d) more than 10 Hz

71. P_1 phsophor material is used for display in CRTs for:
(a) photographic applications
(b) general purpose applications
(c) television applications
(d) all of the above

72. An aquadag is used in a CRO to collect:

(a) primary electrons
(b) secondard emission electrons
(c) both primay and secondary emission electrons
(d) none of the above.

73. A thin aluminium film is usually deposited on the non viewing side fo the phosphor because,
(a) it acts as a heatsink and prevents phoshor burn
(b) the light scatter form the phosphor is reduced
(c) it does not allow the screen to be negatively charged
(d) all of the above

74. During the **retrace time**, the electrons forming the horizontal beam.
(a) move from left to ringht on the screen
(b) move from right to the left on the screen
(c) move from bottom to top of screen
(d) move from top to bottom of screen

75. The bandwith of a CRO is form 0 - 20 MHz. The fastest rise time a sine wave can have to be accurately reproduced by the instrument is
(a) 35 ns (b) 35 μs
(c) 17.5 ns (d) 0.175 μs

76. The horizontal amplifler for a **CRO** should be designed for:
(a) high frequency signals with a fast rise time
(b) high amplitude signals with a slow rise time
(c) high amplitude signals with a fast rise time
(d) low amplitude signals with a fast rise time

77. A vertical amplifler for a **CRO** can be designed for:
(a) Only a high gain
(b) Only a broad bandwidth
(c) a constant gain times band width product
(d) all of the above

78. In an **oscilloscope,** when the unknown signal applierd to the vertical plates is being sychronized with the sweep signal applied to horizontal plates, the pattern seen on the CRO screen moves towards the right. It means that,
(a) the frequency of the signal is lower than that of the sweep signal
(b) the frequency of the signal is greater than that of the sweep signal
(c) the frequency of the signal is equal to the frequecy of sweep signal
(d) none of the above

79. An **oscilloscope** has an input capacitance of 50 pF and a resistance of 2 $M\Omega$ and the voltage divider ratio of 10. The parametrs of a high **impedance probe** are
(a) $C_1 = 5.55\ pF, R_1 = 18\ M\Omega$
(b) $C_1 = 3.33\ pF, R_1 = 9\ M\Omega$
(c) $C_1 = 1.11\ pF, R_1 = 18\ M\Omega$
(d) $C_1 = 5.55\ pF, R_1 = 9\ M\Omega$

80. In a **digital storage oscilloscope**, the input signals are
(a) direcly applied to the oscilloscope
(b) Multilexed, converted to digital form and stored and applied to osilloscope
(c) multiplexed converted to digital form form stored converted to analog form and applied to oscilloscope.
(d) applied to amplifier, stored as analog signals, multiplexed converted to digital form stored in digital form converted to analog form and applied to CRO through an amplifier.

81. The desirable characteristics of a signal source are:
(a) the frequency of the signal should be well known and stable.
(b) the amplitude of a signal source should be controllable from very small to relatively large values.
(c) the output signal should be free from distortion.
(d) all of the above

82. In signal genrators
(a) energy is created
(b) energy is generated
(c) energy is converteed from a simple d.c. source at some specific frequency
(d) all of the above

83. A video frequency oscillator has a frequency range of
(a) 20 Hz to 20 kHz (b) 20 kHz to 30 MHz
(c) d. c. to 5 MHz (d) 2 MHz to 30 MHz

84. It G and H are respectively the gains of forward and feedback paths. Then feedback oscillators have a closed loop gain of
(a) $\frac{G}{1-GH}$ (b) $\frac{G}{1+GH}$
(c) $\frac{G}{1 \pm GH}$ (d) $\frac{H}{1+GH}$

85. In an RC phase shift oscillastor having three stage which have equivalued resistors and capacitors, the amplifer. In order to satisfy, barkhausen criterion should have a minimum gain of
(a) 6 (b) 29
(c) 31 (d) none of these

86. In a **Hartiey oscillator**, the inductor is divided into two parts. The inductor which is on the input of the amplifler has a value of 20 μH while that connected to the out put side of the amplifler has an inductance of 300 μH . If the resistance connected to the input of the amplifler is 15 $k\Omega$, what should be the value of resistance in the feed back path of the amplifler ?
(a) 300 $k\Omega$ (b) 225 $k\Omega$
(c) 1 $k\Omega$ (d) 15 $k\Omega$

87. The pulse rise time is defined as the time taken by the pluse.

(a) to go form 10% to 90% of its amplitude.
(b) to go from 0% to 100% of its amplitude
(c) to go from 0% to 90% of its amplitude
(d) to go from 10% to 100% of its amplitude

88. An Astable meltivibrator
(a) gives one output pulse for every two input pulses
(b) gives a timed output pulse for a trigger input
(c) gives a train of output pulses for a trigger input.
(d) gives four output pulses for a single input pulse.

89. An Astable meltivibrator uses a resitance of 100 $k\Omega$ and capacitance $0.01\,\mu l$. The frequency of square waves genreated by it is.
(a) 924 Hz (b) 593 Hz
(c) 693 Hz (d) 110 Hz

90. A triangular waveshape is obtained:
(a) by integrating a square wave
(b) by differentiating a in wave
(c) by differentiating a square wave
(d) by integrating a sine wave

91. In a **sweep frequecy generator,** two oscillators, one with frequency range of 3 GHz to GHz is heterodyned with a second oscillator having a fisced frequency output frequency varies from.
(a) 0 to 2 GHz (b) 6 GHz to 8 GHz
(c) 0 to 3 GHz (d) 0 to 5 GHz

92. **A random noise generator** produces a signal
(a) whose amplitude varies randomly
(b) which has no periodic frequency
(c) has an unpredictable power spectrum
(d) all of the above

93. **A voltage controlled oscllator** is an instrument.
(a) whose frequency is dependent upon the amplitude of input signal
(b) whose frequency is independent of the amplitude of the input signal
(c) whose frequency is dependent upon the frequency of the input signal
(d) none of the above

94. The **effective reactance** of an inductive coil,
(a) increases because of stray capacitances as the frequency increase
(b) decreases because of stray cpacitances as the frequency increases.
(c) remains the same irrespective of the increase in frequency even if stray capacitances are present.
(d) none of the above.

95. The **effective reactances** of an inductive coil
(a) increases due to stray series inductances as fre quency increases.
(b) decreasses due to stray series indutances as frequency increases.
(c) remaines the same irrespective of the increase in frequency even if stray series inductances are present
(d) none of the above.

96. The **effective resistances** of a coil at high frequences is more than its d.c. resistance on account of
(a) skin effects (b) proximity effects
(c) eddy current losses (d) all of the above

97. A series RLC circuit is used for measurement of inductance of a coil using resonant method. Resonance condition are produced by varyiong the capacitance. A thermelectric type ammeter is used in series with the circuit. Also an elecironic voltmeter is connected across the tuning capacitor.
(a) The conditions of resonance are more reliably indicated when the ammeter reads maximum.
(b) The conditions of resonance are more reliably indicated when the voltmeter connected across the capacitor reads maximum.
(c) The conditions of resonance are the same when both ammeter and voltmeter read maximum.
(d) The conditions of resonance are indicated when both ammeter and voltmeter read minimum

98. If the Q factor of a coil which is $\omega_0 L / R$ is measured by varying the frequency
(a) The plot between Q and frequency is linear
(b) The value of Q intially decreases with increase of frequecy and afterwards increases with increase of frequecy.
(c) The value of Q initially increases with increase of frequecy and after wards decreases with increase of frequecy.
(d) The Q factor remains constant irrespective of the value of frequecy

99. The **Q factor of a coil** at the resonant frequecy 1.5 MHz of an RLC series circuit is 150. The bandwidth is :
(a) 225 MHz (b) 1.06 MHz
(c) 10 kHz (d) none of above

100. The resonant frequency of an RLC series of the circuit is 1.5 MHz with the tuning capacitor set at 150 pF. The band width is 10 KHz. The **effective resistance** of the circuit is:
(a) 29.5 Ω (b) 14.75 Ω
(c) 9.4 Ω (d) 4.7 Ω

101. In a **Q meter**, distributed capaitance of coil is measurement by changing the capacitance of the tuning capacitor. The values of tuning capacitor are C_1 and C_2 for resonant frequencies f_1 *and* $2f_1$ respectively. The value of disturbuted capacitance is:
(a) $\dfrac{C_1 - C_2}{2}$ (b) $\dfrac{C_1 - 2C_2}{3}$
(c) $\dfrac{C_1 - 4C_2}{3}$ (d) $\dfrac{C_1 - 3C_2}{2}$

102. The **true value of Q** of a coil is 245 and the mesured value of 244.5. The ratio of distributed capacitance of self capacitance of the coil is

(a) 2.04×10^{-3} (b) 1.002
(c) 0.997 (d) none of the above

103. In a Q - meter, the value of shunt resistance connected across the oscillator is typically of the order of

(a) Ω (b) $m\Omega$
(c) $\mu\Omega$ (d) $k\Omega$

104. A **temperature** measurment instrument is callbrated between $100°C$ to $600°C$. The **scale span** of the instrument is

(a) $600°C$ (b) $100°C$
(c) $700°C$ (d) $500°C$

105. A **pressure measurment** instrument is callbrated between 50 bar to 150 bar. The **scale range** is
(a) 150 bar (b) 50 bar
(c) 100 bar (d) 200 bar

106. A temperature measuring instrument reads $98.6°F$ and static correction given in the correction curve is $-0.07°F$. Determine ture value of temperature.

(a) $93.53°F$ (b) $98.67°F$
(c) $93.6°F$ (d) none of these

107. An instrument has an input S/N of 50. The input voltage is 3mV. Calculate the noise voltage in mV.
(a) 0.84 (b) 0.42
(c) 0.06 (d) none of these

108. A set of independent current measurement were recorded as $10.10, 10.17, 10.32, 10.51$ and $10.07A$ calcultate (a) the average current and (b) range of error.
(a) 10.23, 0.28 (b) 10.23, 0.16
(c) 10.23, 0.22 (d) 10.23, 0.12

109. In a measurment system '**force**' is an across variable. Its associated power based through variable is
(a) Translational displacement
(b) Translational acceleration
(c) Translational velocity
(d) Rotational displacement

110. A synchro transmitter is used with a synchro reperater for remote Sensing and Amplification and with a control transformer for.
(a) Amplification and addition
(b) Error Detection and amplification
(c) Addition and Error Detection
(d) None of the above

111. To avoid the effect of stray magnetic field in ac bridges we can use.
(a) Magnetic screening
(b) Wagner earthing device
(c) Wave filter
(d) Any of the above

112. The shielding of bridge elements is done to reduce the effect of.
(a) Mutual inductance (b) Self inductances
(c) Stray capacitances (d) Story inductance

113. Wagner's devices are used in bridge circuits for
(a) Eliminating the effects of earth capacitance
(b) Eluiminating the effects of inter - component capacitances.
(c) Eliminating the effects of magnetic field
(d) None of the above.

114. Which one of the following definitions correctly repersents a **data aequisition** systems?
(a) DAS is a group of electronic devices that are connected to perform the measurment and quantization of electrical signals for digital processing.
(b) DAS is a group of devices that are connected to store different signals
(c) DAS is a system to control a process.
(d) DAS is a signal conditioner.

115. Consider the following statements.
1. Low - pass filter
2. Signal transmission medium
3. Amlifier
4. Digital to analog converter
5. Analog to digital converter

Which of the following sequences is the correct sequence for effective signal re - construction in a data acquisition and processing scheme?
(a) 1 - 3 - 5 - 2 - 4
(b) 3 - 1 - 5 - 2 - 4
(c) 1 - 5 - 3 - 2 - 4
(d) 3 - 5 - 1 - 4 - 2

116. Excitation and amplification systems are needed for:
(a) for active transducers only
(b) for passive transducers only
(c) for both active and passive transducers
(d) for both passive and output transducers

117. A .d.c amplifier
(a) needs to have a balanced differential inputs with a high common mode rejection ratio to give very good thermal and long term stability.
(b) easy to calibrate at low frequencies and has ability to recover from overload condition
(c) is prove to drift and low frequency spurious signals come out as data information
(d) all of the above

118. When using a.c. signal conditioning system is normally used for capacitive transducers, the carrier frequencies.
(a) range between 50 Hz and 20 kHz
(b) should be of the order of 0.52 MHz
(c) should be of the order of 20 MHz
(d) none of the above

119. In a carrier system, drift and spurious signals are important,

(a) because they modulate the carrier
(b) because they do not modulate the carrier
(c) because it is easier to achieve a stable carrier than a stabilized d.c. source
(d) none of the above

120. When using a d.c. signal conditioning system, with a carrier of 3 kHz, the data frequency should be limited to:
(a) 1 kHz (b) 5 Hz
(c) 600 Hz (d) 2 MHz

121. An operational amplifier has an open loop gain of $200,000\,\Omega$. Its output exhibits saturation at 10 V. The thereshold differntial voltage of the amplifler is :
(a) $25\,\mu V$ (b) $50\,\mu V$
(c) $0.5\,mV$ (d) $10\,V$

122. An ideal OPAMP has a gain of - 100. The input is connected to inverting end and the input resistance is $1\,k\Omega$. The feedback resistance is,
(a) $100\,k\Omega$ (b) $10\,\Omega$
(c) $100\,\Omega$ (d) $1000\,\Omega$

123. An ideal OPAMP has a gain of - 100. The input is connected to the inverting and the resistance is $1\,k\Omega$. The value of feedback resistance is
(a) $101\,k\Omega$ (b) $99\,k\Omega$
(c) $9.9\,k\Omega$ (d) $1.01\,k\Omega$

124. The properties of an ideal OPAMP are,
(a) it should have zero input impedance
(b) it should have a zero open loop gain
(c) it should have a zero open loop gain
(d) none of the above

125. The closed loop gain of an OPAMP is dependent upon whether the OPAMP is used,
(a) in inverting mode
(b) in non - ivberting mode
(c) is independent of the fact whether the input is connected to inverting or non - liverting terminal
(d) is dependent upon the fact whether the input is connected to inverting or the non - inverting terminal.

126. The differential (open loop)gain of an OPAMP is 20,000. The input resistance is $2\,k\Omega$ and the feedback resistor has a value of $200\,k\Omega$. The closed loop gain of the amplifler is:
(a) -9.9995 (b) -9.9955
(c) -10 (d) $+10$

127. A buffer amplifler has gain of
(a) infinity
(b) zero
(c) unity
(d) dependant upon the circuit parameters

128. An operational amplifier has a bandwidth of 100 kHz. Negative feedback is introduced with a loop gain of 50. The band with feed back is .
(a) 2 kHz (b) 49 kHz
(c) 5.1 MHz (d) 1.96 kHz

129. The order of input resistnace of 741 OPAMP is.
(a) $10\,\Omega$ to $10^3\,\Omega$ (b) $10^3\,\Omega$ to $10^3\,\Omega$
(c) $10^3\,\Omega$ to $10^6\,\Omega$ (d) $10\times10^3\,\Omega$ to $10^9\,\Omega$

130. The order of output resistance of 741 OPAMP is,
(a) $0.1\,\Omega$ to $10\,\Omega$ (b) $10\,\Omega$ to $10^5\,k\Omega$
(c) $10\times10^3\,\Omega$ to $10^9\,\Omega$ (d) $10^3\,\Omega$ to $10^6\,\Omega$

131. The gain of a 741 OPAMP falls off at a low frequency of,
(a) 10 kHz (b) 10 Hz
(c) 100 Hz (d) 1000 Hz

132. An OPAMP having an open loop gain of A_{VOL} is designed with a feed factor β, where $\beta = \frac{R_1}{R_1 + R_f}$ R_1 is the input resistance connected to non - inverting end and R is the resistance in the feedback path, the closed loop gain is :

(a) $-\frac{R_f}{R_1}\left[\frac{1}{1+(1/A_{VOL}\,\beta)}\right]$

(b) $\frac{R_f}{R_1}\left[\frac{1}{1+(1+A_{VOL}\,\beta)}\right]$

(c) $\frac{R_f}{R_1}\left[\frac{1}{1+(1-A_{VOL}\,\beta)}\right]$

(d) $-\frac{R_f}{R_1}\left[\frac{1}{1+(1+A_{VOL}\,\beta)}\right]$

133. An OPAMP with an open loop gain is designed with a loop gain $(A_{VOL}\,\beta)$ where $\beta = \frac{R_1}{R_1 + R_f}$ with R_1 = resistance connected to non - inverting end, and R_f = resistance connected in the feedback path.
If R_{in} = input resistance of amplifler without feed back then the input resistance of the amplifier $(R_{in\,f})$ with feedback is:

(a) $R_{in\,f} = R_{in}\,(1 + A_{VOL}\,\beta)$

(b) $R_{in\,f} = R_{in}\,(1 - A_{VOL}\,\beta)$

(c) $R_{in\,f} = R_{in}\,(1 + \beta)$

(d) $R_{in\,f} = R_{in}\,(1 - \beta)$

134. An OPAMP with an open loop gain is designed with a loop gain $A_{VOL}\,\beta$ where $\beta = \frac{R_1}{R_1 + R_f}$ with R_1 = resistance connected in the feedback path the output resistance of the amplifier, R_f resistance connected in the feedback path, R_0 the output resistance of the amplifier, R_{of} with feed back is

(a) $R_{of} = \frac{R_0}{1 - A_{VOL}\,\beta}$ (b) $R_{of} = \frac{R_0}{1-\beta}$

(c) $R_{of} = \frac{R_0}{1+\beta}$ (d) $R_{of} = \frac{R_0}{1 + A_{VOL}\,\beta}$

135. Offset voltages in OPAMPS are produced because of
(a) variations in the input voltage applied to the amplifer
(b) mismatch between the two differential amplifers which form the IC of the OPAMP chip
(c) mismatch between the input signals applied to the OPAMP
(d) All of the above

136. The typical value of input offset voltage of a 741 OPAMP is:
(a) $2\,\mu$ V (b) 12 mV
(c) 2 V (d) 0.2 V

137. The typical value of input offset current of a 741 OPAMP is:
(a) 10 nA (b) 10 mA
(c) 10 pA (d) 100 mA

138. A 741 OPAMP has,
(a) 6 pins (b) 7 pins
(c) 8 pins (d) 9 pins

139. A 741 OPAMP has an open gain 200,000. The input offset voltage is 2mV. If the input terminals are shorted, the output voltage is:
(a) 0 V (b) ∞
(c) 400 V (d) $\pm 400\,V$

140. In order to eliminate the output voltage on account of input offset voltage,
(a) A nulling potentiometer is connected between pins 1 and 5 and potentiometer is adjusted to give zero output voltage. No voltage input is applied.
(b) A nulling potentiometer is connected between pins 1 and 5 and the sliding contact of the POT is connected to pin 4. The POT is adjusted to give zero. No input voltage is applied.
(c) A blocking capacitor is used at the input of non inverting end.
(d) An external off setting potentiometer is used whose sliding contact is connected to the output terminal. This potentiometer is adjusted to give zero output voltage with no input voltage applied.

141. Ac active half wave voltage follower circuit has an input of $10 \sin \omega t$ V. The input resistance has value $2\,k\Omega$ and feedback resistance is $10\,k\Omega$. The expression for the output voltage is,
(a) $50 \sin \omega t$ V *for* $0 < \omega t < \pi$,
0V for $\pi < \omega t < 2\pi$
(b) $60 \sin \omega t$ V for $0 < \omega t < \pi$,
0 V for $\pi < \omega t < 2\pi$
(c) $60 \sin \omega t$ V for $0 < \omega t < 2\pi$,
(d) $50 \sin \omega t$ V for $0 < \omega t < 2\pi$,

142. The output of an OPAMP is 2 V peak. The slew rate is $5\,V/\mu\,s$. The input sinuesiodal which can be reproduced with no diistortion has the maximum frequency of :
(a) 398 Hz (b) 1592 kHz
(c) $125 \times 10^3\,Hz$ (d) 398 kHz

143. A **differential amplifier** has + 100 mV applied to non - incerting end and + 250 m V applied to inverting end. The output is 1.5 V. The gain of the amplifier is:
(a) 10 (b) 4.29
(c) 0.6 (d) 15

144. A **differential amplifler** having CMRR 50,000 has a difference mode gain of 500. The common mode gain is:
(a) 1 (b) 0.1
(c) 0.01 (d) 0.001

145. Which of the following are data repersentation elements in a genralized measurement system?
1 Analog indication 2 Amplifier
3 A/D converter 4 Digital Display
Select the correct answer using the codes given be low:
(a) 1 and 2 (b) 1 and 4
(c) 2 and 4 (d) 3 and 4

146. Telemetry processes the information from a remote location by means of
(a) Mechanical means (b) Electrical means
(c) CRO (d) All of the above

147. The position telemetering system using synchros is
(a) A pulse telemetering system
(b) A RF telemetering system
(c) A DC telemetering system
(d) An AC telemetering system

148. The bandwidth requirekment of an FM teleetry channel is
(a) Equal to that of an AM telemetry channel
(b) Smaller than that of an AM telemetry channel
(c) About 100 times that of an AM telemetry channel
(d) About 10 times that of an AM telemetry channel

149. If the largest frequency present in the measured signal and number of bits used in the binary code, are respectively 'f_m' and 'n' then minimum bandwidth of pulse code modulation channel used for telemetry would be
(a) $\frac{f_m}{2n}$ (b) $\frac{f_m}{n}$
(c) nf_m (d) $2nf_m$

150. A tempeatuer-sensitive transducer is subjected to a sudden tempeature change. It takes 10 sec for the transducer to reach equilibrium condition (five time constants). How long will it take for the transducer to read half of the tempeature difference?

(a) 1.41 s (b) 1.56 s
(c) 1.39 s (d) 1.81 s

151. The function of reference electrode in a pH meter is to provide.

(a) a constant current
(b) a constant voltage
(c) temperature compensation
(d) both (b) and (c)

152. Measurments of flow, thermal conductivety and liquid level using thermisrtors make use of:

(a) Resistance decrease with temperature
(b) Resistance increase with temperature
(c) Self - heating phenomenon
(d) Change of resistively

153. The electrical transducers used for continous recording and control of humidity are.

(a) resistive hugrometers
(b) electrolytic hygrometers
(c) aluminium oxide hygrometers
(d) all of the above

154. In **hygrometers** the principle of measurment is.

(a) change in resistance of saits with humidity.
(b) change in microways power using kiystron
(c) change in thermal conductivity using thermistor.
(d) none of the above

155. The most suitable transducer for monitoring continuously variations in very fine thickness is

(a) inductive (b) capacitive
(c) ultrasonic (d) nuclear radiation

156. Hydrometer is employed for determination of

(a) Relative humidity (b) bourdon tube
(c) fluid level (d) none of the above

157. The method that can be employed for measuring only fluid level, is

(a) radio - active method (b) bourdon tube
(c) strain gauge (d) bell tybe meter

158. In radio - active method, the receiver measures fluid level by recording the.

(a) Direction of the rays
(b) Number of radioactive particles received.
(c) Time taken by the rays in reaching the receiver.
(d) None of the above.

159. A **differential amplifler** has a comm mode gain of 0.02. It has $+200\ mV$ signals applied to each of its inputs. The amplitude of output signal is :

(a) 0 V (b) 8 mV
(c) 4 mV (d) non of the above

160. The output from a transducer has a frequency of 50 Hz. It modulates a carrier having a frequency of 1 kHz. The output waveshape contains frequencies which are :

(a) 1000 Hz, 1050 Hz (b) 1000 Hz, 950 Hz
(c) 950 Hz and 1050 Hz (d) 2000 Hz $\pm$ 100 Hz

161. The output from a transducer is a sinusoidal wave frequency ω_1, ω_2, ω_3, ω_4,. The carrier frequency is ω_c,. The frequency spectrum of output wave contains frequencies.

(a) $\omega_c \pm \omega_1$
(b) $\omega_c \pm \omega_2$
(c) $\omega_c \pm {}_1$, $\omega_c \pm \omega_2$, $\omega_c \pm \omega_3$
(d) $\omega_c \pm \omega_1$, $\omega_c \pm \omega_2$, $\omega_c \pm \omega_3$, $\omega_c \pm \omega_4$

162. A **low pass filter** has a time constant of τ. Its gain at a frequency ω is,

(a) $\sqrt{1+(\omega\tau)^2}$ (b) $\dfrac{\omega\tau}{\sqrt{1+\omega^2\tau^2}}$
(c) $\dfrac{1}{\sqrt{1+(\omega\tau)^2}}$ (d) none of above

163. A **high pass filter** has a time constant of τ. Its gain at a frequency ω is,

(a) $\sqrt{1+(\omega\tau)^2}$ (b) $\dfrac{\omega\tau}{\sqrt{1+\omega^2\tau^2}}$
(c) $\dfrac{1}{\sqrt{1+(\omega\tau)^2}}$ (d) none of above

164. The following **figure** shows some bode plots. Recognize the Bode plot for a Bandpass filter

(a) (b)
(c) (d) none of above

165. The input circuit of variable resistance trasnducer consists of a ballast resistance in service with the transducer for a particular input is kR_1 where k is a constant. If the input circuit is supplied by an input voltage e_i the sensitivity of the transducer.

(a) $\dfrac{e_i\ R_b}{(R_b + kR_1)}$ (b) $\dfrac{e_i\ R_b}{(R_b + kR_1)^2}$
(c) $\dfrac{e_i\ R_b}{(R_b + kR_1)^3}$ (d) $\dfrac{e_i\ R_b\ kR_t}{(R_b + kR_1)^3}$

166. In a voltage sensitive wheatstone having each are having a resistance, R, the resistance of one of the arms is changed to $R + \Delta R$ where $\Delta R <<< R$. The wheatstone bridge is supplied with an input voltage of e_1. The output voltage on account of unbalance is,

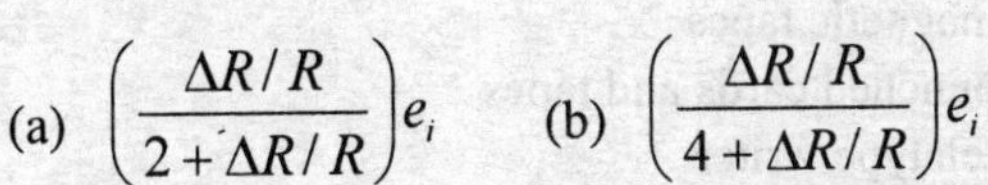

(c) $\left(\dfrac{2\Delta R/R}{4+\Delta R/R}\right)e_i$ (d) $\left(\dfrac{2\Delta R/R}{4+2\Delta R/R}\right)e_i$

167. If L = nominal value of inductance of inductive transducer and L_c = self inductiveof one of the tightly coupled ratio arms. The bridge sensitiviy factor of a Blumlein bridgbe using two inductive transducer in push pull arrangement is,

(a) $\dfrac{2\,L/L}{(1+L_c/L)}$ (b) $\dfrac{4\,L_c/L}{(1+2L_c/L)}$

(c) $\dfrac{4\,L_c/L}{(1+L_c/L)}$ (d) $\dfrac{2\,L_c/L}{(1+2L_c/L)}$

168. In a **Blumlein bridge** using two capacitive transducers in push pull configuration, the bridge sensitivity factor above the resonance frequency for a tightly coupled bridge is,
(a) 2 and is independent of the variations of the bridge frequency and ratio arm inductance.
(b) 2 and is dependent upon the variations of the bridge frequency and ratio arm inductance.
(c) practically constant and is lower than that of bridge with uncoupled arms.
(d) variable and is dependent upon the frequency even though the ratio arms tightly doupled

169. It ω = applied frequency and τ time constant of RC circuit. A low pass RC filter acts as a pure integrator when.
(a) $\omega\tau = 1$ (b) $\omega\tau << 1$
(c) $\omega\tau >> 1$ (d) $\omega\tau = 0$

170. It ω = applied frequency and τ = time constant of RC circuit. A high pass RC filter acts as a pure differentilator when:
(a) $\omega\tau = 1$ (b) $\omega\tau << 1$
(c) $\omega\tau >> 1$ (d) $\omega\tau = 0$

171. An 8 bit converter is used for a d.c. range of $0-10$ V. Find the weight of LSB.
(a) 39 mV (b) 78 mV
(c) 39.2 m V (d) none of the above

172. A 5 bit D/A converter has a current output. The digtal input is 10100 and the output current corresponding to theis is 10 m/A. What will be output current for digital input of 11101?
(a) 10.5 mA (b) 14.5 mA
(c) 29 mA (d) none of the above

173. A successive approximation A/D converter has a resolution of 20 m V. What will be its digital output for an analog input of 2.17 V?
(a) 01101100 (b) 01101101
(c) 01101011 (d) none of the above

174. A analog transducer range 0 - 10V. Calculate bits of an A/D converter if the resolution is 5 m V
(a) 10 (b) 9
(c) 11 (d) none of the above

175. The 0-10 V A/D converter has to have a resolution of 0.025 percent. Find the r.m.s. value of quantization error.
(a) 176 μV (b) 705 μV
(c) 352 μV (d) 1410 μV

176. The tuned circuit of the **oscillator** in a simple A.M. transmitter uses a 100 μ H inductor and 0.5 nF capcitor. If the oscillator output is modulated by audio frequencies upto 20 kHz, what is the range occupled by side frequencies?
(a) 702 to 722 kHz (b) 702 to 732 kHz
(c) 692 to 732 kHz (d) 682 to 722 kHz

177. An A.M. broadcast radio trasnmitter raditaes 10 kW when the modulation percentage is 60, What is the power of the carrier?
(a) 7.69 kW (b) 8.47 kW
(c) 9.17 kW (d) 5 kW

178. In an amplitude modulated system, the total power radiated 600 W. The power of the carrier is 400 W. What is modulation index?
(a) 1 (b) 0.5
(c) 0.75 (d) non of the above

179. In an A.M. system, the modulated output voltage has $E_{min} = 15$ V, the modulation index is :
(a) 2 (b) 1.5
(c) 1 (d) 0.5

180. For an F.M. wave represented by the voltage equation

$$e = 8\sin(6\times10^8 t + 5\sin 4\times10^4\ t)V$$

the carrier and modulating frequencies are :
(a) 127.3 MHz, 9.5 kHz, (b) 95.5 MHz, 6.37 kHz
(c) 190 MHz, 6.37 kHz (d) 95.5 MHz, 12.74 kHz

181. For an F.M. wave represented by the voltage equation

$$e = 8\sin(6\times10^8\ t + 5\sin 4\times10^4 t)V$$

the modulation index and the maximum deviation are:
(a) 5,31,85 kHz (b) 40,31,85 kHz
(c) 5, 95.5 MHz (d) 5, 12.74 kHz

182. The F.M **telemetry** as comparad with A.M. telemetry requires a channel that is
(a) equal to that of A.M.
(b) smaller thanwhat is required for A.M. telemetry
(c) 100 times of tha trequired for A.M. telemetry
(d) 10 times of that required for AM telemetry

183. In F.M. systems operate at
(a) VHF and UHF (b) only VHF
(c) only MF (d) MF and HF

184. F.M. systems as comparad to A.M. systems
(a) are equally effected by noise
(b) are less effected by noise

(c) are more effected by noise
(d) are highly effected by noise sence they operate at VHF and UHF

185. A **typical telemetry** carrier frequency 230 MHZ mau be used with a band width of ± 320 kHz using A.M. A data with a bandwidth of 4k Hz are modulated. The frequency bands utillzed are
(a) 28 kHz to 40 kHz
(b) 28 kHz to 36 kHz, 40 kHz to 48 kHz
(c) 28 kHz to 48 kHz,
(d) 230 KHz $\pm$ 324 KHz

186. Twenty - one data channels, with subcarrier centre frequencies ranging from 400 Hz to 165 kHz have been designated for **telemetry** use. All have a frequency deviation of 75%. If the index of modulation is 5, the intelligence band width of channel 1 for a centre frequency of 3000 Hz is. Assume the frequency multiplexing has been used.
(a) 6 Hz (b) 20 Hz
(c) 60 Hz (d) 45 Hz

187. In **frequency multiplexing systems**, if the subcarrier frequency deviations are proportions to the centre freqency, the scheme is referred to having:
(a) a proportional bandwidth format
(b) a constant bandwidth format
(c) a variable bandwidth format
(d) a consistent bandwidth format

188. Time division multiplexing is used when
(a) constant data transmission
(b) transmission of data samples
(c) transmission of data at random
(d) transmission of data of only one measurand

189. Time division multiplexing is used when
(a) when the data to be transmitted is slow changing
(b) when the data to be transmitted has a small bandwith.
(c) when the data br trasnmitted is slow changing and has a low bandwidth.
(d) when all the data to transmitted is fast changing and has a high bandwidth

190. Modem is an acronym of
(a) modulation
(b) demodulation
(c) modulation and demodulation
(d) all of the above

191. If an information is required to be stored over a short interval of time.
(a) a single number devices should be used
(b) a storage type oscilloscope should be used
(c) a CRO with photographic equipment should be used.
(d) a direct writing recorder or a magnetic tape recorder should be used

192. The machine interpertable outputs from an analog transducer can be had from
(a) magnetic tapes
(b) punched cards and tapes
(c) teletype writer
(d) all of the above

193. Digital instruments have input inpedeance of the order of
(a) Ω (b) $k\Omega$
(c) $M\Omega$ (d) $m\Omega$

194. If four flip - fliops are connected in casecade, they serve as
(a) 2 : 1 scalar (b) 4 : 1 scalar
(c) 8 : 1 scalar (d) 16 : 1 scalar

195. The LEDs, For their display require
(a) a voltage of 1.2 V and a current of 20 m A
(b) a voltage of 25 V and a current of 20 m A
(c) a voltage of 1.2 V and a current of 100 m A
(d) a voltage of 25 V and a current of 100 m A

196. The switching time of LEDs is of the order of
(a) 1 s (b) 1 ms
(c) 1 μs (d) 1 ns

197. LEDs emit light
(a) only in red colour
(b) only in yellow colour
(c) only in green coloue
(d) in red, greed, yellow and amber colour

198. An LCD requires a power of
(a) 20 W (b) 20 m W
(c) 20 μW (d) 20 nW

199. The turn on and turn off times of a LCD are of the order of
(a) 1 s (b) 1 ms
(c) 10 ms (d) 10 ns

200. The power requirement of an LED is
(a) 40 m W per numeral (b) 40 μW per numeral
(c) 10 W per numeral (d) 10 μW per numeral

201. A Nixie tube requires
(a) 10 cathodes (b) 14 cathodes
(c) 15 cathodes (d) 20 cathodes

202. Nixie tubes require
(a) 150 V - 220 V (b) 100 V - 150 V
(c) 150 mV - 220 mV (d) 100 mV - 150 mV

203. In a **digital frequency meter**, the Schmitt trigger is used for
(a) sinusoidal wavefroms into rectaugular pulses
(b) scaling of sinmsiodal wave forms
(c) providing time base
(d) none of the above

204. Pulse of a frequency of 1 MHz are applied to the time base selector of a **digital frequency meter** which consists of 6 frequency dividers, each dividing the incoming frequency by a factor of 10. What is the time base setting at the output of 4th frequency divisor ?
(a) 1 ms (b) 10 ms
(c) 100 ms (d) 1 ms

205. Period measurment is done in frequency meter for achieving high accuracy in the case of

(a) high frequencies (b) medium frequencies
(c) d.c. (d) low frequencies

206. In a **digital frequency meter** the multi period average mode of operation, the main gate is held open for more than one period of the unknown signal
(a) so that the period is extended in order that conters show more digits there by an increased accuracy is obtained.
(b) so that resolution is increased
(c) so that range of th instrument is extended
(d) so that errors on acount of dead time are eliminated

207. A digital voltmeter uses an A/D converter which needs a start pulse uses an analog comparator and has a relatively fixed converslon time independent of the applied voltage. The A/D converter is:
(a) successive apporximation converter
(b) digital ramp converter
(c) divial slope converter
(d) all of the above

208. X - Y recorders
(a) record one quantity with respect to another quantity.
(b) record one quantity on X axis with respect to time on Y axis
(c) record oen qunitity on Y axis with respect to time on X axis.
(d) None of the above

209. In a direct magnetic tape recording system, the reproducting head and amplifier characteristic are
(a) - 6 db/octave, + 6 db/octave
(b) - 6 db/octave, + 20 db/octave
(c) - 6 db/octave, - 6 db/octave
(d) -20 db/decade, + 20 db/decade

210. The advantages of **F.M. magnetic tape recording** are
(a) it can record from d.c. to several kHz
(b) it is fire from dropout effects
(c) it is independent of amplitude varations and accurately reproduces the waveform of input signal.
(d) all of the above

211. An a.c. signal conditioning system is normally used for,
(a) resistive transducer like strain gauges
(b) inductive and capactive transducers
(c) piezoelectric transducers
(d) all of the above

Answers Set-2

1. (c)	2. (b)	3. (d)	4. (d)
5. (a)	6. (a)	7. (b)	8. (b)
9. (a)	10. (c)	11. (d)	12. (d)
13. (c)	14. (c)	15. (c)	16. (a)
17. (c)	18. (a)	19. (a)	20. (d)
21. (d)	22. (b)	23. (b)	24. (b)
25. (a)	26. (b)	27. (c)	28. (d)
29. (c)	30. (b)	31. (b)	32. (a)
33. (d)	34. (c)	35. (c)	36. (b)
37. (a)	38. (c)	39. (c)	40. (b)
41. (d)	42. (d)	43. (c)	44. (c)
45. (d)	46. (d)	47. (b)	48. (d)
49. (a)	50. (a)	51. (b)	52. (a)
53. (d)	54. (b)	55. (d)	56. (b)
57. (c)	58. (d)	59. (d)	60. (c)
61. (d)	62. (d)	63. (c)	64. (c)
65. (c)	66. (b)	67. (a)	68. (b)
69. (b)	70. (c)	71. (b)	72. (b)
73. (d)	74. (b)	75. (c)	76. (b)
77. (c)	78. (a)	79. (a)	80. (d)
81. (d)	82. (c)	83. (c)	84. (a)
85. (b)	86. (b)	87. (a)	88. (c)
89. (c)	90. (a)	91. (a)	92. (a)
93. (a)	94. (b)	95. (a)	96. (d)
97. (b)	98. (c)	99. (c)	100. (d)
101. (c)	102. (a)	103. (b)	104. (d)
105. (a)	106. (a)	107. (b)	108. (c)
109. (c)	110. (a)	111. (a)	112. (c)
113. (a)	114. (a)	115. (b)	116. (d)
117. (e)	118. (b)	119. (a)	120. (c)
121. (b)	122. (a)	123. (b)	124. (d)
125. (d)	126. (a)	127. (c)	128. (c)
129. (d)	130. (b)	131. (b)	132. (a)
133. (a)	134. (d)	135. (b)	136. (b)
137. (a)	138. (c)	139. (d)	140. (b)
141. (a)	142. (d)	143. (a)	144. (c)
145. (d)	146. (b)	147. (c)	148. (a)
149. (c)	150. (b)	151. (b)	152. (c)
153. (d)	154. (a)	155. (d)	156. (b)
157. (a)	158. (c)	159. (c)	160. (c)
161. (d)	162. (c)	163. (b)	164. (c)
165. (b)	166. (d)	167. (b)	168. (a)
169. (c)	170. (b)	171. (a)	172. (b)
173. (a)	174. (c)	175. (b)	176. (c)
177. (a)	178. (a)	179. (d)	180. (b)
181. (a)	182. (d)	183. (a)	184. (b)
185. (b)	186. (d)	187. (a)	188. (b)
189. (c)	190. (c)	191. (b)	192. (d)
193. (c)	194. (d)	195. (a)	196. (d)
197. (d)	198. (c)	199. (c)	200. (a)
201. (c)	202. (a)	203. (a)	204. (b)
205. (d)	206. (a)	207. (a)	208. (a)
209. (c)	210. (d)	211. (d)	

MCQ Set–3

1. Radiation pyrometers are used for the measurement of temperature in the range of
   (a) –200° to 500°C (b) 0°C to 500°C
   (c) 500° to 1200°C (d) 1200°C to 2500°C.
2. Magnetic flux can be measured by
   (a) Capacitor pick-up (b) Inductive pick-up
   (c) Resistance pick-up (d) Half-effect pick-up.
3. A semiconductor based temperature transducer has a temperature coefficient of –2500 μV/°C. This transducer is indeed a
   a. Thmistor
   b. Forward-biased pn junction diode
   c. Reverse-biased pn junction diode
   d. FET.
4. The function of the reference electrode in a pH meter is to
   a. Produce a constant voltage
   b. Provide temperature, compensation
   c. Provide a constant current
   d. Measure average pH value.
5. Pirani gauge is used for the measurement of pressure in the range of
   a. 1^{-8} mm to 10^{-5} mm of Hg
   b. 10^{-3} mm to 10^{-} mm of Hg
   c. 10 mm to 10^{3} mm of Hg
   d. 10^{5} mm to 10^{-} mm of Hg.
6. The most light sensitive transducer for conversion of light into electrical power is the
   (a) Photodiode (b) Solar cell
   (c) Photoconductive (d) Photovoltaic cell.
7. Hot wire anemometer is a device used to measure the
   (a) Pressure in gases (b) Liquid discharge
   (c) Gas velocities (d) Temperature.
8. Consider the following devices:
   1. Anemometer 2. Stroboscope
   3. Accelerometer

   The correct sequence of these devices to measure the rotational speed, vibration and airflow, respectively is
   Codes:
   (a) 2, 3 and 1 (b) 2, 1 and 3
   (c) 1, 3 and 2 (d) 3, 2 and 1.
9. The pH value of solution is defined as
   a. – log (H^+ ion concentration)
   b. log (H^+ ion concentration)
   c. – $\log^{-1}$/(OH – ion concentration)
   d. – log (OH ion concentration).
10. The piezoelectric **crystal voltage sensitivity is** defined as
    a. Voltage developed per unit stress
    b. Field developed per unit stress
    c. Voltage developed per unit force
    d. Field developed per unit force
11. Consider the following statements associated with electrical/electronic transducers:
    1. Mass inertia effects are minimized
    2. These transducers consume very little power
    3. The response time is large.
    4. Transmission and processing the signal for the purpose of measurement are easier.

    Select the correct answer using the codes given below:
    Codes:
    (a) 1, 2 and 3 (b) 2, 3 and 4
    (c) 1, 3 and 4 (d) 1, 2 and 4
12. Consider the following statements:
    A transduction converts
    1. Mechanical energy into electrical energy
    2. Mechanical displacement into electrical signal
    3. One form of energy into another form of energy
    4. Electrical energy into mechanical form.

    Which of these statements is/are correct?
    (a) 1 and 4 (b) 1 and 2
    (c) 3 alone (d) 1 alone.
13. Match List-I (Transducers) with List-II (Measured Quantities) and select the correct answer using the codes given below the lists:

List-I	List-II
A. Capacitive transducer	1. Pressure
B. Thermo couple	2. Torque
C. Load cell	3. Displacement
D. Diaphragm	4. Temperature

Codes:

	A	B	C	D
(a)	2	4	3	1
(b)	3	1	2	4
(c)	2	1	3	4
(d)	3	4	2	1.

14. Gauge factor of a stain gauge is defined as the ratio of a per unit change in the
    (a) Conductivity to the per unit change in applied force acting on the element
    (b) Resistance to the per unit change in the length of the element
    (c) Stress to the per unit change in strain of the element
    (d) Current to the per unit change in the length of the element.

15. Some of the functional building blocks of measurement system are:

Primary Sensing Elements (PSE)

Variable Conservation Element (VCE), or Transducer

Data Transmission Element (DTE)

Variable Manipulation Element (VME)

Data Presentation Elements (DPE)

The correct sequential connection of the functional building blocks for an electronic gauge will be

(a) PSE, VME VCE, DPE, DTE.

(b) PSE, VCE, VME, DTE, DPE.

(c) DTE, DPE, VCE, PSE, VME

(d) PSE, VCE, DTE, DPE, VME.

16. Match List-I (Transducers) with List-II (Measured quantities) and select the correct answer using codes given below the lists:

List-I (Transducers)	List-II (Measured Quantities)
A. Venturi tube	1. Displacement
B. Optical tachometer	2. Pressure
C. Linear variable differential transformer	3. Flow
D. Perini gauge	4. Velocity

Codes:

	A	B	C	D
(a)	1	4	3	2
(b)	3	2	1	4
(c)	1	2	3	4
(d)	3	4	1	2.

17. Rochelle salt is a crystalline material used in producing

a. velocity transducer

b. photoelectric transducer

c. piezoelectric transducer

d. differential transformer transducer.

18. The gauge factor of the material of strain gauge is such that the resistance changes from 1000 ohms to 1009 ohms subjected to a strain of 0.0015. The Poisson's ratio for the material of the gauge wire is

(a) 1.75 (b) 2

(c) 2.5 (d) 6.

19. Consider the following statements:

1. A variable capacitance type transducer gives an output proportional to acceleration.
2. LVDT is a self generating type of transducer.
3. Eddy current type of transducer gives an output proportional to velocity.
4. A piezoelectric transducer cannot be used to measure static variable.

Which of these statements is/are correct?

(a) 1 and 2

(b) Resistance thermometer

(c) 2 and 4

(d) 4 only.

20. A temperature between –200°C and 1000°C may be measured conveniently by

a. Thermistor

b. Resistance thermometer

c. Optical pyrometer

d. Copper constantan thermocouple.

21. A hot-wire anemometer is a device used to measure:

(a) Gas velocities (b) Pressure in gases

(c) Liquid discharge (d) Temperature.

22. Very small displacements are effectively measured using

(a) LVDT (b) Strain gauge

(c) Thermistor (d) Tachogenerator.

23. Which one of the following statements is not correct?

a. The operation of LVDT can be adversely affected by stray magnetic a.c. fields or by the presence of large masses of metal nearby.

b. Fitting a magnetic shield with a longitudinal slot along it, over the transducer can largely be beneficial.

c. Relatively small displacements are required for appreciable differential output.

d. LVDT gives high output and many times, there is no need for amplification. It possesses a high sensitivity, which is typically of about 40 V/mm.

24. Which one of the following defines **piezo-resistive effect?**

a. Production of voltage in a crystal subjected to mechanical strain.

b. Changes in the value of resistivity of a conductor which is strained.

c. Creation of strain in a crystal when electricity is passed through the crystal.

d. Increase in the frictional resistance of a sliding contact under the influence of magnetic field.

25. Consider the following statements relating to an **LVDT** type of transducer.

1. A soft iron core provides the magnetic coupling between a primary coil and two secondary coils, connected in series opposition.

2. The output is proportional to the displacement of the iron core.
3. The device is very sensitive and is linear aver a wide range of motion.
4. This is a variable inductance transducer.

Which of the statements given above are correct?

(a) 1, 2 and 4 (b) 3 and 4
(c) 1 and 2 (d) 1, 2, 3 and 4.

26. Which one of the following is the principal **disadvantage of a piezoelectric transducer?**

a. It can measure force only.
b. It cannot measure static conditions
c. It is too small to handle.
d. It produces only d.c. voltage.

27. Which one of the following **transducers** requires power supply for its operation?

(a) Thermocouple (b) Photovoltaic cell
(c) Piezoelectric crystal (d) Thermistor.

28. Variation of which of the following parameters is related to the **level change in the measurement** of liquid level using two parallel metal rods immersed in a liquid tank?

a. Area of plates
b. Distance between plates
c. Dielectric medium
d. Length of plates immersed in the liquid.

29. A piezoelectric transducer is a voltage source of 10 V with an internal impedance of 10 MΩ. It is connected to a digital oscilloscope with an input impedance of 10 MΩ directly. What is voltage measured by the oscilloscope?

(a) 20 V (b) 10 V
(c) 5 V (d) 2.5 V.

30. Match List-I (Transducer) with List-II (Measured Quantity) and select the correct answer using the code given below the Lists:

List-I (Transducer)	List-II (Measured Quantity)
A. Hall Effect pick up	1. Pressure
B. Piezoelectric pick up	2. Velocity of fluid
C. Hot-wire anemometer	3. Flow rate
D. Rotameter	4. Current

Code:

	A	*B*	*C*	*D*
a.	4	3	2	1
b.	2	1	4	3
c.	4	1	2	3
d.	2	3	4	1.

31. Which one of the following in simple elemental forms is not a **pressure sensor**?

(a) Cantilever beam (b) Bourden tube
(c) Diaphragm (d) Bellows.

32. Which one of the following statements in correct?

a. A piezoelectric pressure transducer can be used for measuring both static and dynamic pressures.
b. A resistance strain gauge type pressure transducer cannot be used for dynamic pressure measurement.
c. Vacuum levels lower than 1 micron can be measured with an ionisation gauge.
d. Accuracy of a manometer is affected by shape or size of the tubes.

33. An inductive pick up is used to measure speed of a shaft on which a 120 tooth wheel is attached. The number of pulses produced per second is 3000. What is the RPM of the shaft?

(a) 1500 (b) 1800
(c) 3000 (d) 3600.

34. Which one of the following is the best method of measurement of temperature of hot bodies radiating energy in visible spectrum?

(a) Thermocouple (b) Thermopile
(c) Optical pyrometer (d) Bolometer.

35. The working principle of a Pirani gauge pressure **transducer** is based on which one of the following?

a. Humidity of the medium
b. Thermal conductivity of the medium
c. Combustibility of the medium
d. Electrical resistivity of the medium.

36. Which one of the following is the **most sensitive device?**

(a) Thermocouple (b) RTD
(c) Hermistor (d) Pyrometer.

37. Which one of the following transducers is the most suitable for the measurement of linear displacement??

(a) Strain gauge (b) LVDT
(c) Piezoelectric crystal (d) Microphone.

38. Match List-I (Type of Hygrometer) with List-II (Related Process/Material) and select the correct answer using the code given below the lists:

List-I (Type of Hygrometer)	List-II (Related Process/Material)
A. Resistive hygrometer	1. Quartz crystal
B. Capacitive hygrometer	2. Lithium chloride crystal
C. Microwave refractometer	3. Change in dielectric constant
D. Crystal hygrometer	4. Change in frequency of oscillator

Codes:

	A	*B*	*C*	*D*
a.	2	3	4	1
b.	4	1	2	3
c.	2	1	4	3
d.	4	3	2	1.

39. Which is the most suitable thermocouple transducer for the measurement of temperature in the range of 1300°C to 1500°C?

a. Chromel-alumel

b. Platinum-rhodium

c. Iron-constantan

d. Chromel-constantan.

40. With a variable tachometer having 60 rotor teeth, the counter 3600 counts per minute. The device speed, in rpm, would be

(a) 36 (b) 60

(c) 360 (d) 3600.

41. In an experiment with a thermocouple with one junction at 0°C, the following data were obtained at θ°C:

θ°C	20°C	100°C	300°C
e (in milivolts)	0.194	0.850	1.650

Assuming parabolic relation, the calculated neutral temperature for this thermocouple will be about

(a) 300°C (b) 333°C

(c) 366°C (d) 399°C.

42. Measurement of a non-electrical quantity involves the following four subsystems:

1. Amplifier Block.
2. Display Block.
3. Instrumentation Block.
4. Transducer Block.

The correct, sequence in which these block occur is

(a) 4, 1, 3, 2 (b) 4, 3, 1, 2

(c) 1, 3, 2, 4 (d) 1, 2, 3, 4.

43. Consider the following statements regarding a phase-sensitive detector:

1. The detector reads zero when the frequencies of input signal and reference signal are different.
2. The detector acts as a linear rectifier.
3. The detector can be used as a vector voltmeter.

Of these statements

(a) 1, 2 and 3 are correct (b) 1 and 2 are correct

(c) 2 and 3 are correct (d) 1 and 3 are correct.

44. Thermistors are made of

(a) Pure metals

(b) Pure insulators

(c) Sintered mixtures of metallic oxides

(d) Pure semiconductor.

45. In a thermocouple, two metal junctions between metals M_1 and M_2 are at temperatures T_1 and T_2. The thermocouple emf is produced because

(a) M_1 and M_2 are similar and T_1 and T_2 also similar

(b) M_1 and M_2 are dissimilar while T_1 and T_2 are similar

(c) M_1 and M_2 are similar while T_1 and T_2 are dissimilar

(d) M_1 and M_2 are similar and T_1 and T_2 also dissimilar.

46. The use of thermocouple meters for ac measurement leads to a meter-scale which is

(a) Linear (b) Square law

(c) Logarithmic (d) Exponential.

47. Match List I with List II and select the correct answer using the codes given below the lists:

List I (Quantity to be measured)	List II (Type of transducer)
A. Strain	1. Magnetostrictive transducer
B. Temperature	2. Magnetoelectric transducer
C. Angular velocity	3. Encoder
	4. Thermocouple

Codes:

	A	B	C
(a)	1	4	3
(b)	2	1	3
(c)	3	4	2
(d)	1	4	2

48. Consider the following statement:
A Linear Variable Differential Transformer (LVDT) has
1. one primary winding
2. two exactly similar secondary windings
3. a toroidal magnetic core

Of these statements
(a) 2 and 3 are correct
(b) 1 and 2 are correct
(c) 1 and 3 are correct
(d) 1,2 and 3 are correct

49. A strain gauge bridge measures the strain in a cantilever where the gauge is fixed, With strain $\in$, the gauge resistance increases form 110 Ω to 110.52 Ω. If the gauge factor is 2.03, then strain in the cantilever will be
(a) 2.06×10^{-3} (b) 3.15×10^{-3}
(c) 3.81×10^{-3} (d) 4.33×10^{-3}

50. Consider the following statements:
The cause of error in the measurement of temperature using a **themistor** are
1. self heating
2. poor sensitivity
3. non–linear characteristics

Of these statements
(a) 1, 2 and 3 are correct
(b) 1 and 2 are correct
(c) 2 and 3 are correct
(d) 1 and 3 are correct

51. For a **thermocouple pair (A, B),** the extension wires (C,D).
(a) Should be identical pair elements
(b) Should have identical temperature–emf relationship
(c) Can be any two dissimilar materials
(d) Should have very small temperature emf sensitivity

52. In a **flux meter**, the controlling torque is
(a) Produced by weights attached to the moving coil
(b) Produced by spring
(c) Not provided at all
(d) Provided by crossed coil mechanism

53. Dummy **strain, gauge** is used in conjunction with the main strain gauge to
(a) Calibrate the system
(b) Compensate temperature effects
(c) Improve sensitivity
(d) Reduce strain on the gauge

54. **Half effect device** can be used to
(a) Multiply two signals
(b) Divide one signal by another on an instantaneous basis
(c) Add two signals
(d) Subtract one signal from another

55. Match List I (Transducer) With List II (Input/Qutput variabiles) and select the correct answer using the codes given below Lists:

List I	List II
A. Electordynamic generator	1. Gas pressure to resistance change
B. Venturimeter	2. Force to displacement
C. Pirani gauge	3. Motion to voltage
D. Spring balance	4. Flow rate to pressure.

Codes:

	A	B	C	D
(a)	2	1	4	3
(b)	2	4	1	3
(c)	3	4	1	2
(d)	3	4	2	1

56. **Doppler shift principle** is used in the measurement of
(a) temperature (b) frequency
(c) speed (d) pressure

57. The emf is measured for a junction temperate different of 20 °C by thermocouples made from materials A, B and C. The pair A-B gives an emf of 165 μV, with the end of A being positive . The pair B-C gives 100 μ V, with the end of C being positive. The pair A-C will give
(a) 265 μV with the end of A being positive
(b) 265 μV with end of C being positive
(c) 65 μV with the end of A being positive
(d) 65 μV with the end of C being positive

58. Consider the following transducers:
1. L.V.DT.
2. Piezoelectric
3. Thermocouple
4. Photovoltaic cell
5. Strain gauge

Which of these are active transducers?
(a) 1, 2 and 5 (b) 1, 3 and 4
(c) 2, 3 and 5 (d) 2, 3 and 4

59. A fixed resistor of suitable value is usually connected across a the thermistor to
a) Decrease its resistance
b) Increase its sensitivity
c) Compensate its self-heating effect
d) Improve linearity

60. A resistance strain gauge with a gauge factor of 3 is fastened to a steel member subjected to a stress of 100 N/mm_2. The modulus of elasticity of steel is approximately 2×10^5 N/mm^2. The percentage change in resistance is
(a) 1.5 (b) 1
(c) 0.15 (d) 0.1

61. Consider the following statements about LVDT as a transducer:

1. The relationship between input displacement and output voltage is almost linear
2. The range of displacement that can be measured is wide
3. It does not from a loading on the mechanical system

Of these statements

(a) 1, 2 and 3 are correct (b) 2 and 3 are correct
(c) 1 and 2 are correct (d) 1and 3 are correct

62. The block diagram of data acquisition system is given. In this context, match List I with List II and select the correct answer using the codes given below the Lists:

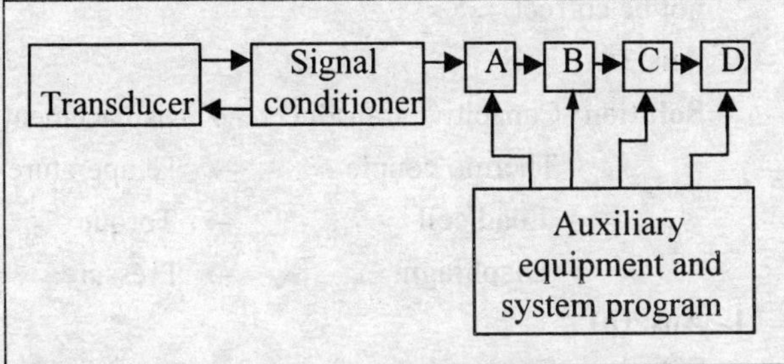

List-I (Block)	List-II (Functional name)
A. Block A	1. A/D converter
B. Block B	2. Signal converter
C. Block C	3. Digital recorder
D. Block D	4. Multiplexer

Codes:

	A	B	C	D
(a)	2	4	1	3
(b)	4	2	3	1
(c)	2	4	3	1
(d)	4	2	1	3

63. A thermocouple ammeter gives full scale deflection at 10 A . When it reads one fifth of the scale , the current will be

(a) 2 A (b) 4 A
(c) 4.47 A (d) 5.78 A

64. The resistance of thermistor is 5000 Ω at 20°C and its resistance temperature coefficient is 0.04/°C. A measurement with a lead resistance of 10 Ω will cause an error of

(a) 0.05°C (b) 0.1°C
(c) 0.4°C (d) 0.8°C

65. Match List I with List II and select the correct answer using codes given below the lists:

List I (transducers)	List II (characteristics)
A. Thermocouple	1. Modulated output
B. Thermistor	2 . Resistance change wit Pressure
C. Strain gauge	3 Negative temperature coefficient
D. LVDT	4. Constant temperature a one end

Codes:

	A	B	C	D
(a)	3	2	4	1
(b)	4	3	2	1
(c)	2	1	4	3
(d)	1	2	3	4

66. A photo-cell is illuminated by a small bright source placed one metre away. When the same source of light is placed two metres away. The electrons emitted by the photocathode.

(a) Each carry one quarter of their previous energy
(b) Each carry one quarter of their previous moments
(c) Are half as numerous
(d) Are one quarter as numerous.

67. A temperature sensitive transducer is subjected to a sudden temperature change. It takes 10 seconds for the transducer to reach equilibrium condition (five time constants). The time taken by the transducer to read half of the temperature difference will be nearly

(a) 1.38 second (b) 5.00 second
(c) 8.62 second (d) 12.4 second

68. Strain gauge, LVDT and Thermocouple may be classified as

(a) active transducers
(b) analogue transducers
(c) primary transducers
(d) none of the above.

69. Which of the following closely represents the temperature (T) – Resistance (R) characteristics of a thermistor.

(a) $R_T = a\,R_o e^{b/T}$ (b) $R_T = a\,R_o e^{-b/T}$
(c) $R_T = -a\,R_o e^{-b/T}$ (d) $R_T = -a\,R_o e^{b/T}$

70. Match List I with List II and select the correct answer by using the Codes given below the Lists:

List-I (Instrument)	List-II (use)
A. Electro- Encephelograph	1. Diagnostic tool for heart ailment
B. Electro- Cardiograph	2. Diagnostic tool for brain element

C. Sphygmo-manometer	3. Instrument used for measuring B.P.
D. Stethoscope	4. Instrument used to hear pulse/heart beat

Codes:

	A	B	C	D
(a)	2	1	3	4
(b)	3	1	2	4
(c)	2	1	4	3
(d)	1	2	3	4

71. In hygrometers the principle of measurements is
(a) change in resistance of salts with humidity
(b) change in microwave power using klystron
(c) change in thermal conductivity using thermistor
(d) none of the above

Answers Set-3

1. Ans. (d)

Solution: Radiation pyrometer measures the radiant heat emitted or reflected by a hot object at higher temperature range of 1200°C to 2500°C.

It is not used for lower temperatures as it exhibit poor sensitivity in the lower temperature ranges.

2. Ans. (d)

Solution: Hall effect pick up can measure the magnetic flux.

3. Ans. (b)

4. Ans. (a)

Solution: The function of the reference electrode in a pH meter is to produce a constant potential regardless of the pH value of the solution under test.

5. Ans. (b)

Solution: Pirani gauge is useful for pressures ranging from 10^{-1} to 10^{-3} mm of Hg.

6. Ans. (a)

7. Ans. (b)

Solution: Hot wire anemometer is a device used to measure the liquid discharge.

8. Ans. (a)

9. Ans. (a)

Solution: pH value of a solution is defined as the negative logarithm of the hydrogen ion concentration

or

$pH = -\log_{10}(H^+)$.

10. Ans. (b)

Solution: Piezoelectric crystal voltage sensitivity (g) is defined as the ratio of electric field intensity to stress

Therefore, $g = \dfrac{\text{electric field}}{\text{stress}}$.

11. Ans. (d)

Solution: Advantage of electronics Transducer

1. When dealing with electrical/electronic signals, the inertia effects are due to electrons which have negligible mass. Thus, the mass inertia effects are minimized
2. These systems can be controlled with a very small power level
3. The electrical output can be easily used, transmitted end processed for the purpose of measurement.

12. Ans. (c)

Solution: Statement (3) is correct other may be or may not be correct.

13. Ans. (d)

Solution: Capacitive transducer → Displacement
Thermo couple → Temperature
Load cell → Torque
Diaphragm → Pressure.

14. Ans. (b)

Solution: Gauge factor of a strain gauge is defined as the ratio of per unit change in resistance to per unit change in length of element i.e.

$$GF = \frac{\nabla R/R}{\nabla L/L}.$$

15. Ans. (b)

16. Ans. (d)

17. Ans. (c)

18. Ans. (b)

Solution: Gauge factor, $G_f = \dfrac{\Delta R/R}{\Delta L/L} = 5$

Relation between Gauge factor (G_f) and Poisson ratio (ρ) is given as

$$G_f = 1 + 2\rho$$

$$5 = 1 + 2\rho$$

$$2\rho = 4 \Rightarrow \rho = \frac{4}{2} = 2.$$

19. Ans. (d)

Solution: The use of Piezo-electric transducer elements is confined primarily to dynamic measurement. The voltage developed by application of strain is not held under static conditions.

20. Ans. (b)

21. Ans. (c)

22. Ans. (a)

23. Ans. (c)

Solution: Relatively large displacement are required for appreciable differential output.

24. Ans. (b)

Solution: If a conductor is stressed or compressed, its resistant changes. This property is called Piezo-resistive effect.

25. Ans. (d)

26. Ans. (b)

Solution: It is confined primarily to dynamic measurements. The voltage developed by application of strain is not held under **static conditions**.

27. Ans. (d)

Solution: Thermistor is a passive transducer, which require power supply.

28. Ans. (c)

29. Ans. (c)

Solution:

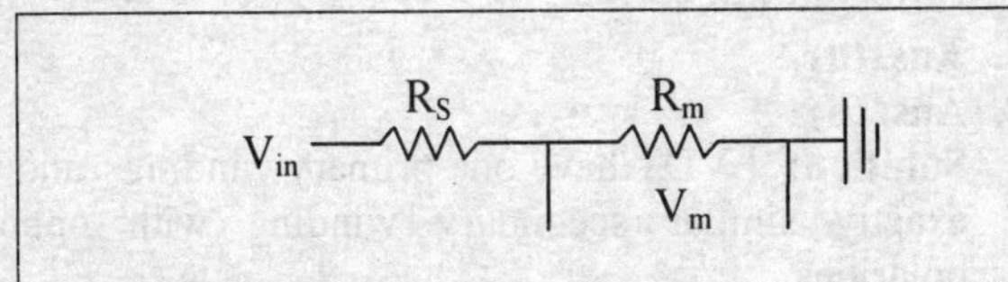

$$V_m = \frac{10}{10+10} \times 10 = 5 \text{ V}.$$

30. Ans. (c)

31. Ans. (a)

32. Ans. (c)

Solution: Pressure that be measured by ionization gauge range from 10^{-3} mm (1 micron) to 10^{-8} mm of Hg.

33. Ans. (a)

Solution: Pulse rate = 3000 pulse/second

$$= 3000 \times 60 \text{ pulse/minute}$$

RPM of the shaft

$$= \frac{\text{pulse/minute}}{\text{No. of teeth}} = \frac{3000 \times 60}{120} = 1500.$$

34. Ans. (c)

Solution: Optical Pyrometer → The radiations from a heated body at high temperature fall within the insides region of the electro-magnetic spectrum. Brightness of the light of a givens colour emitted by a hot source is measured which is an indication of temperature.

The wavelength of radiation accepted is restricted by means of a colour filter and brightness is measured by comparison with a standard lamp.

In can measured up to 3000°C and is used for of the measurement of temperature of furnace metals.

35. Ans. (b)

36. Ans. (c)

37. Ans. (b)

38. Ans. (a)

Solution Hygrometer → It measures the value of humidity directly. Generally the output of the Hygrometer is used to indicate relative humidity.

Types:

(a) Resistance Hygrometer → Some Hygroscopic salts exhibit a *change in resistively* with humidity. **Example:** Lithium chloride crystal.

The higher the relative humidity the more moisture the lithium chloride will absorb.

(b) Capacitance Hygrometer → Some Hydroscopic materials exhibit a *change is dielectric constant* with humidity. These transducers have response time of about 1s.

(c) Microwave Refractometer → A system consisting of 2 cavities each of which is coupled to a Klystron. One contains dry air and the other contain the mixer to be measured.

The change in dielectric constant change the frequency of the oscillator which is a measure of relative humidity is the mixture.

(d) Crystal Hygrometer → Quartz Crystal are used as frequency determination elements in electronic oscillator. Frequency shifts with humidity.

These transducer are useful if a telemetry system is needed.

(e) Aluminum oxide Hygrometer → Change is dielectric constant and change in resistively with a change in humidity.

39. Ans. (b)

Solution: Measurement of temperature

Temperature measuring devices are known as pyrometer

(a) Resistance pyrometer → Pure platinum is used because it has a high temperature coefficient and its resistance change per unit variation in temperature is uniform. Range of temperature covered by platinum pyrometer is

$-200°\text{C}$ to $1450°\text{C}$.

(b) Thermocouple pyrometer → The e.m.f. generated in a circuit containing hot and cold junctions depends upon the temperature difference between the junction.

Copper-constants	Upto 6000°C
Iron-constants	800–1000°C
Aluminium-Chrominum	1000–2000°C
Platinum-Rhodium	Upto 1500°C

(c) Radiation Pyrometer → Radiation energy in the form of heat and light from hot body is focused on to a sensitive thermocouple

Energy radiated from a black body

$$E = K\left(T_1^4 - T_2^4\right) \cong K\,T^4$$

(d) Semiconductor

(e) thermometer → Germanium crystals is used for measurement of cryogenic temperature (below 25° K or –248°C). Semiconductor material have large resistance temperature coefficients. Silicon crystals are used in the ranges –48° C to 252°C.

(f) Thermistor Thermometer → Thermistors are semiconductor which behave as resistance with a high negative temperature. Coefficient.

Thermistor are used in temperature range –100°C to 300°C.

40. Ans. (b)

Solution: Speed in rpm $= \dfrac{\text{Counts per minute}}{\text{Number to teeth}}$

$$= \frac{3600}{60} = 60.$$

41. Ans. (b)

Solution: For thermo couple e.m.f

$$= m_2\left(T_2 - T_1\right)\left(T_n - \frac{T_1 + T_2}{2}\right)$$

where m_2 is a constant and T_n is neutral temperature.

$$\text{Or } 0.194 = 20\, m_2\left(T_n - \frac{20}{2}\right)$$

$$= 20\, m_2\, T_n - 200\, m_2 \qquad (1)$$

$$\text{Similarly } 0.85 = m_2(100)\left(T_n - \frac{100}{2}\right)$$

$$100\, m_2\, T_n - 5000\, m_2 \qquad (2)$$

Solving equations (1) and (2) for m_2, we get

$0.194 - 0.85 = -\ 1000\ m_2 + 5000\ m_2$

or $m_2 = 3 \times 10^{-5}$

Substituting this value in equation (1), we get

$$T_n = \frac{0.194 + 6\times10^{-3}}{20\times10^{-5}} = 333°\text{C}.$$

42. Ans. (a)

43. Ans. (c)

44. Ans. (c)

Solution: A thermistor is generally fabricated a semiconductor material having a large nonlinear temperature coefficient of resistance.

Thermistor are compared of sintered mixture of metallic oxides such as manganese, nickels cobalt copper, iron and uranium.

45. Ans. (b)

46. Ans. (b)

Solution: Deflection of thermo couple instrument

$\theta = K_3 I^2$

$K_3 = a/c_1\ K_2\ R \rightarrow$ constant

Thermocouple indicate

RMS value of current.

47. Ans. (d)

48. Ans. (b)

Solution: LVDT have one primary winding and two exactly similar secondary winding with opposite polarities.

49. Ans. (a)

Solution: Gauge factor $G_f = \dfrac{\Delta R/R}{\text{Strain}}$

$$= \frac{(110.52 - 110)/110}{\text{strain}}$$

$$\therefore \quad \text{Strain} = \frac{\Delta R/R}{G_f} = \frac{0.52}{110.00} \times \frac{1}{2.03}$$

$$= 2.33\times10^{-3} \Rightarrow 2.06\times10^{-3}$$

50. Ans. (d)

51. Ans. (a)

52. Ans. (c)

Solution: In a **flux meter**, the controlling torque is not provided at all

53. Ans. (b)

Solution: Dummy **strain, gauge** is used in conjunction with the main strain gauge to compensate temperature effects

54. Ans. (a)

Solution: Half effect device can be used to multiply two signals

55. Ans. (c)

Solution: The correct match is given below:

Electordynamic generator	Motion to voltage
Venturimeter	Flow rate to pressure.
Pirani gauge	Gas pressure to resistance change
Spring balance	Force to displacement

56. Ans. (c)
57. Ans. (c)
58. Ans. (d)
59. Ans. (d)
60. Ans. (c)

Solution: Gauge factor = $\dfrac{\Delta R/R}{\Delta L/L}$

But, $\dfrac{\Delta L}{L}=\dfrac{100}{2\times10^5}$

Therefore $\dfrac{\Delta R}{R}=\dfrac{3\times100}{2\times10^5}=1.5\times10^{-3}$

Hence percentage change in resistance

$100=\dfrac{\Delta R}{5}\times100 = 1.5\times10^{-3}=0.5$

61. Ans. (d)
62. Ans. (c)
63. Ans. (c)

Solution: Thermocouple reals rms value $\theta\propto I^2$

From $\dfrac{\theta_1}{\theta_2}=\dfrac{I_1^2}{I_2^2}$; $\dfrac{1}{5}=\dfrac{10^2}{I_2^2}$

$I_2^2=\dfrac{100}{5}=20$

$\Rightarrow$ $I_2=4.47\text{ A}$

64. Ans. (a)
Solution: Temperature coefficient

$\alpha_t=\dfrac{\Delta R/R}{\Delta T}\Rightarrow\Delta T=\dfrac{\Delta R/R}{\alpha_t}$

$=\dfrac{10/5000}{0.04}=0.0560\,^{\circ}C$

A change of $1^{\circ}C$ in temperature is reflected as a change of $5000\times0.04=200$ ohms of resistance in thermistor resistance.
A resistance changes of 10 ohm is equivalent to $\dfrac{10}{200}\,^{\circ}C=0.05\,^{\circ}C$

65. Ans. (b)
Solution: The match is given below

Thermocouple	Constant temperature a one end
Thermistor	Negative temperature coefficient
Strain gauge	Resistance change witpressure
LVDT	Modulated output

66. Ans. (d)
Solution: The frequency and hence energy of each of the photons will remain the same . Since I $\propto\dfrac{1}{r^2}$ the number of photons incident on the photocathode will be one fourth of the their initial number. The number of photoelectrons emitted will be accordingly on fourth in number as compared to this first case

67. Ans. (a)
Solution: Time to reach equilibrium condition = 5 τ = 10 seconds
Time constant τ = [10/5] = 2 sec.

Now $\theta=\theta_o\left(1-e^{(-t/\tau)}\right)$

Or $\dfrac{\theta}{\theta_o}=1-e^{-(-t/\tau)}$ Or $\dfrac{1}{2}-e^{-t/\tau}$

Therefore, t = 1.38 second.

68. Ans. (b)
69. Ans. (a)

70. Ans. (a)
Solution: The correct match is given below

Instrument	Use
Electro-Encephalo-graph	Diagnostic tool for brain element
Electro-Cardiograph	Diagnostic tool for heart ailment
Sphygmo-manometer	Instrument used for measuring B.P.
Stethoscope	Instrument used to hear pulse/heart beat

71. Ans. (a)
Solution: Resistive Hygrometers
Dunmore element and the pope cell are the two extensively used percent relative humility (% RH) electrical transducers.
The Dunmore sensor makes use of bifilar-wound inert wire grid on an isolative substrate coated with a lithium chloride (LiCl) solution of a controlled concentration. The hygroscopic nature of this salt makes it to take up water vapour from the surrounding atmosphere.

With the increasing in relative humidity, the salt (lithium chloride) film absorbs more and more water vapour from the air and so its **resistance decreases** markedly. Thus the ac resistance of the transducer is an indication of the prevailing % RH. Bridge type resistance measuring circuitry with an ac excitation is used normally as dc excitation polarizes the transducer, eventually causing loss of calibration. The resistance-relative humidity relation is quite non-linear.

MCQ Set–4

1. The **least suitable transducer** for **static pressure measurement** is
(a) Semi-conductor strain guage
(b) Variable capacitor for transducer
(c) Metal wire strain guage
(d) Piezoelectric transducer

2. A **semi-conductor strain guage**
(a) Has a much higher guage factor then that of a metal wire guage
(b) Employs piezoelectric property of undoped silicon
(c) Does not require temperature compensation
(d) Exhibits very little guage factor variation as compared to that of metal wire gauges

3. The emf developed by a **thermocouple** depends on
(a) The length of wires and temperature difference between the hot and cold junction
(b) Materials used, diameter of wires used and temperature difference between the hot and cold junctions
(c) Materials used, temperature of hot junction and temperature of cold junction
(d) Materials used, shape and size of materials, resistance of the wires and temperature difference between the hot and cold junctions.

4. The operation of a **Pirani guage** is based on
(a) Ionization of gas at low pressures
(b) Variation of volume with pressure
(c) Variation of viscosity with pressure
(d) Variation of thermal conductivity of gas with pressure

5. In a **seismic pick-up for getting** an output proportional to acceleration it is desirable to have
(a) A natural frequency very small in comparison to frequency of input and a damping ratio around 0.7
(b) A natural frequency very large in comparison to frequency of input and a damping ratio around 0.7
(c) A natural frequency equal to the frequency of input signal and a damping ratio around 0.7
(d) A natural frequency very small in comparison to frequency of input and a damping ratio higher that unity.

6. A unity ratio **quarter bridge** strain measuring circuit produces an output of 1 mV for a strain of 500 micro stain when the **bridge excitation** is 4 volts. The gage factor of the element is then
(a) 1 (b) 2
(c) 3 (d) 4

7. The output of the **LVDT** with input mechanical motion of 10 Hz and excitation frequency 400 Hz will contain frequencies
(a) 10 Hz and 400 Hz (b) 400 Hz only
(c) 10 Hz only (d) 390 Hz and 410 Hz

8. **Bellow's expansion** is usually against a spring. The spring is provided to
(a) Increase sensitivity
(b) Increase operating range
(c) Increase linearity
(d) Decrease hysteresis effect

9. A **nozzle flow meter** has a pressure drop of 200 mm of water for a flow rate of 100/min. For a pressure drop of 400 mm of water, the flow rate is
(a) 141 *l*/min (b) 165 *l*/min
(c) 200 *l*/min (d) 362 *l*/min

10. A weight of 4.9 N rests on a rough horizontal floor that has a co-efficient of friction of 0.5. If g = 9.8 m/s^2, the smallest horizontal force that gives the weight an **acceleration** of 4m/s^2 is approximately
(a) 2.0 N (b) 2.5 N
(c) 4.5 N (d) 19.6 N

11. With the **stroboscope**, synchronism has been obtained at the flashing rates of 2.5 and 20 pulses per second. The speed of rotation is
(a) 0.6 rev/min. (b) 50 rev/min.
(c) 100 rev/min. (d) 6000 rev/min.

12. Match the List I instruments with the List II parameters they measure and select the correct answer using the codes:

List I	List II
A. Optical encoder	1. Stress
B. Polaris cope	2. Low pressure
C. Optical Pyrometer	3. Angular displacement
D. Pirani guage	4. Temperature

Codes:

	A	B	C	D
(a)	2	4	3	1
(b)	4	1	3	2
(c)	3	1	4	2
(d)	4	2	1	3

13. Match the List I devices with List II quantities measure and select the correct answer using the codes:

List I	List II
A. Rotameter	1. Relative humidity
B. Hydrometer	2. Density
C. Sling Psychrometer	3. Dynamic pressure
D. Piezoelectric transducer	4. Flow rate

Codes:

	A	B	C	D
(a)	2	4	3	1
(b)	4	1	3	2
(c)	3	1	4	2
(d)	4	2	1	3

14. Match the List I instruments on the left used for the calibration of the instruments List II on the right and select the correct answer using the codes:

List I	List II
A. Inverted bell	1. Slip-guages
B. Kelvin's double bridge	2. Gas flow meter
C. Dead weight tester	3. Pressure guage
D. Interferometer	4. Resistance thermometer

Codes:

	A	B	C	D
(a)	2	4	3	1
(b)	4	1	3	2
(c)	3	1	4	2
(d)	4	2	1	3

15. A **piezoelectric transducer** is directly connected through a cable to an electronic voltmeter. The minimum operating frequency of measurement is 1000 Hz. IF the connecting cable length is doubled., the new minimum operating frequency is

(a) 500 Hz (b) 1000 Hz
(c) 2000 Hz (d) 4000 Hz

16. Assuming uniform **pressure** over the supported area, the friction moment in an axial bearing of a shaft of diameter d for an axial load of F (With co-efficient of friction μ) is given by

(a) [4 / 3] Fμd (b) [1 / 3] Fμd
(c) [2 /3] Fμd (d) Fμd

17. The ideal linear characteristic for transmitting **angular motion** between two misaligned shafts is achieved by

(a) Dog coupling
(b) Universal coupling
(c) Flexible spring coupling
(d) Oldham's coupling

18. Measurement of **viscosity** involves measuring

(a) Frictional force (b) Coriolł's force
(c) Centrifugal force (d) Buoyant force

19. An example of variable area device for **measuring flow** is

(a) Flow nozzle (b) Orifice meter
(c) Venturimeter (d) Rotameter

20. The primary standard for calibrating vacuum is

(a) McLeod gauge (b) Deadweight tester
(c) Thermocouple gauge (d) Kundsen guage

21. When the reading of a **pH meter** changes from 5 to 7, the hydrogen ion concentration of the solution is

(a) Halved (b) Doubled
(c) Increased 100 times (d) Decreased 100 times

22. A **pressure gauge** used to measure vacuum indicates a guage pressure of 5 kPa, If the atmospheric pressure is 100 kPa, the absolute pressure is

(a) 105 kPa (b) 0.05 kPa
(c) 95 kPa (d) 20 kPa

23. A **cantilever** is used as a primary sensing element for the measurement of force. The measured deflection is 1 mm. Suppose each of the dimensions, length width and thickness of the cantilever are doubled for the same material and force, the deflection will be

(a) 0.125 mm (b) 0.25 mm
(c) 0.5 mm (d) 2 mm

24. Match the List I device with List II its working principles and select the correct answer using the codes:

List I	List II
A. Bourdon guage	1. Thermal conductivity
B. Pirani gauge	2. Thermal co-efficient of expansion
C. Semi-conductor strain guage	3. Piezo-resistive effect
D. Bimetallic strip	4. Phote-electric effect
	5. Deformation of element

Codes:

	A	B	C	D
(a)	2	4	3	1
(b)	4	1	3	2
(c)	5	1	3	4
(d)	4	2	1	3

25. Match the List I sensors and the List II properties used and select the correct answer using the codes:

List I	List II
A. Pyroelectric detector	1. Anemometer
B. Heat transfer	2. Magnetostrictive
	3. Electrostatic polarization

Codes:

	A	B
(a)	3	2
(b)	2	1
(c)	3	1
(d)	1	3

26. The natural frequency of a **seismic mass type** pickup is f_n. The pick-up will give the correct motion information, if the exciting frequency f_e is
(a) f_n (b) $[1/\sqrt{2}] f_n$
(c) $>> f_n$ (d) $<< f_n$

27. A **pitol static tube** is used to measure the airflow rate in a squire tube. if p_d is the differential pressure, the flow rate is proportional to
(a) $\sqrt{(P_d)}$ (b) P_d
(c) $[1/\sqrt{(p_d)}]$ (d) p_d^2

28. In a **sensitive spirit level,** the sealed glass tube has a radius of curvature equal to 200 meters. A movement of 2mm of the bubble measures an inclination of the surface, in radians, equal to
(a) 0.40 (b) 0. 10
(c) 10^{-2} (d) 10^{-5}

29. A shaft and bearing have been designated as in fig. 1.25 . The type of the fit is.

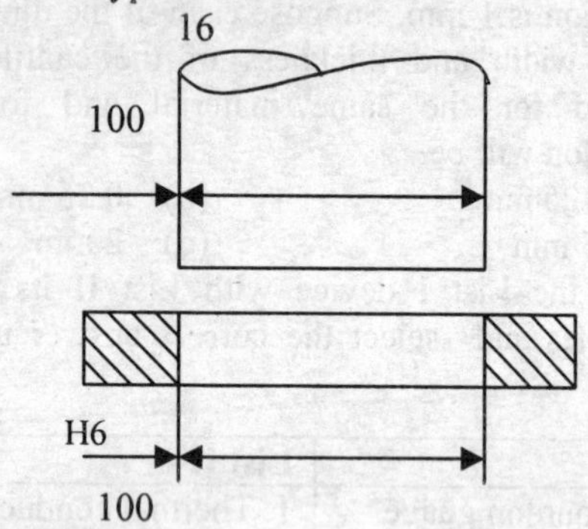

(a) Push fit (b) Interference fit
(c) Transition fit (d) Clearance fit.

30. Match the List I following instruments with the List II measurements and select the correct answer using the codes:

List I	List II
A Rotameter	1 Rotational speed
B. Tachometer	2. Roughness
C. Tracer	3. Temperature
D. Pyrometer	4. Flow rate

Codes:

	A	B	C	D
(a)	2	4	3	1
(b)	4	2	1	3
(c)	3	1	4	2
(d)	2	4	1	3

(Gate Inst.-1997)

31. Match the List I type of the thermocouple with List II its characteristics and select the correct answer using the codes:

List I	List II
A. Type E Chromel–Constantan	1. Low cost
B. Type J Fe – Alumel	2. Good stability
C. Type T Cu – constantan	3. High precision
D. Type S Pt – PtRh	4. High output

Codes:

	A	B	C	D
(a)	2	4	3	1
(b)	4	1	3	2
(c)	3	1	4	2
(d)	4	2	1	3

32. Match the List device with the List II measurement and select the correct answer using the codes:

List I	List II
A. Slip gauges	1. Co – efficient of friction
B. Sine bars	2. Angles
	3. Linear distance

Codes:

	A	B
(a)	3	2
(b)	3	1
(c)	1	2
(d)	2	3

33. Choose the **thermocouple** that can measure a temperature in therange of 1300° C to 1400° C
(a) Iron-constantan
(b) Copper-constantan
(c) Platinum-platinum rhodium
(d) Chromel- alumel

34. The two secondary coils of an LVDT have wrongly been connected as shown in figure below. Then the input-output relationship would be

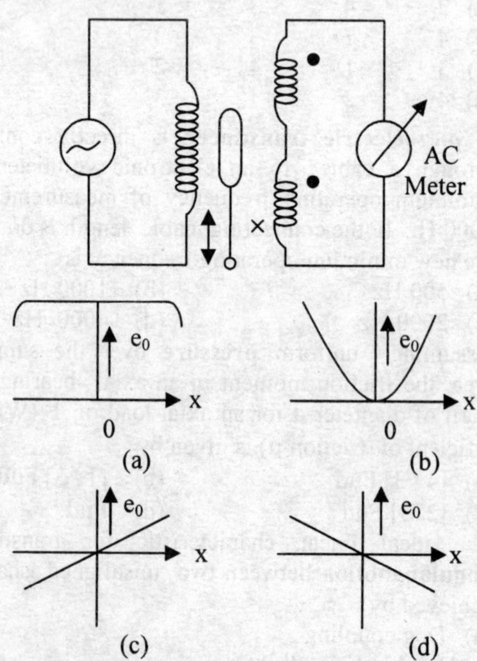

35. A strain gauge has a gauge factor G = -100 . The type of the strain gauge is .
(a) Unbounded metal type
(b) Bonded metal foil type
(c) p – type semi conductor
(d) n – type semi conductor.

36. The extension wires of iron constantan **thermocouple** were improperly wired shown in figure below
The voltmeter calibrated in °C will then read.

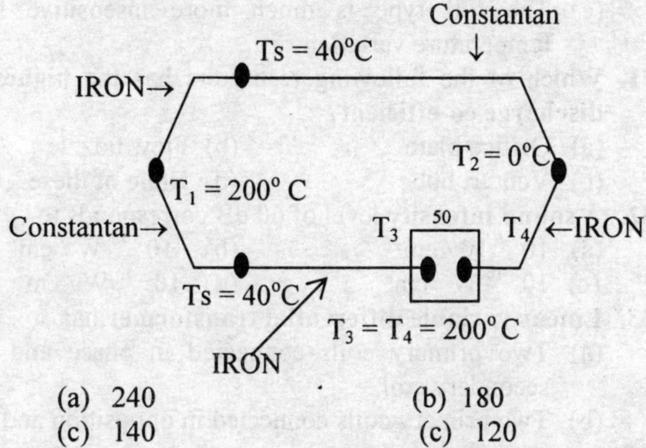

(a) 240 (b) 180
(c) 140 (c) 120

37. Match the List I transducers with the List II materials used and select the correct answer using the codes:

List I	List II
A. Permalloy	1. Piezolectic
B. Advance	2. Magnetostrictive
C. Phosphor bronze	3. Strain gauge
D. Quartz	4. Spring

Codes:

	A	B	C	D
(a)	2	4	3	1
(b)	4	3	2	1
(c)	3	4	1	2
(d)	2	4	1	3

38. Match the List I physical principle with the List II characteristic equation and select the correct answer using the codes:

List I	List II
A. Photonic effect	1. $E = C_1 (T_1 - T_2) + C_2 (T_1^2 - T_2^2)$
B. Absorption of light	2. $E = \sigma T^4$
C. Total radiation pyrometry	3. $E = hc / \lambda$
D. Thermoelectric effect	4. A = abc

Codes:

	A	B	C	D
(a)	3	1	2	4
(b)	4	3	2	1
(c)	3	4	1	2
(d)	2	4	1	3

39. Match the List I **pressure measuring** devices with the List II pressure levels in torr and select the correct answer using the codes:

List I	List II
A. Bourdon tube	1. 10^3
B. Ionization gage	2. 10^{-10}
C. Water manometer	3. 1.0
D. Pirani	4. 10^{-5}

Codes:

	A	B	C	D
(a)	2	4	3	1
(b)	3	4	1	2
(c)	3	1	4	2
(d)	1	2	3	4

40. The **temperature** of fixed points used to define international Temperature Scale are determined by using.
(a) Platinum resistance thermometer
(b) Platinum Platinum-Rhodium thermo – couples.
(c) Vapour – pressure thermometer
(d) Gas thermcmeter to which corrections are applied for non ideal behaviour of the gas.

41. The emf-temperature data for a **thermocouple** with reference junction at 0°C is as follows,

Temperature (°C)	20	180	200
Emf (m V)	1.2	11.8	13.5

The emf developed in mV with the two junctions at 200° C is
(a) 11.8 (b) 12.3
(c) 13.0 (d) 13.5

42. A rotameter with a heavy float for **measuring gas flow** is calibrated with a gas of density 1.2 kg/m^3. It measures the flow rate of a different gas having density of 2 kg/m^3 and indicates a flow rate of 2.2 m^3/min . The actual flow rate in m^3 / min is.
(a) 0.79 (b) 1.32
(c) 1.70 (d) 2.20

43. The **sound pressure level** (SPL) measured in open space (free field), at a distance of 6 m from a noise source is 80 dB. At a distance of 60 m the SPL is.
(a) 80 dB (b) 60 dB
(c) 8 db (d) 1.34 dB

44. An **elastic transducer** is used to measure pressure in a vessel and it indicates a pressure of 3.2 bar. Atmospheric pressure is 1.01 bar. The absolute pressure in the vessel in bar is
(a) 1.01 (b) 2.19
(c) 3.20 (d) 4.21

45. A photo conductive **transducer** works on the principle that when a light beam strikes
(a) The material, its resistance decreases, which is sensed by an external circuit
(b) The barrier between transparent metal layer and a semi-conductor material, a voltage is generated

(c) The barrier between transparent metal layer and a semi-conductor material, a current is generated in the external circuit
(d) The cathode, it releases electrons, which are attracted towards anode, thereby producing electric current in the eternal circuit.

46. The resistance **temperature** characteristic for a temperature transducer is shown in Figure. The transducer is.

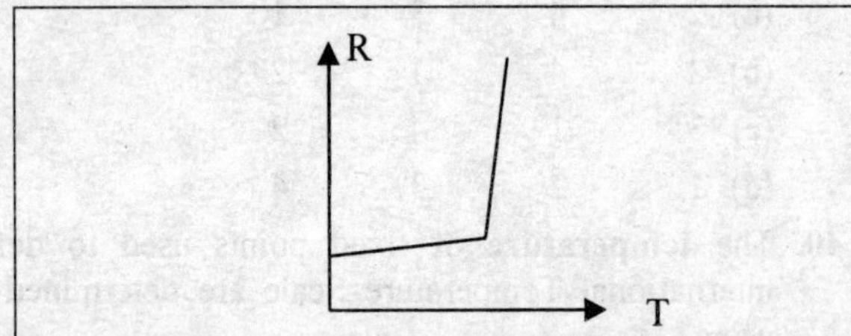

(a) Nicket RTD (b) NTC thermistor
(c) Platinum RTD (d) PTC themistor

47. identify the correct set of matches from the following

List I	List II
A. LVDT	1. Reference junction compensation
B. Piezoelectric crystal	2. Charge amplifier
C. pH probe	3. Phase sensitive detector
D. Thermocouple	4. High input impedance amplifier

	A	B	C	D
(a)	1	3	4	2
(b)	3	2	4	1
(c)	4	1	2	1

(d) None of the above

48. Identify the correct set of matches from the following

List I	List II
A. Pirani gauge	1. Temperature measurement
B. Venturi meter	2. Vacuum pressure measurement
C. Radiation pyrometer	3. Force measurement
D. Load cell	4. Flow measurement

	A	B	C	D
(a)	2	1	4	3
(b)	4	3	2	1
(c)	2	4	1	3

(d) None of the above

49. Which of the following gauges can measure the lowest vacuum pressure?
(a) McLead gauge
(b) Pirani gauge
(c) Ionisation gauge

50. Among the bonded metal **strain gauges**, the foil type is more popular than the wire type because,
(a) Error due to transverse strain is much less in foil type
(b) Gauge factor is much higher in foil type
(c) The foil type is much more insensitive to temperature variation

51. Which of the following restrictors has the **highest discharge co-efficient**?
(a) Orifice plate (b) Flow nozzle
(c) Venturi bube (d) None of these

52. A **sound intensity** level of 60 dB corresponds to.
(a) 10^{-6} W / cm^2 (b) 10^{-10} W / cm^2
(c) 10^{-16} W / cm^2 (d) 10^{-62} W / cm^2

53. **Linear variable differential transformer** has
(a) Two primary coils connected in phase and a secondary coil.
(b) Two primary coils connected in opposition and a secondary coil.
(c) One primary coil and two secondary coils. Connected in phase.
(d) One primary coil and two secondary coils connected in opposition.

54. Value of **pH** of a solution is 4. It indicates that concentration of hydrogen ions is
(a) 10^{-4} g/ litre and the solution is acidic
(b) 10^{-4} g/ litre and the solution is alkaline
(c) 10^{-4} mg/ litre and the solution is acidic
(d) 10^{-4}m g/ litre and the solution is alkaliae.

55. For **flow measurement**, a rotameter can be installed in a pipe line.
(a) Horizontally with flow inlet in a specific direction
(b) Horizontally with flow inlet in any direction
(c) Vertically with flow inlet at the bottom and the outlet at the top.
(d) Vertically with flow inlet at the top and outlet at the bottom.

56. An example of a **positive displacement flow meter** is.
(a) Orifice meter
(b) Rotary vane type meter
(c) Turbine type meter
(d) Ultrasonic flow meter.

57. In a **differential micrometer**, the main screw has pitch of 0.48 mm and the smaller screw has a pitch of 0.40 mm. The thimble has 50 division marked on it . The movement of the spindle per divisions is
(a) 0.0016 mm (b) 0.0176 mm
(c) 0.08 mm (d) 0.88 mm

58. A **strain gauge** is attached to a bar of length 20 cm. which is subjected to a tensile force. The nominal resistance of strain gauge is 100 Ω . The changes in resistance and elongation in the bar measured are 0.35 Ω and 0.2 mm, respectively. The gauge factor of the strain gauge is.

(a) 2 (b) 3.5
(c) 10 (d) 100

59. In the context of **transducers** identify the correct matches using the code:

A. Mean free path	1. Optical pyrometer
B. Humdity	2. Knudsen gauge
C. Heat transfer Coefficient	3. Sling psychrometer
D. Intensity of radiation	4. Hot wire anemometer

Codes:

	A	B	C	D
(a)	1	2	3	4
(b)	2	1	4	3
(c)	2	4	1	2
(d)	2	3	4	1

60. A **seismic vibration sensor**, having natural frequency ω_n and dampijg ratio ζ, used for measuring amplitude of vibration with frequency ω, should have.
(a) $\omega_n << \omega$ and ζ much greater than 1
(b) $\omega_n = \omega$ and ζ slightly less than 1
(c) $\omega_n >> \omega$ and ζ slightly less than 1
(d) $\omega_n << \omega$ and ζ slightly less than 1

61. **A piezoelectric** type 50 MP a full scale pressure sensor with built in electronics has a sensitivity of 100 m v per Mpa. If this sensor is subjected to a static pressure of 10 Mpa then its output will be.
(a) 0 m v (b) 100 mV
(c) $1000 / \sqrt{2}$ m v (d) 1000 mV

62. In the case of a **variable reluctance type inductive transducer**, PUSH – PULL arrangement reduces non linearity by eliminating.
(a) The ODD order harmonics
(b) The EVEN order harmonics
(c) Both ODD and EVEN harmonics
(d) Offset

63. A **seismic type of transducer** has a damping constant of 10. The transducer is designed to measure.
(a) Accelerator (b) Velocity
(c) Displacement (d) Force

64. A **thermocouple** is suddenly immersed in a medium of high temperature. The approximate time taken by the thermocouple to reach 98% of the steady – state value is.
(a) Equal to the time constant of the thermo couple
(b) Equal to twice the value of the time constant of the thermocouple
(c) Equal to four times the value of the time constant of the thermocouple.
(d) Independent of the time constant

65. A **capillary viscometer,** with known dimensions is used for measuring dynamic viscosity of oil .In order to obtain viscosity, it is necessary and sufficient if one measures.
(a) Pressure drop across the capillary
(b) Volume of fluid collected in a given time
(c) Both (a) and (b)
(d) Not only (a) and (b) but also one must ensure that the follow is laminar.

66. The sensing element of a thermocouple at its hot junction is provided with a shielde while taking measurements in a high temperature gas. The principal reason for providing the shield is.
(a) To reduce conduction and convection errors.
(b) To reduce radiation error
(c) To provide temperature compensation to the see beck voltage.
(d) To improve air supply to the sensing element for better response.

67. A **pitot static tube** is used for measuring velocity of a gas, flowing in a duct. The velocity is proportional to.
(a) Square root of the total pressure measured by the tube
(b) The total pressure measured by the tube.
(c) Difference between total and static pressures
(d) Square root of the difference between total and static pressures.

68. A **temperature measuring** system consists of a sensor and a thermal – well. Each of them can be considered as a single capacity system with the capacity of the thermal-well higher than that of the sensor. The overall measuring system will.
(a) Have equal time constants
(b) Have zero damping
(c) Be a non interactive system
(d) Be an interactive system

69. A **quartz crystal** (Young's modulus, $E = 9 \times 10^{10}$ N/m^2) with **piezo electric** properties has a diameter of 10 mm and thickness of 2 mm. Its voltage sensitivity constant is 4500 V / μm. If the voltage output is 127.3 V, the applied load is approximately.
(a) 100 N (b) 200 N
(c) 127.3 N (d) 6.4 N

70. One method of measuring the radius of an arc (R) is to allow a roller of radius (r) to oscillate to and from on the arc and measure the average time per oscillation. T seconds. The roller will then have a linear acceleration of 2 g/3 where g is acceleration due to gravity. m such an experiment, the value of radius of are can be found from the expression :
(a) $(T^2 g / 6\pi^2) + r$ (b) $(T^2 g / 6\pi^2) - r$
(c) $(T^2 g / 6\pi^2)$ (d) $(6r^2 / T^2 g)$

71. In a rotameter, used for measuring flow rate of a fluid P_a : Pressure above the float, P_b: pressure below the float, A : area of float, V : volume of the float , d_1 : density of float material, d_2 : density of

the fluid g: acceleration due to gravity. The following equation describes the equilibrium of the float.
(a) $(P_b - P_a) A = V_g (d_2 - d_1)$
(b) $(P_a + P_b) A = V_g (d_2 + d_1)$
(c) $(P_a - P_b) A = V_g (d_2 - d_1)$
(d) $(P_b - P_a) A = V_g (d_2 + d_1)$

72. The expression for the capacitance (C in pF) of a parallel plate capacitor is given by : $C = 6.94 \times 10^{-3} (d^2 / S)$. The diameter (d) of each plate is 20 mm and the spacing between the plates (S) is 0.25 mm. The **displacement sensitivity** of the capacitor is approximately:
(a) 44.4 pf / mm (b) – 44.4 pf/ mm
(c) 11.1 pF / mm (d) – 11.1 pf/ mm

73. It is reported that an engine is produced a sound **pressure** level of 100 dB, by considering the reference pressure as 2×10^{-5} Pa. The root mean square (r.m.s) of the fluctuating component of the source pressure is given by
(a) 2.0 Pa (b) 9.2 Pa
(c) 20 .0 Pa (d) 2.0m Pa

Answers Set-4

1. Ans. (c)
Solution:
Metal wire strain guage is having lowest gauge factor least suitable for static pressure measurement
Metal wire strain gauge frequently used for dynamic measurement.

2. Ans. (a)
Solution:
Semi conductor strain gauge has higher gauge factor as act ± 130 which is more than the metal wire strain gauge.

3. Ans. (c)
Solution:
Emf developed by the thermocouple mainly depends on metrical used, hot and cold junction temperature.

4. Ans. (d)
Solution:
Pirani gauge operation is based on variation of thermal conductivity of gas with pressure.

5. Ans. (b)
Solution:
In a **seismic pick-up for getting** an output proportional to acceleration it is desirable to have a natural frequency very large in comparison to frequency of input and a damping ratio around 0.7.

6. Ans. (b)
Solution: Gage factor = strain × excitation

$$500 \times 10^{-3} \times 4 = 2000 \times 10^{-3} = 2$$

7. Ans. (d)
Solution: $f = f \pm f_m$
$= 400 \pm 10\ Hz = 390\ Hz\ and\ 410\ Hz$

8. Ans. (d)
Solution:

9. Ans. (c)
Solution: New flow rate =

$$\frac{\text{old flow rate} \times \text{New presure drop}}{\text{old presure drop}}$$

$$\frac{400 \times 100}{200} = 200\ \ell / \text{min}$$

10. Ans. (b)
Solution:

11. Ans. (a)
Solution:

$$\text{Speed} = \frac{20 + 25}{60} = 0.6$$

12. Ans. (c)
Solution:

13. Ans. (d)
Solution:

14. Ans. (a)
Solution:

15. Ans. (a)
Solution: If the length of cable increased then the capacitance increases therefore capacitance become double if the length become double

$$Frequencies \propto \frac{1}{\text{Capacitance}} = \frac{1}{2C}$$

$$\frac{f_{new}}{f_{old}} = \frac{1/2C}{1/C} = \frac{1}{2}$$

$$f_{new} = \frac{1}{2} f_{old} = \frac{1000}{2} = 500\ Hz$$

16. Ans. (d)
17. Ans. (a)
18. Ans. (a)
19. Ans. (d)
20. Ans. (a)
21. Ans. (d)
22. Ans. (c)
Solution:
Absolute pressure
atmospheric presser – guage pressure
$= 100 - 5$
$= 95$ kPa

23. Ans. (b)
Solution:

24. Ans. (c)
Solution:

25. Ans. (d)
Solution:

26. Ans. (c)
Solution:

27. Ans. (a)
Solution:
We know that

$$P_d = \frac{1}{2} \int V^2$$

or $V^2 \propto P_d$

Therefore $V \propto \sqrt{P_d}$

28. Ans. (d)
Solution:
Since radius = 200 mm
Movement of bubble = 2mm = 2×10^{-3} m

$$\text{Inclination of surface} = \frac{2\times10^{-3}}{200} = 10^{-5}$$

29. Ans. (b)
Solution:

30. Ans. (b)
Solution:

31. Ans. (d)
Solution:

32. Ans. (a)
Solution:

33. Ans. (c)
Solution:

34. Ans. (a)
Solution:

35. Ans. (d)
Solution:

36. Ans. (d)
Solution:
$T = 200 - 40 - 40 = 120^\circ$ C

37. Ans. (c)
Solution:

38. Ans. (a)
Solution:

39. Ans. (b)
Solution:

40. Ans. (a)
Solution:
A. Platinum resistance thermometer is used between – 259.34 °C and 630/.74 °C
B. Is used between 630.74 °C and 1064.73 °C

41. Ans. (c)
Solution:

42. Ans. (c)
Solution:

$$W \propto \sqrt{\rho}$$

$$Q\rho \propto \sqrt{\rho}$$

$$Q \propto \frac{1}{\sqrt{\rho}} \quad Q = \frac{k}{\sqrt{\rho}}$$

$$2.2 = \frac{k}{\sqrt{2}}$$

Only $\frac{k}{\sqrt{1.2}}$ will given the correct flow rate

$$2.2 = \frac{k}{\sqrt{2}} \times \frac{\sqrt{1.2}}{\sqrt{2}}$$

or $\frac{k}{\sqrt{2}} = Q_{actual} = 2.2 \times \frac{\sqrt{2}}{\sqrt{1.2}}$

Actual mass flow rate = $2.2 \times \frac{\sqrt{2}}{\sqrt{1.2}} \times \frac{1.2}{2}$

$$= 1.70$$

as mass flow rate = $Q \times \rho$
and not volume flow rate m^3/min

43. Ans. (b)
Solution:

$$SPL_2 - SPL_1 = 20\log\frac{d_1}{d_2} = -20dB$$

$$SPL_2 = 80 - 20 = 60\ dB$$

$$\therefore\ 80 - 20 = 60\ dB$$

44. Ans. (d)
Solution:

45. Ans. (d)
Solution:

46. Ans. (c)
Solution:

47. Ans. (b)
Solution:

48. Ans. (c)
Solution:

49. Ans. (c)
Solution:

50. Ans. (a)
Solution:

51. Ans. (c)
Solution:

52. Ans. (a)
Solution:

$$SIL = 10 \log \frac{W}{W_{rd}} = 60$$

$$\frac{W}{W_{req}} = 10^6 = \omega_{req} = W \times 10^{-6}$$

$$\Rightarrow \; W_{req} = W \times 10^{-6}$$

53. Ans. (d)
Solution:
Linear variable differential transformer has one primary coil and two secondary coils connected in opposition.

54. Ans. (a)
Solution:

55. Ans. (c)
Solution:

56. Ans. (b)
Solution:

57. Ans. (b)
Solution:

58. Ans. (b)
Solution:

$$C_{gauge}\ factor = \frac{\frac{\Delta R}{R}}{\frac{\Delta l}{l}} = \frac{\frac{35}{100}}{\frac{0.02}{20}} = \frac{7}{2} = 3.5\,7$$

59. Ans. (d)
Solution:

60. Ans. (d)
Solution:

61. Ans. (d)
Solution:
Output = Sensitivity × static pressure applied
= (100 mV/MPa) × (10 MPa)
= 1000 mV

62. Ans. (d)
Solution:
In push-pull arrangement, the transducer, are corrected in the adjacent areas and any equal changes such as due to temperature in the two parts of the transducer act against each other and hence are cancelled out.

63. Ans. (a)
Solution:

64. Ans. (c)
Solution:
The output at line t

$$c(t) = c_0\left(1 - e^{-t/T}\right)$$

For, $c(t) = 0.98\,C_0$

We have $0.98\,C_0 = C_0\left(1 - e^{-t/T}\right)$

Or $t \approx 4T$

65. Ans. (d)
Solution:
From, Hagin-Poisuille equal to laminar flow

$$Q = \frac{\pi\, d^4\,(\Delta p)}{128\,ML}$$

The unknown are Δp and Q and applicable only for laminar flow

66. Ans. (b)
Solution:

The sensitive element is required to measure its own temperature change due to radiant energy received relative to the temperature of the immediate surroundings

67. Ans. (b)
Solution:

68. Ans. (a)
Solution:

69. Ans. (a)
Solution:

$E = 9 \times 10^{10}$ N/m^2

g = voltage sensitively = 4500 V/μm

$$= \frac{4500 \times 10^6}{E}\,\frac{V - M}{N}$$

V = voltage output = 127.3 V

We know, that,

$$V = g.t.\frac{F}{A}$$

Where t = thickness
F = applied load,
A = Area

$$\therefore\; F = \frac{VA}{g.t}$$

$$= \frac{127.3 \times \pi \times (0.01)^2 \times 9 \times 10^{10}}{4 \times 4500 \times 0.002 \times 10^6}$$

$$= 99.98\, n \approx 100\, N$$

70. Ans. (a)
Solution:

Time period, $T = 2\pi\sqrt{\frac{l}{2g/3}} = 2\pi\sqrt{\frac{3l}{2g}}$

Where l = distance between centre of rolling to the centre of roll

$$\therefore \quad T^2 = 4\pi^2 \times \frac{3l}{2g}$$

$$\text{or} \quad l = \frac{T^2 g}{6\pi^2}$$

$\therefore$ Radius of arc, $R = l + r = \dfrac{T^2 g}{6\pi^2} + r$

71. Ans. (a)
Solution:
(Difference in pressure) × A = weight of float – Buoyancy force

or $(p_b - p_a)A = V_g(d_2 - d_1)$

72. Ans. (b)
Solution:

$$C = 6.94 \times 10^{-3}\left(\frac{d}{5}\right)^2$$

Displacement sensitivity,

$$\frac{dC}{dS} = -(6.94 \times 10^3)d^2/S^2$$

$$= -(6.94 \times 10^{-3})\frac{20^2}{(0.25)^2} = -44.4\ pF/mm$$

73. Ans. (c)
Solution:

Reader's Notes

4 Analog Electronics

Analog Electronics: Characteristics of diode, BJT, JFET and MOSFET. Diode circuits. Transistors at low and High frequencies, Amplifiers, Single And Multi-Stage. Feedback Amplifiers. Operational Amplifiers, Characteristics And Circuit Configurations. Instrumentation Amplifier. Precision Rectifier. V-To-I and I-To-V Converter. Op-Amp Based Active Filters. Oscillators and Signal Generators.

4.1 CHARACTERISTICS OF DIODE

4.1.1 JUNCTION THEORY

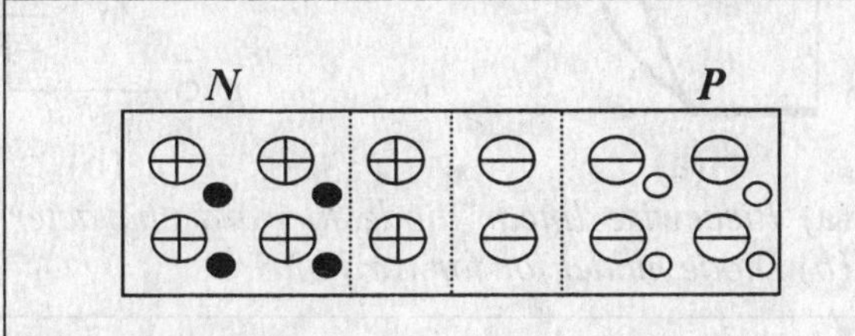

- PN Junction is formed on a single crystal of Semiconductor like *Ge* or *Si* by dropping acceptor impurity on one side and donor impurity on the other side.
- Since P- side is having large hole as compared to electron and N - side is having large electron as compared to hole so due to difference in concentration, diffusion takes place, i.e. hole diffuse to N-side and electron diffuse to P - side.
- Depletion region is completely free from any mobile charge. It is behave just like Capacitor, which is controlled by applied voltage at P and N side.
- Forward bias causes Depletion region to shrink, so increase the value of capacitor.
- Total Depletion region width

$$W = W_n + W_p$$

- $N_D W_n = N_A W_p$
  means for highly dropped region the Depletion region is less in that side.
- Barrier Potential or Build in Potential

$$V_B = V_T \ \ell n \left(\frac{N_A \cdot N_D}{n_i^2}\right) = \frac{KT}{q} \ \ell n \left(\frac{N_A \cdot N_D}{n_i^2}\right)$$

- Current flow in PN junction diode

$$= I_0\left(e^{\frac{nV}{V_T}} - 1\right)$$

Where I_o = Reverse saturation current
n = Factor due to charge generation and recombination.
n = 1 for *Ge*.
= 2 for *Si*

- I_o doubles for every $10^o\,C$ rise in temp

$$I'_o = I_o \ 2^{[(T'-T)/10]}$$

- Diode dynamic resistance

$$r = \frac{dV}{dI} = \frac{1}{dI/dV} = 1/\text{Slop of characteristic of diode}$$

$$= \frac{nV_T}{I} = \frac{26\ (mV)}{I(mA)} \ ; \ n = 1$$

- $I_o = q\left(n_p \dfrac{L_n}{\tau_n} + p_N \dfrac{L_p}{\tau_p}\right)$

Where, L_n and L_p → Diffusion length
τ_n and τ_p → Life time.

4.1.2 CAPACITANCE IN PN JUNCTION DIODE

The depletion region denotes a parallel plate capacitor since the total charge on the 2 side of the depletion region is equal but opposite.

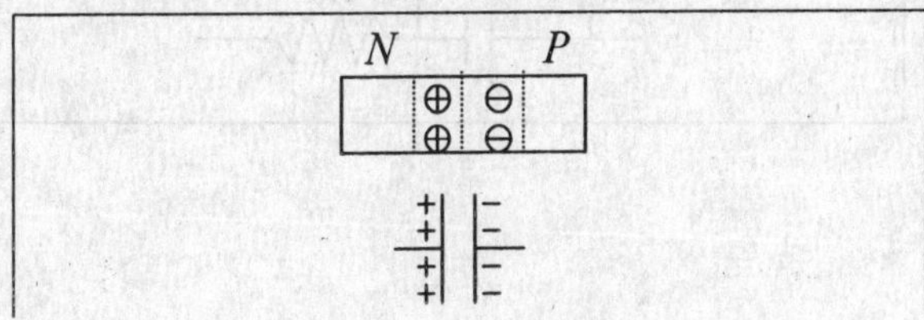

$$C = \frac{\varepsilon A}{d} = \frac{\varepsilon A}{w}$$

Where, ε = permittivity of medium
A= cross section area
d = W=Total width of depletion region.
$= W_N + W_p$

Capacitance are of two type

4.1.3 TRANSITION CAPACITANCE

C_T arises due to reverse bias applied and due to majority charge carrier.

$$C_T(V_r) = \frac{C(o)}{1 + [V_r/V_T]} \quad n$$

V_r = Reverse bias voltage
$C(o)$ = Capacitance at zero bias.
n = [1/2] for step graded/Abrupt junction
= [1/3] for diffused/ linearly graded junction.

Finally,
$C_T = K_1 [V]^{-1/2}$: for abrupt junction
$= K_2 V^{-1/3}$: for linearly graded function
$W_B = K_1 [V_j]^{1/2}$: for abrupt junction
$= K_2 [V_j]^{1/3}$: for linearly graded junction.

Diffusion Capacitance C_D exist for forward bias and is important for minority charge carrier

$$C_D = \frac{\tau_n}{r_n} + \frac{\tau_n}{r_p}$$

Where, τ = Life time
r = Dynamic resistance of region

- Diffusion Capacitance increase rapidly with forward bias i.e. $C_D \propto I$.
- The value of C_D is of the order of several þF and hence it is important when high frequency circuit for transistor are derived.
- Voltage variable diode capacitor are called *Varactor*, Varicap or Volta cap. They are used in frequency modulation circuit.
-

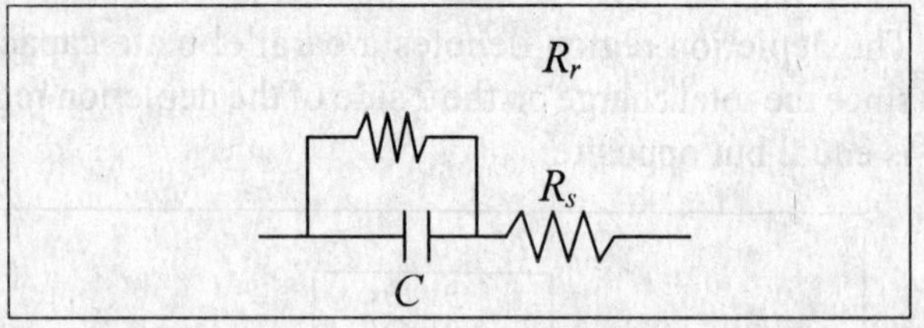

- $L = \sqrt{D\ \tau}$

Where L = Diffusion length
D = Diffusion Co efficient
τ = Life time

The Einstein relation

$$\frac{D}{\mu} = V_T = \frac{KT}{q}$$

4.2 DIFFERENT TYPES OF DIODES AND THEIR CHARACTERISTIC

4.2.1 DIODE

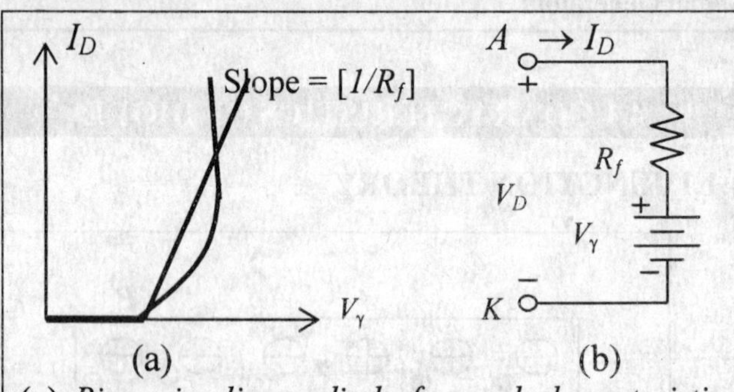

(a) *Piecewise linear diode forward characteristic* (b) *Diode model for forward bias.*

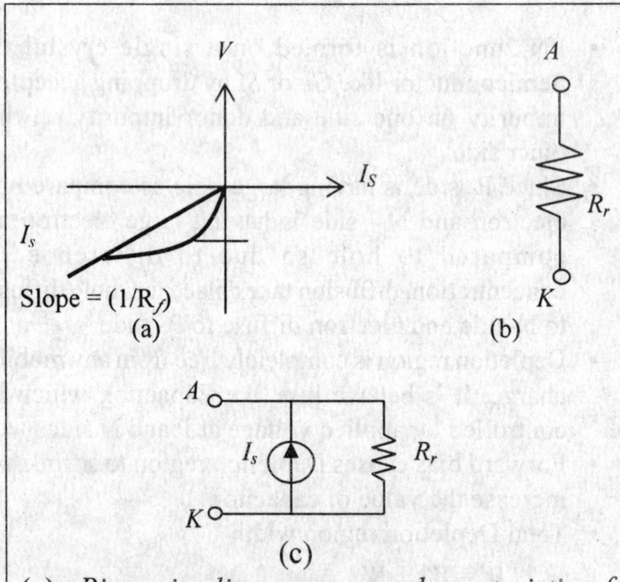

(a) *Piecewise linear reverse characteristic of diode (b) Diode model based on piecewise linear representation. (c) Model to include surface leakage*

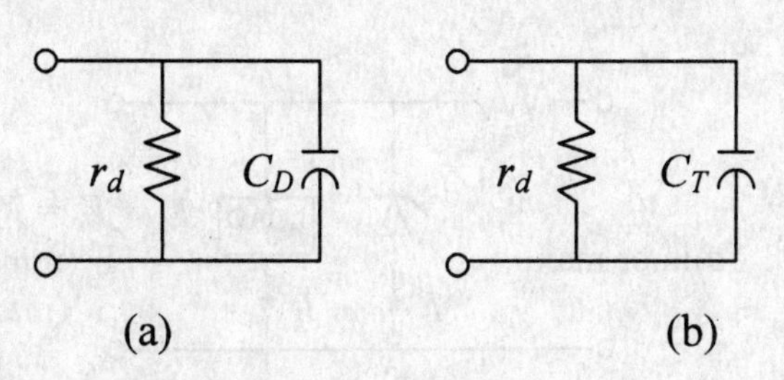

(a) *Small-signal diode models for (a) forward bias* (b) *Reverse bias*

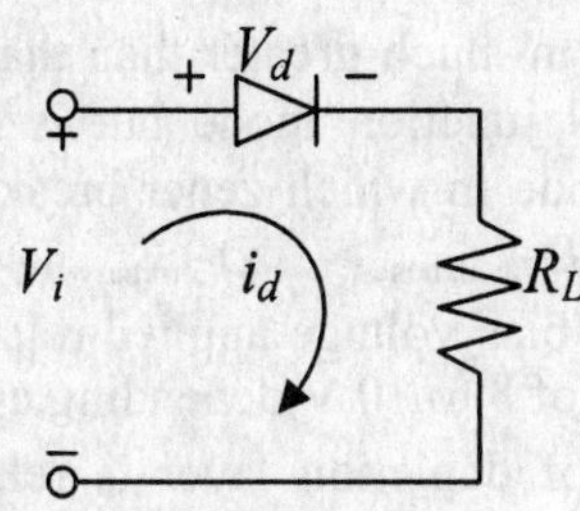

(a) *Diode-resistor circuit*

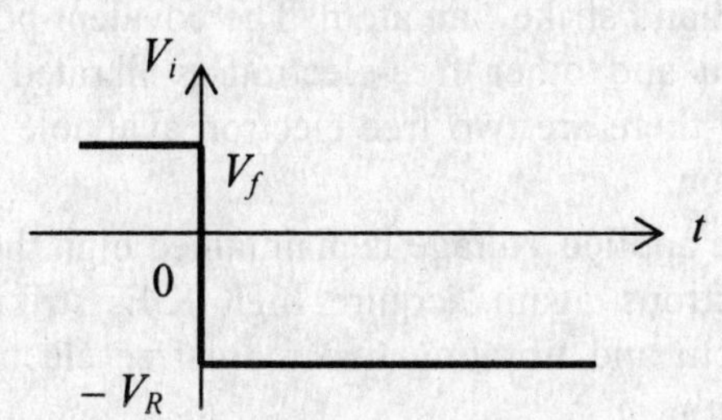

(b) *Input waveform applied to circuit in part figure* (a) *Showing abrupt change from forward to reverse bias.*

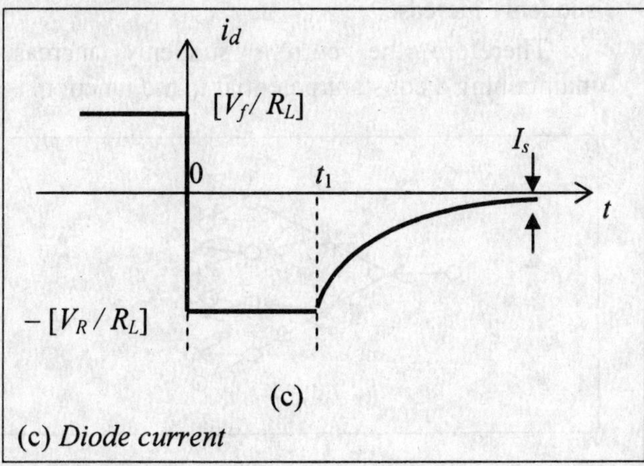

(c) *Diode current*

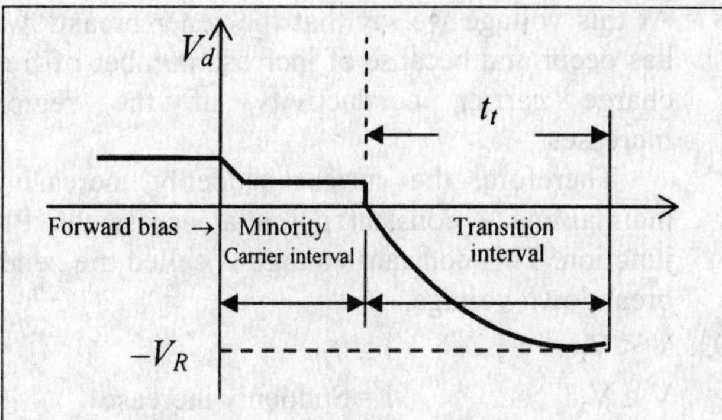

(d) *Diode voltage waveforms displaying storage and transition times.*

4.2.2 ZENER DIODE

It is a PN junction diode with heavy doping Operate under reverse bias condition and for voltage > V_g. (Breakdown Voltage)

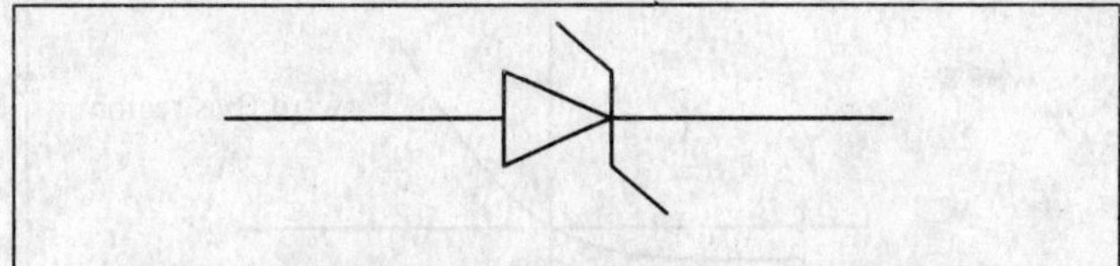

4.2.3 BREAKDOWN MECHANISM

Zener breakdown and Avalanche breakdown

Zener breakdown

1. This phenomena occur in PN junction diode with very high doping.
2. It occur when the diode is reverse bias of voltage approximately 4 to 6V.
3. The junction is narrow since the width of depletion layer is small become of highest doping and relatively low reverse bias voltage.
4. The junction width is of the order of 100A° so that very strong electric field of 10^7 V/m is developed near the junction.
5. Now as the reverse bias Voltage is increased then the current will be only due to minority charge carrier and is of the order of μA or nA.
6. As the reverse voltage is still increased a value of this voltage is reached so that very high electrical field is developed.
7. For this magnitude of the electrical field there will be direct rupture of the covalent bond so that immediately a number of free charge carrier is generated.

8. At this voltage we say that the zener breakdown has occur and because of increase number of free charge carrier conductivity of the region increases.

   Therefore, the current suddenly increasing maintaining a constant potential across the PN junction. This constant voltage is called the zener breakdown voltage.

9. $V < V_z$; $I \cong 0$

   $V \geq V_z$; I = Suddenly increase

10. $V < V_z$

    The dynamic resistance is very large

    $V \geq V_z$

    The dynamic resistance is small.

11. Characteristic

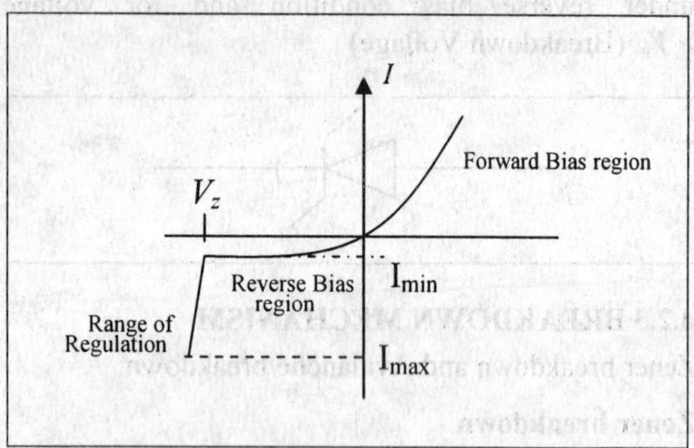

12. In the forward bias condition the V-I characteristic is of the similar nature of ordinary diode.
13. The V-I characteristic of forward bias condition is not practically useful since the same variation we can get using an ordinary PN junction diode.
14. The increase in the current, once the breakdown has occur is limited by the use of a series of resistor along with the zener diode the current varies from I_{min} to I_{max}.

    I_{max} = I (rated current)

15. Practically for designing a circuit, we use I_{min} to the 20% of I (rated).

    $I_{min} \cong 0.2$ I rated

    $I_{max} = 0.8$ I (rated)

16. This device can be utilized as a voltage regulator since the voltage across it remain constant irrespective of current variation through it.

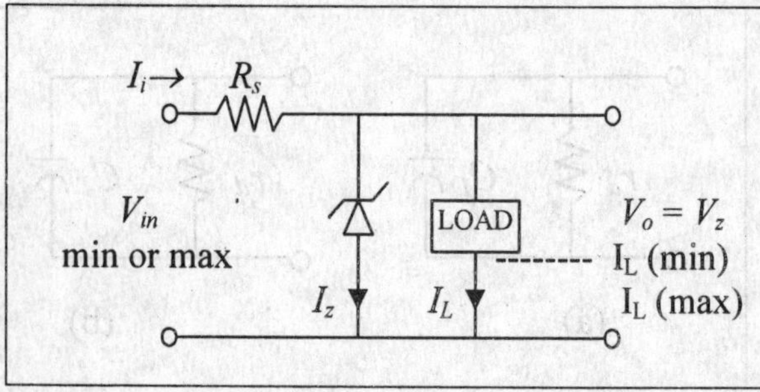

Avalanche Breakdown

1. This breakdown occurs in PN junction diode with reverse bias applied.
2. The doping in much greater than that used in the ordinary PN junction diode but is smaller than that of a diode in which zener breakdown occur, i.e. $D_{zener} > D\text{-}_{avalanche} > D\text{-}_{ordinary}$.
3. The reverse bias voltage applied relatively higher of the order of 8 to 10 V depending upon doping.
4. The width of depletion layer is relatively more than that in the diode with zener breakdown.
5. Since the reverse bias voltage applied is relatively large an free charge carrier say electron acquire high K.E and strike an atom. The covalent bond is broken and other free-electron is librated so that now there are two free electron available for conduction.
6. Since the applied voltage is maintained high there two electrons again acquire high K.E, striking other atom and librating two more free electron and so on.
7. At a given voltage called the avalanche breakdown voltage a number of free charge carrier are available to avalanche multiplication as shown so that the conductivity of the region suddenly increase.

   Therefore the current suddenly increase maintaining a constant potential to the junction.

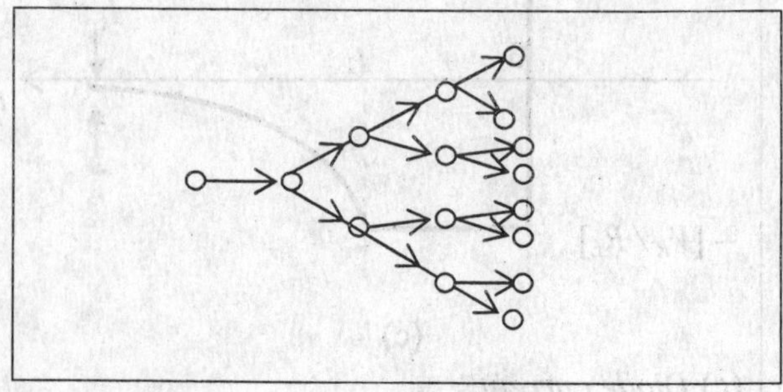

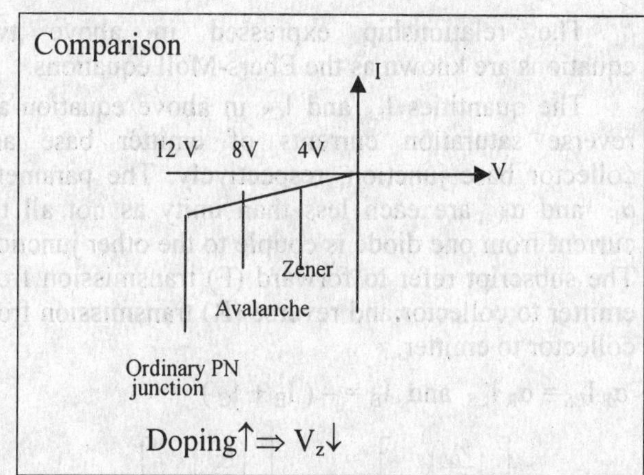

Variation V_z with temperature

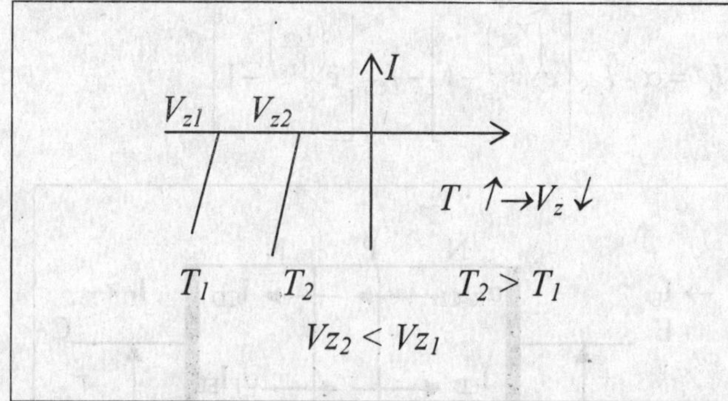

$$\text{Dynamic Resistance } r = \frac{\Delta V}{\Delta I}$$

$= \text{Very large for } V < V_z$

$= \text{Very small for } V > V_z$

8. The VI characteristic is similar to that a diode with zener breakdown.

4.3 CHARACTERISTICS OF BJT

4.3.1 BJT: BIPOLAR JUNCTION TRANSISTOR

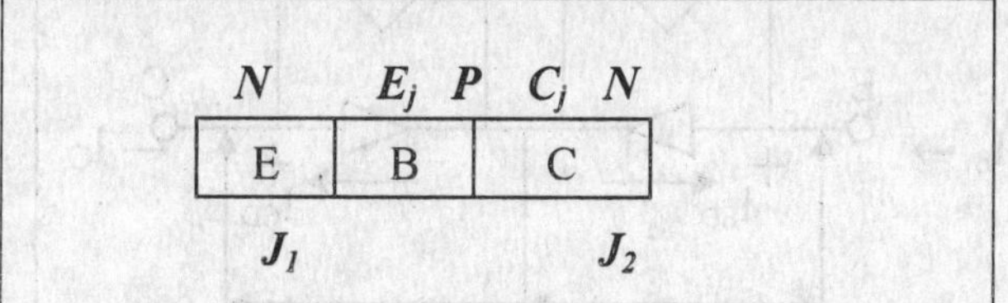

Collector region width is maximum

(a) To collect maximum number of charge carrier.

(b) To reduce heat dissipation.(per unit area surface).

Depletion layer in collector junction base side is maximum Emitter junction. Emitter side is Minimum.

Biasing and application

Input Jn J_1 (EB J_n)	Output Jn J_2 (CB J_n)	Region in Which BJT works	Application
FB	RB	Active	Amplifier
FB	FB	Saturation	BJT is ON Or OFF
RB	RB	Cut off	OFF
RB	FB	Invented (Active)	Amplifier with decreased gain α_I, β_I

- For Normal Biasing (i.e. Emitter junction FB and Collector junction RB) the emitter junction has low resistance.

4.3.2 EARLY EFFECT

For active region operation of transistor collector base region is reversed biased. Due to reverse bias the depletion width in base region increases which causes effective width of base to decrease. This effect is known as early effect.

If the reverse bias VcB is increased beyond a point, breakdown may occur as width of base becomes zero and phenomena is called punch through or Base width modulation.

4.3.3 THE PHYSICAL BEHAVIOR OF BIPOLAR TRANSISTOR

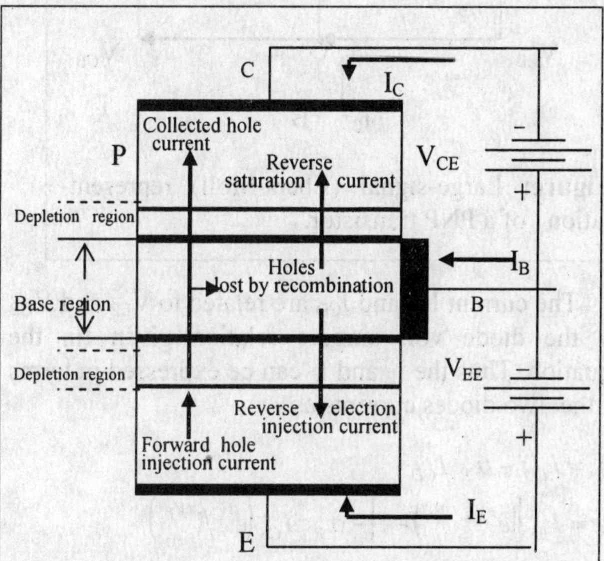

Figure: Transistor carrier components for a forward biased emitter-base junction and reverse biased collector base junction.

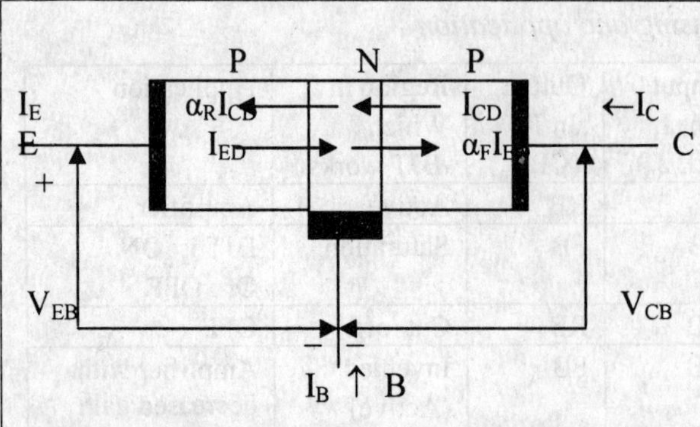

Figure: Current components in PNP transistor

On the basis of the considerations in the preceding paragraph we construct the Ebers-Moll mode in below figure. The two back to back diode (whose cathodes are connected) represent the junctions of the bipolar transistor, whereas the two controlled sources indicate the coupling between junctions.

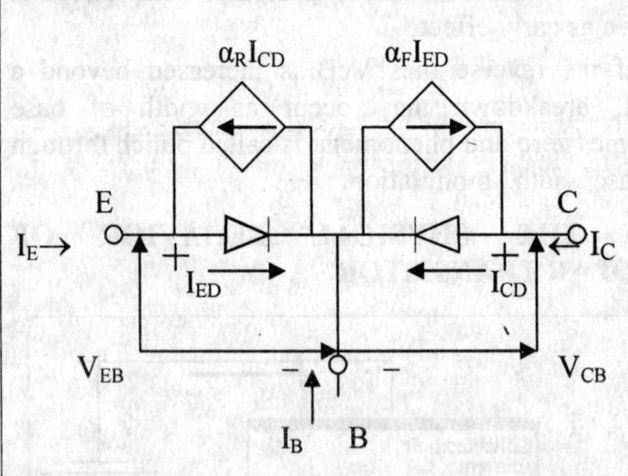

Figure: Large-signal (Ebers-Moll) representation of a PNP transistor.

The current I_{ED} and I_{CD} are related to V_{EB} and V_{CB} by the diode volt ampere relation given tin the equation. Thus the I_E and I_C can be expressed in terms of the two diodes currents as

$$I_E = I_{ED} - \alpha_R\, I_{CD}$$

$$= I_{ES}\left[\left(e^{V_{EB}/V_T}\right) - 1\right] - \alpha_R\;\; I_{CS}\left(e^{V_{CB}/V_T}\right)$$

$$I_C = -\alpha_R\, I_{ED} + I_{CD}$$

$$= -\alpha_F\; I_{ES}\left(e^{V_{EB}/V_T} - 1\right) + I_{CS}\left(e^{V_{CB}/V_T} - 1\right)$$

The relationship expressed in above two equations are known as the Ebers-Moll equations.

The quantities I_{ES} and I_{CS} in above equation are reverse saturation currents of emitter base and collector base junction, respectively. The parameter α_F and α_R are each less than unity as not all the current from one diode is couple to the other junction. The subscript refer to forward (F) transmission from emitter to collector and reverse (R) transmission from collector to emitter.

$\alpha_F\, I_{ES} = \alpha_R\, I_{CS}$ and $I_B = -(I_E + I_C)$

$$I_E = -I_{ES}\left[e^{\left(\frac{V_{EB}}{V_T}\right)} - 1\right] + \alpha_R I_{CS}\left[e^{\left(\frac{V_{CB}}{V_T}\right)} - 1\right]$$

$$I_C = \alpha_F\, I_{ES}\left[e^{\left(\frac{V_{EB}}{V_T}\right)} - 1\right] - I_{CS}\left[e^{\left(\frac{V_{CB}}{V_T}\right)} - 1\right]$$

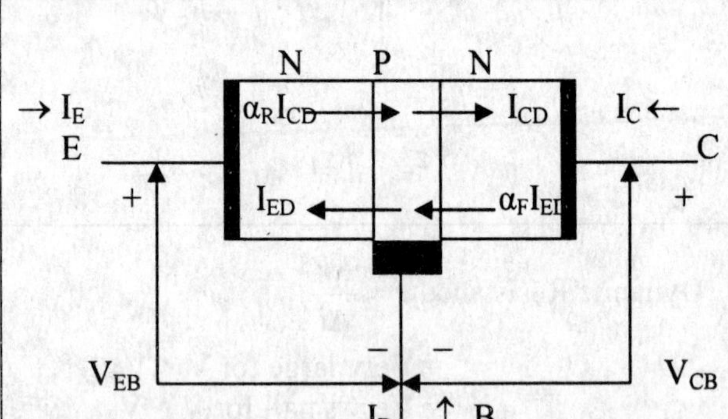

Figure: Current component in NPN transistor

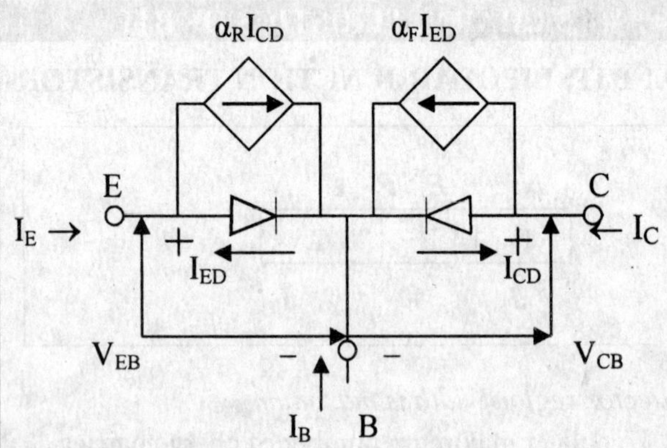

Figure: Large-signal representation of a NPN transistor.

4.4 CHARACTERISTICS OF JFET

4.4.1 TWO TYPES: N CHANNEL AND P CHANNEL

It consists of a semiconductor block which is n type for n channel and p type for p channel. On both sides of n block acceptor impurities p are heavily doped by allowing or by diffusion. The region is called the gate. Ohmic contact are attached to the either ends of the n-block. One end is called Source and other is called Drain. The np region between source and gate is reverse biased. In the vicinity of np junction there is diffusion of majority carriers, electrons and holes leaving uncover immobile positive and negative ions. When RB is applied, the space charge width increases uncovering more ions, until the space charge region from the two end, meets the n type region between the two (P^+) gates is called n channel and when a positive voltage is applied between drain and source, there is a current flow in the channel of majority carriers, (electrons). When a reverse bias is applied between the gate and source, constriction of the channel takes place by widening of the space charge depletion region controlling the flow of electrons to the drain.

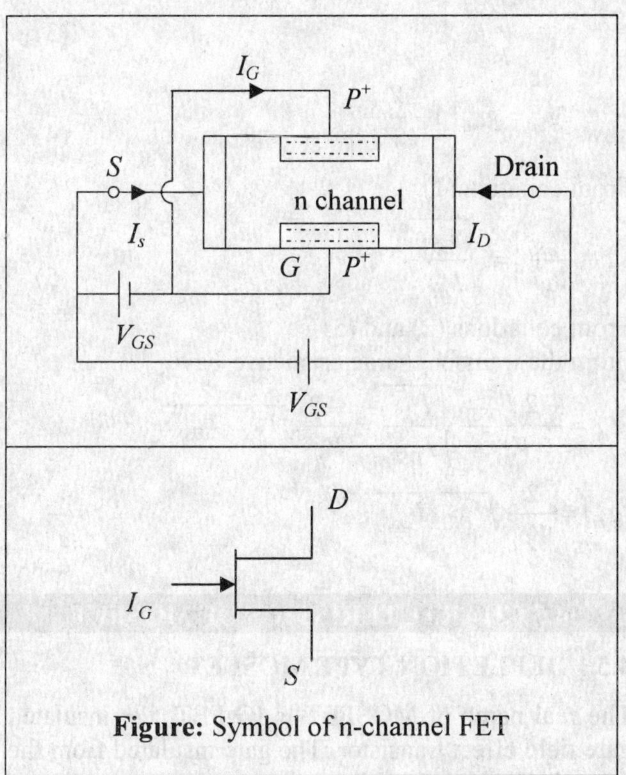

Figure: Symbol of n-channel FET

Note: For a P channel FET the bar is a P type semiconductor and gate is dopped by n^+ donor impurity

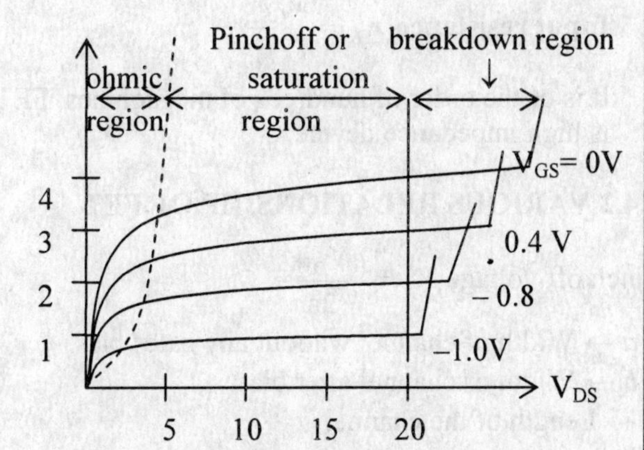

Figure: N-channel FET characteristic

$$I_{DS} = I_{DSS}\left(1 - \frac{V_{GS}}{V_p}\right)^2$$

Figure: FET Transfer characteristic

1. **Drain resistance** $r_d = \left.\frac{V_{DS}}{I_D}\right|_{V_{GS}=\text{Constant}}$

Output resistance of FET and evaluated in constant drain current region of $V_{DS} - I_D$ curve. It is of the order of 10 to 20 kΩs.

2. **Transconductance)** $g_m = \left.\frac{I_D}{V_{GS}}\right|_{V_{DS}=\text{Constant}}$

It will be of the order of 2 k to 6 k micromhos.

3. **Amplification Factor** $\mu = g_m r_d$
4. $= \frac{V_{DS}}{I_D} \times \frac{I_D}{V_{GS}} = \frac{V_{DS}}{V_{GS}}$, It ranges of 40 to 100.
5. **Input resistance** $r_{GS} = \frac{V_{GS}}{I_{GSS}}$

   It is of the order of hundreds of megaohmes. FET is high impedance device.

4.4.2 VARIOUS RELATIONSHIP OF FET

Pinch off voltage $V_p = \frac{eN_D a^2}{2\varepsilon}$

$2a \to$ Width of channel without any gates bias.
$2b \to$ Width of channel after bias.
$L \to$ Length of the channel

$$b = a\left[1 - \left(\frac{V_{GS}}{V_p}\right)^{1/2}\right]; \text{ when } V_{GS} = 0, b = a$$

If $w \to$ channel dimension perpendicular to b direction

$$A = 2bW$$

Drain Current $I_D = AqN_D\mu_n E$

$$= 2bwq\, N_D\mu_n \frac{V_{DS}}{L}$$

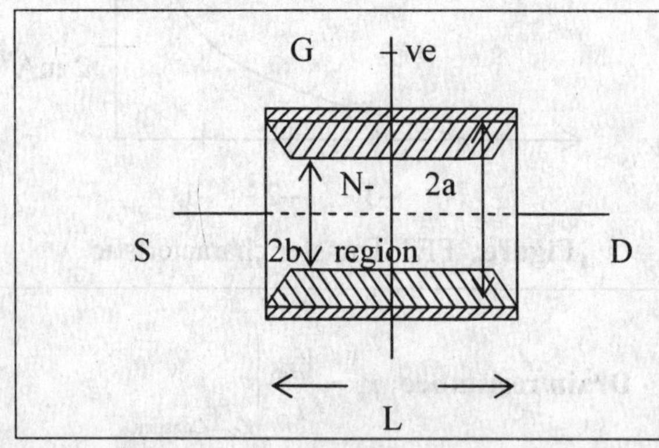

Substituting 'b'

$$I_D = \frac{2awqN_D\mu_n}{L}\left[1 - \left(\frac{V_{GS}}{V_p}\right)^{1/2}\right] V_{DS}$$

The ON resistance $r_d = \frac{V_{ds}}{I_d}$

$$r_{d\cdot ON} = \frac{L}{2awqN_D\mu_n} \text{ for } V_{GS} = 0$$

Transfer Characteristics

In amplifier application, the FETis almost used in the region beyond pinch-off (also called the constant-saturation region).

$$I_{DS} = I_{DSS}\left(1 - \frac{V_{GS}}{V_p}\right)^2$$

Relation between g_m, I_{DDS} and I_{DS}

$$I_{DS} = I_{DSS}\left(1 - \frac{V_{GS}}{V_p}\right)^2 \quad (1)$$

$$\frac{\partial I_{DS}}{\partial V_{GS}} = I_{DSS}\, 2\left(1 - \frac{V_{GS}}{V_p}\right)\left(\frac{-1}{V_p}\right)$$

$$g_m = \left(\frac{-2I_{DDS}}{V_p}\right)\left(1 - \frac{V_{GS}}{V_p}\right) \quad (2)$$

g_{mo} = It is value of g_m for $V_{GS} = 0$ and given by

$$g_{mo} = \left(\frac{-2I_{DSS}}{V_p}\right) \quad (3)$$

now $g_m = g_{mo}\left(1 - \frac{V_{GS}}{V_P}\right)$ (4)

From equation (1)

$$\left(1 - \frac{V_{GS}}{V_p}\right) = \sqrt{\frac{I_{DS}}{I_{DSS}}} \quad (5)$$

From equations (2) and (5)
From the transfer characteristic we have

$$g_m = \frac{-2I_{DSS}}{V_p}\sqrt{\frac{I_{DS}}{I_{DSS}}} = \frac{-2}{V_p}\sqrt{I_{DS}\cdot I_{DSS}}$$

$$|g_m| = \frac{2}{|Vp|}\sqrt{I_{DS}\cdot I_{DSS}}$$

4.5 CHARACTERISTICS OF MOSFET

4.5.1 DEPLETION TYPE MOSFET

The real name of MOSFET is IG FET, i.e. insulated gate field effect transistor. The gate insulated from the substrate by an oxide layer.

In the FET the metal and channel with dielectric in between forms a parallel plate capacitor giving rise to a high input impedance of (10^{10}) to (10^{15}) ohms and is the major difference from JFET. There are two types of MOSFET enhancement and depletion type.

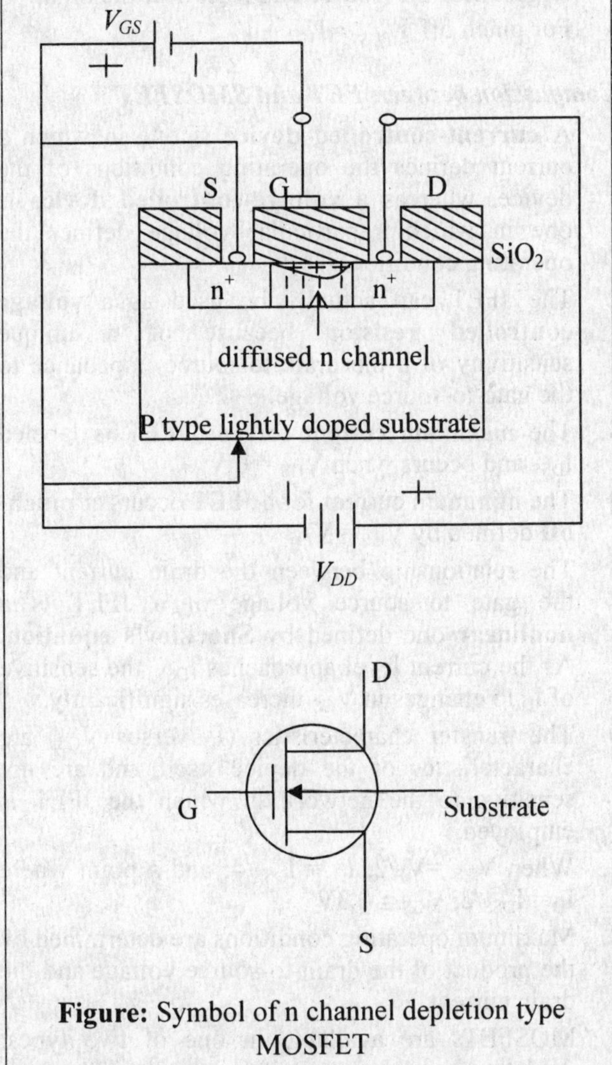

Figure: Symbol of n channel depletion type MOSFET

When a negative voltage V_{GS} is applied to the conducting terminal of the capacitor positive charges are induced in the channel, the other side of the capacitor. These positive charge neutralize the electrons in the n channel and reduce the charge carrier. This reduces the drain current. The depletion region is wider as in JFET near the drain than the near the source, due to voltage drop by the drain current. Unlike JFET when gates is reverse biased, MOSFET there is no pn^+ junction. As the carrier, electron, in n channel are depleted this is called depletion type MOSFET.

The same MOSFET can be made enhancement type when a positive voltage is applied to the gate negative charge are induced in the channel adding to the electron, charge carrier in the channel this enhances the drain current.

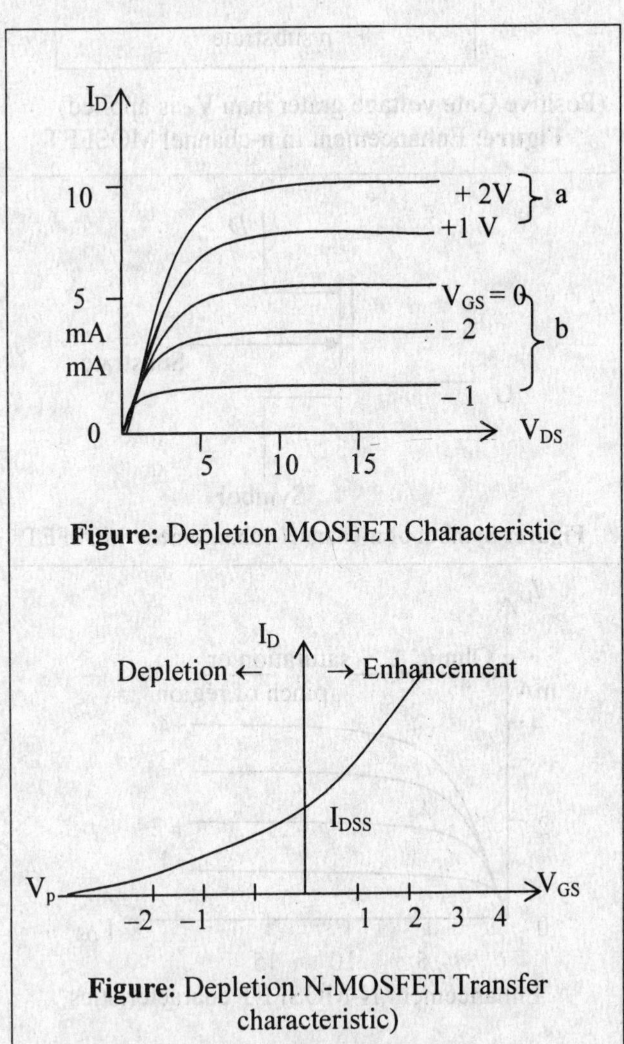

Figure: Depletion MOSFET Characteristic

Figure: Depletion N-MOSFET Transfer characteristic)

4.5.2 ENHANCEMENT TYPE MOSFET

No channel is diffused in the P substrate. So when a positive voltage is applied to the gate, the minority carriers is the P substrate, electrons are drawn towards the dielectric and this forms an effective channel. As the positive voltage on the gate is increased, conductivity increases leading to higher drain current as the negative charge carrier are increased.

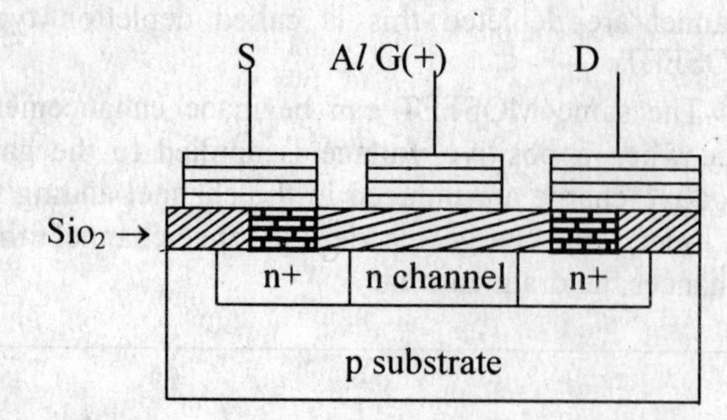

(Positive Gate voltage grater than V_T is applied)
Figure: Enhancement in n-channel MOSFET

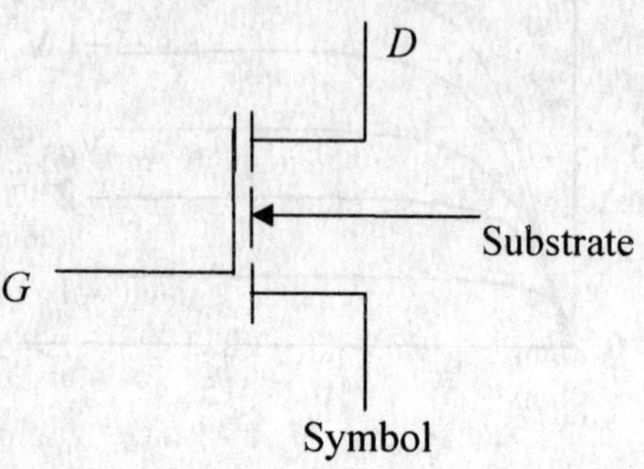

Figure: Symbol of n-channel enhancement MOSFET

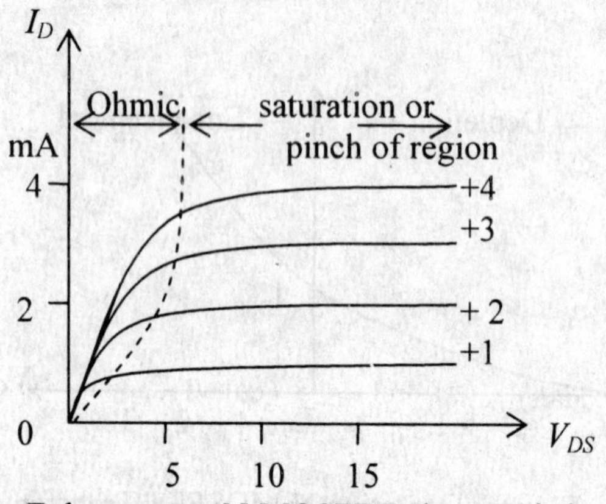

Enhancement N-MOSFET characteristics

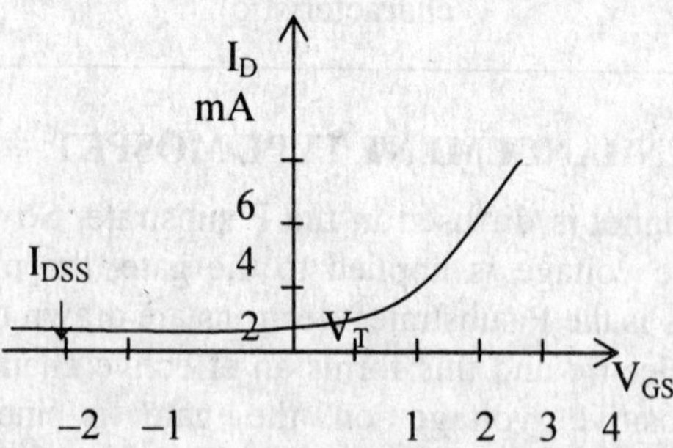

Figure: Transfer characteristic of N-channel E-MOSFET

- We note from the transfer characteristics that a threshold voltage V_{GST} is required for drain current to flow. Without gate voltage, a small current of nanoampere I_D flows in the drain.
- The electric field produced by the gate voltage is largest near the source and least near the drain.
- For pinch off $V_{PDS} = V_{GS} - V_T$

Compassion between FET and SMOSFET

1. A **current-controlled device** is one in which a current defines the operating conditions of the device, whereas a **voltage-controlled device** is one in which a particular voltage defines the operating conditions.
2. The JFET can actually be used as a **voltage controlled resistor** because of a unique sensitivity of a the drain-to-source impedance to the gate-to-source voltage.
3. The maximum current for any JFET is labeled I_{DSS} and occurs when $V_{GS} = 0$ V.
4. The minimum current for a JFET occurs at **pinch-off** defined by $V_{GS} = V_P$.
5. The relationship between the drain current and the gate to source voltage of a JFET is a **nonlinear** one defined by **Shockley's equation.** As the current level approaches $I_{DSS,}$ the sensitive of I_D to changes in V_{GS} increases significantly.
6. The transfer characteristics (I_D versus V_{GS}) are characteristics of the device itself and are not sensitive to the network in which the JFET is employed.
7. When $V_{GS} = V_P/2$, $I_D = I_{DSS}/4$; and a point where $I_D = I_{DSS}/2$, $V_{GS} \cong 0.3V$
8. Maximum operating conditions are determined by the product of the drain-to-source voltage and the drain current.
9. MOSFETs are available in one of two types: depletion and enhancement.
10. The deplention-type MOSFETs has the same transfer characteristics as a JFET for drain currents up to the I_{DSS} level. At this point the characteristics of a depletion point type MOSOFET continue to levels above $I_{DSS.}$ Whereas those of the JFET will end.
11. The **arrow** in the symbol of **n-channel JFETs** or MOSFETs will always point in to the center of the symbol, whereas those of a p-channel device will always point out of the center of the symbol.

12. The transfer characteristics of an enhancement-type MOSFET are not defined by Shockly's equation but rather by a **nonlinear equation** controlled by the gate to source voltage, the threshold voltage, and a constant k defined by the device employed. The resulting plot of I_D versus V_{GS} is one that rises exponentially with increasing values of V_{GS}.
13. Always handle MOSFETs with additional care due to the static electricity that exists in places we might least suspect. Do not remove any shorting mechanism between. The device until it is installed.
14. A **CMOS** (complementary MOSFET) device is one that employs a unique combination of a p-channel an n-channel MOSFT with a single set of external leads. It has the **advantages of a very high input impedance, fast switching speeds, and low operating power levels,** all of which make very useful in logic circuits.

JFET

$$I_D = I_{Dss}\left(1-\frac{V_{GS}}{V_P}\right)^2$$

$$I_D = I_{Dss}\Big|_{V_{GS}=0V} \;;\; I_D = 0mA\Big|_{V_{GS}=V_P}$$

$$I_D = \frac{I_{DSS}}{4}\Big|_{V_{GS}=V_P/2}$$

$$V_{GS} \cong 0.3V_p\Big|_{I_0=I_{DSS/2}} \;;\; V_{GS} = V_P\left(1-\sqrt{\frac{I_D}{I_{DSS}}}\right)$$

$$P_D = V_{DS}I_D$$

$$r_d = \frac{r_0}{(1-V_{GS}/V_r)^2}$$

MOSFET (enhancement):

$$I_D = k(V_{GS}-V_T)^2$$

$$k = \frac{I_{D(on)}}{\left(V_{GS(on)}-V_r\right)^2}$$

Typical values for JFET and MOSFET

Parameter	JFET	MOSFET
Input resistance r_{gs}	$>10^9\,\Omega$	$>10^{13}\,\Omega$
Transconductance g_m (μ mho)	$1K$ to $2.5K$	$1K$ to $20K$
Drain resistance r_d (ohm)	$0.1M$ to $1M$	$1K$ to $50K$
Revere current I_{GSS}	0.1 to $10\,nA$	0.1 to $10\,pA$
Capacitance $C(G \text{ to } D)$	1 to $4\,pF$	0.005 to $1pF$

For D-MOSFET

Characteristic are similar to FET

For E-MOSFET

$$V_{DS\,sat} = V_{GS} - V_T$$

For $V_{GS} << V_T$, the drain current of an enhancement type MOSFET is 0 mA.

For levels of $V_{GS} > V_T$, the drain current is related to the applied gate-to-source voltage by the following nonlinear relationship

$$I_D = K(V_{GS}-V_T)^2$$

Where. K→ constant that is function of the construction of the device, the value of K can be determined from

$$K = \frac{I_{D(on)}}{\left(V_{GS(on)}-V_T\right)^2}$$

$I_{D(on)}$ and $V_{GS(on)}$ are the values for each at a particular point on the characteristic of the device.

Compassion between BJT and FET

BJT	FET
1 High voltage gain	Low voltage gain
2. Low current gain	High current gain
3. Low input and output impedance	High input and output impedance
4. Medium noise generation.	Low noise generation.
5. (Gain ×Band-width) high	(Gain × Band-width) medium
6. Current source	Voltage controlled

4.6 BIASING AND STABILIZATION

The transistor is biased (with the help of external voltage in such a manner so the Q-point is selected in the middle portion the transistor output characteristic. When the ac voltage is superimposed it is ensure that the positive and negative half of input voltage cycle remain it the linear or active region if the transistor characteristic. Under these condition the transistor output voltage is undistorted.

- To fix the Q-point load line curve is needed.
- The dc load line is locus of variation of output voltage and output current.
- The load lines are being drawn for the amplifier chosen in the active region of output of characteristic.
- The Q-point is generally fixed at the middle of load line.
- Ac load line have greater slope than dc load line.
- If the Q-point is chosen near the V_{ce} (Cut-off region or I_c axis (saturation region) than output will be highly distorted.
- If the Q-point is chosen in the saturation region for NPN transistor, it may cause the output to be clipped in the positive half cycle of the input signal.

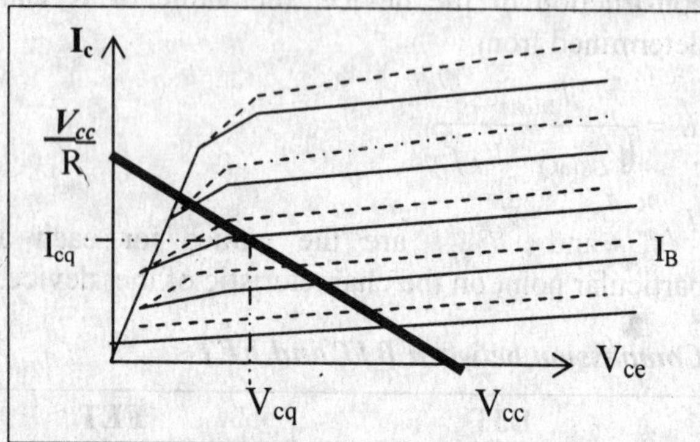

4.6.1 STABILIZATION

Once the Q-point has been fixed due to some biasing arrangement it has to be stabilized against the variation in:

(a) Temperature (T) –I_{co} double for every 10 degree rise in temperature.

(b) I_B (Depends on V_{be}), i.e. V_{be} decrease at the rate of 2.5 mv/degree.

(c) β.

- Due to variation of one or all 0f these parameter I_c current will change which shift the Q-point.
- In stabilization technique resistive biasing circuit is used which permit I_b to vary by an amount so as to keep I_c relatively constant with variation in any of T, I_b,β

Change in Temperature

- We Know that $I_c = \beta I_b + (1+\beta) I_{co}$
- β Varies due to change in temp.
- I_b also changes due to change in temp. As the cut in voltage is increases at 2.5 mv/degree for constant I_c.
- I_{co} changes with temperature doubling with every 10 degree rise in temperature. Increase in I_{co} caused an increase in I_c which further increasing the collector dissipation thereby increasing the collector temperature. This may set in a cumulative effect causing thermal runway burning out the transistor ultimately.

= Si transistor, I_{co} is of the order of nano amperes compared to milliamps in Ge and so *thermal runway* and instability is more problem in Ge transistor.

4.6.2 STABILITY FACTOR

- The merit of a biasing circuit in holding the dc collector current I_c at the operating Q-point may be examine by stability factor

$$S = \frac{\delta I_C}{\delta I_{CO}}\bigg|\, V_{be},\ \beta = \text{constant}$$

$$S_\beta = \frac{\delta I_c}{\delta \beta}\bigg|\, I_{CO}\ \ V_{be} = \text{constant}$$

$$Sv = \frac{\delta I_c}{\delta V_{bo}}\bigg|\, I_{co},\ \beta = \text{constant}$$

- Lower the value of stability factor (S) better is stability of circuit.
- $1 < S < 1+\beta$.
- The stability factor S is most important since it depends upon I_{co} and temperature.

$$S_\beta = \frac{\delta I_c}{\delta I_{\infty}}\bigg|_{V_{be}=\text{Constant}}$$

$$I_c = \beta\, I_b + (1+\beta)\, I_{co} \ \dots \text{ general expression}$$

Differentiate with respect to I_c

$$1 = \beta\, \frac{\delta I_b}{\delta I_c} + (1+\beta)\, \frac{\delta I_{co}}{\delta I_c}$$

$$S = \frac{\delta I_c}{\delta I_{co}} = \frac{(1+\beta)}{(1-\beta)\frac{\delta I_b}{\delta I_c}}$$

- For fixed bias arrangement I_b is independent of I_c, $S = \beta + 1$.
- *Collector to base biasing:*

$$S = \frac{1+\beta}{1+\frac{\beta R_c}{R_b + R_c}}$$

(a) Better than fixed biased circuit.

(b) $\beta R_c >> R_b$ as this will give very small value as stability as near to 1.

Drawback of Collector to Base Bias

- R_b has to be very small but this gives large I_b. This increase in I_b is offset by a decrease in I_b hence there in not much overall improvement in stability.
- Output is feedback to the base via R_b and this reduces the net gain of the transistor.

4.6.3 SELF-BIAS/POTENTIAL DIVIDER/EMITTER BIAS

- AS compared to collector–base bias, if the load of the collector is a transformer, than R_c will be small so $\beta R_c >> R_b$ cannot be achieved. But self-bias is suitable for any type of load resistance.
- Emitter bias is applied by a resistor Rs between the emitter and ground, i.e. Thevenin equivalent of two resistor of self bias.
- When collector current increased the voltage drop across R_e increases and reduce forward bias V_{be} reducing base current I_b and keeping the operating point stable
- Stability factor

$$S = \frac{1+\beta}{1+\frac{\beta R_c}{R_b + R_c}}$$

- $R_e >> R_b$, S =1 Best stability.
- R_e introduces both dc and ac feedback.
- In order to avoid the loss of ac signal gain because of feed back caused by Re. This R_e is bypassed by a leakage capacitor C_e of high value so that is offering little resistance at frequency of interest.

$$S_\beta = \frac{\delta I_c}{\delta \beta} \quad ; \quad S_\beta = \frac{I_{C1} \cdot S_2}{\beta_1 (1+\beta)}$$

$$S_v = \frac{\delta I_c}{\delta V_{be}} \quad ; \quad S_v = \frac{\beta}{R_b + R_c (1+\beta_2)}$$

where S_2 : Stability (S) at $\beta = \beta_2$.

Total increment in I_c is the sum of all three increments due to I_{co}, β, and V_{be}.

$$\Delta I_c = S_1 . \Delta I_{co} + S_v . \Delta V_{be} + S_\beta \Delta\beta$$

4.6.4 COMPENSATION TECHNIQUES

- Major disadvantage of the resistor biasing techniques is the feed back of the output signal to input which causes reduction in gain. To avoid this, compensation technique with diodes, transistor and Thermsitor are employed.
- Diode compensation.

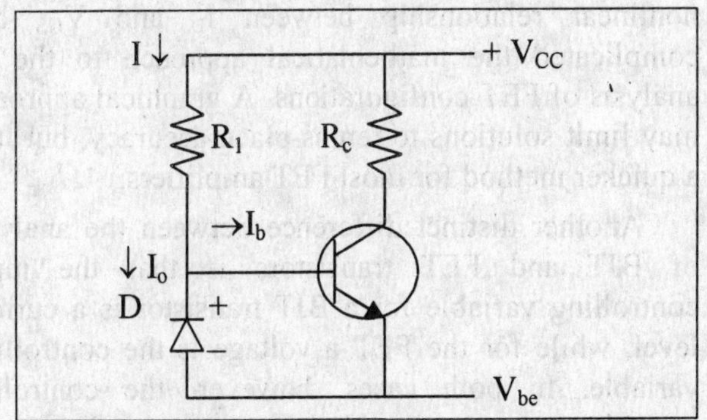

- Reversed biased diode is connected in the base emitter circuit, which is same material as transistor.
- The reverse bias voltage V_{be} allowing the reverse saturation current to I_o to flow through the diode D

$$I = \frac{V_{cc} - V_{be}}{R_I} = \frac{V_{cc}}{R_I}$$

And $I_B = I - I_o$

From collector current expression

$I_c = \beta I_b + (1+\beta) I_{co} = \beta I - \beta I_o + (1+\beta) I_{co}$

$= \beta I - \beta I_o + \beta I_{co}$ if $\beta >> 1$ and I_o of diode and I_{co} of transistor track each other over the desired temperature rang, then I_{co} remains essentially constant.

Reverse saturation current I_o flows through the diode. If I_{co} increases due to temperature, the increased current flows through diode decreasing I_b keeping I_b constant.

4.6.5 THERMISTOR

Thermistors are temperature sensitive element, which has a negative temperature coefficient of resistance.

The thermistor element is place parallel to and across V_{cc} anf E. R_2 of potential divider bias. Increase of temperature reduces resistance and decrease the forward bias reduces I_b and I_c.

4.6.6 SENSISTOR COMPENSATION

It may be placed either in parallel with R_1 or parallel with R_E. It can also placed in placed of R_E rather than in parallel with R_E.

4.6.7 FET/MOSFET BIASING

For the field-effect transistor, the relationship between input and output quantities is nonlinear due to the squared term in Shockley's equation. The nonlinear relationship between I_D and V_{GS} can complicated the mathematical approach to the dc analysis of FET configurations. A graphical approach may limit solutions to tenths-place accuracy, but it is a quicker method for most FET amplifiers.

Another distinct difference between the analysis of BJT and FET transistors is that the input controlling variable for a BJT transistor is a current level, while for the FET a voltage is the controlling variable. In both cases, however, the controlled variable on the output side is a current level that also defines the important voltage levels of the output circuit.

The general relationships that can be applied to the analysis of all FET amplifiers are

$$I_G \cong 0A \quad \text{and} \quad I_D = I_S$$

For JEETS and depletion-type MOSFETs, Shockley's equation is applied to re-late the input and output quantities:

$$I_D = I_{DSS}\left(1 - \frac{V_{GS}}{V_P}\right)^2$$

For Enhancement MOSFET

$$I_D = k\,(V_{GS} - V_r)^2$$

It is particularly important to realize that of the equations above are for the device only! They do not change with each network configurations. So long as the device is in the active region.

Table: FET Bias Configurations

Type/configuration/ pertinent equations	Graphical solution
JFET Fixed-bias V_{DD}, R_D, R_G, V_{GG} $I_G \cong 0$ $V_{GS_Q} = -V_{GG}$ $V_{DS} = V_{DD} - I_D R_S$	I_D, I_{DSS}, Q-point, I_{DQ}, V_P, V_{GG}, 0, V_{GS} V_G SQ
JFET self-bias V_{DD}, R_D, R_G, R_S $V_{GS} = -I_D R_S$ $V_{DS} = V_{DD} - I_D (R_D + R_S)$	I_D, I_{DSS}, $I_{DSS}/2$, Q-point, I_{DQ}, Vp, V'_{GS}, 0, V_{GS} $V_{GS} = -\dfrac{I_{DSS} \cdot R_S}{2}$
JEFT/voltage-divider bias V_{DD}, R_1, R_D, R_2, R_S $V_G = \dfrac{R_2 V_{DD}}{R_1 + R_2}$ $V_{GS} = V_G - I_D R_S$ $V_{DS} = V_{DD} - I_D (R_D + R_s)$	I_D, I_{DSS}, Q-point, $[V_G/R_s]$, V_P, 0, V_G, V_{GS} $V_D = V_{DD} - I_D R_D$ $V_S = I_D R_D$

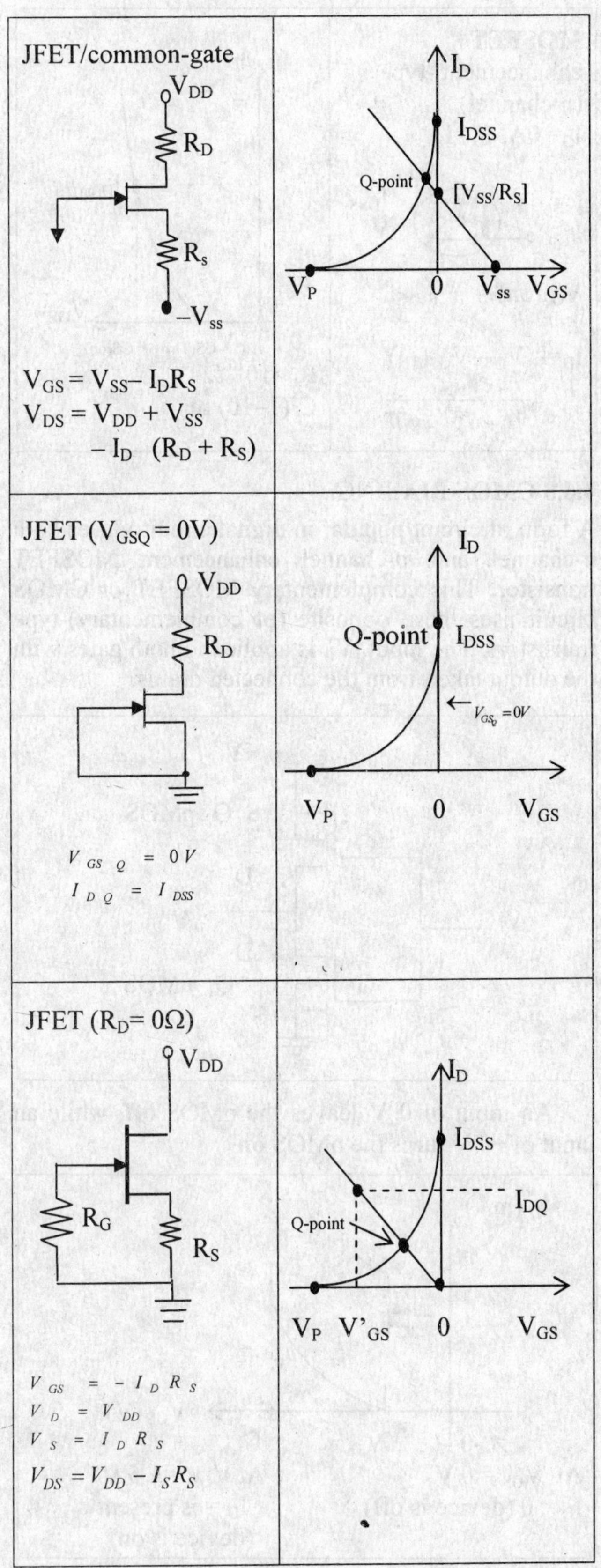
JFET/common-gate
V_{DD}
R_D
R_s
$-V_{ss}$
I_D
I_{DSS}
Q-point
$[V_{SS}/R_S]$
V_P
0
V_{ss}
V_{GS}
$V_{GS} = V_{SS} - I_D R_S$
$V_{DS} = V_{DD} + V_{SS} - I_D (R_D + R_S)$
JFET (V_{GSQ} = 0V)
V_{DD}
R_D
I_D
Q-point
I_{DSS}
$V_{GS_Q} = 0V$
V_P
0
V_{GS}
$V_{GS_Q} = 0V$
$I_{DQ} = I_{DSS}$
JFET (R_D= 0Ω)
V_{DD}
R_G
R_S
I_D
I_{DSS}
I_{DQ}
Q-point
V_P
V'_{GS}
0
V_{GS}
$V_{GS} = -I_D R_S$
$V_D = V_{DD}$
$V_S = I_D R_S$
$V_{DS} = V_{DD} - I_S R_S$

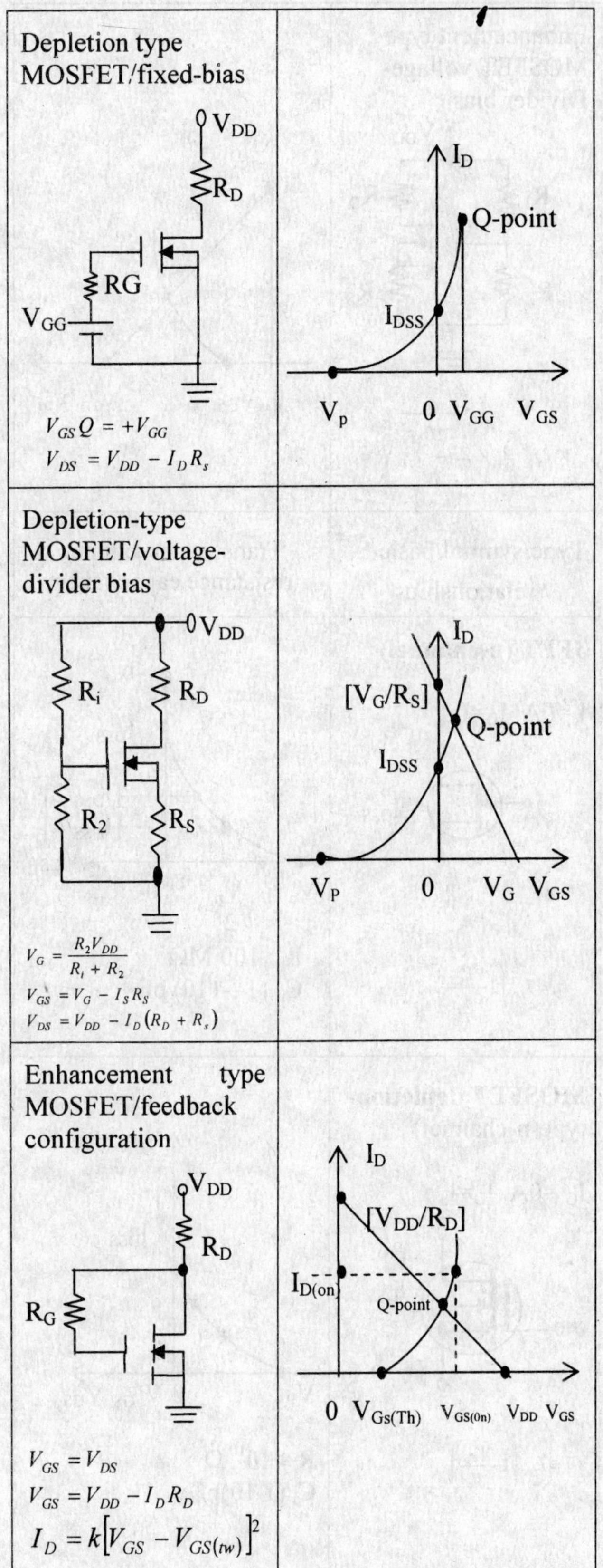
Depletion type MOSFET/fixed-bias
V_{DD}
R_D
RG
V_{GG}
I_D
Q-point
I_{DSS}
V_p
0
V_{GG}
V_{GS}
$V_{GS}Q = +V_{GG}$
$V_{DS} = V_{DD} - I_D R_s$
Depletion-type MOSFET/voltage-divider bias
V_{DD}
R_i
R_D
R_2
R_S
I_D
$[V_G/R_S]$
Q-point
I_{DSS}
V_P
0
V_G
V_{GS}
$V_G = \frac{R_2 V_{DD}}{R_i + R_2}$
$V_{GS} = V_G - I_S R_S$
$V_{DS} = V_{DD} - I_D (R_D + R_s)$
Enhancement type MOSFET/feedback configuration
V_{DD}
R_D
R_G
I_D
$[V_{DD}/R_D]$
$I_{D(on)}$
Q-point
0
$V_{GS(Th)}$
$V_{GS(On)}$
V_{DD}
V_{GS}
$V_{GS} = V_{DS}$
$V_{GS} = V_{DD} - I_D R_D$
$I_D = k[V_{GS} - V_{GS(tw)}]^2$

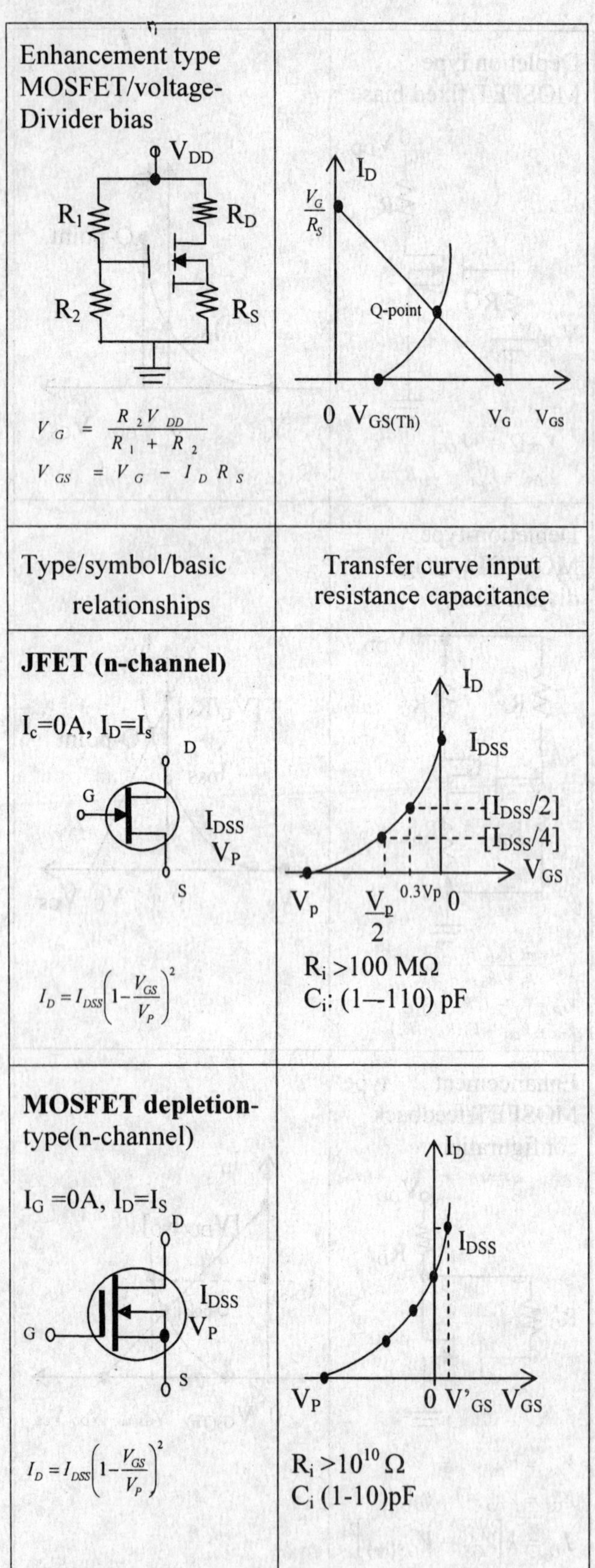

Enhancement type MOSFET/voltage-Divider bias V_{DD}, R_1, R_D, R_2, R_S $V_G = \frac{R_2 V_{DD}}{R_1 + R_2}$ $V_{GS} = V_G - I_D R_S$	I_D, $\frac{V_G}{R_S}$, Q-point, 0, $V_{GS(Th)}$, V_G, V_{GS}
Type/symbol/basic relationships	Transfer curve input resistance capacitance
JFET (n-channel) I_c=0A, I_D=I_s D, G, S, I_{DSS}, V_P $I_D = I_{DSS}\left(1 - \frac{V_{GS}}{V_P}\right)^2$	I_D, I_{DSS}, $[I_{DSS}/2]$, $[I_{DSS}/4]$, V_{GS}, V_p, $\frac{V_p}{2}$, 0.3Vp, 0 R_i >100 MΩ C_i: (1—110) pF
MOSFET depletion- type(n-channel) I_G =0A, I_D=I_S D, G, S, I_{DSS}, V_P $I_D = I_{DSS}\left(1 - \frac{V_{GS}}{V_P}\right)^2$	I_D, I_{DSS}, V_P, 0, V'_{GS}, V_{GS} $R_i > 10^{10}$ Ω C_i (1-10)pF

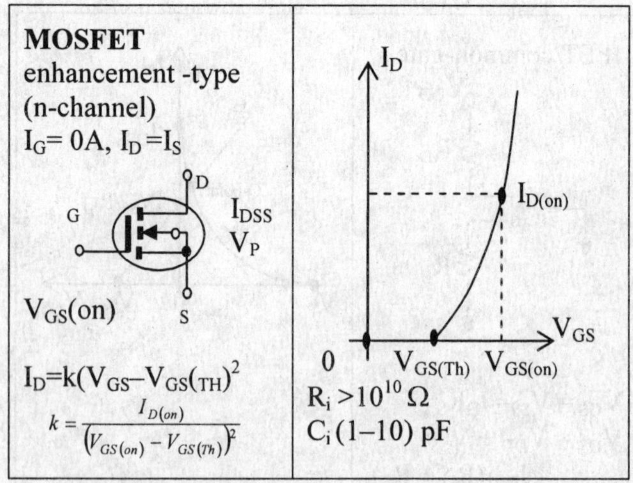

MOSFET enhancement -type (n-channel) I_G= 0A, I_D =I_S D, G, S, I_{DSS}, V_P V_{GS}(on) I_D=k(V_{GS}–$V_{GS(TH)}$)^2 $k = \frac{I_{D(on)}}{\left(V_{GS(on)} - V_{GS(Th)}\right)^2}$	I_D, $I_{D(on)}$, V_{GS}, 0, $V_{GS(Th)}$, $V_{GS(on)}$ $R_i > 10^{10}$ Ω C_i (1–10) pF

4.6.8 CMOS BIASING

A form of circuit popular in digital circuitry uses both n-channel and p-channel enhancement MOSFET transistor. This complementary MOSFET or CMOS circuit uses these opposite (or complementary)-type transistors. The input, V_i, is applied to both gates with the output taken from the connected drains.

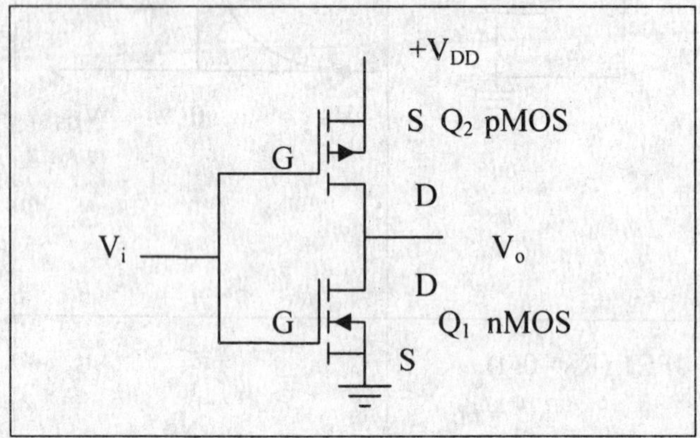

An input of 0 V leaves the nMOS **off**, while an input of +5 V turns the nMOS **on**.

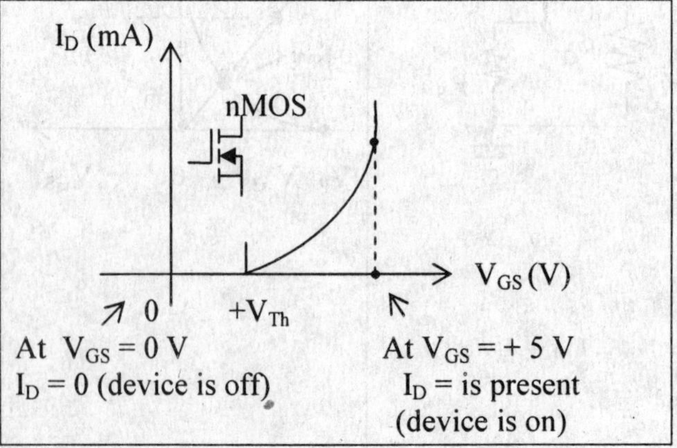

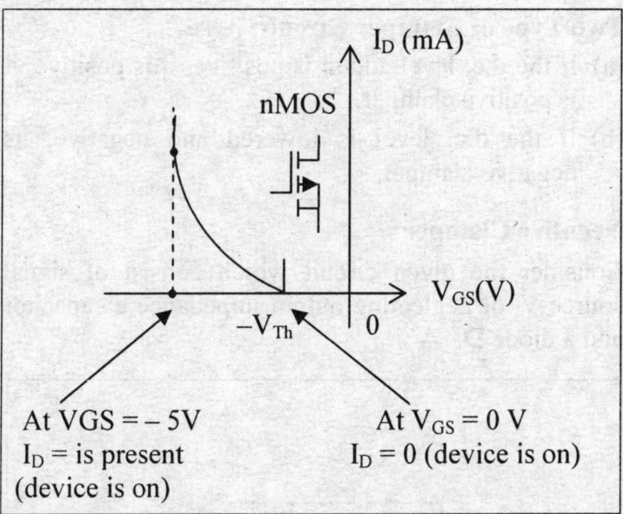

V_{GS} = 0 V leaves pMOS off: VGS = – 5 V turns pMOS on.

Operation of CMOS Circuit			
V_i (V)	Q_1	Q_2	V_o (V)
0	Off	On	+5
+5	On	Off	0

4.7 DIODE CIRCUITS

4.7.1 DIODE CLIPPING CIRCUITS

Clipping circuit are used to select for transmission that part of an arbitrary wave form which lies above or below some particular reference voltage level. These are also called voltage or current limiters. Some typical clipping with input output waveform are shown.

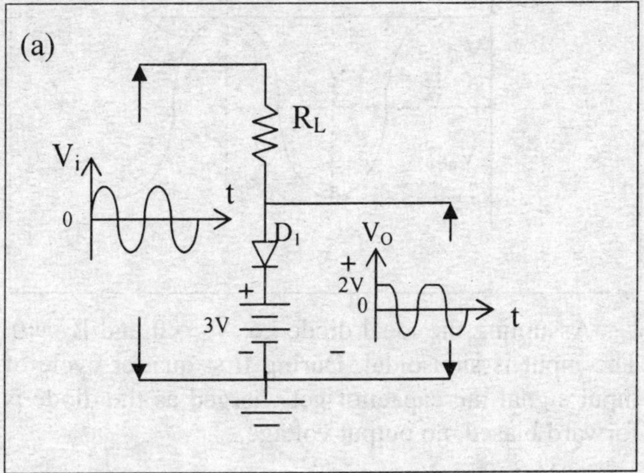

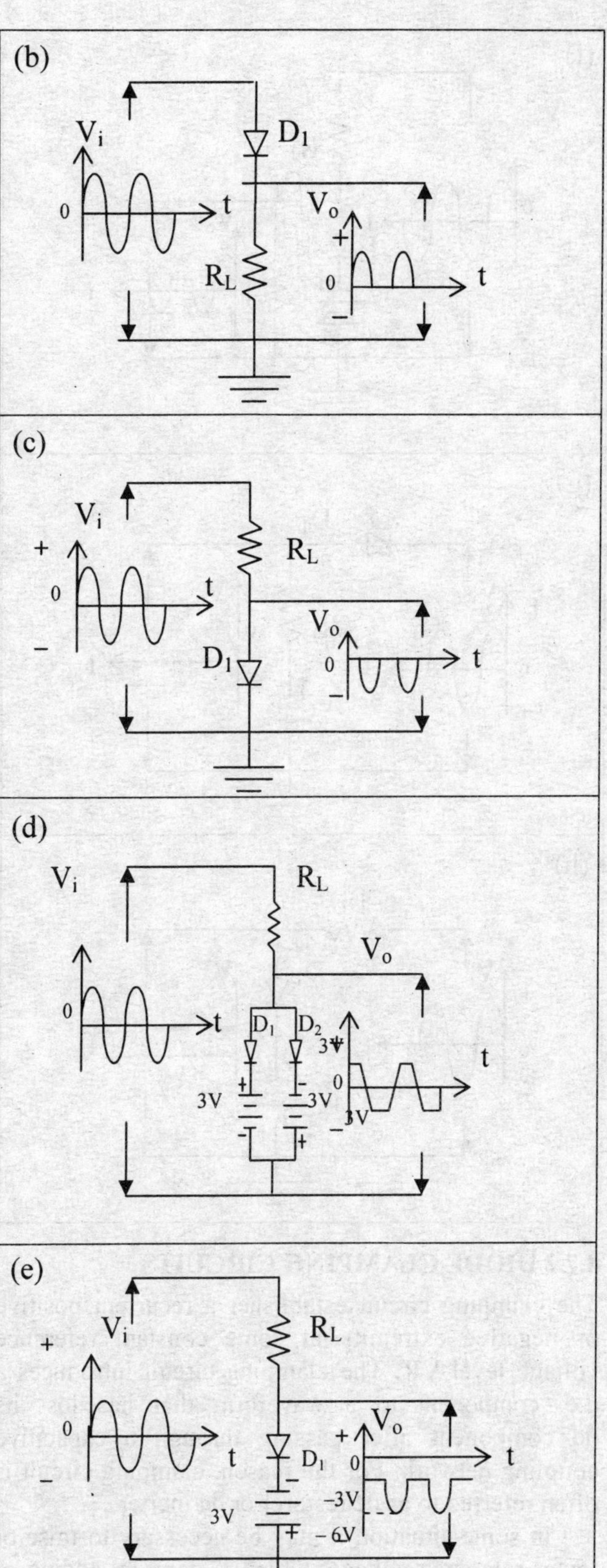

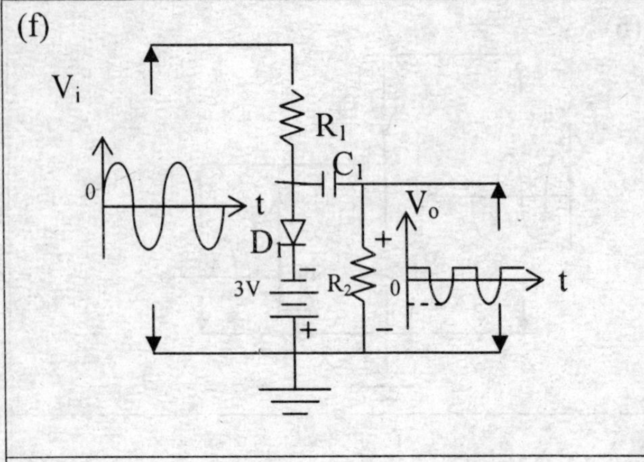

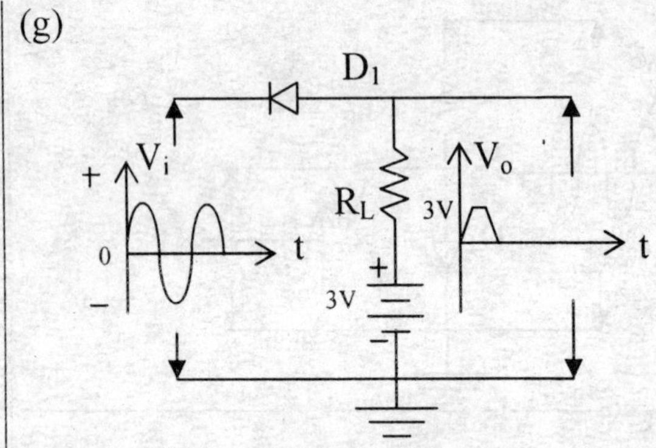

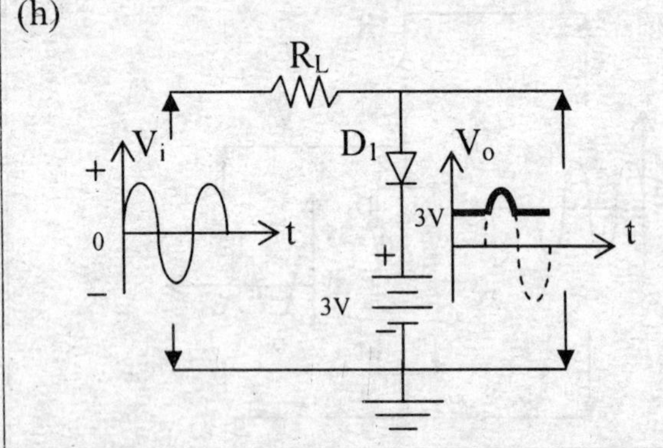

4.7.2 DIODE CLAMPING CIRCUITS

The clamping circuit establisher a recurrent positive or negative extremity at some constant reference voltage level VR. The clamping circuit introduces a d.c. component to a waveform that has lost its dc component after passing through a capacitive coupling network. For the reason, clamping circuit is often referred to as dc restorer or dc inerter.

In some situation it may be necessary to raise or lower reference voltage, which is done by adding or subtracting a dc voltage to the original reference voltage of 0.

Two Type of Clamper Circuit:

(a) If the d.c. level added is positive, it is positive, it is positive clamper.

(b) If the d.c. level is lowered and negative, its negative clamper.

Negative Clamper

Consider the given circuit which consist of signal source V_i of neglecting output impedance a capacitor and a diode D.

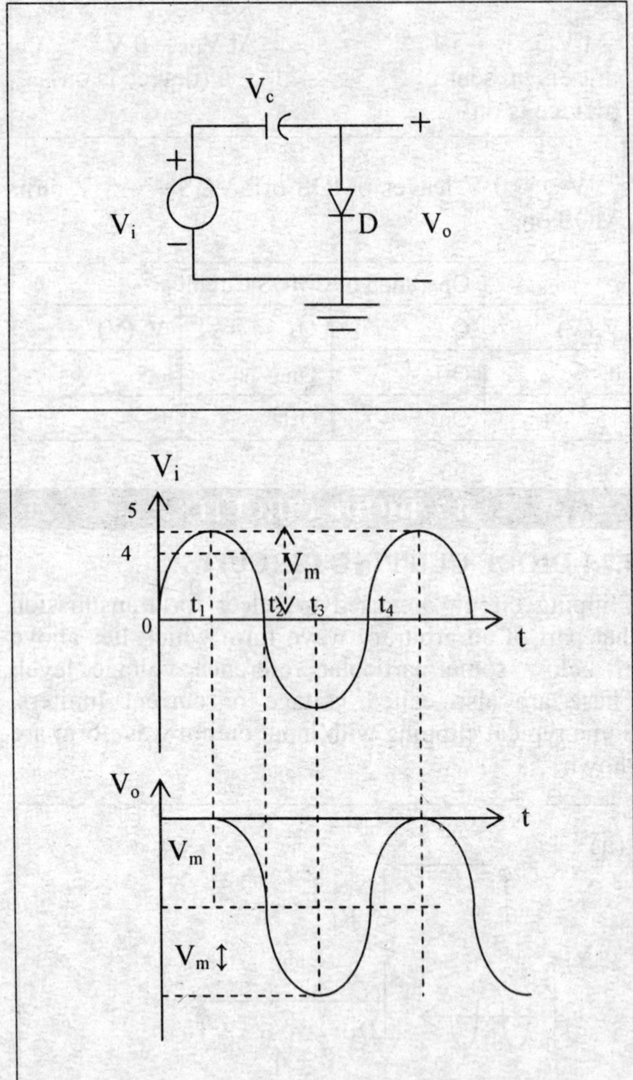

Assuming the ideal diode i.e. $V_y = 0$ and $R_f = 0$. The input is sinusoidal. During first quarter cycle of input signal the capacitor get charged as the diode is forward biased, no output voltage.

$\therefore V_o = 0.$

At the end of first quarter cycle $V_c = V_m$. After the diode comes reverse bias. The input signal reduces but the capacitor is not able to discharge due to diode

$$V_o = V_i - V_m$$

During succeeding cycles cycle the positive excusion of the signal just barely reaches zero. The diode need never again conduct and the positive extremity of the signal has been clamped or restore to zero.

If the amplitude of the input signal increases, the diode will conduct for first quarter cycle. The capacitor will again get charged and positive extremity will get clamped to zero voltage.

However, if the amplitude of input signal decrease. The signal does not remain clamped to zero. The reason being that capacitor does not get path to discharge.

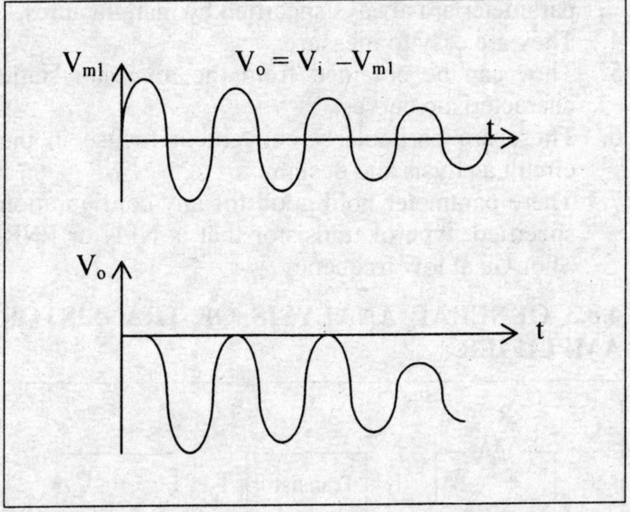

To permit a decrease in capacitor voltage it is necessary to shunt a resistor across c or to shunt a resistor across the diode.

The circuit with shunting diode is given below

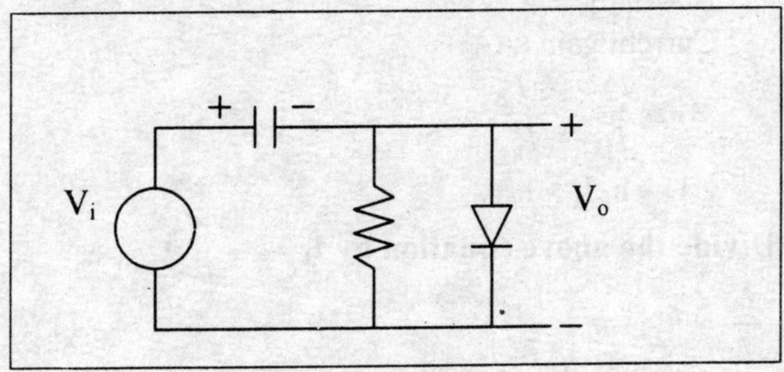

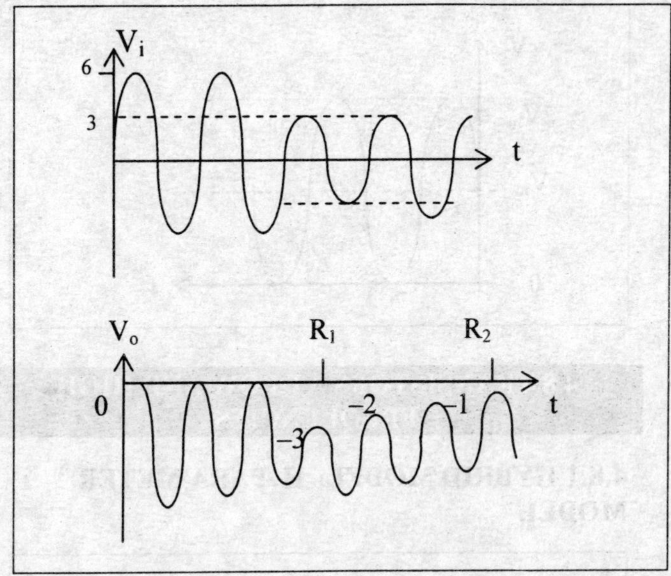

Now the capacitor get a discharge path. When the input suddenly rate.

$$V_o = V_i \rightarrow V_{ml}$$

Even if the input does not decrease to 3V the capacitor continues to discharge and diode get forward bias during the peak of input voltage and capacitor charge during peaks. The RC time constant is very important. It must be very large in comparison with the period of signal. If is become if capacitor loses to much of charge when diode is not conducting, then it may be that it does not charge during the small time internal when diode conduct.

Clamping circuit taking source and diode resistance into account:

The realistic clamping circuit is given below: Here source resistance R_s and diode forward resistance takes into account.

Positive Clamper

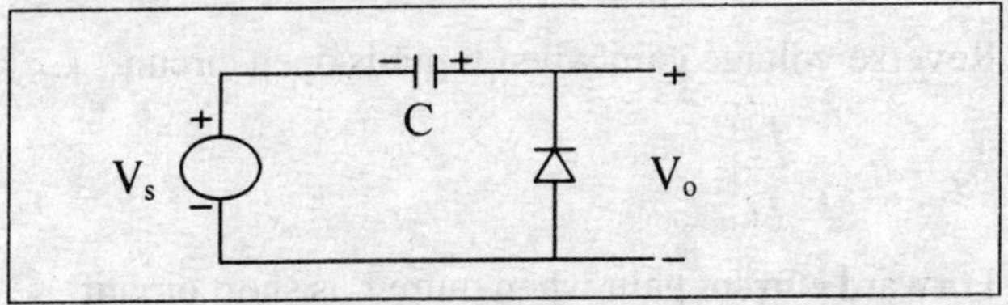

$V_s = V_m$ sin wt, for negative swing D conducts and capacitor charge to peak value V_m and cathode end of capacitor is positive.

$V_0 = V_s + V_c$ (or V_m)

= 2 Vm for positive swing and 0 for negative swing

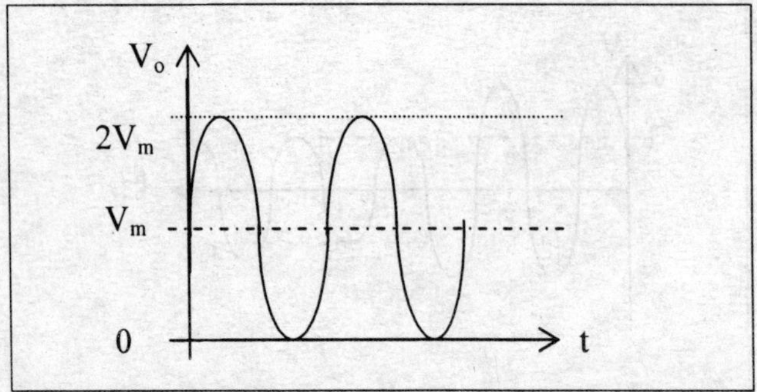

4.8 TRANSISTORS AT LOW AND HIGH FREQUENCIES

4.8.1 HYBRID MODEL: H–PARA METER MODEL

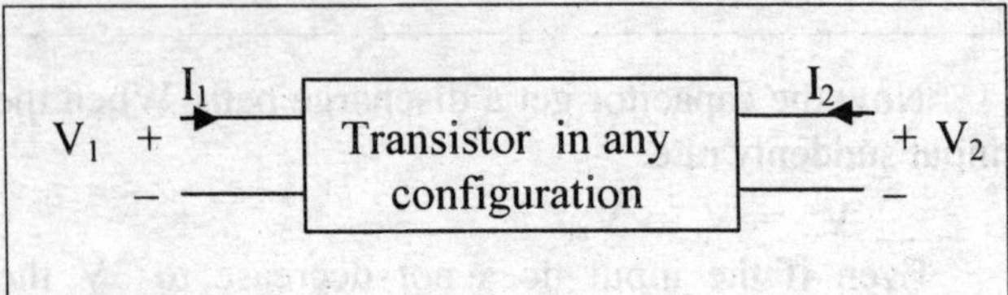

$V_1 = h_{11}\, I_1 + h_{12}\, V_2$

$I_2 = h_{21}\, I_1 + h_{22}\, V_2$

V_1 and I_2 $\rightarrow$ dependent Variable

I_1 and V_2 $\rightarrow$ independent Variable

$$\begin{pmatrix} V_1 \\ I_2 \end{pmatrix} = \begin{pmatrix} h_{11} & h_{12} \\ h_{21} & h_{22} \end{pmatrix} \begin{pmatrix} I_1 \\ V_2 \end{pmatrix}$$

$$h_i = h_{11} = \left.\frac{V_1}{I_1}\right|_{V_2=0}$$

Input impedance when output is short circuit (Ω)

$$h_r = h_{21} = \left.\frac{V_1}{V_2}\right|_{I_1=0}$$

Reverse voltage gain when input is open circuit

$$h_f = h_{21} = \left.\frac{I_2}{I_1}\right|_{V_2=0}$$

Forward current gain when output is short circuit

$$h_o = h_{22} = \left.\frac{I_2}{V_2}\right|_{I_1=0}$$

Output admittance when input is open circuit (s)

$V_1 = h_i\, I_1 + h_r\, V_2 \leftarrow$ KVL

$I_2 = h_f\, I_1 + h_o\, V_2 \leftarrow$ KCL

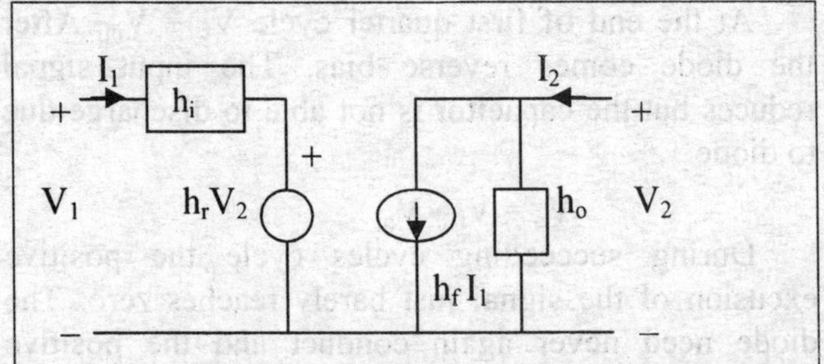

4.8.2 IMPORTANT POINT OF HYBRID MODEL

1. We use h-parameter in the equivalent circuit of a transistor since this parameters are relatively constant (real number) with respect to audio frequency and temperature.
2. We find the h-parameter under different condition, i.e. by open circuit the input or short circuit the output. Since these parameter have mixed units, these are called hybrid parameter
3. For a given configuration of transistor the form h-parameter are always specified by manufactures.
4. They are easy to measure.
5. They can be obtained from the transistor static characteristic curves.
6. These are particularly convenient to use in the circuit analysis and design.
7. There parameter hold good for any configuration specified, type of transistor that is NPN or PNP, Si or Ge at low frequency.

4.8.3 GENERAL ANALYSIS OF TRANSISTOR AMPLIFIER

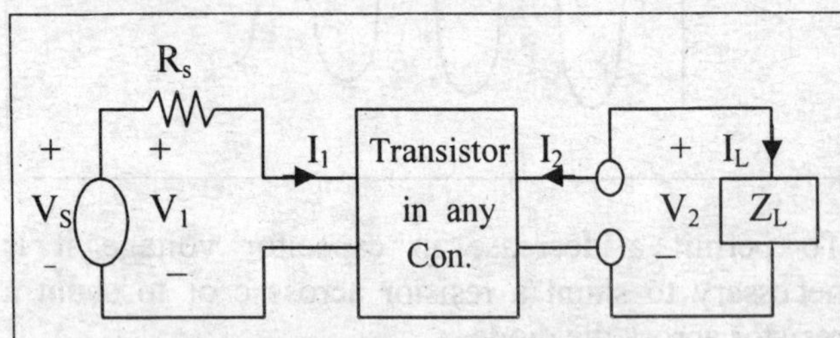

$V_1 = h_i\, I_1 + h_r\, V_2$

$I_2 = h_f\, I_1 + h_o\, V_2$

Current gain (A_I)

$$A_I = \frac{I_L}{I_1} = \frac{-I_2}{I_1}$$

$I_2 = h_f\, I_1 + h_o\, V_2$

Divide the above equation by I_1

$$\frac{I_2}{I_1} = h_f + h_o\left(\frac{V_2}{I_1}\right) \qquad (1)$$

Now $V_2 = I_L Z_L = -I_2 Z_L$ (2)

From equation (1) and (2)

$$\frac{I_2}{I_1} = h_f + h_o\left(\frac{-I_2 Z_L}{I_1}\right)$$

$$\frac{I_2}{I_1}\left(1 + h_o Z_L\right) = h_f$$

$$\frac{I_2}{I_1} = \frac{h_f}{1 + h_o Z_L}$$

$$A_I = -\frac{I_2}{I_1} = \frac{-h_f}{1 + h_o Z_L}$$

Input impedance (Z_{in})

$Z_{in} = [V_1 / I_1]$

$V_1 = h_i I_1 + h_r V_2$

Divide the above equation by I_1

$$\frac{V_1}{I_1} = h_i + h_r \frac{V_2}{I_1} = h_i + h_r\left(\frac{-I_2 Z_L}{I_1}\right)$$

Put the value of V_2 from equation (2)

$Z_{in} = h_i + h_r . A_I Z_L$

Voltage gain (A_v)

$A_v = [V_2 / V_1]$

Put the value from equation (2)

$V_1 = I_1 Z_{in}$

$$A_v = \frac{-I_2 Z_L}{I_1 Z_{in}} \;;\quad A_v = \frac{A_I Z_L}{Z_{in}}$$

Output impedance Z_{out}

$$Z_{out} = \left.\frac{V_2}{I_2}\right|_{V_s = 0}$$

$I_2 = h_f I_1 + h_o V_2$

Divide the above equation by V_2

$$\frac{I_2}{V_2} = h_f \frac{I_1}{V_2} + h_o \quad (3)$$

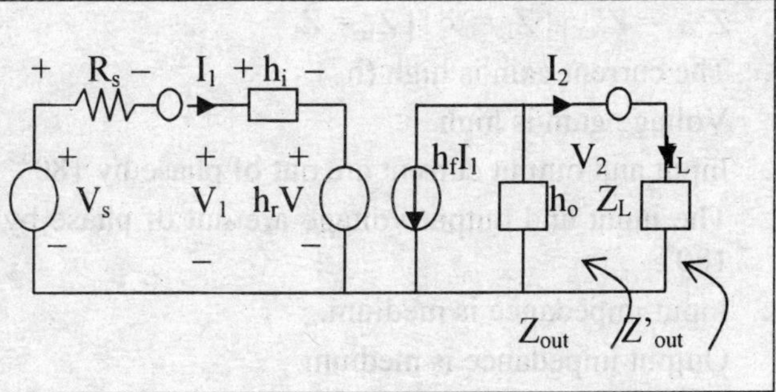

KVL at input:

$-V_s + I_1 (R_s + h_i) + h_r V_2 = 0$

$0 + I_1 (R_s + h_i) = -h_r V_2$

$$\frac{I_1}{V_2} = \frac{-h_f}{R_s + h_i} \quad (4)$$

From equation (3) and (4)

$$\frac{I_2}{V_2} = h_f\left(\frac{-h_f}{R_s + h_i}\right) + h_o$$

$$Z_{out} = \left[h_o - \frac{h_f h_r}{(R_s + h_i)}\right]^{-1} \qquad Z'_{out} = Z_{out} \parallel Z_L$$

Power gain $A_P = A_v . A_I$

Voltage gain taking R_S into account i.e. overall voltage gain

$$A_{vs} = \frac{V_2}{V_s} = \frac{V_2}{V_1} \times \frac{V_1}{V_s} = A_v\left(\frac{V_1}{V_s}\right)$$

$$\frac{V_1}{V_s} = \frac{Z_{in}}{Z_{in} + R_s}$$

$$A_{vs} = A_v\left[\frac{Z_{in}}{Z_{in} + R_s}\right] \quad \Rightarrow \quad A_{vs} < A_v$$

$A_{vs} = A_v \Rightarrow$ when $R_s = 0$

Current gain taking R_S into account i.e. overall current gain

$$A_{IS} = \frac{I_L}{I_S} = \frac{I_L}{I_1} \times \frac{I_1}{I_S} = A_I\left(\frac{I_1}{I_S}\right)$$

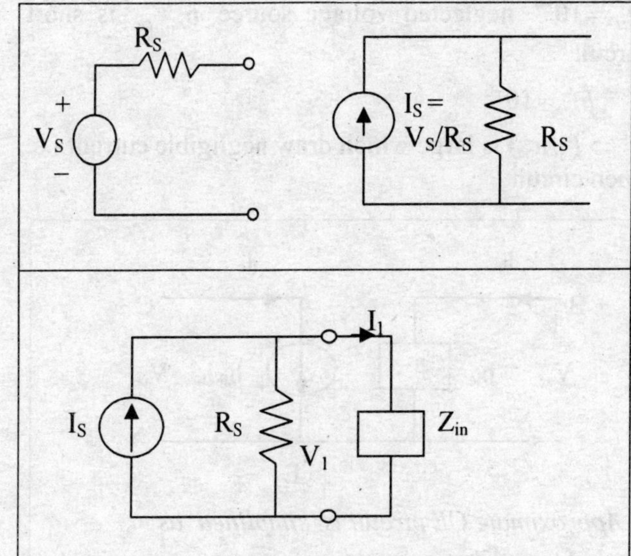

$$\frac{I_1}{I_s} = \frac{R_s}{R_s + Z_{in}}$$

$$A_{IS} = A_1 \left[\frac{R_s}{R_s + Z_{in}} \right] \quad A_{IS} < A_I$$

$A_{IS} = A_I$ when $R_S = \infty$

⇒ ideal current source

Over all power gain

$A_p = A_{VS} \times A_{IS}$.

4.8.4 APPROXIMATE ANALYSIS OF TRANSISTOR

In general for a transistor if

1. $h_{oe} = 10^{-6}$ S
2. $h_{re} = 10^{-4}$
3. $h_{oe} Z_L \leq 0.1$

Then it is seen that the error made in the analysis is < 10%, in such case approximate analysis hold good for CE.

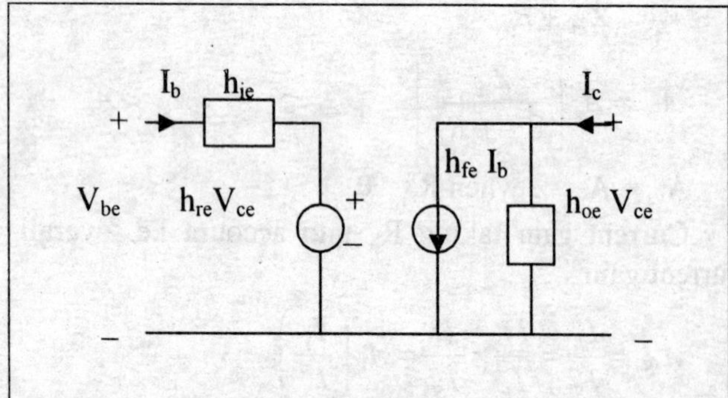

$h_{re} = 10^{-4}$ neglected voltage source $h_{rc}V_{ce}$ is short circuit

$h_{oe} = 10^{-6}$

S ⇒ $(1/h_{oe})$ is large which draw negligible current i.e. open circuit

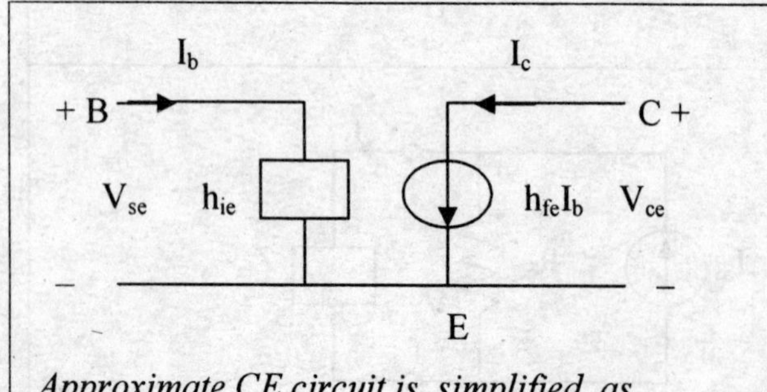

Approximate CE circuit is simplified as

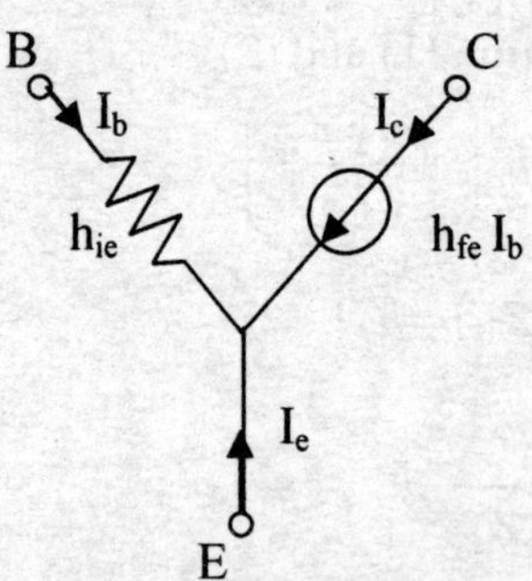

Approximate h-model which may be used for all three configuration

$h_{re} = 10^{-4}$ $h_{oe} = 10^{-4}$ S

$h_{fe} = 100$ $h_{re} = 2$k.

CE Analysis

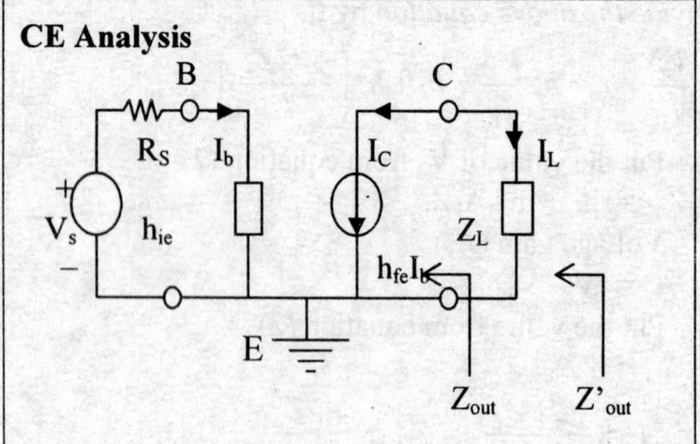

1. $A_I = \frac{I_L}{I_b} = \frac{-I_c}{I_b} = \frac{-h_{fe} I_b}{I_b} = -h_{fe}$ phase inversion
2. $Z_{in} = \frac{V_{be}}{I_b} = \frac{-h_{ie} \cdot I_b}{I_b} = h_{ie} \cong 1\,k\Omega$
3. $A_v = \frac{A_I}{Z_{in}} Z_L = -h_{fe} \frac{Z_L}{h_{ie}}$
4. $Z_{out} = \infty$; due to idle current source

   $Z'_{out} = Z_{out} \,||\, Z_L = \infty \,||\, Z_L = Z_L$

1. The current gain is high (h_{fe}).
2. Voltage gain is high.
3. Input and output current are out of phase by 180°.
4. The input and output voltage are out of phase by 180°.
5. Input impedance is medium.
6. Output impedance is medium.

7. Since voltage gain and current gain both are simultaneously high, this configuration is always used as an amplifier.

CB Analysis

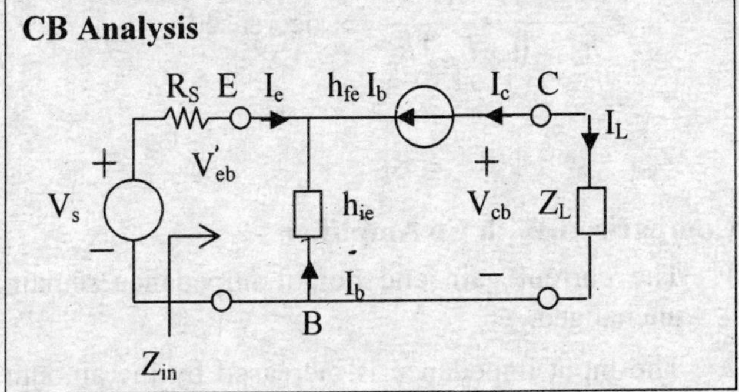

1. Current gain

$$A_I = \frac{I_L}{I_e} = \frac{-I_c}{I_e}$$

$$= \frac{-h_{fe} I_b}{-(1+h_{fe})I_b} = \frac{h_{fe}}{1+h_{fe}} = \frac{\beta}{1+\beta} = <1$$

2. $Z_{in} = \dfrac{V_{eb}}{I_e} = \dfrac{-h_{ie} I_b}{-(1+h_{fe})I_b} = \dfrac{h_{ie}}{1+h_{fe}} \Rightarrow$ very low

3. $A_v = A_I \dfrac{Z_L}{Z_{in}} = \left(\dfrac{h_{fe}}{1+h_{fe}}\right) \times \dfrac{Z_L}{\left(\dfrac{h_{ie}}{1+h_{fe}}\right)} = \dfrac{h_{fe} \cdot Z_L}{h_{ie}}$

4. $Z_{out} = \infty$

$Z_{out} = \infty \| Z_L = Z_L \Rightarrow$ highest

$$Z_{out} = \left.\frac{V_{eb}}{I_C}\right|_{V_s=0} = \frac{V_{eb}}{0} = \infty$$

$V_s = 0$; l $I_C = 0$

Feature

1. Low current gain (α) $\Rightarrow$ always less than 1.
2. High voltage gain (approximately equal CE).
3. Input and output current are in phase.
4. Input and output Voltage are in phase.
5. It has lowest input impedance of all three configurations.
6. It has highest output impedance of all three configurations.
7. Since current gain is approximately equal 1, if can be used as constant current source.
8. Can be used at place where transformation of impedance from a very low value of very high value to desired.
9. Non-inverting amplifier with $A_V > 1$ and $A_I < 1$.

1. $A_I = \dfrac{I_L}{I_b} = \dfrac{-I_e}{I_b} = \dfrac{(1+h_{fe})I_b}{I_b} = 1 + h_{fe}$

2. $Z_{in} = \dfrac{V_{bc}}{I_b} = \dfrac{h_{ie} \cdot I_b + I_L Z_L}{I_b}$

$$= \frac{h_{ie} I_b + Z_L (1+h_{fe}) I_b}{I_b}$$

$Z_{in} = h_{ie} + (1+h_{fe}) Z_L \Rightarrow$ highest

3. $A_v = A_I \dfrac{Z_L}{Z_{in}} = \dfrac{(1+h_{fe})Z_I}{h_{ie} + (1+h_{fe})Z_L} = \dfrac{x}{h_{ie} + x} (<1)$

where $x = (1+h_{fe})Z_L$

4. $Z_{out} = \left.\dfrac{V_{ec}}{I_e}\right|_{V_s=0}$

$$V_{ec} = -h_{ie} I_b - R_s I_b + V_s = -(h_{ie} + R_s) I_b$$

$$I_e = -(1+h_{fe}) I_b$$

$$Z_{out} = \frac{h_{ie} + R_s}{(1+h_{fe})} \Rightarrow \text{lowest}$$

Feature

1. Highest current gain $(1 + h_{fe})$.
2. Voltage gain always less the unity (lowest).
3. Input and output current are in phase.
4. Input and output voltage are in phase.
5. Input impedance is the highest of all the three configurations.
6. Output impedance is the lowest of the there configurations.
7. Used at place where the impedance conversion from a high value to low value is desired. Therefore it is always used as last stage of any audio amplifier where its output impedance is matched to low impedance speaker to transfer maximum amount of power.

CC / EMITTER FOLLOWER

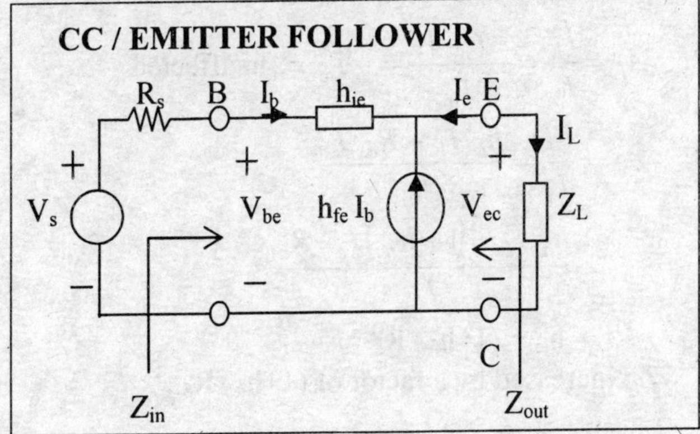

8. Since it is having high input impedance therefore it can be used as a buffer amplifier, which draw negligible current and does not load the pervious stage.
9. Z_{in} depend upon the load.

CE AMPLIFIER WITH EMITTER RESISTANCE R_E.

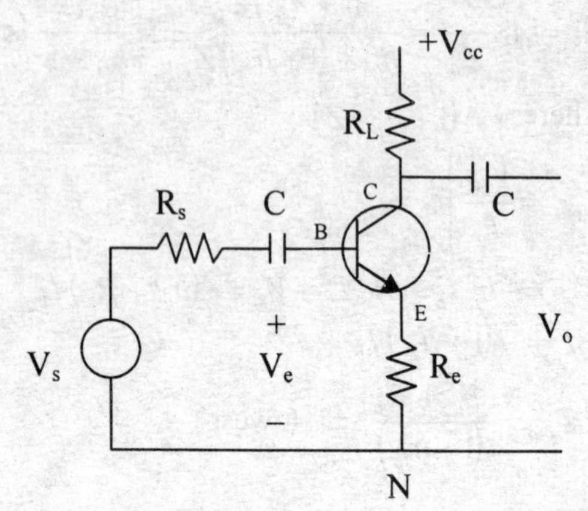

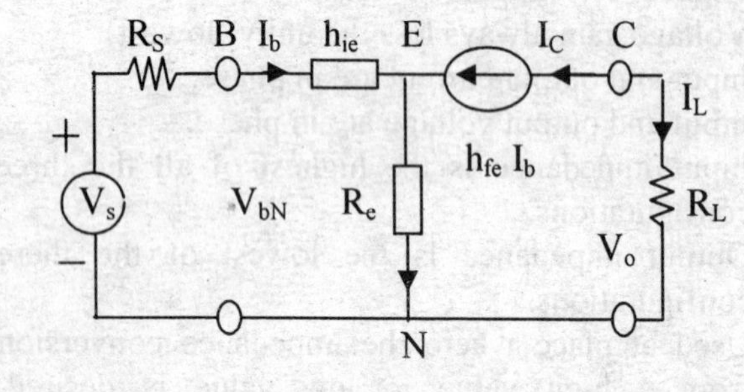

It is very important to stabilize the voltage amplification of an amplifier, i.e. it becomes independent or h-parameter.

$$I = I_b + h_{fe} I_b = (1 + h_{fe}) I_b$$

2. $A_I = \dfrac{I_L}{I_b} = \dfrac{-h_{fe}\, I_b}{I_b} = -h_{fe} \Rightarrow$ unaffected

3. $Z_{in} = \dfrac{V_b}{I_b} = \dfrac{h_{ie}\, I_b + R_e\, I}{I_b}$

$$= \frac{h_{ie}\, I_b + (1 + h_{fe}) I_b \cdot R_e}{I_b}$$

$$= h_{ie} + (1+h_{fe})\, R_e$$

Z_{in} increased by a factor of $(1+h_{fe})\, R_e$

4. $A_v = A_I \cdot \dfrac{Z_L}{Z_{in}}$

$$= \frac{-h_{fe} \cdot R_L}{h_{ie} + (1 + h_{fe}) R_e} \Rightarrow \text{decreased}$$

5. $Z_{out} = \infty$

$Z_{out} = \infty \,||\, R_L = R_L$

Comparison With Ce Amplifier

1. The current gain and output impedance remain unchanged.
2. The input impedance is increased by the amount $(1+h_{fe})\, R_e$ .... this is advantage.
3. The voltage gain decreased by the amount $(1+h_{fe})\, R_e$

$$A_v = \frac{-h_{fe} \cdot R_L}{h_{ie} + (1 + h_{fe}) R_e} \quad \text{as } h_{fe} >> 1$$

then $= \dfrac{-h_{fe} \cdot R_L}{h_{ie} + h_{fe} \cdot R_e} = \dfrac{-h_{fe} \cdot R_L}{h_{fe} \cdot R_e} = \dfrac{-R_L}{R_e}$

Let, $h_{fe}\, R_e >> h_{ie}$ ...... practically

$A_v = -\dfrac{R_L}{R_e}$ i.e. voltage gain is independent of h-parameter.

The h-parameter of transistor vary with respect to temperature so that the voltage gain changes with respect to temperature. Hence since the voltage gain is independent of h-parameter therefore the voltage gain is stabilized against the variation in temperature.

Therefore stability of the circuit increase.

Transistor Parameter value at Low Frequencies

Parameter	CE	CC	CB
h_i	1.1 kΩ	1.1 KΩ	20 Ω
h_r	2.5×10^{-4}	1	3×10^{-4}
h_f	50	– 51	– 0.99
$[1/h_0]$	40 kΩ	40 kΩ	2 MΩ

Approximate values for different quantities using approximate equivalent model are given as follows:

	CB	CC	CE	CE with Re
A_I	$\frac{h_{fe}}{1+h_{fe}}$	$1+h_{fe}$	$-h_{fe}$	$-h_{fe}$
Z_{in}	$\frac{h_{ie}}{1+h_{fe}}$	$h_{ie}+(1+h_{fe})R_L$	h_{ie}	$h_{ie}+(1+h_{fe})R_e$
A_v	$\frac{h_{fe}R_L}{h_{ie}}$	$1-\frac{h_{ie}}{Z_{in}}$	$\frac{-h_{fe}R_L}{h_{ie}}$	$\frac{-h_{fe}R_L}{Z_{in}}$
Z_{out}	∞	$\frac{R_s+h_{ie}}{1+h_{fe}}$	∞	∞
Z_{out}	R_L	$Z_{out} \parallel R_L$	R_L	$\| R_L$

4.8.5 SMALL SIGNAL ANALYSIS FREQUENCY RESPONSE

When a multi frequency signal extending from very low frequency to very high frequency is applied to an amplifier, it should amplify all the signals in the range uniformly. This is known as frequency response. In practice it is not so because the amplifier is constructed out as frequency sensitive element like condenser, which cause attenuation of the signals.

Amplifier may be capable of uniform amplification from low frequency F_L to a high frequency F_H and the range. $(F_H - f_L)$ called b*andwidth.*

Rise Time

The response of amplifier is studied by feeding a step signal.

$$t_r = \frac{2.2}{\omega H} \quad ; \quad t_r = \frac{0.35}{f_h}$$

where, t_r = rise time (time required from 10% to 90%)

f_h = higher cut-off frequency.

4.8.6 TRANSISTOR AT HIGH FREQUENCY

Hybrid π Model/Giacoletto model

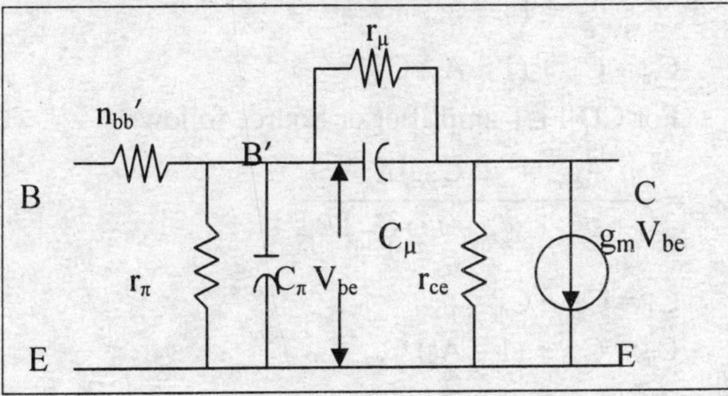

$r_\pi = [1/g_{be}]$; $r_{ce} = [1/g_{ce}]$

$\mathbf{r_{bb'}}$ = It is equivalent resistance due to base spreading Resistance. Base region is very thin in transistor. Current which enter the base from emitter must flow through narrow path to reach the base terminal. the cross section area for current flow in collector or emitter is very much large than in base.

r_π, C_π = euivalent resitance and capacitance due to forward biased junction.

r_π = 1 k ohm , C_π = 100 pF,

as $C = \varepsilon A/d$, FB $\uparrow \rightarrow$ d $\downarrow \rightarrow$ C $\uparrow$

This represent the diffusion capacitance between base and emitter. This is due to excess of minority charge in base.

C_μ, r_μ = Equivalent resistance and capacitance due to reverse biased junction.

r_μ = 4 M ohm, C_μ =3pF, RB $\uparrow \rightarrow$d $\uparrow \rightarrow$C$\downarrow$

4.8.7 SOME IMPORTANT RELATION BETWEEN LOW FREQUENCY AND HIGH FREQUENCY

Transistor Parameter

$g_m = [I_{cq}/V_T] = 50$ mA/V = 50 mS

where, $V_T = T/11600 = [h_{fe}/r_\pi]$

$r_{bb} = h_{ie} - r_\pi$; $\quad C_\pi = g_m/2\pi f_T$

here, f_T is the frequency at $2\pi f_T$ which gain falls to unity.

Short Circuit Current Gain

$$A_I = \frac{I_L}{I_i} = \frac{-hfe}{1+j[f/f_\beta]}; \; f_\beta = \frac{1}{2\pi r_\pi (C_\pi + C_\mu)}$$

where, f_β is the bandwidth of amplifier.

= half power frequency.

Figure of Merit (f_T)

It is the frequency at which the short circuit current gain become unity.

$$A_I = \frac{I_L}{I_i} = \frac{-h_{fe}}{1+j[f/f_\beta]}$$

At $f = f_\beta$; $A_I = 1$

$$f_T = h_{fe} \cdot f_\beta = \frac{h_{fe}}{2\pi r_\pi (C_\pi + C_\mu)}$$

Miller' Theorem

It states that if an impedance Z is connected between the input and output terminal of network between which there is voltage gain Av, the same effect can be held by removing Z and connecting an impedance Z_1 at the input and Z_2 at output.

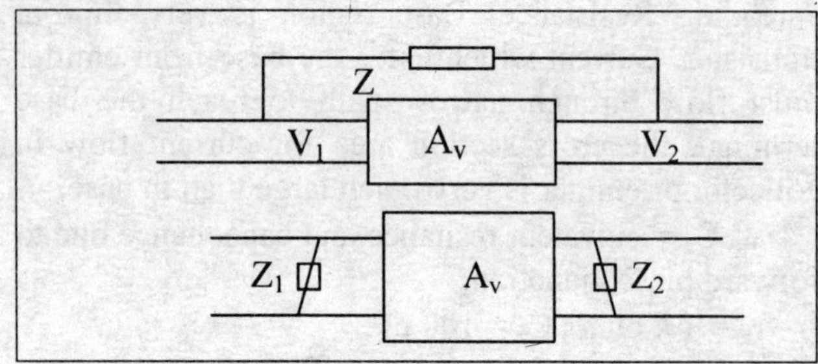

$$Z_1 = \frac{Z}{1 - A_v} \quad ; \quad Z_2 = \frac{Z}{\left[1 - \left(\frac{1}{A_v}\right)\right]}$$

$$Y_1 = Y(1 - A_v) \quad ; \quad Y_2 = [1 - (1/A_v)]$$

4.9 FET AT LOW AND HIGH FREQUENCIES

4.9.1 EQUIVALENT CIRCUIT OF FET AT LOW FREQUENCIES

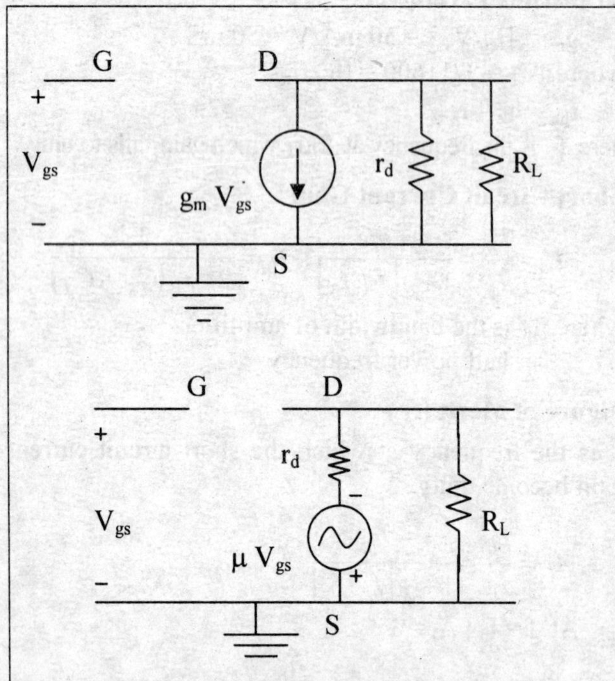

4.9.2 VOLTAGE GAIN AT LOW FREQUENCIES

For CS-FET amplifier

$$Z_{in} = R_1 \| R_2 \; ; \; Z_0 = R_L$$

$$A_v = \frac{-\mu R_L}{r_d + R_L} = -g_m \left(r_d \| R_L\right)$$

For CD-FET amplifier or source follower

$$Z_{in} = R_G$$

$$Z_0 = r_d \left\| R_L \right\| \frac{1}{g_m} = R_L \| g_m$$

$$A_v = \frac{\mu R_L}{r_d + (\mu + 1) R_L} = g_m \frac{r_d \| R_L}{1 + g_m \left(r_d \| R_L\right)}$$

4.9.3 EQUIVALENT CIRCUIT OF FET AT HIGH FREQUENCIES

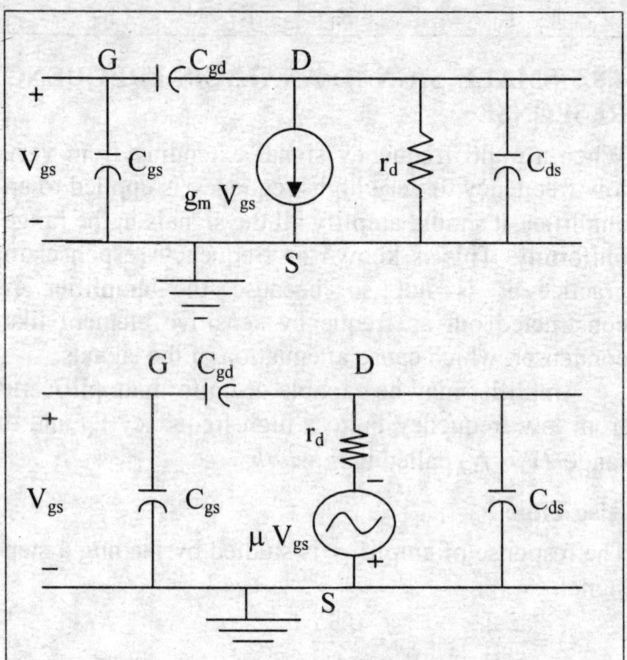

4.9.4 VOLTAGE GAIN AT HIGH FREQUENCIES

For CS-FET amplifier

$$A_v = \frac{-g_m + Y_{gd}}{Y_L + Y_{ds} + g_d + Y_{gd}}$$

$$C_{in} = C_{gs} + (1 - A_v) C_{gd}$$

For CD-FET amplifier or Source follower

$$A_v = \frac{\left(g_m + j\omega C_{gs}\right) R_s}{\left(1 + g_m + g_d + j\omega C_T\right) R_s}$$

$$C_T = C_{gs} + C_{ds}$$

$$C_{in} = C_{gd} + (1 - A_v) C_{gs}$$

4.10 AMPLIFIERS

4.10.1 POWER AMPLIFIER

Power amplifier are meant for large signal amplification power stage is normally the final stage of a cascade amplifier and driver. The energy conversion device i.e. speaker, antenna or transducer etc. The power amplifier does not actually amplifier the power, instead it takes power from the dc power supply connected to the power circuit and converts into use full ac signal power $Z_{in} \rightarrow$ high, $Z_{out} \rightarrow$ low, heat sink are used.

Class A

Here the dc bias is adjusted much less than its cut off value so that the output current flow for full input voltage cycle.

Feature:

1. Minimum distortion.
2. Since the output contains only one frequency component, the output power is small.
3. Efficiency is low.
4. Used as voltage amplifier as modulating amplifier.

Class B

Here the dc bias is adjusted equal to the cut off value so that the output current flow for half of the input voltage cycle. Hence the output dc component will be always equal to zero.

Feature:

1. High distortion
2. High power output
3. Efficiency is large
4. Used as power amplifier as modulated amplifier.

Class AB

Here the dc bias is adjusted slightly less than is cut-off value so that the output current flows for more than half but less than input voltage cycle. This class of amplifier is just a compromise between class A and class B amplifier.

Feature:

1. Output power is medium
2. Medium distortion
3. Medium efficiency
4. Used as modulating amplifier
5. Used with unturned power amplifier.

Class C

Here the dc bias is adjusted greater than its cut-off value so that the output current flow for less than half of the input voltage cycle.

Therefore the output current will be in the from of pulses which will be highly rich in harmonic i.e. frequency multiplier.

Feature:

1. Very high output power due to reach in harmonic
2. Very high distortion
3. Very high conversion efficiency
4. Used as only modulated amplifier when the need is to raise the power level before transmission
5. Used with tuned frequency amplifier.

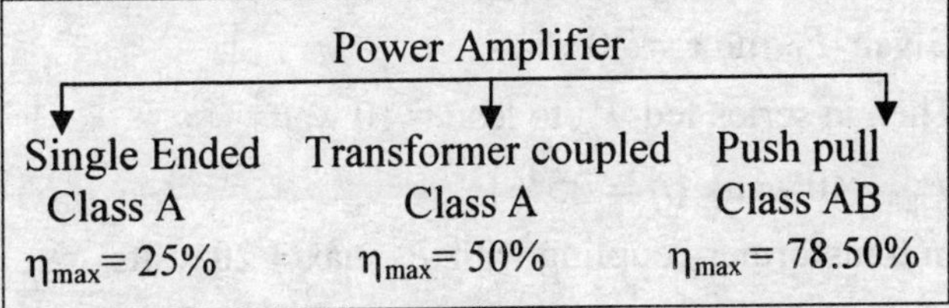

4.10.2 SINGLE ENDED CLASS A AMPLIFIER

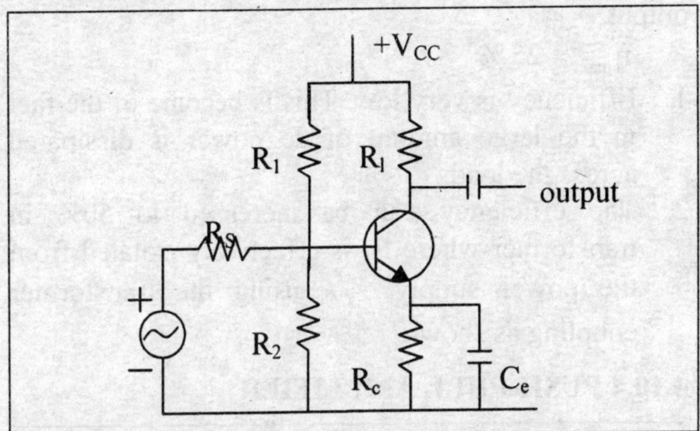

4.10.3 CLASS A TRANSFORMER COUPLED

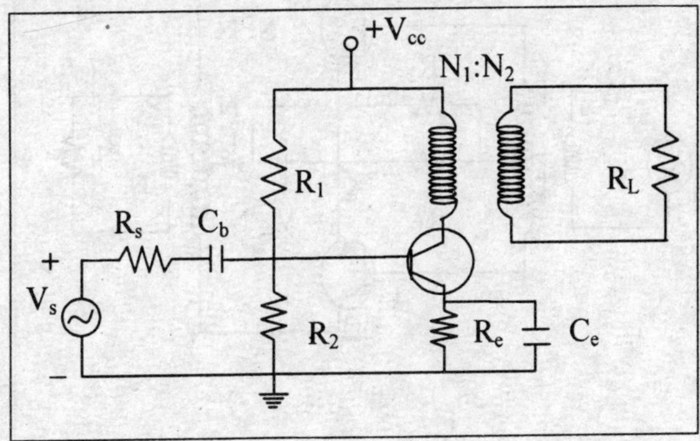

Conversion efficiency (η)

$$= \frac{\text{Signal power delivered to load}}{\text{DC power supplied to output cricuit}} \times 100 = 50\%$$

[DC power supplied $(V_{CC} I_C)$]

$= AC$ power developed in load $\left(\frac{1}{2} V_m I_m\right)$

+ power dissipated in load $R_L \left(I_c^2 R_L\right)$

+ Dissipation in theCollector P_D

With no signal $P_D(\max) = V_{CE} \cdot I_C$

$\therefore P_{ac}$ delivered $= \frac{1}{2} P_D \max$

This statement is true in both series fed and Transformer coupled cases.

Given P_D max = 20 watts

Then in series fed P_{ac} to load = 10 watts

$P_{dc} = 40$ watts $(\eta = 25\%)$

In transformer coupling with P_D max = 20 watts

$P_D = 20$ watts $(\eta = 50\%)$

Since only 25% ac power is useful at the power output.

$\eta_{max} = 25\,\%$

1. Efficiency is very low. This is become of the fact in the large amount of dc power is dissipated across the load.
2. The efficiency can be increased to 50% in transformer where R_L is effectively isolated from the power supply V_{CC} using the transformer coupling as shown.

4.10.4 PUSH-PULL AMPLIFIER

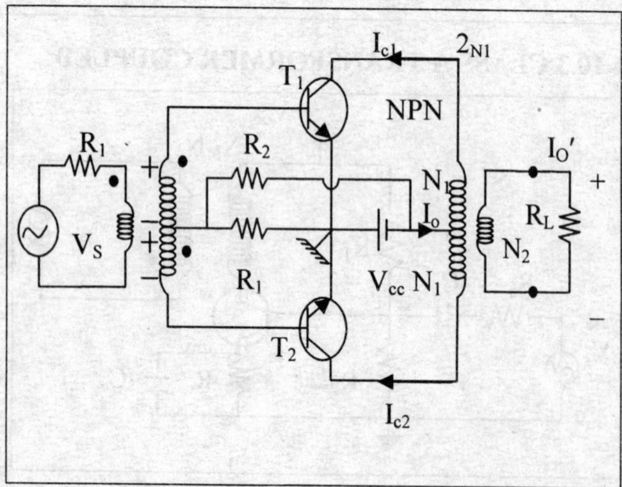

$$R'_L = \left(\frac{2N_1}{N_2}\right)^2 R_L$$

Output load current $= k\left(i_{c1} - i_{c2}\right)$

$= 2k\left(B_1 \cos \omega t + B_1 \cos 3\omega t + ...\right)$

Only odd harmonics are present. The push pull system has balanced out the even harmonics.

→ In class B operation, the operating point is at cut-off so that the signal is amplified by one transistor only for one half of the cycle,

$$\eta = \frac{P_{ac} to\ load}{P_{dc}} = \frac{\pi}{4}\left(\frac{V_{cc} - V_{\min}}{V_{cc}}\right) \times 100$$

theoretical max.η = η max = π / 4 × 100 = 78.5%

also $\frac{P_{D\max}}{P_{ac\max}} = \frac{4}{\pi^2} = 0.4$

Under these maximum values of P_{ac}, P_D per transistor $= 0.2 P_{ac}$, $P_d \max = \frac{1}{5} P_{ac} \max$.

Maximum AC power obtainable in class B push pull $= 5P_D$ (of the transistor).

Characteristic:

1. The even harmonic are absent at the output of a push-pull amplifier i.e.
   $I_C = 2\,I_1 \sin \omega t + 2\,I_2 \sin 3\omega t + \ldots$
2. The amplifier gives more output power per transistor for a given amount of distortion.
3. Less distortion is achieved for a given power output per transistor.
4. The dc components of the output current of the two transistor oppose each other in the transformer. This eliminates the tendency of the output core of the transformer to saturate. Therefore, the losses in the transformer are minimized.
5. The effect of ripple voltage contained in the power supply are just balanced out since the current produced are in the opposite direction.
6. Due to non saturation of core the transformer contains less ferromagnetic material.
7. The output transformer for push pull amplifier circuit are lighter, smaller and less expensive than the transformer of comparable quality used in single ended transform amplifier.

8. If quiescent current is low. Hence efficiency is maximum i.e. $25\pi = 78.5\,\%$

$$\eta = 25\,\pi\left(1 - \frac{V_C(\min)}{V_{CC}}\right)$$

Disadvantage:

1. However push pull does not eliminate power supply hum.
2. Bulk in size become of two transformer.
3. Cross over distortion.

Distortion in the Output Caused by Many Factor

1. If the transistor are not matched in push pull configuration.
2. Collection characteristic many be nonlinear.
3. Input characteristic are non-linear.

Crossover distortion

The input characteristic of transistor are such that no appreciable current flow until the emitter function is forward biased by cut in voltage V_Y. Under there condition, a sinusoidal base voltage excitation does not result in a sinusoidal output current.

To minimize cross over distortion transistor must operate in class AB mode where a small standing current flows in the case of zero excitation. The remedy is class AB push pull amplifier.

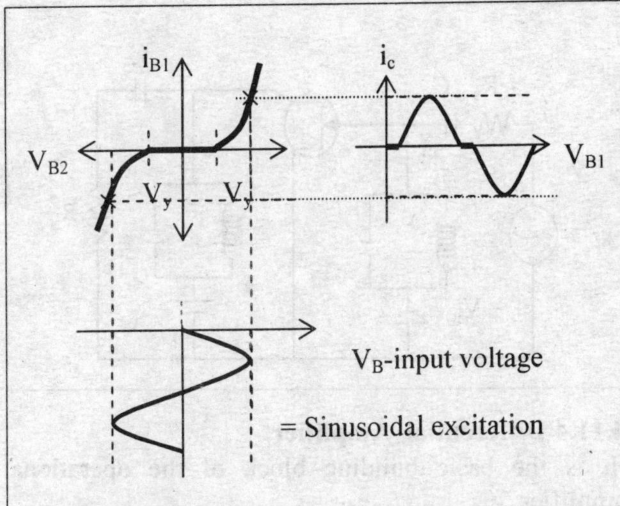

Complementary-Symmetry Connection

In this case two transistors, one NPN and other PNP have their emitter connected together. The Bases are tied and signal is fed to the base.

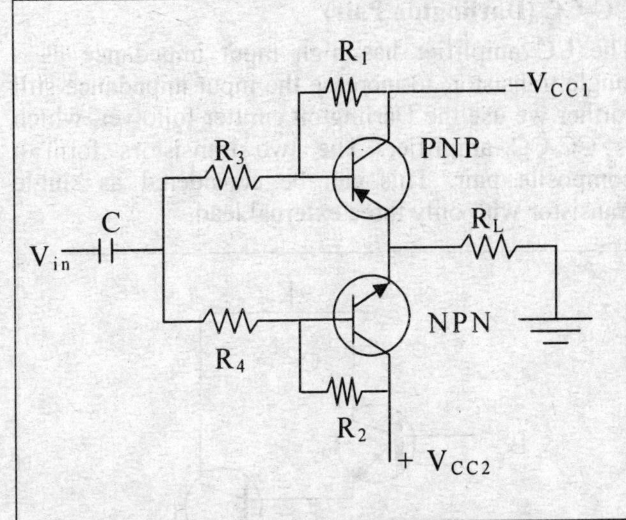

- Bases are tied and signal is fed to the base.
- Both transistors are biased near cut-off.
- This is also known as push-pull and it doubles the dynamic range of the circuit by accommodating a larger input range of the circuit of accommodating a large input signal.

4.11 MULTISTAGE AMPLIFIER

4.11.1 CASCADED AMPLIFIER

- Gain of one stage of amplifier may not be sufficient to obtain the required gain and also the required gain may be worked out in a single stage, the operating condition of the transistors may be impracticable.
- Overall gain is the product of gain of individual stages.
- $A_{VT} = A_{V1} \times A_{V2}$

Frequency Response of Cascaded Stages

For n identical stages:

1. The upper cut-off frequency decreases

$$f_H = f_H\sqrt{\left(2^{1/n} - 1\right)}$$

2. The lower cut off frequency increase

$$f_L = \frac{f_L}{\sqrt{2^{1/n}} - 1}$$

3. Overall bandwidth decreases.
4. The gain bandwidth product remain constant.
5. The Transmission line parameter are used, i.e. A, B, C, D.

CC-CC (Darlington Pair)

The CC amplifier has high input impedance as a single transistor; to increase the input impedance still further we use the Darlington emitter follower, which is CC-CC amplifier. The two transistors form a composite pair. This can be considered as single transistor with only three external lead.

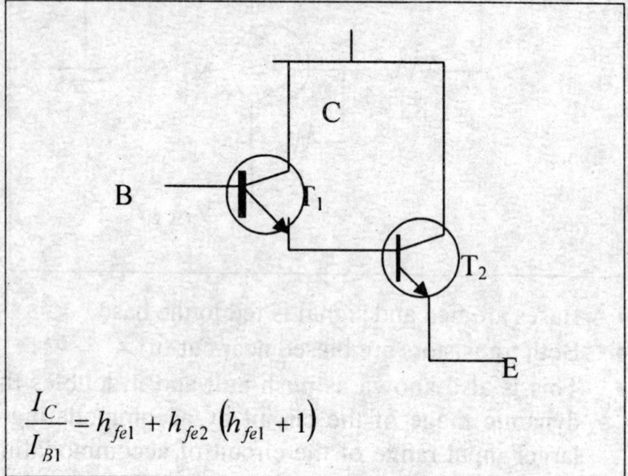

$$\frac{I_C}{I_{B1}} = h_{fe1} + h_{fe2}\left(h_{fe1} + 1\right)$$

Assuming two transistor is having same h_{fe}

$$\frac{I_C}{I_{B1}} = h_{fe} \times h_{fe} = h_{fe}^2 \;; \qquad A_v \cong 1$$

Input Impedance

$R_i = h_{ie1} + (h_{fe1} + 1)\, h_{ie2}$

$R_i = 2h_{ie2}\ (h_{fe1} + 1)$

assuming $h_{ie1} = h_{ie2} = h_{ie}$

4.11.2 CASCODE CONNECTION

This CE-CB connection

- Common emitter stage drives a common base amplifier.
- Gain can be improved by using large value of R_L over a wide range of frequencies.

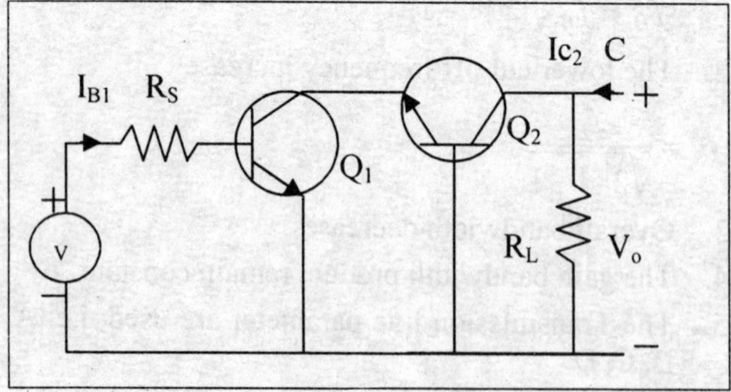

- Overall current gain $A_I - \frac{I_{C2}}{I_{B2}} = \beta_0$
- Voltage gain $= A_v = \frac{-\beta_0 R_L}{(R_s + r_{be})}$
- Output resistance of a CB stage is high, so we can connect a large load resistance R_L.
- This input resistance for the CB stage is low and this is the load resistance for the CE stage. The low input resistance of CB stage makes Q_1 operate into a short-circuit load, so voltage gain of common emitters stage is low. This cause reduces the input capacitances so its Bandwidth gets increased.
- It is used in *wide Band amplifiers* due to its large bandwidth.

4.11.3 TUNED AMPLIFIER

To provide gain at particular frequency tuned circuit are used at the input and output terminal. It is operated on RF frequency. The effective Q of the resonant circuit is of be high enough so the V_o consist only the direct voltage and fundamental component of signal i.e. of to eliminate the harmonics.

Function of the resonant circuit are:

1. To provide the correct load impedance.
2. To reject the unwanted Harmonic.
3. To couple the power to the load.

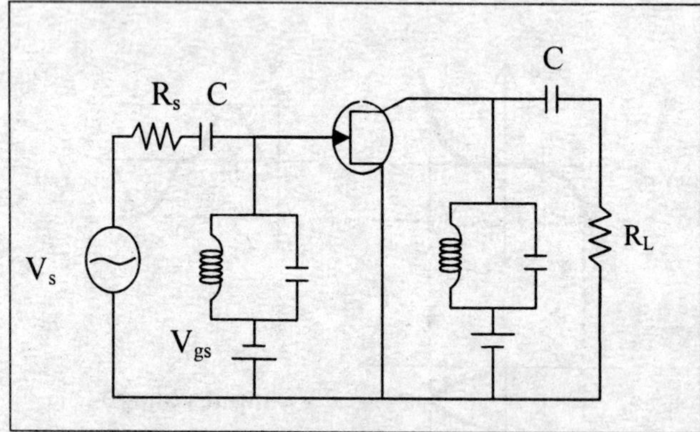

4.11.4 Differential Amplifier

It is the basic building block of the operational amplifier:

- The first stage of an op-amp is Differential amplifier (DA).
- DA consists of two transistor which are emitter coupled and whole characteristic are identical.

There are two type of input signal to the DA:

Differential mode: $V_d = V_1 - V_2$

Common mode: It is average between the input voltage V_2 and V_1

$$V_c = \frac{V_1 + V_2}{2}$$

High Frequency Analysis

$$A_{DM} = \frac{-h_{fe} \cdot R_L}{R_s + r_{b'b} + r_{b'e}} = \frac{-h_{fe} \cdot R_L}{r_{b'e}} = -g_m R_L$$

$$A_{CM} = \frac{-g_m \cdot R_L}{1 + 2\, g_m\, R_E}$$

Low frequency analysis

$$A_d = \frac{-h_{fe} \cdot R_o\, R_L}{h_{ie}(R_o + 2R_L)}$$

$$A_c = \frac{-h_{fe} \cdot R_L}{h_{ie} + 2(1 + h_{fe})R_e}$$

Where ADM and ACM is known as differential mode gain and common mode gain.

Common mode rejection ratio (CMRR)

- $CMRR = \frac{A_{DM}}{A_{CM}}$
- $CMRR = 1 + 2g_m R_E \approx 2g_m.R_E$
- For large value of CMRR, we need large value of R_E, so use current sources having high output resistance. for FET based differential amplifiers.
- $CMRR = \frac{1 + 2\, R_s\, \mu}{r_d} \approx 2g_m\, R_s$

4.12 FEEDBACK AMPLIFIER TOPOLOGIES

There are four basic type:

1. Shunt-Shunt
2. Shunt-Series
3. Series-Shunt
4. Series-Series

These designation corresponds to the input and output port connection respectively between the feedback network and the basic amplifier.

Gain with feedback

$$A_f = \frac{X_o}{X_s} = \frac{A}{1 - \beta \cdot A}$$

where A = transfer function without feedback.

If $|A_f| < |A|$, the feedback is termed as negative.

And if $|A_f| > |A|$, the feedback is termed as positive.

Fundamental Assumption

Three conditions are implicit in the representation of the ideal feedback amplifier:

1. The input signal is transmitted to the output through the amplifier A and not through the feedback network β. Means feedback network is unilateral.
2. The feedback signal is transmitted from output to input through the feedback network only.
3. The transfer ratio β is independent of source and load resistance.

4.12.1 PROPERTY OF NEGATIVE-FEEDBACK AMPLIFIERS

Although negative feedback results in reduced gain, it is used because of following reasons:

- Bandwidth get increased
- Effect of noise get reduced
- Distortion get reduced.

Input Impedance in Feedback Amplifier Input Resistance

If the feedback signal is returned to the input in series with the applied voltage, the input impedance get increased.

$$R_{if} = R_i(1 - \beta A)$$

Where, R_{if} = Input resistance after feedback.

R_i = Input resistance without feedback.

- For negative feedback V_f is 180° out of phase with V_s, V_i is less than it would be if V_f were absent. Hence $I = V_i/R_i$ decreased.

  Causes the ratio V_s/I to increase.

- If shunt connection is used at the input of a negative feedback amplifier, the impedance decreases.

$$R_{if} = \frac{R_i}{(1 - \beta \cdot A)}$$

Output Impedance

When the output of a feedback employs a shunt connection, negative feedback reduces the Output resistance:

$$R_{of} = \frac{R_o}{(1-\beta \cdot A)}$$

Output impedances increased when a negative feedback amplifier employs a series connected output. $R_{Of} = R_o\,(1 - \beta.A)$.

Effect of feedback on Bandwidth

Open loop gain $A_{OL} = \dfrac{A_o}{\left[1+\left(\dfrac{s}{\omega_n}\right)\right]}$

(without feedback)

Return Ratio

$$T(s) = \frac{T_o}{\left[1+\left(\dfrac{s}{\omega_n}\right)\right]}$$

where A_o and T_o are the midband value of A_{OL} and T (s), ω_n is angular frequency of the dominant pole.

$$A(s) = \frac{A_{OL}}{1 + A_{OL}\cdot T(s)} = \frac{A_{FO}}{\left[1+\left(\dfrac{s}{\omega_n}\right)\right]}$$

where, $A_{FO} = \dfrac{A}{(1+T_o)}$

$$\omega_H = (1 + T_o)\,\omega_n$$

Negative feedback has increased the bandwidth by a factor $(1 + T_o)$, the same factor by which A_{OL} is reduced.

Stability

In the design of a feedback amplifier, it must be ascertained that the circuit is stable at all frequencies.

A system is stable if an only if all bounded input signal produce bounded output signals.

- If a pole of transfer function exists with a positive real part, this cause output to increase exponentially with time and hence unbounded.

Stability of Feedback Amplifiers

$$A_F(s) = \frac{A_{OL}(s)}{1+T(s)}$$

Amplifier without feedback is stable if all the pole of AOL lie in the left half of s-plane.

Amplifier with feedback is stable if all the zero of T (s) lie in the left half of s-plane.

4.12.2 FEEDBACK AMPLIFIER CONFIGURATION

Feedback is the process whereby a portion of the output is returned to the input to form part of the system excitation.

Feedback was used to make the operating point of a transistor insensitive to both manufacturing variation in β_1 and temperature.

Effect of negative feedback on amplifier characteristic

	Type of feedback			
	Voltage series	Current series	Current Shunt	Voltage shunt
Z_{out}	Decrease	Increases	Increases	Decreases
Z_{in}	Increases	Increases	Decreases	Decreases
Improves Charac-teristics of	Voltage Amplifier	Transcon ductance amplifier	Current amplifier	Trans-Resistance amplifier
Desen-sitizes	A_{vf}	G_{Mf}	A_{If}	R_{Mf}

4.12.3 TYPE OF FEEDBACK AMPLIFIERS

Voltage Amplifier Or Voltage Series Amplifier

$$R_i >> R_s\ ;\ R_o << R_L$$

R_i = Input resistance of the amplifier.

This circuit provides an output voltage proportional to the input voltage. $R_i >> R_s$ and $R_o >> R_L$.

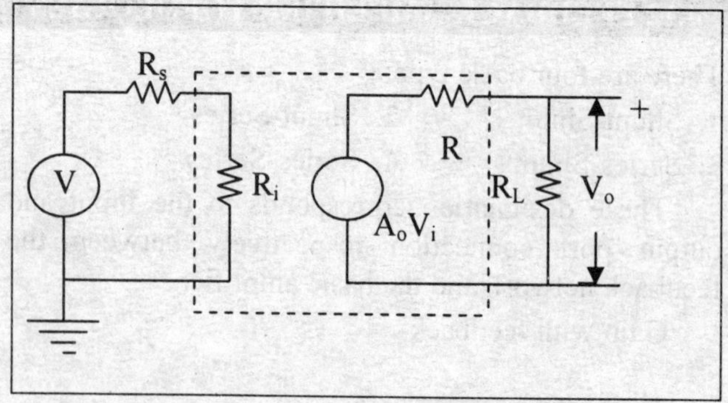

The current Amplifier or Current Shunt Amplifier

An ideal current-controlled source has zero input resistance Ri and has infinite output resistance R_o.

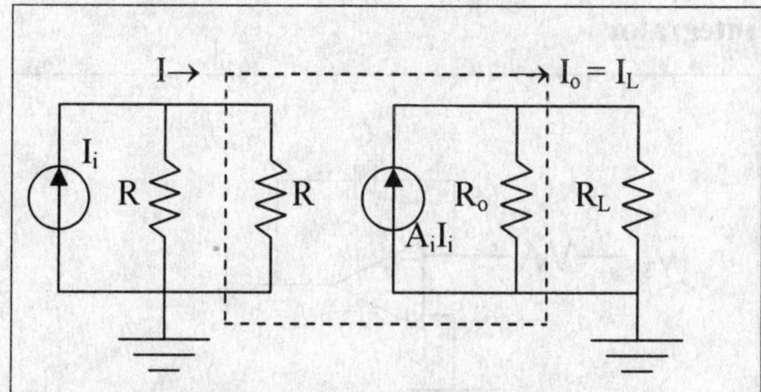

$|R_i| << R_s$ and $|R_o| >> R_L$; $I_o = A_i I_i$

Output current is proportional to input current.

The Voltage-Current Convertor or trans-conductance Amplifier (Voltage Controlled current source) or Current Series Amplifier

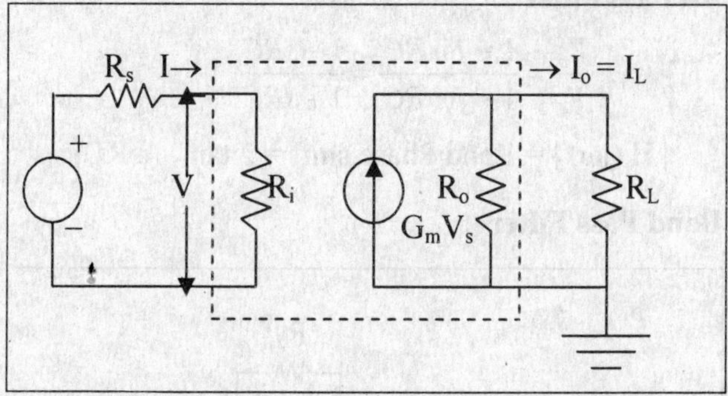

This ideal voltage – controlled current source.

$|R_i| >> R_s$ and $|R_o| >> R_L$

$I_o = G_m V_s$

Current-voltage Converter or Trans-Impedance Amplifier or Voltage Shunt Amplifier (Current Controlled Voltage Source)

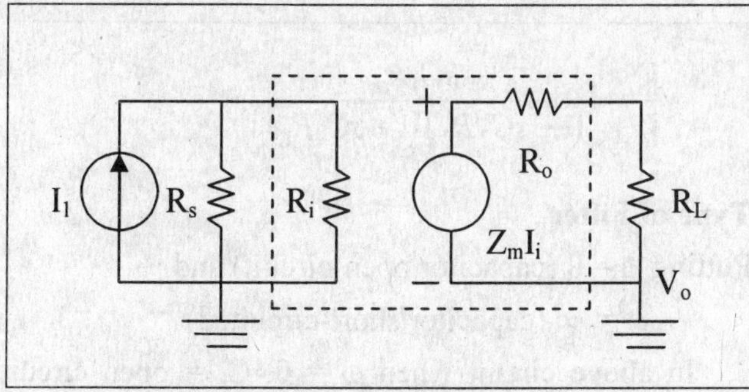

This is ideal current controlled voltage source.

$R_i >> R_s$ and $|R_o| << R_L$

$V_o = Z_m I_s$

4.13 OPERATIONAL AMPLIFIERS

1st Stage : Difference amplifier.

2nd Stage : High gain linear amplifier.

3rd Stage : Emitter follower.

4th Sage : DC level shifter and output amplifier.

Ideal Op-amp Characteristics

1. Input resistance is infinite, $Z_{in} = \infty$, so no current enters the input terminal.
2. A zero output resistance, $Z_{out} = 0$.
3. Infinite voltage gain, $A_v = \infty$.
4. Characteristics do not drift with temperature.

4.13.1 APPLICATION OF OPS AMPLIFIER

Inverting Amplifier

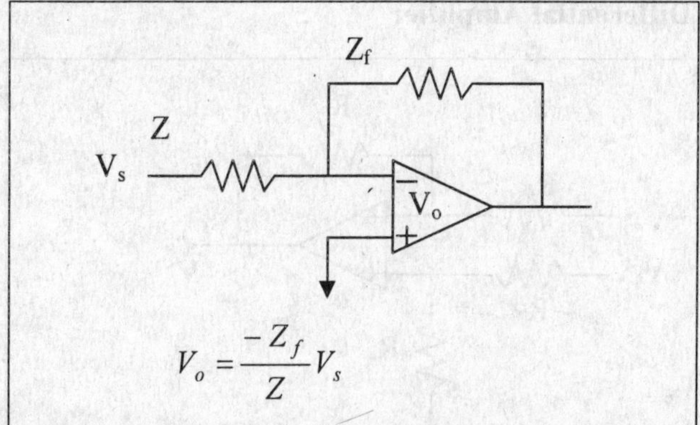

Input Impedance: The input resistance is equal to resistance Z.

Unity Follower

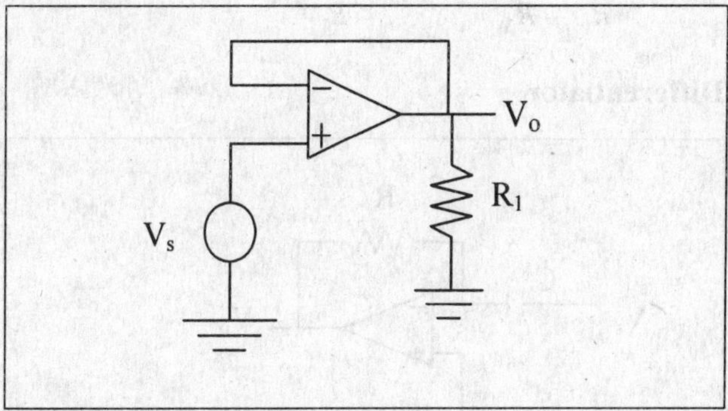

The output signal is in phase with the input signal and of same magnitude.

Input impedance is very high, so it provides very good isolation between stages.

Summing Amplifier

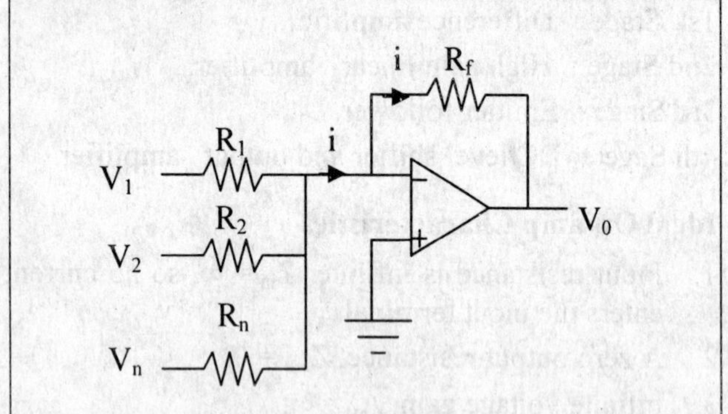

$$V_0 = -\left(\frac{R_f V_1}{R_1} + \frac{R_f V_2}{R_2} + \frac{R_f V_r}{R_n}\right)$$

Differential Amplifier

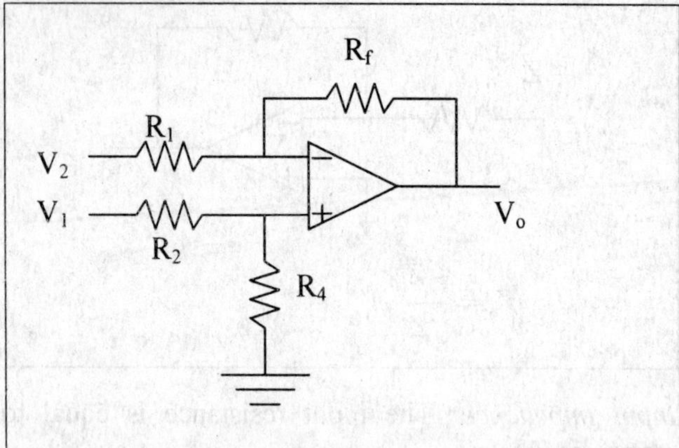

If $\frac{R_2}{R_1} = \frac{R_4}{R_3}$; $V_0 = \frac{R_f}{R_1}[V_1 - V_2]$

Differentiator

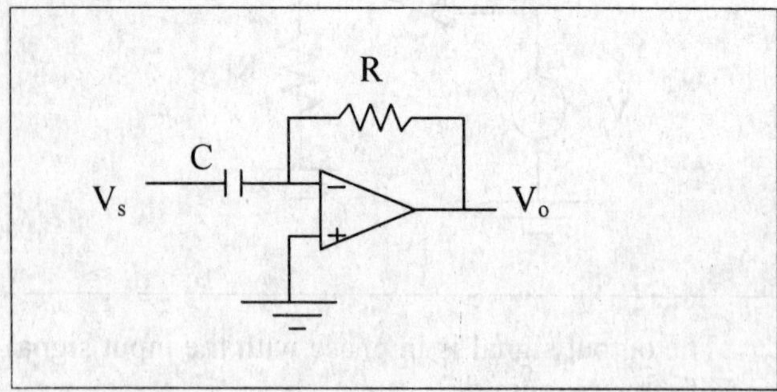

$$\frac{V_o(s)}{V_s(s)} = -sRC$$

Integrator

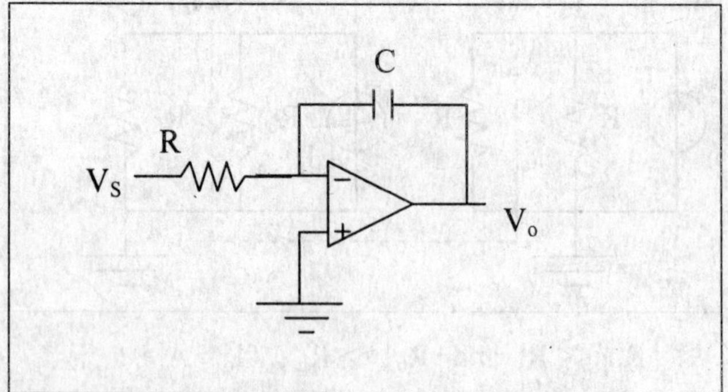

$$\frac{V_o(s)}{V_s(s)} = -\frac{1}{sRC}$$

All Pass Filter

$$\frac{V_o}{V_i} = \frac{1 - j\omega RC}{1 + j\omega RC} = \frac{1 - sRC}{1 + sRC}$$

$| H(j\omega) | = 1$ and Phase shift = $2 \tan^{-1} (\omega RC)$

Band Pass Filter

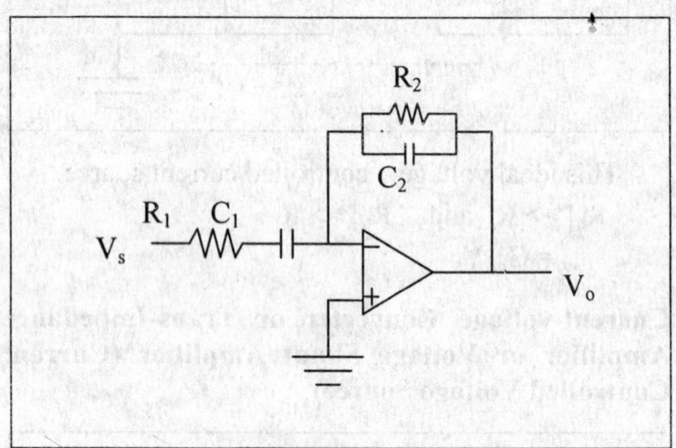

$$\frac{V_o}{V_s} = \frac{-sC_1R_2}{(1 + sC_2R_2)(1 + sC_1R_1)}$$

Type of Filter

Putting $\omega = 0$ (capacitor open-circuit) and

$\omega = \infty$ (capacitor short-circuited)

In above circuit when $\omega = 0$, C_1 = open circuit and $V_0 = 0$ V, similar for $\omega = \infty$ ($C_1 = C_2$) so it does not passes high as well as low frequency so it is band pass.

Non-Inverting Integration

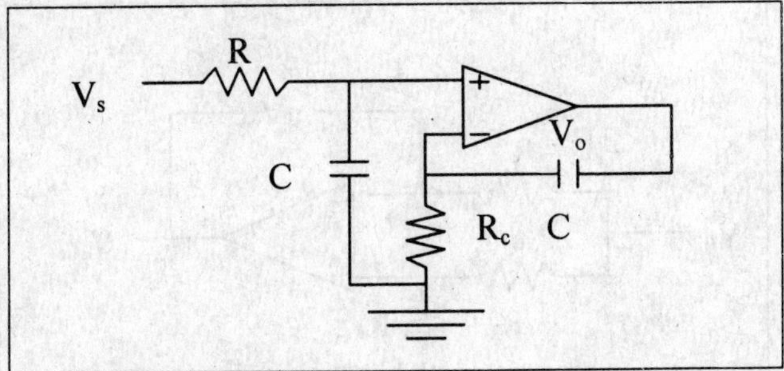

$$V_o = \frac{1}{RC} \int V_s \, dt$$

Double Integration

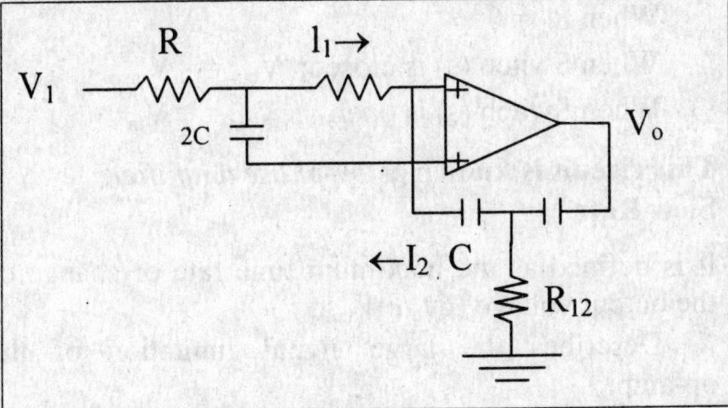

$$\frac{V_o}{V_s} = -\frac{1}{(sRC)^2}$$

Logarithmic Amplifiers

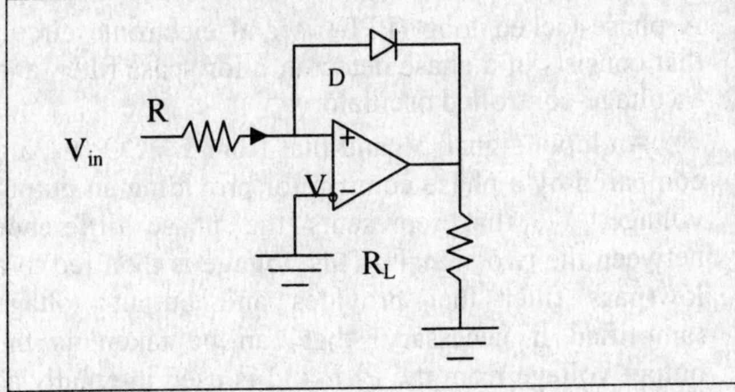

When $V_{in} < 0$; diode is off, Op-amp behaves as comparator. At this time

$V_o = +V_{sat}$

- When $V_{in} > 0$ diode is ON

  $V_0 = -0.7$ V
- For small range of V_{in}'. It behaves as logarithmic amplifier.

$$V_o = 2.3 \frac{KT}{q} \log_{10} \left(\frac{V_{in}}{R_L I_o} \right)$$

Antilogarithmic Amplifier

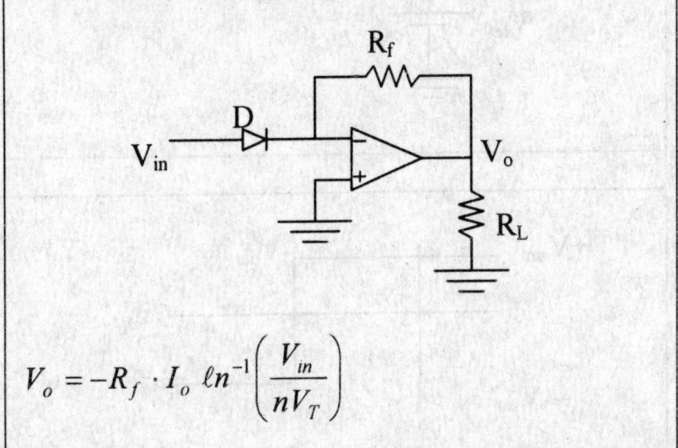

$$V_o = -R_f \cdot I_o \, \ell n^{-1} \left(\frac{V_{in}}{nV_T} \right)$$

Comparator

In a comparator, the level of the applied input voltage is compared with a known reference voltage:

1. As a comparator, the level of the applied input voltage is compared with a Known reference voltage.
2. As a comparator, the Op-amp are operated in the open loop condition without feedback resistor.
3. Since it is operated in open loop condition, the assumption ($V_+ = V_- = V_d = o$) is not applicable.
4. Open loop gain is very high, so output goes to saturation levels ($\pm V_{cc}$) even for very low inputs.
5. If 1 or 2V drops in R_c's of the transistor of the differential stage, the saturation voltage will be (± 13 V) for a supply voltage of (± 15V).

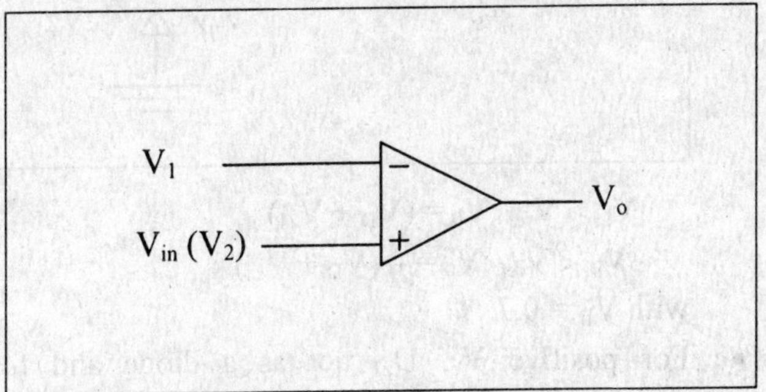

If $(V_2 - V_1) = V_{in} > 0$, $V_o = +V_{sat}$

If $(V_2 - V_1) = V_{in} < 0$, $V_o = -V_{sat}$

Zero Crossing Comparator and Detector

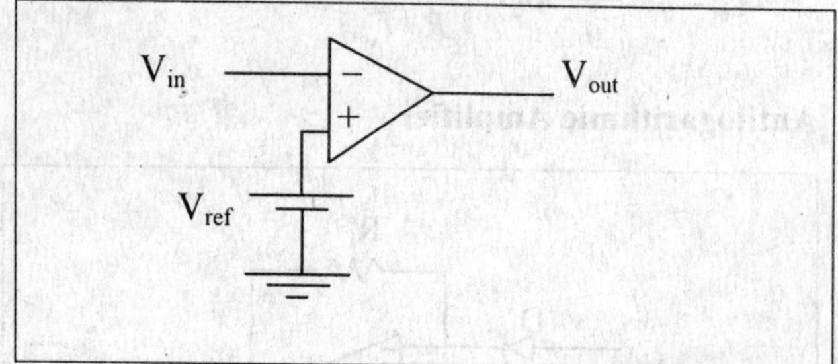

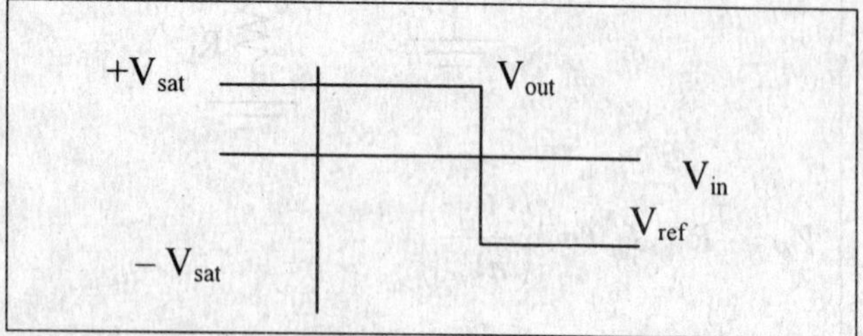

- With sinusoidal, we obtain rectangular pulse waveform at the output.
- Output remains at $+V_{sat}$ or $-V_{sat}$ till the input voltage is above or below V_{ref}.

  $V_{in} > V_{ref}$; $V_{out} = -V_{sat}$

  $V_{in} < V_{ref}$; $V_{out} = V_{sat}$
- Output remains at $\pm V_{sat}$ ($\pm$ 13) is not suitable for TTL operation. We can clamp the output voltage to lowest value by connecting at the output zener diodes back to back.

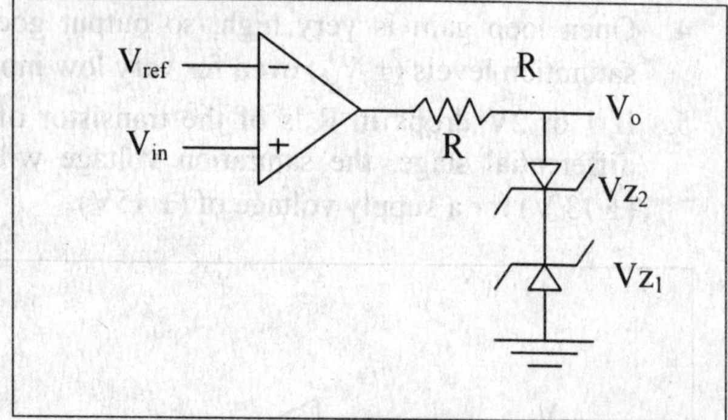

$V_{in} > V_{ref}$; $V_0 = (V_D + V_{Zl})$

$V_{in} < V_{ref}$; $V_0 = -(V_D + V_{Zl})$

with $V_D = 0.7$.

- For positive V_o, D_{z2} act as a diode and for negative V_0, D_{z1} acts as a simple diode.
- R restricts the current to the operating point of zener.

Unity Gain Amplifier Or ± 1 Gain Amplifiers

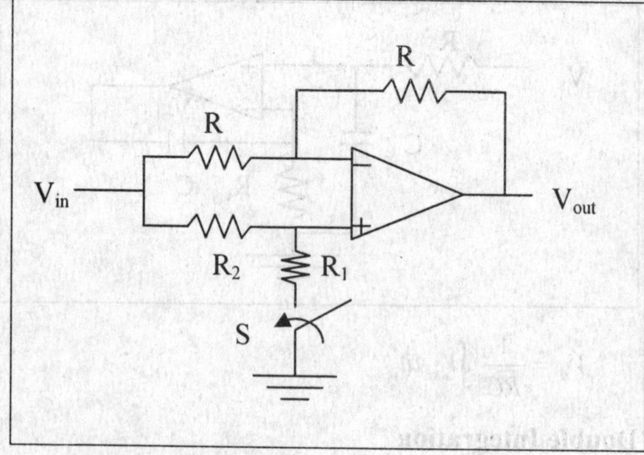

When $R_1 = 0$

When Switch (s) is closed; $V_{out} = -V_{in}$

When Switch (s) is open ; $V_{out} = +V_{in}$

This circuit is known as *by-phase amplifier:*

Slew Rate

It is defined as the maximum time rate of change of the output voltage $(\delta v_o/\delta t)_{max}$.

Describes the large signal limitation of the op-amp.

For faithful reproduction of this sinusoid requires

$$\omega V_m \leq \text{slew rate}$$

where, V_m = peak amplitude of applied signal

ω = frequency of input signal.

Phase-Locked Loop

A phase-locked loop (PPL) is and electronic circuit that consists of a phase detector, a low-pass filter, and a voltage-controlled oscillator

An input signal, V_i and that from a VCO, V_0, are compared by a phase comparator providing an output voltage, V_e, that represents the phase difference between the two signals. This voltage is then fed to a low-pass filter that provides and output voltage (amplified if necessary) that can be taken as the output voltage from the PLL and is used internally as the voltage to modulate the VCO's locked to that of the input signal frequency.

The basic operation of a PLL circuit can be explained using the circuit. We will first consider the *operating in lock* (the input signal frequency and the VCO frequency are the same). When the input signal frequency is the same as that from VCO to the comparator, the voltage, V_d, taken as output is the value needed to hold the VCO in lock with the input signal. The VCO then provides output of a fixed-amplitude square-wave signal at the frequency of the input. Best operation is obtained if the VCO center frequency, f_0, is set with the dc bias voltage midway in its linear operating rang.

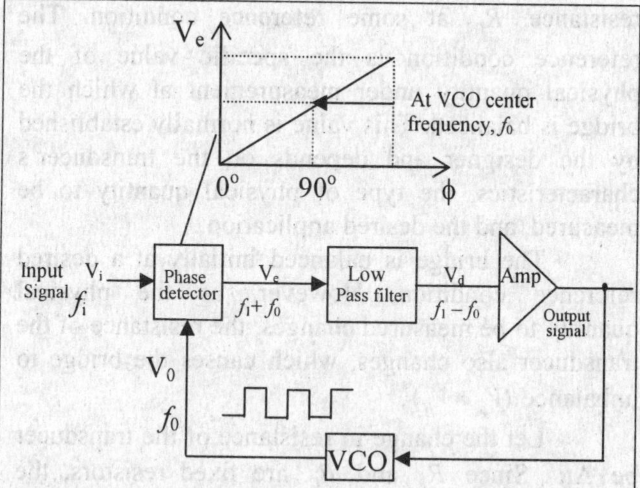

Figure: Block diagram of basic phase-locked loop (PLL)

The amplifier allows this adjustment in lock, the two signals to the comparator are of the same frequency, although not necessarily in phase. A fixed phase difference between the two signal to the comparator results in change in the dc voltage to the VCO. Within a capture–and lock frequency range, the dc voltage will drive the VCO frequency to match that of the input.

While the loop is trying to achieve lock, the output of the phase comparator contains frequency components at the sum and difference of the signals compared. A low pass filter passes only the lower-frequency component of the signals so that the loop can obtain lock between input and VCO signals.

Owing to the limited operating range of the VCO and the feedback connection of the PPL circuit there are two important frequency bands specified for a PPL. The ***capture range*** of a PLL is the frequency range centered about the VCO free-running frequency, f_0, over which the loop can acquire lock with the input signal. Once the PLL has achieved capture, it can maintain lock with the input signal over somewhat wider frequency range called the ***lock range.***

Applications:

Common applications of a PPL include:

1. Frequency synthesizers that provide multiples of a reference signal frequency generated using a PLL.
2. FM demodulation network for FM operation with excellent linearity between the input signal frequency and the PLL output voltage.
3. Demodulation of the two data transmission or carrier frequencies in digital-data transmission used in frequency-shift keying (FSK) operation and
4. A wide variety areas including modems, telemetry receivers and transmitters, tone decoders AM detectors, and tracking filters.

4.14 INSTRUMENTATION AMPLIFIER

Instrument Amplifier

In Instrumentation a transducer is used at the measuring site to obtain the required information easily and safely. The transducer is a device that converts one from of energy into another. For example, a strain gage when subjected to pressure or force undergoes a change in its resistance An instrumentation system is used to measure the output signal produced by a transducer and often to control the physical signal producing it. Figure 6 – 11 shows a simplified form of such system. The input stage is composed of a preamplifier and some sort of transducer, depending on the physical quantity; to be measured. The output stage may use devices such as meters, oscilloscopes, charts, or magnetic recorders.

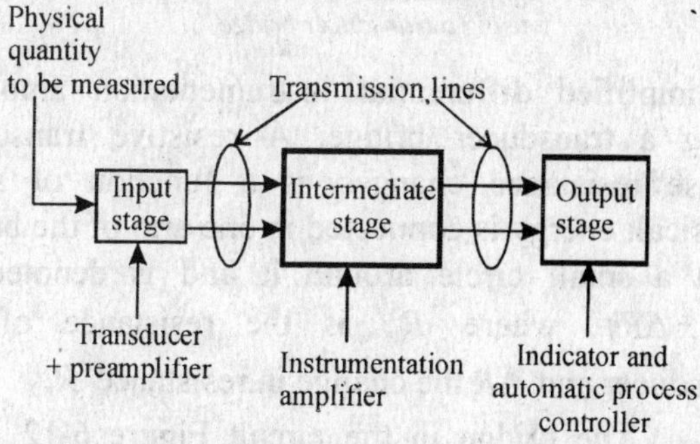

The connecting lines between the blocks represent transmissions lines, used especially when the transducer is at a remote test site monitoring hazardous conditions such as high temperatures or liquid levels of flammable chemicals. These transmission lines depend primarily on the physical quantities to be monitored and on system requirements.

The signal source of the instrumentation amplifier is the output of the transducer. Although some transducers produce outputs with sufficient strength to permit their use directly, many do not. To amplify the low-level output signal of the transducer so that it can drive the indicator or display is the major function of the instrumentation amplifier. In short, the instrumentation amplifier is intended for precise, low-level signal amplification where low noise, low thermal and time drifts, high input resistance, and accurate closed – loop gain are required. Besides, low power consumption, high common – mode rejection ratio, and high slow rate are desirable for superior performance.

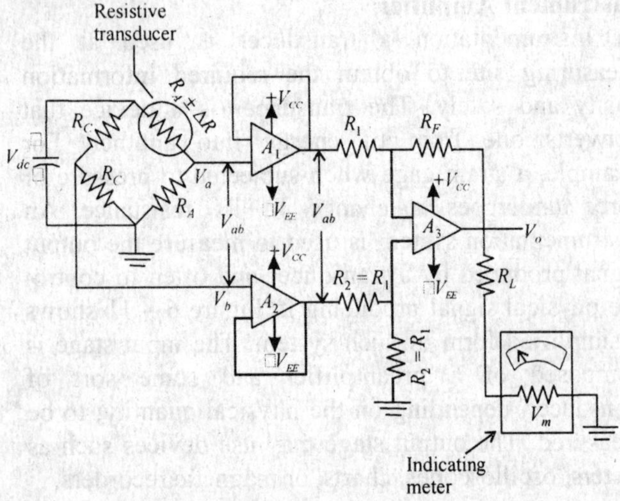

Figure: Differential instrumentation amplifier using a transducer bridge.

A simplified differential instrumentation amplifier using a transducer bridge. A resistive transducer whose resistance changes as a function of some physical; energy is connected in one arm of the bridge with a small circle around it and is denoted by $(R_T \pm \Delta R)$, where R_T is the resistance of the transducer and ΔR the change in resistance R_T.

The bridge in the circuit Figure 6-12 is dc excited but could be as excited as well. For the balanced bridge at some reference condition,

$$V_b = V_a$$

or

$$\frac{R_B(V_{dc})}{R_B + R_C} = \frac{R_A(V_{dc})}{R_A + R_T}$$

That is,

$$\frac{R_C}{R_B} = \frac{R_T}{R_A}$$

Generally, resistors R_A, R_B and R_C are selected so that they are equal in value to the transducer resistance R_T at some reference condition. The reference condition is the specific value of the physical quantity under measurement at which the bridge is balanced. This value is normally established by the designer and depends on the transducer's characteristics, the type of physical quantity to be measured, and the desired application..

The bridge is balanced initially at a desired reference condition. However, as the physical quantity to be measured changes, the resistance of the transducer also changes, which causes the bridge to unbalance $(V_a \neq V_b)$.

Let the change in resistance of the transducer be ΔR. Since R_B and R_C are fixed resistors, the voltage V_b is constant. However, voltage V_a varies as a function of the change in transducer resistance. Therefore, according to the voltage-divider rule,

$$V_a = \frac{R_A(V_{dc})}{R_A + (R_T + \Delta R)}$$

$$V_b = \frac{R_B(V_{dc})}{R_B + R_C}$$

Consequently, the voltage V_{ab} across the output terminals of the bridge is

$$V_{ab} = V_a - V_b$$

$$= \frac{R_A V_{dc}}{R_A + R_T + \Delta R} - \frac{R_B V_{dc}}{R_B + R_C}$$

However, if $R_A = R_B = R_C = R_T = R$, then

$$V_{ab} = \frac{\Delta R(V_{2dc})}{2(2R + \Delta R)}$$

The negative (–) sign in this equation indicates that $V_a < V_b$ because of the increase in the value ΔR.

The output voltage V_{ab} of the bridge is then applied to the differential instrumentation amplifier composed of three op – amps. The voltage followers preceding the basic differential amplifier help to eliminate loading of the bridge circuit. The gain of the basic differential amplifier is $(-R_F/R_1)$; therefore, the output V_a of the circuit is

$$V_a = V_{ab}\left(-\frac{R_F}{R_1}\right) = \frac{(\Delta R)V_{dc}}{2(2R+\Delta R)}\frac{R_F}{R_1}$$

Generally, the change in resistance of the transducer ΔR is very small. Therefore, we can approximate $(2R+\Delta R)=2R$. Thus, the output voltage.

$$V_o = \frac{R_F}{R_1}\frac{\Delta R}{4R}V_{dc}$$

The equation indicates that V_o is directly proportional to the change in resistance ΔR of the transducer.

4.15 PRECISION RECTIFIER

Precision Rectifier

Main defect with Si rectifier is its cut in voltage of 0.7 V and so, voltages less than this value cannot be rectified. This defect is over come in op-amp rectifiers and voltages in the mill volt range can be obtained.

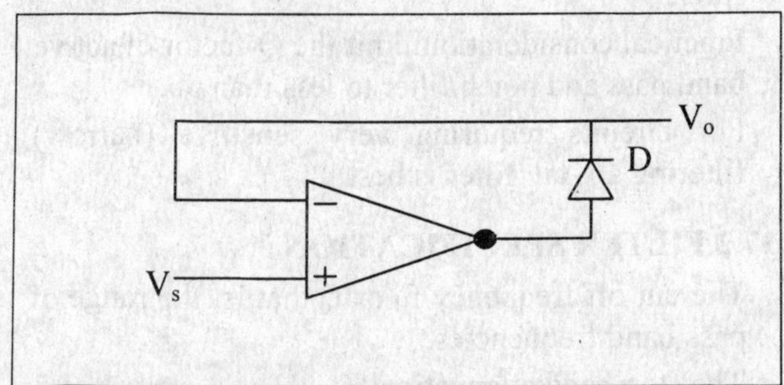

When V_s is +ve diode is ON and feedback loop is established and V_o follows V_s.

When V_s is –ve, diode is OFF and feedback loop is open and $V_o = 0$.

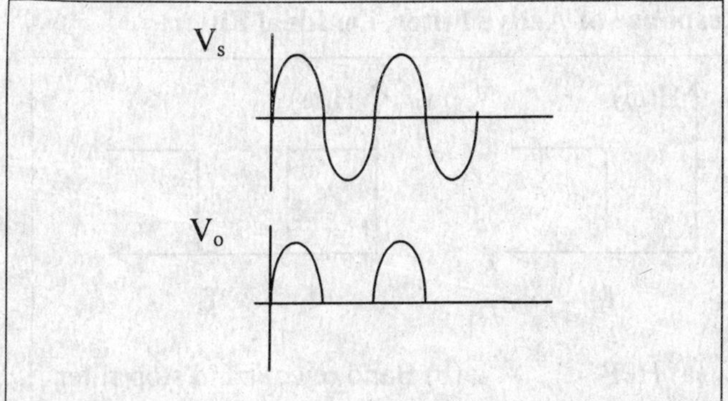

4.16 V TO I AND I TO V CONVERTER

4.16.1 CURRENT-TO-VOLTAGE CONVERTER

Let us reconsider the ideal voltage-gain Equation (3-14) of the inverting amplifier, in given.

$$\frac{V_O}{V_{in}} = \frac{-R_f}{R_1}$$

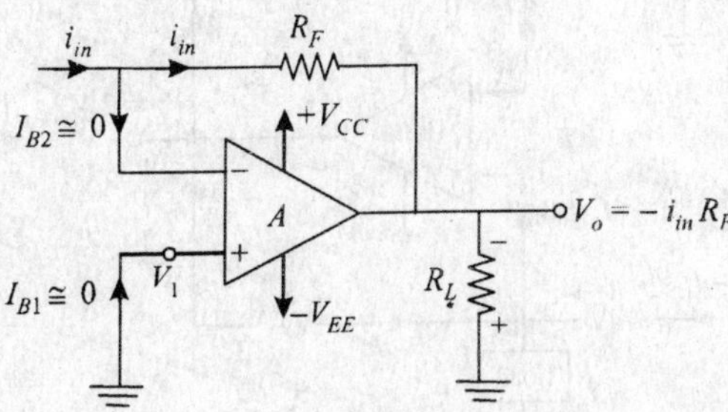

$$v_o = -\left(\frac{v_{in}}{R_1}\right)R_F$$

However, since $v_1 = 0V$ and $v_1 = v_2$

$$\frac{v_{in}}{R_1} = i_{in}$$

And

$$v_o = -i_{in}R_F$$

This means replace the v_{in} and R_1 combination by a current source i_{in} as shown in figure 3-12, the output voltage v_o becomes proportional to the input current i_{in}.

4.16.2 VOLTAGE- TO – CURRENT CONVERTER WITH GROUNDED LOAD

Another version of the voltage-to-current converter is shown in Figure 6-19, In this circuit, on terminal of the load is grounded, and load current is controlled by an input voltage. The analysis of the circuit is accomplished by first determining the voltage V_1 at the no inverting input terminal and then establishing the relation ship between V_1 and the load current.

Writing Kirchhoff's current equation at node V_1.

$$I_1 + I_2 = I_L$$

$$\frac{V_{in}-V_1}{R} + \frac{V_0 - V_1}{R} = I_L$$

$$V_{in} + V_0 + 2V_1 = I_L R$$

Therefore

$$V_1 = \frac{V_{in} + V_o - I_L R}{2} \quad (1)$$

Since the op-amp is connected in the no inverting mode, the gain of the circuit in Figure 6-19 is 1+R/R = 2. Then the output voltage is

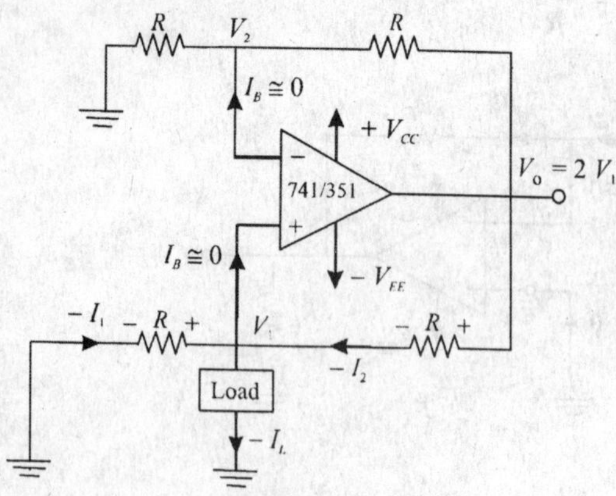

$$V_o = 2V_1$$

$$= V_{in} + V_o - I_L R$$

This is

$$V_{in} = I_L R$$

Or

$$I_L = \frac{V_{in}}{R} \quad (1)$$

This means that the load current depends on the input voltage V_{in} the regretted R.

The voltage to current converted may also be used in testing seen devices as converts and LED s forming a ground load.

4.17 OP-AMP BASED ACTIVE FILTERS

4.17.1 ACTIVE FILTER

1. These in-corporate an amplifier with RC networks to overcome these problem of passive filter. By enclosing a capacitor in feedback loop, the inductor [with all of its LF problems] can be eliminated.
2. Input impedance (Z_{in}) is increased (because of OP-AMP).
3. The load is driven from the output of the OP-AMP, giving on output impedance (Z_{out}) as low as a few ohm.
4. Load is isolated from the frequency determining network, i.e. variation in load does not effects on the active filter characteristic.
5. The amplifier allows to speechify and adjust pass-band gain, pass band ripple, cut off frequency and initial roll off.
6. Because of high Z_{in} of OP-AMP, large value resistor can be used, for reducing, the value (size and cost) of the amplifier.

Limitation of Active Filter

1. High frequency response in limited by the gain, Bandwidth, slew rate of OP-AMP.
2. HF OP-AMP are more expressive, making passive filters a more economical choice for RF application.
3. Active filters requires power supplies (OP-AMP requires two supplies, i.e. positive and Negative). Variation in power supplies may cause variation in output signal from the active filter.
4. Feedback along the power supply lines can cause oscillation unless decoupling technique are rigorously applied.
5. Active filter are much more susceptible to RF interference and ionization than the passive RLC filter.
6. Practical consideration limit the Q-factor of active band pass and notch filter to less than 50.

   For circuits requiring very sensitive (narrow) filtering a x-tal filter is best

4.17.2 FILTER SPECIFICATION

1. The cut off frequency f_H or f_c that is, the range of pass band frequencies.
2. The stop band attenuation $H_0 - H_2$.
3. The stop band frequency range, i.e. f_s is specified.
4. The allowable pass band ripple $\gamma = H_0 - H_1$. If no ripple is permitted, γ -= 0 and $H_1 = H_0$, figure(e).

Response of Active Filter, i.e. Ideal Filter

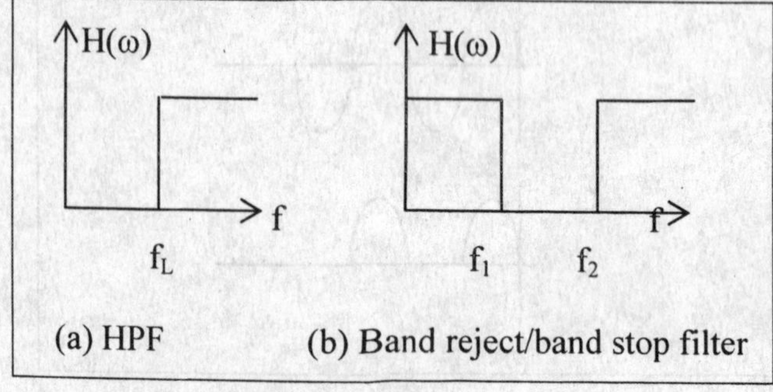

(a) HPF (b) Band reject/band stop filter

Practical Filter

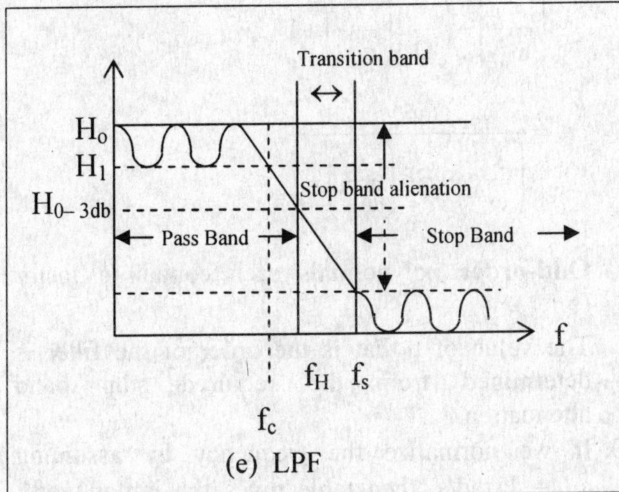

(e) LPF

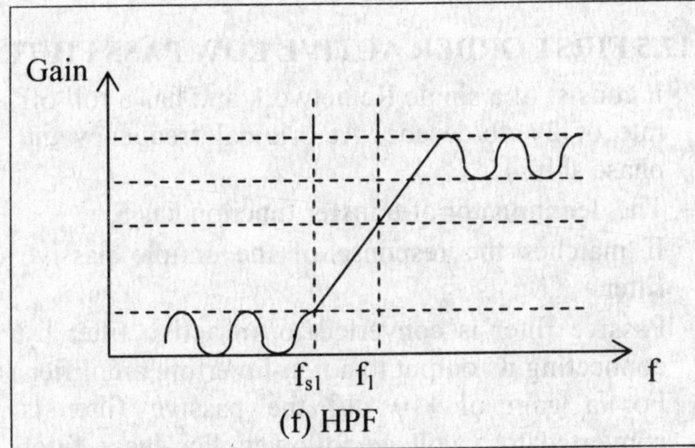

(f) HPF

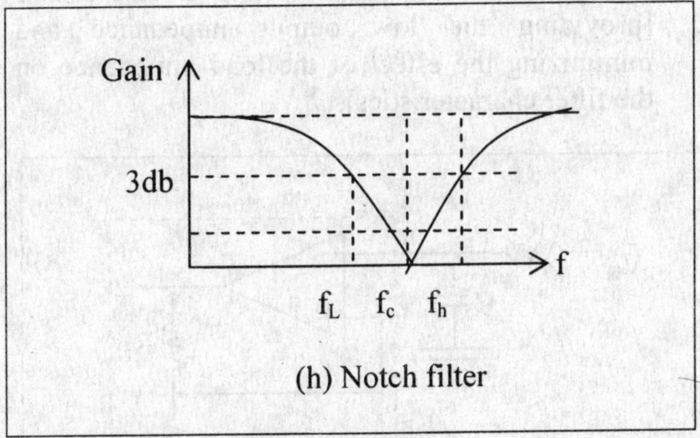

(h) Notch filter

4.17.3 CHARACTERISTIC AND TERMINOLOGY

dB/decade: When the frequency increased by factor of 10, the amplitude or gain roll off at the rate of same dB.

Octave: Frequency has doubled.

Roll off: How rapidly, the gain falls off as the stop band is entered is called Roll off.

dB/decade	dB/octave
20	6
40	12
60	18
80	24
100	24

Note:

1. Decreasing the ratio gain by 10 subtracts 20 db.
2. Increasing by 10 adds 20 dB.
3. Doubling the gain adds 6 dB.
4. Halving the gain subtract 6 dB.
5. Cutting the gain by √2 gives – 3 dB.

Logarithmic Scale

1. On a log scale horizontal axis each decade (factor of 10) increases or decrease moves the same distance along the scale.
2. Starting point is never zero, if we go further left, each increment lower the scale by 10 (i.e. 0.1, 0.01) but zero is never reached.

4.17.4 ALL POLE FILTER

The most common LPF have transfer function that are ratio of a constant to a polynomial. These are called "all pole" filters.

1. Becomes all its poles are finite and zero are at infinite.
2. In general, an nth –order low pass all pole has a transfer function of the form

$$H(s)=\frac{H_0\ a_0}{s^n+a_{n-1}\ s^{n-1}+\ldots a_{1s}+a_o}$$

where H_o is gain of LPF which is defined as that value of transform function at s = 0 and the coefficient a_o, a_1, … a_{n-1} depend on the filter type.

Type of all Pole Filter

The most common types are:

(a) BUTTERWORTH Filter and
(b) CHEBYSHEV Filter

Butter Worth Filter or Maximally Flat LPF

This is a filter which provide **flattest pass band response** and whose magnitude in the nth –order case is defined by

$$H(j\omega)=\frac{H}{\sqrt{1+\left(\frac{\omega}{\omega_o}\right)^{2n}}}$$ where, H_o ... is the gain

ω_o ... is the cut-off frequency

n ... is the number of order or poles

For $H_0 = 1$,

n = 1, 2, 4, ... 6..... the response is

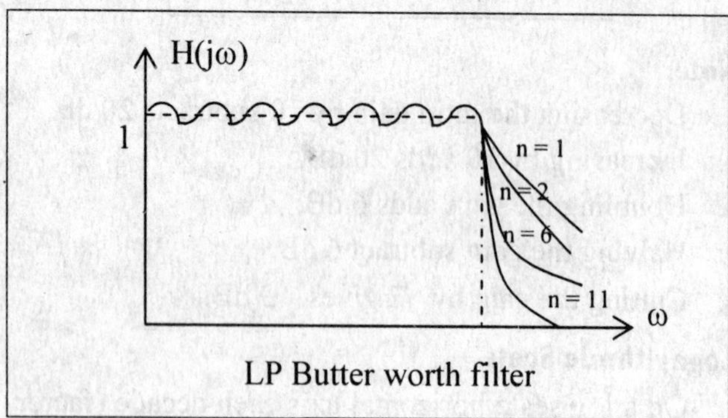

LP Butter worth filter

Feature

1. As the order increase i.e. (higher n) the butter worth magnitude more nearly approaches the ideal case.
2. The magnitude monotonically decrease as the frequency increase.
3. It is flashest in the vicinity of $\omega = 0$ that why it is called **maximally flat filter.**
4. For frequency near the cut-off point, the magnitude is not good.
5. The use of Butter worth polynomials is a common, all pole approximation of the LP characteristic.

   Hence $H(s)=\frac{H_o}{B(s)}$

   where B (s) is the Butter worth polynomials whose magnitude is given by

   $$B^2(\omega)=1+\left(\frac{\omega}{\omega_o}\right)^{2n}$$

6. No pass band ripple exist and response is essentially constant for $\omega < \omega_o$.
7. For **n even**, the polynomials are product of quadratic factors of the form of denominator of following equation.

$$H(s)=\frac{H_o}{\left(\frac{s^2}{\omega_o^2}\right)+\left(\frac{1}{Q}\right)\left(\frac{s}{\omega_o}\right)+1}$$

$$=\frac{\omega_n^2}{\left(\frac{s^2}{\omega_n}\right)+2\delta\left(\frac{s}{\omega_n}\right)+1}$$

8. **Odd-order** polynomials each contain a factor (s + 1).
9. The value of n-that is the order of the filter is determined from the required stop band attenuation.
10. If we normalize the frequency by assuming ω_o = 1 rad/s, then table gives the butter worth polynomial B_n (s) for n.

4.17.5 FIRST ORDER ACTIVE LOW PASS FILTER

1. It consist of a single Rc network and has a roll off rate of 29 db/decade. At original frequency the phase shift 45°.
2. The denominator of transfer function has S.
3. If matches the response of the simple passive filter.
4. Passive filter is converted to an active filter by connecting its output to a non-inverting amplifier. For a gain of [10 db] the passive filter is converted to a voltage follower. For these filter the OP-AMP main purpose is to drive the load [providing the low output impedance and minimizing the effect of the load impedance on the filter characteristics].

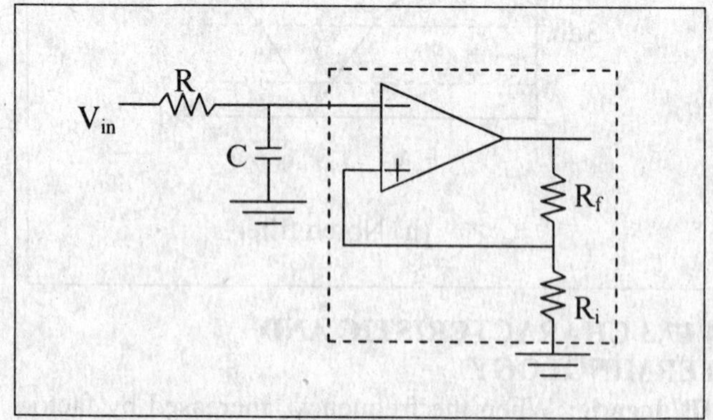

$A_o = 1+\frac{R_f}{R_i}$, Gain > 1

Analysis

For simple RC filter find TF by substituting Laplace function of each variable.

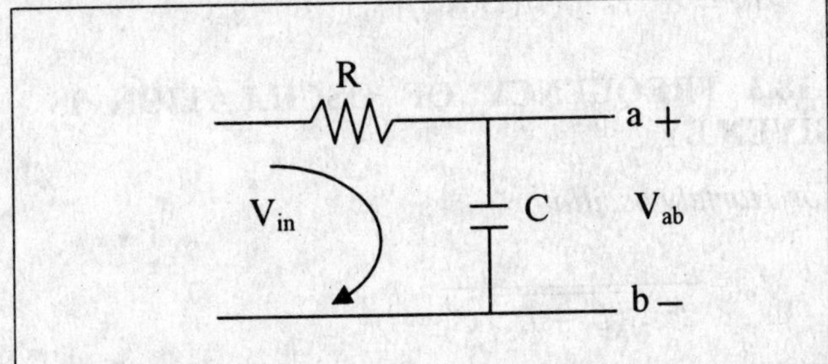

$$V_{ab} = \frac{Z_c}{R + Z_c} . V_{in}$$

$$\frac{V_{out}}{V_{in}} = \frac{A_O}{RCs + 1} \qquad (1)$$

Let, $\omega_o = \frac{1}{RC}$

Then, $\frac{V_{out}}{V_{in}} = \frac{A_o}{\left[\frac{s}{\omega_o}\right] + 1} = \frac{A_o \omega_o}{s + \omega_o}$

$$H(s) = \frac{V_{out}}{V_{in}} = \frac{A_o \omega_o}{s + \omega_o} \qquad (2)$$

Equation (2) is the standard form for the transfer function of 1st order system.

Gain Equation: Put s = jω in equation (1)

$$\frac{V_{out}}{V_{in}} = \frac{A_o}{1 + j\omega RC}$$

$$= \frac{A_o}{1 + \omega^2 R^2 C^3} - \frac{j A_o \omega RC}{1 + \omega^2 R^2 C^3} = \text{Real} + \text{imaginary}$$

$| G | =$ magnitude $= \sqrt{[(\text{Real})^2 + \text{Imaginary})^2]}$

$$= A_o \sqrt{\frac{1}{(1 + \omega^2 R^2 C^2)^2} + \frac{\omega^2 R^2 C^3}{(1 + \omega^2 R^2 C^2)^2}}$$

$$= A_o \frac{\sqrt{1 + \omega^2 R^2 C^2}}{1 + \omega^2 R^2 C^2} + \frac{A_o}{\sqrt{1 + \omega^2 R^2 C^2}} = | G | \qquad (3)$$

The critical frequency (cut-off frequency) f_c is that frequency at which gain drops to 0.707 A_o or (A_o – 3 db).

From equation (3)

$$|G| = 0.707 A_o = \frac{A_o}{\sqrt{1 + \omega_o^2 R^2 C^2}}$$

$$\omega_o^2 = \left(\frac{1}{RC}\right) \quad ; \quad \omega_o = 2\pi f_o$$

$$f_o = \frac{\omega_o}{2\pi} \quad ; \quad f_o = \frac{1}{2\pi RC} \qquad (4)$$

Phase Shift

$\phi_o = -\tan^{-1} \omega_o RC$ from equation (1)

$= -\tan [1/RC] RC = -\text{an}^{-1} 1$

$\phi = -45$

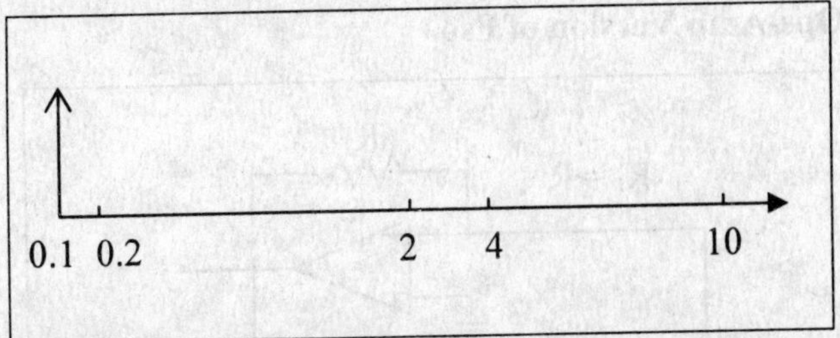

4.18 OSCILLATORS

4.18.1 The Barkhousen Criterion

The condition for sinusoidal oscillation can be expressed as:

$T(j\omega_o) = 1$

$\angle T(j\omega) = -180^o$

1. Two conditions must be satisfied by a circuit to sustain oscillation.
2. The phase shift through the amplifier; and feedback network must be 360° as T includes a
3. minus sign which is equivalent to 180° of phase shift.
4. The magnitude of the gain of the amplifier and feedback network must be unity.

   $T(j\omega) = -1$ known as Barkhausen Criterion.

- Barkhausen Criterion is equivalent to stating that with both the phase margin and gain margin are zero. So the phase and gain, crossover frequencies are equal.
- In every practical oscillator the loop gain is slightly larger than unity and the amplitude of the oscillations is limited by the on-set of non-linearity.

4.18.2 TYPES OF OSCILLATOR

Phase-Shift Oscillator (Pso)

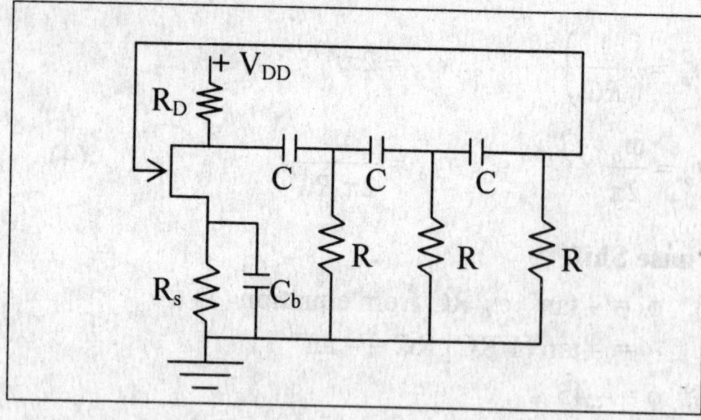

Ops-Amp Version of Pso

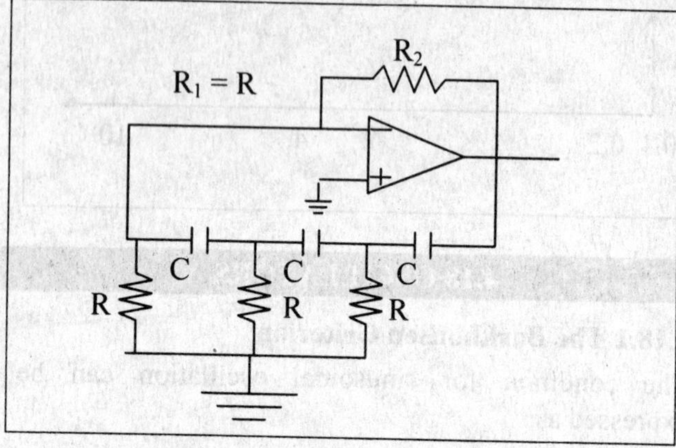

- The oscillation, frequency is same as that of FET, PSO.
- OPS-amp gain $A_v = [-R_2/R_1]$ must be greater then 29.

Wien Bridge Oscillator

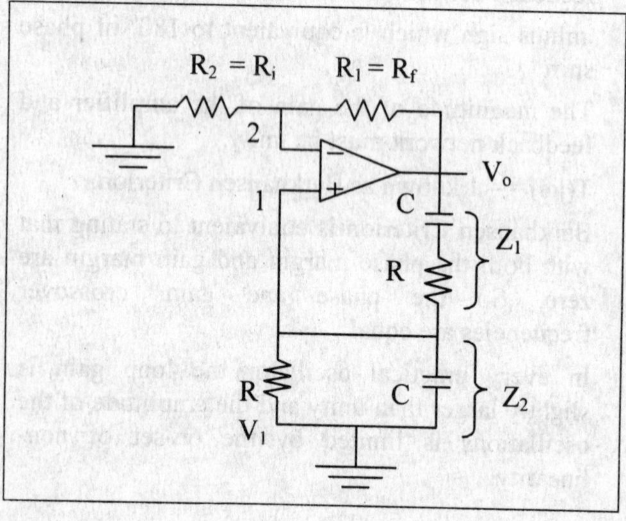

$R_f \rightarrow$ Thermistor , $\dfrac{R_f}{R_i} = 2$

$A_V \geq 3$; $f_o = [1/(2\pi RC)]$

4.18.3 FREQUENCY OF OSCILLATION IS GIVEN BY

For Hartely oscillator

$$f_o = \frac{1}{2\pi\sqrt{(L_1 + L_2)\,C_2}}$$

For Colpitt's oscillator

$$f_o = \frac{1}{2\pi\sqrt{\dfrac{C_1C_2}{C_1 + C_2}\,L_3}}$$

For Clapp oscillator

$$f_o = \frac{1}{2\pi\sqrt{L_3\,C_3}}$$

For Wein bridge oscillator

$$f_o = \frac{1}{2\pi\,RC}$$

$|A_v| > +3$

For RC –Phase shift oscillator

$$f = \frac{1}{2\pi\sqrt{6}\ RC} = \frac{1}{2\pi\ RC\sqrt{6}}$$

$|A_v| > 29$

4.19 SIGNAL GENERATORS

4.19.1 TIMER IC 555

A popular analog-digital integrated is the versatile 555 timer. The IC is made of a combination of linear comparators and digital flip-flops. The entire circuit is usually house in an 8-pin package. A series connection of three resistors sets the reference voltage levels to the two comparators at 2 $V_{CC}/3$ and $V_{CC}/3$, the output of these comparators setting or resetting the flip-flop unit. The output of the flip-flop circuit is then brought out through an output amplifier stage. The flip-flop circuit also operates a transistor inside the IC, the transistor usually being driven low dischargé a timing capacitor.

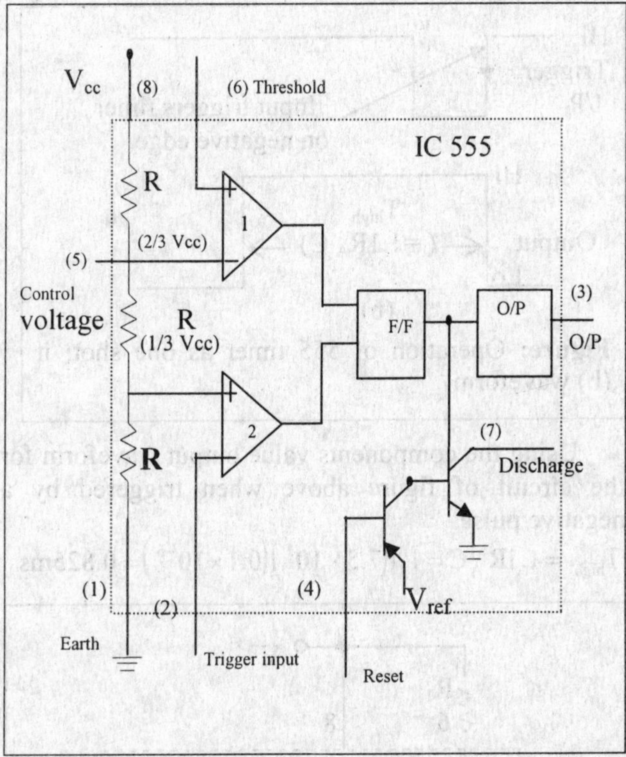

Astable Operation

One popular application of the 555 timer IC is as an astable multivibrator or clock circuit. The following analysis of the operation of the 555 as an astable circuit in cludes details of the different parts of the unit and how the various inputs and outputs are utilized. Figure shows an astable circuit built using an external resistor and capacitor to set the timing interval of the output signal.

Capacitor C charge toward V_{CC} through external Resistor R_A and R_B. the capacitor voltage rises until goes above ($2V_{cc}/3$). This voltage is the threshold voltage at pin 6, which drives comparator 1 to trigger the flip-flop so that the output pin 3 goes low, In addition, the discharge transistor is driven on, causing the output at pin 7 to discharge the capacitor through resistor R_B. The capacitors voltage then decrease until it drops below the triggers level (Vcc/3). The flip-flop is triggered so that output goes back high and the discharge transistor is turned off, so that the capacitor can again charge through resistors R_A and R_B toward V_{cc} figure shows the capacitor and output waveforms resulting form the astable circuit. Calculation of the time intervals during which the output is high and low can be made using the relations.

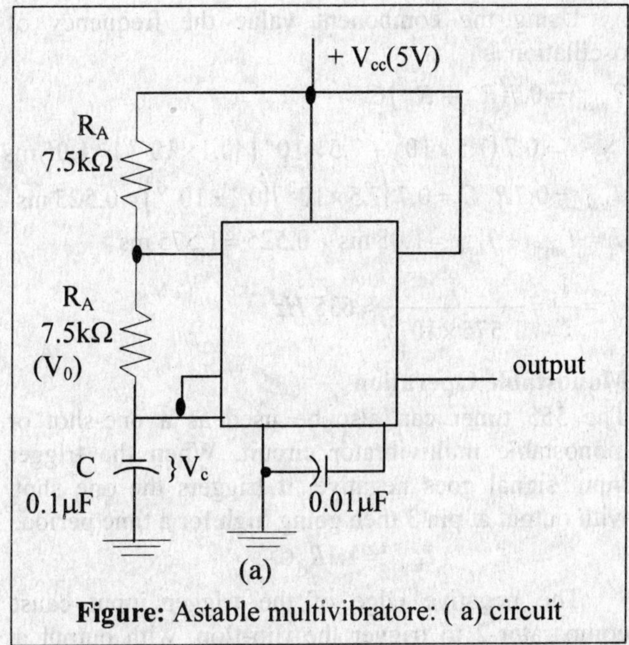

Figure: Astable multivibratore: (a) circuit

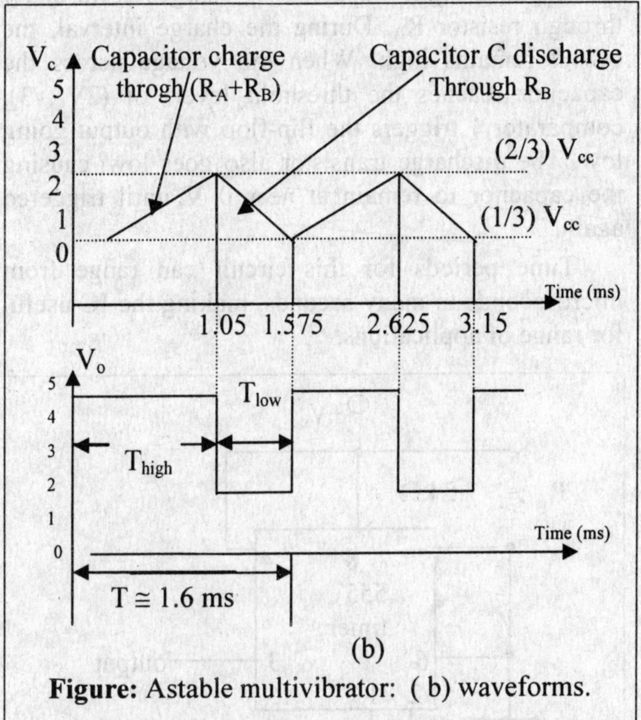

Figure: Astable multivibrator: (b) waveforms.

$$T_{\text{high}} \approx 0.7(R_A + R_B)C \; ; \; T_{\text{low}} \approx 0.7 R_B C$$

The total period is, T = period = $T_{high} + T_{low}$

The frequency of the astable circuit is then calculated using * $f = \frac{1}{T} \approx \frac{1.44}{(R_A + 2R_B)C}$

Using the component value the frequency of oscillation is

$$T_{\text{high}} = 0.7(R_A + R_B)C$$

$$= 0.7\left(7.5\times10^3 + 7.5\times10^3\right)\left(0.1\times10^{-6}\right) = 1.05 \text{ ms}$$

$$T_{\text{low}} = 0.7R_BC = 0.7\left(7.5\times10^3\right)\left(0.1\times10^{-6}\right) = 0.525 \text{ ms}$$

$$T = T_{\text{high}} + T_{\text{low}} = 1.05 \text{ ms} + 0.525 = 1.575 \text{ ms}$$

$$f = \frac{1}{T} = \frac{1}{1.575\times10^{-3}} \approx 635\ Hz$$

Monostable Operation

The 555 timer can also be used as a one-shot or monostable multivibrator circuit, When the trigger input signal goes negative, it triggers the one shot, with output at pin 3 then going high for a time period.

$$T_{\text{high}} = 1.1R_AC$$

The negative edge of the trigger input cause comparator-2 to trigger the flip-flop, with output at pin 3 going high. Capacitor C charges toward V_{cc} through resistor R_A. During the charge interval, the output remains high. When the voltage across the capacitor reaches the threshold levels of ($2V_{CC}/3$), comparator 1 triggers the flip-flop with output going low. The discharge transistor also goes low, causing the capacitor to remain at near 0 V until triggered again.

Time periods for this circuit can range from microseconds to many seconds, making the IC useful for range of applications.

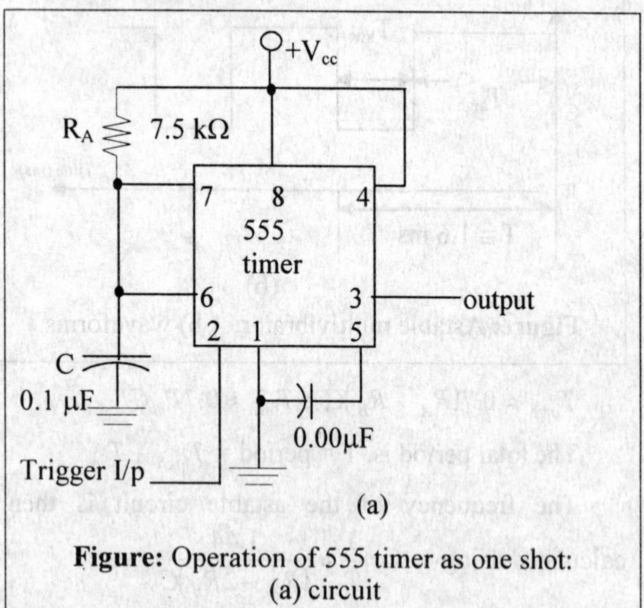

Figure: Operation of 555 timer as one shot: (a) circuit

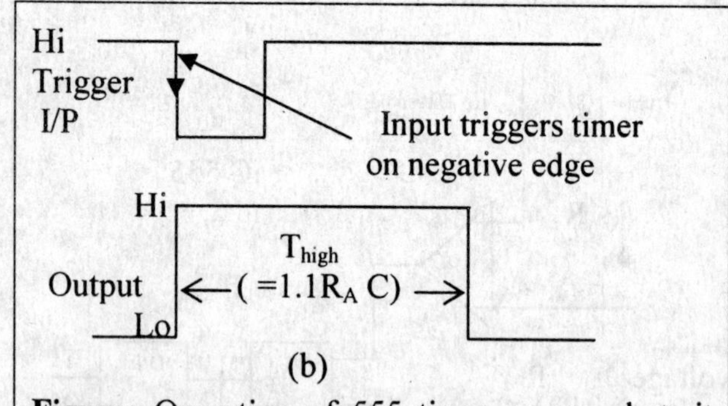

Figure: Operation of 555 timer as one shot: it (b) waveform.

Using the components value output waveform for the circuit of figure above when triggered by a negative pulse.

$$T_{\text{high}} = 1.1R_AC = 1.1\left(7.5\times10^3\right)\left(0.1\times10^{-6}\right) = 0.825\text{ms}$$

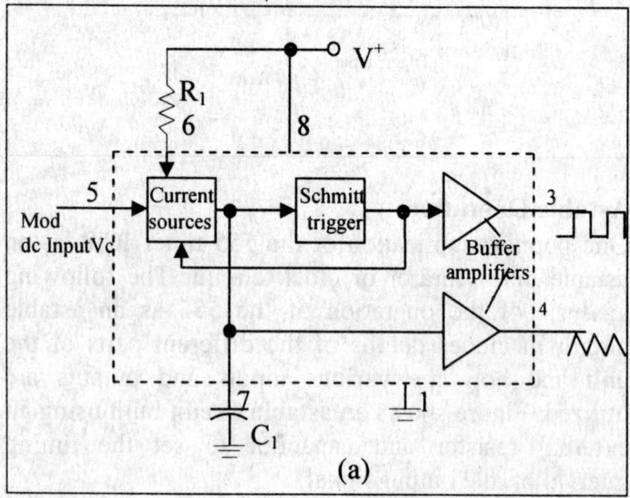

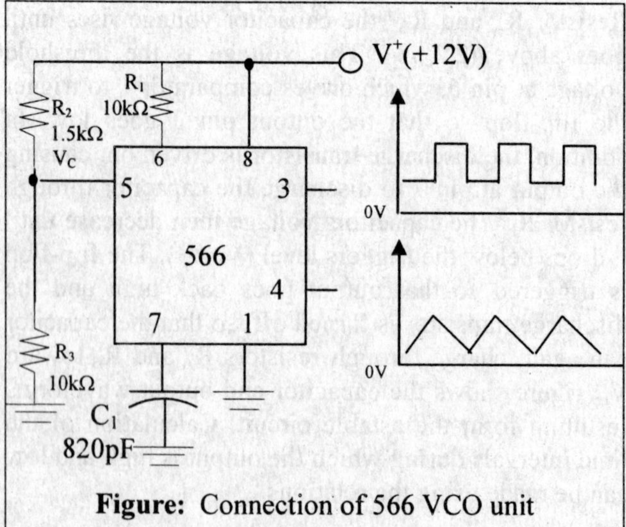

Figure: Connection of 566 VCO unit

Above figure shows an example in which the 566 function generator is used to provide both square-wave and triangular-wave signals at a fixed frequency set by R_1, C_1, and V_c. A resistor divider R_2 and R_3 sets the dc modulating voltage at a fixed value.

$$V_c = \frac{R_3}{R_2 + R_3} V^+ = \frac{10k\Omega}{1.5k\Omega + 10k\Omega}(12V) = 10.4V$$

(which falls properly in the voltage range $0.75V^+ = 9V$ and $V^+ = 12$ V).

$$f_o = \frac{2}{(10\times10^3)(820\times10^{-12})}\left(\frac{12-10.4}{12}\right) = 32.5\ kHz$$

The circuit of below figure shows how the output square-wave frequency can be adjusted using the input voltage, V_c, to vary the signal frequency. Potentiometer R_3 allows varying V_c from about 9 V to near 12V, over the full 10-to-1 frequency range. With the potentiometer wiper set *at the top*, the control voltage is

$$V_c = \frac{R_3 + R_4}{R_2 + R_3 + R_4}(V^+) = \frac{5k\Omega + 18k\Omega}{510\Omega + 5k\Omega + 18k\Omega}(+12V)$$

$$= 11.74V$$

resulting in a lower output frequency of

$$f_o = \frac{2}{(10\times10^3)(220\times10^{-12})}\left(\frac{12-11.74}{12}\right) \approx 19.7\text{kHz}$$

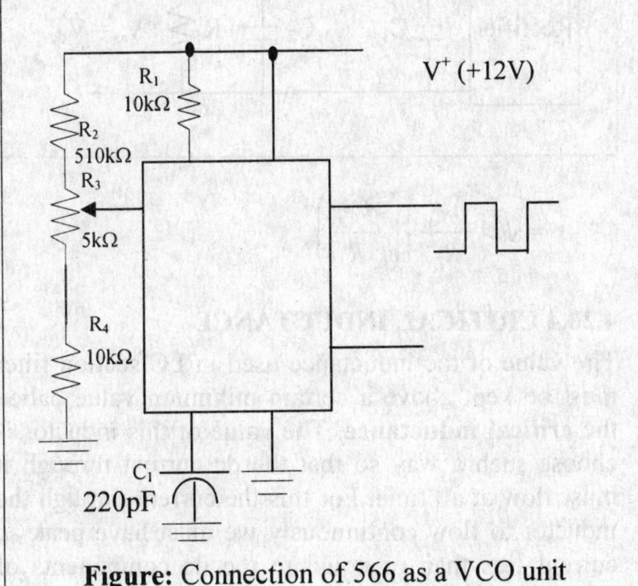

Figure: Connection of 566 as a VCO unit

With the wiper arm of R_3 *set at the bottom*, the control voltage is

$$V_c = \frac{R_4}{R_2 + R_3 + R_4}(V^+) = \frac{18k\Omega}{510\Omega + 5k\Omega + 18k\Omega}(+12V)$$

$$= 9.19V$$

Resulting in an upper frequency of

$$f_o = \frac{2}{(10\times10^3)(220\times10^{-12})}\left(\frac{12-9.19}{12}\right) \approx 212.9\text{kHz}$$

The frequency of the output square wave can then be varied using potentiometer R_3 over a frequency range of at least 10 to 1.

Rather than varying a potentiometer setting to change the value of V_c, an input modulating voltage, V_{in}, can be applied The voltage divider sets V_c at about 10.4 V. An input ac voltage of about 11.8 V, causing the output frequency to vary over about a 10-to-1 range. The input signal V_{in} thus frequency-modulates the output voltage around the center frequency set by the bias value of V_c =10.4 V (f_o =121.2 kHz).

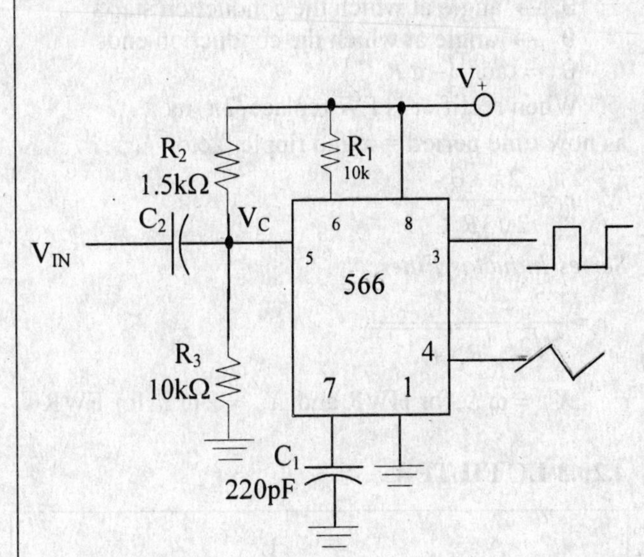

Figure: Operation of VCO with frequency modulating i/p.

20 RECTIFIERS AND POWER SUPPLIES

4.20.1 DIFFERENT TYPES OF RECTIFIERS CHARACTERISTICS

	HW	FW	Bridge	HW (3ϕ)	FW (3ϕ)
No. of diodes	1	2	4	3	6
Secondary voltage	V_m	2 V_m	V_m	V_m	V_m
PIV	V_m	2 V_m	V_m	V_m	V_m
V_{dc}	$\frac{V_m}{\pi}$	$\frac{2V_m}{\pi}$	$\frac{2V_m}{\pi}$	$0.82V_m$	$0.95V_m$
r	1.21	0.482	0.482	0.17	0.055
η_{max}	40.6	81.2	81.2	96.	99.5
f_0	f_i	$2f_i$	$2f_i$	$3f_i$	$6f_i$

4.20.2 Filter

Shunt capacitor filter:
when rectifiers is HW.

$$V_{dc} = \frac{V_m}{2\pi}\sqrt{1+(\omega R_L C)^2}\,(1-\cos\theta_c)$$

where $\theta_c = \theta_2 - \theta_1$

$\theta_1 \rightarrow$ angle at which the conduction starts
$\theta_2 \rightarrow$ angle at which the conduction ends
$\theta_2 = \tan^{-1}(-\omega R_L C)$

When rectifier is FW replace 2π by π as now time period = π and ripple factor

$$r = \frac{2\pi - \theta_c}{2\sqrt{3}R_L C}$$

Series Inductor Filter:

$$r = \frac{2}{3\sqrt{2}}\frac{R_L}{\sqrt{R_L^2 + X_L^2}}$$

$X_L = \omega L$ for HWR and $X_L' = 2\omega L$ for FWR

4.20.3 LC FILTER

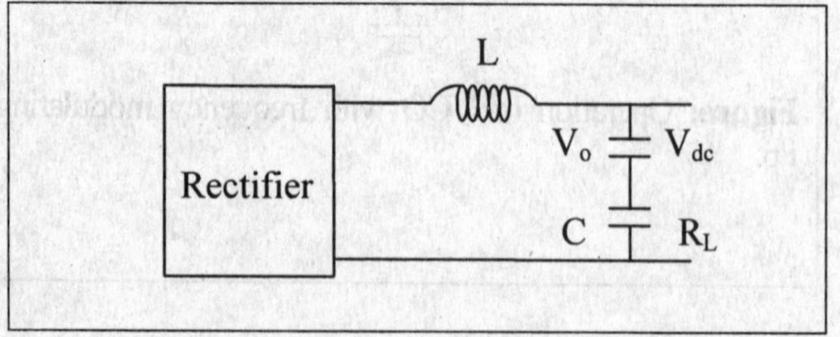

$$r = \frac{\sqrt{2}}{3}\left(\frac{X_C}{X_1}\right)$$ i.e. independent of load resistance

$X_1 = \omega L$ or $2\omega L$ for HW and forward

$X_C = \frac{1}{\omega C}$ or $\frac{1}{2\omega C}$ for HW and forward

Multiple LC Section filter:

$$r = \frac{\sqrt{2}}{3}\left(\frac{X_C}{X_L}\right)^n$$

CLC (π) Filter

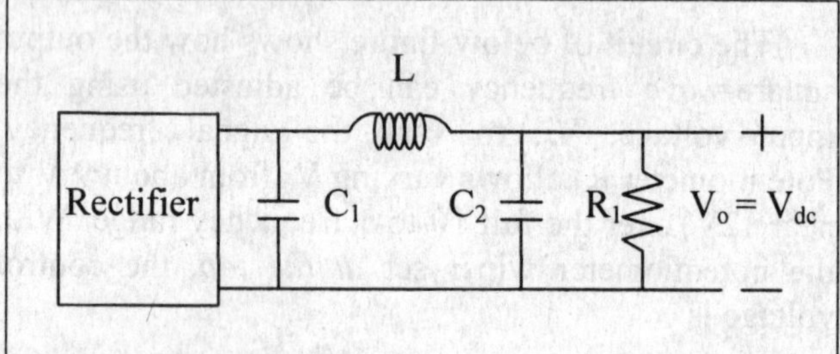

$$r = \sqrt{2}\left(\frac{X_{C_1}}{R_L}\right)\left(\frac{X_{C_2}}{X_L}\right)$$

CRC (π) Filter:

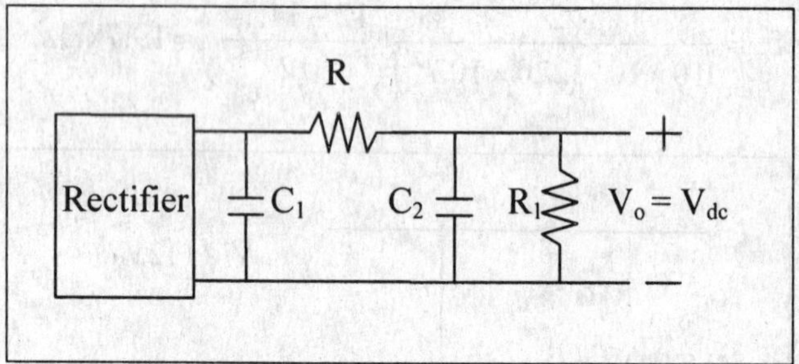

$$r = \sqrt{2}\left(\frac{X_{C_1}}{R_L}\right)\left(\frac{X_{C_2}}{R}\right)$$

4.20.4 CRITICAL INDUCTANCE

The value of the inductance used in LC section filter must be kept above a certain minimum value called the **critical inductance**. The value of this inductor is choose such a way so that the dc current through it must flow at all time. For this the current through the inductor to flow continuously we must have peak ac current less than or equal to the dc components of current.

$$I_{dc} \geq I_{peak}$$

$$\frac{V_{dc}}{V_L} \geq \sqrt{2}\ I'rms$$

$$\frac{V_d}{R_L} \geq \sqrt{2} \cdot \frac{\sqrt{2}}{3}\left(\frac{V_{dc}}{X_L}\right)$$

$$X_L \geq \frac{2}{3} R_L$$

4.20.5 BLEEDER RESISTANCE FUNCTION

1. To maintain the minimum current necessary for optimum inductor operation.
2. Used as a voltage divider in providing output voltage from the supply.
3. It serves as a discharge path for capacitor which may be induced in the circuit so that the voltage does not remain across the output terminal of power supply for a indefinitely long period of time after the load has been disconnected and the current dc-energies.
4. Reduces voltage regulation.

4.20.6 REGULAR

Voltage regulator using zener diode

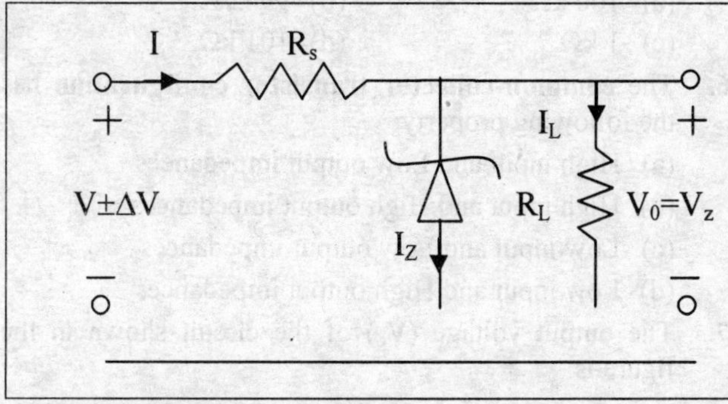

Condition for regulation is $(V + \Delta V) \geq |V_z|$

Case 1:

When $V_{in\,(\max)}$ and $R_{L\,\max}$ (no load)

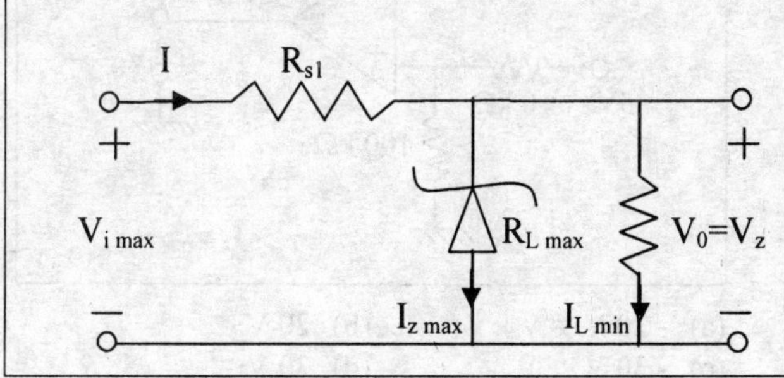

$$R_{\min} = R_{s1} \geq \frac{V_i(\max) - V_z}{I_z(\max) + I_L(\min)}$$

Case 2:

When V_{in} (min) and R_L (min) (maximum load current) this worst case

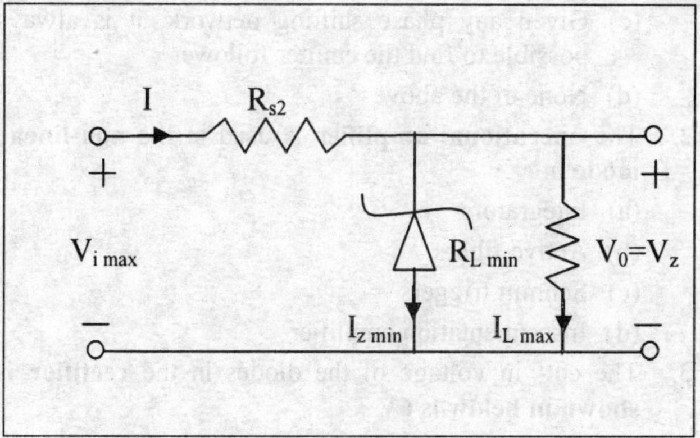

$$R_{\max} = R_{s2} = \frac{V_i(\min) - V_z}{I_z(\min) + I_L(\min)}$$

Practically $I_{z(\min)} = 0.2$ of I_z (rated)

$I_{z(\max)} = 0.8 I_z$ (rated)

- Series regulator is preferable for
  - → High voltage
  - → High output currents
  - → Relatively varying loads.
- Shunt regulator is preferred for:
  1. Medium to low voltages
  2. High currents
  3. Relatively constant loads.
- Emitter follower voltage regulator are limited to preferred for
  1. Low output voltage

  It has poor ripple suppression and poor regulation with respect to variation in the input voltage variation.

4.20.7 REGULATION FACTOR

$$S = \left.\frac{\Delta V_0}{\Delta V_1}\right|_{\Delta I_1 = 0}$$

Constant temperature $\Delta T = 0$ or should be low for good regulation.

MCQ Set–1

1. To realize a **phase shift oscillator**.
   (a) It is possible to use a voltage follower as an amplifier along with a suitable phase shift network
   (b) Only common emitter or common source stages can be used
   (c) Given any phase shifting network, it is always possible to find the emitter follower
   (d) None of the above

2. The **operational amplifier** is used in the **non-linear mode** in
   (a) Integrators
   (b) Active filters
   (c) Schmitt triggers
   (d) Instrumentation amplifier

3. The cut- in voltage of the diodes in the **rectifier is shown in below is 6V**

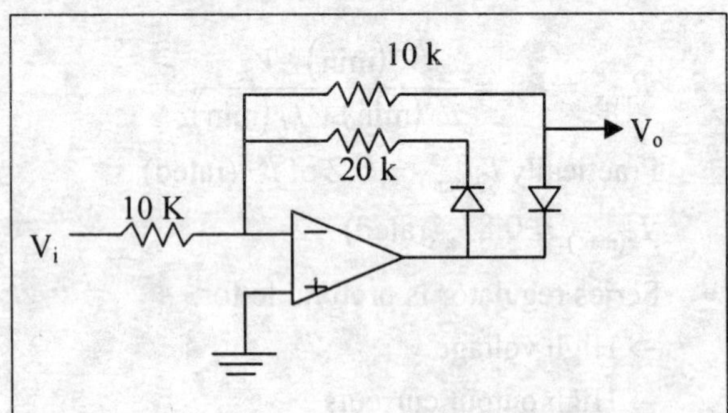

Identify the correct output input characteristic (V_O vs V_i)

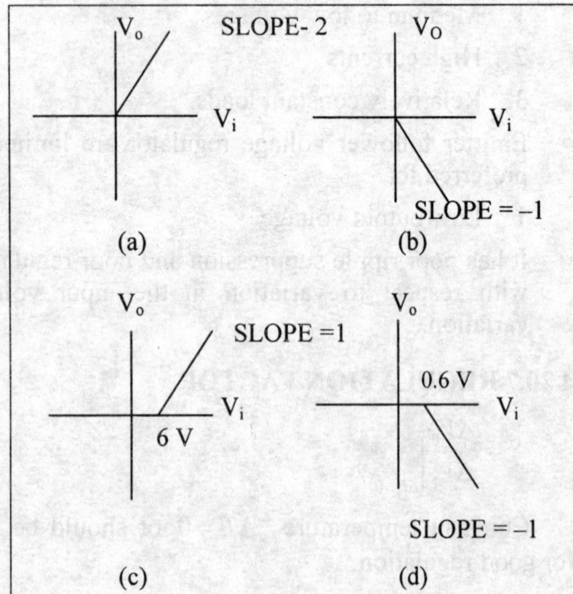

4. The diode circuit shown in figure function as

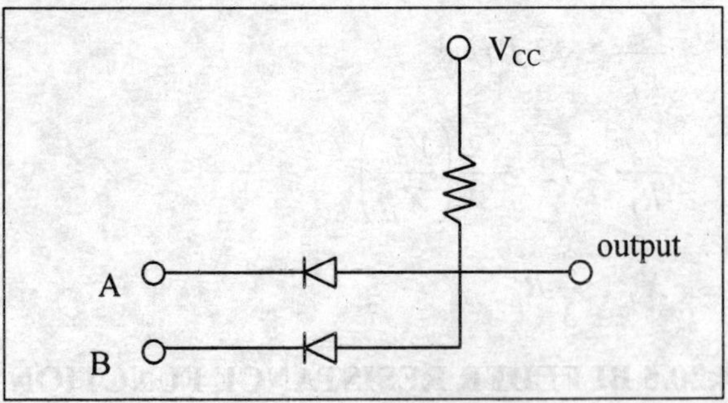

   (a) AND gate (b) OR gate
   (c) NAND gate (d) NOR gate

5. The input resistance of the op-amp circuit shown in figure.

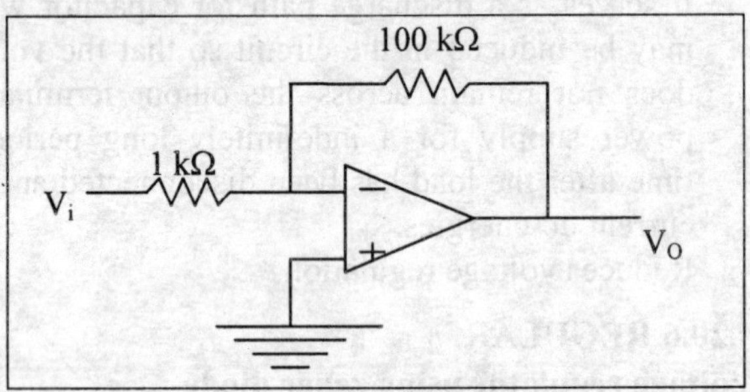

   (a) 100 kΩ (b) 99 kΩ
   (c) 1 kΩ (d) 101 kΩ

6. The **common-collector transistor configuration** has the following property:
   (a) High input and Low output impedances
   (b) High input and High output impedances
   (c) Low input and Low output impedances
   (d) Low input and High output impedances

7. The output voltage (V_o) of the circuit shown in the figure is.

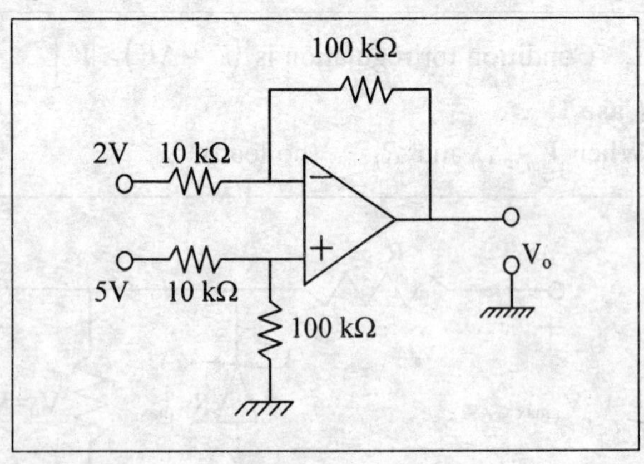

   (a) - 20 V (b) 20 V
   (c) - 30 V (d) 30 V

8. A differential output high impedance signal source is connected to an instrumentation amplifier through two shielded cables. The best way to obtain a high **CMMR** is to connect the cable shields to

(a) Output of the instrumentation amplifier

(b) Mean of the two first stage output in the instrumentation amplifier

(c) Common signal ground

(d) Power supply gourd

9. The signal to noise (S / N) ratio of an amplifier developing an output voltage of 10 V and a noise voltage of 1mV is

(a) – 40 dB (b) 40 dB

(c) 80 dB (d) 100 dB

10. In the circuit shown in figure, if e_i = sin t, the voltage e_0 is

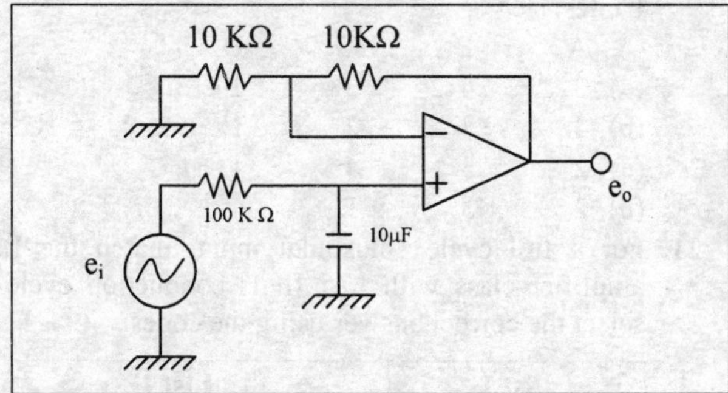

(a) $\sqrt{2}$ sin [t – (π/ 4)]

(b) {1/ $\sqrt{2}$} sin [t + (π/ 4)]

(c) {1/ $\sqrt{2}$} sin [t – (π/ 4)]

(d) $\sqrt{2}$ sin [t + (π/ 4)]

11. The differential input resistance of the circuit shown in figure is

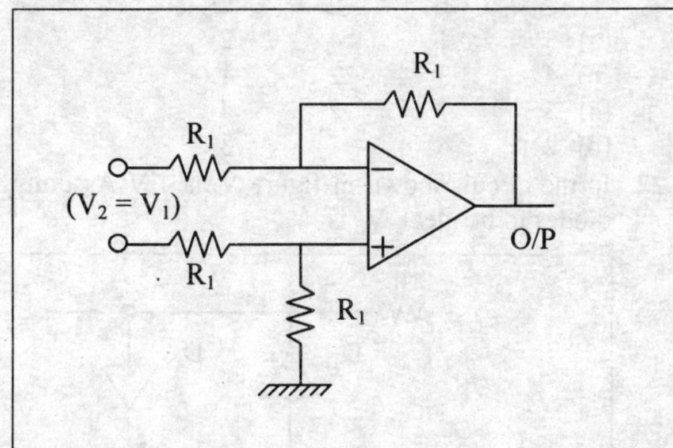

(a) R_1 (b) $R_1 / 2$

(c) $2R_1$ (d) R_f

12. A **common emitter transistor amplifier,** using a collector load of 1 k ohm and operating at room temperature, with a collector current of 1 mA, gives a voltage gain nearly equal to

(a) 25 (b) 40

(c) 250 (d) 1000

13. The **operational amplifiers** use a differential input stage with a constant current source, mainly to obtain

(a) Very low common mode gain

(b) Very high differential gain

(c) Very low input noise

(d) Very high input resistance.

14. n the op – amp circuit shown in figure the output voltage V_o, is equal to

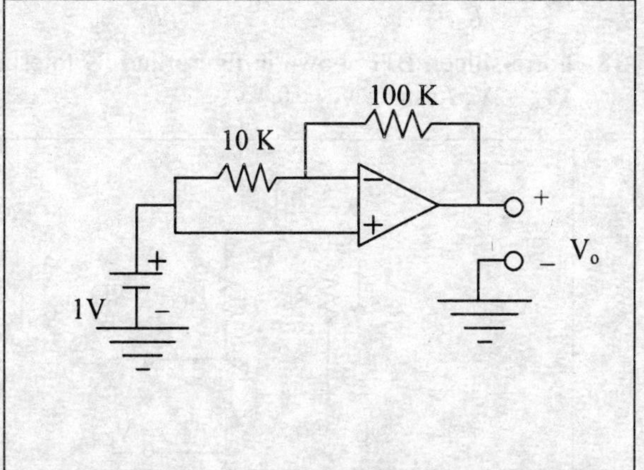

(a) 11 V (b) – 10 V

(c) 9 V (d) 1 V

15. The input v_t to a differentiator consists of a signal voltage v_s = 10 sin (50 t) and a noise voltage v_n =0.1 sin (250 t). The **signal-to -noise** voltage ratio at the output of the differentiator

(a) Decreases by factor of 5.

(b) Decreases by factor of 20.

(c) Increases by a factor of 5

(d) Increases by a factor of 20.

16. A pulse having a rise time of 40 ns is passed through a dc amplifier with a bandwidth of 12 MHz. The **rise time** of the pulse at the output of the amplifier nearly equals.

(a) 30 ns (b) 40 ns

(c) 50 ns (d) 80 ns

17. Pick up the correct relationship for the circuit shown in figure.

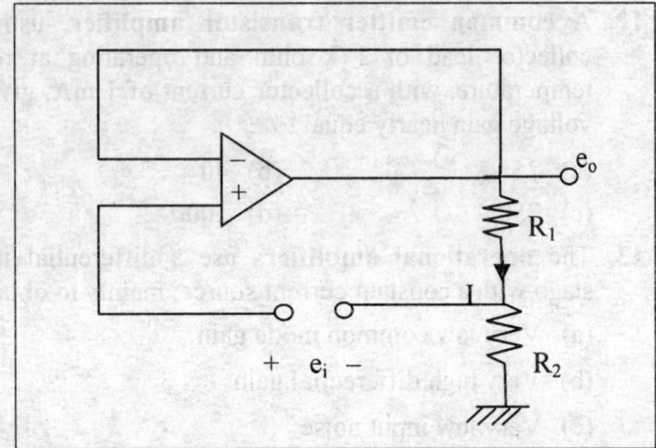

(a) $e_0 = e_i$ (b) $e_0 = \frac{R_2}{R_1} e_i$

(c) $1 = \frac{e_i}{R_2}$ (d) $1 = \frac{e_i}{R_1}$

18. For a silicon BJT shown in figure find R_B to establish V_{CE} =2V, Assume V_{BE} =0.7 V.

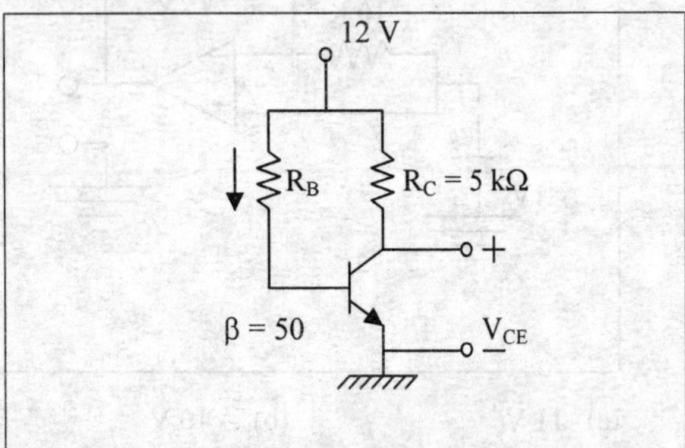

(a) 283 KΩ (b) 107 KΩ
(c) 200 KΩ (c) 242 K Ω

19. A class B push pull complementary symmetry amplifier uses.

a) Two n- p –n transistor
b) One p-n-p and one npn transistor
c) One pre – amplifier of n transistor followed
d) amplifier of two p-n-p transistors.

20. Match the List I figure with the List II device name and select the correct answer using the codes:

List I	List II
A. (G, D, S)	1. Depletion mode MOSFET
B. 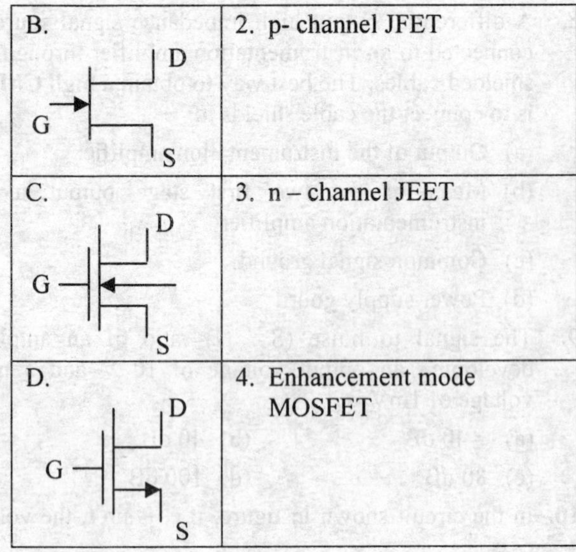	2. p- channel JFET
C.	3. n – channel JEET
D.	4. Enhancement mode MOSFET

Codes:

	A	B	C	D
(a)	2	4	3	1
(b)	4	3	2	1
(c)	2	3	4	1
(d)	2	3	1	4

21. For a full cycle sinusoidal input, match the List I amplifier class with List II its conduction cycle and select the correct answer using the codes:

List I	List II
A. Class A	1. One half cycle.
B. Class B	2. Less than one half cycle
C. Class AB	3. Full cycle
D. Class C	4. More than half cycle but less than full cycle.

Codes:

	A	B	C	D
(a)	3	1	4	2
(b)	4	3	2	1
(c)	3	1	2	4
(d)	2	4	1	3

22. In the circuit shown in figure V_i is 4 V Assuming the diodes to be ideal, V_o is.

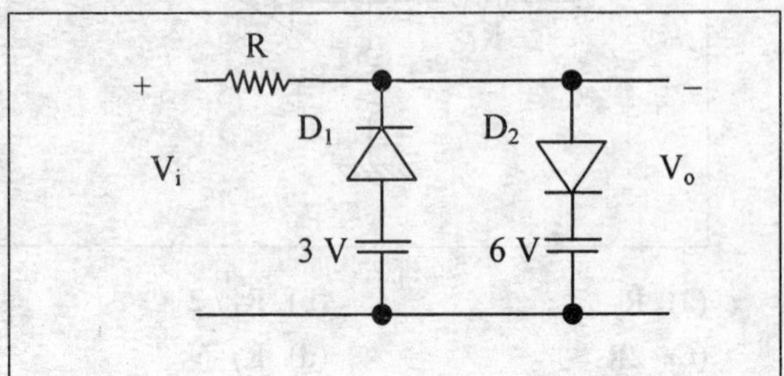

(a) 3 V (b) 4 V
(c) 4.5 V (d) 6 V

23. The input voltage, v_1 is 4 + 3 sin ωt. assuming all elements to be ideal, the average of the output voltage V_0 in figure is.

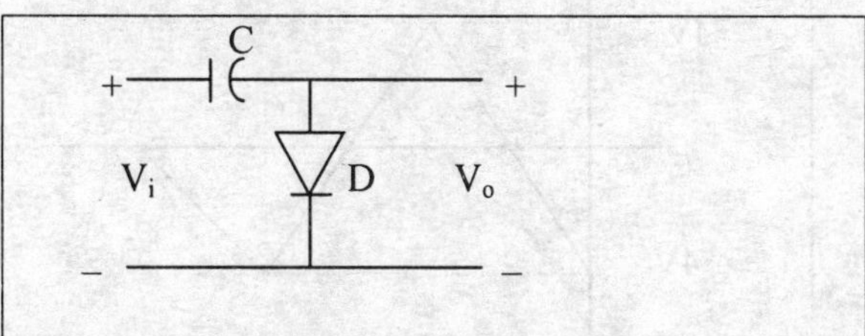

(a) – 3 V (b) + 3 V
(c) – 7 V (d) + 7 V

24. In the JFET amplifier circuit shown in figure the signal outputs V_1 and V_2 are related as.

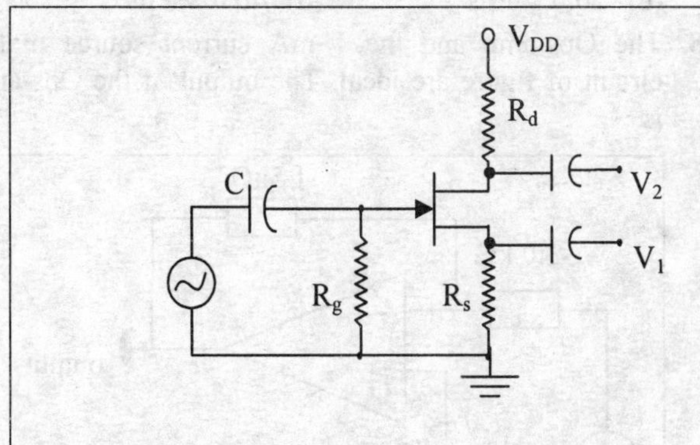

(a) $V_2 = \frac{R_d}{R_c} v_1$ (b) $V_2 = -\frac{R_d}{R_c} V_1$

(c) $V_2 = \frac{R_c}{R_d} V_1$ (d) $V_2 = -\frac{R_c}{R_d} V_1$

25. The op- amp in the amplifier circuit shown in figure has an offset voltage of 10 mV and it is ideal otherwise. If V_1 is zero, the output voltage V_0 is.

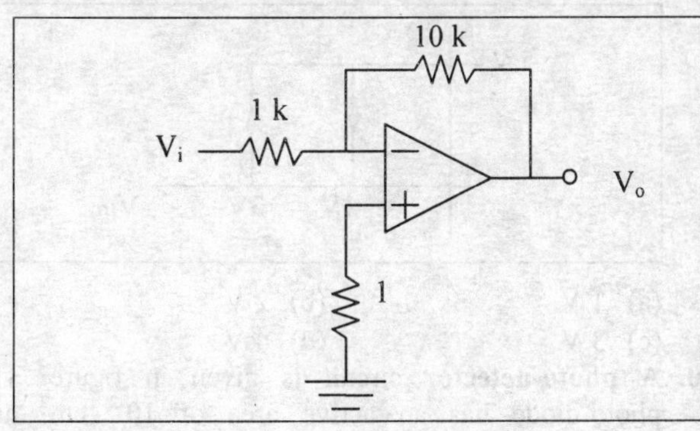

(a) 0 (b) 10 mV
(c) 100 mV (d) 110 m V

26. A phase locked loop can be employed for demodulation of
(a) Pulse amplitude modulated signals
(b) Pulse code modulated signals
(c) Frequency modulated signals
(d) Single side band amplitude modulation signals.

27. Bootstrapping in a buffer amplifier circuit is used for
(a) Increasing the input resistance
(b) Reducing the power consumption
(c) Decreasing the output resistance
(d) Improving the frequency response

28. An op-amp with a **slew rate** of 1 V/μ sec has used to build an amplifier of gain + 10. If the input to the amplifier is a sinusoidal with peak amplitude of 1 V, the maximum allowable frequency of the input signal for undistorted output is
(a) 830 Hz (b) 15.92 Hz
(c) 31.84 kHz (d) 1.0 MHz

29. For a sinusoidal input 50 V amplitude, the circuit shown in figure can be used as

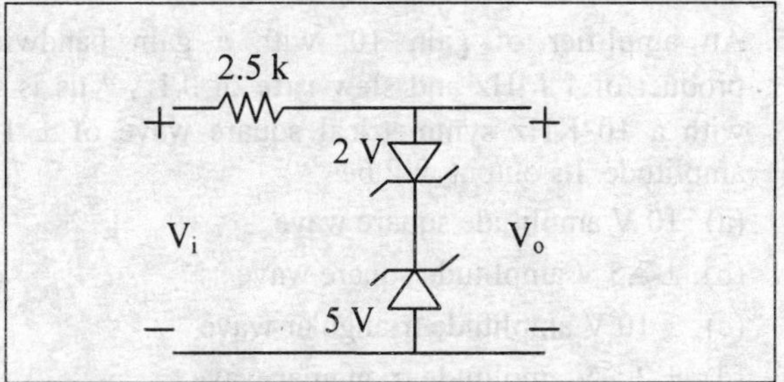

(a) Regulated dc power supply
(b) Square wave generator
(c) Half wave rectifier
(d) Full wave rectifier

30. IF both the junction of a transistor are forward biased, it will be in
(a) Saturation mode (b) Active mode
(c) Cut off mode

31. For a **phase locked loop** identify the correct relationship between the Lock Range (LR) and Capture Rang (CR)
(a) LR = CR (b) LR < CR
(c) LR > CR

32. **A twisted pair of wires** is used for connecting the signal source with the **instrumentation amplifier**, as it helps reducing
(a) The effect of External interference
(b) The error due to bias currents in the amplifier
(c) The loading of the source by the amplifier
(d) The common mode voltage.

33. In figure input offset voltage of the operational amplifier is 2 m V. The output dc-error voltage is

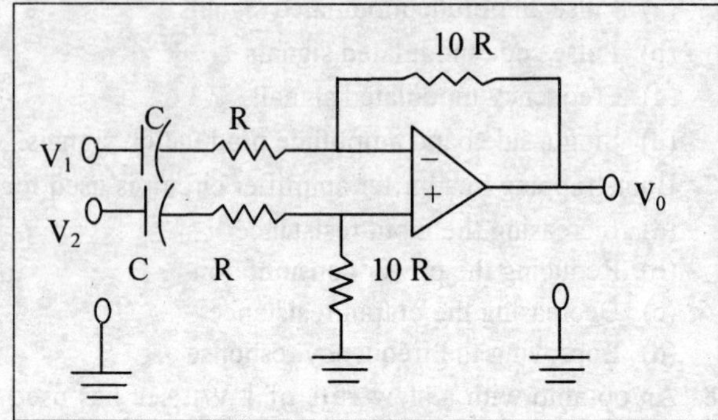

(a) 0 (b) 2 mV
(c) 11 m V (d) 22 m V

34. A unity gain buffer amplifier has a bandwidth of 1 MHz. The output voltage of the amplifier for an input of 2 V sinusoid of frequency 1 MHZ will be.

(a) 2 V (b) $2\sqrt{2}\ V$

(c) $2/\sqrt{2}\ V$ (d) $4/\sqrt{2}\ V$

35. An amplifier of gain 10, with a gain bandwidth product of 1 MHz and **slew rate** of 0.1 v / μs is fed with a 10 KHz symmetrical square wave of ± 1 V amplitude. Its output will be

(a) 10 V amplitude square wave
(b) ± 2.5 V amplitude square wave
(c) ± 10 V amplitude triangular wave
(d) ± 2.5 V amplitude triangular wave.

36. An integrator circuit is shown in figure. The Ops-amp is of type 741 and has an **input offset current** i_{os} of 1 μA. C is 1μF and R is 1 MΩ. If the input V_i is a 1 kHz. Square wave of 1V peak to peak, the output V_o, under steady state condition, will be.

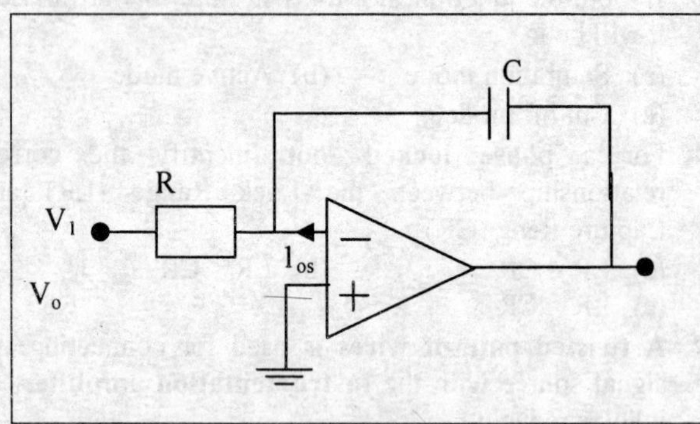

(a) A square wave of 1 V peak to peak.
(b) A triangular wave of 1 V peak to peak
(c) Positive supply voltage + V_{cc}
(d) Negative supply voltage – V_{cc}

37. The output of an **Ops-amp** whose input is a 2.5 MHz square wave is shown in figure. The **slew rate** of the Ops- amp is

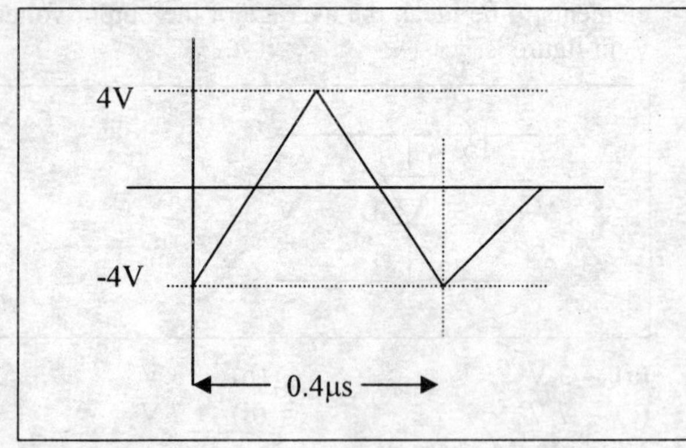

(a) 0.8 V / μs (b) 8.0 V / μs
(c) 20.0 V / μs (d) 40.0 V / μs

38. The Ops-amp and the 1 mA current source in the circuit of figure are ideal. The output of the Ops-amp is.

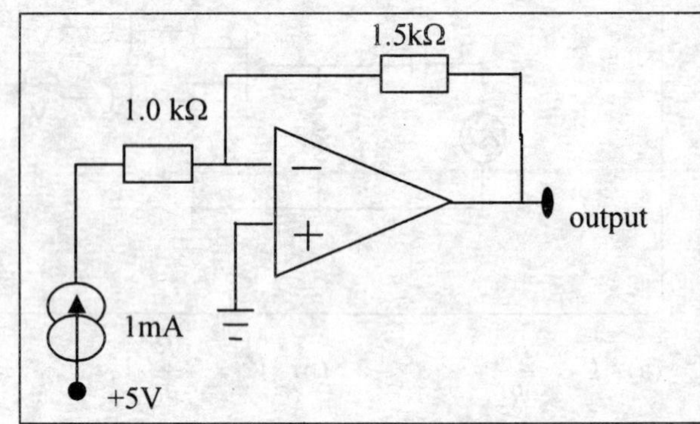

(a) –1.5mA (b) – 1.5 V
(c) –7.5 v (d) + 1.5

39. The input output characteristics of a **Schmitt trigger** is given in figure. The **noise margin of the Schmitt trigger** is.

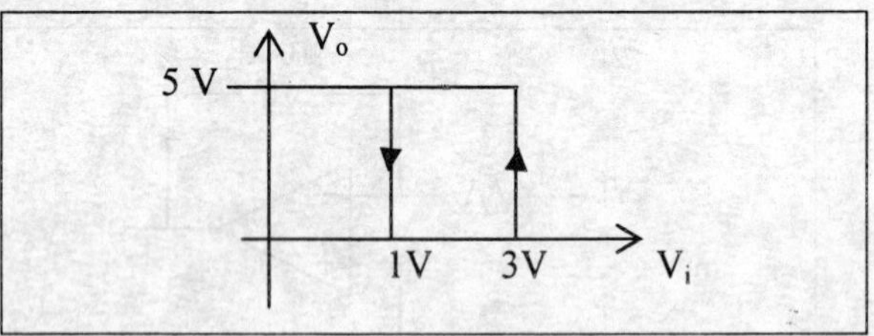

(a) 1 V (b) 2 V
(c) 3 V (d) 4 V

40. A **photo-detector** circuit is given in figure. The photo-diode has an active area of 10^{-2} cm^2 and sensitivity of 0. 55 A/W. When light of intensity 10 mW/cm^2 falls on the photo-diode, the output V_o is

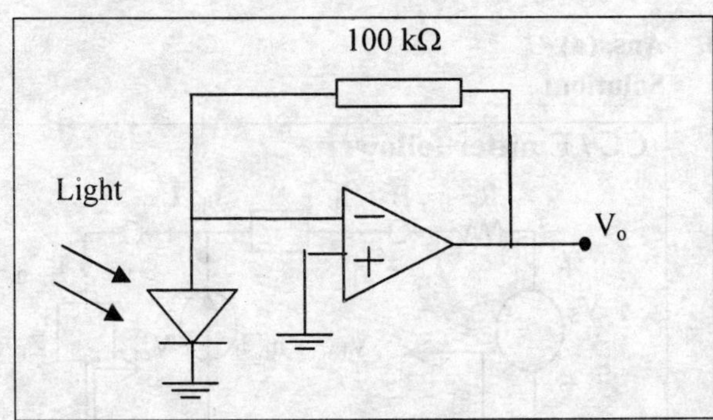

(a) 10.1 V (b) 5.5 V
(c) – 5.5 V (d) –10.1 V

41. A forward biased **silicion diode** when carrying negligible current, has a voltage drop of 0.64 V. When the current is 1 A it dissipates 1 W. The ON resistance of the diode is

(a) 0.36 Ω (b) 0.64 Ω
(c) 0.72 Ω (d) 1.0 Ω

42. A transistor amplifier circuit is shown in figure. The quiescent collector current, rounded off to first decimal, is

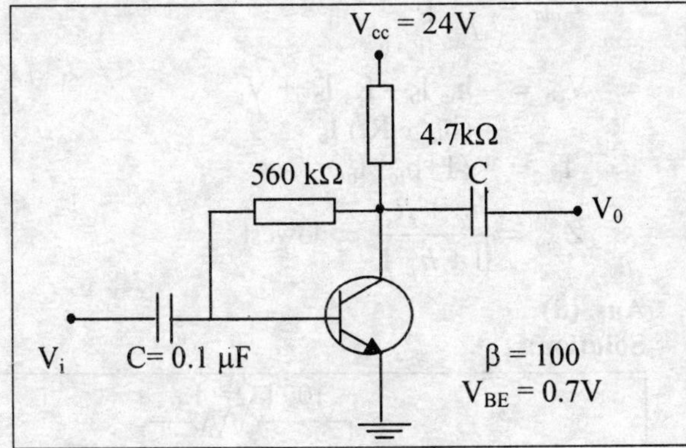

(a) 2.6 m A (b) 2.3 mA
(c) 2.1 m A (d) 2.0 m A.

43. The Ops-amp used in the inverting amplifier shown in figure has an equivalent input offset voltage V_{ios} of 5 mV. The output offset voltage is.

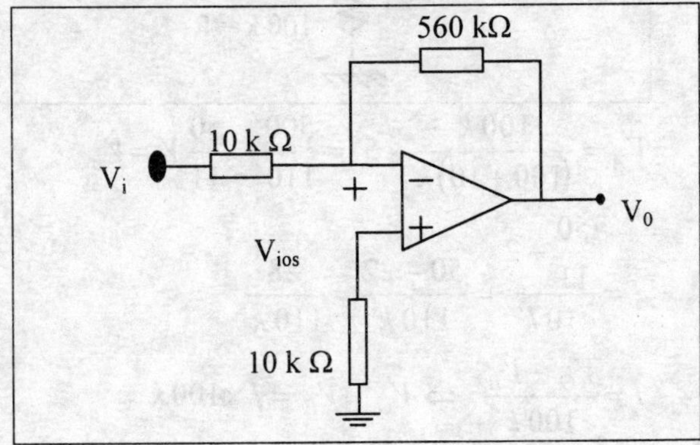

(a) 5 mV (b) 280 mV
(c) 285 mV (d) 560 mV

44. In the circuit shown in figure the Ops-amps used are ideal. The output V_0 is.

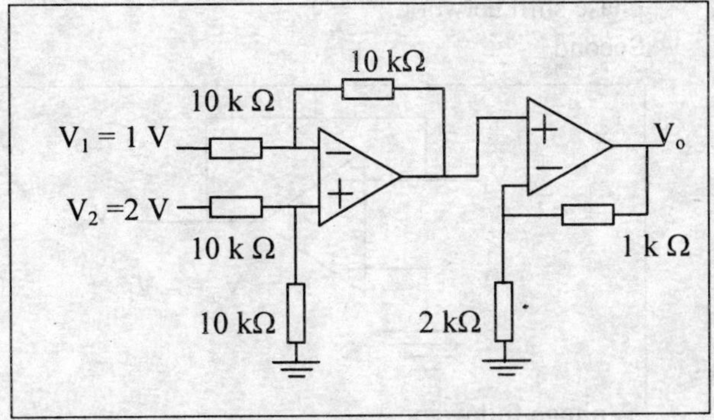

(a) 3.0 V (b) 1.5 V
(c) 1.0 V (d) 0.5 V

45. A 50 W power amplifier has a rated output of 30 V r. m. s and a voltage gain of 40 dB. It is connected to a 40 W loudspeaker having an internal resistance of 10 Ω. Determine the maximum input voltage that can be given to the power amplifier so that neither the power amplifier nor the loudspeaker is overload

(a) 40 V (b) 20 V
(c) 0.4 V (d) 0.2 V

46. The circuit shown in figure is that of a waveform generator. Assuming ideal devices, and ± 12 V supply the output V_o is a

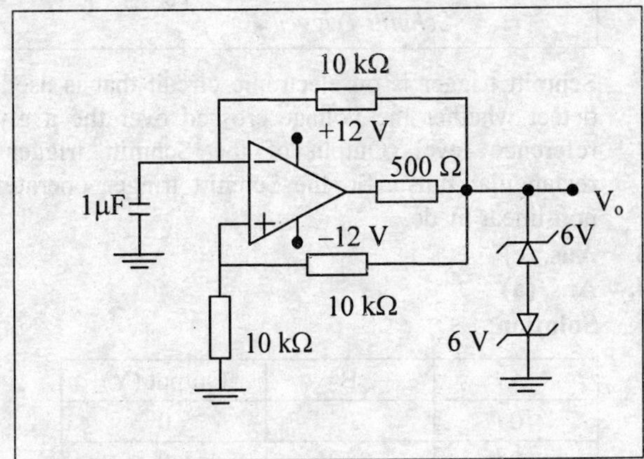

(a) Triangular wave of period 120 ms and amplitude ± 6V.
(b) Square wave of period 60 ms and amplitude ± 6V.
(c) Square wave of period 120 ms and amplitude ± 6V.
(d) Square wave of period 60 ms and amplitude ± 12V

Answers Set-1

1. Ans. (d)

Solution: To realize a phase shift oscillator, collector of the common emitter amplifier connected the R-C phase shift network

Second

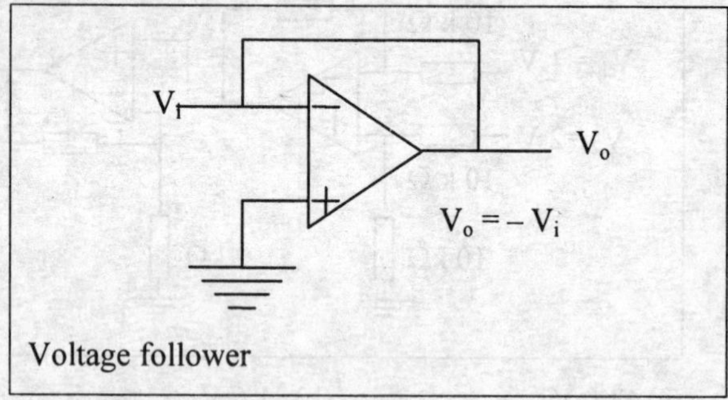

Voltage follower

$V_o = -V_i$ hence 180° phase shift

Phase shift network gives 180° then total become 360°

2. Ans. (c)

Solution:

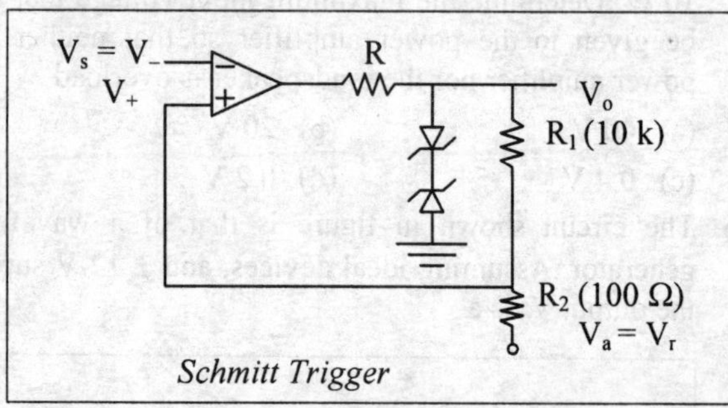

Schmitt Trigger

Schmitt trigger is on electronic circuit that is used to detect whether the voltage crossed over the a given reference level. Output of the Schmitt trigger is rectangular pulse. So the Schmitt trigger operate in non-linear made.

3. Ans. (c)

4. Ans. (a)

Solution:

A	B	Output (Y)
0	0	0
0	1	0
1	0	0
1	1	1

⇓

Y = A . B → AND Gate

5. Ans. (c)

Solution: Since it is inverting amplifier, therefore input impedance $R_i \cong R_1 \cong 1\,k\Omega$

6. Ans. (a)

Solution:

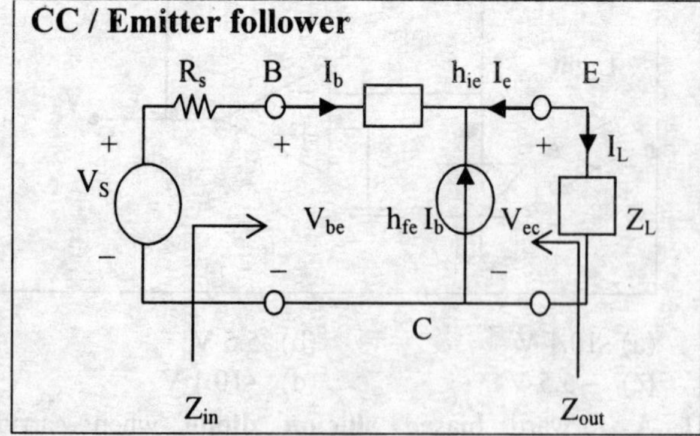

1. $Z_{in} = \dfrac{V_{bc}}{I_b} = \dfrac{h_{ie} \cdot I_b + I_L Z_L}{I_b}$

$= \dfrac{h_{ie} I_b + Z_L (+h_{fe}) I_b}{I_b}$

$Z_{in} = h_{ie} + (1 + h_{fe})Z_L$ .... highest

2. $Z_{out} = \left.\dfrac{V_{ec}}{I_e}\right|_{V_s=0}$

$V_{ec} = -h_{ie} I_b - R_s I_b + V_s$

$= -(h_{ie} + R_s) I_b$

$I_e = -(1+h_{fe}) I_b.$

$Z_{out} = \dfrac{h_{ie} + R_s}{(1 + h_{fe})} \Rightarrow$ lowest

7. Ans. (d)

Solution:

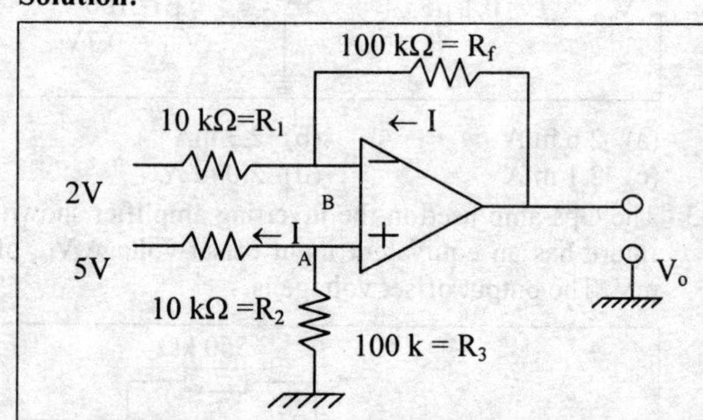

$$V_A = \frac{100\,k}{(100+10)\,k} \times 5 = \frac{500}{110} = \frac{50}{11} V = V_B$$

$$I = \frac{\frac{50}{11} - 2}{10\,k} = \frac{50 - 22}{110\,k} = \frac{28}{110\,k}$$

$$I = \frac{V_O - V_B}{100\,k} \Rightarrow V_O - V_B = I \times 100\,k$$

$$V_O = V_B + I \times 100\,k$$
$$= \frac{50}{11} + \frac{28}{110k} \times 100\,k = \frac{50}{11} + \frac{2800}{110}$$
$$= \frac{50+280}{11} = \frac{330}{11} = 30\,V$$

Another Method

Present case is of differential amplifier

Where $\frac{R_f}{R_1} = \frac{R_3}{R_2}$

Here $R_f = R_3 = 100\,k$

$R_1 = R_2 = 10\,k$

$V_1 = 2\,V,\ V_2 = 5\,V$

Then output $= \frac{-R_f}{R_1}(V_1 - V_2)$

$$= \frac{-100}{10}(2-5) = -10 \times -3 = 30\,V$$

8. **Ans. (b)**
9. **Ans. (c)**

Solution:

$\frac{S}{N}$ ratio = $20 \log_{10} \frac{S}{N}$

$= 20 \log_g \frac{10}{10^{-3}} = 20 \log_g 10^4$

$= 20 \times 4 \log_{10} = 80 \times 1 = 80$ db

10. **Ans. (a)**

Solution:

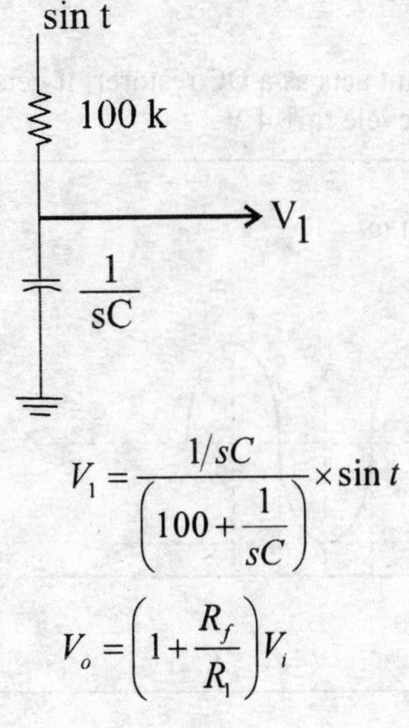

$$V_1 = \frac{1/sC}{\left(100 + \frac{1}{sC}\right)} \times \sin t$$

$$V_o = \left(1 + \frac{R_f}{R_1}\right)V_i$$

$$= (1+1)V_f = 2\,V_1$$

$$= \frac{2 \sin t}{(100\,sC + 1)}$$

$$= \frac{2 \sin t}{100\,k \times 10 \times 10^{-6} s + 1}$$

$$= \frac{2 \sin t}{s+1} = \frac{2 \sin t}{\sqrt{2}\angle 45^\circ}$$

$$= \sqrt{2} \sin\left(t - \frac{\pi}{4}\right)$$

11. **Ans. (c)**

Solution:

We known that gain of the op-amp is very high

$V_t = Lt$

So that Lt is restualy short circuited

$Z_{in} : R_1 + R_1 = 2R_1$

12. **Ans. (d)**

Solution:

$$A_V = \frac{h_{fe} \cdot R_L}{h_{ie}} \rightarrow$$

13. **Ans. (a)**
14. **Ans. (d)**

Solution:

$V_{o1} = 11\,V$

$V_{o2} = -10\,V$

$V = V_{o1} + V_{o2} = 11\,V + (-10\,V) = 1\,V$

15. **Ans. (a)**

Solution:

$$\frac{df(t)}{dt} = \frac{dA \sin \omega t}{dt} \quad A\omega \cos \omega t$$

Initial S/N ratio $= \frac{10}{0.1} = 100$

Final S/N ratio $= \frac{10 \times 50}{0.1 \times 250} = \frac{500}{25} = 20$

Hence S/N decrease 5 times

16. **Ans. (b)**
17. **Ans. (c)**

Solution:

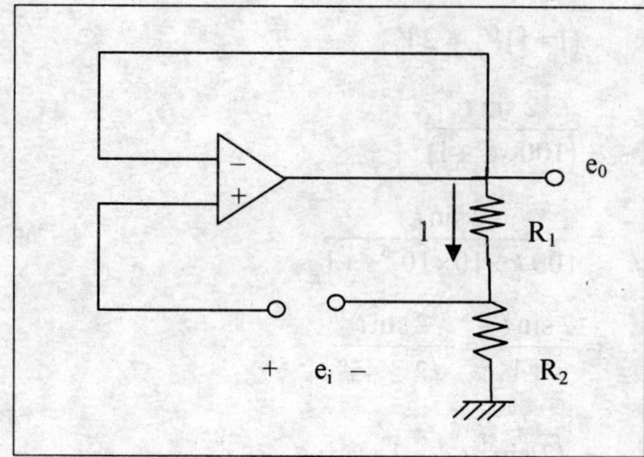

$$e_i = \frac{e_0 \times R_2}{R_1 + R_2} \quad (1)$$

$$e_0 - e_i = IR_1 \quad (2)$$

$$e_0 - \frac{e_0 R_2}{R_1 + R_2} = IR_1$$

$$e_0 \left[1 - \frac{R_2}{R_1 + R_2}\right] = IR_1$$

$$e_0 \left[\frac{R_1}{R_1 + R_2}\right] = IR_1$$

$$e_0 = I(R_1 + R_2) \quad (3)$$

From (2) and (3)

$$I(R_1 + R_2) - e_i = IR_1$$

$$e_i = IR_1 + IR_2 - I\,R_1$$

$$e_i = IR_2$$

$$I = \frac{e_i}{R_2}$$

18. Ans. (a)
Solution:

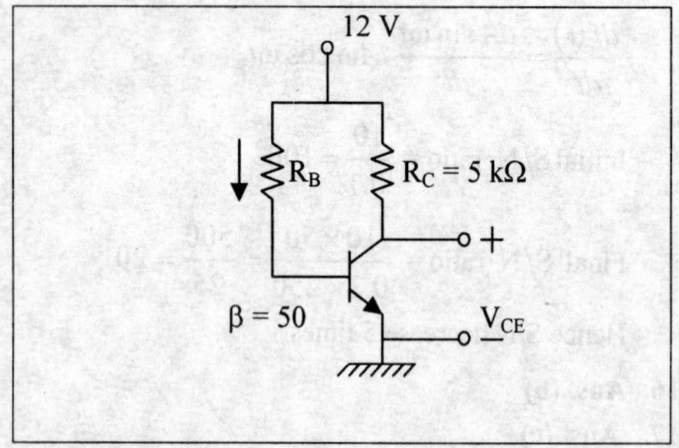

This diagram represent fixed biased

Apply KVL in CE loop

$$V_{CC} = I_C R_C + V_{CE}$$

$$V_{CE} = V_{CC} - I_C R_C$$

$$2 = 12 - I_C \times 5000$$

$$\therefore\ I_C = 2\ mA$$

KVL in V_{CC} – B – E loop

$$V_{CC} = I_B R_B + V_{BE}$$

$$12 = I_B R_B + 0.7$$

$$I_B = \frac{11.3}{R_B} = \frac{I_C}{\beta} = \frac{2\ mA}{50}$$

$$R_B = \frac{11.3 \times 50}{2 \times 10^{-3}} = \frac{11.3 \times 50}{2}\ k\Omega$$

$$R_B = 283\ k\Omega$$

19. Ans. (b)
20. Ans. (c)
21. Ans. (a)
22. Ans. (b)
Solution: For 4 V input both diodes are $V_o = V_i$

23. Ans. (c)
Solution:

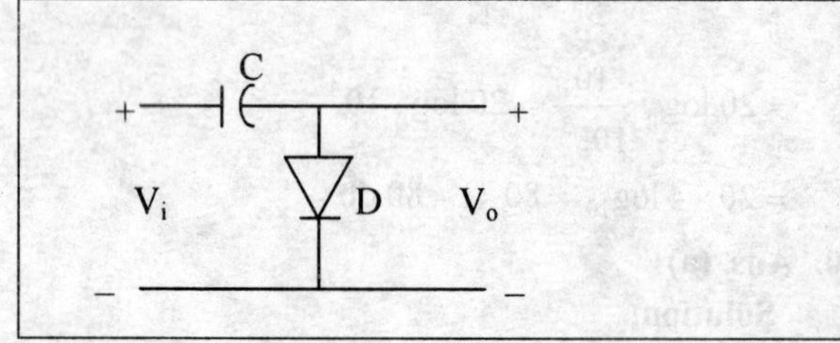

$$V_i = 4 + 3\sin \omega t$$

Capacitor in this circuit acts as a DC restorer, It gets charged posture half cycle to + 4 V.

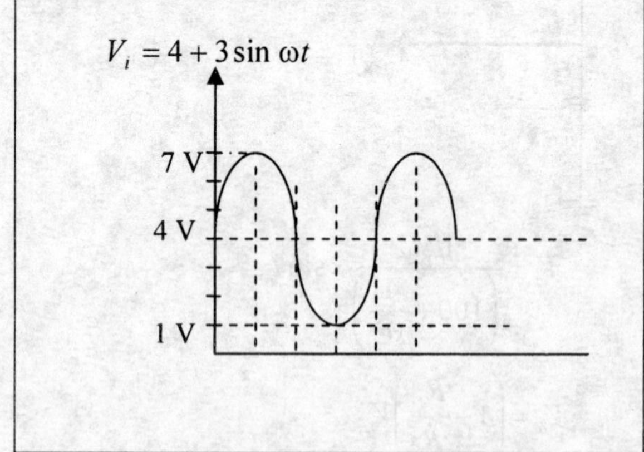

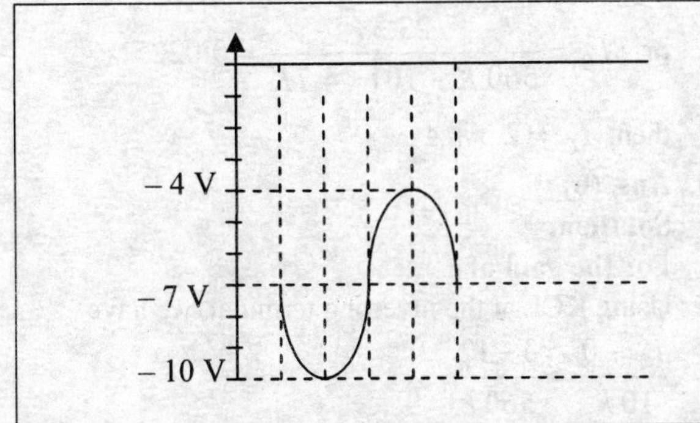

AC signal is using on – 7V. Average of AC over are cycle = 0

∴ Resultant output = – 7 V

24. Ans. (b)

25. Ans. (d)

Solution: $V_{oo} = V_{io} \times \left(1 + \frac{Rf_c}{R_i}\right)$

$= 10\ mV \times \left(1 + \frac{10}{1}\right) = 110$

26. Ans. (c)

27. Ans. (a)

28. Ans. (b)

Solution:

Slew rate $= \frac{d}{dt} V_m \sin \omega_m t\Big|_{\max}$

$= \frac{1V}{\mu s} = V_m \cdot \omega_m$

$= \frac{1V}{\mu s} = V_m \cdot 2\pi f_m$

$= 1 \times 2\pi f_m$

$1 \times 10^6 = 2\pi\ f_m$

$f_m = \frac{10^6}{2\pi} = 15.92\ kHz$

29 Ans. (b)

Solution:

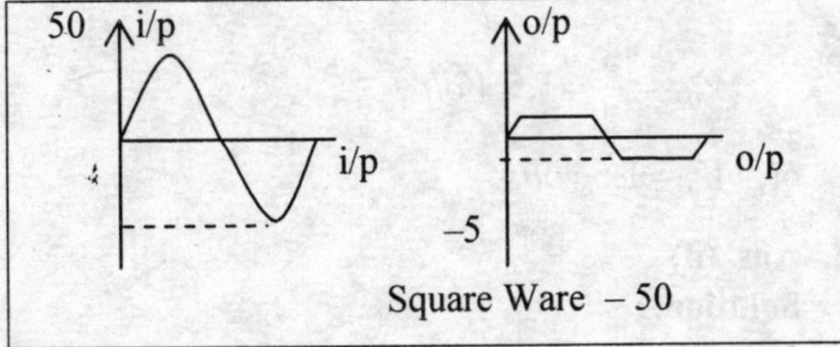

30. Ans. (a)

31. Ans. (b)

32. Ans. (c)

33. Ans. (d)

Solution: $= V_{io}\left(1 + \frac{R_f}{R_1}\right)$

$= 11 \times 2 = 22\ mV$

34. Ans. (a)

Solution: $V_0 = V_i = V_m \sin \omega t$

Then, output voltage $= V_m = 2\ V$

35. Ans. (c)

Solution:

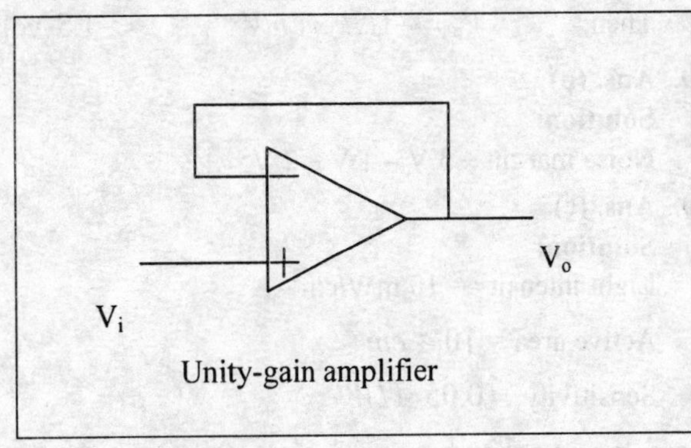

Unity-gain amplifier

Gain BW = constant

Here, Gain A = 10, then BW at output = constant

Amplitude amplified to ± 10 V. Since BW reduces at the amplifier output, therefore shape of output will be triangular

36. Ans. (b)

Solution: Using KVL at the inverting terminal, we have

$$\frac{V_i - 0}{R} = (0 - v_0)\,cS$$

or, $V_o = -\frac{1}{RC}\int_0^t V_i\,dt$

The, given circuit behave as an integrator, therefore integration of square wave result that is triangles

37. Ans. (d)

Solution:

Slew rate $= \frac{dv}{dt} = \frac{8}{T/2} = \frac{8}{0.2 \times 10^{-6}} = 40\ V/\mu s$

38. Ans.: (b)

Solution:

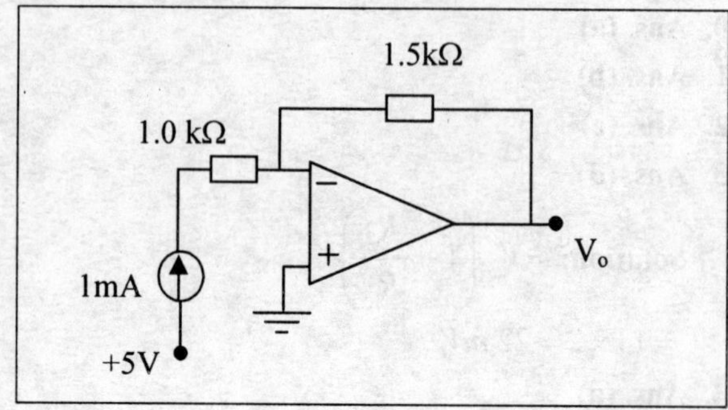

$V_- = V_+ = 0$ volt

This is due to concept virtually grounded

Ideal Ops amp $Z_{in} = \infty$, it has negligible current.

Whole of the current flow through feedback register

Then, $V_0 = -1.5k \times 1\, mV \quad = -1.5$ volts

39. Ans. (b)

Solution:

Noise margin = 3 V – 1 V = 2 V

40. Ans. (c)

Solution:

Light intensity = 10 mW/cm^2

Active area = $10^{-2}\ cm^2$

Sensitivity $= 0.05\ A/W$

$$= \frac{10 \times 10^{-3}\ W}{cm^2} \times 10^{-2}\ cm^2 \times \frac{0.05\ A}{W}$$

The photo-diode current

$= 10 \times 10^{-3}\ 0.55 \times 10^2 \times 10^{-4}$

$= 0.55 \times 10^{-4}$ A

then, $V_o = -(100\ K\Omega) \times 0.55 \times 10^{-4}\ A$

= – 5.5 volt

41. Ans. (d)

Solution:

$P = 1W \ \ dI = 1\,A$

$$r = R_{ON} = \frac{dV}{dI}$$

or $(dI)^2 R_{ON} = P$

or $R_{ON} = 1\,\Omega$

42. Ans. (b)

Solution:

For condition DC analysis is done

Apply KVL in CBE loop

$24 - (I_C + I_B)\, 4.7\ K - I_B\, 560\ K - 0.7 = 0$

as $I_C = \beta I_B$

or $I_B = \dfrac{23.3}{560\,K + 101 \times 4.7K} = 0.0225$

then, $I_C = 2.3\ mA$

43. Ans. (b)

Solution:

For the gain of

Using KCL at the inverting terminal, we have

$$\frac{V_i - 0}{10\,k} = \frac{0 - V_o}{560\,k}$$

or, $A_v = \dfrac{-V_o}{V_i} = 56$

Gain of the inverting amplifier by formula is

$$A_v = \frac{R_f}{R} = \frac{560}{10} = 56$$

then, output offset voltage = $A_v \times v_{ios}$

$= -56 \times 5\ mV$

$V_{oos} = -280\ mV$

44. Ans. (b)

Solution:

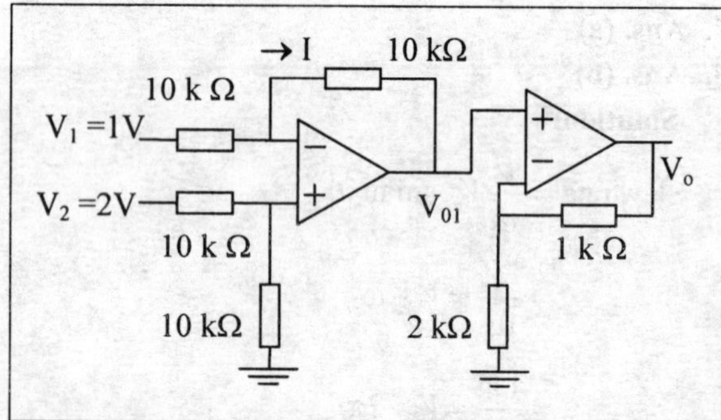

At Ops-amp

Then $I = \dfrac{1-1}{10} = 0$

Using voltage division $\quad V = V_- = V_+ = 1\ volt$

$\therefore V_{01} = 1V$

For Ops-amp, by voltage division

$$V_{01} = 1 = \frac{2}{3} V_{02}$$

$$V_{02} = \frac{3}{2} = 1.5\ Volt$$

or, $V_0 = 1.5\ volts$

45. Ans. (d)

Solution:

46. Ans. (c)
Solution:

Time period, $T = 2RC \ln \dfrac{1+\beta}{1-\beta}$

Here, $\beta = \dfrac{1}{2}$

$\therefore$ $T = 2RC \ln 3$

$= 2 \times 10^4 \times 10^{-6} \ln 3$

$= 0.0219$ sec

The circuit behaves as square wave generator.

One Mark Questions

1. n-type silicon is obtained by doping silicon with
(a) Germanium (b) Aluminum
(c) Boron (d) Phosphorus

2. For an n-channel enhancement type MOSFET, if the source is connected at a higher potential than that of the bulk (i.e $V_{SB} > 0$), the threshold voltage V_T of the MOSFET will.
(a) remain unchanged (b) decrease
(c) change polarity (d) increase

3. The impurity commonly used for realizing the base region of a silicon n-p-n transistor is
(a) Gallium (b) Indium
(c) Boron (d) Phosphorus

4. If for a silicon n-p-n transistor, the base-to-emitter voltage (V_{BE}) is 0.7V and the collector –to –base voltage (V_{CB}) is 0.2 V, then the transistor is operating in the
(a) Normal active mode
(b) Saturation mode
(c) Inverse active mode
(d) Cutoff mode

5. Consider the following statements S_1 and S_2.
S_1: The β of a bipolar transistor reduces if the base width is increased
S_2: The β of a bipolar transistor increases if the doping concentration in the base is increased.
Which one of the following is correct?
(a) S_1 is FALSE and S_2 is TRUE
(b) Both S_1 and S_2 are TRUE
(c) Both S_1 and S_2 are FALSE
(d) S_1 is TRUE and S_2 is FALSE

6. The band gap of Silicon at room temperature is
(a) 1.3 e V (b) 0.7 eV
(c) 1.1 eV (d) 1.4 eV

7. A Silicon PN junction at a temperature of 20°C has a reverse saturation current of 10 pico-Amperes (pA). The reverse saturation current at 40° C for the same bias is approximately.
(a) 30 pA (b) 40 pA
(c) 50 pA (d) 60 pA

8. The primary reason for the widespread use of Silicon in semiconductor device technology is
(a) Aboundance of Silicon on the surface of the Earth
(b) Large band-gap of Silicon in comparison to Germanium.
(c) Favorable properties of Silicon-dioxide (SiO_2)
(d) Lower melting point

9. The values of voltage (V_D) across a tunnel-diode corresponding to peak and valley currents are V_p and V_v respectively . The range of tunnel diode voltage V_D for which the slope of its I - V_D characteristics is negative would be.
(a) $V_D < 0$ (b) $0 \le V_D < V_p$
(c) $V_p \le V_D < V_v$ (d) $V_D \ge V_v$

10. The concentration of minority carriers in an extrinsic semiconductor under equilibrium is
(a) Directly proportional to the doping concentration
(b) Inversely proportional to the doping concentration.
(c) Directly proportional to the intrinsic concentration.
(d) Inversely proportional to the intrinsic concentration.

11. Under low level injection assumption, the injected minority carrier current for an extrinsic semiconductor is essentially the
(a) Diffusion current
(b) Drift current
(c) Recombination current
(d) Induced current

12. The phenomenon known as "Early Effect" in a bipolar transistor refers to a reduction of the effective base – width caused by.
(a) Electron – hole recombination at the base.
(b) The reverse biasing of the base – collector junction.
(c) The forward biasing of emitter – base junction.
(d) The early removal of stored base charge during saturation to cutoff switching

13. The electron and hole concentrations in an intrinsic semiconductor are n_i per cm^3 at 300 K. Now, if acceptor impurities are introduced with a concentration of N_A per cm^3 (where $N_A >> n_i$), the electron concentration per cm^3 at 300 K will be
(a) n_i (b) $n_i + N_A$
(c) $N_A - n_i$ (d) n_i^2 / N_A

14. In a P^+ **n junction diode** under reverse biased, the magnitude of **electric field is maximum** at
(a) The edge of the depletion region of the p-side
(b) The edge of he depletion region on the n-side
(c) The p+n junction
(d) The centre of the depletion region on the n-side

15. The circuit shown in the figure is best described as a

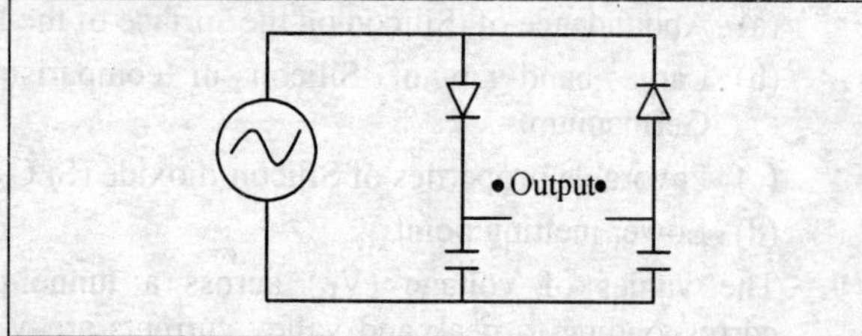

(a) Bridge rectifier
(b) Ring modulator
(c) Frequency discriminatory
(d) Voltage doubler.

16. If the input to the ideal comparator shown in Figure is a sinusoidal signal of 8V (peak to peak) without any DC component, then the output of the comparator has a duty cycle of

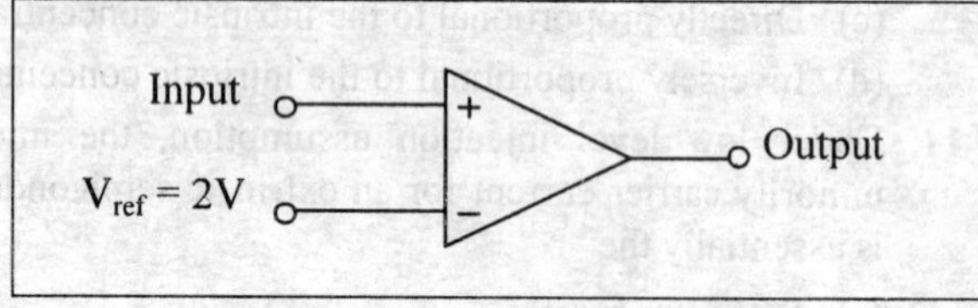

(a) ½ (b) 1/3
(c) 1/6 (d) 1/12

17. If the differential voltage gain and the common mode voltage gain of a differential amplifier are 48 dB and 2dB respectively, then its common mode rejection ratio is:
(a) 23 dB (b) 25 dB
(c) 46 dB (d) 50 dB

18. Generally, the gain of a transistor amplifier falls at high frequencies due to the
(a) Internal capacitance of the device
(b) Coupling capacitor at the input
(c) Skin effect
(d) Coupling capacitor at the output

19. Choose the correct match for input resistance of various amplifier configurations shown below:

Configuration	input resistance
CB: Common Base	LO: Low
CB: Common Collector	MO: Moderate
CE : Common Emitter	HI: High

(a) *CB-LO*, *CC-MO*, *CE-HI*
(b) *CB-LO*, *CC-HI*, *CE-MO*
(c) *CB-MO*, *CC-HI*, *CE-LO*
(d) *CB-HI*, *CC-LO*, *CE-MO*

20. An ideal op-amp is an ideal
(a) Voltage controlled current source
(b) Voltage controlled voltage source
(c) Current controlled current source
(d) Current controlled voltage source

21. Voltage series feedback (also called series-shunt feedback) results in
(a) Increase in both input and output impedances
(b) Decrease in both input and output impedances
(c) Increase in input impedance and decrease in output impedance
(d) Decrease in input impedance and increase in output impedance.

22. The circuit in Figure is a

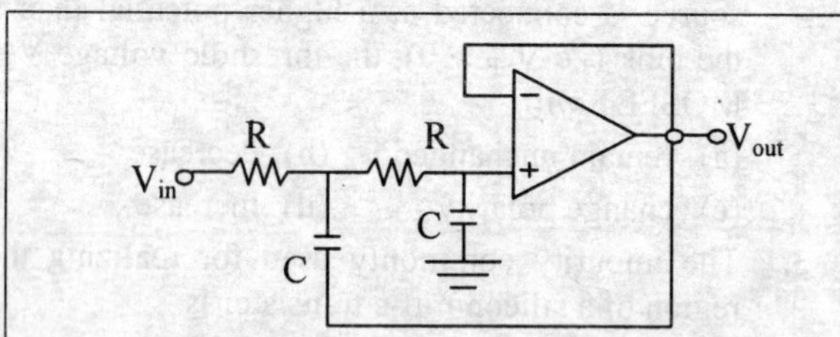

(a) low-pass filter (b) high-pass filter
(c) band-pass filter (d) band-reject filter

23. The OP AMP of figure has a very poor open loop voltage gain of 45 but is otherwise ideal. The gain of the amplifier equals.

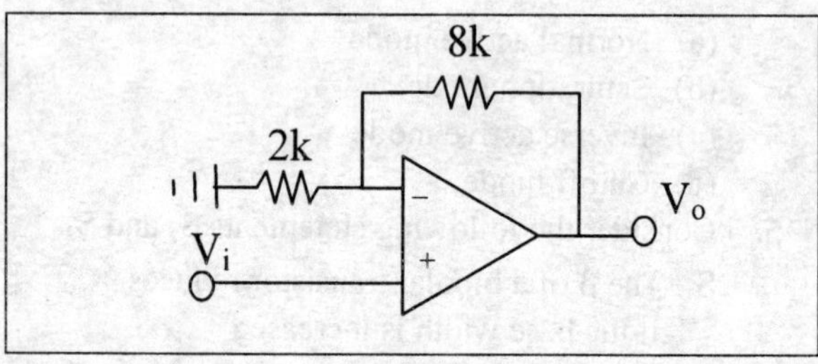

(a) 5 (b) 20
(c) 4 (d) 4.5

24. The effect of current shunt feedback in an amplifier is to
(a) Increase the input resistance and decrease the output resistance
(b) Increase both input and output resistance.
(c) Decrease both input and output resistance.
(d) Decrease the input resistance and increase the output resistance

25. The input resistance R_i of the amplifier shown in the figure is

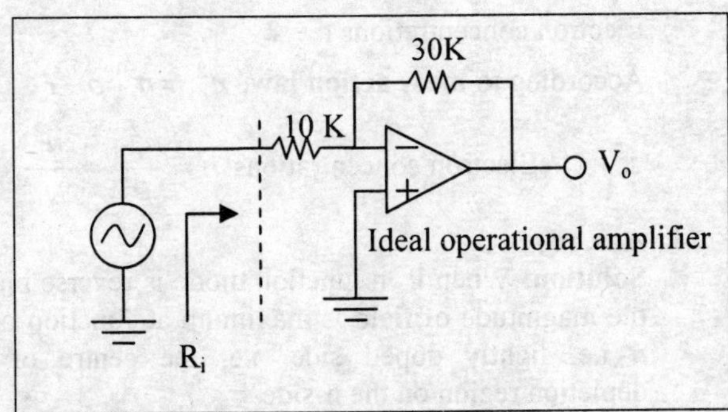

(a) [30 / 4] kΩ (b) 10 kΩ
(c) 40 kΩ (d) infinite

26. The cascode amplifier is a multistage configuration of
(a) CC – CB (b) CE – CB
(c) CB – CC (d) CE – CC

27. The input impedance (Z_i) and the output impedance (Z_o) of an ideal trans conductance (voltage controlled current source) amplifier are:
(a) $Z_i = 0, Z_o = 0$ (b) $Z_i = 0, Z_o = \infty$
(c) $Z_i = \infty, Z_o = 0$ (d) $Z_i = \infty, Z_o = \infty$

28. An n – channel depletion MOSFET has following two points on its I_D - V_{GS} curve
- $V_{GS} = 0$ at $I_D = 12$ mA and
- $V_{GS} = -6$ Volts at $I_D = 0$ mA

Which of the following Q – point will give the highest trans – conductance gain for small signals?
(a) $V_{GS} = -6$ Volts (b) $V_{GS} = -3$ Volts
(c) $V_{GS} = 0$ Volts (d) $V_{GS} = 3$ Volts

29. Consider two transfer function
$G_1(s) = [1 / (s^2 + as + b)]$ and
$G_2(s) = [s / (s^2 + as + b)]$
The 3 dB bandwidths of their frequency responses are, respectively.
(a) $\sqrt{(a^2 - 4b)}$, $\sqrt{(a^2 + 4b)}$
(b) $\sqrt{(a^2 + 4b)}$, $\sqrt{(a^2 - 4b)}$
(c) $\sqrt{(a^2 - 4b)}$, $\sqrt{(a^2 - 4b)}$
(d) $\sqrt{(a^2 + 4b)}$, $\sqrt{(a^2 + 4b)}$

30. The correct full wave rectifier circuit is

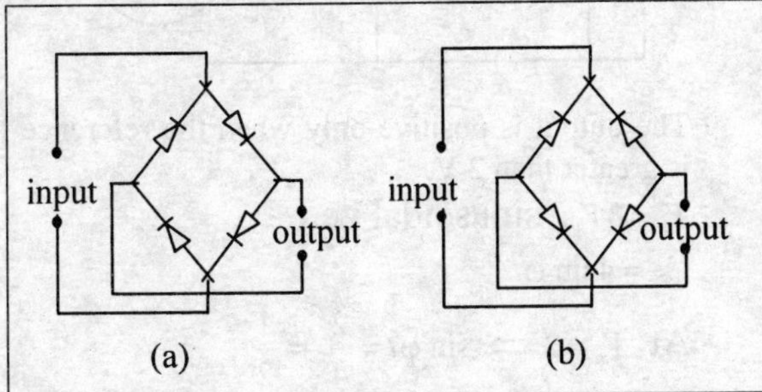

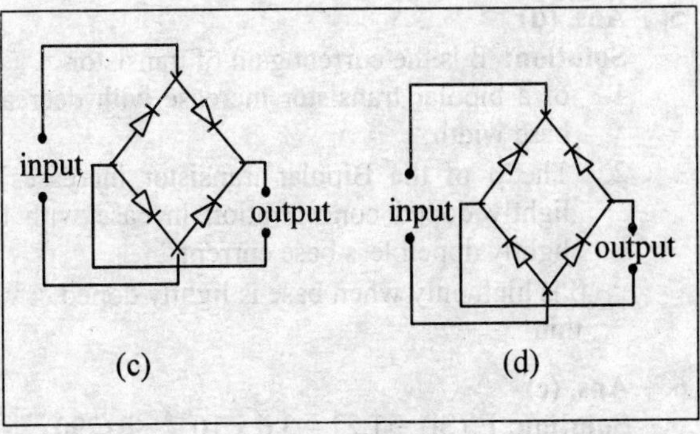

31. In a **tranconductance amplifier**, it is desirable to have
(a) A large input resistance and a large output resistance
(b) A large input resistance and a small output resistance
(c) A small input resistance and a large output resistance
(d) A small input resistance and a small output resistance

Answers
Analog Electronics

1. Ans. (d)
Solution: n type is formed by doping with pentavalent impurity (P).
Others are G(4) and Al (3) B(3)

2. Ans. (a)
Solution:
V_T is independent of circuit parameter, it depends on fabrication process and material used.

3. Ans. (c)
Solution: Boron is an atom of Group III element, Group III elements are used for making p-type series conducts

4. Ans. (a)
Solution:

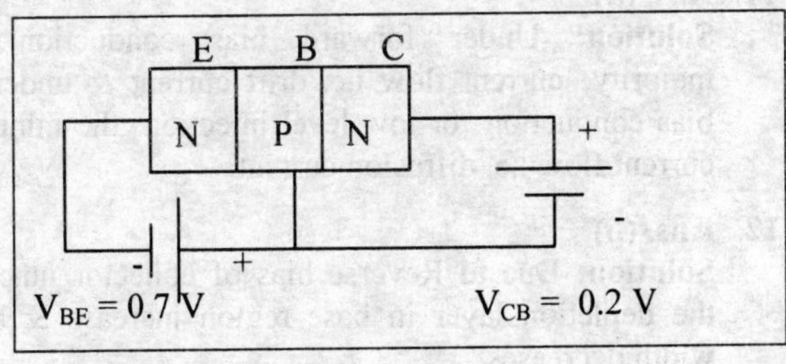

If, $V_{BE} = 0.7$ V - --Forward biased
$V_{CB} = 0.2$ V --- Reverse biased
Transistor operate in normal active mode

5. **Ans. (d)**

Solution: β is the current gain of transistor

1. of a bipolar transistor increase with decrease in base width
2. The β of the Bipolar transistor increases with lightly doped concentration in base with i.e. if lightly doped less base current

∴ β is high only when base is lightly doped & very thin

6. **Ans. (c)**

Solution: $E_g(Si) = 1.21 - 3.6 \times 10^{-4}$ T (290)
= 1.1 eV
E_g (Ge) = $0.785 - 2.23 \times 10^{-4}$ T(290)
= 0.76 eV

7. **Ans. (b)**

Solution: Reverse saturation current becomes doubles for every 10 degree, rise in temperature.

I_s = 10 þA at 20° C ; I_s = 20 þA at 30° C

I_s = 40 þA at 40° C

8. **Ans. (c)**

9. **Ans. (c)**

Solution: The tunnel diode has characteristic

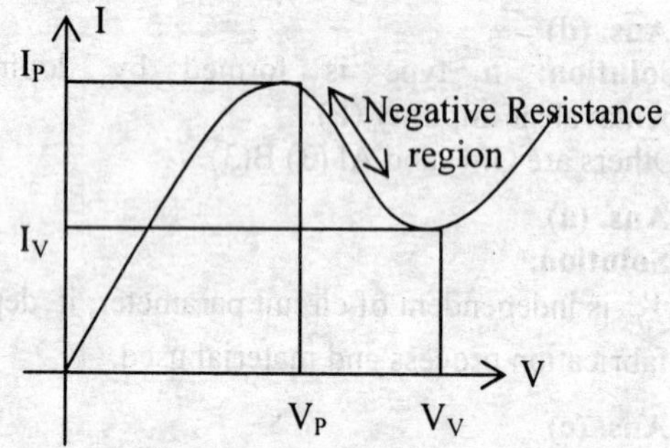

10. **Ans. (b)**

Solution: If it is extrinsic then $n + N_a = N_d + p$

for N-type-----$n = N_d$

So here $n.p = n_i^2$ $p = [n_i^2 / N_d]$

11. **Ans. (a)**

Solution: Under forward bias conduction the majority current flow i.e. drift current & under no bias conduction or low level injection the minority current flow i.e. diffusion current.

12. **Ans. (b)**

Solution: Due to Reverse bias of collector junction the depletion layer in base region increase & base width decreases

13. **Ans. (d)**

Solution: Intrinsic concentrations = n_i/cm^3

Acceptor (Hole) concentrations = N_A/cm^3 i.e.

$\Rightarrow \qquad P \approx N_A$

Electron concentrations n = ?

According to **mass action law,** $n_i^2 = n.\cdot p$

$\therefore \quad$ Electron concentrations $n = \dfrac{n_i^2}{P} = \dfrac{n_i^2}{N_A}$

14. **Ans. (d)**

Solution: When P^+ n junction diode is reverse biased the magnitude of field is maximum at Junction of P^+ n i.e. lightly doped side, i.e. the centre of the depletion region on the n-side

15. **Ans. (d)**

Solution:

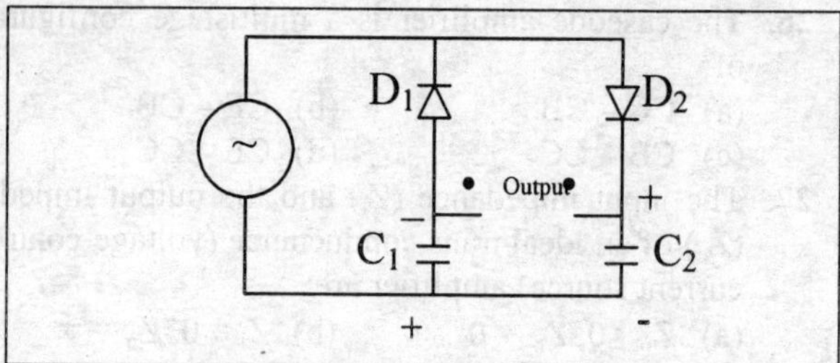

During the positive half cycle D_2 conducts due to C_2 charged to positive V volt. During the negative half cycle C_1 gets charged to Negative V volt. So that output voltage across AB is V + V = 2V Volt i.e. voltage doubler.

16. **Ans. (b)**

Solution:

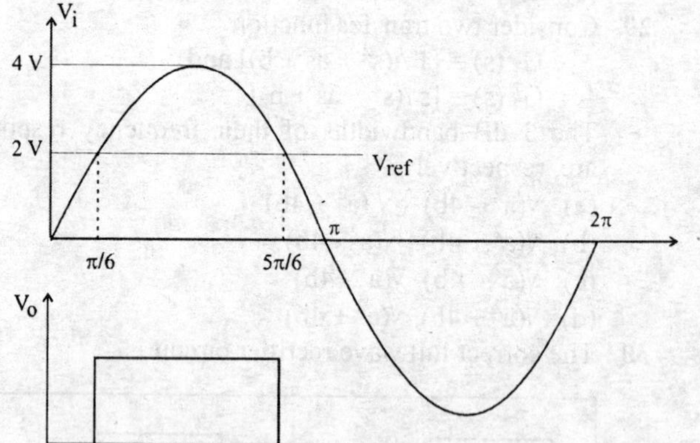

The output is positive only when the reference input is greater than 2 V.

$V_i = 8\, V_{PP}$ sinusoidal i.e.

$= 4 \sin \omega t$

At $V_i = 2 \Rightarrow \sin \omega t = \dfrac{2}{4} = \dfrac{1}{2}$

$\Rightarrow \omega t = \sin^{-1}\frac{1}{2} = \frac{\pi}{6}$

Another cross over at $\pi - \frac{\pi}{6} = \frac{5\pi}{6}$

Duty cycle $= \frac{T_{ON}}{T}$

$= \frac{\frac{5\pi}{6} - \frac{\pi}{6}}{2\pi} = \frac{\frac{4\pi}{6}}{2\pi}$

$= \frac{1}{3}$

17. Ans. (c)

Solution: CMMR = $[A_d / A_c]$, take the log both side

Log (CMMR) = log $[A_d / A_c]$

Or, 20 log (CMMR) = 20 log A_d – 20 log A_c

= 48- 2 = **46 db**

18. Ans. (a)

Solution: At high frequency strong internal capacitor come into existence which have considerable impedance.

19. Ans. (b)

20. Ans. (b) Voltage controlled voltage source

Solution:

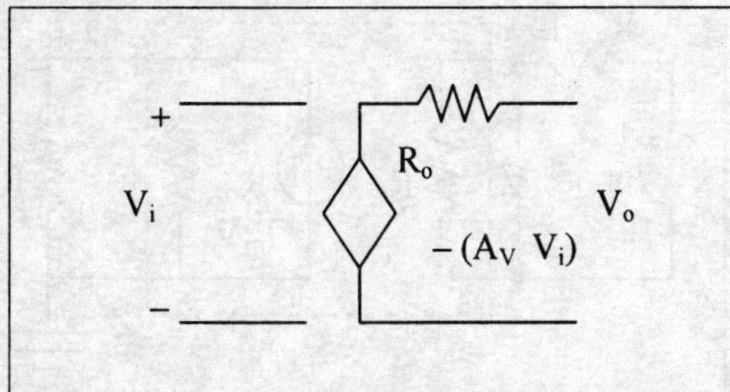

21. Ans. (c)

Solution: For voltage series feedback the input impedance increase by factor of (1 + AB) & output impedance decrease by a factor of [1 / (1 + AB)]

Where, A = Gain of amplifier

B = feedback factor

22. Ans. (a)

Solution: The combined effect of two resistor and two capacitor give a fall off rate of 40 db/decode. If is a second order LPF.

At $\omega = \infty$, $C \to$ short circuit, $\frac{V_o}{V_{in}} = 0$

At $\omega = 0$ c, $C \to$ open circuit, $\frac{V_o}{V_{in}} = 1$

23. Ans. (d)

Solution: If an amplifier has very high gain i.e. $A_V \to \infty$ then

$A_{vf} = [-R_2 / R_1]$ otherwise as below

Close loop Gain $A_{vf} = \frac{V_0}{V_i} = \frac{A_y}{1 + V_v K}$

Where, $K = \frac{R_1}{R_1 + R_f} = \frac{2}{2+8} = \frac{2}{10} = 0.2$

$$A_{Vf} = \frac{45}{1 + 45\,(0.2)} = \frac{45}{10} = 4.5$$

24. Ans. (d)

25. Ans. (b)

Solution:

The input resistance R_i is effective value of parallel combination of $10\,k$ and input impedance of amplifier, i.e.

$R_i = 10\,k \| \infty = 10\,k\Omega$

26. Ans. (b)

Solution:

Cascode connection

This CE-CB connection

- Common emitter stage drives a common base amplifier.
- Gain can be improved by using large value of R_L over a wide range of frequencies.

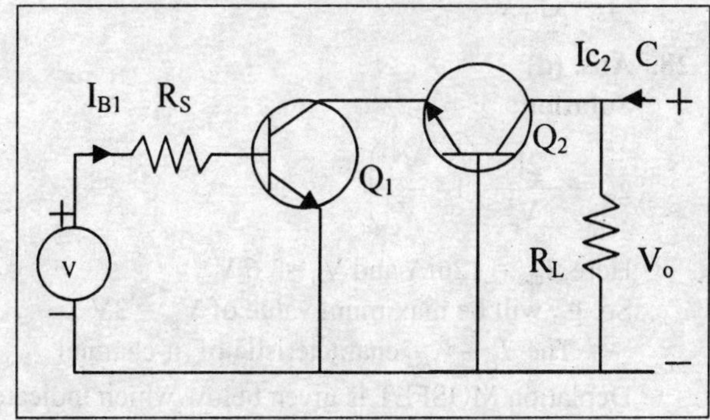

- Overall current gain $A_I - \frac{I_{C2}}{I_{B2}} = \beta_0$
- Voltage gain = $A_v = \frac{-\beta_0 R_L}{(R_s + r_{be})}$
- Output resistance of a CB stage is high, so we can connect a large load resistance R_L.
- This input resistance for the CB stage is low and this is the load resistance for the CE stage. The low input

resistance of CB stage makes Q_1 operate into a short-circuit load, so voltage gain of common emitters stage is low. This cause reduces the input capacitances so its Bandwidth gets increased.

It is used in *wide Band amplifiers* due to its large bandwidth.

27. Ans. (d)

Solution: The voltage - current convertor or trans-conductance Amplifier (Voltage controlled current source).

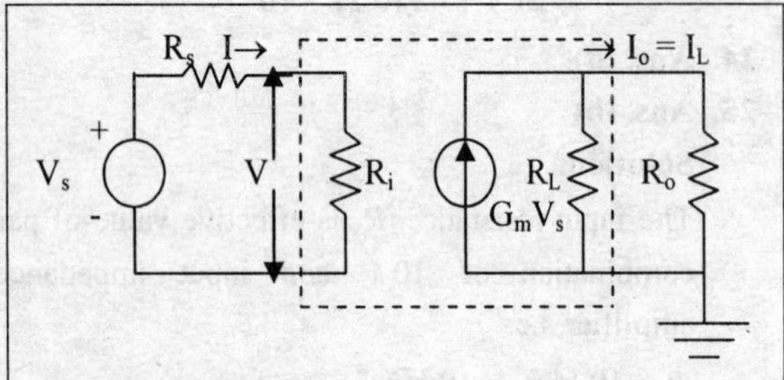

This is ideal voltage-controlled current source.

For a voltage controlled current source input is voltage so input impedance R_i is kept as high as possible i.e. $Z_i = \infty$. Output is current so output impedance R_o is kept high as possible so that max a full current can flow to load i.e. $Z_o = \infty$.

$|R_i| >> R_s$ and $R_L << |R_o|$

It mean, $Z_i = \infty$, $Z_o = \infty$

$I_o = G_m V_s$

28. Ans. (d)

Solution:

$$g_m = \frac{-2I_{dss}}{V_p}\left(1 - \frac{V_{gs}}{V_p}\right)$$

Here, I_{dss} = 12mA and V_p = - 6 V

So, g_m will be maximum value of V_{gs} = 3V

The $I_D - V_{GS}$ characteristic of n-channel Deplation MOSFET is given below which indicate that current is maximum for positive value of V_{GS}

$$g_m = \frac{\Delta I_D}{\Delta V_{GS}}$$

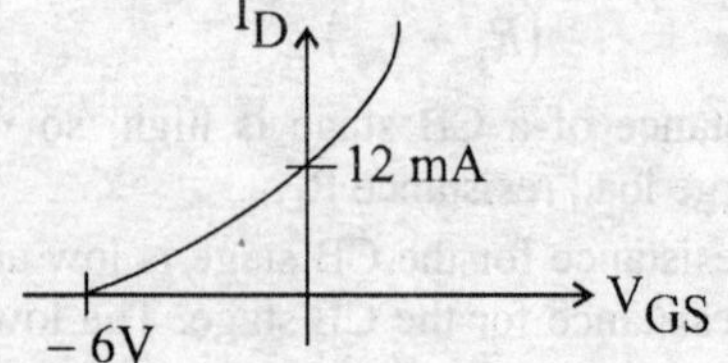

In can be observed from characteristic curve that the slop is maximum for higher positive value of V_{GS}.

Hence trans conductance is maximum at $V_{GS} = 3\,V$

29. Ans. (c)

Solution:

3 db BW is given by determinate of the characteristic equation for G_1 and G_2.

$$BW = \sqrt{a^2 - 4b}$$

30. Ans. (c)

Solution: Full wave rectifier: In all case of supply only two diode conduct and gives terminal A + Ve and terminal B –Ve.

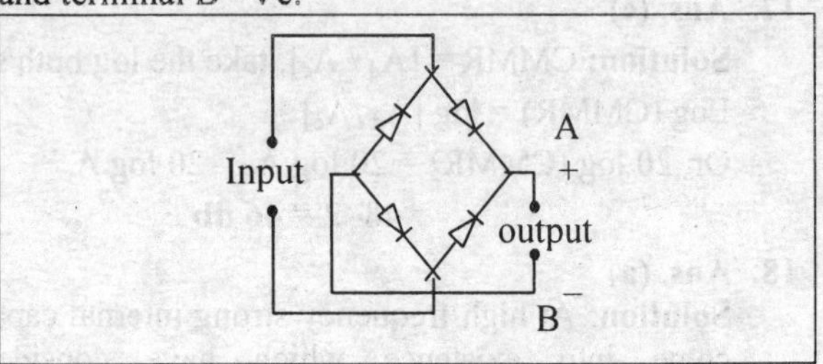

31. Ans. (a)

Solution: The voltage – current convertor or transconductance Amplifier is a Voltage controlled current source.

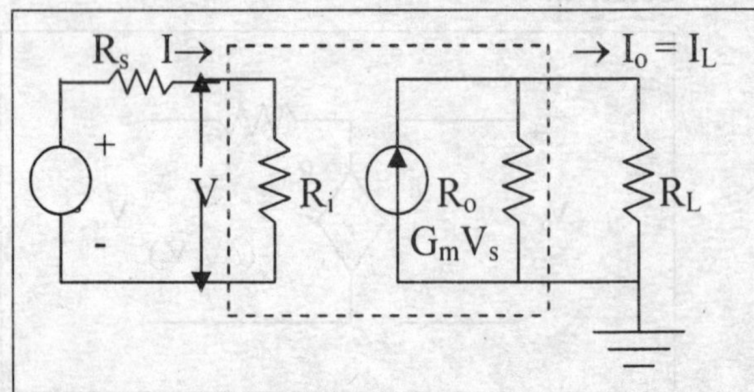

Transductions amplifier has voltage source at input end and current source at output end. Thus input resistance must be high and output resistance must be low.

$|R_i| >> R_s$ and $R_L << |R_o|$

$I_o = G_m V_s$

Two Marks Questions

1. An n-type silicon bar 0.1 cm long and 100 μm^2 in cross sectional area has a majority carrier concentration of 5 x 10^{20}/m^3 and the carrier mobility is 0.13m^2/Vs at 300 K. If the charge of an electron is 1.5 × 10^{-19} coulomb, then the resistance of the bar is:

(a) 10^6 ohm (b) 10^4 ohm
(c) 10^{-1} ohm (d) 10^{-4} ohm

2. Match items in Group 1 with items in Group 2, most suitably.

Group I	Group II
A. LED	1. Heavy doping
B. Avalanche photodiode	2. Coherent radiation
C. Tunnel diode	3. Spontaneous emission
D. LASER	4. Current gain

Codes:

	A	B	C	D
(a)	1	2	4	3
(b)	2	3	1	4
(c)	3	4	1	2
(d)	2	1	4	3

3. A particular green LED emits light of wavelength 5490 A. The energy band gap of the semiconductor material used there is
(Planck's constant= 6.626 x 10^{-34} J-s)
(a) 2.26 eV (b) 1.98 eV
(c) 1.17 eV (d) 0.74 eV

4. If P is Passivation, Q is n-well implant, R is metallization and S is source/drain diffusion, then the order in which they are carried out in a standard n-well CMOS fabrication process, is
(a) P-Q-R-S (b) Q-S-R-P
(c) R-P-S-Q (d) S-R-Q-P

5. The electron concentration in sample of uniformly doped n-type silicon at 300 K varies linearly from $10^{17}/cm^3$ at x = 0 to 6 × $10^{16}/cm^3$ at x = 2 μm. Assume a situation that electrons are supplied to keep this concentration gradient constant with time. If electronic charge is 1.6 x 10^{-19} coulomb and the diffusion constant D_n = 35 cm^2/s, the current density in the silicon, if no election field is present, is
(a) zero (b) – 1120 A/cm^2
(c) + 1120 A/cm^2 (d) – 1120 A/cm

6. At 300 K, for a diode current of 1 mA, a certain germanium diode requires a forward bias of 0.1435 V, whereas a certain silicon diode requires a forward bias of 0.718 V. Under the conditions stated above, the closest approximation of the ratio of reverse saturation current in germanium diode to that in silicon diode is
(a) 1 (b) 5
(c) 4 × 10^3 (d) 8 × 10^3

7. When the gate-to-source voltage (V_{GS}) of a MOSFET with threshold voltage of 400mV, working in saturation is 900mV, the drain current is observed to be 1 mA. Neglecting the channel width modulation effect and assuming that the MOSFET is operating at saturation, the drain current for an applied V_{GS} of 1400 mV is
(a) 0.5 mA (b) 2.0 mA
(c) 3.5 mA (d) 4.0 mA

8. The action of a JFET in its equivalent circuit can best be represented as a
(a) Current Controlled Current Source
(b) Current Controlled Voltage Source
(c) Voltage Controlled Voltage Source
(d) Voltage Controlled Current Source

9. The drain of an n-channel MOSFET is shorted to the gate so that $V_{GS} = V_{DS}$. The threshold voltage (V_T) of MOSFET is 1 V. If the drain current (I_D) is 1 mA for V_{GS} = 2V, then for V_{GS} = 3 V, I_D is
(a) 2 mA (b) 3 mA
(c) 9 mA (d) 4 mA

10. In an abrupt p-n junction, the doping concentrations on the p-side and n-side are N_A = 9 x 10^{16} / cm^3 respectively. The p-n junction is reverse biased and the total depletion width is 3 μm. The depletion width on the p-side is
(a) 2.7 μm (b) 0.3 μm
(c) 2.25 μm (d) 0.75 μm

11. The resistively of a uniform doped n-type silicon sample is 0.5 Ω-cm. If the electron mobility (μ_n) is 1250 cm^2/ V-sec and the charge of an electron is 1.6 x 10^{-19} Coulomb, the donor impurity concentration (N_D) in the sample is
(a) 2 x $10^{16}/cm^3$ (b) 1 x $10^{16}/cm^3$
(c) 2.5 x $10^{15}/cm^3$ (d) 5 x $10^{15}/cm^3$

12. Consider an abrupt p-junction. Let V_{bi} be the built-in potential of this junction and V_R be the applied reverse bias. If the junction capacitance (C_j) is 1 pF for $V_{bi} + V_R$ = 1 V, then for $V_{bi} + V_R$ = 4V, C_j will be
(a) 4pF (b) 2 pF
(c) 0.25 pF (d) 0.5 pF

13. Consider the following statement S1 and S2.
S1: The threshold voltage (V_T) of a MOS capacitor decreases with increase in gate oxide thickness
S1: The threshold voltage (V_T) of a MOS capacitor decreases with increase in substrate doping concentration
Which one of the following is correct?
(a) S1 is FALSE and S2 is TRUE
(b) Both S1 and S2 are TRUE
(c) Both S1 and S2 are FALSE
(d) S1 is TRUE and S2 is FALSE

14. The longest wavelength that can be absorbed by silicon, which has the band-gap of 1.12 eV, is 1.1 μm. If the longest wavelength that can be absorbed by another material is 0.87 μm, then the band-gap of this material is.

(a) 1.416 eV (b) 0.886 eV
(c) 0.854 eV (d) 0.706 Ev

15. The neutral base width of a bipolar transistor, biased in the active region, is o.5 μm. The maximum electron concentration and the diffusion constant in the base are $10^{14}/cm^3$ and D_n = 25 cm^2/ sec respectively. Assuming negligible recombination in the base, the collector current density is (the electron charge is 1.6 x 10^{-19} coulomb).
(a) 800 A/cm^2 (b) 8 A/cm^2
(c) 200 A/cm^2 (d) 2 A/cm^2

16. A Silicon sample A is doped with 10^{18} of Boron. Another sample B of identical dimensions is doped with 10^{18} atoms/cm^3 of Phosphorus. The ratio of electron to hole mobility is 3. The ratio of conductivity of the sample A to B is
(a) 3 (b) [1 / 3]
(c) [2 / 3] (d) [3 / 2]

17. A Silicon PN junction diode under reverse bias has depletion region of width 10 μm. The relative permittivity of Silicon , ε_r = 11.7 and the Permittivity of free space ε_0 = 8.85 x 10^{-12} F/m. The depletion capacitance of the diode per square meter is
(a) 100 μF (b) 10 μF
(c) 1 μF (d) 20 μF

18. For an n-channel MOSFET and its transfer curve shown in the figure, the threshold voltage is

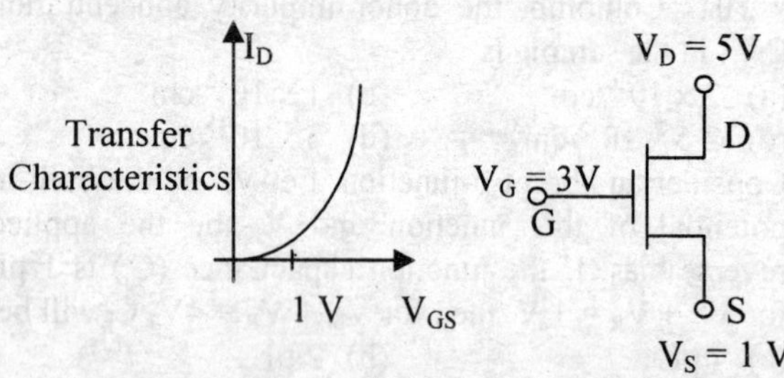

(a) 1 V and the device is in active region
(b) – 1 V and the device is in saturation region
(c) 1 V and the device is in saturation region.
(d) – 1 V and the device is in active region.

19. Find the correct match between Group 1 and Group 2.

Group 1	Group 2
A Varactor diode	1. Voltage reference
B F- PIN diode	2. High frequency switch
C G – Zener diode	3. Tuned circuits
C H- Schottky diode	4. Current controlled attenuator .

(a) A-4, B-2,C-1,D-3 (b) A-2,B-4,C-1,D-3
(c) A-3,B-4,C-1,D-2 (d) A-1,B-3,C-2,D-4

20. A heavily doped n-type semiconductor has the following data.
Hole – electron mobility ratio :0.4
Doping concentration : 4.2 x 10^8 atoms/m^3
Intrinsic concentration : 1.5 x 10^4 atoms /m^3
The ratio of conductance of the n- type semiconductor to that of the intrinsic semiconductor of same material and at the same temperature is given by.
(a) 0.00005 (b) 2,000
(c) 10,000 (d) 20,000

21. Croup I List four different semiconductor devices. Match each device in Group I with its characteristic property in Group II.

Group I	Group II
A. BJT	1. Population inversion
B. MOS capacitor	2. Pinch-off voltage
C. LASER diode	3. Early effect
D. JFET	4. Flat-band voltage

Codes:

	A	B	C	D
(a)	3	1	4	2
(b)	1	4	3	2
(c)	3	4	1	2
(d)	3	2	1	4

22. A P^+n junction has a built-in potential of 0.8 V. The depletion layer width at a reverse bias of 1.2 V is 2 μm. For a reverse bias of 7.2 V, the depletion layer width will be
(a) 4 μm (b) 4.9 μm
(c) 8 μm (d) 12 μm

23. The DC current gain (β) of a BJT is 50. Assuming that the **emitter injection efficiency** is 0.995, the **base transport factor is**.
(a) 0.980 (b) 0.985
(c) 0.990 (d) 0.995

24. In the amplifier circuit shown in figure the values of R_1 and R_2 are such that the transistor is operating at V_{CE} = 3 V and I_C= 1.5 mA when its β is 150. For a transistor with β of 200, the operating point (V_{CE} I_C) is:

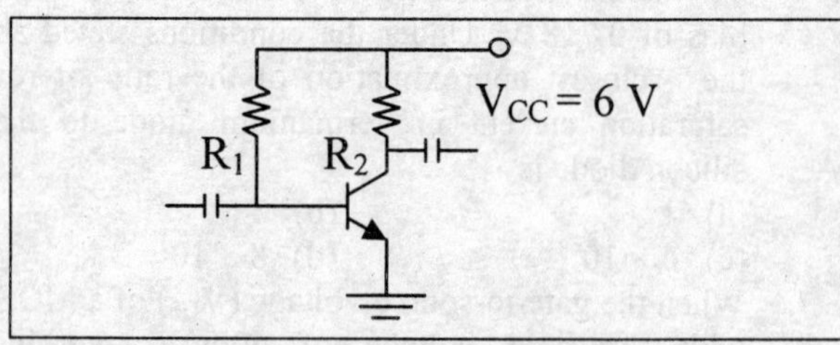

(a) 2V, 2 mA (b) 3V, 2mA
(c) 4V, 2mA (d) 4V, 1mA

25. The oscillator circuit shown in figure has an ideal inverting amplifier. Its frequency of oscillation (in Hz) is:

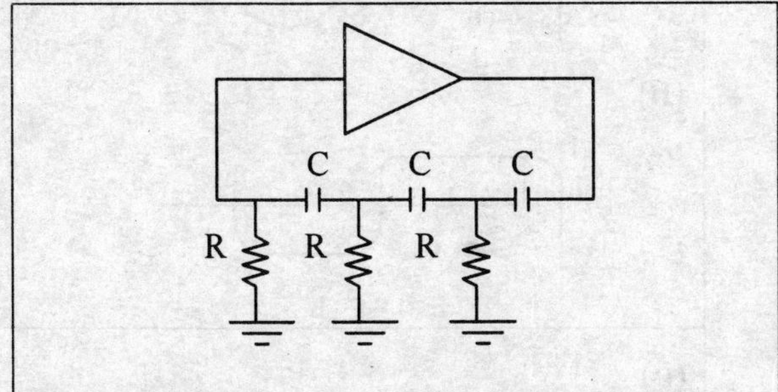

(a) $\frac{1}{(2\pi\sqrt{6}RC)}$ (b) $\frac{1}{(2\pi RC)}$

(c) $\frac{1}{(\sqrt{6}RC)}$ (d) $\frac{\sqrt{(6)}}{(2\pi RC)}$

26. The output voltage of the regulated power supply shown in figure is:

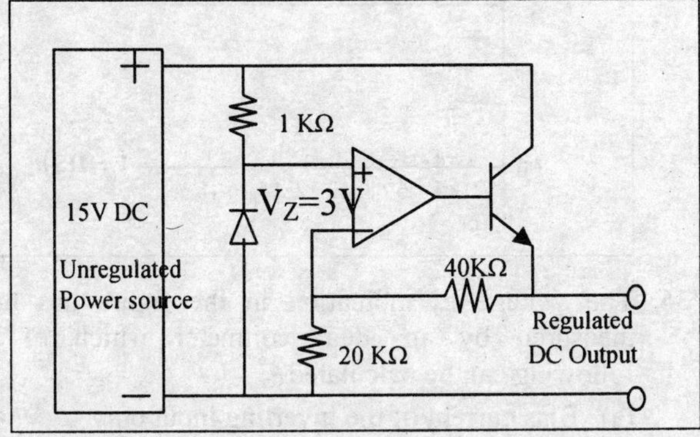

(a) 3 V (b) 6 V

(c) 9 V (d) 12 V

27. If the op-amp in figure is ideal, the output voltage V_{out} will be equal to

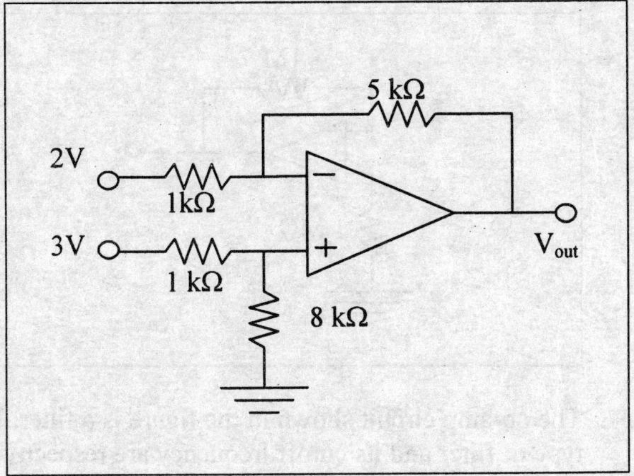

(a) 1 V (b) 6 V

(c) 14 V (d) 17 V

28. Three identical amplifiers with each one having a voltage gain of 50, input resistance of 1 kΩ and output resistance of 250Ω, are cascaded. The open circuit voltage gain of the combined amplifier is

(a) 49 dB (b) 51 dB

(c) 98 dB (d) 102 dB

29. An ideal saw-tooth voltage waveform of frequency 500 Hz and amplitude 3V is generated by charging a capacitor of 2 μF in every cycle. The charging requires.

(a) Constant voltage source of 3 V for 1 ms

(b) Constant voltage source of 3 V for 2 ms

(c) Constant current source of 3 mA for 1 ms

(d) Constant current source of 3 mA for 3 ms

30. A bipolar transistor is operating in the active region with a collector current of 1mA. Assuming that the β of the transistor is 100 and thermal voltage (V_T) is 25 mV, the trans-conductance (g_m) and the input resistance (r_π) of the transistor in the common emitter configuration, are

(a) g_m = 25 mA/V and r_π = 15.625 kΩ

(b) g_m = 40 mA/V and r_π = 4.0 kΩ

(c) g_m = 25 mA/V and r_π = 2.5 kΩ

(d) g_m = 40 mA/V and r_π = 2.5 kΩ

31. The value of C required for sinusoidal oscillations of frequency 1 KHz in the circuit of Figure is:

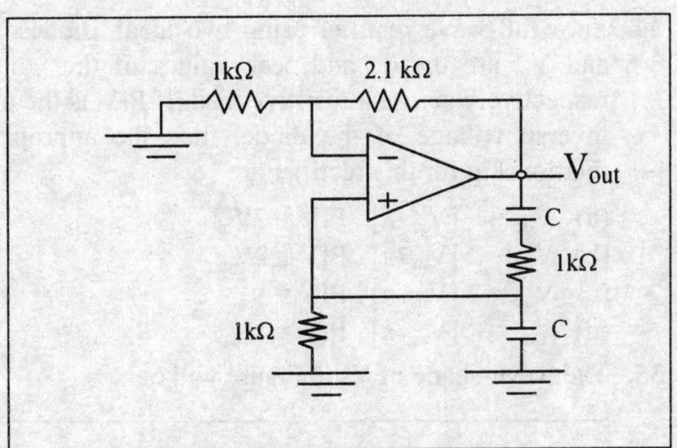

(a) 1 / 2π μF (b) 2π μF

(c) 1 / 2π √6 μF (d) 2π√6 μF

32. Assuming V_{CEsat} = 0.2 V and β = 50, the minimum base current (I_B) required to drive the transistor in Figure to a saturation is:

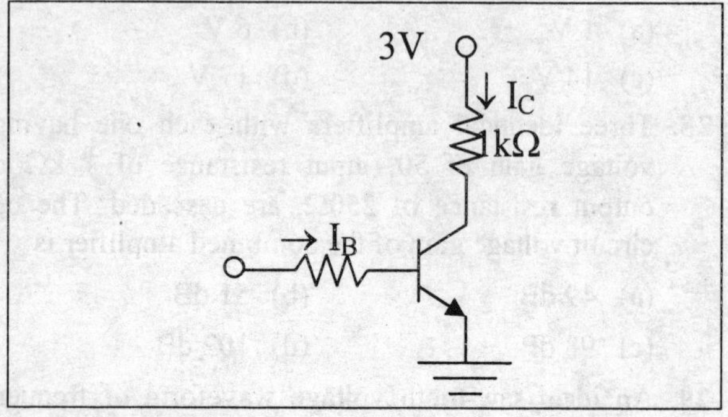

(a) 56 μA (b) 140 mA
(c) 60μA (d) 3mA

33. Assuming that the β of the transistor is extremely large and $V_{BE} = 0.7$ V, I_c and V_{CE} in the circuit shown in figure are:

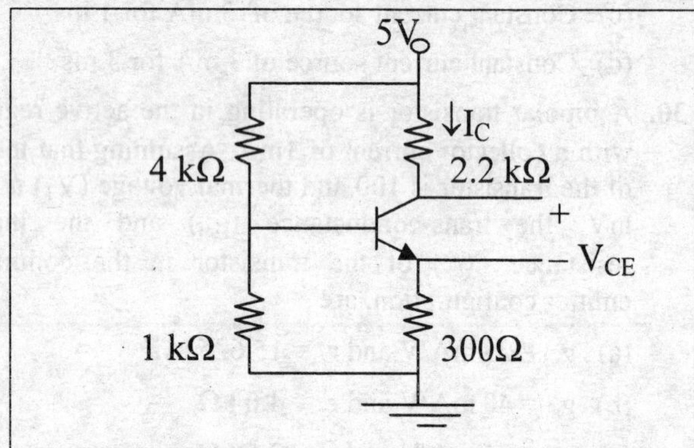

(a) $I_c = 1$ mA, $V_{CE} = 4.7$ V
(b) $I_c = 0.5$ mA, $V_{CE} = 3.75$ V
(c) $I_c = 1$ mA, $V_{CE} = 2.5$ V
(d) $I_c = 0.5$ mA, $V_{CE} = 3.9$ V

34. In a full-wave rectifier using two ideal, diodes, V_{dc} and V_m are the dc and peak values of the voltage respectively across a resistive load. If PIV is the peak inverse voltage of the diode, then the appropriate relationship for this rectifier are.

(a) $V_{dc} = [V_m/\pi]$, PIV = $2V_m$
(b) $V_{dc} = 2[V_m/\pi]$, PIV = $2V_m$
(c) $V_{dc} = 2[V_m/\pi]$, PIV = V_m
(d) $V_{dc} = [V_m/\pi]$, PIV = V_m

35. The wave-shape of V_0 in Figure will be

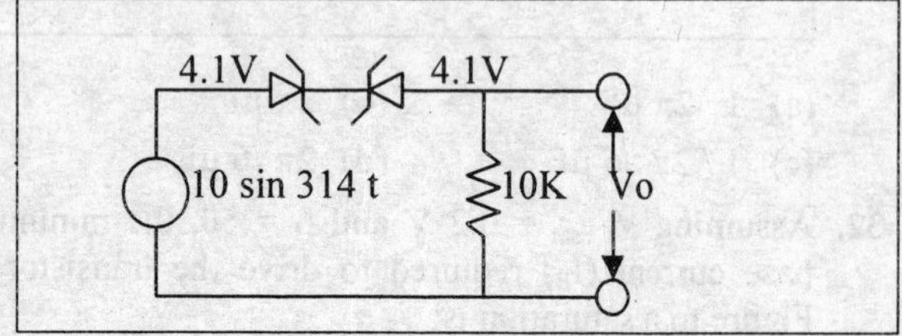

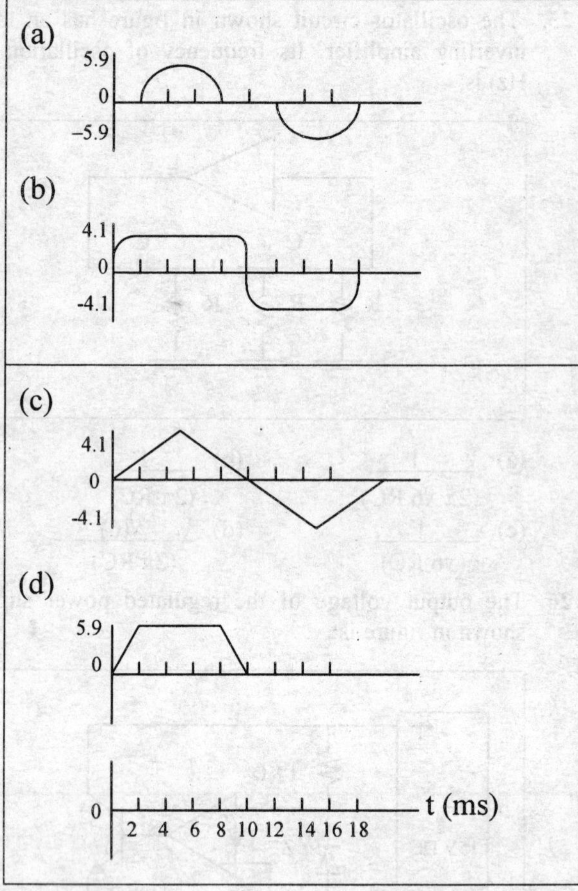

36. The voltage e_0 indication in the figure has been measured by an ideal voltmeter which of the following can be calculated?

(a) Bias current of the inverting input only
(b) Bias current of the inverting and non-inverting inputs only
(c) Input offset current only
(d) Both the bias currents and the input offset current.

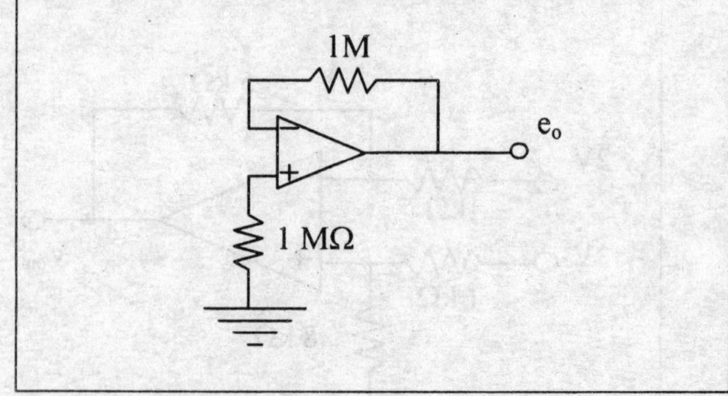

37. The op-amp circuit shown in the figure is a filter. The type of filter and its cutoff frequency are respectively.

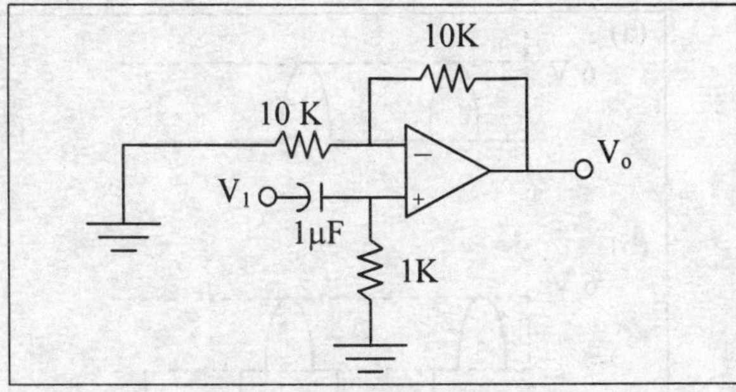

(a) High pass, 1000 rad/sec
(b) Low pass, 1000 rad/sec
(c) High pass, 10000 rad/sec
(d) Low pass, 10000 rad/sec

38. In an ideal differential amplifier shown in the figure, a large value of (R_E).

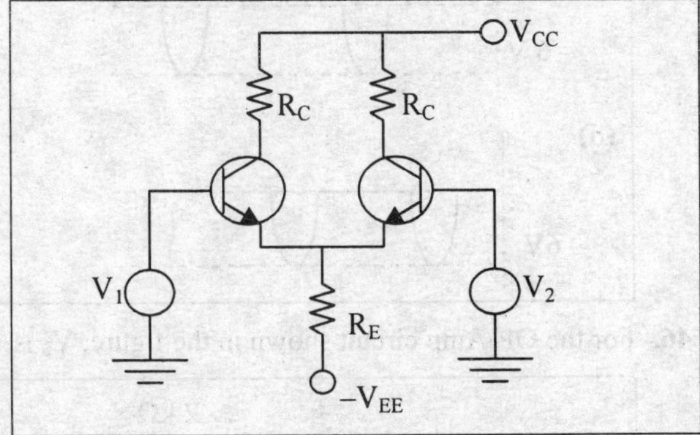

(a) Increase both the differential and common-mode gains
(b) Increases the common-mode gain only
(c) Decreases the differential-mode gain only
(d) Decreases the common-mode gain only

39. The circuit using a BJT with β = 50 and V_{BE} = 0.7 V is shown in the figure The base current I_B and collector voltage by V_C are respectively.

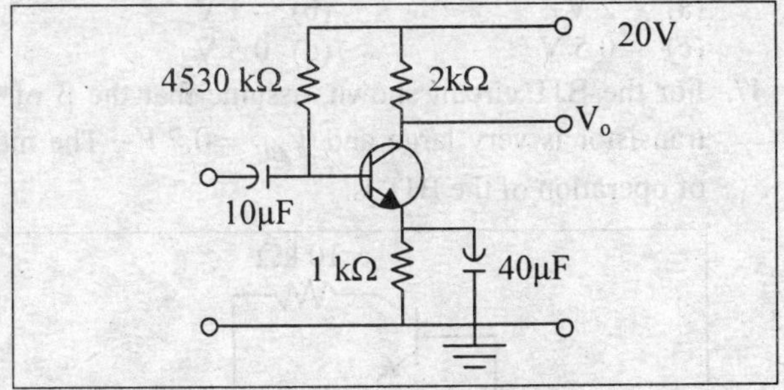

(a) 43 μA and 11.4 Volts (b) 40 μA and 16 Volts
(c) 45 μA and 11 Volts (d) 50 μA and 10 Volts

40. The zener diode in the regulator circuit shown in the figure has a zener voltage of 5.8 Volts and a zener knee current of 0.5 mA. The **maximum load current drawn** from this circuit ensuring proper functioning over the input voltage range between 20 and 30 volts, is:

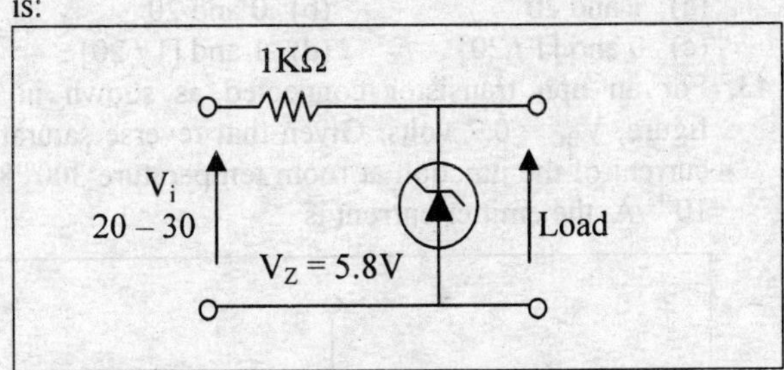

(a) 23.7 mA (b) 14.2 mA.
(c) 13.7 mA (d) 24.2 mA.

41. Given the ideal operational amplifier circuit shown in the figure, indicate the correct transfer characteristic assuming ideal diodes with zero cut-in voltage.

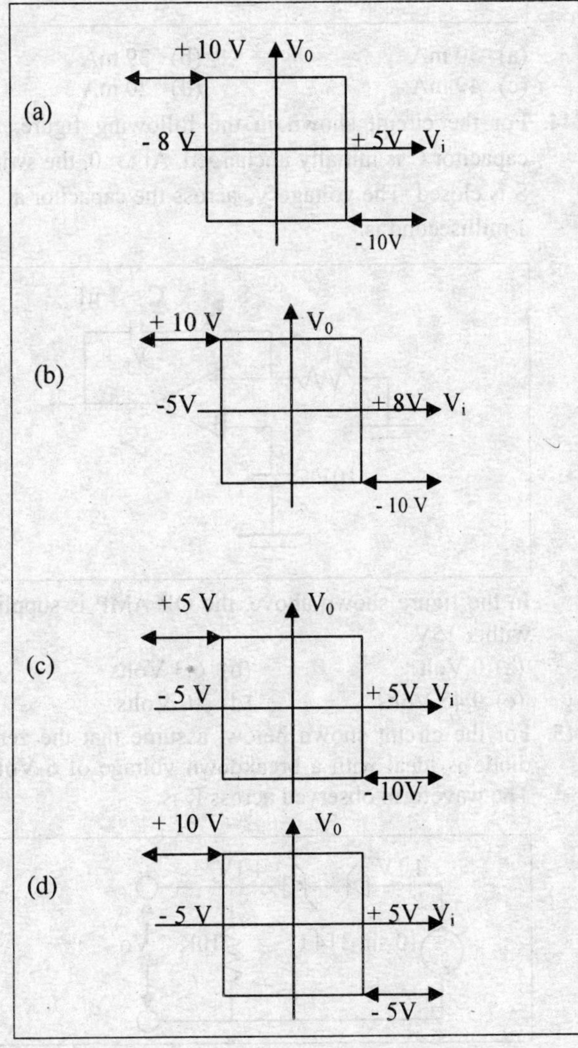

42. A ramp input applied to an unity feedback system results in 5 % steady state error. The type number and zero frequency gain of the system are respectively.

(a) 1 and 20 (b) 0 and 20
(c) 0 and [1 / 20] (d) 1 and [1 / 20]

43. For an npn transistor connected as shown in the figure, V_{BE} = 0.7 volts. Given that reverse saturation current of the junction at room temperature 300° K is 10^{-13} A, the emitter current is

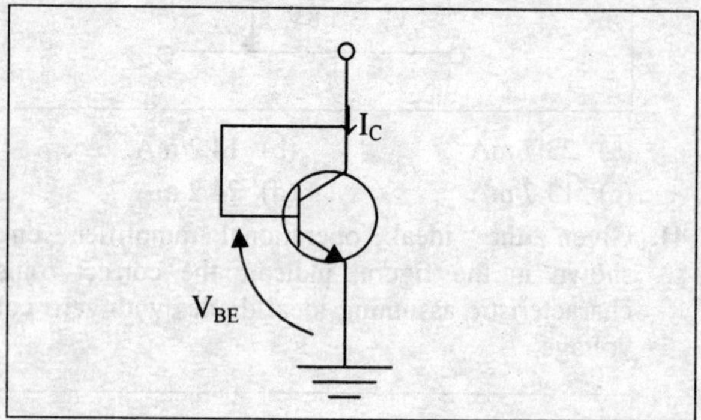

(a) 30 mA (b) 39 mA
(c) 49 mA. (d) 20 mA.

44. For the circuit shown in the following figure, the capacitor C is initially uncharged. At t = 0, the switch S is closed. The voltage V_c across the capacitor at t = 1 millisecond is:

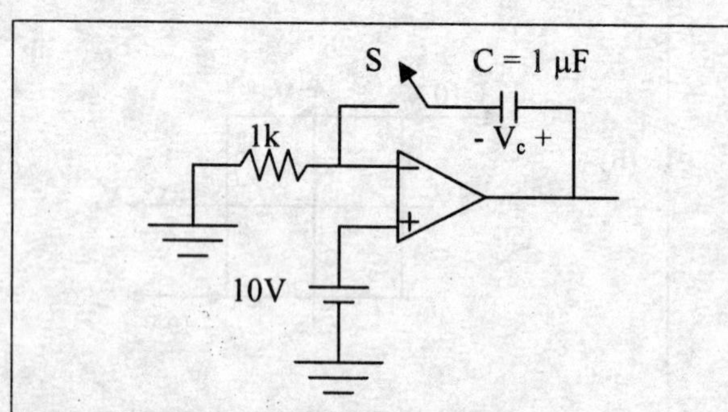

In the figure shown above, the OP AMP is supplied with ± 15V.

(a) 0 Volt (b) 6.3 Volts
(c) 9.45 Volts (d) 10 Volts

45. For the circuit shown below, assume that the zener diode is ideal with a breakdown voltage of 6 Volts. The waveform observed across R is:

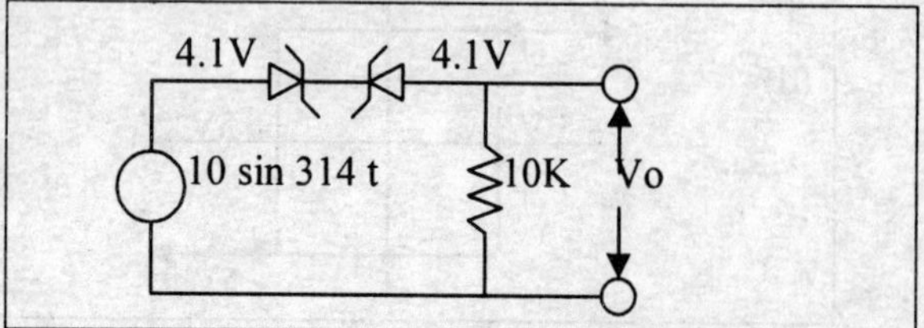

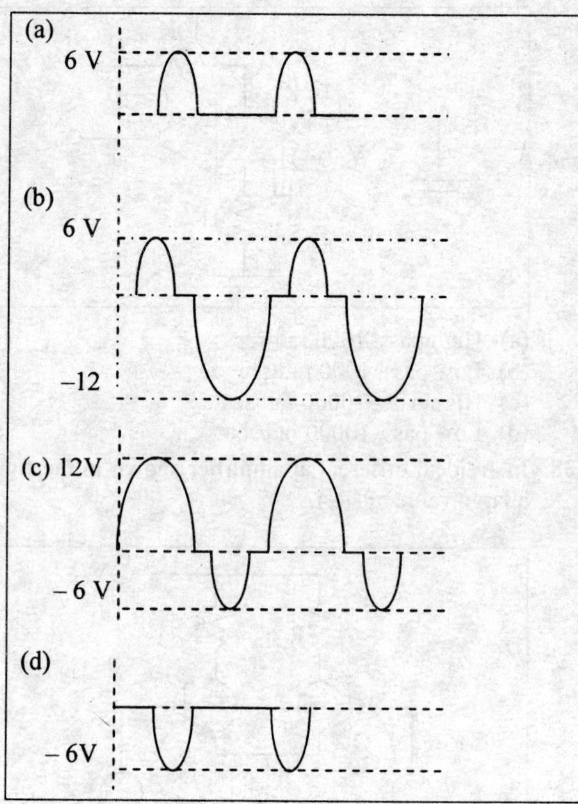

46. For the OP-Amp circuit shown in the figure, V_0 is

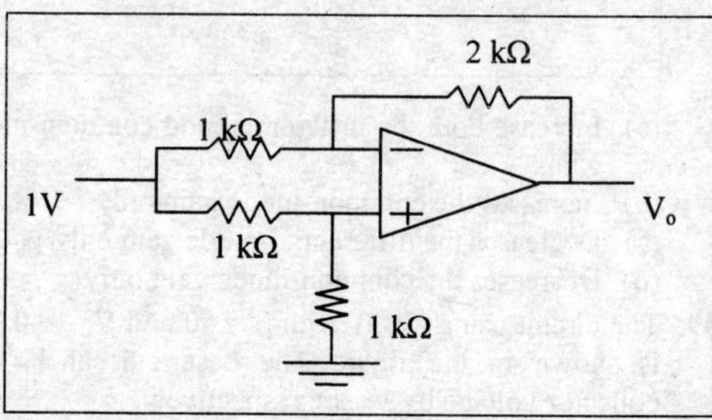

(a) – 2 V (b) – 1 V
(c) – 0.5 V (d) 0.5 V

47. For the BJT circuit shown, assume that the β of the transistor is very large and $V_{BE} = 0.7\,V$. The mode of operation of the BJT is:

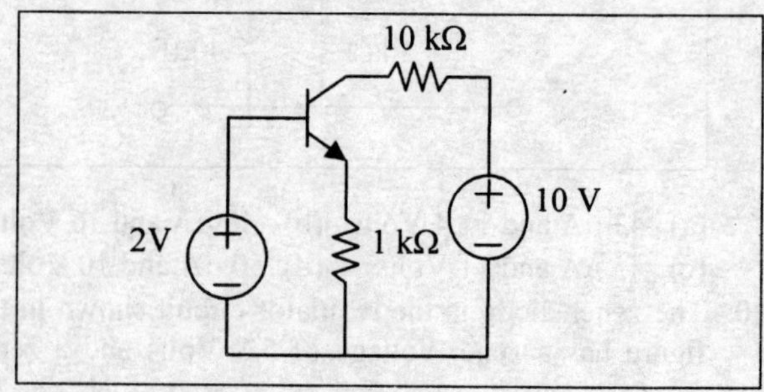

(a) Cut-off (b) Saturation
(c) Normal active (d) Reverse active

48. In the Op-Amp circuit shown, assume that the diode current follows the equations $I = I_s e^{(V/V_T)}$. For $V_i = 2V$, $V_o = V_{01}$, and $V_i = 4V$, $V_o = V_{02}$. The relationship between V_{o1} and V_{o2} is:

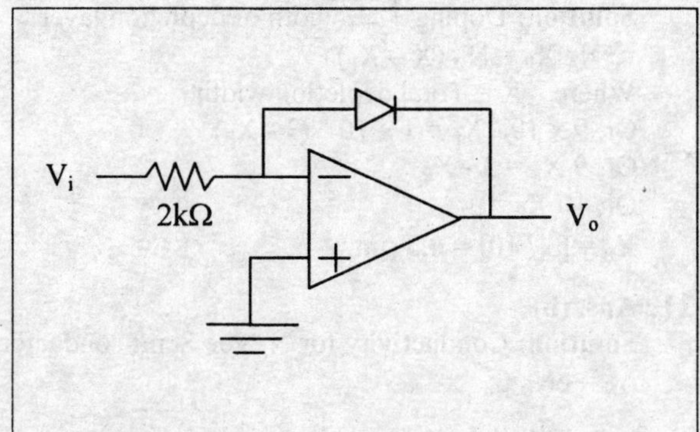

(a) $V_{o2} = \sqrt{2}\, V_{o1}$ (b) $V_{o2} = e^2\, V_{o1}$
(c) $V_{o2} = V_{o1} \ln 2$ (d) $V_{o1} - V_{o2} = V_T \ln 2$

49. In the CMOS inverter circuit shown, if the trans-conductance parameters of the NMOS and PMOS transistors are

$$k_n = k_p = \mu_n C_{ox} \frac{W_n}{L_n} = \mu_p C_{ox} \frac{W_p}{L_p} = 40\ \mu A/V^2$$

and their threshold voltage are $V_{THn} = |V_{THp}| = 1V$, the current I is

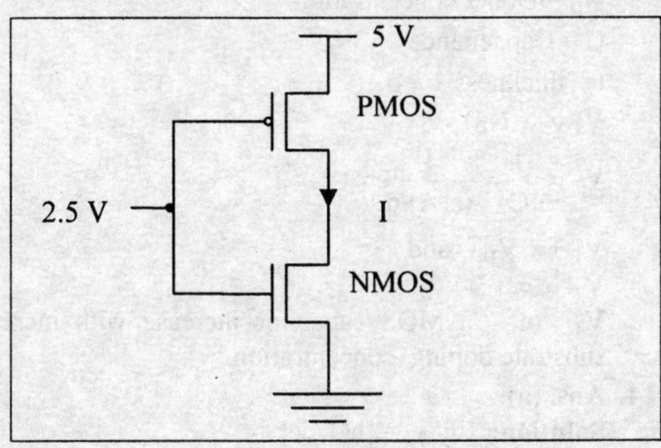

(a) 0 A (b) 25 μA
(c) 45 μA (d) 90 μA

50. For the Zener diode shown in the figure, the Zener voltage at knee is 7 V, the knee current is negligible and the Zener dynamic resistance is 10Ω. If the input voltage (V_i) range is from 10 to 16 V, the output voltage (V_o) ranges from.

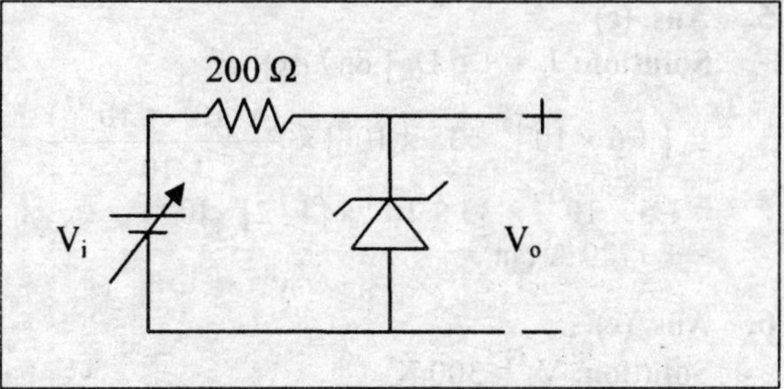

(a) 7.00 to 7.29 V (b) 7.14 to 7.29 V
(c) 7.14 to 7.43 V (d) 7.29 to 12.3

Answers
Analog Electronics

1. Ans. (a)

Solution: Conductivity σ = n e μ$_n$ + p e μ$_p$

For N type σ = n e μ$_n$

Resistively ∫ = [1 / n e μ$_n$]

Resistance of the bar =[ρ ℓ /A] = [ℓ / σ . A]

[ℓ / n e μ$_n$. A]

$$= \frac{10^{-3}}{(5 \times 10^{20} \times 1.6 \times 10^{-19} \times 0.13) \times 100 \times 10^{-6} \times 10^{-6}}$$

= 10^6 ohm

2. Ans. (c)

Solution: The correct match is given below:

LED	Spontaneous emission
Avalanche photodiode	Current gain
Tunnel diode	Heavy doping
LASER	Coherent radiation

3. Ans. (a)

Solution: λ = [C / f] ; f = [C / λ]

$$E = hf = hC/\lambda = \frac{6.625\times10^{-34}\times3\times10^{8}}{5490\times10^{-10}} \text{ Jules}$$

= 3.62 x 10^{-19} Joules.

In eV, the energy Band Gap Energy = [E / e]

$$= \frac{3.62\times10^{-19}}{1.6\times10^{-19}}$$

= 2.26 eV

4. Ans. (b)

Solution:

In n-well CMOS fabrication following are the steps

1. n- well implant
2. Source /drain diffusion
3. Metalization
4. Passivation

5. Ans. (c)

Solution: $J_n = + q D_n [dn / dx] \cong$

$$\cong [1.6 \times 10^{-19} \times 35 \times 10^{-4}] \times \frac{(6\times10^{16} - 10^{+17})}{2\times10^{-6}}$$

$= 1.6 \times 10^{-19} \times 35 \times 10^{-4} \times [4 / 2] \times 10^{22}$

$= + 1120$ A/cm^2

6. Ans. (c)

Solution: $V_T = 300$ K

I = 1 mA

For Si, $I = I_oSi [\exp V / nV_T - 1]$; n= 2 for Si

$$1 \text{ mA} . = I_oSi \left[\exp\frac{0.718}{2\times0.03} - 1\right]$$

$$\text{For, Ge I} = I_oGe \left[\exp\frac{0.1435}{2\times0.03} - 1\right]$$

$$\frac{I_oGe}{I_oSi} = I_oGe \exp\left(\frac{0.1435}{1\times0.03} - 1\right)$$

$$= I_oSi \left(\exp\frac{0.718}{1\times0.03} - 1\right)$$

$$\frac{I_oGe}{I_oSi} = \frac{\exp[0.718/0.06]}{\exp[0.435/0.03]} \cong 4\times10^3$$

7. Ans. (d)

Solution:

Given that,

For $V_T = 400\, mV$

$I_D = 1\, mA$

$V_{gs} = 900\, mV$

For $V_{gs} = 1400\, mV$

$I_D = ?$

Using, $I_D = K(V_{gs} - V_T)^2$

$$\frac{1\times10^{-3}}{I_D} = \left(\frac{900-400}{1400-400}\right)^2$$

$$\frac{1\times10^{-3}}{I_D} = \frac{1}{4}$$

$I_D = 4\, mA$

8. Ans. (d)

Solution:

For a JEET in active region we have

$$I_{DS} = I_{DSS}\left(1 - \frac{V_{GS}}{V_P}\right)^2$$

From above equation it is clear that the action of a JFET is voltage controlled current source.

9. Ans. (c)

Solution: Boron is an atom of Group III element, Group III elements are used for making p-type series conducts.

10. Ans. (b)

Solution: Doping ↑ ⇒ width of depletion layer ↓

$\therefore N_A X_P = N_D (X - X_P)$

Where, X = Total depletion width

Or, $9 \times 10^{16} X_P = 1 \times 10^{16} (3 - X_P)$

Or, $9 X_P = 3 - X_P$

Or, $10 X_P = 3$

$X_P = [3 / 10] =$ **0.3 μm**

11. Ans. (b)

Solution: Conductivity for N-type Semi conductor is

$\sigma_N = eN_D \mu_n$

$$N_D = \frac{\sigma_N}{\mu_n e} = \frac{1}{0.5 \times 1.6 \times 10^{-19} \times 1250}$$

$= \mathbf{1 \times 10^{16} /cm^3}$

12. Ans. (d)

Solution: $C_j \propto [1 / \sqrt{(V_{bi} + V_R)}]$

$= \text{i.e.} \propto [1 / \sqrt{V_j}]$

$[C_2 / C_1] \propto \sqrt{(1 / 4)}$

$\Rightarrow C_2 =$ **0.5 pF**

13. Ans. (c)

Solution: For N-Channel MOSFET

V_T = Threshold voltage

N_D = Doner concentration

C = Capacitance

t = thickness = d

$V_T \propto \sqrt{(N_D)}$

$$V_T \propto \frac{1}{C} = \frac{1}{(\varepsilon_o A)/d} \propto t$$

$V_T \uparrow \propto V_D \uparrow$ and

$V_T \downarrow \propto t \downarrow$

V_T of a MOS capacitor increase with increase substrate doping concentration.

14. Ans. (a)

Solution: $E_{g(ev)} = [hC / \lambda]$

$$E_{g\,(ev)} = \frac{12400}{\lambda\,(A^o)} = \frac{12400}{0.8700 \times 10^{-6} \times 10^{-4}}$$

= 1.416 eV

Second Method

$$E = \frac{hc}{\lambda} = \frac{6.626\times10^{-34} \times 3\times10^8}{0.87\times10^{-6}}$$

$= 22.85\times10^{-20}$ *Joules*

$$E = \frac{22.85\times10^{-20}}{1.6\times10^{-19}} eV = 1.42\, eV$$

15. Ans. (b)

Solution: Diffusion current density

J_n = e. D_n . [dn / dx]

Where, dx =0.5μm = 0.5 x 100 μcm

dn = 10^{14} / cm^3

D_n = 25 cm^2/s

e = 1.6 x 10^{-19} C

J_n = e. D_n . [dn / dx]

$= 1.6 \times 10^{-19} \times 25 \times \frac{10^{14}}{(0.5\times10^{-6})\times10^{2}}$

$= 1.6 \times 50 \times 10^{-1} = 8$ A/m^2

16. Ans. (c)

Solution: E_g(Si) = 1.21 – 3.6 x 10^{-4} T (290)

= 1.1 eV

E_g (Ge) = 0.785 – 2.23 x 10^{-4} T(290)

= 0.76 eV

17. Ans. (b)

Solution: $C = \frac{\in A}{d} = \frac{\in_0 \in_r A}{d}$

$[C / A] = \in_0 \in_r = \frac{11.7\times8.85\times10^{-12}}{10\times10^{-6}}$

[C / A] = 10.35 μF

18. Ans. (c)

Solution:

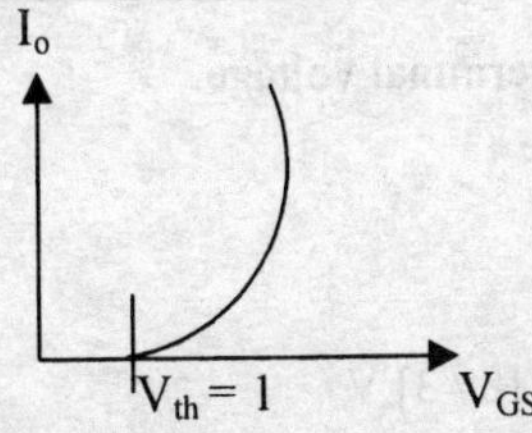

V_{DS} = 5 – 1 = 4 ; V_{GS} = 3 –1 = 2

V_{GS} – V_{th} = 2 – 1 = 1

In present case: V_{DS} >, V_{GS} i.e. 4 > 1

So device is in saturation

19. Ans. (c)

Solution: The correct match is given below

Varactor diode	Tuned circuits
F- PIN diode	Current controlled attenuator.
G- Zener diode	Voltage reference
H-Schottky diode	2High frequency switch

20. Ans. (d)

Solution:

$$\frac{\sigma_{N\text{-type}}}{\sigma_{intrinsic}} = \frac{n_e\, \mu_e}{n_i\mu_i + n_i\mu_h}\left(\frac{\mu_h}{\mu_e} = 0.4\right)$$

$$= \frac{4.2 - 10^8\, \mu_e}{1.5 \times 104\, \mu_e + 1.5\times10^4 \mu_h} = 20,000$$

21. Ans. (c)

Solution:

BJT → Early effect (base width modulation)

MOS Capacitor → Flat-band voltage

Laser diode → Population inversion

JFET → Pinch off voltage

22. Ans. (a)

Solution: P^+n junction built in volt = 0.8 V

for applied 1.2 V reverse bias, width = 2 μm

Diode voltage = $V_a + V_b$

Where, V_a = applied volt

V_b= built in volt

Also $w \propto \sqrt{V_{diode}}$

$$\therefore \frac{w_1}{w_2} = \sqrt{\frac{V_1}{V_2}}$$

$w_1 = 2\, \mu m,\ w_2 = ?$

$V_1 = 1.2 + .8 = 2V$

$V_2 = 7.2 + .8 = 8V$

$$\therefore \frac{2\times10^{-6}}{w_2} = \sqrt{\frac{2}{8}}$$

$\therefore w_2 = 4\, \mu m$

23. Ans. (d/b)

Solution: DC current gain $\beta = 50$

Emitter injection efficiency = 0.995

base transportation factor = ?

$\Rightarrow$ Emitter efficiency $= \frac{I_{PE}}{I_{ETOT}}$

Base transportation factor, $P_T = \frac{I_{PC}}{I_{PE}}$

α = Emitter efficiency × Base transportation factor

$$\therefore \beta_T = \frac{\alpha}{0.995} = \frac{\beta}{1+\beta} \times \frac{1}{0.0995}$$

$$= \frac{50}{51} \times \frac{1}{0.995} = 0.980$$

24. Ans. (a)

Solution: $V_{CC} - I_C R_2 = V_{CE}$

$R2 = \frac{(6-3)}{1.5\text{ mA}} V = 2K\Omega$

$I_C = I_b\,\beta + (1+\beta)\,I_{CO}$

$I_{B1} = \frac{I_{C1}}{\beta}$

$= \frac{1.5\text{ mA}}{150} = 0.01\text{ mA}$

When, $\beta = 200$

I_B remains same, I_C become $= \beta_2\, I_B$

$= 200 \times [1.5 / 150]$ mA

$V_{CC} - I_{C2}\,R_2 = V_{CE}$

$V_{CE2} = 6 - 2\;10^{-3} \times 2 \times 10^3$

$= 6 - 4 = 2V$

This operating point is (2V, 2mA)

25. Ans. (a)

Solution: $F = [1 / 2\pi\, RC\, \sqrt{6}]$

This is RC phase shift AF oscillator

26. Ans. (c)

Solution:

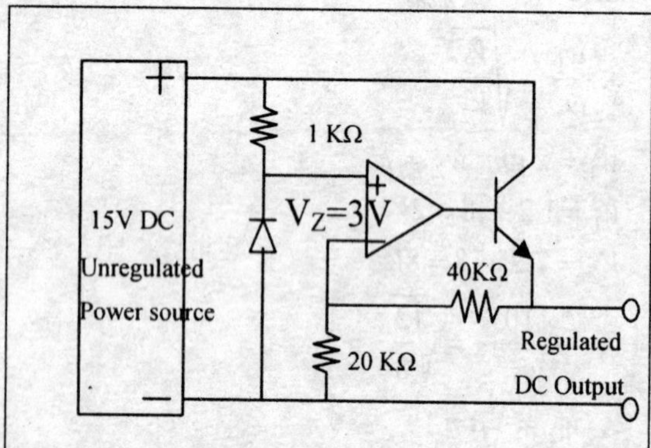

In an amp series regulator the o/p is

$V_o = \left(1 + \frac{R_1}{R_2}\right) Vz$

$= \left(1 + \frac{40}{20}\right) \times 3 = 9V$

Other way, The voltage at op-amp input

$V_+ = V_- = V_z = 3$ Volt

Then, by voltage division rule

$\left(\frac{20}{20+40}\right) V_o = 3,\; V_o = 9V$

Another Approach

As voltage at Non inverting terminal is 3 V due to Zener diode, voltage at inverting terminal will be same as 3 V because of virtual ground concept.

$\therefore$ Current through 20 kΩ resister is

$\frac{3}{20\,k} = \frac{2}{20}\,mA$

$\therefore\; V_o = \frac{3}{20\,k} \times 60\,k = 9\,V$

27. Ans. (b)

Solution:

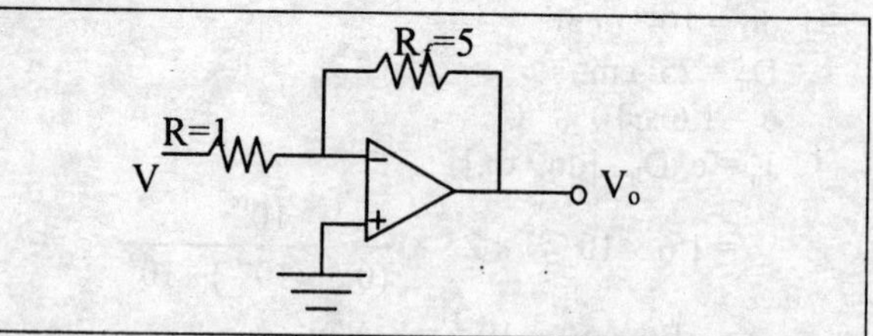

Output due to –Ve terminal voltage V_{01} =

$\frac{R_f}{R_1} \times V = \frac{-5}{1} \times 2 = -10V$

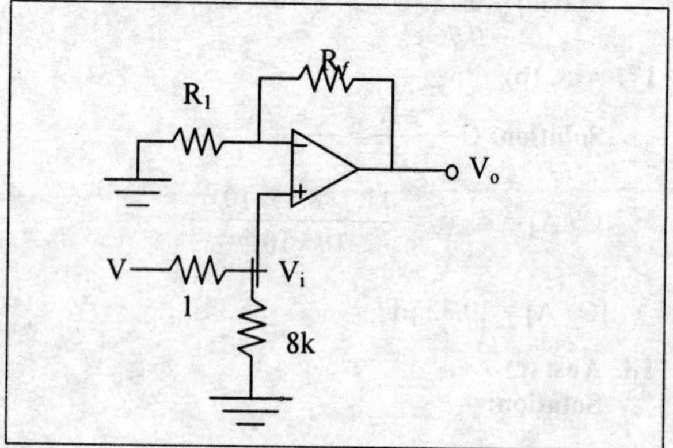

Output due to + Ve terminal voltage

$V_{02} = \left(1 + \frac{R_f}{R}\right) V_i$

in present case,

$V_i = [8 / 8 + 1] \times 3 = [8 / 3]$ V

$V_{02} = \left(1 + \frac{R_f}{R}\right) V_i$

$= [1 + 5 / 1] \times [8 / 3]$

$= 6 \times [8 / 3] = 16V$

V_o (Net o/p) $= V_{02} + V_{01} = 16 - 10 = 6V$.

28. Ans. (d)

Solution: The combined gain of the cascaded amplifier

$A_V = A_{V1} \cdot A_{V2} \cdot A_{V3}$

$= (A_V)^3 = (50)^3$

$[Ad_{db}] = 20 \log A_V = 20 \log (50)^3$

$= 60 \log 50 =$ **102 db**

29. Ans. (d)

Solution:

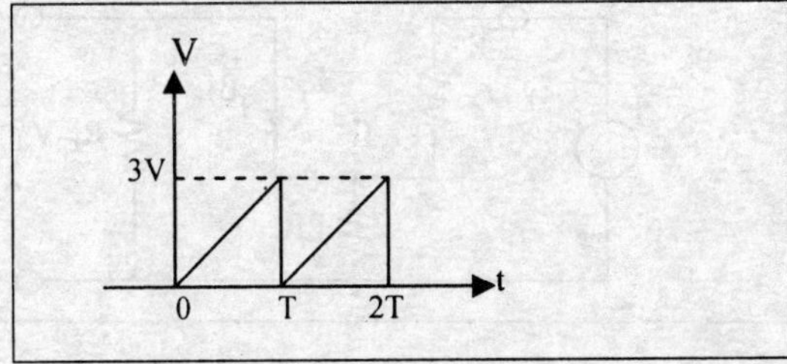

$V_o = \frac{1}{C}\int_0^t i\,dt$: $T = \frac{1}{f} = \frac{1}{500} = 2ms$

If V is constant, then i = c [dV / dt] = 0

dV = 0 → 3V

Otherwise $i = \frac{3 \times 2 \times 10^{-6}}{2 \times 10^{-3}}$ 3mA

Therefore constant current source of 3 mA, for 2ms required.

30. Ans. (d)

Solution:

$$g_m = \frac{|I_C|}{V_T} = \frac{1\ mA}{25 \times (1000)^{-1}} = 40\ mA/V$$

Using relation between low frequency and high frequency parameter

$h_{fe} = g_m r_\pi$

But, $h_{fe} = \beta$

$$\therefore\ r_\pi = \frac{\beta}{g_m} = \frac{100}{40\ mV} = 2.5\ k\Omega$$

31. Ans. (a)

Solution: The given figure represent Wein Bridge oscillator

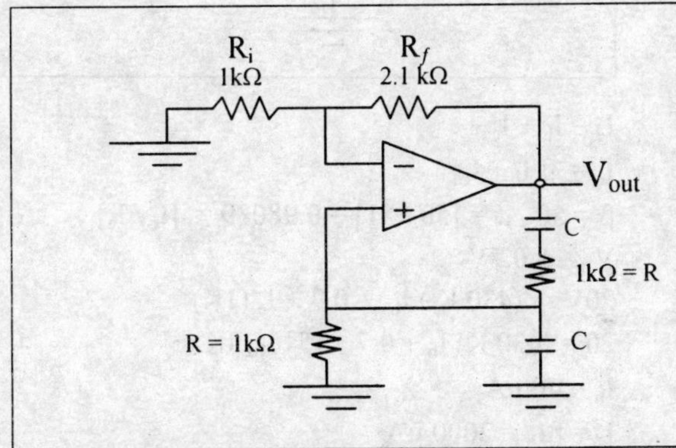

For Wein Bridge oscillator

Frequency of oscillation $f_o = [1 / 2\pi RC]$

The ratio of $[R_f / R_i] \simeq 2$

Gain of amplifier $A_V \geq 3$

$f_o = [1/2\pi RC]$

$$\Rightarrow\ C = \frac{1}{2\pi} \times \frac{1}{1 \times 10^3 \times 1 \times 10^3} = \frac{1}{2\pi}\mu F$$

32. Ans. (a)

Solution: $\beta = h_{fe} = 50$

$V_{CC} = I_C \cdot R_C + V_{CE\ sat}$

$I_C = h_{fe} \cdot I_B$

$$\Rightarrow\ I_C = \frac{3 - 0.2}{1k} = 2.8\ mA$$

$$I_B = \frac{2.8\ mA}{50} = 0.056\ mA = 56\ \mu A$$

33. Ans. (c)

Solution:

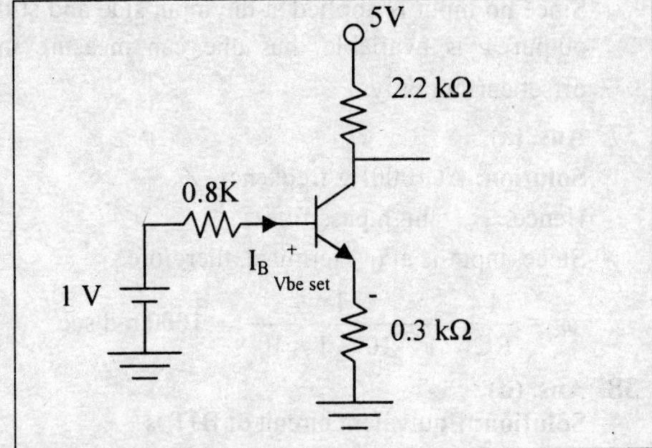

$$R_{th} = \frac{1 \times 4}{1 + 4} = 0.8\ k\Omega$$

V_{th} = [`1 / 5] x 5 = 1 V

KVL in base loop

-1+ 0.8 x I_b + $V_{be\ sat}$ + 0,3 x I_C = 0

I_b → = 0 as β is very large

0.3 I_C = 1 – 0.7 = 0.3

I_C = 1 mA

$V_{CE\ sat} = 5 - 2.2 \times 1 - -0.3 \times 1 = 2.5\ V$

34. Ans. (b)

35. Ans. (a)

Solution:

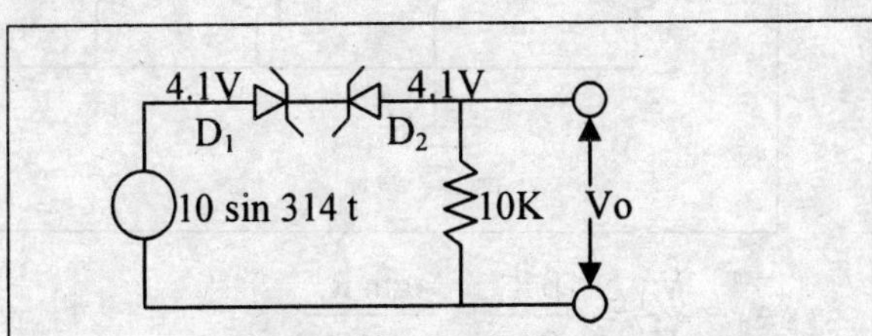

During positive half cycle for input less than 4.1 V

$D_1 \rightarrow$ Forward biased

$D_2 \rightarrow$ Reverse biased

Current through load resistance is zero. Thus $V_o = 0$

During positive half cycle for more than 4.1 V

$D_1 \rightarrow$ Forward biased

$D_2 \rightarrow$ Reverse biased but Zener break down take place and voltage across V_2 is 4.1 Volt.

Thus maximum voltage drop across load resistance is $= 10 - 4.1 = 5.9\,V$

Similarly, During the negative half cycle output is maximum to $-5.9\,V$

36. Ans. (c)

Solution:

Since no input is applied at the input side and still an output e_o is available thus one can measure input offset current only.

37. Ans. (a)

Solution: At 1000Hz frequency, $Z_c \rightarrow \infty$

Hence, high pass filter

Since, input is at (+) terminal, therefore

$$\omega_c = \frac{1}{RC} = \frac{1}{1 \times 10^3 \times 1 \times 10^6} = 1000 \text{ rad/sec}$$

38. Ans. (d)

Solution: Equivalent circuit of BJT is

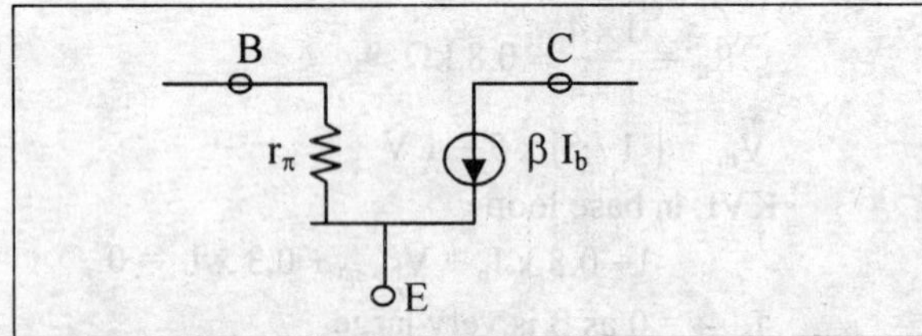

Differential mode gain (A_d):

Small signal model for difference mode

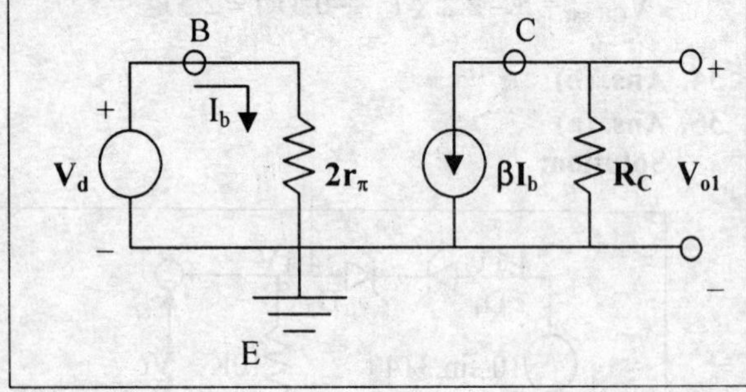

$$A_d = \frac{V_{01}}{V_d} = \frac{-\beta R_C}{2r_\pi} = \frac{-gm\,R_C}{2}$$

$A_d = [-g_m R_C / 2]$

Common mode gain (A_c):

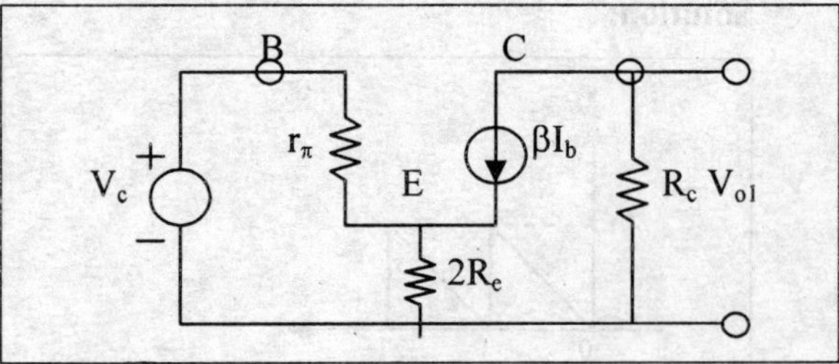

KVL in input loop

$V_c = I_b r_\pi + 2 R_e [1 + \beta] I_b$

Output voltage $V_o = -\beta I_b . R_c$

$$A_c = \frac{V_{01}}{V_c} = \frac{-\beta R_c I_b}{[2(1+\beta) R_e + r_\pi] I_b}$$

$A_c = [-R_c / 2R_e]$ for $\beta >> 1$; $r_\pi << 2R_e$

Decreases A_c with $R_e \uparrow$

CMRR = $[A_d / A_c]$

CMRR = $g_m R_e$

39. Ans. (b)

Solution:

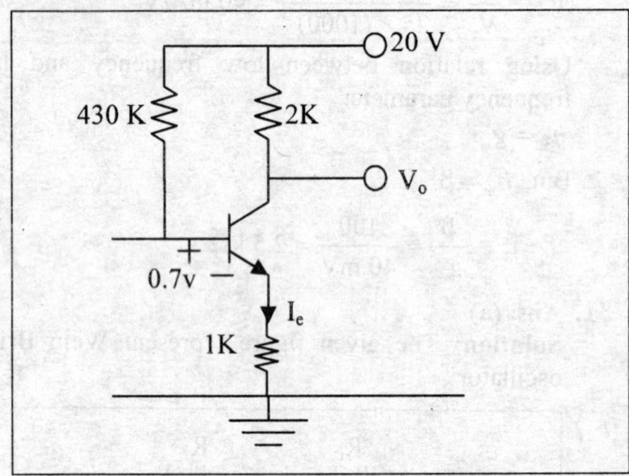

$I_e = I_b + I_c$

$I_c = \beta I_b + I_{co}$

$\beta = 50$, $\alpha = [50 / 51] = 0.98039 = [I_c / I_e]$

$V_{BE} = 0.7V$

$20V = (430\text{ K}) I_b + 0.7 + I_e (1K)$

$20 = (430\text{ K}) I_b + 0.7 + (51 I_b)\text{ IK}$

$I_b = 40\ \mu A$

$I_c = \beta I_b = 2000\ \mu A$

$V_o = 20 - 2\text{ K} \times 2000\ \mu A = 16\text{ V}$

40. Ans. (a)

Solution:

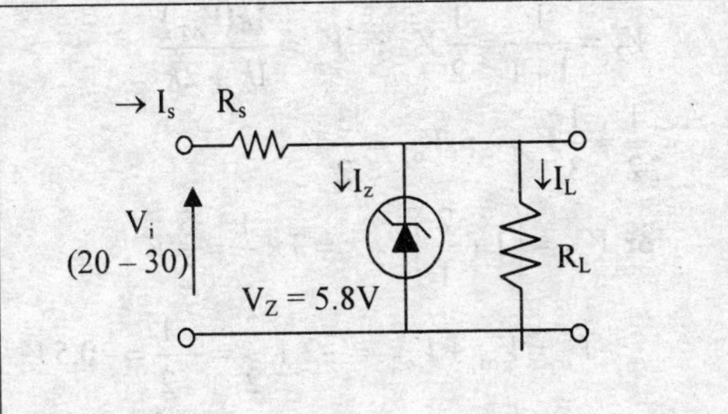

$I_s = I_L + I_z$

$I_L = I_s - I_z$

For design consideration

Load current is maximum when diode is just operate in break down region, As V_{in} increases, the load current remain but zener current increase.

For the zener diode to be operated in the breakdown region, the maximum load current is

$I_L = I_s - I_z$

$$i_{L(max)} = \frac{V_{s(max)} - V_z}{R_s} - I_{z\,(min)}$$

$$= \frac{30 - 5.8}{1\text{ K}} - [1/2]\text{ mA}$$

$i_{L\,(max)} = 23.7$ mA

41. Ans. (b)

Solution: Due to positive feed back circuit is always in saturation

$V_{out} = \pm 10$V; as $V_{supply} = \pm 10$V

When V_o = 10V

Diode D_1 – open circuit, D_2 – Conduct

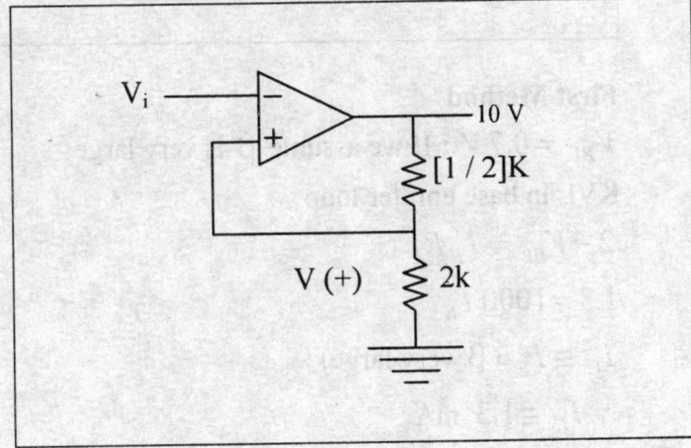

$$V+ = \frac{2K}{[2 + (1/2)]\text{ k}} \times 10$$

$V+ = 8V = V_i$

When V_o = - 10 V,

Diode D_1 → Conduct, D_2 → open circuit

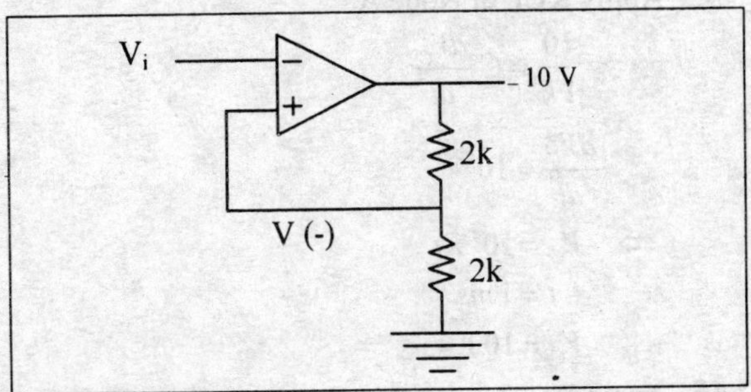

V- = [2k / 4k] x (- 10)

V - = - 5 V = V_i

Hysteresis loop is

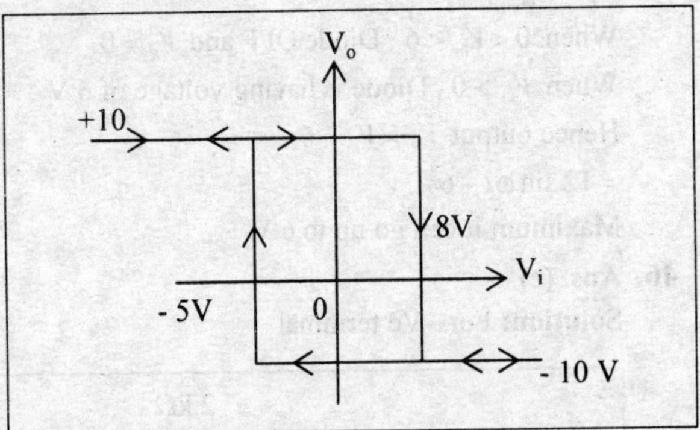

42. Ans. (a)

Solution: Ramp input is applied

$R(s) = [1/s^2]$

$e_{ss} = \text{Lim } sE(s)$

$$sE(s) = \frac{sR(s)}{1 + G(s)} = \frac{s.\,[1s^2]}{1 + G(s)} = \frac{1}{s + s\,G(s)}$$

$$e_{ss} = \lim_{s \to 0} sE(s) = \lim_{s \to 0} \frac{1}{sG(s)}$$

The above expression represent type 1 system therefore, type = 1. e_{ss} for type 1 system is given by

$e_{ss} = [1/K] = 5\,\% = [1/20]$

∴ **K = 20**

43. Ans. (c)

Solution: $I = I_o \;[\exp\{V / nV_T\} - 1]$

44. Ans. (d)

Solution: $V_c^+ = V_c^- = 10\,V$

$$i_c = \frac{10}{1\,k} = 10^{-2}A$$

$$V_c = -\frac{1}{c}\int_0^t i_c\, dt = \frac{1}{10^{-6}} \times i_c \times t$$

$\Rightarrow V_c = 10^4$ t

For t = 1 ms, V_c = 10V

Another way

Apply KCL at Node A

$$\frac{10}{1\,k} = C\frac{dV_C}{dt}$$

$$\frac{dV_C}{dt} = 10^4$$

$$\Rightarrow \quad V_C = 10^4 \times t$$

at $\quad t = 1\,ms$

$$V_C = 10\,V$$

45. Ans. (d)

Solution: When $V_{in} < 0$, Zener diode works as a normal diode in forward biased. Hence output $V_R = V_{in}$

When $0 < V_{in} < 6$, Diode OFF and $V_R = 0$

When $V_{in} > 0$, Diode is having voltage of 6 V.

Hence output $V_R = V_{in} - 6$

$= 12\sin\omega t - 6$

Maximum it can go up to 6 V.

46. Ans. (c)

Solution: For –Ve terminal

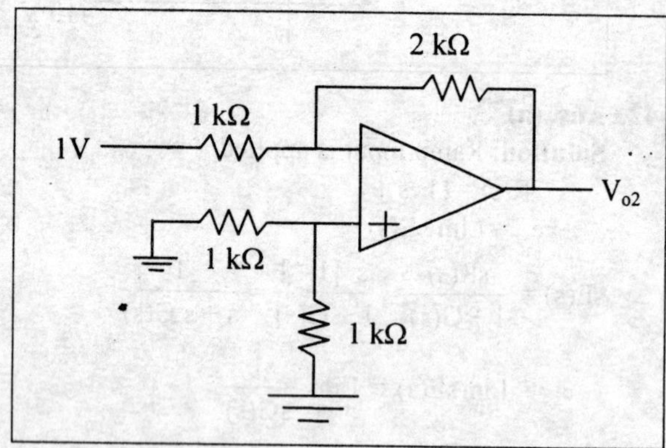

$$V_{o1} = -\frac{2}{1}(1) = -2V \qquad (1)$$

For + Ve terminal

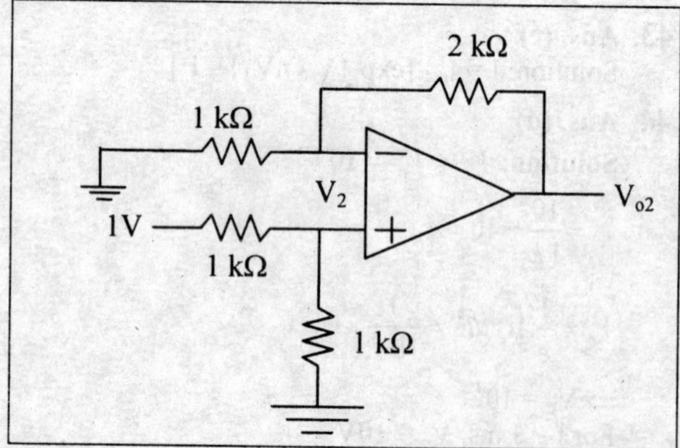

$$V_2 = \frac{1}{1+1} = \frac{1}{2}V \quad ; \quad V_2 = \frac{1k(V_{o2})}{1k + 2k}$$

$$\frac{1}{2} = \frac{1}{3}V_{o2} \quad ; \quad V_{o2} = \frac{3}{2}V$$

$$\text{or } V_{o2} = \left(1 + \frac{2}{1}\right)V_2 \quad = 3 \times \frac{1}{2} = \frac{3}{2}V$$

$$V_o = V_{o1} + V_{o2} \quad = -2 + \frac{3}{2} = -\frac{1}{2} = -0.5\,V$$

Second Method

KCL at Node G,

$V_G = 0.5\,V$

$$\left(\frac{0.5-1}{1}\right) = \frac{(V_o - 0.5)}{2}$$

$$-1 = V_o - 0.5$$

$$V_o = -0.5$$

47. Ans. (b)

Solution:

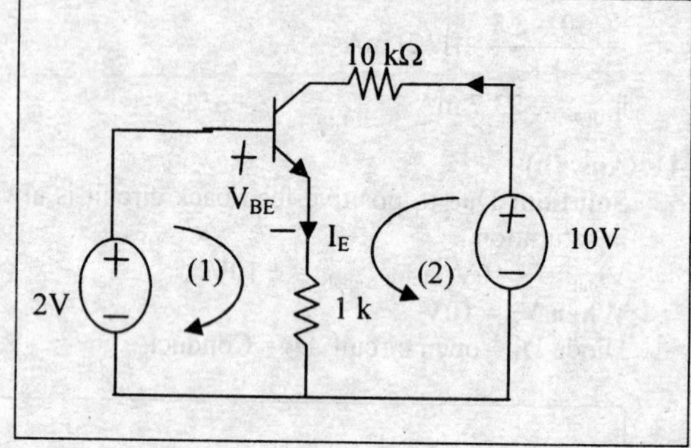

First Method

$V_{BE} = 0.7\,V$; If we assume β is very large

KVL in base emitter loop

$2 - V_{BE} = I_E R_E$

$1.3 = 1000\,I_E$

$I_C \cong I_E$ (β very large)

$\therefore\ I_C \cong 1.3$ mA

KVL in collector emitter loop

$V_{CC} = +I_C\,(R_C + R_E) + V_{CE}$

$V_{CE} = V_{CC} - I_C\,(R_C + R_E)$

$V_{CB} + V_{BE} = 10 - 1.3 \times 10^{-3} \times 11 \times 10^3$

$V_{CB} = -4.6\,V - V_{BE} = -4.6 - 0.7 = -5.3$

Collector junction is reverse bias and emitter junction is already forward bias as $V_{BE} = 2\,V$ applied.

Therefore, BJT is in Saturation

Second Method

Assuming transistor in saturation

$$i_C = \frac{(10-0.2)}{(10+1)}\,mA$$

$$= \frac{9.8}{11}\,mA = 0.8909$$

Since β is very high thus I_B will be very small

For Base emitter loop

$$I_E = \frac{(2-0.7)}{1K} = 1.3\,mA$$

$$I_E = I_B' + I_C$$

$$I_B' = 1.3 - 0.89 = 0.41$$

Thus $I_B' > I_B$ which shows transistor in saturation

48. Ans. (d)

Solution:

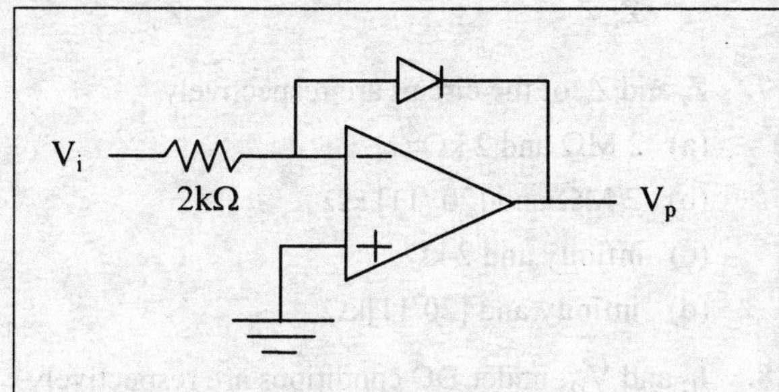

$V_o = V_{o1}$ for $V_1 = 2$ V

$V_o = V_{o2}$ for $V_i = 4$V

Relation between V_{o1} and V_{o2} is?

⇒ For algorithmic Amplifier

$$V_o = -V_T \ln\left(\frac{V_i}{R_o I_o}\right)$$

$$\therefore V_{o1} = -V_T \ln\left(\frac{2}{R_o I_o}\right) = -V_T(\ln 2 - \ln R_o I_o)$$

$$V_{o2} = -V_T \ln\left(\frac{4}{R_o I_o}\right)$$

$$= -V_T(\ln 4 - \ln R_o I_o) = -V_T(2\ln 2 - \ln R_o I_o)$$

$$\therefore V_{o1} - V_{o2} = V_T \ln 2$$

49. Ans. (d)

Solution: For high Gate input (2.5 V) only NMOS will conduct and PMOS switch off

$$\therefore \quad I_d = k_n\left(|V_{gs}| - V_{THn}\right)^2$$

$$= 40\times10^{-6}(2.5-1)^2$$

$$= 40\times10^{-6}\times 2.25 = 90\,\mu A$$

50. Ans. (c)

Solution:

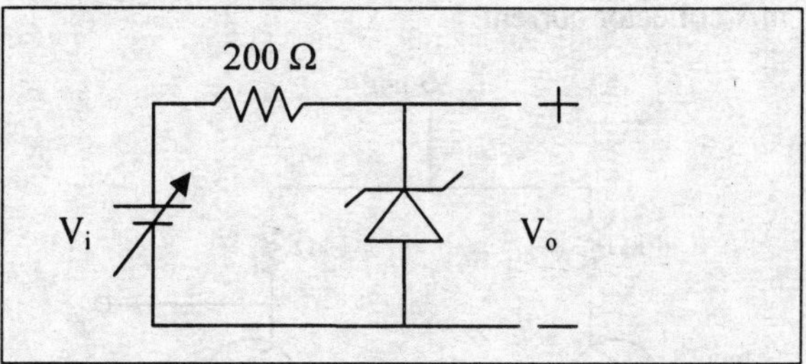

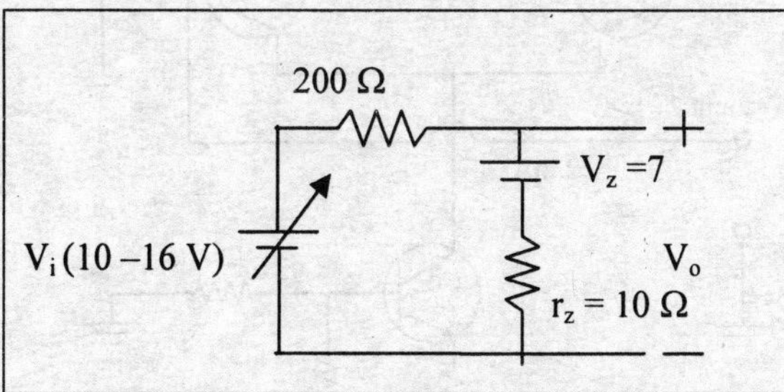

$$V_{r1} = \frac{10-7}{200}\times 10 = 0.15$$

$$V_{r2} = \frac{16-7}{200}\times 10 = 0.45$$

Zener regulator output

= Zener breakdown voltage + voltage drop due to Zener dynamic resistance.

Thus $V_o = (7 + V_{r1})$ to $7 + V_{r2}$ = 7.15 to 7.45 3

Statement for Linked Answers Questions 1 and 2

For Class A, CE amplifier circuit shown in figure, (I_c) Q = 500mA. The collector efficiency of the transistor is 12.5%.

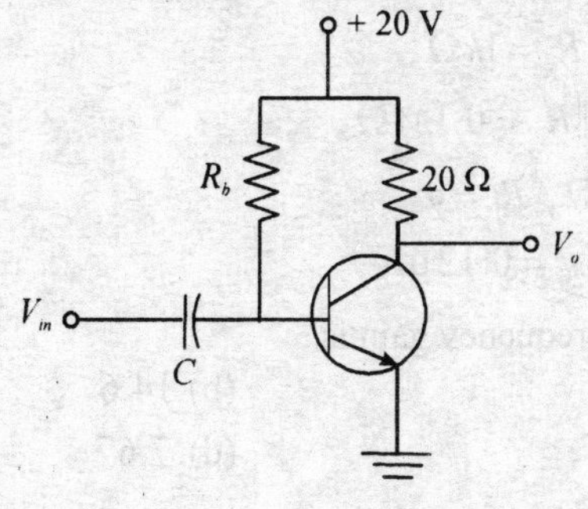

1. The average AC power output is:
(a) 10V (b) 8V
(c) 3V (d) 0.625W
2. The overall efficiency of the amplifier circuit is:
(a) 12.5% (b) 2.8%
(c) 6.25% (d) 30%

Common Data for Questions 3 and 4

The ransistors in the differential amplifier shown in figuer below are identical with $h_{fe} = 100$ and $r_e = 25\ \Omega$ at 1 mA collector current.

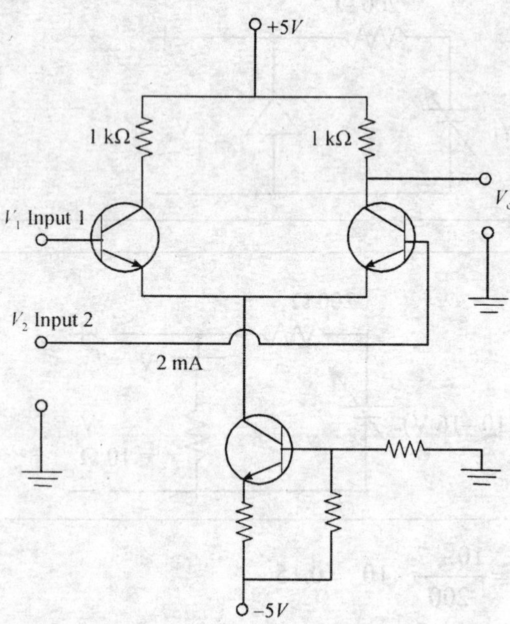

3. What is the differenital gain of the circuit?
(a) −18 (b) −20
(c) −40 (d) −28
4. What is the common mode gain of the circuit?
(a) 1 (b) 0.78
(c) 0.266 (d) 0.92

Statement for Linked Answers Questions 5 and 6

For JEFT amplifier, the data is $g_m = 2.5\ mS$,

$r_d = 200\ k\Omega\ C_{gs} = 12\ pF, C_{gd} = 2pF$ .Circuit

elements $R_s = 1k\Omega$,

$R_G = R_1 \| R_2 = 0.1M\Omega$,

$R_D = 6k\Omega,\ R_1 = \infty$

$C_{C1} = C_{C2} = 0.12\mu F$,

5. The mid frequency gain is:
(a) − 14.6 (b) 14.6
(c) 5.83 (d) 7.67

6. For the above circuit parameter and with $V_1 = 1V$ and element values, predict the upper and lower cut-off frequencies is
(a) 10 Mrad /s,10 rad/s
(b) 10 Mrad /s,100 rad/s
(c) 100 Mard/s,10 rad/s
(d) 100 Mard/s,100 rad/s

Common Data for Questions 7, 8 and 9

Given, r_d = 20 kΩ, I_{DSS} = 10 mA, V_p – = – 8 V

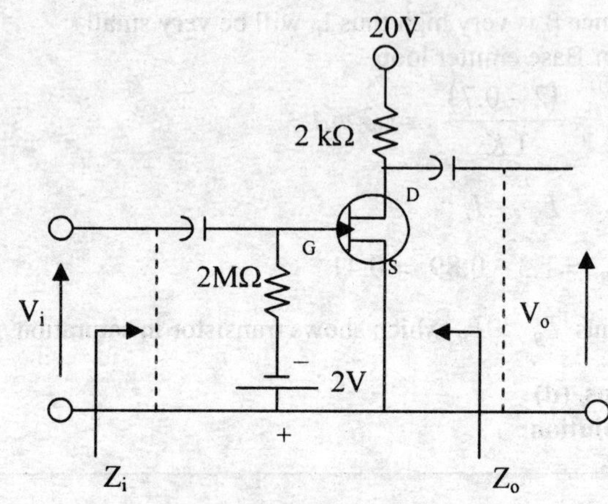

7. Z_i and Z_o of the circuit are respectively
(a) 2 MΩ and 2 kΩ
(b) 2 MΩ and [20 /11] kΩ
(c) infinity and 2 kΩ
(d) infinity and [20/11]kΩ
8. I_D and V_{DS} under DC conditions are respectively
(a) 5.625 mA and 8.75 V
(b) 7.500 mA and 5.00 V
(c) 4.500 mA and 11.00 V
(d) 6.250 mA and 7.50 V
9. Trans-conductance in milli-Siemens (mS) and voltage gain of the amplifier are respectively
(a) 1.875 mS and 3.41
(b) 1.875 mS and – 3.41
(c) 3.3 mS and – 6
(d) 3.3 mS and 6

Common Data for Questions 10, 11 and 12

In the transistor amplifier circuit shown in the figure below, the transistor has following parameters : β_{DC} = 60 , V_{BE} = 0.7 V, $h_{ie} \to \infty$, $h_{fe} \to \infty$. The capacitance C_c can be assumed to be infinite.

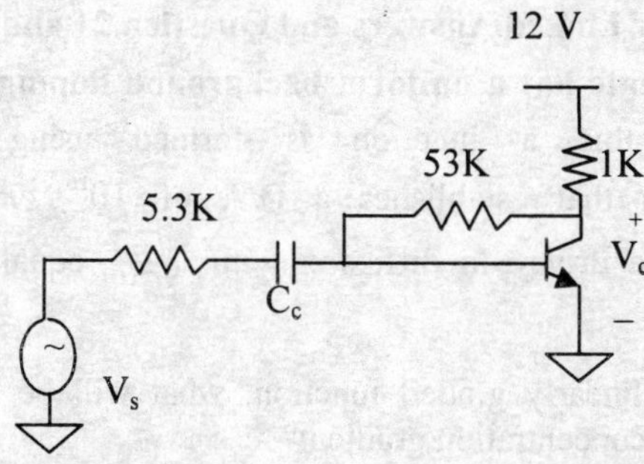

In the figure above, the ground has been shown by the symbol ∇.

10. Under the DC conditions, the collector – to emitter voltage drop is
(a) 4.8 Volts (b) 5.3 Volts
(c) 6.0 Volts (d) 6.6 Volts

11. If β_{DC} is increased by 10%, the collector – to emitter voltage drop .
(a) Increases by less than or equal to 10%
(b) Decreases by less than or equal to 10%
(c) Increases by more than 10%
(d) Decreases by more than 10%

12. The small – signal gain of the amplifier [v_c / v_s] is .
(a) –10 (b) –5.3
(c) 5.3 (d) 10

Statement for linked Answers Questions 13 and 14

A regulated power supply, shown in figure below, has an unregulated input (UR) of 15 Volts and generates a regulated output V_{out} Use the component values shown in the figure.

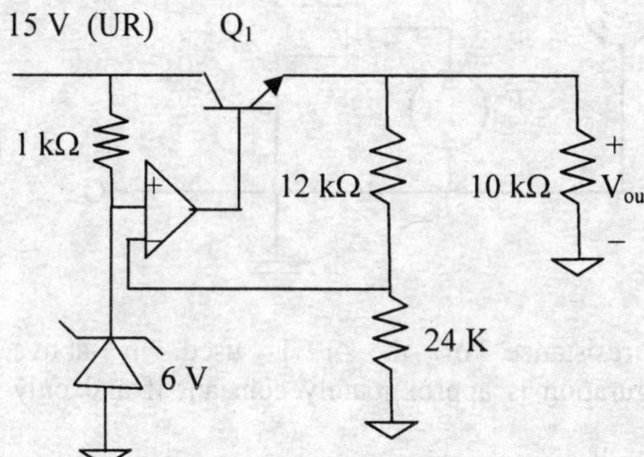

In the figure above, the ground has been shown by the symbol ∇.

13. The power dissipation across the transistor Q_1 shown in the figure is .
(a) 4.8 Watts (b) 5.0 Watts
(c) 5.4 Watts (d) 6.0 Watts

14. If the unregulated voltage increases by 20%, the power dissipation across the transistor Q_1.
(a) Increases by 20%
(b) Increases by 50%
(c) Remains unchanged
(d) Decreases by 20%

Statement for Linked Answer Questions 15 and 16

Consider the Op-Amp circuit shown in the figure

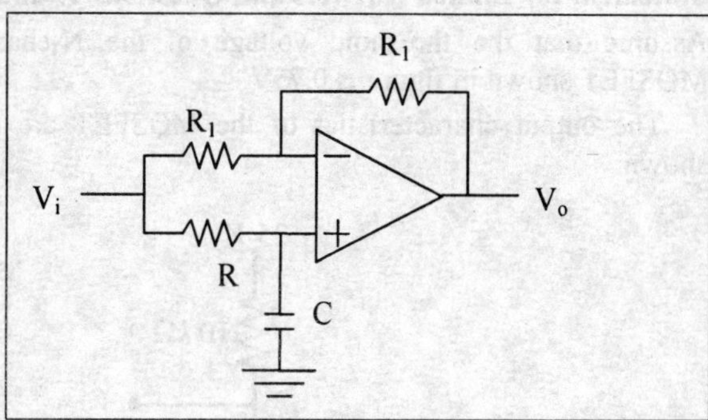

15. The transfer function $V_o(s)/V_i(s)$ is

(a) $\frac{1-sRC}{1+sRC}$ (b) $\frac{1+sRC}{1-sRC}$

(c) $\frac{1}{1-sRC}$ (d) $\frac{1}{1+sRC}$

16. If $V_i = V_1 \sin(\omega t)$ and $V_o = V_2 \sin(\omega t + \phi)$, then the minimum and maximum values of ϕ (in radians) are respectively.
(a) $-\pi/2$ and $\pi/2$ (b) 0 and $\pi/2$
(c) $-\pi$ and 0 (d) $-\pi/2$ and 0

Statement for the Linked Answers Questions 17 and 18

In the Schmitt trigger circuit shown below, the zener diodes have V_z (reverse saturation voltage) = 6 V and V_D (forward voltage drop) = 0.7 V.

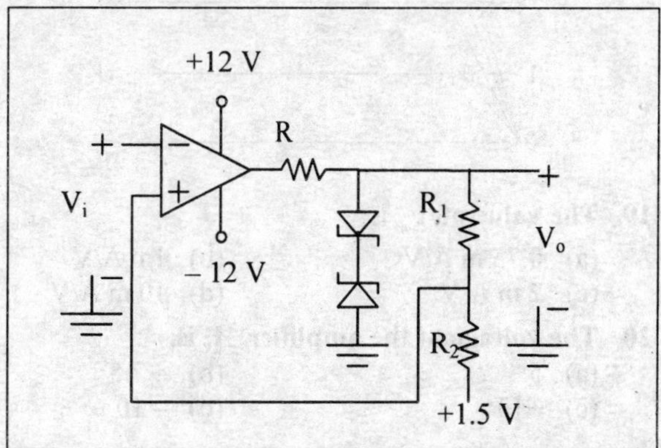

17. If the circuit has the input lower trip point (LTP) = 0V, then value of (R_1/R_2) is given as

(a) 0.223 (b) 2.67
(c) 4.67 (d) ∞

18. In input upper trip point (UTP) of the Schmitt trigger is

(a) 1.5 V (b) 2.1 V
(c) 2.42 V (d) 6.7 V

Statement for Linked Answers and Questions 19 and 20

Assume that the threshold voltage of the N-channel MOSFET shown in figure is 0.75V.

The output characteristics of the MOSFET are also shown.

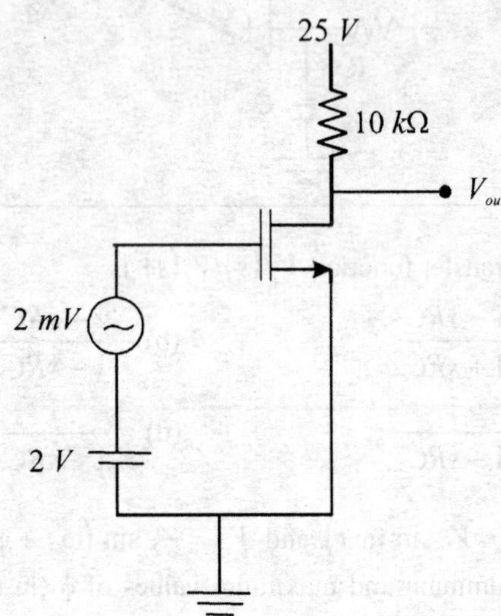

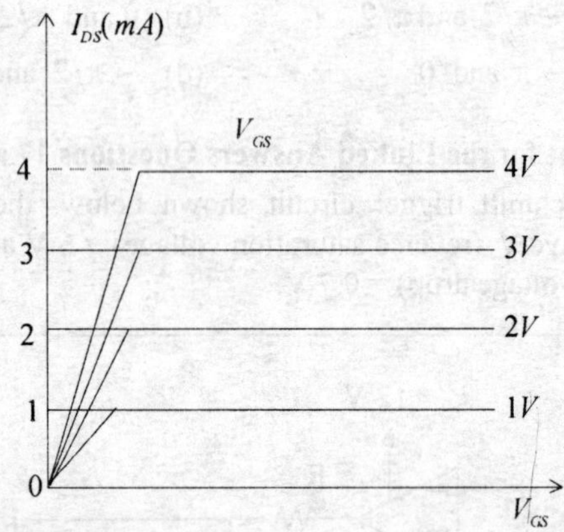

19. The value of g_m is:

(a) 0.75 m A/V (b) 1m A/V
(c) 2 m A/V (d) 10 m A/V

20. The voltage of the amplifier A_v is:

(a) 5 (b) − 7.5
(c) +10 (d) − 10

Statement for Linked Answers and Question 21 and 22

A Si – substrate has a uniform background doping of $10^{16} cm^3$. In this a junction is formed using a predisposition that establishes a Q/A of $10^{16}/cm^{-2}$ followed by a drive –in diffusion with $\sqrt{D_t}$ equal to $2\,\mu m$.

21. Using a linearly graded junction, what will be the impurity concentration gradient?

(a) $1.41\times10^{23}\,cm^{-4}$ (b) $1.41\times10^{20}\,cm^{-4}$
(c) $1.69\times10^{22}\,cm^{-4}$ (d) $2.26\times10^{21}\,cm^{-4}$

22. For the above problem, if the barrier potential is 0.6V, the depletion layer width will be,

(a) 0.629 μm (b) 0.141 μm
(c) 2.96 mm (d) 4.11 mm

Statement for Linked Answers and Questions 23 and 24

A FET Switch is shown in the figure below with $k_n = 10^{-3}$ and $V_T = 3V$. The input voltage varies between −8 to 8V. The gate voltage used to turn the switch ON is $V_G = V_N$ and that to turn the OFF is $V_G = -V$

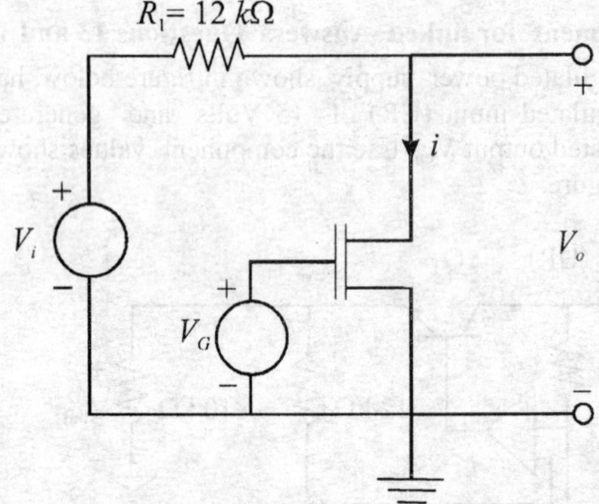

23. The resistance of the FET used in above configuration is approximately constant if and only if,

(a) $v_0 \le 2(V_N - V_T)$ (b) $v_0 \ge (2V_N - V_T)$
(c) $v_0 = 2(V_N - V_T)$ (d) $v_0 = (2V_N - V_T)$

24. For the above problem, find V_F so that the gate can be turned OFF.

(a) 3V (b) 5V
(c) −3V (d) − 5V

Answers
Analog Electronics

1. Ans. (d)
Solution:

$$\left(P_{in}\right)_{dc} = V_{cc} \times \left(I_c\right)_Q = 20 \times 500 \times 10^{-3}$$
$$= 10^4\, mW$$
$$= 10W$$
$$\left(P_{R_C}\right)_{dc} = \left|\left(I_c\right)\right|_Q^2 \times R_C$$
$$(0.5)^2 \times 20 = 5W$$
$$\left(P_{tr}\right)_{dc} = 10W - 5W = 5W$$
$$\left(P_o\right)_{ac} = \left(I_{rms}\right)^2 R_c$$
$$I_{rms} = \frac{0.25}{\sqrt{2}}$$
$$\left(P_o\right)_{ac} = \left(\frac{0.25}{\sqrt{2}}\right)^2 \times 20$$
$$= 0.625\text{ W}$$

2. Ans. (c)
Solurion:.

$$\left(P_c\right)_{dc} = \left(P_{in}\right)_{dc} - \left(P_{RC}\right)_{dc} - (P_O)_{ac}$$
$$= 4.375\, W$$
$$n_{overall} = \frac{\left(P_O\right)_{ac}}{\left(P_{in}\right)_{dc}}$$
$$= \frac{0.625}{10} \times 1000$$
$$= 6.25\%$$

3. Ans. (a)
Solution:
Emitter resistance due to current source is

$$I_c = 2\, mA$$
$$h_{fe} = 100$$
$$I_C = h_{fe} I_b$$
$$I_B = \frac{2 \times 10^{-3}}{100} = 0.02\, mA$$
$$I_E = -(I_C + I_B) = -2.02\ mA$$

Assuming base of current source transisor at ground.

$$R_E = \frac{5 - 0.6}{2.02 \times 10^{-3}} = \frac{4.4}{2.02 \times 10^{-3}}$$
$$= 2.178\, k\Omega \approx 2200\, \Omega$$

Differential gain $A_D = \dfrac{R_C}{2\left(r_e + \dfrac{R_B}{h_{fe}+1}\right)}$

Since, biasing and source resistance are not included in the circuit.

$$\therefore A_D = \frac{R_c}{2r_c} = -\frac{1000}{2 \times 25} = -20$$

4. Ans. (b)
Solution:
Common mode gain is given by,

$$\therefore A_C = -\frac{R_c}{2R_E + r_e + \dfrac{R_B}{h_{fe}+1}} \approx \frac{R_C}{2R_E + r_e}$$
$$= \frac{1000}{2 \times 2200 + 25} = 0.226$$

5. Ans. (a)
Solution:
For midfreqency circuit model, all capacitor is short circuted.

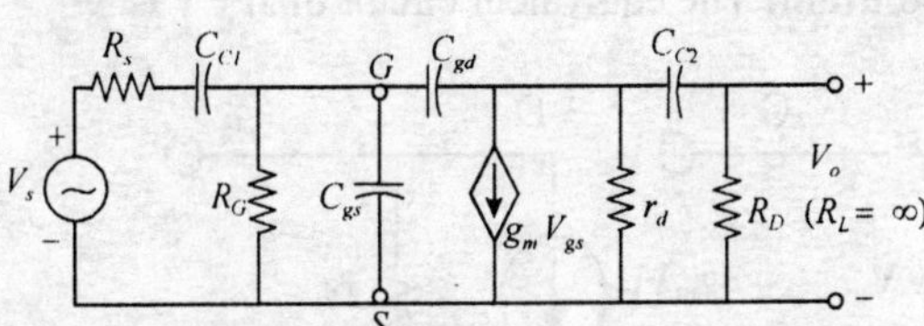

$$R_G = R_1 \| R_2 = 0.1 M\Omega$$
$$V_{GS} = \left(\frac{R_G}{R_S + R_G}\right) V_S$$
$$= \left(\frac{100}{1 + 100}\right) V_S = V_S$$
$$R_0 = r_d \| R_D = \frac{200 \times 6}{200 + 6}$$
$$= 5.083\, k\Omega$$
$$V_0 = -g_m V_{gs} \times (r_d \| R_D)$$
$$= -14.6 V_S$$
$$A_{V0} = \frac{V_0}{V_s} = -14.6$$

6. Ans. (c)
Solution:
For upper frequency region, series (ie circuit) capacitor acts as an open circuit while shunt capacitor (ie inter electrode) has to be considered.

$$V_0 = \frac{1}{50} \times \frac{5\times10^3 \times \frac{10^{12}}{j\omega\times 2}}{(5\times10^3) + \frac{10^{12}}{j\omega\times2}}$$

$$= \frac{1}{50} \times \frac{5\times10^{15}}{j10^4\omega + 10^{12}} = \frac{100}{1 + j\left(\frac{\omega}{10^8}\right)}$$

$\therefore \omega_H = 10^8$ Rad/s or 100 Mard/s

For low frequency gain, shunt capacitor acts as open circuit, while series capacitor is considered.

Taking Thevenin's equivalent

$$V_{TH} = \frac{1}{50} \times 10\times10^3 = 200V$$

$$= V_0 = \frac{200\times10\times10^3}{20\times10^3 + \frac{10^6}{j5\omega}} = \frac{100}{1 - j\frac{10}{\omega}}$$

$\omega_L = 10$ rad/s

7. Ans. (b)

Solution: The equivalent circuit of JFET is

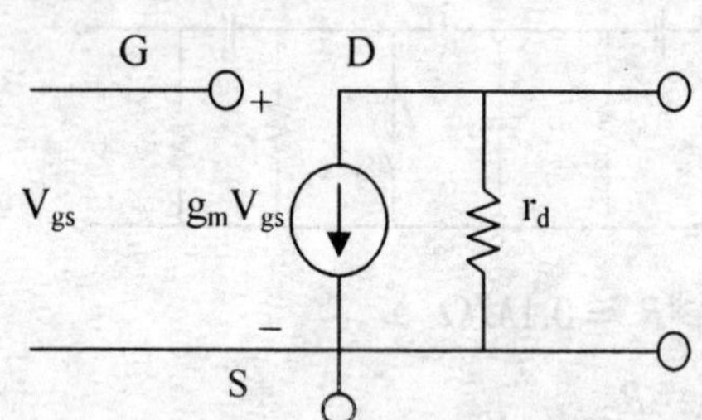

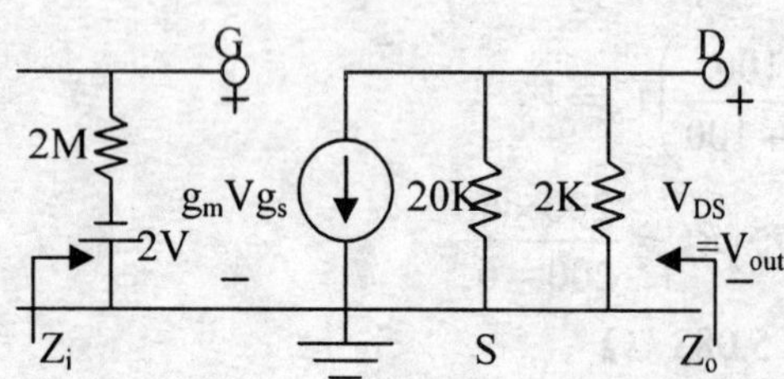

$Z_i = 2M$; $Z_{out} = 20K \parallel 2K$

$$= \frac{20\times2}{20+2} = [20/11]\ k\Omega$$

8. Ans. (a)

Solution:

$$I_D = I_{DSS}\left(1 - \frac{V_{GS}}{V_P}\right)^2 = 10\left(1 - \frac{2}{8}\right)^2 = 5.625$$

$I_D = 5.625$ mA

$V_{DS} = 20 - 2k \times 5.625 = 8.75$ V

$V_{DS} = 8.75$ V

9. Ans. (b)

Solution:

$$g_m = \frac{-2I_{DSS}}{V_p}\left(1 - \frac{VG_S}{V_p}\right)$$

$$= \frac{-2\times10\text{ mA}}{-8}\left(1 - \frac{(-2)}{-8}\right)$$

$g_m = 1.875$ msiemens

For CS amplifier

$A_v = - g_m (r_d \parallel R_L)$

$= - 1.875 \times [20 / 11]$

$[V_0 / V_{gs}] = V_{gain} = - [20 / 11]\ 1.875$

$= - 3.41$

$A_V = - 3.41$

10. Ans. (c)

Solution: $\beta_{DC} = 60$, $V_{BE} = 0.7$ V, $h_{ie} \to \infty$, $h_{fe} \to \infty$

The capacitance C_c can be assumed to be infinite.

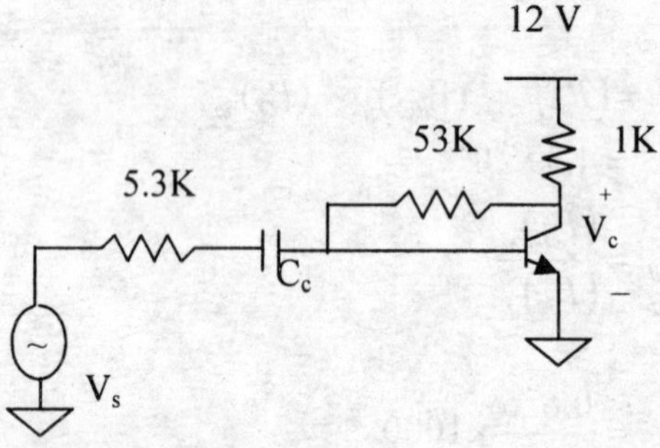

This is collector to Base biasing circuit diagram

$\beta_{dc} = [I_c / I_b]$

$I_b \times \beta_{dc} = I_c$

$$\left(\frac{V_c - 0.7}{53\text{ k}}\right) \times 60 = \left(\frac{12 - V_c}{1\text{ k}}\right)$$

$60\ V_c - 42 = 53 \times 12 - 53V_c$

$V_c = 6.0$ V

$V_{CC} = R_L [I_c + I_b] + R_b I_b + V_{BE}$

$12 = [I_c + I_b] \times 1 + 53.\ I_b + V_{BE}$

$\Rightarrow 12 = 61\ I_b + 53\ I_b + V_{BE}$

$\Rightarrow I_b = .099$ mA

Now $V_{CE} = 53\ I_b + V_{CE} = 5.95$ Volt

11. Ans. (b)

Solution: $V_{CE} = V_{CC} - R_L (I_c + I_b)$

$= V_{CC} - R_L (\beta + 1) I_b$

From above expression if β is increased then V_{CE} is decrease by less than equal to same percentage.

12. Ans. (a)

Solution: $\beta_{DC} = 60$, $V_{BE} = 0.7$ V,
$h_{ie} \to \infty$, $h_{fe} \to \infty$
The capacitance C_c can be assumed to be infinite.

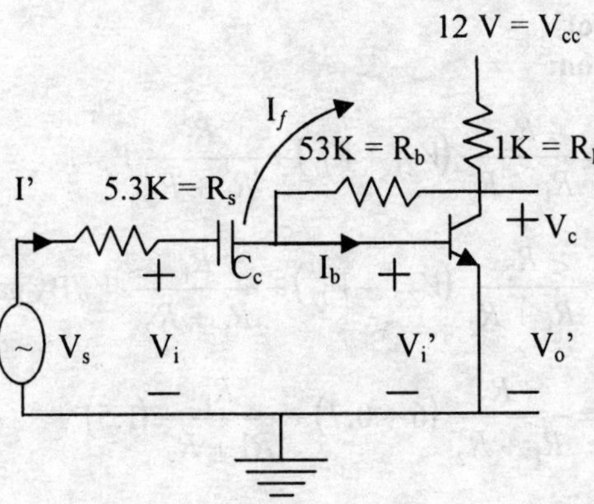

$I' = I_f + I_b$; $V_o >> V_i$

$$If = \frac{V_i - V_o}{R_b} = \frac{-V_o}{R_b} = KV_o$$

Where, $K = [-1/R_b]$

Since feed back current is proportional to output voltage. It represent voltage shunt feed back

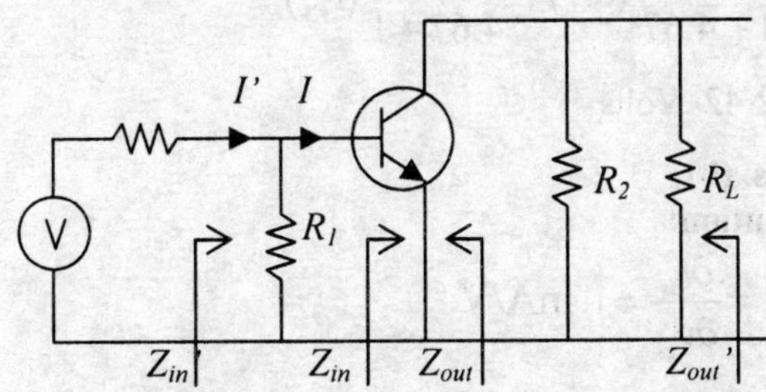

By Millar Theorem,

$$R_1 = \frac{R_b}{-A_v}; \; R_2 = \frac{R_b}{[1/(1/A_v)]} = \frac{R_b A_v}{A_v - 1}$$

In CE, $A_v >> 1$

$$R_1 = \frac{Rb}{-A_v}; \; R_2 \cong R_b$$

For CE: $A_v = \dfrac{-h_{fe} R_L}{h_{ie}}$; $R_1 = \dfrac{R_b \cdot h_{ie}}{h_{fe} \cdot R_L}$

$Z_{in} = h_{ie}$; $Z_{in}' = hi_e \,||\, R_1$

$A_I = -h_{fe}$

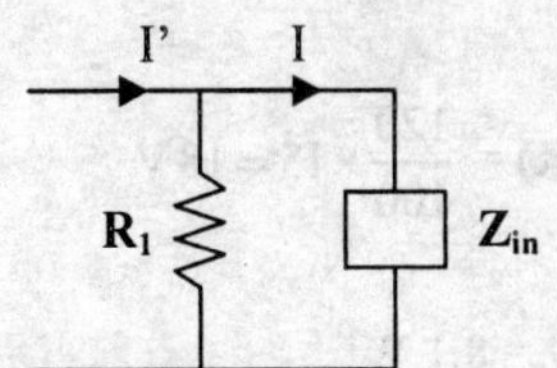

$$\frac{I}{I'} = \frac{R_1}{R_1 + Z_{in}}$$

$$A_{If} = A_I' = \frac{I_L}{I'} = \frac{I_L}{I} \times \frac{I}{I'}$$

$$= AI = \frac{R_1}{R_1 + Z_{in}}$$

$$= \frac{-h_{fe} R_1}{R_1 + Z_{in}} = A_{If}$$

$$A_v = A_I \cdot [Z_L / Z_{in}] = \frac{-h_{fe}(R_2 \,||\, R_L)}{h_{ie}}$$

$Z_{out} = \infty$

$Z_{in} = \infty \,||\, R2 \,||\, R_L$

$= R_2 \,||\, R_L$

$$= \frac{R_b \times R_L}{R_b + R_L}$$

$$A_{vf} = \frac{-h_{fe} \cdot R_b \cdot R_L}{h_{ie}(R_B + R_L)}$$

r_L = 1 k PS53k =- 1k

$A_i = h_{fe}$

$R_i = h_{ie}$

$$A_v = \frac{-h_{fe}}{h_{ie}} rL = -1k$$

$$R_{BI} = \frac{53 \times 100}{1 - (-1000)} = 53\Omega$$

$$A_{vs} = \frac{V_c}{V_s} = 1000 \times \frac{53}{5.3 \times 1000} = 10$$

13. Ans. (c)

Solution:

$V^- = V^+$

$$= \frac{24}{(24+12)} V_{out} = 6$$

$$V_{out} = \frac{6 \times (12+24)}{24} = 6 \times \frac{36}{24} = 9\,V$$

$I_c = I_E = 9 = 0.9$ A

$V_{CE} = 15 - 9 = 6$ V

Power = VI = V_{CE} x I_c = 3 x 0.9 = 5.4

14. Ans. (b)

Solution: $V(UR) = \frac{120}{100} \times 15 = 18\,V$

$V_{CE} = 9$ V

Power = 9 x 0.9 = 8.1 W

$\%\text{ increase} = \frac{8.1-5.4}{5.1} = 50\%$

15. Ans. (a)

Solution: For –Ve terminal Analysis

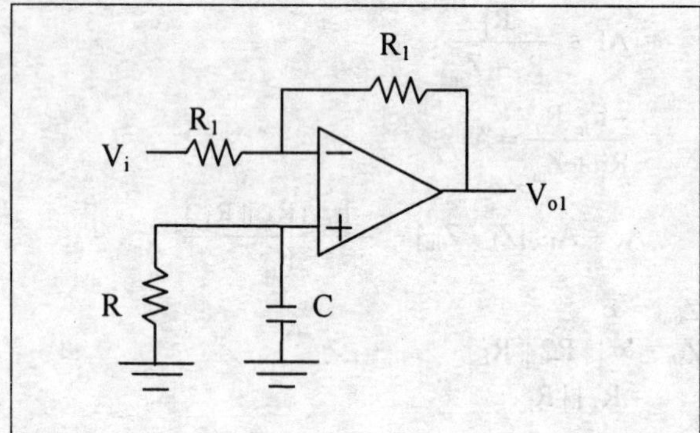

$V_{o1} = -V_i$

For +Ve terminal Analysis

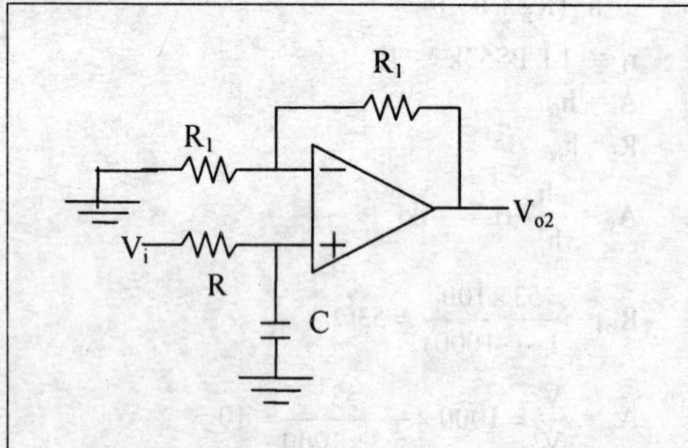

$$V_{o2} = \left(1 + \frac{R_1}{R_1}\right) V_1 = 2\,V_1$$

But $V_1 = \frac{1/sC}{R + \frac{1}{sC}} \cdot V_i = \frac{V_i}{1 + RCs} = \frac{2V_i}{1 + RCs}$

$$V_o = \frac{2V_i}{1 + RCs} - V_i$$

$$\frac{V_o}{V_i} = \frac{1 - RCs}{1 + RCs}$$

16. Ans. (c)

Solution: $T(s) = \frac{V_o(s)}{V_i(s)}$

Argument of $T(s) = -\tan^{-1}\omega RC - \tan^{-1}\omega RC$

$= -2\tan^{-1}\omega RC - \pi$ to 0

17. Ans. (c)

Solution:

$$V_{UT} = \frac{R_2}{R_1 + R_2}(V_{Z1} + V_D) + \frac{R_1}{R_1 + R_2} V_R$$

$$V_{LT} = \frac{R_2}{R_1 + R_2}(V_{Z2} + V_D) + \frac{R_1}{R_1 + R_2} \cdot V_R \text{ By gain}$$

$$\Rightarrow 0 = \frac{-R_2}{R_1 + R_2}(6 + 0.7) + \frac{R_1}{R1 + R_2}(1.5)$$

$$\Rightarrow \frac{R_1}{R_2} = \frac{6.7}{1.5} = 4.67$$

18. Ans. (c)

Solution:

$$V_{UT} = \frac{R_2}{R_1 + R_2}(V_{Z1} + V_D) + \frac{R_1}{R_1 + R_2} \cdot V_R$$

$$= \frac{1}{1 + 4.67}(6.7) + \frac{4.67}{4.67 + 1}(1.5)$$

$$= 2.42 \text{ Volts}$$

19. Ans. (b)

Solution:

$$g_m = \frac{\partial i_{ds}}{\partial v_{gs}} = 1 \text{ mA/V}$$

20. Ans. (d)

Solution:

For Common source amplifier the voltage gain

$A_v = -g_m r_{ds} = -1\,mA/V \times 10\,k\Omega = -10$

The small signal equivalent is:

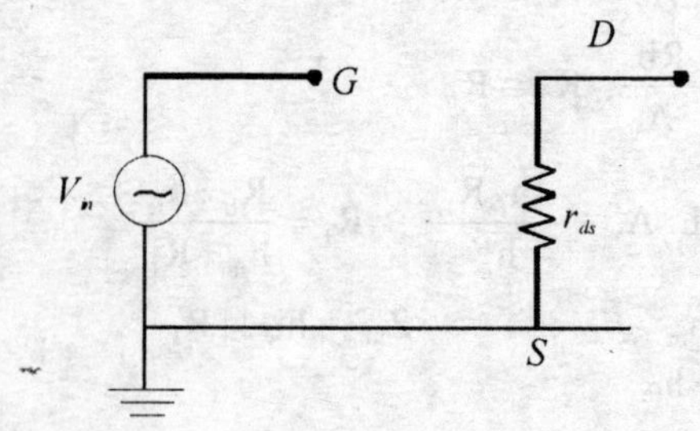

21. Ans. (b)
Solution:

Using the guassian equation,

$$N(x,t) = \frac{Q/A}{\sqrt{\pi D_t}} e^{-x^2/4D_t}$$

Differentiating with respect to x,

$$\frac{dN}{dx} = \frac{x}{2D_t} N(x)$$

At the junction depth x_j,

$N(x) = N_B$ = background doping.

$$\frac{dN}{dx} = \frac{x_j}{2D_t} N_B = m(say)$$

$$\therefore\ x_j = \left[-4D_t\, \ell n \frac{\sqrt{\pi}\, D_t\, N_B}{Q/A} \right]^{\frac{1}{2}}$$

$$= \left[-4 \times 4 \times \ell n \frac{\sqrt{\pi} \times 0.002 \times 10^{16}}{10^{16}} \right]^{\frac{1}{2}}$$

$$= 11.3 \times 10^{-4} cm$$

$$\therefore\ \mathrm{m} = \frac{(11.3 \times 10^{-4})(10+16)}{8 \times 10^{-8}}$$

$$= 1.41 \times 10^{20} cm^{-4}.$$

22. Ans. (a)
Solution:

The depletion layer width W is,

$$\mathrm{W} = \left[\frac{12 t_s V_B}{qm} \right]^{\frac{1}{3}}$$

$$= \left[\frac{12 \times 11.7 \times 55.4 \times 10^4 \times 0.6}{9 \times 1.41 \times 10^{20}} \right]^{\frac{1}{3}}$$

$$= 6.92 \times 10^{-5} cm$$

$$= 0.629\ \mu m.$$

23. Ans. (a)
Solution:

The equivalent circuit for the given FET configuration is,

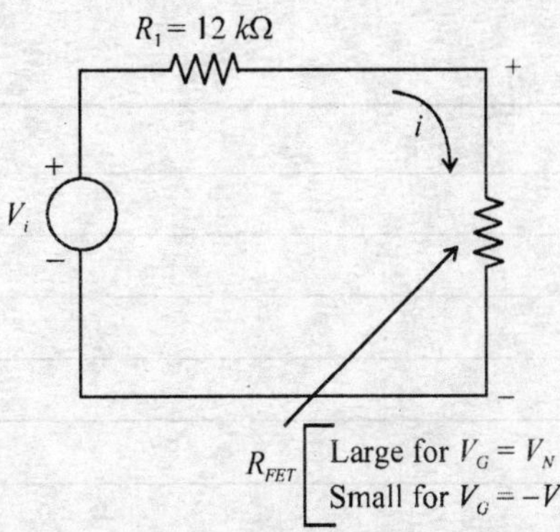

The equation for the current i_{DS} below and above pinch – off is, $i_{DS} = k_N\left[2(V_{GS} - V_{TN})V_{DS} - V_{DS}^2\right]$

From the conditions given,

$i_{DS} = k_N\left[2(V_{GS} - V_{TN})V_o - V_o^2\right]$

Thus, if $v_0 \le 2(V_N - V_T)$, the term v_0^2 can be neglected

24. Ans. (b)
Solution:

$$R_{FET} = \frac{v_0}{i_{DS}} = \frac{1}{2k_N(V_N - V_T)} = \frac{500}{V_N - V_T}$$

Applying voltage divider formula to equivalent circuit,

$$v_0 = v_i \frac{R_{FET}}{R_{FET} + R_1}$$

$$= v_i \frac{\dfrac{500}{V_N - V_T}}{\dfrac{500}{V_N - V_T} + 12 \times 10^3}$$

$$= \frac{v_i}{1 + 24(V_N - V_T)}$$

Since, $(v_i)_{\max} = +8\mathrm{V}$

$$v_{0\,(\max)} = \frac{8}{1 + 24(V_N - V_T)}$$

To ensure that the FET remains OFF for all values of v_i consider the worst condition where $v_i(\min) = -8V$.

$\therefore\ -V_F < V_T + v_{i(\min)}$

$\therefore -V_F < 3 - 8 = -5V$

$\therefore\ V_F \le 5V$.

Reader's Notes

5 Digital Electronics

Digital Electronics: Combinational logic circuits, minimization of Boolean functions. IC families, TTL, MOS and CMOS. Arithmetic circuits. Comparators, Schmitt trigger, timers and mono-stable multi-vibrator. Sequential circuits, flip-flops, counters, shift registers. Multiplexer, S/H circuit. Analog-to-Digital and Digital-to-Analog converters. Basics of number system. Microprocessor applications, memory and input-output interfacing. Microcontrollers.

5.1 COMBINATIONAL LOGIC CIRCUITS

Digital circuit are divided into two broad categories.

1. Combinational circuit and
2. Sequential circuit.

Combinational Circuit

The output at any instant of time depends upon the nput present at the instant of time. No memory in these circuits.

Sequential Circuit

These are circuit in which the output at any time depends upon the present inputs as well as past input/outputs. Memory is essential for these type of circuit.

The Design Requirement of Combinational Circuit

- A set of statements
- Boolean expression
- Truth table

Following methods can be used to simplify the Boolean function:

1. Algebraic method
2. K-map technique
3. Variable entered mapping (VEM) technique and
4. Quine-Mc cluskey method.

5.2 MINIMIZATION OF BOOLEAN FUNCTIONS

Georga Boolean developed rules for manipulation of binary variable, known as Boolean algebra.

Table for Boolean

1.	$0.A = 0$	$A + O = A$
2.	$A.1 = A$	$A + 1 = 1$
3.	$A.A = A$	$A + A = A$
4.	$A.A' = 0$	$A + A' = 1$

Duality is defined for huntingtone system just as it is for other axiomatic system and the dual of a give expression can be derive for simply replacing the "." with "+" and "+" with "." and 0's with 1's and 1's with 0's in the expression.

5.2.1 ALGEBRAIC THEOREMS

1. $A + AB = A$
2. $A + \overline{A}B = A + B$
3. $AB + A\overline{B} = A$
4. $AC + \overline{A}BC = AC + +BC$
5. $AB + AC + \overline{B}C = AB + \overline{B}C$
6. $\overline{A.B.C} = \overline{A} + \overline{B} + \overline{C}$
7. $\overline{A + B + C} = \overline{A}.\overline{B}.\overline{C}$

Realization of Boolean function by digital gate

Universal gates are those which can be used to get desired basic operation (i.e.) AND, OR, NOT, operations.

1. AND using NAND

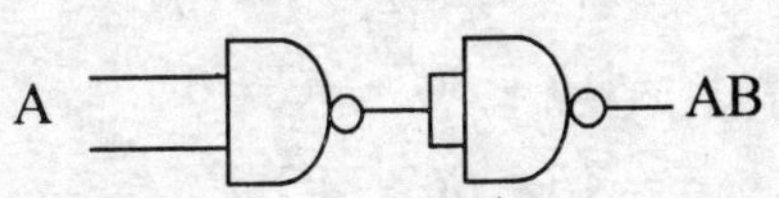

2. OR using NAND

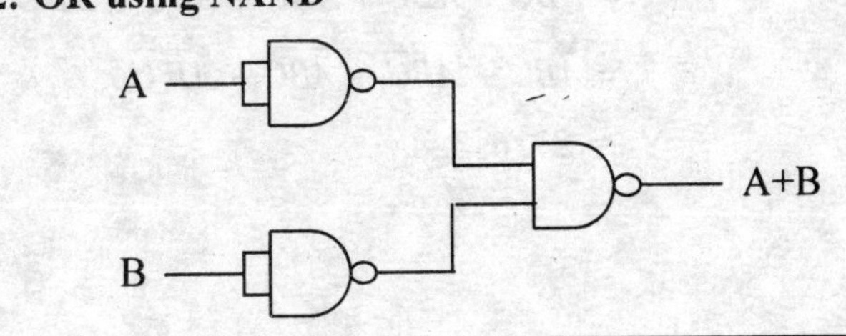

3. NOT using NAND

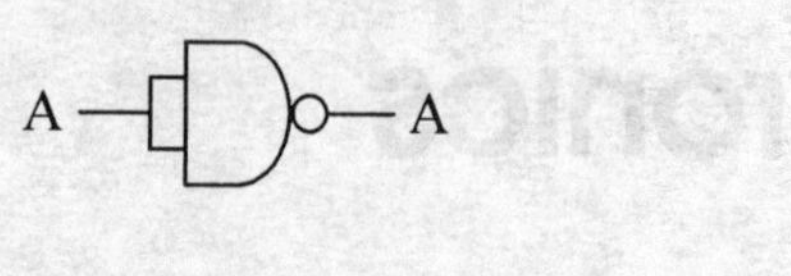

NOR = gate also used to find operation similar to that of AND, OR and NOT.

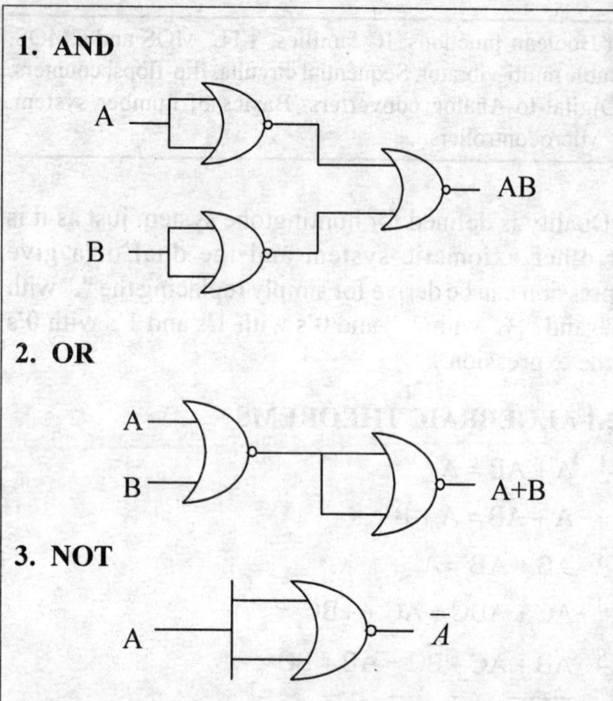

5.2.2 KARNAGUH MAP AND APPLICATIONS

Standard Representations for Logical Functions

Any arbitrary logic function can be expressed in the following forms.

1. Sum-of-product **(SOP)**, and
2. Product-of-sums form **(POS)**

Example: $Y = (A + BC)(B + \overline{AC})$

$$= AB + BC + A\overline{C} = 0$$

$$= AB + BC + A\overline{C}$$

$$= ABC + ABC + ABC + ABC + ABC$$

$$= ABC + AB\overline{C} + ABC + \overline{A}BC + A\overline{B}\overline{C}$$

$$= ABC + AB\overline{C} + \overline{A}BC + A\overline{B}\,\overline{C}$$

$$Y = \Sigma 7, 6, 3, 4$$

Note: If a logical function is specified in terms of minterm/maxterm its maxterm minterm representation can be using this complementary property.

If $y = \Sigma m\,(0, 3, 5, 6, 9, 10, 12, 15)$

Then $\overline{y} = \pi m\,(1, 2, 4, 7, 8, 11, 13, 14)$

Some Special Cases of K-Map

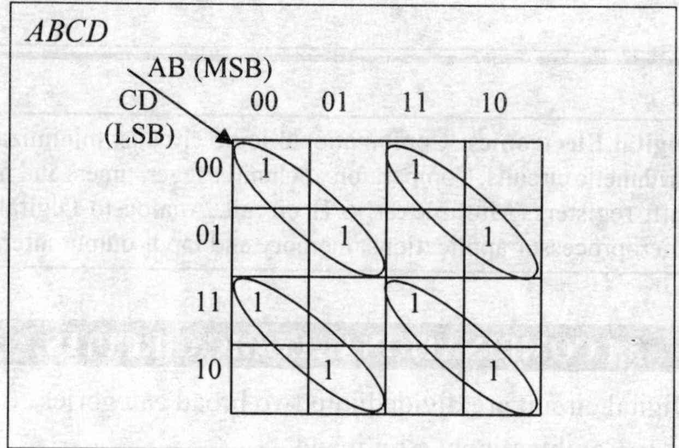

Making pair as shown. To write expression for the pair we follow this technique.

For first pair, we have two term 0000 and 0101. Now we see that (A and C) is same and (B and D) is changing like (00 to 11). This gives **EX-NOR (⊙)** type. If this case like (01 to 10) will give **EX-OR (⊕)** operation. So,

- 1st pair we can write $\overline{A}\ \overline{C}$ [B ⊙ D]
- 2nd pair we have (1100 to1001) = $A\overline{C}$ [B ⊕ D] (as B and D change as 10 to 01)
- 3rd pair we have (0011 to 0110) = $\overline{A}$C[B ⊕ D]
- 4th pair we have (1111 to1010) = AC[B⊙ D]

So, Y = $A\ C$ [B ⊙ D] ± $A\overline{C}$ [B ⊕ D] + $\overline{A}$C[B ⊕ D] +AC [B⊙D]

$$= (A\overline{C} + \overline{A}C)(B \oplus D) + (\overline{A}\overline{C} + AC)(B \odot D)$$

$$= (A \oplus C)(B \oplus D) + (A \odot C)(B \odot D)$$

Similarly we have

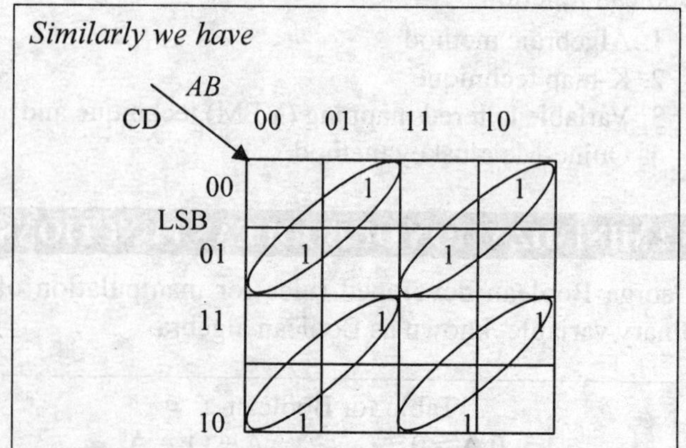

- 1st pair (0100 to 0001) = $\bar{A}\,\bar{C}$ [B ⊕D]
- 2nd pair (1000 to 1101)= A$\bar{C}$ [B⊙D]
- 3rd pair (0111 to 0010) = $\bar{A}$C[B⊙D]
- 4th pair (1011 to 1110 = AC [B⊕D]

$$\text{So, } Y = \bar{A}\,\bar{C}\,[B \odot D] + A\bar{C}\,[B \odot D] + \bar{A}C[B \odot D] + AC\,[B \oplus D]$$
$$= [AC + \bar{A}\,\bar{C}]\,[B \oplus D] + [A\bar{C} + \bar{A}C]\,[B \odot D]$$
$$= [A \odot C]\,[B \oplus D] + [A \oplus C]\,[B \odot D]$$

5.3 IC FAMILIES (TTL MOS AND CMOS

5.3.1 BIPOLAR LOGIC FAMILIES

- Main elements of a bipolar IC (integrated circuit) are resistors, diodes (which are also capacitors) and transistors.
- Two types of operations in Bipolar IC's:
  1. *Saturated*
  2. *Non-Saturated*

1. *In saturated logic* the transistors in the IC are driven to saturation so its operation is slow due to charge storage problem.
2. *In Non saturated* case the transistors are not driven into saturation so its operation is fast.

5.3.2 THE SATURATED BIPOLAR LOGIC FAMILIES ARE

1. Resistor-transistor logic (RTL)
2. Direct-couple transistor logic (DCTL)
3. Integrated-injected logic (I^2 L)
4. Diode-transistor logic (DTL)
5. High threshold logic (HTL)
6. Transistor-transistor logic (TTL)

The Non-Saturated Bipolar Logic Families

1. Schottky TTL
2. Emitter-coupled logic (ECL)

Unipolar Logic Families

MOS devices are unipolar devices and only Enhancement MOSFET'S are employed in MOS logic circuit.

1. PMOS
2. NMOS and
3. CMOS (complementary MOS)

- PMOS uses only p-channel MOSFET'S
- NMOS uses only n-channel MOSFET'S
- CMOS uses both n and p channel MOSFET'S fabricated on the same silicon chip.

IC classification	Equivalent individual Base gates	Numbers of components
SSI	Less than 12	Up to 99
MSI	12 – 99	100-999
LSI	100-999	1000-9999
VLSI	1000-9999	10,000-99,9999
ULSI	10,000 or more	100,000 and above

5.3.3 VARIOUS CHARACTERISTICS

Various characteristics of digital IC's are used to compare their performance are:

1. Speed of operation
2. Power dissipation
3. Figure of Merit
4. Fan out
5. Current and voltage parameters
6. Noise immunity
7. Operating temperature range
8. Power supply requirements
9. Flexibility available.
10. Digital Signal voltage level

Speed of Operation

The speed of a digital circuit is specified in term of the propagation delay times. The input and output waveforms of a logic gate is shown below:

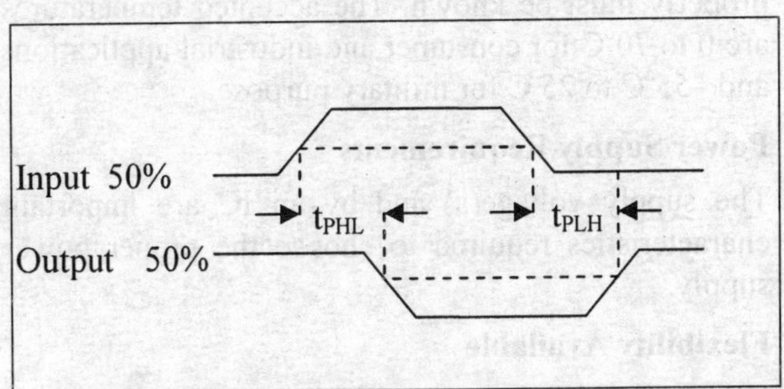

The delay times are measured between the 50% voltage levels of Input Waveform and output waveform.

There are two Delay Time

t_{PHL}: when the output goes from High to Low state.

$t_{PLH:}$ when the out put wave form goes low to high.

The propagation delay time of the logic gate is taken as the average of these two delay times.

Power Dissipation

This is the amount of power dissipated in an IC. It is determined by the current I_{cc}, that it draws from the Vcc Supply, and is given by $V_{cc,X}$ where $I_{cc.}$. I_{cc} is the average value of I_{cc} (0) and I_{cc} (1).

Figure of Merit

The figure of merit of a digital IC is defined as the product of speed and power. A low value of speed-power product is desirable.

Fan Out

This is the number of similar gates which can be driven by gate. High fan out is advantage because it reduces the need for addition drivers to drive more gates.

Current and Voltage, Parameters

Current parameters are I_{OH}, I_{OL}, I_{IH}, $I_{IL.}$ and voltage parameters are V_{OH}, V_{OL}, V_{IH}, V_{IL}.

Noise Immunity

As discussed earlier i.e. NM_H and NM_L, noise may cause the voltage at the input to a logic circuit to drop below V_{IH} or rise above V_{IL} and may produce undesired operation. The circuit ability to tolerate noise signal is referred to as the noise immunity.

Operating Temperature

The temperature range in which an IC functions properly must be known. The accepted temperatures are 0 to 70°C for consumer and industrial applications and –55°C to 25 C for military purpose.

Power Supply Requirements

The supply voltage(s) and by an IC are important characteristics required to choose the proper power supply.

Flexibility Available

Various flexibilities are available in different IC logic families and these must be considered.

- *Wire logic capability*: the output can be connected together to perform additional logic without any extra hardware.
- *Availability of complement output*: This eliminates the need for additional inverters.

Digital Signals Voltage Levels

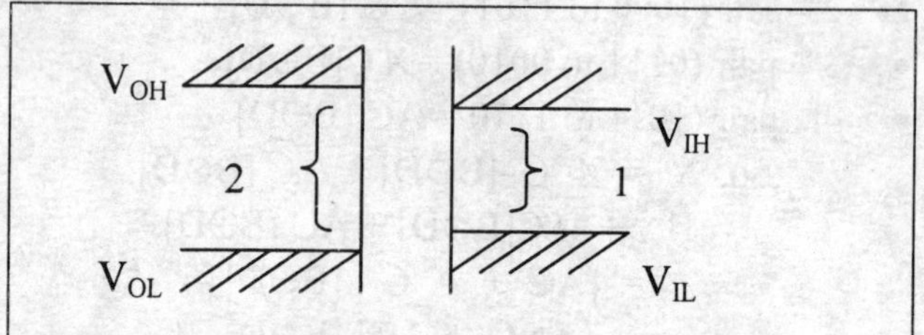

where, V_{OH} = Output voltage at high signal

V_{OL} = Output voltage at low signal

V_{IH} = Input voltage at high signal

V_{IL} = Input voltage at low signal.

(1) and (2) forbidden region.

- If Input voltage to any logic gate is less then V_{IH}, then input may be considered as low (i.e. = 0).
- If input voltage to any logic gate is greater then V_{IH}, then in out may be considered as high (i.e = 1).
- If output voltage of any logic gate is greater then V_{OH}, then output may be considered as high (i.e.= 1).
- If input to any logic gate lies in region (1) then input level is considered as forbidden similarly if output signal of any logic gate lies in region (2) then output is in forbidden region, so we cannot say whether output is high or low.

$NM_L = V_{IL} - V_{OL}$

$NM_H = V_{OH} - V_{IH}$

where, NM_H = Noise margin in high state.

NM_L = Noise margin in low state.

- Large noise margin better logic gate i.e. less effect of noise.
- FOM = Figure of Merit

P_{DP} = Power dissipation

T_D = delay

$FOM = T_D \times P_{DP}$

5.3.4 RESISTORS-TRANSISTOR LOGIC (RTL)

- RTL logic was the most popular form of logic in common use before the development of IC's.
- RTL circuit consists of resistors and transistors.
- The Basic RTL based NOR gate

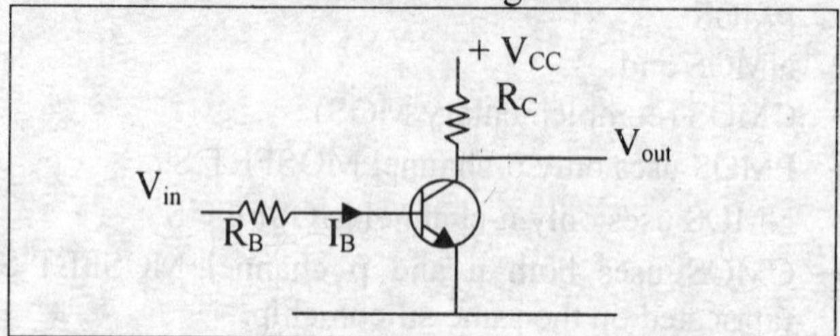

Logic Operation

Input representing the logic levels are voltage corresponding to low level should be low enough to drive the corresponding transistor to cut-off. Similarly, the input voltage corresponding to HIGH level should be high enough to drive the corresponding transistor to saturated.

5.3.5 DIRECT-COUPLED TRANSISTOR LOGIC (DCTL)

In the RTL, if the base resistors R_B are omitted we obtain, DCTL, in which input is directly coupled to Bases.

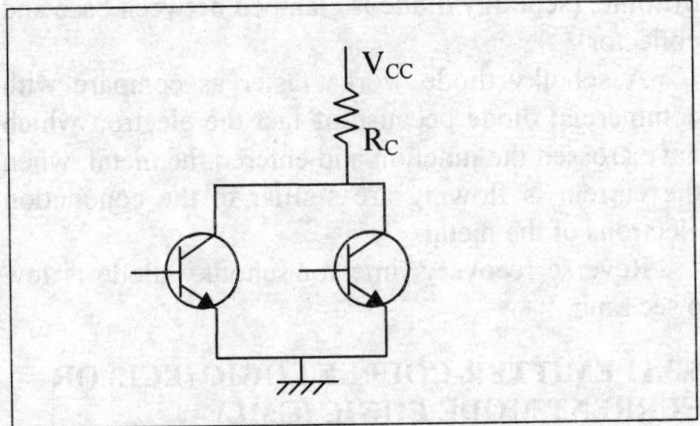

Disadvantage

Voltage corresponding to 1 → 0.8v

Voltage corresponding to 0 → 0.2v

Then logic Swing ; 0.8-0.2 = 0.6v

This gives very low noise margin.

Also, It is having problem of current hogging this problem may not occur if all transistors have same input characteristic but, unfortunately, the input characteristics differ due to the manufacturing tolerance of different IC. Owing to this difference, the saturation voltages of the load transistors may be different.

Let V_{BE} voltages of the transistors corresponding to saturation be 0.78, 0.79, 0.80 V. The first transistor when it enters saturation, will not allow other transistor to enter saturation and will take whole of the current supplies from the driver gate. This is known as current hogging.

5.3.6 INTEGRATED INJECTION LOGIC (I^2L)

- It is simple and uses very small silicon area, consumes very little power.
- Easier and cheaper to fabricate, so suitable for MSI and LSI.
- It uses the concept of merging the components i.e. one semi conductor region is part of two or more device, so also called as *(MTL) merged transistor logic*.

5.3.7 DIODE-TRANSISTOR LOGIC (DTL)

- DTL make use of diode AND or OR gate and transistor invertor in series for making it to be used as NAND and NOR operation.
- DTL is some what more complex than RTL, but due to **high fan out** and **improved noise margin,** it has replaced RTL.
  It is slower speed, poor noise margin.

Propagation Delays

Delays are associated with the turning-on and the turning-off of the output transistor. While turning on, any capacitance shunting the output of the gate discharges rapidly through the low impedance of the output transistor in saturation. On the other hand, at turn-off, the shunt capacitor must charge through the pull-up resistor R_e in addition to the storage time delay. The turn off delay is larger than the turn-on delay.

The propagation delay time of commercially available DTL gates are of the order of 30 to 80n sec.

5.3.8 HIGH-THRESHOLD LOGIC (HTL)

Due to the presence of Electric motors, on-off control circuit, high voltage switching the noise level is quite high and the logic family discussed so far, don't perform the intended function.

For this purpose, DTL gates has been redesigned with a higher supply voltage (15 V).

Diode have been replaced by a zener diode with V_z = 6.9 V and resistances have been increased so that Approximately the same current are obtained as in DTL.

- The propagation delay time is adversely affected due to large resistance value.
- The noise margin obtained with this circuit is typically 7 V.

5.3.9 TRANSISTOR TRANSISTOR LOGIC (TTL)

Difference Between TTL and DTL

The input diode in DTL are replaced by emitter-base Junctions of multiple-emitter transistor, which is easily and economically fabricated IC.

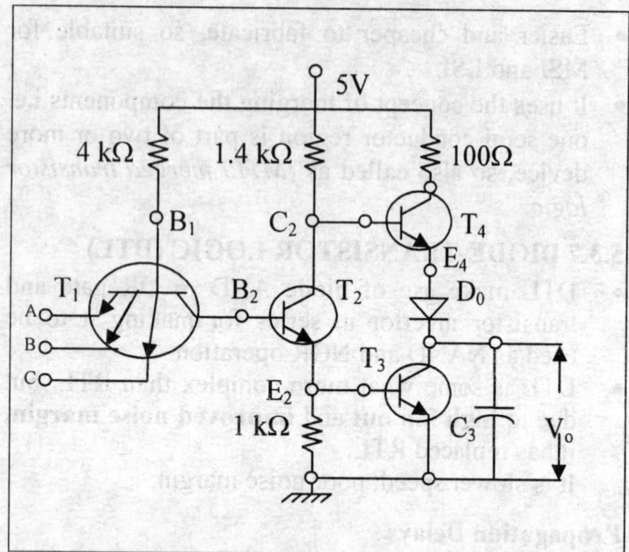

Totem-Pole Transistor

Q_4 is used because they produce low output impedance which act as power driver or power buffer.

TTL Based NAND Gate and its Operation

Diagram of a TTL NAND gate with totem-pole out put. Active pull up is provided by T_4, D_0 and100 ohm resistor.

Condition

At least one input is LOW. The emitter-Base junction. of T_1 corresponding to the input in the low state is forward biased making voltage of (B_1).

$V_m = 0.2 + 0.7 = 0.9$ V

This value of V_m causing T_2 and T_3 to turn off and hence V_{out} charged to V_{cc} so $V_{out} = V_{cc}$.

High Power and Low Power TTL

If we reduce the values of the resistance used it will result in more flow of current and the switching speed of the circuit is increased by factor of 2.

By increasing resistance of the circuit the power dissipation is decreased by factor of 10 and speed by factor of 3 as compared to that of a standard TTL.

5.3.10 SCHOTTKY TTL

Speed limitation of TTL is mainly due to the turn-off time delay involved in transistor while making transition from saturation to cut-off (storage time delay). This can be eliminated by replacing the transistors of TTL gate by schottky transistor.

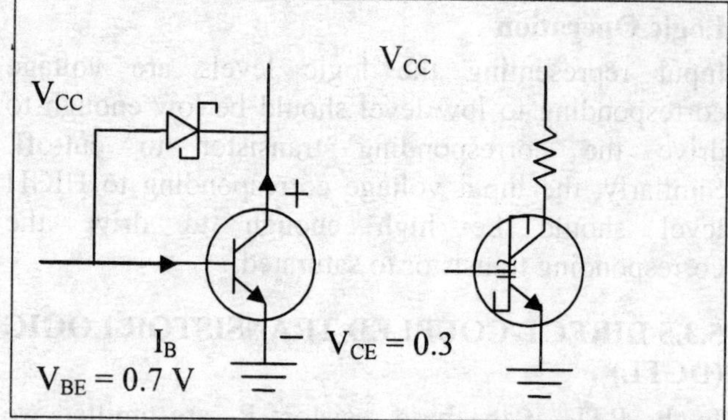

This schottky transistor prevent transistor from entering saturation and hence, there is saving in turn-off time. (schottky diode is clamped between base and collector).

A schottky diode works faster as compare with commercial diode because of fact the electron which have crossed the junction and entered the metal, when the current is flowing are similar to the conduction electrons of the metal.

Reverse recovery time for schottky diode is few p sec rang.

5.3.11 EMITTER-COUPLE LOGIC (ECL) OR CURRENT MODE LOGIC (CML)

ECL is the fastest of all logic families so used in application where very high speed is essential.

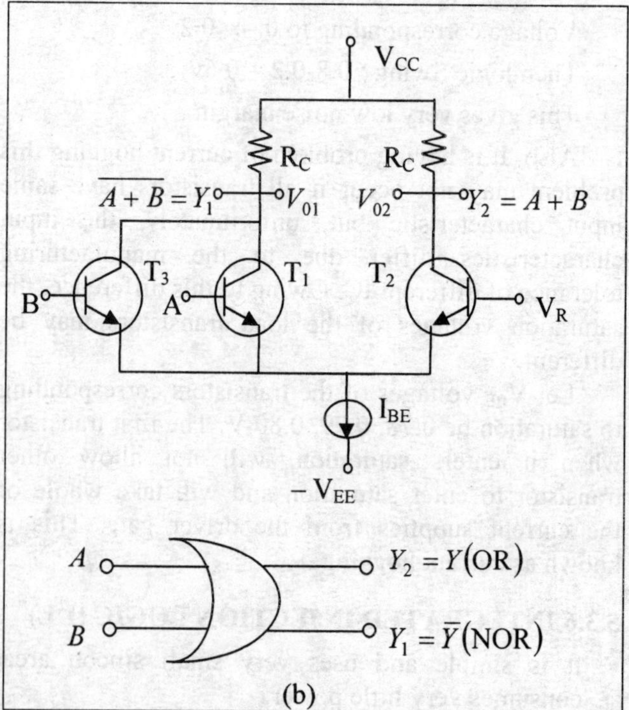

(b)

- High speed is possible in ECL because the transistors are used in difference amplifier either in active or cut-off region and thereby the storage time is eliminated.
- ECL is realized using difference amplifier in which the emitter of the two transistors are connected so called emitter-coupled logic.
- Most of the super fast computer and high speed special purpose computer used these ECL logic families.
- Emitter follower are used for d.c. level shifting of the output. So, that V (0) and V (1) are same for the input and the output.
- Two output Y_1 and Y_2 are available in this circuit Which are complementary. Y_1 correspondent to NOR logic and Y_2 to OR logic.
- In ECL, the positive end of the supply is connected to ground in contrast to other logic families in which negative end of the supply is grounded. This is done to minimize the effect of noise induced in the power supply.
- Fan out is large-why.
- If all the input are low, the input transistors are cut-off. Therefore, the input resistance is very high. On the other hand, if an input if HIGH, the input resistance is that of emitter followers which is also high. So, input impedance is always high.
- The output resistance is either that of an emitter followers or the forward resistance of a diode, which is always low, because of low output impedance and high input impedance, *the fan out is large.*
- The output of two or more ECL gates can be connected to obtain additional logic without using additional hardware.

Unconnected Input

If any input of an ECL gate is left unconnected the corresponding E-B junction of the input transistor will not be conducting, so can be treated as logical 0.

5.3.12 MOS-LOGIC

- High density of fabrication and low power dissipation.
- MOS devices used in logic circuits, are only p or n-channel.
- CMOS lowest power dissipation.

CMOS Based Inverter

This is basic building Block for CMOS circuit:

Circuit diagram

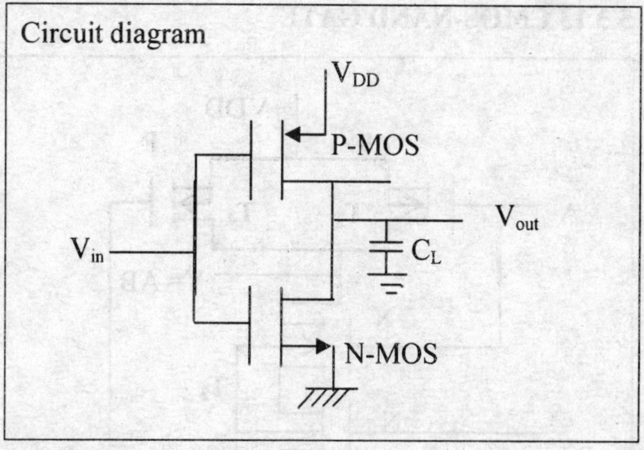

Operation

When $V_{in} > (V_{th})$

$(V_{thn}) \rightarrow$ threshold voltage for NMOS)

NMOS turned ON and V_{out} is '0' as Capacitor discharged to ground.

When V_{in} low which cause PMOS to turn ON and NMOS is turn off, the Capacitor starts charging by supply voltage V_{DD}.

So, $V_{in}(1) \rightarrow V_{out}(0)$

$V_{in}(0) \rightarrow V_{out}(1)$

- No static power dissipation as at a time only one of the MOS, either PMOS or NMOS is turn ON ,so there is no direct path from V_{DD} to ground, so static power consumption is zero (ideally).
- During transition from zero to one (for input signal), there will be voltage level at which Both NMOS and PMOS turned ON, giving direct path from supply to ground so there is power dissipation. This power loss is termed as *Dynamic Power dissipation.*
- *Dynamic Power dissipation* is given by

$$P_d = (1/2)\, C_L\, V_{DD}^2\, f$$

where, P_d = Dynamic power Dissipation

V_{DD} = Supply voltage

f = Switching frequency

C_L = Load Capacitor

- Strongly depends on Supply voltage, so to reduce power dissipation reduce supply voltage, but to that limit so that noise can't effect circuit operation.
- Reduce load capacitance.
- Reduce the frequency of operation but this will reduce frequency at which we want to operate my circuit.

5.3.13 CMOS-NAND GATE

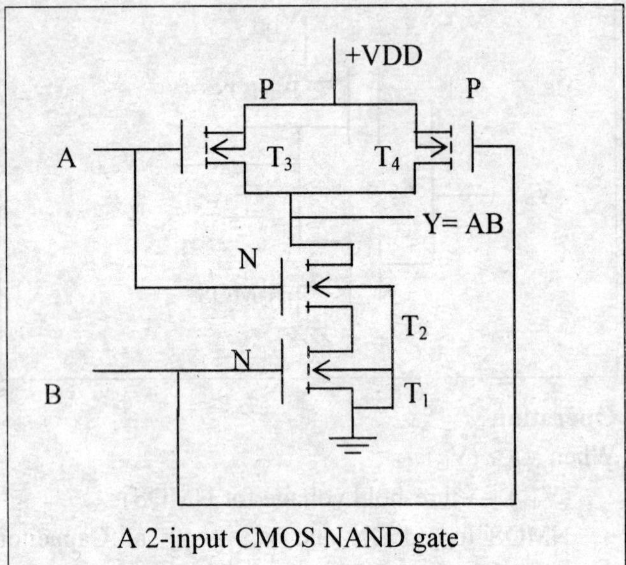

A 2-input CMOS NAND gate

Operation

When Both Signal (A and B) are high, Both NMOS turned ON and output goes to zero. For rest of combination at least one of the NMOS is OFF and at least one PMOS is ON which cause V_{out} to go high.

Realization of Any Boolean Function in CMOS

Realization of Boolean function in CMOS needs following steps.

Complement the function to be realized.

If product term is there, then use two NMOS in series and if add term is there then use to NMOS in parallel.

If NMOS is found then what is series in NMOS, just put them in parallel using PMOS. And what is parallel in NMOS just put them in series using PMOS.

Example: $f = ab + cd$

Step 1: Find complement of this function

$\bar{f} = abcd$

Step 2: There is two term (ab) and (cd) which is product term, it is realized by series NMOS separately.

Then for OR ing ab and cd used two NMOS combination in parallel.

Step 3: To find PMOS Block, just to do manipulation. Whatever there is in NMOS. For example (a and b) NMOS ~~are~~ in series put them in parallel (in PMOS). Similar for (c and d) NMOS's

After this since (a and b) and (c and d) are parallel in NMOS, then put them in series for (PMOS) version.

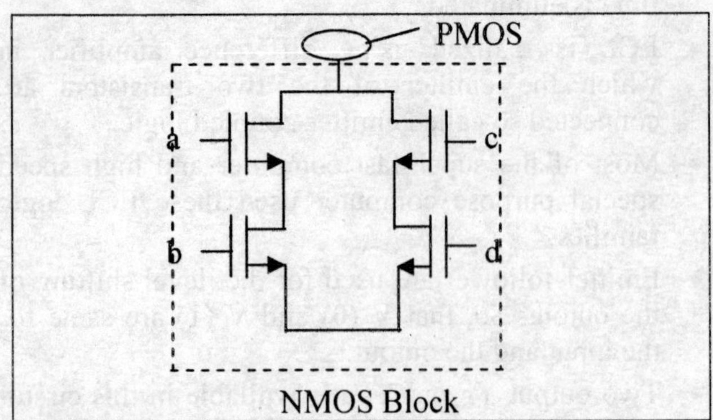

PMOS Block

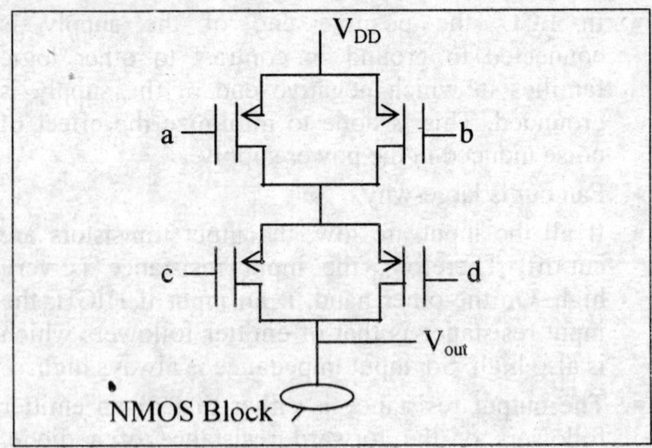

Full Schematic

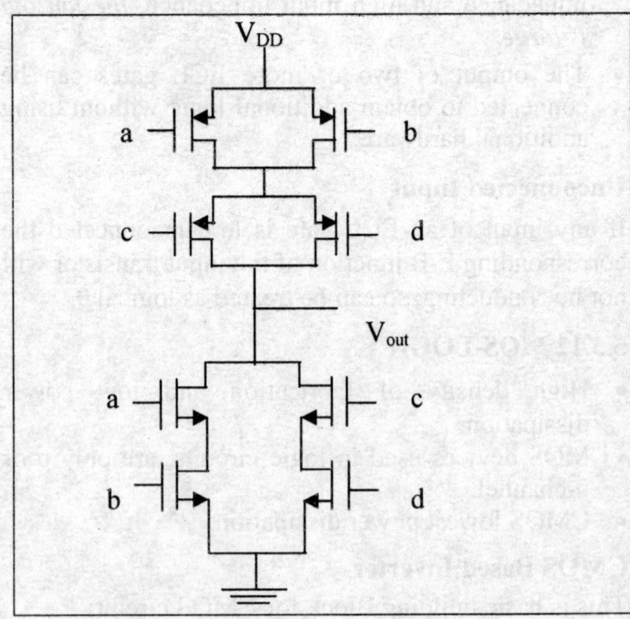

5.3.14 WIRED LOGIC

The output impedance of a CMOS gate is 1 k and connecting gates directly in parallel would draw large current (mA). A special circuitry is devised to achieve wired-OR connection. This is known as **transmission gate.**
Wired logic must not be used for CMOS logic circuit.

CMOS Transmission Gate:

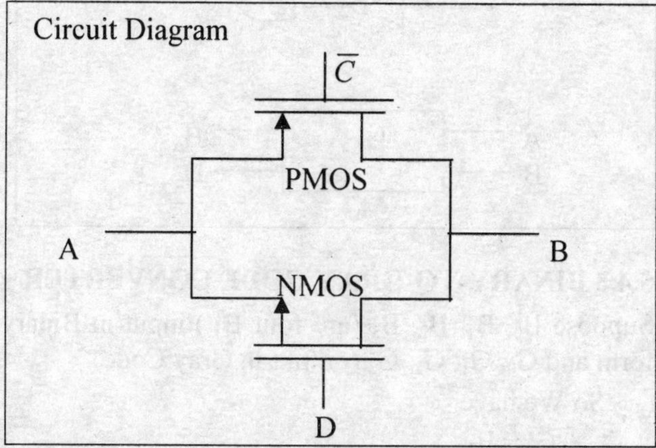

- Controlled by gate voltage C and C'. when C = 1 and C′ = 0 assume A= V(1), then T_1 is OFF and T_2 conducts in the ohmic region, So T_2 behaves as a small resistance connecting the output to the input and B = A = V(1). If A = (0), T_2 is OFF and T_1 is ON, connecting the output to the input and B = A = V (0).
- If C = 0, No transmission is possible.
- **Noise Margin:** Noise margin of CMOS logic IC's is higher than that of TTL.
- **Unconnected Input:** unused input must be connected to either the supply voltage terminal or one of the used input provided that the fan out of the signal source is not exceeded.

5.3.15 LOGIC FAMILIES COMPARISON

Logic	RTL	DTL	HTL	TTL	ECL	MOS	CMOS
Gate	N1	N2	N2	N2	N1 /N2	N2	N1/ N2
Fan-out	5	8	10	10	25	20	50
Power loss mW	12	8	50	12	40	0.2	0.01
Noise	N	G	E	VG	G	N	VG
Propgation nsec	12	30	90	12	4	300	70

* N → Normal, * S → Satisfactory,
* G → Good * VG → Very good
* E → Excellent * N1 → NOR
* N2 → NAND * N3 → OR

5.4 ARITHMETIC CIRCUITS.

5.4.1 HALF ADDER

A logic circuit for the addition of two one-bit number is referred to as an half-adder.

$$S = A \oplus B$$
$$C = AB$$

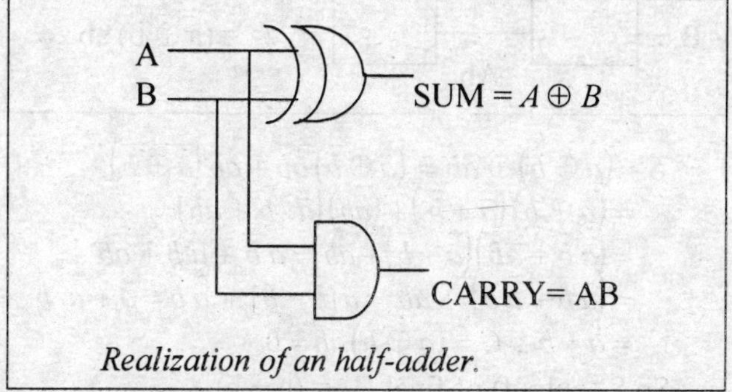

Realization of an half-adder.

5.4.2 FULL ADDER

An half adder has only two inputs and there is no provision to add a carry coming from the lower order bits when multibit addition is performed. For this purpose an extra (third) input is added and this circuit is used to add A_n, B_n and C_{n-1}.

We have $S_n = A_n \oplus B_n \oplus C_{n-1}$

$$C_n = A_n B_n + B_n.C_{n-1} + A_n.C_{n-1}$$

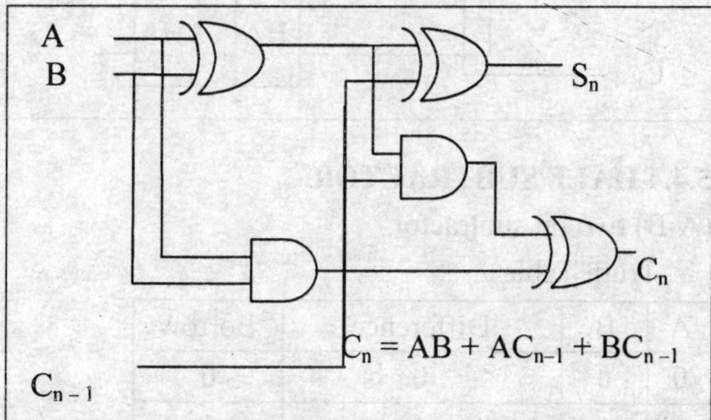

5.4.3 REALIZATION OF FULL ADDER USING HA AND OR GATE

Full adder can be realized with two half Adder (HA) and OR gate.

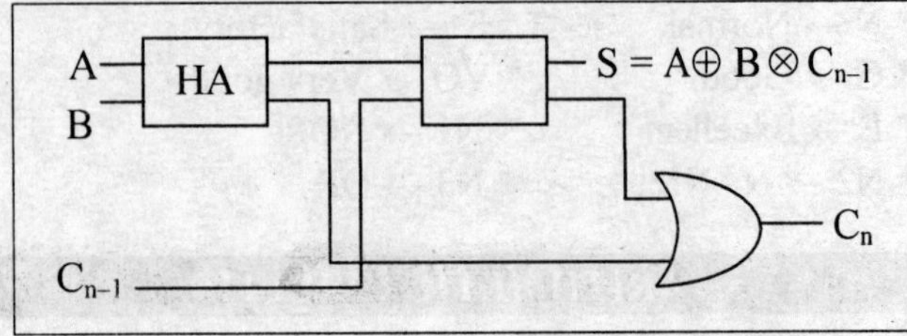

OR gate also be realized by cascading two HA as shown below:

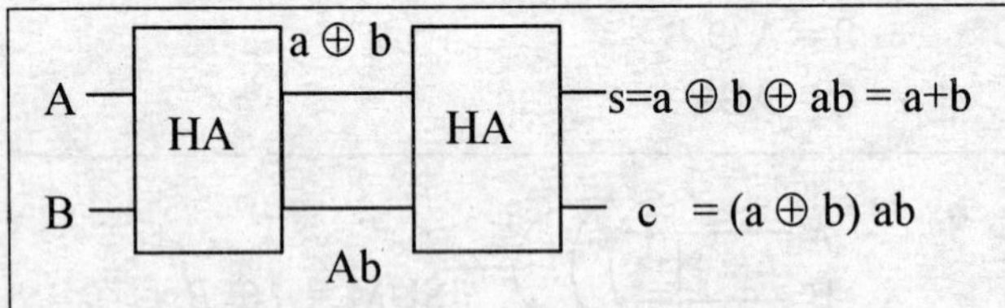

$$\begin{aligned} S &= (a\oplus b)\oplus ab = (a\oplus b)\overline{ab} + ab\left(\overline{a\oplus b}\right) \\ &= (a\oplus b)(\bar a + \bar b) + (ab)(\bar a\cdot\bar b + ab) \\ &= (a\bar b + \bar a b)(\bar a + \bar b) + ab = a\bar b + \bar a b + ab \\ &= (a\bar b + \bar a b) + ab = a[\bar b + b] + \bar a b = a + \bar a\ b \\ &= a + b\ ;\ C = (a\oplus b)\,ab = 0 \end{aligned}$$

So S (output) of final HA= (a + b)

So Cascaded HA behaves as OR gate.

So realization of Full adder only with using HA.

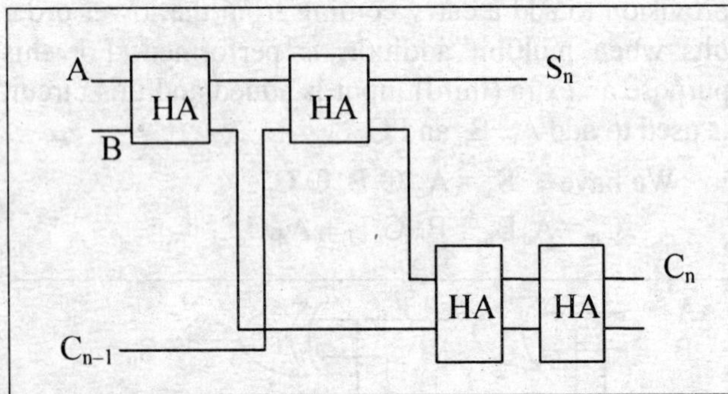

5.4.4 HALF SUBTRACTOR

(A-B) two bit subtractor

Truth Table

A	B	Difference	Borrow
0	0	0	0
0	1	1	1
1	0	1	0
1	1	0	0
		$A\oplus B = \bar A B + \bar B A$	$\bar A\ B$

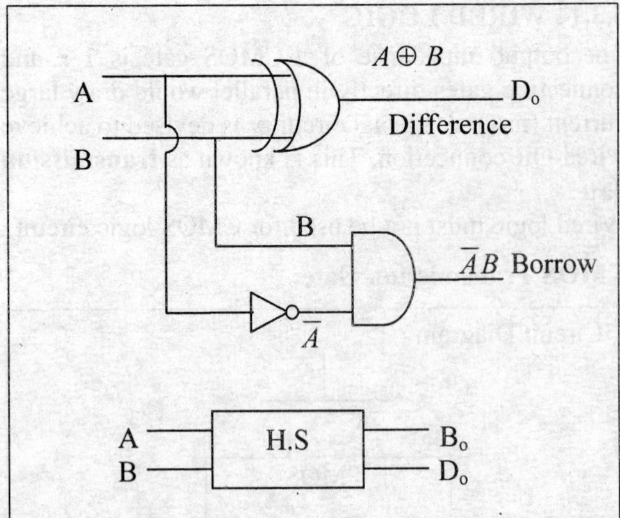

5.4.5 BINARY-TO-GRAY CODE CONVERTER

Suppose B_0, B_1, B_2, B_3 (are four Bit) input in Binary form and G_0, G_1, G_2, G_3 (output in Gray Code).

So We have

$G_3 = B_3$

$G_2 = B_2 \oplus B_3$

$G_1 = B_1 \oplus B_2$

$G_0 = B_0 \oplus B_1$

Example: Binary Input: 1 0 0 0 1 0

Gray : 1 1 0 0 1 1

i.e. MSB remain same and rest of Bit is obtain by EX-OR of Previous and current bit.

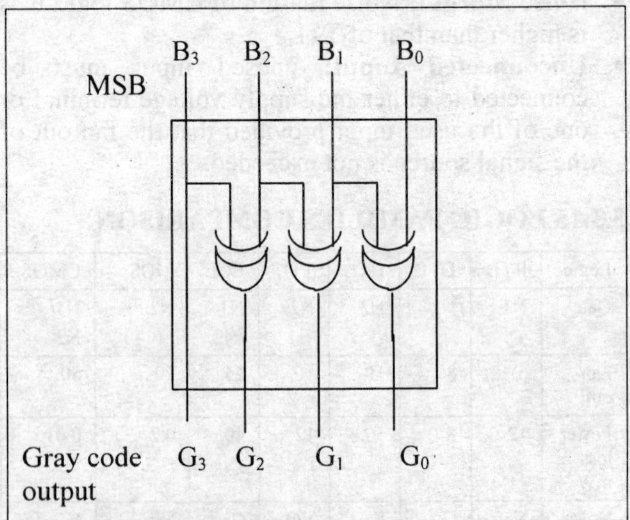

5.4.6 GRAY-TO-BINARY CODE CONVERTER

Gray code input ($G_3\,G_2\,G_1\,G_0$)

Binary output (B_3 B_2 B_1 B_0)

$B_3 = G_3$

$B_2 = G_2 \oplus G_3$

$B_1 = G_1 \oplus G_2 \oplus G_3$

$B_0 = G_0 \oplus G_1 \oplus G_2 \oplus G_3$

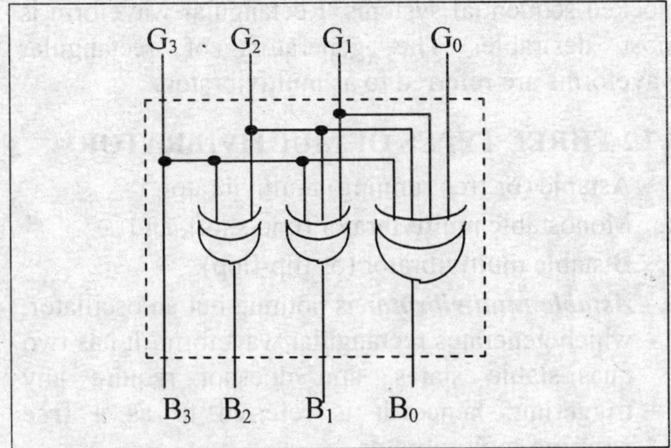

Example:

Gray code: 1 1 0 0 1 1

Binary code: 1 0 0 0 1 0

5.4.7 ERROR DETECTING CODE

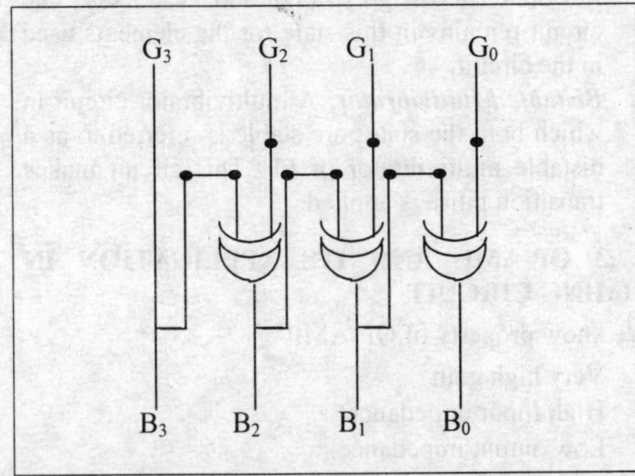

5.5 COMPARATORS

5.5.1 One Bit Comparator

Suppose you have two, one bit number A and B.

A > B only if A = 1 and B = 0 i.e. $C = A\overline{B} = 1$

A < B only if A = 0 and B = 1 i.e. $D = \overline{A}B = 1$

A = B if (A = 1 and B = 1) or (A = 0 and B = 0)

So,

A > B should give high signal when $A\overline{B}$ is high

A < B should give high signal when $\overline{A}B$ is high

A = B should give high signal when $\overline{\overline{A}B + A\overline{B}}$ is high

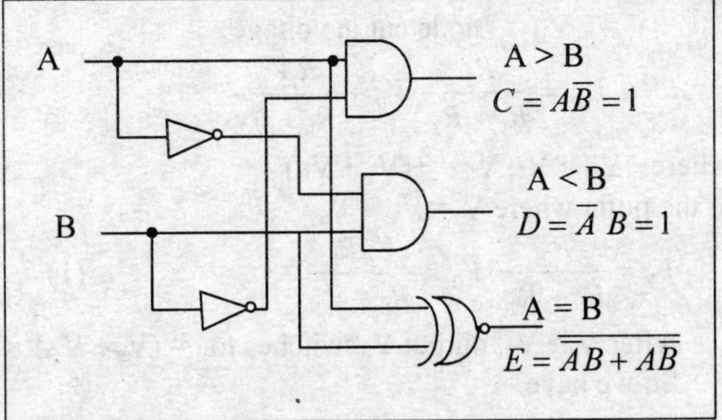

5.5.2 FOUR BIT COMPARATOR

$A = A_3\ A_2\ A_1\ A_0$

$B = B_3\ B_2\ B_1\ B_0$

For A > B

$$C = A_3\overline{B}_3 + E_3A_2\overline{B}_2 + E_3E_2A_1\overline{B}_1 + E_3E_2E_1\ A_0\overline{B}_0$$

For A < B

$$D = \overline{A}_3B_3 + E_3\overline{A}_2B_2 + E_3E_2\overline{A}_1B_1 + E_3E_2E_1\ \overline{A}_0B_0$$

For A = B

$$E = E_3E_2E_1E_0$$

where, $E_n = A_n\overline{B}_n + \overline{A}_nB_n$

5.6 SCHMITT TRIGGER

5.6.1 REGENERATIVE COMPARATOR (SCHMITT TRIGGER)

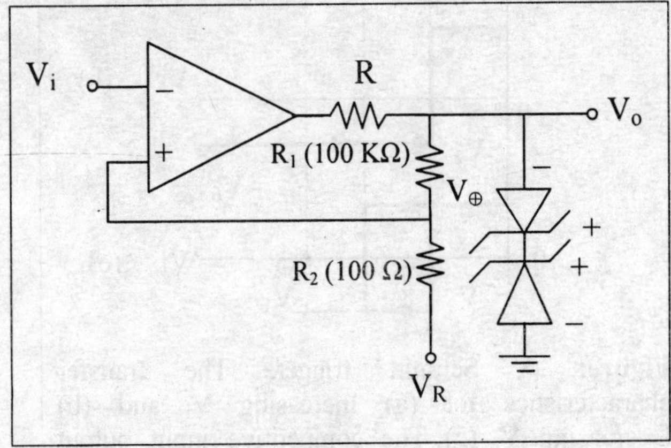

R: used for current limiting.

Output voltage is referred by using two zener diode Back to back.

$$V_0 = (V_z + V_D) \text{ for } V_\oplus > V_i$$

$$V_0 = (V_z + V_D) \text{ for } V_\oplus < V_i$$

where, V_z = Zener breakdown voltage

V_D = Diode cut in voltage

$$V_\oplus = \frac{R_2}{R_1 + R_2} \cdot V_0 + \frac{R_1 V_R}{R_1 + R_2}$$

where, $V_\oplus > V_1$, $V_0 = +(V_Z + V_D)$
at the point where $V_1 = V_\oplus = V_1$

$$V_1 = \frac{R_2}{R_1 + R_2} \cdot V_0 = \frac{V_R R_1}{R_1 + R_2} \quad (1)$$

After $V_1 > V_2$, output V_0 switches to $-(V_z + V_D)$
So we have

$$V_2 = \frac{R_2}{R_1 + R_2} \cdot V_0 = \frac{V_R R_1}{R_1 + R_2} \quad (2)$$

From (1) and (2) we can see that

$$V_2 < V_1$$

Conclusion

For increasing value of V_i output switches from V_0 to $-V_0$ after $V_i > V_1$ and it will continue to be same when Vi is decreasing then output changes from $-V_0$

V_0 at the value of V_2 which is less then that of V_1. Due to the difference in V_1 and V_2, switching point, we referred this situation as hysteresis voltage.

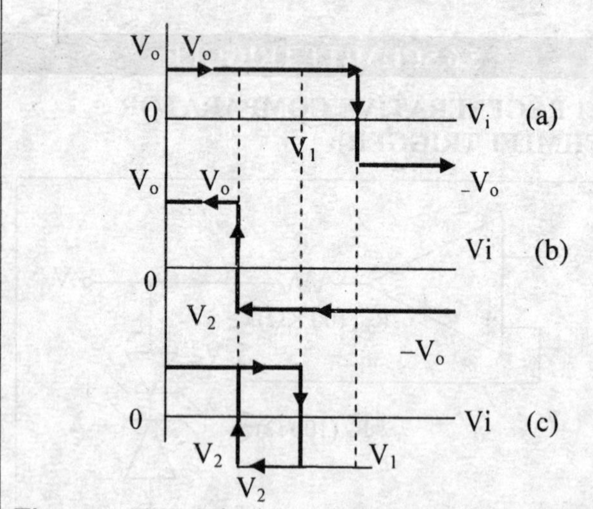

Figure: A Schmitt trigger, The transfer characteristics for (a) Increasing V_i and (b) Decreasing V_i (c) The competitive input output curve

5.7 TIMERS AND MONO STABLE MULTI VIBRATOR

5.7.1 TIMING CIRCUIT

Digital systems require some kind of a timing waveform for example, a source of trigger pulses, for clocked sequential systems. Rectangular waveform is most desirable. The generators of rectangular waveforms are referred to as multivibrators.

5.7.2 THREE TYPES OF MULTIVIBRATOR

1. Astable (or free running) multivibrator.
2. Monostable multivibrator (one-shot), and
3. Bistable multivibrator (or flip-flop).

1. ***Astable multivibrator*** is nothing but an oscillator, which generates rectangular waveform. It has two quasi-stable states, and doesnot require any triggering; hence it is referred to as a **free running** multivibrator.

2. ***Monostable Multivibrator:*** has one stable state i.e. under steady state condition, its output is fixed it is either in the low or the high state. When the circuit is triggered with an externally applied pulse it goes into the other state, i.e. if it was in low state, it will go to high and vice-versa. The circuit remains in this state for the elements used in the circuit.

3. ***Bistable Multivibrator:*** A multivibrator circuit in which both the states are stable is referred to as a bistable multivibrator or FF. This circuit makes transition pulse is applied.

5.7.3 OP-AMP AND ITS APPLICATION IN TIMING CIRCUIT

We know property of OP-AMP

- Very high gain
- High Input impedance
- Low output impedance

Block Diagram of OP-AMP

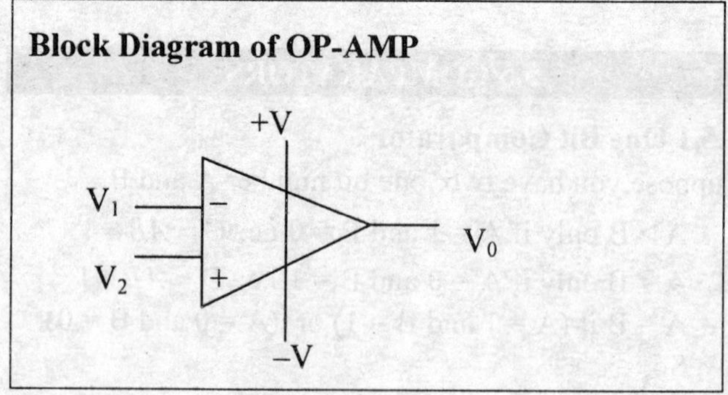

- Labeled negative and positive referred as inverting and non-inverting inputs, respectively.
- The input V_1 is applied between the inverting input terminal and ground, and input V_2 is applied between the non-inverting input terminal and ground.
- The input terminals are called differential input and the output is single-ended.
- The output voltage depends on the difference voltage

$$V_i = V_1 - V_2$$
$$V_0 = A_v.V_i$$

where $A_v \rightarrow$ is the voltage gain of the amplifier.

5.7.4 MONOSTABLE MULTIVIBRATOR

Monotype multivibrator has one stable state and one quasistable state. The circuit remains in its stable state until an external triggering pulse causes a transition to the quasi-stable state. The circuit returns to its stable state after a time T. Hence, it generates a single output pulse in response to an input and is referred to as one-shot or single shot.

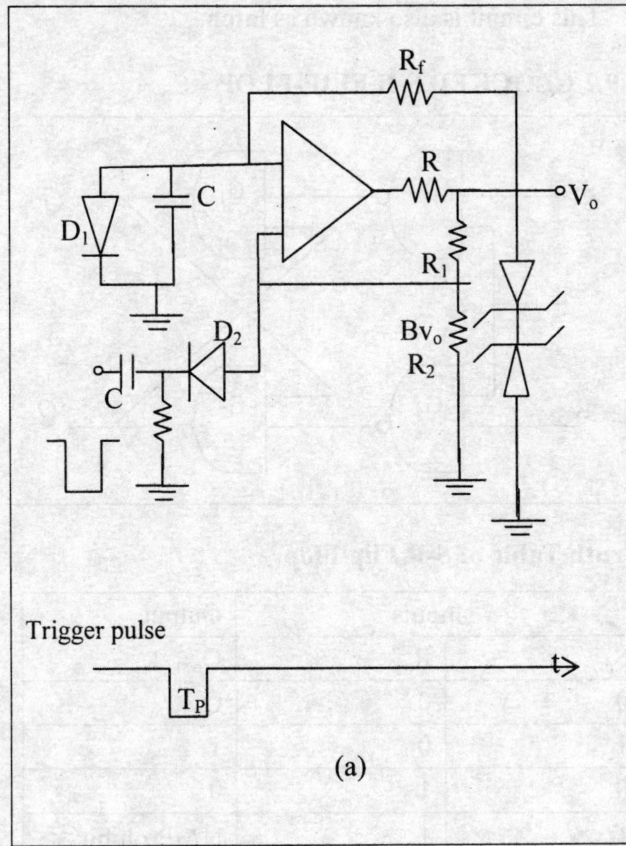

(a)

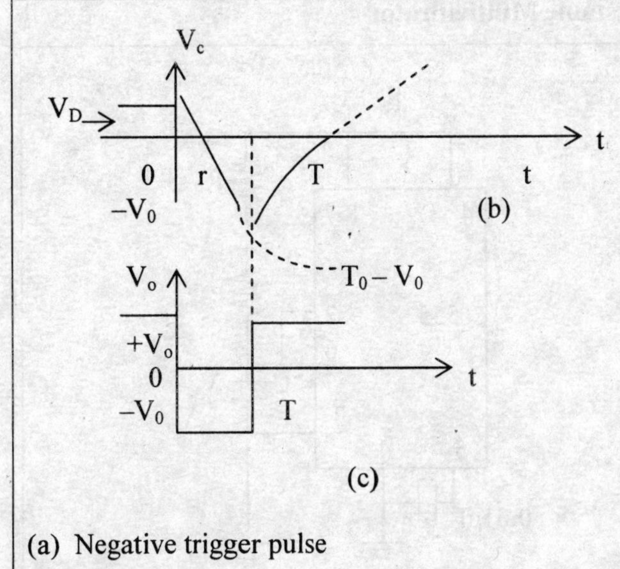

(a) Negative trigger pulse
(b) waveforms of V_c
(c) Vo

$$T = 0.69\ R_f C$$

5.7.5 555 TIMER

Most popular and versatile sequential logic devices which can be used in monostable and astable modes.

Monostable Multivibrator

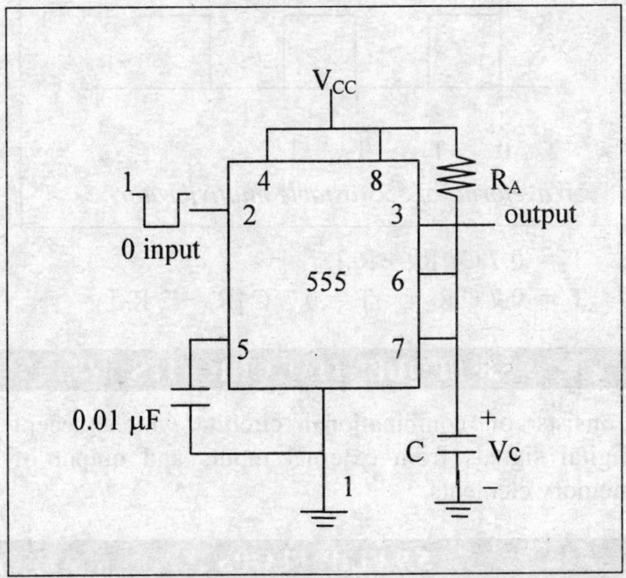

A monostable multivibrator using 555.

$$T = 1.1\ C.R_A$$

Astable Multivibrator

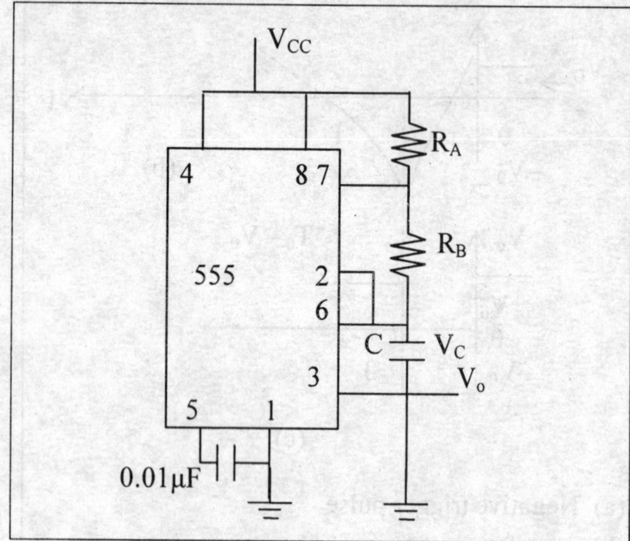

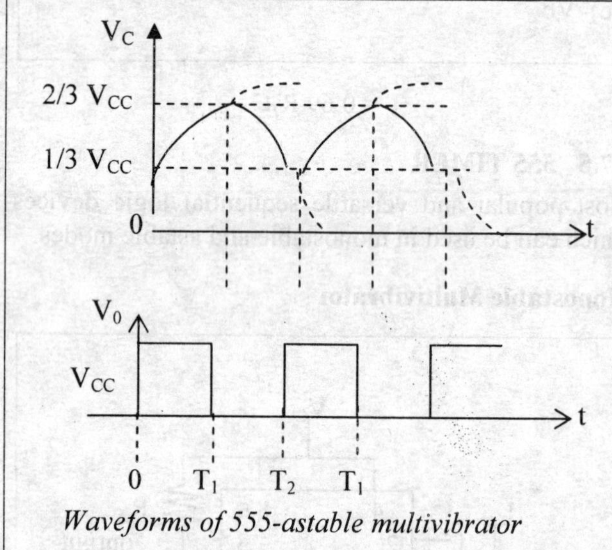

Waveforms of 555-astable multivibrator

$T_1 = 0.7\, C\, [R_A + R_B]$

$T_2 = 0.7\, C\, R_B$; $T = 0.7\, C\, [R_A + 2\, R_B]$

5.8 SEQUENTIAL CIRCUITS

Consists of combinational circuit, which accept digital signals from external inputs and output of memory elements.

5.9 FLIP FLOPS

5.9.1 A 1-BIT MEMORY CELL

The basic digital memory circuit is known as *flip-flop*. It has two stable, 1 state and 0 state.

It consists of two inverters G_1 and G_2 (NAND gates based).

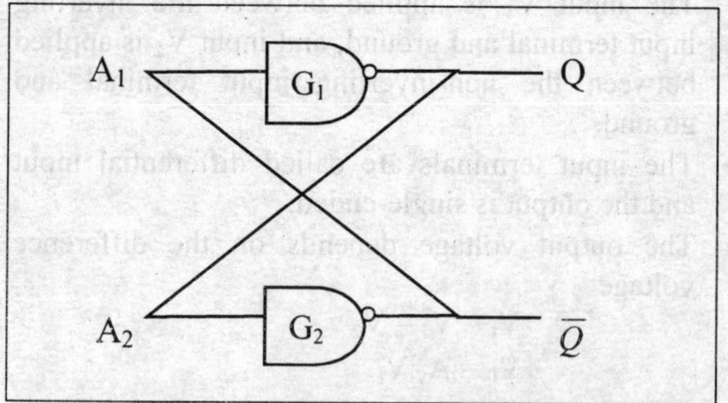

- The output Q and $\overline{Q}$ are complementary.
- The circuits has two states, in one of the stable state Q = 1 which is known as (Set state), where as in the other hand, other stable Q = 0 known as (reset state).
- If the circuit is in state 1, it continues to remain in this state and similarly if it is in 0 state. This property is known as memory, it can store 1-bit of digital information.
- This circuit is also known as **latch.**

5.9.2 CLOCKED S-R FLIP-FLOP

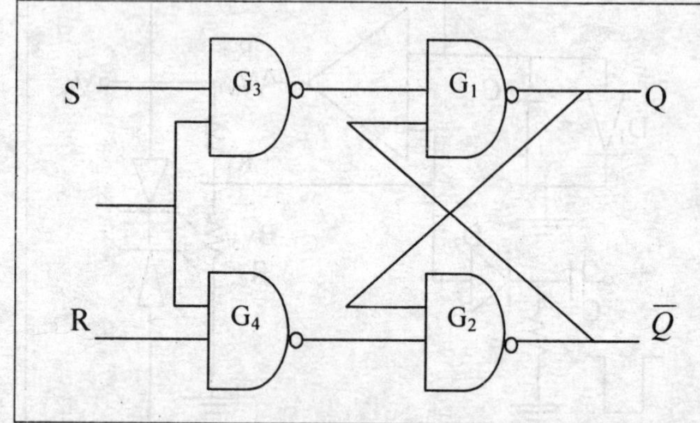

Truth Table of S-R Flip-Flop

Inputs		Output
S_n	R_n	Q_{n+1}
0	0	Q_n
1	0	1
0	1	0
1	1	NA/Prohibit

State Table

Q_n	S_n	R_n	Q_{n+1}
0	0	0	0
	0	1	0
	1	0	1←
	1	1	?
1	0	0	1←
	0	1	0
	1	0	1←
	1	1	?

$$Q_{n+1} = \overline{Q}_n S_n \overline{R}_n + Q_n \overline{S}_n \overline{R}_n + Q_n S_n \overline{R}_n$$

	$\overline{S}_n\overline{R}_n$	$S_n\overline{R}_n$	$\boldsymbol{S_nR_n}$	$\overline{S}_nR_n$
$\overline{Q}_n$		1		
$\boldsymbol{Q_n}$	1	1		

$$Q_{n+1} = S_n \overline{R}_n + \overline{R}_n Q_n$$

Excitation table for S-R

Q_n	Q_{n+1}	S_n	R_n
0	0	0	ϕ
0	1	1	0
1	0	0	1
1	1	ϕ	0

where, ϕ = (don't care)

The Race-Around Condition

The difficulty of both input 1 (S = R = 1) being not allowed in an S-R flip flop is eliminated in J-K flip flop by using the feedback connection from outputs to the input of the gates. When J = K = 1 and Q = 0 and a pulse as shown is applied at the clock input.

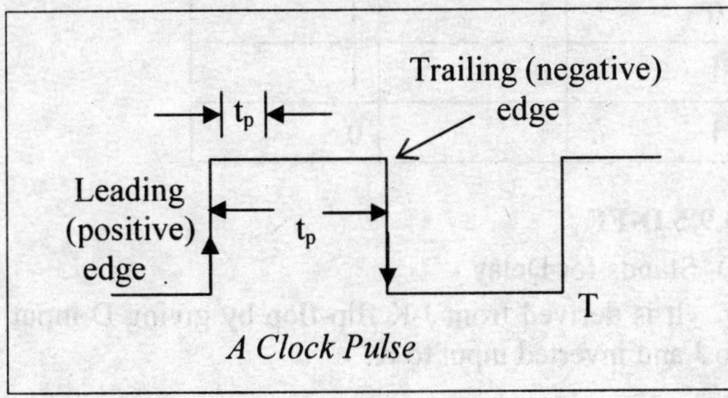

A Clock Pulse

5.9.3 J-K FLIP FLOP

J-K FF is called a **universal flip-flop** because the other FF_s like D, S-R, T can be derived from it. It is a versatile device.

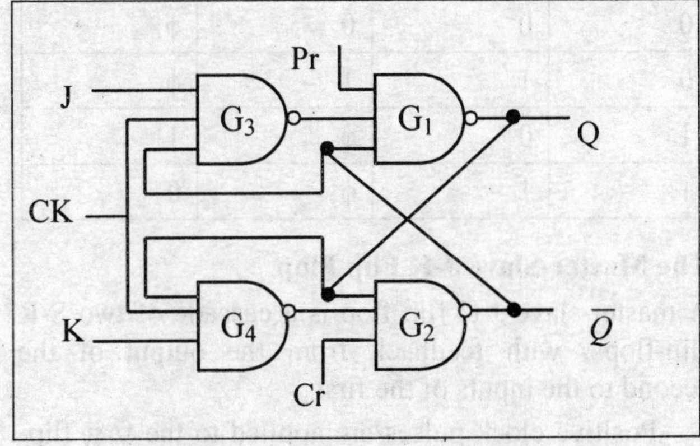

Pr = Present and

C1r = clear (are asynchronous input)

Truth Table for JK

Inputs		Output
J_n	K_n	Q_{n+1}
0	0	Q_n
0	1	0
1	0	1
1	1	$\overline{Q}_n$

State Table for JK

Q_n	J_n	K_n	Q_{n+1}
0	0	0	0
	0	1	0
	1	0	1←
	1	1	1←
1	0	0	1←
	0	1	0
	1	0	1←
	1	1	0

$$Q_{n+1} = \overline{Q}_n J_n \overline{R}_n + \overline{Q}_n J_n k_n + Q_n \overline{J}_n \overline{K}_n + Q_n J_n \overline{K}_n$$

	$\overline{J}_n \overline{K}_n$	$\overline{J}_n K_n$	J_nK_n	$J_n \overline{K}_n$
$\overline{Q}_n$			**1**	**1**
$\boldsymbol{Q_n}$	**1**			**1**

$$Q_{n+1} = J_n \overline{Q}_n + \overline{K}_n Q_n$$

Excitation Table for JK

Q_n	Q_{n+1}	J_n	K_n
0	0	0	φ
0	1	1	φ
1	0	φ	1
1	1	φ	0

The Master-Slave J-K Flip Flop

A master-slave J-K flip flop is a cascade of two S-R flip-flops, with feedback from the output of the second to the inputs of the first.

Positive clock pulses are applied to the first flip-flop (master) and the clock pulses are inverted before these are applied to the second flip flop (slave), master is positive edge triggered and slave is negative edge triggered.

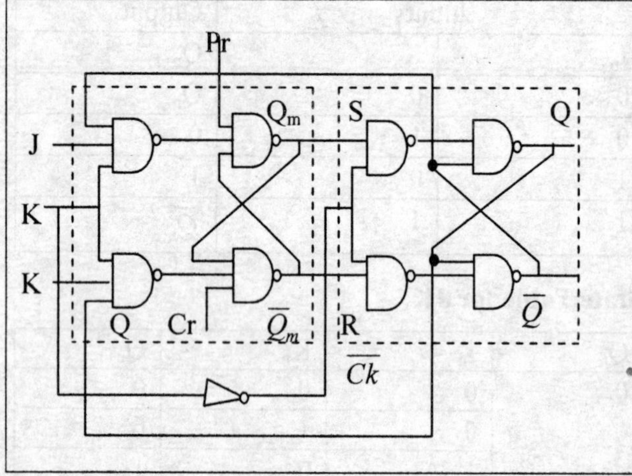

When CK = 1, the first flip flop is enabled and the output Q_m and $\overline{Q}_m$ respond to the input J and K according to the truth table of J-K. At this time, the second flip-flop is inhibited because its clock is low $(\overline{CK} = 0)$, when CK goes low, the first flip-flop is inhibited and the second flip flop is enabled. Therefore, the output Q and $\overline{Q}$ follow the output Q_m and $\overline{Q}_m$. Since the second flip flop simply follows the first one, it is referred to as the slave and first one as the master. Hence this is referred to as Master-Slave (JK) flip-flop.

5.9.4 T- FF

T-stand for Toggle

It is derived from J-K flip-flip by joining J and K input.

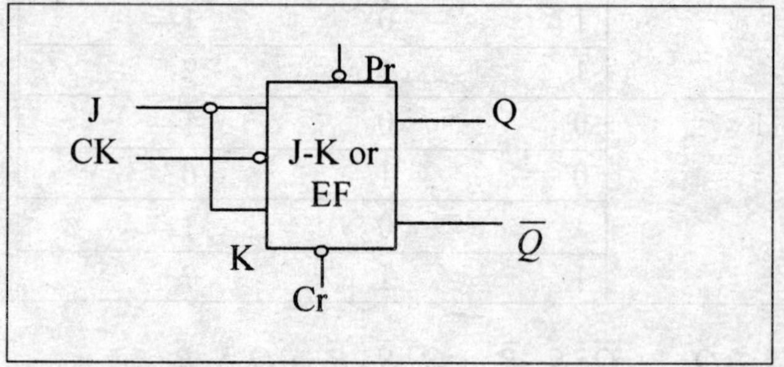

Truth Table

T_n	Q_{n+1}
0	Q_n
1	$\overline{Q}_n$

State Table

Q_n	T_n	Q_{n+1}
0	0	0
	1	1←
1	0	1←
	1	0

$$Q_{n+1} = T_n \overline{Q}_n + \overline{T}_n Q_n = Q_n \oplus T_n$$

Excitation Table For T FF

Q_n	Q_{n+1}	T_n
0	0	0
0	1	1
1	0	1
1	1	0

5.9.5 D-FF

D- Stands for Delay.

It is derived from J-K flip-flop by giving D input to J and inverted input to K.

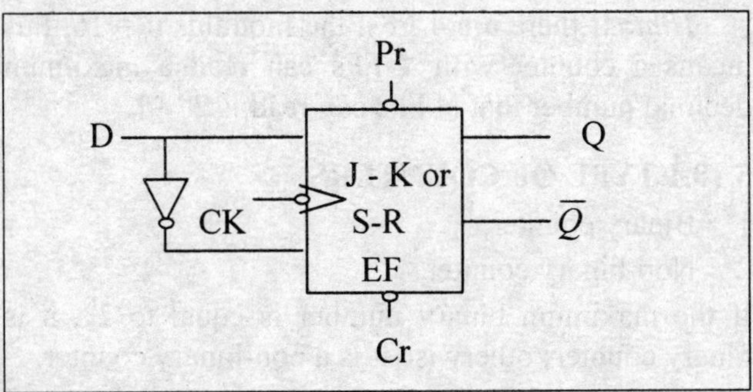

Truth Table for D-FF

D_n	Q_{n+1}
0	0
1	1

State Table

Q_n	D_n	Q_{n+1}
0	0	0
	1	1←
1	0	0
	1	1←

$Q_{n+1} = D_n$

Excitation Table for DFF

Q_n	Q_{n+1}	D_n
0	0	0
0	1	1
1	0	0
1	1	1

5.9.6 CONVERSION OF FFS

Conversion of one FF to another is done by preceding the given FF be a combinational logic circuit in following steps:

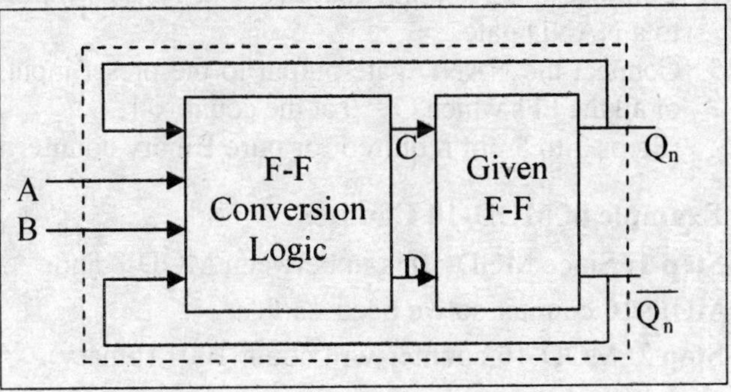

1. Write excitation table for available/given FF and FF to which it is to be converted.
2. Find expression for input C and D of FF in term of output Q_n and input of the target FF.
3. Draw the logic circuit based on expression.

Converting T to D flip flop

Q_n	Q_{n+1}	T_n	D_n
0	0	0	0
0	1	1←	1
1	0	1←	0
1	1	0	1

Expression for T in term of D and Q_n, we get

$$T = \overline{Q}_n D + Q_n \overline{D} \; = D \oplus Q_n$$

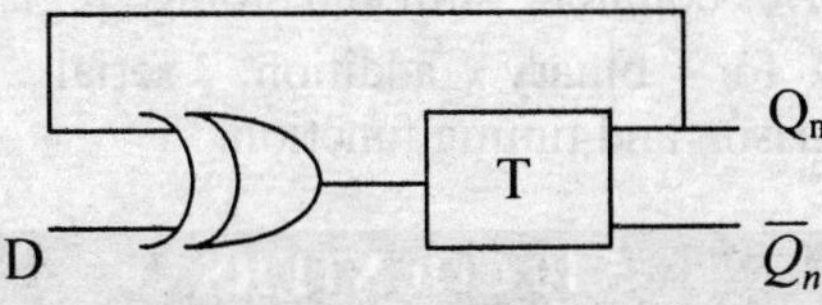

Converting D to T Flip Flop

Q_n	Q_{n+1}	T_n	D_n
0	0	0	0
0	1	1	1←
1	0	1	0
1	1	0	1←

Expression for D in term of T and Q_n we get

$$D = \overline{Q}_n T + Q_n \overline{T} \; = T \oplus Q_n$$

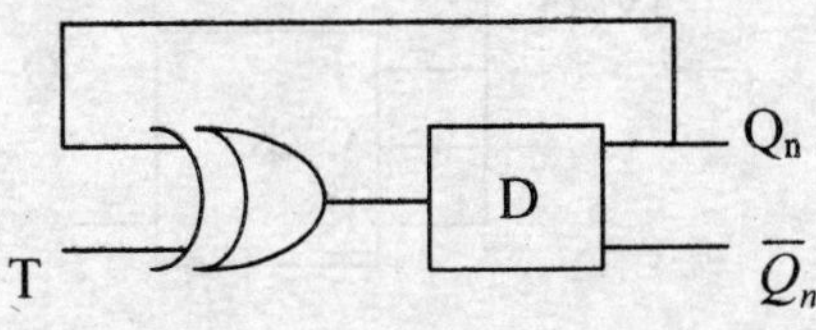

Converting R-S to T Flip Flop

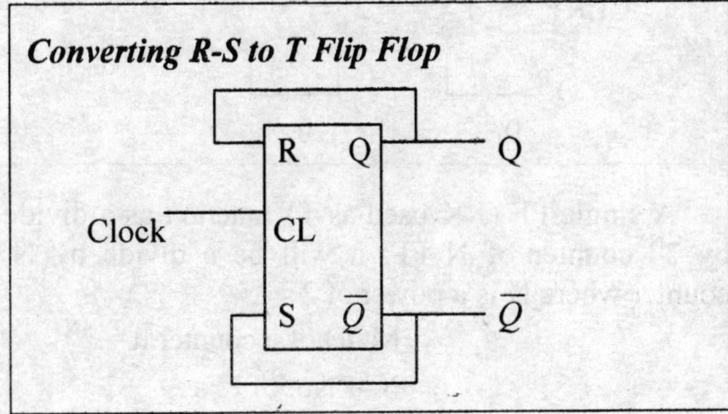

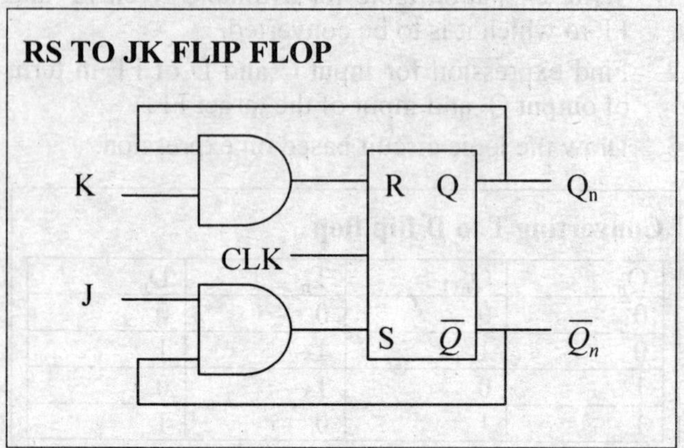

5.9.7 USES OF FFS

- Wide and extensive application in semi conductor memories counters, shift and storage registers.
- Used for binary addition, serial decoder, comparison and timing function.

5.10 COUNTERS

Counters are used for measurement of frequency and its inverse, time period, sequencing of equipment operation, frequency division, and mathematical manipulation.

A clocked J-K flip-flop function as a counter. A J-K flip flop functions as a T-FF with both input J and K tied and kept HIGH(1) and when negative edge triggered toggles to the opposite state $Q_{n+1} = \overline{Q}_n$.

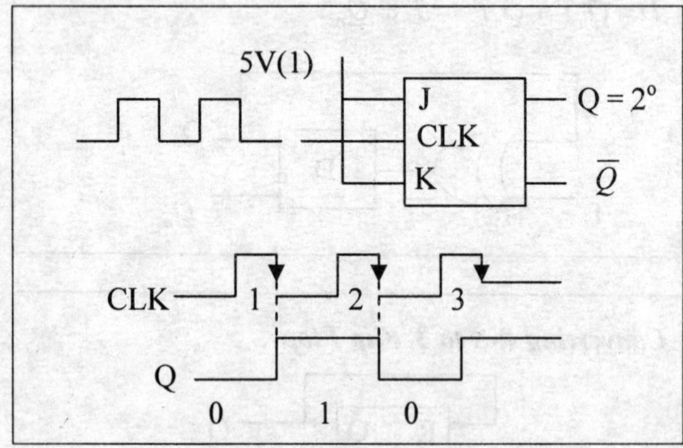

A single FF (J-K used as T) function as a divide by 2^N counter of N FF, it will be a divide by N, counter where N is a power of 2

MOD of a counter $n = 2^N$

N→ No. of FFs

Hint: If there are 4 FFs, the modulus n = 16, this means a counter with 4 FFs can read a maximum decimal number that N FFs can read is $2^N - 1$.

5.10.1 TYPE OF COUNTERS

1. Binary counter
2. Non-binary counter

If the maximum binary number is equal to 2^N, it is binary counter; otherwise it is a non-binary counter. MOD-16 is binary counter, whereas MOD10 is not. Both Binary and non-binary counter are classified as synchronous and asynchronous counter.

5.10.2 ASYNCHRONOUS

If the FFs are connected serially and the output of preceding FF clocks the succeeding FF, it is asynchronous counter as the change of states occurs one after the other. It also known as **ripple counter.**

In a **synchronous counter** all flip flop are clocked simultaneously with the clock input connected in parallel to the clock pulse.

In case of Asynchronous counter clock pulse period should be greater than Nt_{pd}. If in addition there is an interval T_s for **strobing.**

Clock period $T \geq N\, t_{pd} + T_s$

Maximum clock frequency is

$$f \leq \frac{1}{T} \leq \frac{1}{N t_{pd} + T_s}$$

where, N: no of flip flops

t_{pd} : Propagation delay of one FF.

t_s: Strobing interval

Design of Ripple Counter, divide by N

1. Decide the no. Of FFs from the MOD number, from the Basic equation

   Modulus $n = 2^N$, N no. of FFs
2. Connect the FFs serially as a ripple counter.
3. Find the binary number, N – 1.
4. Connect all FF output that are **1** at N–1 as inputs to a NAND gate.
5. Connect the NAND gate output to the preset inputs of all the FFs which Q = 0 at the count N-1.

   Steps 3 to 5 not required for pure Binary counter.

Example of MOD-10 Counter

Step 1: Since MOD-10 is in between MOD-8 and MOD-16 counter so we need 4 FF.

Step 2: MOD-10 counter-will count 1001 (binary).

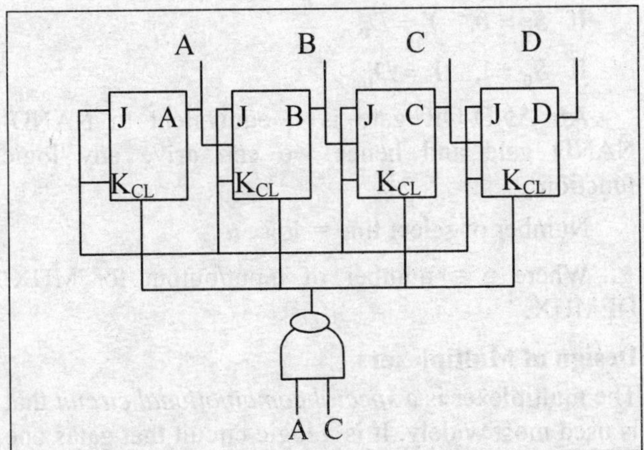

Step 3: whenever 1010 ($Q'_AQ_BQ'_CQ_D$) occur, this should reset all flip flop output to 0000. so connect Q_A and Q_C to NAND gate input and give its output to clear input to all flip flop.

Glitch

It is an unwanted spike. In the decade counter after decimal 9 or binary 1001 the counter does not recycle to 0000 immediately because of t_{pd} of NAND gate. It is unwanted spike for small duration known as Glitch.

5.11 Shift Register

Shift Register

Simplest of the shift register counters is a circulating shift register with output of last FF connected to input of 1st FF. It is also called a RING COUNTER.

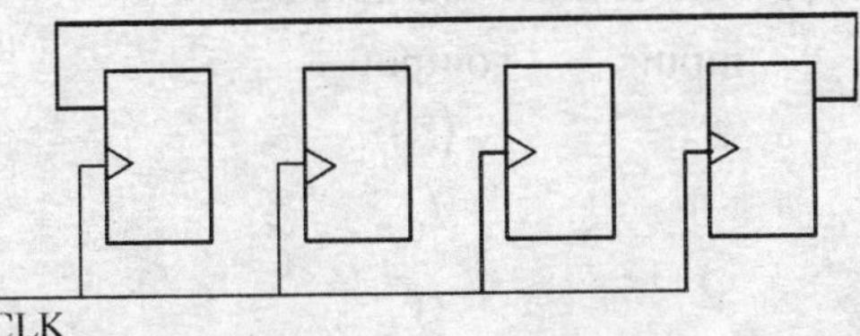

Suppose the initial state of the counter is $Q_3\ Q_2\ Q_1\ Q_0$ = 1000. The state diagram is shown below

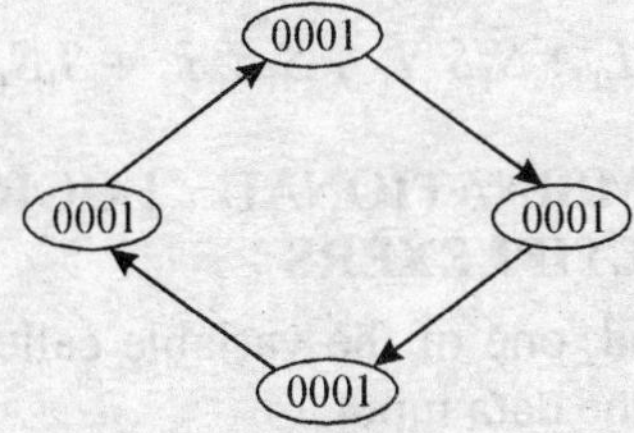

Flip Flops are connected so that information shifts from left to right and back around from $Q0$ to $Q3$.

Q3	Q2	Q1	Q0	Clock
1	0	0	0	0
0	1	0	0	1
0	0	1	0	2
0	0	0	1	3
1	0	0	0	4
0	1	0	0	5
.	.	.	.	.
.	.	.	.	.

This counter functions as a MOD-4 counter

Johnson Counter

It is constructed like a normal ring counter except that the inverted output of the last FF is connected to D input of first FF. Three bit Johnson counter is shown.

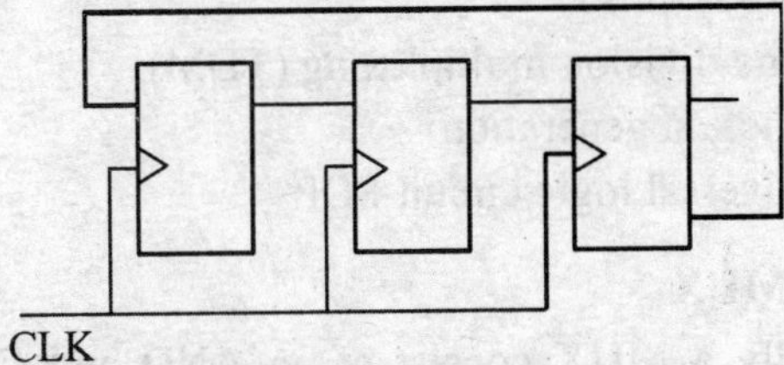

The sequence table and state diagram for Johnson counter are shown.

Q2	Q1	Q0	Clock
0	0	0	0
1	0	0	1
1	1	0	2
1	1	1	3
0	1	1	4
0	0	1	5
0	0	0	6
.			
1	0	0	7
.	.	.	.
.	.	.	.

- This counter has six distinct state. Normally a n-state counter goes through 2n stages i.e., MOD number = 2n.
- Waveform of each FF is square wave at 1/6th of the frequency of the clock.

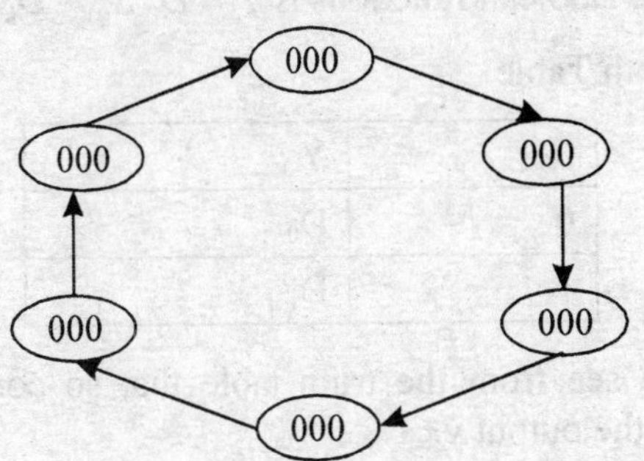

5.12 MULTIPLEXER

A multiplexer selects one out of N data input and feeds it into the single line. It is equivalent to single pole multi position switch. The selection of the input lines is controlled by a set of selection line. Hence MUX is called the data selector.

Parts of Mux

(a) Data or input line
(b) Selection or control line
(c) Output line

Use of MUX:

1. Data selection
2. Time division multiplexing (TDM)
3. Function generation
4. Universal logic circuit-SOP

Input MUX

Basically a MUX consist of an AND and OR gate circuit with an inverter in the select line. There is one AND gate for each input.

$D_0, D_1, D_2 \Rightarrow$ refer to input data
$S_0, S_1, S_2 \Rightarrow$ refer to Control data
$Y_0, Y_1, T_2 \Rightarrow$ refer to output data

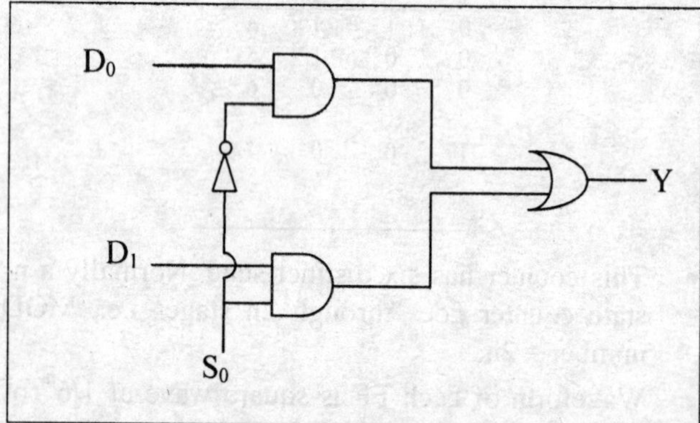

The Boolean functions is $Y = D_0\,\overline{S}_0 + D_1 S_0$

Truth Table

S_0	Y
0	D_0
1	D_1

We see from the truth table that so controls or selects the output y.

If $S_0 = 0, \quad Y = D_0$

If $S_0 = 1, \quad Y = D_1$

An AND-OR gate is equivalent to NAND-NAND gate and hence we can drive any logic function.

Number of select line = $\log_2 n$

Where n = number of input/output for MUX/DEMUX.

Design of Multiplexers

The multiplexer is a *special combinational circuit* that is used most widely. It is a logic circuit that gates one out of several inputs to a single output. The input selected is controlled by a set of select inputs.

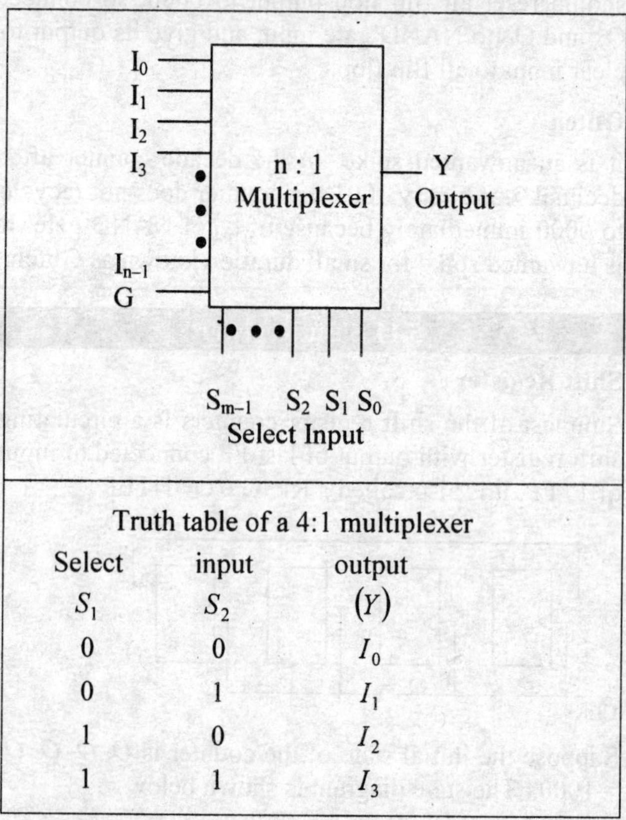

Truth table of a 4:1 multiplexer

Select S_1	input S_2	output (Y)
0	0	I_0
0	1	I_1
1	0	I_2
1	1	I_3

$$y = \overline{S}_1\overline{S}_0 I_0 + \overline{S}_1 S_0 I_1 + \overline{S}_1\overline{S}_0 I_2 + S_1 S_0 I_3$$

5.12.1 COMBINATIONAL LOGIC DESIGN USING MULTIPLEXERS

In this method, one of the variable called the residue is applied to the data input.

Example: f (A,B,C,D) = Σ (0, 1, 5, 7, 10, 14,15)

Solution: This is a 4 variable function.

Number of selection line = 4 – 1 = 3
and input $(2)^3$ = 8 line.

Step 1: The input to the multiplexer are listed I_0, I_1, I_2, I_3…I_7

	I_0	I_1	I_2	I_3	I_4	I_5	I_6	I_7
$\overline{D}$	0	1	2	3	4	5	6	7
D	8	9	10	11	12	13	14	15

where, D(MSB) and A(LSB). 0 to 15 assigned to D' to D such that corresponding to number from 0 to 15, either D = 1 or D = 0. If D =1 then again put it to 2nd row from 8 to 15 other wise from 0 to 7.

Now (ii) circle the representation of function in the above table.

Now, If two column are not circle, apply '0' to the corresponding Mux input.

- If two columns are circle, apply '1' to the Mux input.
- If the top row of column is circled, apply D' to the Corresponding Mux input.
- If the bottom tow of column only is circled, apply 'D' to the corresponding Mux input.

From above we can find that

$I_0 = \overline{D}$, I_2 = D, I_4 = 0, I_6 = D

$I_0 = \overline{D}$, I_3 = 0, $I_5 = \overline{D}$, I_7 =1

So final realization with Mux

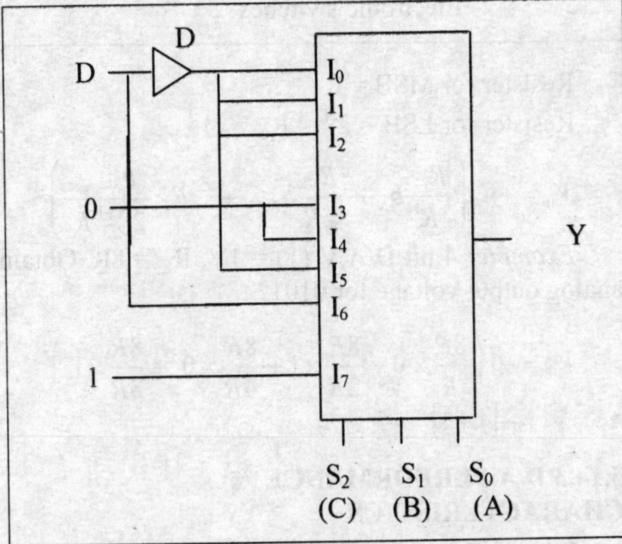

5.12.2 XOR FUNCTION REALIZATION USING (4×1) MUX

$$Y = A\overline{B} + \overline{A}B$$

We have to use A, B for selection line. Whenever

AB' = 1; or A'B = 1,

output is high, otherwise output is zero.

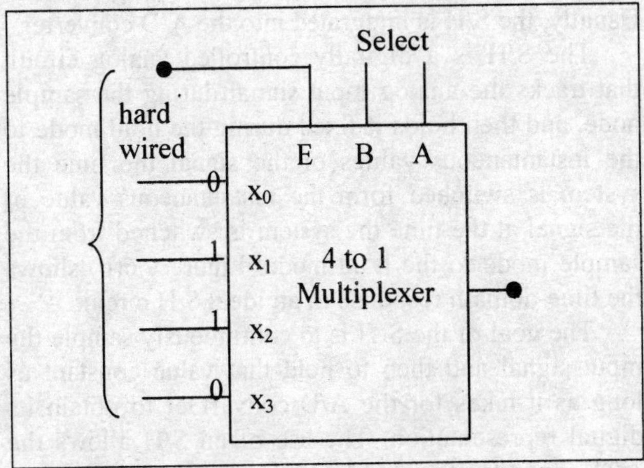

5.12.3 DEMULTIPLEXERS (DATE DISTRIBUTORS)

Demultiplexer functions opposite to that of MUX. It receives information on a single line and distributes it to possible 2n lines where n is the number of selection lines.

5.12.4 DECODERS

A binary word length of n bits can contain upto 2^n distinct elements of coded information on n-lines and feeds them to 2^n unique output lines after conversion.

5.12.5 COMPARISON OF MULTIPLEXER AND DECODER

- Only one decoder is necessary to generate all minterms. However, the number of OR gates required is equal to the number of outputs.
- One multiplexer is required for each output function.
- So if the outputs are small in number, Multiplexers are used. Decoder IC's are used for large output function.

5.13 S/H CIRCUIT

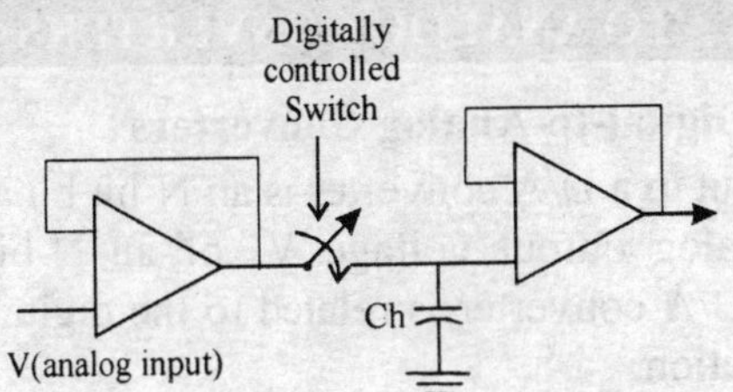

In practice, the sampling of an analog signal is performed by a simple-and-hold circuit. The sampled signal is then quantized and converted to digital form. Usually, the S/H is integrated into the A/D converter.

The S/H is a digitally controlled analog circuit that tracks the analog input signal during the sample node, and then holds it fixed during the hold mode to the instantaneous values of the signal the time the system is switched form the instantaneous value of the signal at the time the system is switched from the sample mode to the hold mode. Figure 9.6(b) shows the time-domain response of an ideal S/H circuit.

The goal of the S/H is to continuously sample the input signal and then to hold that value constant as long as it takes for the A/D convert3er to obtain its digital representation. The use of an S/H allows the A/D converter to operate more slowly compared to the time actually used to acquire the sample. In the absence of A S/H, the input signals must not change by more than one-half of the quantization step during the conversion, which may be an impractical constricting. consequently, the S/H is crucial in high-resolution digital conversion of signals that have large bandwidths.

An ideal S/H introduces no distortion the conversion process and is accurately modeled as an ideal sampler. However, time-related degradations such as errors in the periodicity of the sampling process nonlinear variations in the duration of the sampling aperture, and changes in the collage held during conversion do occur in practical devices.

The A/D converter begins the conversion after it receives a convert command. The time required to complete the conversion should be less than the duration of the hold mode of the S/H. Furthermore, the sampling period T should be larger than the duration of the sample mode and the hold mode.

In the following sections we assume that the S/H introduces negligible errors and we focus on the digital conversion of the analog samples.

5.14 ANALOG TO DIGITAL AND DIGITAL TO ANALOG CONVERTERS

5.14.1 Digital-to-Analog Converters

The input to a D/A converter is an N-bit binary signal. The analog output voltage V_0 of an N-bit straight binary D/A converter is related to the digital input by the equation.

$$V_0 = K\left(2^{N-1} \cdot b_{n-1} + 2^{N-2} \cdot b^{n-2} + \ldots\right)$$

where, K = Proportionality factor.

b_n = 1 if the nth bit of the digital input is 1.

= 0 if the nth of the digital input is 0.

5.14.2 TYPES OF D/A CONVERTERS

1. Weighted-resistor D/A converter and
2. R-2R ladder D/A converter.

Weight-Resister D/A converter

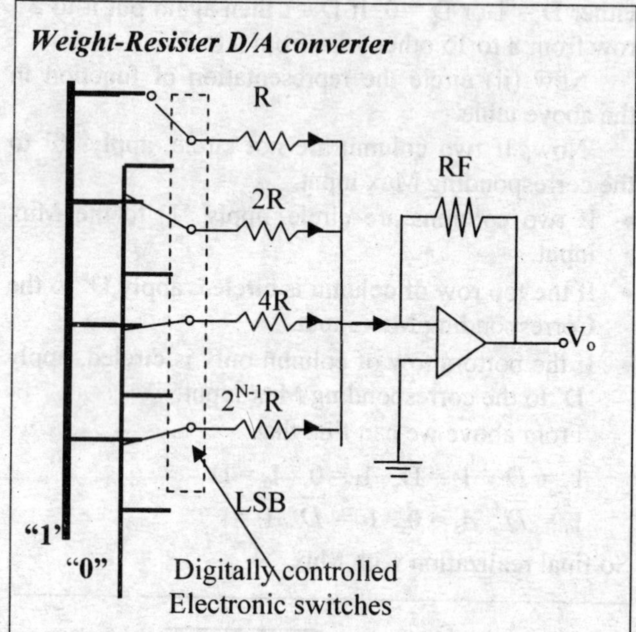

Resister for MSB = R

Resister for LSB = 2^{N-1}.R

$$V_0 = -V_R\left(\frac{R_F}{R}b_N + \frac{R_F}{2R}b_{N-1} + \ldots R_F\frac{b_{N-N}}{2^{N-1}R}\right)$$

Example: 4 bit D/A,V (1) = 1V, R_F = 8R. Obtain analog output voltage for 0101?

$$V_0 = -1\left(\frac{8R}{R}\times 0 + \frac{8R}{2R}\times 1 + \frac{8R}{4R}\times 0 + \frac{8R}{8R}\times 1\right)$$

$$= -[4+1] = -5$$

5.14.3 D/A PERFORMANCE CHARACTERISTICS

1. **FSV (full scale voltage):** It represents maximum output voltage for the DA converter and is obtained when all bits of digital input are 1.

   The output for any bit is $e_0 = V_R/2^n$

   where n = no. of bits.

$FSV = V_R [1 - (1/2^n)]$.

2. **Resolution**

$$\% \text{ Resolution} = \frac{\text{Step size}}{\text{FSV}} \times 100 \quad = \frac{1}{2^{n-1}} \times 100$$

3. *ccuracy:* It is the error between the actual output of DA converter to the expected or theoretical output and expressed as percentage of FSV.
   If the converter has an accuracy of ± 0.1% for FSV of 10 V, the minimum error for any output voltage is [(10/100) × 0.1] = 10 mV Ideally the accuracy should be = ± [1/2]1 LSB.
4. **Monotonicity:** It means that the output increase for an increase of input or decreases for decrease of input. The output should not decrease for increase of input.
5. **Settling Time:** It is the time taken by DA converter to settle within ± [1/2] LSB of its final value.

5.14.4 MAJOR COMPONENT OF THE BINARY WEIGHTED RESISTANCE DAC:

(a) a weighted resistor network, R to $2^{n-1}R$
(b) n switches, one for each bit applied bit input.
(c) A reference voltage V_{ref}
(d) A summing element

Defects: The chief detect of weighted binary DAC is that the resistor values increases in multiples of 2.

If the value of MSB is 2 KΩ. Then for a 10 bit, the value for LSB is 2^9. 2K = 1.024 MΩ.

It is difficult to obtain high precision at very high values. It is also difficult to fabricate high value resistance by integrated circuit methods. This is over come by the ladder type DA converter.

5.14.5 R-2R LADDER NETWORK

In this, only two values of resistors are used.

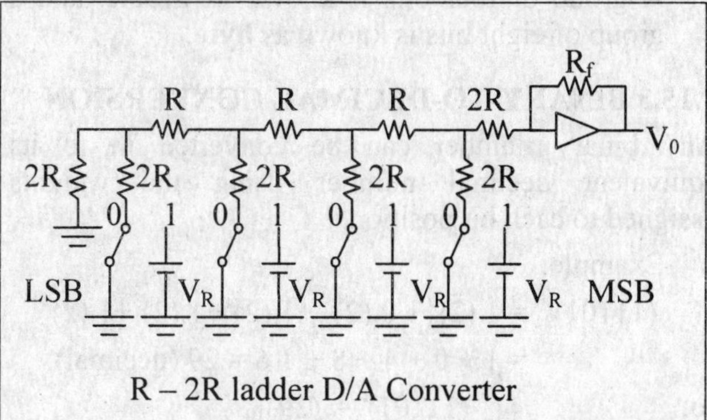

R – 2R ladder D/A Converter

$$V_0 = V_R(2^{N-1} b_{N-1} + 2^{N-2} b_{N-2} + ...b_0)/2^N$$

The number of resistor required for an N-bit ladder D/A converter is 2N in the case of R-2R ladder D/A converter whereas it is only N is the case of weighted-resister D/A converter.

5.14.6 ANALOG-TO-DIGITAL CONVERTERS :TYPES OF ADC

1. *Parallel-comparator/flash/simultaneous A/D*

- Simplest in concept and fastest conversion time.
- Disadvantages are rapid increase in the number of comparators with the number of bits, (2^N-1) comparators are required for an N-bit converter.

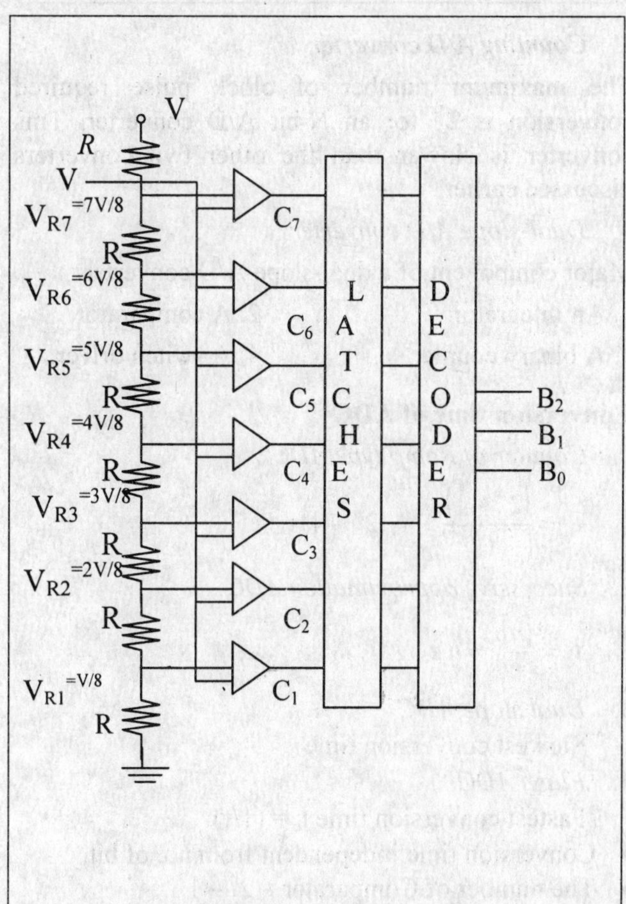

2. *Successive Approximation A/D converter*

- For N-bit converter, the number of clock pulses required would be N.
- Slow than the parallel-comparator A/D
  Offset voltage = 1/2 LSB – 0.5

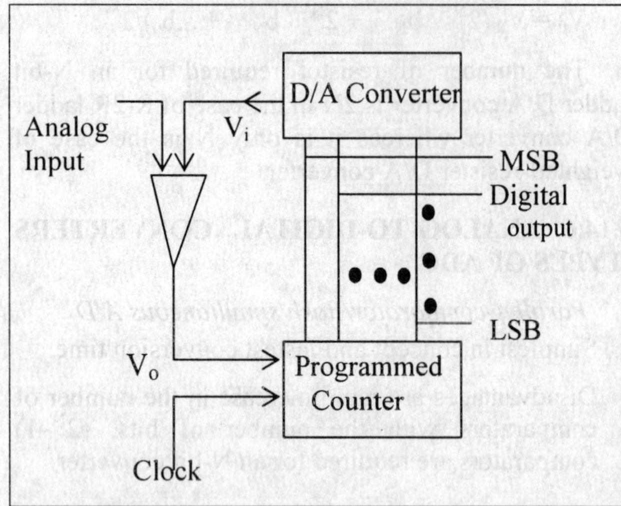

3. *Counting A/D converter*

The maximum number of clock pulse required conversion is 2^N for an N-bit A/D converter. This converter is slower than the other two converters discussed earlier.

4. *Dual-slope A/D converters*

Major component of a dual-slope A/D converter.

1. An integrator
2. A comparator
3. A binary counter
4. A switch driver.

Conversion time of ADC

1. *Counter of Ramp type ADC*

$$t_c = \frac{(2^n - 1)}{f} = (2^n - 1)\tau$$

2. *Successive approximation ADC*

$$t_c = \frac{[n]}{f} = n\,\tau$$

3. *Dual slope ADC*
   Slowest conversion time
4. *Flash ADC*
   Fastest conversion time $t_c = (1/f)$

- Conversion time independent from no. of bit.
- The number of Comparator = $2^n - 1$
  where f = clock frequency.

Aperture Time

Aperture time is called the Maximum allowable conversion time t_c.

$$t_a = \frac{1}{\omega(2^n - 1)}$$

where, ω = angular frequency of signal applied
n = number of bit for ADC.
t_a = aperture time.

5.15 BASICS OF NUMBER SYSTEM

5.15.1 NUMBER SYSTEMS AND CODES

A number system is language system consisting of an ordered set of symbols called digits with rules defined for addition, multiplication, and other mathematical operation. The radix or base of a number system specifies the actual number of digits included in its ordered set. Further, a number system allow for number (collection digits) to have an integer and a fractional part set apart by a radix point.

There are various number system based on radix which are required for various purpose:

1. Decimal number system (i.e. with 0, 1, 2, ... 9), base-10.
2. Binary number system (i.e. with 0, 1), base-2.
3. Trinary number system (i.e. with 0, 1, 2), base-3.
4. Quinaary number system (i.e. with 0, 1, 2, ... 4), base-5.
5. Octal number system (i.e. with 0, 1, 2, ...7), base-8.
6. Hexadecimal number system (i.e. with 0, 1, 2, ... 9, A, B, ... F (15), base-16 (H).

5.15.2 BINARY NUMBER SYSTEM

Almost all digital computers and systems are based on binary (two state) operation, e.g. a switch, magnetic core tape, punched card, tape recorder, transistor action cutoff or saturation.

- The number system with base (or radix) two is known as the binary number system.
- Two symbols are used to represent numbers in this system and there are 0 to 1, known as bits.
- It is potential system, that is every position is assigned a specific weight.
- A group of four bits is known as **nibble**, and a group of eight bits is known as **byte**.

5.15.3 BINARY-TO-DECIMAL CONVERSION

Any binary number can be converted in to its equivalent decimal number using the weights assigned to each bit position.

Example:

$$(11101)_2 = 1\,(2)^0 + 0.(2)^1 + 1.(2)^2 + 1.(2)^3 + 1.(2)^4$$
$$= 1 + 0 + 4 + 8 + 16 = 29 \text{ (decimal)}$$

so, $(11101)_2 = (29)_{10}$

In case we have from like $(111.01)_2$ then we have to manipulate like this

$(111.01)_2 = 1.(2)^0 + 1.(2)^1 + 1.(2)^2 + 0.(2)^{-1} + 1.(2)^{-2}$

$= 1 + 2 + 4 + 0 + (1/4) = (7.25)_{10}$

5.15.4 DECIMAL-TO-BINARY CONVERSION

Any decimal number can be converted into its equivalent binary number. The conversion is obtaining by continuous division by 2 and keeping track of the conversion is affected by continuous multiplication by 2 and keeping track of the integers generated for.

Example: For this $(13)_{10}$

		Carry	
2	13		
2	6	1	LSB
2	3	0	
2	1	1	
	0	1	MSB

So we have $(13)_{10} = (1101)$

For converting $(0.65625)_{10}$ into binary

0.65625 × 2 = 1.131250 (keep integer part, i.e. 1) MSB
0.3125 × 2 = 0.625 (keep integer part, i.e. 0)
0.625 × 2 = 1.25 (keep integer part, i.e. 1)
0.25 × 2 = 0.5 (keep integer part, i.e. 0)
0.5 × 2 = 1.0 (keep integer part, i.e. 1) LSB

so $(0.65625)_{10} = (0.10101)_2$

5.15.5 SIGN-MAGNITUDE REPRESENTATION

- Plus (+) sign is used to denote a positive number and minus (–) sign for denoting a negative number in decimal numbers.
- In digital world, to represent sign we represent the same in term of Bit (0 or 1). An additional bit is used as sign bit, and it is laced as the most significant bit.
  0 = is used to represent a positive number.
  1 = is used to represent a negative number.

Example: 8 bit signed number 01000100 give positive number of value (1000100) 2 = (68) 10. The left most 0 (MSB) indicate that the number is positive.

On the other hand, (11000001) represent negative number of value (–65).

One's Complement Representation

In a binary number, if each 1 is replaced by 0 and each 0 by 1, the resulting number is known as one's Complement of the first numbers.

Both the number are complement of each other.

If one of these number 1's positive, then the other umber will be negative with the same magnitude and vice versa.

(0101) → (+5) whereas
(1101) → (–5)

Two's Complement Representation

If 1 is added to 1's complement of binary number the resulting number is known as the two's complement of the binary number.

Example: Find 2's complement of (0101)

$0101 \xrightarrow{1's\text{ complement}} 1010$

$1010 \xrightarrow{\text{add }1} 1011\,(2's\text{ complement})$

$0101 \longrightarrow (+5)_{10}$

therefore (1011) represent $(-5)_{10}$

2's Representation

If the LSB of a number is 1, its 2's complement is obtained by changing each 0 to 1 and 1 to 0 except the LSB.

If the LSB of a number is 0, its 2's complement is obtained by scanning the number from LSB to MSB bit by bit as they are up to and including the occurrence of the first 1, and complement all other bits.

5.15.6 BINARY ARITHMETIC

Binary Addition

A	B	S(sum)	C(carry)
0	0	0	0
0	1	1	0
1	0	1	0
1	1	0	1

$S = A\overline{B} + \overline{A}B$

where, S→ Sum of Bit A and B

C→ Carry generated after addition of A and B

Binary Subtraction

A	B	D (Difference)	Br (Borrow)
0	0	0	0
0	1	1	1
1	0	1	0
1	1	0	0

$D = A\overline{B} + \overline{A}B$; Br (Borrow) $A\overline{B}$

5.15.7 CODES

It is symbolic representation of information transform. Generally 4 bit codes are used.

Binary Code

If X = Number of element of information to code into binary coded format

J = Number of bits in a code, then following expression must hold

$$X \leq 2^J \quad ; J \geq \log 2^X$$
$$J \geq 3.32 \log_{10}{}^X$$

Self-Complementary Codes

Certain codes have a distinct advantage in that their logical complement is same as the arithmetic complement.

Example are Excess-3, 6311,2421 codes

Reflective Codes

The 9's complement of reflected BCD code word is simply by changing only one o fits bits. A reflected code is characterized by the fact that it is immerged about the center entries with one bit changed.

Unit Distance Code

One bit change in the next or adjacent code word. Independent of the direction has taken in the code.

The unit distance codes have special advantage in that they minimize transition error or flashing.

Excess-3 Code

This is also known as self-complementing code. Each decimal digit is coded into a 4-digit binary code. The code for each decimal digit is obtained by adding decimal 3 to the natural BCD.

It is unweighted code.

It is self complementary code.

Gray Code

In this code, a decimal number is represented in binary from is such a way of that each gray-code number differs from the preceding and the succeeding number by a single bit.

1. It have reflection properties
2. Un weighted code
3. Unit distance code
4. More error free code, used for input/output devices

Decimal	Binary	Gray
0	0000	0000
1	0001	0001
2	0010	0011
3	0011	0010
4	0100	0110
5	0101	0111
6	0110	0101
7	0111	0100
8	1000	1100
9	1001	1101

Alphanumeric Code

ASCII: American standard code for information interchange (8 bit code).

EBCDIC: Extended Binary coded decimal interchange code (8-bit code).

Hollerith Code:**12 bit code used in punched card.**

5.16 MICROPROCESSOR APPLICATIONS

5.16.1 BASIC TERMS

- The microprocessor can be viewed as a **programmable logic device** that can be used to control processes or to turn on/off devices.
- The microprocessor is a programmable integrated device that has **computing and decision-making capability** similar to that of the central processing unit (CPU) of a computer.
- The microprocessor communicates and **operates in the binary numbers 0 and 1** called **bits**.
- Each microprocessor has a fixed set of instructions in the form of binary patterns called a **machine language**.

- Instructions are given abbreviated names called **mnemonics**, which form the **assembly language**.
- A microprocessor is a multipurpose, programmable, clock-driven, register-based electronic device that reads binary instructions from a storage device called memory, accepts binary data as input and processes data according to those instructions, provides results as output.
- The physical components of this system are called hardware.
- A set of instructions written for the microprocessor to perform a task is called a **program**, and a group of programs is called **software**.

5.16. 2 MICROPROCESSOR

The microprocessor is in many ways similar to the CPU, but includes all the logic circuitry, including the control unit, on one chip. The microprocessor can be divided into three segments for the sake of clarity, as shown in below figure arithmetic/logic unit (ALU), register array, and control unit.

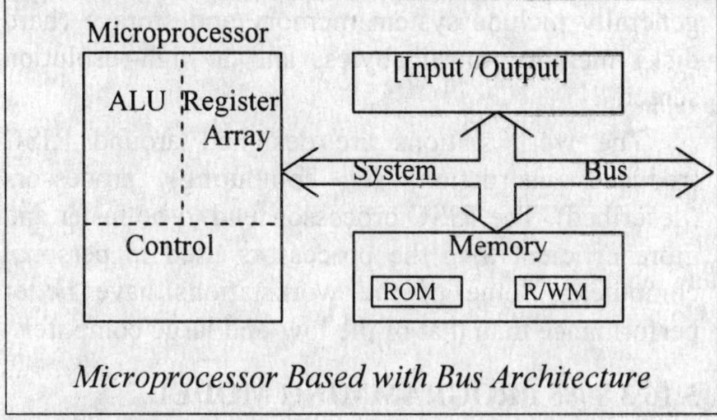

Microprocessor Based with Bus Architecture

Arithmetic/Logic Unit

This is the area of the microprocessor where various computing functions are performed on data. The ALU unit performs such arithmetic operations as addition and subtraction, and such logic operations as AND, OR, and exclusive OR.

Register Array

This area of the microprocessor consists of various registers identified by letters such as B, C, D, E, H and L. These register are primarily used to store data temporarily during the execution of a program and are accessible to the user through instruction.

Control Unit

The control unit provides the necessary timing and control signals to all the operations in the microcomputer. It controls the flow of data between the microprocessor and memory and peripherals.

Memory

- Memory stores such binary information as instructions and data, and provides that information to the microprocessor whenever necessary. To execute programs. The microprocessor reads instructions and data from memory and performs the computing operations in its ALU section. Results are either transferred to the output section for display or stored in memory for later use. The memory block shown in above figure has two sections: **Read-Only memory** (ROM) and **Read/write memory** (R/WM), popularly known as **Random-Access memory** (RAM).
- The ROM is used to store programs that do not need alterations.
- The Read/Write memory (R/WM) is also known as *user memory*: It is used to store user programs and data.

Input/ Output

- **I/O includes two types of devices:** Input and output.
- These **I/O** devices are also known as peripherals. **ASCII** keyboard is found in most microcomputers **(PCs)**.

System Bus

The system bus is a communication path between the microprocessor and peripherals; it is nothing but a group of wires to carry bits.

Instruction Set and Languages

The **word** (or word length) is defined as the number of bits the microprocessor recognizes and processes at a time.

A **byte** is defined as a group of eight bits. The term **nibbles**, which stands for a group of four bits. A Byte has two nibbler.

Each machine has its own set of instructions based on the design of its CPU or of its microprocessor. To communicate with the computer, one must gives instructions in **binary** language **(machine language)**.

Programmers can written program called **assembly language programs** using English like words i.e. mnemonics.

An assembly language is specific to a given machine, programs written in assembly language are not transferable from one machine to another.

Low Level Languages

Machine languages is in binary and the assembly language is in English like words i.e. in mnemonics.

The assembler is a program that translates the mnemonics entered by the ASCII keyboard into the corresponding binary machine codes of the microprocessor .Each microprocessor has its own assembler because the mnemonics and machine codes are specific to the microprocessor being used, and each assembler has rules that must be followed by the programmer.

High Level Languages

Programming languages that are intended to be **machine independent** are called **high level languages**. These include such languages as BASIC, PASCAL, C, C++ and java, all of which have certain sets of rules and draw on symbols and conventions from English Instructions written in these languages are known as **statements** rather than mnemonics.

Compiler or an Interpreter

These programs accept English like statements as their input called the **source code**. The compiler or interpreter then translates the source code into the machine language compatible with the microprocessor being used in the system. This translation in the machine language is called the **object code**. Each microprocessor needs its own compiler or an interpreter for each high level language.

- The compiler reads the entire program first and translates it into the object code that is executed by the microprocessor.
- The interpreter reads one instruction at a time, produces its object code (a sequence of machine actions) and executes the instruction before reading the next instruction.

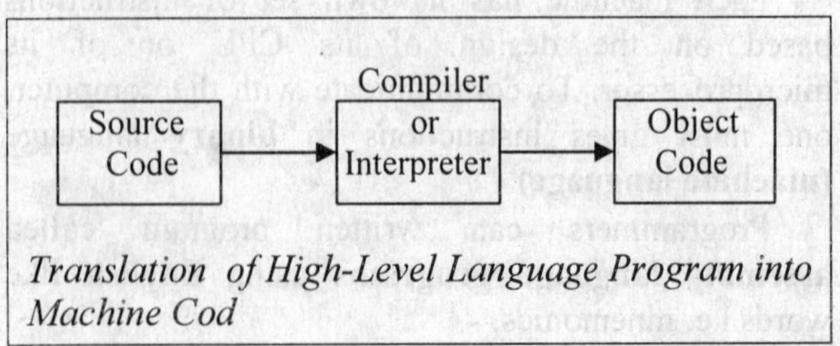

Translation of High-Level Language Program into Machine Cod

The primary advantage of high level languages is in troubleshooting **(debugging)** programs.

Operating Systems

The interaction between the hardware and the software is managed by a set of programs called an **operating system** of a computer; it oversees all the operations of the computer. The operating system is responsible primarily for storing information on the disk and for the communication between the computer and its peripherals.

The Floppy Disk

The floppy disk is a magnetic medium similar to a disk can be accessed randomly using disk drives.

Workstation

These are high-performance cousins of the personal computers. They are used in engineering and scientific applications such as computer-aided design (CAD), computer-aided engineering (CAE), and computer-aided manufacturing (CAM). They generally include system memory and storage (hard disk) memory in gigabytes, and a high-resolution screen.

The work stations are designed around RISC (reduced instruction set computing) processors (described). The RISC processor tend to be faster and more efficient than the processors used in personal computers. Some of the workstations have better performance than that of the low-end large computer.

5.16.3 8085 PROGRAMMING MODEL

The programming model consists of some segments of the ALU and the registers. This model does not reflect the physical structure of the **8085** but includes the information that is critical in writing assembly language programs.

Registers

The 8085 has **six general purpose registers** to store 8 bit these are identified as B, C, D, E, H, and L, They can be combined as register pairs BC,DE, and HL–to perform some 16 bit operations .The programmer can use these registers to store or copy data into the registers by using data copy instructions.

Accumulator

The accumulator is an 8 bit register that is part of the arithmetic/logic unit (ALU). This register is used to store 8 bit data and to perform some 16 bit operations The result of an operation is stored in the accumulator. The accumulator is also identified as register A.

Flags

The ALU include **five flip flops** which are set or reset after an operation according to data conditions of the result in the accumulator and other registers. They are called Zero (Z) Carry (CY) Sign (S), Parity (P) and Auxiliary carry (AC) flags.

These flags have critical importance in the decision making process of the microprocessor. The conditions set or reset) of the flags are tested through software instructions.

The 8085 Flags

The following flags are set or reset after the execution of an arithmetic or logic operation: data copy instruction do not affect any flags.

- *Z-Zero:* The Zero flag is set to 1 when the result is zero: otherwise it is reset
- *CY-Carry:* If an arithmetic operation results in a carry, the CY flag is set: otherwise it is reset.
- *S-Sign:* The Sign flag is set if bit D_7 of the result = 1 : otherwise it is reset
- *P-Parity:* If the result has an even number of 1s, the flag is set: for an odd number of 1s the flag is reset.
- *AC-Auxiliary Carry:* In and arithmetic operation, when a carry is generated by digit D_3 and passed to digit D_4, the AC flag is set. This flag is used internally for BCD (binary-coded decimal) operations: *there is no jump instruction associated with this flag.*

Program Counter (Pc) and Stack Pinter (Sp)

These are two 16 bit registers used to hold memory addresses.

The microprocessor uses the PC register to sequence the execution of the instructions. The function of the program counter is to point to the memory address from which the next byte is to be fetched. When a byte (machine code) is being fetched the program counter is incremented by one to point to the next memory location.

The **stack** pointer is also a 16-bit register used as a memory pointer. It points to a memory location in R/W memory, called the stack. The beginning of the stack is defined by loading a 16-bit address in the stack pointer.

1. **One byte instruction:** A mnemonic followed by a letter (or two letters) representing the registers (Such as A, B, C, D, E, H, L, M, and SP) is a one byte instruction.

   Examples: (a) MOV A, B (b) DCX SP
   (c) RRC
2. **Two byte instruction:** A mnemonic followed by 8-bits (byte) is a two byte instruction.

   Examples: (a) MVI A, 8-bits (b) ADI 8 bits
3. **Three byte instruction:** A mnemonic followed by 16 bit (also terms such as adr or dble) is a three byte instruction.

   Example: (a) LXI B, 16-bit (dble)
   (b) JNZ 16 bit (adr)

Data Transfer (Copy) Instruction

These instructions perform the following six operations:

1. Load an 8 bit number in a register
2. Load 16 bit number in a register pair
3. Copy from register to register
4. Copy between register and memory
5. Copy between I/O and accumulator
6. Copy between register and stack memory.

1. Mnemonics/Operation		**Examples:**
1.1	MVI R,** 8-bit load 8 bit data (byte) in a register	MVI B,4FH
1.2	MOV Rd Rs** Copy data from, source register Rs into destination register Rd	MOV B,A
1.3	LXI Rp,** 16-bits Load 16 bit number in a register pair	LXI B, 2050H
1.4	OUT 8-bit (port address) Send (write) data byte from the accumulator to an output device	OUT 01H
1.5	IN 8-bit (port address) Accept (read) data byte from an input device and place it in the accumulator	IN 07H
1.6	LDA 16 bits Copy the data byte into A from the memory specified by 16 bit address	LDA 2050 H

1.7 STA 16-bit STA 2070H
Copy the data byte from A into the memory specified by 16-bits address

1.8 LDAX R_P LDAX B
Copy the data byte into A from the memory specified by the address in the register pair

1.9 STAX R_P STAX - D
Copy the data byte from A into the memory specified by the address in the register pair

1.10 MOV R.. M MOV B.M
Copy the data byte into register from the memory specified by the address in HL register.

1.11 MOV M. R MOV M. C
Copy the data byte from the register into memory specified by the address in HL register.

2. Arithmetic Instructions: The frequently used arithmetic operations are:

* ADD * Subtract * Increment (Add 1)
* Decrement (Subtract 1)

Mnemonics/Operation Examples

2.1 ADD R ADD B
Add the contents of a register to the register to the contents of A

2.2 ADI 8 – bit ADI 37H
Add 8-bit data to the contents of A

2.3 ADD M ADD M
Add the contents of memory to A; The address of memory is in HL register

2.4 SUB R SUB C
Subtract the contents of a register from the content s of A

2.5 SUI 8-bit SUI 7FH
Subtract 8–bit data from the contents of A content

2.6 SUB M SUB M
Subtract the contents of memory from A : The address o f memory is in HL register

2.7 INR R INR D
Increment the contents of a register

2.9 DCR R DCR E
Decrement the contents of a register

3. Logic and bit Manipulation Instruction: These instructions include the following operations

*AND * OR * X-OR
* Compare * Rotate Bits

Mnemonics/Operation Examples

3.1 ANA R ANA B
Logically AND the contents of a register with the contents of A

3.2 ANI 8-bit ANI 2FH
Compare the contents of register with the contents of A for less than equal to or greater than.

4. **Branch Instructions:** The following instructions change the program sequence.

Mnemonics/Operation Examples

4.1 JMP 16-bit address JMP 2050H
Change the program sequence to the specified 16–bit address.

4.2 JZ 16 –bit address JZ 2080H
Change the program sequence to the specified 16 address if the Zero flag is set.

4.3 JNZ 16 –bit address JNZ 2070H
Change the program sequence to the specified 16-bit address if the Zero flag is reset.

4.4 JC 16 bit address JC 2025 H
Change the program sequence to the specified 16-bit address if the Carry flag is set

4.5 JNC 16 bit address JNC 2025 H
Change the program sequence to the specified 16-bit address if the Carry flag is reset

5. Machine Control Instructions

These instruction effect the operation of the processor

5.1 HLT HLT
Stop processing and wait

5.2 NOP NOP
Do not perform any operation

This set of instructions is a representative sample; it does not include various instructions related to 16-bit data operation, additional jump instructions, and conditional Call and Return instruction.

- The 8085 microprocessor operations are classified into five major groups; data transfer (copy), arithmetic, logic, branch and machine control.
- An Instruction has two parts opcode (operation to be performed) and operand (data to be operated on). The operand can be data (8 or 16-bit), address or register, or it can be implicit. The method of specifying an operand (directly, indirectly, etc.) is called the **addressing mode**:

Microprocessor Architecture and its Operations

This process of data manipulation and communication is determined by the logic design of the microprocessor, called the **architecture**.

- Microprocessor initiated operations
- Internal operations
- Peripheral (or externally initiated) operations

The MPU Performs Primarily Four Operations

Memory Read: Reads data or instructions from memory.

Memory write: Writes data or instructions into memory.

I/O Read: Accepts data from input devices.

I/O Write: Sends data to output devices.

All these operation are part of the communication process between the MPU and peripheral devices (including memory). To communication with a peripheral (or a memory location), The MPU needs to perform the following steps:

Step 1: Identify the peripheral or the memory location (With its address).

Step 2: Transfer binary information (data and instructions).

Step 3: Provide timing or synchronization signals

Address Bus

The address bus is a group of 16 lines generally identified as A_0 to A_{15}. **The address bus is unidirectional:** bits flow in one direction-from the MPU to peripheral devices. The MPU uses the address bus to perform the first function identifying a peripheral or a memory location (Step 1). In a computer system each peripheral or memory location is identified by a binary number called an address, and the address bus is used to carry a 16-bit address.

Data Bus

The data bus is a group of eight lines used for data flow. These lines are **Bi-directional**: Data flow in both directions between the MPU and memory and peripheral devices. The MPU uses the data bus to perform the second function: Transferring binary information (Step 2).

Control Bus

The control bus is comprised of various single lines that carry synchronization signals.

The MPU uses such lines to perform the third function providing timing signals (Step 3). The term bus in relation to the control signals is some what confusing these are indicate an MPU operation. The MPU generates specific control signals for every operation it performs.

5.16.4 INTERNAL DATA OPERATIONS AND THE 8085 REGISTERS

The internal architecture of the 8085 microprocessor determines how and what operations can be performed with the data. These operations are:

1. Store 8-bit data
2. Perform arithmetic and logical operations
3. Test for conditions
4. Sequence the execution of instructions
5. Store data temporarily during execution in the defined R/W memory locations called the stack.

Peripheral or Externally Initiated Operation

External devices (or signals) can initiate the following operations for which individual pins on the microprocessor chip are assigned Reset. Interrupt Ready, Hold.

Reset

When the reset pin is activated by an external key (also called a reset key) all internal operation are suspended and the program counter is cleared (it holds 0000H).

Interrupt

The microprocessor can be interrupted from the normal execution of instructions and asked to execute some other instructions called a service routine.

Ready

The 8085 has a pin called READY. If the signal at this READY pin is low, the microprocessor enters into a Wait state. The signal is used primarily to synchronize slower peripherals with the microprocessor.

Hold

When the **Hold** pin is activated by an external signal the microprocessor **relinquishes** control of buses and allows the external peripheral to use them. For example the **HOLD** signal is used in Direct Memory Access (DMA) data transfer.

Memory Classification

As mentioned earlier memory can be classified into two groups: **Prime** (system or main) **memory** and **storage memory**. The R/WM M and ROM are examples of prime memory this is the memory the microprocessor uses in executing and storing programs. This memory should be able to respond fast enough to keep up with the execution speed of the microprocessor. Therefore, it should be random access memory, meaning that the microprocessor should be able to access information from any resister with the same speed (Independent of its place in the chip). The size of a memory chip is specified in terms of bits for example a 1 K memory chip means it can store 1 K (1024) bits (not bytes). On the other hand, memory in a system such as a PC is specified in bytes. For example, 4M memory in a PC means it has 4 megabytes of memory.

The other group is the storage memory such as magnetic disks and tapes. This memory is used to store programs and results after the completion of program execution.

Flash Memory

This is a variation of **EE-PROM** that is becoming popular. The major difference between the flash memory and **EE-PROM** is in the erasure procedure. The **EE-PROM** can be erased at a register level but the flash memory must be erased either in its entirety or at the sector (block) level.

The Zero Power RAM is a CMOS

Read/write memory with battery backup built internally. It includes lithium cells and voltage-sensing circuitry. When the external power supply voltage falls below 3 V, the power switching circuitry connects the lithium battery, thus. This memory provides the advantages of R/W and Read only memory.

The integrated RAM (iRAM) is a dynamic memory with the refreshed circuitry built on the chip. For the user it is similar to the static R/W memory. The user can derive the advantages of the dynamic memory without have to build the external refresh circuitry.

Input and Output Devices

Input/output devices are the means through which the MPU communicates with the out side world. The MPU accept binary data as input from devices such as keyboards and A/D converters and sends data to output devices such as LED or printers.

The steps in communicating with an I/O device are similar to those in communicating with memory and can be summarized as follows.

1. The MPU places an 8-bits address on the address bus, which is decoded by external decode logic.
2. The MPU sends a control signal (I/O Read or I/O Write) and enables the I/O device.
3. Data are transferred using the data bus.

Logic Devices

Tri-state logic devices have three states: logic 1, logic 0, and high impedance. The term tri-state is a trademark of National Semiconductor and is used to represent three logic states .A tri-state logic device has a third line called enable. When this line is activated the tri state device goes into the high impedance state–as if it were logic 1 states. In the high impedance state, current is drawn from the system.

The **buffer** is a logic circuit that amplifies the current or power. It has one input line and one output line. The logic level of the output is the same as that of the input logic 1. The buffer is commonly used to increase the driving capability of the data bus and the address bus. It is also known as a **line driver** or **line receiver**. This device is commonly used as a driver for the address bus in a bus oriented system.

The **data bus** of a microcomputer system is **bi-directional**; therefore, it requires a buffer that allows data to flow in both directions. Also called an **octal bus** transceiver. This is commonly used as a driver for the data bus.

The 8085 Microprocessor

The 8085 is an 8-bit general purpose microprocessor capable of address 64K of memory. The device has forty pins, requires a +5 V single power supply, and can operate with a 3 MHz signal phase clock.

Address Buss

The 8085 has 16 signal lines (pins) that are used as address bus however, these lines are split into two segments: $A_{15} - A_8$ and $AD_7 - AD_0$. The eight lines, A_{15}-A_8 are unidirectional and used for the most significant bits, called the high-order address, of is 16-bit address. The signal lines AD_7 –AD_0 are used for a dual purpose, as explained in the next section.

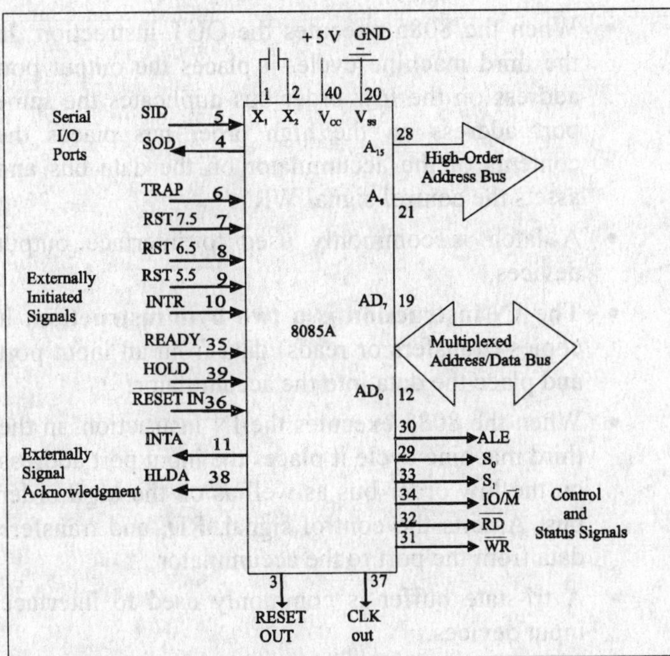

Multiplexed Address/Data Bus

The signal lines, AD_7–AD_0 are bi-directional: they server a dual purpose. They are used as the low-order address bus as well as the low-order address bus. During the later part of the cycle, these lines are used as the data bus. (This is also known as multiplexing the bus). However, the low-order address bus can be separated from these signaled by using a latch.

Control and Status Signals

This group of signals includes two control signals ($\overline{RD}$ and $\overline{WR}$), three status signals ($IO/\overline{M}$, S_1 and S_0) to identify the nature of the operation and one especial signal (ALE) to indicate the beginning of the operation. These signals are as follows.

ALE–Address Latch Enable

This is a positive going pulse generated every time the 8085 begins an operation (machine cycle). It indicates that the bits on $AD_7 - AD_0$ are address bits. This signals is used primarily to latch the low order address from the multiplexed bus and generate a separate set of eight address lines $A_7 - A_0$.

$\overline{RD}$ –Read

This is a Read control signal (active low). This signal indicates that the selected I/O or memory device is to be read and data are available on the data bus.

$\overline{WR}$ -Write

This is a write control signal (active low) this signal indicates that the data on the data bus are to be written into a selected memory or I/O location.

$IO/\overline{M}$

This is a status signal used to differentiate between I/O and memory signals operation.

S_1 and S_0

These status signals similar to IO/M can identify various operations but they are rarely used in small systems.

Power Supply and Clock Frequency

The power supply and frequency signals are as follows:

- V_{CC} : + 5V power supply
- V_{SS}: Ground reference.
- X_1, X_2: A crystal (or RC, LC network) is connected at these two pins. **The frequency is internally divided by two**; therefore, to operate a system at 3 MHz, the crystal should have a frequency of 6 MHz.
- CLK (OUT)–Clock output: This can be used as the system clock for other devices.

8085 Interrupts and Externally Initiated Signals

- INTR (Input) ⇒ *Interrupt Request:* This is used as a general purpose interrupt.
- $\overline{INTA}$ (Output) ⇒ ***Interrupt Acknowledge:* This is used to acknowledge an interrupt.**
- **RST 7.5/RST 6.5/RST 5.5 ⇒ *Restart Interrupts:*** These are vectored interrupts that transfer the program control to specific memory locations. They have higher priorities than the INTR interrupt. Among these three the priority order is 7.5 6.5 and 5.5.
- **TRAP (Input)** ⇒ This is a nonmaskable interrupt and has the highest priority.
- **HOLD (Input)** ⇒ This signal indicates that a peripheral such as a DMA (direct Memory Access) controller is requesting the use of the address and data buses.
- **HLDA (Output)** ⇒ ***Hold Acknowledge:*** This signal acknowledges the HOLD request.

- **READY (Input)** ⇒ This signal is used to delay the microprocessor Read to send or accept data. When this signal goes low the microprocessor waits for an integral number of clock cycles until it goes high.

Various 8085 Operations/Cycle/Timing

Instruction cycle is defined as the time required to complete the execution of an instruction. The 8085-instruction cycle consists of one to six machine cycles or one to six operations.

- ***Machine cycle*** is defined as the time required to complete one operation of accessing memory, I/ O or acknowledging an external request. This cycle may consist of three to six T-states.
- ***T-state*** is defined as one subdivision of the operation performed in one clock period. These subdivisions are internal states synchronized with the system clock, and each T-state is precisely equal to one clock periods. The terms T-state and clock period are often used synonymously.

THE ALU

The arithmetic/logic unit performs the computing function; It includes the accumulator, the temporary register, the arithmetic and logic circuits and five flags. The temporary register is used to hold data during an arithmetic/logic operation. The result is stored in the accumulator, and the flags (flip-flops) are set or reset according to the result of the operation.

The flags are affected by the arithmetic and logic operations in the ALU. In most of these operations, the result is stored in the accumulator. Therefore, the flags generally reflect data condition in the accumulator-with some exception. The descriptions and conditions of the flags are as follows.

Among the five flags the AC flag is used internally for BCD arithmetic the instruction set does not include any conditional jump instructions based on the AC flag. Of the remaining four flags, the Z and CY flags are those most commonly used.

Peripheral Mapped I/O

- The OUT is a two byte instruction. It copies (transfers or sends) data from the accumulator to the addressed port.
- When the 8085 executes the OUT instruction. In the third machine cycle, it places the output port address on the low order bus duplicates the same port address on the high order bus places the contents of the accumulator on the data bus and assets the control signal WR.
- A latch is commonly used to interface output devices.
- **The IN instruction is a two byte instruction.** It copies (transfers or reads) data from an input port and place the data into the accumulator.
- When the 8085 executes the IN instruction, in the third machine cycle it places the input port address on the low order bus as well as on the **high** order bus. Asserts the control signal, RD, and transfers data from the port to the accumulator.
- A tri-state buffer is commonly used to interface input devices.
- To interface an output or an input device the low order address bus A_7 – A_0 (or high order bus A_{15} – A_8) needs to be decoded to generate the device address pus, which must be combined with the control signal IOR (or LOW) to select the device.

Memory Mapped I/O

- Memory related instructions are used to transfer data.
- To interface I/O devices, the entire bus must be decoded to generate the device address pulse, which must be combined with the control signal MEMR (or MEMW) to generate the I/O select pulse device and transfer the data.

Stack

- The **stack** in an 8085 microcomputer system can be described as a set of memory locations in the R/W memory specified by a programmer in a main program. These memory locations are used to store binary information (bytes) temporarily during the execution of a program.
- The beginning of the stack is defined in the program by using the instruction LXI SP, Which loads a 16-bit memory address in the stack pointer register of the microprocessor.
- The stack is shared by the programmer and the microprocessor. The programmer can store and retrieve the contents of a register pair by using

PUSH and POP instructions. Similarly, the microprocessor automatically stores the contents of the program counter when a subroutine is called.

Subroutine

- A subroutine is a group of instructions written separately from the main program to per form a function that occurs repeatedly in the main program. For example, it a time delay is required between three successive event, three delays can be written in the main program. To avoid repetition of the same delay instruction, the subroutine technique is used. Delay instructions are written once, separately form the main program, and are called by main program when needed.
- The 8085 microprocessor has two instruction to implement subroutines: CALL (call a subroutine), and the main program from a subroutine). The CALL in struction is used in the main program to call a subroutine, and the RET instruction is used at the end of the subroutine to return to the main program.

Operating System and Programing Tools

The operating system of a computer is a group of programs that manages or oversees all the operations of the computer. The computers transfer information constantly among peripherals such as a floppy disk, keyboard, and video monitor. It also stores uses programs under file names on a disk. (A file is defined as related records stored as single entity). The operating system is responsible primarily for managing the files on the disk and the communication between the computers and its peripherals. The functional relationship between the operating system and computer's various subsystem is shown in figure. From the user's point of view, the operating system is an access or a pathway to run application programs; it is user interface with the computer.

Operating Systems

Each computer has its own operating system. We focus briefly here on four operating system, known respectively as MS-DOS (Microsoft Disk Operating System), OS/2 (Operating System 2), Windows 95/98/2000, and UNIX.

MS-DOS Operating System

In 16-bit microcomputer, such as IBM PC, XT, and AT, MS-DOS was so widely used that it became the industry standard. Initially, when it was installed on IBM PC, it was known as PC-DOS the terms MS-DOS and PC-DOS are interchangeable. MA-DOS is designed to handle 16-bit data word and 640K system memory and disks with quad (high)-density disk format with memory capacity of 720K and 1.200K. Similarly, it can support a hard disk and includes a hierarchical file directory.

Os/2 (Operating System 2)

OS/2 (and its various versions) is a 32-bit single-user operating system designed by IBM to exploit various powerful features of recent 32-bit (and 64-bit) microprocessor. It has removed the memory limitation of DOS (64k bytes) it can assign 512M bytes of virtual memory space (using storage space form a disk drive) to each application program. Furthermore, it is compatible with DOS and Windows application program. A few of the important features of OS/2 are as follows. It supports:

Windows 95/98/2000 and Me (Millenium)

This is a 32-bit single-user operating system designed by Microsoft, and it is by far the most widely used operating system is personal computers. It has a graphical interface and it supports the use of a mouse, icons, and menus. It is also a multitasking operating system, similar to OS/2. It includes many of the features of OS/2 described above. Windows 98/2000/ME are upgraded versions of Windows 95.

Unix-Based Operating System

Unix is a multi-user, multitasking operating system. Initially, it was designed for mainframe computer, but is now used on various machines ranging from powerful microcomputer to supercomputer. It is independent of any particular hardware structure. It is widely used in engineering, scientific, and research environments and is not limited by any memory constraints. Linux, a variation of UNIX, is gaining popularity in PC system. It is free and its source code is freely distribute. It is ideally suited for Interment and networking environments.

Solaris is a Unix-based operating system designed by Sun Microsystems and is widely used in high-end microcomputer such as workstations and network servers. In the network environment, Solaris is capable of handling 64 computers and is being upgraded to handle 128 computers. One of the major problem with Unix is that there are too many versions in the market; however companies have begun to accept Solaris as a standard operating system for high-end micro-computers.

Tools for Developing Assembly Language Programs

In addition to the operating system of a computer, various programs called **utility programs** are necessary to develop assembly language programs. These utility programs enable the user to perform such functions as copying, printing, deleting, and renaming files, In MS-DOS, commands such as Copy Del Dir are given by using a keyboard and in Windows operating systems, graphical icons and a mouse are used to perform these tasks. The program development utilities enable the user to write, assembler, and assembly language program; they include program such as Editor, Assembler, Linker (or Loader), and Debugger. The descriptions of various programs and the assembler process described below may vary in details depending upon an operating system and its utility programs.

Editor

The Editor is a program that allows the user to enter, modify, and store a group of instruction or text under a file name. The Editor programs can be classified in two groups line editors and full screen editors. Line editors such as EDIT in MS – DOS, work with and manage one line at a time. Full screen editors (also known as word processors) such as MS Word and WordPerfect, manage the full screen or a paragraph at a time.

Assembler

The Assembler is a program that translates source codes or mnemonics into the binary code, called object code, of the micro-processor and generates a file called the object file. This function is similar to manual assembly, where by the user looks up the code for each mnemonic in the listing. In addition to translating mnemonics, the Assembler performs various functions, such as error checking and memory allocations.

Loader

The loader (or Linker) is a program that takes the object file generated by the assembler program and generates a file in binary code called the COM file or the EXE file. The COM (or EXE) file is the only executable file, i.e. the only file that can be executed by the microcomputer.

Debugger

The Debugger is a program that allows the user to test and debug the object file. The user can employ this program to perform the following functions:

- Make changes in the object code.
- Examine and modify the contents of memory.
- Set breakpoints, execute a segment of the program and display register contents after the execution.
- Trace the execution of the specified segment of the program and display register and the memory contents after the execution of each instruction.
- Disassemble a section of the program, i.e. convert the object code into the source code or mnemonics.

RST (Restart) Instructions

The 8085 instruction set includes eight RST (Restart) instructions listed in Section 9.3. These are 1-byte call instruction, that transfer the program execution are executed in a similar way to that of call instruction. The address in the program counter (meaning the address of the next instruction to an RST call instruction) is stored on the stack before the program execution is transferred to the RST call location. When the processor, the program returns to the address that was stored on the stack. In case of a hardware interrupt, we will use an RST instruction to restart the program execution.

8085 Vectored Interrupts

The 8085 has five interrupt inputs one is called INTR. Three are called RST 5.5, 6.5, and 7.5 respectively and the fifth is called TRAP a nonmaskable interrupt. These last four RSTs and TRAP) are automatically vectored (transferred) to specific locations on memory page 00H without any external hardware. They do not require the $\overline{\text{INTA}}$ signal or an input port the necessary hardware is already implemented inside the 8085. These interrupts and their call locations are as follows.

Interrupts	Call Locations
1. TRAP	0024H
2. RST 7.5	003CH
3. RST 6.5	0034H
4. RST 5.5	002CH

The **TRAP has the highest priority** followed by RST 7.5, 6.5, 5.5 and INTR in that order however the TRAP has a lower priority than the Hold signal used for DMA.

TRAP

TRAP a nonmaskable interrupt known as NMI. It has the highest priority among the interrupt signals, it need not be enable and it cannot be disabled. **It is level and edge sensitive** meaning that the input should go high and stay high to be acknowledged. It cannot be acknowledged again until it makes a transition from high to low to high.

When this interrupt is triggered, the program control is transferred to location 002H without any external hardware or the interrupt enable instruction EI. TRAP is generally used for such critical events as power failure and emergency shunt-off.

RST 7.5, 6.5, and 5.5

These mask able interrupts are enabled under program control with two instruction : EI (Enable Interrupt) described earlier, and SIM (Set Interrupt Mask) described below:

Instruction SIM

Set Interrupt Mask. This is a 1-bytes instruction and can be used for three different functions

- One function is to set mask for RST 7.5, 6.5, and 5.5 interrupts. The accumulator and enables or disables the interrupts according to the content of the accumulator. Bit D_3 is a control bit and should = 1 for bits D_0, D_1 and D_2 to be effective. Logic 0 on D_0, D_1 and will enabel the corresponding interrupts, and logic 1 will disable the interrupts.
- The second function is to reset RST 7.5 flip-flop. Bit D_4 is additional control for RST 7.5. If D_4 = 1, RST 7.5 is reset. This is used to override (or ignore) RST 7.5 without servicing it.
- The third function is to implement serial I/O Bits D_7 and D_6 of the accumulator are used for serial I/O and do not affect the interrupt. Bit D_6 = 1 enables the serial I/O and bit D_7 is used to transmit (output) bits. Here we are concerned with RST 7.5, 6.5 and 5.5 interrupts and not with serial I/O. The mnemonic SIM is confusing. The wording Set Interrupt Mask-Implies that the instruction masks the interrupts however the instruction must be executed in order to use the interrupts. The Process required to enable these interrupts can be likened to a switchboard controlling three telephone extensions in a company.

Triggering Levils

These interrupts are sensitive to different types of triggering as listed below.

RST 7.5

This is positive edge sensitive and can be triggered with a short pulse. The request is stored internally by the D flip-flop until the microprocessor responds to the request or until it is cleared by Reset or by bit D_4 in the SIM instruction.

RST 6.5 and RST 5.5

These interrupts are level sensitive meaning that the triggering level should be on until the microprocessor completes the execution of the current instruction. If the microprocessor is unable to respond to these requests immediately, they should be stored or held by external hardware.

Addressing Modes

An instruction format contains an opcode or mnemonic field and an operand or address field. Typically, there may be two source addresses for manipulation of data and the third as destination address. Many times we economies by making one of the source addresses as destination. However we would like to address a large number of memory locations. To achieve this objective, different address techniques are employed. They are:

1. Immediate	2. Direct
3. Indirect	4. Register
5. Register Indirect	6. Indexed
7. Relative	8. Stack

Immediate Addressing

In this The operand or source data is available within the instruction itself and is therefore immediately available.

MVI	A	FE (hex)	(A) (hex)
↓	↓	↓	
Op code	Destination	source data	

Sometimes the destination may be 16 bits register pairs (BC, DE, HL).

LXI H 802D; (H) (L) 802D(hex) Load immediate 802D in HL register pair. Other instructions are ADI SUI, ANI, I indicating immediate.

Use

For initializing counters, obtaining constants in calculations data transfer and manipulation.

Advantage

Fast as the instruction combines the data and easy and convenient to use No need for memory reference and involves only internal processor registers.

Disadvantage

Not flexible having data and instruction together in the same field.

Direct Addressing

In this, the instruction itself contains the address of the memory location.

Equation: LDA 20ADH; (A) ← (20 AD)

It means load accumulator with thecontains the address of the memory Location given by address 20AD please not loaded by 20AD.

STA 20A2H ; (20A2) ← (A)

Store contents of accumulator at memory address 20A2

LHID 20A2 H; (L) ← 20A2
(H) ← 20A3

Here a 16-bit register L pair is used to load contents of memory location address 20A2 at L and the consecutive address 20A3 at H, using single instruction.

Similarly SHLD 20A4 ; (20A4) ← (L).

Content of Register L store at 20A4 and of H 20A5.

Memory addresses are 16 bits long and so two bytes (two 8 bits) are needed to address and this is known as extended direct addressing.

Advantages

Simple to use either for reading from memory and writing into memory.

Disadvantage

Flexibility to process data arrays not available, instruction can access a single fixed address only limiting address space. It also takes longer time to execute an instruction as the number of instruction cycles to fetch and execute are more.

Indirect Addressing

In this method the instruction contains the memory address which refers to the address storing the data. it means that address of the address is part of the instruction.

ADD 20A5 (A) + (20A5) (20AB)

Double parenthesis indicates "contents of contents" of In effect

(A) ← (A) + (20AB)

Advantages

In direct addressing the length of the address field is less than the word length thus limiting the address range. Here for the word length N, the address field possible is 2^N.

It is much more flexible especially when repeated use of the same result is resorted to as in subroutines.

For the same reason it is slower as the instruction execution requires two memory references to fetch the operand one to get its address and another time to get its value.

Register Addressing

Also called Register Direct Addressing this is similar to direct Address respect that the address of the data is in one of the processor registers (B, C, D, E, H, L) and in memory.

MOV B, A (B) ←(A)

Contents of Register A (source) are transferred to B register (destination) contents of A remain unaltered.

XCHG (DE) ↔(HL)

Contents of 16-bits HL registers pair are interchanged with register pair DE.

Advantage

1. As data is to be fetched only from internal registers and not memory it is the fastest addressing method as assess time to register is very much less than to memory.
2. No memory references are required.
3. Instructions can be shorter as there are fewer number of registers compared to memory locations.

Disadvantage

1. This type of addressing holds advantage only if the data address is used a number of times over and over again.

(D) ← (20A5)

ADD D (A) ← (A) + (D)

However if data at 20A5 has to be added a number of times storing in the processor 's' register D is convenient and faster as repeated memory access can be avoided.

Register Indirect Addressing

This is similar to Indirect Addressing except that the address of the data in memory is now stored in the processor register usually HL. It is indirect because the instruction does not contain the address of the memory but has the address in the HL register pair. This is called data/address register.

MOV A, M (A) ← [(H) (L)]. It means load the accumulator with the contents of the data in memory as pointed to by address in HL register pair. So the address of the memory is first transferred to HL register.

Advantages

1. It is faster indirect as the CPU need not fetch the effective address from memory.
2. The register address is shorter than memory address. It has also large address space.

Disadvantages

The advantage is valid only if repeated operations of the same data are resorted to.

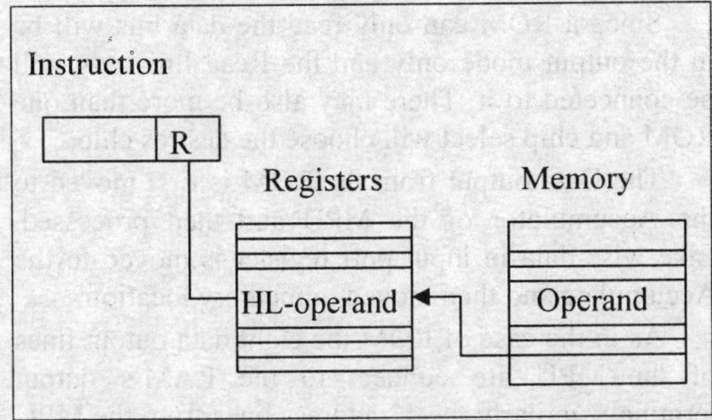

Indexed Addressing

Some 8 bit microprocessors have a separate 16 bit register called Index Register. In this the contents of the Index Register are added by CPU to the address in the instruction creating a new effective address (EA). EA = A + (R). This type of addressing is also called Displacement Addressing.

ADD 100, X (X indicates indexing)

(EA) ← 100 (X)

X is also called offset (A) ← (EA) or distancement.

Equation: LDA 0100 (X) Reference address 0100 and IR holds the value 0002.

EA = RA + (X) = 0100 + 0002 = 0102

So 0102 will be transferred to the A register In this reference address is part of an array, for accessing the next element of array.

Advantage

1. This mode is very useful in accessing arrays and tables.
2. It is much more flexible than direct addressing since the same instructions can handle all the elements in an array or table.

Disadvantage

It is slower than direct addressing as the CPU has to perform an addition in order to obtain the effective address.

Relative Addressing

In this the CPU adds the contents of the PC to the address supplied with the instruction to find the effective address.

Advantages

1. By Relative Addressing it is possible to move programmers to different places in memory without disturbing their relative position. This is called relocating and it given the advantage of occupying any a available area of memory.
2. The results in shorter addresses when the EA is close to current program location. They are used with jump instructions.

Stack Addressing

As already mentioned a stack is a reserved or apportioned are in memory or registers for storing data address. Address are pushed to the top of the stack and a stack pointer (SP) points to the top of the stack.

Stack addressing means the contents of the SP are the SP are the data address. As no explicit addresses are necessary, instructions that use stack addressing are the shortest of all.

Since references are to the address on the top of the stack this is the same as register indirect addressing or ides addressing. This mode of addressing in a form of implied addressing.

Advantages

1. There is no memory reference.
2. Address is updated as part of the execution of instruction.
3. There is no need for an effective address.

In stack operation data address is only pushed down and not destroyed. Hence reentrant programmers (they are executed when the same program has been interrupted or temporarily suspended) and recursive programs (which can use themselves as subroutines) are possible.

Inherent addressing

In this the source of the data and its destination are in the instruction itself. It is a single byte instruction. All actions occur in the CPU and there is no need for retrieving data or addresses from CPU registers memory location or I/O ports.

Example:

STC	- Set Carry Flag
CMA	- Complement (A)
RAR	- Rotate A right through carry
RAL	- Rotate A left Through
NOP	- No operation
HLT	- Halt

Advantage

It is very fast.

Disadvantage

Limited to particular instructions.

5.17 MEMORY AND INPUT OUTPUT INTERFACING

The microprocessor cannot function all by itself. Its main link with the external world is through the buses by which it receives and sends out data and also signals for control and synchronization. Thus, the microprocessor is a part though an important one of a system and its interconnections to other components of the systems such as memory and input/Output ports is called interfacing. A general schematic of a MP's link to the other devices is.

An external device connected to an I/O module is called a peripheral device or peripheral. They are VDU, printers disk and tape systems, sensors and actuators and communication devices such as modems terminals or even another computer. The following activities are associated with transfer of data to and from MPU. *1. Memory read 2. Memory write 3. I/O read 4. I/O write 5. Interrupt 0* or reset handlings sometimes the peripherals can access the main memory directly when the MPU releases control of address and data buses temporarily. *This is called DMA (Direct Memory Access.*

The data bus lines are connected to the output terminal of ROM (D_0-D_7). The single $\overline{RD}$ output line is connected to the output enable $\overline{OE}$ of the ROM chip. This least significant address lines $A_0 - A_{11}$ of the MPU are connected to the 4 K x 8 ROM . The balance 4 address lines are connected to a decoder which selects the particular ROM by activating chip enable $\overline{CE}$ or select $\overline{CS}$. It functions as follows. The chip is first enabled by sending a low (0) to $\overline{CS}$ input of ROM. The required address is sent on the address bus by the MPU. The built in decoder of ROM connects to the required 8 bit address. A short time later a low (0) on the Read line enables the output ($\overline{OE}$) of the ROM. Stored data is placed on the data bus and fed to the MPU. In this one critical time limitation is the Read Access Time, that is, the time it takes to locate the correct memory word (byte) after the chip is enabled, by the ROM's internal decoder. Thus while interfacing the MPU with the ROM the important considerations are addressing and timing.

Since a ROM can only read the data bus will be in the output mode only and the Read line only will be connected to it .There may also be more than one ROM and chip select will choose the desires chip.

The Data output from the ROM is first moved to the Accumulator of the MPU and then processed. Like wise data in input port register is moved to the Accumulate and then moved to memory location.

As in the case of ROM the eight data output lines of the MPU are connect to the RAM's output terminal; similarly the 12 address lines from the MPU to the RAM via. address bus $A_0 - A_{11}$. The high order of the address lines A_{12} to A_{15} are connected to an address decoder and as per the decoder logic sends a low pulse to the chip select ($\overline{CS}$) when it becomes ready to receive read or write signals.

The internal decoder of the RAM chooses the required 8 bit address word. The difference between ROM and RAM is the write line. When the write $\overline{WR}$ of the MPU is Low. $R/\overline{W}$ is enabled and a word can be written at the chosen address. If $\overline{WR}$ line carries a HIGH and $\overline{RD}$ line carries a LOW, the output is enabled and data from the required address is placed on the output terminal.

We have considered SRAMs. Though we have showed a 4 k × 8 SRAM, The same as ROM in practice the number of RAM cells will be less than RAM on account of circuit considerations. Also though they are more difficult to interface on account of the need to refresh memory frequently.

The maximum read access time for SRAM is 250 to 1000 ns. The minimum write cycle time is also of the same order. The minimum address set up time might range from 20 to 200 ns, While the minimum write pulse time might be from 180 to 750 ns.

IN and OUT are the instructions for feeding data into the microprocessor and taking out data from the MPU respectively. The locations or terminals from where data is taken into the MPU are called Input ports and locations to which data flows out of the MPU are called Output ports. Jointly they ate called I/O ports. There are two ways that the I/O port can be connected to the microprocessor so that it can be addressed they are:

1. Program Controlled I/O (PI/O)
2. Memory Mapped I/O

Pi/O

In programmed I/O, data transfer between the external devices (Peripherals) and the MPU takes Place vis. One or more register called I/O ports under the control of instructions stored the in the MPU such as IN and OUT. These are two byte instructions with an op-code and port address. Port address is transferred to the Accumulator in the case of IN and contents of Accumulator are transferred to the port address in the case of OUT.

The I/O port is the connecting medium between the MPU, memory (RAM, ROM) and the wide variety of peripherals which may be mechanical, electro-mechanical or electronic. They may range in speeds from the slowest AD converter providing data every few minutes or seconds to teletypewriter of 100 BPS, to high speed floppy disks of 250,000 BPS Hence sometimes exclusive chips are manufactured to handle I/O function.

Some others, manufacture the I/O ports in the same chip as memory, i.e. I/O with ROM or RAM, e.g. Intel 8155 has 256 bytes of RAM and three I/O ports.

Input To Mpu

This is similar to a memory read cycle and follows the steps given below. The CPU puts out the address on the address bus specifying the particular I/O module and external device. The word length of the I/O ports out the address on the address bus specifying the particular I/O module and external device. The word length of the I/O port is normally the same as that of CPU, that is 8 bits so that $2^8 = 256$ addresses are possible.

1. A CPU waits for the data to become available from the data register.
2. The CPU reads data from the data from the data bus and places it in a register.

These actions takes place under an I/O command called command register. There are four type of commands that an I/O port or module may receive when it is addressed by a CPU. They are control, test, read and write.

1. A control command activates a peripheral and tells it what to do. For instances magnetic tape unit may be asked to rewind or move forward one record. These commands are tailored according to the peripheral deceive.
2. A test command is used to test various status conditions of the I/O module and its peripherals such as being on and available for use after completing a recent I/O operation.
3. A read command causes the I/O module to obtain the data from the peripheral and place it in the data Register. Later the CPU will ask the I/O part to place it on the data bus.
4. Conversely a write command causes the I/O port to take an item of data (byte or word) from the data bus and later transmit it to the peripheral.

Output from MPU

1. The CPU places an address on the address on the address bus.
2. The CPU places data on the data bus.
3. The CPU waits for the data transfer to be successfully completed.

Handshake

From the procedures described above it is clear that the MPU cannot just issue an address on the address bus following it up either by sending data or be ready to receive data from any of the peripherals at any instant of time as it may be already engaged. So a signal is sent to the peripheral to know its status. Common status signals are BUSY and READY.

Synchronization of transfer of data between MPU and external devices is called **Handshaking**. The Handshake control operates as follows:

1. The CPU interrogates the I/O module to check the status of the required device (peripheral).
2. The I/O module replies to the device status.
3. If free the peripheral sends the data and a DATA READY signal to the I/O port.
4. The CPU determines that DATA READY signal is active. A latch holds the DATA READY signal until the CPU reads it.
5. The CPU accepts the data and sends an DATA Accepted Signal (DACC) by way of acknowledgements.
6. DATA READ line is reset to be ready to receive further data.

For sending out data, the peripheral READY signal is sent to I/O module which is detected by the CPU. After sending an OUTPUT READY signal, data is fed is data bus to the peripheral at the completion of which acknowledgment is sent.

Buffering

Another important task of I/O port is buffering that is, matching memory and CPU to the peripheral. The data rates of various devices are:

Device	Data Rates (bytes/sec)
Measuring Instrument	0.1 to 1000
CRT terminals	10 to 2000
Line printer	220 to 88,000
Communication line	30 to 200 k
Magnetic taps	15 k to 3 M
Magnetic Disc	30 k to 10 M
Main memory	1 M to 100 M

High speed data from CPU and main memory are buffered in the I/O port and then sent to the peripheral device at its data rate. In the reverse direction, data are buffered so as not to tie device and memory speeds.

I/O Port Addressing

As stated already there may be more than one I/O ports each characterized by an address and the CPU sends the address of the desired port on the address line.

When the CPU, main memory and I/O port share a common bus, two modes of addressing are possible (1) Standard or Isolated I/O and (2) Memory Mapped I/O.

Isolated I/O

As all the three buses address data and control are common to memory any I/O ports, there should be a means to identify whether the signals and data from MPU are meant for memory or I/O ports. For this there is a special terminal in MPU, $IO/\overline{M}$ control signal. A HIGH on this terminal indicates an I/O operation whereas a LOW indicates that memory is to be operated. Execution of IN and OUT instructions set $IO/\overline{M}$ to HIGH. One advantage is that port addresses can be short and 8 bit addresses are usually sufficient for I/O sections.

Memory Mapped I/O

In this some portion of the available memory address space is reserved for input/output devices, while the rest is for RAM/ROM. One major advantage is that *all the instruction types including arithmetic and logic operation as also the different addressing modes are available to I/O data transfer* in contrast to just two, IN and OUT, instructions of isolated I/O. Also data from any register can be moved to a port or data from a port to moved to any register. In Isolated I/O mode data to and from pots have be routed through the Accumulator.

The disadvantage is that the number of memory addresses available is reduced. The most significant bit A_{15} of the address is typically used to distinguish between memory and I/O. If $A_{15} = 1$, an I/O port is selected; if $A_{15} = 0$ memory is selected. This means that memory locations $7FFF_{16}$ and above will access memory words, and memory locations 8000_{16} and above will access I/O port buffers. This shows that the maximum size of memory has been reduced from 64 k to 32 k bytes. A memory mapped MPU is given below:

Address Range in Hex	Memory Chip	A_{12}	A_{11}	A_{10}
0000-03FF CSI/RAM$_1$	RAM$_1$	0	0	0
0400-07FF CSI/RAM$_2$	RAM$_2$	0	0	1
0800-0BFF CSI/ROM	RAM$_3$	0	1	1
0000-FFF CSI/ROM	ROM	0	1	1

Interrupt I/O

When the microprocessor is asked to communicate with the I/O ports we say that the microprocessor is servicing. When an external device seeks priority in the execution of a program, the current programme is completed, the next instruction in PC is pushed to the stack the interrupt program is executed and the completion of it, the processor returns to the execution or original program popped to the PC. The starting address for interrupt service routine is stored at the two special memory locations called interrupt vector, which is loaded into the PC after the next instruction PC is pushed to the stack.

5.17. 1 DIRECT MEMORY ACCESS (DMA)

In this type of I/O technique the data are transferred to the memory from the external devices by passing the microprocessor. An interface chip called the DMA controller chip is used with the MP for transferring data via. DMA. *DMA is used for transferring blocks of data from the memory to the external device such as a cassette recorder. In this type the transfer of 1 byte of data typically requires 1 μs.*

The DMA controller chip must be able to perform READ and WRITE operation similar to a MPU:

1. The DMA controller chip puts the MPU in a Hold state by means of the HOLD control signal.
2. The MPU then stops executing the program and disconnects the address data and memory control lines from its bus by placing them on a high impedance state. The microprocessor is totally disabled during DMA.
3. The MPU sends a DMA acknowledge signal when the DMA takes control of the address bus and control bus to put data on the RAM from the I/O device.
4. When once the DMA completes transfer of data into or out of memory the DMA controller transfers back the control of the buses to the MPU by releasing the HOLD state.

As may be noted the DMA controller's functions ate similar to those of microprocessor with speeds close to the memory cycle time. DMA chip consists of an address register containing the address of the data to be transferred to or from and a counter containing the length of the data to be transferred. The total number of bytes to be transferred are entered in the counter and is decremented by 1 after each bytes is transferred sequentially When the count register reaches 0, transfer of data is complete and HOLD is released. The DMA chip functions in coordination with the MP and the MP loads the address register and counter.

There are Three Types of DMA

1. Block transfer DMA
2. Cycle stealing DMA
3. Interleaved DMA

1. In this blocks of data are transferred from and to memory by the I/O device over the DMA.
2. On the other hand if the quantum of data to be transferred is not much cycle stealing.
3. DMA is used when data are transferred on a byte by byte basis by sealing a clock cycle of the microprocessor. This is accomplished by stopping the clock signal to the MP. This is done by using the INHIBIT terminal when a LOW on the terminal disables the clock.
4. In this blocks of data are transferred when the microprocessor is not connected to external devices and performs internal operations such as decoding and arithmetic-logic operations. Though this is complex, microprocessor is not stopped. With interleaved DMA, each data transfer includes 1 bytes per instruction cycle.

5.18 MICROPROCESSOR 8086

- The intel 8086 is a 16-bit microprocessor that is intended to be used as the CPU in a microcomputer.
- The term 16-bit means that its arithmetic logic unit, its internal registers, and most of its

instructions are designed to work with 16-bit binary words.

- The 8086 has a 16-bit data bus, so it can read data from or write data to memory and ports either 16-bit or 8 bits at a time.
- The 8086 has a 20-bit address bus, so it can address any one of 2^{20}, or 1,048,576, memory locations.
- Each of the 1,048,576 memory addresses of the 8086 represents a byte-wide location.
- Sixteen-bit words will be stored in two consecutive memory locations.
- If the first byte of a word is at an even address, the 8086 can read the entire word in one operation.
- If the first byte of the word is at an odd address, the 8086 will read the first byte with one bus operation.

The main point here is that if the first byte of a 16-bit word is at an even address, the 8086 can read the entire word in one operation.

5.19 MICROCONTROLLERS

5.19.1 MICROCONTROLLER VERSUS GENERAL – PURPOSE MICROCPROESSOR

What is the difference between a microprocessor and microcontroller? By microprocessor is meant the general – purpose microprocessors such as Intel's x86 family (8086, 80286, 80486, and Pentium) or Motorola's 680x0 family (68000, 68010, 68020, 68030, 68040, etc). These microprocessors contain no RAM, no ROM, and no I/O ports on the chip itself. For this reason, they are commonly referred to as general – purpose microprocessors.

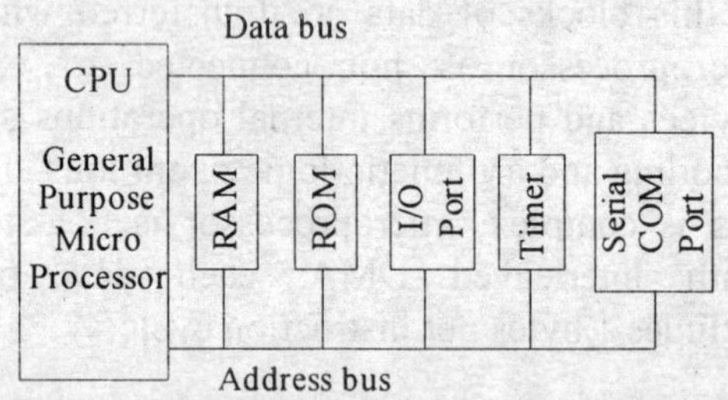

Figure: General-purpose Microprocessor Systems

CPU	RAM	ROM
I/O	Timer	Serial COM Port

Figure: Microcontroller

A system designer using a general-purpose microprocessor such as the Pentium or the 68040 must add RAM, ROM, and I/O ports needed to fit the task at hand. This is not the case with microcontrollers. A microcontroller has a CPU (a microprocessor) in addition to a fixed amount of RAM, ROM, I/O ports, and timer all on a single chip. In other words, the processor, RAM, ROM, I/O ports, and timer are all embedded together on one chip; therefore, the designer cannot add any external memory, I/O, or timer to it. The fixed amount of on-chip ROM, RAM, and number of I/O ports in microcontrollers makes them ideal for many applications in which cost and space are critical. In many applications, for example a TV remote control.

5.19.2 8051 MICROCONTROLLER

The 8051 is the original member of the 8051 family. Intel refers to it as MCS – 51. Table shows the main features of the 8051.

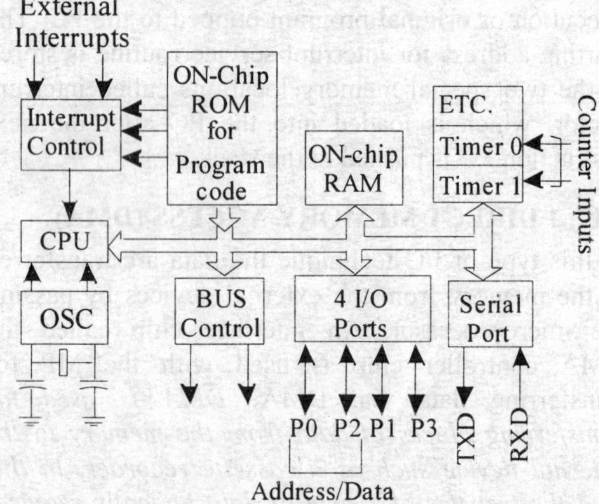

Figure: Microcontroller 8051 block diagram

Table: Feature of the 80 51

Featire	Quantity
ROM	4 K bytes
RAM	128 bytes
Timer	2
I/O pins	32
Serial port	1
Interrupt sources	6

The 8051 architecture consist of the specific Features

1. Eight-bit CPU with registers A and B.
2. Sixteen-bit program counter and data pointer.
3. Eight-bit program status word.
4. Eight-bit stack pointer [SP]
5. Internal ROM of EPROM [8751]of 0 [8031] to 4k [8051]
6. Internal RAM of 128 bytes.
7. Four register banks, each containing eight registers.
8. Sixteen bytes, which may be addressed at the bit level.
9. Eighty bytes of general-purpose data memory.
10. Thirty-two input/ output pins arranged as four 8-bits ports: P0-P3.
11. Two 16-bit timer/counters: T0 and T1.
12. Full duplex serial data receiver/transmitter: SBUF
13. Control registers: TCON, TMOD, SCON, PCON, IP, and IE.
14. Two external and three internal interrupt sources.
15. Oscillator and clock circuits.

Function of 8051 micro controller

Oscillator and Clock

1. The heart of the 8051 is the circuitry that generates the clock pulses by which all internal operations are synchronized pins XTALI and XTAL2 are provided for connecting a resonant network to form an oscillator. Typically, a quartz crystal and capacitors are employed. The crystal frequency is the basic internal clock frequency of the microcontrollers.
2. The time to execute that instruction is then found by multiplying C by 12 and dividing the product by the crystal frequency:

$$T_{inst} = \frac{C \times 12d}{\text{crystal frequency}}$$

Program Counter and Data Pointer

1. The 8051 contains two 16-bit registers: the program counter (PC) and data pointer (DPTR). Each is used to hold the address of a byte in memory.
2. The PC is automatically incremented after every instruction byte is fetched and may also be altered by certain instructions. The DPTR register is made up of two 8-bit registers, named DPH and DPL. Which are used to furnish memory addresses for internal and external code access and external A and CPU register data access.
3. The 8051 contains 34 general-purpose, or working, registers. Two of these registers A and B, hold results of many instruction, particularly math and logical operations, of the 8051 central processing, unit (CPU).

Flags and the program Status word (PSW)

(a) The 8051 has four math flags that respond automatically to the outcomes of math operations and three general-purpose user flags. That can be set to 1 ot cleaved to B by The programmer an desired. The nath flags Include may be used by The programmer to record some enent in The program. Carry (c) Auxiliary Carry Overflow and Parity. User flags are named F0, GF0, and GF1; they are general purpose flags.

(b) The PSW contains the math flags, user program flag F0, and the register select bits that identify which of the four general-purpose register banks is currently is use by the program. The remaining two user flags, GF0 and GF1, are stored in PCON, which is shown in Figure 3.13.

(c)

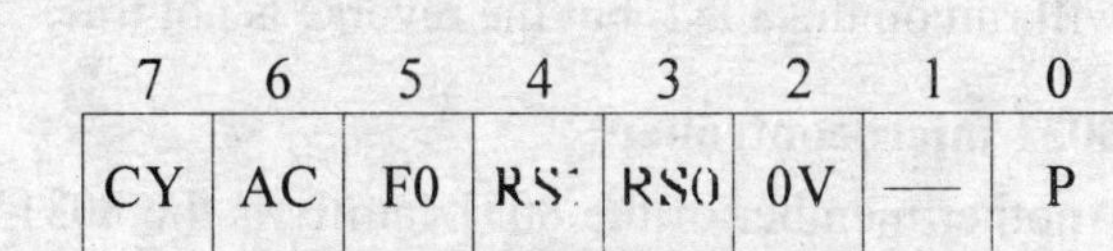

7	6	5	4	3	2	1	0
CY	AC	F0	RS1	RS0	OV	—	P

Figure: PSW Program status word Register

Used in arithmetic, jump, rotate and Boolean instructions.

6 AC Auxiliary Carry flag; used for BCD arithmetic

5 F0 User flag 0

4 RS1 Register bank select bit 1

3 RS0 Register bank select bit 2

RS1	RS0	
0	0	Select register bank 0
0	1	Select register bank 1
1	0	Select register bank 2
1	1	Select register bank3

2 OV Overflow flag; used in arithmetic instructions

1 – Reserved for future use

0 P Parity flag; shows parity of register A:1= Odd Parity

Bit addressable as PSW.0 to PSW. 7

5.19.3 OTHER MEMBERS OF THE 8051 FAMILY

There are two other members in the 8051 family of microcontrollers. They are the 8052 and the 8031.

8052 microcontroller

The 8052 is another member of the 8051 family. The 8052 has all the standard features of the 8051 as well as an extra 128 bytes of RAM and 3 times. It also has 8K bytes of on-chip program ROM instead of 4K bytes. See Table.

Table: Comparison of 8051 Family Members

Feature	8051	8052	8031
ROM (on Chip Program space in bytes)	4 K	8 K	0 K
RAM (bytes)	128	256	128
Timers	2	3	2
I/O pins	32	32	32
Serial port	1	1	1
Interrupt sources	6	8	6

As can be seen from Table 1-4, the 8051 is a subset of the 8052; therefore, all programs written for the 8051 will run on the 8052, but the reverse is not true.

8031 microcontroller

Another member of the 8051 family is the 8031 chip. This chip is often referred to as a ROM – less 8051 since it has 0K bytes of on-chip ROM. To use this chip you must add external ROM to it. This external ROM must contain the program that the 8031 will fetch and execute. Contrast that to the 8051 in which the on-chip ROM contains the program to be fetched and executed but is limited to only 4K bytes of code. The ROM containing the program attached to the 8031 can be as large as 64K bytes. In the process of adding external ROM to the 8031, you lose two ports. That leaves only 2 ports for I/O operations. To solve this problem, you can add external I/O to the 8031.

MCQ Set–1

1. The most suitable coding scheme for coding the successive position in an 8 position **digital shaft encoder** is
   (a) 000, 001, 010, 011, 100, 101, 110, 111
   (b) 000, 010, 100, 110, 001, 011, 101, 111
   (c) 000, 001, 011, 010, 110, 111, 101, 100
   (d) 000, 001, 100, 101, 010, 011, 110, 111

2. In the digital circuit shown in figure, the output f is found to be logic 1 when A is logic '0'. The values of B and C are:

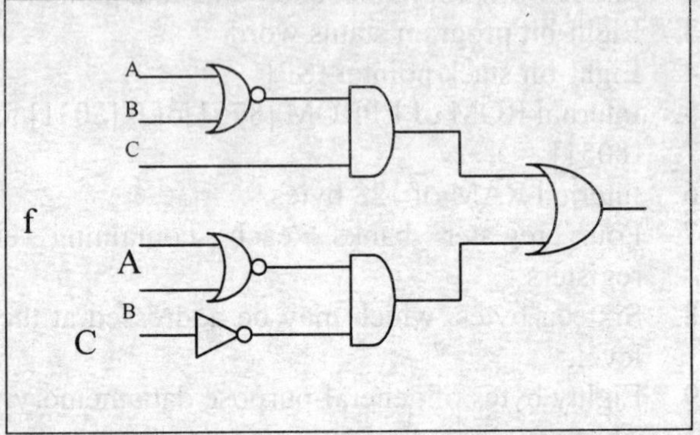

   (a) B = 1, C = 0 (b) B = 0, C = 0 or 1
   (c) B = 1, C = 1 (d) indeterminate

3. The desirable characteristics of a **digital logic family** are:
   1) Large fan-out
   2) Large threshold voltage
   3) Large noise
   4) Margin small threshold voltage
   Choose correct option
   (a) 1, 2 are correct (b) 2, 3 are correct
   (c) 1, 3, 4 are correct (d) 1, 3 are correct

4. The decimal number – 15 is represented in 8 – bit **signed 2's complement notation** as
   (a) 1010001 (b) 11100111
   (c) 11110001 (d) 10001111

5. The Boolean expression $\overline{A+B+C}$ is equal to
   (a) $\overline{A}+\overline{B}+\overline{C}$ (b) $\overline{A}\cdot\overline{B}\cdot\overline{C}$
   (c) $\overline{A}+\overline{B+C}$ (d) A. (B + C).

6. The output Q_{n+1} of a J – K flip-flop for the input J = 1, K = 1 is
   (a) O (b) 1
   (c) Q_n (d) Q_n

7. The minimum number of flip- flop required to design a MOD – 10 counter is
   (a) 3 (b) 10
   (c) 4 (d) 5

8. A combinational circuit has inputs A, B and C and its Karnaugh map is given in figure. The output of the circuit is:

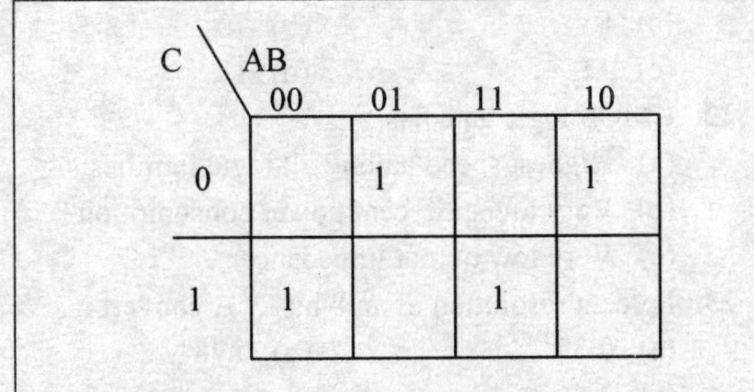

(a) $(\overline{A}\,\overline{B} + AB)\,\overline{C}$ (b) $(\overline{A}B + A\overline{B})\,C$

(c) $A \oplus B \oplus C$ (d) $\overline{A}\,\overline{B}\,\overline{C}$

9. The combinatorial circuit shown in figure employs a 4 to 1 multiplexer. The output Q of the circuit is:

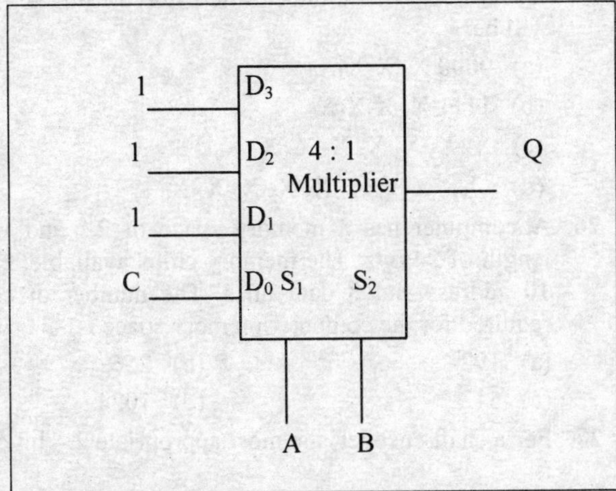

(a) $\overline{A}\,\overline{B}\,C$ (A B C (b) A + B + C

(c) $A \oplus B \oplus C$ (d) $\overline{A}\,\overline{B}\,\overline{C}$

10. The type of A/D converter normally used in a 3 [1/ 2] digit multimeter is:

(a) Dual-slope integrating type

(b) Voltage - to – frequency converter type

(c) Flash (or parallel) type

(d) Successive approximation type

11. If A and B are inputs of a logic gate, then match the logic gate with its output and select the correct answer using the codes:

List I	**List II**
A. NAND	1. $\overline{\overline{A} + \overline{B}}$
B. NOR	2. $\overline{A} + \overline{B}$
C. XNOR	3. $\overline{A}\,\overline{B} + AB$
D. AND	4. $\overline{A}\,\overline{B}$

Codes:

	A	B	C	D
(a)	2	4	3	1
(b)	4	1	3	2
(c)	3	1	4	2
(d)	4	2	1	3

12. A 3- bit resistance ladder D/A (R - 2R network) has resistor of R = 10 kΩ and 20 kΩ V_{REF} equals 8 volts. What is I_{out} for a digital input of 111?

(a) 0.7 mA (b) 0.2 mA

(c) 0.1mA (d) 1 mA

13. The differential input resistance of the circuit shown in figure is:

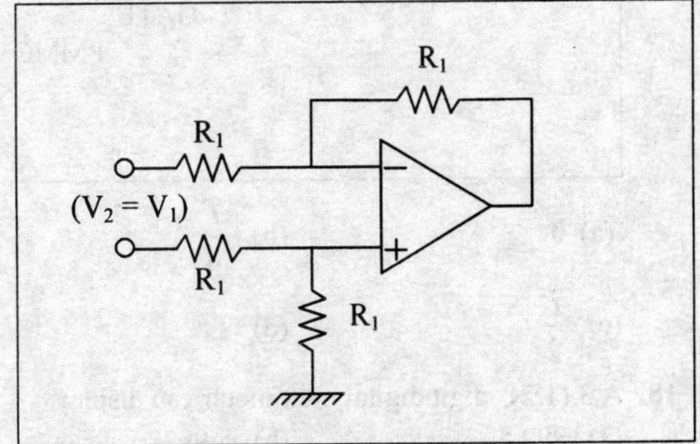

(a) R_1 (b) $R_1 / 2$

(c) $2R_1$ (d) R_f

14. A 4-bit **synchronous counter** with a series carry, uses flip- flops and gates, having a propagation delay of a 30 ns and 10 ns respectively. The maximum time interval required between two successive clock pulses for reliable operation of the counter is:

(a) 10 ns (b) 30 ns

(c) 40 ns (d) 50 ns

15. A **sample and hold circuit** is normally required before the following type of A/ D converter.

(a) Successive approximation

(b) Flash (parallel) converter

(c) Voltage – to frequency converter

(d) Dual slope integrator.

16. Match the List I ADC type with List II applications and select the correct answer using the codes:

List I	**List II**
A. Voltage – to frequency converter	1. High speed
B. Flash (parallel)	2. Telemetry
	3. High precision

Codes:

	A	B
(a)	3	2
(b)	1	3
(c)	3	1
(d)	2	1

17. For the two square inputs in figure the PMMC meter will read maximum when the phase difference between them is:

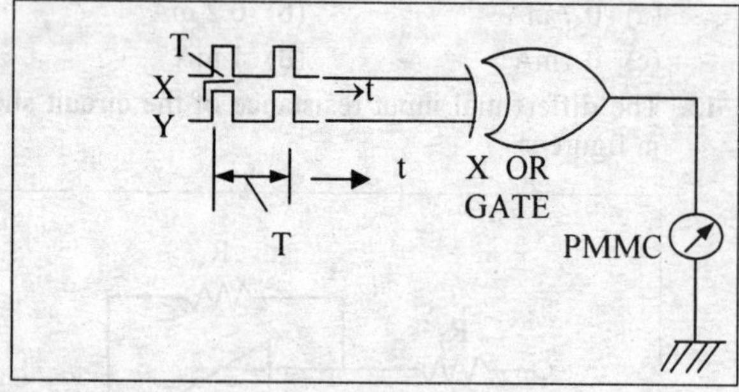

(a) 0 (b) $\frac{T}{4}$

(c) $\frac{T}{2}$ (d) T.

18. A 3 (1/2)- digit digital multimeter can display.

(a) 999.5 (b) 5.999

(c) 5555 (d) 1.999

19. The minimal sum of products form of

$f = A\bar{B}CD + \bar{A}BC + \overline{ABC} + BCD$ is

(a) $\overline{AC} + BD$ (b) $\bar{A}C + CD$

(c) $AC + \bar{B}D$ (d) $A\bar{B} + C\bar{D}$

20. Integrated output waveform for the dual slope ADC is shown in figure. The time T for an 8 bit counter with 4 MHz clock will be:

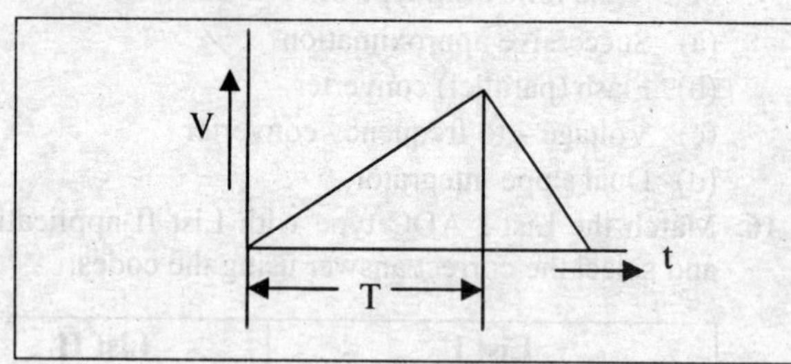

(a) 0.032 ms (b) 0.064 ms

(c) 0. 64 ms (d) 0.024 ms

21. The full scale input voltage to an ADC is 10 V. The resolution required is 0.5 mV. The minimum number of bits required for ADC is.

(a) 8 (b) 10

(c) 11 (d) 12

22. CMOS logic gate has.

(a) Highest speed among all logic families

(b) Very low quiescent power consumption

(c) Very low output impedance.

23. Percent resolution of an 8 bit D/ A convert is

(a) 0.39 (b) 0.78

(c) 2.56

24. The minimum number f flip-flop needed to make mod-2 counter is

(a) 1 (b) 2

(c) 3 (d) 10

25. A number in 4-bit two' complement representation $X_3\ X_2X_1X_0$. This number when stored using 8 – bits will be.

(a) 0000 $X_3\ X_2X_1X_0$

(b) 1111 $X_3\ X_2X_1X_0$

(c) $X_3\ X_3\ X_3\ X_3\ X_3\ X_2\ X_1\ X_0$

(d) $\bar{X}_3\ \bar{X}_3\ \bar{X}_3\ \bar{X}_3\ X_3\ X_2\ X_1\ X_0$

26. A computer has a memory space of 2^{16} and word length of 24 bits. The memory chips available. Have 10 address and 8 data lines. The number of chips required for the computer memory space is.

(a) 192 (b) 256

(c) 512 (d) 1024

27. For a shaft encoder, the most appropriate 2 – bit code is

(a) 11,10,01,00 (b) 11,10,00,01

(c) 01,10,11,00 (d) 01,00,11,10

28. A voltmeter has been connected between the input of a TTL inverting gate and ground. The gate is powered by 5 V. The voltmeter reading will be approximately.

(a) 0 V (b) 2 V

(c) 4 V (d) 5 V

29. The advantage of a dual slope converter over successive approximation converter is that the dual slope converter.

(a) Is faster

(b) Eliminates error due to drift

(c) Can reduce the errors due to power supply

(d) Does not require a stable voltage reference

30. The conversion time of an 8-bit successive approximation converter with a 1 MHz clock is nearly.

(a) 512 μs (b) 256 μs

(c) 128 μs (d) 8 μs

31. The expression $Y = \overline{AB}$ is equivalent to

(a) $\overline{A} + \overline{B}$ (b) AB + A

(c) A + B (d) AB

32. A **sample and hold circuit** has two buffers, one at the input and the other at the output. The primary requirements for the buffers are.

(a) The input buffer should have high slew rate and the output buffer should have low bias current.

(b) The input buffer should have low bias current and the output buffer should have high slew rate

(c) Both the buffers should have low bias currents

(d) Both the buffers should have high slew rate.

33. In a **dual slope type** digital voltmeter, an unknown signal voltage is integrated over 100 cycles of the clock. if the signal has a 50 Hz pick up. The maximum clock frequency can be.

(a) 50 Hz (b) 5 kHz.

(c) 10 kHz (d) 50 kHz.

34. In the circuit shown in figure when inputs A = B = O, the possible logic states of C and D are.

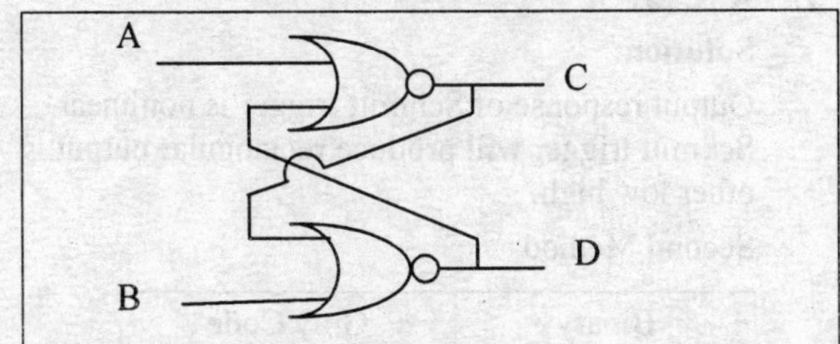

(a) C = 0, D = 1 or C = 1, D = 0

(b) C = 1, D = 1 or C = 0, D = 0

(c) C = 1, D = 0

(d) C = 0, D = 1

35. In the logic circuit shown in figure the output x is

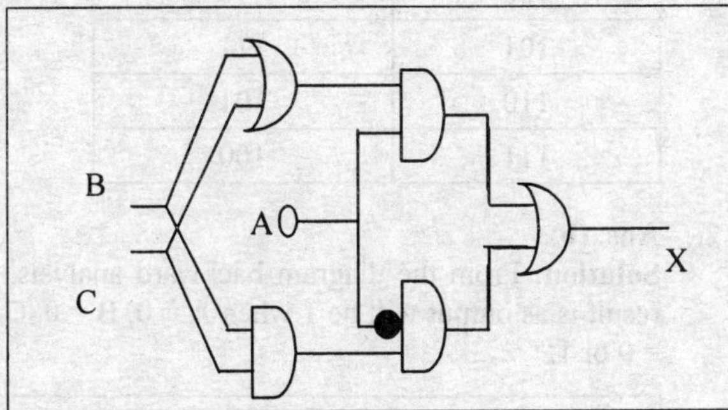

(a) $A\overline{B} + B\overline{C} + C\overline{A}$

(b) $A + B + C$

(c) $AB + BC + CA$

(d) $\overline{AB} + \overline{BC} + \overline{CA}$

36. An 8-bit **DAC** with a reference voltage V, when tested gave the following results.

Input	20H	21H	22H	23H	24H
Output v	0.320	0.330	0.325	0.350	0.360

The DAC is

(a) Monotonic (b) Non monotonic

(c) Bi – polar (d) Quadratic

37. Majority of digital voltmeters are built with a **Dual-slope ADC** because.

(a) Dual slop ADCs are less complex than other types of ADCs.

(b) Dual slope ADCs are faster than other types of ADCs.

(c) Dual slope ADCs can be designed to be insensitive to noise and interference.

(d) Dual slop ADCs provide BCD outputs.

38. Introducing a **schottky diode** between the base and collector of the output transistor in a TTL circuit.

(a) Increases the speed of operation by inhibiting saturation

(b) Decreases the speed of operation by inhibiting saturation

(c) Increases the FAN OUT by enabling saturation

(d) Increases the speed of operation by enabling saturation.

39. The karnaugh map for a four variable Boolean function is given in. The correct Boolean sum of product is.

RS \ PQ	00	01	11	10
00	0	0	0	0
01	1	0	0	1
11	1	0	0	1
10	0	1	0	0

(a) $PQRS + \overline{Q}\ S$ (b) $\overline{P}\ QR\overline{S} + \overline{Q}\ S$

(c) $PQR + Q\overline{S}$ (d) $PQRS + Q$

40. Two 4 – bit 2's complement number 1011 and 0110 are added. The result expressed in 4 – bit 2's complement notation is

(a) 0001

(b) 0010

(c) 1101

(d) Cannot be expressed in 4 – bit 2's complement

41. The logic circuit of figure is a

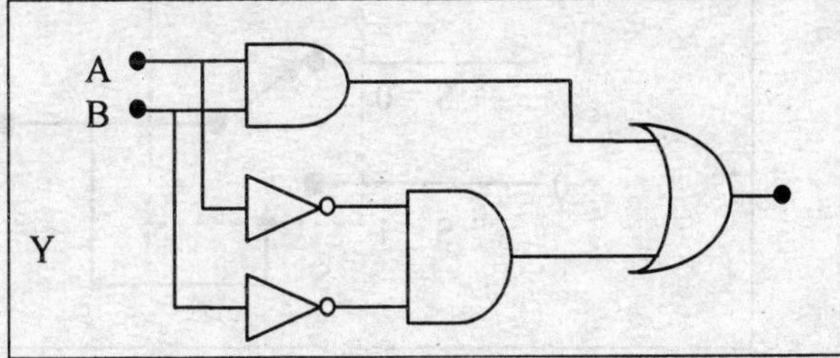

(a) Half adder (b) XOR
(c) Equality detector (d) Full adder

42. The correct cyclic sequence of the output s (Q_0Q_1) for the JK master slave flip flop circuit shown in figure when the input clock is applied is

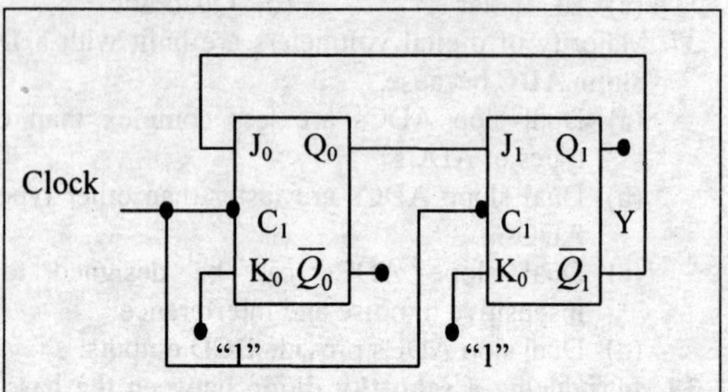

(a) 00, 01, 10, 11, 00, 01,...
(b) 00, 10, 01, 00, 10, 0 1...
(c) 00, 01, 10, 00, 01, 10...
(d) 00, 11, 00, 11, 00, 11...

43. For the CMOS analog switch shown in figure the positive supply is $+V_{DD}$ and the negative supply is $-V_{EE}$. The input V_i is bipolar. The switch will be ON and V_0 will be equal to V_i provided.

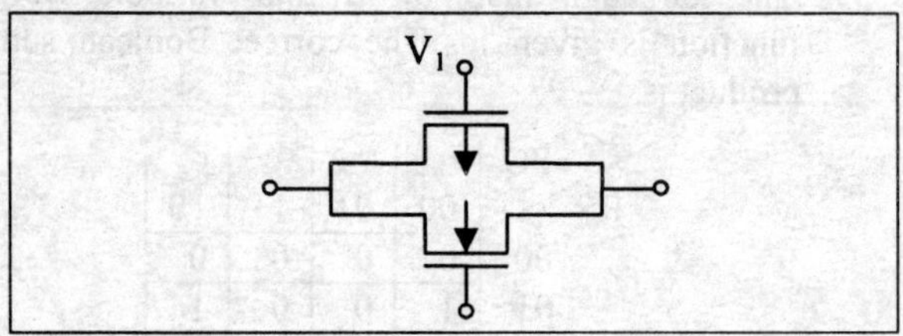

(a) $V_1 = V_2 = +V_{DD}$
(b) $V_1 = V_2 = -V_{EE}$
(c) $V_1 = +V_{DD}$ and $V_2 = -V_{EE}$
(d) $V_1 = -V_{EE}$ and $V_2 = +V_{DD}$

44. In 2 to 1 **digital multiplexer** a switching delay of 1 μs is connected as shown in figure. The output of the multiplexer is tied to its own select input S. The input which gets selected when S = 0 is tied to 1 and the input that get is selected when S = 1 is tied to 0. The output V_0 will be.

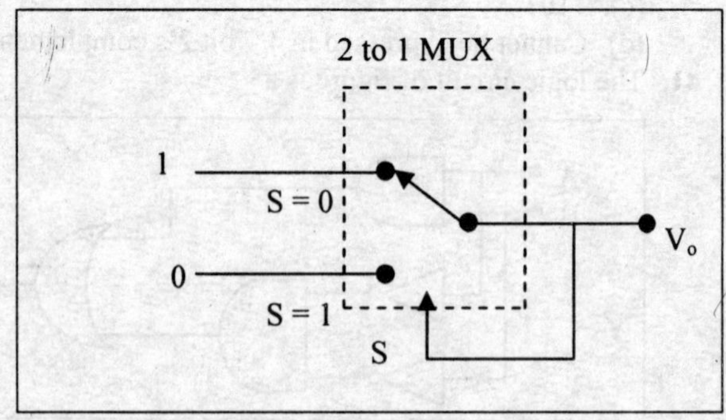

(a) 0
(b) 1
(c) Pulse train of frequency 0.5 MHz.
(d) Pulse train of frequency 1.0 MHz.

45. The square wave C_1 shown in figure is given to the clock input of a 4 – bit binary up/ down counter whose UP/ DN input is fed with pulse train P_u. The counter is a negative edge triggered one. This counter starts with 0000 and will reach 0000 again at the
(a) 15 th clock pulse
(b) 16th clock pulse
(c) 44th clock pulse
(d) 48th clock pulse

46. For an input voltage of 120 mV, the second integration (de – integration) period will be.
(a) 50 ms (b) 60 ms
(c) 100 ms (d) 120 ms

Answers Set-1

1. Ans. (a)

Solution:

Output response of Schmitt trigger is nonlinear Schmitt trigger will produce rectangular output is other low high.

Second Method

Binary	Gray Code
000	00
001	001
010	011
011	010
100	110
101	111
110	101
111	100

2. Ans. (b)

Solution: From the diagram backward analysis. The result is as output will be 1 when A = 0, B = 0, C = X = 0 or 1.

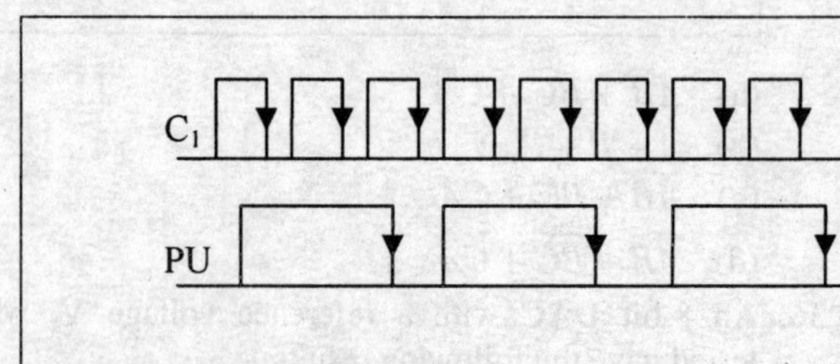

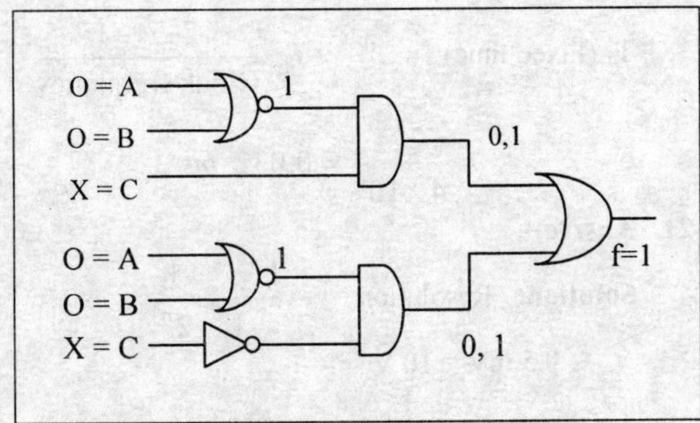

3. **Ans. (a)**
Solution: Digital families should have noiseimmunity

4. **Ans. (c)**
Solution:
Complement of number only taken when it is negative number.

$$-15 \rightarrow 0000\ 1111 \xrightarrow{1's\ complement} 1111\ 0000$$

$$1111\ 0000 \xrightarrow{2's\ complement} 1111\ 0001$$

5. **Ans. (b)**
Solution: Demorgan's theorem
$\overline{A+B+C} = \overline{A}, \overline{B}, \overline{C}$

6. **Ans. (d)**
Solution: For J.K flip –flop when J = 1, K = 1 Then output toggle i.e. $Q_{n+1} = \overline{Q}_n$

7. **Ans. (c)**
Solution: $2^n \geq Mode$ of counter.

8. **Ans. (c)**
Solution:

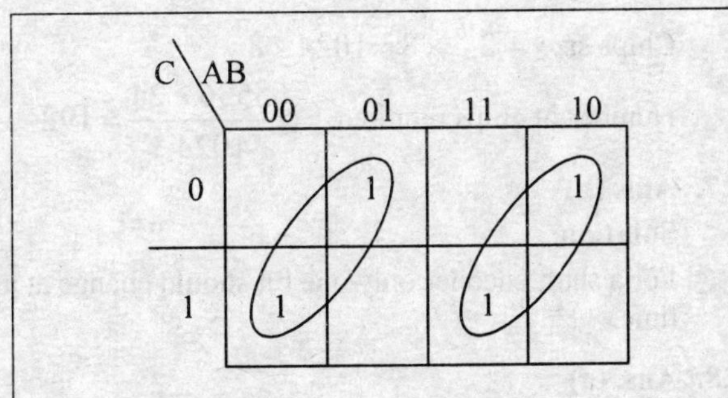

1st Pair (010 to 001) = $\overline{A}\,(B \oplus C)$

2nd pair (100 to 100)= A [B ⊙ C]

$Y = \overline{A}\,(B \oplus C) + A\,[B \odot C]$

$= \overline{A}\,(B \oplus C) + A\left(\overline{B \oplus C}\right)$

= A ⊕ B ⊕ C

9. **Ans. (c)**
Solution:

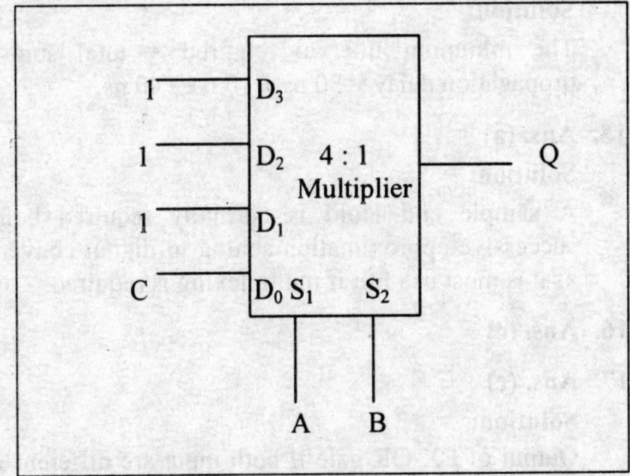

A	B	D_n	
0	0	C	D_0
0	1	1	D_1
1	0	1	D_2
1	1	1	D_3

A = 0, B = 0, D_0 is selected

Therefore, A = 0, B = 0, Q = C

For the rest of A and B. Q = 1

Therefore, Q = A + B + C

10. **Ans. (c)**
Solution: 100 rev/min

Speed of rotation $n = \dfrac{f_1 f_2}{f_2 - f_1}$

$$= \frac{25 \times 20}{25 - 20} = \frac{25 \times 20}{5} = 100$$

11. **Ans. (a)**
Solution: The correct match is given below

A. NAND	1. $\overline{A} + \overline{B} = \overline{A \cdot B}$
B. NOR	2. $\overline{A} \cdot \overline{B} = \overline{A + B}$
C. XNOR	3. $\overline{A}\,\overline{B} + AB$
D. AND	4. $\overline{\overline{A} \cdot \overline{B}} = \overline{\overline{A \cdot B}} = A \cdot B$

12. **Ans. (b)**
Solution:
$28 - 1 : 25\,\pi$

$$\frac{V_{in}}{2^n - 1} = \frac{1.275}{2^6 - 1} = -5\ \text{mV}$$

13. Ans. (a)

14. Ans. (c)

Solution:

The minimum interval required = total sum of propagation delay = 30 ns + 10 ns = 40 ns.

15. Ans. (a)

Solution:

A sample and Hold is normally required before successive approximation analog to digital converter as it is most use full if multiplexing is required.

16. Ans. (c)

17. Ans. (c)

Solution:

Output of EX-OR gate if both input are different i.e. in present case both expect waveforms have a phase difference of T/2 for mean reading of PMMC.

18. Ans. (d)

Solution:

For (1 / 2) digit MSB can be maximum and other full digit gives 9 respectively.

$$3 \text{ and } \frac{1}{2} \text{ digit} = 1.999$$

19. Ans. (b)

Solution:

Apply the Karnaugh's map for given function.

$$f = A\overline{B}CD + \overline{A}BC + \overline{A}\overline{B}C + BCD$$

$$= A\overline{B}CD + \overline{A}BCD + \overline{A}BC\overline{D} + \overline{A}\overline{B}CD + \overline{A}\overline{B}C\overline{D} + ABCD + \overline{A}BCD$$

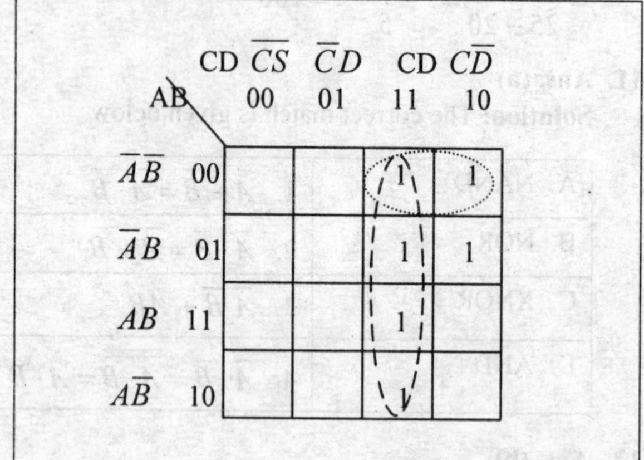

AB \ CD	$\overline{C}\overline{S}$ 00	$\overline{C}D$ 01	CD 11	$C\overline{D}$ 10
$\overline{A}\overline{B}$ 00			1	1
$\overline{A}B$ 01			1	1
AB 11			1	
$A\overline{B}$ 10			1	

$$f = \overline{A}C + CD$$

20. Ans. (a)

Solution:

$$T_1 \text{ (Fixed time)} = 2^{n-1} \times T_C \frac{2^{n-1} - 1}{\text{Clock frequency}}$$

$$= \frac{2^{8-1}}{4 \times 10^6} = 0.032 \ ms$$

21. Ans. (c)

Solution: Resolution $= \dfrac{1}{2^n - 1} \cong \dfrac{1}{2^n}$

0.5 mV = 10 V

$$2^n = \frac{1}{0.5 \times 10^{-3}} = 2 \times 10^3$$

$2^n = 2000$

n = 11

22. Ans. (b)

23. Ans. (a)

Solution:

$$\text{Perent resolution} = \frac{100}{2^n} = \frac{100}{2^8} = 0.39$$

24. Ans. (a)

Solution:

$2^n = 2 \Rightarrow$ **n=1**

25. Ans. (c)

Solution:

When word size is increased copy the original 2's complement representation in least significant bits of the word and make other bits equal to the MSB of the original.

26. Ans. (a)

Solution:

$$\text{Memory bits} = 2^{10} \times 24 = 65536 \times 24$$

$$\text{Chips size} = 2^{10} \times 8 = 1024 \times 8$$

$$\text{Number of chips required} = \frac{65536 \times 24}{1024} = 192$$

27. Ans. (b)

Solution:

For a shaft encoder only one bit should change at a time.

28. Ans. (a)

29. Ans. (c)

30. Ans. (d)

Solution:

Conversion time $t_c = \dfrac{n}{f}$,

Where,

n – Number of bits

f-Clock frequency, $\frac{8}{1\times10^6}=8\ \mu s$

31. Ans. (a)

32. Ans. (c)

33. Ans. (b)

34. Ans. (a)

35. Ans. (c)

Solution:

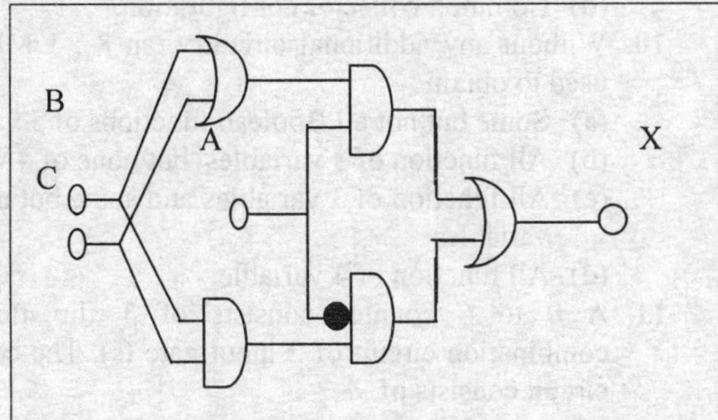

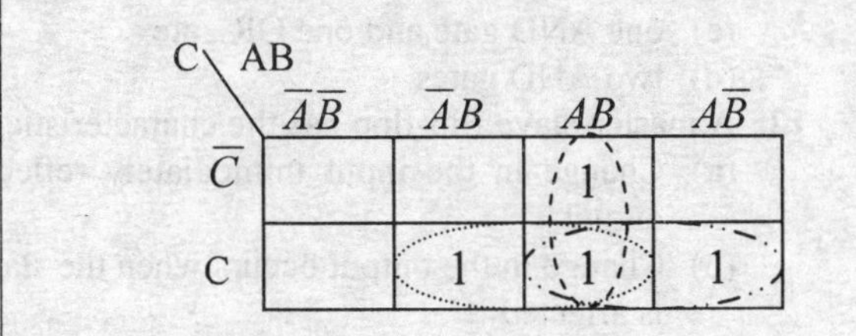

$x = AB + AC + BC$

36. Ans. (a)

37. Ans. (c)

38. Ans. (a)

39. Ans. (b)

Solution:

The given expression have a group of and pair

From the k-map,

Output (Y) $=\overline{Q}\,S+\overline{P}\,Q\,R\,\overline{S}$

40. Ans. (a)

Solution:

2's complement of 1011 (–3) → 0101

2's complement of 0110 (5) → 1010

Addition 2'→ 1111 (1' complement)

Then, in 2's complement it will be 0001

41. Ans. (c)

Solution:

42. Ans. (d)

Solution:

Q_0	Q_1	$J_1=(Q_0)$	$K_1(1)$	$J_0(\overline{Q_1})$	$K_o(1)$
0	0	0	1	1	1
0	1	0	1	0	1
0	0	0	1	1	1
0	1				

43. Ans. (d)

44. Ans. (c)

45. Ans. (d)

46. Ans. (b)

Solution:

De-integration period, $T_2-T_1=\frac{V_a}{V_R}\cdot T_1$

$=\frac{120}{100}\cdot 50=60\ ms$

One Mark Questions

1. Figure shown a 4 to 1 MUX to be used to implement the sum S of a 1-bit full adder with input bits P and Q the carry input C_{in}. Which of the following combination of inputs to I_0, I_1, I_2 and I_3 of the MUX will realize the sum S?

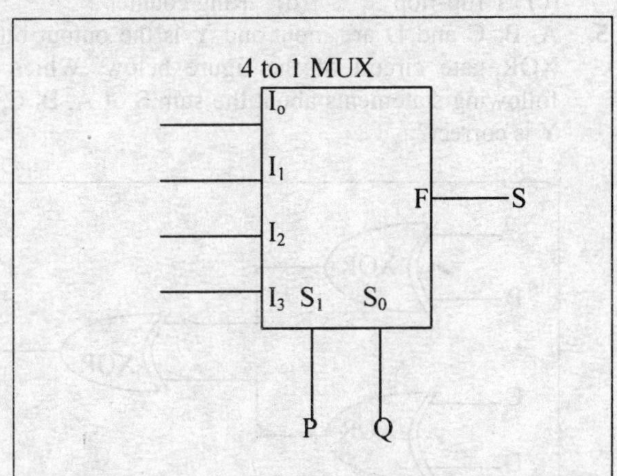

(a) $I_0=I_1=C_{in}$; $I_2=I_3=\overline{C}_{in}$

(b) $I_0=I_1=\overline{C}_{in}$; $I_2=I_3=C_{in}$

(c) $I_0=I_3=C_{in}$; $I_1=I_2=\overline{C}_{in}$

(d) $I_0=I_3=\overline{C}_{in}$; $I_1=- I_2=C_{in}$

2. The digital circuit using two inverters shown in figure will act as.

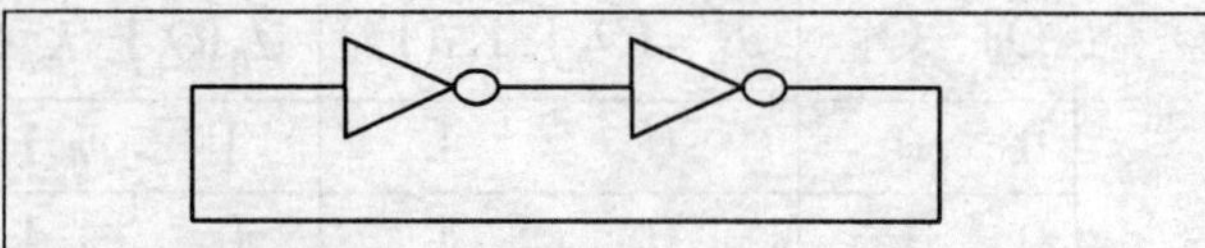

(a) A bistable multi-vibrator
(b) A an astable muti-vibrator
(c) A monostable multi-vibrator
(d) Aan oscillator

3. The voltage comparator shown in figure can be used in the analog-to-digital conversion as.

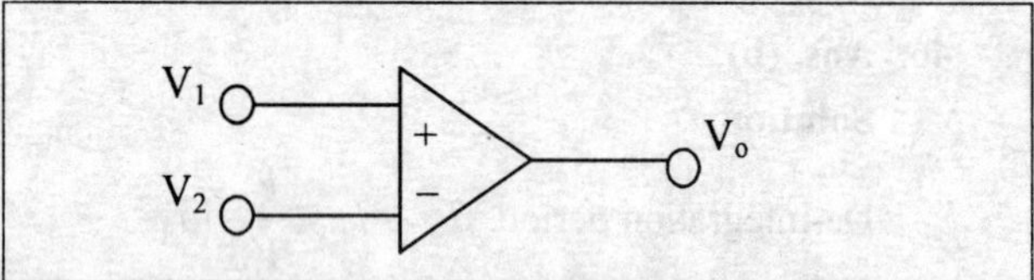

(a) a 1-bit quantizer (b) a 2-bit quantizer
(c) a 4-bit quantizer (d) a 8-bit quantizer

4. The digital circuit shown in the figure works as a

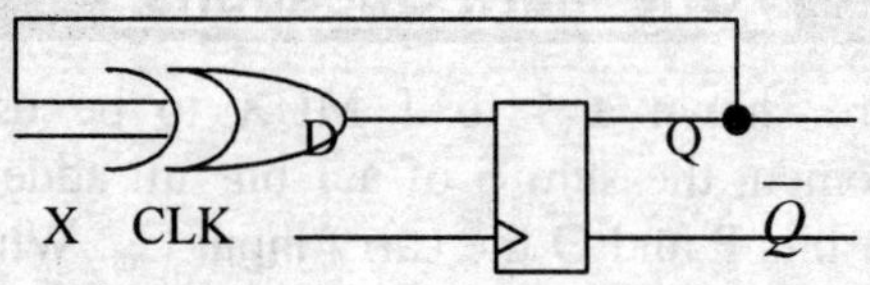

(a) JK flip-flop (b) Clocked RS flip-flop
(c) T flip-flop (d) Ring counter

5. A, B, C and D are input and Y is the output bit in the XOR gate circuit of the figure below. Which of the following statements about the sum S of A, B, C, D and Y is correct?

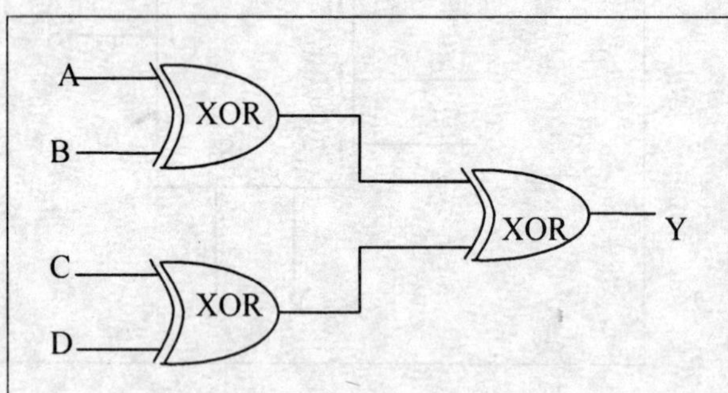

(a) S is always either zero or odd
(b) S is always either zero or even
(c) S = 1 only if the sum of A, B,C and D is even
(d) S = 1 only if the sum of A,B,C and D is odd

6. The octal equivalent of the Hex number AB.CD is
(a) 253.314 (b) 253.632
(c) 526.314 (d) 526.632

7. The number of distinct Boolean expressions of 4 variables is:
(a) 16 (b) 256
(c) 1024 (d) 65536

8. The minimum number of comparators required to build an 8 bit flash ADC is.
(a) 8 (b) 63
(c) 255 (d) 256

9. The output of the 74 series of GATE of TTL gates is taken from a BJT in:
(a) Totem pole and common collector configuration
(b) Either totem pole or open collector configuration
(c) Common base configuration
(d) Common collector configuration.

10. Without any additional circuitry, an 8 : 1 MUX can be used to obtain
(a) Some but not all Boolean functions of 3 variables
(b) All function of 3 variables but none of 4 variables
(c) All function of 3 variables and some but not all of 4 variables
(d) All function of 4 variable.

11. A 0 to 6 counter consists of 3 flip flops and a combination circuit of 2 input gate (s). The combination circuit consists of:
(a) one AND gate
(b) one OR gate
(c) one AND gate and one OR gate
(d) two AND gates

12. A master-slave flip-flop has the characteristic that
(a) Change in the input immediately reflected in the output
(b) Change in the output occurs when the state of master is affected
(c) Change in the output occurs when the state of the slave is affected
(d) Both the master and the slave states are affected at the same time

13. The range of signed decimal number that can be represented by 6-bits 1's complement number is
(a) –31 to + 31 (b) –63 to + 64
(c) –64 to + 63 (d) –32 to + 31

14. A digital system is required to amplify a binary-encoded audio signal. The user should be able to control the gain of the amplifier form a minimum to a maximum in 100 increments. The minimum number of bits required to encode, in straight binary, is
(a) 8 (b) 6
(c) 5 (d) 7

15. Choose the correct one from among the alternatives A, B, C, D after matching an item from Group 1 with the most appropriate item in Group 2.

Group 1	Group 2
P : Shift register	1 : Frequency division
Q : Counter	2 : Addressing in memory chips
R : Decoder	3 : Serial to parallel data conversion

	P	Q	S
(a)	3	2	1
(b)	3	1	2
(c)	2	1	3
(d)	1	2	2

16. Figure show the internal schematic of s **TTL AND-OR – invert (AOI)** gate. For the inputs shown in Figure, the output Y is:

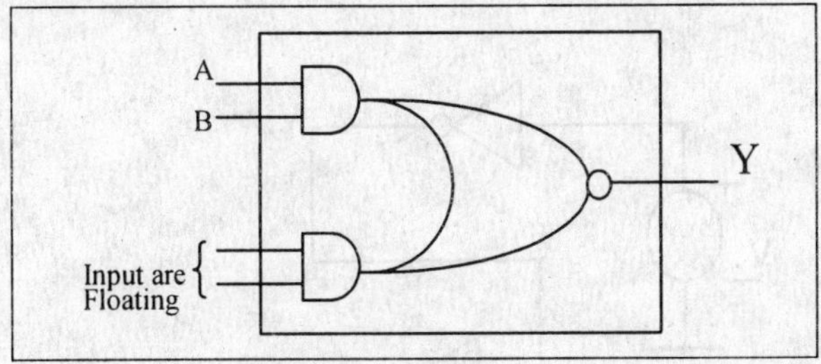

(a) 0 (b) 1

(c) AB (d) $\overline{AB}$

17. Figure is the **voltage transfer characteristic** of

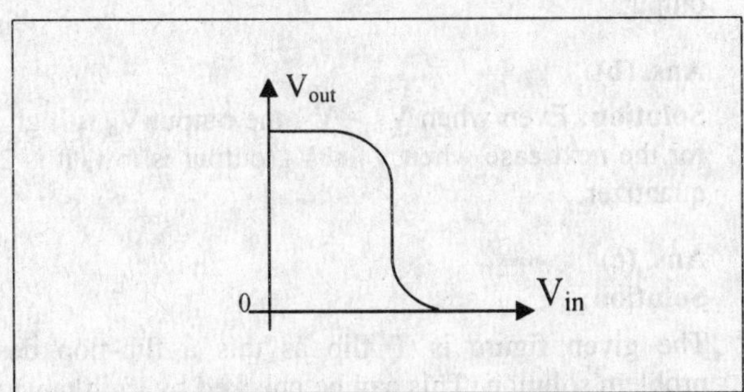

(a) An NMOS inverter with enhancement mode transistor as load
(b) An NMOS inverter with depletion mode transistor as load
(c) A CMOS inverter
(d) A BJT inverter

18. The minimum number of 2-to –1 multiplexes to realize a 4-to 1 multiplexes is:
(a) 1 (b) 2
(c) 3 (c) 4

19. The Boolean expression AC + BĈ is equivalent to

(a) $\bar{A}C + B\bar{C} + AC$

(b) $\bar{B}C + A\bar{C} + BC + \bar{\bar{A}}C\bar{B}$

(c) $AC + B\bar{C} + \bar{B}C + ABC$

(d) $ABC + \bar{A}B\bar{C} + AB\bar{C} + A\bar{B}C$

20. Decimal 43 in Hexadecimal and BCD number system is respectively:
(a) B2, 0100 0011 (b) 2B, 0100 0011
(c) 2B, 0011 0100 (d) B2, 0100 0100

21. The Boolean function *f* implemented in the figure using two input multiplexes is

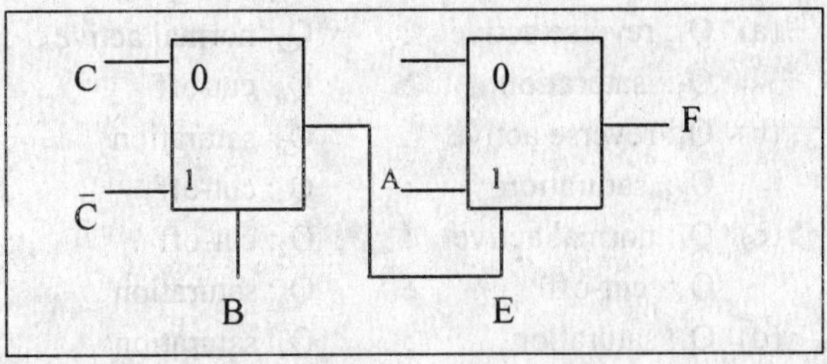

(a) $A\bar{B}C + AB\bar{C}$ (b) $ABC + A\bar{B}\bar{C}$

(c) $\bar{A}BC + A\overline{BC}$ (d) $\overline{AB}C + \bar{A}B\bar{C}$

22. The number of product terms in the minimized sum of product expression obtained through the follwing K – map is (where, "d" denotes denotes don't care states)

1	0	0	1
0	d	0	0
0	0	d	1
1	0	0	1

(a) 2 (b) 3
(c) 4 (d) 5

23. $X = 01110$ and $Y = 11001$ are two 5-bit binary number represented in two's complement format. The sum of X and Y represented in two's complement format using 6 bit is:
(a) 100111 (b) 001000
(c) 000111 (d) 101001

24. The Boolean function $Y = AB + CD$ is to be realized using only 2-input NAND gates. The minimum number of gates required is:
(a) 2 (b) 3
(c) 4 (d) 5

25. The circuit diagram of a0 standard TTL NOT gate is shown in the figure. When $V_i = 2.5\,V$, modes of operation of the transistors will be:

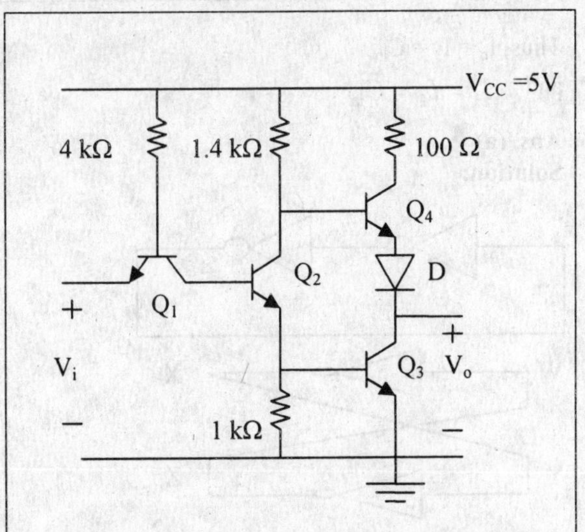

(a) Q_1: reverse active ; Q_2: normal active ;
Q_3: saturation ; Q_4: cut-off ;
(b) Q_1: reverse active ; Q_2: saturation ;
Q_3: saturation ; Q_4: cut-off ;
(c) Q_1: normal active ; Q_2: cut-off ;
Q_3: cut-off ; Q_4: saturation ;
(d) Q_1: saturation ; Q_2: saturation ;
Q_3: saturation ; Q_4: normal active ;

Answers
Digital Electronics

1. **Ans. (c)**
Solution: We draw the Truth-table for full adder is as shown below

S_1 (P)	S_o (Q)	C_{in}	SUM=S
0	0	0	0
0	0	1	1
0	1	0	1
0	1	1	0
1	0	0	1
1	0	1	0
1	1	0	0
1	1	1	1

Implementing the truth-table as for conditions

S_1	S_o	S	
0	0	C_{in}	I_o
0	1	$\overline{C}_{in}$	I_1
1	0	$\overline{C}_{in}$	I_2
1	1	C_{in}	I_3

Thus, $I_o = I_3 = C_{in}$
and $I_1 = I_2 = \overline{C}_{in}$

2. **Ans. (a)**
Solution:

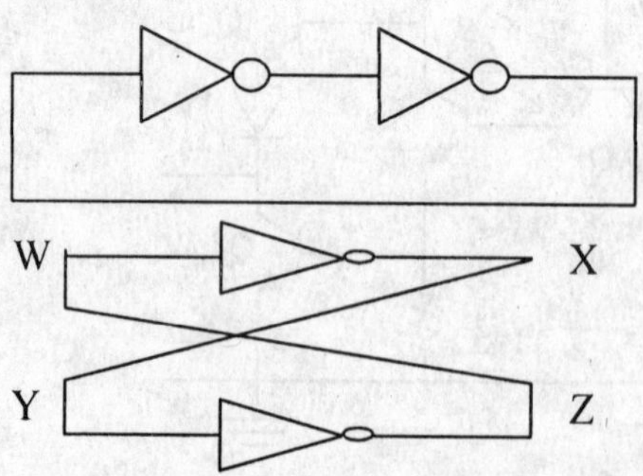

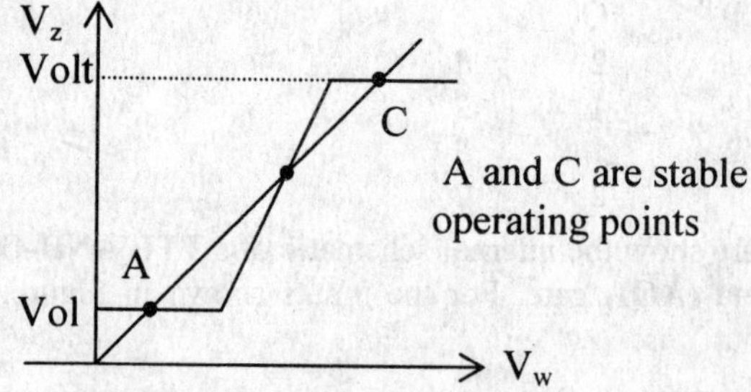

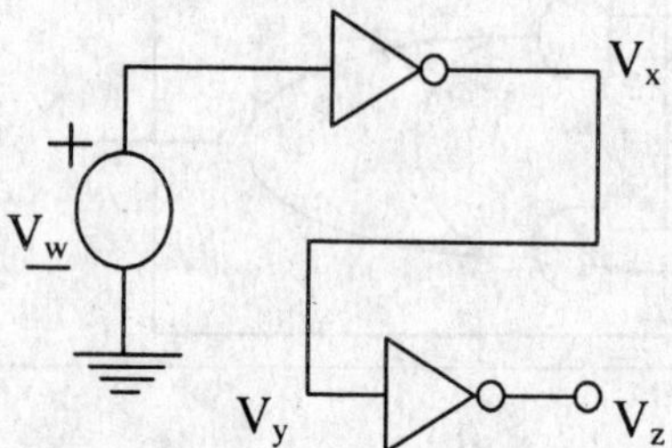

The latch has two stable operating point A and C. Then the latch is a bistable circuit having two complementary output.

3. **Ans. (b)**
Solution: Even when $V_1 > V_2$, the output V_o is high and for the next case when $V_1 < V_2$, output is low, it is 2-bit quantizer.

4. **Ans. (c)**
Solution:
The given figure is T flip as this a flip-flop design problem solution. This can be checked by truth table too. It have one input only.

$Y = A\overline{B} + \overline{A}B$

excitation table for T and D

Qn	Qn + 1	D	T
0	0	0	0
0	1	1←	1
1	0	0	1
1	1	1←	0

Expression for D in term of T

$D = \overline{Q_n}T + Q_n\overline{T}$

5. **Ans. (d)**
Solution:

A	B	C	D	Y	$S = A + B + C + D + Y$
0	0	0	0	0	0
0	0	0	1	1	0
0	0	1	0	1	0
0	0	1	1	0	0

0	1	0	0	1	0
0	1	0	1	0	0
0	1	1	0	0	0
0	1	1	1	1	0
1	0	0	0	1	0
1	0	0	1	0	0
1	0	1	1	1	0
1	1	0	0	0	0
1	1	0	1	1	0
1	1	1	0	1	0
1	1	1	1	0	0

Sum is always zero and zero is a even term therefore, option (b) is correct.

6. **Ans. (b)**
Solution:

$$\begin{array}{cccccc} A & B & \cdot & C & D & \\ \downarrow & \downarrow & & \downarrow & \downarrow & \\ 10 & 11 & \cdot & 12 & 13 & \\ \downarrow & \downarrow & & \downarrow & \downarrow & \\ 1010 & 1011 & \cdot & 1100 & 1101 & \end{array}$$

$$\underbrace{10}_{2}\ \underbrace{101}_{5}\ \underbrace{011}_{3} \cdot \underbrace{110}_{6}\ \underbrace{011}_{3}\ \underbrace{010}_{2}$$

7. **Ans. (d)**
Solution: Let ABCD be 4 variables

⇒ Total number of variables

$= A\bar{A}\ \ B\bar{B}\ \ C\bar{C}\ \ D\bar{D}$ i.e. 8

The total elements in 4 variable K – map $= 2^4 = 16$

∴ Number of distinct Boolean expression

$= 2^{16} = 65536$

8. **Ans. (c)**
Solution: Number of compare for Flash ADC
$= 2^n - 1 = (2^8 - 1) = 255$
Where, n is the number of bit

9. **Ans. (b)**

10. **Ans. (b)**
Solution:
Without any addition circuit i.e. without any input variable only selector expression will pass i.e. expression of 3 variable.
$8 = 2^n$, n = 3 select lines and a function with n + 1 = 4 can be implemented. With addition circuit.

11. **Ans. (b)**
Solution:
0 to 6 counter:

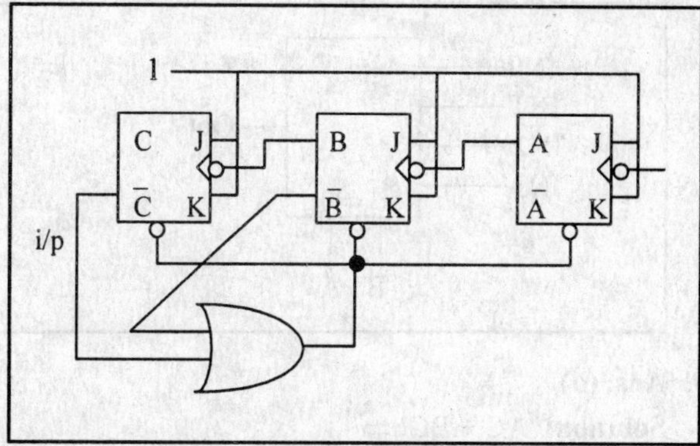

12. **Ans. (c)**
Solution: Change in output occurs when the state of the slave is affected:

13. **Ans. (a)**
Solution: The 1's complement range of numbers is $-2^{n-1} - 1$ to $2^{n-1} - 1$.
When n = 6, so Range = – 31 to + 31

14. **Ans. (d)**
Solution: Range $2^n \geq 100$, $2^7 = 128$
Hence for 100 increase 7 bit are required

15. **Ans. (b)**
Solution:
Shift register → Serial to parallel 7 bits are required
Counter → frequency division
Decoder → Addressing in memory chips

16. **Ans. (a)**
Solution:
For TTL logic floating input represent logic 1

$Y = \overline{(AB + 1)} = \overline{AB} \cdot 0 = 0$

17. **Ans. (c)**

18. **Ans. (c)**
Solution:

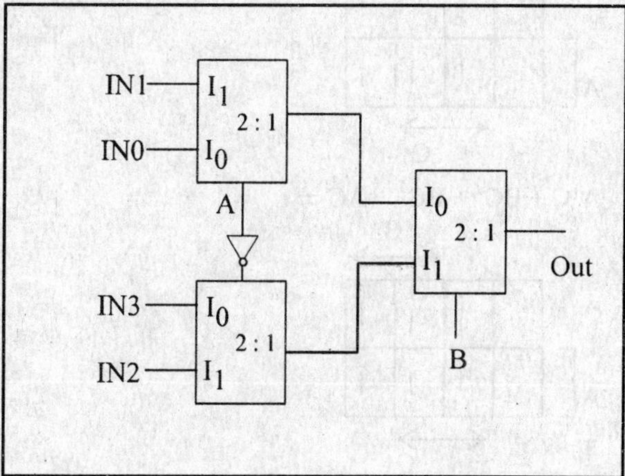

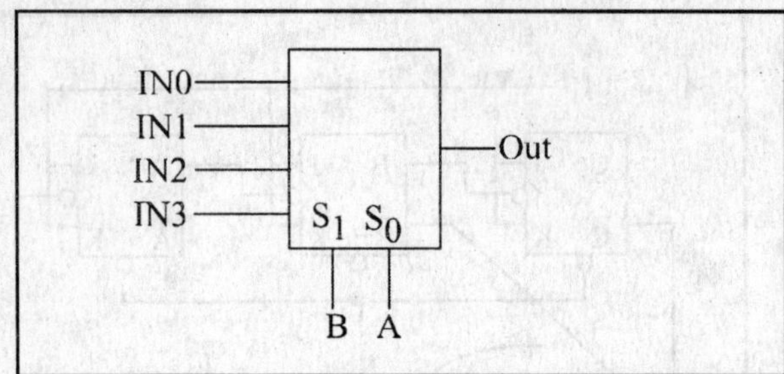

19. Ans. (d)

Solution: AC + BC' ⇒

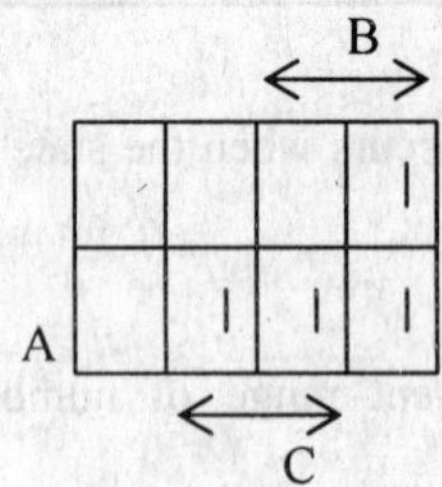

ABC + ĀBC' + ABC' + AB'C ⇒

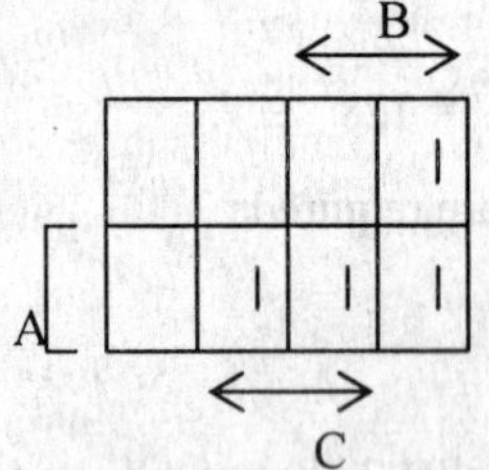

AC + BC' + B'C + ABC ⇒

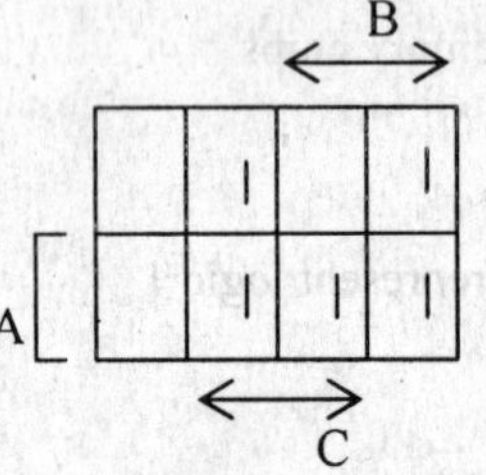

B'C + AC + BC' + A'CB' ⇒

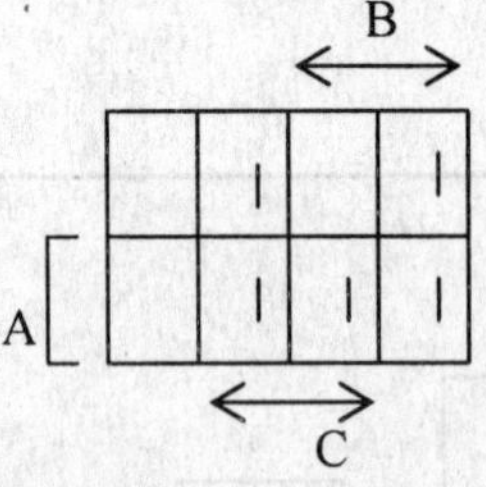

A'C + BC' + BC' + AC ⇒

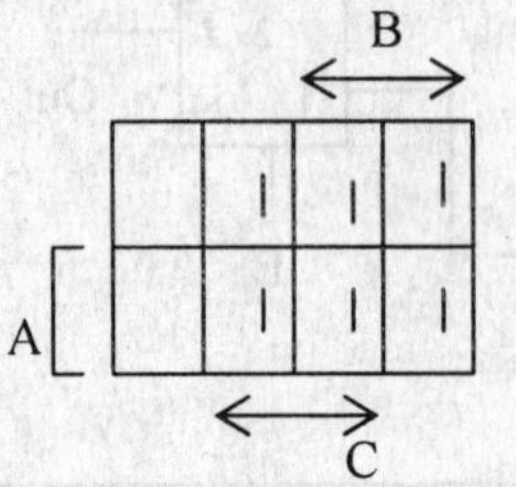

On comparing all, we find that ∞ is correct

20. Ans. (b)

Solution:

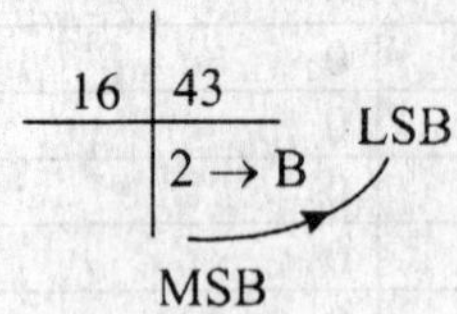

$(43)_{10} = (2B)_H$

$(4 \quad 3)_{10}$

↓ ↓

$0100 \quad 0011 \Rightarrow (0100 \ 0011)_{BCD}$

21. Ans. (a)

Solution: $f = EA$

$E = \bar{B}C + B\bar{C}$

$B = A(\bar{B}C + B\bar{C}) \quad = A\bar{B}C + AB\bar{C}$

22. Ans. (a)

Solution:

	$\bar{A}\bar{B}$	$\bar{A}B$	AB	$A\bar{B}$
$\bar{C}\bar{D}$	1	0	0	1
$\bar{C}D$	0	d	0	0
CD	0	0	d	1
$C\bar{D}$	1	0	0	1

II

I

F = $\bar{B}\,\bar{D} + A\,\bar{B}C$

23. Ans. (c)

Solution: 5 bit number in 2's complement format

$X = 01110 = 14$

$Y = 11001 = -7$, X + Y = 7

$X + Y$ in 2's complement format

$$\begin{array}{rcc} \Rightarrow X + Y = & 0 & 1110 \\ + & 1 & 1001 \\ \hline = \boxed{1} & 0 & 0111 \end{array}$$

↑

ignore it

Note: overflow of sign bit is neglected and sign bit is zero therefore number is 000111.

24. Ans. (b)

Solution:

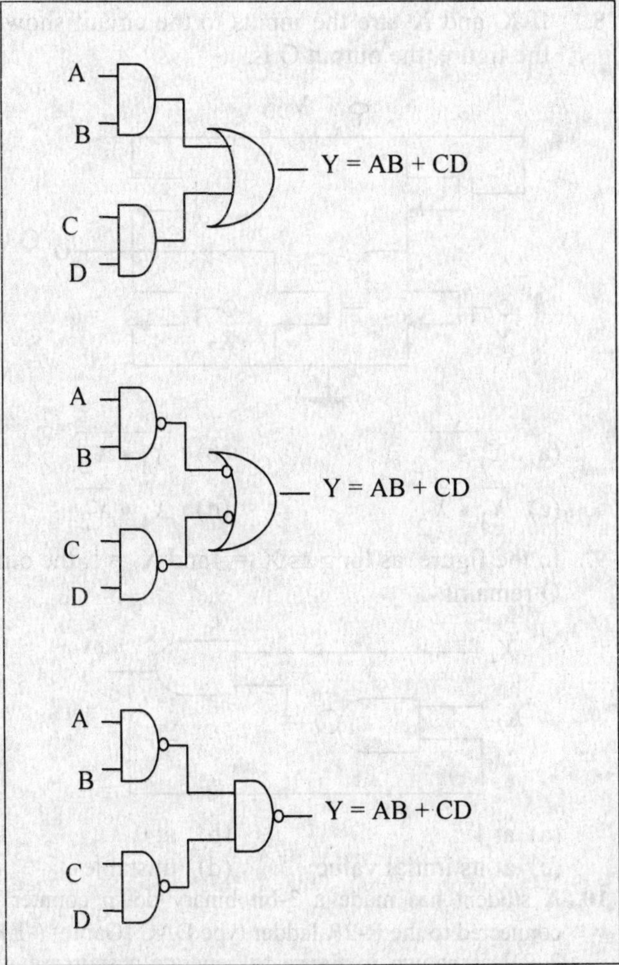

For $Y = AB + CD$ realization using NAND the minimum number of gates required is three

25. Ans. (a)

Solution: As given TTL circuit is NOT gate

V_i = 2.5 V high is applied. Output is low only when the earth is extended to output it means Q_3 is in saturation mode and Q_4 is cut off as Q3 and Q4 always behave negatively. Q_2 work in normal active but Q_1 in reverse active as V_i= 2.5 applied to emitter of Q_1.

Two Marks Questions

1. The Boolean expression XYZ + XYZ + X$\bar{Y}$Z + $\bar{X}$YZ + XYY can be simplified to

   (a) $\bar{X}\,\bar{Z} + XZ + YZ$ (b) $\bar{X}\,Z + \bar{Y}Z + YZ$

   (c) $\bar{X}Y + YZ + XZ$ (d) $\overline{XY} + Y\bar{Z} + \bar{X}Z$

2. The shift register shown in figure is initially loaded with the pattern 1010. Subsequently the shift register is clocked, and with each clock pulse the pattern gets shifted by one bit position to the right. With each shift, the bit at the serial input is pushed to the left most position (msb). After how may clock pulses will the content of the shift register become 1010 again?

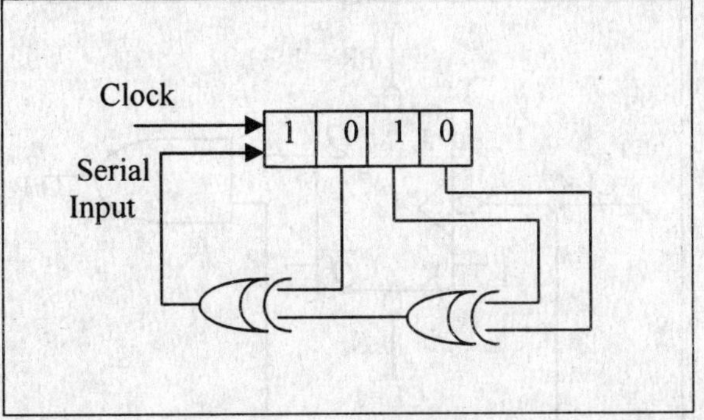

   (a) 3 (b) 7

   (c) 11 (d) 15

3. An X-Y flip flop, whose Characteristic Table is given below is to implemented using a J-K flip flop

X	Y	Q_{n+1}
0	0	1
0	1	Q_n
1	0	$\bar{Q}_n$
1	1	0

   This can be done by making

   (a) J = X, K = $\bar{Y}$ (b) J = $\bar{X}$, K = Y

   (c) J = Y, K = $\bar{X}$ (d) J = $\bar{Y}$, K = X

4. The simplified form of the Boolean expression $Y = \left(\bar{A}.BC + D\right)\left(\bar{A}.D + \bar{B}.\bar{C}\right)$ can be written as

   (a) $\bar{A}.D + \bar{B}.\bar{C}.D$ (b) $AD + B.\bar{C}.D$

   (c) $(\bar{A}. + D)(\bar{B}.C + \bar{D})$ (d) $A.\bar{D} + BC.\bar{D}$

5. A digital circuit which compares two number $A_3A_2A_1A_o$, $B_3B_2B_1B_o$ is shown in figure to get output Y=0, choose one pair of correct input numbers.

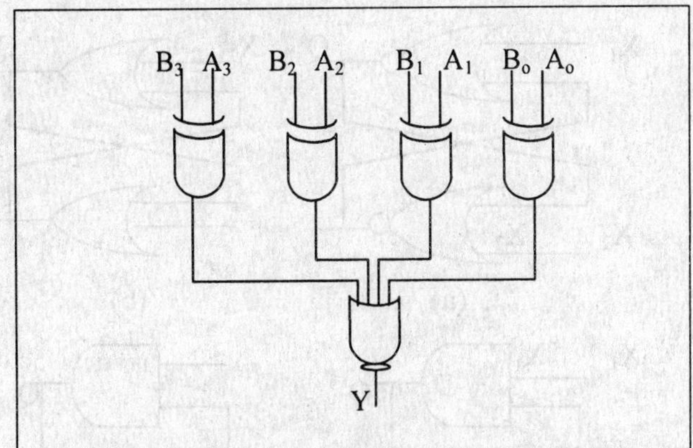

   (a) 1010, 1010 (b) 0101, 0101

   (c) 0010, 0010 (b) 0010, 1011

6. The digital circuit shown in figure generates a modified clock pulse at the output. Choose the correct output waveform from the option given below.

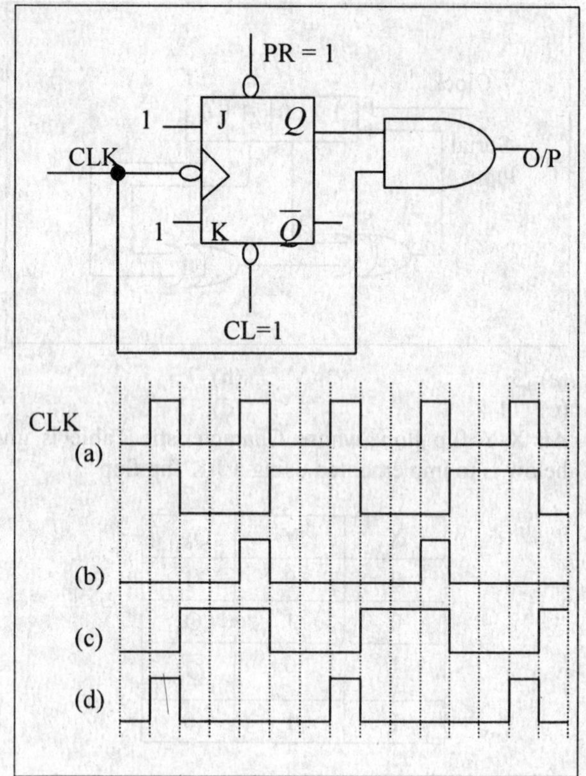

7. Select the circuit which will produce the given output Q for the input signals X_1 and X_2 given in the figure.

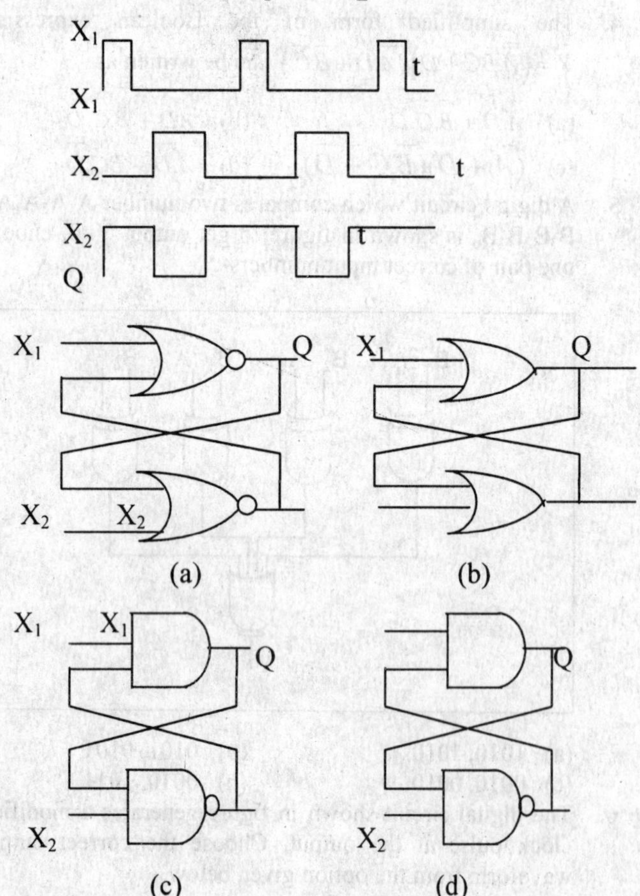

8. If X_1 and X_2 are the inputs to the circuit shown in the figure, the output Q is.

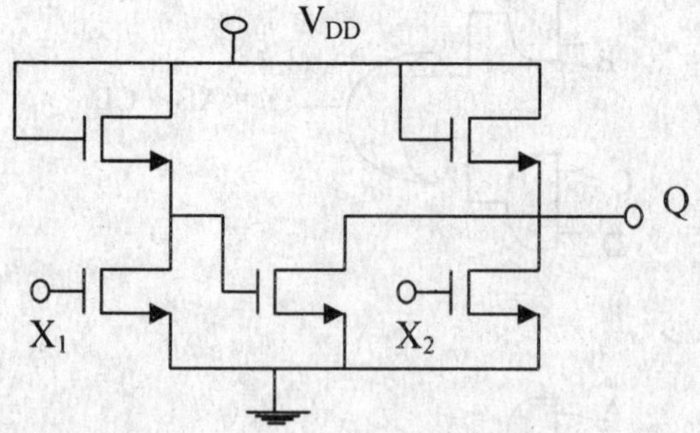

(a) $\overline{X_1 + X_2}$ (b) $\overline{X_1 \bullet X_2}$

(c) $\overline{X_1} \bullet X_2$ (d) $X_1 \bullet \overline{X_2}$

9. In the figure, as long as $X_1 = 1$ and $X_2 = 1$, the output Q remains

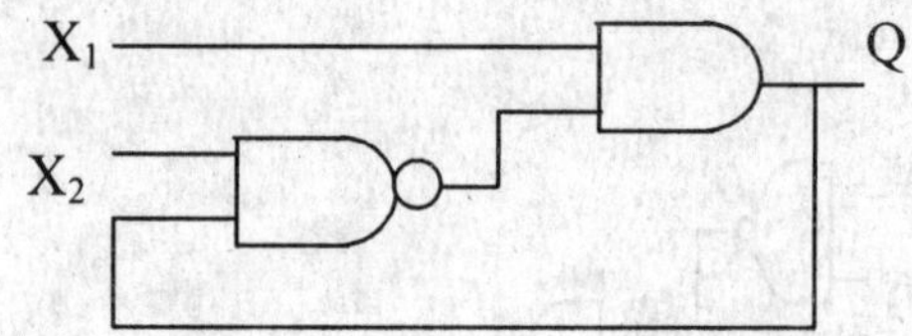

(a) at 1 (b) at 0

(c) at its initial value (d) unstable

10. A student has made a 3-bit binary down counter and connected to the R-2R ladder type DAC [Gain = (-1 KΩ/ 2 R)] as shown in figure to generate a staircase wave form. The output achieved is different as shown in figure. What could be the possible cause of this error?

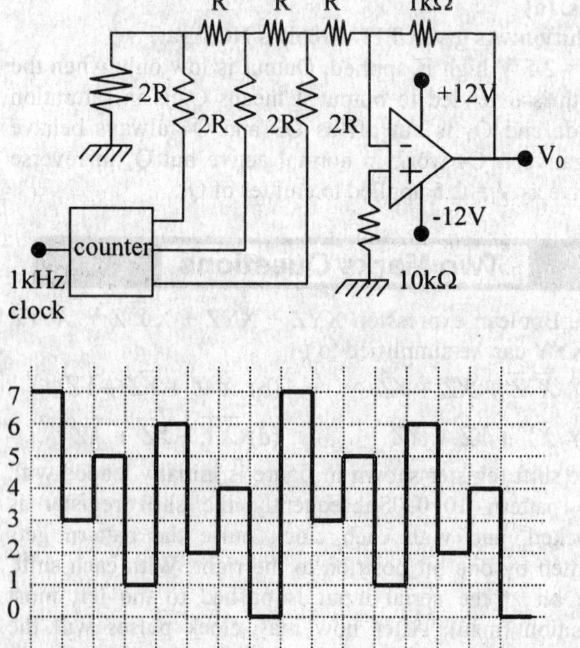

(a) he resistance values are incorrect
(b) The counter is not working properly
(c) The connection from the counter to DAC is not proper
(d) The R and 2R resistances are interchanged

11. A 4 × 1 MUX is used to implement a 3-input Boolean function as shown in figure. The Boolean function F (A, B, C) implemented is

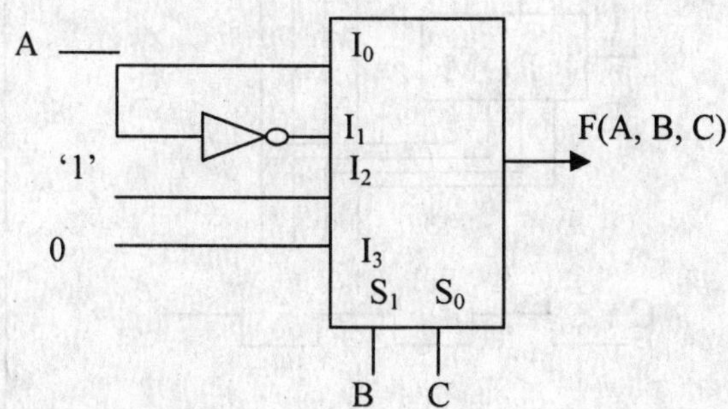

(a) F (A, B, C) = Σ(1, 2, 4, 6)
(b) F (A, B, C) = Σ(1, 2, 6)
(c) F (A, B, C) = Σ(2, 4, 5, 6)
(d) F (A, B, C) = Σ(1, 5, 6)

12. The 8085 assemble language instruction that stores the content of H and L registers into the memory location 2050_H and 2051_H respectively, is
(a) SPHL 2050_H (b) SPHL 2051_H
(c) SPLD 2050_H (d) STAX 2050_H

13. A TTL NOT gate circuit is shown in figure. Assuming V_{BE} = 0.7 V of both the transistors, if $V_i = 3.0\ V$ then the states of the two transistors will be

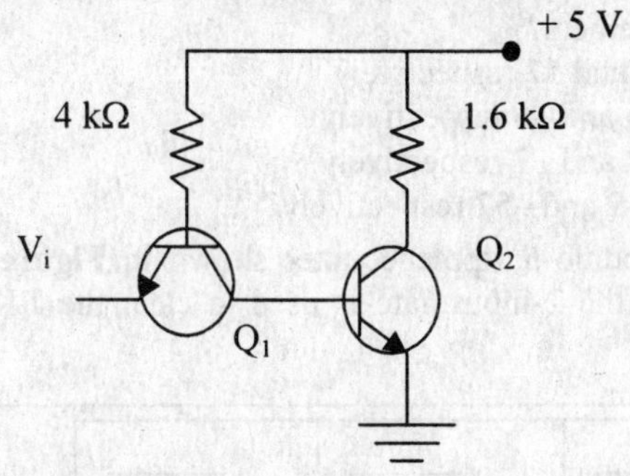

(a) Q_1 *ON and* Q_2 *OFF*
(b) Q_1 *reverse ON and* Q_2 *OFF*
(c) Q_1 *reverse ON and* Q_2 *ON*
(d) Q_1 *OFF and* Q_2 *reverse ON*

14. The octal equivalent of the Hex number AB.CD is
(a) 253.314 (b) 253.632
(c) 526.314 (d) 526.632

15. circuit shown in Figure has 4 boxes each described by inputs P, Q, R and outputs Y, Z with

$$Y = P \oplus Q \oplus R$$
$$Z = RQ + \bar{P}R + Q\,\bar{P}$$

The circuit acts as a

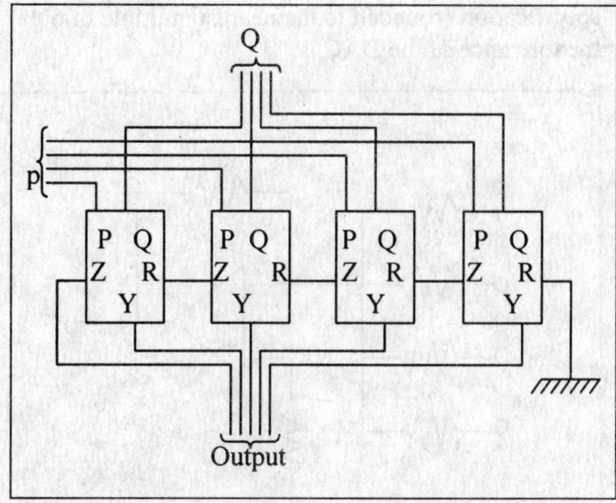

(a) 4 bit adder giving P + Q
(b) 4 bit subtractor giving P – Q
(c) 4 bit subtractor giving Q – P
(d) 4 bit adder giving P + Q + R

16. the functions W, X, Y and Z are as follows.

$$W = R + \bar{P}Q + \bar{R}S$$
$$X = PQ\bar{R}\bar{S} + \bar{P}\bar{Q}\bar{R}\bar{S} + P\bar{Q}\bar{R}\bar{S}$$
$$Y = \overline{RS + \overline{PR + P\bar{Q} + \bar{P}\cdot\bar{Q}}}$$
$$Z = \overline{R + S + PQ + \bar{P}\cdot\bar{Q}\cdot\bar{R} + P\bar{Q}\cdot\bar{S}}$$

Then
(a) W = Z, X = $\bar{Z}$ (b) W = Z, X = Y
(c) W = Y (d) W = Y = $\bar{Z}$

17. A 4 bit ripple counter and a 4 bit synchronous counter are made using flip flops having a propagation delay of 10 ns each. If the worst case delay in the ripple counter and the synchronous counter be R and S respectively, then:
(a) R = 10 ns, S = 40 ns
(b) R = 40 ns, S = 10 ns
(c) R = 10 ns, S = 30 ns
(d) R = 30 ns, S = 10 ns

18. The DTL, TTL, ECL and CMOS families of digital IC_s are compared in the following 4 columns

	(P)	(Q)	(R)	(S)
Fan-out is min.	DTL	DTL	TTL	CMOS
Power consumption Is min.	TTL	CMOS	ECL	DTL
Propagation delay Is min.	CMOS	ECL	TTL	TTL

The correct column is
(a) P (b) Q
(c) R (d) S

19. The circuit shown in Figure is a 4 bit DAC

The input bits 0 and 1 are represented by 0 and 5V respectively. The OP AMP is ideal, but all the resistances and the 5V inputs have a tolerance of ± 10 %. The specification (rounded to the nearest multiple of 5 %) for the tolerance of the DAC is:

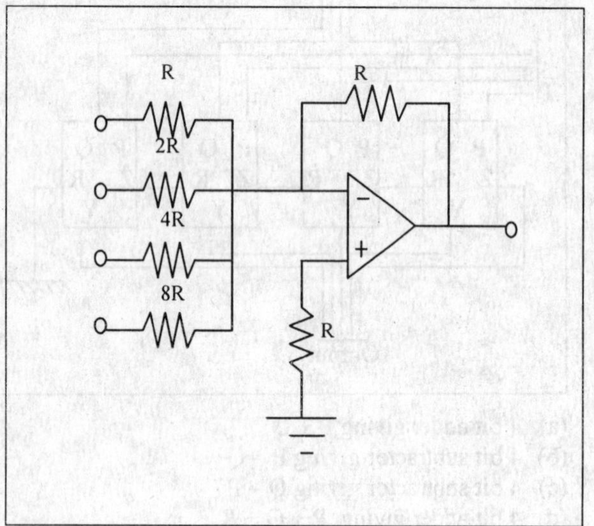

(a) ± 35 % (b) ± 20 %

(c) ± 10 % (d) ± 5 %

20. The circuit shown in figure converts

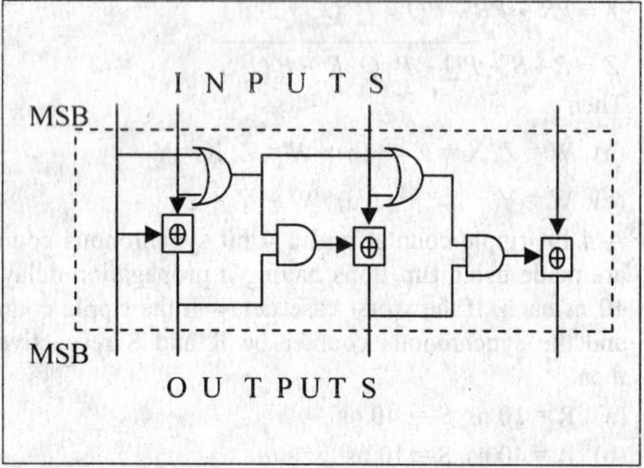

(a) BCD to binary code
(b) Binary to excess –3 code
(c) Excess-3 to Gray code
(d) Gray to Binary code

21. In the circuit shown in figure, A is a parallel-in-parallel-out 4 bit register, which loads at the rising edge of the clock C. The input lines are connected to a 4 bit bus, W. Its output acts as the input to a 16 x 4 ROM whose output is floating when the enable E is 0. A partial table of the contents of the ROM is as follows.

Add	0	2	4	6	8	10	11	14
Data	0011	1111	0100	1010	1011	1000	0010	1000

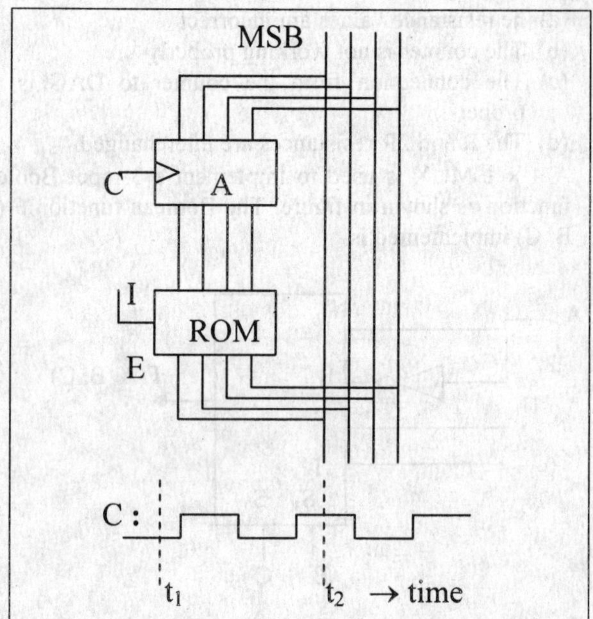

The clock to the register is shown, and the data on the W bus at time t1 is 0110. The data on the bus at time t_2 is

(a) 1111 (b) 1011

(c) 1000 (d) 0010

22. In an 8085 microprocessor the instruction CMP B has been executed while the content of the accumulator is less than that of register B.As a result

(a) Carry flag will be set but zero flag will be reset
(b) Carry flag will be reset but zero flag will be set
(c) Both Carry flag and Zero flag will be reset
(d) Both Carry flag and Zero flag will be set

23. 11001, 1001 and 111001 correspond to the 2's complement representation of which one of the following sets of numbers?

(a) 25, 9 and 57 respectively
(b) –6, -6 and –6 respectively
(c) –7, -7 and –7 respectively
(d) –25, -9 and –57 respectively

24. In the modulo-6 ripple counter shown in **Figure** , the output of the 2-input gate is used to clear the J-K flip-flops:

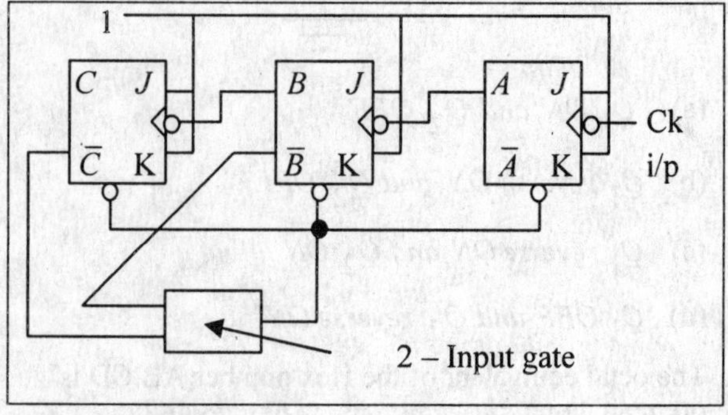

The 2-input gate is

(a) a NAND gate (b) a NOR gate

(c) an OR gate (d) an AND gate

25. A Boolean function f of two variable x and y is defined as follows:

$$f(0,0) = f(0,1) = f(1,1) = 1;\ f(1,0) = 0$$

Assuming complements of x and y are not available, a minimum cost solution for realizing f using only 2-input NOR gates and 2-input OR gate (each having unit cost) would have a total cost of

(a) 1 unit (b) 4 unit
(c) 3 unit (d) 2 unit

26. In figure, the J and K inputs of all the four Flip-Flops are made high. The frequency of the signal at output Y is:

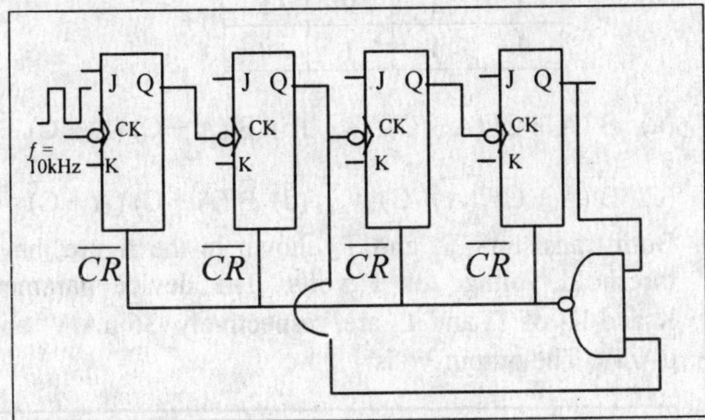

(a) 0.833 kHz (b) 1.0 kHz
(c) 0.91 kHz (d) 0.77 kHz

27. The number of comparators in a 4 bit flash ADC is

(a) 4 (b) 5
(c) 15 (d) 16

28. For a binary half –subtractor having two inputs A and B , the correct set of logical expression for the outputs D (= A minus B) and X (= borrow are)

(a) $D = \bar{A}B + AB,\ \bar{X} = AB$
(b) $D = \bar{A}B + A\bar{B} + A\bar{B},\ X = A\bar{B}$
(c) $D = \bar{A}B + A\bar{B},\ X = \bar{A}B$
(d) $D = AB + \bar{A}\bar{B},\ X = A\bar{B}$

29. The ripple counter shown in figure works as a:

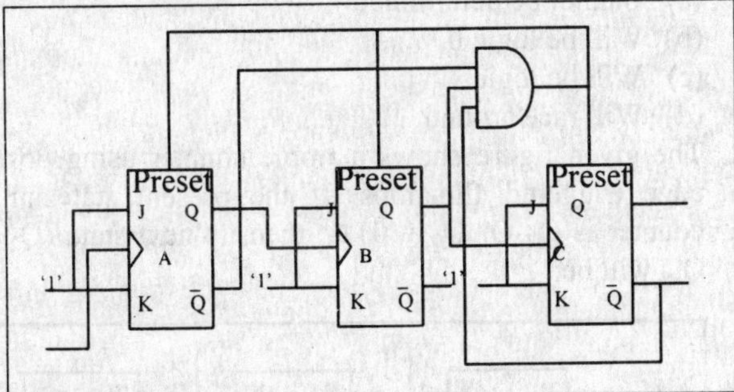

Clock

(a) mod –3 up counter
(b) mod – 5 up counter
(c) mod – 3 down counter
(d) mod – 5 down counter

30. Dual-slope integration type Analog-to-digital converters provide.

1. Higher speeds compared to all other types of A/D converters
2. Very good accuracy without putting extreme requirements on component stability
3. Good rejection of power supply hum
4. Better resolution compared to all other types of A/D converters for the same number of bits.

Codes:

(a) only 1 is correct (b) 2 and 3 correct
(c) 1, 2, 3 and 4 correct (d) 2 and 4 correct

31. The minimized form of the logical expression $(\bar{A}\bar{B}\bar{C} + \bar{A}B\bar{C} + \bar{A}BC + AB\bar{C})$ is

(a) $A\bar{C} + B\bar{C} + \bar{A}B$
(b) $A\bar{C} + BC + \bar{A}B$
(c) $\bar{A}C + \bar{B}C + \bar{A}B$
(d) $A\bar{C} + BC + A\bar{B}$

32. In figure the LED

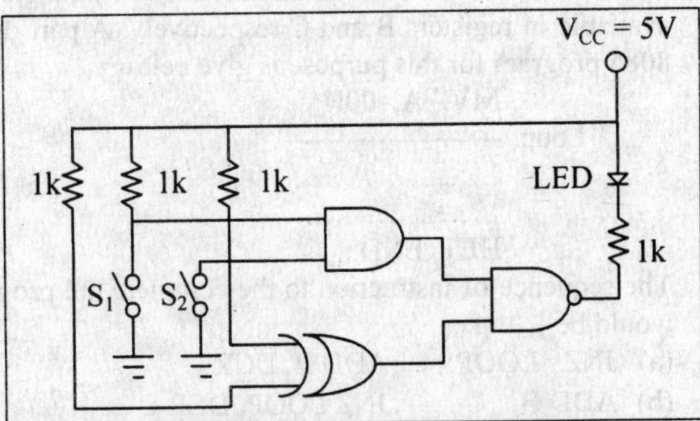

(a) Emits light when both S_1 and S_2 are closed
(b) Emits light when both S_1 ad S_2 are open
(c) Emits light when only of S_1 or S_2 is closed
(d) Does not emit light, irrespective of the switch positions.

33. The 8255 Programmable Peripheral Interface is used as described below

I. An A/D converter is interfaced to a microprocessor through on 8255. The Port C causes data to be strode into Port A.
II. Two computers exchange data using a pair of 8255s. Port A works as bi-directional data port supported by appropriate handshaking signals.

The appropriate modes of operation of the 8255 for (I) and (II) would be

(a) Mode 0 for (I) and Mode 1 for (II)
(b) Mode 1 for (I) and Mode 0 for (II)
(c) Mode 2 for (II) and Mode 0 for (I)
(d) Mode 2 for (I) and Mode 1 for (II)

34. The number of memory cycles required to execute the following 8085 instruction

(I) LDA 3000 H
(II) LXI D, F0F 1 H

would be

(a) 2 for (I) and 2 for (II)
(b) 4 for (I) and 3 for (II)
(c) 3 for (I) and 3 for (II)
(d) 3 for (I) and 4 for (II)

35. Consider the sequence of 8085 instruction given below.
LXI H, 9258,
MOV A, M,
CMA,
MOV M, A
Which one of the following is performed by this sequence
(a) Contents of location 9258 are moved to the accumulator
(b) Contents of location 9258 are compared with the contents of the accumulator
(c) Contents of location 8529 are complemented and stored in location 8529.
(d) Contents of location 5892 are complement and stored in location 5892

36. It is desired to multiply the numbers $0AH$ by $0BH$ and store the result in the accumulator. The numbers are available in registers B and C respectively. A part of the 8085 program for this purpose is give below

```
        MVI A, 00H
Loop; --------------------
        ..................
        ..................
        HLT END
```

The sequence of instruction to the complete the program would be
(a) JNZ LOOP, ADD B, DCR C
(b) ADD B, JNZ LOOP, DCR C
(c) DCR C, JNZ LOOP, ADD B
(d) ADD B, DCR C, JNZ LOOP

37. The minimum number of 2 to 1 multiplexers required to realize a 4 to 1 multiplexers is
(a) 1 (b) 2
(c) 3 (4) 4

38. The transistors used in a portion of th TTL gate shown in the figure have a $\beta = 100$. The base-emitter voltage of is 0.7 V for a transistor in active region and 0.75 V for a transistor in saturation. If the sink current I = 1 mA and the output is at logic 0, then the current I_R will be equal to:

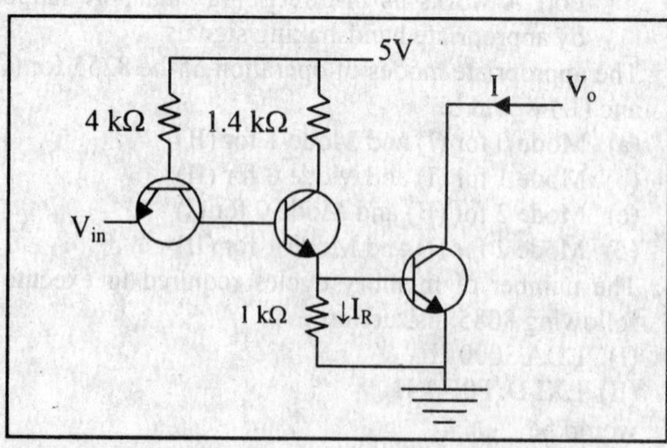

(a) 0.65 mA (b) 0.70 mA
(c) 0.75 mA (d) 1.00 mA

39. The Boolean expression for the truth table shown in

A	B	C	F
0	0	0	0
0	0	1	0
0	1	0	0
0	1	1	1
1	0	0	0
1	0	1	0
1	1	0	1
1	1	1	0

(a) $B(A+C)(\bar{A}+\bar{C})$ (b) $B(A+\bar{C})(\bar{A}+C)$
(c) $\bar{B}(A+\bar{C})(\bar{A}+C)$ (d) $\bar{B}(A+C)(\bar{A}+\bar{C})$

40. Both transistors T_1 and T_2 shown in the figure, have a threshold voltage of 1 volts. The device parameters K_1 and K_2 of T_1 and T_2 are, respectively, 36 $\mu A/V^2$ and 9 $\mu A/V^2$. The output V_0 is:

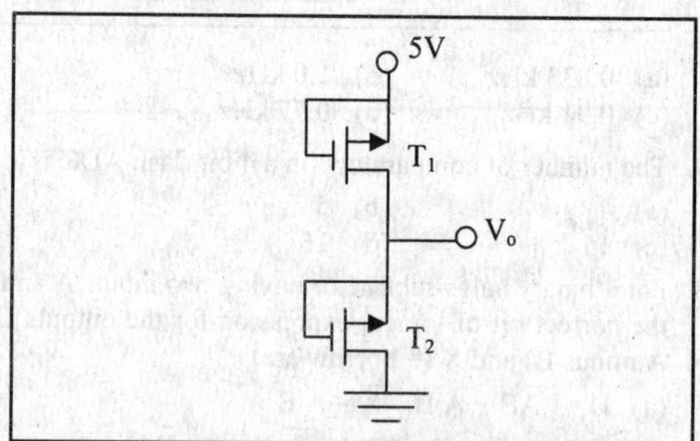

(a) 1 V (b) 2 V
(c) 3 V (d) 4 V

41. The present output Q_n of an edge triggered JK flip-flop is logic 0. If J = 1, then Q_{n+1}.
(a) Cannot be determined
(b) Will be logic 0
(c) Will be logic 1
(d) Will race around.

42. The given figure shows a ripple counter using positive edge triggered flip-flops. If the present state of the counter is $Q_2\ Q_1\ Q_0 = 011$, then its next state ($Q_2\ Q_1\ Q_0$) will be:

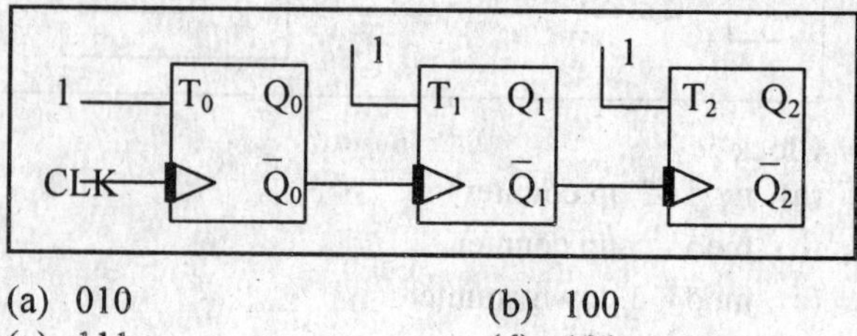

(a) 010 (b) 100
(c) 111 (d) 101

43. What memory address range is NOT represented by chip # 1 and chip # 2 in the figure. A_0 to A_{15} in this figure are the address lines and CS means Chip Select.

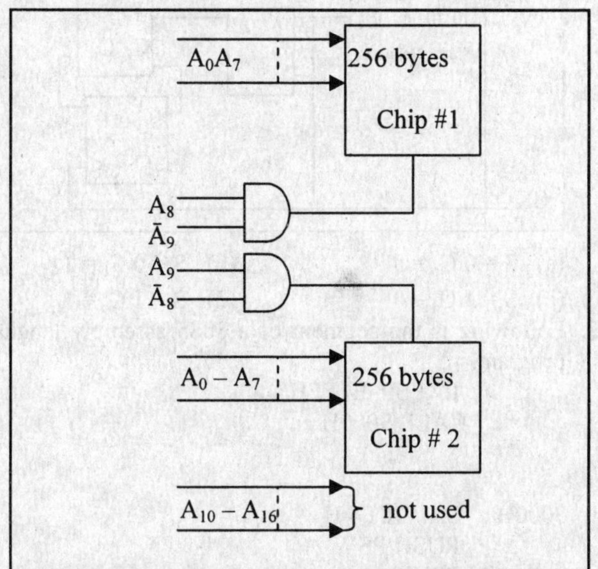

(a) 0100-02FF (b) 1500-16FF
(c) F900-FAFF (d) F800-F9FF

44. Consider an 8085 microprocessor system
The following program starts at location 0100H.

```
LXI SP, 00FF
LXI H, 0701
MVI A, 20H
SUB M
```

The content of accumulator when the program counter reaches 0109 H is
(a) 20 H (b) 02 H
(c) 00 H (d) FF H

45. Consider an 8085 microprocessor system
If in addition following code exists from 0109 H onwards,.

```
ORI 40 H
ADD M
```

What will be the result in the accumulator after the last instruction executed?
(a) 40 H (b) 20 H
(c) 60 H (d) 42 H

46. A new Binary Coded Pentary (BCP) number system is proposed in which every digit of a base – 5 number is represented by its corresponding 3 – bit binary code. For example, the base –5 number 24 will be represented by its BCP code 010100. In this numbering system, the BCP code 100010011001 corresponds to the following number in base – 5 system.
(a) 423 (b) 1324
(c) 2201 (d) 4231

47. A 4 – bit D/A converter is connected to a free – running 3 – bit UP counter, as shown in the following figure. Which of the following waveforms will be observed at V_0?

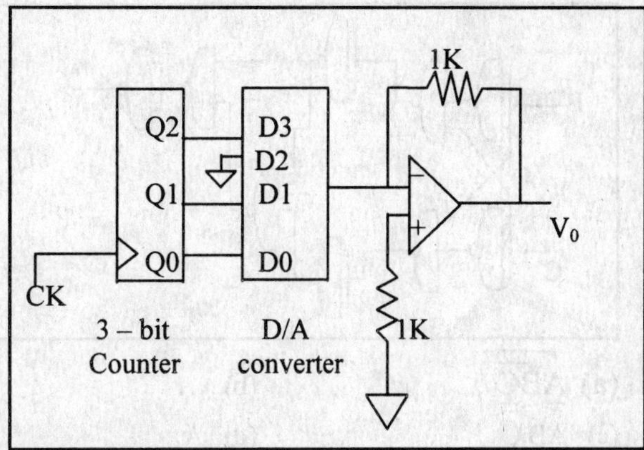

In the figure shown above, the ground has been shown by the symbol ∇.
Figure:

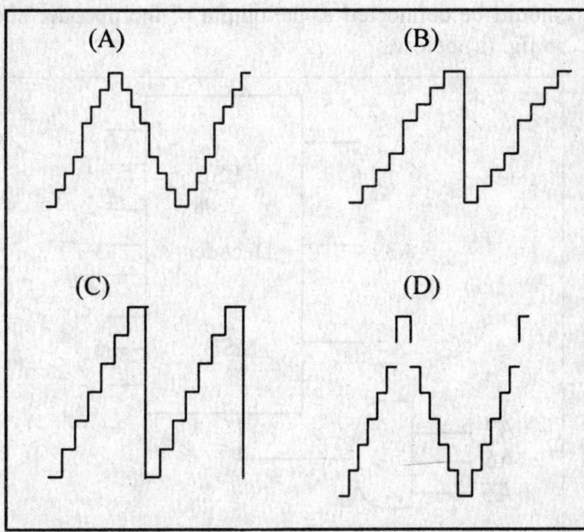

48. Two D – flip – flops. As shown below, are to be connected as a synchronous counter that goes through the following Q_1 Q_0 sequence.
00 → 01 → 11→ 10 → 00 →...........
The inputs D_0 D_1 respectively should be connected as:

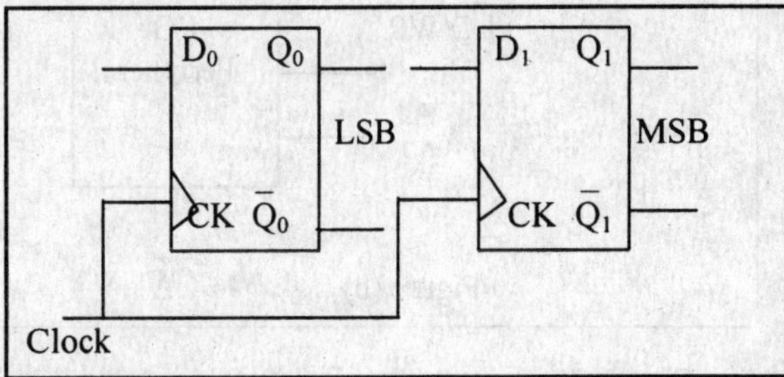

(a) $\bar{Q}_1$ and Q_0 (b) $\bar{Q}_0$ and Q_1
(c) $\bar{Q}_1 Q_0$ and $\bar{Q} Q_0$ (d) $\bar{Q}_1 \bar{Q}$ and $Q_1 Q_0$

49. The point P in the following figure is struck at –1. The output f will be:

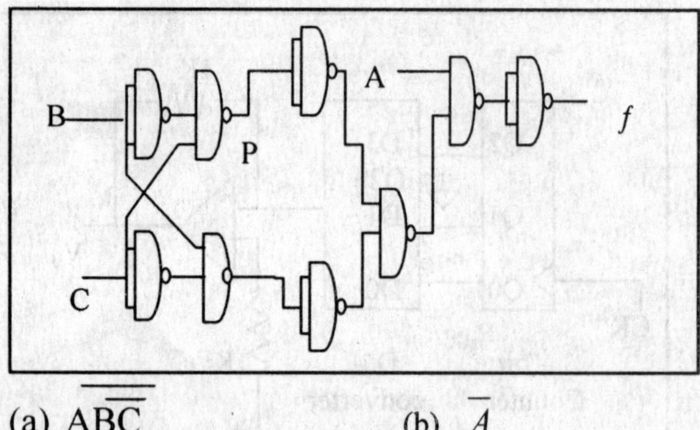

(a) $\overline{A\overline{BC}}$ (b) $\overline{A}$

(c) $AB\overline{C}$ (d) A

50. An I/O peripheral device shown in fig. (b) below is to be interfaced to an 8085 microprocessor. To select the I/O address range D4 H – D7 H, its chip – select ($\overline{CS}$) should be connected to the output of the decoder shown in fig. (a) below.

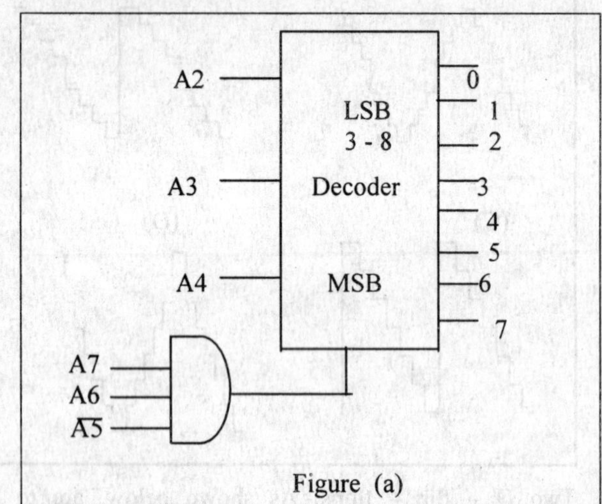

Figure (a)

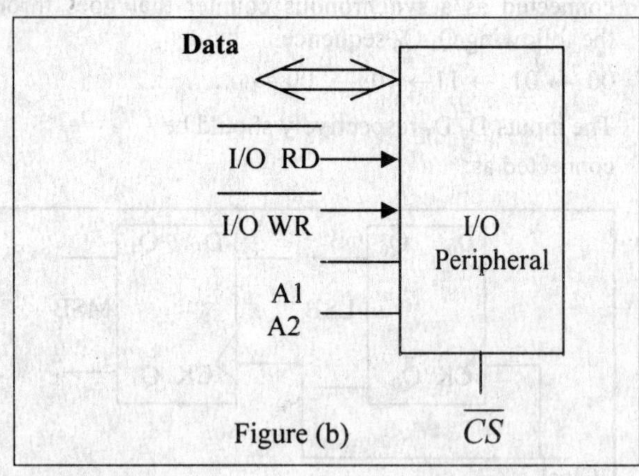

Figure (b)

(a) Output 7 (b) Out put 5

(c) Output 2 (d) Output 0

51. For the circuit shown in figure below, two 4-bit parallel – in serial – out shift registers loaded with the data shown are used to feed the data to a full adder. Initially, all the flip- flops are in clear state. After applying two clock pulses, the outputs of the full – adder should be.

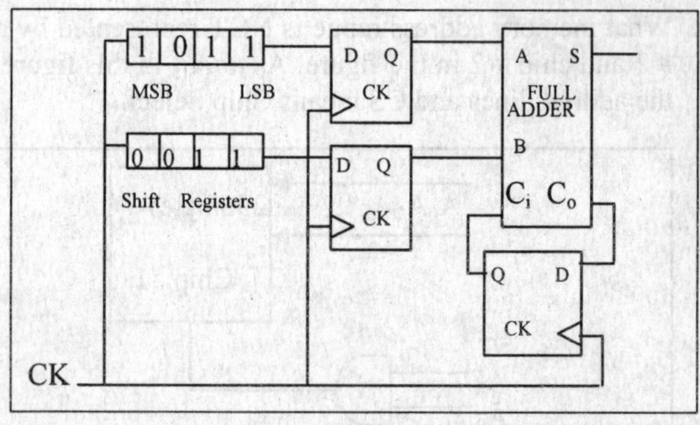

(a) S = 0 C_o = 0 (b) S = 0 C_o = 1

(c) S = 1 C_o = 0 (d) S = 1 C_o = 1

52. Following is the segment of a 8085 assembly language program.

```
        LXI SP, EFFFH
        CALL 3000H
           :
           :
3000H : LXI, 3CF4 H
        PUSH PSW
        SPHL
        POP PSW
        RET
```

On completion of RET execution , the contents of SP is

(a) 3 CFO H (b) 3CF 8 H

(c) EFFD H (d) EFFF H

53. In the following circuit, X is given by:

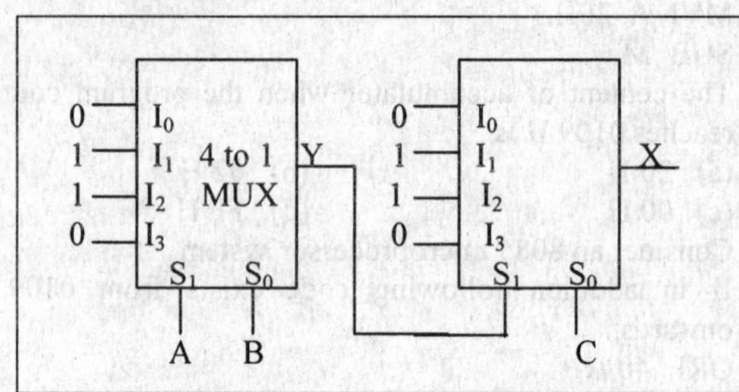

(a) $X = A\overline{B}\,\overline{C} + \overline{A}\,B\overline{C} + \overline{A}\,\overline{B}\,C + A\,B\,C$

(b) $X = \overline{A}\,B\,C + A\overline{B}\,C + A\,B\overline{C} + \overline{A}\,\overline{B}\,\overline{C}$

(c) $X = A\,B + B\,C + A\,C$

(d) $X = \overline{A}\,\overline{B} + \overline{B}\,\overline{C} + \overline{A}\;\overline{C}$

54. The Boolean expression $Y = \overline{A}\,\overline{B}\,\overline{C}\,D + \overline{A}\,B\,C\,\overline{D} + A\,\overline{B}\,\overline{C}\,D + A\,B\,\overline{C}\,\overline{D}$ can be minimized to

(a) $Y = \overline{A}\,\overline{B}\,\overline{C}\,D + \overline{A}\,B\,\overline{C} + A\,\overline{C}\,D$

(b) $Y = \overline{A}\,\overline{B}\,\overline{C}\,D + B\,C\,\overline{D} + A\,\overline{B}\,\overline{C}\,D$

(c) $Y = \overline{A}\,B\,C\,\overline{D} + \overline{B}\,\overline{C}\,D + A\,\overline{B}\,\overline{C}\,D$

(d) $Y = \overline{A}\,B\,C\,\overline{D} + \overline{B}\,\overline{C}\,D + A\,B\,\overline{C}\,\overline{D}$

55. The following binary values were applied to the X and Y inputs of the NAND latch shown in the figure in the sequence indicated below:

X = 0, Y = 1 ; X = 0, Y = 0; X = 1, Y = 1
The corresponding stable P, Q outputs will be:

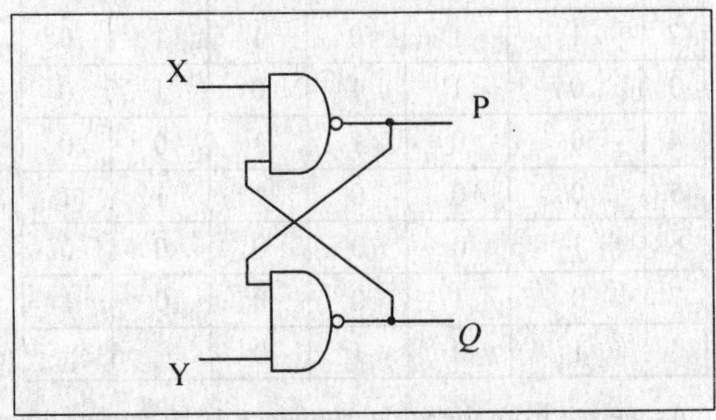

(a) P = 1, Q = 0 ; P = 1, Q = 0 P =1, Q = 0 or P = 0 , Q = 1
(b) P = 1, Q = 0 ; P = 0, Q = 1 or P = 0 , Q = 1 P = 0 , Q = 1
(c) P = 1, Q = 0 ; P = 1, Q = 1 P =1, Q = 0 or P = 0 , Q = 1
(d) P = 1, Q = 0 ; P = 1, Q = 1 P = 1, Q = 1

56. For the circuit shown, the counter state (Q_1,Q_0) follows the sequence.

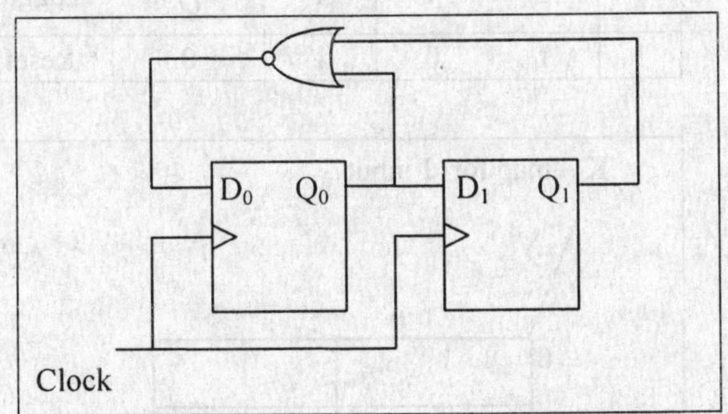

(a) 00, 01, 10, 11, 00... (b) 00, 01, 10, 00, 01...
(b) 00, 01, 11, 00, 01... (d) 00, 10, 11, 00, 10...

57. The circuit diagram of a standard TTL NOT gate is shown in the figure. When $V_i = 2.5\,V$, modes of operation of the transistors will be

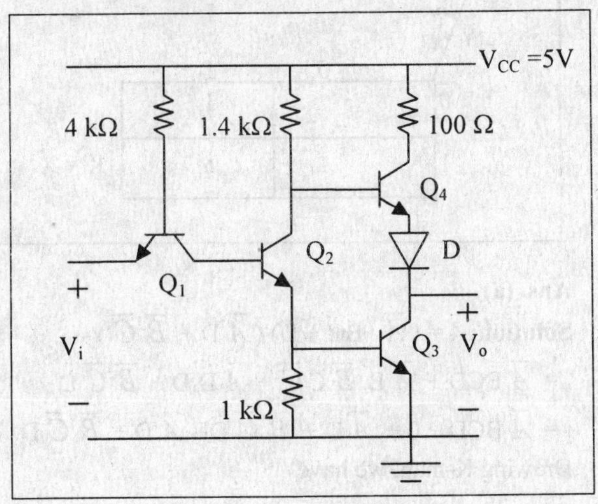

(a) Q_1: reverse active ; Q_2: normal active ; Q_3: saturation ; Q_4: cut-off ;
(b) Q_1: reverse active ; Q_2: saturation ; Q_3: saturation ; Q_4: cut-off ;
(c) Q_1: normal active ; Q_2: cut-off ; Q_3: cut-off ; Q_4: saturation ;
(d) Q_1: saturation ; Q_2: saturation ; Q_3: saturation ; Q_4: normal active ;

58. In the Digital-to-Analog converter circuit shown in the figure below, V_R = 10 V and R = 10 kΩ.

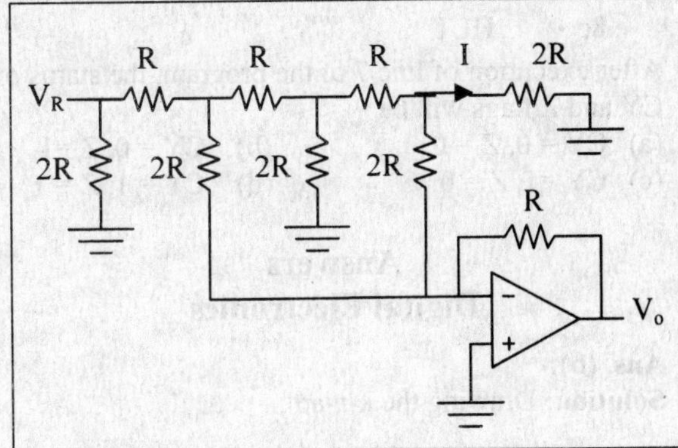

The current I is
(a) 31.25 μ A (b) 62.5 μA
(c) 125 μA (d) 250 μA

59. In the Digital-to-Analog converter circuit shown in the figure below, V_R = 10 V and R = 10 kΩ.

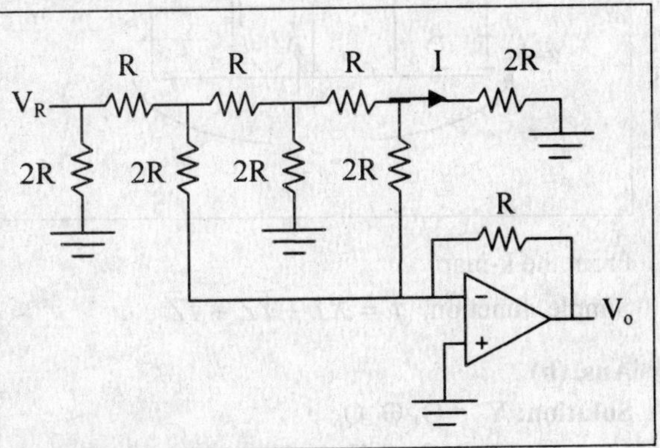

The voltage V_0 is
(a) –0.781 V (b) –1.562 V
(c) –3.125 V (d) –6.250 V

60. An 8085 assembly language program is given below.

```
Line 1:  MVI A,B5H
     2:  MVI B,0EH
     3:  XRI 69H
     4:  ADD B
     5:  ANI 9BH
     6:  CPI 9FH
     7:  STA 3010H
     8:  HLT
```

The contents of the accumulator just after execution of the ADD instruction in line 4 will be

(a) C3H (b) EAH
(c) DCH (d) 69H

61. An 8085 assembly language program is given below.

```
Line 1:  MVI A,B5H
     2:  MVI B,0EH
     3:  XRI 69H
     4:  ADD B
     5:  ANI 9BH
     6:  CPI 9FH
     7:  STA 3010H
     8:  HLT
```

After execution of line 7 of the program, the status of the CY and Z flags will be

(a) CY = 0, Z = 0 (b) CY = 0, Z =1
(c) CY =1, Z = 0 (d) CY = 1, Z = 1

Answers
Digital Electronics

1. Ans. (b)
Solution: Drawing the k-map

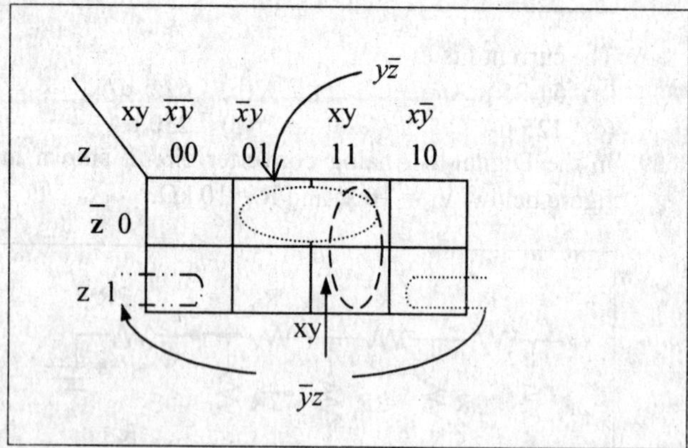

From the k-map,
Simple, function, $f = XY + \bar{Y}Z + Y\bar{Z}$

2. Ans. (b)
Solution: $Y_2 = Q_1 \oplus Q_o$
$Y_1 = Q_2 \oplus Y_2$

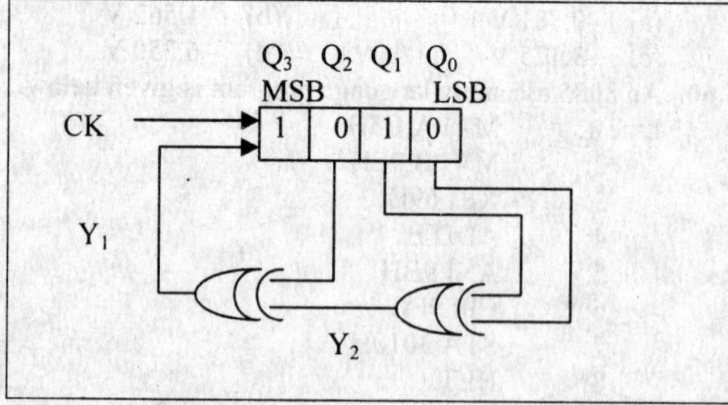

The sequences are as follows.

	Q_3 (Y_1)	Q_2	Q_1	Q_0	Y_2	Y_1
1	1	0	1	0	1	1
2	1	1	0	1	1	0
3	0	1	1	0	1	0
4	0	0	1	1	0	0
5	0	0	0	1	1	1
6	1	0	0	0	0	0
7	0	1	0	0	0	1
	1	0	1	0		

As we see from the table, sequence 1010 repeats after 7 cycles.

3. Ans. (d)
Solution: Finding the JK input for required output XY-FF

X	Y	J	K	Q_{n+1}	
0	0	1	0	1	Set
0	1	0	0	Q_n	Hold
1	0	1	1	$\overline{Q}$	Toggle
1	1	0	1	0	Reset

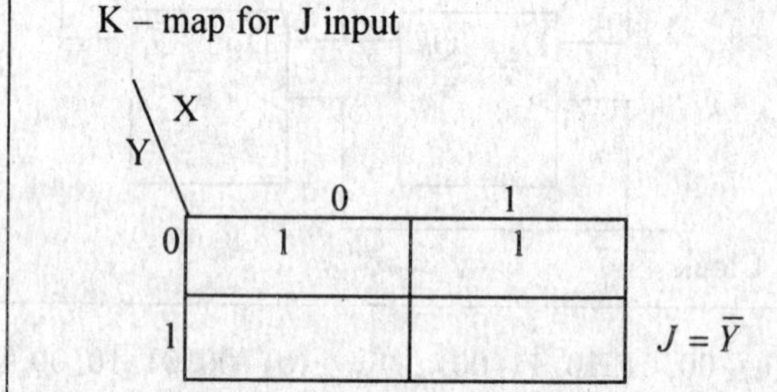

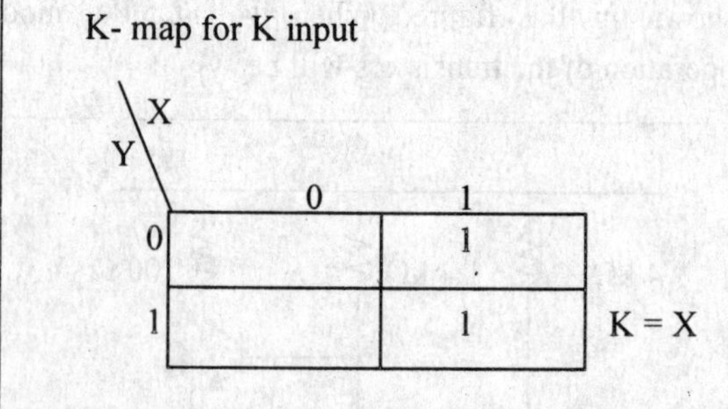

4. Ans. (a)
Solution: $4 = (\bar{A}\,.\,BC + D)(\bar{A}D + \bar{B}\bar{C})$
$= \bar{A}BCD + \bar{A}B.\bar{B}C\bar{C} + \bar{A}D.D + \bar{B}\bar{C}D$
$= \bar{A}BCD + 0 + \bar{A}D + \bar{B}\bar{C}D = \bar{A}D + \bar{B}\bar{C}D.$

Drawing K-map, we have

CD \ AB	$\overline{A}\overline{B}$ 00	$\overline{A}B$ 01	AB 11	$A\overline{B}$ 10
00	1	1		1
01	1	1		
11				
10				

From the K-map, it can be inferred that, simplified expression is

$Y = \overline{A}D + \overline{B}\,\overline{C}D.$

5. **Ans. (d)**
Solution: Exclusive OR gate gives output 1 when the input are different.

6. **Ans. (c)**
Solution:

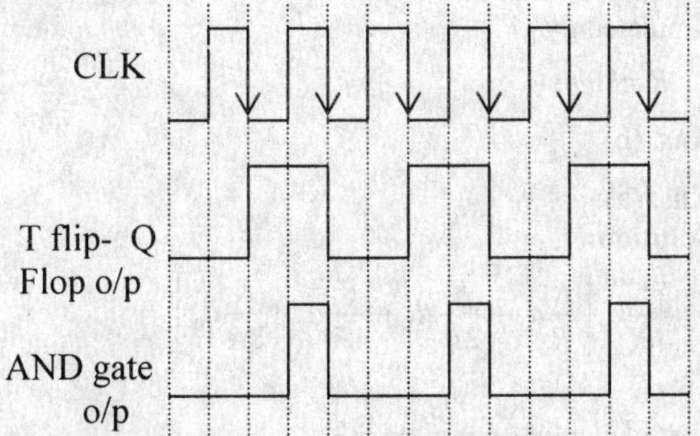

7. **Ans. (b)**
Solution:

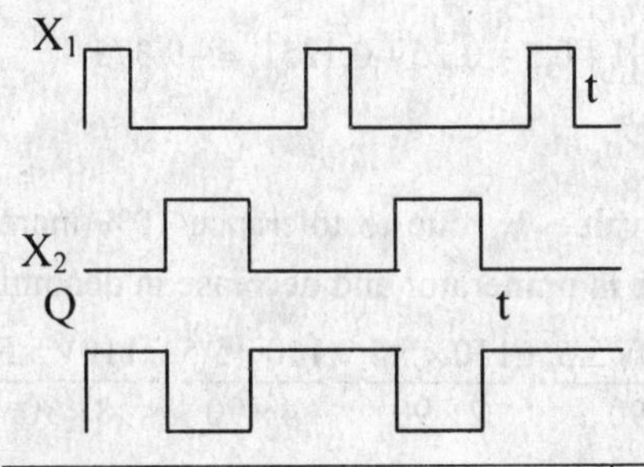

X_4	X_2	Q
1	0	1
0	0	1
0	1	0

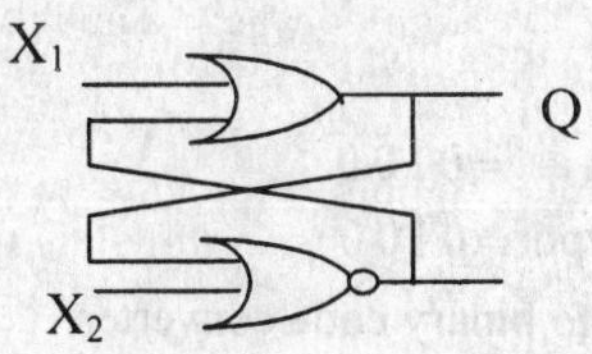

Q_2	X_1	Q	X_2	Q_2
0	1	1	0	0
0	0	0	0	1 →
→1	0	1	0	0
0	0	0	1	0

8. **Ans. (d)**

9. **Ans. (c)**
Solution:

Case 1:-
$X_1 = 1 \rightarrow 1$
$X_2 = 1 \rightarrow 0$ } → 0
$Q = 0$

Case 2:-
$X_1 = 1$ → 1
$X_2 = 1 \rightarrow 1$ } → 1
$Q = 1$

Therefore output remain at its initial value

10. **Ans. (c)**
Solution: The general circuit diagram of the DAC is:

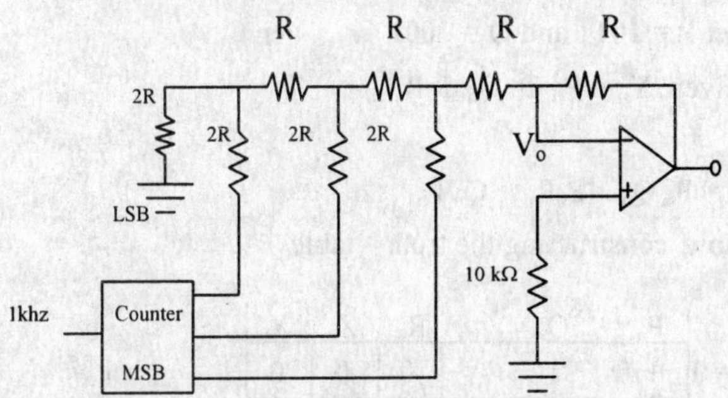

Here LSB and MSB are not correct.

11. **Ans. (a)**
Solution:

A \ BC	00	01	11	10
0		1		1
1	1			1

Decimal	A	B	C	Uni selected	Output
0	0	0	0	I_0	0
1	0	0	1	I1	1←
2	0	1	0	I_2	1←
3	0	1	1	I_3	0
4	1	0	0	I_4	1←
5	1	0	1	I_5	0
6	1	1	0	I_6	1←
7	1	1	1	I_7	0

$F(A,B,C) = \sum (1, 2, 4, 6)$

12. **Ans. (c)**

13. **Ans. (c)**

Solution: Base of Q_2 always at ≤ 0.7 V
SO Q_2 is ON.
Q_1 will conduct in reverse active region as collector potential should always $\geq$ base potential.
$V_{CB} \geq 0$
$V_{CE} > 0$
So, Q_1 is reverse ON

14. **Ans. (b)**

Solution:

$$\begin{array}{cccc} A & B\ \cdot & C & D \\ \downarrow & \downarrow & \downarrow & \downarrow \\ 10 & 11\ \cdot & 12 & 13 \\ \downarrow & \downarrow & \downarrow & \downarrow \\ 1010 & 1011\ \cdot & 1100 & 1101 \end{array}$$

$$\underset{2}{\underline{10}}\ \underset{5}{\underline{101}}\ \underset{3}{\underline{011}}\ .\ \underset{6}{\underline{110}}\ \underset{3}{\underline{011}}\ \underset{2}{\underline{010}}$$

15. **Ans. (b)**

Solution:

Let P = 1011 and Q = 1001

Given, $Y_n = P_n \oplus Q_n \oplus R_n$

$Z = R_n \overline{Q_n} + P_n \overline{R_n} + Q_n P_n$

Now, constructing the truth – table

	P_n	Q_n	R_n	Z_n	Y_n
n = 1	1	1	0	0	0
n = 2	1	0	0	0	1
n = 3	0	0	0	0	0
n = 4	1	1	0	0	0
					0

From the circuited given, we have

$R_{n+1} = Z_n,\ 1 \leq n \leq 3$

And $Z_4 = Y_5$(MSB)

Hence, output is 00010 which show that it is a 4-bit subtractor giving P – Q

16. **Ans. (a)**

Solution:

$$W = R + \overline{P}Q + \overline{R}S = R + S + \overline{P}Q$$

$$Z = R + S(\overline{P} + \overline{Q})(P + Q + R)(\overline{P} + Q + S)$$

$$= R + S + \overline{P}Q + \overline{P}QS + \overline{P}R + \overline{P}RQ + \overline{P}RS +$$

$$\overline{Q}PS + \overline{P}QS + \overline{Q}RS$$

$$= R + S + \overline{P}Q$$

$\therefore$ **W = Z**

$$\overline{Z} = \overline{W} = \overline{R + \overline{R}S + \overline{P}Q}$$

$$\overline{W} = \overline{R + \overline{R}S + \overline{P}Q}$$

$$= \overline{R}(R + \overline{S})(P + \overline{Q}) = \overline{R}\,\overline{S}(P + \overline{Q})$$

$$X = PQ\overline{R}\,\overline{S} + \overline{P}\,\overline{Q}\,\overline{R}\,\overline{S} + P\overline{Q}\,\overline{R}\,\overline{S}$$

$$= P\overline{R}\,\overline{S} + \overline{Q}\,\overline{R}\,\overline{S} = \overline{R}\,\overline{S}(P + \overline{Q})$$

$\therefore$ **X = $\overline{\mathbf{Z}}$**

17. **Ans. (b)**

Solution:

Total propagation delay in 4 bit ripple counter $R = 4\,t_d = 40\ ns$

In synchrous counter all flip flop are given clock simulateonly

$\therefore\ R = 10\ ns$

18. **Ans. (b)**

19. **Ans. (a)**

Solution:

$$V_o = -V_R\left[\frac{R}{R}d_3 + \frac{R}{2R}d_2 + \frac{R}{4R}d_1 + \frac{R}{8R}d_0\right]$$

$$= -V_R\left[\frac{d_3}{1} + \frac{d_2}{2} + \frac{d_1}{4} + \frac{d_0}{8}\right]$$

When $V_R = 5\,V$ and $d_3 = d_2 = d_1 = d_0$

$$V_o = -5[1 + 0.5 + 0.25 + 0.125] = -9.375$$

In worst Case,

Maximum value V_o due to tolerance 10% increase and 5% increase in numerator and decrease in denominator

$$V_o = -\left[\frac{110\times5.5}{90} + \frac{110\times5.5}{2\times90} + \frac{110\times5.5}{4\times90} + \frac{110\times5.5}{8\times90}\right]$$

$$V_o = \frac{-110\times5.5}{90}[1 + 0.5 + 0.25 + 0.125]$$

$$= -12.604$$

Tolerance of $DAC = 12.604 - 9.375$

$= 34.44\% \ = 35\%$

20. **Ans. (a)**

Solution:

Let us take input as $= 0\,1\,0\,0$

output $= 0\,1\,0\,0$

$\therefore$ This is BCD to binary code converter

21. Ans. (c)

Solution: When W has data 0110 i.e. 6 in decimal its data value at address is 1010.

New 1010 i.e. 10 is acting as address at time t_2 and data at that moment is 1000.

22. Ans. (a)

Solution: CMP B: Contents of B and A are compared and result is stared in A that is indicated by flag $A < B$

$\therefore\ CY = 1,\quad Z = 0$

23. Ans. (c)

Solution

1100 ⇒ - 00111 ⇒ - 7

1001 ⇒ - 0111 ⇒ - 7

111001 ⇒ - 000111 ⇒ - 7

24. Ans. (c)

Solution:

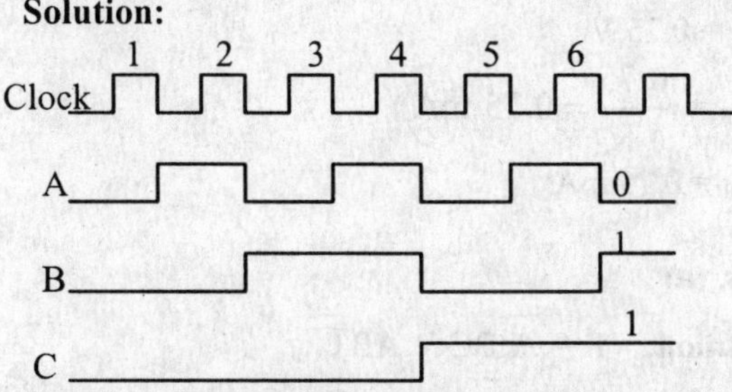

To perform the given ripple counter as module-6, the flip-flop should be cleared at the end of sixth pulse.

Since at 6th pulse, B = C, then $\overline{B} = \overline{C} = 0$

Hence, the 2-input gate, should be OR gate so that Clr = 0 for all the three flip-flops.

The desire gate should generate 0 as clear is active Low when CBA = $(110)_2$

Input to the desire gate is $\overline{C}\,\overline{B}$ i.e. 00

OR gate give 0 when both input are 0.

25. Ans. (d)

Solution:

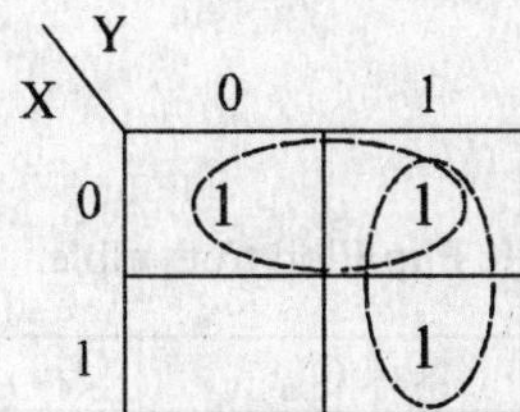

$f = \overline{X}Y$

x	y	F
0	0	1
0	1	1
1	0	0
1	1	1

Since complement are not available

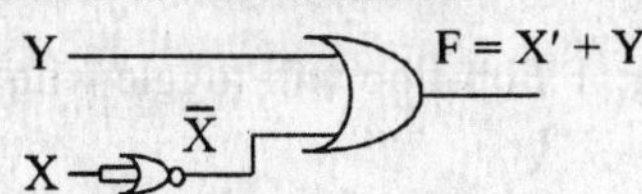

Hence 2 unit are required

26. Ans. (b)

Solution:

The given diagram represent a decade counter

$$\text{Frequency of output} = \frac{10 \times 10^3}{10} 1\ \text{kHz}$$

27. Ans. (c)

Solution: Number of comparators = $2^n - 1$

$= 2^4 - 1 = 15$

28. Ans. (c)

Solution:

A	B	D(difference)	X(boorow)
0	0	0	0
0	1	1←	1
1	0	1←	0
1	1	0	0

$D = \overline{A}B + A\overline{B} = A \oplus B$

$X = \overline{A}B$

29. Ans. (d)

Solution:

3 flip flips are used and it has – ve edge triggering. Hence it is mod 5 down counter

Apply the Trick as below:

- When ever preset is activate it indicate down counter
- When ever clear is activated it indicate up counter

up /down $= CP \oplus Q$, 1 for up and 0 for down

$CP =$ (clock pulse)

$Q = (O/P)$

$0 =$ –ve edge; $Q = 1$

$1 =$ +ve edge; $\overline{Q} = 1$

$= 1 \oplus 1 = 0$ (down counter)

Counting sequence

1 1 1

1 1 0

1 0 1

1 0 0

0 1 1 (preset state) *so* Mod 5

30. Ans. (b)

Solution: Dual slope A/D is

1. Very slow
2. Highly accurate
3. Capable of eliminating hum noise

31. Ans. (a)

Solution:

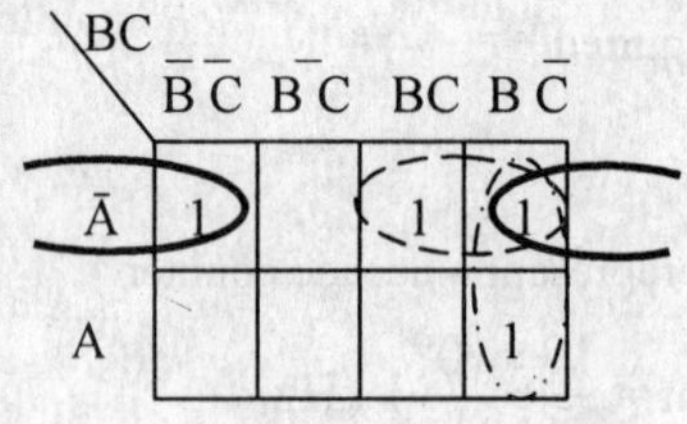

$y = \bar{A}\,\bar{C} + \bar{A}\,B + B\,\bar{C}$

32. Ans. (c)

Solution:

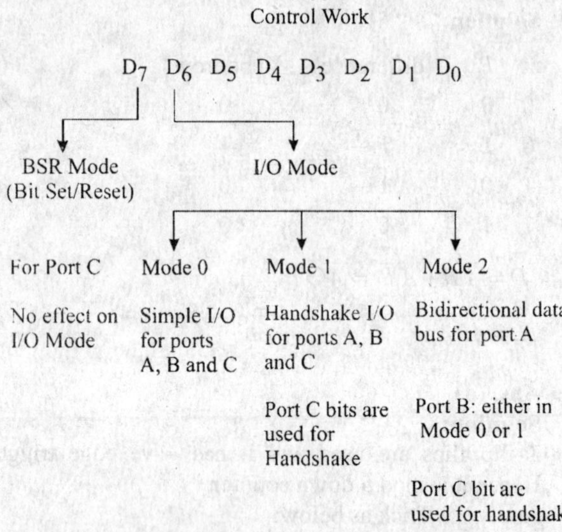

Mode 2: Bidirectional Data Transfer

This mode is used primarily in applications such as data between two computers or floppy disk controller interface, In this mode, port A can be configured as the bidirectional port and port B either in Mode 0 or Mode 1. Port A uses five signal from port C as handshake signals for data transfer. The remaining three signals from port C can be used either as simple I/O or as handshake for Port B.

33. Ans. (b)

Solution:

LDA 3000 H $\rightarrow$ Fetch, (Address) Read, Read,

Read $\Rightarrow 4\,M/C$

LXI D $F0F1H \rightarrow$ Fetch data, read, read $\Rightarrow 3\,M/C$

34. Ans. (c)

Solution:

LXI H 9258 $\Rightarrow$ *HL*. $\leftarrow$ 9258

MOV A M $\Rightarrow$ A $\leftarrow$ contents of Address 9258

CMA $\Rightarrow$ Content of accumulator complemented

MOV M, A $\Rightarrow$ Complemented content of A stored in M(9258)

35. Ans. (d)

36. Ans. (d)

Solution: For LED to emit light, NAND output should be low (o) and input 1, for which AND input to be 1 and ex – oR gate input 0.1 or 1, 0. when S_1 and or S_2 are closed input to the AND Gage is o. When S_1, S_2 are open input to the AND gate is 1 as also to X-OR gate.

37. Ans. (c)

Solution:

Number of MUX is $\frac{4}{2} = 2$ and $\frac{2}{2} = 1$. Thus the total number 3 multiplexer is required

38. Ans. (c)

Solution: V_o is at logic 0. So transistor is in saturation

$V_{BE} = 0.75$ V

$\therefore I_R = \frac{0.75}{1\text{ K}} = 0.75$ mA

$I_R = 0.75$ mA

39. Ans. (a)

Solution: $F = \bar{A}BC + AB\bar{C}$

$= B[\bar{A}C + A\bar{C}]$

$F = B[A + C][\bar{A} + \bar{C}]$

40. Ans. (c)

Solution:

$I_{DS}(T_1) = I_{DS}(T_2)$

$[1/2]\ K_1\ (V_{GS1} - V_t)^2 = [1/2]\ K_2\ (V_{GS2} - V_t)^2$

$36\,(5 - V_o - 1)^2 = 9\,(V_o - 1)^2$

$6\,(4 - V_o) = 3\,(V_o - 1) \Rightarrow 2\,(4 - V_o) = V_o - 1$

$8 + 1 = 3V_o = 9$

$V_o = 3$

41. Ans. (c)

Solution: JK Flip Flop truth table.

Q_n	J	K	Q_{n+1}	Q_{n+1}
0	0	0	Q(n)	0
0	0	1	0	0
0	1	0	1	1
0	1	1	$\overline{[Q(n)]}$	1

42. Ans. (b)

Solution: T Flip-flop will toggle with each trigger pulse clock if T = 1

Q_n	$Q_{n+1} = \overline{Q}_n$
$Q_0 = 1$	0
$Q_1 = 1$	0
$Q_2 = 0$	1

Output $Q_2Q_1Q_0$ =100

43. Ans. (d)

Solution: Address range of chip 1 and chip 2 →

$A_0 - A_7$; $A_9A_8A_7 - A_0$

00 – FF ; 00FF – 00

01FF – 00 } $\bar{A}_9\ A_8$

10FF – 00 } $A_9\ \bar{A}_8$

11FF – 00

$A_8\ \bar{A}_9$ →selecting chip # 1 [Address range is 100 – 1FF]

$\bar{A}_8A_9$ → selecting chip # 2 [Address range is 200 – 2FF]

Address range is 100 – 2 FF

Since A10 – A 15 are not used, the address range can be 100 – 2 FF (or) 500 – 6FF (or) 900 – AFF,

For the address space F800 – F9FF chip # 1 and chip # 2 are not selected

44. Ans. (a)

Solution: MVI A, 20 H puts 20 H in A

When SUB M is encountered, PC reaches 0109H

∴ Value in A remains same.

45. Ans. (c)

46. Ans. (d)

Solution:

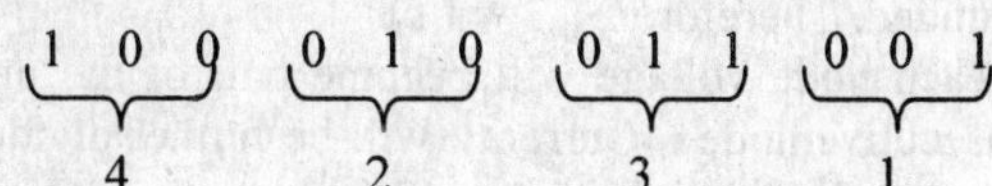

47. Ans. (b)

Solution: In one cycle

Q_2		Q_1	Q_0
D_3	D_2	D_1	D_0
0	0	0	0 → 0
0	0	0	1 → 1
0	0	1	0 → 2
0	0	1	1 → 3 } ***
1	0	0	0 → 8
1	0	0	1 → 9
1	0	0	0 → 10
1	0	1	1 → 11

*** From third level there is abrupt jump to eight level.

48. Ans. (a)

Solution:

Present state		Next state	
Q_1	Q_0	D_1	D_0
0	0	0	1
0	1	1	1
1	1	1	0
1	0	0	0

$D_0 = \overline{Q}_1;\quad D_1 = Q_0$

49. Ans. (d)

50. Ans. (d)

Solution:	A	B	C_i	S	C_0
After I clock	1	1	0	0	1
After II clock	1	1	1	1	1

51. Ans. (c)

Solution: For $\overline{CS}$ × should be 0

i.e. all input should be high for GATE

A_7	A_6	A_5	A_4	A_3	A_2	A_1	A_0
1	1	1	1	1	0	X	X

← F →← 8 →

0	0
0	1
1	0

⇕

A_7	A_6	A_5	A_4	A_3	A_2	A_1	A_0
1	1	1	1	1	1	1	1

← F →← F →

Thus the address range is F8H to FFH

52. Ans. (b)

Solution:

	SP
After SPHL	3 C F 4
After POP PSW	3 C F 6
After RET	3 C F 8

POP & RET are 2 Byte instruction each

53. Ans. (a)

Solution: $Y = I_1 + I_2$

$$Y = \overline{A}B + A\overline{B} = A \oplus B$$

$$X = \overline{Y}C + Y\overline{C} = Y \oplus C = A \oplus B \oplus C$$

$$= \overline{A}B\overline{C} + A\overline{B}\overline{C} + \overline{A}\overline{B}C + ABC$$

54. Ans. (d)

Solution:

$$Y = \overline{A}\ \overline{B}\ \overline{C}\ D + \overline{A}\ B\ C\ \overline{D} + A\ \overline{B}\ \overline{C}\ D + A\ B\ \overline{C}\ \overline{D}$$

$$Y = \overline{B}\ \overline{C}\ D + A\ B\ \overline{C}\ \overline{D} + \overline{A}\ B\ C\ \overline{D}$$

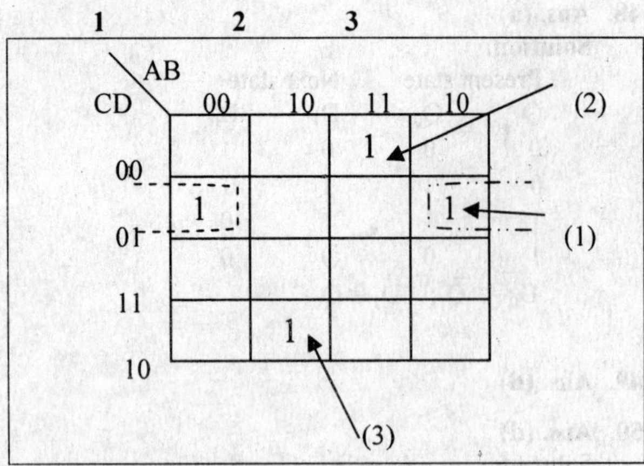

55. Ans. (c)

Solution: For X = 0 ⇒ P = 1

∵ for NAND gate 0X or X0 gives 1 as output

Then Q becomes 0 for 11 input.

The NAND action is represented by table given below

X	Y	P	Q
0	1	1	0
0	0	1	1
1	1	1	0 or 0/1

56. Ans. (b)

Solution:

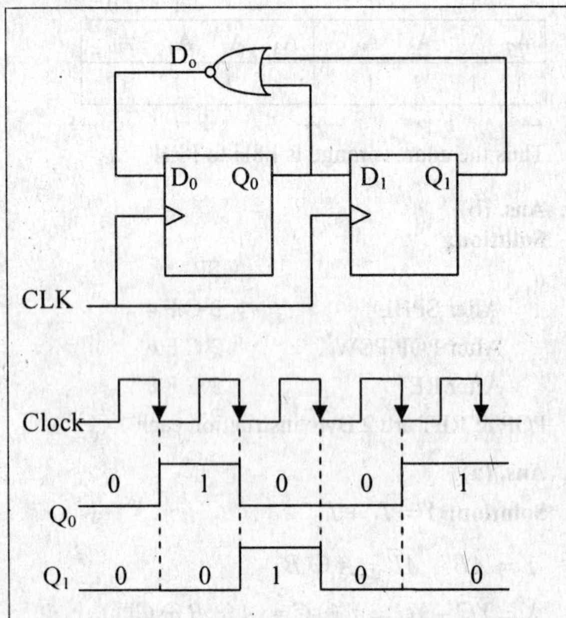

NOR GATE		
Q_0	Q_1	D_0
0	0	1
1	0	0
0	1	0
0	0	1
1	0	0

Sequence is

$\therefore \quad Q_1Q_2 \quad 00,01,10,00,01\ldots$

57. Ans. (a)

Solution: As given TTL circuit is NOT gate

V_i = 2.5 V high is applied. Output is low only when the earth is extended to output it means Q_3 is in saturation mode and Q_4 is cut off as Q3 and Q4 always behave negatively. Q_2 work in normal active but Q_1 in reverse active as V_i = 2.5 applied to emitter of Q_1.

58. Ans. (b)

Solution: In the Digital-to-Analog converter circuit shown in the figure below, V_R = 10 V and R = 10 kΩ.

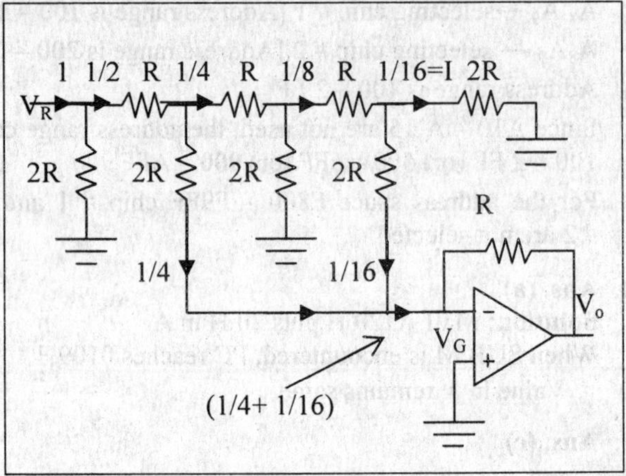

Due to virtual ground each 2 Ω resistor one terminal is grounded. Therefore 2 || 2 = 1 Ω.

At each node voltage will become half of the previous consecutive node . Current will be further divided into half. I = 1/ 16 = 0.0625 A = 62.5 mA.

59. Ans. (c)

Solution: The current through the Ops-Amp resistance R is [1/4 + 1/16] × 10 k = –3.125 V.

60. Ans. (b)

Solution:

1. A → B5 H
2. B → 0EH
3. A XOR 69H ⇒ 10110101

   <u>00111001</u>

   10001100 = 8CH
4. ADDB

   A + B ⇒ 8C + OE = EAH

61. Ans. (c)

Linked Answers Questions 1 and 2

1. The aperture time for a signal V sin ωt using a n-bit ADC where the input sine wave amplitude is equal to the maximum input for ADC will be.

(a) $1/\omega(2^n - 1)$ (b) $1/\omega 2^n$

(c) $1/\omega(2^n + 1)$ (d) $\omega(2^n - 1)$

2. Repeat above problem for a 3 KHz signal using a 6 bit ADC to find aperture time

(a) 816 nsec (b) 842 nsec

(c) 828.93 nsec (d) None of the above.

Statement for Linked Answers Questions 3 and 4

In the Digital-to-Analog converter circuit shown in the figure below, V_R = 10 V and R = 10 kΩ.

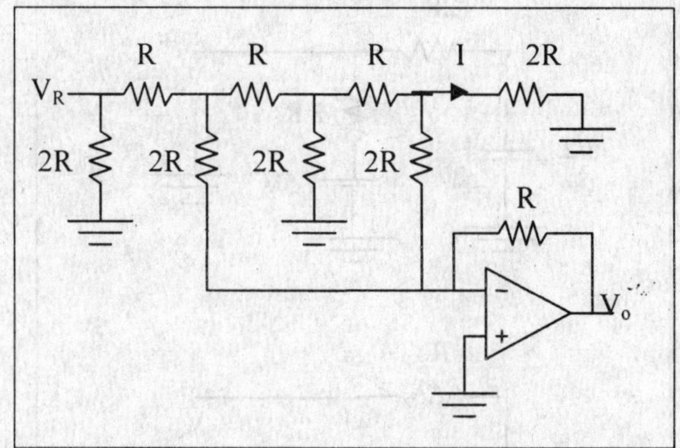

3. The current I is:

(a) 31.25 μ A (b) 62.5 μA

(c) 125 μA (d) 250 μA

4. The voltage V_O is:

(a) –0.781 V (b) –1.562 V

(c) –3.125 V (d) –6.250 V

Statement for Linked Answer Question 5 and 6:

An 8085 assembly language program is given below.

```
Line 1:  MVI A,B5H
     2:  MVI B,0EH
     3:  XRI 69H
     4:  ADD B
     5:  ANI 9BH
     6:  CPI 9FH
     7:  STA 3010H
     8:  HLT
```

5. The contents of the accumulator just after execution of the ADD instruction in line 4 will be.

(a) C3H (b) EAH

(c) DCH (d) 69H

6. After execution of line 7 of the program, the status of the CY and Z flags will be

(a) CY = 0, Z = 0 (b) CY = 0, Z =1

(c) CY =1, Z = 0 (d) CY = 1, Z = 1

Common Data for Questions 7 and 8

An R-2r ladder type DAC is shown below. If a switch status is '0' 0 V is applied and if a switch status is '1', 5 v is applied to corresponding terminals of the DAC.

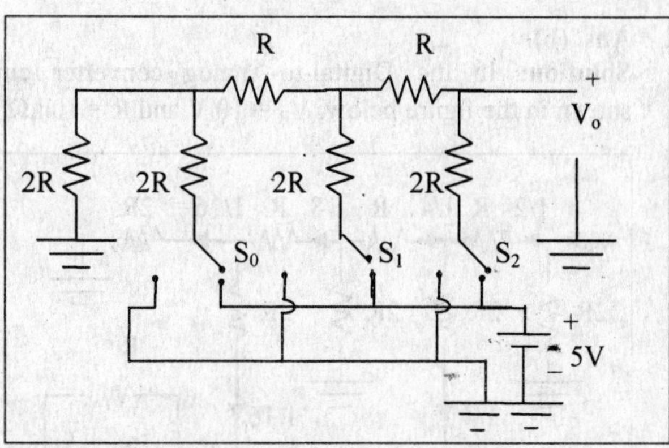

7. What is the output voltage (v_o) for the switch status $S_o = 0, S_1 = 1, S_2 = 1$?

(a) $\frac{5}{4}V$ (b) $\frac{15}{4}V$

(c) $\frac{17.5}{4}V$ (d) $\frac{22.5}{4}V$

8. What is the step size of the DAC?

(a) 0.125 V (b) 0.525 V

(c) 0.625 V (d) 0.75 V

Answers
Digital Electronics

1. Ana. (a)

Solution:

The aperture time is defined as the maximum rate of change of sine wave occurs at zero phase angle and is given by, $\frac{d}{dt}(V\sin\omega t) = \omega V\cos\omega t = \omega V.$

The voltage change during aperture time t_a is $\omega V t_a$

Therefore, $\omega V t_n$ = quantization error

$$= \frac{\frac{V_{pp}}{2}}{2^n - 1}$$

Also, $V = \frac{V_{pp}}{2}$

Therefore, $t_a = \frac{1}{\omega(2^n - 1)}$

2. Ans. (b)

Solution:

Aperture time $t_a = \frac{1}{\omega(2^n - 1)}$

$$= \frac{1}{2\pi f(2^n - 1)}$$

$$= \frac{1}{2\pi(3\times10^3)(2^6 - 1)}$$

= 842 nsec

3. Ans. (b)
Solution: In the Digital-to-Analog converter circuit shown in the figure below, $V_R = 10$ V and R = 10 kΩ

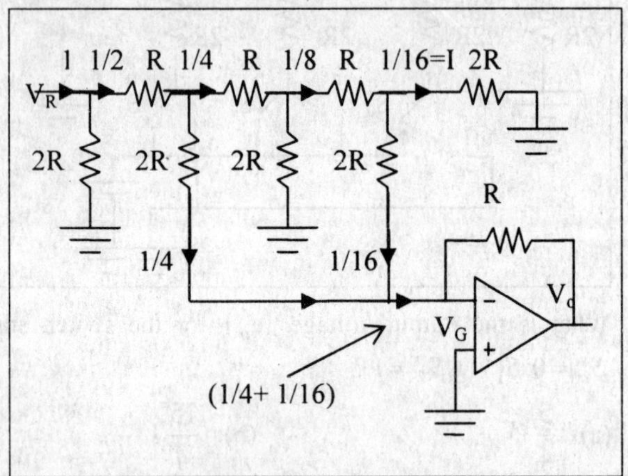

Due to virtual ground each 2 Ω resistor one terminal is grounded. Therefore 2 || 2 = 1 Ω.
At each node voltage will become half of the previous consecutive node . Current will be further divided into half. I = 1/ 16 = 0.0625 A = 62.5 mA.

4. Ans. (c)
Solution: The current through the Ops-Amp resistance R is [1/4 + 1/16] × 10 k = –3.125 V

5. Ans. (b)
Solution:
1. A → B5 H
2. B → 0EH
3. A XOR 69H ⇒ 10110101

$$\frac{00111001}{10001100} = 8CH$$

4. ADDB

A + B ⇒ 8C + OE = EAH

6. Ans. (c)

7. Ans. (b)
Solution:

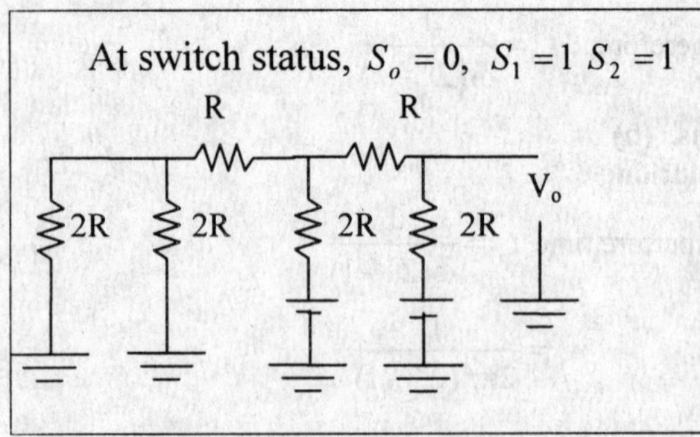

Thevenin equivalent

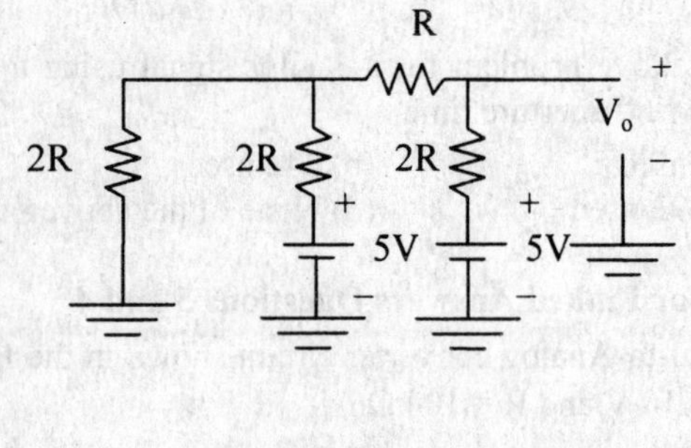

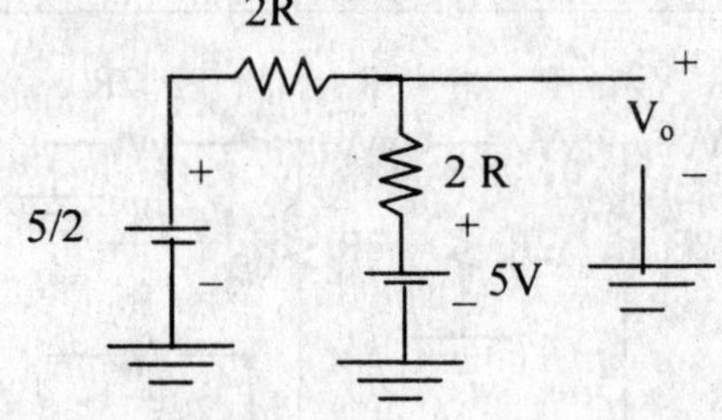

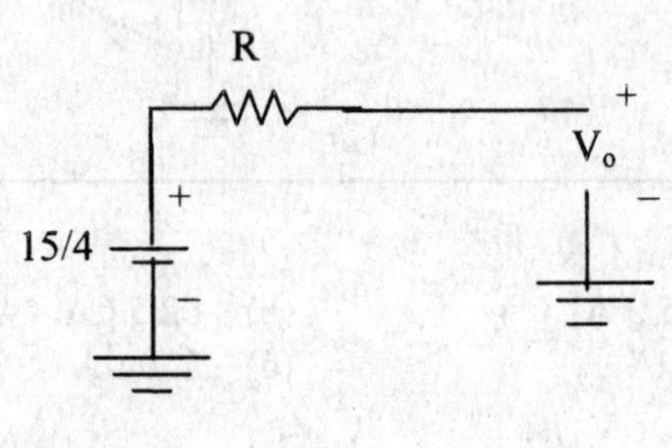

Second Method

$$= \frac{V_{ref}}{2r}\left[D_o \times 2^0 + D_1 \times 2^1 + D_2 \times 2^2\right]$$

$$= \frac{5}{8}[2+4] = \frac{5 \times 6}{8} = \frac{15}{4}$$

8. Ans. (c)
Solution:

$V_{ref} = 5V$; $n = 3 =$ number of bits

$$\text{Step size, } S = \frac{V_{ref}}{2^n} = \frac{5}{2^3} = \frac{5}{8} = 0.625$$

6

Signals, Systems and Communication

Signals, Systems and Communication: Periodic and aperiodic signals. Impulse response, transfer function and frequency response of first- and second order systems. Convolution, correlation and characteristics of linear time invariant systems. Discrete time system, impulse and frequency response. Pulse transfer function. IIR and FIR filters. Amplitude and frequency modulation and demodulation. Sampling theorem, pulse code modulation. Frequency and time division multiplexing. Amplitude shift keying, frequency shift keying and pulse shift keying for digital modulation.

6.1 FUNDAMENTALS OF SIGNALS AND SYSTEMS

6.1.1 SIGNALS

Signal may be describe a wide variety of physical phenomena. It may be represented in many ways, i.e. variation of some form.

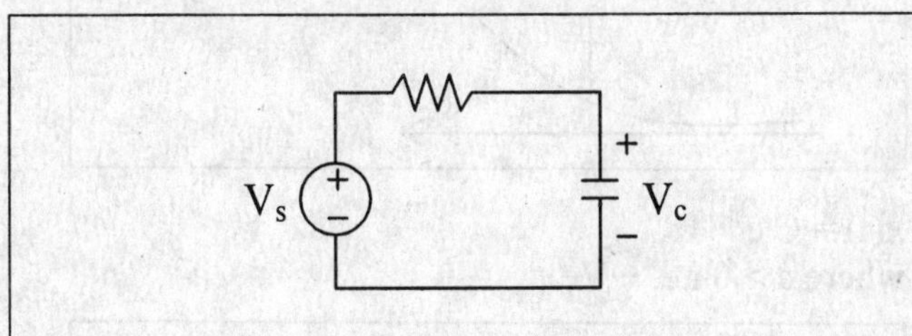

Example

Variation of voltage across the capacitor w.r.t. is the signal.

Signals are represented mathematically as functions of one or more independent variable.

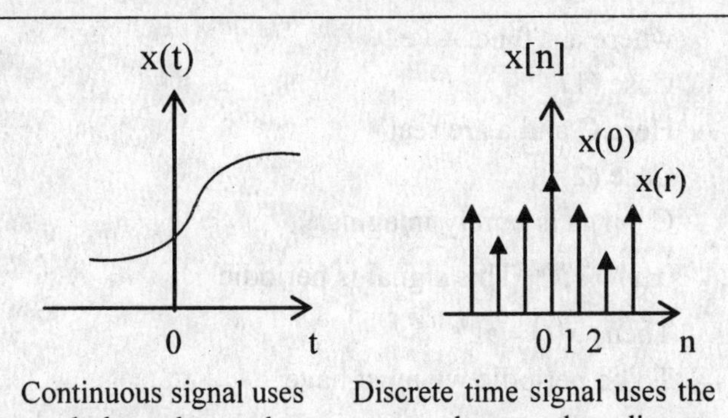

Continuous signal uses symbol t to denote the continuous-time independent variable

Discrete time signal uses the n to denote the discrete time independent variable

Example

Speech signal represented mathematically by acoustic pressure as function of time and picture signal represented by brightness a function of two spatial variable.

$x[n]$ = having a sequence

$x[n]$ = a1 sin t = having one element

Sampling of data gives discrete signal which is use to indicate Aircraft position, velocity, heading.

6.1.2 TRANSFORMATION OF INDEPENDENT VARIABLE

1. Time shifting

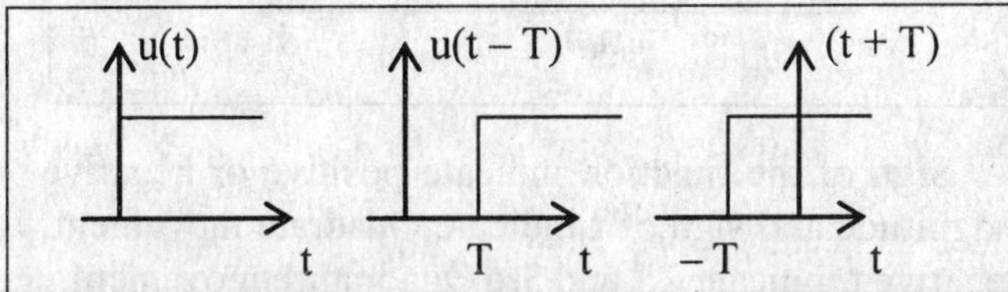

2. Amplitude reversal

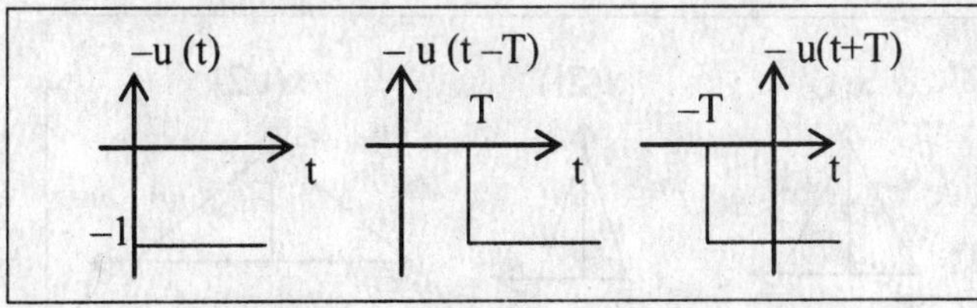

3. Time reversal

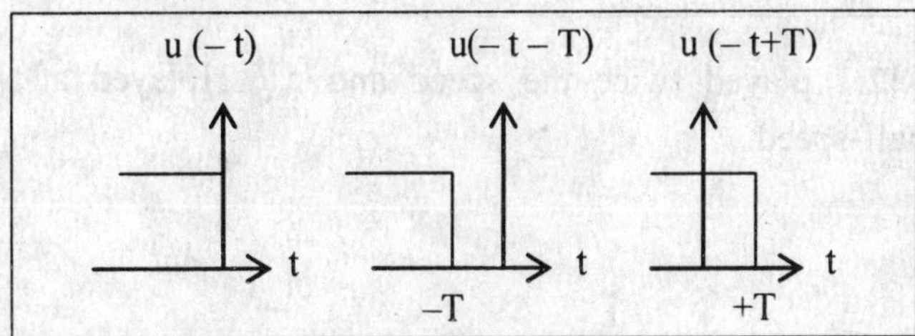

4. Time and amplitude reversal

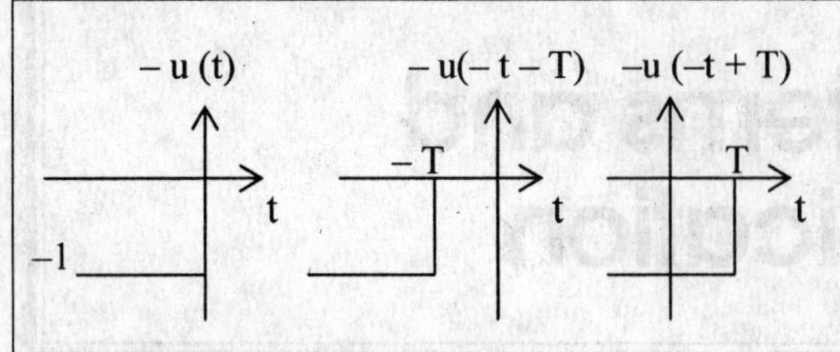

Similar case with the discrete signal are

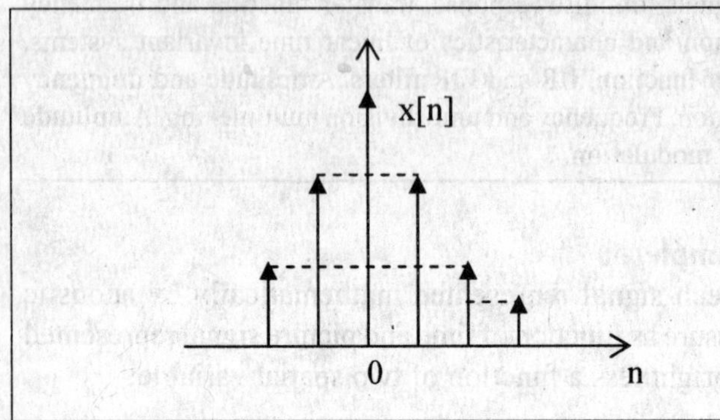

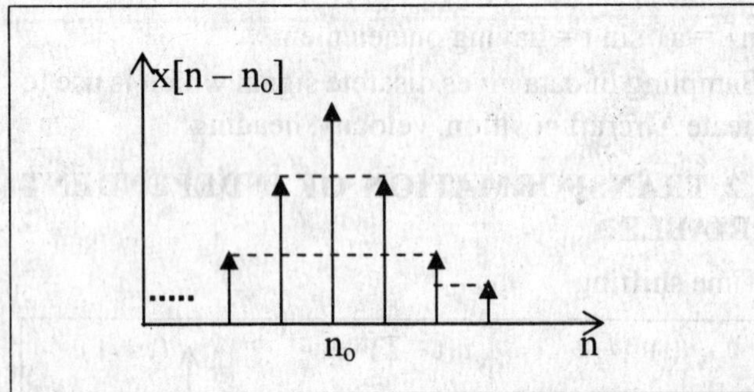

Sign of the function indicate positive or negative magnitude and sign of **t** indicate Quadrate movement, negative t indicate 2nd and 3rd Quadrature movement.

Time Scaling

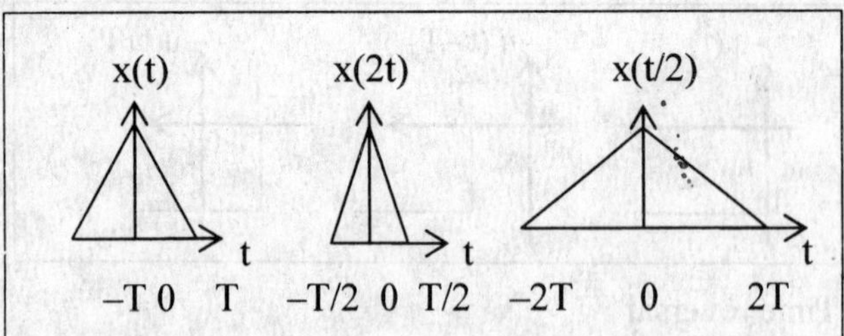

$x(2t)$ played twice the speed and $x(t/2)$ played at half-speed.

Even Signals

A signal x(t) or x[n] is even if is identical to its time-reversed counter part i.e. with its reflection about the origin.

$$x(-t) = x(t) \quad ; x[-n] = x[n]$$

Odd Signal

$x(-t) = -x(t)$; $x[-n] = -x[n]$

odd signal must be 0 at t = 0 or n = 0

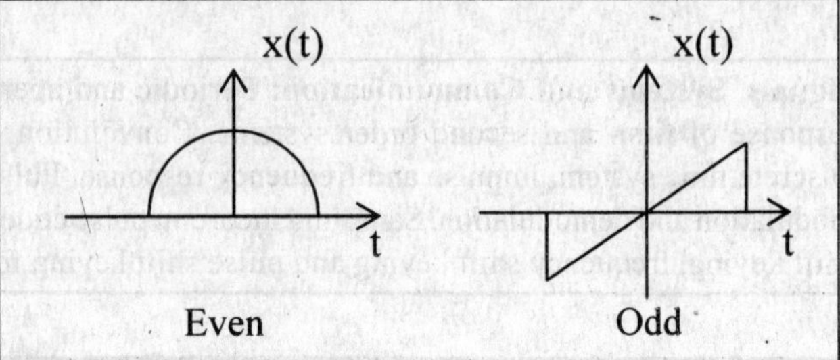

Exponential Signal

$x(t) = C\,e^{at}$

where C and a are complex number.

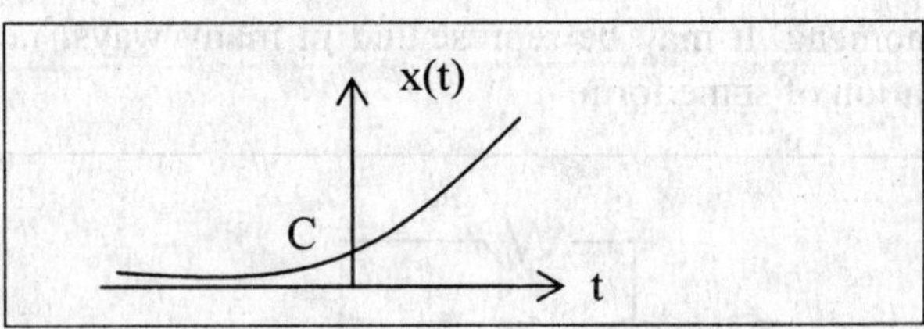

$x(t) = C\,e^{at}$

where a > 0 i.e. + Ve

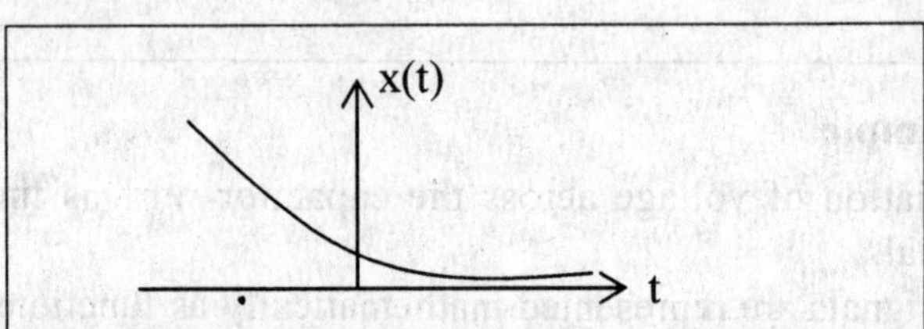

$x(t) = C\,e^{at}$

where a < 0 i.e. – ve

Case (1)

Here C and a are real

Case (2):

C = 1, a is purely imaginary

$x(t) = e^{j\omega_o t}$ This signal is periodic

Then, $x(t) = e^{j\omega_o (t+T)} = e^{j\omega_o t} \cdot e^{j\omega_o T}$

To be periodic we must have

$e^{j\omega_o T} = 1$

If $\omega_o = 0$; $x(t) = 1$

Which is periodic for any value of T

If $\omega_o \neq 0$; $T_o = \dfrac{2\pi}{\omega_o}$

$$\omega_o = \frac{2\pi}{T_o} = 2\pi f_o$$

The complex equation can be written in terms of sinusoidal signals with the same fundamental period.

$$x(t) = e^{j\omega_o t} = \cos\omega_o t + j\sin\omega_o t$$

$$\cos\omega_o t = \frac{e^{j\omega_o t} + e^{-j\omega_o t}}{2}$$

$$\sin\omega_o t = \frac{e^{j\omega_o t} - e^{-j\omega_o t}}{2}$$

$$\cos(\omega_o t + \phi) = \frac{e^{j\omega_o t}e^{j\phi} + e^{-j\omega_o t}e^{-j\phi}}{2}$$

if $x(t) = e^{j2t} + 2^{j3t}$

$$\frac{2+3}{2} = \frac{5}{2} = 2.5$$

$$= e^{j2.5t}(e^{-j0.5t} + e^{j0.5t}) = 2e^{j2.5t}\cos 0.5t$$

$$|x(t)| = 2|\cos 0.5t|$$

For discrete signal

$x[n] = C\alpha^n$

Where C and α are complex number

$= Ce^{\beta n}$ Where $\alpha = e^B$

The real exponential signal $x[n] = C\alpha^n$

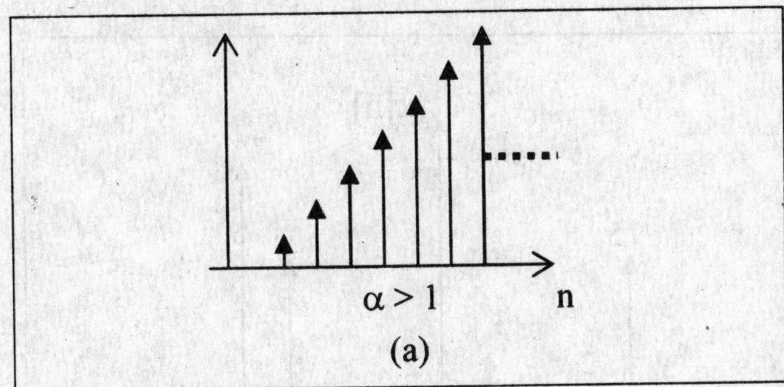

(a)

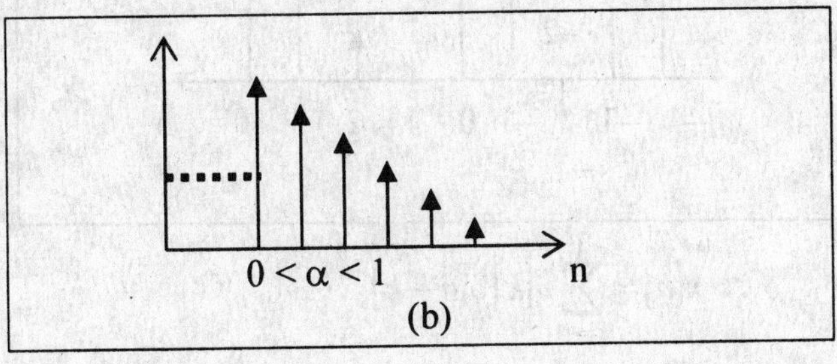

(b)

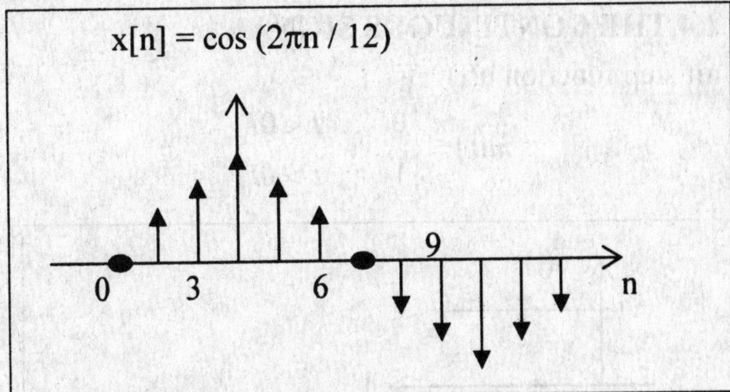

6.1.3 THE UNIT IMPULSE AND STEP FUNCTION

Discrete $\delta[n] = \begin{cases} 0 & ; n \neq 0 \\ 1 & n = 0 \end{cases}$

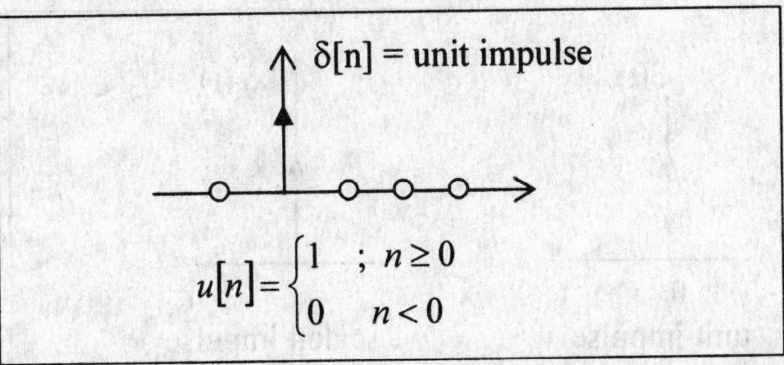

$$u[n] = \begin{cases} 1 & ; n \geq 0 \\ 0 & n < 0 \end{cases}$$

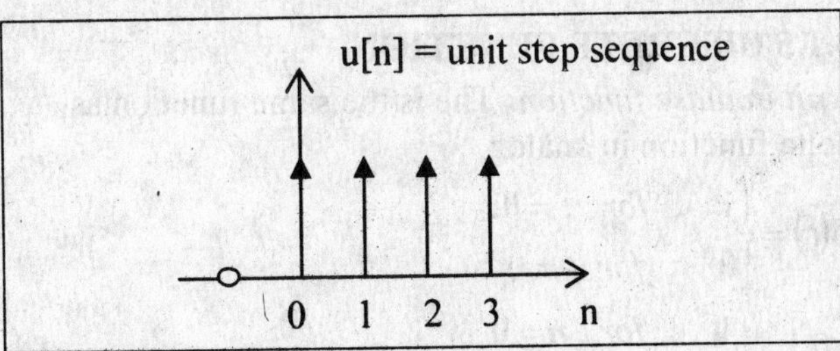

$\delta[n] = u[n] - u[n-1]$

There is a close relationship between the discrete time unit impulse and unit step. In particular the discrete-time impulse is the first difference of the discrete time step.

$$u[n] = \sum_{m}^{n} \delta(m) = \sum_{k=0}^{\infty} \delta(n-k)$$

The discrete-time step is the running sum of the unit sample.

The unit impulse sequence can be used to sample the value of signal n = 0.

$$x[n]\,\delta[n] = x[0]\,\delta[n]$$

$$x[n]\,\delta[n-n_o] = x[n_o]\,\delta[n-n_o]$$

Where $x[0]$ and $x[n_o]$ represent the amplitude of impulse function at $n = 0$ and $n = n_o$.

6.1.4 THE CONTINUOUS SIGNAL

Unit step function u(t)

$$u(t)=\begin{cases}0 & ; \quad t<0\\ 1 & ; \quad t>0\end{cases}$$

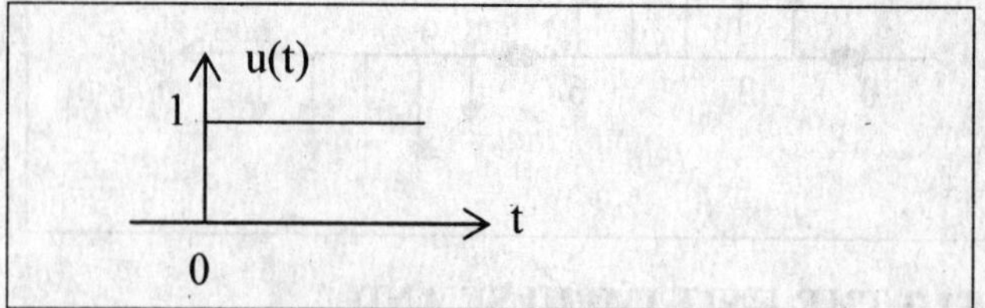

Discontinuous at t = 0

$$\delta(t)=\frac{du(t)}{dt}\ ;\ u(t)=\int_{-\infty}^{t} d(\tau)d\tau$$

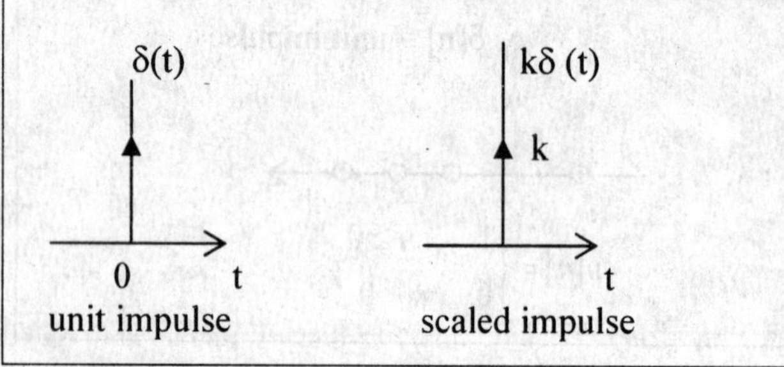

6.1.5 DISCRETE FUNCTION

Unit impulse function: The is the same function as delta function in analog.

$$\delta(t)=\begin{cases}1 & ; \ for \quad t=0\\ 0 & ; \ for \quad t\neq 0\end{cases}$$

$$\delta[n]=\begin{cases}1 & ; \ for \quad n=0\\ 0 & ; \ for \quad n\neq 0\end{cases}$$

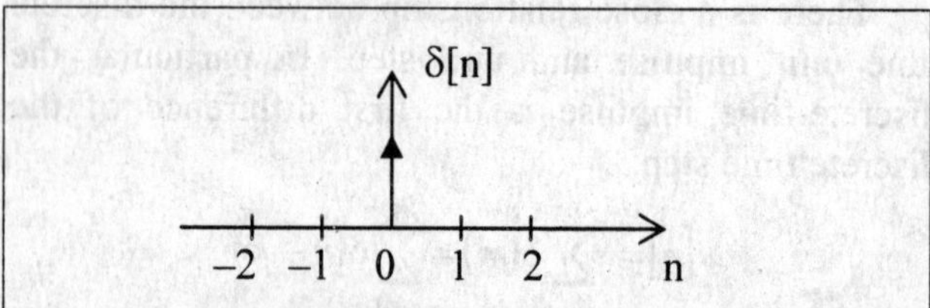

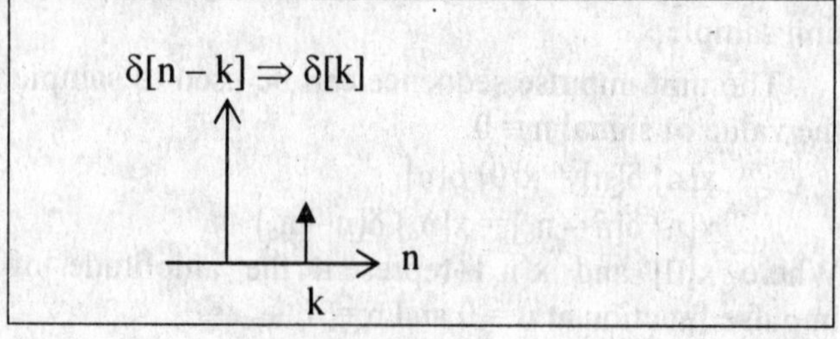

6.1.6 REPRESENTATION OF ANY DISCRETE TIME SIGNAL IN TERM OF IMPULSE

x[n] = x[–3] δ[n + 3] + n[– 2] δ[n + 2]
+ x[– 1] δ[n + 1] + x [0] δ [n]
+ x[1] δ[n – 1] + x[2] δ[n – 2]
+ x [3] δ[n – 3]

Where, x[–3] → is the amplitude value of the sequence at n = [– 3]

x[n] = 2 δ[n + 3] + 1 δ [n + 2]
+ 1 δ [n + 1] + …….. – 0.5 δ[n – 3]

$$\therefore\ x[n]=\sum_{k=-\infty}^{\infty} n[k]\delta[n-k]$$

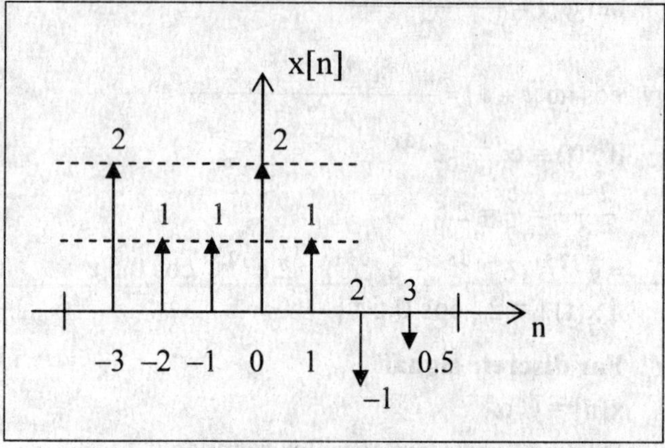

Example: Draw the signal sequence and also write the sequence for x[n] = [5, 2, –1, 3, 2, 0, 1, 5,7]. From n = – 4 to n + 4.

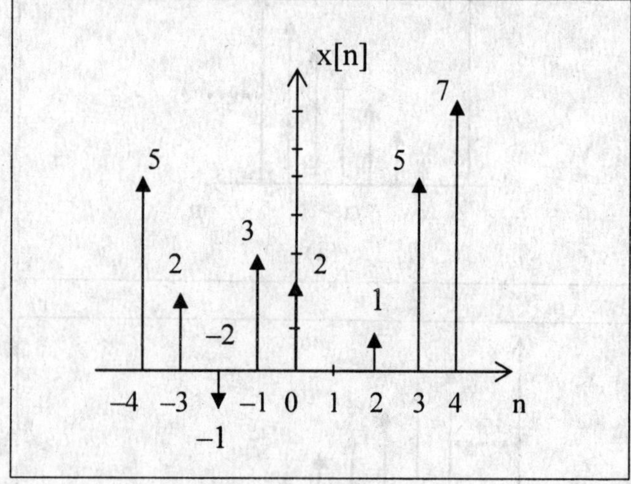

$$\therefore\ x[n]=\sum_{k=-4}^{4} n[k]\delta[n-k]$$

$= 5\,\delta\,[n+3] + 2\,\delta\,[n+3] - 1\,\delta[n+2]$
$+ 3\delta\,[n+1] + 2\,\delta[n] + 0\,\delta\,[n-1]$
$+ 1\,\delta[n-2] + 5\,\delta\,[n-3] + 7\,\delta\,[n-4]$

6.17 SIGNAL ENERGY AND POWER

v(t) and i(t) are continuous signal across R.

Instantaneous Power

$$p(t) = v(t)\cdot i(t) = \left(\frac{1}{R}\right)v^2(t)$$

The total energy $= \int_{t_1}^{t_2} p(t)\,dt = \int_{t_1}^{t_2} (1/R)\,v^2(t)\,dt$

Average power $= \frac{1}{t_2 - t_1}\int_{t_1}^{t_2} p(t)\,dt$

For continuous signal $x(t)$

Total energy $= \int_{t_1}^{t_2} |x(t)|^2\,dt$

For discrete signal x[n]

Total energy $= \sum_{n=n_1}^{n_2} |X[n]|^2$

Total energy of signal for $-\infty < t < +\infty$ or $-\infty < n < +\infty$

Continuous Time

$$E_\infty = \lim_{T\to\infty}\int_{-T}^{T} |x(t)|^2\,dt = \int_{-\infty}^{\infty} |x(t)|^2\,dt$$

Discrete Time $E_\infty = \lim_{T\to\infty}\sum_{n=-N}^{+N} |x[n]|^2 = \sum_{n=-\infty}^{+N} |\,x[n]\,|^2$

Time Average Power $P_\infty = \lim_{T\to\infty}\frac{1}{2T}\int_{-T}^{T} |\,x(t)\,|^2\,dt$

$$P_\infty = \lim_{N\to\infty}\frac{1}{2N+1}\sum_{n-\infty}^{N} |\,x(t)\,|^2$$

6.1.8 SYSTEMS

Physical system in the broadest sense are a inter connection of components, devices, or sub-systems.

A system can be viewed as a process in which input signals are transformed by the system or cause. The system to respond in some way, resulting in other signals as output examples are, signal processor or communication equipment.

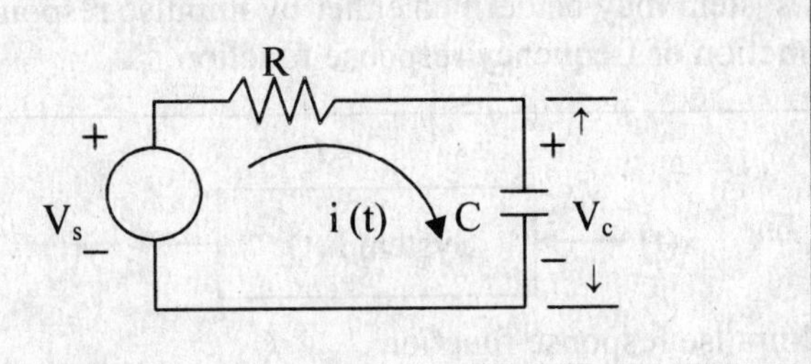

$$i(t) = \frac{V_s(t) - V_c(t)}{R} \tag{1}$$

$$i(t) = \frac{C\,d\,V_c(t)}{dt} \tag{2}$$

From equation, (1) and (2)

$$\frac{dV_c(t)}{dt} + \frac{V_c(t)}{RC} \tag{3}$$

$$= \frac{1}{RC}V_s(t)$$

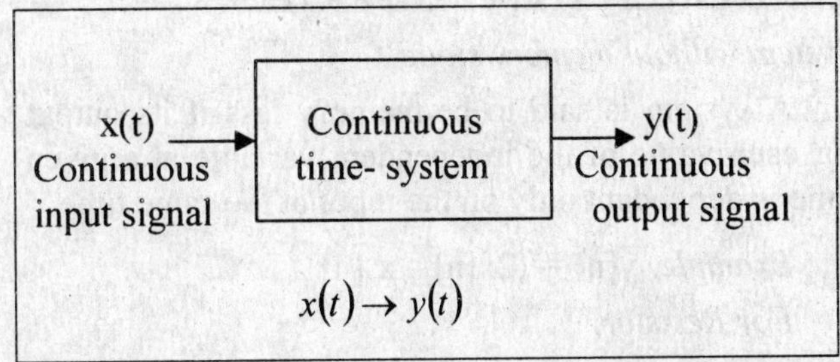

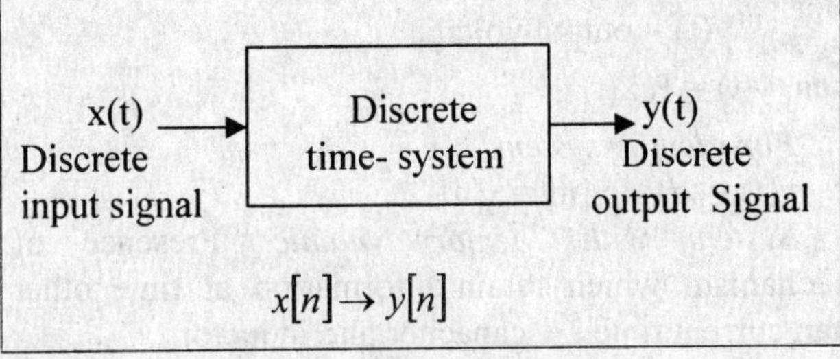

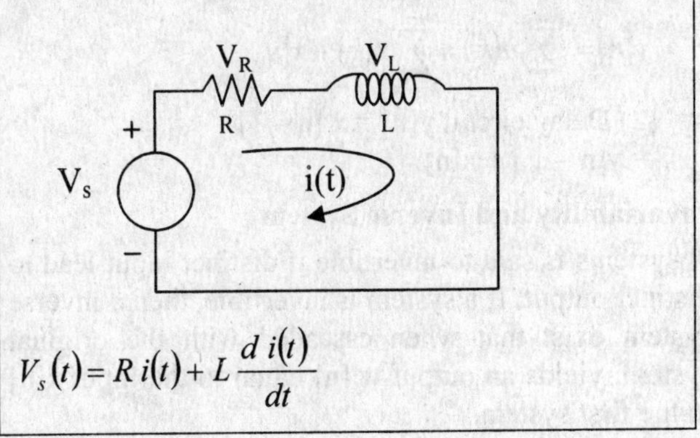

A system may be defined either by impulse response function or frequency response function.

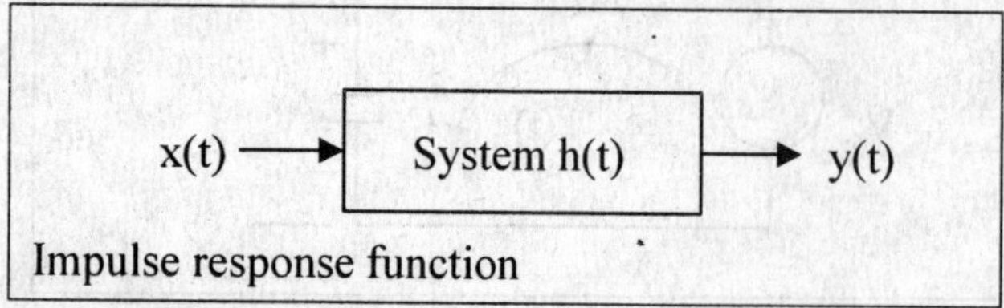

$y(t) = x(t) * h(t)$
Where, * is convolution

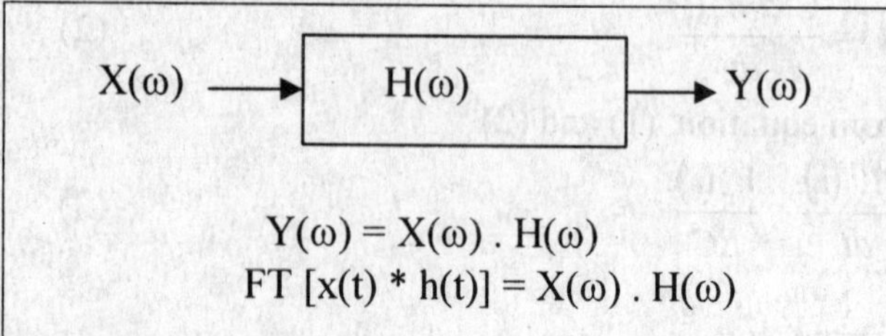

6.1.9 BASIC SYSTEM PROPERTIES

System without memory/static

A system is said to be memory less if its output for each value of the independent variable at a given time is dependent only on the input at the same time.

Example: $y[n] = \{2x[n] - x^2[n]\}^2$

For Resistor,

$x(t)$ = input current

$y(t)$ = output voltage

then, $y(t) = R\,x(t)$

For identity system

$y(t) = x(t)$; $y[n] = x[n]$

System with Memory/dynamic: Presence of mechanism which retain information at time other than current time i.e. capacitor and inductor

Example: accumulator or summer

$$y[n] = \sum_{k=-\infty}^{n} n(k) = \sum_{k=-\infty}^{n-1} x(k) + x[n]$$

(2) Delay circuit $y[n] = x[n-1]$
$= y[n-1] + x[n]$

Invariability and Inverse System

A systems is said to invertible if distinct input lead to distinct output. If a system is invertible, then a inverse system exist that when cascaded with the original system, yields an output w [n] equal to the input x[n] to the first system.

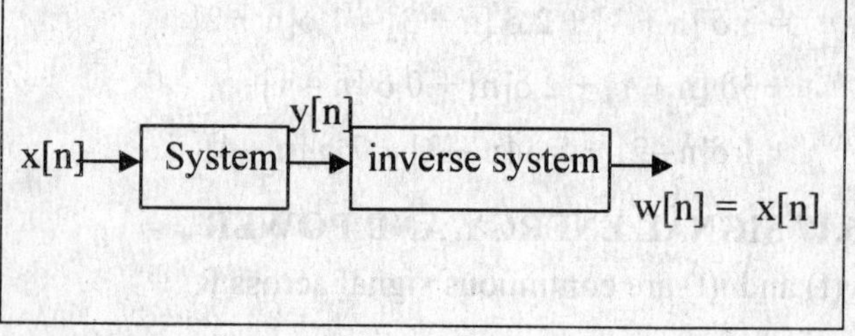

Example:

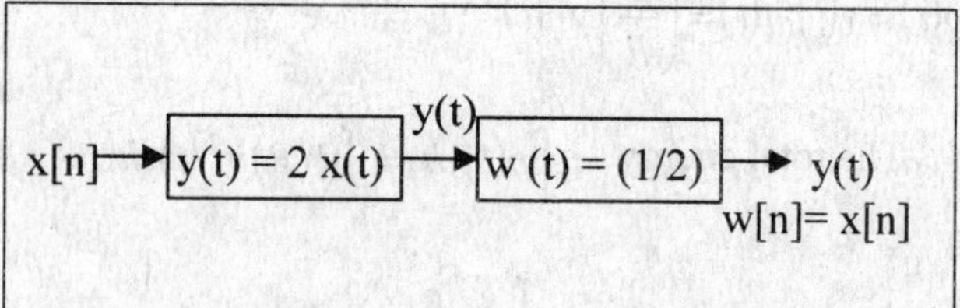

$$H_1 \rightarrow y[n] = \sum_{k=-\infty}^{n} x(k)$$

$H_2 \rightarrow w[n] = y[n] - y[n-1]$

Noninvertible System are

$y[n] = 0$ and $y(t) = x^2(t)$

1. The system that produce zero output sequence for input.
2. In which we cannot determine the sign of the input form knowledge of output.

Use: Encoder must be invertible in communication system.

Causality

A system is casual if the output at any time depends on value of the input at the present time and in the past. Such system is referred as **non - anticipative** as the system does not anticipate future values of the input.

The RC series circuit is casual as the capacitor voltage responds only to the present and past values of source voltage.

Example:

1. The accumulator and capacitor.
2. All memory less system are casual, since the output responds only to the current value of input.

Not casual if

$y[n] = x[n] - x[n+1]$

$y(t) = x(t+1)$

Where, $x[n+1]$ represent future input

Stability

A stable system is one in which small inputs lead to responds that do not diverge.

For the stable system, if the input to a stable system is bounded, then the output must also be bounded.

The stability of physical system generally results from the presence of mechanism that dissipate energy

Examples

1. $S_1 : y(t) = t\,x(t)$
   if $x(t) = u(t)$
   then $y(t) = t \Rightarrow$ unbounded $\Rightarrow$ unstable
2. $S_2 : y(t) = e^{x(t)}$ if $x(t)$ is bounded i.e.
   $|x(t)| < B$

   $\Rightarrow$ bounded or $\Rightarrow$ stable

Time Invariance

A system is time invariant if the behaviour and characteristics of the system are fixed over time.

For example, the RC series circuit is time invariant if the resistance and capacitance value are constant.

A system is time invariant if a time shift in the input signal result in the identical shift in the output signal.

If $x(t)$ input then $y(t)$ output

If $x(t - t_o)$ – input then $y(t - t_o)$ output.

Example

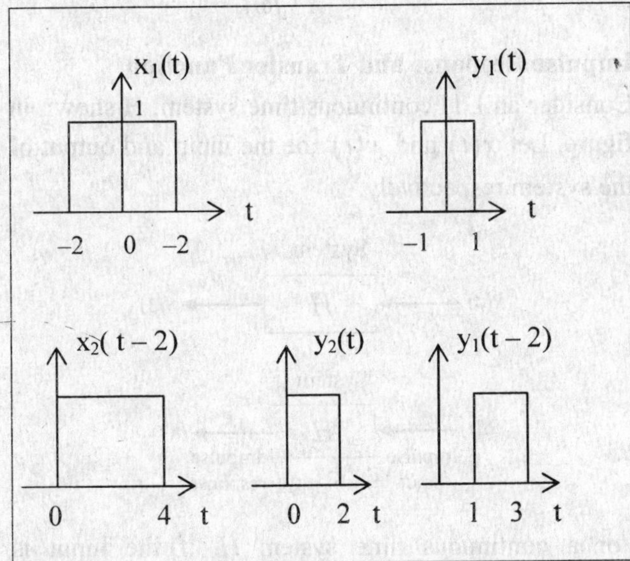

$y_2(t) \neq y_1(t - 2)$ ….. system is not time invariant

Linearity

A linear system in continuous time or discrete time, is a system that possesses the important property of super position, i.e.

1. The response to $x_1(t) + x_2(t)$ is $y_1(t) + y_2(t)$
2. The response to $ax_1(t)$ is a $y_1(t)$ where a is complex constant.

Example: $\left.\begin{array}{l} y(t) = x^2(t) \\ y(t) = \sin[x(t)] \end{array}\right\}$ Not linear

For linear system, an input which is zero for all time results in an output which is zero all time.

6.2 PERIODIC AND APERIODIC SIGNALS

A periodic signal has the property that it is unchanged by a time shift of T.

$x(t) = x(t + T)$ and

$x[n] = x[n + N]$

We can say that $x(t)$ is periodic with time period T.

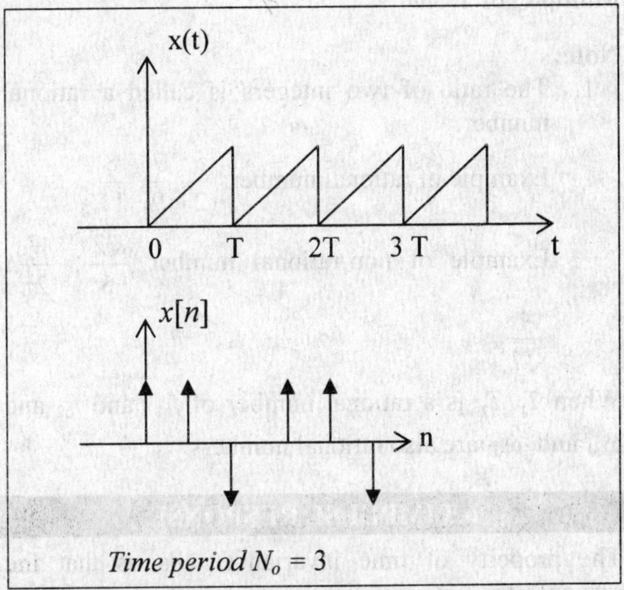

Time period $N_o = 3$

If this property is not satisfy then the signal is called aperiodic signal or non Periodic signal.

PERIODIC AND NON-PERIODIC

A periodic signal will have a definite pattern that repeats again and again over a certain period of time. Therefore the signal which satisfies the condition.

$x(t + T) = x(t)$ is called a periodic signal.

A signal which does not satisfy that condition, $x(t+T)=x(t)$ is called an **aperiodic or nonperiodic** signal.

In periodic signals, then term T is called the fundamental time period of the signal. Hence, inverse to T is called the fundamental frequency, f_o in cycles/sec or Hz, and $2\pi f_o = \omega_o$ is called the fundamental angular frequency in rad/sec.

The sinusoidal signals and complex exponential signals are always periodic with a periodicity of T, where, $T = \frac{1}{f_o} = \frac{2\pi}{\omega_o}$. The proof of this concept is given below.

When A continuous time signal is a mixer of two periodic signals with fundamentals time periods T_1 and T_2, then the continuous time signal will be periodic, if the radio of T_1 and T_2 (i.e. T_1 / T_2) is a rational number. Now the periodicity of the continuous time signal be the LCM (Least Common Multiple) of T_1 and T_2

Note:

1. The ratio of two integers is called a rational number.

   Example of rational number: $\frac{5}{2}, \frac{7}{9}, \frac{8}{11}$

   Example of non-rational number $\frac{\sqrt{2}}{5}, \frac{7}{2\pi}, \frac{4}{\sqrt{7}}$

When T_1/T_2 is a rational number of f_{o1} and f_{o2} and ω_{o1} and ω_{o2} are also rational numbers.

6.3 IMPULSE RESPONSE

The property of time invariance tells us that the response the time-invariant system to the time shifted unit impulse are simply time shifted version of one another.

For any discrete LTI system output is given by

$$y[n] = x[n] * h[n]$$

Where, $x[n] = \sum_{k=-\infty}^{\infty} n(k)\delta[n-k]$

and h[n] → response of δ[n]

$h_k[n]$ → respone of linear system to the shifted unit impulse δ[n – k]

$$\therefore\ y[n] = \sum_{k=-\infty}^{\infty} x(k)\, h_k[n]$$

$$= \sum_{k=-\infty}^{\infty} x(k)\, h[n-k]$$

Note: If x[n] → has n element

h[n] → m element

Then y[n] → [m + n – 1] element

6.4 TRANSFER FUNCTION AND FREQUENCY RESPONSE OF FIRST AND SECOND ORDER SYSTEMS

Transfer Function of LTI Continuous Time System in Frequency Domain

The ratio of fourier transform of output and the Fourier transform of input is called transfer function of LTI continuous time system in frequency domain

Let

$x(t)$ = Input to the continuous time system

$y(t)$ = Output of the continuous time system

$X(j\omega)$ = Fourier transform of $x(t)$

$Y(j\omega)$ = Fourier transform $y(t)$

Now, Transfer function $= \frac{Y(j\omega)}{X(j\omega)}$

Impulse Response and Transfer Function

Consider an LTI continuous time system. H shown in figure. Let $x(t)$ and $y(t)$ be the input and output of the system respectively.

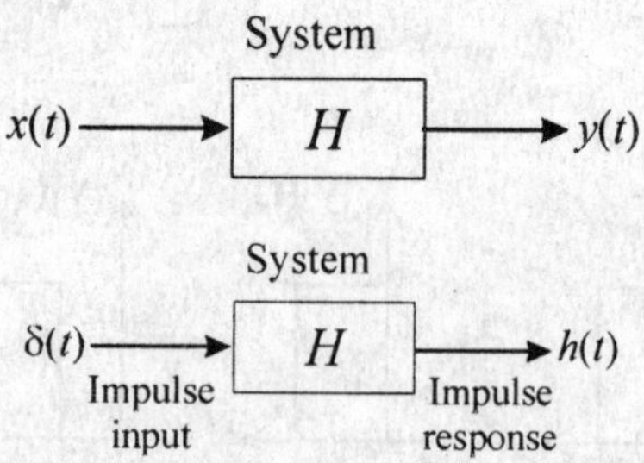

For a continuous time system H, if the input is impulse signal $\delta(t)$ as shown in figure, then the output is called impulse response, which is denoted by $h(t)$.

The important of impulse response s that the response for any input to LTI system is given by convolution of input and impulse response.

$$y(t) = x(t) * h(t)$$

When '*' is the symbol for convolution.

Mathematically, the convolution operation is as,

$$y(t) = x(t) * h(t)$$

$$= \int_{-\infty}^{+\infty} x(\tau) h(t-\tau) d\tau$$

Frequency Response of LTI Continuous Time System

The output $y(t)$ of an LTI continuous time system is given by convolution of $h(t)$ and $x(t)$.

$$y(t) = x(t) * h(t)$$

$$= h(t) * x(t) = \int_{-\infty}^{+\infty} x(\tau) h(t-\tau) d\tau \quad (1)$$

Consider a special class of input (sinusoidal input),

$$Ae^{j\omega t} = A(\cos \omega t + j \sin \omega t)$$

$$x(t) = A\, e^{j\omega t} \quad (2)$$

Where A = Amplitude,

ω = Angular frequency in rad/sec

$$\therefore\ x(t-\tau) = Ae^{j\omega(t-\tau)} \quad (3)$$

On substituting for $x(t-\tau)$ from equation (1) in equation (2) we get,

$$y(t) = \int_{-\infty}^{+\infty} h(\tau)\, A\, e^{j\omega(t-\tau)} d\tau$$

The function $H(j\omega)$ is complex quantity and so it can be expressed as magnitude frequency and phase function.

$$H(j\omega) = |H(j\omega)| \angle H(j\omega)$$

Where, $|H(j\omega)|$ = Magnitude function

$\angle H(j\omega)$ = Phase function

The sketch of magnitude function and phase function with respect to ω will given the frequency response graphically.

$$H(j\omega) = H_r(j\omega) + jH_i(j\omega)$$

Where, $H_r(j\omega)$ = Real part of $H(j\omega)$

$H(j\omega)$ Imaginary part of $H(j\omega)$

The magnitude function is defined as,

$$|H(j\omega)|^2 = H(j\omega)\, H^*(j\omega)$$

$$= [H_r(j\omega) + jH_i(j\omega)][H_r(j\omega) - jH_i(j\omega)]$$

Where, $H^*(j\omega)$ is complex conjugate of $H(j\omega)$

$$\therefore\ |H(j\omega)|^2 = H_r^2(j\omega) + H_i^2(j\omega)$$

$$\Rightarrow |H(j\omega)| = \sqrt{H_r^2(j\omega) + H_i^2(j\omega)}$$

The phase function is defined as

$$\angle H(j\omega) = Arg[H(j\omega)] = \tan^{-1}\left[\frac{H_i(j\omega)}{H_r(j\omega)}\right]$$

From equation ($H(j\omega) = \frac{Y(j\omega)}{X(j\omega)}$)

we can say that the frequency response of an LTI continuous time system is same as transfer function in frequency domain and so, the frequency response is also given by the ratio of Fourier transfer of output to Fourier transform of input.

$$\therefore \text{ Frequency response, } H(j\omega) = \frac{Y(j\omega)}{X(j\omega)}$$

Advantages of Frequency Response Analysis

1. The practical testing of systems can be easily carried with available sinusoidal signal generators and precise measurement equipments.
2. The transfer function of complicated system can be determined experimentally by frequency response tests.
3. The design and parameter adjustment is carried out more easily in frequency domain.
4. In frequency domain, the effects of noise disturbance and parameter variations are relatively easy to visualize and incorporate corrective measures.
5. The frequency response analysis and designs can be extended to certain non linear system.

6.4.1 INITIAL AND FIND CONDITION IN ELECTRONIC NETWORK

Case1: Initial condition i.e. t = 0

(1) Resistance

R

(a)

(2) Capacitor

\+ – ⇒ S· C

(a) (b)

V_o + – ⇒ V_o + – S.C

(3) Inductor

⇒ O.C

I_o ⇒ O.C I_o

Case (2), Final condition i.e. t → ∞ i..e study state

(1) Resistance

R

(2) Capacitor

(a) ⇒ O.C

(b) + – ⇒ O.C + – V_o

(3) Inductor

(a) ⇒ L S.C

I_o ⇒ S.C I_o

Equivalent circuit of L & C using Initial Condition & Laplace Transform

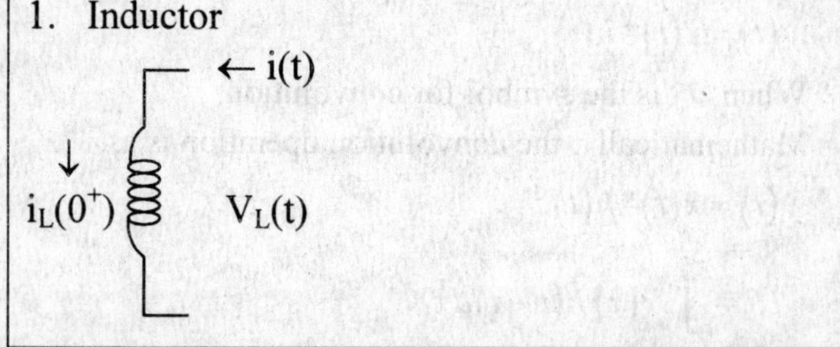

$$V_L(t) = L\frac{di(t)}{dt}$$

$$V_L(s) = L[s\ I(s)\ i_L(0^+)]$$

$$V_L(s) = sL\ \ I(s) - L\ i_L(0^+)] \qquad (1)$$

$$sL\ \ I(s) = V_L(s) + L\ i_L(0^+)$$

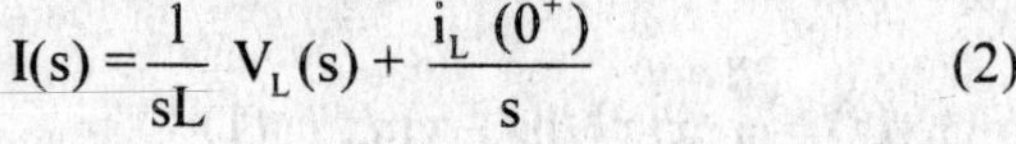

$$I(s) = \frac{1}{sL}\ V_L(s) + \frac{i_L\ (0^+)}{s} \qquad (2)$$

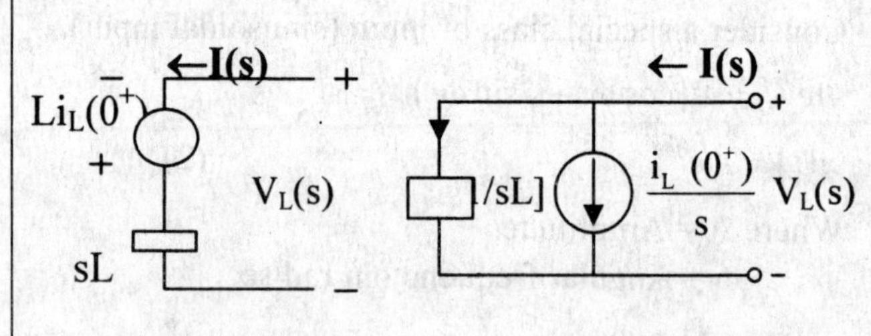

2. Capacitor

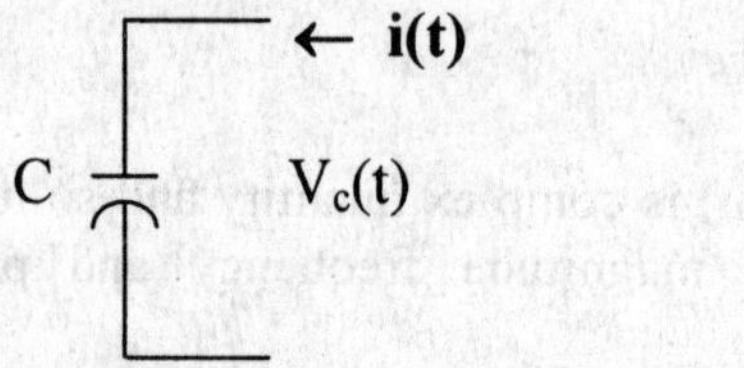

$$i(t) = C\frac{dV_c(t)}{dt}$$

$$I(s) = C\ [s\ V_c(s) - V_c(0^+)]$$

$$I(s) = sC\ V_c(s) - CV_c(0^+) \qquad (1)$$

$$sCV(s) = I(s) + C\ V_c(0^+)$$

$$V(s) = \frac{1}{sC}\ I(s) + \frac{V_c(0^+)}{s} \qquad (2)$$

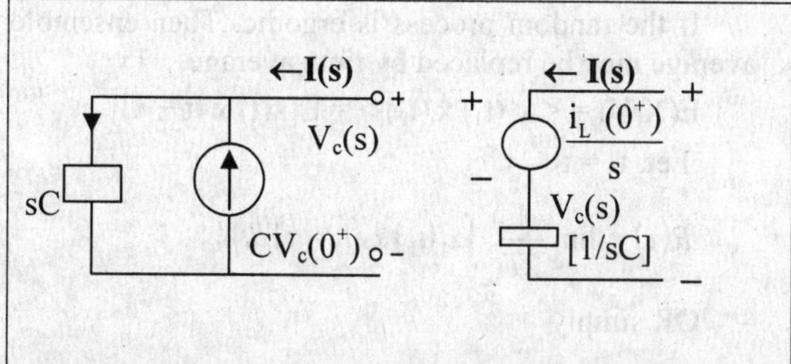

6.5 CONVOLUTION

Convolution is a mathematical operation and is useful for describing the input output relationship in a linear time variant system. It is an important analytical tool for the communication engineers.

The convolution $f(t)$ of two time functions $f_1(t)$ and $f_2(t)$ is defined by the following integral.

$$f(t)=\int_{\tau=-\infty}^{\infty} f_1(\tau)\, f_2(t-\tau)\, d\tau = f_1(t) * f_2(t)$$

FT $[f_1(t) * f_2(t)] = F_1(\omega) \,.\, F_2(\omega)$

Application

FT $[f_1(t) \,.\, f_2(t)] = (1/2\pi)\, [F_1(\omega) * F_2(\omega)]$

Physical Meaning of convolution.

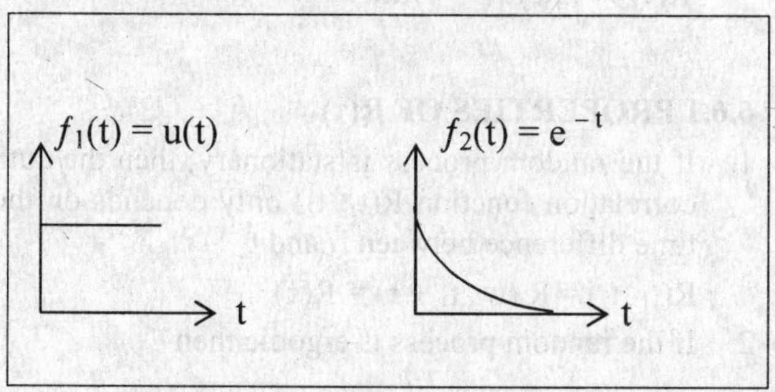

$$f_1(t) * f_2(t) = \int_0^t f_1(t-\tau)\, f_2(\tau)\, d\tau$$

Taking the Mirror image

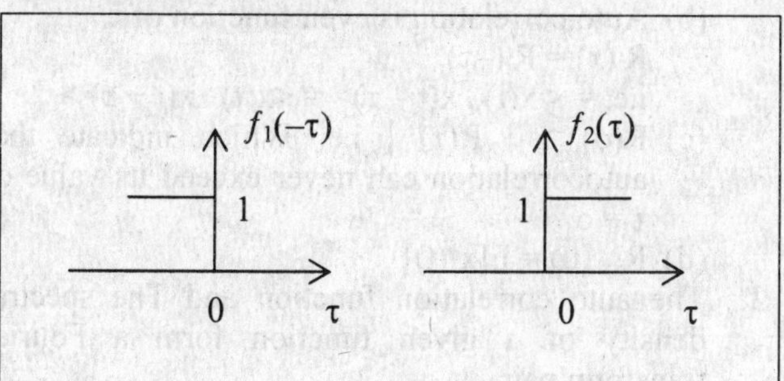

Shifting along the time axis

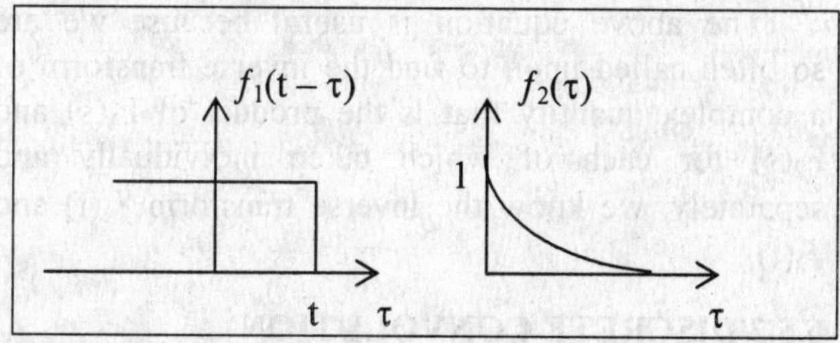

Multiplication

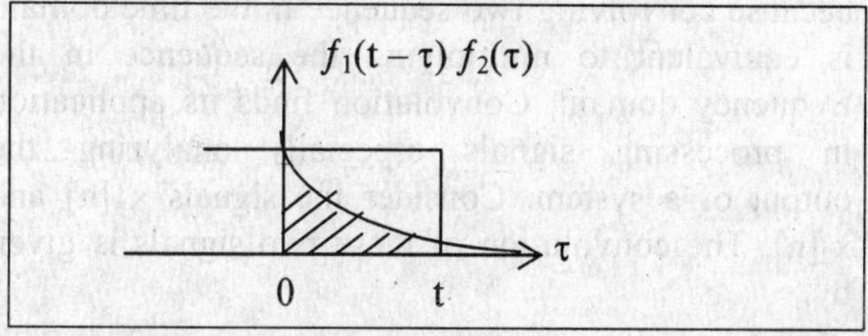

Integration of Shaded Portion of Area From Limit 0 To t

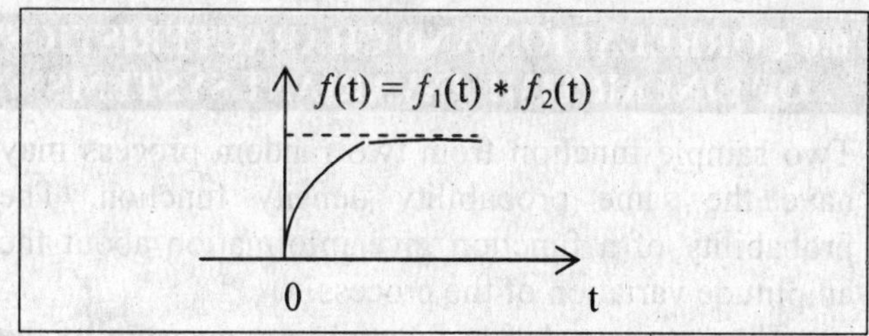

Use:

- The convolution theorem provides a convenient and signal method of finding the inverse laplaace transform of product of Laplace signal.
- This theorem also enable us to evaluate the response of a linear system arbitrary excitation in term of its to a unit impulse.

6.5.1 CONVOLUTION ON LAPLACE TRANSFORM BASIS

$$f_1(t) * f_2(t) = \int_0^t f_1(t-\tau) \cdot f_2(\tau)\, d\tau$$

$$f_1(t) * f_2(t) = \int_0^t f_2(t-\tau) \cdot f_1(\tau)\, d\tau$$

$f_1(t) * f_2(t) = f_2(t) * f_1(t)$

The word convolve means to roll together. The two function $f_1(t)$ and $f_2(t)$ are multiplied together in such a manner that one is continuously moving with time τ relative to the other.

LT $[f_1(t) * f_2(t)] = F_1(s). F_2(s)$

The above equation is useful because we are so often called upon to find the inverse transform of a complex quantity that is the product of $F_1(s)$ and $F_2(s)$ for each of which taken individually and separately, we know the inverse transform $f_1(t)$ and $f_2(t)$.

6.5.2 DISCRETE CONVOLUTION

Convolution is important in digital signal processing because convolving two sequence in the time domain is equivalent to multiplying the sequence in the frequency domain. Convolution finds its application in processing signals especially analyzing the output of a system. Consider the signals x1[n] and $x_2[n]$. The convolution of these two signals is given by.

$$x_3[n] = x_1[n] * x_2[n] = \sum_{k=-\infty}^{\infty} x_1(k)\, x_2(n-k)$$

6.6 CORRELATION AND CHARACTERISTICS OF LINEAR TIME INVARIANT SYSTEMS

Two sample function from two random process may have the same probability density function. The probability of a function give information about the amplitude variation of the process.

The auto correlation function is a measure of the similarities between one probability density function (pdf) with the same pdf when this is shifted by some time along the horizontal time axis.

Quantatively, the auto correlation function is a measure of regularity of random process.

Auto correlation function is defined as

$R(t_1, t_2) = E(XY)$

$= \int\int x\, y\, p(x, y)\, dx\, dy.$

Where x and y are random variable at time t_1, t_2 respectively.

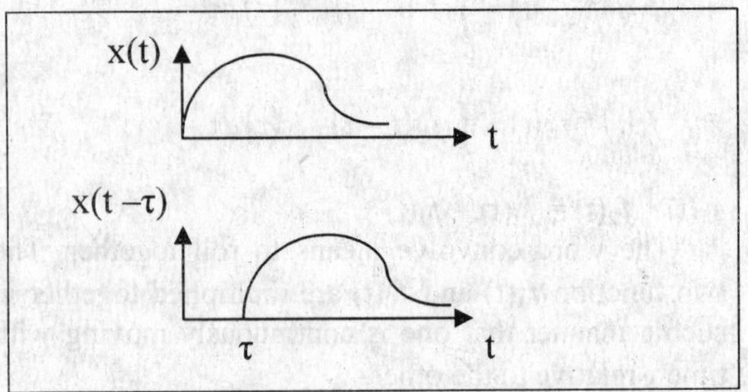

If the random process is ergodic. Then ensemble average may be replaced by time average.

$E(X\,Y) = \langle x_i(t_1)\, x_i(t_2)\rangle = E[x(t)\, x(t+\tau)]$

Let, $t_2 = t_1 - \tau$

$$R(\tau) = \lim_{T\to\infty} \frac{1}{2T} \int_{-T}^{T} x_1(t_1)\, x_i(t_1 - \tau)\, dt_1$$

OR simply

$$R(\tau) = \lim_{T\to\infty} \frac{1}{2T} \int_{-T}^{T} x(t)\, x(t-\tau)\, dt$$

Mathematical definition:

Case 1: Random signal

$$R(\tau) = \lim_{T\to\infty} \frac{1}{T} \int_{-T/2}^{T/2} x(t)\, x(t-\tau)\, dt.$$

Case 2: Periodic signal (Power signal)

$$R(\tau) = \frac{1}{T_o} \int_{-T_o/2}^{T_o/2} x(t)\, x(t-\tau)\, dt$$

Where, T_o = time period

Case 3: Pulse type of signal (Energy signal) i.e. Non periodic

$$R(\tau) = \int_{-\infty}^{\infty} x(t)\, x(t-\tau)\, dt$$

6.6.1 PROPERTIES OF R(τ)

1. If the random process is stationary, then the auto correlation function $R(t_1, t_2)$ only depends on the time difference between t_1 and t_2

   $R(t_1, t_2) = R(t_1, t_1 + \tau) = R(\tau)$
2. If the random process is ergodic then

   (a) $R(\tau) = \lim_{T\to\infty} \frac{1}{T} \int_{-T/2}^{T/2} x(t) \cdot x(t-\tau)\, dt.$

   $= \langle x(t)\, .\, x(t-\tau)\rangle$ --- Time average

   $= R(\tau)$

   (b) Auto correlation is even function of τ

   $R(\tau) = R(-\tau)$

   i.e. $= \langle x(t)\, .\, x(t-\tau)\rangle = \langle x(t)\, .\, x(t+\tau)\rangle$

   (c) $R(o) \geq |R(\tau)|$ i.e. Which indicate that autocorrelation can never exceed its value of $\tau = o$

   (d) $R_{xx}(0) = E[x^2(t)]$
3. The auto correlation function and The spectral density of a given function form a Fourier transform pair.

$R(\tau) \leftrightarrow G(\omega)$ ----Wiener Kintchen Theorem

Let, x(t) be a periodic function of time with time period T

$\therefore$ x(t) Can be written in term of exponential Fourier series

$$x(t) = \sum_{n=-\infty}^{\infty} F_n e^{jn\omega_o t}$$

$$\omega_o = \frac{2\pi}{T}$$

$$\text{Where, } F_n = \frac{1}{T}\int_0^T x(t) e^{-jn\omega_o t}\, dt.$$

Spectral density of periodic signal.

$$G(\omega) = 2\pi \sum_{n=-\infty}^{\infty} |F_n|^2 \delta(\omega - n\omega_o)$$

They are summarized in the following:

1. $\bar{x} = \langle x(t) \rangle$ is equal to the dc level of the signal
2. $[\bar{x}]^2 = \langle x(t) \rangle^2$ is equal to the normalized power in the dc component.
3. $R_{XX}(0) = \langle x^2(t) \rangle$ is equal to the total average normalized power.
4. $\bar{\sigma}x^2 = \langle x^2(t) \rangle$ is equal to the average normalized power in the time-varying or ac component of the signal.
5. $\bar{\sigma}x$ is equal to the rms value of the ac component of the signal.

6.6.2 CORRELATIONS AND POWER SPECTRAL DENSITIES

Autocorrelation R(τ)

The autocorrelation of X(t) is

$$R_{XX}(\tau) = E[X(t)\, X(t+\tau)]$$

Properties of $R_{XX}(\tau)$

1. $R_{XX}(-\tau) = R_{XX}(\tau)$
2. $|R_{XX}(\tau)| \le R_{XX}(0)$
3. $R_{XX}(0) = E[X^2(t)]$

CROSS-CORRELATION $R_{XY}(\tau)$

The cross-correlation of X(t) and Y(t) is

$$R_{XY}(\tau) = E[X(t)Y(t+\tau)]$$

Properties or $R_{XY}(\tau)$

1. $R_{XY}(-\tau) = R_{YX}(\tau)$
2. $|R_{XY}(\tau)| \le \sqrt{[R_{XX}(0)\, R_{YY}(0)]}$
3. $|R_{XY}(\tau)| \le (1/2)\, [R_{XX}(0) + R_{YY}(0)$

Auto-covariance $C_{XX}(\tau)$

The auto-covariance of X(t) is

$$C_{XX}(\tau) = E[\{X(t) - E[X(t)]\}\{X(t+\tau) - E[X(t+\tau)]\}] = R_{XX}(\tau) - m_x^2$$

6.6.3 CROSS-COVARIANCE $C_{XY}(\tau)$

The cross-covariance of X(t) and Y(t) is

$$C_{XY}(\tau) = E[\{X(t) - E[X(t)]\}\{Y(t+\tau) - E[Y(t+\tau)]\}] = R_{XY}(\tau) - m_x m_y$$

Two processes X(t) and Y(t) are called (mutually) orthogonal if $R_{XY}(\tau) = 0$

They are called uncorrelated if $C_{XY}(\tau) = 0$

6.6.4 CROSS SPECTRAL DENSITIES

The cross spectral densities (or cross power spectra) $S_{XY}(\omega)$ and $S_{YX}(\omega)$ of X(t) and Y(t) are defined by

$$S_{XY}(\omega) = \int_{-\infty}^{\infty} R_{XY}(\tau) e^{-j\omega\tau}\, d\tau$$

$$S_{YX}(\omega) = \int_{-\infty}^{\infty} R_{XY}(\tau) e^{-j\omega\tau}\, d\tau$$

The cross-correlations and cross spectral densities thus form Fourier transform pairs. Thus

$$S_{XY}(\tau) = \int_{-\infty}^{\infty} R_{XY}(\omega) e^{j\omega\tau}\, d\omega$$

$$R_{YX}(\tau) = \frac{1}{2\pi}\int_{-\infty}^{\infty} S_{YX}(\omega) e^{j\omega\tau}\, d\omega$$

Substituting the relation

$R_{XY}(\tau) = R_{YX}(-\tau)$

In equation and we, obtain

$S_{XY}(\omega) = S_{YX}(-\omega) = S^*_{YX}(\omega)$

6.6.5 DISCRETE TIME SYSTEM

A **discrete time system** is a device or algorithm that operates on a discrete time signal, called the input or excitation, according to some well defined rule, to propduce another discrete time signal called the output or the response of the system. We can be say that the input signal *x*(*n*) is transformed by the system into a signal *x*(*n*) is transformed by the system into a signal *y*(*n*) and the transformed can be expressed mathematically as shown in equation (1). The diagrammatic representation of discrete time system is shown in figure below

Response $y(n) = H\{x(n)\}$ (1)

Where H denoted the transformation (also called operator.)

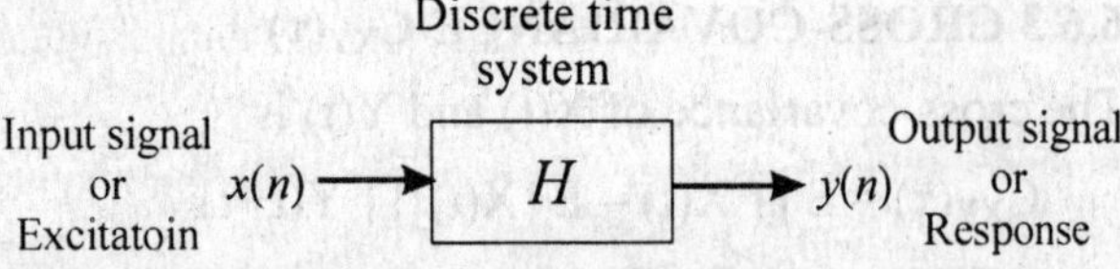

Figure: Representation of discrete time system

A discrete time system is linear if it obeys the principle of superposition and it is time invariant if its input-output relationship do not change with time. When a discrete time system satisfies the properties of linearity and time invariance then it is called an LTI system (Linear Time Invariant system).

Impulse Response

When the input to a discrete time system is a unit impulse $\delta(n)$ then the output is called an impulse response of the system and is denoted by $h(n)$.

$\therefore$ Impulse Response, $h(n) = H\{\delta(n)\}$ (2)

$\delta(n) \longrightarrow \boxed{H} \longrightarrow h(t)$

Figure: Discrete time system with impulse input.

Mathematical Equation Governing Discrete Time System

The response of a discrete time system at any time instant depends on the present input, past inputs or outputs.

Let us consider the response at $n = 0$. Let us assume a related system and so at $n = 0$, there is no past input or output . Therefore the response at $n = 0$, is a function of present input alone.

i.e. $y(0) = f[x(0)]$

Let us consider the response at $n = 1$. Now the present input is $x(1)$, the past input is $x(0)$ and past output is $y(0)$. Therefore the response at $n = 1$, is a function of $x(1)$, $x(0)$, $y(0)$

i.e. $y(1) = f[y(0), x(1), x(0)]$

Let us consider the response at $n = 2$. Now the present input is $x(2)$, the past inputs are $x(1)$ and $x(0)$ and past output are $y(1)$ and $y(0)$. Therefore the response at $n = 2$ is a function of $x(2)$, $x(1)$, $x(0)$, $y(1)$, $y(0)$

i.e. $y(2) = f[y(1), x(0), x(2), x(1), x(0)]$

Similarly , at $n = 3$

$y(3) = f[y(2), y(1), y(0), x(3), x(2), x(1), x(0)]$

at

$$y(4) = f\begin{bmatrix} y(3), y(2), y(1), y(0), \\ x(4), x(3), x(2), x(1) \end{bmatrix} \text{ as so on.}$$

In general , at any time instant n,

$$y(n) = f\begin{bmatrix} y(n-1), y(n-2), y(n-3),... \\ y(1), y(0), x(n), x(n-1), x(n-2) \\ , x(n-3),...x(1), x(0) \end{bmatrix} \quad (3)$$

For and LTI system, the response $y(n)$ can be expressed as a weighted summation of dependent terms. Therefore the equation (3) can be written as.

$$y(n) = \begin{bmatrix} -a_1\, y(n-1) - a_2\, y(n-2) - a_3(y-3) - ... \\ +b_0\, x(n) + b_1 x(n+1) \\ +b_2 x(n-2) + b_3 x(n-3) + ... \end{bmatrix} \quad (4)$$

Where, a_1, a_2, a_3, ... and b_0, b_1, b_2, b_3 , ... are constants.

Note: Negative constants are inserted for output signals, because output signals are feedback from output to input. Positive constants are inserted for input signals, because input signals are feed forward from input to output.

Practically, the response $y(n)$ at any time instant n, may depend on N number of past outputs, present input and M number of past inputs where $M \leq N$. Hence the equation (4) can be written as

$$y(n) = \begin{bmatrix} -a_1\, y(n-1) a_2 y(n-2) - a_3\, y(n-3) - ... \\ -a_N\, y(n-N) + b_0\, x(n) + b_1\, x(n-1) + \\ b_2\, x(n-2) + b_3\, x(n-3) + ... b_M\, x(n-M) \end{bmatrix}$$

$$\therefore\ y(n) = -\sum_{m=1}^{N} a_m\, y(n-m) + \sum_{m=0}^{M} b_m\, x(n-m) \qquad (5)$$

The equation (5) is a constant coefficient difference equation, governing the input –output relation of an LTI discrete time system.

In equation (5) the value of "N" gives the **order** of the system.

Block Diagram and signal flow Graph Representation of Discrete Time system

The discrete time system can be represented diagrammatically by **block diagram** or **signal flow graph**. These diagrammatic representations are useful for physical implementation of discrete time system in hardware or software.

Basic Elements of Block diagram and Signal Flow graph

Element :	Block Diagram/Signal flow graph representation
Adder	$x_1(n)$ → (+) → $x_1(n)+x_2(n)$, with $x_2(n)$ input; signal flow graph: $x_1(n)$ —1→ $x_1(n)+x_2(n)$ ←1— $x_2(n)$
Constant multiplier	$x(n)$ → [a] → $ax(n)$; signal flow graph: $x(n)$ —a→ $ax(n)$

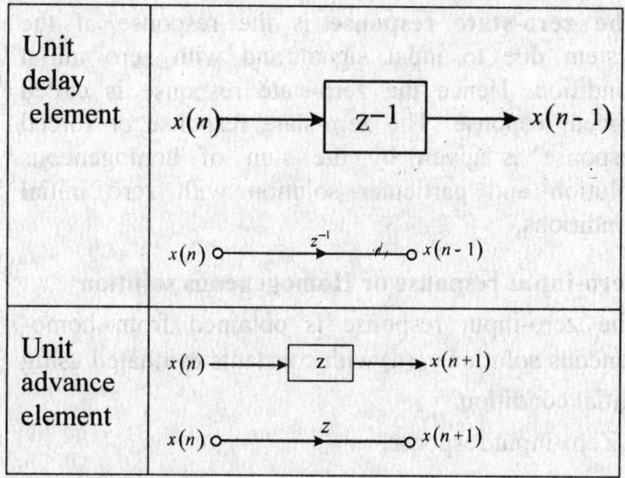

6.6.6 RESPONSE OF LTI DISCRETE TIME SYSTEM IN TIME DOMAIN

The general equation governing an LTI discrete time system is,

$$y(n) = -\sum_{m=1}^{N} a_m y(n-m) + \sum_{m=0}^{M} b_m x(n-m)$$

$$y(n) + \sum_{m=1}^{N} a_m y(n-m) = \sum_{m=0}^{M} b_m x(n-m)$$

$$\sum_{m=0}^{N} a_m y(n-m) = \sum_{m=0}^{M} b_m z(n-m) \text{ with } a_0 = 1 \qquad (1)$$

The solution of the difference equation (1) is the response $y(n)$ of LTI system, which consists of two parts. In mathematics, the two parts of the solution y(n) are homogeneous solution $y_h(n)$ and particular solution $y_p(n)$.

$$\therefore \text{Response, } y(n) = y_h(n) + y_p(n) \qquad (2)$$

The homogeneous solution is the response of the system when there is no input. The particular solution $y_p(n)$ is the solution of difference equation for specific input signal x(n)for $n \geq 0$.

In signal and system, the two part of the solution y(n) are called zero-input response $y_{zi}(n)$ and zero-state response $y_{zs}(n)$.

$$\therefore \text{Response, } y(n) = y_{zi}(n) + y_{zs}(n). \qquad (3)$$

The **zero-input response** is mainly due to initial conditions (or initial stored energy) in the system. Hence zero-input response in also called **free response or natural response** .The zero-input response is given by homogeneous solution with constants evaluated using initial conditions.

The zero-state response is the response of the system due to input signal and with zero initial condition. Hence the zero-state response is called forced response. The zero-state response or forced response is given by the sum of homogeneous solution and particular solution with zero initial conditions.

Zero-input response or Homogeneous solution

The zero-input response is obtained from homogeneous solution $y_h(n)$ with constants evaluated using initial condition.

∴ Zero-input response,

$$y_{zi}(n) = y_h(n)\big|_{\text{with constants evaluated using initial conditions}}$$

The homogeneous solution is obtained when $x(n) = 0$. Therefore the homogeneous solution is the solution of the equation,

$$\sum_{m=0}^{N} a_m y(n-m) = 0 \qquad (4)$$

Let us assume that the solution of equation (4) is in the form of an exponential.

$y(n) = \lambda^n$

On substituting $y(n) = \lambda^n$ in equation (4) we get,

$$\sum_{m=0}^{N} a_m \lambda^{n-m} = 0$$

On expanding the above equation (by taking $a_0 = 1$), we get,

$$\lambda^n + a_1\lambda^{n-1} + a_2\lambda^{n-2} + \ldots + a_{N-1}\lambda^{n-(N-1)}\lambda^{n-N} = 0$$

$$\lambda^{N-n}(\lambda^N + a_1\lambda^{N-1} + a_2\lambda^{N-2} + \ldots + a_{N-1}\lambda + a_N) = 0$$

Now, the characteristic polynomial of the system is given by,

$$\lambda^N + a_1\lambda^{N-1} + a_1\lambda^{N-2} + \ldots + a_{N-1}\lambda + a_N = 0$$

The characteristic polynomial has N roots, which are denoted as $\lambda_1, \lambda_2, \ldots \lambda_N$.

The characteristic polynomial may be distinct real roots, repeated real roots or complex. The assumed solutions for various types of roots are given below.

Distinct Real Roots

Let the roots $\lambda_1, \lambda_2, \lambda_3, \ldots \lambda_N$ be distinct real roots. Now the homogeneous solution will be in the form,

$$y_h(n) = \begin{bmatrix} (C_1 + C_2 n + C_3 n^2 + \ldots + C_p n^{p-1})\lambda_1^n \\ +C_{p+1}\lambda_{p+1}^n \quad + \ldots C_N\lambda_N^n \end{bmatrix}$$

Where, $C_1, C_2, C_3, \ldots C_N$ are constants that can be evaluated using initial conditions.

Repeated Real Roots

Let one of the real roots λ_1 repeats p times and the remaining (N–p) roots are distinct real roots. Now, the homogeneous solution is in the form,

$$y_h(n) = \begin{bmatrix} (C_1 + C_2 n + C_3 n^2 + \ldots + C_p n^{p-1})\lambda_1^n \\ +C_{p+1}\lambda_{p+1}^n \quad + \ldots + C_N\lambda_N^n \end{bmatrix}$$

Where, $C_1, C_2, C_3, \ldots C_N$ are constants that can be evaluated using initial conditions.

Complex Roots

Let the characteristic polynomial has a pair of complex roots λ and λ^* and the remaining (N-2) roots be distinct real roots. Now, the homogeneous solution will be in the form,

$$y_h(n) = r^n\left[C_1\cos n\theta + C_2\sin\theta\right] + C_3\lambda_3^n + C_4\lambda_4^n + \ldots + C_N\lambda_N^n$$

$$\lambda = a + jb,\ \lambda^* = a - jb$$

$$r = \sqrt{a^2 + b^2},\ \theta = \tan^{-1}\frac{b}{a}$$

$C_1, C_2, C_3 \ldots C_N$ are constants that can be evaluated using initial conditions.

Particular Solution

The particular solution, $y_p(n)$ is the solution of the difference equation for specific input signal x(n) for $n \geq 0$. since the input signal may have different form, the particular solution depends on the form or type of the input signal x(n).

If x(n) is constant, then $y_p(n)$ is also a constant.

Zero-State Response

The zero-state response or forced response is obtained from the sum of homogeneous solution and particular solution and evaluating the constants with zero initial conditions.

∴ Zero-state response,

$$y_{zs}(n) = y_h(n) + y_p(n)\big|_{\text{with constanst } C_1, C_2, \ldots C_N \text{ evaluated with zero initial conditions}}$$

Total response

The total response of discrete time system can be obtained by the following two methods.

Method-1

The total response is given by sum of homogeneous solution and particular solution.

$\therefore$ Total response, $y(n) = y_h(n) + y_p(n)$

6.6.7 ANALYSIS OF LTI DISCRETE TIME SYSTEM USING DISCRETE TIME FOURIER TRANSFORM (DTFT)

Transfer Function of LTI Discrete Time System in Frequency Domain (DTFT)

The ratio of Fourier transform of output and the Fourier transform of input is called transfer function of LTI discrete time system in frequency domain.

Let, $x(n)$ = Input to the discrete time system

$y(n)$ = Output of the discrete time system

$\therefore X(e^{j\omega})$ = Fourier transform of x(n)

$Y(e^{j\omega})$ = Fourier transform of y(n)

Now, Transfer function $= \dfrac{Y(e^{j\omega})}{X(e^{j\omega})}$ (1)

The transfer function of an LTI discrete time system in frequency domain can be obtained from the difference equation governing the input-output relation of the LTI discrete time system given below

$$y(n) = -\sum_{m=1}^{N} a_m y(n-m) + \sum_{m=0}^{M} b_m x(n-m)$$

On taking Fourier transform of above equation and rearranging the resultant equation as a ratio of $Y(e^{j\omega})$ and $X(e^{j\omega})$, the transfer function of LTI discrete time system in frequency domain is obtained.

Impulse Response and Transfer Function

Let, $x(n)$ = Input of an LTI discrete time system

$y(n)$ = Output/Response of the LTI discrete time system for the input x(n)

$h(n)$ = Impulse response (I.e., response for impulse input)

Now, the response the y(n) of the discrete time system is given by convolution of input and impulse response,

$$y(n) = x(n) * h(n) = \sum_{m=-\infty}^{+\infty} x(m)\, h(n-m) \quad (2)$$

Let, $FT\{y(n)\} = Y(e^{j\omega})$

$FT\{x(n)\} = X(e^{j\omega})$

$FT\{h(n)\} = H(e^{j\omega})$

Now by convolution theorem of Fourier transform,

$$FT\{x(n) * h(n)\} = X(e^{j\omega})\, H(e^{j\omega}) \quad (3)$$

Using equation (2), the equation (3) can be written as,

$FT\{y(n)\} = X(e^{j\omega})\, H(e^{j\omega})$

$Y(e^{j\omega}) = X(e^{j\omega})\, H(e^{j\omega})$

$$H(e^{j\omega}) = \frac{Y(e^{j\omega})}{X(e^{j\omega})} \quad (4)$$

From equation (1) and (4) we can say that the **transfer function** of a discrete time system in frequency domain is also given by discrete time Fourier transform of impulse response.

Response of LTI Discrete Time System using Discrete Time Fourier Transform (DTFT)

Consider the transfer function of LTI discrete time system. $H(e^{j\omega}) = \dfrac{Y(e^{j\omega})}{X(e^{j\omega})}$

Now, response in frequency domain,

$$Y(e^{j\omega}) = X(e^{j\omega})\, H(e^{j\omega}) \quad (5)$$

On taking inverse Fourier transform of equation (5) we get,

$$Y(n) = FT^{-1}\left[X(e^{j\omega})H(e^{j\omega})\right] \quad (6)$$

From the equation (6) we can say that the output y(n) is given by the inverse Fourier transform of the product of $X(e^{j\omega})$ and $H(e^{j\omega})$.

Since the transfer function is defined with zero initial conditions, the response of discrete time system.

Frequency Response of LTI Discrete Time System

The output $y(n)$ of LTI system is given by convolution of $h(n)$ and $x(n)$.

$y(n) = x(n)*h(n)$

$= h(n)*x(n)$

$$= \sum_{m=-\infty}^{+\infty} h(m)x(n-m) \quad (7)$$

Consider a special class of input (sinusoidal input),

$$x(n) = A\,e^{j\omega n}\,;\ -\infty < n < \infty \quad (8)$$

Where,
A = Amplitude
ω = Arbitrary frequency in the interval $-\pi$ to $+\pi$

$$x(n-m) = Ae^{j\omega(n-m)} \quad (9)$$

On substituting for $x(n-m)$ form equation (8) in equation (7) we get,

$$y(n) = \sum_{m=-\infty}^{+\infty} h(m)\,Ae^{j\omega(n-m)}$$

$$= \sum_{m=-\infty}^{+\infty} h(m)\,Ae^{j\omega n}e^{-j\omega m}$$

$$= A\,e^{j\omega n}\sum_{m=-\infty}^{+\infty} h(m)\,e^{-j\omega m} \quad (10)$$

By the definition of Fourier transform,

$$H(e^{j\omega}) = FT\{h(n)\}$$

$$= \sum_{m=-\infty}^{+\infty} h(n)\,e^{-j\omega m} \quad (11)$$

Using equation (8) and (11), the equation (10) can written as, $y(n) = x(n)\,H(e^{j\omega})$

From equation (12) we can say that if a complex sinusoidal signal is given as input signal to an LTI system, then the output is a sinusoidal of the same frequency modified by $H(e^{j\omega})$. Hence $H(e^{j\omega})$ is the called the **frequency response** of the system.

The $H(e^{j\omega})$ produces a change in the amplitude and phase of the input signal .An LIT system is characterized in the frequency domain by its frequency response. The function $H(e^{j\omega})$ is a complex quantity and so it can be expressed as magnitude function and phase function.

$$\therefore H(e^{j\omega}) = \left|H(e^{j\omega})\right| \angle H(e^{j\omega})$$

Where, $\left|H(e^{j\omega})\right|$ = Magnitude function

$\angle H(e^{j\omega})$ = Phase function

The sketch of magnitude function and phase function with respect to ω will give the frequency response graphically.

Let, $H(e^{j\omega}) = H_r(e^{j\omega}) + jH_i(e^{j\omega})$

Where $H_r(e^{j\omega})$ = Real part of $H(e^{j\omega})$

$H_r(e^{j\omega})$ = Imaginary part of $H(e^{j\omega})$

The magnitude function is defined as,

$$\left|H(e^{j\omega})\right|^2 = H(e^{j\omega})H^*(e^{j\omega})$$

$$= [H_r(e^{j\omega}) + jH_i(e^{j\omega})][H_r(e^{j\omega}) - jH_i(e^{j\omega})]$$

Where, $H^*(e^{j\omega})$ is complex conjugate of $H(e^{j\omega})$

$$\therefore \left|H(e^{j\omega})\right|^2 = H_r^2\,(e^{j\omega}) + H_i^2\,(e^{j\omega})$$

$$\Rightarrow \left|H(e^{j\omega})\right| = \sqrt{H_r^2\,(e^{j\omega}) + H_i^2\,(e^{j\omega})}$$

The phase function is defined as,

$$\angle H(e^{j\omega}) = Arg[H(e^{j\omega})] = \tan^{-1}\left[\frac{H_i(e^{j\omega})}{H_r(e^{j\omega})}\right]$$

The drawback in frequency response analysis using Fourier transform is that the frequency response is a continuous function of ω and so it cannot be processed by digital system. This drawback is overcome in Discrete Fourier Transform (DFT)

From equation (97) we can say that the frequency response $H(e^{j\omega})$ of an LTI system is same as transfer function in frequency domain and so, the frequency response is also given by the ratio of Fourier transform of output to Fourier transform of input. i,e

Frequency response $H(j\omega) = \dfrac{Y(e^{j\omega})}{X(e^{j\omega})}$

Properties of Frequency Response

1. The frequency response is periodic function of ω with a period of 2π.
2. If $h(n)$ is real, then the magnitude of $H(e^{j\omega})$ is symmetric and phase of $H(e^{j\omega})$ is antisymmetric over the interval $0 \le \omega \le 2\pi$.
3. If $h(n)$ is complex, then the real part of $H(e^{j\omega})$ is symmetric and the imaginary part of $h(n)H(e^{j\omega})$ is antisymmetric over the interval $0 \le \omega \le 2\pi$
4. The impulse $h(n)$ is discrete, whereas the frequency response $H(e^{j\omega})$ is continuous function of ω.

Frequency Response of First Order Discrete Time System

A first order discrete time system is characterized by the difference equation,

$$y(n) = x(n) + a(n-1) \quad (14)$$

On taking Fourier transform of equation (14) we get

$$y(e^{j\omega}) = x(e^{j\omega}) + a\, e^{-j\omega}\, Y(e^{j\omega})$$

or, $y(e^{j\omega}) - a\, e^{-j\omega}\, Y(e^{j\omega}) = X(e^{j\omega})$

$$\therefore\ Y(e^{j\omega})\left[1 - a\, e^{-j\omega}\right] = X(e^{j\omega})$$

or, $H(e^{j\omega}) = \dfrac{Y(e^{j\omega})}{X(e^{j\omega})} = \dfrac{1}{1 - ae^{-j\omega}} \quad (15)$

The equation (15) is the frequency response of first order system. The frequency response can be expressed graphically as two functions: Magnitude function and Phasé function.

Frequency Response of Second Order Discrete Time System

A second order discrete time system is characterized by the different equation

$$y(n) = \begin{bmatrix} 2r\cos\omega_o(n-1) - r^2\, y(n-2) \\ + x(n) - r\cos\omega_o\, x(n-1) \end{bmatrix}$$

Let $a = -r\cos\omega_o$

$\alpha = -2r\cos\omega_o$,

$\beta = r^2$

$$y(n) = \begin{bmatrix} -\alpha\, y(n-1) - \beta\, y(n-2) \\ +x(n) + a\, x(n-1) \end{bmatrix} \quad (16)$$

On taking Fourier transform of the equation (16) we get,

$$Y(e^{j\omega}) = \begin{bmatrix} -\alpha\, e^{-j\omega}\, Y(e^{j\omega}) - \beta\, e^{-j2\omega}\, Y(e^{j\omega}) \\ +X(e^{j\omega}) + a\, e^{-j\omega}\, X(e^{j\omega}) \end{bmatrix}$$

$$Y(e^{j\omega}) = \begin{bmatrix} Y(e^{j\omega}) + \alpha\, e^{-j2\omega}\, Y(e^{j\omega}) \\ +\beta e^{-j2\omega} Y(e^{j\omega}) = X(e^{j\omega}) + a\, e^{-j\omega}\, X(e^{j\omega}) \end{bmatrix}$$

$$Y(e^{j\omega})\left[1 + \alpha\, e^{-j\omega} + \beta\, e^{-j2\omega} = X(e^{j\omega})(1 + a\, e^{-j\omega})\right]$$

$$H(e^{j\omega}) = \frac{Y(e^{j\omega})}{X(e^{j\omega})}$$

$$= \frac{1 + a\, e^{-j\omega}}{1 + \alpha\, e^{-j\omega} + \beta e^{-2j\omega}}$$

The equation (16) is the frequency response of second order system. The frequency response can be expressed graphically as two function: Magnitude function and phase function.

LINEAR TIME-INVARIANT (LTI)

Continuous-Time and Discrete-Time system

A system which possesses two properties namely linearity and time invariance is called linear time invariant (LTI) system. These two properties, linearity and time invariance, play a fundamental role in signal and system analysis. Many physical processes have these properties and thus can be modeled as linear time invariant (LTI) system. LTI systems can be analyzed in adequate detail with the help of a set of powerful tools which are already available. These tolls are the basis of signal and system analysis. LTI systems are of two types viz., continuous time and discrete-time. One of the primary reasons why linear time invariant (LTI) systems are amenable to analysis is that any such system has the super position property. LTI systems are characterized by their impulse response in both continuous time and discrete time and these are represented as linear combinations of delayed impulses. The complete characterization of any LTI system in terms of its impulse response is performed by convolution integral in the continuous time case and by the convolution sum indiscrete time case. A continuous time LTI system is described by linear constant–coefficient differential equations. Similarly a described- time (LTI) system is described by linear constant coefficient difference equations.

Continuous time linear time invariant system

A system of which both input and output are continuous time signals and possesses two properties namely linearity and time invariance is referred to as continuous time linear time invariant (LTI) system.

These systems are characterized either by their impulse response h(t) or by their transfer function H(s).

Continuous time signals are represented mathematically as the superposition of scaled and shifted continuous time unit impulse functions δ (t).

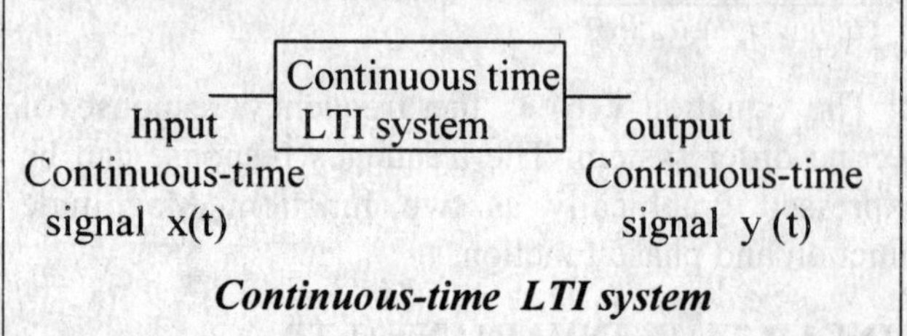

Continuous-time LTI system

Continuous time input signal x(t) is represented in terms of δ(t) as

$$x(t) = \int_{\tau=-\infty}^{\infty} x(\tau)\,\delta(t-\tau)\,d\tau$$

Continuous time impulse response h(t) is the output of a continuous time system when applied input x(t) is 0 initially relaxed. For an initially relaxed system all the initial conditions are zero.

Unit Impulse Response for Continuous Time LTI Systems

Output of a continuous time LTI system is called unit impulse response when we apply an input equal to unit impulse function δ (t).

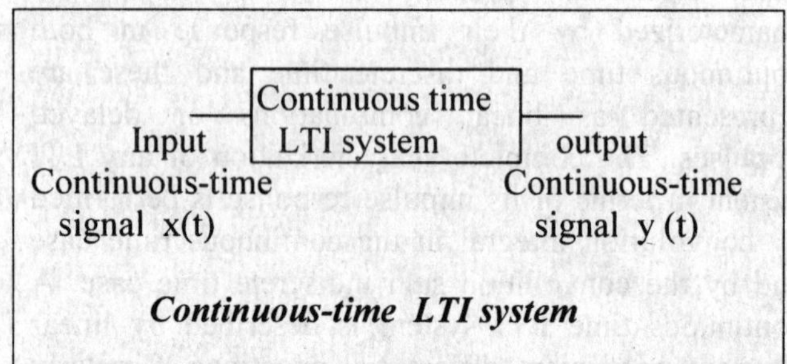

Continuous-time LTI system

If input x(t) = δ (t) is unit impulse function then output y(t) = h(t) is unit impulse response where unit impulse function δ(t) is given by

$$\delta(t) = \begin{cases} 1, & t = 0 \\ 0, & t \neq 0 \end{cases}$$

Transfer function of continuous time LTI system is equal to Laplace transform of unit impulse response of a continuous time LTI system.

Transfer function of a continuous time LTI system

H(s) = Laplace transform of h(t).

$$= LT\,[h(t)] = \int_{-\infty}^{\infty} h(t)\,e^{-st}\,dt$$

Where s is a complex variable and $s = \sigma + j\omega$, σ real part and ω imaginary part.

Properties of continuous time and Discrete time LTI systems

We know that continuous time and discrete time systems are characterized by their unit impulse responses. Response of continuous time LTI system is determined by using convolution integral.

While the response of discrete time LTI system is determined by using convolution sum.

Now we will discuss various properties of continuous time and discrete time LTI systems.

Convolution integral for continuous time LTI is given by $y(t) = x(t) * h(t) = \int_{\tau=-\infty}^{\infty} x(\tau)\,h(t-\tau)d\tau$

and convolution sum for discrete time LTI system is given by. $y[n] = x[n] * h[n] = \sum_{K=-\infty}^{\infty} x(k)\,h(n-k)$

Both continuous time and discrete time LTI systems are completely characterized by their impulse responses. These systems are also completely characterized by their transfer function. For continuous time LTI system, the transfer function is the ratio of Laplace transform of output to Laplace transform of input to the system when system is initially relaxed. It is also defined as Laplace transform of impulse response of a continuous time LTI system.

For discrete time LTI system, transfer function is the ratio of z-transform of output to z-transform of input when initially system is relaxed. It is also defined. as z-transform of unit impulse response of discrete time LTI system.

Causality for LTI Systems

The output of a causal system depends only on the present and past values of the input to the system.

A continuous tie LTI system is called causal system if its impulse response h(t) is zero for t < 0.

For a causal continuous time LTI system, convolution integral is given by

$$y(t) = x(t) * h(t) = \int_{\tau=-\infty}^{\infty} x(\tau)\,h(t-\tau)d\tau$$

$$= h(t) * x(t) = \int_{\tau=0}^{\infty} h(\tau)\, x(t-\tau)\, d\tau$$

A discrete time LTI system is called is called causal system if its impulse response h[n] is zero for n < 0.

For a causal discrete time LTI system, convolution sum is given by.

$$y[n] = x[n] * h[n] = \sum_{k=-\infty}^{n} x(k)\, h(n-k)$$

$$= h[n] * x[n] = \sum_{k=0}^{\infty} h(k)\, x(n-k)$$

Causality of an LTI system is equivalent to its impulse response being a causal signal. Accumulator with impulse response.

$$h[n] = u[n] = \begin{cases} 0 & n < 0 \\ 1 & n < 0 \end{cases}$$

is a causal discrete time LTI system.

A pure time shift with unit impulse response $h(t) = \delta(t - t_o)$ is a causal continuous time LTI system for $t \geq 0$. In this case time shift is called a delay.

A pure time shift is non causal continuous time LTI system for $t < 0$. In this case time shift is called an advance.

Stability for LTI Systems

A stable system produces bounded output from every bounded input.

Condition of stability for continuous time LTI system: Consider an input x(t) that is bounded in magnitude.

$| x(t) | < M$ for all values of t (1)

Now we apply this input to an continuous time LTI system with unit impulse response h(t).

Output of this LTI system is determined by convolution integral and is given by.

$$y(t) = \int_{\tau=-\infty}^{\infty} h(\tau)\, x(t-\tau)\, d\tau \qquad (2)$$

Magnitude of output y(t) is given by

$$|y(t)| = \left| \int_{\tau=-\infty}^{\infty} h(\tau)\, x(t-\tau) d\tau \right|$$

$$= \int_{\tau=\infty}^{\infty} | h(\tau) |\, | x(t-\tau) |\, d\tau \qquad (3)$$

[Substituting the value $|s(t-\tau)| < M$ for all values of τ and t].

$$\text{or } |y(t)| \leq \int_{\tau=-\infty}^{\infty} |h(\tau)| \,.\, M.\, d\tau$$

$$\leq M \int_{\tau=-\infty}^{\infty} | h(\tau) |\, d\tau \text{ for all values of t} \qquad (4)$$

From equation we can conclude that if the impulse response h(t) is absolutely integerable then output of a continuous time LTI system is bounded in magnitude, and hence, the system is bounded input bounded- output (BIBO) stable.

A sufficient and necessary condition to stability of a continuous time LTI system is given by.

$$S = \int_{\tau=-\infty}^{\infty} |h(\tau)|\, d\tau < \infty$$

Condition to stability of a discrete time LTI system

Consider an input x[n] that is bounded in magnitude.

$| x[n] | < M$ for all values of n (1)

Now we apply this input to a discrete time LTI system with unit impulse response h(n). Output of discrete time LTI system is determined by convolution sum and is given by $y[n] = \sum_{K=-\infty}^{\infty} h(k)\, s(n-k)$ (2)

Magnitude of y (n) is given by.

$$|y[n]| = \left| \sum_{K=-\infty}^{\infty} h(k)\, x(n-k) \right|$$

$$= \sum_{K=-\infty}^{\infty} |h(k)|\, |x(n-k)| \qquad (3)$$

[Substituting the value of $|x(n-k)| < M$ for all value of k and n].

$$\text{or } |y[n]| \leq \sum_{K=-\infty}^{\infty} |h(k)|\, M$$

$$\leq M \sum_{K=-\infty}^{\infty} | h(k) | \qquad (4)$$

for all values of k.

From Equation we can conclude that if the impulse response h (n) is absolutely summable then output of a discrete time LTI system is bounded in magnitude, and hence, the system is bounded input bounded output (BIBO) stable.

A sufficient and necessary condition of stability of a discrete time LTI system is given by.

$$S = \sum_{K=-\infty}^{\infty} |h(k)| < \infty.$$

6.7 IMPULSE AND FREQUENCY RESPONSE (LTI System)

6.7.1 IMPULSE RESPONSE

The unit sample response or simply the impulse response h[n] of a linear time invariant system is the system's response to a unit impulse input signal {δ[n]} located at n = 0, when the initial conditions of the system are zero.

$$\delta[n] = \begin{cases} 1 & n = 0 \\ 0 & n \neq 0 \end{cases}$$

δ[n]

n x[n] = δ[n] → LTIC System → y[n] = h[n]

0

y[n] = h[n]

e^{-x}

n

Impulse Response of linear Time Invariant Causal (LTIC) System

If the input to the system is the time shifted version of the δ(t) function, the output is a time shifted version of the impulse response. The result shown in Figure is due to the impulse response. The result shown in figure is due to the time invariance of the system. In the linear system, application of αδ (n – k) at the input will produce α h (n – k) at the output as shown in figure.

x[n] αδ(n–k) → LTIC System → y[n] αh [n–k]

Time Shifted Impulse Response

The shifted impulse sequence is given by

$$\delta[n-k] = \begin{cases} 1 & n = k \\ 0 & n \neq k \end{cases}$$

Where k is an integer. The graphs of δ[n] and δ (n – 2) are shown in figure.

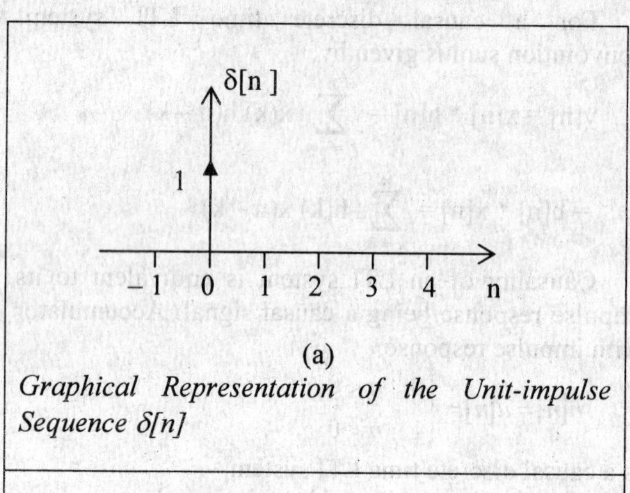

Graphical Representation of the Unit-impulse Sequence δ[n]

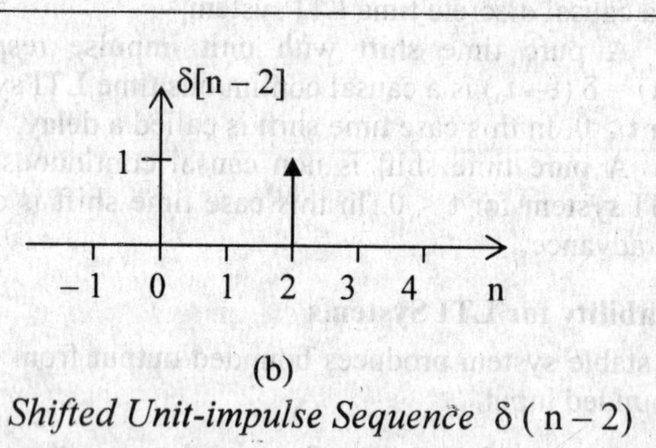

The Shifted Unit-impulse Sequence δ (n – 2)

The response of the LTI system to the unit impulse sequence δ[n] is represented by h[n], i.e., h[n] = F{δ [n]} and hence h(n – k) = F_T [δ (n – k)]

Therefore,

$$y[n] = F\, x[n] = F \sum_{k=-\infty}^{\infty} x(k)\, \delta(n-k)$$

$$= \sum_{k=-\infty}^{\infty} x(k)\, F[\delta(n-k)] = \sum_{k=-\infty}^{\infty} x(k)\, h(n-k) = x[n] * h[n]$$

$$y[n] = \sum_{k=-\infty}^{\infty} h(k)\, x(n-k) = h[n] * x[n]$$

This expression gives the output response y[n] of the LTI system as a function of the input signal x[n] and the unit impulse (sample) response h[n] and is referred to as a **convolution sum**.

Unit Step Response {u[n]}

The unit step sequence u[n] is defined by

$$u[n]=\begin{cases}0 & n<0\\ 1 & n\geq 0\end{cases}$$

The shifted unit step sequence u(n – k) is given by

$$u[n-k][n]=\begin{cases}0 & n<0\\ 1 & n\geq 0\end{cases}$$

The graphical representations of u[n] and u(n –2) are below shown in figure.

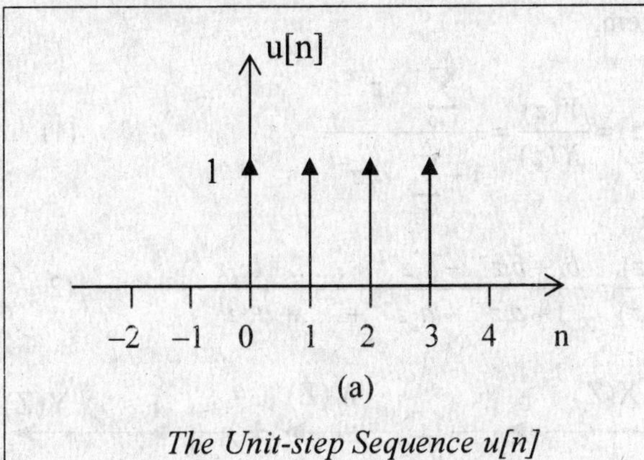

(a)

The Unit-step Sequence u[n]

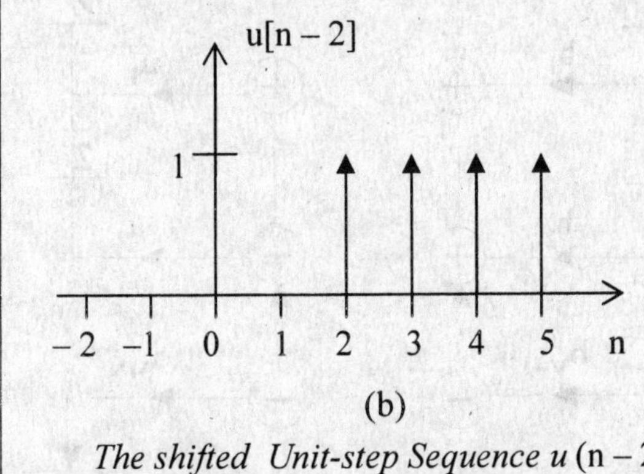

(b)

The shifted Unit-step Sequence u (n – 2)

The step response can be obtained by exciting the input of the system by a unit-step sequence, i.e., x[n] = u[n]. Hence, the output response y[n] is obtained by using the convolution formula as.

$$y[n]=\sum_{m=0}^{n} h(k)\,u(n-k)$$

Relation between the Unit Sample and the Unit-step sequence

The unit sample sequence δ[n] and the unit-step sequence u[n] are related as

$$u[n]=\sum_{m=-\infty}^{\infty}\delta[m]$$

$$\delta[n] = u[n] - u[n-1]$$

6.7.2 FREQUENCY RESPONSE

Frequency response is a complex function that describes the magnitude and phase shift of a filter over a range of frequencies.

Properties of the Frequency Response

If h[n] is a real sequence, the frequency response has the following properties:

(a) $H(e^{j\omega})$ takes on values for all ω, i.e. on a continuum of ω.

(b) $H(e^{j\omega})$ is periodic in ω with period 2π.

(c) The magnitude response $|H(e^{j\omega})|$ is an even functions of ω and symmetrical about π.

(d) The phase response $\angle H(e^{j\omega})$ is an odd function of ω and antisymmetrical about π.

The impulse function is a derivative of step function *u*(*t*) with respect to times. Conversely, the step function *u*(*t*) is the integral of the impulse with respect to time. The continuous-time unit impulse function is commonly denoted by δ(*t*) and is defined as

$$\delta(t)=0,\ t\neq 0 \tag{1}$$

and

$$\int_{-\infty}^{\infty}\delta(t)\,dt=1 \tag{2}$$

Equation (1) says that the impulse δ(*t*) is zero everywhere except at the origin and equation (2) says that the area under the unit impulse is unity. The impulse function is also known as Dirac delta function. The impulse function is illustrated s figure (a).

The discrete-time impulse function is commonly denoted by δ[*n*] and is defined as

$$\delta[n]=\begin{cases}1, & n=0\\ 0, & n\neq 0\end{cases} \tag{4}$$

Equation (4) says that the impulse δ[*n*] is zero everywhere except at the origin, whose magnitude is unity. The impulse function is illustrated in figure (b).

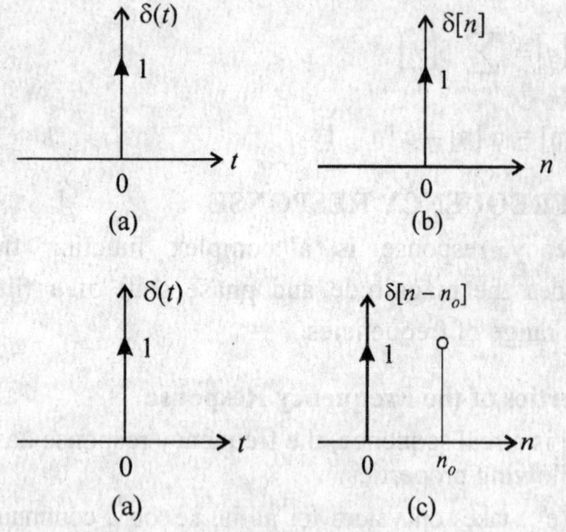

Impulse signal

Figure: (a) Continuous-time impulse (b) Discrete-time impulse

6.9 IIR AND FIR FILTERS

6.9.1 INFINITE IMPULSE RESPONSE FILTER

The bock diagram approach gives more information on implementation of the system. It also explains the relation between input and output.

Block diagram consists of an interconnection of three elementary operation on signals.

1. Scalar multiplication

x(t) / x[n] ⟶ y(t) / y[n]

$y(t) = ax(t) \Rightarrow$ continuous - time system

$y[n] = ax[n] \Rightarrow$ discrete - time system

where constant, which is a scalar.

2. Addition

$y(t) = x(t) + w(t) \Rightarrow$ continuous – time system

$y[n] = x[n] + w[n] \Rightarrow$ discrete – time system

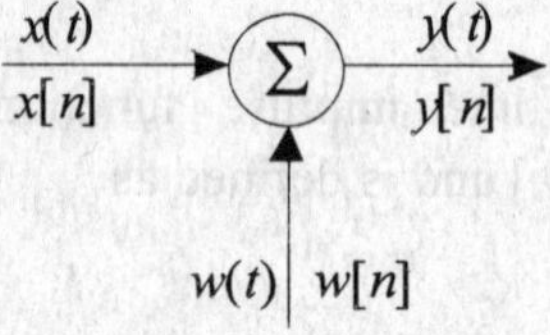

w(t) | w[n]

3. Shift operator

x[t] ⟶ ∫ ⟶ y[t]

x[n] ⟶ Z ⟶ x[n 1]

"integration" for continuous - time system

"Time - shift" for discrete - time system

Following are the various configuration in which a given system can be realized.

1. Direct form - I
2. Direct form - II

System Realization Using Direct Form-I

Let us consider an LTI recursive discrete - time system,

$$H(z) = \frac{Y(z)}{X(z)} = \frac{\sum_{k=0}^{M} b_k z^{-k}}{1 + \sum_{k=1}^{N} a_k z^{-k}} \quad (1)$$

$$\frac{Y(z)}{X(z)} = \frac{b_0 + b_1 z^{-1} + b_2 z^{-2} + \ldots\ldots + b_M z^{-M}}{1 + a_1 z^{-1} + a_2 z^{-2} + \ldots\ldots + a_N z^{-N}} \quad (2$$

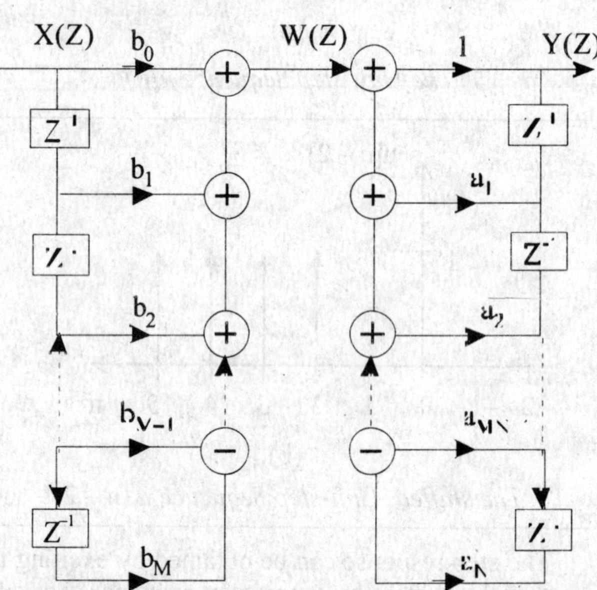

Figure: Complete Realization of system Using Direction From-I

System Realization Using Direct Form - II

Let us consider the system function,

$$H(Z) = \frac{Y(Z)}{X(Z)} = \frac{\sum_{k=0}^{M} b_k Z^{-k}}{1 + \sum_{k=1}^{N} a_k Z^{-k}} \quad (12.9)$$

$$H(z) = H_1(z)H_2(R)$$

$$\frac{W(Z)}{X(Z)} = \frac{1}{1+\sum_{k=1}^{N} a_k Z^{-k}}$$

$$H(z) = \frac{Y(z)}{W(z)}$$

$$= \sum_{k=0}^{M} b_k z^{-k}$$

Figure: Realization of I / R system Direction From-II

6.9.2 FINITE IMPULSE RESPONSE FILTER

The unit sample response of the FIR system is identical to the coefficient, i.e.

$$h[n] = \begin{cases} b_k, & 0 \le n \le m-1 \\ 0, & \text{otherwise} \end{cases}$$

Following are the various configuration in which a given in which a given system can be realized.

1. Parallel from
2 Cascade form

System Realization Using Parallel Form

In parallel realization, the system function is expressed as an addition of several subsystems functions. Each subsystem in Parallel structure relized in direct form II. The order of each subsystem may be one or two or more.

$$H(Z) = C + \sum_{k=1}^{N} \frac{C_k}{1-P_k Z^{-1}} \qquad (12.16)$$

where poles of the system.

$$H_1(Z) = \frac{C_1}{1-P_1 Z^{-1}}$$

$$H_2(Z) = \frac{C_2}{1-P_2 Z^{-1}}$$

where p_1, p_2 are The poles of the sub system transfer function $H_1(z), H_2(z)$ respect $c_1\ c_2$ are the contents of partial fraction expansion.

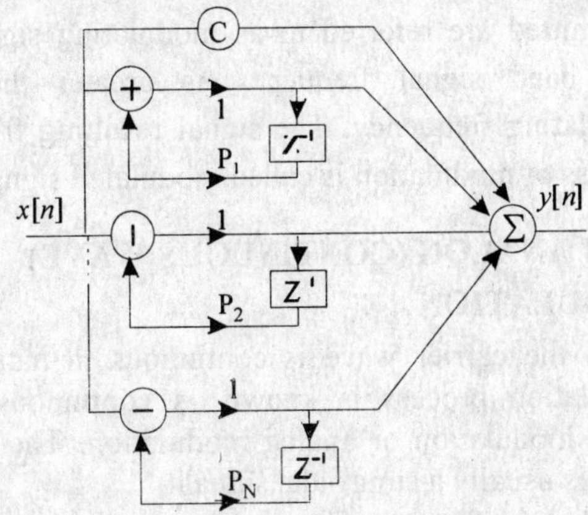

Figure: Parallel Structure Realization

Direct Form Realization of FIR System

Direct form is one of the simplest structure. The direct form realization can be obtained from the no recursive difference equation given in equation i.e.

$$H(z) = \sum_{k=0}^{m-1} b_k z^{-k}$$

$$H(Z) = \sum_{n=0}^{M-1} h[n] Z^{-n}$$

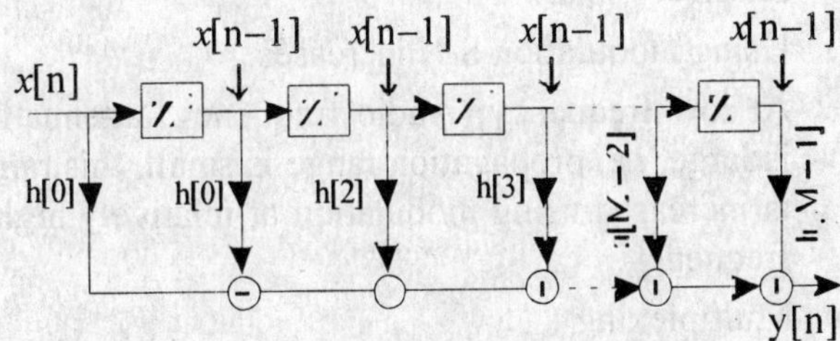

Figure: Direct form Realization of FIR Filter

The major advantages of FIR system is that

1. Better system stability.
2. Better linear phase can be achieved.
3. Structure can be realized in both recursively and no recursively.

6.10 AMPLITUDE AND FREQUENCY MODULATION AND DEMODULATION

Modulation is a fundamental requisite of communication system. It is defined as the process by which some characteristics of signal called carrier is varied in accordance with the instantaneous value of another signal called modulating signal. Signals containing information or intelligence to be transmitted are referred to as modulating signals or base band signal frequency is greater than the modulating frequency. The signal resulting from the process of modulation is called modulated signal.

6.10.1 ANALOG (CONTINUOUS WAVE) MODULATION

When the carrier wave is continuous in nature the modulation process is known as continuous wave (CW) modulation or analog modulation. The carrier wave is usually a sinusoidal signal.

Modulating signal $\rightarrow m(t)$ or $f(t) \rightarrow$ Audio signal 15Hz – 15 kHz

Carrier signal $\rightarrow \sin \omega_c t$ or $\cos \omega_c t$

ω_o or ω $\rightarrow$ Carrier frequency

$f(t)$ or $m(t)$ $\rightarrow$ Band limited signal limited to frequency ω_m

$\rightarrow$ Frequency spectrum of this signal extends upto ω_m.

Advantage of Modulation

1. Using modulation; length of receiving antenna is relatively small.
2. Using modulation S/N increases.
3. At low frequency (Audio frequency) attenuation is large, i.e. propagation range is small, this range is increased using modulation at relatively higher frequency.
4. Multiplexing
5. Narrow banding
6. Audio signal 20 Hz to 20 kHz, when transmitted as usual mix-up inseparably. Broadcasting station alone would blanket the air, very less number of channel can be used.
7. Common processing (Tx and Rx).

Amplitude Modulation

Here the amplitude of the carrier is varied in accordance with the instantaneous value of the amplitude of the modulating signal. Keeping frequency and phase of the carrier to be fixed.

$$f(t) = v_m(t) = V_m \sin \omega_m (t) \rightarrow \text{mod. signal}$$

$$v_c(t) = V_c \sin \omega_c (t) \rightarrow \text{carrier}$$

Modulated Signal (AM Signal)

$$V(t) = [V_c + K_a V_m(t)] \sin \omega_c t$$

$$= V_c \sin \omega_c t + K_a V_m, 2 \sin \omega_c t \sin \omega_m t$$

$$= V_c \sin \omega_c t + K_a \frac{V_m}{2} \begin{bmatrix} \cos(\omega_c - \omega_m)t \\ -\cos(\omega_c + \omega_m)t \end{bmatrix}$$

Modulation Index (m_a = 0.4 to 0.6)

The extent of amplitude variation in AM about an un-modulated carrier amplitude is measured in terms of a factor called modulation index. It is define as:

$$m_a = \frac{V_m\,(\max)}{V_c} = \frac{K_a V_m}{V_c} = \frac{V_m}{V_c} \text{ (if } K_a = 1)$$

$$V_m = \frac{m_a V_c}{K_a} \text{ or } K_a V_m = m_a V_c$$

This factor m_a is also known as: (i) depth of modulation or (ii) degree of modulation or (iii) modulation factor.

Power Contained in the AM Signal

$$V(t) = V_c \sin \omega_c t$$

$$= \frac{m_a V_c}{2} [\cos(\omega_c - \omega_m)t - \cos(\omega_c + \omega_m)t]$$

Total Tx power

= Power in the carrier + Power USB + Power LSB

= Power (Carrier) + 2 Power (SB)

Power in Carrier $\propto V_c^2 = P_c$

$$\text{Power (USB)} \propto \left(\frac{m_a V_c}{4}\right)^2 = \frac{m_a^2}{4} P_c$$

$$\text{Power (LSB)} \propto \left(\frac{m_a V_c}{2}\right)^2 = \frac{m_a^2}{4} P_c$$

Power in AM signal, i.e. AM–DSB/FC

$$P_t = P = P_c + \frac{m_a^2}{4} P_c + \frac{m_a^2}{4} P_c$$

Power $= P_c \left(1 + \frac{m_a^2}{2}\right) \rightarrow$ Only for the sinusoidal type of modulating signal.

Example: Power saving in different AM system.

Let $m_a = 100$ %, i.e. $m_a = 1$

Case 1: *AM-DSB/FC*

Power $P_t = P_c \left(1 + \frac{m_a^2}{2}\right) = \frac{3 P_c}{2}$

Power required to transmit AM-DSB/FC

Case 2: AM-DSB/SC

Power saving $= P_c$

Power transmitted = Power in 2 SB

$$= P_c \left(\frac{m_a^2}{2}\right) = \left(\frac{P_c}{2}\right)$$

% Power saving w.r.t. AM-DSB/FC

$$= \frac{P_c}{(3P_c / 2)} \times 100 = \frac{200}{3} = 67\%$$

Case 3: AM-SSB/FC

Power saving $= P_c \frac{m_a^2}{4} = \frac{P_c}{4}$

Power transmitted $= P_c \left(1 + \frac{m_a^2}{2}\right) = \frac{3P_c}{4}$

% Power saving w.r.t. AM-DSB/FC

$$= \frac{(P_c / 4)}{(3P_c / 2)} \times 100 = \frac{50}{3} = 17\%$$

Case 4: AM – SSB/SC

Power saving $= P_c + \frac{m_a^2 P_c}{4}$

$$= P_c \left(1 + \frac{m_a^2}{2}\right) = \frac{5P_c}{4}$$

Power transmitted $= P_c \frac{m_a^2}{4} = \frac{P_c}{4}$

% Power saving w.r.t. AM –DSB/FC

$$= \frac{(5P_c / 4)}{(3P_c / 2)} \times 100 = 83\%$$

Efficiency of Transmission AM-DSB/FC

$$\eta = 33.3\%$$

For *broadcasting* AM. DSB/FC is always preferred since the receiver configuration is the simplest.

6.10.2 AM-DSB/SC

Modulation signal $= f(i)$

Carrier $= \cos \omega_c t$

Frequency components.

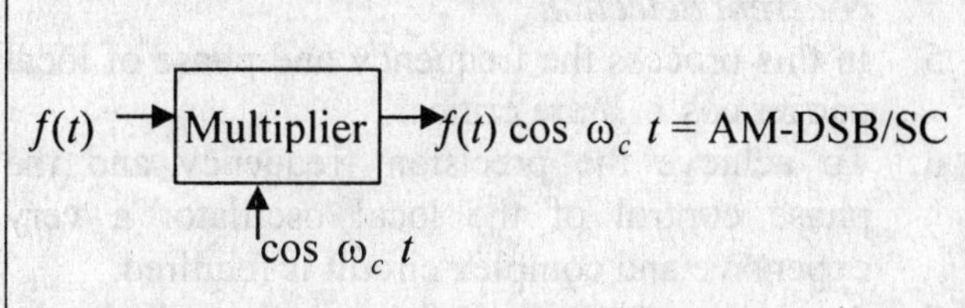

$$f(t) \cos \omega_c t = f(t) \frac{\left(e^{j\omega_c t} + e^{-j\omega_c t}\right)}{2}$$

$$= FT\left[1/2\, f(t) e^{j\omega_c t}\right] + FT\left[1/2\, f(t) e^{-j\omega_c t}\right]$$

$$= (1/2)\left[F(\omega + \omega_c)\right] + F(\omega - \omega_c)$$

Principle of Demodulator

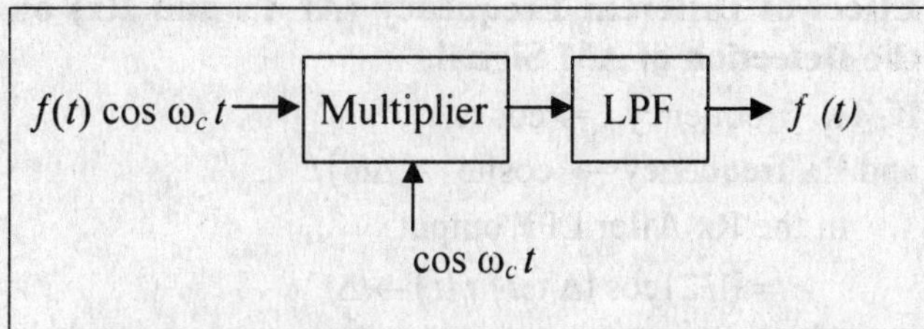

After the multiplier

$$= f(t) \cos^2 \omega_c t$$

$$= (1/2)\, f(t)\left[1 + \cos 2\omega_c t\right]$$

$$= (1/2)\, f(t) + (1/2)\, f(t) \cos 2\omega_c t$$

After LPF

(1/2) $f(t)$ is passed and (1/2) $f(t)$ cos $2\omega_c t$ is attenuated

$\therefore$ $f(t)$ is recovered

$$f(t) \cos^2 \omega_c t$$

$$= (1/2)\, f(t) + (1/2)\, 2 \cos 2\omega_c t$$

$$= (1/2)\, F(\omega) + (1/4)\left[F(\omega + 2\omega_c) + F(\omega - 2\,\omega_c)\right]$$

Basic Principle of Synchronous Detection

1. Multiplication of $f(t)$ by $\cos \omega_c t$ translate its Spectrum by $\pm \omega_c$.
2. The new spectrum can be retranslated to the original position by another translation of $\pm \omega_c$. This is accomplished by multiplication of modulated signal by $\cos \omega_c t$.
3. The process at the receiver end is therefore exactly same as that required at the transmitting end.
4. Hence this method of recovery the original signal is called the *synchronous detection or coherent detection.*
5. In this process the frequency and phase of local carrier $\cos \omega_c t$ are critical.
6. To achieve the precision frequency and the phase control of the local oscillator a very expensive and complex circuit is required.
7. A very small amount of the carrier called *pilot carrier* is transmitted along with the modulated signal.
8. At the receiver end the pilot carrier is separated by an appropriate filter and amplified. This weak carrier is then used to phase lock the local oscillator which generates the strong carrier of the same frequency as that at the transmitter.

Effect of Different Frequency (At Tx and Rx) on the Detection of AM Signal

If, Tx frequency $\rightarrow \cos \omega_c t$

and Rx frequency $\rightarrow \cos(\omega_c + \Delta\omega) t$

In the Rx After LPF output

$$= (1/2)\cos(\Delta \omega t) f(t) \rightarrow \Delta f$$

Instead of getting the fixed amplitude signal, we receive the signal *f(t)* whose amplitude varies with respect to time t.

6.10.3 AM-DSB/FC or (AM -DSB)

$$= \text{AM SDB/SC} + \text{Carrier}$$

$$V(t) = f(t)\cos \omega_c t + A\cos \omega_c t$$

AM-DSB/SC carrier

$$V(t) = \cos \omega_c t + [A + f(t)]$$

where A $\rightarrow$ DC component

$$FAM(\omega) = (1/2)[F(\omega + \omega_c) + F(\omega - \omega_c)] + \pi A[\delta(\omega + \omega_c) + \delta(\omega - \omega_c t)]$$

(i) $|A| \geq [f(t)]_{\max} \rightarrow$ Envelop detection

(ii) $|A| = [f(t)]_{\max} \rightarrow$ Synchronous detection.

Envelop Detection

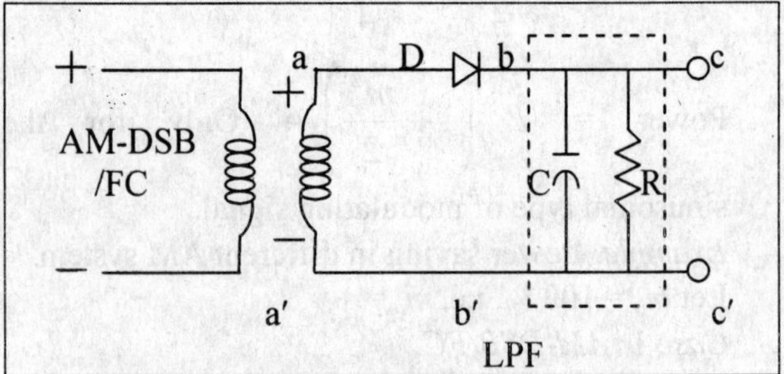

Case 1: $T_2 = RC$ is large

Result *Diagonal Clipping.* It may occur due to higher modulating frequency.

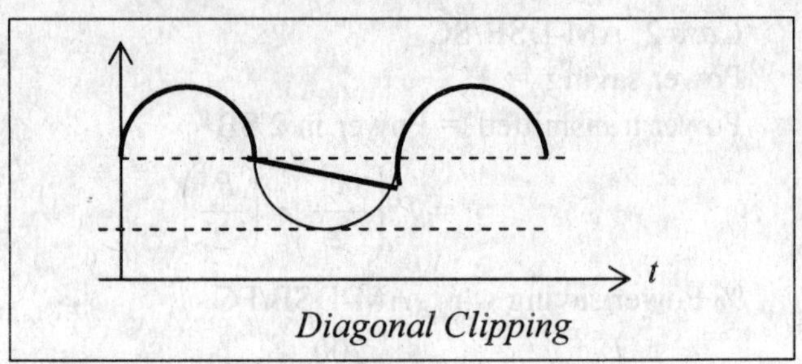

Diagonal Clipping

Case 2:

$T_2 = RC$ is small. Result *Negative peak clipping.* It may be due to large modulation index.

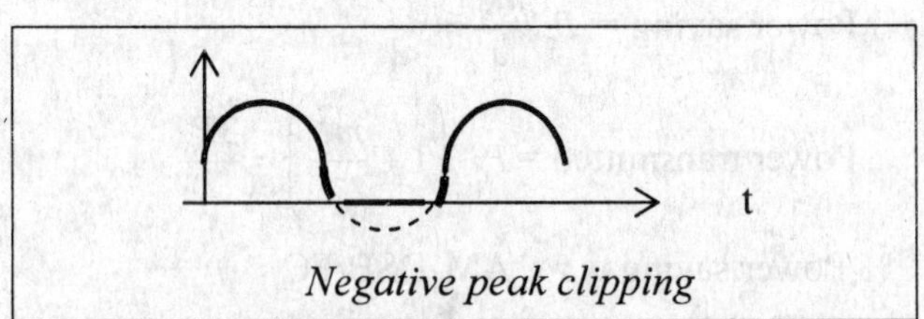

Negative peak clipping

Case 3: $T_3 = RC$ is medium

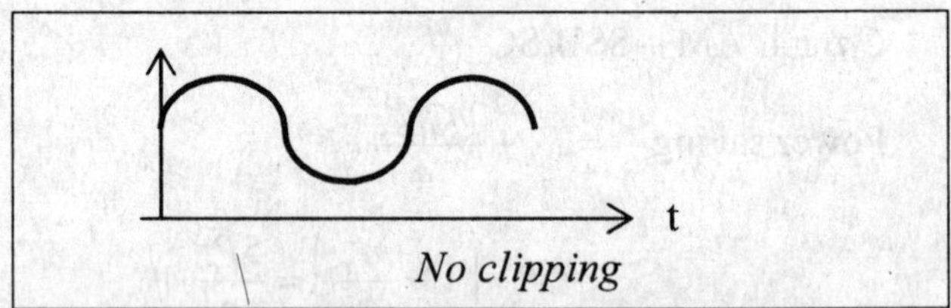

No clipping

The T_2 is adjusted so that exponential decay of capacitor voltage during discharge follows the envelope.

Discharge time constant *RC* should be such that the rate of capacitor discharge = Slop of envelope of modulated signal. Then the output capacitor voltage follow the envelope and f(t) is detected with high S/N

$$\frac{1}{RC} \geq \frac{\omega_m \ m_a}{\sqrt{1 - m_a^2}} \rightarrow \text{for no clipping}$$

Typically $m_a = 0.4$, $m_a^2 = 0.16$

$$\frac{1}{RC} \geq m_a \omega_m$$

$$\frac{1}{RC} \geq 2\pi f_m \ m_a \rightarrow \text{approximately}$$

Representation of AM

$$V(t) = V_c \sin \omega_c t + \left(\frac{K_a V_m}{2}\right) \cos(\omega_c - \omega_m) t$$

$$-\left(\frac{K_a V_m}{2}\right) \cos(\omega_c + \omega_m) t$$

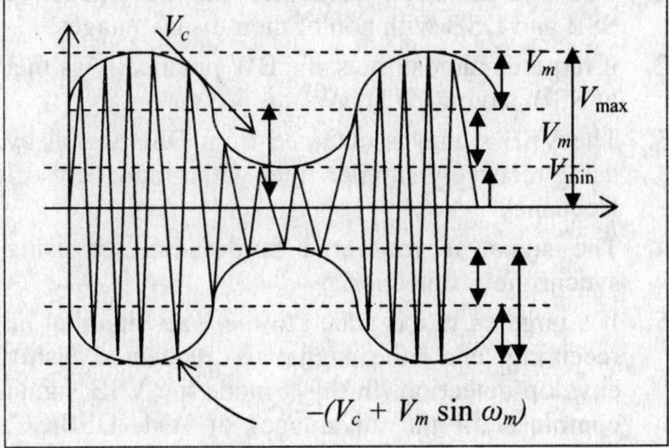

$$V_m = \frac{V_{max} - V_{min}}{2}$$

$$V_c = V_{max} - V_m = \frac{V_{max} + V_{min}}{2}$$

Modulation index

$$m_a = [V_m / V_a] = \frac{[(V_{max} - V_{min}) / 2]}{[(V_{max} + V_{min}) / 2]} = \frac{[V_{max} - V_{min}]}{[V_{max} + V_{min}]}$$

6.10.4 AM – SSB/SC

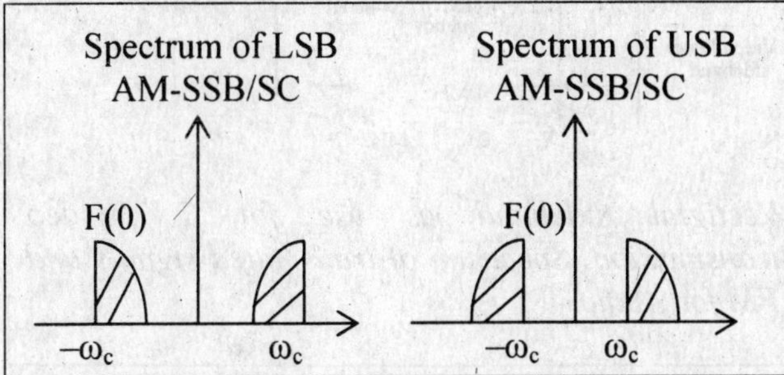

There are three method to Generate SSB/SC

1. The Filter System
2. The "Third" Method or Weaver Method

Phase Shift Method of Generating AM-SSB/SC Signal

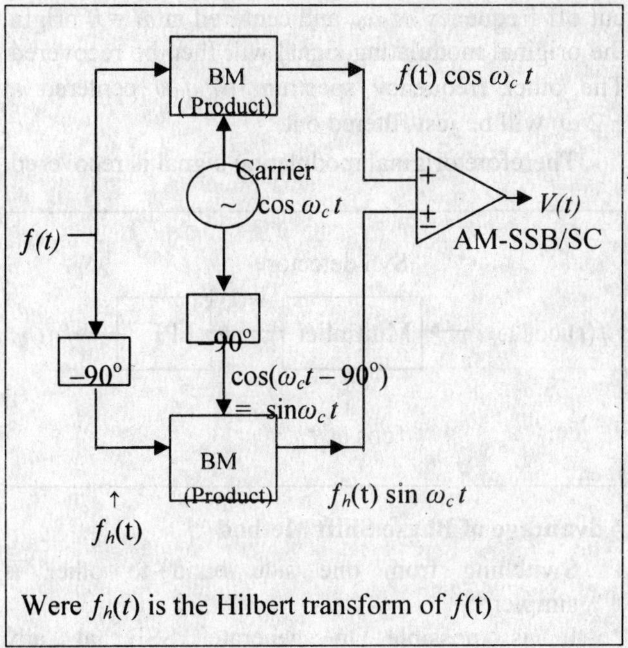

Were $f_h(t)$ is the Hilbert transform of $f(t)$

Special Case:

Let $f(t) = \cos \omega_m t$

→ It holds good for any general signal

$$\therefore \ V(t)\big|_{SSB/SC} = \cos \omega_m t . \cos\omega_c t + \sin \omega_m t . \sin\omega_c t$$

$$= \cos(\omega_c - \omega_m) t$$

= LSB is generated

$$V(t)\big|_{SSB/SC} = f(t) \cos \omega_c t + f_h(t) \sin \omega_c t$$

→ LSB is generated

$$= f(t) \cos \omega_c t - f_h(t) \sin \omega_c t$$

→ USB is generated

In general representation of AM.SSB/SC

$$V(t) = f(t) \cos \omega_c t \pm f_h(t) \sin \omega_c t$$

Detection of AM-SSB/SC Signal

Basic Principle

$$V(t) = f(t) \cos \omega_c t \pm f_h \ \omega_i t - AM - SSB/SC$$

For synchronous detection

$$V(t) = V(t) \ \cos \omega_c t$$

$$= f(t) \cos^2 \omega_c t \pm f_h(t) \sin \omega_c t \cos \omega_c t$$

$$=\frac{1}{2}[f(t)(1+\cos 2\,\omega_c t)] \pm \frac{1}{2}[f_h(t)\sin 2\,\omega_c t]$$

$$=\frac{1}{2}\,f(t)+\frac{1}{2}[f(t)(1+\cos 2\,\omega_c t)] \pm [f_h(t)\sin 2\,\omega_c t]$$

When this signal is passed through a LPF having cut off frequency of ω_m and centered at $\omega = 0$ origin, the original modulating signal will then be recovered. The other frequency spectrum of $f(t)$ centered at $\pm 2\,\omega_c$ will be just filtered out.

Therefore original modulating signal is recovered.

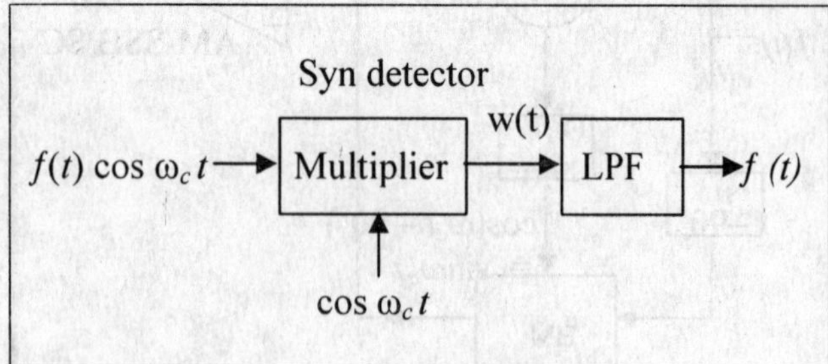

Advantage of Phase Shift Method

1. Switching from one side band to other is simpler.
2. It is possible to generate SSB at any frequency.
3. SSB with lower Audio frequency can be generated.

Generation and Detection of AM-SSB/FC

$$V(t)=[f(t)\cos\omega_c t \pm f_h(t)\sin] + A\cos\omega_c t$$

AM SSB/SC

Generation is similar to AM-SSB/SC and then carrier is added.

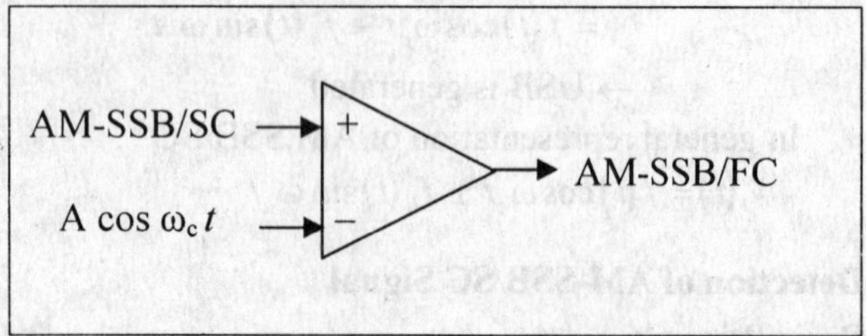

Note: Video signal in television broadcast transmitted by AM-SSB/FC.

Detection of AM-SSB/FC

→ Envelop detection is done

6.10.5 VESTIGIAL SIDEBAND TRANSMISSION

→ Compromise between DSB and SSB.

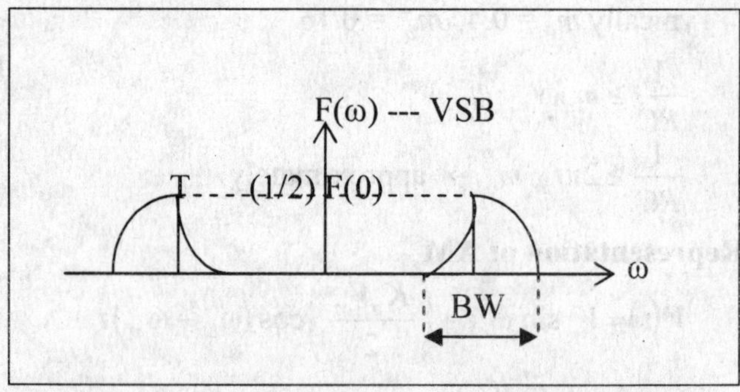

The Basic Feature of VSB Transmission

1. VSB transmission combines the advantage of SSB and DSB with non of their disadvantages.
2. It requires almost the same BW practically as that of SSB, saving (1/3) BW.
3. The VSB signal is obtained from DSB signal by using relatively simpler filter with gradual cut off frequency.
4. The signal is recovered at the detector using synchronous detection.
5. If a large carrier is added to the VSB signal at he receiving end, the signal is recovered using envelop detection. In this mode the VSB signal combines all the advantages of AM, DSB/SC, SSB/SC.
6. In the public TV system the Video signal are transmitted using this type of transmission.

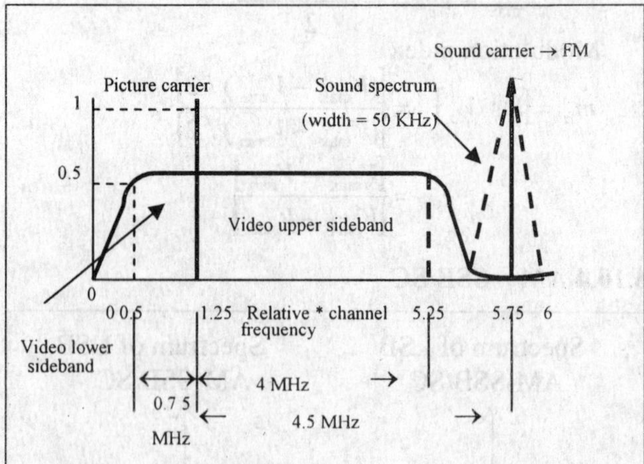

Vestigial sideband as use for T.V video transmission. Spectrum of transmitted signals and FM for Audio.

6.10.6 ANGLE MODULATION

A general modulated carrier signal can be represented mathematically as

$$x_c(t) = A(t)\cos[\omega_c t + \phi(t)]$$

where, $A(t)$ and $\phi(t)$ are called the instantaneous amplitude and phase angle of the carrier. When $A(t)$ is linearly related to the message signal $f(t)$ the result is amplitude modulation. If $\phi(t)$ or its derivative is linearly related to $f(t)$, then we have *phase or frequency* modulation.

FM: Here the frequency of the carrier is varied in accordance with the instantaneous value of the amplitude of the modulating signal.

PM: Here the phase of the carrier is varied in accordance with the instantaneous value of the amplitude of the modulating signal.

Let, modulating signal $= f(t)$

$$\text{Carrier} = A\cos\omega_c t = A\cos\theta(t)$$

For angle modulation, the modulated carried is represented by

$$x_c(t) = A\cos[\omega_c t + \phi(t)]$$

where, A and ω_c are constant and the phase angle $\phi(t)$ is function of the message signal $f(t)$. If we rewrite above equation

$$x_c(t) = A\cos\theta(t)$$

where $\theta(t) = [\omega_c t + \phi(t)]$

Then we can define the *instantaneous radian frequency* of $x_c(t)$ defined by ω_i as

$$\omega_i = \frac{d\,\theta(t)}{dt} = \omega_c + \frac{d\,\phi(t)}{dt}\ \omega_c + \omega\Delta$$

The function $\phi(t)$ and $[d\,\phi(t)\,/\,dt]$ are known as the instantaneous phase deviation and instantaneous frequ-ency deviation of $x_c(t)$. The quantity $\Delta\omega = \delta$ defined by

$$\Delta\omega = |\,\omega_i - \omega_c\,|_{max}$$

Is called the maximum (or peak) radian frequency deviation of the angle modulated signal

where, $\theta(t) = [\omega_c t + \phi(t)]\ \ \omega_c\ \ 1 \rightarrow$ frequency

$$\frac{d\,\theta(t)}{dt} = \omega_c + \frac{d\,\phi(t)}{dt}\ ;\ \phi(t) \rightarrow \text{Phase}$$

$$\theta(t) = \int\left[\omega_c + \frac{d\,\phi(t)}{dt}\right]dt$$

FM: Frequency is varied with $f(t)$

$\omega'_c = \omega_c + k_f\, f(t) \rightarrow$ instantaneous frequency

$$\omega'_c = \left[\omega_c + \frac{d\phi(t)}{dt}\right]$$

$$K_f\, f(t) = \frac{d\,\phi(t)}{dt}\ ; \qquad \phi(t) = \int K_f\, f(t)\,dt$$

where K_f = the frequency deviation constant expressed in radian/sec for FM

$$\theta'(t) = \int\omega'_c\,dt = \int[\omega_c + K_f\, f(t)]\,dt$$

$$= \int\left[\omega_c + \frac{d\,\phi(t)}{dt}\right]dt$$

FM signal $f(t) = A\cos\theta'(t)$

$$= A\cos\int[\omega_c + K_f\, f(t)]\,dt$$

$$f_{FM}(t) = A\cos\omega_c t + K_f\int f(t)\,dt$$

PM: The instantaneous phase deviation of the carrier is proportional to the message signal; that is

$$\phi(t) = K_p\, f(t)$$

$$\theta(t) = [\omega_c t + \phi(t)]$$

$$\theta'(t) = |\omega_c t + K_p\, f(t)|$$

$\omega'_c = \omega_c + K_p\, d\dfrac{f(t)}{dt} \rightarrow$ instantaneous frequency

where, K_p is the phase deviation constant expressed in radian /unit of $f(t)$ for PM.

PM signal $= A\cos\theta'(t)$

$$f_{PM}(t) = A\cos\left[\omega_c t + K_p\, f(t)\right]$$

$$f_{FM}(t) = A\cos\left[\omega_c t + K_p\int f(t)\,dt\right]$$

6.10.7 RELATIONSHIP BETWEEN PM AND FM

PM and FM are closely related in the sense that the net effect of both is variation in total angle. In PM the phase angle varies linearly with $f(t)$ where as in FM phase angle varies linearly with the integral of $f(t)$.

1. *Indirect Method*

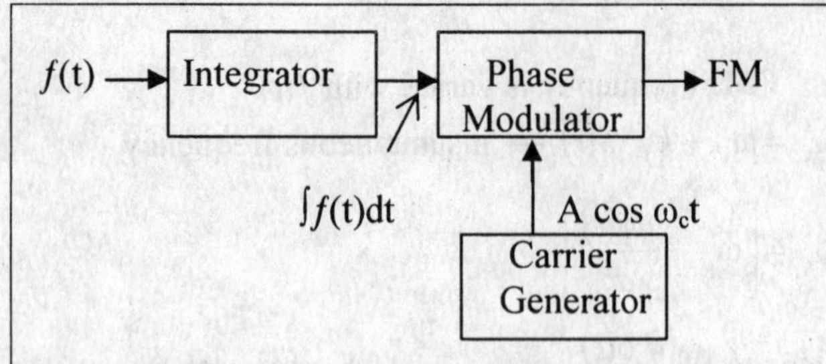

2. *Direct method* of frequency Modulation

$\rightarrow$ Reactance Modulation

Definition of Modulation Index

For *FM* deviation is proportion to modulating voltage E_m, i.e.

$$[\Delta\omega = \delta] \propto E_m \quad ; \quad \left(\frac{\delta}{E_m}\right) = K_f$$

$$\delta = K_f \cdot E_m \; Hz \quad \text{or} \quad \delta = K_f \cdot E_m \quad \text{rad}$$

$$m_f \cdot \omega_m = \Delta\omega = \delta = K_f \cdot E_m$$

Definition of Modulation Index for FM

$$m_f = \frac{K_f \,.\, E_m}{\omega_m}$$

For PM

$$m_p = (K_p \,.\, E_m)$$

In general, ***modulation index*** *β for angle modulation is defined as*

$$\beta = \begin{cases} K_P \; E_m & \text{for } PM \\ \dfrac{K_f \, E_m}{\omega_m} & \text{for } FM \end{cases}$$

Relation Between Δ, f_m, M_f

$\delta \rightarrow$ Frequency deviation

$m_f \rightarrow$ Modulation index

$f_m \rightarrow$ Modulation frequency.

$$\delta = m_f \cdot f_m$$

In general angle modulation index is given by

$$\beta = \frac{\Delta\omega}{\omega_m}$$

Main Point

1. In PM the angle is varied linearly with the modulating signal where as in FM the angle varies linearly with the integral of the modulating signal.
2. If we integrate the modulating signal $f(t)$ then Phase modulation it, we get FM signal.
3. If we differentiate $f(t)$ first and use it for Frequency modulation, the carrier we get the phase modulation signal.
4. PM and FM are inseparable since, any variation in phase of the carrier result in the variation in frequency and vice versa.
5. Relation between frequency deviation (δ), modulation index (m_f) and modulating frequency (f_m) is given by

$$\delta = m_f \cdot f_m \rightarrow \text{Hz}$$
$$\delta = m_f \cdot \omega_m \rightarrow \text{rad/sec}$$

6. The frequency deviation controls the Bandwidth required to transmit FM signal

$$\delta \uparrow \rightarrow BW \uparrow$$

7. The modulation index m_f controls the number of side Band to be included in FM signal.

$$m_f \uparrow \rightarrow \text{number of SB} \uparrow$$

8. In contrast to AM where only 2SB are present, FM signal has infinite number of SB, Therefore ideal, Bandwidth required to transmit FM is infinite.
9. If the power contribution due to SB after a specified number is negligible, the BW is restricted to a finite value.

$$BW\big|_{FM} \cong 2(\delta + f_m)$$

This is known as *Carson's* rule

1. For a narrow Band FM, $m_f < 0.6$
2. For wide Band FM, $m_f > 0.6$
3. For angle modulation amplitude is always constant
4. The power of FM carrier is same as modulated carrier. By proper choice of m_f the efficiency of transmission can be raised to 100%.

6.10.8 REACTANCE MODULATOR

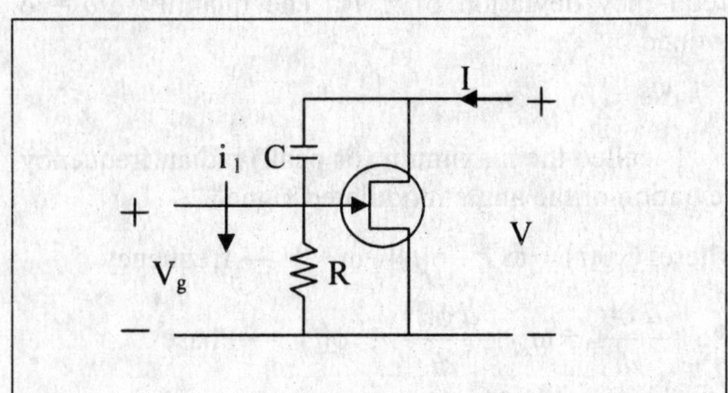

$$Z_o = \frac{1}{j\omega C_{eq}}$$

where, $C_{eq} = g_m RC$

Therefore, the output impedance of the given circuit is capacitive

If $X_c >> R$

Let, $\left(\frac{X_c}{R}\right) = n$ and $(n >> 1)$

$$Z_o = \frac{-j X_c}{g_m R} = \frac{1}{j g_m} \frac{X_c}{R}$$

$$Z_o = \left(\frac{n}{jg_m}\right) \quad ; \quad C_{eq} = \left(\frac{g_m}{n\omega}\right)$$

Basic Principle of Reactance Modulator

1. The modulating signal is applied at the gate terminal of the FET.
2. The trans conductance g_m of FET varies w.r.t to the input voltage V_g or V_m

$$f = \frac{1}{2\pi\sqrt{(LC)}}$$

3. This FET circuit has output impedance which is purely Capacitive in nature, which is equal to $C_{eq} = g_m RC$
4. The C_{eq} is connected effectively in parallel with a fixed frequency carrier oscillator The frequency of carrier oscillator is given by

$$f = \frac{1}{2\pi\sqrt{(LC)}}$$

where effect of C_{eq} is taken then

$$C' = C_{eq} + C_o$$

Frequency at the output

$$= \frac{1}{2\pi\sqrt{[L_o\,(C_o + C_{eq})]}}$$

5. Hence the frequency of the Carrier is effectively varied in accordance with the amplitude of input modulating signal $V_m = V_g$.
6. The result is a frequency modulated signal.

Disadvantage

1. The carrier frequency changes with the reactance parameter variation. The parameter change with the time.
2. Separate circuit is needed for phase modulation.

6.10.9 ARMSTRONG METHOD

FM Signal Range = 88 MHz to 108 MHz $\cong f_c$

Frequency deviation $\delta = \pm$ 75 kHz

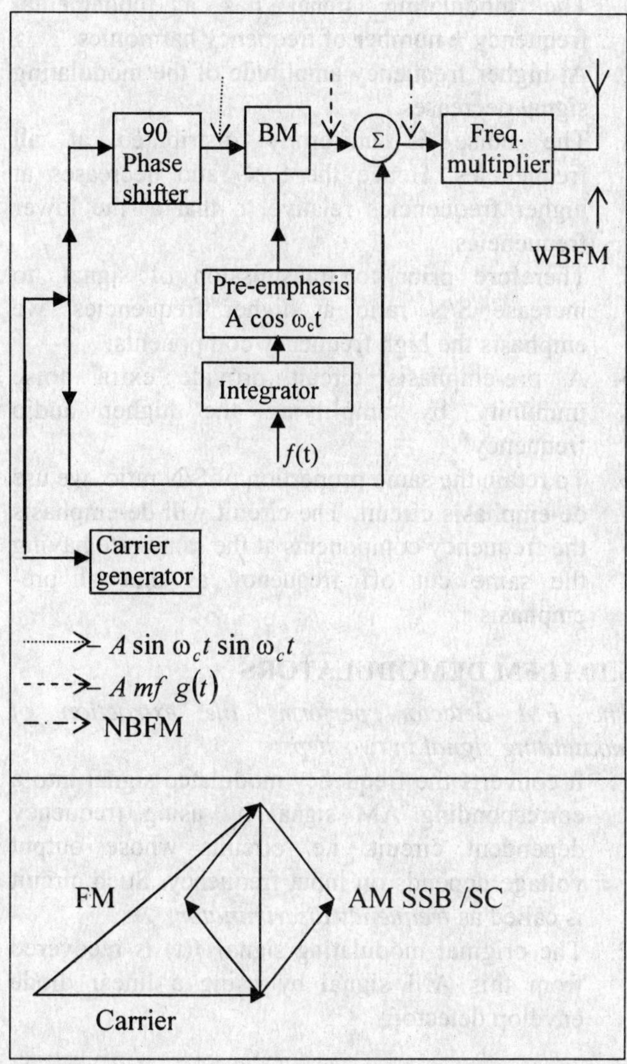

Generation of narrow Band FM using Balance Modulator.

Narrow Band FM is multiplied than the result is multiplication of frequency and modulation index which result as a wide Band FM and known as

$$\phi_{FM}(t) = A\cos\omega_c t - A\,m_f\; g(t)\sin\omega_c t$$

$$\phi_{PM}(t) = A\cos\omega_c t - A\,m_f\; f(t)\sin\omega_c t$$

Note: The expression for FM and PM are very much similar to the expression of AM signal, with only slight modification. Hence the narrowband FM is same as AM.

AM + NBFM → SSB with carrier.

6.10.10 PRE-EMPHASIS CIRCUIT (HPF) AND DE-EMPHASIS CIRCUIT [L P F]

Need of Pre-Emphasis Circuit

1. The modulating signal has a fundamental frequency + number of frequency harmonics.
2. At higher frequency amplitude of the modulating signal decreases.
3. The noise is uniformly distributed at all frequencies. Hence the S/N ratio decreases at higher frequencies relative to that at the lower frequencies.
4. Therefore prior to transmission of signal, to increase S/N ratio at higher frequencies, we emphasis the high frequency components.
5. A pre-emphasis circuit provide extra noise immunity by amplifying the higher audio frequency.
6. To retain the same proportion of S/N ratio, we use de-emphasis circuit. The circuit will de-emphasis the frequency components at the same rate having the same cut off frequency as that of pre-emphasis.

6.10.11 FM DEMODULATORS

The FM detector performs the extraction of modulating signal in two steps:

1. It converts the frequency modulated signal into a corresponding AM signal by using frequency dependent circuit, i.e. circuits whose output voltage depends on input frequency. Such circuit is called as *frequency discriminator*.
2. The original modulating signal f(t) is recovered from this AM signal by using a linear diode envelop detector.

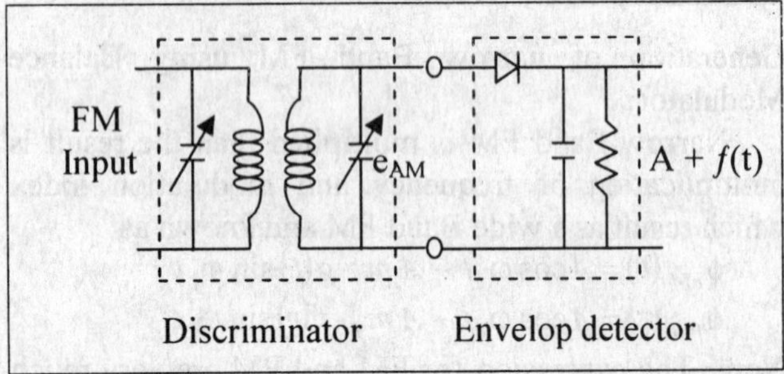

Feature of Foster-Seeley discriminator (phase discriminator)

1. Foster Seeley discriminator is used in FM receiver AFC system. AFC system consist in.
   a. Picking up voltage between the junction of load and the reference terminal.
   b. Apply this Dc voltage as the controlling voltage to the reactance modulator which is placed in shunt with the tuned circuit of local oscillator automatically in such a way that IF center
   c. Frequency remains constant. Separating out the DC component with the help of suitable filter.
2. The discriminator curve is almost linear over a range of ± 200 kHz. The performance is much better than the other type of FM Detector.

Disadvantage

Any variation in the amplitude of the input FM signal due of noise modifies the discriminator characteristic. The distortion is reduced using a *limiter*circuit which precede the discriminator stage.

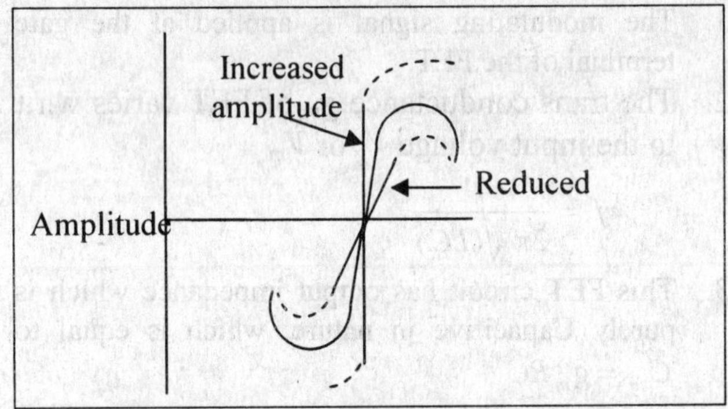

Advantage of Foster-Seeley

Much easier Alignment

Better linearity

Fewer tuned circuit

Modified version of Foster Seely Discriminator Ratio Detector

Ratio detector is an improvement over the Foster-seely discriminator and is widely used. As it does not respond to amplitude variation, a limiter is not needed.

Feature:

1. It gives an excellent noise free output.
2. Unlike foster seely discriminator, it requires no limiter, i.e. Greater limiting.
3. Relatively lesser components are required.
4. Since no limiter is required small signal AM rejection is considerably improved. The input required in ratio detector for noise free output is considerably smaller than that required in foster seely discriminator.

5. Although no separate limiter is required in the ratio detector, AM rejection is little inferior to that of foster seely discriminator with limiter.
6. The ratio detector does not provide protection against long period variation in signal strength and hence AGC is required with this type of detector.
   The dc voltage developed against output or stabilizing capacitor is generally used as AGC control voltage.

Advantage of Slope Detector

Greater limiting.

6.10.12 COMPARISON BETWEEN AM AND FM

1. *Noise performance* (Better noise immunity is provided):
   a. Wideband FM has greater S/N ratio than AM. Greater the BW, better the noise performance.
   b. Narrow band FM S/N ratio as AM.
   c. In large noise case AM is better than FM.
2. *Common channel Interference:* FM is better than AM because of *Capture effect*. The interfering week signals received in FM from the neighboring channels are almost completely suppressed, but produce distortion.
3. *Externally generated noise pulses:* FM receiver responds slightly to noise pulse because of its wide band RF and IF circuit.
4. *Channel BW*: The wideband FM has a larger BW as compared to AM as FM produces a large number of sidebands. For AM BW = 15 kHz Where as for FM BW = 150 kHz.
5. *Operating carrier frequency*: Transmitted power is more useful. FM utilizes higher carrier frequency (88 – 108 MHz) because of its higher BW. If FM is used at the lower frequency (HF or MF). The number of channels available will be too less.
6. *Transmission efficiency:* FM has more/better efficiency than AM, but not better than AM-SC.
7. Less modulating power is required.

Disadvantage

1. Transmitter and receiver circuits are complex.
2. Limited range, i.e. range limited to line of sight (LOS) communication.
3. Bandwidth is large.

6.10.13 FREQUENCY SPECTRUM OF FM WAVE

Let, $f(t) = V_m \cos \omega_m t$

carrier $= A \sin \omega_c t$

$$f_{FM}(t) = A\cos\left(\omega_c t + m_f \int f(t)\,dt\right)$$

$$f_{FM}(t) = A\sin\left(\omega_c t + m_f \sin \omega_m t\right)$$

→ it is sin of a sin function

→ the solution for this function involve the use of Bassel function.

$$f_{FM}(t) = A\sin\left(\omega_c t + m_f \sin\omega_m t\right)$$
$$= A\,J_o(m_f)\sin\omega_c t$$
$$+ J_1(m_f)\left(\left[\sin\omega_c t + \omega_m\right]t - \sin\left(\omega_c - \omega_m\right)t\right)$$
$$+ J_2(m_f)\left(\left[\sin\omega_c t + 2\omega_m\right]t - \sin\left(\omega_c - \omega_m\right)t\right)\ldots$$

It is seen that the output consist of a carrier and apparently infinite sidebands each preceded by J coefficient. These are Bessel function.

Main Points

1. Unlike AM where there ore only three frequencies, FM has an infinite number of sidebands.
2. The J coefficient eventually decrease in value as n increases but not in any manner.
   The modulation index thus determines how many sideband components have significant amplitude.
3. The sidebands at equal distances from f_c have equal amplitudes.
4. In AM increased depth of modulation increases the sideband power and therefore the total transmitted power.

In FM the total transmitted power always remains constant but with increased depth of modulation, the required sideband is increased.

6.11 SAMPLING THEOREM

Modulating signal = $f(t)$; $m(t)$

Sampling signal carrier = $f_s(t)$, i.e. Pulse train or impulse train (for pulse modulation).

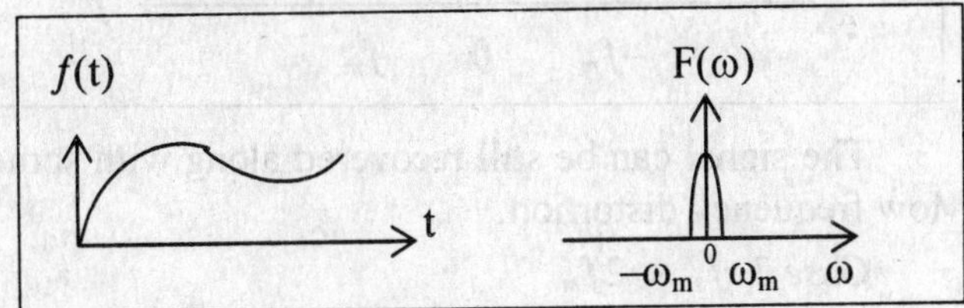

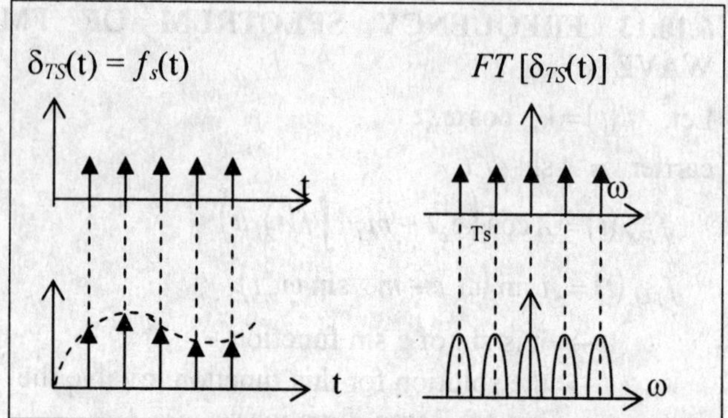

$f(t) \rightarrow$ Modulating signal is a band limited signal with highest modulating frequency component of frequency or ω_m.

$$f(t) = \delta_{TS}(t) = \sum_{n=-\infty}^{\infty} \delta(t - nT_S)$$

$$= P_{TS}(t) = \sum_{n=-\infty}^{\infty} P(t - nT_S)$$

Recovery of f(t)

Case 1. $f_s > 2f_m$

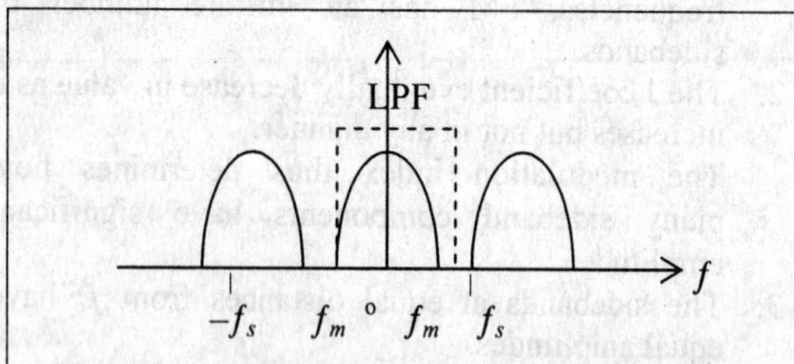

The signal can then always be recovered from its sample value with a LPF whose cut off frequency is f_m. With minimum amount ofnoise.

Case 2. $f_s = 2f_m$

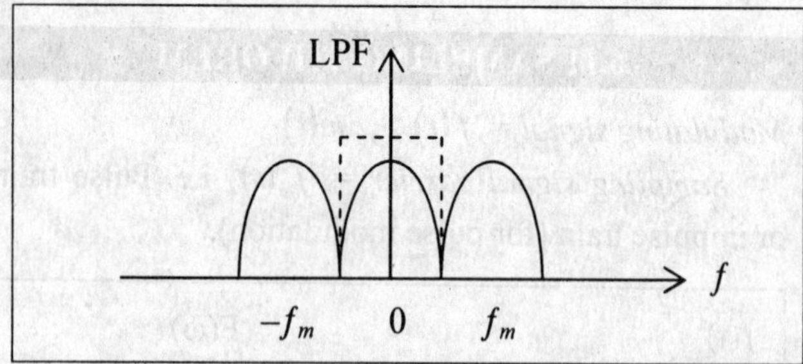

The signal can be still recovered along with some low frequency distortion.

Case 3. $f_s < 2f_m$

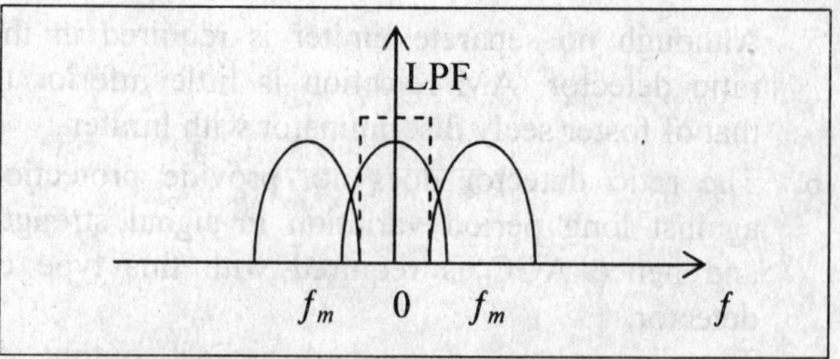

When $f_s < 2f_m$ then the spectral of $f(t)$ at frequency 0, f_s, $2f_s$ ...etc will overlap each other so that the resultant spectrum of the sampled signal all together changed. Therefore by passing through any LPF having any cut off frequency the original signal be never to recovered from its sample value.

Hence in order to recover the signal from its sample version we must have $f_s \geq 2f_m$.

$f_{s(\min)}$ ***Nyquist rate*** $= 2f_m = 2\omega_m$

or $\quad T = (1/2f_m)$

The maximum internal of Sampling is known as ***Nyquist interval.***

Statement of Sampling Theorem

1. *Frequency domain statement:* A Band limited signal of finite energy which has no frequency components higher than f_m Hz is completely described by specifying the value of signal at the instant of time separated by.

$$Ts = \left(\frac{1}{2f_m}\right) = \left(\frac{1}{2\omega_m}\right) \text{sec}$$

Sampling Theorem Mathematical Proof

The $f(t)$ is band limited signal which is limited to frequency $\pm\,\omega_m$.

Let, $f(t) \leftrightarrow F(\omega)$

Also $\delta_{TS}(t)$ = Sampling signal

$$F_s(\omega) = \omega_o \sum_{-\infty}^{\infty} F(\omega - n\,\omega o)$$

Aliasing Effect

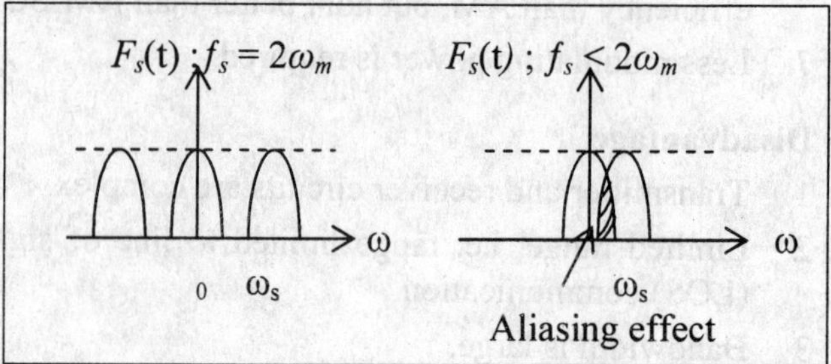

The phenomena of high frequency component in the Spectrum of the original signal f(t) effectively taking on the identity of lower frequency component in the spectrum of its sampled version is called the ***Aliasing effect***.

This effect occurs when the sampling is done at a rate lower than f_s i.e. $f_s < 2f_m$. Under this condition the original signal f(t) can never be recovered from its sampled value.

To minimize Aliasing effect:

1. The filtered signal is sampled at a rate slightly higher than the Nyquist rate $\omega_s > 2\omega_m$.
2. Prior to sampling a LP pre Aliasing filter is used to attenuate those high frequency components of that lie out side the Band of interest.

6.12 PULSE CODE MODULATION

Basic Principle of PCM

The essential processes of PCM are sampling, quantizing and encoding as shown in Figure.

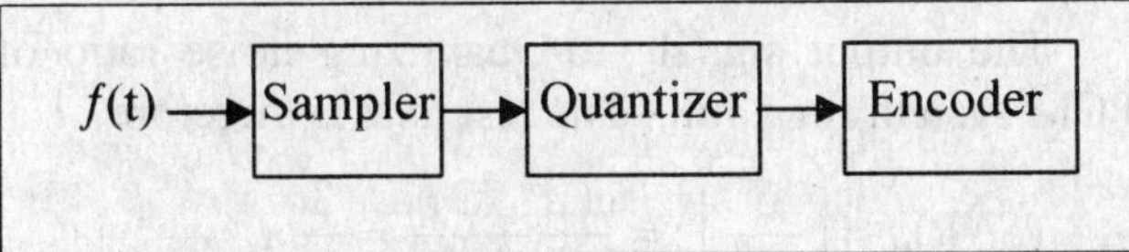

Sampling is the process in which a continuous-time signal is sampled by measuring its amplitude t discrete instants. Representing the sampled values of the amplitude by a finite set of levels is called quantizing. Designating each quantized level by a code is called encoding.

While sampling converts a continuous time signal to a discrete-time signal, quantizing converts a continuous amplitude sample to a discrete–amplitude sample. Thus sampling and quantizing operations transform an analog signal to a digital signal.

The quantizing and encoding operations are usually performed in the same circuit, which is called analog-to-digital (A/D) converter. The combined use of quantizing and encoding distinguishes PCM from analog pulse modulation techniques.

In the following sections, we discuss the operations of sampling, quantizing, and encoding:

A. Uniform Quantizing

An example of the quantizing operation is shown in figure. We assume that the amplitude of *m(t)* is confined to the range ($-m_p$, m_p). As illustrated in figure, this range is divided in L.

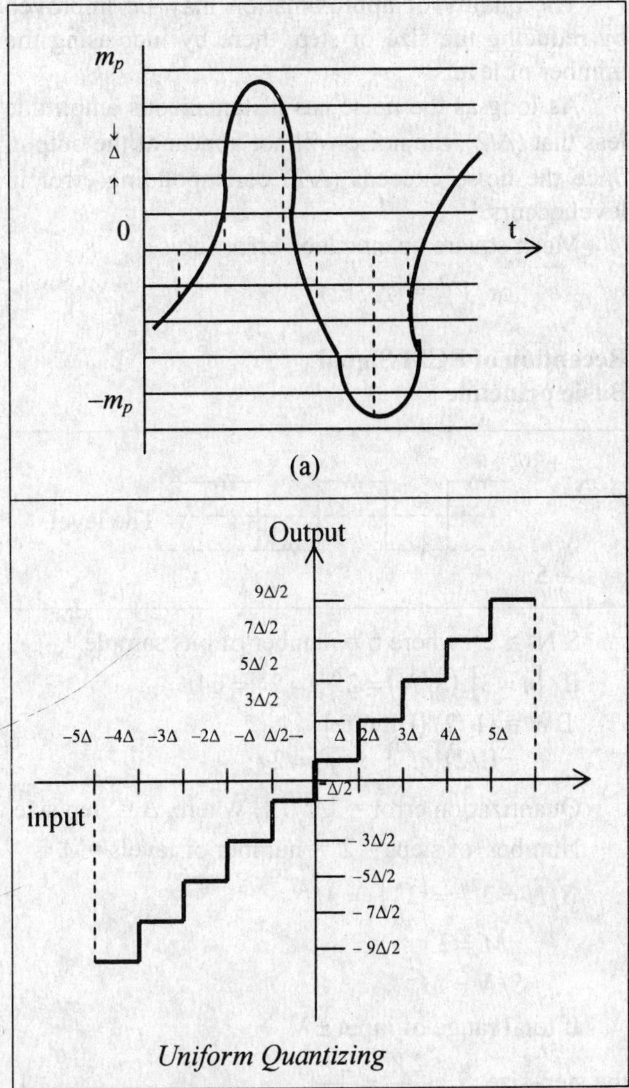

Uniform Quantizing

Uniform Quantizing

Zones, each of step size (Δ = s) given by

$$\Delta = \frac{2m_p}{L} \qquad (1)$$

where s = Δ and L = M.

A sample amplitude value is approximated by the midpoint of the interval in which it lies. The input-output characteristics of a uniform quantizer are shown in above figure.

Quantization of Signal

To eliminate the noise, we equalize a signal m(t) to new m_q(t) which approximated equal to m(t).

However the signal $m_q(t)$ has the great advantage that it is large measure separable from this additive noise.

The quality of approximation may be improved by reducing the size of step, there by increasing the number of levels.

As long as the noise has instantaneous amplitude less that $(\Delta/2)$, the noise will not appear at the output, once the noise exceeds $(\Delta/2)$ corresponding error in level occurs.

Mean square quantization error (noise)

$\overline{e^2}$ or $V_n^2 = (\Delta^2/12)$

Reception of PCM Signal

Basic principle

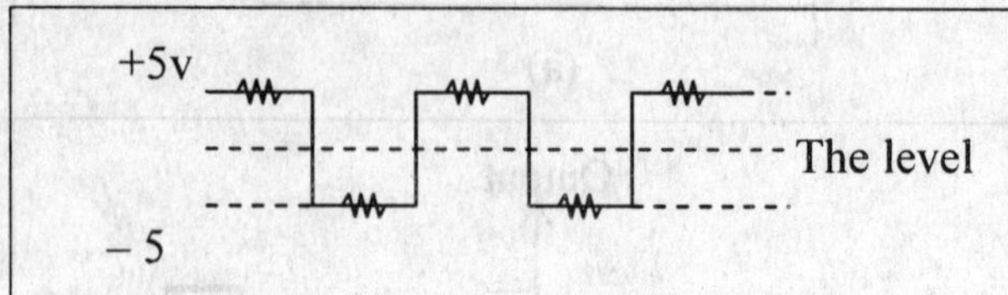

S/N ≅ 2^{2n} where n = number of bits sample

If $[n=3]:(S/N)=2^{2(3)}=2^6=64$

BW = (1 /2) [Bit rate]

$=(1/2)[nf_s]$; $f_s = 2f_m$

Quantization error = $(\Delta^2/12)$ where, Δ = step size

Number of steps = 2^n = number of levels =M

$S/N = 2^{2n} = (2^n)^2 = M^2$

$\therefore \quad M = 2^n$

$S/N = M^2$

If total range of input ± V

the step-size $S = \frac{2V}{2^n} = \frac{2V}{M}$

if n ↑ → S / N ↑

→ BW ↑

→ Design of quantizer becomes complex

→ Quantization error ↓

Quantizing Noise

The difference between the input and output signals of the quantizer becomes the quantizing error, or quantizing noise. It is apparent that with a random input signal, the quantizing error q_e varies randomly within the interval (s = Δ).

$$\frac{-\Delta}{2} \le q_e \le \frac{\Delta}{2} \quad (2)$$

Assuming that the error is equally likely to lie anywhere in the range (–Δ/2, Δ/2), the mean-square quantizing error (q_e^2) is given by

$$q_e^2 = 1\int_{-\Delta/2}^{\Delta/2} q_e^2\, dq_e = \frac{\Delta^2}{12} \quad (3)$$

Substituting Equation (1) into Equation (3) we have

$$(q_\epsilon^2) = \frac{m_p^2}{3L^2} \quad (4)$$

The peak-to-peak excursion of the quantizer input is 2A. Form Equation. The quantizer step size is

$$\Delta = \left(\frac{2S}{L}\right)$$

Then from Equation. The average Quantizing noise power is

$$N_q = q_e^2 = \frac{\Delta^2}{12} = \left(\frac{2A}{L}\right)^2 \times 1 = \frac{A^2}{3L^2}$$

Signal power $S_o = \left(\frac{A^2}{2}\right)$

The output signal –to-Quantizing noise ratio of PCM system for a full-scale test tone is therefore

$$(SNR)_0 = \left(\frac{S}{N_q}\right)_0 = \frac{(A^2/2)}{A^2/(3L^2)} = \frac{3}{2}L^2$$

Expressing this in decibels, we have

$$\left(\frac{S}{N_q}\right) dB = 10\log\left(\frac{S}{N_q}\right)_0 = 1.76 + 20\log L$$

$$= 1.76 + 20\log 2^n \; [L = 2^n]$$

$$= 1.76 + 6.02\, xndB$$

Where L = M is the number of Quantizing levels.

Above equation indicates that each bit in the code word of a binary PCM system contributes 6 dB to the output signal-to-Quantizing ratio. This is called the **6 dB rule**.

C. Non-uniform Quantizing and Companding

The same result can be achieved by first compressing signal samples and then using a uniform quantizing. The input-output characteristic of a compressor are shown in figure. The horizontal axis is the normalized input signal (that is, m/m_p), and the vertical axis is the output signal y. The compressor maps input signal increment Δm into large increment Δy for small input signals and small increments for large input signals. Hence by applying the compressed signal to a uniform quantizer, a given interval Δm contains a larger number of steps (or smaller step size) when m is small.

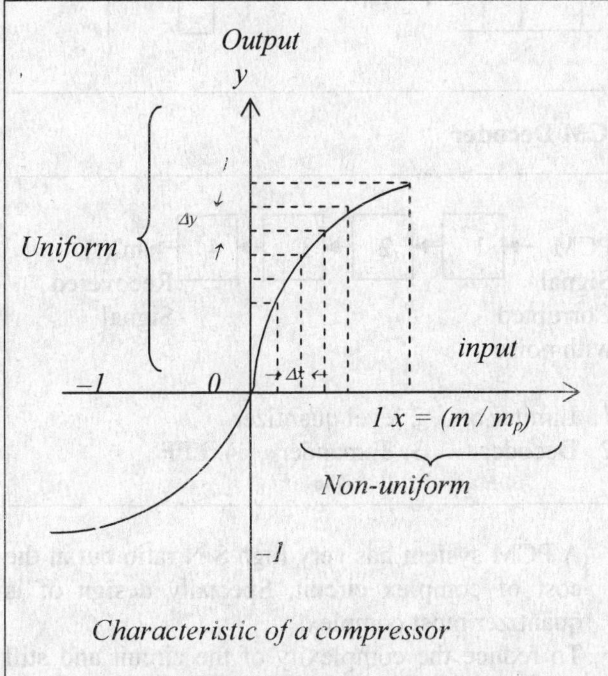

Characteristic of a compressor

A particular form of compression law that is used in practice (in North America and Japan) is the so-called μ law, defined by

$$y = \frac{\ell n\left(1 + \mu \left|m/m_p\right|\right)}{\ell n\,(1+\mu)} \operatorname{sgn}(m)\ ; \left|\frac{m}{m_p}\right| \le 1$$

where μ is a positive constant and

$$\operatorname{sgn}(m) = \begin{cases} +1 & m > 0 \\ 1 & m < 0 \end{cases}$$

or $|V| = \dfrac{\log\,(1 + \mu\,|\,m\,|)}{\log\,(1+\mu)}$

where = m and V the normalized input & output voltage and μ is positive complex.

The μ-law is neither strictly linear not strictly logarithmic, but it is approximately linear at low input level corresponding to μ |m| << 1, and approximately logarithmic at high input levels corresponding to μ | m | >> 1.

Another compression law that is used in practice (in Europe) is the so-called **A law**, defined by.

$$y = \begin{cases} \dfrac{A}{1 + \ell n\,A}\left(\dfrac{m}{m_p}\right) & \left|\dfrac{m}{m_p}\right| \le \dfrac{1}{A} \\[2ex] \dfrac{\left(1 + \ell n\,A\,|m|/m_p\right)}{1 + \ell n\,A} \operatorname{sgn}(m) & \dfrac{1}{A} \le \left|\dfrac{m}{m_p}\right| \le 1 \end{cases}$$

For the **μ law**, μ = 255 is used in digital telephone systems in North America. For the **A law**, A = 87.6 is used in European system.

To restore the signal samples to their correct relative level, an expander with a characteristic complementary to that of the compressor is used in the receiver. The combination of compression and expansion is called ***companding***.

Compressing used in T x or PCM decoder	Expanding used in Rx or PCM decoder

Compressing

1. For small input voltage level the signal strength is small so that the signal power is small.
2. Since the noise is informally distributed noise power is relatively same at all different voltage levels.
3. Therefore for small signal level the S/N ratio is small relative to that at higher voltage level. Therefore Prior to transmission the PCM signal we compress. The higher voltage level relative to lower +voltage level.
4. This is done by the use of compressor.
5. At the receiver in order to maintain the same relative S/N ratio we use the expanding technique where we expand the higher voltage level relative to smaller one.

This is done by the use of expander in Rx.

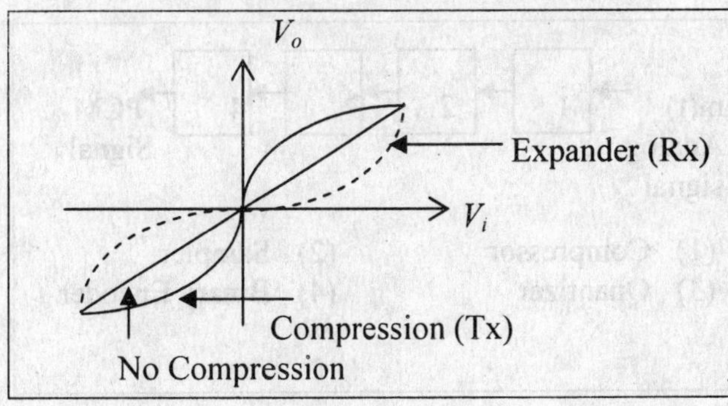

Encoding

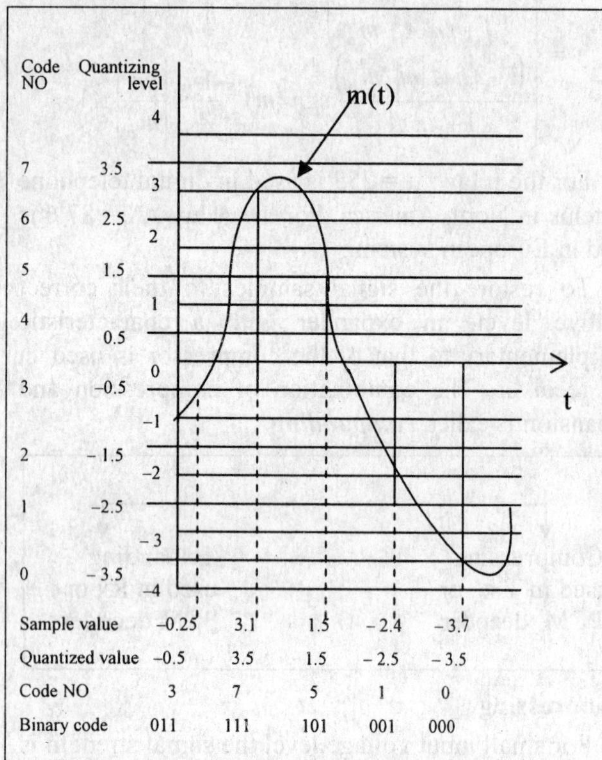

An encoder in PCM translates the quantized sample into a code number. Usually the code number is converted to its representation in binary sequence. The binary sequence is converted to a sequential string of pulses for transmission. In this case the system is referred to as binary PCM. The essential features of binary PCM are shown in Figure.

Bandwidth Requirements of PCM

$$2\,f_{PCM} = nf_s$$

or, $$f_{PCM} = \frac{n\,f_s}{2} \geq n\,f_m \ Hz$$

PCM Encoder

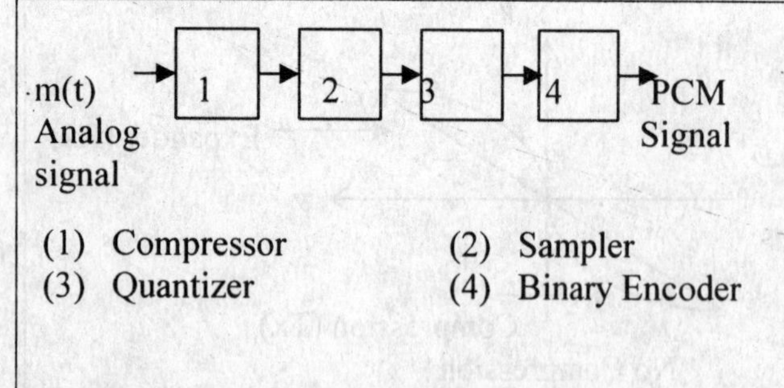

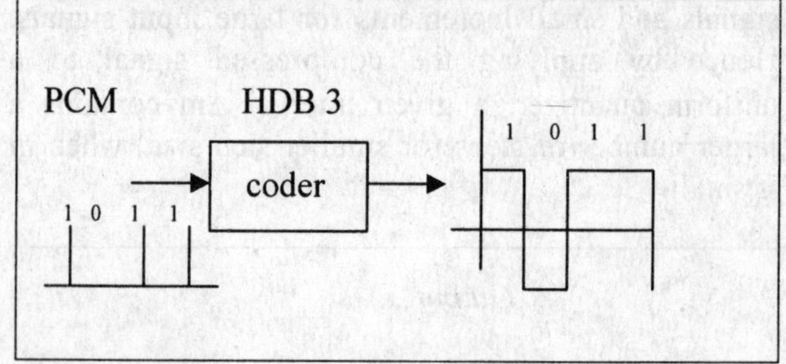

PCM Decoder

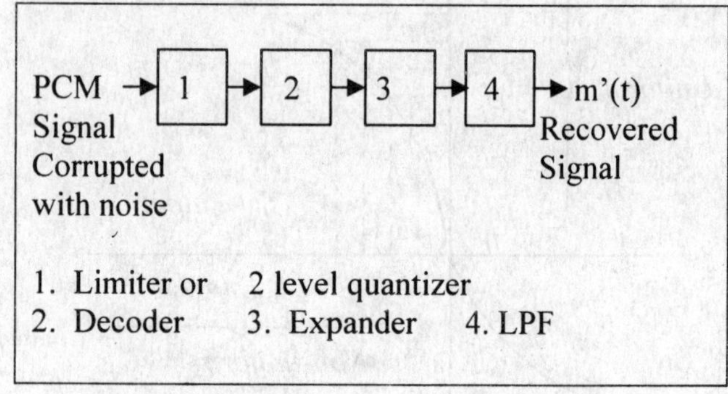

1. A PCM system has very high S/N ratio but at the cost of complex circuit. Specially design of is quantizer most complex.
2. To reduce the complexity of the circuit and still maintain same S/N ratio we use another coded modulation called *delta modulation.*

Advantages of PCM

1. PCM offers a considerable increase in S/N ratio at the receiver.
2. The decision by receiver is to take pulse is available or not; nothing to be known concerning its amplitude width etc.
3. PCM permits amplification of encoded signal without significant distortion.
4. PCM system promot flexibility and increasing system utilization design for one system can be used for the other too.
5. Using many repeater station PCM communication range can be increased.
6. PCM/TDM system applicable.
7. PCM-AM is common.

Disadvantage

PCM requires large bandwidth compare to analog system.

6.13 FREQUENCY AND TIME DIVISION MULTIPLEXING

6.13.1 Frequency–Division Multiplexing

An important signal processing operation is multiplexing, whereby a number of independent signals can be combined into a composite signal suitable for transmission over a common channel. Voice frequencies transmitted over telephone over telephone systems, for example, range form 300 to 3400Hz. To transmit a number of these signals over the same channel, the signals must be kept apart so that they do not interfere with each other, and thus they can be separated at the receiving end. This is accomplished by separating the signals either in frequency or in time. The technique of separating the signals is referred to as frequency-division multiplexing (FDM), whereas the technique of separating the signals in time is called time-division multiplexing (TDM).

A block diagram of an FDM system is shown in figure. The incoming message signals are assured to be of the low pass type, but their spectra do not necessarily have nonzero values all the way down to zero frequency. Following each signal input, a low pass filter, is designed to remove high frequency components that do not contribute significantly to signal representation but are capable of disturbing other massage signals that share the common channel. These low-pass filters may be omitted only if the input signals are sufficiently band limited initially. The filtered signals are applied to modulators that shift the frequency ranges of the signals so as to occupy mutually exclusive frequency intervals. The necessary carrier frequencies needed perform theses frequency translations are obtained form a carrier supply. The most widely used method of modulation in frequency-division multiplexing is single sideband modulation, which, in the case of voice signals, requires a bandwidth that is approximately equal to that of the original voice signals. Each voice input is usually assigned a bandwidth of 4kHz. The band pass filters following the modulation are used to restrict the band of each modulated wave its prescribed rang. The resulting band pass filter outputs are next combined in parallel to form the input to the common channel, At the receiving terminal, a bank of band pass filters, with their inputs connected in parallel, is used to separate the message signals on a frequency occupancy basis,. Finally, the original message signals are recovered by individual demodulations. Note that the FDM system shown in figure operates in only one direction To provide for two-way transmission, as in telephony, have to completely duplicate the multiplexing facilities, with the components connected in reverse order and with the signal waves proceeding from right to left.

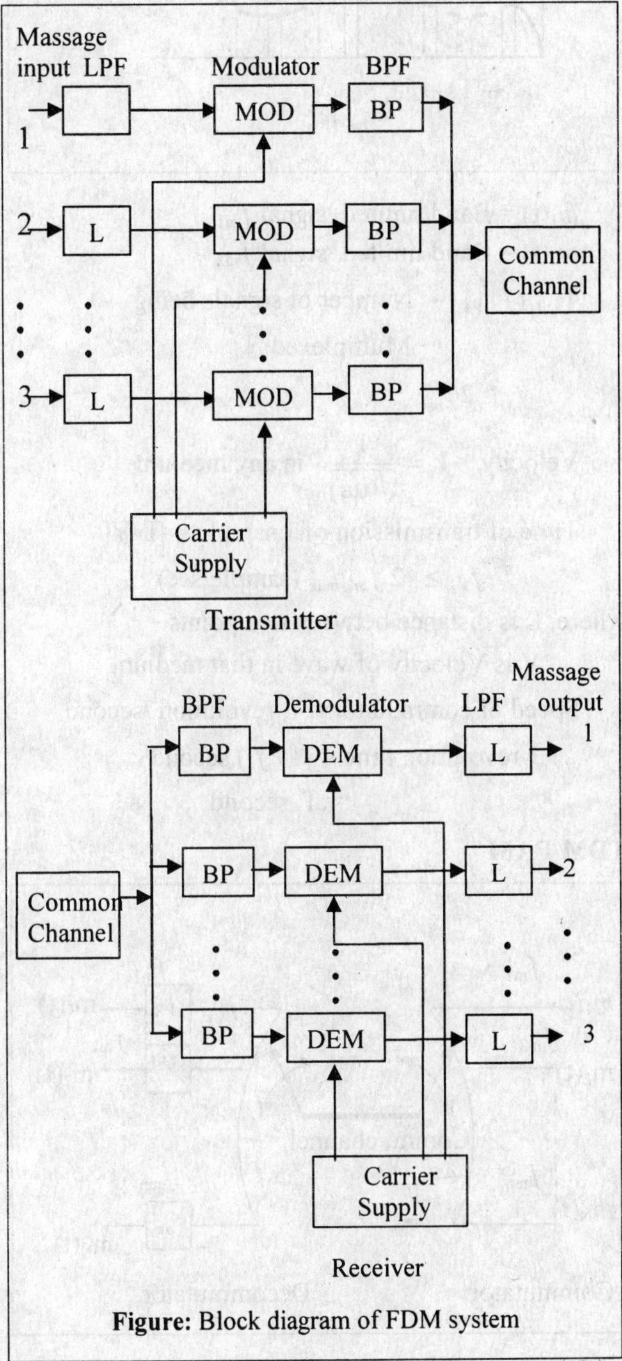

Figure: Block diagram of FDM system

6.13.2 Time Division Multiplexing

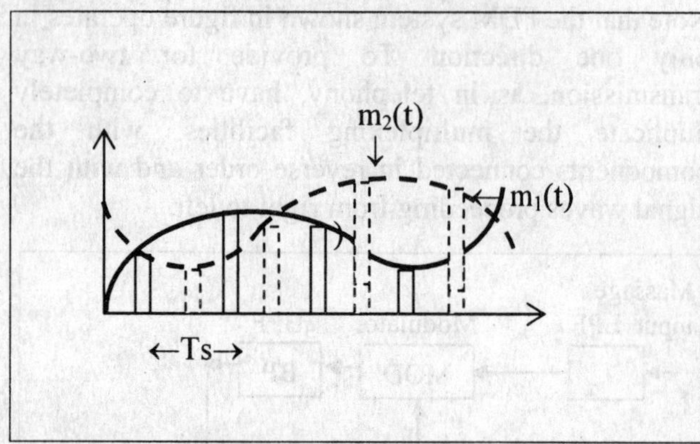

$m_1(t)$ = Band limited signal f_{m1}

$m_2(t)$ = Band limited signal f_{m2}

$T_s \uparrow (f_s \downarrow) \rightarrow$ Number of signals being Multiplexed. ↑

$f_s \geq 2 f_m$

Velocity $V = \frac{1}{\sqrt{(\mu\varepsilon)}}$ in any medium

Time of transmission of 1 sample = [L/V[

$f_s \geq 2 f_{m\ \max.}$ (sample/sec)

where, L is distance between two points

V is Velocity of wave in that medium

Speed of commutator = f_s revolution /second

$\therefore$ 1 revolution time = $(1 / f_s)$ second

= T_s second

TDM/PAM

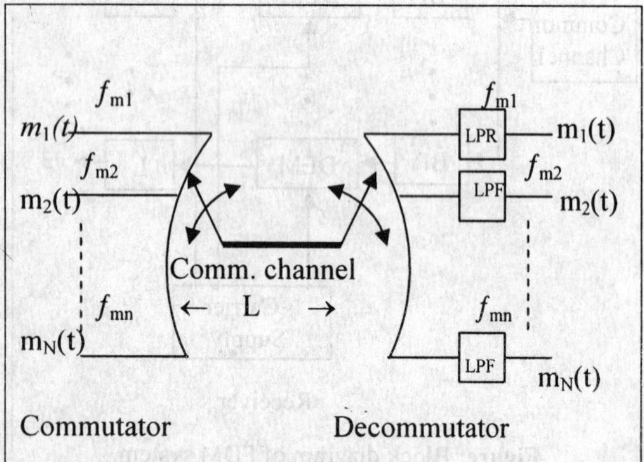

Main Points

1. The speed of commutator will be f_s revolution/ sec which is equal or greater, than 2 time the maximum modulating frequency component i.e. $f_s \geq 2 f_m$.
2. Lower the value of f_s, higher is T_s and higher is number of signals to be multiplexed.
3. The number of multiplexed signals will increase as the width of each sample decrease.
4. The commutator and Decommutator switches must be synchronized with some time delay between them, depending on the length of channel.
5. The cut off frequency of LPF at the Decommutator is in general different, corresponding to each signal.
6. Therefore by using PAM/TDM, number of signals can be transmitted simultaneously & still maintaining high S/N ratio compared to that of analog modulation.
7. Multiplexing of several PAM signal is possible because of various signals are kept distinct and are separately recovered by virtue of the fact that they are sampled at different times. Hence this is an example of TDM.
8. BW $= nf_s$

   $= 2\,nf_m$ number of channels.

   where n = Number of channels.

6.13.3 Comparison between FDM and TDM

There are two methods of multiplexing. viz FDM (frequency Division Multiplexing) TDM (Time Division Multiplexing). The basic deference between them in FDM the frequency scale is shared by different signals; in TDM the time scale is shared by different signals, as shown in figure. The bandwidth requirement of AM/SSB and PAM is n_{fm} Hz, and that of AM/DSB and PAM/AM is $2n_{fm}$ Hz. Thus it can be said that the bandwidth requirement of the FDM and TDM system is the same TDM is superior to FDM in the following ways:

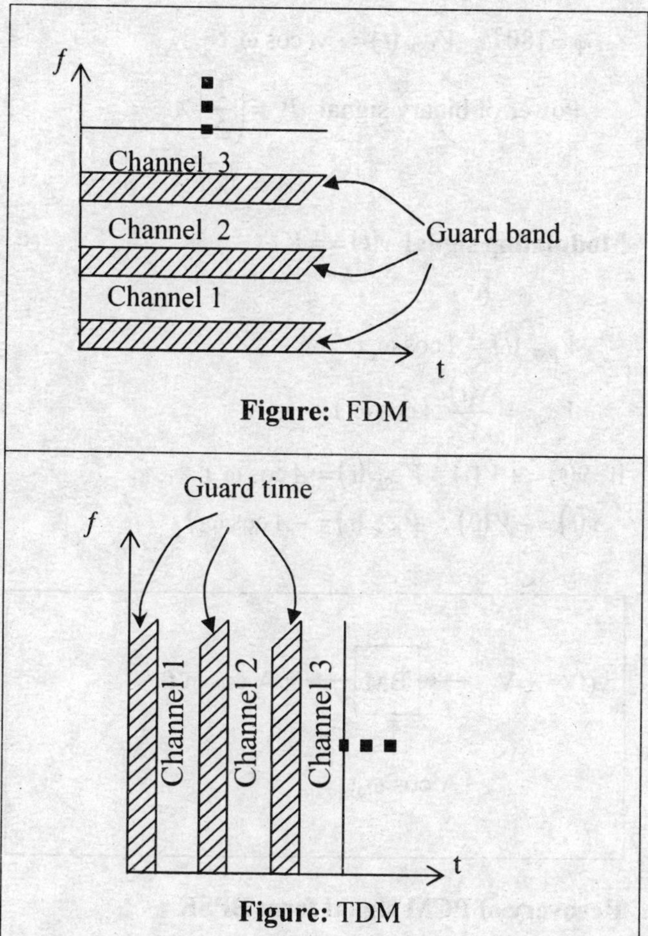

Figure: FDM

Figure: TDM

1. In the FDM system, different carriers are to be generated for different channels. Also, as each channel occupies a different frequency band, different band pass filters are required. On the other hand, in TDM system, all the channels require identical circuits, consisting of simple synchronous switches, gates, and a low pass filter. Thus the circuitry needed in the TDM system is much simple than the one needed in the FDM system.
2. The non-linearties in the various amplifiers of an FDM system produce harmonic distortion and, hence, the introduce interference within the channels. Therefore, the non-linearity requirements of an FDM system are much more stringent than those for a single channel. On the other hand, in a TDM system, the signals from different channels are allotted different time slots and they are not applied to the system simultaneously. Hence the non linearity requirements of a TDM system are the same as those for a single channel. Thus the TDM system is relatively immune to interference within the channels (inter channel cross-talk) as compared to the FDM system.

6.14 DIGITAL MODULATION

6.14.1 AMPLITUDE-SHIFT KEYING (ASK)

In ASK, the modulated signal can be expressed as

$$x_c(t) = \begin{cases} A\cos\omega_o t & \text{symbol 1} \\ 0 & \text{symbol 0} \end{cases}$$

Note that the modulated signal is still an on-off signal. Thus ASK is also known as on-off keying.

ASK (Amplitude shift keying)

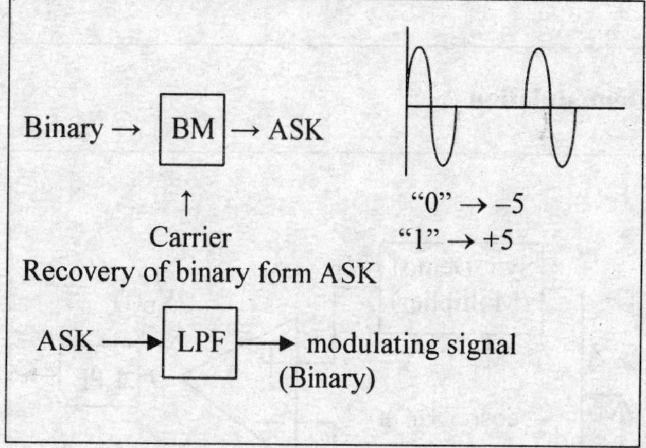

Not in use due to fading & does not give a satisfactory value for probability of error.

6.14.2 FREQUENCY SHIFT KEYING AND PULSE SHIFT KEYING FOR DIGITAL

Frequency-Shift Keying (FSK)

In FSK, the modulated signal can be expressed as

$$x_c(t) = \begin{cases} A\cos\omega_1 t & \text{symbol 1} \\ A\cos\omega_2 t & \text{symbol 0} \end{cases}$$

FSK (Frequency shift keying)

$$= A\cos\left(\omega_c' + d(t)\,\omega'\right)t$$

$$V_{PSK} = A\cos(\omega_o \pm \omega')t$$
$$= A\cos(\omega_o + \omega')t \rightarrow \text{"1"}$$
$$= A\cos(\omega_o - \omega')t \rightarrow \text{"0"}$$
$$\omega_H = \omega_o + \omega' \quad ; \quad \omega_L = \omega_o - \omega'$$

d(t)	$P_H(t)$	$P_L(t)$
+ 1	+1 V	0 V
–1	0 V	+ 1 V

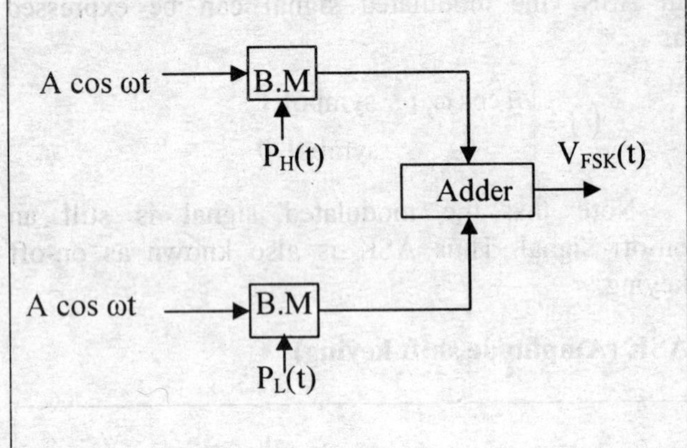

Demodulation

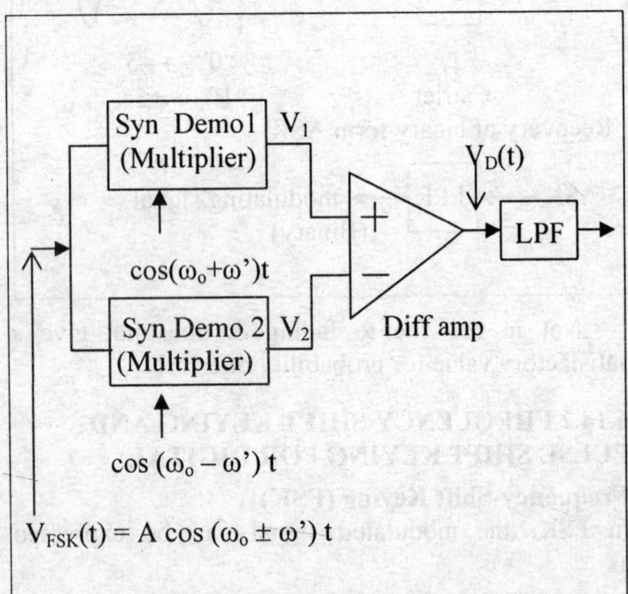

6.14.3 BPSK (binary phase shift keying)

$$V_{PSK}(t) = A\cos(\omega_c t + \phi)$$

let $\phi = 0^o$; $V_{PSK}(t) = A\cos\omega_c t$

$\phi = 180^o$; $V_{PSK}(t) = -A\cos\omega_c t$

Power of binary signal $P_s = \left(\frac{1}{2}\right)A^2$

$$A = \sqrt{2P_s}$$

Modulating signal $v(t) = \pm V$

$$\phi = 0^o$$
$$V_{PSK}(t) = A\cos\omega_c t$$
$$V_{PSK} = \frac{v(t)}{V}A\cos\omega_c t$$

if $v(t) = +V(1)$ $V_{PSK}(t) = A\cos\omega_c t$

$v(t) = -V(0)$; $V_{PSK}(t) = -A\cos\omega_c t$

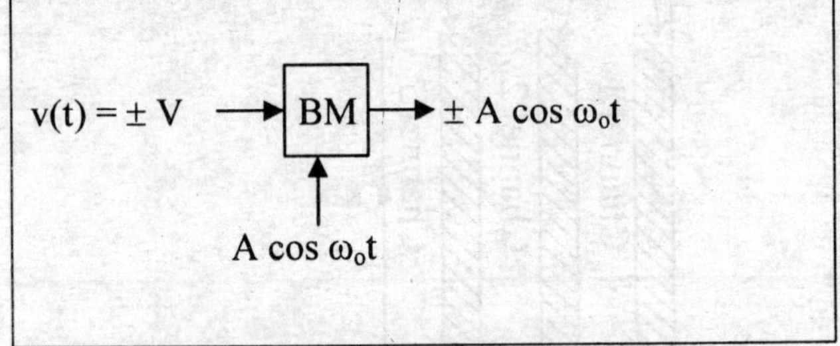

Recovery of PCM signal form BPSK

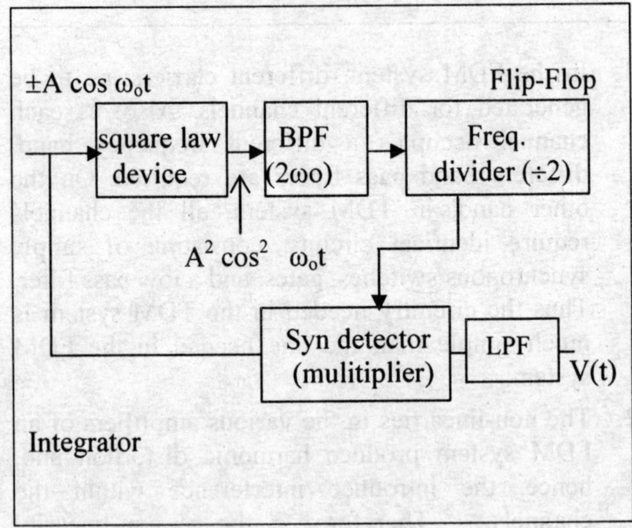

Advantage of PSK

1. Smaller Bandwidth.
2. Signaling speed fastest up to 2400 baud.
3. It is better for working under noisy condition.

Disadvantage of PSK

We need a synchronous detector and the associated circuit so that the receiver configuration of PSK system becomes more complex. To avoid this we use differential PSK, i.e. (*DPSK*) system which does not need synchronous detector.

6.14.4 DPSK (Differential Phase shift keying)

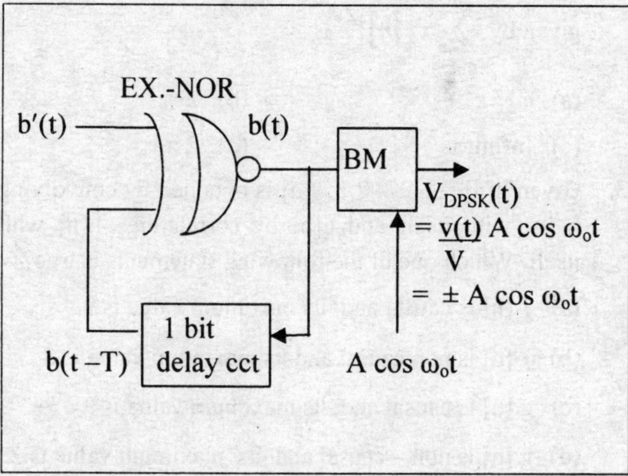

b'(t) = input modulating bit stream

b(t) = arbitrary generated bit stream.

ω_o = carrier frequency

T = delay equal to period of 1 bit

Demodulation of DPSK

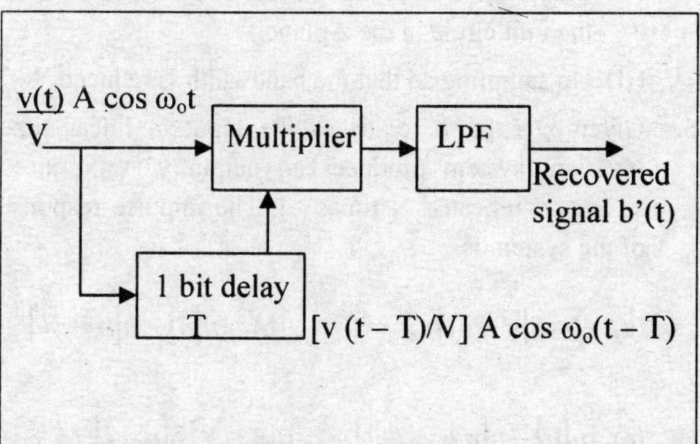

Characteristic of DPSK

1. The DPSK system avoids the use of synchronous detector. This is possible by generation of arbitrary waveform b(t) at the modulator. The generation is simple.
2. The error rate at DPSK is more so that S/N ratio is less than PSK system. *Bit error occur in pair or single.*

6. 14. 5 Quadrature PSK (QPSK)

[+ 45°+135°– 135°– 45°]

We have seen that when a data stream whose bit duration is T_b is to be transmitted by PSK the channel bandwidth must be normally $2f_b$ where $f_b = (1/T_b)$.

QPSK allows bits to be transmitted using the half bandwidth. required by PSK.

$$V(t) = b_e(t) A \cos \omega_o t + b_o(t) A \sin \omega_o t$$

for PSK BW = $2 \times (1/T_b)$

for QPSK BW = $2 \times (1/2T_b)$

PSK BW = (1/2) FSK BW

QPSK BW = (1/2) PSK BW

= (1/4) FSK BW

when $b_o = +1$; $S_o(t) = A \sin \omega_o t$

$b_o = -1$; $S_o(t) = -A \sin \omega_o t$

and $b_e(t) = +1$; $S_e(t) = A \cos \omega_o t$

$b_e(t) = -1$; $S_e(t) = -A \cos \omega_o t$

These four signals are represented as phase

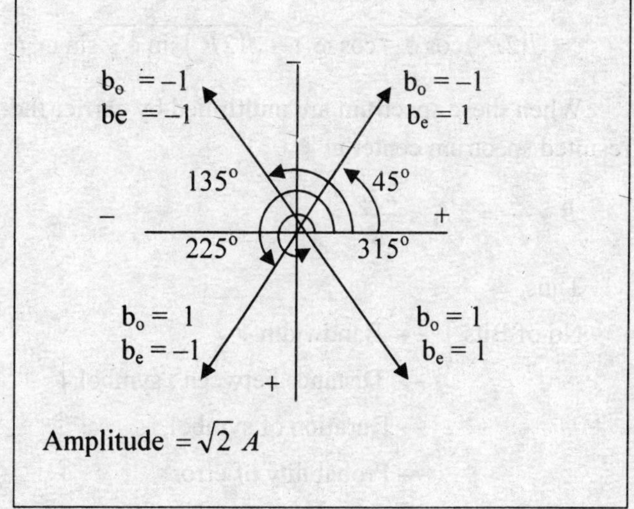

Note: After each T_b its phase charge by 90°

The bit stream $b_e(t) = (\pm 1V)$ is superimposed on a carrier A cos ω_o t and the b_o (t) = (±1) is superimposed on a carrier A sin ω_o t by the use of two multiplier (BM) to general $S_e(t)$ and $S_o(t)$. These signals are then adder to generate the T x R output signal $V_m(t)$

$V_m(t) = A\, b_e(t) \sin \omega_o t + A\, be(t) \cos \omega_o t.$

M-ARRAY-PSK

Single bits: In BPSK-Sinusoid phase

Difference = [2π/2] =180°

Two bits: In QPSK–Sinusoid phase difference

= [2π/4] = 90°

where, QPSK = Two bit symbol

N bit in a symbol, extending over the time NT_b

$2^N = M$ = Possible symbol

each symbol by N T_b = T_s which differ from on another by [2π/M]

In M-ary PSK the waveform used to identify the symbols are

$$V_m(t) = \sqrt{(2P_s)} \cos(\omega_o t + \phi_m)$$

where, m = 0, 1, 2 → (M – 1)

With the symbol phase angle given by

$$\phi_m = (2m+1)\left[\frac{\pi}{M}\right]$$

$$V_m(t) = \sqrt{(2P_s)} \cos(\omega_o t + \phi_m)$$

$$= \sqrt{(2P_s)} \cos \phi_m \cos \omega_o t - \sqrt{(2P_s)} \sin \phi_m \sin \omega_o t$$

When these spectrum are multiplied by carrier the resulted spectrum center at

$$B = \frac{2}{T_s} = 2f_s = \frac{2f_b}{N}$$

Thus,

No of Bits ↑ → Bandwidth ↓

→ Distance between symbol ↓

→ Duration of symbol ↑

→ Probability of error ↑

Except N = 1 and N = 2

MCQ Set–1

1. A real function f (t) has a Fourier transform F(ω). The Fourier transform of [f (t) – f (–t] is .
   (a) Zero (b) Real
   (c) Real and odd (d) Imaginary

2. Given $x[n] = \frac{\sin \omega_c n}{\pi n}$, the energy of the signal given by $\sum_{n=-\infty}^{\infty} x|[n]|^2$ is
   (a) ω_c/π (b) $\pi\omega_c$
   (c) Infinite (d) $2\pi\omega_c$

3. Given h[n] = [1, 2, 2,] , *f*[n] is obtained by convolving h [n] with itself and g[n] by correlating h[n] with itself. Which one of the following statements is true?
   (a) *f* [n] is causal and its maximum value is 9.
   (b) *f* [n] is non causal and its maximum value is 8
   (c) g [n] is causal and its maximum value is 9
   (d) g [n] is non – causal and its maximum value is .

4. **Bilinear transformation** avoid the problem of aliasing encountered with the use of impulse invariance through.
   (a) Mapping the entire imaginary axis of the s- plane on to the unit circle in the z - plane.
   (b) Pre-filtering the input signal to impose bank limitedness.
   (c) Mapping zeros of the left half of the s-plane inside the unit circle in the z-plane.
   (d) Up sampling so that the bandwidth is reduced.

5. Given x = [a, b, c, d] as the input, a linear time invariant system produces an output y = [x, x, x, x,........< repeated N times >]. The impulse response of the system is.
   (a) $\sum_{i=0}^{N-1} \delta[n-4i]$ (b) $u[n]-u[n-N]$
   (c) $u[n]-u[n-N-1]$ (d) $\sum_{i=0}^{N-1} \delta[n-i]$

6. Direct form implementation of the FIR structure shown in figure (a) is also implemented in figure (b) lattice form. The coefficients k_1 and k_2 are respectively.

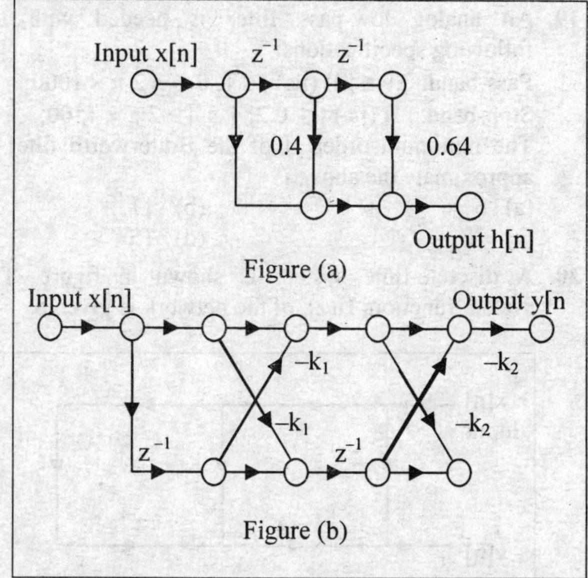

Figure (a)

Figure (b)

(a) 0.640 and –0.244 (b) 0.640 and 0.244
(c) 0.244 and 0.640 (d) – 0.244 and –0.640

7. The circuit in figure is a

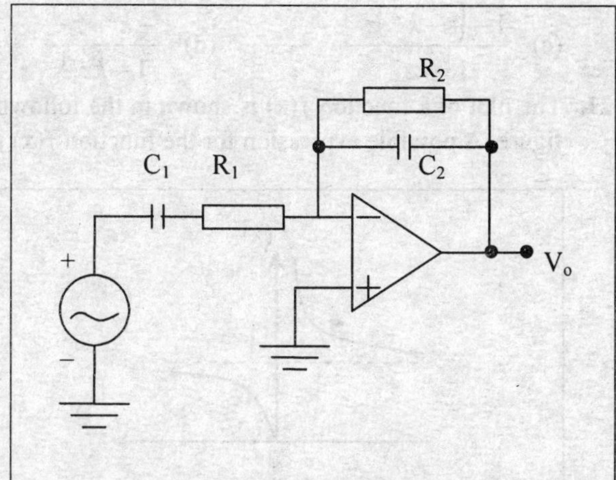

(a) Band-pass filter with lower cut-off $\omega_L = \dfrac{1}{R_1C_1}$ and higher cut off $\omega_H = \dfrac{1}{R_2C_2}$

(b) Band-reject filter with lower cut-off $\omega_L = \dfrac{1}{R_1C_1}$ and higher cut off $\omega_H = \dfrac{1}{R_2C_2}$

(c) Band-pass filter with lower cut-off $\omega_L = \dfrac{1}{R_2C_2}$ and higher cut off $\omega_H = \dfrac{1}{R_1C_1}$

(d) Band-pass filter with lower cut-off $\omega_L = \dfrac{1}{R_2C_2}$ and higher cut off $\omega_H = \dfrac{1}{R_1C_1}$

8. Given $X(z) = \dfrac{\frac{1}{2}}{1 - a z^{-1}} + \dfrac{\frac{1}{3}}{1 - b z^{-1}}$, $|a|$ and $|b|$ $< 1$with the ROC specified as $|a| < |z| < |b|$, x[0] of the corresponding sequence is given by

(a) 1/3 (b) 5/6
(c) 1/2 (d) 1/6

9. A direct form implementation of an LTI system with $H(z) = \dfrac{1}{1 - 0.7z^{-1} + 0.13z^{-2}}$ is shown in figure. The value of a_0, a_1 and a_2 are respectively

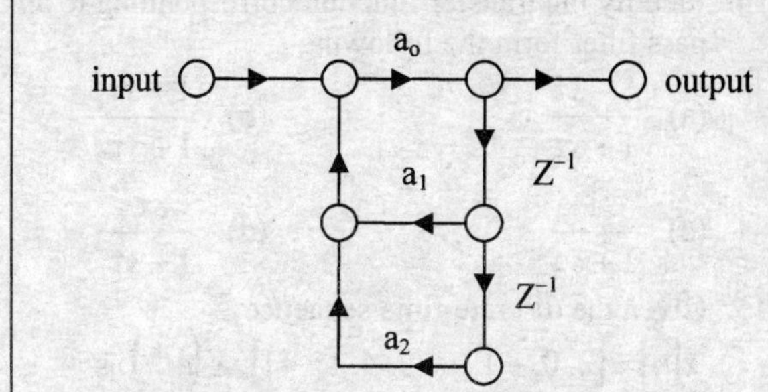

(a) 1.0, 0.7 and – 0.13 (b) – 0.13, 0.7 and 1.0
(b) 1.0, – 0.7 and 0.13 (c) 0.13, – 0.7 and 1.0

10. Alow-pass filter has sampling frequency $f_T = 10\,kHz$. The passband (f_p) and stop band (f_s) edge frequencies are 2 kHz and 3 kHz respectively. The frequency axis is pre-warped before applying bilinear transformation to get new angular frequencies, $\tilde{\Omega}_P$ and $\tilde{\Omega}_3$. The ratio of the warped edge frequencies $\left(\tilde{\Omega}_s / \tilde{\Omega}_P\right)$ is

(a) – 1.213 (b) 1.213
(c) 1.575 (d) 1.894

11. A casual, analog system has a transfer function $H(s) = \dfrac{a}{s^2 + a^2}$. Assuming a sampling lime of T seconds, the poles of the transfer function H(z) for an equivalent digital system obtained using impulse invariance method are at

(a) $\left(e^{aT}, e^{-aT}\right)$ (b) $\left(j\dfrac{a}{T}, -j\dfrac{a}{T}\right)$
(c) $\left(e^{jaT}, e^{-jaT}\right)$ (d) $\left(e^{aT/2}, e^{-aT/2}\right)$

12. A discrete-time signal x[n], suffered a distortion modeled by an LTI system with $H(z) = (1 - az^{-1})$, a is real and $|a| > 1$. The impulse response of a stable system that exactly compensates the magnitude of the distortion is

(a) $\left(\frac{1}{a}\right)^n u[n]$ (b) $-\left(\frac{1}{a}\right)^n u[-n-1]$

(c) $a^n u[n]$ (d) $a^n u[-n-1]$

13. A digital notch filter with a notch frequency of 60 Hz is to be transformed into one operating at a new notch frequency of 120 Hz. The sampling frequency is 400 Hz. A lowpass to lowpass transformation $z^{-1} = \frac{1-\alpha \bar{z}}{\bar{z}-\alpha}$ is used on the first to obtain the new notch filter. Then the value of α should be

(a) -0.95 (b) 0.95

(c) -0.46 (d) 0.46

14. Identify the transfer function corresponding to an all-pass filter form the following

(a) $\frac{1-s\tau}{1+s\tau}$ (b) $\frac{1+s\tau_1}{1+s\tau_2}$

(c) $\frac{1}{1+s\tau}$ (d) $\frac{s\tau}{1+s\tau}$

15. Given the discrete-time sequence $x[n] = [2, 0, -1, \; -3, 4, 1, \; -1], X(e^{j\pi})$ is

(a) 8 (b) 6π

(c) 8π (d) 6

16. A voltage signal $V(t) = A + B\sin \omega t$ is fed to an **oscilloscope** with the coupling mode set to DC, other setting being appropriate. The displayed wave can be expressed as

(a) $A + B\sin \omega t$ (b) $B\sin \omega t$

(c) A (d) $|A + B\sin \omega t|$

17. The continuous-time signal $x(t) = \frac{1}{a^2+t^2}$ has the Fourier transform $\frac{\pi}{a} e^{(-\alpha|\omega|)}$. The signal $x(t)$ cos bt has the Fourier transform

(a) $\frac{\pi}{2a}\left[e^{(-a|\omega-b|)} + e^{(-a|\omega+b|)}\right]$

(b) $\frac{\pi}{2a}\left[e^{(-a|\omega|)} + e^{(-a|\omega|)}\right]$

(c) $\frac{\pi}{2a}\left[e^{(-a|\omega|)} \cos b\omega\right]$

(d) $\frac{\pi}{2a}\left[e^{(-a|\omega-b|)} - e^{(-a|\omega+b|)}\right]$

18. The **fundamental period** of the sequence x[n] = 2 sin (1.3 πn + 0.5π) + 5 sin (1.2 πn) is

(a) 20 (b) $\frac{2\pi}{1.3\pi}$

(c) $\frac{2\pi}{1.2\pi}$ (d) 10

19. An analog low-pass filter is needed with the following specifications:
Pass-band: $0.9 \le |H(j\omega)| \le 1, 0 \le \omega \; 2\pi \times 1000$;
Stop-band: $|H(j\omega)| \le 0.2, \; |\omega| \ge 2\pi \times 1500$;
The minimum order N of the Butterworth filter to approximate the above is

(a) 6 (b) 11

(c) 12 (d) 15

20. A discrete-time system is shown in figure. The system function, H(z), of the network is given by.

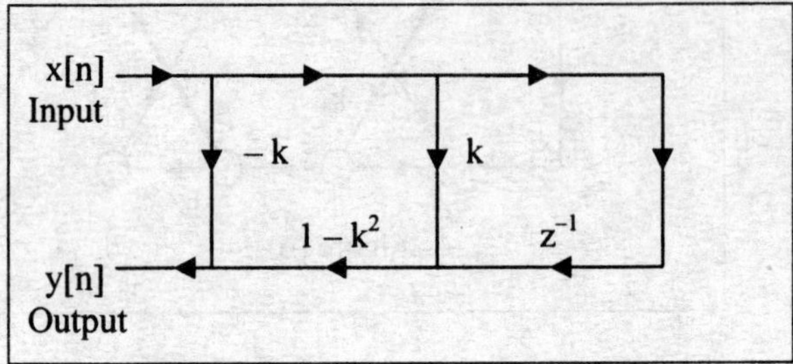

(a) $\frac{-k+z^{-1}}{1-(1-k^2)z^{-1}}$ (b) $\frac{(1-k^2)+z^{-1}}{1-kz^{-1}}$

(c) $\frac{1-(1-k^2)z^{-1}}{1-kz^{-1}}$ (d) $\frac{-k+z^{-1}}{1-kz^{-1}}$

21. The plot of a function $f(x)$ is shown in the following figure. A possible expression for the function $f(x)$ is

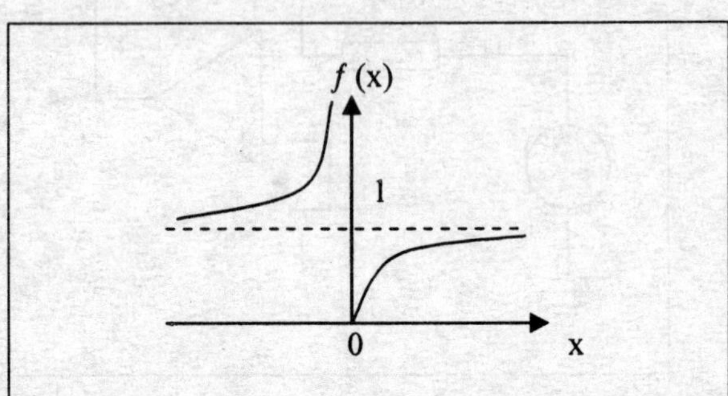

(a) $e^{(|x|)}$ (b) $e^{\left(\frac{-1}{x}\right)}$

(c) $e^{(-x)}$ (d) $e^{\left(\frac{1}{x}\right)}$

22. Given $x(t) * x(t) = t\, e^{(-2t)} u(t)$ the function x (t) is

(a) $e^{(-2t)} u(t)$ (b) $e^{(-t)} u(t)$

(c) $t\, e^{(-t)} u(t)$ (d) $0.5\, t\, e^{(-t)} u(t)$

23. The Fourier series for a periodic signal is given as. $x(t) = \cos(1.2\,\pi t) + \cos(2\,\pi t) + \cos(2.8\,\pi t)$

The fundamental frequency of the signal is

(a) 0.2 Hz (b) 0.6 Hz

(c) 1.0 Hz (d) 1.4 Hz

24. If the waveform shown in the following figure, corresponds to the second derivative of a given function $f(t)$, then the Fourier transform of $f(t)$ is

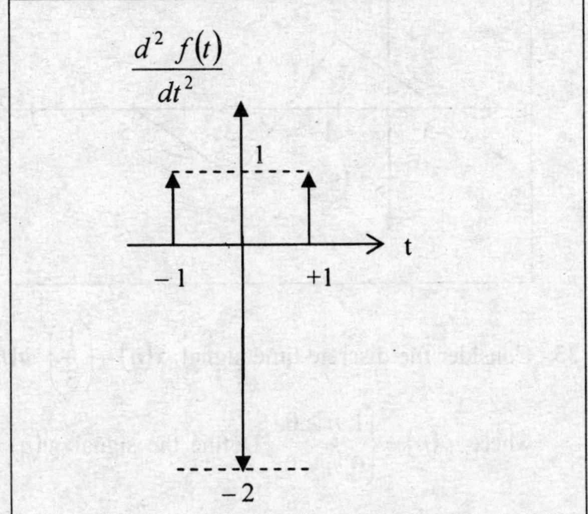

(a) $1 + \sin \omega$ (b) $1 + \cos \omega$

(b) $\frac{2(1 - \cos \omega)}{\omega^2}$ (d) $\frac{2(1 + \cos \omega)}{\omega^2}$

25. The Fourier transform of a function g(t) is given as.

$$G(\omega) = \frac{\omega^2 + 21}{\omega^2 + 9}$$

Then the function g(t) is given as

(a) $\delta(t) + 2 \exp(-3|t|)$ (b) $\cos 3\omega t + 21 \exp(-3t)$

(c) $\sin 3\omega t + 7 \cos \omega t$ (d) $\sin 3\omega t + 21 \exp(3t)$

26. The spectrum of a bandlimited singal after sampling is shown below. The value of sampling interval is

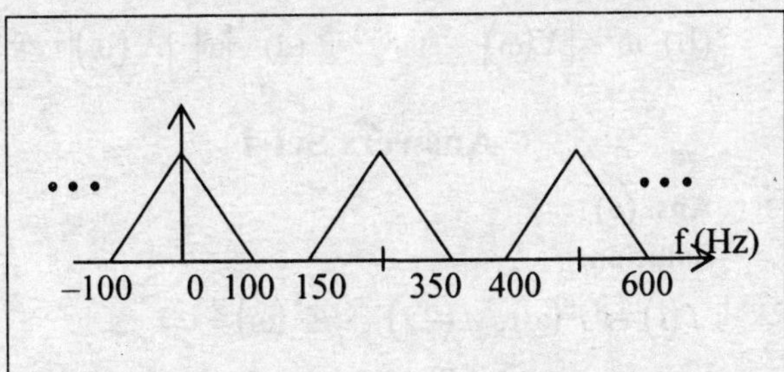

(a) 1 ms (b) 2 ms

(c) 4 ms (d) 8 ms

27. The magnitude of Fourier transform X(ω) of a function x(t) is shown below in figure (a). The magnitude of Fourier transform Y (ω) of another function y(t) is shown below in Fig. (b).The phases of X (ω) and Y(ω) are zero for all ω.The magnitude and frequency units are identical in both the figures. The function y (t) can be expressed in terms of x(t) as

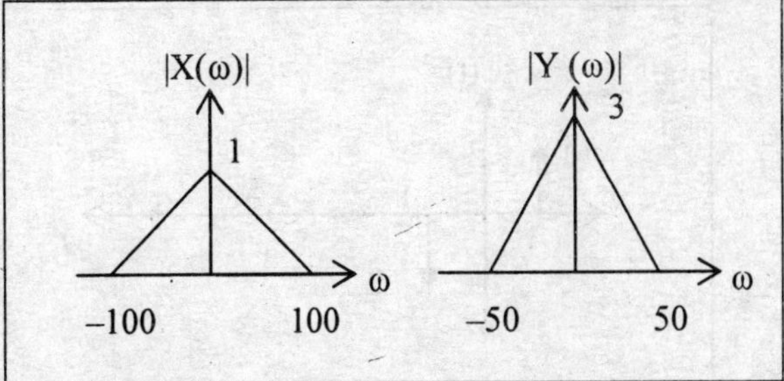

(a) $\frac{2}{3} x\left(\frac{t}{2}\right)$ (b) $\frac{3}{2} x(2t)$

(c) $\frac{2}{3} x(2t)$ (d) $\frac{3}{2} x\left(\frac{t}{2}\right)$

28. In the IIR filter below, a is a variable gain. For which of the following cases, the system will transit form stable to unstable condition?

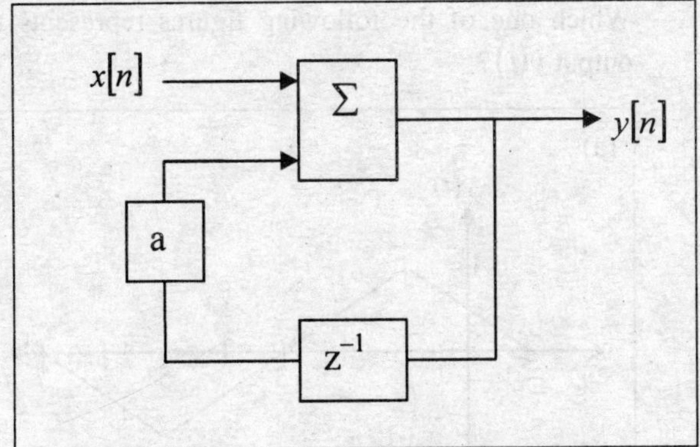

(a) $0.1 < a < 0.5$ (b) $0.5 < a > 1.5$

(c) $1.5 < a < 2.5$ (d) $2 < a < \infty$

29. A digital filter has the transfer function.

$$H(z) = \frac{z^2 + 1}{z^2 + 0.81}$$

If this filter has to reject a 50 Hz interference from the input, then the sampling frequency for the input signal should be

(a) 50 Hz (b) 100 Hz

(c) 150 Hz (d) 200 Hz

30. Let x(t) be a continuous-time, real-valued signal band limited to F Hz. The Nyquist sampling rate, in Hz, for $y(t) = x(0.5t) + x(t) - x(2t)$ is

(a) F (b) 2 F

(c) 4 F (d) 8 F

31. Consider the periodic signal

$x(t) = (1 + 0.5 \cos 40\pi t) \cos 200\pi t$, where t is in seconds. Its fundamental frequency, in Hz is

(a) 20 (b) 40

(c) 100 (d) 200

32. The signal $x(t)$ and $h(t)$ shown in the figure are convolved to yield $y(t)$.

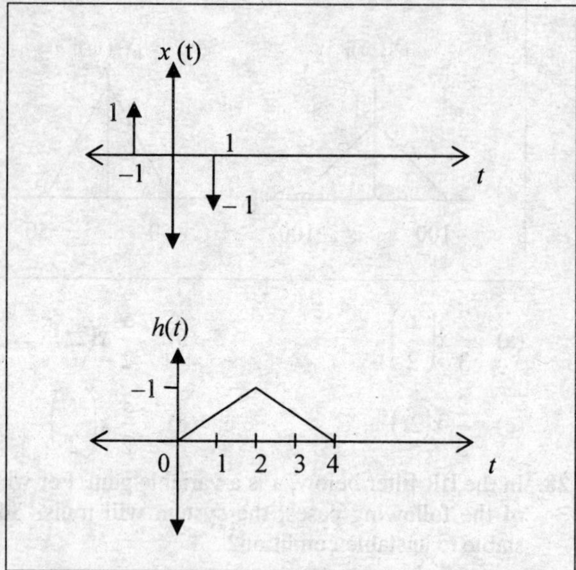

Which one of the following figures represents the output $y(t)$?

(a)

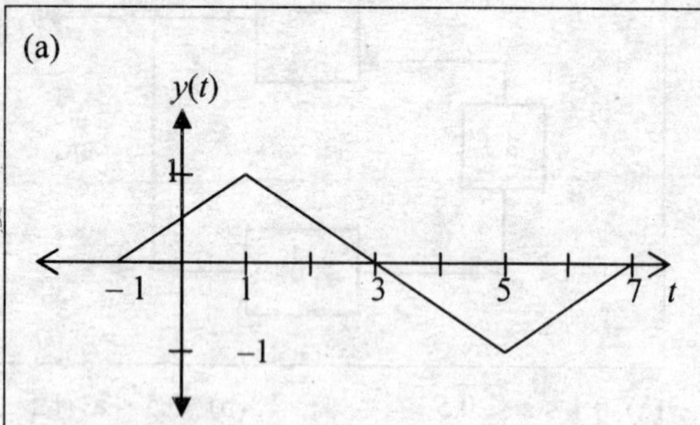

(b)

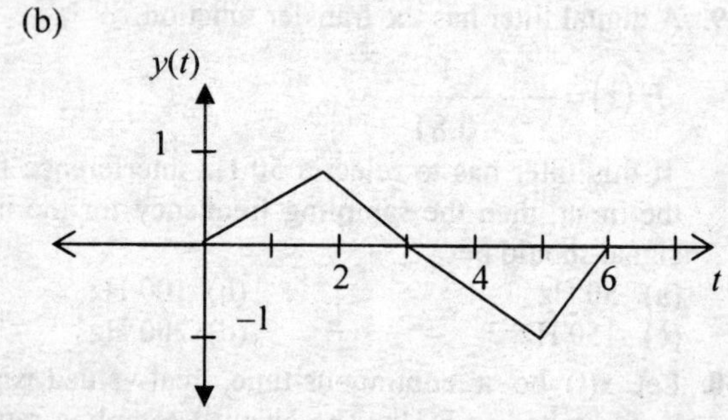

(c)

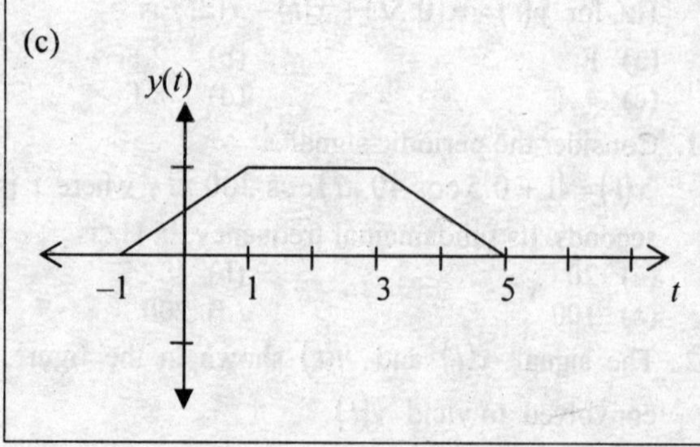

(d)

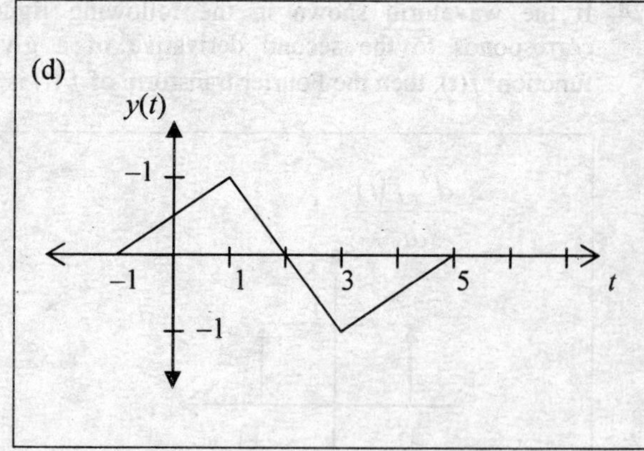

33. Consider the discrete-time signal $x(n)=\left(\frac{1}{3}\right)^n u(n)$, where $u(n)=\begin{cases}1, n\geq 0\\ 0, n<0\end{cases}$. Define the signal $y(n)$ as $y(n)=x(-n), -\infty<n<\infty$. Then $\sum_{n=-\infty}^{\infty} y(n)$ equals

(a) $-\frac{2}{3}$ (b) $\frac{2}{3}$

(c) $\frac{3}{2}$ (d) 3

34. Let the signal $x(t)$ have the Fourier transform $X(\omega)$. Consider the signal $y(t)=\frac{d}{dt}[x(t-t_d)]$ where t_d is an arbitrary delay. The magnitude of the Fourier transform of $y(t)$ is given by the expression.

(a) $|X(\omega)|\cdot|\omega|$ (b) $|X(\omega)|\cdot\omega$

(b) $\omega^2\cdot|X(\omega)|$ (d) $|\omega|\cdot|X(\omega)|\cdot e^{-j\omega t_d}$

Answers Set-1

1. Ans. (d)

Solution:

$f(t)\to F(\omega),\ f(-t)\to F^*(\omega)$

F^* is conjugate of $F(\omega)$

The Fourier transform pair

$f(t)-f(-t)=F(\omega)-F^*(\omega)$

it must be imaginary

Then, $F^*(\omega)$ as difference of $F(\omega)$ and $F^*(\omega)$

2. Ans. (d)

Solution:

The given signal is sampling signal and it Fourier Transform is gate function.

The Fourier transform of x[n] is shown below.

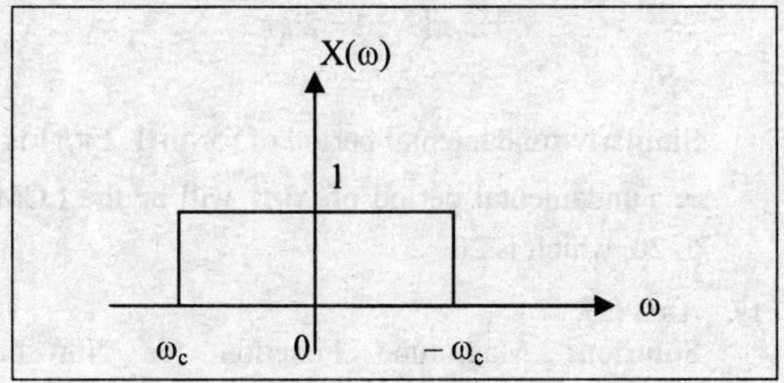

$$\text{Energy} = \pi \int_{-\infty}^{\infty} X(\omega).\, d(\omega)$$

$$= \pi \int_{-\omega_c}^{\omega_g} 1.\, d(\omega) = 2\pi\, \omega_c$$

3. **Ans. (d)**
   Solution:
4. **Ans. (a)**
   Solution:
5. **Ans. (d)**
   Solution:
6. **Ans. (a)**
   Solution:
   From Figure (a)
   $y[n] = x[n] + 0.4\, x[n-1] + 0.6\, x[n-2]$ (1)
   From figure (b)
   $y[n] = x[n] + [-k_1] \cdot xx[n-1] + k_1 \;\; k_2$
   $x[n-1] - \;k_2 \;\; x[n-2]$ (2)
   $= x[n] + x[n-1](k_1 k_2 - k_1) - k_2\, x[x-2]$
   Comparing equation (1) and (2), we have
   $k_2 = -0.640$
   and $k_1(k_2 - 1) = 0.40$
   or $k_1 = -0.244$
7. **Ans. (a)**
   Solution:

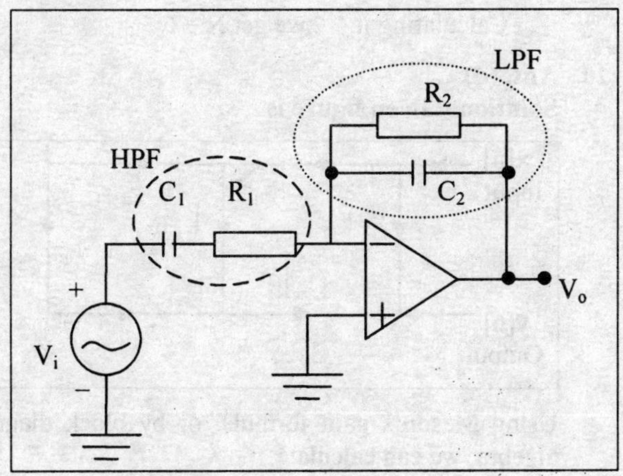

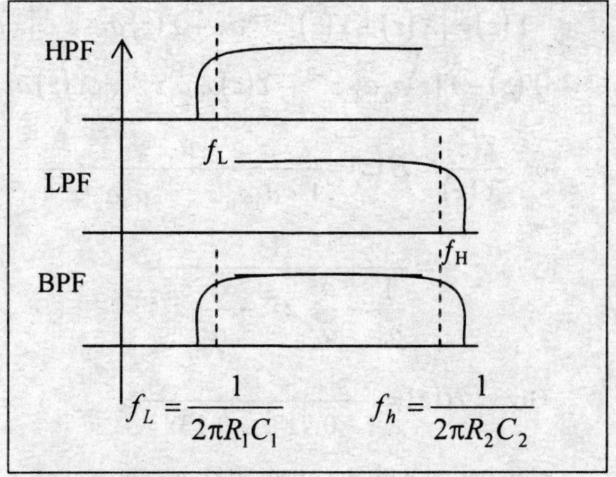

Second Method

KCL at the inverting terminal

$$\frac{V_i - 0}{R_1 + \frac{1}{C_1 s}} = \frac{0 - V_0}{\frac{R_2}{R_2 C_2 s + 1}}$$

$$\text{or, } A = \frac{V_0}{V_i} = \frac{-R_2 C_1 s}{(R_2 C_2 s + 1)(R_1 C_1 s + 1)}$$

$$\text{or, } A_v(f) = \frac{-R_2/R_1}{1 - j(f_L/f)} \cdot 1 + j(f/f_H)$$

$$= [-A_L(jf) \cdot A_H(jf)]$$

$$\text{Where, } f_L = \frac{1}{2\pi R_1 C_1},\; f_H = \frac{1}{2\pi R_2 C_2}$$

8. **Ans. (d)**
   Solution:

$$a^n u[n] \leftrightarrow \frac{1}{1 - a^{z-1}}\;; \qquad |a| < z$$

$$b^n u[-n] \leftrightarrow \frac{-1}{1 - b^{z-1}}\;; \qquad |b| > z$$

$$x[n] = \frac{1}{2} a^n - \frac{1}{3} a^n$$

$$x[0] = \frac{1}{2} - \frac{1}{3} = \frac{1}{6}$$

9. **Ans. (a)**
   Solution:

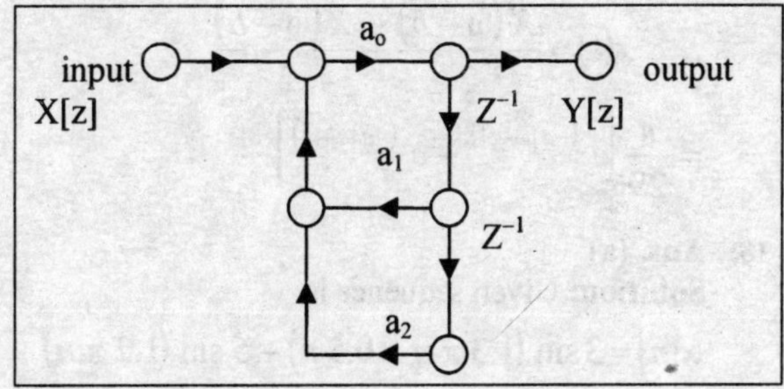

$$Y(z) = \left[X(z) + Y(z)\cdot z^{-2}a_2 + Y(z)a_1 \cdot z^{-1}\right] \times a_0$$

$$Y(z) - Y(z)a_0a_2\, z^{-2} - Y(z)a_0a_1z^{-1} = X(z)\, a_0$$

or $$\frac{Y(z)}{X(x)} = H(z) = \frac{a_0}{1 - a_0a_1\, z^{-1} - a_0a_2\, z^{-2}}$$

$$= \frac{1}{\frac{1}{a_0} - \frac{a_1}{a_0} z^{-1} - \frac{a_2}{a_0} z^{-2}}$$

Given $H(z) = \frac{1}{1 - 0.7z^{-1} + 0.13z^{-2}}$

Comparing with the given H(z), we get

$$\frac{1}{a_0} = 1, \quad -\frac{a_1}{a_0} = 0.7, \quad -\frac{a_2}{a_0} = 0.13$$

$$a_0 = 1, \qquad a_1 = 0.7, \qquad a_2 = -0.13$$

10. Ans. (c)
Solution:

11. Ans. (c)
Solution:

12. Ans. (b)
Solution:

13. Ans. (a)
Solution:

14. Ans. (a)
Solution: Since, pole zero plot of all pass filter is symmetrical about jω axis, hence only possible answers is

$$\frac{1 - s\tau}{1 + s\tau}$$

15. Ans. (c)
Solution:

16. Ans. (c)
Solution: Since, DC mode has been set, so it will read only the DC voltage applied at probe.

$$V(t) = A + B\sin\omega t$$

17. Ans. (a)
Solution:

$$x(t)\cos bt = x(t)e^{jbt} + \frac{x(t)}{2}e^{-jbt}$$

$$\therefore \quad x(t) = \frac{X(\omega - b)}{2} + \frac{X(\omega - b)}{2}$$

$$= \frac{\pi}{29}\left[e^{(-a|\omega - b|)} + e^{(-a|\omega + b|)}\right]$$

18. Ans. (a)
Solution: Given sequence is

$$x[n] = 3\sin(1.3\,\pi\, n + 0.5\,\pi) + 5\sin(1.2\,\pi\, n)$$

Fundamental period of $3\sin(1.3\,\pi n + 0.5\,\pi)$

$$\Rightarrow \qquad 1.3\,\pi(N_o) = 2\,k\pi$$

$$\Rightarrow \qquad N_o = 20$$

Similarly fundamental period of $5\sin(1.2\,\pi h)$ is 5

∴ Fundamental period of $x[n]$ will be the LCM of 5, 20, which is 20

19. Ans. (a)
Solution: Magnitude Function for Nth order Butterworth filter with passband edge ω_p is given by

$$|T(j\omega)| = \frac{1}{\sqrt{1 + \varepsilon^2\left(\frac{\omega}{\omega_p}\right)^{2N}}}$$

At $\omega = \omega_p$, $|T(j\omega_p)| = \frac{1}{\sqrt{1 + \varepsilon^2}}$

Here, $\omega_p = 2\pi \times 1000$

$$\frac{1}{\sqrt{1 + \varepsilon^2}} = 0.9$$

$$\Rightarrow \qquad 1 + \varepsilon^2 = \frac{1}{0.81}$$

$$\Rightarrow \qquad \varepsilon = 0.4843$$

$$\omega_s = 2\pi \times 1500\, rad\, s^{-1}$$

$$|T(j\omega)| = \frac{1}{\sqrt{1 + 0.23\left(\frac{3}{2}\right)^{2N}}}$$

$$\Rightarrow \qquad 1 + 0.23\left(\frac{3}{2}\right)^{2N} = \frac{1}{|T(j\omega)|^2}$$

At stop band edge $|T(j\omega)| = 0.2$

$$\therefore \; 1 + 0.23\,(1.5)^{2N} = \frac{1}{0.04}$$

Calculating it, we get N = 6

20. Ans. (d)
Solution: Given figure is

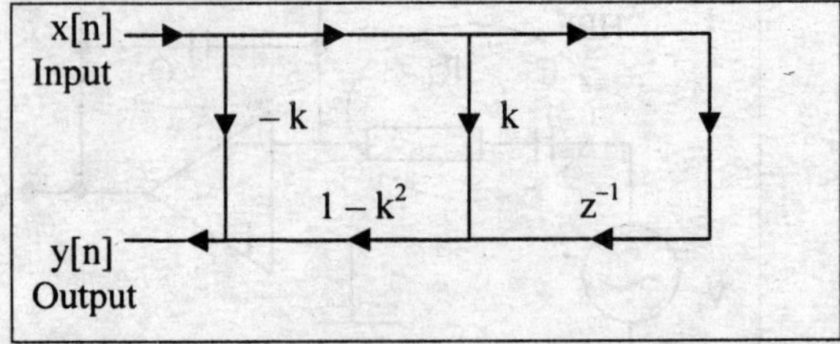

Using Mason's gain formula, or by block diagram algebra, we can calculate.

$$\frac{Y(z)}{X(z)} = -k + \frac{(1-k^2)z^{-1}}{1-k^{z-1}} = \frac{-k+z^{-1}}{1-kz^{-1}}$$

21. Ans. (b)
Solution:

22. Ans. (a)
Solution: $x(t) * x(t) = t\, e^{(-2t)} \cdot u(t)$
Taking Laplace Transform both sides,

$$X(s).X(s) = \frac{1}{(s+2)^2}$$

$$\Rightarrow \qquad X(s) = \frac{1}{s+2}$$

$$x(t) = e^{(-2t)}.u(t)$$

23. Ans. (b)
Solution:

$$\omega = 2\pi f \; : \; f = \frac{\omega}{2\pi f} = \frac{1.2\pi}{2\pi} = 0.6\ Hz$$

Fourier series represented by sinusoidal function and its harmonic component. Therefore, all terms frequency must be multiple of fundamental frequency. In present case that is only possible with 0.2 Hz.

24. Ans. (c)

Solution: $\dfrac{d^2(t)}{dt^2} = \delta(t-1) + \delta(t+1) - 2\,\delta(t)$
Taking Fourier transform both sides

$$\Rightarrow \quad (j\omega^2)F(j\omega) = e^{-j\omega} + e^{j\omega} - 2$$

$$\Rightarrow \quad F(j\omega) = \frac{2\,(1-\cos\omega)}{\omega^2}$$

25. Ans. (a)
Solution: $g(t) = \delta(t) + 2\,e^{(-3|t|)}$
Taking Lapace transform both sides

$$G(\omega) = 1 + 2\int_{-\infty}^{0} e^{(3t)}\, e^{(-j\omega t)}.\,dt$$

$$+ 2\int_{0}^{\infty} e^{(-3t)}\, e^{(-j\omega t)}.\,dt$$

$$= 1 + 2\int_{-\infty}^{0} e^{(3-j\omega)t}\, dt + 2\int_{0}^{\infty} e^{(-3-j\omega t)t}.\,dt$$

$$= 1 + \frac{2}{3-j\omega} + \frac{2}{3+j\omega}$$

$$\frac{\omega^2 + 9 + 6 + 2j + 6\cdot 2j}{\omega^2 + 9}$$

$$G(\omega) = \frac{\omega^2 + 21}{\omega^2 + 9}$$

26. Ans. (c)
Solution: $f_s \geq 2f_m$
In present case
$f_s \geq 200\ Hz \quad f_s = 250\ Hz$ u

$$\text{or, } T_s = \frac{1}{250} = 4ms$$

27. Ans. (b)
Solution: $x(t) \Leftrightarrow \times(\omega)$

$$x(at) \Leftrightarrow \frac{1}{|a|} \times \left(\frac{\omega}{a}\right)$$

28. Ans. (a)
Solution:

29. Ans. (b)
Solution:

30. Ans. (c)
Solution:

31. Ans. (a)
Solution:
$\omega = 40\,\pi = 2\pi f$
$f = 20Hz$

32. Ans. (d)
Solution:

$$x(t) = \delta(t+1) - \delta(t-1)$$

$$h(t) = \frac{t}{2}[U(t) - U(t-2)] + \frac{(4+2)}{2}$$

$$[U(t-2) - U(t-4)]$$

$$y(t) = h(t) * x(t)$$

$$[\text{using } x(t) - \delta(t-t_0) = x(t_0)]$$

We have

$$y(t) = \frac{(t+2)}{2}[U(t+1) - U(t-1)] - \frac{(t-1)}{2}$$

$$[U(t-1) - U(t-3)]$$

$$+ \frac{(3-t)}{2}[U(t-1) - U(t-3)] + \frac{(5-t)}{2}$$

$$[U(t-3) - U(t-5)]$$

and the graph would be as shown below

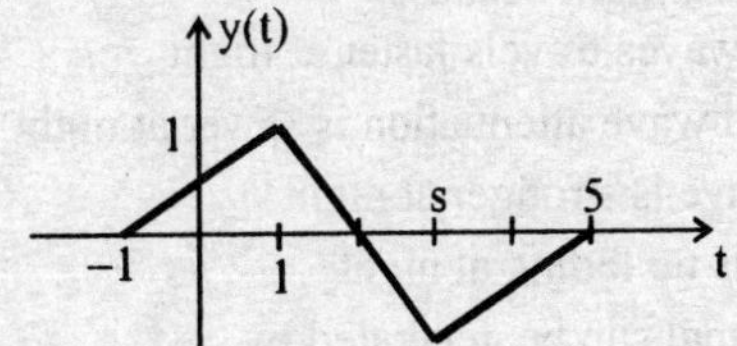

33. Ans. (d)
Solution:

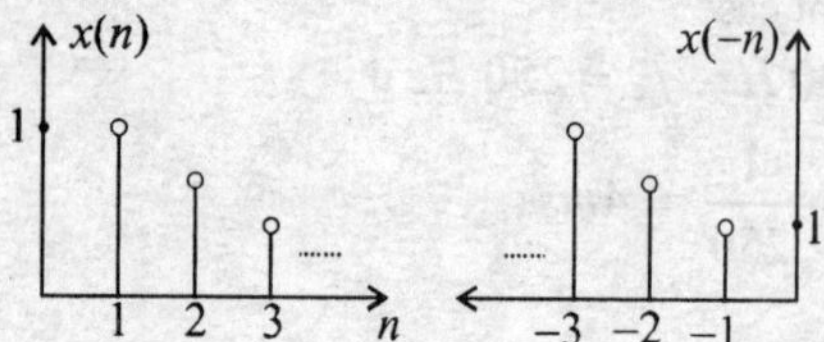

Then

$$S\sum_{n=-\infty}^{\infty} x(y)+\sum_{n=\infty}^{\infty} x(n)$$

$$=\sum_{n=-\infty}^{0}\left(\frac{1}{3}\right)^{-n} U(n)+S\sum_{n=\infty}^{\infty}\left(\frac{1}{3}\right)U(n)$$

$$=1+\frac{1}{3}+\frac{1}{9}+......+1+\frac{1}{3}+\frac{1}{9}+$$

$$=\frac{1}{1-1/3}+\frac{1}{1-1/3}=3$$

34. Ans. (d)
Solution:

Using

$$x'(t)=\frac{d}{dt}x(t)\leftrightarrow j\omega X(\omega)$$

MCQ Set–2

1. Which of the following pulse communication system is inherently immune to noise?
(a) PPM (b) PCM
(c) PWM (d) PAM
2. PDM is genertad by using.
(a) Monostabel muitivibrator
(b) Free runing multivibrator
(c) Bistable multivibrator
(d) JK flip flop
3. If large amount of information is to be transmitted in a small amount of time, we require.
(a) Low - frequency signals
(b) Narrow - band signals
(c) Wide - band signals
(d) high - frequency signals
4. Which one of the following circuits transmit two massages simultaneously in one direction?
(a) Duplex (b) Diplex
(c) Simplex (d) Quadruplex
5. Medium wave radio signals can be received at for off distance at night because.
(a) Radio waves travels faster at night
(b) Ground wave attenuation is lower at night
(c) Sky wave is stronger at night
(d) There is no fading at night
6. A PWM signal can be generated by
(a) An astable muitivibrator
(b) A monostable multivibrator
(c) Integrating a PPM signal
(d) Differentiating a PPM signal
7. In PCM
(a) Noise is removed by regeneration pulses at each repeater station.
(b) Noise free signal is transmitted at each repeater station.
(c) Only noise of the link between repeater stations is to be considered.
(d) All of the above
8. Which one of the following statement is correct? Quantising noise is produced in?
(a) All pulse modulation systems.
(b) PCM
(c) All modulation systems
(d) PDM
9. Pulse code modulation is commonly used in telemetry because.
(a) It ensures immunity from noise during transmission
(b) The band width requirement of the channel is reduced
(c) It eliminates quantization error.
(d) It permits lower rate of sampling than what is required under shannon's theorem.
10. Quantising noise can be reduced by increasing the.
(a) Number of standard quantum levels.
(b) Sampling rate
(c) Band width
(d) All of the above
11. Which of the following multiplexing technique involves signal composed of light beam?
(a) CDM (b) FDM
(c) TDM (d) WDM
12. Time division multiplexing needs
(a) Transmission of data samples
(b) Transmission of data at random
(c) Constant data at random
(d) Trabsmission of data of only one measuarand
13. In TDM, non-essential frequency components of modulating signal are removed by.
(a) Sampler (b) Attenuator
(c) Pre-allas filter (d) Modulator
14. Which one of the following systems offers the best trade off between bandwidth and S/N ratio.
(a) PAM (b) PDM
(c) PPM (d) PCM
15. Companding is used
(a) To overcome quantizing noise in PCM.
(b) In PWM recelvers to reduce impulse noise

(c) To protect small signal in PC from quantizing noise
(d) None of the above

16. In PCM, the conditional provability of error is proportial to.
(a) Noise power (no) (b) Peak signal energy
(c) $\sqrt{N_0}$ (d) $\sqrt{E_{max}}$

17. Time divison multiplexing is used when the data to be transmitted is
(a) Slow changing
(b) Of small bandwidth
(c) Slow changing and has a small bandwidth
(d) Fast changing and has a wide band width

18. In case of data transmission, which one of following systems will give the maximum probability error?
(a) ASK (b) FSK
(c) PSK (d) DPSK

19. A handshake signal in a data transfer is transmitted
(a) Along with the data bits
(b) Before the data transfer
(c) After the data transfer
(d) Either along the data bits or after the data transfer.

Answers Set-2

1. (b)	2. (a)	3. (b)	4. (b)
5. (d)	6. (b)	7. (d)	8. (b)
9. (a)	10. (a)	11. (d)	12. (a)
13. (c)	14. (d)	15. (c)	16. (d)
17. (d)	18. (a)	19. (b)	

One Mark Questions

1. If the Laplace transform of the voltage across a capacitor of value of (1/2) F is

$$V_c(s) = \frac{s+1}{s^3+s^2+s+1}$$

The value of the current through the capacitor at t = 0^+ is
(a) 0 A (b) 2 A
(c) (1/2) A (d) 1 A

2. The Fourier series expansion of a real periodic signal with fundamental frequency f_o is given by

$$g_p(t) = \sum_{n=-\infty}^{\infty} C_n e^{j2\pi n f_o t}$$

It is given than $C_3 = 3 + j5$. Then C_{-3} is
(a) 5 + j3 (b) –3 – j5
(c) –5 + j3 (d) 3 – j5

3. Let x(t) be the input to a linear, time-invariant system. The required output is 4x(t - 2). The transfer function of the system should be
(a) $4\,e^{j4\pi f}$ (b) $2\,e^{-j8\pi f}$
(c) $4\,e^{-j4\pi f}$ (d) $2\,e^{j8\pi f}$

4. A sequence x(n) with the z-transform $X(z) = z^4 + z^3 - 2z + 2 - 3z^{-4}$ is applied as an input to a linear, time-invariant system with the impulse response $h(n) = 2\delta(n-3)$ where
δ (n) = 1 ; n = 0
0 ; otherwise
The output at n = 4 is
(a) – 6 (b) zero
(c) 2 (d) –4

5. The impulse response h[n] of a linear time-invariant system is given by
h[n] = u[n + 3] + u[n – 2] – 2u[n – 7]
where u[n] is the unit step sequence. The above system is
(a) Stable but not causal
(b) Stable and causal
(c) Causal but unstable
(d) Unstable and not causal

6. The z-transform of a system is $H(z) = \dfrac{z}{z-0.2}$ If the ROC is | z | < 0.2, then the impulse response of the system is
(a) $(0.2)^n$ u[n] (b) $(0.2)^n$ u[– n – 1]
(c) – $(0.2)^n$ u[n] (d) –$(0.2)^n$ u[- n –1]

7. The Fourier transform of a conjugate symmetric function is always
(a) Imaginary (b) Conjugate anti-symmetric
(b) Real (d) Conjugate symmetric

8. Choose the function f(t) ; - ∞ < t < ∞, for which a Fourier series cannot be defined.
(a) 3 Sin (25 t)
(b) 4 Cos (20t + 3) + 2 Sin (710 t)
(c) exp (- | t |) Sin (25 t)
(d) 1

9. The function x(t) is shown in the figure. Even and odd parts of a unit-step function u(t) are respectively.

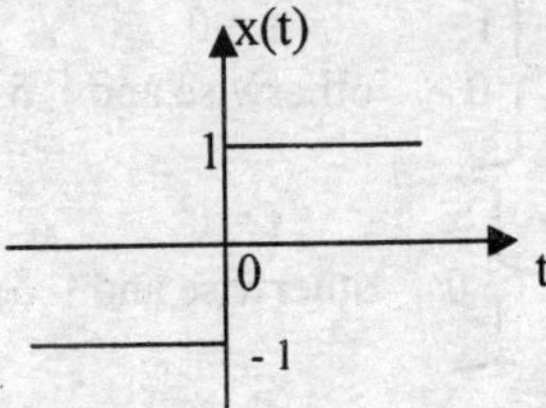

(a) $\frac{1}{2}, \frac{1}{2}$ x (t) (b) $-\frac{1}{2}, \frac{1}{2}$ x (t)
(c) $\frac{1}{2}, -\frac{1}{2}$ x (t) (b) $-\frac{1}{2}, -\frac{1}{2}$ x (t)

10. The region of con convergence of Z-transform of sequence $[5/6]^n$ u(n) - $[6/5]^n$ u (- n - 1) must be
(a) | z | < [5 / 6] (b) | z | > [6 / 5]
(c) [5 / 6] < | z | < [6 / 5] (d) [6 / 5] < | z | < ∞

11. Which of the following can be impulse response of a causal system?

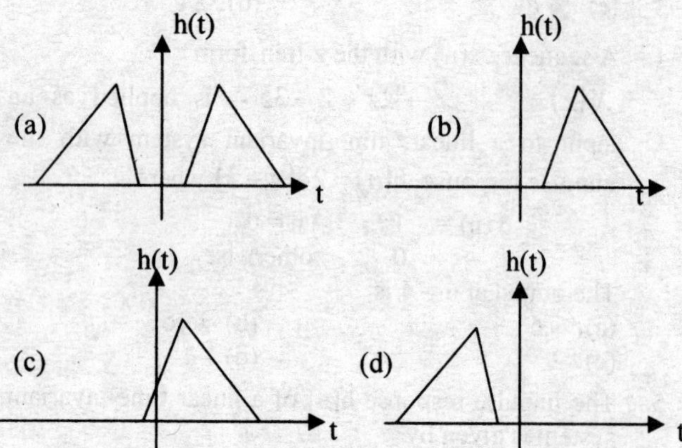

12. Let $x(n) = [1/2]^n u(n)$, $y(n) = x^2(n)$ and $Y(e^{j\omega})$ be the Fourier transform of y (n). Then $Y(e^{j0})$ is.

(a) [1 / 4] (b) 2
(c) 4 (d) [4 / 3]

13. The power in the signal $s(t) = 8\cos[20\pi t - (\pi/2)] + 4\sin(15\pi t)$ is

(a) 40 (b) 41
(c) 42 (d) 82

14. Let $x(t) \leftrightarrow X(j\omega)$ be Fourier Transform pair . The Fourier Transform of the signal x(5t – 3) in terms of X (j ω) is given as.

(a) $(1/5)\exp(-j3\omega/5)\,X(j\omega/5)$
(b) $(1/5)\exp(j3\omega/5)\,X(j\omega/5)$
(c) $(1/5)\exp(-j3\omega)\,X(j\omega/5)$
(d) $(1/5)\exp(-j3\omega/5)\,X(j\omega/5)$

15. The Dirac delta function δ (t) is defined as

(a) $\delta(t) = \begin{cases} 1 & t = 0 \\ 0 & \text{otherwise} \end{cases}$

(b) $\delta(t) = \begin{cases} \infty & t = 0 \\ 0 & \text{otherwise} \end{cases}$

(c) $\delta(t) = \begin{cases} 1 & t = 0 \\ 0 & \text{otherwise and } \int_{\infty}^{\infty} \delta(t)\,dt = 1 \end{cases}$

(d) $\delta(t) = \begin{cases} \infty & t = 0 \\ 0 & \text{otherwise and } \int_{\infty}^{\infty} \delta(t)\,dt = 1 \end{cases}$

16. If the region of convergence of $x_1[n] + x_2[n]$ is $[1/3] < |z| < [2/3]$, then the region of convergence of $x_1[n] - x_2[n]$ includes.

(a) $[1/3] < |z| < 3$ (b) $[2/3] < |z| < 3$
(c) $[3/2] < |z| < 3$ (d) $[1/3] < |z| < [2/3]$

17. In the system shown below, x(t) = (sin t) u(t) . In steady state, the response y (t) will be.

x(t) → $\frac{1}{s+1}$ → y (t)

(a) $(1/\sqrt{2})\sin[t - (\pi/4)]$
(b) $(1/\sqrt{2})\sin[t + (\pi/4)]$
(c) $(1/\sqrt{2})\,e^{-1}\sin t$
(d) sin t – cos t

18. If the Laplace of a signal $y(t)$ is $Y(s) = \dfrac{1}{s(s-1)}$, then its final value is

(a) – 1 (b) 0
(c) 1 (d) Unbounded

19. The frequency spectrum of a signal is shown in the figure. If this signal is ideally sampled at intervals of 1 ms, then the frequency spectrum of the sampled signal will be.

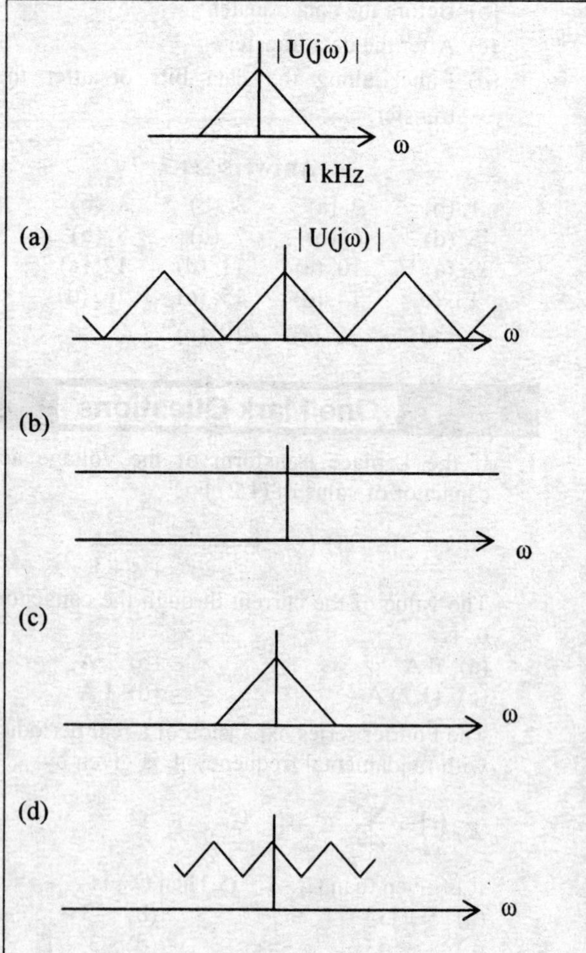

20. The input to a coherent detector is DSB-SC signal plus noise. The noise at the detector output is.
(a) The in-phase component
(b) The Quadrature-component
(c) Zero
(d) The envelope

21. The noise at the input to an ideal frequency detector is white. The detector is operating above threshold. The power spectral density of the noise at the output is.
(a) raised-cosine (b) flat
(c) parabolic (d) Gaussian

22. At a given probability of error, binary coherent FSK is inferior to binary coherent PSK by.
(a) 6 dB (b) 3 dB
(c) 2 dB (d) 0 dB

23. In PCM system, if the code word length is increased from 6 to 8 bits, the signal to quantization noise ratio improves by the factor.
(a) [8 / 6] (b) 12
(c) 16 (d) 8

24. An AM signal is detected using an envelope detector. The carrier frequency and modulating signal frequency are 1 MHz and 2 kHz respectively. An appropriate value for the time constant of the envelope detector is.
(a) 500 μsec (b) 20 μsec
(c) 0.2 μsec (d) 1 μsec

25. An AM signal and a narrow-band FM signal with identical carriers, modulating signals and modulation indices of 0.1 are added together. The resultant signal can be closely approximated by.
(a) broadband FM (b) SSB with carrier
(c) DSB-SC (d) SSB without carrier

26. In the output of a DM speech encoder, the consecutive pulses are of opposite polarity during time interval $t_1 \le t \le t_2$. This indicates that during this interval
(a) The input to the modulator is essentially constant
(b) The modulator is going through slope overload
(c) The accumulator is in saturation
(d) The speech signal is being sampled at the Nyquist rate

27. Find the correct match between group 1 and group 2.

Ground I

P – $\left[1+km(t)\right]A\sin(\omega_c t)$

Q – $km(t)A\sin(\omega_c t)$

R – $A\sin\left[\omega_c' + k\right]_{-\infty}^{t} m(t)\,dt$

S – $A\sin\left[\omega_c t + k\int_{-\infty}^{t} m(t)\,dt\right]$

Group II

W – Phase modulation
X – Frequency modulation
Y – Amplitude modulation
Z – DSB-SC modulation

	P	Q	R	S
(a)	Z	Y	X	W
(b)	W	X	Y	Z
(c)	X	W	Z	Y
(d)	Y	Z	W	X

28. Which of the following analog modulation scheme requires the minimum transmitted power and minimum channel bandwidth?
(a) VSB (b) DSB - SC
(c) SSB (d) AM

29. A low – pass filter having a frequency response H (jω) = A (ω) $^{j\phi(\omega)}$ does not produce any phase distortion if.
(a) $A(\omega) = C\omega^2$, $\phi(\omega) = k\omega^3$
(b) $A(\omega) = C\omega^2$, $\phi(\omega) = k\omega$
(c) $A(\omega) = C\omega$, $\phi(\omega) = k\omega^2$
(d) $A(\omega) = C$, $\phi(\omega) = k\omega^{-1}$

30. If $R(\tau)$ is the autocorrelation function of a real, wide-sense stationary random process, then which of the following is **NOT true?**
(a) $R(\tau) = R(-\tau)$
(b) $|R(\tau)| \le R(0)$
(c) $R(\tau) = -R(-\tau)$
(d) The mean square value of the process is $R(0)$

31. If $S(f)$ is the power spectral density of a real, wide-sense stationary random process, then which of the following is **ALWAYS true?**
(a) $S(0) \ge S(f)$ (b) $S(f) \ge 0$
(c) $S(-f) = -S(f)$ (d) $\int_{-\infty}^{\infty} S(f)\,df = 0$

32. If E denote expectation, the variance of a random variable X is given by
(a) $E[X^2] - E^2[X]$ (b) $E[X^2] + E^2[X]$
(c) $E[X^2]$ (d) $E^2[X]$

Answers
Signals, Systems and Communication

1. Ans. (c)

Solution: $C = \frac{1}{2}F$

$$I_c(s) = \frac{V(s)}{Z(s)} = \frac{V_c(s)}{1/sC}$$

$$= \frac{(s+1)}{\frac{1}{sC}\left(s^3+s^2+s+1\right)} = \frac{sC(s+1)}{\left(s^3+s^2+s+1\right)}$$

$$= \frac{s}{\left(s^2+1\right)\cdot 2}$$

Apply initial value theorem

$$I_c(0^+) = \underset{s\to\infty}{Lt}\ sI_c(s)$$

$$= \underset{s\to\infty}{Lt}\ sI_c(s) \times \frac{s}{\left(s^2+1\right)2}$$

$$= \underset{s\to 0}{Lt}\ \frac{s^2}{\left(s^2+1\right)2} = \underset{s\to\infty}{Lt}\ \frac{1}{2+\frac{2}{s^2}} = \frac{1}{2}$$

2. **Ans. (d)**

Solution: For real Periodic signal, Fourier Series coefficient are complex conjugate of each other

$C_{-K} = \overline{C}_K$

$\therefore\quad C_{-3} = \overline{C}_3 = \overline{3+5j} = 3-5j$

3. **Ans. (c)**

Solution: $V_g(t) \Leftrightarrow G(f)$

$g(t - t_o) \Leftrightarrow \exp[-j\,\omega_o t_o]\,G(f)$; $\omega = 2\pi f$

The output, $y(t) = 4x(t-2)$

FT of output, $Y(f) = 4X(f)\,e^{-j4\pi f}$

The transfer function = $[Y(f)/X(f)] = 4e^{-j4\pi f}$

$= \mathbf{4\exp[-j\,4\,\pi\,f]}$

4. **Ans. (b)**

Solution: $h[n] = 2\delta[n-3]$

$H(z) = 2z^{-3}$

The output, $Y[z] = H(z)\cdot X[z]$

$= 2z^{-3}\left[z^4 + z^2 - 2z + z - 3z^4\right]$

$= 2\left[z + z^{-1} - 2z^{-1} + z^{-3} - 3z^{-7}\right]$

Taking inverse Z -transform we have

$$y[n] = 2\begin{bmatrix}\delta[n+1] + \delta[n-1] - 2\delta[n-2] + 2\delta[n-3] \\ -3\,\delta[n-7]\end{bmatrix}$$

At, $n = 4$. $y[n] = 0$

5. **Ans. (a)**

Solution:

$$\sum_{L=-\infty}^{+\infty} h(K) =$$

$$\sum_{K=-3}^{\infty} u(K+3) + \sum_{K=-2}^{\infty} u(K-2) - 2\sum_{K=7}^{\infty} u(K-7)$$

$$= \sum_{K=-3}^{6} 1 + \sum_{K=2}^{6} 1 = 10 + 5 = 15 < \infty$$

Therefore, the system is stable

Also, the three system described by h[n] is non-causal because, the response depends upon the future value of the input signal, namely u[n + 3].

Second Method

$$h(n) = u(n+3) + u(n-2) - 2u(n-7)$$

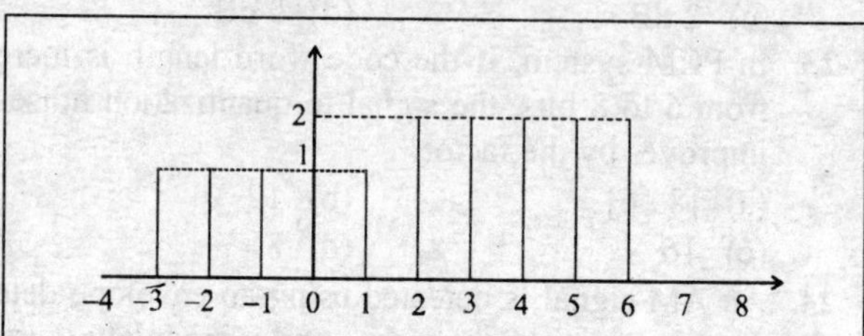

$$\sum_{-\infty}^{\infty} h(n) = 1+1+1+1+1+2+2+2+2+2 = 15$$

We can see that $\sum_{-\infty}^{\infty} h(n)$ is finite hence system is stable.

Since $h(n)$ depends on $u(n+3)$ i.e. future values, so it is not a causal system.

6. **Ans. (d)**

Solution: Using the following Z-transform pair,

$$-A_k(dk)^n\, u[-n-1] \xrightarrow{z} \frac{A_k}{1-dk\,z^{-1}}$$

with ROC 1 | Z | < dk

Comparing with above equation

$$\frac{z}{z-0.2} \longleftrightarrow -(0.2)^n\, u[-n-1]$$

$$H(z) = \frac{z}{z-0.2}$$

$$= \frac{1}{1-0.2\,z^{-1}}\ ;\ ROC: |z| < 0.2$$

7. **Ans. (c)**

Solution: Fourier transform of a conjugate symmetric function is always real.

8. **Ans. (c)**

Solution: Fourier series is defined for a periodic function.

Here $e^{-|t|}$ is not a periodic function

9. Ans. (d)

Solution: Even part = $\frac{u(t) + u(-t)}{2} = [1/2]$

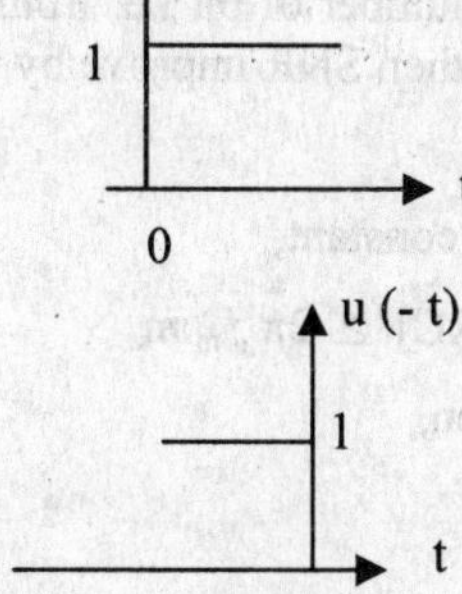

odd part = $\frac{u(t) - u(-t)}{2} = \frac{x(t)}{2}$

10. Ans. (c)

Solution: The given sequence is a 2 sided sequence so its ROC is a circle

ROC for $[5/6]^n \ \mu(n)$ is $|z| > [5/6]$

ROC for $[5/6]^n \ \mu(-n-1)$ is $|z| > [5/6]$

$[5/6] < |z| < [5/6]$

11. Ans. (b)

Solution: For causal system,

h(t) = 0 , for t < 0

12. Ans. (d)

Solution: $x(n) = [1/2]^n u(n)$; $x^2(n) = [1/2]^{2n} u(n)$

$y(n) = [1/4]^n u(n)$

$Y(e^{j\omega}) = \frac{1}{1 - [1/4] e^{-j\omega}} \Rightarrow Y(e^{j\omega}) = \frac{1}{1 - [1/4]}$

$Y(e^{j\omega}) = [4/3]$

13. Ans. (a)

Solution: $s(t) = 8 \cos [20 \pi t - \{\pi/2\}] + 4 \sin (15\pi t)$

$= 8 \sin (20\pi t) + 4 \sin (15 \pi t)$

Power in the signal = $[8/\sqrt{2}]^2 + [4/\sqrt{2}]^2$

$= [64/2] + [16/2]$

$= [80/2] = \mathbf{40}$

14. Ans. (A)

Solution: By scaling Property

$f(t) \rightarrow F(\omega)$

$f(at) \rightarrow \frac{1}{|a|} F(\omega/a)$

By shifting property

$f(t - T) \rightarrow F(\omega) e^{-j\omega T}$

By shifting and scaling

$f(at - T) \rightarrow \frac{1}{|a|} F(\omega/a) e^{-j\omega T}$

15. Ans. (d)

Solution: The Dirac delta function δ (t) is defined as

$$\delta(t) = \begin{cases} \infty & ; \quad t = 0 \\ 0 & ; \quad \text{otherwise} \end{cases}$$

and $\int_{\infty}^{\infty} \delta(t)\, dt = 1$

16. Ans. (d)

Solution: ROC, remains same in both cases.

17. Ans. (a)

Solution:

System [1 / (s + 1)] gives change in output as

$\frac{1}{\sqrt{(1+1)}} \angle -\tan^{-1}\frac{1}{1}$

$\frac{1}{\sqrt{2}} \angle -45°$

Therefore output = $[1/\sqrt{2}] \ \sin [t - (\pi/4)]$

18. Ans. (d)

Solution: $Y(s) = \frac{1}{s(s-1)}$

$y(t) =: LT^{-1}\left(\frac{-1}{s} + \frac{1}{s-1}\right) = \left(-1 + e^{-t}\right)u(t)$

$y(\infty) = -1$

This result is unbounded

19. Ans. (d)

Solution:

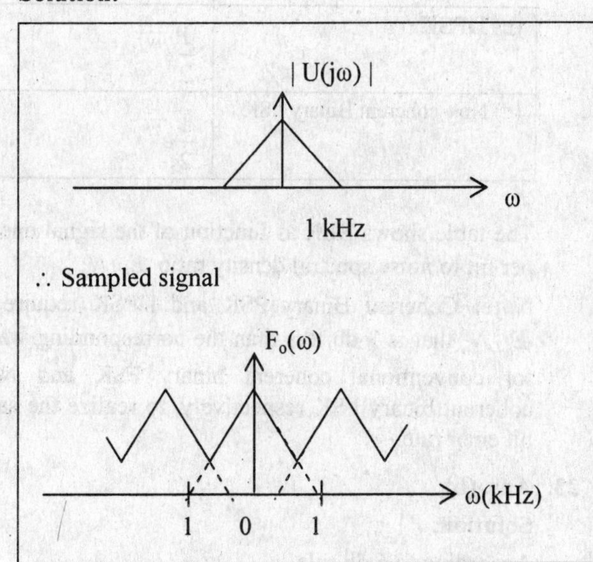

Given $f_s = 1$ kHz

Because $f_s = f_m$ or $f_s < 2f_m$

Therefore, Aliasing effect occurs

Also sampled signals $= \omega_o \sum_{-\infty}^{\infty} (\omega - n\omega_s)$

Due to ω_o factor the amplitude is increased

20. Ans. (a)

Solution: The noise at (DSB - SC) detector output is only **in phase component.**

21. Ans. (d)

Solution:

By experimental result and central limit theorem noise is Gaussian in nature, As input is an independent white noise, so power spectral density is Gaussian.

22. Ans. (b)

Solution: In FSK $P_e = \text{erfc}\,[\sqrt{(E / n_o)}]$

$= \sqrt{(n_o / 2\pi E)}\, e^{-E/2n_o}$

In PSK $P_e = \text{erfc}\,[\sqrt{(E / n_o)}]$

$= [1/2]\,\sqrt{(n_o / \pi E)}\, e^{-E/n_o}$

Coherent PSK given minimum P_e and supperier to FSK by **3 db.**

Signaling scheme	Bit error rate or, probability of error
(a) Coherent Binary PSK Coherent QPSK Coherent MSK	$\frac{1}{2} erfc\sqrt{\frac{E_b}{N_o}}$
(b) Coherent Binary FSK	$\frac{1}{2} erfc\sqrt{\frac{E_b}{2N_o}}$
(c) DPSK	$\frac{1}{2} e^{\frac{-E_b}{N_o}}$
(d) Non-coherent Binary FSK	$\frac{1}{2} e^{\frac{-E_b}{2N_o}}$

The table shows BER as function of the signal energy per bit to noise spectral density ratio E_b/N_o

Note: Coherent Binary PSK and DPSK require an E_b/N_o that is 3 db less than the corresponding, value for conventional coherent binary FSK and Non-coherent binary FSK respectively, to realize the same bit error rate.

23. Ans. (b)

Solution:

According to 6 db rule

$SNR|_{db} = 1.76 + 6 \cdot n\ db$

Where $n =$ number of bits.

For 1 bit increase SNR by 6 db

Hence for change in number of bit i.e. from 6 to 8 bit increase in bit is by 2 then SNR improve by 12 db.

24. Ans. (b)

Solution: RC = Time constant,

For No Clipping $[1 / RC] \geq 2\pi f_m m_a$

Typically $m_a = 0.4$ then,

$$\frac{1}{2\pi f_m m_a} \geq RC$$

or $$\frac{1}{2\pi \times 2 \times 10^3 \times 0.4} \geq RC$$

or $10^{-3} \geq RC$

or, $4 \times 10^{-4} \geq RC$

$RC \leq 10^{-4}$

$$= \frac{4}{10^4} \times \frac{100}{100} = \frac{400}{10^6} = 400\ \mu s$$

The time constant of Envelop detector should be less than time period of modulating signal but grater than time period of carrier i.e. $\frac{1}{f_c} < RC < \frac{1}{f_m}$

Where $\frac{1}{f_c} \Rightarrow$ to avoid the fluctuation at recoved output

$\frac{1}{f_m} \Rightarrow$ to avoid diagonal clipping

Hence $\frac{1}{1\,MHz} < RC < \frac{1}{2\,kHz}$

$1\mu s < RC < 0.5\ ms$

$1\mu s < RC < 500\ \mu s$

The best choice is 20 μs

25. Ans. (b)

Solution:

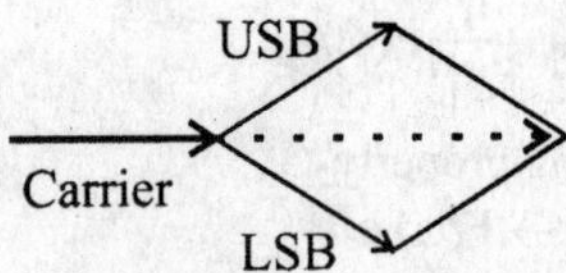

Figure (a) AM vector diagram

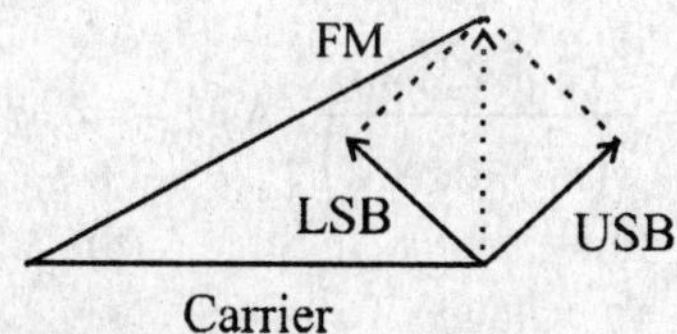

Figure (b) FM vector diagram

When we add both the vector diagram. Then one SB is cancelled out remaining with carrier and one side band i.e. AB-SSB with full carrier.

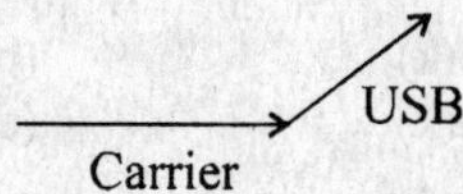

Here $m_f = m_a = 0.1$

Modulating signal $f(t) = \cos\omega_m t$

$$s(t)_{FM} = A\cos\omega_c t - A\,m_f\,g(t)\sin\omega_c t$$

$$= A\cos\omega_c t + \frac{A\,m_f}{2}\cos(\omega_c + \omega_m)t - \frac{Am_f}{2}\cos(\omega_c - \omega_m)t$$

$$s(t)_{AM} = A\cos\omega_c t + Am_a\,f(t)\cos\omega_c t$$

$$= A\cos\omega_c t + \frac{A\,m_a}{2}\cos(\omega_c + \omega_m)t + \frac{Am_a}{2}\cos(\omega_c - \omega_m)t$$

$$s(t)_{FM} + s(t)_{AM} = 2A\cos\omega_c t + \left(\frac{A\,m_f}{2} + \frac{Am_a}{2}\right)\cos(\omega_c + \omega_m)$$

= Carrier + *USB*

⇒ SSB with carrier

26. Ans. (a)

Solution:

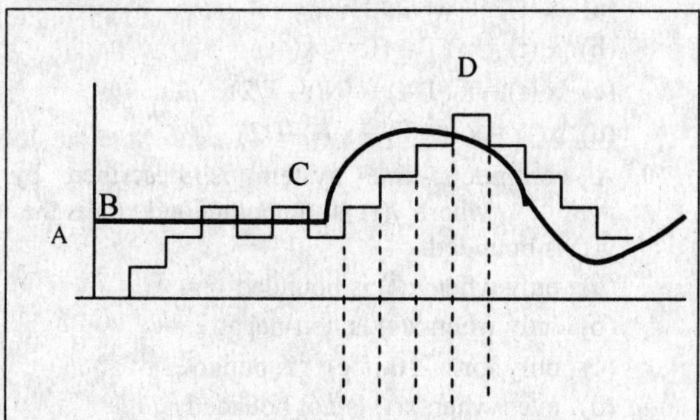

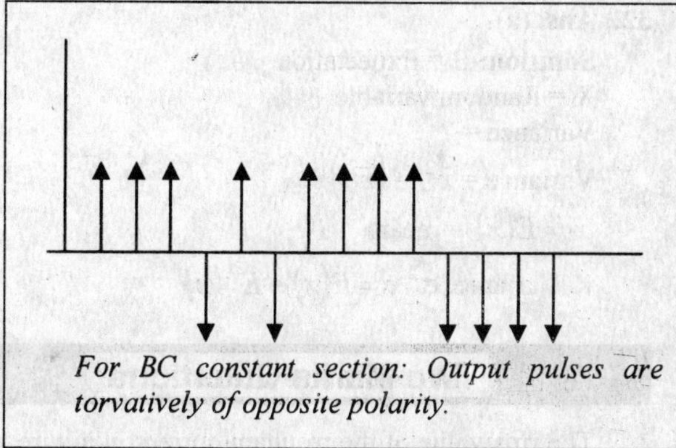

For BC constant section: Output pulses are torvatively of opposite polarity.

27. Ans. (d)

Solution: The correct match is given below.

$[1 + km(t)]A\sin(\omega_c t)$	Amplitude modulation
$km(t)A\sin(\omega_c t)$	DB-SC modulation
$A\sin[\omega_c' + k]_{-\infty}^{t} m(t)dt$	Phase modulation
$A\sin\left[\omega_c t + k\int_{-\infty}^{t} m(t)dt\right]$	Frequency modulation

28. Ans. (c)

Solution:

VSB → $f_m + f_c$

DBS -SC → $2f_m$

SSB → f_m

AM → $2f_m$

Thus SSB has minimum bandwidth and it required minimum power i.e. 17% as compared to AM.

29. Ans. (c)

Solution: Maximum frequency of signal f_m = 3000 Hz

So, sampling rate = 2 f_m = 6000 Hz

30. Ans. (c)

Solution: Because autocorrelation function $R(\tau)$ is even function in which R(0) ≥ R(τ) then $R(\tau) = -R(-\tau)$ is not true.

31. Ans. (b)

Solution: If $S(f)$ is the power spectral density of a real, wide-sense stationary random process, then $S(f) \geq 0$ **Always true**

32. Ans. (a)

Solution: E = Expectation

X = Random variable

Variance = ?

Variance = $\overline{x}^2 - \left(\overline{x}^2\right)$

$\overline{x} = E(x)$ = mean

$\therefore$ Variance $\sigma^2 x = E(x) - E^2(x)$

Two Marks Questions

1. The rms value of the resultant current in a wire which carries a dc current of 10 A and a sinusoidal alternating current of peak value 20 A is.

(a) 14.1A (b) 17.3 A

(c) 22.4 A (d) 30.0 A

2. The rms value of the periodic waveform given in figure is.

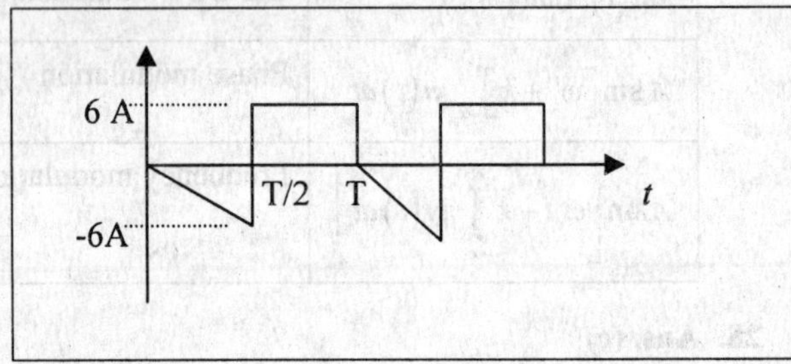

(a) $2\sqrt{6}\ A$ (b) $6\sqrt{2}\ A$

(c) $\sqrt{\frac{4}{3}}\ A$ (d) 1.5 A

3. For the triangular wave form shown in the figure, the RMS value of the voltage is equal to.

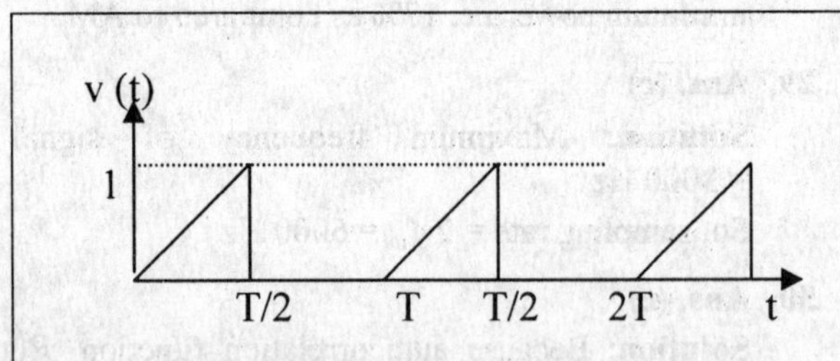

(a) $\sqrt{\frac{1}{6}}$ (b) $\sqrt{\frac{1}{3}}$

(c) $\frac{1}{3}$ (d) $\sqrt{\frac{2}{3}}$

4. The Laplace transform of a function f(t) if.

$F(s) = \frac{5s^2 + 23s + 6}{s\left(s^2 + 2s + 2\right)}$. As $t \to \infty$, $f(t)$ approaches

(a) 3 (b) 5

(c) $\frac{17}{2}$ (d) ∞

5. If u(t) is the unit step and δ (t) is the unit impulse function, the inverse z-transform of $F(z) = \frac{1}{z+1}$ for k > 0 is.

(a) $(-1)^k\ \delta(k)$ (b) $\delta(k) - (-1)^k$

(c) $(-1)^k\ u(k)$ (d) $u(k) - (-1)_k$

6. The Fourier series for the function $f(x) = \sin^2 x$ *is*

(a) sin x + sin 2x (b) 1 –cos 2x

(c) sin 2x+cos 2x (d) 0.5 – 0.5 cos 2x

7. Which of the following is true.

(a) A finite signal is always bounded

(b) A bounded signal is always possesses infinite energy

(c) A bounded signal is always zero outside the interval $\left[-t_0, t_0\right]$ for some t_0

(d) A bounded signal always finite

8. A discrete real allpass system has a pole at $z = 2\angle 30^o$: it, therefore.

(a) Also has a pole at $\frac{1}{2}\angle 30^o$

(b) Has a constant phase response over the z- plane: $\arg\ |H(z)| = \text{constant}$

(c) Is stable only if it is anticausal

(d) Has a constant phase response over the unit circle: $\arg\ |H(e^{j\Omega})| = \text{const.}$

9. x (t) is a real valued function of a real variable with period T. its trigonometric Fourier Series expansion contains no terms of frequency $\omega = 2\pi\ (2K)/T; K = 1,2,.......$ Also, no sine terms are present. Then x (t) satisfies the equation

(a) x (t) = –x(t – T)

(b) x (t) = x (T – t) = – x(–t)

(c) x (t) = x (T–t)= – x(t –T/2)

(d) x(t) = x (t –T) = x (t –T/2)

10. A continuous-time system is described by $y(t) = e^{-|x(t)|}$, where y(t) is the output and x(t) is the input. y(t) is bounded.

(a) only when x(t) is bounded

(b) only when x(t) is non-negative

(c) only for t ≥ 0 if x (t) is bounded t ≥ 0

(d) even when x(t) is not bounded

11. x[n]=0; n<-1, n>0, x[-1]=-1, x[0]=2 is input and y[n]=0; n<-1, n>2, y[-1]=-1=y[1], y[0]=3, y[2]=-2 is the output of a discrete-time LTI system. The system impulse response h[n] will be.
    (a) h[n] = 0; n <0, n > 2, h[0] = 1, h[1] = h[2] = -1
    (b) h[n] = 0; n <-1, n>1, h[-1] =1, h[0] = h[1] = 2
    (c) h[n] =0; n <0, n>3, h[0]= -1, h[1] =2, h[2] = 1
    (d) h[n] = 0; n <-2, n> 1, h[-2] = h[1] = h[-1] = H (0) = 3

12. The discrete-time signal.

$$x[n] \leftrightarrow X(z) = \sum_{n=0}^{\infty} \frac{3^n}{2+n} z^{2n}$$

where $\leftrightarrow$ denotes a transform-pair relationship, is orthogonal to the signal.

(a) $y_1[n] \leftrightarrow Y_1(Z) = \sum_{n=0}^{\infty} \left(\frac{2}{3}\right)^n z^{-n}$

(b) $y_2[n] \leftrightarrow Y_2(Z) = \sum_{n=0}^{\infty} (5^n - n) Z^{-(2n+1)}$

(c) $y_3[n] \leftrightarrow Y_3(z) = \sum_{n=-\infty}^{\infty} 2^{-|n|} z^{-n}$

(d) $y_4[n] \leftrightarrow Y_4(z) = 2z^{-4} + 3z^{-2} + 1$

13. The running integratot, given by y(t) $= \int_{-\infty}^{H} x(t')\,dt'$
    (a) has no finite singularities in its double sided Laplace Transform Y (s)
    (b) produces a bounded output for every causal bounded input
    (c) produces a bounded output for every anticausal bounded input
    (d) has no finite zeroes in its double sided Laplace Transform Y(s)

14. A signal $x(t)$ is given by.

$$x(t) = \begin{cases} 1, & \frac{-T}{4} < t \le \frac{3T}{4} \\ -1 & \frac{3T}{4} < t \le \frac{7T}{4} \\ -x(t+T) & \end{cases}$$

Which among the following gives the fundamental Fourier term of $x(t)$?

(a) $\frac{4}{\pi}\cos\left(\frac{\pi t}{T} - \frac{\pi}{4}\right)$ (b) $\frac{\pi}{4}\cos\left(\frac{\pi t}{2T} + \frac{\pi}{4}\right)$

(c) $\frac{4}{\pi}\sin\left(\frac{\pi t}{T} - \frac{\pi}{4}\right)$ (d) $\frac{4}{\pi}\sin\left(\frac{\pi t}{2T} + \frac{\pi}{4}\right)$

15. A signal is processed by a causal filter with transfer function $G(s)$. For a distortion free output signal waveform $G(s)$ must.
    (a) Provide zero phase shift for all frequency
    (b) Provide constant phase shift for all frequency
    (c) Provide linear shift that is proportional to frequency
    (d) Provide a phase shift that is inversely properly proportional to frequency

16. $G(z) = \alpha z^{-1} + \beta z^{-3}$ is a low-pass digital filter with a phase characteristics same as that of the above question if.
    (a) $\alpha = \beta$ (b) $\alpha = -\beta$
    (c) $\alpha = \beta^{1/3}$ (d) $\alpha = \beta^{-(1/3)}$

17. If $u(t), r(t)$ denote the unit step and unit ramp function respectively and $u(t) * r(t)$ their convolution, then the function $u(t+1) * r(t-2)$ is given by.
    (a) (1/2) (t – 1) (t–2) (b) (1/2) (t – 1) (t–2)
    (c) (1/2) $(t-1)^2$ u(t–1) (d) None of the above

18. Consider the **discrete-time system** shown in the figure where the impulse response of $G(z)$ is $g(0) = 0$, $g(1) = g(2) = 1$, $g(3) = g(4) = \ldots = 0$

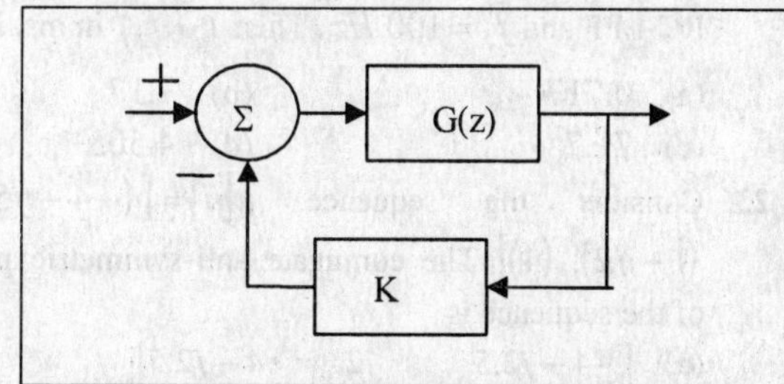

The system is stable for range of values of K
    (a) [–1, 1 /2] (b) [–1, 1]
    (c) [–1/2, 1] (d) [–1/2, 2]

19. $X(z) = 1 - 3z^{-1}, Y(z) = 1 + 2z^{-1}$ are Z-transforms of two signals $x[n], y[n]$ respectively. A **linear time invariant system** has the impulse response $h[n]$ defined by these two signals as.

$$h[n] = x[n-1] * y[n]$$

where * denotes discrete time convolution. Then the output of the system for the input $\delta[n-1]$
    (a) Has Z-transform $Z^{-1}X(z)Y(z)$
    (b) Equals $\delta[n-2] - 3\delta[n-3] + 2\delta[n-4] - 6\delta[n-5]$
    (c) Has Z-transform $1 - 3z^{-1} + 2z^{-2} - 6z^{-3}$
    (d) Does not satisfy any of the above three

20. Let P be linearity, Q be time-invariance, R be causality and S be stability. A discrete-time system has the input-output relationship,

$$y(n) = \begin{cases} x(n) & n \geq 1 \\ 0 & n = 0 \\ x(n+1) & n \leq -1 \end{cases}$$

where x(n) is the input and y(n) is the output. The above system has the properties.

(a) P, S but not Q, R (b) P, Q S but not R
(c) P, Q, R, S (d) Q, R, S but not P

21. The system under consideration is an RC low-pass filter (RC-LPF) with $R = 1.0\ k\Omega$ and $C = 1.0\ \mu F$

Let $H(f)$ denote the frequency response of the RC-LPF. Let f_1 be the highest frequency such that $0 \leq |f| \leq f_1, \left|\frac{H(f_1)}{H(0)}\right| \geq 0.95$. Then f_1 (in Hz) is

(a) 327.8 (b) 163.9
(c) 52.2 (b) 104.4

22. The system under consideration is an RC low-pass filter (RC-LPF) with $R = 1.0\ k\Omega$ and $C = 1.0\ \mu F$

Let $t_g(f)$ be the group delay function of the given RC-LPF and $f_2 = 100\ Hz$. Then $t_g(f_2)$ in ms, is

(a) 0.717 (b) 7.17
(c) 71.7 (d) 4.505

23. Consider the sequence $x[n] = [(-4 - j5), (1 + j2), (4)]$ The conjugate anti-symmetric part of the sequence is

(a) $[-4 - j2.5 \quad j2 \quad 4 - j2.5]$
(b) $[-2.5 \quad 1 \quad j2.5]$
(c) $[-j5 \quad j2 \quad 0]$
(d) $[-4 \quad 1 \quad 4]$

24. A causal LTI system is described by the difference equation.

$$2y[n] = \alpha y[n-2] - 2x[n] + \beta x[n-1]$$

The system is stable only if

(a) $|\alpha| = 2$, $|\beta| < 2$
(b) $|\alpha| > 2$, $|\beta| > 2$
(c) $|\alpha| < 2$, any value of β
(d) $|\beta| < 2$, any value of α

25. The impulse response h[n] of a linear time invariant system is given as.

$$h[n] = \begin{cases} -2\sqrt{2} & n = 1, -1 \\ 4\sqrt{2} & n = 2, -2 \\ 0 & \text{otherwise} \end{cases}$$

If the input to the above system is the sequence $e^{j\pi n/4}$, then the output is

(a) $4\sqrt{2}\ e^{j\pi n/4}$ (b) $-4\sqrt{2}\ e^{j\pi n/4}$
(c) $4\ e^{j\pi n/4}$ (d) $-4\ e^{j\pi n/4}$

26. Let x(t) and y(t) (With Fourier transforms X(*f*) and Y(*f*) respectively) be related as shown in Figure.

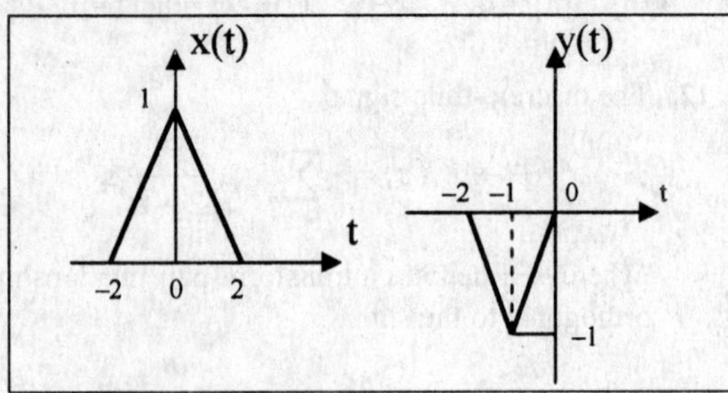

Then Y(f) is

(a) $-[1/2]\, X(f/2)\, e^{-j2\pi f}$
(b) $-[1/2]\, X(f/2)\, e^{j2\pi f}$
(c) $-X[f1/2]\, e^{j2\pi f}$
(d) $-X(f/2)\, e^{-j2\pi f}$

27. The output y(t) of a linear time invariant system is related to its input x(t) by the following equation:
$y(t) = 0.5x(t - t_d + T) + x(t - t_d) + 0.5x(t - t_d - T)$.
The filter transfer function H(ω) of such a system is given by

(a) $(1 + \cos\omega T)e^{-j\omega t_d}$
(b) $(1 + 0.5\cos\omega T)e^{-j\omega t_d}$
(c) $(1 + \cos\omega T)e^{j\omega t_d}$
(d) $(1 - 0.5\cos\omega T)e^{-j\omega t_d}$

28. Match the following and choose the correct combination.

Group-1

E – continuous and a periodic signal
F – continuous and periodic signal
G-Discrete and a periodic signal
H – Discrete and periodic signal

Group-2

1. Fourier representation is continuous and aperiodic
2. Fourier representation is discrete and aperiodic
3. Fourier representation is continuous and periodic
4. Fourier representation is discrete and periodic

	E	F	G	H
(a)	3	2	4	1
(b)	1	3	2	4
(c)	1	2	3	4
(d)	2	1	4	3

29. A signal (n) = sin (ω_0 n + ϕ) is the input to a linear time-invariant system having a frequency response H($e^{j\omega}$). If the output of the system Ax (n – n_0), then the most general form $\angle$H ($e^{j\omega}$) will be

(a) – $n_0\omega_0$ + β for any arbitrary real β.

(b) – $n_0\omega_0$ + 2πk for any arbitrary integer k.

(c) $n_0\omega_0$ + 2πk for any arbitrary integer k.

(d) –$n_0\omega_0$ ϕ

30. For a signal x(t) the Fourier transform is X(f). Then the inverse Fourier transform of X (3 f + 2) is given by

A sequence x(n) has non-zero values as shown in the figure

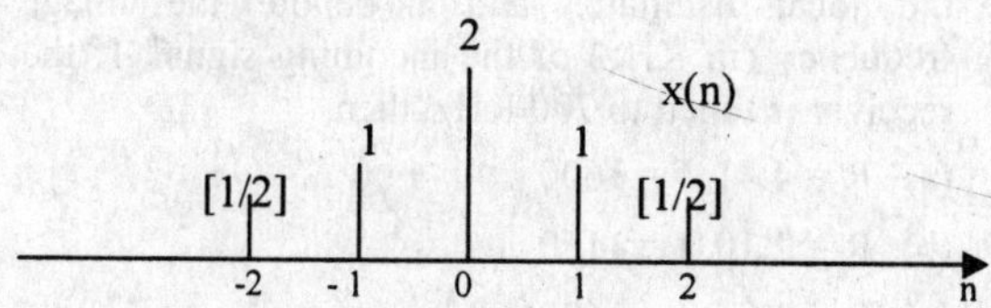

(a) [1 / 2] x (t / 2) $e^{j3\pi t}$

(b) [1 / 3] x (t / 3) $e^{-j4\pi t/3}$

(c) 3x(3t) $e^{-j4\pi t}$

(d) x (3t + 2)

31. The sequence

$$y(n) = \begin{cases} x\left(\dfrac{n}{2}-1\right) & \text{for n even} \\ 0 & \text{for n odd} \end{cases}$$

(a)

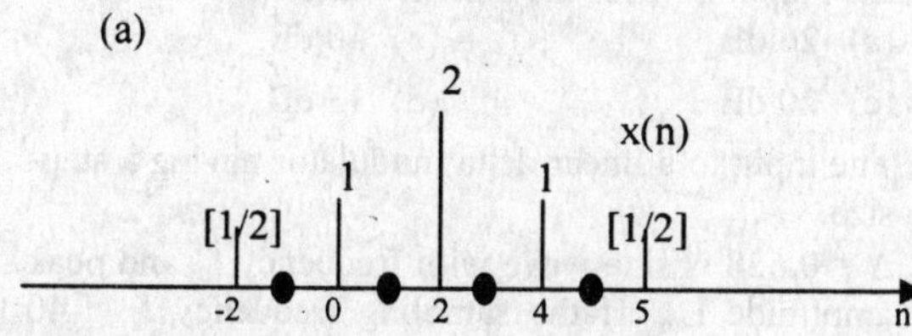

(b)

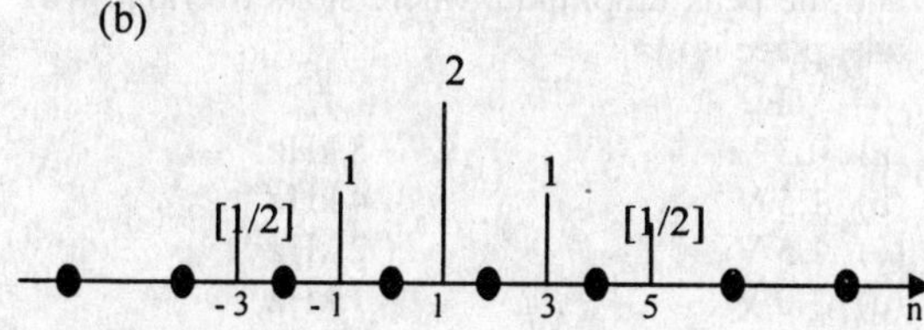

(c)

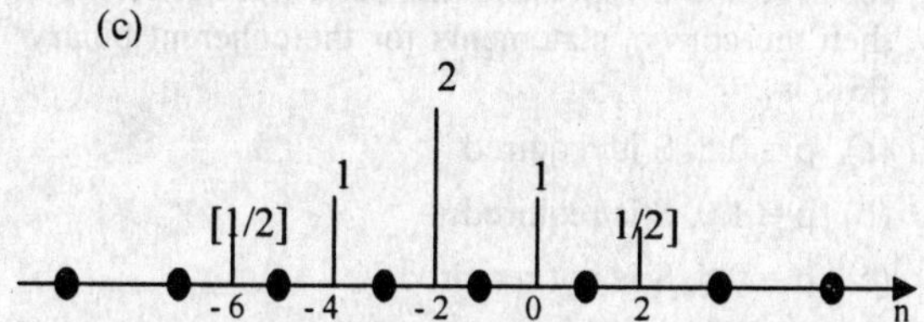

(d)

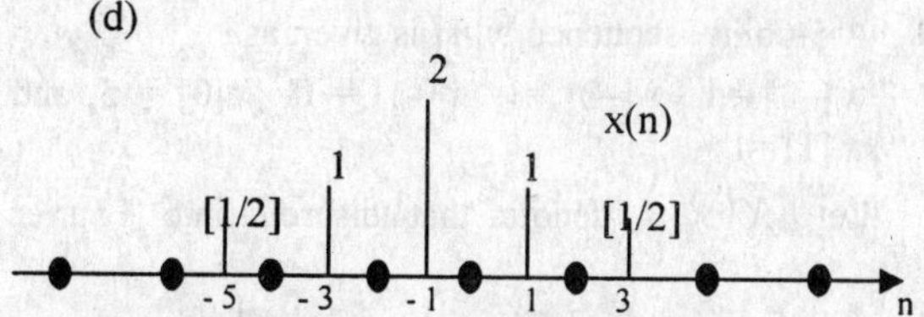

32. The Fourier transform of y(2n) will be

(a) $e^{-j2\omega}$ [Cos 4 ω + 2 Cos 2ω + 2]

(b) Cos 2 ω + 2 Cos ω + 2]

(c) $e^{-j\omega}$ [Cos 2 ω + 2 Cos ω + 2]

(d) $e^{-j2\omega}$ [Cos 2 ω + 2 Cos 2ω + 2]

33. Consider the function f (t) having Laplace transform

F (s) = [ω_0 / (s^2 + ω_0^2)] Re [s] > 0

The final value of f (t) would be:

(a) 0 (b) 1

(c) -1 $\leq$ f (∞) $\leq$ 1 (d) ∞

34. A system with input x [n] and output y[n] is given as y[n] = [sin (5 / 6)π n] x(n).

The system is.

(a) Linear, stable and invertible

(b) Non – linear, stable and non – invertible

(c) Linear, stable and non – invertible

(d) Linear, unstable and invertible

35. The unit – step response of a system starting from rest is given by.

c(t) = 1-e^{-2t} for t $\geq$ 0

The transfer function of the system is.

(a) [1 / (1 + 2s)] (b) [2 / (2 + s)]

(c) [1/ (2 + s)] (d) [2s / (1 + 2s)]

36. The unit impulse response of a system is.

h(t) = e^{-t} , t $\geq$ 0

For this system, the steady- state value of the output for unit step input is equal to.

(a) –1 (b) 0

(c) 1 (d) ∞

37. The 3-dB bandwidth of the low-pass signal $e^{-t}u(t)$, where u(t) is the unit step function, is given by

(a) $\dfrac{1}{2\pi}$ Hz (b) $\dfrac{1}{2\pi}\sqrt{\sqrt{2}-1}\ Hz$

(c) ∞ (d) 1 Hz

38. A Hibert transformer is a

(a) Non-linear system (b) Non-causal system

(c) Time-varying system (d) Low-pass system

39. The frequency response of a **linear, time-invariant** system is given by $H(f) = \dfrac{5}{1 + j10\pi f}$. The step response of the system is

(a) $5\left(1 - e^{-5t}\right)u(t)$ (b) $5\left(1 - e^{-t/5}\right)u(t)$

(c) $\dfrac{1}{5}\left(1 - e^{-5t}\right)u(t)$ (d) $\dfrac{1}{5}\left(1 - e^{-t/5}\right)u(t)$

40. A 5-pointy sequence $x[n]$ is given as $x[-3]=1$, $x[-2]=1$, $x[-1]=0$, $x[0]=5$, and $x[1]=1$

Let $X(e^{j\omega})$ denote that discrete-time Fourier transform of $x[n]$. The value of $\int_{-\pi}^{\pi} X(e^{j\omega})d\omega$ is

(a) 5 (b) 10π
(c) 16π (d) $5+j10\pi$

41. The z-transform X[z] of a sequence $x[n]$ is given by $X[z]=\frac{0.5}{1-2z^{-1}}$. It is given that the region of convergence of $X[z]$ includes the unit circle. The value of $x[0]$ is

(a) -0.5 (b) 0
(c) 0.25 (d) 0.5

42. A DSB-SC signal is to be generated with a carrier frequency f_c = 1 MHz using a non-linear device with the input –output characteristic

$$v_0 = a_0 v_i + a_1 v_i^2$$

where a_0 and a_1 are constants. The output of the non-linear device can be filtered by an appropriate band pass filter.

Let $v_i = A_c' \cos(2\pi f'_c t) + m(t)$

Where m(t) is the message signal.

Then the value of f'_c (in MHz) is

(a) 1.0 (b) 0.333
(c) 0.5 (d) 3.0

43. Let $m(t) = \cos(4\pi \times 10^3)t$ be the message signal and $c(t) = 5\cos(2\pi \times 10^6)t$ be the carrier c(t) and m(t) are used to generate an AM signal. The modulation index of the generated AM signal is 0.5. Then the quantity [Total sideband power/ Carrier power] is.

(a) (1/2) (b) (1/4)
(c) (1/3) (d) (1/8)

44. Let $m(t) = \cos(4\pi \times 10^3)t$ be the message signal and $c(t) = 5\cos(2\pi \times 10^6)t$ be the carrier c(t) and m(t) are used to generate an FM signal. If the peak frequency deviation of the generated FM signal is three times the transmission bandwidth of the AM signal, then the coefficient of the term $\cos[2\pi(1008 \times 10^3 t)]$ in the FM signal (in terms is the Bessel coefficients) is.

(a) $5J_4(3)$ (b) $(5/2)\,J_8(3)$
(c) $(5/2)\,J_8(4)$ (d) $5J_4(6)$

45. Match List I with List II and select the correct answer using the codes given below the Lists:

	List I		List II
A	Ring modulator	1.	Clock recovery
B	VCO	2.	demodulation of FM
C	Foster-Seely discriminator	3.	Frequency Conversion
D	Mixer	4.	Summing the two inputs
		5.	Generation of FM
		6.	Generation of DSB-SC

Codes:

	A	B	C	D
(a)	1	3	2	4
(b)	6	5	2	3
(c)	6	1	3	2
(d)	5	6	1	3

46. A super-heterodyne receiver is to operate in frequency range 550 KH-1650 KHz, with the intermediate frequency of 450 KHz. Let R = C_{max}/C_{min} denote the required capacitance ratio of the local oscillator and I denote the image frequency (in KHz) of the incoming signal. If the receiver is tuned to 700 KHz, then.

(a) R = 4.41, I = 1600
(b) R = 2.10, I = 1150
(c) R = 3.0, I = 1600
(d) R = 9.0, I = 1150

47. A sinusoidal signal with peak-to-peak amplitude of 1.536 V is Quantized into 128 levels using a mid-rise uniform quantizer. The quantization-noise power is.

(a) 0.768 V (b) 48×10^{-6} V^2
(c) 12×10^{-6} V^2 (d) 3.072 V

48. If E_b, the energy per bit of a binary digital signal, is 10^{-6} watt-see and the one sided power spectral density of the white noise, $N_0 = 10^{-5}$ W/Hz, then the output SNR of the matched filter is.

(a) 26 dB (b) 10 dB
(c) 20 dB (d) 13 dB

49. The input to a linear delta modulator having a step-size.

Δ = 0.628 is sine wave with frequency f_m and peak amplitude E_m. If the sampling frequency f_s = 40 KH, the combination of the sine-wave frequency and the peak amplitude, where slops overload will take place is

	E_m	f_m
(a)	0.3 V	8 kHz
(b)	1.5 V	4 kHz
(c)	1.5 V	2 kHz
(d)	3.0 V	1 kHz

50. If S represents the carrier synchronization at the receiver and ρ represents the bandwidth efficiency, then the correct statements for the coherent binary PSK is.

(a) ρ = 0.5. S is required
(b) ρ = 1.0, S is required
(c) ρ = 0.5, S is not required
(d) ρ = 1.0, S is not required

51. A signal is sampled at 8 KHz and is quantizer using 8-bit uniform quantizer. Assuming SNR_q for a sinusoidal signal, the correct statement for PCM signal with a bit rate of R is

(a) R = 32 kbps, SNR_q = 25.8 dB
(b) R = 64 kbps, SNR_q = 49.8 dB
(c) R = 64 kbps, SNR_q = 55.8 dB
(d) R = 32 kbps, SNR_q = 49.8 dB

52. A 1 mW video signal having a bandwidth of 100 MHz is transmitted to a receiver through a cable that has 40 dB loss. If the effective one-sided noise spectral density at the receiver is 10^{-20} Watt/Hz, then the signal to noise ratio at the receiver is

(a) 50 dB (b) 30 dB
(c) 40 dB (d) 60 dB

53. A 100 MHz carrier of 1 V amplitude and a 1 MHz modulating signal of 1 V amplitude are fed to a balance modulator. The output of the modulator is passed through and ideal high-pass filter with cut-off frequency of 100 MHz. The output of the filter is added with 100 MHz signal of 1 V amplitude and 90° phase shift as shown in figure. The envelope of the resultant signal is

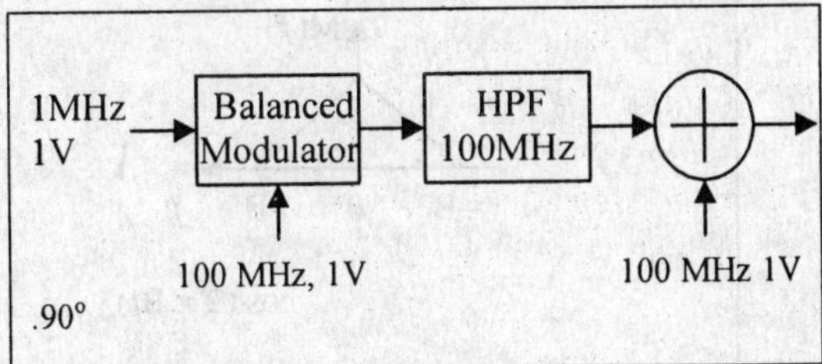

(a) Constant
(b) $\sqrt{[1 + \sin (2\pi \times 10^6 t)]}$
(c) $\sqrt{[5/4 - \sin (2\pi \times 10^6 t)]}$
(d) $\sqrt{[5/4 + \cos (2\pi \times 10^6 t)]}$

54. Two sinusoidal signal of same amplitude and frequencies 10 KHz and 10.1 KHz are added together. The combined signal is given to an ideal frequency detector. The output of the detector is

(a) 0.1 kHz sinusoid
(b) 20.1 KHz sinusoid
(c) a linear function of time
(d) a constant

55. Consider a binary digital communication system with equally likely 0's and 1's. When binary 0 is transmitted the voltage at the detector input can lie between the levels –0.25 V and + 0.25 V with equal probability: when binary 1 is transmitted, the voltage at the detector can have any value between 0 and 1V with equal probability. If the detector has a threshold of 2.0 V (i. e,. if the received signal is greater than 0.2V, the bit is taken as 1), the **average bit error** probability is

(a) 0.15 (b) 0.2
(c) 0.05 (d) 0.5

56. A random variable X with uniform density in the interval 0 to 1 is Quantized as follows:

If $0 \le X \le 0.3$, $x_q = 0$
If $0.3 \le X \le 1$, $x_q = 0.7$

Where X_q is the quatized value of x

The root mean square value of the quantization noise is

(a) 0.573 (b) 0.198
(c) 2.205 (d) 0.266

57. Choose the correct one from among the alternatives A, B, C, D after matching an item from Group 1 with the most appropriate item in Group 2.

Group 1	Group 2
1 : FM	P : Slope overload
2 : DM	Q : μ- Law
3 : PSK	R : Envelope detector
4 : PCM	S : Capture effect
	T : Hilbert transform
	U : Matched filter

(a) 1 - T, 2-P, 3-U, 4-S
(b) 1 - S, 2U, 3-P, 4-T
(c) 1 - S, 2-P, 3U, 4-Q
(d) 1-U, 2-R, 3-S, 4-Q

58. Three analog signals, having bandwidth 1200 Hz, 600 Hz and 600 Hz, are sampled at their respective Nyquist rates, encoded with 12 bit words, and time division multiplexed. The bit rate for the multiplexed signal is

(a) 115.2 kbps (b) 28.8 kbps
(c) 57.6 kbps (d) 38.4 kbps

59. A device with input x(t) and output y(t) is characteristic by: y (t) = x^2(t). An FM signal with frequency deviation of 90 KHz and modulation signal bandwidth of 5 KHz is applied to this device. The bandwidth of the output signal is

(a) 370 KHz (b) 190 KHz
(c) 380 KHz (d) 95 KHz

60. A signal as shown in the figure is applied to a matched filter. Which of the following does represent the output of this matched filter?

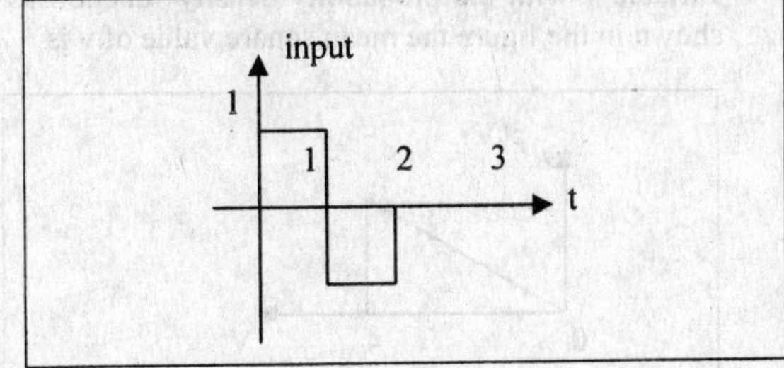

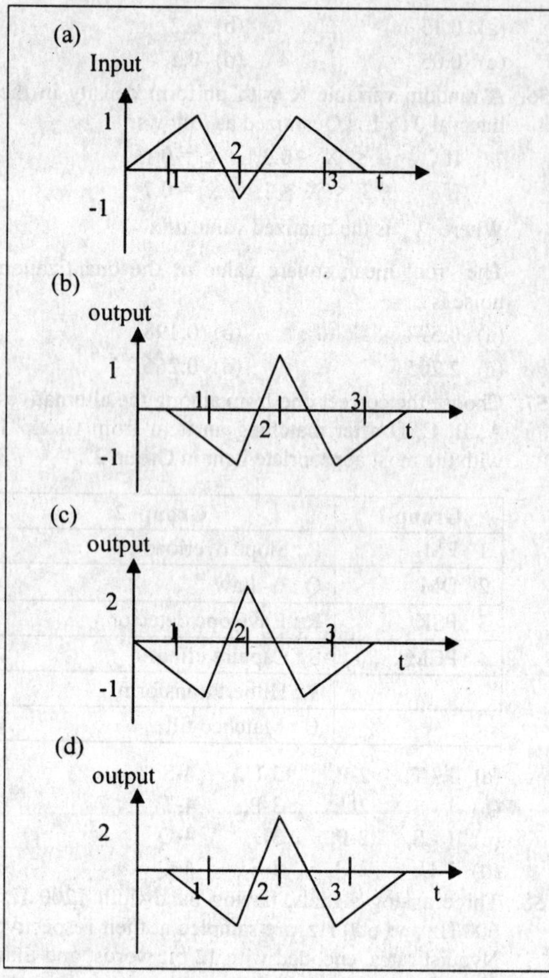

61. Noise with uniform power spectral density of N_0 W/Hz is passed through a filter H (ω) = 2 exp (-jωt$_d$) followed by an ideal pass filter of bandwidth B Hz. The output noise power in Watts is

(a) $2 N_0B$ (b) $4 N_0B$
(c) $8 N_0 B$ (d) $16 N_0B$

62. A carrier is phase modulated (PM) with frequency deviation of 10 KHz by a single tone frequency of 1 KHz. If the single tone frequency is increased to 2 KHz, assuming that **phase deviation remains unchanged**, the bandwidth of the PM signal is

(a) 21 kHz (b) 22 kHz
(c) 42 kHz (d) 44 kHz

63. An output of a communication channel is a random variable v with the probability density function as shown in the figure the mean square value of v is

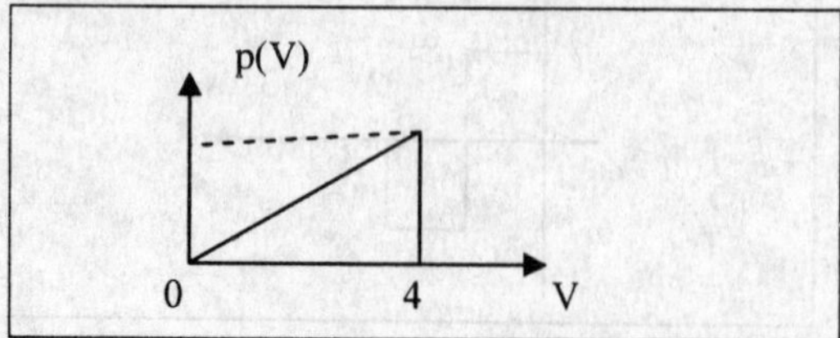

(a) 4 (b) 6
(c) 8 (d) 9

64. In **delta modulation**, the slope overload distortion can be reduced by.

(a) Decreasing the step size
(b) Decreasing the granular noise
(c) Decreasing the sampling rate
(d) Increasing the step size

65. The **raised cosine pulse** p(t) is used for zero ISI digital communications. The expression for p(t) with unity **roll-off factor** is given by $p(t)=\dfrac{\sin 4\pi \omega t}{4\pi \omega t(1-16\omega^2 t^2)}$. The value of $p(t)$ at $t=\dfrac{1}{4\omega}$ is

(a) – 0.5 (b) 0
(c) 0.5 (d) ∞

66. In the following scheme, if the spectrum $M(f)$ of $m(t)$ is as shown, then the spectrum $Y(f)$ of $y(t)$ will be

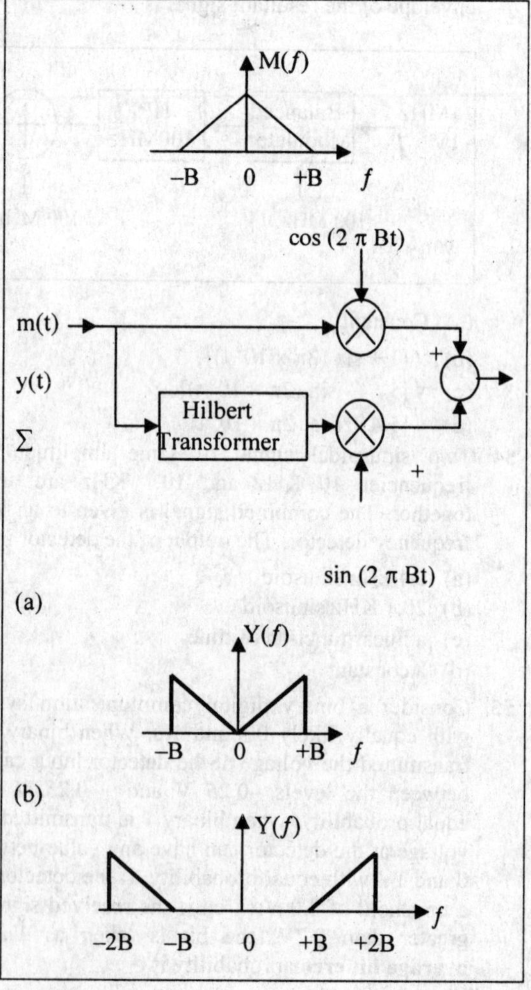

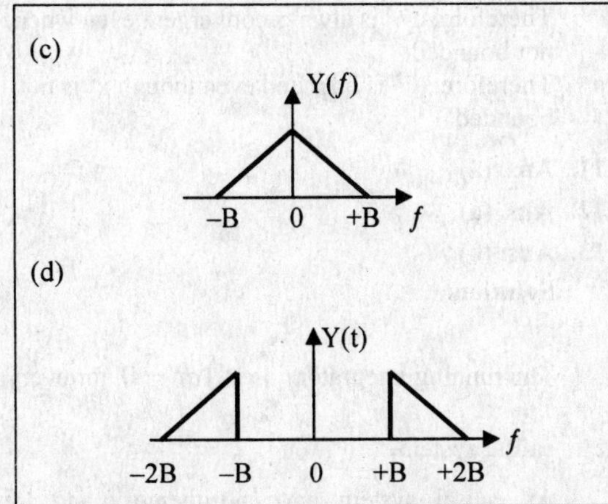

67. During transmission over a certain binary communication channel, bit errors occur independently with probability p. The probability of AT MOST one bit in error in a block of n bit is given by.

(a) p^n

(b) $1 - p^n$

(c) $np(1-p)^{n-1} + (1-p)^n$

(d) $1-(1-p)^n$

68. In a **GSM system,** 8 channels can co-exist in 200 kHz bandwidth using TDMA. A GSM based cellular operator is allocated 5 MHz bandwidth. Assuming a frequency reuse factor of $1/5$, i.e. a five-cell repeat pattern, the maximum number of simultaneous channels that can exist in one cell is

(a) 200 (b) 40

(c) 25 (d) 5

69. In a **Direct Sequence CDMA system** the chip rate is 1.2288×10^6 chips per second. If the processing gain is desired to be LEAST 100, the data rate

(a) Must be less than or equal to 12.288×10^3 bits per sec

(b) Must be grater than 12.288×10^3 bits per sec

(c) Must be exactly equal to 12.288×10^3 bits per sec

(d) Can take any value less than 122.88×10^3 bits per sec

Answers
Signals, Systems and Communication

1. Ans. (b)

Solution:

i = A sin ωt ; ω = 2π/T

i(t) = 10 +20 sin t

$$I_{rms} = \left[\frac{1}{2\pi}\int_o^{2\pi} (10+20\sin t)^2 .dt\right]^{1/2}$$

$$= \left[\frac{1}{2\pi}\int_o^{2\pi} (100+400\sin^2 t+400\sin t).dt\right]^{1/2}$$

$$= \left[\frac{1}{2\pi}\int_o^{2\pi} (100+200(1-\cos 2t)+400\sin t).dt\right]^{1/2}$$

$$= \left[\frac{1}{2\pi}\int_o^{2\pi} (300-200\cos 2t+400\sin t).dt\right]^{1/2}$$

$$= \left[\frac{1}{2\pi}\int_o^{2\pi} 300.dt\right]^{1/2} = \sqrt{300} = 17.3\ A.$$

2. Ans. (a)

Solution:

Straight line equation, y = mt + C

When m = slop of line = tan θ.

then, $m = \dfrac{-6}{T/2} = \dfrac{-12}{T}$ and C = 0

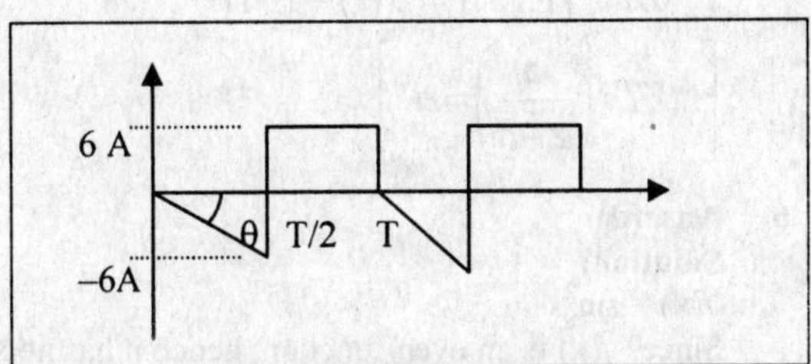

f(t) = mt ; 0≤ t ≤ T/2

= 6 ; T/2 ≤ t ≤ T

rms value for any waveform = $\sqrt{\dfrac{1}{T}\int_o^T f^2(t)\cdot dt}$

$$= \left[\frac{1}{T}\left(\int_o^{T/2} (mt)^2\cdot dt + \int_{T/2}^T 6^2\cdot dt\right)\right]^{1/2}$$

$$= \left[\frac{1}{T}\left(\frac{144}{T^2}\times\frac{T^3}{3\times 8} + 36\times\frac{T}{2}\right)\right]^{1/2}$$

$$= [6+18]^{1/2} = \sqrt{24} = 2\sqrt{6}$$

3. Ans. (a)

Solution:

$$V(t) = \frac{2t}{T},\ 0 \le t \le T/2$$

$$= 0\ ; T/2 \le t \le T$$

$$V(t)_{rms} = \sqrt{\frac{1}{T}\int_o^T |V(t)|^2 dt}$$

$$= \sqrt{\frac{1}{T}\int_o^{T/2} \frac{4}{T^2} t^2 dt}$$

$$= \sqrt{\frac{1}{T}\cdot\frac{4}{T^2}\frac{t^3}{3}\Big|_o^{T/2}} = \sqrt{\frac{4}{T^3}\cdot\frac{T^3}{8\times 3}} = \sqrt{\frac{1}{6}}$$

4. Ans. (a)
Solution:
Apply find value theorem

$$x(\infty) = \lim_{t\to\infty} f(t) = \lim_{s\to 0} sF(s)$$

$$= \lim_{s\to 0}\frac{s(5s^2+23s+6)}{(s^2+2s+2)s} = 3$$

5. Ans. (b)
Solution:

$$F(z) = \frac{1}{z+1} = \frac{z+1-z}{z+1}$$

$$= 1 - \frac{z}{z-(-1)}$$

$$\therefore\ ZT^{-1}[F(z)] = \delta(t) - (-1)^n$$

$$\left[\therefore ZT^{-1}\left(\frac{z}{z-a}\right) = a^n\right]$$

6. Ans. (d)
Solution:
$f(x) = \sin^2 x$
Since, $f(x)$ is an even function, hence it has no sine terms.

$$a_o = \frac{2}{7}\int_o^T f(x)dx$$

$$a_0 = \frac{2}{2\pi}\int_o^{2T} f(x)dx = 1$$

$$\therefore\ f(x) = \frac{a_o}{2} + \sum_{n=1}^{\infty} a_n \cos nx$$

f(x) = 0.5 + terms containing cosine

7. Ans. (c)

8. Ans. (d)

9. Ans. (d)
Solution:

This is care of

Half wave and even symmetry

a_o = Find out.

$a_n = 0$ n even

$$a_n = \frac{8}{T}\int_0^{T/4} f(t)\cos(n\,\omega_o t)\,dt \quad \text{n odd}$$

$b_n = 0$ all n

10. Ans. (d)
Solution: $y(t) = e^{-|x(t)|}$

Therefore, e^{-x} is always convergent even when x is not bounded.
Therefore e^{-x} is bounded even though x^2 is not bounded

11. Ans. (a)

12. Ans. (a)

13. Ans. (b)
Solution:

The running integrator $\int_{-\infty}^{t} x(t')\,dt' = 0$ for every causal system.

As causal system have no memory and initial values, is zero the output is followed by input so, $y(t)$ will always be bounded if this function is causal bounded system.

14. Ans. (c)

15. Ans. (c)
Solution:

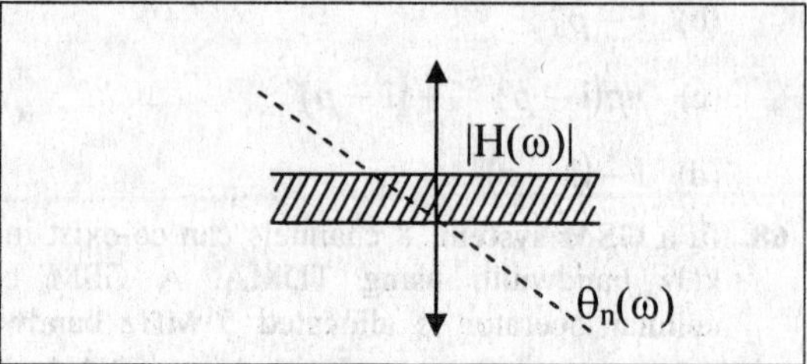

For Distortion Less filter
$Y(j\omega) = A(j\omega).X(j\omega)$
$H(j\omega) = A\,e^{-j\omega}$
Magnitude → constant
Phase → linear

16. Ans. (d)

17. Ans. (c)

Solution: $LT[u(t+1)] = \frac{1}{s}e^{s}$

$$LT[r(t-2)] = \frac{1}{s^2}.e^{-2s}$$

$$u(t) * r(t) \to U(s)\cdot R(s)$$

$$u(t-1) * r(t-2) \to \frac{e^s}{s}\cdot\frac{e^{-2s}}{s^2}$$

$$\to \frac{e^{-s}}{s^3}$$

Then function output $= LT^{-1}\left(\frac{e^{-s}}{s^3}\right)$

$$= \frac{1}{2}(t-1)^2\,u(t-1)$$

18. Ans. (b)

Solution: Since the output is 1 only for $g(1)=g(2)=1$ and o for other then

$G(z)=z^{-1}+z^{-2}$

Characteristic equation, $1+KG(z)=0$

or $1+K\left(z^{-1}+z^{-2}\right)=0$

or $z^2+Kz+K=0$

Using Jusy's stability criteria

$|K|<1$ i.e $-1<K<1$

19. Ans. (d)

Solution: $X(z)=1-3z^{-1}$; $Y(z)=1+2z^{-1}$

$h[n]=x(n-1)*y(n)$

$ZT\,[x(n-1)]=z^{-1}X(z)=z^{-1}\left(1-3z^{-1}\right)$

$=z^{-1}-3z^{-2}$

$\therefore\ H(z)=\left(z^{-1}-3z^{-2}\right)\left(1+2z^{-1}\right)$

$=z^{-1}-3z^{-2}+2z^{-2}-6z^{-3}$

$=z^{-1}-z^{-2}-6z^{-3}$

For $u[n]=\delta[n-1]$

$U(z)=z^{-1}$

$Y(z)=H(z)\times U(z)$

$=\left(z^{-1}-z^{-2}-6z^{-3}\right)\left(z^{-1}\right)$

$=z^{-2}-z^{-3}-6z^{-4}$

20. Ans. (b)

Solution:

- $y(n)=x(n)\Rightarrow$ Thus it is linear $\rightarrow$ P
- $y(n)=x(n+1)\Rightarrow$ Thus it is not causal $\rightarrow$ not R
- **If** $x(n)$ is bounded, output also going to be bounded $\Rightarrow$ Thus it is stable system $\rightarrow$ S
- $y(n)$ has different value for the same time shift in input i.e. $y(n)$ is not shifted by same amount as $x(n)\Rightarrow$ Thus it is time variant $\rightarrow$Q

21. Ans. (c)

Solution:

For RC low pass filter

$$H(f)=\frac{1}{j2\pi fRC+1}$$

$H(o)=1$

Given $\dfrac{H(f_1)}{H(o)}\geq 0.95$

$$\frac{1}{\sqrt{4\pi^2 f_1^2R^2C^2+1}}\geq 0.95$$

$$\Rightarrow\frac{1}{4\pi^2R^2C^2f_1^2+1}>0.9025$$

$4\pi^2R^2C^2f_1^2\leq 0.108$

$$f_1^2<\frac{0.108}{4\pi^2R^2C^2}$$

$$f_1<\frac{0.329}{2\pi RC}\ f_1<\frac{0.329}{2\pi RC}<\frac{0.329}{2\pi\times10^3\times10^{-6}}$$

$=52.2$

22. Ans. (d)

Solution:

$$H(f)=\frac{1}{j2\pi fRC+1}$$

$\phi(f)=-\tan^{-1}(2\pi fRC)$,

Group delay $\tau_g=-\dfrac{d\phi(f)}{d(f)}$

$$=-\frac{1}{(2\pi fRC)^2+1}\times 2\pi RC$$

$f=1000\,Hz$, $R=1.0\,k\Omega$, C

$C=1\,\mu F$ Putting f, R and C in this equation,

Group delay $\tau_g=$

$$\frac{2\times3.14\times1\times10^3\times10^{-6}}{4\times3.14\times3.14\times100\times100\times1\times10^3\times10^3\times10^{-6}\times10^{-6}+1}$$

$=\dfrac{6.28}{1.39}=4.51\,ms$

23. Ans. (c)

Solution:

$\dfrac{s[n]-s*[n]}{2}\Rightarrow$ as conjugate antisymmetirc

$$a_i=\frac{-4-5i-(-4+5i)}{2}=-5i$$

$$a_j=\frac{(1+j2)-(1-i2)}{2}=\frac{4i}{2}=2j$$

$$a_k = \frac{(4+0)-(4-0)}{2} = 0$$

Thus $(-5j, j2, 0)$

24. Ans. (c)

Solution:

$$H(z) = \frac{Y(z)}{X(z)} = \frac{b_0 + b_1 Z^{-1} + b_2 Z^{-2}}{1 + a_1 Z^{-1} + a_2 Z^{-2}}$$

(1)

$$= q[n+1] = \begin{vmatrix} -a_1 & 1 \\ -a_2 & 0 \end{vmatrix} q[n] + \begin{vmatrix} b_1 & -a_1 b_0 \\ b_2 & -a_2 b_0 \end{vmatrix} x[n] \quad (2)$$

$y[n] = [1 \quad 0]\, q[n] + b_0\, x[n]$.

Given $2Y[n] = \alpha Y[n-2] - 2x[n] + \beta x[n-1]$

$$y[n] = \frac{\alpha}{2} y[n-2] - x[n] + \frac{\beta}{2} \times [n-1]$$

z transforming.

$$Y[z]\left(1 - \frac{\alpha}{2}\right) z^{-2} = X[z]\left[-1 + \left(\frac{\beta}{2}\right) z^{-1}\right]$$

$$H(z) = \frac{Y(z)}{X(z)} = \frac{-1 + \left(\frac{\beta}{2}\right) z^{-1}}{1 - \left(\frac{\alpha}{2}\right) z^{-2}}$$

Comparing with (1)

$$b_0 = -1,\ b_1 = \left(\frac{\beta}{2}\right),\ a_1 = 0,\ a_2 = -\left(\frac{\alpha}{2}\right)$$

From (2)

$$= q[n+1] = \begin{pmatrix} 0 & 1 \\ \alpha/2 & 0 \end{pmatrix} q[n] + \begin{pmatrix} (\beta/2) \\ -(\alpha/2) \end{pmatrix} x[n]$$

$$c(\lambda) = [\lambda_1 - A] = \begin{vmatrix} \lambda & -1 \\ (\alpha/2) & \lambda \end{vmatrix}$$

$$\text{Eigen value} = \lambda^2 - \left(\frac{\alpha}{2}\right) = 0 \Rightarrow \lambda^2 = \left(\frac{\alpha}{2}\right)$$

$$\Rightarrow \lambda = \pm\sqrt{\frac{\alpha}{2}}$$

For stability $\lambda < \pm 1$, $\therefore\ \alpha < 2$

β does not figure in stability

Method 2

Given system differential equation is

$$2y(n) = \alpha y(n-2) - 2x(n) + \beta x(n-1)$$

Taking Z-transform.

$$2Y[z] = \alpha z^{-2} Y[z] - 2X[z] + \beta z^{-1} X[z]$$

$$Y[z]\left[2 = \alpha z^{-2}\right] = X[z]\left[-2 + \beta z^{-1}\right]$$

$$H[z] = \frac{Y[z]}{X[z]} = \left[\frac{-2 + \beta z^{-1}}{2 - \alpha z^{-2}}\right]$$

$$= \frac{-2 + \beta z^{-1}}{2\left(1 - \frac{\alpha}{2} z^{-2}\right)}$$

For stable system it ROC should include unit circle $H[z]$ i.e. its pole must lie. The unity circle.

$$1 - \frac{\alpha}{2} z^{-2} > 0$$

$$\text{or} \quad 1 > \frac{\alpha}{2z^{-2}}$$

$$\text{or} \quad z^2 > \frac{\alpha}{2}$$

$$\text{or} \quad 1 > \frac{\alpha}{2}$$

Thus $\alpha < 2$

β does not appear in the analysis.

25. Ans. (d)

Solution: Given input $x[n] = e^{[j\pi/4]n}$

By definition output y(n) is given by

$y(n) = x(n) * h(n)$

$$= \sum_{k=-\infty}^{\infty} x(k)\, h(n-k)$$

$= x(n-1)\,h(1) + x(n+1)\,h(-1) +$

$x(n-2)\,h(2) + x(n+2)\,h(-21) + 0$

$= e^{j\pi/4\{n-1\}}(-2\sqrt{2}) + e^{j\pi/4\{n+1\}}(-2\sqrt{2})$

$+ e^{j\pi/4\{n-2\}}(4\sqrt{2}) + e^{j\pi/4\{n+2\}}(4\sqrt{2})$

$= \sqrt{2}\, e^{\{j\pi/4\}n}\,[-2\{e^{-j\pi} + e^{j\pi/4}\}]$

$+ 4[\{e^{-j\pi/2} + e^{j\pi/2}\}]$

$= \sqrt{2}\, e^{\{j\pi/4\}n}\,[-2 \times 2\cos\{\pi/4\} + 4 \times 2\cos[\pi/2]$

$= \sqrt{2}\, e^{\{j\pi/4\}n}\,[-4/\sqrt{2} + 0] \quad = \mathbf{-4\, e^{\{j\pi/4\}n}}$

26. Ans. (b)

Solution:

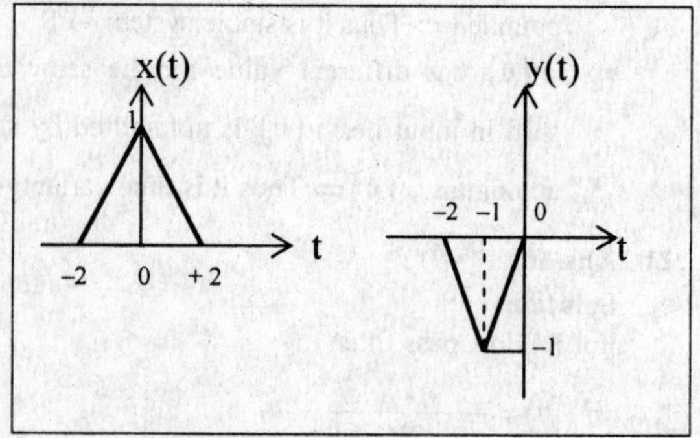

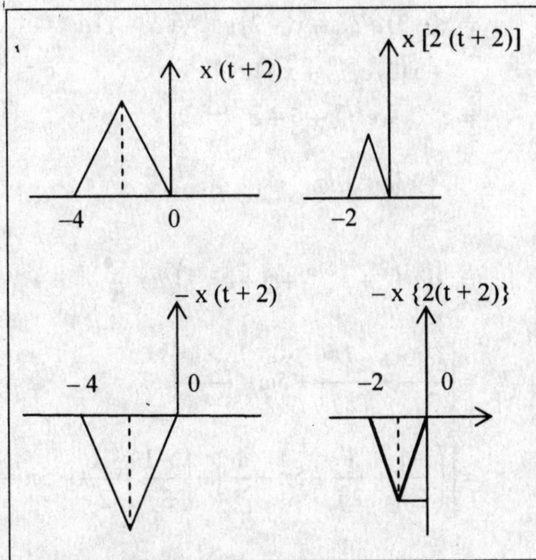

$x(t - t_o) \rightarrow X(f)\, e^{-j2\pi f\, t_o}$

$X(a t) \rightarrow [1/|a|]\ X(f/a)$

$y(t) = -x[2(t+2)] \ \rightarrow -[1/2]\, X(f/2)\, e^{-j2\pi f \cdot 2/2}$

$\rightarrow -\frac{1}{2} \times \left(\frac{f}{2}\right) e^{j2\pi f}$

27. Ans. (a)

Solution: Fourier transform relation for shifted function.

$x(t - t_o) \leftrightarrow e^{-j\omega t_o}\, x(\omega)$

$y(t) = 0.5x(t - t_d + T) + x(t - t_d) + 0.5x(t - t_d - T).$

$Y(\omega) = 0.5e^{-j\omega(td - T)}\, X(\omega) + e^{-j\omega td}\, X(\omega) + 0.5e^{-j\omega c(td + T)}\, X(\omega)$

$= 0.5e^{-j\omega(td - T)}\, X(\omega) + 0.5e^{-j\omega c(td + T)}\, X(\omega) + e^{-j\omega td}\, X(\omega)$

$= 0.5\, X(\omega)\, [e^{-j\omega(td - T)} + e^{-j\omega c(td + T)}] + e^{-j\omega td}\, X(\omega)$

$= e^{-j\omega td}\, X(\omega)\, [1 + 0.5\,(2 \cos\{\omega T\})]$

$\frac{Y(\omega)}{X(\omega)} = e^{-j\omega td}\, [1 + \cos \omega T] = H(\omega)$

28. Ans. (c)

29. Ans. (b)

Solution: $x(n) = \text{Sin}(\omega_o n + \phi)$

$X(n - n_o) = \text{Sin}[\omega_o (n - n_o) + \phi]$

$= \text{Sin}[\omega_o n - \omega_o n_o + \phi]$

$y(n) = Ax(n - n_o)$

$Y(e^{j\omega}) = A \exp[-j\omega n_o]\, X(e^{j\omega})$

$H(e^{j\omega}) = [Y(e^{j\omega}) / X(e^{j\omega})]$

$= A \exp[-j\omega n_o] = |H(e^{j\omega})| \quad \angle H(e^{j\omega})$

$\angle H(e^{j\omega}) = -\omega n_o$

$\angle H(e^{j\omega}) = -\omega_o n_o + 2\pi k$

General solution; k is a arbitrary integer

30. Ans. (b)

Solution: $f(t) \rightarrow F(\omega)$

$f(at) \rightarrow [1/|a|]\ F[\omega/a]$

$f(t - T) \rightarrow F(\omega)\, e^{-j\omega T}$

$f(at - T) \rightarrow [1/|a|]\, F(\omega/a)\, e^{-j\omega T}.$

If $\quad x(t) \Leftrightarrow X(f)$

Then $x(at) \Leftrightarrow X[1/|a|]\ [f/(|a|)]$

Now, X has $3f$

$\therefore |a| = [1/3]$

Inverse Fourier transform of

$X(af + b) \leftrightarrow [1/a]\ e^{-j2\pi[b\backslash a]t}\ x\ [t/a]$

$a = 3, b = 2 \leftrightarrow [1/3]\ x[t/3] \exp -j(4\pi t/3)$

By scaling property

31. Ans. (a)

Solution: $y(2) = x(0) = 2$

$y(4) = x(1) = 1$

32. Ans. (c)

Solution:

$$y(2n) \left\{ \frac{1}{2}, \underset{n=0}{\overset{1}{\uparrow}}, 2, 1, \frac{1}{2} \right\}, \ldots$$

Fourier transform of y(n) is $\sum_{n=-\infty} y(n)\, e^{-j\omega n}$

$y(2n)\ [1/2]\, e^{j\omega} + 1 + 2e^{-j\omega} + e^{j2\omega} + [1/2]\, e^{-j3\omega}$

$y(2n)\ e^{j\omega}\, [\text{Cos } 2\omega + 2 \text{ Cos } \omega + 2]$

33. Ans. (c)

Solution: $F(s) = \frac{\omega_0}{S2 + \omega_0^2}$; Re(s) . 0

$f(t) = \sin \omega_o t$

Final value lie between – 1 and 1.

34. Ans. (c)

35. Ans. (b)

Solution: $c(t) = 1 - e^{-2t}$; $r(t) = u(t)$; $R(s) = [1/s]$

$$C(s) = \frac{1}{s} - \frac{1}{s+2} = \frac{2}{s(s+2)}$$

$$H(s) = \frac{C(s)}{R(s)} = \frac{2}{(s+2)}$$

36. Ans. (c)

Solution: $h(t) = e^{-t}$; $H(s) = \frac{1}{s+1}$

$$Y(s) = H(s)\, X(s) = \frac{1}{s+1} \times \frac{1}{s}$$

$y(t) = 1 - e^{-t}$

In steady state $t \rightarrow \infty$; $y(\infty) = 1$

37. Ans. (a)

Solution: 3dB bandwidth of low pass signal

$$e^{-t}\ u(t) \rightarrow \frac{1}{s+1}$$

For series RL

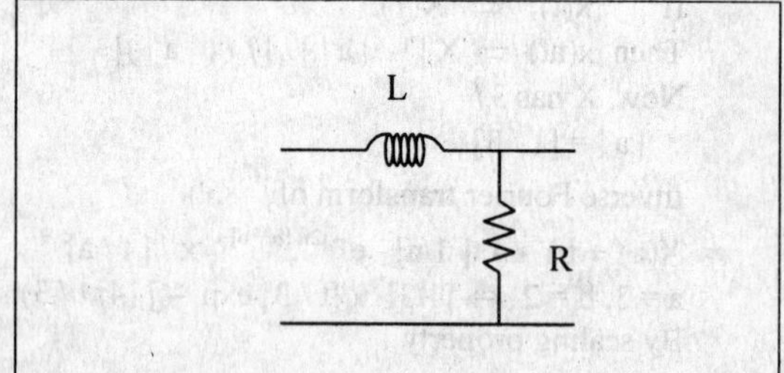

$$H(s)=\frac{R}{R+sL} = \frac{R/L}{s+R/L}$$

Comparing to given

$$\frac{R}{L}=1$$

Bandwidth of RL circuit = $\frac{R}{L}=1$

38. Ans. (a)

Solution: For any signal g(t) its Hilbert transform is

$$\hat{g}(t)=\frac{1}{\pi}\int_{-\infty}^{\infty}\frac{g(\tau)}{(t-\tau)}d\tau$$

System is non-liner because for

$t=0 \quad \hat{g}(t)\neq 0$

39. Ans. (b)

Solution: For step response Y(*f*), $x(t)$ is u(t)

$$X(f)=\frac{1}{1+j10\,\pi f}$$

$$Y(f)=X(f)\times H(f)$$

$$=\left(\frac{1}{j2\pi f}\right)\left(\frac{5}{1+j2\pi f\,(5)}\right)$$

$$=5\left(\frac{1}{j2\pi f}-\frac{1}{\frac{1}{5}+j2\pi f}\right)$$

$$y(t)=5\left(1-e^{-t/5}\right)u(t)$$

$$\therefore\ j2\pi f=S$$

40. Ans. (b)

Solution: $X\left(e^{j\omega}\right)=\sum_{-\infty}^{\infty}x[n]\,e^{-j\omega_n}$

$$=x(-3)\,e^{3j\omega}+x(-2)\,e^{2j\omega}+x(-1)\,e^{j\omega}$$
$$+\,x(0)\,e^{-j0}+x(1)\,e^{-j\omega}$$
$$=e^{3j\omega}+e^{2j\omega}+5+e^{-j\omega}$$

$$\therefore\ \int_{-\pi}^{\pi}X\left(e^{j\omega}\right)d\omega$$

$$=\int_{-\pi}^{\pi}\left(e^{3j\omega}+e^{2j\omega}+5+e^{-j\omega}\right)d\omega$$

$$=\left|\frac{e^{3j\omega}}{3j}+\frac{e^{2j\omega}}{2j}+5\omega+\frac{e^{-j\omega}}{-j}\right|_{-\pi}^{\pi}$$

$$=\left[\left(\frac{-1}{3j}+\frac{1}{2j}+5\pi+\frac{1}{j}\right)-\left(\frac{-1}{3j}+\frac{1}{2j}-5\pi+\frac{1}{j}\right)\right]$$

$$=2\,(5\pi)=10\,\pi$$

41. Ans. (b)

Solution:

$$X[z]=\frac{0.5}{1-2z^{-1}}$$

For right sided sequence ROC = $|Z|>2$

$\therefore\ x(n)$ is anti causal, ROC is $|Z|<2$

$$x[n]=-\frac{1}{2}2^{n}\ U[-n-1]$$

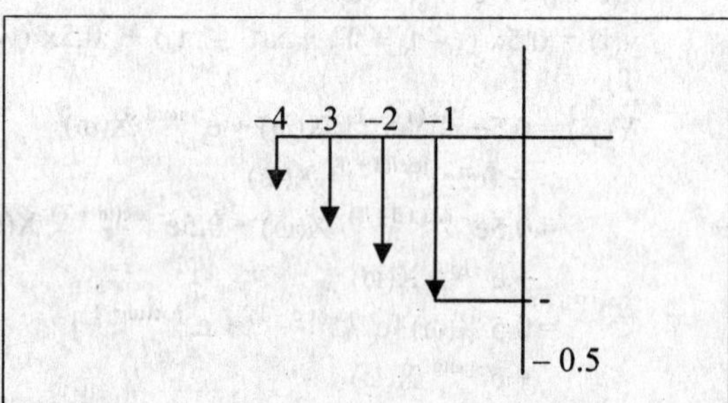

$$\therefore\ x(0)=\frac{-1}{2}Z(0)U(-1)$$

$$=-\frac{1}{2}\times 1\times -(2)^{-1}\ \frac{1}{2}\times 1\times\frac{1}{2}=\frac{1}{4}=0.25$$

42. Ans. (a)

Solution: Modulated AM – DSB /SC,

s(t) = A_c m (t) Cos ω_ct

= 2 m(t) A_c' cos (2π f_c't)

carrier frequency

s(*f*) = A_c [M (*f* + f_c))] + M (*f* + f_c)]

f_c' = f_c + f_m = **1 MHz**

i.e. since carrier frequency is 1 MHz, so $f_e'=1\,\text{MHz}$

43. Ans. (d)

Solution: Modulation index = 0.5 = [1/2] = m_a

$$P_t = P_c\left(1+\frac{m_a^2}{2}\right) = P_c + P_{SB}$$

$$\frac{P_t}{P_c} = 1+\frac{m_a^2}{2} = 1+\frac{0.5}{2}$$

$$= 1+\frac{0.25}{2}$$

Side Band power = P_t - P_c

Then, $\frac{P_t - P_c}{P_c} = \frac{1.125}{1} = \frac{0.125}{1} = \frac{1}{8}$

Second Method

$$P_t = P_c\left(1+\frac{m_a^2}{2}\right) = P_c + P_{SB}$$

$$\frac{P_{SB}}{P_c} = \frac{\frac{m_a^2}{2}P_c}{P_c} = \frac{m_a^2}{2} = \frac{(0.5)^2}{2}$$

$$= \frac{0.25}{2} = \frac{1}{8}$$

44. Ans. (d)

Solution: Audio Frequency = ω_m

AM BW = 2 ω_m

For FM, $\Delta\omega = \delta = 3 \times 2\,\omega_m = 6\,\omega_m$

$\omega_m = 4\pi \times 10^3$

$\omega_c = 2\pi \times 10^6$

$$m_f = \beta = \frac{\Delta\omega}{\omega_m} = 6$$

$$X_{FM}(t) = A\sum_{n=-\infty}^{\infty} J_n(\beta)\cos(\omega_c + n\omega_m)t$$

$\omega_c + n\,\omega_m = 2\pi\,(1008 \times 10^3)$

or, $2\pi \times 10^6 + n \,.\, 4\pi \times 10^3 = 2\pi\,(1008 \times 10^3)$

or, $2\pi \times 1000 \times 10^3 + n \,.\, 4\pi \times 10^3 = 2\pi\,(1008 \times 10^3)$

$n\,4\pi \times 10^3 = 8 \times 2\pi\,10^3$

$n = \frac{8 \times 2\pi}{4\pi}$ or, n = 4

A $J_n(m_f)$ = **5 J_4 (6)**

45. Ans. (b)

Solution:

Ring modulation → Generation of DSB / SC

VCO → Generation of FM

Foster seely Disc → Detection of FM

Mixer → Frequency. Conversion.

46. Ans. (a)

Solution:

Frequency range 550 KHz ----1650 KHz, IF = 450 KHz

Rx is tuned to 700 KHz

$f = [1/2\pi\sqrt{(LC)}$

For Rx capacitance ratio, $\frac{C_{max}}{C_{min}} = \left(\frac{1650}{550}\right)^2 = 9.1$

For Local oscillator ratio, $\frac{C_{max}}{C_{min}} = \left(\frac{1650+450}{550+450}\right)^2$

$$= \left(\frac{2100}{1000}\right)^2 = 4.41$$

As the local oscillator is timed to higher frequency (f_s + IF).

Image frequency = f_s + 2 IF = 700 + 2 x 450

= **1600 kHz**

47. Ans. (c)

Solution: Step size $S = \frac{V_{PP}}{M} = \frac{1.536}{128} = 0.012$

Quantization Noise Power $= \frac{S^2}{12} = 0.012 \times \frac{0.012}{12}$

$$= \frac{0.00014}{12}$$

$= (12 \times 10^{-3})^2$

= **12×10^{-6} V²**

48. Ans. (d)

Solution:

For matched filter Peak pulse signal to Noise ratio.

$$(SNR) = \frac{2\,E_b}{N_o} = \frac{2\times10^{-5}}{10^{-6}} = 20$$

$$SNR|_{db} = 10\log_{10}20 = 13db$$

Since spectral density of white noise is constant w.r.t. to frequency i.e. one sided or two sided.

49. Ans. (b)

Solution: For slope over load noise to take place.

$$\left[\frac{dx(t)}{dt}\right]_{max} > f_s \cdot S \quad x(t) = A\sin 2\pi f_m t$$

$|x'(t)|_{max}$ = A. $2\pi f_m$

= 2 x 3.14 A. f_m

f_s. S = 40 x 10^3 x 0.628

= 25.12 KHz

= 6.28 A. f_m

6.28 × 1.5 × 4 = 37. 68 --- > --- **25.12**

50. Ans. (a)

Solution: Bandwidth efficiency defined as the ratio of data rate in bits per second to the effectively utilized channel Bandwidth:

$\rho = [R_b / B]$ bit /s/ Hz

$= [\log_2 M/2]$

where M = Number of changing in coding i.e. M array PSK

For Binary PSK, M = 2

The $\rho = [\log_2 2 / 2] = [1 / 2] =$ **0.5**

51. Ans. (b)

Solution:

$f_s = 8$ kHz, n = 8 bits

Bit rate R $= nf_s$

$=$ (n bits/sample) x (f_s sample/sec)

R = 8 x 8

R = 64 kbps

$$SNR|_{db} = 1.76 + 6 \times n$$
$$= 1.76 + 6 \times 8$$
$$= 1.76 + 48 = 49.8$$

52. Ans. (a)

Solution:

$$P_n = \int_{-\infty}^{\infty} G(\omega)\, d\omega = \int_0^{100 \times 10^6} 10^{-20}\, d\omega$$

$$= 10^{-20}\ \omega \Big|_0^{100 \times 10^6}$$

$$= 10^{-20} \times 100 \times 10^{-6} = 10^{-12} \text{ watt}$$

$$\frac{S}{N} = \frac{1\,m\,watt}{10^{-12}} = \frac{1 \times 10^{-3}}{10^{-12}} = 10^9$$

$$= 10 \log 10^9 = 90\ db$$

Considering loss of 40 db, the S/N ratio at the receiver is

= (90 – 40) db = 50 db.

53. Ans. (c)

Solution: cos [2 π x 10^8 t] cos [2π x 10^6 t]

= [1 / 2] cos [2π (10^8 + 10^6)t] + cos [2π (10^8 - 10^6)t

After HPF

$f(t)$ = [1 / 2] cos [2π (10^8 + 10^6)t]

After adding another signal at the output one sets

Y(t) = f_{Add} (t) = [1/2] cos [2π (10^8 + 10^6)t]

+ cos (2π x 10^8 t)

Its Hilbert transform gives

$\hat{y}(t)$) = [1 / 2] sin [2π(10^8 + 10^6)t +sin(2π x 10^8)t]

Pre. Envlop Y_t (t)=Y(t) + j $\hat{y}(t)$

= [1/2] $e^{j e\pi}$ (10^8 + 10^6) + j2

Second Method

$$\frac{1}{2}\cos 2\pi\left(101 \times 10^6 t\right) + \sin 2\pi\left(100 \times 10^6 t\right)$$

$$= \frac{1}{2}\cos 2\pi\left(10^8 t\right)\cos 2\pi\left(10^6 t\right)$$
$$-\frac{1}{2}\sin 2\pi\left(100 \times 10^6 t\right)\sin 2\pi 10^6 t + \sin \pi\left(100 \times 10^6 t\right)$$
$$\frac{1}{2}\sin 2\pi\left(10^8 t\right)\cos 2\pi\left(10^6 t\right) - \left(1 - \frac{7}{2}\sin 2\pi\left(10^8 t\right)\right)$$

So envelope detector output

$$\left[\frac{1}{4}\cos^2(\omega_m t) + \left(1 - \frac{1}{2}\sin \omega_m t\right)^2\right]^{1/2}$$

$$= \sqrt{\frac{5}{4} - \sin\left(2\pi \times 10^6\right)}$$

Third Method

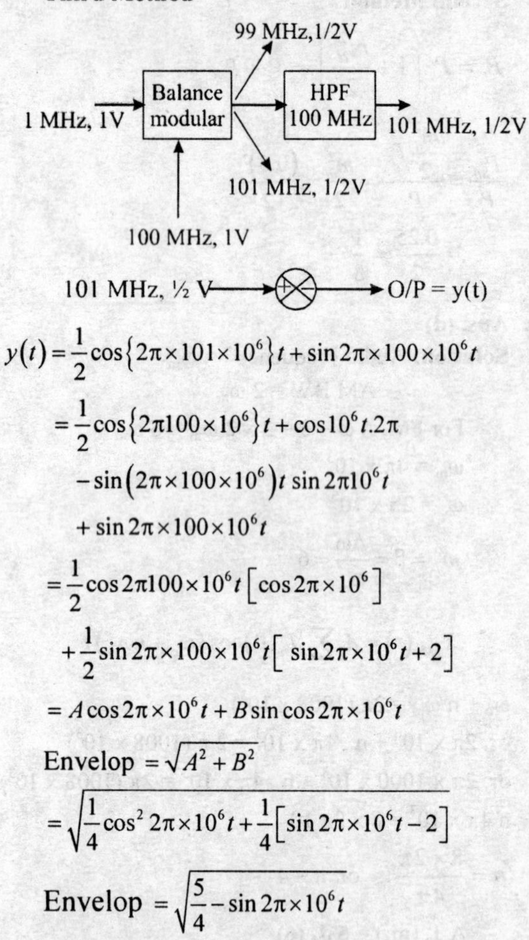

$$y(t) = \frac{1}{2}\cos\left\{2\pi \times 101 \times 10^6\right\}t + \sin 2\pi \times 100 \times 10^6 t$$
$$= \frac{1}{2}\cos\left\{2\pi 100 \times 10^6\right\}t + \cos 10^6 t.2\pi$$
$$- \sin\left(2\pi \times 100 \times 10^6\right)t \sin 2\pi 10^6 t$$
$$+ \sin 2\pi \times 100 \times 10^6 t$$
$$= \frac{1}{2}\cos 2\pi 100 \times 10^6 t\left[\cos 2\pi \times 10^6\right]$$
$$+ \frac{1}{2}\sin 2\pi \times 100 \times 10^6 t\left[\sin 2\pi \times 10^6 t + 2\right]$$
$$= A\cos 2\pi \times 10^6 t + B\sin\cos 2\pi \times 10^6 t$$

$$\text{Envelop} = \sqrt{A^2 + B^2}$$
$$= \sqrt{\frac{1}{4}\cos^2 2\pi \times 10^6 t + \frac{1}{4}\left[\sin 2\pi \times 10^6 t - 2\right]}$$
$$\text{Envelop} = \sqrt{\frac{5}{4} - \sin 2\pi \times 10^6 t}$$

54. Ans. (a)

Solution:

Ideal frequency detector work as mixer whose output is $(f_2 + f_1)$, f_2, f_1, $(f_2 - f_1)$

Detector output is connected with LPF. Hence only $(f_2 - f_1)$ signal through it i.e. $(16.1 - 10) = 0.1\,kHz$ sinusoidal.

Another way

$$f(t) = A\cos\omega_1 t + A\cos\omega_2 t$$

$$\omega = 2\pi f = \frac{2\pi}{T}$$

$$T_1 = \frac{1}{10\,k} = 100\,\mu s$$

$$T_2 = \frac{1}{10.1\,k} = 99\,\mu s$$

$f(t)$ will be periodic with time period LCM of T_1 and T_2

$\therefore$ Frequency of detector $= 0.1\,kHz$

55. Ans. (a)

Solution:

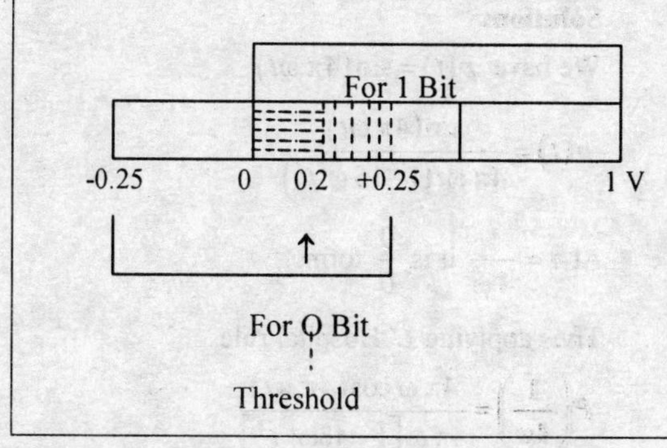

Shaded area shows the probability of error

Average $BEP = P(0)\,P(1/0) + P(1)\,P(0/1)$

$$P(0/1) = P_e(0) = \frac{\text{error}}{\text{total}}$$

$$= \frac{0.25-0.2}{0.25-(0.25)} = \frac{0.05}{0.50}$$

And $P(0/1) = P_e(1) = \dfrac{0.2-0}{1-0} = 0.2$

$$P(0) = P(1) = \frac{1}{2}$$

$$BEP = \frac{1}{2}(0.2+0.1) = 0.15$$

56. Ans. (b)

Solution:

Since it is uniform as

$x_q = 0$ in the range $0 \le x \le 0.3$

$x_q = 0.7$ in the range $0.3 \le x \le 1$

The square mean value is

$$\sigma^2 = \int_{-\infty}^{\infty} \left(x - x_q\right)^2 f(x)\,dx$$

$$= \int_0^1 \left(x - x_q\right)^2 f(x)\,dx$$

$$= \int_0^{0.3} (x-0)^2 f(x)\,dx + \int_{0.3}^{0.1} (x-0.7)^2 f(x)\,dx$$

$$= \left[\frac{x^3}{3}\right]_0^{0.3} + \left[\frac{x^3}{3} + 0.49\,x - 1.4\right]_{0.3}^{1}$$

or $\sigma^2 = 0.039$

The root mean square quantization noise

$$\text{RMS} = \sqrt{\sigma^2}$$

$$= \sqrt{0.039} = 0.198$$

57. Ans. (c)

Solution:

FM --- Capture effect ----- Receives only strong signal

DM --- Slop over load Noise

PSK --- Matched filter

PCM - μ law – Non linear quantization by using Companding with a law

$$V = \frac{\log\left(1+\mu|M|\right)}{\log\left(1+\mu\right)}$$

58. Ans. (c)

Solution: The three analog Signals having BW 1200 Hz, 600 Hz and 600 Hz are sampled at their respective Nyquist rate i.e. at 2400, 1200, 1200 sample/sec respectively.

The total of (2400 + 1200 + 1200)

= 4800 sample/sec

The Bit rate $= n.\,f_s =$ (4800 sample/sec) x 12

= 57. 6 Kbps

Where n = number of bit in a symbol

59. Ans. (a)

Solution: In present case

$\Delta f = 90;\ f_m = 5$

$\beta = [\Delta f / f_m] = [90/5] = 18$

FM equation

A cos $[\omega_c t + \beta \sin \omega_m t]$

= A cos $[\omega_c t + 18 \sin \omega_m t]$

$y(t) = x^2(t) = A^2 \cos^2 [\omega_c t + 18 \text{ Sin } \omega_m t]$

Note: $\text{Cos}^2\,\theta = [1 + \text{Cos}2\theta]/2$

If there is change in frequency the modulation index also changes in same ratio

$y(t) = A^2 [(1/2) + (1/2) \text{ Cos } \{2\omega_c t + 36 \text{Sin } \omega_m t\}]$

$y(t) = [(A^2/2) + (A^2/2) \text{ Cos } \{2\omega_c t + 36 \text{Sin } \omega_m t\}]$

After the device,

$\beta_{(new)} = 36 = [\Delta f_{(new)} / f_m]$

$\Delta f_{(new)} = 36 \times 5 = 180$

By carson's rule

Bandwidth = 2 $(\Delta f + f_m)$

= 2 (180 + 5)

Bandwidth = 370 kHz

60. Ans. (c)

Solution: x(t) = u(t) – 2 u(t - 1) + u(t - 2) for x (t)

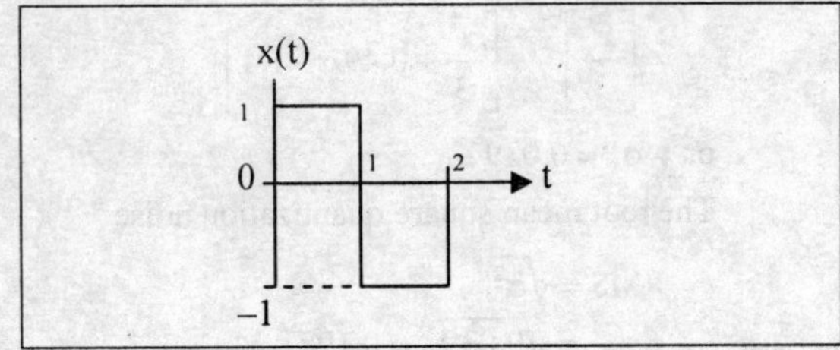

The impulse of a matched filter is delayed version of the input.

Now delay not given, so assumed to be zero.

∴ y (t) = x (t)* x(t) [n(t) = x(t)] where * represent convolution.

y(t) = r(t) - 4r (t – 1) + 6r (t - 2) – 4r (t -3) + r (t - 4)

Hence u (t) * u (t) = r(t) Ramp function.

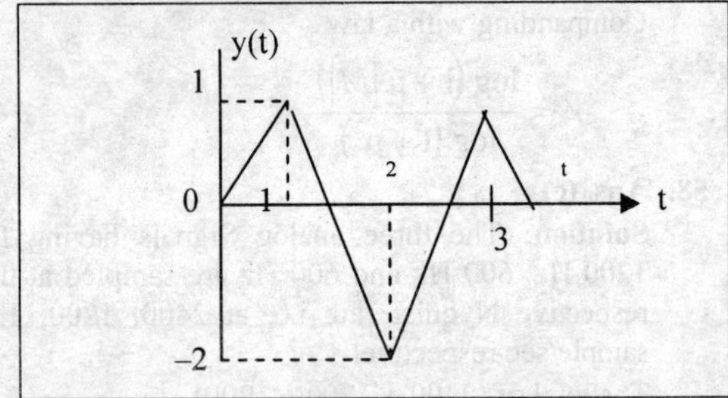

61. Ans. (a)

Solution:

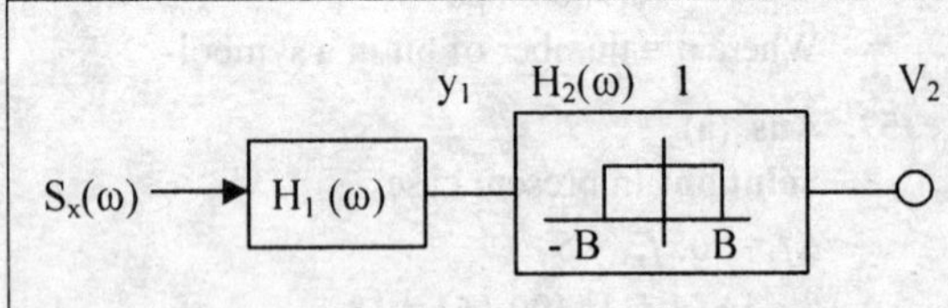

PSD of input $G_x(\omega) = N_o$ W/ Hz. $H_1(\omega) = 2\,e^{-j\omega td}$

$G_{y1}(\omega) = |H_1(\omega)|^2 |G_x(\omega) = 4\,N_o$ W / Hz

$G_{y2}(\omega) | = H_2(\omega) | G_{y1} x(\omega)$

$= 4\,N_o$ W / Hz , $-B \le \omega \le B$

Noise power $= \int_{-B}^{B} G_{y2}(\omega)\, dW = 2 \times 4\,N_oB = 8\,N_oB$

62. Ans. (d)

Solution: Δf = 10 KHz $f_{m\,(new)}$ = 2 KHz

f_m = 1 KHz

By carson's Rule

BW = 2 $(\Delta f + f_m)$ = 2 (10 + 1) = 22 KHz

$\Delta f \propto f_m$

$\Delta f_{(new)}$ = 2 x 10 = 20

$BW_{(new)}$ = 2 (20 + 2) = **44 kHz**

63. Ans. (b)

Solution:

Mean noise power = Area under the PSD curve

$$= 4\left[\frac{1}{2} \times B \times \frac{N_o}{2}\right]$$

$$= BN_o$$

The ratio of average sideband power to mean noise power is

$$\frac{\text{Side Band Power}}{\text{Noise Power}} = \frac{6.25}{N_oP} = \frac{25}{4N_oB}$$

64. Ans. (d)

Solution: In **delta modulation**, the slope overload distortion can be reduced by Increasing the step size

65. Ans. (c)

Solution:

We have $p(t) = \sin(4\pi\,\omega t)$

$$p(t) = \frac{\sin(4\pi\,\omega t)}{4\pi\,\omega t\left(1 - 16\,\omega^2 t^2\right)}$$

At $t = \dfrac{1}{4\,\omega}$ it is $\dfrac{0}{0}$ form.

Thus applying L' Hospital rule

$$P\left(\frac{1}{4\omega}\right) = \frac{4\pi\,\omega\cos(4\pi\,wt)}{4\pi\,\omega\left[1 - 48\omega^2 t^2\right]}$$

$$= \frac{\cos(4\pi\,\omega t)}{1 - 48\omega^2 t^2} = \frac{\cos\pi}{1 - 3} = 0.5$$

66. Ans. (a)

Solution: This diagram represent SSB generation by phase shift method.

For USB, r(t) = m(t) cos π Bt – m′(t) sin 2 π Bt

For LSB, r(t) = m(t) cos π Bt + m′(t) sin 2 π Bt

Since y(t) is sum of both.

∴ It is LSB generator and its Fourier transform $Y(f)$ is given as:

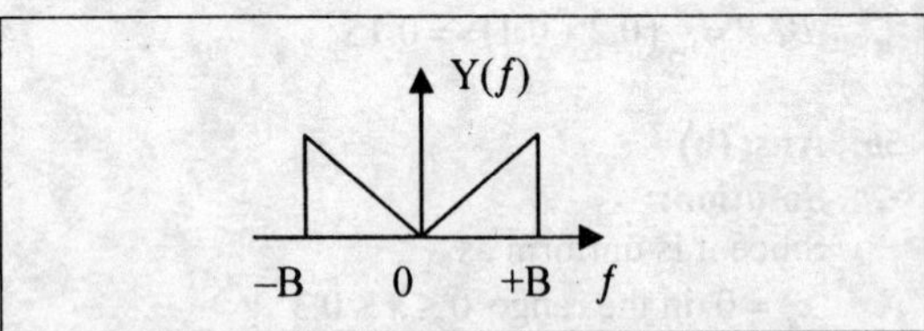

67. Ans. (c)

Solution:

By Binomial distribution the probability of error is

$$p_e = nC_r p^r (1 - p)^{n-r}$$

Probability of at most one error

= Probability of no (zero bit) error

+ Probability of one bit error

$= {}^nC_0\left(p^0\right)\left(1-p\right)^{n-0} + {}^nC_1\left(p^1\right)\left(1-p\right)^{n-1}$

$= \left(1-p\right)^n + np\left(1-p\right)^{n-1}$

68. Ans. (b)

Solution:

Bandwidth allocated for 1 channel = 5 MHz

Average bandwidth for 1 channel $\frac{5}{5} = 1\,MHz$

Total Number of Simultaneously channel

$= \frac{1\,M \times 8}{200\,K} = 40$ **channel.**

69. Ans. (b)

Solution:

Chip Rate $R_C = 1.2288 \times 10^6$ chips/sec

Data Rate $R_b = \frac{R_C}{G}$

Since the processing gain G must be at least 100, thus for $G_{\min}$ we get

$$R_{b\max} \geq \frac{R_C}{G_{\min}} = \frac{1.2288 \times 10^6}{100}$$

$$\geq 1.2288 \times 10^3$$

Common Data for Questions 1 and 2

The system under consideration is an RC low-pass filter (RC-LPF) with $R = 1.0\,k\Omega$ and $C = 1.0\,\mu F$

1. Let $H(f)$ denote the frequency response of the RC-LPF. Let f_1 be the highest frequency such that $0 \leq |f| \leq f_1, \left|\frac{H(f_1)}{H(0)}\right| \geq 0.95$. Then $f_1(\text{in } Hz)$ is

(a) 327.8 (b) 163.9
(c) 52.2 (b) 104.4

2. Let $t_g(f)$ be the group delay function of the given RC-LPF and $f_2 = 100\,Hz$. Then $t_g(f_2)$ in ms, is

(a) 0.717 (b) 7.17
(c) 71.7 (d) 4.505

Common Data for Questions 3 and 4

3. A signal is processed by a causal filter with transfer function $G(s)$. For a distortion free output signal waveform $G(s)$ must.

(a) Provide zero phase shift for all frequency
(b) Provide constant phase shift for all frequency
(c) Provide linear shift that is proportional to frequency
(d) Provide a phase shift that is inversely properly proportional to frequency.

4. $G(z) = \alpha z^{-1} + \beta z^{-3}$ is a low-pass digital filter with a phase characteristics same as that of the above question if.

(a) $\alpha = \beta$ (b) $\alpha = -\beta$
(c) $\alpha = \beta^{1/3}$ (d) $\alpha = \beta^{-(1/3)}$

Statement for Communication Linked Answers Questions 5 and 6

The message signal with bandwidth 10 kHz is transmitted via a channel with 80 dB attenuation and additive noise with power spectral density $S_n(\text{f}) = \frac{N_0}{2} = 10^{-12}$ W/Hz.

5. The required transmitted power is, if the channel bandwidth to achieve a SNR of at least 50 dB at the modulator output, if the modulating scheme is DSB AM.

(a) 400 kW (d) 200 W
(c) 300 W (d) 200 kW

6. If the modulating index of the system is 40% than side band power is

(a) 16 kW (b) 14.82 kW
(c) 25.52 kW (d) 22.30 kW

Statement for Linked Answers Questions 7 and 8

An amplitude modulated voltage in volts is given by $v = 10(1 + 0.5\sin 1000t)\,\sin(6.28 \times 10^6)t$.

7. The rms value of the un-modulated carrier voltage in volts is

(a) 10 (b) $\frac{10}{\sqrt{2}}$
(c) 20 (d) $\frac{20}{\sqrt{2}}$

8. The upper sideband frequency of the modulated signal is

(a) 1000 kHz (b) 1000.159 kHz
(c) 999.841 kHz (d) 1000.318 kHz

Common Data for Questions 9 and 10

Let m(t) = cos $(4\pi \times 10^3)$ t be the message signal and c(t) = 5 cos $(2\pi \times 10^6)$ t be the carrier.

9. c(t) and m(t) are used to generate an AM signal. The modulation index of the generated AM signal is 0.5. Then the quantity [Total sideband power / Carrier power] is

(a) (1/2) (b) (1/4)
(c) (1/3) (d) (1/8)

10. c(t) and m(t) are used to generate an FM signal. If the peak frequency deviation of the generated FM signal is three times the transmission bandwidth of the AM signal, then the coefficient of the term cos [2 π (1008 x 10^3t)] in the FM signal (in terms is the Bessel coefficients) is.

(a) $5J_4(3)$ (b) $(5/2)\ J_8\ (3)$
(c) $(5/2)\ J_8\ (4)$ (d) $5J_4(6)$

Statement for Linked Answer Questions 11 and 12:
Consider the following Amplitude Modulated (AM) signal, where $f_m < B$:
$x_{AM}(t) = 10\,(1 + 0.5 \sin 2\pi f_m t) \cos 2\pi f_c t$.

11. The average side band power for the AM signal given above is
(a) 25 (b) 12.5
(c) 6.25 (d) 3.125

12. The AM signal gets added to a noise with power spectral density $S_n(f)$ given in the figure below. The ratio of average sideband power to mean noise would be

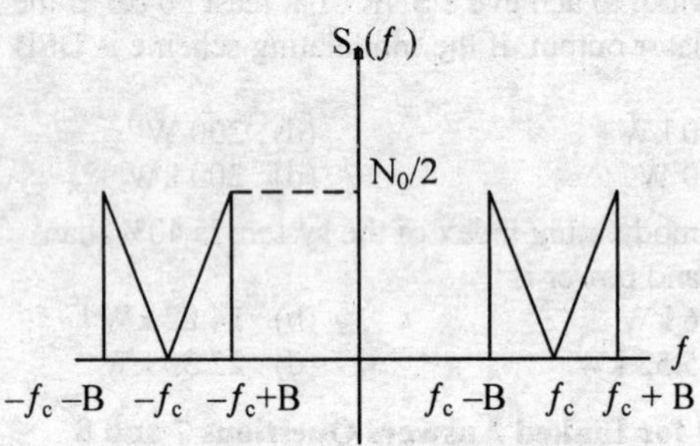

(a) $25/(8N_0B)$ (b) $25/(4N_0B)$
(c) $25/(2N_0B)$ (d) $25/(N_0B)$

Answers
Signals, Systems and Communication

1. Ans. (c)
Solution:
For RC low pass filter

$$H(f)=\frac{1}{j2\pi fRC+1}$$

$$H(o)=1$$

Given $\frac{H(f_1)}{H(o)} \geq 0.95$

$$\frac{1}{\sqrt{4\pi^2 f_1^2 R^2C^2+1}} \geq 0.95$$

$$\Rightarrow \frac{1}{4\pi^2 R^2C^2 f_1^2+1} > 0.9025$$

$$4\pi^2 R^2C^2 f_1^2 \leq 0.108$$

$$f_1^2 < \frac{0.108}{4\pi^2 R^2C^2}$$

$$f_1 < \frac{0.329}{2\pi RC}\ f_1 < \frac{0.329}{2\pi RC} < \frac{0.329}{2\pi\times 10^3\times 10^{-6}}$$

$$= 52.2$$

2. Ans. (d)
Solution:

$$H(f)=\frac{1}{j2\pi fRC+1}$$

$$\phi(f)=-\tan^{-1}(2\pi fRC),$$

Group delay $\tau_g = -\frac{d\phi(f)}{d(f)}$

$$=-\frac{1}{(2\pi fRC)^2+1}\times 2\pi RC$$

$f = 1000\ Hz,\ R = 1.0\ k\Omega,$
$C = 1\ \mu F$ Putting f, R and C in this equation,

Group delay $\tau_g =$

$$\frac{2\times 3.14\times 1\times 10^3\times 10^{-6}}{4\times 3.14\times 3.14\times 100\times 100\times 1\times 10^3\times 10^3\times 10^{-6}\times 10^{-6}+1}$$

$$=\frac{6.28}{1.39} = 4.51\ ms$$

3. Ans. (c)
Solution:

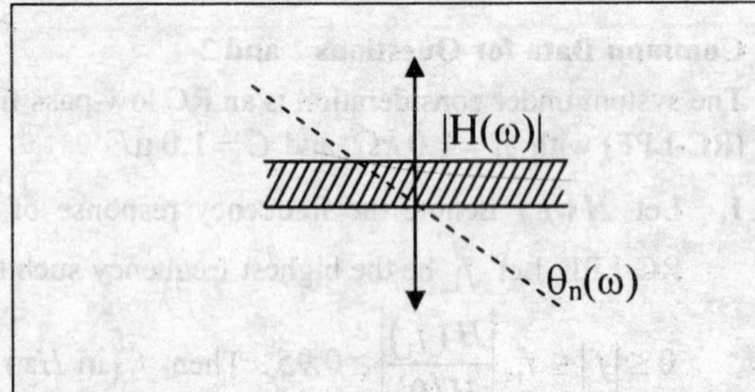

For Distortion Less filter
$Y(j\omega) = A(j\omega) \cdot X(j\omega)$
$H(j\omega) = A e^{-j\omega}$
Magnitude → constant
Phase → linear

4. Ans. (d)
Solution:

$$G(z)\Big|_{z=e^{j\omega}} = \alpha e^{-j\omega} + \beta e^{-3j\omega}$$

For linear phase characteristic $\alpha = \beta$.

5. Ans (d)
Solution:

$$\left(\frac{S}{N}\right)_b = \frac{P_R}{N_o\times BW}$$

$$= \frac{P_R}{2\times 10^{-12}\times 10^4}$$

$$=\frac{10^8 P_R}{2}$$

Hence the channel attenuation is 80 dB, the ratio of transmitted power to received power P_R is,

$$10\log\left(\frac{P_T}{P_R}\right)=80$$

$$\Rightarrow \quad P_R = 10^8 P_T$$

Therefore, $$\left(\frac{S}{N}\right)_b = \frac{10^8 \times 10^{-8} P_T}{2}$$

$$=\frac{P_T}{2}$$

$$\left(\frac{S}{N}\right)_0 = \left(\frac{S}{N}\right)_b$$

$$\approx 50\,\text{dB}$$

$$=10^5$$

Therefore, $$\frac{P_T}{2} = 10^5$$

Therefore, $$P_T = 2\times 10^5 \text{ watt}$$

$$= 200\,\text{kW}$$

6. **Ans. (b)**
Solution:

$$P_T = P_C\left(1+\frac{m^2}{2}\right)$$

Where $m_a = 0.4$

Hence, $P_C = 185.19\,\text{kW}$

Therefore, Sideband power

$$=\frac{m_a^2 \ P_C}{2}$$

$$=14.82\,kW$$

7. **Ans. (b)**
Solution:
RMS value of un-modulated carrier voltage is $=\frac{10}{\sqrt{2}}$

8. **Ans. (b)**
Solution:
Let f_c = carrier frequency

$$=10^6\,\text{Hz} = 1000\,\text{kHz.}$$

From the equation of modulated signal

$$v = 10(1+0.5\sin 1000t)\sin\left(6.28\times 10^6\right)t.$$

The modulating signal freq

$$f_m = \frac{10^3}{2\pi} \quad = \frac{10^3}{6.25}\text{Hz}$$

$$=159Hz \ = 0.159kHz.$$

So the upper sideband frequency is

$$(f_e + f_m) = 1000 + 0.159$$

$$= 1000.159\,kHz.$$

9. **Ans. (d)**
Solution: Modulation index = 0.5
= [1/2] = m_a

$$P_t = P_c\left(1+\frac{m_a^2}{2}\right) = P_c + P_{SB}$$

$$\frac{P_t}{P_c} = 1+\frac{m_a^2}{2} = 1+\frac{0.5}{2}$$

$$=1+\frac{0.25}{2}$$

Side Band power = P_t - P_c

Then, $$\frac{P_t - P_c}{P_c} = \frac{1.125}{1} = \frac{0.125}{1} = \frac{1}{8}$$

Second Method

$$P_t = P_c\left(1+\frac{m_a^2}{2}\right) = P_c + P_{SB}$$

$$\frac{P_{SB}}{P_c} = \frac{\frac{m_a^2}{2}P_c}{P_c} = \frac{m_a^2}{2} = \frac{(0.5)^2}{2}$$

$$=\frac{0.25}{2} = \frac{1}{8}$$

10. **Ans. (d)**
Solution: Audio Frequency $= \omega_m$

AM BW = 2 ω_m

For FM, $\Delta\omega = \delta = 3\times 2\,\omega_m = 6\,\omega_m$

$\omega_m = 4\pi \times 10^3$

$\omega_c = 2\pi\times 10^6$

$$m_f = \beta = \frac{\Delta\omega}{\omega_m} = 6$$

$$X_{FM}(t) = A\sum_{n=-\infty}^{\infty} J_n(\beta)\cos\left(\omega_c + n\omega_m\right)t$$

$\omega_c + n\,\omega_m = 2\pi\,(1008 \times 10^3)$

or, $2\pi\times 10^6 + n.\,4\pi\times 10^3 = 2\pi\,(1008\times 10^3)$

or, $2\pi\times 1000\times 10^3 + n\,.\,4\pi\times 10^3$
$= 2\pi\,(1008\times 10^3)$

$n\,4\pi\times 10^3 = 8\times 2\pi\,10^3$

$$n = \frac{8\times 2\pi}{4\pi} \quad \text{or, } n = 4$$

A J_n (m_f) = **5 J_4 (6)**

11. Ans. (c)

Solution:

We have $x_{AM}(t) = 1(1 + 0.5\sin 2\pi f_m t)\cos 2\pi f_c t$

Carrier power $P_c = \frac{(10)^2}{2} = 50$

Side band power $P_c = \frac{m^2 P_c}{2}$

$$= \frac{(0.5)^2 (50)}{2} = 6.25$$

12. Ans. (b)

Solution:

Mean noise power = Area under the PSD curve

$$= 4\left[\frac{1}{2} \times B \times \frac{N_o}{2}\right]$$

$$= BN_o$$

The ratio of average sideband power to mean noise power is

$$\frac{\text{Side Band Power}}{\text{Noise Power}} = \frac{6.25}{N_o P} = \frac{25}{4N_o B}$$

7 Electrical and Electronic Measurements

Electrical and Electronic Measurements: Bridges and potentiometers, measurement of R,L and C. Measurements of voltage, current, power, power factor and energy. A.C and DC current probes. Extension of instrument ranges. Q-meter and waveform analyzer. Digital voltmeter and multi-meter. Time, phase and frequency measurements. Cathode ray oscilloscope. Serial and parallel communication. Shielding and grounding.

7.1 BRIDGES

AC Bridges

- A.C bridge are used for measurement of inductance, capacitance storage factor, loss factor.
- An A.C bridge , consists of four arms, a source of excitation and a balance detector.
- In A.C bridge, each of the four arms is an impedance and the battery and the galvanometer of the Wheatstone bridge are replaced respectively by an a.c. source and a detector.
- Also used for measurement of phase shifting, providing feedback paths for oscillators and amplifiers, filtering out undesirable signal and measuring the frequency of audio signals.

Detectors: Commonly used for a.c. bridges are

1. Head phone
2. Vibration galvanometers
3. Tunable Amplifier Detector.

General Equation for Bridge Balance,

$$Z_1 Z_4 = Z_2 Z_3$$

7.2 POTENTIOMETERS

Potentiometer is basically used to measure an unknown voltage by comparing it with a known voltage. They can be used to measure both d.c. and a.c. voltage. The basic d.c. potentiometer is also known as *slide wire potentiometer* as shown in the figure contains a standard cell which gives the known voltage.

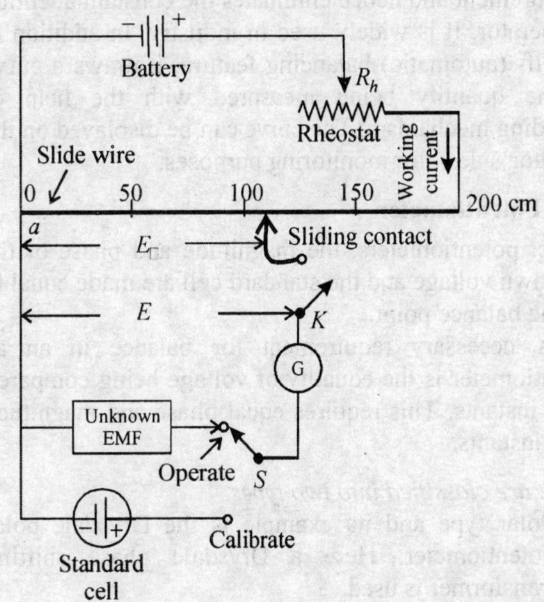

Figure: Circuit diagram of a basic slide wire potentiometer

The side wire is provided with a scale and by siding the contact the unknown e.m.f. is adjusted with the standard cell. When no current flows through the galvanometer, the proportional length of the slide wire is calculated. The scale of slide wire is calibrated directly in term of measured e.m.f. This method of direct reading is called standardization of potentiometers.

Brooks Deflection Potentiometer:

This potentiometer is used for application where the voltage to be measured is continuously changing. With conventional type of potentiometer it becomes impossible to obtain exact balance even when changes are slow since it is difficult to follow the changes by manipulation of the dials.

To measure higher voltage ranges, a precision potential divider called a *volt-ratio box* is used. It provides multiple ranges.

Potentiometers are also used for calibration of voltmeter and ammeter measurement of resistance and power.

Measurement using comparison method are capable of a high degree of accuracy because the result obtained does not depend upon the actual deflection of a pointer, as is the case in deflectional methods, but only upon the accuracy with which the voltage of the reference source is known.

Self-balancing Potentiometer: In industries, a continuous measurement of non – electrical quantity is required as the process is dynamic and a value of quantity is changing with time. A constant attention of an operator is a must in such situation. A self – balancing potentiometer helps in continuous measurement and hence eliminates the constant attention of operator. It is widely used in industry. In addition to its self- (automatic) balancing feature, it draws a curve of the quantity being measured with the help of recording mechanism. The curve can be displayed on the operator's desk for monitoring purposes.

A.C. Potentiometer

In a.c. potentiometers, the magnitude and phase of the unknown voltage and the standard cell are made equal to get the balance point.

A necessary requirement for balance in an ac potentiometer is the equality of voltage being compared at all instants. This requires equal phase and magnitude at all instants.

These are classified into two types:

1. Polar type and its example is the Drysdale polar potentiometer. Here a Drysdale phase shifting transformer is used.

   Polar-type Potentiometer: It measures the unknown e.m.f. in polar form, i.e. terms of its magnitude and phase.

2. Co-ordinate type and its example is the Gall Tinsley AC potentiometer.

Rectangular co-ordinate type Potentiometer: It measures the unknown e.m.f. in terms of its regular co-ordinates. In this potentiometer, balance is obtained in terms of two voltages which are in quadrature (i.e. have a phase difference of 90°). Each of these voltages is varied in magnitude and a means is usually provided to adjust or check the quardrature phase relation.

Application of AC potentiometer:

1. Ammeter calibration
2. Wattmeter and Energy meter testing
3. Voltmeter calibration
4. Measurement of self reactance of coil

7.3 MEASURMENT OF RESISTANCE (R)

1. Measurement of Medium resistance:
   (a) Ammeter-voltmeter-method.
   (b) Substitution method.
   (c) Wheatstone bridge method
   (d) Ohmmeter method.
2. Measurement of low resistance:
   (a) Kelvin's double bridge method.(KDB).
   (b) Potentiometer method.

Kelvin Double Bridge

It is one of the best available for the precise measurement of low resistances. It is only a development of the Wheatstone Bridge by which the error due to contact and lead resistance are eliminated. The scheme of connections is as shown below

X = unknown resistance

S = standard resistance of same order of magnitude as X

r = connecting link of low resistance to balance the error due to contact and lead

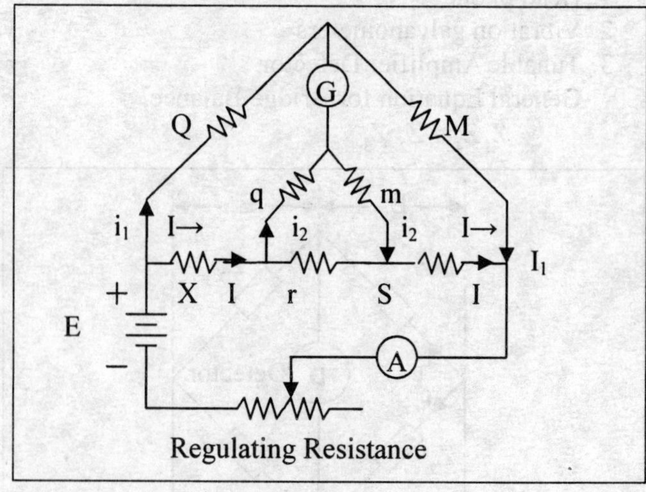

Q M and q m are two pairs of known non-inductive resistance, and one pair out of Q q or $M.m$ is variable (for purposes of balancing the bridge).

With large currents as mentioned above is often difficulty in ensuring that essentially all the voltage drop occurs along of the specimen, and not at the contacts carrying current in and out of the specimen. It is very difficult even with soldered leads to avoid appreciable resistance in the contacts between a specimen and the leads connected to it. An arrangement as shown in often adopted where, essentially, the voltage measurement is isolated from the current leads and only small currents flow through the voltmeter. The arrangement helps also to fix the points between resistance is measured.

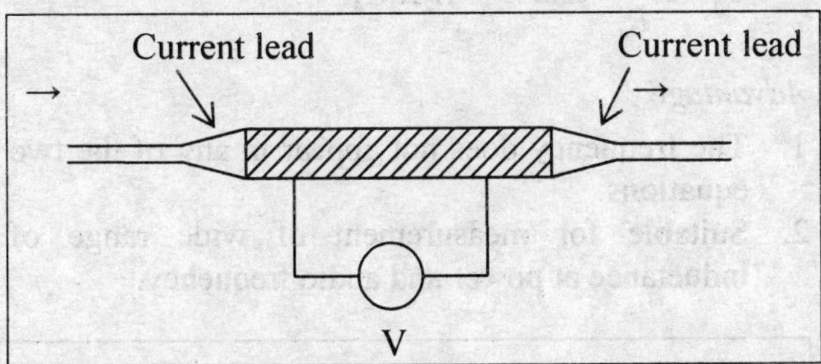

Reducing the delta of q, m and r to a star circuit if a, b, c as shown below, one can write the balance equation of the resulting Wheatstone Bridge.

$$a = \frac{qr}{q+m+r} \quad ; \quad b = \frac{qm}{q+m+r}$$

$$c = \frac{mr}{q+m+r}$$

and at balance , $\dfrac{Q}{M} = \dfrac{X+a}{c+S}$

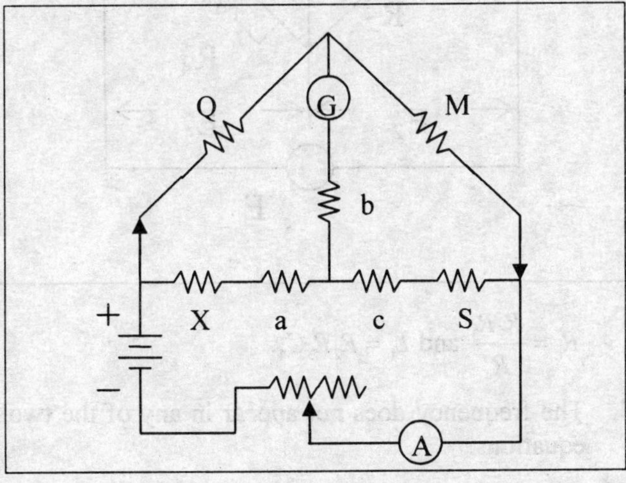

or $X = \dfrac{Q}{M}\cdot(S+c) - a$

$$= \frac{Q}{M}S + \frac{Q}{M}\cdot\frac{mr}{q+m+r} - \frac{qr}{q+m+r}$$

$$X = \frac{Q}{M}\cdot S + \frac{mr}{q+m+r}\left[\frac{Q}{M} - \frac{q}{m}\right]$$

where for 'balancing' the bridge we adjust $\dfrac{Q}{M} - \dfrac{q}{m}$ as closely as possible. Also by marking r of very low value of resistance $\dfrac{mr}{q+m+r}$ can be made very small indeed. Thus $X \approx \dfrac{Q}{M}S$ at balance.

The balance condition is actually

$$\frac{X}{S} = \frac{Q}{M} = \frac{q}{m}$$

That is why the bridge is called double bridge.

Measurement of High Resistance

High resistance includes.

1. Insulation resistance of components.
2. Resistance of high resistance circuit elements like in vacuum tube circuit.
3. Leakage resistance of capacitors.
4. Surface Resistivity.

Methods for measurement of high Resistance

1. Direct deflection method.
2. Loss of charge method.
3. Megohm bridge.
4. Megger.

Loss of charge method: In this method the insulation resistance R to be measured is connected in parallel with a capacitor C and an electrostatic voltmeter. Capacitor is charge to some suitable voltage, by means of a battery having voltage V, and is then allowed to discharge through the resistance.

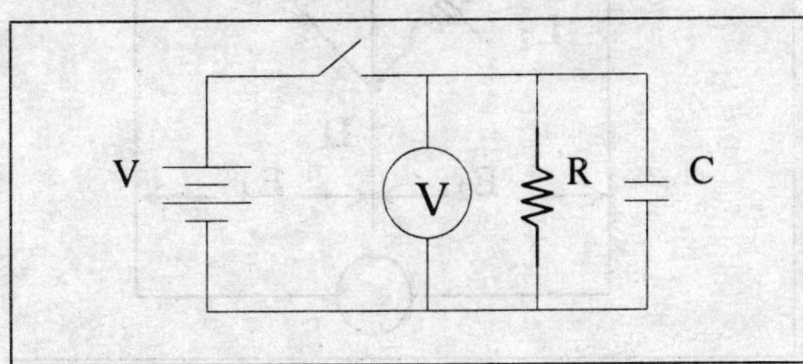

- Voltage across capacitor at any instance t after the application of voltage is

$$V = V_o \; e^{(-t/RC)}$$

- Insulation resistance:

$$R = \frac{0.4343 \cdot t}{C \log_{10}\left(\frac{V}{V_0}\right)}$$

- If the resistance R very large, the time for an appreciable fall in voltage is very large and this process is time consuming.
- Applicable for high resistance, but it requires a capacitor of very high leakage resistance. More accurate result when we take resistance of electrostatic voltmeter and the leakage resistance of the capacitor into account we have

$$R' = \frac{0.4343 \cdot t}{C \log_{10}\left(\frac{V}{V}\right)} \qquad R' = \frac{R \cdot R_1}{R + R_1}$$

where, R = Resistance to be measured
R_1 = Leakage resistance of capacitor.

7.4 MEASUREMENT OF SELF-INDUCTANCE (L)

Maxwell's Inductance Bridge

This bridge circuit measure an inductance by comparison with a variable standard self-inductance.

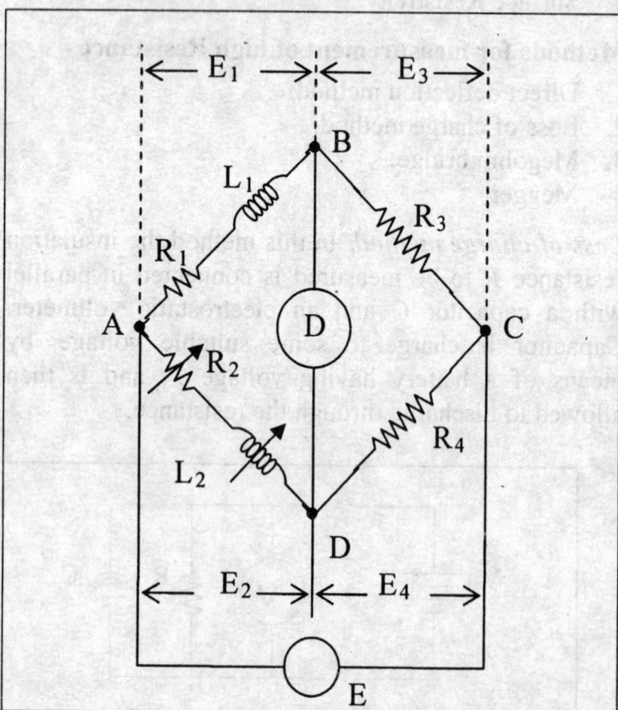

where,
L_1 = Unknown inductance of resistance R_1
L_2 = Variable inductance of fixed resistance
R_2 = Variable resistance connected in series in with inductor L_2,
R_3 and R_4 = Known and non-inductive resistance.

$$L_1 = \frac{R_3}{R_4} \cdot L_2.$$

Maxwell's Inductance-Capacitance Bridge

In this bridge, an inductance is measured by comparison with a standard variable capacitance.

$$R_1 = \frac{R_2 R_3}{R_4} \text{ and } L_1 = R_2 R_3 C_4.$$

Advantage

1. The frequency does not appear in any of the two equations.
2. Suitable for measurement of wide range of inductance at power and audio frequency.

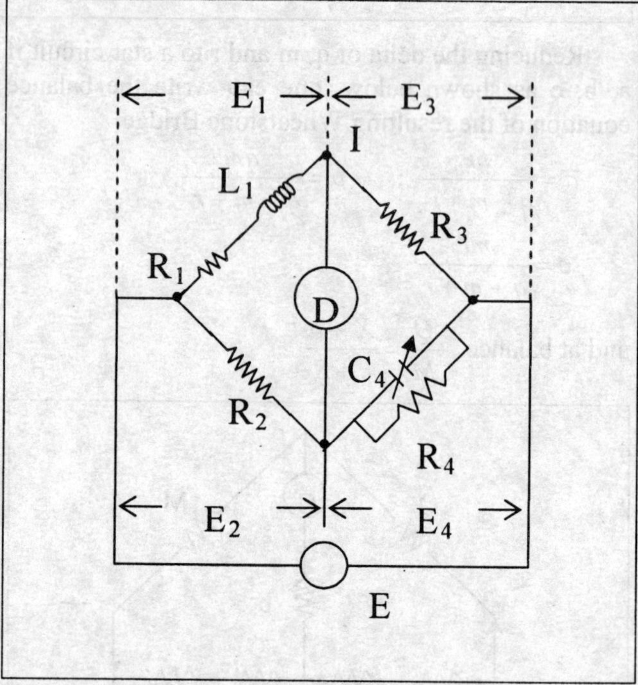

$$R_1 = \frac{R_2 R_3}{R_4} \text{ and } L_1 = R_2 R_3 C_4.$$

3. The frequency does not appear in any of the two equations.

4. Suitable for measurement of wide range of inductance at power and audio frequency.

Disadvantages

1. Requires variable standard capacitor, which is expensive, for high degree of accuracy.
2. Either varying R_2 and R_4 and since R_2 appears in both the Balance equations, the Balance adjustment difficult.

Hay's Bridge

1. Modified Maxwell's bridge.
2. Used resistance in series, which uses a resistance in parallel with the capacitor.

- Advantage (*i*)

$$L_1 = \frac{R_2 R_3 C_4}{1 + \omega^2 C_4^2 R_1^2}$$

gives simple Expression for high Q coil.

(*ii*) Quality $Q = \dfrac{1}{\omega C_4 R_4}$

- For high Q coil, R_4 should have small value. So it requires low value of resistor for R_4 whereas the Maxwell's bridge requires a parallel resistor, R_4, of a very high value.

Disadvantage

Hay's bridge is suited for the measurement of high Q inductor, $Q > 10$, for low value of Q, $\dfrac{1}{Q^2}$ can't be neglected and in that case use of maxwell's bridge is more suited.

Some more Bridge for measurement of Inductances

(a) Anderson's Bridge
(b) Owen's Bridge.

Measurement of Mutual Inductance

Heaviside Campbell Equal Ratio Bridge:
It is a modified and more updated version of Heaviside bridge used for mutual inductance measurement. Mutual inductance is measured in terms of self inductance.

7.5 MEASUREMENT CAPACITANCE (C)

De Sauty's bridge simplest method of comparing two capacitances.

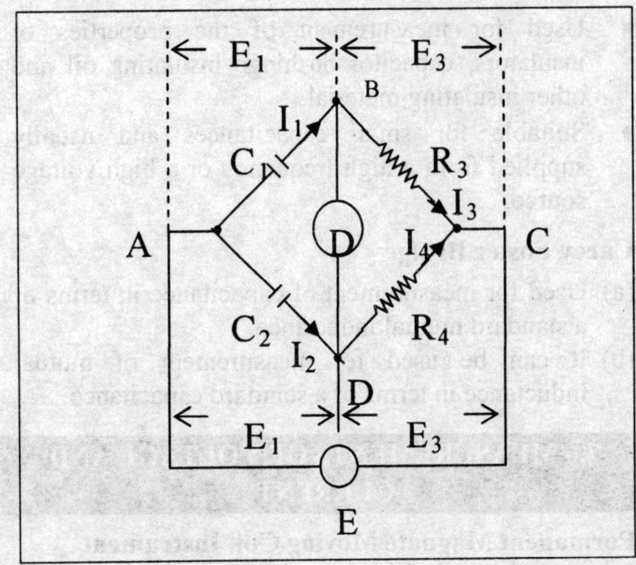

$$C_1 = \frac{C_2 R_4}{R_3}$$

C_1 = Capacitor whose value is to be measured
C_2 = a standard capacitor
R_3 and R_4 = non-inductive resistance.

It is impossible to obtain balance if both the capacitor are not free from dielectric loss. So this is applicable only for loss-less capacitor like air capacitor.

Shering Bridge

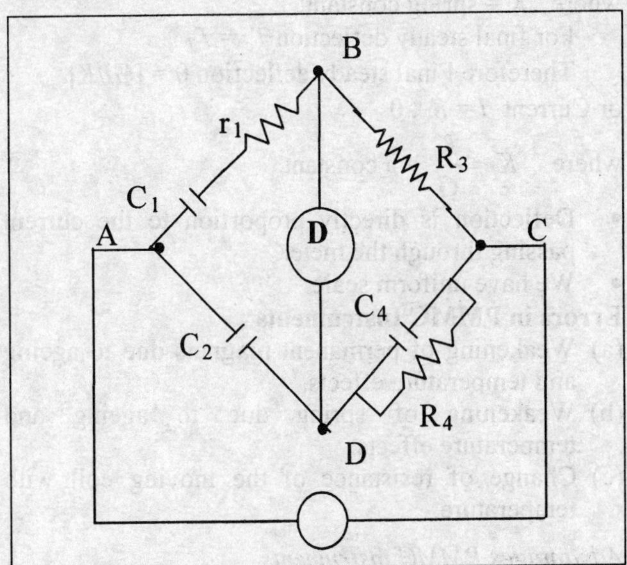

$$r_1 = \frac{R_3 C_4}{C_2} \text{ and } C_1 = C_2\left(\frac{R_4}{R_3}\right).$$

- Used for measurement of the properties of insulators, capacitor bushings insulating oil and other insulating material.
- Suitable for small capacitances and usually supplied from a high frequency or a high voltage source.

Carey Foster Bridge

(a) Used for measurement of capacitance in terms of a standard mutual inductance.
(b) It can be used for measurement of mutual inductance in terms of a standard capacitance

7.6 MEASUREMENTS OF VOLTAGE AND CURRENT

Permanent Magnate Moving Coil Instrument

It is most accurate for measurement of DC.

Construction of PMMC Instruments:

(a) Moving coil
(b) Magnetic system
(c) Control
(d) Damping
(e) Pointer and scale.

Torque Equation

Defection torque is $T_d = NB\,l\,dI = GI$

where G = a constant = $NB\,ld$

The spring control provides a restoring (controlling) torque $T_c = K\theta$

where, K = spring constant.

For final steady deflection $T_c = T_d$

Therefore Final steady deflection $\theta = [GI/K]$

or Current $I = K_i\,.\,\theta$

where $K_i = \frac{K}{G}$ = a constant.

- Deflection is directly proportion to the current passing through the meter.
- We have uniform scale.

Errors in PMMC Instruments

(a) Weakening of permanent magnets due to ageing and temperature effects.
(b) Weakening of spring due to ageing and temperature effects.
(c) Change of resistance of the moving coil with temperature.

Advantages PMMC instruments

1. The scale is uniform.
2. The power consumption is very low as 2.5 μW to 200 μW.
3. The Torque-weight ratio is high which gives a high accuracy.
4. A single instrument may be used for many different current and voltage range by using different values for shunts and multipliers.

Disadvantages

1. Used only for d.c. measurement.
2. Cost is higher than that of moving iron instrument.

Sensitivity of PMMC voltmeters and their loading Effects

The sensitivity of a dc voltmeter is an important factor when selecting a meter for a certain voltage measurement. A low resistance voltmeter may give correct readings when measuring voltage in low resistance circuit but is certain produce very unreliable reading in high resistance circuits.

Moving Iron (M.I) Instruments

- At power frequencies, we are using moving iron instruments.
- Deflecting torque
$$T_d = [1/2]\,I^2\,[dL/d\theta\,]$$
- Controlling torque
$$T_e = K\theta$$
- Deflection
$$\theta = \frac{1}{2}\cdot\frac{I^2}{K}\frac{dL}{d\theta}$$
- Deflection is proportional to square of the r.m.s value of the operating current.
- Scale is non-uniform.
- For a liner scale product θ, $[dL/d\theta]$ should be a constant.

Attraction Type

When the current flows through the coil, a magnetic field is produced and moving iron moves from the weaker field outside the coil to the strong field inside or the moving iron is attracted in.

Repulsion Type

In the repulsion types, there are two vanes inside the coil, one fixed and other movable. These are magnetized when the current flows through the coil and there is a force or repulsion between the two vanes resulting in the movement of the moving vane.

Errors in Moving Iron Instruments

1. Errors with Both DC and AC

(a) Hysteresis error
(b) Temperature error
(c) Stray magnetic field.

2. Error with A.C only.

Frequency Errors

Changes in frequency may cause error due to changes of reactance of the working coil. The change of reactance of the instrument coil is important in case of voltmeter where an additional resistance R_s is used in series with instrument coil.

C
I L R R_s
Coil Series Resistance
V

Frequency error can be eliminated by using Capacitor in parallel with series resistance R_s.

$$C = 0.41\frac{L}{R_s}.$$

Electrodynamics Type Instruments

A transfer instrument is one that may be calibrated with a DC source and then used without modification to measure AC This needs the same accuracy for both DC and AC.

Sinusoidal Currents

If currents i_1 and i_2 are sinusoidal and are displaced by a phase angle ϕ, i.e.

$$i_1 = I_{m1}\sin\omega t \text{ and } i_2 = I_{m2}\sin(\omega t - \phi)$$

$$\text{Deflection } \theta = \frac{I_1 I_2}{K}\cos\phi\cdot\frac{dM}{d\theta}$$

It follows from the above that for sinusoidal alternating currents the deflecting torque and deflection are determined by the product of the rms values of the coil currents and cosine of the phase angle between them.

Electrodynamometer Ammeters

In this case the fixed and moving coils are connected in series and, therefore, carry the same current i.e. $I_1 = I_2$ and $\phi = 0$.

$$\text{Therefore, Deflecting torque } T_d = I^2\frac{dM}{d\theta}$$

$$\text{Therefore, Deflection } \theta = \frac{I^2}{K}\frac{dM}{d\theta}.$$

Electrodynamometer Voltmeters

The electrodynamometer movement is used as a voltmeter by connecting the fixed and moving coils in series with a high non-inductive resistance.

$$\text{Deflection torque } T_d = I_1 I_2 \cos\phi\frac{dM}{d\theta}$$

$$\text{In this case } I_1 = I_2 = \frac{V}{Z} \text{ and } \phi = 0$$

$$T_d = \frac{V}{Z}\cdot\frac{V}{Z}\frac{dM}{d\theta} = \frac{V^2}{Z^2}\frac{dM}{d\theta}$$

where V = Voltage across the instrument,
Z = Impedance of the instrument circuit.

$$\text{Deflection } \theta = \frac{V^2}{KZ^2}\frac{dM}{d\theta}$$

Electrodynamometer type voltmeter is the most accurate type of AC voltmeter. But the sensitivities are low in comparison with the DC instruments and range from 10 to 30 Ω/V.

Electrostatic Instruments

In electrostatic instruments, the deflecting torque is produced by action of electric field on charged conductors.

$$T_d = \frac{\varepsilon_o r^2}{d}V^2$$

where

r is the radius

ε_o is the permittivity of air

d is the distance between the plates.

Electrostatic instruments can measure both ac and dc. Their deflection is directly proportional to square of the r.m.s voltage.

Hot Wire Instruments

The current to be measured is passed through a fine *platinum iridium wire.* On passage of current the wire gets heated and it expands. This expansion of wire causes the pointer to deflect.

Thermocouple Instrument

When two metals having different work function are placed together, a voltage is generated at the junction which is nearly proportional to the temperature. This junction is called a thermocouple temperature. This principal is used to convert heat energy to electrical energy at the junction of two conductors.

$$E = a(T_1 - T_2) + b\,(T_1 - T_2)^2 = a\,.\Delta\theta + b\,(\Delta\theta)^2$$

where a, b are constants whose value depends upon metals used.

$\Delta\theta$ = difference of temperature of hot and cold junction = $T_1 - T_2$.

These methods are popular for ac measurement at high frequency.

Induction Instruments

The working principle of induction instruments is similar to that of single phase induction, motor. A torque is developed due to the reaction between two magnetic fields with a phase difference.

The deflecting torque is given by,

$$T_d \propto \frac{I^2 f}{Z}$$

where,

I is the current in the coil

Z is the impedance

f is the frequency.

Induction instruments are used only for ac measurements.

The full scale deflection produced by these instruments is about 300°.

The main errors occur due to frequency and temperature variations.

Rectifier Instruments

Rectifier instruments are used for measurements of ac voltages. A rectifier elements is used which converts the ac to a unidirectional dc and then using a PMMC meter the value of the rectified dc is measured. Rectifier instruments are highly sensitive with a very rugged construction.

Extension of range of rectifier instruments as voltmeter.

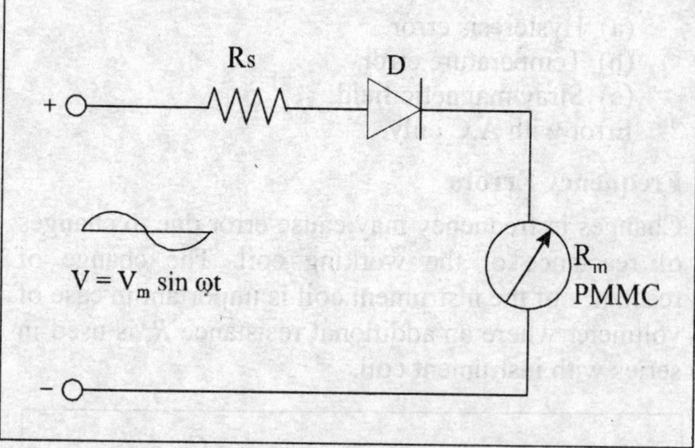

$$R_s = 0.45\,S_{de}V - R_m R_d \text{ (For half wave rectification)}$$

$$R_s = 0.9\,S_{de}V - R_m - 2R_d \text{ (For full wave rectification).}$$

7.7 MEASUREMENTS OF POWER

Power measurement if both the current and voltage waves are sinusoidal, the current lagging in phase by an angle ϕ, then

$$v = V_{max} \sin \omega t$$

and $$i = I_{max} \sin(\omega t - \phi)$$

$$P = \frac{V_{max} V_{max}}{4\pi} \cos\phi$$

or $$P = V \times I \cos\phi$$

Where, P = true power in watts

V = rms value of circuit voltage in volt

I = rms value of current in amp

ϕ = phase angle between V and I

$\cos\phi$ = power factor of the circuit

Due to the fact of of involvement of power factor ($\cos\phi$) in the expression for power, a wattmeter is used for the measurement of power in a. c. circuit instead of voltmeter and ammeter methods, because latter method does not take any account of the power factor.

Power is measured by wattmeter. Most of the wattmeters are electrodynamometer type instruments. Dynamometer consists of two types of coil one moving coil and two fixed coils. Here the fixed coils are connected in series with the load and carries the current in the circuit. So they are also called '**current coils**'. On the other hand the moving coil carries current proportional to the voltage and it is called '**pressure coil' or 'voltage coil**'.

In this type of wattmeters **spring control is used** and the damping used is **air friction damping**

The deflecting torque is given by,

$$T_d = \frac{VI}{R_p}\cos\phi\frac{dM}{d\theta}$$

where

$R_P \rightarrow$ Resistance of pressure coil

$M \rightarrow$ mutual inductance between the current coils and pressure coil

$\phi \rightarrow$ Phase angle which the current lags with voltage in current coil circuit.

Control torque, $T_c = K\phi$

where

$K \rightarrow$ Spring constant

$\phi \rightarrow$ Final steady deflection.

Various widely used **power-measuring** techniques are as follows:

- Bolometer method
- Calorimeter method
- Wattmeter method

Bolometer Method

This method unknown power to be measured is absorbed in a specially constructed bolometer element of resistive material. The resultant temperature rise is then detected by measuring the change in bolometer resistance by mean of an auxiliary bridge circuit. Power raising from a few microwatts to a fraction of a watt are commonly measured in this method.

The word bolometer encompasses two sections, namely, (a) barretters, and (b) thermistors, Barretters are normal resistance, while thermistors are compounded of metallic oxide materials possessing a negative temperature coefficient.

Advantages/Applications:

- Both the barretter and thermistor are very sensitive elements, and can be used for **measuring very small powers** (as small as tens of microwatts).
- The barrerter and thermistor are more stable with respect to time and exposure to overload than are detectors for very high frequencies, such as the silicon crystal, and in most circuits can be replaced without recalibration.
- They can also be used for monitoring/measure large amounts of power by inserting a **directional coupler** between the bolometer element and the main ratio-frequency system.

Calorimeter Method

In this method the radio-frequency energy is converted into heat. This heat is absorbed in a fluid (usually water) that flows through the system, and then the temperature rise of the fluid is measured. The **radio-frequency power** may be absorbed directly in the calorimeter fluid, or the fluid may be used as a coolant for a resistive load of solid material. It is essential that all of the radio -frequency energy be transferred to the fluid, i.e., the system must have no radio-frequency leakage, either by radiation or in lossy joints. Likewise, the heat radiation from the fluid must be minimized until after the point in the system where the temperature measurement is made.

The power dissipated in a calorimeter can be calculated directly from the temperature rise, the specific heat of the fluid, and the rate of flow of the fluid, according to the relation ship given as

$$P = 4.18 \times m \times s_P \times \Delta t$$

Where,

P = power in watts

m = flow of fluid in grams per second (g/s)

s_p = specific heat in calories per °C

Δ_t = temperature rise in °C

Alternative, the relationship between temperature rise and power dissipated can be determined experimentally by dissipating known amounts of ac or dc power in the calorimeter system.

Advantage/Applications:

The calorimeter method is well suited for laboratory purposes.

The calorimeter method of measuring power is specially for measurement of large power.

Wattmeter Method

Wattmeters are indicating-type instruments, generally used for power measurement of the electrical circuit.

In ac circuit,

deflecting torque T_m given by

$T_m \propto VI\cos\theta$

$\propto$ True power

Damping: Air damping is provided in the dynamometer wattmeter.

Advantages:

1. Such instruments can be made to give a very **high degree of accuracy**. Hence, they are used as a standard for calibration purposes.
2. They are equally accurate on dc as well as ac measurements.
3. Scales are more or less uniform because the deflection is proportional to the average power
4. It can be used on both ac and dc supply for any wave-form of voltage and current, and is not restricted to sinusoidal wave-forms.

Disadvantages/Errors

At power factor, the inductance of the voltage coil causes serious error unless special precautions are taken to reduce this effect.

1. The inductance of the moving (or voltage) coil causes errors, which can be avoided to some extent by connecting a high non-inductive resistance series with the coil.
2. Errors due to voltage drop in the circuit.
3. Errors due to the current taken by the voltage coil

Methods of Wattmeter Connection

There are two methods of connecting a wattmeter in electrical circuit for power measurement, as shown in figure.

In the first method, the voltage (or pressure) coil is connected on the supplied side of the current coil. In this connection, the voltage applied to the voltage coil is higher than that of the load on account of the voltage coil in the current coil. The wattmeter measures the power I^2R_c lost in the current coil.

In the second method, the current coil carries the small current taken by the voltage coil, in addition to the load current. In this case, the wattmeter measures the power lost in the voltage coil as well the power in the load.

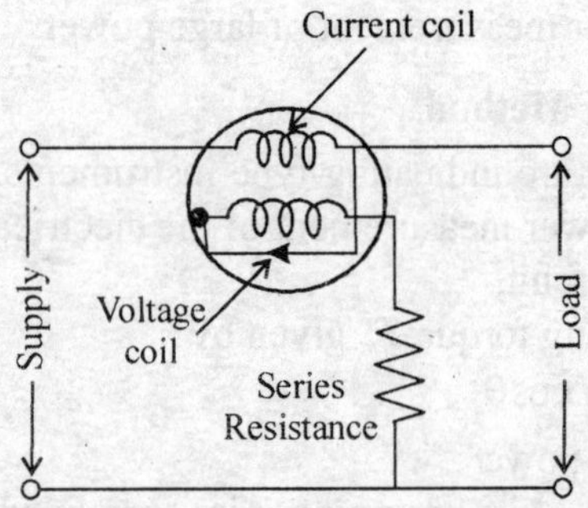

Figure: Alternate wattmeter connections

If the load current is small, the voltage drop in the current coil is small, so that the first method of connection introduces a very small error.

On the other hand, if the load current is large, the power lost in the voltage coil will be small compared with the power in the load, and the second method of connection is better.

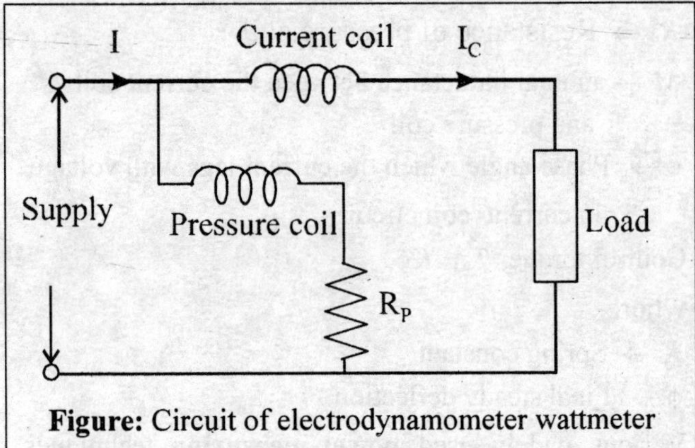

Figure: Circuit of electrodynamometer wattmeter

The possible error sources are voltage coil, capacitance and inductance, eddy current and mutual inductance between the coils, temperature error etc.

Measurement of power in 3-phase circuit is done by 'Three Wattmeter method' 'Two Wattmeter method and also' One wattmeter method'.

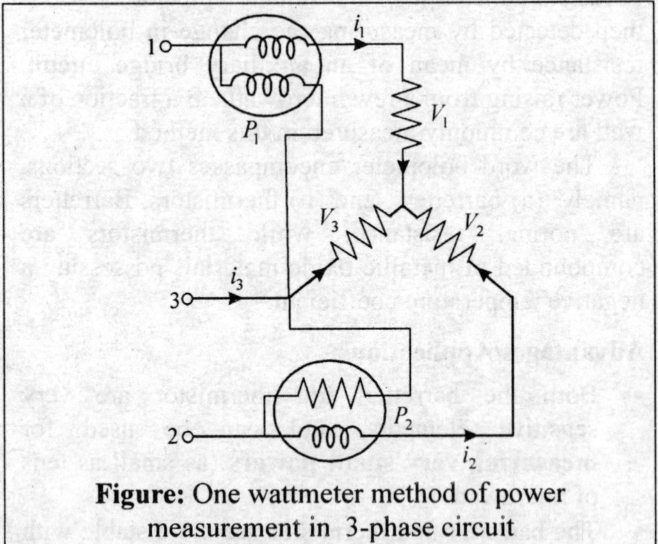

Figure: One wattmeter method of power measurement in 3-phase circuit

The most common method of power measurement in three-phase circuit is the two wattmeter method. The arrangement is shown in figure.

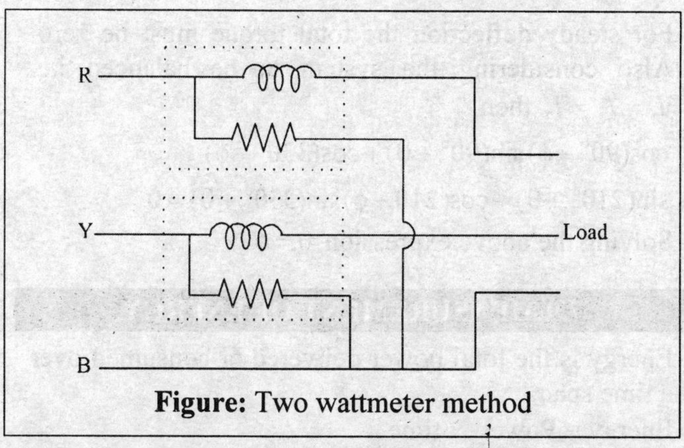

Figure: Two wattmeter method

7.8 MEASUREMENT OF POWER FACTOR

On measuring current, voltage and power in an ac circuit, its power factor can be determined from the relationship $\cos\phi = \frac{P}{VI}$, where ϕ is phase angel.

Power factor measurement is done is similar way as power is measured. These instruments also work mostly in electro-dynamic principle containing voltage and current coils.

But there are two moving coils instead of one along with two fixed coils. The two moving coils produces two deflecting torques.

$$T_{d1} \propto VI\cos\phi\sin\theta$$

$$T_{d2} \propto VI\sin\phi\sin\theta$$

$\theta \rightarrow$ angle of deflection

$\phi \rightarrow$ phase angle of the circuit

In equilibrium, $T_{d1} = T_{d2}$

$$\therefore \quad \theta = \phi$$

Thus, deflection becomes equal to phase angle. However, power factor meters indicates directly, by a single reading, the power factor of the circuit to which they are connected.

Power factor meters like wattmeters have a current circuit and pressure circuit.

The current circuit carries the current in the circuit whose power factor is to be measured.

The pressure circuit is connected across the circuit whose power factor is to be measured and is usually split up into two parallel paths: one inductive and other non – inductive. The deflection of the instrument depends upon the phase difference between the main current and currents in the two paths of the pressure circuit i.e. upon the phase angle or power factor circuit. The deflection is indicated by a pointed.

Two Types of Power Factor Meter

(a) Electrodynamometer type

(b) Moving Iron type.

Advantages and disadvantages of moving Iron P.F. meters

Advantages

(a) The working forces are very large as compared with those in electrodynamometer type.

(b) All the coils in a moving iron instrument are fixed are hence use of alignments is eliminated.

(c) Simple and robust in construction and cheap also.

(d) The scale extends over 360°.

Disadvantages

(a) Errors are introduced in these meters owing to losses in iron parts.

(b) The calibration of these instruments is appreciably affected by variation in supply frequency voltage and waveform.

There are two types of power factor meter

1. Single phase electrodynamic type *power factor meter:*

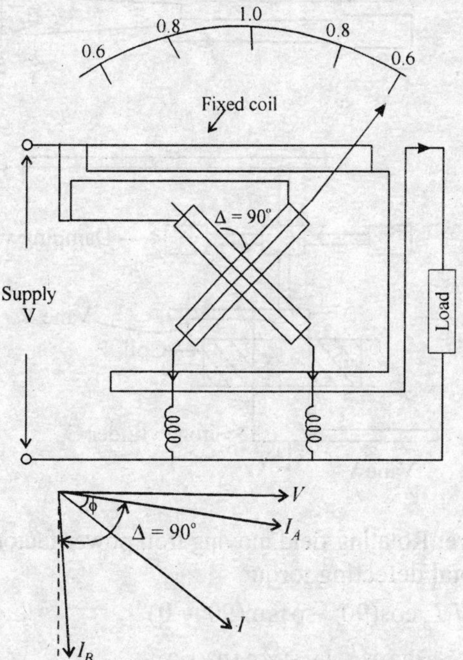

Figure: Construction of single phase electrodynamics type power factor meter

For equilibrium of the circuit

$KVIM_{max} \cos\phi \sin\theta = KVIM_{max} \sin\phi \cos\theta$

or $\theta = \phi$

Where,

θ = angular deflection from the phase reference

M_{max} = Maximum value of mutual inductance between the two coils.

2. Three phase Moving iron *Power factor meter:*

This instrument may be divided into two categories according to whether the operation of the instrument depends upon a rotating magnetic field or a number of alternating fields.

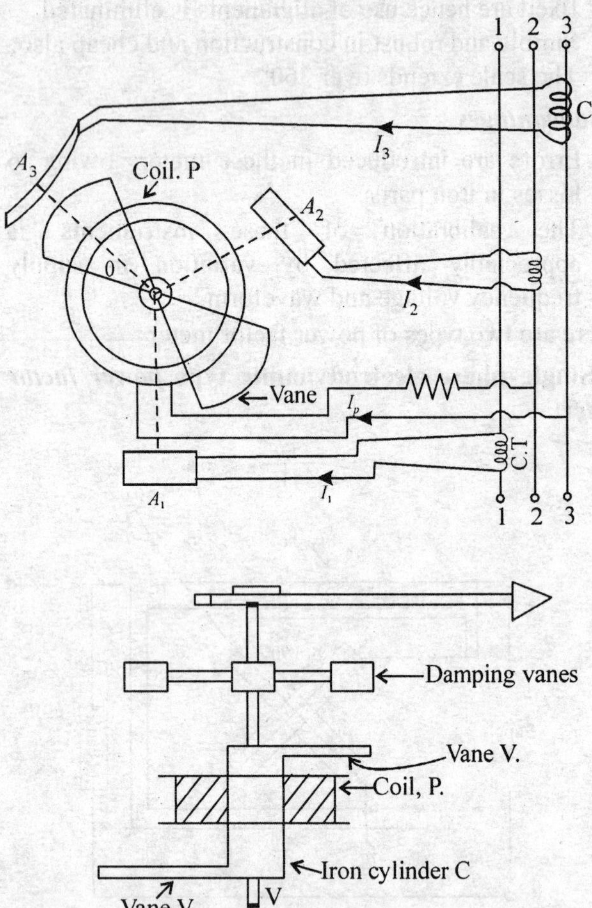

Figure: Rotating field moving iron power factor meter

The total defecting torque

$$T_d \propto [I_1 I_p \cos(90° - \phi)\sin(90° + \theta)$$
$$+ I_2 I_p \cos(330° - \phi)\ \sin(210° + \theta)$$
$$+ I_3 I_p \cos(210° - \phi)\sin(330° + \theta)]$$

For steady deflection the total torque must be zero. Also considering the system to be balanced i.e. $I_1 = I_2 = I_3$ then

$$\cos(90° - \phi)\sin(90° + \theta) + \cos(330° - \phi)$$
$$\sin(210° + \theta) + \cos(210 + \phi)\sin(330° + \theta) = 0$$

Solving the above expression $\theta = \phi$

7.9 MEASUREMENT OF ENERGY

Energy is the total power delivered or consumed over a time span.

Energy = Power × time

$$W = \int_0^t vi\, dt.$$

Motor meter are a class of energy meter which measure the total energy in kilowatt hour (KWh) or the quantity of electricity in ampere-hour. Energy meters are integrating instruments. Motor meters are used for measurement of energy in both dc and ac circuit.

In a motor meter the moving system revolves continuously. The speed of rotation is proportional to the power in the circuit. The total number of revolution made by the meter in a given interval of time is proportional to the energy supplied. In this context the term meter constant comes.

Meter constant is defined as the number of revolutions made per Kilo-Watt-hour (KWh).

The speed of the moving system is controlled by braking system. It is done with help of a permanent magnet.

Energymeters for ac circuit are generally induction type.

Testing: The energy meters are tested generally at 5%, 50%, 100% and 120% of its rated current both for unity and 0.5 power factors.

Single-phase Energy Meter

There are used to measure electric energy in ac single-phase circuits.

In a single-phase energy meter, rotation of the disc is produced by the eddy current due to the combined fluxes of two electromagnets. The fluxes of these two magnets are kept 90° apart to make the rotating field.

Creep

In some meters a slow but continuous rotation is

obtained even when there is no current flowing though the current coil and only pressure coil is energized. This is called *creeping*.

Reason of Creeping

Due to over-compensation for friction. If the friction compensation device is adjusted to give a driving torque to compensate for statice friction which is larger then the running friction, there is a tendency for the disc to run over when there is no current.

Excessive voltage across the potential coil.

Prevention of Creeping

In order to prevent creeping two diametrically opposite hole are drilled in the disc.

7.10 A.C & DC CURRENT PROBES

Features & Benefits

- Easy-to-use and Accurate AC/DC current measurements.
- Connects Directly to DPO 7000 and DPO 4000 series Oscilloscope using the new Tex *VPI™* Probe interface.
- Provides automatic units scaling and Readout on the Oscilloscope Display
- Split – core construction Allows Easy Circuit Connection.
- DC to > 120 MHz Bandwidth.
- 30 A RMS Maximum Current Capability
- 50 A Peak Pulse current capability
- High Accuracy with Typically Less than 1% DC Gain
- Accurately Measures Current Levels as low as 1mA
- Low Noise and DC drift
- Easy to Degauss and Auto zero
- Setup Controls and Probe Status/Diagnostic indicators are Provided on Both the Probe Hardware, and though as Easy – to – Acces Oscilloscope UI Display Menu
- Safely Certified for U.S. and Canada

Applications

- Power Supplies
- Semiconductor Devices
- Power inverters/Converters
- Electronic Ballasts
- Industrial/Consumer Electronics
- Mobile Communications
- Motor Drives

Characteristics

- Bandwidth – DC to ≥ 120 MHz
- Rise Time – ≤ 2.92 *ns*
- Max RMS Current – 30 A
- Max Peak Pulse Current – 50 A
- Minimum Sensitivity
- 1 mA AC coupling
- Current Ranges – 5 A and 30 A
- Max Bare – wire Voltage – Use insulated wire only.

Simple Logic Probe

This sample logic probe has both LEDs on with no signal at the input but due to the nor gates connected to the prove, in dates correctly when a high or low signal is present. It also works correctly for pulse trains. Normally both LEDs are forward biased and therefore on, powered by the 12V supply. When a logic "high" is present at the probe, ICI a's output goes low sending ICI b's output high. This turns off LED 1 but turning LED 1 on and LED2 off.

7.11 EXTENSION OF INSTRUMENT RANGES

Ammeter Shunts

A DC ammeter is PMMC D'Arsonval galvanometer.

- Multiplying power

$$m = \frac{I}{I_m} = I + \frac{R_m}{R_{sh}}$$

Resistance of shunt

$$R_{sh} = \frac{R_m}{(m-1)}$$

where I = Current to be measured.

$I_{fs} = I_m$ = full-scale deflection current of meter.

R_m = internal resistance of meter.

R_{sh} = resistance of the shunt.

Requirement for Shunt

1. The temperature co-efficient of shunt and instrument should be low and should be equal as nearly as possible.
2. The resistance of shunts should not vary with time.
3. They should carry the current without excessive temperature rise.

4. They should have a low thermal automotive force.
5. Manganin is used for shunt of DC instrument.

Effect of temperature changes in Ammeters

The temperature error can be eliminated when the shunt and the moving-coil are made of the same material.

Use of ***swamping resistance of manganin*** having resistance 20 to 30 times the coil resistance is connected in series with the coil and a shunt.

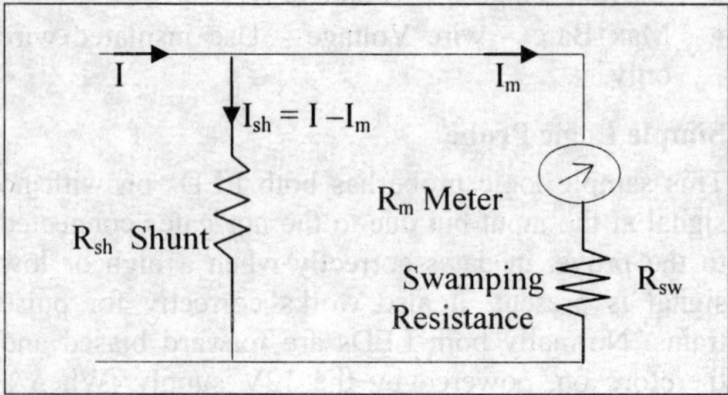

Voltmeter Multipliers

Arsonval basic meter movement is converted into a voltmeter by connecting a series resistance with it.

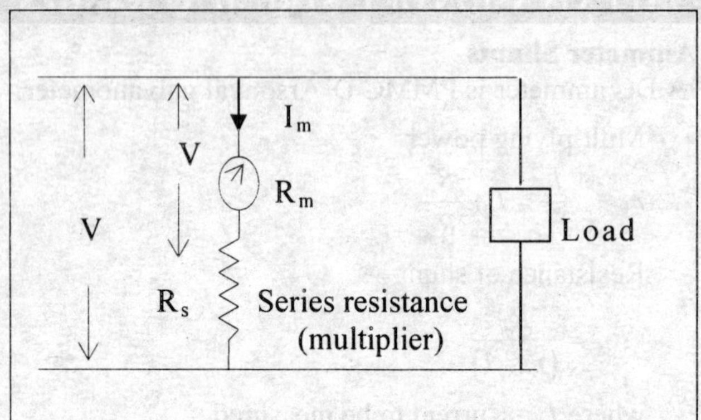

$$V = I_m\left(R_m + R_s\right); \quad \frac{V}{I_m} = \left(R_m + R_s\right)$$

$$\frac{V}{I_m} - R_m = R_s; \qquad R_s = \frac{V}{I_m} - R_m$$

where, R_s = multiplier resistance.

$I_{fs} = I_m$ = full scale deflection current of meter

R_m = internal resistance of meter

V = full range voltage of instrument.

Multiplying factor for multiplier

$$\frac{V}{v} = \frac{I_m\left(R_m + R_s\right)}{I_m\, R_m} = \left[1 + \frac{R_s}{R_m}\right]$$

$$M = \frac{V}{v} = \left(1 + \frac{R_s}{R_m}\right).$$

Resistance of multiplier $R_s = (m - 1)\, R_m$.

7.12 Q METER

The determination of the storage factor Q is one of the most widely used means in the laboratory for testing radio frequency coils (RFC), inductors and capacitors. The storage factor is equal to $Q = \omega_o\, L/R$ where ω_o is the resonant angular frequency, L is the inductance and R is the effective resistance of a coil. The effective resistance R, is never determined directly since its value depends upon the value of frequency. As is well known, the value of effective resistance with AC differs from its DC value on account of skin effect and eddy current losses. Since the value of R varies in a complex manner with frequency, it is indirectly determined by measuring the value of Q.

The Q meter is an instrument which is designed to measure the value of Q directly and as such is very useful in measuring the characteristics of coils and capacitors.

The principle of working of this useful laboratory instrument is based upon the well-known characteristics of a resonant series R, L, C, circuit shown a coil of resistance R and inductance L in series with a capacitor C.

Principle of Working

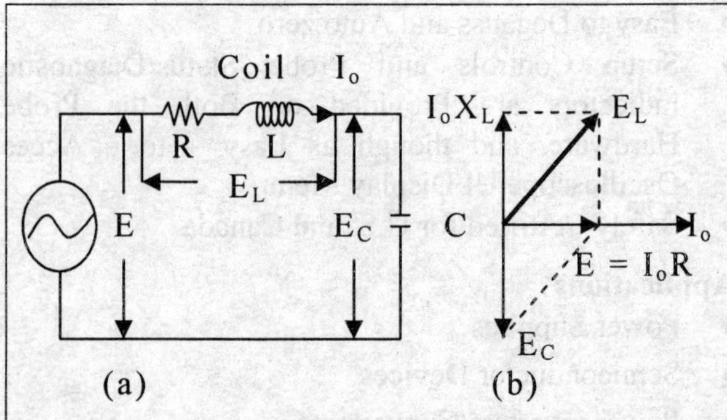

Figure: A resonant RLC circuit and its phasor diagram

At resonant frequency f_o we have

$X_C = X_L$

where capacitive reactance

$$X_C = \frac{1}{2\pi f_o C} = \frac{1}{\omega_o C}$$

and inductive reactance

$X_L = 2\pi f_o L = \omega_o L$

Resonant frequency

$$f_o = \frac{1}{2\pi\sqrt{LC}}$$

and current $I_o = \frac{E}{R}$.

Voltage across capacitor,

$$E_C = I_o X_C = I_o X_L = I_o \omega_o L$$

Input voltage $E = I_o R$

$$\therefore \quad \frac{E_C}{E} = \frac{I_o \omega_o L}{I_o R} = \frac{\omega_o L}{R} = Q$$

$$E_C = QE.$$

Thus, the input voltage E is magnified Q times.

If the input voltage E is kept constant, the voltage appearing across the capacitor is Q times E and a voltmeter connected across the capacitor can be calibrated to read the value of Q directly.

Practical Circuit

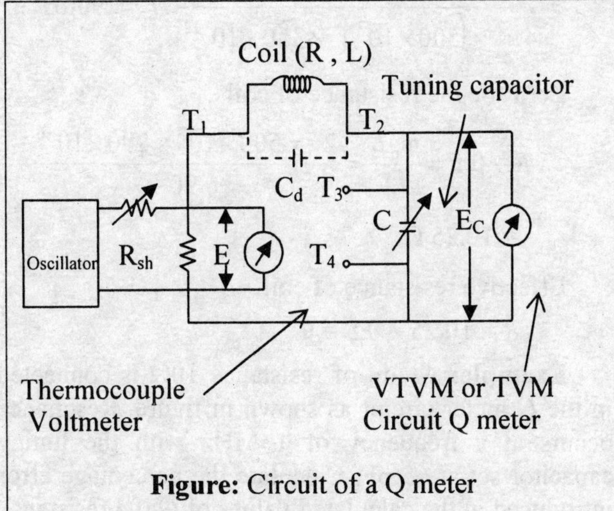

Figure: Circuit of a Q meter

A practical Q meter circuit is shown in figure it consists of a self contained variable frequency RF oscillator. This oscillator delivers current to a low value shunt resistance R_{sh}. This low resistance is of the order of a hundredths of an ohm; a typical value may be 0.02Ω. Through this resistance a small value of voltage E is injected into the resonant circuit. This voltage is measured by a thermocouple voltmeter. Since the value of shunt resistance is very low, it introduces almost no resistance into the oscillatory circuit and therefore represent a voltage source of magnitude E with a vary small internal resistance. A calibrated standard variable capacitor C is used for resonating the circuit. A VTVM or a TVM is connected across this capacitor. The coil under test is connected to terminals T_1 and T_2.

Measurement of Q

The circuit for measurement of Q shown in figure. The oscillator is set to the desired frequency and then the tuning capacitor is adjusted for maximum value of E_C. As discussed earlier under these conditions $[Q = (E_C/E)]$ and if the voltage is kept constant, the voltmeter connected across the capacitor may be calibrated to read the value of Q directly. This measured value of Q is commonly regarded as the Q of the coil under test. However, there is an error. The measured value of Q is the Q of whole the circuit and not of the coil. There are errors caused on account of the shunt resistance and also due to the distributed capacitance of the circuit.

Correction for shunt resistance:

$$\text{Measured value, } Q_{meas} = \frac{\omega_o L}{R + R_{sh}}$$

$$\text{True value, } Q_{ture} = \frac{\omega_o L}{R}$$

$$\therefore \quad Q_{ture} = Q_{meas}\left(1 + \frac{R_{sh}}{R}\right).$$

Thus, the measured value of Q is smaller than the true. Now if coils of high resistance (low Q coil) are being measured, the difference between the two may be negligible. But when measurement are done on low resistance (high Q) coils, the error caused on this account may be serious.

Correction for Distributed Capacitance

The distributed capacitance or self-capacitance of measuring circuit modifies the true value of Q. the measured value of Q is less than the true value by a factor that upon the value of the distributed (self) capacitance and the tuning capacitor. It can be shown that

$$\therefore\ Q_{true} = Q_{meas}\left(1 + \frac{C_d}{C}\right)$$

C_d = distributed or self-capacitance and

C = tuning capacitance.

Measurement of Inductance

The value of inductance is given by

$$L = \frac{1}{4\pi^2 f_o^2 C}$$

The value f_o and C are known and therefore the value of inductance may be calculated.

Measurement of Effective Resistance

The value of the effective resistance may be computed from the relation $R = \frac{\omega_o L}{Q_{ture}}$.

Measurement of Self Capacitance

The self-capacitance is measured by making two measurement at different frequencies. The capacitor is set to a high value, and the circuit is resonated by adjustment of the oscillator frequency. The Resonance is indicated by the circuit Q meter. Let the values of tuning capacitor be C_1 and that of frequency be f_1 under these conditions. Therefore

$$f_1 = \frac{1}{2\pi\sqrt{L(C_1 + C_d)}}.$$

The frequency is now increased to twice its initial value and the circuit is resonated again this time with the help of tuning capacitor. Let the value of tuning capacitor C_2 and that of frequency be f_2 under these conditions, Thus,

$$f_2 = \frac{1}{2\pi\sqrt{L(C_2 + C_d)}}.$$

Now $f_2 = 2f_1$

$$\therefore\ \frac{1}{2\pi\sqrt{L(C_1 + C_d)}} = 2 \times \frac{1}{2\pi\sqrt{L(C_1 + C_d)}}$$

or distributed capacitance, $C_d = \frac{C_1 - 4C_2}{3}$.

Measurement of Bandwidth

The measurement of bandwidth with a Q meter is similar to the one used for measurement of effective resistance by variation of frequency.

Measurement of Capacitance

For measurement of capacitance a dummy coil is connected across terminal T_1 and T_2. The circuit is resonated by varying the value of the tuning capacitor. Let this value be C_1.

This capacitor under test C_T, is now connected across terminals T_3 and T_4. This puts the test capacitor in parallel which tuning capacitor. The circuit is resonated again by varying the value of tuning capacitor. Let this value be C_2. Therefore value of capacitance under test is $C_T = C_1 - C_2$.

Example: A circuit consisting of a coil, a resistance and a variable capacitor connected in series is tuned to resonance using a Q meter. If the frequency is 500 kHz, the resistance 0.5Ω and the variable capacitor set to 350 pF. Calculate the effective inductance and resistance of the coil, if the Q meter indicates 90.

Solution: At resonance Inductance

$$L = \frac{1}{4\pi^2 f_o^2 C}$$

$$= \frac{1}{4\pi^2 \times (500 \times 10^3)^2 \times 350 \times 10^{-12}} H = 290\mu H.$$

Let R be the resistance of coil

$$\therefore\ R + 0.5 = \frac{\omega_o L}{Q} = \frac{2\pi \times 500 \times 10^3 \times 290 \times 10^{-6}}{90}$$

$$= 10.25\ \Omega$$

Effective resistance of coil

$$R = 10.25 - 0.5 = 9.75\Omega.$$

Example: A coil of resistance 10Ω is connected in the Q meter circuit as shown in figure. Resonance occurs at a frequency of 1 MHz with the tuning capacitor set at 65 pF. Calculate the percentage error introduced in the calculated values of Q if a resistance of 0.2 Ω is used across the oscillator circuit.

Solution: True value of Q is:

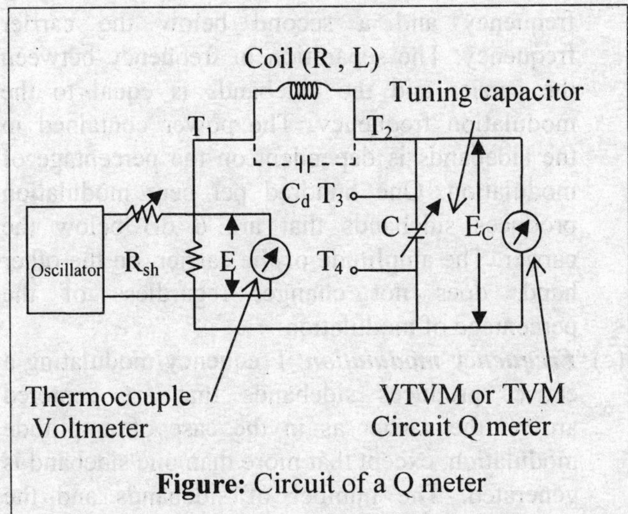

Figure: Circuit of a Q meter

$$Q_{ture} = \frac{1}{\omega CR} = \frac{1}{(2\pi \times 10^6) \times (45 \times 10^{-12}) \times 10.02} = 245$$

Measured value of

$$Q_{meas} = \frac{1}{\omega C(R + 0.02))}$$

$$= \frac{1}{(2\pi \times 10^6) \times (45 \times 10^{-12}) \times 10.02} = 244.5$$

$$\therefore \ \% \text{ age error } = \frac{245 - 244.5}{245} \times 100 = 0.2\ \% \text{ high.}$$

7.13 WAVEFORM ANALYZER

Spectrum Analyzer

In a spectrum analyzer, we obtain a visual display on a CRT at one instant a range or sweep of frequencies. The variation on range is done electronically at a rapid rate. (*This method is distinct from the discrete method. In the discrete method of the frequency response an amplifier is to be checked, from 50 Hz to 15 kHz, spot frequencies from an audio oscillator at 50 Hz, 100 Hz, 400 Hz, 1 k, 5 k, 10 k and 15 k. Constant level are fed to the amplifier and its output measured on a voltmeter probably calibrated in dB or dBm frequency after frequency. This is a time consuming process*). In a spectrum analyzer, the frequency is swept, rapidly from 50 Hz to 15 kHz and output checked on the CRT calibrated logarithmically i.e., in dB. The output of the amplifier is fed to the vertical plates and the sweep is fed a AC voltage proportional to the output frequency. The curve traced on the CRT is the frequency vs gain of the test circuit. One defect of the system is the generation of harmonics in the analysis circuit on account of on-linearly of transistor characteristics. So filters are introduced in the circuit to pass the fundamental and also check outputs a harmonics of fundamental.

The spectrum analyzer is similar to converting super-heterodyne receiver. This block diagram of an analyzer is

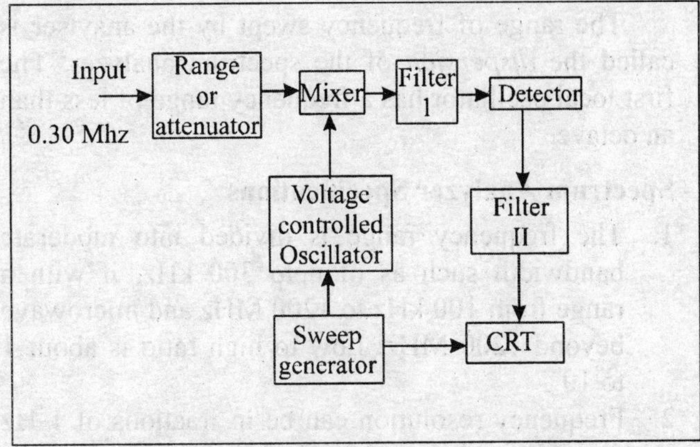

The sweep generator produces the saw tooth waveform which is fed to the X plates of CRT and also to the VCO. The VCO is a varactor diode whose capacitance varies with the applied signal. The output of the VCO is swept over a predetermine range and then sent to the mixer. An oscillator feeds the circuit under test whose output is connected to the ranger or attenuator so that the mixer is fed correct level without over loading. The mixer heterodynes the two frequencies creating a sum and difference. The LP filter passes the difference. A first IF of 400 MHz is heterodyned again to obtain a lower 2nd IF of 21.4 MHz. By bringing the frequencies down from 400 MHz is heterodyned again to obtain a lower 2nd IF of 21.4 MHz. By bringing the frequencies down from 400 MHz to 21.4 MHz it is possible to attain good selectivity with crystal filter as it is difficult to obtain good selectivity at very high frequencies. The 2nd IF is fed to detector which removes the negative half of the signal, changing the signal to varying DC. The varying DC is smoothened by the integrator circuit in filter 2 and fed to the vertical palates of CRT. We obtain a frequency vs amplitude display on the CRT.

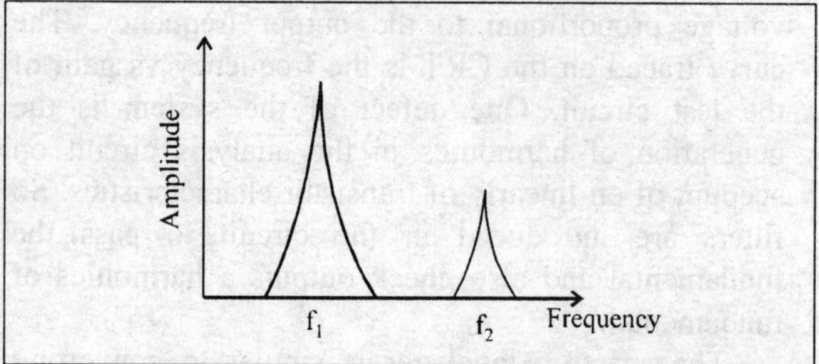

If we want a display in time domain i.e. time vs amplitude an ADC – Memory-DAC circuit is introduced before the mixer.

The range of frequency swept by the anaylser is called the *dispersion* of the spectrum analyzer. The first local oscillator has a frequency range of less than an octave.

Spectrum Analyzer Specifications

1. The frequency range is divided into moderate bandwidth such as of upto 300 kHz, if with a range from 100 kHz to 1200 MHz and microwave beyond 1200 MHz. Low to high ratio is about 1 to 10^6.
2. Frequency resolution can be in fractions of 1 Hz for audio and about 100 Hz for microwave.
3. Accuracy of the frequency varies between 2% and 10%.
4. Input attenuation range is specified in dBm i.e., for 1 mW reference. It is from 50 to 70 dBm i.e. from a signal as low as –60 dBm. Input impedance is 50Ω.
5. The amplitude accuracy is ± 1 dB, i.e. the response is fed for the range between ± 1 dB.

Spectrum Analyser Application

The spectrum analyzer such as the unit shown in figure is a powerful tool and has many applications. To those who have never used the instrument, these application may not be readily apparent. To illustrate some of the applications, the following signals, their descriptions, and the spectrum analyzer display, as shown in will be presented.

(a) ***Pure sinusoid with no modulation or harmonic distortion:*** The signal is characterized by a single spectral line regardless of what the dispersion of the spectrum analyzer is or the IF filter bandwidth

(b) ***Amplitude modulation***: When a carrier is modulated with amplitude modulation, two sidebands are generated, one above the carrier frequency and a second below the carrier frequency. The separation in frequency between the carrier and the sidebands is equal to the modulation frequency. The power contained in the sidebands is dependent on the percentage of modulation. One hundred per cent modulation produces sidebands that are 6 dB below the carrier. The amplitude of the carrier, on the other hand, does not change, regardless of the percentage of modulation.

(c) ***Frequency modulation***: Frequency modulating a carrier produces sidebands that are centered around the carrier as in the case of amplitude modulation, except that more than one sideband is generated. The number of sidebands and the amplitude of those sideband is described by complex formulas based on the Bessel functions. The sidebands are all multiples of the modulating frequency, and the amplitude of the carrier is affected by the amount of modulation supplied. The precise amount of frequency modulation can be determined if the modulation is adjusted so that the amplitude of the carrier or other sidebands goes to zero.

(d) ***Asymmetrical spectra***: The generation of a spectrum that is not symmetrical about the carrier is usually an indication that both frequency and amplitude modulation are occurring simultaneously. This could occur in a FM system where the passband of an amplifier is not flat and the frequency modulation is introducing amplitude modulation. Likewise, amplitude modulation applied to a carrier that also causes frequency instabilities, which is a common problem with phase-locked loops, will cause a similar spectrum.

(e) ***Harmonic distortion:*** Harmonics appear as additional signals in the spectrum analyzer display at multiples of the carrier frequency. It is often required that the harmonic content of a signal be kept low, on the order of 60 or more dB below the carrier. As an example, this may be required so that a transmitter operating at an assigned frequency will not interfere with other radio services at twice the assigned frequency that may be located near the

(f) ***Pulse modulation:*** Examining pulse modulation was the first application of the spectrum analyzer. Determining the pulse modulation of radar transmitters was a difficult task in the early development of radar, and the spectrum analyzer was used to evaluate the quality of the pulse modulation. The spectrum of a rectangular amplitude pulse is shown in figure. The structure of the sidebands shows the rise-and fall-times of the pulse modulation, and the symmetry indicates the presence or absence of frequency modulation, which is a problem with modulated oscillators such as those used with high-power radar transmitters.

Other applications include components linearity, pulse shape analysis and noise measurement. Under component linearity is measured filter characteristics such as in 3 dB attenuation.

Distortion Analyzer

An amplifier to which a pure sine wave is fed, may have at its output signals other than the fundamental though the magnitude may be at much lower intensity. This is because of the non-linearity of components like valves and transistors. The output voltage can be stated as

$$V_o = K_o + K_1 V_{in} + K_2 V_{in}^2 + K_3 V_{in}^3 + \ldots$$

The first 2 term are linear but the third term and above are of higher degree. Taking the 3rd term for a sine wave, the output is

$$K_2 V_{in}^2 = K_2 \left(A \sin \omega t\right)^2 = \frac{K_2 A}{2}\left(1 - \cos 2\omega t\right)$$

We see that the output contains a 2nd harmonic of the input frequency at half the amplitude. So the presence of frequencies other the fundamental or the desired frequency is called distortion.

The distortion analyzer mainly consists of a Wein circuit which tunes out the input frequency to the circuit. What remains are the spurious frequencies whose voltage is measured on a thermocouple or square law voltmeter calibrated in percentage. This measurement is called THD, total harmonic distortion.

$$\text{THD} = \sqrt{D_2^2 + D_3^2 + D_4^2 + \ldots D_n^2}$$

$$\text{Where } D_2 = \frac{A_2}{A_1},\ D_3 = \frac{A_3}{A_1} \ldots D_n = \frac{A_n}{A_1}$$

Where A_2, A_3 etc are the effective magnitudes of 2nd, 3rd harmonic frequencies and A1 the effective magnitude of the fundamental.

Block diagram of a measuring system is

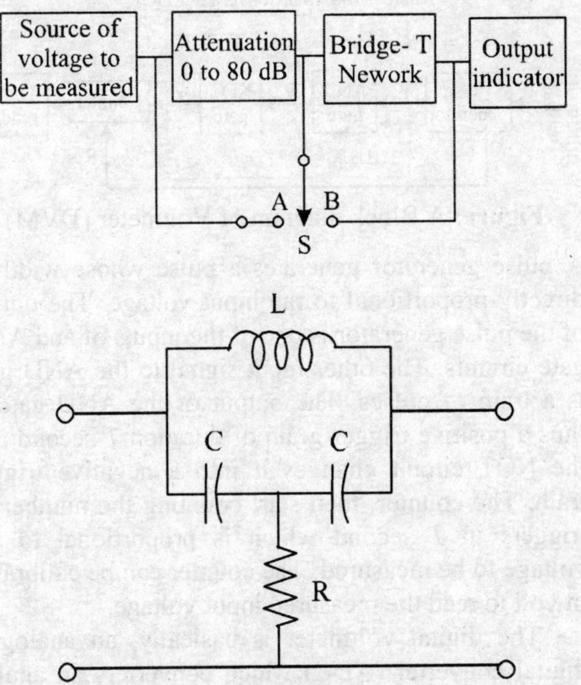

The bridged-T filters out the fundamental with the switch S in position A, the meter indication being minimum. Switch S is thrown to position B and attenuator adjusted to give the same output as in A. The attenuator reading is the rms distortion in % or db below the fundamental.

Wave Analyzer

It is designed to measure relative amplitudes of single frequency components in a complex waveform. It acts as a frequency selective voltmeter which is tuned to the frequency of one signal while rejecting the other signal components.

Applications of the wave analyzer are found in the fields of electrical measurements and sound and vibration analysis. For example, harmonic distortion of an amplifier can readily be measured, and the contribution of each harmonic to the total distortion figure can be determined.

7.14 DIGITAL VOLTMETER

A Digital electronics voltmeter, also called digital voltmeter (DVM), measures and displays dc or ac voltages as discrete numberals instead of a pointer deflection on a continuous scale as in analog electronics instruments.. This meter scaled inputs and than digitized them, showing the results in a numeric digital display.

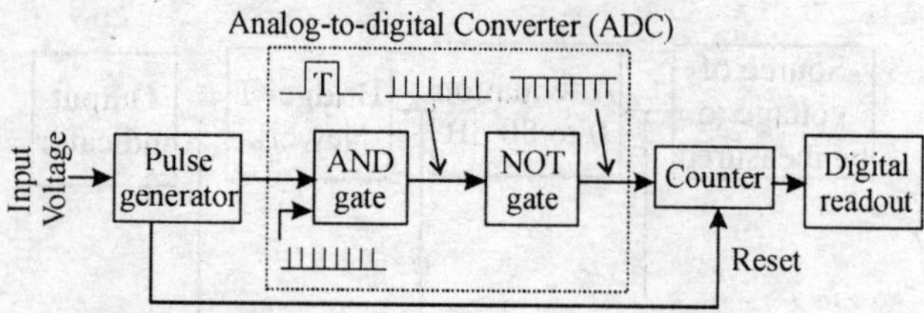

Figure: A Block diagram of Voltmeter (DVM)

A pulse generator generates a pulse whose width is directly proportional to the input voltage. The output of the pulse generator is one of the inputs of and AND gate circuits. The other input signal to the AND gate is a train of pulses. The output of the AND gate is thus a positive trigger grain of duration *T* second and the NOT circuit changes it into a negative trigger train. The counter, then start counting the number of triggers in *T* second which is proportional to the voltage to be measured. The counter can be calibrated in volt to read the measured input voltage.

The digital voltmeter is basically, an analog-to digital converter (ADC) which converters an analog signal into a train of pulses, the number of which is proportional to the magnitude of the input voltage. With appropriate signal conditional of the input voltage, DVM can be used to measure many electrical and physical quantities such as ac voltage dc and ac current, resistance, temperature, pressure, etc.

Digital voltmeters can be classified into the following broad categories on the basis of analog-to-digital converter (ADC) used:

1. Ramp-type DVM
2. Integrating-type DVM
3. Continuous-balance or servo-balance DVM
4. Successive-approximation DVM

Advantages of Continuous-balance or Servo-balance DVM

1. It is low cost instrument that provides excellent performance.
2. The accuracy of this DVM is quite satisfactory, usually of the order of 0.1 of its input dc voltage
3. It has high input impedance, of about 10 MΩ

Analog meters require no power supply, they give a better visual indication of trends and changes, suffer less from electric noise and isolation problems and are simple and inexpensive.

Digital meters offer higher accuracy and input impedance, unambiguous readings at greater viewing distances, smaller size and a digital electrical output in addition to visual readout.

Three major classes of digital meters are

1. Panel meter
2. Bench type meters
3. System meters

- All employ some type of A/D converter and have a visible readout that displays the converter output.
- Usually panel meters are dedicated to a signal function
- On the other hand bench and system meters often are multimeters.

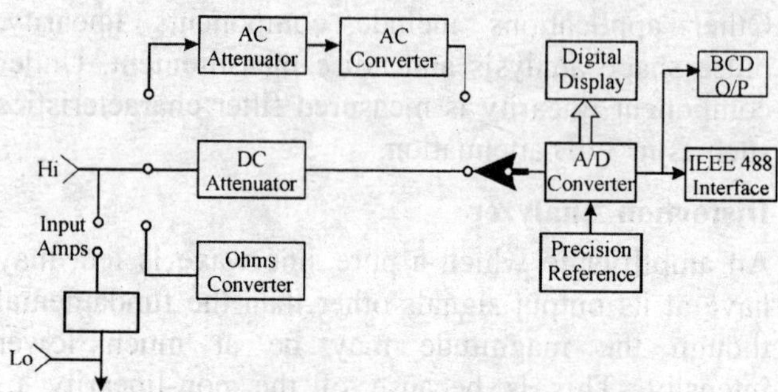

Figure: Digital Millimeter configuration

Ramp Type DVM

The basic principle of this DVM is to measure the time that a linear ramp voltage takes to change from level of input voltage to zero voltage (or vice versa). This time interval is measured with an electronic time interval counter and the count is displayed as a number of digits on electronic indicating tubes of the output readout of the voltmeter. *The conversion of a voltage value to a time interval* is shown in figure.

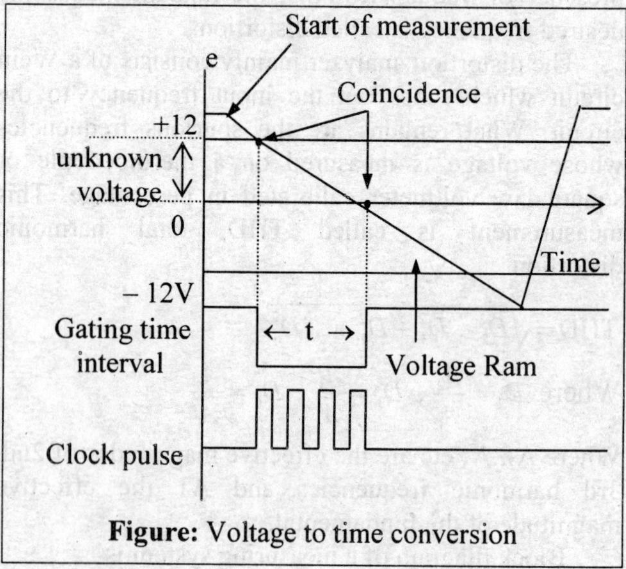

Figure: Voltage to time conversion

At the start of measurement a ramp voltage is initiated. The negative going ramp is shown in figure but a positive going ramp may also be used. The ramp voltage value is continuously compared with voltage being measured (unknown voltage). At the instant the value of ramp voltage is equal to that of unknown voltage coincidence circuit, called an input comparator, generates a pulse which opens a gate as shown in figure. At this instant another comparator called ground comparator generates a pulse and close the gate.

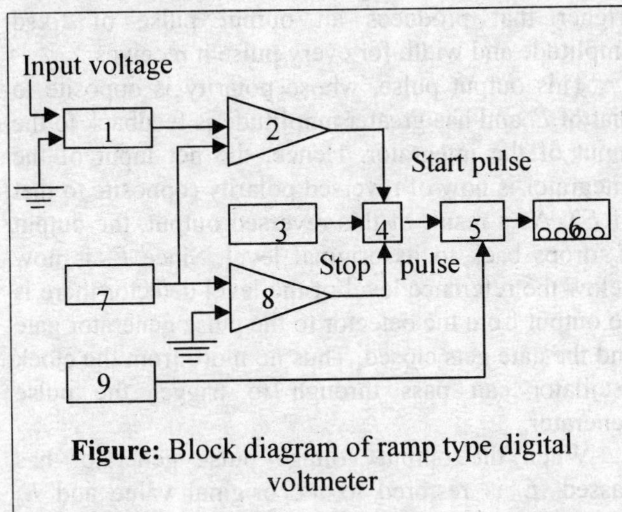

Figure: Block diagram of ramp type digital voltmeter

Block Name

1. Ranging and attenuator
2. Input comparator
3. Clock oscillator
4. Gate
5. Counter
6. Readout
7. Ramp generator
8. Ground comparator
9. Sample rate.

The time elapsed between opening and closing of the gate is t as shown in Fig. During this time interval pulse from clock pulse generator pass through the gate and are counted and displayed. The decimal number is indicated by the readout is a measure of the value of input voltage. The sample rate of multivibrator determines the rare at which the measurement cycles are initiated. The sample rate circuit provides an initiating pulse generator to start its next ramp voltage. At the same time it sends a pulse to the counters which sets all of them to zero.

Advantages of Ramp-type DVM

1. The ramp-type circuit is easy to design and its cost is low.
2. The output pulse can be transmitted over long distance

Disadvantages of Ramp-type DVM

1. The single ramp requires excellent characteristics regarding linearity of the ramp and time measurement
2. Large error are possible when noise is superimposed on the input signal input filters are usually required for this type of converter.

Integrating Type DVM

This voltmeter measures the true average value of the input voltage over a fixed measuring period. This voltmeter employs an integration technique which uses a *voltage to frequency conversion* (V/F). The V/F converter functions as a feedback control system which governs the rate of pulse generation in proportion to the magnitude of input voltage.

In V/F conversion technique, a train of pulses, whose frequency depends upon the voltage being measured, is generated. Then the number of pulse appearing in a definite interval of time is counted. The basic requirement of this technique is operation integrator, which is a modification of operational amplifier.

(a) Circuit diagram

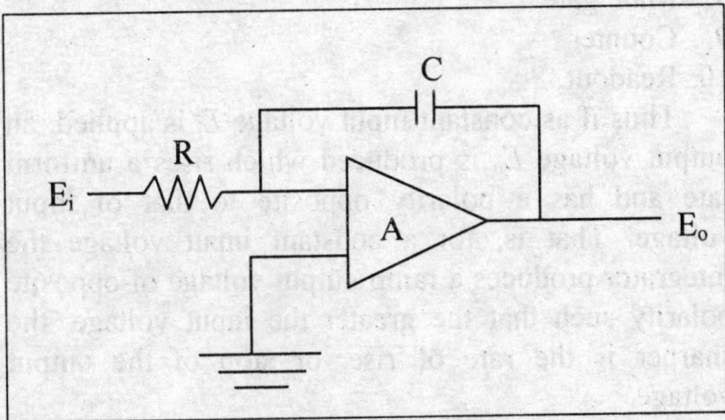

(b) Symbolic representation

$$E_i \circ\!\!-\!\!- \boxed{\frac{1}{RC}\int E_i\,dt} -\!\!-\!\!\circ E_o$$

If the input voltage is constant, we have

$$E_e = -\frac{E_i}{RC}t$$

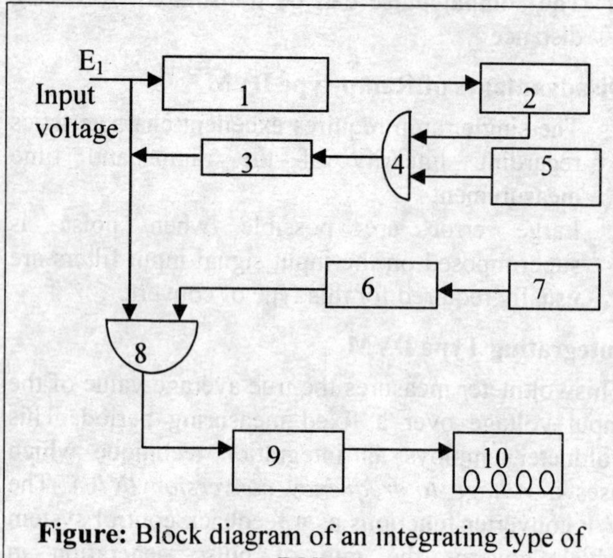

Figure: Block diagram of an integrating type of DVM

Block Name

1. Integrator
2. Level detector
3. Pulse generator
4. Pulses generator gate
5. Clock oscillator
6. Start stop gate
7. Time base selector
8. Main gate
9. Counter
10. Readout.

Thus if as constant input voltage E_i is applied, an output voltage E_o is produced which rises a uniform rate and has a polarity opposite to that of input voltage. That is, for a constant input voltage the integrator produces a ramp output voltage of opposite polarity such that the greater the input voltage, the sharper is the rate of rise, or stop of the output voltage.

The unknown voltage, E_i is applied to the input of integrator as shown in Fig. and the output voltage E_o starts to rise. The slope of output voltage E_o is determined by the value of input voltage E_i. This voltage is fed to a level detector, and when E_o reaches a certain reference level, the detector sends pulse to the generator gate.

The level detector is a device similar to a voltage comparator. The output voltage from integrator E_o is compared with the fixed voltage of an internal reference source and when E_o reaches that level detector produces an output pulse. As already stated, greater the value of input voltage E_i, the sharper will be the slope of output voltage E_o and quicker the output voltage E_o will reach its reference level.

The output pulse of the level detector opens the pulse level gate, permitting pulses from a fixed frequency clock oscillator to pass through pulse generator. This pulse generator is similar to Schmitt trigger that produces an output pulse of fixed amplitude and width for every pulse it receives.

This output pulse, whose polarity is opposite to that of E_i and has greater amplitude is feedback to the input of the integrator. Hence, the net input of the integrator is now of reversed polarity (opposite to that of E_i). As a result of this reversed output, the output E_o drops back to its original level. Since E_o is now below the reference level of the level detector there is no output from the detector to the pulse generator gate and the gate gets closed. Thus no more from the clock oscillator can pass through to trigger the pulse generator.

When the output voltage pulse generator has passed, E_i is restored to its, original value and E_o starts its rise again. When it reaches the level of reference voltage again, the pulse generator gate is opened. The pulse generator is triggered by a pulse from the clock generator and the entire cycle is repeated again. Thus, the waveform of E_o is a saw tooth wave whose rise time is dependent upon the value of the output pulse from the pulse generator.

Thus the frequency of the saw tooth E_n is a function of the value of E_i the voltage being measured. Since one pulse from the pulse generator is produced for each cycle of the saw tooth wave, the number of pulses produced in a given time interval and hence the frequency of saw tooth wave is an indicator of the value of voltage being measured. The frequency of saw tooth wave may be measured by counting the number of pulses in a given interval of time.

Pulses from clock oscillator are applied to a time base selector. The first pulse passes through start-stop gate, producing an output which is applied to the main gate, thus opening the gate. As a result of this, the same output pulses from the pulse generator (that are applied to integrator) also pass through the main gate.

The next pulse from the time base closes the start-stop gate and also the main gate. Thus no more pulse generator pulses can pass through. Hence the counters and their associated readout indicate the number of pulses that have passed during a known interval of time. This count is an indication of the voltage being measured. In order to make the counter read directly the terms of voltage the amplitude and width of pulse generator can be suitably adjusted.

Advantages of Integrating-type DVM

Integrating-type DVM is capable of providing *accurate measurement in the presence of large amounts of superimposed noise.*

Disadvantages of Integrating-type DVM

Integrating-type DVM requires excellent characteristics in linearity of the ramp.

Potentiometric type DVM

This employs *voltage comparison technique.* In this the unknown voltage is compared with a reference voltage whose value is fixed by the setting of the calibrated potentiometer. The potentiometer setting is changed to obtain balance, i.e. null condition. When null conditions are obtained the value of the unknown voltage, is indicated by the dial setting of the potentiometer.

In this DVM the balance is not obtained manually but is arrived at automatically. Thus it is a self balance potentiometer, the only exception is that a potentiometer DVM is provided with a readout which displays the voltage being measured.

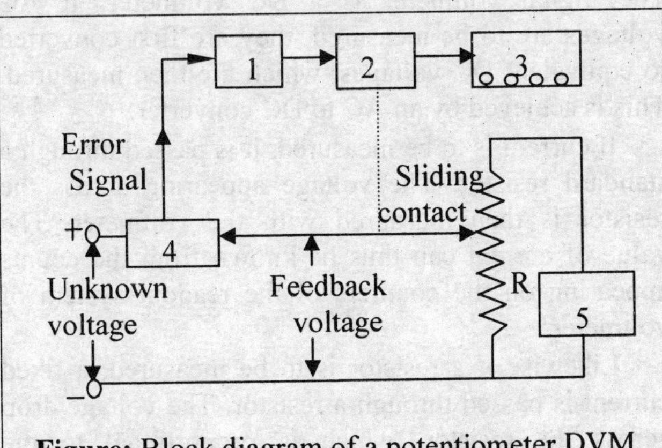

Figure: Block diagram of a potentiometer DVM

Block Name

1. Error amplifier.
2. Potentiometer adjustment device
3. Read out
4. Comparator
5. Reference voltage.

The block diagram of basic circuit of potentiometric DVD is shown in Fig. The unknown voltage is filtered and attenuated to a suitable level. This input voltage is applied to a comparator (also known as error detector). This error detector may be chopper. The reference voltage is obtained from fixed voltage source. This voltage is applied to a potentiometer *R*. The value of the feedback voltage depends upon the position of the sliding contact. The feedback voltage is also applied to the comparator.

The unknown voltages and the feedback voltages are compared in the comparator. The output voltage of the comparator is the difference of the above two voltages, often called as error signal. The error signal is amplified and is fed to a potentiometer adjustment device, which moves the sliding contact of the potentiometer. The magnitude by which the sliding contact moves depends upon the magnitude of the error signal. The direction of movement of slider depends upon whether the feedback voltage equals the unknown voltage. In this case, there will not be any error voltage and hence there will come to rest.

The condition of the potentiometer adjustment device at the point is indicated in numerical form on the digital readout device associated with it. Since the position at which no voltage appears at potentiometer adjustment device is the one where the unknown voltage equals the feedback voltage, the reading of readout device indicates the value of unknown voltage. The potentiometer adjustment device i.e. the device which moves the sliding contact is a two phase servomotor. The reference voltage source must be extremely stable and generally consists of standard cells or Zener diode sources. A practical potentiometric type DVM does not consist of a single resistor *R*, but consists of a number of decade resistors connected in the Kelvin Varley side system.

Successive-approximation DVM

Digital voltmeter capable of 1000 readings per second or more are now commercially available. These instruments generally use successive approximation converters to perform the digitization (analog-to-digital conversion).

At the beginning or the measurement cycle, a start pulse is applied to the start–stop multivibrator. This sets a 1 in the most significant digit (MSB) of the control register and a 0 in all the bits of less significance. Assuming an 8-bit control register, its reading would then be 1000,0000. This initial setting of the control register causes the output of the D/A converter to be one-half the reference supply voltage ½V. The converter output is compared to the unknown input by the comparator produces an output that causes the control register to retain the 1 setting in its MSB, and the converter continues to supply its reference output voltage of 1/2V.

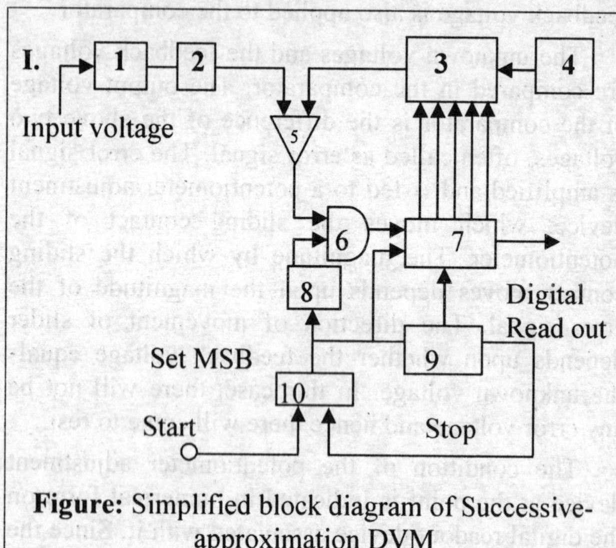

Figure: Simplified block diagram of Successive-approximation DVM

Block Name

1. Input alternator and amp.
2. S/H circuit
3. D/A converter
4. Ref supply
5. Cooperator
6. Gate
7. Control register
8. Delay
9. Ring counter
10. Start/stop.

The ring counter next advances one count, shifting a 1 in the second MSB of the control register, and its reading becomes 1100,0000. This causes the D/A converter to increase its reference output by one increment, to [½ V + ¼V], and another comparison with the unknown input voltage takes place. If in this case, the accumulated reference voltage exceeds the unknown voltage, the comparator produces an output that causes the control register rest its second MSB to 0. The converter output then returns to its previous level of ½ V awaits another input from the control register for the next approximation. When ring counter advances another count, the third MSB of the control register is set to 1 and the converter output rises by the next increment to [½V + ⅛V]. The measurement cycle thus proceeds through a series of successive approximation, retaining or rejecting the converter output in the manner described. Finally, when the ring counter reaches its last count, the measurement cycle stops, and the digital output of the control register represents the final approximation of the unknown input voltage.

With input voltages other than DC, the input level changes during digitization and decisions made during conversion are nor consistent. To avoid this conversion error, a sample land hold (S/H) circuit is placed in the input, directly following the input form S/H circuit can be represented by a switch and a capacitor as shown in Fig. in the sample mode the switch is switch closed and the capacitor holds the voltage. In the hold mode the switch is opened and the capacitor holds the voltage that it had at the instant the switch was opened. If the switch drive is synchronous with the ring converter pulse, the actual measurement and conversion take place when S/H circuit is in the old mode.

Other Measurements wits Digital Voltmeter

The digital voltmeter is a DC voltmeter. If AC voltages are to be measured, they are first converted to equivalent DC voltages, which are then measured. This is achieved by an AC to DC converter.

If current is to be measured, it is passed through a standard resistor. The voltage appearing across the resistor is then measured with the voltmeter. The value of current can thus be known from the counts appearing on the counters of the readout system of voltmeter.

Likewise if a resistor is to be measured, a fixed current is passed through a resistor. The voltage drop across the resistor, which is proportional to the resistance is measured by the voltmeter. The number of counts appearing on the digital voltmeter is the resistance other physical quantities can be measured with digital voltmeter by using suitable transducers to feed the input.

7.15 DIGITAL MULTIMETER

Digital Multimeter (DMM)

Analog multimeter dc and ac voltage, resistance and dc current. These meter were known as volt-ohm-milliammeters or VOMs. After commercial manufacturing of digital voltmeter digital measurement and display technologies developed in parallel with digital integrated circuit technology portable digital versions of the VOM apperated. This form of meter has since been known as the digital multimeter (DMM).

Construction and working

In this shows the basic flow of information from the analog input signal through the various analog signal conversion circuits which convert the measured quantity to a d.c. voltage equivalent. Then the ADC translates this d.c. signal to digital form, and the display system shows the resultant value with appropriate annunciation, such as measurement units of the original input signals. Thus, a DMM is made up of following three basic elements:

- Signal conditioning
- Analog-to-digital conversion (ADC)
- Numeric digital display

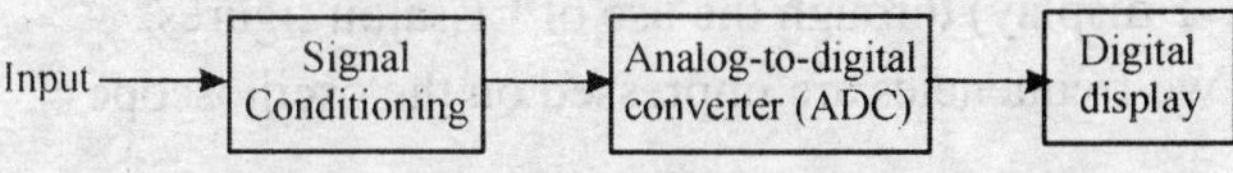

Figure: Basic block diagram of digital multimeter (DMM)

A.C. or d.c. voltage, A.C. or d.c. current, and resistance are the most common parameters to be measured by DMM. Now DMMs are also being used to measure additional functions such as frequency, temperature. etc. To measure these parameters, first these inputs must be switched and conditioned to present an equivalent d.c. voltage of the appropriate range to the analog-to-digital converter or a scaled input to the counting circuit.

For d.c. voltage measurement by DMM, the a wide range of d.c. voltage inputs is scaled to the limited range of the ADC. A resistive divider and switching generally used for this function.

For a.c. voltage measurement by DMM, the signal is scaled and then converted to a d.c. equivalent value before sending it to the ADC.

For d.c. and a.c. current measurements by DMM, it is necessary to convert the current at the input to a voltage for use by the ADC. This is done through a series of switched resistors, called shunts. A small – valued resistor in the input is placed in series with the current to be measured. Ohm's law $(V = IR)$ defines the small voltage proportional to the input current which is then measured. AC current measurement uses the same method except that the voltage across the shunt is routed through an a.c. to d.c. voltage converter before going to the ADC.

For resistance measurement, It is necessary to create a voltage proportional to the resistance because the ADC measures only d.c. volts. The DMM input circuit must provide a d.c. current flowing through the resistor and then measure the resulting voltage. Many DMMs provide a calibrated constant current source for this purpose. Many others use reference resistors in series with the resistor being measured, and then measure the unknown and reference voltage drops. When combined with the known reference resistor voltage drops. When combined with the known reference resistor value, this voltage – ratio technique also yields the desired resistance value

- DMMs offer high measurement accuracy.
- These instruments have a high input impedance.
- They are smaller in size.
- These meters eliminate observational, parallax and approximation errors.
- The measured outputs of these instruments can be directly fed to a computer for further analysis and use

With the additional capability to measure AC/DC voltage and current, and resistance, the digital multimeter (DMM) offers increased versatility. In the A.C. voltage mode, the applied input is fed through a calibrated, compensated attenuator, to a precision full-wave rectifier circuit followed by a ripple-reduction filter. The resulting DC is fed to ADC and the subsequent display system.

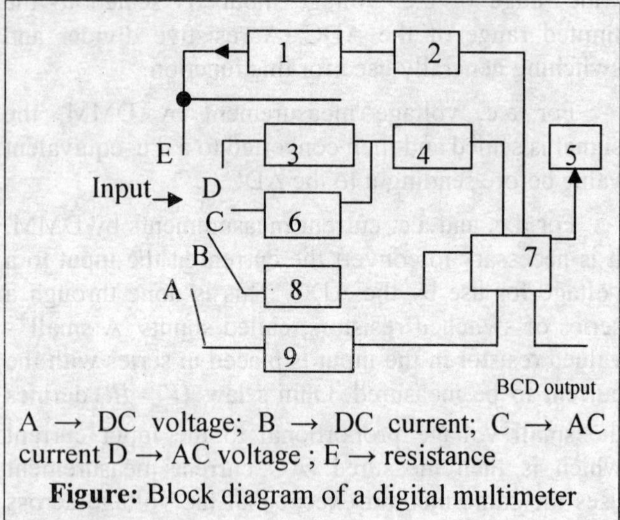

A → DC voltage; B → DC current; C → AC current D → AC voltage ; E → resistance

Figure: Block diagram of a digital multimeter

Block Name

1. Current source
2. Buffer AMP
3. Calibrated attenuator
4. Precision AC/DC converter
5. Display
6. Shunt
7. A/D converter
8. Shunt
9. Calibrated attenuator.

For current measurements, the drop across an internal calibrated shunt is measured, directly by the ADC in the DC. conversion in the A.C. current mode. This drop is often in the range of 200 mV. Due to the lack of precision in the AC-DC conversion, the accuracy in the A.C. range is in general of the order of 0.2 to 0.5%. In addition, the measurement range is often limited to about 50 Hz at lower frequency end due to the ripple in the rectified signal becoming a non- negligible percentage of the display and hence resulting in fluctuation of the displayed number. At the higher frequency end, detoriation of the performance of the AC/DC converter limits the accuracy. The A.C. measurement range is often average reading, RMS calibrated

In the resistance range the DMM operates by measuring the voltage across the externally connected resistance, resulting from a current forced through it from a calibrated internal current source. The accuracy and stability of the resistance measurement is of the order of 0.1 to 0.5% depending on the accuracy and stability of the internal current sources. The accuracy may be poorer in the highest range, which is often a bout 10 to 20 MΩ. In the lowest range, the full scale may≈200 ohms with a resolution of about 0.01 ohms for a 4(1/2) digit DMM. The effect of the lead resistance will have to be carefully considered in this range of resistance measurement.

7.16 TIME, PHASE AND FREQUENCY MEASUREMENTS

Frequency and time period

Frequency is the number of recurrences of a phenomenon or series of events during as given time internal. The unit of frequency is hertz (Hz) which equals one-cycle per second. The time taken for one cycle of events is known as periodic time T (second). Apparently the frequency and the periodic time are related to each other by the expression $f = \frac{1}{T} Hz$.

The concept of frequency and periodic time has been illustrated in figure for the pulse and sinusoidal wave forms.

Lissajou Figures

The frequency of sine-wave signals and the phase shift can be determined by an oscilloscope (capable of *X-Y* display) through the use of Lissajou figures.

Two frequencies are impressed on the oscilloscope

1. A known variable frequency is varied until applied to the X-channel (horizontal input).
2. The unknown signal is applied to the Y-channel (vertical input).

The oscillator signal is varied until a stable figure known as Lissajou figure appears on the oscilloscope screen. A suitable size of the figure can be obtained with some adjustment of the X and Y amplifier gains. They shapes of the Lissajou figure depends on the oscillator frequency and also on the phase shift between the two sine waves.

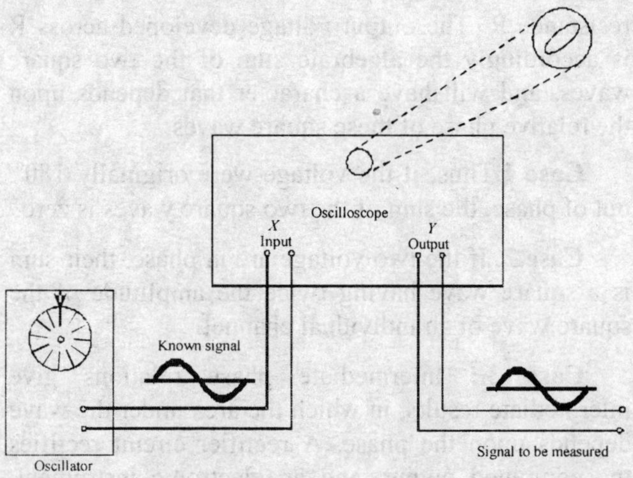

Figure: Schematic arrangement of an oscilloscope for obtaining Lissajous pattern

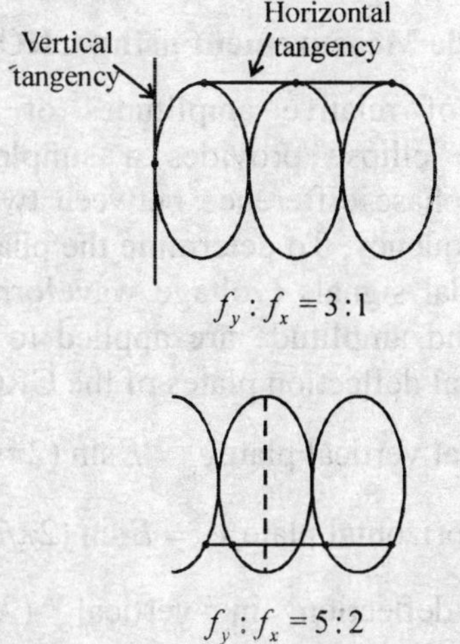

Figure: Frequency estimation from Lissajou pattern

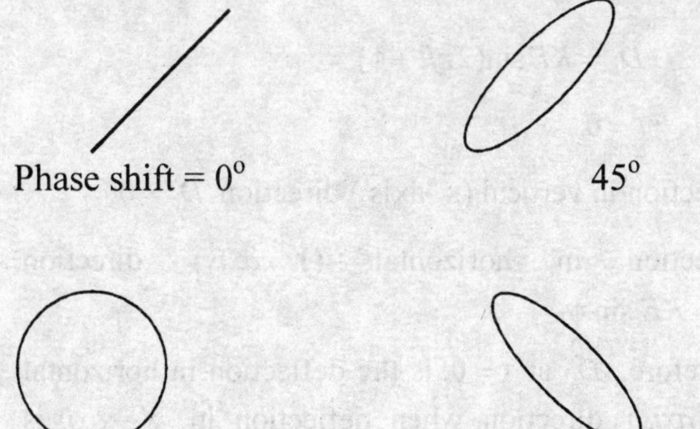

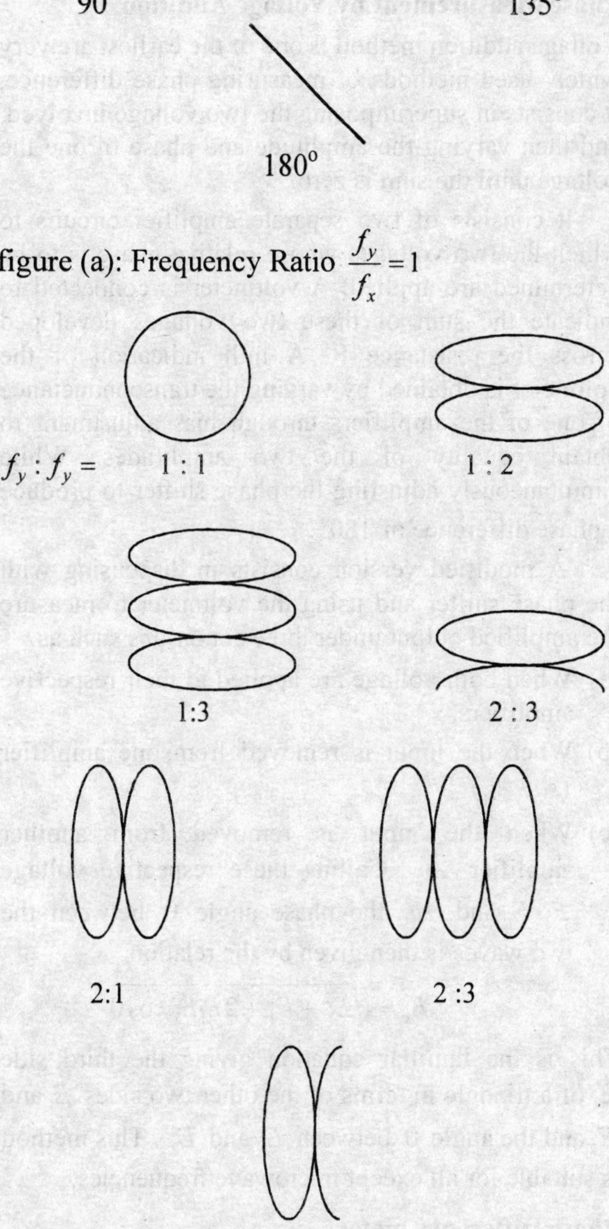

Figure: Lissajou patterns for different frequency ratios and phase shifts

7.17 MEASUREMENT OF PHASE

Phase angle measurement

Phase angle between two signals, or the power factor of the circuit, is an important to be measured for various control purpose.

Phase Measurement by Voltage Addition

Voltage addition method is one of the earliest are very widely used methods of measuring phase difference. It consists in superimposing the two voltage involved, and then varying the amplitude and phase of one the voltage until the sum is zero.

It consists of two separate amplifier circuits to which the two voltages whose relative phase is to be determined are applied. A voltmeter is connected to indicate the sum of these two voltages developed across the resistance R. A null indication of the voltmeter is obtained by varying the transconductance of one of the amplifiers through bias adjustment to obtain equality of the two amplitudes. While simultaneously adjusting the phase shifter to produce a phase difference of 180°.

A modified version consists in dispensing with the phase shifter and using the voltmeter to measure the amplified output under three conditions such as

(a) When both voltage are applied to their respective amplifiers.

(b) When the input is removed from one amplifier $(say\ A_{v1})$

(c) When the input is removed from another amplifier A_{v2}. Calling these respective voltage E_s, E_1, and E_2, the phase angle θ between the two waves is then given by the relation.

$$E_s = \sqrt{E_1^2 + E_2^2 + 2E_1E_2 \cos\theta}$$

This is the familiar equation giving the third side E_s of a triangle in terms of the other two sides E_1 and E_2 and the angle θ between E_1, and E_2. This method is suitable for all except microwave frequencies.

Phase Difference meter

There are a number of circuits available for measuring phase difference between two waves directly on a meter. The voltage whose phase difference is to be determined are applied to separate amplifier channels, which are so designed and operated that they develop square wave outputs.

By using the same clipping level in both channels, the square waves thus produced not only have amplitudes independent of the strength of the applied voltages, but also have the same amplitude in each of the two channels.

These two square waves are applied to separate amplifiers $A_{v1} and\ A_{v2}$. Which have a common load resistance R. The output voltage developed across R is accordingly the algebraic sum of the two square waves, and will have a character that depends upon the relative phase of these square waves.

Case 1.Thus, if the voltage were originally 180° out of phase, the sum of the two square waves is zero,

Case 2. If the two voltage are in phase, their sum is a square wave having twice the amplitude of the square wave of an individual channel.

Case 3.: Intermediate phase relations give intermediate results, in which the area under the wave depends upon the phase. A rectifier circuit rectifies the combined output, and an electronic instrument, which can be calibrated in degrees, indicates the resulting rectified current. It is to be noted that the indication is independent of frequency.

Phase – angle Measurement using CRO

Regardless of relative amplitudes of the applied voltage, the ellipse provides a simple means of finding the phase difference between two signals of the same frequency, To determine the phase angle the two sinusoidal signals (voltage waveforms) of same frequency and amplitude are applied to the vertical and horizontal deflection plates of the CRO.

Let, voltage at vertical plate $e_v = E\sin(2\pi f)t$

Voltage at horizontal plate $e_h = E\sin(2\pi ft + \phi)$

Then the deflection in vertical $(X-axis)$ and horizontal $(Y-axis)$ direction will be

$$D_v = KE\sin(2\pi f)t$$

And $\quad D_h = KE\sin(2\pi ft + \phi)$

For $\quad t = 0,$

deflection in vertical (x -axis) direction $D_v = 0$

deflection in horizontal $(Y-axis)$ direction $D_h = KE\sin\phi$.

Therefore, D_h at $t = 0$ is the deflection in horizontal $(Y-axis)$ direction when deflection in $X-axis$ is zero, i.e. the intercept on $Y-axis$ intercept Y_{int} and the maximum vertical deflection Y_{peak}. It is given as

$$\sin\phi = \frac{Y_{int}}{Y_{peak}} \quad (1)$$

$$\phi = \sin^{-1}\frac{Y_{int}}{Y_{peak}} \quad (2)$$

For convenience, the gains of the vertical and horizontal amplifiers are adjusted so that the ellipse fits exactly within a square as marked by the coordinate lines on the graticule.

Consider the characteristics of patterns that appear on the screen of a CRT when sinusoidal voltages are simultaneously applied to horizontal and vertical plates. These patterns are called 'Lissajous Patterns'.

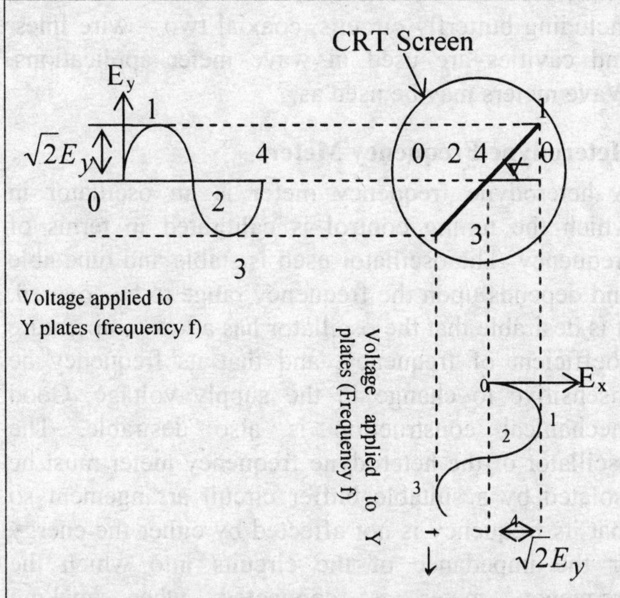

Figure: Lissajous pattern with equal frequency voltages and zero phase shift

When two sinusoidal voltage of equal frequency which are in phase with each other are applied to the horizontal and vertical deflection plates, the pattern appearing on the screen is a straight line.

When two equal voltages of equal frequency but with a phase shift ϕ (not equal to 0° or 90°) are applied to a CRO we obtain an ellipse. An ellipse is also obtained when unequal voltage of same frequency are applied to the CRO.

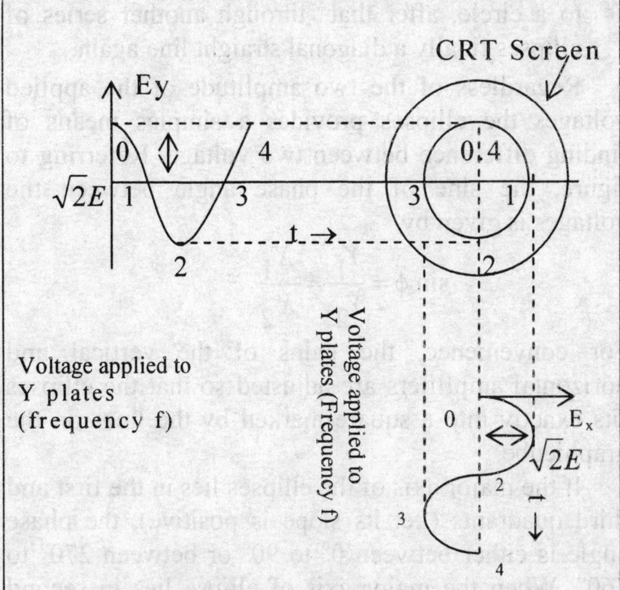

Figure: Lissjous pattern with equal voltages of equal frequency and a phase shift of 90°

A number of conclusions can be drawn. When two sinusoidal voltage of same frequency are applied:

1. A straight line results when the two voltages are equal and are either in phase with each other or 180° out of phase with each other. The angle formed with the horizontal is 45° when the magnitudes of voltages are equal. An increase in the vertical deflection voltage causes the line to have an angle greater than 45° with the horizontal. On the other a greater horizontal voltages makes the angle less than 45° with the horizontal.
2. Two sinusoidal waveforms of the same frequency produce a Lissajous pattern, which may be a straight line, a circle or an ellipse depending upon the phase and magnitude of the voltage

   A circle can be formed only when the magnitude of the two signals are equal and the phase difference between them is either 90° or 270°. However, if the two voltages are not equal and/or out of phase an ellipse is formed. If the Y voltage is larger, an ellipse with vertical major axis is formed while of the X plate voltage has greater magnitude the major axis of the ellipse lies along horizontal axis.
3. It is clear that for equal voltages of same frequency progressive variation of phase voltages cause the pattern to vary from a straight diagonal line to ellipses of different eccentricities and then

to a circle, after that through another series of ellipses finally a diagonal straight line again.

Regardless of the two amplitude of the applied voltages the ellipses provides a simples means of finding difference between two voltage. Referring to figure, the sine of the phase angle between the voltages is given by

$$\sin\phi = \frac{Y_1}{Y_2} = \frac{X_1}{X_2}$$

For convenience, the gains of the vertical and horizontal amplifiers are adjusted so that the ellipses fits exactly into a square marked by the lines on the graphicube.

If the major axis of the ellipses lies in the first and third quadrants (i.e. its slope is positive), the phase angle is either between 0° to 90° or between 270° to 360°. When the major axis of ellipse lies in second and fourth quadrants i.e. when its slope is negative, the phase angle is either between 90° and 180° or between 180° and 270°.

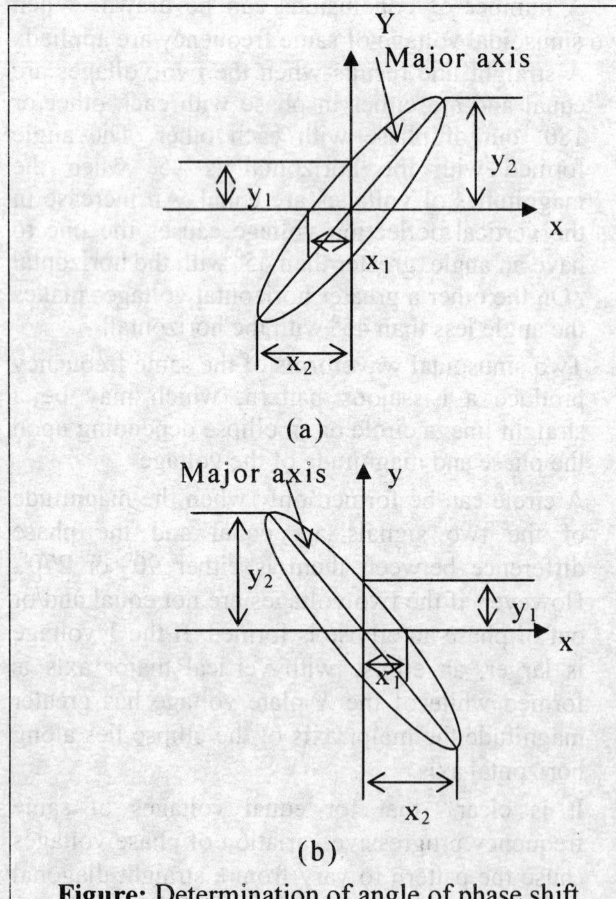

Figure: Determination of angle of phase shift

7.17 FREQUENCY MASUREMENTS

Frequency Measurement

Frequency is an important parameter, which is required to be measured for wide applications, particularly in industrial applications.

Wave meters

A wave meter is one of the earliest devices used to *measure resonant frequency*. It is an adjustable resonant circuit provided with a calibration that gives the resonant frequency in terms of the setting of the tuning adjustment. Wave meters are used to measure frequency when the higher accuracy of a primary or secondary standard is not required, and where simplicity and portability are important. A wave meter may employ any type of resonant circuit that is convenient for the frequency range to be covered. Thus, resonant circuits based on lumped constants, including butterfly circuits, coaxial two – wire lines, and cavities are used in wave meter applications. Wave meters may be used as.

Heterodyne Frequency Meters

A heterodyne frequency meter is an oscillator in which the tuning control is calibrated in terms of frequency. The oscillator used is stable and tune able and depends upon the frequency range to be covered. It is desirable that the oscillator has a low temperature coefficient of frequency, and that its frequency be insensitive to change in the supply voltage. Good mechanical construction is also desirable. The oscillator of the heterodyne frequency meter must be isolated by a suitable buffer circuit arrangement so that its frequency is not affected by either the energy or the impedance of the circuits into which the frequency meter is connected when making measurements. When a heterodyne frequency meter is to be used for interpolation purposes, it is desirable that the frequency be linearly proportional to the angle of rotation of the tuning dial to a high degree of precision.

Frequency Monitors

Frequency monitors are devices for continuously observing and indicating the frequency of a radio transmitter through comparison with a reference frequency generated by the monitor. Such arrangements are used to ensure that radio transmitter are maintaining their assigned frequency within the assigned limits by government regulation. While carefully engineered crystal oscillators determine the frequency of modern ratio transmitters, the use of a monitor based on a crystal oscillator designed with the highest possible frequency stability, is more dependable.

Measurement of Audio Frequency

Audio frequencies can be measured by comparing the unknown frequency with a known frequency, by balance depend upon frequency, or by some form of cycle-counting device.

There are various methods of measuring audio frequencies, such as

- Comparison method
- Cycle – counting method
- Frequency – selective wave analyzer

In comparison method of audio – frequency measurement, the standard frequencies are obtained by frequency division of the audio frequency generated by a standard. Electrically driven tuning forks represent another source of standard audio frequencies. Where known radio frequencies are available, it is always possible to adjust a low – frequency oscillator in a known relation to the high frequency by the use of harmonics, etc. When one or more audio frequencies are available, it is possible to calibrate completely an audio – frequency oscillator by the usual frequency – comparison methods

Frequency Measurement using Counters

The basic principle for the precise determination of frequency using counters (also called frequency counters) operates on the principle of gating the input frequency into counter for a predetermined time. For example, if an unknown frequency were gated into the counter for exactly 1 second, the number of counts allowed into the counter would be precisely the frequency of the input singnal. The term 'gated' stems from the fact that an AND or an OR gate is used to allow an unknown input into the counter to be accumulated.

Frequency Measurement Using CRO

The frequency of periodic signals are easily measured with a CRO. The frequency of a signal to be measured is compared with a known frequency. A sinusoidal voltage of known frequency is applied to the horizontal plates and the sinusoidal voltage of unknown frequency (the frequency of which is to measured) is applied to the vertical plates. If the ratio of two frequencies is an integer a clearly defined and distinctive pattern appears on the screen. These patterns are known as Lissajous figures.

Let us assume that two sinusoidal voltages applied to the vertical and horizontal plates of the CRO have equal frequencies and same phase, and let that voltage applied to the plates be given as

Voltage at vertical late

$$e_v = E_v \sin(2\pi f)t \qquad (1)$$

Voltage at horizontal plate

$$e_h = E_h \sin(2\pi f)t \qquad (2)$$

Where,

E_v = amplitude of voltage applied at vertical plate

E_h = amplitude of voltage applied at horizontal plate

Then, the deflections of the spot (on CRT screen) on the vertical and horizon direction are given respectively by

Deflection at vertical plate

$$D_v = k\, e_v = E_v \sin(2\pi f)t \qquad (6.29)$$

Deflection at horizontal plate

$$D_h = k\, e_h = E_h \sin(2\pi f)t \qquad (6.30)$$

Where, K is a constant.

Hence, the ratio,

$$\frac{D_v}{D_h} = \frac{E_v}{E_h} = m \quad \text{(a constant)}$$

$$D_v = mD_h$$

When the frequencies of the two voltages are not equal then to obtain the Lissajous patterns is not simple. Lissajous patterns may be divided into two general classes, the *closed and the open Lissajous patterns*. A closed Lissajous pattern is continuous and has no break, i.e. it has no loose or free ends, as shown in Fig 6.34(a). The open Lissajous patterns are simply steady traces and have free ends.

Lissajous patterns may be used for accurate measurement of frequency. The signal, whose frequency is to be measured, is applied to the *Y* plates. An accurately calibrated standard variable frequency source is used to supply voltage to the *X* plates, with the internal sweep generator switched off. The standard frequency is adjusted until the patterns appears as circle or an ellipse, indicating that both signals are of the same frequency. Where it is not possible to adjust the standard signal frequency to the exact frequency of the unknown signal, the standard is adjusted to a multiple or submultiples of the frequency of the unknown source so that the pattern appears stationary.

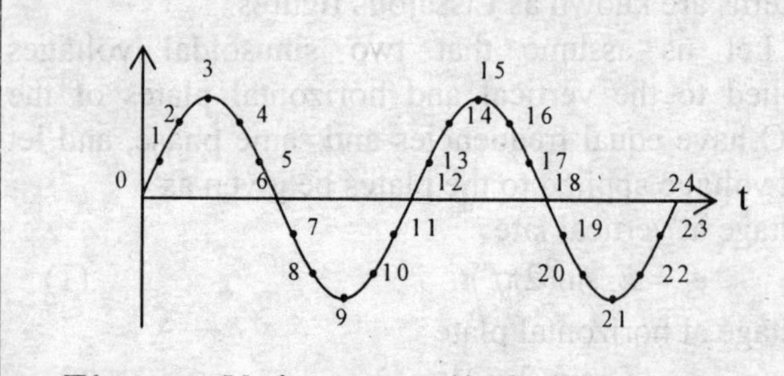

Figure: Voltage applied to Y plates (frequency = 2f)

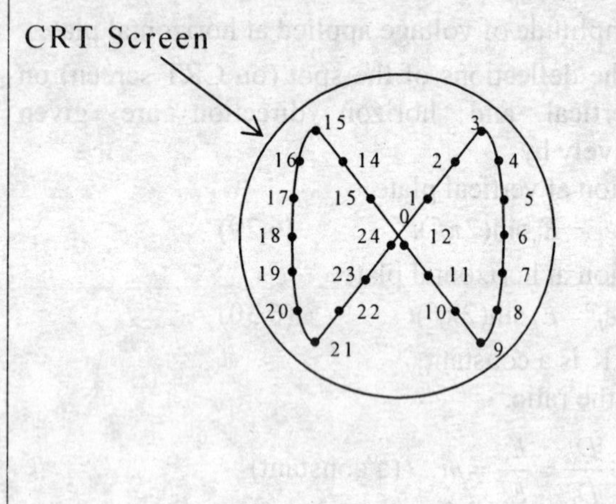

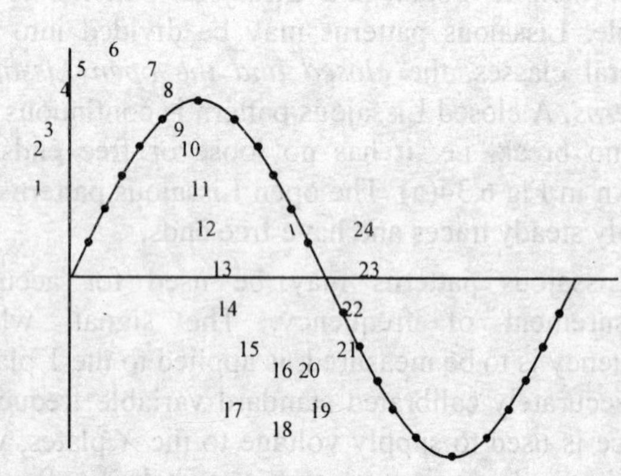

Voltage applied to X plates (frequency = f)

Figure: Lissajous pattern with frequency ratio 2 : 1

Let us consider an example suppose sine wave are applied to *X* and *Y* plates. Let the frequency of wave applied to *Y*-plates is twice that of the voltage applied to *X*-plates. This means that the CRT spot travels two complete cycles in the vertical against one in the horizontal direction.

The two waves start at the same instant. Lissjous pattern may be constructed in the usual way and a 8 shaped pattern with two loops is obtained. If the two waves do not start at the same instant we get different patterns for the same frequency ratio. The Lissajous patterns for other frequency ratios can be similarly drawn.

$$\frac{f_y}{f_x} = \frac{\text{number of times tangent touches top or bottom}}{\text{number of times tangent touches either side}}$$

$$= \frac{\text{number of horizontal tangencies}}{\text{number of vertical tangencies}}$$

f_y = frequency of signal applied to *Y* plates

f_x = frequency of signal applied to *X* plates

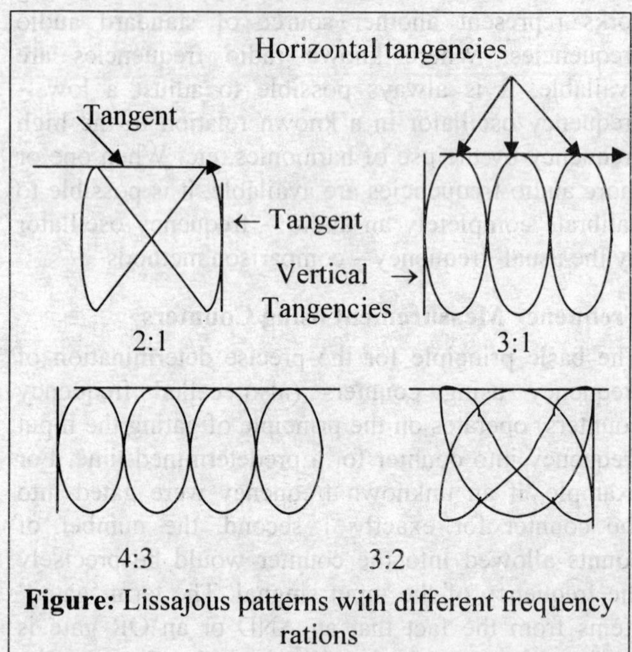

Figure: Lissajous patterns with different frequency rations

The above rule, however, does not hold for the Lissajous patterns with free ends. The simple rule mentioned above needs following modifications.

Two lines are drawn, one horizontal and the other vertical so that they doe not pass through any intersections of the different parts of the Lissajous curve. The number of intersections of the horizontal and the vertical line with the Lissajous curve are individually counted. The frequency ratio is given by

$$\frac{f_y}{f_x} = \frac{\text{number of intersections of the horizontal line with the curve}}{\text{number of intersections of the vertical line with the curve}}$$

The modified rule is applicable in all cases whether the Lissajous pattern in open or closed.

The ratio of frequencies when open ended Lissajous patterns are obtained can also be found by treating the open ends as half tangencies as shown in figure.

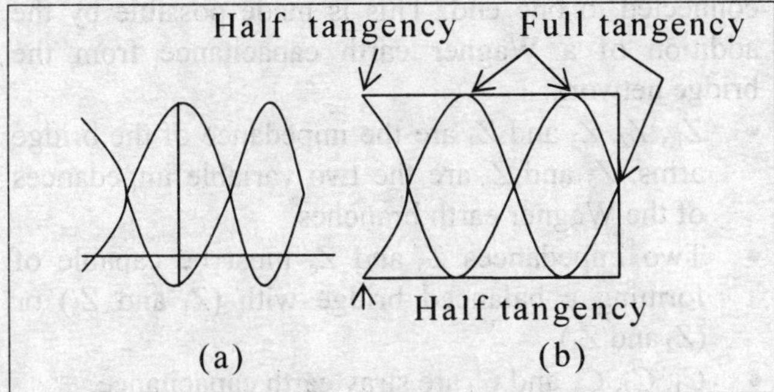

$$\frac{f_y}{f_x} = \frac{\text{number of horizontal tangencies}}{\text{number of vertical tangencies}}$$

$$= \frac{2 + [1/2]}{1}$$

$$= \frac{5}{2}$$

There are some restrictions on the frequencies which can be applied to the deflection plates. One obviously, is the CRO must have the bandwidth required for these frequencies. The others restriction is that the ratio of the two frequencies should not be such as to make the pattern complicated otherwise determination of frequency would become difficult. As a rule rations as high as 10 : 1 and as low as 10 : 9 can be determined comfortably.

7.18 CATHODE RAY OSCILLOSCOPE

- It is a very useful and versatile laboratory instrument used for display, measurement and analysis of waveforms.
- Normal form of a CRO uses a horizontal input voltage, which is an internally generated ramp voltage, called *'time base'*. This move the luminous spot periodically in a horizontal direction from left to right over screen.
- Vertical input to the CRO is the voltage under investigation. This moves luminous spot up and down with current value of the voltage.
- CRO operates on voltage. But it is possible to convert current, strain, acceleration, and pressure into voltage with the help of transducers.
- Used to investigate waveforms, transient phenomena, for very low frequency range to the radio frequencies.

Cathode Ray Tube (CRT)

CRT is heart of CRO.

Main part of CRT

(a) Electron gun assembly.
(b) Deflection plate assembly.
(c) Fluorescent screen.
(d) Glass envelope.

- *Electron gun* assembly produces a sharp focused beam of electron, which are accelerated to high velocity. This focused beam strikes the fluorescent screen with sufficient energy.
- *Deflection plate* assembly moves beam vertically up and down and the voltage applied to the other pair or plates move the beam horizontally from one side to another
- *Screens*: made up of phosphorescence material like phosphors.

Aquadag

The bombarding electrons stiking the screen, releases secondary emission electron. These are collected by an aqueous solution of Graphite called 'Aquadag'. The collection of secondary electron is necessary to keep the CRT in a state of electrical equilibrium.

- Short persistence phosphors require more frequent refreshes whereas long persistence phosphors may result in character fading slowly, giving a trace after glow or trails.
- Short persistence phosphors (P_1, P_2, P_{11}, P_{31},) long persistence phosphors (P_{19}, P_{26}, P_{33}).

7.19 SHIELDING AND GROUNDING

Shielding of Bridge elements

Capacitance between the bridge elements and between the elements and ground effect the accuracy of ac bridge measurement. An effecting way of controlling these capacitance consist of enclosing the element in a shield, thereby placing the capacitance at a position in circuit where it can do no harm.

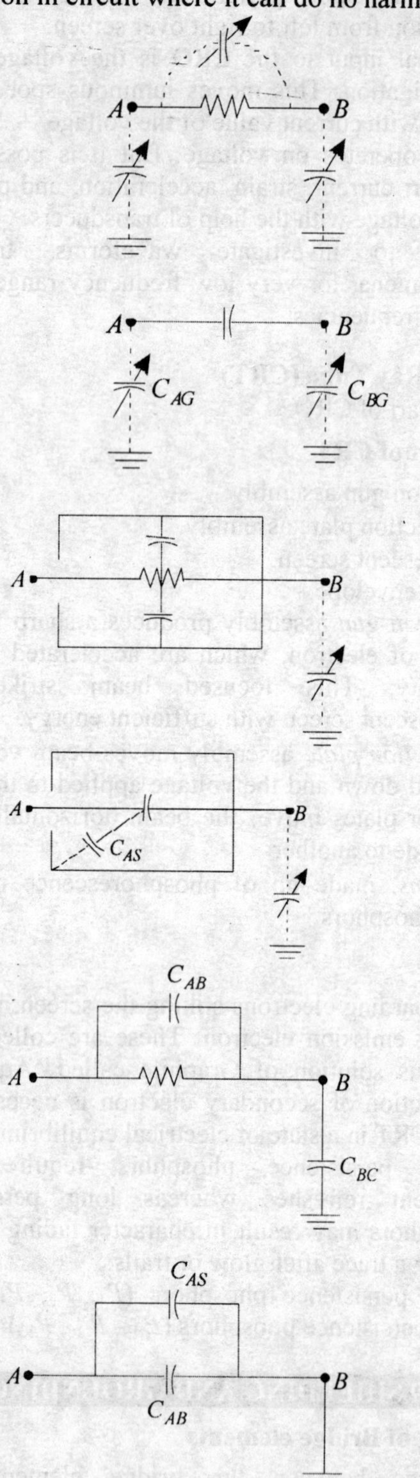

1. Shielding is effecting in preserving the true magnitude of the component of a bridge.
2. It is also important in keeping noise out of measuring system.
3. Both magnetic and electrostatic shielding are necessary to keep stray electro-magnetic and electro static field from generating unwanted emf in the bridge circuit.
4. Provide only one ground connection to ac bridge. Otherwise ground loop produces series noise result in accuracy.

Wagner Earthing Device

For having accuracy in measurement, each component in a bridge has a defining screen connected to one end. This is made possible by the addition of a Wagner earth capacitance from the bridge network.

- Z_1,, Z_2, Z_3 and Z_4 are the impedance of the bridge arms. Z_5 and Z_6 are the two variable impedances of the Wagner earth branches.
- Two impedances Z_5 and Z_6 must be capable of forming a balanced bridge with (Z_1 and Z_3) or (Z_2 and Z_4).
- C_1, C_2, C_3 and C_4 are stray earth capacitance.
- Putting such in *d* position and adjusting Z_2 and Z_4 until minimum sound is obtain. Then putting switch to *e* and adjusting Z_5 and Z_6 to give minimum sound. Then again connect switch to *d* to do same. The process is repeated until silence is obtained with the switch on d and one. Then all three point *b*, *d* and *e* are at earth potential.

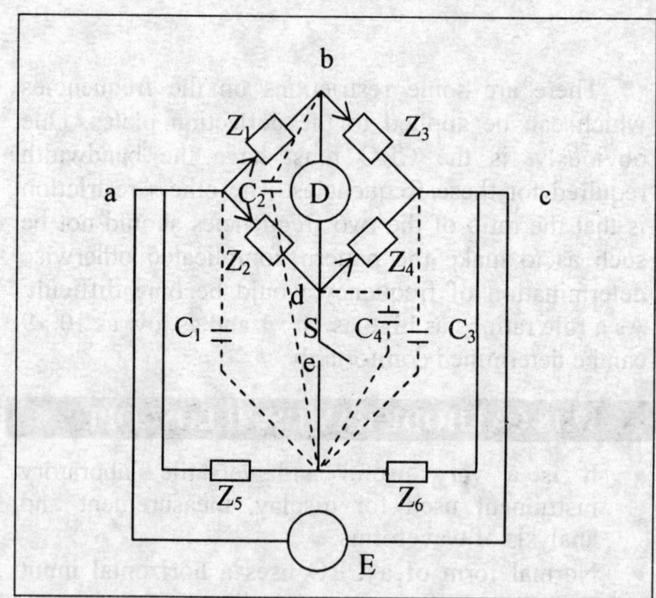

MCQ Set–1

1. The **threshold** of an instrument is normally defined
   (a) as the smallest measurable input change (non-zero value) which can be detected.
   (b) As the smallest measurable input which can be detected.
   (c) In terms of linearity of scale.
   (d) As a function of drift.
2. Disc in an **induction type energy-meter** rotates in the opposite direction when.
   (a) A large capacitor is connected on the load side
   (b) A large inductor is connected on the load side
   (c) Current coil and voltage coil are wrongly connected
   (d) Either current coil or voltage coil is wrongly connected
3. The **ratio error** of a **current transformer** depends mainly on.
   (a) Type of the ammeter used on the secondary
   (b) Eddy current loss of the core
   (c) Magnetising current of the core
   (d) Power factor of the load
4. The **deflection sensitivity** of a Cathode ray tube is
   (a) Directly proportional to the deflection voltage
   (b) Directly proportional to the square of the deflection voltage
   (c) Inversely proportional to the deflection voltage
   (d) Independent of the deflection voltage.
5. All **override control** is used
   (a) To improve stability
   (b) For better accuracy
   (c) To protect the equipment
   (d) To have flexibility in design
6. The emf developed by a thermocouple depends on.
   (a) The length of wires and temperature difference between the hot and cold junction
   (b) Materials used, diameter of wires used and temperature difference between the hot and cold junctions
   (c) Materials used, temperature of hot junction and temperature of cold junction
   (d) Materials used, shape and size of materials, resistance of the wires and temperature difference between the hot and cold junctions.
7. Below figure shows a balanced ac bridge excited by a voltage source of fixed frequency.

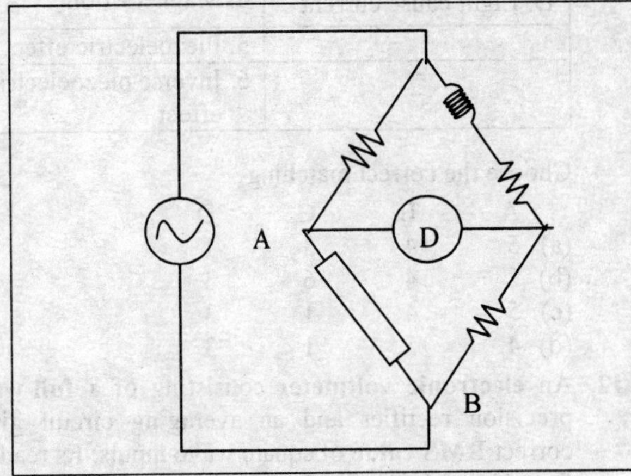

   (a) An R and L in series
   (b) An R and C in series
   (c) An R and C in parallel
   (d) An R and L in parallel
8. Care should be taken not to open-circuit
   (a) The secondary of a **voltage transformer** when the primary is energized at rated voltage
   (b) The secondary of a **current transformer** when the primary is energized at rated current
   (c) The primary of a **current transformer** when the secondary is connected to rated burden
   (d) The primary of a **voltage transformer** when the secondary is connected to rated burden.
9. A [3 (1 / 2)] digit **DVM** has a lowest measuring range of 200 mV. Hence it can be concluded that
   (a) Its best resolution is 0.1 mV
   (b) Its poorest resolution is 0.2 mV
   (c) Its accuracy is at least 0.05 %
   (d) The maximum voltage that can be measured in this range is 199.9 mV.
10. A **heterodyne type wave analyzer** with a band pass filter centered at 5 KHz is used to measure the 11th harmonic amplitude of a 50 Hz waveform. Which of the following local oscillator frequencies will provide the desired output?
    (a) 550 Hz (b) 4450 Hz
    (c) 5000 Hz (d) 5550 Hz
11. Match the following

Phenomenon	Name
A. Applied force causes emf	1. Seebeck effect
B. Applied voltage causes Vibrations	2. Hall effect
C. Temperature difference causes emf	3. Photoelectric effect

D. Light cause current	4. Magnetostion
	5. Piezoelectric effect
	6. Inverse piezoelectric effect

Choose the correct matching

	A	B	C	D
(a)	5	4	3	1
(b)	2	4	6	3
(c)	5	4	1	3
(d)	4	5	1	3

12. An **electronic voltmeter** consisting of a **full-wave** precision rectifier and an averaging circuit gives correct **RMS value** of square wave inputs. Its reading for a 2V peak- to – peak sinusoidal input will be

(a) $[2/\pi]$ V
(b) $[1/\pi]$ V
(c) $[2/\sqrt{(2\pi)}]$ V
(d) $[\sqrt{(2)}/\pi]$ V

13. An amplitude modulated signal can be observed on a CRO by applying the following waveform to the "external trigger" input:

(a) The modulated waveform itself
(b) derivative of the modulated waveform
(c) the modulating waveform
(d) the carrier waveform

14. In the a.c. bridge circuit shown in figure at balance the perameters of the coil are calculated to the 50 mH and 20Ω. Another coil is connected in parallel to the previous coil, and balance is restored by changing R_2 to 2.5 kΩ. The parameters of the new coil are than.

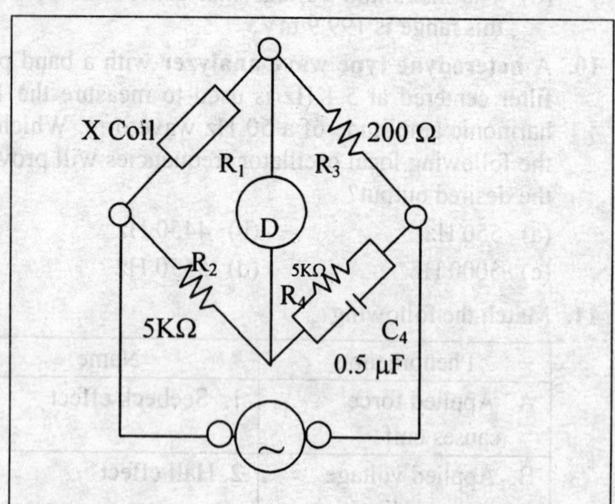

(a) 25 mH, 20 Ω
(b) 50 mH, 10Ω
(c) 50 mH, 20 Ω
(d) 100 mH, 40 Ω

15. In the potentiometric circuit shown in figure balance is obtained. The unknown E_x is then

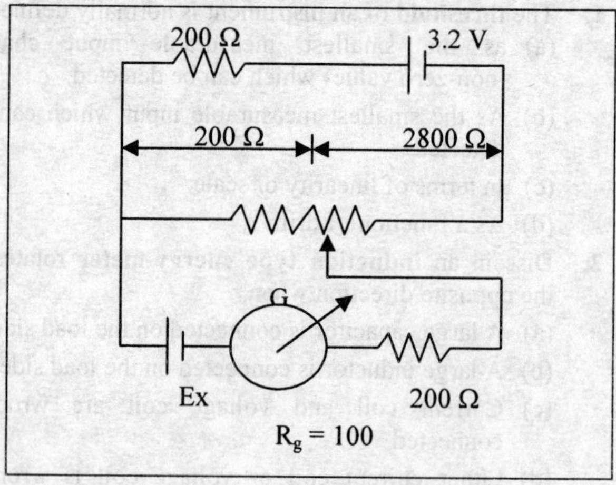

(a) 20 mV
(b) 2.8 V
(c) 3.0 V
(d) 400 mV

16. The equations under balance conditions for a bridge are:

$R_1 = [(R_2R_3)/R_4]$ and $L_1 = R_2 R_3 C_4$, where R_1 and L_1 respectively unknown resistance and inductance.

In order to achieve faster balance which of the following variables are chosen:

(a) R_2 and R_3
(b) R_2 and C_4
(c) R_4 and C_4
(d) R_3 and C_4

17. The relationship between T_S = rise time of the signal, T_O = rise time of the oscilloscope, is T_D (rise time of the signal observed on the oscilloscope) equal to

(a) $T_s + T_O$
(b) $[T_s^2 + T_O^2]^{1/2}$
(c) $T_s - T_O$
(d) $[T_s^2 - T_O^2]^{1/2}$

18. In the bridge circuit shown in Figure, when $[X_C / R] = 1$, the voltmeter reads.

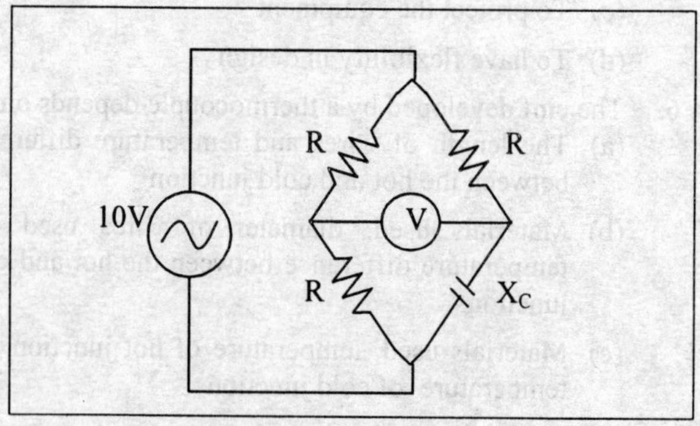

(a) 5 V
(b) 0.0 V
(c) 2.5 V
(d) 10 V

19. The degree of damping most desirable for an analog indicating instrument is:

(a) zero
(b) slightly less than critical
(c) critical
(d) slightly more than critical

20. The instrument that does not have any restoring torque is:
(a) D' Arsonval galvanometer
(b) Flux meter
(c) Ballistic galvanometer
(d) Moving iron instrument

21. A voltage drop across a non-linear element is given as $v = i + 2i^2$. The power loss across it for a current of sin t is
(a) 1.0 W (b) 2.0W
(c) 3.0 W (d) 0.5 W

22. Match the terms and their definitions the ability to display the same reading while measuring a quantity

List I	List II
A. Accuracy	1.smallest perceptible change in the output
B. Sensitivity	2. Error between the measured value and the absolute value.
C. Repeatability	3. Closeness of the measurement to the mean value
D. Resolution	4. Ratio of change in the instrument reading to the change in the measured quantity

Codes:

	A	B	C	D
(a)	1	2	3	4
(b)	3	5	2	1
(c)	3	4	5	1
(d)	3	2	4	5

23. The following type of error is regularly repetitive in nature.
(a) Random error (b) Environmental error
(c) Error of observation (d) Systematic error

24. The SI unit of length, the metre, was defined in 1960, in term of
(a) The length of Earth's meridian passing through Barcelona and Dunkirk
(b) The distance between two marks on a platinum-iridium bar kept in France
(c) The wavelength of light radiation from krypton - 86
(d) The frequency of a quartz crystal.

25. An ammeter with input resistance of 50Ω gives full scale deflection for 50 μA current. The input resistance of a 0-1 mA ammeter obtained by connecting a shunt across the 0-50 μA meter will be
(a) 50 Ω (b) 25 Ω
(c) 5 Ω (d) 2.5 Ω

26. Match the List I parameter with the List II type of bridge used for measurement and select the correct answer using the codes:6

List I	List II
A. Low value of R	1. Schering bridge
B. High value of Q	2. Maxwell-Wien bridge
C. Inductance L	3. Kelvin double bridge
D. Capacitance C	4. Hay bridge

Codes:

	A	B	C	D
(a)	2	4	3	1
(b)	4	1	3	2
(c)	3	4	1	2
(d)	3	4	2	1

27. Match the List I term with List its description and select the correct answer using the codes:

List I	List II
A. Tolerance	1. Difference between the actual and nominal values
B. Deviation	2. Difference between the high and low values
	3. Half the difference between the high and low values.

Codes:

	A	B
(a)	3	1
(b)	1	2
(c)	1	3
(d)	2	1

28. For the circuit given in figure the unbalance voltage e_0 is:

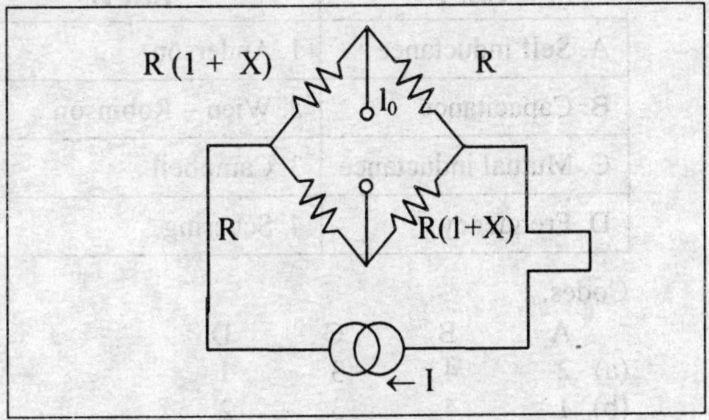

(a) e_0 = IRX (b) e_0 = [IRX / 2]
(c) e_0 = 2 IRX (d) e_0 = IR (1 + X)

29. The extension wires of iron constantan thermocouple were improperly wired shown in figure
The voltmeter calibrated in °C will then read.

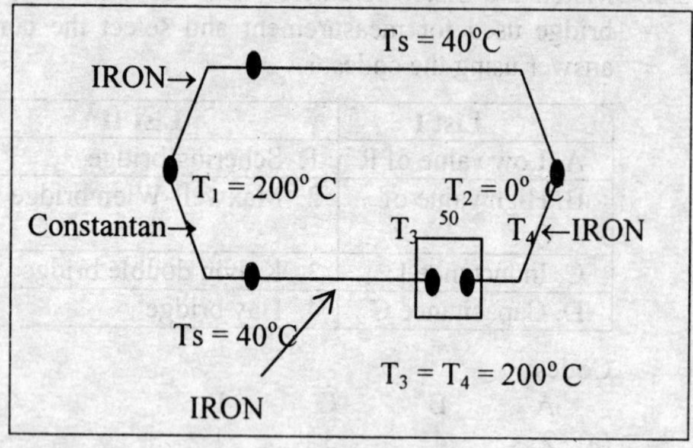

(a) 240 (b) 180
(c) 14C (c) 120

30. The a.c bridge in figure remains balanced if Z comprises of:

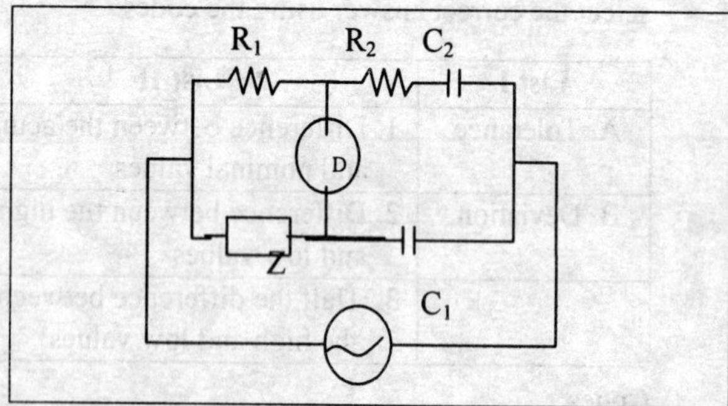

(a) Resistance only
(b) Capacitance only
(c) Resistance and capacitance in parallel
(d) Resistance and inductor in series.

31. Match the measured variables with the bridges and select the correct answer using the codes:

List I	List II
A. Self inductance	1. Anderson
B. Capacitance	2. Wien – Robinson
C. Mutual inductance	3. Campbell
D. Frequency	4. Schering

Codes:

	A	B	C	D
(a)	2	4	3	1
(b)	1	4	3	2
(c)	3	1	4	2
(d)	2	4	1	3

32. Ann oscilloscope in put impedance consists of 1 MΩ in parallel with 100pF. A compensated 20:1 attenuator is obtained by connecting a parallel combination of:
(a) 19 MΩ and [100 / 19] pF
(b) 20 MΩ and [00 / 20] pF
(c) 19 MΩ and 1900 pF
(d) 20 MΩ and 2000 pF.

33. The bridge circuit in figure is balanced. The magnitude of current I is:

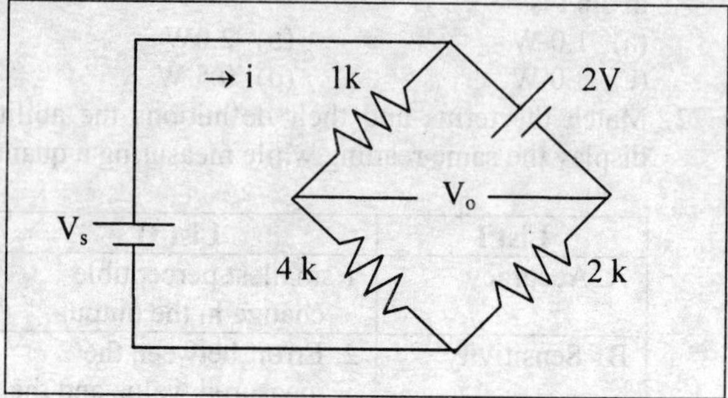

(a) 2 mA (b) 4 mA
(c) 5 mA (d) 6 mA

34. The term 'precision' used in instrumentation means.
(a) Gradual departure of the measured value from the calibrated value.
(b) Smallest increment in the measurand that can be detected by the instrument.
(c) Maximum distance or angle through which any part of a mechanical system may be moved in one direction without causing motion of the next part.
(d) The ability of the instrument to give output readings close to each other, when the input is constant.

35. A resistance potentiometer has a total resistance of 10000Ω and is rated 4 W. If the range of potentiometer is 0 to 100 mm, then its sensitivity in V/mm is.
(a) 1.0 (b) 2.0
(c) 2.5 (d) 25

36. The temperature of fixed points used to define international Temperature Scale are determined by using.
(a) Platinum resistance thermometer
(b) Platinum Platinum-Rhodium thermo – couples.
(c) Vapour – pressure thermometer
(d) Gas thermcmeter to which corrections are applied for non ideal behaviour of the gas.

37. The emf-temperature data for a thermocouple with reference junction at 0°C is as follows,

Temperature (°C)	20	180	200
Emf (m V)	1.2	11.8	13.5

The emf developed in mV with the two junctions at 200° C is

(a) 11.8 (b) 12.3
(c) 13.0 (d) 13.5

38. Attenuation of a narrow monochromic X-ray beam in a metal plate of thickness 'd' is given by the equation.

(a) $I = I_0 \exp(-\mu d)$
(b) $I = I_0 \exp(-\mu d^2)$
(c) $I = I_0 \exp(-\mu /d)$
(b) $I = I_0 \exp(-\mu/d^2)$

39. The elliptical Lissajous pattern shown in figure was observed from two sinusoidal signal sources of same frequency on a CRO

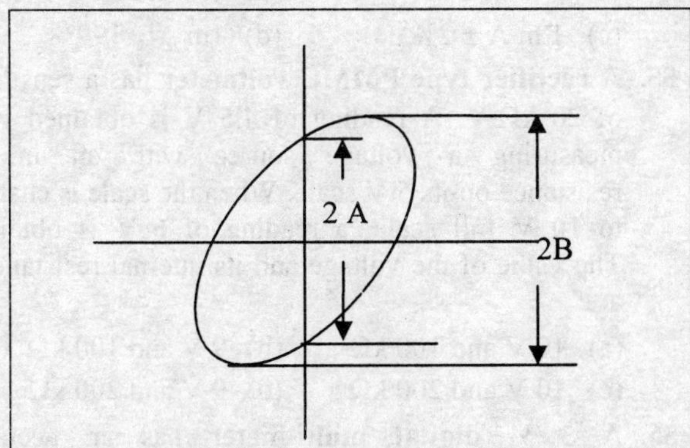

The phase angle between the signals is.

(a) $\cos^{-1}\left(\frac{A}{B}\right)$ (b) $\tan^{-1}\left(\frac{A}{B}\right)$

(c) $\sin^{-1}\left(\frac{A}{B}\right)$ (d) $\left(\frac{A}{B}\right)$

40. The scale of a full wave rectifier type average sensing meter is calibrated to read the r.m.s. for ideal sine wave signal (form factor 1.11). Determine the multiplying factor of the meter if it is used to measure the r.m.s. value of square waves

(a) 1.11
(b) 1/1.11
(c) 1.0
(d) Depends on the frequency of the square wave.

41. In the dc millivoltmeter circuit shown in figure the input for full scale deflection is:

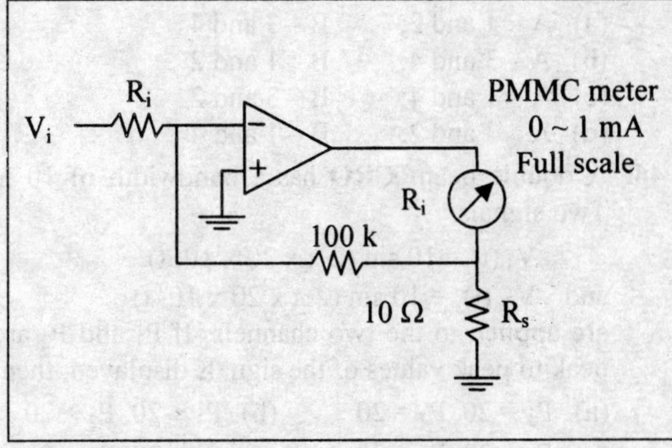

(a) 10 V (b) 1 V
(c) 100 mV (d) 10 mV.

42. Which one of the following conditions must be satisfied for the two wattmeter method of three phase power measurement?

(a) Load must be balanced
(b) Source must be balanced
(c) System should be 3-phase 3-wire type

43. Threshold of a measurement system is

(a) The smallest change in input which can be detected
(b) A measure of linearity of the system.
(c) The smallest input which can be detected
(d) A measure of precision of the system.

44. In the context of low resistances, identify the INCORRECT statement.

(a) Standard low resistances have four terminal construction.
(b) In Kelvin bridge for measuring low resistance, precision depends on the detector sensitivity.
(c) A Pair of ratio arms in kevin bridge for measuring the low resistance eliminates the error due to thermo emf.
(d) Low resistances used for ammeter shunts are usually made with a suitable number of plates of large area and all the plates connected in parallel.

45. In the context of a PMMC instrument, identify the correct matches.

A. A pair of springs	1. To provide controlling Torque
B. Aluminium former	2. To provide damping Torque
	3. To act as a base for the coil
	4. To provide current into and a out of the moving coil

(a) A – 1 and 2; B – 3 and 4
(b) A – 3 and 4; B – 1 and 2
(c) A – 1 and 4; B – 3 and 2
(d) A – 3 and 2; B – 1 and 4

46. A double beam **CRO** has a bandwidth of 10 MHz. Two signals.

$$V_1(t) = 10 \sin (2 \pi \times 20 \times 10^3 t)$$

and $V_2(t) = 10 \sin (2 \pi \times 20 \times 10^6 t)$

are applied to the two channels. If P_1 and P_2 are the peak to peak values of the signals displayed, then.

(a) $P_1 = 20, P_2 = 20$ (b) $P_1 < 20, P_2 > 20$
(c) $P_1 = 20, P_2 < 20$ (d) $P_1 > 20, P_2 < 20$

47. **Kelvin double bridge** is best suited for the measurement of:

(a) Resistances of very low value
(b) Low value capacitance
(c) Resistance of very high value
(d) High value capacitance

48. A **counter timer** has a basic clock of 16 MHz. The count value displayed is in error by 1 count. The frequency at which the error in the displayed value is the same whether the counter timer is used in the frequency mode of operation or period mode of operation is:

(a) 15 MHz (b) 10 MHz
(c) 8 MHz (d) 4 MHz.

49. **Wien bridge** is best suited for the measurement of

(a) Frequency (b) Capacitor
(c) Inductor (d) Resistor

50. A common practice of reducing **hysterisis error** in the output for a given value of input is to maintain.

(a) High rate of change of input
(b) Low rate of change of input
(c) Taking observations either in the ascending or in the descending order.
(d) Taking observations both in the ascending and descending order and take average value of the output.

51. A **thermocouple** is suddenly immersed in a medium of high temperature. The approximate time taken by the thermocouple to reach 98% of the steady – state value is:

(a) Equal to the time constant of the thermo couple
(b) Equal to twice the value of the time constant of the thermocouple
(c) Equal to four times the value of the time constant of the thermocouple.
(d) Independent of the time constant

52. The sensing element of a **thermocouple** at its hot junction is provided with a shielde while taking measurements in a high temperature gas. The principal reason for providing the shield is.

(a) To reduce conduction and convection errors.
(b) To reduce radiation error
(c) To provide temperature compensation to the see beck voltage.
(d) To improve air supply to the sensing element for better response.

53. A 3 ½ **digit voltmeter** has an internal reference of 200 mV. Its full scale count and resolution are:

(a) 2000 and 1 mV (b) 4000 and 0.1 mV
(c) 2000 and 0.1 mV (d) 4000 and 1 m V

54. Three DC current I_1, I_2 and I_3 meet at a node with I_1 entering and I_2 and I_3 leaving the node. I_1 and I_2 are measured as 100 m A and 99 mA with a ± 1% accuracy. Then the value of I_3 and the accuracy of I_3 are:

(a) 1 m A ± 2% (b) 199 m A ± 2%
(c) 1 m A ± 2% (d) 1m A ± 199%

55. A **rectifier type PMMC voltmeter** has a sensitivity of 20 kΩ/V. A reading of 4.5 V is obtained when measuring a voltage source with an internal resistance on its 5 V scale. When the scale is changed to 10 V full scale, a reading of 6 V is obtained. The value of the voltage and its internal resistance R are:

(a) 10 V and 100 kΩ (b) 9 V and 100 kΩ
(c) 10 V and 200 kΩ (d) 9 V and 200 kΩ

56. A 3 ½ **digital multi-meter** has an accuracy specification of 0.5% of reading plus 5 counts. The value of an unknown resistance is read as 50 .0Ω the 200 Ω scale of the meter. The value of the resistance is:

(a) 50.0 ± 0.25 Ω (b) 50.0 ± 0.5 Ω
(c) 50.0 ± 0.75 Ω (d) 50.0 ± 1.0 Ω

57. A 5 A, 110 V **electrodynamics** type **wattmeter** has a scale having 110 divisions. Its pressure coil is fed a voltage of [110 √2 cos (314 t) + √2 sin (942t)] V and its current coil carries a current of [5 √2 cos (314 t + 60) + 2 √2 sin (628t + 90) + √2cos (942t + 90)] A. The needle of the watt meter will move to

(a) 110 divisions (b) 50 divisions
(c) 54 divisions (d) 55 divisions

58. The loop transfer function of a system is given by $G(s)H(s) = \frac{10\,e^{-Ls}}{s}$ the phase cross – over frequency is 5 rad/s. The value of the dead time L is

(a) $\pi / 20$ (b) $\pi / 10$
(c) $-\pi / 20$ (d) Zero

59. A large number of 230 Ω resistors are obtained by combining 120 Ω resistors with a standard deviation of 4.0 Ω and 110 Ω resistors with a standard deviation of 3.0 Ω .The **standard deviation** of the 230 Ω resistors thus formed will be:
(a) 3.5 Ω (b) 5.0 Ω
(c) 7.0 Ω (d) 12.0 Ω

60. A 3 ½ digit multimeter has an accuracy specification of (± 0.5 % of reading ± 5 counts) If the meter reads 2.00 mA, on a full scale of 20 mA, the worst case error in the reading is:
(a) 0.5 % (b) 2.5%
(c) 3.0% (d) 5.5%

61. Which one of the following is a derived unit (Not a basic unit) in SI system?
(a) Candela (b) Coulomb
(c) Kelvin (d) mol

62. A shaft encoder attached to a d c motor has sensitivity of 500 pulses per revolution. A frequency meter connected to the output of the encoder indicates the frequency to be 5500 Hz. The speed of the motor in RPM is.
(a) 110 (b) 220
(c) 550 (d) 660

63. The current coil of a 200 V, 5A, **electrodynamometer** type LPF **wattmeter carries** a current of $\sqrt{2}\cos(100\pi t)$ A. The voltage across the pressure coil is $\sqrt{2}\, 200 \sin(100\pi t)$ V. The meter will indicate
(a) 0 W (b) 100 W
(c) 200 W (d) 400 W

64. The clock frequency of a **timer counter** is 10 MHz. The timer counter is used in the period mode and the input to the timer counter is a square wave of frequency 2 kHz. The display of the timer counter will show a value.
(a) 200 (b) 2000
(c) 5000 (d) 50000

65. The insulation level for a 5 A full scale **rectifier type ammeter** is specified as 2.5 k V. It then safe to use the meter with a potential difference of upto 2.5 kV between:
(a) The terminals of the meter
(b) Case and the ground
(c) Both the terminals and the case
(d) One of the terminals and the ground

66. An **oscilloscope** screen displays a line inclined at 45°. Its Y-input is a sine wave of frequency f then the X-input should be:
(a) Sine wave of frequency f and 0° phase shift with the Y-input.
(b) Sine wave of frequency f and 45° phase shift with the Y-input
(c) Sine wave of frequency f and 90° phase shift with the Y-input.
(d) Saw tooth wave of frequency f

67. An RTD is assumed to be represented in the range 0 °C to 100 °C by a linear model as $R_T = R_0(1 + 0.004\,T)$:
Where T is the temperature in °C. R_0 = 100 Ω with a variation of ± 2 Ω. The true model of the RTD is.
$R_T = R_0(1 + 0.004T + 6 \times 10^{-7}\,T^2)$.
The worst case error magnitude in the resistance value that will be introduced, if the linear model is used over a range of 0 to 100 °C, is:
(a) 2 Ω
(b) 2.8 Ω
(c) 3.0 Ω
(d) 3.4 Ω

68. The thermocouple pair that gives the maximum sensitivity around 273 K is:
(a) Platinum – Constantan
(b) Nichrome – Constantan
(c) Nickel – Constantan
(d) Copper - Nickel

69. Two copper-constantan thermocouples are connected such that the two constantan wires are joined together. The two copper wires are connected to the input of a low noise chopper stabilized differential amplifier having a gain of 1000. One of the thermo-couple junctions is immersed in a flask containing ice and water in equal proportion. The other thermo-couple is at a temperature T. If the output of the amplifier is 2.050 V, the temperature T is:
(a) 205°C (b) 102.5 °C
(c) 51.25 °C (d) 50 °C

70. For the **bridge circuit** in figure the voltages are:

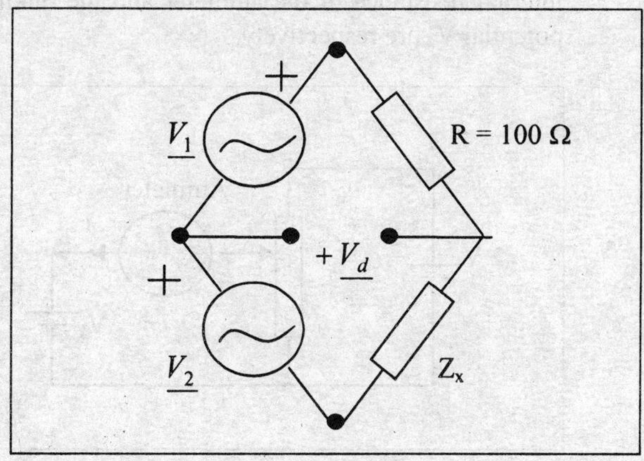

$\underline{V}_1 = \sqrt{2} \cos 1000\, t\, V,$

$\underline{V}_2 = 2 \cos \left(1000\ t + 45^{o} V\right)$, and

$V_d = 0$

The value of R= 100Ω Then Z_x will be

(a) 100 Ω resistor in series with 100 mH inductor

(b) 100Ω resistor in series with 100 μF inductor

(c) 100Ω resistance in parallel with 100 mH inductor

(d) 100Ω resistance in parallel with 100μF capacitor.

71. An oscilloscope is powered from 230 V mains having a nominal frequency of nearly 50 Hz. The triggering mode of the oscilloscope is set to LINE. A square wave from a function generator is fed to the Y input and a stable display (properly triggered) is obtained when the frequency of the function generator is 297 Hz. The actual frequency of the power supply is:

(a) 50 Hz (b) 49.5 Hz

(c) 49.0 Hz (d) 48.5 Hz

72. The primary of a 200/1 A, 5 VA, 0.8 PF current transformer (CT) carriers 100 A. The secondary current is 0.495 mA. The ratio error of the CT is

(a) – 1.0% (b) – 0.5%

(c) + 0.5% (d) + 1.0%

73. A very low loss coil is tested with a **Q-meter** and the distributed (self) capacitance of the coil is found to be 820 pF. Resonance occurred at an angular frequency of 10^6 rad/s with a capacitance of 9.18nF. The inductance of the coil is:

(a) 100 pH (b) 100 μH

(c) 100 nH (d) 100 mH

74. A **potentiometer** circuit is shown in figure. The jockey of the potentiometer is kept exactly at the mid point and the ammeter reads the current, I, as 0.2 mA When V_x is reversed the ammeter reads 3.8 mA. The internal resistance of the ammeter and the unknown potential V_x are respectively.

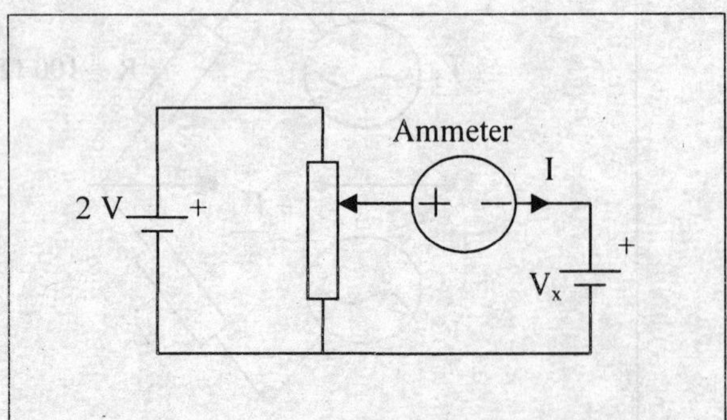

(a) 500Ω, 1.1 V (b) 50Ω, 1.1 V

(c) 500Ω, 0.9 V (d) 50Ω, 0.9 V

Answers Set-1

1. Ans. (a)

Solution: The threshold of an instrument is normally defined (a) As the smallest measurable input charge (non zero value) which can be detectable.

If the instrument input is increased very gradually form zero there will be some minimum value below which no output change can be detected. This minimum value defines the threshold of the instrument.

2. Ans. (b)

Solution:

3. Ans. (b)

Solution: Ratio error of current transformer

$$= \frac{\text{norminal ratio-actual ratio}}{\text{actual ratio}} = \frac{R_n - R_a}{R_a}$$

$$K = n + \frac{I_m \sin\delta + I_c \cos\delta}{I_s}$$

So ratio error of current transformer depends upon the I_m magnetize current and I_c eddy current

4. Ans. (d)

Solution: Deflection of CRT screen is

$$D = \frac{L\,\ell_d\, E_d}{2d\, E_a}$$

Deflection sensitivity of CRT is defined as deflection of screen per unit deflection voltage

$$S = \frac{D}{E_d} = \frac{L\ell_d E_d}{2d\, E_a \times E_d} = \frac{L \cdot \ell_d}{2d\, E_a}$$

5. Ans. (c).To protect the equipment

Solution: One example of over ride control system is the protection of boiler system usually the steam pressure in a boiler is controller through the use of a pressure control loop on the discharge line (loop 1). At the same time the water level in the boiler should not fall below a lower limit which is necessary to keep the heating coil immersed in water and thus prevent its burning out. Here over ride control system is using a low switch selector (LSS). Whenever the liquid level falls below allow limit, LSS switches the control action from pressure control to level control (loop 2) and closes the value on discharge line.

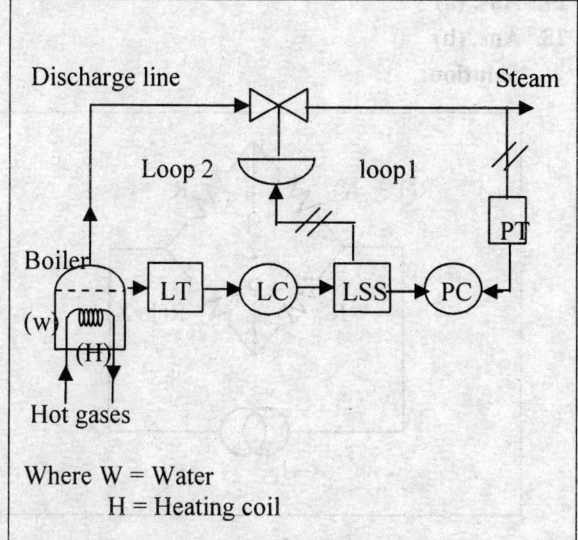

6. **Ans. (c)**

Solution: The emf developed by a thermo couple depends on Material used, temperature of hot junction and cold junctions.

The emf produced in thermocouple is

$$E = a(\Delta\theta) + b(\Delta\theta)^2$$

Where $\Delta\theta$ = difference in temperature between the hot thermocouple junction and reference junction

a, b = Constants

a is generally very large that b

$$a \approx 40\,\mu V/^{o}C \text{ to } 50\,\mu V/^{o}C$$

$$b \approx 0.1\,\mu V/^{o}C^2 \text{ to } 0.01\,\mu V/^{o}C^2$$

Generally metals used are pair of copper, iron, constantan, Chrome, Platinum Rhodium tungsten, iridium etc.

The point where temperature is to be measured is called HOT JUNCTION and other end is called COLD JUNCTION (Which has to be kept ideally at 0°C) but cold junction compensation is required in case cold junction not kept at 0°C.

7. **Ans. (d)**

Solution: As per the rule of bridge to obtain the balance if one arm contains a series combination then its opposite arm has to be a parallel combination of the same components in other arm.

8. **Ans. (b)**

Solution: Adding Integral Mode

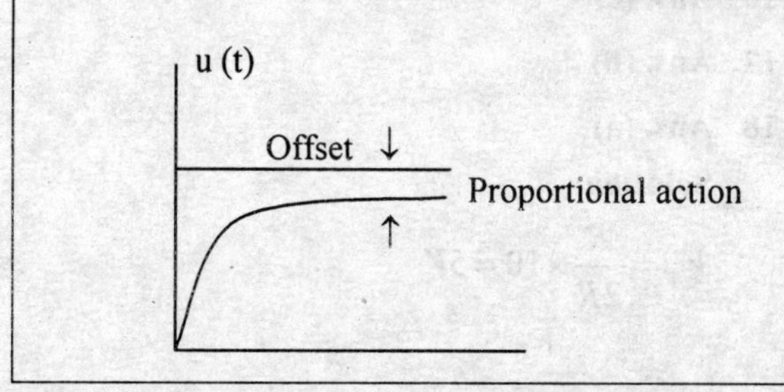

- This offset can be removed by using integral action.
- Integral action also improves steady state accuracy of the system.
- Response becomes slower with adding Integral
- Response can become faster by adding gain but oscillations increase and it becomes unstable

9. **Ans. (b)**

10. **Ans. (d)**

Solution:

Local oscillator frequency is always higher than the in coming signal frequency.

$$11 \times 50\ Hz + 5\ k\ Hz = 5550\ Hz$$

11. **Ans. (c)**

12. **Ans. (a)**

Solution:

Here V_m = 1 V or V_{pp} = 2 V

Voltmeter reading = rms value of square wave

= average value of sine wave

$$= \frac{2V_m}{\pi} = \frac{2}{\pi}$$

13. **Ans. (a)**

Solution:

14. **Ans. (c)**

Solution:

15. **Ans. (a)**

Solution:

In balanced condition current through G (Galvanometer) is zero, therefore voltage across 200 Ω resistance.

$$V = \frac{200}{2800 + 200 + 200} \times 3.2$$

$$= \frac{200 \times 3.2}{3200} = 200\text{ mV}$$

16. Ans. (c)

17. Ans. (b)

18. Ans. (a)

Solution:

$$V_A = \frac{R}{2R} \times 10 = 5V$$

$$V_B = \frac{\frac{1}{j\omega C}}{R + \frac{1}{j\omega C}} \times 10 = \frac{1}{j\omega RC + 1} \times 10 = \frac{10}{1 + j}$$

Since $\frac{X_c}{R} = 1 \Rightarrow \frac{1}{j\omega RC} = 1$

$$V = V_A - V_B = 5 - \frac{10}{1+j} = \frac{5 + 5j - 10}{1+j}$$

$$= \frac{5(j-1)}{1+j} = 5 \text{ V}$$

19. Ans. (b)

20. Ans. (d)

21. Ans. (d)

22. Ans. (c)

23. Ans. (a)

24. Ans. (c)

25. Ans. (d)

Solution:

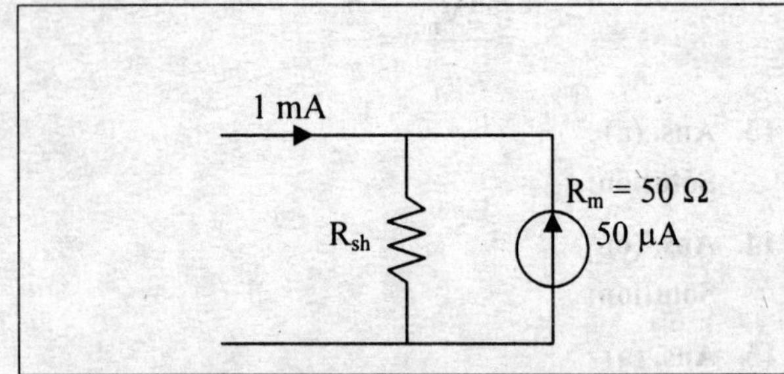

Voltage across meter

$$V = 50 \times 50 \times 10^{-6} = R_{sh}(1000 - 50) \times 10^{-6}$$

$$2.5 \times mV = R_{sh}(950 \times 10^{-6})$$

$$R_{sh} = \frac{50 \times 50}{950} = \frac{250}{950}$$

$$= 2.63\ \Omega$$

26. Ans. (d)

27. Ans. (a)

28. Ans. (b)

Solution:

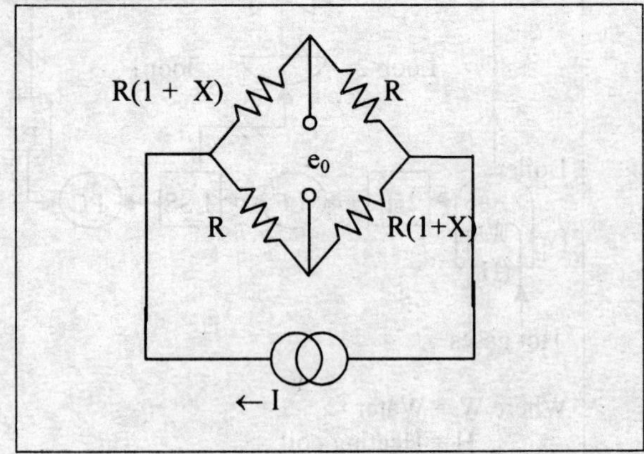

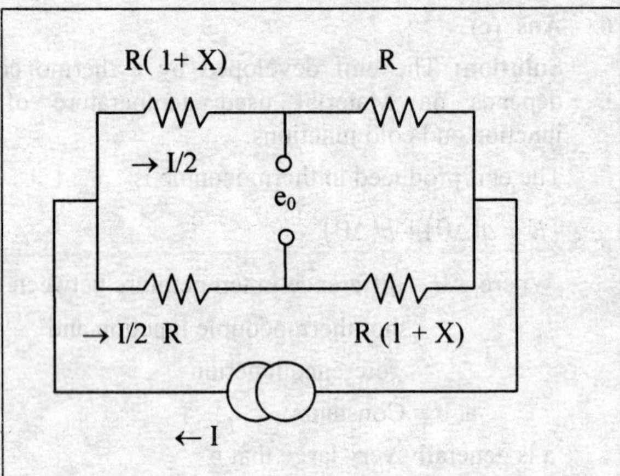

I divides equally as resistance in two parallel branches are same

$$e_0 = \frac{I}{2} \times R(1 + X) - \frac{I}{2} R$$

$$e_0 = \frac{I}{2}[R - RX - R]$$

29. Ans. (d)

Solution: T = 200 – 40 – 40 = 120° C

30. Ans. (c)

31. Ans. (b)

32. Ans. (a)

Solution: $\frac{R_i}{R_i + R_p} = \frac{1}{20}$

$$R_1 + R_2 = 20$$

$R_p = 20 - 1 = 19\, M\Omega$

$$\frac{X_{ci}}{X_{cp}} = \frac{R_i}{R_p}$$

$$\frac{C_p}{C_i} = \frac{R_i}{R_p}$$

$$C_p = 100 \times \frac{1}{19} = \frac{100}{19}\, pF$$

33. **Ans. (d)**
Solution:

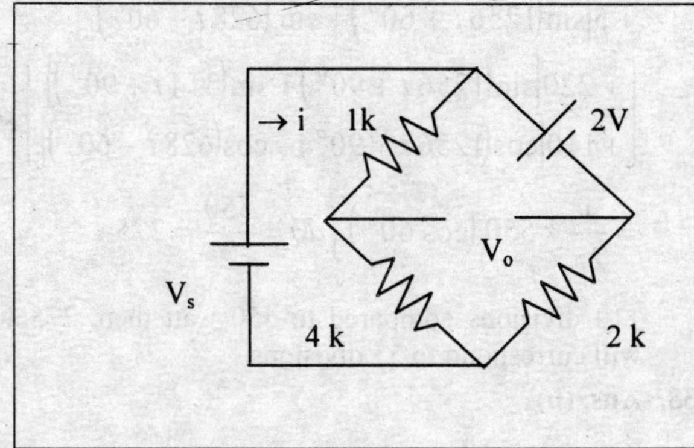

In balanced condition

$V_{AB} = V_{AB} = 2\, r = I \times 1\, K$

$$I = \frac{2}{1\, K} = 2\, mA$$

$V_{AD} - V_{DC} = 2\, V + 2 \times 2\, K = 2 + 4 = 6\, V$

34. **Ans. (d)**

35. **Ans. (b)**

36. **Ans. (a)**
Solution: Platinum resistance thermometer is used between – 259.34 °C and 630/.74 °C
A. Is used between 630.74 °C and 1064.73 °C

37. **Ans. (c)**

38. **Ans. (a)**
Solution: $E_y = E_0 e^{-\alpha x}$
$E_y = E_0 e^{-\alpha x}\, e^{-1\ \beta x}$
(is terns of attenuations)

39. **Ans. (c)**

40 **Ans. (b)**

41. **Ans. (d)**
Solution: $A_d = 1$
Full scale deflection = 1 mA
$V_0 = 1\, mA \times 10 = 10\, mV$
$V_2 = V_0 = 10\, mV$

42. **Ans. (c)**

43. **Ans. (a)**

44. **Ans. (c)**

45. **Ans. (c)**

46. **Ans. (d)**

47. **Ans. (a)**

48. **Ans. (a)**

49. **Ans. (a)**

50. **Ans. (b)**

51. **Ans. (c)**
Solution: The output at time t

$$c(t) = c_0\left(1 - e^{-t/T}\right)$$

For, $c(t) = 0.98\, C_0$

We have $0.98\, C_0 = C_0\left(1 - e^{-t/T}\right)$

Or $t \approx 4T$

52. **Ans. (b)**
Solution:
The sensitive element is required to measure its own temperature change due to radiant energy received relative to the temperature of the immediate surroundings

53. **Ans. (a)**
Solution:

$$\text{Resolution} = \frac{1}{10^n} = \frac{1}{10^3} = 0.001\, V = 1\, mV$$

There are three full digits in this DVM, and thus meter will be able to cont from zero to 999 the fourth digit placed to the left of active digits indicates either 0 or 1.Thus count goes from above from 999 to 1999, thereby making total 2000 counts

54. **Ans. (c)**
Solution: $-I_1 + I_2 + I_3 = 0$ (Using KCL at the node)
$I_3 = I_1 - I_2 = 100\, mA - 99\, mA = 1\, mA$

$$\frac{dI_3}{I_3} = \pm\frac{dI_1}{I_1} \pm \frac{dI_2}{I_2}$$

$= \pm 1 \pm 1 = \pm\, 2\,\%$

55. Ans. (b)

Solution:

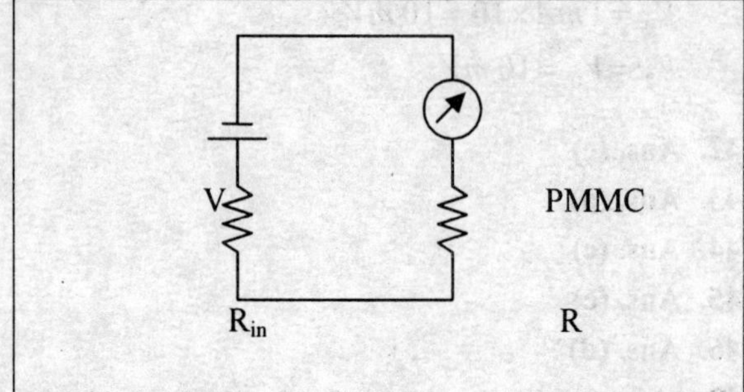

Let R_{in} be the internal resistance of the unknown voltage source V.

On 5 V scale

$$R_n = \text{Sensitivity} \times \text{Scale range} = (20)5 = 100\ k\Omega$$

then Voltage across meter $= \dfrac{V_{in} 100\ k}{R_{in} + 100\ k} = 4.5$ (1)

On 10 V scale,

$$R_m = \text{Sensitivity} \times \text{Scale range} = (20\ k\Omega) = 10\ V$$

$$= 200\ k\Omega$$

then voltage across the meter $= \dfrac{V_{in}.200\ k}{R_{in} + 200\ k} = 6$ (2)

From equations (1) and (2) we get

$V = 9\ V$ and $R_{in} = 100\ k\Omega$

56. Ans. (c)

Solution:

Error in reading 0.5 % of 50 Ω

Then, reading error $= \dfrac{0.5}{100} \times 50 = 0.25\ \Omega$

5 count will contribute error of $5 \times 0.1\ \Omega = 0.5\ \Omega$

Thus, total error = 0.75 Ω

and the value of resistance = 50 ± 0.75 Ω

57. Ans. (d)

Solution:

Watt meter reading, $P = \dfrac{1}{2\pi} \int_0^{2\Pi} vi d\theta$

or $P = \dfrac{1}{2\pi} \int_0^{2\pi} \left[110\sqrt{2}\cos(314t) + \sqrt{2}\cdot\sin(942t)\right]$

$$\begin{bmatrix} 5\sqrt{2}\cos(314t + 60^o) + 2\sqrt{2}\sin(628t + 90^o) \\ + \sqrt{2}\cos(942t + 90^o). \end{bmatrix} d\theta$$

$$p = \frac{1}{2\pi} \int_0^{2\pi} 500\left[\left(\cos 628t + 60^o\right) + \cos 60^o\right]$$

$$+ 220\left[\sin(942t) + \sin(314t + 90^o).\right]$$

$$+ 110\left[\cos(1156t + 90^o) + \cos(628t + 90^o)\right]$$

$$+ 5\left[\sin(1256t + 60^o) + \sin(628t - 60^o).\right]$$

$$\begin{bmatrix} + 220\left[\sin(1256t + 90^o) + \sin(314t + 90^o)\right] \\ + 110\left[\cos(1256t + 90^o) + \cos(628t - 60^o)\right] \end{bmatrix} .d\theta$$

$$= \frac{1}{2\pi} \times 550\left(\cos 60^o\right) \int_0^{2\pi} d\theta = \frac{550}{2} = 275$$

110 divisions compared to 550 watt then, 275 watt will correspond to 55 divisions

58. Ans. (b)

Solution: $e^{jx} = \cos x + j\sin x$

$$\phi = \tan^{-1}\frac{\sin x}{\cos x} = x$$

$$\angle G(s)H(s)\big|_{s=j\omega} = -90^o - \frac{(\omega L)180^o}{\pi}$$

At phase cross-over frequency, $\angle GH = -180^o$

Then, $-90^o - \dfrac{(\omega L)180^o}{\pi} = -180^o$

At $\omega = 5$

Or $90^o = \dfrac{5(L)\cdot 180^o}{\pi}$

$\pi = 10\ L$

Or $L = \dfrac{\pi}{10}$

59. Ans. (b)

Solution: When the number of observation is less than 20, then the standard deviation of a finite number of of data having (Standard deviation) as d_1, d_2, d_3, … d_n is expressed as

$$SD = \sqrt{\frac{d_1^2 + d_2^2 + \ldots + d_n^2}{n-1}}$$

Here, n = 2, and d_1 = 4 Ω, d_2 = 3 Ω

$$\therefore \quad SD = \sqrt{\frac{4^2+3^2}{1}} = 5\,\Omega$$

60. Ans. (c)

Solution:

0.5 percent of reading = 0.005 × 20 = 0.1 mA

Resolution

$$= \frac{1}{10^n} = \frac{1}{10^3} = 10^{-3} = 0.001 = 0.1\%$$

Then worst-case error $= 0.001 + 0.101 \times 20$

$$= 0.03\, mA = 3\%$$

61. Ans. (b)

Solution: The fundamental units is SI system are

Unit	Name
Length	Meter
Mass	kg
Time	Second
Intensity of electric current	Ampere
Temperature	Kelvin
Luminous intensity	Candela
Amount of substance	Mole

62. Ans. (d)

Solution:

Pulses/min = 5500 × 60

Then, speed in RPM

$$= \frac{5500 \times 60}{500}\left(\frac{pulse}{\min}\right)\left(\frac{revolutions}{pulse}\right)$$

= 660 revolution/minutes

63. Ans. (a)

Solution: Average deflection torque

$$T_d = \frac{1}{T}\int_0^T T_i \; d(\omega t)$$

Where, T_i (instantaneous torque)

$$= i_p i_v \frac{dM}{d\theta} = \left(\frac{V_p}{R_p}\right) i_v \frac{dM}{d\theta}$$

$$T_d = \frac{1}{T}\int_0^T \frac{\sqrt{2}\cos(100\pi t) . 200\sqrt{2}\sin(100\pi t)}{R_p}$$

$$\cdot \frac{dM}{d\theta} \cdot d(100\,\pi\, t)$$

$$= \frac{100\pi}{T}\int_0^T \frac{200 \sin 200\,\pi\, t}{R_p} \cdot \frac{dM}{d\theta} \cdot dt$$

$$= \frac{100\pi}{R_p . T} \cdot \frac{dM}{d\theta} \int_0^T 200 \sin 200\,\pi\, t . dt$$

= 0 [cosine average over a cycle is zero]

64. Ans. (c)

65. Ans. (d)

66. Ans. (b)

67. Ans. (c)

68. Ans. (b)

Solution: For Nickel-Constantan, sensitivity = 60 μV/k is highest

69. Ans. (d)

Solution:

For copper –constants, sensitivity = 41 μV/k

Then, $\quad emf = 41(T - 0) \times 10^{-6}\ volt$

Now $\quad 41\,T \times 10^{-6} \times 1000 = 2/050$

Or $\quad T = 50^o\ C$

70. Ans. (b)

Solution: Shorting the voltage source, we have

$$R_{12} = [20 \parallel 30] + [25 \parallel 25] k\Omega\, \text{t}$$

$$= 24.5\, k\Omega$$

71. Ans. (b)

Solution:

72. Ans. (a)

Solution:

$$R = n + \frac{I_p}{I_s} = n + \frac{1}{0.495} = 202.02,$$

$k = 200$

then, $= \dfrac{k - R}{R} = \dfrac{200 - 202.02}{202.02} \times 100 = -1.0\%$

73. Ans. (b)
Solution:

$$\omega = \frac{1}{\sqrt{L(C + C_d)}}$$

$$\text{or } L = \frac{1}{\omega 2(C + C_d)} = \frac{1}{10^{12}(0.820 + 9.18) \times 10^{-9}}$$

$$= 100\ \mu H$$

74. Ans. (b)
Solution:

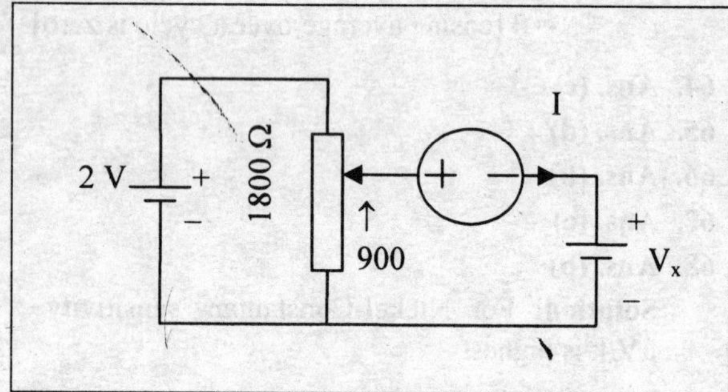

$$V_{AB} = V_{BC} = 1\ volt$$

$$R_{BC} = R_{AB} = 900\ \Omega$$

MCQ Set–2

1. When a voltage $V_0 \sin \omega_0 t$ is applied to the pure inductor, the ammeter shown in the figure reads I_0. If the voltage applied is $-V_0 \sin \omega_0 t + 2V_0 \sin 2\omega_0 t + 3V_0 \sin 3\omega_0 t + 4V_0 \sin 4\omega_0 t$, the ammeter reading would be

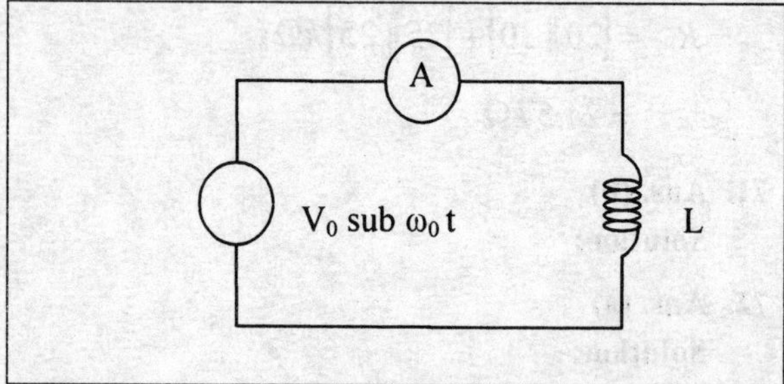

(a) 0 (b) $10\ I_0$
(c) $\sqrt{(4^3 + 3^3 + 2^2 + 1)}\ I_0$ (d) $2I_0$.

2. In two watt meters method of measurements if one of the wattcmeters reads zero, then what will be the power factor
(a) Zero (b) Unity
(c) 0.5 (d) 0.866.

3. What will be the power consumed by the voltage source, current source and resistance respectively?

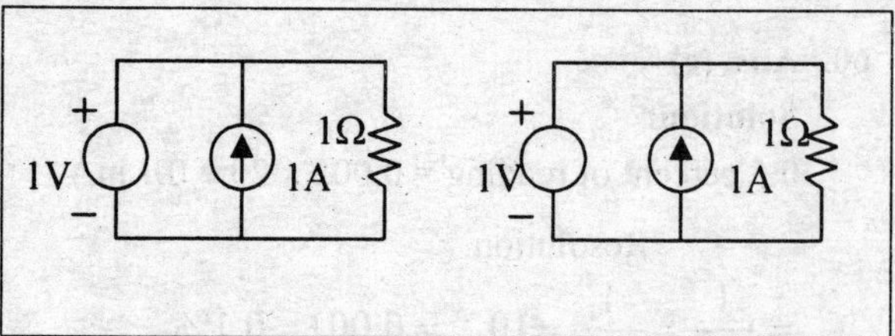

(a) 1 W, 1 W, 2 W (b) 0 W, 1 W, 1 W
(c) 1 W, 0 W, 1 W (d) 0 W, 0 W, 0 W.

4. A 200 V, 5 A wattmeter is tested on full load using the circuit shown in the figure. The power consumed for testing is

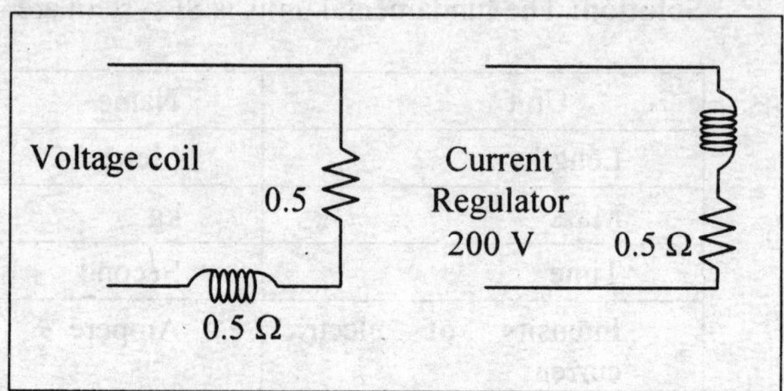

(a) 8 W (b) 25 W
(d) 33 W (d) 1000 W.

5. An ac current of 5A and dc current of 5 A flow simultaneously through a circuit. Which of the following statements is true?
(a) An ac ammeter will read less than 10 A but more than 5 A
(b) An ac ammeter will read only 5 A
(c) A dc ammeter will read 10 A
(d) A dc ammeter will read zero.

6. An engineer uses an ohmmeter to measure the direct current resistance of a transformer winding. Measure. The engineer notices that when once of the ohmmeter lead is disconnected from the transformer, a spark results. What accounts for this phenomenon?
(a) Ohm's law (b) Faraday's law
(c) Thomson effect (d) Seebeck effect.

7. An inductive coil connected to a 200 V, 50 Hz ac supply with 10 A current flowing through it dissipates 1000 watts. Which of the following will have least value in ohms?
(a) Resistance, R (b) Reactance, X_L
(c) Impedance, Z (d) None of the above.

8. Which of the following bridges can be uses for the measurement of dielectric loss of a capacitor?

(a) Schering bridge
(b) Heaviside Campbell equal ratio voltage
(c) Owen bridge
(d) Anderson bridge.

9. At a frequency below the resonant frequency…circuit is capacitive and …circuit is inductive.
(a) series, parallel (b) parallel, series
(c) parallel, parallel (d) series, series.

10. Reactive current through the capacitive load produces…field and reactive current through the inductive load process ... field:
(a) magnetic, electric (b) electric, magnetic
(c) magnetic, magnetic (d) electric, electric.

11. Shering bridge measures
(a) Capacitance, dielectric loss
(b) Inductance
(c) Resistance
(d) Mutual inductance.

12. Match List-I with List-II and select the correct answer by using the codes given below the lists:

List-I	List-II
Resistance	1. $M^{-1} L^{-2} T^4 I^2$
Inductance	2. $ML^2 T^{-2} I^{-2}$
Capacitance	3. $ML^2 T^{-3} I^{-2}$

Codes:

	A	B	C
(a)	1	2	3
(b)	2	1	3
(c)	2	3	1
(d)	3	2	1.

13. Which of the following methods can be used for measuring power without using wattmeter?
1. One voltmeter, one ammeter
2. Two voltmeter, two ammeter
3. Three voltmeters
4. Three ammeters.

(a) 1 and 2 only
(b) 2 and 4 only
(c) 1 and 3 only
(d) 4 only.

14. A dc-supply of 35 V is connected across 600 Ω resistance in series with an unknown resistance. The voltmeter having a resistance of 1.2kΩ is connected across 600Ω resistance which reads 5 V. Then the unknown resistance should be
(a) 50 Ω (b) 1.7 Ω
(c) 7.2 Ω (d) 2.4 Ω.

15. Consider the following statements in connection with DAC:
1. The resolution of a 10-bit DAC is nearly equal to 0.1 of its full-scale range.
2. The linearity of a DAC depends principally on the accuracy of the resistors used.
3. The R – 2R ladder type DAC requires less number of resistor compared to weighted–resistor type of DAC
4. The output of a DAC should be monotonic.

Which of these statements are correct?
(a) 1, 2 and 3 (b) 1, 2 and 4
(c) 2, 3 and 4 (d) 1, 3 and 4.

16. Match List-I, List-II and select correct answer by using the Codes given below the lists:

List-I	List-II
A. Digital Counter	1. Measurement of harmonic
B. Scherring bridge	2. Measurement of frequency
C. Megger	3. Measurement of loss angle of a dielectric
D. Spectrum Analyser	4. Measurement of insulation resistance

Codes:

	A	B	C	D
(a)	1	2	3	4
(b)	2	3	4	1
(c)	4	3	2	1
(d)	4	1	3	2.

17. Match List-I (ADCs) with List-II (Characteristics) and select the correct answer using codes given below the lists:

List-I (ADCs)	List-II (Characteristics)
A. Parallel comparator	1. Null balancing type
B. Successive approximation	2. Fastest converter
C. Dual-slope	3. Voltage-dependent conversion time
D. Counter-ramp	4. Intergrading type

Codes:

	A	B	C	D
(a)	2	1	3	4
(b)	2	1	4	3
(c)	1	2	4	3
(d)	1	2	3	4.

18. A digital voltmeter has read-out range from 0 to 9999 counts. When full scale reading is 9.999 V, the resolution of the full scale reading is

(a) 0.001 (b) 1000
(c) 2 digit (d) 1 mV.

19. The input impedance of a CRO is nearly

(a) Zero (b) Around 10 ohms
(c) Around 100 ohms (d) Around one megaohm.

20. When all the resistance in the circuit are of one each, the equivalent resistance across the points A and B will be

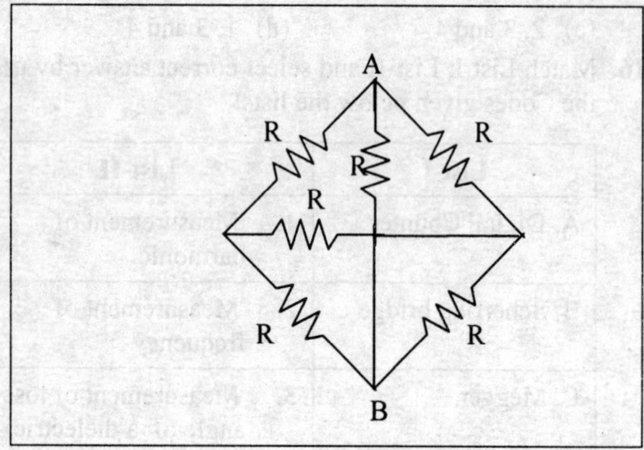

(a) 1 R (b) 0.5 R
(c) 2 R (d) 1,5 R.

21. The meters X and Y requires 40 mA and 50 mA respectively for full scale deflection. Then

(a) X is more sensitive
(b) Y is more sensitive.
(c) Both are equally sensitive
(d) Not possible to determine from the given data.

22. The common mode error voltage in a DVM can be eliminated by using at its input

(a) a differential amplifier
(b) a wide band amplifier
(c) a tuned amplifier
(d) a low-pass filter.

23. The source in the circuit shown is a sinusoidal source. The voltage across various elements are marked in the figure. The input voltage is

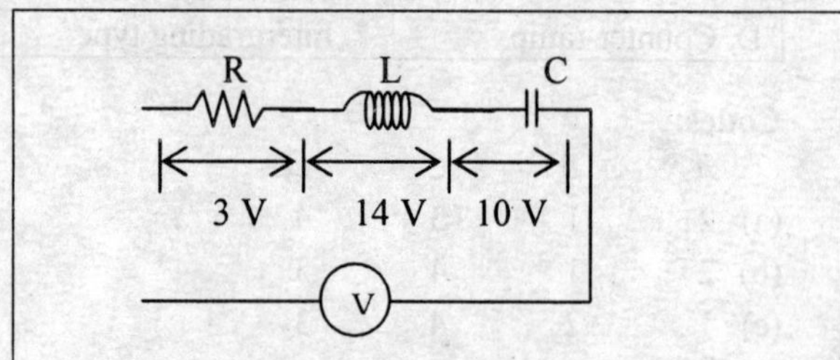

(a) 10 V (b) 5 V
(c) 27 V (d) 24 V.

24. Match List-I with List-II select the correct answer using the codes gives below the lists:

List-I	List-II
Resistance	1. $M^{-1} L^{-2} T^{2} I^{-2}$
Inductance	2. $ML^{2} T^{2} I^{2}$
Capacitance	3. $M^{-1} L^{-2} T^{4} I^{2}$
Reluctance	4. $ML^{2} T^{-3} I^{2}$

Codes:

	A	*B*	*C*	*D*
(a)	1	2	3	4
(b)	4	2	3	1
(c)	2	1	3	4
(d)	4	2	3	1.

25. If the **transient during** switching of a power supply are to be studied, which oscilloscope will be preferred?

(a) An ordinary oscilloscope with high frequency sweep generator.
(b) Dual beam oscilloscope
(c) Dual trace oscilloscope
(d) Storage oscilloscope.

26. Which of the following does not have the same units as the others? The symbols have their usual meanings.

(a) L/R (b) RC
(c) $\sqrt{(LC)}$ (d) $[1/\sqrt{(LC)}]$.

27. A signal of 10mV at 75 MHz is to be measured. Which of the following instrument can be used?

(a) VTVM
(b) Cathode ray oscilloscope
(c) Moving iron voltmeter
(d) Digital multimeter.

28. The quantity $[1/\sqrt{\varepsilon_0 \mu_0}]$ is SI units has the

(a) Value 33om/s (b) Value 1.73×10^{4}
(c) Dimensions LT^{-1} (d) None of the above.

29. Which of the following statement is true about two wattmeter method for power measurement in three phase current?

(a) Power can be measured using two wattmeter method only for star connected three phase circuits.
(b) When two meters show identical readings, m the power factor is 0.5.
(c) When power factor is unit, one of the wattmeter reads zero.
(d) When the readings of the two wattmeter are equal but of opposite sign, the power factor is zero.

30. An alternating current source having voltage E = 110 sin ($\omega t + \pi/3$) is connected in an a.c circuit. If the current drawn from the circuit varies as I = 5 sin ($\omega t - \pi/3$), impedance of the circuit will be
(a) 22 ohms (b) 16 ohms
(c) 30.8 ohms (d) None of the above.

31. A force with which the plates of a parallel plate capacitor having charge Q and area of each plate A, attract each other is
1. Directly proportional to Q
2. Directly proportional to Q^2
3. Directly proportional to A
(a) 1 and 2 only (b) 2 and 3 only
(c) 1 and 3 only (d) 1, 2 and 3.

32. Consider the following statements in connection with measurements of frequency/time interval using a digital frequency counter:
1. Period measurement are preferred over frequency measurements at lower frequencies due to 'gating error' of ± 1 count ambiguity.
2. The error in time-interval and period measurement due to trigger level uncertainty can be reduced with large signal amplitudes and fast rise times.
3. Short-term frequency stability errors can be minimized by taking frequency measurements over long gate times.
4. Long-term frequency stability errors are generally negligible since they tend to a average out.

Which of these statements are correct?
(a) 1, 2 and 3 (b) 1, 3 and 4
(c) 2, 3 and 4 (d) 1, 2 and 4.

33. The energy spent in moving a charge of 10 coulomb form one point 'a' to another point 'b' is 50 joules. The potential difference between points 'a' and ' b' is
(a) 2 volts (b) 5 volts
(c) 10 volts (d) 100 volts.

34. In a series R-L-C circuit excited by a voltage, e = E sin ωt where $LC < 1/\omega^2$
(a) Current lags the applied voltage
(b) Current leads the applied voltage
(c) Current is in phase with the applied voltage
(d) Voltage across L and C are equal.

35. Consider the following statements regarding a moving coil instrument:
1. The sensitivity of a moving coil voltmeter is specified in terms of ohms per volt.
2. A higher range moving coil voltmeter has higher sensitivity.
3. A higher current moving coil instrument has higher sensitivity.
4. Higher sensitivity meters give more reliable results.

Which of these statements are correct?
(a) 1, 2 and 3 (b) 1, 3 and 4
(c) 1, 2 and 4 (d) 2, 3 and 4.

36. In CRT aquadag carries
(a) Aqueous solution of graphite
(b) Sweep voltage
(d) Secondary emission electrons
(d) None of the above.

37. In the context of ac bridge measurements, the term "Wagner Ground" means
(a) A special RC connection that eliminates stray magnetic effects
(b) Any unwanted and unintended ground connection
(c) A large metal plate buried in the ground and connection to one corner of the bridge
(d) A special RC connection that eliminates stray capacitance effects.

38. Any two-port network having a 6 dB loss will give
(a) An output power which is one-quarter of the input power
(b) An output power which is one-half of the input power
(c) An output voltage which is 0.707 of the input voltage
(d) An output power is 0.707 of the input power.

39. In a set-up for the determination of self-capacitance of a coil, the standard variable capacitor has a value of 460 pF at 2 MHz and a value of 100 pF at 4 MHz. The coil self-capacitance is
(a) 2 pF (b) 20 pF
(c) 60 pF (d) 200 pF.

40. The coil of a moving iron voltmeter has a resistance of 200 hms and an inductance of 0.3 H. It is connected in series with a swamping resistance of 2000 ohms. The capacitor connected in shunt with the swamping resistance for making the meter read correctly at dc as well as at 50Hz ac will have a value of
(a) 0.0125 μF (b) 0.0311 μF
(c) 0.0521 μF (d) 0.0622 μF.

41. The arrangement shown in the given figure represent an RC potentiometer for measuring ac voltage. What should be the value of C so that V_o/V_{in} is independent of the frequency of the input signal?

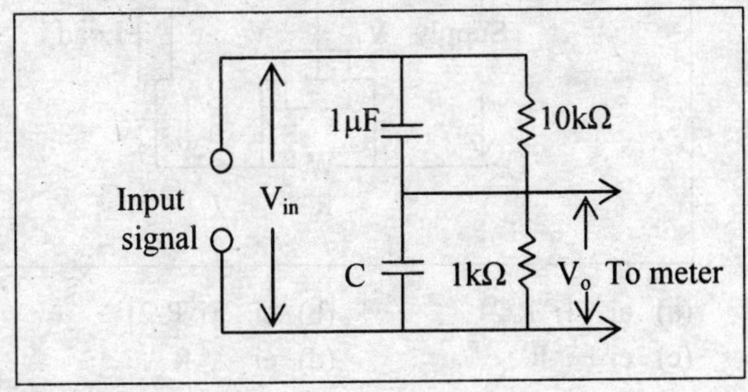

(a) 10 μF (b) 11 μF
(c) 0.10 μF (d) 0.09 μF.

42. Consider the circuit shown in the given figure for the measurement of resistance X using voltmeter-ammeter method.
Ammeter resistance, $R_a = 0.1\Omega$
Voltmeter resistance, $R_v = 5000\Omega$
If X_1 and X_2 are the measured values of X for switch positions 1 and 2 respectively, them the arithmetic mean of $X_1$and X_2 will be

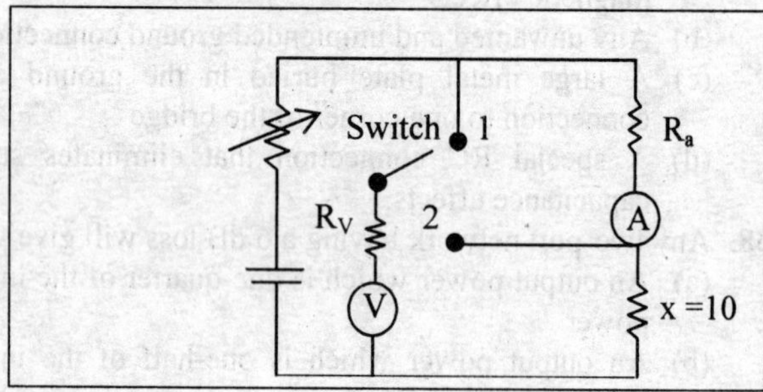

(a) 9.98 Ω (b) 10.00 Ω
(c) 1004 Ω (d) 10.10 Ω.

43. It is desired of convert a 0–1000μA meter moment, with an internal resistance of 100 ohms, into a 0–100 mA meter. The required value of shunt resistance is about
(a) 1 ohm (b) 10 ohms
(c) 99 ohms (d) 100 ohms.

44. A moving coil ammeter has a fixed shunt of 0.02-ohm resistance. If the coil resistance of the meter is 1000 ohms, a potential difference of 500 mV is required across it for full-scale deflection. Under this condition, the current in the shunt would be
(a) 2.5 A (b) 25 A
(c) 0.25 A (d) 0.025 A.

45. In the circuit of an electrostatic wattmeter given below, the instantaneous torque is proportional to

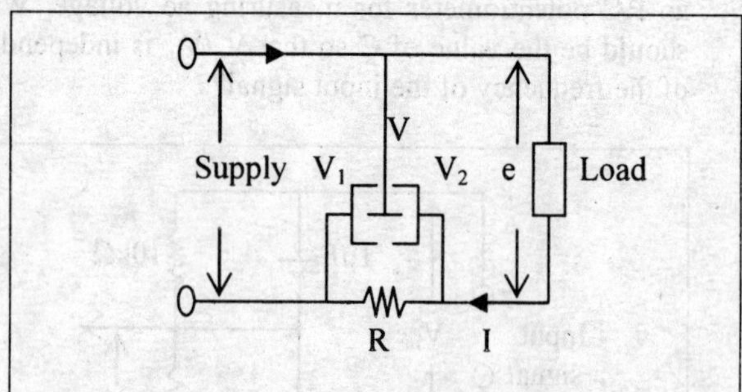

(a) ei + [i^2 R/2] (b) ei – [i^2 R/2]
(c) ei + i^2R (d) ei – i^2 R.

46. In the Wheatstone bridge shown in the given figure, if the resistance in each arm is increased by 0.05% then the value of V_{out} will be

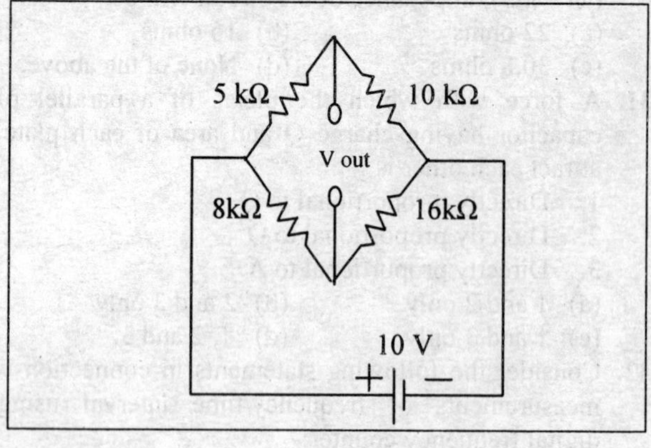

(a) 50 mV (b) 5 mV
(c) 0.1V (d) zero.

47. The resolution of a logic analyser is
(a) The maximum number of input channels
(b) The maximum duration of the glitch it can capture
(c) Its internal clock period
(d) The minimum amplitude of input signal it can display.

48. The internal resistance of a –25-0-25μA meter can be safely measure by using a laboratory multimeter whose sensitivity is
(a) 1 k ohm/volt (b) 10 k ohm/volt
(c) 100 k ohm/volt (d) 20 k ohm/volt.

49. While transporting a **sensitive galvanometer**
(a) The terminals are kept shorted
(b) Critical **damping resistance** is connected across the terminals
(c) The terminals are kept open-circuited
(d) It does not matter as to what is connected across the terminals.

50. The total capacitance across points 'a' and 'b' in the given figure is

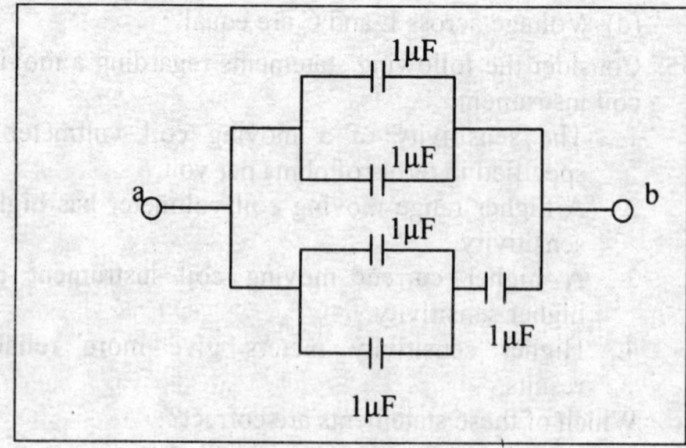

(a) 1.66 μF (b) 2.66 μF
(c) 3.5 μF (d) 4.5 Mf.

51. Consider the following statements regarding the use of **electrodynamometer** type indicating instruments: It is used for measuring
1. direct current 2. alternating current
3. direct voltage 4. alternating voltage
Of these statements
(a) 1 and 2 are correct
(b) 3 and 4 are correct
(c) 1 and 3 are correct
(d) 1, 2, 3, and 4 are correct.

52. **Harmonic distortion** for each frequency can be obtained by harmonic analyzer of the
(a) heterodyne type
(b) tuned circuit type
(c) fundamental suppression type
(d) bridge circuit type.

53. A voltage waveform shown in the given figure is applied to an ideal full-wave rectifier voltmeter. The voltmeter is calibrated to read the RMS values of sine waves. The voltmeter will read

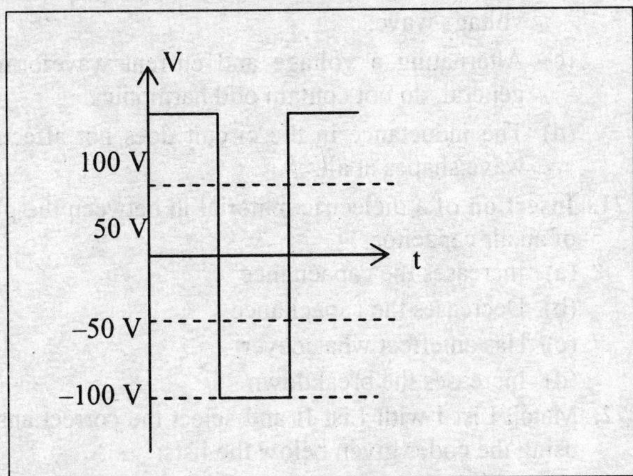

(a) $50/(2)^{1/2}$ (b) $100/(2)^{1/2}$ V
(c) 100 V (d) 111 V.

54. The equivalent circuit of a **wire-wound resistor** can be represented as

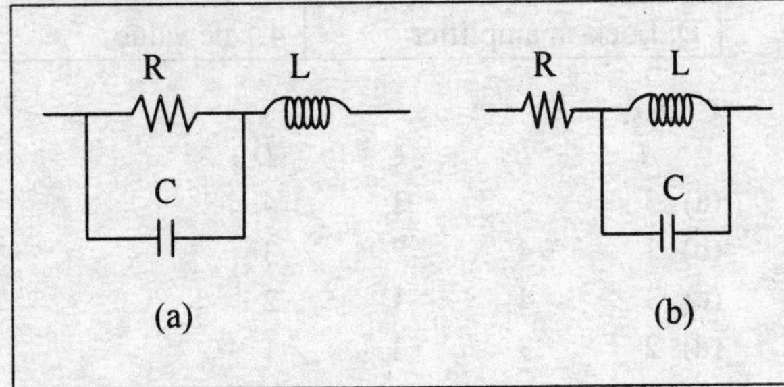

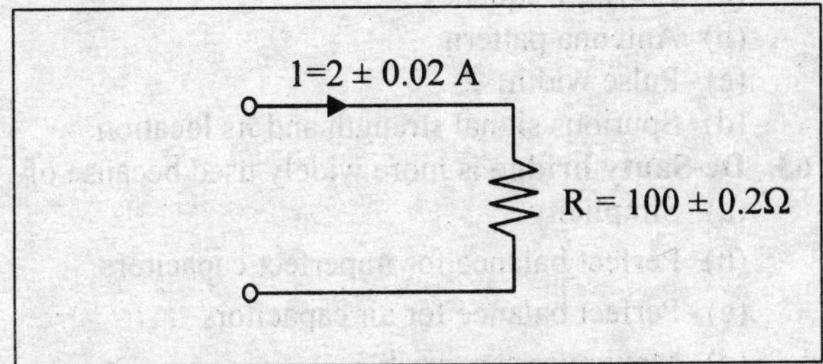

55. The current flowing through the resister R is as indicated in the given figure. The computed value of power is

I = 2 ± 0.02 A

R = 100 ± 0.2Ω

(a) 400 ± 0.42 W (b) 400 ± 4.60 W
(c) 400 ± 8.85 W (d) 400 ± 10.65 W.

56. The most serious source of error in a.c. bridge measurement is
(a) eddy currents
(b) leakage currents
(c) residual imperfectness
(d) stray fields.

57. **A peak-responding ac voltmeter** employing a half-wave precision rectifier has been calibrated to give a reading of 1 V for 1 V rms sinusoidal input. For 1 V dc input, the meter reading will be
(a) 0.707 V (b) 0.707 V or 0 V
(c) 1.414 V (d) 1.414 V or 0 V.

58. The **Lussajous pattern** observed on the screen of a CRO is a straight line inclined at an angle of 45^0 to the x-axis. If the voltage applied to the x-plate is 2 sin ωt, then the voltage applied to the Y-plates will be
(a) 2 sin ωt (b) 2 sin (ωt + 45°)
(c) 2 sin (ωt – 45°) (d) $2\ (2)^{1/2}$ sin (ωt + 45°).

59. An **oscilloscope** has an input impedance consisting of 1 MΩ and 20 pF in parallel. A high impedance probe connected to the input of this oscilloscope has a 10 MΩ series resistance
(a) Need not be shunted
(b) Should be shunted by a 2 pF capacitor
(c) Should be shunted by a 20 pF capacitor
(d) Should be shunted by a 200 pF capacitor.

60. In a **digital measuring device,** if the input electrical signal is in the frequency range dc to f_{max} Hz, then it must be sampled at a rate of
(a) f_{max} time/sec (b) $f_{max}/2$ times/sec
(c) every $2f_{max}$ sec (d) 2 f_{max} times/sec.

61. If the secondary **burden** of a **current transformer** is 15 VA and the secondary current is 5 A, then the impedance to the connected load will be
(a) 0.6 ohms (b) 5 ohms
(c) 6 ohms (d) 10 ohms.

62. Without a **spectrum Analyser,** it is not possible to determine
(a) Carrier frequency
(b) Antenna pattern
(c) Pulse width
(d) Spurious signal strength and its location.

63. De-Sauty bridge is more widely used because of
(a) Simplicity
(b) Perfect balance for imperfect capacitors
(c) Perfect balance for air capacitors
(d) Maximum sensitivity.

64. The three impedance of an bridge shown in given figure are:

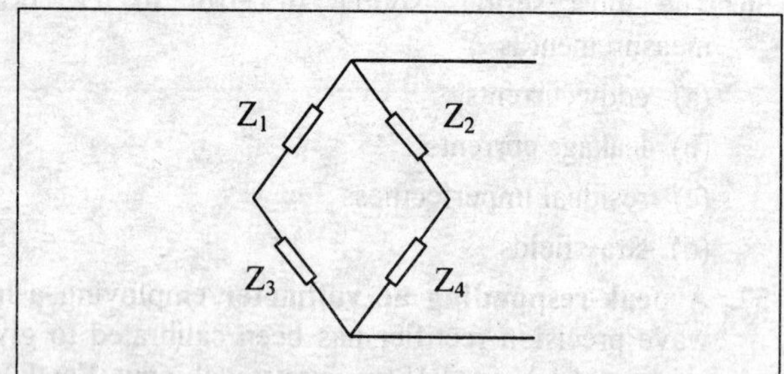

(a) 150 Ω ∠30° (b) 400 Ω ∠–90°
(c) 300 Ω ∠90° (b) 600 Ω ∠–150°.

65. A **diaphragm,** has a natural frequency of 30 kHz. If both is diameter and thickness are halved, then the natural frequency will be
(a) 15 kHz (b) 60 kHz
(c) 120 kHz (d) 240 kHz.

66. The **equivalent series resistance of a capacitor** (C) in terms of its power factor and operation frequency (*f*) is given by

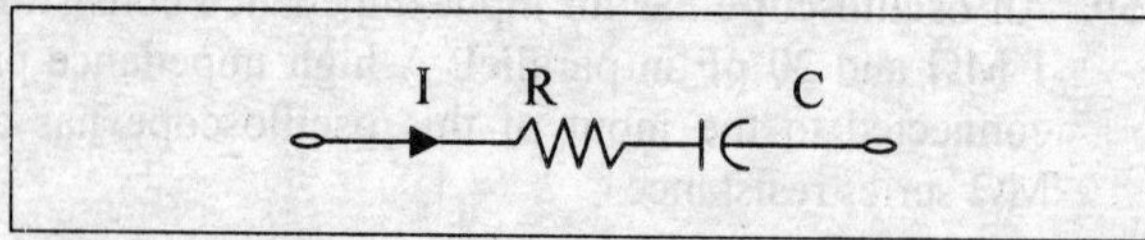

(a) Power factor/$2\pi f C$
(b) $2\pi f C$/Power factor
(c) $2\pi f C$ (Power factor)
(d) $1/2\pi f C$ (Power factor).

67. A straight conductor 1 meter long is moved at right angles to a uniform field of 1 Wb/m^2 at a speed of 50 cm per second. The **emf inducted** in the conductor is
(a) 0.5 V (b) 0.1 V
(c) 1 V (d) 5 V.

68. The phenomenon of **creeping** occurs in
(a) Energy meters (b) Ammeters
(c) Watt meters (d) Voltmeters.

69. Potentiometer method of **dc voltage measurement** is, in principle, more accurate than direct measurement of a voltage using voltmeter because,
(a) It loads the circuit to the maximum extent
(b) It loads the circuit moderately
(c) It does not load the circuit at all
(d) It uses zero center galvanometer instead of a voltmeter.

70. Which one of the following statements is true?
(a) Alternating voltage and current waveforms contain even and odd harmonics in general.
(b) The inductance in a circuit reduces the amplitude of the harmonic in the current weave and makes it more nearly to the sinusoidal wave shape than the voltage wave.
(c) Alternating a voltage and current waveforms in general, do not contain odd harmonics.
(d) The inductance in the circuit does not affect, the wave shapes at all.

71. Insertion of a dielectric material in between the plates of an air capacitor
(a) Increases the capacitance
(b) Decreases the capacitance
(c) Has an effect whatsoever
(d) Increases the breakdown.

72. Match List-I with List-II and select the correct answer using the codes given below the lists:

List-I	List-II
A. Hot wire instrument	1. Only voltage
B. PMMC instrument	2. phase
C. Electrostatic instrument	3. True rms value
D. Lock-in amplifier	4. dc value

Codes:

	A	*B*	*C*	*D*
(a)	3	2	1	4
(b)	1	4	2	3
(c)	3	4	1	2
(d)	2	3	4	1.

73. **Pointing vector wattmeter** uses
(a) See beck effect (b) Ferranite effect
(c) Induction effect (d) Hall effect.

74. An average responding a voltage, employing a full-wave precision rectifier has been calibrated to given a reading of 1 V for 1 V rms sinusoidal input. For 1V peak-to-peak square wave input, the meter reading will be
(a) $[\pi/4\sqrt{(2\text{ V})}]$ (b) zero
(c) $[\pi/2\sqrt{2}]$ (d) $\pi\sqrt{2}$ V.

75. A digital frequency counter can be connected to a DVM by addition of a stage of suitable
(a) Voltage controlled oscillator
(b) D/A converter to it
(c) Power amplifier to it
(d) Operational amplifier to if.

76. Match List-I with List-II and selected the correct answer using the codes given below the lists:

List -I	List-II
A. Moving coil gaivanometer	1. High inertia soft spring
B. Ballistic galvanometer	2. No control, high damping
C. Vibration Galvanometer	3. Soft spring control, critical damping
D. Moving coil Flux meter	4. Low inertia, tuned spring control

Codes:

	A	*B*	*C*	*D*
(a)	4	2	3	1
(b)	1	3	2	4
(c)	2	1	4	3
(d)	3	1	4	2.

77. Consider the following statements with reference to analog to digital converters:
1. The dual-slope integration type has high accuracy and immunity from noise.
2. The successive approximation type a higher conversion time than the dual-slope converter.
3. The counter-type does not use a D/A converter unlike the dual-slope type.

Of these statements
(a) 1, 2 and 3 are correct (b) 1 and 2 are correct
(c) 1 alone is correct (d) 2 and 3 are correct.

78. A permanent magnet moving coil measures the
(a) True rms value (b) dc value
(c) Peak value (d) rms plus dc value.

79. Consider the following statements about a moving iron instrument:
1. It is an un-polarised meter.
2. It has a permanent magnet
3. It has a fixed coil
4. It has two iron vanes.

Of these statements
(a) 1, 2 and 3 are correct (b) 1, 2 and 4 are correct
(c) 1, 3 and 4 are correct (d) 2, 3 and 4 are correct.

80. The coil of a moving coil ammeter of 100 turns is 40mm long and 30 mm wide. The control torque is 240×10^{-6} Nm on full scale. If the magnetic flux density in the air-gap is 1 wb/m^2, what will be the range of the ammeter?
(a) 1 mA (b) 2 mA
(c) 3 mA (d) 4 mA.

81. To measure low resistance four-terminal approach is preferred because it
(a) Eliminates the effect of thermoelectric emf
(b) Minimizes the effect of parasitic capacitances
(c) Reduces the effect of parasitic inductances
(d) Eliminates the effect of lead and contact resistance.

82. A 1 mA ammeter has a resistance of 100 ohms. It is to be converted into a 1A ammeter. The value of shunt resistance required is
(a) 0.001 Ω (b) 0.1001 Ω
(c) 100 Ω (d) 10000 Ω.

83. Consider the following statements about secondary standards of mutual inductances:
1. They shall remain constant under all conditions of use.
2. Their construction should be such that the mutual inductance varies as little as possible with change in temperature.
3. As far as possible, their inductance should be independent of excitation frequency.
4. They usually consist of two coils wound on a bobbin of hard magnetic material, coils being separated by a flange.

Of these statements
(a) 2, 3 and 4 are correct
(b) 1, 2 and 3 are correct
(c) 1, 3 and 4 are correct
(d) 1, 2 and 4 are correct.

84. A dual trace oscilloscope usually offers two modes, chop and alternate. The alternate mode can be used for displaying.

(a) Any two waveforms
(b) Two waveforms of relatively high frequency
(c) Two waveforms of relatively low frequency
(d) One low frequency and one high frequency waveform.

85. An average responding rectifier type electronic ammeter has its scale calibrated in terms of the rms value of a sine wave. If wave voltage of peak magnitude 100 V is measured using this voltmeter will be reading indicated by the meter?
(a) 11 V (b) 100 V
(c) 90.90 V (d) 70.7 V.

86. Wagner earthing device eliminates the effects of
(a) All stray capacitances in the bridge network
(b) Stray capacitance between components
(c) Stray capacitance between the detector terminals to the ground
(d) The mutual coupling between components.

87. Kelvin double bridge is chosen to measure low resistance because
(a) It has high sensitivity
(b) Thermoelectric emf's can be taken
(c) Resistance variation due to temperature can be accounted for
(d) Resistance variation due of contacts of leads can be eliminated.

88. In the bridge shown in the given figure, the value of R_x and C_x at balance will be

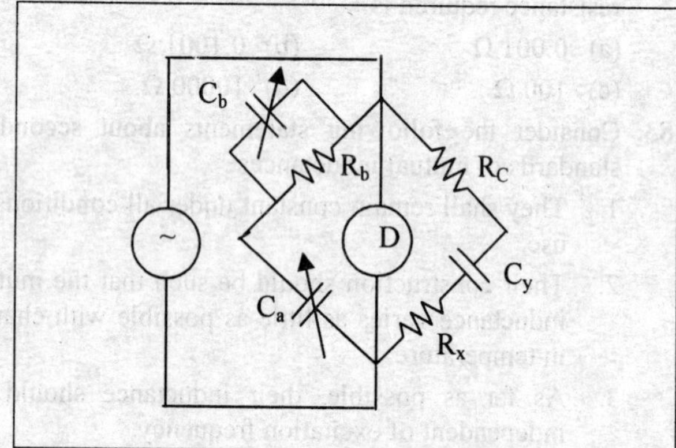

(a) $R_x = \frac{C_b}{C_a} R_c$; $C_x = \frac{R_b}{R_c} C_a$

(b) $R_x = \frac{C_a}{C_b} R_c$; $C_x = \frac{R_c}{R_b} C_a$

(c) $R_x = \frac{C_a}{C_b} R_c$; $C_x = \frac{R_b}{R_c} C_a$

(d) $R_x = \frac{C_b}{C_a} R_c$; $C_x = \frac{R_c}{R_b} C_a$.

89. The circuit shown in the given figure is that of a half-bridge which is often used if only two identical resistive transducers are employed, for example two strain gauges. For $\delta < 1$ the output V_0 will be

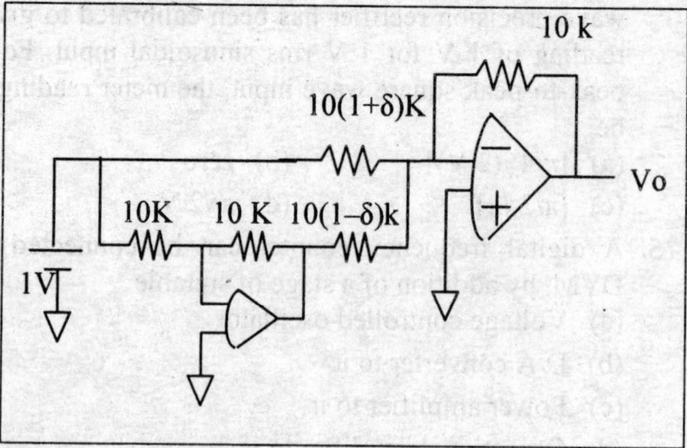

(a) δ (b) 2δ
(c) -2δ (d) $-\delta$.

90. An attenuator drops a 10 V signal to 50 mV in an experiment. The loss in decibels is
(a) –40 dB (b) –46 dB
(c) –55 dB (d) –60 dB.

91. A moving has a resistance of 10 ohms and takes 40 mA to produce full-scale deflection. The shunt resistance required to convert this instrument for use as an ammeter of range 0 to 2 A is
(a) 0.1021 Ω (b) 0.2041 Ω
(c) 0.2561 Ω (d) 0.4210 Ω.

92. Which one of the following measuring instruments consume the LOWEST power from the source during measurement
(a) Permanent magnet moving coil
(b) Electronic multi meter
(c) Electrostatic instrument
(d) Moving iron instrument.

93. A **dynamometer type of wattmeter** is connected in an ac circuit. The power indicated by the wattmeter is the
(a) rms power (b) average power
(c) peak power (d) instantaneous power.

94. In terms of LMTQ system of dimensional parameters, the **dimension of 'permittivity'** can be expressed as
(a) $L^{-3} M^{-1} T^2 Q^2$ (b) $L^{-1} M^{-1} T^2 Q^2$
(c) $L^2 MT^{-1} Q^{-1}$ (d) $L^{-2} M^{-1} T^2 Q^2$.

95. The D' Arsonal meter movement can be converted into audio frequency ac ammeter by adding to it a
(a) Thermocouple (b) Rectifier
(c) Chopper (d) Transducer

96. The time base of a high frequency CRO whose screen diameter is 10 cm is set at 10 μ sec/cm. The lowest frequency of the signal that can be fully displayed an this CRO for measurement and analysis with great accuracy is
(a) Greater than 100 kHz (b) Equal to 100 kHz
(c) Equal to 10 kHz (d) Less than 10 kHz.

97. In the circuit shown in the figure, if the voltmeter and ammeter are interchanged, it i likely to result in damage to

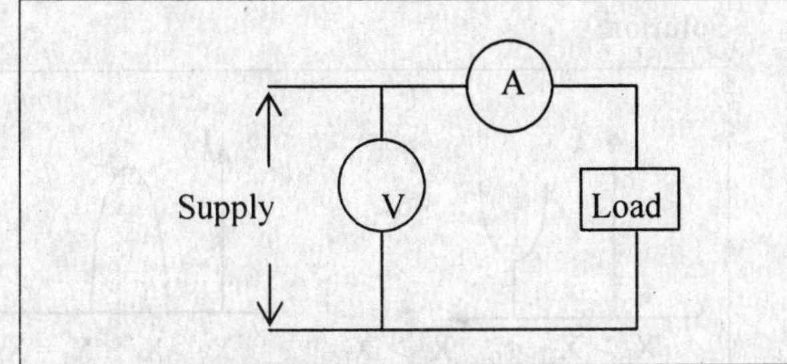

(a) Both the instruments
(b) The ammeter
(c) The voltmeter
(d) Neither of less instruments.

98. A special voltmeter can be devised which can measure the amplitude of a signal at two points in a circuit and simultaneously **measure the phase difference** between the voltage waveforms at these points. Such a meter would be a
(a) Phase meter (b) Waveform meter
(c) Vector voltmeter (d) Digital voltmeter.

99. An instrument needs an amplifier to amplify pulse of **one microsecond** duration. This amplifier must have a bandwidth of at least
(a) 10 kHz (b) 10 MHz
(c) 1 kHz (d) 1 MHz.

100. A **diode peak reading VTVM** and a thermocouple meter are connected across the output of an amplitude modulator circuit. In the absence of modulation, both the meters read 10 V. When a sinusoidal AM is applied, the VTVM reads 15 V. The reading of the thermocouple meter in this case will be
(a) 10 V (b) 12.5 V
(c) 14.1 V (d) 15 V.

101. The input impedance of a CRO is 1MΩ is parallel with 10 pF. If CRO is required to display pulses using a 10:1 **attenuator**, the attenuator will have to use a
(a) 9 M ohm resistor
(b) 1.11 pF capacitor
(c) parallel combination of 9 M ohm resistor and 1.11 pF capacitor
(e) series combination of 9 M ohm resistor and 1.11 pF capacitor.

102. Which one of the following detectors is generally used in ac bridges for audio frequency range?
(a) AC voltmeter (b) CRO
(c) Headphones (d) Vibration galvanometer

103. The temperature coefficient of resistance for a thermistor is
(a) Low and negative (b) Low and positive
(c) High and negative (d) High and positive.

104. A variable reluctance type tachometer has 150 teeth on the rotor. The counter records 13,500 pulses per second. The rotational speed is
(a) 4800 rpm (b) 5400 rpm
(c) 6000 rpm (d) 7200 rpm.

105. A linear displacement transducer of the digital type generally uses
(a) Straight binary code (b) BCD
(c) Gray code (d) Hexadecimal code

Answers Set-2

1. Ans. (d)

Solution: The impedance of the inductor for n[th] harmonic is n times the impedance for the fundamental frequency i.e.

$$X_L = \omega L,$$

so as impedance increase the current decreases but since the nth harmonic voltage is n times the fundamental frequency signal, each harmonic signal gives the same current I_0.

∴ Ammeter reading

$$= \sqrt{I_0^2 + I_0^2 + I_0^2 + I_0^2}$$

$$= \sqrt{4I_0^2} = 2I_0.$$

2. Ans. (c)

Solution: In two wattmeter method of power measurement in balanced, three phase Network, the reading. R_1 and R_2 are proportional to

cos (θ – 30) and cos (θ + 30),

where θ is the power factor angle of the load, therefore, it θ = 60°, i.e. if load power factor is 0.5 one of the wattmeter shall read zero.

3. Ans. (b)

Solution: Since the voltage across the resistor of 1Ω is also to be 1 V, the current flow through the resistor only (= 1 A). The voltage source carries no current. Hence the corresponding Power are 0, 1 Watt, 1 Watt.

4. Ans. (c)

Solution:

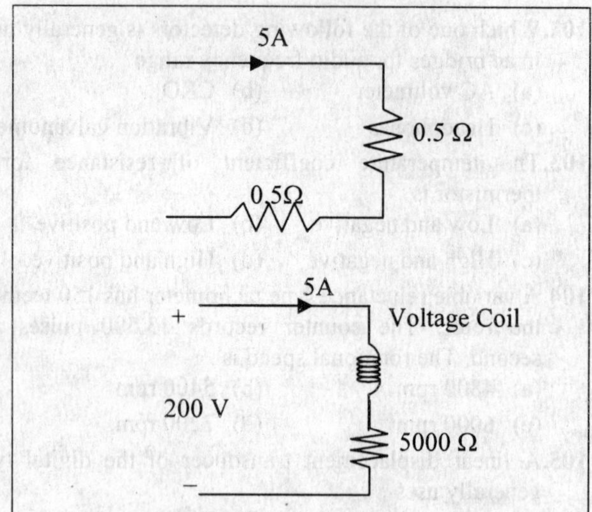

Watt-meter current coil circuit resistance

$= (0.5 + 0.5) = 1\ \Omega$

Watt meter voltage coil circuit resistance

$= 5000\ \Omega$

The watt-meter is tested for full load, i.e. the current in current coil is 5 A and voltage a cross voltage coil is 200 V.

The associated power consumed are

I^2 R and $[V^2/R]$

$= 5^2 \times 1$ and $[(200)^2/5000] = 25 + 8 = 33$ watt.

5. **Ans. (a)**

Solution: Ammeter reads Rms value

$$\frac{5}{\sqrt{2}} + 5 = \frac{5}{1.732} + 5$$

5 A < I < 10 A.

6. **Ans. (b)**

Solution: Faradays law: Whenever the magnetic flux linked with as electric circuit is changed, an emf is induced in the circuit. By disconnecting the instrument, the flux gets charged and hence the spark results.

7. **Ans. (a)**

Solution: Power dissipated is given by

I^2 R = 1000 Watt

$$\text{Hence } R = \frac{1000}{(10)^2} = 10\ \Omega.$$

Impedance Z = [V/I] = [200/10] = 20 Ω

$$Z = \sqrt{R^2 + Z_L^2}$$

$$\text{Inductive reactance } X_L = \sqrt{Z^2 - R^2}$$

$$= \sqrt{(20)^2 - (10)^2}$$

$$= \sqrt{400 - 100} = \sqrt{300}$$

$$= 17.4\ \Omega.$$

8. **Ans. (a)**

Solution: Only shearing bridge can be used for measurement of dielectric loss of a capacitance. Other is DE-Sauty's Bridge. Which can measure Dielectric loss.

9. **Ans. (a)**

Solution:

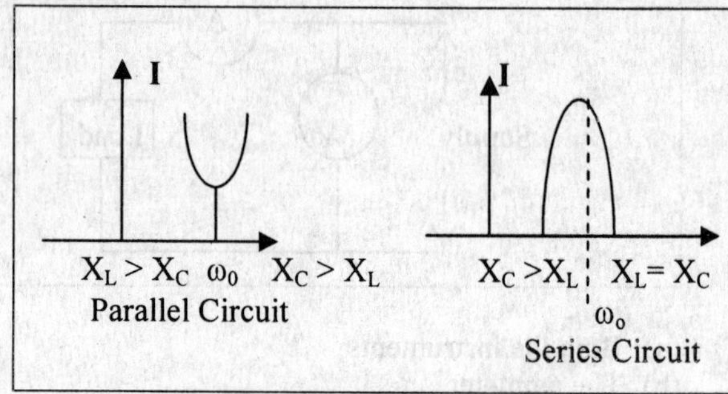

10. **Ans. (b)**

Solution: Capacitor store energy in the form of electric field and inductor store energy in the form of magnetic field. $E_E = [1/2]\ V^2$. $E_M = [1/2]\ L\ I^2$, Due to current magnetic field exist i.e. Ampere law.

11. **Ans. (a)**

Solution: Shering bridge is widely use for the measurement of capacitance and dissipation factor.

12. **Ans. (d)**

Solution: Resistance $= I^{-2} M L^2 T^{-3}$

Capacitance $= I^2 M^{-1} L^{-2} T^4$

Inductance $= I^{-2} M L^2 T^{-2}$.

13. **Ans. (a)**

Solution: Since power = V × I, i.e. Minimum one voltmeter and one ammeter is required to measure the power.

14. **Ans. (d)**

Solution:

1.2 K ← R = 400

600 Ω R = ?

35V

Since the continuation of voltmeter Resistance 1.2 K and $600 = 1200 \| 600$

$$= \frac{1200 \times 600}{1200 + 600}$$
$$= 400\ \Omega$$

Voltage across unknown resistance R = 35 – 5 = 30 V

Then Unknown resistance value $R = 30 \times \left[\frac{400}{5}\right]$

= 2.4 kΩ.

Second Method

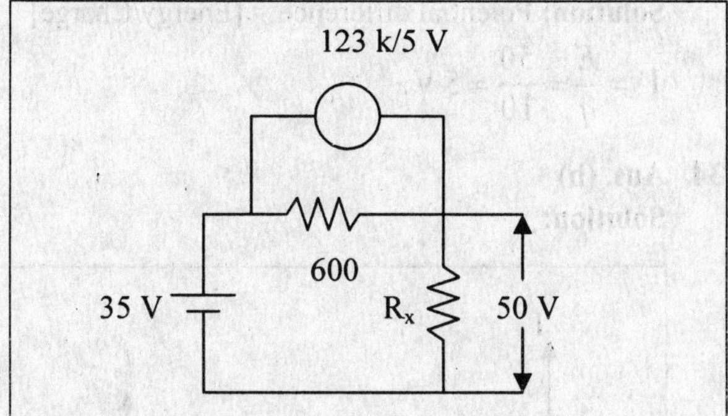

Parallel combination of series resistance and meter resistance

$$= \frac{600 \times 1200}{600 \times 1200} = 400\ \Omega$$

Voltage across the unknown resistance is 30 V.
By voltage division rule

$$V_x = \frac{R_x}{R_x + R_s} \times V_s$$

$$30 = \frac{R_x}{R_x + 400} \times 35$$

$30\,[R_x + 400] = R_x \times 35$
$30 \times R_x + 12000 = R_x \times 35$
$5R_x = 12000$
$R_x = 2.4$ kΩ.

15. Ans. (b)

16. Ans. (b)

Solution: The correct match is given below:

Digital Counter	Measurement of frequency
Schering bridge	Measurement of loss angle of dielectric
Megger	Measurement of insulation resistance
Spectrum Analyzer	Measurement of harmonic

17. Ans. (b)

18. Ans. (d)

Solution: Resolution $= \frac{1}{\text{Number of steps}} \times \text{FSV}$

$$= \frac{\text{FSV}}{\text{Number of steps}}$$

$$= \frac{1}{9999} \times 9.999\ \text{V} = 10^{-3} = 1\ \text{mV}.$$

19. Ans. (d)

20. Ans. (b)

Solution: When bridge is balance
R + R = 2R
R + R = 2R
2R || 2R = R
R || R = **0.5R.**

21. Ans. (a)

Solution: Sensitivity is deflection per unit current
Sensitivity = $[1/I_{FSD}]$

$$S_x = \frac{1}{40\ mA} = \frac{1000}{40} = 25$$

$$S_y = \frac{1}{50\ mA} = \frac{1000}{50} = 20$$

Assuming full scale angle of deflection to be same for the two meter, then we conclude **X is more sensitive.**

22. Ans. (a)

23. Ans. (b)

Solution: $V_L = 14$

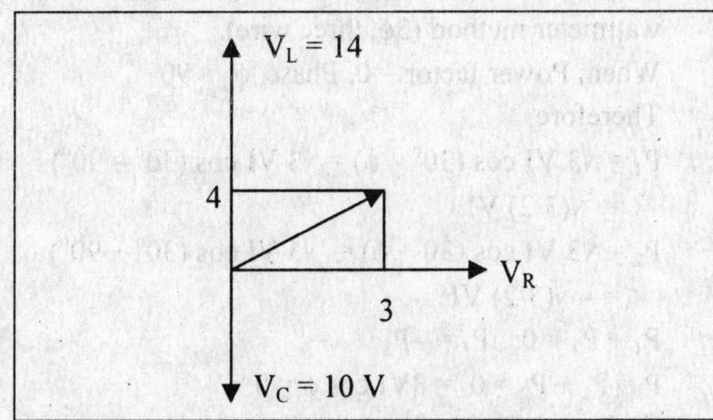

Net voltage across reactive element taken together
$V_X = (V_L - V_C) = (14 - 10) = 4$ V

Supply voltage $= \sqrt{V_R^2 + V_X^2} = \sqrt{3^2 + 4^2} = 5$ V.

24. Ans. (b)

Solution: Resistance $ML^2\,T^{-3}\,I^{-2}$
Inductance $ML^2\,T^{-2}\,I^2$

Capacitance $M^{-1} L^{-2} T^4 I^2$
Reluctance $M^{-1} L^{-2} T^2 I^{-2}$.

25. Ans. (d)
Solution: Storage oscilloscope are capable of retaining the image on screen for longer time than that possible with conventional high persistence phosphors. Example of these are capture storage of transient and steady display of very low frequency signals.

26. Ans. (d)
Solution: $RC = ML^2 T^{-3} I^{-2} \times M^{-1} L^{-2} T^4 I^2 = T$

$$\frac{L}{R} = \frac{ML^2T^{-2}I^2}{ML^2T^{-3}I^{-2}} = T\,I^4 = T$$

$LC = ML^2 T^{-2} I^2 \times M^{-1} L^{-2} T^4 I^2 = T^2 I^4$.

27. Ans. (b)
Solution: Cathode ray oscilloscope are available in the range of which can be measured upto 1 GHz signal.

28. Ans. (c)
Solution: $\mu_o \rightarrow$ Permeability of medium
$= 4\pi \times 10^{-7} = 1.257 \times 10^{-6}$
$\varepsilon_o \rightarrow$ Electric Permittivity of medium F/m
$\varepsilon = [1/36\pi] \times 10^9 = 8.85 \times 10^{-12}$

Velocity of light $C = \dfrac{1}{\sqrt{\mu_o \varepsilon_o}}$

$$= \frac{1}{\sqrt{1.257 \times 8.884 \times 10^{-6} \times 10^{-12}}}$$

$= 3 \times 10^8$ m/s------ unit LT^{-1}.

29. Ans. (d)
Solution: When power is measured with two wattmeter method (3ϕ, three wire).
When, Power factor = 0, Phase, $\phi = 90$
Therefore,
$P_1 = \sqrt{3}\,VI \cos(30° - \phi) = \sqrt{3}\,VI \cos(30° - 90°)$
$= \sqrt{(3/2)}\,VI$
$P_2 = \sqrt{3}\,VI \cos(30 + \phi) = \sqrt{3}\,VI \cos(30° + 90°)$
$= -\sqrt{(3/2)}\,VI$
$P_1 + P_2 = 0;\ \ P_1 = -P_2$
$P = P_1 + P_2 = 0 = 3VI \cos \phi = 0$
With zero Power factor readings of tow wattmeter are equal but opposite sign.

30. Ans. (a)
Solution: Maximum voltage $V_{max} = 110$ V
Maximum Current $I_{max} = 5A$
Impedance $(Z) = \dfrac{110}{5} = 22\,\Omega$.

31. Ans. (b)

Solution: $F = \dfrac{Q_1Q_2}{4\pi\varepsilon d^2} = \dfrac{Q^2}{4\pi\varepsilon d^2}$

$$F = \frac{1}{2}V^2 = \frac{dC}{dx}$$

where $C = \dfrac{\varepsilon A}{x} = \dfrac{dC}{dx} = -\dfrac{\varepsilon A}{x^2} = -\dfrac{1}{2}\dfrac{V^2 \varepsilon A}{X^2}$.

32. Ans. (a)

33. Ans. (b)
Solution: Potential difference = [Energy/Charge]

$$V = \frac{E}{q} = \frac{50}{10} = 5\text{ V}.$$

34. Ans. (b)
Solution:

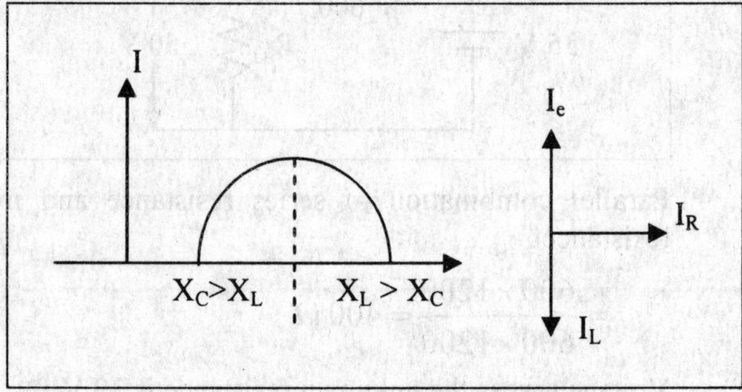

When frequency is less than resonating frequency, circuit is capacitive, therefore I_C lead the I_R.

Second Method

$$LC < \left[\frac{1}{\omega^2}\right];\ \omega^2 < \left[\frac{1}{LC}\right];\ f < \frac{1}{2\pi\sqrt{LC}}$$

When $f < f_r \rightarrow$ Series resonance circuit and as capacitive circuit. So current leads the applied voltage.

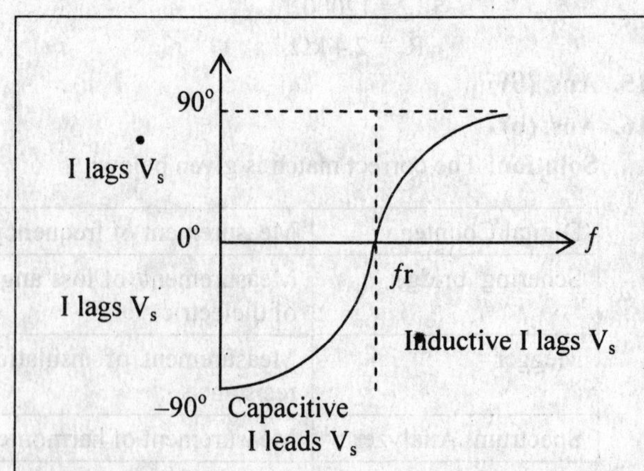

35. Ans. (b)

Solution: Sensitivity of moving coil instrument

$S_V = [1/I_{fs}] = [1/I_m]\ \Omega/V$

I_{fs} = Current required for full scale deflection

Moving coil instrument with higher the ohm per volt or with high or sensitivity gives more reliable results. More current passing through the moving coil it give the more defections.

36. Ans. (a)

Solution: The bombardment of electron on screen given rise of secondary emission of election. The secondary electrons are collected by an aqueous solution of graphite called aquadag coating. Collecting of secondary electron is necessary to keep screen in a state of electrical equilibrium.

37. Ans. (d)

Solution: Each component in a A.C bridge has defining screen connected to one end, very high accuracy in measurement is made possible by additional Wagner earthy device, it remove all the earth capacitance from bridge network.

38. Ans. (a)

Solution: $10 \log_{10}\left(\frac{P_{out}}{P_{in}}\right) = 6$

$$\left(\frac{P_{out}}{P_{in}}\right) = 10^{0.6} = \left(\frac{1}{4}\right).$$

39. Ans. (b)

Solution: $C_d = \frac{C_1 \times f_1 - C_2 \times f_2}{f_2 - f_1}$

$$= \frac{469 \times 1 - 4 \times 100}{3} = 20\ pF.$$

40. Ans. (d)

Solution:

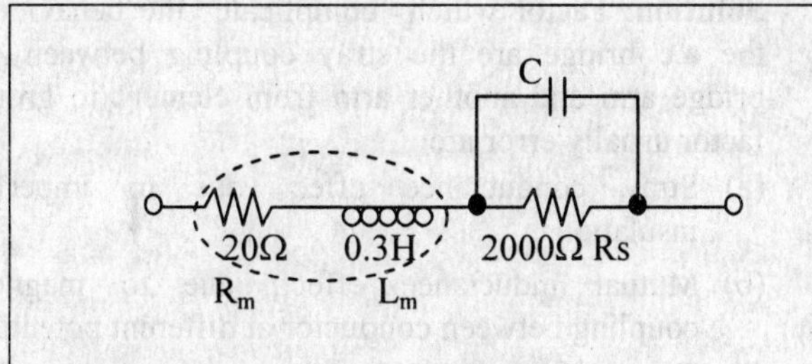

$R_T = R_s + R_m$

$= 2000 + 200 = 2200$

The value of Capacitance $C \cong \left(\frac{L}{R^2}\right)$

$$= \left(\frac{0.3}{2200^2}\right)$$

$= 0.062\ \mu F.$

41. Ans. (a)

Solution:

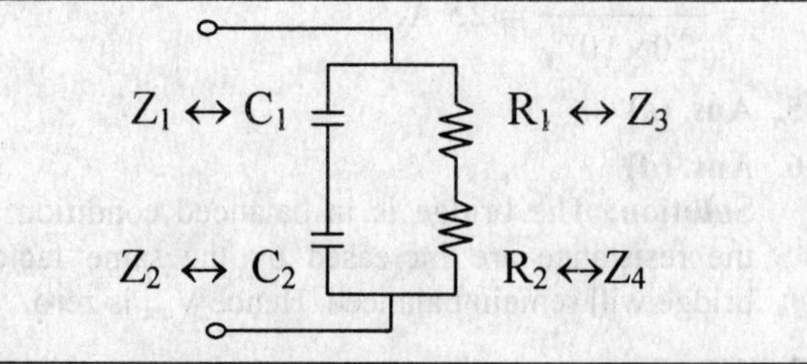

For the voltage division to be independent of frequency, the impedance ratio is equal and independent of frequency i.e.

$$\frac{Z_1}{Z_2} = \frac{Z_3}{Z_4}$$

or $Z_1 \times Z_4 = Z_2 \times Z_3$

$$\frac{1}{j\omega C_1} R_4 = \frac{1}{j\omega C_2} R_3$$

or $C_2 R_4 = C_1 R_3$

where, $C_1 = 1\ \mu F$, $R_3 = 10\ k$, $R_4 = 1\ \Omega$

$$C_2 = \frac{C_1 R_3}{R_4} = \frac{1\ \mu F \times 10\ k}{1\ k} = 10\ \mu F.$$

42. Ans. (c)

Solution: $X_1 = (10 + 0.1) = 10.1\Omega$

$$X_2 = \frac{5000 \times 10}{5000 + 10} = \frac{50000}{5010} = 9.88\ \Omega$$

----Parallel with voltmeter

Arithmetic mean $= \frac{10.1 + 9.98}{2} = 10.04$ ohms.

43. Ans. (a)

Solution: Shunt resistance, $S = \left(\frac{R_m}{m - 1}\right)$ o

where, R_m = movement resistance 100 ohm

m = multiplying multiplying factor

$$= \left(\frac{I}{i}\right) = \frac{100 \times 10^{-3}}{1000 \times 10^{-6}} = \frac{1}{10} \times 1000 = 100$$

I = total current
I = current through the movement

Hence, $S = \dfrac{100}{100-1} \approx 1$ ohm.

44. Ans. (b)
Solution: Current in shunt

$$= \frac{\text{Potential difference across the shunt coil}}{\text{Shunt resistance}}$$

$$= \frac{500\times10^{-3}}{20\times10^{-3}} = 25\text{ A.}$$

45. Ans. (a)

46. Ans. (d)
Solution: The bridge is in balanced condition. If all the resistance are increased by the same factor the bridge will remain balanced. Hence V_{out} is zero.

47. Ans. (d)

48. Ans. (d)
Solution: Sensitivity

$$= \frac{1}{(25+25)\,\mu A} = 200\ k\Omega/\text{volt}$$

$$= \frac{1}{\text{Current range of meter}}$$

Second Method
Smallest increment in input which can be detected by an instrument.

49. Ans. (b)

50. Ans. (b)
Solution: In parallel

$$C = C_1 + C_2$$

In series $= \dfrac{C_1\,C_2}{C_1 + C_2}$

2μF
a b
2μF 1μF
⇕
2μF
a b
[2/3] μF

In present case total capacitance

$$= 2 + \left(\frac{2}{3}\right) = 2.66\ \mu F.$$

51. Ans. (d)
Solution: Electrodynamometer type of instrument are used both an a.c as well d.c. Their calibration for both d.c and a.c is the same.

52. Ans. (a)

53. Ans. (d)
Solution: A full wave factor rectifier type meter for measuring sinusoidal current or voltage indicates 1.11 times the means value. I.e. form factor = $[V_{rms}/V_{mean}]$. For the this given wave the mean value is 100 V itself. Therefore, the meter will indicate, RMS value 100 × 1.11 = 111 V.

54. Ans. (d)

55. Ans. (c)
Solution: $P = I^2R$

$$\frac{\Delta I}{I} = \frac{0.02}{2}\times 100\,\% = 1\%$$

$$\frac{\Delta R}{R} = \frac{0.2}{100}\times 100\,\% = 0.2\%$$

$$P = I^2R$$

$$\Delta P = 2\Delta IR + 2\Delta I^2$$

$$\frac{\Delta P}{P} = \frac{2\Delta IR}{I^2R} + \frac{\Delta RI^2}{I^2R} = \frac{2\Delta I}{R} + \frac{\Delta R}{I}$$

$$\therefore\quad \frac{\Delta P}{P} = \frac{2\Delta I}{I} + \frac{\Delta R}{R} = 2\times 1 + 0.2 = 2.2\,\%$$

Or, $\Delta P = (2^2\times 100)\times 2..2\% = 400\times\dfrac{2.2}{100} = 8.83$

Hence $P = (400 \pm 8.8\,W)$.

56. Ans. (d)
Solution: Factor which complicate the behavior of the a.c bridge are the stray coupling between one bridge arm and another arm from element to ground factor usually error are:

(a) Stray conductance effect due to imperfect insulation.
(b) Mutual inductance effect, due to magnetic coupling between conductor at different potential.
(c) Stray capacitance effect due to electro static field between conductor at different potential.
(d) Residue component.

57. Ans. (b)

Solution: Since it is a peak responding meter, for $(1/\sqrt{2})$ volt peak value it indicates 1 V and 1 V DC input, peak values is 1 V itself or would be taken as zero for negative polarity.

Consequently reading would be

$$\frac{1}{\sqrt{2}} \times 1 \text{ or } 1 \times \frac{1}{\sqrt{2}} \times 0 \text{ i.e. } 0.707 \text{ V or } 0 \text{ V}.$$

58. Ans. (a)

Solution: When on X and Y plate signal of same frequency it represent straight line.

For such a trace, both the signals are identical.

Since x-plate signal is 2 sin ωt, the same is the y-plate signal.

Second Method

When two sinusoidal signal of equal frequency which are in phase with each other are applied to the horizontal and vertical deflection plate, pattern appear on screen is straight line. So applied to X-plate is 2sin ωt then the voltage applied y–plate also 2 sin ωt.

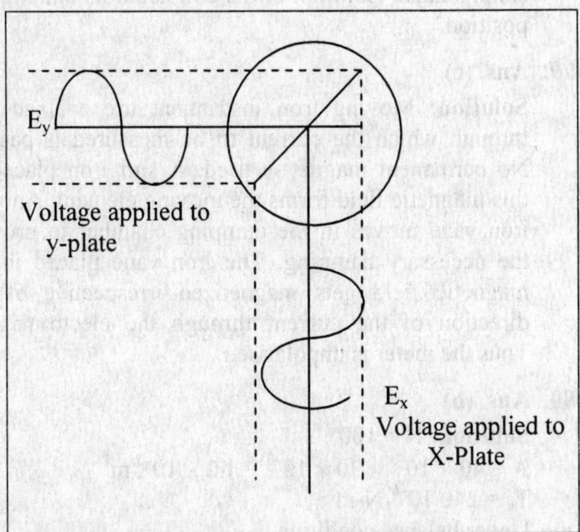

59. Ans. (b)

Solution: Time constant $CR = P = 10\ M\Omega \times 2\ \text{pF}$.

60. Ans. (d)

Solution: Nyquist sampling i.e. $f_s \geq 2 f_{max}$.

61. Ans. (a)

Solution: Burden = V × I = 15 VA and I = 5

Therefore V = 3 volt and the impedance = [3/5] = 0.6 ohm.

Method (2)

Burden = V × I = 15 VA

Current (I) = 5 A so, V = [15/5] = 3 V

Impedance = [V/I] = [3/5] = 0.6 Ω.

62. Ans. (a)

63. Ans. (a)

Solution: Dc Sauty's bridge is the simplest method of comparing two capacitors, capacitance is measured by only varying the value of resistance.

64. Ans. (b)

Solution: $$Z_4 = \frac{Z_2 Z_3}{Z_1} = \frac{(400 \angle -90^\circ)\cdot(300 \angle 0^\circ)}{(200 \angle 60^\circ)}$$

$$= \frac{400 \times 300}{200} \angle -90^\circ - 60^\circ$$

$$= 600 \angle -150^\circ.$$

Second Method

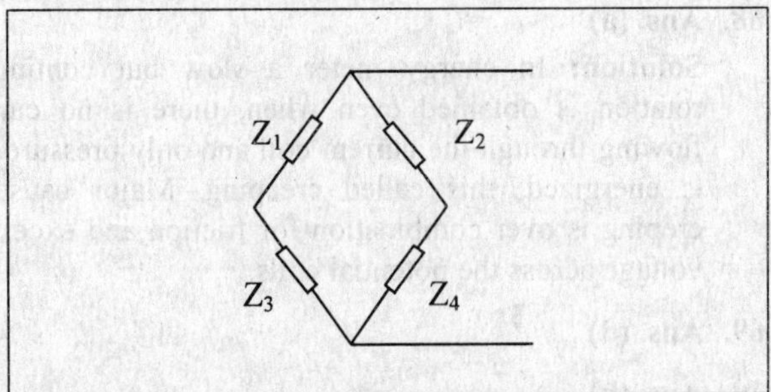

$Z_1 = 200\ \Omega \angle 60^\circ$

$Z_2 = 400\ \Omega \angle -90^\circ$

Value of Z_4 for bridge to be balanced

$$Z_4 = \frac{Z_2 Z_3}{Z_1} = \frac{(400 \angle -90^\circ)\cdot(300 \angle 0^\circ)}{(200 \angle 60^\circ)}$$

$$= (600 \angle -90^\circ \angle -60^\circ) = (600 \angle -150^\circ).$$

65. Ans. (d)

Solution: $$\omega_n = \left(\frac{2\Delta t}{D^2}\right)\sqrt{\frac{E}{3P(1-\nu^2)}} \text{ rad/s}$$

ω_n = 30 kHz

$$\text{New } \omega_n = \frac{2\Delta t/2}{D^2/4} = \frac{4\Delta t}{D^2}$$

= 2 × 30 kHz = **60 kHz.**

66. Ans. (a)

Solution:

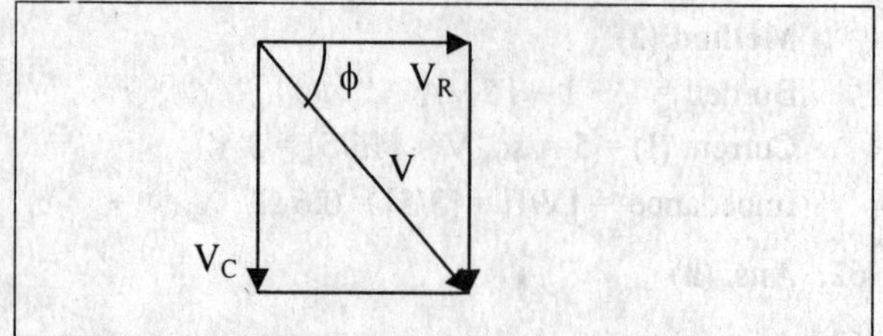

$$\cos\phi = \frac{V_R}{V_C} = \frac{IR}{1/\omega C} = RC\,\omega$$

$$\therefore\ R = \frac{\cos\phi}{C\omega} = \frac{\cos\theta}{2\pi/C}.$$

R_4 is largest of all. Hence (d) is the correct choice.

67. Ans. (a)

Solution: $L = 1$ m; $B = 1$ Wb/m^2, $v = 50$ cm/sec.

$= 50 \times 10^{-2}$ m/sec

Hence $e = LB\,v \sin\theta = 1 \times 1 \times 50 \times 10^{-2} \times 2 \sin 90^\circ$

$= 0.5$ V.

68. Ans. (a)

Solution: In energy meter a slow but continuous rotation is obtained even when, there is no current flowing through the current coil and only pressure coil is energized, this called creeping. Major cause of creping is over combination for friction and excessive voltage across the potential coils.

69. Ans. (d)

70. Ans. (b)

Solution: Impedance of an inductance is directly proportional to the frequency. Hence harmonic frequency components are more attenuated as compared to the fundamental frequency.

71. Ans. (a)

Solution: C = [$\varepsilon_o\ \varepsilon_r$A/d]. On inserting a dielectric material (εr > 1), C increases.

72. Ans. (c)

73. Ans. (c)

74. Ans. (c)

75. Ans. (b)

76. Ans. (d)

Solution: *Ballistic galvanometer*: High inertial soft spring have moment of inertial "S" of moving system should be large and control constant of suspension "K" Should be small.

Moving coil flux meter: No control high damping flux meter is special type of ballistic galvanometer in which controlling torque is very small e electromagnetic damping is heavy.

Vibration galvanometer: Low inertia, tuned spring control damping in these galvanometer is very small in order to obtain sharp resonance.

Moving coil galvanometer: Soft spring control, critical damping in these galvanometer have short periodic time and nearly critical damping.

77. Ans. (c)

Solution: Dual slope integration type A/D, not reducing actual conversion time, but improve accuracy conversion in successive conversion type A/D multiplexing of input in counter-type A/D converter used the D/A converter.

78. Ans. (b)

Solution: In a permanent magnet moving coil (PMMC) meter, the current to be measured is passed through a movable coil place in the magnetic field of a permanent magnet. The interaction between steady magnetic and the magnetic field due to current in the coil causes angular movement of the coil. Resulting angular deflection is directly proportional to the dc current through the coil. A sinusoidal current will simply cause vibration of the coil about its undeflected position.

79. Ans. (c)

Solution: Moving iron instrument use a fixed coil through which the current to be measured is passed. No permanent magnet is used. A soft iron placed in this magnetic field forms the moving element. Another iron vane moves in the damping chamber to provide the necessary damping. The iron vane placed in the magnetic field gets magnetized irrespective of the direction of the current through the electromagnet. Thus the meter is unpolarised.

80. Ans. (b)

Solution: N = 100

A = $40 \times 10^{-3} \times 30 \times 10^{-3} = 1.2 \times 10^{-3}$ m^2

$T_d = 240\ 10^{-6}$ N-m

Under balance condition,

$$T_c = T_d = NI\,AB$$

or $240 \times 10^{-6} = 100 \times \mathbf{I} \times 1.2 \times 10^{-3} \times 1$

Hence I = 0.002 A = 2 mA.

81. Ans. (d)

Solution: Kelvin diode bridge is used to measure the low resistance. The Kelvin double bridge in corporate idea of second set of ratios–hence name double bridge the use of four terminal resister for low resistance arms. Shown in figure, first of ratio arm P and Q. The second set of ratio arms P and q is used to connect the galvanometer to point d at appropriate potential between point in m and n eliminate the affect of connecting load of resistance r between known resistance R and standard resistance S.

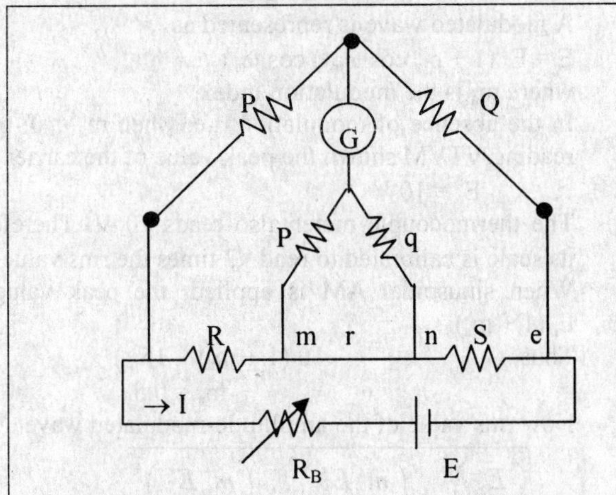

82. Ans. (b)

Solution: $m = \frac{I}{I_m} = \frac{1\,A}{1\,mA} = 1000$

$R_s(m-1) = R_m \times I$

$R_s = \frac{R_m \times I}{m-1} = \frac{100 \times 1}{999}$

$R_s \times 999 = 1 \times 100$

Hence $R_s = \frac{100}{999} = 0.1001$ ohm.

83. Ans. (b)

84. Ans. (b)

Solution: In Dual trace oscilloscope in alternate mode provide the sweep speed is much greater than the of CRT phosphor. The alternate mode con not be use for displaying low frequency signals it used for high frequency signals.

85. Ans. (a)

Solution: Form factor

$= \frac{RMS}{\text{Average}} = \frac{0.707\, I_m}{0.637\, I_m} = 1.11$

$I_{av} = [2I_m/\pi]$

RMS value $= 1.11 \times$ average value

$I_{rms} = [I_m/\sqrt{2}] = 1.11 \times 100 = 111$ V.

86. Ans. (c)

Solution: Wagner earth connection is a method of connection used in c bridge circuit which minimizes the admittance between detector terminals and the ground.

87. Ans. (d)

88. Ans. (a)

Solution:

$$\left(R_x + \frac{1}{j\omega C_x}\right)\left(\frac{R_b}{1 + j\omega R_b C_b}\right) = R_c \cdot \frac{1}{j\omega Ca}$$

or $R_b\left(1 + j\omega R_x C_x\right)$

$= R_c \frac{C_x}{C_a}\left(1 + j\omega R_x C_b\right)$

Equating the real parts: $C_x = R_b \cdot \left(\frac{C_a}{R_c}\right)$

Equating the imaginary parts

$R_x C_x = R_c\left(C_x/C_a\right)C_b$

or $R_x = R_c \cdot \left(C_b/C_a\right)$.

89. Ans. (b)

Solution: $V_1 = -\frac{1}{10\,k\Omega} \times 10\,k\Omega = -1$ Volt

$$V_0 = \left(\frac{-1}{10(1+\delta)} + \frac{1}{10(1-\delta)}\right) \times 10 = +2\delta$$

$(\delta << 1)$.

90. Ans. (b)

Solution: $\text{Loss} = 20\log\frac{50 \times 10^{-3}}{10}$

$= 20 \log 5 \times 10^{-3} = -46$ dB.

91. Ans. (b)

Solution:

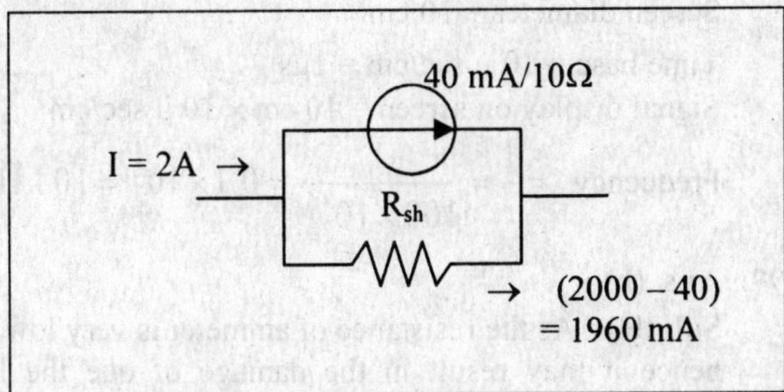

$40 \times 10 = 1960\, R_{sh}$

$R_{sh} = [40/196] = 0.2041$ ohm.

Second Method

I → Current to measurement

I_m → Full scale deflection current movement

$$m = \left(\frac{I}{I_m}\right)$$

$$R_{sn} = \frac{R_m}{m-1} = \frac{10}{\left(\frac{2}{40\times10^{-3}}\right)=1} = \frac{10}{49} = 0.2041\Omega.$$

92. Ans. (c)

Solution: Electrostatic instrument draw negligible power from mains initial current drawn by instrument which connected cross the d.c mains is charging current which is dies down gradually depending upon the insulation resistance of meter which is order of 10^{12} Ω. The final current with d.c operation to zero on a.c operation current draw depends upon the impedance of meter circuit. The current draws on a.c fairly small.

93. Ans. (a)

94. Ans. (a)

Solution: $F = \frac{1}{4\pi\varepsilon_0} \times \frac{Q^2}{r^2}$

$$4\pi\varepsilon = \left(\frac{Q^2}{Fr^2}\right)$$

$$= \frac{Q^2}{M\cdot L\cdot T^{-2}\cdot L^2} = M^{-1}L^{-3}T^{+2}Q^{+2}.$$

95. Ans. (b)

96. Ans. (c)

Solution: Time base = 100 μs/cm

Time period = 10 μs/cm × 10 cm = 100 μs

Frequency = [1/T] = [1/(100 × 10^{-6})]

= 10 kHz

Second Method

Screen diameter = 10 cm

Time base = 10 μ sec/cm = 1 μs

Signal display on screen = 10 cm × 10 μ sec/cm

$$\text{Frequency} = \frac{1}{t} = \frac{1}{100\times10^{-6}} = 0.1\times10^6 = 10\text{ kHz}.$$

97. Ans. (b)

Solution: As the resistance of ammeter is very low and hence it may result in the damage of due the high current.

98. Ans. (c)

99. Ans. (d)

Solution: $f = \left(\frac{1}{T}\right) = \left(\frac{1}{1\times10^{-6}}\right) = 1\text{ MHz}.$

100. Ans. (a)

Solution: Diode peak reading VTVM deflects proportional to the peak value of the applied signal; and a thermocouple meter deflects proportional to the rms value of the applied signal.

A modulated wave is represented as

$E = E_c (1 + m_a \cos \omega_m t) \cos \omega_c t$

where m_a is the modulation index.

In the absence of modulation, i.e. when $m_a = 0$; peak reading VTVM shown the peak value of the carrier

∴ $E_c = 10$ V

The thermocouple meter also reads 10 V. Therefore, its scale is calibrated to read √2 times the rms value.

When sinusoidal AM is applied, the peak value is $E_c (1 + m_a)$.

Thus $10 (1 + m_a) = 15$

∴ $m_a = 0.5$

Now rms value of the amplitude-modulated wave

$$= \sqrt{\left(\frac{E_c}{\sqrt2}\right)^2 + \left(\frac{m_aE_c}{2\sqrt2}\right)^2 + \left(\frac{m_aE_c}{2\sqrt2}\right)^2}$$

$$= \frac{E_c}{\sqrt2}\sqrt{1^2 + 2\frac{m_a^2}{4}} = \frac{10}{\sqrt2}\sqrt{1+\frac{0.5^2}{2}} = \frac{1}{\sqrt2}\times10.61$$

Hence the thermocouple mete will read.

$$\sqrt2\times\frac{1}{\sqrt2}\times10.61 = 10.61\text{ V}.$$

101. Ans. (a)

102. Ans. (c)

Solution: Headphone used in frequency range upto 3 or 4 kHz most sensitive for this frequency range.

103. Ans. (c)

Solution: Thermistor: A resistor, usually fabricated from semiconductor material, that a large nonlinear negative temperature coefficient of resistance. An electrolyte, such as a viscous solution of water glass, is some time used as the temperature-sensitive element. The thermistor is usually shaped as a rod, bead or disc and named accordingly. Applications include compensation for temperature variations in other components, used as a nonlinear circuit element, and for temperature and power measurements.

104. Ans. (b)

Solution: Rotational speed

$$= \frac{13500}{150} = 90\text{ rps}$$

= 90 × 60 rpm = 5400 rpm.

105. Ans. (c)

MCQ Set–3

1. Consider the following statements:
DC Potentiometer is the means available for measurement of dc voltages because
   1. The precision in measurement is independent of the type of detector used
   2. It is base on null-balance technique
   3. It is possible to standardize before a measurement is undertaken
   4. It is possible to measure dc voltages ranging in value from millivolts to hundreds of volts.

   Of these statements
   (a) 2 and 3 are correct
   (b) 1and 4 are correct
   (c) 2 and 4 are correct
   (d) 3 and 4 are correct

2. The bridges suitable for the measurement of an unknown inductance in terms of a known capacitance would include
   (a) Maxwell and Hay
   (b) Maxwell and Schering
   (c) Hay and Schering
   (d) Maxwell, Hay and Schering

3. In the **Maxwell bridge** as shown in given figure, the value of resistance R_x and inductance L_x of a coil are to be calculated after balancing the bridge. The component values are shown in the figure at balance. The value of R_x and L_x will respectively be

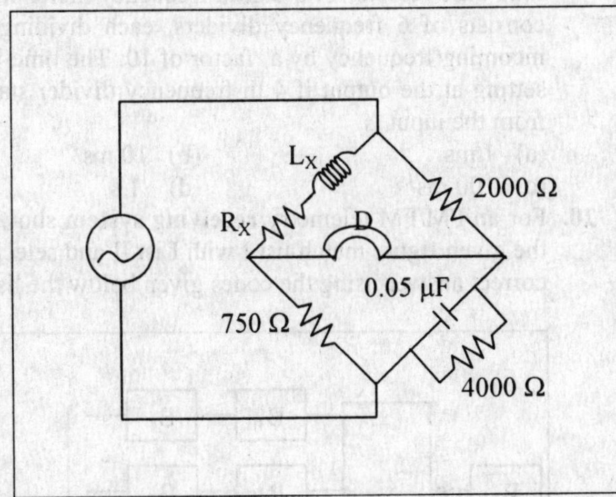

   (a) 375 Ω, 75 mH (b) 75 Ω, 150 mH
   (c) 37.5 Ω, 75 mH (d) 75 Ω, 75 mH

4. In the bridge given in figure, the reading of the high impedance voltmeter is

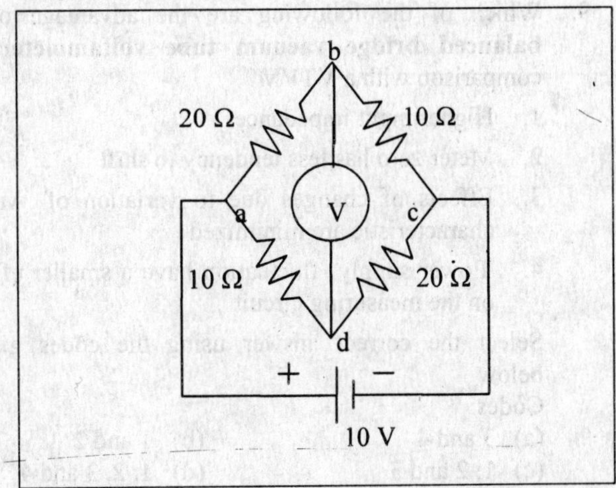

   (a) Zero (b) 6.66 V
   (c) 4.20 V (d) 3.33 V

5. Rectifier Moving coil instruments respond to
   (a) Peak value, irrespective of the nature of the waveform
   (b) Average value, for all waveforms
   (c) rms value for all waveforms
   (d) rms value , for symmetrical square waveforms

6. An average response rectifier type electronic ac voltmeter has a dc voltage of 10 V applied to it. The meter reading will be
   (a) 7.1 V (b) 10.0 V
   (c) 11.1 V (d) 22.2 V

7. Consider the following statements regarding the source of error in a Q-meter:
   1. If a coil with a resistance R is connected in the direct measurement mode and if the residual resistance of the Q-meter is 0.1 R, then the measured Q of the coil would be 1.1 times the actual Q.
   2. IF the inductance to be measured is less than 0.1μH , the error due to the presence of residual inductance cannot be neglected
   3. The presence of distributed capacitance in a coil modifies the effective Q of the coil

   Of these statements
   (a) 1, 2 and 3 are correct
   (b) 1 and 2 are correct
   (c) 2 and 3 are correct
   (d) 1 and 3 are correct

8. An advantage which a VTVM has over a non-electronic voltmeter is
   (a) Lower power consumption
   (b) Lower input impedance
   (c) The ability to measure wider ranges of voltage and resistance
   (d) Grater portability

9. Which of the following are the advantage of a **balanced bridge vacuum tube voltammeter** in comparison with a VTVM?

1. Higher input impedance
2. Meter zero has less tendency to shift
3. Effects of changes due to variation of valve characteristic are minimized
4. Power supply fluctuation have a smaller effect on the measuring circuit

Select the correct answer using the codes given below

Codes:

(a) 3 and 4 (b) 1 and 2
(c) 1, 2 and 3 (d) 1, 2, 3 and 4

10. A meter has full-scale deflection of 90° at a current of 1 A. The response of the meter is square-law. Assuming spring control, the current for a deflection 45° will be

(a) 0.25 A (b) 0.50 A
(c) 0.67 A (d) 0.707 A

11. A permanent magnet moving coil type ammeter and moving iron type ammeter are connected in series in a resistive circuit fed from output of a half wave rectifier voltage source. If the moving iron type instrument reads 5 A, the permanent magnet moving coil type instrument is likely to read.

(a) Zero (b) 2.5 A
(c) 3.18 A (d) 5 A

12. A **compensated wattmeter** has its reading corrected for error due to

(a) The frequency
(b) Friction
(c) Power consumed in current coil
(d) Power consumed in potential coil

13. While measuring power in a three-phase load by two-wattmeter method, the reading of the two wattmeter will be equal and opposite when

(a) Power factor is unity
(b) Load is balanced
(c) Phase angle is between 60° and 90°
(d) The load is purely inductive

14. The coils having self-inductances of 10 mH and 15 mH have an effective inductance of 40 mH, when connected in series-aiding. What will be the equivalent inductance if we connect them in series-opposing?

(a) 20 mH (b) 10 mH
(c) 5 mH (d) Zero

15. A **polarized electrolytic capacitor** can be used

(a) Only sinusoidal circuits
(b) In dc and sinusoidal ac circuits
(c) Only in dc circuits
(d) In circuits having any arbitrary waveform of current

16. The Lissajous pattern obtained on CRO screen is shown in the given figure:

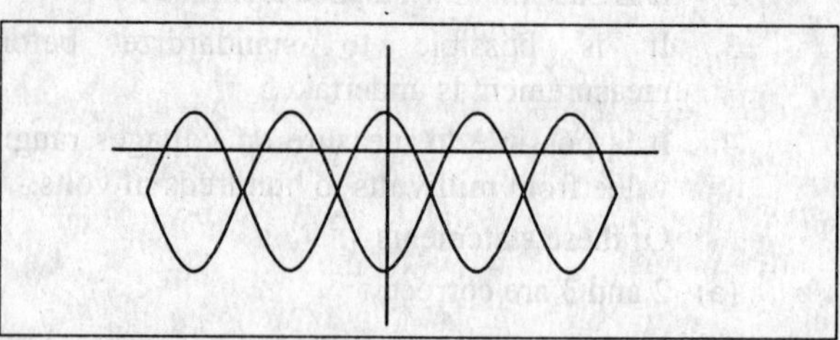

The frequency ration of the vertical signal to the horizontal one is

(a) 3 : 2 (b) 5 : 1
(c) 1: 5 (d) 2 : 3

17. The major cause of **creeping** in an energy meter is

(a) Over-compensation for friction
(b) Mechanical-vibrations
(c) Excessive voltage across the potential coil
(d) Stray magnetic fields

18. In a **digital voltmeter,** the oscillator frequency is 400 kHz and the ramp voltage falls from 8 V to 0 V in 20 m sec. The number of pulses counted by the counter is

(a) 800 (b) 2000
(c) 4000 (d) 8000

19. Pulses of a frequency of 1 MHz are applied to the time base selector of a digital frequency meter which consists of 6 frequency dividers, each dividing the incoming frequency by a factor of 10. The time-base setting at the output if 4 th frequency divider staring from the input is

(a) 1 ms (b) 10 ms
(c) 100 ms (d) 1 s

20. For an FM/FM telemetry **receiving system** shown in the given figure match list I with List II and select the correct answer using the codes given below the list

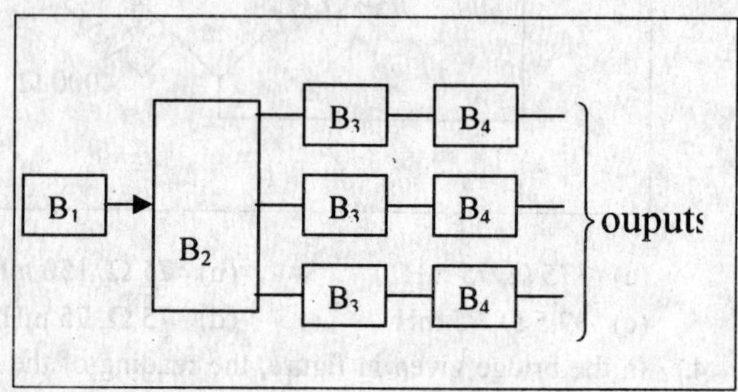

List I Black Box	List II Circularly
A B 1	1. Band-pass filter
B. B 2	2. Receiver
C. B 3	3. Low-pass filter
D. B 4	4. Amplifier discriminator
	5. Sub carrier discriminators

Codes:

	A	B	C	D
(a)	3	2	4	1
(b)	2	3	5	1
(c)	2	1	5	3
(d)	1	2	4	5

21. Match List I with List II and select the correct answer using the codes given below the lists:

List I	List II
A. Display device (lowest power)	1. Light emitting diode
B. High speed display device	2. Nixie Tube
C. Bright perfect numeral display	3. Liquid crystal display
D. Perfect light transmitting device	4. Optical fibre

Codes:

	A	B	C	D
(a)	1	2	3	4
(b)	2	4	3	1
(c)	3	1	2	4
(d)	4	3	2	1

22. In two–wattmeter method of power measurement, one of the watt meters will show negative reading when the load pf angle is strictly

(a) less than 30° (b) less than 60°
(c) greater than 30° (d) greater than 60°

23. Which of the following are the necessary conditions for an entire three–phase system to be balanced?

1. The line voltages are equal in magnitude
2. The phase difference between successive line voltages are equal
3. The impedance in each of the phase are identical

Select the correct answer using the codes given below:

Codes:

(a) 1,2 and 3 (b) 1 and 3
(c) 1 and 2 (d) 2 and 3

24. The closed circuit shown in the given figure is linked by a steady magnetic flux density B and is in two parts. Initially, when the switch 'S' is closed, V, a centre zero voltmeter, reads zero. If 'S' is suddenly opened, the voltmeter pointer will.

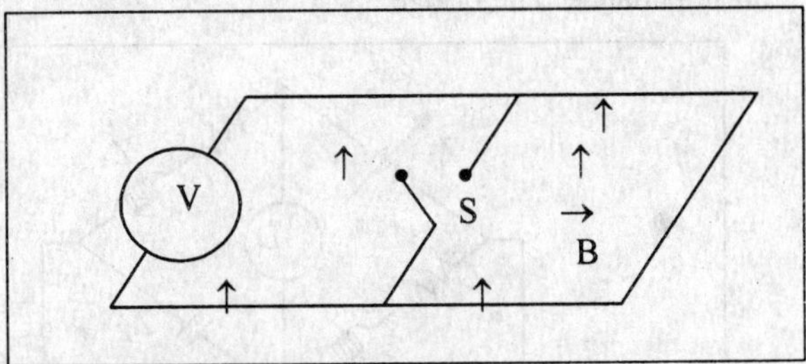

(a) Remain at zero
(b) Momentarily move to the left
(c) Momentary move to the right
(d) Show a damped oscillation about zero

25. Consider the following statements: ·The **absolute value of current** can be measured by

1. dynamometer type moving coil ammeters
2. tangent galvanometers
3. current balance instruments

Of these statements

(a) 1, 2 and 3 are correct
(b) 1 and 2 are correct
(c) 1 and 3 are correct
(d) 2 and 3 are correct

26. A slide wire potentiometer has 10 wires of 1 m each. With the help of a standard voltage source of i.018 V it is standardized by keeping the jockey at 101.8 cm. If the resistance of the potentiometer wires is 1000 ohm, then the value of the working current is

(a) 0.1 mA (b) 0.5 mA
(c) 1 mA (d) 10 mA

27. Match List I (Bridges) with List II (Quantity measured) and select the correct answer using the codes given below the lists:

List I	List II
A. Anderson Bridge	1. Frequency
B. Kelvin Double Bridge	2. Resistance
C. Schering Bridge	3. inductance
D. Wien Bridge	4. Capacitance

Codes :

	A	B	C	D
(a)	3	2	1	4
(b)	3	2	4	1
(c)	2	1	3	4
(d)	2	3	4	1

28. In the ac bridge shown in the given figure, the impedances Z_1, Z_2, Z_3 and Z_4 in ohms at the supply frequency are as indicated. Z_3 is a variable impedance. The bridge.

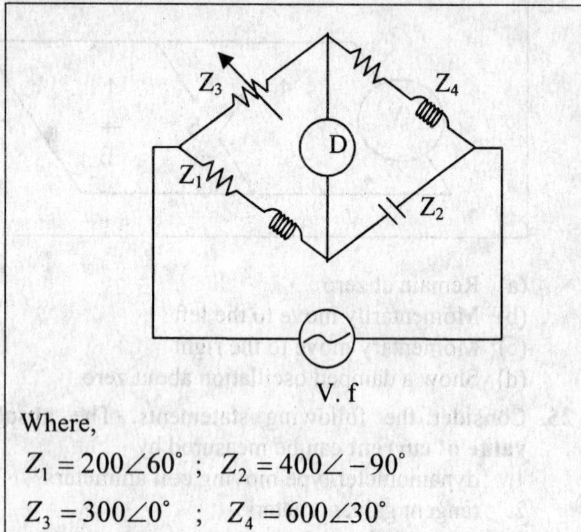

Where,

$Z_1 = 200\angle 60^\circ$; $Z_2 = 400\angle -90^\circ$

$Z_3 = 300\angle 0^\circ$; $Z_4 = 600\angle 30^\circ$

(a) is balanced with the indicated vlaues of impedances
(b) can be balanced if $Z_3 = 400\angle 0^\circ$
(c) can be balanced if $Z_4 = 600\angle 60^\circ$
(d) cannot be balanced with the given configuration

29. A PMMC zero to 10 A ammeter is not provided with any controlling mechanism and the moving part are free to rotate. If a current of 1 A dc is passed through the moving coil, then the instrument.
(a) will read 1 A
(b) will read 10 A
(c) pointer will continually rotate
(d) pointer will remain stationary

30. Match List I (Expression for torque) with List II (Type of instrument) and select the correct answer using the codes given below the lists :

List I	List II
A. $\phi_1 \phi_2 \cos\phi$	1. Ferrodynamic instruments
B. $I_1 I_2 \cos\phi$	2. D' Arsonval instruments
C. ki	3. Electrodynamometer instruments

Codes :

	A	B	C
(a)	2	1	3
(b)	3	2	1
(c)	1	3	2
(d)	3	1	2

31. A PMMC instrument uses a coil of resistance 0.2 ohm and produces full–scale deflection with a steady current of 10 A. If this instrument is used as an ammeter without the addition of any external resistance then the range of the instrument will be zero to.
(a) 0.2 A (b) 1 A
(c) 2 A (d) 10 A

32. Two voltages V_1 and V_2 are connected to the X and Y plates of a CRO as shown in the given figure. To get a circular pattern on the CRO screen , R should be adjusted to a value of

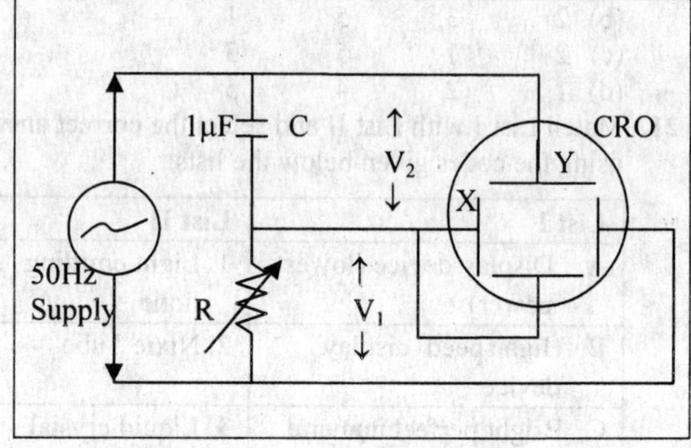

(a) 1 kΩ (b) 16 kΩ
(c) 2.35 kΩ (d) 3.18 kΩ

33. If Q_e is the effective Q of the coil, C is the resonance capacitance and C_d is the distributed capacitance then the true Q in a Q –meter will be
(a) $Q_e[(C + C_d)/C]$ (b) $Q_e[C/(C + C_d)]$
(c) $Q_e[C_d/(C + C_d)]$ (d) $Q_e[(C + C_d)/C_d]$

34. A Q–meter is supplied with an oscillator having a 500 mV output voltage. While testing an unknown inductor, the voltage across the variable capacitance of the Q-meter, measured by a VTVM, is obtained as 10 V. The Q–factor of the inductor is
(a) 5 (b) 10
(c) 20 (d) 100

35. A heterodyne spectrum analyzer has several mixer stages and covers the frequency range form 10 kHz to 1.3 GHz. The first IF stage has a center frequency of 1.45 GHz. The first stage local oscillator should
(a) Be at the fixed frequency of 2.75 GHz
(b) Sweep the frequency range from 0.15 GHz to 1.45 GHz
(c) Sweep the frequency range from 1.45 GHz to 2.75 GHz
(d) Be at the fixed frequency of 0.15 GHz

36. The voltmeter of choise for measuring the emf of a 100 V dc source would be

(a) 100 V, 1mA (b) 100 V, 2 mA
(c) 100 V, 10kΩ/V (d) 100 V, 100 Ω/V

37. An electrodynamometer type wattmeter is connected (as shown in the given figure) in a 3–phase supply and having a 3–phase balanced load. E and I are the values of phase voltage and current and ϕ is the phase angle between them. The wattmeter reading will be

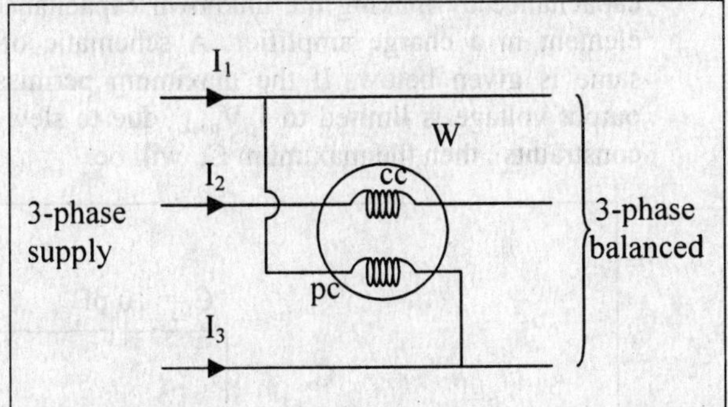

(a) Proportional to EI sin ϕ
(b) Proportional to EI cos ϕ
(c) Proportional to EI tan ϕ
(d) Zero

38. The natural frequency of vibration of a reed used in a vibrating reed frequency meter, can be in increased by
(a) Increasing the gap between the reed and the electromagnet
(b) Decreasing the gap between the reed and the electromagnet
(c) Increasing the length of the reed
(d) Decreasing the length of the reed

39. A single–phase energy meter is operating on 230 V, 50 Hz supply with a load of 20 A for two hours at upf. The meter makes 1380 revolutions in that period. The meter constant is
(a) 695 rev/kWh (b) 150 rev/kWh
(c) 0.15 rev/kWh (d) 1/150 rev/kWh

40. In **noise–level measurement**, the output of the microphone pick –up is weighted suitably by electrical filters so as to
(a) Provide linearised response at different frequencies
(b) Provide frequency spectra at different frequencies
(c) Filter the radio frequency interference
(d) Provide a response similar to that of the human ear

41. In a digital voltmeter, ' **over–ranging** 'imples that
(a) The next four digits are switched on
(b) 1 /2 digit is switched off
(c) 1 /2 digit is switched on
(d) An over–range indicator starts glowing

42. Which of the following features determine the **precision** of an **integrating type** digital voltmeter
1. Time constant of the integrator
2. Input impedance of the integrator
3. Reference voltage of the comparator
4. Opening time of the gate before the counter

Select the correct answer using the codes given below:
Codes :
(a) 1,2 and 3 (b) 1,2 and 4
(c) 1,3 and 4 (d) 2,3 and 4

43. In a **telemetry system** used for measuring inaccessible non–electrical quantity, the following subsystems are used at the measuring end :
1. Transmitter 2. Transducer
3. Amplifier 4. Modulator

The correct sequence in which these sub–systems are placed is
(a) 2,3,14 (b) 2,3,4,1
(c) 3,2,14 (d) 3,2,4,1

44. Galvanometer type recorders use
(a) Vibration galvanometer
(b) Ballistic galvanometer
(c) D' Arsonval galvanometer
(d) Tangent galvanometer

45. Horizontal deflection in a CRO is due to E sin ωt while vertical deflection is due to E $\sin(\omega t+\theta)$ with a positive θ. Consider the following patterns obtained in the CRO.
The correct sequence of these patterns in increasing order of the values of θ is

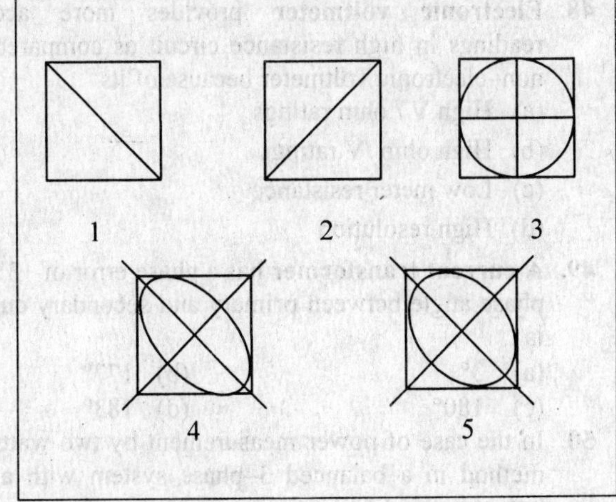

(a) 3, 2, 5, 1, 4 (b) 3, 2, 4, 5, 1
(c) 2, 3, 5, 1, 4 (d) 2, 3, 5, 4, 1

46. Which of the following units are present in a **spectrum analyzer**?
1. Mixer
2. Saw–tooth generator
3. Local oscillator

Select the correct answer using the codes given below:

Codes:
(a) 1, 2 and 3 (b) 1 and 2
(c) 1 and 3 (d) 2 and 3

47. For the voltmeter circuit shown in the given figure, the basic D' Arson Val meter used has full-scale current of 1 mA and meter resistance (R_m) of 100 ohms. The values of the series resistance R_1 and R_2 required for 10 V range and 50 V range will be respectively.

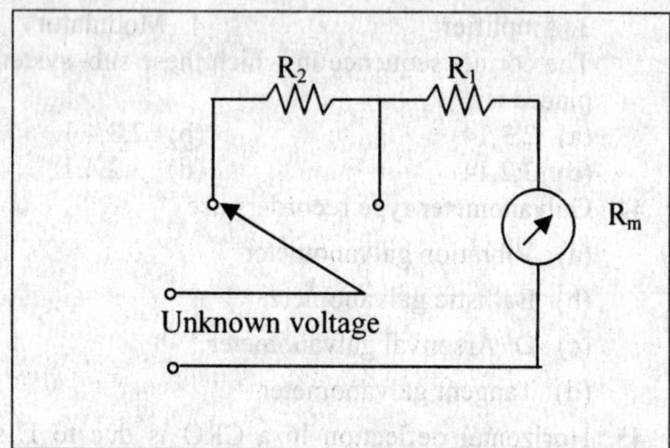

(a) $9.9 k\Omega$ and $40 k\Omega$
(b) $10 k\Omega$ and $50 k\Omega$
(c) $20 k\Omega$ and $30 k\Omega$
(d) $200 k\Omega$ and $250 k\Omega$

48. Electronic voltmeter provides more accurate readings in high resistance circuit as compared to a non–electronic voltmeter because of its
(a) High V / ohm ratings
(b) High ohm /V ratings
(c) Low meter resistance
(d) High resolution

49. A current transformer has a phase error of +3°. The phase angle between primary and secondary currents is
(a) 3° (b) 177°
(c) 180° (d) 183°

50. In the case of power measurement by two wattmeter method in a balanced 3–phase system with a pure inductive load.
(a) Both the wattmeter's will indicate the same value but of opposite sign
(b) Both the wattmeter will indicate zero
(c) Both the wattmeter will indicate the same value and of the same sign
(d) One wattmeter will indicate zero and other will indicate some non–zero value

51. Many modern capacitance meters measure capacitance by making the unknown capacitance an element in a charge amplifier. A schematic of the same is given below. If the maximum permissible output voltage is limited to 1 V_{peak} due to slew rate constraints , then the maximum C_x will be.

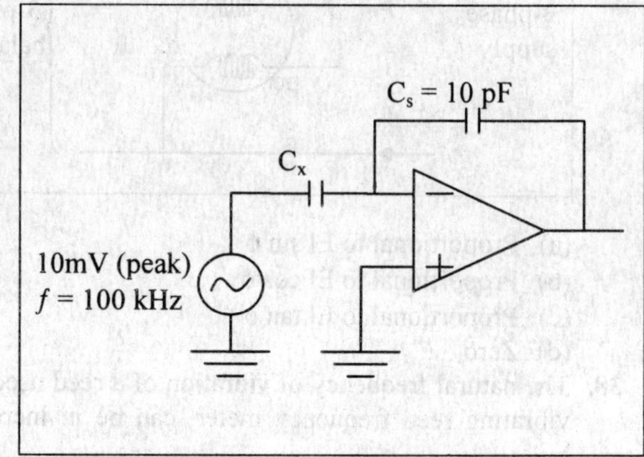

(a) 10 pF (b) 100 pF
(c) 1000 pF (d) 10 nF

52. A 230 V,10 A, single-phase energy meter makes 90 revolutions in 3 minutes at half load rated voltage and unity pF. If the meter constant is 1800 revolutions /kWh, then its error at half load will be.
(a) 13.04% slow (b) 13.04% fast
(c) 15% slow (d) 15% fast

53. A piezoelectric transducer has the following parameter values:

Crystal capacitance $= 10^{-9}\ F$

Cable capacitance $= 2\times10^{-10}\ F$

Charge sensitivity $= 4\times10^{-6}$ Coulomb/cm

If the oscilloscope used for read-out has an input resistance of 1 MΩ in parallel with $C = 4\times10^{-10}\ F$ then the voltage sensitivity constant will be
(a) 2500 V/cm (b) 3334 V/cm
(c) 4000 V/cm (d) 4500 V/cm

54. A-7 bit successive approximation DVM will convert the analog input into digital output in a period of
(a) 7 clock pulses
(b) 8 clock pulses

(c) $\dfrac{\text{input signal amplitude}}{\text{full-scale output amplitude}} \times 7 \text{ clock pulses}$

(d) $\dfrac{\text{input signal amplitude}}{\text{full-scale output amplitude}} \times 8 \text{ clock pulses}$

55. Match List I with List II and select the correct answer using the codes given below the Lists:

List I	List II
A. Low pass filter	1. $\dfrac{T_s}{1+T_s}$
B. High pass filter	2. $\dfrac{1}{1+\left(T_s+\dfrac{1}{T_s}\right)}$
C. Band pass filter	3. $\dfrac{T_s+\dfrac{1}{T_s}}{1+\left(T_s+\dfrac{1}{T_s}\right)}$
D. Band reject filter	4. $1+\dfrac{1}{1+T_s}$

56. Data buffering means.
(a) A buffer stock of data already stored in a computer unit
(b) An electronic circuit which is used to maintain level of data
(c) That data are first collected in an external unit and then processed into the computer
(d) Application of buffer (unit gain amplifier) at interface to the computer

57. A circuit draws a current I when a single phase ac voltage V is applied to it. If the power factor is cos ϕ, then the dimensions of VI cos ϕ would be
(a) ML^3T^{-2} (b) ML^2T^3
(c) ML^3T^3 (d) ML^2T^{-3}

58. Swamping resistance is a resistance which is added to the moving coil of a meter to
(a) Reduce the full –scale current
(b) Reduce the temperature error
(c) Increase the sensitivity
(d) Increase the filed strength

59. The dimensional equation of resistance is
(a) $L^2MT^{-2}\ I^{-4}$ (b) $L^2MT^{-2}\ I^{-2}$
(c) $L^2M^2T^{-3}\ I^{-2}$ (d) $L^2MT^{-3}\ I^{-2}$

60. A high frequency as signal is applied to a PMMC instrument. If the rms value of the ac signal is 2 V, then the reading of the instrument will be
(a) zero (b) 2 V
(c) $2\sqrt{2}\,V$ (d) $4\sqrt{2}\,V$

61. Three resistance of a shunt for a precision grade **ammeter** can be best measured by
(a) De Sauty bridge
(b) Scherring bridge
(c) Maxwell bridge
(d) Kelvin double bridge

62. Which one of the following has the highest accuracy?
(a) Standard resistance
(b) Standard inductance
(c) Standard capacitance
(d) Standard mutual inductance

63. In the circuit shown in the figure, if ammeter indicates 1 A, and the voltmeter having and internal resistance of 1 kΩ indicated 100 V, then the value of R would be.

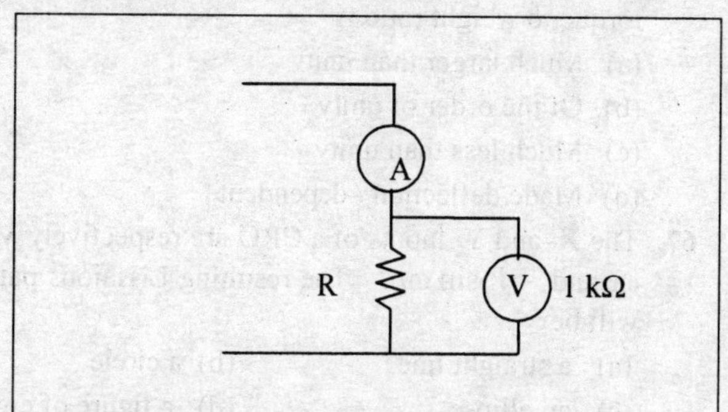

(a) 111.11 Ω (b) 105.2 Ω
(c) 100 Ω (d) 90.9 Ω

64. In the balanced Wheatstone bridge shown in the figure, if the value of R_6 is increased, the current I_2

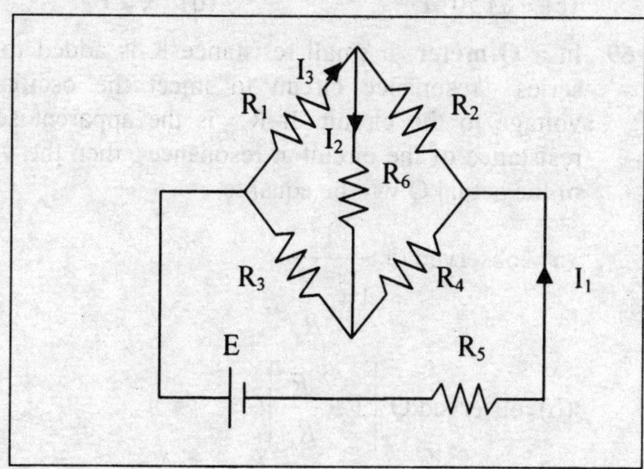

(a) will increase (b) will decrease
(c) will remain unchanged

65. Match list I with List II and select the correct using the codes given below the Lists:

List I	List II
A. Low value of R.	1. Scherring bridge
B. High–Q inductor	2. Maxwell bridge
C. Low–Q inductor	3. Kelvin double bridge
D. High voltage capacitors	4. Hay bridge

Codes:

	A	B	C	D
(a)	1	2	4	3
(b)	1	2	4	3
(c)	3	2	4	1
(d)	3	4	2	1

66. An **indicting instrument** is more **sensitive** if its torque to weight ratio is
(a) Much larger than unity
(b) Of the order of unity
(c) Much less than unity
(d) Made deflection –dependent

67. The X–and Y–inputs of a CRO are respectively V sin ωt and $-V\sin\omega t$. The resulting Lissajous pattern will be
(a) a straight line (b) a circle
(c) an ellipse (d) a figure of eight

68. A current $i=(10+10\sin t)$ amperes is passed through an ideal **moving iron type ammeter**, its reading will be
(a) zero (b) 10 A
(c) $\sqrt{150}\,A$ (d) $\sqrt{2}\,V$

69. In a **Q-meter**, a small resistance R is added to the series resonance circuit to inject the oscillatory voltage to the circuit. If R_s is the apparent series resistance of the circuit at resonance , then the value of the actual Q will be equal to

(a) observed Q $\dfrac{1}{1+\dfrac{R}{R_s}}$

(b) observed Q $\left[1+\dfrac{R}{R_s}\right]$

(c) observed Q $\left[\dfrac{1}{1+\dfrac{R_s}{R}}\right]$

(d) observed Q $\left[1+\dfrac{R_s}{R}\right]$

70. In a **two –wattmeter method** of measuring power, one of the wattmeters is reading zero watts. The power factor of the circuit is
(a) zero (b) 1
(c) 0.5 (d) 0.8

71. In **distortion factor meter,** the filler is used is to suppress
(a) dc component (b) odd harmonics
(c) even harmonics (d) fundamentals

72. Which of the following measurements can be made with the help of a **frequency counter?**
1. Fundamental frequency of input signal
2. Frequency components of the input signal at least upto third harmonic
3. Time interval between two pulses
4. Pulse width

Select the correct answer using the codes given below
Codes:
(a) 1, 3 and 4 (b) 1, 2 and 3
(c) 2 and 4 (d) 1 and 2

73. The bandwidth requirement of an FM telemetry channel is
(a) Equal to that of an AM telemetry channel
(b) Smaller than that of an AM telemetry channel
(c) About 100 times of an AM telemetry channel
(d) About ten times that of an AM telemetry channel.

74. The recording head in a magnetic tape responds to
(a) Electrical signal and creates a magnetic signal
(b) Thermal signal and creates a magnetic signal
(c) Magnetic signal and creates a electron signal
(d) Thermal signal and creates a electron signal

75. In the SI system, the dimension of emf is
(a) $M\,L^3\,T^{-3}\,I$ (b) $M\,L^3\,T^{-3}\,I^{-1}$
(c) $M\,L^2\,T^{-3}\,I^{-1}$ (d) $M^{-1}\,L^2\,T^3\,I$

76. Which one of the following techniques is 'adopted to reduce the self-inductance of a wire-wound standard resistor?

(a) Bifilar winding
(b) Negative impedance converter
(c) Force balance type flux compensation
(d) Providing additional potential terminals

77. The voltage of a standard cell is monitored daily over a period of one year. The mean value of the voltage for every month shows a standard deviation of 0.1 mV. The **standard deviation** of the set constituted by the monthly mean values will be

(a) Zero (b) $\frac{0.1}{12}$

(c) $\frac{0.1}{\sqrt{12}}$ (d) 0.1

78. A bridge is shown in the given figure. If the resistance R_s is increased from 2kΩ to 2.5 kΩ, the current I_2 will.

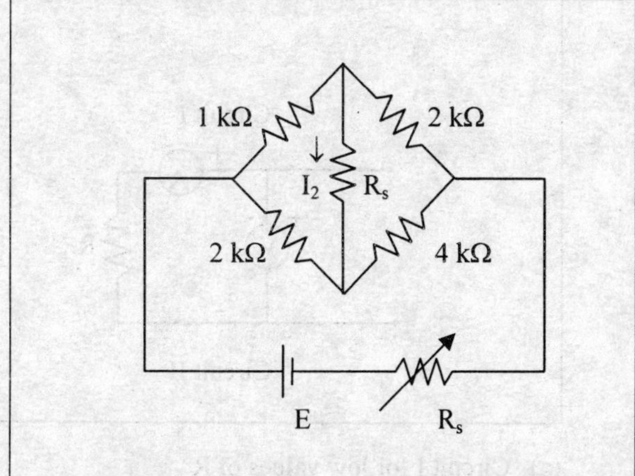

(a) Increase
(b) Decrease
(c) Not change
(d) Increase or decrease depending on the polarity of E

79. When a current i(t) =5 + 10 cos 314 t is measured by an **electrodynamics ammeter,** the meter will read

(a) $\sqrt{5^2+10^2}$ (b) $\sqrt{5^2+\frac{10^2}{2}}$

(c) $\sqrt{5^2+\left(\frac{10}{2}\right)^2}$ (d) $\sqrt{\frac{5^2+10^2}{2}}$

80. Two equal voltage of same frequency applied to the X and Y plates of a CRO, produces a circle on the screen. The phase difference between the two voltage is

(a) 30° (b) 60°

(c) 90° (d) 150°

81. In a Q-meter, an inductor tunes to 2 MHz with 450 pF and to 4 MHz with 90 pF. The distributed capacitance of the inductor is

(a) 30 pF (b) 45 pF
(c) 90 pF (d) 360 pF

82. Consider the network shown in the given figure. If a voltmeter of internal resistance 10 kΩ read V_1 and V_2 respectively when connected across 5 kΩ and 2 kΩ in turn, then

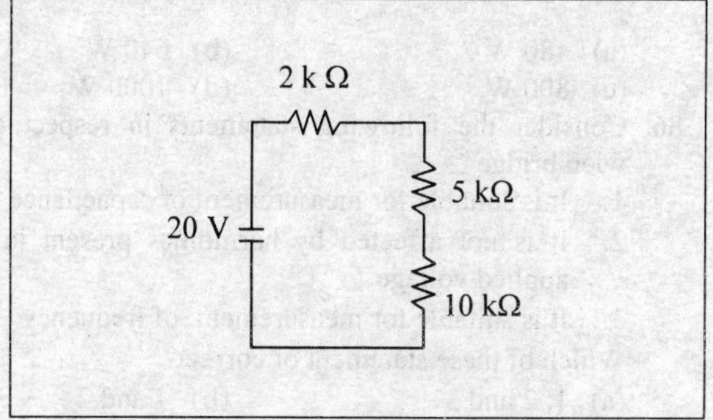

(a) $V_1 > \frac{V_2}{2}$ (b) $V_1 < \frac{V_2}{2}$

(c) $V_1 = \frac{V_2}{2}$ (d) $V_1 = 6.67\ V$

83. Consider the following statements
A current transformer is used for measurement of large current to

1. Isolate the instrument from a high voltage bus bar
2. Increase the accuracy of measurement
3. Decrease the cost of measuring arrangements
4. Extend the range of measurement of a conventional ammeter on ac

Which of these statements are correct?

(a) 1 and 2 (b) 2 and 3
(c) 3 and 4 (d) 1 and 4

84. If the reading of the two wattmeters are equal and positive in two-wattmeter method, the load pf in a balance 3-phase 3-wire circuit will be

(a) Zero (b) 0.5
(c) 0.866 (d) Unity

85. In the circuit shown in the given figure, the wattmeter reading will be

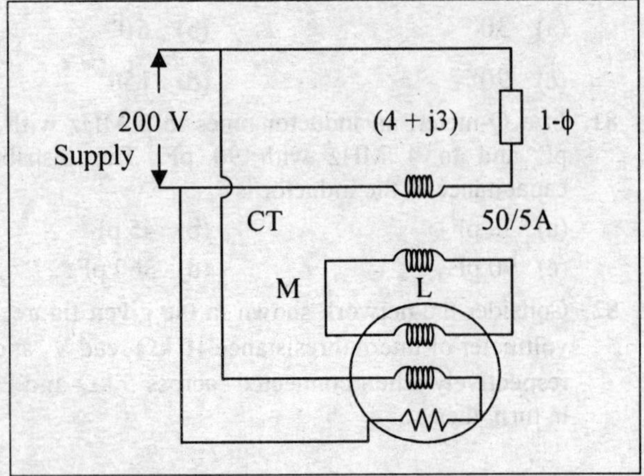

(a) 480 W (b) 640 W
(c) 800 W (d) 1000 W

86. Consider the following statements in respect of a wien bridge
1. It is suitable for measurement of capacitance
2. It is not affected by harmonics present in the applied voltage
3. It is suitable for measurement of frequency

Which of these statement of correct?
(a) 1, 2 and 3 (b) 1 and 2
(c) 2 and 3 (d) 1 and 3

87. Consider the following A/D converters used commonly in digital instruments
1. Successive approximation type
2. Flash type
3. Dual slope type

The correct sequence in increasing order of their conversion is
(a) 1, 2, 3 (b) 2, 3, 1
(c) 2, 1, 3 (d) 3, 1, 2

88. A shunt type ohmmeter is shown in the figure. With R_x disconnected, the meter reads full scale, 'S' represents the meter current as a fraction of full scale current R_x connected such that, $S = \frac{R_x}{R_x + R_p}$

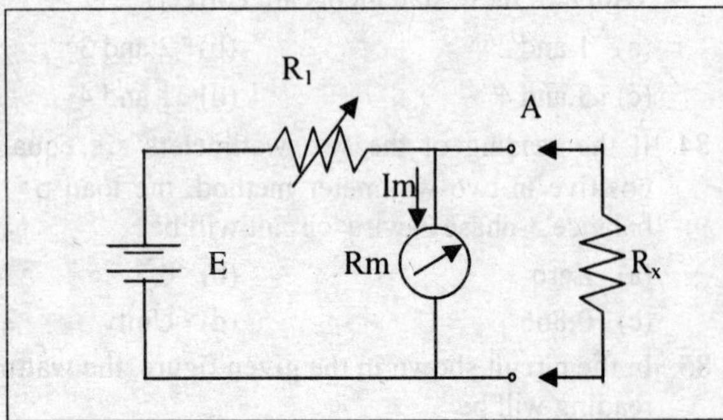

The value of R_p is given by
(a) R_m (b) $R_1 + R_m$
(c) $\frac{R_1 R_x}{(R_1 + R_m)}$ (d) $\frac{R_1 R_m}{(R_1 + R_m)}$

89. The power in a resistor R is estimated by measuring the voltage and current using the voltmeter-ammeter method. Two different arrngements can be used as shown in circuits I and II. Less erroneous results are obtained by adopting.

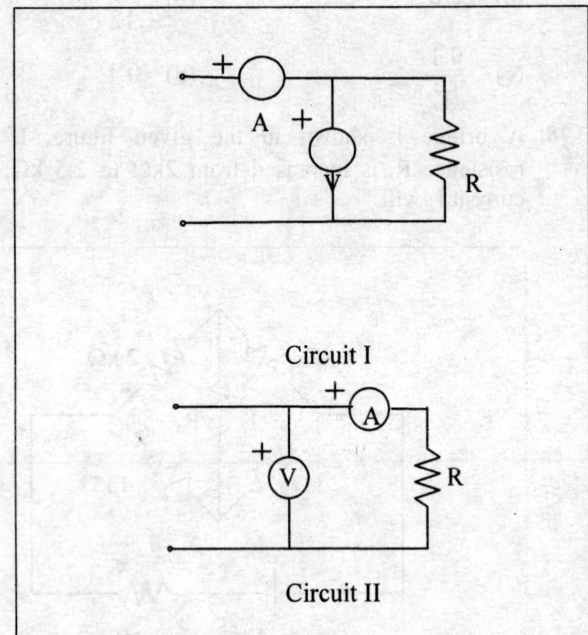

(a) Circuit I for low values of R
(b) Circuit II for low values of R
(c) Circuit I for high values of R
(d) Circuit II for low and high values of R

90. A simple dc potentiometer is to be standardised voltage by keeping the slidewire setting at 1.0183 V. If by mistake, the setting is at 1.0138 V and the standardisation is made to obtain a source voltage of 1.0138 V, then the reading of the potentiometer will be
(a) 1.0138 V (b) 1.0183 V
(c) $\frac{(1.0138)^2}{1.0183} V$ (d) $(1.0138)^2$ V

91. In the bridge shown in the figure
$Z_1 = 450\Omega$,
$Z_2 = (300 - j\,600)\ \Omega$
$Z_3 = (200 + j\,100)\ \Omega$

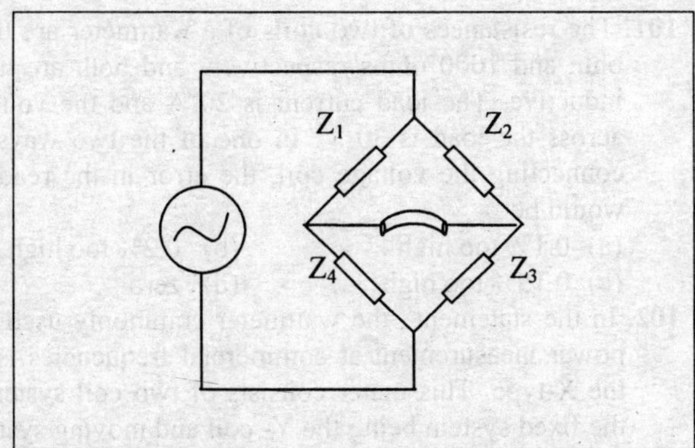

The value of unknown impedance Z_4 at balance is

(a) $(100 + j\,150)\ \Omega$ (b) $(100 - j\,150)\ \Omega$
(c) $(300 + j\,100)\ \Omega$ (d) $(0 + j\,150)\ \Omega$

92. Which of the following conditions are to be satisfied so that the common variable shaft of resistance R_1 and R_3 can be graduated in frequency to measure the frequency of E under balanced condition?

(a) $R_1 = R_3$ (b) $C_1 = C_3$
(c) $R_2 = 2R_4$ (d) $R_2 = R_4$

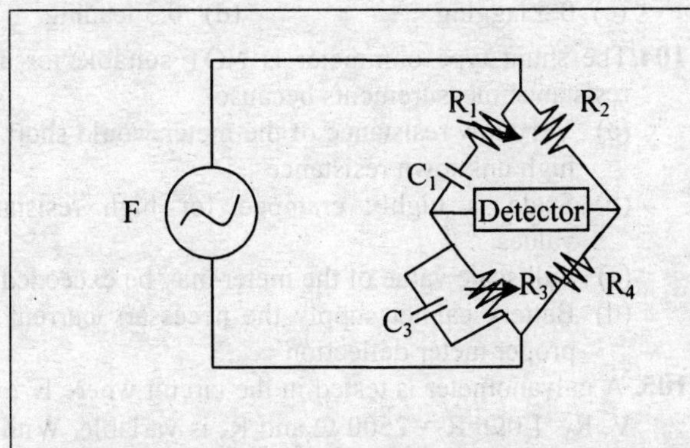

Select the correct answer using the codes below:

(a) 1and 4 (b) 1 and 2
(c) 2 and 4 (d) 1, 2 and 3

93. In the multimeter circuit shown in the figure for ac voltage measurement, the function of diode D_1

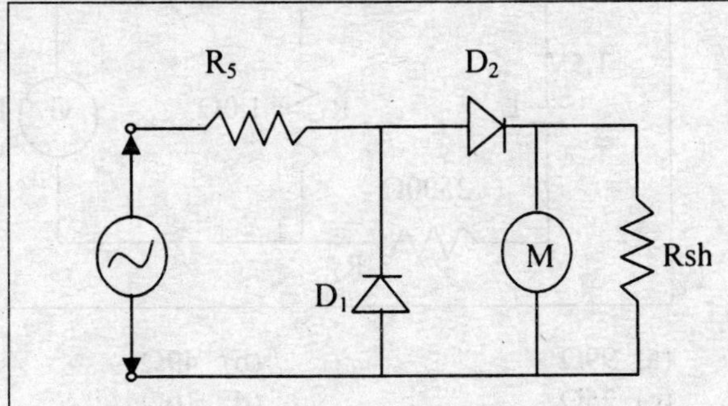

(a) Provide half-wave rectification
(b) Make the rectifier D_2 perform full-wave rectification
(c) By-pass reverse leakage current of D_2 in the negative cycle of the input
(d) Short circuit over-range voltages

94. The main advantage in using a three OP-AMP instrumentation amplifier over a single OP-AMP differential amplifier lies in

(a) Higher values of CMRR
(b) Lower noise figure
(c) Elimination of the need for accurate matching of resistors
(d) Simplicity of gain adjustment

95. Two sine waves of the same frequency are impressed on the X and Y plates of a CRO and the Lissajous figure seen is shown in the diagram. The phase difference between the signals is

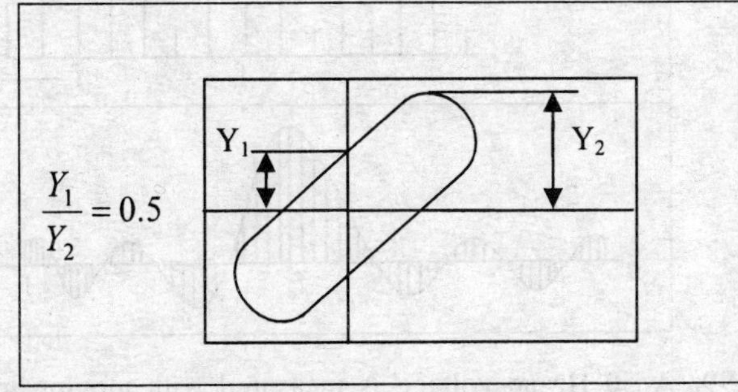

(a) 30° or 330° or 150° or 210°
(b) 30° or 330° or 150°
(c) 30° or 330°
(d) 30°

96. When the horizontal deflection plates of CRO are kept at the ground potential and a 30 volt dc is applied to the vertical deflecting plates, the bright spot moves 1 cm away from the centre. If with the same setting, a 30 volt ac is applied to the vertical deflecting plates, then the picture observed on the screen would be

(a) A spot approximately 3 cm away from the center
(b) A vertical line 2 cm long
(c) A vertical line approximately 3 cm long
(d) Two spots 2 cm vertically above each other

97. When testing a coil having a resistance of 10 ohms, resonance occurred when the oscillator frequency was 10 MHz and the rotating capacitor was set at 500/2pF. The effective value of Q of the coil is

(a) 200 (b) 254
(c) 314 (d) 542

98. The resultant spectrum of a carrier, amplitude modulated with a rectangular pulse will be (f_c is the carrier frequency).

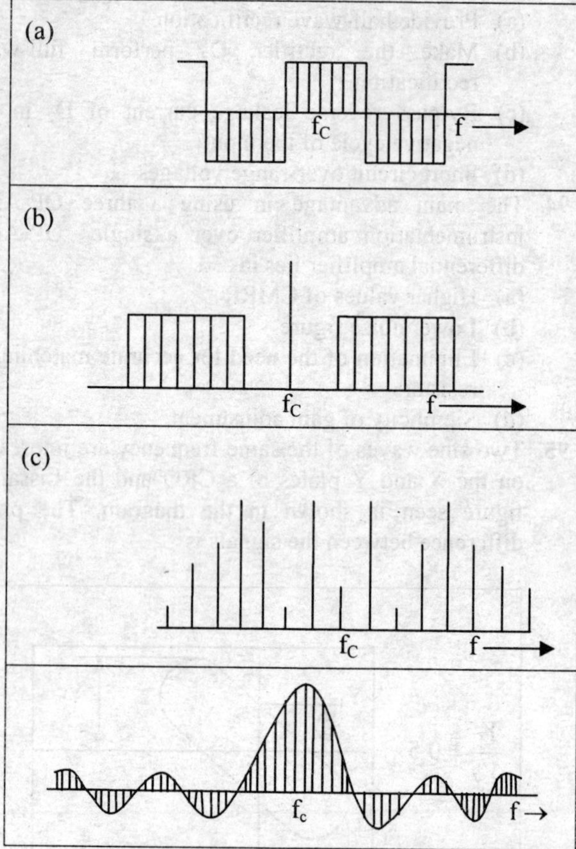

99. A 50 Hz ac voltage is measured with moving iron voltmeter and a rectifier type ac voltmeter connected in parallel. If the meter readings are V_1 and V_2 respectively and the meters are free from calibration errors, then the form factor of the ac voltage may be estimated as

(a) $\frac{V_1}{V_2}$ (b) $1.11\frac{V_1}{V_2}$

(c) $2\frac{V_1}{V_2}$ (d) $\pi\frac{V_1}{2V_2}$

100. Consider the following statements regarding the causes of error in current transformers:

1. Some exciting mmf is required by the primary winding to produce a flux.
2. The flux density in the core is not a linear function of the magnetizing force.
3. There is some magnetic leakage in the secondary winding
4. There is significant power consumption in the metering current of these statement

(a) 1 and 2 are correct
(b) 1, 2 and 3 are correct
(c) 2 and 4 are correct
(d) 1, 3 and 4 are correct

101. The resistances of two coils of a wattmeter are 0.01 ohm and 1000 ohms respectively and both are non-inductive. The load current is 20 A and the voltage across the load is 30 V. In one of the two ways of connecting the voltage coil, the error in the reading would be

(a) 0.1% too high (b) 0.2% too high
(c) 0.15% too high (d) zero

102. In the statement "the wattmeter commonly used for power measurement at commercial frequencies is of the X-type. This meter consists of two coil systems, the fixed system being the Y- coil and moving system being the Z-coil " X, Y and Z stand Respectively for

(a) Dynamometer, voltage and current
(b) Dynamometer, current and voltage
(c) Induction voltage and current
(d) Induction, current and voltage

103. Two-wattmeter method is employed measure power in a 3-phase balanced system with the current coils connected in the A and C lines. The phase sequence is ABC. If the wattmeter with its current coil in A-phase line reads zero, then the power factor of the 3-phase load will be

(a) zero lagging (b) zero leading
(c) 0.5 lagging (d) 0.5 leading

104. The shunt type ohmmeter is NOT suitable for high resistance measurements because

(a) Very low resistance of the meter would short the high unknown resistance
(b) Scale is highly cramped for high resistance values
(c) Full scale value of the meter may be exceeded
(d) Battery cannot supply the necessary current for proper meter deflection

105. A galvanometer is tested in the circuit where E = 1.5 V, R_1=1.0Ω, R_2= 2500 Ω and R_3 is variable. With R_3 set at 450Ω the galvanometer deflection is 140 mm and with R_3 set at 950Ω, the galvanometer deflection is 70mm. The resistance of the galvanometer is

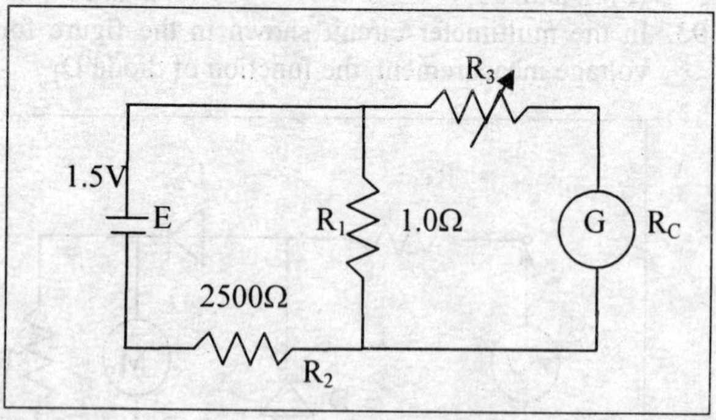

(a) 99Ω (b) 49Ω
(c) 25Ω (d) 10Ω

106. The bridge network shown in the figure is connected up with a view estimate the increase in the resistance of a coil due to skin effect. The procedure involves the following steps, not necessarily in that order

1. Switch on S_2 with S_1 off
2. Switch on S_1 with S_2 off
3. Adjust R_4
4. Adjust R_5

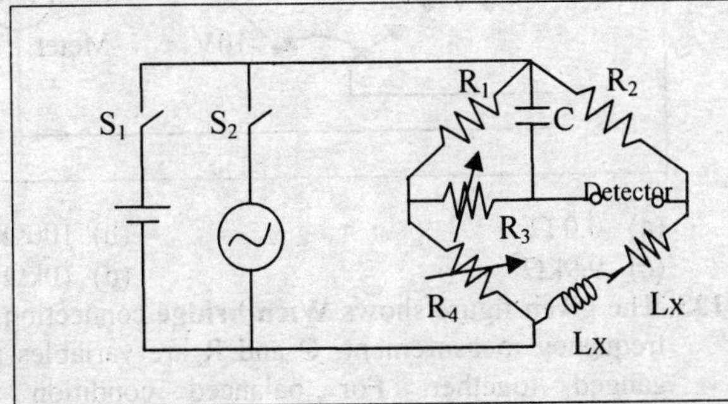

The correct sequence of these steps (a step could be repeated) is

(a) 1, 3, 2, 4, 3 (b) 2, 4, 1, 4, 3
(c) 1, 4, 2, 3, 4 (d) 2, 3, 1, 3, 4

107. The measurement of very low and very high frequencies is invariably done using a frequency counter/timer in the

(a) Frequency measurement mode only
(b) Period measurement mode only
(c) Frequency and period measurement modes respectively
(d) Period and frequency measurement modes respectively

108. In a particular from of frequency meter, a 1 μF capacitor is connected across a symmetrical square wave signal of 1 volt peak value. If the average value of the current taken by the capacitor, after full wave rectification is measured as 2 mA, then the frequency of the signal will be

(a) $1000/\pi$ Hz (b) 500 Hz
(c) 1000 Hz (d) $1000\,\pi$ Hz

109. The meter constant of a single-phase 240 V induction watt-hour meter is 400 revolutions per kWh. The speed of the meter disc for a current 0 to 10 amperes at 0.8 p.f. lagging will be

(a) 12.8 rpm (b) 16.02 rpm
(c) 18.2 rpm (d) 21.1 rpm

110. A digital voltmeter uses a 10 MHz clock and has a voltage controlled generator which provides a width of 10 μ sec per volt of unit signal. 10 volt of input signal would correspond to a pulse count of

(a) 500 (b) 750
(c) 1000 (d) 1500

111. An analog voltage signal whose highest significant frequency is 1 kHz is to be digitally coded with a resolution of 0.01% covering the voltage range of a to 10 volts. To avoid loss of information the MINIMUM number of bits in the digital code should be

(a) 4 (b) 8
(c) 10 (d) 14

112. The A to D converter used in a digital instrument could be

1. Successive approximation converter type
2. Flash converter type
3. Dual slope converter type

The correct sequence of the increasing order of conversion time taken by these type is

(a) 1, 2, 3 (b) 2, 1, 3
(c) 3, 2, 1 (d) 3, 1, 2

113. Which on of the following terms is NOT associated with a sample-and-hold circuit?

(a) Conversion time (b) Acquisition time
(c) Aperture time (d) Sample mode

114. A 35 V dc supply is connected across a combined resistance of 600 ohms and an unknown resistance of R ohms in series. A voltmeter having a resistance of 1. 2 kΩ is connected across 600 ohms resistor and reads 5V. the resistance R will be

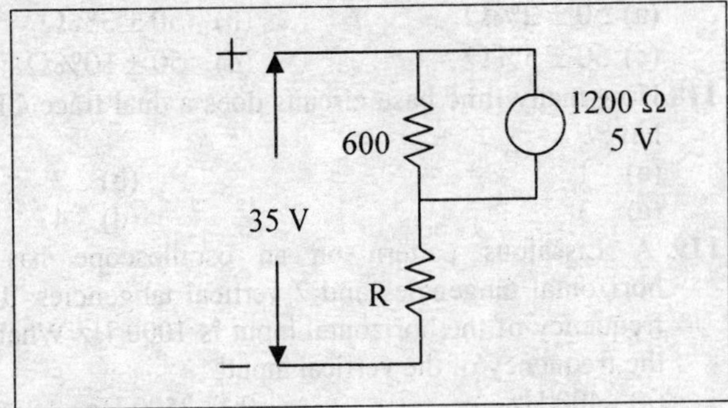

(a) 120 ohms (b) 500 ohms
(c) 1.7 k ohms (d) 2.4 k ohms

115. If two 300 V full –scale voltmeters V_1 and V_2 having sensitivities of 100 kΩ/V and 150 k Ω/V are connected in series to measure 500 V, then

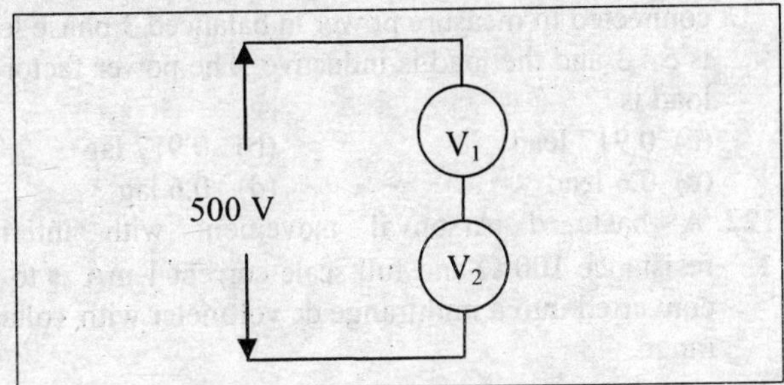

(a) V_1 and V_2 will read 250 V each
(b) V_1 will read 200 V and V_2 will read 300 V
(c) V_1 will read 300 V and V_2 will read 200 V
(d) V_1 and V_2 will read 0 V each

116. If the practical units of voltage and current were each made 20 times as large as they are at present, what would be the consequent alteration in the size of the unit of capacitance?
(a) 200 times (b) 60 times
(c) 20 times (d) Nil

117. The arms of a Wheatstone bridge are shown in the given figure. For the balanced condition, the least tolerance value of R_4 will be

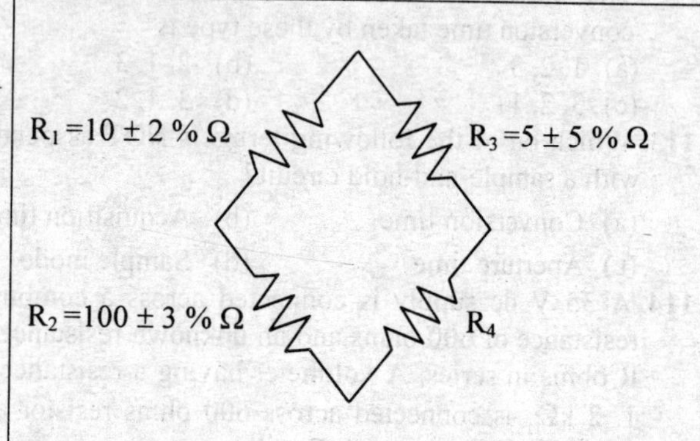

(a) $50 \pm 2\%\Omega$ (b) $50 \pm 3\%\Omega$
(c) $50 \pm 5\%\Omega$ (d) $50 \pm 10\%\Omega$

118. How many time base circuits does a dual trace CRO have?
(a) 1 (b) 2
(c) 3 (d) 4

119. A Lissajous pattern on an oscilloscope has 5 horizontal tangencies and 2 vertical tangencies. The frequency of the horizontal input is 1000 Hz What is the frequency of the vertical input?
(a) 400 Hz (b) 2500 Hz
(c) 4000 Hz (d) 5000 Hz

120. In measurements made using a Q-meter , high impedance element should preferably be connected in
(a) star (b) delta
(c) series (d) parallel

121. The ratio of the readings of two wattcmeters connected to measure power in balanced 3 phase load is 5 : 3 and the load is inductive .The power factor of load is
(a) 0.917 lead (b) 0.917 lag
(c) 0.6 lead (d) 0.6 lag

122. A basic d Arsonval movement with internal resistance 100 Ω and full scale current 1 mA is to be converted into a multirange dc voltmeter with voltage range. 0-10V, 0-30V, 0-100 V and 0-300 V. The value of R_1 is

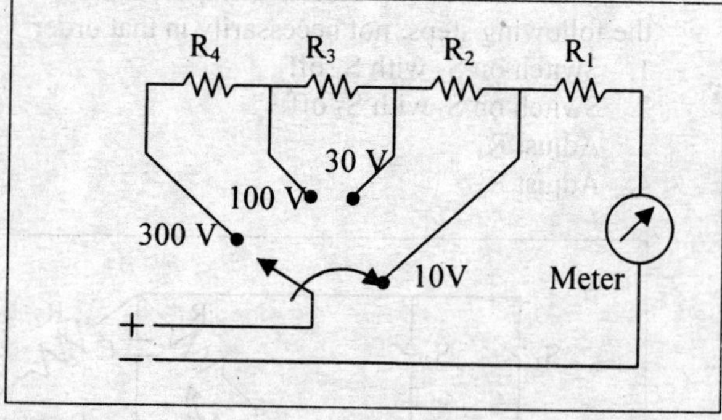

(a) 0.0 Ω (b) 100Ω
(c) 9.9kΩ (d) 10kΩ

123. The given figure shows **Wien bridge** connection for frequency measurement. C and R are variables and ganged together. For balanced condition the expression for frequency is $f = \dfrac{1}{2\pi CR}$.

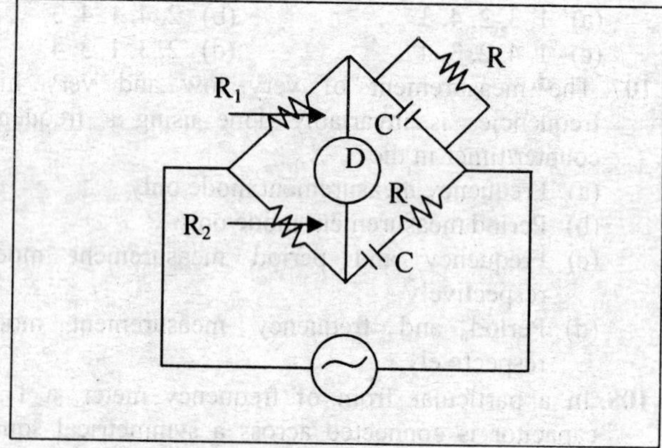

(a) $R_1 = R_2$ (b) $R_1 = 2R_2$
(c) $R_1 = \dfrac{R_2}{2}$ (d) $R_1 = 3R_2$

124. In modern electronic multi meters a FET or MOSFET is preferred over BJT because
(a) Its input resistance is high
(b) Its input resistance is high and does not vary with the change of range
(c) Its input resistance is low
(d) It is cheaper

125. The number of flip flops required in a decade counter is
(a) 2 (b) 3
(c) 4 (d) 10

126. Holes are drilled on the opposite sides of disc of an induction type energy meter to

(a) Avoid creep on no load
(b) Balance the disc
(c) Dissipate heat energy due to eddy currents
(d) Increase the deflecting torque

127. In a digital voltmeter, the oscillator frequency is 400 kHz, the ramp voltage falls from 8 V to 0 V in 20 msec. The number of pulses counted by the counter is

(a) 8000 (b) 4000
(c) 3200 (d) 1600

128. A digital voltmeter has $4\frac{1}{2}$ digit display. The 1 volt range can read upto

(a) 1.0000 (b) 1.1111
(c) 1.9999 (d) 19999

129. Match List I with List II and select the correct answer using the codes given below the lists:

List I	List II
A. LED	1. Optical scattering
B.LCD	2. Glow discharege
C. NIXIE	3. Electromagnetic radiation emitter

Codes:

	A	B	C
(a)	1	3	2
(b)	3	1	2
(c)	2	1	3
(d)	3	2	1

130. The **zero-suppression in recorders** implies

(a) Recording signals with reference to a point other than zero
(b) Removing the static component so that rest of the signal is displayed with more expansion
(c) Providing inertialess components to improve transient response
(d) Designing the recorder for zero error.

131. Suppose the potential and current coil a wattmeter are changed accidentally and supply is given. Then,

(a) Wattmeter will not read
(b) Current coil will burn out
(c) Potential coil will burn out
(d) Nothing will happen.

132. A battery is connected to a resistance causing a current of 0.5 A in the circuit. The current drops to 0.4 A when an additional resistance of 5Ω is connected in series. The current will drop to 0.2A when the resistance is further increased by

(a) 10 Ω (b) 15 Ω
(c) 25 Ω (d) 40 Ω

133. The ratio of the charges stored by two metallic spheres raised to the same potential is 6. The ratio of the surface areas of the sphere is

(a) 6 (b) 1/6
(c) 36 (d) $1/\sqrt{6}$

134. A wheatstone bridge requires a change of 6 ohms in the unknown arm of the bridge to produce a change in deflection of 3mm of the galvanometer. The sensitivity of the instrument is:

(a) 0.5 percent (b) 2.0
(c) 0.5mm/ohm (d) 2.0 ohm/mm.

135. Which of the following does not have the same units as others? The symbols have their usual meanings.

(a) $\frac{L}{R}$ (b) RC
(c) $\sqrt{LC}$ (d) $\frac{1}{\sqrt{LC}}$

136. Mach List-I with List-II and select the correct answer by using the Codes given below the lists:

List-I	List-II
A. Digital Counter	1. Measurement of harmonics
B. Schering bridge	2. Measurement of frequency
C. Megger	3. Measurement of loss angle in a dielectric
D. Spectrum Analyser	4. Measurement of insulation resistance

Codes:

	A	C	B	D
(a)	1	2	3	4
(b)	4	3	2	1
(c)	2	3	4	1
(d)	4	1	2	3

137. Which of the following statements about impulse voltage is true?

(a) An impulse voltage is a unidirectional voltage
(b) In chopped impulse voltage, flashover does not occur
(c) Time taken to rise is exactly equal to the time taken to fall
(d) RMS value of impulse voltage is a always less than 50% of average value.

138. The digital voltmeter has a read-out range from 0 to 99999 counts. When full scale reading is 9.999 v, the resolution of the full scale reading is

(a) 0.001 (b) 1000
(c) 3 digit (d) 1mV

139. When C_1, C^2, C_3, C = Capacitance

L = Self inductance
M_1, M_2, M = Mutual inductances
ω = angular frequency
r, R, R_1, R_2, R_3, R_4 = resistances.
Which of the following is dimensionally incorrect?

(a) $C = \dfrac{1}{\omega^2 M}$

(b) $r = \dfrac{R\omega^2 M_1 M_2}{R^2 + \omega^2 L^2}$

(c) $R_4 = \dfrac{R_2\left(1+\omega^2 R^2 C_3{}^2\right)}{\omega^2 R_1 R_2 C_3}$

(d) $I = \dfrac{V\,\omega\, M}{\left[(\omega\, M + R_2)^2 + \omega^2 L_1 L_2 R_1{}^2\right]^{1/2}}$

140. Which of the following statement is true about two wattmeter method for power measurement in three phase circuit.
(a) Power can be measured using two wattmeter method only for star connected three phase circuits
(b) When two meters show identical readings, the power factor is 0.5
(c) When power factor is unity, one of the wattmeter reads zero
(d) When the readings of the two wattmeter's are equal but of opposite sign, the factor is zero.

141. In CRT aquadag carries
(a) Aqueous solution of graphite
(b) Sweep voltage
(c) Secondary emission electrons
(d) None of the above.

142. A dc supply of 35 V is connected across 600 Ω resistance in series with an unknown resistance. The voltmeter having a resistance of 1.2 kΩ is connected across 600 Ω, resistance which reads 5 V. Then the unknown resistance should be
(a) 50 Ω (b) 1.7 kΩ
(c) 7.2Ω (d) 2.4 kΩ

143. Two meters X and Y required 40 mA and 50mA respectively for full-scale deflection. Then,
(a) X is more sensitive
(b) Y is more sensitive
(c) Both are equally sensitive
(d) Not possible to determine from given data

144. What are the other methods of measuring power in a three phase circuit without using wattmeter.
1. one voltmeter and one ammeter
2. Two voltmeter and two ammeters
3. Three voltmeters
4. Three ammeters

(a) 1 and 2 only (b) 3 and 4 only
(c) 1 and 3 only (d) 4 only

145. Mach List-I with List-II and select the correct answer by using the Codes given below the lists:

List I	List II
A. Resistance	1 $M^{-1}L^{-2}T^{2}I^{-2}$
B. Inductance	2 $ML^{2}T^{-2}I^{2}$
C. Capacitance	3 $M^{-1}L^{-2}T^{4}I^{2}$
D. Reluctance	4 $ML^{2}T^{-3}I^{-2}$

Codes:

	A	B	C	D
(a)	1	2	3	4
(b)	2	1	3	4
(c)	4	2	3	1
(d)	4	3	2	1

146. The input impedance of a CRO is nearly
(a) Zero
(b) Around 10ohms
(c) Around 100 ohms
(d) Around one mega-ohm

147. If the transients during switching of a power supply are to be studied, which oscilloscope will be preferred?
(a) All ordinary oscilloscope with high frequency sweep generator
(b) Dual beam oscilloscope
(c) Dual trace oscilloscope
(d) Storage oscilloscope.

148. In a **dual beam oscilloscope**
(a) There are two separate vertical input and two separate horizontal inputs
(b) There are two separate vertical inputs and there is only one set of horizontal deflection plates.
(c) There is only one vertical input but there are two separate horizontal deflection plates.
(d) There is only vertical and one horizontal input.

149. Which of the following is **not an advantage** of semi-conductor strain gauges as compared to conventional strain gauges?
(a) Excellent hysteresis characteristics
(b) Least sensitive to temperature changes
(c) High fatigue life
(d) Smaller size.

150. Which of the following instrument is **free from hysteresis and eddy current errors**?
(a) Moving iron instrument
(b) Electrostatic instrument
(c) Moving coil–permanent magnet type instrument
(d) Moving coil dynamometer type instrument

151. The term **'artificial aging'** in instrument is associated with
(a) springs (b) permanent magnets
(c) controlling torques (d) damping

152. Two voltmeter of 0–300 V range are connected in parallel to a ac circuit. One voltmeter is moving iron type and reads 200 V. If the other instrument is moving coil type, its reading will be
(a) 200 √3 volts
(b) 200 × 1.41volts
(c) slightly less than 200V
(d) zero

153. Which voltmeter would you select for measuring 50000 V direct current?
(a) Moving coil voltmeter
(b) Hot wire voltmeter
(c) Electro dynamo meter
(d) Electrostatic voltmeter

154. A signal of 10 m v at 75 MHz is to be measured. Which of the following instrument can be used?
(a) VTVM
(b) Cathode ray oscilloscope
(c) Moving iron voltmeter
(d) Digital multimeter.

Answers Set-3

1. **Ans. (a)**
2. **Ans. (a)**
3. **Ans. (a)**
Solution: Under balance condition

$$Z_1 Z_4 = Z_2 Z_3$$

or $$(R_1 + j\omega L)\times\left[\frac{R_4 \times \dfrac{1}{j\omega C_4}}{R_4 + \dfrac{1}{j\omega C_4}}\right] = R_2 R_3$$

Equating real and imaginary part,

$$R_1 = \frac{R_2 R_3}{R_4} \Rightarrow R_X = \frac{750\times 2000}{4000} = 375\,\Omega$$

$$L_1 = R_2 R_3 C_4 \Rightarrow L_X = 750\times 2000\times 0.05\times 10^{-6}$$

$$= 75\ mH$$

4. **Ans. (d)**
Solution: Voltmeter is high impedance or open circuit meter. Impedance in each branch abc and adc are equal Current in each branch abc and adc is

$$I = \frac{10}{30} = \frac{1}{3}A$$

$$V_{bd} = -\left(20\times\frac{1}{3}\right)+10\times\frac{1}{3} = -\frac{10}{3}V$$

$$= -3.33\ V$$

5. **Ans. (b)**
6. **Ans. (c)**
Solution: Rectifier type instrument calibrated to rms Value
Hence reading $= 10\times 1.11 = 11.1\,V$
7. **Ans. (c)**
Solution: Q-METER
The determination of the storage factor Q is one of the most widely used means in the laboratory for testing radio frequency coils (RFC), inductors and capacitors. The storage factor is equal to $Q = \omega_o L/R$ where ω_o is the resonant angular frequency, L is the inductance and R is the effective resistance of a coil. The effective resistance R, is never determined directly since its value depends upon the value of frequency. As is well known, the value of effective resistance with AC differs from its DC value on account of skin effect and eddy current losses. Since the value of R varies in a complex manner with frequency, it is indirectly determined by measuring the value of Q.
The Q meter is an instrument which is designed to measure the value of Q directly and as such is very useful in measuring the characteristics of coils and capacitors. $Q = \dfrac{\omega L}{R}$
8. **Ans. (a)**
Solution: Advantage of electronic meter
   - High input impedance and low power consumption due to power supplied from auxiliary source.
   - Detection o f low level signal of the order of less than 50 μA and less than 5 mV due to amplifying properties.
   - High frequency range (10–100 MHz) some electronic voltmeter have range from dc to hundred of MHz. It is suitable for RF measurement.
9. **Ans. (d)**
Solution: Advantage of balance bridge VTVM
   - The effect of change and variation in the tube characteristic are minimized.
   - Fluctuation in the power supply have a small effect on measuring circuit.

10. Ans. (d)

Solution: $\theta_x \propto I^2$

$$\frac{\theta_2}{\theta_1} = \frac{I_2^2}{I_1^2} \Rightarrow I_2 = I_1\sqrt{\frac{\theta_2}{\theta_1}} = 1\sqrt{\frac{1}{2}} = 0.707\ A$$

11. Ans. (c)

Solution: Moving iron reals I_{rms}

I_{rms} for half wave rectifier is, $I_{rms} = \frac{I_m}{\pi}$

$\Rightarrow I_m = 2 \times I_{rms} = 2 \times 5 = 10$

Moving coil reals I_{av}

$$I_{av} = \frac{I_m}{\pi} = \frac{10}{\pi} = 3.18$$

12. Ans. (d)

13. Ans. (d)

Solution: $W_1 \propto \cos(30 + \phi)$

$W_2 \propto \cos(\phi - 30)$

14. Ans. (b)

Solution: Coil series aiding

$L_{eq} = L_1 + L_2 + 2M$

$2M = L_{eq} - L_1 - L_2$

$= 40 - 10 - 15 = 15$

Coil Series opposing

$L_{eq} = L_1 + L_2 - 2M$

$= 10 + 15 - 15 = 10\ mH$

15. Ans. (c)

16. Ans. (b)

Solution:

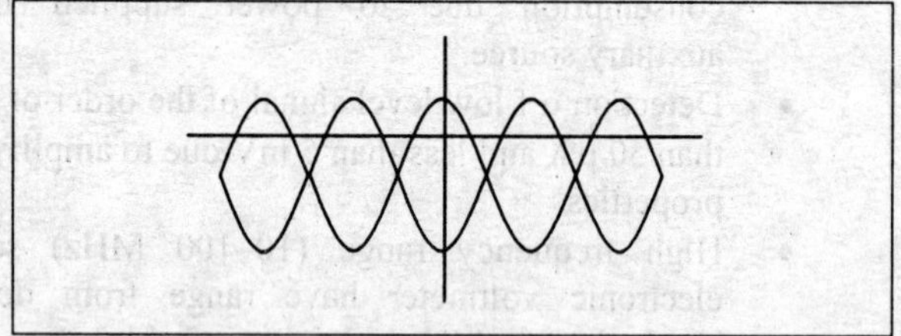

$$\frac{f_v}{f_h} = \frac{\text{No of cutting the Lissajous patterns by H-line}}{\text{No of cutting the Lissajous patterns by V-line}}$$

$$= \frac{10}{2} = 5:1$$

17. Ans. (a)

18. Ans. (d)

Solution: Time period $= \frac{1}{f} = \frac{1}{400 \times 10^{-3}} = 2.5\ \mu s$

No. of pulse $= \frac{\text{Total Time}}{\text{Time period}} = \frac{20\ ms}{2.5\ \mu s}$

$= 8000$

19. Ans. (b)

Solution:

4 th stage output frequency $f = \frac{1 \times 10^6}{1 \times 10^4} = 100$ Hz

Hence Time period $T = \frac{1}{f} \Rightarrow T = \frac{1}{100} = 10\ ms$

20. Ans. (c)

21. Ans. (c)

Solution: The correct match is given below

Display device (lowest power)	Liquid crystal Display
High speed display device	Light emitting diode
Bright perfect numeral display	Nixie Tube
Perfect light transmitting device	Optical fibre

22. Ans. (d)

Solution: One of the wattmeter reading $\alpha \cos(30^o + \phi)$

23. Ans. (a)

24. Ans. (a)

25. Ans. (a)

Solution: The **absolute measurement of current** involves the measurement of force.

The force may be exerted in two ways

- By the current in a solenoid upon a suspended magnetic needle-as in a tangent or sine galvanometer.
- By the current in one part of the circuit upon another part of the circuit in series with it and carrying the same current-as in an electro-dynamometer or current balance.

Absolute measurements of current, are therefore, carried out with instrument so carefully constructed that the current can be calculated from their dimensions and the measured effect produced.

The current can be determined by any one of the methods given below:

1. Tangent galvanometer method.
2. Electro dynamometer
3. Current balance method

Galvanometer methods of determination of current suffer from the following disadvantage

1. Horizontal component of the magnetic field of earth is required to be determined separately with great accuracy.
2. There is always some uncertainty about the exact position of the poles of the magnetic needle used.

In electro-dynamometer method, the current is measured in terms of the torsion of suspension wire or of a bifilar suspension, therefore this method is also not satisfactory.

Hence the method, which make use of some form of **current balance, are the most satisfactory** and therefore, are commonly used. One of the current balance method.

$$F = I^2 \frac{dM}{dx}$$

Where $\frac{dM}{dx}$ is the rate of change of M with distance from the coils.

If the three coils are so placed that the moving coil is at a distance half their radius from each of the fixed coils, the values of $\frac{dM}{dx}$ becomes dependent only on the radio of the effective radial of the two coils, a ratio that can be determined by electrical means.

The advantage of this method of measurement of current are that accurate measurement of dimensions, measurement of earth magnetic field horizontal component and determination of the tension constant of a suspension are not required.

26. Ans. (d)

Solution: Standard resistance

$$R_s = \frac{101.8}{10 \times 100} \times = 101.8\,\Omega$$

Standard Current $I_s = \frac{1.018}{101.8} = 10\ mA$

27. Ans. (b)

Solution: The correct match is given below

Anderson Bridge	Inductance
Kelvin Double Bridge	Resistance
Schering Bridge	Capacitance
Wien Bridge	Frequency

28. Ans. (d)

Solution: For bridge at balance $Z_1 Z_4 = Z_2 Z_3$

But $200\angle 60^\circ \times 600\angle 30^\circ \neq 400\angle 90^\circ$

i.e. $120000\angle 90^\circ \neq 1200\angle -90^\circ$

29. Ans. (b)

Solution: Without controlling torque pointer strike to beyond the scale for Half circle scale. For full circle scale rotate freely

30. Ans. (c)

Solution: The correct match is given below

$\phi_1 \phi_2 \cos\phi$	Ferrodynamic instruments
$I_1 I_2 \cos\phi$	Electrodynamometer instruments
Ki	D' Arsonval instruments

For D Arsonval instruments $T_d \propto KI$

For Electrodynamometer type $T_d \propto I_1 I_2 \sin\phi$

For inductor type, $T_d \propto \phi_1 \phi_2 \cos\phi$

31. Ans. (d)

Solution: External resistance of low value are connected in parallel to meter to increases the range of meter. If no resistance are used then it reads maximum value.

32. Ans. (d)

Solution: For circular pattern in CRO

$$R = X_C = \frac{1}{\omega C} = \frac{1}{2\pi f C}$$

$$= \frac{1}{2\pi \times 50 \times 1 \times 10^{-6}}$$

$$= \frac{1000}{100\,\pi} \times 10^3 = \frac{10}{3.14} \times 10^3$$

$$= 3.18\,k\Omega$$

33. Ans. (a)

Solution: Correction for Distributed Capacitance

The distributed capacitance or self-capacitance of measuring circuit modifies the true value of Q. the measured value of Q is less than the true value by a factor that upon the value of the distributed (self) capacitance and the tuning capacitor. It can be shown that,

$$\therefore \ Q_{true} = Q_{meas}\left(1+\frac{C_d}{C}\right)$$

C_d = distributed or self-capacitance and

C = tuning capacitance

34. Ans. (c)

Solution: $Q \cdot F = Q_L = \frac{V_L}{V_R} = \frac{V_C}{V_R} = \frac{10\,V}{500\,mV} = 20$

35. Ans. (c)

Solution: $f_{Lo} = f_s + f_{IF}$

$= (10\,kHz + 0.13\,GHz) + 1.45\,GHz$

$= 10\,kHz + 1.45\,GHz$ to $01.3\,GHz + 1.45\,GHz$

$= 1.45\,GHz$ to $02.75\,GHz$

36. Ans. (c)

Solution: Always used high sensitivity (high input impedance) meter to avoid the loading effect

37. Ans. (b)

38. Ans. (d)

Solution: The natural frequency of vibration of a reed used in a vibrating reed frequency meter, can be in increased by decreasing the length of the reed

39. Ans. (b)

Solution: Energy consumed

$= \frac{230 \times 20 \times 1 \times 2}{1000}\ kWh$

$= 9.2\ kWh$

Revolution $= 1380$

$$\text{Meter constant} = \frac{\text{Revolution}}{\text{Energy consumed}}$$

$$= \frac{1380}{9.2} = 150 \text{ revs/kWh}$$

40. Ans. (a)

41. Ans. (d)

Solution: In a digital voltmeter, ' **over–ranging** 'impels that an over–range indicator starts glowing

42. Ans. (b)

43. Ans. (b)

44. Ans. (a)

Solution: Galvanometer Recorders

In these recorders a pen assembly mouthed on the end of the pointer of a rugged **D'Arsonval movement** is used. While measuring a quantity by such recorders, the restraining springs of the movements provide a counterforce which balance the force produced by the quantity under measurement. Such recorders usually employ strip charts.

A basic diagram of such a recorder D'Arsonval galvanometer is some what different form the ordinary D'Arsonval galvanometer. As the frication torque in a recorder is high due to larger weight of the moving system and frication between the pen and the chart, so that large deflecting torque is required. Obviously the controlling torque required will be also large. So in such recorders large moving coil and strong magnetic field are employed. Galvanometer uses in a recorder is also made critical damped to avoid overshoot but this result in the slow response time.

45. Ans. (d)

Solution: $x = k_x \sin \omega t$

$y = k_y \sin(\omega t + \theta)$

Since $k_x = k_y = k$, therefore $x = k \sin \omega t$,

$y = k \sin(\omega t + \theta)$

Angle which the position vector of the spot on the screen makes the x-axis at any time t

$$\tan^{-1}\left(\frac{y}{x}\right) = \tan^{-1}\frac{\sin(\omega t + \theta)}{\sin(\omega t)}$$

For $\theta = 0^o$, angle of the trace $= \tan^{-1}(1) = 45^o$

For $\theta = 180^o$, angle of the trace $= \tan^{-1}(1) = -45^o$

or 135°

For $\theta - 90^o$, $x = k \sin \omega t$, and $y = \cos \omega t$ which will trace out a circle.

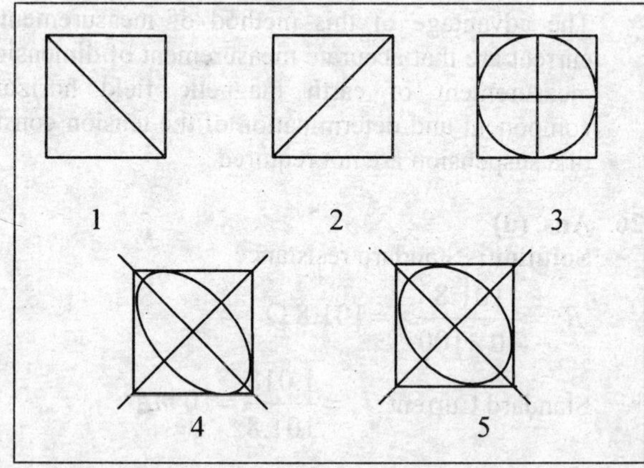

Thus figure 2 is for $\theta = 0$ further greater than 90° but still less than 180°, and figure 1 is for $\theta = 180^o$. Thus the sequence is 2, 3, 5, 4, 1

46. Ans. (a)

47. Ans. (a)

Solution: Full scale meter current = 1 mA

For 10 v range: $10^{-3} = \dfrac{10}{(100 + R_1)}$

or $R_1 = 9.9 \times 10^3 \ \Omega$

For 50 V range:

$$10^{-3} = \frac{50}{(100 + R_1 + R_2)}$$

or $R_1 + R_2 = 49.9 \times 10^3$

or $R_2 = (49.9 - 9.9) \times 10^3 = 40 \ k\Omega$

Another Method

For 10 V Range

$$m_1 = \frac{V_1}{V_m} = \frac{10\,V}{1\,mA \times 100\,\Omega} = \frac{10}{0.1} = 100$$

Series resistance $R_{s2} = (m_1 - 1) R_m$

$$= (100 - 1)\,100\,\Omega$$

$$= 9900\,\Omega = 9.9\,k\Omega = R_1$$

For 50 V range

$$m_2 = \frac{V_2}{V_m} = \frac{50\,V}{0.1} = 500$$

$$= (500 - 1)100$$

$$= 49900 = 49.9\,k\Omega$$

$$= R_1 + R_2$$

$$R_2 = R_{s2} - R_1$$

$$= 49.9\,k\Omega - 9.9\,k\Omega$$

$$= 40.8\,k\Omega$$

48. Ans. (b)

Solution: Advantage of electronic meter

- High input impedance and low power consumption due to power supplied from auxiliary source.
- Detection o f low level signal of the order of less than 50 μA and less than 5 mV due to amplifying properties.
- High frequency range (10–100 MHz) some electronic voltmeter have range from dc to hundred of MHz. It is suitable for RF measurement.

49. Ans. (b)

Solution:

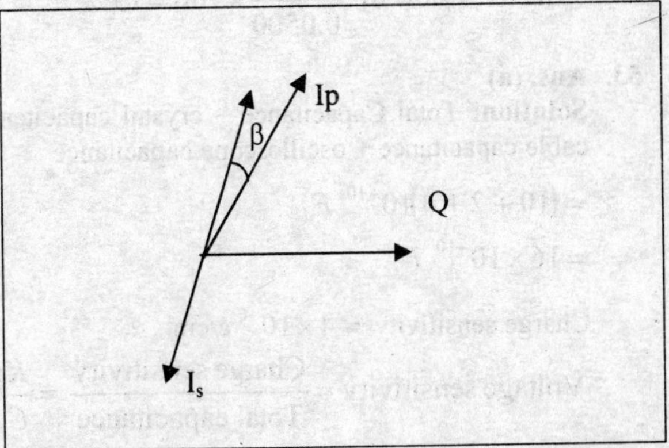

Phase error $\beta = 3^o$

∴ angle between the primary and secondary currents

$$= 180^\circ - \beta = 180^\circ - 3^\circ = 177^\circ$$

50. Ans. (a)

Solution: Two wattmeter reading are

$$W_1 = V_L I_L \cos(30 - \phi)$$

$$V_L I_L \cos(30^o - 90^o) = V_L I_L \cos 60^o$$

$$W_2 = V_L I_L \cos(30 - \phi)$$

$$V_L I_L \cos 120^o = -V_L I_L \cos 60^o$$

51. Ans. (c)

Solution: For Ops amplifier gain

$$\frac{V_o}{V_{in}} = \frac{X_S}{X_X} = \frac{1/\omega C_S}{1/\omega C_X} = \frac{C_X}{C_S}$$

$$C_X = \frac{V_o}{V_{in}} C_S = \frac{1}{10 \times 10^{-3}} \times 10 \times 10^{-12}$$

$$= 10^{-9}\,F$$

$$= 1000\,pF$$

52. Ans. (c)

Solution: Energy consumed as indicated by energy meter = Revolution/meter constant

$$= \frac{90}{1800} 10^{-3} = 0.05\,kwh$$

Actual energy consumed

$$= 230 \times \left(\frac{10}{2}\right) \times 1 \times \frac{3}{60} \times 10^{-3}\,kwh$$

$$= 0.0575\,kwh$$

Error $= 0.0500 - 0.0575 = -0.0075\ kwh$

$\therefore$ meter is slow by $\dfrac{0.0075}{0.0500} \times 100 = 15\%$

53. Ans. (a)
Solution: Total Capacitance = crystal capacitance + cable capacitance + oscilloscope capacitance

$= (10 + 2 + 4)10^{-10}\ F$

$= 16 \times 10^{-10}\ F$

Charge sensitivity $= 4 \times 10^{-6}$ c/cm

Voltage sensitivity $= \dfrac{\text{Charge sensitivity}}{\text{Total capacitance}} = \dfrac{Kq}{C_T}$

$= \dfrac{4 \times 10^{-6}}{16 \times 10^{-10}} = 2500$ V/cm

54. Ans. (a)
Solution: Conversion time of successive
$ADC = nT$
Where n = number of bit
T = time period of clock

55. Ans. (b)

56. Ans. (c)

57. Ans. (d)

58. Ans. (b)
Solution: The temperature error can be eliminated when the shunt and the moving-coil are made of the same material.

Use of ***swamping resistance of manganin*** having resistance 20 to 30 times the coil resistance is connected in series with the coil and a shunt.

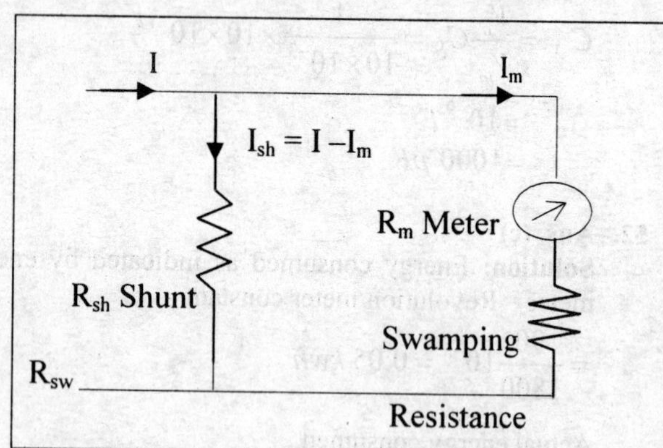

59. Ans. (d)

Solution: $R = \dfrac{V}{A} = \dfrac{Joules/Coulomb}{Ampere}$

$= \dfrac{ML^2T^{-2}}{IT} \times \dfrac{1}{I}$

$= ML^2\, T^{-3}\, I^{-2}$

60. Ans. (a)
Solution: PMMC instrument reals dc average value only, therefore reading is zero.

61. Ans. (d)
Solution: Ammeter shunt are of very low value, for low value resistance measurement Kelvin double bridge is best suitable.

62. Ans. (a)

63. Ans. (a)
Solution:

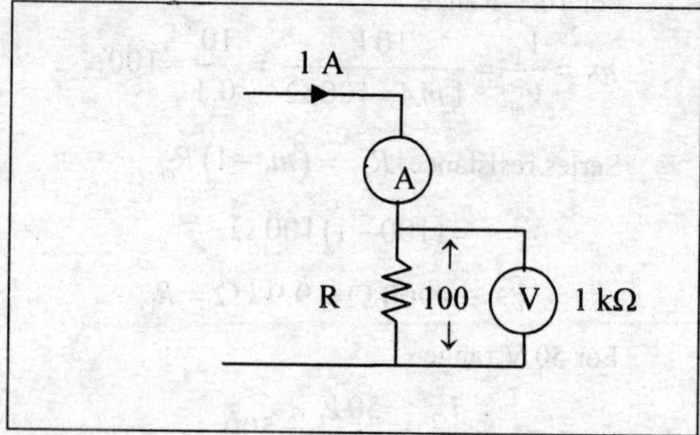

Current through voltmeter

$= \dfrac{100}{1000} = 0.1\ A$

$\therefore$ Current through resistor $= 1 - 0.1 = 0.9$

Hence $R = \dfrac{100}{0.9} = 111.11\ \Omega$

64. Ans. (c)
Solution: If the bridge is balanced then current $I_2 = 0$.

65. Ans. (d)
Solution: The match is given below

Low value of R.	Kelvin double bridge
High–Q inductor	Hay bridge
Low–Q inductor	Maxwell bridge
High voltage capacitors	Scherring bridge

66. Ans. (a)

67. Ans. (a)
Solution:

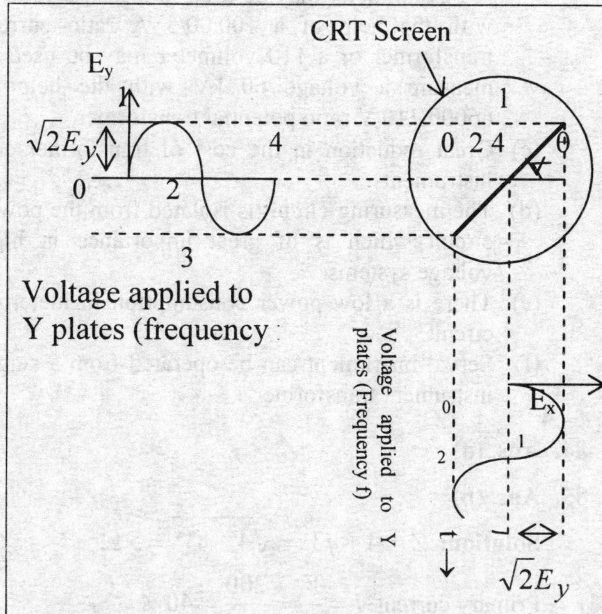

Figure: Lissajous pattern with equal frequency voltages and zero phase shift

68. Ans. (c)
Solution: Moving iron type meter measures rms value

$$\text{Reading} = \sqrt{10^2 + \left(\frac{10}{\sqrt{2}}\right)^2}$$

$$= \sqrt{100 + \frac{100}{2}} = \sqrt{100 + 50} = \sqrt{150}$$

69. Ans. (d)
Solution: Measurement of Q

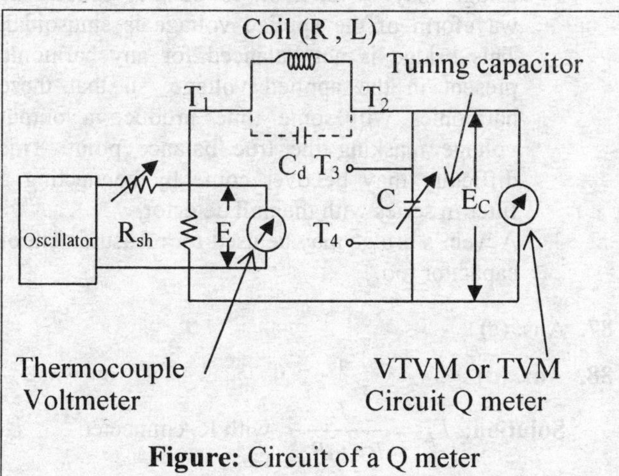

Figure: Circuit of a Q meter

The circuit for measurement of Q shown in figure. The oscillator is set to the desired frequency and then the tuning capacitor is adjusted for maximum value of E_C. As discussed earlier under these conditions $[Q = (E_C / E)]$ and if the voltage is kept constant, the voltmeter connected across the capacitor may be calibrated to read the value of Q directly. This measured value of Q is commonly regarded as the Q of the coil under test. However, there is an error. The measured value of Q is the Q of whole the circuit and not of the coil. There are errors caused on account of the shunt resistance and also due to the distributed capacitance of the circuit.

Correction for shunt resistance:

$$\text{Measured value, } Q_{meas} = \frac{\omega_o L}{R + R_{sh}}$$

$$\text{True value, } Q_{ture} = \frac{\omega_o L}{R}$$

$$\frac{Q_t}{Q_m} = \frac{\omega_o L/R}{\omega_o L/(R + R_{sh})}$$

$$= \frac{R + R_{sh}}{R} = 1 + \frac{R_{sh}}{R}$$

$$\therefore Q_{ture} = Q_{meas}\left(1 + \frac{R_{sh}}{R}\right)$$

Thus, the measured value of Q is smaller than the true. Now if coils of high resistance (low Q coil) are being measured, the difference between the two may be negligible. But when measurement are done on low resistance (high Q) coils, the error caused on this account may be serious.

70. Ans. (c)
Solution: In a **two –wattmeter method** of measuring power, one of the wattmeters is reading zero watts. The power factor of the circuit is 0.5.

71. Ans. (d)

72. Ans. (a)
Solution: Measurements can be made with the help of a **frequency counter** fundamental frequency of input signal and frequency components of the input signal at least upto third harmonic.

73. Ans. (d)
Solution: The bandwidth requirement of an FM telemetry channel is about ten times that of an AM telemetry channel.

74. Ans. (a)
Solution: The recording head in a magnetic tape responds to electrical signal and creates a magnetic signal.

75. Ans. (c)

Solution: In the SI system, the dimension of emf is $M L^2 T^{-3} I^{-1}$.

76. Ans. (a)
Solution: Bifilar winding techniques is 'adopted to reduce the self-inductance of a wire-wound standard resistor.

77. Ans. (c)

78. Ans. (c)
Solution: Bridge is under balanced conditions as

$Z_1 Z_4 = Z_2 Z_3$

i.e. $1 \times 4 = 2 \times 4$

then no current flow through Rs

79. Ans. (b)

80. Ans. (c)

81. Ans. (a)

Solution: $C_d = \dfrac{C_1 - n_2 C_2}{n^2 - 1}$

Here $n = \dfrac{4}{2} = 2$

$C_1 = 4500\ pF$

$C_2 = 90\ pF$

$$C_d = \frac{C_1 - 4C_2}{2^2 - 1} = \frac{450 - 4 \times 90}{3}$$

$$= \frac{450 - 360}{3} = \frac{90}{3} = 30\ pF$$

82. Ans. (a)
Solution: Since 5 kΩ and 2 kΩ connected in series then same current flow through it then $V_1 = 5\,kI$, $V_2 = 2\,kI$

$$\therefore \quad V_1 > \frac{V_2}{2}$$

83. Ans. (d)
Solution: Advantage of Transformers

(a) When instruments are used with instrument transformer, their readings does not depend upon their constant viz. R, L and C as is the case with shunts and multipliers.

(b) Very cheap moderate rating instruments may be used to measure large currents and high voltages. A 5 A ammeter may be used to measure 1000 A with the help of a 100.00/5 A ratio current transformer or a 110 voltmeter may be used to measure a voltage 60 kV with the help of 66000/110 V ratio potential transformer.

(c) Great reduction in the cost of transformer and instruments.

(d) The measuring circuit is isolated from the power circuit which is of great importance in high voltage systems.

(e) There is a low power consumption in metering circuit.

(f) Serval instrument can be operated from a single instrument transformer.

84. Ans. (d)

85. Ans. (b)

Solution: $Z = 4 + j3 = \sqrt{4^2 + 3^2} = 5\,\Omega$

Primary current $I = \dfrac{V}{R} = \dfrac{200}{5} = 40\ A$

Power factor $\cos\phi = \dfrac{R}{Z} = \dfrac{4}{5} = 0.8$

Secondary current $= \dfrac{5}{50} \times 40 = 4\ A$

Wattmeter reading $= VI \cos\phi$

$= 400 \times 4 \times 0.8 = 640$ watt

86. Ans. (d)
Solution: Wien's Bridge Characteristic

1. This bridge is suitable for measurement form 100 Hz to 10 kHz. It is possible to obtain an accuracy of 0.1 to 0.5 %.
2. Because of its frequency sensitivity. The wien's bridge may be difficult to balance unless the waveform of the applied voltage is sinusoidal. This bridge is not balanced for any harmonic present in the applied voltage, so that these harmonics will some time produce a output voltage masking the true balance point. This difficulty may be over come by connecting a fitter in series with the null detector.
3. A wein's bridge may be used for measurement of capacitor too.

87. Ans. (c)

88. Ans. (d)

Solution: $I_{fs} = \dfrac{E}{R_i + R_m}$, with R_x connected

$$I_m = \frac{R_x}{R_x + R_m} \frac{E}{R_1 \dfrac{R_m R_x}{R_m + R_x}}$$

$$= \frac{E}{R_x + R_m} \cdot \frac{R_x}{R_1 \dfrac{R_m R_x}{R_m + R_x}}$$

$$= \frac{ER_x}{R_1(R_m + R_x) + R_m R_x} = \frac{ER_x}{R_x(R_1 + R_m) + R_1 R_m}$$

$$= \frac{E}{(R_1 + R_m)} \frac{R_x}{R_x \dfrac{(R_1 R_m)}{R_1 + R_m}} = I_{fs} \frac{R_x}{R_m + R_p}$$

Where

$$R_p = \frac{R_1 R_m}{(R_1 + R_m)}$$

89. Ans. (a)

90. Ans. (c)

Solution: Setting of the potentiometer is at 1.0138 instead of 1.0183. Therefore of working current in the slide wire is more that what it would have been if the setting was at 1.0183, i.e. the correct value for applied voltage of 1.0183 V,

$$\frac{I}{I_C} = \frac{1.0183}{1.0138}$$

If the setting is at 1.0138 and the applied voltage is also 1.0138, the standardization is then correct.

91. Ans. (d)

Solution: Apply balance condition we have

$Z_1 Z_3 = Z_2 Z_4$

Or $450\,(200 + j\,100) = (300 - j600)\,(R_4 + j \times X_4)$

$90000 + j\,45000 = (300\,R_4 + 600\,X_4)$

$+ j(300\,X_4 - 600R_4)$

$$\text{or } X_4 = \frac{90000 - 300R_4}{600}(150 - 0.5R_4)$$

Hence $300(150 - 0.5R_4) - 600R = 45000$

Or $R_4 = 0$

Therefore $X_4 = 150$ or $Z_4 = j\,150\,\Omega$

92. Ans. (d)

Solution: Form bridge balance equations

$$\left(R_1 + \frac{1}{jC_1\omega}\right) R_4 = R_2 \left(\frac{R_3 + \dfrac{1}{jC_2\omega}}{R_3 + \dfrac{1}{jC_3\omega}}\right)$$

$$= \frac{R_2 R_3}{1 + jR_2 C_3 \omega}$$

$$R_1 + \frac{1}{jC_1\omega}\left(1 + j\,R_3 C_3 \omega\right) = \frac{R_2 R_3}{R_4}$$

It gives $R_1 + R_3 \dfrac{C_3}{C_1} = \dfrac{R_2 R_3}{R_4}$ (1)

And $R_1 R_3 C_3 \omega = \dfrac{1}{C_1\,\omega}$ (2)

Therefore $\omega = \dfrac{1}{\sqrt{R_1 R_3 C_1 C_3}}$

If $C_1 = C_3$, equation (1) becomes

$$R_1 + R_3 = \frac{R_2 R_3}{R_4}$$

Further if $R_2 = 2R_4$, the condition to be maintained to satisfy the balance equation is $R_1 = R_3$ and common variable shaft of resistance R_1 and R_3 can be graduate in frequency to measure the frequency

$$\omega \propto \frac{1}{R_1}$$

Thus, conditions to be satisfied are

$C_1 = C_3$, $R_2 = 2$ R and $R_1 = R_3$

93. Ans. (c)

Solution: If D_1 were not there in the negative half cycle of the applied voltage, there would be some small reverse through D_2 (in the reverse direction and the meter, thereby giving errorneous mean value and hence of the applied voltage).

With D_1 in the circuit, the current in the negative half cycle flows through R_5 and D_1 and there is no voltage across D_2 and hence no current through D_2.

94. Ans. (c)

95. Ans. (c)

Solution: When major area of the Lissajous figure lies in the Ist and 3rd quadrant, the angle is within 0 and ± 90°.

In the present case the angle is given by

$$\theta = sin^{-1}\frac{Y_1}{Y_2}$$

$= \sin^{-1} 0.5 = 30°$

Therefore the phase difference is 30° or –30°, i.e. 30° or 330°.

Y_1 is the deflection on the Y-plates when X-plates defection is zero, i.e. when $V_x(t) = 0$ and $V_y(t) = V_{my} \sin \phi$ and Y_2 is the deflection due to V_{my}.

96. Ans. (c)

Solution: Total length observed ∝ 2 V, i.e. $2\sqrt{2} \times 30$. Since 30 V dc causes I cm deflection from the centre point, thus, length observed, = $2 \times \sqrt{2}$ cm ≈ 3 cm

97. Ans. (c)

Solution: $Q = \frac{L\omega_o}{R} = \frac{1}{C\omega_o R}$

$$= \frac{1}{\frac{500}{20} \times 10^{-12} \times 2\pi \times 10^7 \times 10} = 341$$

Answer would be been (a) if it were 20 instead of 200.

98. Ans. (c)

99. Ans. (b)

Solution: Form factor = $\frac{\text{Rms value of the wave}}{\text{Mean value of the wave}}$

Moving iron instrument will shown RMS value. Rectifier voltmeter is calibrated to read RMS of sinusoidal voltage, i.e., with form factor of 1.11. Therefore, Mean value of the applied voltage = $\left(\frac{V_2}{1.11}\right)$

Hence Form factor = $\frac{V_1}{V_2/1.11} = \frac{1.11V}{V_2}$

100. Ans. (d)

101. Ans. (c)

Solution: Given $r_{cc} = 0.01\ \Omega$ and $r_{vc} = 1000\ \Omega$

There are two ways in which the current coil and voltage can be connected. The wattmeter reads high by an amount equal to power loss in current coil in one case and in voltage coil in the other.

The two possible power losses are

(1) $I_c^2 \cdot r_{cc} = 20^2 \times 0.01 = 4\text{W}$

(2) $\frac{V^2}{r_{ve}} = \frac{30^2}{1000} = 0.9\text{W}$

Percentage errors are

(1) $\frac{4}{100} \times 100 = 0.66$

(2) $\frac{0.9}{600} \times 100 = 0.15$

102. Ans. (b)

103. Ans. (c)

Solution: For the phase sequence and connections specified, the reading in the wattmeter of line A ∝ (30 + θ),

Where θ = lagging load power factor angle

Since it reads zero, θ = 60°

Or load p.f. = cos 60° = 0.5 lagging

104. Ans. (b)

Solution: Meter current, $I_m = I_{fs}\frac{R_x}{R_x + R_p}$

Where Rx = unknown, and

$R_p = \frac{R_1 R_m}{R_1 + R_m}$, which is quite low

Where R_x is in high value range, variation in the value of R_x causes hardly any variation in I_m and thus the scale is highly cramped for high resistance values

105. Ans. (b)

Solution: $I_c = \frac{1}{1 + R_3 + R_G} \cdot \frac{E}{2500 + \frac{1(R_3 + R_G)}{1(R_3 + R_G)}}$

From the given data

$$\frac{1}{1 + 450 + R_G} \cdot \frac{E}{2500 + \frac{450 + R_G}{1 + 450 + R_G}}$$

$$= \frac{1}{1 + 950 + R_G} \cdot \frac{2E}{2500 + \frac{950 + R_G}{1 + 950 + R_G}}$$

or 2[2500 (451 + R_G)] + 450 + R_G]

= 2500 (951 + R_G) + 50

or 2501 RG = 2500 (951 – 902) + 50

Therefore, $R_G = 49 + \frac{1}{2500} \approx 49\ \Omega$

106. Ans. (d)

Solution: First dc voltage is applied with ac off. Then to balance the bridge (under dc condition), adjustment of R3 has no effect. Therefore only R4 need be adjusted. After that dc is switched off, and ac is switched on. Bridge is adjusted by adjusting R_4 and R_5 alternately.

107. Ans. (c)

108. Ans. (a)

Solution: $f = \frac{2}{2\pi\sqrt{10^{-6}}} = \frac{1000}{\pi}$

109. Ans. (a)

Solution: 400 revolutions → kWh

When the power is 1 kW, the speed $= \dfrac{400}{60}$

$= \dfrac{20}{3}$ rpm

When the power $= \dfrac{240 \times 10 \times 0.8}{1000} = 19.2$

Then, $= \dfrac{20}{3} \times 1.92 = 12.8$ rpm

110. Ans. (c)

111. Ans. (b)

112. Ans. (c)

113. Ans. (a)

114. Ans. (d)

Solution: $600\,\Omega \| 1200\,\Omega$

$$= \frac{600 \times 1200}{1800} = 400\,\Omega$$

$$\frac{R}{35-5} = \frac{400}{5}$$

$\therefore$ R = 2400 $= 2:4\,k\Omega$

115. Ans. (b)

Solution: In series current is same voltage develop in proportion to resistance value

$$\frac{R_1}{R_2} \propto \frac{V_1}{V_2} = \frac{100 \times 10^3 \times 300}{150 \times 10^3 \times 300} = \frac{1}{1.5} \Rightarrow$$

$$V_2 = 1.5\,V_1$$

$$V_1 + V_2 = 500\,V \Rightarrow 2.5\,V_1 = 500\,V$$

$\therefore$ $V_1 = 200\,V$ and $V_2 = 300\,V$

116. Ans. (d)

Solution: Capacitance, $C = \dfrac{\text{Charge } Q}{\text{Voltage } V} = \dfrac{It}{V}$

Since I and V unit are both changed by the same factor and time unit is not changed; capacitance unit is not affected.

117. Ans. (d)

Solution:

When **Bridge is at balance then** $R_1\,R_4 = R_2\,R_3$

$$R_4 = \frac{R_2 R_3}{R_1} = \frac{5 \times 100}{10} = 50 \text{ ohm}$$

Relative limiting error of unknown resistance is,

$$\frac{\delta R_x}{R_x} = \pm\left(\frac{\delta R_2}{R_2} + \frac{\delta R_3}{R_3} + \frac{\delta R_1}{R_{1x}}\right)$$

$$= \pm(3+5+2) = \pm 10\,\%$$

118. Ans. (a)

Solution: Dual trace oscilloscope user a single electron gun and produces multiple traces by switching the y-deflection plates from one input signal to another It (this means the y-channel is shared by may signal.

119. Ans. (b)

Solution: $\dfrac{f_V}{f_H} = \dfrac{T_H}{T_V}$

$\therefore$ $f_V = \dfrac{5}{2} \times 1000 = 2500$ Hz.

120. Ans. (d)

Solution: Measurement of High impedance Components:

High impedance components, such as high value resistors, inductors and low-value capacitors are measured by connecting them in parallel with the measuring circuit, as shown is figure.

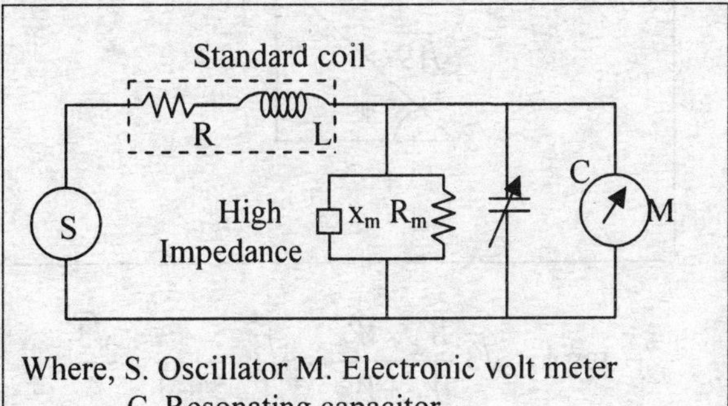

Where, S. Oscillator M. Electronic volt meter
C. Resonating capacitor

The unknown impedance is measured in terms of its parallel components X_m and R_m. Firstly the measurement is carried out without connecting unknown impedance and reading of resonating capacitor is noted with the standard coil in the circuit. Q-factor of the circuit is also calculated. Let their values be C_1 and Q_1 respectively.

Second measurement is carried out after inserting the unknown impedance in the circuit. New value of the resonating capacitor is obtained and the Q-factor of the circuit is determined. Let the corresponding values be C_2 and Q_2.

In the first case when unknown impedance is not connected in the circuit, under resonant condition

$$\omega_L = \frac{1}{\omega C_1}$$

$$Q = \frac{\omega_L}{R} = \frac{1}{\omega C_1 R}$$

$$X_m = \frac{1}{\omega C_m}$$

$$\text{or } C_m = \frac{1}{\omega X_m} = C_1 - C_2$$

$$\text{or } R_m = \frac{Q_1 Q_2}{\omega C_1 (Q_1 - Q_2)}$$

and the Q-factor of the unknown impedance

$$Q_m = \frac{R_m}{X_m} = \frac{(C_1 - C_2) Q_1 Q_2}{C_1 (Q_1 - Q_2)}$$

121. Ans. (b)

Solution:

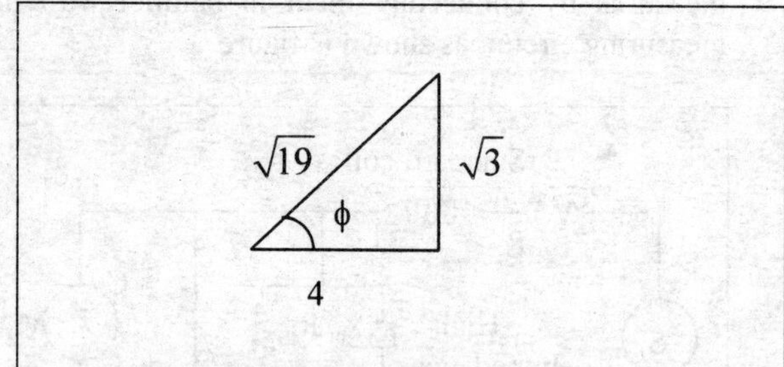

$$\tan\phi = \sqrt{3}\,\frac{W_1 \times W_2}{W_1 + W_2} = \sqrt{3}\,\frac{5-3}{5+3} = \frac{\sqrt{3}}{4}$$

$$\therefore \quad \cos\phi = \frac{4}{\sqrt{19}} = 0.917$$

(lagging due to inductive load)

122. Ans. (c)

Solution: Maximum meter current

$$I_m = \frac{V}{R_m + R_s}$$

$$\frac{10}{R_1 + 100} = 10^{-3}$$

$$\therefore \quad R_1 + 100 = 1000$$

$$\text{Or} \quad R_1 = 9900 = 9.9\ k\Omega$$

123. Ans. (c)

Solution: At balance $Z_1 Z_4 = Z_2 Z_3$

$$R_1\left(R + \frac{1}{j\omega C}\right) = R_2\left[\frac{R\dfrac{1}{j\omega C}}{R + \dfrac{1}{j\omega C}}\right]$$

$$\text{or} \quad \frac{R_1(j\omega CR + 1)}{j\omega C} = \frac{R_2 R}{(j\omega CR + 1)}$$

$$\text{or} \quad R_1\left(-C^2\omega^2 R^2 + 1 + 2j\omega CR\right) = jR_2\, C\omega R$$

Compare imaginary and real terms then

$$2C\,\omega RR_1 = C\,\omega RR_2$$

$$\text{or} \quad R_1 = \frac{R_2}{2} \text{ and } -C^2\omega^2 R^2 + 1 = 0$$

$$\text{or} \quad \omega = \frac{1}{RC}$$

$$\text{or} \quad f = \frac{1}{2\pi RC}$$

124. Ans. (b)

125. Ans. (c)

Solution: Number of flip flop for a counter is $n \geq \log_2 N$

126. Ans. (a)

127. Ans. (a)

Solution: Time period of clock frequency

$$T = \frac{1}{400 \times 10^{-3}} = 0.0025\ ms$$

$$\text{Number of pulse} = \frac{\text{Total time}}{\text{Time period}} = \frac{20}{0.0025}$$

$$= 8000$$

128. Ans. (c)

129. Ans. (b)

Solution: The correct match is given below

LED	Electromagnetic radiation emitter
LCD	Optical scattering
NIXIE	Glow discharege

130. Ans. (b)

131. Ans. (b)
Solution: when a high voltage is impressed across current coil which is of very low resistance, the coil current will immediately burn out unless some protective device operates.

132. Ans. (c)
Solution: Let the e.m.f. of the battery be E and its internal resistance be r and assume $r \to 0$

Then $\frac{E}{R} = 0.5$ and $\frac{E}{R+5} = 0.4$

Therefore $\frac{R+5}{R} = \frac{0,5}{0.4} = 1.25.$

Hence, R = 20 Ω

Therefore E = 20 × 0.5 V = 10 V

If R_1 = is the desired resistance

$\frac{10}{R_1+25} = 0.2$; $\frac{R_1+25}{10} = 5$

$\Rightarrow R_1 = 25\ \Omega$

133. Ans. (c)

Solution: For a sphere, $V_1 = \frac{Q}{4\pi\varepsilon_o R_1}$

$V_2 = \frac{Q_2}{4\pi\varepsilon_o R_2}$

Then, $\frac{V_1}{V_2} = 6$

or $\frac{Q_1}{Q_2} = \frac{R_1}{R_2} = 6$

Hence, $\frac{A_1}{A_2} = \frac{R_1^2}{R_2^2} = 36$

134. Ans. (c)
Solution:

$$\text{Sensitivity} = \frac{\text{magnitude of output response}}{\text{magnitude of input}}$$

$$= \frac{3.0\ mm}{6.0\ ohm} = 0.5\ mm/ohm$$

135. Ans. (d)
Solution:

$\frac{L}{R} \to$ Time constant

$RC \to$ Time constant

$\sqrt{LC} \to$ Time constant

$\frac{1}{\sqrt{LC}} = \omega_o \to$ Radians

136. Ans. (c)
Solution: The correct match is given below

Digital Counter	2. Measurement of frequency
Schering bridge	3. Measurement of loss angle in a dielectric
Megger	4. Measurement of insulation resistance
Spectrum Analyser	1. Measurement of harmonics

137. Ans. (b)
Solution: AN impulse voltage rise rapidly to its maximum value and falls more or less rapidly to zero. If an impulse voltage develops without causing a false over or puncture, it is called a full impulse voltage. However, if a flash over or a puncture occur, It is called a chopped impulse voltage.

138. Ans. (d)
Solution: The resolution of this instrument is 1 count in 99999.

Therefore, Resolution $= \frac{1}{9999}$

$\text{Count} = \frac{1}{99999} \times 9.999\ V = 10^{-4} = 10\ mV$

139. Ans. (a)

140. Ans. (d)
Solution: The wattmeter method can be used for star as well as data connected circuits Readings of two wattmeters.

(1) Equal unity
(2) One reads zero other reads troal poser 0.5
(3) Equal but of opposite sign zero

141. Ans. (a)
Solution: The bombarding electrons, striking the screen, releases secondary emission electrons. The secondary electrons are collected by an aqueous solution of graphite called aquatic which s connected

to the Second electrode. Collection of electrons is necessary to keep the screen in a state of electrical equilibrium.

142. Ans. (d)

Solution: $600\ \Omega \| 1200\ \Omega$

$$= \frac{600 \times 1200}{1800} = 400\ \Omega$$

$$\frac{R}{35-5} = \frac{400}{5}$$

$\therefore$ R = 2400 = 2:4 $k\Omega$

143. Ans. (a)

Solution: Sensitivity $S = \frac{Q_{FSD}}{I_{fsd}}$

$$S_1 = \frac{Q_{FSD}}{40\ mA} = 25$$

$$S_2 = \frac{1}{50\ mA} = 20$$

Taken less current for FDD
Sensitivity is deflection per unit current. Assuming full-scale angle of defection to be same for the meters one concludes that X is more sensitive.

144. Ans. (a)

145. Ans. (c)
Solution: The correct match is given below

Resistance	$ML^2T^{-2}I^{-2}$
Inductance	$M^{-1}L^{-2}T^4I^2$
Capacitance	$ML^2T^{-2}I^2$
Reluctance	$M^{-1}L^{-2}T^2I^{-2}$

146. Ans. (d)
Solution: Measuring instrument have high input impedance a CRO too have higher impedance of order of 1MΩ.

147. Ans. (d)
Solution: The storage type CRO is rapidly becoming one of the most useful tools in the presentation of very slowly swept signal and finds many application in the mechanical and biomedical fields. In the conventional CRT the persistence of the phosphor ranges from microsecond to perhaps seconds. In application where the persistence of the screen in smaller than the rate at which the signal sweeps across the screen, the start of the display will have disappeared before the end of the display is written.

148. Ans. (b)

Solution: Multiple Beam Oscilloscopes: In many cases it becomes necessary to compare one signal with another. In such cases **Multiple Beam Oscilloscopes** are used. Multiple Beam Oscilloscopes enclose in the signal tube several beam producing system each with is vertical pair or plates, but mostly with a common time-base. Each Y-channel has its own amplifier. The synchronization or triggering is done form the input of a desired Y-channel or form an external input voltage.

Double beam Oscilloscopes use two electron guns with in the same cathode ray tube. The electron beams of the two channel of the two channel are completely independent of each other. The same effect may be produced by a single electron gun, the output it being split into two independently controllable electron beams.

149. Ans. (b)

150. Ans. (b)

Solution: Advantage of electronic meter

- High input impedance and low power consumption due to power supplied from auxiliary source.
- Detection o f low level signal of the order of less than 50 μA and less than 5 mV due to amplifying properties.
- High frequency range (10–100 MHz) some electronic voltmeter have range from dc to hundred of MHz. It is suitable for RF measurement.

151. Ans. (a)

Solution: The essential qualities , for materials used for springs are that they should be non-magnetic proof against mechanical fatigue, which can be achieved by proper **annealing** and **pre-aging** (artificial aging) during manufacturing and the particular of low resistively and low temperature coefficient specially when it is used as lead for current flowing to the moving systems.

152. Ans. (d)

Solution: Moving coil does not reals ac

153. Ans. (d)

Solution: Very high voltage are measured by electrostatic voltmeter.

154. Ans. (b)

Solution: VTVM are used for V measurement but giver poor frequency response. CRO is suitable for VHF range and mV.

MCQ Set–4

1. In measurement systems, which of the following static **characteristice are disirabel**:
(a) Accuracy (b) Sensitivity
(c) Reproducibility (d) All of the above

2. In measurement systems, which of the following are undesirable static characteristics:
(a) Sensitivity and accuracy
(b) Drift, statoc error anjd dead zone
(c) Reproducibility and non - linearity
(d) Drift, static error, dead zone and non - linearity.

3. A **pressure measurement instrument** is callbrated between 10 bar and 250 bar. The scale span of the instrment is
(a) 10 bar, (b) 250 bar,
(c) 240 bar, (d) 260 bar,

4. The output of an instrument under particular enviromental conditions is given by the relationship $q_0 = a + b\, q_i$ = input and a and b are constants $b \neq 1$. The theoretical input - output relationship is repesented by a stralght passing through zero. This is a case of.
(a) zero drift only
(b) sensitivity drift only
(c) zero and sensitivity drift together
(d) none of the above

5. A reading is recorded as $23.90\, C^{o}$. The reading has significant figures of
(a) three (b) five
(c) four (d) none of the above

6. The scale of a $0 - 500\, V$ voltmeter is divided into ten large divisions representimg $50\, V$ each and each large division is further subdivided into 10 small divisions, each representing $5\, V$. It is used for measurment of output voltage of a **potpentiometer** which can be varied from 0 to 500 V. It is observed that when the sliding contact is moved from its zero position, there is no perceptible movement of pointer of the voltmeter till then sliding contant reaches a position where the output voltage should be $5\, V$. Therefore, it can be concluded that.
(a) The threshold of the voltmeter is $5\, V$.
(b) The resolution of the voltmeter is $5\, V$.
(c) The sensitivity of the voltmeter is $5\, V$.
(d) None of the above.

7. In the centre zero **analog ammeter** having a range of $-10\, A$ to $+10\, A$, there is a detectable change of the pointer from its zero position on either side of the scale only if the current reaches a value of 1A. The ammeter has a
(a) resolution of 1 A (b) dead zone of 1A
(c) dead zone of 2 A (d) sensitivity of 1A

8. A **pressure gauge** is calibrated from $0 - 50\, kN/m^2$. It has a uniform scale with 100 scale divisions. One fifth of a scale division can be read with certainty. The gaguge has a:
(a) resolution of $0.1\, kN/m^2$
(b) threshold of $0.1\, kN/m^2$
(c) dead zone of $0.2\, kN/m^2$
(d) a resolution of $0.5\, kN/m^2$

9. A d.c. circuit can be represented by an internal voltage source of 50 V with an output resistance of $100\, k\Omega$. In order to achieve 99 percent **accuracy** for voltage measurement across its terminals, the voltage measuring device should have a resistance of
(a) atleast $10\, M\Omega$ (b) $100\, k\Omega$
(c) atleast $10\, \Omega$ (d) none of the above.

10. A d.c. circuit can be represented by a voltage source of 10 V. in series with as output resistance of $1\, k\Omega$. An ammeter of $50\, \Omega$ resistance is connected to the source terminals for measurement of current. The accuracy of measurement is nearly.
(a) – 4.8 percent (b) + 4.8 percent
(c) 99 percent (d) 95.2 percent

11. In a.c. circuits, the connection of measuring instruments cause **loading errors** which may effect.
(a) only the magnitude of quantity being measured.
(b) only the phase of the quantity being measured

(c) both the magnitude and phase of the quantity being measured
(d) magnitude, phase and also the waveform of the quantity being measured.

12. The input resistance of a **cathode ray oscilloscope** is of the order of.
(a) tens of ohm` (b) megohm
(c) kilohm (d) fraction of an ohm

13. A resistor, at room temperature of $290^\circ K$ has a noise voltage of $2\,\mu V$ for a 50 kHz bandwidth. The **temperature** at which the noise voltage is 20% of its value at room temperture is,
(a) $58^\circ K$ (b) $11.6^\circ K$
(c) $241.7^\circ K$ (d) None of these

14. An amplifier has an input signal voltage of $25\,\mu V$ and a notise $2.5\,\mu V$. The signal to noise ratio is
(a) 10 (b) 100
(c) 10 db (d) none of the above

15. Johonson noise is:
(a) caused by thermal agitation of free electrons carrying current thereby modulating the current.
(b) noise carried into a circuit through conductors
(c) noise of an electromagnetic origin that is radiated into a circuit.
(d) shot noise which results from randoim emission of electrons across a PN junction

16. In a measurement system if a q_1 is an across variable and its associated throuhg variable is q_2 (power passed) the static impedance of the system is:
(a) $q_1 \times q_2$ (b) q_1 / q_2
(c) q_2 / q_1 (d) q_1^2 / q_2

17. In a measurement system 'force' is an across variable. Its associates.
(a) translational displacement
(b) translational
(c) translational velocity
(d) rotational displacement

18. The **power can be controlled** in a circuit by:
(a) controlling the voltage applied to load.
(b) controlling the current level in the circuit.
(c) using pulse modulation
(d) all of the above

19. The value of **static** stiffness and **compliance** in a measurement system determine the amount of.
(a) power drain from a system.
(b) current drain from a system.
(c) energy drain from a system.
(d) none of the above.

20. A **wheatstone bridge** is balanced with all the four resistance equal to $1\,k\Omega$ each. The bridge supply voltage is 100 V. The value of one of the resistance is changed to $1010\,\Omega$. The output voltage is measured with a voltage measuring device of infinite resistance. The **bridge sensitivity** is:
(a) $10\,V/\Omega$ (b) $25\,mV/\Omega$
(c) $2.5\,mV/\Omega$ (d) None of the above.

21. **Digital multimeters** will be the types of
1. panel meters 2. bench type meters
3. system meters
Select the correct answer uisng the codes given below
(a) only 1 (b) 2 and 3
(c) 1, 2 and 3 (d) 1 and 2

22. In **digital meters** we mostly use:
(a) Dual slope integrating type A/D converter
(b) R - 2R, ladder circuit type A/D converter
(c) Any kind to D/A converter
(d) Flash converter

23. **Bench meters** are intended mainly for
(a) Visual reading and parallelly at least an electrical binary coded decimal output.
(b) Stand alone operation
(c) Stand alone operation and visual reading
(d) None of these

24. For **ohm measurement**, what should be added with the digital multimeter.
(a) Precision high voltage source, i.e. applied series with the unknown resistor.
(b) A resistance pot parallel with the unknown resistance .
(c) Precision low - current source i.e. applied across the unknown resistance.
(d) Resistance pot that can tally with the unknown resistnace.

25. For **digital panel meters,** read outs range.
(a) From basic 3 digit to high precision $4\frac{3}{4}$ digit.
(b) From basic $4\frac{1}{2}$ digit to high precision $8\frac{1}{2}$ digit
(c) From basic 2 digit to high precision 4 digit
(d) None of these

26. These meters can be obtained with Tri - state binary coded decimal outputs - About which types of types of **digital multimeter** is the above statements talking?
(a) Digital panel meter and bench type digital meter.
(b) Digital panel meter

(c) Bench meter and system meter.
(d) System meter.

27. Nano voltmeters (nVM) and **digital multimeters** (DMM) have input impedance about.
(a) $10\,T\Omega$ (b) $200\,M\Omega$
(c) $50\,K\Omega$ (d) $10\,G\Omega$

28. For **high current measurment** in the range of few amperes, which digital meter is not suitable.
(a) DMMS
(b) SMU
(c) Source Meter Instruments
(d) Electrometer

29. For D.C. voltage measuring capability which **digital meter** shows minimum ranges for voltage measurment?
(a) SMU
(b) DMMS
(c) Nano voltmeter
(d) Plug in data acqusion meer

30. Source measure units (SMU) combine.
(a) An electrometer, **DMM,** voltage source and current source.
(b) An electrodes, voltage source in parallel with the unknown resistance.
(c) An electrometer, potentiometer and current source.
(d) Combinations of voltage source and current source.

31. A **phase sequence indicator** rotates clockwise for phase sequence of RYB. If the phase sequence is changed to BRY it will rotate.
(a) anti - clockwise
(b) clockwise
(c) clockwise and antoclockwise
(d) none of these

32. An **electrodynamometer type** of instrument is used in Weston type synchroscope while synchronizing an incoming machine to bus bars.
(a) For checking the power factor of two circuits
(b) For checking the voltages of two circuits
(c) For checking whether the incoming machine is fast or slow
(d) For checking the phase sequency of the two circuits

33. **Phase sequence indicators** are two types, one is static type, another is
(a) Dynamic type (b) Rotating type
(c) Sequential type (d) None of these

34. In **ratio type frequency meter** which one is true ?
(a) Ratio meter gives a linear relationship between the current ratio and the deflection.
(b) Ratio meter gives a linear relationship between the voltage ratio and the deflection.
(c) Ratio meter gives a non - linear relationship among deflection, power dissipation and current ratio.
(d) Ratio meter gives a blinking indicator for over loading.

35. Which is not a **electrical resonance freqency meter**.
(a) Ferro dynamic type frequency meter.
(b) Electridynamometer type frequency meter
(c) Weston frequency meter
(d) All are electrical resonance frequency meter.

36. **Power factor meter** are two types, they are
(a) Electrodynamometer type and moving coil type.
(b) Moving coil type and moving iron type.
(c) Electrodynamometer type and moving iron type.
(d) None of these

37. Among the below which is **not a frequency meter**.
(a) Mechanical resonance type.
(b) electrical resonance type.
(c) electromagnetic type
(d) electro dynamometer type

38. Which type of instrument is working on the principle of 3 - phase induction moter.
(a) Rotating type phase sequence indicator.
(b) Electro dynamometer type synchroscope.
(c) Saturable core frequency meter.
(d) None of these.

39. If $f_o =$ frequency of crystal oscillator for 1's sampling periodd, this frequency f_0 is given by.
(a) $f_0 = \sqrt{f_c}$ (b) $f_0 = \frac{f_c}{2}$
(c) $f_0 = \frac{f_c^2}{2}$ (d) $f_0 = \frac{3f_c}{2}$

40. The most convenient and widely utilized instrument for accurate measurment of freqnecy and time interval is the.
(a) Electronic counter timer
(b) Electronic encoder
(c) JK Filp flop
(d) None of these

41. From **Lissajous figure** we can measure.
(a) frequency (b) phase angle
(c) both (d) None of these

42. There will be serious errors, If power factor of non - sinusoidal waveforms is measured by **electrodynamometer power factor meters.** This is true if.
(a) Only single phase meter
(b) Only 3 phase meter
(c) both
(d) none of these

43. In a **read type frequency** meters all the reads.
(a) Have the same natural frequency
(b) Have different natural frequency
(c) Have different natural frequencies but the difference in natural frequences of adjacent reeds is ±0.5 Hz.
(d) None of these

44. The controlling torque in **signal phase power factor meters is provided** by
(a) Spring control
(b) Gravity control
(c) Striffness of suspension
(d) None of these

45. Practical Q - meter circuit consists of a oscillator of type.
(a) Hartley
(b) Col pit oscillator
(c) Variable frequency RF oscillator
(d) None of these

45. The principle for working of Q - meter is based upon characteristics of a.
(a) Resonant series RLC circuit
(b) Reasonant parallel RLC circuit
(c) R and L are in series and C in parallel
(d) None of these

46. If C_d = distributed capacitance.
C_1 = value of turning capacitor to set frequency f.
C_2 = value of turning capacitor to set frequency 2f.
For measurment of self capacitance by Q - meter which relationship is true ?
(a) $C_d = \frac{C_1 - 4C_2}{3}$ (b) $C_d = \frac{C_1 - 3C_2}{4}$
(c) $C_d = \frac{C_1 - 2C_2}{4}$ (d) $C_d = \frac{C_1 - C_2}{4}$

48. Which of the below is not the part of CRT?
(a) Glass envelope
(b) Photon gun assembly
(c) Deflection plate assembly
(d) Horizontal deflection plate

49. The particular function of focusing anode is:
(a) accelerating the electron beam
(b) focusing the electron beam
(c) focusing the image
(d) none of these

50. If D = Deflection on the screen.
E_d = Potential between deflecting plates.
E_0 = Voltage of Pre - accelerating anode.
Deflection sensitivity of CRT is defined as
(a) $S = \frac{D}{E_d}$ (b) $S = \frac{E_d}{D}$
(c) $S = \frac{D}{E_0}$ (d) $S = \frac{E_0}{D}$

51. "Aquadag" is a
(a) Aqueous solution of graphite
(b) Acidic solution of graphite
(c) Basic solution of graphite
(d) none of these

52. If the bandwidth of an oscilloscope is given as direct current to 10 MHz, what is the fastest rise time a sine wave can have to be accurately reproduced by the instrument?
(a) 35 ns (b) 40 ns
(c) 80 ns (d) 25 ns

53. Synchornization of wave form can be accomplished in CRO by.
(a) Free Running sweep only
(b) Free Runing sweep and Triggred sweep
(c) Vertical sweep and horizontal sweep
(d) Alternator sweep and amplifier sweep.

54. When two equal voltage of equal frequency is applied but with 90° phase shift is applied to a CRO, the trace on the screen is a/an:
(a) ellipse (b) circle
(c) straight line (d) oval

55. To examine the very fast signals we use.
(a) Dual beam osciolloscope
(b) Dual trance oscilloscope
(c) Sampling oscilloscope
(d) None of these

56. Among the following which is not a **potentiometer**.
(a) Brooks deflection potentiometer
(b) Duo range type potenitiometer
(c) Vernier potentiometer
(d) Screw gauge potentiometer

57. D.C. **potentiometer** is used to callbrate.
(a) Voltmeter (b) Ammeter
(c) Watt meter (d) All of these

58. Which detector is usually used as a detector in a.c. **potentiometers**?
(a) Bridge detector
(b) Vibration galvanometer
(c) LVDT
(d) Phase sensitive detector

59. For D.C. current measurment capability, which digital voltage and current instrument is having range from little below. If A to serval mA?
(a) DMMS

(b) 6430 Remote source meter
(c) Electrometer
(d) Pico ammeter

60. Successive apporximation type and integrating type instruments are types of.
(a) Digital voltmeter (b) Digital multimeters
(c) Analog voltmeter (d) None of these

61. The potentiometric type DVM does not consist of
(a) comparator
(b) error amplifler
(c) potentiometric adjancent device
(d) pulse generator

62. By **Digital voltmeters** we can only measure.
(a) a.c. voltage
(b) d.c. voltage
(c) both (a) and (b)
(d) integrating type is used to measure the d.c. only and the rest are used to measure a.c. and d.c.

63. A 53 Hz **reed type frequecy meter** is pilarized with d.c. the new range of frequency is
(a) 106 Hz (b) 26.5 Hz
(c) 53 Hz (d) None of these

64. For a 3 - phase moving iron synchroscope
f_2 = frequency of incoming machine
f_1 = frequency of bus bar
At equilibrium $\theta = \pm 2\pi f't$
Where $f' = f_2 - f_1$
If f' is positive:
(a) The incoming machine is faster
(b) The incoming machine is slower
(c) The incoming machine can be synchronized to the bus bars
(d) None of these

65. An alternator is being synchronized to the bus bars whose frequency 50 Hz. The bulb of the synchroscope tickers at frequency of 10 Hz. The frequency of alternator is:
(a) 60 Hz (b) 40 Hz
(c) 60 Hz to 40 Hz (d) None of these

66. For a synchroscope the correct instant of synchronizing is when the bus bar and the incomding machine voltage.
1. are equal in magnitude
2. are in phase
3. have the same frequency
(a) 1and 2 are correct (b) 1 and 3 are correct
(c) 2 and 3 are correct (d) 1, 2, 3 are correct

67. For **saturable core frequency meter**, which part is not a part is not a part of core assembly.
(a) An outer core made by magnetic material
(b) A leg A made of non - magnetic meterial
(c) A leg B made of magnetic material
(d) A bar joining leg A and B, made of paramagnetic material.

68. Which one is true?
(a) The working force are large for electrodynamom eter type P.F. meter compared with those in mov ing iron P.F. meter.
(b) Electrical resonance type is a classification of synchroscope.
(c) For moving iron P.F. meter, the scale extends over 360°.
(d) All are true.

69. For **measurment of power factor** in 3 - phase unblanced systems a two element power factor meter has to be used. Two element power factor meter consists of.
(a) Two sets of fixed coil and two sets of moving coils mounted on the spindle.
(b) One sets of fixed coil and one set of moving coil mounted on the spindle.
(c) Two sets of fixed coil mounted on the spindle.
(d) Two sets of moving coils mounted one the spindle.

70. For a **vibrating reed type frequency meter**, If the natural frequency of first reedis 47 Hz, then what will be natural frequeny of fifth reed?
(a) 51 Hz (b) 49 Hz
(c) 47 Hz (d) 44 to 50 Hz

71. A **saturable core frequency** meter does not consist of
(a) bridge rectifier
(b) phase sensitive detector
(c) PMMC meter
(d) magnetic legs

72. In a **static type phase sequence indicator** which uses two identical lamps of pure resistance in the circuit of the phases R and Y and inductor in phase B having inductive resistance = resistance of bulb with circuit being connected in star. The reactance is equal to the bulb resistance. If bulb connected to phase R glows brightly and lamp connected to phase Y glows dimly, the phase sequence is.
(a) RBY (b) RYB
(c) BRY (d) None of these

73. For **saturable core frequency meter** the deflection θ is:
(a) Directly proportional to frequency (f)
(b) Directly proportional to current (I_{av})
(c) Directly proportional to $(1/\mathit{frequency})$
(d) Directly proportional to $(1/\mathit{curent}(I_{av}))$

74. During the measurment of a low resistance using a potentiometer the following readings were obtained: Voltage drop across the low resistance under test $= 0.4221\text{ V}$.

Voltage drop across the $0.1\,\Omega$ standard resistance $= 1.0235\text{ V}$.

Calulate the value of unknown resistance?

(a) $0.01\,\Omega$ (b) $0.041208\,\Omega$
(c) $0.101180\,\Omega$ (d) $0.051018\,\Omega$

75. Calculate the inductance of a coil from the following measurment on an a.c. potentiometer.

Voltage drop across a $0.1\,\Omega$ standard register connected with the coil $= 0.613\angle 12°6$

Voltage across the test coil through a $100/1$ voltage ratio box $= 0.781\angle 50°48'V$ Frequency is 50 Hz.

(a) $L = 25.3\,mH$ (b) $L = 7.95\,mH$
(c) $L = 9.94\,mH$ (d) $L = 50.8\,mH$

76. Calculate the maximum velocity of the beam of electron in a CRT having a cathode voltage of 800 V. Assume that the electrons to leave the cathode with zero velocity. Change of electron $= 1.6\times10^{-19}C$ and mass of electron $= 9.1\times10^{-31}\,m/s$ kg.

(a) $16.8\times10^{3}\,m/s$ (b) $17.5\times10^{3}\,m/s$
(c) $1000\,m/s$ (d) $9.1\times10^{3}\,m/s$

77. A CRT has an anode voltage of 2000V and parallel deflecting plates 2 cm long and 5 mm apart. The centre of the plates. Find the i/p voltage required to deflect the beam through 3 cm. The i/p voltage is applied to the deflecting plates throgh amplifiers having an overall gain of 100.

(a) 2 V (b) 1000 V
(c) 10 V (d) 1 V

78. An electrically deflected CRT has final anode voltage of 2000 V and parallel deflecting plates 1.5 cm long and 5 mm from the centre of deflecting plates, find (i) beam speed (ii) the deflection sensitvity of the tube?

(a) (i) 26.5 m/s, (ii) 0.375 mm/V
(b) (i) 25.6 m/s, (ii) 0.735 mm/V
(c) (i) 52.0 m/s, (ii) 0.51 mm/V
(d) (i) 25 m/s, (ii) 0.5 mm/V

79. An electrically deflected CRT has final anode voltage of 2000 V and parallel deflecting plates 1.5 cm long and 5 mm from the centre of deflecting plates, find (i) beam speed (ii) the deflection sensitvity of the tube? determine Deflection factor (G).

(a) 2.66 V/mm (b) 2 V/mm
(c) 1.98 V/mm (d) 1.31 V/mm

80. If the vertical amplifier of an oscilloscope has a bandwidth of 15 MHz. What is the rastest rise time that an input may have to be displayed without distortion?

(a) 23.33 nS (b) 23.33 μS
(c) 23.33 nS (d) 0.33 μS

81. If the Q factor of a coil which is $\frac{\omega_0 L}{R}$ is measured by varying the frequency.

(a) The plot between Q and frequency is linear
(b) The value of Q intially decrease with increase of frequency and afterwards increase with increases of frequency.
(c) The value of Q initially increases with increase of frequency and afterwards decreases with increase of frequency.
(d) The Q factor remains constant irrespective of the value of frequency.

82. The true value of Q of a coil is 245 and measured value is 244.5. The ratio of distributed capacitance of self capacitance of the coil is:

(a) 2.04×10^{-3} (b) 1.002
(c) 0.997 (d) None of these

83. In a Q-meter, the value of shunt resistance is typically of the order of

(a) Ω (b) $m\Omega$
(c) μΩ (d) $k\Omega$

84. The value of capacitance of a capactor is specified as $1\,\mu F \pm 5\%$ by the manufacture. Find the limits between which the value of capacitance is guaranteed?

(a) 0.95 *to* $1.05\,\mu F$ (b) 0.90 *to* $1.00\,\mu F$
(c) 0.92 *to* $1.02\,\mu F$ (d) 0.96 *to* $1.06\,\mu F$

85. By plug in data acquisiton types of instrument, we can measure voltage in range of

(a) serval μ volts to serval volts
(b) serval μ volts to several m volts
(c) serval μ volts to serveral K volts
(d) none of these

86. Which current meters cannot be used to measure less than 1 pA current?

(a) Electrometer (b) Pico meters
(c) DMMS (d) SMU

87. Suppose we have to measure a very high resistance of $1T\Omega = (10^{12}\,\Omega)$ range which instrument we can use.

(a) DMM
(b) Electrometer
(c) Nano voltmeter
(d) Source meter instrument

88. **Level detector** is used in integrated type DVM circuit. It works much like as a/an.
(a) Voltage comparator (b) pulse geneator
(c) Counter (d) None of these

89. For **weston frequency**, the moving element is a
(a) Soft iron needle (b) Soft iron core
(c) Moving coil (d) None of these

90. Iron bar and bus bar is a part of
(a) Rotating type phase sensitivedetector
(b) Naider Lipman types synchroscope
(c) Weston type synchroscope
(d) None of these

91. These insturments consists of a fixed coil divided into to parts. The fixed coil is designed to carry a small current and is connected in series with resistance across busbars
The statements is talking about
(a) Electrodynamometer type insturment
(b) Saturable core frequendy meter
(c) Electrodynamometer type frequency meter
(d) Electrodynamometer type sychroscoope

92. For **rotating field power factor meter** A_1, A_2 and A_3 are three coils are connected with their axes displaced.
(a) 120° from each other
(b) 60°, 60° and 240° from each other
(c) 90°, 90° and 180° from each other
(d) None of these

93. If a deflection frequnecy miter working on the princial resonance there are two parallel circuit each consisting of an inductance and capacitance in series. One circuit has $C_1 = 1\mu F$ and is tuned to a frequency $f_1 = 60 Hz$. The other has $C_1 = 1\mu F$ has is tunedb to a frequency $f_1 = 60 Hz$. The other has $C_2 = 1.5\mu s$ and is turned to a frequency f_2 below $50 Hz$. The resistance of each circuit $R_1 = R_2 = 1000\Omega$.
What must be the inductance of the second circuit in order that the current in both the circuits shall be same at a frequency of 50 Hz.
(a) 8.82 H (b) 9.82 H
(c) 971 H (d) 100 H

94. If a deflection frequnecy miter working on the princial resonance there are two parallel circuit each consisting of an inductance and capacitance in series. One circuit has $C_1 = 1\mu F$ and is tuned to a frequency $f_1 = 60 Hz$. The other has $C_1 = 1\mu F$ has is tunedb to a frequency $f_1 = 60 Hz$. The other has $C_2 = 1.5\mu s$ and is turned to a frequency f_2 below $50 Hz$. The resistance of each circuit $R_1 = R_2 = 1000\Omega$.
For the above question in what frequency it must be taken?
(a) 42.15 H (b) 41.15 Hz
(c) 40.15 H (d) 39.15 Hz

95. A circuit consisting of a coil, a resistance and a variable capacitor connected in series is tuned to responce using a Q - meter. If the frequency is 500 KHz, the resistance 0.5Ω and the variable capacitor set of 350pF. Calculate the effective value of L and R.
(a) $L = 290\mu H,\ R = 10.25\,\Omega$
(b) $L = 290\mu H,\ R = 9.75\Omega$
(c) $L = 270\mu H,\ R = 10.25\Omega$
(d) None of these

96. A simple **slide wire** is used for measure of current in a circuit. The voltage drop across a standard resistor of 0.1Ω is balanced at 50 cm. Find the magnitude of the current if the standard cell emf of 1.45 V is balanced at 50 cm.
(a) 21.75 A (b) 217.5 A
(c) 2.17A (d) None of these

97. In the measurment of power by a **polar potentiometer**, the following readings were obtained:
Voltage across a 0.2Ω standard resistance in series with load $= 1.46\angle 32°$ Voltage across a 200:1 potential divider across the line $= 1.37\angle 56° V$
Estimate the power factor of the load?
(a) 0.9135 (leading) (b) 0.9135 (lagging)
(c) 0.5125 (leading) (d) 0.5125 (lagging)

98. The law of deflection of a galvanometer is $I = \frac{K\theta}{\cos\theta}$, where I is the current, K is a constant and θ is the deflection. If the angle of deflection θ is known to be whthin $\pm 1°$ (standard deviation) of $15°$, what is the percent standard deviation of current, I?
(a) ±76.5% (b) ±7.65%
(c) ±0.765% (d) 1%

99. In **weston frequency meter** which one is true?
(a) It consists of one coil
(b) It consists of two coil parallel to each other

(c) It consists of two coils perpendicular to each other.
(d) It consists of three coils two of them parallel to each other and the rest one is perpendicular to those.

100. The measurement of a quantity is an act of comparison of an unknown quantity with
(a) another quantity.
(b) a known quantity whose sccuracy may be known or may not be knowm
(c) a Pedefined acceptable standard which is accurately knowm.
(d) none of the above.

101. In present day measurement systems.
(a) direct methods are commonly used.
(b) use of direct methods is limited but indirect methods are commonly used.
(c) both direct and indirect methods are commonly used.
(d) all of the above.

102. Purely **mechanical instruments** cannot be used for dunamic measurements because they have:
(a) high inertia (b) large time constant
(c) higher response time(d) all of the above

103. The usage of **electronic Instrument** is becoming more extensive because they have.
(a) A high sensitiveity and reliability.
(b) A fast response and compatibility with digital computers.
(c) The capability to respond to signals from remote places.
(d) All of the above.

104. A **null type of instrument** as compared to a deflection type instrument has
(a) a higher accuracy (b) a lower sensitivity
(c) a faster reponse (d) all of the above

105. In an **intelligent information system** used for the measurement of a physical quantity.
(a) the digital computer must perform all or nearly allthe signal and information processing.
(b) the observer must perform all or nearly all the signal and information processing.
(c) the digital computer is not required.
(d) none of the above.

106. In a **generalized measurement system**, the function of the signal manipulating element is to
(a) change the quantity under measurement to an analogous signal.
(b) change the magnitude of the input signal while retaining its identity.
(c) to perform linear operations like addition and multiplication.
(d) none of above.

107. A modifying input to a measurement system can be defined as an input which change the input-output relationship for
(a) desired inputs only.
(b) desired as well as interfering inputs.
(c) interfering inputs only.
(d) none of the above.

108. In a measurement system, the open loop transfer function is 10 and changes by 10 percent. If a closed loop system is adopted where in the feedback system has a transfer function of 100. What will be the change in output in the case of open - loop and in closed loop systems.
(a) 10 percent, 10×10^{-3} percent.
(b) 10 percent 1×10^{-3} percent
(c) 100 percent 10×10^{-3} percent
(d) 100 percent 1×10^{-3} percent

109. In a **permanent magent moving** coil type of ammeter, the current is,

$$I=\frac{K\theta}{NBA}$$

where
θ = deflection,
K = spring constant,
N = number of turns of moving coil,
A = area of moving coil and
B = flux density of field produced permanent magnet.
Temperature is an interfering input to the system. It is known that strength of magnets reduces by 0.02 percent per $^\circ C$. Which method should be adopted to apply corrections for the interfering input ?
(a) Method of inherent insensitivity.
(b) Method of high gain feed back.
(c) Method of calculated output corrections.
(d) Method of opposing inputs.

Answers Set-4

1. (d)	2. (d)	3. (c)	4. (c)
5. (c)	6. (a)	7. (c)	8. (a)
9. (a)	10. (d)	11. (d)	12. (b)
13. (b)	14. (b)	15. (a)	16. (b)
17. (c)	18. (d)	19. (c)	20. (b)
21. (c)	22. (a)	23. (c)	24. (c)
25. (a)	26. (b)	27. (d)	28. (d)
29. (a)	30. (a)	31. (b)	32. (c)
33. (b)	34. (a)	35. (c)	36. (c)
37. (c)	38. (a)	39. (a)	40. (a)

41. (b)	42. (a)	43. (c)	44. (d)
45. (c)	46. (a)	47. (a)	48. (b)
49. (b)	50. (a)	51. (a)	52. (a)
53. (b)	54. (b)	55 (c)	56. (d)
57. (d)	58. (b)	59. (b)	60. (a)
61. (d)	62. (c)	63. (a)	64. (a)
65. (c)	66. (d)	67. (d)	68. (c)
69. (a)	70. (b)	71. (b)	72. (a)
73. (b)	74. (b)	75. (a)	76. (a)
77. (d)	78. (a)	79. (a)	80. (a)
81. (c)	82. (a)	83. (b)	84. (a)
85. (c)	86. (c)	87. (b)	88. (a)
89. (a)	90. (b)	91. (a)	92. (a)
93. (b)	94. (b)	95. (b)	96. (a)
97. (b)	98. (c)	99. (c)	100. (c)
101. (b)	102. (d)	103. (d)	104. (a)
105. (a)	106. (b)	107. (b)	108. (a)
109. (c)			

One Mark Questions

1. A Manganin swamp resistance is connected in series with a moving coil ammeter consisting of a milli-ammeter and a suitable shunt in order to
   (a) Minimum the effect of temperature variation
   (b) Obtain large deflection torque
   (c) Reduce the size of the meter
   (d) Minimize the effect of stray magnetic fields

2. The effect of stray magnetic fields on the actuating torque of a portable instrument is maximum when the operating field of the instrument and the stray fields are:
   (a) Perpendicular (b) Parallel
   (c) Inclined at 60° (d) Inclined at 30°

3. A reading of 120 is obtained when a standard inductor was connected in the circuit of a Q meter and the variable capacitor is adjusted to a value of 300 pF. A loss-less capacitor of unknown value C_x is then connected in parallel with the variable capacitor and the same reading was obtained when the variable was obtained when the variable capacitor is readjusted to a value of 200 pF. The value of C_x in pF is
   (a) 100 (b) 200
   (c) 300 (d) 500

4. A dc potentiometer is designed to measure up to about 2 V with a slide wire of 800 mm. A standard cell of emf 1.18 V obtains balance at 600 mm. A test cell is seen to obtain balance at 680 mm. The emf of the test cell is:
   (a) 1.00 V (b) 1.34 V
   (c) 1.50 V (d) 1.70 V

5. The circuit in figure is used to measure the power consumed by the load. The current coil and the voltage coil of the wattmeter have 0.02 W and 1000 W resistance respectively. The measured power compared to the load power will be
   (a) 0.4 % less (b) 0.2 % less
   (c) 0.2 % more (d) 0.4 %more

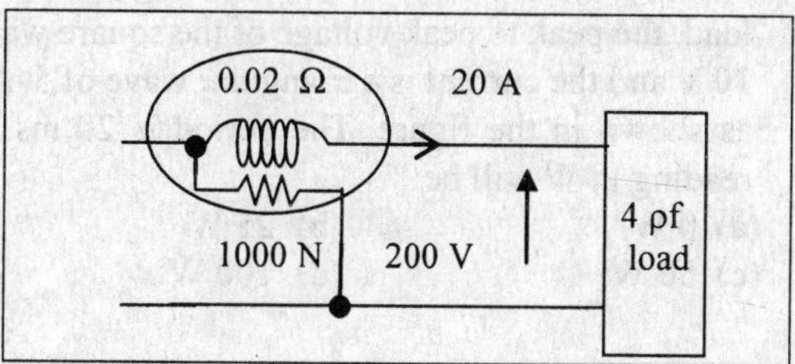

6. A galvanometer with a full scale current of 10 mA has a resistance of 1000W. The multiplying power (the ratio of measured current to galvanometer current) of a 100 W shunt with this galvanometer is
   (a) 110 (b) 100
   (c) 11 (d) 10

7. The Q – meter works on the principle of
   (a) mutual inductance (b) self inductance
   (c) series resonance (d) parallel resonance

8. A PMMC voltmeter connected across a series combination of a DC voltage source V_1 = 2V and an AC voltage source V_2 = (t) 3 sin (4t) V. The meter reads
   (a) 2V (b) 5 V
   (c) $\left(2+\sqrt{3/2}\right)V$ (d) $\left(\sqrt{17/2}\right)V$

9. A digital-to-analog converter with a full-scale output voltage 3.5 V has a resolution close to 14 m V. its bit size is
   (a) 4 (b) 8
   (c) 16 (d) 32

10. The time/div and voltage/div axes of an oscilloscope have been erased. A student connects a 1 kHz, 5 V p-p square wave calibration pulse to channel 1 of the scope and observe the screen to be as shown is the upper trace of the figure. An unknown signal is connected to channel 2 (lower trace) of the scope. It the time/div and V/div on both channel are the same, the amplitude (p-p) and period of the unknown signal are respectively.
    (a) 5 V, 1 ms (b) 5 V, 2ms
    (c) 7.5 V, 2ms (d) 10 V, 1ms

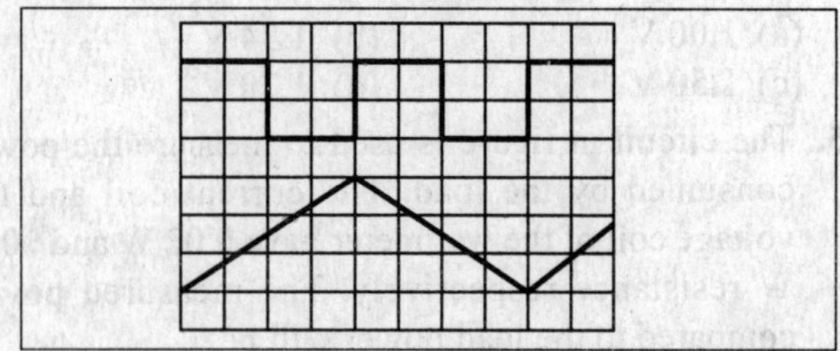

11. A sampling wattmeter (that computes power from simultaneously sampled values of voltage and current) is used to measure the average power of a load. the peak to peak voltage of the square wave is 10 V and the current is a triangular wave of 5 A p-p as shown in the figure. The period is 20 ms. The reading in W will be:

(a) 0 W (b) 25 W
(c) 50 W (d) 100 W

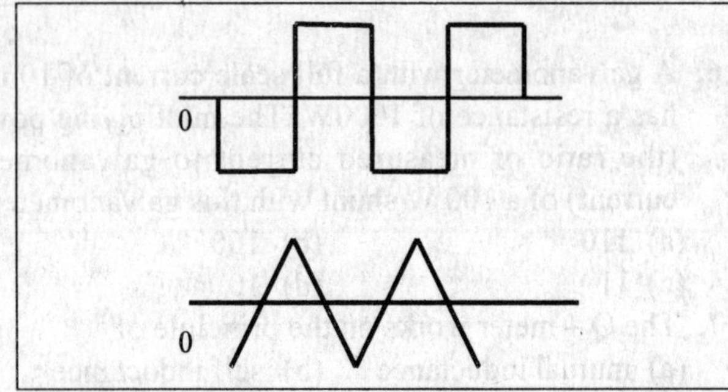

12. The probes of a non-isolated, two-channel oscilloscope are clipped to points A, B and C in the circuit of the adjacent figure. V_{in} is a square wave of a suitable low frequency. The display on Ch_1 and Ch_2 are as shown on the right. Then the "Signal" and "Ground" probes S_1, G_1 and S_2, G_2 of Ch_1 and Ch_2 respectively are connected to points:

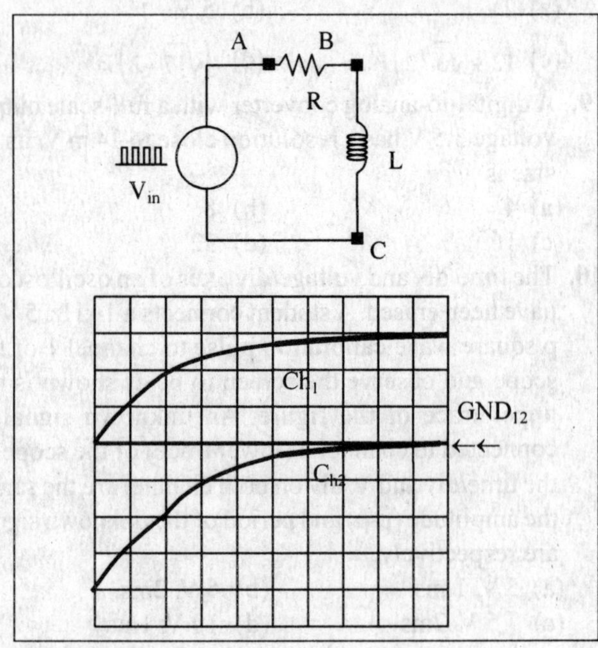

(a) A,B,C,A (b) A,B,C,B
(c) C,B,A,B (d) B,A,B,C

Answers

Electrical and Electronic Measurements

1. Ans. (a)

2. Ans. (b)

3. Ans. (a)

Solution: Let $C_V = 200$ pF be the readjusted value and $C = 300$ pF.

Quality factor is given by $(Q) = \frac{1}{RX_C}$

According to given condition

$$\frac{1}{\omega RC} = \frac{1}{\omega R(C_V + C_X)}$$

or $C = C_V + C_x$

or $C_x = C - C_V = 300 - 200 = 100$ pF

4. Ans. (b)

Solution: For the d.c. potentiometer, we have

$$\frac{E_1}{E_2} = \frac{l_1}{l_2}$$

or, Emf of the test cell,

$$E_2 = E_1 \cdot \frac{l_2}{l_1} = (1.18) \times \frac{680}{600} = 1.34\text{V}.$$

5. Ans. (c)

Solution: Power indicated by wattmeter,

= Power consumed by load + Power loss in current coil

$= 200\times20+20^2\times0.02 = 4008$ W

Load power = 4000 W

Then %increase $= \frac{4008-4000}{4000} \times 100 = 0.2\%$ more

6. Ans. (c)

Solution: For galvanometer,

$$\text{Multiplying Power} = \frac{I_m}{I_g}$$

$$\frac{I_g}{I_{sh}} = \frac{100}{1000}$$

$I_{sh} = 10\ I_g$

$I_m = I_g + I_{sh} = I_g + 10\ I_g = 11$ g;

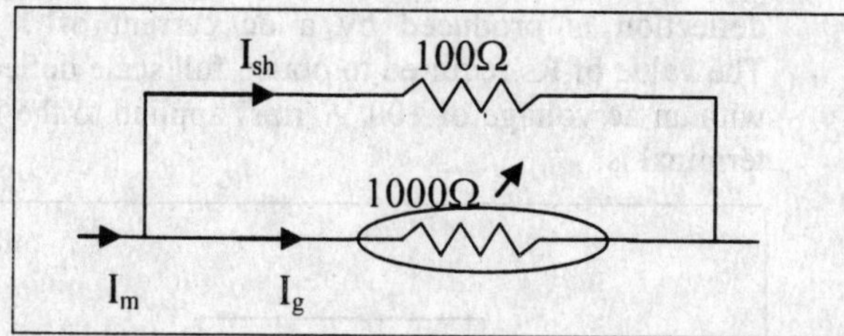

$$\frac{I_m}{I_g} = 11$$

7. Ans. (c)

Solution: Q-Meter

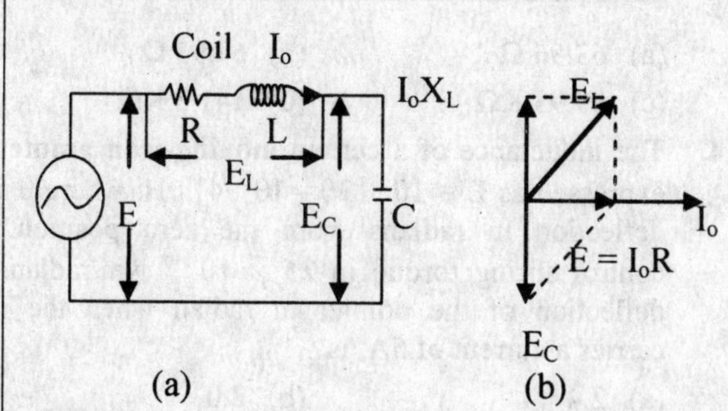

Figure: A resonant RLC circuit and its phasor diagram.

The Q-Meter works on the principle of series resonance.

8. Ans. (a)

Solution: PMMC instruments measure DC value only.

9. Ans. (b)

Solution:

$$\text{Resolution} = \frac{\text{FSV}}{2^n}$$

Where, n = number of bits.

$$\text{Then } 2^n = \frac{3.5}{1.4 \times 10^{-3}}$$

N = 8.

10. Ans. (c)

Solution: Upper trace represent setting of CRO

Lower trace is display value to be measured

2 divisions → 5 V P – P

$$3 \text{ divisions} \rightarrow \frac{3 \times 5}{2} = 7.5 \text{ P} - \text{P}$$

$\left.\begin{array}{l}\text{Frequency} = 1 \text{ kHz, T} = 1 \text{ ms} \\ \text{Division on x} - \text{axis} = 4\end{array}\right\}$ represent setting

Result, 8 division on x-axis = 2ms

11. Ans. (a)

Solution:

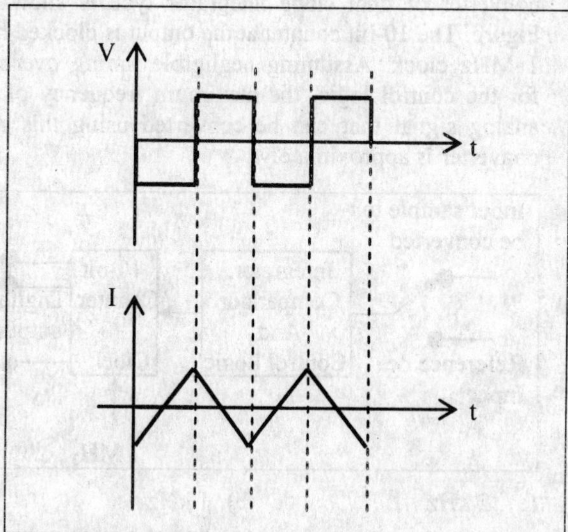

If we consider both the waves we, can see positive power and negative power in each cycle is equal.

Therefore, Net power = 0.

12. Ans. (b)

Solution:

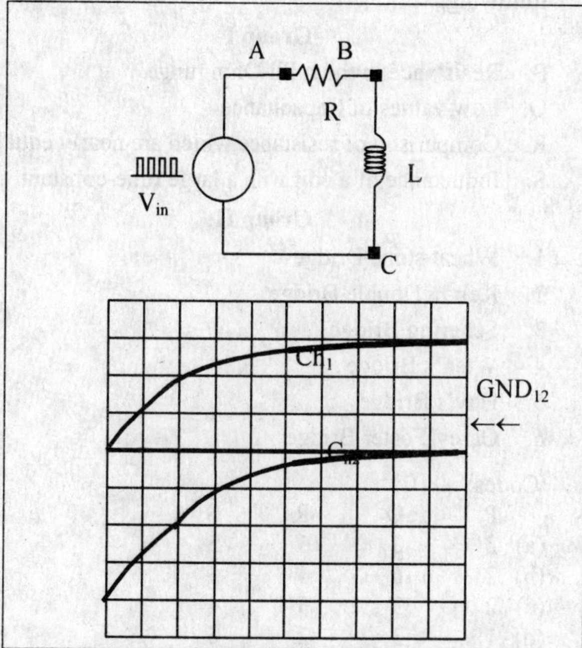

Since, Ch_2 shows a maximum step ramp curve so it is used against inductor.

As current through L is $= \frac{1}{L} \int V \, dt$

Integral of square is ramp.

Two Marks Questions

1. The simplified block diagram of a 10-bit A/D converter of dual slope integrator type is shown in Figure. The 10-bit counter at the output is clocked by a 1 MHz clock. Assuming negligible timing overhead for the control logic, the maximum frequency of the analog signal that can be converted using this A/D converter is approximately.

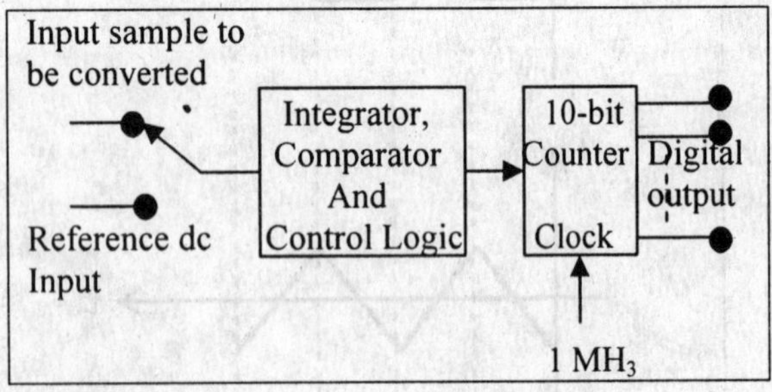

(a) 2 kHz (b) 1 kHz
(c) 500 Hz (d) 250 Hz

2. The items in group I represent the various types of measurements to be made with a reasonable accuracy using a suitable bridge. The items in Group-II represent the various bridges available for this purpose. Select the correct choice of the item in Group II for the corresponding item in Group I form the following.

Group I

P. Resistance in the milli-Ohm range
Q. Low values of Capacitance
R. Comparison of resistance which are nearly equal
S. Inductance of a coil with a large time-constant

Group II

1. Wheat stone Bridge
2. Kelvin Double Bridge
3. Schering Bridge
4. Wien's Bridge
5. Hay's Bridge
6. Carey-Foster Bridge

Codes:

	P	Q	R	S
(a)	2	3	6	5
(b)	2	6	4	5
(c)	2	3	5	4
(d)	1	3	2	6

3. A rectifier type ac voltmeter consists of a series resistance R_s, and ideal full-wave rectifier bridge and PMMC instrument as shown in figure. The internal resistance of the instrument is 100 Ω a full scale deflection is produced by a dc current of 1 mA. The value of Rs required to obtain full scale deflection with an ac voltage of 100 V (rms) applied to the input terminal is:

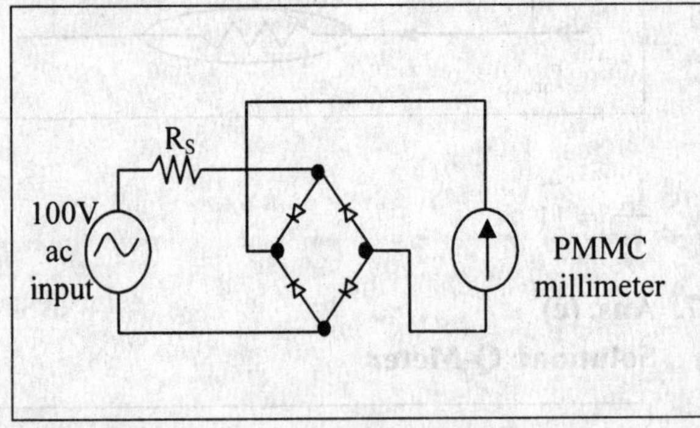

(a) 63.56 Ω (b) 69.93 Ω
(c) 89.93 KΩ (d) 141.3 KΩ

4. The inductance of a certain **moving-iron ammeter** is expressed as $L = 10 + 3\theta - [\theta^2/4]$ μH, where θ is the deflection in radians from the zero position. The control spring torque in 25×10^{-6} Nm/radian. The deflection of the pointer in radian when the meter carries a current of 5A, is

(a) 2.4 (b) 2.0
(c) 1.2 (d) 1.0

5. The voltage-flux adjustment of a certain 1-phase 220 V induction watt-hour meter is altered so that the phase angle between the applied voltage and the flux due to it is 85° (instead of 90°). The errors introduced in the reading of this meter when the current is 5 A at power factors of unity and 0.5 lagging are respectively

(a) 3.8 mW, 77.4 mW
(b) – 3.8 mW, – 77.4 mW
(c) – 4.2W, – 85.1 W
(d) 4.2W, 85.1 W

6. Group I represents the figure obtained on a CRO screen when the voltage signals $V_x = V_{xm} \sin \omega t$ and $V_y = V_{ym} \sin(\omega t + \Phi)$ are given to its X and Y plates respectively and Φ is changed. Choose the correct value of Φ from Group I to match with the corresponding figure of Group II.

Group I
P. $\Phi = 0$
Q. $\Phi = \pi/2$
R. $\pi < \Phi < 3\pi/2$
S. $\Phi = 3\pi/2$

Group II

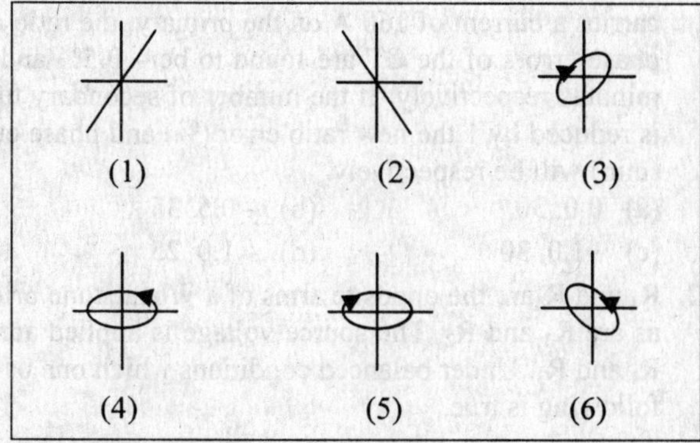

Codes:

	P	Q	R	S
(a)	1	3	6	5
(b)	2	6	4	5
(c)	2	3	5	4
(d)	1	5	6	4

7. A CRO probe has an impedence of 500 kΩ in parallel with a capacitance of 10 pF. The probe is used to measure the voltage between P and Q as shown in figure. The measured voltage will be

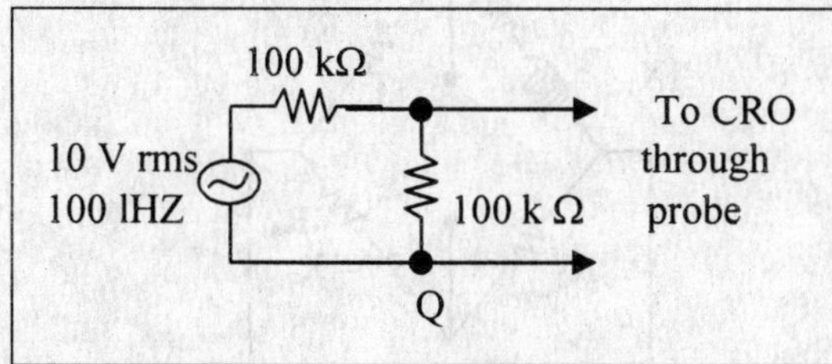

(a) 3.53 V (b) 4.37 V
(c) 4.54 V (d) 5.00 V

8. A moving coil of a meter has 100 turns, and a length and depth of 10 mm and 20 mm respectively. It is positioned in a uniform radial flux density of 200 mT. The coil carrier a current of 50 mA. The torque on the coil is

(a) 200 μNm (b) 100 μNm
(c) 2 μNm (d) 1 μNm

9. A dc A-h meter is rated for 15 A, 250 V. the meter constant is 14.4 A-sec/rev. The meter constant at rated voltage may be expressed as

(a) 3750 rev/kWh (b) 3600 rev/kWh
(c) 1000 rev/kWh (d) 960 rev/kWh

10. A moving iron ammeter produces a full scale torque of 240 μNm with a deflection of 120° at a current of 10 A. The rate of change of self inductance (μH/radian) of the instrument at full scale is

(a) 2.0 μH/radian (b) 4.8 μH/radian
(c) 12.0 μH/radian (d) 114.6 μH/radian

11. A single-phase load is connected between R and Y terminals of a 415 V, symmetrical, 3-phase, 4 wire system with phase sequence RYB. A wattmeter is connected in the system as shown in figure the power factor of the load is 0.8 lagging. The wattmeter will read.

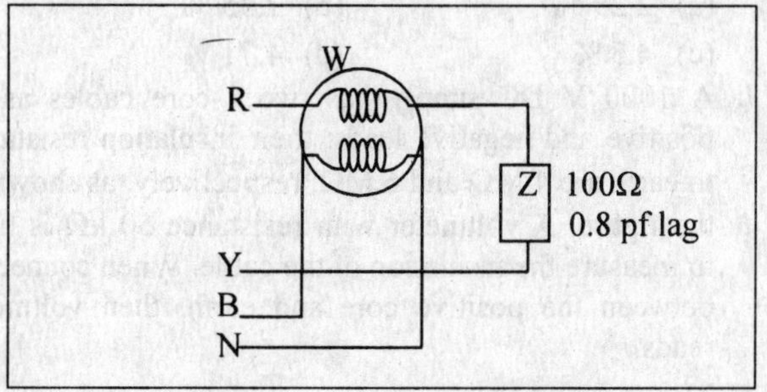

(a) – 795 W (b) – 597 W
(c) + 597 W (d) + 795 W

12. A 50 Hz, bar primary CT has a secondary with 500 turns. The secondary supplies 5 A current into a purely resistive burden of 1Ω. The magnetizing ampere-turns is 200. The phase angle between the primary and secondary current is

(a) 4.6° (b) 85.4°
(c) 94.6° (d) 175.4°

13. The core flux in the CT of Prob. Q. 18, under the given operating conditions

(a) 0 (b) 45.0 μWb
(c) 22.5 mWb (d) 100.0 mWb

14. The simultaneous application of signals x(t) and y(t) to the horizontal and vertical plates, respectively, of an oscilloscope, produces a vertical figure-of-8 display. If P and Q are constants, and x(t)=P sin (4t+30), then y(t) is equal to

(a) Q sin (4t – 30) (b) Q sin (2t + 15)
(c) Q sin (8t + 60) (d) Q sin (4t + 30)

15. A DC ammeter has a resistance of 0.1 Ω and its current range is 0-100 A. if the range is to be extended to 0-500 A, then meter requires the following shunt resistance

(a) 0.010 Ω (b) 0.011 Ω
(c) 0.025 Ω (d) 1.0 Ω

16. The set-up in the figure is used to measure resistance R. The ammeter and voltmeter resistances are 0.01 Ω and 2000 Ω, respectively. Their readings are 2 A and 180 V, respectively, giving a measured resistance of 90 Ω. The percentage error in the measurement is

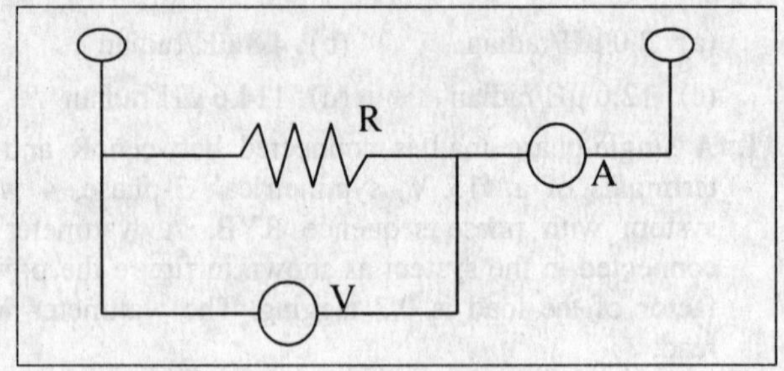

(a) 2.25 % (b) 2.35 %
(c) 4.5 % (d) 4.71 %

17. A 1000 V DC supply has two 1-core cables as its positive and negative leads: their insulation resistance to earth are 4 MΩ and 6 MΩ, respectively, as shown in the figure. A voltmeter with resistance 50 kΩ is used to measure the insulation of the cable. When connected between the positive core and earth, then voltmeter reads.

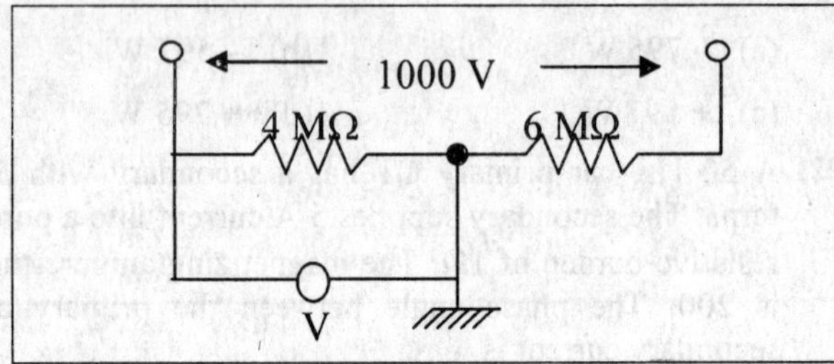

(a) 8 V (b) 16 V
(c) 24 V (d) 40 V

18. Two wattmeters, which are connected to measure the total power on a three – phase system supplying a balanced load, read 10.5 kW and – 2.5 kW, respectively. The total power and the power factor, respectively, are
(a) 13.0 kW, 0.334 (b) 13.0 kW, 0.684
(c) 8.0 kW, 0.52 (d) 8.0 kW, 0.334

19. A current of $-8+6\sqrt{2}\cdot\left(\sin \omega t+30^0\right)$ A is passed through three meters. They are a center zero I'MMC meter, a true rms meter and a moving iron instrument. The respective readings (in A) will be
(a) 8, 6, 10 (b) 8, 6, 8
(c) – 8, 10, 10 (d) –8, 2, 2

20. A variable w is related to three other variables x, y, z as w=xy/z. The variables are measured with meters of accuracy ±0.5% reading, ±1 % of full scale value and ± 1.5% reading. The actual readings of the three meters are 80, 20 and 50 with 100 being the full scale value for all three. The maximum uncertainty in the measurement of w will be
(a) ± 5% rdg (b) ± 6.5% rdg
(c) ± 6.7% rdg (d) ± 7.0 rdg

21. A 200/1 Current transformer (CT) is wound with 200 turns on the secondary on a toroidal core. When it carries a current of 160 A on the primary, the ratio and phase errors of the CT are found to be – 0.5% and 30 minutes respectively. If the number of secondary turns is reduced by 1 the new ratio error (%) and phase error (min) will be respectively.
(a) 0.0, 30 (b) – 0.5, 35
(c) –1.0, 30 (d) – 1.0, 25

22. R_1 and R_4 are the opposite arms of a Wheatstone bridge as are R_3 and R_2. The source voltage is applied across R_1 and R_3. Under balanced conditions which one of the following is true.
(a) $R_1 = R_3\, R_4 / R_2$
(b) $R_1 = R_2\, R_3 / R_4$
(c) $R_1 = R_2\, R_4 / R_3$
(d) $R_1 = R_2 + R_3 + R_4$

23. A bridge circuit is shown is shown in the figure below. Which one of the sequence given below is most suitable for balancing the bridge?

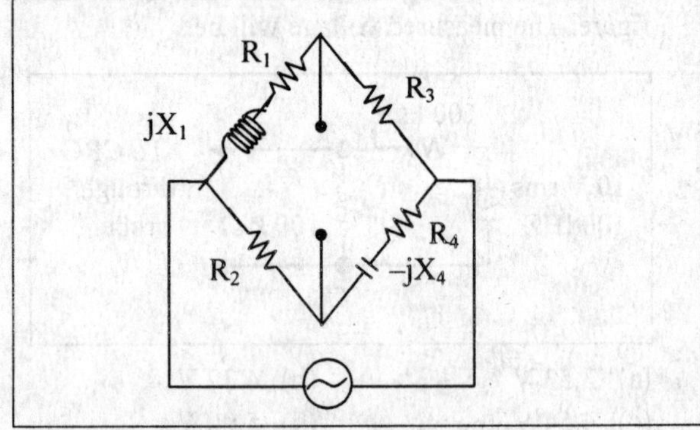

(a) First adjust R_4, and then adjust R_1
(b) First adjust R_2, and then adjust R_3
(c) First adjust R_2, and then adjust R_4
(d) First adjust R_4, and then adjust R_2

Answers

Electrical and Electronic Measurements

1. **Ans. (b)**
Solution:

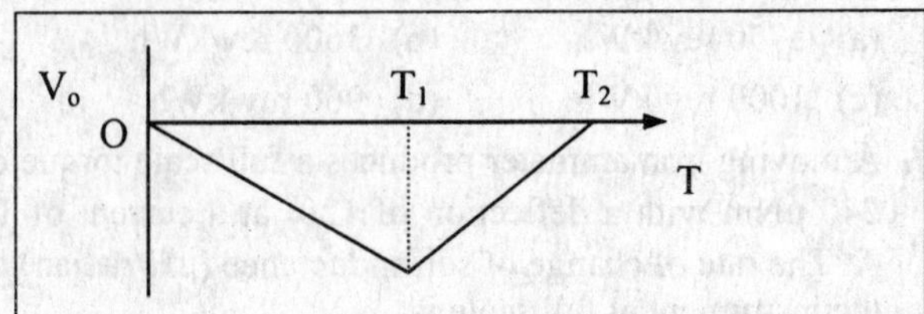

The waveform for dual slope integrater is shown. The maximum frequency can be attained when $T_2 - T_1 = 0$ and, as

$T_1 = 2^n T_C$

where T_C = clock period and

n = number of bits.

$$\therefore f_{max} = \frac{1}{2^n . T_C} = \frac{f_c}{2^n} = \frac{10^6}{1024} \approx 1\,kHz$$

2. **Ans. (a)**
3. **Ans. (c)**

Solution: Output of full wave rectifier.

$$V_{dc} = \frac{2V_m}{\pi} = \frac{2\sqrt{2}}{\pi} V_{rms} = \frac{2\sqrt{2} \times 100}{\pi} = 90.03$$

For full deflection

$$R_m + R_s = \frac{90.03}{1 \times 10^{-3}}$$

$$R_s = 90.03 \times 10^3 - 0.1 \times 10^3 = 89.93\ k\Omega$$

Method 2

d.c. sensitivity is, $S_{dc} = \frac{1}{I_{fs}} = \frac{1}{1mA} = 10^3\,\Omega / V$

For full-wave rectifier a.c. sensitivity,

$S_{ac} = 0.9\ S_{dc} = 0.9 \times 10^3 = 900\ \Omega/V$

Resistance of multiplier,

$R_s = S_{ac}\ V - R_m - 2\ R_d$

Since, diodes are ideal, $R_d = 0$

The, $R_s = 900 \times 100 - 100 = 89.90\ k\Omega$

4. **Ans. (b)**

Solution: $T_d = \frac{1}{2} I^2 \frac{dL}{d\theta}$,

$$L = 10 + 3\theta - \frac{\theta^2}{4}$$

$$\frac{dL}{d\theta} = 3 - \frac{2\theta}{4}\ \mu H$$

$$25 \times 10^{-6} = \frac{1}{2} \times 5^2 \times \left(3 - \frac{\theta}{2}\right) \times 10^{-6}$$

$$\frac{\theta}{2} = 3 - 2 = 1, \theta = 2$$

5. **Ans. (c)**

Solution: At $\Delta = 85°$, $\phi = 60°$ (pf = 0.5)

Energy registered = V 1 sin $(\Delta - \phi)$

Then error = V 1 sin $(\Delta - \phi)$ – V 1 cos ϕ

= 200 × 5 (sin 85° – cos 60°)

= – 85.1 W

At $\Delta = 80°$, $\phi = 0°$ (pf = 0°)

error = V I sin $(\Delta - \phi)$ – VI cos ϕ

= V I (sin85° – cos 0°)

= – 4.2 W

6. **Ans. (d)**
7. **Ans. (a)**

Solution: $f = 100$; $\omega = 2\pi f$; C = 10pF

Here, $X_c = \frac{1}{\omega C} = 159K$

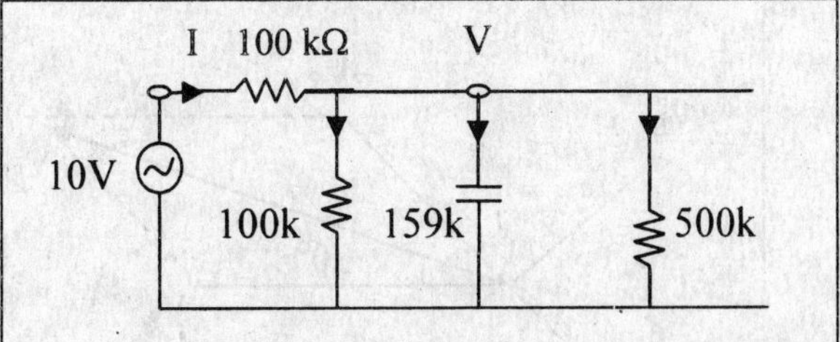

Apply the KCL

$$\therefore \frac{10 - V}{100} = V\left(\frac{1}{100} + \frac{1}{159} + \frac{1}{500}\right) = 0.018V$$

$\Rightarrow$ 10 – V = 1.82V

$$V = \frac{10}{2.82} = 3.53V.$$

8. **Ans. (a)**

Solution: Deflecting torque

T_d = NbldI

= (10)(200 × 10^{-3})(100 × 10^{-3})

(20 × 10^{-3})(50× 10^{-3})

= 200 μNm.

9. **Ans. (c)**

Solution: Meter constant is 14.4 A sec per rev.

i.e., $\frac{14.4}{3600} Ahr = \frac{14.4 \times 250}{3600}$ Whr = 1 Whr per rev.

hence 1rev/Whr = 1000 rev/kWhr.

10. **Ans. (b)**

Solution: At full scale position, $\frac{1}{2} I^2 \frac{dL}{d\theta} = T_c$

or $\frac{1}{2} \cdot (10)^2 \cdot \frac{dL}{d\theta} = 240 \times 10^{-6}$

or $\frac{dL}{d\theta} = 4.8\,\mu H / radian$

11. **Ans. (b)**

Solution: From the figure given in the question, we get.

$V_{RY} = 415\ \angle 30°$; $V_{BN} = \dfrac{415}{\sqrt{3}} \angle 120°$

Now, current through current coil is,

$$I_C = \frac{V_{RY}}{Z} = \frac{415\angle 30°}{100\angle 36.87°} = 4.15\angle -6.87°$$

S = VI* = 415/√3 × 4.15 ∠126.87

= 994.34 ∠126.57

Hence P = –597 W.

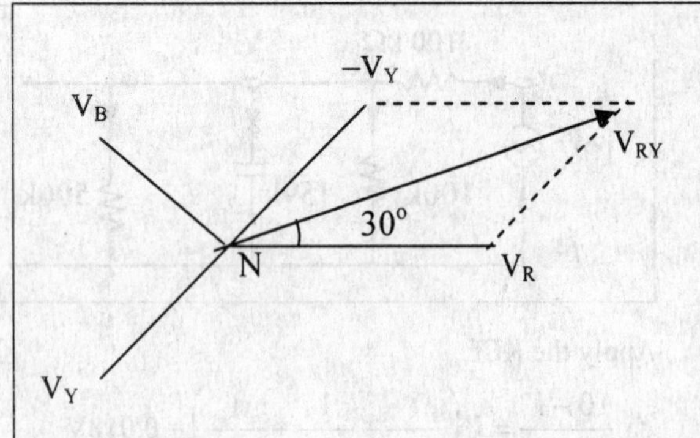

12. Ans. (a)

Solution: Phase angle between the I_s and I_p is,

$$\theta \approx \text{deg}ree \approx \left(\frac{180}{\pi}\right)\cdot\left(\frac{200.1}{500.5}\right) \approx 4.6^o$$

13. Ans. (b)

Solution: Voltage induced in the secondary,

$E_s = I_s \times Z_s = 5$ Volts; Now $E_s = 4.44 f \phi_m N_s$.

$$\therefore\ \phi_m = \frac{E_s}{4.44 fN_s} = \frac{5}{4.44 \times 50 \times 500} = 45\ \mu Wb.$$

14. Ans. (b)

Solution: $\dfrac{f_y}{f_x} = \dfrac{x - peak}{y - peak}$

Here, x-peak = 1, and y-peak = 2

$$\therefore\ f_y = \frac{f_x}{2}$$

Since x(x) = P sin (4t + 30)

$$\therefore\ y(t) = Q\sin\left(\frac{4t+30}{2}\right) = Q\sin(2t+15)$$

15. Ans. (c)

Solution: $I_m = 100$ A, $R_m = 0.1\Omega$

$$I_m = \frac{V_m}{R_m} = \frac{V_m}{0.1} = 100,\ \Rightarrow V_m = 10\text{ V}$$

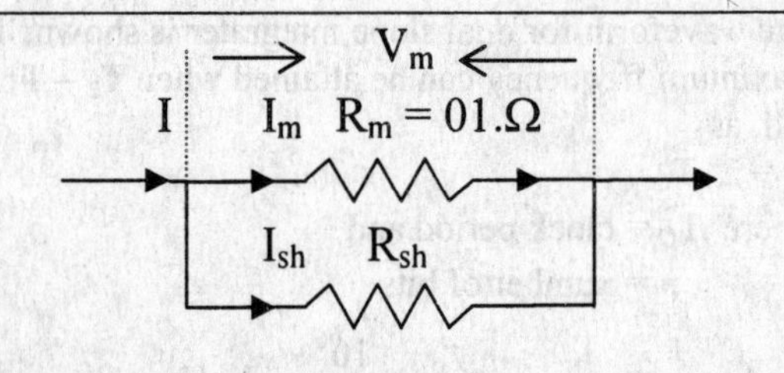

Meter current remain same, $V_m = \left(\dfrac{R_m.R_{sh}}{R_m + R_s}\right) I$

$$\text{Now, } 10V = 500 \times \frac{0.1R_{sh}}{0.1 + R_{sh}}$$

$\Rightarrow R_{sh} = 0.025\ \Omega$

16. Ans. (d)

Solution:

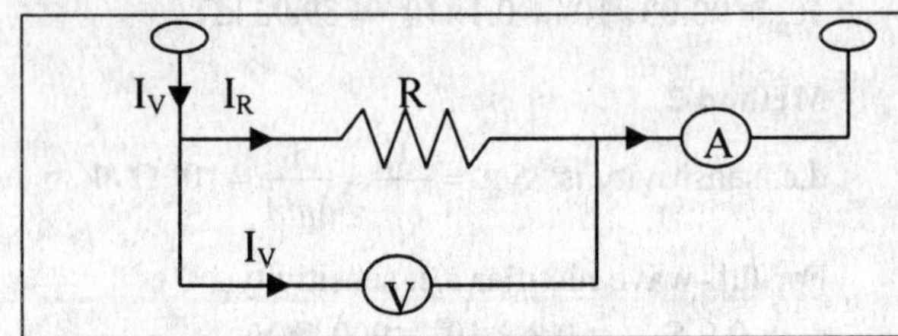

$R_A = 0.01\Omega$; $R_V = 2000\ \Omega$

$I_R = 2A$; $V_R = V_V = 180$ V

$I_A = I_R + I_V$; $I_R = I_A - I_V$

$$\text{Since, } I_V = \frac{180}{2000}$$

$$\therefore I_R = 2 - \frac{9}{100} = 1.91\,A$$

Since, 1.91R = 180

∴ R = 94.24 Ω

$$\text{Error} = \frac{\text{True value} - \text{measured value}}{\text{measured value}}$$

$$\therefore Error = \frac{94.24 - 90}{90} \times 100 = 4.71\%$$

17. Ans. (a)

Solution: Since, $R_T = \dfrac{4 \times 0.05}{4.05} + 6 = 6.05\Omega$

$$\therefore\ I_T = \frac{1000}{6.05} A$$

$$\therefore\ V_T = \frac{1000}{6.05} \times 0.05 = 8V$$

18. Ans. (d)

Solution: Power factors by two wattmeter method is given by

$$\tan\phi = \frac{\sqrt{3}\,(w_1 - w_2)}{w_1 + w_2}$$

Where w_1 and w_2 are the reading of two wattmeter.

$$\therefore\ \phi = \tan^{-1}\left(\frac{\sqrt{3}\,(w_1 - w_2)}{w_1 + w_2}\right)$$

Given $w_1 = 10.5$ w,

$w_2 = -2.5$ w

$\therefore \cos\phi = 0.334$

19. Ans. (c)

Solution: PMMC instrument reads only DC value and since it is centre zero type so it will give – 8 value RMS meter will read rms value of the current, so

$$\text{RMS} = \sqrt{8^2 + 6^2} = 10\text{A}$$

Moving iron also reads rms value, so its reading will also be 10 A.

Hence ultimately it will be – 8, 10, 10.

20. Ans. (a)

Solution: Given, $\omega = [xy / z]$

Taking log, we get, log ω x + log y – log z

Differentiating with respect to ω, we get

$$\frac{d\omega}{\omega} = \left[\frac{dx}{x} + \frac{dy}{y} - \frac{dz}{z}\right] \times 100$$

$$dx = \frac{0.5 \times 80}{100} = 4$$

$$dy = \frac{1 \times 20}{100} = 0.2$$

$$dz = \frac{1.5 \times 50}{100} = 0.75$$

$$\frac{d\omega}{\omega} = \left[\frac{4}{80} + \frac{0.2}{20} - \frac{0.75}{50}\right] \times 100$$

$$= \frac{20 + 4 - 6}{400} \times 100$$

$$= \frac{18}{4} = 4.5 = 5\%$$

21. Ans. (b)

Solution: Ratio error $= \frac{1}{200} \times 100 = 0.5\,\%$ and will be negative because one turn is reduced. Reduction in turn has no effect on phase error. So, net

Ratio error $= -(0.5\,\% + 0.5\,\%) = -1\,\%$

Phase error $= 30$ min (no charge)

22. Ans. (b)

Solution:

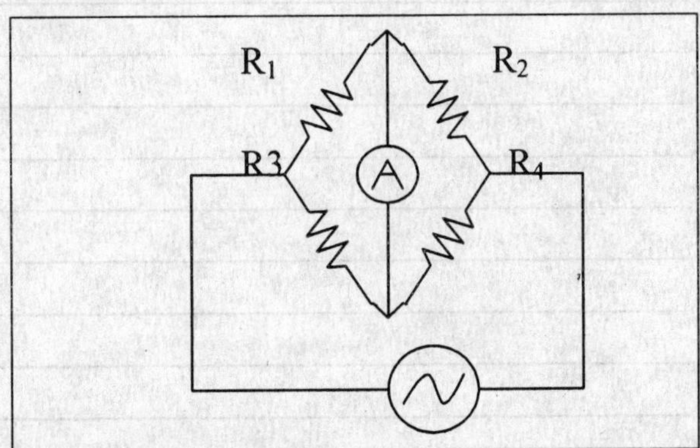

When the bridge is balanced,

$$\frac{R_1}{R_2} = \frac{R_3}{R_4} \Rightarrow R_1 = \frac{R_2 R_3}{R_4}$$

23. Ans. (d)

Reader's Notes

8 Control Systems and Process Control

Control Systems and Process Control: Feedback principles. Signal flow graphs. Transient Response, steady-state-errors. Routh and Nyquist criteria. Bode plot, root loci. Time delay systems. Phase and gain margin. State space representation of systems. Mechanical, hydraulic and pneumatic system components. Synchro pair, servo and step motors. On-off, cascade, P, P-I, P-I-D, feed forward and derivative controller, Fuzzy controllers.

8.1 FEEDBACK PRINCIPLES

System

System is an arrangement of physical components connected or related in such a manner to from or acts as entire unit.

Control

Control means to regulate, direct or command.

Control System

A control system is a combination of components that are inter connected in a configuration design to provide a specified system in such a manner so as to command direct or regulate itself or an other system.

Feed Back Control System

A control system that tends to maintain a prescribed relationship of one system variable to another by comparing function of, these variable and using the difference as a means of control.

Figure given below show block diagram of an automatic control system.

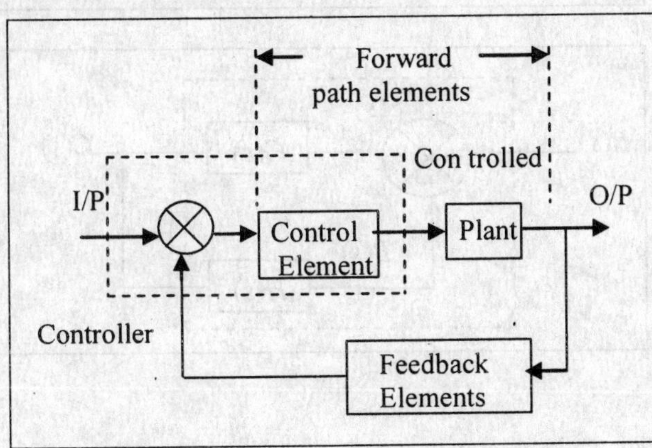

- Error detector compares a signal obtained through feedback elements, with the reference input.
- Any difference between these two signals constitutes an error or actuating signal, which actuates the control elements.
- Control element in turn alters the condition in such a manner as to reduce the original error.

Open Loop Control System

1. In which the output has no effect upon the control action. The output is neither measured nor feed back for comparison with the input. Any system which operate on time basis is an open loop system. Examples are Traffic lights, washing machine, toaster operation.
2. Feed back not applied
3. Response is pre calibrated
4. Stability is not a major problem.
5. Comparatively cheap.

Close Loop Control System

1. In which the output signal has direct effect upon the control action.
2. It is feed back system.
3. It has the tendency to oscillate.
4. Response relatively insensitive to external disturbances because of fed back.
5. Stability of the system is always a major problem.

Adaptive Control System

It is a system, which adjusts itself against environ-ment, aging deterioration and even engineering errors.

Multivariable Control System

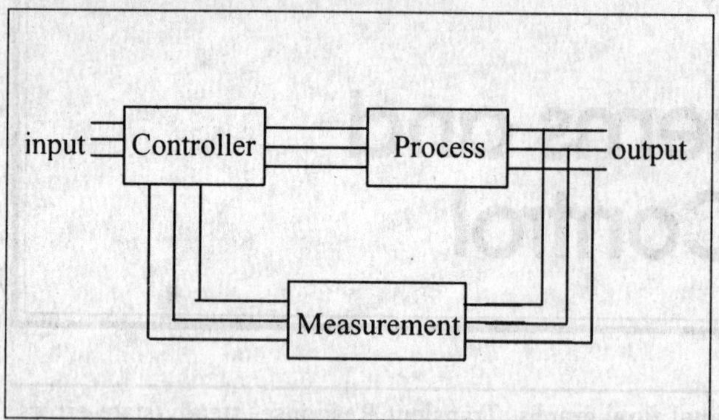

With time the systems have become more complex the interrelationship of many-controlled variable must be considered in the control scheme. Block diagram depicts a multivariable control system.

System with more than one inputs and one or more outputs are called multivariable system. Examples are automobile control, Autopilot system, Engine.

Typical Autopilot System

The state of motion of variable (velocity) acceleration fed to by means of motion errors is Gyros accelerometer. It tells the direction of vehicle on command from ground generates signals which controls the hydraulic actuators and in turn move the engine.

Multiple Input and Multiple Output System

When multiple input are present in a linear system, each input can be treated independently of the others. Complete output of the system can then be obtained by superposition, i.e. output corresponding to each input alone are added together.

$$C_i(s) = \sum_{j=1}^{Y} G_{ij}(s) R_1(s) \quad i = 1, 2, \ldots\ldots n$$

Regulating System

It is feed back control system is which the reference input or the desired output is either constant or slowly varying with time and in which the primary task is to maintain output at the desired value in the presence of disturbance.

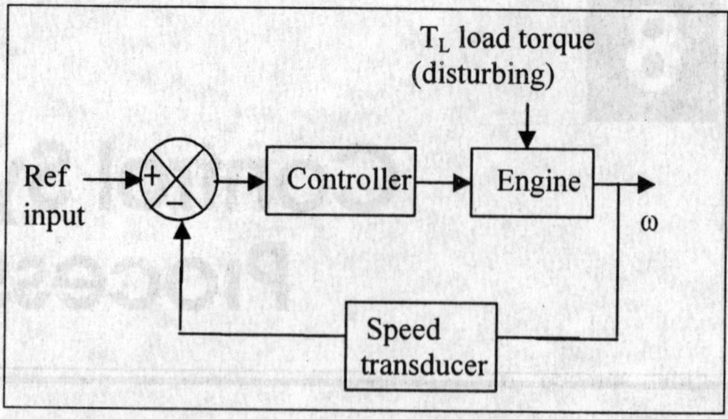

Example: Voltage control of generator, speed control of motor.

Servomechanism

A servomechanism is a feed back control system in which the output is some mechanical position, velocity or acceleration i.e. in which one or more signal represent mechanical motion.

According to the nature of controlled signal flowing in the system, they are classified.

1. Containers data feedback control system
2. Relay type data feedback control system
3. Sampled data feedback control system
4. Optimal data feedback control system
5. Adoptive data feedback control system

Block Diagram Retention Technique

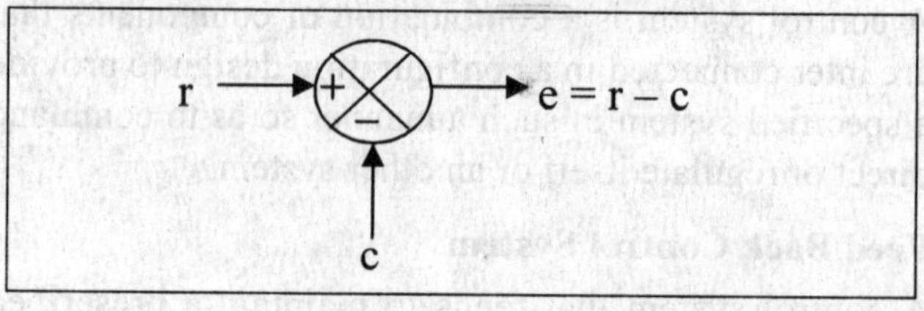

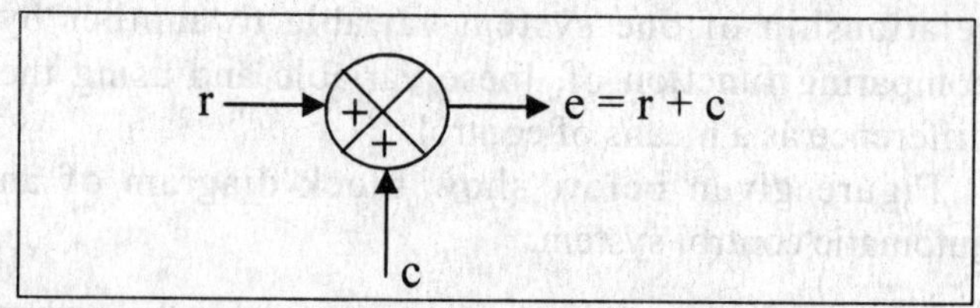

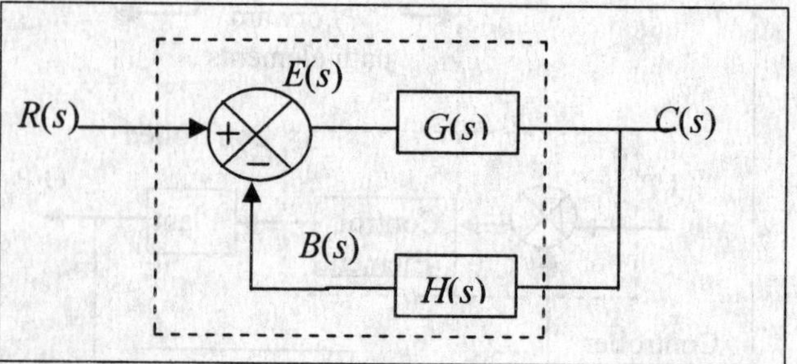

$$G(s)=\frac{C(s)}{E(s)} \quad ; \quad H(s)=\frac{B(s)}{C(s)}$$

Term Used

$r(t), R(s) \rightarrow$ reference input

$c(t), C(s) \rightarrow$ output signal or controlled variable

$b(t), B(s) \rightarrow$ feed back signal

$e(t), E(s) \rightarrow$ actuating signal or error signal

$G(s)=\frac{C(s)}{E(s)}$ forward path T.F or open loop T.F

$H(s)=$ feed back path Transfer function

$G(s)H(s)=$ loop transfer function or loop gain

$M(s)=\frac{C(s)}{R(s)}$ closed loop transfer function

$$C(s) = E(s)\,G(s) \tag{1}$$

$$B(s) = H(s)\,C(s) \tag{2}$$

$$E(s) = R(s) - B(s) \tag{3}$$

From equations (2) and (3), we eliminate B(s) and get

$$E(s)=R(s)-H(s)C(s) \tag{4}$$

Put the value of $E(s)$ in equation (1) and we set

$$C(s)=G(s)[R(s)-H(s)C(s)] \tag{5}$$

$$C(s)=G(s)R(s)-G(s)H(s)C(s)$$

$$C(s)+C(s)G(s)H(s)=G(s)R(s)$$

$$C(s)[1+G(s)H(s)]=G(s)R(s)$$

$$\frac{C(s)}{R(s)}=\frac{G(s)}{1+G(s)H(s)}=M(s)$$

This is the close loop transfer function for negative feedback control system.

System can be reduced to a single block shown below:

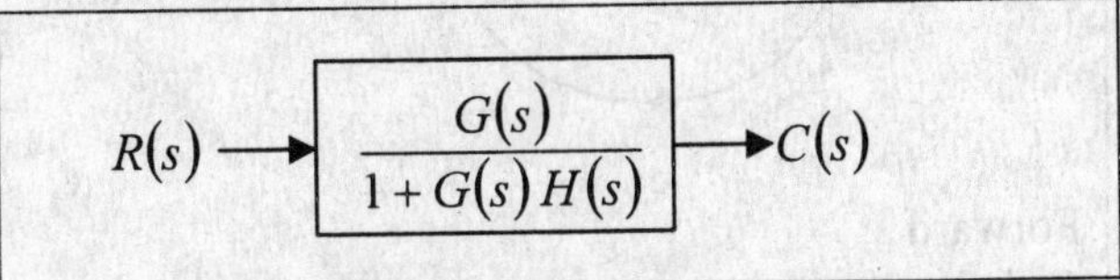

Block-Diagram Reduction

A complex block diagram configuration can be simplified by certain re-arrangement.

Some Rules are Given Below

1. Combining block in cascade

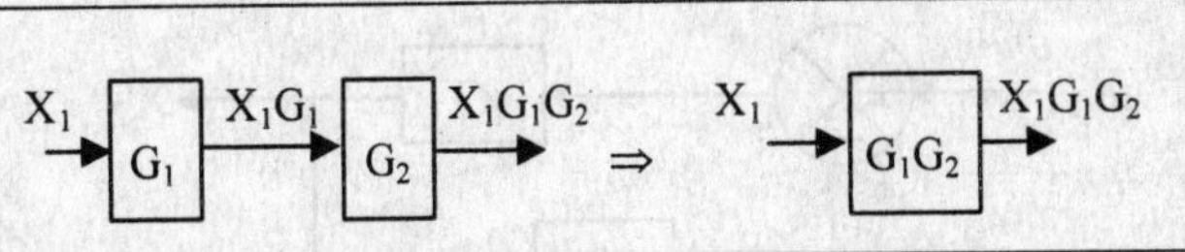

2. Moving a summing point after a block

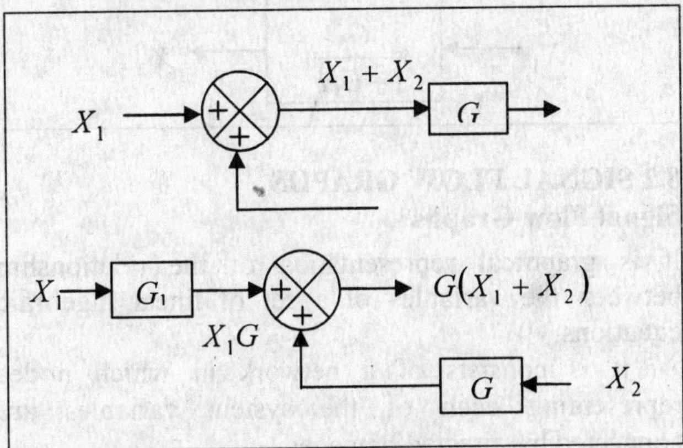

3. Moving a take off point after a block

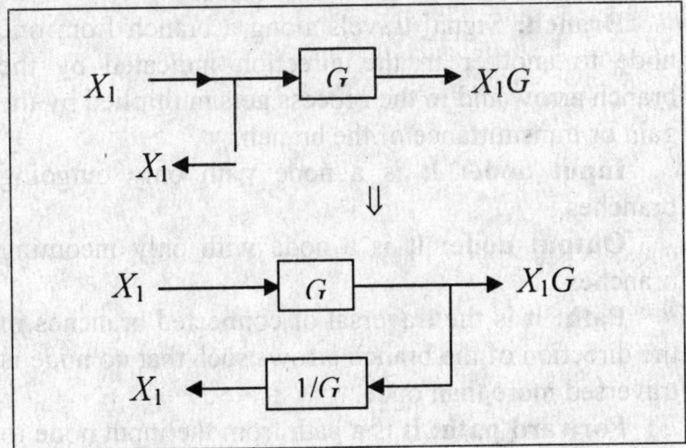

4. Moving a take off point ahead of a block

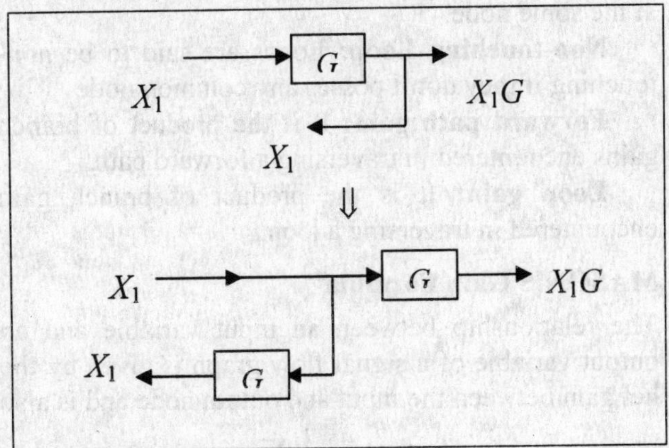

5. Eliminating a feedback loop

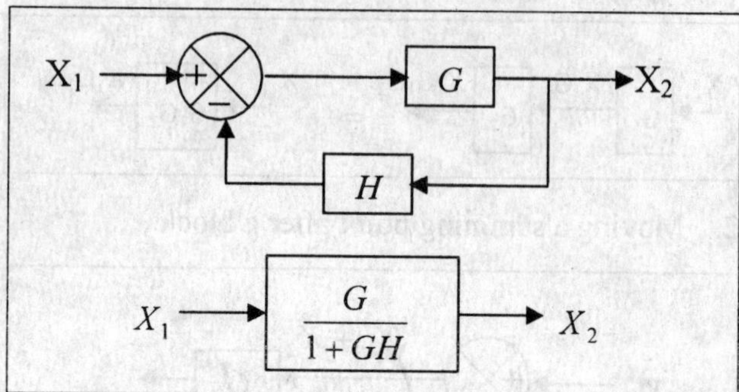

8.2 SIGNAL FLOW GRAPHS

Signal Flow Graphs

It is graphical representation of the relationship between the variables of a set of linear algebraic equations.

It is consists of a network in which nodes representing each of the system variables are connected by directed branches

Node: A system variable, which is equal to the sum of all incoming signals at the node.

Branch: Signal travels along a branch from one node to another in the direction indicated by the branch arrow and in the process gets multiplied by the gain or transmittance of the branch.

Input node: It is a node with only outgoing branches.

Output node: It is a node with only incoming branches.

Path: It is the traversal of connected branches in the direction of the branch arrows such that no node is traversed more than once.

Forward path: It is a path from the input node to the output node.

Loop: It is path, which originates and terminates at the some node.

Non-touching Loop: Loops are said to be non-touching if they don't posses any common node.

Forward path gain: It is the product of branch gains encountered in traversing a forward path.

Loop gain: It is the product of branch gain encountered in traversing a loop.

MASON'S Gain Formula

The relationship between an input variable and an output variable of a signal flow graph is given by the net gain between the input and output node and is also known as overall gain $M = \frac{1}{\Delta}\sum_k P_k \cdot \Delta_k$

where, M = overall system gain

P_k = path gain of kth forward path.

Δ = determinant of the graph

= 1 – (sum of loop gains of all individual loop) + (sum of gain products of all possible combination of non touching loop) – ....

$$\Delta = 1 - \sum_m m_1 + \sum_m m_2 - \sum_m m_3 + \ldots\ldots$$

where, P_{mr} = gain product of mth possible combination of non-touching loop

Δ_k = The value of Δ for that part of the graph not touching the kth forward path.

Example

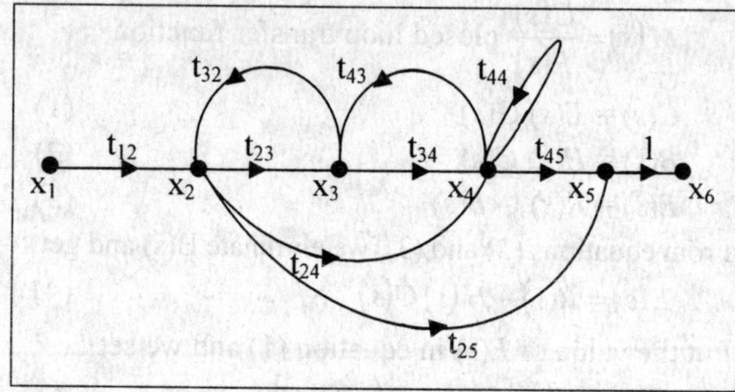

Solution:

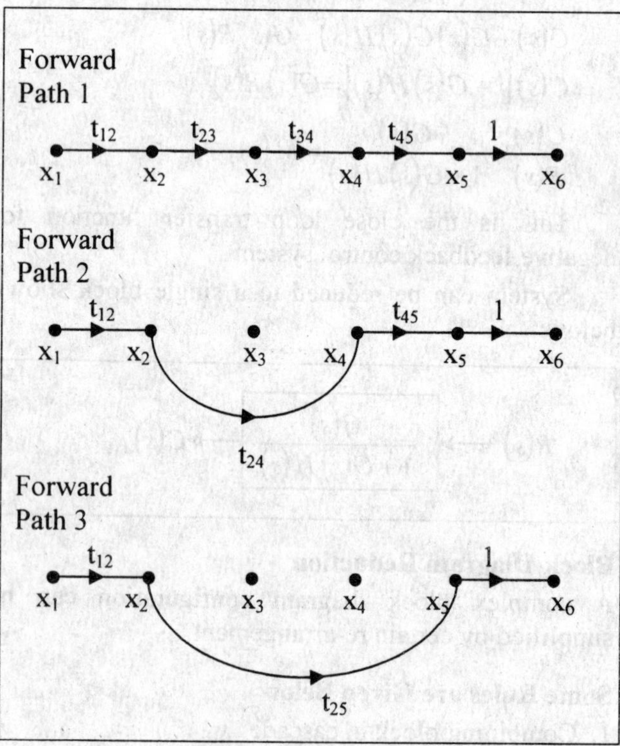

Individual loop

(1)

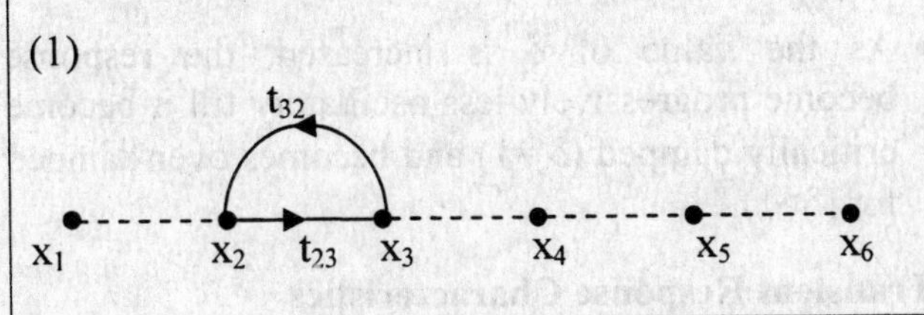

(2)

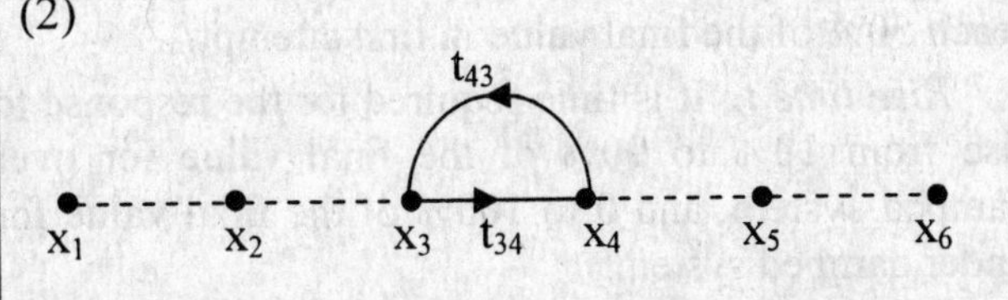

(3)

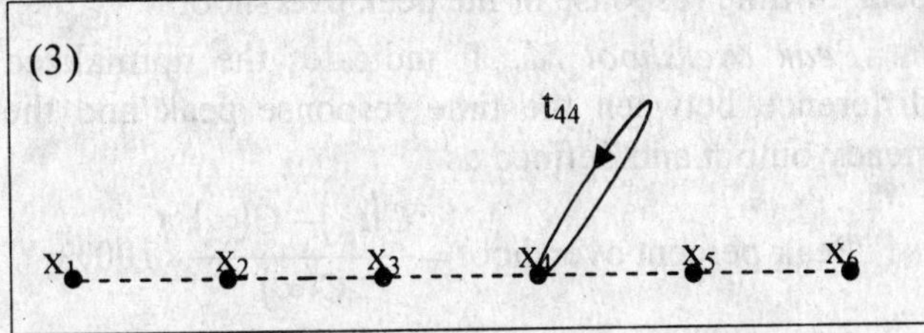

(4)

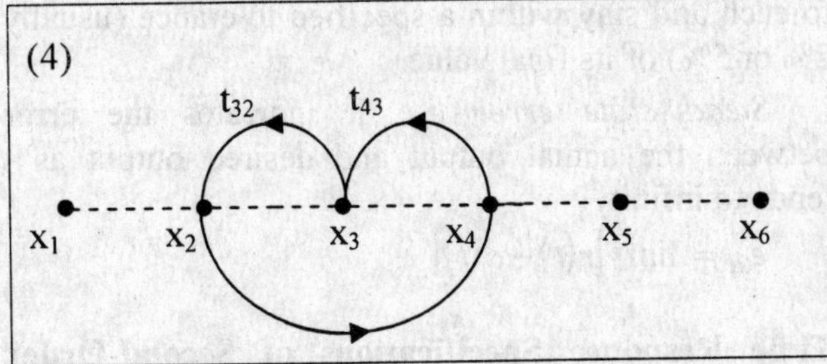

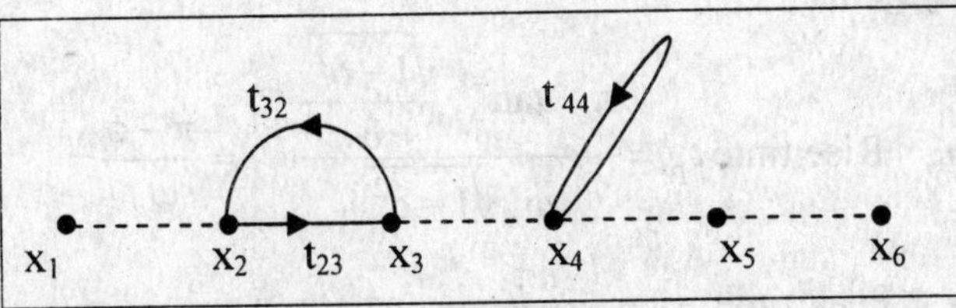

Forward Path Gain

$$M_1 = t_{12} \cdot t_{23} \cdot t_{34} \cdot t_{45}$$

$$M_2 = t_{12} \cdot t_{24} \cdot t_{45}$$

$$M_3 = t_{12} \cdot t_{25}$$

Individual Loop Gain

$$P_{11} = t_{23} \cdot t_{32}$$

$$P_{21} = t_{34} \cdot t_{43}$$

$$P_{31} = t_{44}$$

$$P_{41} = t_{24} \cdot t_{43} \cdot t_{32}$$

$$\Delta = 1 - [t_{23} . t_{32} + t_{34} . t_{43} + t_{44} + t_{24} . t_{43} .t_{32}]$$

$$\Delta = 1 - \left[t_{23} \cdot t_{32} + t_{34} \cdot t_{43} + t_{44} + t_{24} \cdot t_{43} \cdot t_{32}\right] + \left[t_{23} \cdot t_{32} \cdot t_{44}\right] - [\text{non}]$$

Path 1 $\Delta_1 = 1 - [0]$

Path 2 $\Delta_2 = 1 - [0]$

Path3 Δ_3 non-touching graph for path 3

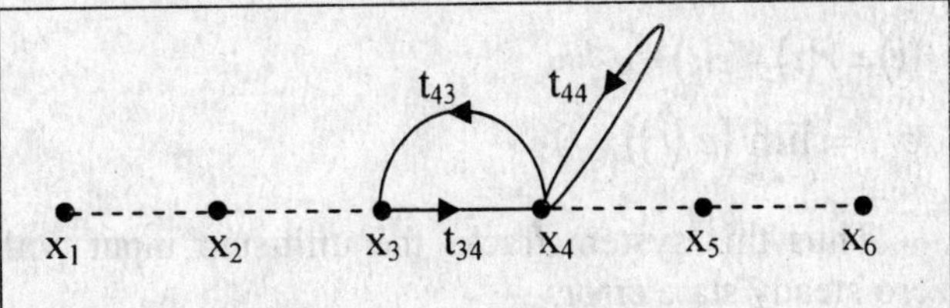

$$\Delta_3 = 1 - (t_{34} \cdot t_{43} + t_{44}) + (0)$$

$$M = \frac{M_1\Delta_1 + M_2\Delta_2 + M_3\Delta_3}{\Delta}$$

$$= \frac{(t_{12} \cdot t_{23} \cdot t_{34} \cdot t_{45}) \cdot 1 + (t_{12} \cdot t_{24} \cdot t_{45}) \cdot 1 + t_{12} \cdot t_{25} \cdot (1 - t_{34} \cdot t_{43} - t_{44})}{1 - (t_{23} \cdot t_{32} + t_{34} \cdot t_{43} + t_{44} + t_{24} \cdot t_{43} \cdot t_{32}) + (t_{23} \cdot t_{32} \cdot t_{44} \cdot)}$$

8.3 TRANSIENT RESPONSE

Standard Test Signal

(a) Step input $u(t) = 0$ for $t < 0 = 1$ for $t \geq 0$

(b) Impulse input $\delta(t) = 0$ for $t \neq 0 = 1$ for $t = 0$

- $\int_{-\infty}^{\infty} \delta(t)\, dt = 1$ i.e. area under impulse function is unity
- $\int_{-\infty}^{\infty} x(t)\,\delta(t)\, dt = x(0)$

(c) Ramp function $r(t) = Kt\ u(t)$ for $t \geq 0$
$= 0$ for $t < 0$

- It is the integral of the unit step.

(d) Parabolic function:
$a(t) = [K t^2 u(t))/2]$ for $t \geq 0$
$= 0$ for $t < 0$

Time Response of First-Order System

Suppose we have transfer function

$$\frac{C(s)}{R(s)} = \frac{1}{1+sT}$$

1. Response to the Unit-step input.

$$c(t) = 1 - e^{-t/T}$$

- Slope at t = 0 ; $\left.\frac{\partial c(t)}{\partial t}\right|_{t=0} = \frac{1}{T}$

where, T = time constant

Steady State Error

$$e(t) = r(t) - c(t) = e^{-t/T}$$

$$e_{ss} = \lim_{t \to \infty} (c\,(t)) = 0$$

Thus this system tracks the unit-step input with zero steady state error.

Response Due To Unit-Ramp Signal

$$c(t) = t - T\left(1 - e^{-t/T}\right)$$

Error signal $e(t) = T\left(1 - e^{-t/T}\right)$

$$e_{ss} = \lim_{t \to \infty} (e\,(t)) = T$$

Thus the first order system under consideration will track the unit-ramp input with a steady state error = T.

Response of 2nd Order System to Unit-step

$$\frac{C(s)}{R(s)} = \frac{\omega_n^2}{s^2 + 2\delta\omega_n \cdot s + \omega_n^2}$$

where, δ = damping factor

ω_n = undamped natural frequency.

Characteristic Polynomial

$q(s) = 0$ i.e.

$$s^2 + 2\delta\omega_n \cdot s + \omega_n^2 = 0$$

$\omega_d = \omega_n\sqrt{1-\delta^2}$ known as damped natural frequency.

With Unit Step Signal

$$c(t) = 1 - \frac{e^{-\delta\omega_n t} \cdot \sin}{\sqrt{1-\delta^2}}\left(\left(\omega_n\sqrt{1-\delta^2}\right)t + \tan^{-1}\frac{\sqrt{1-\delta^2}}{\delta}\right)$$

- Time response is under damped for (δ< 1).
- For δ = 0, it breaks into continuous oscillation.
- $c(t) = 1 - \cos(\omega_n t)$
- As the value of δ is increased, the response become progressively less oscillatory till it become critically damped (δ =1) and becomes over damped for (δ>1).

Transient Response Characteristics

Delay Time t_d: It is time required for the response to reach 50% of the final value in first attempt.

Rise time t_r: It is time required for the response to rise from 10% to 90% of the final value for over damped system, and 0 to 100% of the final value for under damped system.

Peak time t_p: It is the time required to reach the peak of time response or the peak overshoot.

Peak overshoot M_p: It indicates the normalized difference between the time response peak and the steady output and defined as

$$\text{Peak percent overshoot} = \frac{C(t_p) - C(\infty)}{C(\infty)} \times 100\%$$

Setting time t_s: It is time required for the response to each and stay within a specified tolerance (usually 2% or 5%) of its final value.

Steady-state error e_{ss}: It indicates the error between the actual output and desired output as t tends to infinity.

$$e_{ss} = \lim_{t \to \infty} [r(t) - c\,(t)]$$

Time Response Specifications of Second-Order Systems

- Rise time $t_p = \dfrac{\pi - \tan^{-1}\dfrac{\sqrt{1-\delta^2}}{\delta}}{\omega_n\sqrt{1-\delta^2}} = \dfrac{\pi - \phi}{\omega_d}$
- Peak time = $\dfrac{n\pi}{\omega_n\sqrt{1-\delta^2}}$ where (n = 1, 2, …
- Peak time corresponds to first over shoot.

$$t_p = \frac{\pi}{\omega_n\sqrt{1-\delta^2}}$$

- Peak overshoot M_p :

$$M_p = c(t_p) - 1 = \exp\left(\frac{-\pi\delta}{\sqrt{1-\delta^2}}\right)$$

Settling Time t_s

(a) For 2% tolerances, $t_s = \frac{4T}{\delta\omega_n}$

(b) For 5% tolerances, $t_s = \frac{3T}{\delta\omega_n}$

8.4 TEADY-STATE-ERRORS

Steady State Error and Error Constants

If R(s) is input and G(s) open loop gain then steady state error *for unity feedback* system is given by

$$e_{ss} = \lim_{s\to\infty} e(t) = \lim_{s\to 0} sE(t) = \lim_{s\to 0} \frac{sR(s)}{1+G(s)}$$

where, $e(t) =$ error signal $= r(t) - c(t)$

$r(t) =$ input signal

c(t) = output of unity feedback system.

1. Unit Step Input $R(s) = (1/s)$

$$e_{ss} = \lim_{s\to 0} \frac{s(1/s)}{1+G(s)} = \frac{1}{1+G(0)} = \frac{1}{1+K_p}$$

where, $K_p = G(0)$ is defined as the *positional error constant.*

2. Unit-Ramp (Velocity) Input

$$R(s) = \frac{1}{s^2}$$

$$e_{ss} = \lim_{s\to 0} \frac{s(1/s^2)}{1+G(s)} = \frac{1}{K_p}$$

where $K_p = \lim_{s\to 0} sG(s)$ is known as ***velocity error constant.***

3. Unit-Parabolic (Acceleration) Input

$$R(s) = \frac{1}{s^3}$$

- $$e^{ss} = \lim_{s\to 0} \frac{s(1/s^3)}{1+G(s)} = \frac{1}{\lim_{s\to 0}(s^2G(s))} = \frac{1}{K_p}$$

where $K_p = \lim_{s\to 0} s^2G(s)$ is defined *acceleration error constant.*

Classification of Feed Back System by Type

Since the error constant for a given system with unity feed back depend on the type of G (s). G(s) are classified as.

$$G(s) = \frac{K(1+S\tau_1)(1+S\tau_2)}{S^n(1+S\tau_a)(1+S\tau_b)} \quad ; \quad n = 0,1,2$$

where K and τ are constant.

The type of control system referred to the order of the pole of G(s) at s = 0. Therefore the system that is describe by G(s) is of type j where n = 0, 1, 2…

n = 0, Type 0 system

n = 1, Type 1 system

- The open loop transfer function having s^n term in the denominator with corresponds to number of integration in the system.

Type	K_p	K_v	K_a	e_{ss}= R/1+K_p	e_{ss}= R/K_V	e_{ss}= R/K_a
0	K	0	0	R/1+K_p	∞	∞
1	∞	K	0	0	R/K_v	∞
2	∞	∞	K	0	0	R/K
3	∞	∞	∞	0	0	0

$e_{ss} = 0$ means output like input with zero actuating signal.

$e_{ss} = \infty$, means actuating signal invariance with time

The steady state error because of an input signal can easily be found by the use of supper position theorem, taking one input at a time.

- Error constant K_p, K_v, K_a describe the ability of a system to reduce or eliminate steady-state error.
- As the type of system become higher (i.e. increasing no. of integration), more steady state error are eliminated
- Although it appears that there is no limit to the number of integration, type –0, 1 and 2 are the most commonly employed system in practice.
- System of type higher than 2 i.e. with more that two integration is not employed in practice as.
  (a) These are more difficult to stabilize
  (b) The dynamic errors for such system tend to be large than those for type-0, 1 and 2.
- One of the disadvantages of error constant is that don't give information on the steady state error. When inputs are other than the three basic types.
- It fails to indicate the exact manner in which error function changes with time.

8.5 ROUTH AND NYQUIST CRI0TERIA

This criterion is based on ordering the co-efficient of the characteristic equation into an array, called the routh-array.

$$q(s) = a_o s_n + a_1 s^{n-1} + \ldots + a_{n-1} s + a_n = 0$$

Routh array:

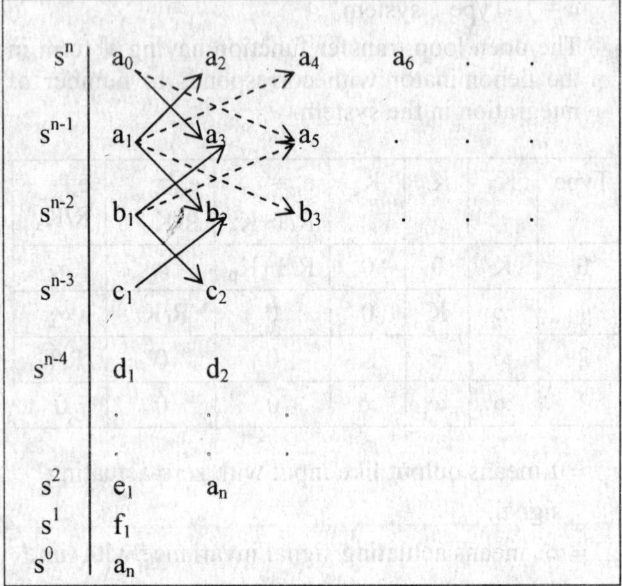

The co-efficient b_1, b_2 ... are evaluated as follow

$$= b_1 = \frac{a_1 a_2 - a_0 a_3}{a_1} \; ; \; b_2 = \frac{a_1 a_4 - a_0 a_5}{a_1}$$

$$= c_1 = \frac{b_2 a_3 - a_1 b_2}{b_1} \; ; \; c_2 = \frac{b_1 a_5 - a_1 b_3}{b_1}$$

For a system to be stable, it is necessary and sufficient that each term of first column of Routh array of its characteristic equation be positive if $a_0 > 0$. If this condition in not met, the system is unstable and number of sign changes of the terms of the first column of the Routh array corresponds to the no of roots of the characteristic equation in the right half of the s-plane.

Special Case: In applying the Routh stability criterion, certain difficulties arise causing the breakdown of the Routh's test.

Problem 1: When the first term in any row of the Routh array is 0 while rest of the row has at least one non-zero term. Due to this, the term in the next row becomes infinite and test breaks out. It can be solved by

(a) Put a small positive number ε for the zero and proceed to evaluate the rest of the Routh array. Then examine the signs of the first column of Routh array by letting $\varepsilon \to 0$.

(b) Modify the original characteristic equation by replacing S by (1/Z) and than apply the Routh test on the modified equation in term of Z.

Problem 2: When all the elements in any one row of the Routh array are zero

- This condition indicates that there are symmetrically located roots in the S-plane.
- The polynomial whose coefficient are the elements of the row just above the row of zero in the Routh array is called an *auxiliary polynomial.* This gives the number and location of root pairs of the characteristic equation.

8.6 NYQUIST PLOT

The frequency-domain analysis of a linear control system is based on the closed-loop transfer function, such as $M(s) = \frac{C(s)}{R(s)} = = \frac{G(s)}{1+G(s)H(s)}$.In the analysis problem we are again involved in determining the performance characteristics of the system, and the one of the most important is stability. Thus, we are interested in the location of the characteristic equation roots which are the poles of the closed-loop transfer function.

Thus far, two methods of determining stability by investigating the location of the roots of the characteristic equation have been indicated:

1. The roots of the characteristic equation are actually solved. These roots can actually be plotted as a function of some system parameter in the form of a root locus diagram. Thus, it is possible to study not only the absolute stability but also the relative stability of a closed-loop system through the location of the characteristic equation roots in the s-plane.
2. The relative position of the characteristic equation roots with respect to the imaginary axis of the z-plane are determined by means of the *Routh-Hurwitz criterion.*

The basic difference between these two methods is that the root locus is generally more versatile, and is

useful for analysis as well as design problems, whereas the Routh-Hurwitz criterion gives only an absolute answer on stability.
Another important difference to realize is that the starting point of the root locus method is the loop transfer function of the closed-loop system, whereas for the Routh-Hurwitz test it is the characteristic equation.

The Nyquist criterion is a frequency-domain method which is based on *polar plot* and has the following ***features*** that make it desirable for the analysis as well as the design of control system:

1. It provides the same amount of information on the absolute stability of a control system as does the Routh-Hurwitz criterion.
2. In addition to absolute system stability, the Nyquist criterion indicates the degree of stability of a stable system and gives an indication of how the system stability may be improved, if needed
3. It gives information on the frequency-domain response of the system.
4. It can be used for a stability study of systems with time delay.
5. It can be modified for nonlinear system.

We can formulate the Nyquist criterion by referring to the closed-loop transfer function $M(s)=\dfrac{C(s)}{R(s)}=$ $=\dfrac{G(s)}{1+G(s)\,H(s)}$ which may represent any linear time-invariant system.

The stability of the system can be studied by investigating the poles of the closed-loop transfer function matrix M (s). For asymptotic stability, all the poles of the transfer function matrix M(s) must be located in the left-half s-plane.

In short, the Nyquist criterion is a graphical method of determining the stability of a closed-loop system by investigating the properties of the frequency-domain plots of $G(s)\,H(s)$.

Encircled: If point is found to be inside the path, the point is said to be encircled by the closed path. Figure shows that point X is encircled by the closed path where the point Y is not encircled by the closed path.

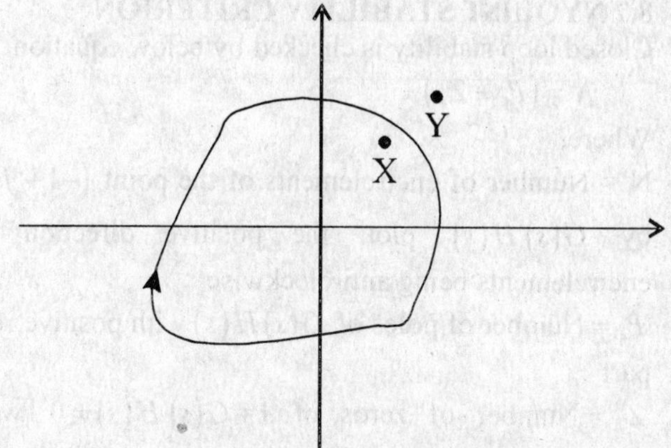

Figure: Point X is encircled

Enclosed: If a path is traversed in the clockwise direction and the point is found to lie to the right of the path, the point is said to be enclosed by the path. Figure shows that point X is enclosed by the path, whereas point Y is enclosed by the path in figure.

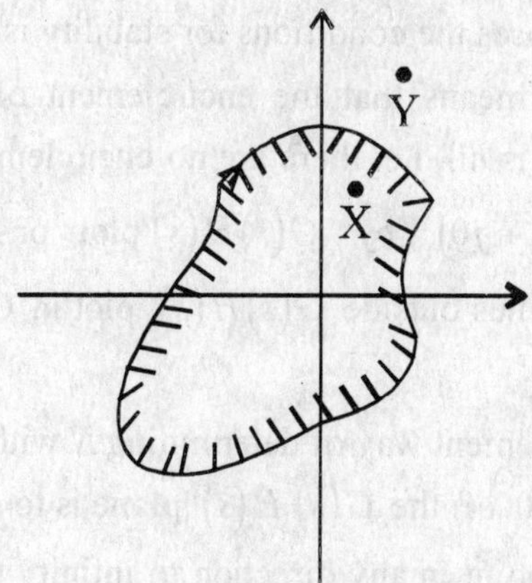

(a) Point X is enclosed

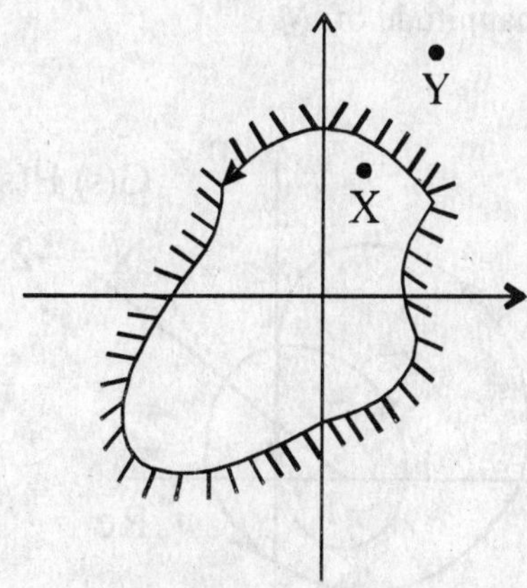

(b) Point Y is enclosed

Figure: Definition of enclosed

8.7 NYQUIST STABILITY CRITERION

Closed loop stability is checked by below equation

$$N = (P_+ - Z_+)$$

Where,

N = Number of encirclements of the point $(-1 + j0)$ by $G(s)\,H(s)$ plot. the positive direction of encirclements being anti-clockwise

P_+ = Number of poles of $G(s)H(s)$ with positive real part

Z_+ = Number of zeros of $1 + G(s)\,H(s) = 0$ with positive real part.

Condition for stability $N = P_+ - Z_+$

For a stable control system $Z_+ = 0$, therefore, the condition for a control system to be stable is $N = (P_+ - 0)$ or, $N = P_+$

For most of the control system $P_+ = 0$, therefore, for such cases the conditions for stability is $N = 0$.

Which means that the encirclement of the point $(-1 + j0)$ is nil, i.e. there are no encirclements of the point $(-1 + j0)$ by $G(s)H(s)$ plot or the point $(-1 + j0)$ lies outside $G(s)H(s)$ plot in $G(s)H(s)$-plane.

A convenient way of determining *N* with respect to the $-1 + j0$ of the $G(s)\,H(s)$ plane is to draw a line from the point in any direction to infinity the number of net intersections of this line with $G(s)\,H(s)$ locus gives the magnitude of *N*.

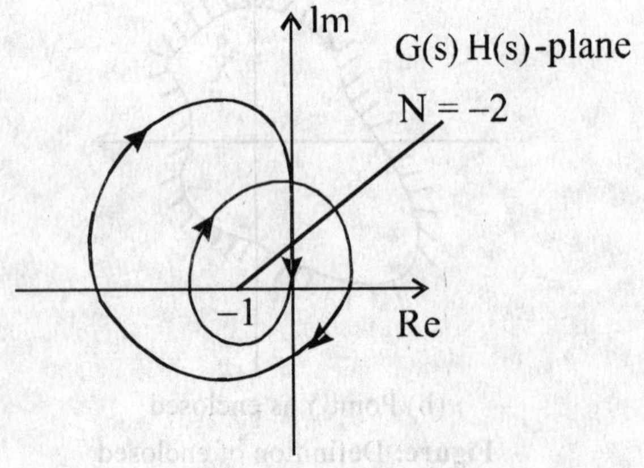

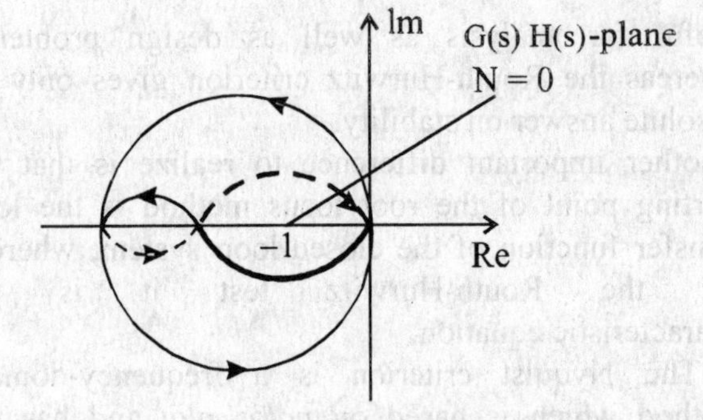

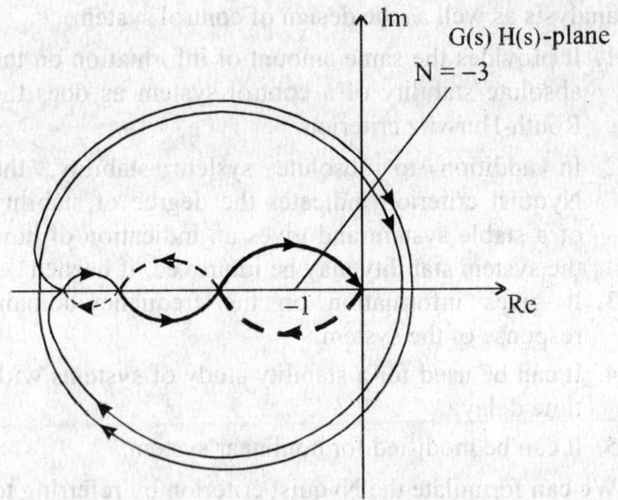

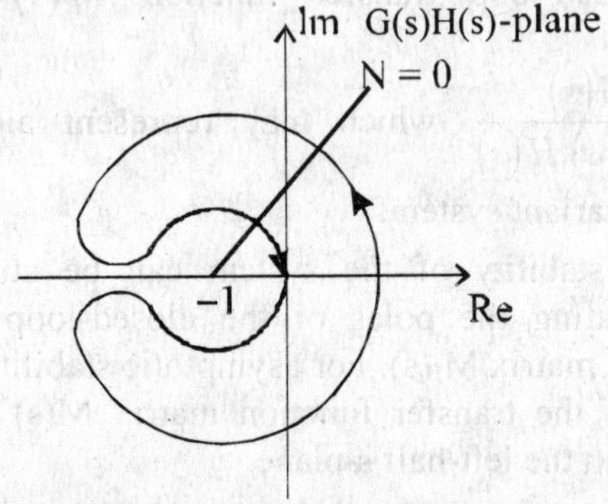

Figure: Examples of the determination of N in the $G(s)\,H(s)$ plane.

8.8 PROCEDURE FOR DRAWING OF A NYQUIST STABILITY PLOT

For the convenience of analysis the Nyquist path is divided into a minimum of three sections

Step 1: $G(s)\,H(s)$ plot for $\omega = +0$ to $\omega = +\infty$ i.e. polar plot of a function.

Step 2: $G(s)H(s)$ plot for $\omega = -0$ to $\omega = -\infty$ i.e. mirror image of polar plot of a function.

Step 3: $G(s)H(s)$ plot for $\omega = -0$ to $+0$. It is drawn with radius of infinite magnitude through an angle of $-n\pi$ (clockwise direction) where n is type of system.

Step 4: In most of the case $\omega = +\infty$ and $\omega = -\infty$ lies at the same point, i.e. origin.

Step 5: Connect all the curves.

Step 6: Note the number of clockwise encirclements about $(-1, 0)$.

Step 7: Decide stability using $N = (P_+ - Z_+)$

8.9 BODE PLOT

BODE plot of a transfer function

The Bode plot contain two graphs, one with the magnitude of G (jω) plotted in decibels Vs log w or w and the other with the phase of G (jω) in degrees as a function of log w or w. The Bode plot is also known as the **corner plot** or the **logarithmic plot** of G (jω)

Advantage:

1. Since the magnitude of G (jω) in the Bode plot is expressed in decibels, the product and division factor in G (jω) be comes addition and subtraction respectively. The phase relation are also added and subtracted from each other in natural way.
2. The shape of the Bode plot for the most commonly function is servo system make it possible to represent approximately the exact function plot by straight-line asymptotes. i.e. the magnitude plot of he Bode plot of most G(jω) faction encountered in control system may be approximated by straight line segments. This makes the construction of Bode plot very simple.
3. Since the Bode plot are easy to Construct the data necessary for the Construction of polar plot and the magnitude is phase plot can be obtained directly form the Bode plot.

Let open loop transfer function be

$$G(s) = \frac{K(1+T_1 s)(1+T_2 s)}{s(1+T_a s)(1+j2\delta\mu - \mu^2)}$$

where, K, T_1, T_2, δ, μ, are real coefficient.

It is assume that second order polynomial $1 + j2\delta\mu - \mu^2$, $\mu = \omega/\omega_n$ has two complex conjugate.

The magnitude of |G(jω)| db = 20 log | G (jω) |

$$G(j\omega) \text{ db} = 20\log K + 20\log(1 + j\omega T)$$
$$+ - 20\log(j\omega) - 20\log(1 + j\omega T_a) \quad (1)$$
$$\angle G(j\omega) = \angle K + \angle(1 + \angle j\omega T_1) + \angle(1 + j\omega T_2)$$
$$- \angle(j\omega) - \angle(1 + j\omega T_a) \quad (2)$$

Equations 1 and 2 indicate that addition tern in the magnitude and phase expression. So the basic method of construction of Bode plot would be the same.

Any Transfer faction will have form type of factors

1. Constant factor: K
2. Poles or zero at the origin: $(j\omega)^{\pm n}$
3. Poles or zero not at $\omega = 0$; $(1 + j\omega T)^{\pm n}$
4. Quadratic poles and zero:

$$\left(1 + \frac{2\delta(j\omega)}{\omega_n} - \frac{(j\omega)^2}{\omega_n^2}\right)$$

Unique characteristic of the Bode plot is such that each of the four type of factor listed may be considered as a separate plot: The individual plots are then added or subtracted accordingly to yield the total magnitude in db and phase of G (jω).

Constant K

$$K_{db} = 20\log_{10} K = \text{constant}$$
$$\angle K = 0 \quad ; \quad K < 0$$
$$= 180 \quad ; \quad K < 0$$

Bode plot on semi log co-ordinate.

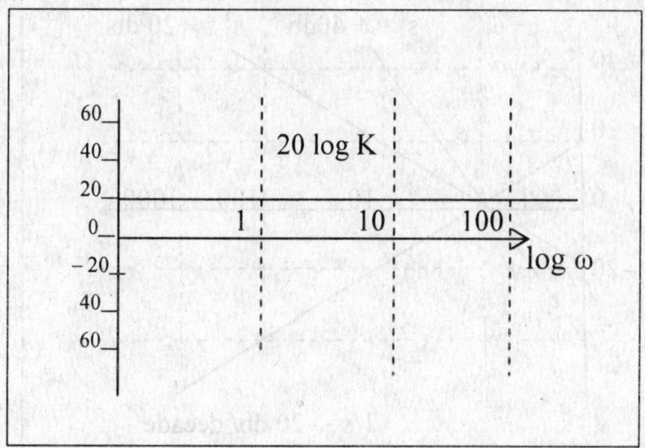

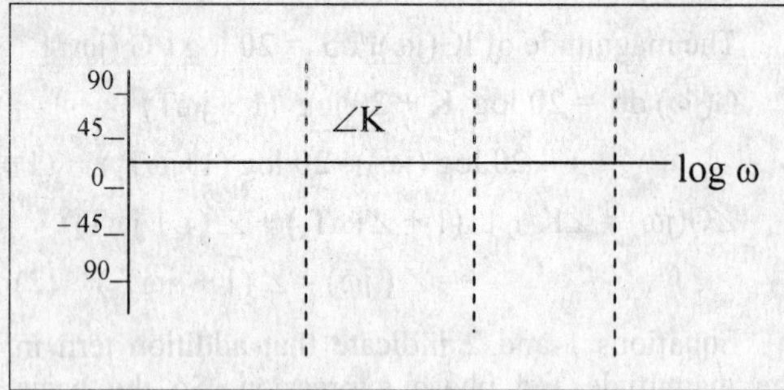

Pole and zero at the origin

$= (j\omega)^{+n}$

magnitude of $(j\omega)^{\pm n}$

$= 20 \log_{10}(j\omega)^{\pm n}$

$= \pm 20 n \log \omega$ db (1)

The above equation represent the straight line i.e. either semi log co-ordinate or rectangular co-ordinate for $\omega \geq 0$.

The slop of the straight line w.r.t $\log_{10} \omega$ will be

$$\frac{d(\pm 20n \log_{10} \omega)}{d \log_{10} \omega} = \pm 20\, n\, db$$

Hence in the rectangular coordinate a unit change in $\log_{10}\omega$ will corresponded to a change of ± 20n db in magnitude. Further more a unit charge in $\log_{10}$ w is equivalent to a change of 1 to 10, 10 to 100 in log scale.

Then the slop of these straight lines are describe by ± 20n db for decade of frequency.

Slop (m) = ± 20 n db/ decade of frequency

These lines all passes through 0 db point at $\omega = 1$.

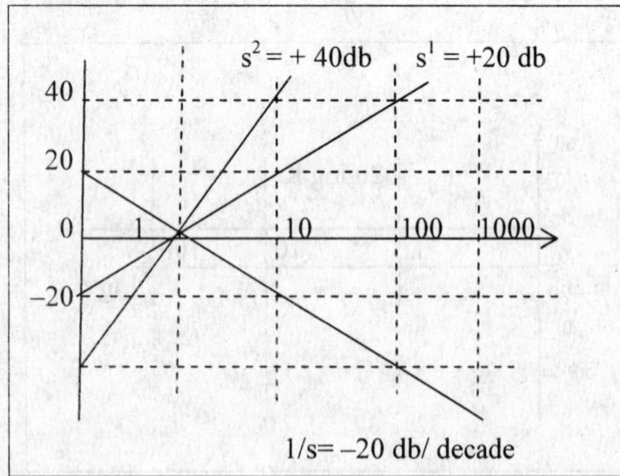

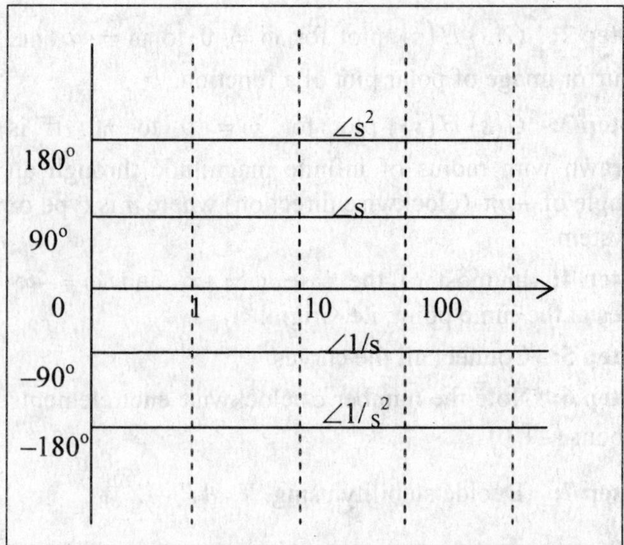

Instead of using decade, some time the unit **octave** is used to represent the separation of frequency.

$$\text{Number of octave} = \frac{\log_{10}(\omega_2/\omega_1)}{\log_{10} 2} \qquad \frac{\omega_2}{\omega_1} = 2$$

Between, ω_2 and ω_1

$$\text{No. of decade} = \frac{\log_{10}(\omega_2/\omega_1)}{\log_{10} 10} = \log_{10} \frac{\omega_2}{\omega_1}$$

Between, ω_2 and ω_1

$$\text{No of octaves} = \frac{\log_{10}(\omega_2/\omega_1)}{\log_{10} 2} = \frac{1 \text{ decade}}{0.301}$$

± 20 n db/decade = 20 × 0.301 db/octave

= ± 6 n db/octave

± 20 n db / decade = ± 6 n db/octave

Simple pole and zero $(1 + j\omega T)^{\pm 1}$

Let $|G(j\omega)| = 1 + j\omega T$

$|G(j\omega)|$ db $= 20 \log \sqrt{(1 + \omega^2 T^2)}$

$\phi = \tan^{-1} \omega T$

Magnitude, $\omega T \ll 1$; $|G| \approx 0$ db

$\omega T \gg 1$; $|G| \approx 20 \log \omega T$

This represents a straight line with slop of +20 db/decade of frequency.

The intersection of this straight line with 0 db axis is found by equity above equation to zero. Which gives.

$0 \log \omega T = 0$; $\log \omega T = 0 = \log 1$

$\omega T = 1$, $\omega = 1/T$ --- (*corner frequency*)

This frequency is also the intersection of high frequency approximate plot and low frequency plot which is 0 db line.

OUT LINE

(a) Locate the corner frequency $\omega = 1/T$

(b) Draw 20 db/decade or 6 db/octave line and horizontal line at 0 db, with two line intersecting at $\omega = 1/T$

Phase of $G(j\omega) = 1 + j\omega T$

$\angle G(j\omega) = \tan^{-1} \omega T$

Since the phase of G (jω) varies from 0 to 90^0 we may draw straight line 0^0 at 1 decade below the corner frequency to $+90^0$ at 1 decade above the corner frequency.

Max deviation of straight line is 6°

$\phi = \tan^{-1} \omega T$; When $\omega T = 1$; $\phi = 45°$

$\omega T = 10$; $\phi = 90°$

$\omega T = 0$; $\phi = 0°$

when $G(j\omega) = (1 + j\omega T)^{-1}$; magnitude is negative and phase is negative angle.

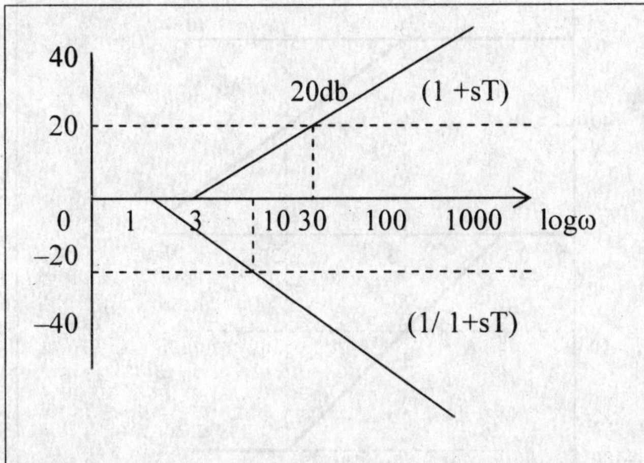

Note 1:

The error between the actual magnitude curve and the straight line asymptotes is symmetrical with respect to he corner freq 1/T. Further more it is useful to remember that the *error is 3db at the corner frequency* and 1 db at 1 octave above (2/T) and 1 octave below (0,5/T) the corner frequency.

Note 2:

Similar to the magnitude curve, a straight line approximation can be made for the phase curve, since the phase of G(jω) varies from 0^0 to 90^0 we may draw a line 0^0 at 1 decade below the corner frequency to $+90°$ at 1 decade above the corner frequency.

The deviation of straight line approximation from the actual curve is less than 6°

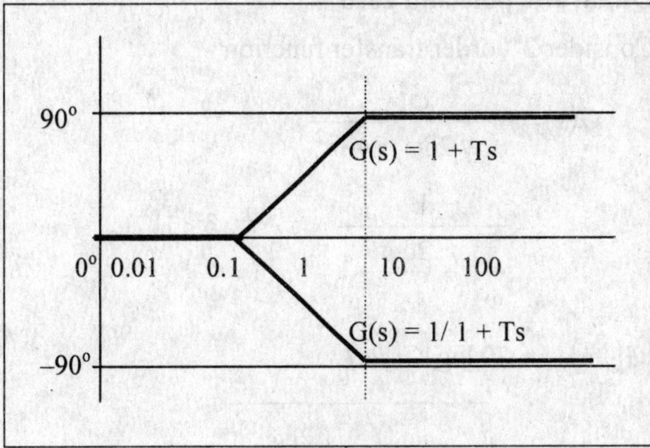

Example: $$G(s) = \frac{10(s+10)}{s(s+2)(s+5)}$$

$$G(j\omega) = \frac{10(1+0.1j\omega)}{j\omega(1+j0.5\omega)(1+j0.2\omega)}$$

This equation show that G(jω) has the corner frequency ω = 10, 2, and 5 rad /sec.

The pole at the origin gives a magnitude curve that is a straight line with a slop of –20 db/decade and passing wrought ω = 1 rad/sec point on the w axis at 0 db.

For K =10

$K_{db} = 20 \log_{10} 10 = 20$ db.

$\angle K = 0°$ for $K > 0$

For $(j\omega)^{-1}$

Slop of line = – 20 db/decade and

passing through 0 db axis at ω = 1

$\angle (j\omega)^{-1} = -1 \times 90° = -90°$

For (1+j 0.1ω)

$|G(j\omega)|db = 20 \log \sqrt{(1 + (0.1)^2 \omega^2)}$

= + 20 db/decade straight line

corner frequency, $\omega = (1/T) = (1/0.1) = 10$

For $(1+ j\,0.5\omega)^{-1}$

$|G(j\omega)|$ db = –20 db/decade straight line

Corner frequency, $\omega = (1/0.5) = 2$

For $(1+j\,0.2\omega)^{-1}$

$|G(j\omega)|$ db = –20 db/decade straight line

Corner frequency, $\omega = (1/0.2) = 5$

Quadratic pole and zero

Consider 2nd order transfer function

$$G(s)=\frac{\omega_n^2}{s^2+2\delta\omega_n s+\omega_n^2}$$

$$=\frac{1}{\frac{s^2}{\omega_n^2}+\frac{2\delta\omega}{\omega_n}+1} \quad ; \quad (\because \delta<1)$$

$$|G(j\omega)|_{db}=20\log|G(j\omega)|$$

$$=-20\log=\sqrt{\left(1-\frac{\omega}{\omega_n}\right)^2+\left(\frac{2\delta\omega}{\omega_n}\right)^2}$$

$$\angle G(j\omega)=-\tan^{-1}\frac{2\delta\,\omega/\omega}{\left(1-\left(\frac{\omega}{\omega_n}\right)^2\right)}$$

where, ω_n--- natural frequency

ω-----operating frequency

Case 1: when $\omega/\omega_n << 1$

$G(j\omega)$ db = 0 (1)

Straight line on 0 db axis

Case 2: where $\omega/\omega_n >> 1$ i.e. at least 10 times

$$G(j\omega)\,db=-20\log\left(\frac{\omega}{\omega_n}\right)^2=-40\log\left(\frac{\omega}{\omega_n}\right)^2 \quad (2)$$

Straight line with slop of–40 db/decade

Intersection of –40 db/decade line with 0 db line i.e.

$$-40\log\left(\frac{\omega}{\omega_n}\right)=0$$

or, $$\frac{\omega}{\omega_n}=1$$

Or, $\omega = \omega_n$ --------corner frequency.

The actual $G(j\omega)$ db curve is is differ from the asymptotic line become $G(j\omega)$db and $\angle G(j\omega)$ curve depends on corner frequency ω_n and damping ratio

Phase

$$\angle G(j\omega)=-\tan^{-1}\frac{2\delta\,\omega/\omega_n}{\left(1-\left(\frac{\omega}{\omega_n}\right)^2\right)}$$

Case 1: When, $\omega/\omega_n << 1$ at least 10 time

$\angle G(j\omega) = 0$

Case 2: When $\omega/\omega_n = 1$ i.e. $\omega = \omega_n$ .. corer freq.

$\angle G(j\omega) = -90$

Case 3: When $\omega/\omega_n >> 1$ at least 10 time

$\angle G(j\omega) = -180^0$

Note:

For quick solution to $\frac{\omega_n^2}{s^2+2\delta\omega_n s+\omega_n^2}$

1. If $\delta < 1$ i.e. below 0.8 , then
2. Slop will be – 40 db/decade and
3. Angle 0^0 at 1 decade below the corner frequency and -180^0 at 1 decade above the corner frequency
4. Corner frequency = ω_n

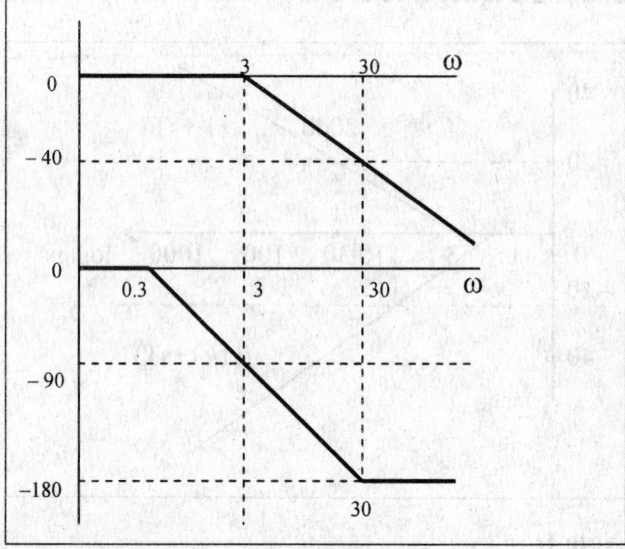

Example:

$$G(s)=\frac{64(s+2)}{s(s+0.5)(s^2+3.25+64)}$$

Solution: Convert the problem in the form of $(1+j\omega T)^{\pm n}$ and $\frac{\omega_n^2}{s^2+2\delta\omega_n s+\omega_n^2}$

i.e. $$\frac{(s+2)\times(1/2)\times 2}{(1/2)[s(s+0.5)\times 2]}\times\frac{64}{s^2+3.2s+64}$$

or $\frac{4(1+s/2)}{s(1+2s)} \times \frac{64}{s^2+3.2s+64}$

The rearrangement of the function in the time-constant form gives.

$$G(s)=\frac{4(1+s/2)}{s(1+2s)(1+0.05s+s^2/64)}$$

Therefore, the sinusoidal transfer function in the time-constant form is given by

$$G(s)=\frac{4(1+j\omega/2)}{j\omega(1+2j\omega)\left[1+j0.4(\omega/8)-(\omega/8)^2\right]}$$

Here, $3.2\,s = 2\delta\,\omega_n s$

$= 2 \times 8 \times \delta s$

$\delta = \frac{3.2}{16} = 0.2$

Corner frequency are 0.5, 2 and $\omega_n = 8$

For quadratic section,

Amplitude curve slop = – 40 db/decade

Phase curve Angle 0° ---- one decade below corner frequency and –180° --- one decade above corner frequency.

This factors of this transfer function in order of their occurrence as frequency increases, are

1. Constant gain, K = 4
2. Pole at origin, $1/j\omega$
3. Pole at s = –0.5 ; corner frequency $\omega_1 = 0.5$
4. Zero at s = –2 ; corner frequency $\omega_2 = 2$
5. Pair of complex conjugate poles with $\delta = 0.2$, $\omega = 8$; corner frequency $\omega_3 = 8$

,

$$G(s)=\frac{4(1+j\omega/2)}{j\omega(1+2j\omega)\left[1+j0.4(\omega/8)-(\omega/8)^2\right]}$$

Factor	Corne Freq.	Asymptotic log-magnitude characteristic	Phase angle
$4/j\omega$	None	Straight line of constant slope-20 db/ decade, assign through 20 log 4 =12 db point at $\omega = 1$	Constant-90°
$1/(1+j2\omega)$	$\omega_1 =$ 0.5	Straight line of the 0 db for $\omega < \omega_1$, straight line of slope –20 db/decade for $\omega > \omega_1$	Phase angle varies, from 0 to –90°, angle at $\omega_1 = -45°$
$1 + j0.5\omega$	$\omega_2 = 2$	Straight line of 0 db for $\omega < \omega_2$ straight line of slope + 20 db/decade for $\omega > \omega_3$	Phase angle varies from 0 to 90°, angle at $\omega_3 = 45^0$
$\left(1+\frac{j0.4\omega}{8}-\left(\frac{\omega}{8}\right)^2\right)$	$\omega_3 = 8$ (δ= 0.2)	Straight line of 0 db for $\omega < \omega_3$ straight line of slope +40 db/ decade for $\omega > \omega_3$	Phase angle varies from 0 to 180°, angle at $\omega_3 = -90^0$

All-Pass and Minimum-Phase Systems

The cases we have considered so far are termed as *minimum-phase transfer functions, i.e. those with all poles and zeros in the left that of the s-plane.*

Consider now the special class of transfer function having a pole-zero pattern which is antisymmetric about the imaginary axis, i.e. for every pole in the left half plane, there is a zero in the mirror-image position. A common example of such a transfer function is

$$G(j\omega)=\frac{1-j\omega T}{1+j\omega T}$$

Whose pole-zero configuration is shown in figure (c). it has a magnitude of unity at all frequencies and a phase angle ($-2\tan^{-1}\omega T$) which varies from 0° to 180° as ω is increased from 0 to ∞. *The property of unit magnitude at all frequencies applies to all transfer function with ant symmetric pole-zero pattern. Physical system with this property, are called all-pass systems.*

Consider next the case *where the transfer* function has poles in the left half s-plane and zero in both the left-and right-half s-plane are not permitted to lie in the right-half s-plane because such a system would be unstable. Consider for example the transfer function.

$$G_1(j\omega)=\frac{1-j\omega T}{(1+j\omega T_1)(1+j\omega T_2)}$$

Whose pole-zero pattern is shown in Figure (a). This transfer function may be rewritten as

$$G(j\omega) = \frac{(1+j\omega T)}{(1+j\omega T_1)(1+j\omega T_2)} \frac{(1-j\omega T)}{(1+j\omega T)}$$

$$G_2(j\omega)G(j\omega)$$

Which now becomes the product of two transfer functions, one $G_2(j\omega)$ having no poles or zero in the right half of s-place figure. It is evident that $G_1(j\omega)$ and $G_2(J\omega)$ have identical curves of magnitude versus frequency but their phase versus frequency curves are different as shown in figure with $G_2(j\omega)$ having a smaller range of phase angle that $G_1(j\omega)$. It means that for $G_1(j\omega)$, there is no unique relationship between magnitude and phase, as it is always possible to alter the phase curve without affecting the associated magnitude curve by the addition of an all-pass transfer function. *A transfer function which curve has one or more zero in the right half s-plane is known as* ***non-minimum phase transfer function.***

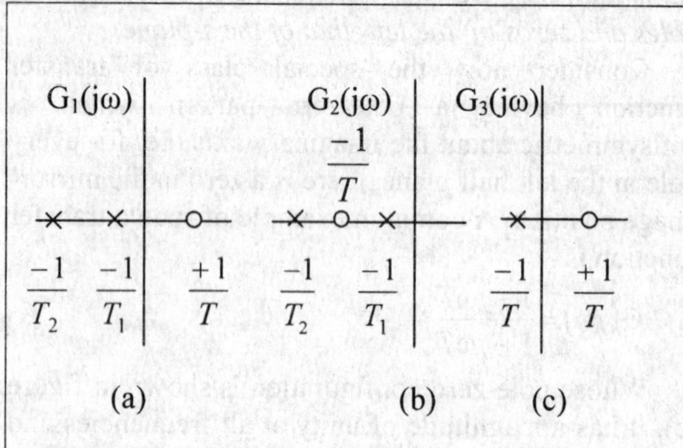

Figure: Pole zero patterns for

(a) Non minimum-phase function;
(b) Minimum phase function;
(c) All pass function

In general, if a transfer function has any zero in the right half a-plane transfer function in a manner shown in above figure each time this is done, the magnitude curve remains unaltered but the phase lag is reduced, until eventually we are left with a function, which contains no zero in the half s-plane. Such a transfer function has the least (minimum) phase angle range for a given magnitude curve and is called a minimum-phase function. It has a unique relationship between its phase and magnitude curves.

Typical phase angle characteristics of minimum and no minimum-phase transfer function are shown in below figure. It will be seen that the lager the phase lags present in a system, the more complex are its stabilization problems.

Therefore for control system, elements with non-minimum-phase transfer functions are avoided as far as possible.

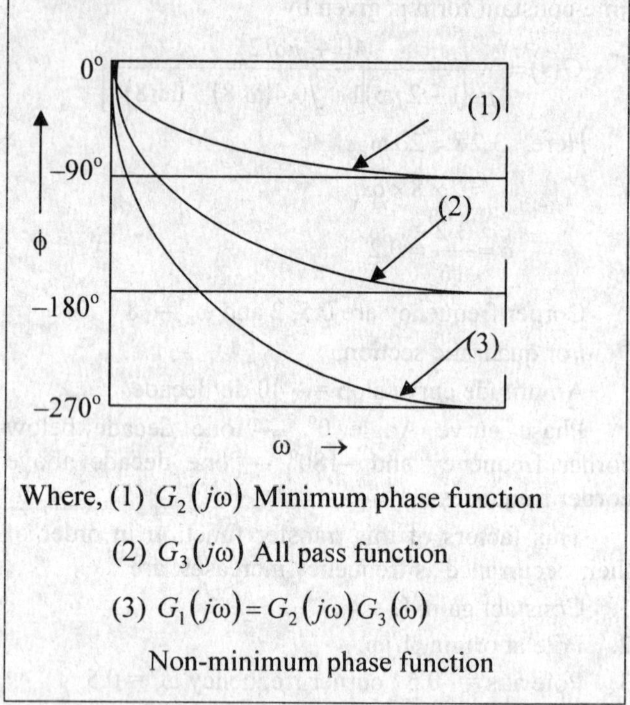

Where, (1) $G_2(j\omega)$ Minimum phase function

(2) $G_3(j\omega)$ All pass function

(3) $G_1(j\omega) = G_2(j\omega)G_3(\omega)$

Non-minimum phase function

A common example of a non minimum phase element is transportation lag which has the transfer function.

$$G(j\omega) = e^{-j\omega T}$$

$$= 1\angle -\omega T \text{ rad} = 1\angle -57.3\,\omega T \text{ deg}$$

The phase angle characteristics where a non minimum-phase transfer can arise are shown in below figure.

Other possible situations where a non minimum phase transfer function can arise are, when more than one possible signal paths are variable between input and output and output in addition to a conductive path.

Experimental Determination of Transfer Functions

The Bode plots are of great value in situations where the transfer function of a given system in unknown. In such cases we proceed to obtain the frequency response data experimentally in the frequency range of interest, the system transfer function within a certain degree of accuracy can then be obtained by fitting an asymptotic log-magnitude plot the experimental data as per the procedure outlined below.

1. The experimental data is used to plot the exact log-magnitude and phase angle versus frequency curves on a semi log graph sheet.

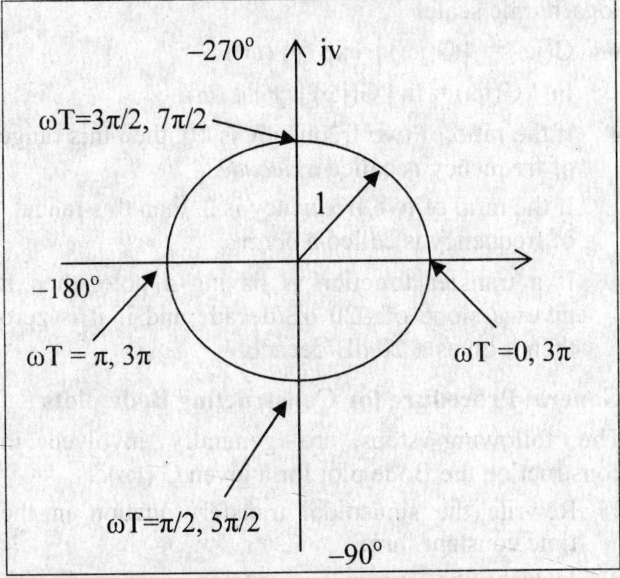

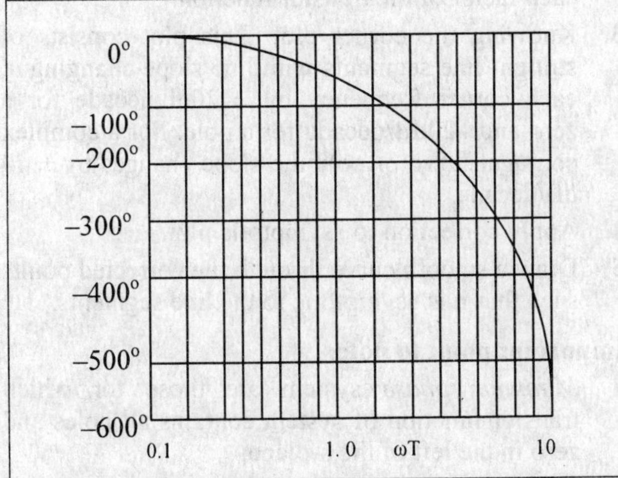

2. Asymptotes are then drawn on the log-magnitude curve keeping in view the fact that the slopes of these asymptotes must be multiples of ±20 db/decade. The corner frequencies are so adjusted that the db value at the corner frequency on the asymptotic plot differs from the actual log-magnitude plot be amount which is in close agreement the db correction of the kind of factor revealed.

3. If the slope of asymptotic log-magnitude curve obtained in (2) above, changes by –20n db/decade at $\omega = \omega_1$ it indicates that the factor $1/(1 + j\omega/\omega_2)^n$ exist in the transfer function.

4. If the slope changes by +20n db/decade at $\omega = \omega_2$, it indicates the presence of the factor $(1 + j\omega/\omega_2)^n$ in the transfer function.

5. The change of slope by – 40 db/decade at $\omega = \omega_3$ indicates that either a double pole (i.e., n = 2) or pair of complex conjugate pole is present. If the error between the asymptotic and actual curve is about –6 db, then a factor of the form $(1 + j\omega/\omega_3)^2$ is present and if the error is positive then a quadratic factor of obtained with the help of the error graph.

6. In the low frequency range, the plot is dominated by a factor of the form $\dfrac{K}{(j\omega)^r}$.

In most practical system r equal 0, 1 or 2. The value r is determined as follows:

(a) If the low frequency asymptote is a horizontal line at x db, it indicates that the transfer function represents a type-0 system with a gain k given by

$$20 \log k = x$$

$$k = (1/20) \log^{-1} x$$

(b) If the low frequency asymptote has a slope of – 20 db/decade, it indicates the presence of a factor of the form K/jω in the transfer function (type-1 system). The frequency at which the asymptote (extended if necessary) intersects the 0-db line numerically represent the value of k. Also the asymptote (extended) if necessary has as gain of 20 k at ω = 1.

(a) If the low frequency asymptote has a slope of – 40 db/decade, then the transfer function has a factor of the form $K/(j\omega)^2$ (types-2 system). The frequency at which this asymptote (extended if necessary) intersect the 0-db line numerically equal to $\sqrt{k}$. Also the asymptote (extended if necessary) has a gain of 20 log k at ω = 1.

The log-magnitude curves of type-0, type-1 and type-2 system are shown in figures a, b and c respectively.

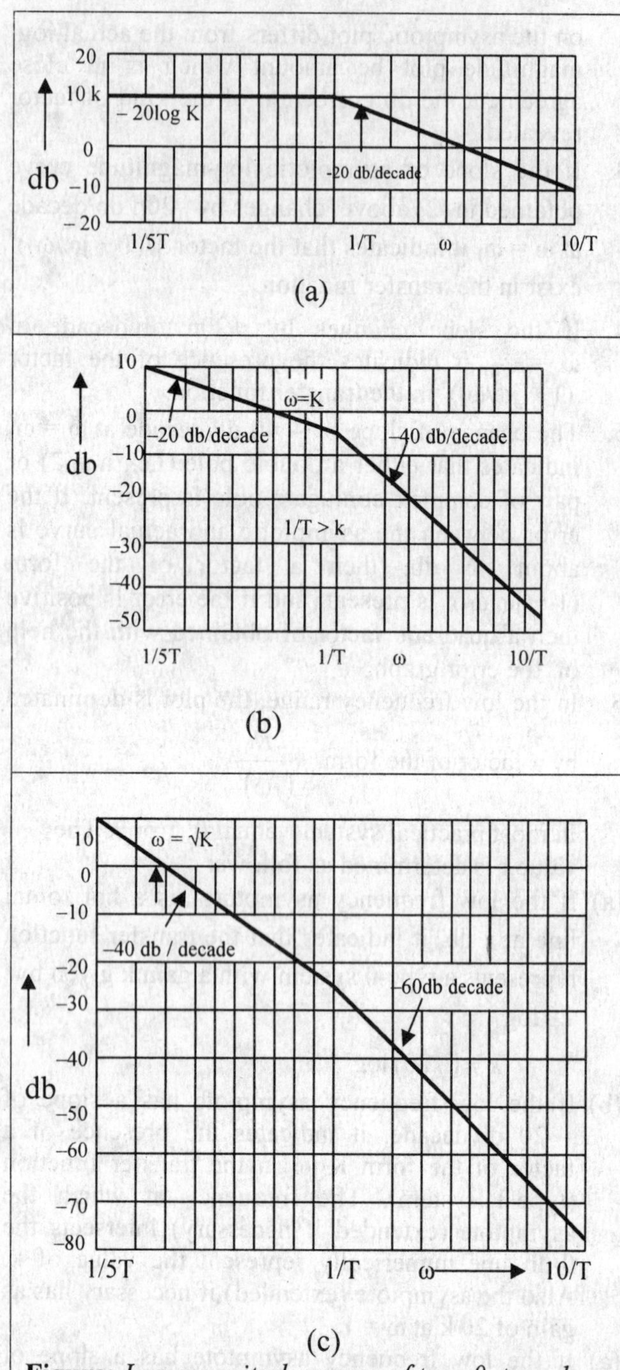

Figure: *Log-magnitude curves of type-0, type-1 and type- 2 systems.*

After obtaining the transfer function form the log-magnitude curve, the phase angle curve is constructed from it an s then compared with the one obtained experimentally. If the two curve are in fair agreement and if the phase angles of both the curves at very high frequency tend to –90° (q – p) where p and q are degrees of the numerator and denominator polynomials respectively of the computed phase angle is 180° less negative than the one obtained experimentally, then the transfer function is of non minimum –phase type and one of the zeros of the transfer function lies in the right half-s-plane.

Important point on Bode Plots: This is logarithmic plot which consists of two graphs one giving the logarithm of $|G(j\omega)|$ and the other phase angle of $G(j\omega)$ both plotted against frequency in logarithmic scale.

- $G(j\omega) = |G(j\omega)| \exp(j\phi(\omega))$
  $\ln G(j\omega) = \ln|G(j\omega)| + j\phi(\omega)$
- If the ratio of two frequency is 10, then this range of frequency is called a *decade.*
  If the ratio of two frequency is 2, then this range of frequency is called a *octave.*
- If a transfer function is having a pole, then it gives a slope of –20 dB/decade and if it is zero then it gives + 20 dB/decade.

General Procedure for Constructing Bode plots

The following steps are generally involved in construction the Bode plot for a given G (jω).

1. Rewrite the sinusoidal transfer function in the time constant form.
2. Identify the corner frequencies associated with each factor of the transfer function.
3. Knowing the corner plot. This plot consists of straight line segments with line slope changing at each corner frequency by ± 20db/decade for a zero and –20dB/decada for a pole. For a complex conjugate zero or pole the slope changes by ± 40 dB/decade.
4. Apply correction to asymptotic plot.
5. Draw a smooth curve through the corrected points such that it is asymptotic to the line segments.

Important point to note:

- *Minimum phase* system are those for which transfer function of system contains all poles and zero in the left of the s-plane.
- If pole –zero pattern which is ant symmetric about the imaginary axis exist, then these types of transfer function known as *All-phase system* for all pass system, magnitude is unity at all frequency.

- A transfer function which has one or more zero in the right half S-plane is known as *non minimum-phase transfer function.*

8.10 ROOT LOCI

Root Locus

In the design and analysis of control system we often need to investigate a system's performance when more than one parameter vary over a given range. Our purpose may be to select the appropriate value of a system parameters, such as gain, or we may want to study parameter variation due to aging of the system components or environment changes.

Close loop transfer function is expression as

$$\frac{C(s)}{R(s)} = \frac{G(s)}{1+G(s)H(s)}$$

Characteristic equation of close loop control system $1+G(s)H(s)=0$ (1)

- The Conventional **root locus plots** includes the loci or the roots of characteristic equation when the gain K is varied from zero to infinity.
- **Root Contours (RC)** loci of roots when more, then one parameter varies.
- **Complementary root loci (CRL):** the portion of the root loci when K varies form −∞ to 0; i.e., K is negative.
- **Root Locus (RL):** are plots of the variation of poles of the closed loop function (as obtained from characteristic equation), with variation in the open loop gain (Changed from zero to infinity).

Properties of Root Locus

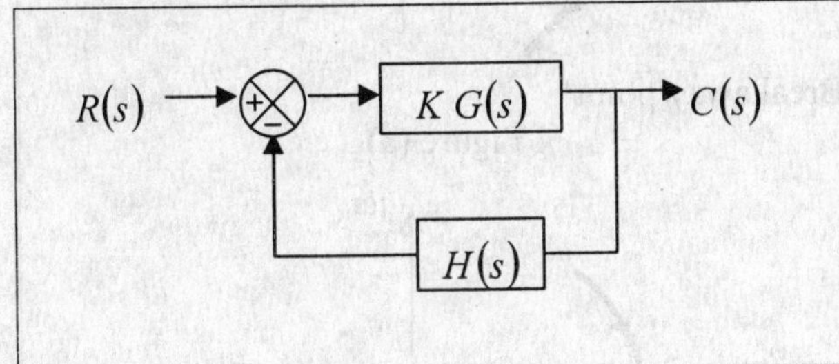

$$\frac{C(s)}{R(s)} = \frac{K}{1+K\,G(s)H(s)}$$

Characteristic equation can be written as

$1+G(s)H(s)=0$

$$\sum_{i=1}^{P}(s-p_i)+K\sum_{j=1}^{Z}(s-z_i)=0$$

Construction of Root Locus

K = 0, is the starting point of RL so poles of G (s) H(s) provides the starting point for the root locus.

1. K = ∞, is the end point of locus, so zero of G (s) H (s) provide the termination of root locus.
2. Number of Separate Root Loci.

   The total number of root loci is equal to the order of equation $F(s)=1+G(s)H(s)=0$

   N = Number of branch of RL

   P = Number of finite Pole of GH

   Z = Number of finite zero of GH

   N = Z if Z > P

   = P if P > Z

3. **Symmetry of the Root Loci**

   1. Root locus is symmetrical to the real axis of the s-plane.
   2. In general the complete root loci are symmetrical with respect to the axis of symmetry of the poles and zero of G (s) H(s).

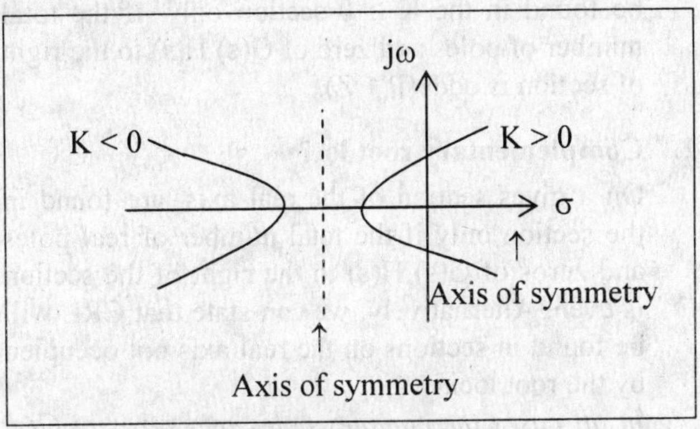

4. **Asymptotes of the complete root loci**

   (i.e. behavior of root loci at s = ∞)

   For large value of s, the RL (K ≥ 0) are asymptotic to straight lines with angle value

$$\theta_q = \frac{(2q+1)\pi}{P-Z}$$

where, q = 0, 1, 2-------[| (P − Z) | − 1]

P and Z are total number of **poles and zeros of** GH.

Number of asymptotes for

$R_L = P - Z$

Number of asymptotes for

$CRL = P - Z$

Total number of asymptotes of complete $R_L = 2(P - Z)$

For CRL, $(K \le 0)$

$$\theta_q = \frac{2q\pi}{P - z}$$

where, q = 0, 1, 2-------[|(P – Z)| – 1]

Intersection of the Asymptotes (Centriod)

1. The intersection of the | P–Z | asymptotes lies only on the real axis in the s-plane.
2. The intersection of the asymptotes is given by

$$\sigma = \frac{b_1 - a_1}{P - Z}$$

$$= \frac{\sum \text{finite poles of GH} - \sum \text{finite zores of GH}}{\text{Number of finite poles of GH} - \text{Number of finite zero GH i.e. } (P - Z)}$$

1. **Root locus on the real axis**

   On a given suction of real axis the root loci may be found in the K ≥ 0 section only. If the total number of poles and zero of G(s) H(s) to the right of section is odd (P + Z).

2. **Complementary root loci**

   On a gives section of the real axis, are found in the section only if the total number of real poles and zeros of G(s) H(s) to the right of the section is even. Alternatively, we can state that CRL will be found in sections on the real axis not occupied by the root loci.

   In all cases the complex poles and zeros of G(s) H(s) do not affect the existence properties of the root loci on the real axis.

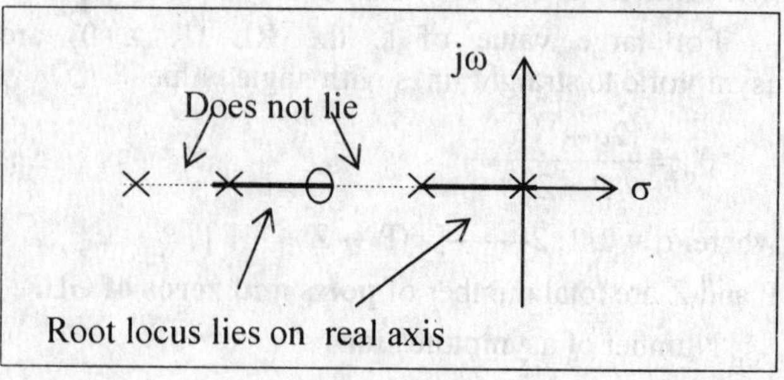

Angle of Departure (form Pole) and Angle of Arrival (at zero) of the RL

The angle of departure (arrival) of the complete root - locus a pole (zero) of G(s) H(s) denotes the behavior of the root loci that pole (zero). For the RL (K ≥ 0) these angles can be determined by use of equation.

$$(\angle G_1(s)H_1(s)) = \sum_{i=1}^{m}(\angle s + z_i) - \sum_{j=1}^{n}(\angle s + p_i)$$

$$\Sigma\angle\theta(Z) - \Sigma\angle\theta(P) = (2q + 1)\pi$$

Angle of departure from complex pole is given by the $\theta_D = +180 + \angle G'(s)$ where G′(s) is phase angle of G′(s) counted at complex pole but dropping the contribution of that particular pole. *θ_D is measured in the counter clock wise direction from the Horizontal (σ axis) in the s – plane.*

Inter section of RL with Imaginary axis

The point where the complete Root Loci intersect the imaginary axis of s-plane, and the corresponding value of K may be determined by means of Routh Hurwitz criterion.

Break away point (saddle point) on the complete Root Loci

Breakaway points or saddle points on the root loci of an equation correspond to multiple-order root of the equation.

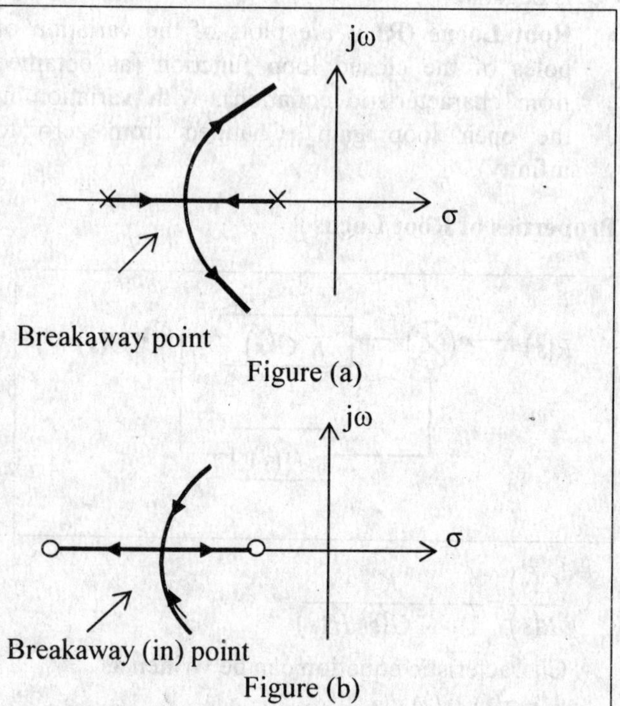

Figure (a)

Figure (b)

Breakaway point computed by analytical method

The breakaway point on the complex root loci of $1 + KG_1(s)H_1(s) = 0$ must satisfy.

$$\frac{dG_1(s)H_1(s)}{ds} = 0 \quad \text{or,} \quad \left(\frac{dK}{ds}\right) = 0$$

It is important to point out that the condition for the breakaway point given by above equation is necessary but not sufficient. In other words, ***all breakaway points must satisfy above equation, but not all solution of above equation are breakaway point***. To be a breakaway point, the solution of above equation. must be a factor of characteristic equation for some real K.

Break away point not on the real axis

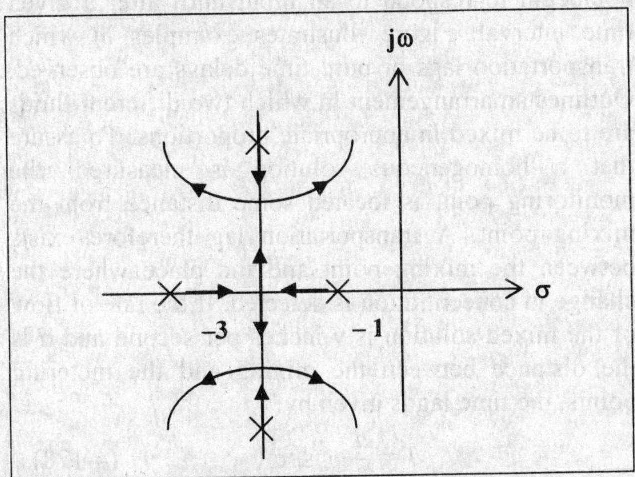

Calculation of K on RL

Once the root loci from the characteristic equation have been constructed, the value of K at any point S_i on the loci can be determined by the equation.

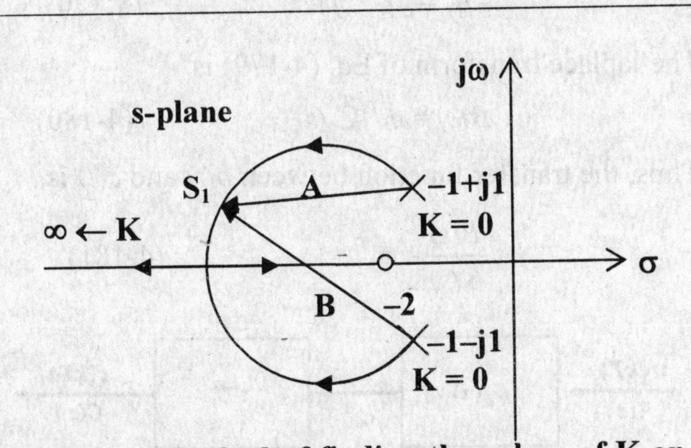

Graphical method of finding the values of K on the root loci.

$$|K| = \frac{\text{Product of length of vector draw from poles to } S_1}{\text{Product of length of vector draw from zeros to } S_1}$$

$$K = \frac{AB}{C}$$

Some Important Aspect of Root Loci

The analyst or designer may conduct a quick study of the system in the s- plane by making a sketch of the root loci using some or all rules of construction.

Effect of Adding Poles and Zeros to G(s) H(s)

1. Addition of Poles to G(s) H(s) reduce the relative stability of the close loop system.

Let, $G(s)H(s) = \dfrac{K}{s(s+a)}$;

$a > 0$

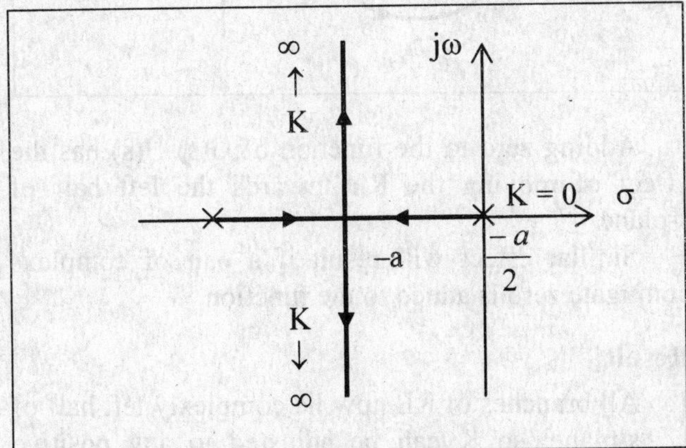

Let us introduce a pole at $s = -b$

$$G(s)H(s) = \frac{K}{s(s+a)(s+b)} \; ; \; b > a$$

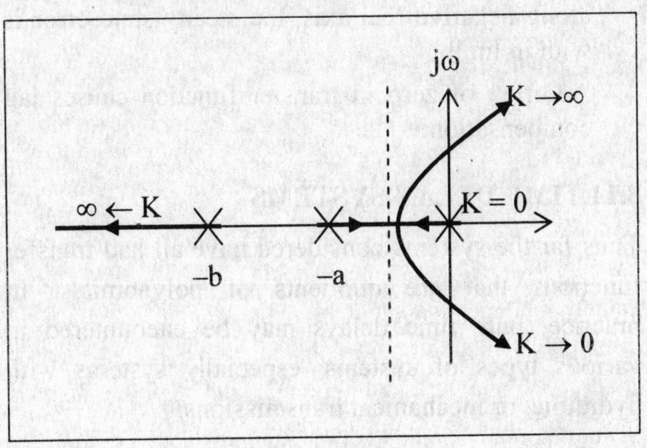

In general we may state the adding a pole to the function G(s) H(s) in the left half of s-plane has the effect pushing the original root loci towards the right half of s-plane.

Addition of zero to G(s) H(s) has the tendency to stabilize the close loop system.

Addition of zeros to above function result as

$$G(s)H(s)=\frac{K(s+b)}{s(s+a)}\ ;\ b>a$$

Its Root locus is as below:

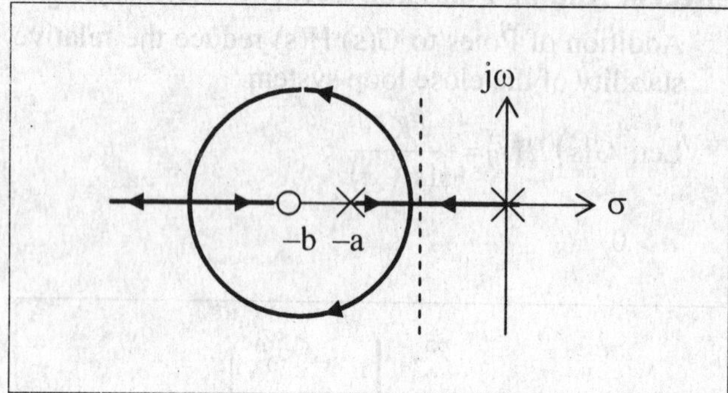

Adding zero to the function of G(s) H(s) has the effect of moving the RL towards the left half of s-plane.

Similar effect will result if a pair of complex-conjugate zero is added to the function

Result

1. All branches of RL now lie complexly left half of s-plane, so K can be adjusted to any positive value without causing in stability.
2. The complete R.L are now bent away to LHS of s-plane from jω axis.
3. By suitable adjustment of zero (compensating) along negative real axis, the steady state error is with in limit.
4. Addition of zero in transfer function causes lag compensation.

8.11 TIME DELAY SYSTEMS

Thus far the systems considered have all had transfer functions that are quotients of polynomials. In practice, pure time delays may be encountered in various types of systems, especially systems with hydraulic, or mechanical transmissions.

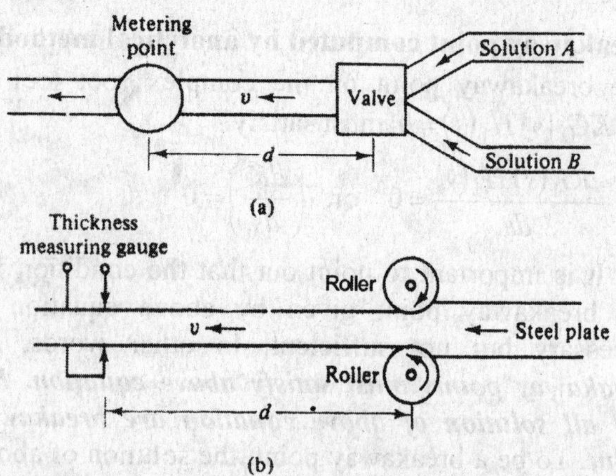

Systems with computer control also have time delays, since it takes time for the computer to execute numerical operations. In these systems the output will not begin to respond to an input until after a given time interval. Figure illustrates examples in which transportation lags or pure time delays are observed. Outlines an arrangement in which two different fluids are to be mixed in appropriate proportions. To assure that a homogeneous solution is measured, the monitoring point is located some distance from the mixing point. A transportation lag therefore exists between the mixing point and the place where the change in concentration is detected. If the rate of flow of the mixed solution is *v* inches per second and *d* is the distance between the mixing and the metering points, the time lag is given by:

$$T=\frac{d}{v}\ \text{sec} \qquad (4\text{-}178)$$

If it is assumed that the concentration at the mixing point is c(*t*) and that it is reproduced without change T seconds later at the monitoring point, the measured quantity is

$$b(t)=c(t-T) \qquad (4\text{-}179)$$

The laplace transform of Eq. (4-179) is

$$B(s)=e^{-Ts}C(s) \qquad (4\text{-}180)$$

Thus, the transfer function between *b(t)* and *c(t)* is

$$\frac{B(s)}{C(s)}=e^{-Ts} \qquad (4\text{-}181)$$

r(kT) R(z) → $G_c(z)$ → z^{-N} → c(kT) C(z)

Digital processor

The arrangement shown in figure may be thought of as a thickness control of the rolling of steel plates. As in the case above, the transfer functions between the thickness at the rollers and the measuring point is given by Eq. (4-181).

Other examples of transportation lags are found in human beings as control systems where action and reaction are always accompanied by pure time delays. The operation of the sample-and-hold device of a sampled-data system closely resembles a pure time delay; it is sometimes approximated by a simple time-lag term, e^{-Ts}.

$$\frac{dx(t)}{dt} = Ax(t) + Bu(t) \qquad (4\text{-}182)$$

A general state description of a containing time lags is given by the following matrix differential-difference equation:

$$\frac{dx(t)}{dt} = \sum_{i=1}^{p} A_i x(t - T_i) + \sum_{j=1}^{q} B_j u(t - T) \qquad (4\text{-}183)$$

Where T_i and T_j are fixed time delays. In this case Eq. (4-183) represents a general situation where time delays may exist on both the inputs as well as the states.

Figure shows the block diagram of a digital control system. The block with the transfer function z^{-N} represents pure time delay resulting from the time required to execute the digital processing. In this case, N is an integer, so that the time delay is approximated as an integral multiple of the sampling period T.

8.12 PHASE AND GAIN MARGIN

Gain Margin

The gain margin is defined as the reciprocal of open loop transfer function evaluated at the frequency ω_π at which the phase angle is $-180°$

Gain margin = $1/|GH(j\omega)|$

$\phi = \angle GH(j\omega) = -180° = -\pi$ radian

ω_π = *Phase cross over frequency,* that is the frequency ω_p at which the polar plot crosses the negative real axis.

Gain margin gives the relative distance between $(-1 + j\,0)$ point and the p(jω) or G(jω) H(jω) plot.

If the phase cross over point A is too near $(-1 + j0)$ (small G_m) will know how much to reduce the gain and if too far away (large G_m) how much to increase.

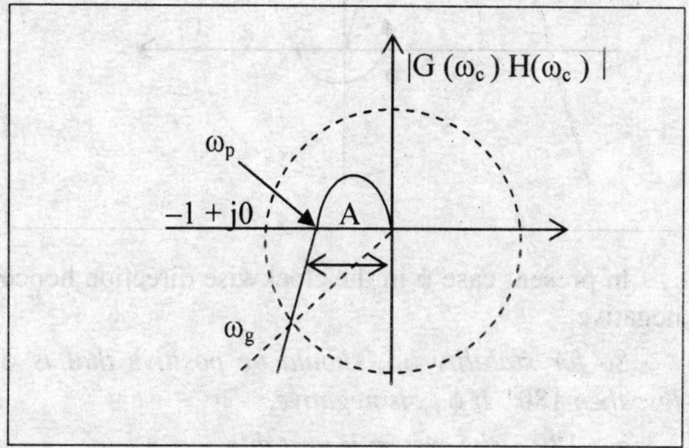

A = Phase cross over point
ω_p = Phase cross over frequency
ω_g = Gain cross over frequency

Phase Margin (ϕ_{pm})

It is the angle by which the polar plot is rotated to cause it to pass through $(-1, j\,0)$ point ϕ_{pm} is the angle between the negative read axis and the radius vector joining the origin to the gain cross over frequency ω_g. The radius vector is $|G(j\omega)\,H(j\omega)| = 1$.

The gain cross over frequency w_g is the frequency at which $|G(j\omega)\,H(j\omega)| = 1$, that is the point of intersection of polar plot and (–1, jo) circle.

Phase margine $\phi_{Pm} = 180° + \phi$

where, $\phi = \angle G(j\omega)\,H(j\omega)$ and

$G(j\omega)\,H(j\omega)\ | = 1$

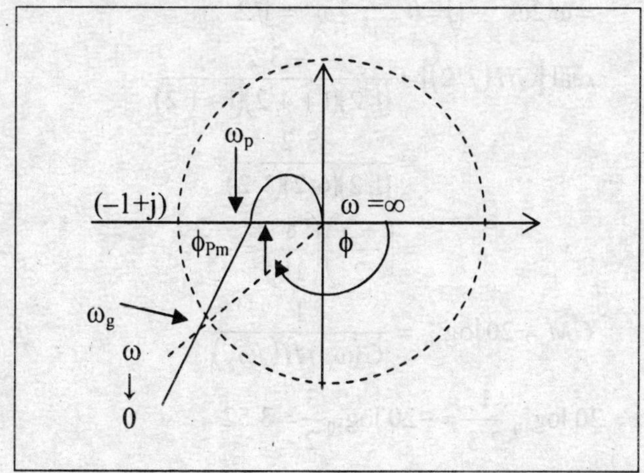

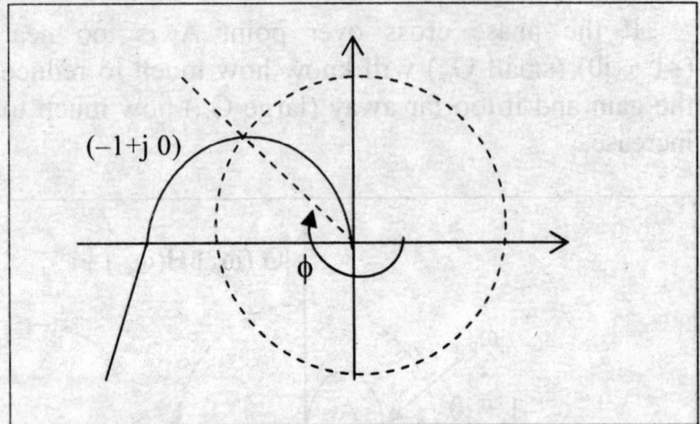

In present case ϕ in the clockwise direction hence negative.

So for stability ϕ_{Pm} *should be positive that is* ϕ *less then* 180°. *If* ϕ_{Pm} *is negative,*

$\phi > 180^{o}$, *the system is unstable.*

Example: A unity feed system has open-loop transfer function.

$$G(s) = \frac{1}{s(2s+1)(s+1)}$$

Find (a) Gain margin (b) Phase margin

Solution: The intersection with the real axis give the phase cross over frequency, at this

Im [G(jω)] = 0

$$= \frac{j\omega(-2j\omega+1)(-j\omega+1)}{\omega^2(1+4\omega^2)(1+\omega^2)}$$

$$= \frac{-j\omega(-2j\omega^2-3j\omega+1)}{\omega^2(1+4\omega^2)(1+\omega^2)} = \frac{2j\omega^2-3j\omega^3-j2}{\omega^2(1+4\omega^2)(1+\omega^2)}$$

$$\text{Im}\,[G(j\omega)] = \frac{2\omega^3-\omega}{Dr} = 0$$

$$= \omega(2\omega^2-1) = 0 \quad ; \quad \omega^2 = 1/2$$

$$\text{real}\,[GH(j/2)] = \frac{-3/2}{(1/2)(1+4/2)(1+1/2)}$$

$$= \frac{-3/2}{(1/2)(6/2)(3/2)}$$

$$= \left(\frac{-3}{2}\right)\left(\frac{8}{18}\right) = \frac{-2}{3}$$

$$GM = 20\log_{10} = \frac{1}{|G(\omega_c)\,H(j\omega_c)|}$$

$$= 20\log_{10}\frac{1}{2/3} = -20\log_{10}\frac{3}{2} = 3.52$$

$$G(j\omega) = \frac{1}{j\omega(2j\omega+1)(j\omega+1)}$$

$$\phi = -90 - \tan^{-1}\omega - \tan{-}12\omega$$

$$|G(j\omega)| = 1$$

$$\omega^2(1+\omega^2)(1+4\omega^2) = 1$$

put $\omega^2 = x$; $x(1+x)(1+4x) = 1$

By trail and error; $x = 0.33$

The equation ; $\omega^2 \;= 0.33$

$\omega = \sqrt{0.33} = 0.574$ rad/sec

Phase Margin $(\phi_{Pm}) = 180 + \phi$

$= 180 - 90\tan^{-1} 0.574 - \tan^{-1} 2 \times 0.374$

$= 180 - 90 - 29.68 - 48.94 = 11.38^{o}$

Relative Stability Using Nyquist Criterion

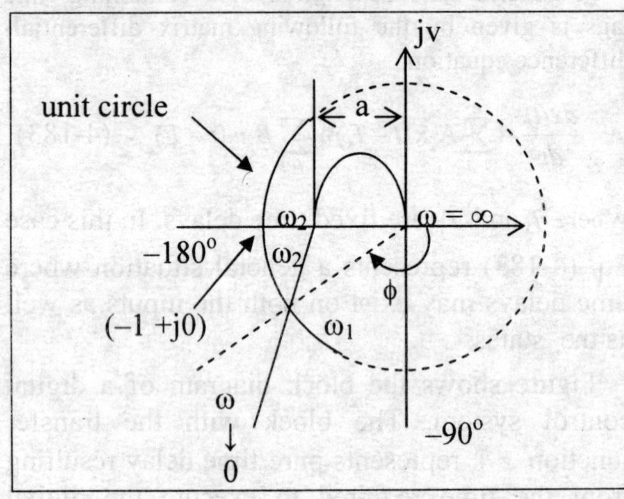

- G(jω) H (jω) which crosses the negative real axis at a frequency $\omega = \omega_2$ with a intercept of a.
- Let a unit circle center at origin intersect the G(jω) H(jω) make an angle of ϕ with the negative real axis measured positively in counted-clockwise direction.
- As the value of G(jω) H(jω) approaches (−1 + j0), the relative stability reduces i.e. as a approaches unity and that of ϕ tends to zero.

Gain Margin and Phase margin:

- Gain margin: It is the factor by which the system gain can be increased to drive it to the verge of instability.

  at $\omega = \omega_2$, $\angle G(j\omega)H(j\omega) = 180^{o}$ and

  $|G(j\omega)\,H(j\omega)| = a$,if gain is increased by (1/a) then $|G(j\omega)\,H(j\omega)|$ becomes = 1 and hence the G(jω) H(jω) – plot will pass thought (−1 + j0).

- *The gain margin* (GM) may be defined as the reciprocal of the gain at the frequency at which the phase angle becomes 180°
- The frequency at which the angle is 180°, is called the *phase cross over frequency*.

  GM = 1/a

  where a = | G(jω) H(jω) | at

  $\omega = \omega_2$

Phase margin:

- The frequency at which | G(jω) H(jω) | = 1, is called the *gain cross-over frequency.*
- It is given by the intersection of the G(jω) H(jω) plot and a unit circle, centered at the origin.
- This gives phase angle.

  G(jω)H(jω) = – 180° + ϕ
- The phase margin is thus defined as the amount of additional phase-lag at the gain cross-over frequency required to bring the system to the verge of instability.
- Phase margin is always positive for stable feedback systems.

  Phase Margin $\phi = \angle|\,(j\omega)\ H(\omega)\big|_{\omega=\omega_c} + 180^o$

  $= \omega_c + 180^o$

  where ω_1 the gain cross over frequency.
- A large gain margin or a large phase margin indicates a very stable feedback but usually a very sluggish one.
- A GM close to unity or PM close to zero corresponds to a highly oscillatory system.

8.13 STATE SPACE REPRESENTATION OF SYSTEMS

- Transfer function model provides us with simple and powerful analysis and design technique, but it suffers from certain drawback i.e. it is defined only under zero initial condition.
- Also it may sometimes be necessary and advantageous to provide a feedback proportional to some of the internal variable of a system, rather than the output alone, for the purpose of stabilizing and improving the performance.
- We need more general mathematical representation of a system which, along with output, yields information about the state of the system variable at some predetermined points along the flow of signal.
- *State Variable:* This approach is a direct time domain approach which provides as basis for modern control theory and system optimization.
- It is very powerful technique for the analysis and design of linear and non linear, time-invariant or time-varying, multi-input –multi-output system.
- The state of a dynamical system is a minimal set of variable (state variables) such that the knowledge of these variables at $t = t_o$ together with the knowledge of the inputs for $t \geq t_o$, completely determines the behavior of the system for $t > t_o$.

State Model of Linear System

In vector matrix form the sate model of linear time invariant system is

$$X'(t) = A\,x(t) + B\,u(t)$$
$$Y(t) = C\,x(t) + D\,u(t)$$

where,

x(t)	→	State vector
u(t)	→	Input vector
A	→	System matrix
B	→	Input matrix
Y(t)	→	Output vector
C	→	Output matrix
D	→	Transmission matrix

Block Diagram Representation of State Model

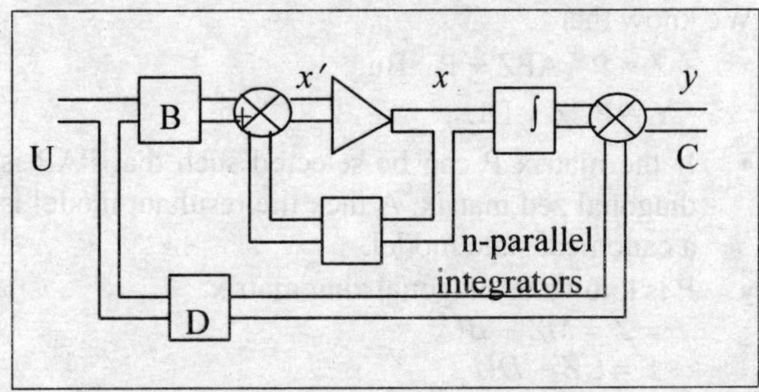

- State model of a system is not unique, but the number elements in the state vector is equal and minimal.
- If $x = Pz$ where P is any (n × n) non singular\r matrix

  $x = Pz$

  So $Pz = APZ + Bu$

  $Z = P^{-1}\ \underline{APZ} + P^{-1}\ Bu$

  $= \Delta + Bu$

where $\Delta = P - AP$ and $B = \overline{P^{-1}B}$

- $Y = CPZ + Du = \overline{C}\,CZ + Du$

where $\overline{C} = CP$

So (1) and (2) gives another state model for a given system.

Derivation of Transfer Function Form State Model

$$T(s) = \frac{Y(s)}{U(s)} = C(SI - A)^{-1}B + D$$

$$= \frac{C\ adj\ (sI - A)}{\det(SI - A)} B + D$$

- Where the state model is non unique, the transfer function of the system is unique, i.e. the transfer function must out to be the same irrespective of which particular state model is used to described the system.

Digitalization:

- Some of the state models presented employed physical variable, phase variable and canonical variables.
- In canonical state model wherein A is in diagonal form is most suitable for this purpose.
- Digitalization is a techniques for transforming a general state model into a canonical one.

Method:

We know that

$Z = P^{-1}APZ + P^{-1}Bu$

$Y = PCZ + DU$

- If the matrix P can be selected such that PAP is diagonal zed matrix. A then the resultant model is a canonical state model.
- P is known as diagonal zing matrix.

$= Z \equiv MZ + BU$

$Y = CZ + DU$

where M is a diagonal matrix.

$B = P^{-1}B$; $C = PC$

- Determination of the diagonal zing matrix is vacillated by the use of eigenvector.

Special Cases:

$$\text{when } A = \begin{vmatrix} 0 & 1 & 0\ldots\ldots & 0 \\ 0 & 0 & 1\ldots\ldots & 0 \\ 0 & 0 & 0\ldots\ldots & 1 \\ -a_n & -a_{n-1} & \ldots\ldots & a_1 \end{vmatrix}$$

then digitalization matrix can be directly written as

$$V = \begin{vmatrix} 0 & 2\ldots\ldots & 1 \\ \lambda_1 & \lambda_2\ldots\ldots & \lambda_n \\ \lambda_1^2 & \lambda_2^2\ldots\ldots & \lambda_n^2 \\ \lambda_1 & \lambda_2\ldots\ldots & \lambda_n \end{vmatrix}$$

where λ_1, λ_2 -----λ_n are eigenvector of matrix-A solution of state equation:

- $x(t) = e^{AT}\ x(0) + \int_0^t e^{-A(t-t_o)}\ Bu(t)\,dt$
- If the initial state is known at $t = t_o$ then
- $x(t) = e^{At}\,x(t_o) + \int_{to}^{t} e^{A(t-t)}\,Bu(t)\,dt$

Properties of the State Transition Matrix

- state transition matrix $\phi(t) = e^{At}$

Properties of state transition matrix $\phi(t)$

1. $\phi(0) = eA_o = 1$
2. $\phi(t) = e^{At} = e^{(-At)-1} = [\phi(-t)]^{-1}\ \phi^{-1}(t) = \phi(-t)$
3. $\phi(t_1 + t_2) = \phi(t_1)\,\phi$
4. $\phi(t_2) = \phi(t_2)\,\phi(t_2)$

Computation of State transition matrix:

1. Computation by Lap lace transformation:

$X(s) = (SI - A)^{-1} \times (0)$.Taking inverse Lap lace transform

$x(t) = LT^{-1}[(SI - A)^{-1}] \times (0)$

$\phi(t) = LT^{-1}[(SI - A)^{-1}] = \rho^{-1}\sigma(s)$

where $\phi(s) = [(SI - A)^{-1}$ is called the resolving matrix.

$x(t) = \phi(t)\,x_0 + \delta^{-1}\,\phi(s)\,Bu(s)$

Concepts of Controllability and Observability

Controllability

A system is said to be controllable at time t_0 if it is possible by means of an unconstrained control vector to transfer the system from any initial state x (t_0) to any other state in a finite interval of time.

Observability

A system is said to be observable at time t_0 if, with the system in state x (t_0), it is possible to determine this state from the observation of the output over a finite time interval.

Kalman's Test for Controllability:

$X' = AX + BU$

This state variable is completely controllable if and only if the rank of the composite matrix is same as that of order of the matrix

$Q = [B \mid AB \mid \text{----} \mid A^{n-1}B \mid]$

i.e. The system is completely state controllable if and only if the vector B, AB, A^2 B, A^3 B, ..., A^{n-1} B are linearly independent or the n × n matrix

$[B : AB : A^2B : \ldots A^{n-1}B]$ is the rank of N

Observability

In canonical state variable representation, for complete Observability non-of the column of the matrix c is zero.

Kalman's test of Observability

$Q_C = [C^T \mid A^T C^T \mid \text{-----} \mid (A^T)^{n-1} C^T]$

For completely observable system Q_C must have rank of n.

Example: $\begin{vmatrix} x_1' \\ x_2' \end{vmatrix} \begin{vmatrix} 1 & 1 \\ 0 & -1 \end{vmatrix} \begin{vmatrix} x_1 \\ x_2 \end{vmatrix} + \begin{vmatrix} 1 \\ 0 \end{vmatrix} u$

where, $x' = Ax + Bu$

$[B : AB] = \begin{vmatrix} 1 & 1 \\ 0 & 0 \end{vmatrix} = 0 =$ singular

System is not completely controllable

Example: $\begin{vmatrix} x_1' \\ x_2' \end{vmatrix} \begin{vmatrix} 2 & 1 \\ 2 & -1 \end{vmatrix} \begin{vmatrix} x_1 \\ x_2 \end{vmatrix} + \begin{vmatrix} 0 \\ 1 \end{vmatrix} u$

Solution: $[B : AB] = \begin{vmatrix} 0 & 1 \\ 1 & -1 \end{vmatrix} =$ non singular

System completely state controllable

Example: $x_1'(t) = -2\, x_1(t) + 4\, x_2(t)$

$x_2'(t) = 2\, x_1(t) - x_2(t) + 4(t)$

Check controllable and stable

Solution: (a) $\begin{vmatrix} x_1'(t) \\ x_2'(t) \end{vmatrix} = \begin{vmatrix} -2 & 4 \\ 2 & -1 \end{vmatrix} \begin{vmatrix} x_1(t) \\ x_2(t) \end{vmatrix} + \begin{vmatrix} 1 \\ 1 \end{vmatrix} u(t)$

$$A = \begin{vmatrix} -2 & 4 \\ 2 & -1 \end{vmatrix} \quad B = \begin{vmatrix} 0 \\ 1 \end{vmatrix}$$

$$AB = [A][B] = \begin{vmatrix} -2 & 4 \\ 2 & -1 \end{vmatrix} \begin{vmatrix} 0 \\ 1 \end{vmatrix} = \begin{vmatrix} -2\times 0 & + & 4\times 1 \\ 2\times 0 & + & -1\times 1 \end{vmatrix}$$

$$= \begin{vmatrix} 0 + 4 \\ 0 - 1 \end{vmatrix} = \begin{vmatrix} 4 \\ -1 \end{vmatrix}$$

$$[B : AB] = \begin{vmatrix} 0 & 4 \\ 1 & -1 \end{vmatrix} = -4 \neq 0 \text{ system controllable}$$

(b) Characteristic equation $[S\,I - A] = 0$

$\begin{vmatrix} s+2 & -4 \\ -2 & s+1 \end{vmatrix} = 0$; $(s + 2)(s + 1) - 8 = 0$

or, $s^2 + 3s - 6 = 0$; $s = [-3 \pm \sqrt{(9 + 2\,4)}]/2$

$s = (-3 \pm 5.7)/2 = -4.35, 1.35$

The roots are in the right side of s-plan .therefore system is unstable.

Example: $\begin{vmatrix} x_1' \\ x_2' \end{vmatrix} \begin{vmatrix} -1 & 0 \\ 0 & -2 \end{vmatrix} \begin{vmatrix} x_1 \\ x_2 \end{vmatrix} + \begin{vmatrix} 0 \\ 1 \end{vmatrix} u$

$Y = [1 \quad 2] \begin{vmatrix} x_1 \\ x_2 \end{vmatrix}$

Check for controllable, observable

Solution: (a)

$$A = \begin{vmatrix} -1 & 0 \\ 0 & -2 \end{vmatrix} ;\ B = \begin{vmatrix} 0 \\ 1 \end{vmatrix} ; C = 0 ; D = [1 \quad 2]$$

$$[AB] = \begin{vmatrix} -1 & 0 \\ 0 & -2 \end{vmatrix} \begin{vmatrix} 0 \\ 1 \end{vmatrix} = \begin{vmatrix} 0 \\ -2 \end{vmatrix}$$

$$|B \quad AB| = \begin{vmatrix} 0 & 0 \\ 1 & -2 \end{vmatrix} = 0$$

system is not controllable.

Solution : (b)

$$D = [1 \quad 2]; [DA] = [1 \quad 2] \begin{vmatrix} -1 & 0 \\ 0 & 2 \end{vmatrix} [-1 \quad -4]$$

Hence $D = \begin{vmatrix} 1 & 2 \\ -1 & -4 \end{vmatrix} = -2 \neq 0$

System is completely observable.

Eigen values of n × n matrix A

The eigen value of a n × n matrix A are the roots of the characteristic equation

$[\lambda I - A] = 0$

The eigen values are called the ***characteristic roots***

If $A = \begin{vmatrix} 0 & 1 & 0 \\ 0 & 0 & 1 \\ -6 & -11 & -6 \end{vmatrix}$

The characteristic equation is

$$[\lambda I - A] = \begin{vmatrix} \lambda & -1 & 0 \\ 0 & \lambda & -1 \\ 6 & 11 & \lambda+6 \end{vmatrix}$$

$$= \lambda^3 + 6\lambda^2 + 11\lambda + 6$$

$$= (\lambda+1)(\lambda+2)(\lambda+3) = 0$$

The eigen values of A are the roots of characteristic equation or –1, –2, –3.

8.14 MECHANICAL HYDRAULIC AND PNEUMATIC SYSTEM COMPONENTS

Mechanical System

Hydraulic Syste:

Advantage of hydraulic system;

- Size to power ratio is high.
- Rapid action as compared to electrical system.

Disadvantage

- Problem of leakage

Application:

- Power steering
- Hydraulic breaks

Components:

(1) Hydraulic pumps and motors
Mechanical power → input
Hydraulic power→ output

(2) Hydraulic valves

- *Hydraulic valve introduces non-linearity in the system but it is fast in response*

$$G(s) = \frac{K_m}{s(sT_m + 1)}$$

(3) Hydraulic liner actuator
These are the piston devices

$$G(s) = \frac{K_m}{s(sT_m + 1)}$$

(4) Hydraulic liner amplifier

- *It is obtained by providing negative feedback in liner actuator*

Pneumatic Systems

As different from hydraulic fluid, air medium (or other gases in special situations) is used in pneumatic control system.

Advantages:

- Theses are not inflammable
- Negligible viscosity

Disadvantage:

Because of compressibility of air the response of pneumatic system is slower than hydraulic system.

Components:

(1) Pneumatic bellows.
(2) Pneumatic flapper valve (action is similar to spring).
Input is air supply
Output is pressure
(1) Pneumatic relays: works like an amplifier. D
(2) Pneumatic actuator: it responds to change in the input pressure.

8.15 MATHEMATICAL MODELING

Electrical System

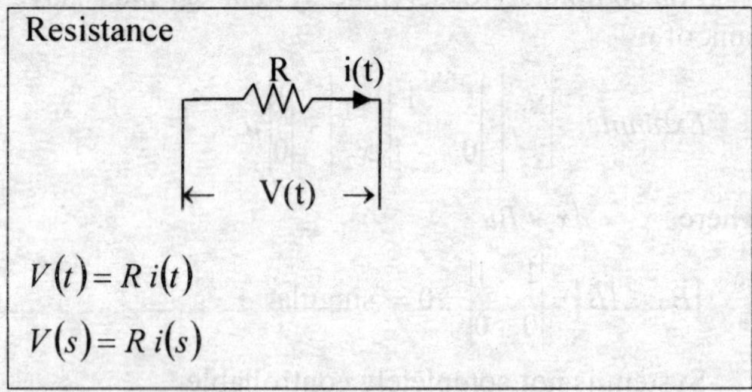

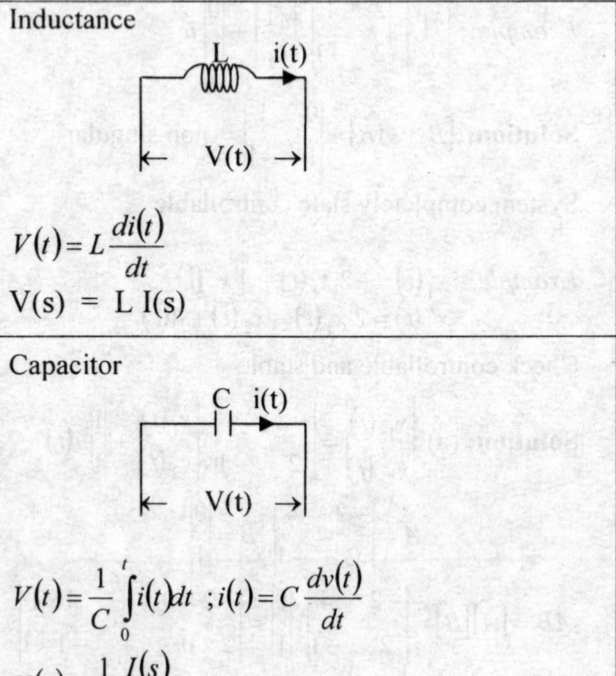

Example: Series RLC Circuit

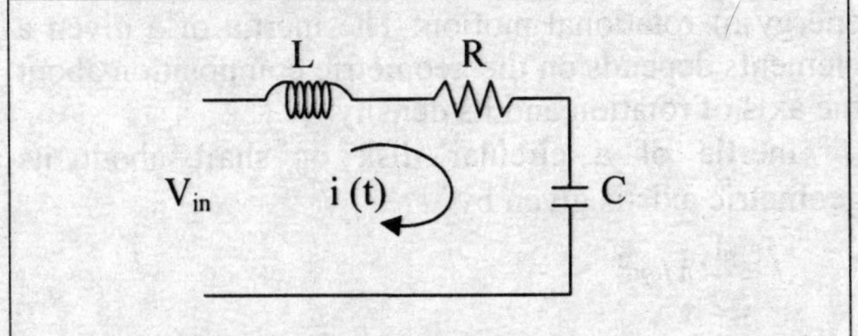

Apply KVL in loop

$$iR + \frac{Ldi}{dt} + \frac{1}{C}\int idt = V_{in}$$

$$but \quad i = \frac{dq}{dt}$$

$$L\frac{d^2q}{dt^2} + R\frac{dq}{dt} + \frac{q}{C} = V_{in}$$

Example: Parallel RLC Circuit

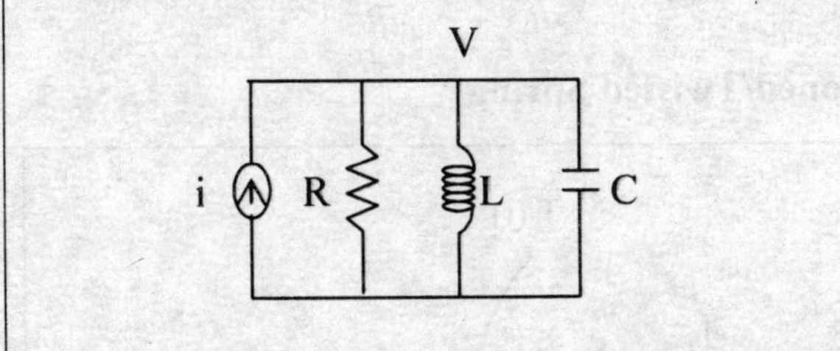

Applying KCL at node

$$\frac{V}{R} + \frac{1}{L}\int Vdt + C\frac{dV}{dt} = i$$

$$V = \frac{d\phi}{dt}, \text{ where } \phi = \text{magnetic flux}$$

$$C\frac{d^2\phi}{dt^2} + \frac{1}{R}\frac{d\phi}{dt} + \frac{\phi}{L} = i$$

Duality

$C \rightarrow L$; $R \rightarrow 1/R$; $\phi \rightarrow q$

Mechanical System

Translation system

Three basic components of a mechanical system are mass, elasticity (idealized by a helical spring) and friction (idealized by a damper). These are for transnational system.

(a) Mass → L

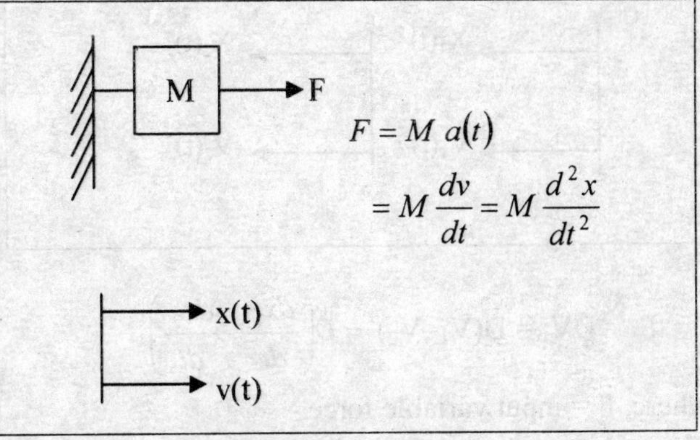

$$F = M\,a(t) = M\frac{dv}{dt} = M\frac{d^2x}{dt^2}$$

where x(t) represents displacement
v(t) represent velocity
a(t) represent acceleration.

All referred in the direction of force.

(b) Elasticity (helical spring) → C

Considered to be element which store potential energy.

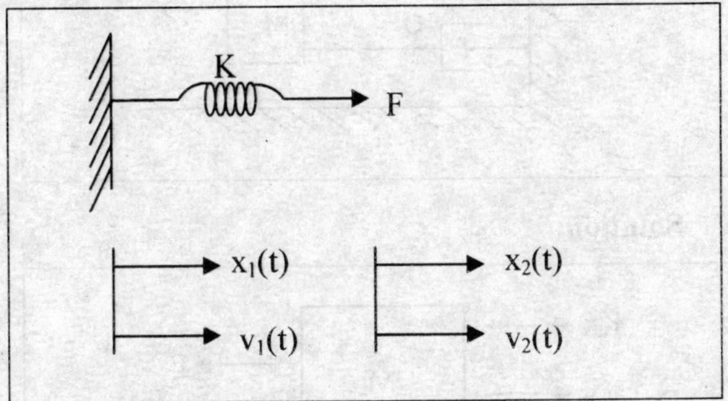

$$F = K\,x(t) = K(x_1 - x_2) = K\int_{-\infty}^{t}(v_1 - v_2)dt = K\int_{-\infty}^{t}v(t)dt$$

where, K = spring constant
F = input variable force
X = output variable displacement

Friction (damper) → R

When ever there is motion or tendency of motion between two elements, friction force exists.

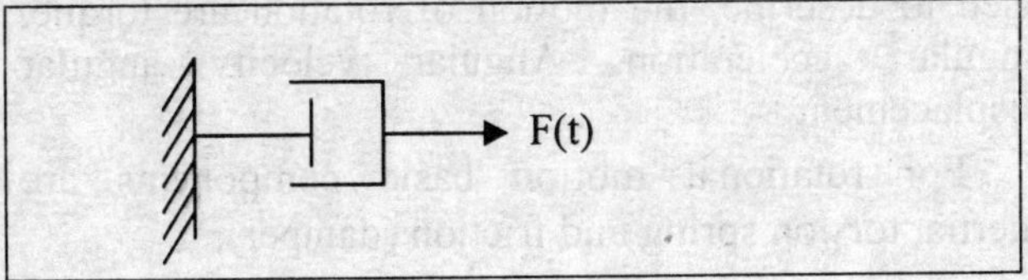

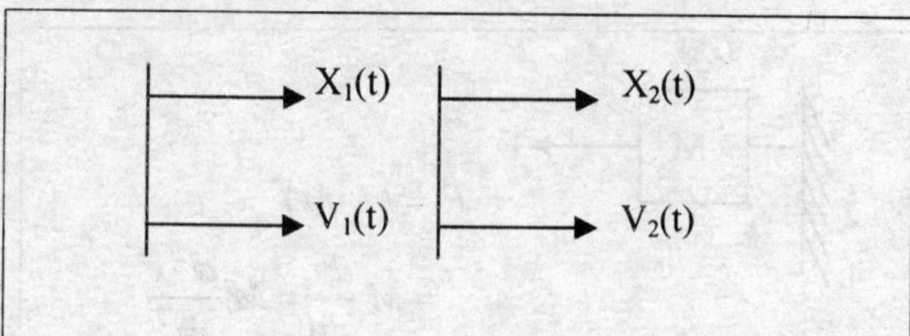

$$F = DV = D(V_1 - V_2) = D\left(\frac{dx_1}{dt} - \frac{dx_2}{dt}\right)$$

where, F = input variable force
V = output variable displacement
D = frictional constant

Example: Derive the mathematical model for the system.

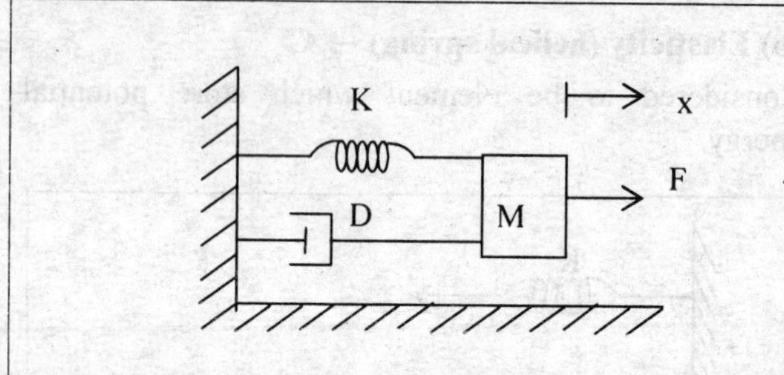

Solution:

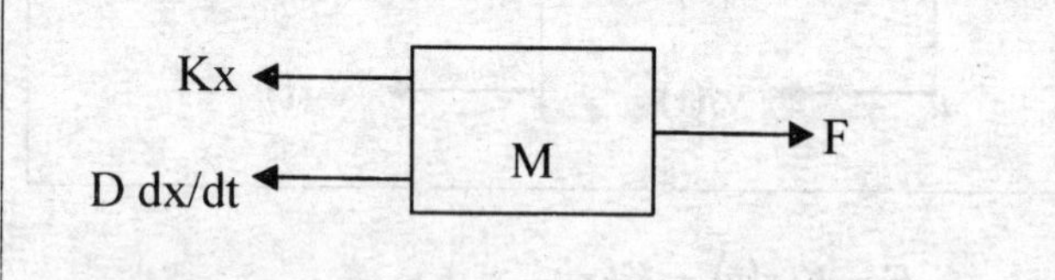

At balance

$$F - Kx - D\frac{d^1x}{dt} = M\frac{d^2x}{dt^2}$$

Mathematical model

$$\therefore\ M\frac{d^2x}{dt^2} + D\frac{dx}{dt} + Kx = F$$

Rotational System

Motion about a fixed axis. The variable generally used to describe, the motion of rotation are torque, angular acceleration, Angular velocity angular displacement.

For rotational motion basic components are inertia, torsion spring and friction (damper).

Movement of Inertia (J) → mass

Property of an element which stores the kinetic energy of rotational motion. The inertia of a given a elements depends on the geometric composition about the axis of rotation and its density

Inertia of a circular disk or shaft about its geometric axis is given by

$$J = \frac{1}{2}Mr^2$$

where, M = mass of the disk
f = radius

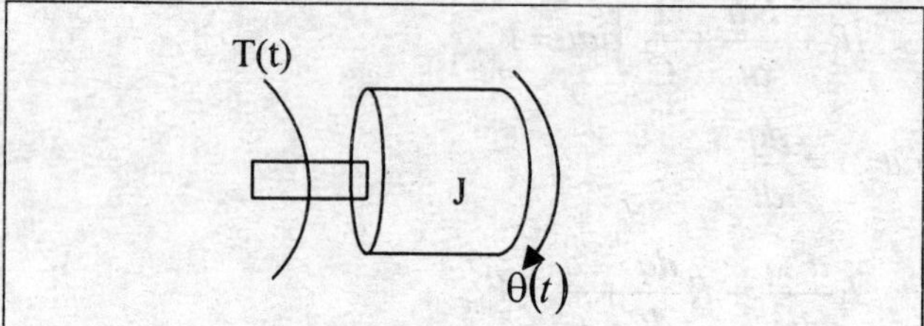

$$T(t) = J_d(t) = J\frac{d\omega(t)}{dt} = j\frac{d^2\theta(t)}{dt^2}$$

Tortioned/Twisted Spring

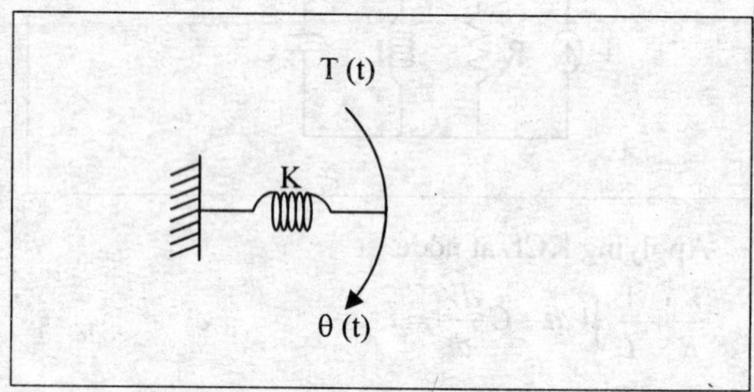

$$T(t) = K\theta(t) = K\int_{-\infty}^{t}\omega(dt)$$

Friction (damper)

$$T(t) = D\frac{d\theta(t)}{dt} = D\omega(t)$$

Example: Derive the mathematical model for the given system.

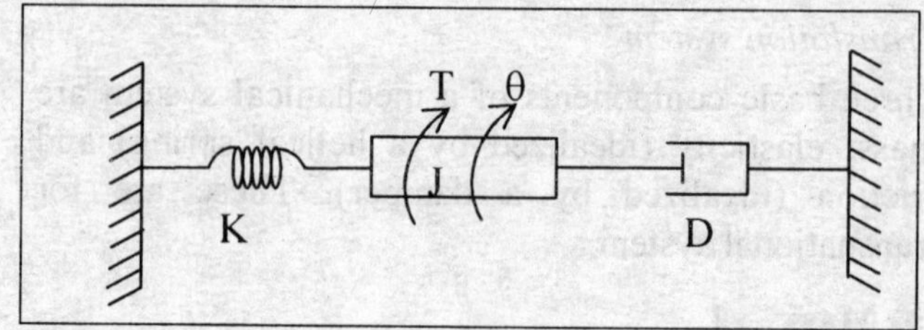

Solution: Mathematical model

$$J\frac{d^2\theta(t)}{dt^2}+D\frac{d\theta(t)}{dt}+K\theta(t)=T$$

Example: Find the transfer function for given diagram.

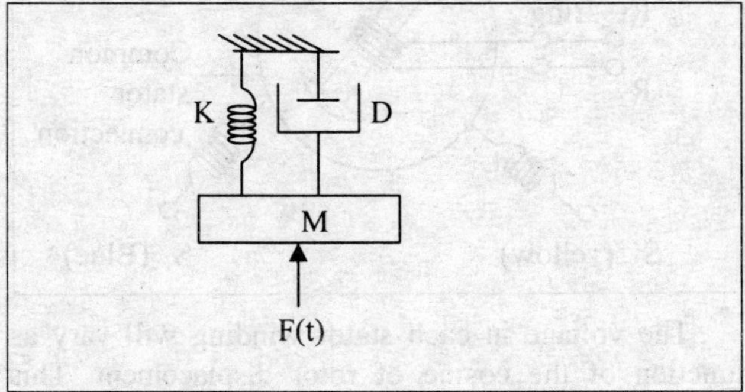

Solution:

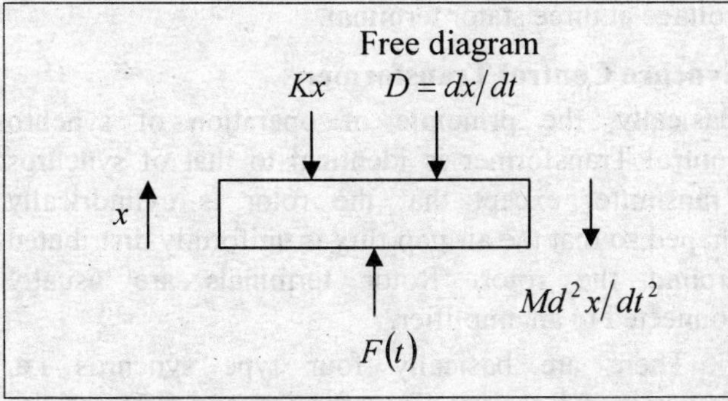

The applied force F is balanced by the acceleration of the mass and resistive force of spring and damper.

$$F(t)=M\frac{d^2x}{dt^2}+D\frac{dx}{dt}+K\,x(t)$$

This is a 2nd order differential equation, apply the LT to the above equation.

$$F(s)=M[s^2\,X(s)-s\,x(0^+)+x(0^+)]$$
$$+D\,(s\,X(s)-x\,(0^+)]\ \ +K\,X(s)$$

initial condition zero $x(0^+)\to 0$

$$F(s)=M\,s^2\,X(s)+Ds\,X(s)+kX(s)$$

$$\text{The transfer function}=\frac{\text{LT of output}}{\text{LT of input}}$$

$$G(s)=\frac{X(s)}{Fs}=\frac{1}{Ms^2+Ds+K}$$

Example: Find the transform function for given diagram.

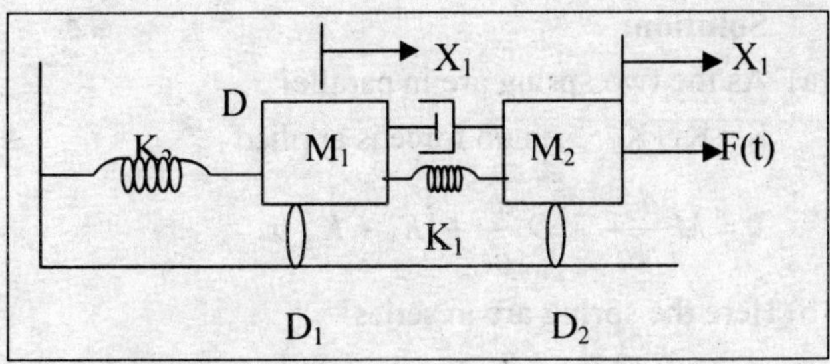

Solution: For mass M_2

$$F(t)=M_2\frac{d^2x_1}{dt^2}+D_2\frac{dx_1}{dt}$$
$$+D\left(\frac{dx_1}{dt}-\frac{dx_2}{dt}\right)+K1(x_1-x_2)$$

No force is applied on mass M_1 so

$$0=M_2\frac{d^2x_2}{dt^2}+D_1\frac{dx_2}{dt}+K_2x_2+D\left(\frac{dx_2}{dt}-\frac{dx_1}{dt}\right)$$
$$+K1(x_2-x_1)$$

$(x_1 - x_2)$ is the displacement of spring form equilibrium

$$F(s)=M_2s^2X_1(s)+D_2sX_1(s)+Ds[X_1(s)X_2(s)]$$
$$+K_1[X_1(s)-X_2(s)]$$
$$0=M_1s^2X_2(s)+D_1sX_2(s)+K_2X_2(s)$$
$$+Ds[X_2(s)-X_1(s)]$$
$$+K_1[X_1(s)-X_2(s)]\ M_1s^2X_2(s)$$
$$+D_1sX_2(s)+K_2X_2(s)$$
$$+DsX_2(s)+K_1X_1(s)$$
$$=DsX_2(s)+K_1X_1(s)$$

By substituting the value of $X_2(s)$ in equation for F(s) and rearrange again we will get the transfer $X_1(s)/F(s)$ If will be 4th order **s** equation.

Example: Write the differential equation for the systems.

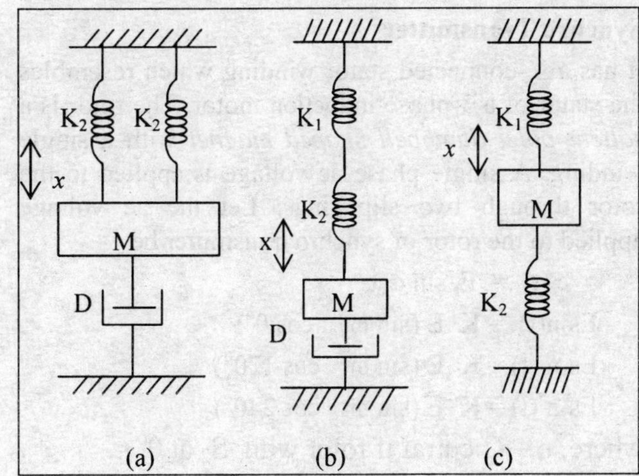

Solution:

(a) As the two spring are in parallel

$K = K_1 + K_2$ As no force is applied

$$0 = M\frac{d^2x}{dt} + D\frac{dx}{dt} + (K_1 + K_2)x$$

(b) Here the spring are in series

$$K = \frac{K_1K_2}{K_1 + K_2}$$

$$0 = M\frac{d^2x}{dt} + D\frac{dx}{dt} + \frac{(K_1K_2)}{(K_1 + K_2)}x$$

(c) Hence spring are in opposite side of mass m, which result in parallel configuration of spring

$$0 = M\frac{d^2x}{dt^2} + (K_1 + K_2)x$$

8.16 SYNCHRO PAIR, SERVO AND STEP MOTORS

Synchros

A synchros is an electromagnetic transducer commonly used to convert an angular position of shaft into electric signal. It is known as *selsyn or antosym, Telesyn.*

Basically a synchros is a rotary device that operates on the same principal as transformer and produces a correlation between angular position and a voltage or set of voltage.

- Used to determine the magnitude and the direction of angular displacement.
- Synchros are used widely in control systems as detectors and encoders because of their ruggedness in constriction and high reliability.

Synchro Transmitter

It has a Y-connected stator winding which resembles the stator of a 3-phase induction motor. The rotor is a *salient pole, dumbbell shaped magnet* with a single winding. A single phase ac voltage is applied to the rotor through two slip rings. Let the ac voltage applied to the rotor of synchro transmitter be

$e_r(t) = E_r \sin \omega_c t$

$Es_2n(t) = K\ E(\sin \omega t \quad \cos 0^\circ)$

$Es_1n(t) = K\ E(\sin \omega t \quad \cos 120^\circ)$

$Es_3n(t) = K\ E(\sin \omega t \quad \cos 240^\circ)$

where, n → neutral if rotor with S_2 at 0°

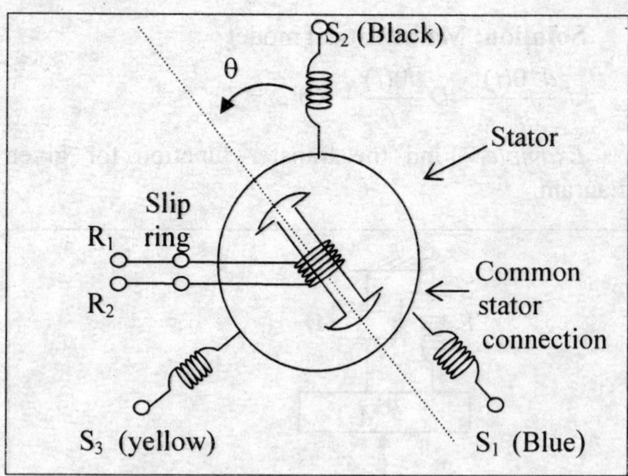

The voltage in each stator winding will vary as function of the cosine of rotor displacement. This head to the use of synchros Transmitter to identify angular position by measuring and identify the set of voltage at three stator terminal.

Synchro Control Transformer

Basically, the principle of operation of synchro control Transformer is identical to that of synchros Transmitter except that the rotor is cylindrically shaped so that the air gap flux is uniformly distributed around the rotor. Rotor terminals are usually connected to an amplifier.

There are basically four type synchros i.e. Transmitter, Receiver, Transformers and differentials.

Application of synchros as error detector

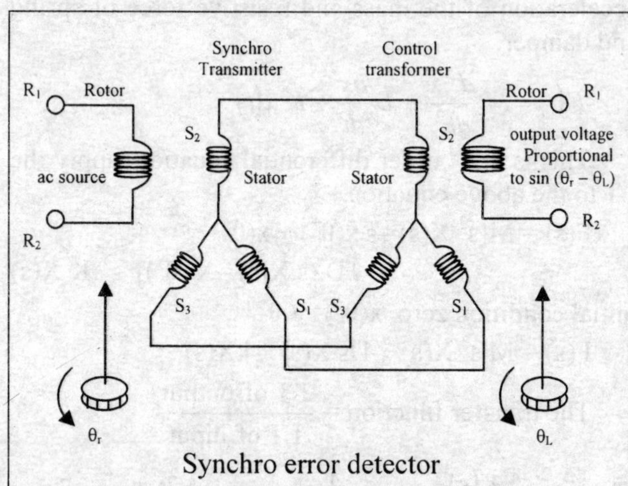

Synchro error detector

When the rotor position of two synchro are in perfect alignment, the voltage generated across the terminal of output rotor winding is zero. When the rotor shaft are not align, the rotor voltage of control Transformer is approximately a sine function of difference between the two shaft angle.

Servo Motor

A motor used in high accuracy position or in speed control system, must is general satisfy the following specification:

1. It should have high stall torque and low inertia resulting in high initial acceleration.
2. Internal damping should be constant over the operating range.
3. Torque Vs speed characteristic should be parallel and equally spaced for equal increments of the control variable.
4. Reversible operation should be possible.

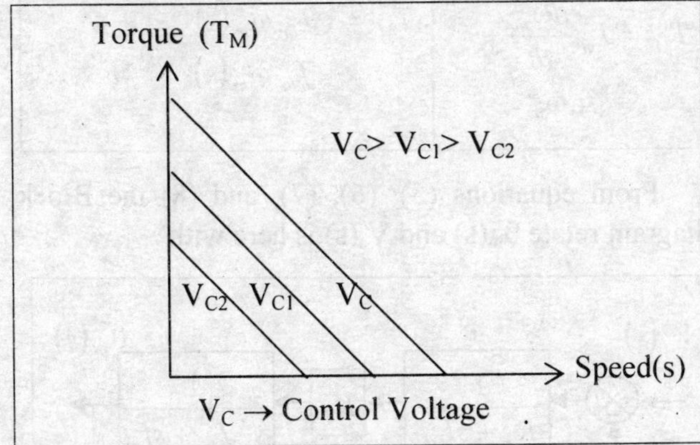

DC Motor

It is an important transducer that converts electrical energy to mechanical energy that is dc voltage is converted into rotational or linear motion by gear driving a helical gear. They can be continuously and accurately controlled over a wide range of speed and loads.

Field Controlled DC Motor

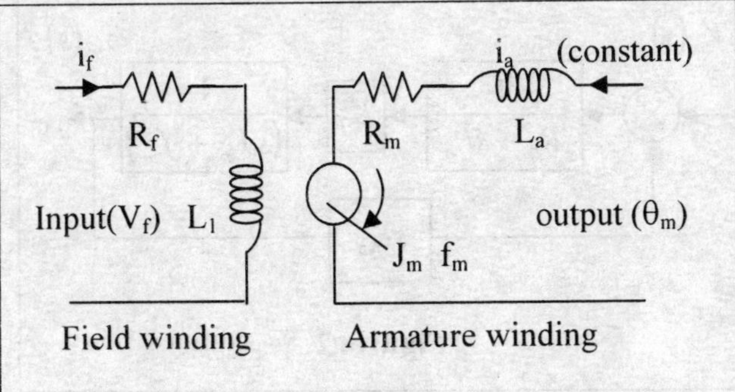

The input voltage V_f is applied to the field winding, which has a resistance R_f and inductance L_f.

The armature current i_a supplied to the armature is kept constant and thus the motor shaft is control by the input voltage V_f.

The field current i_f produces a flux in the machine which in turn produces a torque at the motor shaft.

The moment of inertia and the coefficient of viscous friction at the motor shaft are J_m (K_g m^2) and f_m [Nm/ (rad/sec)] respectively.

The angular shift resulting in the motor shaft being θ_m (radian) and the corresponding angular velocity being ω_m (rad/sec).

Since the armature current i_a is kept constant, the relationship between the developed motor Torque T_M and the field current i_f given by

$$T_M \propto i_f \quad ; \; T_M = K_f \cdot i_f$$

where $k_f \rightarrow$ motor torque constant Nm/A

The related Dynamic Equation	Corresponding LT equation
$V_f = R_f i_f + L_f \, d i_f / dt$	$V_f(s) = R_f I_f(s) + s L_f I_f(s)$ (1)
$T_M = K_f \, i_f$	$T_M(s) = K_f I_f(s)$ (2)
$T_M = J_m \dfrac{d^2\theta_m}{dt^2} + f_m \dfrac{d\theta_m}{dt}$	$T_M(s) = s^2 J_m \theta_m(s) + s f_m \theta_m(s)$ (3)
$T_M = J_m \dfrac{d\omega_m}{dt} + f_m \omega_m$	$T_M(s) = sJ_m \omega_m(s) + f_m \omega_m(s)$ (4)

From equations (1) to (3) the block diagram relating the output $\theta_m(s)$ and the input $V_f(s)$ is drawn.

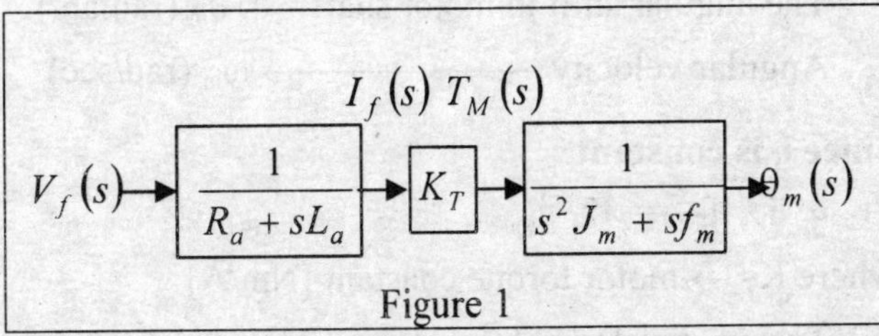

Figure 1

Form equation (1), (2) and (4) the block diagram

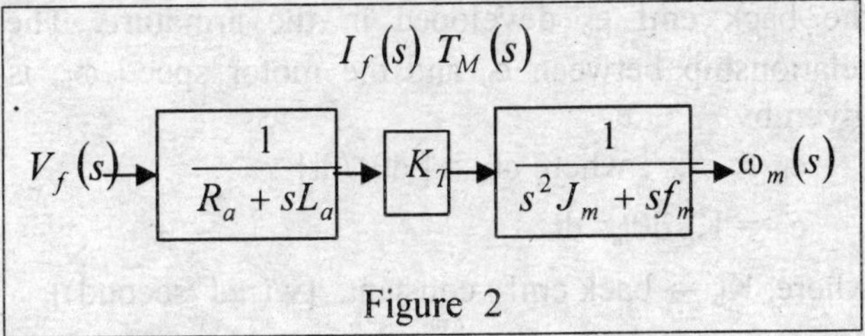

Figure 2

Transfer function

$$\frac{\theta_m(s)}{V_f(s)}=\frac{K_f}{s(R_f+sL_f)(sJ_m+f_m)}$$

$$\frac{\omega_m(s)}{V_f(s)}=\frac{K_f}{(R_f+sL_f)(sJ_m+f_m)}$$

Armature controlled DC motor

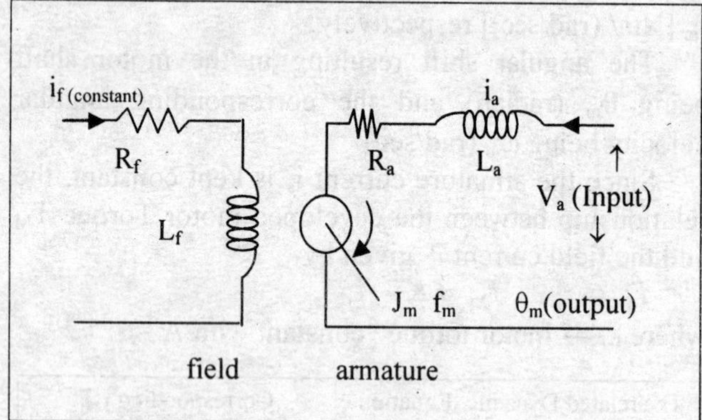

In this system

Input voltage ⟶ V_a (V)

Armature resistance ⟶ R_a (Ω)

Armature inductance ⟶ L_a (H)

Armature current ⟶ i_a (A)

Field current ⟶ i_f(constant)

Moment of inertia at motor shaft → J_m (kg–m^2)

Coefficient of viscous friction at motor shaft ⟶ f_m [Nm/(rad /sec)]

The angular shift in motor shaft ⟶ θ_m (radian.)

Angular velocity ⟶ ω_m (rad/sec)

Since i_f is constant

$T_M \propto i_a$; $T_M = K_T\, i_a$

where K_T → motor torque constant [Nm/A]

Torque developed by motor → T(Nm)

The applied input voltage V_a is being opposed by the back emf e_b developed in the armature. The relationship between e_b and the motor speed ω_m is given by

$e_b \propto \omega_m$; where ω_m = [$d\theta_m/dt$]

$e_b = K_b\, d\theta_m / dt$

where, K_b → back emf constant, [v/(rad /second)]

Dynamic equation	Corresponding LT equation
$V_a - e_b = R_a i_a + L_a \dfrac{di_a}{dt}$	$V_a(s) - E_b(s) = R_a I_a(s) + s L_a I_a(s)$ (5) $V_a(s) - Ea(s)$
$e_b = K_b\left(\dfrac{d\theta_m}{dt}\right)$	$E_b(s) = sK_b\, \theta_m(s)$ (6)
$e_b = K_b\, \omega_m$	$E_b(s) = K_b \omega_m(s)$ (6a)
$T_m = K_T\, i_a$	$T_M(s) = K_T\, I_a(s)$ (7)
$T_M = J_m \dfrac{d^2\theta_m}{dt} + f_m \dfrac{d\theta_m}{dt}$	$T_M = s^2 J_m \theta_m(s) + sf_m\, \theta_m(s)$ (8)
$T_M = J_m \dfrac{d\omega_m}{dt} + f_m \omega_m$	$T_M = s\, J_m\, \omega_m(s) + f_m\, \omega_m(s)$

From equations (5), (6), (7), and (8) the Block diagram relate $\theta_m(s)$ and $V_a(s)$ is here with

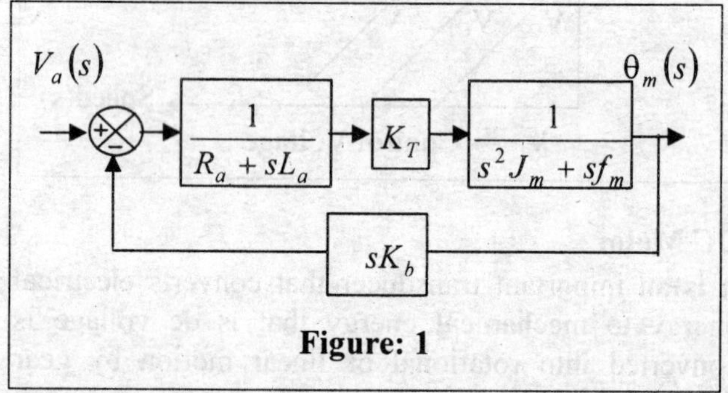

Figure: 1

From equation (5), (6a), (7) and (8a) Block diagram which relate output as wm(s) $\theta_m(s)$ and $V_a(s)$ is with

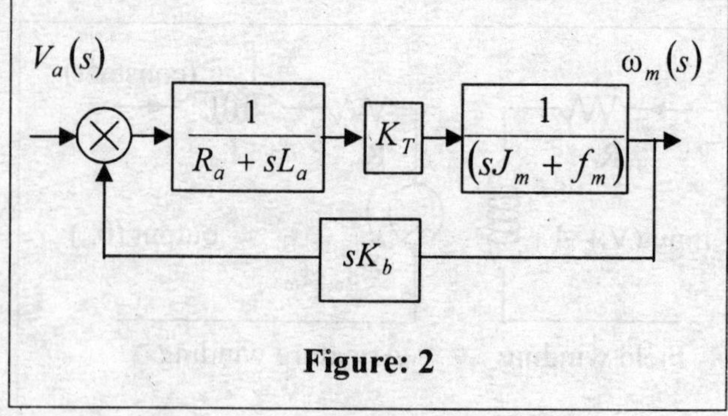

Figure: 2

From the block diagram figure (2) the forward path Transfer function is

$$G(s)=\frac{K_T}{s(R_a+sL_a)(sJ_m+f_m)}$$

And the feed back path

$$H(s)=sK_b$$

Hence the over all transfer function relating the output $\theta_m(s)$ and input $V_a(s)$ is given by the relation.

$$\frac{\theta_m(s)}{V_a(s)}=\frac{G(s)}{1+G(s)H(s)}$$

$$\frac{\theta_m(s)}{V_a(s)}=\frac{\dfrac{K_T}{s(R_a+L_a)(sj_m+f_m)}}{1+\left\{\dfrac{K_T}{s(R_a+L_a)(sj_m+f_m)}\right\}\times sK_b}$$

$$=\frac{K_T}{s(R_a+sL_a)(sJ_m+f_m)+sK\tau K_b}$$

if armature inductance is neglected i.e. $L_a \to o$

$$\frac{\theta_m(s)}{V_a(s)}=\frac{K_T}{sR_a(sJ_m+f_m)+sK_T K_b}$$

$$=\frac{K_T}{s(sR_aJ_m+R_a f_m+K_T K_b)}$$

$$\frac{\theta_m(s)}{V_a(s)}=\left(\frac{K_T}{R_a f_m+K_T K_b}\right)\div s\left(\frac{sR_aJ_m+1}{R_a f_m+K_T K_b}\right)+1$$

$$\frac{\theta_m(s)}{V_a(s)}=\frac{K_m}{s(1+sT_m)}$$

where $K_m=\dfrac{K_T}{R_a f_m+K_T K_b}$

= Motor gain constant

and $T_m=\dfrac{R_aJ_m}{(R_a f_m+K_TK_b)}$ = Motor time constant.

The block diagram representing the transfer function is

$V_a(s) \rightarrow \boxed{\dfrac{K_m}{s(1+sT_m)}} \rightarrow \theta_m(s)$

Since $\omega_m(s) = s\,\theta_m(s)$

$$\frac{\omega_m(s)}{V_a(s)}=\frac{K_m}{(1+sT_m)}$$

Comparative study of armature controlled dc servomotor over field controlled dc servomotor

Field control

$V_f(s) \rightarrow \boxed{\dfrac{K_f}{R_f+sL_f}} \rightarrow \boxed{\dfrac{1}{s^2J_m+sf_m}} \rightarrow \theta_m(s)$

- Open loop structure result in poor performance.
- The inductacne of field L_f is not neglegible. Which result 3rd order type 1 system.
- Improved damping is not possible.
- Low power application.
- Efficiency is low.

Armature Controlled

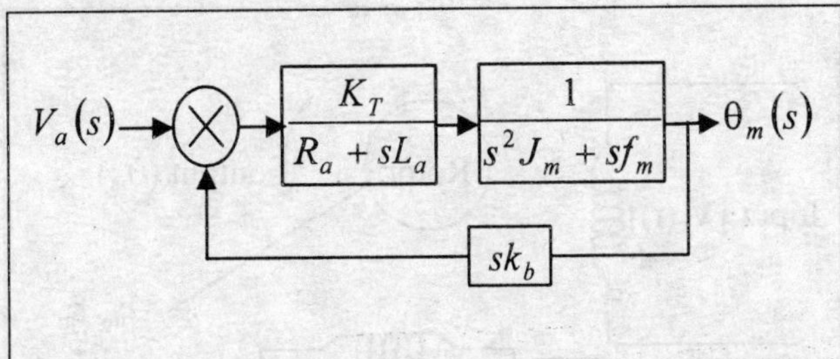

- Close loop structure results better performance.
- The inductance of armature is small and neglegible. Hence reduce the order of the system equation which result 2nd order type 1 system.
- Extra damping due to back emf is obtained. Increased damping improves the transient response of the system.
- High power application.
- Efficiency is high.

However, it may be noted that the amplifier required to supply the large armature current.

Stepper Motor

A stepper motor is an electromechanical device which actuates a train a of step angular (or linear) movement in response to a train of input pulses one to one basis, one step actuation for each pulse input.

A stepper motor is the actuator element of incremental motion control system. Computer peripherals like printers, tape drives, capstan drive, machine tool and process control system.

The mostly used type are:

(1) Variable reluctance motor

(2) Permanent magnet motor

Step size of the stepper motor is given by

$$\alpha = \left(\frac{360}{nT}\right)$$

where n is number of stake and T is number of teeth of rotor.

Advantage over conventional Motor

There are two advantages in using stepper motors. One is their compatibility with digital systems and secondly no sensors are needed for position and speed sensing as these are directly obtained by counting input pulses and periodic counting if speed information is needed.

AC servomotor

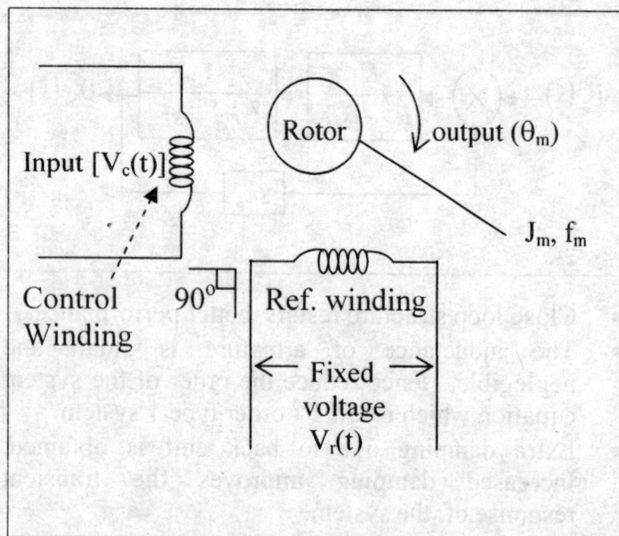

A two phase ac servomotor is similar to two phase induction motor. Rotor of the AC servomotor is generally made up of Aluminium. It can be either squirrel cage type or drag cup type. For low inertial application drag cup type is preferred.

The control voltage V_c (t) is applied to the control winding and a fixed voltage having a phase difference of 90^0 w.r.t control winding voltage is applied to the reference winding.

The control voltage result in the development of the motor torque T_M. The torque speed characteristic of motor shown here.

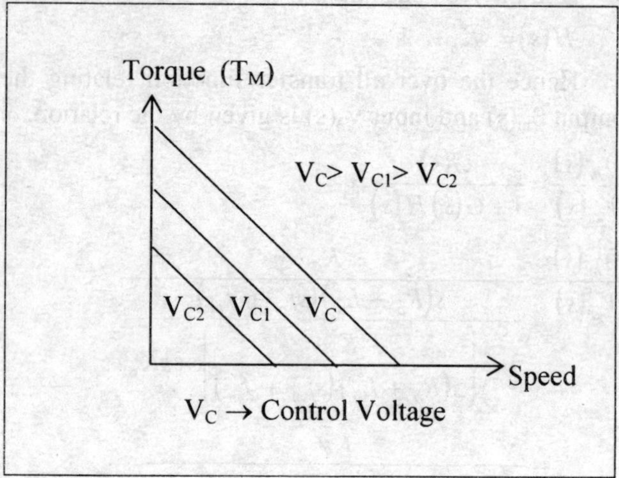

In this system

Moment of Interia ⟶ J_m (kg.m^2)

Coefficient of friction ⟶ f_m [N_m/(rad/sec)]

Angular shift ⟶ θ_m [radian]

Angular velocity ⟶ ω_m [rad/sec]

Form the torque speed characteristics the dynamic equation relating the motor torque and the speed is formed below:

$$T_M = K_m\,\omega_m + K_c V_c(t)$$

$$T_M = K_m \frac{d\theta_m}{dt} + K_c V_c(t)$$

If T_o is the stalling torque (initial torque), ω_o is the no load speed. The parameter K_m are K_c are determined in term of T_o and ω_o.

1. When the speed is zero the torque T_o and this stalling torque is proportional to the control voltage V_c

2. $T_o = K_c V_c$ i.e.

$$K_c = \frac{T_c}{V_c} \quad \text{[Nm/v]}$$

3. The slop of the torque speed characteristic

$$K_m = \frac{-T_o}{\omega_o} \quad \text{[Nm/ (rad/sec)]}$$

Dynamic equation	Corresponding L.T equation
$T_M = K_m \frac{d\theta_m}{dt} + K_c V_c(t)$	$T_M(s) = sK_m\theta_m(s) + K_c V_c(s)$ (1)
$T_M = J_m \frac{d^2\theta_m}{dt^2} + f_m \frac{d\theta}{dt}$	$T_M(s) = s^2 J_m\ \theta_m(s) + f_m\ \theta(s)$ (2)

From equation (1) and (2) the block diagram is drawn.

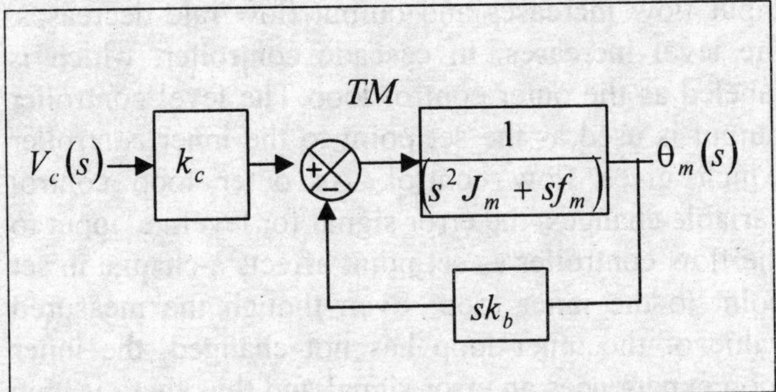

where $K_n = \frac{K_c}{f_m - K_m}$ = motor gain constant

$T_m = \frac{J_m}{f_m - K_m}$ = motor time constant

Since $\omega_m(s) = s\,\theta_m(s)$ i.e. $\frac{\theta_m(s)}{V_c(s)} = \frac{K_n}{s(1+sT_m)}$

$$\frac{\omega_m(s)}{V_c(s)} = \frac{K_n}{s(1+sT_m)}$$

Advantages of AC Servo Motor Over DC Servo Motor

1. It is having low brush contacts ∴ lesser losses.
2. Lighter in weight.
3. Ac servomotor is generally used for low power application.
4. Almost instantaneous starting, stopping and reversal.
5. Long life with high reliability.
6. Compatibility with ac signals of synchros, LVDTs and other ac transducers without the need for demodulation as in the case of dc motors.

Torque speed characteristic of Ac servomotor:

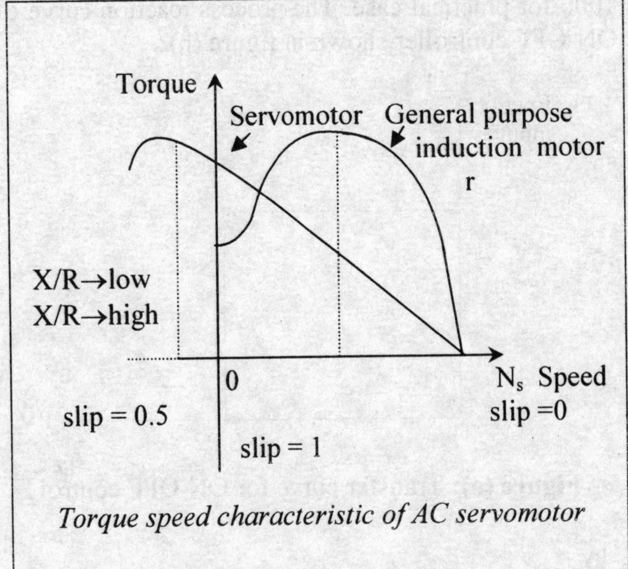

Torque speed characteristic of AC servomotor

R → resistance of rotor

X→ Inductance of rotor

Difference between general purpose Induction motor and AC servomotor:

1. $\left(\frac{X}{R}\right)_{servomotor} < \left(\frac{X}{R}\right)_{General\ purpose\ induction\ motor}$
2. Torque speed characteristics is linear for A.C. servomotor whereas it is non-linear for general purpose induction motor.
3. Rotor is generally drag cup type or squirrel cage type.
4. Ac servomotor is an example of type 1 order 2 system.

8.17 ON-OFF Controller

For a control of an ON-OFF controller is active only when process variable crosses the set point in the positive or negative direction. When the process variable is greater than set point the controller is OFF and when *PV* is less than set point the controller is ON. The transfer curve for the controller is given in figure(a).

The figure suggests that correcting action is taken by the controller only when manipulated or error i.e.

(PV – set point), is non zero. The gain of these types of controller is infinity for ideal case and greater than 1000 for practical case. The process reaction curve of ON-OFF controller shown in figure (b).

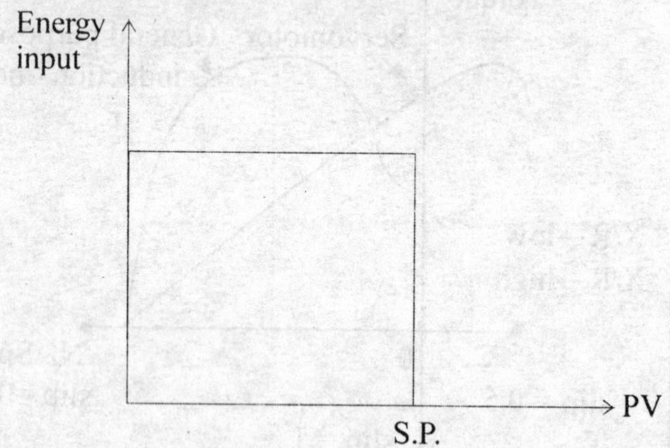

Figure (a): Transfer curve for ON-OFF control

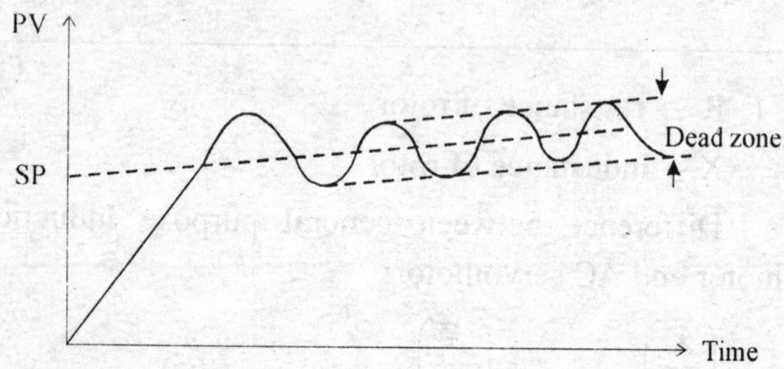

Figure (b): Process reaction curve for ON-OFF control

For ideal ON-OFF controller the dead zone indicated in figure (b) is 0, whereas, in practical based the dead zone is 1% of the set point.

8.18 CASCADE, P

The inherent interaction which occurs between two interactive control variables like flow rate and level in a level control experiment by input flow and output flow and controlling the input flow valve i.e. in a compound control loop, sometimes helps the user with better and efficient control scheme. One way to realize the efficient control scheme is to provide set point in one control loop where the set point is determined by the measurement of other interactive variable. Figure 6 illustrates the essential feature of a cascade control.

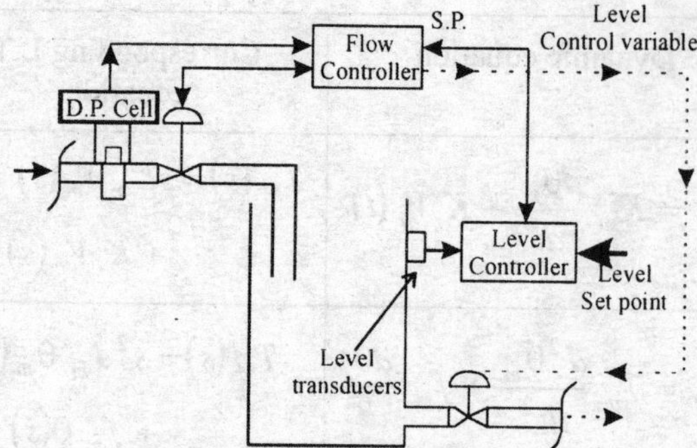

Figure: Cascade control scheme

Hence level and flow rate are two distinct process variables which are interactive in the true sense i.e. if input flow increases and output flow rate decreases, the level increases. In cascade controller, which is labeled as the outer control loop The level controller output is used as the set point to the inner controller which gives flow control. As outer loop control variable changes, the error signal for level i.e. input to the flow controller as set point affects a change in set point to the inner loop. Even though the measured value of the inner loop has not changed, the inner loop experiences an error signal and thus there is new output due to change in set point. Cascade control provides better control of outer variable then the single variable loop system. The block diagram of cascade control is shown in fig .7.

Advantages of Cascade control system over direct control

It is possible to control the level of the bath by applying direct control of level and controlling the input valve by the controlling signal from flow controller. In this direct control, the change in upstream load causes change in flow rate and with input flow rate change, level changes. Hence, level change is a second stage effect and the system cannot respond until the level has actually been changed. This is the problem of direct control of level. Whereas in Cascade control upstream load change i.e. input flow rate change, is never transparent to the level controller and flow control system regulates such changes before they appear as substantial change in level. Hence cascade control loop is taken and more efficient than direct control. In a different version, a disturbance in the outer control loop is faster and more efficient than direct control. In a different version, disturbances in the outer control loop is taken care by the inner controller before it can influence.

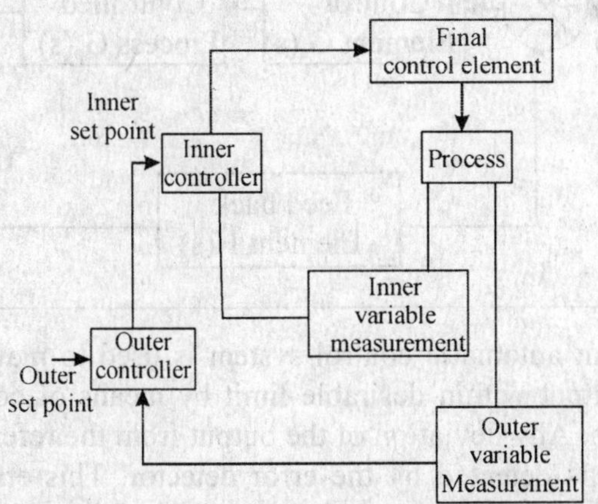

Figure: Block diagram of generalized cascade control system

The outer variable particularly in the case where flow loop is more sluggish than level loop. To add, in a sluggish process, time constant of PV is large where as the bandwidth is small. Thus, inner loop reduces the response time of the outer loop.

Though, it is a disadvantage of Cascade control that, Cascade control only shall be applied to control due to the presence of suitable intermediate variable.

Feed forward control: Drum boiler control

Drum boiler control is one type of feed forward control. In particular, feed forward control may be tried where (1) the disturbance is easily measurable (2) Band width of the disturbance shall not be too large with respect to bandwidth of the control system i.e. there shall be matching of bandwidth between the two.

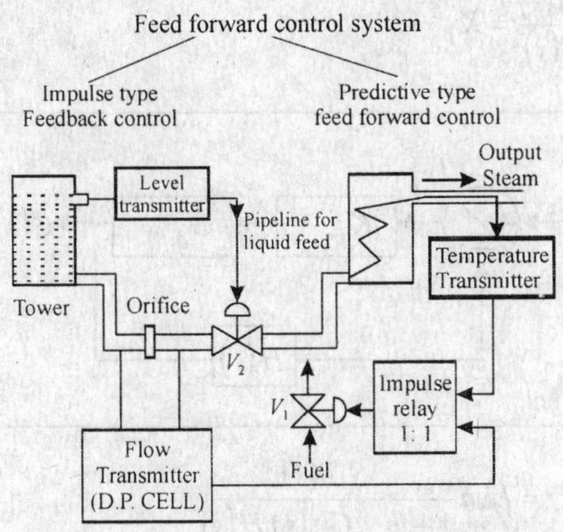

Figure: Impulse type feedback control

Figure shows a control experiment where the Reoile heater heats up the liquid flow from the tower and the objective is to control the flow rate of liquid from the tower and as good as the temperature of the output steam. As the liquid flow from tower to reoiled of depend on the level of liquid in the tower, thus, the change in level in the tower acts as an disturbance in the process.

Hence, the process variables to be measured are level in the tower, flow from tower to reboiler and the temperature of the reboiler output, where as, the variable to be controlled is the fuel flow, to reboiler heater.

With the inclusion of the feed forward control, the output of temperature is fed to the impulse relay (1:1) whose other input is the feed forward signal from flow transmitter. The feed forward impulse relay, output controls the fuel flow to reboiler heater. The level transmitter output directly controls the valve V_2 which controls the liquid flow from tower to teboiler.

The scheme described is a completely an open loop feed forward system i.e. not a feedback closed loop control, but it shows a tight control on any disturbance is the liquid level in tower as any change in level is reflected to impulse transformer which controls the fuel flow to reboiler heater.

Predictive feed forward control

Three element control for drum boiler.

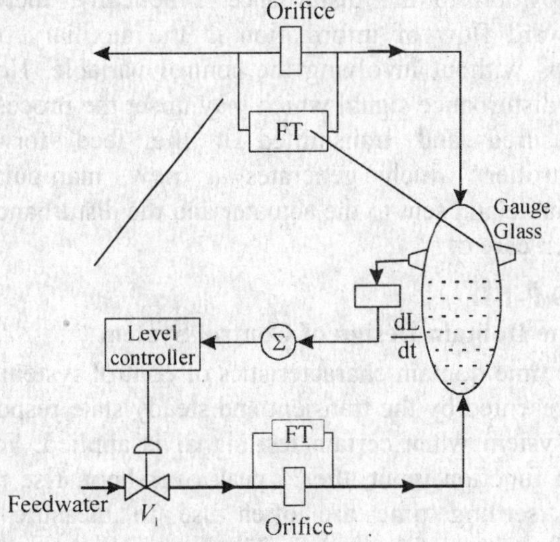

FT: Flow transmitter. LT: Level transmitter.

Figure: Drum boiler contorl

The drum boiler control system is a three element control system where feed water flow, Level in drums and steam flow and the three elements. When the steam pressure is higher then feed water pressure the level bugler where as the steam pressure in higher. Then feed water pressure the level shrinks. Hence feed forward signal is generated from the rate of change of drum boiler level $\left(\frac{dl}{dt}\right)$ i.e. the disturbance, and the disturbance is taken care of only when the disturbance reaches the output by controlling the feed water flow. The flow transmitters in the feed water line and steam line give a signal proportional to differential pressure which are proceed with level disturbance to output the control signal to value v.

This is also an open loop control system but The proven disturbance is taken care of by the feed toward signal from measurement of disturbance of level which is reflected in the control signal and corrective action is taken.

Advantages of feed forward control

Thought feed forward control is an open loop control, it consists of measurement sensors to measure of disturbances and they feed forward controller in some time in called the load compensator. The sense of the output from feed forward compensator should be in opposition to the disturbance. Practically, there is forward flow of information in the auxiliary open loops without involving the control variable, Hence the disturbance signal which may upset the process is measured and transmitted it the feed forward controller which generates a new manipulated variable and sent to the actuator and the disturbance is taken care of.

8.19 P-I-D, Time Domain Design of Control System

The time domain characteristics of control system are represented by the transient and steady state response of system when certain test signal is applied. For a step function input, the % peak overshoot, rise time and settling time are often use to measure the performance of the system. The damping ratio and the natural un-damped frequency may be use to indicate the relative sterility of the system.

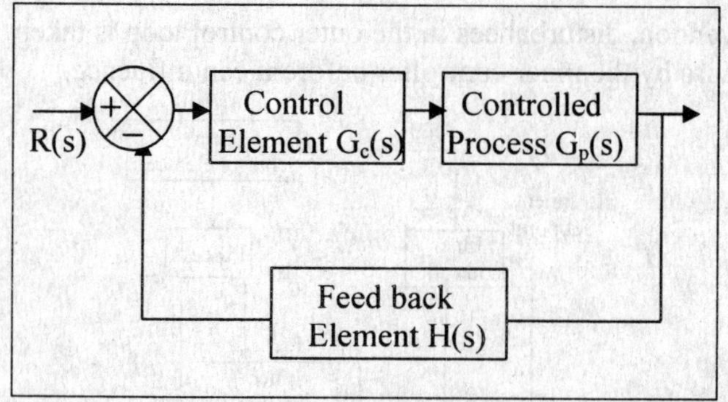

An automatic control system is used to maintain its output within desirable limit by means of control action. Any deviation of the output from the reference input is detected by the error detector. This error is used a actuating signal for control action through. The controller signal for control action through the controller. Various control actions are used for improving transient and steady state response of control system.

Proportional Control Action

There is a continuous linear relationship between the output and the input of the control element.

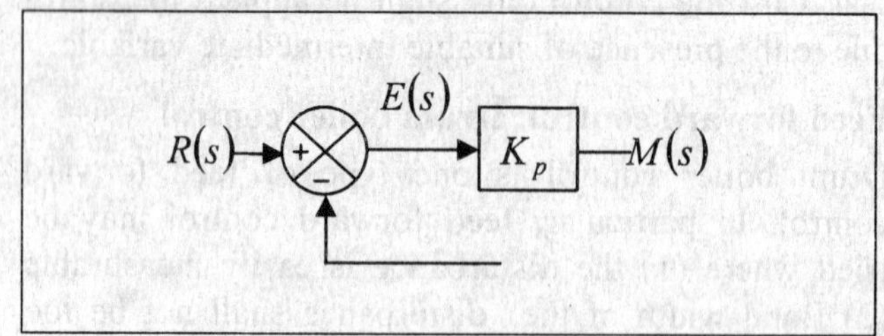

$$m(t) = K_p\, e(t)$$

$$\frac{M(s)}{E(s)} = K_p$$

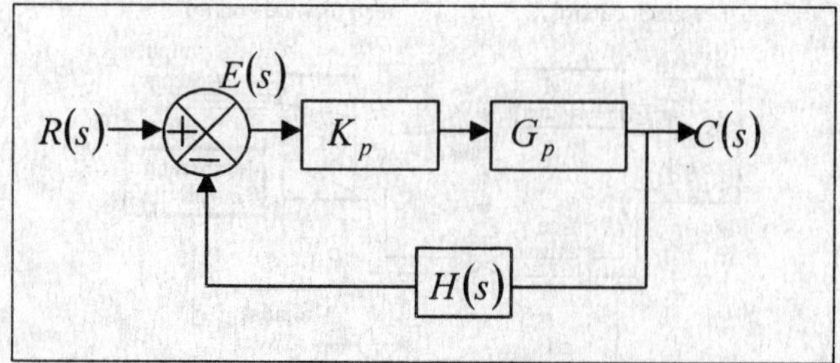

$$E_{ss} = \lim_{s \to 0} \frac{S\, R(s)}{1 + K_p\, G_p(s) H(s)}$$

- Increase in forward path gain reduces the steady state error.
- It has very simple implementation.
- If forward path gain is increase then *system is made faster,* steady state error educes but maximum over shoot i.e. offset is increased.
- By proper choice of forward path gain adjustments can be made between maximum overshoot and see.

Derivative Control Action

In this the control action the output is proportional to the time rate of change of input.

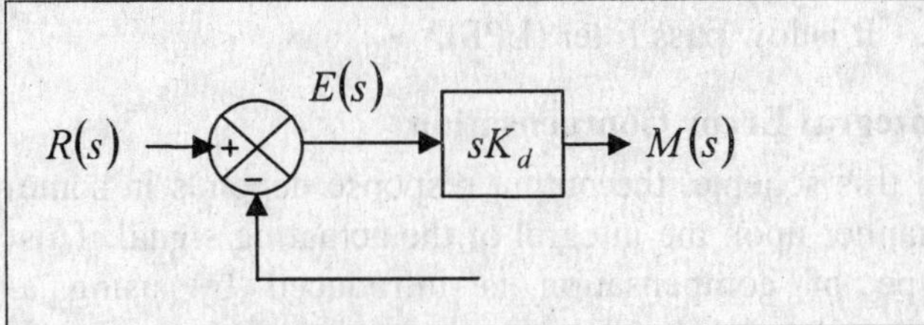

$$m(t) = K_d \frac{de(t)}{dt}$$

$$\frac{M(s)}{E(s)} = sK_d$$

- Damping increases and setting time decrease.

Integral Control Action

In this control action the output is proportional to time integral of the input.

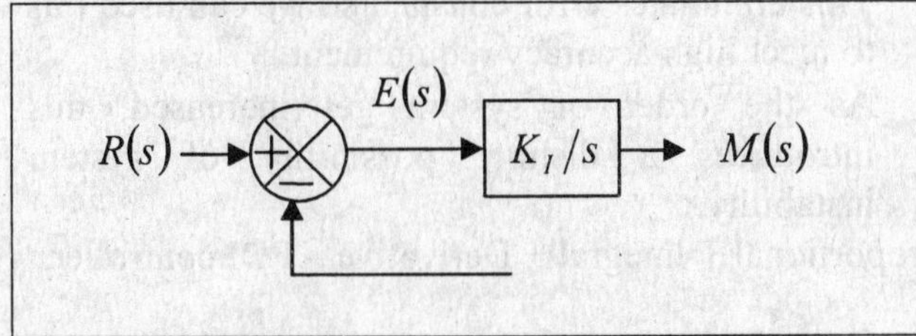

$$m(t) = K_I \int e(t)\,dt$$

$$\frac{M(s)}{E(s)} = \frac{K_I}{s}$$

- Highly steady state accurate.

Proportional plus Derivative controller (PD)

The output of the controlling system is proportional to a linear rate change.

$$\frac{M(s)}{E(s)} = K_p\,(1 + s\,\tau_d)$$

where, τ_d = delay time

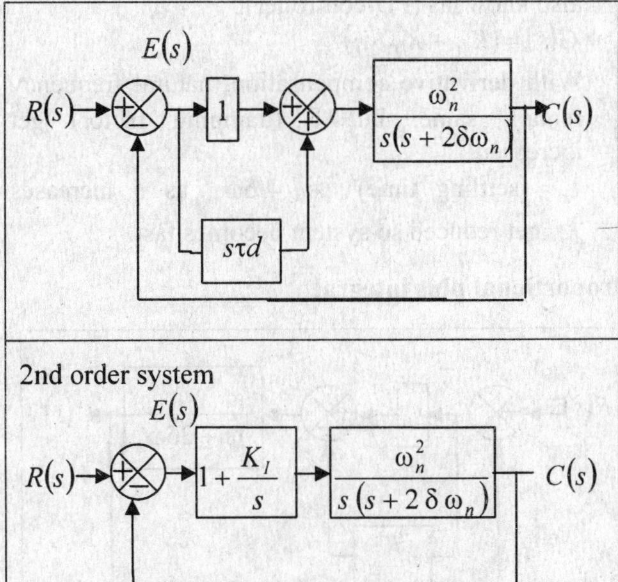

$$\frac{C(s)}{E(s) + E(s)\,s\tau_d} = \frac{\omega_n^2}{s(s + 2\delta\,\omega_n)}$$

$$\frac{C(s)}{E(s)} = \frac{(1 + s\tau_d)\,\omega_n^2}{s(s + 2\delta\,\omega_n)} = G(s)$$

$$\frac{C(s)}{R(s)} = \frac{G}{1+G} = \frac{(1 + s\tau_d)\,\omega_n^2}{s^2 + (2\delta\,\omega_n + \tau_d\,\omega_n^2)s + \omega_n^2}$$

By PD controller one zero gets added up in the transfer function.

Characteristic equation of new system

$$s^2 + (2\delta\,\omega_n + \tau_d\,\omega_n^2)s + \omega_n^2$$

1. SSE is not affected
2. It increases damping
3. It decrease overshoot
4. Rise time faster
5. Proportional band unaffected
6. It is a high pass filter HPF.

1. Derivative error compensation

A system is said to posses derivative error compensation when the generation of its output depends in some way on the rate of change of actuating signal.

- This type of compensation is easily introduced by using an containing two term, one proportional to the derivative of the actuating signal and the other proportional to the actuating signal itself. This is also know as (PD-controller).
- $G(s)=(K_A+K_D\cdot s)$
- With derivative compensation, natural frequency remain same, but δ (damping factor) get increased.
- t_s = (settling time) = $4/\delta\omega_n$ as δ increases t_s get reduced so system becomes fast.

Proportional plus integral

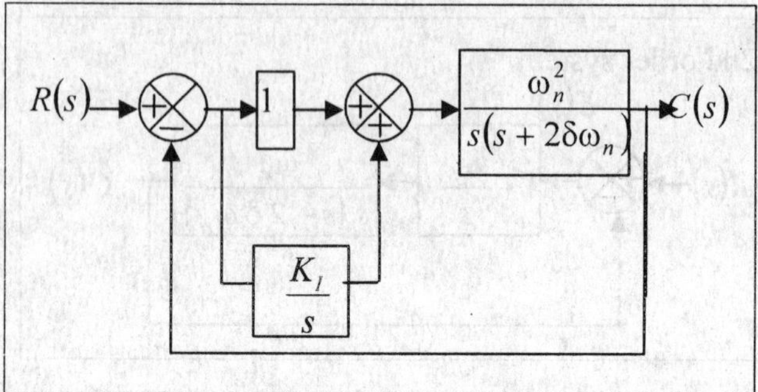

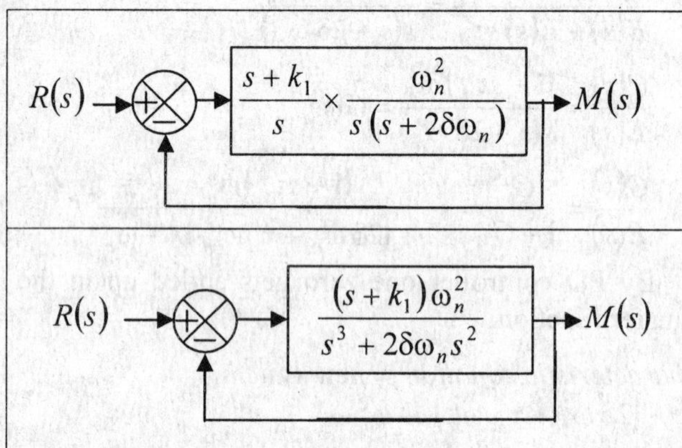

Close loop Transfer frequency

$$\frac{C(s)}{R(s)}=\frac{(1+K_I)\omega_n^2}{s^3+2\delta\omega_n s^2+\omega_n^2+K_I\,\omega_n^2}=G(s)$$

$$E(s)=\frac{R}{1+G}$$

$$E(s)=\frac{s^2(s+2\delta\,\omega_n)\cdot R(s)}{s^3+\left(s+2\delta\omega_n+\omega_n^2\,(s+K_I)\right)}=G(s)$$

$$e_{ss}=\lim_{s\to 0} s.\,E(s)$$

$$=\lim_{s\to 0}\frac{s\cdot s^2(s+2\delta\omega_n)\cdot R(s)}{\left[s^2(s+2\delta\omega_n)+\omega_n^2(s+K_I)\right]}$$

Case 1: $R(s)=$ Step function

$=1/s$; $e_{ss}=0$

Case 2: $R(s)=$ Ramp function

$=1/s^2$; $e_{ss}=0$

Case 3: $R(s)=$ Accelerating function

$=1/s^3$; e_{ss} = some constant

1. Steady state error can be reduced by in corporating integral control.
2. The order of the system increases and higher order system are more likely to unstable.
3. It is low pass filter (LPF).

Integral Error Compensation

In this scheme, the output response depends in some manner upon the integral of the actuating signal. This type of compensation is introduced by using a controller which produces an output signal consisting of two terms, one proportional to the actuating signal and the other proportional to its integral. This is know as proportional plus integral controller **(PI controller)**.

- In this system, we have two blocks one is proportional and other is integral block

  $G(s)=\frac{(K_p+K_1)}{s}$u
- Integral error compensation changes a second order system into a third-order.
- This eliminates error constant so we can used this to meet high accuracy requirement.
- As the order of system get increased, this introduces a distinct possibility of system instability.

Proportional + Integral + Derivative – PDI controller.

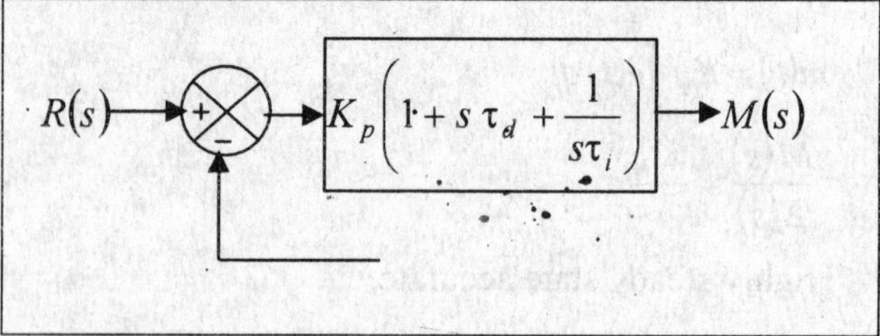

$$G(s)=K_p+K_d s+\frac{K_I}{s}$$

Combines all the advantage of proportional derivative and integral control action. It is BPF or BSF depends upon controller parameter.

8.20 FEED FORWARD AND DERIVATIVE CONTROLLER

Derivative (rate) Feed back control

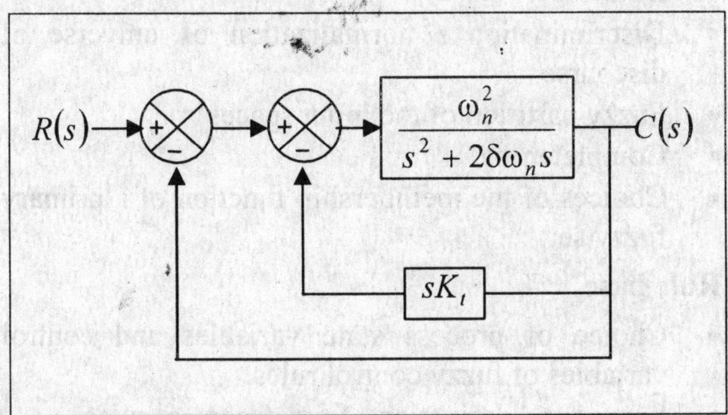

$$\frac{C(s)}{E(s)}=\frac{\omega_n^2/s(s+2\delta\omega_n)}{1+\dfrac{\omega_n^2(sK_t)}{s(s+2\delta\omega_n)}}$$

$$=\frac{\omega_n^2}{s^2+2\delta\omega_n s+K_t\,\omega_n^2 s}=G(s)$$

$$\frac{C(s)}{R(s)}=\frac{\omega_n^2}{s^2+\left(2\delta\omega_n+K_t\omega_n^2\right)s+\omega_n^2}$$

Characteristic equation:

$$s^2+\left(2\delta\omega_n+K_t\omega_n^2\right)s+\omega_n^2=0$$

If $R(s)=\frac{1}{s^2}$

$$e_{ss}=\frac{\omega_n\,K_t+2\delta}{\omega_n}$$

Steady state error is increased by incorporating feed back control.

Actual damping with feed back control is increased which decreases the overshoot.

8.21 FUZZY CONTROLLERS

Fuzzy Logic Control

Fuzzy logic is a very powerful method of reasoning when mathematical models are not available and input data are imprecise. Their application, mainly to control is being studied throughout the world by control engineers.

Fussy Sets Vs, Crisp Sets

The main objective behind fuzzy logic is to represent and reason with some particular form of knowledge expressed in a linguistic form. However, when using a language-oriented approach for knowledge representation, one has to build a conceptual framework to tackle its inherent vagueness. So let's start with a brief introduction to fuzzy-set concepts.

The traditional sets are based on a two-valued logic: objects are either members or not members. Every individual object is assigned a member shi value μ of either 1 or 0 that discriminates between members and non-members of the crisp set.

Fuzzy Set Theory and Operations

In this section, we shall dwell in brief on some of the basic elements of fuzzy set theory that is required in fuzzy control applications. Consider a universe of discourse. μ, that consists of elements which shall be denoted by x. A fuzzy set A can be defined in μ as follows.

Operation

1. Set Union: The union of two fuzzy sets A and B in μ is a fuzzy set $A\cup B$, is defined as follows,

$$A\cup B=\{(\mu_{A\cup B}(x),x)\,1\,x\in\mu\}$$

2. Set Intersection: The intersection of two fuzzy sets A and B in μ is a fuzzy set $A\cap B$, is defined as follows.

$$A\cap B=\{(\mu_{A\cap B}(x),x)\,1\,x\,\varepsilon\mu\}\text{ where}$$

$$\mu_{A\cap B}(x)=\min\,(\mu_A(x),\mu_B(x)\,)\;\forall x\varepsilon\,\mu$$

Set complement. The complement of fuzzy sets A in μ is a fuzzy set $\overline{A}$, and is defined as follows.

$$\overline{A}=\{(\mu_{\overline{A}}(x),x)|x\,\varepsilon\mu\}\text{ b}$$

Where,

$$\mu_{\overline{A}}(x)=1-\mu_A(x)\;\forall\;x\,\varepsilon\,\mu$$

Fuzzy Relations

If x and y are universal sets, then a fuzzy relation R on $x\times y$ is defined as follows,

$$R=\{(\mu_R(x,y),(x,y))|(x,y)\varepsilon x\times y\}$$

Composition: The composition of relations is the net effect of applying one relation after another, For the case of two binary relation P and Q, the composition of their relations as the binary relation R is given by,

$$R(A,C) = Q(A,B) \text{ O } P(B,C)$$

Where
R (A, C) is relation between A and C.
Q (A, B) is relation between A and B.
P (B, C) is a relation between B and C
A, B, C are fuzzy sets and o is the composition operator. The relation of R is the same as applying R first, followed by Q. In terms of membership function.

$$R = \{(\mu_R(a,c),(a,c)) | a \,\varepsilon\, A, c \,\varepsilon\, C\}$$

Where $\mu_R(a,c) = \max(\min(\mu_P(b,c)))\, b\varepsilon\, B$

Thus the composition is commonly defined by the min – mix matrix product or simply minmax.

A fuzzy Logic based controller (FLC) is developed using the concept of prevailing conventional controllers. Consider of example an ordinary PID regulator. Its linear behavior can be described by the model:

$$u(t) = k\left[e(t) + \frac{1}{\tau_i}\int^t e(t)dt + \tau_d\left(\frac{de(t)}{dt}\right)\right]$$

Where u(t) is the control signal and $e(t)$ the error signal. PID control can be understood very well from this linear equation. To obtain a good PID regulator, it is also necessary to consider operator interfaces, operational issues likes switching smoothly between manual and automatic operations, transients due to parameter changes, the effects of non – linear actuators, wind up of the integral term, maximum and minimum selectors, etc. An operational industrial PID regulator consists of an implementation of the above model and heuristic logic that takes care of these issues.

It can thus be concluded that practical solution to such mundane problems likes care of these are not done by theory alone but that heuristics play an import role. The discrete form of equation can be equation can be represented as

$$u(k) = u(k-1) + \delta u(k)$$

Where $\delta u(k) = f(e(k), \delta e(k)...)$

I.e.. Incremented controller output is some function of error and other terms related to present and past status of error.

Once the system behavior is thoroughly studied. Comprehensive performance objectives of the controller can be speculated. Then the heuristics rules that de-scribe the equation can be formulated by the help of intelligent trial and error science as well as self institution learned from experience. This forms the basis for the design of a FLC.

The principal design parameters for an FLC are the following:

- Fuzzification strategies and the interpretation of a fuzzification operation operator
- Data base:
- Discrimination / normalization of universe of discourse.
- Fuzzy partition of the input spaces.
- Completeness.
- Choices of the membership function of a primary fuzzy set;

Rule base

- Choice of process state variables and control variables of fuzzy control rules.
- Source and derivation of fuzzy control rules.
- Types of fuzzy control rules.
- Consistency interactivity, completeness of fuzzy control rules;
- Fuzzy inference mechanism.
- Defuzzification strategies and the interpretation of a defuzzification operator.

The section deals with the design of components of a FLC. The above listed components are necessary for the construction of any FLC. Fuzzification implies the process of transforming the crisp values of the inputs to a controller, to the fuzzy domain. The fuzzified values now fire the rule base and generate a so called fuzzified output. As a final step, the fuzzified output is defuzzified to yield the crisp controller output. figure gives an idea of the construction of a fuzzy logic based controller. Now, let's deal with each of the components in detail.

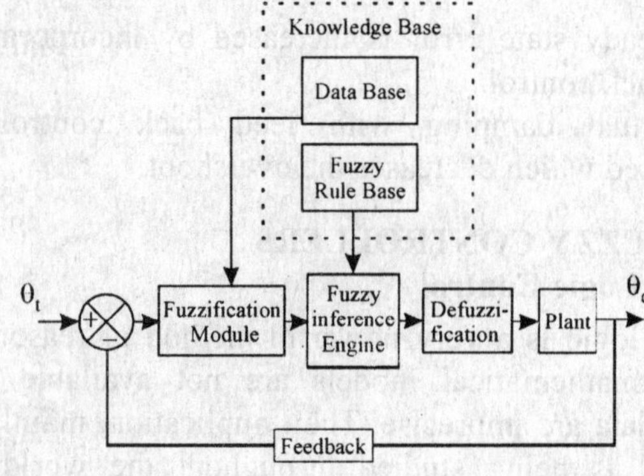

Figure: Fuzzy logic based control system

Fuzzification Module

The fuzzification module performs the following function

- Measure the values of input variables.
- Performs a scale transformation which maps the physical values of the current process state variable into a normalized universe of discourse. It also maps the normalized value of the control output variable onto its physical domain is used then there is no need for FM. A normalized universe implies the knowledge of input/output space via appropriate scale mappings. Due to normalization, a well – formed terms set can be achieved. If this is not the case, or a non normalized universe is used, the terms could be asymmetrical and unevenly distributed in the universe.
- Performs the so called fuzzification which converts a point – wise current value of a process state variable into a fuzzy set, in order to make it compatible with the fuzzy set representation of the process state variable in the rule – antecedent.

Defuzzification Module or Technique

The function of a defuzzification module are as follows:

- Performs the so called defuzzification which converts the set of modified control output values into a single point- wise value of the control output onto its physical domain. This step is not needed if non-normalized fuzzy sets are used.

A defuzzification strategy is aimed a producing a non – fuzzy control action that best represents the possibility of an inferred fuzzy control action.

There many procedures outlined in the literature of defuzzification which are, center of gravity area, center of sums, center of largest area, first of maxima, middle of maxima, and height. Of these the center of gravity is the most efficient in that it gives a defuzzified output which conveys the real meaning of the action that had to be taken at then instant.

The COG strategy generates the center of gravity of the possibility distribution of a control action action. In the case of a discrete universe, this method yields.

$$F_o = \frac{\sum_{i=l}^{m} \mu(i).\omega_i}{\sum_{i=l}^{m} \mu(i)}$$

Where m is the number of quantization levels of the output, $\mu(i)$ represents the "truth" of the ith rule and w_i represents the action that i^{th} rule would dictate.

In the temperature controller example considered in the Section 14.2.4, Rule No. 13 would make zero change, while Rule No. 14 would make +2.5% change. The actual output is an average of the active rule outputs weighted by their corresponding truths. Thus the output is given by

$$F_o = \frac{0\times 0.2 + 2.5\times 0.8}{0.2+0.8}$$
$$= 2\%$$

For this particular example, the sum of the truths of activated tules. Rule 13 and 14 is unit: $0.8 + 0.2 = 1.0$. This is neither a necessary, nor a desirable, not common event. It is simply the happenstance of this example.

MCQ Set–1

1. If the open loop transfer function of negative feedback control system is:

$$G(s)H(s) = \left(\frac{K}{(s+1)^3}\right)$$

then the gain K for a closed loop pole to be at $\left(-\frac{1}{2}+j\frac{\sqrt{3}}{2}\right)$ is:

(a) 1 (b) 2

(c) 3 (d) 4

2. In **sizing a control valve**

(a) A constant flow through the valve is assumed

(b) A constant pressure drop across the valve is assumed

(c) An equal percentage opening of the valve is assumed

(d) A constant pressure downstream and a constant flow through the valve is assumed

3. A **temperature control** system is usually very sluggish to improve its dynamics if:

(a) A PI controller can be used
(b) An I controller can be used
(c) A PID controller with large I and a negligible D action can be used
(d) A PD controller can be used

4. In a **seismic pick-up for getting** an output proportional to acceleration it is desirable to have:
(a) A natural frequency very small in comparison to frequency of input and a damping ratio around 0.7
(b) A natural frequency very large in comparison to frequency of input and a damping ratio around 0.7
(c) A natural frequency equal to the frequency of input signal and a damping ratio around 0.7
(d) A natural frequency very small in comparison to frequency of input and a damping ratio higher that unity.

5. A **First order system with a proportional controller** exhibits an offset to step input. The offset can be reduced by:
(a) increasing the gain
(b) adding integral mode
(c) adding derivative mode
(d) decreasing the gain.

6. A plant is controlled by a **proportional controller**. If a **time delay element** is introduced in the loop then
1. Gain margin remains the same
2. Phase margin remains the same
3. Gain margin decreases
4. Decreasing the gain.

Chose the correct option
(a) 1, 2 are correct (b) 2, 3 are correct
(c) 2, 3 are correct (c) 3, 4 are correct

7. A **synchro transmitter** is used with a synchro receiver for the purpose of
(a) Amplification (b) Error detection
(c) Remote sensing (d) Addition

8. A current of 5 mA in a standard process controller indicate while zero current output indicates.
(a) Live zero (b) Dead zero
(c) Full scale (d) Threshold

9. A **cascade controller** is used when the process
(a) Gain is too small
(b) Gain is too large
(c) Has widely different time constant
(d) oscillation at the output is not permitted

10. The minimum number of states required to describe the network shown in Figure.

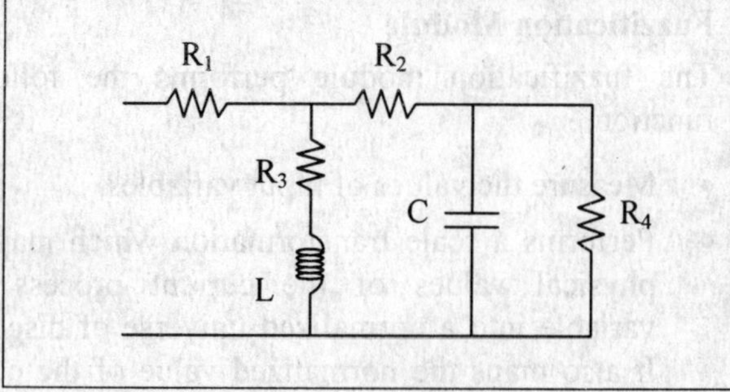

(a) 1 (b) 2
(c) 3 (d) 4

11. The root, locus plot of a unity feedback system is shown in figure. The system transfer function has the form:
(a) $\frac{K}{s(s+4)(s+8)}$ (b) $\frac{K(s+8)}{s(s+4)}$
(c) $\frac{K(s+4)}{s(s+8)}$ (d) $\frac{Ks}{(s+4)(s+8)}$

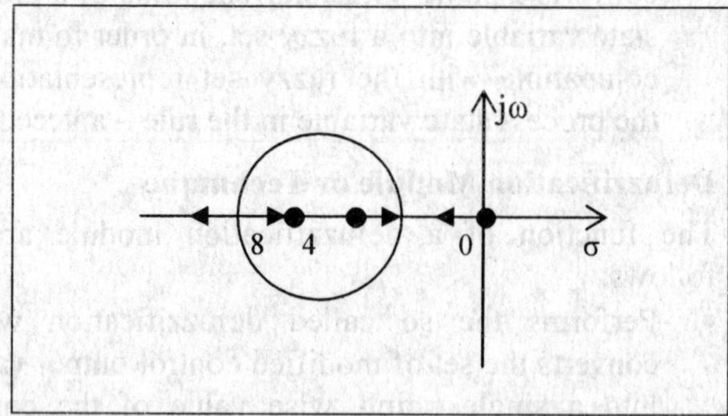

(a) $\frac{K}{s(s+4)(s+8)}$ (b) $\frac{K(s+8)}{s(s+4)}$
(c) $\frac{K(s+4)}{s(s+8)}$ (d) $\frac{Ks}{(s+4)(s+8)}$

12. A unity feedback closed loop second order system has a transfer function $[81 / (s^2 + 0.6\,s + 9)]$ and is excited by a step input of 10 units. The **steady state error** of the output is:
(a) 10 (b) 0.0
(c) 1.0 (d) 0.1

13. Match the List I elements and the List II transfer function and select the correct answer using code

List I	List II
A. Pneumatic actuator	1. e^{-sT}
B. Transport delay	2. K

C. A servo-moter	3. $\left(\dfrac{K}{as^2+bs+c}\right)$
D. A nozzle-flapper	4. $\left(\dfrac{K}{s(1+as)}\right)$

Codes:

	A	B	C	D
(a)	1	2	3	4
(b)	3	1	4	2
(c)	1	2	4	3
(d)	4	1	2	3

14. The break way point of the root locus from the real axis for a closed-loop system with a loop gain

$$G(s)\,H(s) = \frac{K\,(s+10)}{(s+2)\,(s+5)} \text{ lies}$$

(a) Between – 2 and origin

(b) Between – 2 and –5

(c) Between – 10 and – ∞

(d) at – ∞

15. For a unit step input, a system with a closed-loop transfer function of $G(s) = \dfrac{20}{s^2+2s+5}$ has a steady –state output of

(a) 10 (b) 5
(c) 2 (d) 4

16. The condition for suability of a closed-loop system with a characteristic equation $s^3 + bs^2 + cs + 1 = 0$ with positive co-efficient is:

(a) $b + c > 1$ (b) $bc > 1$
(c) $b = c$ (d) $b > c$

17. Of the following transfer functions, the function that has a minimum phase characteristic is:

(a) $\dfrac{1+s}{1-s}$ (b) $\dfrac{1-s}{1+s}$

(c) $\dfrac{e^{-3}}{s}$ (d) $\dfrac{s}{(s+1)\,(s+2)}$

18. For a unity feedback system with open loop gain $G(s) = [9 / s(s+3)]$, the closed loop system will have a damping ratio of

(a) 0.33 (b) 0.5
(c) 1.0 (d) 3

19. Match the List I characteristic with the List II system transfer function and select the correct answer using the code:

List I	List II
A. All pass	1. $e^{-0.5s}$
B. Transport delay	2. $\dfrac{1-s}{1+s}$
C. Lead compensator	3. $\dfrac{1}{s^2+2s-3}$
D. Unstable	4. $\dfrac{s}{s^2+s+100}$
	5. $\dfrac{10(1+0.3s)}{1+0.1s}$

Codes:

	A	B	C	D
(a)	1	2	5	3
(b)	5	2	4	3
(c)	2	1	5	3
(d)	5	1	2	3

20. The transfer function of a control system is given by $\dfrac{Cs}{(Rs)} = \dfrac{25}{s^2+6s+25}$.

The first maximum value of the response occurs at a time t_{max} given by

(a) [π/8] (b) [π /4]
(c) [π/2] (d) π

21. The open – loop transfer function of a feedback Control system is $[1/(s+1)^3]$. The gain margin of the system is:

(a) 16 (b) 8
(c) 4 (d) 2

22. The transfer function of a basic PD controller is given by (K's are real constants).

(a) $K_0 + [K_1/s] + K_2s$ (b) $K_0 + K_2s$
(c) $K_1s + K_2s$ (d) $K_0 + [K_1/s]$

23. The two secondary coils of an LVDT have wrongly been connected as shown in figure. Then the Input-output relationship would be:

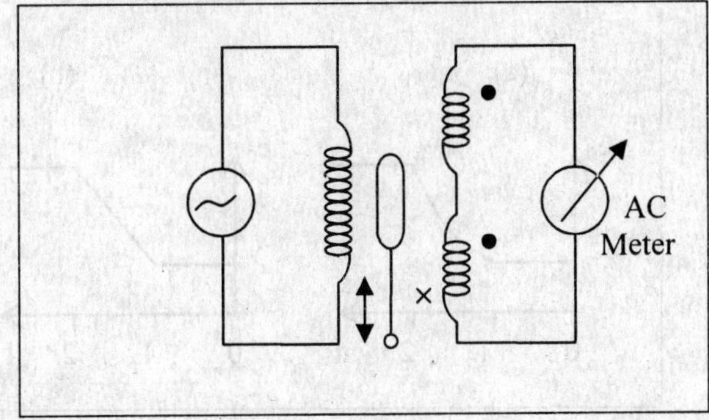

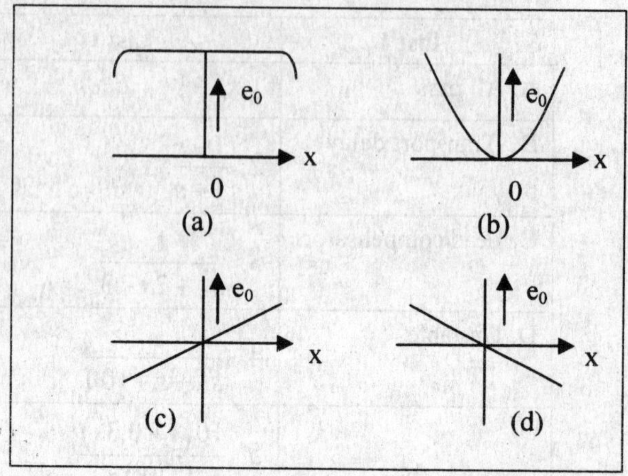

24. The phase margin of the system for which loop gain $GH(j\omega) = 1/(j\omega + 1)^3$ is

(a) $-\pi$ (b) π
(c) 0 (d) $\pi/2$

25. The number of roots, lying outside the unit circle, of the polynomial

$F(z) = 4z^2 - 8z^2 - z + 2$ is

(a) 3 (b) 2
(c) 1 (d) 0

26. The input error voltage to a PD-controller, e_i is shown in figure. The output voltage of the controller e_o is:

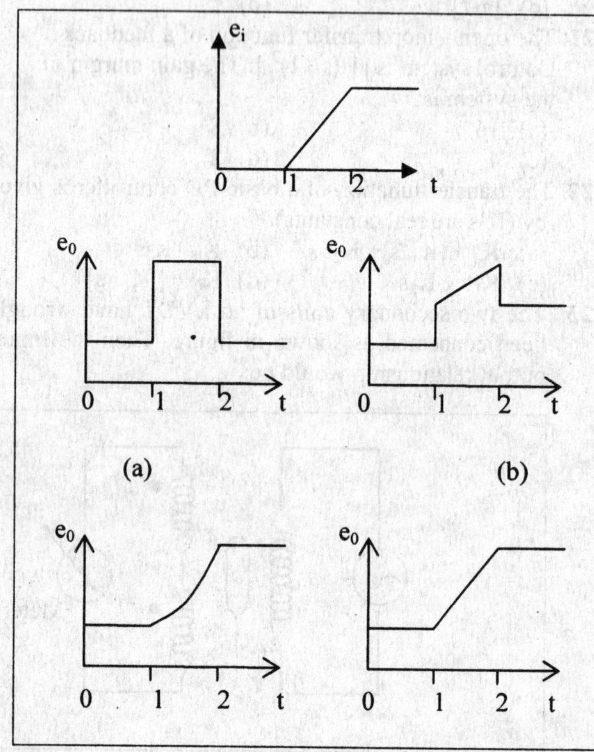

27. Match the List I transfer functions with the List II systems and select the correct answer using the codes:

List I	List II
A. All pass filter	1. e^{-as}
B. Transport delay	2. $\dfrac{1-s}{1+s}$
C. Lag network	3. $\dfrac{1+as}{1+bs}, a<b$
D. Servomotor	4. $\dfrac{K}{s(1+as)}$

Codes:

	A	B	C	D
(a)	2	1	3	4
(b)	4	3	2	1
(c)	3	1	4	2
(d)	2	3	4	1

28. The transfer function between y_2 and y_1 in figure.

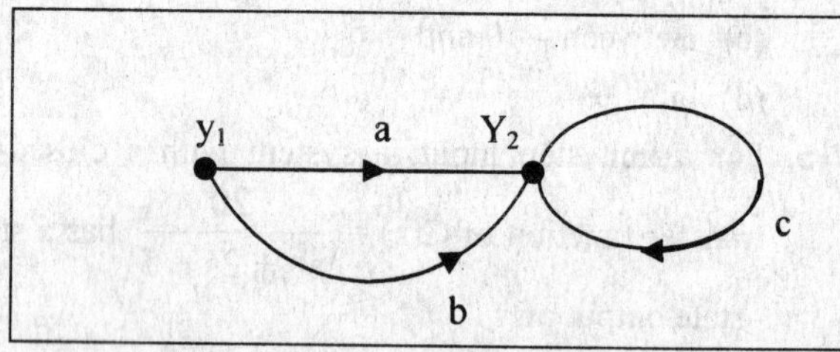

(a) a + b (b) (a + b) c
(c) [(a + b) / (1- c)] (d) [(a + b) / (1 + c)]

29. In control system design, gain and phase margins are usually provided to.

(a) Account for the uncertainties in the system.
(b) Make the system respond fast.
(c) Reduce the overshoot in step response
(d) Reduce the steady state offset.

30. First order system with a proportional controller in the negative feedback loop has an offset to a step/input. This offset can be eliminated by

(a) Adding a derivative mode to the controller
(b) Adding an integral mode to the controller
(c) Decreasing the magnitude of the gain of the proportional controller
(d) Adding a delay in the controller loop

31. For a unity feedback system with open loop transfer function $G(s) = \dfrac{9}{s(s+2)}$, the damping ratio is

(a) 1/3 (b) 1/2
(c) 1 (d) 2

32. The asymptotic Bode plot for the gain magnitude of a minimum phase system G(s) is shown in figure:

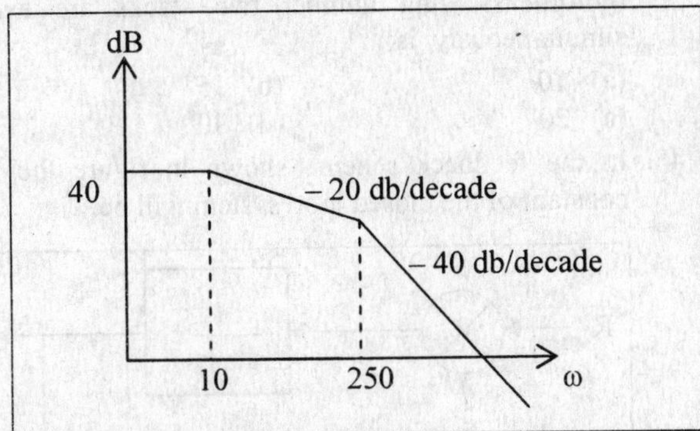

The transfer function is G(s) =

(a) $\dfrac{100}{\left(1+\dfrac{s}{10}\right)\left(1+\dfrac{s}{250}\right)}$

(b) $\dfrac{40}{s\,(s+250)}$

(c) $\dfrac{100}{(s+10)\,(s+250)}$

(d) $\dfrac{100\,s}{(s+1)\,(s+250)}$

33. The proportional gain of a PID controller can be expressed in terms of its proportional band (PB) as,

(a) PB (b) $\dfrac{100}{PB}$

(c) $\dfrac{PB}{100}$ (d) $\dfrac{10}{PB}$

34. Linear variable differential transformer has

(a) Two primary coils connected in phase and a secondary coil.
(b) Two primary coils connected in opposition and a secondary coil.
(c) One primary coil and two secondary coils. connected in phase.
(d) One primary coil and two secondary coils connected in opposition.

35. A second order feedback system is found to be: oscillating with a high frequency. The oscillations

(a) Can be reduced by increasing the proportional action
(b) Can be reduced by increasing the integral action
(c) Can be reduced by increasing the derivative action.
(d) Cannot be reduced.

36. A process in a feedback loop with a proportional controller with ultimate gain k_u = 10 is oscillating at a frequency of P_u = 8Hz. The *Ziegler-Nichols* setting for the proportional controller is

(a) 8 (b) 5
(c) 1.25 (d) 10

37. An **integral control** is used to

(a) Improve the transient response
(b) Reduce the offset
(c) Eliminate the offset
(d) Reduce the settling time

38. A unity feed back system has a forward path transfer function $G(s)=\dfrac{10}{s(s+5)}$. It is subjected to a unit ramp input. The **integral of the squared** error (ISE) is.

(a) Zero (b) Infinity
(c) Always negative (d) 0.5

39. The loop transfer function of a system has a pole in the right half of the s – plane. The **Nyquist plot** makes one clock wise encirclement of (-1, 0) point. The closed loop system is.

(a) Unstable
(b) Stable
(c) Marginally stable
(d) Asymptotically stable

40. The loop transfer function of a system is given by

$$G(s)\,H(s)=\frac{10e^{-0.1s}}{s(s+5)(s^2+4s+16)}$$

When the system is subjected to a **parabolic input** of $t^2/2$, the **steady state error** is.

(a) Zero (b) 0.1
(c) Infinity (d) $e^{-0.1}/0.1$

41. The partial **Routh array** of the characteristic equation of a system is given by.

S^4	1	a	8
S^3	3	12	

The system oscillates with a frequency of 2 rad/s. The value of the parameter'a' of the system is.

(a) 6 (b) 2
(c) 8 (d) 12

42. The transfer function of a PID Controller is given by $G(s)=4\left(1+\dfrac{1}{2s}+0.5s\right)$ as ω tends to infinity

(a) Magnitude of G (jω) tends to zero and phase angle of G (jω) tends to infinity
(b) Magnitude of G (jω) tends to infinity and phase angle of G (jω) tends to zero.

(c) Magnitude tends of G (jω) to infinity and phase angle of G(jω) tends to –90°.
(d) Magnitude tends of G (jω) to zero and phase angle of G(jω) tends to + 90°.

43. The loop transfer function of a system is given by $G(s)H(s)=\frac{10\,e^{-Ls}}{s}$ the phase cross – over frequency is 5 rad/s. The value of the dead time L is

(a) $\pi/20$ (b) $\pi/10$
(c) $-\pi/20$ (d) Zero

44. The forward path transfer function of an unity feedback system is given by.

$$G(s)=\frac{(1+5s)(1+10s)(1+2s)}{(1+s)(1+8s)(1+20s)}$$

If e(t) is the error to a unit impulse input the value of the performance index

$J=\int_0^{\infty} e(t)\,dt$ is equal to

(a) Zero (b) Infinity
(c) –12 (d) 0.5

45. In a **synchro pair,** the control transmitter excites the three stator windings of the control transformer .The stator winding voltages will have:

(a) Equal magnitudes but different phases
(b) Different magnitudes and different phases
(c) Equal magnitudes and phase
(d) Different magnitudes but equal phases

46. The **steady state error** for the feedback control system shown in figure when subjected to a unit step input is

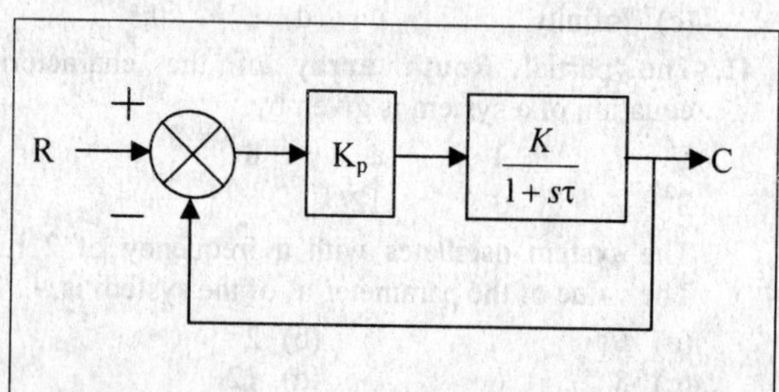

(a) Zero (b) $\frac{1}{1+KK_p}$
(c) $\frac{KK_p}{1+KK_p}$ (d) Infinite

47. A three stack **variable reluctance stepper motor** has 12 numbers of rotor and stator teeth on each stack. The resolution of angular rotation for the stepper motor, assuming number two stacks are excited simultaneously, is:

(a) 10° (b) 15°
(c) 20° (d) 30°

48. In the feedback scheme shown in figure the time constant of the closed loop system will be.

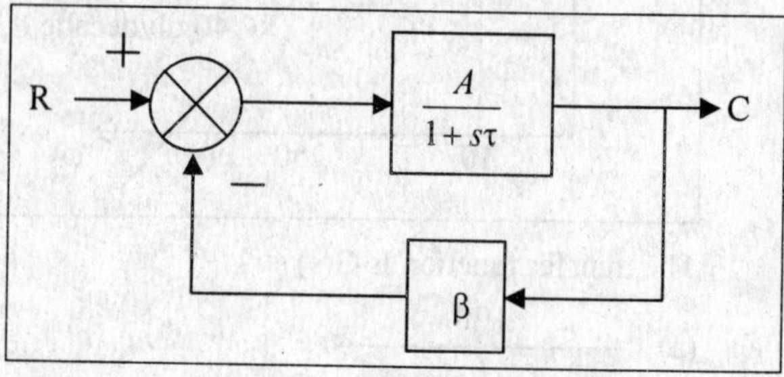

(a) $A\beta\tau$ (b) $[(1+A\beta)\tau]$
(c) τ (d) $\tau/(1+A\beta)$

49. The transfer function of a system is $\frac{A}{(s^2+\omega^2)}$. The **steady state gain** of the system to a *unit step input* is

(a) A/ω^2
(b) 0
(c) ∞
(d) Not possible to be determined

50. The closed loop control system shown in figure has $\tau_1 > 0$. The system will remain stable for all τ_1 in the range

(a) $\tau_1 < 0.5$ (b) $0.5 < \tau_1 < 1.0$
(c) $\tau_1 < 1.0$ (d) $\tau_1 > 1.0$

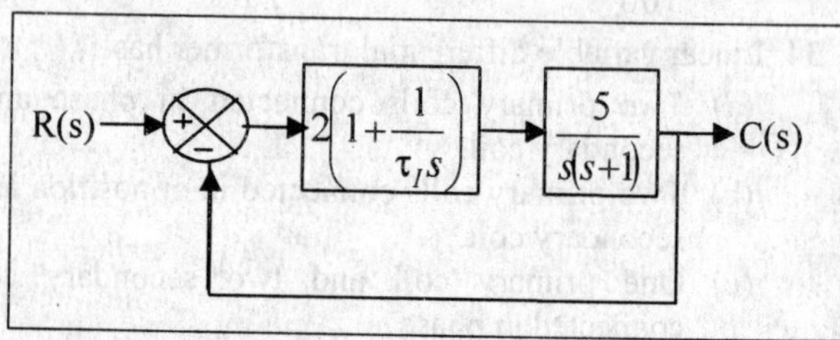

51. In figure K_1 and K_2 are proportional controllers. For the control, system to remain stable, the upper limits of K_1 and K_2 are

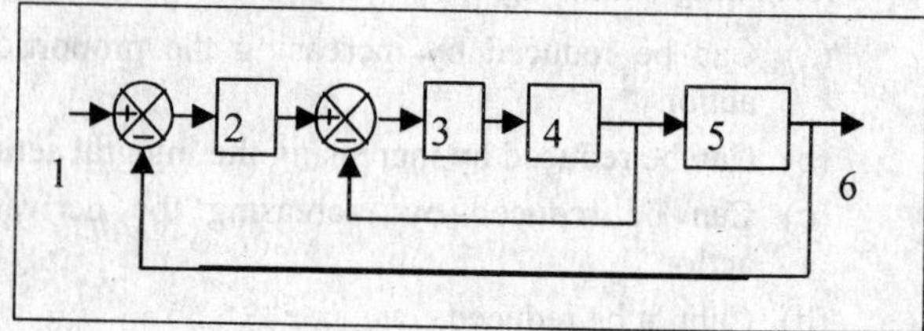

Where, 1 = x(t), 2 = K_2, 3. = K_1, 4. = $\frac{1}{s+1}$

5. = $\frac{8}{(s+2)(s+4)}$ 6. = y(t)

(a) ∞ , ∞ (b) 1, ∞
(c) ∞ , 10 (d) 10, 1

52. The response of a system to a unit impulse is $y(t) = e^{-5(t-1)}$. Which one of the following is the correct statement about the system?
(a) The system is nonlinear
(b) The system is unstable in open loop.
(c) The steady - state gain of the system for a unit step input is 0.2.
(d) The steady state gain of the system for a unit step input is 1.0

53. A PID controller has the transfer function $2 + \frac{0.4}{s}$ with the unit of time expressed in minutes. The parameters **proportional band and reset time** for the above controller are respectively:
(a) 200% and 0.4 min
(b) 50% and 0.4 min.
(c) 200% and 5 min
(d) 50% and 5 min.

54. For a dynamic system, various matrices for a state space model are given by

$$A = \begin{pmatrix} 0 & 1 \\ -2 & -3 \end{pmatrix} \quad B = \begin{pmatrix} 0 \\ 1 \end{pmatrix}$$

and C = [1 0]. The system poles are located at.
(a) –2, –1 (b) –2, –3
(c) –3, –1 (d) –1, 0

55. The transfer function of a system is given by $\frac{Y(s)}{X(s)} = \frac{e^{-0.1s}}{1+s}$. If x(t) is 0.5 sin t, then the phase angle between the output and the input will be:
(a) – 39.27° (b) – 45°
(c) – 50.73° (d) – 90°

56. The transfer function of second order band pas filter, having a centre frequency of 1000 rad./s, selectivity of 100 and a gain of 0dB at the center frequency, is

(a) $\frac{10s}{s^2+10s+10^6}$ (b) $\frac{s}{s^2+s+10^6}$

(c) $\frac{100s}{s^2+100s+10^7}$ (d) $\frac{100s}{s^2+100s+10^6}$

57. In a second order instrument, 'K' is the spring constant, 'M' is the mass of the instrument, and 'B' is the damping constant. The damping constant necessary to the just prevent overshoot in the step response is
(a) equal to √(KM) (b) less than √(KM)
(c) greater than √(KM) (d) equal to 2 √(KM)

Answers Set-1

1. Ans. (a)

Solution: 1 + GH = 0
Substitute S = – [1 / 2] + j [√3 / 2]

Second Method
Characteristic equation

$$C(S)H(S) = \frac{K}{(s+1)^3}$$

$$1 + G(S)H(S) = 0$$

$$= \frac{(s+1)^3 + K}{(s+1)^3} = 0$$

$$(s+1)^3 + K = 0$$

or $-K = (s+1)^3$

Put and $= \left(\frac{-1}{2} + j\frac{\sqrt{3}}{2}\right)$

Now, $-K = (s+1)^3$

$$= \left(\frac{-1}{2} + j\frac{\sqrt{3}}{2} + 1\right)^3$$

or $K = -\left(\frac{1}{2} + j\frac{\sqrt{3}}{2}\right)^2$

$$= \frac{1}{8} - j\frac{3\sqrt{3}}{8} + \frac{3\times 1}{2} \times j\frac{\sqrt{3}}{2}\left(\frac{1}{2} + j\frac{\sqrt{3}}{2}\right)$$

$$= \frac{1}{8} - j\frac{3\sqrt{3}}{8} + j\frac{3\sqrt{3}}{8}\left(\frac{1+j}{2} + \frac{\sqrt{3}}{2}\right)$$

$$= \frac{1}{8} - j\frac{3\sqrt{3}}{8} + j\frac{3\sqrt{3}}{8} - \frac{9}{8} = \frac{1}{8} - \frac{9}{8} = \frac{-8}{8} = -1$$

$-K = -1$; $K = 1$

2. Ans. (c)

3. Ans. (d)

Solution: A temperature control system is usually Very sluggish, To improve its dynamics a PD controller can be used.

An integral system always makes a system sluggish (slow).The response can be made faster b y increasing the gain but oscillation will increase which will tend the system towards instability. This is not desirable. Hence to improve the dynamics, an derivative controller is always introduced.

4. Ans. (b)

Solution:

In a **seismic pick-up for getting** an output proportional to acceleration it is desirable to have a natural frequency very large in comparison to frequency of input and a damping ratio around 0.7

5. Ans. (b)

Solution:

Integration reduces the offset caused by proportional controller action. It integrates area under curve thus minimizing the offset.

6. Ans. (a)

Solution:

When proportional controller action is used and time delay is introduce then gain margin is not affected and phase margin also remain the saem.

7. Ans. (b)

Solution: Selsyn used for error detection

8. Ans. (a)

9. Ans. (d)

10. Ans. (b)

Solution:

Number of state variable depends on the number of storage element i.e. Total number of L and C is 2.

11. Ans. (b)

Solution:

Root locus start from poles and terminate at zero of GH. In present case pole are at s = 0 and s = – 4 and zero at s = – 8.

12. Ans. (d)

Solution:

Steady state error for second order system is given by

$$\frac{2\delta}{\omega_n} =$$

From given function $\omega_n = 3$; $2\delta\,\omega_n = 0.6 \rightarrow$

$$\delta = \frac{0.6}{2\times 3} = 0.1$$

$$\text{Steady state error} = \frac{2\times 0.1}{3} = 0.15$$

13. Ans. (b) nagrath

14. Ans. (b)

Solution:

$$G(s)X(s) = \frac{K(s+10)}{(s+2)(s+5)}$$

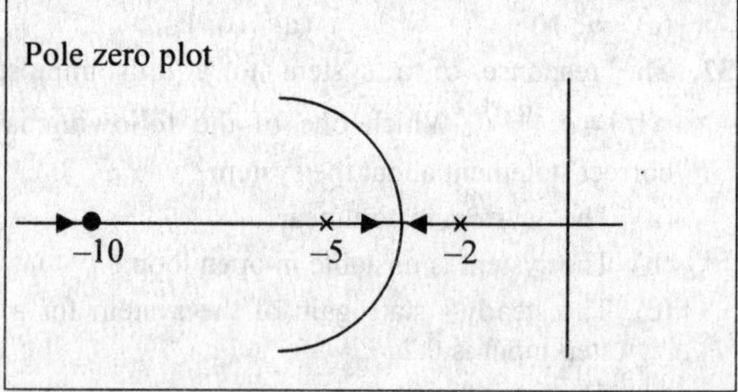

The characteristic equation:

$$1 + G(s)H(s) = 0$$

$$1 + \frac{K(s+10)}{(s+2)(s+5)} = 0$$

$$\text{or } (s+2)(s+5) + K(s+10) = 0$$

$$\text{or } s^2 + 7s + 10 + K(s+10) = 0$$

$$\Rightarrow \quad K = \frac{-(s^2 + 7s + 10)}{(s+10)}$$

For breakaway point, put $\frac{dK}{ds} = 0$

$$\frac{dK}{ds} = -\left[\frac{(s+10)(2s+7) - (s^2 + 7s + 10)(1)}{(s+10)^2}\right]$$

$$= -\left[\frac{2s^2 + 7s + 20s + 70 - s^2 - 7s + 0}{(s+10)^2}\right]$$

$$= -\left[\frac{s^2 + 20s + 60}{(s+10)^2}\right]$$

$$\Rightarrow \quad -\frac{(s^2 + 20s + 60)}{(s+10)^2} = 0$$

$$s^2 + 20s + 60 = 0$$

$$= s + -20 \pm \sqrt{\frac{(20)^2 - 4(1)(60)}{2}}$$

$$= \frac{-20 \pm \sqrt{160}}{2} = -3.6, 16.4$$

15. Ans. (d)

Solution: $F_o(s) = \frac{1}{s}\left(\frac{20}{s^2 + 2s + 5}\right)$

$$f(\infty) = \underset{s \to 0}{Lt}\ sF_o(s) = \underset{s \to 0}{LT}\ s \times \frac{1}{s} \times \frac{20}{(s^2 + 2s + 5)}$$

$$= \underset{s \to 0}{Lt} \frac{20}{s^2 + 2s + 5}$$

$$= \frac{20}{5} = 4$$

16. Ans. (b)

Solution: The characteristic equation is

$s^3 + bs^2 + cs + 1 = 0$

Apply the RH criterion

s^3	1	c
s^2	b	1
s^1	$\frac{bc-1}{b}$	0
s^0	1	

For stability

$$\frac{bc-1}{b} \geq 0 \Rightarrow c - \frac{1}{b} \geq 0$$

$$c \geq \frac{1}{b}; \quad cb \geq 1$$

17. Ans. (c)

18. Ans. (b)

Solution:

Close loop gain = $\frac{G(s)}{1 + G(s)} = T(s)$

1 + G(s) gives the stability criteria

$$G(s) = \frac{9}{s(s+3)}$$

$$\frac{C(s)}{R(s)} = \frac{9}{9 + s(s+3)}$$

or $s^2 + 3s + 9 = 0$

Therefore $\omega_n^2 = 9$

i.e. $\omega_n = \sqrt{9} = 3$

Now $2\delta_e \cdot \omega_n = 3$

Now $2\delta_e \cdot \omega_n = 3$

i.e. $\delta_e = \frac{3}{2 \times 3} = \frac{1}{2} = 0.5$

19. Ans. (c)

20. Ans. (b)

Solution:

$$\frac{C(s)}{R(s)} = \frac{25}{s^2 + 6s + 25} \cong \frac{\omega_n^2}{s^2 + 2\delta\omega_n s + \omega_n^2}$$

Here, $\omega_n = 5$; $2\delta\omega_n = 6$

$$\delta = \frac{6}{2 \times 5} = 0.6$$

$$t_{peak} = \frac{\pi}{\omega_n\sqrt{1-\delta^2}} = \frac{\pi}{5\sqrt{1-(0.6)^2}}$$

$$= \frac{\pi}{5\sqrt{1-0.36}} = \frac{\pi}{5\sqrt{0.64}}$$

$$= \frac{\pi}{5 \times 0.8} = \frac{\pi}{4}$$

21. Ans. (b)

Solution:

The inter section with the real axis gives the phase cross over frequency, at this $I_m[G(j\omega)H(j\omega)] = 0$

OLTP, $G(s)H(s) = \frac{1}{(s+1)^3}$

Putting s = jω, we have

$$G(j\omega)H(j\omega) = \frac{1}{(j\omega+1)^3}$$

$$= \frac{(1-j\omega)^3}{(j\omega+1)^3(1-j\omega)^3} = \frac{(1-j\omega)^3}{(1+\omega^2)^3}$$

$$= \frac{(1-j\omega)(1-\omega^2-2j\omega)}{(1+\omega^2)^3}$$

$$= \frac{(1-j\omega-\omega^2+j\omega^3-2j\omega-2\omega^2)}{(1+\omega^2)^3}$$

$$=\frac{1-3\omega^2+j\left(\omega^2-3\omega\right)}{\left(1+\omega^2\right)^3}$$

Equation imaginary part to zero, we have

$$\left(\omega^2-3\omega\right)=0$$

or $\omega^2=3$

or $\omega_c=\pm\sqrt{3}$

Gain margin, $GM=\left|\frac{1}{G(j\omega_c)H(j\omega_c)}\right|_{\omega=\sqrt{3}}$

$$=\left|\frac{\left(1+\omega_c^2\right)^3}{1-3\omega_c^2}\right|_{\omega_c=\sqrt{3}}$$

$$=\left|\frac{(1+3)^3}{-8}\right|=\frac{4^3}{8}=8$$

22. Ans. (b)

Solution: $\frac{E_o(s)}{E_i(s)}=K_1+K_2 s$

23. Ans. (a)

24. Ans. (b)

Solution:

$$GH=\frac{1}{(j\omega+1)^3}$$

$$\phi=-3\tan^{-1}\omega$$

$$|G(j\omega)|=1$$

$$\left[1+\omega^2\right]^3=1$$

Put $\omega^2=x$

$$(1+x)^3=1$$

$$x=0$$

$$\omega=0$$

$$\phi=-3\tan^{-1}0=0$$

$$PM=180+\phi=180+0=180°$$

25. Ans. (c)

26. Ans. (d)

Solution:

Proportional will give the output as the error signal and derivative will output as a step for a the ramp, so result is figure:

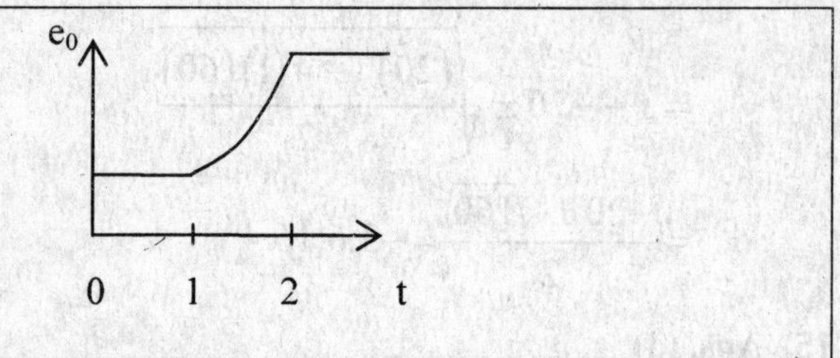

27. Ans. (a)

28. Ans. (c)

Solution:

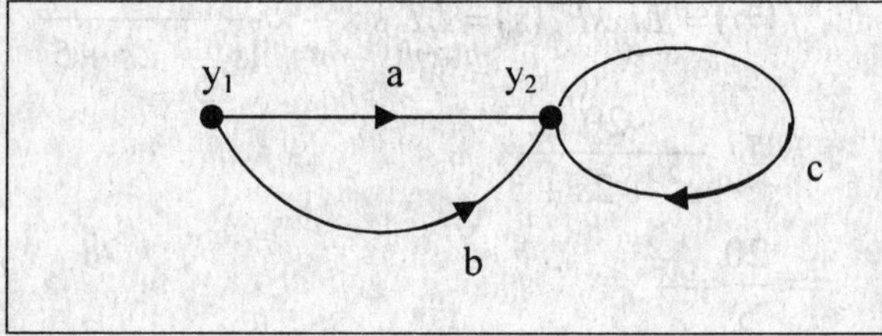

$$y_2=(a+b)y_1+cy_2$$

i.e. $y_2(1-c)=(a+b)y_1$

i.e. $\frac{y_2}{y_1}=\frac{a+b}{1-c}$

29. Ans. (a)

30. Ans. (a)

31. Ans. (a)

Solution: $\frac{C(s)}{R(s)}=\frac{9}{s^2+2s+9}$

$$\cong\frac{\omega_n^2}{s^2+2\delta_n+\omega_n^2}$$

$$\omega_n=\sqrt{9}=3$$

$$2\delta\omega_n=2$$

$$\Rightarrow 2\times 3\,\delta=2$$

32. Ans. (a)

Solution: $20\log_{10}K=40$

$$log_{10}K=2$$

$$\therefore K=100$$

Corner frequencies = 10, 250

$$\therefore=\frac{100}{(1+j\omega T_1)(1+j\omega T_2)}$$

$$G(s)=\frac{100}{\left(1+\frac{s}{10}\right)\left(1+\frac{5}{250}\right)},$$

33. Ans. (b)

34. Ans. (d)

Solution:

Linear variable differential transformer has one primary coil and two secondary coils connected in opposition.

35. Ans. (b)

36. Ans. (b)

Solution:

For proportional controller $K_p = 0.5$ Ku

$$K_u = 0.5 \times 10 = 5$$

37. Ans. (a)

38. Ans. (b)

Solution:

$$G(s)=\frac{10}{s(s+5)}$$

$$H(s=1)$$

$$G(s)\,H(s)=\frac{10}{s(s+5)}$$

This is type 1 system

Ramp error constant $k_v = \underset{s\to 0}{Lt}\,(s..G(s)\,H(s))$

$$= \underset{s\to 0}{Lt}\left(s.\times\frac{10}{s(s+5)}\right)=\frac{10}{5}=2$$

$$=\frac{R}{k_v}=\frac{1}{2}=0.5$$

39. Ans. (a)

Solution:

For the closed loop to be stable, contour τ_{GH} of $G(s)\,H(s)$ does not encircle $(-1+j0)$ point, i.e. the net encirclement is zero

40. Ans. (c)

Solution:

$$e_{ss} = \underset{s\to 0}{Lt}\, sE(s)= \underset{s\to 0}{Lt}\,\frac{sR(s)}{1+G(s)\,H(s)}$$

For $r(t)=\frac{t^2}{2}$, $R(s)=\frac{1}{s^3}$, $H(s)=1$

Then, $e_{ss} = \underset{s\to 0}{Lt}\,\frac{1}{s^2+\frac{10se^{-0.1s}}{(s+5)(s2+45+16)}}=\infty$

41. Ans. (a)

Solution: Constructing Routh-array

s^4	1	a	8
s^3	3	12	0
s^2	$\frac{3a-12}{3}$	8	
s^1			
s^0			

42. Ans. (b)

Solution:

For $\omega\to\infty$ take higher order term of s and for $\omega\to\infty$ take the lower order term of s.

In present case one s is in denominator and other in numerator which given phase angle of zero.

43. Ans. (b)

Solution: $GH=\frac{10e^{-LS}}{s}$

$$e^{jx} = \cos x + j\sin x$$

$$\phi = \tan^{-1}\frac{\sin x}{\cos x}=x$$

$$\angle G(s)\,H(s)\,\big|_{s=j\omega}=-90^o-\frac{(\omega L)180^o}{\pi}$$

At phase cross-over frequency,

$$\angle GH=-180^o$$

Then, $\quad -90^o-\frac{(\omega L)180^o}{\pi}=-180^o$

At $\omega=5$

Or $\quad 90^o=\frac{5(L)\cdot 180^o}{\pi}$

$$\pi=10\,L$$

Or $\quad L=\frac{\pi}{10}$

44. Ans. (a)

Solution: $e(t)=\underset{s\to 0}{Lt}$

$$S\cdot E(s)$$

$$=\underset{s\to 0}{Lt}\,S\cdot\frac{R(s)}{1+GH}$$

$$= \underset{s\to 0}{Lt}\ S \cdot \frac{1}{1+\frac{(1+5s)(1+10s)(1+2s)}{(1+s)(1+8s)(1+20s)}}$$

$$= \underset{s\to 0}{Lt}\ S \times \frac{(1+s)(1+8s)(1+20s)}{(1+s)(1+8s)(1+20s) + (1+5s)(1+10s)(1+2s)} = 0$$

$$J = \int_0^\infty e(t)\,dy = \int_0^\infty 0\ dt = 0$$

45. Ans. (a)

46 Ans. (b)

Solution: $G(s)H(s) = \frac{K K_p}{1+s\tau}$

Positional error constant

$$K_{pec} = \underset{s\to 0}{Lt} \frac{K K_p}{1+s\tau} = K K_p$$

Steady state error e_{ss}

$$= \frac{1}{1+K_{pec}} = \frac{1}{1+K K_p}$$

47. Ans. (a)

Solution: Resolution, $\alpha = \frac{360^o}{nT}$,

Where, n = number of stacks

T = number of teeth

$$\therefore \alpha = \frac{360^o}{12\times 3} = 10^o$$

48. Ans. (d)

Solution:

$$T(s) = \frac{C(s)}{R(s)} = \frac{G(s)}{1+G(s)H(s)} \frac{A}{1+s\tau+A\beta}$$

$$= \frac{A}{s\tau + A\beta + 1} = \frac{A/(A\beta+1)}{\frac{s\tau}{(A\beta+1)}+1}$$

or $\quad T(s) = \frac{\alpha}{\tau' s + 1}$

$$\alpha = \frac{A}{A\beta+1} \text{ and } \tau' = \frac{\tau}{(1+A\beta)}$$

Thus, time constant of $T(s), \tau' = \frac{\tau}{(1+A\beta)}$

49. Ans. (a)

Solution:

$$H(s) = \frac{A}{\omega^2 + s^2}$$

$$R(s) = \frac{1}{s}$$

$$V_o(s) = \frac{A}{s(s^2+\omega^2)}$$

$$V_o(\infty) = \underset{s\to 0}{Lt}\ s\left[\frac{A}{s(s^2+\omega^2)}\right] = \frac{A}{\omega^2}$$

50. Ans. (d)

Solution:

The characteristic equation is,

$$1 + G(s) = 0$$

or $\quad 1 + \frac{10(\tau_1 s + 1)}{\tau_1 s^2 (s+1)} = 0$

or $\quad \tau_1 s^3 + \tau_1 s^2 + 10\tau_1 s + 10 = 0$

Constructing routh-array,

s^3	τ_1	$10\tau_1$
s^2	τ_1	10
s^1	$10(\tau_1 - 1)$	0
s^0	10	

For stability, $\tau_1 - 1 \geq 0$

or $\quad \tau_1 > 1$

51. Ans. (d)

52. Ans. (a)

53. Ans. (c)

54. Ans. (a)

Solution:

$$T(s) = C|sI - A|^{-1} B$$

$$= [1\ \ 0]\begin{bmatrix} s & -1 \\ 2 & s+3 \end{bmatrix}^{-1} \begin{bmatrix} 0 \\ 1 \end{bmatrix}$$

$$= [1 \quad 0]\begin{bmatrix} s+3 & 1 \\ -2 & s \end{bmatrix}\begin{bmatrix} 0 \\ 1 \end{bmatrix}$$

$$\overline{s^2 + 3s + 2}$$

$$= \frac{1}{s^2 + 3s + 2}$$

The system poles are given by

$$s^2 + 3s + 2 = 0$$

or $(s+1)(s+2) = 0$

or $s = -1,\ s = -2$

55. Ans. (c)

Solution:

$$\text{Phase angle} = -\tan^{-1}\omega - \frac{0.1\omega \times 180^o}{\pi}$$

For $\omega = 1$

$$\angle Y(j\omega) - \angle X(j\omega) = -\tan^{-1} 1 - \frac{0.1\omega \times 180^o}{\pi}$$

$$= -50.73^o$$

56. Ans. (a)

Solution:

Transfer function second order band-pass filter

$$A_v(s) = \frac{(\omega_0/Q)A_0 s}{s^2 + (\omega_0/Q)s + \omega_0^2}$$

Given, Q = 100, A_0 = 1, ω0 = 1000, we get

$$A_v(s) = \frac{10s}{s^2 + 10s + 10^6}$$

57. Ans. (d)

Solution:

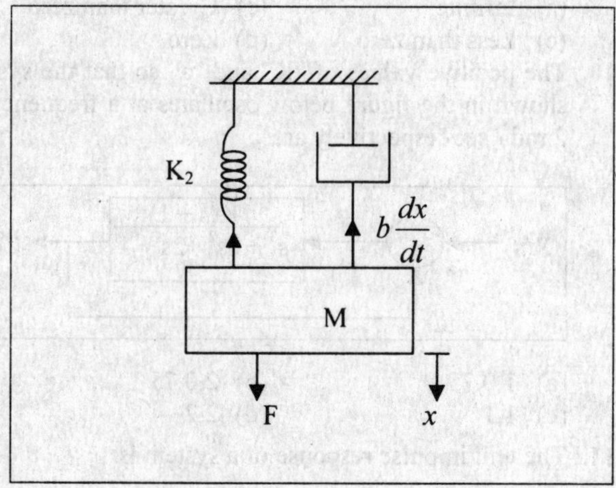

$$F - \frac{b\,dx}{dt} - K_x = \frac{Md^2x}{dt^2}$$

$$F = \frac{M\,d^2x}{dt^2} + b\frac{dx}{dt} + K_x$$

$$F(s) = Ms^2X(s) + bs\,X(s) + K\,X(s)$$

$$I.F = \frac{X(s)}{F(s)} = \frac{1}{Ms^2 + bs + K}$$

$$Ms^2 + bs + K = 0$$

$$s^2 + 2\delta\omega_n s + \omega_n^2$$

$$s^2 + \frac{b}{M}s + \frac{K}{M} = 0$$

$$\omega_n = \sqrt{\frac{K}{M}} rad/\sec$$

$$2\delta\omega_n = \frac{b}{M}$$

$$\delta = \frac{b}{2M\cdot\omega_n}$$

$$\delta = \frac{b}{2M\cdot\sqrt{\frac{K}{M}}} = \frac{b}{2\sqrt{KM}}$$

$$\delta = \frac{b}{2\sqrt{KM}}$$

If $b = 2\sqrt{KM}$

Then $\delta = 1$ critically damped system.

Overshoot is observed if $\delta < 1$ i.e. under damped system.

This is n overshoot in $\delta = 1$ (critically damped system) a for $\delta > 1$ (over-damped system.

$\therefore$ $b = 2\sqrt{KM}$ to just prevent overshoot in step response.

One Mark Questions

1. Figure shows the Nyquist plot of the open-loop transfer function G(s) H(s) of a system. If G(s)H(s) has one right-hand pole, the closed-loop system is:

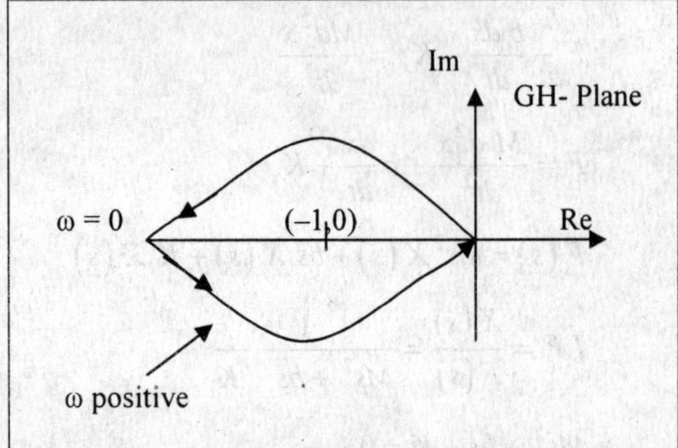

(a) Always stable
(b) Unstable with one closed-loop right hand pole
(c) Unstable with two closed-loop right poles
(d) Unstable with three closed-loop right hand poles

2. A PD controller is used to compensate a system Compared to the un-compensated system the compensated system has
(a) A higher type number
(b) Reduced damping
(c) Higher noise amplification
(d) Larger transient overshoot

3. The gain margin for the system with open-loop transfer function $G(s)\,H(s) = \dfrac{2\,(1+s)}{s^2}$ is
(a) ∞ (b) 0
(c) 1 (d) – ∞

4. Given $G(s)H(s) = \dfrac{K}{s(s+1)(s+3)}$, the point of intersection of the asymptotes of the root loci with the real axis is
(a) – 4 (b) 1.33
(c) –1.33 (d) 4

5. A linear system is equivalently represented by two sets of state equations.

$X' = AX + BU$ and $W = CW + DU$

The eigenvalues of the representations, are also computed as (λ) and (μ). Which one of the following statements is true?
(a) [λ] = [μ] and X = W
(b) [λ] = [μ] and X ≠ W
(c) [λ] ≠ [μ] and X = W
(d) [λ] ≠ [μ] and X ≠ W

6. Which one of the following polar diagrams corresponds to a lag network?

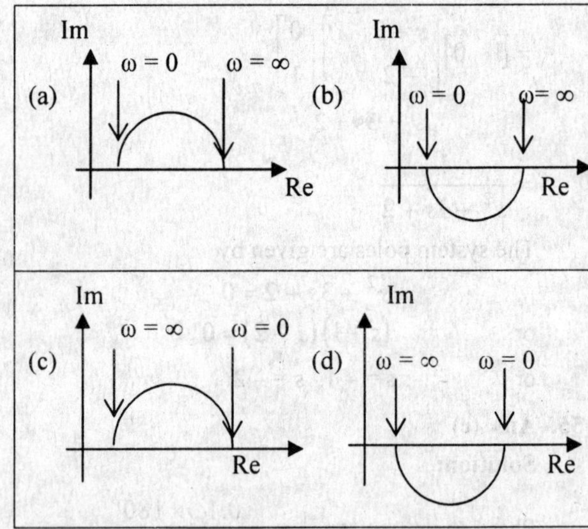

7. Despite the presence of negative feedback, control systems still have problems of instability because the
(a) Components used have non-linearities.
(b) Dynamic equations of the subsystem are not known exactly
(c) Mathematical analysis involves approximations
(d) System has large negative phase angle at high frequencies.

8. The open - loop transfer function of a unity – gain feedback control system is given by

$$G(s) = \frac{K}{(S+1)(S+2)}$$

The gain margin of the system in dB is given by.
(a) 0 (b) 1
(c) 20 (d) ∞

9. The Nyquist plot of G (jω) H (jω) for a closed loop control system, passes through (-1, j0) point in the GH plane. The gain margin of the system is dB is equal to:
(a) Infinite (c) Greater than zero
(c) Less than zero (d) Zero

10. The positive values of "K" and "a" so that the system shown in the figure below oscillates at a frequency of 2 rad / sec respectively are.

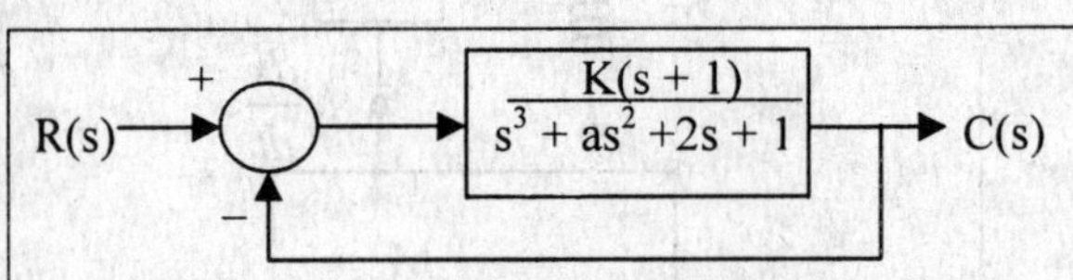

(a) 1,0.75 (b) 2,0.75
(c) 1,1 (d) 2,2

11. The unit impulse response of a system is:

$h(t) = e^{-t},\ t \geq 0$

For this system, the steady- state value of the output for unit step input is equal to:

(a) –1 (b) 0

(c) 1 (d) ∞

12. The transfer function of a phase lead compensator is given by.

$G_c(s) = [(1 + 3Ts) / (1 + Ts)]$ where $T > 0$

The maximum phase – shift provided by such a compensator is:

(a) $\pi / 2$ (b) $\pi / 3$

(c) $\pi / 4$ (d) $\pi / 6$

13. A linear system is described by the following state equation.

$X'(t) = AX(t) + BU(t)$; $A = \begin{bmatrix} 0 & 1 \\ -1 & 0 \end{bmatrix}$

The state – transition matrix of the system is

(a) $\begin{pmatrix} \cos t & \sin t \\ -\sin t & \cos t \end{pmatrix}$ (b) $\begin{pmatrix} -\cos t & \sin t \\ -\sin t & -\cos t \end{pmatrix}$

(c) $\begin{pmatrix} -\cos t & -\sin t \\ -\sin t & -\cos t \end{pmatrix}$ (d) $\begin{pmatrix} \cos t & -\sin t \\ \cos t & \sin t \end{pmatrix}$

14. Consider two transfer function

$G_1(s) = [1 / (s^2 + as + b)]$ and

$G_2(s) = [s / (s^2 + as + b)]$

The 3 dB bandwidths of their frequency responses are, respectively .

(a) $\sqrt{(a^2 - 4b)}$, $\sqrt{(a^2 + 4b)}$

(b) $\sqrt{(a^2 + 4b)}$, $\sqrt{(a^2 - 4b)}$

(c) $\sqrt{(a^2 - 4b)}$, $\sqrt{(a^2 - 4b)}$

(d) $\sqrt{(a^2 + 4b)}$, $\sqrt{(a^2 + 4b)}$

15. Consider a unity gain feedback control system whose open loop transfer function is

$G(s) = [(as + 1) / s^2]$

The value of "a" so that the system has a phase margin equal to $\pi / 4$ is approximately equal to

(a) 2.40 (b) 1.40

(c) 0.84 (c) 0.74

16. Consider a unity gain feedback control system whose open loop transfer function is

$G(s) = [(as + 1) / s^2]$

With the value of set for a phase margin of $[\pi / 4]$, the value of unit impulse response of the open loop system at t = 1 second is equal to .

(a) 3.40 (b) 2.40

(c) 1.84 (d) 1.74

17. If the closed-loop transfer function of a control system is given as:

$T(s) = \dfrac{s-5}{(s+2)(s+3)}$, then it is

(a) An unstable system

(b) An uncontrollable system

(c) A minimum phase system

(d) A non-minimum phase system

18. A control system with a PD controller is shown in the figure. If the velocity error constant $K_v = 1000$ and the damping ratio $\delta = 0.5$, then the values of K_p and K_D are:

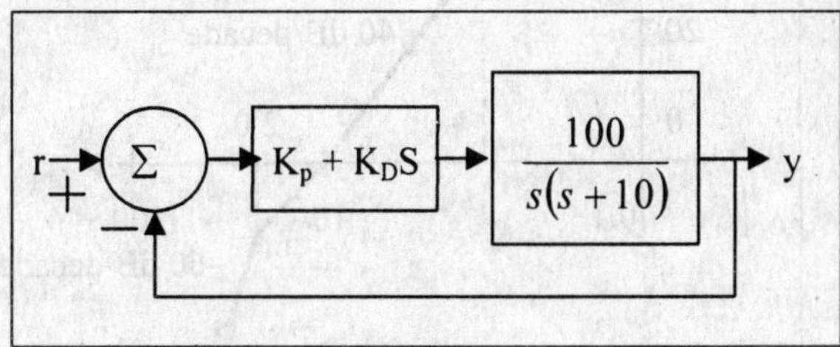

(a) $K_P = 100, K_D = 0.09$

(b) $K_P = 100, K_D = 0.9$

(c) $K_P = 10, K_D = 0.09$

(d) $K_P = 10, K_D = 0.9$

19. The transfer function of a plant is $T(s) = \dfrac{5}{(s+5)(s^2+s+1)}$. The second-order approximation of T(s) using **dominant pole concept** is

(a) $\dfrac{1}{(s+5)(s+1)}$ (b) $\dfrac{5}{(s+5)(s+1)}$

(c) $\dfrac{5}{s^2+s+1}$ (d) $\dfrac{1}{s^2+s+1}$

20. The open-loop transfer function of a plant is given as $G(s) = \dfrac{1}{s^2 - 1}$. If the plant is operated in a unity feedback configuration , then the *lead compensator* that can stabilize this control system is:

(a) $\dfrac{10(s-1)}{s+2}$ (b) $10\dfrac{(s+4)}{s+2}$

(c) $\dfrac{10(s+2)}{s+10}$ (d) $\dfrac{2(s+2)}{s+10}$

21. A unity feedback control system has an open-loop transfer function $G(s) = \dfrac{K}{s(s^2+7s+12)}$. The gain K for which s = –1 + j1 will lie on the root locus of this system is:

(a) 4 (b) 5.5

(c) 6.5 (d) 10

22. The asymptotic Bode plot of a transfer function is as shown in the figure .The transfer function G(s) corresponding to this Bode plot is:

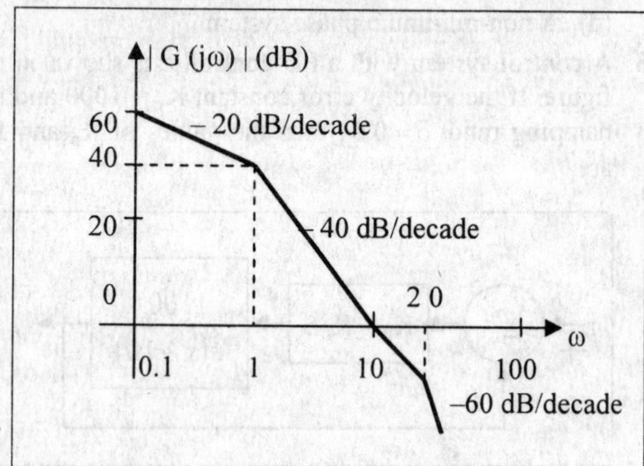

(a) $\frac{1}{(s+1)(s+20)}$ (b) $\frac{1}{s(s+1)(s+20)}$

(c) $\frac{100}{s(s+1)(s+20)}$ (d) $\frac{100}{s(s+1)(1+0.05s)}$

23. The **state space** representation of a separately excited DC servo motor dynamics is given as:

$$\begin{bmatrix} \frac{d\omega}{dt} \\ \frac{di_o}{dt} \end{bmatrix} = \begin{bmatrix} -1 & 1 \\ -1 & -10 \end{bmatrix} \begin{bmatrix} \omega \\ i_a \end{bmatrix} + \begin{bmatrix} 0 \\ 10 \end{bmatrix} u$$

where ω is the sped of the motor, i_a is the armature current and u is the armature voltage. The transfer function $\frac{\omega(s)}{U(s)}$ of motor is

(a) $\frac{10}{s^2+11s+11}$ (b) $\frac{1}{s^2+11s+11}$

(c) $\frac{10s+10}{s^2+11s+11}$ (d) $\frac{1}{s^2+s+1}$

24. Consider a linear system whose space representation is $x(t)=Ax(t)$. If the initial state vector of the system is $x(0)=\begin{bmatrix} 1 \\ -2 \end{bmatrix}$, then the system response is $x(t)=\begin{bmatrix} e^{-2t} \\ -2e^{-2t} \end{bmatrix}$. If the initial state vector of the system changes to $x(0)=\begin{bmatrix} 1 \\ -1 \end{bmatrix}$, them the system response becomes $x(t)=\begin{bmatrix} e^{-t} \\ e^{-t} \end{bmatrix}$.

The eigenvalue and eigenvector pairs (λ_i, v_i) for the system are

(a) $\left(-1, \begin{bmatrix} 1 \\ -1 \end{bmatrix}\right)$ *and* $\left(-2, \begin{bmatrix} 1 \\ -2 \end{bmatrix}\right)$

(b) $\left(-1, \begin{bmatrix} 1 \\ -1 \end{bmatrix}\right)$ *and* $\left(-1, \begin{bmatrix} 1 \\ -2 \end{bmatrix}\right)$

(c) $\left(-1, \begin{bmatrix} 1 \\ -1 \end{bmatrix}\right)$ *and* $\left(2, \begin{bmatrix} 1 \\ -2 \end{bmatrix}\right)$

(d) $\left(-2, \begin{bmatrix} 1 \\ -1 \end{bmatrix}\right)$ *and* $\left(1, \begin{bmatrix} 1 \\ -2 \end{bmatrix}\right)$

25. Consider a linear system whose space representation is $x(t)=Ax(t)$. If the initial state vector of the system is $x(0)=\begin{bmatrix} 1 \\ -2 \end{bmatrix}$, then the system response is $x(t)=\begin{bmatrix} e^{-2t} \\ -2e^{-2t} \end{bmatrix}$. If the initial state vector of the system changes to $x(0)=\begin{bmatrix} 1 \\ -1 \end{bmatrix}$, them the system response becomes $x(t)=\begin{bmatrix} e^{-t} \\ -e^{-t} \end{bmatrix}$.

The system matrix A is

(a) $\begin{bmatrix} 0 & 1 \\ -1 & 1 \end{bmatrix}$ (b) $\begin{bmatrix} 1 & 1 \\ -1 & -2 \end{bmatrix}$

(c) $\begin{bmatrix} 2 & 1 \\ -1 & -1 \end{bmatrix}$ (d) $\begin{bmatrix} 0 & 1 \\ -2 & -3 \end{bmatrix}$

26. A control system is defined by the following mathematical relationship

$$\frac{d^2x}{dt^2} + 6\frac{dx}{dt} + 5x = 12(1 - e^{-2t})$$

The response of the system as $t \to \infty$ is

(a) x = 6 (b) x = 2

(c) x = 2.4 (d) x = –2

27. A lead compensator used for a closed loop controller has the following transfer function

$$\frac{K(1+s/a)}{(1+s/b)}$$

For such a lead compensator

(a) $a < b$ (b) $b < a$
(c) $a > Kb$ (d) $a < Kb$

28. A second order system starts with an initial condition of $\begin{bmatrix}2\\3\end{bmatrix}$ without any external input. The state transition matrix for the system is given by $\begin{bmatrix}e^{-2t} & 0\\0 & e^{-t}\end{bmatrix}$. The state of the system at the end of 1 second is given by

(a) $\begin{bmatrix}0.271\\1.100\end{bmatrix}$ (b) $\begin{bmatrix}0.135\\0.368\end{bmatrix}$
(c) $\begin{bmatrix}0.271\\0.736\end{bmatrix}$ (b) $\begin{bmatrix}0.135\\1.100\end{bmatrix}$

29. The Nyquist plot of loop transfer function G(s) H(s) of a closed loop control system passes through the point (-1, j0) in the G(s) H(s) plane. The phase margin of the system is

(a) 0° (b) 45°
(c) 90° (d) 180°

30. Consider the function, $F(s) = \frac{5}{s(s^2+3s+2)}$

Where F(s) is the Laplace transform of the function f(t). The initial value of f(t) is equal to

(a) 5 (b) $\frac{5}{2}$
(c) $\frac{5}{3}$ (d) 0

31. For a tachometer, if θ(t) is the rotor displacement in radians, e(t) is the output voltage and K_t is the tachometer constant V/rad/sec, then the transform function, $\frac{E(s)}{Q(s)}$ will be

(a) $K_t s^2$ (b) $\frac{K_t}{s}$
(c) $K_t s$ (d) K_t

32. A system with zero initial conditions has the closed loop transfer function. $T(s) = \frac{s^2+4}{(s+1)(s+4)}$. The system output is zero at the frequency

(a) 0.5 rad/sec (b) 1 rad/sec
(c) 2 rad/sec (d) 4 rad/se

33. Figure shows the root locus plot (location of poles not given) of a third order system whose open loop transfer function is:

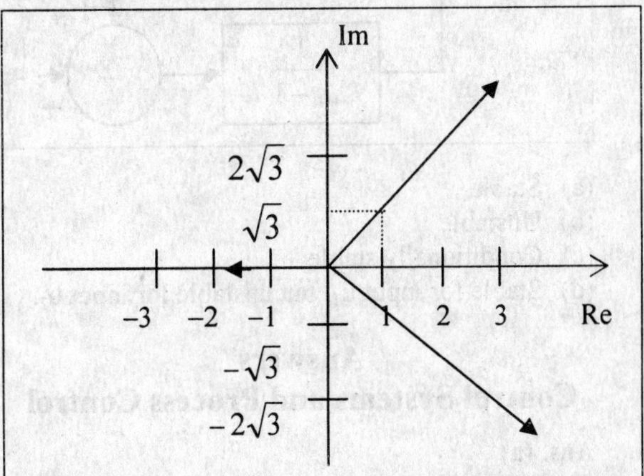

(a) $\frac{K}{s^3}$ (b) $\frac{K}{s^2(s+1)}$
(c) $\frac{K}{s(s^2+1)}$ (d) $\frac{K}{s(s^2-1)}$

34. The gain margin of a unity feedback control system with the open loop transfer function $G(s) = \frac{(s+1)}{s^2}$ is

(a) 0 (b) $\frac{1}{\sqrt{2}}$
(c) $\sqrt{2}$ (d) ∞

35. For a system with the transfer function $H(s) = \frac{3(s-2)}{4s^2-2s+1}$, the matrix A in the state space form x′ = Ax + Bu is equal to

(a) $\begin{bmatrix}1 & 0 & 0\\0 & 1 & 0\\-1 & 2 & -4\end{bmatrix}$ (b) $\begin{bmatrix}0 & 1 & 0\\0 & 1 & 0\\-1 & 2 & -4\end{bmatrix}$

(c) $\begin{bmatrix}0 & 1 & 0\\3 & -2 & 1\\1 & -2 & 4\end{bmatrix}$ (d) $\begin{bmatrix}1 & 0 & 0\\0 & 0 & 1\\-1 & 2 & -4\end{bmatrix}$

36. The system shown the figure is:

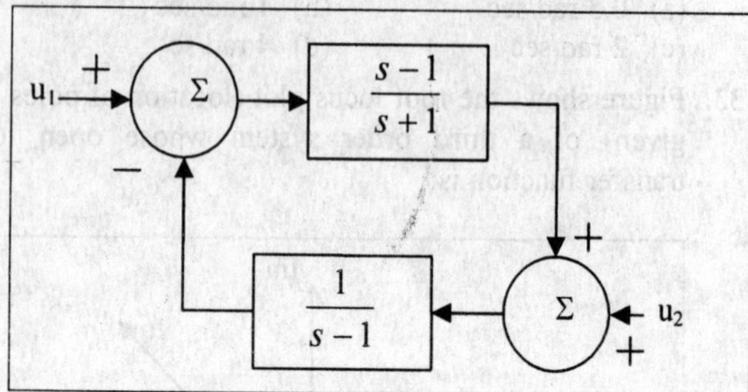

(a) Stable
(b) Unstable
(c) Conditionally stable
(d) Stable for input u_1, but unstable for input u_2.

Answers

Control Systems and Process Control

1. **Ans. (a)**

Solution: The encirclement of critical point $(-1, j0)$ is once in anticlockwise direction

$\therefore N = 1$

$G(s)H(s)$ has one right side pole

$\therefore P = 1$

$Z = P - N = 1 - 1 = 0$

No zero in the right hand side of s-plane

So system is stable

2. **Ans. (c)**

Solution: Differentiator amplify the noise

3. **Ans. (b)**

Solution:

$GH(s) = \frac{2(1+s)}{s^2}$

$GH(j\omega) = \frac{2(1+j\omega)}{(j\omega)^2}$

$= \frac{2(1+j\omega)}{-\omega^2}$

The interaction with real axis gives the phase cross over frequency.

$Im(GH(j\omega)) = 0 \Rightarrow \omega = 0$

$Real\ GH(j0) = [2/-0] = -\infty$

$$GM = \frac{1}{|Real\ GH(j\omega)|} = \frac{1}{\infty} = 0$$

Another way

$\angle G(s)H(s) = -180 + \tan^{-1}\omega$

For phase cross over frequency

$-180 + \tan^{-1}\omega = -180$

$\omega = 0$

$$|G(s)H(s)|_{\omega=0} = \frac{2\sqrt{1+\omega^2}}{\omega^2} = \infty = a$$

$$GM = \frac{1}{|a|} = \frac{1}{\infty} = 0$$

In present case GM is not asked into db. Hence correct option is (b) only.

4. **Ans. (c)**

Solution: Point of interaction of Asymptotes with real axis

$$\text{Centroid, } (-\sigma_A) = \frac{\sum_{i=1}^{m} P_i - \sum_{i=1}^{n} Z_j}{m-n}$$

Where m and n are the number of poles and zero respectively.

$$= \frac{-4-0}{3} = -1.33$$

5. **Ans. (b)**

Solution:

We have

$\dot{X} = AX + BU$ Where λ is set of Eigen values

$\dot{W} = CW + DU$ Where μ set of Eigen values

If a linear system is equivalently represented by two set of state equation, then for both sets, states will be same but their sets of Eigen values will not be same i.e.

$X = W$ but $\lambda \neq \mu$

6. **Ans. (d)**

Solution: Lag network is an RC network

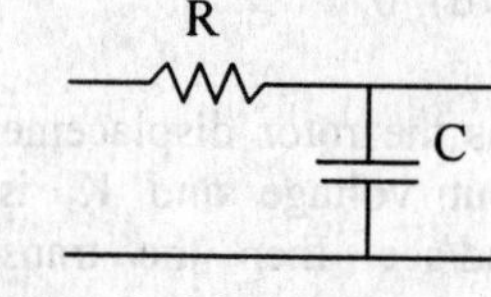

$G(s) = [1/(1+sT)]$

$$G(j\omega) = \frac{1}{1+j\omega T} = \frac{1}{\sqrt{1+\omega^2T^2}} \angle -\tan^{-1}\omega T$$

$= M\angle\phi$

When $\omega = 0$, $M = 1$ and $\phi = 0$

The phasor at $\omega = 0$ has unit length & lies along the positive real axis. As ω increases M decreases ε^1 phase angle increases negatively.

When $\omega = [1/T]$, $M = [1/\sqrt{2}]$, $\phi = -45$

$\omega \to \infty$, $M \to 0$, $\phi = -90$

7. **Ans. (a)**

Solution:

Despite the presence of negative feedback, control system still have problems of instability because components used have nonlinearities. There are always some variation as compared to ideal characteristics

8. **Ans. (d)**

Solution: $\phi = -\tan^{-1}\dfrac{\omega}{1} - \tan^{-1}\dfrac{\omega}{2}$

So either $\omega = 0$ or $\omega = \infty$ for ϕ to be -180°

But $\omega = \infty$ will given G.M zero and in dB it will ∞ db

9. **Ans. (d)**

Solution:

As the system Nyquest plot passes through $-1 + j0$,

$|G(s)H(s)| = 1$

So, G.M = 1, but in dB this will be zero

10. **Ans. (b)**

Solution: Characteristic equation = 1 + GH = 0

$= s^3 + as^3 + 2s + 1 + k(s+1) = 0$

$= s^3 + as^2 + (2+k)s + (1+k) = 0$

Routh Hurwitz array

s^3	1	$2+k$
s^2	a	$1+k$
s^1	$\dfrac{(1+k)-a(2+k)}{a}$	0
s^0	$1+k$	

For oscillating at 2 rad/sec.

$$\frac{2+k}{1} = \frac{1+k}{a} = 2^2$$

$\Rightarrow k = 2$ and $a = 0.75$

11. **Ans. (c)**

Solution: $h(t) = e^{-t}$; $H(s) = \dfrac{1}{s+1}$

$Y(s) = H(s)\,X(s) = \dfrac{1}{s+1} \times \dfrac{1}{s}$

$y(t) = 1 - e^{-t}$

In steady state $t \to \infty$; $y(\infty) = 1$

12. **Ans. (d)**

Solution: $\alpha = [1/3]$

$$\theta_m = \sin^{-1}\left(\frac{1-\alpha}{1+\alpha}\right) = \sin^{-1}\left(\frac{1}{2}\right) = \frac{\pi}{6}$$

Another way

Maximum phase shift $\phi_m = \angle G_c(s)$

$\phi = \tan^{-1} 3\omega T - \tan^{-1}\omega T$

For max phase shift $\dfrac{d\phi}{dt} = 0$

$$\text{or,}\quad \frac{3T}{1+(3T\omega)^2} - \frac{T}{1+(T\omega)^2} = 0$$

$$\text{or,}\quad \frac{3T}{1+(3T\omega)^2} = \frac{T}{1+(T\omega)^2}$$

$$\text{or}\quad 3\left[1+(T\omega)^2\right] = 1+(3T\omega)^2$$

$$\text{or,}\quad 3+3T^2\omega^2 = 1+9T^2\omega^2$$

$$\text{or}\quad 2 = 6(\omega T)^2$$

$$(\omega T)^2 = \frac{1}{3}$$

$$\omega T = \frac{1}{\sqrt{3}}$$

$$\phi_{max} = \tan^{-1} 3\times\frac{1}{\sqrt{3}} - \tan^{-1}\frac{1}{\sqrt{3}}$$

$$= \frac{\pi}{3} - \frac{\pi}{6} = \frac{\pi}{6}$$

13. **Ans. (a)**

Solution: State transition matrix = $[SI - A]^{-1}$

$$\phi(s) = \begin{pmatrix} s & -1 \\ 1 & -s \end{pmatrix}^{-1} \frac{1}{s^2+1}\begin{pmatrix} s & 1 \\ -1 & s \end{pmatrix}$$

Another way

$$\phi(t) = LT^{-1}[SI - A]^{-1}$$

$$= LT^{-1}\left[\begin{pmatrix} s & 0 \\ 0 & s \end{pmatrix} - \begin{pmatrix} 0 & 1 \\ -1 & 0 \end{pmatrix}\right]$$

$$= LT^{-1}\left[\begin{pmatrix} s & -1 \\ 1 & s \end{pmatrix}\right]^{-1}$$

$$= \frac{LT^{-1}}{s^2+1}\begin{pmatrix} s & 1 \\ -1 & s \end{pmatrix}$$

$$= \begin{pmatrix} \cos t & \sin t \\ -\sin t & \cos t \end{pmatrix}$$

14. **Ana. (b)**

15. **Ans. (c)**

Solution:

We gave $G(s)=\dfrac{as+1}{s^2}$

$\angle G(j\omega)=\tan^{-1}(\omega a)-\pi$

or $\dfrac{\pi}{4}=\pi+\tan^{-1}(\omega_g a)-\pi$

or $\dfrac{\pi}{4}=\tan^{-1}(\omega_g a)$

or $a\omega_g=1$

At fain crossover frequency $|G(j\omega_g)|=1$

Thus $\dfrac{\sqrt{1+a^2\omega_g^2}}{\omega_g^2}=1$

or $\sqrt{1+1}=\omega_g^2$

or $\omega_g=(2)^{1/4}$

Now $a=\dfrac{1}{2^{1/4}}=0.84$

16. Ans. (c)

Solution: Y(s) = G(s) U(s) = $\dfrac{as+1}{s^2}$

U(s) LT $\delta(t)=1$

Output $\xrightarrow{LT}$ $\dfrac{as+1}{s^2}=\dfrac{a}{s}+\dfrac{1}{s^2}$

$\Rightarrow$ au (t) + r (t) = a u (t) + t . u (t)

$\Rightarrow$ a + 1 $\Rightarrow$ 1.84

17. Ana. (d)

Solution: $T(s)=\dfrac{s-5}{(s+2)(s+3)}$, then system is

$\Rightarrow$ Non minimum phase
because of 1 zero on RHS and 2 poles on LHS
i.e. all poles and zero does not lie in LHS of s-plane

18. Ana. (b)
Solution:

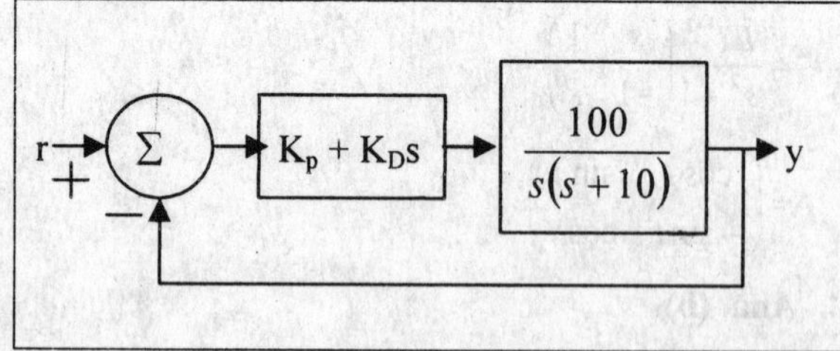

Given that,

$K_V=1000, \quad \delta=0.5$

$T(s)=\dfrac{100(K_P+K_D s)}{s^2+(100K_D+10)s+K_P 100}$

$1+G(s)H(s)=s^2+(100K_D+10)s+100K_P$

$\therefore\ \omega_n^2=100K_P$

$2\delta\omega_n=2\times0.5\omega_n=\omega_n=100K_D+10$

$\therefore\ \sqrt{K_P 100}=100K_D+10$

$\sqrt{K_P}=10K_D+1 \quad (1)$

Also $K_V=1000$

$K_V=\underset{s\to0}{Lt}\ sG(s)H(s)$

$1000=\underset{s\to0}{Lt}\ \dfrac{s\,100(K_P+sK_D)}{s(s+10)}$

$1000=10K_P$

$K_P=100$

From (1)

$\sqrt{K_P}=10K_D+1 \Rightarrow 10K_D=\sqrt{100}-1=9$

$K_D=0.9$

19. Ana. (d)

Solution: $T(s)=\dfrac{5}{(s+5)(s^2+s+1)}$

Dominant poles are poles near origin. or imaginary axis in LHS of s-plane which give rise to transient response that decay relatively slow.
Thus s + 5 nearly equal to 5.

Which result as $T(s)=\dfrac{1}{(s^2+s+1)}$

Another way

Dominant poles are poles near origin. So eliminate $(s+5)$ but in dominant pole concept, the factor that has to be eliminated should be in time domain form

$$=\frac{5}{(s+5)(s^2+s+1)}=\frac{5}{5\left(1+\frac{s}{5}\right)(s^2+s+1)}$$

$$=\frac{1}{s^2+s+1}$$

20. Ana. (a)
Solution:

$$G(s)=\frac{1}{s^2-1}=\frac{1}{(s+1)(s-1)}$$

Uncompensated system have pole in the Right side of s-plane i.e. unstable system. The right side pole is to be neutralized by compensator of $\frac{10(s-1)}{s+1}$.

$$G_C(s) = \frac{1}{(s+1)(s-1)} \times \frac{10(s-1)}{(s+2)}$$

$$= \frac{10}{(s+1)(s+2)} \text{ is stable system}$$

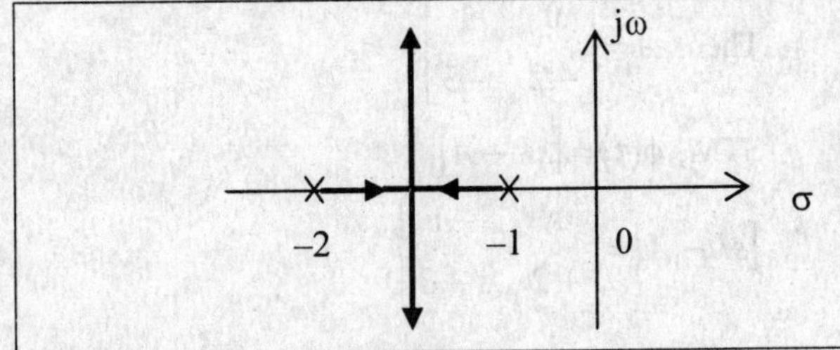

21. **Ana. (d)**

Solution: We known that characteristic equation is

$1 + G(s)H(s) = 0$

or $G(s)\ H(s) = -1$

or $|G(s)\ H(s)| = 1$

unity feed back system

Hence put $H(s) = 1$ in above equation then

$$\therefore \left|\frac{K}{s(s^2+7s+12)}\right|_{s=-1+j} = 1$$

$$\Rightarrow \left|\frac{K}{(-1+j)(5+5j)}\right| = 1$$

$$\Rightarrow \left|\frac{K}{-5+5j-5j-5}\right| = 1$$

$$\because (-1+j)\left[(-1+j)^2 + 12 + 7(-1+j)\right]$$

$$= (-1+j)(5+5j)$$

$$\left|\frac{K}{-10}\right| = 1$$

$$K = 10$$

22. **Ana. (d)**

Solution: At ω = 0.1 gain is 60 db that is due to K and [1/s].

[1/s] give gain = 20 db at 0.1 and 40 db due to K

Therefore, 20 log K = 40 ⇒ K = 100

$$T_1 = \frac{1}{1}$$

$$T_2 = \frac{1}{20} = 0.05$$

$$G(s) = \frac{K}{s(1+sT_1)(1+sT_2)}$$

$$= \frac{100}{s(1+s)\ (1+0.05s)}$$

Another way

$$G(s) = \frac{K}{s(s+1)(s+20)} = \frac{K}{s(s+1)\,20(1+0.05\,s)}$$

as there is slop in the beging of plot

$$G(s) = \frac{K/20}{s(s+1)\left(1+\frac{s}{20}\right)} = \frac{K^1}{s(s+1)(1+0.5\,s)}$$

As the term are in the $(1+sT)$ form

$$20\log \left.\frac{K'}{\omega}\right|_{\omega=0.1} = 60\ db$$

$$= \left.\frac{K'}{\omega}\right|_{\omega=0.1} = (10)^3$$

or $$\frac{K'}{0.1} = (10)^3$$

or $$K' = 100$$

$$G(s) = \frac{100}{s(s+1)(1+0.5\,s)}$$

23. **Ans. (a)**

Solution: Transfer function

$$\frac{\omega(s)}{U(s)} = C(sI-A)^{-1}\ B + D$$

Where $x' = Ax + Br(t)$

$$y = [1\quad 0]\begin{bmatrix}\omega \\ i_a\end{bmatrix} = CX$$

$$\frac{\omega(s)}{U(s)} = [1\quad 0]\begin{bmatrix}\frac{s+10}{s^2+11s+11} & \frac{+1}{s^2+11s+11} \\ \frac{-1}{s^2+11s+11} & \frac{s+1}{s^2+11s+11}\end{bmatrix}\begin{bmatrix}0 \\ 10\end{bmatrix}$$

$$= \begin{bmatrix} \dfrac{s+10}{s^2+11s+11} & \dfrac{1}{s^2+11s+11} \end{bmatrix} \begin{bmatrix} 0 \\ 10 \end{bmatrix}$$

$$= \frac{10}{s^2+11s+11}$$

Where, STM $\phi(x) = [sI - A]^{-1}$

$$= \frac{\begin{bmatrix} s+10 & -1 \\ 1 & s+1 \end{bmatrix}}{s^2+11s+11}$$

24. Ans. (a)

Solution: If $x(0) = \begin{bmatrix} 1 \\ -2 \end{bmatrix}$

Then response $x(t) = \begin{bmatrix} e^{-2t} \\ -2e^{-2t} \end{bmatrix} \rightarrow \begin{bmatrix} 1/(s+2) \\ -2/(s+2) \end{bmatrix}$

If $x(0) = \begin{bmatrix} 1 \\ -1 \end{bmatrix}$

Then response $x(t) = \begin{bmatrix} e^{-2t} \\ -e^{-2t} \end{bmatrix} \rightarrow \begin{bmatrix} 1/(s+2) \\ -1/(s+2) \end{bmatrix}$

Let, $A = \begin{bmatrix} A_0 & A_1 \\ A_2 & A_3 \end{bmatrix}$

Then, $x(t) = [sI - A]^{-1} x(0)$

Multiply both sides by (sI – A) to above equation

$[sI - A]x(t) = x(0)$

Case 1

$$\begin{bmatrix} s - A_0 & -A_1 \\ -A_2 & s - A_3 \end{bmatrix} \begin{bmatrix} 1/(s+2) \\ -2/(s+2) \end{bmatrix} = \begin{bmatrix} 1 \\ -2 \end{bmatrix}$$

$$\frac{s - A_o}{s+2} + \frac{2A_1}{s+2} = 1$$

or, $s - A_o + 2A_1 = s + 2$

or, $-A_o + 2A_1 = 2$

or, $2A_1 - A_o = 2$

Finally,

$2A_1 - A_0 = 2$ (1)

$A_2 - 2A_3 = 4$ (2)

Case 2

$$\begin{bmatrix} s - A_0 & -A_1 \\ -A_2 & s - A_3 \end{bmatrix} \begin{bmatrix} 1/(s+1) \\ -1/(s+1) \end{bmatrix} = \begin{bmatrix} 1 \\ -1 \end{bmatrix}$$

$A_1 - A_0 = 1$ (3)

$A_2 - A_3 = 1$ (4)

From (1), (2), (3) and (4)

$A_o = 0 \quad ; \quad A_1 = 1$

$A_2 = -2 \quad ; \quad A_3 = -3$

Then, $A = \begin{bmatrix} 0 & 1 \\ -2 & -3 \end{bmatrix}$

STM, $\phi(\tau) = [sI - A]^{-1}$

$[sI - A] = \begin{bmatrix} s & -1 \\ +2 & s+3 \end{bmatrix}$

For eigen values

$|(sI - A)| = 0$

$s(s+3) + 2 = 0$

$s^2 + 3s + 2 = 0$

$(s+2)(s+1) = 0$

$s = -1, -2$

Eigen vectors can be find out from

$[sI - A]x = 0$

If for eigen value s = –1 then eigen vector

$$\begin{bmatrix} -1 & -1 \\ 2 & 2 \end{bmatrix} \begin{bmatrix} x_1 \\ x_2 \end{bmatrix} = 0$$

$-x_1 - x_2 = 0$

$2x_1 + 2x_2 = 0$

$$\frac{x_1}{1} = \frac{-x_2}{1}$$

Vector = $\begin{bmatrix} 1 \\ -1 \end{bmatrix}$

If Eigen value then eigen vector s = – 2

$$\begin{bmatrix} -2 & -1 \\ +2 & 2 \end{bmatrix} \begin{bmatrix} x_1 \\ x_2 \end{bmatrix} = 0$$

$-2x_1 - x_2 = 0$

$$\frac{x_1}{1} = \frac{-x_2}{1}$$

Vector = $\begin{bmatrix} 1 \\ -2 \end{bmatrix}$

25. Ans. (b)

Solution: It can be verified with options, Option (d) gives two eigen value – 2 and – 1

26. Ans. (c)

Solution: By applying Laplace-transform on both-sides, we get

$$\left[s^2 X(s) - sx(0) - x'(0)\right] + 6\left[s X(s) - x(0)\right] + 5X(s)$$

$$= 12\left(\frac{1}{s} - \frac{1}{s+2}\right)$$

It ignition condition not given assume zero

$$[s^2 + 6s + 5]X(s) = 12\left(\frac{1}{s} - \frac{1}{s+2}\right) = \frac{24}{s(s+2)}$$

or $$X(s) = \frac{24}{s(s^2 + 6s + 5)(s+2)}$$

then, $$\lim_{t\to\infty} x(t) = \lim_{s\to\infty}\left[sX(s)\right] = \lim_{s\to 0}\frac{24}{(s^2 + 6s + 5)(s+2)}$$

$$= \frac{24}{5 \times 2} = 2.4$$

27. Ans. (a)

Solution: Phase of circuit

$$\theta = \tan^{-1}\frac{\text{Im}}{\text{Re}} = \frac{\tan^{-1}\omega/a}{\tan^{-1}\omega/b}$$

$$\theta = \tan^{-1}\frac{\omega}{a} - \tan^{-1}\frac{\omega}{b}$$

For lead compensator, $\theta > 0$

Then, $b > a$

28. Ans. (a)

Solution: The state of the system at time t is given by formula

$x(t) = \phi(t)\, X(0)$

$= LT^{-1}[sI - A]^{-1}\, x(0)$

$$= \begin{bmatrix} e^{-2t} & 0 \\ 0 & e^{-t} \end{bmatrix}\begin{bmatrix} 3 \\ 2 \end{bmatrix} = \begin{bmatrix} 2e^{-2t} \\ 2e^{-t} \end{bmatrix}$$

At, t = 1, $$x(1) = \begin{bmatrix} 2e^{-2} \\ 3e^{-1} \end{bmatrix} = \begin{bmatrix} 0.271 \\ 1.100 \end{bmatrix}$$

29. Ans. (d)

Solution:

Apply initial value theorem = $f(0) = \lim_{t\to 0} f(t)$

$$= \lim_{s\to\infty} sF(s) = \lim_{s\to\infty} = \frac{s\,5}{s(s^2 + 3s + 2)} = 0$$

30. Ans. (c)

Solution:

Transfer function is derived as, $e(t) = K_t \frac{d\theta}{dt}$

E(s) = K_t sQ(s) ; $\frac{F(s)}{Q(s)} = s\,K_t$

31. Ans. (b)

Solution: Characteristic equation

$s^3 - 4s^2 + s + 6$

Constructing Routh Harwitz array

s^3	1	1	0
s^2	→	6	0
	–4		
	→		
s^1	2	0	
s^0	6		

Number of sign changes in the first column is two, therefore the number of roots in the right half s-pane is 2, total roots = 3.

∴ Roots in left half = 3 – 2 = 1

32. Ans. (a)

Solution:

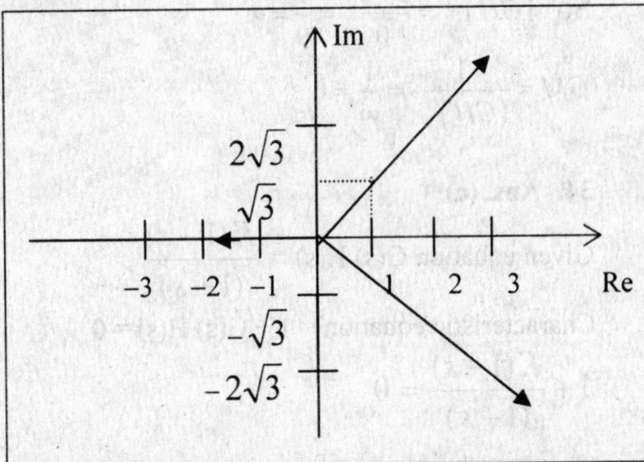

The given figure shows that all the roots are at origin or Root locus breaking point is at origin. Therefore the correct option is k./s^3.

$$G(s)H(s)=\frac{K}{s^3}$$

$$\frac{dK}{ds}=0$$

$$3s^2=0$$

$$s=0,0$$

Thus breaking points are at origin.

33. Ans. (a)

Solution:

Given function $G(s)=\frac{(s+1)}{s^2}$

$$G(j\omega)=\frac{(j\omega+1)}{(j\omega)^2}=\frac{1+j\omega}{-\omega^2}$$

For gain margin, take G(jω) complex part as zero i.e.

$$I_m G(j\omega)=\frac{\omega}{Dr}=0$$

$$\omega=0$$

$$\left|G(j\omega_p)\right|=\frac{0+1}{0_2}=\frac{1}{0}=\infty$$

$$\left|G(j\omega_p)\right|=20\log\frac{1}{|G(j\omega)|}=20\log\frac{1}{\infty}=0$$

Another way

$$GH(s)=\frac{s+1}{s^2}$$

Phase cross over frequency

$$-180=-180+\tan^{-1}\omega_p$$

$$\tan^{-1}\omega_p=0$$

$$\omega_p=0$$

$$\text{So } (GH)_{\omega_p}=\frac{1+0}{0}=\frac{1}{0}=\infty$$

$$GM=\frac{1}{(GH)_{\omega_p}}=\frac{1}{\infty}=0$$

34. Ans. (c)

Given equation G(s) H(s) = $\frac{K(1-s)}{(1+s)}$

Characteristic equation = 1 + G(s) H(s) = 0

$$1+\frac{K(1-s)}{(1+s)}=0$$

or (1 + s) + K(1 – s) = 0

or s – Ks + K + 1 = 0

or (1 – k)s + (K + 1) = 0

RH Table

s^1	1 – K
s^0	1 + K

For stability 1 – K > 0

1 + K > 0 Thus | K| <|

35. Ans. (a)

Solution: Given $H(j\omega)=\frac{10^4\,(1+j\omega)}{(10+j\omega)(100+j\omega)^2}$

Convert the given equation is standard form

i.e. $\frac{K(1+j\omega T_1)\ldots\ldots}{(j\omega)^n(1+j\omega T_a)(1+j\omega T_b)^2}$

$$=\frac{10^4\,(1+j\omega)}{10\left(1+\frac{j\omega}{10}\right)(100)^2\left(1+\frac{j\omega}{100}\right)^2}$$

$$=\frac{0\cdot 1\,(1+j\omega)}{\left(1+\frac{j\omega}{10}\right)\left(1+\frac{j\omega}{100}\right)^2}$$

Value of K_{db} = + 20 log $_{10}$ (0.1) = – 20 db.

So plot will start from – 20 db ft.

In other way this problem can be seen as

When ω → 0

Gain → H(0) = $\frac{10^4}{10\times 100}$ = 10

20 log [1 / 10] = – 20 dB

36. Ans. (b)

Solution: $G(s)=\frac{1}{s(s+1)(s+2)}$

$$G(j\omega)=\frac{1}{j\omega(1+j\omega)(2+j\omega)}$$

$$=\frac{-j(1-j\omega)(2-j\omega)}{\omega(1+\omega^2)(4+\omega^2)}$$

$$=\frac{-j(2-3j\omega-\omega^2)}{\omega(1+\omega^2)(4+\omega^2)}$$

$$=\frac{-3\omega-j(2-\omega^2)}{\omega(1+\omega^2)(4+\omega^2)}$$

$$=\frac{-3}{(1+\omega^2)(4+\omega^2)}-j\frac{(2-\omega^2)}{\omega(1+\omega^2)(4+\omega^2)}$$

$$\therefore\ X=-\frac{3}{(1+\omega^2)(4+\omega^2)},\ y=-\frac{(2-\omega^2)}{\omega(1+\omega^2)(4+\omega^2)}$$

at ω → 0, x → [– 3 / 4]; y → ∞

Two Marks Questions

1. A control system with certain excitation is governed by the following mathematical equation

$$\frac{d^2x}{dt^2}+\frac{1}{2}\frac{dx}{dt}+\frac{1}{18}x=10+5e^{-4t}+2e^{-5t}$$

The natural time constants of the response of the system are

(a) 2s and 5s (b) 3s and 6s
(c) 4s and 5s (d) 1/3s and 1/6s

2. The block diagram shown in figure gives a unity feedback closed loop control system. The steady state error in the response of the above system to unit step input is

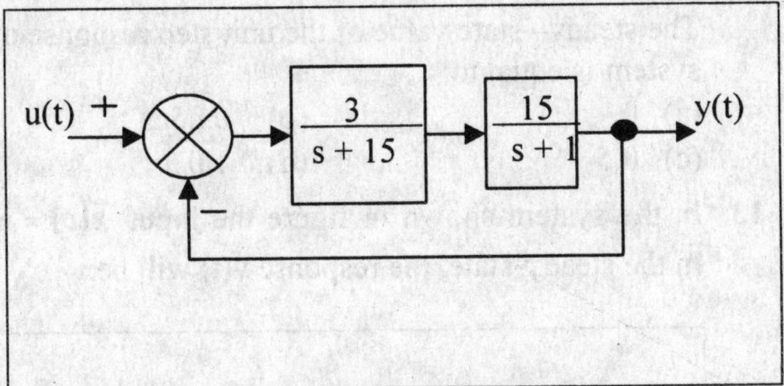

(a) 25 % (b) 0.75 %
(c) 6 % (d) 33 %

3. The roots of the closed loop characteristic equation of the system shown in figure are

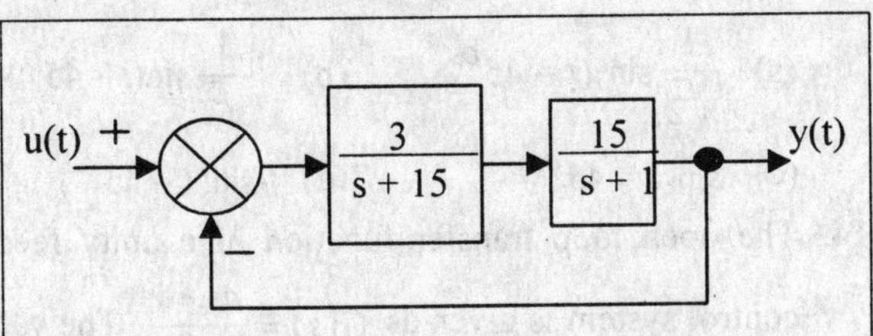

(a) – 1 and –15 (b) 6 and 10
(c) – 4 and –15 (d) – 6 and –10

4. The following equation defines a separately excited dc motor in the form of a differential equation

$$\frac{d^2\omega}{dt^2}+\frac{B}{J}\frac{d\omega}{dt}+\frac{K^2}{LJ}\omega=\frac{K}{LJ}V_a$$

The above equation may be organized in the state-space form as follows

$$\begin{bmatrix}\frac{d^2\omega}{dt^2}\\ \frac{d\omega}{dt}\end{bmatrix}=P\begin{bmatrix}\frac{d\omega}{dt}\\ \omega\end{bmatrix}+QV_a$$

Where the P matrix is given by

(a) $\begin{bmatrix}-\frac{B}{J} & -\frac{K^2}{LJ}\\ 1 & 0\end{bmatrix}$ (b) $\begin{bmatrix}-\frac{K^2}{LJ} & -\frac{B}{J}\\ 0 & 1\end{bmatrix}$

(c) $\begin{bmatrix}0 & 1\\ -\frac{K^2}{LJ} & -\frac{B}{J}\end{bmatrix}$ (d) $\begin{bmatrix}0 & 1\\ -\frac{B}{J} & -\frac{K^2}{LJ}\end{bmatrix}$

5. The loop gain GH of a closed loop system is given by the following expression $\frac{K}{s(s+2)(s+4)}$

The value of K for which the system just becomes unstable is

(a) K = 6 (b) K = 8
(c) K = 48 (d) K = 96

6. The asymptotic Bode plot of the transfer function [K / 1 + (s/a)] is given in figure. The error in phase angle and dB gain at a frequency of ω = 0.5 a are respectively

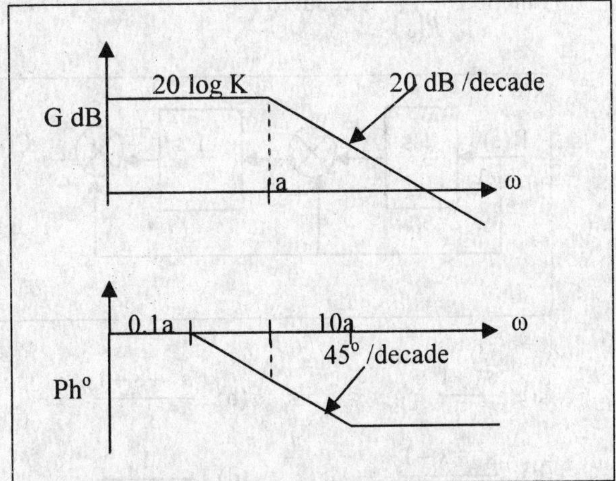

(a) 4.9°, 0.97 dB (b) 5.7°. 3 dB
(c) 4.9°, 3 dB (d) 5.7°, 0.97 dB

7. The block diagram of a control system is shown in figure. The transfer function G(s) = Y(s) / U(s) of the system is:

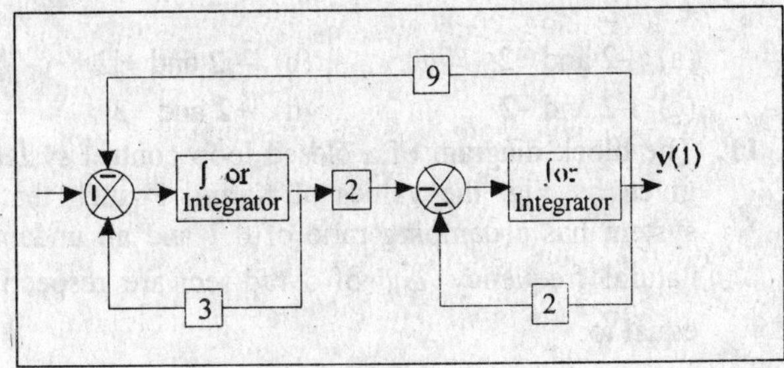

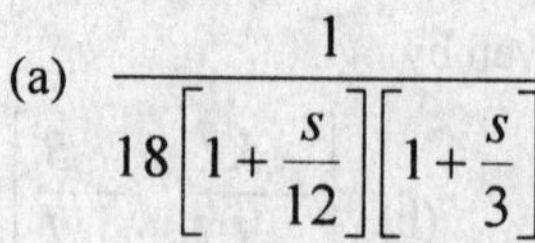

(a) $\dfrac{1}{18\left[1+\dfrac{s}{12}\right]\left[1+\dfrac{s}{3}\right]}$

(b) $\dfrac{1}{27\left[1+\dfrac{s}{6}\right]\left[1+\dfrac{s}{9}\right]}$

(c) $\dfrac{1}{27\left[1+\dfrac{s}{12}\right]\left[1+\dfrac{s}{9}\right]}$

(d) $\dfrac{1}{27\left[1+\dfrac{s}{9}\right]\left[1+\dfrac{s}{3}\right]}$

8. For the equation, $s^3 - 4s^2 + s + 6 = 0$ The number of roots in the left half of s-plane will be

(a) zero (b) one
(c) two (d) three

9. For block diagram shown in figure the transfer function $\dfrac{C(s)}{R(s)}$ is equal to

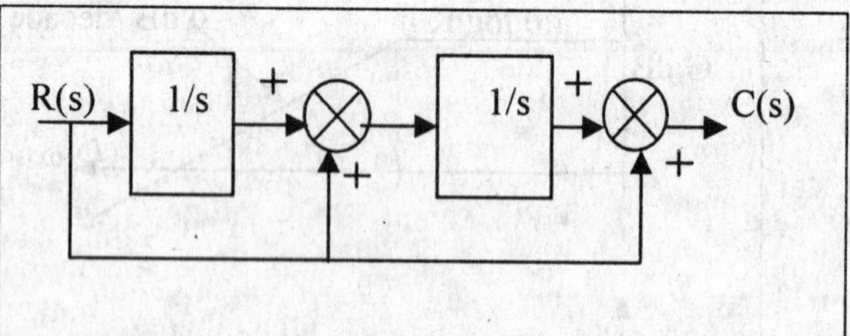

(a) $\dfrac{s^2+1}{s^2}$ (b) $\dfrac{s^2+s+1}{s^2}$

(c) $\dfrac{s^2+s+1}{s^2}$ (d) $\dfrac{1}{s^2+s+1}$

10. The state variable description of a linear autonomous system is, X'=AX, where X is the two dimensional state vector and A is the system matrix given by A= $\begin{bmatrix} 0 & 2 \\ 2 & 0 \end{bmatrix}$. The roots of the characteristic equation are

(a) −2 and +2 (b) −j2 and +j2
(c) −2 and −2 (d) +2 and +2

11. The block diagram of a closed loop control system is given by Fig. the values of K and P such that the system has a damping ratio of 0.7 and an undamped natural frequency ω_n of 5 rad/sec, are respectively equal to:

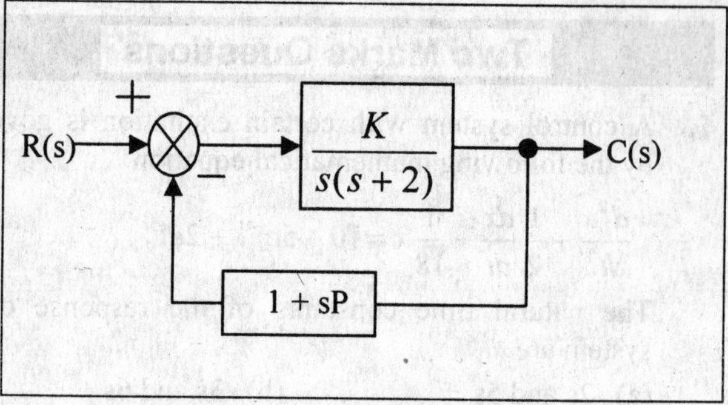

(a) 20 and 0.3 (b) 20 and 0.2
(c) 25 and 0.3 (d) 25 and 0.2

12. The unit impulse response of a second order under-damped system starting from rest is given by
$C(t) = 12.5\,e^{-6t} \sin 8t,\ t \ge 0$
The steady –state value of the unit step response of the system is equal to

(a) 0 (b) 0.25
(c) 0.5 (d) 1.0

13. In the system shown in figure the input $x(t) = \sin t$. In the steady-state, the response y(t) will be:

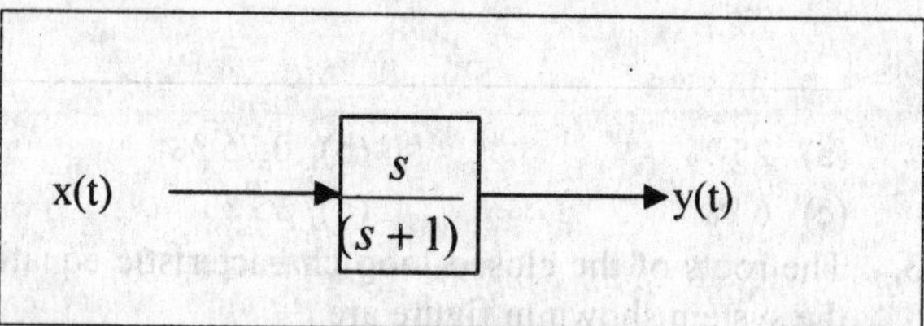

(a) $\frac{1}{\sqrt{2}} \sin(t - 45^o)$ (b) $\frac{1}{\sqrt{2}} \sin(t + 45^o)$

(c) $\sin(t - 45^o)$ (d) $\sin(t - 45^o)$

14. The open loop transfer function of a unity feedback control system is given as $G(s) = \dfrac{as+1}{s^2}$. The value of 'a' to give a phase margin of 45° is equal to

(a) 0.141 (b) 0.441
(c) 0.841 (d) 1.141

15. A unity feedback system, having an open loop gain $G(s)H(s) = \dfrac{K(1-s)}{(1+s)}$, becomes stable when

(a) $|K| > 1$ (b) $K > 1$
(c) $|K| < 1$ (d) $K < -1$

16. When subjected to a unit step input, the closed loop control system shown in the figure will have a steady state error of:

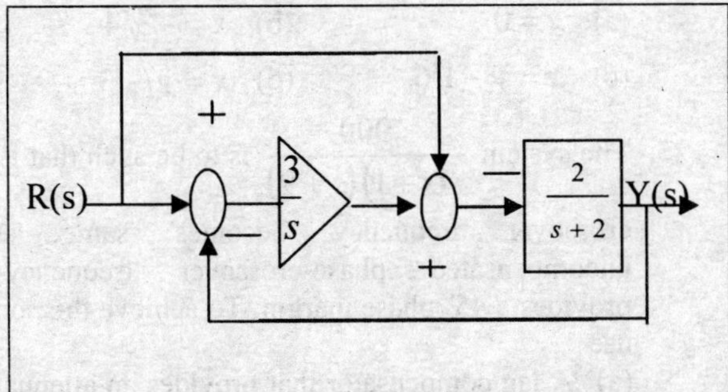

(a) −1.0 (b) −0.5
(c) 0 (d) 0.5

17. In the GH(s) plane, the Nyquist plot of the loop transfer function $G(s)H(s) = \dfrac{\pi e^{-0.25s}}{s}$ passes through the negative real axis at the point

(a) (0.25, j0) (b) (-0.5, j0)
(c) (1, j0) (d) (-2, j0)

18. If the compensated system shown in the figure has a phase margin of 60° at the crossover frequency of 1 rad/sec, then value of the gain K is:

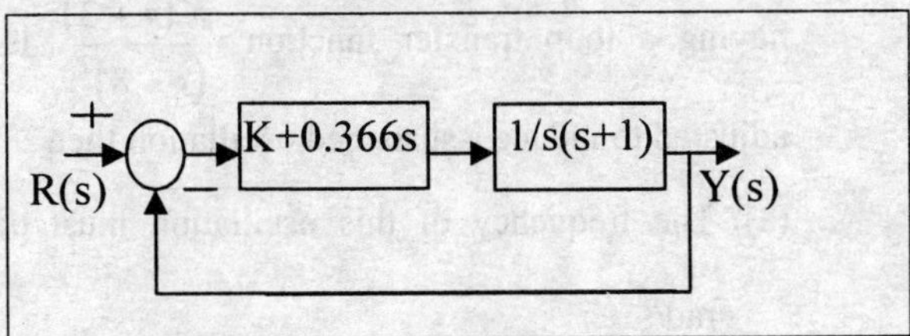

(a) 0.366 (b) 0.732
(c) 1.366 (d) 2.738

19. A state variable system $X'(t) = \begin{bmatrix} 0 & 1 \\ 0 & -3 \end{bmatrix} X(t) + \begin{bmatrix} 1 \\ 0 \end{bmatrix}$ u (t), with the initial condition $X(0) = [-1 \;\; 3]^T$ and the unit step input u(t) has

The state transition matrix

(a) $\begin{bmatrix} 1 & \frac{1}{3}\left(1 - e^{-3t}\right) \\ 0 & e^{-3t} \end{bmatrix}$ (b) $\begin{bmatrix} 1 & \frac{1}{3}\left(e^{-t} - e^{-3t}\right) \\ 0 & e^{-t} \end{bmatrix}$

(c) $\begin{bmatrix} 1 & \frac{1}{3}\left(e^{-t} - e^{-3t}\right) \\ 0 & e^{-3t} \end{bmatrix}$ (d) $\begin{bmatrix} 1 & \left(1 - e^{-t}\right) \\ 0 & e^{-t} \end{bmatrix}$

20. A state variable system $X'(t) = \begin{bmatrix} 0 & 1 \\ 0 & -3 \end{bmatrix} X(t) + \begin{bmatrix} 1 \\ 0 \end{bmatrix}$ u (t), with the initial condition $X(0) = [-1 \;\; 3]^T$ and the unit step input u(t) has

The state transition equation

(a) $X(t) = \begin{bmatrix} t - e^{-t} \\ e^{-t} \end{bmatrix}$ (b) $X(t) = \begin{bmatrix} t - e^{-t} \\ 3e^{-3t} \end{bmatrix}$

(c) $X(t) = \begin{bmatrix} t - e^{-3t} \\ 3e^{-3t} \end{bmatrix}$ (d) $X(t) = \begin{bmatrix} t - e^{-3t} \\ e^{-t} \end{bmatrix}$

21. The Bode magnitude plot of

$$H(j\omega) = \frac{10^4 (1 + j\omega)}{(10 + j\omega)(100 + j\omega)^2} \text{ is}$$

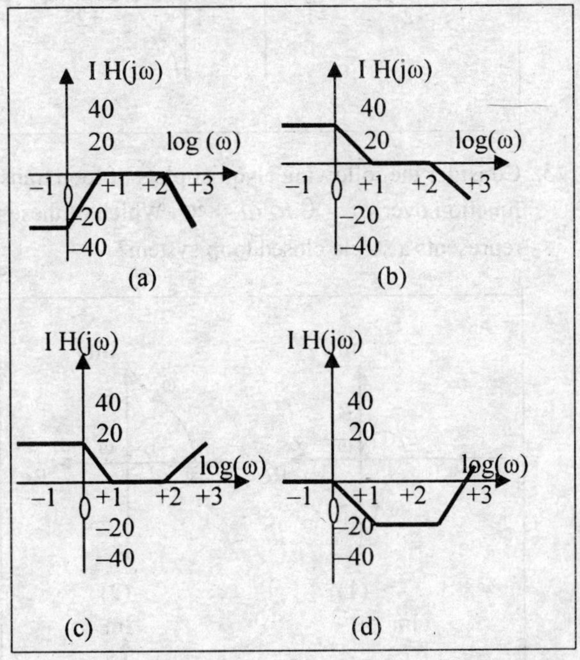

22. A closed-loop system has the characteristic function. $(s^2 - 4)(s + 1) + K(s - 1) = 0$ Its root locus plot against K is

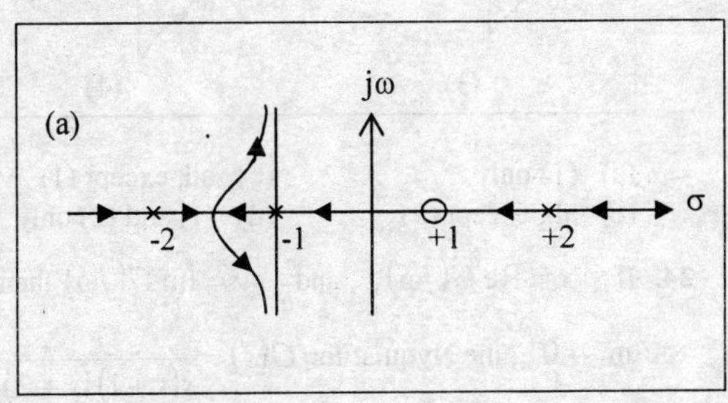

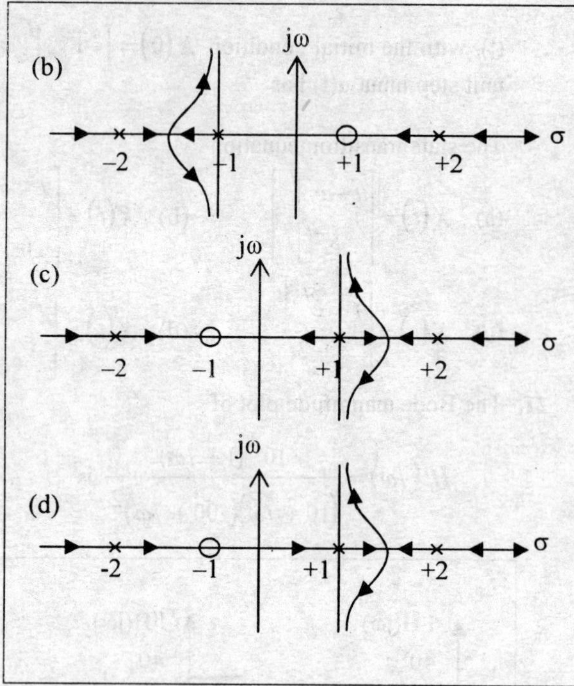

23. Consider the following Nyquist plots of loop transfer function over $\omega = 0$ *to* $\omega = \infty$. Which of these plots represents a stable closed loop system?

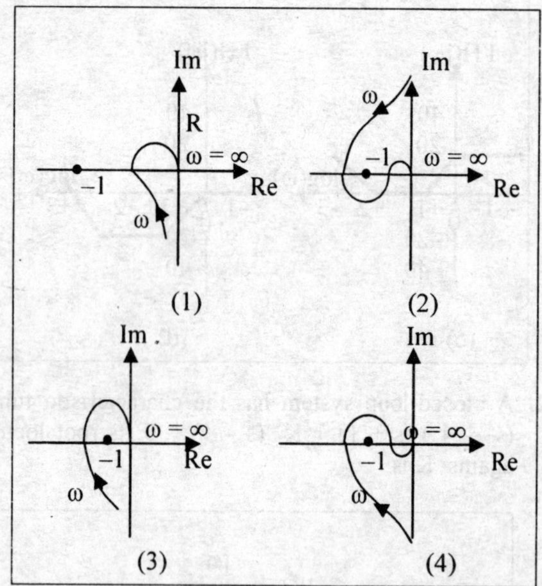

(a) (1) only
(b) all, except (1)
(c) all, except (3)
(d) (1) and (2) only

24. If $x = \text{Re}\, G(j\omega)$, and $y = \text{Im}\, G(j\omega)$ then for $\omega \to 0^+$, the Nyquist for $G(s) = \dfrac{1}{s(s+1)(s+2)}$

(a) $x = 0$
(b) $x = -3/4$
(c) $x = y - 1/6$
(d) $x = y/\sqrt{3}$

25. The system $\dfrac{900}{s(s+1)(s+9)}$ is to be such that its gain-crossover frequency becomes same as its uncompensated phase-crossover frequency and provides a 45° phase margin. To achieve this, one may use

(a) A lag compensator that provides an attenuation of 20 dB and a phase lag of 45° at the frequency of $3\sqrt{3}$ rad/s
(b) A lead compensator that provides an amplification of 20 dB and a phase laed of 45° at the frequency of 3 rad/s
(c) A lag--lead compensator that provides an amplification of 20 dB and a phase lag of 45° at the frequency of $\sqrt{3}$ rad/s
(d) A lag-lead compensator that provides an attenuation of 20 dB and a phase lead of 45° at the frequency of 3 rad/s

26. If the loop gain K of a negative feedback system having a loop transfer function $\dfrac{K(s+3)}{(s+8)^2}$ is to be adjusted to induce a sustained oscillation then

(a) The frequency of this oscillation must be $\dfrac{4}{\sqrt{3}}$ rad/s
(b) The frequency of this oscillation must be 4 rad/s
(c) The frequency of this oscillation must be 4 or $\dfrac{4}{\sqrt{3}}$ rad/s
(d) Such a K does not exist

27. The system shown in the figure below:

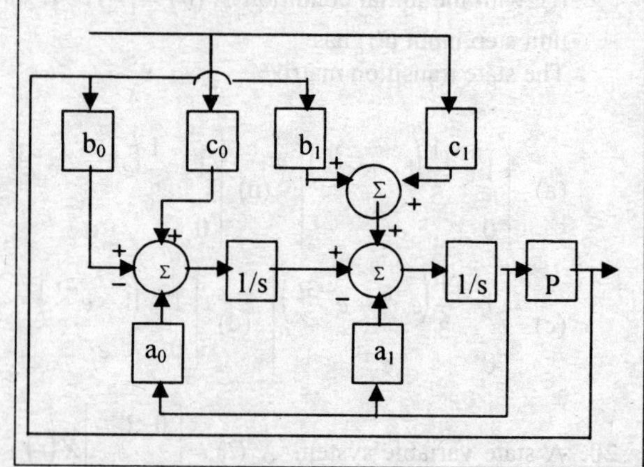

can be reduced to the form

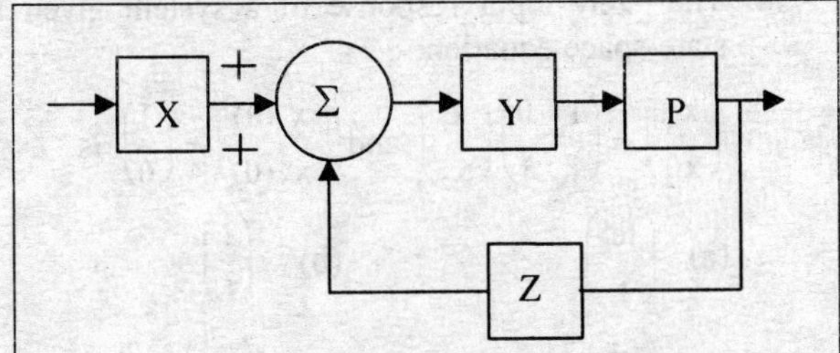

With

(a) $X = c_0 s + c_1,\ Y = \dfrac{1}{(s^2 + a_0 s + a_1)},\ Z = b_0 s + b_1$

(b) $X = 1,\ Y = \dfrac{(c_0 s + c_1)}{(s^2 + a_0 s + a_1)},\ Z = b_0 s + b_1$

(c) $X = c_1 s + c_0,\ Y = \dfrac{(b_1 s + b_0)}{(s^2 + a_1 s + a_0)},\ Z = 1$

(d) $X = c_1 s + c_0,\ Y = \dfrac{1}{(s^2 + a_1 s + a_0)},\ Z = b_1 s + b_0$

28. Consider the feedback control system shown below which is subjected to a unit step input. The system is stable and has the following parameters $k_p = 4$, $k_i = 10, \omega = 500$ and $\xi = 0.7$

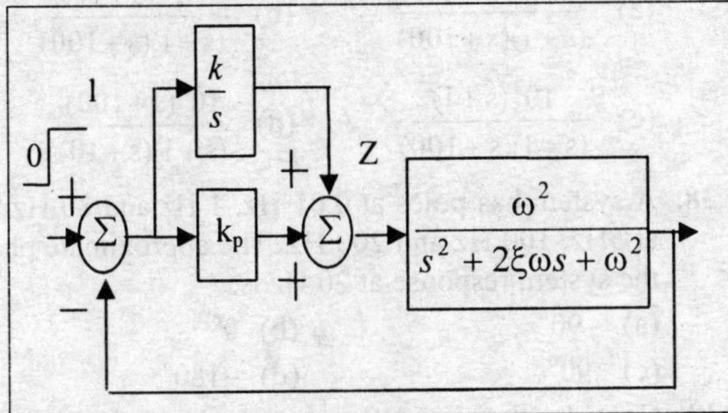

The steady state value of z is

(a) 1 (b) 0.25

(c) 0.1 (d) 0

29. Consider the R-L-C circuit shown in figure

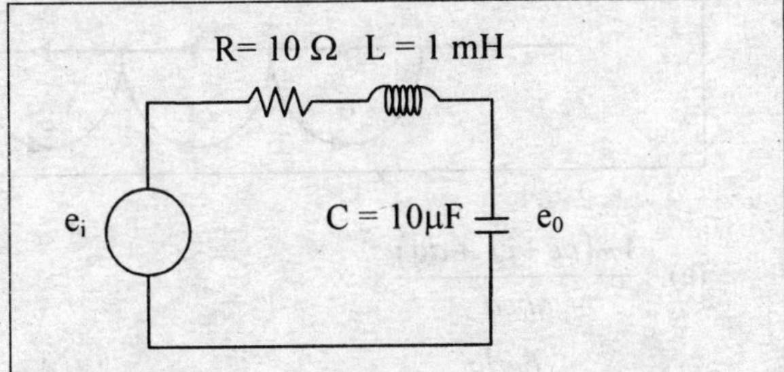

For a step-input e_i , the overshot in the output e_o will be

(a) 0, since the system is not under-damped

(b) 5 %

(c) 16 %

(d) 48 %

30. Consider the R-L-C circuit shown in figure:

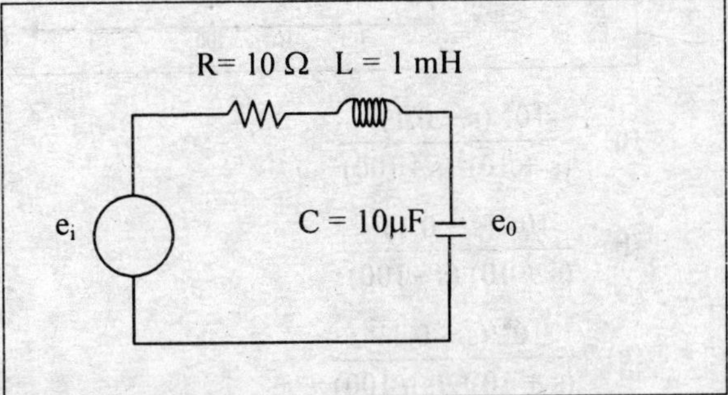

If the above step response is to be observed on a non-storage CRO, then it would be best to have the e_i as a

(a) Step function

(b) Square wave of frequency 50 Hz

(c) Square wave of frequency 300 Hz

(d) Square wave of frequency 2.0 kHz

31. The signal flow graph of a system is shown in figure. The transfer function [C(s)/R(s)] of the system is:

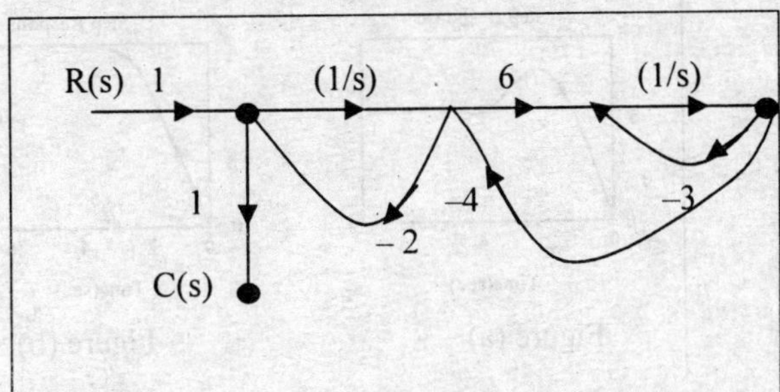

(a) $\dfrac{6}{s^2 + 29s + 6}$ (b) $\dfrac{6s}{s^2 + 29s + 6}$

(c) $\dfrac{s(s+2)}{s^2 + 29s + 6}$ (d) $\dfrac{s(s+27)}{s^2 + 29s + 6}$

32. The root locus of the system $G(s)H(s)=[K/s(s+2)(s+3)]$ has the break-away point located at:

(a) (-0.5, 0) (b) (–2.548, 0)

(c) (-4, 0) (d) (–0.784, 0)

33. The approximate Bode magnitude plot of minimum-phase system is shown in Figure. The transfer function of the system is:

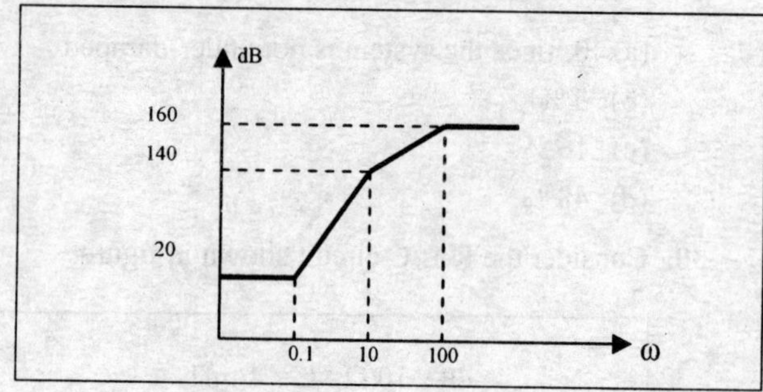

(a) $\dfrac{10^8\,(s+0.1)^3}{(s+10)^2(s+100)}$

(b) $\dfrac{10^7\,(s+0.1)^3}{(s+10)\,(s+100)}$

(c) $\dfrac{10^8\,(s+0.1)^2}{(s+10)^2\,(s+100)}$

(d) $\dfrac{10^9\,(s+0.1)^3}{(s+10)\,(s+100)^2}$

34. A second-order system has the transfer function

$$\frac{C(s)}{R(s)} = \frac{4}{s^2+4s+4}$$

with r(t) as the unit-step function, the response c(t) of the system is represented by:

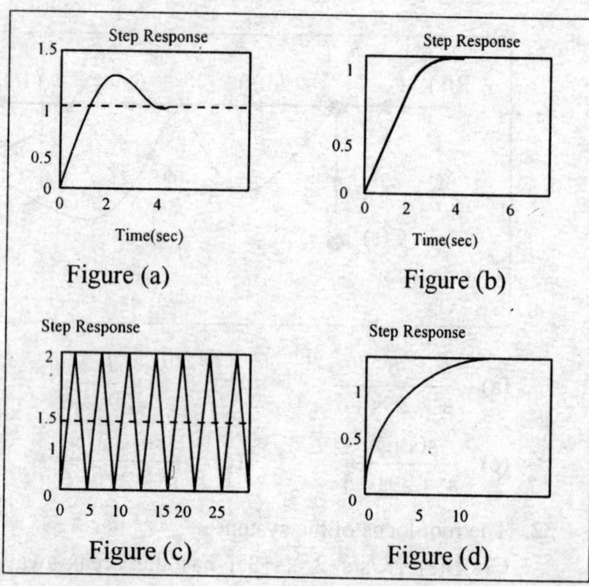

(a) Figure (a) (b) Figure (b)
(c) Figure (c) (d) Figure (d)

35. The gain margin and the phase margin of a feedback system with G(s) H(s) = are

(a) 0 dB, 0° (b) ∞, ∞
(c) ∞, 0° (d) 88.5 dB, ∞

36. The zero-input response of a system given by the state-space equation:

$$\begin{pmatrix} x'_1 \\ x'_2 \end{pmatrix} = \begin{pmatrix} 1 & 0 \\ 1 & 1 \end{pmatrix}\begin{pmatrix} x_1 \\ x_2 \end{pmatrix} \text{ and } \begin{pmatrix} x_1(0) \\ x_2(0) \end{pmatrix} = \begin{pmatrix} 1 \\ 0 \end{pmatrix} \text{ is}$$

(a) $\begin{pmatrix} te^t \\ t \end{pmatrix}$ (b) $\begin{pmatrix} \\ t \end{pmatrix}$

(c) $\begin{pmatrix} e^t \\ te^t \end{pmatrix}$ (d) $\begin{pmatrix} \\ te^t \end{pmatrix}$

37. Consider the Bode magnitude plot shown in figure. The transfer function H(s) is:

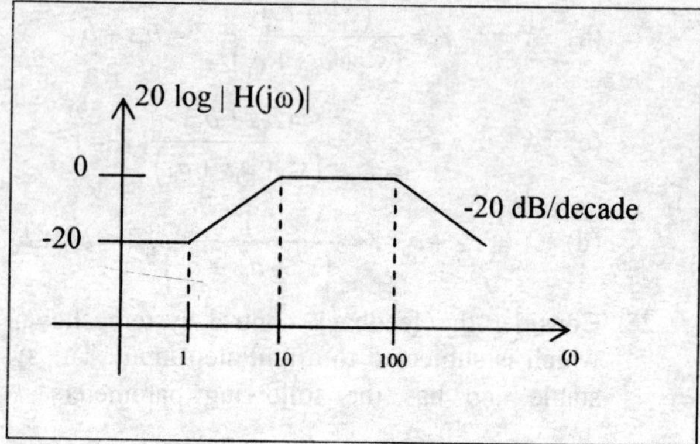

(a) $\dfrac{(s+10)}{(s+1)(s+100)}$ (b) $\dfrac{10(s+1)}{(s+1)(s+100)}$

(c) $\dfrac{10^2(s+1)}{(s+1)(s+100)}$ (d) $\dfrac{10^3(s+100)}{(s+1)(s+10)}$

38. A system has poles at 0.01 Hz, 1 Hz and 80 Hz; zeros at 5Hz. 100 Hz and 200 Hz. The approximate phase of the system response at 20 Hz is:

(a) −90° (b) 0°
(c) 90° (d) −180°

39. Consider the signal flow graph shown in Figure. The gain $[x_5/x_1]$ is:

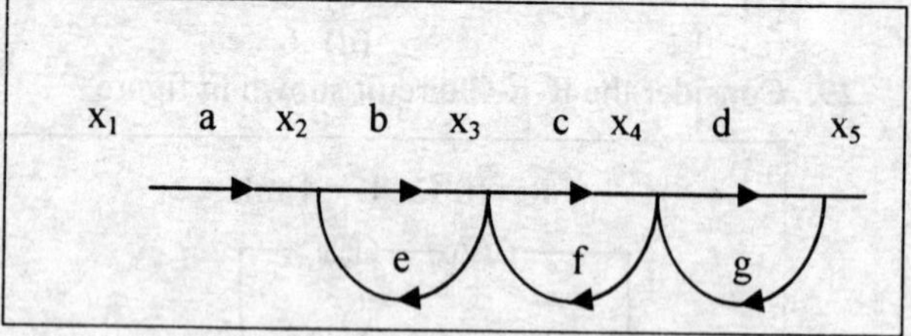

(a) $\dfrac{1-(be+cf+dg)}{abcd}$

(b) $\dfrac{bedg}{1-(be.+cf+dg)}$

(c) $\dfrac{abcd}{1-(be+cf+dg)+bedg}$

(d) $\dfrac{1-(be+cf+dg)+bedg}{abcd}$

40. If $\begin{pmatrix} A = & -2 \\ 1 & -3 \end{pmatrix}$ 2, then sin At is

1 -3

(a) $\begin{pmatrix} 1\sin(-4t)+2\sin(-t) & [2\sin(-4t)+2\sin(-t)] \\ 3\ \sin(-4t)+\sin(-t) & [2\sin(-4t)+\sin(-t)] \end{pmatrix}$

(b) $\begin{pmatrix} \sin(-2t) & \sin(2t) \\ \sin(t) & \sin(-3t) \end{pmatrix}$

(c) $\begin{pmatrix} l\sin(4t)+2\sin(t) & [2\sin(-4t)-2\sin(-t)] \\ -\sin(-4t)+\sin(t) & [2\sin(4t)+\sin(-t)] \end{pmatrix}$

(d) $\begin{pmatrix} l\cos(-t)+2\cos(t) & [2\sin(-4t)-2\sin(-t)] \\ 3-\sin(-4t)+\sin(-t) & [2\sin(-4t)+\sin(-t)] \end{pmatrix}$

41. The open-loop transfer function of a unity feedback system is $G(s) = \dfrac{k}{s(s2+s+2)(s+3)}$

The range of K for which the system is stable is

(a) $[21/4] > K > 0$
(b) $13 > K > 0$
(c) $[21/4] < K < \infty$
(d) $-6 < K < \infty$

42. For the polynomial $P(s) = s^5 + s^4 + 2s^3 + 2s^2 + 3s + 15$, the number of roots which lie in the right half of the s-plane is:

(a) 4 (b) 2
(c) 3 (d) 1

43. Given $A = \begin{bmatrix} 1 & 0 \\ 0 & 1 \end{bmatrix}$, the state transition matrix e^{At} is given by

(a) $\begin{pmatrix} 0 & e^{-t} \\ e^{-t} & 0 \end{pmatrix}$ (b) $\begin{pmatrix} e^{-t} & 0 \\ 0 & e^{t} \end{pmatrix}$

(c) $\begin{pmatrix} e^{-t} & 0 \\ 0 & e^{-t} \end{pmatrix}$ (d) $\begin{pmatrix} 0 & e^{t} \\ e^{t} & 0 \end{pmatrix}$

44. The state variable equations of a system are:

$x_1' = -3x_1 - x_2 + u;$

$x_2' = 2x_1$ and

$y = x_1 + u$

The system is:

(a) controllable but not observable
(b) observable but not controllable
(c) neither controllable nor observable
(d) controllable and observable

45. The polar diagram of a conditionally stable system for open gain K = 1 is shown in the figure. The open loop transfer function of the system is known to be stable. The closed loop system is stable for:

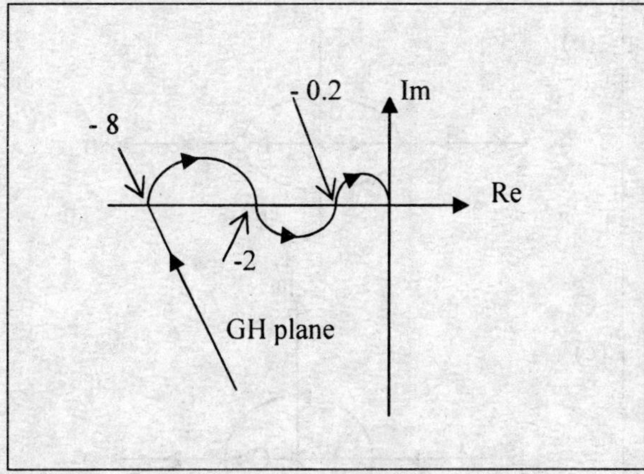

(a) $K < 5$ and $[1/2] < K < [1/8]$.
(b) $K < [1/8]$ and $[1/2] < K < 5$.
(c) $K < [1/8]$ and $5 < K$.
(d) $K > [1/8]$ and $K < 5$.

46. In the derivation of expression for peak percent overshoot, $M_P = \exp[-\pi\delta / \sqrt{(1-\delta^2)}] \times 100\ \%$ which one of the following condition is NOT required?

(a) System is linear and time invariant.
(b) The system transfer function has a pair of complex conjugate poles and no zeros.
(c) There is no transportation delay in the system.
(d) The system has zero initial conditions

47. A double integrator plant, $G(s) = [K/s^2]$, $H(s) = 1$ is to be compensated to achieve the damping ratio $\zeta = 0.5$, and an un-damped natural frequency, $\omega_n = 5$ rad/s. Which one of the following compensator $G_c(s)$ will be suitable?

(a) $\dfrac{s+3}{s+9.9}$ (b) $\dfrac{s+9.9}{s+3}$

(c) $\dfrac{s-6}{s+8.33}$ (b) $\dfrac{s-6}{s}$

48. An unity feedback system is given as,

$G(s) = \dfrac{K(1-s)}{s(s+3)}$

Indicate the correct root locus diagram

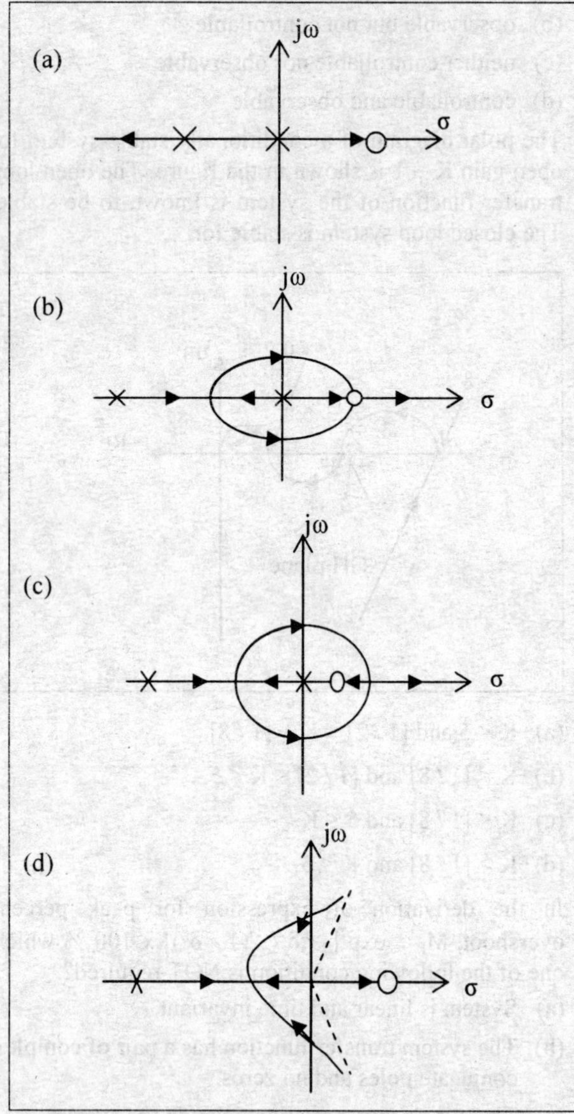

49. A ramp input applied to an unity feedback system results in 5 % steady state error. The type number and zero frequency gain of the system are respectively.

(a) 1 and 20 (b) 0 and 20
(c) 0 and [1 / 20] (d) 1 and [1 / 20]

50. The open loop transfer function of a unity feedback system is given by

$$G(s) = \frac{3e-2}{s(s+2)}$$

The gain and phase crossover frequencies in rad/sec are, respectively

(a) 0.632 and 1.26 (b) 0.632 and 0.485
(c) 0.485 and 0.632 (d) 1.26 and 0.632

51. The open loop transfer function of a unity feedback system is given by:

$$G(s) = \frac{3e-2}{s(s+2)}$$

Based on the above results, the gain and phase margins of the system will be

(a) – 7.09 and 87.5°
(b) 7.09 and 87.5°
(c) 7.09 dB and – 87.5°
(d) – 7.09 dB and – 87.5°

Answers
Control Systems and Process Control

1. Ans. (b)

Solution: Natural response is when differential equation is zero i.e. forcing function is zero.

$$\left(s^2+\frac{1}{2}s+\frac{1}{18}\right)=0.$$

$$s_{1,2}=\frac{-\frac{1}{2}\pm\sqrt{\frac{1}{4}-\frac{4}{18}}}{12}=\frac{1}{4}\pm\frac{1}{12}=\frac{1}{3},\frac{1}{6}$$

The time constant $T_1 = 3$ sec

$T_2 = 6$ sec

Second Method

$$T(s)=\frac{C(s)}{R(s)}=\frac{1}{s^2+\frac{s}{2}+\frac{1}{18}}=\frac{18}{18s^2+9s+1}$$

$$=\frac{18}{(3s+1)(6s+1)}$$

As, T(s) have the form, $\frac{A}{(T_1s+1)(T_2s+1)}$

The time – constant are $T_1 = 3$ sec, $T_2 = 6$ sec.

Third Method

For natural response

$$\frac{dx^2}{dt^2}+\frac{1}{2}\frac{dx}{dt}+\frac{x}{18}=\delta(t)$$

$$X(s)=\frac{1}{s^2+\frac{s}{2}+\frac{1}{18}}=\frac{1}{\left(s+\frac{1}{3}\right)\left(s+\frac{1}{6}\right)}$$

Thus $s=-\frac{1}{3}$ and $s=-\frac{1}{6}$

Natural time constant is 3 second and 6 second.

2. Ans. (a)

Solution: $e_{ss}=\lim_{s\to 0}\frac{s.R(s)}{1+G(s)H(s)}$ as $R(s)=\frac{1}{s}$

$$= \lim_{s\to 0} \frac{s \times \frac{1}{s}}{1+\left(\frac{3}{s+15}\right)\left(\frac{15}{s+1}\right)}$$

$$= \lim_{s\to 0} \frac{s^2+16s+15}{s^2+16s+60} = \frac{15}{60}$$

$$= \frac{1}{4}\times 100 = 25\%$$

3. **Ans. (d)**

Solution: Characteristic equation = 1 + G(s)H(s) = 0

$$= s^2 + 16s + 60 = 0$$

$$s = \frac{-16 \pm \sqrt{16^2 - 240}}{2} = \frac{-16 \pm 4}{2} = -6, -10$$

4. **Ans. (a)**

Solution: $x_1 = \omega, x_1' = x_2 = \frac{d\omega}{dt}, x_2' = x_1'' = \frac{d^2\omega}{dt^2}$

$$\frac{d^2\omega}{dt^2} = -\frac{B}{J}\frac{d\omega}{dt} - \frac{K^2}{LJ}\omega + \frac{K}{LJ}V_a$$

$$x_2' = -\frac{B}{J}x_2 - \frac{K^2}{LJ}x_1 + \frac{K}{LJ}V_a$$

$$\begin{bmatrix} x_2' \\ x_1' \end{bmatrix} = \begin{bmatrix} -\frac{B}{J} & \frac{K^2}{LJ} \\ 1 & 0 \end{bmatrix}\begin{bmatrix} x_2 \\ x_1 \end{bmatrix} + \begin{bmatrix} \frac{K}{LJ} \\ 0 \end{bmatrix} V_a$$

5. **Ans. (c)**

Solution: The characteristic equation is,

$$1 + GH = 0 \text{ or } 1+\frac{K}{s(s+2)(s+4)} = 0 \text{ or } 1+\frac{K}{s(s^2+6s+8)} = 0$$

or $s^3 + 6s^2 + 8s + K = 0$

Apply Routh-Hurwitz array

$$\begin{array}{c|cc} s^3 & 1 & 8\ 0 \\ s^2 & 6 & K \\ s^1 & K-48-0 & \\ s & K & \end{array}$$

The system will be just if first column of all Row have same sign i.e. $K - 48 \geq 0$ or $K \geq 48$.

6. **Ans. (a)**

Solution: $\angle GH\big|_{\omega=0.5a} = -\tan^{-1}\frac{\omega}{a}$

$$= -\tan^{-1} 0.5 = 26.56^o$$

From the plot,

At ω = a, $\angle GH = 22.5^o$

Then error = $-22.50^o - (-26.56^o) = 4.06^o$

At ω = 0.5 a, $|GH|_{dB} = 20 \log K - 20 \log |1 + j0.5|$

$= 20 \log K - 0.97$

From the plot, at ω = 0.5 a $|GH|_{dB}$

$= 20 \log |GH| = 20 \log K$

Then, error in dB gain = 0.97 dB

7. **Ans. (b)**

Solution:

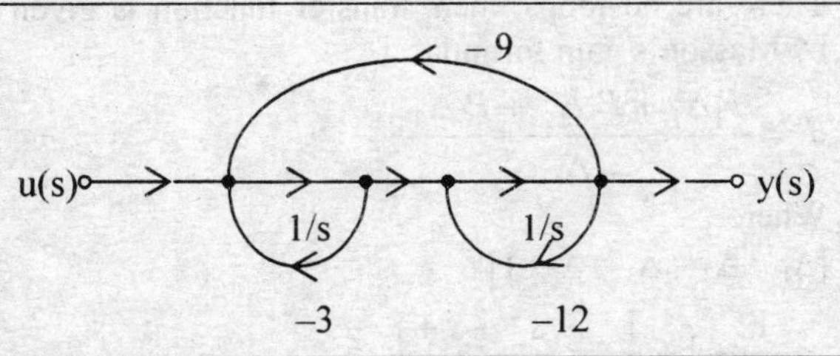

Forward path gain, $P_1 = \frac{2}{s^2}$

Individual loop gains = $\left(\frac{-3}{s}\right), \left(\frac{-12}{s}\right), \left(\frac{-18}{s^2}\right)$

P_{12} (Non-touching loop gain) = $36/s^2$

$$\frac{Y(s)}{U(s)} = T(s) = \frac{P_1\Delta_1}{1 - P_{11} - P_{21} - P_{23} + P_{12}}$$

$$= \frac{2/s^2}{1+\frac{15}{s}+\frac{18}{s^2}+\frac{36}{s^2}} = \frac{2}{s^2+15s+54}$$

$$= \frac{2}{(s+6)(s+9)} = \frac{\frac{2}{(6\times 9)}}{\frac{\{(s+6)(s+9)\}}{(6\times 9)}}$$

$$= \frac{1}{27\left(1+\frac{s}{6}\right)\left(1+\frac{s}{9}\right)}$$

8. **Ans. (a)**

Solution: At point (–1, j0), $\angle G(j\omega).H(j\omega)\big|_{\omega=\omega_1} = 180^o$

Phase margin, $\phi = \angle G(j\omega)H(j\omega)\big|_{\omega=\omega_1} + 180^o$

$= -180^o + 180^o = 0^o$

9. **Ans. (b)**

Solution: Constructing signal-flow graph as shown here,

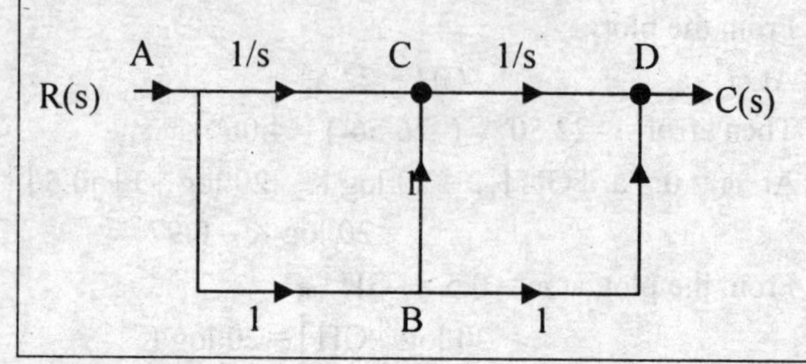

Number of forward paths is 3 i.e. ACD, ABD, ABCD, having path gains,

$P_1 = 1/s^2$, $P_2 = 1$, $P_3 = 1/s$

There are no loops, then, transfer function is given a by Masson's gain formula

$$T = \frac{P_1\Delta_1 + P_2\Delta_2 + P_3\Delta_3}{\Delta}$$

Where,

$[\Delta_1 = \Delta_2 = \Delta_3 = \Delta = 1]$

$$= \frac{1}{s^2} + 1 + \frac{1}{s} = \frac{s^2 + s + 1}{s^2}$$

10. Ans. (a)

Solution: Characteristic equation for the system is, given by

$|sI - A| = 0$

$$\text{or} \quad \begin{vmatrix} s & -2 \\ -2 & s \end{vmatrix} = 0$$

or $s^2 - 4 = 0$ or $s = \pm 2$.

11. Ans. (d)

$$\textbf{Solution: } T(s) = \frac{C(s)}{R(s)} = \frac{k/s(s+2)}{1 + \dfrac{k(1+sP)}{s(s+2)}}$$

$$= \frac{k}{s^2 + (2 + Pk)s + k} \qquad (1)$$

Standard 2nd order equation

$$= \frac{\omega_{n^2}}{s^2 + 2\delta\omega_n s + \omega_{n^2}} \qquad (2)$$

Compare equation (1) and (2)

Now, $k = \omega_n^2$, and $(Pk + 2) = 2\delta\omega_n$

Then, $k = 25$ and $25P + 2 = 2 \times .7 \times 5 = 7$

$25\,P = 5$

$P = 5/25 = 1/5 = 0.2$

12. Ans. (d)

$$\textbf{Solution: } H(s) = C(s) = \frac{12.5 \times 8}{(s+6)^2 + 8}$$

$$= \frac{100}{s^2 + 12s + 100}$$

Then, output for unit step response

$$C_u(s) = \frac{1}{s} \cdot \frac{100}{s^2 + 12s + 100}$$

and steady-state value $= \lim_{s \to 0} sC_u(s)$

$$= \lim_{s \to 0} \frac{100}{s^2 + 12s + 100} = 1$$

13. Ans. (b)

$$\textbf{Solution: } H(s) = \frac{s}{(s+1)}$$

$\phi = 90 - 45 = 45°$

$$|H(s)| = \frac{1}{\sqrt{1+1}} = \frac{1}{\sqrt{2}}$$

Therefore this transfer function circuit will give amplitude change of $1/\sqrt{2}$ and phase 45°

Hence, in steady state output is $1/\sqrt{2}\,\sin(t + 45°)$

Second Method

$x(t) = \sin t$; $x(s) = \dfrac{1}{s^2 + 1}$; $Y(s) = X(s)\dfrac{s}{s+1}$

$$Y(s) = \frac{1}{(s^2+1)} \cdot \frac{s}{s+1} = \frac{A}{(s+j)} + \frac{B}{(s-j)} + \frac{C}{(s+1)}$$

$$= \frac{1}{2(+j)}\left(\frac{1}{s-j}\right) + \frac{1}{2(1-j)(s+j)} - \frac{1}{2(s+1)}$$

Taking inverse Laplace transformation, we have

$$y(t) = \frac{1}{2(1+j)}e^{jt} + \frac{1}{2(1-j)}e^{-jt} - \frac{1}{2}e^{-t}$$

$$= \frac{1}{2}(\sin t + \cos t) - \frac{1}{2}e^{-t}$$

$$= \frac{1}{\sqrt{2}}\sin(t + 45°) - \frac{1}{2}e^{-t}$$

Then, steady state value $= \lim_{t \to \infty} y(t)$

$$\frac{1}{\sqrt{2}}\sin(t + 45^o)$$

14. Ans. (c)

Solution: For Phase Margin, find the cut off frequency for gain = 1

The value of ω at which $|G(j\omega)| = 1$ is given by,

$$1 = \frac{\sqrt{(\omega_c a)^2 + 1}}{\omega_c^{\,2}} \qquad (1)$$

Also $\angle G(j\omega)\,|_{\omega=\omega c} + 180^\circ = 45^\circ$
Or $\tan^{-1}(\omega_c a) - 180^\circ + 180^\circ = 45^\circ$
Or $\omega_c a = 1$
From equation (i), we have,

$$\omega_c^{\;2} = \sqrt{(\omega_c a)^2 + 1}$$

$\omega_c^{\;4} = (\omega_c a)^2 + 1$
$\omega_c^{\;4} = 1 + 1 = 2$
$\omega_c = (2)^{1/4}$
$\omega_c = 2^{1/4} = 1.189$ rad/sec.
Then, a = 0.841.

15. Ans. (c)
Solution:

$$T(s) = \frac{s^2 + 4}{(s+1)\,(s+4)};$$

At the frequency ω, gain is zero i.e.

$$T(j\omega) = \frac{\left|(j\omega)^2 + 4^1\right|}{\left|(j\omega+1)\,(j\omega+4)\right|} = 0$$

$-\omega^2 + 4 = 0$
$\omega = 2$ rad/sec.

16. Ans. (c)
Solution:

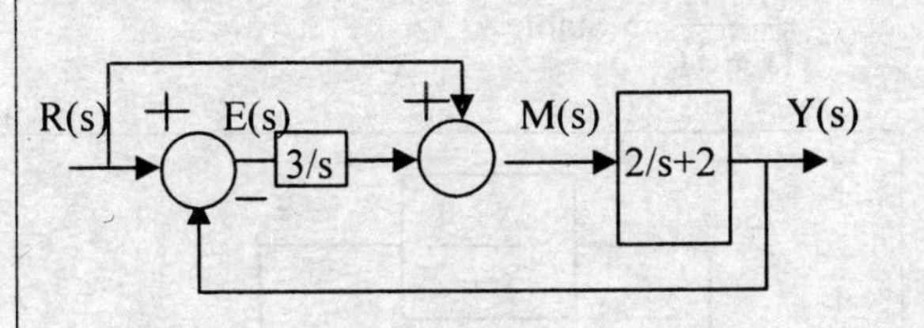

R(s) = 1/s

$$M(s) = R(s) + [R(s) - Y(s)]\frac{3}{s}$$

$$\text{And } Y(s) = \frac{2}{s+2}\left[R(s)\left(1+\frac{3}{s}\right) - \frac{3}{s}Y(s)\right]$$

$$\text{or } Y(s) + \frac{6}{s(s+2)}Y(s) = R(s)\frac{(s+3)}{s}\frac{2}{(s+2)}$$

$$\text{or } Y(s)\left[\frac{s(s+2)+6}{s(s+2)}\right] = R(s)\frac{2(s+3)}{s(s+2)}$$

$$\frac{Y(s)}{R(s)} = \frac{2(s+3)}{s(s+2)} \times \frac{s(s+2)}{s^2+2s+6}$$

$$\therefore \frac{Y(s)}{R(s)} = \frac{2(s+3)}{s^2+2s+6}$$

$$E(s) = R(s) - Y(s) = R(s)\left[1 - \frac{2(s+3)}{s^2+2s} + 6\right]$$

$$E(s) = R(s)\frac{s^2}{s^2+2s+6} = \frac{s}{s^2+25+6}$$

$$e_{ss} = \lim_{s\to 0} sE(s) = 0$$

17. Ans. (b)

Solution: $G(s)\,H(s) = \dfrac{\pi\, e^{-0.25s}}{s}$

$$G(j\omega)H(j\omega) = \frac{\pi[\cos(0.25\omega) - j\sin(0.25\omega)]}{j\omega}$$

$$= \frac{-\pi}{\omega}\sin(0.25\omega) - j\frac{\pi}{\omega}\cos(0.25\omega)$$

To get the interception on real axis put imaginary part = 0

$$\frac{\pi}{\omega}\cos(0.25\omega) = 0;\ \cos\frac{\omega}{4} = \cos\pi/2$$

$$\frac{\omega}{4} = \frac{\pi}{2};\ \omega = 2\pi$$

intercept on Negative real axis.

$$\therefore \left|G(j\omega)H(j\omega)\right|_{\omega=2\pi} = \left|-\frac{\pi}{2\pi}\sin\left(\frac{2\pi}{4}\right)\right| = \left|-\frac{1}{2}\right| = 0.5$$

18. Ans. (c)
Solution:

Here, $= G(s) = \dfrac{K + 0.366s}{s(s+1)}$, in present case, unity feed back is applied

Since, Phase Margin = 180° + ∠G(jω) H(jω)

$$60^o = 180^o + \tan^{-1}\left(\frac{0.366\omega}{K}\right) - 90^o - \tan^{-1}\omega$$

$$\left[\tan^{-1}A - \tan^{-1} = \tan^{-1}\left\{\frac{A-B}{1+A\cdot B\cdot}\right\}\right]$$

$$\Rightarrow \tan^{-1}\left(\frac{\frac{0.366\omega}{K} - \omega}{1 + \frac{0.366\omega^2}{K}}\right) = -30^o$$

$$\Rightarrow \frac{0.366\omega - \omega K}{K + 0.366\omega^2} = -\frac{1}{\sqrt{3}}$$

Solving we get, K + 0.366 = 1.732 k − 0.634
∴ k = 1.366

19. Ans. (a)

Solution:

$$(sI - A)^{-1} = \begin{bmatrix} s & +1 \\ 0 & s+3 \end{bmatrix} = \frac{Adj(sI - A)}{|sI - A|}$$

$$= \frac{\begin{bmatrix} s+3 & -1 \\ 0 & s \end{bmatrix}}{\begin{bmatrix} s & 1 \\ 0 & s+3 \end{bmatrix}} = \frac{\begin{bmatrix} s+3 & -1 \\ 0 & s \end{bmatrix}}{s(s+3)} = \begin{bmatrix} \frac{1}{s} & \frac{-1}{s(s+3)} \\ 0 & \frac{1}{s+3} \end{bmatrix}$$

$$\therefore\ \phi(t) = LT^{-1}(sI - A)^{-1} = \begin{bmatrix} 1 & \frac{1}{3}(1 - e^{-3t}) \\ 0 & e^{-3t} \end{bmatrix}$$

20. Ans. (c)

Solution:

Zero state response = $L_T^{-1}\phi(s)\,B\,U(s)$

$$= LT^{-1}\begin{bmatrix} \frac{1}{s} & \frac{-1}{s(s+3)} \\ 0 & \frac{1}{s+3} \end{bmatrix}\begin{bmatrix} 1 \\ 0 \end{bmatrix}\frac{1}{s} = LT^{-1}\begin{bmatrix} \frac{1}{s^2} \\ 0 \end{bmatrix} = \begin{bmatrix} t \\ 0 \end{bmatrix}$$

State transition equation

= Zero input response + zero state response.

$\therefore\quad x(t) = \phi(t)\,x(0) + t$

$$\begin{bmatrix} 1 & \frac{1}{3}(1 - e^{-3t}) \\ 0 & e^{-3t} \end{bmatrix}\begin{bmatrix} -1 \\ 3 \end{bmatrix} + \begin{bmatrix} t \\ 0 \end{bmatrix}$$

$$= \begin{bmatrix} -1 + 1 - e^{-3t} \\ 0 + 3e^{-3t} \end{bmatrix} + \begin{bmatrix} t \\ 0 \end{bmatrix} = \begin{bmatrix} t - e^{-3t} \\ 3e^{-3t} \end{bmatrix}$$

21. Ans. (b)

22. Ans. (b)

Solution: Characteristics equation

$(s^2 - 4)(s + 1) + (s - 1) = 0$ i.e. $1 + GH = 0$

$$\text{i.e. } 1 + \frac{(s+1)}{(s^2 - 4)(s + 1)} = 0$$

zero of GH = 1

Poles of GH = –1; ± 2

Therefore the Root locus plot is as

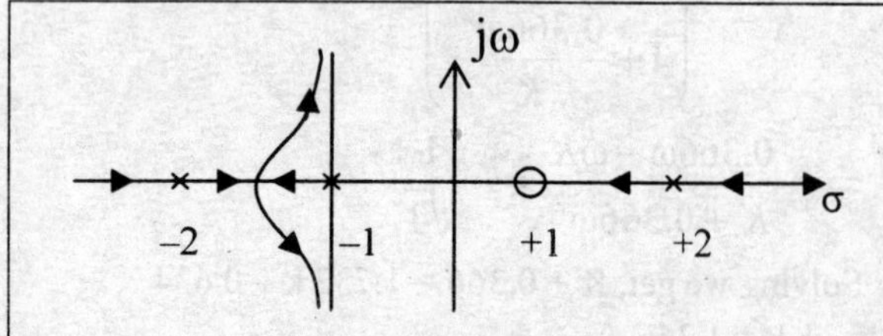

23. Ans. (a)

Solution: If the GH plot in circles (–1, j0) that is the critical point, then the system becomes unstable. So in option (1) that does not enclose the (-1, j0) other all are in circling the critical point.

24. Ans. (d)

Solution: For U_1 only

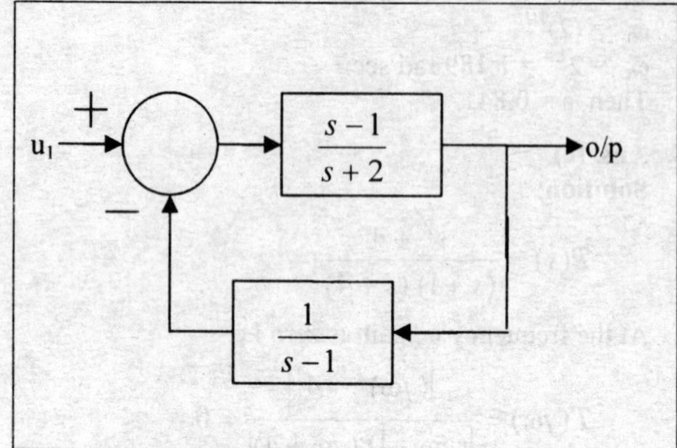

For U_1 transfer function

$$TF = \frac{(s-1)/(s+2)}{1 + \frac{(s-1)}{(s+2)} \cdot \frac{1}{(s-1)}} = \frac{(s-1)}{s+2+1}$$

$$= \frac{(s-1)}{(s+3)} \Rightarrow \text{Stable}$$

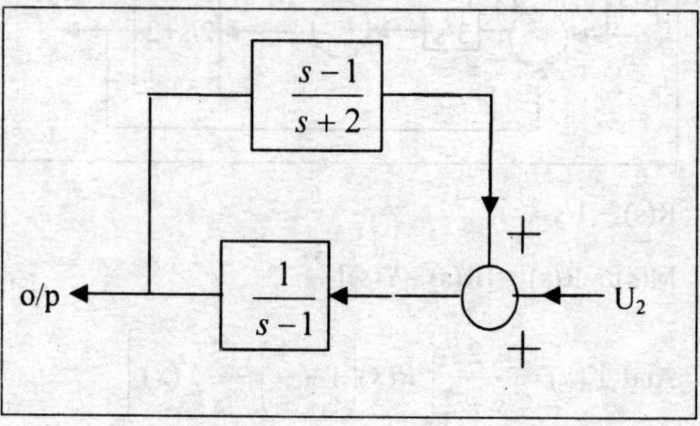

For U_2 transfer function

$$TF = \frac{\frac{1}{s-1}}{1 - \frac{1}{(s-1)} \times \frac{(s-1)}{(s+2)}} = \frac{1/(s-1)}{\frac{s+2-1}{s+2}}$$

$$= \frac{1/(s-1)}{(s+1)/(s+2)} = \frac{s+2}{(s+1)/(s+2)}$$

$$= \frac{s+2}{(s-1)(s+1)} \Rightarrow \text{Unstable}$$

25. Ans. (c)

26. Ans. (b)

Solution: Characteristics equation.

$$1+\frac{K(s+3)}{(s+8)^2}=0$$

or $s^2+(16+K)s+3K+64=0$

Routh array

S^2	1 3K + 64
S^1	(16 + K)
S^0	(3K + 64)

$16+K=0$; $K=-16$

$3K+64=0$; $K=\frac{-64}{3}=-21$

$K>-16$

Put the value $K=-16$ which both the condition in auxilay equation,

$s^2+3(-16)+64$

$s^2+16=0$

$s=\pm j4 = \pm j\omega$

$s=\pm j4 = \pm j\omega$

$\omega = 4\ rad/\sec$

27. Ans. (d)

Solution:

Step: 1 Solve loop (1)

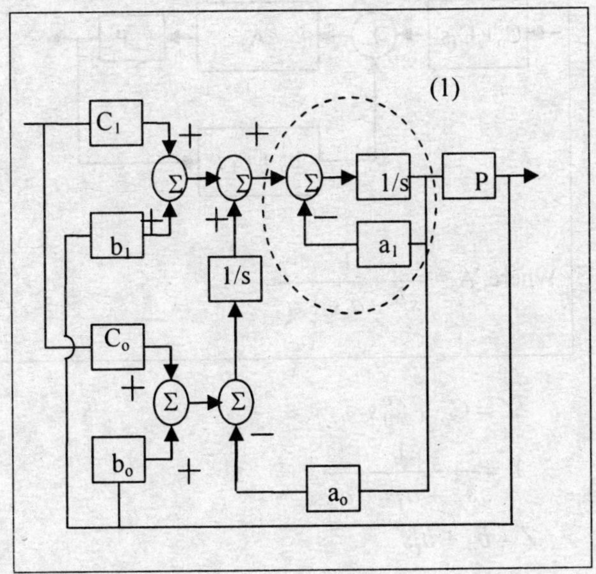

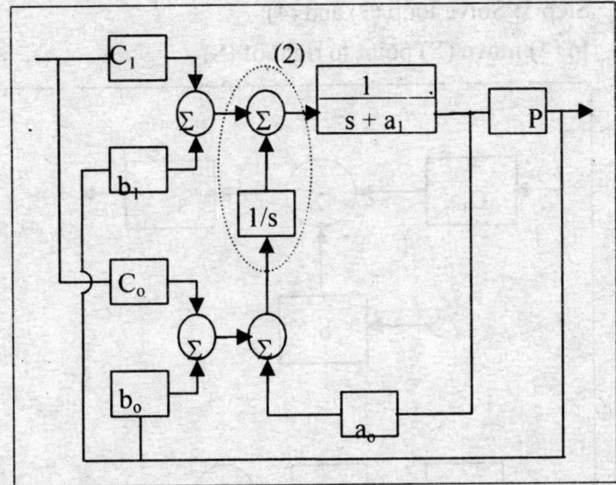

Step 2: Solve loop (2) bring summing point before (1/s).

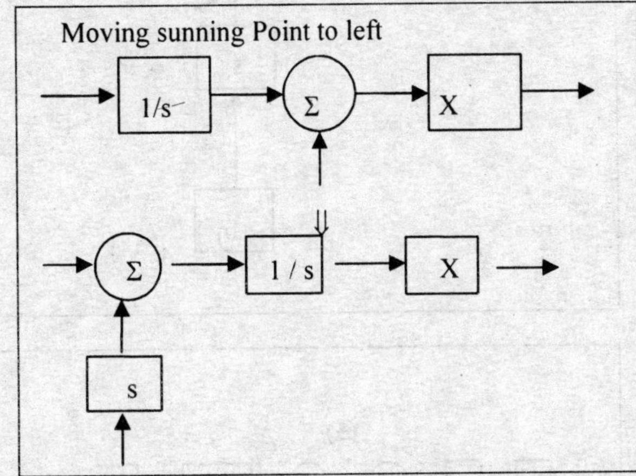

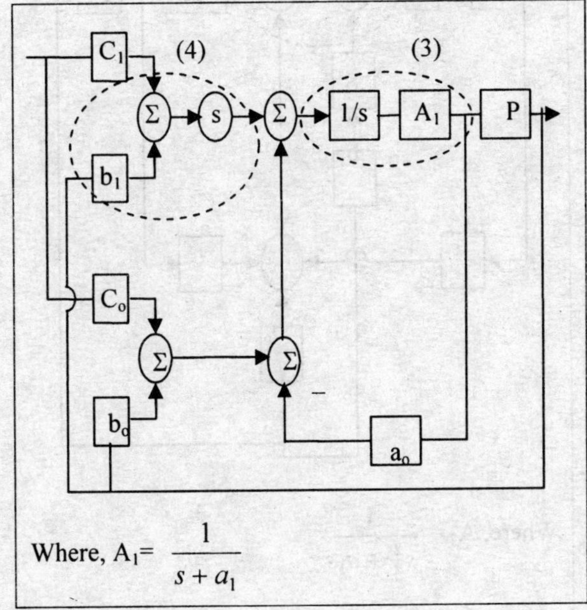

Where, $A_1 = \frac{1}{s+a_1}$

Step 3: Solve loop (3) and (4)

In (4) move (Σ) point to right of (S)

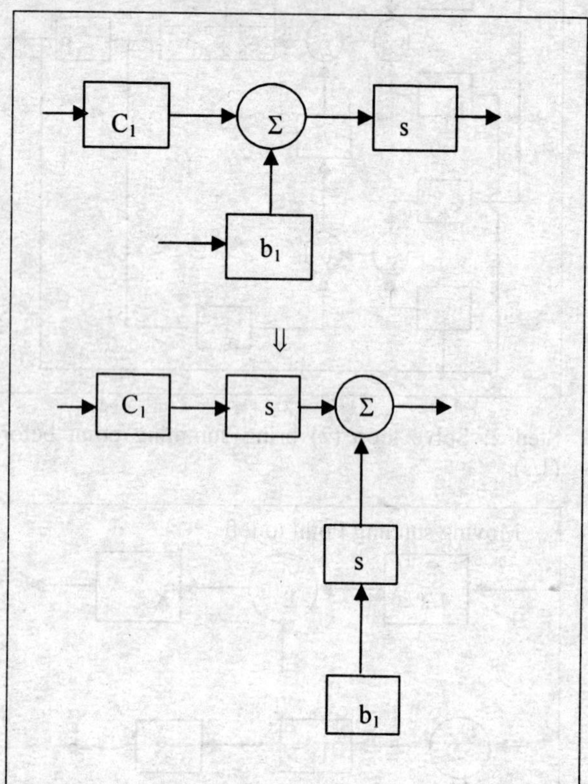

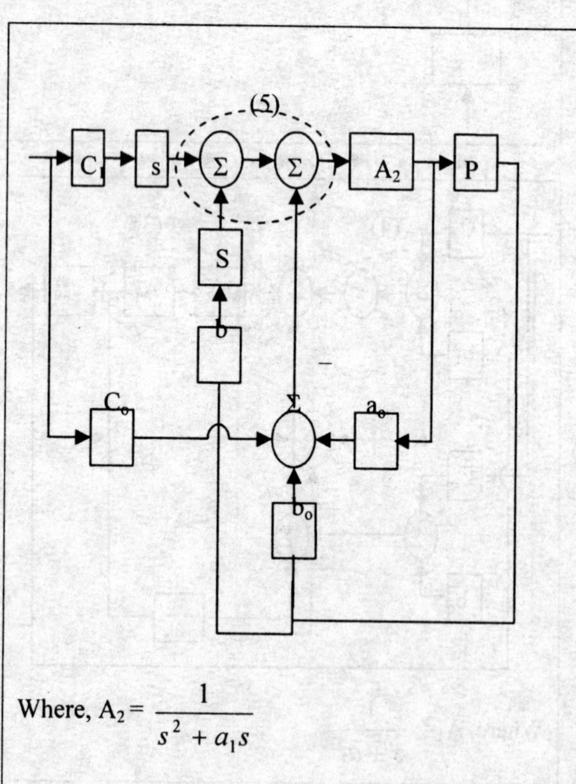

Where, $A_2 = \dfrac{1}{s^2 + a_1 s}$

Step 4: Interchange summing Point in loop 5.

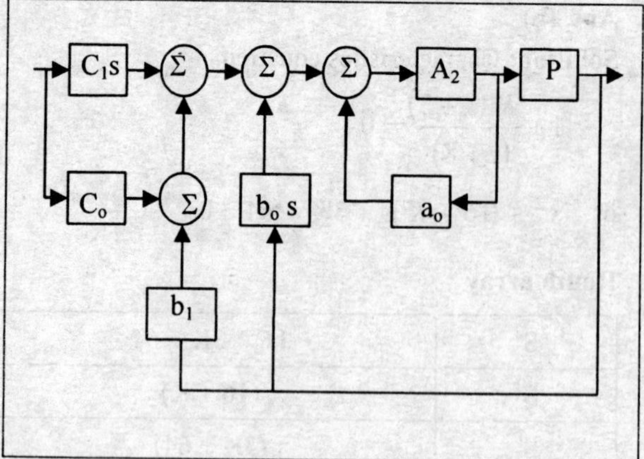

Step 5

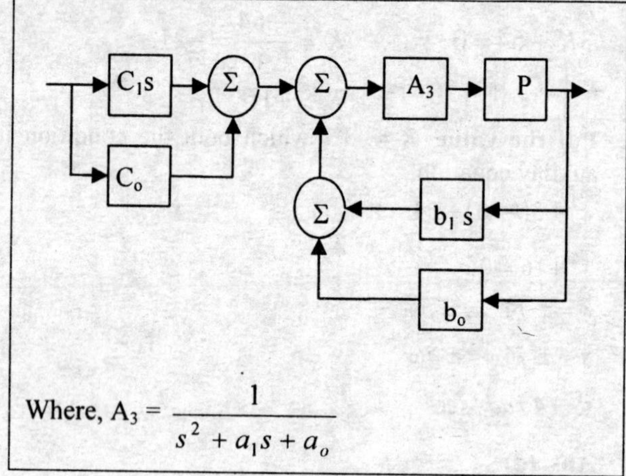

Where, $A_3 = \dfrac{1}{s^2 + a_1 s + a_o}$

Step 6:

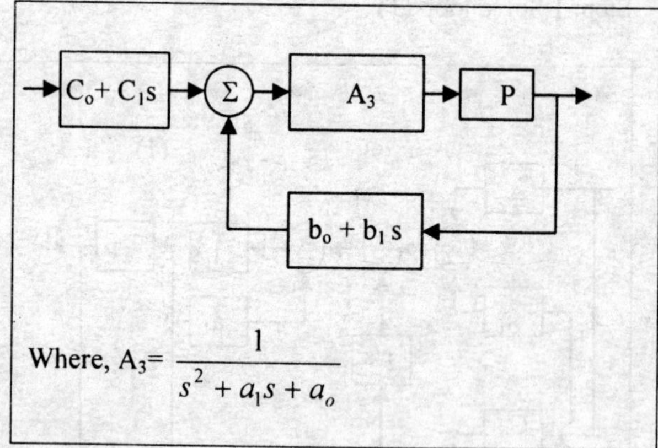

Where, $A_3 = \dfrac{1}{s^2 + a_1 s + a_o}$

$$X = C_o + C_1 s$$

$$Y = \frac{1}{s^2 + a_1 s + a_o}$$

$$Z = b_o + b_1 s$$

Another method
Transform the block diagram into signal flow graph, then use Masson's gain formula.

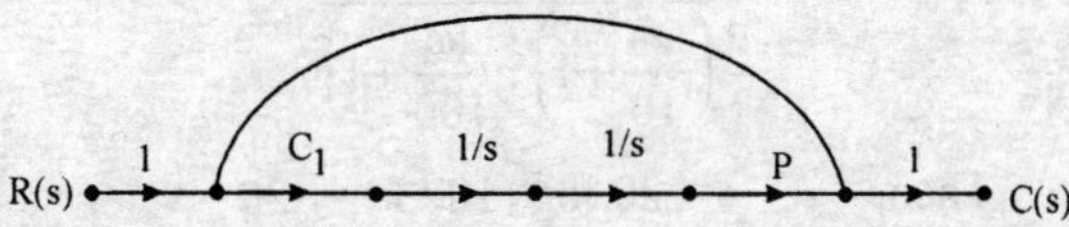

28. Ans. (a)
Solution:

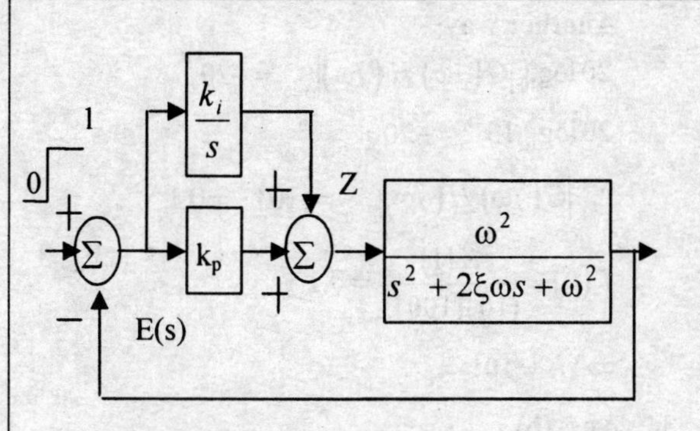

$$E(s)=\frac{R(s)}{1+G(s)H(s)}$$

$r(t) \to$ Unit step

$R(s) \to (1/s)$

The steady state of z is

$$z(\infty)=\underset{s\to 0}{Lt}\; s\,Z(s)=\underset{s\to 0}{Lt}\; sE(s)\frac{K_i}{s}$$

$$=\underset{s\to 0}{Lt}\frac{s.\frac{1}{s}.\frac{K_i}{s}}{1+\left(K_P+\frac{K_i}{s}\right)\left(\frac{\omega^2}{s^2+2s\omega_r s+\omega^2}\right)}$$

$$=\underset{s\to 0}{Lt}\frac{\frac{K_i}{s}}{1+\left(K_P+\frac{K_i}{s}\right)\left(\frac{\omega^2}{s^2+2s\omega_r s+\omega^2}\right)}$$

$$\underset{s\to 0}{Lt}\frac{K_i}{s\left(1+\left(K_P+\frac{K_i}{s}\right)\left(\frac{\omega^2}{s^2+2s\omega_r s+\omega^2}\right)\right)}=1$$

29. Ans. (c)
Solution: Characteristic equation

$$LC\,s^2+RCs+1=0$$

$$s^2+\frac{R}{L}s+\frac{1}{LC}=0$$

$$s^2+2\delta\omega_n s+\omega_n^2=0$$

$$\omega_n=\frac{1}{\sqrt{LC}}=\frac{1}{\sqrt{10^{-3}\times 10\times 10^{-6}}}=\frac{1}{\sqrt{10^{-8}}}$$

$$=\sqrt{10^8}=10^4\;;\;2\delta\omega_n=\frac{R}{L}$$

$$2\times\delta\times 10^4=\frac{10}{10^{-3}}=10^4$$

$$\delta=\frac{1}{2}=0.5$$

$$\text{overshoot}==e^{-\frac{n\pi}{\sqrt{1-\delta^2}}}=0.16\Rightarrow 16\,\%$$

30. Ans. (c)

31 Ans. (d)
Solution:

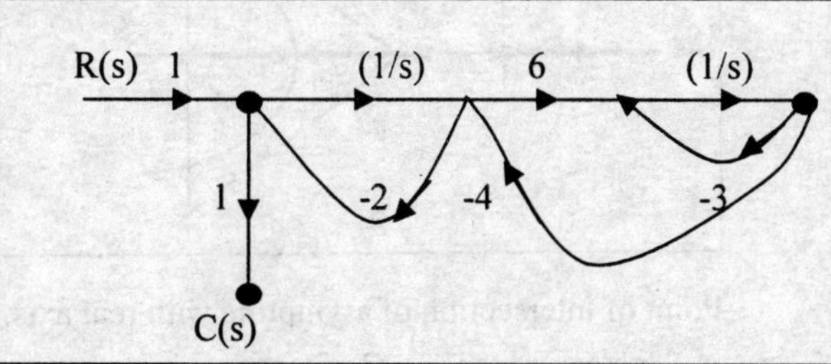

Loop gain l_1 = - [2 / S]

l_2 = - [3 / S]

l_3 = - [24 /S]

Non touching loops

[loop$_1$ x loop$_2$] = - [2 / S] x [-3 / S] = [6 / S^2]

$$G(s)=\frac{C(s)}{R(s)}=\Sigma\frac{\Delta_k M_k}{\Delta}$$

Δ = 1 – (Sum of loop gain) + (Product of two Non touching loop)

$$=\left(1-\frac{-2}{S}-\frac{3}{S}-\frac{24}{S}\right)+\frac{6}{S^2}=\frac{S^2+29S+6}{S^2}$$

$M_1 = 1$

For path No touching = 1 + [3 / S] + [24 /S]
= [(S+ 27) / S]

$$C(s)=\frac{M_1\Delta_1}{\Delta}=1\times\frac{(S+27)/S}{\Delta}$$

= [(S + 27)/S] = S (S + 27)

$$\frac{S^2+29+6}{S^2} \quad S^2+29S+6$$

32. Ans. (d)

Solution: The Characteristic equation:

1 + GH = 0 or, $\{1 + [K/\{S(S+2)(S+3)\}] = 0$

$K = -(S^3 + 5S^2 + 6S)$

For break away point [dK /dS] = 0

$-(3S^2 + 10S + 6) = S^2 + 3.33S + 2 = 0$

$$S = \frac{-3.33 \pm \sqrt{[(3.33)^2 - 04.2]}}{2}$$

$$= \frac{-3.33 \pm \sqrt{(12-8)}}{2}$$

$$= \frac{-3.33 \pm 2}{2}$$

$S = -0.784$ or -2.55

RL for $G(s)H(s) = \dfrac{K}{s(s+7)(s+3)}$ is

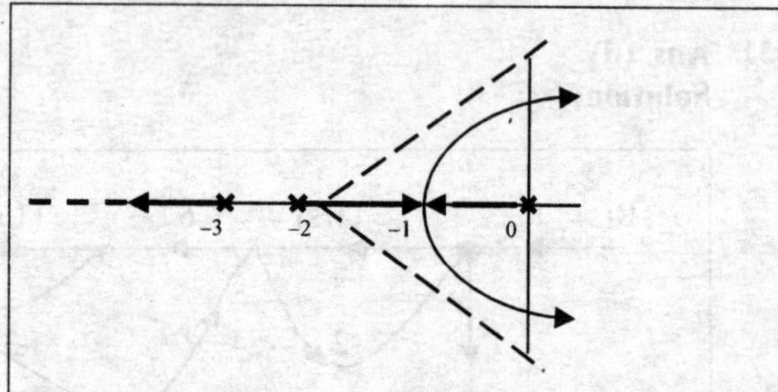

Point of intersection of asymptote with real axis,

$$= \frac{\Sigma\, pole - \Sigma\, zero}{|P-z|} = \frac{-2-3}{3} = -1.66$$

$$\text{Angle of asymptote} = \frac{(2K+1)\pi}{|P-z|} = \frac{\pi}{3}, \pi, \frac{5\pi}{3}$$

As root locus does not lie on -2.55

Hence breakaway point is -0.784

Note: As there are two poles i.e. even number of pole and zero beyond left to point -2.55. Hence RL does not exist between -2 and -3. So there is no breakaway point at -2.55

33. Ans. (c)

Solution: At ω = 1, slope changes form odB/decade to 20 dB/decade, and, at ω = 10, slope changes from 20 dB/decade to 0dB/decade and at ω = 100, slope changes form odB/decade to – 20 dB/.decade.

Hence, there is a zero at ω = 1 and poles at ω = 10,100.

Now, $$G(s) = \frac{k(s+1)}{(s+10)(s+100)}$$

$$G(j\omega) = \frac{k(j\omega+1)}{(j\omega+10)(j\omega+100)}$$

$$= \frac{k(j\omega+1)10^{-3}}{\left(\frac{j\omega}{10}+1\right)\left(\frac{j\omega}{100}+1\right)}$$

Then, $20 \log (10^{-3} k) = -20$

Or, $k = 10^2$

Hence, $$G(s) = \frac{10^2(s+1)}{(s+10)(s+100)}$$

Another way

$$20\log_{10}\left|G(j\omega)H(j\omega)\right|_{\omega=0} = -20$$

$$20\log_{10}10^{-1} = -20$$

$$\therefore \left|G(j\omega)H(j\omega)\right|_{\omega=0} = (10)^{-1} = 0.1$$

$$T(0) = \frac{K(1)}{(10)(100)} = 0.1$$

$$\Rightarrow K = 10^2$$

34. Ans. (b)

Solution:

$2\delta\, \omega_n = 4$

$\delta\, \omega_n = 2$

and $\omega_n = 2$

then, $\delta = 1$ (critically-damped system)

and t_{ss} (Setting time) $= \dfrac{4}{\delta\omega_n} = 2\,\text{sec}$

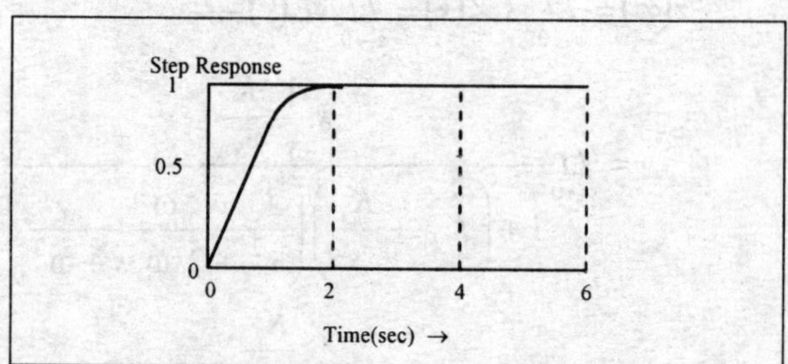

Thus, only response is as shown in Figure (b). Figure (d) is also critically-damped system, but its setting time is about 5 sec. Hence, the only choice is (b).

35. Ans. (d)

Solution:

$$G(s)H(s) = \frac{s}{(s+100)^3}$$

$$G(j\omega)H(j\omega) = \frac{j\omega}{(100+j\omega)^3} = \frac{j\omega\,(100-j\omega)^3}{(100^2+\omega^2)^{3/2}}$$

$$= \frac{j\omega\,[(100)^3 - 3\times100^2\,j\omega - 300\,\omega^2 + j\omega^3]}{(100^2+\omega^2)^{3/2}}$$

$$= \frac{-\omega^4 + 3\times100^2\omega^2 + j\,(100^3\omega - 3\times100\omega^2)}{(100+\omega^2)^{3/2}}$$

Intersection with the real axis gives the phase cross over frequency.

i.e. Im [G(jω) H (jω)] = 0

$100^3\omega$ - 3 x $100\omega^2$ = 0

$(100)^3$ - 3 x $100\omega^3$ = 0

$$\omega^2 = \frac{(100)^3}{3\times100} = \frac{(100)^2}{3}$$

$\omega = [100 / (\sqrt{3})]$

Real GH [j 100 / √3]

$$= \frac{-[100/\sqrt{3}]^4 + 3\,(100)^2 \times [1003\sqrt{3}]^2}{\{(100)\}^2 + (100/\sqrt{3})^2\}^3}$$

$$\frac{\dfrac{-(100)^4}{9} + \dfrac{3\times(100)^4}{3}}{\{(100)^2 + [(100)^2/3]\}^3}$$

$= [8 \text{ x } (100)^4 / 9]$

$[4 \text{ x } (100)^2 / 3]^3$

$$= \frac{8\times(100)^4}{9} \times \frac{27}{64\times(100)^6}$$

$= [3 / \{8 \text{ x } (100)^2\}] = [3 / 80000]$

$$\left(\text{Gain margin} = 20\log_{10}\frac{1}{|G(j\omega)|}\right)$$

$-20\log[3/80{,}000] =$ **88.5 dB**.

Plot must cross some where in - Ve real axis for Gain margin definition.

In present case plot is crossing form origin which indicate Gain margin infinite

36. Ans. (c)

Solution: If the function is Non Homogenous i.e.

X' (t) = A X (t) + B u(t) ; X(o) = X_o

The solution of this equation is given by

$$X(t) = e^{At}X(0) + \int_0^t e^{-At}Bu(d\tau)$$

$$X(t) = \phi(t)\,X(0) + \int_0^t \phi(t-\tau)\,Br(\tau)\,d\tau$$

Where, $\phi(t) = LT^{-1}[SI - A]^{-1}$

Where ϕ (t) is STM.

In present case

$$A = \begin{pmatrix}1 & 0\\ 1 & 1\end{pmatrix} \text{ and } X(t_o) = \begin{pmatrix}1\\0\end{pmatrix}$$

$$B = \begin{pmatrix}0\\0\end{pmatrix}$$

$x(t) = \exp[A(t - t_o)]\; X(t_o)$

$$+ \int_{t_0}^{t} \exp[A(t-t_o)]\,X(t_o)$$

$= \exp[A(t - t_o]\,X(t_o) + 0$

$$(sI - A) = \begin{pmatrix}s & 0\\ 0 & s\end{pmatrix} - \begin{pmatrix}1 & 0\\ 1 & 1\end{pmatrix}$$

$$(sI - A) = \begin{pmatrix}s-1 & 0\\ -1 & s-1\end{pmatrix}$$

$$(sI - A)^{-1} = \frac{1}{(s-1)^2}\begin{pmatrix}s-1 & 0\\ 1 & (s-1)\end{pmatrix}$$

$$= \begin{pmatrix}[1/(s-1)] & 0\\ [1/(s-)^2] & [1/(s-1)]\end{pmatrix}$$

$$LT^{-1}[(sI-A)^{-1}] = e^{At} = \begin{pmatrix}e^t & 0\\ te^t & e^t\end{pmatrix}$$

$x(t) = e^{At} \text{ x } [0]$

$$= \begin{pmatrix}e^t & 0\\ te^t & e^t\end{pmatrix}\begin{pmatrix}1\\0\end{pmatrix}$$

$$= \begin{pmatrix}e^t\\ te^t\end{pmatrix}$$

37. Ans. (c)

Solution: At ω = 1, slope changes form odB/decade to 20 dB/decade, and, at ω = 10, slope changes from 20 dB/decade to 0dB/decade and at ω = 100, slope changes form odB/decade to – 20 dB/.decade.

Hence, there is a zero at ω = 1 and poles at ω = 10,100.

Now, $$G(s) = \frac{k(s+1)}{(s+10)\,(s+100)}$$

$$G(j\omega) = \frac{k(\omega+1)}{(j\omega+10)\,(s+100)}$$

$$= \frac{k(j\omega+1)}{(j\omega+10)\,(j\omega+100)}$$

$$= \frac{k(j\omega+1)\,10^{-3}}{\left(\dfrac{j\omega}{10}+1\right)\left(\dfrac{j\omega}{100}+1\right)}$$

Then, $20 \log (10^{-3} k) = -20$

Or, $k = 10^2$

Hence, $G(s) = \dfrac{10^2 (s+1)}{(s+10)(s+100)}$

Another way

$$20\log_{10}|G(j\omega)H(j\omega)|_{\omega=0} = -20$$

$$20\log_{10}10^{-1} = -20$$

$$\therefore |G(j\omega)H(j\omega)|_{\omega=0} = (10)^{-1} = 0.1$$

$$T(0) = \frac{K(1)}{(10)(100)} = 0.1$$

$$\Rightarrow K = 10^2$$

38. Ans. (a)

Solution: Transfer function =

$$G(s) = \frac{[(s/5)+1][(s/100)+1][(s/200)+1]}{[(s/0.1)+1][(s/1)+1][(s/80)+1]}$$

$\angle G(j\omega) =$

$= \tan^{-1}[\omega/5] + \tan^{-1}[\omega/100] + \tan^{-1}[\omega/200]$

$-\tan^{-1}[\omega/.01] - \tan^{-1}\omega - \tan^{-1}[\omega/80]$

At $\omega = 20$, transfer function. $\angle G(j\omega)$

$= \tan^{-1}(4) + \tan^{-1}[1/5] + \tan^{-1}[1/10]$

$-\tan^{-1}(2000) - \tan^{-1} 20 - \tan^{-1}[1/4] = -90°$

39. Ans. (c)

Solution: Forward path gain, P_1 = abcd

And, as it touches all the three loops, $\Delta_1 = 1$.

Now, Three individual loops with loop gains

P_{11} = be, P_{21} = cf, P_{31} = dg

And combination of two non-touching loops with loop gains products,

P_{12} = bedg

Then, $\dfrac{C}{R} = \dfrac{x_5}{x_1} = \dfrac{P_1\Delta_1}{1-(P_{11}+P_{21}+P_{31})+P_{12}}$

$$= \frac{abcd}{1-(be+cf+dg)+be.\,Dg}$$

40. Ans. (b)

Solution:

We have $A = \begin{bmatrix} -2 & 2 \\ 1 & -3 \end{bmatrix}$

Characteristic equation is

$$|\lambda I - A| = 0$$

or $\begin{vmatrix} \lambda+2 & -2 \\ -1 & \lambda+3 \end{vmatrix} = 0$

or $(\lambda+2)(\lambda+3) - 2 = 0$

or $\lambda^2 + 5\lambda + 4 = 0$

Then $\lambda_1 = -4$ and $\lambda_2 = -1$

Eigen value are -4 and -1

Eigen vectors for $\lambda_1 = -4$

$$(\lambda_1 I - A)X_1 = 0$$

or $\begin{bmatrix} \lambda_1+2 & -2 \\ 1 & \lambda_1+3 \end{bmatrix}\begin{bmatrix} x_{11} \\ x_{21} \end{bmatrix} = 0$

$$\begin{bmatrix} -2 & -2 \\ -1 & -1 \end{bmatrix}\begin{bmatrix} x_{11} \\ x_{21} \end{bmatrix} = 0$$

or $-2x_{11} - 2x_{21} = 0$

or $x_{11} + x_{21} = 0$

We have only one independent equation $x_{11} = -x_{21}$

Let $x_{21} = K$, then $x_{11} = -K$, the Eigen vector will be

$$\begin{bmatrix} x_{11} \\ x_{21} \end{bmatrix} = \begin{bmatrix} -K \\ K \end{bmatrix} = K\begin{bmatrix} -1 \\ 1 \end{bmatrix}$$

Now Eigen vector for $\lambda_2 = -1$

$$(\lambda_2 I - A)X_2 = 0$$

or $\begin{bmatrix} \lambda_2+2 & -2 \\ -1 & \lambda_2+3 \end{bmatrix}\begin{bmatrix} x_{12} \\ x_{22} \end{bmatrix} = 0$

or $\begin{bmatrix} 1 & -2 \\ -1 & 2 \end{bmatrix}\begin{bmatrix} x_{12} \\ x_{22} \end{bmatrix} = 0$

We have only one independent equation $x_{12} = 2x_{22}$

Let $x_{22} = K$, then $x_{12} = 2K$. Thus Eigen vector will be

$$\begin{bmatrix} x_{12} \\ x_{22} \end{bmatrix} = \begin{bmatrix} 2K \\ K \end{bmatrix} = K\begin{bmatrix} 2 \\ 1 \end{bmatrix}$$

Diagonalizing matrix

$$M = \begin{bmatrix} x_{11} & x_{12} \\ x_{21} & x_{22} \end{bmatrix} = \begin{bmatrix} -1 & 2 \\ 1 & 1 \end{bmatrix}$$

Now $M^{-1} = \left(\dfrac{-1}{3}\right) = \begin{bmatrix} 1 & -2 \\ -1 & -1 \end{bmatrix}$

Now Diagonal matrix of sin At is D where

$$D=\begin{bmatrix}\sin(\lambda_1 t) & 0\\ 0 & \sin(\lambda_2 t)\end{bmatrix}$$

$$=\begin{bmatrix}\sin(-4t) & 0\\ 0 & \sin(-t)\end{bmatrix}$$

Now matrix $B=\sin At$

MDM^{-1}

$$=\left(-\frac{1}{3}\right)\begin{bmatrix}-1 & 2\\ 1 & 1\end{bmatrix}\begin{bmatrix}\sin(-4t) & 0\\ 0 & \sin(-t)\end{bmatrix}\begin{bmatrix}1 & -2\\ -1 & -1\end{bmatrix}$$

$$=-\frac{1}{3}\begin{bmatrix}-\sin(-4t) & 2\sin(-t)\\ \sin(4t) & \sin(-t)\end{bmatrix}\begin{bmatrix}1 & -2\\ -1 & -1\end{bmatrix}$$

$$=-\left(\frac{1}{3}\right)$$

$$\begin{bmatrix}-\sin(-4t)-2\sin(-t) & 2\sin(-4t)-2\sin(-t)\\ \sin(-4t)-\sin(-t) & -2\sin(-4t)-\sin(-t)\end{bmatrix}$$

$$=\left(\frac{1}{3}\right)\begin{bmatrix}\sin(-4t)+2\sin(-t) & -2\sin(-4t)+2\sin(-t)\\ -\sin(-4t)+\sin(-t) & 2\sin(-4t)+\sin(-t)\end{bmatrix}$$

Hence (a) is correct option.

41. Ans. (a)

Solution: The characteristic equation is, 1 + G(s) = 0

or, $s^4 + 4s^2 + 5^2 + 6s + K = 0$

Apply Routh – Hurwitz array, we have

s^4	1	5	K
s^3	4	6	0
s^2	(7 / 2)	K	
s^1	(21-4K) (2/7)	0	
s^0	K		

For the system to be stable

21 – 4 K > 0 and K > 0; [21 / 4] > k & k > 0

On combining we get

[21 / 4] > k > 0

42. Ans. (b)

Solution: $P(s) = s^5 + s^4 + 2s^3 + 2s^2 + 3s + 15 = 0$

Apply RH Criterion

S^5	1	2	3
S^4	1	2	15
S^3	0	–15	

This test Method fails then apply another method

Put S = [1/ p] , & test it again with RH table

$[1/p]^5 + [1/p]^4 + 2[1/p]^3 + 2[1/p]^2 + 3[1/p] + 15 = 0$

$15p^4 + 3p^4 + 2p^3 + 2p^2 + p^1 + 1 = 0$

There are two sign change in the first column of RH table which indicate two roots in the RHS of s plan.

R.H table

p^5	15	2	1	0
p^4	3	2	1	0
p^3	([6 – 30) / 3] = [-24 / 3]	[(3 – 15) / 3]		
p^3	= - 8	= - 4	0	
p^2	[(-16 + 12) –8]	1	0	
p^2	= [1 /2]			
p^1	$\dfrac{-2+8}{1/2}$	0		
p^1	= 12			
p^0	1			

43. Ans. (b)

Solution:

$$[sI - A]^{-1} = \begin{pmatrix}s-1 & 0\\ 0 & s-1\end{pmatrix}^{-1}$$

STM, $\phi(t)$ = $LT^{-1}[sI - A]^{-1}$

$$= LT^{-1}\begin{Bmatrix}[1/(s-1)] & 0\\ 0 & [1/(s-1)]\end{Bmatrix}$$

$$=\begin{pmatrix}e^t & 0\\ 0 & e^t\end{pmatrix}$$

44. Ans. (d)

Solution:

Rewriting state equations in matrix form, we get

$$\begin{bmatrix}X_1'\\ X_2'\end{bmatrix}=\begin{bmatrix}-3 & -1\\ 2 & 0\end{bmatrix}\begin{bmatrix}X_1\\ X_2\end{bmatrix}+\begin{bmatrix}1\\ 0\end{bmatrix}u(t)$$

$$Y=\begin{bmatrix}1 & 0\end{bmatrix}\begin{bmatrix}X_1\\ X_2\end{bmatrix}+\begin{bmatrix}1\\ 0\end{bmatrix}u(t)$$

Then,

$$A=\begin{bmatrix}-3 & -1\\ 2 & 0\end{bmatrix};\ B=\begin{bmatrix}1\\ 0\end{bmatrix};\ D=\begin{bmatrix}1 & 0\end{bmatrix}$$

Now, $Q_c = [B : BA]$

$$=\begin{bmatrix}-3 & -1\\ 0 & 20\end{bmatrix}$$

$$\det Q_c = \begin{bmatrix} -3 & -1 \\ 0 & 2 \end{bmatrix} \neq \text{ i.e. controllable}$$

$$Q_o = \begin{bmatrix} D^T & D^T & A^T \end{bmatrix}$$

$$\begin{bmatrix} D \\ DA \end{bmatrix} = \begin{bmatrix} -3 & -1 \\ 0 & 20 \end{bmatrix}$$

$$\det Q_c = \begin{bmatrix} 1 & -3 \\ 0 & -1 \end{bmatrix} \neq \text{ i.e. observable}$$

i.e. controllable

Hence the system is controllable and observable

45. Ans. (b)

Solution: The closed loop system is stable for

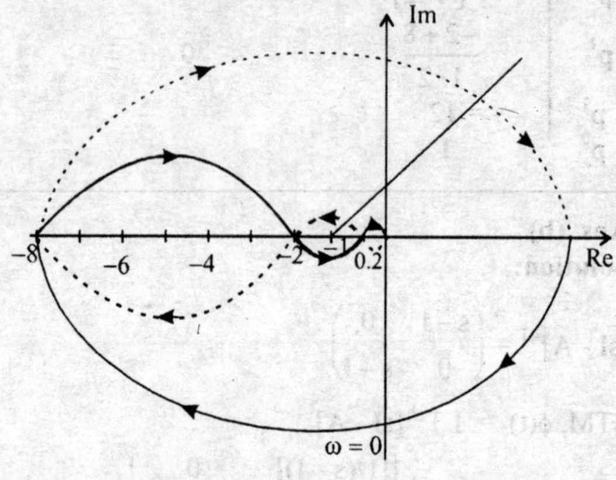

System is stable in region -0.2 to -2 and on the left side of -8 as the number of encirclement is zero.

$$0.2\,K < 1$$

$$\Rightarrow \quad K < 5$$

$$2K > 1$$

$$\Rightarrow K > 0.5$$

$$\therefore \quad 0.5 < K > 5$$

and $8K < 1$

$$K < \frac{1}{8}$$

46. Ans. (c)

Solution:

The peak percent overshoot is determined for LTI second order closed loop system with zero initial condition. It's transfer function is:

$$T(s) = \frac{\omega_n^2}{s^2 + 2\delta\omega_n s + \omega_n^2}$$

Transfer function has a pair of complex conjugate poles and zeros.

47. Ans. (a)

Solution:

$$\delta = 0.5$$

$$\omega_n = 5\, rad/\text{sec}$$

$$G(s)H(s) = \frac{K}{s^2}$$

$$\cos^{-1}\delta = \theta$$

$$\cos^{-1} 0.5 = 60^\circ$$

$$\theta = 60^\circ$$

With $\theta = 60^\circ$ and $\omega_n = 5\, rad/\text{sec}$ we get a point on the Root locus with coordinate $-2.5 + j\,4.33$.

$$\angle G(s) = \left.\frac{K}{s^2}\right|_{s=-2.5+j4.33}$$

$$= -2\tan^{-1}\frac{4.33}{-2.5} = 120^\circ$$

$\therefore$ For compensated system $D = 180 - 120^\circ = 60^\circ$

Hence option (b) and (c) are lag network

For compensating lag network $\frac{K}{s^2}$, we need a lead network

Putting $s = -2.5 + j4.33$ in option (a) network gives

$$\left.\frac{K(s+3)}{s^2(s+9.9)}\right|_{s=-2.5+j4.33}$$

$$= \frac{0.5 + j4.33}{7.4 + j4.33} = 53^\circ \triangleq 60^\circ$$

48. Ans. (c)

Solution: The Rough sketch of the root locus is a circle

{d / ds} [G (s) H (s) = 0;

The breakaway point is

$$\frac{-s(s+3) - (1-s)(2s+3)}{s\,(s+3)} = 0$$

s = 3 or – 1

49. Ans. (a)

Solution: Ramp input is applied

$$R(s) = [1/s^2] \quad ; e_{ss} = \lim_{s\to 0} sE(s)$$

$$sE(s) = \frac{sR(s)}{1+G(s)} = \frac{s.[1/s^2]}{1+G(s)} = \frac{s.[1/s^2]}{1+G(s)} = \frac{1}{s + s\,G(s)}$$

$$e_{ss} = \lim_{s\to 0} sE(s) = \lim_{s\to 0}\frac{1}{sG(s)}$$

The above expression represent type 1 system therefore, type = 1. e_{ss} for type 1 system is given by

$e_{ss} = [1 / K] = 5\,\% = [1 / 20]$

$\therefore \quad K = 20$

50. Ans. (d)

Solution: Gain crossover frequency gain of

$$\frac{3e^{-2s}}{s\,(s+2)} = \left|\frac{3e^{-2s}}{s\,(s+2)}\right| = 1$$

$$e^{-2s} = \cos(2\omega) - j\,\text{Sin}\,2\omega \quad ; \quad |e^{-2s}| = 1$$

$$\left|\frac{3}{j\omega\,(2 + jGw2)}\right| = 1; \quad \left|\frac{3}{-\omega^2 + j^2\omega}\right| = 1$$

ω = 1.26

Phase crossover frequency

Phase angle = -90 – $\tan^{-1}$ [ω / 2] – 2 ω [180 / π]

Phase = -180 ⇒ - 90 – $\tan^{-1}$ [ω / 2] - 2ω [180 / π]

= - 180

ω= 0.632

51. Ans. (d)

Solution:

Gain margin

$$A = \text{Gain} = \left|\frac{3e^{-2s}}{S\,(s+2)}\right|_{\omega=0.632} = 2.2631$$

GM = 20 log [1 / a] = - 7.094 dB

Phase margin =

180 + [-90 – $\tan^{-1}$(ω / 2) – 2 ω (180 / π)]

ω = 1.26

Phase margin = – 86.59°

Statement for Linked Answers Questions 1 and 2

The open loop transfer function of a unity feedback system is given by

$$G(s) = \frac{3\,e^{-2}}{s\,(s+2)}$$

1. The gain and phase crossover frequencies in rad/sec are, respectively
   (a) 0.632 and 1.26 (b) 0.632 and 0.485
   (c) 0.485 and 0.632 (d) 1.26 and 0.632
2. Based on the above results, the gain and phase margins of the system will be
   (a) – 7.09 and 87.5°
   (b) 7.09 and 87.5°
   (c) 7.09 dB and – 87.5°
   (d) – 7.09 dB and – 87.5°

Statement for Linked Questions 3 and 4

Consider a unity gain feedback control system whose open loop transfer function is:

$$G(s) = [(as + 1) / s^2]$$

3. The value of "a" so that the system has a phase margin equal to π / 4 is approximately equal to
   (a) 2.40 (b) 1.40
   (c) 0.84 (c) 0.74
4. With the value of set for a phase margin of [π / 4], the value of unit impulse response of the open loop system at t = 1 second is equal to .
   (a) 3.40 (b) 2.40
   (c) 1.84 (d) 1.74

Statement for Linked Answers Questions 5 and 6

Consider a linear system whose space representation is $x(t) = Ax(t)$. If the initial state vector of the system is $x(0) = \begin{bmatrix} 1 \\ -2 \end{bmatrix}$, then the system response is $x(t) = \begin{bmatrix} e^{-2t} \\ -2e^{-2t} \end{bmatrix}$. If the initial state vector of the system changes to $x(0) = \begin{bmatrix} 1 \\ -1 \end{bmatrix}$, them the system response becomes $x(t) = \begin{bmatrix} e^{-t} \\ -e^{-t} \end{bmatrix}$.

5. The eigenvalue and eigenvector pairs (λ_i, v_i) for the system are
   (a) $\left(-1, \begin{bmatrix} 1 \\ -1 \end{bmatrix}\right)$ *and* $\left(-2, \begin{bmatrix} 1 \\ -2 \end{bmatrix}\right)$
   (b) $\left(-1, \begin{bmatrix} 1 \\ -1 \end{bmatrix}\right)$ *and* $\left(-1, \begin{bmatrix} 1 \\ -2 \end{bmatrix}\right)$
   (c) $\left(-1, \begin{bmatrix} 1 \\ -1 \end{bmatrix}\right)$ *and* $\left(2, \begin{bmatrix} 1 \\ -2 \end{bmatrix}\right)$
   (d) $\left(-2, \begin{bmatrix} 1 \\ -1 \end{bmatrix}\right)$ *and* $\left(1, \begin{bmatrix} 1 \\ -2 \end{bmatrix}\right)$
6. The system matrix A is:
   (a) $\begin{bmatrix} 0 & 1 \\ -1 & 1 \end{bmatrix}$ (b) $\begin{bmatrix} 1 & 1 \\ -1 & -2 \end{bmatrix}$
   (c) $\begin{bmatrix} 2 & 1 \\ -1 & -1 \end{bmatrix}$ (d) $\begin{bmatrix} 0 & 1 \\ -2 & -3 \end{bmatrix}$

Date for Questions 7 and 8 are given below. Solve the problems and choose the correct answers.

A state variable system $X'(t) = \begin{bmatrix} 0 & 1 \\ 0 & -3 \end{bmatrix} X(t) + \begin{bmatrix} 1 \\ 0 \end{bmatrix} u(t)$, with the initial condition $X(0) = [-1 \;\; 3]^T$ and the unit step input u(t) has

7. The state transition matrix

(a) $\begin{bmatrix} 1 & \frac{1}{3}(1 - e^{-3t}) \\ 0 & e^{-3t} \end{bmatrix}$ (b) $\begin{bmatrix} 1 & \frac{1}{3}(e^{-t} - e^{-3t}) \\ 0 & e^{-t} \end{bmatrix}$

(c) $\begin{bmatrix} 1 & \frac{1}{3}(e^{-t} - e^{-3t}) \\ 0 & e^{-3t} \end{bmatrix}$ (d) $\begin{bmatrix} 1 & (1 - e^{-t}) \\ 0 & e^{-t} \end{bmatrix}$

8. The state transition equation

(a) $X(t) = \begin{bmatrix} t - e^{-t} \\ e^{-t} \end{bmatrix}$ (b) $X(t) = \begin{bmatrix} t - e^{-t} \\ 3e^{-3t} \end{bmatrix}$

(c) $X(t) = \begin{bmatrix} t - e^{-3t} \\ 3e^{-3t} \end{bmatrix}$ (d) $X(t) = \begin{bmatrix} t - e^{-3t} \\ e^{-t} \end{bmatrix}$

Statement for Linked Answers Questions 9 and 10

Consider the R-L-C circuit shown in figure.

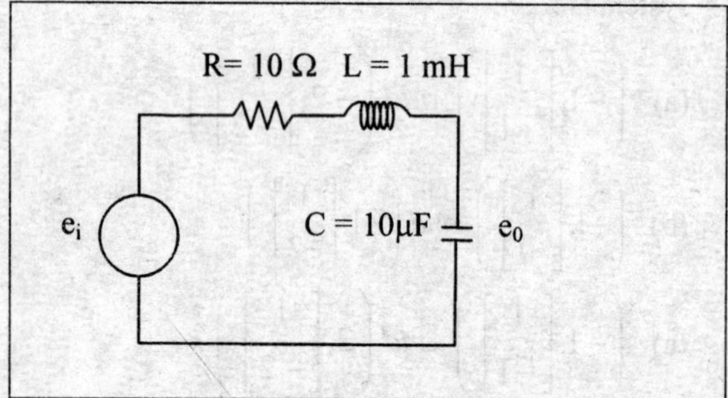

9. For a step-input e_i, the overshot in the output e_o will be
   (a) 0, since the system is not under-damped
   (b) 5 %
   (c) 16 %
   (d) 48 %

10. If the above step response is to be observed on a non-storage CRO, then it would be best to have the e_i as a
   (a) Step function
   (b) Square wave of frequency 50 Hz
   (c) Square wave of frequency 300 Hz
   (d) Square wave of frequency 2.0 kHz

Data given below should be used answering questions 11 and 12

The root locus plot of a system with transfer function $\frac{2}{s(s+1)(s+2)}$, is shown in figure. A unity feedback proportional system is built using this system.

11. The maximum possible controller gain, for which the unity feedback system is stable, is approximately.

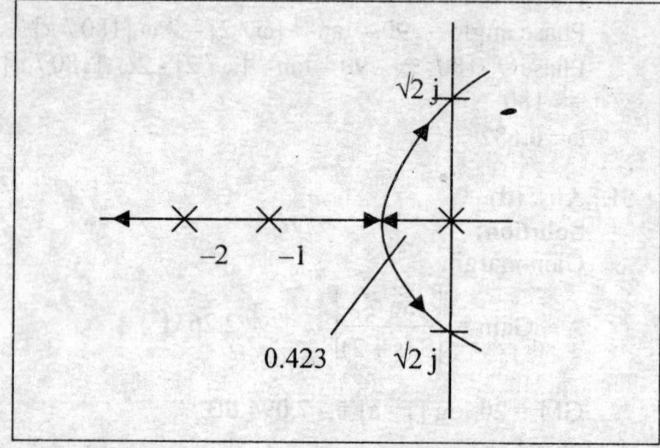

   (a) 6.0 (b) 3.0
   (c) 0.4 (d) 0.2

12. The maximum possible controller gain, for which the unity feedback system exhibits a non-oscillatory response to a unit step input, is
   (a) 6.4 (b) 3.0
   (c) 0.4 (d) 0.2

Statement for Linked Answers Questions 13 and 14

A system of linear simultaneous equations is given as Ax = b where

$$A = \begin{bmatrix} 1 & 0 & 1 & 0 \\ 0 & 1 & 0 & 1 \\ 1 & 1 & 0 & 0 \\ 0 & 0 & 0 & 1 \end{bmatrix} \text{ and } b = \begin{bmatrix} 0 \\ 0 \\ 0 \\ 1 \end{bmatrix}$$

13. The rank of matrix a is
   (a) 1 (b) 2
   (c) 3 (d) 4

14. Which of the following statements is true?
   (a) x is a null vector
   (b) x is unique
   (c) x does not exist
   (d) x has infinitely many values

Statement for Linked Answers Questions 15 and 16

A transfer function with unity DC gain has three-poles at – 1, –2 and – 3 and no finite zeros. A plant with this transfer function is connected with a proportional controller of gain K in the forward path, in a unity feedback configuration

15. The transfer function is

(a) $\frac{s}{(s-1)(s-2)(s-3)}$

(b) $\frac{6}{(s+1)(s+2)(s+3)}$

(c) $\frac{s}{(s+1)(s+1)(s+3)}$

(d) $\frac{6}{(s-1)(s-2)(s-3)}$

16. If the **root locus plot** of the closed loop system passes through the points $\pm j\sqrt{11}$, the maximum value of K for stability of the unity feedback closed loop system is

(a) $\sqrt{11}$ (b) 6

(c) 10 (d) $6\sqrt{11}$

Answers
Control Systems and Process Control

1. Ans. (d)

Solution: Gain crossover frequency gain of

$$\frac{3e^{-2s}}{s(s+2)} = \left|\frac{3e^{-2s}}{s(s+2)}\right| = 1$$

$e^{-2s} = \cos(2\omega) - j\,\text{Sin}\,2\omega$; $|e^{-2s}| = 1$

$$\left|\frac{3}{j\omega(2+j\omega 2)}\right| = 1 \;;\; \left|\frac{3}{-\omega + j2\omega}\right| = 1$$

ω = 1.26

Phase crossover frequency

Phase angle = $-90 - \tan^{-1}[\omega/2] - 2\omega[180/\pi]$

Phase = -180 ⇒ $-90 - \tan^{-1}[\omega/2] - 2\omega\,[180/\pi]$

= - 180

ω= 0.632

2. Ans. (d)

Solution:

Gain margin

$$A = \text{Gain} = \left|\frac{3e^{-2s}}{S(s+2)}\right|_{\omega=0.632} = 2.2631$$

GM = 20 log [1 / a] = - 7.094 dB

Phase margin =

$180 + [-90 - \tan^{-1}(\omega/2) - 2\omega(180/\pi)]$

ω = 1.26

Phase margin = - 86.59°

3. Ans. (c)

Solution:

We gave $G(s) = \frac{as+1}{s^2}$

$\angle G(j\omega) = \tan^{-1}(\omega a) - \pi$

or $\frac{\pi}{4} = \pi + \tan^{-1}(\omega_g a) - \pi$

or $\frac{\pi}{4} = \tan^{-1}(\omega_g a)$

or $a\omega_g = 1$

At fain crossover frequency $|G(j\omega_g)| = 1$

Thus $\frac{\sqrt{1+a^2\omega_g^2}}{\omega_g^2} = 1$

or $\sqrt{1+1} = \omega_g^2$

or $\omega_g = (2)^{1/4}$

Now $a = \frac{1}{2^{1/4}} = 0.84$

4. Ans. (c)

$$= \frac{as+1}{s^2}$$

Solution: Y(s) = G(s) U(s)

U(s) LT δ(t) = 1

Output $\xrightarrow{LT} \frac{as+1}{s^2} = \frac{a}{s} + \frac{1}{s^2}$

⇒ au (t) + r (t) = a u (t) + t . u (t)

⇒ a + 1 ⇒ 1.84

5. Ans. (a)

Solution: If $x(0) = \begin{bmatrix} 1 \\ -2 \end{bmatrix}$

Then response $x(t) = \begin{bmatrix} e^{-2t} \\ -2e^{-2t} \end{bmatrix} \rightarrow \begin{bmatrix} 1/(s+2) \\ -2/(s+2) \end{bmatrix}$

If $x(0) = \begin{bmatrix} 1 \\ -1 \end{bmatrix}$

Then response $x(t) = \begin{bmatrix} e^{-2t} \\ -e^{-2t} \end{bmatrix} \rightarrow \begin{bmatrix} 1/(s+2) \\ -1/(s+2) \end{bmatrix}$

Let, $A = \begin{bmatrix} A_0 & A_1 \\ A_2 & A_3 \end{bmatrix}$

Then, $x(t) = [sI - A]^{-1}\, x(0)$

Multiply both sides by (sI – A) to above equation

$$[sI - A]x(t) = x(0)$$

Case 1

$$\begin{bmatrix} s - A_0 & -A_1 \\ -A_2 & s - A_3 \end{bmatrix}\begin{bmatrix} 1/(s+2) \\ -2/(s+2) \end{bmatrix} = \begin{bmatrix} 1 \\ -2 \end{bmatrix}$$

$$\frac{s - A_o}{s+2} + \frac{2A_1}{s+2} = 1$$

or, $s - A_o + 2A_1 = s + 2$

or, $-A_o + 2A_1 = 2$

or, $2A_1 - A_o = 2$

Finally,

$$2A_1 - A_0 = 2 \qquad (1)$$

$$A_2 - 2A_3 = 4 \qquad (2)$$

Case 2

$$\begin{bmatrix} s - A_0 & -A_1 \\ -A_2 & s - A_3 \end{bmatrix}\begin{bmatrix} 1/(s+1) \\ -1/(s+1) \end{bmatrix} = \begin{bmatrix} 1 \\ -1 \end{bmatrix}$$

$$A_1 - A_0 = 1 \qquad (3)$$

$$A_2 - A_3 = 1 \qquad (4)$$

From (1), (2), (3) and (4)

$A_o = 0$; $A_1 = 1$

$A_2 = -2$; $A_3 = -3$

Then, $A = \begin{bmatrix} 0 & 1 \\ -2 & -3 \end{bmatrix}$

STM, $\phi(\tau) = [sI - A]^{-1}$

$$[sI - A] = \begin{bmatrix} s & -1 \\ +2 & s+3 \end{bmatrix}$$

For eigen values

$$|(sI - A)| = 0$$

$$s(s+3) + 2 = 0$$

$$s^2 + 3s + 2 = 0$$

$$(s+2)(s+1) = 0$$

$$s = -1, -2$$

Eigen vectors can be find out from

$$[sI - A]x = 0$$

If for eigen value $s = -1$ then eigen vector

$$\begin{bmatrix} -1 & -1 \\ 2 & 2 \end{bmatrix}\begin{bmatrix} x_1 \\ x_2 \end{bmatrix} = 0$$

$$-x_1 - x_2 = 0$$

$$2x_1 + 2x_2 = 0$$

$$\frac{x_1}{1} = \frac{-x_2}{1}$$

Vector $= \begin{bmatrix} 1 \\ -1 \end{bmatrix}$

If Eigen value then eigen vector $s = -2$

$$\begin{bmatrix} -2 & -1 \\ +2 & 2 \end{bmatrix}\begin{bmatrix} x_1 \\ x_2 \end{bmatrix} = 0$$

$$-2x_1 - x_2 = 0$$

$$\frac{x_1}{1} = \frac{-x_2}{1}$$

Vector $= \begin{bmatrix} 1 \\ -2 \end{bmatrix}$

6. **Ans. (b)**

Solution: It can be verified with options, Option (d) gives two eigen value – 2 and – 1

7. **Ans. (a)**

Solution:

$$(sI - A)^{-1} = \begin{bmatrix} s & +1 \\ 0 & s+3 \end{bmatrix} = \frac{Adj(sI - A)}{|sI - A|}$$

$$= \frac{\begin{bmatrix} s+3 & -1 \\ 0 & s \end{bmatrix}}{\begin{bmatrix} s & 1 \\ 0 & s+3 \end{bmatrix}} = \frac{\begin{bmatrix} s+3 & -1 \\ 0 & s \end{bmatrix}}{s(s+3)} = \begin{bmatrix} \frac{1}{s} & \frac{-1}{s(s+3)} \\ 0 & \frac{1}{s+3} \end{bmatrix}$$

$$\therefore\ \phi(t) = LT^{-1}(sI - A)^{-1} = \begin{bmatrix} 1 & \frac{1}{3}(1 - e^{-3t}) \\ 0 & e^{-3t} \end{bmatrix}$$

8. **Ans. (c)**

Solution:

Zero state response = $L_T^{-1}\phi(s)\, B\, U(s)$

$$= LT^{-1}\begin{bmatrix} \frac{1}{s} & \frac{-1}{s(s+3)} \\ 0 & \frac{1}{s+3} \end{bmatrix}\begin{bmatrix} 1 \\ 0 \end{bmatrix}\frac{1}{s} = LT^{-1}\begin{bmatrix} \frac{1}{s^2} \\ 0 \end{bmatrix} = \begin{bmatrix} t \\ 0 \end{bmatrix}$$

State transition equation

= Zero input response + zero state response.

$\therefore \quad x(t) = \phi(t)\, x(0) + t$

$$\begin{bmatrix} 1 & \frac{1}{3}(1-e^{-3t}) \\ 0 & e^{-3t} \end{bmatrix}\begin{bmatrix} -1 \\ 3 \end{bmatrix} + \begin{bmatrix} t \\ 0 \end{bmatrix}$$

$$= \begin{bmatrix} -1+1-e^{-3t} \\ 0+3e^{-3t} \end{bmatrix} + \begin{bmatrix} t \\ 0 \end{bmatrix} = \begin{bmatrix} t-e^{-3t} \\ 3e^{-3t} \end{bmatrix}$$

9. **Ans. (c)**

Solution: Characteristic equation

$$LC\,s^2 + RCs + 1 = 0$$

$$s^2 + \frac{R}{L}s + \frac{1}{LC} = 0$$

$$s^2 + 2\delta\omega_n s + \omega_n^2 = 0$$

$$\omega_n = \frac{1}{\sqrt{LC}} = \frac{1}{\sqrt{10^{-3}\times 10\times 10^{-6}}} = \frac{1}{\sqrt{10^{-8}}}$$

$$= \sqrt{10^8} = 10^4 \;;\; 2\delta\omega_n = \frac{R}{L}$$

$$2\times\delta\times 10^4 = \frac{10}{10^{-3}} = 10^4$$

$$\delta = \frac{1}{2} = 0.5$$

$$\text{overshoot} = = e^{-\frac{n\pi}{\sqrt{1-\delta^2}}} = 0.16 \Rightarrow 16\,\%$$

10. **Ans. (c)**

11. **Ans. (b)**

Solution:

At s = 0 $\pm\sqrt{2}$ j, the system will have maximum controller gain and is,

$$k = \left|\frac{s(s+1)(s+2)}{2}\right|_{s=\sqrt{2}j}$$

$$= \frac{\left|\left(\sqrt{2}j\right)\left(1+\sqrt{2}j\right)\left(2+\sqrt{2}j\right)\right|}{2} = 3$$

12. **Ans. (d)**

Solution:

At s = – 0.425, the system have maximum non-oscillatory response, because after the system will become oscillatory since roots will be become complex.

Then

$$k = \left|\frac{s(s+1)(s+2)}{2}\right|_{s=-0.425} = 0.2$$

13. **Ans. (d)**

Solution:

14. **Ans. (d)**

Solution:

15. **Ans. (b)**

Solution:

16. **Ans. (c)**

Solution:

Characteristic equation

$$1+G(s)H(s)=0$$

$$1+\frac{6K}{(s+1)(s+2)(s+3)}=0$$

$$s^3+6s^2+11s+6+6K=0$$

Apply RH criterion

$s^3 \quad 1 \quad 11$

$s^2 \quad 6 \quad 6+6K$

s^1

s^0

Root locus intersection with imaginary axis is obtained from 2nd order simple auxiliary equation i.e.

According to question

$$6s^2+6+6K=0$$

$$s=\left(\pm j\sqrt{11}\right)$$

$$\Rightarrow j\sqrt{11}$$

$$6\left(j\sqrt{11}\right)^2+6+6K=0$$

$$-66^2+6+6K=0$$

$\therefore$ K = 10

Reader's Notes

9

Analytical, Optical and Biomedical Instrumentation

Analytical, Optical and Biomedical Instrumentation: Mass spectrometry. UV, visible and IR spectrometry. X-ray and nuclear radiation measurements. Optical sources and detectors, LED, laser, Photo-diode, photo-resistor and their characteristics. Interferometers, applications in metrology. Basics of fiber optics. Biomedical instruments, EEG, ECG and EMG. Clinical measurements. Ultrasonic transducers and Ultrasonography. Principles of Computer Assisted Tomography.

9.1 ANALYTICAL INSTRUMENTATION

Mass spectrometry. UV, visible and IR spectrometry. X-ray and nuclear radiation measurements.

Principle of Specroscopy

Spectroscopy is the measurement and interpretation of electromagnetic radiation absorbed or emitted, when the molecules or atoms, or ions of a sample move from are energy state to another. Every atom, ion or molecule the area of spectroscopy that steam resulting form every changes that arise when a sample is placed in a magnetic or electric field are susceptible to spectroscopic studies. Nuclear magnetic resonance and electrons spin resonance are two such studies.

The Nature of Electromagnetic Radiation

A beam of radiation may be regarded as an electromagnetic waveform disturbance or photon of energy propagated at the speed of High. The Heisenberg uncertainty principle shown that it is not important or even possible to measure both the wave and particle properties of a photon simultaneously.

A photon originating at a point in space radiates from the point in a spherical wave characteristic by elasticized by electrical field vectors which have maxima perpendicular to the direction of propagation. Wavelength and frequency are related to the energy of a photon, E, by Planck's constant h.

$$\Delta E = h\nu = \frac{hc}{\lambda}$$

Where c is the velocity of light in vacuum, $3.00 \times 10^8\ m\ \sec^{-1}$

Only frequency is truly characteristic of a particular radiation. When the light of a particular wavelength enters matter, its velocity decreases but its frequency remains constant.

Wave numbers, $\bar{\nu}$, are sometimes used to express frequency. They are calculated as follows:

$$\bar{\nu} = \frac{1}{\lambda}$$

Wave numbers, in unit of cm^{-1}, express the number of waves that occur per centimeter this number is directly proportional to the frequency $\nu = c\,\bar{\nu}$.

The Electromagnetic Spectrum

The interaction of matter and radiation takes place throughout the entire electromagnetic spectrum, the name given to broad range of radiations extending from cosmic rays with wave lengths as short as 10^{-9} nm all the way u to radio waves, which have wavelength in excess of 1000 km. Encompassed with in these extremes and moving from short to long wave lengths are gamma rays, middle and near ultraviolet rays, the visible light portion of the spectrum, infrared rays, and microwaves. The nature of all radiations is the same, and all move with the speed of light. They differ only in frequency and wavelength and in the effects which they can produce.

The chemical and physical effects of various types f radiations are quite different and these difference can be understood in terms of the differing energies of the photons. In the radio frequency range. The energy of one photon is very low, and the energy transition are concerned with reorientation of nuclear spin state so substance in a magnetic field. In the slightly-energy microwave region, there are change in infrared region absorption with unpaired electrons when in a magnetic field. In the infrared region absorption causes changes is vibration energy accompanied by changes in rotational energy. Changes in the visible and ultraviolet regions involve the electron energy of atoms, or molecules; for, molecules this is accompanied by changes in vibrational and rotational energy. These changes in electronic energy involve the most loosely held or so-called electrons. Finally, at the high energy end, X-rays will raise inner electrons to excited states.

Atomic Energy Level

Atoms can exist only at discrete potential energy level. The potential energy of an atom depends on the electrons configuration and transition of outer electrons between fixed energy level to emit or absorb radiation at discrete energies. The frequency of radiation absorbed or emitted is proportional to the change in potential energy involved, and is given by $\Delta E = h\nu$. Thus atomic spectra involve only transitions of electrons from one electronic energy level to anther. Each transition accouter for the presence of a specific frequency of light and hence the presence of a spectral line either in absorption or in emission.

Molecular Electronic Energy levels.

The electrons of a molecule can also be excited to higher-energy states and the radiation that is absorbed in this process, or the energy emitted in the return to ground state, can be studied. The energy involve are generally large, 50–150 kcal mol^{-1}; consequently, electronic spectra of molecules are usually found in the ultraviolet or visible region of the electromagnetic spectrum. In molecular spectra, transition between electronic states may be accompanied by transition between rotational and vibration energy level which impart fine structure on the electronic absorption bands. As a result the spectra of molecules are much more complicated than those of atoms. But this means that information may be obtained about molecular vibration and rotations that revel a great deal about molecular structures.

Vibrational Energy Levels

Vibrational energies are usually of an order of magnitude smaller than the electronic energy. The simplest case involves the vibrational motion of the atoms a diatomic molecule. In a diatomic molecule the only vibration is the stretching of the bond between two atoms. The allowed energies for a diatomic molecule, as given by the quantum-mechanical vibrational energy level are

$$e_{vib} = \left(v + \frac{1}{2}\right)\frac{h}{2\pi}\sqrt{\frac{K}{u}} \quad v = 0, 1, 2, \ldots \qquad (1)$$

Where K is the force constant and measures the force required to streach a bond by a given distance (that is, the stiffness of the chemical bond) and u is the reduced mass for the two atoms, m_1 and m_2

$$u = \frac{m_1\ m_2}{m_1 + m_2} \qquad (2)$$

Source of Light and their Spectrum

Light is an electromagnetic radiation. Hence emission of light from any source result from the acceleration of electric charges. These electric charges involved in the emission of visible and ultraviolet light are the negative electrons is the outer pat of the atom. BY assuming that vibratory or orbital motions of these electrons cause radiation, many of the characteristic of different light sources can be explained.

Light Sources

Source of light may be divided into two main classes: (1) thermal source in which the radiation is the result of high temperature and (2) sources depending on the electrical charges through gases. The sun, with it's surface temperature of 5000 °C to 6000 °C is an important example of the first class, but here must also be added such important sources as filament lamps, the various electric arcs at atmospheric pressure and the flame. Under the second class high voltage sparks, the glow discharge to vacuum tubes of low pressure, and certain low pressure arcs like the mercury arc.

Coherence and Uses of Light Waves

Conventional light sources produces incoherent light. This means the light is a mixture of wave at various frequencies the reinforce of cancel each other in a random fashion A coherent wave is an orderly wave of one frequency where the whole beam is spatially in phase. There are actually two facets to this coherent wave, one time and the other spatial. Time coherence comes about when cycle of the wave takes exactly the same time to pass a point in space. This really means that the frequency of the wave is not varying and is given the name monochromatic. The prefix mono means "one" while chromatic means "colour" – the word monochromatic mans one colour, which is a single non-varying frequency in the light region of the frequency spectrum.

Classification of Spectra

There are two principal classes of spectra, known a emission spectra and absorption spectra.

1. Continuous emission spectra
2. Continuous absorption spectra
3. Line emission spectra
4. Line absorption spectra
5. Band emission spectra
6. Band absorption spectra

Emission spectra are obtained when the light coming directly form a source is examined with a spectroscope. Absorption spectra are obtained when the light from a source showing a continuous emission spectra is passed through an absorbing material and thence in to the spectroscope.

Soilds and liquids, with a few rare exception, give only continuous emission and absoption section , in which a wide rage of wave without any sharp discontinuities, is covered. Discontinuous spectra (line and band) are obtained with gases.

Continuous Spectra

The most common source of continuous emission spectra are solids at high temperature. One can observe in the continuous spectrum the distribution indifferent wavelengths of the energy. According to Kirchhoffs law, this depends on the ability of the surface to absorb light of different wavelengths.

Line Spectra

When the slit of a prism or grating spectroscope is illuminated with the light from mercury are, several lines of different colours are seen in the eyepiece. It is important to realize that line spectra derive their name from the fact a slit is customarily used in the collimator the line. If a point, a disk, or any other forms of aperture wave used in the collimator, the spectrum lines would become points, disks etc. as the case may be. In this case, each "line" in the spectrum has the shape of the source.

The most intense sources of line spectra are metallic arcs and sparks, although vacuum tubes containing hydrogen or one of the rare gases are vary suitable.

Spectroscopic Analysis

Many a times it is essential to know the composition and the percentage of constituents of a substance correctly. For analyzing any material for its compositional and percentage of constituents, it is necessary to use properties or qualities not possessed by other materials or compounds being ayalysed at the same time.

Optical analysis is now a days finding extensive use and may employ different phenomena and a fairly wide portion of the electro-magnetic spectrum ranging from the ultraviolet to the infrared region. The former have a grater sensitivity because of the greater amount of ultraviolet energy then the methods employing the infrared energies. However, the methods employing infrared techniques are more selective because the absorption bands in the infrared regions are much sharper than the corresponding bands used in ultraviolet spectroscopy.

Infrared analysis can be used in two forms-dispersive and the non-dispersive. The dispersive system does not employ an optical system other then polished reflecting surface which causes the energy to be reflected from surface to surface until it finally reaches the absorbing cell . The non-dispersive system, using the more optical methods, reflects and refracts the beams as it causes them to actuate the detector-amplifier system.

In gas chromatography which is batch type of analysis is sent to an absorption column, the different constituents are then removed in a definite sequence by the carrier gas. The time of releases of the material indicates its composition. The carrier gas thus carries but one variable through the analyzer at may one time.

The property of refractive index of a material is also many times utilized to determine the composition and connection of the material as it proportional to both. By measuring the index of refraction, the degree of connect ration can be determined. When dirty or turbid materials are to be analyzed the critical angle is more generally employed.

Certain gases can be analyzed by untilsing the property of their different densities. The analysis is made by measuring the forces of deceleration involved when different gasses undergo the same change in speed.

Methods of Composition Analysis

One of the greatest advances in the field of instrumentation in past years has been in the application of direct and quantitative methods of composition measurement in the continuous processing inductees. The newer methods of composition analysis constitute positive methods as they are based on atomic or molecular or crystalline characteristic. We shall now review some of the measurement techniques used in composition analysis.

Spectroscopic Analysis

The fundamental properties of a substance are identified and molecular spectra. As we all know, atoms are composed of a nucleus, having a net positive electric charge, and a relatively large mass, surrounded by number of negatively charged electrons of small mass. The atomic of the element is determined by the total number of the electrons surrounding the nucleus and therefore by the nuclear charge.

The electrons surround the nucleus is so called with a discrete number in each shell. The number of elections in shell is limited, and the energy state of each electron corresponds to a particular configuration of electrons. If the atom is in a ground state, it has the lowest possible energy, and a given electron configuration is determined. If the atom is excited to a higher energy state, as by light radiation or any other means, the electrons absorb energy by shifting to a new configuration.

In spectroscopic analysis, the atoms of a substance are excited, thereby their configuration is changed and they either absorb or radiate energy at unique frequencies. Therefore, a means becomes available for identifying the existence of a particular atom of a substance. The above could be further clarified by considering an example. Say a light beam containing all frequencies is passed through a then film of sodium and then through a refracting prism. The prism breaks the light beam into a spectrum which can be recorded on a photographic plate. From studying this spectrum it will be observed that certain frequencies will be missing from the light, which obviously must have absorbed by the sodium. Each of these frequencies corresponds to particular changes of energy-level within the atom. Thus, these frequencies corresponds to particular these particular lines in an absorption spectrum.

Line spectra are caused by emission or absorption by atoms of discrete wavelength of radiation, thus causing lines. If, however we obtain an absorption spectrum from a molecule or a molecular compound, a band spectrum is observed. That is many difference wavelengths are absorbed but in varying degree.

The absorption spectroscopy is based on the principle that when infrared, ultraviolet or X-ray radiation is passed through the sample of unknown martial; certain frequencies of the radiation are absorbed. This is determined by separating the radiation into a spectrum and measuring the intensity of radiation at each frequency. Absorption line are dark lines on a light background on a photographic negative.

In emission spectroscopy the unknown substance is placed in a flame or an arc. This causes emission of radiation which is characteristic of each substance. By collecting the emitted radiated and separating it by refraction or diffraction, the intensity or radiation at each frequency can be measured. Emission liens are bright lines bright lines on a dark background on a photographic negative.

In fluorescence spectroscopy which actually is a from of emission spectroscopy, the photoluminescence or fluorescence, resulting from excitation of light, is employed. The excitation of substance causes an emission of light which, by measurement of its spectral distribution, can be related to the composition of the substance. This called method can be applied to only those substance which undergo fluorescence.

In the method of mass spectroscopy, the sample of unknown substance to be identified (usually a vapouror gas) is bombarded by an electron beam in an evacuated chamber. The bombardment produces ions by "Knocking off" an electron from each of the various kinds of atoms present in the substance. These ions are then accelerated into a circular path, the radius of which varies with the mass of ions and thus the ions are thus the ions are stored into beams. By simply measuring the strength of each beam, all the various kinds of atoms can be counted are identified, thereby enabling the composition of the substance to be identified. A broad classification of spectrometer can be done as follows:

Absorption Spectroscopy

The absorption type spectrometers can be further classified as follows depending upon the wave length range of electromagnetic spectrum.

1. Infrared absorption spectrometers operating from radiation in the range from 1 to 25 microns wavelength
2. Ultraviolet absorption spectrometer operating in the range from 0.2 to 0.8 micro actually includes the visible band.
3. X-ray absorption spectrometers operating in the wavelength range from 0.4 to 1.5 A.

9.1.1 Mass Spectrometer

In a mass spectrometer, the sample to be analyzed is first bombarded with an electron beam to produce ionic fragments of the original molecule. These ions are then sorted out by accelerating them through electric and magnetic fields, according to their mass/charge ratio. A record of the numbers of different kinds of ions is called the mass spectrum. The uniqueness of the molecular fragmentation is the basis for identification of different molecules in a complex mixture, as no two molecules will be fragmented and ionized in exactly the same manner. Very small sample sizes, which may be of the order of a few tenths of a milligram, are generally sufficient as long as the material is able to exist in the gaseous state at the temperature and pressure existing in the ion source. Several other useful applications of mass spectrometry include the direct determination of the molecular weight, the placement of functional groups into certain areas of the molecules and their interconnection and investigation of reaction mechanism.

A significant advantage of mass spectra is its suitability for data storage and library retrieval, since the positions (masses) of the peaks in the spectrum of a given compound are fixed. Another asset of mass spectrometry is its capability in handling complex mixtures, currently obtained via asset of mass spectrometry is its capability in handling complex mixtures, currently obtained via the has chromatograph-mass spectrometer.

The procedure for analyzing a substance by mass spectrometry starts by converting the substance into a gaseous state by chemical means. The gas is introduced into the highly evacuated spectrometer tube, where it is ionised by means sof an electron beam. The positive ions thus formed and deflected and focused by means of suitable magnetic and electric fields. For a given accelerating voltage, only positive ions of a specific mass pass through a slit and reach the collecting plate. The ion currents thus produced are measured by using a sensitive electrometer tube. By varying the accelerating voltage, ions from other mass species may be collected, and the ion current measured in such cases would be proportional to the amount of the given mass species present. A great advantage of mass spectrometry over other direct methods is that the impurities of masses different from the one being analyzed, do not interfere with the results.

Principle of Operation

Figure 9.1 shown the principle of operation of a mass spectrometer. The molecules in the gas sample (*A*) to be analyzed are bombarded with electrons to produce ions (*B*). These ions are accelerated in a high vacuum onto a magnetic field (*C*), which deflects them into circular paths (*D*). Since the deflection for light ions is greater than that for heavy ions, the ion stream separates into beams of different molecular weight. A suitable placed slit (*F*) allows a beam of a selected molecular weight to pass through to a collection electrode (*E*). As the accelerating voltage on the ion source is gradually reduced, ion beams of successively greater mass pass through the slit. When these ions fall on the collector electrodes, they produce minute electric currents, which may be measured father suitable amplification. Their amplitude will indicate the number of ions in each beam. The proportion of molecules of different masses in the gas sample may thus be found and a complete analysis of the gas sample may be made; provided all the constituent gases have a different molecular weight. This is usually the case in respiratory gas analysis work.

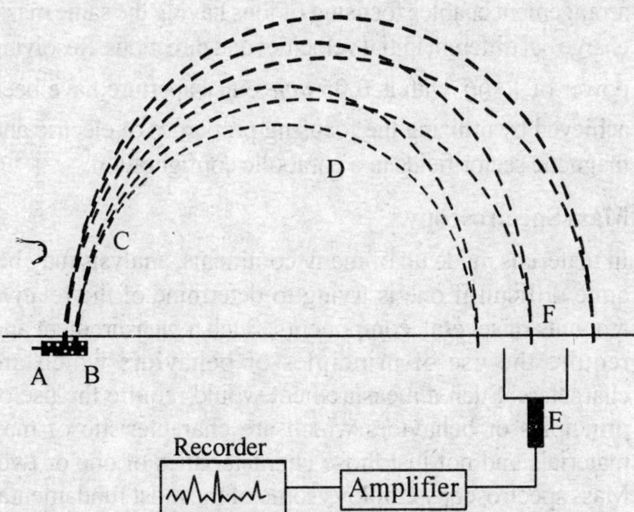

Figure: Principle of operation of a mass spectrometer

It is apparent that there is a close analogy between mass spectrometry and optical spectroscopy. This is shown in Figure 9.2.

In optical spectroscopy, the source is a visible light, which is composed of individual colours (different wavelengths of light) that are present at different intensities. A prism (or grating) is used to separate the light into its different wavelengths followed by a slit which selects the wavelength reaching the detector. The different wavelengths are then scanned across the detector slit and the light intensity is measured as a function of wavelenght.

In scanning mass spectrometry, the starting point is a mixture of ions having different mass-to-charge ration. A slit is used to select which mass-to-charge ratio reaches the detector. The different mass-to-charge rations are then scanned across the detector slit and the ion current is recorded as a function of time (mass).

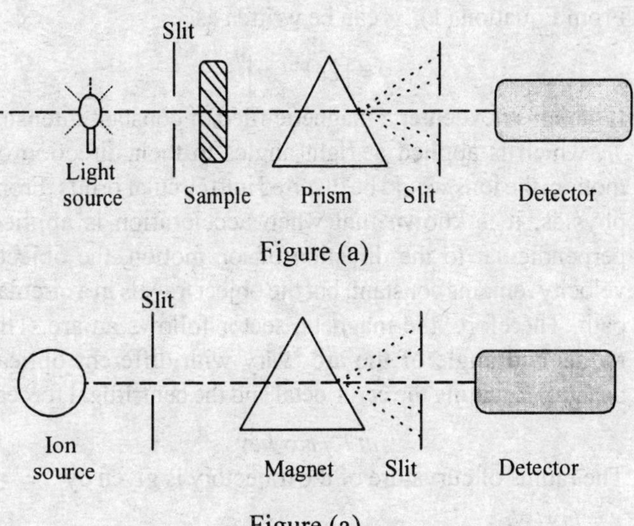

Figure: Analogy between optical spectroscopy and mass spectrometry

Magnetic Deflection Mass Spectrometer

Figure 9.4 shown the essential parts of a typical magnetic deflection mass spectrometer tube due to Nier (1940) . The heated tungsten filament produces an electron beam, which passes between

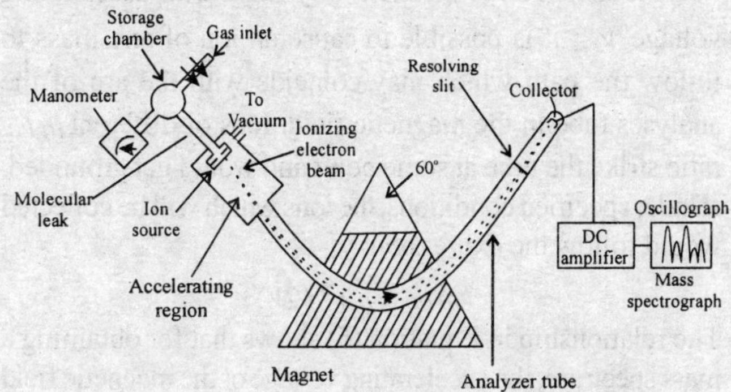

Figure: Schematic diagram of a Nier 60° sector mass spectrometer

plates *A* and *B*. *A* difference in electrical potential between *A* and *B* pulls ions out of the beam, so that they pass through the slit in *B* into the region between *B* and *C*. The potential difference between *B* and *C* is adjustable from 0 up to several thousand volts. The ion beam then enter the space between two trapezoid shaped magnet poles, where it is deflected through an angle of $60, 90, 120$ or 180.

When ions of mass m and charge e pass through an accelerating electric field, they would attain a celocity V which can be expressed in terms of the accelerating voltage V as follows:

$$1/2mv^2 = eV,$$

where $1/2mv^2$ is the kinetic energy of the ion as it leaves the electric field.

From Equation (1), v can be written as

$$v = [2eV/m]^{1/2}$$

If the ions next enter a magnetic filed of constant intensity h, which is applied at right angles to their direction of motion, the ions would be diverted into circular orbits. From physics, it is known that when acceleration is applied perpendicular to the direction of ion motion, the objects velocity remains constant, but the object travels in a circular path. Therefore, the magnetic sector follows an arc. The radius and angle of the arc vary with different optical designs. Equating the centripetal and the centrifugal forces:

$$mv^2/r = hev$$

The radius of curvature of the trajectory is given by:

$r = mv/eh$

Substituting for v from Equation (2), we get :

$$r = [2eV/m]^{1/2} \times m/eh$$

$$= \left[2eV/m \times m^2/e^2h^2\right]^{1/2}$$

$$= \left[2V/h^2 \times m/e\right]^{1/2}$$

In practice, all the quantities of Equation (5) are kept constant with the exception of m and V. By varying the accelerating voltage V, it is possible to cause an ion of any mass to follow the path which may coincide with the arc of the analyses tube in the magnetic field. Ions of different m/e ratio strike the tube at some point and would get grounded.

Under specified conditions, the ions which will be collected would follow the expression:

$$m/e = h^2r^2/2V$$

The relationship in Equation (6) shows that for obtaining a mass spectrum, the accelerating voltage or the magnetic field strength can be varied. Usually, it is the magnetic field which is kept constant and the voltage is adjusted to bring to a focus specific m/e. The magnetic field must be very uniform over a large area. If electromagnets are employed, they would require large power supplies capable of providing several Kilowatts of power and must be highly regulated. For a particular instrument, the angle of defection (radius) is fixed for a given analyzer tube $(180°, 90°, 60°)$. Therefore, Equation (6) can be written as:

$$mV = \text{constant}$$

It this constant is, say 2000, then as V is set at $100\,V$, particles of mass 20 can reach the collector plate, while to direct particles of mass 25 to it, V must be set at $80\ V$. Direction focusing is achieved by deflecting the ion beam along a $180°$ trajectory through the magnetic filed. A large magnet is required for an $180°$ mass spectrometer. When the ion source is in the uniform magnetic field, the gap between the poles must be large enough to contain the ion source. In order to overcome this difficulty, Nier (1940) employed the direction focusing properties of a $60°$ sector magnetic field. In this arrangement, the magnetic field dos not envelop the ion source; a separate source magnet is required. With sector type instruments, a mass resolution of 200 to 6000 mass units can be obtained.

The use of simultaneous direction and velocity focusing results in a high in a resolution mass spectrograph. This is done by placing an electrostatic analyzer between the ion accelerating slits and the magnetic field. After the usual acceleration of the beams in an electrostatic field, the ions are deflected through a tandem arrangement of an electrostatic analyzer and then a magnetic analyzer. The arrangement enables focusing of ions having the same mass/charge but different initial velocities and directions. Resolving power of 8500 with a $0.05\,mm$ entry aperture have been achieved by utilizing the focusing properties of electric and magnetic sector fields in a parabolic configuration.

Mass Spectroscopy

In materials made up of many continents, analysis may be, quite difficult if one is trying to determine of the relative amounts of several components. Such a measurement and require the use of principles or behaviors which are characters. Such a measurement would require the use of principles or behaviors which are characteristic of may materials and not just those characteristics of one or two. Mass spectroscopy employs some of the most fundamental characteristic of materials (a) For a given accelerating force, the acceleration in inversely proportional to the mess (b) As the ions of various materials have different weight, they are accelerated at different rates. (c) These ions follow a curved path when they are caused is pass through a magnetic field. For a given voltage and magnetic field, the radius depends upon the mass. The radius of curvature which ions possess in a uniform magnetic field is given by the relation.

$$x = \frac{144\sqrt{mV/e}}{B}$$

Where r = radius

m = mass, atomic-weight units

V = accelerating potential, volts

B = magnetic field strength, gauss

e = number of electronic charges

(d) Materials that have equal kinetic energies but different masses will have unequal speeds and will necessarily take different time to traverse the same distance (e) If a radio-frequency field is applied to particles that have different speeds. Then energy can be imparted to some is far grater measure than it can be imparted to others. Consequently, there is a selective accelerating process (f) The angular velocity of a charged particle moving in a magnetic field is dependent upon the following relationship:

$$f = \frac{1}{2\pi} \cdot \frac{B}{mc}$$

Where f = frequency, cps

e = charges of the electron, electrostatic units

m = mass of the particle

B = magnetic field strength, gauss

c = speed of light, cm/sec

Thus it is obvious that the path followed by an ion depends upon the accelerating voltage the strength of the magnetic field and the mass of the ion. For practical design, varying the applied can cause all of the ions to be counted at some common point, thereby providing a relationship companion and voltage.

9.1.2 Ultraviolet Spectroscopy

Ultraviolet Absorption Spectrometers

These are widely used for nearly all types of analysis and as already indicated, these operate in the range from 0.2 to 0.8 micron. The optical path of a widely used ultraviolet spectrometer (photoelectric quarts spectrometer) is shown in the figure 16(a). An image of the light source (tungsten lamp for use is the visible region from 0.3 to 1.0 micros and a hydrogen-discharge lamp for uses in the ultraviolet region from 0.22 to 0.35 micron) is focused by the condensing mirror and by flat mirror on the entrance slit, from where it proceeds to a collimating mirror and through a quartz prism and the collimating mirror, the portion of the spectrum mirror through a quartz prism and the collimating mirror, the portion of the spectrum passes through the exit slit and then collimating mirror, the portion of the spectrum passes through the exit slit and then through the sample cell and falls on the phototube. The intensity of the beam is measured by a suitable amplifier connected to an automatic balance potentiometer instrument. The particular wavelength passed through the sample cell is determined by the angle of the angle of setting of the prism. The angle of the quartz prism determines which part of the refracted beam falls on the exit slit. The sample cell can be.

The ultraviolet absorption spectroscopy in the other respect is similar to infrared spectroscopy. The instrument may be used either for complete analysis samples by running a complete spectrogram or for continuous analysis of process steams by operating at a fixed wavelength critical to the analysis being made.

9.1.3 Visible Absorption Spectroscopy

Ultraviolet and visible (UV-Vis) absorption spectroscopy is the measurement of the attenuation of a beam of light after it passes through a sample or after reflection from a sample surface in the ultraviolet and visible range of the electromagnetic spectrum. Absorption measurements can be done at a single wavelets or over an extended spectral range. Ultraviolet and visible light are energetic enough to promote outer electrons to higher energy levels. UV-Vis spectroscopy is usually applied to molecules or enraging complexes in solution. The UV-Vis spectra have broad features that are of limited use for sample identification but are very useful for quantitative measurement. The concentration of an annulate in solution can be determined by measuring the absorbance at some wavelength and applying the Beer Lambert Law.

Since the UV-Vis range spans the range of human visual acuity of approximately $400-750\,nm$, UV-Vis spectroscopy is useful for characterizing the absorption, transmission and reflexivity of a variety to technologically important materials such as pigments, coatings, windows and filters .This more qualitative application usually required recording at last a portion of the UV-Vis spectrum for characterization of the optical or electronic properties of materials (Manning, 1969).

The UV-Vis spectral range is approximately 190 to 900 nm, as defined by the working range of typical commercial UV-Vis spectrophotometers. The short-wavelengths limit for simple UV-Vis spectrometers is the absorption of ultraviolet wavelengths of less than 180 nm by atmosphericgases. Purging a spectrometer with nitrogen gas extends this limit to 175 nm. For working beyond 175 nm requires a vacuum spectrometer and a suitable UV light source. The long-wavelength limit is usually determined by the wavelengths response of the director in the spectrometer.

High-end commercial UV-Vis spectrophotometers extend the measurable spectral range into the NIR region to as far as 3300 nm.

The light source is usually a deuterium discharge lamp for UV measurements and a tungstenhalogen lamp for visible and NIR measurements. The instruments automatically swap lamps when scanning between the UV and visible regions. The wavelengths of these continuous light sources are typically dispersed by a diffraction/ holographic grating in a single or double monochromator or spectrograph. The spectral pampas is than determine by the monochromator slit width or by the array-element width in array-detects spectrometers. Spectrometer designs and optical components are optimized to reject stray light, which is one of the limiting factors in quantitative adsorbates measurement. The detector in single-detector instruments is a photodiode, phototube, or photomultiplier tube (PMT). UV-Vis-NIR spectrometers utilize a combination of a PMT and a Platier-cooled PbS IR detector. The light beam is redirected automatically to the appropriate detector when scanning between the visible and NIR regions. The diffraction gyration and instrument parameters such as slit width can also be changed.

Most commercial UV-Vis absorption spectrometers use one of the following three overall optical designs:

- A fixed or scanning spectrometer with a single light beam and sample holder,
- A scanning spectrometer with dual light beams and dual sample holders for simultaneous measurement of I and I_o, or
- A non-scanning spectrometer with an array detector for simultaneous measurement of multiple wavelengths.

The infrared region extends from 0.8 to 200 μ in the electromagnetic spectrum. However, most of the commercial instruments are available in the region from 0.8 to 50 μ. The position of subshrub-tion bands in the infrared spectrum is expressed in both wavelength as well as wave numbers. The wavelengths λ is generally measured in microns (μ). The wavenmber (v) is the number of wavelength per centimeter and is given by:

$$v(\text{in cm}^{-1}) = 10^4/\lambda(\text{in } \mu)$$

Infrared spectrophotometers using prisms produce spectra which are spread linearly with wavelength, whereas instruments fitted with gratings generally deliver spectra that are spread linearly with wave number. However, the results are preferably reported in wave numbers in either case, since these are proportional to molecular properties like frequency and energy, whereas the wavelengths is a property of the radiation only.

Each of the motions of a molecule occurs with a certain frequency, which is characteristic of the molecule and of the particular vibration. The energy involved in a particular vibration is characterized by the amplitude of the vibration, so that the higher the vibrational energy, the larger is the amplitude of the motion. A series of energy levels or states is associated with each of the vibrational motions of the molecule. The molecule may be made to ascend from one unearth level to a higher one by absorption of a quantum of electromagnetic radiation. While undergoing such a transition, the molecule gains vibrational energy, which is manifested in an increase in the amplitude of the vibration. The frequency of light required to cause a transition for a particular vibration os equal to the frequency of that vibration, so that we may measure the vibrational frequency by measuring the frequencies of light which are absorbed by the molecule.

Since most vibrational motions in molecules occur at frequencies of about 10^{14} $\sec^{-1}$, then the light of the wavelength

$$\lambda = c/f = 3\times10^{10} \; cm/\sec/10^{14}\sec^{-1} = 3\times10^{-4} cm = 3$$

microns will be required to cause transitions. As it happens, the light of this wavelength lies in the so-called infrared region of the spectrum. IR spectroscopy, then, deals with transitions between vibrational energh levels in molecules, and is therefore also called vibrational spectroscopy. An IR spectrum is generally displayed as a plot of the energy of the infrared radiation (expressed either in microns or wave numbers) versus the percentage of light transmitted by the compound. This is indicated schematically in Figure 3.1. Here, sample transmittance is usually presented linearly on the vertical axis in an IR spectrum and absorption bands are generally presented pointing down wards. Transmittance, in this case, is defined as the radiant power of the radiation, which is incident on the sample, divided by the radiant power transmitted by the sample.

The energy acquired by a molecule can be utilized in three ways, namely electronic excitation, vibrational change and rotational change, and the various molecular energy transformations that may take place are quantized. The amount of energy required to cause the various types of transitions generally corresponds to definite regions of the electromagnetic spectrum. For a molecule, the average energy involved in electronic excitations is 5 eV. For molecules in a particular electronic state, the average energy involved in a vibrational excitation is 0.1 eV and a rotational exciton involves about 0.005 eV.

In infrared spectroscopy, we are interested manual in the vibrations and rotations induced in a molecule by the absorption of radiation. The infrared spectrum based on these absorption properties constitutes a powerful tool for the study of molecular structures and identification. Chemical identification is based in the empirical correlations of vibrating groups, with specific absorption bands and the quantitative estimations are dependent on the intensity immurements.

Infrared spectrophotometry is applicable to solids, liquids and gases. It is fast and requires small sample sizes. It can also differentiate between subtle structural differences, On the other hand, interpretation of infrared spectra is highly empirical and requires huge libraries of reference spectra. The interpretation of spectra is a skilled art; nevertheless, it is now a part of the repertoire of many chemists. Infrared spectrophotometry provides the means for monitoring many common atmospheric pollutants such as ozone, oxides of nitrogen, carbon monoxide, sulphur dioxide and others.

The region of the infrared spectrum which is of greatest interest to organic chemists is the wavelength range 2.5 to ≈ 15 micrometers (μ). In practise, units proportional to frequency , (wavenumber in units of cm^{-1}) rather than wavelength, are commonly used and the region 2.5 to $\approx 15\ \mu$ corresponds to approximately 4000 to $600\ cm^{-1}$.

9.1.4 Infrared Spectrophotometers

Infrared spectrophotometers are produced by several instrument manufacturers. Almost all the commercial designs employ a double-beam system, wherein the radiant energy passes alternately through the sample and then through a reference to a single detector. They incorporate a low-frequency chopper (5 to $13\ Hz$) to modulate the output radiation from the source. The various sources and detectors described above are incorporated in one or more of the commercial instruments.

The double-beam system offers the advantage that scans are not disturbed by absorption bands produced by atmospheric water vapour and carbon dioxide. Also, fluctuations in source output detector sensitivity and gain have no influence on the record.

There are two types of arrangements for recording the infrared spectra. These are the ratic recording and the optical null methods.

Optical Null Method

In this method, the infrared radiation is passed simultaneously through two separate shingles, one containing the sample, and the other, the reference. The two beams are recombined into a common axis and are alternately focused on the detector. If the intensities of the sample and reference beams are exactly equal, then no alternating intensity radiation goes through the slip. If the sample absorbs some radiation, an alternating intensity radiation is observed by the detector, which produces an ac signal. This ac signal can be selectively amplified in a tuned amplifier. As the amplifier is tuned to the chopping frequency of one light beam, the signals of frequency that are different from the chopping frequency are not amplifier. The alternating signal from the detector is used to drive a serve-motor, which is mechanically coupled to an epactal wedge or a fine-toothed comb attenuator. The movement of the comb occurs when a difference in power of the two beams is sensed by the detector. The motor will stop when the reference and the sample beam intensities are exactly equal. The movement of the motor is synchronized with the recorder pen so that its position gives a measure of the relative power of the two beams and thus the transmittance of the sample. The record obtained therefore is of the sample absorption as a function of spectral frequency. The teeth of the attenuator comb are accurately cut so that a linear relationship exists between the lateral movement of the comb and the decrease in power of the beam.

Figure 3.13 shows the diagram of a typical double-beam optical-null principle infrared spectrophotometer. Emanating from the source, the two light beams pass through the sample and reference cells. With a switching frequency of $12.5\ Hz$, the rotary sector mirror alternately conducts the radiation of the sample and reference channels into the monochromator. Here, the light of the respective wavenumber is reflected and passed on to the thermocouple, which delivers a $12.5\ Hz$ ac voltage signal, when the intensities of the two optical paths are not equal. After amplification in the pre-amplifier and main amplifier, and after phase-sensitive rectification in the demodulate to, this ac voltage signal controls the servo-motor, which adjusts the attenuator diaphragm in the reference channel, until the intensities of both beams are equal and the signal disappears. The servo-motor also drives the recording pen-which records the sample's transmittance corresponding to the position of the attenuator diaphragm.

Ratio Recording Method

In the direct ratio recording system, radiation from the source is passed altermately though two separate channels,, one passing through the sample cell and the other through the reference cell. The two signals from the detector are amplified, rectified and recorded separately. The sample signal is displayed as a proportion of the reference signal. In this method, attempts are not made to achieve a physical equalization of the sample and the reference beam intensities. The reference signal is used to drive the slit width control so that the energy level reaching the detector is constant. This system is, however, not favoured in infrared instruments, as it requires a very stable amplifier system and chopping system.

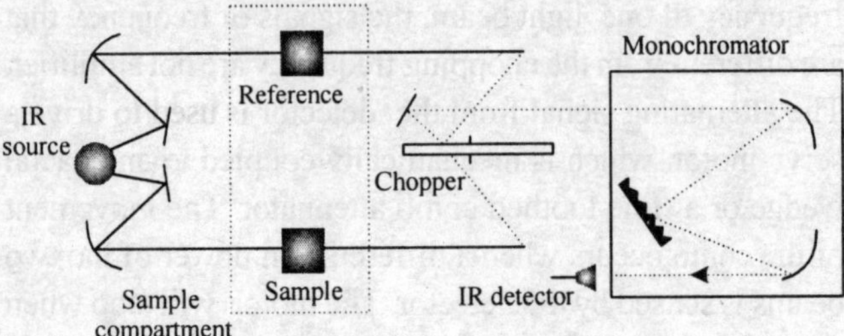

Figure: Block diagram of a double-beam infrared spectrophotometer

9.1.5 X-Ray Insturment

The purpose of radiography is to show the presence and nature of defets of other structural discontinuities in the interior of the materials under examination. The principle of radiographic examination is shown in figure below.

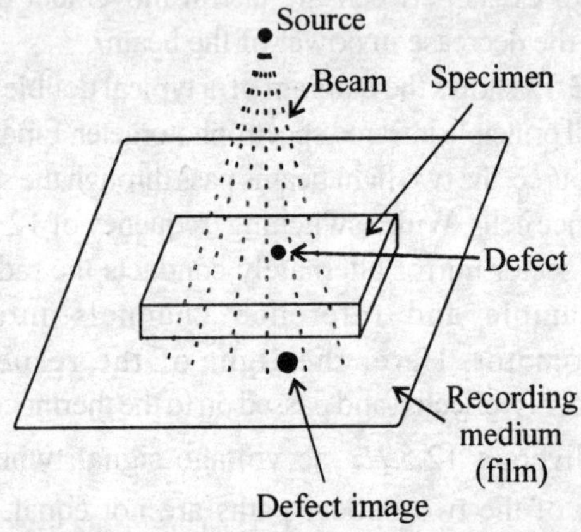

Figure:Principle of radiographic examination

This technique makes use of the ability of short wavelength electromagnetic rediations, such as X-rays or gamma rays, to penetrate objects. In general the shorter the wavelength, the greater is the penetrating power. The radiation that enters through the material, some being absorbed in the material itself and amount of absorption is a function of the density and thickness of the material. Should there be a cavity or discontinuty in the interior of the material, the bram of radiation will have less material to pass through than in solid material. Consequently, there is a variation in the absorption of the rays by the material in the defective area. The variation, if measured or recoded on a film sensitive to X-or gamma radiation, produces an image that will indicate the presence of the defect, The image is an X-ray shadow of the interior of the material. Thus, radiography is essentially based on the principle of shadow projection and such a shadow picture is called a radiograph. Variation in the darkness may be interpreted to provide information concerning the internal strcture of the material. The basic setup essenitally consists of a source of radiation, the object to be radiographed and a detector which is normally a sheet of photographic film.

X-Ray Source

In the widely used conventional X-radiography, the source of radiation is an X-ray tube. The X-ray tube consists of a glass bulb under vacuum, enclosing a positive electrode or anode and a negative electrode or cathode. The cathode comprises a filament which, when brought to incandescence by a current of a few amperes emits electrons. Under the

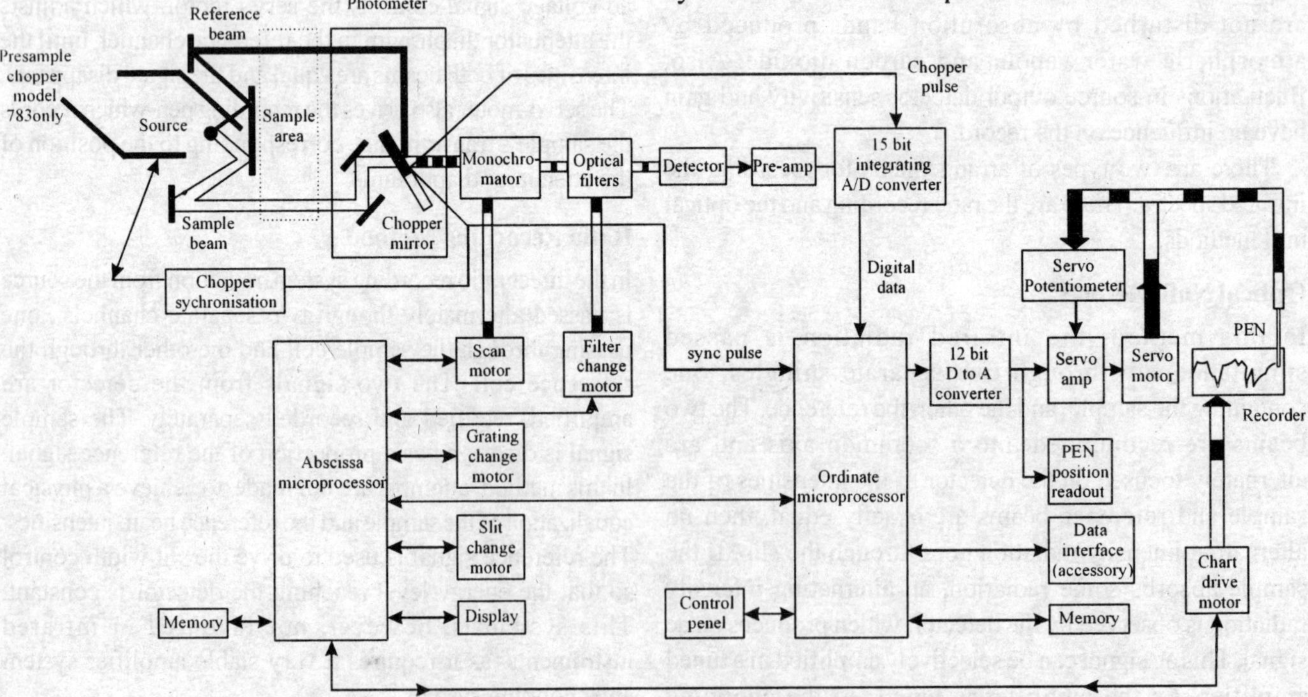

Figure: Block Diagram of a scanning infrared ratio recording infrared spectrophotometer

effect of electrical tension set up between the anode and cathode, these electrons from the cathode are attracted to the anode. This stream of electrons is concentrated in a beam by a cylinder or a focusing cup. The anti-cathode is a slip of metal with high melting point recessed into the anode at the place where it is struck by the beam of electrons. It is by impinging on the anticathode that fast moving electrons give rise to X-rays. Figure below shows the layout of a typical X-ray tube built by Collidge in 1913 which also used a heated filament to produce electrons.

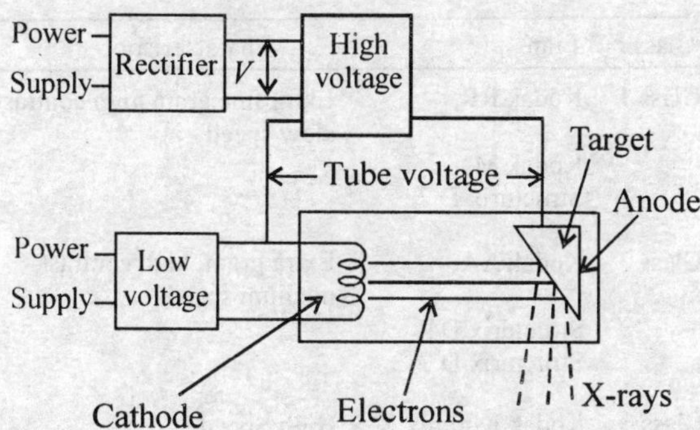

Figure: Schematic setup of X-ray tube and circuit

The development of electronics has led to the availability of constant potential units which give stable operating conditions. The replacement of the glass tubes by metal ceramic ones has led to an extended tube life. X-ray machines are charactersed by the operating voltage and current which determine the penetrability and intensity of the radiation produced. Modern X-generators are available up to 450 kV and 15 mA. X-ray equipment with dual focal spot sizes and ultra-small focal spot and portable $(\sim 15kg)$ equipment with an output voltage of 200 kV and 3 mA current are also available. Highly automated self propelled X- ray mint-crawlers which travel within pipelines are used to take radiographs of pipelines/welds from inside.

The area of the anti-cathode which is struck by the electron flux is called the "focalspot" or "TARGET". It is essential that area should be sufficiently large in order to avoid local overheating which might damage the anti-cathode and to allow rapid dissipation of heat. The projection of the focal spot on a surface perpendicular to the axis of the beam of X-rays is termed as the "*optical focus" or "focus*". This focus has to be as small as possible in order to achieve maximum sharpness in the radiographic image. The size of the focal spot is measured by the *pinhole imaging technique*. Here, a pinhole made in a suitably thick material with high density (lead) and having dimension one order less than the expected focal sport size is used. The pinhole is aligned parallel to the tube axis and perpendicular to X-ray beam and a radiographic image of the focal spot is obtained. By scanning this image using a **microdensitometer**, the focal spot size is obtained.

EFFECT OF RADIATION ON FILM

The radiation source required to produce a certain blackening on the file depends on the energy of radiation. Low energy radiation needs smaller dose to ensure a certain file density as compared with hard radiation. Various isotopic sources with different radiation energy will cause different film ionisation. For example,

Cobalt $-60 = 1$ Iridium $-192 = 2.35$
Caesium $-137 = 5.5$ Thulium $-170 = 4-4.5$

Inherent Unsharpness

The inherent unsharpness is the results of the interaction of high energy radiation with emulsion on the film. During the interaction, the electrons are dislodged from the silver halide emulsion after gaining excess energy in the form of kinetic energy. The electronswith hihg kinetic energy tend to fly off causing ionisation in the adjacent silver halide grains. Thus the bounaries of the exosed areas will show an unsharpness of the image which is called inherent unsharpness or film unsharpness (U_i). Inherent unsharpness value is about 0.1 mm without lead screen and it is 0.2 mm when a film is used with a lead screen.

RADIOGRAPHIC IMAGING

The appearance of a distinguishable image on a radiograph is dependent on several factors, the most important of which is the difference in radiation intensity at various locations in the image plane. The difference in intensity, then must be a function of the abnormality within the part being inspected.

Geometrical Factors

True focal spot for sources used in conventional radiography is not point source but rather is of a few *millimetres* in size. Due to the finite source size, the image projected on to the film is enlarged leading to geometric unsharpness shown figure below. Geometric unsharpness (U_g) is given by

$$U_g = \frac{Ft}{L_o}$$

where $F =$ focal spot size,
$t =$ distance from the object to the receiving plane,
$L_o =$ distance from the source to the object.

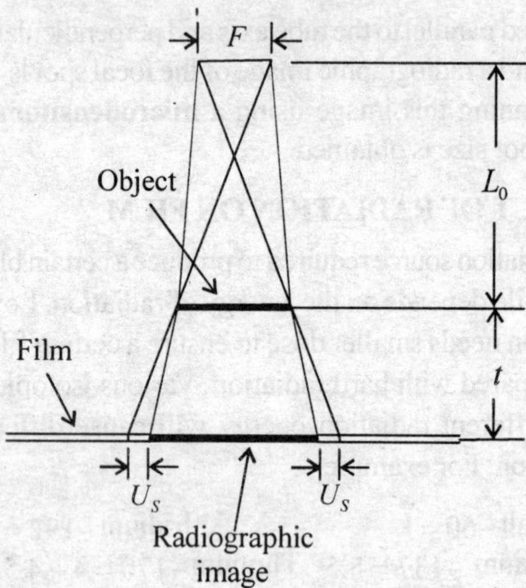

Figure: Geometric unsharpness in radiographic examination

General rules for optimum results:

1. Source or focal spot size should be as small as practicable. Ideal sorce will be a point source.
2. Source to object distance should be as large as possible.
3. The film should be in close contact with object.
4. Source loation should be such that radiation passes through the object thickness.

Radiographic Film

Radiographic film is similar to photographic film in that there is a central carrier called the film base that is made of thin sheet of polyester type material. This is normally transparent and serves only as the carrier for the chemically reactive material that forms the emulsion. Emulsion consisting of a silver halide recording medium with a bider (gelatin) is applied to both sides of the base. Additionally, a protective layer may be applied over the emulsion. The silver halide is a granular material and its grain size has a significant effect on the exposure as well as the resolution ability of the film for defect detail. When radiation strikes the emulsion a change takes place in the emulsion. This change is referred to as the "*latent image*" (latent simply means 'not available', or hidden). Upon processing, the grains that have been exposed will be darkened. The silver halide is removed from the unexposed grains during film processing leaving a transparent area. Thus the darkness or density of the film is directly a function of the exposure of the grains on the film to the radiation.

The film speed is another important film parameter. A film is called high speed film when its grains would begin reacting to the exposure considerably sooner than other films. Realizing that exposure is the product of time and intensity, the effect of film speed is rather significant. For a constant intensity, for, e.g. the grains of high speed film would produce the required density before the grains of slow speed film. It is to be noted that, the faster speed films have larger grains and, therefore may not be able to produce the minute detail. Grain size in a film affects quality and time of exposure. Slow films have extra-fine grain or find grain and give better quality even though the exposure time is longer of the many films available, the films that are referred in many standards and their classification are given in the Table below:

Table: Various Class of Films and Their Characteristic

Class	Film	Characteristic
Class 1	Kodak RR, Kodak M, Structurix D 2	Extra fine grain high contrast, slow speed
Class 2	Kodak AA, Structurix D4, Structurix D 7	Extra grain, high contrast, medium speed
Class 3	Kodak Kodirex X Structurix D 10	High Speed
Class 3	They are screen type films used in conjunction with fluorescent screen These films are not recommended in industrial radiography	

INSPECTION TECHNIQUES

With the various te hniques available, the choice of appropriate one is made on the basis of geometry, size, sensitivity requirements, in-situ space availability etc. The techniques followed for various engineering components for radiographic inspection are given below:

Single Wall Single Image Technique

This technique is used when both the sides of the specimen are accessible. This is used for plates, cylinders, shells

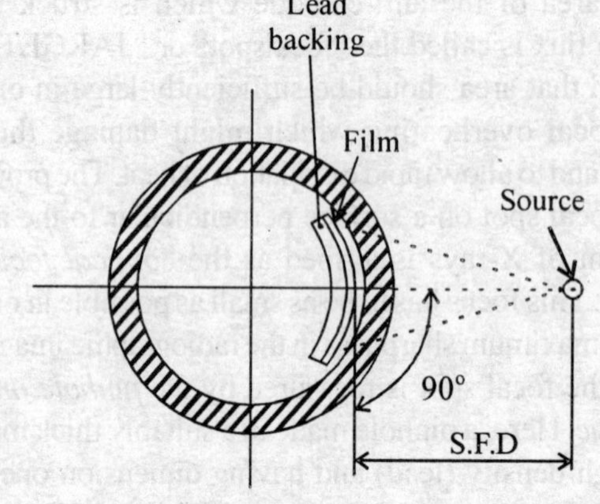

Figure (a) Source outside film inside

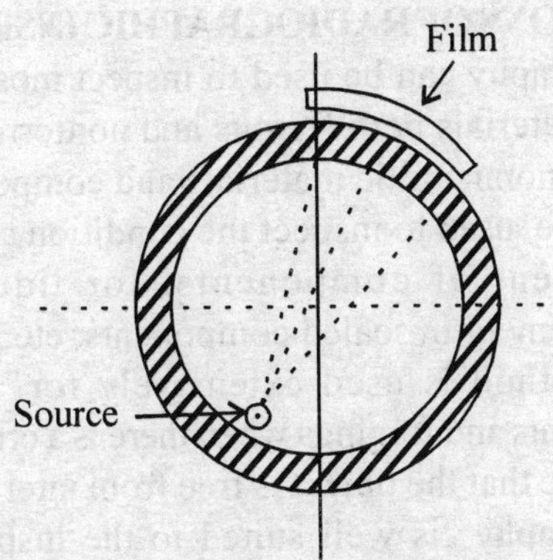

Figure: (b) Source indise film inside

Figuer: Radiographic examination of pipes with single wall penetration

and large diameter pipes. This technique is illustrated in figure below. The source is kept outside and the film inside or vice versa and the weld is exposed part by part (a smaller length of weld).

Panoramic technique: In this technique, the radiation source is kept in the centre of the pipe and the film is fixed around the weld on the outer surface of the pipe. The total circumferential weld length is exposed at a time. This technique reduces the examination time considerably. It can be effectively employed only when the source to film distance is sufficient enough to ensure the proper sensitivity. The required IQI, as per the governing code, can be placed either on source or film side as the case may be.

Double wall penetration technique

The double wall penetration technique can be effectively adopted, in three different methods, based on the prevailing pipe diameter and site restriction. They are:

(a) Double wall single image, (b) Double will double image and (c) Double wall superimposing image.

These techniques are used where the inside surface of the pipe is not accessible. The source of radiation and the film are kept outside. The radiation penetrates both the walls of the pipe.

(b) *Double wall single image:* The radiation source generally is kept on the pipe or very near to the OD, and just near the weld so that the source side weld is not falling on the film side weld. This technique is employed for the pipes with diameter more than 90 mm OD. The IQI is placed on the film side. Here film side weld only can be interpreted. As the interpretable weld length is being small, this technique requires number of exposures to cover the entire weld length, depending upon the pipe diameter.

(b) *Double wall double image*: This technique is specially suited for the smaller diameter pipes upto 90 mm OD. The radiation source is kept at a distance (SFD) with an offset from the axis of the weld, to avoid the superimposing of the source side weld over the film side weld and to obtain an elliptical image on the film. The IQI is positioned on the source side. In this both the source and the film side weld can be interpreted from the image. This requires minimum of two exposures, perpendicular to each other, to cover the entire circumference of the weld.

(c) *Superimposing technique:* This technique is attempted whenever the requried offset to obtain double image could not be possible due to site restrictions for the pipes with diameter upto 90 mm OD. The source is kept at a distance (SFD) without offset, thereby the source side weld is superimposed on the film side weld on the film. The IQI is positioned on the source side. This requries minimum of 3 expsoures at 120 each, to cover the entire length of the weld.

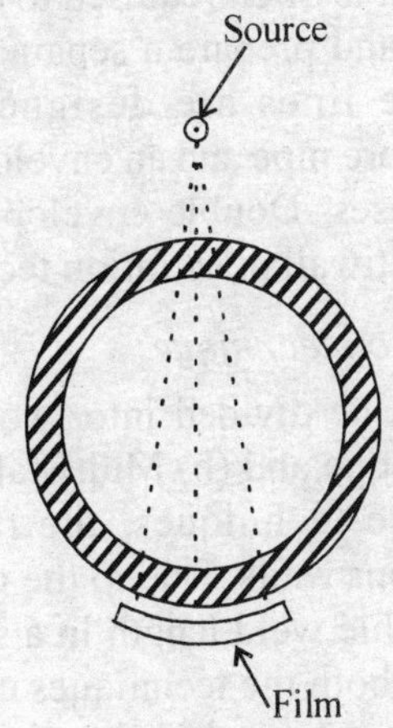

Figure(a): Double wall single image

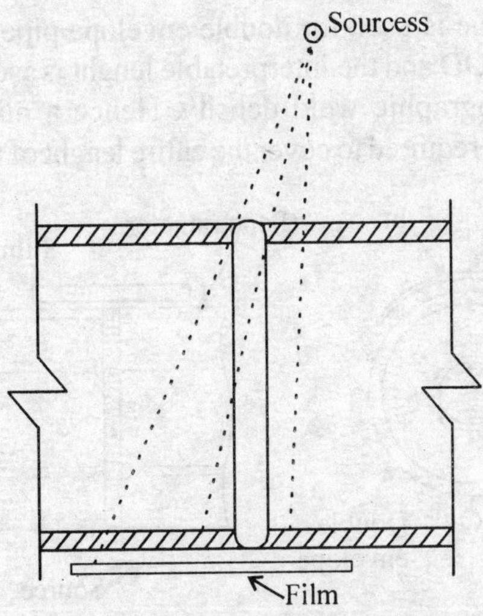

Figure (b): Double wall double image

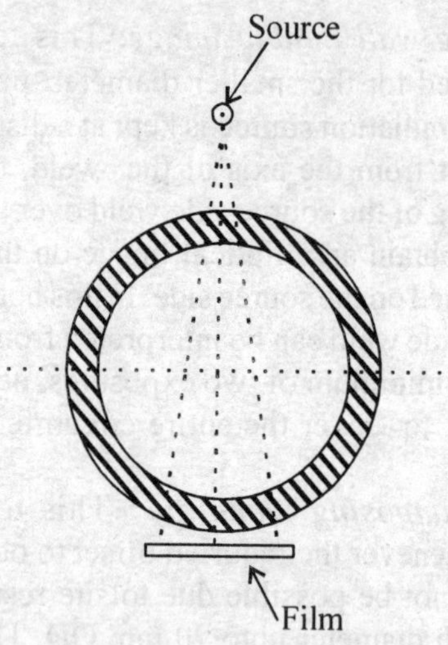

(c): Super imposing image

Figuer: Radiographic examination of pipes with double wall penetration

Special Techniques

In a complex part, it is often required to consider certain areas individually and prepare a separate technique for each . Some pipe lines are designed to have the configurations of core pipe and an envelope pipe to meet the intended purposes. Double envelope welds can be tested by using multiwall penetration technique.

Multiwall penetration technique:

This technique can be divided into two: (a) Multiwall single image technique and (b) Multiwall double image technique. In these techniques, the radiation beam penetrates all the four walls. Due to the geometry of the joint, the interpretable weld length in a single exposure is much reduced in both the techniques compared to the techniques normally employed for the same diameter pipe.

(a) Multiwall single image technique

The technique is used for double envelope pipe of more than 90 mm OD and the interpretable lenght is ascertained by the radiographic weld density. Hence a number of expoures are required to cover the entire lenght of the weld.

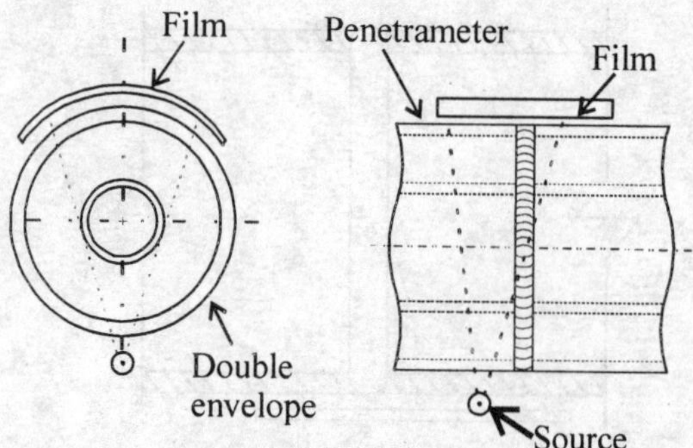

Figure: Multiwall single image technique

APPLICATIONS OF RADIOGRAPHIC INSPECTION

a. Radiography can be used to inspect most types of solid materials both ferrous and nonferrous alloys as well nonmetallic materials and composites.
b. It can be used to inspect the condition and proper placement of components. for liquid level measurement in sealed components, etc.
c. The method is used extensively for castings, weldments and forgings when there is a critical need to ensure that the object is free from internal flaws.
d Radiography is well suited to the inspection of semiconductor devices for detection of cracks, broken wires, unsoldered connections, foreign material and misplaced components, whereas other methods are limited in ability to inspect semiconductor devices.

9.1.6 NUCLEAR READIATION MEASURMENTS NEUTRON RADIOGRAPHY

Neutron radiography (NR) is a valuable NDT technique that is identical in principle to the conventional radiography but complementary with respect to the nature of the information obtained. The history of neutron radiography with thermal neutrons can be traced shortly after the discovery of neutron by Chadwick, in 1932. Thewlis and Derbyshire produced the first reactor based neutron radiographs in 1956, using a reactor beam of the 8 MW BEPO reactor at Harwell. Commercial interest in neutron radiography began in mid 1960's and today this technique, apart from its applications in nuclear industry, finds extensive applications in aerospace and other industries.

Principles of Neutron Radiography

Neutron radiography extends the ability to image the internal structure of a specimen beyond what can be accomplished with photon (X-ray & Gamma) radiation. Similarities as well as obvious differences exist when neutron radiography is compared to photon radiography.

The basic principle underlying in this technique is due to the penetrating nature of neutron radiation and its differential absorption by the material to obtain details of the internal structure. Just as in conventional radiography, in neutron radiography also the object to be extamined is placed in a colimated neutron beam. Neutrons on passing through the object are differentially absorbed. This absorption depends on the atomic number, thickness of the materials, homogeneity and composition. The geometric pattern of the transmitted neutron intensity is recorded using a suitable detector and visualzied. Figuer below gives the schematic sketch of the above description.

It is observed that the mass absorption coefficient for neutron abruptly change and present a random pircture when plotted aganist regularly increasing atomic number of abosrbed. On the other hand the X-ray mass absorption cofficients increase with atomic number in a regular fashion. the differences in absorption coefficents between the neutrons and X-rays suggest a number of possible application for NR, such as:

a. Examination of dense materials like uranium, lead etc.
b. Detection of light materials enveloped in denser materials
c. Differentiating between isotopes of same elements and
d. Examination of radioactive material due to availability of image detection methos, which are not sensitive to the assoicated gamma rays

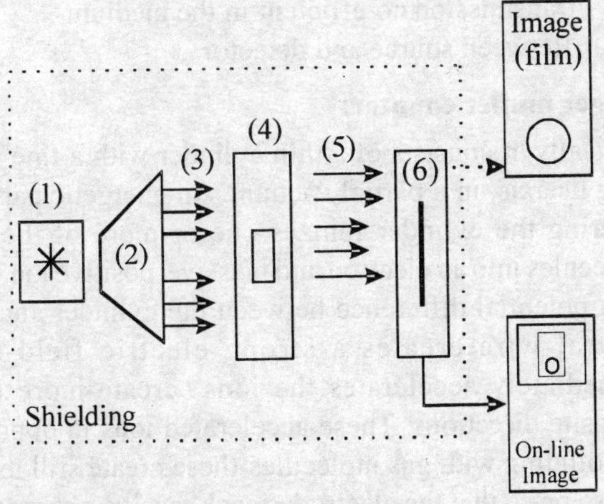

Where

1 → Radiation source
2 → Collimater (moderator)
3 → Directed beam
4 → Object
5 → Attenuated beam
6 → image detector or recorder

Moderator and Collimator

Practically all neutron radiogrphy work is performed using thermal neutrons. Whether it is a reactor, or accelerator or isotopic source, the primary neutrons are high energy neutrons. their average energy is mederated to thermal range using a moderator such as water or other hydrogenous materials.

For neutron radiogpraphy, a useful beam of neutron is requierd. The extraction of the neutron beam from the reactor core is achieved through the insertion of probe (beam) tube or colliator into the moderator. This permits only those neutrons having a direction to that of the tube axis to pass through. The divergent type collimator is widely used since a unifrom beam can be projected over a large inspection area. This improtant geometric factors for a neutron collimator are the total lenght (L) from inlet apertuer to detector and effective dimensiion of the inlet collimator (D). This is usually expressed as L/D ratio. Higher this ratio, better would be the resolution.

Neutron Detection

Neutrons are not directly ionizing radiation and hence have no effect on the conventional films used in industrial radiography. Hence, the detection system for neutorns consists of a latent image recorded in close contant with a thin sheet of material (referred to as converter screen) , which absorbs the neutrons and converts them into a form of secondary radiation of which the image recorder respends. The converters screens are often metallic foils. the emissions from these foils can be either charged particles of elecromagnetic radiation, which produce the image on the film/screen. the technique used for imaging can be classified as direct and indirect. Apart from the industrial X-ray film, there are also etchable plastic films, image intensifiers and imaging plates for displaying the image.

(a) Direct Technique

A foil of **gadolinium** is used before the film. Gadolinium atoms in the foil absorb a neutron and promptly emit other radiation such as electron. Alternatively, a scintillator screen containing a mixture of lithium-6 and zinc sulphide can be used. On observing a neutron a lithium atoms emits an alpha particle and this then strikes the zinc sulphide screen, which in turn emits a light photon. As the above processes are continuous reactions this type of foil and scintillator screen can be used with low neutron fluxes and long integrating exposures. Also, as the film is in contact with the converter during the neutron exposure all the forward emitted radiations takes part in the exposure of the film. Thus, *the direct technique is fast, the scintillator screens being 30 to 100 times faster than metal foils.*

(b) Indirect Techique

This is also referred to as *transfer technique*. this method relies on the build up of radioactivity in the produced by neutron absorption. In this way an activation image is formed in the foil and this is subsequently transferred to a photographic film in contact and allowing the decay radiations from the foil to produce the letent image on the film. The method is useful for nuclear applications since the process of activation and film exposure are two independent processes. However, the technique is much slower compared to direct one.

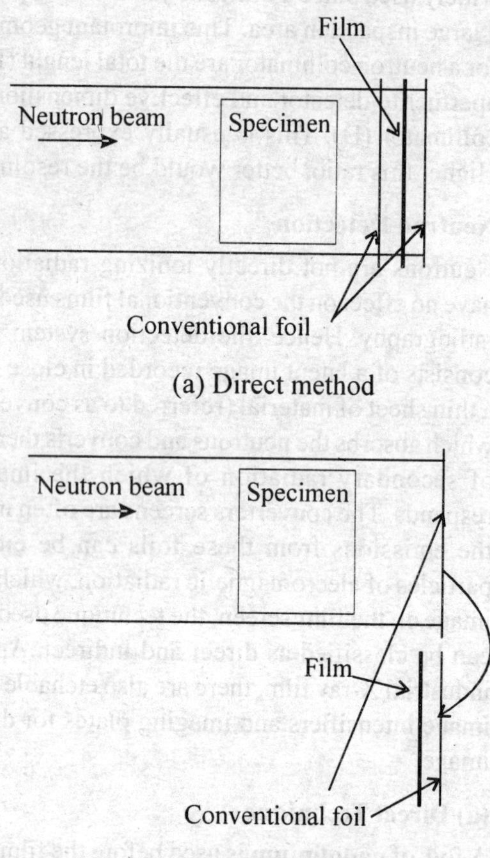

Figure: Detection techniques for neutrons

(c) Track-etch Technique

For neutron radiography of radioactive objects, nitrocellulose film is used as a neutron detechor. This is the dielectric matrerial which can detect charged particles by the radiation damage caused in it. The charged particles are produced by an α emitting converter. The radiation damage is made visible by etching the film in hot sodium hydroxide solution. Higher definition and very good spatial resolution can be obtained by this technique.

Nuclear Radiation Measurements

Ionization detectors: These detectors are devices, in which the electric charge of the ion-pairs formed through radiation interaction is collected and electronically amplified.

Scintillation detector: The passage of radiations through certain inorganic materials produces excited atomic state, whereas in certain organic materials molecular vibrational states get excited. In both cases, the light that is emitted during de-excitation in converted into electric and fed to measuring system. The detectors based on this principle are called scintillation detectors.

Counting efficiency:

1. Intrinsic efficiency of a detector

$$= \frac{\text{No. of pulses recorded}}{\text{No. of particles incident}}$$

$\leq 100\%$ for electrons

$< 100\%$ for photons and neutrons

2. Absolute efficiency $= \frac{\text{No. of pulses detected}}{\text{No. of particles emitted}}$

$$= E_{\text{int}} \frac{d\Omega}{4\pi} T$$

$d\Omega =$ soils angle substended by detector

T = transmission co-efficient in the medium Between source and detector

Geiger muller counter:

Basically, it consists of a thin cylinder with a fine wire long its axis, in a partial vacuum. An energetic particle entering the cylinder ionizes one or more of the gas molecules into an electron and massive positive ion. The high potential difference between the cylinder and the central wire creates a strong electric field that immediately accelerates the ions create more ions opposite directions. These accelerated ions in opposite by colliding with gas molecules-these create still more, and so on so that the electrodes each receive a shower of electric charges. The resulting small surge of current through the external circuit is amplified to operate a counter, a rate circuit is amplified to operate a counter, a rate meter (which shows the number of surges occurring per second) or a loudspeaker that produces a click for each particle entering the tube.

Semi-conductor detectors:

These detectors can be made very small and robust. Silicon and germanium crystals have been employed mainly for counting alpha and beta particle. They function in a manner similar to that of gas ionization chamber. On obsorption of radiation in crystal, electrons and positive holes are formed which towards opposite electrodes under the influences of applied potential. The resulting current is proportional to energies of ionizing radiation.

Pattern (montage) as well as the position of scalp sites with electrode-to-skin contact. Individual channel control settings for gain and filter positions can be displayed on the video monitor for immediate review. Therefore, a setting can be changed by a simple push button operation while looking at the display.

Electrometer

This simple instrument operates, in principle, exactly like an electroscope and is *used for detecting radioactive radiations*. The Braun type electrometer consists of a light weight metal pointer which is pivoted a little bit above its centre of gravity and is insulated from its ring support. When charged either positively or negatively, the needle rises due to mutual repulsion and the pointer indicates the acquired potential on a recalibrated scale.

When a charged particle passes through the electrometer. it ionizes the air thereby producing ion-electron pairs. If the electrometer is negatively-charged. It attractive positive ions and is discharged. Hence, the pointer returns to zero. The electrons are repelled to the 'grounded' walls where they aready neutralized.

Ionization Chamber

It is *used for the detection of a moving charged particles.* It is based on the principle that charged particles in motion produces ionization in a gas which varies with their nature and velocity, But estimating the number of ions produced, one can calculate the nature and energy of the ionizing particle.

Ionization chambers are of the following two types:

- Non-integrating (or pulse) type in which each ionization particle is recorded separately
- Integrating (or current measuring) type in which current pulses are not separated; rather they are allowed to combine into a steady flow of current.

The saturated current from an ingegating type chamber is directly proportional to the amount of radiation pressent. When air is the gas, the response is proportional to the air density, which is affected by atmospheric pressure and temperature. The approximate output current of are ionzation chamber can be calculated from the following expression.

$$\text{Output current } I = k\frac{VBP}{1.08}\times10^{-7} \quad (3)$$

where k is proportional constant and depends on the types of gas used (for air $k = 1$), V is the volume of chamber in m^3, P is gas pressure in atmosphere pressure and R is the radian intensity in *R/h*.

Ion chambers may be operated with several atmoshperes of gas pressrue, Gases employed for an ion chamber filling include air, nitrogen, argon, krypton, xenon and boron trifluoride. Sealed ion chambers can be filled with a gas other than air under pressure, but the untility of the devices for measuring low-energy x-rays and beta radiation dimishers as the well thickness must be sufficient to withstand the internal pressure.

The advantges of gas-filled ion chambers are smaller size, low weight, and rugged construction, low sensitivity, low sensitivity to humidity, fast response, no requirement for warm-up and not correction for temperature or pressure.

The major advantage of air ion chambers is that they directly measure the exposure rate over a wide range of photion energies. the major disadvantage is assoicated with attempts to measure a voltge drop across a high megohm resistor, especially at low exposure rate, when the ionization current is extremely small. Moisture condensate indise the chamber can cause erroneous readings and sometimes make the device ineffective.

Ionizzation chambers are mainly for detection measurement of alpha particles.

Basically, the difference between the two lies in the value of the time constant (or response time) of the system relative to frequency of arrival of the ionizing particles. The pulse-counting ionization chamber has a very short time constant where as the integrating type has a comparatively long time constant.

Scintillation Counters

A sointillation counter is a very sensitive device *used in nuclear physics study for the detection and measurement of high-energy atomic radiations*. It has assumed great importance because of are sensitive to $\beta-$and $\gamma-$rays and the development of highly efficient photo-multiplier tube.

Working Principle

The working of a sointillation counter is based on the fact that when high-energy charged particles pass through certain transparent materials, florissant light is produced all along the path. As the fast-moving particle collides with the atoms and molecules of the material, electrons are raised to excited energy levels and in returning to their ground states, they emit light (called scintillation). For many crystals and particles, this fluorescent light is blue or violet in colors where as for many others, it is ultraviolet or infra- red. This tiny flash of light is made to fall on the photocathode of a very sensitive photo-electric cell called *photomultiplier tube* where it ejects photoelectrons. These photoelectrons are accelerated by the successive dynodes which are kept at progressively higher voltages. These primary photoelectrons knock off secondary electrons as they travel from one anode to another. Ultimately, the number of electrons is so much increased in successive stage that a measurable current pulse is obtained. This current pulse is recorded by a counter-rate meter.

Scintillation counter have the following advantages:

1. Counting rater is very fast so pulses of as short a duration as 2 ns may easily be meaured.
2. These are capable of detecting radiations of lower levels becuase of their high sensitivity.
3. Scintiliation counter produces a flash of light when struck by x-rays.

Cloud Chamber

It is one of the most important instruments for basic research radiations like detecting and measuring moving charged particles, studying the behaviour of individual atoms, photographing the actual paths of ionizing radiations and analyzing the complicated inter sections which take place between charged particles and individual atoms.

Basic Principle

The working of a cloud chamber is based on the fact that supercooled vapour needs some discontinuity or nuclei to initials its condensation. In the absence of such nuclei, a super cooled vapour remains in the vapour phase and does not condense. As an example, imagine a certain volume of air which is saturated with water vapour its room temperature. If this air is subjected to an adiabatic expansions, its temperature is lowered so that its saturation limit is decreased. Hence, air now contains more water vapour is necessary for saturation at this temperature. If any particles of dust amoke are present in air, they will act as condensation nuclei and the excess of water vapour will separate out as fine droplets of liquid the from of a cloud or mist. If on the other hand, there are dust or smoke particles, no condensation is possible so that air will become supersaturated will water vapour. In Wilson cloud chamberions act as nuclei for condensation of supersaturated water vapour. These ions are produced due to the passage of a high-energy particle like $\alpha-$ or $\beta-$ particle etc. Through the chamber which leading the ionization of the air or gas contained therein.

Bubble Chamber

This instrument has assumed great important in the study of minute details of high-energy nuclear events. It contains superheated liquid instead of supersaturated vapors as in condensation chambers of both types. In this instrument, the tracks consist of a series of closely-spaced bubbles which develop on the ions produced by the passage of a charged participles through the superheated liquid contained in the chamber.

Basic Principle

It is well- known that boiling point of a liquid can be realized by increasing pressure on its surface. For example water boils at 100°*C* at standard atmospheric pressure. If pressure on its surface in increased, as in a pressure cooker, boiling does not start till a highest temperature is reached. If pressure is then suddenly decreased, boiling begins with sudden formation of tiny bubbles that grow quickly in size. If at that exact moment, a charged particle passes through the liquid, bubbles grow on the ions produced by it and form bubble track which can be photographed.

It consists of a box having thick glass walls which is filled with and liquid and is also connected to a pressure heated to a pre- determined energy particles enter the liquid through a thin window.

Spark Chamber

This chamber is used for the study of cosmic and high-energy particles produced by accelerators.

Basic Principle

Its working depends on the of a spark through a gas that has been ionized by nuclear radiations. Usually, a high voltage a maintained between two plates of electrodes situated in a gas such that the electric field is not strong enough to produce and spark in the was under normal conditions. If an ionizing particle enters the gas between the plates, and spark will pass which would tend to follow the with of the ion-pairs produced by the particle. In this way, the track of the ionizing particle can be delineated by the spark and recorded photograpically by means of stereoscopic cameras for three-dimensional representation.

Photographic Emulsions

This technique is employed for detecting high-speed particles. When such as ionizing particle travels through the sensitive emulation of a photographic plate, it leaves & track consisting of a number of silver bromide grain which turn into black silver upon being developed by a regular film-developing process. Special nuclear emulsions have been manufactured for this purpose. Their silver bromide content grain size and emulsion thickness in much greater as oompared to the optical film. The study of detailed structure of tracks under a microscope yields information about the mass, charger and energy of the ionizing particle.

Application of Radioactive Radiations

- Many radiation sources such as cobalt – 60. (a γ-ray emitter) have been used for *industrial radiography* i.e. for *investigating the interiors of metallic castings* for detecting any flows or defects.
- Radioactive nuclides emitting a α or β particles have been used for the production of electric power by thermo-electric conversion. When α or β particles are absorbed in matter. The energy of the radiation is converted into heat which can be utilized for production of electric power. Such radionuclide power generators are particularly suitable for space vehicles because of their light weight, long life and absolute reliability.
- *Radio nuclides* have been used as compact sources of heat energy because of which they find many space-related applications. Radio nuclides of promethium-147 polonium- 210 and plutonium -238 have been used to heat the propellants gas (hydrogen) in low-throws rockets.
- Chemical changes initiated by radiations from radioactive substances is the basic of a newly-developed subject called radiation chemistry which has been applied in the production and modification of plastics and in the production of irradiated wood – plastic combined nations.
- Nuclear radiations like $\gamma - rays$ have been utilized for the *preservation of food*. The irradiation of food-stuff mainly meat, poultry, fish and fruits is achieved by exposure to $\gamma - rays$ from cobalt 60 or cesium - 137. A dose of 2 to 5 million rads is sufficient to destroy almost at bacteris in food. This increases the sheif-life of these articles without regeneration. Radiations have also been employed for insect disinfestations of wheat and flour.
- Gamma radiations from cobalt-60 are *used in hospital for sterilization of materials like dressings, hypodermic syringes and surgical sutures.*
- Population of insects which causes considerable damages to both plant crops and livestock can be controlled by irradiating the male members of these insects so that they become sterile.
- Radiation mutations in plants have been practiced the produce new varieties of these plants.
- Self-luminous paints for use on instruments and watch dials have been made by adding a natural $\propto$ – emitting radioactive substance to phosphor.
- The ionization produced by β- particles has been widely used for the elimination of static electricity which constitutes a serious fire and explosion hazard in the paper, textile, rubber and plastic industries.
- The phenomenon of spontaneous emission of powerful radiations exhibited by heavy elements is called radioactivity.
- Radioactive radiations consists of α – rays, β – rays and γ – rays.
- The two laws of radioactive disintegration are :
- Atoms of all radioactive elements undergo spontaneous disintegration to form fresh radioactive products with the emission of α, β and γ-rays.
- The rate of radioactive disintegration is not affected by environmental factors but depends on the number of the atoms of the original kind present at any time.

The decay is given by $N = N_0 e^{-\lambda t}$

Also, $$M = M_0 e^{-\gamma t}$$

- The decay constant λ may be defined as the reciprocal of the time during which the number of atoms of a radioactive substance falls to 37 percent of its original value.
- Half-life (T) of a radioactive element is equal to the time during which a given amount of that element is reduced by disintegration to half its initial amount.

$$T = 0.693 / \lambda$$

- The average life expectancy of a radioactive atom is equal to the reciprocal of the decay constant.

$$l = 1/\lambda$$

- The rate of disintegration or activity of a radioactive material is given by

$$dN / dt = \lambda t$$

- Curie (Ci) is the unit of activity, i.e. ratio of disintegration. One curie is that amount of radioactive materials which gives 3.7×10^{10} disintegrations per second.

Solid State Detectors

The solid state detector may be considered as a parellel plate ion chamber in which the plates separate out in proporion to the applied voltage. the sensitive volume of the detector is called the deplection laver which is essenstially the extent of the region at the junction of the semiconductor materials produced by an external influence, such as nuclear radiation, is quickly removed with negligible recombination. The thickness of the deplection layer is given by the expresion.

$$\text{Deplection layer thickness } x = \frac{10}{3}\sqrt{\rho V}\ \mu m$$

Where, ρ is the conductivity of the semiconductor material in Ω-m and V is the reverse bias voltage applied to the junction in volts.

For measurement of the energy of ionizing particles, particularly alpha and protons, it is essential that the radiation particles dissipate negligible energy before entering the deplection layer. Two techniques are employed. A very then layer of P doped semiconductor may be superimposed on a n N-type semiconductor, thus forming a junction very near one outer surface; or a surface is used to provide a high-potential-gradient region. The surface is coated with a thin layer of gold to provide a conductive contact.

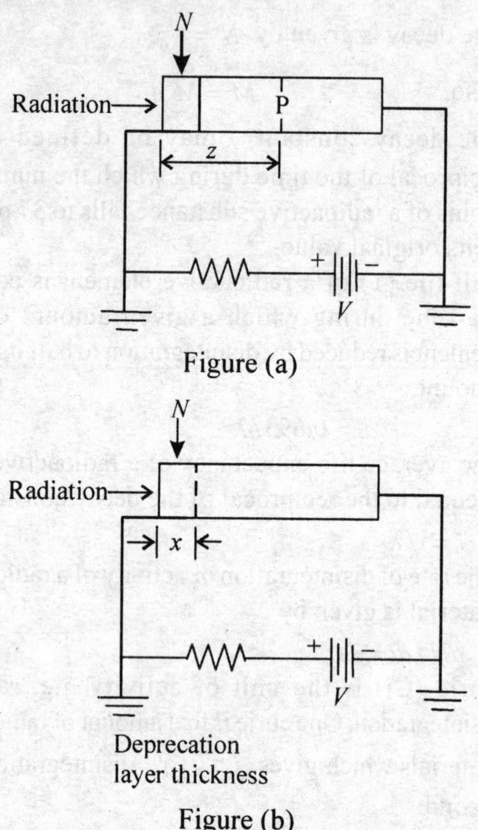

Figure: (a) Junction diode and (b) surface barrier

The advantages of solid-state detectors are that

1. They are of small size
2. Have high specific ionization

The specific iionization, the number of ion pairs per eV or its reciprocal, the number of elctron volts of energy lost by the ionizing particles to create one ions pair, is 10 times more favourable for solid-state detectors (approximate 3eV per ion pair) than for gas chamber detectors.Thus the ionization intensity is an order of magnitude higher in the semiconductor for a given ionizing particle and as a result the energy resolution of solid-state detectors is grater than that of ion chambers.

For integrated current measurement, the solid-state detector does not appear as favourable as the gas ionizatin chamber.

Neutron Dectection

Neutrons are very difficult to detect as they do not carry any change. The neutron is detected by means of a secondary charged particle which to releases in passing through matter, or by means of the radioactivity which the neutron induces in stable elements. Protons may be projected by collisins with neutrons in hydrogenous material and the ionizatior from the protron; can be measured in an ionization chamber. Also fission may be used for the detection of neutrons by placing fissionable material inside the ionization chamber and observing the ionization produced by the fission fragments.

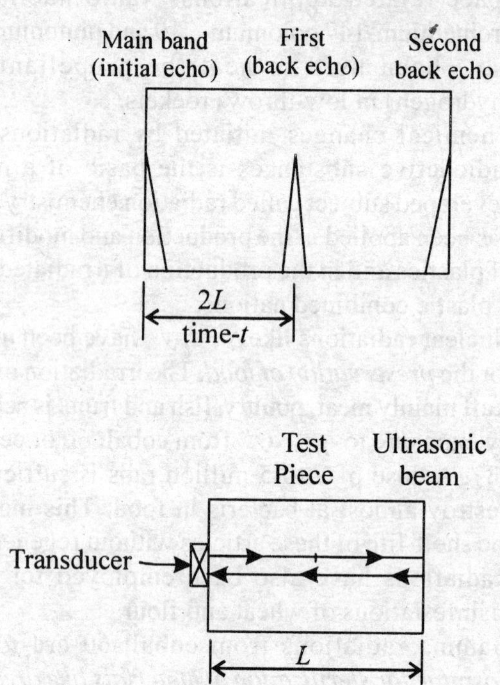

Figure: Pulse echo inspection of bar of lenght L

Normal incident Through-Transmission Testing

Many a time, pulse-echo technique may not provide required test information. This may occur when a flaw or other anomaly does not provide a suitable reflection surface or where the orientation or location of the flaw which is not favourable for detection using single probe. Also, highly attenuating materials are often tested with through-transmission technique. **Figure 7.4** shown the probe arrangement for through-transmission technique. The technique is often used in large casting and highly attenuating materials.

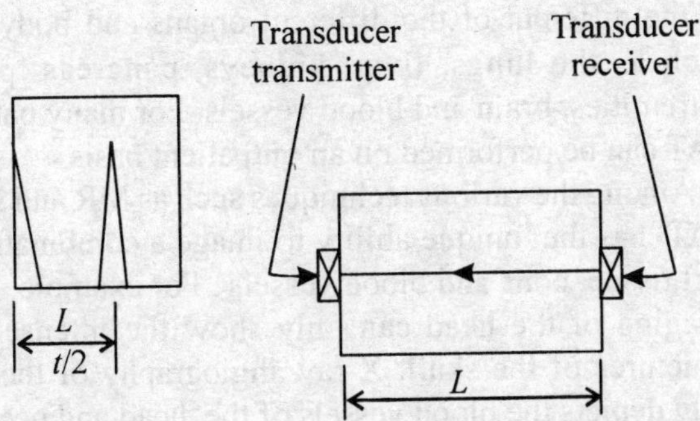

Figuer: Throuth-trnasmission inspection of bar of lenght L

Angle Beam Pulse-Echo Testing

Angle beam transducers provide access to areas that are inaccessible to normal beam probes. **Figure 7.5** shows the angle beam test arrangement. Angle beam inspection is accomplished with shear wave probes.

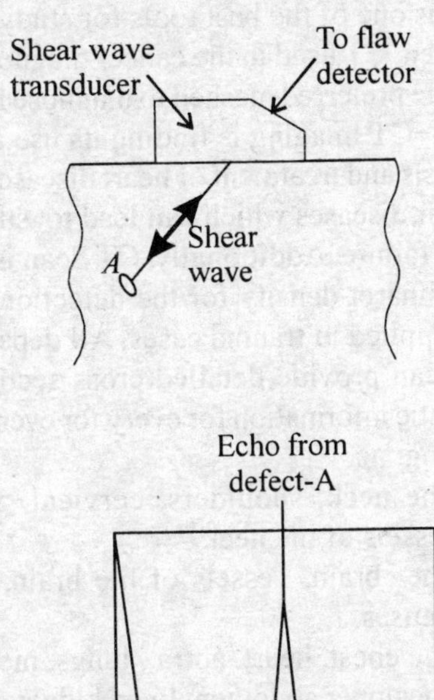

Figure: Anagle bteam test arrangement

9.3.6. ULTRASONOGRAPHY

Principles of Ultrasonic Measurement

Ultrasonic is sonic energy at frequencies above audible range (greater than 20 kHz). It use in medical diagnosis dates back to the period following World War II and is a directly outgrowth of the military development of sonar.

Properties of Ultrasound

Like other forms of sonic energy, Ultrasound exists as a sequence of alternate compressions and rarefactions of a suitable medium (air, water , bone, tissue, etc.) and is propagated through that medium at some velocity. Its behavior also depends on the frequency (wavelength) of the sonic energy and the density mechanical compliance of the medium through which is travel. At the frequencies normally used in diagnostic application. Ultrasound can be focused into a beam and obeys the laws of refection and refraction.

Whenever a beam of ultrasound phases from one medium to another, a portion of the sonic energy is reflected and the remainder is refracted as shown in figure (9.7). The amount of energy reflected depends on the difference in density between the angle at which the transmitted beam strikes the medium. The grater the difference in media, the grater will be the amount reflected. Also the nearer the angle of incidence the beam and the interface is to 90° the greater will be the reflected portion.

Figure 9.7: Reflection and refraction out ultrasound at an interface between media of different densities.

The velocity of sound propagation through a medium varies with the density of the medium and its elastic properties. It also varies with temperature.

Every material has an acoustic impedance, which is a ratio of the acoustic pressure of the applied ultrasound to the resulting particle velocity in the material. Since acoustic impedance is a complex value, consisting of both resistive and reactive components, a simpler term called characteristic impedance, in more often used. The characteristic impedance of a material is the product of its density and the velocity of the sound through it.

As ultrasound travels through the material , some of the energy is absorbed and the wave is attenuated a curtains amount for each centimeter through which it travel it travels The amount of attenuation is a function of both the frequency of the ultrasound and the characteristics of the materials. The attenuation constant, α , is defined by the equation.

$$\frac{\text{amplitude at point } X}{\text{amplitude at } X+1 \text{ unit distance}} = \beta$$

$$\alpha \text{ (per cm)} = cf_{\beta}$$

Wherec = proportionality constant

f = ultrasound frequency

β = exponential term determined by the properties of the material

This formula shows that attenuation increases with some power of the frequency, which means that the higher the frequency, the less distance it can penetrate into the body with a given amount of ultrasonic energy. For this reason, lower ultrasound frequencies are used for deeper penetration. However, lower frequencies are incapable of refection small objects. As a rule, a solid object surrounded by water or saline must be a least a quarter-wave thick in order to cause a usable reflection. Thus, for finer resolution, higher frequencies must be used. Ultrasound frequencies of 1 to 15 Mhz are usually used for diagnostic purposes. At 2 MHz, distinct echoes can be recorded from interface 1 mm apart. Higher-frequency ultrasound is also more subject to scattering than ultrasound at lower frequencies. However, the high-frequency ultrasound beam can be focused for greater resolution at a given depth.

Another useful way of assessing the attenuation of ultrasound as it penetrates the body is the half-value layer of the medium given in Table 9.2. The half-value layer is the depth of penetration at which the ultrasound energy is attenuated to that the applied amount.

Table 9.2 Ultrasound absorption

Type of Tissue	Frequency (MHz)	Half-value Layer (cm)
Blood	1.0	35.0
Bone	0.8	0.23
Fat	0.8	3.3
Music	0.8	2.1

A well-known characteristic of ultrasound frequently utilized in biomedical instrumentation is the Doppler effect, in which the frequency of the reflected ultrasonic energy is increased or decreased by a moving interface. The amount of frequency shift can be expressed in the formula:

$$\Delta f = \frac{2V}{\lambda}$$

Where f = shift in frequency of the reflected wave

V = velocity of the interface

λ = wavelength of the transmitted ultrasound

The frequency increases when the interface moves the transducer and decreases when it moves away. With an ultrasound frequency of 3 MHz, the shift is about 40 Hz for each cm/sec of interface velocity.

9.3.7 PRINCIPLE OF COMPUTER ASSISTED TOMOGRAPHY

(1) Use of CT Scan: Computed Tomography (CT) imaging, also known as "CAT scanning" (Computed Axial Tomography), combines the use of a digital computed together with a rotating with a rotating X-ray device to create exposed X-radiation detailed cross sectional images of the different of the different organs and body parte such as the lungs, liver, kidneys, pancreas, pelvis, extremities, brain and blood vessels. For many patients, CAT can be performed on an outpatient basis.

Among the various techniques such as MR and X-ray, CAT has the unique ability to image a combination of soft tissue, bone and blood vessels . For example, X-ray imaging of the head can only show the intense bone structures of the skull. X-ray angiography of the head only depicts the blood vessels of the head and neck and not the soft brain-tissue. Magnetic resonance (MRI) imaging does an excellent job of showing soft tissue and blood vessels, but MRI does not give as much detail of bony structures such as the skull. CAT images of the head allow physicians to see soft-tissue a structures like the brain's ventricles or gray and white matter. Example selectively "window" the digital CT images on the computer monitor to look at tissue, then the bone and then the blood vessels.

CT is one of the best tools for studying the lungs and abdomen. CT used in the cancer diagnosis process and is often the preferred meshed to diagnose liver and pancreas cancer . CT imaging is finding its use in the detraction, diagnosis and treatment of heart disease, acute stroke and vascular diseases which can lead to stroke, gangrene or kidney failure. Additionally, CT Scan is used to measure bone minaret density for the detection of osteoporosis. CT is applied in trauma cases. All department.

CT can provide detailed cross sectional images and diagnostic information for every for every part of the body including:

- The neck, shoulders, cervical spine and blood vessels of the neck.
- The brain, vessels of the brain, eyes, inner are, sinuses
- The chest, heart, aotra , lungs, mediastinum
- The upper abdomen, liver, kidney, spleen, pancreas and other abdomin
- The thoracic and lumbar spine
- The skeletal system including bones of the hands, feed ankles, legs and jaws
- The pelvis and hips, male and female reproductive system, bladder.

Working Principle CT Scan

Computed Tomography is based on the X-ray principal: as X-rays pass through the body they are absorbed or attenuated at differing levels creating a matrix or profile of X-ray beams of different strength. This X-ray profile is registered on film, thus creating an image in a computer. In the case of CT, the film is replaced by a banana shaped detector which measures the X-ray profile.

A CT scanner looks like a big, square. The patent aperture (opening) is $60\,cm$ to $70\,cm(24" \,to\, 28")$ in diameter. Inside the covers of the CT scanner is a rotating frame which has an X-ray tube mounted on one side and detector mounted on the opposite side. A fan beam of X-ray is created as the rotating frame spins the X-ray tube and detector around the patient (Figure. 36.20.). Each time the X-ray tube and detector mark a $360°$ rotation, an image has been acquired. This image is collimated (focused) to a thickness between *I mm* and 10 *mm* using lead shutters in front of the X-ray tube and X-ray detector.

As the X-ray tube and detector mark this $360°$ rotation, the detector takes numerous snapshots (called profiles) of the attenuated X-ray beam. Typically, in one 360 lap, about 1,000 profiles are sampled. Each profile is subdivided spatially by the detectors and fed into about 700 individual channels. Each profile is then backward projected by a dedicated computer into a two-dimensional image of the slice that was scanned.

Applications and Clinical Benefits of CT Imaging

Unlike other medical imaging techniques, such as conventional X-ray imaging (radiography), CT enables direct imaging and differentiation of soft tissue structures, such as liver, lung tissue, and fat. CT is especially useful in searching for large space occupying lesions, tumours and metastasis and can not only reveal their presence, but also the size, spatial location and extent of a tumour.

CT imaging of the head and brain can detect tumours, show blood clots blood vessel defects, show enlarged ventricles (caused by a build up of cerebrospinal fluid) and image other abnormalities such as those of the nerves or muscles of the eye.

Due to the short scan times of 500 milliseconds to a few seconds. CT can be used for all anatomic regions, including those susceptible to patient motion and breathing. For example, in the thorax CT can be used for visualization of modular structures, infatuations of fluid, fibrosis (for example from asbestos fibres), and effusions (filling of an air space with fluid).

CT has seen the basis for interventional work like CT guided biopsy and minimally invasive therapy. CT images are also used as basis for planning radiotherapy cancer treatment. CT is also often used to follow the course of cancer treatment to determine how the tumour id responding to treatment.

For many computer tomography examinations patients may be asked to take a special contrast agent (orally , rectally or via injection). Intravenous, oral and rectal CT contrast are pharmaceutical agents (liquids) and are sometimes referred to as "dye". CT contrast is used to make specific organs, blood vessels and/or tissue type "stand out" with more image contrast to better show the presence of disease or injury. Thus CT contrast highlights specific areas of the resultant CT image or "dyes" it.

There are four types of contrast agents used in CT:

1. The type that is given via intravenous (through a vein) injection
2. The type id given orally
3. The type that is given orally
4. A much less common type of contrast used in CT is inhaled as a gas and used for special lung and brain imaging. This technique (called Xenon CT) is only available at a small number of locations throughout the world and is only performed for rare cases.

Use of Computers in CT Scan

Multiple computers are used to control the entire CT system. The main computer that combines the operation of the entire system is called the host computer". There is also a dedicated computer that reconstructs the "raw CT data" into an image. A workstation with a mouse, keyboard and other dedicated controls allows the technologist to control and monitor the exam. The CT gantry and table have multiple microprocessors that control the rotation of the gantry, movement of the table (up/down and in/out), tilting of the gentry for angled images, and other functions such as turning the X-ray beam on and off.

Magnetic Resonance Imaging (MRI)

Magnetic resonance imaging was, developed from knowledge gained in the study of nuclear magnetic resonance. The original name for the medical technology is nuclear magnetic resonance imaging (NMRI), but the word nuclear is almost universally dropped. This is done to avoid the negative connotations of the word nuclear, and to prevent patients from associating the examination with radiationn exposure, which is not one of the safety concerns for MRI.

Difference between MRI & CT

A CT scanner uses ionizing radiation, X-rays, to acquire its images, making it a good tool for dense tissue (bone) exams ,MRI, on the other hand, used radio frequency signals to acquire its images, and is best suited for soft (non-calcified) tissue exams.

9.2 OPTICAL INSTRUMENTATION

Optical sources and detectors, LED, laser, Photo-diode, photo-resistor and their characteristics. Interferometers, applications in metrology. Basics of fiber optics.

9.2 OPTICAL INSTURMENTATION

The optical source is often considered to be the active component in an optical fiber communication system. Its fundamental function is to convert electrical energy in the form of a current into optical energy in an efficient manner which allows the light output to be effectively launched or coupled into the optical fiber. Three main types of optical light source are available. These are:

a. Wideband 'continues spectra' sources (incandescent lamps)
b. Monochromatic incoherent sources (light emitting diodes, LEDs) ;
c. Monochromatic coherent sources (lasers).

To a large extent these two sources fulfil the major requirements for an optical fiber emitter which are outlined below:

1. A size and configuration compatitable with launching light into an optical fiber. Ideally, the light output should be highly directional.
2. Must accurately track the electrical input signal to minimize distortion and noise. Ideally, the source should be linear.
3. Should emit light at wavelengths where the fiber has low losses and low dispersion and where the detector are efficient.
4. Preferably capable of simple signal modulation over a wide bandwidth extending form audio frequencies to beyond the gigathertz range.
5. Must couple sufficient optical power to overcome attenuation in the fiber plus addition connector losses and leave adequate power to drive the detector.
6. Should have a very narrow spectral bandwidth in order to minimize dispersion in the fiber.
7. Must be cable of maintaining a stable optical output which is largely unaffected by changes in ambient conditions (e.g. tempeature)
8. It is essential that the source is comparatively cheap and highly reliable in order to compete with conventional transmission techniques.

9.2.1 LASER

LASER as anacronym for light ampplificatio bye Stimulated emission or radiation. Lasers, however, are seldom used as amplifiers since there are practical difficulties in relation to the achievement of high gain whilst avoiding oscillation from the required energy feedback. Thus the practical realization of the **laser is as an optical oscillator**. The operation of the device may be described by the formation of an electromagnetic standing wave within a cavity (or optical resonator) which provides an output of monochromators highly coherent radiation.

By contrast the LED provides optical emission without an inherent gain mechansim. This result in incoherent light output.

Absorption and emission of radiation

The interaction of light with matter takes place in discrete packets of energy or quanta, called photons. The quantum theory suggests that atoms exist only in certain discrete energy states such that absorption and emission light causes them to make a transition from one discrete energy state to the difference in energy E between the higher energy state E_2 and the lower energy state E_1 by the expression:

$$E = E_2 - E_1 = hf \quad (1)$$

Where $h = 6.626 \times 10^{-34}\,Js$ is Planck's constnat.

Below fingure Illustrates a two energy state or level atomic system where an atom is initially in the lower energy state E_1. Where a photon with energy $(E_2 - E_1)$ is incident on the atom it may be excited into the higher energy state E_2 through absorption the photon. This process is sometimes referred to as **stimulated absorption**. Alternatively, when the atom is initially in the higher energy state E_2 it can be make a transition to the lower state E_1 providing the emission of a photon at a frequency corresponding to equation (1) This emission process can occur in two ways:

a. by spontaneous emission in which the atom returns to the lower energy state in an entirely random manner;
b. by stimulated emission when a photon having an energy equal to the energy difference between the two states $(E_2 - E_1)$ interacts with the atom in the upper energy state causing it to return to the lower state the creation of a second photon.

These two emission processes are illustrated in figure (b) and (c) respectively.

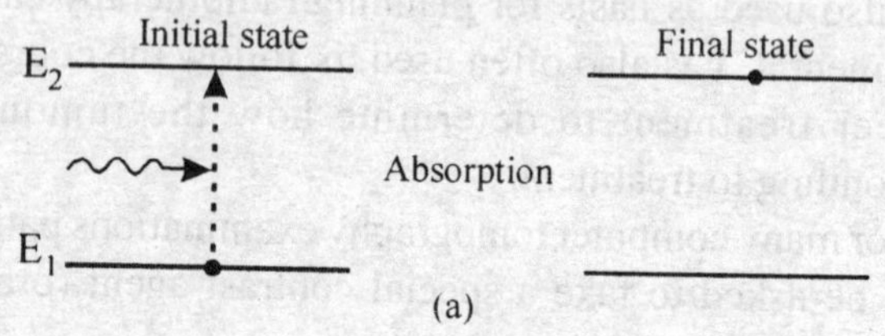

(a)

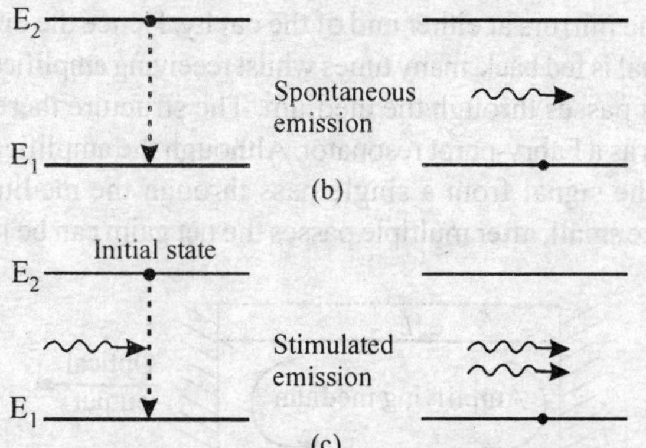

Figure: Energy state diatram showing (a) absorption, (b) spontaneous emission,(c) stimulated emission. the Black dot indicates the state of the atom beforeand after a transition takes place.

Firstly, the photon proudced by stimulated emission is generally of an identical energy tothe one which cause it and hence the lihg associated with them is of the same frequency. Secondaly the light associates with the simulating and stimulated photon is in phase and has the same polarization.

Einstein domesticated that the rates of the three transition processes of absorption, spontaneous emission and stimulated emission were related mathematically. He achieved this by considering the atomic system to be in thermal equilibrium such that the rate olf the upward transition must equal the rate of the downward transition. The population of the two energy leaves of such a system are described by Boltzmann statistics which give:

$$\frac{N_1}{N_2} = \frac{g_1 \exp(-E_1/KT)}{g_2 \exp(-E_2/KT)} = \frac{g_1}{g_2} \exp(E_2 - E_1/KT)$$

$$= \frac{g_1}{g_2} \exp(hf/KT) \qquad (2)$$

where N_1 and N_2 represent the density of atoms in energy levels E_1 and E_2, respectively with g_1 and g_2 being the corresponding degeneracies of the levels, K is Boltzmann's constant and T is the absolute temperature.

As the density of atoms in the lower of ground energy state E_1 is N_1, the rate of upward transition of absorption is proportional to both N_1 and the spectral density ρ_f of the radiation energy at the transition frequency f. Hence, the upward transition rate R_{12} may be written as :

$$R_{12} = N_1 \rho_f B_{12} \qquad (3)$$

where the constant of proportionality B_{12} is known as the Einstein coefficient olf absorption.

By contrast atoms in the higher or excited energy state can undergo electron transitions from level 2 to level 1 either spontaneously or throwing stimulation by the radiation field. For spontaneous emission the average time an electron exists in the excited state before a transition occurs is known as the spontaneous lifetime τ_{21}.

If the density of atoms within the system with energy E_2, is N_2, then the spontaneous emission rate is givenby the product of N_2 and $\frac{1}{\tau_2}$. This may be written as $N_2 A_{21}$ where A_{21}, the Einstein coefficient of spontaneous emission, is equal to the reciprocal of the spontaneous lifetime.

The rate of stimulate downward transition of an electron form level 2 to level 1 may be obtained in a similar manner to the rate of stimulated upward transition. Hence the rate of stalemated emission is given by $N_2 \rho_f B_{21}$, where B_{21} is the Einstein coefficient of stimulated emission. The total transition rates from level 1, R_{21}, is the sum olf the spontaneous and stimulated contributions.

When the degeneracies of the two levels are equal. then the probabilities of absorption and stimulated emission are equal. Furthermore, the ratio of the stimulated emission rate to the spontaneously emission rate is given by:

$$\frac{\text{Stimulated emission rate}}{\text{spontaneous emission rate}} = \frac{B_{21}\rho_f}{A_{21}} = \frac{1}{\exp(hf/KT) - 1} \qquad (4)$$

Population Inversion

Under the conditions of thermal equilibrium given by the Boltzmenn distribution the lower energy level E_1 of the two level atomic system contains more atoms than the upper energy level E_2. This situation, which is normal for structures at room temperature. However, to achieve optical amplification it is necessary to create a nonequilibrium distribution of atoms such that the population of the upper energy level is greater then that of the lower energy level. This condition, which is knows as population inversion.

In order to achieve population inversion it is necessary to excite atoms into the upper energy level E_2 and hence obtain a nonequilibrium distribution. This process is achieved using an external energy source and is referred to an '**pumping**'. A common method used for pumping involves the application of intense radiation. In the former case atoms are excited into the higher energy state through stimulated absorption. However, the two level system discussed above does not lend it self to suitable population inversion. From Einstein relation $B_{12} = \left(\frac{g2}{g1}\right) B_{21}$ when the two levels are equally degenerate, then $B_{12} = B_{21}$. Thus the probabilities of absorption and stimulated emission are equal, providing at best equal populations in the two levels.

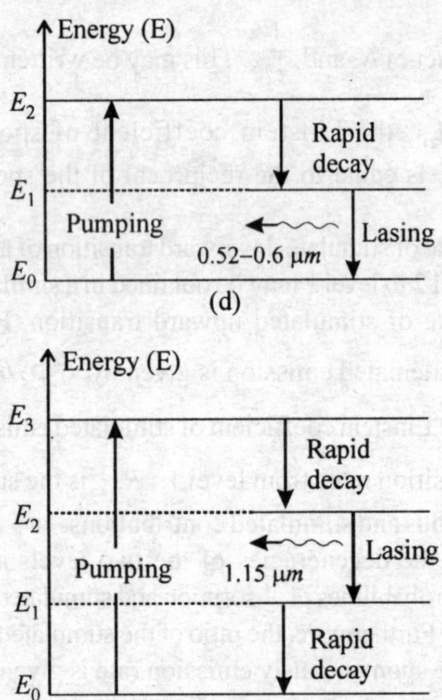

Figure: Energy level diagrams showing population inversion and lasing for two nonsemiconductor laser (d) three level system -ruby (crystal) laser (e) four level system - He-Ne (gas) laser.

Optical feedback and laser oscillation

Light amplification in the laser occurs when a photon colliding with an atom in the excited energy state causes the stimulated emission of a second photon and then both these photons release two more. Continuation of this process effectively creates **avalanche multiplication**, and when the electromagnetic waves associated with these photons are in phase, amplified **coherent emission** is obtained. To achieve this laser action it is necessary to contain photons with in the laser medium and maintain the conditions for coherence. This is accomplished by placing or forming mirrors at either end of the amplifying medium, as illustrated in figure (f). The optical cavity formed is more analogous to an oscillator than an amplifier as it provides positive feedback of the photons by reflection at the mirrors at either end of the cavity. Hence the optical signal is fed back many times whilst receiving amplification as it passes through the medium. The structure therefore acts as a Fabry-perot resonator. Although the amplification of the signal from a single pass through the medium is quite small, after multiple passes the net gaim can be large.

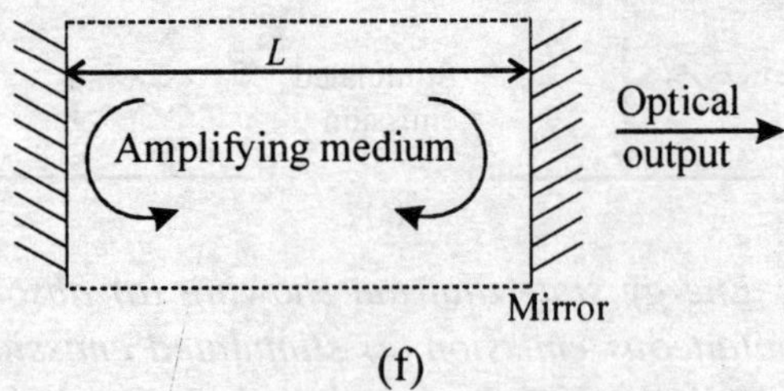

Figure: (f) The basic laser structure incorporting plane mirrors

Furthermore, if one mirror is made partially transmitting, usful radiation may escape from the **cavity**.

A stable output is obtained at saturation when the optical gain is exactly matched by the losses incurred in the amplifying medium. The major losses result from factor such as absorption and scattering in the amplifying medium, absorption, screaming and diffraction at the mirrors and nonuseful transmission through the mirrors.

Oscillations occur in the laser cavity over a small range of frequencies where the cavity gain is sufficient to overcome the above losses. Hence the device is not a perfectly monochromatic source but emits over a narrow spectral band. The central frequency of this spectral band is determined by the mean energy level difference of the stimulated emission transition. Other oscillation frequencies within the spectral band result from frequency variations due to the thermal motion of atoms within the amplifying medium (known as Doppler broadening) and by atomic collision. Hence the amplification within the laser medium results in a broadened laser transition or gain curve over a finite spectral width, as illustrated in figure (g)the spectral emission from the device therefore lies with in the frequency range dictated by this gain curve.

Since the structure forms a resonant cavity, when sufficient population inversion exists in the amplifying medium the radiation builds up and becomes established as standing waves between the mirrors. These is an intrgral number of half wavelengths. Thus when the optical spacing between the mirrors is L the resonance condition along the axis olf the cavity is given by:

$$L = \frac{\lambda q}{2\pi} \quad (5)$$

Figure: (g) The relative amplification in the laser amplifying medium showing the broadened laser transition line or gain curve

where λ is the emission wavelength,
n is the refractive index of the amplifying medium and
q is an integer. Alternatively, discrete emission frequencies f are defined by

$$f = \frac{qc}{2nL} \tag{6}$$

Where c is the velcoity of light

$$\delta f = \frac{c}{2n\,L} \tag{7}$$

and
$$\delta\lambda = \frac{\lambda^2}{2nL} \tag{8}$$

Optical emission from semiconductors

For a semiconductor in thermal equilibrium the energy level occupation is described by the Fermi-Dirac distribution function (rather than the Boltzmann). Consequently, the probability P(E) that an electron gains sufficient thermal energy at an absolute temperature T such that it will be found occupying a particular energy level E, is given by the Fermi-Dirac distribution.

$$P(E) = \frac{1}{1 + e^{(E-E_F)/KT}}$$

where K is Boltzmann's constant and E_F is known as the Fermi energy of Fermi level. The Fermi level is only mathematical parameter but it gives an indication of the distribution of carriers within the material. This is shown in figure for the intrinsic semiconductor where the Fermi level is at the centre of the bandgap, indication that there is a small probability of electrons occupying energy levels at the bottom of the conduction band and a corresponding number of holes occupying energy levels at the top of the valence band.

To create an extrinsic semiconductor the material is doped with impurity atoms which either create more free electrons (donor impurity) or holes (acceptor impurity). These two situations are shown in figure where the donor impurities form energy level just below the conduction band whilst acceptor impurities form energy level just above the valence band.

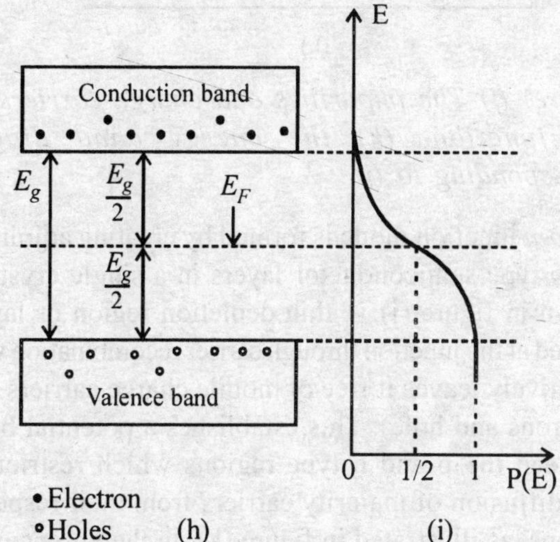

figure: (h) The energy band structure of an intrinsic semiconductor at a tempeature above absolute zero, showing an equal number of electrons and hold in the conduction band and the valence band respectively. (i) The Fermi-Dirac probability distribution corresponding to (a)

When donor impurities ae added, thermally excited electrons from the donor levels are raised into the conduction band to create an excess of negative charge carriers and the semiconductor is said to be n-type with the majority carriers being electrons. The Fermi level corresponding to this carrier distribution is raised to a position above the centre of the bandgap. When acceptor impurities are added, thermally excited electrons are raised from the valence band to the acceptor impurity levels leaving an excess of positive charge carriers in the valence band and creating a *p*-type semiconductor where the majority carriers are holes. In this cases Fermi level is lowered below the centre of the bandgap.

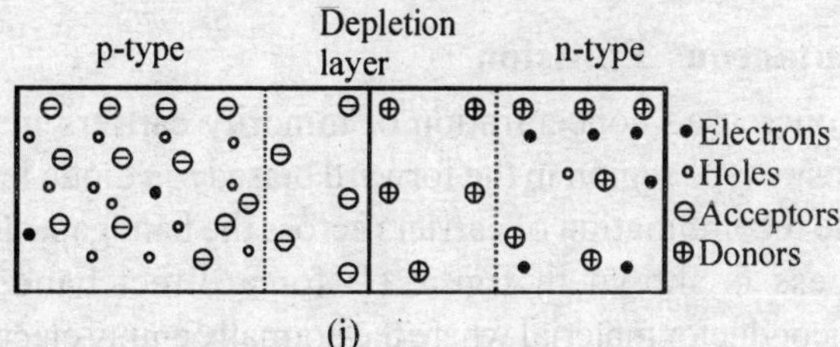

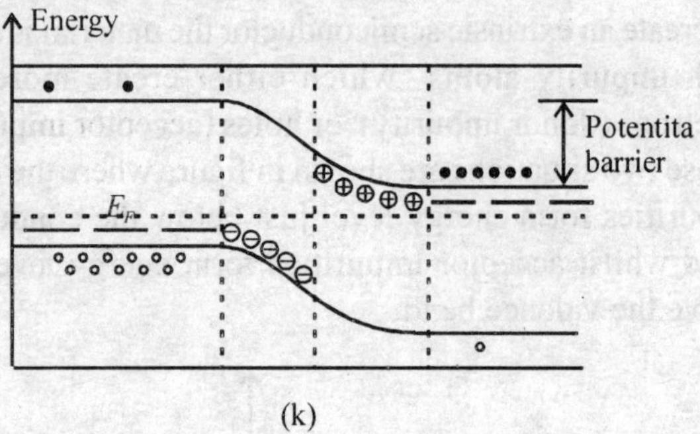

Figure: (j) The impurities and charge carriers at a p-n junction. (k) the energy band diagram corresponding to (j)

The *p-n* junction diode is formed by creating adjoining *p* and *n* type semiconductor layers in a single crystal, as shown in figure (j). A thin depletion region or layer is formed at the junction through carrier recombination which effectively leaves it free of mobile charge carriers (both electrons and hole). This establishes a potential barrier between the p and n type regions which restricts the interdiffusion of majority carriers from their respective regions, as illustrated in figure (k). In the absence of an external applied voltage no current flows as the potential barrier prevents the net flow of carriers from one region to another. When the junction is the same equilibrium state the fermi level for the *p* and *n* type semiconductor is the same as shown figure (k).

The width of the depletion region and thus the magnitude of the potential barrier is dependent upon the carrier concentration (doping) in the *p* and *n* type regions, and any external applied voltage. When an external positive voltage is applied to the *p* type region with respect to the *n* type, both the depletion region width and the resulting potential barrier are reduced and the diode is said to be forward biased. Electrons from the *n* type region and holes from the p type region can flow more readily across the junction into the opposite type region. These minority carriers are effectively injected across the junction by the application of the external voltage and form a current flow through the device as they continuously diffuse away from the interface. However, this situation in suitable semiconductor materials allows carrier recombination with emission of light.

Spontaneous Emission

The increased concentration of minority carriers in the opposite type region in the forward biased *p-n* diode leads to the recombination of carriers across the bandgap. This process is shown in figure (*l*) for a direct bandgap semiconductor material where the normally empty electron

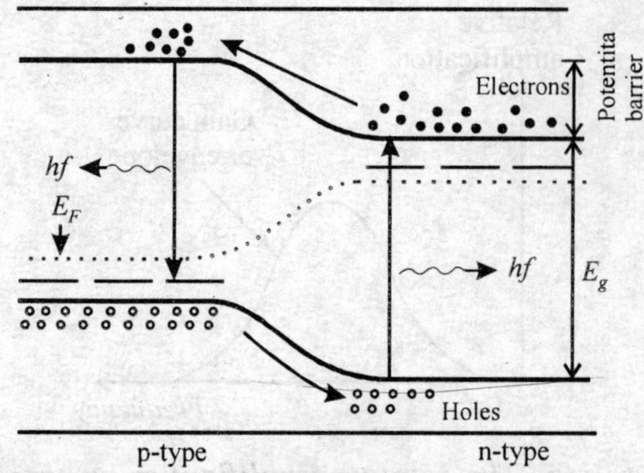

Figure: (l) The p-n junction with forward bias giving spontaneous emission of photons

states in the conduction band of the *p* type material and the normally empty hole state in the valence band of the *n* type material populated by injected carriers which recombine across the bandgap. The energy relased by this eelctron-hole recombination is approximately equal to the bandgap energy E_g. Excess carrier population is therefore decreased by recombination which may be radiative or nonradioactive.

In nonradioactive recombination the energy released is dissipated in the form of lattice vibration and thus heat. However, the band to band radiative recombination the energy is released with the creation of a photon figure (*l*) with a frequency following equation (1) where the energy is approximately equal to the bandgap energy E_g and therefore:

$$E_g = hf = \frac{hc}{\lambda} \tag{10}$$

Where c is the velocity of light in a vacuum and λ is the optical wavelength. Substituting the appropriate values for h and c in Eq (9) and rearranging gives:

$$\lambda = \frac{1.24}{E_g} \tag{11}$$

Where λ is written in μm and E_g in eV.

This spontaneous emission of light from within the diode structure is know as electroluminescence. The light is emitted at the site of carrier recombination which is primarily closed to the junction, although recombination may take place through hole diode structure as carriers diffuse away from the junction region figure (m). However, the amount of radiative recombination and the emission area within the structure is dependent upon the semiconductor materials used and the fabrication of the device.

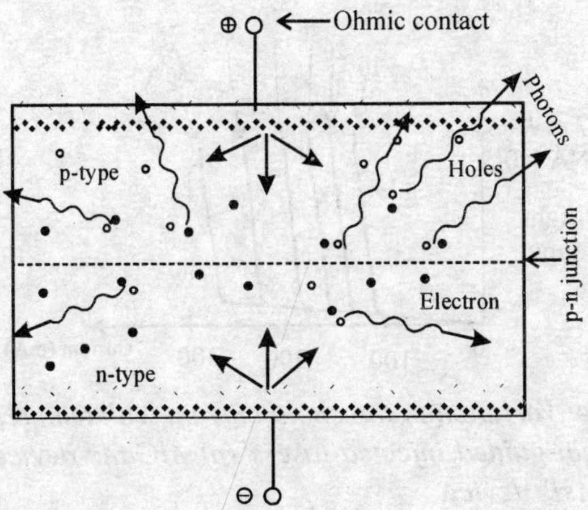

Figure: (m)An illustration of carrier recombination giving spontaneous emission of light in a p-n junction diode

Heterojunctions

The preceding sections has considered the photoemissive properties of a single *p-n* junction fabricated from a single crystal semiconductor material. This is known as a **homojunction**. However, the radiative properties of a junction diode may be improved by the use of heterojunction. A heterojunction is an interface between two adjoining single crystal semiconductors with **different bandgap energies**. Devices which are fabricated with heterojunctions are said to have heterostructure.

Heterojunctions are classified into either an isotype (*n-n or p-p*) or an anisotype (*p-n*). The isotype heterojunction provides a potential barrier within the structure which is useful for the confinement of minority barrier within the structure which is useful for the confinement of minority carriers to a small active region (carrier confinement). It effectively reduces the carrier diffusion length the thus the volume within the structure where radiative recombination may take place. This technique is widely used for the fabrication of injected lasers and highradiance LEDs. Isotype heterjunctions are are also extensively used in LEDs to provide a transparent layer close to the active region which substantially reduces the absorption of light emitted from the structure.

The semiconductor injection laser

The electroluminescent properties of the forward biased *p-n* junction diode have been considered in the preceding section. Stimulated emission by the recombination of the injected carriers is encouraged in the semiconductor injection laser (often called the injection laser diode (ILD) or simply the injection laser) by the provision of an optical cavity in the crystal structure in order to provide the feedback of photons. This gives the injection laser several major advantages over other semiconductor source (e.g. LEDs) that may be used for optical communications. These are as follows:

1. High radiance due to the amplifying effect of stimulated emission. Injunction lasers will generally supply milliwatts of optical output power.
2. Narrow linewidth of the order 1 nm (10 $\mathring{A}$) or less which is useful in minimizing the effect of material dispersion.
3. Modulation capabilities which at present extend up into the gigahertz range and will undoubtedly be improved upon.
4. Relative temporal coherence which is consider essential to allow heterodyne (coherent) detection in high capacity system, but at present is primarily of use in single-mode systems.
5. Good spatial coherence which allows the output to be focused by a lens into a spot which has a greater intensity than the dispersed unfocused emission. This permits efficient coupling of the optical power into the fiber even for fibers with low numerical aperture. The spatial fold matching to the optical fiber which may be obtained with the laser source is not possible with an incoherent emitter and, consequently, coupling efficiencies are much reduced.

These advantages, together with the compatibility of the injection laser with optical fiber (e.g. size) led to the early developments of the device in the 1960s. Early injection lasers had the form of a Fabry-Perot cavity often fabricated in **gallium arsenide** which was the major III-V compound semicouductor with electro-luminescent properties at the appropriate wavelengths for first generation systems. The basic structure of this homojunction device is shown in figure (n), where the cleaved ends of the crystal act as partial mirros in order to encourage stimulated emission in the cavity when electrons are injected into the p-type region. However as mentioned previously these devices has a high threshold current emission in the cavity when electrons are injected into the p type region. However, as mentioned previously these devices had a high threshold current density (greate than 10^4 A cm-2) due to their lack of carrier containment and proved inefficient light sources. The high current densities required dictated that these devices when operated at 300 *K* were largely utilized in a pulsed mode in order to minimize the junction temperature and thus avert damage.

Improved carrier containment and thus lower threshold current densities (around 10^3 A com^{-2}) are achieved using heterojunction structures. The double heterojunction injection laser fabricated from matched III-V alloys provided both carriers and optical confinement on both sides of the *p-n* junction, giving the injection laser a greatly enhanced performance. This enabled these devices with the appropriate heat sinking to be operated in a continuous wave (CW) mode at 300 *K* with obvious advantages for optical communication (e.g. analog transmission). However, in order to provide reliable CW operation of the DH injection laser it was necessary to provide further carrier and optical confinement discussion of this subtraction, however, it is useful to consider the efficiency of the semiconductor injection laser as an optical source.

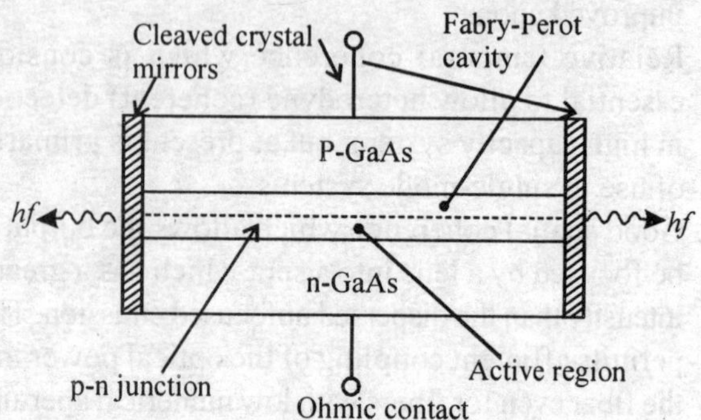

Figure: (n) Schematric diagram of a GaAs homojunciton injection laser laser with a Fabry-Perot cavity

Threshold current tempeature depedence

Figuer (p) and (q) shows the variation in threshold current with tempeature for two gain-guided (oxide insulated stripe) injection lasers . Both devices that stripe widths of aproximately 20 μm but wre fabricated from different material systems for emission at wavelengths of 0.85 μm and 1.55 μm (AlGaAs) and InGaAsP devices respectively).

In general terms the threshold current tends to increase with tempeature, the temperature dependence of te thresphold current density j_{th} being approximately exponential for msot common structures. It is given by:

$$J_{th} \propto e^{\frac{T}{T_0}}$$

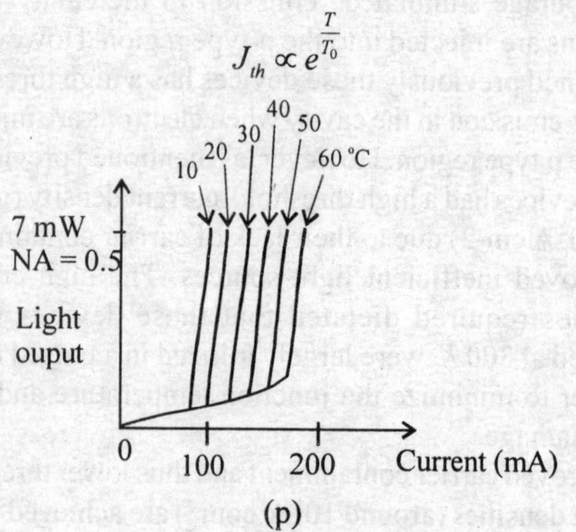

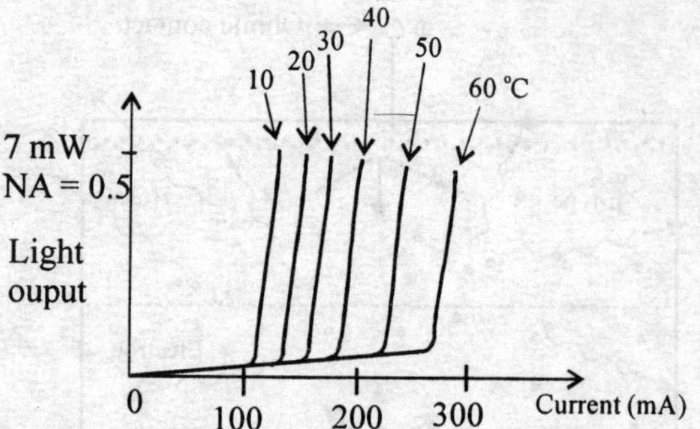

Figure: Variation in threshold current with temperature for gain-guided injected lasers (p) AlGaAs device, (q) InGaAsP device

Where T is the device absolute temperture and T_0 is the threshold temperature coefficient which is a charactersitic temperature describing the quality of the material.

9.2.2 LED

Spontaneous emission of radiation in the visible and infrared regions of the spectrum from a forward biased p-n junction was discussd in section 6.3.2. The normally empty conduction based of the semiconductor is populated by electrons injected into it by the forward current through the junction, and light is generated when these electrons recombine with hole in the valence band to emit a photon. This is the mechanism by which light is emitted from from an LED, but stimulated optical cavity and mirror facets to provides feedback of photons.

The LED can therefore operate at lower current densities than the injection laser, but the emitted photons has random phases and the device is an incoherent optical source, Also, the energy of the emitted photons is only rughly equal to the bandgap energy of the seimcounductor meateris, which gives a much wider spectral linewidht (possibly by a factor of 100) thanthe injection laser.

Add odd File

1. *Simpler fabrication:* There are no mirror facets and in some structures no striped geometry.
2. *Cost*: This simpler construction of the LED leads to much reduced cost which is always likely to be maintained.
3. Reliability. The LED does not exhibit catasrophic degradation and has proved far less sensitive to gradual degradation than the injection laser. It is also immune to self pulsation and modal noise problems.
4. *Generlly less temperature dependence*: The light output aginst current characteristic is less affected by temperature than the corresponding characteristic for the injected laser. Furthermore,

the LED is not threshold devices and therefore raising the temperature doest nto increase the threshold current above the operating piont and hence halt operation.

5. *Simpler drive circuitry:* This is due to the generally lower drive current and reduced temperature depedence which makes temperature compensation circuits unnecessary.
6. *Linearity* : Ideally, the LED has a linear light output against current characteristic, unlike the injection laser. This can prove advantageous where modulation is concerned.

RADIGRAPHY

The LED internal quantum efficiency η_{int}, Which can be defined as the ratio of the radiative recombination rate to the total recombination rate, following:

$$\eta_{int} = \frac{r_r}{r_t} = \frac{r_r}{r_r + r_{nr}} \quad (1)$$

$$= \frac{R_r}{R_t} \quad (2)$$

where R_r is the total number of radiative recombinations per second.

$$R_r = \eta_{int} \frac{i}{e} \quad (7.9)$$

$$\eta_{int} = \frac{1}{1+(r_{nr}/r_r)} = \frac{1}{1+(\tau_r/\tau_{nr})} \quad (7.12)$$

Furthermore, the total recombination lifetime τ can be written as $\tau = \Delta n / r_t$ which, givens:

$$\frac{1}{\tau} = \frac{1}{\tau_r} + \frac{1}{\tau_{nr}} \quad (7.13)$$

Hence Eq. (7.12) becomes

$$\eta_{int} = \frac{\tau}{\tau_r}$$

The double heteroxanthin LED

The principal of operation of the DH LED is illustrated in Figure 7.2. The device shown consists of a *p* type GaAs layer sandwiched between a *p* type AlGaAs and an n type AlGaAs layer. When a forward bias is applied electrons form the *n* type layer are injected through the *p-n* junction into the *p type GaAs* layer where they become minority carriers. These minority carriers diffuse away from the junction [Ref. 9] recombining with majority

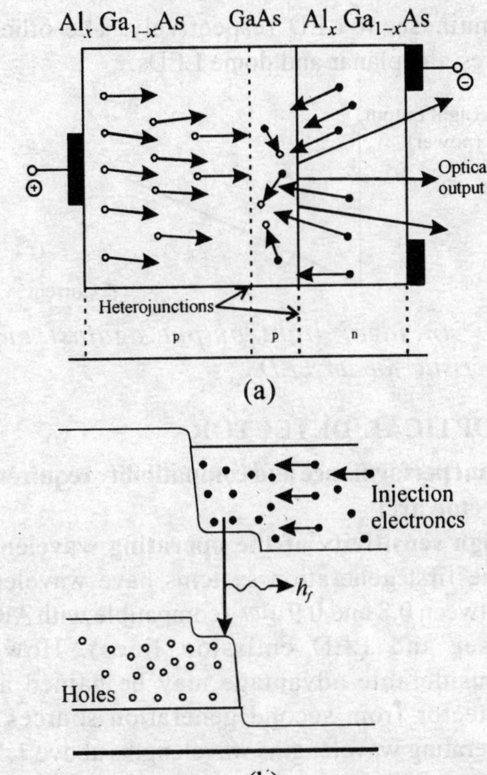

Figure: The double heterojunction LED (a) the layer structure, shown with an applied forward bias, (b) the corresponding energy band diagra

carriers (holes) as they do so. Photons are therefore produced with energy corresponding to the bandgap energy of the *p* type AlGaAs layer. The injected electrons are inhibited form diffusing into the *p* type AlGaAs layer because of the potential barrier presented by the $p-p$ heterojunction. Hence, electroluminescence only occurs in the GaAs junction layer, providing both good internal quantum efficiency and high radiance emission. Furthermore, light is emitted from the device without reabsorption because the bandgap energy in the AlGaAs lauer is large in comparison with that in in GaAs. The DH structure is therefore used to provide the most efficient incoherent sources for application within optical fiber communications. Nevertheless, these devices generally exhibit the previously discussed constraints in relation to coupling efficiency to optical fibers. This and other LED structures are considered in greater detail in the following section.

LED structure

There are five major types of LED structure and although only two have found extensive use in optical fiber communications a third is becoming of increasing interest. These are the surface emitter, the edge emitter and the superluminescent LED respectively. The other two structures, the planar and dome LEDs.

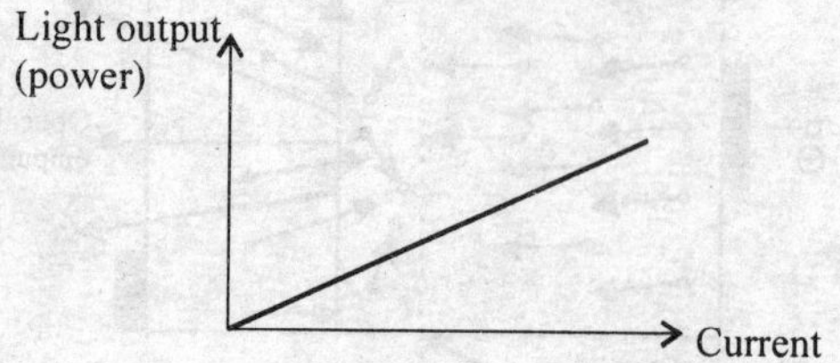

Figure: An ideal light output against current characteristic for an LED

9.2.3 OPTICAL DETECTOR

Important performance and compatibility requirements for detector are:

1. **High sensitivity at the operating wavelengths:** The first generation systems have wavelengths between 0.8 and 0.9 μm (compatible with AlGaAs laser and LED emission lines). However, considerable advantage may be gained at the detector from second generation sources with operating wavelengths wavelengths above 1.1 μm as both fiber attenuation and dispersion are reduced. There is much research activity at present in this longer wavelength region, especially concerning wavelengths around 1.3 μm where attenuation and material dispersion can be minimized. In this case semiconductor materials are currently under investigation in order to achieve good sensitivity at normal operating.
2. **High fidelity:** To reproduce the received signal waveform with fidelity, for analogy transmission the response of the photodetector must be linear with regard to the optical signal over a wide range.
3. **Large electrical response to the received optical signal:** The photodetector should produces a maximum electrical signal for a given amount of optical power, i.e. the quantum efficiency should be high.
4. **Short response time to obtain a suitable bandwidth:** Present systems extend into the hundreds of megahertz. However, It is apparent that future systems will operate in the gigahertz range, and possibly above.
5. **A minimum noise introduced by the detector:** Dark currents, leakage currents and shunt conductance must be of low. Also the gain mechanism within either the detector of associated circuitry must be of low noise.
6. **Stability of performance characteristics:** Ideally, the performance chatacteristics of the detector should be independent of changes in ambient conditions However, the detectors currently favoured have chatacteristics which vary with temperature , and therefore compensation for temperature effects is often necessary.
7. **Small size**: The physical size of the detector must be small for efficient coupling to the fiber and to allow easy packaging with the following electronics.
8. **Low bias voltages:** Ideally the detector should not require excessive bias voltage or currents.
9. **High reliability:** The detector must be capable of continuous stable operation at room temperature for many year.

Optical detection principles

The basic detection process in an intrinsic absorber is illustrated in figure (a) which shows a p-n photodiode. This device is reverse biased and the electric field developed across the p - n junction sweeps mobile carriers to their respective majority sides. A deplertion region or layer is therefore created on either side of the junction. This barrier has the effect of stopping the majority carriers crossing the junction in the opposite direction to the field. However, the field accelerates minority carriers from both sides to the opposite side of the junction, forming the reverse leakage current of the diode. Thus intrinsic conditions are created in the depletion region.

A photon incident in or near the depletion region of this device which has an energy greater than or equal to the bandgap energy E_g of the fabricating material will excite an electron form the valence band into the conduction band . This process leaves an empty hole in the valence band is known as the photogeneration of an electron-hole pair, as shown in figure (a). Carrier pairs so generated near the junction are separated and swept under the influence of the the electric field to produce a displacement by current in the external circuit in excess of any reverse leakage current. Photo generation and the separation of a carrier pair in the depletion region of this reverse.

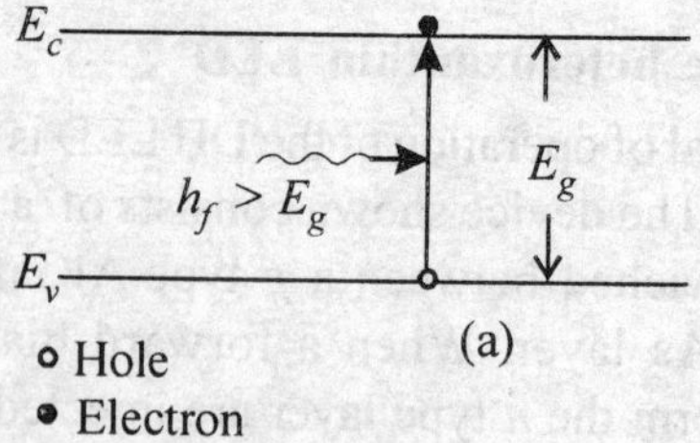

biased $P-n$ junction is illustrated in Figure 8.1 (c) The depletion region must be sufficiently thick to allow a large fraction of the incident light to be absorbed in order to achieve maximum carrier-pair generation. However, since long carrier drift times in the depletion region restrict the speed of operation of the photodiode it is necessary to limit its width. Thus there is a trade off between the number of photons absorbed and the speed of response.

Absorption

The absorption of photons to produce carrier pairs and thus a photocurrent, is dependent on the absorption coefficient $\alpha_0$of the light in the semiconductor used to fabricate the device. At a specific wavelength and assuming only bandgap transition the photocurrent I_P produced by incident light of optical power P_o is given by

$$I_P = \frac{P_o e(1-r)}{hf}[1-e^{(-\alpha_o d)}] \qquad (8.1)$$

Quantum efficiency

The quantum efficiency η is defined as the fraction of incident photons which are absorbed by the photodetector and generate electrons which are collected at the detector terminals:

$$\eta = \frac{number\ of\ electrons\ collected}{number\ of\ incident\ photons} \qquad (8.2)$$

Hence,

$$\eta = \frac{r_e}{r_P} \qquad (8.3)$$

where r_P is the incident photon rate and

r_e is the corresponding electron rate.

Responsivity

The expression for quantum efficiency does not invole photon energy and therefore the reponsivity R is often of more use when characterizing the performance of a photodetector. It is defined as:

$$R = \frac{I_p}{P_o}(AW^{-1})$$

where I_p is the output photocurrent in amperes and P_o is the incident optical power in watts. The responsivity is a useful parameter as it gives the transfer characteristic of the dector (i.e. photocurent per unit incident optical power).

The relationship for responsivity may be developed to include quantum efficiency as follows. Considering the energy of a photon $E = hf$. Thus the incident photon rate r_p may be written in term of incident optical power and the photon energy as:

$$r_p = \frac{P_o}{h_f}$$

In equation the electron rate is given by

$$r_e = \eta r_p$$

Substituting from equation we obtain

$$r_e = \frac{\eta P_o}{h_f}$$

where e is the charge on an electron. Thus from equation the respeonsivity may be written as:

$$R = \frac{\eta e}{hf}$$

9.2.4 $p-n$ PHOTODIODE

Figure shown a reverse biased $p-n$ pohotodiode with both the depletion and diffusion regions . The depletion region is formed by immobile positively charged donor atoms in the n type semiconductor material and immobile negatively charged acceptor atoms in the p type material, when the mobile carriers are swept to their majority sides under the influence of the electric filed. The width of the depletion region is therefore dependent upon the doping, concentration for a given applied reverse bias (i.e. the lower the doping the wider the depletion region).

Photons may be the absorbed in both the depletion and diffusion regions, as indicated by the absorbed regions figure. The absorption region's position and width depends upon the energy of the incident photons and on the material form which the photodiode is fabricated. Thus in the case of the weak absorption of photons, the absorption region may extend completely throughout the device. Electron-hole pairs are therefore generated in both the depletion and diffusion regions. In the depletion region the carrier parts separate and drift under the influence of the electric field, whereas outside the region the hole diffuses process is very slow compared to drift and thus limits the response of the photodiode.

It is therefore important that the photons are absorbed in the depletion region. Thus it is made as long as possible by decreasing the doping in the n type material. The depletion region width in a $p-n$ photodiode is normally 1 to 3 μm and is optimized for the efficient detraction of light at a given wavelength. For silicon.

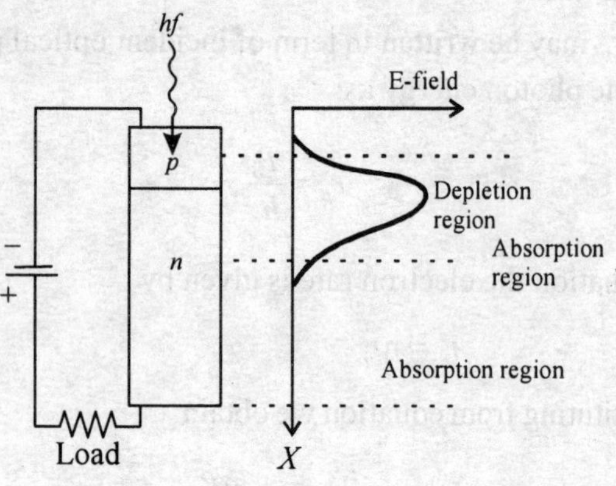

Figure: p-n photodiode showing depletion and diffusion regions

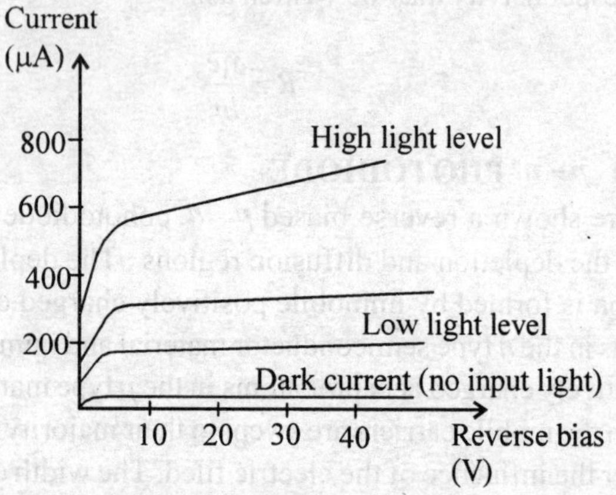

Figure: Typical p-n photodiode output characteristics

$p-i-n$ Photodiode

In order to allow operation at longer wavelengths where the light penetrates more deeply into the semiconductor material a wider depiction region is necessary. To achieve this the *n* type material is doped so lightly that it can be considered intrinsic, and to make a low resistance constant a highly doped *n* type (n^+) layer is added. This creates a $p-i-n$ structure, as many be seen in figure where all the absorption takes places in the depletion region.

The top entry device shown is Figure 8.8(a) is the simplest structure with the light being introduced through the 0upper p^+ layer. However, a drawback with this structure is a quantum efficiency penalty which results form optical absorption in the undepleted p^+ region. In addition, there is a limit to how small such a device can be fabricated as both light access and metallic contact are required on the top. To enable smaller devices with lower capacitances to be made a substrate entry technique is employed.

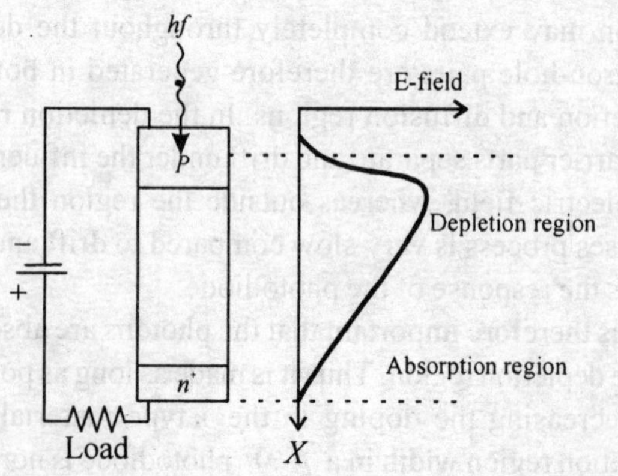

Figure: p-i-n photodiode showing combined absorption and deplecttion region

A substrate entry $p-i-n$ photodiode is shown in Figure 8.9(a). This device incorporates a p^+ – In GaAsP layer to provide a heterojunction structure which improves quantum efficiency. Moreover, it is fabricated as a mesa structure which reduces parasitic capacitances. Unfortunately. Charge trapping can occur at

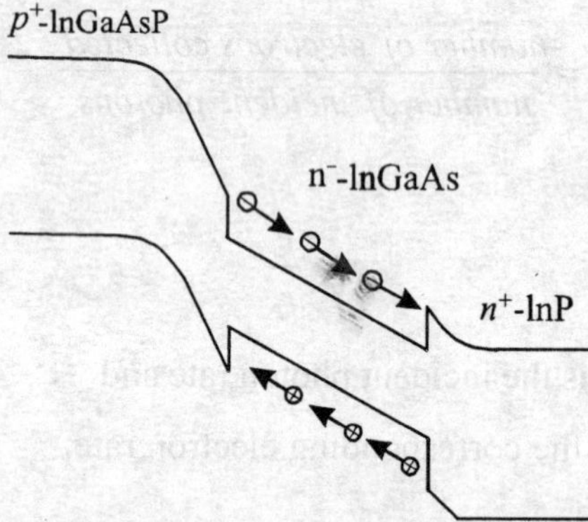

Figure: Plane lnGaAs p-i-n photodiode (a) structure (b) energy band diagram shwowing homojunction assicated with the conventional p-i-n structure

Speed of Response

Three main factors limit the speed of response of a photodiode.

(a) Drift time of carriers through the depletion region. The speed of response of a photodiode is fundamentally limited by the time it takes photogenerated carriers to drift across the depletion region. When the field in the depletion region exceeds a saturation value then the carriers may be assumed to travel at a constant drift velocity v_d. The longest transit time, d_{rift} is for carriers which must traverse the full depletion layer width w and is given by

$$t_{drift} = \frac{w}{vd}$$

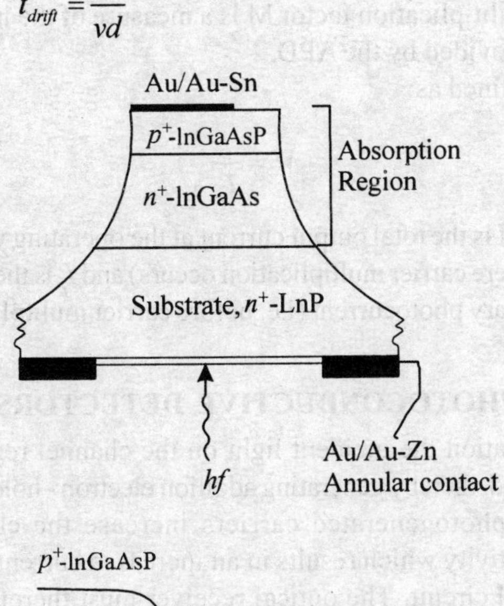

Figure: Subsrate enety lnGaAs p-i-n photodiode (a) sturcture (b)energy band diagram illustrating the heterojunction and charge trapping

(b) Diffusion time of carriers generated outside the depletion region

Carrier diffusion is a comparatively slow process where the time taken, t_{diff}, for carriers to diffuse a distance d may be written as

$$t_{diff} = \frac{d^2}{2D_c}$$

where D_c is the minority carrier diffusion coefficient. For example, the hole diffusion time through 10 μm of silicon is 40 *ns* where as the electrum diffusion time over a similar distance is around 8 ns.

(c) Time constant incurred by the captaincies of the photodiode with its load. A reversed based photodiode exhibits a voltage dependent capacitance caused by the variation in the stored charge at the junction. The junction capacitance C_j is given by.

Photodiode Bandwidth

$$C_j = \frac{\varepsilon_s A}{w} \qquad (8.17)$$

where ε_s is the permittivity of the semiconductor material and A is the diode junction area.

$$B_m = \frac{1}{2\pi t_{drift}} = \frac{vd}{2\pi w}$$

Semiconductor photodiodes with internal gain

The second major type of optical communication detector is the **avalanche photodiode**. This has a more sophisticated structure than the $p-i-n$ photodiode in order to create an extremely high electric field region as may be seen in figure (a) Therefore, as well as the depletion region where most of the photons are absorbed and the primary carrier pairs generated there is a high field region in which holes and electrons can acquire sufficient energy to excite new electron-hole pairs. This process is known as Impact ionization and is the phenomenon that leads to avalanche break down in ordinary reverse biased diodes. It often requires high reverse bias voltage in order that the now carriers created by impact lonization can themselves produce additional carriers by the same mechanism as shown if figure (b). More recently, however, it should be noted that devices which will operated at much lower bias voltage (15 tol 25 V) have become available.

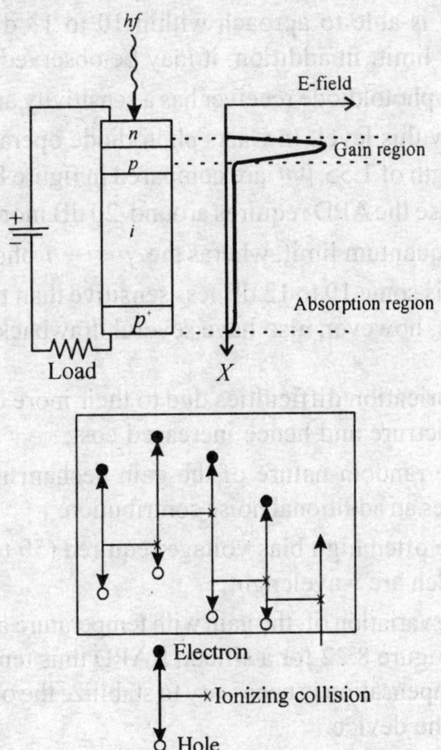

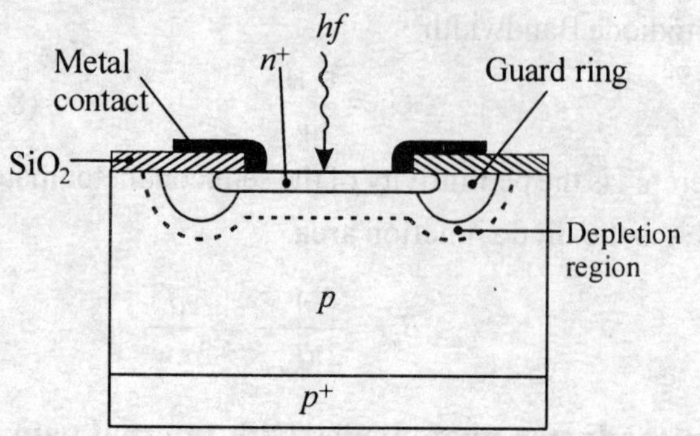

Benefits and drawbacks with the avalanche photodiode.

APDs have a distinct advantage over photodiodes without internal gain for the detection of the very low light levels often encountered in optical fiber communications.

They generally provide an increase in sensitivity of between 5 and 15 dB over $p-i-n$ photodiodes whilst often giving a wider dynamic range as a result of their gain varication with response time and reverse bias.

The optimum sensitivity improvement of APD receivers over $p-i-n$ photodiode devices is illustration in the characteristics shown in Figure 8.21. The characteristics display the minimum detectable optical power for direct detection versus the transmitted bit rate in order to maintain a bir error rate of 10^{-9} in the shorter and longer wavelenmght regions. Figure 8.21 (a) compares silicon photodiode operating at a wavelength of 0.82 μm wher the APD is able to aproach within 10 to 13 dB of the quantum limit. In addition, it may be observed that the $p-i-n$ photoldiode receiver has a sensitivity around 15 dB below this level. In GaAs photodiode operating at a wavelength of 1.55 μm are compared in figure 8.21 (b). In this case the APD requires around 20 dB more power than the quantum limit, wheras the $p-i-n$ photodiode receiver is some 10 to 12 dB less sensitive than the APD.

APDs, however, also have several drawbacks which inculde:

a. Fabrication difficulties due to their more complex structrure and hence increased cost;
b. The random nature of the gain nechanism which gives an additional noise contribution;
c. The often high bias voltage required (50 to 400V) which are wavelenght;
d. The variation of. the gain with temperature as shown in Figure 8.22 for a silicon RAPD thus temperture compensation is necessary to stabilize the operation of the device.

Multiplication Factor

The mulitiplication factor M is a measure of the internal gain provided by the APD.
It is defined as:

$$M = \frac{I}{I_p}$$

Where I is the total output current at the operating voltage (i.e. where carrier multiplication occurs) and I_p is the initial or primary photocurrent (i.e. before carrier mulitplication occurs).

9.2.5 PHOTOCONDUCTIVE DETECTORS

In operation the incident light on the channel region is absorbed, thereby generating addition electron - hole pairs. These photogenerated carriers increase the channel conductivity which results in an increased current in the external circuit. The optical receiver must therefore be sensitive to very small changes in resistance induced by the incident light. Furthermore, once the carriers have been generated the electrons will be swept by the applied electric field towards the anode whilst the holes move towards the cathode. In general, however, the mobility of the holes is considerably smaller than that of the electrons in $III-V$ alloys such as GaAs and In GaAs. Thus the electrons, being the fastest charges carriers, provide the minimum time for detection and hence the limitation on the speed of response of the photoconductive detector.

In addition, whilst the fast electrons are collected at the anode, the corresponding holes are still processing across the channel. This creates an absence of electrons figure and hence a net positive charge in the channel region. However, the excess charge is immediately compensated by the injection further electrons form the cathode into the channel. Thus further electrons may be generated from the absorption of a single photon. This factor creates what is known as the photoconductive gain G which may be defined as the ratio of the slow carrier transit time t_s to the fast carrier transit time t_f Hence:

$$G = \frac{t_s}{t_f}$$

Moreover, the photocurrent I_P produced by the photodetector can be written as

$$I_P = \frac{\eta P_o e}{hf} G$$

where η is the device quantum efficiency which in a similar manner to the absorption process follows an exponential distribution, P_o is the incident optical power and e is the charge on an electron.

Since the current response in the photodetector remains for a time t_s after the end of an incident optical pulse and exhibits an exponential decay with a time constant equal to this slow carrier transit time, then the maximum 3 dB bandwidth B_m for the device is given by:

$$B_m = \frac{1}{2\pi t_s}$$

Equations (8.32) and (8.34) can be combined to provide an expression for the gain - bandwidth product of a photoconductor as:

$$G.B_m = \frac{1}{2\pi t_f}$$

It may be noted that an implication of Eq. (8.35) is that when t_f is fixed photoconductive gain can only be obtained at the expense of the maximum bandwidth permitted by the device. There is therefrom a trade - off between gain and speed of response.

9.2.6 INTERFEROMETERS

An interferometer is apparatus that can be used *measure the wavelength* of light by utilizing interference between 2 light waves.

As a rough guide, the following table serves to give the wavelengths of the colours:

Colour	Wavelength
Violet	400–450 nm
Blue	450–500 nm
Green	500–570 nm
Yellow	570–590 nm
Orange	590–620 nm
Red	620–750 nm

Young's experiment:

For a bright fringe, $= m\lambda = x_m \frac{d}{D}$

For a dark fringe $= (2m+1)\frac{\lambda}{2} = X_m \frac{d}{D}$

Fringe width, $X = \lambda \frac{D}{d}$

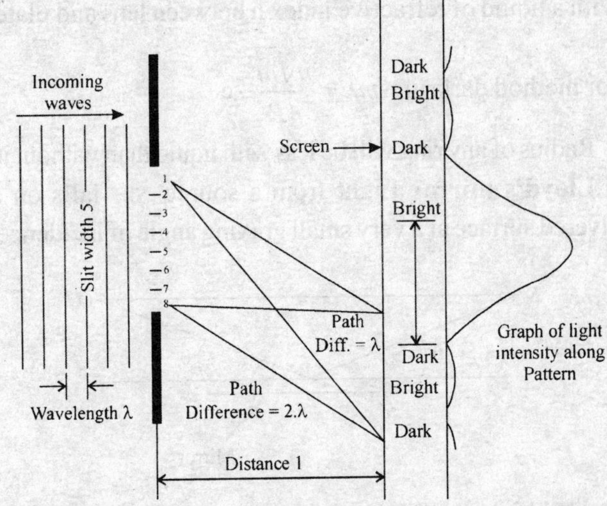

Figure: Interference pattern produced by a single slit

Newton's rings (For determining wavelength of light). A poano-convex lens of layer radius of curvature R is illuminated from top with a parallel beam of monochromatic light. Some of the light is reflected from the upper surface of glass plate and some from lower surface of lens. Interference occurs by division of amplitude.

The fringes are localized with air gap between the lens and plate.

h = distance between lens and plate as the point Which is at a distance r from axis of length

For a bright ring viewed by reflection,

$$(2m+1)\frac{\lambda}{2} = \frac{r_m^2}{R}$$

For dark viewed by reflection. $m\lambda = \frac{r_m^2}{R}$

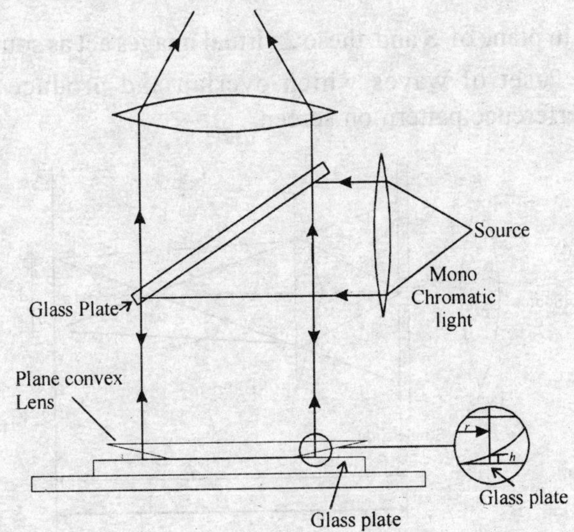

With a liquid of refractive index n between lens and plate:

For method darkring $m\lambda = \frac{n\sqrt{m^2}}{R}$ c

Radius of any ring will be less with liquid than without it.

Lloyd's mirror Light from a source S_0 falls on a silvered surface at a very small graxing angle of incidence.

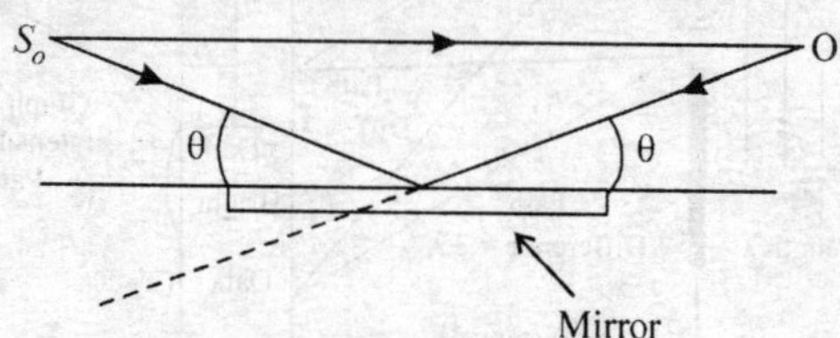

A virtual image of S_0 is formed at S_1.

Interference occurs between direct beam from S_0 to observer O and reflected beam.

The Zeroth fringe will be black because of phase change due to reflection at surface.

Fringe width $= \frac{D}{2d}\lambda$

Conditions of maxima and minima in interference pattern interchange,

$$\Delta = \frac{2n\lambda}{2} \text{ .... Minima}$$

$$\Delta = (2n+1)\lambda/2 \text{.....Maxima}$$

Where D = distance of screen from plane of slits;

2d = distance between S_0 S_1

Fresnel's biprism: Monochromatic light from a narrow slit S falls on a baptism the axis of which must be in line with slit. The refracting angle of biprism are very small $(0.25°)$. The prism forms 2 Virtual images of slit S_1 and S_2 in plane of S and these 2 virtual images act as source for 2 set of waves which overlap and produce an interference pattern on screen.

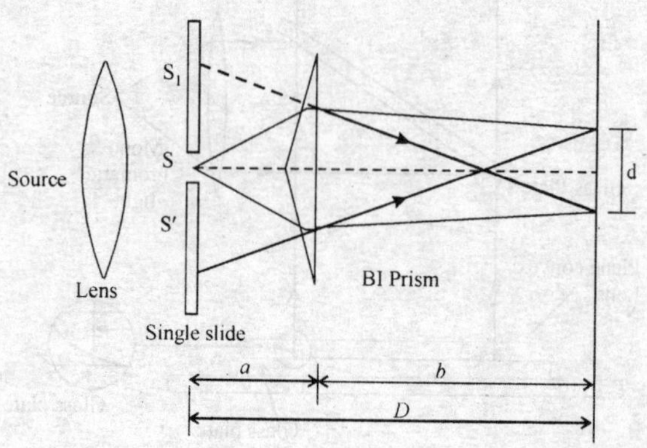

Fringe width (distance between 2 consecutive bright or dark fringes) $= \frac{D\lambda}{2d}$

Michelson Interferometer: Michelson Interferometer used to measure the wavelength of light by utilizing interference between 2 light waves.

Observer views combination of waves A and F through the telescope depending on constructive or destructive interferences.

Fabry-Perot interferometer Condition of maxima in transmitted beam,

$2t \cos\theta = n\lambda$

Where, t = separation between plates

θ = angle of incidence of beam on surface

n = order of fringe.

MICHELSON INTERFEROMETR

The problem of light setting reflected into the light source and distortion in the interference pattern due to slight tilts in the moving mirror while making displacement measurements is solved by using corner cube reflector

Three dimensional corner cube has the characteristic that any light beam is reflected parallel to itself regardless of its angle of incidence. When corner cube reflects during motion. The propagation direction of reflected beam remains unchanged. Second mirror is also replaced by a corner cube reflects during motion, the propagation direction reflected beam remains unchanged. Second mirror is also replaced by a corner cube due of symmetry reasons.

The transversal displacement of reflector induces a transversal displacement of beam. This is eliminated by inserting a flat mirror.

This arrangement leads to tilt as well as transversal shift invariance with respect to moving reflector and incident beam is always reflected upon itself.

This system is used for velocity and displacement measurements. Also the double transit increases the sensitivity by a factor of two.

DOUBLE – BEAM INTERREFOMETERS

- The Mach – Zehnder interferometers. Here the mirrors are turned to direct the light waves into a second beam splitter which act as a beam combiner. Both output beams exit the system. Also two interfering waves travel parallel routes rather than two orthogonal routes as in ordinary. *This interferometer is used in the analysis of phase objects while the interferometer set- up remains fixed.*

- **Ring interferometer** Here the beam combiner is replaced by a 45° mirror. The two waves circulate the device in opposite directions returning to be recombined in the beam splitter. This interferometer is since it is a common path interferometer. Both interfering waves traverse exactly the same path. but in opposite direction. There is no phase difference between the interfering waves except, if a non – reciprocal effect exists. This interferometer is used in *rotation rate sensing.*
- **The jamin interferometer** Here two parallel plates incorporate a beam splitter and a mirror. The beam splitter and mirrors are implemented by Fresnel reflections and transmissions with no need for any coating. It is very stable as the two interfering waves use the same components. Although it is not a common path interferometer. Additional waves are produced since each wave incident on an interface is partially reflected and partially transmitted.
- **Beam splitter** Here the splitting layer is along the meridian of an equilateral triangle. The beam splitter cube has the splitting surface at 45° ti its outer surface and is used is the same wave as the regular beam splitter. The Ko steers prism is different in that both output beams are deviated from the original propagation erection. The Ko steers prism is illuminated by a light beam normal to the side of the triangle. Part of the beam is reflected and other part is transmitted. Both beams are totally reflected from the sides of the two 30° prisms and emerge as two parallel beams.

Prism interferometer It is more flexible and easier to align.

A reflecting surface is used when operating in reflecting mode. It is used in high accuracy profile measurement. By moving the reflecting surface is a transverse plane, the surface profile can be measured.

Fibry-Perot interferometer Here the amplitude is slit by partially transparent mirrors. It consists of two high quality flat mirrors, M_1 and M_2 having respective amplitude transmissions t_1 and t_2 and reflectance, r_1 and r_2. A plane wave of complex amplitude E, incident from the left is split by the first mirror into transmitted wave E_{t1}' and reflected wave E_{r1}' and reflected wave E_{r1}. The transmitted wave arrives at the second mirror at a distance, L and $E_{t1}'\ r_2 - e^{jkL}$. Where $k = n\ k_0$ with n denoting refracting index of the medium inside interferometer and k_0 is the wave number in free space outside the interferometer. At the second mirror the wave is split into transmitted wave $E_{t1}'t_2 e^{JKL}$ and reflected wave is $E_{t1}'r_2 - e^{JKL}$. The reflected wave arrives at the first mirror where it is again split into a transmitted wave a transmitted wave and a reflected wave. In this way we obtain a superposition of all these partial waves.

Condition of maxima is transmitted beam

$2t\cos\theta = n\lambda$

Where,

t = Separation between plates,

θ = Angle of incidence of beam on surface.

n = Order of fringe

Two beam interferometry vs. Multiple beam interferometry

The fiber optic version of Mach- Zender, Michelson and signal interferometer fails under the 2-beam category. In all of these interferometers. Light is divided into two paths by means of a fiber directional couple. Either of these two paths could be considered as a reference aim and the other as the signal arm. Light propagating through these two paths are recombined by a second directional couple (DC) or by the same DC to beat at a detector. Depending on the phase retardation introduced to the propagating light between the two paths, the light intensity could vary at the detector end and hence the detector output will contain signatures of the phase difference between the two arms of the interferometer and the information about can retrieved. Typical phase sensitivities achievable in fiber optical interferometers operating at 850 nm are:

- 10 rads/ microstrain meter. 10 rads/bar/meter, 100 rad/ °C/ meter for M-Z interferometer.
- 0.1 tad/ sec/ meter sq. for signal interferometer.

The major applications of these 2 – beam interferometer sensors lies in 3 broad categories: *hydro phones, magnetometers and gyroscopes.*

Multibeam interferometry is used in situation which require extremely high measurement resolutions e.g. fiber optic version of Fabry Perot interferometer (FPI). In this interferometer, intensity variations with phase induced by a measured is more rapid than a 2 – beam case. These can be constructed in miniature sizes with refractive coatings at each fiber end and can be fusion spliced to a long length of low loss fiber which serves to address the sensor from a remote point. Calorimeters have been constructed as heat transfer gauges in transient wind tunnel experiments by embedding such FPI sensors in test models. Multiple interferometry in a fiber firm can be implemented to configure a ring resonator which not only find applications in sensing, but also as a resonant cavity in fiber lasers.

9.2.7 BASIC OF OPTICAL FIBER

For optical fiber communication the system shown in Figure 1.2 (a) may be considered in slightly greater detail, as given in Figure 1.2 (b). In this case the information source provides an electrical signal to a transmitter comprising an electrical stage which drives an optical source to give modulation of the light wave carrier. The optical source which provides the electrical-optical conversion may be either a semiconductor laser or light emitting diode (LED). The transmission medium consists of an optical fiber cable and the receiver consists of an optical detector which drives a further electrical stage and hence provides demodulation of the optical carrier. Photodiodes (*p-n, p-i-n* or avalanche) and, in some instances, phototransistors and photoconductors are utilized for the detection of the optical signal and the optical – electrical conversion. Thus there is a requirement for electrical interfacing at either end of the optical link and at present the signal processing is usually performed electrically.

The optical carrier may be modulated using either an analog or digital information signal. In the system shown in Figure 1.2 (b) analog modulation involves the variation of the light emitted from the optical source in a continuous manner. With digital modulation, however, discrete changes in the light intensity are obtained (i.e on-off pulse). Although often simpler to implement, analog modulation with an optical fiber, communication system is less efficient, requiring a far higher signal to noise ratio at the receiver than digital modulation. Also the linearity needed for analog modulation is not always provided by semiconductor optical sources, especially at high modulation frequencies. For these reasons, analog optical fiber communication links are generally limited to shorter distances and lower bandwidths than digital links.

Figure 1.3 shows a block schematic of a typical digital optical fiber link. Initially the input digital signal from the information source is suitably encoded for optical transmission. The laser drive circuit directly modulates the intensity of the semiconductor laser with the encoded digital signal. Hence a digital optical signal is launched into the optical fiber cable. The avalanche photodiode (APD) detector is followed by a front-end amplifier and equalizer or filter to provide gain as well as linear signal processing and noise bandwidth reduction. Finally, the signal obtained is decoded to give the original digital information. The various elements of this and alternative optical fiber system configurations are:

Advantages of optical fiber communication

(a) Enormous potential bandwidth: The optical carrier frequency in the range 10^{13} to 10^{16} *Hz* (generally in the near infrared around 10^{14} *Hz* or 10^{5} GHz) yields a far greater potential transmission bandwidth than metallic cable systems (i.e. coaxial cable bandwidth up to around 500 MHz) or even millimeter wave radio systems (i.e systems currently operating with modulation bandwidths of 700 MHz).

(b) Small size and weight: Optical fibers have very small diameters which are often no greater than the diameter of a human hair. Hence, even when such fibers are covered with protective coatings they are far smaller and much lighter than corresponding copper cables.

(c) Electrical isolation: Optical fibers which are fabricated from glass, or sometimes a plastic polymer, are electrical insulators and therefore, unlike their metallic counterparts, they do not exhibit earth loop and interface problems.

(d) Immunity to interference and crosstalk: Optical fibers form a dielectric waveguide and are therefore free from electromagnetic interference (EMI), radiofrequency interference (RFI), or switching transients giving electromagnetic pulses (EMP). Hence the operation of and optical fiber communication system is unaffected by transmission through an electrically noisy environment and the fiber cable requires no shielding form EMI.

(e) Signal security: The light from optical fibers does not radiate significantly and therefore they provide a high degree of signal security. Unlike the situation with copper cables, a transmitted optical signal cannot be obtained from a fiber in a noninvasive manner (i.e. without drawing optical power from the fiber).

(f) Low transmission loss: The development of optical fibers over the last twenty years has resulted in the production of optical fiber cables which exhibit very low attenuation or transmission loss in comparison with the best copper conductors. Fibers haves been fabricated with losses as low as 0.2 dB km^{-1} (see Section 3.3.2) and this feature has become a major advantage of optical fiber communication.

(g) Ruggedness and flexibility: Although protective catechize are essential, optical fivers nay be manufactured with very high tensile strengths (see Section 4.7). perhaps surprisingly for a glassy substance. The fibers may also be bent to quite small radii or twisted without damage.

(h) System reliability and ease of maintenance: These features primarily stem from the low loss property of optical fiber cables which reduces the requirement for intermediate repeaters or line amplifiers to boost the transmitted signal strength.. Hence with fewer repeaters, system reliability is generally enhanced in comparison with conventional electrical conductor systems.

(i) Potential low cost: The glass which generally provides the fiber transmission medium is made from sand–not a scarce resource. So in comparison with copper conductors, optical fibers offer the potential for low cost line communication.

Ray theory transmission

Total internal reflection

To consider the propagation of light within an optical fiber utilizing the ray theory model it is necessary to take account of the refractive index of the dielectric medium. The refractive index of a medium is defined as the ratio of the velocity of light in a vacuum to the velocity of light in the medium.

It may be observed that total internal reflection occurs at the interface between two dielectrics of differing refractive indices when light is incident on the dielectric of lower index from the dielectric of higher index, and the angle of incidence of the ray exceeds the critical value. This is the mechanism by which light at a sufficiently shallow angle (less than $90°-\phi_C$) may be considered to propagate down an optical fiber with low loss the transmission of a light ray in an optical fiber via a series of total internal reflections at the interface of the silica core and the slightly lower refractive index silica cladding .The ray has an angle of incidence ϕ at the interface which is greater than the critical angle and is reflected at the same angle to the normal.

The light ray passes through the axis of the fiver core is known as a *meridional ray* as it This type of ray is the simplest to describe and is generally used when illustrating the fundamental transmission properties of optical fivers. It must also be noted that the light transmission assumes a prefect fiver, and that any discontinuities or imperfections at the core-cladding interface would probably result in refraction rather than total internal reflection , with the subsequent loss of the light ray into the cladding.

Acceptance Angle

The geometry concerned with launching a light ray into an optical fiber illustrates a meridional ray A at the critical angle ϕ_C within the fiber at the core – cladding interface. It may be observed that this ray enters the fiber core at an angle θ_a to the fiber axis and is reforested at the air – core interface before transmission to the core–cladding interface at an the critical angle. Hence any rays which are incident into the fiber core at angle. greater than θ_a will be transmitted to the core – core cladding interface at an angle less than ϕ_c, and will not be totally internally reflected. If where the incident ray B at an angle greater than θ_a is refracted into the cladding and eventually lost by radiation. Thus for rays to be transmitted by total internal reflection within the fiber core they must be incident on the fiver core within an acceptance cone defined by the conical half angle θ_a. Hence θ_a *is the maximum angle to the axis at which light may enter the fiber in order to be propagated, and is often referred to as the acceptance angle for the fiber.*

Numerical Aperture

It is possible to continue the ray theory analysis to obtain to obtain a relationship between the acceptance angle and the refractive indices of the three media involved , namely the core, cladding and air. This leads to the definition of a more generally use term, the numerical aperture (NA) of the fiber. it must be noted that within this analysis, as with the preceding discussion of acceptance angle , we are concerned with meridional rays within the fiber. A light ray incident on the fiber core at an angle θ_1 to the fiber axis which is less than the acceptance angle for the fiber θ_a. The ray enters the fiber from a medium (air) of refractive index n_0 and the fiber core has a refractive index n_1 which is slightly greater than the cladding refractive index n_2. Assuming the entrance face at the fiber core to be normal to the axis, then considering the refraction at the air-core interface and using Snell' s law given by equation.

$$n_o \sin\theta_1 = n_1 \sin\theta_2 \quad (1)$$

Considering the right –angled triangle ABC indicated in Figure 2.5, then;

$$\phi = \frac{\pi}{2} - \theta_2 \quad (2)$$

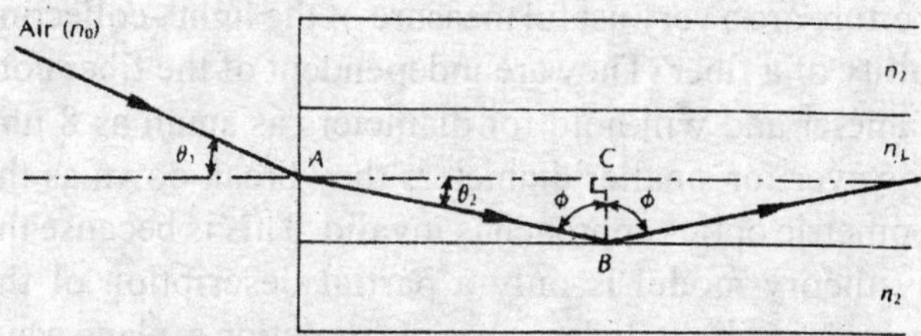

The ray path for a meridional ray launcher into an optical fiber in air at an input angle less than the acceptance angle for the fiber

Where ϕ is greater than the critical angle at the core–cladding interface. Hence Eq(1) becomes

$$n_o \sin\theta_1 = n_1 \cos\phi \quad (3)$$

Using the trigonometrically relationship

$\sin^2\phi + \cos^2\phi = 1, Eq$ (3) may be written in the form

$$n_0 \sin\theta_1 = n_1\left(1 - \sin^2\phi\right)^{\frac{1}{2}}$$

When the limiting case for total internal reflection is considered , ϕ becomes equal to the critical angle for the core–cladding interface. Also in this limiting case θ_1 becomes the acceptance angle for the fiber θ_a Combining these limiting cases into eq. (4) gives:

$$n_0 = n_o \sin\theta_a = \left(n_1^2 - n_2^2\right)^{\frac{1}{2}}$$

Equation (3), apart from relating the acceptance angle to the refractive indices, serves as the basis for the definition of the important optical fiber parameter, the numerical aperture aperture (NA). Hence the NA is defined as:

$$NA = n_0 \sin\theta_a = \left(n_1^2 - n_2^2\right)^{\frac{1}{2}} \quad (6)$$

Since the NA is often used with the fiber in air where n_o is unity, it is simply equal to $\sin\theta_a$. It may also be noted that incident meridional rays over the range $0 \le \theta_1 \le \theta_a$ will be propagated within the fiber. The numerical aperture may also be given in terms of the relative refractive index difference Δ between the core and the cladding which is defined as:

$$\Delta = \frac{n_1^2 - n_2^2}{2n_1^2}$$

$$\cong \frac{n_1 - n_2}{n_1}$$

For $\qquad \Delta << 1 \qquad (7)$

Hence combining Eq.(6) with Eq.(7) we can write:

$$NA = n_1(2\Delta)^{\frac{1}{2}} \quad (8)$$

The relationships given in Eq.(6) and (8) for the numerical aperture are a very useful measure of the light–collecting ability of a fiber .They are independent of the fiber core diameter and will hold for diameters as small as 8 μm. However for smaller diameters they break down as the geometric optics approach is invalid. This is because the ray theory model is only a partial description of the character of light. It describes the direction a plane wave component takes in the fiber but does not take into account interference between such components. When interference phenomena are considered it is found that only rays with certain discrete characteristics propagate in the fiber core. Thus the fiber will only support a discrete number of guided modes. This becomes critical in small core diameter fibers which only support one or a few modes.

Goos–Haenchen Shift

The phase change incurred with the total internal reflection of a light beam on a planar dielectric interface may be understood from physical observation. Careful examination shows that the reflected beam is shifted laterally from the trajectory predicted by simple ray theory analysis, as illustrated in figure below. This lateral displacement is know as the Goos–Haenchen shift,

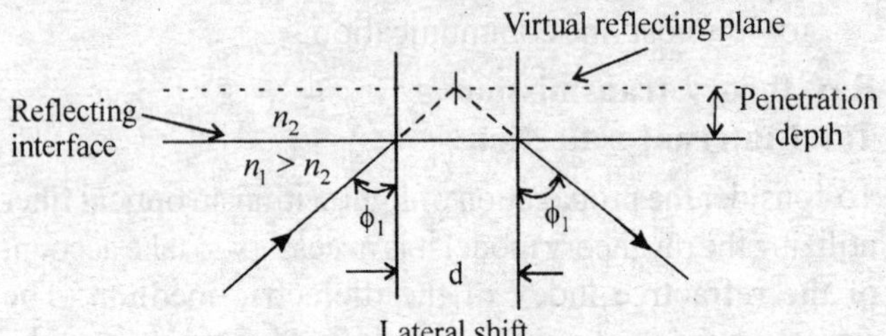

Figure: The lateral displacement of a light beam on reflection at a dielectric interoffice (Goos–Henchen Shift)

The geometric reflection appears to take place at a virtual reflecting plane which is parallel to the dielectric interface in the lower index medium. Utilizing wave theory it is possible to determine this lateral shift.

Step Index Fibers

The optical fiber considered with a core of constant refractive index n_1 and a cladding of a slightly lower refractive index n_2 is know as step index fiber. This is because the refractive index profile fore this type of fiber makes a step change at the core –cladding interface, as indicated in figure below which illustrates the two major types of step index fiber. The refractive index profile may be defined as:

$$n(r) = \begin{cases} n_1 & r < a \quad (\text{core}) \\ n_2 & r \ge a \quad (cladding) \end{cases} \quad (9)$$

in both cases.

Figure (a) shows a multimode step index fiber with a core diameter of around 50 μm or greater, which is large enough to allow the propagation of many modes within the fiber core. This is illustrated in Figure (a) by the many different possible ray paths through the fiber. Figure (b) shows a single mode or monomode step index fiber which allows the propagation of only one transverse electromagnetic mode (typically HE_{11}) ,and hence the core diameter must be of the order of 2 to 10 μm. The propagation of a single mode is illustrated in figure (b) as corresponding to a single ray path only (usually shown as the axial ray) through the fiber.

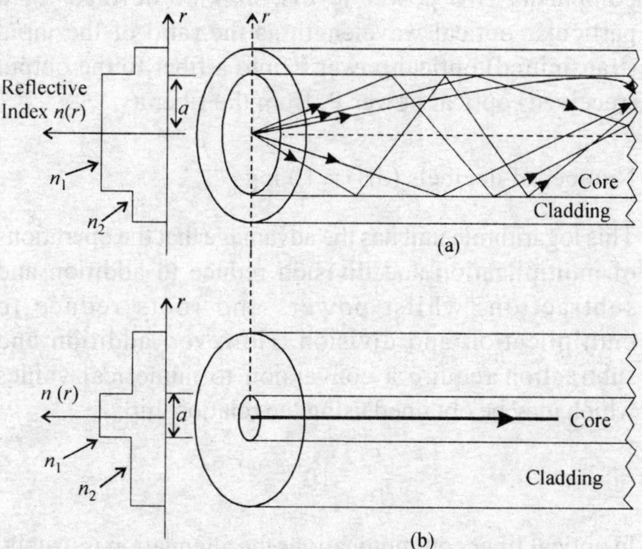

Figure: The refractive index profile and ray transmission in step index fibers: (a) multimode step index finer; (b) single-mode step index fiber

The single-mode step index fiber has the district advantage of low intermodal dispersion (broadening of transmitted light pulses), as only one mode is transmitted, where as with multimode step index finer considerable dispersion may occur due to the differing group velocities of the propagating modes. This in turn restricts the maximum bandwidth attainable with multimode step index figure, especially when compared with single-mode fibers. However, for lower bandwidth applications multimode fibers have sacral advantages over single-mode fibers. These are:

(a) The use of spatially incoherent optical sources (e.g. most light emitting diodes) which cannot be efficiently coupled to single mode fibers;
(b) Larger numerical apertures, as well as core diameters, facilitating easier coupling to optical sources:
(c) Lower tolerance requirements on fiber connectors.

Multimode step index fibers allow the propagation of a finite number of guided modes along the channel. The numbers of guided modes is dependent upon the physical parameters (i.e. relative refractive index deference, core radius) of the fiber and wavelengths of the transmitted light which are included in the normalized frequency V for the fiber.

Graded Index Fibers

Graded index fibers do not have a constant refractive index in the core* but a decreasing core index n(r) with radial distance from a maximum value of n_1 at the axis to a constant value n_2 beyond the core radius a in the cladding. This index variation may be represented as:

$$n(r) = \begin{cases} n_1\left(1-2\Delta(r/a)^{\alpha}\right)^{\frac{1}{2}} & r < a\,(\text{core}) \\ n_1(1-2\Delta)^{\frac{1}{2}} = n_2 & r \le a\,(\text{cladding}) \end{cases} \quad (10)$$

where Δ is the relative refractive index difference and α is the profile parameter which gives the characteristic refractive index profile of the fiber core. Equation (10) which is a convenient method of expressing the refractive index profile of the fiber core as a variation of α allows representation of the step index profile when $\alpha = \infty$ a parabolic profile when $\alpha = 2$ and a triangular profile when $\alpha = 1$. This range of refractive index profile is illustrated in figure below:

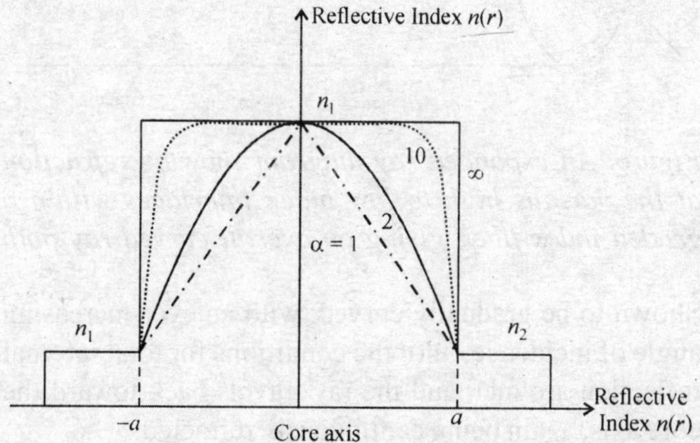

Figure: Possible fiber refractive index profiles for different values of α (given in Eq.(2)

The graded index profiles which at present produce the best results for multimode optical propagation have a near parabolic refractive index profile core with $\alpha = 2$. Figure with such core index profiles are well established and consequently when the term 'graded index' is used without qualification it usually refers to a fiber with this profile. For this reason in this section we consider the wave guiding properties of graded index fiber with a parabolic refractive index profile core.

A multimode graded index fiber with a parabolic profile core is illustrated in figure below It may be observed that the meridional rays shown appear to follow curved paths through the fiber core. Using the concepts of geometric optics, the gradual decrease in refractive index from the centre of the core creates much refraction of the rays as they are effectively incident on a large number of high to low index interfaces. This mechanism is illustrated in figure below where a ray is:

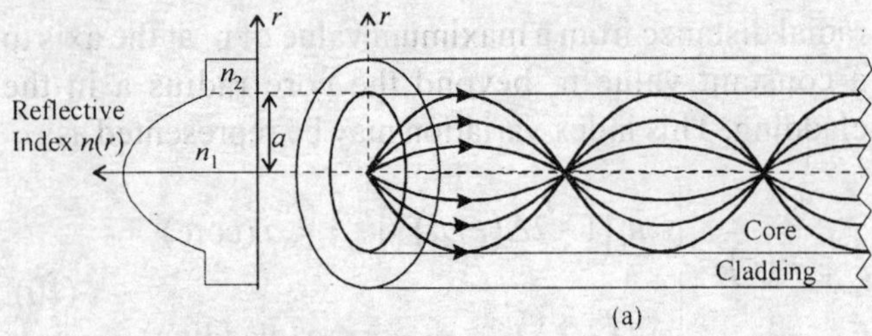

Figure: The refractive index profile and ray transmission in a multimode graded index fiber

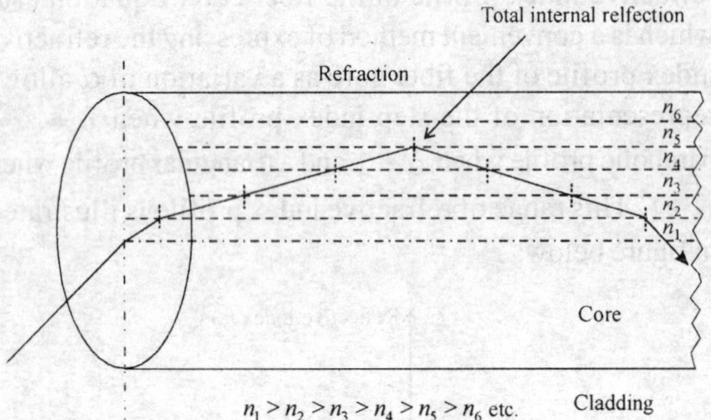

Figure: An expanded ray diagram showing refraction at the various high to low index interfaces within a graded index fiber, giving an overall curved ray path

Shown to be gradually curved, with an ever-increasing angle of incidence, until the conditions for total internal reflection are met, and the ray travels back toward the core axis, again being continuously refracted.

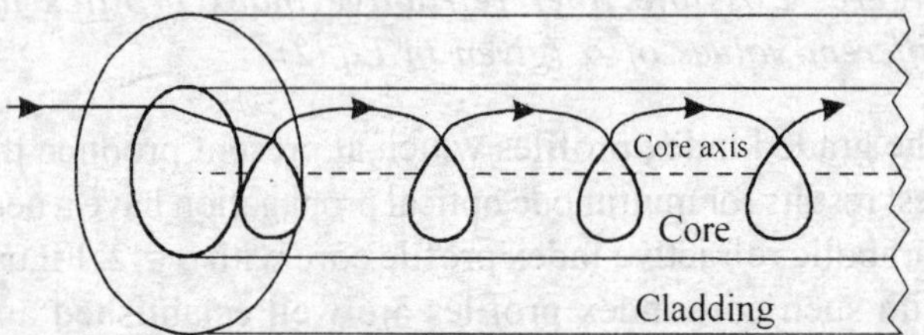

Figure: A helical skew ray path within a graded index fiber

Transmission Characteristic

Attenuation

The attenuation or transmission loss of optical fibers has proved to be one of the most important factors in bringing about their wide acceptance in telecommunications. As channel attenuation largely determined the maximum transmission distance prior to single restoration, optical fiber communications become especially attractive when the transmission losses of fibers were reduced below those of the competing metallic conductors (less than 5 dB km^1). Signal attenuation within optical fibers, as with metallic conductors, is usually expressed in the logarithmic unit of the decibel. The decibel, which is used for comparing two power levels, may be defined for a particular optical wavelength as the ratio of the input (transmitted) optical power P_i into a fiber to the output (received) optical power P_0 from the fiber as:

$$\text{Number of decibels (dB)} = 10 \log_{10} \frac{P_i}{P_o}$$

This logarithmic unit has the advantage that the operations of multiplication and division reduce to addition and subtraction, whilst powers and roots reduce to multiplication and division. However, addition and subtraction require a conversion to numerical values which may be obtained using the relationship:

$$\frac{P_i}{P_o} = 10^{(dB/10)}$$

In optical fiber communications the attenuation is usually expressed in decibels per unit length (i.e. dB km^{-1}) following:

$$\alpha_{dB} L = 10 \log_{10} \frac{P_i}{P_o}$$

Where α_{dB} the signal attenuation per unit length in decibels and L is is the fiber length

Material absorption losses in silica glass fibers

Material absorption is a loss mechanism related to the material composition and the fabrication process for the fiber, which results in the dissipation of some of the transmitted optical power as heat in the waveguide. The absorption of the light may be intrinsic (caused by the interaction with one or more of the major components of the glass) or extrinsic (caused by impurities within the glass).

Linear Scattering Losses

Linear scattering mechanism cause the transfer of some or all of the optical power contained within one propagating mode to be transferred linearly (proportionally to the mode power) into a different mode. This process tends to result in attenuation of the transmitted light as the transfer may be to a leaky or radiation mode which does not continue to propagate within the fiber core, but is radiated from the fiber. It must be noted that as with all linear processes there is no change of frequency on scattering. Linear scattering may be categorized into two major types: Rayleigh and Mie scattering. Both result from the no ideal physical properties of the manufactured fiber which are' difficult and, in certain cases, impossible to eradicate at present.

Rayleigh Scattering

Rayleigh scattering is the dominant intrinsic loss mechanism in the low absorption window between the ultraviolet and infrared absorption tails. It results from inhomogeneities of random nature occurring on a small scale compared with the wavelength of the light. These inhomogeneities manifest theme selves as refractive index fluctuations and arise from density and compositional variations which are frozen into the glass lattice on cooling .The compositional variations may be reduced by improved fabrication but the index fluctuations caused by the freezing in of density in homogeneities are fundamental and cannot be avoided. The subsequent scattering due to the density fluctuations which is in almost all directions produces attenuations to $1/\lambda^4$ following the Raleigh scattering formula. For a single component glass this is given by:

$$\gamma_R = \frac{8\pi^3}{3\lambda^4} n^8 p^2 \beta_c K T_F$$

where, $\gamma_R \rightarrow$ is the Raleigh scattering coefficient

$\lambda \rightarrow$ the optical wavelength

n $\rightarrow$ the refractive index of the medium

p $\rightarrow$ the average photo elastic coefficient coefficient

$\beta_c \rightarrow$ the isothermal compressibility at a fictive temperature

T_F and $K \rightarrow$ Bolt Mann's constant.

The fictive temperature is defined as the temperature at which the glass can reach a state of thermal equilibrium and is closely related to the anneal temperature. Furthermore, the Raleigh scattering coefficient is related to the transmission loss factor (transmissivety) of the fiber following the relation.

$$P = e^{-\gamma_v L}$$

where L is the length of the fiber. It is apparent from Eq. that the fundamental component of Rayleigh scattering is strongly reduced by operating at the longest possible wavelength.

Mie Scattering

Linear scattering may also occur at in homogeneity which are comparable in size to the guided wavelength. These result from the no perfect cylindrical structure of the waveguide and may be caused by fiber imperfections such as irregularities in the core–cladding interface, core –cladding refractive index differences along the fiber length, diameter fluctuations, strains and bubbles. When the scattering in homogeneity size is greater than $\lambda/10$, the scattered intensity which has an angular dependence can be very large.

The scattering created by such in homogeneities is mainly in forward direction and is called Mie scattering. Depending upon the fiber material, design and manufacture, Mie scattering can cause significant losses. The in homogeneities may be reduced by:

1. Removing imperfections due to the glass manufacturing process;
2. Carefully controlled extrusion and coating of the fiber;
3. Increasing the fiber guidance by increasing the relative refractive index difference.

By these means it is possible to reduce Mie scattering to insignificant levels.

Nonlinear Scattering Losses

Optical waveguides do not always behave as completely linear channels whose increase in output optical power is directly proportional to the input optical power. Several nonlinear effects occur, which in the case of scattering cause disproportionate attenuation, usually at high optical power levels. This nonlinear scattering causes the optical power from one mode to be transferred in either the forward or backward direction to the same, or other modes, at a different frequency. It depends critically upon the optical power density within the fiber and hence only becomes significant above threshold power levels.

The most important types of nonlinear scattering within optical fibers are stimulated Brillouin and Ramam scattering both of which are usually only observed at high optical power densities in long single-mode fibers. These scattering mechanisms in fact give optical gain but with a shift in frequency, thus contributing to attenuation for light transmission at a specific wavelength. However ,it may be noted that such nonlinear phenomena can also be used to give optical amplification in the context of integrated optical techniques.

Stimulated Brillouin Scattering

Stimulated Brillouin Scattering (SBS) may be regarded as the modulation of light through thermal molecular vibraitions within the fiber. The scattered light appears as upper and lower sidebands which are separated from the incident light by the modulation frequency .The incident photon in this scattering process produces a phonon of acoustic frequency as well as a scattered photon. This produces an optical frequency shift which varies with the /scattering angle because the frequency of the sound wave varies with acoustic wavelength. The frequency shift is a maximum in the backward direction reducing to zero in the forward direction making SBS a mainly backward process.

Brillouin scattering is only significant above a threshold power density. Assuming that the polarization state of the transmitted light is not maintained threshold power P_B is given by:

$$P_B = 4.4\times10^{-3} d^2\lambda^2\alpha_{dB} V \text{ watt}$$

where

d → the fiber core diameter and the operating wavelength

λ → respectively, both measured in micrometers

α_{dB} → the fiber attenuation in decibels per kilometers

v → is the source bandwidth (i.e. injection laser) in gigahertz.

The expression given in equation allows the determination of the threshold optical power which mist be launched into a single-mode optical fiber before SBS occurs.

Stimulated Rama Scattering

Stimulated Raman Scattering (SRS) is similar to stimulated Brillouin scattering except that a high frequency optical phonon rather than an acoustic phonon is generated in the scattering process. Also SRS can occur in both the forward and backward directions in an optical fiber and may have an optical power threshold of up to three orders of magnitude higher than the Brillouin threshold in a particular fiber.

Using the same criteria as those specified for the Brillouin scattering threshold the threshold optical power for SRS P_R in a long single mode fiber is given by:

$$P_R = 5.9\times10^{-2} d^2\lambda\alpha_{dB} \quad watts$$

Fiber Bend Loss

Optical fibers suffer radiation losses at bends or curves on their paths. This is due to the energy in the evanescent field at the bend exceeding velocity the velocity of light in the cladding and hence from the fiber. An illustration of this situation is shown in figure below. The part of the mode which is on the outside of the bend is required to travel faster than that on the inside so that a wave front perpendicular to the direction of propagation is maintained. Hence ,part of the mode in the cladding needs to travel faster than the velocity of light in that medium. As this is not possible ,the energy associated with this part of mode is lost through radiation. The loss can generally be represented by a radiation attenuation coefficient which has the from:

$$\alpha_r = C_1 e^{(-C_2 R)}$$

where R is the radius of curvature of the fiber bend c_1, c_2 are constants which are independent of R.

Furthermore large bending losses tend to occur in multimode fibers at a critical radius or curvature R_c which may estimated from:

$$R_C = \frac{3_1^2\lambda}{4\pi\left(n_1^2 - n_2^2\right)^{\frac{1}{2}}}$$

It may be observed from the expression given in eq. that potential macro bending losses may be reduced by:

(a) Designing fibers with large relative refractive index differences;

(b) Operating at the shortest wavelength possible.

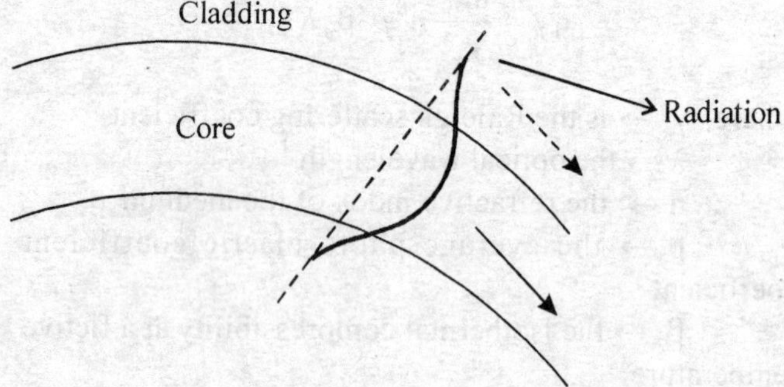

Figure: An illustration of the radiation loss at a fiber bend. The part of the mode in the cladding outside the dashed arrowed line may be required to travel faster than the velocity of light in order to maintain a plane wave front. Since it cannot do this, the energy contained in this part of the mode is radiated away

The above criteria for the reduction of bend losses also apply to single mode fibers One theory [Ref. 21] based on the concept of a single quasi-guided mode provides an expression from which the critical radius of curvature for a single-mode fiber R_{CS} can be estimated as:

$$R_{CS} = \frac{20\lambda}{(n_1 - n_2)^{\frac{1}{2}}}\left(2.748 - 0.996\frac{\lambda}{\lambda_c}\right)^{-3}$$

where λ_c is the cutoff wavelength for the single mode fiber .Hence again, for a specific single mode fiber (that is, a fixed relative index difference and cutoff wavelength), the critical wavelength of the radiated light becomes progressively shorter as the bend radius is decreased.

Polarization

Cylindrical optical fibers do not generally maintain the polarization state of the light input for more than a few meters, and hence for many applications involving optical fiber transmission some from of intensity modulations of the optical source is utilized. The optical signal is thus detected by a photodiode which is insensitive to optical polarizations have been investigated.

9.3 BIOMEDICAL INSTRUMENTATION

EEG, ECG and EMG. Clinical measurements. Ultrasonic transducers and Ultrasonography. Principles of Computer Assisted Tomography.

9.3.1 ELECTROCARDIOGRAPH (ECG)

The electrocardiograph (ECG) is an instrument, which records the electrical activity of the heart. Electrical signals from the heart characteristically precede the normal mechanical function and monitoring of these signals has great clinical significance. ECG provides valuable information about a wide range of cardiac disorders such as the presence of an inactive part (infarction) or an enlargement (cardiac hypertrophy) of the heart muscle. Electrocardiographs are used in catheterization laboratories, coronary care units and for routine diagnostic applications in cardiology.

Although the electric field generated by the heart can be best characterized by vector quantities, it is generally convenient to directly measure only scalar quantities, i.,. voltage difference of mV order between the given points of the body. The diagnostically useful frequency range is usually accepted as mV mV to $150\ Hz$ (Golden *al* 1973). The amplifier and writing part should faithfully reproduce signals in this range. A good low frequency response is essential to ensure stability of the baseline. High frequency response is a compromise of several factors like isolation between a useful ECG signal from other signals of biological origin (myographic potentials) and limitations of the direct writing pen recorders due to mass, inertia and friction. The interference of non- biological origin can be handled by using modern differential amplifiers, which are capable of providing excellent rejection capabilities CMRR of the order if $100-120\ dB$ with $5\ k\Omega$ unbalance in the leads is a desirable feature of ECG machines. In addition to this, under specially adverse circumstances, it becomes necessary to include a notch filter tuned to $50\ Hz$ reject hum due to power mains. The instability of the baseline, originating from the changes of the contact impedance, demands the application of the automatic baseline stabilizing circuit. A minimum of two paper speeds in necessary (25 and $50\ mm$ per sec) for ECG recording.

Block Diagram Description of an Electrocardiograph

Figure 5.1 shows the block diagram of an electrocardiograph machine. The potentials picked up by the patient electrodes are taken to the lead selector switch. In the lead selector, the electrodes are selected two by two according to the lead program. By means of capacitive coupling, the signal is connected symmetrically to the long-tail pair differential preamplifier. The preamplifier is usually a three or four stage differential amplifier having a sufficiently large negative current feedback, from the end stage to the first stage, which gives a stabilizing effect. The amplified output signal is picked up single-ended and is given to the power amplifier. The power amplifier is generally of the push-pull differentical type. The base of one input transistor of this amplifier is driven by the preamplified unsymmetrical signal. The base of the other transistor is driven by the feedback signal resulting from the pen position and connected via frequency selective network. The output of the power amplifier is single-ended and is fed to the pen motor, which deflects the writing arm on the paper. A direct writing recorder is usually adequate since the ECG signal of interest has limited bandwidth. Frequency selective network is an $R-C$ network, which provides necessary damping of the pen motor and is preset by the manufacturer. The auxiliary circuits provide a $1\ mV$ calibration signal and automatic blocking of the amplifier during a change in the position of the lead switch. It may include a speed control circuit for the chart drive motor.

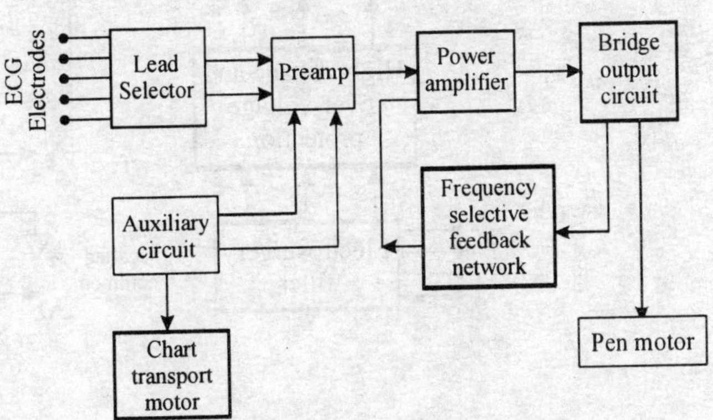

Figure: Block diagram of an ECG machine

A, stand by' mode of operation is generally provided on the electrocardiograph. In this mode, the stylus moves in response to input signals, but the paper is stationary. This mode allows the operator to adjust the gain and baseline position controls without wasting paper.

Electrocardiograms are almost invariably recorded on graph paper with horizontal and vertical lines at 1 mm intervals with a thicker line at 5 mm intervals. Time measurements and heart rate measurements are made horizontally on their electrocardiogram. For routine work, the paper recording speed is 25 mm/s. Amplitude measurements are made vertically in millivolts. The sensitivity of an electrocardiograph is typically set at 10 mm/mV.

Isolated Preamplifier: It has been traditional for all electrocardiographs to have the right leg (RL) electrode connected to the chassis, and from there to the ground. This provided a ready path for any ground seeking current through the patient and presented an electrical hazard. As the microshock hazard became better understood, particularly when intracardiac catheters are employed, the necessity of isolating the patient from the ground was stressed. The American Heart Association guidelines state that the leakage current should not be greater than 10 microamperes when measured from the patient's leads to the ground or through the main instrument grounding wire with the ground open or intact. For this, patient leads would have to be isolated from the ground for all line operated units.

Figure shows a block diagram of an isolation preamplifier used in modern electrocardiographs. Difference signals obtained from the right arm (RA), left arm (LA) and right leg (RL) is given to a low-pass filter. Filtering is required on the input leads to reduce interference caused by electrosurgery and radio frequency emissions and sometimes from the 50 kHz current used for respiration detection. The filter usually has a cut off frequency higher then 10 kHz multistage filter in needed to achieve a suitable reduction in high frequency signal.

The filter circuit is followed by high voltage and over voltage protection circuits so that the amplifier can withstand large voltages during defibrillation. However, the price of this protection is a relatively high amplifier noise level arising from the high series resistance in each lead.

The lead selector switch is used to derive the requite lead configurations and give it to a dc-coupled amplifier. A dc level of 1 mV is obtained by dividing down the power supply, which can be given to this amplifier through a push button for calibration of the amplifier. Isolation of the patient circuit is obtained using a low capacitance transformer whose primary winding is driven from a 100 kHz oscillator. The transformer recondary is used

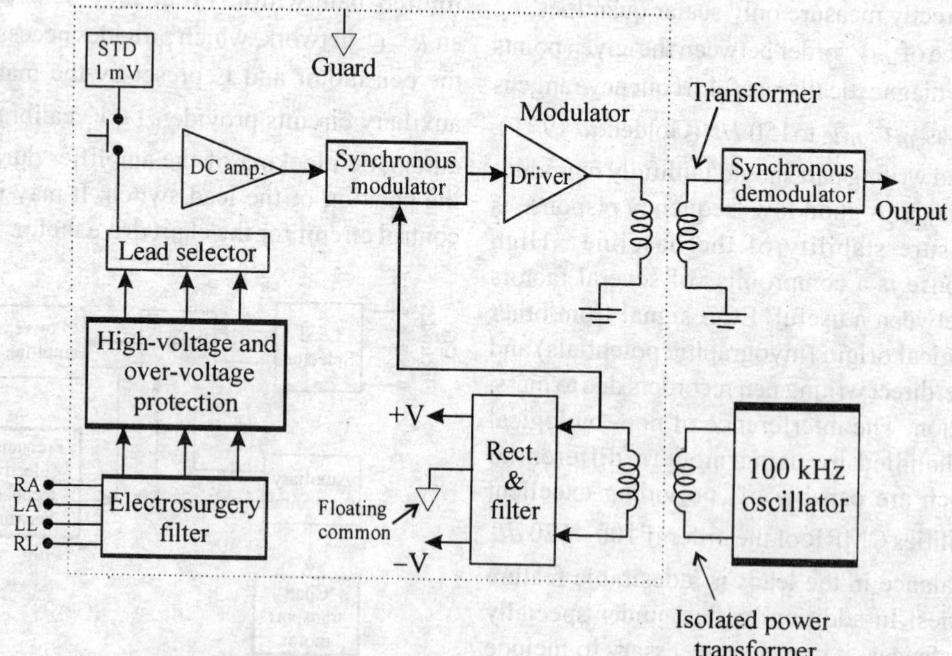

Figure: Block diagram of an isolation pre-amplifier (transformed-coupled) commonly used in modern ECG machines

to obtain an isolated power supply of $\pm 6\ V$ for operating the devices in the isolated portion of the circuit and to drive the synchronous modulator at $100\ kHz$, which linearly modulates an ECG signal given to it. The oscillator frequency of $100\ kHz$ is chosen as a compromise so that reasonable size transformers (higher the frequency the smaller the transformer) could be used and that the switching time is not too fast, so that inexpensive transistors and logic circuity can be utilized. A square wave is utilized to minimize the power requirements of the driven transistors. A synchronous demodulator is chosen to give low noise performance utilizing switching FET's.

Isolation of the patient preamplifier can also be obtained using an optical isolator. The high common-mode rejection of the amplifier is obtained by proper shielding. The effective capacitance from the input leads to the earth is made negligible. The preamplifier circuitry should is preferably be shielded in a separate case.

To minimize the common-mode signal between the body of the patient and the gloating ground, a right leg drive circuit is used. The common-mode signals after amplification in a preamplifier are inverted and fed back to the right leg electrode, reducing the common mode voltage on the input with respect to the floating ground. Winter and Webster (1983) examined optimal design parameters for a driven-right-leg circuit.

The presence of stray capacitance at the input of the preamplifier causes common-mode currents to flow in LA and RA, resulting in a voltage drop at the electrode resistors. An imbalance of the stray capacitance or the electrode resistors causes a difference signal. This difference signal can be almost eliminated, in that the common-mode currents of stray capacitances are not allowed to flow through the electrode resistors but are neutralized by currents delivered to stray capacitances from the common-mode rejection amplifier. In other words, the potentials at A, B and C are equalised through an in-phase component of the common-mode voltage, which the amplifier delivers via C_1 and `C_2 ti LA and RA. As a result, the potentials at A, B and C are kept equal, in dependent of the imbalance in the electrode resistors and stray capacitance.

The ECG Leads

Bipolar Leads: In bipolar leads, ECG os recorded by using two electrodes such that the final trace corresponds to the difference of electrical potentials existing between them. They are called standard leads and have been universally adopted. They are sometimes also referred to as Einthoven leads (Figure. 5.4 (a))

In standard lead I, the electrodes are placed on the right and the left arm (RA and LA). In lead II, the electrodes are placed on the right arm and the left leg and in lead III,

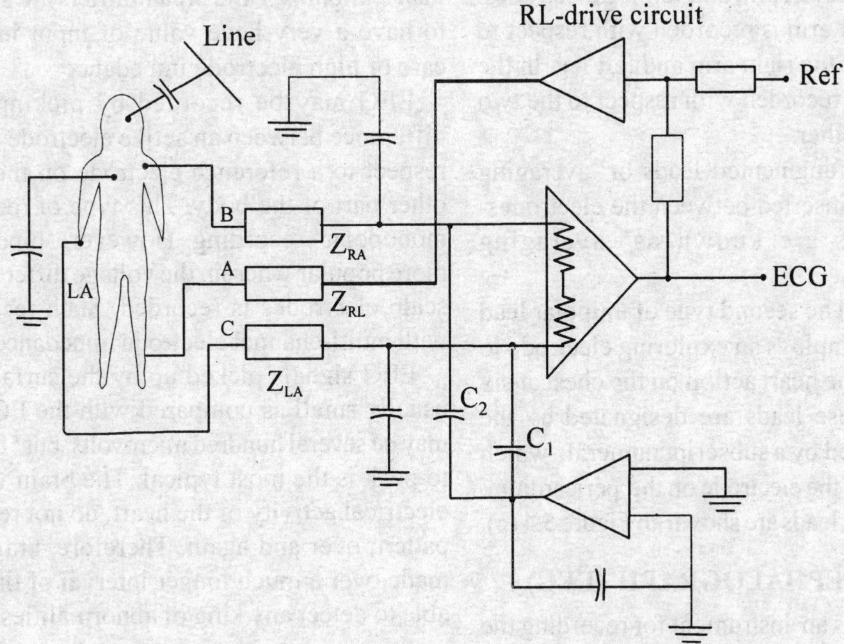

Figure: Improvement in CMRR using right leg drive (Courtesy: Hewlett Packard, USA)

they are placed on the left arm and the left leg. In all lead connections, the difference of potential measured between two electrodes is always with reference to a third point on the body. This reference point is conventional taken as the "right leg". The records are, therefore, made by using three electrodes at a time, the right leg connection being always present.

Unipolar Leads (V Leads): The standard leads record the difference in electrical potential between two points on the body produced by the heart's action. Quite often, this voltage will show smaller changes than either of the potentials and so better sensitivity an be obtained if the potential of a single electrode is recorded. Moreover, if the electrode is placed on the chest close to the heart, higher potentials can be detected than normally available at the limbs. This lead to the development of unipolar leads introduced by Wilson in 1894. In this arrangement, the electrooculogram is recorded between a single exploratory electrode and the central terminal, which has a potential corespondent to the centre of the body. In practice, the reference electrode or central terminal is obtained by a combination of several electrodes tied together at one point. Two types of unipolar leads are employed which are: (i) limb leads, and (ii) precordial leads.

(1) *Limb leads* In unipolar limb leads (Figure. 5.4(b)), two of the limb leads are tied together and recorded with respect to the third limb. In the lead identified as AVR, the right arm is recorded with respect to a reference established by joining the left arm and left leg electrodes. In the AVL lead, the left arm is recorded with respect to the common junction of the right arm and left leg. In the AVF lead, the left leg is recorded with respect to the two arm electrodes ties together.

They are also called augmented leads or' averaging leads'. The resistances inserted between the electrodes-machine connections are known as' averaging resistances'.

(2) *Precordial leads* The second type of unipolar lead is a precordial lead. It employs an exploring electrode to record the potential of the heart action on the chest at six different positions. These leads are designated by the capital letter 'V' followed by a subscript numeral, which represents the position of the electrode on the pericardium. The positions of the chest leads are shown in Figure 5.4 (c).

9.3.2 ELECTROENCEPHALOGRAPH (EEG)

Electroencephalograph is an instrument for recording the electrical activity of the brain, by suitably placing surface electrodes on the scalp. EEG, describing the general function of the brain activity, is the superimposed wave of neuron potentials operating in a non-synchronized manner in the physical sense. Its stochastic nature originates just from this, and the prominent signal groups can be empirically connected to diagnostic conclusions.

Monitoring the electroencephalogram has proven to be an effective method of diagnosing many neurological illnesses and diseases, such as epilepsy, tumour, cerebrovascular lesions, ischemia anaesthetize and to establish the integrity of the anaesthetized patient's nervous system. This has become possible with the advent of small, computer-based EEG analyzers. Consequently, routine EEG monitoring in the operating room and intensive care units is becoming popular.

Several types of electrodes may be used to record EEG. These include: Peel and Stick electrodes, Silver plated cup electrodes and Needle electrodes.

EEG electrodes are smaller in size than ECG electrodes. They may be applied separately to the scalp or may be mounted in special bands, which can be placed on the patient's head. In either case, electrode jelly or paste is used to improve the electrical contact. It the electrodes are intended to be used under the skin of the scalp, needle electrodes are used. They offer the advantage of reducing movement artefact. EEG electrodes give high skin contact impedance as compared to ECG electrodes. Good electrode impedance should be generally below 5 kilohms. Impedance between a pair of electrodes must also be balanced or the difference between them should be less than 2 kilohms. EEG preamplifiers are generally designed to have a very have value of input impedance to take care of high electrode impedance.

EEG may be recorded by picking up the voltage difference between an active electrode on the scalp with respect to a reference electrode on the ear lode or any other part of the body. This type of recording is called' monopolar' recording. However, 'bipolar' recording is more popular wherein the voltage difference between two scalp electrodes is recorded. Such recordings are dons with multi-channel electrode impedance.

EEG signals picked up by the surface electrodes are usually small as compared with the ECG signals. They may be several hundred microvolts, but 50 microvolts peak-to-peak is the most typical. The brain waves, unlike the electrical activity of the heart, do not represent the same pattern over and again. Therefore, brain recordings are made over a much longer interval of time in order to be able to detect any king of abnormalities.

Block Diagram Description of Electroencephalograph

The basic block diagram of an EEG machine with both analog and digital components is shown in Figure. 5.11.

Montages: A pattern of electrodes on the head and the channels they are connected to is called a montage. Montages are always symmetrical. The reference electrode is generally placed on a non-active site such as the forehead or earlobe. EEG electrodes are arranged on the scalp according to a standard known as the 10/20 system, adopted by the American EEG Society (Barlow et, *al* 1974) Traditionally, there are 21 electrode locations in the 10/20 system.

Electrode Montage Selector: EEG signals are tensimeter from the electrodes to the head box, which is ladled according to the 10–20 system, and then to the montage selector. The montage selector on analog EEG machine is a large panel containing switches that allow the user to select which electoral pair will have signals subtracted from each other to create an array of channels of output called a montage. Each channel is created in the from of the input from one electrode minus the input from a second electrode.

Montages are either bipolar (made by the subtraction of signals from adjacent electrode pairs) or referential (made by subtracting the potential of a common reference electrode from each electrode on the head). In order to minimize noise, a separate reverence is often chosen for each side of the head e.g. the ipsilateral are Bipolar and referentially montages contain the same basic information that is transformable into either format by simple substraction as long as all the electrodes, including reference, are included in both montages and linked to one common reference.

Preamplifier: Every channel has an individual, multistage, ac coupled, very sensitive amplifier with differential input and adjustable gain in a wide range. Its frequency response can be selected by single-stage passive filters. A calibrating signal is used for controlling and documenting the sensitivity of the amplifier channels. This supplies a voltage step of adequate amplitude to the input of the channels. A typical value of the calibration signal is $50\, uV/cm$.

The preamplifier used in electroencephalograph must have high gain and low noise characteristics because the EEG potentials are small in amplitude. In addition, the amplifier must have very high common-mode rejection to minimise stray interference signals from power lines and other electrical equipment. The amplifier must be free from drift so as to prevent the slow movement of the recording pen from its centre position as a result of changes in temperature, etc.

EEG amplifiers must have high gain in the presence of unbalanced source resistances and dc skin potentials at least up to $100\, mV$. Noise performance is crucial in EEG work because skin electrodes couple brain waves of only

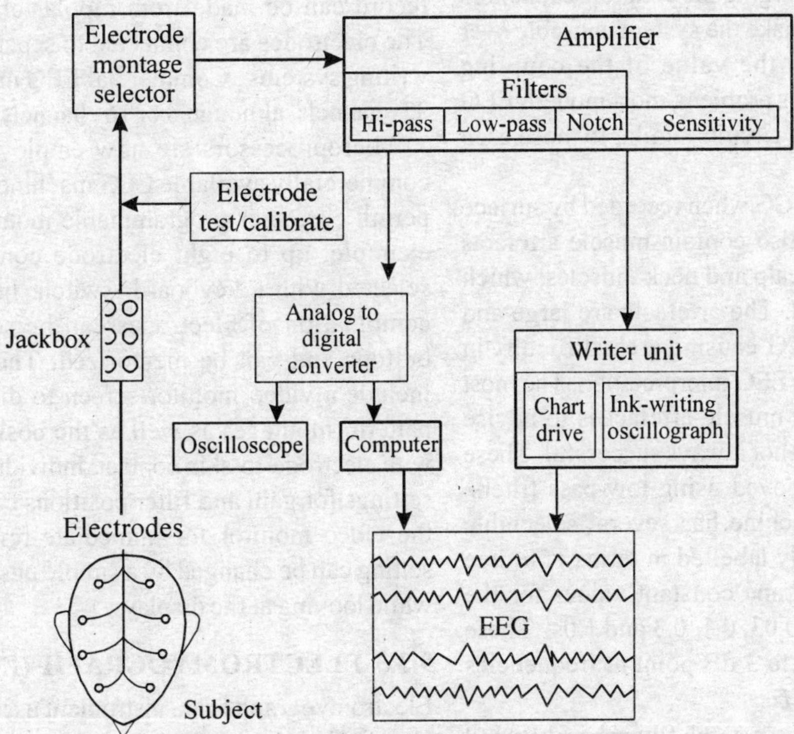

Figure: Schematric diagram of an EEG machine (after Isley et a1, 1998)

a few microvolts to the amplifier. Each individual EEG signal should be preferably amplified at the bedside. Therefore, a specially designed connector box, which can be mounted near the patient, is generally with EEG machines. This ensures the avoidance of cable or switching artefacts. The use of electrode amplifiers at the site also eliminates undesirable cross-talk effects of the individual electrode potentials. The connector box also carries arrangement for measuring the skin contact impedance of electrodes with ac. Then, poor electrode-to-skin contacts above a predetermined level can be easily spotted out.

Sensitivity Control: The overall sensitivity of an EEG machine is the gain of the amplifier multiplied by the sensitivity of the writer. Thus, if the writer sensitivity is $1\, cm/V$, the amplifier must have an overall gain of $20{,}000$ for a $50\, \mu V$ signal. The various stages are capacitor coupled. An EEG machine has two types of gain controls. One is continuously variable and it is used to equalize the sensitivities of all channels. The other control operates in steps and is meant to increase or reduce the sensitivity of a channel by known amounts. This control is usually calibrated in decibels. The gain of amplifiers is normally set so that signals of about $200\, \mu V$ deflect the pens over their full linear range. Artefacts, several times greater than this, can cause excessive deflections of the pen by charging the coupling capacitors to large voltage. This will make the system unusable over a period depending upon the value of the coupling capacitors. To overcome this problem, most modern EEG machines have be-blocking circuits similar to those used in ECG machines.

Filters: Just like in an ECG when recorded by surface electrodes, an EEG may also contain muscle artefacts due to contraction of the scalp and neck muscles, which overlie the brain and skull. The artefacts are large and sharp, in contrast to the ECG, causing great difficulty in both clinical and automated EEG interpretation. The most effective way to eliminate muscle artefact is to advise the subject to relax, but it is not always successful. These artefacts are generally removed using low-pass filters. This filter on an EEG machine has several selectable positions, which are usually labelled in terms of a time constant. A typical set of time constant values for the low-frequency control are 0.03, 0.1, 0.3 and 1.0s. These time constants correspond to 3 dB point at frequencies of 5.3, 1.6, 0.53 and $0.16\, Hz$.

Some EEG machines have a notch filter sharply tuned at $50\, Hz$ so as to eliminate mains frequency interference. These however have the undesirable property of 'ringing' i.e. they produce a damped oscillatory response to a square wave calibration waveform or a muscle potential. The use of notch filters should preferably be restricted to exceptional circumstances when all other methods of eliminating interference have been found to be ineffective.

The high frequency response of an EEG machine will be the resultant of the response of the amplifier and the writing part. However, the figure mentioned on the high frequency filter control of most EEG machines generally refers to the amplifier. The typical frequency range of standard EEG machines is from $0.1\, Hz$ to $70\, Hz$, though newer machines allow the detection and filtering of frequencies up to several hundred Hertz. This may be of importance in some intracranial recordings.

Noise: EEG amplifiers are selected for minimum level, which is expressed in terms of an equivalent input voltage. Two microvolts is often stated as the acceptable figure for EEG recording. Noise contains components at all frequencies and because of this, the recorded noise increases with the band width of the system. It is therefore important to restrict the bandwidth to that required for faithful reproduction of the signal. Noise level should be specified as peak-to-peak value as it is seen on the record rather than rms value, which could be misleading.

Channels: An electroencephalogram is recorded simultaneously from an array of many electrodes. The record can be made from bipolar or monopolar leads. The electrodes are connected to separate amplifiers and writing systems, Commercial EEG machines have up to 32 channels, although 8 or 16 channels are more common.

Microprocessors are now employed in most of the commercially available EEG machines. These machines permit customer programmable montage selection; for example, up to eight electrode combinations can be selected with a keyboard switch. In fact, any desired combination of electrodes can be selected with push buttons and can be memorized. These machines also include a video monitor screen to display the selected pattern (montage) as well as the position of scalp sites with electrode-to-skin contact. Individual channel control settings for gain and filter positions can be displayed on the video monitor for immediate review. Therefore, a setting can be changed by a simple push button operation while looking at the display.

9.3.3 ELECTROMYOGRAPH (EMG)

Electromyograph is an instrument used for recording the electrical activity of the muscles to determine whether the muscle is contracting or not; or for displaying on the

CRO and loudspeaker the action potential spontaneously present in a muscle or those induced by voluntary contractions as a means of detecting the nature and location of motor unit lesions; or for recording the electrical activity evoked in a muscle by the stimulation of its nerve. The instrument is useful for making a study of several aspects of neuromuscular function, neuromuscular condition, extent of nerve lesion, reflex responses, etc.

EMG measurements are also important for the myoelectric control of prosthetic devices (artificial limbs). This use involves picking up EMG signals from the muscles at the terminated nerve endings of the remaining limb and using the signals to activate a mechanical arm. This is the most demanding requirement from an EMG since on it depends the working of the prosthetic device.

EMG is usually recorded by using surface electrodes or more often by using needle electrodes, which are inserted directly into the muscle. The surface electrodes may be disposable, adhesive types or the ones which can be used repeatedly. A ground electrode is necessary for providing a common reference for measurement. these electrodes pick up the potentials produced by the contracting muscle fibres. The signal can then be amplified and displayed on the screen of a cathode ray tube. It is also applied to an amplifier connected to a loudspeaker. A trained EMG interpreter can diagnose various muscular disorders by listening to the sounds produced when the muscles potentials are fed to the loudspeaker. The block diagram (Figure. 5.15) shown a typical set-up for EMG recordings. The oscilloscope displays EMG waveforms. The tape recorder is included in the system to facilitate playback and study of the WMG sound waveforms at a later convenient time. The waveform can also be photographed from the CRT screen by using a synchronized camera.

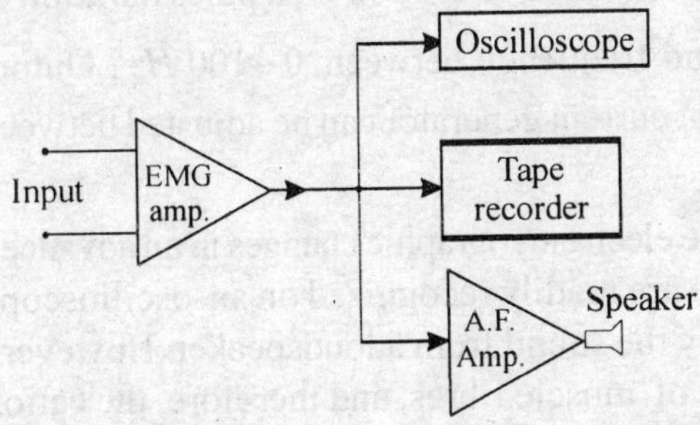

Figure: Block diagram of a typical set-up for EMG recording

The amplitude of the EMG signals depends upon various factors, e.g. the type and placement of electrodes used and the degree of muscular exertions. The needle electrode in contact with a single muscle fibre will pick up spike type voltages whereas a surface electrode picks up many overlapping spikes and therefore produces an average voltage effect. A typical EMG signal ranges from 0.1 to $0.5\, mV$. The may contain frequency components extending up to $10\, kHz$. Such high frequency signals cannot be recorded on the conventional pen recorders and therefore, they are usually displayed on the CRT screen.

Preamplifier: The preamplifiers used for EMG are generally of differential type with a good bandwidth. The input impedance of the amplifier must be greater than $2 \times 50\ M\Omega$. Present day electronic devices easily provide input impedances of the order of 10^{12} ohms in parallel with 5 picofarads. It is preferable to mount the preamplifiers very near the subject using very small electrode leads, in order to avoid the undesirable effects of stray capacitance between connecting cables and the earth. Also, any movement of the cable from the output of the electrode will not generate significant noise signals in the cable, which feeds into the subsequent amplifier. The preamplifier provides an output with low impedance and, therefore, the high frequencies do not get attenuated even if long cables are used to connect the preamplifier and the rest of the machine. The common-mode rejection should be greater than $90\, dB$ up to $5\, kHz$. A calibrating square wave signal of $100\, \mu V$ (peak-to-peak) at a frequency of $100\, Hz$ is usually available. The main amplifier has control for gain adjustment from $5\, mV$ / div to $10 mV$ / div for selecting the sensitivity most appropriate to the incoming signal from the patient.

The two ICs in the input stage act as village followers, which present the desired high input impedance to the electrodes. They are coupled via C_1 and R_5 to provide a high differential signal gain. Capacitor C_1 determines the low frequency performance of the circuit. It also eliminates the effects, st the output, of any dc offsets due to IC_1 and IC_2 or any ombalance in electrode potentials. The second stage IC_3 provides further differential signal gain, while rejecting common-mode signals. The overall gain of the amplifier is 1000.

Input impedance of the amplifier must be higher by several orders than the electrode impedance. Also selections of the electrode type without the knowledge of the amplifier's input resistance results in distorted records and considerable error. The larger the surface of the electrode, the less input resistance is allowable. For example. For example, a needle, electrode with a surface of 15,000 μm^2 may need an amplifier with input impedance of 5 $M\Omega$, while, a needle electrode with a surface of 500 μm^2 will ensure a record with acceptable distortion by means of an amplifier with minimum input impedance of 100 $M\Omega$.

The capacitance present parallel with the input resistance of the amplifier reduces the frequency response of the amplifier as well as lowers the common-mode rejection at higher frequency. Owing to these, the electrode cable, the extension cable and the input stage of the amplifier require careful designing. Generally, shielded cables are used which reduce the disturbing signals but at the same time, the parasite capacitance will increase. By careful design, a capacitance value of 50 pF or less can be achieved for the input stage. McRobbie (1990) illustrate a rapid recovery EMG preamplifier without AC coupling capacitors.

Low Frequency and High Frequency Filters: These are used to select the passband of the incoming signal and to modify the progressive reduction in voltage output which occurs at either end of the frequency spectrum roll-off. The low frequency 3 dB point may be selected over the range of 0.016 to 32 Hz while the high frequency 6 dB point can be selected over the range 16 Hz to 32 kHz. Thus, the passband may be varied over a very wide range but is normally made as narrow as possible, subject to the requirements of the particular application in order to restrict displayed noise.

Signal Delay and Trigger Unit: Sometimes, it is necessary to examine the signals from individual fibres of muscle tissue. For this purpose, special needles are available with a 25 micron diameter electrode surface and up to 14 pick-up surfaces down the side of one needle . These 14 points are scanned sequentially to determine which point is acquiring the largest the largest signal. This point is then considered as the reference and its signal is used to trigger the sweep. Signals from the remaining 13 point are then scanned squintingly and recorded with respect to the reference signal. To examine these signals, it is necessary to trigger the sweep from the signal and to delay the signal so that the whole of its leading edge is displayed. The delay is a achieved by passing the digitized signal through the shift register or random access memory into the recirculate mode to obtain a non-fade display of a transient phenomenon.

Integrator: The integrator is used for quantifying the activity of a muscle. Lippold (1952) established that a linear relationship exists between the integrated EMG and the tension produced by a muscle. The integrator operates by rectifying an incoming EMG signal, i.e. by converting all negative potentials to identical positive deflections so that the EMG pattern consists of positive deflections only. The area under the rectified potentials is a accumulated using a low-pass filter so that the module output , at any time , represents the total area summed from a selected starting time. The integrator indicates the EMG activity either as a variable frequency saw tooth waveform or as a steady deflection. In the former case, the output curve is a measure of the total electrical activity per second, recorded from a muscle during voluntary contraction within the analysis time. The slope of this curve, measured as the number of resets per second, can be used to select changes in the number of motor units firing over a period of time. The steady deflection or mean voltage mode is used in plotting mean voltage v. isometric tension curves of muscle interference patterns during voluntary contraction, to show changes in muscle activity due to neuro-muscular disease such as muscular dystrophy, poliomyelitis, etc. Different time constants determine the amount of smoothing applied to the output signal. When rapid changes in the signal have to be followed, the shortest time constant provides maximum smoothing of the signal and most easily read mean value.

Stimulators: The stimulators incorporated in the EMG machines are used for providing a single or double pulse or a train of pulses. Stimulus amplitude, duration, repecition and delay are all adjustable and facilities are provided for external type. The output is either of the constant voltage type or of the constant current type. The constant voltage type stimulator provides square wave pulses with amplitudes in the range of $0 - 500\ V$, a pules duration of $0.1 - 3$ ms and frequency between $0 - 100\ Hz$. Output of the constant current generator can be adjusted between 0 to 100 mA.

Usually, the electromyographic changes in an advanced diseased state are readily recognized on an oscilloscope display and by the sound from a loudspeaker. However , since the loss of muscle fibres, and therefore, the cation potential changes are relatively are relatively small in early or mild disease states, changer in the EMG signals may be obscured by the usual variability or action potentials. Quantitative analysis is thus necessary to determine when the waveforms have changed beyond the normal range. Quantities measured for such analysis include zero-crossing rate, peak rate, negative wave duration and wave rise time. These time-domain techniques are somewhat different from the classical frequency spectrum and correlation function methods, but are much simpler to implement with electronic techniques. Fusfeld (1978) details out circuits used to implement quantitative analysis of the electromyogram.

9.3.4 CLINICAL INSTRUMENTS

Most hospitals have a laboratory area separate from patient areas that is used solely for chemical analyses and measurements of body fluids and tissues. Typically, analysis is done on **blood** and **urine** and on body **tissue**. These measurements are made to aid physicians in the diagnosis of disease states and to help them monitor the effects of therapy.

Operation of The Clinical Laboratory

The clinical laboratory performs tests on a variety of specimens of diverse origin. These may be samples of normal body fluids (blood, cerebrospinal fluid, etc), abnormal body fluids (pleural or peritoneal effusions, cyst contents, etc), abnormal products (purulent secreations, exudates, etc) products of normal excretory functions (feces, urine, etc) and others. Such specimens and samples may be obtained directly laboratory by nursing personnel.

The results of a test may be expressed in one or more of the following ways:

- Quantitatively, as a value followed by a unit of measure, such as serum cholesterol, 150 mg percent.
- Quantitatively, as an indication of the presence or absence of a particular substance or element in a specimen, such as albumin in a urinalysis as 'trace' occult blood in stools as '3 +' platelets in a blood sample as' ade l quate or a blood culture as 'negative'
- In narrative form, such as 'no malignant tells were found in a report of a biopsy specimen' (cytology).
- In interpretive form, such as 'further laboratory evidence of myocardial damage' in one of a series of daily sequential reports of a cardiac profile.

Table 10.1

Test Name	S	L	H	U	C/L	C/H
Chemistry						
Random serum glucose	M-F	70	120	Mg/dl	50	300
Serum sodium	M-F	135	145	Mg/dl	120	150
Serum potassium	M-F	3.5	5.0	Mg/dl	3	5.9
Serum calcium	M-F	8.5	10.5	Mg/dl	7	1.3
Cholesterol	M-F	150	300	Mg/dl	-	-
Uric acid	M-F	2.5	8.0	Mg/dl	-	-
Triglycerides	M-F	30	150	Mg/dl	-	-
Haematology						
Haemoglobin	M-F	14	18	gm%	10	-
	F	12	16	gm%	10	-
Heamatocrit	M-F	42	52	%	-	-
	M-F	37	47	%	-	-
White blood cell coun		4800	10800	Cells/ mm^3	-	-
Reticulocyte count		0.5	1.5	%	-	-
Platelet count		1,50,000	3,50,00	Cell/ mm^3	50000	-

Where S $\rightarrow$ Sex, L $\rightarrow$ Low, H $\rightarrow$ High, U $\rightarrow$ Units, C/L $\rightarrow$ Critical low, C/J $\rightarrow$ Critical high

Chemical Electrodes

Chemical electrodes produce a potential that depends on the ionic concentration of fluids under test. These potentials are determined by the Goldman equation or the Nernst equation. These equations show that these potentials are directly proportional to temperature. Ions that are of particular interest in body fluid analysis include calcium. Potassium, sodium, lithium and the chlorides.

Chemical electrodes range in complexity from a simple solid in contact with the solution under test to multiple – element structures. A chloride membrane electrode, for example, consists of silver chloride membrane on the tip of a glass tube. A wire from one side of the membrane connects to a high – impedance voltmeter. The membrane potential is a function of the chloride content of perspiration of the skin in contact with the electrode.

BLOOD GAS ANALYSER

A major function of blood is to carry oxygen to the cells and carbon dioxide to the lungs for expiration. These gases mix with the blood to form a partial pressure in the blood.

A blood gas analyzer is used to measure the partial pressure of oxygen, $^PO_2$*,z The partial pressure of carbon dioxide,* $^PCO_2$ *and the blood pH.* The fundamental measurement from which the others are derived is the pH measurement. In normal blood this must be maintained between 7.34 and 7.44 slightly basic. pH reading from 0 to 7 are acidic and readings from 7 to 14 are basic. Since pH measurements are very much dependent on temperature that parameter is care fully regulated to normal body temperature, 37° C. A pH electrode, may be used in a blood gas analyzer.

PHOTOMETERS AND COLORIMETERS

The chemical content of biological substance can be determined by measuring how they either absorb or emit visible light. *The colorimeter uses light absorption to determine blood proteins and iron levels.* In order to enhance the colour of these substances in blood serum, It is necessary to mix it with reagents. Measurement of light emitted by irons, such as sodium or potassium in serum or urine, excited by heat are made with a flame photometer.

A colorimeter consists of a light source broken into its spectrum of colours by a prism or diffraction grating. The individual colours are then passed through the sample. The amount of each colour absorbed is measured to determine the type and concentration of substances in the sample. Since these measurements are highly temperature dependent, control of this parameter is necessary.

Blood Cell Counter

Knowledge about the number of particles in blood is important data from which a physician can diagnose disease. These particles are commonly known as red blood cells (RBCs), white blood cells (WBCs) and platelets (PLTs). They can be distinguished from each other by virtue of their size and density, as table shown . Because of these distinguishing features it is possible to use a microscope to count the particles or to use electronic circuitry to do the task automatically. Notice in the table that the RBCs are more numerous but smaller than the WBCs. Table shows that most of the particles in blood are RBCs. To determine relative proportion of blood volume made up by cell particles, known as the *hematocrit* (*ht*), a centrifuge is used (the centrifuge is sometimes called a hematocrit also). The blood sampleis placed on a test tube, which is spun so that the cells are packed on the bottom under centrifugal force.

The ht value equals the height of the packed cells divided by the height of the blood in the tube. This is typically 45 per cent.

Table Blood Particles

Type	Density(millions/μl)	Individual size (μm)
RBC	4.26-6.2	6.8–7.5
PLT	0.15-0.40	2–4
WBC	0.004–0.011	6–18

The function of the red blood cell is to carry heamoglobin to the cells of the body. It is important to know the mean volume of each red blood cell (MCV) in litres. MCV is defined as:

$$MCV = \frac{ht}{RBC} \text{(in litres/cell)}$$

Where RBC represents the density of red blood cells in cells per litre. A measure of haeoglobin is made by destroying the red blood cells with an acid and releasing the red colour haemoglobine into solution 1 in a process called *lysing*. The haeomglobin is then separated and weighed. A measure of the means cell haemoglobin (MCH) is then computed by the definition

$$MCH = \frac{\text{Haemoglobin (g/litre)}}{\text{RBC(litre}^{-1})}$$

the units of MCH are grams per cell. The concentration of haemoglobin in the blood is then given by the mean cell haemoglobin concentration (MCHC) in grams per litre.

$$MCHC = \frac{\text{Haemoglobin (g/litre)}}{ht}$$

9.3.5 SCINTILLATION COUNTERS

Scintillation is a word used to denote a process similar to that which gererates light on the screen of a cathode ray tube. When a radiation particle strikes an atom of certain phosphorous mterials, its kinetic energy may be added to the energy of the orbitial electrons. When an electron is thus excited, it jumps to a higher energy level. It is, however, unstable in this high-energy state and will soon fall back to its normal or ground state. But energy must be conserved and just as it attained extra energy from the radiation, it must lose energy as it returns to its normal state. This energy is released in the from of light energy called photons. Quite simply then, a scintillator is a device that allows the allows the counting of the light flashes produced by scintillation.

9.3.6 SEMI-CONDUCTOR RADIATION DETECTORS

The elements of semi-conductor theory form the basis for the operation of diodes and transistors and this should be somewhat familiar to you. If a radiation particle strikes the semi-conductor crystal, its kinetic energy can dislodge an electron and through the normal semi-conductor mechanism, a current is generated. This current can be accelerated by an external electric field applied across the crystal, creating an even greater current pulse. The pulse amplitude from this type of detector is substantial and it is nearly linear for a wide range of particle kinetic energy (usually 20 ke V to 200 MeV). Pulse durations down to the nanosecond range can be achieved.

9.3.7 ULTRASONIC TRANSDUCERS

Ultrasonic waves can be generated and detected in a number of ways. The one which is most commonly used in Non Destruction testing (DST) is described here. Quartz and some other crystals have a lattice structure such that if plate is cut out of the crystal with a certain orientation with respect to the crystallographic axes and subjected to an electric field in the right direction. it will change its dimensions: it will contract or expand according to the polarity of the field. Conversely, when a similar deformation of the plate is brought about by an external mechanical force, electric charges appear on its opposite surfaces. This phenomenon is known as piezoelectric effect. The materials which exhibit this property are known as piezoeiectric materials.

Among the various naturally occurring piezoelectric materials, quartz is the most important one, because it combines reasonably good piezoelectric properties with excellent mechanical and dielectric strength and stability .X-cut quartz plate is used for generating and receiving longitudinal waves. Y-cut plate is used for generating transverse and surface waves in solids. Quartz transducers can be operated at high temperatures up to 773K.

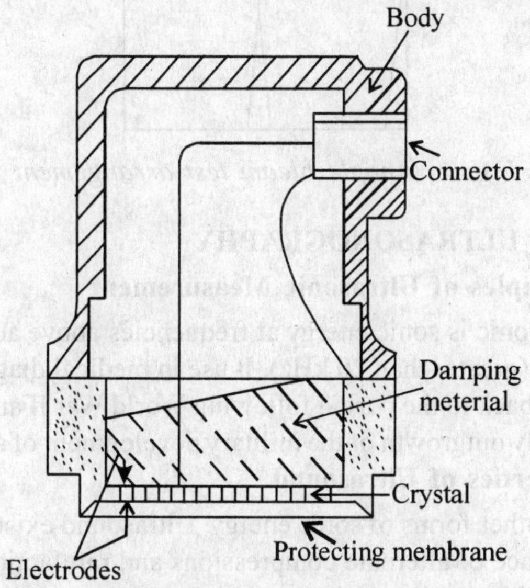

Figure (a): Normal proble

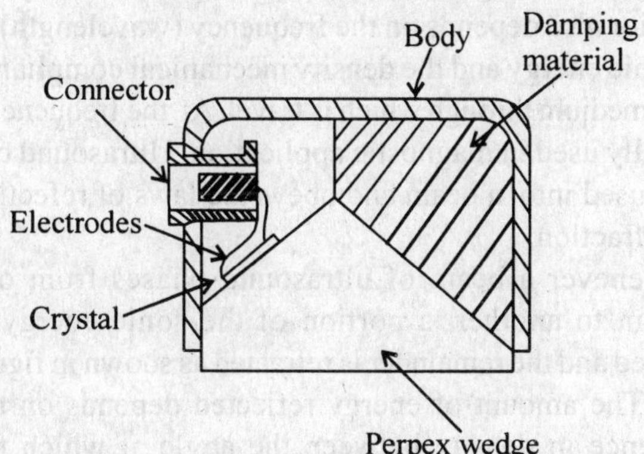

Figure (b): Angle probe

Figure: Types of ultrasonic probes

Barium titanate and lead zirconate titanate (PZT) are used for ultrasonic flaw detection. These transducer materials are mechanically less resistant. Lithium sulphate is the most sensitive but barium titanate is the best transmitter. Because of its higher acoustic impedance, the matching of barium titanate is always unsatisfactory and its sensitivity cannot be fully exploited. Lead metaniobate and lithium sulphate are superior in this respect. A gain because of their low acoustic impedance and high intrinsic internal damping, they are most suited to produce short pulses as is required in pulse-echo technique.

The transducers (piezoelectric crystals) can not be used on their own, but have to be mounted as suitable probes. The role of the probe is to protect the operator from electric shock, to protect the transducer from mechanical damage, and to make the transducer more suitable for the job. Various types of probes **Figure7.2** are made for different applications. Normal beam transducers are used for testing by using waves at normal incidence. For under-water testing, the probe, especially the cable, must be waterproof. For good performance, the transducer impedance should be matched to that of the water. For very short range operation, a twin probe is needed with separate transmitter and receiver probes built into one housing and acoustically isolated from each another. There in an acoustic delay rod, also called as stand-off, in front of both.

Normal Incident Pulse-Echo Inspection

The ultrasonic energy is coupled to the component being inspected through a couplant (usually oil or grease or glycerine) that transmits the ultrsound between the face of the transducer and the surface of the compoent. When ultrasonic energy travels through a test sample and strikes a discontinuity, part of the energy will be reflected back because ultresonic cannot propogate throuhg space and at discontinuties the medium is empty thus no meidum to propogate and hence reflected back and the remaining part propagaytes in the material in the forward direction. Also a refracted beam from the discontinuity is available for interpretation. Ultrasonic energy that is reflected and returned to the probe is the source of the (find)defect indications shown on the instrument screen **Figure 7.3.** The sound energy that travelled completely through the test piece will be reflected at the end, giving the large back-echo indication.Once excited in the material, the ultrasonic pulse will continue to reflect from the parallel surfaces, creating a multiple echo display when the time base is sufficiently large. As there is continuous attenuation of the sound energy in the material the amplitude of the echoes decreases when timeis sufficienly large. As there is continous of sound energy in the material the amplitude of echoes decreases.

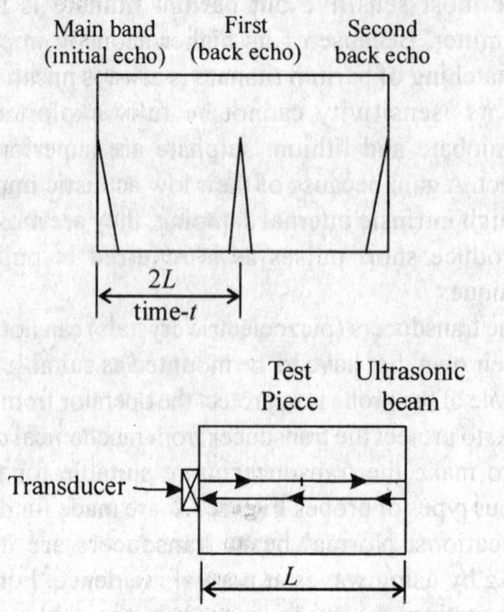

Figure: Pulse echo inspection of bar of lenght L

Normal Incident Through-Transmission Testing

Many a time, pulse-echo technique may not provide required test information. This may occur when a flaw or other anomaly does not provide a suitable reflection surface or where the orientation or location of the flaw which is not favourable for detection using single probe. Also, highly attenuating materials are often tested with through-transmission technique. **Figure 7.4** shown the probe arrangement for through-transmission technique. The technique is often used in large casting and highly attenuating materials.

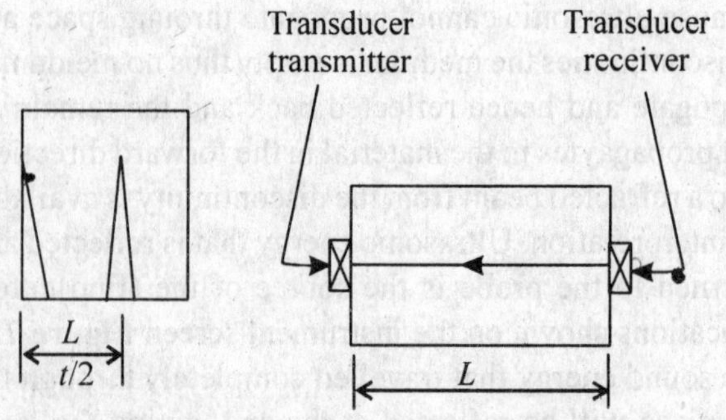

Figuer: Throuth-trnasmission inspection of bar of lenght L

Angle Beam Pulse-Echo Testing

Angle beam transducers provide access to areas that are inaccessible to normal beam probes. **Figure 7.5** shows the angle beam test arrangement. Angle beam inspection is accomplished with shear wave probes.

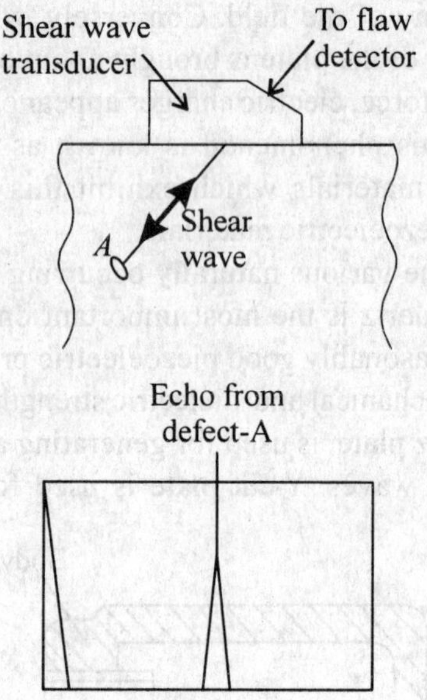

Figure: Anagle bteam test arrangement

9.3.6. ULTRASONOGRAPHY

Principles of Ultrasonic Measurement

Ultrasonic is sonic energy at frequencies above audible range (greater than 20 kHz). It use in medical diagnosis dates back to the period following World War II and is a directly outgrowth of the military development of sonar.

Properties of Ultrasound

Like other forms of sonic energy, Ultrasound exists as a sequence of alternate compressions and rarefactions of a suitable medium (air, water , bone, tissue, etc.) and is propagated through that medium at some velocity. Its behavior also depends on the frequency (wavelength) of the sonic energy and the density mechanical compliance of the medium through which is travel. At the frequencies normally used in diagnostic application. Ultrasound can be focused into a beam and obeys the laws of refection and refraction.

Whenever a beam of ultrasound phases from one medium to another, a portion of the sonic energy is reflected and the remainder is refracted as shown in figure (9.7). The amount of energy reflected depends on the difference in density between the angle at which the transmitted beam strikes the medium. The grater the difference in media, the grater will be the amount reflected.

Also the nearer the angle of incidence the beam and the interface is to 90° the greater will be the reflected portion.

Figure 9.7: Reflection and refraction out ultrasound at an interface between media of different densities.

The velocity of sound propagation through a medium varies with the density of the medium and its elastic properties. It also varies with temperature.

Every material has an acoustic impedance, which is a ratio of the acoustic pressure of the applied ultrasound to the resulting particle velocity in the material. Since acoustic impedance is a complex value, consisting of both resistive and reactive components, a simpler term called characteristic impedance, in more often used. The characteristic impedance of a material is the product of its density and the velocity of the sound through it.

As ultrasound travels through the material, some of the energy is absorbed and the wave is attenuated a curtains amount for each centimeter through which it travel it travels The amount of attenuation is a function of both the frequency of the ultrasound and the characteristics of the materials. The attenuation constant, α, is defined by the equation.

$$\frac{\text{amplitude at point } X}{\text{amplitude at } X+1 \text{ unit distance}} = \beta$$

α (per cm) $= cf_{\beta}$

Where c = proportionality constant

f = ultrasound frequency

β = exponential term determined by the properties of the material

This formula shows that attenuation increases with some power of the frequency, which means that the higher the frequency, the less distance it can penetrate into the body with a given amount of ultrasonic energy. For this reason, lower ultrasound frequencies are used for deeper penetration. However, lower frequencies are incapable of refection small objects. As a rule, a solid object surrounded by water or saline must be a least a quarter-wave thick in order to cause a usable reflection. Thus, for finer resolution, higher frequencies must be used. Ultrasound frequencies of 1 to 15 Mhz are usually used for diagnostic purposes. At 2 MHz, distinct echoes can be recorded from interface 1 mm apart. Higher-frequency ultrasound is also more subject to scattering than ultrasound at lower frequencies. However, the high-frequency ultrasound beam can be focused for greater resolution at a given depth.

Another useful way of assessing the attenuation of ultrasound as it penetrates the body is the half-value layer of the medium given in Table 9.2. The half-value layer is the depth of penetration at which the ultrasound energy is attenuated to that the applied amount.

Table 9.2: Ultrasound Absorption

Type of tissue	Frequency (MHz)	Half-value Layer (cm)
Blood	1.0	35.0
Bone	0.8	0.23
Fat	0.8	3.3
Music	0.8	2.1

A well-known characteristic of ultrasound frequently utilized in biomedical instrumentation is the Doppler effect, in which the frequency of the reflected ultrasonic energy is increased or decreased by a moving interface. The amount of frequency shift can be expressed in the formula:

$$\Delta f = \frac{2V}{\lambda}$$

Where f = shift in frequency of the reflected wave

V = velocity of the interface

λ = wavelength of the transmitted ultrasound

The frequency increases when the interface moves the transducer and decreases when it moves away. With an ultrasound frequency of 3 MHz, the shift is about 40 Hz for each cm/sec of interface velocity.

9.3.8 PRINCIPLE OF COMPUTER ASSISTED TOMOGRAPHY

(1) Use of CT Scan: Computed Tomography (CT) imaging, also known as "CAT scanning" (Computed Axial Tomography), combines the use of a digital computed together with a rotating with a rotating X-ray device to create exposed X-radiation detailed cross sectional images of the different of the different organs and body parte such as the lungs, liver, kidneys, pancreas, pelvis, extremities, brain and blood vessels. For many patients, CAT can be performed on an outpatient basis.

Among the various techniques such as MR and X-ray, CAT has the unique ability to image a combination of soft tissue, bone and blood vessels . For example, X-ray imaging of the head can only show the intense bone structures of the skull. X-ray angiography of the head only depicts the blood vessels of the head and neck and not the soft brain-tissue. Magnetic resonance (MRI) imaging does an excellent job of showing soft tissue and blood vessels, but MRI does not give as much detail of bony structures such as the skull. CAT images of the head allow physicians to see soft-tissue a structures like the brain's ventricles or gray and white matter. Example selectively "window" the digital CT images on the computer monitor to look at tissue, then the bone and then the blood vessels.

CT is one of the best tools for studying the lungs and abdomen. CT used in the cancer diagnosis process and is often the preferred meshed to diagnose liver and pancreas cancer . CT imaging is finding its use in the detraction, diagnosis and treatment of heart disease, acute stroke and vascular diseases which can lead to stroke, gangrene or kidney failure. Additionally, CT Scan is used to measure bone minaret density for the detection of osteoporosis. CT is applied in trauma cases. All department.

CT can provide detailed cross sectional images and diagnostic information for every for every part of the body including:

- The neck, shoulders, cervical spine and blood vessels of the neck.
- The brain, vessels of the brain, eyes, inner are, sinuses
- The chest, heart, aotra , lungs, mediastinum
- The upper abdomen, liver, kidney, spleen, pancreas and other abdomin
- The thoracic and lumbar spine
- The skeletal system including bones of the hands, feed ankles, legs and jaws
- The pelvis and hips, male and female reproductive system, bladder

Working Principle CT Scan

Computed Tomography is based on the X-ray principal: as X-rays pass through the body they are absorbed or attenuated at differing levels creating a matrix or profile of X-ray beams of different strength. This X-ray profile is registered on film, thus creating an image in a computer. In the case of CT, the film is replaced by a banana shaped detector which measures the X-ray profile.

A CT scanner looks like a big, square. The patent aperture (opening) is $60\,cm$ to $70\,cm(24"\ to\ 28")$ in diameter. Inside the covers of the CT scanner is a rotating frame which has an X-ray tube mounted on one side and detector mounted on the opposite side. A fan beam of X-ray is created as the rotating frame spins the X-ray tube and detector around the patient (Figure. 36.20.). Each time the X-ray tube and detector mark a 360° rotation, an image has been acquired. This image is collimated (focused) to a thickness between $I\ mm$ and $10\ mm$ using lead shutters in front of the X-ray tube and X-ray detector.

As the X-ray tube and detector mark this 360° rotation, the detector takes numerous snapshots (called profiles) of the attenuated X-ray beam. Typically, in one 360 lap, about 1,000 profiles are sampled. Each profile is subdivided spatially by the detectors and fed into about 700 individual channels. Each profile is then backward projected by a dedicated computer into a two-dimensional image of the slice that was scanned.

Applications and Clinical Benefits of CT Imaging

Unlike other medical imaging techniques, such as conventional X-ray imaging (radiography), CT enables direct imaging and differentiation of soft tissue structures, such as liver, lung tissue, and fat. CT is especially useful in searching for large space occupying lesions, tumours and metastasis and can not only reveal their presence, but also the size, spatial location and extent of a tumour.

CT imaging of the head and brain can detect tumours, show blood clots blood vessel defects, show enlarged ventricles (caused by a build up of cerebrospinal fluid) and image other abnormalities such as those of the nerves or muscles of the eye.

Due to the short scan times of 500 milliseconds to a few seconds. CT can be used for all anatomic regions, including those susceptible to patient motion and breathing. For example, in the thorax CT can be used for visualization of modular structures, infatuations of fluid, fibrosis (for example from asbestos fibres), and effusions (filling of an air space with fluid).

CT has seen the basis for interventional work like CT guided biopsy and minimally invasive therapy. CT images are also used as basis for planning radiotherapy cancer treatment. CT is also often used to follow the course of cancer treatment to determine how the tumour id responding to treatment.

For many computer tomography examinations patients may be asked to take a special contrast agent (orally , rectally or via injection). Intravenous, oral and rectal CT contrast are pharmaceutical agents (liquids) and are sometimes referred to as "dye". CT contrast is used to make specific organs, blood vessels and/or tissue type "stand out" with more image contrast to better show the presence of disease or injury. Thus CT contrast highlights specific areas of the resultant CT image or "dyes" it.

There are four types of contrast agents used in CT:

1. The type that is given via intravenous (through a vein) injection
2. The type id given orally
3. The type that is given orally
4. A much less common type of contrast used in CT is inhaled as a gas and used for special lung and brain imaging. This technique (called Xenon CT) is only available at a small number of locations throughout the world and is only performed for rare cases.

Use of Computers in CT Scan

Multiple computers are used to control the entire CT system. The main computer that combines the operation of the entire system is called the host computer". There is also a dedicated computer that reconstructs the "raw CT data" into an image. A workstation with a mouse, keyboard and other dedicated controls allows the technologist to control and monitor the exam. The CT gantry and table have multiple microprocessors that control the rotation of the gantry, movement of the table (up/down and in/out), tilting of thc gentry for angled images, and other functions such as turning the X-ray beam on and off.

Magnetic Resonance Imaging (MRI)

Magnetic resonance imaging was, developed from knowledge gained in the study of nuclear magnetic resonance. The original name for the medical technology is nuclear magnetic resonance imaging (NMRI), but the word nuclear is almost universally dropped. This is done to avoid the negative connotations of the word nuclear, and to prevent patients from associating the examination with radiationn exposure, which is not one of the safety concerns for MRI.

Difference between MRI & CT

A CT scanner uses ionizing radiation, X-rays, to acquire its images, making it a good tool for dense tissue (bone) exams ,MRI, on the other hand, used radio frequency signals to acquire its images, and is best suited for soft (non-calcified) tissue exams.

Example 9.1: A p-n junction LED has an infection efficiency of 60% and a light extraction efficiency of 50%. Calculate t_r provided $\tau_{nr} = 10^{-8}$ sec. and the overall efficency is 1.5%.

Solution: Overall efficency

$$= \frac{1.5}{100} = \frac{50}{100} \times \frac{60}{100} \times \eta_r$$

η_r = radiative recombination efficiency.

$$\eta_r = \frac{1.5 \times 100}{60 \times 50} = 0.05 \text{ or } 5\%$$

τ_r = 19 τ_{nr}= $\mathbf{1.9 \times 10^{-7}}$ **sec.**

Example 9.2: Calculate the divergence angle of a $H_e - N_e$ Gaussian beam whose spot size is 25 um. Operating wavelength 633 nm.

Solution: Divergence angle = $\frac{2\lambda}{\pi w}$,

λ = operating wavelength

w = beam diameter.

$$= \frac{(2 \times 633)}{3.14 \times 25} = 0.016 \text{ radians} = 0.92°.$$

Example 9.3: When a LED has 2V applied to its terminals, it draws 100 mA and produces 2 mw of optical power. What is the LED's convesion efficiency from electrical to optical power?

Solution: Conversion efficiency

$$= \frac{(P_0)\text{ optical power}}{(P_E)\text{ electricalpower}} \times 100\%$$

P_0 = 2×10^{-3} watts.

$P_E = IV = 100 \times 10^{-3} \times 2$ watts.

$$\eta = \frac{2 \times 10^{-3}}{100 \times 10^{-3} \times 2} \times 100\% = \mathbf{1\%}.$$

Example 9.4: Calculate the efficiency of a PIN silicon photo-detector if the responsivity is 0.58 A/W at 800 nm.

Solution: $\eta = \frac{\text{Responsivity}}{\text{e}} \times \frac{hc}{\lambda}$

$$= \frac{0.58(6.63 \times 10^{-34}\text{ J.S}) \times (3 \times 10^8 \text{ m/s})}{(1.6 \times 10^{-19}\text{ C})(800\text{nm})}$$

= 85%.

Example 9.5: The dark current of an InGaAs PIN photo-diode is 5nA, responsivity 0.86 A/W at 1300 nm. The received optical power is –27 dBm. If it is operated with a 4 kΩ bias resistor, calculate the thermal noise in a 1 Hz. bandwidth.

Solution: $i_{th}^2 = \frac{4(1.38 \times 10^{-23}) \times 300}{4000}$

$= 4.14 \times 10^{-24}$ A²/Hz

$i_{th} = 2.04 \times 10^{-12}$ A/ $\sqrt{\text{Hz}}$.

Example 9.6: In continuation of the problem 2 if detected optical powe is 2 uw, calculae the quantum shot noise current and NEP.

Solution: I_{Nsh}^2 = 2e ($R_a \times P_d$) × I_d

= 2 (1.6×10^{-19}c) (0.86 A/w × 2 u w × 5 nA)

So I_{Nsh} = 7.42×10^{-13} A $\sqrt{\text{Hz}}$.

NEP = I_{Nsp}/R_a = 7.42 × 10–13/86

= 0.865 pW/ $\sqrt{\text{Hz}}$

Example 9.7: A photemissive material has a work function of 1.8 eV. Calculate its cutoff wavelength.

Solution: $\lambda = \dfrac{1.24}{\text{Work function}} = \dfrac{1.24}{1.8}$

$= 0.689$ um $= 689$ nm.

Example 9.8: Calculate the responsivity of a detector having a quantum efficency of 15% at 0.8 um.

Solution: $= \dfrac{0.015\times(1.6\times10^{-19})(0.8\times10^{-6})}{(6.63\times10^{-34}(3\times10^{8})}$

$= 9.6 \times 10^{-3}$ A/W.

Example 9.9: A detector has a load resistor of 50 Ω and the optical power absorbed by the detector is l uW. The detector has a quantum efficiency of 1% at the operating wavelength of 800 nm. Calculate the woltage across the load.

Solution: The detector current

$i_d = (6.4 \times 10^{-3})\ 10^{-6}) = 6.4$ nA

V_2 (Voltage across the load)

$= (6.4 \times 10^{-9})\ (50)$

$= 320$ nV

Example 9.10: Calculate the 3dB bandwidth and the resistor that can be used withour significantly increasing the rise time. The PIN diode has a capactance of 5 pf and a trasit-timelimited rise time of 2 ns.

Solution: $f_{3-dB} = \dfrac{1.24}{2\times10^{-9}} = 175$ MHz.

$f_{-3dB} = \dfrac{0.35}{t_r}\quad t_r \times 2 \times 10^{-9}$ sec.

Largest resistance

$R_{Lmax} \le \dfrac{0.5}{2.19\,C_d}\quad \because t_r = 2.19\ R_L$ Cd.

for $\quad R_{Lmax}\ 2.19\ R_{Lmax}\ C_d \le \dfrac{t_r}{4}$

$R_{Lmax} = 4652$

Example 9.11: A soap film 5×10^{-5} cm thick is viewed at an angle of 35° to the normal. Find the wavelength of light in the visible spectrum which will be absent from the reflected light (μ = 1.33).

Solution : Let i be the angle of incidence and r be the angle of refraction. Then

$\mu = \dfrac{\sin i}{\sin r}$ or $1.33 = \dfrac{\sin 35°}{\sin r}$

or r = 25.5° and

cos r = 0.90

The condition for destructive interference is $2\mu\, t \cos r = m\lambda$.

Using different values for m in the above relation, we get following values for wavelengths.

When m = 1,

$\lambda_1 = 2 \times 1.33 \times 5 \times 10^{-5}$ cm $\times 0.90$

$= 12.0 \times 10\ 10^{-5}$ cm $=$ **120 μm.**

When m = 2,

$\lambda_2 = (2 \times 1.33 \times 5 \times 10^{-5}$ cm $\times 0.90) \div 2$

$= 6.0 \times 10^{-5}$ cm $=$ **6000 Å.**

When m = 3,

$\lambda_3 = (2 \times 1.33 \times 5 \times 10^{-5}$ cm $\times 0.90) \div 4$

$= 3.0\ 10–5$ cm $=$ **4000 Å.**

When m = 4,

$\lambda_4 = (2 \times 1.33 \times 5 \times 10^{-5}$ cm $\times 0.90) \div 4$

$= 3.0 \times 10–5$ cm $=$ **3000 Å.**

Out of the above wavelengths, $\lambda_2 = 6000$ Å, and $\lambda_3 = 4000$ Å lie in the visible region. Therefore, these two wavelengths are absent in the reflceted light.

Example 9.12: A glass wedge of angle 0.01 radian is illuminated by monochromatic light of wavelength 6000 Å falling normally on it. At what distance from the edge of the wedge will the 10th fringe be observed by reflected light?

Solution: Given that $\theta = 0.01$ rad, m = 10, $\lambda = 6000 \times 10^{-8}$ cm.

The condition for dark fringer is $2\,t = m\lambda$ (1)

The angle of the wedge $\theta = \dfrac{t}{x}$

From equation (1) and (2)

or $t = \theta\, x$ (2)

$\therefore\quad 2\,\theta\, x = m\,\lambda$

or $\quad x = \dfrac{m\lambda}{2\theta} = \dfrac{10\times6000\times10^{-8}\text{ cm}}{2\times0.01} =$ **3 mm**

Example 9.13: A beam of monochromatic light of wavelength 5.82×10^{-7} m falls normally on a glass wedge with the wedge angle of 20 seconds of an arc. If the refractive index of glass is 1.5, find the number of dark fringes per cm of the wedge length.

Solution : Given wedge angle θ = 20"

$= \dfrac{20\times\pi}{60\times60\times180}$ radians,

$\lambda = 5.82 \times 10^{-7}$ m, μ = 1.5.

Fringe width $\beta = \dfrac{\lambda}{2\mu\theta}$

$$= \frac{5.82\times10^{7}\ \text{m}\times60\times60\times180}{2\times1.5\times20\times\pi} = 2\ \text{mm}.$$

Number of fringes per cm = $\dfrac{1}{0.2\ \text{cm}}$ = **5 per cm.**

Example 9.14: A thin equiconvex lens of focal length 4 m and refractive index 1.50 rests on and in contact with an optical flat, and using light of wavelength 5460Å, Newton's rings are viewed normally by reflection. What is the diameter of the 5th bright ring?

Solution : We know that

$$\frac{1}{f} = (\mu - 1)\left[\frac{1}{R_1} - \frac{1}{R_2}\right].$$

Here $R_1 = R$ and $R_2 = -R$

Therefore $\dfrac{1}{f} = (\mu - 1)\left[\dfrac{2}{R}\right]$

$\therefore \quad \dfrac{1}{4\text{m}} = 0.5\left[\dfrac{2}{R}\right]$

or R = 4 m.

The diameter of the mth bright ring is given by

$$D_m = \sqrt{2(2m-1)\lambda R}$$

$$= \sqrt{2(2\times5-1)\times5460\times10^{-10}\ \text{m}\times4\ \text{m}}$$

= **6.2 mm**

Example 9.15: Newton's rigns are observed in reflected light of $\lambda = 5.9\times10^{-5}$ cm. diameter of the 10th dark ring is 0.5 cm. Find the radius of curvature of the lens and the thickness the air film.

Solution : Given that $\lambda = 5.9\times10^{-5}$ cm, m = 10.

The radius of mth dark ring is given by

$$\therefore \quad R = \frac{r^2}{m\lambda} = \frac{(0.5\,\text{cm})^2}{10\times5.9\times10^{-5}\ \text{cm}}$$

= 106 cm

= **1.06 cm**

The thickness of air film is given by

$$\therefore \quad t = \frac{m\lambda}{2} = \frac{10\times5.9\times10^{-5}\ \text{cm}}{2}$$

= **2.95μm.**

Example 9.16: In a Newton's rigns experiment, the diameter of 10th dark ring due wavelength 6000 Å in air is 0.5 cm. Find the radius of curvature of the lens.

Solution : Radius of curvature,

$$R = \frac{(D/2)^2}{m\lambda} = \frac{(0.5\times10^{-2}/2)^2\ \text{m}^2}{10\times6000\times10^{-10}\ \text{m}}$$

= **1.04m**

Example 9.17: In a Newton's rings experiment the diameter of the 15th ring was found to be 0.59 cm and that of the 5th ring was 0.336 cm. If the radius of the plano-convex lens is 100 cm calculate the wavelength of light used.

Solution: $\lambda = \dfrac{D_{m+p}^2 - D_m^2}{4pR} = \dfrac{D_{15}^2 - D_5^2}{4\times10\times R}$

$$= \frac{(5.9-3.36)^2\times10^{-6}\ \text{m}^2}{4\times10\times1\text{m}} = \mathbf{5880\ Å.}$$

Example 9.18: In a Newton's rings experiment the diameter of 10th ring changes from 1.40 to 1.27 cm when a drop of liquid is introduced between the lens and the glass plate. Calculate the refractive index of the liquid.

Solution: $\mu = \dfrac{(D_m^2)_{air}}{(D_m^2)_{liq}} = \dfrac{(1.40\,\text{cm})^2}{(1.27\,\text{cm})^2}$

= **1.215**

Example 9.19: In a Michelson interferometer 200 fringes cross the field of view when the movable mirror is moved through 0.0589 mm. Calculate the wavelength of ligh used.

Solution : $\lambda = \dfrac{2d}{m} = \dfrac{2\times0.0589\times10^{-3}\ \text{m}}{200}$

= **5890 Å**

Example 9.20: In an experiment for determining the refractive index of gas using Michelson interferometer a shift of 140 fringes is observed, when all the gas is removed from the tube. If the wavelength of ligh used is 5460 Å and the length of the tube is 20 cm, calculate the refractive index of the gas.

Solution: Given that $n = 140$,

$\lambda = 5460$ Å $= 5460\times10^{-10}$ m,

$l = 20$ cm = 0.2 m.

$$\mu = 1 + \left(\frac{n\lambda}{2l}\right) = 1 + \left[\frac{140\times5460\times10^{-10}\ \text{m}}{2\times.02\,\text{m}}\right]$$

= **1.00019.**

Example 9.21: A glass microscope lens ($\mu = 1.50$) film to increase the transmission of normally incident yellow ligh ($\lambda = 5800$ Å). What minimum film thickness should be deposited on the lens?

Solution: Given that $\mu_g = 1.50$, $\mu_f = 1.38$, $\lambda = 5800$ Å.

Minimum film thickness $t_{min} = \dfrac{\lambda}{4\mu_f}$

$$= \frac{5800}{4 \times 1.38} \text{ Å} = \mathbf{1050 \text{ Å}}$$

Example 9.22: Find the separation of two points on the moon that can be resolved by a 500 cm telescope. The distance of the moon is 3.8×10^5 km. The eye is most sensitive to light of wavelength 5500 Å.

Solution: The limit of resolution of a telescope is given by $d\theta = \dfrac{1.22\lambda}{a}$

Here $\lambda = 5500$ Å $= 5500 \times 10^{-8}$ cm,

a = 500 cm

Hence $d\theta = \dfrac{1.2 \times 5500 \times 10^{-8} \text{ cm}}{500 \text{cm}}$

or $d\theta = 13.42 \times 10^{-8}$

Let the distance between the two points be x.

$$\therefore \quad d\theta = \frac{x}{R}$$

Here $R = 3.8 \times 10^{10}$ cm

$x = R.\, d\theta$

$= 3.8 \times 10^{10} \text{ cm} \times 13.42 \times 10^{-8}$

$=$ **51.0 m.**

Example 9.23: Two pin holes 1.5 mm apart are placed in front of a source of light of wavelength 5.5×10^{-5} cm and seen through a telescope with its objective stopped down to a diameter of 0.4 cm. Find the maximum distance from the telescope at which the pin holes can be resolved.

Solution: Here, $\lambda = 5.5 \times 10^{-5}$ cm; a = 0.4 cm; x = 1.5 mm = 0.15 cm

Now, $d\theta = \dfrac{1.22\,\lambda}{a}$

$$d\theta = \frac{x}{d}$$

Hence $d = \dfrac{x\,a}{1.22\,\lambda}$

$$= \frac{0.15 \times 0.4}{1.22 \times 5.5 \times 10^{-5}} \text{ cm} = 894.2 \text{ cm}$$

= 8.9 m

Example 9.24: Calculate the useful magnifying power of a telescope of 10 cm objective, assuming that the limit of resolution of the eye is 2 minutes of an arc. Wavelength of light used is 6000 Å.

Solution: Here diameter of the objective $D = 10$ cm, $\lambda = 6000$ Å $= 6 \times 10^{-5}$ cm.

Limit of resolution of the telescop, $d\theta$

$$= \frac{1.22\,\lambda}{D} = \frac{1.22 \times 6 \times 10^{-5}}{10} \text{ cm}$$

$= 7.32 \times 10^{-6}$ radian

Limit of resolution of the eye, $d\theta$ = 2 minutes of an arc

$$= \frac{2}{60} \times \frac{22}{7 \times 180} = 582 \times 10^{-6} \text{ radian}$$

Useful manifying power of the telescope

$$= \frac{d\theta}{d\theta'} \times \frac{582 \times 10^{-6}}{\times 7.32 \times 10^{-6}} = \mathbf{97.5}$$

Example 9.25: Sodium light of wavelength 5890 Å is used to view an object under a microscope. The aperture of the objective has a diameter of 0.9 cm.

(a) Calculate the limiting angle of resolution

(b) Using violet light, what is the maximum limit of resolution for this microscope?

Solution: (a) Limiting angle of resolution,

$$\theta_m = 1.22\left(\frac{\lambda}{d}\right)$$

Here, $\lambda = 5890$ Å $= 5.89 \times 10^{-7}$ m,

$d = 0.9$ cm $= 9 \times 10^{-3}$ m

$$\theta_m = 1.22\left(\frac{5.89 \times 10^{-7}}{9 \times 10^{-3}}\right)$$

$= \mathbf{7.89 \times 10^{-5}}$ **radiam**

(b) The wavelength of violet light is 4000 Å.

The maximum limit of resolution for the microscope corresponds to the smallest angle

Here, $\lambda = 4000$ Ã $= 4 \times 10^{-7}$ m

$$\theta_m = 1.22\left(\frac{\lambda}{d}\right)$$

$$\theta_m = 1.22\left(\frac{4 \times 10^{-7}}{9 \times 10^{-3}}\right)$$

$= \mathbf{5.42 \times 10^{-5}}$ **radian**

Example 9.26: Calculate the minimum thickness of the base of a prism which will just resolve the D_1 and D_2 lines of sodium. The refractive index of glass is 1.6545 for $\lambda = 6563$ Å and 1.6635 for $\lambda = 5270$ Å.

Solution: Resolving power of the prism

$$= \frac{\lambda}{d\lambda} = t\frac{d\mu}{d\lambda}$$

$$\frac{d\mu}{d\lambda} = \left[\frac{1.6635 = 1.6545}{(6563 - 5270)\times 10^{-10}\text{ m}}\right]$$

and $\frac{\lambda}{d\lambda} = \frac{5893}{6}$

now $t = \frac{\lambda / d\lambda}{d\mu / d\lambda}$

$$\therefore \quad t = \frac{5893 \times 1293 \times 10^{-10}\text{ m}}{6 \times 0.0090} = \mathbf{14\ mm}$$

Example 9.27: Light is incident normally on a grating of total ruled width 5×10^{-3} m with 2500 lines in all. Calculate the angular separation of the two sodium lines inthe first order spectrum. Can they be seen distinctly?

Solution: (*i*) Here, $N = 2500$,
width of ruling $= 5 \times 10^{-3}$ m, $n = 1$,
$\lambda_1 = 5890 \times 10^{-10}$ m, $\lambda_2 = 5896 \times 10^{-10}$ m

$$\text{Number of lines per meter} = \frac{2500}{5\times 10^{-3}} = 5\times 10^{5}$$

$$\text{Hence } (a+b) = \frac{1}{5\times 10^{5}} = 2\times 10^{-6}\text{ m}$$

For the first ordr ($n = 1$),

$$\sin\theta_1 \frac{n\lambda_1}{(a+b)} = \frac{\lambda_1}{(a+b)}$$

$$= \frac{5890\times 10^{-10}\text{ cm}}{2\times 10^{-6}\text{ cm}} = 0.2945$$

$\theta_1 = 17°8'$

and $\sin\theta_2 = \frac{n\lambda_2}{(a+b)} = \frac{\lambda_2}{(a+b)}$

$$= \frac{5896\times 10^{-10}\text{ cm}}{2\times 10^{-6}\text{ cm}} = 0.2948$$

$\therefore \quad \theta_2 = 17°9'$

(ii) The resolving power of the grating,

$$\frac{\lambda}{d\lambda} = \frac{5890 + 10^{-10}\text{ cm}}{6\times 10^{-10}\text{ cm}} = 982$$

As the total number of lines on the grating is 2500 which is more than 982, the lines can beseen distinctly.

Example 9.28: Unpolarized light falls on two polarizing sheets placed on top of the other. What must be the angle between the characteristic directions of the sheets if the intensity of the transmitted light is one third intensity of the incident beam?

Solution:

Intensity of the light transmitted through the first polarizer $I_1 = I_0/2$,

where I_0 is the intensity of the incident uupolarized light.

Intensity of the light transmitted through the second polarizer is $I_2 = I_1 \cos^2\theta$ where θ is the angle between the characteristic directions of the polarizersheets.

But $I_2 = I_0/3$ (given)

$$I_2 = I_1 \cos^2\theta/2 = I_0/3$$

$$\cos^2\theta = 2/3$$

$$\theta = 0.8165$$

$$\theta = \mathbf{35.3°}$$

Example 9.29: If the plane of vibration of the incident beam makes an angle of 30° with the optic axis, compare the intensities of extraordinary and ordinary light.

Solution: Intensity of the extraordinry ray.

$$I_e = E^2 \cos^2\theta$$

Intensity of the ordinary ray $I_0 = E^2 \sin^2\theta$

$$\frac{I_e}{I_0} = \frac{\cos^2\theta}{\sin^2\theta} = \frac{\cos^2 30°}{\sin^2 30°} = 3$$

Example 9.30: A plane-polarized light is incident perpendicularly on a quartz plate cut with faces parallel to optic axis. Find the thickness of quartz plate, which introduces phase difference of 60° between e-and o-rays.

Solution:

The path difference between the waves is given by
$\Delta = (\mu_e - \mu_0)d$

The phse dfference between the waves s given by

$$\delta = \frac{2\pi}{\lambda}\Delta$$

$$\Delta = \frac{60°}{360°}.\lambda = \frac{\lambda}{6}$$

$$d = \frac{\lambda}{6(\mu_e - \mu_0)}$$

$$d = \frac{5400\text{ Å}}{6(1.553 - 1.544)} = \frac{0.54}{0.054}\ \mu\text{m} = \mathbf{10\ \mu m}$$

Example 9.31: Calculate the thicness of double refracting plate capable of producing a path difference of $\lambda/4$ between extraordinary nd ordinary waves. Given : $\lambda = 5890$ Å,

$\mu_0 = 1.53$, $\mu_e = 1.54$.

Solution:

$$d = \frac{\lambda}{4(\mu_0 - \mu_e)} = \frac{5890 \text{ Å}}{4(1.54 - 1.53)}$$

$$= \frac{0.589}{0.04} \mu\text{m} = \mathbf{14.7\ \mu m}$$

Example 9.32: Calculatae the least thickness of a calcite plate which would convert plane polarized light into circularly light. Given $\mu_0 = 1.658$, $\mu_e = 1.486$ and wavelength of light is 5890Å.

Solution:

Plane polarize light gets converted into circularly polarized light by a suitably oriented quarter wave plate. Its thickness is given by

$$d = \frac{\lambda}{4(\mu_0 - \mu_e)} = \frac{5890 \text{ Å}}{4(1.65 - 1.486)}$$

or d = **0856μm.**

Example 9.33: Plane polarized light passes through a calcite plate with its optic axis parallel to the faces. Calculate the least thickness of the plate for which the emergent beam will be plante polarized. Given μ_0 1.6584, $\mu_e = 1.4864$ and wavelength of light is 5000Å.

Solution:

When a plane-polarized beam is incident on a half wave plate, the emergent beam will also be plane polarized but the plane of polarization undergoes a rotation through an angle 2θ.

$$d = \frac{\lambda}{2(\mu_0 - \mu_e)} = \frac{5 \times 10^{-7} \text{ m}}{2(1.6584 - 1.4864)}$$

$= \mathbf{1.45\ \mu m}$

Example 9.34: The rotation in the plane of polarization in a certain substance is 10°/cm. Calculate the difference between the refractive indices for right and left circularly polarized light in the substance. Given λ = 5893 Â.

Solution:

$$\delta = \frac{2\pi}{\lambda}[\mu_R - \mu_L]d$$

$$\therefore [\mu_R - \mu_L] = \frac{\delta\lambda}{2\pi d}$$

It is given that $\frac{\delta}{d} = 10° = \frac{10 \times 2\pi}{360°} = \frac{\pi}{18}$ radian/cm and λ = 5890 Å = 5893 × 10^{-8} cm.

$$\mu_R - \mu_L = \frac{\pi}{18} \cdot \frac{5893 \times 10^{-8} \text{ cm}}{2\pi} = 1.6 \times 10^{-6}.$$

Example 9.35: A 200 mm long tube containing 48 cm³ of sugar solution produces an optical rotatin of 11° 2hen placed in a saccharimeter. If the specific rotation of sugar solution is 66°, calculate the quantity of sugar contained in the tube in the form in the form of a solution.

Solution: It is given that θ = 11°, l = 200 mm = 20 cm, S = 66°, and V = 48 cm³.

$$C = \frac{10\theta}{lS} = \frac{10 \times 11°}{20 \text{ cm} \times 66°} = 0.0833 \text{ g/cm}^3$$

Mass of sugar in solution M = CV = 0.0833 g/cm³ × 48 cm³ = **4 grams.**

Example 9.36: About 0.1% of electrical energy supplied to a laboratory mercury vapour lamp of 80 watt is converted into UV light of wavelenth 2500 Å. Calculate the number of UV photons emittd per second by the lamp.

Solution: Number of UV photons emitted per second

$$= \frac{\text{Energy converted into UV light}}{\text{Energy carried by one UV photon}}$$

$$= \frac{80J/s \times (0.100) \times 6.24 \times 10^{18} eV/J)}{12400\ eV/2500}$$

$= \mathbf{10^{17}}$ **photons/s.**

Example 9.37: X-rays of 0.5 Å are scattered by free electrons in a block of carbon through 90°. FInd the velocity of recoil electrons.

Solution:

K.E. of the recoil electrons = $h\nu_i - h\nu_f$

$$= 12400 \left(\frac{1}{\lambda_i} - \frac{1}{\lambda_f}\right) \text{ eV } (1.602 \times 10^{-19} \text{ J/eV})$$

$$\lambda_f = \lambda_i + \frac{h}{m_0 c}(1 - \cos\theta)$$

$$= (0.5 + 0.02426) \text{ Å} = 0.5243 \text{ Å}$$

$$\text{K.E.} = 12400 \left(\frac{1}{0.5} - \frac{1}{0.5243}\right)(1.602 \times 10^{-19})$$

$= \mathbf{1.84 \times 10^{-16}\ J}$

$$= \left[\frac{2(K.E.)}{m}\right]^{1/2} = \left[\frac{2(1.84\times10^{-16}}{9.11\times10^{-31}\,Kg.}\right]^{1/2}$$

$$= \left\{4.04\times10^{14}\,\frac{kg.m^2/s^2}{kg}\right\}^{1/2}$$

$\therefore$ **ν = 2 × 107 m/s.**

Example 9.38: Calculate de Broglie wavelength of an electron moving with velocity 10^7 m/s.

Solution:

$$\lambda = \frac{h}{m\upsilon} = \frac{6.626\times10^{-34}\,J.s}{9.11\times10^{-31}\,kg\times10^7\,m/s}$$

$$= \frac{6.626\times10^{-34}\,kg.m^2/s^2.(s^2)}{kg.m}$$

$= 7.28 \times 10^{-11}\, m$

$\therefore$ **λ = 0.72 Å**

Example 9.39: Compute the minimum uncertainty in the location of a mass of 2.0 gm moving with a speed of 1.5 m/s and the minimum uncertainty in the location of an electron moving with a spped of 0.5 × 10^8 m/s. Given that the uncertainty in the momentum *p* for both is $\Delta p = 10^3$ p.

Solution: $p = m\upsilon = 2 \times 10^{-3}$ kg × 1.5 m/s

$= 3 \times 10^{-3}$ kg m/s.

$\therefore \Delta p = 10^{-3}\, p = 3 \times 10^{-6}$ kg m/s

$$\therefore \Delta x = \frac{h}{2\pi\Delta p} = \frac{6.63\times10^{-34}\,Js}{2\pi\times3\times10^{-6}\,kg.m/s}$$

= 3.5 × 10^−19 Å

For an electron, $p = m\upsilon$

$= (9.11 \times 10^{-31}$ kg) $(0.5 \times 10^8$ m/s)

$= 4.55 \times 10^{-25}$ kg. m/s

$$\therefore \Delta x = \frac{h}{2\pi.\Delta p} = \frac{6.63\times10^{-34}\,Js}{2\pi\times4.55\times10^{-26}\,kg.m/s}$$

= 23 Å

Example 9.40: Find the ratio of populations of the two states in a He-Ne laser that produces light of wavelength 6328 Å at 27° C.

Solution: Thr ratio of population is given by $\frac{N_2}{N_1} = e^{-}$

(E2–E1)/KT.

$$E_2 - E_1 = \frac{12400}{6328}\text{ eV} = 1.96\text{eV}$$

$$\therefore \frac{N_2}{N_1} = \exp\left[\frac{-1.96\text{eV}}{(8.61\times10^{-5}\text{ eV})(300\text{ K})}\right]$$

$= e^{-75.88} = 1.1\times10^{-33}$

Example 9.41: The wavelength of emission is 6000 Å and the coefficient of spontaneous emission is 106/s. Determine the coefficient for the stimulated emission.

Solution: The ratio of population is given by

$$B_{21} = \frac{c^3}{8\pi h\nu^3\mu^3}A_{21} = \frac{\lambda^3}{8\pi h}A_{21} \text{ (Taking } \mu = 1)$$

$$B_{21} = \frac{\left(6000\times10^{-10})^3\right)m^3\left(10^6/s\right)}{8\pi\times6.626\times10^{-34}\,Js}$$

$= 1.3 \times 10^{19}$ m/kg.

Example 9.42: At what temperature are the rates of spontaneous and stimulated emission equal? Assume λ = 5000 Å.

Solution: If the rates of spontaneous and stimulated emission are equal, then

$$R_1 = \left[\frac{1}{e^{h\nu/kT}-1}\right] = 1$$

or $e^{h\nu/kT} = 2$

As λ = 5000Å, ν = c/λ = 6 × 10^{14} Hz and

$$\frac{h\nu}{kT} = \frac{6.626\times10^{-34}\,Js\left(6\times10^{14}/s\right)}{\left(1.38\times10^{-23}\,J/K\right)T}$$

$$= \frac{28.8\times10^3}{T}K$$

$$e^{h\nu/kT} = \exp\left[\frac{28.8\times10^3}{T}\right] = 2$$

or $\frac{28.8\times10^3}{T}$ K = ln 2 = 0.693

$$\therefore T = \frac{28.8\times10^3}{0.693}\text{ K}$$

= 41,558 K.

Example 9.43: The length of a laser tube is 50 mm and the gain factor of the laser material is 0.0005/cm. If one of the cavity mirrors reflects 100% light that is incident on it, what is the required reflectance of the other cavity mirror?

Solution: $\gamma_{th} = \frac{1}{2L}\ln\frac{1}{r_1 r_2}$

$$\therefore\ r_2 = \frac{1}{r_1 e^{2L\gamma}} = \frac{1}{1\times e^{2\times5\times0.0005}} = 0.985$$

It means that the second mirror should have a reflectance of 98.5%.

Example 9.44: The half-width of the gain profile of a He-Ne laser material is 2×10^{-3} nm. If the length of the cavity is 30 cm, how many longitudinal modes can be excited? The emission wavelength of He-Ne laser is 6328 Å.

Solution: The separation between successive longitudinal modes is given by

$$\Delta\lambda = \frac{\lambda^2}{2L} = \left(\frac{\left(6328\times10^{-10}\,m\right)^2}{2\left(30\times10^{-2}\,m\right)}\right)$$

$= 0.663 \times 10^{-3}$ nm

Number of modes

$$N = \frac{\delta\lambda}{\Delta\lambda} = \frac{2\times10^{-3}\,nm}{0.66\times10^{-3}\,nm} = \mathbf{3.}$$

Example 9.45: In an optical fibre, the core material has refractive index 1.43 and refractive index of clad material is 1.4. Find the propagation angle.

Solution: $\cos\theta c = \dfrac{n_2}{n_1} = \dfrac{1.40}{1.43} = 0.979$

Therefore, propagation angle

$\theta c = \cos^{-1}(0.979) = 11.8^\circ$

Example 9.46: In an optical fibre, the core material has refractive index 1.6 and refractive index of clad material is 1.3. What is the value of critical angle? Also calculate the value of angle of acceptance cone.

Solution: Critical angle is given by

$\sin\phi_c$

$= \dfrac{n_2}{n_1} = \dfrac{1.3}{1.6} = 0.8125$

$\therefore\ \phi_c = \mathbf{54.3^\circ}$

Acceptance angle

$$\theta_0 = \sin^{-1}\left[\sqrt{n_1^2 - n_2^2}\right] = \sin^{-1}\left[\sqrt{1.6^2 - 1.3^2}\right]$$

$= \sin^{-1}(0.87)$

$= \mathbf{60.5^\circ}$

Angle of acceptance cone $= \theta_0$

$= \mathbf{121^\circ}$

Example 9.47: Calculate the numerical apeture and acceptance angle of an optical fibre from the following data:

μ_1 (core) = 1.55 and μ_2 (cladding) = 1.50

Solution: $NA = \sqrt{n_1^2 - n_2^2}$

$= \sqrt{1.55^2 - 1.50^2} = \sqrt{0.153}$

$= \mathbf{0.391^\circ}$

Acceptance angle

$$\theta_0 = \sin^{-1}\left[\sqrt{n_1^2 - n_2^2}\right] = \sin^{-1}\left[\sqrt{1.55^2 - 1.50^2}\right]$$

$= \mathbf{23.02^\circ}$

Example 9.48: What is the numerical aperture of an optical fibre cable with a clad index of 1.378 and a core index of 1.546?

Solution: $NA = \sqrt{n_1^2 - n_2^2}$

$= \sqrt{1.546^2 - 1.378^2}$

$= \sqrt{0.491} = 0.70$

Example 9.49: A fibre cable has an acceptance angle of 30° and a core index of refraction of 1.4. Calculate the refractive index of the cladding.

Solution: $\sin\theta_0 = \sqrt{n_1^2 - n_2^2}$

$\therefore\ \sin\theta_0 = n_1^2 - n_2^2$

$n_2^2 = n_1^2 - \sin\theta_0 = (1.4)^2 - \sin 30^\circ$

$= 1.96 - 0.25 = 1.71$

$n_2 = \mathbf{1.308}$

Example 9.50: Calculate the angle of acceptance of a given optical fibre, if the refractive indices of the core and the cladding are 1.563 and 1.498 respectively.

Solution: $\sin\theta_0 = \sqrt{n_1^2 - n_2^2}$

$= \sqrt{(1.563)^2 - (1.498)^2}$

$= 0.4461$

$\theta_0 = \sin^{-1}(0.4461) = \mathbf{26.49^\circ}$

Example 9.51: Calculate the fractional index change for a given optical fibre if the refractive indices of the core and the cladding are 1.563 and 1.498 respectively.

Solution: Fractional index change $\Delta = \dfrac{n_1 - n_2}{n_1}$

$$= \frac{1.563 - 1.498}{1.563} = \frac{0.065}{1.563} = \mathbf{0.0415}$$

Example 9.52: Calculate the refractive indices of the core and the cladding material of a fibre from the following data:

Numerical aperture (NA) = 0.22 and Δ = 0.012 where Δ is the fractional refractive index change.

Solution: NA = $n_1\sqrt{2\Delta}$

$0.22 = n_1\sqrt{2\times0.012} = 0.155\, n_1.$

$\therefore\ n_1 = \frac{0.22}{0.155} = \mathbf{1.42}$

$\Delta = \frac{n_1 - n_2}{n_1}\ \therefore\ \frac{1.42 - n_2}{1.42}$

$= 0.012$

$\therefore\ n_2 = 1.42 - 1.42 \times 0.012 = \mathbf{1.403}$

Example 9.53: Find the fractional refractive index and numerical aperture for an optical fibre with refractive indeices of core and cladding as 1.5 and 1.49 respectively.

Solution: $\Delta = \frac{n_1 - n_2}{n_1} = \frac{1.5-1.49}{1.5}$

$= \mathbf{0.0067}$

$\text{NA} = n_1\sqrt{2\Delta} = 1.5\sqrt{2\times0.0067}$

$= \mathbf{4.049}$

Example 9.54: A step-index fibre is made with a core of refractive index 1.52, a diameter of 29 μm and a fractional difference index of 0.0007. It is operated at a wavelength of 1.3 μm. Find the V-number and the number od modes that the fibre will support.

Solution: $V = \frac{\pi d}{\lambda_0} n_1\sqrt{2\Delta}$

$= \frac{3.143\times29\times10^{-6}\,m}{1.3\times10^{-6}\,m}\times1.52\sqrt{2\times0.0007}$

$= \mathbf{4.049}$

$\therefore$ Number of modes, $N = \frac{1}{2}V^2 = \frac{1}{2}(4.049)^2$

$= \mathbf{8\ modes}$

Example 9.55: A step-index fibre is with a core of refractive index 1.55 and cladding of refractive index 1.51. Compute the intermodal dispersion per kilometer of length of the fibre and the total dispersion in a 15 km length of the fibre.

Solution: $\Delta t = \frac{n_1 L}{c}\left[\frac{n_1}{n_2} - 1\right]$

$= \frac{1.55\times10^3\,m}{3\times10^8\,m/s}\left[\frac{1.55}{1.51} - 1\right]$

$= \mathbf{138\ ns/km.}$

Total dispersion for 15 km length = $\Delta t \times$ 15 km
= (138 ns/km) × 15 km = **2.07 μs.**

Example 9.56: Optical power of 1 m W is launched into an optical fibre of length 100 m. If the power emerging from the other end is 0.3.mW, calculate the fibre attenuation.

Solution: Attenuation, $\alpha = \frac{10}{L}\log\frac{P_i}{P_0}$

$= \frac{10}{0.1km} = \log\frac{1mW}{0.3mW}$

$= \mathbf{52.3\ dB/km}$

Example 9.57: What is the attenuation in dB/km, if 15% of the power fed at the launching end of a $\frac{1}{2}$ km fibre is lost during propagation?

Solution: Attenuation, $\alpha = \frac{10}{L}\log\frac{P_i}{P_0}$

$= \frac{10}{0.5km}\log\frac{1}{0.15} = 16.48\ \text{dB/km}$

MCQ Set–1

1. Operating wavelength window of 1550 nm is not yet in use with fibre optic system because
   (a) loss in fibre is higher at 1300 nm
   (b) suitable Laser devices have not yet been obtained
   (c) wavelength multiplexing is not possible at this wavelength
   (d) both (b) and (c)
2. An optical fibre consists of a cylindrical dielectric rod of refractive index μ_1 surrounded by another dielectric of refractive index u_2, where $u_2 < \mu_1$. If a ray is incident from air at an angle i to the axis, then it udergoes total internal reflection at the interface AB if:
   (a) $i \geq \sin^{-1}\sqrt{\mu_1^2 - \mu_2^2}$
   (b) $i = < \sin^{-1}\sqrt{\mu_1 - \mu_2}$
   (c) $i = < \sin_{-1:}\sqrt{\mu_1^2 - \mu_2^1}$
   (d) $i = < \sin^{-1}\sqrt{\mu_1 - \mu_2}$
3. Total internal reflections can take place when light travels from

(a) diamond to glass
(b) water to glass
(c) air to water
(d) air to glass

4. Loss in fibre is due to (indicate false statement)
(a) Impurities
(b) microbending
(c) attenuation in fibre
(d) stepped index operation

5. The colour of a LED can be changed by
(a) using different band gap semiconductor
(b) by changing the doping level of the semiconductor
(c) by increasing applied voltage
(d) none of the above

6. Q-Switched Laser is a
(a) continuous Laser
(b) short burst Laser
(c) Laser produced by Switching operation
(d) low powered laser

7. Optical communication is feasible only in IR range because
(a) it offers minimum dispersion
(b) it offers minimum attenuation
(c) both (a) and (b)
(d) none of the above

8. Silicon photodetector can only used to detect wavelength in the
(a) 1st window of optical communication
(b) 2nd window of optical communication
(c) 3rd window of optical communication
(d) both 2nd and 3rd window.

9. Material dispersion of an optial fibre vansihes if:
(a) refractive index of the core varies lineraly with wavelength
(b) refractive index of core remains constant with wavelength
(c) both (a) and (b)
(d) neither (a) nor (b)

10. Invention of optical windows are:
(a) Five (b) three
(c) two (d) four

11. In comparison to LED, Laser has
(a) no tuning arrangements
(b) higher emission efficiency
(c) narrow spectral width
(d) provision for confinement

12. The cable that offers lowest dispersion is
(a) SMSI (b) MMSI
(c) MMGI (d) None of these

13. A graded index profile in optical fibre cable provides:
(a) Less waveguide dispersion than SI profile
(b) Less material dispersion than SI profile
(c) Less attenuation in SI profile
(d) Less modal dispersion than SI profile

14. Normalised frequency (v) for a single mode cable is
(a) $v \leq 2.4$ (b) $2.4 \leq n \leq 3.98$
(c) $n \leq 2.4$ (d) $0 \leq n \leq 3.8$

15. Which of the semiconductor can be used to fabricate a LED.
(a) Si (b) Ge
(c) GaAs (d) none of these

16. Which of the following condition should be obeyed by every Laser
(a) rate stimulated emission should be greater than the rate of spontaneous emission
(b) there should be no spontaneous emission
(c) rate of stimulated emission should be equal to the rate of spontaneous
(d) there should be no stimulated emission

17. If BW is the bandwidth of an optical fibre and NA is the numerical aperture then
(a) $BW \propto NA$ (b) $BW \propto \frac{1}{NA}$
(c) $BW \propto \frac{1}{(NA)^2}$ (d) $BW \propto \frac{1}{(NA)^3}$

18. In a dispersion shifted fibre desired wavelength of poeration is
(a) 1.33 μm (b) 1.55 μm
(c) 0.85 (d) none of the above

19. Noise equivalent power of a photodetector is:
(a) inversely proportional to the dark current
(b) proportional to the square root of the dark current
(c) directly proportional to the dark current
(d) none of the above

20. To be used as photodetector PIN diode should be connected
(a) forward bias (b) reverse bias
(c) zero bias (d) either (a) or (c)

21. SJT is used as
(a) transmitter
(b) receiver
(c) both transmitter and Receiver
(d) none of these

22. Which of the following is not applicable for Laser
(a) spatial coherence
(b) temperal cohorence
(c) both (a) and (b)
(d) none of these

23. Which of the following is not applicable for Laser
(a) no tuning arrangement
(b) higher emission efficiency
(c) narrow spectral width
(d) provision for confinement

24. The configuration utilized by a line communication system capable of transmitting a data at the rate of 100 Mb/s is
(a) coaxial cable system
(b) open line system
(c) optical fibre system
(d) twisted wire system

25. Which of the following is the transmission frequency in optical fibre.
(a) 10^9Hz (b) 10^{11}Hz
(c) 10^{14}Hz (d) None

26. Optical fibre was invented in
(a) 1950 (b) 1960
(c) 1970 (d) 1975

27. Light propagates along optical fibre by
(a) total internal reflection
(b) total internal refraction
(c) both (a) and (b)
(d) either (a) or (b).

28. Optical fibre can be satisfactorily operated if
(a) RI of the core is lesser than that of cladding
(b) RI of the core is greater than that of cladding
(c) RI of the core is equal to the that of cladding
(d) None of the above

29. RI of the core of an optical fibre is higher than that of cladding because of
(a) better confinement of light
(b) maximum distance operation
(c) easy to handle
(d) higher life time of the material

30. In the fabrication of optical fibre silica is used because
(a) it is cheap and easily avialable
(b) it is obtained in abundance
(c) it is obtained in purest form
(d) none of above

31. Attenuation in optical can be measure in
(a) dB/km (b) dB/m
(c) Kdb/m (d) dBm/m.

32. Transmission of signal through optical fibre is of the form of
(a) sound (b) electricity
(c) light (d) speed

33. For extremely low transmission losses in optical fibre, operating wavelength is in the range of
(a) (0.1 to 1) μm (b) (0.01 to 2) μm
(c) (0.2 to 8) μm (d) (0.0001 to 2) μm.

34. The material used for optical fibre for least losses is
(a) SiF_4 (b) NaF_4
(c) ZrF_4 (d) $NaSiF_4$

35. In communication repeaters are used to
(a) amplify the signal
(b) reshape the signal
(c) either (a) or (b)
(d) both (a) and (b)

36. The ability of a receiver used in optical fibre communication is to achieve a certain performance levels depends on
(a) type of the photo-detector
(b) effect of noise in the system
(c) the characteristics of the successive amplifying stages
(d) all the three mentioned above

37. Which of the folliwing is more sensitive
(a) PIN diode
(b) APD
(c) neither (a) or (b)
(d) either (a) or (b)

38. At critical angle of incidence
(a) light ray in air is perpendicular to the glass surface
(b) light ray in air is parallel to the glass surface
(c) there is no such relation
(d) none of the above

39. Which of the following cannot be used for multimode fibre optic communication
(a) stepped index
(b) graded index
(c) either (a) or (b)
(d) none of (a) and (b)

40. Refractin of light ray at the interface is a result of
(a) difference in speed of light in two material having same refractive indices
(b) same speed of light in two materials having difference in refractive indices
(c) Difference in spped of light in two materials having different refractive indices
(d) none of the above

41. Which of the following is having the highest refractive index
(a) diamond (b) air
(c) water (d) glass

42. An optical repeater consists of
(a) series of receivers
(b) series of transmitter
(c) A receiver and a transmitter placed back to back
(d) none of the above

43. Function of receiver in optical fibre is to
(a) reshape the degraded signal only
(b) only amplify of degraded signal
(c) both amiplify and reshaped the degraded signla
(d) none of the above

44. Photo detector is a
(a) traingular device
(b) square law device
(c) linear device
(d) both (a) and (b)

45. Photodetectors used in optical fibre is
(a) PIN, APDs
(b) PIN, Gunn diodes
(c) APD, Gunn diodes
(d) none of the above

46. Photo-detector converts
(a) optical power into electric voltage
(b) optical power into electric current
(c) electrical currents into optical power
(d) none of the above

47. In the wavelength range (11–6) μm, the principal optical source material is
(a) In GaAsBr
(b) In GaAsF
(c) In GaAsP
(d) None of (a), (b) and (c)

48. Source of light for optical fibre is
(a) PIN diode
(b) Photodiode
(c) LED
(d) none of these

49. Square-law device in optical source
(a) Linear variation in drive voltage results in a corresponding linear change in the optical output power.
(b) Linear variation in the drive current results in a corresponding linear change in the output power.
(c) Non-linear variation in the drive current result in a coreesponding linear change in the optical output power.
(d) None of the above

50. In the first window of optical fibre, light sources are generally.
(a) GaAIP (b) GaABr
(c) GaAIAs (d) GeAIAs

51. Distance bandwidth product is maximum in
(a) step index multimode fibre
(b) GI multimode fibre
(c) single mode fibre
(d) None of the above

52. Dispersion in silica fibrees is minimum at λ =
(a) 850 nm (b) 1300 nm
(c) 1550 nm (d) 1800 μm

53. Band gap energy of a material is 1.24 eV: wavelength of peak emission is
(a) 0.75 μm (b) 1.0 μm
(c) 1.24 μm (d) 2.24 μm

54. – 23 dBM indicates a power of
(a) 23 mwatts (b) 10 mwatts
(c) 5 mwatts (d) 100 mwatt

55. Responsivity of a photodiode is specified as
(a) amps/watts (b) watts/amp
(c) volts/watt (d) watt/volt

56. Graded index fibre with the following value of index profile, (α), results in minimum dispersion.
(a) 1.0 (b) 2.0
(c) 3.0 (d) 4.0

57. The V-parameter for an optical fibre is 50. The number of modes in that fibre is approximately.
(a) 50 (b) 1250
(c) 2500 (d) 1400

58. Spectral width of LED is typically.
(a) 20–100 nm (b) 1–10 nm
(c) 500–1000 nm (d) 100–500 nm

59. Scattering loss in optical varies with wavelength as
(a) $\frac{1}{\lambda^2}$ (b) $\frac{1}{\lambda^3}$
(c) $\frac{1}{\lambda^4}$ (d) $\frac{1}{\lambda}$

60. The high impedance front-end has the advantage of:
(a) equalisation is not required
(b) low noise
(c) wide dynamic range
(d) none of the above

Answers Set-1

1. (b)	2. (a)	3. (a)	4. (d)
5. (a)	6. (b)	7. (b)	8. (a)
9. (c)	10. (b)	11. (c)	12. (a)
13. (c)	14. (a)	15. (c)	16. (a)
17. (d)	18. (b)	19. (b)	20. (b)
21. (c)	22. (c)	23. (a)	24. (c)
25. (b)	26. (b)	27. (a)	28. (b)
29. (a)	30. (c)	31. (a)	32. (c)
33. (c)	34. (c)	35. (d)	36. (d)
37. (b)	38. (b)	39. (b)	40. (c)
41. (a)	42. (c)	43. (c)	44. (b)
45. (a)	46. (c)	47. (c)	48. (c)
49. (b)	40. (c)	51. (c)	52. (b)
53. (b)	54. (c)	55. (a)	56. (b)
57. (b)	58. (a)	59. (c)	60. (b)

MCQ Set–2

1. The **optical spectrum** has wavelengths ranging from:
   (a) 10 nm to 10^6 nm (b) 10 nm to 770 nm
   (c) 370 m to 770 nm (d) 370 nm to 10^6 nm
2. The **visible spectrum** has wavelengths ranging from:
   (a) 10 nm to 10^6 nm (b) 10 nm to 770 nm
   (c) 370 nm to 770 nm (d) 370 nm to 10^6 nm.
3. A certain emissive material has a work function of 2.24 e V, the radiations emitted from it will have **threshold length** of approximately:
   (a) 900 nm (b) 600 nm
   (c) 533.6 nm (d) 527.6 nm
4. In order that **photoelectric emission** to be possible over whole of the visible spectrum of 370 nm - 770 nm, the work function of the photoemission surface should be aproximately:
   (a) 1.61 to 3.35 eV (b) 1.3 to 2.25 eV
   (c) 1.5 to 4.206 eV (d) 0.231 to 5.62 eV
5. When light beam is made incident on a surface which is rough as compared with the wavelength of light.
   (a) Specular reflection takes place
   (b) Diffused reflection takes place
   (c) Absorbtion takes place
   (d) Transmittance take place
6. **Radiometry** refers to measurement of optical rediations.
   (a) from a wavelength of 370 nm - 770 nm
   (b) from a wavelength of 10^{-12} nm to 10^6 *nm*
   (c) from a wavelength of 10^2 nm to 10^6 nm
   (d) from a wavelength of 10 nm to 10^6 nm
7. The responce of human eye under normal light conditions is called.
   (a) stoppoic vision and it peaks at awavelength of 555 nm.
   (b) stoppic vision an it peaks at a wvelength of 507 nm.
   (c) photopic vision and it peaks at a wevalength of 507 mn
   (d) photopic vision and it peaks of 555 nm
8. A **light source** may be considered as a point soutce if its diameter is.
   (a) less than 10% of distance it and the detector.
   (b) less than 1% of the distance between it and the dejector.
   (c) more less 10% of the distance between it and the detector.
   (d) more than 2% of the distance between it and the detector.
9. The units of luminous Intensity are:
   (a) lux (b) Cd
   (c) W/sr (d) Cd/sr
10. The units of lumuance are:
    (a) nit (b) lm/sr - m^2
    (c) Cd/m^2 (d) all of the above.
11. The errors invovled in **photometric measurements** are of the order of the order of:
    (a) 1% (b) 1–5%
    (c) 10-20 % (d) 40–50%
12. The desirable properties of a **laser** beam are:
    (a) it has a high radiant energy
    (b) it can be polarized
    (c) it is monochromatic and coherent
    (d) all of the above
13. The current through a **photoemissive tube depends** upon:
    (a) intensity of light incident upon it.
    (b) colour and wave length of incident light
    (c) voltage applied between cathode and anode of tube
    (d) all of the above
14. The **luminous sensitivity** of gas filled photelectric cells varies between:
    (a) 10 to 100 $\mu A/lm$
    (b) 40 to 150 $\mu A/lm$

(c) 0.002 to 0.1 $\mu A/\mu W$

(d) 0.01 to 0.15 $\mu A/\mu W$

15. A **photomultiplier tube** uses 10 stages. The sensitivity of each dynnode of the tube is 100 $\mu A/lm$ the overall sensitivity of the photomultiplier tube, if the gain of each stage is 2.7, is approximately:

(a) 2.06 A/lm (b) 270 $\mu A/lm$

(c) 2700 $\mu A/lm$ (d) None of above

16. A **photoconductive cell** may be called as:

(a) photoresistive cell

(b) photo resistor

(c) light dependent resistor

(d) all of the above

17. The resistance of a **photo resistive cell**

(a) increases with increase in light intensity

(b) remains constant irrespective of the light inten sity incident upon it

(c) decreases with increase in light intensity in a lin ear manner

(d) decreases with increase in light intensity in an expenential manner.

18. Photodiodes

(a) do not have any response in darkness

(b) have a reponse even in darkness

(c) have a leakage current even in darkness which is called dark current that doubles for every $10^{o}C$ increase in the ambient temperature that which decrease the signal to noise ratio.

(d) have a leakage current even in darkness which is called current that doubles for every increase of $5^{o}C$ in temperture and which increase the signal to noise ratio.

19. The sensitivity of a photovaltaic cell

(a) is almost constant when it is short circuited.

(b) decrease with increase of load resistance is abouT 1000 W

(c) is most negligible when the load resistance

(d) all of the above

(e) none of the above

20. In case of measurement of Intensity of a test source is used. The photometric head used when comparing two lights to different colours, is:

(a) Bunsen photometer head

(b) Constast type Lummer Brodhun photometer head

(c) Comparion type Lummer Brodhum photometeR head.

(d) Flicker photometer head.

21. One of the combination of materials used for **optical fibers** is:

(a) Copper core and glass cladding

(b) Glass core and aluminium cladding

(c) Glass core and plastic cladding

(d) Plastic core and glass cladding

22. For **fiber optical sensors**:

(a) single mode transmission is used.

(b) multimode transmission is used.

(c) both single and multimode transmissions are commonly used

(d) none of the above

23. In **optical fibres**, if the total reflection is to takes place, the following conditions must be satisfied.

(a) $n_2 < n_1$ *and* $\phi_1 > \cos^{-1}(n_2/n_1)$

(b) $n_1 < n_2$ *and* $\phi_1 < \cos^{-1}(n_2/n_1)$

(c) $n_2 < n_1$ *and* $\phi < \cos^{-1}(n_2/n_1)$

(d) $n_2 > n_1$ *and* $\phi_1 > \cos^{-1}(n_2/n_1)$

24. In a **graded index fibre**, the total reflected light takes a

(a) straight line path (b) parabolic path

(c) cliptical pathe (d) circular path

25. The optical sensors may be classifled as:

(a) Intensity and intrinsic types

(b) Interferometric and extrinsic types

(c) Intensity or Instrinsic type and interfermetric or extrimsic types

(d) all of the above

Answers Set-2

1. (a)	2. (c)	3. (c)	4. (a)
5. (b)	6. (d)	7. (d)	8. (a)
9. (b)	10. (d)	11. (c)	12. (d)
13. (d)	14. (b)	15. (a)	16. (d)
17. (d)	18. (c)	19. (d)	20. (d)
21. (c)	22. (a)	23. (a)	24. (b)
25. (c)			

MCQ Set–3

1. Korotkoff sound are used in:
   (a) Blood flow measurements
   (b) Blood pressure measurements
   (c) Heart valve functioning
   (d) As a reference for sound level measurement.
2. CMRR in an ECG amplifier when input impedance is 2 MW and difference of electrode skin contant as 1 $k\Omega$ is
   (a) 1000 (b) 2000
   (c) 3000 (d) 5000
3. Intensity of an X-ray beam of intensity I_o when passed through a substance of thickness X reduces to (μ = absorption co-efficient of substance).
   (a) $I = I_0\, e^{ux}$ (b) $I = I_0\, e^{ux^2}$
   (c) $I = I_0\, e^{-ux^2}$ (d) $I = I_0\, e^{-ux}$
4. Absorption co-efficient μ of a substant when X-ray falls on it is (where, Z = atomic number absorbed atom, λ = wavelength of X - ray, ρ = Lenght of absorber, C = Constants).
   (a) $Z^3\, \lambda^2 \rho$ (b) $C\, Z^4\, \lambda^3 \rho$
   (c) $Z^2\, \lambda^2 \rho$ (d) $Z\, \lambda^2 \rho^2$
5. A graphic record of heart sounds is known
   (a) Electrocardiogram (b) Phonocardiogram
   (c) Vectorcardiogram (d) Electroencephatogram
6. Instrument measuring volume changes in any part of the body due to pulsations of blood occuring with each heart beat is:
   (a) Plethysmograph (b) Arithmetic unit.
   (c) Ventilator (d) Cardiac pacemaker
7. The triangle whose sides represent the line along which 3 projections of ECG are measured is known a
   (a) Galvani tringle (b) Einthoven triangle
   (c) Abbott triangle (d) Korotkoff triangle
8. Sequence, direction and distribution of forces generated by heart are showm by:
   (a) E C G (b) Vectorcardiogram
   (c) Electroencephalogram(d) Rheoencephalogram
9. Heart sounds range in the frequency form
   (a) 2 kHz to 6 kHz (b) 6 kHz to 10 kHz
   (c) 5 to 2 kHz (d) 5 to 100 kHz
10. To produce a magnified image atleast distance of distinct vision with the help of a simple microscope, the object should be placed at:
    (a) focus
    (b) a point far away form focus
    (c) a point between focus and optical centre
    (d) none of the above
11. If the least distance of distinct vision is 25 cm, then a convex lens of focal length 5 cm acts as a magnifier of magnifying power.
    (a) 5 (b) less then 5
    (c) 6 (d) more then 6
12. The final image formed by a microscope is:
    (a) real and magnified (b) real and diminished
    (c) virtual and magnified (d) virtual and diminished
13. When the length of a microscope tube is increased, its maginfying power.
    (a) decreases (b) increases
    (c) does not change (d) cannot be predicted
14. To increase the magnifying power of a telescope, one should increase
    (a) the focal length of the objective
    (b) the focal lenght of eye piece
    (c) aperture of the objective
    (d) aperture of the eye piece
15. A telescope in normal adjustment has a distance between the objective and eye-piece equal to 100 cm. If the magnifying power of the telescope is 24, the focal length of the two lenses are
    (a) 96 cm, 4 cm (b) 124 cm, 100 cm
    (c) 100 cm, 76 cm (d) 62 cm, 38 cm
16. An observer looks at a tree of height 15 m with a telecope of magnifying power 10. To him, tree appears.
    (a) 10 times taller (b) 15 times taller
    (c) 10 time nearer (d) 15 times nearer
17. Magnifying power of a telescope is *m*. If the focal length of the eye-peice is doubled, then its magnifying power will become.
    (a) 2 m (b) m/2
    (c) $\sqrt{2m}$ (d) 3 m
18. A film projector lense magnifies 10 cm square film strip on a screen. If the linear magnification is 4, the area of the magnified film on the screen is:
    (a) 1600 cm^2 (b) 400 cm^2
    (c) 800 cm^2 (d) 400 cm^2
19. The resolving power of telescope depends upon.
    (a) the focal length of the eye lens
    (b) the focal length of the object lens

(c) the length of the telescope
(d) the diameter of the object lens.

20. A fly is sitting on the objective of a telescope pointed to the moon. What effect is expected in the photograph of the moon taken through the telescope.
(a) The entire yield of vision is blocked.
(b) There is an image of the fly on the photograph
(c) There is no effect at all
(d) the image of the moon is of lesser intensity.

21. For observing a circket match, binocular is preferred to terrestrial telescope for the reason that
(a) the binocular is very easy to handle
(b) the binocular provides three dimenison vision
(c) the binocular produces image free chromatic aberration
(d) the binocular produces erect image

22. CT number (Hounsfield unit) H of a tissue when u is attenuation co-efficient of X-ray and μ water attenuation co-efficient of water (0.205) is given by

(a) $\frac{\mu_{water}}{\mu_{water}-\mu}\times 1000$ (b) $\frac{\mu}{\mu_{water}-\mu}\times 1000$

(c) $\frac{\mu-\mu_{water}}{\mu_{water}}\times 1000$ (d) $\frac{\mu_{water}-\mu}{\mu}\times 1000$

23. In NMR blood flowmeter, **larms** frequencies is given as:

(a) $\frac{2\pi}{v}$ (b) $\frac{r}{B_o}$

(c) rB_0 (d) $\frac{v}{B_0}$

24. In nuclear instrumentation, heaviest and flow moving particles are:
(a) Beta particles (b) Alpha particles
(c) Gamma rays (d) Protons

25. Angular magnitication of a telescope if α = angle subtended by object at eye and β = angle subtended by image at eye.

(a) $\frac{\alpha}{\beta}$ (b) $\alpha\beta$

(c) $\frac{\beta}{\alpha}$ (d) $\frac{\alpha}{\beta^2}$

26. Coherence length of He-Ne laser beam is 120 cm. Its conherence time in seconds is:
(a) 2×10^{-9} secs (b) 4×10^{-9} secs
(c) 6×10^{-9} secs (d) 8×10^{-9} secs

27. Resolving power of a telescope is proportional to:

(a) $\frac{\lambda}{a}$ (b) λa

(c) $\frac{a}{\lambda}$ (d) $\frac{\lambda}{a^2}$

28. When light passes form a medium of refractive index n_1 into a medium of refractivbe index n_2, reflection at boundary.

(a) $(n_2-n_1)^2$ (b) $\left(\frac{n_2-n_1}{n_2+n_1}\right)^2$

(c) $\frac{1}{(n_2-n_1)^2}$ (d) $(n_2+n_1)^2$

29. Angle of minimum deviation for a prism of refractive index 1.5 is equal to the angle of prism. The angle of prism is:
(a) 62° (b) 41°
(c) 82° (d) 31°

30. The refracting angle of a prism is A and the retractive index of the prism is cot A/2. The angle of minimum deviation is:
(a) $180°-3A$ (b) $180°+2A$
(c) $90°-A$ (d) $180°-2A$

31. An equilateral prism is placed on the prism table of a spectrometer in the position of minmum deviation. If the angle of incidence is $60°$, the angle of deviation the ray is:
(a) 90° (b) 60°
(c) 45° (d) 30°

32. A ray is incident at an angle of incidence is on one surface of a prism of small angle A and emerges normally from the opposite surface. If the refractive index of the material of the prism is μ, the angle of incidence *i* is nearly equal to:
(a) A/μ (b) $A/2\mu$
(c) μA (d) $\mu A/2$

33. The photo diode as compared to a photo transistor has:
(a) faster switching time
(b) lower sensitivity

(c) higher size for the same value of output current
(d) all of the above

34. An achromatic combination of lenses produces
(a) images in black and white
(b) cololured images
(c) images unaffected by variation of refractive index with wavelength.
(d) highly enlarged images.

35. The refractive indices of glass for violet and red light are 1.54 and 1.52 respectively. If the angle of prism is $10°$, the angular dispersion is
(a) 0.02° (b) 0.2°
(c) 3.06° (d) 30.6°

36. Which of the following phenomenon is not explained by Huygen's construction of wavefront?
(a) Refraction (b) Reflaction
(c) Diffraction (d) Origin of spectra.

37. At sunrise or sunset, the sun looks reddish, because
(a) the sun is coldest at these times
(b) of the effects of reflection and refraction.
(c) the sun is hottest at these times
(d) of the scattrering of light.

38. The sky appears blue because.
(a) red light is absorbed
(b) blue light is scattered the most.
(c) blue light is absorbed.
(d) it is its natural colour

39. One cannot see through fog, because
(a) fog absorbs light
(b) the refractive index of fog is infinity
(c) light suffers total internal reflection at the droplet in a fog.
(d) light is scattered by the droplets in fog

40. An object is placed at a distance of $f/2$ from a convex lens. The image will be
(a) at $3f/2$, real and inverted
(b) at $2f$, virtual and erect
(c) at f, real and inverted
(d) at one of the foci, virtual and double its size.

41. A film projector magnifies a 100 cm^2 film strip on a screen. If the lineat magnification is 4, the area of the magnified film on the screen is:
(a) 1600 cm^2 (b) 400 cm^2
(c) 800 cm^2 (d) 200 cm^2

42. A converging lens is used to form an image on a screen. When the upper half of the lens is convered by an opaque screen.
(a) half the image will disappear
(b) complete image will be formed
(c) intensity of image will decrease
(d) intensity of image will increase

43. If in a plano convex lens radius of curvature of convex surface is 10 cm and the focal length of the lens is 30 cm, the refractive index of the material of the lens will be:
(a) 1.5 (b) 1.66
(c) 1.33 (d) 3

44. A thin symmertrical convex lens of refractive index 1.5 and radius of curvature 0.3 m is put in water of regractive index $\frac{4}{3}$. Its focal length is
(a) 0.15 m (b) 0.30 m
(c) 0.60 m (d) 1.20 m

45. Focal length of a convex lens of refractive eindex 1.5 and 2 cm. Focal length of lens when immersed in a liquid of refractifve index of 1.25 will be
(a) 10 cm (b) 7.5 cm
(c) 5 cm (d) 2.5 cm

46. A convex lens of glass is immersed in water. Compared to its poower in air, its power in water will
(a) diminish for red light (b) remain the same
(c) increase (d) diminish

47. A lens behaves as a converging lens in air and a diverging lens in water. The refractive index of the material of the lens is:
(a) equal to unity
(b) equal to 1.33
(c) between unity and 1.33
(d) greater than 1.33

48. A double convex lens of refractive index μ_1 is immersed in a liquid of refractive index μ_2. The lens will act as
(a) diverging lens, if $\mu_1 > \mu_2$
(b) diverging lens, If $\mu_1 < \mu_2$
(c) converging lens, if $\mu_1 > \mu_2$
(d) converging lens, if $\mu_1 < \mu_2$

50. Two lense having powers + 6D and $-4D$ are placed in contact. The power of the combination is:

(a) $-2D$ (b) $-4D$

(c) $+4D$ (d) $+2D$

51. Two coherent sources are virtual in:

(a) Young's double slit experiment

(b) Leoyd's mirror

(c) Fresnel's prism

(d) All of the above

52. A photo diode function as:

(a) photo conductive device while working with re verse voltage.

(b) photo voltaic devices while working without a reverse voltage.

(c) photo voltaic devces while working with or without reverse voltage

(d) none of the above.

53. Resting potential of normal living cell is about.

(a) + 20 mV (b) + 0.1 mV

(c) - 10 m V (d) - 210 m V

54. When ECG is recorded by connecting 2 - electrodes, One in left arm and other in left leg, recording is made in.

(a) Lead I configureation only

(b) Lead II configurations only

(c) Both Lead I and II configuration only

(d) Lead II configuration.

55. The lung volume expired during maximal forced expiration following a maximal inspiration is known as:

(a) vital capacity

(b) total functional residual capacity

(c) total lung capacity

(d) total ventillatory capacity

56. In diagnostic applications, ultrasound is both generated and detected by the piezoelectric effect in the

(a) MF and HF ranges

(b) LF range

(c) VLF range

(d) EHF range

57. Normal human heart, at rest,pumps blood at a rate of about (in letres/mt).

(a) 3 (b) 5

(c) 7 (d) 11

58. Which of the following materials may be used to produce polaroids

(a) Tourmaline

(b) Calcite

(c) Quartz

(d) Crystals of quinine iodosulphate

59. When a polaroid is rotated, the intensity of light varies but never reduces to zero. It shows that the incident light is:

(a) Completely plane polarised

(b) Partially plane polarised

(c) Unpolasrised plane polarised

(d) None of the above.

60. A convex mirror froms an image which is $\frac{1}{n}$ times the size of the object. I the focal length of the mirror is f, then the distance of the object is:

(a) nf (b) f/n

(c) $\frac{n-1}{f}$ (d) $(n+1)f$

61. Focal length of concave mirror produicing magnification m is:

(a) $\frac{(m-1)f}{m}$ (b) $\frac{(m+1)f}{m}$

(c) $\frac{(m-1)}{f}$ (d) $\frac{(m+1)}{f}$

62. When light travels form one medium to another, which are separated by a sharp boundry, the characteristic which does not change is:

(a) velocity (b) wavelength

(c) frequency (d) amplitude

63. Green light of wacelength 5460 incident on a air-glass interface. It refractive index of glass is 1.5 wavelength of light in glass would be

(a) $3460\,\mathring{A}$ (b) $5460\,\mathring{A}$

(c) $4861\,\mathring{A}$ (d) none of these

64. Air has refractive index 1.0003 thickness of air column, which will have more wavelength of yellow light (6000) than in the same thickness of vacuum is:

(a) 2 mm (b) 2 cm

(c) 2 m (d) 2 km

65. Time taken by the sum light to pass through window of thickness 4 mm whose refractive index is 1.5 is

(a) $2\times10^{-8}\ s$ (b) $2\times10^{8}\ s$

(c) $2\times10^{-11}\ s$ (d) $2\times10^{11}\ s$

66. The refractive index of material varies with wave length as:
(a) $\mu \propto \lambda$ (b) $\mu \propto \frac{1}{\lambda}$
(c) $\mu \propto \lambda^2$ (d) $\mu \propto \frac{1}{\lambda^2}$

67. The speed of light in air is $3\times10^8\ ms^{-1}$ ms. What will be its speed in diamond where refractove Index is 2.4?
(a) $3\times10^8\ ms^{-1}$ (b) $322\ ms^{-1}$
(c) $1.25\times10^8\ ms^{-1}$ (d) $7.2\times10^8\ ms^{-1}$

68. A beam monochrolmatic light is refracted from vacuum into a medium of refractive index 1.5. The wavelength of refracted light will be.
(a) same
(b) dependent on intersity of refracted light
(c) larger
(d) smaller

69. A ray of light is travelling throught a medium of regractive index $\sqrt{2}$ w.r.t. air. When it is incident on the suface at 45° which of the following will take place?
(a) Angle of refraction will be 45°
(b) Angle of refraction will be 90°
(c) The ray will be internally reflected
(d) None of the above

70. The cirtical angle for right, when it passes from glass to air is minimum for
(a) red (b) green
(c) yellow (d) violet

71. A fish looking up through the water sees the outside world contained in a circular harizon.If the refractive index of water is 4/3 and fish is 12 cm below the surface of water, the radiuos of the cirucular horizon is:
(a) $36\sqrt{5}\ cm^2$ (b) $4\sqrt{5}\ cm^2$
(c) $36\sqrt{7}\ cm^2$ (d) $36\sqrt{7}\ cm^2$

72. The twinking of stars is due to
(a) periodic brust of light from the star
(b) interference between light form the sum and star
(c) partial asbsorption of light in the atmosphere.
(d) refractive index fluctuation in the atmosphere.

73. The sun become visible before the actual sunrise and remains visible even after the actual sunset. It is because of:
(a) scattering of light (b) diffraction of light
(c) refraction of light (d) dispersion of light

74. Hygen's principle of secondary wave is used to
(a) Obtain new position of wave front geometrically
(b) Explain the principle of superposition of waves
(c) Explain the interference phenomenon
(d) Explain the polarisation

75. Two points are situated out the same distance from a source of light but in opposite direction. The phase diference between the light waves from the 2-point will be
(a) π (b) zero
(c) $\pi/2$ (d) none of the above

76. Which of the following produce a plane wave front
(a) Point source (b) Line source
(c) Extended soutce (d) None of the above

77. As a plane wave front propagates, its radius of curvature.
(a) decreases
(b) increases
(c) first increases then decteases
(d) remains infinity

78. True about a plane wave front propagating in a medium.
(a) it propagates parallel to itself
(b) it cannot propagate in medium
(c) it changes to spherical wavefront
(d) it changes to cylindrical wavefront

79. Doppler effect in sound and light are
(a) exactly identical to each other
(b) both synmmetrical
(c) both asymmertical
(d) different from each other

80. Correct statement about doppler effect
(a) Doppler effect in light merely depends upon relative velocity of source and observer
(b) Doppler effect in light is saymmetrical
(c) Doppler shift (Δv) is more when source moves with the velocity v w.r.t. stationary observer than that what it is when observer moves with velocity v w.r.t. stationary source
(d) Doppler shift (Δv) is more when source moves with the velocity w.r.t. stationary source than that what it is when source moves with velocity v w.r.t. stationary observer.

81. 6000 Å spectral line emitted by a star is found to be red shift by 20 Å. The speed with which star is recording is given by

(a) $3\times10^8\ m/s$ (b) $3\times10^5\ m/\sec$

(c) $10^6\ m/\sec$ (d) none of these

82. Interference phenomenon deffers from diffraction in that unlike diffraction.

(a) Interference fringes are of varying intensity

(b) Interference cannot be observed with white light

(c) Interference minima may be perfectly dark.

(d) Interference fringes are of unequal width.

83. Width of diffraction fringes varies

(a) Directly as distance between slit and screen

(b) Inversely as wavelength

(c) Directly as width of slit

(d) None of the above

84. Replacing red light by blue light in a single slit diffraction pattern.

(a) No change in diffraction pattern

(b) Diffraction pattern disappears

(c) Diffraction fringes become narrower and crowded together.

(d) Diffraction fringes become broader and farther about

85. Scattering of light is a direct consequence of.

(a) Interference (b) Dispersion

(c) Diffraction (d) Reflection

86. A ray of light is indent on a glass plate angle 60° Reflacted and regracted rays found to be mutually perpendicular. The refractive index of glass.

(a) $\frac{2}{\sqrt{3}}$ (b) $\sqrt{3}$

(c) 1.5 (d) 1.62

85. Incorrect statement

(a) Doppler effect in sound is symmetrical

(b) Doppler effect in light is symmetrical

(c) Red shift in doppler effect in light predicts expanding mature of universe

(d) A spectral line shifts towards the blue end, if source of light moves towards the oberver.

86. An isotropic source of 2-candela produces light flux equal to

(a) 2π lumen (b) 4π lumen

(c) 6π lumen (d) 8π lumen

87. The tangent of angle of prolarisation is equal to refractive index of the material is called

(a) Malu's law (b) Brewster's law

(c) Bragg's law (d) None of the above

89. Which of the following types of wave cannot be polarised.

(a) Longitudinal waves in gas

(b) Transverse waves in string

(c) Radio waves

(d) X-rays

90. In double slit experiment, the 2- slits are illuminated with 2-induendent identical sources of light. What happens to interference pattern?

(a) No interference pattern is produced

(b) Intensity of bright fringes gets doubled

(c) Intensity of bright fringes becomes 4 times

(d) Two sets olf interference patterns are produced

91. In Youngs double slit experiment the fringe width is

(a) directly proportional to distance between the sources.

(b) inversely proportional to distance between the sources

(c) inversely proportional to distance of screen form the soruces

(d) inversely proportional to wavelength of light used.

92. Intensity of light issuing out of 2-slits in Young's experiment is in the ratio 1:4. The intensitt of the minimum to maximum will be in the ratio

(a) 1 : 2 (b) 1 : 4

(c) 1 : 9 (d) None of the above

Answers Set-3

1. (b)	2. (a)	3. (d)	4. (b)
5. (b)	6. (a)	7. (b)	8. (b)
9. (c)	10. (c)	11. (c)	12. (c)
13. (b)	14. (a)	15. (a)	16. (c)
17. (b)	18. (a)	19. (d)	20. (d)
21. (b)	22. (c)	23. (c)	24. (b)
25. (c)	26. (b)	27. (c)	28. (b)
29. (c)	30. (d)	31. (b)	32. (c)
33. (d)	34. (c)	35. (b)	36. (d)
37. (d)	38. (b)	39. (d)	40. (d)
41. (a)	42. (c)	43. (c)	44. (d)
45. (d)	46. (d)	47. (c)	48. (c)
50. (d)	51. (d)	52. (c)	53. (d)
54. (c)	55. (a)	56. (b)	57. (d)
58. (b)	59. (c)	60. (b)	61. (c)

62. (a)	63. (a)	64. (c)	65. (d)
66. (c)	67. (d)	68. (b)	69. (d)
70. (d)	71. (d)	72. (c)	73. (a)
74. (b)	75. (d)	76. (d)	77. (a)
78. (d)	79. (a)	80. (c)	81. (c)
82. (a)	83. (b)	84. (c)	85. (b)
86. (a)	87. (b)	88. (b)	89. (a)
90. (a)	91. (b)	92. (c)	

MCQ Set–4

1. What is the Dispersive power of a prism to separate 2-lines of wavelengths 5997 Å and 6003 Å
(a) 0.001 (b) 0.002
(c) 0.003 (d) 0.004

2. What is the refractive index of glass (5 μm thick) introduced in the path of one of the interfering beams in a prism with λ of light used as 589 nm and central bright fringe shifts a distance equal to width of 5 fringes.
(a) 1.589 (b) 1.329
(c) 1.129 (d) 1.789

3. What is the separation on photographic plate in the velocity selector of **Bain bridge mass spectrocope** with E = 200 V/cm and H = 0.2 Teasla. H in main chamber as 0.3 Teasla. Mixture of signaly charged ions of carbon isotopes C^{12} and C^{13} is analysed.
(a) 0.69 cm (b) 0.59 cm
(c) 0.49 cm (d) 0.39 cm

4. What is the μ of air with a shift of 150 fringes of wave lenght 4000 Å in a **Michelson interferometer** when all air is removed tube length is 20 cm.
(a) 1.00000 (b) 1.00015
(c) 1.00050 (d) 1.00075

5. Refractive index of a liquid intoduced below the convex sourface when diameter of 5th dark ring is redcued to (1/2) of its value is
(a) 2 (b) 4
(c) 6 (d) 8

6. Wavelength of light used in **Newton's rings experiment** when radius of curvaure of plano-convex lens is 1 m and diameters of nth and (n + 14) th rings are 4.2 mm and 7.0 mm.
(a) 3600 A (b) 4600 A
(c) 5600 A (d) 6600 A

7. In **Young's interference experiment** two slits are illuminated with orange light of wavelength 6000 Å. The interferece pattern is observed on a screen very far from the slits. If the central bright fringe is numbered zero. What must be the path difference for the light from the two slits at the fouth-right fringe?
(a) 1.4 μm (b) 2.4 μm
(c) 3.4 μm (d) None of the above

8. Two wavelengths of light λ_1 and λ_2 are sent through a **Young's double-slit apparatus** simulrancously. What most be true concerning λ_1 and λ_2 if the third - order λ_1 bright fringe is to coincide with the fourth - order λ_2 fringe?
(a) $\lambda_1 = \lambda_2$ (b) $\lambda_1 = 2\lambda_2$
(c) $\lambda_1 = 3\lambda_2$ (d) $\lambda_2 = \frac{3}{4}\lambda_1$

9. Two wavelengths λ_1 and λ_2 are used in the **double-slit experiment**. If one is 430 nm, what value must the other have for the fourth-order bright fringe of one to fall on the sixth-order bright fringe of the other?
(a) $2\lambda_1 = \lambda_2$ (b) $3\lambda_1 = 2\lambda_2$
(c) $4\lambda_1 = 6\lambda_2$ (d) $\lambda_1 = 3\lambda_2$

10. Two identical radiators have a separartion $d = \lambda/8$, where λ is the wavelength of the waves emitted by either source. The phase diffrence $\Delta\phi$ of the sources is $\pi/4$. Find the **intensity distribution** in the radiatioin field as a function of the angle θ which specifies the direction from the radiators to the distant observation point P.
(a) $I_0 \sin^2\left[\frac{\pi}{8}(\sin\theta + 1)\right]$
(b) $4 I_0 \cos^2\left[\frac{\pi}{8}(\sin\theta + 1)\right]$
(c) $I_0 \cos^2\left[\frac{\pi}{8}(\sin\theta + 1)\right]$
(d) $4 I_0 \sin^2\left[\frac{\pi}{8}(\sin\theta + 1)\right]$

11. A flake of glass of index of refraction 1.5 is placed over one of the openings of a **double-slit apparatus**. There is a displacement of the interference pattern

through seven successive maxima toward the side where the flake was placed. If the wavelenght of the diffracted $\lambda = 600\ nm$, how thick is the flake?

(a) 7400 *nm* (b) 8400 *nm*
(c) 9400 *nm* (d) 1040 *nm*

12. What is the wavelenght of light when inital and final reading of a **Michelson interferometer** screw are 10.7347 mm and 10.7051 mm with 100 fringes pan.

(a) 4920 $\overset{o}{A}$ (b) 5920 $\overset{o}{A}$
(c) 6920 $\overset{o}{A}$ (d) 7920 $\overset{o}{A}$

13. What is the wavelenths in the region 4000 $\overset{o}{A}$ to 6000 Å for which interface minimum exists in a **Fabry Periot interferometer** with a plate separation of 2×10^{-6} and normal incidence of light

(a) 5714 $\overset{o}{A}$, 5000 $\overset{o}{A}$, 4444 $\overset{o}{A}$
(b) 5714 $\overset{o}{A}$, 6000 $\overset{o}{A}$, 4444 $\overset{o}{A}$
(c) 5714 $\overset{o}{A}$, 6000 $\overset{o}{A}$, 5555 $\overset{o}{A}$
(a) 5714 $\overset{o}{A}$, 7000 $\overset{o}{A}$, 5555 $\overset{o}{A}$

14. A thin parallel film of thickness 0.2 mm is formed using a liquid of refractive index 1.25. It is illuminated with a light of wavelength ranging from 4000 to 5000 $\overset{o}{A}$. It is observed at an angle of 35°. Determine the wavelenghs, absent in reflected light.

(a) 1442 *A* (b) 2442 *A*
(c) 3442 *A* (d) 4442 *A*

15. A film of liquid of refractive index 1.39 is formed and illuminated with white light. When it is observed at an angle of 30°, two consecutive dark bands are observed corresponding to wavelengths 5125 *A* and 5000 *A*. Determine the thickness of the film.

(a) $5.9\times10^{-4}cm$ (b) $6.9\times10^{-4}cm$
(c) $7.9\times10^{-4}cm$ (d) $8.9\times10^{-4}cm$

16. A wedge-shaped film of refractive index 1.4 is formed. Its is illuminated by yellow light of wavelength 5893 *A*. When viewed normally 10 bands were observed in 9 mm. Determine the anlge of wedge.

(a) 0.0134° (b) 0.0144°
(c) 0.0154° (d) 0.0164°

17. Determine the thickness of wedge-shaped air film at fifth bright band from the edg when the film is illuminated by red light of wavelength 6500 $\overset{o}{A}$ and observed normally.

(a) $1.4\times10^{-1}\ cm$ (b) $1.4\times10^{-2}\ cm$
(c) $1.4\times10^{-3}\ cm$ (d) $1.4\times10^{-4}\ cm$

18. Wavelength of light with shift of 400 fringes when movable mirror of **Michelson interferometer** moves a distance of 0.1 mm.

(a) 4000 $\overset{o}{A}$ (b) 5000 $\overset{o}{A}$
(c) 6000 $\overset{o}{A}$ (d) 7000 $\overset{o}{A}$

19. Through what distance does the mirror in **Michelson interferometer** is to be moved between 2-positions of disapperance of frings with a source which consists of 2-wavelenghts 4882 $\overset{o}{A}$ and 4886 $\overset{o}{A}$.

(a) 0.049 mm (b) 0.059 mm
(c) 0.069 (d) 0.079 mm

20. Thickness of a thin glass plate ($\mu = 15$) introduced in one of the arms of a **Michelson interferometer** using 5890 $\overset{o}{A}$ with a shift of 10 fringes.

(a) 4.89 μm (b) 5.89 μm
(c) 6.89 μm (d) 7.89 μm

21. In a certain **Young's double-slit experiment** for which $D = 1.00m$ and 0.10 cm. bright fringes are 0.5 mm a part. What wavelength of light is being used?

(a) 100 mm (b) 200 mm
(c) 300 mm (d) 500 mm

22. A Young's double-slit exeriment is set up using a light source of wavelength 500 *nm* and placing slits 2 m from an observer. Given that the angular resolution of the observer's eye is 1 in (0.000291rad), how far a part are the two slits. If he can just distinguish between the interference frings

(a) 1.32 mm (b) 1.42 mm
(c) 1.62 mm (d) 1.72 mm

23. **Laser** light (630 mm) incident on a pair of slits produces an interference pattern in which the bright fringes are separated by 8.3 mm. A second light produces an interference pattern in which the bright fringes are separated by 7.6 mm. What is the wavelength of this second light?

(a) 377 mm (b) 477 mm
(c) 577 mm (d) 677 mm

24. In a Young's double-slit experiment, the slits are 2 mm apart and are illuminated with a mixture of two wavelengths, $\lambda = 750nm$ and $\lambda' = 900nm$. At what minimum distance from the common central bright

fringe on a screen 2 m from the slits will a bright fringe from one **interference pattern** coincide with a bright fringe from the other?

(a) 1.5 mm (b) 2.5 mm
(c) 3.5 mm (d) 4.5 mm

25. Find the wave length of light incident on a convex lens of 350 cm (R) placed on a flat plate and illuminated by monochromatic light and dia of 6th bright ring as 0.68 cm.

(a) 3080 Å (b) 4080 Å
(c) 5080 Å (d) 6080 Å

26. Find the angle of wedge when light of $\lambda = 6000$ Å incident normally on a wedge shaped film ($\mu = 1.35$) and fringes are 2 mm apart.

(a) 0.0032° (b) 0.0064°
(c) 0.0048° (d) 0.0080°

27. Find the Fringe separation between 2 consecutive fringes when a wedge shaped film is illuminated by light of wavelength 4650 Å angle of wedge = 40°.

(a) 0.09 cm (b) 0.12 cm
(c) 0.15 cm (d) 0.18 cm

28. 40 fringes are seen with a biprism with sodium light, $\lambda = 5892$. Find the number of fringes obtained when Na lamp is replaced by mercury lamp $\lambda = 4358$ Å.

(a) 34 (b) 44
(c) 54 (d) 64

29. In Fresnels Biprism experiment on putting a thin plate of glass in path of one of interfering beams, centeral bright fringe shifts into posiotion pereviously occupied by 5th bright fringe, $\lambda = 6000$ Å, refractive index of glass plate = 1.5. Thickness of plate is:

(a) 4 μm (b) 5 μm
(c) 6 μm (d) 7 μm

30. In a **Doppler - shift flow velocity meter**,. If f is the transmitted frequency,
C is velocity of sound in blood and
θ is angle of inclination of incident wave to direction of blood flow
Δf is doppler shift,
the velocity of blood cells is given as.

(a) $\dfrac{2f\,\Delta f\,C}{\cos\theta}$ (b) $\dfrac{2f\,C}{\Delta f\cos\theta}$
(c) $\dfrac{\Delta f\,C}{2f\cos\theta}$ (d) $\dfrac{2f}{\Delta f C\cos\theta}$

31. Find the time spent by a photon in cavity before it is absorbed/ emitted ina **GaAs optical cavity** (lenght 200 mm)and reflectivity 0.33. Absorption loss in cavity is 10 cm^{-1}

(a) 0.31 ps (b) 0.41 ps
(c) 0.51 ps (d) 0.61 ps

32. If refractive index of water is $\eta = 1.333$ A person looks into a swimming pool at the 5-ft-deep level. How deep does it look to the person?

(a) 2.75 ft (b) 3.75 ft
(c) 4.75 ft (d) 5.75 ft

33. Light enters a glass prism having a **refracting angle** of 60°. If the angle of incidnece is 30° (incident ray is parallel to base) and the index of refraction of the glass is 1.50, What is the angle the ray leaving the prism makes with the normal?

(a) 35° (b) 44°
(c) 55° (d) 77°

34. A ray of light is incident at 30° on a prism with apex angle 55° and index of refraction 1.50. Calculate the angles the ray makes with the sides of the prism.

(a) 15.5° (b) 25.5°
(c) 35.5° (d) 45.5°

35. In **interference** pattern due to **young's double slit arrangement**, maximas are located by

(a) $y = \dfrac{D\lambda}{d}$ (b) $y = \dfrac{m\,D\lambda}{d}$
(c) $y = \dfrac{m\lambda}{d}$ (d) $\dfrac{md}{\lambda}$

36. The **interference pattern** of two identical slits separated by a distance $d = 0.25mm$ is observed on a screen at a distance of 1 m from the plane of the slits. The slits are illuminated by monochromatic light of wavelength 589.3 mm travelling perpendicular to the plane of the slits. Bright bands are observed on each side of the central maximum. The separation between adjacent bright bands is:

(a) 1.86 mm (b) 1.96 mm
(c) 2.16 mm (d) 2.36 mm

37. Find the wavelength of *X*-rays emitted when electrons acceletatyed through 30 kV strike a target

(a) 0.031 nm (b) 0.041 nm
(c) 0.051 nm (d) 0.061 nm

38. Find the atomic spacing when a beam of X-rays of wavelength 0.3 nm is incident on a crystal and gives a first-order maximum when the glancing angle is 9.0 degrees.

(a) 0.76 nm (b) 0.86 nm
(c) 0.96 nm (d) 1.06 nm

39. Find the wavelength of primary *X*- radiation if Comptom scattering of this radiation graphite at an angle of $60°$, the wavelength of **scattered radiation** is 2.54×10^{-11} m.

(a) $0.232 \mathring{A}$ (b) $0.242 \mathring{A}$
(c) $0.252 \mathring{A}$ (d) $0.262 \mathring{A}$

40. Find the Angle at which 3rd coder reflection take place from crystal when *X*- rays from a undergo first coder reflection at a gane angle of 12° from the face of a calcuater crystal. Given grating space of calcite crying as 3.04×10^8 cm.

(a) 19.4° (b) 22.9°
(c) 25.4° (d) 27.3°

41. Find the minimum applied potential required produce *X*- rays of 1 $\mathring{A}$ wavelength.

(a) 11.7 kV (b) 12.4 kV
(c) 13.5 kV (d) 15.6 kV

42. Find the speed of electrons with which they strike the target when potential different applied to 12.4 kV.

(a) 3.2×10^7 m/sec (b) 4.5×10^7 m/sec
(c) 6.6×10^7 m/sec (d) 7.5×10^7 m/sec

43. The bottom of a glass vessel is a thick plane plate that has an index of **refraction** of 1.50 The vessel is filled with water. What is the largest angle of incidence in the water for which a ray can pass through the glass bottom and into the air below? How does this compare with the critical angle of incidence for a water-air interface which it 48.6° ? What is the **angle of refraction** in the glass?

(a) 38.6° (b) 48.6°
(c) 58.6° (d) 68.6°

44. Light is passing from air into a liquid and is deviated 19° when the **angle of incidence** is 52° Under what conditions will total reflection occur at this interface?

(a) 23.7° (b) 33.7°
(c) 43.7° (d) 53.7°

45. Find the Angle of refraction in water, when the angle of incidence in air is 40° and a beam of light is passing through air to oil to water.

(a) 14.9° (b) 18.9°
(c) 22.9° (d) 28.9°

46. Wavelength of line in the target of $Z = 42$ when wavelengrh of line in the target of $z = 24$ is $2\mathring{A}$. Taking m as 1.

(a) $0.43 \mathring{A}$ (b) $0.53 \mathring{A}$
(c) $0.63 \mathring{A}$ (d) $0.73 \mathring{A}$

47. Find the wavelength of scattered *X*-rays in **Compton scattering** of $\lambda\, 3\mathring{A}$ and scattered at 30° angle w.r.t. incident beam.

(a) $2.005678 \mathring{A}$ (b) $3.005678 \mathring{A}$
(c) $4.005678 \mathring{A}$ (d) $5.005678 \mathring{A}$

48. Find the shortest λ of **X - ray produced** in a *X*-ray tube with accelerating potential as 70 kV.

(a) 0.017 (b) 0.018
(c) 0.019 (d) 0.015

49. A beam of sodium light passes from air into water and then into flint glass, all with parallel surfaces. If the **angle of incidence** in the air is 45°. What is the angle of refraction in the glass? (n = 1333 and 1.63 for water and flint glass, respectively).

(a) 12° (b) 18°
(c) 23° (d) 26°

50. A layer of benzene (index of refraction = 1.50) floats on water. If the **angle of incidence** of the light entering the benzene from air is 60° What is the angle the light makes with the vertical in the benzene.

(a) 15° (b) 25°
(c) 35° (d) 45°

51. Find the C_T number (House field unit) H of heart muscle ($\mu = 0.212$) and μ Water = 0.205 is

(a) $\frac{205}{7000}$ (b) $\frac{405}{7000}$
(c) $\frac{7000}{205}$ (d) $\frac{7000}{405}$

52. Sensitivity of a **LVDT** ($\mu V/mm$) when produces an rms voltage of 2 volt for displacement of 500×10^{-6} mm.
(a) 2×10^3 (b) 4×10^3
(c) 6×10^3 (d) 8×10^3

53. In **Newton's rings experiment** with air film the diametr of the fifth brignt ring is 0.5 cm. The wavelength of light used is 5893 Å. The light enters the air film with the angle of 30°. Determine the radius of curvature of lens.
(a) 104 cm (b) 144 cm
(c) 184 cm (d) 204 cm

54. Find the Width of central band if slit width 0.6 mm when wavelength of light used is 6000 Å in a **Fraunhofer diffraction** due to single slit screen is 1 m away from slit.
(a) 1 nm (b) 2 nm
(c) 3 nm (d) 4 nm

55. A diffraction grating has 6000 lines per cm. The diffracted second oder spectral line is observed at 50°. Calculate the wavelength of radiation.
(a) 4383 Å (b) 5383 Å
(c) 6383 Å (d) 6383 Å

56. A transmission grating has 15000 LPI. What is the **maximum order of diffraction** that can be observed when a light of wavelength 6000 Å is used?
(a) $n=1$ (b) $n=2$
(c) $n=3$ (d) $n=4$

57. Calculate the ratio of **intensity** of central maxima to frist secondary maxima.
(a) 0.025 (b) 0.035
(c) 0.045 (d) 0.055

58. The light of wavelength 6400 Å is used to obtain **single-slit Fraunhofer diffraction patron**. The slit width is 0.3 mm. The screen is at a distance of 110 cm from the slit. Calculate the distance between the centre of the central maximum and the third dark fringe (third piont zer intensity).
(a) 5 mm (b) 6 mm
(c) 7 mm (d) 8 mm

59. In an ***X*-ray** tube, accelerating potential 70 kV. Determine the shortest wavelength X-ray produced
(a) 0.0177 nm (b) 0.0133 nm
(c) 0.0144 nm (d) 0.0166 nm

60. In an **X-ray spectrum**, the wavelength of certain line is 2 Å for the target of $z=24$. Calculate the wavelength of the same line the target of $z=42$. Take a as unity. From Moseley's we can write.
(a) 0.33 A (b) 0.43 A
(c) 0.43 A (d) 0.63 A

61. In compton scatterijng, **X- ray** wavelength 3 Å are Scattered at an angle 40° with respect to incident beam. Calculate the wavelength of the scattered X-rays.
(a) 1 Å (b) 2 Å
(c) 3 Å (d) 4 Å

62. The **optical fibre** is 2 m long and has a diameter of 20 μm. If a ray of light is incident on one end of the fiber at an angle $\theta_1 = 40°$. How many reflections does it make before emerging from the other end (The index of refraction of the fibre is 1.30)?
(a) 27000 (b) 37000
(c) 47000 (d) 57000

63. In practice, **optical fibres** have a coating of glass $(n_3 = 1.512)$ to protect the optical surface of the fibre. If the fibre itself has an index of refraction $n_2 = 1.700$, what is the **critical angle for** total reflection of a ray inside the fibre?
(a) 42.7° (b) 52.7°
(c) 62.7° (d) 72.7°

64. The index of refraction (for sound) of water $(v=1498)$ when index of refraction for sound is defined as ratio of speed in air (343 m/s) to speed of sound in medium.
(a) 0.119 (b) 0.229
(c) 0.339 (d) 0.449

65. Find the critical angle θc for total reflection of sound from water if the index of refraction (for sound) of water $(v=1498)$ when index of refraction for sound is defined as ratio of speed in air (343 m/s) to speed of sound in medium.
(a) 7.2° (b) 9.2°
(c) 11.2° (d) 13.2°

66. Calculate the total radiant power output W of an LED chip dia 0.2 nm viewed from a distance of 1m, $\lambda = 550\ nm$ $\eta_e = 0.1\%$ operated at 2 V and 50 mA.

(a) 1.13×10^{-1} watts (b) 1.13×10^{-2} watts

(c) 1.13×10^{-3} watts (d) 1.13×10^{-4} watts

67. Coherence length given that n^{w} at 0.8 μm 1.5019 and n^{2w} at 0.4 μm is 1.4802 in KDP

(a) $10^{-1}m$ (b) $10^{-2}m$

(c) $10^{-3}m$ (d) $10^{-4}m$

68. Superimposing $y_1 = 20\sin\ \omega t$ and $y_2 = 20$ is sin $(\omega t + 60^o)$ we get

(a) 20 cos ωt (b) 20 sin ωt

(c) 20 sin $(\omega t + 60^o)$ (d) 34.6 sin $(\omega t + 30^o)$

69. Superimposing $X_1 = 30\sin\omega t$ and $y_2 = 30\cos\omega t$.

(a) 30 sin ωt

(b) 30 cos ωt

(c) 90 sin ωt cos ωt

(d) 42.4 sin $(\omega t + 45^o)$

70. Light from a sodium vapour lamp (589 nm) forms an **interference pattern** on a screen 0.8m from a pair of slits. The bright fringes is the pattern are 0.35 cm apart. Slit separatic is:

(a) 0.115 mm (b) 0.125 mm

(c) 0.135 mm (d) 0.145 mm

71. A ray of light is incident on the left vertical face of a glass cube of **refractive index** n_2. The plane of incidence is the plane of the page, and the cube is surrounded by liquid (n_1). What is the largest angle of incidnece θ_1 for which total internal reflection occurs at the top surface?

(a) $\sqrt{\frac{n_1}{n_2}-1}$ (b) $\sqrt{\frac{n_1}{n_2}+1}$

(c) $\sqrt{\left(\frac{n_1}{n_2}\right)^2-1}$ (d) $\sqrt{\left(\frac{n_2}{n_1}\right)^2-1}$

72. A ray of light incident at an angle θ_1 on one end of an **optical fiber**. Its angle of refraction is θ_2 and it strikes the side of the fiber at an angle ϕ_2. If the index of refraction of the fibre is 1.30, what is the largest angle of incidence θ_2 that a ray can have and still be totally reflected from the side of the fibre?

(a) 36.2° (b) 46.2°

(c) 56.2° (d) 66.2°

73. Find the **Bragg angle** for a acousts-optic modulator length = 50 mm, diffraction efficieny 70% acoustic wavelenght 4.3×10^{-5} m acoustic velocity 3500m/sec.

(a) 0.22° (b) 0.32°

(c) 0.42° (d) 0.52°

74. Find the minimum change in refractive index of material if **Bragg angle** for a acousts-optic modulator length = 50 mm, diffraction efficieny 70% acoustic wavelenght 4.3×10^{-5} m acoustic velocity 3500m/sec.

(a) 1.27×10^{-1} (b) 1.27×10^{-2}

(c) 1.27×10^{-5} (d) 1.27×10^{-4}

75. Find the maximum irradiance of four coherent and in coherent sources.

(a) $16I, 2I$ (b) $16I, 4I$

(c) $8I, 8I$ (d) $8I, 4I$

76. Find the radiated power from a source of area $10^{-5}m^2$ at a temperature of 2000 K. Given the emissivity of surface is 0.7.

(a) 4.35 W (b) 5.35 W

(c) 6.35 W (d) 7.35 W

77. Find the ionization energy of hydriogen atom $(z = 1)$ from $n = 1$ to $n = \infty$

(a) 7.6 eV (b) 9.6 eV

(c) 11.6 eV (d) 13.6 eV

78. Find the reflectance at Ga As surface with refractive index 3.6 of Ga As.

(a) 0.12 (b) 0.22

(c) 0.32 (d) 0.42

79. Find the overall power efficiency of a low power **He-Ne laser** operating with a current of 10 mA at a dc voltage of 2500 V and given an optical output of 5 mV.

(a) 0.01 % (b) 0.02 %

(c) 0.03 % (d) 0.04 %

80. Energy separation in a GaAs laser with a cavity length of 200 μm.

(a) 0.63 m e V (b) 0.73 m e V
(c) 0.83 m e V (d) 0.93 m e V

81. Find the Cavity length at which absorption loss and mirror loss become equal in a GaAs **Fary perot laser cavity** when absorption loss in cavity length is given by an absorption co-efficient of 20/cm.

(a) 254 μm (b) 354 μm
(c) 474 μm (d) 574 μm

82. Ratio of rates of spontaneous and stimulated emission for tungesten filament lamp operating at a temperature of 2000 K.

(a) 0.5×10^3 (b) 1×10^3
(c) 1.5×10^3 (d) 1.5×10^5

83. If the speed of light in ice is $2.3\times10^8\,m/s$. what is its **index of refraction**? What is the **critical angle** if incidence for light going from ice to air.

(a) 1.1 and 20° (b) 1.304 and 30°
(c) 1.3 and 40° (d) 1.304 and 50°

84. The index of refraction of a flint glass is 1.64. Find the speed of light in the glass, the **angle of refraction** in glass if light is incident from water at an angle of 44°, and the cirtical angle of incidence for a glass-water interface.

(a) $1.83\times10^8\,m/s$ and 14.4°
(b) $1.38\times10^8\,m/s$ and 24.4°
(c) $1.83\times10^8\,m/s$ and 34.4°
(d) $1.38\times10^8\,m/s$ and 44.4°

85. Find the Angle of incidence when a ray of light shines down vertically on the mirror which lies face up making an angle of 15° with horizontal.

(a) 10° (b) 15°
(c) 20° (d) 25°

86. Find Angle between reflected ray, and the horizontal in above situation.If **Find the Angle of incidence** when a ray of light shines down vertically on the mirror which lies face up making an angle of 15° with horizontal.

(a) 60° (b) 90°
(c) 120° (d) 150°

87. Frequency separation in a GaAs laser with a cavity length of 200 μm.

(a) $2\times10^5\,Hz$ (b) $2\times10^7\,Hz$
(c) $2\times10^9\,Hz$ (d) $2\times10^{11}\,Hz$

88. Responsivity R_λ with a quantum efficiency $\eta=0.25$ at $\lambda=0.5\,\mu m$.

(a) $0.1\,A/W$ (b) $0.2\,A/W$
(c) $0.3\,A/W$ (d) $0.4\,A/W$

89. Minimum detectable signal for **pohotomultipiler** with bandwith 1 Hz.

(a) $2\times10^{-16}\,W$ (b) $4\times10^{-16}\,W$
(c) $6\times10^{-16}\,W$ (d) $8\times10^{-20}\,W$

90. An inductive pick-up operating from a 120 tooth wheel is used with a digital frequency meter to measure the speed of rotation of shaft on which the wheel is mounted, The gating period is set to $10^4\,\mu s$, and a reading of 0.030 is obtained on the four digit display. Determine the **shaft speed** in r.p.m.

(a) 10 rpm (b) 15 rpm
(c) 20 rpm (d) 25 rpm

91. A disc mounted on the shaft of a machine has 12 pattern points. The number of flasher projected on the disc by a **stroboscope** is 6000 in a minute. Find the speed of the machine if the disc apperas stationary and has singls image of 12 points.
If the disc apperas to move forward in the direction of rotation at 10 r.p.m., find the **speed of the disc.**

(a) 310 rpm (b) 410 rpm
(c) 510 rpm (d) 610 rpm

92. Smallerst value of internal angle θ_c in a fiber in air with $n_1=1.53$, $n_2=1.50$ and $n_0=1$.

(a) 58.6° (b) 78.6°
(c) 68.6° (d) none of above

93. Numerical aperture if fiber in air with $n_1=1.53$, $n_2=1.50$ and $n_0=1$.

(a) 0.101 (b) 0.202
(c) 0.303 (d) 0.404

94. Fibre acceptance angle if internal angle θ_c in a fiber in air with $n_1=1.53$, $n_2=1.50$ and $n_0=1$.

(a) 15.6° (b) 16.6°
(c) 17.6° (d) 18.6°

95. What Maximum pulse broadening in a **step index fiber** with NA as 0.2, core refractivbe index as 1.45 over a distance 1 km is:

(a) 26 ns/km (b) 36 ns/km
(c) 40 ns/km (d) 46 ns/km

96. What material dispersion over 1 km of silica based fiber when a Ga As LEO operating at 850 nm and with a 50 nm line width is used as a source is

(a) $1.3\times10^{-9} s$ (b) $2.3\times10^{-9} s$

(c) $3.3\times10^{-9} s$ (d) $4.3\times10^{-9} s$

97. What is the **Rayleigh scattering** for silica fibre at $\lambda = 1\,\mu m$, $n = 1.46$, $p = 0.286$, $\beta = 7\times10^{-11}$ m^2/N, TF = 1400 K is:

(a) $1.64\times10^{-4}/m$ (b) $1.74\times10^{-4}/m$

(c) $1.84\times10^{-4}/m$ (d) $1.94\times10^{-4}/m$

98. Light is **incident at angle** θ_1 on an isosceles triangular prism of apex angle a and refractive index $n > 1$. Obtain an expression for the minimum angular deviation, δ_{min}, produced by the prism.

(a) $\frac{\sin \alpha + \delta_{min}}{2} = n \sin\frac{\alpha}{2}$

(b) $\frac{\sin \alpha - \delta_{min}}{2} = n \sin\frac{\alpha}{2}$

(c) $\frac{\cos \alpha + \delta_{min}}{2} = n \cos\frac{\alpha}{2}$

(d) $\frac{\cos \alpha - \delta_{min}}{2} = n \cos\frac{\alpha}{2}$

99. Find the Wavelength in water of a blue light whose wavelength in air is 420 nm.
Given : $n = 1.33$ for water

(a) 116 nm (b) 216 nm

(c) 316 nm (d) 416 nm

100. Find frequency of light of a certain colour has 1800 waves to the millimeter in air (h = 1.52 for glass).

(a) $5.4\times10^{7}\,Hz$ (b) $5.4\times10^{9}\,Hz$

(c) $5.4\times10^{11}\,Hz$ (d) $5.4\times10^{14}\,Hz$

101. Number of waves per millimeter when the light is travelling in water if frequency of light of a certain colour has 1800 waves to the millimeter in air (h = 1.52 for glass).

(a) 1399 mm^{-1} (b) 1499 mm^{-1}

(c) 1799 mm^{-1} (d) 2399 mm^{-1}

102. A plane harmonic infrared wave traveling through a transparent medium is given by:

$$E_x = (y,t) = E_{ox} \sin 2\pi\left(\frac{y}{5\times10^{-7}} - 3\times10^{14} t\right)$$

in SI as usual. Determine the **refractive index** of the medium at that frequency?

(a) 1 (b) 2

(c) 3 (d) 4

103. Light from a sodium lamp ($\lambda_0 = 589\,nm$) passes through a tank of glycerin (refractive index 1.47) 20 m long in a time t_1. If it takes a time t_2 to traverse the same tank when filled with carbon disulphide (index 1.63). Determine the difference $t_2 - t_1$.

(a) $1.07\times10^{-5} s$ (b) $1.07\times10^{-6} s$

(c) $1.07\times10^{-7} s$ (d) $1.07\times10^{-8} s$

104. A beam of 436-nm light is incident perpendicular to a glass plate that is 2.0 cm thick and has $n = 1.66$. How long does it take a point on the beam to pass through the plate?

(a) 0.57 ns (b) 0.67 ns

(c) 0.87 ns (d) 0.111 ns

105. How many wavelengths thick is the plate is terms of beam within it if A beam of 436-nm light is incident perpendicular to a glass plate that is 2.0 cm thick and has $n = 1.66$.

(a) 53127 (b) 43717

(c) 53417 (d) 71647

106. What is the Critical angle of indicence when a ray light passes from crown glass to water A beam of 436-nm light is incident perpendicular to a glass plate that is 2.0 cm thick and has $n = 1.66$.

(a) 41.3° (b) 51.3°

(c) 61.3° (d) 71.3°

107. What is the angle of refraction when angle of incident in glass is 55° .A beam of 436-nm light is incident perpendicular to a glass plate that is 2.0 cm thick and has $n = 1.66$.

(a) 59° (b) 69°

(c) 79° (d) 89°

108. What is **Critical angle** if the sphere is immersed is air ($n_2 = 1$) if index of refraction of a glass sphere, $n_1 = 1.76$.

(a) 14.6° (b) 24.6°

(c) 34.6° (d) 44.6°

109. To some extent, the brilliance of diamonds is attributable to total internal reflection. Calcuate the **critical angle** for a diamond-air surface. (For dimond $\eta = 2.42$).

(a) 12° (b) 15°
(c) 18° (d) 24°

110. At what angle must a ray of light be incident on acetone to be refracted into the liquid at 25°? $(n = 1.358)$.
(a) 15° (b) 25°
(c) 35° (d) 45°

111. Find the angle of regraction of light incident on a water surface at an angle of 48° to the normal. $(n = 1.333)$.
(a) 23.9° (b) 33.9°
(c) 43.9° (d) 53.9°

112. The **index of refraction of** n-propy alcohol is 1.39. Find angle of refraction if light comes from air with an angle of incidence of 55°.
(a) 32° (b) 36°
(c) 40° (d) 44°

113. Light passes from air into a liquid and is deviated 19° when the angle of incidence is 52°. Find the refreactive index of liquid.
(a) 1.15 (b) 1.25
(c) 1.35 (d) 1.45

114. At what **angle of incidence** should a beam of light strike the surface of a still pond if the angle between the reflected ray and the refracted ray is to be 90°? (n = 1.33 for water).
(a) 13° (b) 23°
(c) 33° (d) 53°

115. Find the sine of the crictical angle of incidence for an air-water interface.
(a) 0.550 (b) 0.650
(c) 0.750 (d) 0.850

Answers Set-4

1. Ans. (a)
Solution:

$$\lambda_1 = 5997 \text{ Å}$$

$$\lambda_2 = 6003 \text{ Å}$$

Disersive power = ratio of wavelength

$$= \frac{6003}{5997} \times 1000 = 1 \times 10^{-3} = 0.001$$

2. Ans. (a)
Solution:
Shift in central fringe.

$$\Delta = \frac{D}{2d}(\mu - 1)t$$

Given that $\frac{D}{2d}(\mu - 1)\, t = \frac{D}{2d} 5\lambda$

or $(\mu - 1) = 5\lambda = 5 \times 589 \times 10^{-9}$

or $\mu = 1.589$

3. Ans. (a)
Solution:

$$V = \frac{\overline{E}}{B} = \frac{2 \times 10^4}{0.2} = 10^5\, m/s$$

$$\text{Separation } = 2\,(R_2 - R_1) = \frac{2\,V\,(m_2 - m_1)}{qB}$$

$$= \frac{(2 \times 10^5)(13 - 12)(1.67 \times 10^{-27})}{1.6 \times 10^{-19} \times 0.3}$$

$$= 0.69 \text{ cm.}$$

4. Ans. (b)
Solution:

$t = 20\ cm$

$= 0.2\ m$

$$\Delta = t(\mu_a - 1) = \frac{n\lambda}{2}$$

$$\mu_a = \frac{150 \times 4000 \times 10^{-10}}{2 \times 0.2} = 1.00015$$

5. Ans. (b)
Solution:
Diameter of nth dark ring with liquid

$$D_n^2 = \frac{4n\lambda R}{\mu}$$

$$D_{na}^2 = 4n\,\lambda\,R \text{ (with air)}$$

Given that $\frac{D_{na}}{D_n} = 2$

Hence, $\mu = \frac{D_{na}^2}{D_n^2} = 4$

6. Ans. (c)
Solution:

$$\lambda = \frac{D_{n+p}^2 - D_n^2}{4\,pR}$$

Here $D_n = 4.2\ mm = 0.42\ cm$

$D_{n+p} = D_{n+14} = 7\ mm = 0.7\ cm$

$P = 14$

$R = 1$ m = 100 cm

Hence $\lambda = 5600\ \text{Å}$

7. Ans. (b)

Solution:

Path difference $= 4\lambda$

$= 4\ (6000\ \text{Å})$

$= 24000\ \text{Å}$

$= 2.4\ \mu m$.

8. Ans. (d)

Solution:

The distance along the screen for the third order fringe

is $x_1 = [(3\lambda_1)D]d$:

for λ_2 it is $x_2 = [(4\lambda_2)D]/d$.

Since they fall at the same location,

$\therefore\ x_1 = x_2$

Hence $\lambda_2 = \frac{3}{4}\lambda_1$.

9. Ans. (c)

Solution:

The condition for constructive interference twice is $(4\lambda_1)/d = x/D$

and $(6\lambda_2)/d = x/d$,

so that $4\lambda_1 = 6\lambda_2$.

If $\lambda_1 = 430\ nm$, then $\lambda_2 = 287\ nm$.

10. Ans. (c)

Solution:

Since $\Delta r = (\lambda/8)\ \sin\theta$,

the phase difference at P is

$$\phi = \frac{2\pi}{\lambda}\Delta r + \Delta\phi = \frac{\pi}{4}(\sin\theta + 1)$$

and $I(\theta) = 4I_0 \cos^2\frac{\phi}{2}$

$$= 4I_0 \cos^2\left[\frac{\pi}{8}(\sin\theta + 1)\right]$$

11. Ans. (b)

Solution:

$$t = \frac{\sum n\lambda}{\eta - 1} = \frac{7(600)nm}{0.5} = 8400\ nm$$

12. Ans. (b)

Solution:

$X_2 - X_1 = N\lambda/2$

$$\lambda = \frac{2(X_2 - X_1)}{N}$$

$$= \frac{2\,(10.7347\times10^{-3} - 10.7051\times10^{-3})}{100}$$

$= 5920\ \text{Å}$

13. Ans. (a)

Solution:

For a fabry Periot interferometer the conduct of maxima in transimted beam is given by

$2t\cos\theta = n\lambda$

t = Sepration between heater

For normal incidence, $\theta = 0$

$\therefore\quad 2t = n\lambda$

or, $n = \frac{2t}{\lambda}$

For $4000\ \text{Å}$, $n = \frac{2\times2\times10^{-6}}{400\times10^{-10}} = 10$

For $6000\ \text{Å}$, $n = \frac{2\times2\times10^{-6}}{6000\times10^{-10}} = 6.67$

Hence intermediate wavelengths orders will be 7, 8, 9, Corresponding wavelengths are

$\lambda_1 = \frac{2t}{7} = 5714\ \text{Å}$

$\lambda_2 = \frac{2t}{8} = 5000\ \text{Å}$

$\lambda_3 = \frac{2t}{9} = 4444\ \text{Å}$

14. Ans. (d)

Solution:

The condition for the wavelength absent in the reflected light is

$2\mu t\cos r = \lambda,\ 2\lambda\ 3\lambda, \ldots\ldots, etc$

also $\mu = \dfrac{\sin i}{\sin r}$

Therefore $r = \sin^{-1}\left(\dfrac{\sin i}{\mu}\right)$

$= \sin^{-1}\left(\dfrac{\sin 35°}{1.25}\right) = 27°18$

$\lambda_1 = 2\mu t \cos r$

$= 2\times 0.2\times 10^{-6}\times 1.25\times \cos 27°18$

$= 4442\ \text{Å}$

15. Ans. (c)

Solution:

$r = \sin^{-1}\left(\dfrac{\sin 30°}{1.39}\right) = 21.082524°$

Therefore, cos r $= 0.9330633$

$2\mu t \cos r = n\lambda$ and

$2\mu t \cos r = (n+1)\lambda$

Therefore, n (5125) = (n + 1) 5000

or $n = 40$

Therefore, $t = \dfrac{40\times 512510^{-8}}{2\times 1.39\times 0.9330633}$

$= 7.9031\times 10^{-4}\ cm$

16. Ans. (a)

Solution:

$x = \dfrac{\lambda}{2\mu \tan\beta \cos(\beta + r)}$

$= \dfrac{\lambda}{2\mu \tan\beta \cos\beta} = \dfrac{\lambda}{2\mu \sin\beta}$

Therefore, $\beta = \sin^{-1}\dfrac{\lambda}{2\mu x}$

$= \sin^{-1}\dfrac{5893\times 10^{-8}}{2\times 1.4\times 0.09} = 0.0134°$

17. Ans. (a)

Solution:

Given that

n = 4, $\mu = 1$,

$2\mu t = \left(n + \dfrac{1}{2}\right)\lambda$

Therefore, $t = \dfrac{4.5\times 6500\times 10^{-8}}{2}$

$= 1.4625\times 10^{-4}\ cm.$

18. Ans. (b)

Solution:

Distance moved for shift of 1 fringe

$= \lambda/2$

Displacement for n-fringe, $x = n\lambda/2$

Given that $x = 0.1\ mm,\ n = 400$

Hence, $\lambda = \dfrac{2\times 0.1\times 10^{-3}}{400} = 500\ \text{Å}$

19. Ans. (b)

Solution:

$\Delta\lambda = \lambda_1 - \lambda_2 = \dfrac{\lambda_a^2}{2d}$

$\lambda_a = \dfrac{4882 + 4886}{2} = 4884\ \text{Å}$

$\Delta\lambda = 4886 - 4882 = 4\ \text{Å}$

$\therefore d = \dfrac{(4884\times 10^{-10})}{4\times 10^{-10}} = 0.059\ mm$

20. Ans. (b)

Solution:

$2(\mu - 1)t = N\lambda$

$t = \dfrac{N\lambda}{2(\mu - 1)}$

Given that $N = 10$,

$\lambda = 590\times 10^{-10}\ m$

$\mu = 1.5$

Hence, $t = \dfrac{10\times 5890\times 10^{-10}}{2\times 0.50}$

$= 5.89\ \mu m$.

21. Ans. (d)

Solution:

$\lambda = \dfrac{d\Delta r}{D}$.

22. Ans. (d)

Solution:

Bright fringes are located at

$$\sin\theta_m = \theta_m = \frac{m\lambda}{d}$$

the angular separation of adjacent fringes is λ/d. Thus the condition for seeing adjacent fringes as distinct is

$$\frac{\lambda}{d_{max}} = 0.000291$$

or $$d_{max} = \frac{500\times10^{-9}}{0.000291} = 1.72\ mm$$

23. Ans. (c)

Solution:

From $\lambda = (d/D)\Delta x$, Where d and D are fixed,

$$\lambda' = \frac{\Delta x'}{\Delta x}\lambda = \frac{7.6}{8.3}\times 630 = 577 nm.$$

24. Ans. (d)

Solution:

The mth bright fringe of the λ pattern and the mth bright fringe of the λ' pattern are located at

$$y_m = \frac{mD\lambda}{d} \text{ and } y_m = \frac{m'D\lambda'}{d}$$

Equating these distances gives

$$\frac{m}{m'} = \frac{\lambda'}{\lambda} = \frac{900}{750} = \frac{6}{5}$$

Hence, the first position at which overlappling occurs is

$$y_6 = y_5' = \frac{(6)(2)(750\times10^{-9})}{2\times10^{-3}} = 4.5mm$$

25. Ans. (c)

Solution:

$D_n^2 = 2(2n+1)\lambda R/\mu,$

Given that $\mu = 1$ (Air)

$n = 6,\ D = 0.68\ cm,\ R = 350\ cm$

Hence, $$\lambda = \frac{D_n^2}{2(2n+1)R}$$

$$= \frac{(60)^2}{2\times(2\times6+1)\times350} = 5080\ \mathring{A}$$

26. Ans. (b)

Solution:

Fringe width

$$\beta = \frac{\lambda}{2\mu\tan\alpha\cos(r+\alpha)}$$

α = angle of wedge

For normal indience r = 0

$$\beta = \frac{\lambda}{2\mu\left(\frac{\sin\alpha}{\cos\alpha}\right)\cos\alpha} = \frac{\lambda}{2\mu\sin\alpha}$$

or, $$\sin\alpha = \frac{\lambda}{2\mu\beta}$$

Given that $\lambda = 6000\times10^{-10}\,m$

$\mu = 1.35$

$\beta = 2\ mm = 2\times10^{-3}\,m$

$$\alpha = 1.11\times10^{-4}\times\frac{180}{\pi}$$

$\alpha = 0.00064°$

27. Ans. (b)

Solution:

Fringe separation, $\beta = \dfrac{\lambda}{2a}$

$$a = 40 = \left(\frac{40}{3600}\right)^0$$

$$\beta = \frac{4650\times10^{-10}}{2\times1.9\times10^{-4}} = 0.12\ cm$$

28. Ans. (c)

Solution:`

Fringe width $= \dfrac{D\lambda}{2d}$

$$W_1 = \frac{D\lambda_1}{2d} = \frac{D\eta_1\lambda}{2d}$$

$$W_2 = \frac{D\lambda_2}{2d_2} = \frac{D\eta_2\lambda}{2d}$$

$$\therefore \quad \frac{W_1}{W_2} = \frac{\lambda_1}{\lambda_2} = \frac{\eta_1}{\eta_2}$$

Also $\eta_2 W_1 = 40\,W_2$

$$\eta_2 = \frac{40\times5996}{4358} = 54$$

29. Ans. (c)

Solution:

Displacement of fringe,

$$\frac{\beta}{\lambda}(\mu-1)t = 5\beta$$

$$\therefore \quad t = \frac{5\lambda}{\mu-1} = 6\times10^{-6} = 6\ \mu m$$

30. Ans. (c)

Solution:

When blood is moving towards transmitter, the apparent Frequency, f_1

$$f_1 = \frac{f(C - v\cos\theta)}{C}$$

If incident and scattered radiations are both inclined at θ to the direction of flow.

$$f_2 = f_1\left[\frac{C}{C + v\cos\theta}\right]$$

Resultant doppler shift,

$$\Delta f = f - f_2 = f.\ \frac{f(C - v\cos\theta)}{C(C + C\cos\theta)}$$

Since C >> V

$$\therefore \quad v = \frac{\Delta f C}{2f\cos\theta}$$

31. Ans. (c)

Solution:

Loss co-effcient for phon

$$a_{\omega t} = a_{loss} + a_R = a_{loss}$$

$$= \frac{1}{2}\ell n\, R$$

$$= 65.43/cm$$

$$\tau_{ph} = \frac{1}{v\, a_{\omega t}}$$

$$= \frac{3.6}{3\times10^6\times65.43} = 0.51\ ps$$

32. Ans. (b)

Solution:

We find that, n = actual depth/apparent depth, so $1.333 = (5 ft)/y$ and $y = 3.75\ ft$.

33. Ans. (d)

Solution:

Analyze the refraction in sequence, from the left

$$n = \frac{\sin\theta_1}{\sin\theta_2}$$

$$\text{or } 1.50 = \frac{\sin 30^o}{\sin\theta_2}$$

$$\sin\theta_2 = \frac{0.5}{1.50} = 0.3333 \text{ and}$$

$$\theta_2 = 19.5^o$$

$$\theta_2 + \theta_3 + 120^o = 180^o$$

$$\text{or } \theta_2 + \theta_3 = 60^o$$

The prism angle. Then $19.5^o + \theta_3 = 60^o$ and

$$\theta_3 = 40.5^o$$

$$n = \frac{\sin\theta_4}{\sin\theta_3},\ 1.50 = \frac{\sin\theta_4}{\sin 40.5^o}$$

$0.6494(1.50) = \sin\theta_4$ and $\theta_4 = 77^o$

34. Ans. (a)

Solution:

The angles are obtained from snell's law

$$\theta_2 + \phi_2 + (180 - A) = 180^\circ$$

$$\theta_2 + \phi_2 = A$$

$$\sin\theta_2 = \frac{n_1}{n_2}\sin\theta_1$$

$$= \frac{\sin 30^o}{1.50} = 0.333$$

$$\theta_2 = 19.5^o$$

$$\phi_2 = A - \theta_2 = 55^o - 19.5^o = 35.5^o$$

35. Ans. (b)

Solution:

Where the two slits act as in - phase source of equal instensities and we assume $d << D$, intensity maxima are located by (for any integer m).

$$\Delta r = m\lambda$$

$$\text{or } \quad d\sin\theta = m\lambda$$

$$\sin\theta = \frac{m\lambda}{d}$$

$$\frac{y}{D} = \frac{m\lambda}{d}$$

or $$y = \frac{mD\lambda}{d}$$

36. Ans. (d)
Solution:
For the first bright band, $n = 1$.
The path difference is one wavelength i.e.

$$n\lambda = \frac{dy}{D}$$

$$5839 \times 10^{-10} = \frac{0.25 \times 10^{-3}\, y}{1}$$

$$y = \frac{5839 \times 10^{-10}}{0.25 \times 10^{-3}}$$

$$= .2.36 \times 10^{-3} m = 2.36\ mm$$

37. Ans. (b)
Solution:

$$\frac{1}{2}mv^2 = eV = hf$$

Hence frequency

$$f = \frac{eV}{h} = \frac{1.6 \times 10^{-19} \times 3 \times 10^4}{6.63 \times 10^{-34}}$$

$$\therefore \quad \lambda = \frac{C}{f} = 0.41 \times 10^{-10} m = 0.041\ nm.$$

38. Ans. (c)
Solution:
From Bragg's law,
$2d \sin\theta = m\lambda$

$$\therefore \quad d = \frac{\lambda}{2\sin\theta} = \frac{0.3}{2 \times 0.156} = 0.96\ nm$$

39. Ans. (b)
Solution:

$$\Delta\lambda = 0.024\,(1 - \cos\theta)\ \mathring{A}$$

$$= 0.024(1 - \cos 60)\ A$$

$$= 0.012\ \mathring{A}$$

Now,

$$\lambda + \Delta\lambda = 0.254 \times 10^{-11} m$$

$$\therefore \quad \lambda = 0.254 - 0.012$$

$$= 0.242\ \mathring{A}$$

40. Ans. (b)
Solution:
Given : $\theta = 12^\circ$, $n = 1$,
$d = 3.04 \times 10^{-10} m$

$$2d \sin\theta = n\lambda$$

$$\therefore \quad 2 \times 3.04 \times 10^{-10} \sin 12 = 1 \times \lambda$$

$$\lambda = 0.79 \times 10^{-10}$$

For 3rd order reflection,

$$2 \times 3.04 \times 10^{-10} \sin\theta_3 = 3 \times 0.79 \times 10^{-10}$$

$$\sin\theta_3 = 0.3898$$

Hence $\theta_3 = 22.9^\circ$

41. Ans. (b)
Solution:

$$\lambda_{\min} = \frac{12400}{V}\mathring{A}$$

$$\therefore \quad V = \frac{12400}{\lambda_{\min}} = 12.4\ kV.$$

42. Ans. (c)
Solution:

$$V = 5.93 \times 10^5 \sqrt{V}$$

$$= 5.93 \times 10^5 \sqrt{12.4\ kV}$$

$$= 6.6 \times 10^7\ m/\text{sec}.$$

43. Ans. (b)
Solution:
For largest angle of incident $\theta_a = 90^o$ for light from glass. Then, for angle in glass.

$$1 \sin 90^\circ = 1.50 \sin\theta_{gt}$$

$$\sin\theta_{gt} = 0.66667$$

$$\sin\theta_{gt} = 41.8^\circ$$

Then for anghle in water

$$1.50 \sin\theta_{gl} = 1.333 \sin\theta_w$$

$$\sin\theta_W = 0.7500$$

$$\theta_W = 48.6^\circ$$

44. Ans. (c)

Solution:

When light travels from the liquid to the air at angles of incidence greater than the crical angle, we have total internal reflection at the interface. To find the critical angle, we need n_1, the index of refeaction in the liquid. Now for the liquid.

$$\theta_1 = 52^\circ - 19^\circ = 33^\circ$$

and $1 \sin 52^\circ = n_1 \sin 33^\circ$

so, $0.7889 = 0.54464\, n_1$

$$n_1 = 1.447$$

Then for the ciritical angle.

$$2 \sin 90^\circ = 1.447 \sin \theta_d$$

$$\sin \theta_d = 0.69116$$

and $\theta_d = 43.7^\circ$

45. Ans. (d)

Solution:

At the air oil interface.

$\sin 40^\circ = 1.45 \sin \theta_{oil}$

At the oil- water interface

$1.45 \sin \theta_{oil} = 1.33 \sin \theta_w$

$$\theta_w = \sin^{-1}\left[\frac{\sin 40^\circ}{1.33}\right] = 28.9^\circ$$

46. Ans. (c)

Solution:

From Moseley's law

$$\lambda_2 = \lambda_1 \frac{(Z_1 - 1)^2}{(Z_2 - 1)}$$

$$= \frac{2(24-1)^2}{(42-1)} = 0.63\ A$$

47. Ans. (b)

Solution:

$$\Delta\lambda = \frac{h}{mC}(1 - \cos\theta)$$

$$= \frac{(6.626 \times 10^{-34})(1 - \cos 40)}{9.11 \times 10^{31} \times 3 \times 10^8}$$

$$= 0.005678\ \text{Å}$$

$$\therefore \quad \lambda = \lambda_1 + \Delta\lambda = 3 + 0.005678$$

$$= 3.005678\ \text{Å}$$

48. Ans. (a)

Soution:

$$\lambda_{min} = \frac{hC}{eV} = \frac{6.626 \times 10^{-34} \times 3 \times 10^8}{1.6 \times 10^{-19} \times 70 \times 10^3}$$

$$= 0.017\ nm$$

49. Ans. (d)

Solution:

Use snell's law for the air - water inter surface.

$$n_w = \frac{\sin \theta_1}{\sin \theta_2}$$

$$1.333 = \frac{\sin 45^\circ}{\sin \theta_2} = \frac{0.7071}{\sin \theta_2}$$

$$\sin \theta_2 = \frac{0.7071}{1.333} = 0.5305$$

and $\theta_2 = 32^\circ$

Now use snell's law for the water - glass inter surface.

$$n_{BA} = \frac{n_g}{n_w} = \frac{\sin \theta_2}{\sin \theta_3}$$

$$\frac{1.63}{1.333} = \frac{0.5305}{\sin \theta_3}$$

$$\sin \theta_3 = \frac{0.5305\,(1.333)}{1.63} = 0.4338$$

and $\theta_3 = 26^\circ$

50. Ans. (c)

Solution:

Here the light rays bend inward in the benzene and outward in the water. Use Snell's law first at the air - benzene surface and then at the benzene - water surface.

$$n_b = \frac{\sin \theta_1}{\sin \theta_2}$$

$$1.50 = \frac{\sin 60^\circ}{\sin \theta_2}$$

$$\sin \theta_2 = \frac{0.8660}{1.50} = 0.5773$$

and $\theta_2 = 35$

51. Ans. (c)
Solution:

$$H = \frac{\mu - \mu_{water}}{\mu_{water}} \times 1000$$

$$\mu = 0.212$$

$$\mu_{water} = 0.205$$

$$H = \frac{0.212 - 0.205}{0.205} \times 1000$$

$$= \frac{0.007 \times 1000}{205} = \frac{7000}{205}$$

52. Ans. (b)
Solution:

$$\text{Sensitivity} = \frac{2}{500 \times 10^{-6}} V/mm$$

$$= 4 \times 10^{-9} V/mm$$

$$= 4 \times 10^{-9} \mu V/mm.$$

53. Ans. (d)
Solution:

$$2\mu t \cos r = \left(n + \frac{1}{2}\right)\lambda \text{ and } t = \frac{r_n^2}{2R}$$

Therefore, $R = \frac{d_n^2 \mu \cos r}{2(2n+1)\lambda}$

$$= \frac{(0.5)^2 \cos 30^\circ}{2(2 \times 4 + 1)5093 \times 10^{-8}}$$

$$= 204.1 \text{ cm.}$$

54. Ans. (b)
Solution:

$$2x_1 = \frac{2D\lambda}{W}$$

$$= \frac{2 \times 1 \times 6000 \times 10^{-10}}{0.6 \times 10^{-3}} = 2mm.$$

55. Ans. (c)
Solution:

$$\lambda = \frac{d \sin\theta}{n} = \frac{1}{6000}$$

$$\frac{\sin 50^\circ}{2} = 6383.7 \text{ Å}$$

56. Ans. (b)
Solution:

$$n = \frac{d \sin\theta}{\lambda} = \frac{1}{6000 \times 6000 \times 10^{-8}}$$

$$= \frac{100}{36} = 2.78 < 3$$

Therefore, $n = 2$.

57. Ans. (c)
Solution:

$$\frac{I}{I_0} = \left(\frac{\sin B}{B}\right)^2$$

$$= \left(\frac{\sin(3\pi/2)}{3\pi/2}\right)^2 = 0.045$$

58. Ans. (c)
Solution:

$$Sin\,\theta = \frac{x}{D} = \frac{n\lambda}{d}$$

Therefore, $x = \frac{n\lambda D}{d}$

$$= \frac{3 \times 6400 \times 10^{-10} \times 1.1}{0.3 \times 10^{-3}}$$

$$= 7.04 \ mm$$

59. Ans. (a)
Solution:

$$\lambda_{min} = \frac{hC}{eV} = \frac{6.6}{1.6 \times 10^{-19} \times 70 \times 10^3}$$

$$= 0.0177$$

60. Ans. (d)
Solution:

$$\lambda_2 = \frac{\lambda_1 (Z_1 - a)^2}{(Z_2 - a)^2}$$

$$= \frac{2(24-1)^2}{(42-1)^2} = 0.63 \text{ Å}$$

61. Ans. (c)
Solution:

$$\Delta\lambda = \frac{h}{mC}(1 - \cos\theta)$$

$$= \frac{(6.626\times10^{-34})(1-\cos 40^\circ)}{9.11\times10^{-31}\times3\times10^{8}}$$

$$= 0.005678 \text{ Å}$$

$$\lambda = \lambda_1 + \Delta\lambda = 3 + 0.005678$$

$$= 3.005678 \text{ Å}$$

62. Ans. (d)

Solution:

It the condition for total internal reflection is fulfilled. Then,

$$\sin\theta_2 = \frac{n_1}{n_2}\sin\theta_1$$

$$= \frac{\sin 40^\circ}{1.3} = 0.495$$

and $\theta_2 = 29.7^\circ$

The length along fiber to the first reflection is

$$l = \frac{d}{\tan\theta_2} = \frac{2\times10^{-5} m}{0.570} = 3.51\times10^{-5} m$$

By the law of reflection, l also represents the length along the fiber between successive reflections, the number of reflections is

$$\frac{L}{l} = \frac{2m}{3.51\times10^{-5} m} = 57000$$

63. Ans. (c)

Solution:

The critical angle is determined by the interface between fibre and coating.

Then,

$$\sin\theta_c = \frac{n_3}{n_2} = \frac{1.512}{1.700}$$

$$= 0.899$$

$$\theta_c = 62.7^\circ$$

64. Ans. (b)

Solution:

$$\eta_{water} = \frac{V_{air}}{V_{water}} = \frac{343\ m/s}{1498\ m/s} = 0.229$$

65. Ans. (d)

Solution:

$$\sin\theta_c = \frac{\eta_{water}}{\eta_{air}} = 0.229$$

$$\theta_c = 13.2^\circ$$

66. Ans. (d)

Solution:

$$W = \frac{hC}{\lambda}\times\eta_e\times \text{Photon emission rates}$$

$$= \frac{hc}{550\times10^{-9}}\times0.001\times\frac{50\times10^{-3}}{e}$$

$$= 1.13\times10^{-4} \text{ watts.}$$

67. Ans. (d)

Solution:

$$\lambda_c = \frac{\lambda_0}{4(n^w - n^{2w})}$$

$$= \frac{0.8\times10^{-6}}{4\times0.00217} = 10^{-5} m.$$

68. Ans. (d)

Solution:

$$y = 20\sin\omega t + 20\sin(\omega t + 60^\circ).$$

We can think of y_1 and y_2 as projections on the y-axis of vectors of length 20 rotating with angular velocity ω and making angles ωt and $\omega t + 60^\circ$, repectively, with the x – axis. The projection of the vector sum of these two vectors on the y axis is then our result $y = y_1 + y_2$. $A^2 = 20^2 + 20^2 + 2(20)\cos 60^\circ$

$$= 3(20^2);$$

$A = 34.6$; and from the isosceles triangle $2\phi = 60^\circ$;

$\phi = 30^\circ$.

Thus, $y = 34.6\sin(\omega t + 30^\circ)$.

69. Ans. (d)

Solution:

$\cos\omega t = \sin\left(\omega t + \frac{\pi}{2}\right)$. Thus y_2 "leads" y_1 by 90°.

$A^2 = 30^2 + 30^2 = 2(30^2)$; $A = 42.4$. For the right isosceles tringle, $\phi = 45^\circ$, and

$y = y_1 + y_2 = 42.4\sin(\omega t + 45^\circ)$.

70. Ans. (c)

Solution:

With $y = x_n$ the position of the nth maximum on the screen, measured from the central maximum. The distance between adjacent fringes is:

$$\Delta x = x_{n+1} - x_n = \frac{D\lambda}{d}(n+1) - \frac{D\lambda}{d}n = \frac{D\lambda}{d}$$

So, $$d = \frac{D\lambda}{\Delta x} = \frac{(0.8m)(589\times10^{-9}m)}{0.35\times10^{-2}m}$$

$$= 1.35\times10^{-4}m = 0.135\ mm$$

71. Ans. (d)

Solution:

Appluing Snell's law at A.

$$n_1 \sin\theta_1 = n_2 \sin\theta_2 \quad (1)$$

But $\theta_2 = 90° - \theta_c$,

and so $$\cos\theta_2 = \sin\theta_c = \frac{\eta_1}{\eta_2} \quad (2)$$

Elimination of θ_2 from equation (1) and (2) gives

$$\sin\theta_1 = \sqrt{\left(\frac{n_2}{n1/}\right)^2 - 1} \quad (3)$$

72. Ans. (c)

Solution:

$$n_1 = 1$$

$$\sin\theta_1 = \sqrt{1.30^2 - 1} = 0.831$$

$$\therefore \quad \theta_1 = 56.2°$$

73. Ans. (c)

Solution:

$$\sin\theta = \frac{m\lambda}{2d}$$

$$\phi = \sin^{-1}\left(\frac{1\times4.3\times10^{-5}}{2\times50\times10^{-3}}\right)$$

$$= 0.42°$$

74. Ans. (c)

Solution:

$$\phi = 2\sin^{-1}\sqrt{\eta} = 2\sin^{-1}\sqrt{0.7}$$

$$= 113.6°$$

$$\Delta n = \frac{\phi\lambda\cos\theta_s}{2\pi}$$

$$= 1.27\times10^{-5}$$

75. Ans. (b)

Solution:

For coherent sources $I\max = n^2 I = 16I$

For incoherent sources $I\max = nI = 4I$

76. Ans. (c)

Solution:

$$W = 0.7\times5.67\times10^{-8}\times10^{-5}\times(2000)$$

$$= 6.35\ W$$

77. Ans. (d)

Solution:

$$E_{ion} = \frac{9.1\times10^{-31}\times(1.6\times10^{-19})^4}{8\times(6.6\times10^{-34}\times8.85\times10^{-12})^2}$$

$$= 13.6\ eV$$

78. Ans. (c)

Solution:

$$R = \left(\frac{n_2 - n_1}{n_2 + n_1}\right)^2 = \left(\frac{3.6-1}{3.6+1}\right)^2 = 0.32$$

79. Ans. (b)

Solution:

Overall power efficiency

$$= \frac{5\times10^{-3}}{2500\times1\times10^{-2}} = 0.02\%$$

80. Ans. (c)

Solution:

$$hf = \frac{6.6\times10^{-34}\times2\times10^{11}}{1.6\times10^{-19}}$$

$$= 0.83\ me\ V$$

81. Ans. (b)

Solution:

$$a_{loss} = -\frac{1}{L}\ \ell n\ R$$

$$L = -\frac{1}{20}\ In\ 0.33$$

82. Ans. (d)

Solution:

Ratio of rates of spontanous and stimulated emission

$$R = \exp^{(h\gamma/kT)} - 1$$

$$= \exp\left(\frac{6.6\times10^{-34}\times0.5\times10^{14}}{1.38\times10^{-23}\times2000}\right) - 1$$

$$= e^{12} - 1$$

$$= 1.5\times10^5$$

83. Ans. (d)

Solution:

From the data we can determine the index of refraction

of ice, $n = \frac{c}{v} = \frac{(3\times10^8)}{(2.3\times10^8)} = 1.304$,

Then for the critical angle,

$$1 \sin 90^\circ = 1.304 \sin i_c$$

and $\sin i_c = 0.7667$

$$i_c = 50.1^\circ$$

84. Ans. (c)

Solution:

Speed of light in glass

$$v_g = \frac{c}{n} = \frac{3\times10^8}{1.64} = 1.83\times10^8 \, m/s$$

Then, $\frac{\sin 44^\circ}{\sin \theta_g} = \frac{v_w}{v_g}$

or $1.333 \sin 44^\circ = 1.64 \sin \theta_g$

$$= 1.64 \sin \theta_g = \frac{v_w}{v_g}$$

and $\sin \theta_g = 0.56462$

$$\theta_g = 3.44^\circ$$

85. Ans. (b)

Solution:

The angle of incident is the angle made with the normal to mirror. Thus $i = 15^\circ$.

86. Ans. (a)

Solution:

Angle of reflection is also $15 (i = r)$. Hence the reflected ray makes angles of 30° with vertical and 60° with horixontal.

87. Ans. (d)

Solution:

$$\frac{C}{2n_r L} = \frac{3\times10^{10}}{2(3.66)(200\times10^{-4})}$$

$$= 2\times10^{11} \, Hz$$

88. Ans. (a)

Solution:

$$R_\lambda = \frac{\eta e \lambda}{hc}$$

$$= \frac{0.25\times1.6\times10^{-19}\times0.5\times10^{-4}}{6.6\times10^{-34}\times3\times10^8}$$

$$= 0.1 \, A/W.$$

89. Ans. (c)

Solution:

$$W_{min} = \frac{(2\times2\times10^{-14}\times1.6\times10^{-19}\times1)^{1/2}}{0.1}$$

$$= 8.10 \, W$$

90. Ans. (d)

Solution:

Number of pulses per second

$$= \frac{\text{Reading of digital meter}}{\text{Gating period}}$$

$$= \frac{0.030}{10\times10^{-6}} = 300$$

$$= \frac{3000}{120} = 25 \, r.p.m.$$

Therefore, speed

91. Ans. (c)

Solution:

Speed of the machine when pattern become stationary is given by

$$s = \frac{f}{m} = \frac{6000}{12} = 500 \, r.p.m.$$

Now the pattern move in the direction of rotation as a speed of 10 r.m.p. There fore

The speed of machine will become 500 + 10

$$= 510 \, rpm$$

92. Ans. (b)

Solution:

$$\sin \theta_c = \frac{1.50}{1.53}$$

$$\therefore \quad \theta_c = 78.6^\circ$$

93. Ans.(c)

Solution:

$$NA = 1.53\times\sqrt{2\times0.0196}$$

$$= 0.303$$

94. Ans. (c)

Solution:

$$\alpha_{max} = \sin^{-1}(0.303) = 17.6^\circ$$

95. Ans. (d)

Solution:

Maximum pulse broadening

$$= \frac{d \times NA}{2 \times C \times \eta}$$

$$\Delta_C \ step = \frac{10^3 (0.2)^2}{2 \times 3 \times 10^8 \times 1.45} S \ km^{-1}$$

$$= 46 \ ns / km$$

96. Ans. (d)

Solution:

Material dispersion

$$\Delta C = \frac{10^3 \times 850 \times 10^{-9} \times 3 \times 10^{10} \times 50 \times 10^{-9}}{3 \times 10^8}$$

$$= 4.3 \times 10^{-9} s$$

97. Ans. (c)

Solution:

$$a_R = \left(\frac{8\pi^2}{3(10^{-6})^4} \times (1.46)^6 \times)(0.286)^2 \times 7 \times 10^{-11} \times 1.38 \times 10^{-23} \times 1400 \right)$$

$$= 1.84 \times 10^{-4} / m$$

98. Ans. (a)

Solution:

Because the prism is symmetrical and light paths are reversible, it is easy to see that the deviation δ will be a minimum when gives at the two inter faces.

$$\sin K = n \sin \theta_2$$

and $n \sin (a - \theta_2) = \sin K$

which imply that $\theta_2 = a/2$.

Then, $\sin K = n \sin \frac{a}{2}$

But, $\delta_{\min} = \theta_1 + \theta_3 - \alpha = 2k - \alpha$

or, $K = \frac{\alpha + \delta_{min}}{2}$

Hence, $\sin \frac{\alpha + \delta_{\min}}{2} = n \sin \frac{a}{2}$

99. Ans. (c)

Solution:

$$\lambda' = \frac{\lambda}{n} = \frac{420 \ nm}{1.33} = 316 \ nm$$

100. Ans. (d)

Solution:

$$\lambda = \frac{1 \ mm}{1800} = 5.556 \times 10^{-7} m$$

$$= 556 \ nm$$

$$v = \frac{3 \times 10^8}{5.556 \times 10^{-7}} = 5.4 \times 10^{14} \ Hz$$

101. Ans. (d)

Solution:

No of waves per millimeter in glass.

$$1800 \times 1.333 = 2399 \ mm^{-1}$$

102. Ans. (b)

Solution:

The phase is familiar in the form $k(y - vt)$; accordingly, we rewrite the above as $E_{0x} \sin \phi$ where

$$\phi = \frac{2\pi}{5 \times 10^{-7}} (y - 15 \times 10^7 t)$$

Clearly, $\lambda = 5 \times 10^{-7} m$

and $v = 1.5 \times 10^8 m/s$

Thus, $n = c/v = 2$.

103. Ans. (d)

Solution:

Since $v = \frac{c}{n}$

$$t_1 = \frac{20}{c/n} = \frac{20(1.7)}{e}$$

and $t_2 = \frac{20(1.63)}{c}$

Accordingly, $t_2 - t_1 = \frac{20}{c}(1.63 - 1.47)$

$$= 10.7 \times 10^{-8} \ s.$$

104. Ans. (d)

Solution:

v within the platc $= \frac{c}{n}$

$$= \frac{c}{1.66} = 1.807 \times 10^8 m/s$$

$$t = \frac{\theta}{d} = \frac{0.020}{v} = 0.111 \ ns$$

105. Ans. (d)

Solution:

$$\lambda' = \frac{436 \ nn}{1.65}.$$

No of complete waves $= \frac{d}{\lambda'} = 71647$

106. Ans. (c)
Solution:

$1.52 \sin \theta_c = 1.333 \sin 90$

$\sin \theta_c = 0.877\ \theta_c = 61.3^\circ$

107. Ans. (b)
Solution:

$1.52 \sin 55 = 1.33 \sin \theta_w$

$\theta_w = 69^\circ$

108. Ans. (c)
Solution:

$$\sin C_c = \frac{n_2}{n_1}\theta_c$$

$$= \text{arc} \sin \frac{1}{1.76} = 34.6^\circ$$

109. Ans. (d)
Solution:

$$\theta_c = \theta_2 = \text{arc} \sin \frac{n_A}{n_B} = \text{arc} \sin \frac{1}{2.42}$$

$$= \text{arc} \sin 0.4312 = 24^\circ$$

110. Ans. (c)
Solution:

We use snell's law : $n_1 \sin \theta_1 = n_2 \sin \theta_2$ with measured from normal. These $= n_2 \sin \theta_2$ with measured from normal. These

$$\sin \theta_1 = \frac{n_2 \sin \theta_2}{n_1}$$

$$= \frac{(1.358)(\sin 25^\circ)}{1.00} = 0.574$$

$\theta_1 = 35^\circ$

111. Ans. (b)
Solution:
From snell's law,

$1 \sin 48^\circ = 1.333 \sin \theta_w$

$\sin \theta_w = 0.5575$;

$\theta_w = 33.9^\circ$

112. Ans. (b)
Solution:

$$\frac{\sin 55^\circ}{\sin \theta_{al}} = 1.39$$

$\sin \theta_{al} = 0.58932$

$\theta_{al} = 36.1^\circ$

113. Ans. (d)
Solution:

$\theta_2 = 52^\circ - 19^\circ = 33^\circ$

$$n = \frac{\sin 52^\circ}{\sin 33^\circ} = \frac{0.7880}{0.5446} = 1.45$$

114. Ans. (d)
Solution:

For angle of incidence and refraction θ, ϕ we must have $\theta + \phi = 90^\circ$. or $\theta = \phi - 90^\circ$.

From snell's law, $\sin \theta = n \sin (90^\circ - \theta)$

$= n \cos \theta$

or $\theta = \tan^{-1} n = \tan^{-1} 1.33 = 53^\circ$

115. Ans. (c)
Solution:
For light passing from water to air the angle of refraction is greater than the angle of incidence, so for angle of incidence greater than $i_e < 90^\circ$, the refraction angle will be greater than 90°, implying no refraction. The angle i_c is the critical angle and corresponds to a refraction angle of exactly 90°. Then,

$1 \sin 90^\circ = 1.333 \sin i_c$

$\sin i_c = 0.750$

MCQ Set–5

1. CT sean was invented by
   (a) Godfrey Hounsfield (b) Enc Storz
   (c) John Snow (d) Taka Shita Koba
2. On usual CT scale of densities (–1000 to + 1000) brain tissue measures.
   (a) -10 to $+10$ (b) 0 to $+20$
   (c) $+20$ to $+45$ (d) $+50$ to $+100$

3. Ultrasound wavws are produced by means of a crystal
(a) Ferromagnetic (b) Piezoelectric
(c) Ferroelectric (d) Paramagnetic
4. Increased density on x-ray film denotes
(a) A black shadow
(b) White shadow
(c) Increased radiolucency
(d) None of the avove
5. Diagnostic ultrasound is based on
(a) Slit beam (b) Singel x-rey source
(c) Pulse echo principle (d) None
6. The optimum distence used in radiographic exam is
(a) 90-100 cm (b) 150-160 cm
(c) 180-190 cm (d) 200-210 cm
7. Amongst the following phases of iodinated con trast medium enhancement, CT scan makes use of
(a) Vascular enhancement
(b) Tissue opacification
(c) Opacification of urinary tract
(d) Any of the above
8. Acoustic shadow is ultrasound means
(a) Reflector of sound waves
(b) Absorption of sound waves
(c) Reflection of x-reys
(d) All of the above
9. Maximum radiation exposures is caused by
(a) Sonography (b) CT scan
(c) Chest x-rey (d) MRI
10. The fields is used in MRL in clinical practicl practice is tesla
(a) 0.15 to 1.5 (b) 1.5 to 2.5
(c) 2.5 to 3
11. The fluid on MR appears as
(a) Black on T_1 W images
(b) White on T_2 W images
(c) Intermediate on T_1 and T_2 W images
(d) A, B are true
12. MR imaging is not useful for
(a) CNS evaluation
(b) Soft tissue evaluation
(c) Clacification evaluation
(d) Abd imaging
13. The substance most widely used in medical ultra sound is zirconate titanate.
(a) Silver (b) Sodium
(c) Magnesium (d) Lead
14. Ultrasound is not useful in
(a) Neonatal patients
(b) Obstetric patients
(c) Obese patients
(d) Adult patients
15. Structures seen white on US image are
(a) Calculi (b) Air
(c) Fat (d) All
16. Xeroradiography is mainly used for-cancer
(a) Lung (b) Breast
(c) Thyroid (d) Stomach
17. MRI should be used with caution in patients with
(a) Acute stroke
(b) Brainstem lesions
(c) Demyelinating dz.
(d) Cardiac pacemaker
18. Least resolution is observed in
(a) Conventional x-ray (b) US
(c) CT (d) MR
19. Sialography term is applied to x-ray demonstration by injection of a radiopaque media of which of the following.
(a) Lymph glands (b) Axillary glands
(c) Salivary glands (d) Adrenal glands
20. Most sensitive test for metastatic deposit is
(a) Isotope scan (b) CT scan
(c) Skeletal survey (d) Any of the above
21. Hydatid cyst of the lung in a chest x-ray
(a) is seen as a calcified ring shadow
(b) shows speckled calcification
(c) will not calcify
(d) a and b are correct
22. Water lily appearance in chest x-ray is suggestive of
(a) Bronchiectasis
(b) Bronchopleural fistula
(c) Hydatid cyst
(d) Sequestration cyst of lung
23. Cause of homogenous opacity on a chest x-ray is all except.
(a) Pleural effusion
(b) Diaphragmatic hernia
(c) Massive consolidation
(d) Emphysema
24. Characteristic of benign lesion of lung in chest x-ray is
(a) Size > 5 cm in diameter
(b) Cavitation
(c) Peripheral location
(d) concentric dense calcification
25. Multiple translucent cysts on x-ray are found in the chest in all, except.
(a) congenital diaphragmatic hernia
(b) congenital adenomatoid malformation

(c) lobar agenesis
(d) bilateral multiple cysts

26. The imaging modality for brest, in a patient of 20 years.
(a) Mammography (b) Ultrasound
(c) MRI (d) Thermography

27. The modality of choice for early detection of breast cancer is:
(a) Mammography (b) Ultrasound
(c) MRI (d) Thermography

28. Mammography is:
(a) Infrared imaging
(b) Metabolic imaging
(c) X-ray imaging
(d) Ultrasounic imaging

29. Thermography is:
(a) Infrared imaging
(b) Metabolic imaging
(c) X-ray imaging
(d) Uitrasonic imaging

30. Radiation exposure is associated with which of the fllowing modalities.
(a) Thermography (b) Mammography
(c) Ultrasound (d) MRI

31. Which modality is commonly used to differentiate a solid mass from the cystic one.
(a) Thermography (c) Mammography
(b) Ultrasound (d) MRI

32. Which of the following is an analytical tool
(a) Tharmography (b) Mammography
(c) Ultrasound (d) MR spectroscopy

33. The advantage of ultrasound
(a) Economical
(b) Can differentiate cystic vs solid
(c) No radiation
(d) All of the obove

34. The advantage of mammography
(a) detects non-palapable of breast
(b) Can differentiate cystic vs solid
(c) No radiation
(d) All of the above

35. Features of a benign lesion on ultrasound.
(a) Well-defind margins
(b) Round or oval shape
(c) Homogeneous echopattem
(d) all of the above

36. Following is the only imaging modality capable of detecting microcalcifications.
(a) Mammography (b) Ultrasund
(c) MRI (d) Thermography

37. Isotope used in myocardila perfusion scan is:
(a) Technetium
(b) Thallium
(c) Stannous pyrophosphate
(d) Gallium

38. Left atrial hypertrophy is seen radiologically as:
(a) Double cardiac silhouette
(b) Left bronchial elevation
(c) Straightening of left heart border
(d) All of the above

39. Left to right shunt is usually demonstrated in chest x-ray by
(a) Increase in pulmonary venouse marking.
(b) Increase in size of pulmonary arteries.
(c) Increase in no. of pulmonary arteries.
(d) Increase in pulmonary vascuiar markings.

40. Frequency of sound waves used for transabdomainal ultrasonography is
(a) 2.5 – 3.5 MHz (b) 3.5 – 5.0 MHz
(c) 5.0 – 7.5 MHz (d) 7.5 – 10.0 MHz

41. Chronc Budd-Chiary syndrome is usually assocated with
(a) Hypertrophied caudate lobe
(b) Right lobe hypertrophy
(c) Caudate lobe atrophy
(d) Left lobe hypertrophy

42. Which of the following is false concerning fatty infiltration of the liver.
(a) May occur as a focal abnormality
(b) MRI spin and echo sequences are sensitive to fatty infiltration
(c) Ultrasound is often helpfull in making the diagnosis
(d) With fatty infiltration, liver is of lower attenuation than the spleen on CT

43. Which statement is not true reharding hepatk amoebic abscess.
(a) Male predpminant disorder
(b) Usually multuple lesion seen by imaging
(c) Liver involvement more common than lung or brain
(d) Infection spreads to liver via mesenteric wiens

44. Most useful and safest diagnostic method for amoebic liver abscess is:
(a) Needie aspiration
(b) Chemotherapeutic trial
(c) Ultrasound
(d) Endoscopy

45. Ultrasound is not useful in
(a) CBD stone at the distal end of the CBD
(b) Breast cyst

(c) Ascites
(d) Full bladder

46. Investigation of choice in a case of obstaructive jaundice is:
(a) Ultrasonography (b) E R C P
(c) CT scan (d) Plain x-ray

47. An ultrasound exmaination shows dilated intrahepatic biliary channels with asmall gall bladder. The most likely daignosis is:
(a) Gall bladder stone
(b) Pancreatic calculus
(c) Common bile duct stone
(d) Head of pancreas

48. The method of choice in preliminary evaluation of biliary obstructoin with clinical jaundice is
(a) OCG (b) ERCP
(c) CT scan (d) Ultrasonography

49. Organ that appears most dense on a CT nonecontrast scan is:
(a) Pancrease (b) Kidney
(c) Liver (d) Gall bladder

50. Diagnosis of early lesion of pancreas is best made by
(a) Barium meal (b) Ultrasound
(c) CT Scan (d) Duodenoscopy

51. Dilated jejunal loops on plain X-ray abdomen are identified by
(a) Haustration
(b) Valvulae conniventes
(c) Characterless bowel
(d) Air fluid levels

52. X-ray findings suggestive of achalasia cardiac are all, except.
(a) Megaesophagus
(b) Tortuous esophagus
(c) Fluid levels in esophagus
(d) Diverticulum in esophagus

53. Minimal ascites can be best deceted by
(a) X-ray (b) CT
(c) USG (d) MRI

54. X-ray findings characteristic of small intestinal malabsorption syndrome are all, axcept.
(a) Flocculation of bearium
(b) Dilatation of bowel
(c) Increased transit time
(d) Mucosal atrophy

55. Abdominal organ best suited for CT scanning.
(a) Liver (b) Gall bladder
(c) Pancresa (d) Kidney

56. Colon is identified on X-ray by
(a) Haustra (b) Valva conniventis
(c) Characterless (d) None

57. In contrast to USG which adrenal is better visulized on CT scan
(a) Left
(b) Right
(c) Both are equally visualized
(d) None of the above

58. On CT scan which structure is confused with adrenal mass
(a) Upper pole of left kidney
(b) Gastric diverticulum
(c) Accessory spleen
(d) All of the above

60. The preliminary investigation of choice in a patient of cryptorchidism is
(a) X-ray pelvis (b) US
(c) CT scan (d) MR

61. On US, the undescended testis has one of the following characteristics
(a) Smaller and less echogenic than the normal testis
(b) Simillar in size and echogenicity to the normal testis
(c) Larger in size and more echogenic then the normal testis
(d) Larger in size and less echogenic than the normal tastis

62. Investigation of choice for varicocele is
(a) B-mode US
(b) Colour Doppler imaging
(c) CT
(d) MR

63. The preferred imaging modality for tubal patency.
(a) Hysterosalpingography
(b) Sonosalpingography
(c) CT
(d) MRI

64. The imaging modalities of choice for evaluation of ovarian malignancy include
(a) X-ray + US (b) US + CT
(c) CT + MR (d) US + MR

65. Enndometrial carcinoma is best evaluated with
(a) Transabdominal US (b) CT
(c) MR (d) Scintigraphy

66. The imaging modality that is most specific for differentiating benign from malignant ovarian tumour
(a) Transabdominal US (b) Endovaginal US
(c) Colour Doppler (d) CT

67. Modality of choice for detecting hyperacute infarct

(a) Noncontrast CT
(b) Contrast enhanced CT
(c) Contrast enhanced MR
(d) Diffusion MR

68. Modality of choice for diagnosing acute subarachnoid hemorrhage
(a) Noncontrast CT
(b) Contrast enhanced CT
(c) Contrast enhanced MR
(d) Diffusion MR

69. Investigation of choice for multiple sclerosis
(a) CT (b) MR
(c) Angiography (d) Myelography

70. The imaging modality of choice for aortic dissection
(a) X-ray (b) US
(c) CT (d) CT angiography

71. Imaging modality of choice for diaphragmatic rupture
(a) X-ray (b) US
(c) CT (d) None of the above

72. Spinal fractures are best evaluated by:
(a) X-ray (b) CT
(c) MR (d) Scintigraphy

73. Criculating (mature) red blood cells:
(a) contain ribosomes and mitochondria.
(b) continue to synthesize glycolytic enzymes throughout their life span.
(c) are larger in size in venous blood than in arterial blood
(d) normally do not cotain carbamino haemoglo bin in the arterial blood

74. In the criculating blood:
(a) life span of a neutrophil is 8–12 days.
(b) neutrophils are actively phagocytic in the blood stream.
(c) megakaryocytes are the largest cell type.
(d) haemoglobin acts as an important acid-base buffer.

75. In nuclear instrumentation, which of the following is the heaviest and slow moving particle ?
(a) Beta particies (b) Gamma Rays
(c) Alpha Particles (d) Neutrons

76. The wave length of gamma particles is on the range of
(a) 3 to 6 angstrom (b) 0.03 to angstrom
(c) 6 to 9 angstrom (d) none of the above

77. Mass of alpha particles is around.
(a) $62\times10^{-27}\ Kg$ (b) $9\times10^{-31}\ Kg$
(c) $90.1\times10^{-17}\ Kg$ (d) None of the above

78. Enviromental pollution is not caused by:
(a) Atmospheric air and water pollution
(b) High humidity content in atmospheric air
(c) Noise pollution
(d) Redioactive fallout.

79. Scintillation counters have the advantages of:
(a) Very fast counting rate
(b) Capability of detecting of radiations of lower levels.
(c) Capability of detecting of X-rays
(d) All of the above

80. The activity of a radioactive sample goes down to 12.5% in a time of X days. The half life of this sample is nearly.
(a) $12\times$ (b) $8\times$
(c) $\times/3$ (d) $3\times$

81. Geiger counter has the advantages of:
(a) Being relatively inexpensive, reliable and rugged in construction.
(b) Producing easily countable large pulses
(c) Having response independent of variations in humidity temperature and atmospheric pressure.
(d) All of the above.

82. The method used for analysis of gases and their mixtures is:
(a) thermal conductivity (b) Eletrical conductivity
(c) relative humidity (d) Specific gravity

83. C_1 and C_2 are the acitivities of Ions on the two sides of the membrans. The Nernst potential developed across the membrane is proportional to.
(a) $\frac{C_1}{C_2}$ (b) $\frac{C_1^2}{C_2^2}$
(c) $\log e\frac{C_1}{C_2}$ (d) $\exp(C_1/C_2)$

84. IR spectroscopy.
(a) has a useful range of radiation from 2.5 to 5 microns.
(b) is unsuitable for analysis of maxiture of metals.
(c) is unsuitable for analysis of organic gases.
(d) uses bolo meter as one of the detectors.

85. The Geiger Counter has high quantum efficiency in the wavelength (A°) range.
(a) 0.2 to 0.3 (b) 0.4 to 0.5
(c) 1.5 to 2.1 (d) 6 to 7

86. The Duane - Hunt law represents the minimum wavelength $\lambda(A^\circ)$ of X-rays generated for an applied voltage V (Volts) and given by:
(a) $V=\frac{12400}{\lambda}$ (b) $V=12400\,\lambda$
(c) $V=\frac{12400}{\lambda^2}$ (d) $V=12400\,\lambda^2$

87. Attenuation of a narrow monochromatic X-ray beam in a metal platc of thickness 'd' is given by the equaltion.

(a) $1 = 1_o \exp(-\mu d)$ (b) $1 = 1_o \exp(-\mu d^2)$

(c) $1 = 1_o \exp(-\mu / d)$ (d) $1 = 1_o \exp(-\mu / d^2)$

88. The mathematical basis for nuclear magnetic resonance (NMR) states that, If μ the unclear magnetic moment moment of the molecule, the gyro metric ration proportional to:

(a) μ (b) $\frac{1}{\mu}$

(c) $\sqrt{\mu}$ (d) μ^2

89. When the reading of a pH meter changes from 5 to 7, the hydrogen ion concentration of the solution is:

(a) Halv ed (b) Doubled

(c) Increased 100 times (d) Decreased 100 times

90. In a spectrophotometer, the monochromator must be able to resolve two wavelengths 599.9 nm and 600.1 nm.

The required resolution is:

(a) 100 (b) 1000

(c) 3000 (d) 5000

91. The transmettance of a coloured solution is 0.5. The absorbance of the solution is:

(a) 0.3 (b) 0.69

(c) 3.16 (d) 1.5

92. In a Gas chromatograph, the gas which can not be used as a carrier gas is:

(a) Helium (b) Nitrogen

(c) Ammonia (d) None of the above

93. In a perticular sample, the absorbance is 0.6. For a molar concentration of the solution of 1.0×10^4 M and a 2.0 cm path length, the molar absorptivity is:

(a) 1.2 (b) 1200

(c) 3000 (d) 3×10^{-4}

94. The most used He-Ne gas laser has a wave length of:

(a) 0.5 μm (b) 0.633 μm

(c) 0.70 μm (d) 0.80 μm

95. Para magnetism can be used for analysis of:

(a) O_2 (b) SO_2

(c) Co_2 (d) CO_2

96. If the half life of a radiatiom source is 1 year , it's decay constant per year is:

(a) 1 (b) 2

(c) 0.693 (d) 63.2

97. A frequency established He Ne laser with a wavelength of 6328 A^o has a bandwidth of 1 MHz. Its Cohereence length is:

(a) 0.3 m (b) 3 m

(c) 30 m (d) 300 m

Answers Set-5

1. (a)	2. (c)	3. (b)	4. (b)
5. (c)	6. (a)	7. (d)	8. (a)
9. (b)	10. (a)	11. (d)	12. (c)
13. (d)	14. (c)	15. (d)	16. (b)
17. (d)	18. (a)	19. (c)	20. (a)
21. (c)	22. (c)	23. (d)	24. (d)
25. (c)	26. (b)	27. (a)	28. (c)
29. (a)	30. (b)	31. (c)	32. (d)
33. (d)	34. (a)	35. (d)	36. (a)
37. (b)	38. (d)	39. (d)	40. (b)
41. (a)	42. (b)	43. (b)	44. (c)
45. (a)	46. (a)	47. (b)	48. (d)
49. (a)	50. (c)	51. (b)	52. (d)
53. (b)	54. (c)	55. (c)	56. (a)
57. (a)	58. (d)	60. (b)	61. (a)
62. (b)	63. (b)	64. (d)	65. (c)
66. (c)	67. (d)	68. (a)	69. (b)
70. (d)	71. (c)	72. (b)	73. (c)
74. (d)	75. (c)	76. (b)	77. (a)
78. (b)	79. (d)	80. (c)	81. (d)
82. (a)	83. (c)	84. (d)	85. (d)
86. (b)	87. (a)	88. (a)	89. (d)
90. (c)	91. (a)	92. (d)	93. (c)
94. (c)	95. (b)	96. (c)	97. (d)

MCQ Set–6

1. In a **photo multiplier**.

(a) Gain is independent of stray magnetic fields

(b) High frequency response is improved by increasing the number of dynodes (emitting surfaces)

(c) Secondary emission is used for amplification of low level photo-current

(d) The electrons are directed to the anode by applying a strong magnetic field

2. A compound microscope has an optical tube length of 160 mm, an objective of focal length 8 mm and eyepiece of focal length 25mm . For an eye with the least distance of distinct vision of 250 mm, the magnification is:

(a) 500 (b) 200

(c) 160 (d) 10

3. The Geiger Counter has high **quantum efficiency** in the wavelength ($\overline{A}$) range.
(a) 0.2 to 0.3 (b) 0.4 to 0.5
(c) 1.5 to 2.1 (d) 6 to 7

4. The Duane-Hunt law represents the minimum wavelength λ. (A') of X-rays generated for an applies voltage V (volts) and give by:
(a) V = [12400 / λ] (b) V = 12400 λ
(c) V= [12400 / λ^2] (d) V = 12400λ^2

5. Match the List I lasers with List II their wavelengths and select the correct answer using the codes:

List I	List II
A. Nd Yag	1. 514.5 nm
B. CO_2	2. 514.5 nm
C. Argon ion	3. 1.06 μm
D. He Ne	4. 1.06 μm

Codes:

	A	B	C	D
(a)	2	4	3	1
(b)	4	1	3	2
(c)	3	1	4	2
(d)	4	2	1	3

6. For a glass of **refractive index** 1.732, the angle of incidence of ordinary light, for which the reflected beam is completely polarized is:
(a) 30 degrees (b) 60 degrees
(c) 45 degrees (d) 75 degrees

7. A **step index optical fiber**, whose refractive indices of the core and cladding are 1.44 and 1.40 respectively, is surrounded by air. Its **numerical aperture** is:
(a) 0.12 (b) 0.75
(c) 0.06 (d) 0.34

8. The mathematical basic for nuclear magnetic resonance (NMR) states, that, if μ is the nuclear magnetic moment of the molecule, the gyromagnetic ratio is proportional to:
(a) μ (b) [1/μ]
(c) √μ (d) μ^2

9. In a **spectrophotometer**, the monochromator must be able to resolve two wavelengths 599.9 nm and 600 1nm. The required resolution is:
(a) 100 (b) 1000
(c) 3000 (d) 5000

10. A **laser** beam (λ = 632.8 nm) having a divergence of 1 milli-radian is directed towards the moon at a distance of approximately 4 × 10^5 km. The beam would have spread to a diameter of:
(a) 400 km (b) 200 km
(c) 100 km (d) 4 km

11. Match the List I source and List II their spectral bandwidth and select the correct answer using the codes:

List I	List II
A. Nd Y A G laser	1. 10^9 Hz
B. He Ne laser	2. 10^{13} Hz
C. Semiconductor laser	3. 10^{11} Hz
D. LED	4. 10^{12} Hz
	5. 10^5 Hz

Codes:

	A	B	C	D
(a)	2	4	3	1
(b)	2	4	1	5
(c)	3	1	4	2
(d)	4	2	1	3

12. Match the List I name of the equation with List II the formula of the equation and select the correct answer using the codes:

List I	List II
A. Bragg equation	1. A = a b c
B. Duane Hunt equation	2. N = N_o $e^{-\lambda t}$
C. Radioactive decay equation	3. λ = [hc / eV]
D. Beer Lambert Law	4. (μ – 1) t = nλ
	5. 2d Sin θ = nλ

Codes:

	A	B	C	D
(a)	5	4	2	1
(b)	4	1	3	2
(c)	3	1	4	2
(d)	4	2	1	3

13. Match the List I instruments with List II the parameters they measurer and select the correct answer using the codes:

List I	List II
A. Autocollimator	1. Radiation scattered
B. Hygrometer	2. Angular error
C. Nephelometer	3. Molecular weight
D. Mass spectrometer	4. Humidity
	5. Density

Codes:

	A	B	C	D
(a)	2	4	3	1
(b)	4	1	3	2
(c)	3	1	4	2
(d)	2	4	1	3

14. **Light** from a mains powered fluorescent lamp is falling on a gear rotating clockwise. The gear wheel has one tooth painted white. The white tooth appears to be rotating in the anti – clockwise direction with a speed of 60 r.p.m. if the mains frequency is 50 Hz, the true speed of the gear wheel is:

(a) 2940 r.p.m. (b) 3060 r.p.m.
(d) 5940 r.p.m. (d) 5960 r. p.m.

15. When **light** passes from a medium of refractive index n_1 into a medium of refractive index n_2, the reflection at the boundary is expressed as:

(a) $(n_2 - n_1)^2 b)$ (b) $[(n_2 - n_1)/(n_2 + n_1)]^2$
(c) $[1/(n_2 - n_1)^2]$ (d) $(n_2 + n_1)^2$

16. Two crossed polarizes are placed in the part of a light beam. The **light** output is:

(a) Zero (b) plane polarized
(c) Circularly polarized (d) Elliptically polarized.

17. The **coherence** length of the He–Ne laser beam is 120 cm. Its coherence time in seconds is:

(a) 4×10^{-1} (b) 4×10^{-3}
(c) 4×10^{-5} (d) 4×10^{-9}

18. The resolving power of a telescope is proportional to:

(a) λ/a (b) λa
(c) a/λ (d) λ/a^2

19. In a particular sample, the absorbance is 0.6 for a molar concentration of the solute of 1.0×10^{-4} M and a 2.0 cm path length, the molar absorptivity (per mol per cm) is:

(a) 1.2 (b) 1200
(c) 3000 (d) 3×10^{-4}

20. Match the List I electromagnetic radiation with the List II wavelength range and select the correct answer using the codes:

List I	List II
A. X-rays	1. 2.5 μm to 25 μm
B. Vacuum ultraviolet	2. 400 mm to 700 mm
C. Visible	3. 0.1 mm to 1mm
D. Infrared	4. 10 mm to 100 nm

Codes:

	A	B	C	D
(a)	2	4	3	1
(b)	4	3	2	1
(c)	3	4	2	1
(d)	2	4	1	3

21. Match the List I instrument to the List II characteristic and select the correct answer using the codes:

List I	List II
A. Interferometer	1. Wavelength dispersion
B. Monochromator	2. Magnifying power
	3. Flatness measurement.

Codes:

	A	B
(a)	3	2
(b)	1	2
(c)	3	1
(d)	2	1

22. The most used He – Ne gas laser has a wave length of .

(a) 0.5 μm (b) 0.633 μm
(c) 0.70 μm (d) 0.80 μm

23. For a change in elevation level of 1.5 mm between an **optical** flat and a test surface, a light wave of wavelength 0.5 μm produces fringes. The number of fringes is:

(a) 3 (b) 6
(c) 9 (d) 12

24. Paramagnetism can be used for analysis of:

(a) O_2 (b) SO_2
(c) CO (d) CO_2

25. If the half life of a radiation source is 1 year, its decay constant per year is:

(a) 1 (b) 2
(c) 0.693 (d) 63.2

26. Match the List I sources and the List II nature of radiations and select the correct answer using the codes:

List I	List II
A. Tungsten lamp	1. Laser
B. Cobalt – 60	2. IR
C. He – Ne	3. UV
D. Discharge lamp	4. γ - ray

Codes:

	A	B	C	D
(a)	2	4	1	3
(b)	4	3	2	1
(c)	3	1	4	2
(d)	3	4	1	2

27. Match the List I detector with the List II spectroscopy and select the correct answer using the codes:

List I	List II
A. Bolometer	1. UV
B. Photomultiplier	2. IR
C. G.M Counter	3. y – ray
D. Photographic plate	4. X – ray

Codes:

	A	B	C	D
(a)	2	4	3	1
(b)	2	1	3	4
(c)	3	1	4	2
(d)	2	4	1	3

28. An **optical fibre** has refractive index of 1.641 for the core and 1.422 for the cladding. The critical angle above which a ray will be totally internally reflected is:

(a) 37° (b) 41°
(c) 45° (d) 60°

29. A frequency stablised HeNe laser with a **wavelength** of 6328. A Has a bandwidth of 1 MHz. Its coherence length is:

(a) 0.3 m (b) 3 m
(c) 30 m (d) 300 m

30. Attenuation of a narrow monochromic **X-ray** beam in a metal plate of thickness 'd' is given by the equation.

(a) $I = I_0 \exp(-\mu d)$ (b) $I = I_0 \exp(-\mu d^2)$
(c) $I = I_0 \exp(-\mu/d)$ (b) $I = I_0 \exp(-\mu/d^2)$

31. For alignment and testing of two surfaces at right angles, a constant deviation prism, also called an optical square, is used. It is a

(a) Right angled prism (b) square prism
(c) Pentagonal prim (d) hexagonal prism

32. C_1 and C_2 are the activities of the ions on the two sides of a membrane. The Nernst potential developed across the membrane is proportional to:

(a) $[C_1/C_2]$ (b) $[C_1^2 / C_2^2]$
(c) $\log_e [C_1/C_2]$ (d) $\exp(C_1/C_2)$

33. The bandwidth of an electrocardiogram (ECG) amplifier is:

(a) dc to 0.01 Hz (b) 0.05 to 500 Hz
(c) 550 to 1500 Hz (d) 2000 to 10000 Hz

34. Korokoff sound s are used.

(a) As a reference for sound level measurement
(b) For studying heart muscle functioning
(c) For blood pressure measurement
(d) For study of heart value functioning

35. In an electromagnetic blood flow meter, the induced voltage is directly proportional to the

(a) Blood flow rate
(b) Square root of the blood flow rate
(c) Square of the blood flow rate
(d) Logarithm of the blood flow rate

36. The **relative permittivity** (ε_r) of an optical medium is 2.5. Its refractive index is:

(a) 0.4
(b) 1.25
(c) 1.58
(d) There is no interrelation between the two parameters

37. An interference filter has been made with a dielectric of magnesium fluoride (refractive index $\eta = 1.38$) for a nominal wave length of 455 nm. The required thickness of the dielectric for first order interference is:

(a) 8.2 nm (b) 16.5 nm
(c) 82 nm (d) 16.5 nm

38. The transmittance of a coloured solution on 0.5. The **absorbance** of the solution is:

(a) 0.3 (b) 0.69
(c) 3.16 (d) – 1.5

39. Identify the correct set of matches from the following:

List I	List II
A. Beer's law	1. X-ray diffraction
B. Bragg's law	2. Thermo-emf
C. Peltier effect	3. Electro-chemical cell
D. Nernst equation	4. Absorption of light

Codes:

	A	B	C	D
(a)	4	1	3	2
(b)	3	2	1	4
(c)	4	1	2	3

(d) None Of the above

40. Identify the correct set of matches from the following:

List I	List II
A. Electron capture detector	1. IR spectroscopy
B. Scintillation counter	2. Oxygen analyser
C. Thermopile	3. Gas chromatography
D. Zirconia probe	4. Nuclear radiation spectrosocopy

	A	B	C	D
(a)	4	1	2	3
(b)	2	4	1	3
(c)	3	1	4	3

(d) None of the above

41. **Brewster angle** is the angle of incidence of monochromatic at the interface of two media, for which:
(a) The incidence beam would be totally internally reflected
(b) A polarized beam would become unpolarised after refection
(c) An unpolarised beam would become polarised after reflection

42. Which of the following instruments uses an ultrasonic transducer?
(a) Echo-cardiograph
(b) Electro-cardiograph
(c) Electro-encephalograph

43. Among the electrodes which are used to record ECG signals, the one used for ground referencing is placed at,
(a) Left arm (b) right leg
(c) left leg

44. The refractive indices of glass and water are 1.50 and 1.33, respectively. If the glass is immersed in water its relative **refractive** index is..
(a) 0.89 (b) 1.13
(c) 1.99 (d) 2.83

45. Light is incident on a glass plane of **refractive index** 1.732. It is plane polarized on reflection when the incidence angle is:
(a) 30° (b) 45°
(c) 60° (d) 75°

46. A **gas chromatograph** is used for.
(a) Measuring flow rate of a gas
(b) Measuring the temperature of a gas
(c) Measuring the pressure of a gas
(d) Analysing the composition of a gas

47. In a spirit level, 2.5 mm of movement of the bubble corresponds to a tilt angle of 25 seconds. The readius of curvature of the tube of the spirit level is:
(a) 52.1 m (b) 34 . 4 m
(c) 26 .3 m (d) 15.6 m

48. An optical flat is tested against a standard reference flat in a Newton's interferoscope. If the optical flat is as good as the reference, the fringes observed are a set of:
(a) Concentric rings
(b) Elliptical rings
(c) Hyperbolic fringes
(d) Straight and parallel fringes.

49. Identify the correct matches using the code:

A. Spirometer	1. Electricity activity of the heart.
B. Sphygmomanomete	2. Respiratory volume measurement
C. Plethysmograph	3. Measurement of change in volume of body part
D. Electrocardiograph	4. Blood pressure measurement

Codes:

	A	B	C	D
(a)	1	4	3	2
(b)	2	3	4	1
(c)	4	1	2	3
(d)	2	4	3	1

50. The typical maximum output powers associated with the lasers are given below identify the correct matches.

A. He – Net laser	1. 10 m W
B. Argon ion laser	2. 100 mW
C. Carbon dioxide laser	3. 10 W
D. Diode laser	4. 500 W

Codes:

	A	B	C	D
(a)	1	2	3	4
(b)	2	1	4	2
(c)	3	4	1	2
(d)	2	3	4	1

51. The following method is widely accepted to determine oxides of Nitrogen in an **automobile emission**.
(a) Orsat analysis
(b) Gas – chromatography
(c) Chemi luminescence
(d) Flame ionization detection

52. The minimum number of lines that must be drawn on a grating needed to resolve the Sodium Doublet having nominal wavelengths of 589.0 nm and 589.6 nm is:
(a) 600 (b) 1200
(c) 982 (d) 300

53. A parallel beam is incident normally on a air glass interface. If the refractive index of glass is 1.720 them the loss of intensity through reflection is:

(a) 7% (b) 4%
(b) 2% (d) 1%

54. In a He – Ne **laser** operating at a wavelength of 632 8nm. The actual laser transition takes place between the energy levels of:
(a) Helium (b) Neon
(c) Helium and Neon (d) None of the above

55. To a swimmer viewing a Sodium lamp beside the pool. The color of the lamp seen from underneath the water appears.
(a) Yellow
(b) Yellow with a reddish tinge
(c) Yellow with a greenish tinge
(d) White

56. In a Gas Chromatograph, the gas which can not be used as a carrier gas is:
(a) Helium (b) Nitrogen
(c) Ammonia (d) None of these

57. A He-Ne **laser** beam having a wavelength of 632.8 mm is expanded to a diameter of 1 meter and then pointed towards the moon. If the distance of the moon from earth is 3.8×10^5 km, the diameter of the central maxima on the surface of the moon is:
(a) 1 m (b) 100 m
(c) 587 m (d) 720 m

58. An unpublicized He - Ne **laser** beam of wavelength 632.8 nm is reflected from the flat face of a dielectric see figure. A detector analyzes the reflected beam. It is found that when the angle between the incident and reflected beam is 115°–34' the light is completely plane polarized. The refractive index of the dielectric at this wavelength is:

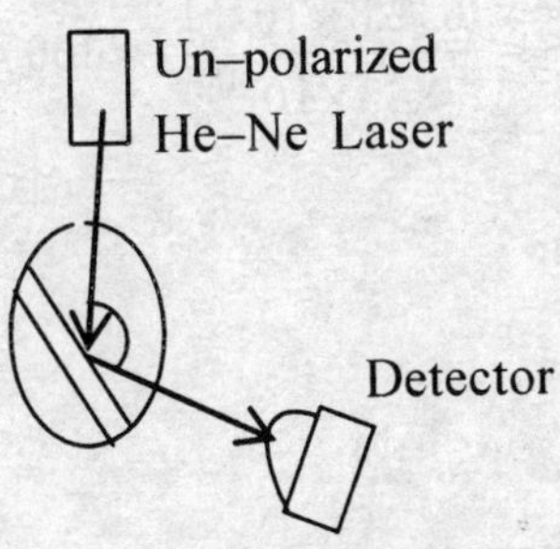

(a) 1.521 (b) 1.587
(c) 1.621 (d) 1.571

59. Two narrow straight slits, o.25 mm apart are illuminated by an unknown monochromatic source of light. If the fifth bright fringe is 12 mm away from the central fringe when the screen Constance is 0.8 the wave length of the unknown source is:
(a) 540 nm (b) 600 nm
(c) 700 nm (d) 750 nm

60. A **microscope** uses a micro-objective 10 X, numerical aperture 0.25 and an eyepiece of focal length of 25 mm. The magnification of the microscope is:
(a) 25 (b) 50
(c) 100 (d) 125

Answers Set-6

1. Ans. (c)
Solution: In a photo multiplier secondary emission is used for amplification of low level photo current.

2. Ans. (d)
Solution:

3. Ans. (d)
Solution:

4. Ans. (b)
Solution:

5. Ans. (b)
Solution:

6. Ans. (a)

Solution: Angle of refraction, $\tan\theta = \frac{\mu_1}{\mu_2} = \frac{1}{1.752}$

or $\theta = \tan^{-1}\left(\frac{1}{1.752}\right) = 30^\circ$

7. Ans. (d)

Solution: Numerical aperture $= \sqrt{\mu_1^2 - \mu_2^2}$

$$= \sqrt{\mu_1 + \mu_2 \times (\mu_1 - \mu_2)}$$

$$= \sqrt{2.84\ (1.44 - 1.4)}$$

$$= \sqrt{2.84 \times 0.04}$$

$$= \sqrt{0.1136} = 0.34$$

8. Ans. (a)
Solution:

9. Ans. (d)
Solution:

$$\text{Resolution} = \frac{\lambda_2 + \lambda_1}{2(\lambda_2 - \lambda_1)} = \frac{1}{2} \times \frac{(600.1 + 599.9)}{600.1 - 599.9}$$

$$= \frac{2 \times 600}{0.02 \times 2} = \frac{6000}{2} = 3000$$

10. Ans. (a)
Solution:

11. Ans. (b)
Solution:

12. Ans. (a)
Solution:

13. Ans. (d)
Solution:

14. Ans. (a)
Solution:
Speed = 50 × 60 rpm = 3000 rpm
One tooth moving in anti-clock wise direction then the true speed of gear wheel
= 3000 – 60
= 2940 rpm

15. Ans. (b)
Solution:

16. Ans. (b)
Solution:

17. Ans. (d)
Solution:

$$\lambda = \frac{c}{f} = cT = 3\times10^8 \times T$$

or $$T_o = \frac{1.2}{3\times10^8} = 4\times10^{-9} \text{ sec}$$

18. Ans. (c)
Solution:

19. Ans. (c)
Solution:
Absorbance = Molar can centration of the solute mc × Path length × molar absorbance.

or $$0.6 = 10^{-4}\times 2 \text{ cm}\times \text{x}$$

$$= \frac{0.6}{10^{-4}\times 2}$$

$$\text{x} = 3000$$

20. Ans. (b)
Solution:

21. Ans. (d)
Solution:

22. Ans. (b)
Solution:

23. Ans. (a)
Solution:

No. of fringes

$$= \frac{\text{Change in elevation level}}{\text{Wave length fo light producting fringes}}$$

$$= \frac{1.57\ \mu m}{0.574\ \mu m} = 3$$

24. Ans. (a)
Solution:
Oxygen is the only gas which shows Paramagnetism

25. Ans. (c)
Solution:

26. Ans. (d)
Solution:

27. Ans. (b)
Solution:

28. Ans. (d)
Solution:

$$\cos\phi_c = \sqrt{\frac{n_1^2 - n_2^2}{n_1}}$$

$$= \sqrt{\frac{1.641^2 - 1.422^2}{1.641}} = 0.499,\ \phi_c = 60^o$$

29. Ans. (d)
Solution:

$$\Delta\lambda = \frac{\lambda^2\Delta f}{c} = \frac{632.8\times10^{-9}\times10^6}{3\times10^8} = 1.33\times10^{-15}$$

$$\ell_c = \frac{\lambda^2}{\Delta\lambda} = \frac{(632.8\times10^{-9})^2}{1.33\times10^{-15}} = 300\ m$$

30. Ans. (a)
Solution:

$$E_y = E_0 e^{-\alpha x}$$

$$E_y = E_0 e^{-\alpha x}\ e^{-1\ \beta x}$$

(is terns of attenuations)

31. Ans. (c)
Solution:

32. Ans. (c)
Solution:

33. Ans. (b)
Solution:

34. Ans. (c)
Solution:

35. Ans. (a)
Solution:

36. Ans. (c)

Solution: $\eta = \sqrt{\varepsilon_r} = 2.5$
$= 1.58$

37. Ans. (d)
Solution:

38. Ans. (a)
Solution:

39. Ans. (c)
Solution:

40. Ans. (d)
Solution:

41. Ans. (a)
Solution:

42. Ans. (a)
Solution:

43. Ans. (c)
Solution:

44. Ans. (a)
Solution:

$$\mu_{rg} = \frac{\mu_w}{\mu_g} - \frac{1.33}{1.7} = 0.886$$

45. Ans. (c)
Solution:

$$\mu = \frac{\sin i}{\sin r} - \frac{\sin 60}{\sin 30} = \sqrt{3}$$

46. Ans. (d)
Solution:

47. Ans. (b)
Solution:

$d = 2.5\ m$

$Q = 25$ second

$$d = \frac{h}{\tan\theta} = \frac{2.5}{\tan 25^o}$$

$$= \frac{2.5}{0.466} = 3.437$$

48. Ans. (a)
Solution:

49. Ans. (d)
Solution:

50. Ans. (d)
Solution:

51. Ans. (b)
Solution:

52. Ans. (c)
Solution:

53. Ans. (a)
Solution:

Reflectivity, $$R = \left(\frac{n_2 - n_1}{n_2 + n_1}\right)^2 = \left(\frac{1.720 - 1}{1.720 + 1}\right)^2$$

$= 0.07$

Thus, loss of intensity through refraction is 7 %

54. Ans. (c)
Solution:

55. Ans. (a)
Solution:

56. Ans. (c)
Solution:

57. Ans. (c)
Solution:

58. Ans. (b)
Solution:

59. Ans. (d)
Solution:
The distance of the nth bright fringe from the central fringe.

$$Y_n = \frac{n\lambda D}{d}$$

where d = 0.25 mm D = 0.8m = 0.8 × 10^3 mm

$Y_n = 12\ mm$

for n = 5,

$$y_n = 12 = (5), \lambda..(0.8 \times 10^3)$$

or $$\lambda = 0.75 \times 10^{-3}\ \text{mm}$$

60. Ans. (c)
Solution:

MCQ Set–7

1. PH scale was introduced by:
(a) Sorensen (b) Rutherford
(c) Berthelot (d) Johnson

2. Which one of the following is used as reference electrode?
(a) Glass electrode
(b) Hydrogen electrode
(c) Quinhydrone electrode
(d) Saturated calomel electrode

3. Which one of the following is incorrect for hydrogen electrode?
(a) can be used over the entrie range of pH
(b) exhibits no salt error
(c) can be used in presence of strong about glass electrode is incorret?
(d) has low internal resistance

4. Which of the following statements about glass electrode is incorret?
(a) can be used in strongly alkaline solutions
(b) can be used for coloured solutions
(c) immediate equilibrium is established
(d) can be used in 0–12 pH range

5. Which of the expressions can be used for calculating pH using glass electrode
(a) $pH = \frac{-0.009 - E_{cell}}{0.059}$
(b) $pH = \frac{0.454 - E_{cell}}{0.0591}$
(c) $pH = \frac{E_G - E_{cell}}{0.0591}$
(d) $pH = \frac{E_{cell} - 0.0246}{0.0591}$

6. pH of the blood is
(a) 7.4 (b) 8.3
(c) 4.2 (d) 11.5

7. The pH of the solutions containing reducible ions or substance having high positive electrode potential can be determined using
(a) glass electrode
(b) antimony electrode
(c) quinhydrode electrode
(d) hydrogen electrode

8. A solution having pH = 0 is
(a) Neutral (b) acidic
(c) alkaline (d) None of these

9. Which of the following statements about X-ray is incorrect?
(a) X-rays are more penetrating than cathode rays.
(b) X-rays are deflected by electrical and magnetic field.
(c) X-rays are electromagnetic waves of short wave length.
(d) X-rays produce ionisation in gass.

10. X-rays were discovered by
(a) Mosley, (b) Rontgen,
(c) Bragg, (d) Chadwick

11. Bragg's equation is
(a) $\sqrt{v} = k\,(z - \sigma)$,
(b) $n\lambda = 2d \sin\theta$.
(c) $hv_{max} = e.v.$,
(d) $R = Ew/l$

12. A systematic study of X-ray spectra was first made by
(a) Mosley, (b) Rutherford
(c) Bohr (d) Bragg

13. The wave length at which a suden change in absorption occurs is known as
(a) absorption ocefficient
(b) Critical absorption wave length
(c) Moley's wave length
(d) absorption edge

14. X-ray emission analysis cannot be used to study
(a) Elements below atomic number 11,
(b) Metals with high atomic number
(c) non-metals with high atomic number
(d) lead in gasoline.

15. Which of these methods can not be used to determine crystal structure?
(a) Laue photographic method
(b) Goug's method
(c) Oscillating crystal method
(d) Powder method

16. Rotating crystal method for the determination of structure of crystals was given by
(a) Laue
(b) M. Polanyi
(c) P. Debyl, P. Scherrer and A. W. Hill.
(d) W. L. Bragg

17. The potential required to effect the fragmentation is called
(a) fragmentation potential
(b) appearance potential
(c) ionization potential
(d) oxidation potential

18. Which of the following is not the unit of a common mass spectrometer?

(a) ion source (b) mass analyser
(c) ion deterctor (d) monochromater

19. Which of the following is not the ion source in mass spectrometer?
(a) proton source
(b) electron impact ion source
(c) spark source
(d) photons.

20. A peak at mass T_9 always indicates the presence of
(a) benezene mg (b) butanol
(c) chloropropane (d) propionic acid

21. Alkyl chloride has a tendency to eliminate
(a) e^- (b) H^+
(c) HCl (d) Cl_2

Answers Set-7

1. (a)	2. (d)	3. (c)	4. (a)
5. (c)	6. (a)	7. (d)	8. (b)
9. (b)	10. (b)	11. (b)	12. (a)
13. (d)	14. (a)	15. (b)	16. (b)
17. (b)	18. (d)	19. (a)	20. (a)
21. (c)			

Reader's Notes

General Ability

Verbal Ability

- Type A : English Grammar
- Type B : Sentence Completion
- Type C : Verbal Analogies
- Type D : Word Groups
- Type E : Instructions/Comprehension
- Type F : Critical reasoning
- Type G : Verbal Deduction

Numerical Ability

- Type A : Numerical Computation
- Type B : Numerical Estimation
- Type C : Numerical Reasoning
- Type D : Data Interpretation

General Ability

VERBAL ABILITY

TYPE A: ENGLISH GRAMMAR

Vocabulary

Entry	Meaning	Synonyms	Antonyms
altercation (*n*)	a noisy argument or disagreement squabbling, wrangle	bickering, dispute, embroilment, fuss, hassle, quarrel, spat,	agreement, concord, harmony, eace, union, unity
anathema (*n*)	a thing or an idea which you hate because it is the opposite of what you believe	ban, censure, commination, condemnation, curse, damnation	benediction, blessing
appease (*v*)	to make sb calmer or less angry by giving them what they want	allay, alleviate, assuage, conciliate, mitigate, mollify pacify, placate, soothe	aggravate, annoy, antagonize, exacerbate, irritate, provoke
arduous (*adj*)	involving a lot of effort and energy, especially over a period of time	ball-buster, burdensome, exhausting, fatiguing, formidable, rigorous, strenuous, toilsome, trying, uphill	easy, facile, painless, simple
assidusous (*adj*)	working very hard and taking great care that everything is done as well as it can be	attentive, diligent, industrious, laborious, scrupulous, untiring	casual, lax, lazy, laid-back, neglectful, neglignet
attrition (*n*)	the act or process of wearing away or grinding down by friction; loss of personnel in an organization in the normal course of events, as by retirement	abrasion, attenuation, debilitation, depreciation, erosion, grinding, rubbing, wear	building, fortification, reinforcement
badinage	friendly joking between people	teasing, desision, jocking, persiflage, banter	seriousness
banal (*adj*)	(disapproving) very ordinary and containing nothing that is interesting, or important	bland, bromidic, cliched, common, conventional, insipid, ordinary, stereotyped, trite	fresh, new, original, rich
beguile (*v*)	to trick sb into doing sth, especially by being nice to them	bluff, cheat, deceive, manipulate, mislead, trick	advise, guard, protect
besmirch (*v*)	(formal) to damage the opinion that people have of sb/sth	defame, denigrate, slander	commend, praise
bluster (*v*)	to talk in an aggressive or threatening way, but with little effect	badger, boast, brag, brazen, browbeat	leave alone
broach (*v*)	to begin talking about a subject that is difficult to discuss, especially because it is embrrassing or because people disagree about it	advance, approach, hint	suppress

Contd...

Contd...

Entry	Meaning	Synonyms	Antonyms
buoyant (*adj*)	cheerful and feeling sure that things will be successful	animated, effervescent, jovial	depressed, somber
cadaverous (*adj*)	(of a person) extremely pale, thin and looking ill/sick	emaciated, pallid, wasted	lifelike, ruddy
cajole (*V*) nice to	To make to do by talking to them	tantalize, tempt, urge, wheedle	bettle, criticize them and being very
callow (*adj*)	young and without experience	guileless, inexperienced, naive, puerile	mature, worldly
candor (*n*)	sharp honesty or frankness in expressing oneself	diectness, forthrightness, impartiality	indirectness, unfairness
capricious (*adj*)	tending to change abruptly and without apparent reason	fickle, impulsive, inconstant, mutable, unpredictable, whimsical	dependable, predictable
carping (*adj*)	tending to carp, or find fault	deprecatory, fault-finding, hypercritical quibbling	complimentary, forgiving
censure (*n*)	a condemning as wrong; strong disapproval	admonition, condermnation, disapproval, repriment	approval, blessing
chicanery (*n*)	the use of clever but tricky talk or action to deceive, evade, etc. as in legal dealings	artfice, chicane, fraud, skulduggery, trickery	honesty, integrity
circumspect (*adj*)	careful to consider all related crcumstances before acting, judging, or deciding	cagey, canny, careful, discreet,	careless, incautious
cogent (*adj*)	forceful and to the point, as a reason or argument	compelling, convictive, forcible, justified, potent, powerful, significant, weighty	impotent, ineffective
collate (*v*)	to compare (texts, data, etc.) critically in order to consolidate, note similarities and differences, etc.	adduce, analogize, compare, examine, relate	disorder, disorganize
commensurate	corresponding in extent or degree	appropriate, compatible, consistent, fitting, symmetrical	disproportionate, inappropriate
commodious (*adj*)	offering plenty of room	capacious, roomy	confined, cramped
concatenate (*v*)	link together	artifulate, couple, integrate	disconnect
conjecture (*n*)	an opinion or idea that is not based on definite knowledge and is formed by guessing	assumption, guess, hypothesis, presumption, surmise	certainty, fact
construe (*v*)	to understand the meaning of a word, a sentence, or an action in a particular way	decipher, interpret, translate	complicate, obscure
convoke (*v*)	to call together for a meeting; summon to assemble	congregate, convene, summon	disband, scatter
cower (*v*)	to shrink and tremble, as from someone's anger, threats, or	blench, crouch, truckle, wince	intimidate, strut

Contd...

Contd...

Entry	Meaning	Synonyms	Antonyms
cursory (*adj*)	done quickly and without giving enough attention to details	haphazard, hasty, random, sloppy, swift	meticulous, unknon
debauch (*v*) defection (*n*)	to lead astray morally abandonment of loyalty, duty, or principle	corrupt, deprave, seduce alienation, desertion, disloyalty, repudiation, separation	upgrade, value loyalty, staunchness
delusion (*n*)	a false belief or opinion about yourself or your situation	apparition, fallacy, illusion, misbelief	reality
denizen (*n*)	a person, an animal or a plant that lives, grows or is often found in a particular place	dweller, habitant, native, occupant	alien, foreigner
detriment (*n*)	the act of causing harm or damage	handicap, impairment, prejudice, spoiling	advantage, benefit
diaphanous (*adj*)	so fine or gauzy in texture as to be transparent or translucent	clear, delicate, filmy, gossamer, pellucid	opaque, thick
diligent (*adj*)	persvering and careful in work	assiduous, careful, industrious, laborious	dilatory, inactive, lazy, slack
disinterested (*adj*)	not influenced by personal interest or selfish motives	aloof, impartial, unbiased, unprejudiced, dispassionate	biased, concerned, passionate
dissemble (*v*)	to conceal under a false appearance	camouflage, conceal, disguise	expose, reveal
distend (*v*)	to expand; make or become swollon	augment, bulge, clilate expand, inflate	compress, constrict
dotage (*n*)	feeble and childish state due to old age	decrepitude, elderliness, imbecility senility, weakness	adolescence, youth
duress (*n*)	the use of force of threats	bondage, coercion, compulsion, incarceration	free will
eclectic (*adj*)	selecting from various systems, doctrines, or sources	assorted, inclusive, multifarious, wide-ranging	incomprehensive, limited, narrow, particular, specialized
effulgent (*adj*)	shining brightly	beaming, dazzling, luminous, radiant	dark, dim, dull
egregious (*adj*)	outstanding for undesirable qualities; remarkably bad	arrant, deplorable, flagrant, heinous, nefarious, notorious,	measly, minor, petty
emasculate (*v*)	to destroy the strength or force of	castrate, debilitate, devitalize, impoverish, mutilate, weaken	energize, invigorate, strengthen, vitalize
empirical (*adj*)	relying or based solely on experiment and observation rather than theory	experiential, experimental, observational, pragmatic	hypothetical, impractical theoretical
ennui (*n*)	weariness and dissatisfaction resulting from inactivity or lack of interest	dejection, depression, lassitude, boredom	enthusiasm, liveliness
emphemeral (*adj*)	short-lived	brief, fleeting, transient, transitory, volatile	eternal, everlasting
equitable (*adj*)	characterized by equity; said of actions, results of actions, etc.	candid, fair, honest, just, stable	biased, partial

Contd...

Contd...

Entry	Meaning	Synonyms	Antonyms
espouse (*v*)	to take up, support, or advocate (some cause, idea, etc.	advocate, defend, imbrace, uphold	abandon, deny
exacerbate (*v*)	to make more intense or sharp	aggravate, enrage, exasperate, inflame, intensify, provoke	appease, mitigate, placate
excogitate (*v*)	to think out carefully and fully	consider, contemplate, educe, ponder, ruminate	ignore, neglect
exhort (*v*)	to urge earnestly byt advice, warning, etc. (to do what is proper or required)	admonish, beseech, goad, stimulate, warn	discourage, impede
extirpate (*v*)	to destroy or remove completely	annihilate, deracinate, exterminate, extinguish	breed, generate
facile (*adj*)	produced without effort or careful thought	adroit, dexterous, glib, proficient, skillful, voluble	difficult, laborius
fathom (*v*)	to understand or find an explanation for sth	apprehend, cognize, discern, interpret, penetrate, perceive, savvy	confound, confuse, misunderstand
flippant (*n*)	frivolous and disrespectful	impertinence, impudence, rudeness, waggishness	seriousness
fortuitous (*adj*)	hapening by chance	accidental, casual, incidental, unforeseen	predictable, unlucky
furtive (*adj*)	behaving in a way that shows that you want to keep sth secret and do not want to be noticed	artful, circuimspect, conspiratorial, cunning, surreptitious, undercover	aboveboard, open
garrulous (*adj*)	talking much or too much, esp. about unimportant things	loquacious, verbose, voluble	concise, taciturn
germane (*adj*)	truly relevant	applicable, apposite, pertinent	inappropriate
gossamer (*adj*)	a very thin, soft, filmy cloth	diaphanous, sheer, transparent	opaque
grandiloquent (*adj*)	using high-flown, pompous, bombastic words and expressions	magniloquent, pompous	unpretentious
gullible (*adj*)	easily cheated or tricked	believing, credulous, innocent, naive, unsophisticated	suspicious, untrusting
harangue (*n*)	a long, blustering, noisy, or scolding speech	address, diatribe, oration, spouting, tirade	panegyric
headstrong (*adj*)	determined not to follow orders, advice, etc. but to do as one pleases	determined, imprudent, intractable, refractory, uncontrollable, unruly	tractable
heresy (*n*)	any opinion (in philosophy, politics, etc.) opposed to official or established views or doctrines	heterodoxy, iconoclasm, nonconformity, unorthodoxy	otherodoxy
horrendous (*adj*)	extermely shocking; extremely unpleasant and unacceptable	abhorrent, detestable, grisly, loathsome, terrible	agreeable, delightful
humdrum (*adj*)	lacking variety	boring, dull, monotonous, repetitious, wearisome	exciting, remarkable

Contd...

Contd...

Entry	Meaning	Synonyms	Antonyms
imperative (*adj*)	absolutely necessary	compulsory, exigent, indispensable, obligatory, urgent, compelling	optional, unimoportant
impolitic (*adj*)	not wise	brash, imprudent, indiscreet, inexpendient, injudicious	prudent, wise
import (*n*)	to mean; signify	consequence, importance, substance, value	insignificance
incarcerate (*v*)	to put sb in prison or in another place from which they cannot escape	bastille, confine, immure	free, release
Incontrovertible	that is true and cannot be disagreed with or denied	authentic, incontestable, indisputable, irrefutable, undeniable	questionable
indelible (*adj*)	that canot be erased, blotted out, eliminated, etc	ineradicable, inexpungible, inextirpable, permannent	erasable
indomitable (*adj*)	not easily discouraged, defeated, or subbued	insurmountable, invincible, unbeatable, unyielding	submissive, weak
infirmity (*n*)	weakness or illness over a long period	affliication, decrepitude, disorder, feebleness, sickliness	health, wellness
innocuous (*adj*)	that does not injure or harm	banal, harmless, inane, insipid, unobjectionable	harmful
insidious (*adj*)	operating in a slow or not easily apparent manner; more dangerous than seems evident	crafty, deceitful, intriguing, perfidious, secret, subtle, surreptitious	straightforward
integrity (*n*)	the quality or state of being of sound moral principle	honesty, incorruptibility righteousness, sincerity, uprightness	dishonesty
intrepid (*adj*)	very brave; not afraid of dnger or difficulties	courageous, dauntless, gallant unafraid, valorous	cowardly, fearful
invidious (*adj*)	unpleasant and unfair; likely to offened sb or make them jealous	abominable, defamatory, detestable, , envious, repugnant, slanderous	desirable, pleasant
jaunty (*adj*)	showing that you are feeling confident and pleased with yourself	animated, breezy, buoyant, hilarious, showy, vivacious	depressed
juvenile (*adj*)	of characteristic of, or suitable for children or young persons	adolescent, babyish, infantile pubescent	adult
kindred (*adj*)	of like nature consanguine, relqated, similar	affiliated, akin, alike, cognate,	unrelated
lackadaisical (*adj*)	showing lack of interest or spirit	apathetic, disinterested, indolent, languid, lethargic, listless	caring, enthusiastic
latitude (*n*)	freedom from narrow restrictions; freedom of opinion conduct or action	independence, laxity, leeway, unrestrictedness	constraint, restriction
limpid (*adj*)	perfectly clear; not cloudy or turbid	bright, lucid, pellucid, transparent, unambiguous	muddy, opague
ludicrous (*adj*)	so absurd, ridiculous, or exaggerated as to cause or merit laughter	comical, crazy, funny, laughable, ridiculous, risible	sensible

Contd...

Contd...

Entry	Meaning	Synonyms	Antonyms
magnanimous (*adj*)	kind, generous and forgiving, especially towards an enemy or a rival	benevolent, charitable, unselfish	mean - spirited, selfish
malaise (*n*)	a vague feeling of physical discomfort or uneasiness, as early in an illness	anxiety, decrepitude, feebleness, lassitude, weakness	health, well - being
mendacious (*adj*)	not truthful; lying or false	deceitful, deceptive, fallacious, insincere, perfidious, wrong	truthful
meteoric (*adj*)	like a meteor; momentarily	brief, fleeting, momentary, dazzling or brillaiant, flashing, swift	gradual, slow spectacular, swift, transient
misgiving (*n*) mortify (*v*)	a disturbed feeling of fear, doubt, apprehension, etc. to make sb feel very ashamed or embarrassed	anxiety, apprehension premonition, reservation, unease abase, abash, chagrin, humiliate, ridicule	certainty, confidence be pleased, be proud
munificent (*adj*)	very gnerous in giving	benevolent, charitable, lavish, philanthropic	mean , miserly
nefarious (*adj*)	very wicked	atrocious, heinous, iniquitous, opprobrious, villainous	admirable, good
noisome (*adj*)	having a bad odor, foul-smelling	noxious, offensive, putrid, repulsive, sickening, stinking	moral, pleasant
novice (*n*)	a person new to a particular occupation, activity, etc.	amateur, apprenticer, beginner, rookie, tyro	expert, veteran
obliterate (*v*)	to remove all signs of sth, either by destroying or covering it completely	annihilate, delete, eliminate, squash	create
officious (*adj*)	too ready to tell people what to do or to use the power you have to give orders.	dictatorial, impertinent, interfering, meddlesome, opinionated, pushy	self - effacing
onerous (*adj*)	needing great effort; causing trouble or worry	arduous, burdensome, demanding, laborious, vexatious	easy, effortless
ornate (*adj*)	heavily ornamented or adorned, often to excess	adorned, uscious, ornamented, ostentatious, variegated	plain
overt (*adj*)	not hidden	apparent, manifest, observable, open, undisguised	covert, hidden
palatable (*adj*)	pleasant or acceptable to the taste; fit to be eaten or drunk	acceptable, appetizing, flavorsome, luscious, relishing	disagreeable
parochial (*adj*)	restricted to a small area or scope	biased, limited, narrow, provincial, sectarian	broad-minded, cosmopolitan
pejorative (*adj*) penitent (*adj*)	disparaging or derogatory truly sorry for having sinned or done other wrong and willing to atone	debasing, derogatory, irreverent attritional, contrite, penitential, regretful, repentant, sorry	complimentary unashamed, unrepentant
personable (*adj*) perturb (*v*)	having a pleasing appearance and personality to make sb worried or anxious	amiable, attractive, charming likable, sociable alarm, bewilder, bother, disconcert, dismay, perplex, pester	unfriendly, unpleasnat reassure

Contd...

Contd...

Entry	Meaning	Synonyms	Antonyms
placate (*v*)	to stop from being angry	appease, assuage, conciliate, reconcile, tranquilize	anger, provoke
pragmatic (*adj*)	concerned with actual practice, everyday affairs, etc. not with thory or speculation	logical, philistine, practical, realistic	idealistic, impractical, unrealistic
pretentious	trying to appear important, intelligent, etc. in order to impress other people; trying to be sth that you are not, in order to impress	assuming, imposing, magniloquent, ostentatious, pompous	natural, unaffected
puerile (*adj*)	suitable for a child rather than an adult	babyish, inane, infantile, juvenile, silly	mature, sensible
petulant (*adj*)	bad-tempered and unreasonable, especially because you cannot do or have what you want	fractious, irritable, peevish, querulous, waspish	affable
platonic (*adj*)	(of a relationship) friendly but not involving sex	idealistic, quixotic, spiritual, utopian	physical, sexual
profligate (*adj*)	extremely wasteful	extravagant, lavish, prodigal, squandering	frugal, thrifty
quaint (*adj*)	attractive in an unusual or old-fashioned way	antiquated, baroque, charming, fanciful, picturesque	modern

Idioms

Idioms	Meaning	Uses
a bean counter	an impolite way of describing someone who is responsible the financial decisions within a company	When decisions that affects people's lives are in the hands of bean counters, it's bad news.
be a bit much	if you say something is a bit much, you think that it is not fair or that it is more than you can deal with	I think it's much a bit much to except her to play 3 matches in one day.
break the ice	to make people who have not met before feel more relaxed with each other if you use a	We played a couple of party games to break the ice.
carrot and stick	carrot-and-stick method to make someone do something, you both offer rewards and threaten punishments	I've had to take the carrot-and-stick approach to disciplining my kids. The harder they work, the more money they get
clear the air	if an argument or disscusiion clears the air, it causes bad feelings between people to disappear	The meeting didn't solve anything, but at least it cleared the air.
give sb the cold shoulder	to bahave towards someone in a way that is not at all friendly, sometimes for reasons that this person does not under stand.	What have I done to him? He gave me the cold shoulder the whole evening at the party.
have a cow	to be very worried, upset, or angry about something	I though he was going to have a cow when I told him I'd lost his key.
as the crow files	if the distance between two places is measured as the crow files, it is measured as a straight line between the two places	'How far is it from Cambridge to london?' 'About 50 miles as the crow files.

Contd...

Contd...

Idioms	Meaning	Uses
die hard	if a habit, custom, or belief, dies hard, it takes a long time to change or end it	After a successfully 30-years career, he no longer has any need to work-but old habits die hard.
donkey's years	a very lone time	I've been doing this job for donkey's years.
the halcyon days	a very happy or successful period in the past	That was in the halcyon days of the 1980's when the economy was boming.
a litmus test	something that shows clearly what someone's opinions or intentions are	His views on abortion are effectively a litmus test of his views on women's rights.
get it in the neck	to be punished or criticized for something that you have done	It always seems to be the chairman of these football clubs who gets it in the neck when the team does badly.
a red herring	something that takes people's attention away from the main subject being talked or written about	About halfway through the book it looked as though the butler was the murderer, but that turned out to be a red herring.
talk turkey	to discuss a problem in a serious way with a real intention to solve it	It the two sides in the dispute are to meet, they must be prepared to talk turkey
from top to be	on every part of a person's body	He gazed at her across the room, dressed in black leather from top to toe.
wax and wane	to grow bigger and stronger and then to become smaller or weaker again	Their influence waxes and wanes depending on which party is in power.
a witch-hunt	an attempt to find and punish people who have opinions that are believed to be dangerous	Senator McCarthy led a withhunt against suspected communists during the 1950's
be writ large wax and wane	to be very obvious to grow bigger and stronger	Anger was writ large in his face Their influence wazse

TYPE B: SENTENCE COMPLETION

Directions: There is a blank space in each of the following sentences. Fill in the blank bty choosing the correct word or group of words from the alternative (a), (b), (c) and (d).

1. Are you ... your lessons?
(a) busy for preparing (b) busy preparing
(c) busy to prepare (d) busy of preparing

2. He is more intelligent ... in the class.
(a) than all the boys (b) than any others
(c) than any other boys (d) to all the boys

3. I admitted that I was ...
(a) on the wrong (b) on wrong
(c) in wrong (d) in the wrong

4. She was much ... at your indecent remarks.
(a) cut short (b) cut up
(c) cut in (d) cut off

5. She ... in the negative.
(a) said to him (b) replied to him
(c) replied him (d) told him

6. I shall have ... his services.
(a) to dispense off (b) to dispense with
(c) to dispense (d) to dispense of

7. I am delayed; now I ... .
(a) take you leave (b) take leave of you
(c) take leave from you (d) take leave with you

8. Would you mind ... me that pen?
(a) giving (b) for giving
(c) to give (d) in giving

9. Tell me whether she burst ... tears.
(a) into (b) of into
(c) in (d) out in

10. At the last election I ... voting.
(a) abstained in (b) abstained from
(c) abstained to (d) refrained

11. .....Cloudy day, we did not go out for a walk.
(a) Being (b) Being a
(c) It being (d) It being a

12. He ... from Tokyo yesterday.
(a) has returned (b) had returned
(c) was returning (d) returned

13. Everything depends on ... on time.
(a) them to come (b) them coming
(c) they coming (d) their coming

14. That road is under ... .
(a) repairs (b) repair
(c) the repair (d) the repairs

15. Know it for certain that it is no use ... everyone.
(a) to find fault with (b) finding fault with
(c) finding fault at (d) to find fault at

16. They reiterated the demand ... elections.
(a) on holding (b) for holding
(c) by holding (d) of holding

17. It is my firm belief that she ... .
(a) shall come (b) will come
(c) should come (d) must come

18. ...... Foresight and prudence, he set up an industrial empire.
(a) Blessed with (b) Conferred with
(c) Endowed with (d) Blessed in

19. Do you enjoy ...?
(a) to playing cricket
(b) to play cricket
(c) playing cricket
(d) himself to play cricket

20. Can you ...?
(a) mix oil by water (b) mix oil with water
(c) mix oil into water (d) mix oil in water

21. Bear in mind that it is not always easy to put ... an air of innocence.
(a) up (b) on
(c) in (d) down

22. As he shouted for a long time, his voice had become...
(a) coarse (b) hoarse
(c) hose (d) course

23. A judge should be ... boundless patience.
(a) skilled in (b) capable of
(c) able of (d) averse to

24. What sort of ...?
(a) girl is she (b) girl she is
(c) the girl is she (d) a girl is she

25. Did you take ... an insurance policy on attaining the age of 30?
(a) out (b) over
(c) up (d) in

26. I want to know what your ... are in this line.
(a) prospects (b) future prospects
(c) prospectives (d) future prospectives

27. She is my ... neighbour.
(a) close (b) immediate
(c) next door's (d) next-door

28. I shall spend ... life in Calcutta.
(a) my remainder of (b) my remaining
(c) my life's rest (d) the rest of my

29. I can always ... her help.
(a) count on (b) count with
(c) count (d) made for

30. Do not interfere ... my affairs.
(a) at (b) by
(c) with (d) in

31. She exampted me ... taking the test.
(a) of (b) from
(c) with (d) by

32. Her mother ... of influenza yesterday.
(a) has died (b) will die
(c) died (d) dies

33. The ... were beaten by the students in the examination hall.
(a) teachers (b) tutors
(c) investigators (d) invigilators

34. It has been years since I last ... her.
(a) seen (b) have seen
(c) had seen (d) saw

35. She was ... for her misdeeds.
(a) famous (b) noted
(c) notorious (d) celebrated

36. He is ... of becoming a lawyer.
(a) disirous (b) desirable
(c) desired (d) desiring

37. The bomb ... and killed several persons.
(a) dropped (b) broke
(c) excited (d) exploded

38. He has developed prejudice ... me.
(a) against (b) for
(c) from (d) upon

39. Are you very fond of ... jokes?
(a) throwing (b) boosting
(c) suggesting (d) cracking

40. I was provoked ... by her remarks.
(a) with (b) at
(c) by (d) on

41. He is not blind ... the faults of his sons
(a) up (b) with
(c) to (d) on

42. She was born ... wealthy parents.
(a) from (b) with
(c) to (d) at

43. His father passed ... last month.
(a) off (b) on
(c) out (d) away

44. State of living unmarried is ... .
(a) virginity (b) bachelorhood
(c) celibacy (d) single

45. A person who always thinks about himself and what is best for himself is an ... .
(a) egoist (b) selfish
(c) self-centred (d) egotist

46. The custom of having more than one husband at the same time is ...
(a) polyandry (b) polyandrous
(c) No word

Answers

1. (b)	2. (c)	3. (d)	4. (b)	5. (b)
6. (b)	7. (b)	8. (a)	9. (a)	10. (b)
11. (d)	12. (d)	13. (d)	14. (b)	15. (b)
16. (b)	17. (b)	18. (c)	19. (c)	20. (b)
21. (b)	22. (b)	23. (b)	24. (a)	25. (a)
26. (a)	27. (d)	28. (d)	29. (a)	30. (d)
31. (b)	32. (c)	33. (d)	34. (d)	35. (c)
36. (a)	37. (d)	38. (a)	39. (d)	40. (c)
41. (c)	42. (c)	43. (d)	44. (c)	45. (a)
46. (a)				

Passage 1

The group of tired dusty riders arrived at a fork in the road. Their leader immediately sprang to the ground after first throwing his rein to one of the others, and began to examine minutely the sandy track. The problem was simple; if the fleeing enemy had taken the left turning, there remained little hope of catching them, since he knew that it led back to a small settlement of native huts where they should be sheltered by the friendly inhabitants. If, on the other hand, they had branched to the right, they would have before them the open desert, not a flat expanse of sand such as they had just crossed, but a country broken by a series of ridges, behind any of which a whole army could hide.

He turned to his companies to see if he could read any solution in their faces. But they, were too occupied by their aching limbs and several were taking a quick drink from the flasks which hung at their belts. He realised every minute's delay lessened their chances of overtaking their adversaries. So, with a rapid glance at the sun, he jumped once more into the saddle and with his whip indicated the way they were to go.

Questions

1. Why did the leader look at the sun?
(a) To estimate what daylight still remained
(b) He sent a silent prayer to the sun-god to show the way
(c) To estimate chances of their overtaking their adversaries
(d) To estimate their position in the desert

2. What, if any, was the difference in the nature of the terrain they had already traversed and the one lying ahead?
(a) The country-side already traversed was ups and downs
(b) There was hardly any difference as it was a vast expanse of unending desert
(c) The terrain hitherto was dusty and tiring, the road ahead led back to a habitation
(d) What they had crossed was a flat expanse of sand and what they lay ahead to the right was a series of ridges of sand

3. What difficulties did they expect if they took the branch to the right?
(a) The open desert ahead promised no oasis to shelter them at night
(b) They were absolutely in the dark about what they lay ahead of them
(c) Ahead lay a country full of ridges of sand providing cover to the feeling enemy
(d) They would have before them the open desert obstructing pursuit

4. What was the aim of the group of riders?
(a) They were fleeing from a pursuing enem
(b) They had a small settlement of native huts as their goal
(c) They were exploring a desert
(d) They were pursuing a fleeing enemy

5. What difficulties did the pursuers expect if they took to the left hand track?
(a) It led through a trackless expanse of unending desert
(b) It led to a settlement where the residents would surely shelter the enemy they were after
(c) The settlement it led to was hostile to them
(d) The leader anticipated a possible revolt among his followers, foot-sore, dust covered and tired as they already were.

6. Where and why did they stop?
(a) At a fork in the road to examine the sandy track for foot-prints of the fleeing enemy
(b) At the huts of some friendly inhabitants in order to rest their weary limbs
(c) At a three-ways crossing where the leader wanted to study the position of the sun in the sky
(d) At a fork in the road in order to decide whether they should turn left or right

Passage 2

Scarcity not created by war, it is a permanent characteristic of all human society and is the basis of the problem that faces, and always had faced, the human race whatever its form of organisation. It springs from the fact that the material resources of the world are limited and that our ability to make use of those resources is even more limited by our ignorance.

Everything that we need to satisfy our wants has to be derived finally from two sources- the natural resources that are available and the human ability to make use of them. As our knowledge grows and we increase our skill, we can exploit more and more of the opportunities that nature offers to us. The increase in communications, for example, has made accessible many minerals from depths below the earth's surface that could not be reached by earlier generations. But whatever the rate of development may be, there is, at any one time, a limit to the total of what can be produced. Here and now, there is only so much of the material resources of the world accessible to us and there are only so many people capable of work and endowed with the capacities and skills that they have developed. No doubt we waste a lot by stupidity even if the very best possible use is made of all the resources we can reach, there is still a limit to what can be produced out of them, and it is out of that total that we must all satisfy our wants.

Questions

1. Some of the material resources are:
   (a) wasted by man's stupidity
   (b) neglected by man
   (c) accessible to man
   (d) stored for the future
2. The first sentence of the passage tells us that scarcity is:
   (a) created by war
   (b) caused by human beings
   (c) not permanent
   (d) present in all human societies
3. The main idea of the passage is:
   (a) the cause of scarcity
   (b) the need to work hard
   (c) how to overcome scarcity
   (d) the need to increase our knowledge
4. The cause of scarcity is:
   (a) limited natural resources and still more limited knowledge of how to use them
   (b) the limitation of man's knowledge
   (c) human ignorance of natural resources
   (d) the limitation of various natural resources

Passage 3

In our boyhood we beheld the dying rays of that intimate sociability which was characteristic of the last generation. Neighbourly feelings were then so strong that the informal gatherings were a necessity, and those who could contribute its amenities were held in great respect. People nowadays call on each other on business, or as a matter of social duty, but not to foregather by way of informal gatherings. They have not the time, nor are there the same intimate relations. What goings and comings we used to see, how merry were the rooms and varandahs with the hum of conversation and the snatches of laughter. The faculty our predecessors had of becoming the centre of groups and gatherings, of starting and keeping up animated and amusing gossip, has vanished. Man still come and go, but those same verandahs and rooms seem empty and deserted. In those days everything from furniture to festivity was designed to be enjoyed by the many, so that whatever of pomp or magnificence there might have been did not savour of hauteur. These appendages have since increased in quantity, but they have become unfeeling and know not the art of making high and low alike feel at home. The barebodied, the indigently clad, no longer have the right to use and occupy them, without a permit, on the strength of their smiling faces alone. Those whom we nowadays seek to imitate in our housebuilding and furnishing, they have their own society, with its wide hospitality. The mischief with us is that we have lost what we had but have not the means of building up afresh on the European standard, with the result that our home-life has become joyless. We still meet for business or political purposes, but never for the pleasure of simply meeting one another. We have ceased to contrive opportunities to bring men together simply because we love our fellow-men. I can imagine nothing more ugly than this social miserliness, and, when I look back on those whose ringing laughter, coming straight from their hearts, used to lighten for us the burden of household cares, they seem to have been visitors from some other world.

Questions

1. Which of the following is social miserliness according to the author?
   (a) Diminishing intimate sociability
   (b) Increasing social inequality
   (c) Bare-bodied, half-starved masses
   (d) We don't wish to incur expenditure merely to get people together
2. Why does the author feel that the rooms and verandahs have lost their charm?

(a) The rooms and verandahs have been deserted by the people
(b) They have been crowded by gatherings of gossiping people
(c) The conversational quality our predecessors had has now vanished
(d) Unlike in the past, no facilities for arranging meetings are available

3. People whom do which of the following now would look like visitors from some other world?
(a) Eradication of poverty and establishing social equalit
(b) Strengthening the ties of love and affection through recurring informal meetings
(c) Reducing the burden of household cares by extending monetary assistance
(d) Enhancing business opportunities with European countries

4. How is the meeting of people of present generations different from the meeting of the people of the past?
(a) Intimate sociability has crept in
(b) Neighbourly feelings have now been strengthened
(c) Informal gossiping is not considered ethical these days
(d) People now do not meet merely for the pleasure of meeting one another

5. The approach of the author seems to be:
(a) extremely critical about our indifference towards the past generation
(b) emphatic about the need for informal social interaction
(c) cynical about our blindly imitating the Western style
(d) pessimistic about our business and political avenues

6. Why, according to the author, has our home life become joyless?
(a) We have lost the means of building up houses on the European standards
(b) Complete imitation of the West is not possible owing to different climatic conditions
(c) We have lost interpersonal intimacy and sociability. We do not have the means to build up afresh on the European standards
(d) Most people are bare-bodied and half-starved

7. Which of the following statements is definitely true in the context of the passage?
(a) Poor people in the past generation did not have the right to enjoy anything without permission
(b) We have ceased to bring people together solely for the pleasure of meeting them
(c) Dying rays of intimate sociability was the characteristic of the last generation
(d) The gap between different strata of the present society is on the increase

Passage 4

A recent report in New York Times says that in American colleges, students of Asian Origin outperform not only the minority group students but the majority Whites as well. Many of these students must be of Indian origin, and their achievement is something we can be proud of. It is unlikely that these talented youngsters will come back to India, and that is the familiar brain drain problem. However, recent statements by the nation's policy makers indicate that the perception of this issue is changing. Brain bank and not brain drain is the more appropriate idea, they suggest, since the expertise of Indians abroad is only deposited in other places and not lost.

This may be so, but this brain bank, like most of ther banks, is one that primarily serves customers in its neighbourhood. The skills of the Asians now excelling in America's colleges will mainly help the US(a) No matter how significant, what Non-Resident Indians do for India and what their counterparts do for other Asian counteries is only a by-product. But it is also necessary to ask, or be reminded, why Indians study more fruitfully when abroad. The Asians whose accomplishments New York Times records would have probably had a very different fate if they had studied in Indi(a) In America they found elbow room, books and facilities not available and not likely to be available here. The need to prove themselves in their new country and the competition of an international standard they faced there must have cured mental and physical laziness. But other things helping them in America can be obtained here if we achieve a change in social attitudes, especially towards youth.

We need to learn to value individuals and their unique qualities more than conformity and respectability. We need to learn the language of encouragement to add to out skill in flattery. We might also learn to be less liberal with blame and less tight-fisted with appreciation, especially to those showing signs of independence.

Questions

1. Among the many groups of students in American colleges, Asian students:
(a) are often written about in newspapaers like New York Times
(b) are the most successful academically

(c) have proved that they are as good as the Whites
(d) have only a minority status like the Blacks

2. The students of Asian origin in America include:
(a) a fair number from India
(b) a small group from India
(c) persons from India who are very proud
(d) Indians who are the most hardworking of all

3. In general, the talented young Indians studying in America:
(a) have a reputation for being hardworking
(b) have the opportunity to contribute to India's development
(c) can solve the brain drain problem because of recent changes in policy
(d) will not return to pursue their careers in India

4. There is talk now of the brain bank. This idea;
(a) is a solution to the brain drain problem
(b) is a new problem caused partly by the brain drain
(c) is a new way of looking at the role of qualified Indians living abroad
(d) is based on a plan to utilise foreign exchange remittances to stimulate research and development.

5. The brain bank has limitations like all banks in the sense that:
(a) a bank's services go mainly to those near it
(b) small neighbourhood banks are not viable in this age of multinationals.
(c) only what is deposited can be withdrawn and utilised
(d) no one can be forced to put his assets in a bank.

6. The author feels that what Non - Resident Indians do for India:
(a) Will have many useful side - effects
(b) will not be their main interest and concern
(c) can benefit other Asian countries, as a lay - product
(d) can help American colleges be of service of the world community.

Answers

Passage 1

1. (a) 2. (d) 3. (c) 4. (d) 5. (c)
6. (d)

Passage 2

1. (c) 2. (d) 3. (c) 4. (a)

Passage 3

1. (d) 2. (c) 3. (d) 4. (d) 5. (b)
6. (c) 7. (b)

Passage 4

1. (b) 2. (a) 3. (d) 4. (c) 5. (d)
6. (c)

TYPE C: VERBAL ANALOGIES

Exercise 1

Directions: (Question 1 to 11): In each of the following questions, four pairs of words are given, out of which the words in three pair bear a certain common relationsip. Choose the pair in which the words are differently related.

1. (a) See: Eye (b) Hear: Ears
(b) Smell: Nose (d) Tongue : Taste

2. (a) Lion: Roar (b) Snake: Hiss
(c) Frog : Bleat (d) Bees : Hum

3. (a) Ornithology : Birds
(b) Mycology : Fungi
(c) Biology : Botany
(d) Phycology : Birds
(e) Entomology : Insects

4. (a) Cat : Paw (b) Lizard : Pad
(c) Horse : Hoof (d) Man : Leg

5. (a) Tree : Branch (b) Hand : Finger
(c) Table : Chair (d) Room : Floor

6. (a) Chaff : Wheat (b) Grit : Pulses
(c) Grain : Crop (d) Dregs : Wine

7. (a) Proteins : Marasum (b) Sodium : Rickets
(c) Iodine : Goitre (d) Iron : Anaemia

8. (a) Jaundice : Liver (b) Rickets : Bones
(c) Tuberculosis : Lungs (d) Dengue : Fever

9. (a) Bronze : Zinc
(b) Duralumin : Aluminium
(c) Steel : Iron
(d) Brass : Copper

10. (a) Flower : Petal (b) Chair : Leg
(c) Circle : Arc (d) Cover : Page

11. (a) Iran : Asia (b) Canberra : Australia
(c) Norway : Europe (d) Algeria : Africa

Answers

Exercies 1

1. **Ans. (d)**
Solution: In all other pairs, first denotes the function performed by the second.

2. **Ans. (c)**
Solution: In all other pairs, second is the noise produced by the first.

3. **Ans. (c)**
Solution: In all other pairs, first is the study of second.

4. **Ans. (d)**
Solution: In all other pairs , second is the name given to the foot of the first.
5. **Ans. (c)**
Solution: In all other pairs, second is a part of the first.
6. **Ans. (c)**
Solution: In all other pairs, first is the waste obtained from the second.
7. **Ans. (b)**
Solution: In all other pairs, second is the disease caused by the deficiency of first.
8. **Ans. (d)**
Solution: In all other pairs, second denotes the part affected by the disease mentioned in first.
9. **Ans. (a)**
Solution: In all other pairs, first is an alloy of the second.
10. **Ans. (d)**
Solution: In all other pairs, second os a part of the first.
11. **Ans. (b)**
Solution: In all other pairs, second is the continent to which the country denoted by the first belongs.

Exercise 2

Directions: In each of the following questions, certain pairs of words are given, out of which the words in all pairs except one, bear a certain common relationship. Choose the pair in which the words are differently related.

1. (a) Water : Thirst (b) Talent : Education
(c) Food : Hunger (d) Air :Suffocation
2. (a) Broad : Wide (b) Light : Heavy
(c) Tiny : Small (d) Big : Large
3. (a) Shopkeeper : Customer
(b) Doctor : Patient
(c) Lawyer : Client
(d) Clerk : File
4. (a) Metal :Ore
(b) Ghee : Milk
(c) Gasoline : Petroleum
(d) Soil : Earth
5. (a) Beans : Pulses (b) Rice : Cereals
(c) Tea : Beverages (d) Legumes : Nodules
6. (a) Principal : School (b) Soldier : Barrack
(c) Artist : Troupe (d) Singer : Chorus
7. (a) Broom : Sweep (b) Jupiter : Planet
(c) Nut : Crack (d) Soap : Bathe
8. (a) Army : General (b) Team : Captain
(c) Crache : Infant (d) Meeting : Chairman
9. (a) Newspaper : Editor (b) Film : Director
(c) Stamps : Philatelist (d) Book : Author
10. (a) Mulder : Proteins
(b) Curie : Radium
(c) Becquerel : Radioactivity
(d) Einstein : Television
11. (a) Fan : Blades (b) Lamp : Bulb
(c) Clock : Alarm (d) Bicycle : Pedal
12. (a) Sahara : Africa (b) Thar : India
(c) Kalahari : America (d) Gobi : Mongolia

Answers

Exercies 2

1. **Ans. (b)**
Solution: In all other pairs, lack of first causes the second.
2. **Ans. (b)**
Solution: The words in all other pairs are aynonyms.
3. **Ans. (d)**
Solution: In all other pairs, second is the person for whom the first works to earn money.
4. **Ans. (d)**
Solution: In all other pairs, first is a product extracted from the second.
5. **Ans. (d)**
Solution: In all other pairs, second denotes the class to which the first belongs.
6. **Ans. (a)**
Solution: In all other pairs, second is a collective group of the first.
7. **Ans. (c)**
Solution: In all other pairs, second denotes the purpose for which the first is used.
8. **Ans. (c)**
Solution: In all other pairs, second is the head of the first.
9. **Ans. (c)**
Solution: In all other pairs, first is prepared by the second.
10. **Ans. (d)**
Solution: In all other pairs, first is the name of the scientist who discovered the second.
11. **Ans (c)**
Solution: In all other pairs, second is an essential part which serves the main purpose of the first.
12. **Ans. (c)**
Solution: In all other pairs, first is the name of a desert which is situated in the country denoted by the second.

Verbal Analogy

'Analogy' means 'correspondence'. In questions based on analogy, a particular relationship is given and another similar relationship has to identified from the alternatives provided. Analogy tests are, therefore, meant to test a candidate's overall knowledge, power of reasoning and ability to think concisely and accurately.

Exercise 3

1. Moon : Satellite : : Earth : ?
(a) Sun (b) Planet
(c) Solar System (d) Asteroid

2. Coconut : Shell : : Letter : ?
(a) Letter-box (b) Stamp
(c) Mail (d) Envelope

3. Smoke : Pollution : : War : ?
(a) Destruction (b) Treaty
(c) Victory (d) Peace

4. Commodore : Navy : : Brigadier : ?
(a) Captain (b) Commander
(c) Air-Force (d) Army

5. Square : Diamond : : Circle : ?
(a) Smooth (b) Round
(c) Oval (d) Ball

6. Man : Machine : : Master : ?
(a) Worker (b) Manager
(c) House (d) Slave

7. Scrap : Food : : Lees : ?
(a) Bread (b) Tea
(c) Wine (d) Rice

8. Rupee : India : : Yen : ?
(a) Turkey (b) Bangladesh
(c) Japan (d) Pakistan

9. Amnesia : Memory : : Paralysis : ?
(a) Movement (b) Limbs
(c) Handicapped (d) Engine

10. Meningitis : Brain : : Cirrhosis : ?
(a) Lungs (b) Brain
(c) Liver (d) Heart

11. Influenza : Virus : : Typhoid : ?
(a) Bacillus (b) Parasite
(c) Protozoa (d) Bacteria

12. Radio : Listener : : Film : ?
(a) Producer (b) Actor
(c) Viewer (d) Director

13. Mountain : Kapile : : Tree : ?
(a) Ground (b) Leaf
(c) Forest (d) Shrub

14. Doctor : Patient : : Politician : ?
(a) Masses (b) Voter
(c) Power (d) Chair

15. Hot : Oven : : Cold : ?
(a) Ice cream (b) Air conditioner
(c) Snow (d) Editor

16. Malaria : Disease : : Spear : ?
(a) Wound (b) Sword
(c) Weapon (d) War

17. Coins : Mint : : Bricks : ?
(a) Kiln (b) Foundry
(c) Cemetery (d) Furnace

18. Antiseptic : Germs : : Antidote : ?
(a) Allergy (b) Poison
(c) Wound (d) Infection

19. Moth : Insect : : Monuse : ?
(a) Cat (b) Mole
(c) Rodent (d) Rat

20. Muslims : Mosque : : Thermometer : ?
(a) Golden Temple (b) Medina
(c) Fire Temple (d) Gurudwara

21. Eye : Myopia : : Teeth : ?
(a) Pyorrhoea (b) Cataract
(c) Trachoma (d) Eczema

Answers

Exercies 3

1. **Ans. (b)**
Solution: Moon is a satellite and earth is a planet.

2. **Ans. (d)**
Solution: First is enclosed inside the second.

3. **Ans. (a)**
Solution: Second is the result of the first.

4. **Ans. (d)**
Solution: Commodore is a rank in Navy, while Brigadier is an equivalent rank in Army.

5. **Ans. (c)**
Solution: Second is a shape obtained by pulling the opposite ends of the first.

6. **Ans. (d)**
Solution: Machine is made to work by a man. Similarly, slave works under the instructions of his master.

7. **Ans. (c)**
Solution: First is the left over of the second .

8. **Ans. (c)**
Solution: Rupee is the currency of Ondia. Similarly, Yen is the currency of Japan.

9. **Ans. (a)**
Solution: First is the name given to the state of less of the second.
10. **Ans. (c)**
Solution: First a disease which affevts the second.
11. **Ans. (d)**
Solution: First is a disease caused by the second.
12. **Ans. (c)**
Solution: First is meant for the second.
13. **Ans (d)**
Solution: First is a bigger form of the second.
14. **Ans. (a)**
Solution: First works for the second.
15. **Ans. (d)**
Solution: An over is an appliance to keep food items hot. Similarly, a refrigerator keeps food items cold.
16. **Ans. (c)**
Solution: Second denotes the class to which the first belongs.
17. **Ans. (a)**
Solution: Coins are cast and stamped in a mint. Similarly, bricks are baked in a kiln.
18. **Ans. (b)**
Solution: First conters the effect of the second.
19. **Ans. (c)**
Solution: Second denotes the class to which the first belings.
20. **Ans. (d)**
Solution: Second is the place of worship for the first.
21. **Ans. (a)**
Solution: Second is a disease which affects the first.

Exercise 4

1. Race : Fatigue : : Fast : ?
(a) Food (b) Appetite
(c) Hunger (d) Weaknessnd
2. Earth : Sun : : Moon : ?
(a) Orbit (b) Sky
(c) Star (d) Earth
3. Misogamy : Marriage : : Misogyny : ?
(a) Children (b) Husband
(c) Relations (d) Women
4. House : Garbaye : : Ore : ?
(a) Rubbish (b) Gangue
(c) Sand (d) Dregs
5. Calendar : Dates : : Dictionary : ?
(a) Vocabulary (b) Language
(c) Words (d) Book
6. Elegance : Vulgarity : : Doctor : ?
(a) Awkward (b) Comely
(c) Asperity (d) Dirty
7. Nurture : Neglect : : Denigrate : ?
(a) Reveal (b) Extol
(c) Recognize (d) Calumniate
8. Pesticide : Crop : : Antiseptic : ?
(a) Wound (b) Infection
(c) Bandage (d) Bleeding
9. Harp : Drum : : Flute : ?
(a) Violin (b) Bugle
(c) Harmonium (d) Piano
10. Jeopardy : Peril : : Jealousy : ?
(a) Envy (b) Insecurity
(c) Minerals (d) Sin
11. Head : Cap : :Finger : ?
(a) Glove (b) Thimble
(c) Nail (d) Thumb
12. Cellls : Tissues : : Atoms : ?
(a) Elements (b) Molecules
(c) Electrons (d) Organs
13. Book : Critic : : Building : ?
(a) Engineer (b) Contractor
(c) Weigher (d) Appraiser
14. Hill : Mountain : : Stream : ?
(a) River (b) Canal
(c) Glacier (d) Avalanche
15. Chromite : Chromium : : Ilmenite : ?
(a) Glass (b) Porcelain
(c) Cement (d) Powder
16. Skeleton : Body : : Grammar : ?
(a) Language (b) Sentence
(c) Meaning (d) Education
17. Archipelage : Islands : : Massif : ?
(a) Mountains (b) Caves
(c) Forests (d) Hillocks
18. Deuce : Fear : : Yawn : ?
(a) Football (b) Cricket
(c) Hockey (d) Polo
19. Current : Circuit : : Earth : ?
(a) Planet (b) Solar system
(b) Orbit (d) Path

Answers

Exercies 4

1. **Ans. (c)**
Solution: First causes the second.
2. **Ans. (d)**
Solution: First revolves around the second.

3. **Ans. (d)**
Solution: First is a hatred for the second.

4. **Ans. (b)**
Solution: The waste of the house is called garbage. Similarly, the impurities in the ore are called gangue

5. **Ans. (c)**
Solution: One looks for the second in the first.

6. **Ans. (a)**
Solution: The words in each are antonyms of each other.

7. **Ans. (b)**
Solution: The words in each pair are antonyms of each other.

8. **Ans. (a)**
Solution: Pesticide protects crops from insects and antiseptic protects wound from germs.

9. **Ans. (b)**
Solution: The voice of harp is shrill as compared to that of drum. Flute produces a shrill sound as compared to that of bugle.

10. **Ans. (a)**
Solution: First is a more intense form of the second.

11. **Ans. (b)**
Solution: First is covered by the second.

12. **Ans. (b)**
Solution: Cells constitute tissues and atoms constitute molecules.

13. **Ans. (d)**
Solution: Second evaluates and comments on the first.

14. **Ans. (a)**
Solution: Seeond is a bigger form of the first.

15. **Ans. (b)**
Solution: Chromite is a mineral of chromium and ilmenite is a mineral of titanium.

16. **Ans. (a)**
Solution: Just as skeleton froms the frame work of the body, so also Grammar lays down the essentials of language.

17. **Ans. (a)**
Solution: First is a group of the second.

18. **Ans. (b)**
Solution: First is a term associated with the second.

19. **Ans. (c)**
Solution: Current flows through electric circuit. Similarly, earth moves in its orbit in its orbit round the Sun.

Exercise 5

Directions: In each of the following questions, there is a certain relation between two given words on one side of : : and one word is given on another side of : : while onother word id to be found from the given alternatives, having the same relation with this word as the words of the given pair bear. Choose the best alternative.

1. Man : Biography : : Nation : ?
(a) Health (b) People
(c) Geography (d) History

2. Taxonomy : Classification : : Pedology : ?
(a) Nature (b) Farming
(c) Soil (d) Mountain

3. Tectonics : Building : : Taxidermy : ?
(a) Classification (b) Conserving
(c) Stuffing (d) Collecting

4. Steel : Rails : : Alnoce : ?
(a) Aircraft (b) Machinery
(c) Silver ware (d) Magnets

5. Cardiology : Heart : : Zoology : ?
(a) Insects (b) Animals
(c) Plants (d) Germs

6. Connoisseur : Art : : Gourmet : ?
(a) Food (b) Money
(c) Drink (d) Flesh

7. Heart : Pericardium : Brain : ?
(a) Bones (b) Head
(c) Skull (d) Cranium

8. Oxygen : Burn : : Carbon dioxide : ?
(a) Isolate (b) Foam
(c) Extinguish (d) Explode

9. Dark : Fear : : Honesty : ?
(a) Personality (b) Money
(c) Treachery (d) Trust

10. Mattock : Dig : : Shovrl : ?
(a) Break (b) Push
(b) Scoop (d) Whittle

11. Carbon : Diamond : : Corundum : ?
(a) Pearl (b) Pukhraj
(c) Garnet (d) Ruby

12. Pyrophobia : Fire : : Ochlophobia : ?
(a) Horses (b) Crowd
(c) Light (d) Foreigners

13. Vestry : Monkey : : Cataclysm : ?
(a) Labortory (b) Hospital
(c) School (d) Monastery

14. Jungle : Zoo: Sea : ?
(a) Trees (b) Theatre
(c) Flowers (d) Tapestry

Answers

Exercies 5

1. **Ans. (d)**
   Solution: Second contains the stroy of the first.
2. **Ans. (c)**
   Solution: Taxonomy is the science dealing with classification. Similarly, pedology deals with study of soils.
3. **Ans. (c)**
   Solution: Tectonics is the science dealing with the art of building. Similarly, taxidermy is the art of stuffing animals.
4. **Ans. (d)**
   Solution: First is an alloy used to make the second.
5. **Ans. (b)**
   Solution: Cardiology is the branch of science which deals with the study of heast. Similarly, zoology is the branch of science which deals with the study of animals.
6. **Ans. (a)**
   Solution: First has a good taste for the second .
7. **Ans. (d)**
   Solution: Second is a digger form of the first.
8. **Ans. (c)**
   Solution: Oxygen supports burning while carbon dioxide extinguishes fire.
9. **Ans. (d)**
   Solution: First leads to the second.
10. **Ans. (c)**
    Solution: Mattock is a tool to dig hard ground. Similarly, shovel is a tool to scoop.
11. **Ans. (b)**
    Solution: Second is a form of the first.
12. **Ans. (b)**
    Solution: Pyrophobia is a morbid fear of fire. Similarly, ochlophobia is a morbid fear of crowds.
13. **Ans. (b)**
    Solution: Second is a higher institution than the first.
14. **Ans. (c)**
    Solution: First is an art associated with the second.

Exercise 6

1. *Orthopaedist* is related to *Bones* in the same way as *Chiropodist* is related to ....?........
   (a) Nails (b) Sounds
   (c) Feet (d) Heart
2. *Tapeworm* is related to *Taeniasis in* the same way as *Plasmodium* is related to ....?..........
   (a) Malaria (b) Constipation
   (c) Diphtheria (d) Diarrhoea
3. *Hygienic* is related to *Polluted* in the same way as *Knowing* is related to ....?...........
   (a) Ingenuous (b) Uncommon
   (c) Secret (d) Sagacious
4. *Doctor* is rlated to *Diagnosis* in the same way as *Judge* is related to......?........
   (a) Crime (b) Guilt
   (c) Allegation (d) Punishment
5. *Librarian* is related to *Books* in the same way as *Banker* is related to .....?.........
   (a) Customers (b) Bank
   (c) Money (d) Creditors
6. *Museum* is related to *Curator* in the same way as *Prison* is related to .... ? ....
   (a) Manager (b) Monitor
   (c) Jailor (d) Earden
7. *Aflatoxin* is related to *Food Poisoning* in the same way as *Histamine* is related to.........?.........
   (a) Allergy (b) Headache
   (c) Anthrax (d) Contamination
8. *Helm* is related to *Rudder* in the same way as *Brain* is related ro ....?..........
   (a) Heart (b) Ribs
   (c) Limds (d) Body
9. *Hamlet* is related to *Village* in the same way as *Metropolis* is related to..........?..........
   (a) Urban (b) City
   (c) District (d) Place
10. *Ballworm* is related to *Cotton* in the same way as *Ghundibug* is related to ....?.............
    (a) Wheat (b) Rice
    (c) Millet (d) Tomato
11. *Rung* is related to *Ladder* in the same way as *Twig* is related to .... ? ....
    (a) Leaf (b) Flower
    (c) Tree (d) Bud

Answers

Exercies 6

1. **Ans. (c)**
   Solution: First is a specialist of the second.
2. **Ans. (a)**
   Solution: Second is a disease caused by the first.
3. **Ans. (a)**
   Solution: The words in each pair areantonyms of each other.
4. **Ans. (d)**
   Solution: The function of a doctor is to diagnose a disease and that of a judge is to give judgement.

5. **Ans. (c)**
 Solution: First feeds on the first.
6. **Ans. (c)**
 Solution: First is looked after and managed by the second.
7. **Ans. (ca)**
 Solution: First causes the second.
8. **Ans. (b)**
 Solution: First comprises of the second.
9. **Ans. (b)**
 Solution: First is a type of the second.
10. **Ans. (b)**
 Solution: First is a pest that damages the crop of the second.
11. **Ans. (c)**
 Solution: First is a part of the second.

Exercise 7

1. Ase*el* is to poultry as *Salmon* is to ....?......
 (a) Cow (b) Camel
 (c) Fish (d) Horse
2. *Flower* is to *Bouquet* as *Mainister* is to ....?..........
 (a) Department (b) Constituecy
 (c) Voter (d) Cabinet
3. *Sulphur* is to *Vulcanisation* as *Chloring* is to......?......
 (a) Extraction (b) Bleaching
 (c) Metallurgy (d) Allotropy
4. *Liquid* is to *Fluidity* as *Comedian* is to.........?..........
 (a) Ridicule (b) Humour
 (c) Solemnity (d) Companion
5. *Kilometre* is to *Distance* as *Poundal* is to ......?......
 (a) Density (b) Acceleration
 (c) Momentum (d) Force
6. *Preface* is to *Book* as *Overture* is to ....?.......
 (a) Opera (b) Ballad
 (c) Novel (d) Symphony
7. *Sweet* is to *Chocolate* as *Book* is to.......?.......
 (a) Distionary (b) Library
 (c) Encyclopaedia (d) Atlas

Answers
Exercies 7

1. **Ans. (c)**
 Solution: Aseel is a breed of poultry and Salmon is a breed of fish.
2. **Ans. (d)**
 Solution: First is a unit/member of the second.
3. **Ans. (b)**
 Solution: Sulphur is used for vulcanisation of rubber. Similarly, chlorine is used for bleaching.
4. **Ans (b)**
 Solution: Second is the defining characteristic of the first.
5. **Ans (d)**
 Solution: Kilometre is a unit of distance and Poundal is a unit of force.
6. **Ans (a)**
 Solution: The first is an opening comment on the second.
7. **Ans. (c)**
 Solution: Secons is an enlarged form of the first.

CHOOSING THE ANALOGOUS PAIR

In this type of questions, a pair of words is given, followed by four pairs of words as alternatives. The candidate is requied to choose the pair in which the words bar the same relationship to each other as the words of the given pair bear.

Exercise 8

Directions: The following questions consist of two words each that have a certain relationship to each other, followed by four lettered pairs of words. Select the lettered pair that has the same relationship as the original pair of words.

1. Train : Track
 (a) Idea : Brain (b) Bullet : Barrel
 (c) Water : Boat (d) Fame : Television
2. Sea : Ghore
 (a) Train : Rails (b) Road : Footpath
 (c) Ocean : Bay (d) River : Delta
3. Sphere : Ball
 (a) Circle : Disc (b) Cube : Dice
 (c) Line : Rope (d) Triangle : Cone
4. Hymn : Praise
 (a) Dirge : Grief
 (b) Prayer : Gongregation
 (c) Liturgy : Rite
 (d) Lullaby : Child
5. Blister : Skin
 (a) Sore : Toe (b) Sty : Eye
 (c) Ball : Pitcher (d) Wound : Arm
6. Injury : Pain
 (a) Matter : Labour (b) Rotate : Churning
 (c) Thunder : Lightning (d) Grades : Merit
7. Signal : Traffic
 (a) Dam : River (b) Lens : Light
 (c) Door : House (d) Operation : Doctor

8. Paper : Ream
   (a) Eggs : Ship (b) Books : Pile
   (c) Twigs : Bush (d) Food : Packet
9. Geology : Earth
   (a) Architect : Building
   (b) Biology : Science
   (c) Aquarium : Fish
   (d) Archaeology : Artifacts
10. Cream : Cosmetics
   (a) Tiger : Forest (b) Magazine : Editor
   (c) Teak : Wood (d) Mountain : Valley
11. Flag : Nation
   (a) Emblem : Prosperity (b) Insignia : Rank
   (c) Wealth : Prestige (d) Honour : Status
12. Auger : Carpenter
   (a) Cement : Mason (b) Awl : Cobbler
   (c) Apron : Chef (d) Seam : Seamstress
13. Threat : Insecurity
   (a) Challenge : Fight (b) Thunder : Lightning
   (c) Clouds : Rain (d) Disease : Death

Answers

Exercies 8

1. **Ans. (b)**
   Solution: A train moves on tracks. Similarly, a bullet travels through the barrel of the gun.
2. **Ans. (b)**
   Solution: First has the second on either sides of it.
3. **Ans. (b)**
   Solution: First is followed the three-dimensinal shape of the second.
4. **Ans. (a)**
   Solution: Hymn is a song praise sung during religious evebts. Similarly, Dirge is a song of grief sung at a funeral.
5. **Ans. (b)**
   Solution: Blister is an infection of the skin. Similarly, aty is an infection of the eye.
6. **Ans. (c)**
   Solution: First causes the second.
7. **Ans. (a)**
   Solution: First controls the flow or movement of the second.
8. **Ans. (a)**
   Solution: Papers are bought in reams. Similarly, eggs are bought in dozens.
9. **Ans. (d)**
   Solution: First is the branch of science which deals with the study of second.
10. **Ans. (c)**
   Solution: Second denotes the calass to which the first belongs.
11. **Ans. (b)**
   Solution: Flag is the symbol of nation. Similarly, insignia is the symbol of rank.
12. **Ans. (b)**
   Solution: First is a tool used by the second.
13. **Ans. (a)**
   Solution: First leads to the second.

Exercise 9

1. Doctor : Hospital
   (a) Plumber : Wrench (b) Chef : Kitchen
   (c) Water : Reservoir (d) Farmer : Village
2. Bridge : Cards
   (a) Dam : River (b) Gamble : Money
   (c) Image : Mirror (d) Fencing : Sword
3. Vaseline : Petrol
   (a) Tea : Leaves (b) Plant : Tree
   (c) Butter : Ghee (d) Cream : Milk
4. Hierarchy : Ranked
   (a) Infinity : Fixed
   (b) Equation : Solved
   (c) Chronology : Sequential
   (d) Critique : Biased
5. Grain : Salt
   (a) Chip : Glass (b) Blades : Grass
   (c) Shred : Wood (d) Shord : Pottery
6. Spring : Summer
   (a) Adolescence : Youth (b) Fight : Battle
   (c) Stagger : Walk (d) Read : Learn
7. Diva : Opera
   (a) Thespian : Play (b) Conductor : Bus
   (c) Director : Drama (d) Producer : Theatre
8. Read : Legible
   (a) Hear : Audible
   (c) Qualify : Eligible
   (c) See : Illegible
   (d) Require : Admissible
9. Coal : Thermal
   (a) Power : Energy (b) Bulb : Light
   (c) Air : Atmosphere (d) Water : Hydel
10. Mork : Monastery
   (a) Noble : House (b) Peasant : Village
   (c) School : Teacher (d) Nun : Concent
11. Behead : Guilloting
   (a) Polish : Nail (b) Perfect : Picture
   (c) Thief : Prison (d) Hang : Gallows

12. Indigent : Wealth
(a) Smug : Complacency
(b) Contented : Happiness
(c) Aristocratic : Stature
(d) Emaciated : Nourishment

13. Abrupt : Gradual
(a) Barren : Fertile (b) Reverse : Backward
(c) Motion : Forward (d) Agile : Calm

14. Prologue : Dull
(a) Epilogue : Oration
(b) Intermezzo : Symphony
(c) Overture : Opera
(d) Chapter : Novel

Answers
Exercies 9

1. **Ans. (b)**
Solution: Second is the working place of the forst.
2. **Ans. (d)**
Solution: The game of bridge is played by cards. Similarly, the game of fencing is played by swords.
3. **Ans. (d)**
Solution: First is extracted from yhr second .
4. **Ans. (c)**
Solution: Hierarchy is a ranked order of individuals. Similarly, chronology is a sequential order of events.
5. **Ans. (a)**
Solution: Salt consists of grains and glass is made up of chips.
6. **Ans. (a)**
Solution: First is followed by the second.
7. **Ans. (a)**
Solution: Diva (woman singer) plays a leading role in an opera (musical play). Similarly, thespian (actor) plays a leading role in a play.
8. **Ans. (a)**
Solution: Legible means able to be read. Similarly, audible means able to be heard.
9. **Ans. (d)**
Solution: Coal produces thermal energy. Similarly, water produces hydel energy.
10. **Ans. (d)**
Solution: Second is the dwelling place of the first.
11. **Ans. (d)**
Solution: Second is the structure where an accused is given the punishment denoted by the first .
12. **Ans. (d)**
Solution: First needs the second badly.
13. **Ans. (a)**
Solution: The words in each pair are antonyms of each other.
14. **Ans. (c)**
Solution: Prologure is a freliminary speech given at the beginning of a play. Similarly, overture is an orchestral composition forming the introduction to an opera.

Exercise 10

1. Defunct : Life
(a) Stagnant : Motion (b) Orderly : Pattern
(c) Arid : Desert (d) Obese : Weight

2. Wan : Colour
(a) Enigmatic : Puzzle (b) Pallid : Complexion
(c) Insipid : Flavour (d) Copulent : Weight

3. Appointment : Ability
(a) Transfer : Punishment
(b) Business : Money
(c) Promotion : Merit
(d) Examination : Success

4. Legendray : Myth
(a) Witty : Amicable (b) Lenght : Sermon
(c) Comic : Epic (d) Didactic : Fable

5. Confine : Prisoner
(a) Impeach : President (b) Detain : Suspect
(c) Ambush : Sentry (d) Trace : Fugitive

6. Dexterous : Clumsy
(a) Sheer : Transparent (b) Browse : Look
(c) Blame : Criticise (d) Pessimist : Hopeful

7. Cliche : Commonplace
(a) Annoy : Destroy (b) Maxim : Terse
(c) Serious : Play (d) Bitter : Bitter : Pill

8. Arc : Circle
(a) Segment : Line (b) Fraction : Percentage
(c) Pie : Slice (d) Number L Count

9. Season : Summer
(a) Chapter : Book (b) Week : Seven
(c) Hot : Temperature (d) Month : December

10. Judge : Adijudicate
(a) Researcher : Emendate
(b) Mediator : Reconcile
(c) Applellamt : Implore
(d) Advocate : Jury

11. Implicate : Incriminate
(a) Involement : Malpractice
(b) Exonerate : Acquit
(c) Embezzlement : Charge
(d) Perjury : Fugitive

12. Basement : Attic
(a) Nadir : Zenith (b) Zenith : Apex
(c) Zenith : Root (d) Apex : Pinnacle

13. Court : Justice
(a) Police : Crime (b) Teacher : Study
(c) Doctor : Sikness (d) Auditor : Accuracy

14. Man : Humaniity
(a) Frame : Piture (b) Scholar : Books
(c) Flower : Fragrance (d) Drop : Ocean

15. Marathon : Stamina
(a) Relay : Independence (b) Jog : Weariness
(c) Sprint : Celerity (d) Hurdle : Perseverance

Answers

Exercies 10

1. Ans. (a)
Solution: The words in each pair are antonyms of each other.

2. Ans. (b)
Solution: Wan refers to pale colour. Similarly, pallid refers to pale complexion.

3. Ans. (c)
Solution: First depensd to form the second.

4. Ans. (d)
Solution: First denotes the nature of the second.

5. Ans. (b)
Solution: A prisoner is confined for punishment. Similarly, a suspect is detained for inquiry.

6. Ans. (d)
Solution: The words in each pair are antonyms of ach other.

7. Ans. (b)
Solution: The words in each pair are synonyms of each other.

8. Ans. (a)
Solution: First is a fragment of the second.

9. Ans. (d)
Solution: First denotes t5he class to which the second belongs.

10. Ans. (b)
Solution: Second denotes the function of the first.

11. Ans. (b)
Solution: The words in each pair are synonyms of each second.

12. Ans. (a)
Solution: Nadir (lowest point) and zenith (highest point) can be compared to basement and attic respectively.

13. Ans. (d)
Solution: The duty of court is to provide justice rain. Similarly, the duty of an auditor is to provide accuracy in financial matters.

14. Ans. (c)
Solution: Second is the quality for which the first is liked or praissed.

15. Ans. (c)
Solution: Second is essentially required for the first.

1. Canvas : Painter
(a) Leather : Shoe (b) Chisel : Wood
(c) Marble : Sculptor (d) hammer : Carpenter

2. Rank :Badges
(a) Person : Status (b) Office : Designation
(c) Flag : Nation (d) Medal : Honour

3. Hero : Accolade
(a) Clodhopper :Grace
(b) Mentor : Advice
(c) Laughingstock : Ridicule
(d) Defendant : Indictment

4. Pusillanimous : Dastard
(a) Optimistic : Pessimist
(b) Magnanimous : Misanthrope
(c) Pledeian : Entrepreneur
(d) Impecunious : Mendicant

5. Surplus : Deficit
(a) Famine : Scarcity
(b) Monsoon : Drought
(c) Need : Wage
(d) Breadth : Width

6. Aphorism : Sententions
(a) Maxim : Allegorical
(b) Adage : Symbolic
(c) Eulogh : Laudatory
(d) Criticism : Redundant

7. Encomium : Laudatory
(a) Prologue : Conclusive
(b) Critique : Insolent
(c) Monologue : Lengthy
(d) Tirade : Abusive

8. Vindictive : Mercy
(a) Skeptical : Trustfulness
(b) Modest : Dignity
(c) Insightful : Hope
(d) Avaricious : Greed

9. Kick : Football
(a) Throw : Ring (b) Wash : Dishes
(c) Mop : Sweep (d) Dust : Rag

10. Candle : Wick
    (a) Bicycle : Wheel (b) Oven : Fire
    (c) Hammer : Nail (d) Light : Bulb
11. Intimidate : Wheedle
    (a) Resolute : Impetuous
    (b) Coordinate : Disinter
    (c) Defile : Rebuke
    (d) Extol : Disparage

Answers
Exercies 11

1. **Ans. (c)**
   Solution: Second wroks on the first.
2. **Ans. (b)**
   Solution: Second is indicative of the postion/ status of a person in terms of the first.
3. **Ans. (d)**
   Solution: First is worthy of the second for his actions.
4. **Ans. (d)**
   Solution: The words in each pair are synonyms of each other.
5. **Ans. (b)**
   Solution: The words in each pair are antonyms of each other.
6. **Ans. (c)**
   Solution: Second describes the nature of the first.
7. **Ans. (d)**
   Solution: Second describes the nature of the first.
8. **Ans. (a)**
   Solution: Second id quite contrary to the first.
9. **Ans. (a)**
   Solution: First is a way of playing with/using the second.
10. **Ans. (a)**
    Solution: Second is the most essenial working part of the first.
11. **Ans. (d)**
    Solution: The words in each pair are antonyms of each other.

Double Analogy

Exercise 12

Directions: (Questions 1 to 8) In each of the following questions, a group of three/four inter-related is given. Choose a word from the given altrernatives, that is similar to the given words and belongs to the same group.

1. Peas : Chram : Pulses
   (a) Rice (b) Barley
   (c) Beans (d) Coconut
2. Sodium : Cod : Eel
   (a) Sulphur (b) Calcium
   (c) Water (d) Ioding
3. Deisy : Pansy : Rose
   (a) Butter (b) Plant
   (c) Violet (d) Red
4. Edge : Corner : Tip
   (a) Snow (b) Hill
   (c) Brink (d) Cap
5. Which of the following is the same as **Rabbit, Rat, Mole?**
   (a) Mongoose (b) Frog
   (c) Earthworm (d) Ant
6. Whish of the following is the same as **Ode, Ballad , Lyric ?**
   (a) Prose (b) Epic
   (c) Sonnet (d) Precis
7. Which of the following is the same as **Emancipate , Free, Release ?**
   (a) Liderate (b) Quit
   (c) Pardon (d) Ignore
8. Which of the following is same as **Bhilai, Rourkela, Durgapur ?**
   (a) Chandigarh (b) Baroda
   (c) Lucknow (d) Bokaro

Answers
Exercies 12

1. **Ans. (c)**
   Solution: All are rich in proteins.
2. **Ans (b)**
   Solution: All are metals .
3. **Ans. (c)**
   Solution: All are flowers.
4. **Ans (c)**
   Solution: All are synonyms.
5. **Ans. (a)**
   Solution: All are rodents and live in holes.
6. **Ans. (c)**
   Solution: All are forms of poery.
7. **Ans (a)**
   Solution: All are synoyms.
8. **Ans. (d)**
   Solution: All are industrial towns famous for steel plants.

Verbal Analogy

The word verbal is defined as pertaining to words rather than things. Verbal reasoning tests use words, letters and numbers, and require logical reasoning and a reasonable knowledge of the English language. It is also necessary to be familiar with simple manipulations with figure, like addition, subtraction, division and multiplication. The problems of numbers in test of reasoning will not require any advanced knowledge of moths. Instead, they will test how logical you are, that is, how well you reason and think while carrying out simple arithmetic manipulation.

Like non-verbal reasoning tests, verbal reasoning also includes three broad categories, namely Series Completion, Classification (finding the odd man out) and Analogical Relationships, in addition to other types of logical and reasoning question.

Type 1

In such questions, you will be presented with a pair of words, letters, groups of letters, numbers or groups of numbers. From the answer choices, you need to select the choice that best matches the relationship existing between the first two words, letters or numbers.

The given pair of words, letters of number is set off by a colon or ratio sign (:). The colon is merely a kind of code or a symbol which signifies that word A is related to word B. The double colon (: :) that separate the given pair stands for "in the sane way as".

For example: Dark : Light : : ?

(a) Novel : Writer (b) Stout : Strong
(c) Germs : Disease (d) Hot : Cold
(e) Candle : Bulb

Ans. (d)

Solution: In the above example, the answer is (d) Hot : Cold because Hot is related to Cols in the same way as Dark is related to Light. In other words, the two words are antonyms.

Exerice 1

Directions: In each of the following questions, you are given a related pair of words (question pair) in capital letters. Each capialised pair is followed by five pairs of words. Choose the pair from the answer choises. (A-E) that best expresses the relationship similar to that expressed by the question pair.

1. EDITOR : MAGAZINE
(a) Nivel : Writer (b) Poem : Poet
(c) Chair : Carpenter (d) Director : Film
(e) Psychiatrist : Neurotic

2. DETERIORAION : RUST
(a) Iron : Water
(b) Emaciation : Dediliation
(c) Depression : Unemployment
(d) Recession : Inefficiency
(e) Promulgation : Legislation

3. INTELLIGENSIA : ELITIST
(a) I.Q. : Intelligent
(b) Outrshell : Sea-shell
(c) Rabble : Plebeian
(d) Gentry : Public
(e) Commonality : Common class

4. TEDIOUS : EXALTED
(a) Promoted : Excellence
(b) Pressing : Crushing
(c) Poor : Poverty
(d) Incongruous : Consistent
(e) Indefatigable : Untiring

5. ELEVATED : EXALTED
(a) Promoted : Excellence
(b) Raise : Commensurate
(c) Dirty : Filthy
(d) Disorferly : Unfaithful
(f) Purified : Hygienic

6. TICKET : ADMISSION
(a) School : Book
(b) Strong : Moral
(c) Neck : Collar
(d) Coat : Pocket
(e) Money : Luxuries

7. MUSIC : HARMONIUM
(a) Novel : Writer
(b) Words : Word-Processor
(c) Author : Book
(d) Water : Tank
(e) Guitar : Sound

8. ENERGY : DISSIPATE
(a) Charge : Battery
(b) Food : Temperature
(c) Money : Squandor
(d) Power : Generator
(e) Atom : power

9. COMMANDER : COMMANDS
(a) Surgeon : Diagnosis
(b) Senator : Iegislates
(c) Teacher : Checks
(d) Aviator : Aerates
(e) Specific : Responses

10. FORMULA : CONSTITUENT
(a) Verdict : Sentence
(b) Rocket : Pilot

(c) Carburettor : Mixture
(d) Binomial : Monocular
(e) Equation : Term

Answers

Exercies 1

1. **Ans. (d)**
Solution: The editor works on the magazine to bring it into ahape for presention to readers. The director works on the film to make it ready for release to the viewers. Here is the relationship of association. In other choices, the sequence is not matching the question pair.

2. **Ans. (c)**
Solution: As rust can be caused by deterioration, unemployment can be caused by depression.

3. **Ans. (c)**
Solution: Intelligensia is characterised as elitist; rabble (common class of people) may be characterised as pledeian (belonging to a lower social class of people).

4. **Ans. (e)**
Solution: Tedious is a synonym for boring and indefatigable is a synonym for Untiring.

5. **Ans. (c)**
Solution: Elevated (promoted/advancement) and exalted both have the same meaning but the latter is less in degree or intensity. Similarly, dirty is a lower degree of uncleanliness.

6. **Ans. (e)**
Solution: A ticket enables admission and money enables one to buy luxuries.

7. **Ans. (b)**
Solution: Music will be produced from a harmonium only if a skilled person plays it. Similarly words will be processed if a skilled operator uses the word processor. Choice (e) could have been correct if words were in reverse order to match the question pair sequence.

8. **Ans. (c)**
Solution: Energy can be wasted or dissipated just as money can be wasted or squan dered.

9. **Ans. (b)**
Solution: The primary function of the commander is to command. Similarly a senator legislates. There is a functional relationship.

10. **Ans. (e)**
Solution: Constituent is one of the components of a formula. Similarly, yerm is a constuent of equation. This is a whole and part relationship.

Type 2

For Example: Composer : Song : :? : ?
(a) Buiding : Architect
(b) Poem : Post
(c) Writer : Book
(d) Chair : Carpenter

Ans. (c)
Solution:

1. Establish the relationship between the given pair, before looking at the answer choices. After you have found the relationship, look at the answer choices to find a similar relationship amongst them
The primary relationship between the given pair of items Composer : Song is that the composer creates or composes a song. The same relationship exists in choice (c) where a Writer crates a book.
2. The sequence of items in the answer choices contain a person and the aricle he creates or produces. In the question pair Composer : Song, the person comes first, follwed by the article he creates. Therefore, in the correct answer choice, the person must come first followed by the article the creaters. The onle answer choice that satisfies this condition is (c) Writer : Book. It is, therefore, necessary to compare the sequence of items in the question pair with answwes that may match the primary relationship.

Exerice 2

1. LUGUBRIOUS : SORROWFUL
(a) Unhappy : Gloomy
(b) Lustrous : Luscious
(c) Credible : Incredible
(d) Euphoric : Cheerful
(e) Frenzied : Excited
2. GLOVE : HAND
(a) Neck : Collar
(b) Tie : Shirt
(c) Shoe : Lace
(d) Coat : Pocket
(e) Socks : Feet
3. STUDENT : MARK
(a) Teacher : Class
(b) Pen : Nid
(c) Scholar : Book
(d) Discipline : Student
(e) Waiter : Tip
4. KNOB : DOOR
(a) Show : Socks
(b) Belt : Trousers

(c) Nose : Face
(d) Necklace : Neck
(e) Ring : Ear

5. PRIMEVAL : MEDIEVAL
(a) Snow : Ice
(b) Thorn : Rose
(c) Evolution : Revelation
(d) Gorilla : Soldier
(e) Dinosaur : Dragon

6. DEPLETE : ENERVATE
(a) Stregthen : Weak
(b) Invigorate : Tired
(c) Exhaust : Enfeeble
(d) Tighten : Loosen
(e) Weariness : Restless

7. LIGHT : CANDLE
(a) Exercise : Strenght
(b) Dieting : Overweight
(c) Power : Battery
(d) Heat : Coil
(e) Automobile : Engine

8. EMBARRASS : HUMILIATE
(a) Difficult : Painful
(b) Annoy : Exasperate
(c) Gamble : Investment
(d) Embezzle : Peculate
(e) Enquire : Ask

9. OPERATION THEATRE : SURGERY
(a) Judge : Lawyer
(b) Doctor : Disease
(c) Settlement : Client
(d) School : Classes
(e) Couet : Litigation

10. OVERLOOK : ABERRATION
(a) Coniviction : Criminal
(b) Error : Omission
(c) Condone : Offense
(d) Careless : Loyalty
(e) Mitigate : Penitence

Answers

Exercies 2

1. **Ans. (c)**
Solution: Lugubrious is a person more gloomy and a sorrowful person is only gloomy. Similarly, a euphoric person has more feelings of well being than a cheerful person. The words have the sane meaning, the difference being only one of degree.

2. **Ans. (e)**
Solution: Just as socks are made to fit and cover the feet a glove is made to fit and cover the hand.

3. **Ans. (e)**
Solution: As a student gets marks for goods work in a test or examination, a waiter gets a tip for giveng good service.

4. **Ans. (c)**
Solution: A knob must be on a door, a nose must be on a face.

5. **Ans. (e)**
Solution: Primeval period is associated with the dinosaur and medieval period with the dragon.

6. **Ans. (c)**
Solution: Deplete means to exhaust and enervate means to become weak. Exhaust and enfeeble also mean the same. The relationship is analogous with a difference of degree only.

7. **Ans. (c)**
Solution: A candle is a source of light as a bettery is a source of power.

8. **Ans. (b)**
Solution: Embarrass and himiliate are similar in meaning but the latter is stronger in degree. The same relation exists between annoy and exasperate.

9. **Ans. (e)**
Solution: Operation theatre is a place where surgery is performed and court is a place for litigation.

10. **Ans. (c)**
Solution: One can overlook an error or an aberration; one can also pardon (condone) an offence.

Type 3

There may be questions on analogy in different formats. Some illustrations are given below:

Exerice 3

Directions: Which makes the best comparison ?

1. Foot is to have as leg is to
(a) Elbow (b) Tow
(c) Finger (d) Arm
(e) Nails

2. Love is to valour is to
(a) Courage (b) Scurity
(c) Cowardice (d) Anger
(e) Terror

3. Daughter is to father as niece is to
   (a) Nephew (b) Cousin
   (c) Uncle (d) Mother
   (e) Brother
4. Bark is to treeas scales are to
   (a) Gills (b) Elephant
   (c) Fish (d) Skiing
   (e) Tape
5. Pillow is to pillowcase as arm is to
   (a) Body (b) Sleeve
   (c) Hand (d) Glove
   (e) Ring

Answers
Exercies 3

1. **Ans. (d)** Arm
   Solution: A foot is attached to a leg, a hand is attached to an arm.
2. **Ans. (c)** Cowardice
   Solution: Love is the opposite of hate. Valour is the opposite of cowardice.
3. **Ans. (c)** Uncle
   Solution: Daughter is the female child of father, niece (daughter of one's brother or sister) has a similar relationship with uncle
4. **Ans. (c)** Fish
   Solution: Bank is the outer protective cover of the trunk of a tree, scales are the outer protective cover of fish.
5. **Ans. (b)** Sleeve
   Solution: Pillow fits inside the *pollowcase*, an *arm* fits inside a *sleeve*.

Type 4

Letters and Number Analogies

As word analogy questions illustrated above, there may also be alphabets and/or number analogies. Some examples are given below.

Exerice 4

Directions: Find the correct letter pair from the alternatives given in given in answer choices below, to locate the same relationship among the group of letters as established between the sets in the question.

1. ZA : YB : : XC :?
   (a) YZ (b) NM
   (c) BC (d) OP
   (e) WD
2. ABCD : WXYZ : : EFGH : ?
   (a) STUV (b) STOU
   (c) STUE (d) TSUV
   (e) STUV
3. AD : BE : : CF : ?
   (a) DE (b) EC
   (c) DG (d) FG
   (e) GD
4. ACDE :OGHI : : ESTU ?
   (a) EPQU (b) ABCD
   (c) OGHK (d) XYZE
   (e) ZXZE
5. BAC : DEF : : ? NOP
   (a) GHI (b) JHI
   (b) GH (d) HJI
   (e) HIJ

Answers
Exercies 4

1. **Ans. (e)**
   Solution: Each pair contains a letter each from backward sequence (Z-A) and forward sequence (A-Z) following the natural alphabetical sequence.
2. **Ans. (e)**
   Solution: The first term in the question pair consists of four letters in forward sequence (A-Z) and the second term cosists of four letters from the end of the alphabetie order and the sequence remairs forward in both cases (A to Z).
3. **Ans. (c)**
   Solution: Each term keeps the alphabetic sequence A (bc) D, B (cd) E, and D (ef) G.
4. **Ans. (a)**
   Solution: Each group of letters has one preseding and one final vowel with two consonants in between.
5. **Ans. (e)**
   Solution: Letters ib each term term follow the alphabetic sequence with an intrvening vowel.

Type 5

1. Which numbers in choices (a), (b), (c) and (d) makes the best comparison?
   YY ZZZ Y ZZ Y is to 22 III 2 II 2 as YY ZZ YZZ Y is to
   (a) 12211212 (b) 211221121
   (c) 22112112 (d) 21211221
2. Which nakes the best comparison?
   LEAP is to PEAL as 8326 is to?

(a) 2368 (b) 6283
(c) 6328 (d) 2683

Answers

1. **Ans. (c)**
Solution: Substitute numbers for letters : Y=2, Z=1

2. **Ans. (c)**
Solution: Substitute numbers for letters : L = 8, E = 3,A = 2, P = 6, PEAL = 6328

Exerice 5

Directions: In each of the questions below, a related pair of words in capital in capital letters is followed by five pairs of words (a-e). Select that lettersd pair that expresses the relationship that is most similar to that the capitalised pair.

1. AGREEMENT : DISSENT
(a) Contract : Clause
(b) Schism : Diverge
(c) Impasse : Concede
(d) Touchdown : Penalty
(e) Latitude : Resistance

2. LOATH : COERCION
(a) Detest : Craessing
(b) Irritate : Caressing
(c) Irate : Antagonism
(d) Reluctant : Prsuasion
(e) Contemplative : Meditative

3. APPEAL : REFUSAL
(a) Obesity : Over-eating
(b) Deny : Affirmation
(c) Try : Falilure
(d) Struggle : Victory
(e) Examination : Passing

4. WEIGHT : KILOGRAM
(a) Pint : Liquid
(b) Distance : Kilometre
(c) Mile : Lenght
(d) Pound : Weight
(e) Bushel : Corn

5. ERITER : PEN
(a) Needle : Tailor
(b) Artist : Brush
(c) Paint : Painter
(d) Tacher : Class
(e) Saw : Carpenter

6. VANDALISM : PROPERTY
(a) Implication : Crime
(b) Embezzlement : Fraud
(c) Perjury : Testimony
(d) Malpracticicing : Cheating
(e) Testify : Reputation

7. FLOW-SHEET : MANUFACTURER
(a) Formula : Product
(b) Blue print : Architect
(c) Cameraman : Film
(d) Scenario : Writer
(e) Script : Actor

8. GOOD :EXCELLENT
(a) Bad : Immoral
(b) Caution : Careless
(c) Hill : Mountain
(d) Jealousy : Respect
(e) Sickess : Medicines

9. PAIN : MISERY
(a) Disease : Poverty
(b) Despair : Loneliness
(c) Ignornce : Confusion
(d) Superstition : Poverty
(e) Ignore : Greet

10. PAINTING : ARTIST
(a) Sing : Singer
(b) Musician : Music
(c) Author : Novel
(d) Cook : Mean
(e) Poem : Post

11. BARREL : VIAL
(a) Length : Height
(b) Low : High
(c) Brochure : Compiler
(d) Book : Pamphlet
(e) Book : Reader

12. CABIN : SHIP
(a) Cockpit : Plane
(b) Whel : Chair
(c) Room : Bungalow
(d) Kitchen : Store room
(e) Aeroplane : Pilot

13. RESTING : FATIGUE
(a) Poverty : Disease
(b) Over -Obesity
(c) Gourmet : Underweight
(d) Race : Exercise :
(e) Dieting : Over-weight

14. IMPLICATE : ALUMINIUM
(a) Pejury : Lead
(b) Embezzlement : Charging
(c) Exonerate : Acquittal
(d) Involve : Crime
(e) Involvement : Malpractice

15. EXEMPT : OBLIGED
(a) Affluent : Fluent
(b) Immune : Susceptible
(c) Valiant : Puissant
(d) Steadfast : Putaive
(e) Cavalier : Presumption

16. JUDGE : ADJUDICATE
(a) Advocate : Jury
(b) Mediator : Reconcile
(c) Lawyer : Client
(d) Appellant : Implore
(e) Researcher : Emendate

Answers
Exercies 5

1. **Ans. (c)**
Solution: An agreement may end if one of the parties to it begins to dissent. An impasse may end if one of the parties will concede a point or points to the other.

2. **Ans. (d)**
Solution: If someone is loath (unwilling) to do something coercion (use of force) may be required to change his mind. Similarly, if a person is reluctant, he may be persuded to chenge his mind.

3. **Ans. (c)**
Solution: When one appeals, he expects an acceptance and not refusal. When one tries he expects success and not failure. The relationship is one of negative response to an action.

4. **Ans. (b)**
Solution: Measurement of weight is expressed in kilogram as measurement of distance is expressed in kilometre. In answer choices (a) and (c) could have ben acceptable if the sequence of the words would have been reverse in order to match the question pair.

5. **Ans. (b)**
Solution: A writer needs a pen for his work and an artist needs a brush for his work. Other answer choices have deen rejected as they do not follow the sequence of question pair.

6. **Ans. (c)**
Solution: Vandalism is a malicious crime related with property. Perjury (false statement) is a crime connected with testimony.

7. **Ans. (b)**
Solution: A flow-sheet is prepared by a manufacturer for the guidance of those involved in the manufacturing operations. A blue-print is prepared by the architect to guide the construction people.

8. **Ans. (c)**
Solution: Excellent is greater in degree than good. Similarly, a mountain is higher than a hill.

9. **Ans. (c)**
Solution: Pain causes misery and ignorance causes confusion.

10. **Ans. (e)**
Solution: A painting is made by an artist and a poem is created by a poet. In other choices, the sequence of words is not in line with the question pair.

11. **Ans. (c)**
Solution: The living quarter in a ship is a cabin, similarly living place in a bungalow is a room.

12. **Ans. (c)**
Solution: The living quarter in a ship is a cabin, similarly living place in a bungalow is a room.

13. **Ans. (e)**
Solution: Resting is necessary for fatigue, as dieting is necessary for overweight.

14. **Ans. (c)**
Solution: Incrimination means to charge someone with a crime. To implicate someone in a crime is to connect him with the crime which would lead to his incrimination. To exonerate a person charged with a crime would free him and lead to his acquittal.

15. **Ans. (b)**
Solution: Exempt means free from obligation or liability, whereas obliged means subject to comply or perform. There is an opposite meaning relatioship. Immune means proteced from and susceptible means vulnerable to which also have as apposite meaning relationship.

16. **Ans. (b)**
Solution: A judge adjicates and a mediator help in reconciliation. The second term in both cases is major function of the first term.

TYPE-6: VERBAL ANSLOGY-WORD ANALOGIES

Type 6

1. UNCLE : AUNT : : FATHER
(a) Affection (b) Son
(c) Daughter (d) Mother

2. ORANGH : FRUIT : : BRINJAL : ?
(a) Vehetable (b) Potato
(c) Flower (d) Cucumber

3. Cloud : rain

(a) Flower : Smell (b) Man : Work
(c) Sun : Light (d) Mountain : River

4. POINT : LINE
(a) Centre : Circumference
(b) Line : Circle
(c) Triangle : Square
(d) Word : Newspaper

5. MUSIC : TUNE
(a) Game : Player (b) Poetry : Poet
(c) Dance : Rhythm (d) Religion : God

6. SCULPTOR : STONE
(a) Essagyist : Words
(b) Painter : Turnpentine
(c) Logger : Timber
(d) Etcher : Acid

7. CALCUTTA : WEST BENGAL
(a) Pune : Maharahtra
(b) Gwalior : Madhya Pradesh
(c) Jaipur : Rajasthan
(d) Kampur : Uttar Pradesh

8. PLANT : FUNGUS
(a) Police : Robbers (b) Sailor : Pirate
(c) Lion : Lamb (d) Major : Sepoy

9. CONFERENCE : CHAIRMAN : : NEWSPAPER :?
(a) Reporter (b) Distributor
(c) Printer (d) Editor

10. SKIRMISH : WAR : : DISEASE : ?
(a) Epidemic (b) Infection
(c) Patient (d) Medicine

11. WYE : BRAIN : : LENS : ?
(a) Camera (b) Mind
(c) Vision (d) Telescope

12. EXERCISE : FITNESS
(a) Concern : Care (b) Intimidation : Fear
(c) Sensitivity : Poetry (d) Retain : Sustain

13. BOUQUET : FLOWERS
(a) Jail : Thieves (b) Garland : String
(c) Necklace : Pearls (d) Party : Cake

14. PLANT : DRINK
(a) Bush : Forest (b) Star : Galaxy
(c) Child : Adult (d) Young : Sprightly

15. WINE : GRAPES : :Vodka : ?
(a) Apples (b) Rye
(c) Orange (d) Flour

16. TASTE : TONGUE : : Walk : ?
(a) Panement (b) Cruteh
(c) Legs (d) Walking Stick

17. BOOK : LIBRARY : : ANIMAL : ?
(a) Domestic (b) Hunter
(c) Wild (d) Zoo

18. HEALTH : SICKNESS : : HAPPINESS : ?
(a) Medicine (b) Treatment
(c) Misery (d) Cheerfulness

19. ARC : CIRCLE
(a) Square : Rectengle
(b) Straightline :Triangie
(c) Side : Angle
(d) None of these

20. ALPHABET : WORD : : SENTENCE
(a) Word (b) Paragraph
(c) Syllables (d) Phrase

21. MEAT : VEGETARIAN : LIQUOR : ?
(a) Insane (b) Teetotaller
(c) Introvert (d) Irrational

22. CELL : CYTOLOGY : : BIRDS : ?
(a) Odontoligy (b) Mycology
(c) Ornithology (d) Etymology

23. ABANDON : KEP
(a) Abhor : Hate
(b) Cloudy :IIInminating
(c) Compree : Squeeze
(d) Confers : Grant

Answers
Exercies 6

1. **Ans. (d)**
Solution: Aunt is the wife of Uncle . Similarly, Light are available by Sun.

2. **Ans. (a)**
Solution: Orange is come under Fruit. Similarly. Brinjal is come under Vegetable.

3. **Ans. (c)**
Solution: First word is main source of second.

4. **Ans. (d)**
Solution: Point is smallest part of the Lien. Similarly, word is smallest part of the Newspaper.

5. **Ans. (c)**
Solution: As tune is in music ; Rhythm is in Dance.

6. **Ans. (a)**
Solution: Sculptor works with stone, similarly essayist works with works.

7. **Ans. (c)**
Solution: As Calcutta is capital of West Bengal, Jaipur is capital of Rajasthan.

8. Ans. (b)
Solution: Fungus is considered negative of Plant similarly, Pirate is considered negative of sailor.

9. Ans. (d)
Solution: Conference is headed by Chairman and Newspaper by Editor.

10. Ans. (a)
Solution: Skirmish full-fledged was, juat as some disease into.

11. Ans. (c)
Solution:

12. Ans. (a)
Solution: As Exercise enhance Fitness, same way Concern enhance care.

13. Ans. (c)
Solution: As Bouquet is of flowers, similarly Necklace is of Pearl.

14. Ans. (c)
Solution: As Bouquet is of flowers, similarly Necklace if of Pearl.

15. Ans. (b)
Solution: Wine is produced from Grapes, similarly, Vodka is produced from Rye.

16. Ans. (c)
Solution:

17. Ans. (d)
Solution: Library stack Books similarly Zoo stocks Animals.

18. Ans. (c)
Solution: Health is opposite of sickness same way Happiness is opposite of Misery.

19. Ans. (b)
Solution: Arc is a part of circle, straight line is a part of triangle.

20. Ans. (a)
Solution: A set of Alphabet is called word. A set of word is called sentence.

21. Ans. (b)
Solution: Vegetarian never eat mat similarly teeyoealer nerver takes Liquor.

22. Ans. (c)
Solution: Study of Cell os called Cytologh, study of Birds is called Ormithology

23. Ans. (b)
Solution: Abandon and keep are antonymous to each other. Similarly, cloudly and IIluminating are antonymous to each other.

Type 7

Word Analogy

Exerice 7

Directions: There is a blank space in each of these questions in which only one of the four altenatives given under qyestion satisfies the same relationship as is found between two terms on the other side of the sign (: :). Find the correct alternative to fill in the blank space.

1. Condine : Offence : : ______ : ______
   (a) Overlook : Aberration
   (b) Error : Omission
   (c) Mitigate : Penitence
   (d) Conviction : Criminal
2. Drill : Bore : : ______ : ______
   (a) Painter : Brush (b) Sieve : Sift
   (c) Helmet : Head (d) Mason : Wall
3. Philatelist : Stamps : : ______ : ______
   (a) Carpenter : Saw
   (b) Runner : Sueakers
   (c) Numismatist : Coins
   (d) Astrologer : Predictions
4. Myth : Legendary : : ______ : ______
   (a) Sermon : Lengthy (b) Eqic : Comic
   (c) Fable : Didactioc (d) Anecdote : Witty
5. Deepening : Filling : : ______ : ______
   (a) Arctic Circle : Antarctic circle
   (b) Tropic of Cancer : Tropic of Capricon
   (c) Ridge of High Pressure : Through of Low Pressure.
   (d) Dead Vally : Dry Valley
6. Manager : Office : : ______ : ______
   (a) Doctor : Patient
   (b) Curator : Mseum
   (c) Bank : Account
   (d) Fruit : Seed
7. Brave : Valour : : ______ : ______
   (a) Look : See (b) Cynic : Doubt
   (c) Man : Management (d) Theif : Victim
8. Spasm : Pain : : ______ : ______
   (a) Touch : Feel (b) Fast : Quick
   (c) Flash : Light (d) Water : Wet
9. Artist : Painting : : ______ : ______
   (a) Lawyer : Penal code
   (b) Fisherman : Herring

(c) Caroenter : Saw
(d) Auther : Book

10. Obliquity : Straight forwardness: :____ : ____
(a) Affable : Persuasive
(b) Diression : Conciseness
(c) Honest : Offensive
(d) Speech : Audience

11. Sky : Sea : : ____ : ____
(a) Terrace : Parking Lot (b) Solider : Rifle
(c) Hat : Shoe (d) Roof : Top

12. Soup : Dessert : : ____ : ____
(a) Tendulkar : Kumble (b) River : Island
(c) Sunlight : Twilight (d) Bow : Arrow

13. Sip : Gulp : :____ : ____
(a) Soup : Water (b) Tent : Hut
(c) Touch : Push (d) Cup : Glass

14. Liquor : Intoxication : : ____ : ____
(a) Whisky : Alcoholic (b) Bottle : Cock
(c) Medicine : Cure (d) Engine : Fuel

15. Joke : Laugh : : ____ : ____
(a) Humar : Relation
(b) Gluttory : Food
(c) Pill : Headache
(d) Cracker : Explosion

16. Facade : Buliding : : ____ : ____
(a) Fence : Garden (b) Neck : Bottle
(c) Dial : Watch (d) Page : Back

17. Open : Secretive : : ____ : ____
(a) Forth : right (b) Tunnel : Toil
(c) Mystery : Detective (d) Better : Best

18. Bouquet : Flower : : ____ : ____
(a) Chain : Link (b) Skin : Body
(c) Product : Factory (d) Page : Book

19. Ogle : Observe : : ____ : ____
(a) Haggle : Outbid (b) Clamour : dispute
(c) Discern : Perceive (d) Flaunt : Display

20. Sap : Vitality : : ____ : ____
(a) Persevere : Fortitude
(b) Bore : Tedium
(c) Drain : Resovle
(d) Enhance : Allure

21. Karma : Life : : ____ : ____
(a) Race : Goal (b) Concert : Finale
(c) Design : Buliding (d) Painting : Canvas

Answers

Exercies 7

1. **Ans. (a)**
Solution:

2. **Ans. (b)**
Solution:

3. **Ans. (c)**
Solution:

4. **Ans. (c)**
Solution:

5. **Ans. (c)**
Solution:

6. **Ans. (b)**
Solution:

7. **Ans. (b)**
Solution:

8. **Ans. (c)**
Solution:

9. **Ans. (d)**
Solution:

10. **Ans. (b)**
Solution:

11. **Ans. (a)**
Solution:

12. **Ans. (c)**
Solution:

13. **Ans. (c)**
Solution:

14. **Ans. (c)**
Solution:

15. **Ans. (d)**
Solution:

16. **Ans. (a)**
Solution:

17. **Ans. (a)**
Solution:

18. **Ans. (a)**
Solution:

19. **Ans. (d)**
Solution:

20. **Ans. (b)**
Solution:

21. **Ans. (d)**
Solution:

Type 8
Verbal Analogy-word Analogy
Exerice 8

1. MATHEMATICS is related to NUMEROLOGY in related to
   (a) Science (b) Astrology
   (c) Philosophy (d) Planets
2. FLUTE is related to SEXAPHONE in the same way as VIOLIN is related to
   (a) Opera (b) Orchestra
   (c) Band (d) Bass
3. SUGAR is related to SACCHARIN in the same way as BUTTER is related to
   (a) Salt (b) Margaine
   (c) Pepper (d) Spice
4. MONEY is related to GREED in the same way as FOOD is related to
   (a) Noracity (b) Slavery
   (c) Desire (d) Greed
5. LITMUS is related to ACID in the same way as THERMOMETER is related to
   (a) Climate (b) Wheather
   (c) Cock (d) Heat
6. PAW is related to CAT in the same way as HOOF is related to
   (a) Lion (b) Cow
   (c) Dog (d) Horse
7. DISCONTENT is related to STRIKE in the same way as FRICTION is related to
   (a) Science (b) Astrology
   (c) Philosophy (d) Planets
8. LETTUCE is related to LEAF in the same way as ONION is related to
   (a) Stem (b) Thorn
   (c) Bulb (d) Vine
9. DAM is elated to WATER in the same way as EMBARGO is related to
   (a) Over (b) Trade
   (c) Stream (d) River
10. CALIBER is related to RIFLE in the same way as GUAGE is related to
    (a) Infantry (b) Army
    (c) Rails (d) Blast
11. PUBLICATION is related to LIBEL in the same way as SPEECH is related to
    (a) Libility (b) Stander
    (c) Attack (d) Information
12. BEHAVLATE is related to IMPROPRIETY in the same way as WORD is related to
    (a) Malapropism (b) Usage
    (c) Morality (d) Honesty

Answers
Exercies 8

1. **Ans. (b)**
   Solution: Mathematics is stidy of number and numderology is occult science of numbers; just as Astrinimy is study of celestiall bodies and astrology is occult science of celestial bodies,
2. **Ans. (d)**
   Solution: All are musical instruments. This one is a little tougher one as such musical instruments are not in common knowledge. Flute is smaller while sexaphone is larger wind musical instrument; similarly violir is small version of a bass.
3. **Ans. (b)**
   Solution: Saccharin is substitute of sugar, while margarine is substitute of butter.
4. **Ans. (a)**
   Solution: Insatiable devise of money is greed and that of food is voracity.
5. **Ans. (d)**
   Solution: Acidity is checked by litmus peper while heat by thermometer.
6. **Ans. (d)**
   Solution: Paw is a limb of cat, and hoof is that of horse.
7. **Ans. (a)**
   Solution: 'Discontent' among labour force makes way to 'strike', while 'friction' can also grow into full-fledged war.
8. **Ans. (c)**
   Solution: Leaf of lettuce and bulb of onion are eatable by humans.
9. **Ans. (b)**
   Solution: A 'dani' obstract flow of 'water' and an 'emdaryo' obstructs 'trade'.
10. **Ans. (c)**
    Solution: 'Guns' has 'Caliber' as standard measurement, while 'gauge' is standard measurement of railways (Broad gauge, narrow gauge etc.)
11. **Ans. (b)**
    Solution: Libel is written defamation while slander is oral defamation.

12. Ans. (a)
Solution: Impropriety is a derohatory from of behaviour; and an incorrect use of word is 'Malapropism'.

TYPE D: WORD GROUPS

Classifications

'Classification' means 'to assort the items of a given group on the basis of a certain common quality they possess and then spot the stranger or odd one out'.

In this test, you are given a group of certain items, out of which all except one are similar to one another in some manner. The candidate is required to choose this one item which does not fit into the given group.

Exercise 3A

1. (a) Shehnai (b) Bagpipe (c) Flute (d) Sitar (e) Harmonica
2. (a) Tortoise (b) Duck (c) Snake (d) Whale (e) Crow
3. (a) Raid (b) Attack (c) Assault (d) Defence (e) Ambush
4. (a) Engineer (b) Architect (c) Mechanic (d) Mason (e) Blacksmith
5. (a) Shoulder (b) Foot (c) Elbow (d) Arm (e) Fingers
6. (a) Poland (b) Greece (c) Spain (d) Italy (d) Korea
7. (a) Sketch (b) Diagram (c) Poster (d) Chart (e) Paper
8. (a) Spade (b) Spanner (c) Shovel (d) Rake (e) Pick-axe
9. (a) Giraffe (b) Hyena (c) Deer (d) Rhinoceros (e) Zebra
10. (a) Arrow (b) Axe (c) Knife (d) Dagger (e) Sword
11. (a) Brick (b) Heart (c) Bridge (d) Spade (e) Club
12. (a) Guava (b) Litchi (c) Papaya (d) Watermelon (e) Jackfruit
13. (a) Month (b) Year (c) Fortnight (d) Season (e) week
14. (a) Oyster (b) Whelk (c) Scallop (d) Mussel (e) Clam
15. (a) Sweep (b) Wipe (c) Scrub (d) Wash (e) Stain
16. (a) Cot (b) Sheet (c) Quilt (d) Pillow (e) Blanket
17. (a) Tsland (b) Coast (c) Harbour (d) Oasis (e) Peninsula
18. (a) Tall (b) Huge (c) Thin (d) Sharp (e) Small
19. (a) Beaker (b) Glass (c) Mug (d) Saucer (e) Cup
20. (a) Hangar (b) Platform (c) Dock (d) Park (e) Bus stand
21. (a) Chameleon (b) Crocodile (c) Alligator (d) Locust (e) Salamander
22. (a) Cabbage (b) Papaya (c) Gourd (d) Cucumber (e) Brinjal

Answers
Exercies 3A

1. Ans. (d)
Solution: All except *Sitar* are wind instruments, while sitar is a string instrument.

2. Ans. (d)
Solution: All except *Whale* lay eggs.

3. Ans. (d)
Solution: All except *Defence* are forms of attack.

4. Ans. (c)
Solution: All except *Mechanic* help in building a house.

5. Ans. (b)
Solution: All except *Foot* are parts of hand.

6. **Ans. (e)**
   Solution: All except *Korea* are European countries, while Korea is an Asian country.
7. **Ans. (e)**
   Solution: All others can be drawn on paper.
8. **Ans. (b)**
   Solution: All except *Spanner* are tools used by a gardener, while a spanner is used by a carpenter.
9. **Ans. (b)**
   Solution: *Hyena* is the only flesh-eating animal in the group.
10. **Ans. (a)**
    Solution: All except *Arrow* are used while holding in hand.
11. **Ans. (a)**
    Solution: All except *Brick* are suits of cards.
12. **Ans. (d)**
    Solution: All except *Watermelon* grow on trees, while watermelon grows on creepers.
13. **Ans. (d)**
    Solution: All except *Season* are precise measurements for days.
14. **Ans. (c)**
    Solution: All except *Scallop* live in shells.
15. **Ans. (e)**
    Solution: All except *Stain* are terms related to cleaning.
16. **Ans. (a)**
    Solution: All except *Cot* are parts of bed-spread.
17. **Ans. (d)**
    Solution: All except *Oasis* are features related to sea, while oasis is related to desert.
18. **Ans. (d)**
    Solution: All except *Sharp* are related to dimension.
19. **Ans. (d)**
    Solution: All except *Saucer* are used to contain liquids.
20. **Ans. (d)**
    Solution: All except *park* are halting places of various transport means.
21. **Ans. (d)**
    Solution: All except *Locust* are reptiles, while locust is an insect.
22. **Ans. (b)**
    Solution: All except *Papaya* are vegetables, while papaya is a fruit.

Exercise 3B

Directions: In each of the following questions, five words nave been given, out of which four are alike in some manner, while the fifth one is different. Choose out the odd one.

1. (a) Collection (b) Compilation
   (c) Cluster (d) Assemblage
   (e) Assortment
2. (a) Anger (b) Grief
   (c) Humorous (d) Kindness
   (e) Joy
3. (a) Crow (b) Pigeon
   (c) Parrot (d) Butterfly
   (e) Peacock
4. (a) Pupil (b) Iris
   (c) Cornea (d) Medulla
   (e) Retina
5. (a) Table (b) Cupboard
   (c) Chair (d) Sofa
   (e) Paper weight
6. (a) Radio (b) Television
   (c) Computer (d) Telephone
   (e) X-ray
7. (a) Telescope (b) Periscope
   (c) Microphone (d) Bioscope
   (e) Kaleidoscope
8. (a) Verse (b) Rhyme
   (c) Couplet (d) Rhetoric
   (e) Stanza
9. (a) Venus (b) Saturn
   (c) Earth (d) Mercury
   (d) Neptune
10. (a) Fern (b) Moss
    (c) Algae (d) Fungi
    (e) Grass
11. (a) Honey (b) Pearl
    (c) Coral (d) Wax
    (e) Resin
12. (a) Cabbage (b) Cauliflower
    (c) Ladyfinger (d) Radish
    (e) Peas
13. (a) Konark (b) Madurai
    (c) Ellora (d) Khajuraho
    (e) Dilwara
14. (a) Mars (b) Sun
    (c) Saturn (d) Mercury
    (e) Pluto
15. (a) Tiger (b) Dolphin
    (c) Zebra (d) Lion
    (e) Crocodile

16. (a) Garlic (b) Chilli
(c) Ginger (d) Potato
(e) Sugar beet

17. (a) Radium (b) Thorium
(c) Sodium (d) Polonium
(e) Uranium

18. (a) Diarrhoea (b) Jaundice
(c) Measles (d) Cholera
(e) Typhoid

19. (a) Wheat (b) Barley
(c) Rice (d) Pea
(e) Mustard

20. (a) Mercury (b) Mars
(c) Earth (d) Jupiter
(e) Neptune

21. (a) Iron (b) Potassium
(c) Sodium (d) Chlorine
(e) Iodine

Answers
Exercies 3B

1. **Ans. (c)**
Solution: All except *Cluster* denote collections of selected items.

2. **Ans. (c)**
Solution: All except *Humorous* are nouns, while *Humorous* is an adjective.

3. **Ans. (d)**
Solution: All except *Butterfly* are dirds, while, butterfly is an insect.

4. **Ans. (d)**
Solution: All except *Medulla* are parts of the ex-Presidents of India.

5. **Ans. (e)**
Solution: All except *Paper weight* are items of furniture.

6. **Ans. (e)**
Solution: All except *X-ray* are means of mass media.

7. **Ans. (c)**
Solution: All except *Microphone* are instruments based on theories and principles of light.

8. **Ans. (d)**
Solution: All except Rhetoric are terms associated with poetry.

9. **Ans. (c)**
Solution: All except *Earth* denote Roman or Greek gods and goddesses.

10. **Ans. (e)**
Solution: All except *Grass* are non-flowering plants.

11. **Ans. (d)**
Solution: *Resin* is the only product in the group which is obtained from trees.

12. **Ans. (c)**
Solution: All excep*t Ladyfinger* are vegetables which grow best in winter.

13. **Ans. (c)**
Solution: All except *Ellora* are famous for temples, while Ellora is famous for caves.

14. **Ans. (b)**
Solution: All except *Sun* are planets, while Sun is a star.

15. **Ans. (e)**
Solution: All except *Crocodile* are mammals, while crocodile is a reptile.

16. **Ans. (b)**
Solution: All except *Chilli* are modified stems.

17. **Ans. (c)**
Solution: All except *Sodium* are radio-isotopes, while sodium is a metal.

18. **Ans. (c)**
Solution: All except *Measles* are diseases which spread through infected food and water.

19. **Ans. (c)**
Solution: All except *Rice* are rabi crops, while rice is a kharif crop.

20. **Ans. (a)**
Solution: *Mercury* is the only planet in the group which does not have any moon.

21. **Ans. (a)**
Solution: All except Iron are very volatile.

Exercise 3C

Directions: In each of the following questions, four words have been given, out of which three are alike in some manner and the fourth one is different. Choose out the odd one.

1. (a) Mushroom (b) Yeast
(c) Mould (d) Smut

2. (a) Rectangle (b) Square
(c) Cube (d) Triangle

3. (a) Sandstone (b) Basalt
(c) Granite (d) Pumice

4. (a) Copper (b) Silver
(c) Gold (d) Platinum

5. (a) Crimson (b) Scarlet
(c) Vermillion (d) Red

6. (a) Feldspar (b) Bauxite
(c) Azurite (d) Cryolite

7. (a) Albatross (b) Ostrich
(c) Pelican (d) Penguin
8. (a) Honest (b) Intelligent
(c) Traitor (d) Wise
9. (a) Brook (b) Bridge
(c) Stream (d) Canal
10. (a) Day (b) Fortnight
(c) Calendar (d) Month
11. (a) Tortoise (b) Alligator
(c) Snail (d) Oyster
12. (a) Advise (b) Counsel
(c) Suggest (d) Direct

Answers
Exercies 3C

1. **Ans. (d)**
Solution: All except *Smut* are forms of fungi.
2. **Ans. (c)**
Solution: All except *Cube* are two-dimensional plane figures.
3. **Ans. (a)**
Solution: All except *Sandstone* are igneous rocks formed by cooling down of magma, while sandstone is a sedimentary rock.
4. **Ans. (a)**
Solution: All except *Copper* are metals used to make jewellery.
5. **Ans. (d)**
Solution: All the rest are shades of red.
6. **Ans. (c)**
Solution: All except *Azurite* are ores of aluminium.
7. **Ans. (b)**
Solution: All except *Ostrich* are water birds.
8. **Ans. (c)**
Solution: All except *Traitor* denote good qualities.
9. **Ans. (b)**
Solution: All except *Bridge* contain water.
10. **Ans. (c)**
Solution: All others are parts of a calendar.
11. **Ans. (b)**
Solution: All except *Alligator* have protective shells.
12. **Ans. (d)**
Solution: All except *Direct* are synonyms.

Exercise 3D

Directions: In each of the following questions, four words have been given, out of which three are alike in some manner and the fourth one is different. Choose out the odd one.

1. (a) Censure (b) Admonish
(b) Ieebuke (d) Retrieve
2. (a) Mosses (b) Lichens
(c) Ferns (d) Cacti
3. (a) Lake (b) Brook
(c) Stream (d) River
4. (a) Electricity (b) Telephone
(c) Telegram (d) Post
5. (a) I1lusion (b) Delusion
(c) Identification (d) Hallucination
6. (a) Panaji (b) Bhopal
(c) Pune (d) Shillong
7. (a) Herb (b) Flower
(c) Tree (d) Shrub
8. (a) Bomdila (b) Lhasa
(c) Nathula (d) Shipkila
9. (a) Saw (b) Axe
(c) Hammer (d) Screw-driver
10. (a) Leader (b) Rubber
(c) Chief (d) Boss

Answers
Exercies 3D

1. **Ans. (d)**
Solution: All others are synonyms.
2. **Ans. (d)**
Solution: All except *Cacti* grow in polar regions .
3. **Ans. (a)**
Solution: All except *Lake* contain moving water.
4. **Ans. (a)**
Solution: All except *Electricity* are means of communication.
5. **Ans. (c)**
Solution: All except *Identification* are synonyms.
6. **Ans. (c)**
Solution: All except *Pune* are capitals of states of India.
7. **Ans (b)**
Solution: All except *Flower* are types of plants.
8. **Ans. (b)**
Solution: All except *Roof* are dwelling places of different animals.
9. **Ans. (b)**
Solution: All except *Axe* are tools used by a carpenter.
10. **Ans. (b)**
Solution: All except *Robber* are names given to the Head.

TYPE E: INSTRUCTIONS/COMPREHENSION

Deriving Conclusions from Passages

In this section of logical deduction, the question consists of a brief passage (usually a report containing certain data regarding some social or economic problem), followed by certain inferences based on it. The candidate is required to analyze the contents (or data) of the passage and grasp the desirable facts from it. Then, he has to consider each of the given inferences in the context of the given passage, decide upon its degree of truth or falsity and then choose the best alternative provided accordingly.

Question 1 to 5: Wind is an inexhaustible source of energy and an aerogenerator can convert it into electricity. Though not much has so far been done in this field, the survey shows that there is a vast potential- direction, duration speed and distribution. On this basis U.P. hill areas have been found an ideal place for setting up aerogenerators . In U.P. hills alone, as many as 58 sites have been identified.

1. Only the hilly areas of U.P. were surveyed for setting up aerogenerators.
(a) Data inadequate (b) Definitely true
(c) Probably false (d) Definitely false
(e) Probably true

2. The survey was conducted under the government of U.P.
(a) Definitely true (b) Probably true
(c) Data inadequate (d) Definitely false
(e) Probably false

3. Wind, as a source of energy, can replace exhaustible sources of energy.
(a) Definitely false (b) Data inadequate
(c) Probably true (d) Probably false
(e) Definitely true

4. Energy by wind is a comparatively new emerging field.
(a) Probably true (b) Probably false
(c) Definitely true (c) Data inadequate
(e) Definitely false

5. 58 sites identified in U.P. did not have electricity.
(a) Definitely true (b) Definitely false
(c) Data inadequate (d) Probably true
(e) Probably false

Questions 6 to 10: Indian granite industry is in peril in the absence of a uniform policy from the State Governments, despite the thrust given by liberalization policies of the Union Government in the last two years. Compared to the remarkable progress in the field during the last three years, the absence of matching policies by State Governments had put granite quarry owners and others involved in the industry on the verge of collapse in the international market. The policies differed from state to state, had created problems as far as loyalty, dead rent and duration of lease were concerned.

6. The granite production is largely controlled by individuals.
(a) Data inadequate (b) Definitely true
(c) Probably true (d) Probably false
(e) Definitely false

7. The granite produced in India dose not match with the quality of international level.
(a) Definitely false (b) Definitely true
(c) Probably false (d) Probably true
(e) Data inadequate

8. The Union Government's liberalization policy became applicable to granite industry only during the last two years.
(a) Definitely true (b) Probably true
(c) Data inadequate (d) Probably false
(e) Definitely false

9. Each state having granite quarry has set up its own rules which are contrary to the interest of the industry.
(a) Probably true (b) Data inadequate
(c) Probably false (d) Definitely false
(e) Definitely true

10. Till three years ago, granite production in India was not profitable.
(a) Data inadequate (b) Probably true
(c) Definitely false (d) Probably false
(e) Definitely true

Questions 11 to 15: Dryland farming is the only way to not only combat recurring drought but also meet the increasing food requirements of India. About 45% of India's total crop production now comes from drylands. By the end of this century, this will have to increase to 60% if India is to provide adequate food for projected population of one billion by the turn of the century.

11. Dryland farming is important for India.
(a) Data inadequate (b) Definitely true
(c) Probably true (d) Probably false
(e) Definitely false

12. The per acre crop production is more in drylands than others.
(a) Definitely false (b) Definitely true
(c) Probably false (d) Probably true
(e) Data inadequate

13. India is self-sufficient in food production.

(a) Definitely true (b) Probably true
(c) Data inadequate (d) Probably false
(e) Definitely false

14. At present India gets larger food production from wetlands.
(a) Probably true (b) Data inadequate
(c) Probably false (d) Definitely false
(e) Definitely true

15. In India, the rate of growth of population is 15 percent per year.
(a) Data inadequate (b) Probably true
(c) Definitely true (d) Probably false
(e) Definitely false

Answrs

1. Ans. (c)
Solution: It is mentioned in the in the passage that the wind survey showed that there is a vast potential for harnessing wind energy. This means that an overall survey must have been made and not only of the hilly areas of U.P.

2. Ans. (b)
Solution: It may be true that the survey was conducted under the U.P. government as the areas of U.P. were also surveyed.

3. Ans. (e)
Solution: According to the passage, wind is an inexhaustible source of energy and efforts are being made to develop wind as an alternative source of energy. This clearly implies the given fact.

4. Ans. (c)
Solution: In the passage, it is given that not much has been done in the field of wind energy and efforts are on. Thus, it is clear that wind energy is a comparatively new emerging filed.

5. Ans. (c)
Solution: In the passage , it is not mentioned whether the 58 sites identified in U.P. had the facility of electricity or not.

6. Ans. (b)
Solution: It is mentioned in the passage that the absence of matching policies by State Governments has 'granite quarry owners' on the verge of collapse. This implies the given fact.

7. Ans. (e)
Solution: Nothing about the quality of granite produced in India is mentioned in the passage.

8. Ans. (a)
Solution: It is mentioned in the passage that the liberalization policies of Union Government gave a thrust to the granite industry in the last two years. This clearly implies the given fact.

9. Ans. (e)
Solution: The given fact directly follows from the last sentence of the passage.

10. Ans. (a)
Solution: Nothing about profitability of granite industry three years ago is mentioned in the passage.

11. Ans. (b)
Solution: The given statement can be derived from the first sentence of the passage.

12. Ans. (e)
Solution: Nothing about the per acre crop production in drylands is mentioned in the passage.

13. Ans. (c)
Solution: The fact is not mentioned in the passage.

14. Ans. (e)
Solution: At present, India gets 45% of its total crop production from drylands. Obviously the rest is obtained from wetlands and is more.

15. Ans. (a)
Solution: The annual rate of growth of population is not mentioned in the passage.

Theme Detection

In this type of questions, a paragraph is given followed by certain statements which may or may not be inferred from the passage. The candidate is required to choose that statement which contains the gist or the theme of the passage, i.e. the idea that it conveys.

1. The only true eduction comes through the stimulation of the child's powers by the demands of the social situations in which he finds himself. Through these demands he is stimulated to act as a member od a unity, to emerge from his original narrowness of action and feeling, and to conceive himself from the standpoint of the welfare of the group to which he belongs.

The passage best supports the statement that real education:
(a) will take place if the children imbibe action and feeling
(b) will tale place if the children are physically strong.
(c) is not provided in oue schools today.
(d) comes through the interaction with social situations.
(e) comes from the seld-centred approch of the students.

2. Emerson said that the poet was landlord, sealord, airlord. The flight of imagination made the post master of land, sea and air. But a poet's dream of yesterday becomes today an actual achievement and a reality for all men. Even those who invented, improved and perfected the aeroplane could hardly have dreamt of the possibility of flight into outer space.

 The passage best supports the statement that:
 (a) seemingly impossible imaginations make one a good poet.
 (b) all imaginations become a reality some day.
 (c) what man imagined has never been impossible; he has always turned it a reality through his conception of ideas and sheer hard labour.
 (d) man has reached the climax of technological development with his exploration into outer space.

3. To forgive an injury is often considered to be a sign of weakness; it is really a sign of strength. It is easy to allow oneself to be carried away by resentment and hate into an act of vengeance ; but it takes a strong character to restrain those natural passions. The man who forgives an injury proves himself to be the superior of the man who wronged himself and puts the wrong-doer to shame.

 The passage best supports the statement that:
 (a) the sufferer alone knows the intensity of his sufferings.
 (b) people tend to forgive the things happened in the past.
 (c) natural passions are difficult to suppress.
 (d) mercy is the noblest form of revenge.
 (e) a person with calm and composed nature has depth of thought and vision.

4. Throughout the ages the businessman has helped build civilisation's great cities, provided people with luxuries and artists with patronage, and lift his fellow citizens to understand the standard of living . In the last few centuries the businessman has seeded the Industrial Revolution around the wo,ld.

 The passage best supports the statement that the businessman:
 (a) is accountable to the society.
 (b) lives luxurious and comfortable fife.
 (c) is the beneficiary of the Industrial Revolution.
 (d) is capable of raising his standard of living.
 (e) has contributed to the growth of civilisation.

5. Satisfaction with co-workers, promotion opportunities, the nature of work, and pay goes with high performance among those with strong growth needs. Among those with weak growth needs, no such relationship is present - and, in fact , satisfaction with promotion opportunities goes with low performance.

 This passage best supports the statement that:
 (a) satisfaction is an inevitable orgainisational variable.
 (b) job satisfaction and performance are directly and closely related.
 (c) relationship between job satisfaction and performance is moderated by growth need.
 (d) every organisation has few employees having weak growth need.
 (e) high performance is essential for organisational effectiveness.

6. The consumption of harmful drugs by the people can be prevented not only by banning their sale in the market but also by instructing users about their dangerous effects which they must understand for their safety. Also the drug addicts may be provided with proper medical facilities for their rehabilitation. This will help in scaling down the use of drugs.

 The passage best supports the statement that consumption of harmful drugs:
 (a) are on increase in the society.
 (b) can always be reduced.
 (c) are due to take of medical facilities.
 (d) can be eliminated with the help of banning their sale.
 (e) may be channelised through proper system.

7. The school has always been the most important means of transferring the wealth of tradition from one generation to the next. This applies today in an even higher degree than in former times for, through the modern development of economy , the family as bearer of tradition and education has become weakened.

 This passage best supports the statement that for transferring the wealth of tradition from one generation to the next:
 (a) there are means other than the school.
 (b) several different sources must be tried.
 (c) economic development plays a crucial role.
 (d) modern technology must be put to use.
 (e) family, as ever, is the most potent means.

8. The paragraph above also supports the statement that:
 (a) there are individual differences in industrial productivity.
 (b) there are numerous side - effects of technology.
 (c) size of the pay and social status contradict each other.

(d) labour force is treated well in modern day industies.
(e) division of labour based on caste was justified.

Answrs

1. **Ans. (c)**
   Solution:
2. **Ans. (d)**
   Solution:
3. **Ans. (d)**
   Solution:
4. **Ans. (e)**
   Solution:
5. **Ans. (c)**
   Solution:
6. **Ans. (d)**
   Solution:
7. **Ans. (c)**
   Solution:
8. **Ans. (a)**
   Solution:

TYPE F: CRITICAL REASONING

Cubes and Dice

Counting the number of cubes/blocks in the given figure.

When the number of cubes (or blocks/cuboids) in a figure are to be counted, the procedure to be adopted is as described in the following examples:

Example 1: Count the number of blocks in the given figure.

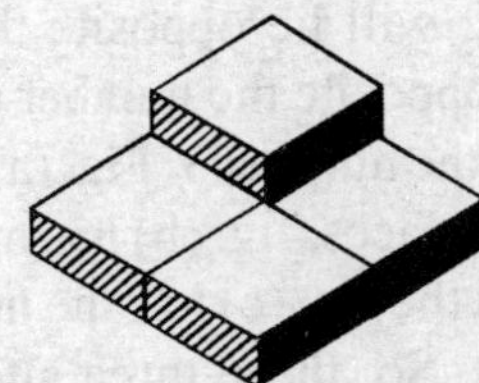

(a) 3 (b) 4
(c) 5 (d) 6

Solution: In the figure there are – 1 column containing 2 blocks and 3 columns containing 1 block each.

Therefore, total number of cubes = (1 × 2) + (3 × 1) = 5.

Example 2: Count the number of cubes in the given figure.

(a) 14 (b) 12
(c) 10 (d) 8

Solution: Clearly, in the figure there are – 1 column containing 3 cubes, 2 column containing 2 cubes each and 3 columns containing 1 cube each.

Number of cubes in columns of 3 cubes = 1 × 3 = 3;
Number of cubes in columns of 2 cubes = 2 × 2 = 4;
Number of cubes in columns of 1 cubes = 3 × 1 = 3;
Therefore, total number of cubes = 3 + 4 + 3 = 10.
Hence, the answer is (c).

Example 3: The figure below is a drawing of pile of blocks. When taken apart, how many blocks would there be?

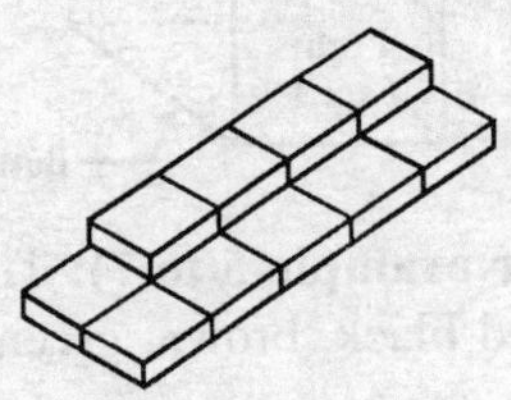

(a) 10 (b) 14
(c) 18 (d) 20

Solution: Clearly, there are 4 columns containing 2 cubes each and 6 columns containing 1 cube each.

Number of cubes in columns of 2 cubes = 4 × 2 = 8;
Number of cubes in columns of 1 cube = 6 × 1 = 6.
Therefore, total number of cubes = 8 + 6 = 14.
Hence, the answer is (b).

Example 4: A cube whose two adjacent faces are coloured is cut into 64 identical small cubes. How many of these small cubes are not coloured at all?

(a) 60 (b) 48
(c) 36 (d) 24

Solution: As shown in the adjoining figure, let the upper face and the RHS face of the cube be coloured. Then, clearly, the rows of smaller cubes (formed by cutting the large cube into 64 parts) which are indicated by dots, have none of their sides coloured.

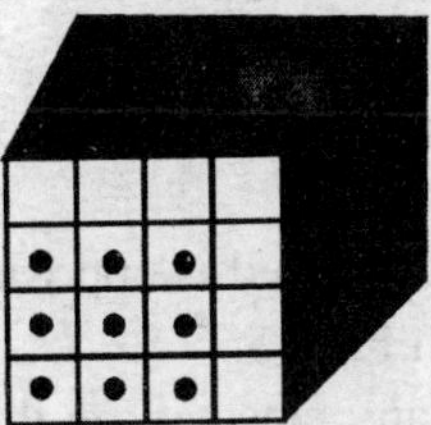

Since there are 9 such rows and each row consists of 4 cubes so there are 9 × 4 = 36 cubes which are not coloured at all.

Hence, the answer is (c).

Colouring the Six Faces of a Cube

In this type of questions, certain specifications are provided for colouring the six faces of a cube and then based upon these specifications, we have to analyse the colour structure of the cube. Then, we have to answer the questions based upon it.

Note: Here we will label the six faces of a cube as:
Front face → Face No. 1; RHS face → Face No. 2
Rear face → Face No. 3; LHS face → Face No. 4
Top face → Face No. 5; Bottom face → Face No. 6

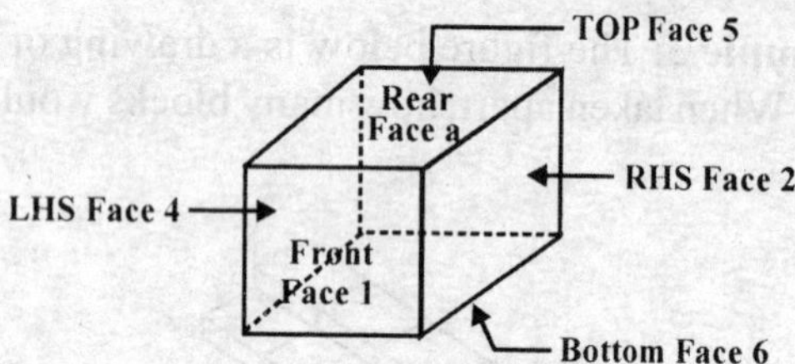

Direction (for example 5 to 7): The six faces of a cube are coloured black, brown, green, red, white and blue, such that

(i) Red is opposite black
(ii) Green is between red and black
(iii) Blue is adjacent to white
(iv) Brown is adjacent to blue
(v) Red is at the bottom.

Answer the following questions based on this information.

Example 5: Which colour is opposite brown?

(a) White (b) Red
(c) Green (d) Blue

Example 6: The three adjacent colours are

(a) Black, Blue, Red (b) Black, Brown, White
(c) Black, Blue, White (d) Brown, Red, White

Example 7: Which of the following can be deduced from (i) and (v)?

(a) Black is on the top
(b) Blue is on the top.
(c) Brown is on the top
(d) Brown is opposite Black.

Solution: From statement (v): Red is at the bottom i.e. on Face 6.

From statement (i): Red is opposite Black i.e. Black is on the top i.e. on Face 5.

Let Blue be on any face say, on the front face i.e. on Face 1.

From statement (iii) and (iv): Brown and White lie on either sides of Blue. We can take Brown on face 4 and White on Face 2.

Then, clearly Green lies on Face 3. This satisfies statement (ii) also.

Thus, the cube will be coloured as indicated in the figure below:

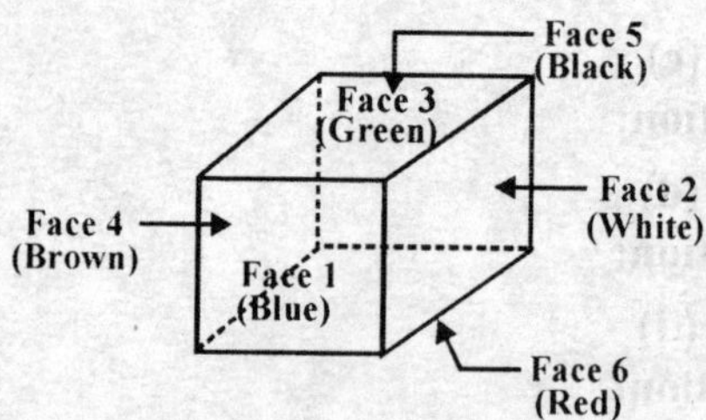

Solution 5: From the figure it is clear that White is opposite Brown. So, the answer is (a).

Solution 6: Since, out of the three colours Black, Blue and White, not two colours lie on opposite faces, hence these three are adjacent colours. Hence, the answer is (c).

Solution 7: From (v) it is clear that Red colour lies at the bottom face and from (i) it is derived that since Red is opposite Black, hence Black lies on the top.

The answer is (a).

Example 8: Select from the alternatives, the box that can be formed by folding the sheet shown in figure (X):

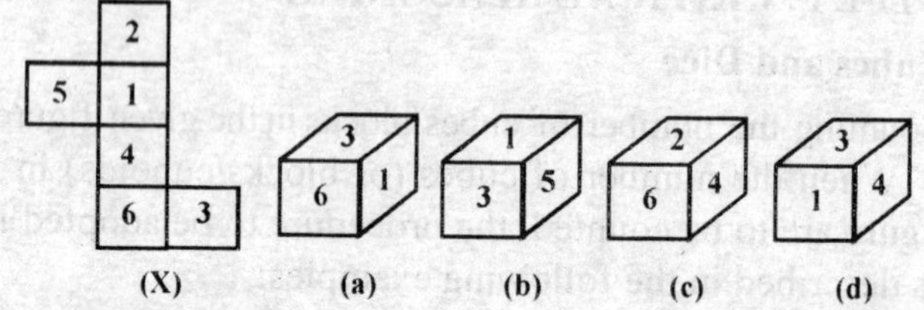

Solution: The fig.(X) is similar to Form III. Hence, when the sheet in fig. (X) is folded to form a box (cube) then the number 2 will lie opposite the number 4; the number 1 will be opposite the number 6 and the number 5 will be opposite the number 3. Fig. (a) has the numbers 1 and 6 on adjacent faces, fig. (b) has number 3 and 5 on adjacent faces and the fig. (c) has the number 2 and 4 on the adjacent faces. So, these three alternatives are not possible. Since, the number 1, 3 and 4 can appear on adjacent faces, so fig. (d) is possible.

Hence, only the box shown in fig. (d) can be formed by folding fig. (X).

Problems on Dice

Sometimes we are given figures showing the same dice (may be in the form of a cube or a cuboid) in various positions. After observing these figures, we have to find the number opposite a given number on the dice.

Activity 1: In solving such problems, we shall label the six faces of a cube in the following manner:

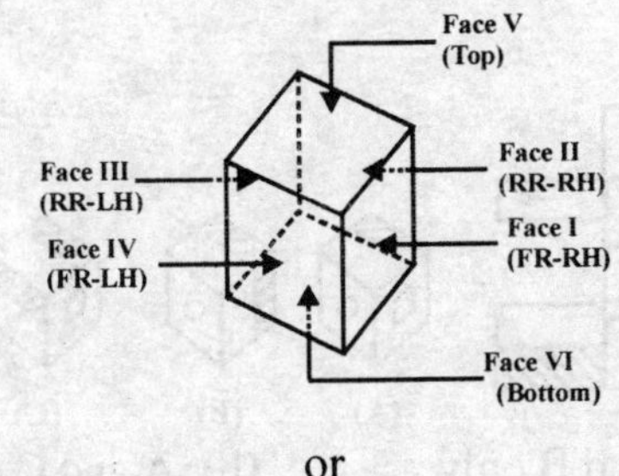

or

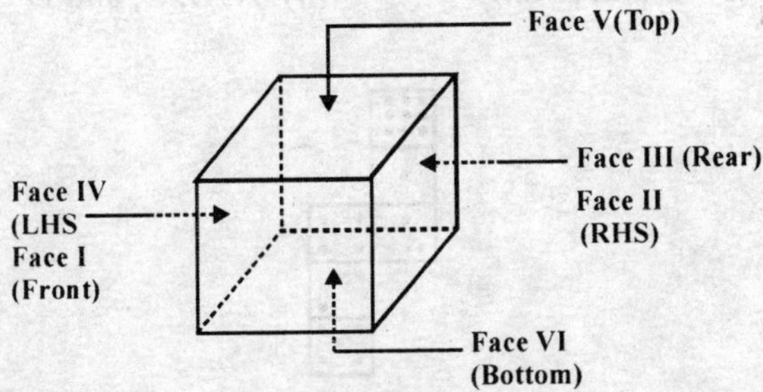

Example 9. A dice is thrown three times and its three different positions are given below. Find the number on the face opposite the face showing 3.

Solution: The number 3 occurs most often in the given three figures. From these three figures it is clear that 1, 2, 5 and 4 lie adjacent to 3. Clearly, 6 lies opposite the face showing 3.

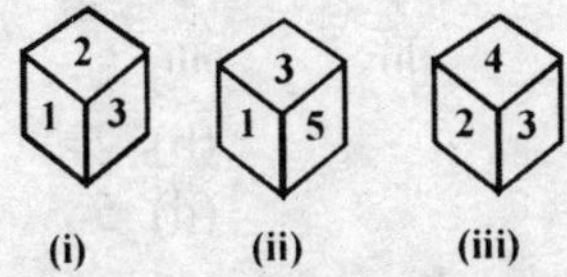

The answer is (d).

Exercise

1. Count the number of cubes in the given figure.

(a) 6 (b) 8
(c) 10 (d) 12

2. **Directions:** Count the number of cubes in each of the following figures:

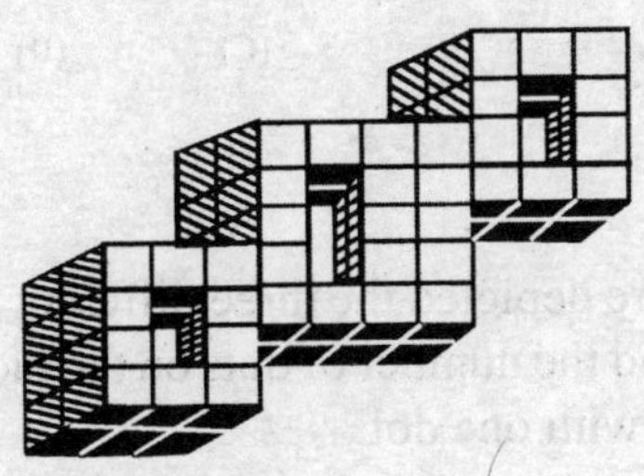

(a) 64 (b) 66
(c) 68 (d) 70

3. A cube is painted blue on all faces and is then cut into 125 cubes of equal size.
How many cubes are not painted on any face?
(a) 8 (b) 16
(c) 27 (d) 36
(e) 54

Directions: (for Question 4 to 9) A cube is painted red on two adjacent faces, yellow on the two faces opposite to the red faces and green on the remaining faces. It is then cut into 64 smaller cubes of equal size. Answer the following questions based on this information:

4. How many cubes are painted on all faces?
(a) 16 (b) 8
(c) 4 (d) 0

5. How many cubes are there which have no face painted?
(a) 24 (b) 16
(c) 8 (d) 4

6. How many cubes are painted yellow on one face only?
(a) 4 (b) 8
(c) 16 (d) 32

7. How many cubes have three faces painted?
(a) 16 (b) 12
(c) 8 (d) 6

8. How many cubes have one face green and one of the adjacent faces red or yellow?
(a) 24 (b) 20
(c) 16 (d) 12

9. How many cubes are painted on two faces only and with the same colour?
(a) 16 (b) 8
(c) 4 (d) 0

Directions: (For Question 10 to 14) A painter is given a task to paint a cubical box with six different colours for different faces of the cube. The detailed account of it was given as:

(i) Red face should lie between Yellow and Brown faces.
(ii) Green face should be adjacent to the Silver faces.
(iii) Pink face should lie adjacent to the Green face.
(iv) Yellow face should lie opposite to the Brown one.
(v) Brown face should face down.
(vi) Silver and Pink faces should lie opposite to each other.

10. The face opposite to Red is
(a) Yellow (b) Green
(c) Pink (d) Silver

11. The upper face is
(a) Red (b) Pink
(c) Yellow (d) Silver

12. The faces adjacent to Green are
(a) Yellow, Pink, Red, Silver
(b) Brown, Pink, Red, Silver
(c) Red, Silver, Yellow, Brown
(d) Pink, Silver, Yellow, Brown

13. The face opposite to Silver is
(a) Pink (b) Brown
(c) Red (d) Green

14. Three of the faces adjacent to Red face are
(a) Silver, Green, Brown (b) Silver, Brown, Pink
(c) Silver, Pink, Green (d) Yellow, Pink, Green

Directions: (for Question 15 to 19) The sheet of paper shown in the figure (X) given on the left hand side, in each problem, is folded to form a box. Choose from amongst the alternatives (a), (b), (c) and (d), the boxes that are similar to the box that will be formed.

15.

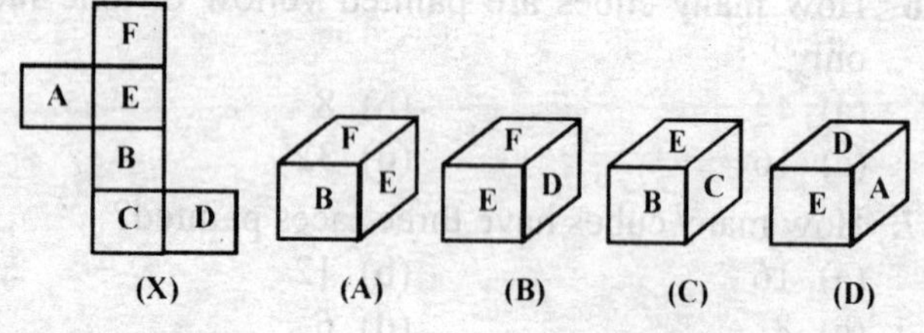

(a) A only (b) B only
(c) A and C only (d) A, B, C and D

16.

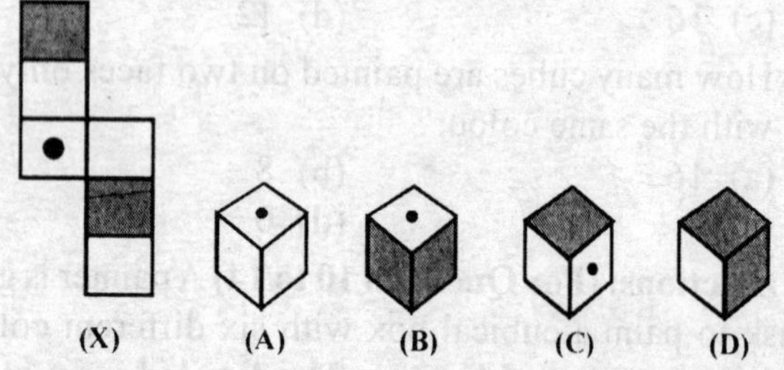

(a) B and C only (b) A , C and D only
(c) B and D only (d) A and D only

17.

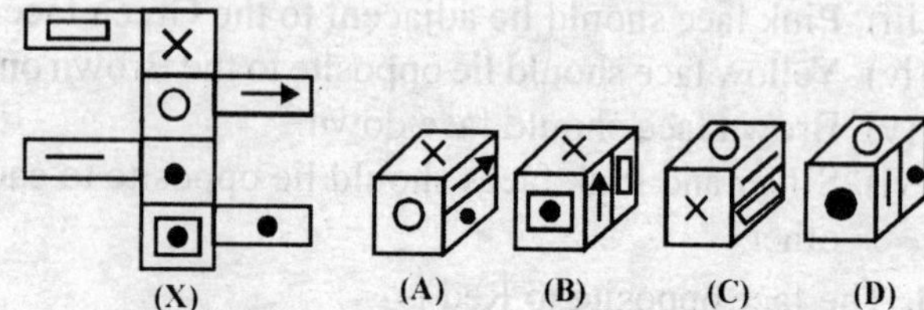

(a) A and B only (b) B and C only
(c) B and D only (d) A, B, C and D

18.

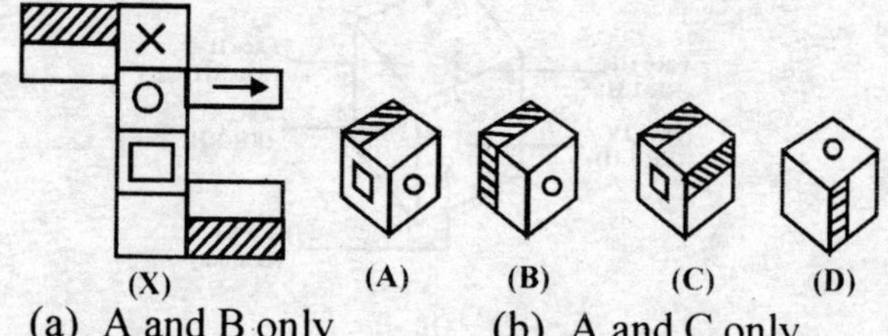

(a) A and B only (b) A and C only
(c) C and D only (d) A, B, C, and D

19.

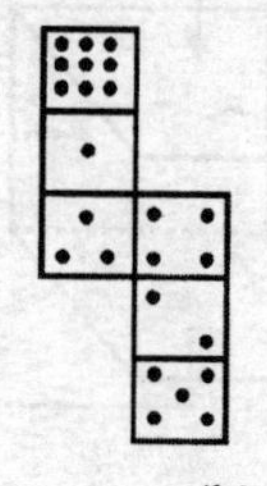

(a) 2 (b) 4
(c) 5 (d) 6

20. Show below are four different positions of the same dice. Find the number on the face opposite the face showing 6.

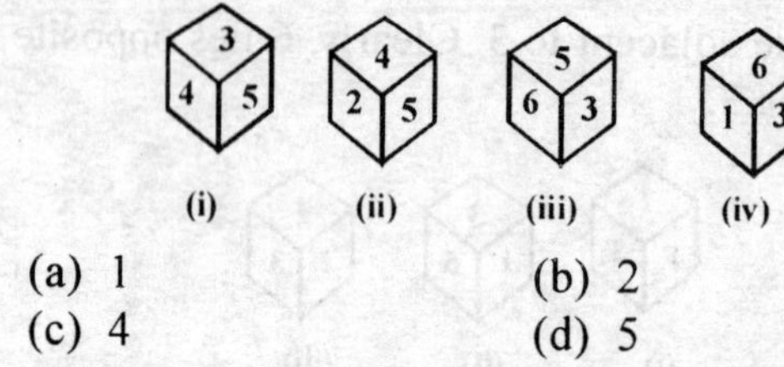

(a) 1 (b) 2
(c) 4 (d) 5

21. If the total number of dots on opposite faces of a cubical block is always 7, find the figure which is correct.

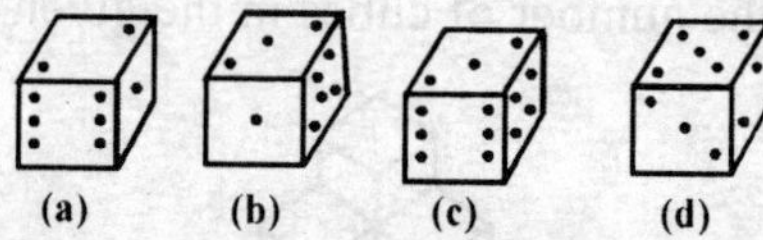

Directions: Questions 22 are based on the following illustrations, which are four views of a cube.

22. The symbol opposite the face having the symbol '=' is

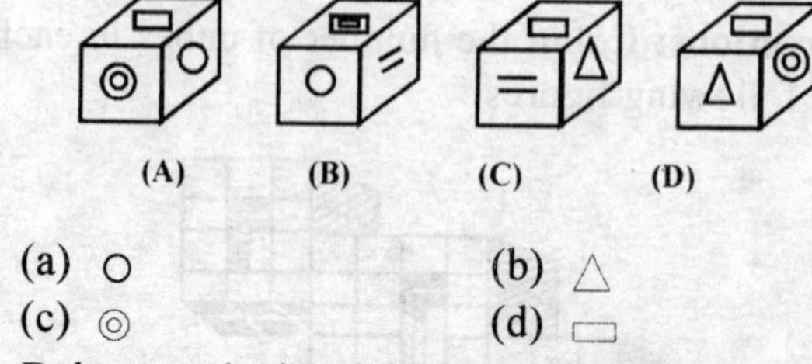

(a) ○ (b) △
(c) ◎ (d) ▭

23. Below are depicted the three different positions of a dice. Find the number of dots on the face opposite to the face with one dot.

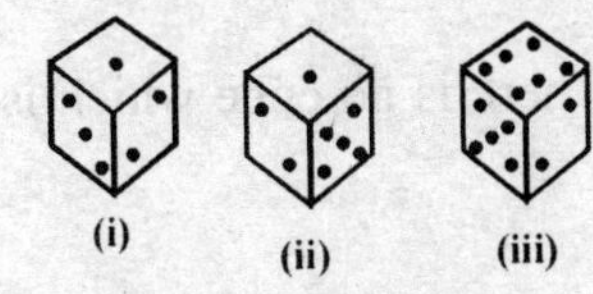

(a) 2 (b) 3
(c) 4 (d) 6

24. Two positions of a dice are shown. When 4 is at the bottom, what number will be on the top?

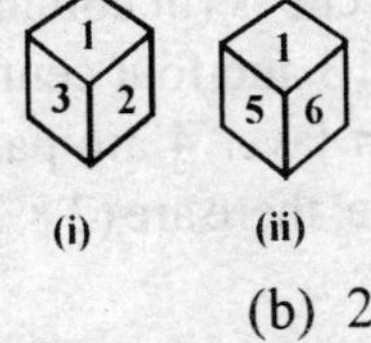

(a) 1 (b) 2
(c) 5 (d) 6

25. Two positions of a cube are shown below. When the number 4 will be at the bottom, then which number will be at the top?

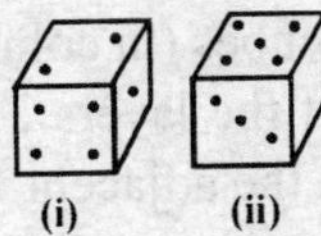

(a) 3 (b) 5
(c) 6 (d) Cannot be determined

26. Two positions of a dice are shown below. When there are two dots at the bottom, the number of dots at the top will be

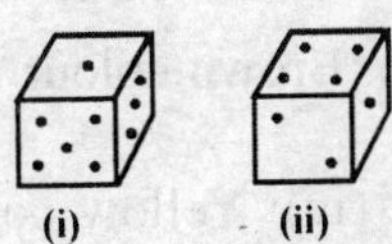

(a) 3 (b) 5
(c) 6 (d) Cannot be determined

27. Two positions of a block are shown below:

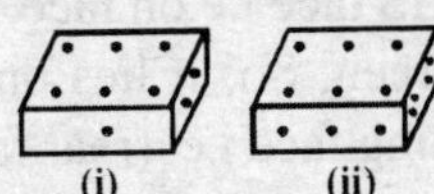

When six is at the bottom, what number will be at the top?

(a) 1 (b) 2
(c) 4 (d) 5

Directions: The questions from 28 to 30 are based on a dice numbered 1 to 6 in different ways as indicated:

28. If 1 is opposite to 2 and adjacent to 3 and 4, then which of the following statements is necessarily true?

(a) 2 is adjacent to 6 (b) 3 is adjacent to 4
(c) 4 is opposite to 6 (d) 3 is opposite to 5

Statement for Question 29 and 30: Six dice with their upper faces erased are as shown:

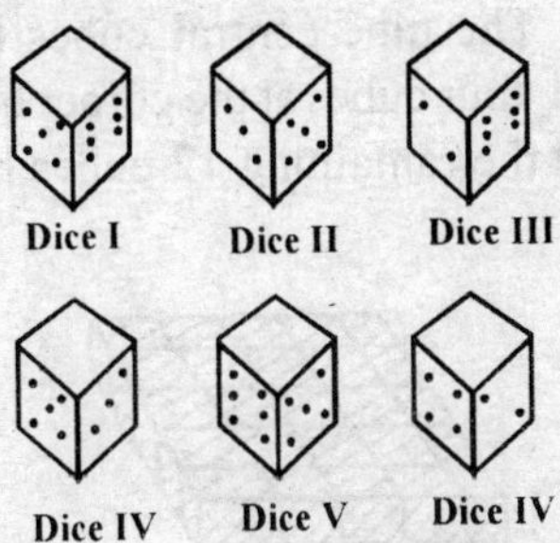

The sum of the numbers of dots on the opposite faces in 7.

29. If the dice I, II and III have even number of dots on their bottom faces, then what would be the total number of dots on the top faces?

(a) 7 (b) 12
(c) 14 (d) 21

30. If odd numbered dice have odd number of dots on their bottom faces, what would be the total number of dots on the top faces of these dice?

(a) 4 (b) 6
(c) 10 (d) 12

Answers

1. **Ans. (c)**

Solution: Clearly, there are 1 column containing 3 cubes, 2 columns containing 2 cubes each and 3 columns containing 1 cube each.
Hence, there are $(1 \times 3) + (2 \times 2) + (3 \times 1) = 3 + 4 + 3 = 10$ cubes in the given figure.

2. **Ans. (c):**

Solution: In the figure there are 34 columns containing 2 cubes each.
Total number of cubes $= (34 \times 2) = 68$.
The method of solving questions based upon painting of selected sides of a stack of cubes (or blocks) will be clear form the following examples.

3. The following figure shows the cubes painted blue on all faces and divided into 125 smaller cubes:

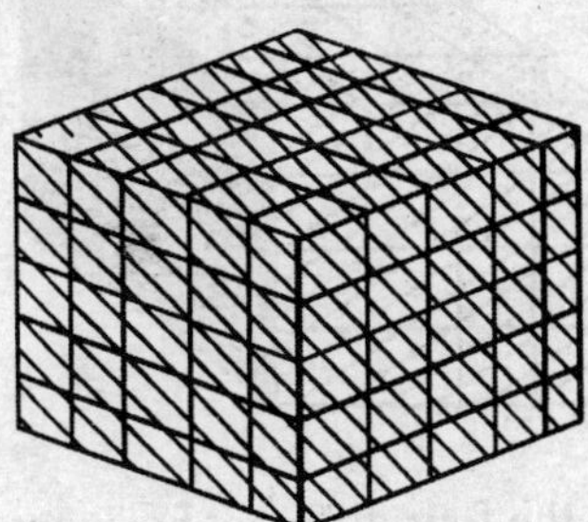

The figure may be analysed by dividing it into five horizontal layers:

In layer I: The nine central cubes have only one face painted, four cubes at the corner have three faces painted and the ramaining 12 cubes have two faces painted.

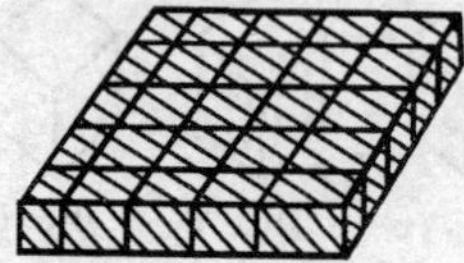

In each of the layers II, III and IV: The nine central cubes have no face painted, the four cubes at the corner have two faces painted and the remaining 12 cubes have one face painted.

In layer V: The 9 central cubes have only one face painted, four cubes at the corner have three faces painted and the remaining 12 cubes have two faces painted.

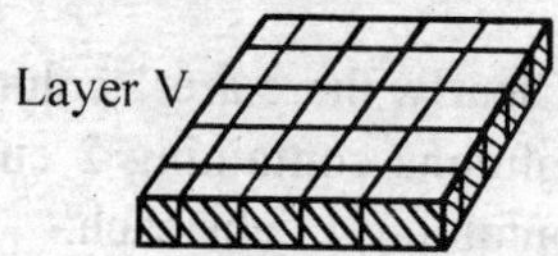

(1) (c): There are 9 central cubes in each of the layers. II, III, and IV which have no face painted. Thus, there are $9 \times 3 = 27$ such cubes.

(Statement for answers 4 to 9)

Let the upper and the LHS faces of the cube be painted red. Then, the lower and the RHS faces are painted yellow and the front and the rear faces are painted green, as indicated in the following figure, which also shows the division of this cube into 64 smaller cubes.

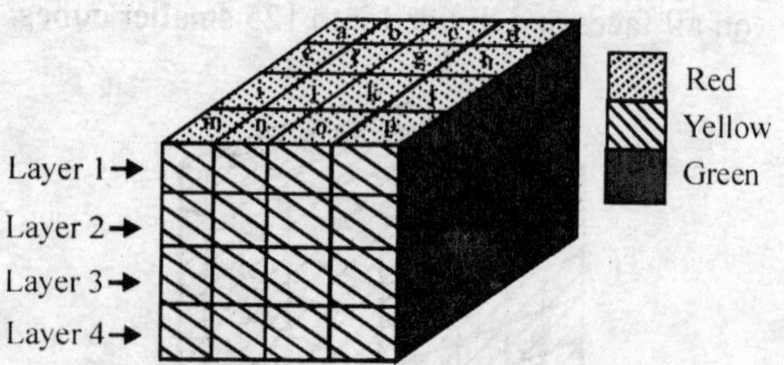

We shall label the four different horizontal layers (of 16 cubes each) as layers 1 to 4. The 16 columns (of 4 cubes each) may also be labelled from a to p as shown in the figure.

4. **Ans. (d)**
   Solution: There is no cube which is painted on all faces.

5. **Ans. (c)**
   Solution: Four central cubes (in columns f, g, j and k) in each of the layers 2 and 3 have no face painted. Thus, there are $(4 \times 2) = 8$ such cubes.

6. **Ans. (b)**
   Solution: Two cubes (in column h and l) in each of the layers 2 and 3 and four central cube (in columns f, g, j and k) in layer 4 are painted yellow on one face only. Hence, there are $(2 \times 2) + 4 = 8$ such cubes.

7. **Ans. (c)**
   Solution: Four corner cubes (in columns a, d, m and p) in each of the layers 1 and 4 have 3 faces painted. Thus, there are $(4 \times 2) = 8$ such cubes.

8. **Ans. (a)**
   Solution: Eight cubes (in columns a, b, c, d, m, n, o and p) in each of the layers 2 and 3 have one face green and one of the adjacent faces red or yellow.

9. **Ans. (c)**
   Solution: Two cubes (in columns e and i) in layer 1 and two cubes (in columns h and l) in layer 4 have only two faces painted and with the same colour.

(Statement for answer 10 to 14)

From statement (v): Brown colour is at the bottom i.e. on face 6.

From statement (iv): Yellow colour is opposite to brown. So, it lies on the top i.e. on face 5.

Let Green colour be in front i.e. on face 1.

From statement (ii) and (iii): Silver colour lies adjacent to Green say on LHS face i.e.on face 4; then Pink colour lies adjacent to Green. So, it lies on RHS face i.e. on face 2. The remaining face i.e. (rear face) face 3 gets the Red colour. This also satisfies statement (i).

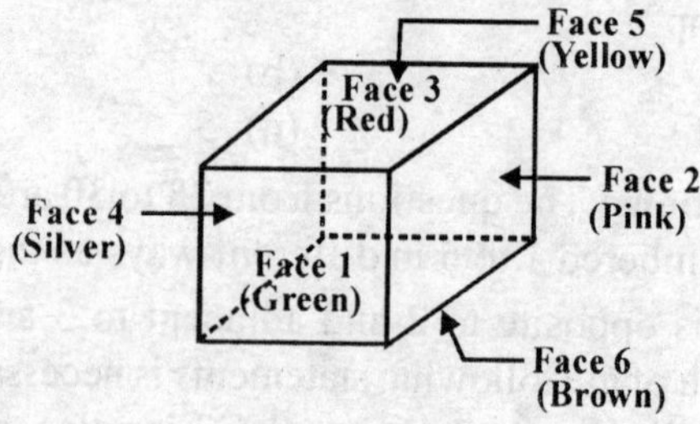

10. **Ans. (b)**
    Solution: Red colour is on face 3. The face opposite to the Red face is face 1 which has Green colour.

11. Ans. (c)

Solution: The upper face i.e. face 5 has Yellow colour.

12. Ans. (d)

Solution: From the figure, it is clear that Red face lies opposite to the Green face. So, the face adjacent to Green are Pink, Silver Yellow and Brown.

13. Ans. (a)

Solution : Silver colour lies on face 4. The face opposite to this face is face 2 which has Pink colour.

14. Ans. (b)

Solution: From the figure, it is clear that Green face (face 1) lies opposite to the Red face (face 3). Hence, Green face cannot lie adjacent to Red face. Out of the given alternatives, the alternatives (b) has coloured Silver, Brown and Pink which lie adjacent to Red face.

15. Ans. (b)

Solution: The fig. (X) is similar to the Form III. So, when the sheet in fig. (X) is folded to form a cube, then 'F' appers opposite 'B', 'E appears opposite 'C' and 'A' appears opposite 'D'. Therefore, the cube in fig. (A) which shows 'F' adjacent to 'B', the cube in fig. (C) which shows 'E' adjacent to 'C' and the cube in fig. (D) which shows 'A' adjacent to 'D' cannot bhe formed. Hence, only the cube in fig. (B) can be formed.

16. Ans. (b)

Solution: The fig. (X) is similar to the Form V. So, when the sheet in fig. (X) is folded to form a cube, then the face bearing a dot lies opposite to one of the shaded faces. Therefore, the cube shown in fig. (B) which has both the shaded faces adjacent to the face bearing the dot, cannot be formed. Hence, the cubes shown in figures (A), (B) and (D) can be formed.

17. Ans. (b)

Solution: The fig. (X) is similar to the Form VIII. So, when a cube is formed by folding the sheet shown in fig. (X), then [→ / •] and [□ / –] are the two faces and these two faces lie opposite to each other. Also, the face bearing the 'x' sign lies opposite to the face bearing the black circle and the face bearing the white circle lies opposite to the face bearing the square (having a dot inside it). Now, the cubes in figures (B) and (D) consist of faces which are not formed when the sheet in fig. (X) is folded. Hence, these two cubes are not formed. Therefore, only the cubes in figures (A) and (C) are formed.

18. Ans. (a)

Solution: The fig. (X) is similar to the Form II. So, when the sheet shown in fig. (X) is folded to form a cube then the two hald-shaded faces lie opposite to each other, the face bearing a square lies opposite to one of the two blank faces and the face bearing a circle lies opposite to the other blank face. Therefore, the cubes shown in figures (b) and (c) which have the two half-shaded faces adjacent to each other, cannot be formed by folding the sheet shown in fig. (X). Also, though the cube shown in fig. (d) has faces that can appear adjacent to each other but the cube formed by folding the sheet in fig. (X) cannot be rotated to form the cube in fig. (d). Hence, only the cube in fig. (a) can be formed.

19. Ans. (d)

Solution: The given figure is similar to Form V. Therefore, when this figure is folded to form a cube then the face bearing six dots will lie opposite the face bearing three dots.

20. Ans. (c)

Solution: From figures (i), (ii) and (iii), we conclude that 3, 4, 2 and 6 lie adjacent to 5. Therefore, 1 must lie opposite 5.

From figures (i), (iii) and (iv), we conclude that 4, 5, 6 and 1 lie adjacent to 3. Therefore, 2 must lie opposite 3. Now, we have 1 opposite 5 and 2 opposite 3. Hence, 4 must lie opposite 6.

21. Ans. (b)

Solution: Since the total number of dots on opposite faces is always 7, therefore, 1 dot appears opposite 6 dots, 2 dots appear opposite 5 dots and 3 dots appear opposite 4 dots.

22. Ans. (c)

Solution: From figures (ii) and (ii), we conclude that the symbols ▭, ○, and △ lie adjacent to the symbol =. Therefore, the symbol ◎ must lie opposite to the face having the symbol =.

23. Ans. (d)

Solution: From figures (i), (ii) and (iii), we conclude that 1, 3, 5 and 6 dots appear adjacent to the face with 2 dots. Therefore, 4 dots will appear opposite to 2 dots. Now, from figures (i) and (ii), we conclude that 2, 3, and 5 dots appear adjacent to 1 dot. Therefore, either 4 or 6 dots will appear opposite to

1 dot. But since, 4 dots appear opposite to 2 dots it follows that 6 dots will appear opposite 1 dots.

24. Ans. (a)

Solution: From figures (i) and (ii), we conclude that 2, 3, 5 and 6 lie adjacent to 1. Therefore, 4 lies opposite 1. Hence, when 4 is at the bottom, then 1 must be on the top.

25. Ans. (a)

Solution: Number 1 is common to both the positions of the dice. We assume the dice in fig. (ii) to be rotated so that 1 remains on the RHS face (i.e. face II as per activity 1) and the numbers 5 and 3 move to the faces hidden behind 2 and 4 respectively [in fig. (i)]. Then, clearly 3 lies opposite 4. Hence, when 4 is at the bottom then 3 must lie on the top.

26. Ans. (a)

Solution: Number 1 is common to both the positions of the dice. We assume the dice in fig. (ii) to be rotated so that 1 dot moves to the top face (i.e. face V as per activity 1) i.e. to the same position as in fig. (i) and 2 and 4 dots move to the faces hidden behind the faces with 3 and 5dots respectively. Thus, the combined figure will have 1 dot on the Top (i.e. on face V), 5 dots on Front face (i.e. on face I), 3 dots on RHS face (i.e. face II), 4 dots on the Rear face (i.e. face III) and 2 dots on the LHS face (i.e. face IV). Clearly, 3 dots lie on the face opposite the face having 2 dots. Therefore, when there are 2 dots at the bottom, the number of dots at the top will be 3.

27. Ans. (d)

Solution: From figures (i) and (ii) we conclude that the number 1, 2, 3 and 4 appear adjacent to 6. Thus, the number 5 will appear opposite 6. Therefore, when six is at the bottom, then 5 will be at the top.

28. Ans. (a)

Solution: If 1 is opposite to 2, then 2 cannot lie opposite to any of the numbers 3, 4, 5 or 6. Hence, 2 necessarily lies adjacent to 6.

29. Ans. (b)

Solution: If the Dice I, II and III hve even number of dots on their bottom faces, then the number of dots on their bottom are 416 and 4. Therefore the number of dots on their top faces are 3, 1 and 3, the requiring sum = 3 + 1 + 3 = 7 .

30. Ans. (d)

Solution: If odd numbered dice i.e. Dice, I, III and V have odd number of dots on their bottom faces, then the number of dots on their bottom faces are 3, 3 and 3 respectively. Therefore, the numbers of dots on their top faces are 4, 4 and 4.

Mirror-Images

Mirror Image: The image of an object, as seen in a mirror reflection or mirror image.

In such an image, the right side of the object appears on the left side and vice-versa. A mirror-image is therefore said to be laterally inverted and the phenomenon is called Lateral Inversion.

Mirror-images of capital letters

Letters	Mirror Images	Letters	Mirror Images	Letters	Mirror Image
A	A	J	Ⴑ	S	Ƨ
B	ᗺ	K	ꓘ	T	T
C	Ɔ	L	⅃	U	U
D	ᗡ	M	M	V	V
E	Ǝ	N	И	W	W
F	ꟻ	O	O	X	X
G	Ә	P	ꟼ	Y	Y
H	H	Q	Ọ	Z	Ƹ
I	I	R	Я		

Remark : The letters which have their mirror images identical to the letter itself are:

A, H, I, M, O, T, U, V, W, X, Y

Examples: Mirror-images of certain words are given below:

1. FUN : ИUꟻ
2. STOP : ꟼOTƧ
3. ZEBRA : AЯᗺƎƸ
4. GOLKONDA : AᗡИOꓘ⅃OӘ
5. XYLOPHONE : ƎИOHꟼO⅃YX

Mirror-images of small letters

Letters	Mirror Images	Letters	Mirror Images	Letters	Mirror Image
a	ɒ	j	į	s	ƨ
b	d	k	ʞ	t	ɟ
c	ɔ	l	l	u	u
d	b	m	m	v	v
e	ɘ	n	n	w	w

f	ɟ	o	o	x	x
g	ɡ	p	q	y	ʏ
h	ɖ	q	p	z	ƨ
i	i	r	ɿ	–	–

Examples: Mirror-images of certain words are given below:

1. a r p i t : tiqɿa
2. b l a d e : ebald
3. d e t e r m i n e : enimɿeted

Mirror-images of numbers

Letters	Mirror Images	Letters	Mirror Images	Letters	Mirror Image
1	I	4	ᔭ	7	Ⲋ
2	Ƨ	5	ꙅ	8	8
3	Ɛ	6	∂	9	୧

Examples: Mirror-Images of certain combinations of alphabets and numbers are given below:

1. alpha348mzl : lzm8ᔭƐahqla
2. BMC49JN2317 : 7ƖƐƧИႱ୧ᔭƆMЯ
3. 15bg82XQh : hϘX28gdꙅ1

Exercise

Directions: (Question 1 to 5) In each of the following questions, your are given a combination of alphabets and/ or numbers followed by four alternatives (a), (b), (c) and (d). Choose the alternative which most closely resembles the mirror-image of the given combination.

1. SUPERVISOR
(a) ЯOƧIVЯƎPUƧ (b) ƧUPEЯVIƧOЯ
(c) ЯƧOUPƧEЯIV (d) ƧUPEЯVIOƧЯ

2. MALAYALAM
(a) MALAYALAM (b) MA⅃AYA⅃AM
(c) MALAYALAM (d) MAΓAYAΓAM

3. MAGAZINE
(a) MAGAZIИƎ (b) ƎNIZAGAM
(c) MAGAZIИƎ (d) ƎИIZAGAM

4. TARA1N1014A
(a) AᔭI0IИIAЯAT (b) IA0IᔭИIAЯAT
(c) AI0IᔭTAЯAIИ (d) AᔭI0IИIAЯAT

5. Nu56p7uR
(a) ИuƧ∂qⲊuЯ (b) ЯuⲊq∂ꙅuИ
(c) ЯuⲊqdƧuИ (d) ЯuⲊb∂ƧuИ

Mirror-images of Clocks

There are certain questions in which the positions of the hour - hand and the minute -hand of a clock as seen in a mirror are given. On the basis of the time indicated by the mirror-image of the clock we have to detect the actual time in the clock. In the solution of such questions we use the fact that if an object A is the mirror - image of another object B then Bis the mirror - image of A. The method of solving such questions would be clear from the following example.

Example: Looking into a mirror, the clock shows 9 : 30 as the time. The actual time is

(a) 2 : 30 (b) 3 : 30
(c) 4 : 30 (d) 6 : 30

Solution: In the mirror the clock appears as

Fig : A

This is the mirror image of the clock shown below:

Fig : B

(Fig. B is obained by taking the mirror image of fig. A) Clearly, this clock shows the time 2 : 30. Therefore, the actual time is 2:30. Hence, the answer is (a).

Exercise

Directions: (Question 6to 10): In each of the following questions, choose the correct mirror-image fo the Fig. (X) from amongst the four alternativs (a), (b), (c) and (d) given along with it.

6.

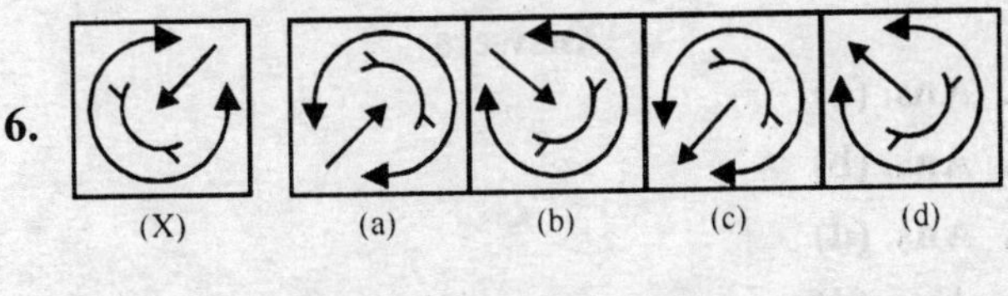

7.

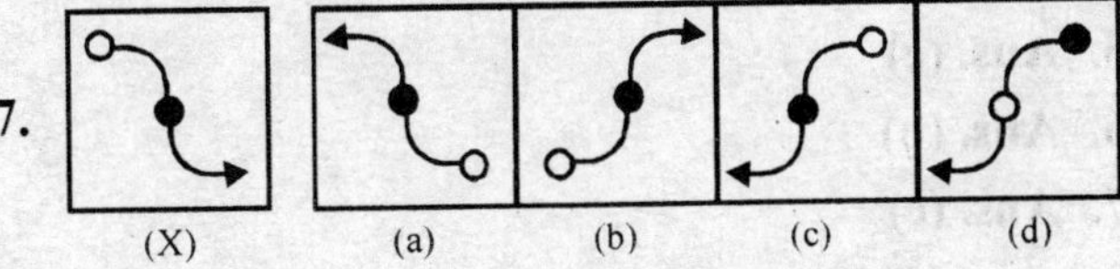

8\. 9. 10.

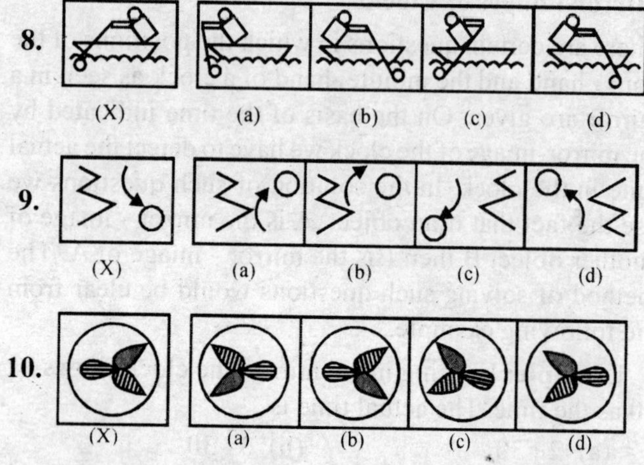

Directions: (Questions 11 to 12): In each of the following questions, choose the correct mirror-image of the Fig. (X) from the four alternatives (a), (b), (c) and (d) given along with it. The mirror may be represented by a line MN or M_1M_2.

11\.

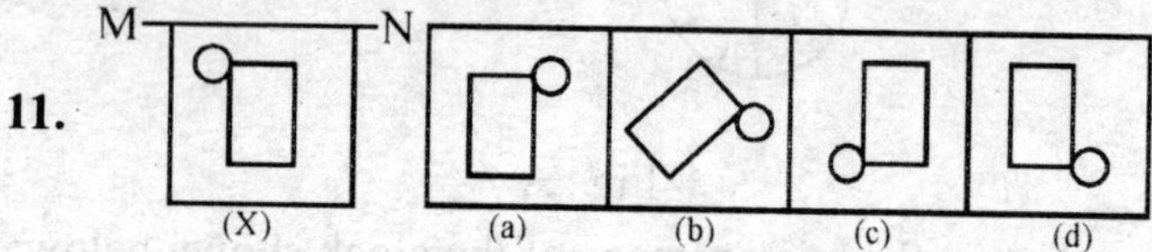

12\.

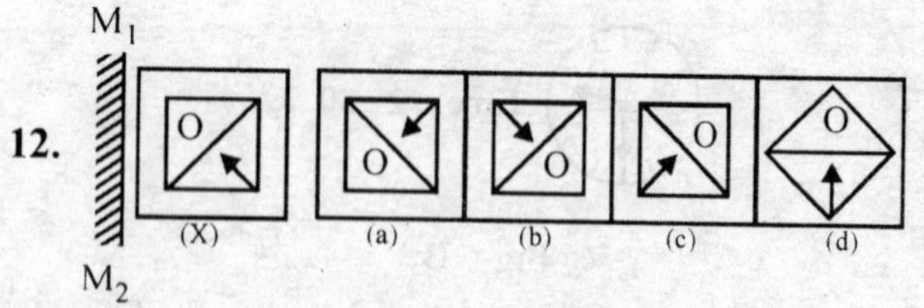

13. By looking in a mirror, it appears that it is 6 : 30 in the clock. What is the real time?

(a) 6 : 30 (b) 5 : 30
(c) 6 : 00 (d) 5 : 30

14. A clock seen through a mirror shows quarter to three. What is the correct time shown by the clock?

(a) 8 : 15 (b) 9 : 12
(c) 8 : 17 (d) 9 : 15

Answers

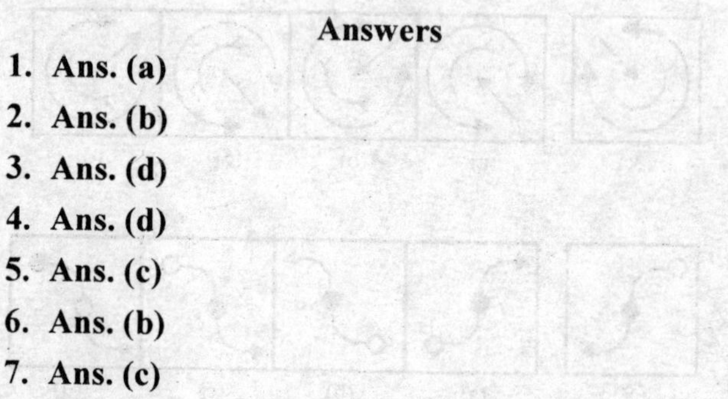

1. **Ans. (a)**
2. **Ans. (b)**
3. **Ans. (d)**
4. **Ans. (d)**
5. **Ans. (c)**
6. **Ans. (b)**
7. **Ans. (c)**
8. **Ans. (d)**
9. **Ans. (c)**
10. **Ans. (c)**
11. **Ans. (c)**
12. **Ans. (c)**
13. **Ans. (b)**

Solution:

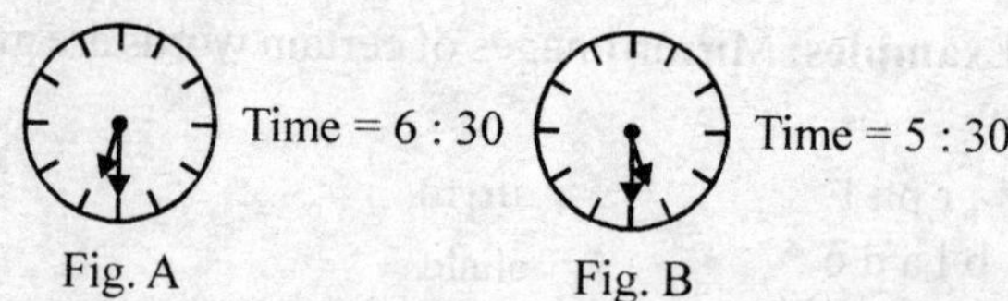

Fig. A Fig. B

Clearly, fig.(A) shows the time (6 : 30) in the clock as it appears in a mirror.
Then, its mirror-image, i.e. Fig.(B) shows the actual time in the clock i.e. 5 : 30.

14. **Ans. (d)**

Solution:

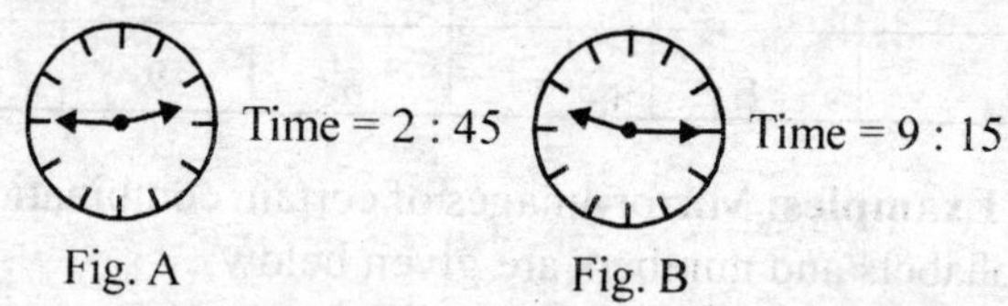

Fig. A Fig. B

Clearly, fig.(A) shows the time (2 : 45) in the clock as it appears in a mirror.
Then, its mirror-image, i.e. Fig. (B) shows the actual time in the clock i.e. 9 : 15.

Logical Venn Diagram

This section deals with questions which aim at analysing a candidate's ability to relate a certain given group of items and illustrate it diagrammatically.

1. If the items evidently belong to three different groups, the Venn diagram representing it would be as shown alongside.
   Ex. Doctors, Engineers, Lawyers

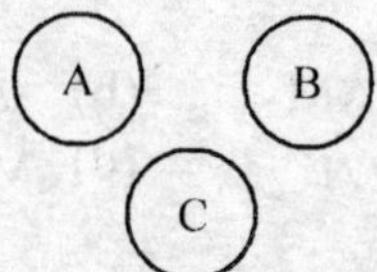

2. If one item belongs to the class of the second and the second belongs to the class of third, then the representation is in the form of three concentric circles, as shown in Fig.2.
   Ex. Seconds, Minutes, Hours

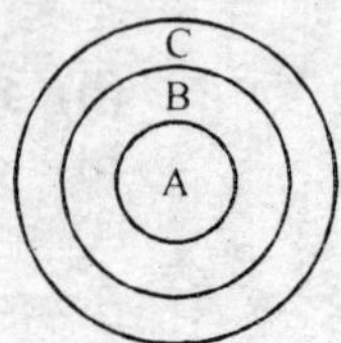

3. If two separate items belong to the class of the third, they are represented by two disjoint circles inside a bigger circle as shown in Fig.3.
Ex. Table, Chair, Furniture

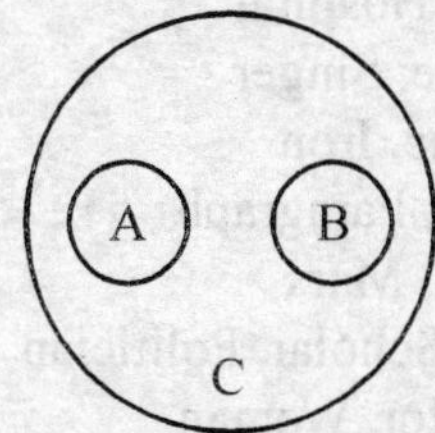

4. If two items belong to the class of the third such that some items of each of these two groups are common in relationship, then they are represented by two intersecting circles enclosed within a bigger circle.
Ex. Males, Fathers, Brothers

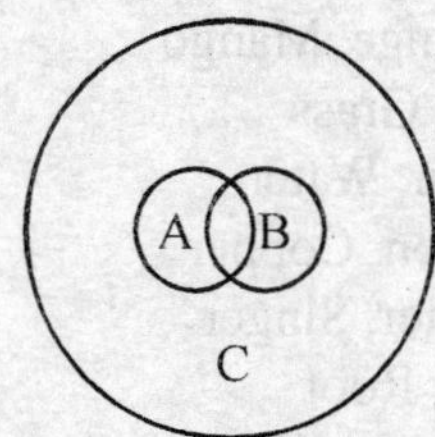

5. If two items are partly related to the third, and are themselves independent of each other they are represented by three intersecting circles in a line.
Ex. Dogs, Pets, Cats

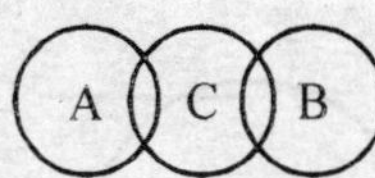

6. If the three items are partly related to each other, they are represented as shown in the adjoining figure.
Ex. Clerks, Government Employees, Educated Persons

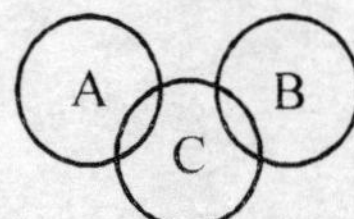

7. If one item belongs to the class of second while third item is entirely different from the two, then they may be represented by the adjoining diagram.
Ex. Engineers, Human Beings, Rats

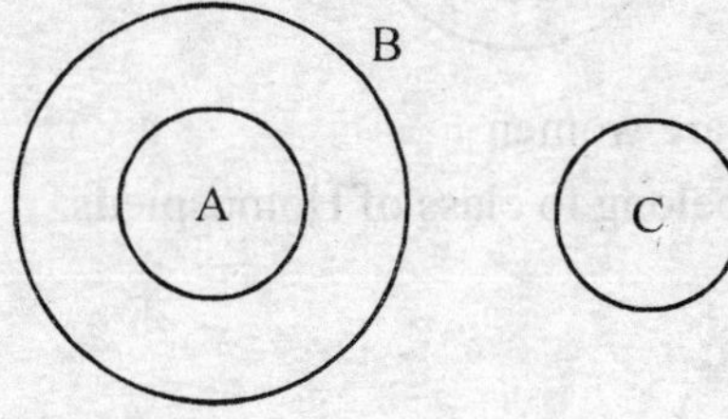

8. If one item belongs to the class of second and the third item is partly related to these two, they are represented as shown alongside.
Ex. Females, Mothers, Doctors

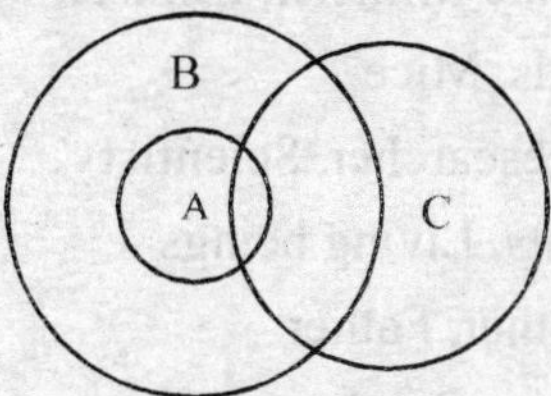

9. If one item belongs to the class of second and the third item is partly related to the second, they are represented as shown alongside.
Ex. Males, Fathers, Children

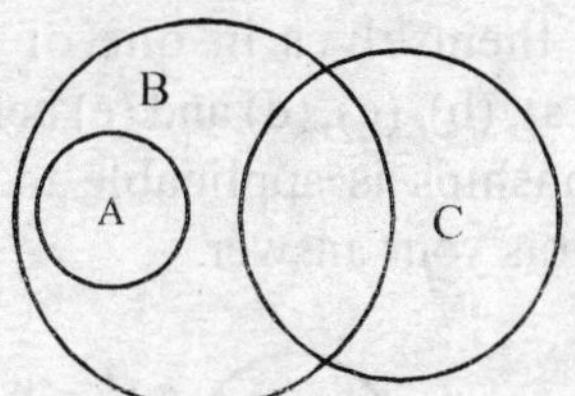

10. If two item are partly related to each other and the third item is entirely different from the two, they are represented as shown alongside.
Ex. Professor, Author, Children

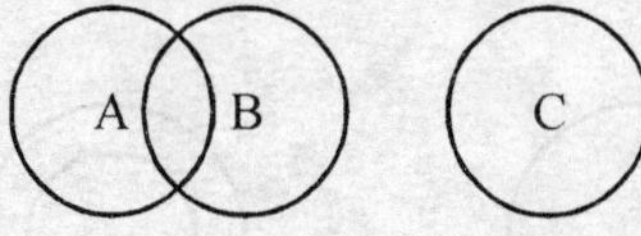

Type 1

Directions: (Questions 1 to 20): In each of these questions, three words are related in some way. The relationship among the words in the question can best be represented by one of the five diagrams (a), (b), (c), (d) and (e) given below, Mark your answer accordingly.

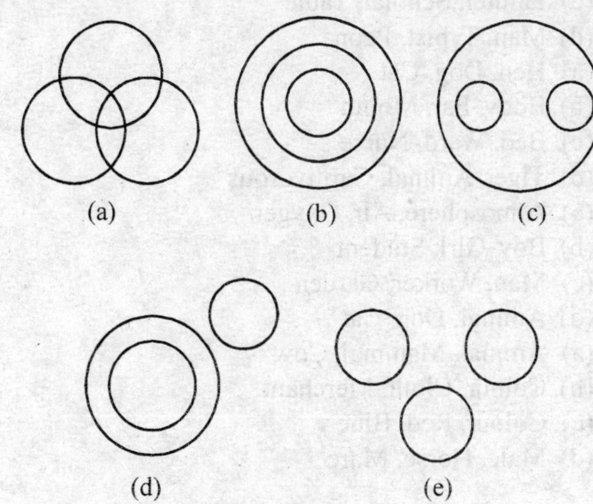

1. Teacher, College, Student
2. Mother, Homosapien, Woman
3. Iron, Lead, Nitrogen
4. Cabinet, Home Minister, Minister
5. Parrots, Birds, Mice
6. Professor, Researcher, Scientist
7. Men, Rodents, Living beings
8. Parents, Mother, Father
9. English, Latin, Greek
10. Nitrogen, Ice, Air

Directions: (Questions 11 to 125): Of the four alternatives in each of the following questions, three alternatives are such that the three words in each are related among themselves in one of the five ways represented by (a), (b), (c), (d) and (e) below while none of these relationships is applicable to the remaining alternative. That is your answer.

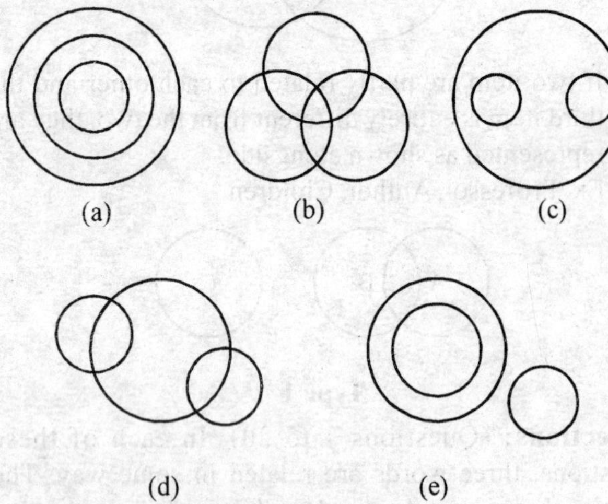

11. (a) Army, General, Colonel
    (b) Boy, Student, Player
    (c) Painter, Scholar, Table
    (d) Man, Typist, Peon
12. (a) Hen, Dog, Cat
    (b) Body, Ear, Mouth
    (c) Bed, Ward, Nurse
    (d) Tiger, Animal, Carnivorous
13. (a) Atmosphere, Air, Oxygen
    (b) Boy, Girl, Student
    (c) Man, Worker, Garden
    (d) Animal, Dog, Cat
14. (a) Animal, Mammal, Cow
    (b) Colour, Cloth, Merchant
    (c) Colour, Red, Blue
    (d) Male, Horse, Mare
15. (a) Metal, Sodium, Copper
    (b) Mammel, Nurse, Woman
    (c) Cereal, Wheat, Rice
    (d) Males, Cousins, Nephews
16. (a) Bed, Ward, Hospital
    (b) Girl, Athlete, Singer
    (c) Copper, Zinc, Iron
    (d) Book, Page, Paragraph
17. (a) Star, Moon, Mars
    (b) Professor, Scholar, Politician
    (c) Maid, Doctor, Woman
    (d) Swimmer, Carpenter, Singer
18. (a) Periodical, Weekly, Book
    (b) Mineral, Copper, Wood
    (c) Doctors, Human beings, Married people
    (d) Army, Doctor, Engineer
19. (a) Director, Engineer, Musician
    (b) Apple, Orange, Mango
    (c) Fruit, Pear, Grass
    (d) Oxygen, Air, Water
20. (a) Mineral, Iron, Copper
    (b) Dean, Painter, Singer
    (c) Seed, Leaf, Root
    (d) Piston, Engine, Wheel

Answers

Type 1

1. **Ans. (a)**
   Solution:

Teacher and Student are entirely differnt. But, both are parts of colege.

2. **Ans. (b)**
   Solution:

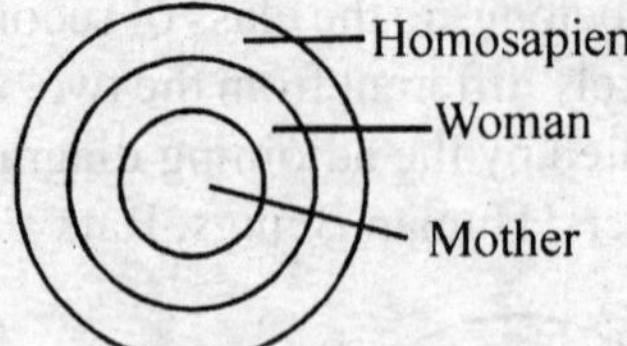

All mothers are women
All women belong to class of Homospiens.

3. **Ans. (e)**
   Solution:

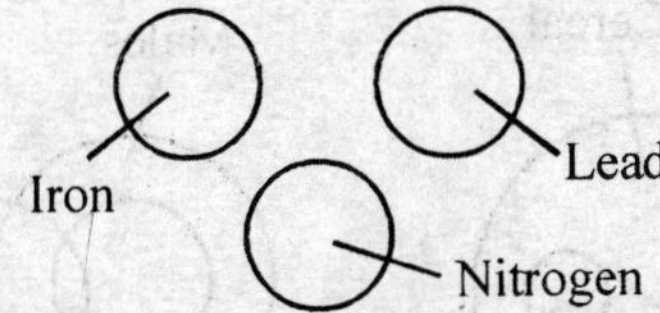

Iron, Lead and Nitrogen are entirely differnet form each other.

4. **Ans. (b)**
Solution:

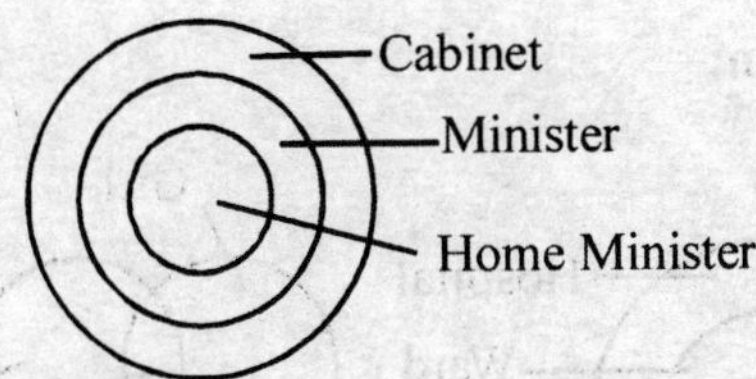

Home Ministr is a minister.
Minister is a part of Cabinet.

5. **Ans. (d)**
Solution:

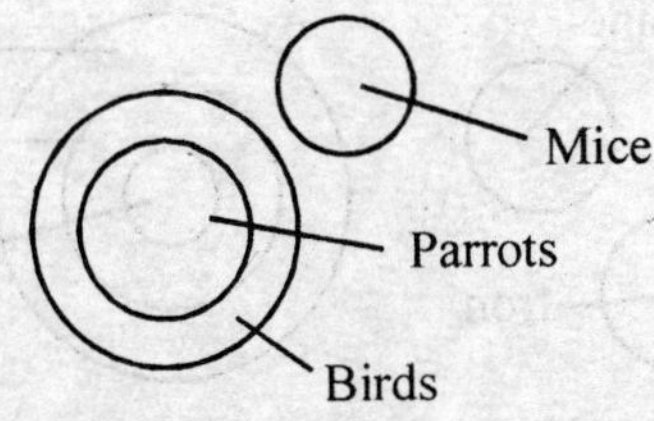

All parrots are birds. But mice is entirely differnet.

6. **Ans. (a)**
Solution:

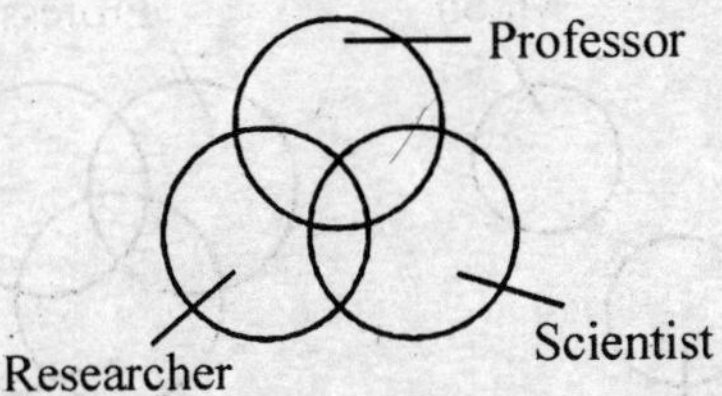

Some professional may be scientists or researchers. Some scientists may be researchers.

7. **Ans. (c)**
Solution: Men and Rodents are entirely differnet. But, both are livin

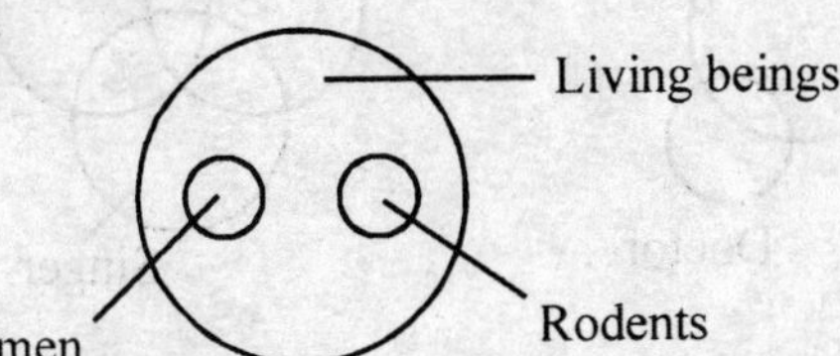

8. **Ans. (c)**
Solution: Mother and Father are entirely differnet. But, both are parents.

9. **Ans. (e)**
Solution:

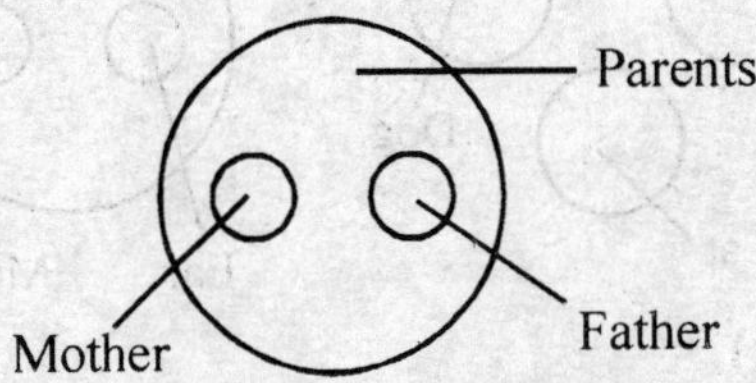

English, Lattin and Greek are entirely different from each other.

10. **Ans. (d)**

Solution:

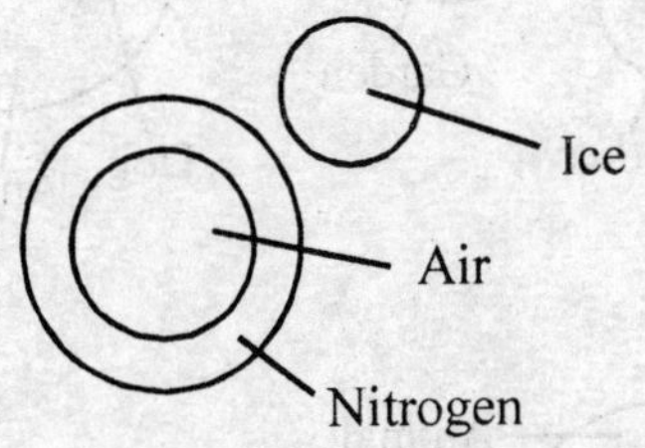

Nitrogen is a constitutent of air. But, Ice is different

11. **Ans. (c)**
Solution:

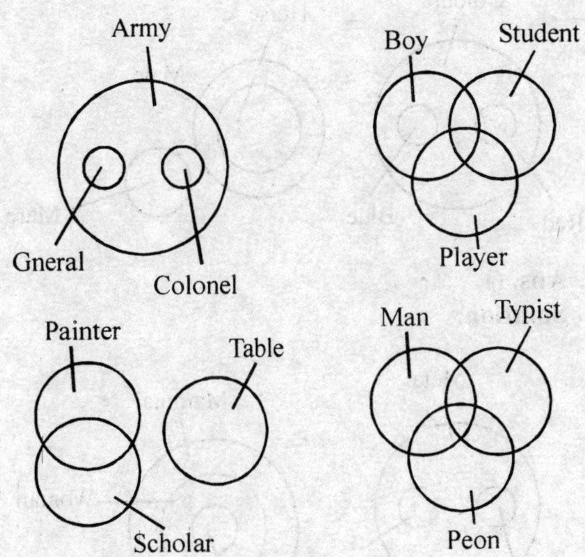

12. **Ans. (a)**
Solution:

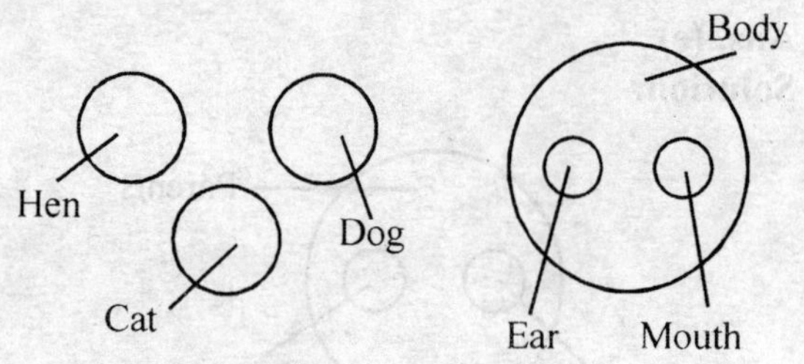

13. Ans. (c)
Solution:

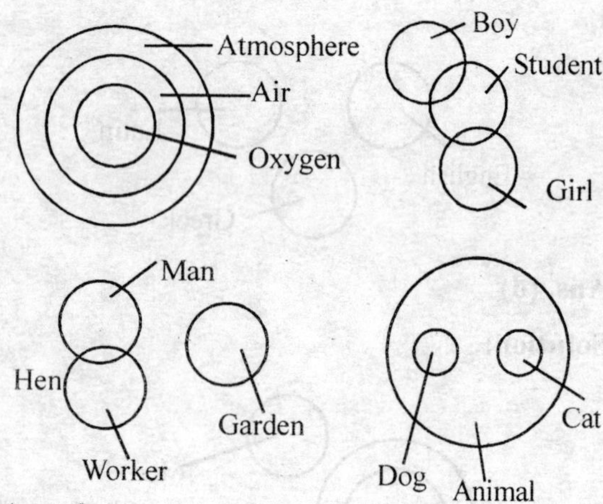

14. Ans. ()
Solution:

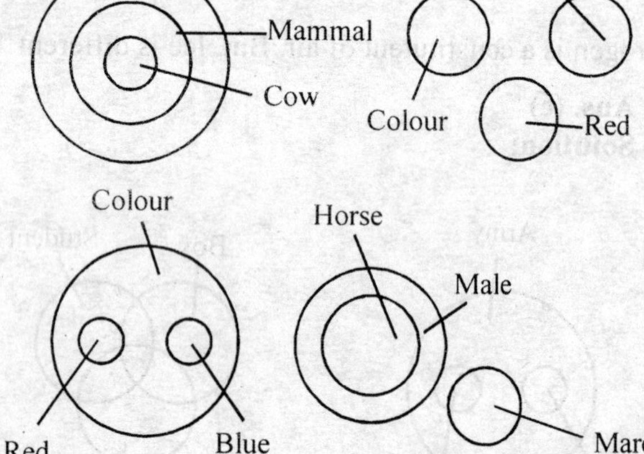

15. Ans. ()
Solution:

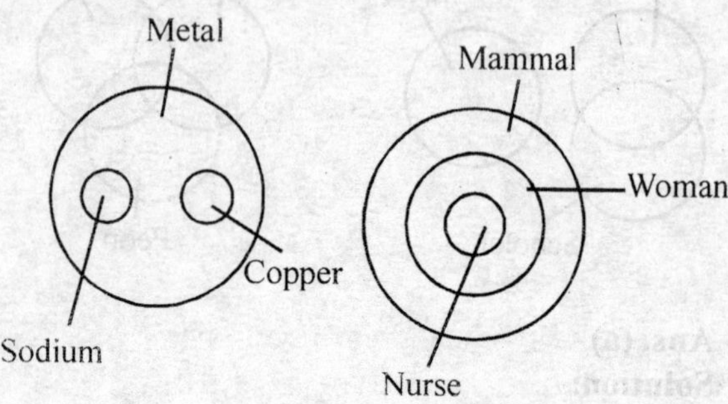

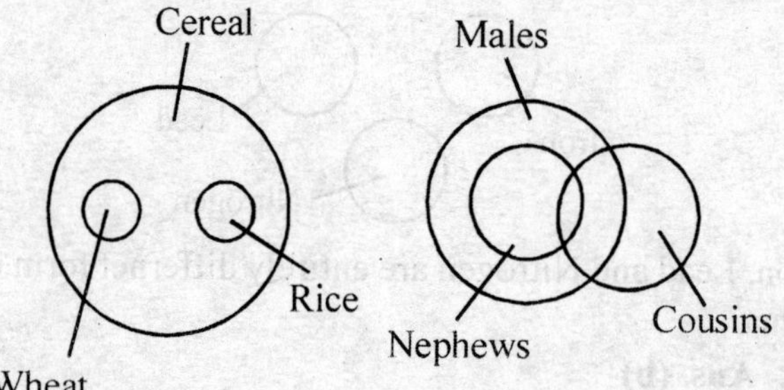

16. Ans. (c)
Solution:

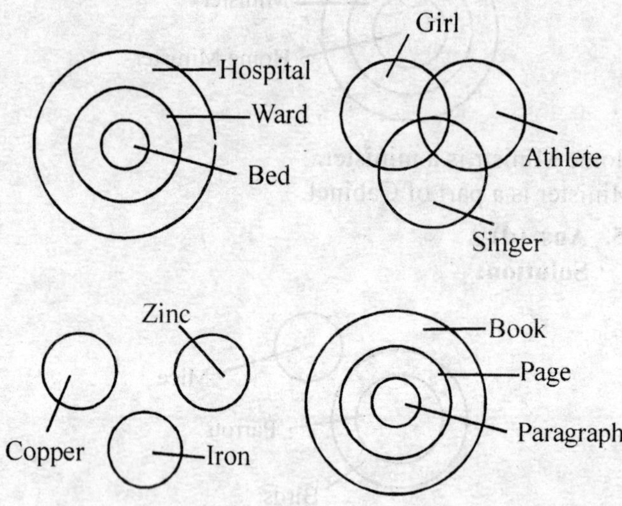

17. Ans. (a)
Solution:

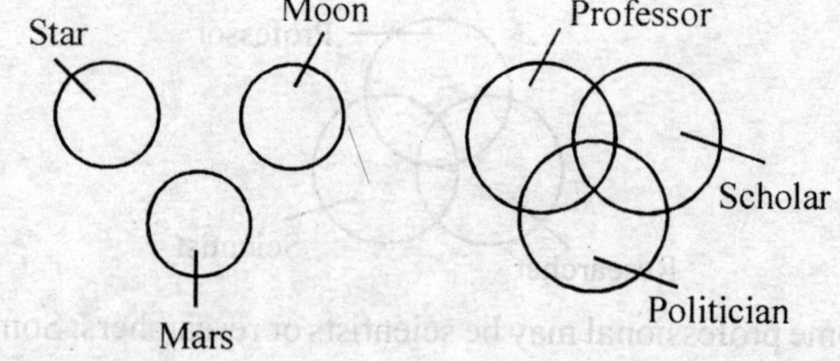

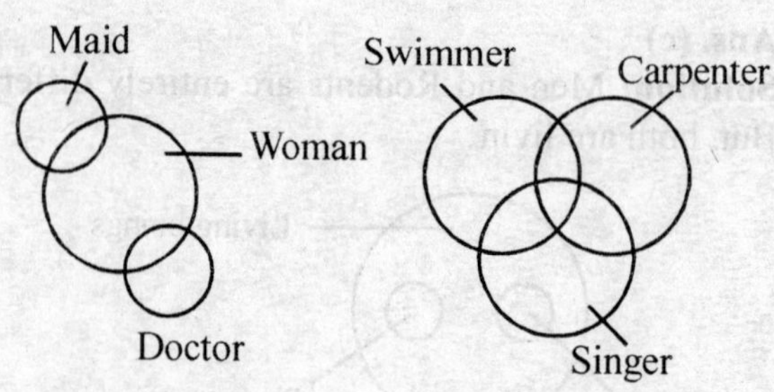

18. Ans. (c)
Solution:

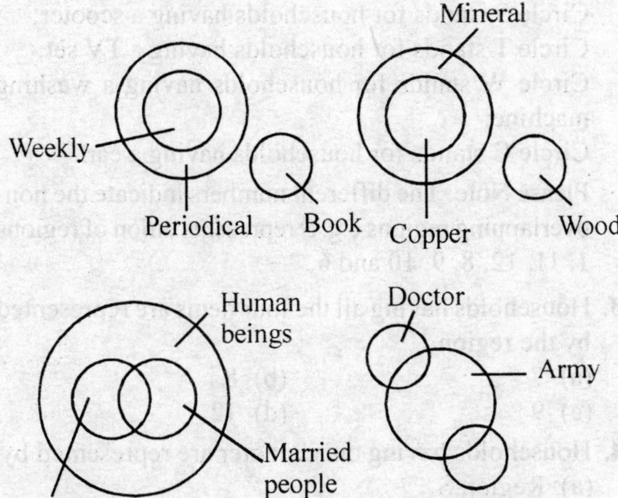

19. Ans. (b)
Solution:

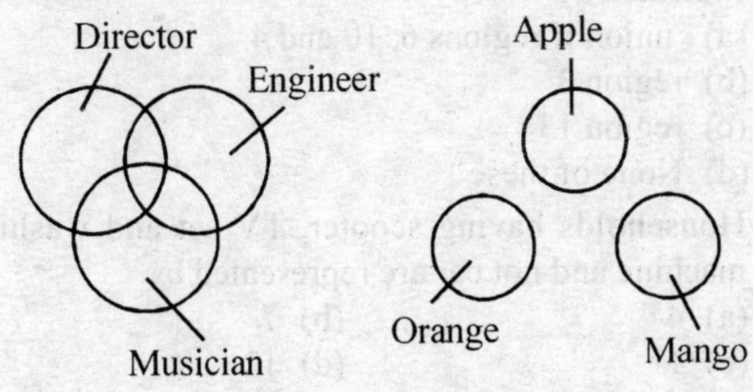

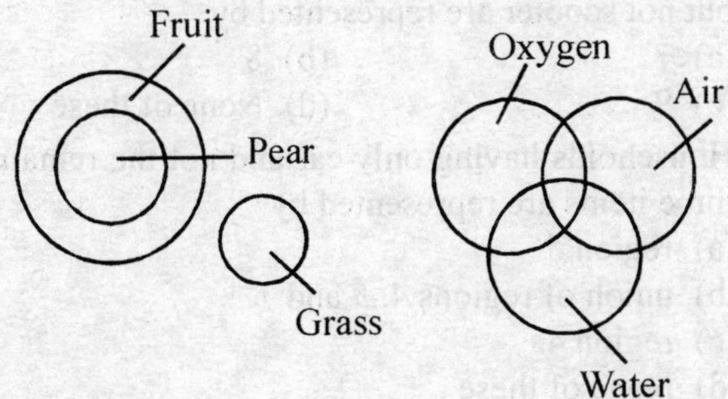

20. Ans. (c)
Solution:

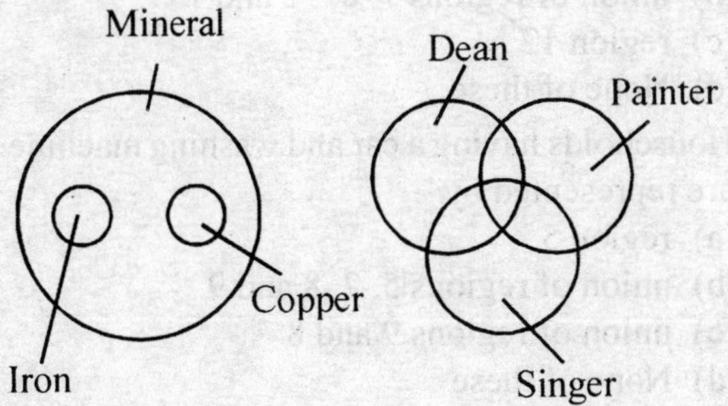

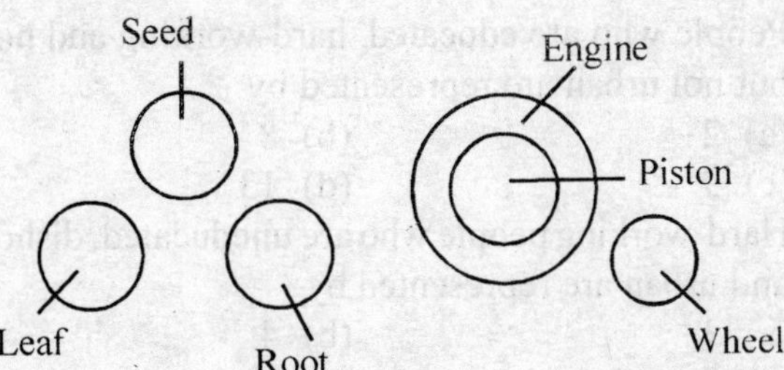

Type 2

In this type of questions, generally a Venn diagram is given. Each geometrical figure in the diagram represents a certain class. The candidate is required to study and analyse the figure carefully and then answer certain questions regarding the given data.

Example 1. In the following diagram, the square represents girls, the circle tall persons, the triangle is for tennis players and the rectangle stands for the swimmers.

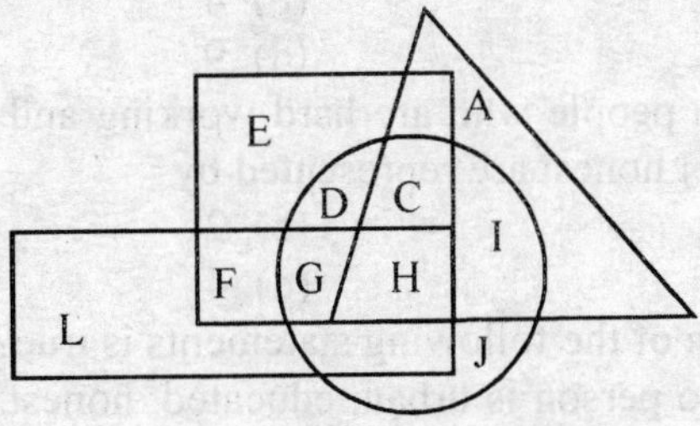

On the basis of the above diagram, answer the following questions.

1. Which letter represents tall girls who are swimmers but don't play tennis?

(a) C (b) D
(c) G (d) H

2. Which letter represents girls who are swimmers, play tennis but are not tall?

(a) B (b) E
(c) F (d) None of these

3. Which letter represents tall girls who do not play tennis and are not swimmers?

(a) C (b) D
(c) E (d) G

4. Which letter represents tall persons who are gents and swimmers but do not play tennis?

(a) I (b) J
(c) K (d) L

Directions: (Questions 5 to 12): In the given diagram, the circle stands for educated, square for hard-working, triangle for urban and the rectangle for honest people. Different regions in the diagram are numbered from 1 to 12. Study the diagram carefully to choose the correct answer in the following questions.

5. People who are educated, hard-working and honest but not urban are represented by
(a) 2 (b) 8
(c) 9 (d) 13

6. Hard-working people who are uneducated, dishoenst and urban are represented by
(a) 3 (b) 4
(c) 8 (d) 9

7. Non-urban, educated people who are neither honest nor hard-working are represented by'
(a) 12 (b) 11
(c) 10 (d) 8

8. Hard-working , non-urban people who are neither educated nor honest are indicated by
(a) 9 (b) 10
(c) 11 (d) 12

9. Urban educated people who are neiter honest nor hard-working are indicated by
(a) 2 (b) 4
(c) 6 (d) 9

10. Urban people who are hard-working and educated but not honest are represented by
(a) 1 (b) 2
(c) 3 (d) 4

11. Which of the following statements is true?
(a) No person is urban, educated, honest and hard working.
(b) Some honest people are also hard-working and educated.
(c) All educated people are urban.
(d) Uneducated people are either honest or hard working.

12. Region 4 is best described as consisting of
(a) people who are non-urban, honest uneducated and hard-working
(b) people who are uneducated, urban, honest and hard-working
(c) people who are uneducated, urban, hard-working and dishonest
(d) people who are urban, hard-working, honest and educated

Directions: (Questions 13 to 22): To answer these questions, study the following diagram carefully:

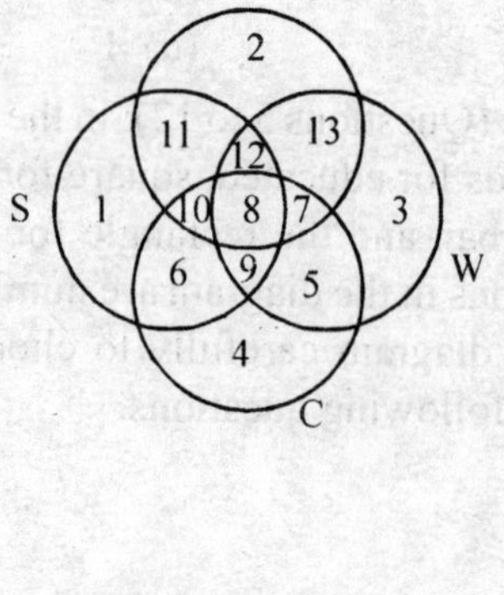

Circle S stands for households having a scooter;
Circle T stands for households having a TV set;
Circle W stands for households having a washing machine;
Circle C stands for households having a car.

Please Note: The different numbers indicate the non-overlapping regions e.g. S represents union of regions 1, 11, 12, 8, 9, 10 and 6.

13. Households having all the four items are represented by the region
(a) 7 (b) 8
(c) 9 (d) 12

14. Households having only scooter are represented by
(a) Region S
(b) Union of regions 1, 6 and 9
(c) Region 1
(d) None of these

15. Households having car but not washing machine are indicated by
(a) union if regions 6, 10 and 4
(b) region 3
(c) region 13
(d) None of these

16. Households having scooter, TV set and washing machine and not car are represented by
(a) 4 (b) 7
(c) 11 (d) 12

17. Households having a car, washing machine, TV set but not scooter are represented by
(a) 7 (b) 8
(c) 9 (d) None of these

18. Households having only car and not the remaining three items are represented by
(a) region 5
(b) union of regions 4, 5 and 7
(c) region 4
(d) None of these

19. Households having only a TV set and washing machine are given by the
(a) region 13
(b) union of regions 7, 8, 12 and 13
(c) region 12
(d) None of these

20. Households having a car and washing machine only are represented by
(a) region 5
(b) union of regions 5, 7, 8 and 9
(c) union of regions 9 and 8
(d) None of these

21. Households having TV set but neither scooter and washing machine are given by the region
(a) 2 (b) 11
(c) 12 (d) 13

22. Households having neither scooter nor car are represented by
(a) region 2
(b) union of regions 2, 13 and 3
(c) region 3
(d) None of these

Answers
Type 2

1. **Ans. (c)**
Solution: Tall girls, who are swimmers are represented by the region common to the square, circle and the rectangle i.e., G and H. But according to the given conditions, the girls shouldn't be tennis players. So, the required region should not be a part of the triangle i.e., H should be excluded. Thus, the region representing the persons satisfying the given conditions is G. Hence, the answer is (c).

2. **Ans. (d)**
Solution: Girls who are swimmers and play tennis are represented by the region common to the square, triangle and rectangle i.e., H. But, it is given that the girls shouldn't be tall. So, the required region should not be a part of the circle. Since H is part of the circle, so the answer is (d).

3. **Ans. (b)**
Solution: Tall girls are represented by the region common to the square and the circle i.e., D, C, G and H. But, according to the given conditions, the girls are neither tennis players nor swimmers. So, the required region should be neither a part of the retangle nor the triangle. G lies inside the rectangle, C inside the triangle and H is common to both. So, the answer is (b).

4. **Ans. (c)**
Solution: The Tall persons are represented by regions inside the circle i.e., C, D, G, H, I, J and K. Since the persons are not girls and do not play tennis, so the region should not be a part of either the square of the triangle. Thus, C, D, G, H should be executed. Also, according to the given conditions, the persons should be swimmers. So, the required region should be a part of the rectangle and such a region is K. Hence, the answer is (c).

5. **Ans. (b)**
Solution: The required set of people is represented by the region which lies outside the triangle and is common to the circle, square and rectangle i.e. 9.

6. **Ans. (a)**
Solution: The required set of people is represented by the region which is common to the triangle and the square but lies outside the circle and rectangle i.e. 3.

7. **Ans. (d)**
Solution: The required set of people is represented by the region which lies outside the triangle, inside the circle but outside the rectangle and the square i.e. 7.

8. **Ans. (d)**
Solution: The required set of people is denoted by the region which lies inside the square but outside the triangle, circle and rectangle i.e. 12.

9. **Ans. (c)**
Solution: The required set of people is denoted by the region which is common to the triangle and the circle, but is not a part of either the rectangle or the square, i.e. 6.

10. **Ans. (b)**
Solution: The required set of people is represented by the region which is common to the triangle, square and circle but is not a part of the rectangle i.e. 2.

11. **Ans. (b)**
Solution: For (a) to be true, there should be no region common to all the four figures. But such a region exists, which is 1. So (a) is false.

For (b) to be true, there should be region common to rectangle, circle and square, which is 9. So, (b) is true.

For (c) to be true, the circle should lie inside the triangle. But this is not so. So, (c) is false.

It follows from the figure that uneducated people are honest, hard-working or urban or may possess a combination of these qualities. So, (d) is false.

12. **Ans. (b)**
Solution: 4 lies outside the circle and is common to the triangle, square and rectangle. So, 4 represents people who are uneducated, urban, honest and hard-working.

13. **Ans. ()**
Solution:

14. **Ans. (c)**
Solution: The required region is the one which lies inside circle S but outside circles T, W and C i.e. 1.

15. **Ans. (a)**
Solution: The required region is the one which lies inside circle C but outside circle W i.e. 6, 10 and 4.

16. Ans. (d)

Solution: The required region is the one which is common to circles S, T and W but lies outside circle C i.e. 12.

17. Ans. (a)

Solution: The required region is the one which is common to circles C, W and T but lies outside circle S i.e. 7.

18. Ans. (c)

Solution: The required region is the one which lies inside circle C but outside circles S, T and W i.e. 4.

19. Ans. (a)

Solution: The required region is the one which is common to circles T and W but lies outside circles S and C i.e. 13.

20. Ans. (a)

Solution: The required region is the one which is common to circles C, and W but lies outside circles S and T i.e. 5.

21. Ans. (a)

Solution: The required region is the one which lies inside circle T but outside circles S and W i.e. 2.

22. Ans. (b)

Solution: The required region is the one which lies outside circles S and C i.e. 2, 13 and 3.

TYPE G: VERBAL DEDUCTION

Set 1

Directions: (Questions 1 to 6): Each question given below has a set of three or four statements. Each set of statements is further divided into three segments. Choose the alternative where the third segment in the statement can be logically deduced using both the preceding two, but not just from one of them.

1. A : All mammals are viviparous. Some fish are vi viparous. Some fish are mammals.

B : All beds are oviparous. Some fish are not oviparous. Some fish are birds.

C : No mammal is oviparous. Some creatyres are oviparous. Some creatures are not mammals.

D : Some creatures are mammals. Some creatures are viviparous. Some mammals are viviparous.

(a) A only (b) B only
(c) C only (d) D only

2. A : All good people are knights. All warriors are good people. All kinghts are warriors.

B : No footballs are ministers. All footballers are tough. Some ministers are players.

C : All pizzas are snacks. Some meals are pizzas. Some meals are snacks.

D : Some barkers are musk deer. All barkers are sloth bears. Some sloth bears are musk deer.

(a) A only (b) B and C
(c) C only (d) D and D

3. A : All earthquakes cause havoc . Some landslides cause havoc. Some earthquakes cause land slides.

B : All glass things are transparent. Some curios are gless thing. Some curios are transparent.

C : All caly objects are brittle. All X are clay objects. All X are brittle.

D : No criminal is a patriot. Rohit is not a patriot. Rohit is a criminal.

(a) A only (b) B only
(c) B and C (d) D only

4. A : Few professors are scientists. Some scientists eat grass. A few professors eat grass.

B : Ajay is a tree. Some trees are made of iron. Ajay is made of iron.

C : No cows eat grass. All bulls eat grass. No bulls are cows.

D : Some squirrels are donkeys. All squirrels are wolves. Some wolves are donkeys.

(a) A and C (b) B only
(c) C only (d) C and D

5. A : Some substances are crystalline. Marble is crystalline Marble a sub stance.

B : All greyhounds are dogs. Some dogs are cows. Some greyhounds are dogs.

C : All locks are keys. Some keys do not open. Some locks do not open.

(a) Aonly (b) B and C
(c) A and C (d) None of these

6. A : Citizens of X Isalnds speak only the truth. Citi zens of X Islands are young people. Young people speak only the truth.

B : Citizens of X Islands speak only the truth. Some X Islands are in Atlantic . Some Citizens of X Islands are in the Atlantic.

C : Citizens of X Islands speak only the truth. Some young people are citizens of X Islands. Some young people speak only the truth.

D : Some people only the truth. Some citizens of X Islands speak only the truth. Some people who speak only the truth are citizens of X Islands.

(a) A only (b) B only
(c) C only (d) D only

Directions: (Questions 7 to 14): Each question given below consists below consists of five or six statements followed by options consisting of three statements put together in a specific order. Choose the option which indicates a valid argument containing logically related statements that is, where the third statements is a conclusion drawn from the preeceding two statements.

7. A : All cushions are pillows.
   B : Some pillows are sheets.
   C : Some sheets are cushions.
   D : Some pillows are not cushions.
   E : All pillows are sheets.
   F : No pillows are sheets.
   (a) ABC (b) BCD
   (c) EAC (d) EDC

8. A : All heroines are pretty.
   B : Some heroines are popular.
   C : Shweta is pretty.
   D : Shweta is a popular. heroine.
   E : Some popular girls are pretty.
   F : No pillows are sheets.
   (a) ABE (b) ACD
   (c) DCA (d) EDC

9. A : Ann ostrich lays eggs.
   B : All birds lay eggs.
   C : Some birds can fly.
   D : An ostrich cannot fly.
   E : An ostrich is a bird.
   F : An ostrich cannot swim.
   (a) ABE (b) BEA
   (c) DEC (d) ECB

10. A : Some women are smart.
    B : All smart women are good-looking.
    C : Monika is a woman.
    D : Monika is smart.
    E : Monika is good - looking
    F : All women are good-looking
    (a) AFC (b) BDE
    (c) DCB (d) FBE

11. A : Some crows are flies
    B : Some flies are mosquitoes.
    C : All mosquitoes are flies.
    D : Some owls are flies.
    E : All owls are mosquitoes.
    F : Some mosquitoes are not owls.
    (a) ABC (b) ACF
    (c) BDF (d) FDA

12. A : MBAs are in great demand.
    B : Shashi and Mansi are in great demand.
    C : Shashi is in great demand.
    D : Mansi is in great demand.
    E : Shashi and Mansi are MBAs.
    (a) ABE (b) ADE
    (c) CEF (d) EDB

13. A : Six is five.
    B : Five is not four.
    C : Some five is ten.
    D : Some six is twelve.
    E : Some twelve is five.
    F : Some ten is four.
    (a) ABC (b) ADE
    (c) AEF (d) EDC

14. A : Some pins are made of tin.
    B : All tin is made of copper.
    C : All sopper is used for pins.
    D : Some tin is copper.
    E : Some pins are used for tin.
    F : Some copper is used for tin.
    (a) ABC (b) ABE
    (c) CDA (d) CEf

Answers
Set 2

1. **Ans. (c)**
   Solution:
2. **Ans. (d)**
   Solution:
3. **Ans. (c)**
   Solution:
4. **Ans. (d)**
   Solution:
5. **Ans. (d)**
   Solution:
6. **Ans (c)**
   Solution:
7. **Ans (c)**
   Solution:
8. **Ans. (a)**
   Solution:
9. **Ans. (b)**
   Solution:
10. **Ans. (b)**
    Solution:
11. **Ans. (d)**
    Solution:
12. **Ans. (b)**
    Solution:
13. **Ans. (b)**
    Solution:
14. **Ans. (c)**
    Solution:

Reader's Notes

General Ability

NUMERICAL ABILITY

TYPE A: NUMERICAL COMPUTATION

GEOMETRY

Mensuration

Mensuration is the science of measurement of lengths of lines, areas of surfaces and volumes of solids. In this chapter we will deal with the area of plain figures.

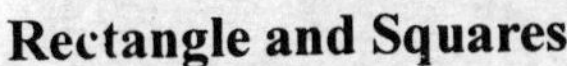

Rectangle and Squares

Perimeter of a rectangle $= 2(length + breadth)$

Perimeter of a rectangle $4 \times length$

Area of a rectangle $= (length \times breadth)$

Area of a square

$= l^2$ (*As in square, length = breadth*)

Note: Area is expressed in square units

Measurements of area in the Metric System

100 square millimeters	= 1 square centimeter
100 square centimeters	= 1 square decimeter
100 square decimeter	= 1 square meter
100 square meter	= 1 square decameter
100 square decameters	= 1 square hectameter
100 squae hectameters	= 1 square kilometer

Diagonal of a Rectangle

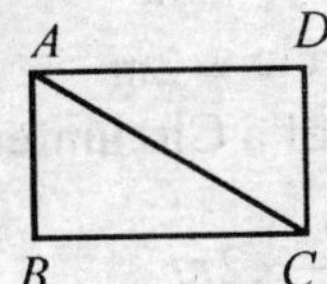

$AC = \sqrt{AB^2 + BC^2}$

Diagonal of a Square

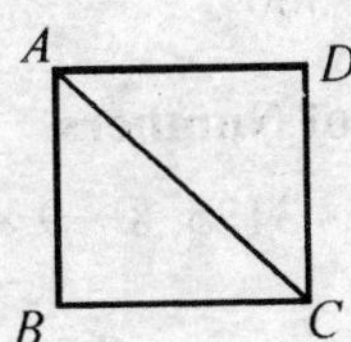

$AC = AB\sqrt{2}$

Triangle

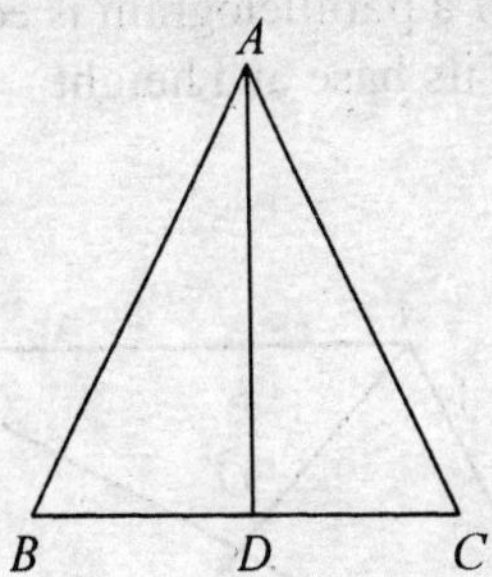

Area of a triangle $= \frac{1}{2}$ Base • Height

$= \frac{1}{2} \times BC \times AD$

Where AD must be a perpendicular to BC

Area of a triangle $= \sqrt{s(s-a)(s-b)(s-c)}$

where a, b, c are the sides of a triangle while

$s = \frac{1}{2}(a + b + (c))$ or sei perimeter

A triangle whose all the three side are equal is called an equilateral triangle.

In an equilateral triangle ABC.

$a = b = c$

$s = \frac{3a}{2}$

$$\text{area} = \sqrt{\frac{3a}{2} \times \frac{1}{2}a \times \frac{1}{2}a \times \frac{1}{2}a} = \frac{\sqrt{3}}{4}a^2$$

$$\text{and Height} = \frac{2\ Area}{Base} = \frac{2 \times \frac{\sqrt{3}}{4}a^2}{a} = \frac{\sqrt{3}}{2}a$$

Parallelogram

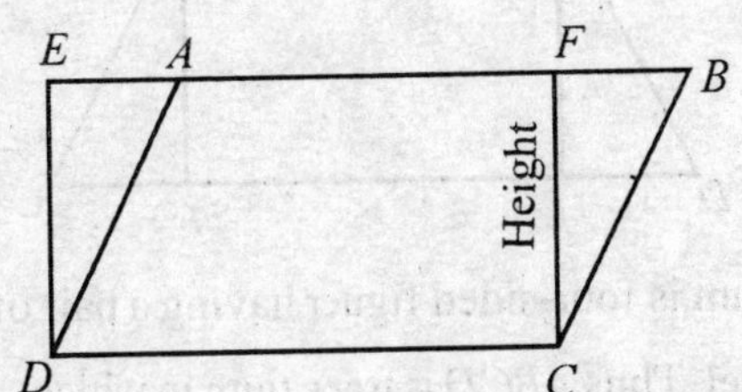

Area of parallelogram = Base Hight

Let $CDEF$ be a rectangle on the same base DC and of the same height FC. Then since parallelograms on the same base and of the same height are equal in area.

Paralellogram ABCD = Reactangle $CDEF$

$= CD \times FC$

= Base × Height

Hence the area of a parallelogram is equal to the product of its base and height

Rhombus

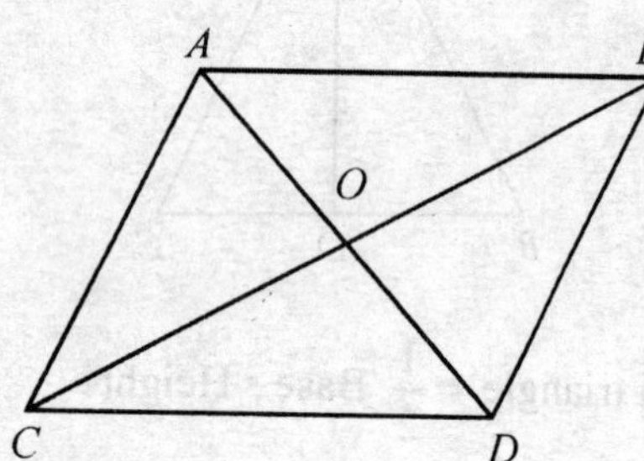

A rhombus is a prarallelogram all of whose side are equal. In a rhombus the diagonals bisect each other at right angles.

$AB = BD = CD = DA$ and $AO = OC$, $BO = OD$

$\angle BOA = 1$ rt. angle

Area of rhombus $ABCD = 2$ area of $\Delta\ BCD$

$= 4$ area of $\Delta\ BOA$

$$= 4 \times \frac{1}{2} BO \times OA$$

$$= 2 \times \frac{BC}{2} \times \frac{AD}{2}$$

$$= \frac{1}{2} BC \times AD$$

Hence the area of a rhombus is equal to half the product of its diagonals.

Trapezium

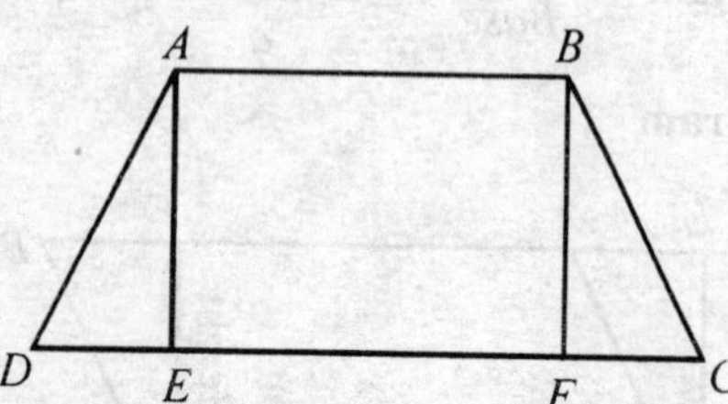

A trapezium is four-sided figuer having a pair of opposite side parallel. Thus $ABCD$ is trapezium in which $AB \parallel DC$

Draw AE and BF perpendiculars from A and B to DC

Area of trapezium $ABCD$

$$= \Delta\ SADE + \text{Rect.}\ ABFE + \Delta\ BFC$$

$$= \frac{1}{2} DE \cdot AE + AE \cdot EF + \frac{1}{2} BF \cdot FC \left(BF = AE\right)$$

$$= \frac{1}{2} AE\left[DE + 2EF + FC\right]\left(EF = A(B)\right)$$

$$= \frac{1}{2} AE\left[DE + EF + FC + AB\right]$$

$$= \frac{1}{2} AE\left[DC + AB\right]$$

$= \frac{1}{2}$ Height sum of parllel sides

Regular Polygon

Regular polygon is a sided enclosed figuer where all the sides are equal.

Area of hexagon $= \dfrac{3\sqrt{3}a^2}{2}$

Area of a OCtagon $= 2a^2\left(\sqrt{2} + 1\right)$

Area of the Regular Polygon of n sides having given the lenght of one side is and the radius of circumscirbed circle.

$$= \frac{n}{2} a \sqrt{R^2 - \left(\frac{a}{2}\right)^2}$$

Circle

Area of Circle $= \pi r^2$

Circumference of Circle $= 2\pi r$

Since, arc is the part of a Circumference, therefore

Length of an arc $= \dfrac{\theta}{360} \times 2\pi r$

Since, segment is the part of Area, therefore

Area of a segment $= \dfrac{\theta}{360} \times \pi r^2$

H.C.F. and L.C.M. of Numbers

1. H.C.F. of 4 × 27 × 3125, 8 × 9 × 25 × 7 and 16 × 81 × 5 × 11 × 49 is

(a) 180 (b) 360

(c) 540 (d) 1260

2. The L.C.M. of 22, 54,108,135 and 198 is:

(a) 330 (b) 1980

(c) 5940 (d) 11880

3. The H.C.F.of 0.54,1.8 and 7.2 is:
(a) 1.8 (b) 0.18
(c) 0.018 (d) 18

4. The H.C.F.of two numbers is 12 and their difference is 12.The numbers are:
(a) 66,78 (b) 70,82
(c) 94,106 (d) 84 96.

5. The L.C.M.and H.C.F.of two numbers are 84 and 21 respectively. If the ratio of the two numbers is $1:4$, then the larger of the two numbers is:
(a) 12 (b) 48
(c) 84 (d) 108.

6. The L.C.M.of three different numbers is 120. Which of the following can not be their H.C.F.?
(a) 8 (b) 12
(c) 24 (d) 35

7. Three different containers contain 496 litres, 403 litres and 713 litres of mixtures of milk and water respectively. What biggest measure can measure all the different quantities exactly?
(a) 1 litre (b) 7 litre
(c) 31 litres (d) 41 litres.

8. Which of the following fractions is the largest ?
(a) $\frac{7}{8}$ (b) $\frac{13}{16}$
(c) $\frac{31}{40}$ (d) $\frac{63}{80}$

9. The least number which when increased by 5 is divisible by each one of 24, 32,36 and 54, is:
(a) 427 (b) 859
(c) 869 (d) 4320

10. Find the least number which when divided by 16,18,20, and 25 leaves 4 as remainder in each case, but when divided by 7 leaves no remainder.
(a) 17004 (b) 18000
(c) 18002 (d) 18004

11. H.C.F. of 2^3, 3^2, 4 and 15 is:
(a) 2^3 (b) 3^2
(c) 1 (d) 360

12. H.C.F. of $\frac{1}{2}, \frac{2}{3}, \frac{3}{4}, \frac{4}{5}$ is:
(a) 1 (b) 12
(c) $\frac{4}{5}$ (d) $\frac{1}{60}$

13. Which of the following is in ascending order
(a) $\frac{5}{7}, \frac{7}{8}, \frac{9}{11}$ (b) $\frac{5}{7}, \frac{9}{11}, \frac{7}{8}$
(c) $\frac{7}{8}, \frac{5}{7}, \frac{9}{11}$ (d) $\frac{9}{11}, \frac{7}{8}, \frac{5}{7}$

14. H.C.F. of three numbers is 12. If they be in the ratio 1: 2: 3: the numbers are:
(a) 12, 24, 36 (b) 10, 20, 30
(c) 5, 10 , 15 (d) 4, 8, 12

15. The sum of two numbers is 216 and their H.C.F. is 27. The numbers are:
(a) 54, 162 (b) 108, 108
(c) 27, 189 (d) None of these

16. The least perfect square number which is divisible by 3, 4, 5, 6 and 8, is :
(a) 900 (b) 1200
(c) 2500 (d) 3600

17. The largest natural number which exactly divides the product of any four consecutive natural numbers, is:
(a) 6 (b) 12
(c) 24 (d) 120

18. The least number which when divided by 16, 18 and 21 leaves the remainders 3, 5, and 8 respectively, is:
(a) 893 (b) 992
(c) 995 (d) 1024

19. The greatest number by which if 1657 and 2037 are divided the remainders will be 6 and 5 respectively, is:
(a) 127 (b) 235
(c) 260 (d) 305

20. The total number of prime factors of the product $(8)^{20}$, $(15)^{24} \cdot (7)^{15}$ is:
(a) 59 (b) 98
(c) 123 (d) 138

Answers
H.C.F. and L.C.M. of Numbers

1. **Ans. (a)**
Solution:
Write the number into smaller number multiples.
$4 \times 27 \times 3125 = 2^2 \times 3^3 \times 5^5$
$8 \times 9 \times 25 \times 7 = 2^3 \times 3^2 \times 5^2 \times 7$
$16 \times 81 \times 5 \times 11 \times 49 = 2^4 \times 3^4 \times 5 \times 11 \times 7^2$
$\therefore$ H.C.F $= 2^2 \times 3^2 \times 5 = 180$

2. Ans. (a)
Solution:

2	22 , 54 , 108 , 135 , 198
3	11 , 27 , 54 , 135 , 99
3	11 , 9 , 18 , 45 , 33
3	11 , 3 , 6 , 15 , 11
11	11 , 1 , 2 , 5 , 11
	1 , 1 , 2 , 5 , 1

L.C.M $= 2\times3\times3\times3\times11\times2\times5 = 5940$

3. Ans. (b)
Solution:
Given numbers are 0.54, 1.80 and 7.20. H.C.F. of 54. 180 and 720 is 18.
$\Rightarrow$ H.C.F. of given numbers = 0.18

4. Ans. (d)
Solution:
From the given numbers, the two with H.C.F 12 and difference 12 are 84 and 96.

5. Ans. (c)
Solution:
Let the numbers be x and $4x$.
Then, $x\times 4x = 84\times 21$

$\Rightarrow\ x^2 = \left(\dfrac{84\times 21}{4}\right)$

$\Rightarrow\ x = 21$

Therefore larger number $= 4x = 84$

6. Ans. (d)
Solution:
Since H.C.F. is always a factor of L.C.M. There on con not have three numbers with H.C.F. 35 and L.C.M. 120.

7. Ans. (c)
Solution:
Required measurment
= (H.C.F. of 496, 403,713) liters
= 31 litres.

8. Ans. (a)
Solution:
Convert these number with equal denominator
L.C.M. of 8, 16, 40 and 80 = 80

$\dfrac{7}{8}=\dfrac{70}{80}$; $\dfrac{13}{16}=\dfrac{65}{80}$; $\dfrac{31}{40}=\dfrac{62}{80}$

Since, $\dfrac{70}{80}>\dfrac{63}{80}>\dfrac{65}{80}>\dfrac{62}{80}$,

So, $\dfrac{7}{8}>\dfrac{63}{80}>\dfrac{13}{16}>\dfrac{31}{40}$

So, $\dfrac{7}{8}$ is the largest

9. Ans. (b)
Solution:
Least number = (L.C.M. of 24, 32, 36, 54) - 5
= 864 − 5 = 859

10. Ans. (d)
Solution:
L.C.M of 16, 18, 20, 25 = 3600.
Required number is of the form of 3600 k + 4.
Least value of k for which $(3600k+4)$ is divisible by 7 is $k = 5$
$\Rightarrow$ Required number $= (3600\times5+4) = 18004$

11. Ans. (c)
Solution: 1 is the highest common factor of 2^3, 3^2, 4 and 15.

12. Ans. (d)
Solution:

$$H.C.F. = \frac{H.C.F.\text{ of }1,2,3,4}{L.C.F.\text{ of }2,3,4,5} = \frac{1}{60}$$

13. Ans. (b)
Solution:

$\dfrac{5}{7}=0.714$, $\dfrac{7}{8}=0.875$ and $\dfrac{9}{11}=0.818$.

Now, $0.714<0.818<0.875$ i.e. $\dfrac{5}{7}<\dfrac{9}{11}<\dfrac{7}{8}$.

14. Ans. (a)
Solution:
Let the numbers be x, $2x$ and $3x$.
Then, their H.C.F $= x$.
So, $x = 12$
$\therefore$ The numbers are 12, 24, 36.

15. Ans. (c)
Solution:
Let the number be 27a and 27b

Then, $27a+27b=216$ or $27(a+b)=216$

or $a+b=\dfrac{216}{27}=8$

$\therefore$ Values of Co-primes (with sum 8) are (1, 7) and (3, 5)

So, the numbers are $(27\times1, 27\times7)$ i.e. (27×189)

16. Ans. (d)
Solution:

3	3 − 4 − 5 − 6 − 8
2	1 − 4 − 5 − 2 − 8
2	1 − 2 − 5 − 1 − 4
	1 − 1 − 5 − 1 − 2

L.C.M of 3, 4, 5, 6, 8 $= 3 \times 2 \times 2 \times 5 \times 2$.

Required number $= (3 \times 3 \times 2 \times 2 \times 5 \times 5 \times 2 \times 2)$

$= 3600$.

17. Ans. (c)

Solution: $1 \times 2 \times 3 \times 4 = 24$

$\therefore$ Required number = 24

18. Ans. (c)

Solution: Here $(16 - 3) = 13$ $(18 - 5) = 13$

and $(21 - 8) = 13$.

So, required number

= (L.C.M. of 16, 18, 21) −13

$= (1008 - 13) = 995$

19. Ans. (a)

Solution: Required number

= H.C.F of $(1657 - 6)$ and $(2037 - 5)$

= H.C.F. of 1651 and 2032 = 127

20. Ans. (c)

Solution:

Since 2, 3, 5, 17 are prime numbers and the given expression is $(2^3)^{20} \times (3 \times 5)^{24} \times (17)^{15}$ i.e. $2^{60} \times 3^{24} \times 5^{24} \times 17^{15}$, so the number of prime factors is the given expression is (60 + 24 + 24 +15) = 123.

Linear Equations

Main Points

1. A pair of linear equations in two variables x and y can be represented algebraically as follows:

$a_1x + b_1y + c_1 = 0$

$a_2x + b_2y + c_2 = 0$

where $a_1, a_2, b_1, b_2, c_1, c_2$ are real numbers such that

$a_1^2 + b_1^2 \neq 0, a_2^2 + b_2^2 \neq 0$.

2. Graphically or geometrically a paif of linear equations

$a_1x + b_1y + c_1 = 0$

$a_2x = b_2y + c_2 = 0$

in two variables represents a pair of straight lines which are

1. intersecting,if $\frac{a_1}{a_2} \neq \frac{b_1}{b_2}$

2. parallel, if $\frac{a_1}{a_2} = \frac{b_1}{b_2} \neq \frac{c_1}{c_2}$

3. coincident,if $\frac{a_1}{a_2} = \frac{b_1}{b_2} = \frac{c_1}{c_2}$

3. A pair of linear equations in two variables can be solved by the:
   1. Graphical method
   2. Algebraic method.
4. To solve a pair of linear equations in two variables by graphical method, we first draw the lines represented by them.
   1. If the pair of lines intersect at a point, then we say that the pair is consistent and the coordinates of the point provide us the unuque solution.
   2. If the pair of lines are parellel then the pair has no solution and is called consistent pair of equations.
   3. If the pair of lines are coincident then it has infinitely many solutions each point on the line being of solution. In this case, we say that the pair of linear equations is consistent with infinitely many solutions.
5. To solve a pair of linear equation in two variables algebraically,we have followings methods:
   1. Substitution method
   2. Elimination method
   3. Cross-multiplication method.
6. If $a_1x + b_1y + c_1 = 0$

$a_2x + b_2y + c_2 = 0$

is a pair of linear equations in two variabes x and y such th at

1. $\frac{a_1}{a_2} \neq \frac{b_1}{b_2}$, then the pair of linear equations is consistent with a unique solution.

2. $\frac{a_1}{a_2} = \frac{b_1}{b_2} \neq \frac{c_1}{c_2}$, then the pair of linear equations is inconsistent.

3. $\frac{a_1}{a_2} = \frac{b_1}{b_2} = \frac{c_1}{c_2}$, then the pair of linear equations is consistent with infinitely many solutions.

Linear Equations

Mark the correct alternative in each of the following :

1. The value of k for which the system of equations

   $kx - y = 2$

   $6x - 2y = 3$

   has a unique solution, is

   (a) $= 3$ (b) $\neq 3$

   (c) $\neq 0$ (d) $= 0$

2. The value of k for which the system of equations

   $2x + 3y = 5$

   $4x + ky = 10$

   has infinite number of solution,is

   (a) 1 (b) 3

   (c) 6 (d) 0

3. The value of k for which the system of equations $x + 2y - 3 = 0$ and $5x + ky + 7 = 0$ has no solution, is

   (a) 10 (b) 6

   (c) 3 (d) 1

4. The value of k for which the system of equations

   $3x + 5y = 0$ and

   $kx + 10y = 0$

   has a non-zero solution, is

   (a) 0 (b) 2

   (c) 6 (d) 8

5. If the system of equations

   $2x + 3y = 7$

   $(a + b)x + (2a - b)y = 21$

   has infinitely many solutions, then

   (a) $a = 1, b = 5$ (b) $a = 5, b = 1$

   (c) $a = -1, b = 5$ (d) $a = 5, b = -1$

6. If the system of equations

   $3x + 2y = 1$

   $(2k - 1)\ x + (k - y)\ y = 2k + 1$

   is consistent ,then k=

   (a) 1 (b) 0

   (c) –1 (d) 2

7. If am $\neq bl$, then the system of equations

   $ax + by = c$

   $lx + my = n$

   (a) has a unique solution

   (b) has no solution

   (c) has infinitely many solutions

   (d) may or may not have a solution.

8. If the system of equations

   $2x + 3y = 7$

   $2ax + (a + b)\ y = 28$

   has infinitely many solutions,then

   (a) $a = 2b$ (b) $b = 2a$

   (c) $a + 2b = 0$ (d) $2a + b = 0$

9. The value of k for which the system of equations

   $x + 2y = 5$

   $3x + ky + 15 = 0$

   has no solution is

   (a) 6 (b) –6

   (c) $\frac{3}{2}$ (d) None of these

10. If $2x - 3y = 7$ and $(a+b)x - (a+b-3)y = 4a + b$ coincident lines, then a and b satisfy the equation

    (a) $a + 5b = 0$ (b) $5a + b = 0$

    (c) $a - 5b = 0$ (d) $5a - b = 0$

Answers

Linear Equations

1. **Ans. (b)**
2. **Ans. (c)**
3. **Ans. (a)**
4. **Ans. (c)**
5. **Ans. (b)**
6. **Ans. (d)**
7. **Ans. (a)**
8. **Ans. (b)**
9. **Ans. (a)**
10. **Ans. (c)**

Areas Related Circles

Main Point

1. For a circle of a radius r, we have

   (a) Circumference $= 2\pi r$

   (b) Area $= \pi r^2$

   (c) Area of semi- circle $= \frac{\pi r^2}{2}$

   (d) Area of a quadrant $= \frac{\pi r^2}{4}$

2. If R and r are the radii of two concentric circles such taht R >r then. Area enclosed by the two circles

   $= \pi R^2 - \pi r^2$

   $= \pi(R^2 - r^2)$

3. If a sector of circle of radius r contains an angle of θ°.Then,

   (a) Length of the are of the sector

   $= \frac{\theta}{360} \times 2\pi r$

$=\frac{\theta}{360}\times(\text{Circumference of the circle})$

(b) Perimet^r of the sector $=2r+\frac{\theta}{360}\times 2\pi r$

(c) Area of the sector

$=\frac{\theta}{360}\times \pi r^2$

$=\frac{\theta}{360}\times(\text{Area of the circle})$

(d) Area of the segment
= Area of the correponding sector – Area of the corresponding triangle

$=\frac{\theta}{360}\times \pi r^2 - r^2\sin\frac{\theta}{2}\cos\frac{\theta}{2}$

$=\left(\frac{\pi\theta}{360}-\sin\frac{\theta}{2}\cos\frac{\theta}{2}\right)r^2$

Areas Related Circles

1. If the circumference and the area of a circle are numerically equal, then diameter of the circle is
(a) $\frac{\pi}{2}$ (b) 2π
(c) 2 (d) 4

2. The difference between the circumference and radius of a circle is 37 cm. The area of the circle is
(a) $111\,cm^2$ (b) $148\,cm^2$
(c) $154\,cm^2$ (d) $259\,cm^2$

3. A wire can be bent in the from of a circle of radius 56 cm. If it is bent in the form of a square, then its area will be
(a) $3520\,cm^2$ (b) $6400\,cm^2$
(c) $7744\,cm^2$ (d) $8800\,cm^2$

4. If a wire is bent into the shape of a square, then the area of the square is $81\,cm^2$. When wire is bent into a semi-circular shape, then the area of the semi-circle will be
(a) $22\,cm^2$ (b) $44\,cm^2$
(c) $77\,cm^2$ (d) $154\,cm^2$

5. A circular park has a path of uniform width around it. The difference between the outer and inner circumferences of the circular path is 132 m. Its width is:
(a) 20 m (b) 21 m
(c) 22 m (d) 24 m

6. The radius of a wheel is 0.25 m. The number of revolutions it will make to travel a distance of 11 km will be
(a) 2800 (b) 4000
(c) 5500 (d) 7000

7. The ratio of the outer and inner perimeters of a circular path is 23 : 22. If the path is 5 metres wide, the diameter of the inner circle is
(a) 55 m (b) 110 m
(c) 220 m (d) 230 m

8. The circumference of a circle is 100 cm. The side of a square inscribed in the circle is
(a) $50\sqrt{2}\,cm$ (b) $\frac{100}{\pi}\,cm$
(c) $\frac{50\sqrt{2}}{\pi}\,cm$ (d) $\frac{100\sqrt{2}}{\pi}\,cm$

9. The area of the incircle of an equilateral triangle of side 42 cm is
(a) $22\sqrt{3}\,cm^2$ (b) $231\,cm^2$
(c) $462\,cm^2$ (d) $924\,cm^2$

10. The area of incircle of an equilateral triangle is $154\,cm^2$. The perimeter of the triangle is
(a) 71.5 cm (b) 71. 7 cm
(c) 72.3 cm (d) 72 .7 cm

11. The area of the largest triangle that can be inscribed in a semi-circle of radius r, is:
(a) r^2 (b) $2r^2$
(c) r^3 (d) $2r^3$

12. The perimeter of a triangle is 30 cm and the circumference of its incircle is 88 cm .The area of the triangle is
(a) $70\,cm^2$ (b) $140\,cm^2$
(c) $210\,cm^2$ (d) $420\,cm^2$

13. The area of a circle is $220\,cm^2$ .The area of a square inscribed in it is:
(a) $49\,cm^2$ (b) $70\,cm^2$
(c) $140\,cm^2$ (d) $150\,cm^2$

14. If the circumference of a circle increases from 4π to 8π, then its area is
(a) Halved (b) Doubled
(c) Tripled (d) Quadrupled

15. If the radius of a circle is diminished by 10 % , then its area is diminished by

(a) 10% (b) 19%
(c) 20% (d) 36%

Answers
Areas Related Circles

1. Ans. (d)
2. Ans. (c)
3. Ans. (c)
4. Ans. (c)
5. Ans. (b)
6. Ans. (d)
7. Ans. (c)
8. Ans. (c)
9. Ans. (c)
10. Ans. (d)
11. Ans. (a)
12. Ans. (c)
13. Ans. (c)
14. Ans. (d)
15. Ans. (b)

Surface Areas and Volumes

Main Point

Surface area and volumes

1. The diameter of a sphere is 6 cm. It is melted and drawn into a wire of diameter 2 mm. The length of the wire is
(a) 12 m (b) 18 m
(c) 36 m (d) 66 m
2. A metallic sphere of radius 10 .5 cm is melted and then recast into small cones, each of radius 3.5 cm and height 3 cm. The number of such cones is
(a) 63 (b) 126
(c) 21 (d) 130
3. A solid is hemispherical at the bottom and conical above. If the surface areas of the two parts are equal, then the ratio of its radius and the height of its conical part is
(a) 1 : 3 (b) $1:\sqrt{3}$
(c) 1 : 1 (d) $\sqrt{3}:1$
4. A solid sphere of radius r is melted and cast into the shape of a solid cone of height r, the radius of the base of the cone is
(a) 2r (b) 3r
(c) r (d) 4r
5. The material of a cone is converted into the shape of a cylinder of equal radius .If height of the cylinder is 5 cm, then height of the cone is
(a) 10 cm (b) 15 cm
(c) 18 cm (d) 24 cm
6. A circus tent is cylindrical to a height of 4 m and conical above. If its diameter is 105 m and its slant height is 40 m, the total area of the canvas required in m^2 is
(a) 1760 (b) 2640
(c) 3960 (d) 7920
7. The number of solid spheres, each of diameter 6 cm that could be moulded to form a solid metal cylinder of height 45 cm and diameter 4 cm, is
(a) 3 (b) 4
(c) 5 (d) 6
8. A sphere of radius 6 cm is dropped into a cylindrical vessel partly filled with water. The radius of the vessel is 8 cm. If the sphere is submerged completely, then the surface of the water rises by
(a) 4.5 cm (b) 3 cm
(c) 4 cm (d) 2 cm
9. If the radii of the circular ends of a bucket of height 40 cm are of lengths 35 cm and 14 cm, then the volume of the bucket in cubic centimeters, is:
(a) 60060 (b) 80080
(c) 70040 (d) 80160
10. If a cone is cut into two parts by a horizontal plane passing through the mid. point of its axis, the ratio of volumes of the upper part and the cone is
(a) 1 : 2 (b) 1 : 4
(c) 1 : 6 (d) 1 : 8
11. The height of a cone is 30 cm . A small cone is cut off at the top by a plane parallel to the base. If its volume be $\frac{1}{27}$ of the volume of the given cone, then the height above the base at which the section has been made, is
(a) 10 cm (b) 15 cm
(c) 20 cm (d) 25 cm
12. A solid consists of a circular cylinder with an exact fitting right circular cone placed at the top. The height of the cone is h. If the total volume of the solid is 3 times the volume of the cone, then the height of the circular cylinder is
(a) $2h$ (b) $\frac{2h}{3}$
(c) $\frac{3h}{2}$ (d) 4 h
13. A reservoir is in the shape of a frustum of a right circular cone. It is 8 cm across at the top and 4 m across at the bottom. If it is 6 m deep, then its capacity is

(a) $176\,m^3$ (b) $196\,m^3$

(c) $200\,m^3$ (d) $110\,m^3$

14. Water flows at the rate of 10 metre per minute from a cylindrical pipe 5 mm in diameter. How long will it take to full up a conical vessel whose diameter at the base is 40 cm and depth 24 cm?
(a) 48 minutes 15 sec (b) 51 minutes 12 sec
(c) 52 minutes 1 sec (d) 55 minutes

15. A cylindrical vessel 32 cm high and 18 cm as the radius of the base, is filled with sand. This bucket is emptied on the ground and a conical heap of sand is formed. If the height of the conical heap is 24 cm, the radius of its base is
(a) 12 cm (b) 24 cm
(c) 36 cm (d) 48 cm

16. The curved surface area of a right circular cone of height 15 cm and base diameter 16 cm is
(a) $60\pi\, cm^2$ (b) $68\pi\, cm^2$
(c) $120\pi\, cm^2$ (d) $136\pi\, cm^2$

17. A right triangle with sides 3 cm, 4 cm and 5 cm is rotated about the side of 3 cm to form a cone. The volume of the cone so formed is
(a) $12\pi\, cm^3$ (b) $15\pi\, cm^3$
(c) $16\pi\, cm^3$ (d) $20\pi\, cm^3$

18. The curved surface area of a cylinder is 264 m^2 and its volume is $924\,m^3$ .The ratio of its diameter to its heights is
(a) 3 : 7 (b) 7 : 3
(c) 6 : 7 (d) 7 : 6

19. A cylinder with base radius of 8 cm and hight of 2 cm is melfted to form a cone of height 6 cm. The radius of the cone is
(a) 4 cm (b) 5 cm
(c) 6 cm (d) 8 cm

20. The volumes of two spheres are in the ratio 64 : 27.The ratio of their surface areas is
(a) 1 : 2 (b) 2 : 3
(c) 9 : 16 (d) 16 : 9

21. If three metallic spheres of radii 6 cm , 8 cm and 10 cm are melted to form a single sphere, the diameter of the sphere is
(a) 12 cm (b) 24 cm
(c) 30 cm (d) 36 cm

22. The surface area of a sphere is same as sthe curved s;urface area of a right circular cylinder whose height and diameter are 12 cm each. The radius of the sphere is
(a) 3 cm (b) 4 cm
(c) 6 cm (d) 12 cm

23. The volume of the greatest sphere that can be cut off from a cylindrical log of wood of base radius 1 cm and height 5 cm is
(a) $\frac{4}{3}\pi$ (b) $\frac{10}{3}\pi$
(c) 5π (d) $\frac{20}{3}\pi$

24. A cylindrical vessel of radius 4 cm contains water . A solid sphere of radius 3 cm is lowered into the water until it is completely immersed. The water level in the vessel will rise by
(a) $\frac{2}{9}\,cm$ (b) $\frac{4}{9}\,cm$
(c) $\frac{9}{4}\,cm$ (d) $\frac{9}{2}\,cm$

25. 12 spheres of the same size are made from melting a solid cylinder of 16 cm diameter and 2 cm height. The dimameter of each sphere is
(a) $\sqrt{3}\,cm$ (b) 2 cm
(c) 3 cm (d) 4 cm

26. A solid metallic spherical ball of diameter 6 cm is melted and recast into a cone with diameter of the base as 12 cm .The height of the cone is
(a) 2 cm (b) 3 cm
(c) 4 cm (d) 6 cm

27. A hollow sphere of internal and external diameters 4 cm and 8 cm respectively is melted into a cone of base diameter 8 cm. The height of the cone is
(a) 12 cm (b) 14 cm
(c) 15 cm (d) 18 cm

28. A solid piece of iron of dimensions $49 \times 33 \times 24\, cm$ .is moulded into a sphere. The radius of the sphere is
(a) 21 cm (b) 28 cm
(c) 35 cm (d) None of these

Answers

Surface Areas and Volumes

1. **Ans. (c)**
2. **Ans. (b)**
3. **Ans. (b)**
4. **Ans. (a)**
5. **Ans. (b)**
6. **Ans. (d)**

7. **Ans. (c)**
8. **Ans. (a)**
9. **Ans. (b)**
10. **Ans. (d)**
11. **Ans. (c)**
12. **Ans. (b)**
13. **Ans. (a)**
14. **Ans. (b)**
15. **Ans. (c)**
16. **Ans. (d)**
17. **Ans. (a)**
18. **Ans. (b)**
19. **Ans. (d)**
20. **Ans. (d)**
21. **Ans. (b)**
22. **Ans. (c)**
23. **Ans. (a)**
24. **Ans. (c)**
25. **Ans. (d)**
26. **Ans. (b)**
27. **Ans. (b)**
28. **Ans. (a)**

LOGARITHMS

Important facts and Formulae

1. **Logarithm:** If a is positive real number,other than 1 and $a^m = x$, then we write: $m = \log_a x$ and we say that the value of log x to the base a is m.

   Example:

   (i) $10^3 = 1000 \Rightarrow \log_{10} 1000 = 3$

   (ii) $3^4 = 81 \Rightarrow \log_3 81 = 4$

   (iii) $2^{-3} = \frac{1}{8} \Rightarrow \log_2 \frac{1}{8} = -3$

   (iv) $(.1)^2 = .01 \Rightarrow \log_{(.1)} .01 = 2.$

2. **Properties of Logarithms:**

   (i) $\log_a (xy) = \log_a x + \log_a y$

   (ii) $\log_a \left(\frac{x}{y}\right) = \log_a x - \log_a y$

   (iii) $\log_x X = 1$

   (iv) $\log_a 1 = 0$

   (v) $\log_a (x^p) = p(\log_a x)$

   (vi) $\log_a x = \frac{1}{\log_x a}$

   (vii) $\log_a x = \frac{\log_b x}{\log_b a} = \frac{\log x}{\log a}$.

   Remeber: When base is not mentioned, it is taken as 10.

3. ***Common Logrithms:*** Logrithms to the base 10 are known as common logarithms.

4. The logarithm of a number contains two parts, namely characteristic and mantissa.

   Characteristic: The integral part of the logarithm of a number is called its Characteristic.

   Case I: When the number is greater than 1.

   In this case, the characteristic is one less than the number of digits in the left of the decimal point in the given number.

   Case II: When the number is less than 1.

   In this case, the characteristic is one more than the number of zeros between the decimal point and the first significant digit of the number and it is negative. Instead of –1, –2, etc. we write, $\bar{1}$ (one bar), $\bar{2}$ (two bar), etc.

Examples

1. **Simplify**

$$\left[\frac{1}{\log_{xy}(xyz)} + \frac{1}{\log_{yz}(xyz)} + \frac{1}{\log_{zx}(xyz)}\right]$$

Solution: Given expression

$$= \log_{xyz}(xy) + \log_{xyz}(yz) + \log_{xyz}(zx)$$

$$= \log_{xyz}(xy \times yz \times zx) = \log_{xyz}(xyz)^2$$

$$= 2\log_{xyz}(xyz) = 2 \times 1 = 2.$$

$$\left[\because \log_a x = \frac{1}{\log_x a}\right]$$

2. If $\log_{10} 2 = 0.30103$, find the value of $\log_{10} 50$.

Solution: $\log_{10} 50 = \log_{10}\left[\frac{100}{2}\right]$

$= \log_{10} 100 - \log_{10} 2 = 2 - 0.3103 = 1.69897.$

3. If log 2= 0.30103, find the number of digits in 2^{56}.

Solution: $\log\left(2^{56}\right) = 56 \log 2$

$= (56 \times 0.30103) = 16.85768.$

Its characteristic is 16.Hence, the number of digits in 2^{56} is 17.

Exercise

1. Which of the following statements is not correct?

(a) $\log_{10} 10 = 1$

(b) $\log(2+3) = \log(2 \times 3)$

(c) $\log_{10} 1 = 0$

(d) $\log(1+2+3) = \log 1 + \log 2 + \log 3$

2. If $\log_4 x + \log_2 x = 6$, then X is equal to:

(a) 2 (b) 4

(c) 8 (d) 16

3. The value of $(\log_9 27 + \log_8 32)$ is:

(a) $\frac{7}{2}$ (b) $\frac{19}{6}$

(c) 4 (d) 7

4. $(\log_b a \times \log_c b \times \log_a c)$ is equal to:

(a) 0 (b) 1

(c) abc (d) a +b + c

5. If $a = b^x$, $b = a^y$ and $c = a^z$, then the value of xyz is equal to:

(a) −1 (b) 0

(c) 1 (d) abc

6. If $\log_{10} 2 = 0.3010$, then $\log_2 10$ is equal to:

(a) $\frac{699}{301}$ (b) $\frac{100}{301}$

(c) 0.3010 (d) 0.6990

7. If log (0.57) = $\bar{1}.756$, then the value of log 57 + log $(0.57)^3$ + log $\sqrt{0.57}$ is:

(a) 0.902 (b) $\bar{2}.146$

(c) 1.902 (d) $\bar{1}.146$

Answers

1. Ans. (b)
Solution:

(a) Since $\log_a a = 1$, so $\log_{10} 10 = 1$.

(b) log (2 + 3) = log 5 and log (2×3) = log 6

$\therefore \log(2+3) \neq \log(2\times3).$

(c) Since $\log_a 1 = 0$, so $\log_{10} 1 = 0$

(d) log (1 + 2 + 3) = log 6 = log $(1\times2\times3)$

= log 1 + log 2 + log 3.

2. Ans. (d)
Solution:

$$\log_4 x + \log_2 x = 6 \Leftrightarrow \frac{\log x}{\log 4} + \frac{\log x}{\log 2} = 6$$

$$\Leftrightarrow \frac{\log x}{2\log 2} + \frac{\log x}{\log 2} = 6 \Leftrightarrow 3\log x = 12\log 2$$

$$\Leftrightarrow \log x = 4\log 2$$

$$\Leftrightarrow \log x = \log\left(2^4\right) = \log 16$$

$$\Leftrightarrow x = 16.$$

3. Ans. (b)
Solution:

let $\log_9 27 = x$. Then, $9^x = 27$

$$\Leftrightarrow \left(3^2\right)^x = 3^3 \Leftrightarrow 2x = 3$$

$$\Leftrightarrow x = \frac{3}{2}.$$

let $\log_8 32 = y$. Then, $8^y = 32$

$$\Leftrightarrow \left(2^3\right)^y = 2^5$$

$$\Leftrightarrow 3y = 5 \Leftrightarrow y = \frac{5}{3}.$$

$$\therefore \log_9 27 + \log_8 32 = \left[\frac{3}{2} + \frac{5}{3}\right] = \frac{19}{6}.$$

4. Ans. (b)
Solution:

Given exspression

$$= \left[\frac{\log a}{\log b} \times \frac{\log b}{\log c} \times \frac{\log c}{\log a}\right] = 1.$$

5. Ans. (c)
Solution:

$$a = b^x, b = c^y, c = a^z$$

$$\Rightarrow x = \log_b a, y = \log_c b, z = \log_a c$$

$$\Rightarrow xyz = (\log_b a) \times (\log_c b) \times (\log_a c)$$

$$\Rightarrow xyz = \left[\frac{\log a}{\log b} \times \frac{\log b}{\log c} \times \frac{\log c}{\log a}\right] = 1.$$

6. Ans. (b)
Solution:

$$\log_2 10 = \frac{1}{\log_{102}} = \frac{1}{0.3010} = \frac{10000}{3010} = \frac{1000}{301}$$

7. Ans. (a)
Solution:

$$\log(0.57) = \bar{1}.756 \Rightarrow \log 57 = 1.756$$

$$[\because \text{mantissa will remain the same}]$$

$$\therefore \log 57 + \log(0.57)^3 + \log\sqrt{0.57}$$

$$= \log 57 + 3\log\left(\frac{57}{100}\right) + \log\left[\frac{57}{100}\right]^{1/2}$$

$$= \quad \log 57 + 3\log 57 - 3\log 100$$

$$+\frac{1}{2}\log 57 - \frac{1}{2}\log 100$$

$$= \frac{9}{2}\log 57 - \frac{7}{2}\log 100$$

$$= \frac{9}{2} \times 1.756 - \frac{7}{2} \times 2 = 7.902 - 7 = 0.902.$$

PERMUTATIONS AND COMBINATIONS

Factorial Notation: Let n be a positive integer. Then, factorial n, denoted by $\underline{|n}$ or n! is defined as; $n! = n(n-1)(n-2).....3.2.1$

Examples: (1) 5! = (5 × 4 × 3 × 2 × 1) = 120;
(2) 4! = (4 × 3 × 2 ×1) = 24 etc.
we define, 0! = 1

Permutations: The different arrangmeents of a given number of things by taking some or all at a time, are called permutations.

1. All permutations (or arrangements) made with the letter a, b, c by taking two at a time one (ab, ba, ac, ca, bc, cb).
2. All permutations mde with the letters a,b, c taking all at a time are: (abc, acb, bac, bca, cab, cba). Number of permutations: Number of all permutations of n things, taken r at a time, is given by:

$${}^nP_r = n(n-1)(n-2)......(n-r+1) = \frac{n!}{(n-r)!}$$

3. ${}^6P_2 = (6 \times 5) = 30.$
4. ${}^7P_3 = (7 \times 6 \times 5) = 210$

Cor. number of all permutations of n things, taken all at a time = n!

An important Result: If there are n objects of whcih p_1 are alike of one kind; p_2 are alike of another kind; p_3 are alike of third kind and so on and p_r are alike of rth kind, such that $(p_1 + p_2 + ... + p_r) = n$.
Then, number of permutations of these n objects is:

$$\frac{n!}{(p_1!).(p_2!).....(p_r!)}$$

Combinations: Each of the different groups or selections whcih can be formed by taking some or all of a number of objects, is called a combination.

5. Suppose we want to select two out of three boys A, B, C Then, possible selections are AB, BC and CA. Note that AB and BA represent the same selection.
6. All the combinations formed by a, b,c taking two at a time are ab, bc, ca.
7. The only combination that can be formed of three letters a, b, c ataken all at a time is abc.
8. Various groups of 2 out of four persons A, B, C and D are
   AB, AC, AD, BC, BC, CD
9. Note that ab and ba are two different permutations but they represent the same combiantion.
   Number of combinations: The number of all combinations of n things, taken r at a time is:

$${}^nC_r = \frac{n!}{(r!)(n-r)!} = \frac{n(n-1)(n-2).....to\ r\ factors}{r!}$$

Note that: ${}^nC_n = 1 \, and \, {}^nC_0 = 1$

An important: ${}^nC_n = 1 \, and \, {}^nC_0 = 1$

10. $${}^{11}C_4 = \frac{(11 \times 10 \times 9 \times 8)}{(4 \times 3 \times 2 \times 1)} = 330$$

11. $${}^{16}C_{13} = {}^{16}C_{(16-13)} = {}^{16}C_3 = \frac{16 \times 15 \times 14}{3!}$$

$$= \frac{16 \times 15 \times 14}{3 \times 2 \times 1} = 560$$

Example: How many words can be formed by using all the letters of the word 'DAUGHTER' so that the vowels always come together?

Solution: Given word contains 8 diffeerent letter. When the vowels AUE are always together, we may suppose them to form an entity, treated as on letter. Then, the letters to be arranged are DGHTR (AUE). These 6 letters to be arranged in ${}^6P_6 = 6! = 720$ ways The vowels in the group (AUE) may be arranged in 3! = 6ways.

$\therefore$ Required number of words $= (720 \times 6) = 4320.$

Example: How many words can be formed from the letters of the words 'DIRECTOR' so that the vowels are always together?

Solution: In the given word, we treat the vowels IEO as one letter.

Thus, we have DRCTR (IEO).

This group has 6 letters of which R occurs 2 times and others are different.

Number of ways of arranging these letters

$= \frac{6!}{2!} = 360.$

Now 3 vowels can be arranged among themselves in 3! = 6 ways.

$\therefore$ Required number of ways $= (360 \times 6) = 2160.$

Exercise

1. In how many different ways can the letters of the word 'MACHINE' be arranged so that the vowels may occupy only the odd positions?
   (a) 210 (b) 576
   (c) 144 (d) 1728
2. In how many ways can a group of 5 men and 2 women be made out of a total of 7 men and 3 women?
   (a) 63 (b) 90
   (c) 126 (d) 45
3. In how many ways a committee, consisting of 5 men and 6 women can be formed from 8 men and 10 women?
   (a) 266 (b) 5040
   (c) 11760 (d) 86400
4. A box contains 2 white balls, 3 black balls and 4 red balls. In how many ways can 3 balls be drawn from the box, if at least one black ball is to be included in the draw?
   (a) 32 (b) 48
   (c) 64 (d) 96
5. How many 3-digit numbers can be formed from the digits 2,3,5,6,7 and 9, which are divisible by 5 and none of the digits is repeated?
   (a) 5 (b) 10
   (c) 15 (d) 20
6. Out of 7 consonants and 4 vowels, how many words of 3 consonants and 2 vowels can be formed?
   (a) 210 (b) 1050
   (c) 25200 (d) 21400

Answers

1. **Ans. (b)**

Solution:

There are 7 letters in the given word, out of which there are 3 vowels and 4 consonants.

Let us mark the positions to be filled up as follows:

$[^{1}][^{2}][^{3}][^{4}][^{5}][^{6}][^{7}]$

Now, 3 vowels can be placed at any of the three places, out of the four marked 1, 3, 5, 7.

$\therefore$ Number of ways of arranging the vowels

$= {}^4P_3 = (4 \times 3 \times 2) = 24.$

Also, the 4 consonants at the remaining 4 positions may be arranged in

$= {}^4P_4 = 4! = 24$ *ways.*

$\therefore$ Required number of ways $= (24 \times 24) = 576.$

2. **Ans. (a)**

Solution:

Required number of ways

$= ({}^7C_5 \times {}^3C_2) = ({}^7C_2 \times {}^3C_1)$

$= \left[\frac{7 \times 6}{2 \times 1} \times 3\right] = 63.$

3. **Ans (c)**

Solution:

Required number of ways $= ({}^8C_5 \times {}^{10}C_6)$

$= ({}^8C_3 \times {}^{10}C_4)$

$= \left[\frac{8 \times 7 \times 6}{3 \times 2 \times 1} \times \frac{10 \times 9 \times 8 \times 7}{4 \times 3 \times 2 \times 1}\right] = 11760.$

4. **Ans (c)**

Solution:

We may have (1 black and 2 non-black) or (2 black and 1non-black) or (3 black).

$\therefore$ Required number of ways

$= ({}^3C_1 \times {}^6C_2) + ({}^3C_2 \times {}^6C_1) + ({}^3C_3)$

$= \left[3 \times \frac{6 \times 5}{2 \times 1}\right] + \left[\frac{3 \times 2}{2 \times 1} \times 6\right] + 1$

$= (45 + 18 + 1) = 64.$

5. **Ans. (d)**

Solution:

Since each desired number is divisible by 5, so we must have 5 at the unit place. So, there is 1 way of doing it.

Tens place can be filled by any of the remaining 5 numbers.

So, there are 5 ways of filling the tens place.

The hundreds place can now be filled by any of the remaining 4 digits. So, there are 4 ways of filling it.

$\therefore$ Required number of numbers

$= (1 \times 5 \times 4) = 20.$

6. Ans. (c)
Solution:
Number of ways of selecting (3 consonants out of 7) and (2 vowels out of 4)

$$= ({}^7C_3 \times {}^4C_2) = \left[\frac{7\times6\times5}{3\times2\times1} \times \frac{4\times3}{2\times1}\right] = 210.$$

Number of groups, each having 3 consonants and 2 vowels =210.
Each group contains 5 letters.
Number of ways of arranging 5 letters among themselves

$= 5! = (5\times4\times3\times2\times1) = 120.$

$\therefore$ Required number of words

$= (210 \times 120) = 25200.$

TYPE B: NUMERICAL ESTIMATION

Time and Work

Work is defined as the amount of job assigned or the assigned or the amount of job actually done.

Work is always considered as a whole or 1.

- If A can be a piece of work in 'a' number of days, then in one day $\frac{1}{a}th$ of the work is done. Conversely, if a man does $\frac{1}{a}th$ of a work in 1 day, then he can complete the work in $\frac{1}{a}th$ days.
- If A is 'x' times as good a working as B, then he will take $\frac{1}{x}th$ of the time taken by B to do the same work.
- If A and B can do a piece of work in 'x' and 'y' days respectively, then working together, they will take $\frac{xy}{x+y}th$ part of the work.
- To compare the work done by different people, frost find the amount of work each can do in the same time.
- If the number of men to do a job is changed in the ratio a: b, then the time required to de the work will be in the ratio b: a, assuming the amount of work done by each of then in the given time is the same or they are identical.
- If two men A and B together can finish a job in 'x' days and if A working alone takes 'a' days more than A and B working together and B working alone takes 'b' days more then A and B working together, then $x = \sqrt{ab}$.
- To do a piece of work, the number of men employed and the number of days required to do the work are in inverse proportion. Also, the number of men employed and the hours working per day are in inverse proportion.

Efficiency

When we say that A can complete a work in 10 day we put 1 day work i.e. 1/10, but we have to work on it as, an efficiency which is inversely proportional to number of days in which a person can complete a piece of work.

$\therefore$ Fraction of work done by A

$$= \frac{\text{Number of days for which A did the work}}{\text{Number of days in which A can complete the work}}$$

If it is known that 3 persons are completing one work, then

Fraction of the work completed by 1st + Fraction of the work completed by 2nd + Fraction of the work completed by 3rd = 1.

Ratio and Proportion

Ratio is the relation which on quantity bears to another of the same kind. The ratio of A to B is usually written: A: B. The first term is called the antecedent and the second term the consequent.

$$\frac{a}{b} = \frac{ma}{mb}$$

i.e., the value of a ratio remains unaltered if the antecedent and the consequent are multiplied or divided by the same quantity.

e.g. A and B have income in the ratio of 5 : 4 does not mean that the incomes are Rs 5 and Rs 4, but but it means any income which is a multiple of 5 and 4, i.e. 5x and 4x, where x is an variable

- When the ratio a : b is compounded with it self the resulting ratio is $a^2 : b^2$ and is called the duplicate ratio of a:b. Also $a^{\frac{1}{2}} : b^{\frac{1}{2}}$ is called the sub duplicate ratio of a:b
- The duplicate ratio of 2a: 3b is $4a^2 : 9b^2$.
- The sub duplicate ratio of 49 : 25 is 7 : 5.
- The triplicate ratio of $2x:1$ is $8x^3:1$.

Ratio of Equality, Greater Inequality and Less inequality

Ratio of Equality, Greater inequality, of less inequality, of less inequality, or of equality, according as the antecedent is greater than less than or equal to the consequent.

- A ratio of greater inequality is diminished and a ratio of less inequality is increased, by adding the same quantity to both its terms.
- When a series of fractions are equal, each of them is equal, each of them is equal to the sum of all the numerators divided by the sum of all the denominators.

An Important Theorem

If $\frac{a_1}{b_1}, \frac{a_2}{b_2}, \frac{a_3}{b_3}, \ldots\ldots\ldots, \frac{a_n}{b_n}$, be unequal fractions, of which the denominators are all of the same sign, then the fraction

$\frac{a_1 + a_2 + a_3 + \ldots + a_n}{b_1 + b_2 + b_3 + \ldots + b_n}$ lies in magnitude between the greatest and least of them.

Proportion

When two ratios are equal, the four quantities composing then are said to be proportional. Thus if $\frac{a}{b} = \frac{c}{d}$, and a, b, c, d, are proportional. This is expressed by saying that a is to be as c is to d, and the proportion is written a : b : : c : d

The terms a and d are called the extremes, b and c the means.

- Let a, b, c, d be the proportionals

Then by definition

$\frac{a}{b} = \frac{c}{d} \quad \Rightarrow ad = bc$

Hence if any three terms of a proportion are given, the fourth may be find out.

Thus if a, c, d are given, then $b = \frac{ad}{c}$.

Quantities are said to be in continued proportion when the first is to the second, as the second is to the third. If three quantities a, b, c, are in continued proportion, then

$a:b = b:c$ and $ac = b^2$

If three quantities be proportional, the first is to the third is the duplicate ratio of the first to the second.

Let three quantities be a, b, c, then $\frac{a}{b} = \frac{d}{c}$

Now $\frac{a}{c} = \frac{a}{b} \times \frac{b}{c} = \frac{a}{b} \times \frac{a}{b} = \frac{a^2}{b^2}$

i.e. $a:c = a^2 : b^2$

If $a:b = c:d$, and $e:f = g:h$,

Then $ae:bf = cg:dh$.

If four quantities a, b, c, d form a proportion, many other proportions may be deduced by the properties of fractions.

- If $a:b = c:d$, then $b:a = d:c$
- If $a:b = c:d$, then $a:c = b:d$
- If $a:b = c:d$, then $a+b:b = c+d:d$.
- If $a:b = c:d$, then $a+b:b = c+d:d$.
- If $a:b = c:d$, then $a+b:a-b = c+d:c-d$

Ratio and Proportion

1. If $2A = 3B$ and $4B = 5C$, Then $A:C$ is
   (a) $4:3$ (b) $8:15$
   (c) $15:8$ (d) $3:4$
2. If $5x^2 - 13xy + 6y^2 = 0$ then, $x:y$ is
   (a) $(2:1)$ only (b) $(3:5)$ only
   (c) $(5:3)$ or $(1:2)$ (d) $(3:5)$ or $(2:1)$
3. Rs. 1210 were divided among A, B, C so that $A:B = 5:4$ and $B:C = 9:10$. Then, C gets:
   (a) Rs. 340 (b) Rs. 400
   (c) Rs. 450 (d) Rs. 475
4. A and B together have Rs. 1210. If $\frac{4}{15}$ of A's amount is equal to $\frac{2}{5}$ of B's amount how much amount does B have?
   (a) Rs. 460 (b) Rs. 484
   (c) Rs. 550 (d) Rs. 664
5. Ratio of the earnings of A and B is 4 : 7. If the earnings of A increase by 50% and those of B decreases by 25%, the new ratio of their earnings becomes $8:7$. What are A's earnings?
   (a) Rs. 21,000 (b) Rs. 26,000
   (c) Rs. 28,000 (d) Data inadequate
6. The prices of a scooter and a T.V. are in the ratio $7:5$. If the scooter costs Rs. 8000 more than a T.V. set, then the price of a T.V. set is

(a) Rs. 20,000 (b) Rs. 24,000
(c) Rs. 28,000 (d) Rs. 32,000

7. 85 kg of a mixture contains milk and water in the ratio 27 : 7. How much more water is to be added to get a new mixture containing milk and water in the ratio 3 : 1?
(a) 5 kg (b) 6.5 kg
(c) 7.25 kg (d) 8 kg

8. x varies inversely as square of y. Givern that $y = 2$ for $x = 1$. The value of x for $y = 6$ will be equal to:
(a) 3 (b) 9
(c) $\frac{1}{3}$ (d) $\frac{1}{9}$

9. Which of the following ratios is greatest ?
(a) $7:15$ (b) $15:23$
(c) $17:25$ (d) $21:29$

10. If $2A = 3B = 4C$ then $A:B:C$ is
(a) $2:3:4$ (b) $4:3:2$
(c) $6:4:3$ (d) $3:4:6$

12. Two number are in the ratio $3:5$. If each number is increased by 10, the ratio becomes $5:7$. The numbers are
(a) 3, 5 (b) 7, 9
(c) 13, 22 (d) 15, 25

13. Out of the ratios $7:15$, $15:23$, $17:25$ and $21:29$ the smallest one is
(a) $17:25$ (b) $7:15$
(c) $15:23$ (d) $21:29$

14. A fraction bears the same ratio to $\frac{1}{27}$ as $\frac{3}{7}$ does to $\frac{5}{9}$. The fraction is:
(a) $\frac{7}{45}$ (b) $\frac{1}{35}$
(c) $\frac{45}{7}$ (d) $\frac{5}{21}$

15. The weight of a 13 m long iron rod is 23.4 kg. The weight of 6 m long such rod will be
(a) 7.2 kg (b) 12.4 kg
(c) 10.8 kg (d) 18 kg

16. If a carton containing a dozen mirrors is dropped, which of the following cannot be the ratio of broken mirrors to unbroken mirrors?
(a) $2:1$ (b) $3:1$
(c) $3:2$ (d) $7:5$

17. A bag, contains 25 paise, 10 paise and 5 paise coins in the ratio $1:2:3$. If their total value is Rs. 30, the number of 5 paise coins is
(a) 50 (b) 100
(c) 150 (d) 200

18. In a class, the number of boys is more than the number of girls by 12 % of the total strength. The ratio of boys to girls is
(a) $11:14$ (b) $14:11$
(c) $25:28$ (d) $28:25$

19. A right cylinder and right circular cone have the same radius and the same volume. The ratio of the height of the cylinder to that of the cone is
(a) $3:5$ (b) $2:5$
(c) $3:1$ (d) $1:3$

20. The prices of a scooter and a television set are in the ratio $3:2$. If a scooter costs Rs. 6000 more than the television set, the price of the television set is
(a) Rs. 6000 (b) Rs. 10000
(c) Rs.12000 (d) Rs. 18000

21. Rs. 5625 is divided among A, B and C so that A may receive $\frac{1}{2}$ as much as B and C together receive and B receives $\frac{1}{4}$ of what A and C together receive. The share of A is more than that of B by
(a) Rs. 750 (b) Rs. 775
(c) Rs. 1500 (d) Rs. 1600

22. The students in three classes are in the ratio $2:3:5$. If 20 students are increased in each class, the ratio changes to $4:5:7$. The total number of student before the increase were
(a) 10 (b) 90
(c) 100 (d) None of these

23. The incomes of A and B are in the ratio $3:2$ and their expenditures in the ratio $5:3$. If each saves Rs. 1000, A's income is
(a) Rs. 3000 (b) Rs. 4000
(c) Rs. 6000 (d) Rs. 9000

24. The speeds of three cars in the ratio $3:4:5$. The ratio between times taken by them to travel the same distance is
(a) $3:4:5$ (b) $5:4:3$
(c) $12:15:20$ (d) $20:15:12$

25. Rs. 680 has been divided among A, B, C such that A gets $\frac{2}{3}$ of what B gets $\frac{1}{4}$ of what C gets. Then, B's share is
(a) Rs. 60 (b) Rs. 80
(c) Rs. 120 (d) Rs.160

26. Rs. 385 has been divided among A, B, C is such a way that A receives $\frac{2}{9}$ the of what B and C together receive. Then, A's share is
(a) Rs. 70 (b) Rs. 77
(c) Rs. 82.50 (d) Rs. 85

27. A and B are two alloys of gold and copper prepared by mixing metals in proportions $7:2$ and $7:11$ respectively. If equal quantities of the alloys are melted to form a third alloy C, proportion of gold and copper in C will be
(a) $5:9$ (b) $5:7$
(c) $7:5$ (d) $9:5$

28. 729 ml of a mixture contains milk and water in the ration $7:2$. How much more water is to be added to get a new mixture containing milk and water in the ratio of $7:3$
(a) 60 ml (b) 70 ml
(c) 81 ml (d) 90 ml

29. Rs. 1050 is divided among P, Q and R> The shared of P is $\frac{2}{5}$ of the combined share of Q and R. Thus, P gets
(a) Rs. 200 (b) Rs. 300
(c) Rs. 320 (d) Rs. 420

30. If $a:b = c:d$, then $\frac{ma+nc}{mb+nd}$ is equal to
(a) $m:n$ (b) $na:mb$
(c) $a:b$ (d) $md:nc$

31. One-fourth of the boys and three-eighth of the girls in a school participated in the annual sports. What fractional part of the total student population of the school participated in the annual sports?
(a) $\frac{4}{12}$ (b) $\frac{5}{8}$
(c) $\frac{8}{12}$ (d) Data inadequate

32. Vinay got thrice as many marks in Maths as in English. the proportion of his marks in Maths and History is $4:3$. If his total marks in Maths, English and History are 250, what are his marks in English ?
(a) 120 (b) 90
(c) 40 (d) 80

33. A's money is to B's money as $4:5$ and B's money is to C money $2:3$. If A has Rs. 800, C has
(a) Rs. 1000 (b) Rs. 1200
(c) Rs. 1500 (d) Rs. 2000

34. In a mixture of 60 liters, the ratio of milk and water is $2:1$. What amount of water must be added to make the ratio $1:2$?
(a) 42 liters (b) 56 litres
(c) 60 litres (d) 77 litres

35. A man has some hens and cows. If the number of heads be 48 and number of feet equals 140, the number of hens will be
(a) 22 (b) 23
(c) 24 (d) 26

36. The ratio between the ages of Kamla and savitri is $6:5$ and the sum of their ages is 44 years. The ratio of their ages after 8 years will be
(a) $5:6$ (b) $7:8$
(c) $8:7$ (d) $14:13$

37. The ages of Vivek and Sumit are in the ratio $2:3$. After 12 years, their ages will be in the raio $11:15$. The age of Sumit is
(a) 32 years (b) 42 years
(c) 48 years (d) 56 years

Answers

Ratio and Proportion

1. Ans. (c)
Solution:

$2A = 3B$ and $4B = 5C$

$\Rightarrow \frac{A}{B} = \frac{3}{2}$ and $\frac{B}{C} = \frac{5}{4}$

$\Rightarrow \frac{A}{C} = \left(\frac{A}{B} \times \frac{B}{C}\right) = \frac{3}{2} \times \frac{5}{4} = \frac{15}{18}$

$\Rightarrow A:C = 15:8$

2. Ans. (d)
Solution:

$\Leftrightarrow 5x^2 - 13xy + 6y^2 = 0$

$\Leftrightarrow 5x^2 - 10xy - 3xy + 6y^2 = 0$

$\Leftrightarrow 5x(x-2y) - 3y(x-2y) = 0$

$$\Leftrightarrow (x-2y)(5x-3y)=0$$
$$x=2y \text{ or } 5x=3y$$
$$\Leftrightarrow \frac{x}{y}=\frac{2}{1} \text{ or } \frac{x}{y}=\frac{3}{5}$$

3. Ans. (b)
Solution:

$$A:B=5:4,\ C=9:10=\left(9\times\frac{4}{9}\right):\left(10\times\frac{4}{9}\right)$$
$$=4:\frac{40}{9}$$
$$\therefore\ A:B:C=5:4:\frac{40}{9}=45:36:40$$

Sum of ratio terms $=(45+36+40)=121$

C's Share $= Rs.\left(1210\times\frac{40}{121}\right)=Rs.400$

4. Ans. (b)
Solution:

$$\frac{4}{15}A=\frac{2}{5}B \text{ Û}$$
$$A=\left(\frac{2}{5}\times\frac{15}{4}\right)B$$
$$A=\frac{3}{2}B$$
$$\frac{A}{B}=\frac{3}{2}$$
$$A:B=3:2$$

$\Leftrightarrow$ B's share $= Rs.\left(1210\times\frac{2}{5}\right)=Rs.\,484$

5. Ans. (d)
Solution:
Let the original earnings of A and B be Rs. $4x$ and Rs. $7x$.
New earnings of $A=150\%$ of $4x$.

$$= Rs.\left(\frac{150}{100}\times 4x\right) = Rs.6x$$

New earnings of $B=75\%$ of $Rs.7x$

$$= Rs.\left(\frac{75}{100}\times 7x\right) = Rs.\frac{21x}{4}$$

$$6x:\frac{21x}{4}=8:7 \Leftrightarrow \frac{6x\times 4}{21x}=\frac{8}{7}$$

This does not give x. So, the given data is inadequate.

6. Ans. (c)
Solution:
Let the prices of a scooter and a T.V. set be Rs. $7x$ and Rs. $5x$ respectively. Then

$$7x-5x=8000 \Leftrightarrow 2x=8000$$
$$x=4000$$

$\therefore$ Price of T.V. set

$$=Rs(7\times 4000)=Rs.\,28000.$$

7. Ans. (a)
Solution:

Milk in 85 kg of maximum $=\left(85\times\frac{27}{34}kg\right)=\frac{135}{2}kg$

Water in it $=\left(85-\frac{135}{2}\right)kg=\frac{35}{2}kg$

Let x kg of water be added to it.

Then, $\dfrac{\frac{135}{2}}{\frac{35}{2}+x}=\frac{3}{1}$

$$\frac{135}{35+2x}=\frac{3}{1}\quad 105+6x=135$$
$$6x=30\quad x=5$$

$\therefore$ Quantity of water to be added $=5\ kg$

8. Ans. (d)
Solution:

Given $x=\frac{k}{y^2}$, where k is a constant.

Now, $y=2$ and $x=1$ gives $k=4$

$$\therefore\ x=\frac{4}{y^2} \Rightarrow x=\frac{4}{6^2}$$

When $y=6 \Rightarrow x=\frac{4}{36}=\frac{1}{9}$

9. Ans. (d)
Solution:

$$\frac{7}{15}=0.466,\ \frac{15}{23}=0.652,\ \frac{17}{23}=0.68 \text{ and } \frac{21}{29}=0.724$$

0.724 is greatest and therefore, $21:29$ is greatest.

10. Ans. (c)
Solution: Let $2A=3B=4C=x$.

$A = \frac{x}{2}, B = \frac{x}{3}$ *and* $C = \frac{x}{4}$.

$A:B:C = \frac{x}{2}:\frac{x}{3}:\frac{x}{4} = 6:4:3$

Hence, $A:B:c = 6:4:3$.

11. Ans. (d)

Solution: Let the numbers be 3 x and 5 x.

then, $\frac{3x+10}{5x+10} = \frac{5}{7}$ *or* $7(3x+10) = 5(5x+10)$

$4x = 20$ or $x = 5$.

So, the numbers are 15,25

12. Ans. (b)

Solution:

$\frac{7}{15} = 0.466; 15:23 = \frac{15}{23} = 0.652;$

$17:25 = \frac{17}{25} = 0.68$ & $21:29 = \frac{21}{29} = 0.724$

$\therefore$ *The smallest one is* $7:15$.

13. Ans. (b)

Solution:

$x:\frac{1}{27} = \frac{3}{7}:\frac{5}{9} \quad \frac{5}{9}x = \frac{1}{27}\times\frac{3}{7}.$

$\frac{5}{9}x = \frac{1}{63}$ *or* $x = \left(\frac{1}{63}\times\frac{9}{5}\right) = \frac{1}{35}.$

14. Ans. (c)

Solution:

Let $0.2:0.12::0.3:x$

Then, $0.2x = 0.12\times 0.3$

or, $x = \frac{0.12\times 0.3}{0.2} = 0.18$

15. Ans. (c)

Solution: Less length, less weight 13 : 6 : : 23 . 4:x.

so, 13x = 6×23.4 *or* $x = \frac{6\times 23.4}{13} = 10.8$

16. Ans. (c)

Solution: For dividing 64 into two whole numbers, the sum of the terms of the ratio must be a factor of 12. so, they can not be in the ratio 3 : 2.

17. Ans (c)

Solution: Ratio of their values

$= \frac{1}{4}:\frac{2}{10}:\frac{3}{20} = 5:4:3:$

$\therefore$ *Value of* $5-$ *paise coins* $= \frac{750}{5} = 150$

No. of 5 paise cons $= \frac{750}{5} = 150$

18. Ans. (b)

Solution: Let the number of boys and girls be x and y respectively.

$(x-y) = 12\%$ *of* $(x+y)$ *or* $x - y = \frac{3}{25}(x+y)$

$\therefore 25x - 25y = 3x + 3y$ *or* $22x = 28y$ *or* $\frac{x}{y} = \frac{28}{22} = \frac{14}{11}$

19. Ans. (d)

Solution: Let the height of the cylinder and cone be h and H respectivelt.Then,

$\pi r^2 h = \frac{1}{3}\pi r^2 H$ *or* $\frac{h}{H} = \frac{1}{3}$

So, their heights are in the ratio 1:3

20. Ans. (c)

Solution: Let the price of scooter be Rs.3x and that of a television set be Rs. 2x.

Then,*3x-2x =6000 or x=6000*

$\therefore$ Cost of a television set *=2x=Is.12000.*

21. Ans. (a)

Solution:

$A = \frac{1}{2}(B+C)$ *or* $B + C = 2A \Rightarrow A + B + C = 3A.$

Thus, $3A = 5625$ *or* $A = 1875.$

$\therefore$ *Again,* $B = \frac{1}{4}(A+C) \Rightarrow 4B \Rightarrow A + B + C = 5B$

$\therefore 5B = 5625$ *or* $B = 1125.$

22. Ans. (c)

Solution: Let the number of students be *2x,3x and5x*

$(2x+20):(3x+20):(5x+20) = 4:5:7$

Then, *So,* $\frac{2x+20}{4} = \frac{3x+20}{5} = \frac{5x+20}{7}$

$\therefore 5(2x+20) = 4(3x+20)$ *or* $x = 10.$

Hence, total number of students before increase $=10x = 100$

23. Ans. (c)

Solution: Let their incomes be *3x,2x* and expenditure*5y,3y* respectively.

Then,

3x-5y = 1000 and 2x-3y =1000

Solving this equations, we get $x = 2000, y = 1000$

$\therefore$ A's income = $3x$ = Rs.6000

24. Ans. (d)

Solution: Ratio of time taken

$$=\frac{1}{3}:\frac{1}{4}:\frac{1}{5}=20:15:12$$

25. Ans. (c)

Solution:

Suppose C gets Re. 1. Then, B gets Rs. $\left(\frac{1}{4}\right)$

A gets = Re. $\left(\frac{2}{3}\times\frac{1}{4}\right)$ = Re. $\frac{1}{6}$

$$\therefore\ A:B:C=\frac{1}{6}:\frac{1}{4}:1=2:3:12$$

Hence, B's share $= Rs.\left(680\times\frac{3}{17}\right) = Rs.120$

26. Ans. (a)

Solution:

$$A:(B+C)=2:9$$

$\therefore$ A's share $= Rs.\left(385\times\frac{2}{11}\right) = Rs.70$

27. Ans. (c)

Solution:

Gold in $C=\left(\frac{7}{9}+\frac{7}{18}\right)=\frac{21}{18}=\frac{7}{6}$

Copper in $C=\left(\frac{2}{9}+\frac{11}{18}\right)=\frac{15}{18}=\frac{5}{6}$

$\therefore$ Gold : Copper $=\frac{7}{6}:\frac{5}{6}=7.5$

28. Ans. (c)

Solution:

Milk $=\left(729\times\frac{7}{9}\right)=567\ ml$

Water $=\left(729\times\frac{2}{9}\right)=162\ ml$

$$\therefore\ \frac{567}{162+x}=\frac{7}{3}$$

or, $7(162+x)=3\times567$

$\therefore\ 7x=1701-1134$

or, $x=\frac{567}{7}=81\ ml$

29. Ans. (b)

Solution:

$$P:(Q+R)=2:5$$

$\therefore$ P's share $= Rs.\left(1050\times\frac{2}{7}\right) = Rs.300$

30. Ans. (c)

Solution: Let

$$\frac{a}{b}=\frac{c}{d}=k.$$

Then, $a=bk$ *and* $c=dk$.

$$\therefore\ \frac{ma+nc}{mb+nd}=\frac{mbk+ndk}{mb+nd}$$

$$=k\left(\frac{mb+nd}{mb+nd}\right)=k=\frac{a}{b}.$$

31. Ans. (d)

Solution:

$$\therefore\ 19w+9w\times x=(1+x)\times15w$$

Thus 4w = 6w x or $x=\frac{4w}{6w}=\frac{4}{6}=\frac{2}{3}.$

So, the ratio is 1: $\frac{2}{3}$ *i.e.* 3 : 2.

32. Ans. (c)

Solution: M=3E AND

$$\frac{M}{H}=\frac{4}{3}.$$

$$\therefore H=\frac{3}{4}M=\frac{3}{4}\times3E=\frac{9}{4}E.$$

Now, $M+E+H=250 \Rightarrow 3\,E+E+\frac{9}{4}E=250$

$\therefore$ 25E =1000 OR E=40.

33. Ans. (c)

Solution: A : B = 4 : 5 = 8 : 10 & B : C = 2 : 3 = 10:15.

$\therefore$ A : B : C = 8 : 10 : 15.

IF A Has Rs. 8, C has Rs. 15

If a has Rs. 800,C has Rs. $\left(\frac{15}{8}\times 800\right)$ = Rs. 1500

34. Ans. (c)

Solution: Milk = $\left(60\times\frac{2}{3}\right)$ Litres = 40 litres.

Water = (60 – 40) litres = 20 litres.

$\therefore \frac{40}{20+x}=\frac{1}{2}\Rightarrow 20+x=80\ or\ x=60$

Water to be added = 60 liters

35. Ans. (d)

Solution: Let the number of hens =*x* & number of coys = *y*.

Then, *x + y = 48* and *2x + 4y =140.*

Solving these equations, we get , *2y = 44 or y = 22.*

So, x = (48–22) = 26.

∴ Number of hens = 26.

36. Ans. (c)

Solution: Let their ages be *6x* and *5x* years.

$\therefore\ 6x+5x=44\ or\ x=4$

So, their present ages are 24 years & 20 years.

Ratio of their ages after 8 years = 32 : 28 = 8:7.

37. Ans. (c)

Solution: Let their ages be *2x* and *3x* years.

$\frac{2x+12}{3x+12}=\frac{11}{15}\ or\ 15(2x+12)=11(3x+12)$

$or\ 3x=48\ or\ x=16$

∴ Age of sumit = 3x = 48 years.

TIME, SPEED AND DISTANCE

In this chapter the relative motion is most important concept. There is only one formula applicable

$$D=S\times T$$

Some Important Conversions:

- 1 hour = 60 minutes = 60×60 seconds.
- 1 Kilometer = 1000 metres
- 1 kilometer = 0.6514 mile;
  1 mile = 1.609 kilometre, i.e.
  8 kilometeers ≈ 5 mile
- 1 Yard = 3 feet
- $\frac{km}{hr}=\frac{5m}{18s}$
- $\frac{m}{s}=\frac{18\ km}{5\ hr}$
- $\frac{km}{hr}=\frac{5\ miles}{8\ hr}$
- $\frac{miles}{hr}=\frac{22\ ft}{15\sec}$

- Average Speed = $\frac{\text{Total distance travelld}}{\text{Total time taken}}$

Relative Motion can be difined in the floowing ways:

1. Dependent as in the case of boats and streams.
2. Independent as in the case of trains.

When one body which is moving inside another moving body, the motion is known as dependent, e.g. When a boat is moving inside a stream, the speed of boat depends on the speed of stream. If boat is going in the same direction as the stream, then the boat will move faster thean the speed at which boat is moving in still water, but when the boat is moving in opposite direction, the speeed of boat is slower than speed of boat in still water.

Boats and Streams

We consider four different types of speed:

1. S_d = speed downstream or speed of boat with the stream.
2. S_{up} = speed upstream or seed of boat against the stream.
3. S_b = speed of boat in still water.
4. S_s = speed of stream or water or river or current.

$$S_d=S_b+S_a$$
$$S_d=S_b-S_a$$

Adding equation (i) and (ii), we get

$$2S_b=S_d+S_{up}$$

$$S_b=\frac{S_b+S_{up}}{2}$$

If we subtract equation (ii) from equation (i), we get

$$2S_s=S_d-S_{up}$$

$$\Rightarrow\quad S_8=\frac{S_d-S_{up}}{2}$$

In above 4 equation there are 3 variables each, in every, question 2 variable are known and 3rd, you have to find out

But in certain question you have two variable unknown and you can make 2 equation to solve these 2 variables.

Questions on Trains

Other questions are related to trains or independent motion, i.e. a policeman runining after a thief, a car moving after another car, a dog runing after a hare and so on.

- Conversion of units is very important whereever necessary. $1\ km=\frac{5}{18}m/\sec$

- Let two trains or two bodies are moving in opposite directions at u *kimph* and v *kmph*, then their relative speed is $(u+v)$ *kmph*.
- Time taken by a train x meters long in passing a point or a pole or a man = Time taken by the train to cover X meters.
- Time taken a train x meters long in passing a stationary object of length y meters e.g. passing a plarform, passing a bridge or another train = Time taken by the train to cover $(x+y)$ meter.
- If two trains of length x *meters* and y *meters* are moving in same directions at u *kmph* and v *kmph*, then time takenn by faster trains to cross slower train $= \frac{x+y}{(u-v)} \times \frac{5}{18}$

Note: The relative speed $(u-v)$ is multiplied by $\frac{5}{18}$ so as to convert the speed given in kmph to m/sec.

- If two trains of length x meters and y meters are moving in opposite directions at *u kmph* and *v kmph*, then time taken by faster train to cross slower train $= \frac{x+y}{(u+v)} \times \frac{5}{18}$.

Note: The relative speed $(u+v)$ is multiplied by $\frac{5}{18}$ so as to convert the speed given in *kmph* to m/sec.

- If two trains at the same time from two points A and B towards each other and after crossing they take a and b hours in reaching B and A respectively. Then, A's speed : B's speed $= \sqrt{b} : \sqrt{a}$.

  In the example below distance convered by two bodies are same because both starts at the same place and one catches other at the same place. So we can say.

General Questions

- If the ratio of the speeds of A and B is $a:b$, then the ratio of the time taken by them to conver the same distance is $\frac{1}{a}:\frac{1}{b}$ or $b:a$.
- Suppose a man covers a certain distance at *x kmph* and an equal distance at *y kmph*. Then, average speed during the whole journey $= \left(\frac{2xy}{x+y}\right)$ *kmph*.
- Suppose a man covers a certain distance at x *kmph*, equal distance at y *kmph* and same distance at z *kmph*, then average speed during the whole journey $= \left(\frac{3xyz}{xy+yz+zx}\right)$.

Average is a very simple but effective way of representing an entire group by a single value Average" of a group is defined as.

$$\text{Average} = \frac{\text{Sum of all items in the group}}{\text{Number of items in the group}}$$

"Sum of all the items in the group" means "sum of the values of all the items is the group".

- If the value of each items is increased by the same value p, then the average of the group or items will also increase by p.
- If the value of each time is decreased by the sme value p, then the average of the group or items will also be mutiplied by p.
- If the value of each items is multiplied by the same value p, then the average of the group or items will also multiplied by p.
- If the value of each item is divided by the same value p $(p \neq 0)$, then the average of the group or items will also be divided by p.
- The average of a group of items will always lie between the smallest value in the group and largest value in the group, i.e. the average will be greater than the smallest value and less than the largest value in the group.

Alligation Rule

The alligation rule states that "when different quanitiesw of the same of differnt ingredients, of different consts are mixed together to produce a mixture of a mean cost the ratio of their quantities are incersely proportional to the difference in their cost from the mean cost".

$$\frac{\textit{Quantity of cheap}}{\textit{Quantity of Dear}} = \frac{\textit{price of Dear} - \textit{Mean price}}{\textit{Mean price} - \textit{price of Cheap}}$$

Problems on Ages

1. Present ages of X and Y are in the ratio 5:6 respectivrly.Seven years hence this ratio will become 6 : 7 respectively. What is X's present age in years?

   (a) 35 (b) 42
   (c) 49 (d) Can not be determined

2. Hitesh is 40 years old and Ronnie is 60 years old. How many years ago was the ratio of their ages 3 : 5?
(a) 5 years (b) 10 years
(c) 20 years (d) 37 years

3. The ratio between the present ages of A and B is 5:3 respectively. The ratio between A's age 4 years ago and B's age 4 years hence is 1 : 1. What is the ratio between A's age 4 years hence and B's age 4 years ago?
(a) 1:3 (b) 2:1
(c) 3:1 (d) 4:1

4. A person,s present age is two-fifth of the age of his mother. After 8 years, he will be one-half of the age of his mother. How old is the mother at present?
(a) 32 years (b) 36 years
(c) 40 years (d) 48 years

5. One yesr ago, Promila was four times as old as her daughter Sakshi. Six years hence, Promila,s age will exceeds her daughter's age by 9 years. The ratio of the present ages of Promila and her daughter is:
(a) 9 : 2 (b) 11 : 3
(c) 12 : 5 (d) 13 : 4

6. The age of a man is three times the sum of the ages of his two sons. Five years hence, his age will be double the sum of the ages of his sons. The father's present age is:
(a) 40 years (b) 45 years
(c) 50 years (d) 55 years

7. Father is aged three times more than his son Ronit. After 8 years, he would be two and a half times of Ronit's age. After further 8 years, how many times would he be of Ronit's age?
(a) 2 times (b) $2\frac{1}{2}$
(c) $2\frac{3}{4}$ (d) 3 times

8. In 10 years, A will be twice as old as B was 10 years ago. If A is now 9 years older than B, the present age of B is:
(a) 19 years (b) 29 years
(c) 39 years (d) 49 years

9. In 10 years, A will be twice as old as B was 10 years ago. If A is now 9 years older than B, the presaent age of B is
(a) 19 years (b) 29 years
(c) 39 years (d) 49 years

10. The sum of the ages of a father and son is 45 years. Five years ago the product of their ages was 4 times the fathers age at that time. The present ages of the father and son, respectively are
(a) 25 years, 10 years (b) 36 years, 9 years
(c) 39 years, 6 years (d) None of these

11. Five years ago Viney's age was onethird of the age of Vikas and now Viney's age in 17 years. What is the present age of Vikas?
(a) 9 years (b) 36 years
(c) 41 years (d) 51 years

12. Pushpa is twice as old as Rita was two years ago. If the difference between their ages be2 years, how old is Pushpa today?
(a) 6 years (b) 8 years
(c) 10 years (d) 12 years

13. The age of Arvind's father is 4 times his age. If 5 years ago, father's age was 7 times of the age of his son at that time, what is Arvind's father's present age?
(a) 35 years (b) 40 years
(c) 70 years (d) 84 years

14. 10 years ago, Chandravati's mother was 4 times older than her daughter. After 10 years, the mother will be twice older than the daughter. The present age of Chandravati is:
(a) 5 years (b) 10 years
(c) 20 years (d) 30 years

15. Kamla got married 6 yeras ago. Today her age is $1\frac{1}{4}$ times age at the time of marriage. Here son's is $\frac{1}{10}$ times her age. Here son's age is:
(a) 2 years (b) 3 years
(c) 4 years (d) 5 years

16. Ratio of Ashok's age to Pradeep's age is equal to $4:3$. Ashok will be 26 years old after 6 years. How old is Pradeep now?
(a) 12 years (b) 15 years
(c) $19\frac{1}{2}$ years (d) 21 years

17. The ratio of the ages of father and son at present is 6 : 1. After 5 years the ratio will become 7 : 2. The present age of the son is:
(a) 5 years (b) 6 years
(c) 9 years (d) 10 years

18. The ratio between the ages of A and B at present is $2:3$. Five years hence the ratio of their ages will be $3:4$. What is the present age of A?

(a) 10 years (b) 15 years
(c) 25 years (d) data inadequate

19. One year ago the ratio between Samir and Ashok's age was 4 : 3. One year hence the ratio of their age will be 5 : 4. What is the sum of their present ages in years?
(a) 12 years (b) 15 years
(c) 16 years (d) Cannot be determined

20. Jayesh is as much younger to Anil as he is older to Prashant. If the sum of the ages of Anil and Prashant is 48 years, What is the age of Jayesh ?
(a) 20 years (b) 24 years
(c) 30 years (d) cannot be determined

21. The ratio of Vimal's age and Aruna's age is $3:5$ and sum of their ages is 80 years. The ratio of their ages after 10 years will be:
(a) $2:3$ (b) $1:2$
(c) $3:2$ (d) $3:5$

Answers

Problems and Ages

1. Ans. (a)
Solution:
Let the present ages of X and Y be $5x$ years and $6x$ years respectively

Then, $\frac{5x+7}{6x+7}=\frac{6}{7}$

$7(5x+7)=6(6x+7)$

$x=7$

$\therefore$ X's present age $=5x=35$ years

2. Ans. (b)
Solution:
Suppose, the ratio was $3:5$, x years ago.

Then $\frac{40-x}{60-x}=\frac{3}{5}$

$5(40-x)=3(60-x)$

$2x=20$

$x=10$

3. Ans. (c)
Solution:
Let the present ages of A and B $5x$ and $3x$ years respectively.

Then, $\frac{5x-4}{3x+4}=\frac{1}{1}$

$\Rightarrow 5x-4=3x+4$

$\Rightarrow 2x=8$

$\Rightarrow x=4$

$\therefore$ Required ratio $=(5x+4):(3x+4)$

$=24:8=3:1$

4. Ans. (c)
Solution:
Let the mother's present age be x years. Then, the person's present age $=\left(\frac{2}{5}x\right)$ years.

$\therefore \left(\frac{2}{5}x+8\right)=\frac{1}{2}(x+8)$

$\Rightarrow 2(2x+40)=5(x+8)$

$\Rightarrow x=40$

5. Ans. (d)
Solution:
Let the ages of Promila and Sakshi 1 year ago be $4x$ and x years respectively.

Then, $(60-x)-6=5(x-6)$

$\Rightarrow 54-x=5x-30$

$\Rightarrow 6x=84$

$\Rightarrow x=14$

Son's age after 6 years $=(x+6)=20$ years

6. Ans. (b)
Solution:
Let the sum of present ages of the two sons be x years.
Then, father's present age $=3x$ years.

$\Rightarrow (3x+5)=2(x+10)$

$\Rightarrow 3x+5=2x+20$

$\Rightarrow x=15$

Hence, father's present age = 45 years.

7. Ans. (a)
Solution:
Let Ronit's present age be x years. Then father's present age $=(x+3x)$ years $=4x$ years.

$\Rightarrow (4x+8)=\frac{5}{2}(x+8)$

$\Rightarrow 8x+16=5x+40$

$\Rightarrow 3x=24$

$\Rightarrow x=8$

Hence, required ratio $= \frac{(4x+16)}{(x+16)} = \frac{48}{24} = 2$

8. Ans. (c)

Solution:

Let B's present age $= x$ years.

Then A's present age $= (x+9)$ years

$\therefore \quad (x+9)+10 = 2(x-10)$

$\Rightarrow \quad x+19 = 2x-20$

$\Rightarrow \quad x = 39$

9. Ans. (c)

Solution:

Let the present ages of B and A be x years and $(x+9)$ years respectively

Then, $(x+9+10) = 2(x-10)$

or $\quad x = 39$.

10. Ans. (b)

Solution:

Let the present ages of father and son be x years and $(45-x)$ years.

Then $(x-5)(45-x-5) = 4(x-5)$

or $\quad -x^2 + 41xx - 180 = 0$

or $\quad x = 36$

$\therefore$ The present case ages of father and son are 36 yeares and 9 years respectively.

11. Ans. (c)

Solution:

Let the age present age of Vikas be x years. Then.

$17-5 = \frac{1}{3}(x-5)$

or $\quad x-5 = 36$

or $\quad x = 41$

12. Ans. (b)

Solution:

Let Rits's age 2 years ago be x years.

Pushpa's present age $= (2x)$ years.= 8 years

13. Ans. (b)

Solution:

Let Avind's age be x years

Then, his father's age $= 4x$ years

$\therefore \quad (4x-5) = 7(x-5)$

or $\quad x = 30$

or $\quad x = 10$

14. Ans. (c)

Solution:

Let Chandravati's age 10 years ago be x years

Mother's age 10 years ageo $= (4x)$ years

$2(x+20) = (4x+20)$

$\Rightarrow \quad x = 10$

Present age of Chandravati $= (x+10) = 20$ years

15. Ans. (b)

Solution:

Let Kmals's age 6 years ago be x years.

Then, Kamla's present age $= (x+6)$ years

$\therefore \quad x+6 = \frac{5}{4}x$

or $\quad 4x+24 = 5x$

or $\quad x = 24$

So, Kamla's present age $= 30$ years

Son,s present age $= \left(\frac{1}{10} \times 30\right) = 3$ years

16. Ans. (c)

Solution:

Let Ashok's age $= 4x$ and Pradeep's age $= 3x$ years.

$\therefore \quad 4x+6 = 26$

$\Rightarrow \quad x = 5$

$\therefore \quad$ Pradeep's age $= 3x = 15$ years

17. Ans. (b)

Solution:

$\frac{6x+5}{x+5} = \frac{7}{2}$

$\Rightarrow \quad 2(6x+5) = 7(x+5)$

$\Rightarrow \quad x = 5$

$\therefore \quad$ Son's present age $= 5$ years

18. Ans. (b)

Solution:

Let the ages of A and B be $2x$ and $3x$ years

$\frac{2x+5}{3x+5} = \frac{3}{5}$

$\Rightarrow \quad 4(2x+5) = 3\,(3x+5)\; x = 5$

$\therefore \quad$ A's present age $= 2x = 10$ years

19. Ans. (c)

Solution:

Let their ages one year ago be $4x$ and $3x$ years.

$\frac{2x+8}{5x+8} = \frac{1}{2}$

$2(2x+8)=(5x+8)\ x=8$

Difference of their ages $=(5x-2x)=3x=24$ years

20. Ans. (b)

Solution:

Let Anil's age $=x$ years

Then, Prashant's age $=(48-x)$ years

Let the age of Jayesh be p years. Then,

$p-(48-x)=x-p\ 2\ p=48$

or $p=24$

21. Ans. (a)

Solution:

$3x+5x=80$

$x=10$

Ratio of their ages after 10 years

$=(3x+10:5x+10)=40:60=2:3$

Alligation or Mixure

Alligation: It is the rule that enables us to find the ratio in which two or more ingredients at the given price must be mixed to produce a mixture of a desired price.

Mean price: Thecost price of a unit quantity of the mixture is called the mean price.

Rule of alligation: If two ingredients are mixed, then

$$\left(\frac{\text{Quantity of cheaper}}{\text{Quwntity of dearer}}\right)$$

$$=\frac{(\text{C.P. of dearer})-(\text{Mean price})}{(\text{Mean price})-(\text{C.P. of cheaper})}.$$

We present as under :

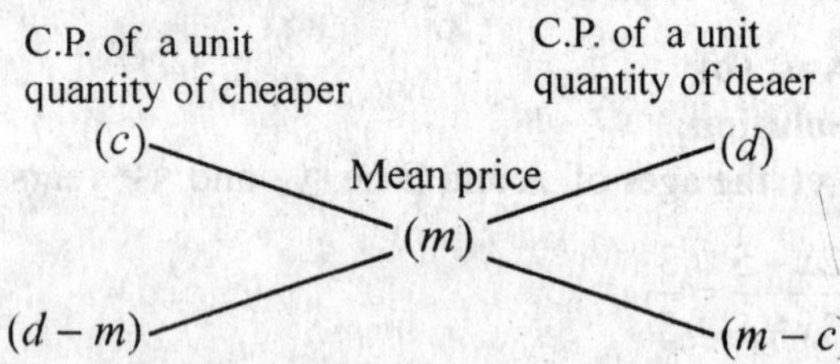

Therefore,

(Cheapre quantity): (Dearer quantity)

$=(d-m):(m-c)$.

Suppose a container contains x units of liquid from which y units are taken out and replaced by water. After n operations, the quantity of pure liquid $=\left[x\left(1-\frac{y}{x}\right)^n\right]$ units.

Example 1

In what ratio must water be mixed with milk to gain 20% by selling the mixture at cost price?

Solution:

Let C.P. of milk be Rs. 1 per litre.

The, S.P. of 1 litre of mixture = Rs. 1

Gain obtained = 20%

Therefore C.P. of 1 litre mixture

$$=Rs.\left(\frac{100}{120}\times 1\right)$$

$$=\text{Re.}\frac{5}{6}.$$

By the rule of allegation, we have:

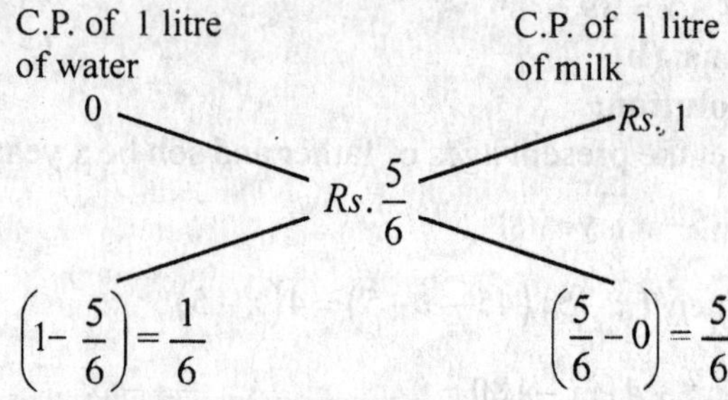

Therefore, Ratio of water and milk $=\frac{1}{6}:\frac{5}{6}=1:5$.

Example 2

How many kgs. of wheat costing Rs. 8 per kg must be mixed with 36 kg of rice costing Rs. 5.40 per kg so that 20% gain may be obtained by selling the mixture at Rs. 7.20 per kg?

Solution:

S.P. os 1 kg mixture $=Rs.\,7.20$,

Gain $=20\%$

Therefore, C.P. of 1 kg mixture

$$\left(\frac{100}{120}\times 7.20\right)=Rs.\,6$$

By the rule of alligation,we have:

C.P. of 1 kg wheat of 1st kind

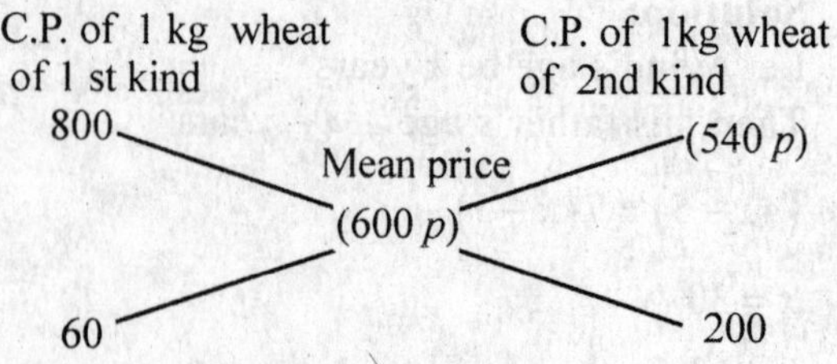

wheat of 1st kind : Wheat of 2nd kind

= 60 : 200 = 3 : 10.

Let x kg of wheat of 1st kind be mixedwith 36 kg of wheat of 2nd kind.

Then, 3 : 10 = x : 36

or 10x = 3 × 36

or x = 10.8 kg.

Excercise

1. The cost of Type 1 rice is Rs. 15 per kg and Type 2 rice is Rs. 20 per kg. If both Type 1 and Type 2 are mixed in the ratio of 2 : 3, then the price per kg of the mixed vatiety of rice is:
   (a) Rs. 18 (b) Rs. 18.50
   (c) Rs. 19 (d) Rs. 19.50
2. A milk vendor has 2 cans of milk. The first contains 25% water and the rest milk. The second contains 50% water. How much milk should he mix from each of the containers so as to get 12 litres of milk such that the ratio of water to milk is 3 : 5?
   (a) 4 liters, 8 litres (b) 6 litres, 6 litres
   (c) 5 litres, 7 litres (d) 7 litres, 5 litres
3. Tea worth Rs. 126 per kg and Rs.135 per kg are mixed with a third variety in the ratio 1 : 1 :2. If the mixture is worth Rs. 153 per kg,the price of the third variety per kg will be:
   (a) Rs. 169.50 (b) Rs. 170
   (c) Rs. 175.50 (d) Rs. 180
4. A jar full of whisky contains 40% alcohal. A part of this whisky is replaced by another containing 19% alchol and now the percentsge of alcohol was found to be 26%. The quantity of whisky replaced is:
   (a) $\frac{1}{3}$ (b) $\frac{2}{3}$
   (c) $\frac{2}{5}$ (d) $\frac{3}{5}$
5. A can contains a mixture of two liquids A and B in the ratio 7 : 5. When 9 litres of mixture are drawn off and the can is filled with B, the ratio of A and B becomes 7 : 9. How many litres of liquid A was contained by the can initially ?
   (a) 10 (b) 20
   (c) 21 (d) 25

Answers

1. **Ans. (a)**
   Solution:
   Let the price of the mixed variety be Rs. x per kg. By the rule of alligation, we have:

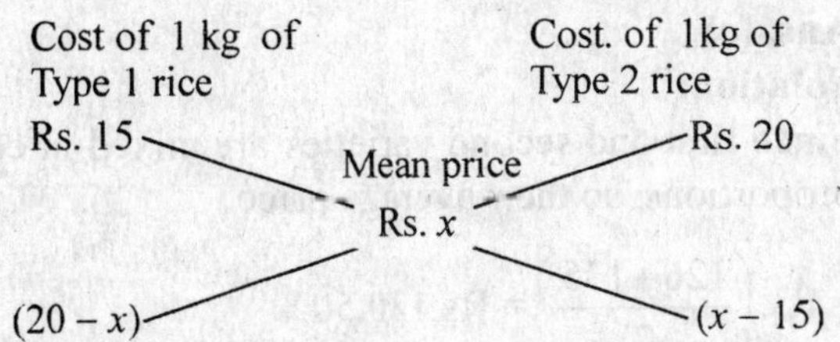

Therefore, $\frac{(20-x)}{(x-15)} = \frac{2}{3}$

$\Rightarrow 60-3x$

$= x2-30$

$\Rightarrow 5x = 90$

$\Rightarrow x = 18$.

2. **Ans. (b)**
   Solution:
   Let cost of 1 litre milk be Rs. 1.

Milk in 1 liter mix. in 1st can $= \frac{3}{4}$ litre,

C.P.of 1 litre mix.in 1st can = Rs. $\frac{3}{4}$.

Milk in 1 litre mix. in 2nd can $= \frac{1}{2}$ litre,

C.P. of 1 litre mix. in 2nd can = Rs. $\frac{1}{2}$.

Milk in one litre of final mix. $= \frac{5}{8}$ litre,

Mean price = Rs. $\frac{5}{8}$.

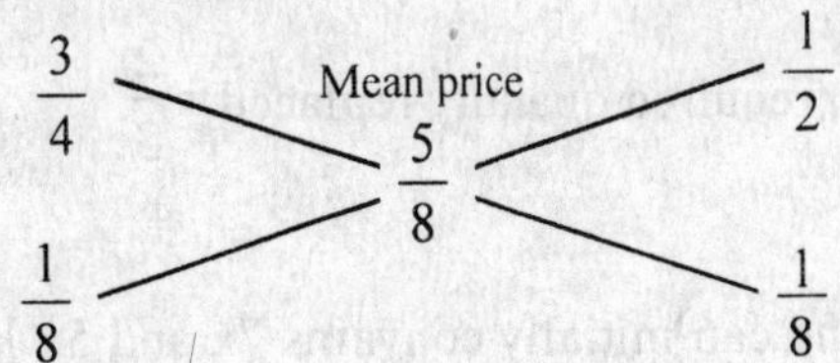

Therefore,

Ratio of two mixtures $\frac{1}{8} : \frac{1}{8} = 1:1$

Hence,

quantity of mixture taken from each can $= \left(\frac{1}{2} \times 12\right) = 6$

3. Ans. (c)

Solution:

Since first and second varieties are mixed in equal proportions, so their average price

$= \text{Rs.}\left(\frac{126+135}{2}\right) = \text{Rs.}130.50.$

So, the mixture is formed by mixing two varieties, one at Rs.130.50 per kg and the other at say, Rs. x per kg in the ratio 2 : 2, i.e. 1 : 1 we have to find x.

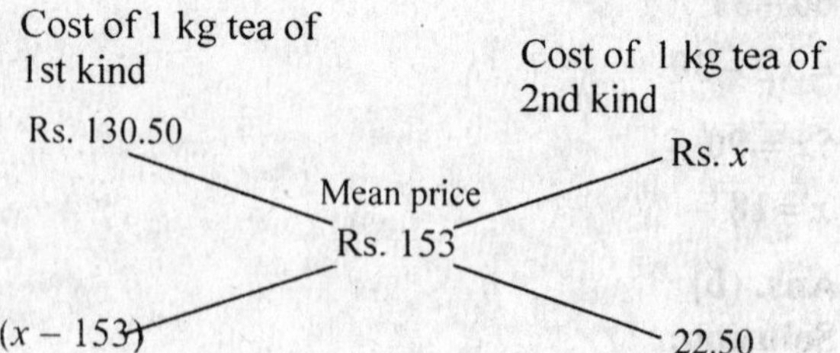

Therefore, $\frac{x-153}{22.5} = 1$

or $x - 153 = 22.50$

or $x = 175.50$.

Hence, price of the thirdvariety = Rs.175.50 per kg.

4. Ans. (b)

Solution:

By the rule of alligation, we have:

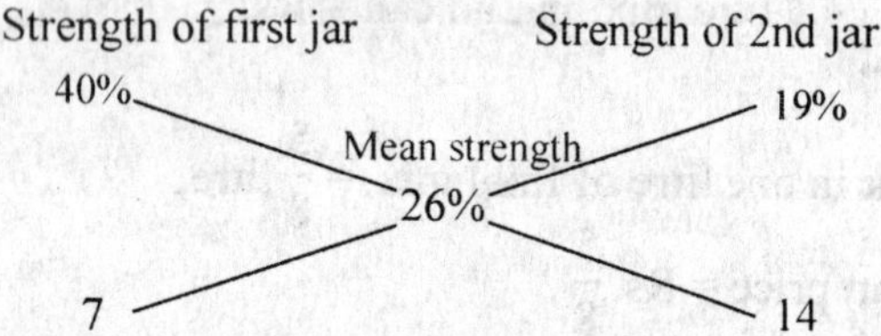

Hence, ratio of 1st and 2nd quantities = 7

14 = 1 : 2.

Therefore, required quantity replaced $= \frac{2}{3}$.

5. Ans. (c)

Solution:

Suppose the can initially conyains 7x and 5x litres of mixtures of A and B respectively.

Quantity of A in mixture left

$= \left(7x - \frac{7}{12}\times 9\right)\text{litres} = \left(7x - \frac{21}{4}\right)\text{litres.}$

Quanitity of B in mixture

left $= \left(5x - \frac{5}{12}\times 9\right)\text{litres} = \left(5x - \frac{15}{4}\right)\text{litres.}$

Therefore,

$$\frac{\left(7x - \frac{21}{4}\right)}{\left(5x - \frac{15}{4}\right) + 9} = \frac{7}{9}$$

or, $\frac{28x - 21}{20x + 21} = \frac{7}{9}$

or, $252x - 189 = 140x + 147$

or, $112x = 336$

or, $x = 3$.

Hence, the can contained 21 litres of A.

TYPE C: NUMERICAL REASONING

Percentage

Percent means out of hundred. For example if we say 50% that means $\frac{50}{100}$, 25% means $\frac{25}{100}$, Or we can say percentage is a fraction in which denominator is 100.

First learn how to change percentage into fraction or fraction into percentage.

- 25% means or in conversion of % into fraction divide the number by 100.
- In conversion of fraction into percentage multiply by 100.

$1/4 \times 100 = 25\%$

Normally a student gets questions based on percentages given in percentage table.

Note: If a fraction is given more and you have to find for less percentage, then decrease the fraction and a lower fraction is the answer or vice-versa.

Use of Percentage Table

Type I

Example 1. If the salary of A is 20% more than B, then by what %, B's salary is less than A?

Solution: If we see the percentage table $20\% = \frac{1}{5}$.

We have to find a fraction less than this, i.e. the answer is.

Example 2. If the rate has reduced by 10%, by what % the consumption has to be increased, so that expenditure remains the same?

Solution: In percentage table. We have to find increased fraction, i.e. or % is the answer.

Example 3. If length of a rectangle is reduced by , then by what % the breath has to be increased so that area remains the same?

Solution: As 40% is not given kin percentage table, we are asking for more, so denominater must be less.

Type II

There are three different cases
We have to solve in the example:

$A + x\%$ of $A = B$

Exercise

1. 860% *of* 50 + 50% *of* 860 = ?
   (a) 430 (b) 516
   (c) 860 (d) 960
2. What percent of 7.2 kg is 18 gms?
   (a) 0.025% (b) 0.25%
   (c) 2.5% (d) 25%
3. Rajeev buy goods worth Rs. 6650. He gets a rebate of 6% on it. After getting the rebate, he pays sales takes @ 10%. Find the amount he will have to pay for the goods.
   (a) Rs. 6876.10 (b) Rs. 6999.20
   (c) Rs. 6654 (d) Rs. 7000
4. 4598 is 95% of?
   (a) 4800 (b) 4840
   (c) 4850 (d) 4880
5. 15% of (?) 582 = 17.46
   (a) 2 (b) 10
   (c) 20 (d) none of these
6. Subtracting 40% of a number from the number, we get the result as 30. The number is:
   (a) 28 (b) 50
   (c) 52 (d) 70
7. A number when 35 is subtracted from it, reduces to its 80 percent. What is four-fifth of that number?
   (a) 70 (b) 90
   (c) 120 (d) 140
8. If one number is 80% of the other and 4 times the sum of their squares is 656, then the numbers are:
   (a) 4,5 (b) 8,10
   (c) 16,20 (d) none of these
9. In a competitive examination in state A,6% candidates got selected from the total appeared candidates. State B had an equal number of candidates appeared and 7% candidates got selected with 80 more candidates got selected than A. What was the number of candidates appeared from each state?
   (a) 7600 (b) 8000
   (c) 8400 (d) data inadequate
10. In an examination, 35% of the students passed and 455 failed. How many students appeared for the examination?
    (a) 490 (b) 700
    (c) 845 (d) 1300
11. At an election involving two candidates,68 votes were declared invalid. The winning candidate secures 52% and wins by 98 votes. The total number of votes polled is:
    (a) 2382 (b) 2450
    (c) 2518 (d) none of these
12. A student secures 90%, 60% and 54% marks in test paper with 100, 150 and 200 respectively as maximum marks. The percentage of his aggregate is:
    (a) 64 (b) 68
    (c) 70 (d) None of these
13. Two students appeared at an examination. One of them secured 9 marks more than the other and his marks was 56% of the sum of their marks. The marks obtained by them are:
    (a) 39,30 (b) 41,32
    (c) 42,33 (d) 43,34
14. If x% of y is 100 and y% of z is 200, then find a relation between x and z.
    (a) $z = \frac{x}{2}$ (b) $z = 2x$
    (c) $z = \frac{x}{4}$ (d) $z = 4x$
15. A man spends 35% of his income on food, 25% on childrens education and 80% of the remaining on house rent. What percent of his income he is left with?
    (a) 8% (b) 10%
    (c) 12% (d) 14%
16. The sum of the number of boys and girls in a school is 150. If the number of boys is x, then the number of girls becomes x% of the total number of students. The number of boys is:
    (a) 40 (b) 50
    (c) 60 (d) 90
17. In an examination, there are three papers and a candidate has to get 35% of the total to pass. In one paper, he gets 62 out of 150 and in the second 35 out of 150. How much must he get, out of 180, in the third paper to just qualify for a pass?
    (a) 60.5 (b) 68
    (c) 70 (d) 71
18. In a city, 40% of the people are illiterate and 60% are poor. Among the rich, 10% are illiterate. What percentage of the poor population is illiterate?

(a) 36% (b) 40%
(c) 60% (d) none of these

19. The boys and girls in a college are in the ratio 3:2. If 20% of the boys and 25% of the girls are adults, the percent of students who are not adults is:
(a) 58% (b) 67.5%
(c) 78% (d) 82.5%

20. A number is decreased by 10% and then increased by 10%. The number so obtained is 10 less than the original number. What was the original number?
(a) 1000 (b) 1050
(c) 1500 (d) 2000

21. What percent decrease in salaries would exactly cancel out the 20 percent increase?
(a) $16\frac{2}{3}$ (b) 18
(c) 20 (d) $33\frac{1}{3}$

22. The income of a broker remains unchanged though the rate of commission is increased from 4% to 5%. The percentage of slup in business is:
(a) 1% (b) 8%
(c) 20% (d) 80%

23. If the price of sugar rises from Rs. 6 per kg to Rs. 7.50 per kg, a person, to have no increase in his expenditure on sugar, will have to reduce his consumption of sugar by
(a) 15% (b) 20%
(c) 25% (d) 30%

24. The population of a town 2 yerars ago was 62,500. Due to migration to big cities, it decrease every year at the rate of 4%. The present population of the town is:
(a) 56,700 (b) 57,600
(c) 58,800 (d) 60,000

25. The population of a variety of tiny bush in an experimental field increased by 10% in the first year, increased by 8% in the second year but decreased by 10% in the third year. If the present number of bushes in the experimental field is 26730, then the number of bushes in the beginning was:
(a) 25000 (b) 27000
(c) 28000 (d) 24600

26. The population of a town increases 4% annualy but is decreased by emigration annualy to the extent of (1/2)%. What will be the increase present in 3 years?
(a) 9.8 (b) 10
(c) 10.5 (d) 10.8

27. P is six times as large as q. The present that q is less than p is:
(a) $16\frac{2}{3}$ (b) 60
(c) $83\frac{1}{3}$ (d) 90

28. Peter earned 40% more money than Albert. Albert earned 20% less than Michael peter earned more than Michael by:
(a) 10% (b) 12%
(c) 20% (d) 25%

29. A large watermelon weights 20 kg with 96% of its weight being water. It is allowed to stand in the sun and some of the water evaporates so that only 95% of its weight is water. Its reduced weight will be:
(a) 16 kg (b) 16.5 kg
(c) 17 kg (d) 18 kg

30. One type of liquid contains 20% water and the second type of liquid contains 35% of water. A glass is filld with 10 parts of first liquid and 4 parts of second liquid. The percentage of water in the new mixture in the glass is :
(a) 20% (b) $24\frac{2}{7}$%
(c) 37% (d) 40%

31. A bag contains 600 coins of 25 p denomination and 1200 coins of 50 p denomination. If 12% of 25 p coins 24% of 50 p coins are removed, the percentage of money removed from the bag is nearly:
(a) 15.6% (b) 17.8%
(c) 21.6% (d) 30%

32. In an examination, 34% of the students failed in mathematics and 42% failed in english. If 20% of the students failed in both the subjects,then the percentage of students who passed in both the subject was:
(a) 44 (b) 50
(c) 54 (d) 56

33. In a certain office, 72% of the workers prefer tea and 44% prefer coffee. If each of them prefers tea or coffee and 40 like both, the total number of workers in the office is:
(a) 200 (b) 240
(c) 250 (d) 320

34. $0.02 = (\cdots?\cdots)\%$
(a) 20 (b) 2
(c) 0.02 (d) 0.2

35. ? % of 130 = 10.4
(a) 80 (b) 8
(c) 0.8 (d) 0.08

36. ? % of 250 + 25% or 68 = 67
(a) 10 (b) 15
(c) 20 (d) 25

37. The number 0.05 is how many per cent of 20:
(a) 25 (b) 0.025
(c) 0.25 (d) 2.5

38. $\frac{30\% \text{ of } 80}{?} = 24$
(a) $\frac{3}{10}$ (b) $\frac{3}{17}$
(c) 1 (d) 2

39. If x is 90 % of y, then what per cent of x is y?
(a) 90 (b) 190
(c) 101.1 (d) 111.1

40. Which number is 60 % less than 80?
(a) 48 (b) 42
(c) 32 (d) 12

41. Which is greatest $33\frac{1}{3}\%$, $\frac{4}{15}$ or 0.35
(a) $33\frac{1}{3}\%$ (b) $\frac{4}{15}$
(c) 0.35 (d) Can not be compared

42. If 0.5 % of $x = 85$ paise, then the value of x is:
(a) Rs. 170 (b) Rs. 17
(c) Rs. 1.70 (d) Rs. 4.25

43. 30 quintals is what per cent of 2 metric tonnes:
(a) 15% (b) 1.5%
(c) 150% (d) 30%

44. If $37\frac{1}{2}\%$ of a number is 900, then $62\frac{1}{2}\%$ of the number is:
(a) 1200 (b) 1350
(c) 1500 (d) 540

45. The price of an article is cut by 10%. To restore it to the former value, the new price must be increased by:
(a) 10% (b) $9\frac{1}{11}\%$
(c) $11\frac{1}{9}\%$ (d) 11%

46. Rakesh credits 15 % of his salary in his fixed deposit account and spends 30% of the remaining amount on groceries. If the cash in hand is Rs. 2380, what is his salary?
(a) Rs. 3500 (b) Rs. 4000
(c) Rs. 4500 (d) Rs. 5000

47. A number increased by $37\frac{1}{2}\%$ gives 33. The numbers is:
(a) 22 (b) 24
(c) 25 (d) 27

48. 96% of the population of a village is 23040. The total population of the village is:
(a) 32256 (b) 24000
(c) 24936 (d) 25640

49. If 70% of the students in a school are boys and the number of girls be 504, the number of boys is:
(a) 1176 (b) 1008
(c) 1208 (d) 3024

50. In an examination, 65% of the total examinees passed. If the number of failures is 420, the total number of examinees is:
(a) 567 (b) 693
(c) 1000 (d) 1200

51. In mathematics exam. a student scored 30% marks in the first paper, out of a total of 180. How much should he score in second paper out of a total of 150, if he is to get an overall average of at least 50 %:
(a) 74% (b) 76%
(c) 70% (d) 80%

52. In an examination, it is required to get 36% of maximum marks to pass. A student got 113 marks and declared failed by 85 marks. The maximum marks are:
(a) 500 (b) 550
(c) 640 (d) 1008

53. A student who secures 20% marks in an examination fails by 30 marks. Another student who secures 32% marks gets 42 marks more than those required to pass. The percentage of marks required to pass is:
(a) 20 (b) 25
(c) 28 (d) 30

54. In an election between two candidates, the candidate who get 30% of the votes polled is defeated by 15000 votes. The number of votes polled by the winning candidate is:
(a) 11250 (b) 15000
(c) 26250 (d) 37500

55. The price of cooking oil has increased by 25 %. The per centage of reduction that a family should effect in the use of cooking oil so as not to increase the expenditure on this account is:
(a) 15% (b) 20%
(c) 25% (d) 30%

56. Water Tax is increased by 20% but its consumption is decreased by 20%. Then, the increase or decrease in the expenditure of the money is :
(a) No charge (b) 5% decrease
(c) 4% increase (d) 4% decrease

57. In a vocational course in a college, 15% seats increase annually. If there were 800 students in 1992, how many students will be there in 1994?
(a) 920 (b) 1040
(c) 1058 (d) 1178

58. A papaya tree was planted 2 years ago. It increase at the rate of 20% every year. If at present, the height of the tree is 540 cm, what was it when the tree was planted?
(a) 324 cms (b) 375 cms
(c) 400 cms (d) 432 cms

59. A man's basic pay for a 40 hour week is Rs. 20. Overtime is paid for at 25% above the basic rate. In a certain week he worked overtime and his total wage was Rs.25. He therefore worked for a total of:
(a) 45 hours (b) 47 hours
(c) 48 hours (d) 50 hours

60. A man's wages were decreased by 50%. Again the reduced wages were increased by 50%. He has a loss of:
(a) 0% (b) 0.25%
(c) 2.5% (d) 25%

61. The price of an article has been reduced by 25%. In order to restore the original price, thenew price must be increased by:
(a) $33\frac{1}{3}\%$ (b) $11\frac{1}{9}\%$
(c) $9\frac{1}{11}\%$ (d) $66\frac{2}{3}\%$

62. The boys and girls in college are in ratio 3 : 2. If 20% of the boys and 25% of the girls are adults, the percentage of students who are not adults is :
(a) 58% (b) 67.5%
(c) 78% (d) 82.5%

63. The price of rice has increased by 60%. In order to restore to the original price, the new price must be reduced by:
(a) $33\frac{1}{3}\%$ (b) $37\frac{1}{2}\%$
(c) 40% (d) 45%

64. In measuring the side of a square, an error of 5 % in excess is made. The error per cent in the calculated area is :
(a) 10% (b) 10.25%
(c) 10.5% (d) 25%

65. The length of a rectangle is increased by 60%. By what pe cent would the width have to be decreased to maintain the same area:
(a) $37\frac{1}{2}\%$ (b) 60%
(c) 75% (d) None of these

66. For a sphere of radius 10 cms, the numerical value of the surface area is how many percent of the numerical value of its volume
(a) 24% (b) 26.5%
(c) 30% (d) 45%

Answers

1. Ans. (c)
Solution:

$$\text{Given expression} = \left(\frac{860}{100}\times 50+\frac{50}{100}\times 860\right)$$
$$= 430+430 = 860$$

2. Ans. (d)
Solution:
Reqquired Percentage
$$= \left(\frac{18}{7200}\times 100\right)\% = \frac{1}{4}\% = 0.25\%$$

3. Ans. (a)
Solution:

$$\text{Rebate} = 6\% \text{ of } Rs.6650 = Rs.\left(\frac{6}{100}\times 6650\right)$$
$$= Rs.399$$
Sales tax
$$= 10\% \text{ of } (6650-399) = Rs.\left(\frac{10}{100}6251\right)$$
$$= Rs.625.10$$
∴ Final amount
$$= Rs.(6251+625.10) = Rs.6876.10$$

4. Ans. (b)
Solution: Let 95% of $x = 4598$.

Then, $\frac{95}{100} \times x = 4598$

or $x = \left(4598 \times \frac{100}{95}\right) = 4840$

5. Ans. (c)

Solution: Let 15% of x% of $582 = 17.46$

Then, $\frac{15}{100} \times \frac{x}{100} \times 582 = \frac{1746}{100}$

$x = \left(\frac{1746}{100} \times \frac{100 \times 100}{15 \times 582}\right) = 20$

6. Ans. (b)

Solution: Let the number b x. Then, $x - 40\%$ of $x = 30$

$\Leftrightarrow x = \frac{40}{100} x = 30 \Leftrightarrow x - \frac{2}{5} x = 30$

$\Leftrightarrow \frac{3x}{5} = 30 \Leftrightarrow x = \left(\frac{12 \times 100}{15}\right) = 80$

7. Ans. (d)

Solution: Let the number of x.

Then, $x - 35 = \frac{80}{100} x$

$\Rightarrow x = \frac{80}{100} x = 35$

$\Rightarrow x = \frac{35 \times 100}{20} = 175$

$\Rightarrow \frac{4}{5} x = 140$

8. Ans. (b)

Solution:

Let the number = x

Then, other number = 80% of $x = \frac{4}{5}$

$$4\left[x^2 + \left(\frac{4}{5}x\right)^2\right] = 656$$

$\therefore x^2 + \frac{16}{25} x^2 = 164$

$\Rightarrow \frac{41}{25} x^2 = 164$

$\Rightarrow x^2 = \left(\frac{164 \times 25}{41}\right) = 100$

$\Rightarrow x = 100$

So, the numbers are 10 and 8.

9. Ans. (b)

Solution: Let the number of candidates appeared from each state be x.

Then, 7% of x – 6% of $x = 80$

1% of $x = 80$ $x = 80 \times 100 = 8000$

10. Ans. (b)

Solution: Let the number of students appeared be x.

Then, 65% of $x = 455$

$\Rightarrow \frac{65}{100} x = 455$ $x = \left(\frac{455 \times 100}{65}\right) = 700$

11. Ans. (c)

Solution: Let the number of valid votes be x.

Then, 52% of x – 48% of $x = 98$ 4% of $x = 98$

$\frac{4}{100} x = 98$

$x = 98 \times 25 = 2450$

$\therefore$ Total number of votes polled $= (2450 + 68) = 2518$

12. Ans. (a)

Solution: Total marks secured

$= (90\%$ of $100 + 60\%$ of $150 + 54\%$ of $200)$

$= \left(\frac{90}{100} \times 100 + \frac{60}{100} \times 150 + \frac{54}{100} \times 200\right)$

$= (90 + 90 + 108) = 288$

Total maximum marks $= (100 + 150 + 200) = 450$

Aggregate percentage $= \left(\frac{288}{450} \times 100\right)\% = 64\%$

13. Ans. (c)

Solution: Let their marks be $(x+9)$ and x.

Then, $x + 9 = \frac{56}{100}(x + 9 + x)$

$\Rightarrow 25(x+9) = 14(2x+9)$

$\Rightarrow 3x = 99$

$\Rightarrow x = 33$

So, their marks are 42 and 33.

14. Ans. (b)

Solution:

$\Rightarrow\ y\%\text{ of } z = 2(x\%\text{ of } Y)$

$\Rightarrow\ \frac{yz}{100} = \frac{2xy}{100}$

$\Rightarrow\ z = 2x$

15. Ans. (a)

Solution: Let the total income be x.

Then, income left

$=(100-80)\%\text{ of }[100-(35+25)]\%$

of $x = 20\%$ of 40% of x.

$=\left(\frac{20}{100}\times\frac{40}{100}\times 100\right)\%$ of $x = 8x$ of x.

16. Ans. (c)

Solution: We have $x + x\%$ of $150 = 150$

$\Rightarrow\ x + \frac{x}{100}\times 150 = 150$

$\Rightarrow\ \frac{5}{2}x = 150$

$\Rightarrow\ x = \left(\frac{150\times 2}{5}\right) = 60$

17. Ans. (d)

Solution: Let the marks rquired be x.

Then, $(62+35+x) = 35\%$ of $(150+150+180)$

$97 + x = \frac{35}{100}\times 480$

$x = 168 - 97 = 71$

18. Ans. (c)

Solution: Let the total population be x. Then

Poor population $= \frac{60}{100}x = \frac{3}{5}x$.

Illiterate population $= \frac{40}{100}x = \frac{2}{5}x$

Illiterate rice $= 10\%$ of $(100-60)\%$

of $x = \left(\frac{10}{100}\times\frac{40}{100}\times x\right) = \frac{x}{25}$

Illiterater poor $= \left(\frac{2}{5}x - \frac{x}{25}\right) = \frac{9x}{25}$

Required percentage $= \left(\frac{9x}{25}\times\frac{5}{3x}\times 100\right)\% = 60\%$

19. Ans. (c)

Solution:

Let the number of boys and girls be $3x$ and $2x$ respectively.

Then, No. of students who are not adults

$=\left(\frac{80}{100}\times 3x\right)+\left(\frac{75}{100}\times 2x\right)=\left(\frac{12}{5}+\frac{3x}{2}\right)=\frac{39x}{10}$

Required percentage $= \left(\frac{39x}{10}\times\frac{1}{5x}\times 100\right)\%$

$= 78\%$

20. Ans. (a)

Solution: Let the original number be x.

Final number obtained $= 110\%$ of $(90\%$ of $x)$

$=\left(\frac{110}{100}\times\frac{90}{100}\times x\right) = \frac{99}{100}x$

$\Rightarrow\ x - \frac{99}{100}x = 10$

$\Rightarrow\ \frac{1}{100}x = 10$

$\Rightarrow\ x = 10\times 100 = 1000$

21. Ans. (a)

Solution:

Let original slary $= Rs.100$. New salary $= Rs.120$

Decrease on $120 = 20$.

decrease on $100 = \left(\frac{20}{120}\times 100\right)\% = 16\frac{2}{3}\%$

22. Ans. (c)

Solution:

Suppose the business value changes form x to y. 4% of $x = 5\%$ of y.

$\frac{4}{100}x = \frac{5}{100}y$

$y = \frac{4}{5}x$

$=\left(x - \frac{4}{5}x\right) = \frac{x}{5} = \left(\frac{x}{5}\times\frac{1}{x}\times\frac{1}{100}\right)\% = 20\%$

23. Ans. (b)

Solution:

Let the original consumption $=100$ kg and new consumption $=x$ kg

so, $100\times6=x\times7.50$ $x=80$ kg

$\therefore$ Reduction in consumption $=20\%$

24. Ans. (b)

Solution:

Present population $=62500\times\left(1-\frac{4}{100}\right)^2$

$=\left(62500\times\frac{24}{25}\times\frac{24}{25}\right)=57600$

25. Ans. (a)

Solution: Number of bushes in the beginning

$=\frac{26730}{\left(1+\frac{10}{100}\right)\left(1+\frac{8}{100}\right)\left(1-\frac{10}{100}\right)}$

$=\left(26730\times\frac{10}{11}\times\frac{25}{27}\times\frac{10}{9}\right)=25000$

26. Ans. (d)

Solution:

The original population $=100$

Population after 3 years $=100\times\left(1+\frac{3\frac{1}{2}}{100}\right)^3$

$=100\times\frac{207}{200}\times\frac{207}{200}\times\frac{207}{200}=110.87$

Increase $=(110.87-100\%)$

$=10.87\%=10.8\%$

27. Ans. (c)

Solution:

$p=6q$. So, q is less than p by 5q.

Requried percentage $=\left(\frac{5q}{p}\times100\right)\%$

$=\left(\frac{5q}{6p}\times100\right)\%=83\frac{1}{3}\%$

28. Ans. (b)

Solution:

$P=\frac{140}{100}A=\frac{140}{100}\left(\frac{80}{100}M\right)$

$=\left(\frac{140}{100}\times\frac{80}{100}\times100\right)\%$

$M=112\%$ of M.

29. Ans. (a)

Solution: Let the reduced weight be x kg. The quantity of pulp reminas the same in both the cases.

So, $(100-96)\%$ of $20\,kg=(100-95)\%$ of $x\,kg$

$\Rightarrow$ 4% of $20\,kg=5\%$ of $x\,kg$

$\Rightarrow$ $x=\left(\frac{4}{5}\times20\right)kg=16\,kg$

30. Ans. (b)

Solution: Required percentage

$=\left(\frac{20\%\text{ of }10+35\%\text{ of }4}{10+4}\times100\right)\%$

$=\left(\frac{3.4}{14}\times100\right)\%=24\frac{2}{7}\%$

31. Ans. (c)

Solution:

Total money $=Rs.\left(600\times\frac{25}{100}+1200\times\frac{50}{100}\right)$

$=Rs.750$

25 paise coins removed $=\left(\frac{12}{100}\times600\right)=72$

25 paise coins removed $=\left(\frac{24}{100}\times1200\right)=288$

Money removed $=Rs.\left(72\times\frac{25}{100}+288\times50\right)$

$=Rs.162$

$\therefore$ Required percentage $=\left(\frac{162}{750}\times100\right)\%$

$=21.6\%$

32. Ans. (a)

Solution:

$n(A)=34$, $n(B)=42$, $n(A\cap B)=20$

So, $n(A\cup B)=n(A)+(B)-n(A\cap B)$

$34+42-20=56$

$\therefore$ Percentage failed in either or both the subject $=56$

Hence, percentage passed $=(100-56)\%=44\%$

33. Ans. (c)

Solution: Let the total number x. Then

$n(A)=\frac{72}{100}x=\frac{18x}{25}$, $n(B)=\frac{44}{100}x=\frac{11x}{25}$ and

$n(A\cap B)=40$

$n(A\cup B)=n(A)+n(B)-n(A\cup B)$

$x=\frac{18x}{25}+\frac{11x}{25}-40$

$\Rightarrow \frac{29}{25}-x=40$

$\Rightarrow \frac{4x}{25}=40$

$\Rightarrow x=250$

Percentage

34. Ans. (b)

Solution:

$0.2=\left(\frac{2\%}{100}\times 100\right)\%=2\%$

35. Ans. (b)

Solution: Let $x\%$ of $130=10.4$

Then $\frac{x}{100}\times 130=10.4$

or $x=\frac{10.4\times 100}{130}=8$

36. Ans. (c)

Solution: Let $x\%$ of $250+25\%$ of $68=67$.

Then, $\frac{x}{100}\times 250+\frac{25}{100}\times 68=67$

or, $\frac{5x}{2}=50$

$\therefore x=\left(\frac{50\times 2}{5}\right)=20$

37. Ans. (c)

Solution: Let $x\%$ of $20=0.05$

Then, $\frac{x}{100}\times 20=0.05$ or $x=0.25$

38. Ans. (c)

Solution: Let $\frac{30\% \text{ of } 80}{x}=24$

Then, $24x=\left(\frac{30}{100}\times 80\right)$ or

$x=\left(\frac{30\times 80}{24\times 100}\right)=1$

39. Ans. (d)

Solution:

$x=90\%$ of $y\Rightarrow x=\frac{90}{100}y$

Let $z\%$ of $x=y$. Then, $\frac{z}{100}\times x=y$

or $\frac{z}{100}\times\frac{90}{100}y=y$ or $z=\frac{100\times 100}{90}-111.1$

40. Ans. (c)

Solution:

Required number $=(80-60\%\text{ of } 80)$

$=\left(80-\frac{60}{100}\times 80\right)=32$

41. Ans. (c)

Solution:

$33\frac{1}{3}\%=\left(\frac{100}{3}\times\frac{1}{100}\right)=\frac{1}{3}=0.33$

$\frac{4}{15}=0.26$

Clearly $0.35>0.33>0.26$.

$\therefore$ 0.35 is greatest.

42. Ans. (a)

Solution:

$\frac{0.5}{100}$ of $x=\frac{85}{100}$ or $x=Rs.\left(\frac{85}{0.5}\right)=Rs.170$

43. Ans. (c)

Solution:

Required percent $=\left[\frac{30}{2\times 10}\times 100\right]\%$

$=150\%$

44. Ans. (c)

Solution: Let $37\frac{1}{2}\%$ of $x = 900$.

Then, $\frac{75x}{2\times100} = 900$

$x = \frac{900\times2\times100}{75} = 2400$

So, $62\frac{1}{2}\%$ of $x = \left(\frac{125}{2}\times\frac{1}{100}\times2400\right) = 1500$

45. Ans. (c)

Solution: Let, original price = Rs. 100. Then, new price = Rs. 90.

Increase on Rs. 90 = Rs.10

Increase % $= \left(\frac{10}{90}\times100\right)\% = 11\frac{1}{9}\%$.

46. Ans. (b)

Solution: Let salary be Rs. x. Then,

$x - 15\%$ of $x - 30\%$ of 85% of $x = 2380$

or $x - \frac{15x}{100} - \frac{30\times85\times x}{100\times100} = 2380$

or $200x - 30x - 51x = 2380\times200$

or $119x = 2380\times200$ or

$x = \frac{2380\times200}{119} = 4000$

47. Ans. (b)

Solution: $137\frac{1}{2}\%$ of $x = 33$.

$\therefore \frac{275}{2}\times\frac{1}{100} = 33$ or $x = \frac{33\times2\times100}{275} = 24$

48. Ans. (b)

Solution: 96% of $x = 23040$

$\therefore x = \frac{23040\times100}{96} = 24000$

49. Ans. (a)

Solution: Let total number of student be x.

Then, 30% of $x = 504$ or

$x = \frac{504\times100}{30} = 1680$

$\therefore$ Number of boys $= (1680 - 504) = 1176$

50. Ans. (d)

Solution:

35% of $x = 420$ or $x = \frac{420\times100}{35} = 1200$

51. Ans. (a)

Solution: 30% of $180 + x\%$ of 150

$= 50\%$ of $(180+150)$

or $54 + \frac{x}{100}\times150 = 165$

or $\frac{3x}{2} = 111$ or $x = \frac{111\times2}{3} = 74$

52. Ans. (b)

Solution: 36% of $x = (113+85)$

or $x = \frac{100\times198}{36} = 550$

53. Ans. (b)

Solution: 20% of $x + 30 = 32\%$ of $x - 42$

or 12% of $x = 72$, $x = \frac{72\times100}{12} = 600$

Pass marks $= 20\%$ of $600 + 30 = 150$

Pass percentage $= \left(\frac{150}{600}\times100\right)\% = 25\%$

54. Ans. (c)

Solution:

Let the votes polled by the winning candidate be x.

Then, $(x - 15000) = 30\%$ of $[x + (x - 15000)]$ or

$x = 26250$

55. Ans. (b)

Solution:

Reduction in consumption $= \left(\frac{25}{125}\times100\right)\%$

$= 20\%$

56. Ans. (d)

Solution: Let Tax = Rs. 100 and Consumption = 100 units.

Orginal Expenditure = Rs. (100×100) = Rs. 10000.

New Expenditures = Rs. (120×80) = Rs. 9600.

$\therefore$ Decrease in expenditure $= \left(\frac{400}{10000} \times 100\right)\%$

$= 4\%$

57. Ans. (c)
Solution:
Required number

$$= 800 \times \left(1 + \frac{15}{100}\right)^2 = \left(800 \times \frac{23}{20} \times \frac{23}{20}\right) = 1058$$

58. Ans. (b)
Solution:

$$540 = x\left(1 + \frac{20}{100}\right)^2 \text{ or } x = \left(540 \times \frac{5}{6} \times \frac{5}{6}\right) = 375$$

59. Ans. (c)
Solution:

Basic rate per hour $= \text{Re.}\left(\frac{20}{40}\right) = \text{Re.}\frac{1}{2}$

Overtime per hour $= 125\%$ of $\text{Re.}\frac{1}{2} = \text{Re}\frac{5}{8}$

Suppose he worked x hours overtime.

Then, $20 + \frac{5}{8}x = 25$ or $\frac{5}{8}x = 5$

$\therefore \quad x = \frac{5 \times 8}{5} = 8$ hours

So, he worked in all for $(40 + 8)$ hours $= 48$ hours

60. Ans. (d)
Solution: Let original wages = Rs. 100.
Reduced wages = Rs. 50.
Increased wages = 150 % of Rs. 50

$= \left(\frac{150}{100} \times 50\right) =$ Rs. 75.

$\therefore$ Loss = 25 %

61. Ans. (c)
Solution: Let, original price = Rs. 100
Reduced price = Rs 75.
Increase on Rs. 75 = Rs. 25

Increase on Rs. 100 $= \left(\frac{25}{75} \times 100\right)\%$

$= 33\frac{1}{3}\%$

62. Ans. (c)
Solution: Suppose boys $3x$ and girls $= 2x$
Not adults $= (80\% \text{ of } 3x) + (75\% \text{ of } 2x)$

$$= \left(\frac{12x}{5} + \frac{3x}{2}\right) = \frac{30x}{10}$$

Required percentage $= \left(\frac{39x}{10} \times \frac{1}{5x} \times 100\right)\%$

$= 78\%$

63. Ans. (b)
Solution: Let, original price = Rs. 100
Increased price = Rs 160.
Decrease on Rs. 160 = Rs. 60

Increase on Rs. 100 $= \left(\frac{60}{160} \times 100\right)\%$

$= 37\frac{1}{2}\%$

64. Ans. (b)
Solution: Let actual side = 100 cm
Measured length = 105 cm
Error in area $= (105)^2 - (100)^2$

$= (105 + 100)(105 - 100) = 1025$

Error $\% = \left(\frac{1025}{10000} \times 100\right)\% = 10.25\%$.

65. Ans. (a)
Solution: Let length = 100 m, breadth = 100 m
New length = 160 m, new breadth $= x$ meters.
Then, $160 \times x = 100 \times 100$

or $x = \frac{100 \times 100}{160} = \frac{125}{2}$

Decrease in breadth $= \left(100 - \frac{125}{2}\right)\% = 37\frac{1}{2}\%$

66. Ans. (c)
Solution:

Surface are $= 4\,\pi \times (100)^2 = (400\,\pi)\,cm^2$

Volume $= \frac{4}{3}\pi \times (10)^3 = \left(\frac{4000\pi}{3}\right) cm^3$

$\therefore$ Required percentage

$$= \left(400\pi \times \frac{3}{4000\pi} \times 100\right)\% = 30\%$$

Profit and Loss

When one person enters into a transaction, he buys the article or manufactures the article at a certain price known as Cost Price. Then he tries to sell at prices more than the cost price, the price at which he sells the article is known as selling price.

If he sells the article at more than the cost price, he gets the profit.

Profit = selling price – cost price

But if he sells the article at lower price than the const prices than he gets the loss.

Loss = Cost price – selling price

Same way when person decides to sell the article above the selling price or the price at which he wants to sell it. The price that he marks is known as **Marked Price**. He writes more price on the article because everybody asks for decreasing the price, which is written or marked. So decrease in price is known as discount.

Discount = Marked price – selling price

At times 2 customers come for the same article; the shopkeeper sells the article above the marked price in known as premium.

Premium = selling price – marked price

Profit % and loss%

Profit is the difference between selling price and cost price and is taken on the cost price.

Similarly, if a person gets some loss, then

$$Loss\% = \frac{Loss}{Cost\ price} \times 100$$

Discount% and premium%

Profit on the cost price is the same as premium on the Marked price and loss on the cost price is same as Discount on the Marked priced, i.e.

- A man purchases an article for Rs 2000 and sells for Rs 2500 gets a premium of Rs 500.
- A man purchases an article of Rs 2000 and sells for Rs 1600 gets a loss of Rs 400. So

Premium % = Premium/Marked price × 100 and
Discount % = Discount/Marked price × 100

There is no direct relationship between const price and Marked price. But both are related to selling price in some way.

$$CP + profit = SP$$

$$CP + \left(\frac{P\%}{100} \times CP\right) = SP$$

$$\left(100 + \frac{P\%}{100}\right) \times CP = SP$$

Same way if it is loss than

$$\left(100 - \frac{L\%}{100}\right) CP = SP$$

Selling price and marked price are related as

MP + Premium = SP

or $$\left(100 + \frac{Premium\%}{100}\right) MP = SP$$

or $$\left(100 - \frac{Discount\ \%}{100}\right) MP = SP$$

Exercise

1. Sam purchased 20 dozen of toys at the rate of Rs. 375 per dozen . He sold each one of them at the rate of Rs. 33.What his percentage profit?
(a) 3.5 (b) 4.5
(c) 5.6 (d) 6.5

2. A shopkeeper expects a gain of $22\frac{1}{2}\%$ on his cost price. If in a week, his sale of Rs. 392, what was his profit?
(a) Rs.18.20 (b) RS. 70
(c) Rs.72 (d) Rs. 88.25

3. When a plot is sold for Rs. 18,700, the owner loses 15%.At what price must the plot be sold in order to gain 15%?
(a) Rs. 21,000 (b) Rs. 22,500
(c) Rs. 25,300 (d) Rs. 25,800

4. A man gain 20% by selling an articale for a certain price. If he sells it at double the price, the percentage of profit will be :
(a) 40 (b) 100
(c) 120 (d) 140

5. If loss is $\frac{1}{3}$ of S.P., the loss percentage is:
(a) $16\frac{2}{3}$ (b) 20%
(c) 25% (d) $33\frac{1}{3}\%$

6. The cost price of 19 articles is equal to the selling priceof 16 articles. Gain percent is :
(a) $3\frac{9}{17}\%$ (b) $15\frac{15}{19}\%$
(c) $18\frac{3}{4}\%$ (d) 20%

7. On selling 17 balls at Rs.720, there is a loss equal to the cost price of a ball is:
(a) Rs. 45 (b) Rs. 50
(c) Rs. 55 (d) Rs. 60

8. A man purchased a box full of pencil at the rate of 7 for Rs. 9 and sold alll of them at the rate of 8 Rs. 11. In this transaction, he gained Rs. 10. How many pencils did the box contain?
(a) 100 (b) 112
(c) 114 (d) 115

9. A trader mixes 26 Kg of rise at Rs. 20 per kg with 30 kg of rise of other variety at Rs. 36 per kg and sells the mixture at Rs. 30 per kg. His profit percent is:
(a) NO profit,no loss (b) 5%
(c) 8% (d) 10%

10. A dishonest dealer uses a scale of 90 cm instead of a meter scale and claims to sell at cost price. His profit is:
(a) 9% (b) 10%
(c) 12% (d) None of these

11. A fair price shopkeeper takes 10% profit on his goods. He lost 20% goods during theft. His loss percent is :
(a) 8 (b) 10
(c) 11 (d) 12

12. An acticle was sold for Rs. 144. If the percentage of profit was numerically equal to the cost price, the cost of the article was
(a) 72 (b) 80
(c) 90 (d) 100

13. If a man reduces the selling price of a fan from Rs. 400 to Rs. 380, his loss increases by 2%. The cost price of the fan is
(a) Rs. 480 (b) Rs. 500
(c) Rs. 600 (d) None of these

14. The each difference between the selling prices of an article at a profit of 4% and 6% is Rs. 3. The ratio of the two selling prices is
(a) 51:52 (b) 52:53
(c) 51:53 (d) 52:55

15. A cloth merchant sold half of his cloth at 20% profit, half of the remaining at 20% loss and the rest was sold at the cost price. In the total transaction, his gain or loss will be.
(a) Neither loss nor gain
(b) 5% loss
(c) 5% gain
(d) 10% gain

16. Two-thrid of a consignment was sold at a profit of 5% and the remainder at a loss of 2%. If the total profit was Rs. 400, the value of the consignment (in Rs.) was:
(a) 10,000 (b) 12,000
(c) 15,000 (d) 20,000

17. On selling a chair at 7% loss and a table at 17% gain, a main gain Rs. 296. If he sells the chair at 7% gain and the table at 12% gain, then he gain Rs. 400. The actural price of the table is:
(a) Rs. 1600 (b) Rs.1800
(c) Rs. 2200 (d) Rs.2400

18. An article was sold for Rs. y after giving a discount of x%. Then, its list price is:
(a) $\frac{100y}{100-x}$ (b) $\frac{100y}{1-x}$
(c) $\frac{100y}{1-(x/100)}$ (d) None of these

19. Find the selling price of an article if a shopkeeper allows two successive discount of 5% each on the marked price of Rs. 80.
(a) Rs. 70.10 (b) Rs. 7020
(c) Rs. 72 (d) Rs. 70.20

20. The price of an article is raised by 30% and then two successvie discount of 10% each are allowed. Ultimately, the price of the article is.
(a) 1 (b) 2
(c) 3 (d) 3

21. At what percent above the cost prime must a shopkepeer mark his goods so that he gains 20% even after giving a discont of 10% on the marked price.
(a) 25% (b) 30%
(c) $33\frac{1}{3}$% (d) $37\frac{1}{2}$%

22. Kunal bought a suitcase with 15% discount on the labelled price. He sold the suitcase for Rs. 2880 with 20% profit on the labelled price. At what price did he buy the suitcase?
(a) Rs. 2040 (b) Rs. 2400
(c) Rs. 2640 (d) None of these

23. A man boys 10 articles for Rs. 8 and sells them at the rate of Rs. 1.25 per article. His gain is
(a) 20% (b) 50%
(c) $19\frac{1}{2}$% (d) $56\frac{1}{4}$%

24. There would be 10% loss if a toy is sold at Rs. 10.80 per piece. At what price should it be sold to earns a profit of 20%

(a) Rs. 12 (b) Rs. 12.96
(c) Rs. 14.40 (d) None of these

25. There would be 10% loss if rice is sold at Rs. 5.40 per kg. At what price per kg shold it be sold to earn a profit of 20%?
(a) Rs. 6 (b) Rs. 6.48
(c) Rs. 7.02 (d) Rs. 7.20

26. A man sells 320 mangoes at the cost porce of 400 mangoes. His gain per cent is:
(a) 10% (b) 15%
(c) 20% (d) 25%

27. If books bought at prices ranging from Rs.200 of Rs.350 are sold at prices ranging from Rs. 300 to Rs. 425, what is the greatest possible profit that might be made in selling 8 book?
(a) Rs. 400
(b) Rs. 600
(c) Cannot be determined
(d) None of these

28. If 1 purchased 11 books for Rs. 10 and sold all the books at the rate of 10 books for Rs. 11, the profit per cent is
(a) 10% (b) 11%
(c) 21% (d) 100%

29. By selling 36 oranges, a vendor loses the selling price of 4 oranges. His loss per cent is:
(a) $12\frac{1}{2}\%$ (b) $11\frac{1}{9}\%$
(c) 10% (d) None of these

30. The C.P. of an artice is 40% of the S.P. The per cent that the S.P. is of C.P. is
(a) 40 (b) 60
(c) 240 (d) 250

31. Aloke bought 25 kg of rice at the rate of Rs. 6 per kg and 35 kg of rice at the rate of Rs. 7 per kg. He mixed the two and sold the mixture at the rate of the Rs. 6.75 per kg. What was is gain or loss inthe transaction?
(a) Rs. 16 gain (b) Rs. 16 loss
(c) Rs. 20 gain (d) None of these

32. A dealer professing to sell at cost price, uses a 900 gms weight for a kilogram. His gain per cent is:
(a) 9 (b) 10
(c) 11 (d) $11\frac{1}{9}$

33. By selling 12 oranges for one rupee, a man loses 20%. How many for a rupee should he sell to get a gain of 20%?
(a) 5 (b) 8
(c) 10 (d) 15

34. Jimmy bought paper sheets for Rs. 7200 and spent Rs. 200 on transport. Paying Rs. 600 he had 330 boxes made, which he sold at Rs. 28 each. What is his profit percentage?
(a) 15.5 (b) 40 toffees
(c) 60 (d) None of these

35. A owns a hous worth Rs. 10000. He sells it to B at a profit of 10% based on the worth of the house. B sells the house back to A at a loss of 10%. In this transaction A gets
(a) No profit no loss
(b) Profit of Rs. 1000
(c) Profit of Rs. 1100
(d) Profit of Rs. 2000

36. When the price of a toy was increased by 20%, the number of toys sold wes decreased by 15% what was the effect on the sales of the shop?
(a) 4% increase (b) 4% decrease
(c) 2% increase (d) 2% decrease

37. When the price of fans was reduced by 20%, the number of fans sold increased by 40%. What was the effect on the sales in rupees?
(a) 12% increase (b) 12% decrease
(c) 30% increase (d) 40% increase

38. A man sold two houses for Rs. 7.81 lakhs each. On one he gained 5% and on the other he lost 5%. What per cent is the effect of the sale on the whole?
(a) 0.25% loss (b) 0.25% gain
(c) 25% loss (d) 25% gain

39. A horse and a cow were sold for Rs. 12000 each. The horse was sold at a loss of 20% and the cow at a gain of 20%. The entire transaction resulted in
(a) No loss or gain (b) Loss of Rs. 100
(c) Gain of Rs 1000 (d) Gain of Rs 2000

40. Profit after selling a commodity for Rs. 425 is same as loss after selling it for Rs.355. The cost of the commdity is
(a) Rs. 385 (b) Rs. 390
(c) Rs. 395 (d) Rs. 400

41. If two mixers and one T.V. cost Rs. 7000, while two T.V. and one mixer cost Rs. 9800, the value of one T.V. is
(a) Rs. 2800 (b) Rs. 2100
(c) Rs. 4200 (d) Rs. 8400

42. Bhajan Singh purchased 120 reams of paper at Rs. 80 per ream. He spent Rs. 280 on transportation, paid octroi at the rate of 40 paise per ream and paid

Rs. 72 to the coolie. If he wants to have a gain of 8% what must be the selling price per ream

(a) Rs. 86 (b) Rs. 87.48
(c) Rs. 89 (d) Rs. 90

43. The cost price of an article,which on being sold at a gain of 12% yields Rs. 6 more than when it is sold a loss of 12%, is

(a) Rs. 30 (b) Rs. 25
(c) Rs. 24 (d) Rs. 20

44. A man sells an article at at gain of 15 %. If he had bought it at 10% less and sold it for Rs. 4 less, he would have gined 25%. The C.P. is the article is

(a) Rs. 140 (b) Rs. 150
(c) Rs. 160 (d) Rs. 180

45. A radio dealer sold a radio at a loss of 2.5%. Had he sold it for Rs.100 more, he would have gained $7\frac{1}{2}\%$. In order to gain $12\frac{1}{2}\%$, he should sell it for

(a) Rs. 850 (b) Rs. 925
(c) Rs. 1080 (d) Rs. 1125

46. Subhas purchased a taperecorder at $\frac{9}{10}$th of its selling price and sold it at 8% more than its S.P. His gain is

(a) 8% (b) 10%
(c) 18% (d) 20%

47. A discount series of 10%, 20% and 40% is equal to a single discount of

(a) 50% (b) 56.8%
(c) 60% (d) 70.28%

48. The ratio of the prices of three different types of cars is $4:5:7$. If the difference between the costliest and the chepest cars is Rs. 60000, the proce of the car of modest price is

(a) Rs. 80000 (b) Rs. 100000
(c) Rs. 140000 (d) Rs. 120000

49. Kabir buys an article with 25% discount on its marked price. He makes a profit of 10% selling it at Rs. 660. The marked price is

(a) Rs. 600 (b) Rs. 700
(c) Rs. 800 (d) Rs. 685

50. A retailer buys a sewing machine at a discount of 15% and sells it for Rs. 1955. Thus, he makes a profit of 15%. The discount is

(a) Rs. 270 (b) Rs. 290
(c) Rs. 300 (d) None of these

51. A shopkeeper earns a profit of 12% on selling a book at 10% discount on the printed price. The ratio of the cost price and the printed price of the book is

(a) $45:56$ (b) $50:61$
(c) $99:125$ (d) None of these

52. While selling a watch, a shopkeeper gives a discount of 5%. If he gives a discount of 7%, he earns Rs. 15 less as profit. The maked price of the watch is

(a) Rs. 697.50 (b) Rs. 712.50
(c) Rs. 787.50 (d) None of these

53. A trader lists his articles 20% above C.P. and allows a discount of 10% on cach payment. His gain percent is

(a) 10% (b) 8%
(c) 6% (d) 5%

54. Tarun bought a T.V. with 20% discount on the labelled price. Had he bought it with 25% discount,he would have saved Rs. 500. At what price did he buy the T.V?

(a) Rs. 5000 (b) Rs. 10000
(c) Rs. 12000 (d) Rs. 16000

55. A dealer marks his goods 20% above cost price. He then allows some discount on it and makes a profit of 8%. The rate of discount is

(a) 12% (b) 10%
(c) 6% (d) 4%

Answers

1. Ans. (c)
Solution:

$$\text{C.P of 1 toy} = Rs.\left(\frac{375}{12}\right) = Rs.31.25.$$

$$\text{S.P of 1 toy} = Rs.33$$

$$\therefore \quad \text{Profit \%} = \left(\frac{1.75}{31.25}\times 100\right)\%$$

$$= \frac{28}{5}\% = 5.6\%$$

2. Ans. (c)
Solution:

$$\text{C.P} = Rs.\left(\frac{100}{122.50}\times 392\right) = Rs.\left(\frac{100}{1225}\times 392\right)$$

$$= Rs.320$$

$$\therefore \quad \text{Profit \%} = Rs.(392-320) = Rs.72$$

3. Ans. (c)
Solution:

$$85:18700 = 115 = 115:x$$

or $x = \left(\frac{18700 \times 115}{85}\right) = 25300$

Hence S.P $= Rs. 25,300$

4. Ans. (d)
Solution:
Let C.P $= Rs. x$ Then, S.P $= Rs(120\% \text{ of } x)$

$= Rs.\frac{6x}{5}$

So, S.P. $= Rs.\left(2 \times \frac{6x}{5}\right) = Rs\frac{12x}{5}$. Profit

$= Rs.\left(\frac{12x}{5} - x\right) = Rs.\frac{7x}{5}$

$\therefore$ Profit% $= \left(\frac{7x}{5} \times \frac{1}{x} \times 100\right)\% = 140\%$

5. Ans. (c)
Solution: Let S.P. $= Rs. x$.

Then, $= Rs.\frac{x}{3}$. C.P. $= Rs.\left(x + \frac{x}{3}\right) = Rs.\frac{4x}{3}$

$\therefore$ Loss% $= \left(\frac{x}{3} \times \frac{3}{4x} \times 100\right)\% = 25\%$

6. Ans. (c)
Solution:
Let C.P. of each article be Re. 1
Then, C.P. of 16 articles $= Rs.16$, S.P of 16 articles $= Rs.19$

$\therefore$ Gain % $= \left(\frac{3}{16} \times 100\right)\% = 18\frac{3}{4}\%$

7. Ans. (d)
Solution:
(C.P. of 17 balls) - (S.P. of 17 balls) = (C.P. of 5 balls)
$\Rightarrow$ C.P of 12 balls = S.P. of 17 balls $= Rs.720$

C.P. of 1balls $= Rs.\left(\frac{720}{12}\right) = Rs.60$

8. Ans. (b)
Solution:
Suppose, number of pencils bought = LCM of 7 and 8 = 56

C.P. of 56 pencils $= Rs\left(\frac{9}{7} \times 56\right) = Rs.72$

S.P. of 56 pencils $= Rs.\left(\frac{11}{8} \times 56\right) = Rs.77$

Now, Rs. 5 are gained on 56 pencils

So, Rs. 10 are gained on $\left(\frac{56}{5} \times 10\right) = 112$

9. Ans. (b)
Solution:
C.P. of 56 kg rice $= Rs.(26 \times 20 + 30 \times 36)$
$= Rs.(520 + 1080) = Rs.1600$

S.P of 56 kg rice $= Rs.(56 \times 30) = Rs.1680$

$\therefore$ Gain $= \left(\frac{80}{1600} \times 100\right)\% = 5\%$

10. Ans. (d)
Solution:

Gain% $= \left(\frac{10}{90} \times 100\right)\% = 11\frac{1}{9}\%$

11. Ans. (d)
Solution:
Suppose he has 100 items. Let C.P. of each item be Re. 1.
Total cost = Rs. 100. Number of items left after theft = 80
S.P of each item = Rs. 1.10
$\therefore$ Total sale $= Rs.(1.10 \times 80) = Rs.88$

Hence, Loss% $= \left(\frac{12}{100} \times 100\right)\% = 12\%$

12. Ans. (b)
Solution:
Let C.P. $= Rs. x$, Profit % $= x\%$ and S.P. $= Rs.144$

$\therefore$ $x = \left[\frac{100}{(100 + x)} \times 144\right]$

$\Rightarrow x^2 + 100x = 14400$

$\Rightarrow x^2 + 100x - 14400 = 0$

$\Rightarrow x^2 + 180x - 80x - 14400 = 0$

$\Rightarrow (x + 180)(x - 80) = 0$

$\Rightarrow x = 80$

13. Ans. (d)
Solution:
Let C.P. be $= Rs. x . (400 - 380) = 20$
Then, 2%

$\Rightarrow \quad \frac{x}{50} = 20$

$\Rightarrow \quad x = 1000$

14. Ans. (b)

Solution:

Let C.P. of the article be $Rs. x$.

Then, Required ratio $= \frac{104\% \text{ of } x}{104\% \text{ of } x} = \frac{104}{106} = \frac{52}{53}$

$= 52:53$

15. Ans. (c)

Solution:

Let C.P. of whole be $Rs. x$. C.P. of $\frac{1}{2}$ stock $= Rs.\frac{x}{2}$,

C.P. of $\frac{1}{4}$ Stock $= Rs.\frac{x}{4}$

Total S.P.

$= Rs.\left[\left(120\% \text{ of } \frac{x}{2}\right)+\left(80\% \text{ of } \frac{x}{4}\right)+\frac{x}{4}\right]$

$= Rs.\left(\frac{3x}{5}+\frac{x}{5}+\frac{x}{4}\right) = Rs.\frac{21x}{20}$

Gain $= Rs.\left(\frac{21x}{20}-x\right) = Rs.\frac{x}{20}$

$\therefore$ Gain % $=\left(\frac{x}{20}\times\frac{1}{x}\times 100\right)\% = 5\%$

16. Ans. (c)

Solution:

Let the total value be Rs. x. Value of $\frac{2}{3}rd = Rs.\frac{2x}{3}$,

value of $\frac{1}{3}rd = Rs.\frac{x}{3}$

Total $Rs.\left(105\% \text{ of } \frac{2x}{3}\right)+\left(98\% \text{ of } \frac{x}{3}\right)$

$= Rs.\left(\frac{210x}{300}+\frac{98x}{300}\right) = Rs.\frac{308x}{300}$

$= Rs.\frac{308x}{300} - x = 400 = \frac{8x}{300} = 400$

$x = \left(\frac{400\times 300}{8}\right) = 15000$

17. Ans. (d)

Solution:

Let C.P. of the chair be $Rs. x$ and that of the table be $Rs. y$.

Then, 17% of $y - 7\%$ of $x = 296$

$\Rightarrow 17y - 7x = 29600$

And, 12% of $y + 7\%$ of $x = 400$

$\Rightarrow 12y + 7x = 40000$

Solving (1) and (2), we get

$y = 2400$ and $x = 1600$

$\therefore$ C.P. of table $= Rs. 2400$

18. Ans. (a)

Solution:

Let the list price be Rs. z

$(100-x)\%$ of $z = y$

$\Rightarrow \left(\frac{100-x}{100}\right)\times z = y$

$\Rightarrow z = \left(\frac{100y}{100-x}\right)$

19. Ans. (d)

Solution:

S.P. $= 95\%$ of 85% of

$Rs. 80 = Rs.\left(\frac{95}{100}\times\frac{95}{100}\times 80\right) = Rs. 72.20$

20. Ans. (c)

Solution:

Let the original price be Rs. 100. Then marked price = Rs. 130.

Final price = 90% of 90% of Rs. 130

$= Rs.\left(\frac{90}{100}\times\frac{90}{100}\times 130\right) = Rs. 105.30$

21. Ans. (c)

Solution:

Let C.P. $= Rs. 100$.

Then, S.P $= Rs. 120$

Let marked price be Rs x.

Then, 90% $x = 120$

$x = \frac{120\times 100}{90} = 133\frac{1}{3}$

$\therefore$ Marked price $= 33\frac{1}{3}\%$ above C.P.

22. Ans. (a)
Solution:
Let the labelled price be. x. Then, 120% of $x = 2880$

$$x = \left(\frac{2880 \times 100}{120}\right) = 2400$$

$$C.P = 85\% \text{ of } = Rs.2400$$

$$= Rs.\left(\frac{85}{100} \times 2400\right) = Rs.2040$$

23. Ans. (d)
Solution:
C.P. of articales $= Rs.8$

S.P of 10 articles $= Rs.(1.25 \times 10) = Rs.12.50$

$$\therefore \text{ Gain} = \left(\frac{4.5}{8} \times 100\right)\% = 56\frac{1}{4}\%$$

24. Ans. (c)
Solution:
$90 : 10.80 :: 120 : x$

$$\therefore \; = \frac{10.80 \times 120}{90} = 14.40$$

Hence, the toy must be sold for Rs. 14.40.

25. Ans. (d)
Solution:
$90 : 5.40 :: 120 : x$

$$\therefore \; x = \frac{5.40 \times 120}{90} = 7.20$$

So, the rice must be sold at Rs. 7.20 per kg

26. Ans. (d)
Solution:
Let C.P. of each mango $= Rc.1$

C.P. of 320 mangoes $= Rs.320$

S.P. of 320 mangoes $= Rs.400$

$$\text{Gain} = \left(\frac{80}{320} \times 100\right)\% = 25\%$$

27. Ans. (d)
Solution:
Profit is maximum when C.P. is minimum and S.P. is maximum.

Thus, C.P. $= Rs.(200 \times 8) = Rs.1600$

S.P. $= Rs.(425 \times 8) = Rs.3400$

$\therefore$ Gain $= Rs.1800$

28. Ans. (c)
Solution:
Suppose I purchansed 110 books.

$$C.P. = Rs.\left(\frac{10}{11} \times 110\right) = Rs.100$$

$$S.P. = Rs.\left(\frac{11}{10} \times 110\right) = Rs.121$$

$\therefore$ Gain% $= 21\%$

29. Ans. (c)
Solution:
Loss = (C.P. of 36 oranges) - (S.P. of 36 oranges)
(S.P. of 4) = (S.P. of 36) - (S.P. of 36)
(S.P. of 40) = (C.P. of 36)
Let C.P. of each $=$ Re1
C.P. of $40 = Rs.40$, S.P. of $40 = Rs.36$

$$\text{Loss} \left(\frac{4}{40} \times 100\right)\% = 10\%$$

Let S.P. $= Rs.\, x$
Then, $400 + 20\%$ of $x = x$

$$\text{or } 400 + \frac{x}{5} = x$$

$$\text{or } \frac{4x}{5} = 400 \text{ or } x = \frac{400 \times 5}{4} = 500$$

$\therefore$ S.P. $= Rs.\, 500$

30. Ans. (d)
Solution:
Let S.P. $= Rs.100$
Then, C.P. $= Rs.\, 40$

$$\therefore \text{ Required per cent} = \left(\frac{100}{40} \times 100\right)\% = 250\%$$

31. Ans. (d)
Solution:
C.P. of 60 kg mix $= Rs.(25 \times 6 + 35 \times 7) = Rs.\, 395$

S.P. of 60 kg mix $= Rs.(60 \times 6.75) = Rs.\, 405$

Gain $= Rs.(405 - 395) = Rs.\, 10$

32. Ans. (d)
Solution:
Let the error be x grms. Then

$$\frac{x}{1000 - x} \times 100 = \frac{300}{47}$$

$$\text{or } \frac{x}{100 - x} = \frac{3}{47}$$

33. Ans. (b)

Solution:

Suppose he buys 12 oranges. Then, S.P. = Re.1

Now, $80:1::120:x$

$$x = \frac{1\times120}{80} = \frac{3}{2}$$

For $Rs.\frac{3}{2}$, organges sold $=12$

For Re.1, oranges sold $= \left(12\times\frac{2}{3}\right) = 8$

So, he must sell them at 8 for a rupee.

34. Ans. (a)

Solution:

C.P. of 330 boxes $= Rs.(7200+200+600)$

$= Rs.8000$

S.P. of 330 boxes $= Rs.(330\times28) = Rs.9240$

$\therefore$ Gain% $= \left(\frac{1240}{8000}\times100\right)\% = 15.5\%$

35. Ans. (c)

Solution:

C.P. of B $=110\%$ Rs. 10000 $= Rs. = 110\%$

Loss of $B = 10\%$

S.P. of $B = 90\%$ of $B = 90\%$

of $Rs.11000 = Rs.9900$

Thus, C.P. of $A = Rs.9900$

So, A gets

$[(10\% \text{ of } Rs.10000) + (10000-9900)]$

$= Rs.1100$

36. Ans. (c)

Solution:

Let original cost of each toy be Rs. 100 and number originally sold be 100

Original sale Proceed

$= Rs.(100\times100) = Rs.10000$

New sale proceed $= Rs.(120\times85) = Rs.10200$

$\therefore$ Increase% $= \left(\frac{200}{10000}\times100\right)\% = 2\%$

37. Ans. (a)

Solution:

Let original cost of each be Rs. 100 and number originally sold be 100.

Original sale proceed $= Rs.(100\times100)$

$= Rs.10000$

New sale proceed $= Rs.(80\times140) = Rs.11200$

$\therefore$ Increase % $= \left(\frac{1200}{10000}\times100\right) = 12\%$

38. Ans. (a)

Solution:

$$\text{Loss\%} = \left(\frac{5}{10}\right)^2 = \frac{1}{4}\% = 0.25\%$$

39. Ans. (b)

Solution:

$$\text{Loss\%} = \left(\frac{20}{10}\right)^2 = (2)^2 = 4\%$$

40. Ans. (b)

Solution:

Let C.P. $= Rs.x$. Then,

$425 - x = x - 355$

or $2x = 780$

or $x = 390$

41. Ans. (c)

Solution:

$2x + y = 7000$ (1)

$x + y = 9800$ (2)

Solving (1) and (2), we get $y = 4200$

42. Ans. (d)

Solution:

C.P. of 120 reams

$= Rs.(120\times80 + 280 + 72 + 120\times40)$

$= Rs.10000$

C.P. of 1 ream $= \left(\frac{10000}{120}\right) = Rs.\left(\frac{250}{3}\right)$

S.P. of 1 ream $= Rs.\left(\frac{108}{100}\times\frac{250}{3}\right) = Rs.90$

43. Ans. (b)

Solution:

Let C.P. $= Rs.x$

Then, $\frac{112}{100}x - \frac{88}{100}x = 6$

or $24x = 600$

or $x=\frac{600}{24}=25$

44. Ans. (c)

Solution:

$(115\% \text{ of C.P}) - (125\% \textit{ of } 90\% \text{ of C.P.}) = 4$

or, $\frac{115}{100}x - \frac{125}{100}\times\frac{90}{100}x = 4$

or, $\frac{23x}{20} - \frac{9x}{8} = 4$

$x = 160$

$\therefore$ C.P. $= Rs.160$

45. Ans. (d)

Solution: It is clear that

$(2.5\% \text{ of } C.P. + 75\% \text{ of } C.P.) = 100$

or 10% of C.P $=100$ or $\frac{1}{10}$ of C.P. $=100$

C.P. $= Rs.1000$

Now, gain $= 12\frac{1}{2}\%$

So, S.P. $112\frac{1}{2}\%$ of C.P.

$= Rs.\left(\frac{225}{200}\times 1000\right) = Rs.1125.$

46. Ans. (d)

Solution:

Then, C.P. paid by Subhash $= Rs.\frac{9x}{10}$

S.P. received by subhash $= (108\% \text{ of } Rs.x)$

$= Rs.\frac{27x}{25}$

$\therefore$ Gain $= Rs.\left(\frac{27x}{25} - \frac{9x}{10}\right) = Rs.\frac{9x}{50}$

Hence gain % $= \left(\frac{9x}{50}\times\frac{10}{9x}\times 100\right)\% = 20\%$

47. Ans. (b)

Solution:

Let original price = Rs. 100

Price after first discount = Rs. 100

Price after seocnd discount $= Rs.\left(\frac{80}{100}\times 90\right)$

= Rs. 72

Price after third discount $= (100 - 43.20) = 56.8\%$

$\therefore$ Single discount $= (100 - 43.20) = 56.8\%$

48. Ans. (b)

Solution:

Let the prices be $4x, 5x$ and $7x$ rupees

Then, $7x - 4x = 60000$

$x = 20000$

Required price $= 5x = Rs.100000$

49. Ans. (c)

Solution:

Let the original price be Rs. x.

C.P $= (x - 25\% \text{ of } x) = \frac{3x}{4}$

S.P. $= \left(\frac{3x}{4} + 10\% \text{ of } \frac{3x}{4}\right) = \frac{33x}{40}$

$= \frac{33x}{40} = 660$

$x = 800$

50. Ans. (c)

Solution:

Let the marked price be Rs. x.

Discount availed by the retailer $= 15\%$ of Rs. x.

C.P. of the machine by retailer

$= (x - 15\% \text{ of } x) = Rs.\frac{17x}{20}$

So, 15% of $\frac{17x}{20} = 1955 - \frac{17x}{20}$

$\frac{51x}{400} + \frac{17x}{20} = 1955$

or $x = 2000$

Discount received by retailer

$= (15\% \text{ of } Rs.2000) = Rs(300)$

51. Ans. (a)

Solution:

Let the printed price of the book be Rs. 100.

After a discount of 10%, S.P. $= Rs.90$

Profit earned $= 12\%$

C.P. of the book $= Rs.\left(\frac{110}{112}\times 90\right) = Rs.\frac{1125}{14}$

Hence, (C.P.): (Printed Price) $=\frac{1125}{14}:100$

or $45:56$

52. Ans. (d)

Solution:

Let the marked price be Rs. *x*.

Then $(7\% \text{ of } x)-15=15\% \text{ of } x$.

or $\frac{7x}{100}-\frac{5x}{100}=15$ or $x=750$

53. Ans. (b)

Solution:

Let C.P $=Rs.100$. Then market price $=Rs.120$

S.P. $=90\%$ of $Rs.120=Rs.108$

54. Ans. (b)

Solution:

Let the abelled price be Rs. 100

S.P. in 1st case $=Rs.80$, S.P. in 2nd case $=Rs.75$

If saving is Rs. 5, labelled price = Rs. 100.

If saving a is Rs. 500, labelled price

$=Rs.\left(\frac{100}{5}\times 500\right)=Rs.10000$

55. Ans. (b)

Solution:

Let C.P. $=Rs.100$

Marked Price $=Rs.120$, S.P. $=Rs.108$

$\therefore$ Discount $=\left(\frac{12}{120}\times 100\right)\%=10\%$

Statistics

Main point

1. Three measures of central value are:
   (i) Mean (ii) Median and
   (iii) Mode
2. Mean is computed by following methods:
   (i) Direct method
   (ii) short-cut method
   (iii) step-deviation method
3. If a variate X takes value $x_1, x_2, ..., x_n$ with corresponding frequencies $f_1, f_2, ..., f_n$ respectively, then the arithmetic mean of these values is given by

$$\bar{x}=\frac{1}{N}\sum_{i=1}^{n} f_i x_i \text{ Where, } N=\sum_{i=1}^{n} f_i$$

Also, $\bar{x}=A+\frac{1}{n}\sum_{i=1}^{n} f_i d_i$, *where* $d_i=x_i-A$.

the number A is called the assumed mean.

If $u_i=\frac{x_i-A}{h}, i=1,2,...,n$. *Then,*

$$\bar{X}=A+h\left\{\frac{1}{n}\sum_{i=1}^{n} f_i u_i\right\}$$

4. The median is the middle value of a distribution, i.e.median of a distribution is the value of the variable which divides it into two equal parts.

   The median of a grouped or continuous frequency distribution may be computed by using the following formula:

$$\text{Median}=\frac{1+\frac{n}{2}-f}{f}\times h, \text{ where}$$

I = Lower limit of the median class

f = frequency of the median class

h = Width of the median class

f = cumulative frequency of the class preceding the median class and, N= $\sum_{i=1}^{n} f_i$

5. Mode is the value of the variable which has the maximum frequency. The mode of a continuous of grouped frequency distribution may be computed by using the following formula:

$$\text{Mode}=1+\frac{f-f_1}{2f-f_1-f_2}\times h, \text{ where}$$

l = Lower limit of the modal class

f = Frequency of the middle class

h = width of the middle class

f_1 = Frequency of the class preceding the modal class

f_2 = Frequency of the class following the modal class.

6. Three measures of central value are connected by the following relation:
   Mode = 3 Median -2 Mean
7. Ogive (s) can be used to find the median of a frequency distribution.

Exercise

1. Which of the following is not a measure of central tendency?

(a) Mean (b) Median
(c) Mode (d) Standard deviation

2. The algebraic sum of the deviations of a frequency distribution from its mean is
(a) Always positive (b) Always negative
(c) 0 (d) A non- zero number

3. The arithmetic mean of 1, 2 , 3 ...., n is
(a) $\frac{n+1}{2}$ (b) $\frac{n-1}{2}$
(c) $\frac{n}{2}$ (d) $\frac{n}{2}+1$

4. For a frequency distribution, mean, median and mode are connected by the relation
(a) Mode = 3 Mean – 2 Median
(b) Mode = 2 Median – 3 Mean
(c) Mode = 3 Median – 2 Median
(d) Mode = 3 Median + 2 Mean

5. Which of the following canont be determined graphically?
(a) Mean (b) Median
(c) Mode (d) None of these

6. The meadian of a given frequency distribution is found grphically with the help of
(a) Histogram (b) Frequency curve
(c) Frequency polygon (d) Ogive

7. The mode of a frequency distribution can be determined graphically from
(a) Histogram (b) Frequency polygon
(c) Ogive (d) Frequency curve

8. Mode is
(a) least frequent value (b) middle most value
(c) most frequent vlaue (d) None of these

9. The mean of n observation is $\bar{x}$. If the first term is increased by 1, second by 2 and so on,then the new mean is
(a) $\bar{x}+n$ (b) $\bar{x}+\frac{n}{2}$
(c) $\bar{x}+\frac{n+1}{2}$ (d) None of these

10. One of the methods of determining mode is
(a) Mode = 2 Median – 3 Mean
(b) Mode = 2 Median + 3 Mean
(c) Mode = 3 Median – 2 Mean
(d) Mode = 3 Median + 2 Mean

11. If the mean of the following distributiion is 2.6, then the value of y is

Variable (x):	1	2	3	4	5
Frequency:	4	5	y	1	2

(a) 3 (b) 8
(c) 13 (d) 24

12. The relationship between mean, median and mode for a moderately skewed distribution is
(a) Mode = 2 median – 3 mean
(b) Mode = Median – Mean
(c) Mode = 2 Median – Mean
(d) Mode = 3 Median –2 mean

13. The mean of a discrete frequency distribution $\frac{x_i}{f_i}; i=1,2,\cdots,n$ is given be
(a) $\frac{\sum f_i x_i}{\sum f_i}$ (b) $\frac{1}{n}\sum_{i=1}^{n} f_i x_i$
(c) $\frac{\sum_{i=1}^{n} f_i x_i}{\sum_{i=1}^{n} x_i}$ (d) $\frac{\sum_{i=1}^{n} f_i x_i}{\sum_{i=1}^{n} i}$

Answers

1. Ans. (b)
2. Ans. (a)
3. Ans. (a)
4. Ans. (c)
5. Ans. (b)
6. Ans. (c)
7. Ans. (c)
8. Ans. (d)
9. Ans. (b)
10. Ans. (b)
11. Ans. (c)
12. Ans. (b)
13. Ans. (d)

SIMPLE INTEREST AND COMPOUND INTEREST

Interset is money paid to the lender by the borrower for using the money for a specified period of time. Various terms and their general representation.

Interset

Money paid by borrower for using the money. Denoted by I.

Principal

The original sum borrowed. Denoted by P.

Time

Time for which money is borrowed. Denoted by n. (n is expressed in number of periods, which is normally one year).

Rate of Interest

Rate at which interest is calculated on the original sum. Denoted by r and is expressed as a percentage or decimal fraction.

Amount

Sum of principal and interest. Denoted by A.

SIMPLE INTEREST

When interest is calculated every year on the oringinal principal, i.e. sum of the beginning of frist year, such interest is called simple Interest.

$$\text{Simple Interest} = \frac{Pnr}{100}$$

$$\text{Total Amount } A = P + \frac{Pnr}{100} = P\left[1 + \frac{nr}{100}\right]$$

COMPOUND INTEREST

Under Compound Interest, the interest is add to the principal at the end of each period to arrive at the new principal for the next period.

In other words, the amount at the end of first year will become the principal for the second year and the Amount at the end of second period become the principal for the third period and so on.

If P denotes the principal at the beginning of Period 1, then.

$$\text{P at the beginning of period } 2 = P\left[1 + \frac{r}{100}\right] = PR$$

Amount at the end of peroid 1.

$$\text{P at the beginning of perod } 3 = P\left[1 + \frac{r}{100}\right]^2 = PR^2$$

$$\text{where } R = \left\{1 + \left(\frac{r}{100}\right)\right\}$$

$$\text{Amount after } n \text{ years} = PR^n = A$$

$$\text{Interest} = I = A - P = P\left[R^n - 1\right]$$

Note:

1. The time period after which the interest is added each time to form a new principal is called conversion period.
2. In case interest is paid semi annually N is the number of half years and R is the rate percent per half year i.e.

   $Number\ of\ year \times 2 = Number\ of\ half\ years$

   $Rate\ percent\ per\ annum + 2$

   $= Rate\ percent\ per\ half\ year$
3. In case interest is paid quarterly, N is the number of quarters and R is the rate percent per quarter.

   $Number\ of\ years \times 4$

   $= Number\ of\ quarters$

   $Rate\ percet\ per\ quarter + 4$

   $=$ Rate percent per quarter
4. When rates are different for different years, say $R_1, R_2, R_3,$ percent for 1st, 2nd, and 3rd year respectively, then

   Amount

   $$= P\left(1 + \frac{R_1}{100}\right) P\left(1 + \frac{R_2}{100}\right) P\left(1 + \frac{R_3}{100}\right)$$
5. In case the time is a fraction of a year, say $y\frac{x}{z}$ year, then

   $$\text{Amount} = P\left(1 + \frac{R}{100}\right)^y \times \left(1 + \frac{\frac{x}{z} \times R}{100}\right)$$

Exercise

1. How much time will it take for an amount of Rs. 450 to yeild Rs. 81 as interest at 4.5% per annum of simple interest?

(a) 3.5 years (b) 4 years
(c) 4.5 years (d) 5 years

2. What is the present worth of Rs. 132 due in 2 years at 5% simple interest per annum?

(a) Rs. 112 (b) Rs. 118.80
(c) Rs. 120 (d) Rs. 122

3. In how many years, Rs. 150 will produce the same interest @ 8% as Rs. 800 produce in 3 years @ $4\frac{1}{2}$

(a) 6 (b) 8
(c) 9 (d) 12

4. What will be the ratio of simple interest earned by certain amount at the same rate of interest for 6 years and that for 9 years?
(a) 1:3 (b) 1:4
(c) 2:3 (d) Data inadequate

5. A sum of money at simple interest amounts to Rs. 815 in 3 years and to Rs. 854 in 4 years. The sum is:
(a) Rs. 650 (b) Rs. 690
(c) Rs. 698 (d) Rs. 700

6. At what rate percent of simple interest will a sum of money double itself in 12 years ?
(a) $8\frac{1}{4}\%$ (b) $8\frac{1}{3}\%$
(c) $8\frac{1}{2}\%$ (d) $9\frac{1}{2}$

7. The simple interest on a sum of money at 8% per annum for 6 years is half the sum. The sum is:
(a) Rs. 4800 (b) Rs. 6000
(c) Rs. 8000 (d) Rs. Data inadeuate

8. Smple interest on a certain amount is (9/16) of the principal. If the numbers representing the rate of interest in percent and time in years be equal, then times, for which the principal is lent out, is:
(a) $5\frac{1}{2}$ years (b) $6\frac{1}{2}$ years
(c) 7 (d) $7\frac{1}{2}$

9. A man invests a certain sum of money at 6% p.a. simple interest and another sum at 7% p.a. simple interest. His income from interest after 2 years was Rs. 354. One-fourth of the first sum is equal to one-fifth of the second sum. The total sum invested was:
(a) Rs. 2600 (b) Rs.2700
(c) Rs. 2880 (d) Rs.2900

10. The difference between the simple interest received from two different sources on Rs. 1500 for 3 years is Rs. 13.50. The different between their rates of interest is:
(a) 0.1% (b) 0.2%
(c) 0.3% (d) 0.4%

11. If the annual rate of simple interest increases from 10% to $12\frac{1}{2}\%$, a man's yearly income increases by Rs. 1250. His principal (in Rs.) is:
(a) 45,000 (b) 50,0000
(c) 60,000 (d) 65,000

12. A sum of Rs. 10 is lent to be returned in 11 monthly instalment of Re. 1 each, interest being simple. The rate of interest is:
(a) $9\frac{1}{11}\%$ (b) 10%
(c) 11% (d) $21\frac{9}{11}\%$

13. A man lends Rs. 10,000 in four parts. If he gets 8% on Rs. 2000; $7\frac{1}{2}\%$ on Rs. 4000 and $8\frac{1}{2}\%$ on Rs. 1400; What percent must he ger for the remainder, if his average annual interest is 8.13%?
(a) 7% (b) 9%
(c) $9\frac{1}{4}\%$ (d) $10\frac{1}{2}\%$

14. A person invested in all Rs. 2600 at 4%,6% and 8%, per annum simple interest. At the end of the year, he got the some interest in all the three cases. The money invested at 4% is:
(a) Rs. 200 (b) Rs. 600
(c) Rs. 800 (d) Rs. 1200

15. A sum of money will double itself in 16 years at simple interset with yearly rate of:
(a) 10% (b) $6\frac{1}{4}\%$
(c) 8% (d) 16%

16. Rs. 800 amounts to Rs. 920 in 3 years at simple interst. If the interest rate is increased by 3%, it would amout to how much?
(a) Rs. 1056 (b) Rs. 1112
(c) Rs. 1182 (d) Rs. 992

17. A sum of money at simple interest amount to Rs. 2240 in 2 years and Rs. 2600 in 5 years. The sum is:
(a) Rs. 1880 (b) Rs. 2000
(c) Rs. 2120 (d) data inadequate

18. A certain sum of money at simple interest amounts to Rs. 1260 in 2 years and to Rs. 1350 in 5 yeras. The rate per cent annum is
(a) 2.5% (b) 3.75%
(c) 5% (d) 7.5%

19. A lent Rs. 600 to B for 2 yerars and Rs. 150 to C for 4 year and received altogether from both Rs. 90 as simple interest. The rate of interest is:
(a) 4% (b) 5%
(c) 10% (d) 12%

20. At simple interest, a sum doubles after 20 years. The rate of interest per annum is:
(a) 5% (b) 10%
(c) 12% (d) data inadequta

21. A certain sum of money at simple interest amount to Rs. 1012 in $2\frac{1}{2}$ years and to Rs. 1067.20 in 4 years. The rate per annm is:
(a) 2.5% (b) 3%
(c) 4% (d) 5%

22. The rate of interest on a sum of money is 4% per annnum for the first 2 years, 6% per annum for the next 4 years and 8% per annum for the period beyond 6 years. If the simple interest accrued by the sum for a totla period of years is Rs. 1120, what is the sum?
(a) Rs. 1500 (b) Rs. 2000
(c) Rs. 2500 (d) Rs. 4000

Answers

1. Ans. (b)
Solution:

$$\text{Time} = \left(\frac{100\times 81}{450\times 4.5}\right) \text{years} = 4 \text{ years}$$

2. Ans. (c)
Solution: Let the present worth be Rs. x.

Then, S.I. $= Rs\,(132 - x)$

$$\therefore \left(\frac{x\times 5\times 2}{100}\right) = 132 - x$$

$$\Leftrightarrow 10x = 13200 - 100x \Leftrightarrow 110x = 13200$$

$$\Leftrightarrow x = 120$$

3. Ans. (c)

Solution: $P = Rs.800$, $R = 4\frac{1}{2}\% = \frac{9}{2}\%$

$T = 3$ years,

Then, $S.I. = Rs\left(800\times\frac{9}{2}\times\frac{3}{100}\right) = Rs.108$

Now, $P = Rs.150$, S.I. $= Rs.1-108$, $R = 8\%$

$$\therefore \text{Time} = \left(\frac{100\times 108}{150\times 8}\right) \text{years} = 9 \text{ yeras}$$

4. Ans. (c)
Solution:
Let the principal be p and rate of interest be R %.

$$\text{Required Ratio} = \left(\frac{\frac{P\times R\times 6}{100}}{\frac{P\times R\times 9}{100}}\right) = \frac{6\,PR}{9PR} = \frac{6}{9} = 2:3$$

5. Ans. (c)
Solution:
S.I. for 1 year $= Rs.(854 - 815) = Rs.39$

S.I. for 3 year $= Rs.(39\times 3) = Rs.117$

$\therefore$ Principal $= Rs.(815 - 117) = Rs.698$

6. Ans. (b)
Solution: Let sum $= x$. Then , S.I. $= x$.

$$\therefore \text{Rate} = \left(\frac{100\times S.I.}{P\times T}\right) = \left(\frac{100\times 3x}{x\times 12}\right)\% = \frac{25}{3}\%$$

$$= 8\frac{1}{3}\%$$

7. Ans. (d)
Solution:

Let Sum $= x$. Then S.I. $= \frac{x}{2}$

$\therefore \frac{x}{2} = \frac{x\times 8\times 6}{100}$. Clearly, data is inadequate.

8. Ans. (d)
Solution:

Let sum $= x$. Then $S.I. = \frac{9}{16}x$

Let rate $= R\%$ and time $= R$ years

$$\therefore \left(\frac{x\times R\times R}{100}\right) = \frac{9x}{16}$$

$$\Leftrightarrow R^2 = \frac{900}{16} \Leftrightarrow R = \frac{30}{4} \Leftrightarrow R = 7\frac{1}{2}$$

Hence time $= 7\frac{1}{2}$ years

9. Ans. (b)
Solution: Let the sums be x and y.

$$\frac{x\times 6\times 2}{100} + \frac{y\times 7\times 2}{100} = 354 \quad (1)$$

or $6x + 7y = 17700$

Also, $\frac{x}{4} = \frac{y}{5}$

or $5x-4y=0$ (2)

Solving (1) and (2), we get : $x=1200$ and $y=1500$

$\therefore$ Total sum = $Rs.2700$

10. Ans. (c)

Solution:

$$\left(\frac{1500\times R_1\times 3}{100}\right)-\left(\frac{1500\times R_2\times 3}{100}\right)=13.50$$

$$\Leftrightarrow\ 4500(R_1-R_2)=1350$$

$$\Leftrightarrow\ R_1-R_1=\frac{1350}{4500}=0.3\%$$

11. Ans. (b)

Solution: Let the sum be Rs. x.

$$\text{Then }\left(x\times\frac{25}{2}\times\frac{1}{100}\right)-\left(\frac{x\times 10\times 1}{100}\right)=1250$$

$$\Leftrightarrow\ 25x-20x=250000$$

$$\Leftrightarrow\ 5x=250000$$

$$\Leftrightarrow\ x=50000$$

12. Ans. (d)

Solution:

$Rs.10+S.I.$ on $Rs.10$ for 11 months

$=Rs.11+S.I.$ on $=$ Re.1

for $(1+2+3+4+\cdots+10)$ months

$Rs.10+S.I.$ on Re.1 for 110 months

$\Rightarrow\ =Rs.11+S.I.$ on Re.1 for 55 months

$\Rightarrow\ S.I.$ on Re.1 for 55 months $=Rs.1$

$$\therefore\ \text{Rate }=\left(\frac{100\times 12}{1\times 55}\right)\%$$

$$=21\frac{9}{11}\%$$

13. Ans. (b)

Solution: Let the required rate be R. Then

$$\left[\left(\frac{20000\times 8\times 1}{100}\right)+\left(4000\times\frac{15}{2}\times\frac{1}{100}\right)+\left(1400\times\frac{17}{2}\times\frac{1}{100}\right)+\left(2600\times R\times\frac{1}{100}\right)\right]$$

$$=\left(\frac{813}{10000}\times 10000\right)$$

$$\Leftrightarrow\ 160+300+119+26R=813$$

$$\Leftrightarrow\ R=9$$

14. Ans. (d)

Solution:

Let the part be x, y and $\left[2600-(x+y)\right]$.

$$\text{Then, }\frac{x\times 4\times 1}{100}=\frac{y\times 6\times 1}{100}$$

$$=\frac{\left[2600-(x+y)\right]\times 8\times 1}{100}$$

$$\frac{y}{x}=\frac{4}{6}=\frac{2}{3}\text{ or }y=\frac{2}{3}x$$

$$\text{So, }\frac{x\times 4\times 1}{100}=\frac{\left(2600-\frac{5}{3}x\right)\times 8}{100}$$

$$\Leftrightarrow\ 4x=\frac{(7800-5x)\times 8}{3}=52x$$

$$\Leftrightarrow\ x=(7800\times 8)$$

$$\Leftrightarrow\ x=\left(\frac{7800\times 8}{52}\right)=1200$$

Money invested at $4\%=Rs.1200$

15. Ans. (b)

Solution:

Let, Principal $=P$. Then, S.I. $=P$

$$\text{Then, Rate }=\frac{100\times P}{P\times 16}=\frac{100}{16}=6\frac{1}{4}\%$$

16. Ans. (d)

Solution:

Principal $=Rs.800$, $S.I.=Rs(920-800)$

$=Rs.120$ and Time $=3\,year$

$$\therefore\ \text{Original Rate }=\frac{100\times 120}{800\times 3}=5\%$$

New rate $=8\%$

$$\text{Now, }S.I.=Rs.\left(\frac{800\times 8\times 3}{100}\right)=Rs.192$$

$\therefore$ Amount $=Rs.992$

17. Ans. (b)

Solution:

S.I. for 3 years $=Rs.(2600-2240)=Rs.360$

S.I. for 2 years $= Rs.\left(\frac{360}{3} \times 2\right) = Rs.240$

Sum $= Rs.(2240 - 240) = Rs.2000$

18. Ans. (a)
Solution:

S.I. for 3 year $= Rs.(1350 - 1260) = Rs.90$

S.I. for 2 year $= Rs.\left(\frac{90}{3} \times 2\right) = Rs.60$

$\therefore$ Sum $= Rs.(1260 - 60) = Rs.1200$

$\therefore$ Rate $= \frac{100 \times 60}{1200 \times 2} = 2.5\%$

19. Ans. (b)
Solution:

Let, rate $= x\%$ per annum.

Then, $\frac{600 \times x \times 2}{100} + \frac{150 \times x \times 4}{100} = 90$

or $18x = 90$ or $x = 5$

20. Ans. (a)
Solution:

Let, sum $= P$. Then S.I. $= P$ and Time $= 20$

$\therefore$ Rate $= \frac{100 \times P}{P \times 20} = 5\%$ per annum.

21. Ans. (c)
Solution:

S.I. for $\frac{3}{2}$ years $= Rs.(1067.20 - 1012)$

$= Rs.55.20$

S.I. for $\frac{5}{2}$ year $= Rs.\left(55.20 \times \frac{2}{3} \times \frac{5}{2}\right) = Rs.92$

Sum $= Rs.(1012 - 92) = Rs.920$

Hence, Rate $= \left(\frac{100 \times 92 \times 2}{920 \times 5}\right) = 4\%$

22. Ans. (b)
Solution:

Let, sum $= Rs.\ x$. Then,

$\frac{x \times 4 \times 2}{100} + \frac{x \times 6 \times 4}{100} + \frac{x \times 8 \times 3}{100} = 1120$

$56x = 112000$ or $x = \frac{112000}{56} = 2000$

Compound Interest

Compound Interest: Sometimes it so happens that the borrower and the lender agree to fix up a certain unit of time, say yearly or half-yearly or quarterly to settel the previous account. In such cases, the amount after first unit of time becomes the principal for the second unit, the amount after second unit becomes the principal for the third unit and so on.

After a specified period, the difference between the amount and the money borrowed is called the Com pound Interest (abbreviated as C.I) for that period.

Important Points

Let Principal = P,
Rate = R% per annum,
Time = n years.

1. When interest is compound annually:

$$\text{Amount} = P\left(1 + \frac{R}{100}\right)^n$$

2. When interest is copounded half-yearly:

$$\text{Amount} = P\left[1 + \frac{(R/2)}{100}\right]^{2n}$$

3. When interest is compounded Quarterly:

$$\text{Amount} = P\left[1 + \frac{(R/4)}{100}\right]^{4n}$$

4. When interest is compounded Annually but time is in

fraction, say $3\frac{2}{5}$ years.

$$\text{Amount} = P\left(1 + \frac{R}{100}\right)^3 \times \left(1 + \frac{\frac{2}{5}R}{100}\right)$$

5. When rates are different for different years, say $R_1\%, R_2\% R_3\%$ for 1st, 2nd and 3rd year respectively. Then,

$$\text{Amount} = P\left(1 + \frac{R_1}{100}\right)\left(1 + \frac{R_2}{100}\right)\left(1 + \frac{R_3}{100}\right).$$

6. Present worth of Rs. x due n years hence is given by:

$$\text{Present worth} = \frac{x}{\left(1 + \frac{R}{100}\right)^n}$$

Example 1

Find compound interest on Rs. 8000 at 15% per annum for 2 years 4 months, compounded annually.

Solution:

Time = 2 years 4 months $= 2\frac{4}{12}$ years

$= 2\frac{1}{3}$ years.

Amount =

$$\text{Rs.}\left[8000\times\left(1+\frac{15}{100}\right)^2\times\left(1+\frac{\frac{1}{3}\times 15}{100}\right)\right]$$

$$= \text{Rs.}\left(8000\times\frac{23}{20}\times\frac{23}{20}\times\frac{21}{20}\right)$$

= Rs. 11109.

C.I. = Rs. (11109 − 8000) = Rs. 3109.

Example 2

If Rs. 500 amounts to Rs. 583.20 in two years compounded annually, find the rate of interest per annum.

Solution:

Principal = Rs. 500; Amount = Rs. 583.20; Time = 2 years.
Let the rate be R% per annum . Then,

$$\left[500\left(1+\frac{R}{100}\right)^2\right] = 583.20$$

or $\left(1+\frac{R}{100}\right)^2 = \frac{5832}{5000} = \frac{11664}{10000}$

Therefore, $\left(1+\frac{R}{100}\right)^2 = \left(\frac{108}{100}\right)^2$

or $1+\frac{R}{100} = \frac{108}{100}$

or $R = 8$

Hence, rate = 8% p.a.

Example 3

The difference between the compound interest and simple interest and simple interest on a certain sum at 10% per annum for 2 years is Rs. 631. Find the sum.

Solution:

$$\text{C.I.}\left(1+\frac{10}{100}\right)^2 - x = \frac{21x}{100}$$

$$\text{S.I.} = \left(\frac{x\times 10\times 2}{100}\right) = \frac{x}{5}.$$

$$\therefore \quad (\text{C.I.})-(\text{S.I.}) = \left(\frac{21x}{100}-\frac{x}{5}\right) = \frac{x}{100}.$$

$$\therefore \quad \frac{x}{100} = 631$$

or, $x = 63100$

Hence, the sum is Rs. 63,100.

Excercise

1. Sam invested Rs. 15,000 @ 10% per annum for one year. If the interest is compounded half-yearly, then the amount receved by Sam at the end of the year will be.
(a) Rs. 16,500 (b) Rs. 16,537.50
(c) Rs. 16,537.50 (d) Rs. 18,150
2. If the simple interest on a sum of money for 2 years at 5% per annum is Rs.50, what is the compound interest on the same sum rate and for the same time?
(a) Rs. 51.25 (b) Rs. 52
(c) Rs.54.25 (d) Rs. 60
3. The present worth of Rs. 169 due in 2 years at 4% per annum compound interest is:
(a) Rs. 150.50 (b) Rs.154.75
(c) Rs.156.25 (d) Rs. 158
4. The compound interest on a sum of money for 2 years is Rs. 832 and the simple interest on the same sum for the same period is Rs. 800. The difference between the compound interest and the simple in terest for 3 years will be.
(a) Rs. 48 (b) Rs. 66.56
(c) Rs. 98.56 (d) None of these
5. The difference between compound interest and simple interest on a sum for 2 years at 10% per annum, when the interest is compounded annually is Rs. 16. If the interest were compounded half-yearly, the difference in two interest would be.
(a) Rs. 24.81 (b) Rs.26.90
(c) Rs. 31.61 (d) Rs. 32.40
6. Mr. Due invested money in two schemes A and B offering compound interest @ 8 p.c.p.a. and 9 p.c.p.a. respectively. If the total amount of interest accrued

through two schemes together in two years was Rs. 4818.30 and the total amount invested was Rs. 27,000,what was the amount invested in Scheme A?

(a) Rs. 12,000 (b) Rs.13,500
(c) Rs. 15,000 (d) None of these

7. What annual payment will discharge a debt of Rs. 1025 due in 2 years at the rate of 5% compound interest?

(a) Rs. 550 (b) Rs. 551.25
(c) Rs. 560 (d) Rs. 560.75

Answers

Compound Interest

1. **Ans. (c)**

Solution:

P Rs. 15000; R = 10% p.a. = 5% per half-year; T = 1 year = 2 half-years.

$$\therefore \text{ Amount} = Rs.\left[15000\times\left(1+\frac{5}{100}\right)^2\right]$$

$$= Rs.\left(15000\times\frac{21}{20}\times\frac{21}{20}\right)$$

$$= Rs.16537.50.$$

2. **Ans. (a)**

Solution:

$$\text{Sum} = Rs.\left(\frac{50\times100}{2\times5}\right) = Rs.500.$$

$$\text{Amount} = Rs.\left[500\times\left(1+\frac{5}{100}\right)^2\right]$$

$$= Rs.\left(500\times\frac{21}{20}\times\frac{21}{20}\right)$$

$$= Rs\,551.25.$$

3. **Ans. (b)**

Solution:

$$\text{Present worth} = Rs.\left[\frac{169}{\left(1+\frac{25}{100}\right)^2}\right]$$

$$= Rs.\left(169\times\frac{25}{26}\times\frac{25}{26}\right)$$

$$= Rs.156.25.$$

4. **Ans. (c)**

Solution:

Difference in C.I. and S.I. for 2 years = Rs. 32.

S.I. for one year = Rs. 400.

$\therefore$ S.I.on Rs. 400 for one year = Rs. 32.

$$\text{So, Rate} = \left(\frac{100\times32}{400\times1}\right)\% = 8\%$$

Hence, difference in C.I. and S.I. for 3rd year

$$= \text{S.I. on } Rs.832 = Rs.\left(\frac{832\times8\times1}{100}\right)$$

$$= Rs.66.56.$$

Total. difference = Rs. (32 + 66.56) = Rs. 98.56

5. **Ans. (a)**

Solution:

For first year, S.I. = C.I.

Now, Rs. 16 is the S.I on S.I.for one year.

Rs. 10 is S.I. on Rs. 100.

$$\therefore \text{Rs. 16 is S.I.on Rs.}\left(\frac{100}{10}\times16\right) = Rs.160$$

Amount for 2 years compounded half yearly

$$= Rs.\left[1600\times\left(1+\frac{5}{100}\right)^4\right]$$

$$= Rs.1944.81.$$

$$\therefore \quad C.I. = Rs.(1944.81-1600) = Rs.344.81.$$

$$\text{S.I.} = Rs.\left(\frac{1600\times10\times2}{100}\right) = Rs.24.81$$

$$\therefore \quad (\text{C.I.})-(\text{S.I.}) = \text{Rs.}(344.81-320)$$

$$= Rs.24.81.$$

6. **Ans. (a)**

Solution:

Let the investment in schema A be Rs. x.

Then, investment in schema B = Rs. $(27000-x)$.

$$\therefore \left[x\times\left\{\left(1+\frac{8}{100}\right)^2-1\right\}+(27000-x)\left\{\left(1+\frac{9}{100}\right)^2-1\right\}\right]$$

$$= 4818.30$$

or, $\left(x \times \frac{104}{625}\right) + \frac{1881(27000 - x)}{10000} = \frac{481830}{100}$

or, $1664x + 1881(27000 - x) = 48183000$

or, $(1881x - 1664x)$

$= (50787000 - 48183000)$

or, $217X = 2604000$

or, $x = \frac{2604000}{217} = 12000.$

7. **Ans. (b)**
Solution:
Let each instalment be Rs. x. Then,

$$\frac{x}{\left(1+\frac{5}{100}\right)} + \frac{x}{\left(1+\frac{5}{100}\right)} = 1025$$

or, $\frac{20x}{21} + \frac{400x}{441} = 1025$

or, $820x = 1025 \times 441$

or, $x = \left(\frac{1025 \times 441}{1275}\right) = 551.25.$

So, value of each instalment = Rs. 551.25.

Number Series

This chapter of questions in which in which series of numbers of alphabetical or combiations of both are given, which are generally called the trems of the series. These terms follow a certain pattern throughout the series. The candidate is required to study the given series, identify the pattern throughout the series. The candidate is required to study the given series, identify the pattern followed in the series and either complete the given series with the môst suitable alternative or find the wrong term in the series.

Exercise

Directions: In each of the following questions, a number series is given with one term missing. Choose the correct alternative that will continue the same pattern and replace the question mark in the given series:

1. 2, 5, 9, ?, 20, 27
(a) 14 (b) 16
(c) 18 (d) 24

2. 4832, 5840, 6848?
(a) 7815 (b) 7846
(c) 7856 (d) 7887

3. 1, 5, 13, 25, 41?
(a) 51 (b) 57
(c) 61 (d) 63

4. 0.5, 0.55, 0.65, 0.8,?
(a) 0.9 (b) 0.82
(c) 1 (d) 0.95

5. 1, 1, 4, 8, 9, 27, 16,?
(a) 32 (b) 64
(c) 81 (d) 256

6. 5760, 960, ?, 48, 16, 8
(a) 120 (b) 160
(c) 192 (d) 240

7. 9, 27, 31, 155, 161, 1127,?
(a) 316 (b) 1135
(c) 1288 (d) 2254

8. 4, 10, ?, 82, 244, 730
(a) 24 (b) 28
(c) 77 (d) 218

9. 2, 3, 8, 27, 112?
(a) 226 (b) 339
(c) 452 (d) 565

10. 24, 60, 120, 210?
(a) 300 (b) 336
(c) 420 (d) 525

11. 2, 8, 16, 128?
(a) 2042 (b) 2046
(c) 2048 (d) 2056

12. 66, 36, 18?
(a) 3 (b) 6
(c) 8 (d) 9

13. 20, 20, 19, 16, 17, 13, 14, 11, ?, ?
(a) 10, 10 (b) 10, 1
(c) 13, 14 (d) 13, 16

14. 11, 10, ?, 100, 1001, 1000, 10001
(a) 101 (b) 110
(c) 111 (d) None of these

15. 90, 180, 12, 50, 100, 200, ? , 3, 50, 4, 25, 2, 6, 30, 3
(a) 150 (b) 175
(c) 225 (d) 250

16. In the series 3, 9, 15, ..., what will be the 21st term?
(a) 117 (b) 121
(c) 123 (d) 129

Answers

Number Series

1. **Ans. (a)**
Solution: The pattern is + 3, + 4, + 5, + 6, ...
So, missing term $= 9 + 5 = 14$.

2. Ans. (c)
Solution: The pattern is + 1008
So, missing term $= 6848 + 1008 = 7856$

3. Ans (c)
Solution: The pattern is + 4, + 8, + 12, + 16, ...
So, missing term $= 41 + 20 = 61$.

4. Ans. (c)
Solution: The pattern is + 0.05, + 0.10, + 0.15, ...
So, missing term $= 0.8 + 0.20 = 1$.

5. Ans (b)
Solution: The series consists of squares and cubes of consecutive natural numbers *i.e.* 1^2, 1^3, 2^2, 2^3, 3^2, 3^3, 4^2, ...
So, missing term $= 4^3 = 64$

6. Ans. (c)
Solution: The pattern is $\div 6, \div 5, \div 4, \div 3, \div 2$
So, missing term $= 960 \div 5 = 192$.

7. Ans. (b)
Solution: The pattern is $\times 3, +4, \times 5, +6, \times 7$, ...
So, missing term $= 1127 + 8 = 1135$.

8. Ans. (b)
Solution: Each number in the series is 2 less than thrice the preceding number.
So, missimg number $= (10 \times 3) - 2 = 28$.

9. Ans. (d)
Solution: The pattern is $\times 1 + 1$, $\times 2 + 2$, $\times 3 + 3$, $\times 4 + 4$
So, missing term $= 112 \times 5 + 5 = 565$

10. Ans. (b)
Solution: The pattern is + 36, +60, + 90.....*i.e.* $+[6 \times (6+0)]$, $+[6 \times (6+4)]$, $+[6 \times (4+5)]$, ...
So, missing term $= 210 + [6 \times (9+6)] = 210 + 126 = 336$

11. Ans. (c)
Solution: Each term in the series, except the first two terms, is the product of the preceding two terms.
So, missing trem $= 16 \times 128 = 2048$.

12. Ans. (c)
Solution: Each term in the series is the product of the digits of the preceding term.
So, missing term $= 1 \times 8 = 8$.

13. Ans. (a)
Solution: Let the missing terms of the series be x_1 and x_2.
Thus the sequence 20, 20, 19, 16, 17, 13, 14, 11, x_1, x_2 is a comdination of two series:
I. 20, 19, 17, 14, x_1 and
II. 20, 16, 13, 11, x_2
The pattern in I is – 1, – 2, – 3, ...
So, missing term, $x_1 = 14 - 4 = 10$
The pattern in (II) is – 4, – 3, – 2, ...
So, missing term, $x_2 = 11 - 1 = 10$.

14. Ans. (a)
Solution: The pattern is –1 ×, 10 + 1, – 1, × 10 + 1, – 1, × 10 + 1, ...
So, missing term = 10 × 10 + 1 = 101.

15. Ans. (a)
Solution: Clearly, 90 = 30 × 3, 180 = 6 × 30, 12 = 2 × 6, 50 = 25 × 2, 100 = 25 × 4, 200 = 50 × 4.
So, missing term = 3 × 50 = 150.

16. Ans (c)
Solution:
Clearly, 3 + 6 = 9, 9 + 6 = 15, ...
So, the series is an A.P. in which $a = 3$ and $b = 6$.
∴ 21st term
$= a + (21-1)d = a + 20d = 3 + 20 \times 6 = 123$.

Exercise

Directions: (Questions 1 to 40): In each of the following questions, one term in the number series is wrong. Find out the wrong term.

1. 121, 143, 165, 186, 209
(a) 143 (b) 165
(c) 186 (d) 209

2. 325, 259, 202, 160, 127, 105, 94
(a) 94 (b) 127
(c) 202 (d) 259

3. 24576, 6144, 1536, 386, 96, 24
(a) 96 (b) 386
(c) 1536 (d) 6144

4. 3, 7, 15, 39, 63, 127, 255, 511
(a) 15 (b) 39
(c) 63 (d) 127

5. 445, 221, 109, 46, 25, 11, 4
(a) 25 (b) 46
(c) 109 (d) 221

6. 1, 5, 5, 9, 7, 11, 11, 15, 12, 17
(a) 11 (b) 12
(c) 17 (d) 15

Answer

1. **Ans (c)**
Solution: Each term in the series obtained by adding 22 to the preceding term.
So, 186 is wrong and must be replaced by (165 + 55) *i.e.* 187.

2. **Ans. (c)**
Solution: The correct pattern is – 66, – 55, – 44, – 33, – 22, – 11.
So, 202 is wrong and must be replaced by 259 – 55 *i.e.* 204.

3. **Ans. (b)**
Solution: Each term of the series is obtained by dividing the preceding term by 4.
So, 1536 is wrong and must be replaced by $(1536 \div 4)$ *i.e.* 384

4. **Ans. (b)**
Solution: The correct pattern is × 2 + 1.
So, 39 is wrong and must be replaced by (15 × 2 + 1) *i.e.* 31.

5. **Ans. (b)**
Solution: 3 is subtracted from each number and the result divided by 2 to obtain the next number of the series.
So, 46 is wrong and must be replaced by $\left(\frac{109-3}{2}\right)$ *i.e.* 53.

6. **Ans. (b)**
Solution: The given sequence is a combination of two series.
I. 1, 5, 7, 11, 12 and II. 5, 9, 11, 15, 17
The pattern in both I and II is + 4, + 2, + 4, + 2.
So, 12 is wrong and must be replaced by (11 + 2) *i.e.* 13.

TYPE D: DATA INTERPRETATION

Data Sufficiency

This section comprises of questions in which certain data regarding common disciplines as production over a period of a few years: imports, exports, incomes of employees in a factory, students applying for and qualifying a certain field of study, etc. are given in the form of a table. The candidate is required to understand the given information and thereafter answer the given questions on the basic of comparative analysis of the data.

Thus, here the data collected by the investigator are arranged in a systematic form in a table called the tabular form. In order to avoid some heads again and again, tables are made consisting of horizontal lines called rows and vertical lines called columns with distinctive heads, known as captions. Units of measurements are given with the captions.

Ex.2. Study the following table carefully and answer these questions:

Number of candidates appeared and qualified in a competitive examination from different states over the years.

1. Combining the states P and Q together in 1998, what is the percentage of the candidates qualified to that of the candidates appeared?
(a) 10.87% (b) 11.49%
(c) 12.35% (d) 12.54%
(e) 13.05%

2. The percentage of the total number of qualified candidates to the total number of appeared candidates among all the five states in 1999 is:
(a) 11.49% (b) 11.84%
(c) 12.21% (d) 12.57%
(e) 12.73%

3. What is the percentage of candidates qualified from State N for all the years together, over the candidates appeared from State N during all the years together?
(a) 12.36% (b) 12.16%
(c) 11.47% (d) 11.15%
(e) None of these

Year/State	1997		1998		1999		2000		2001	
	App	Qual	App	Qual	App	Qual	App	Qual	App	Qual
M	5200	720	8500	980	7400	850	6800	775	9500	1125
N	7500	840	9200	1050	8450	920	9200	980	8800	1020
P	6400	780	8800	1020	7800	890	8750	1010	9750	1250
Q	8100	950	9500	1240	8700	980	9700	1200	8950	995
R	7800	870	7600	940	9800	1350	7600	945	7990	885

4. What is the average of candidates who appeared from State Q during the given years?
(a) 8700 (b) 8760
(c) 8810 (d) 8920
(e) 8990

5. In which of the given years the number of candidates appeared from State P has maximum percentage of qualified candidates?
(a) 1997 (b) 1998
(c) 1999 (d) 2000
(e) 2001

6. Total number of candidates qualified from all the states together in 1997 is approximately what percentage of the total number of candidates qualified from all the states together in 1998?
(a) 72% (b) 77%
(c) 80% (d) 83%
(e) 86%

Answers

1. **Ans. (c)**
Solution: Required percentage

$$=\left[\frac{(1020+1240)}{(8800+9500)}\times 100\right]\%.$$

$$=\left(\frac{2260}{18300}\times 100\right)\% = 12.35\%$$

2. **Ans. (b)**
Solution: Required percentage

$$=\left[\frac{(850+920+890+980+1350)}{(7400+8450+7800+8700+9800)}\times 100\right]\%$$

$$=\left(\frac{4990}{42150}\times 100\right)\% = 11.84\%.$$

3. **Ans. (d)**
Solution: Required Percentage

$$=\left[\frac{(840+1050+920+980+1020)}{(7500+9200+8450+9200+8800)}\times 100\right]\%$$

$$=\left(\frac{4810}{43150}\times 100\right)\% = 11.15\%$$

4. **Ans. (e)**
Solution: Required average

$$=\frac{8100+9500+8700+9700+8950}{5}=\frac{44950}{5}=8990.$$

5. **Ans. (e)**
Solution: The percentages of candidates qualified to candidates appeared from State P during different years are:

For 1997 $=\left(\frac{780}{6400}\times 100\right)\% = 12.19\%;$

For 1998 $=\left(\frac{1020}{8800}\times 100\right)\% = 11.59\%;$

For 1999 $=\left(\frac{890}{7800}\times 100\right)\% = 11.41\%;$

For 2000 $=\left(\frac{1010}{8750}\times 100\right)\% = 11.54\%;$

For 2001 $=\left(\frac{1250}{9750}\times 100\right)\% = 12.82\%.$

$\therefore$ Maximum percentage is for the year 2001.

6. **Ans. (c)**
Solution: Required percentage

$$=\left[\frac{(720+840+780+950+870)}{(980+1050+1020+1240+940)}\times 100\right]\%$$

$$=\left(\frac{4160}{5230}\times 100\right)\% = 79.54\% \approx 80\%.$$

Directions: (Questions 1 to 21): The following table shows the number of new employee added to different categories of employees in a Company and also the number of employees from these categories who left the company every year since the foundation of the Company in 1995.

1. During the period between 1995 and 2000, the total number of operators who left the company is what percent of the total number of operators who joined the company?
(a) 19% (b) 21%
(c) 27% (d) 29%
(e) 32%

2. For which of the following categories the percentage increase in the number of employees working in the company from 1995 to 2000 was the maximum?
(a) Managers (b) Technicians
(c) Operators (d) Accountants
(e) Peons

3. What is the difference between the total number of Technicians added to the Company and the total number of Accountants added to the Company during the years 1996 to 2000?

(a) 128 (b) 112
(c) 96 (d) 88
(e) 72

4. What was the total number of Peons working in the Company in the year 1999?
(a) 1312 (b) 1192
(c) 1088 (d) 968
(e) 908

5. What is the pooled average of the total number of employees of all categories in the year 1997?
(a) 1325 (b) 1285
(c) 1265 (d) 1235
(e) 1195

Answers

1. **Ans. (d)**
Solution: Total number of operators who left the company during 1995–2000
= (104 + 120 + 100 + 112 + 144) = 580.
Total number of operators who joined the company during 1995–2000
= (880 + 256 + 240 + 208 + 192 + 248) = 2024.
∴ Required percentage

$$= \left(\frac{580}{2024} \times 100\right)\% = 28.66\% \approx 29\%.$$

2. **Ans. (a)**
Solution: Number of managers working in the company:
In 1995 = 760.
In 2000 = (760 + 280 + 179 + 148 + 160 + 193) – (120 + 92 + 88 + 72 + 96) = 1252.
∴ Percentage increase in the number of managers

$$= \left[\frac{(1252-760)}{760} \times 100\right]\% = 64.74\%$$

Number of technicians working in the company:
In 1995 = 1200.
In 2000 = (1200 + 272 + 240 + 236 + 256 + 288) – (120 + 128 + 96 + 100 + 112) = 1936.
∴ Percentage increase in the number of technicians

$$= \left[\frac{(1936-1200)}{1200} \times 100\right]\% = 61.33\%.$$

Number of operators working in the company:
In 1995 = 880.
In 2000 = (880 + 256 + 240 + 208 + 192 + 248) – (104 + 120 + 100 + 112 + 144) = 1444.
∴ Percentage increase in the number of Operators

$$= \left[\frac{(1444-880)}{880} \times 100\right]\% = 64.09\%$$

Number of accountants working in the company:
In 1995 =1160.
In 2000 = (1160 + 200 + 224 + 248 + 272 + 260) – (100 + 104 + 96 + 88 + 92) = 1884.
∴ Percentage increase in the number of accountants

$$= \left[\frac{(1884-1160)}{1160} \times 100\right]\% = 62.41\%$$

Number of peons working in the company:
In 1995 = 820.
In 2000 = (820 + 184 + 152 + 196 + 224 + 200) – (96 + 88 + 80 + 120 + 104) = 1288.
∴ Percentage increase in the number of peons

$$= \left[\frac{(1288-820)}{820} \times 100\right]\% = 57.07\%$$

Clearly, the percentage increase is maximum in case of managers.

3. Required difference = (272 + 240 + 236 + 256 + 288) – (200 + 224 + 248 + 272 + 260) = 88.

4. Total number of peons working in the company in 1999 = (820 + 184 + 152 + 196 + 224) – (96 + 88 + 80 + 120) = 1192.

5. Total number of employees of various categories working in the company in 1997 are:
Managers = (760 + 280 + 179) – (120 + 92) = 1007
Technicians = (1200 + 272 + 240) – (120 + 128) = 1464
Operators = (880 + 256 + 240) – (104 + 120) = 1152
Accountants = (1160 + 200 + 224) – (100 + 104) = 1380
Peons = (820 + 184 + 152) – (96 + 88) = 972
∴ Pooled average of all the five categories of employees working in the company in 1997

$$= \frac{1}{5} \times (1007 + 1464 + 1152 + 1380 + 972)$$

$$= \frac{1}{5} \times 5975 = 1195.$$

Bar Diagram

This section comprises of questions in which the data collected in a particular discipline are represented in the form of vertical or horizontal bars drawn by selecting a particular scale. One of the parameters is plotted on the horizontal axis and the other on the vertical axis. The candidate is required to understand the given information and thereafter answer the given questions on ther basic of data analysis.

Directions: (Questions 1 to 6): The bar graph provided below gives the data of the production of paper (in lakh tonnes) by three different companies X,Y, and Z over the years. Study the graph and answer the questions that follow.

Production of paper (in lakh tonnes) by three companies X,Y, and Z over the years

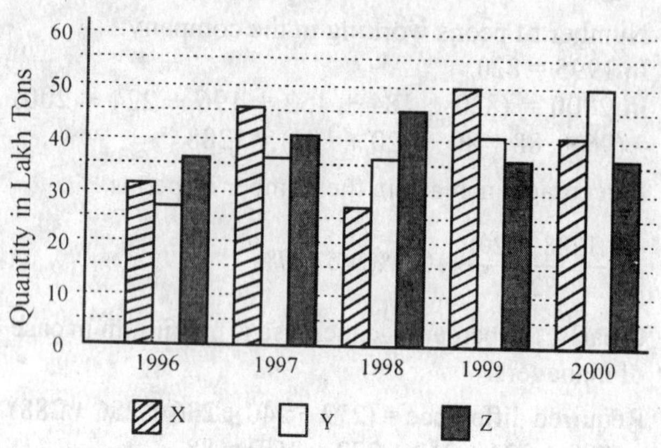

1. What is the difference between the production of Company Z in 1998 and Company Y in 1996?
   (a) 2,00,000 tons (b) 20,00,000 tons
   (c) 20,000 tons (d) 2,00,00,000 tons
   (e) None of these
2. What is the ratio of the average production of Company X in the period 1998-2000 to the average production of Company Y in the same period?
   (a) 1:1 (b) 15:17
   (c) 23:25 (d) 27:29
   (e) None of these
3. What is the percentage increase in the production of Company Y from 1996 to 1999?
   (a) 30% (b) 45%
   (c) 50% (d) 60%
   (e) 75%
4. The average production for five years was maximum for which Company?
   (a) X (b) Y
   (c) Z (d) X and Y both
   (e) X and Z both
5. For which of the following years, the percentage rise fall in production from the previous year is the maximum for Company Y?
   (a) 1997 (b) 1998
   (c) 1999 (d) 2000
   (e) 1997 and 2000
6. In which year was the percentage of production of Company Z to the production of Company Y the maximum?
   (a) 1996 (b) 1997
   (c) 1998 (d) 1999
   (e) 2000

Directions: (Question 7 to 11): Out of the two bar graphs provided below, one shows the amounts (in lakh Rs.) invested by a Company in purchasing raw materials over the years and the other shows the values (in lakh Rs.) of finished goods sold by the Company over the years. Study the two bar graphs and answer the questions based on them.

Amount invested in raw materials and the value of sales of finished goods for a company over the years.

Amount Invested in Raw Materials (Rs. in Lakhs)

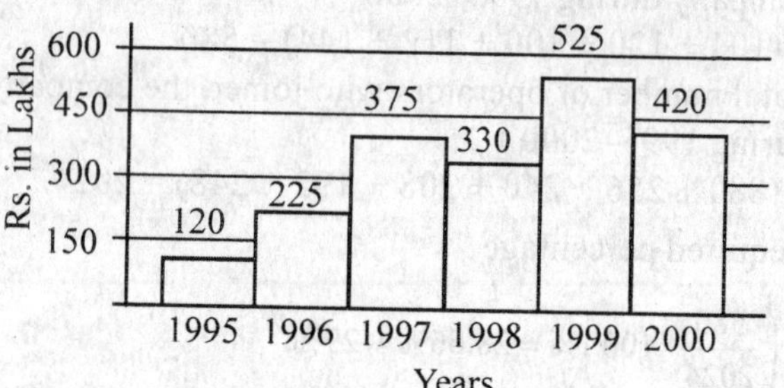

Value of Sales of Finished Goods (Rs. in Lakhs)

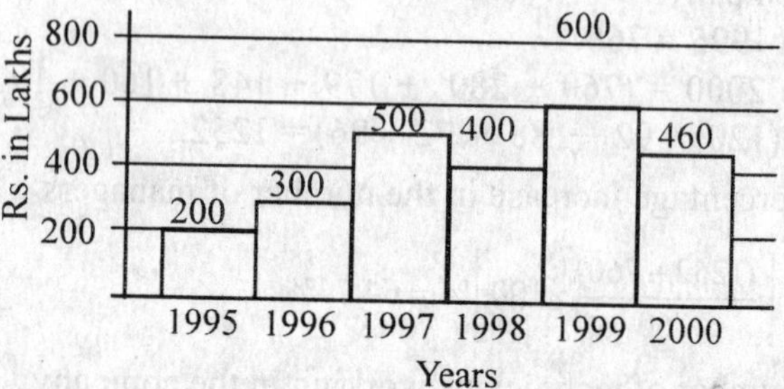

7. In which year, there has been a maximum percentage increase in the amount invested in raw materials as compared to the previous year?
   (a) 1996 (b) 1997
   (c) 1998 (d) 1999
   (e) 2000
8. In which year, the percentage change (compared to the previous year) in the investment on Raw Materials is the same as that in the value of sales of finished goods?
   (a) 1996 (b) 1997
   (c) 1998 (d) 1999
   (e) 2000

9. What was the difference between the average amount invested in Raw Materials during the given period and the average value of sales of finished goods during this period?
(a) Rs. 62.5 lakhs (b) Rs. 68.5 lakhs
(c) Rs. 71.5 lakhs (d) Rs. 77.5 lakhs
(e) Rs. 83.5 lakhs

10. The value of sales of finished goods in 1999 was approximately what percent of the average amount invested in Raw Materials in the years 1997, 1998 and 1999?
(a) 33% (b) 37%
(c) 45% (d) 49%
(e) 53%

11. The maximum difference between the amount invested in Raw Materials and the value of sales of finished goods was during the year:
(a) 1995 (b) 1996
(c) 1997 (d) 1998
(e) 1999

Directions: (Questions 12 to 16): A soft-drink company prepares drinks of three different flavours -X, Y, and Z. The production of the three flavours over a period of six years has been expressed in the bar-graph provided below. Study the graph and answer the questions based on it.

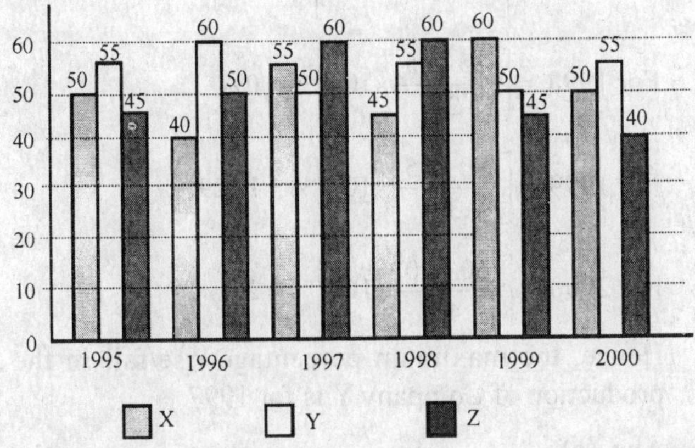

12. For which of the following years the percentage of rise/fall in production from the previous year is the maximum for the flavour Y?
(a) 1996 (b) 1997
(c) 1998 (d) 1999
(e) 2000

13. For which flavour was the average annual production maximum in the given period?
(a) X only (b) Y only
(c) Z only (d) X and Y
(e) X and Z

14. The total production of flavour Z in 1997 and 1998 is what percentage of the total production of flavour X in 1995 and 1996?
(a) 96.67% (b) 102.25%
(c) 115.57% (d) 120%
(e) 133.33%

15. What is the different between the average production of flavour X in 1995, 1996 and 1997 and the average production of flavour Y in 1998, 1999 and 2000?
(a) 50,000 bottles (b) 80,000 bottles
(c) 2,40,000 bottles (d) 3,30,000 bottles
(e) 5,00,000 bottles

16. What was the approximate decline in the production of flavour Z in 2000 as compared to the production in 1998?
(a) 50% (b) 42%
(c) 33% (d) 25%
(e) 22.5%

Directions: (Questions 17 to 21): The bar-graph given below shows the percentage distribution of the total production of a car manufacturing company into various models over two years. Study the graph carefully and answer the questions that follow.

Percentage of six different types of cars manufactured by a company over two years.

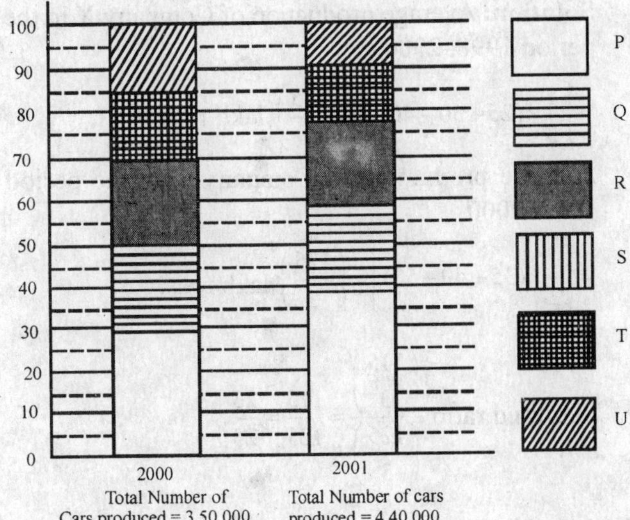

17. Total number of cars of models P, Q and T manufactured in 2000 is:
(a) 2,45,000 (b) 2,27,500
(c) 2,10,000 (d) 1,92,500
(e) 1,57,500

18. For which model the percentage rise/fall in production from 2000 to 2001 was minimum?

(a) Q (b) R
(c) S (d) T
(e) U

19. What was the difference in the number of Q type cars produced in 2000 and that produced in 2001?
(a) 35,500 (b) 27,000
(c) 22,500 (d) 17,500
(e) 16,000

20. If the percentage production of P type cars in 2001 was the same as that in 2000, then the number of P type cars produced in 2001 would have been:
(a) 1,40,000 (b) 1,32,000
(c) 1,17,000 (d) 1,05,000
(e) 97,000

21. If 85% of the S type cars produced in each year were sold by the Company, how many S type cars remained unsold?
(a) 7650 (b) 9350
(c) 11,850 (d) 12,250
(e) 13,350

Answers

1. Ans. (b)

Solution: Required difference $=[(45-25)\times 1,00,000]$ tons = 20,00,000 tons.

2. Ans. (c)

Solution: Average production of Company X in the period 1998–2000

$$=\left[\frac{1}{3}\times(25+50+40)\right]=\left(\frac{115}{3}\right) \text{ lakh tons.}$$

Average production of Company Y in the period 1998–2000

$$=\left[\frac{1}{3}\times(35+40+50)\right]=\left(\frac{125}{3}\right) \text{ lakh tons.}$$

$$\therefore \text{ Required ratio } \frac{\left(\frac{115}{3}\right)}{\left(\frac{125}{3}\right)}=\frac{115}{125}=\frac{23}{25}.$$

3. Ans. (d)

Solution: Percentage increase in the production of Company Y from 1996 to 1999

$$=\left[\frac{(40-25)}{25}\times 100\right]\%$$

$$=\left(\frac{15}{25}\times 100\right)\% = 60\%.$$

4. Ans. (e)

Solution: Average production (in lakh tons) in five years for the three companies are:

For Company $X=\left[\frac{1}{5}\times(30+45+25+50+40)\right]$

$$=\frac{190}{5}=38$$

For Company $Y=\left[\frac{1}{5}\times(25+35+35+40+50)\right]$

$$=\frac{185}{5}=37$$

For Company $Z=\left[\frac{1}{5}\times(35+40+45+35+35)\right]$.

$$=\frac{190}{5}=38$$

$\therefore$ Average production of five years is maximum for both the Companies X and Z.

5. Ans. (a)

Solution: Percentage change (rise/fall) in the production of Company Y in comparison to the previous year, for different years are:

For 1997 $=\left[\frac{(32-25)}{25}\times 100\right]\% = 40\%$

For 1998 $=\left[\frac{(35-35)}{25}\times 100\right]\% = 0\%$

For 1999 $=\left[\frac{(40-35)}{35}\times 100\right]\% = 14.29\%$

For 2000 $=\left[\frac{(50-40)}{40}\times 100\right]\% = 25\%$

Hence, the maximum percentage rise/fall in the production of Company Y is for 1997.

6. Ans. (a)

Solution: The percentages of production of Company Z to the production of Company Z for various years are:

For 1996 $=\left[\frac{35}{25}\times 100\right]\% = 140\%$;

For 1997 $=\left[\frac{40}{35}\times 100\right]\% = 114.29\%$;

For 1998 $=\left[\frac{45}{35}\times 100\right]\% = 128.57\%$;

For 1999 $= \left[\frac{35}{40} \times 100\right]\% = 87.5\%$;

For 2000 $= \left[\frac{35}{50} \times 100\right]\% = 70\%$.

Clearly, this percentage is highest for 1996.

7. Ans. (a)

Solution: The percentage increase in the amount invested in raw-materials as compared to the previous year, for different years are:

For 1996 $= \left[\frac{(225-120)}{120} \times 100\right]\% = 87.5\%$

For 1997 $= \left[\frac{(375-225)}{225} \times 100\right]\% = 66.67\%$

For 1998 there is a decrease.

For 1999 $= \left[\frac{(525-330)}{330} \times 100\right]\% = 59.09\%$

For 2000 there is a decrease.

$\therefore$ There is maximum percentage increase in 1996.

8. Ans. (b)

Solution: The percentage change in the amount invested in raw-materials and in the value of sales of finished goods for different years are:

Year	Percentage change in Amount invested in raw - material	Percentage change in value of sales of finished goods
1996	$\left[\frac{225-120}{120} \times 100\right]\%$ $= 87.5\%$	$\left[\frac{300-200}{200} \times 100\right]\%$ $= 50\%$
1997	$\left[\frac{375-225}{225} \times 100\right]\%$ $= 66.67\%$	$\left[\frac{500-300}{300} \times 100\right]\%$ $= 66.67\%$
1998	$\left[\frac{330-375}{375} \times 100\right]\%$ $= -12\%$	$\left[\frac{400-500}{500} \times 100\right]\%$ $= -20\%$
1999	$\left[\frac{525-330}{330} \times 100\right]\%$ $= 59.09\%$	$\left[\frac{600-400}{400} \times 100\right]\%$ $= 50\%$
2000	$\left[\frac{420-525}{525} \times 100\right]\%$ $= -20\%$	$\left[\frac{460-600}{600} \times 100\right]\%$ $= -23.33\%$

Thus, the percentage difference is same during the year 1997.

9. Ans. (d)

Solution: Required difference = Rs.

$$\left[\frac{1}{6} \times (200+300+500+400+600+460) - \frac{1}{6} \times (120+225+375+330+525+420)\right] \text{ lakhs}$$

$$= \text{Rs.} \left[\left(\frac{2460}{6}\right) - \left(\frac{1995}{6}\right)\right] \text{ lakhs}$$

= Rs. (410-332.5) lakhs = Rs. 77.5 lakhs.

10. Ans. (d)

Solution: Required percentage

$$= \left[\frac{600}{(375+330+525)} \times 100\right]\% = 48.78\% \approx 49\%.$$

11. Ans. (c)

Solution: The difference between the amount invested in raw material and the value of sales of finished goods for various years are:

For 1995 = Rs. (200–120) lakhs = Rs. 80 lakhs.

For 1996 = Rs. (300–225) lakhs = Rs. 75 lakhs.

For 1997 = Rs. (500–375) lakhs = Rs. 125 lakhs.

For 1998 = Rs. (400–330) lakhs = Rs. 70 lakhs.

For 1999 = Rs. (600–525) lakhs = Rs. 75 lakhs.

For 2000 = Rs. (460–420) lakhs = Rs. 40 lakhs.

Clearly, maximum difference was during 1997.

12. Ans. (b)

Solution: The percentage rise/fall in production from the previous year for flavour Y during various years are:

In 1996 $= \left[\frac{(60-55)}{55} \times 100\right]\% = 9.09\%$ (increase)

In 1997 $= \left[\frac{(60-50)}{60} \times 100\right]\% = 16.67\%$ (decrease)

In 1998 $= \left[\frac{(55-50)}{55} \times 100\right]\% = 10\%$ (increase)

In 1999 $= \left[\frac{(55-50)}{55} \times 100\right] = 9.09\%$ (decrease)

In 2000 $= \left[\frac{(55-50)}{55} \times 100\right]\% = 10\%$ (increase)

$\therefore$ Maximum change is decrease of 16.67% during 1997.

13. Ans. (b)

Solution: Average annual productions over the given period for various flavours are:

For flavour $X = \left[\frac{1}{6}\times(50+40+55+45+60+50)\right]$ lakh bottles = 50 lakh bottles.

For flavour $Y = \left[\frac{1}{6}\times(55+60+50+55+50+55)\right]$ lakh bottles = 54.17 lakh bottles.

For flavour $Z = \left[\frac{1}{6}\times(45+50+60+60+45+40)\right]$ lakh bottles = 50 lakh bottles.

∴ Maximum average production is for flavour Y.

14. Ans. (e)

Solution: Required percentage

$$=\left[\frac{(60+60)}{(50+40)}\times 100\right]\% = \left(\frac{120}{90}\times 100\right)\% = 133.33\%.$$

15. Ans. (e)

Solution: Average production of flavour X in 1995, 1996 and

$$1997 = \left[\frac{1}{3}\times(50+40+55)\right] = \left(\frac{145}{3}\right) \text{ lakh bottles.}$$

Average production of flavour Y in 1998, 1999 and 2000

$$= \left[\frac{1}{3}\times(55+50+55)\right] = \left(\frac{160}{3}\right) \text{ lakh bottles.}$$

$$\therefore \text{ Difference } = \left(\frac{160}{3}-\frac{145}{3}\right) = \frac{15}{3}$$

= 5 lakh bottles = 5,00,000 bottles.

$$= \left(\frac{160}{3}-\frac{145}{3}\right) = \frac{15}{3}$$

16. Ans. (c)

Solution: Percentage decline in the production of flavour Z in 2000 as compared to the production in 1998

$$= \left[\frac{(60-40)}{60}\times 100\right]\% = \left(\frac{20}{60}\times 100\right)\% = 33.33\% \approx 33\%.$$

17. Ans. (c)

Solution: We shall first determine the number of cars of each model produced by the Company during the two years:

In 2000: Total number of cars produced = 3,50,000.

P = (30–0)% of 3,50,000 = 30% of 3,50,000 = 1,05,000

Q = (45–30)% of 3,50,000 = 15% of 3,50,000 = 52,500

R = (65–45)% of 3,50,000 = 20% of 3,50,000 = 70,000

S = (75–65)% of 3,50,000 = 10% of 3,50,000 = 35,000

T = (90–75)% of 3,50,000 = 15% of 3,50,000 = 52,500

U = (100–90)% of 3,50,000 = 10% of 3,50,000 = 35,000.

In 2001: Total number of cars produced = 4,40,000.

P = (40–0)% of 4,40,000 = 40% of 4,40,000 = 1,76,000

Q = (60–40)% of 4,40,000 = 20% of 4,40,000 = 88,000

R = (75–60)% of 4,40,000 = 15% of 4,40,000 = 66,000

S = (85–75)% of 4,40,000 = 10% of 4,40,000 = 44,000

T = (95–85)% of 4,40,000 = 10% of 4,40,000 = 44,000

U = (100–95)% of 4,40,000 = 5% of 4,40,000 = 22,000.

Now, we shall solve the questions.

Total number of cars of models P, Q and T manufactured in 2000 = (105000 + 52500 + 52500) = 2,10,000.

18. Ans. (b)

Solution: Using the above calculation, the percentage change (rise/fall) in production from 2000 to 2001 for various model is:

$$\text{For P} = \left[\frac{(176000-105000)}{105000}\times 100\right]\% = 67.62\%, \text{ rise.}$$

$$\text{For Q} = \left[\frac{(88000-52500)}{52500}\times 100\right]\% = 67.62\%, \text{ rise.}$$

$$\text{For R} = \left[\frac{(70000-66000)}{70000}\times 100\right]\% = 5.71\%, \text{ fall.}$$

$$\text{For S} = \left[\frac{(44000-35000)}{35000}\times 100\right]\% = 25.71\%, \text{ rise.}$$

$$\text{For T} = \left[\frac{(52500-44000)}{52500}\times 100\right]\% = 16.19\%, \text{ fall.}$$

$$\text{For U} = \left[\frac{(35000-22000)}{35000}\times 100\right]\% = 37.14\%, \text{ fall.}$$

∴ Minimum percentage rise/fall in production is in the case of model R.

19. Ans. (a)

Solution: Required difference = 88000-52500 = 35500 (Using calculations in the Solution of Q. 21)

20. Ans. (b)

Solution: If the percentage production of P type cars in 2001 = percentage production of P type cars in 2000 = 30%

Then, number of P type cars produced in 2001 = 30% of 440000 = 132000.

21. Ans. (c)

Solution: Number of S type cars which remained unsold in 2000 = 15% of 35000 and number of S type cars which remained unsold in 2001 = 15% of 44000

∴ Total number of S type cars which remained unsold $= 15\%$ of $(35000 + 44000) = 15\%$ of $79000 = 11850$.

Pic Chart

The pie-chart or a pie - graph is a method of representing a given numerical data in the form of sectors of a circle.

The sectors of the circle are constructed in such a way that the area of each sector is proportional to the corresponding value of the component of the data.

From geometry, we know that the area of the sector of a circle is proportaional to the central angle.

So, the central angle of each sector must be proportional to the corresponding valeu of the component.

Since the sum of all the central angles is 360°, we have central angle of the component

$$= \left(\frac{\text{Value of the component}}{\text{Total value}} \times 360\right)^0$$

Exercise

Directions: (Question 1 to 9): The following pie-chart shows the percentage distribution of the expenditure incurred in publishing a book. Study the pie-chart and answer the question based on it.

Various expenditures (in percentage) incurred in publishing a book

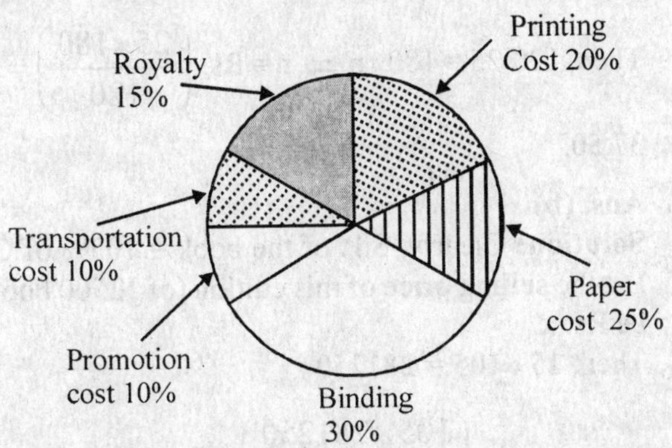

1. What is the central angle of the sector corresponding to the expenaiture incurred on Royalty?
   (a) 15° (b) 24°
   (c) 54° (d) 48°
2. Which two expenditures together have a central angle of 108°?
   (a) Binding Cost and Transportation Cost
   (b) Printing cost and Paper Cost
   (c) Royalty and Promotion Cost
   (d) Binding Cost and Paper Cost
3. If the difference between the two expenditures are represented by 18° in the pie-chart, then these expenditures possibly are:
   (a) Binding Cost and Promotion Cost
   (b) Paper Cost and Royalty
   (c) Binding Cost and Printing Post
   (d) Paper Cost and Printing Cost
4. If for an edition of the book, the cost of paper is Rs. 56250, then find the promotion cost for this edition.
   (a) Rs. 20,000 (b) Rs. 22,500
   (c) Rs. 25,500 (d) Rs. 28,125
5. If for a certain quantity of books, the publisher has to pay Rs. 30,600 as printing cost, then what will ble the amount of royalty to be paid for these books ?
   (a) Rs. 19,450 (b) Rs. 21,200
   (c) Rs. 22,950 (d) Rs. 26,150
6. The price of the book is marked 20% above the C.P. If the marked price of the book is Rs. 180, then what is the cost of the paper used in a single copy of the book?
   (a) Rs. 36 (b) Rs. 37.50
   (c) Rs. 42 (d) Rs. 44.25
7. For an edition of 12,500 copies, the amount of Royalty paid by the publisher is Rs. 2,81,250. What should be the selling price of the book if the publisher desires a profit of 5% ?
   (a) Rs. 152.50 (b) Rs. 157.50
   (c) Rs. 162.50 (d) Rs. 167.50
8. If 5500 copies are published and the transportation cost on them amounts to Rs. 82,500, then what should be the selling price of the book so that the publisher can earn a profit of 25%?
   (a) Rs. 187.50 (b) Rs. 191.50
   (c) Rs. 175 (d) Rs. 180
9. Royalty on the book is less than the printing cost by:
   (a) 5% (b) $33\frac{1}{3}\%$
   (c) 20% (d) 25%

Directions: (Question 10 to 14): The circle-graph given here shows the spendings of a country on various sports during a particular year. Study the graph carefully and answer the questions given below it.

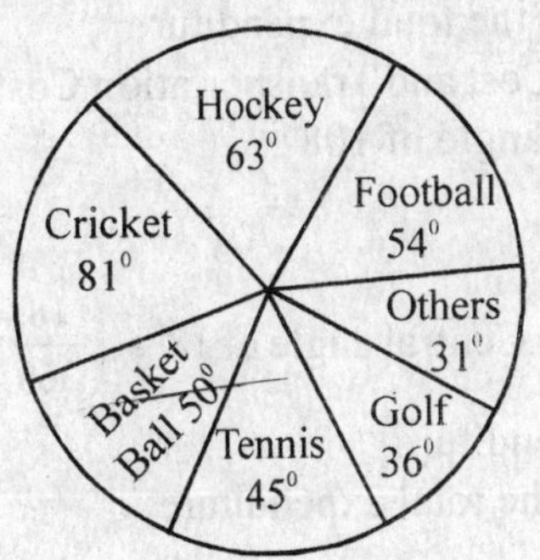

10. What percent of the total spendings is spent on Tennis?

(a) $12\frac{1}{2}\%$ (b) $22\frac{1}{2}\%$

(c) 25% (d) 45%

11. How much percent more is spent on Hockey than that on Golf?

(a) 27% (b) 35%

(c) 37.5% (d) 75%

12. How much percent less is spent on Football than that on Circket ?

(a) $22\frac{2}{9}\%$ (b) 27%

(c) $33\frac{1}{3}\%$ (d) $37\frac{1}{2}\%$

13. If the total amount spent on sports during the year was RS. 2 crores, the amount spent on Cricket and Hockey together was:

(a) Rs. 8,00,000 (b) Rs. 80,00,000

(c) RS. 1,20,00,000 **(d) Rs. 1,60,00,000**

14. If the total amount spent on sports during the year be Rs. 1,80,00,000, the amount spent on Basketball exceeds that on tennis by :

(a) Rs. 2,50,000 (b) Rs. 3,60,000

(c) Rs. 3,75,000 (d) Rs. 4,10,000

Answers

1. Ans. (c)

Solution: Central angle corresponding to royalty = $(15\% \text{ of } 360)^\circ = 54^\circ$

2. Ans. (a)

Solution: Central angle of $108^\circ = \left(\frac{108}{360}\times 100\right)\%$ of the total expenditure

= 30% of the total expenditure.

(from the pie-chart it is clear that)

Bindng Cost + Transportation Cost

= (20% + 10%) of the total expenditure

= 30% of the total expenditure.

∴ Binding Cost and Transportation Cost together have a central angle of 108°.

3. Ans. (d)

Solution: Central angle of $18^\circ = \left(\frac{18}{360}\times 100\right)\%$ of the total expenditure

= 5% of the total expenditure.

From the pie-chart it is clear that:

Out of the given combinations, only in combination (d) the difference is 5%, i.e. Paper Cost - Printing Cost = (25% – 20%) of total expenditure

= 5% of total expenditure.

4. Ans. (b)

Solution: Let the Promotion Cost for this edition be Rs.p.

Then, 25 : 10 = 56250:

$$p \Rightarrow p = \text{Rs.}\left(\frac{56250\times 10}{25}\right) = \text{Rs.}22500.$$

5. Ans. (c)

Solution: Let the amount of Royalty to be paid for these books be Rs. r.

Then, 20 : 15= 30600 :

$$r \Rightarrow r = \text{Rs.}\left(\frac{30600\times 15}{20}\right) = \text{Rs. } 22950.$$

6. Ans. (b)

Solution: Clearly, marked price of the book = 120% of C.P.

Also, cost of paper = 25% of C.P.

Let the cost of paper for a single book be Rs. n.

Then, $120{:}25 = 180 : n \Rightarrow n = \text{Rs.}\left(\frac{25\times 180}{120}\right) = \text{Rs.}$

37.50.

7. Ans. (b)

Solution: Clearly, S.P. of the book = 105% of C.P.

Let the selling price of this edition (of 12500 books) be Rs. x.

Then, 15 : 105 = 281250 :

$$x \Rightarrow x = \text{Rs.}\left(\frac{105\times 281250}{15}\right) = \text{Rs. } 1968750.$$

∴ S.P. of one book

$$= \text{Rs.}\left(\frac{1968750}{12500}\right) = \text{Rs. } 157.50.$$

8. Ans. (a)

Solution: For the publisher to earn a profit of 25%, S.P. = 125% of C.P.

Also Transportation Cost = 10% C.P.

Let the S.P. of 5500 books be Rs. x.

Then, 10 : 125 = 82500:

$x \Rightarrow x = \text{Rs.}\left(\frac{125\times 82500}{10}\right) = \text{Rs. } 1031250.$

$\therefore$ S. P. of one book =

Rs. $\left(\frac{1031250}{5500}\right)$ = Rs. 187.50.

9. Ans. (d)

Solution: Printing cost of book = 20% of C.P.

Royalty on book = 15% of C.P.

Difference = (20% of C.P.) – (15% of C.P.) = 5% of C.P.

$\therefore$ Percentage difference

$= \left(\frac{\text{Difference}}{\text{Printing Cost}}\times 100\right)\%$

$= \left(\frac{5\% \text{ of C.P.}}{20\% \text{ of C.P.}}\times 100\right)\% = 25\%$

Thus, Royalty on the book is 25% less than the Printing Cost.

10. Ans. (a)

Solution: Percentage of money spent on Tennis =

$\left(\frac{45}{360}\times 100\right)\% = 12\frac{1}{2}\%.$

11. Ans. (d)

Solution: Let the total spendings on sports be Rs. x. Then, Amount spent on Golf

$= \text{Rs.}\left(\frac{36}{360}\times x\right) = \text{Rs. } \frac{x}{10}.$

Amount spent on Hockey

$= \text{Rs.}\left(\frac{63}{360}\times x\right) = \text{Rs. } \frac{7}{40}x.$

$\text{Difference} = \text{Rs.}\left(\frac{7}{40}x - \frac{x}{10}\right) = \text{Rs. } \frac{3x}{40}.$

$\therefore$ Required Percentage

$= \left[\left(\frac{3x/40}{x/10}\right)\times 100\right]\% = 75\%.$

12. Ans. (c)

Solution: Let the total spendings on sports be Rs. x. then, Amount spent on Cricket

$= \text{Rs.}\left[\frac{81}{360}\times x\right] = \text{Rs.}\left[\frac{9}{40}x\right].$

Amount spent on Football

$= \text{Rs.}\left[\frac{54}{360}\times x\right] = \text{Rs.}\left[\frac{3}{20}x\right].$

$\text{Difference} = \text{Rs.}\left[\frac{9}{40}x - \frac{3}{20}x\right] = \text{Rs.} \frac{3}{40}x.$

$\therefore$ Required percentage

$= \left[\left(\frac{3x/40}{9x/40}\right)\times 100\right]\% = 33\frac{1}{3}\%.$

13. Ans. (b)

Solution: Amount spent on Cricket and Hockey together

$= \text{Rs.}\left[\frac{(81+63)}{360}\times 2\right]$ crores

= Rs. 0.8 crores = Rs. 8000000.

14. Ans. (a)

Solution: Amount spent on Basketball exceeds that on Tennis by :

$\text{Rs.}\left[\frac{(50-45)}{360}\times 18000000\right] = \text{Rs. } 250000$

Test Paper General Ability

One Mark Questions

1. Choose the most appropriate word from the options given below to the complete the following sentence: His rather casual remarks on politics_______ his lack of seriousness about the subject.
   (a) masked (b) belied
   (c) betrayed (d) suppressed
2. Which of the following options is closest in meaning to the word Circuitous.
   (a) Cyclic (b) Indirect
   (c) Confusing (d) Crooked
3. Choose the most appropriate word from the options given below to complete the following sentence: If we manage to _______ our natural resources, we would leave a better planet for our children.
   (a) uphold (b) restrain
   (c) cherish (d) conserve

4. 25 persone are ina room. 15 of them play hockey, 17 of them play football and 10 of them play both hockey and football. Then the number of persons playing neither hockey nor football is:

(a) 2 (b) 17
(c) 13 (d) 3

5. The question below consists of a pair of related words followed by four pairs of words. Select the pair that best expresses the relation in the original pair. **wordrelation**

Unemployed : Worker

(a) follow : land (b) unaware : sleeper
(c) wit : jester (d) renovated : house

6. If Log (P) = (1/2)Log (Q) = (1/3) Log (R), then which of the following options is **TRUE**?

(a) $P^2 = Q^3R^2$ (b) $Q^2 = PR$
(c) $Q^2 = R^3P$ (d) $R = P^2Q^2$

7. Choose the most appropriate word (s) from the options given below to complete the following sentence. **I contemplated __________ Singapore for my vacation but decided against it.**

(a) to visit (b) having to visit
(c) visiting (d) for a visit

8. Choose the most appropriate word from the options given below to complete the following sentence.

If you are trying to make a strong impression on your audience, you cannot do so by being understated, tentative or __________.

(a) hyperbolic (b) restrained
(c) argumentative (d) indifferent

9. Choose the world from the options given below that is most neary opposite in meaning to the given word:

Amalgamate

(a) merge (b) split
(c) collect (d) separate

10. Which of the following options is the closest in the meaning to the word below:

Inexplicable

(a) Incomprehensible (b) Indelible
(c) Inextricable (d) Infallible

Two Marks Questions

11. If $137 + 276 = 435$ how much is $731 + 672$

(a) 534 (b) 1403
(c) 1623 (d) 1513

12. Hari (H), Gita (G) Irfan (I) and Saira (S) are siblings (i.e. brothers and sisters). All were born on 1st january. The age difference between any two successive siblings (that is born one after another) is less than 3 years. Given the following facts:

1. Hari's age + Gita's age > Irfan's age + Saira's age
2. The age difference between Gita and Saira is 1 year. However, Gita is not the oldest and Saira is not the youngest.
3. There are no twins

In what order were they born (oldest first)?

(a) HSIG (b) SGHI
(c) IGSH (d) IHSG

13. Modern warfare has changed from large scale clashes of armies to suppression of civilian populations. Chemical agents that do their work silenty appear to be suited to such warfare; and regretfully, there exist people in military establishments who think that chemical agents are useful tools for their cause.

Which of the following statements best sums up the meaning of the above passage:

(a) Modern warfare has resulted in civil strife.
(b) Chemical agents are useful in modern warfare.
(c) Use of chemical agents in warfare would be undesirable
(d) People in military establishments like to use chemical agents in war.

14. 5 skilled workers can build a wall in 20 days : 8 semi-skilled workers can build a wall in 25 days; 10 unskilled workers can build a wall in 30 days. If a team has 2 skilled, 6 semi-skilled and 5 unskilled worker, how long will it take to build the wall?

(a) 20 days (b) 18 days
(c) 16 days (d) 15 days

15. Given digits 2, 2, 3, 3, 3, 4, 4, 4,4 how many distinct 4 digit numbers greater than 3000 can be formed?

(a) 50 (b) 51
(c) 52 (d) 54

16. P.Q.R and S are four types of dangerous microbes recently found in a human habitat. The area of each circle with its diameter printed in brackets represents

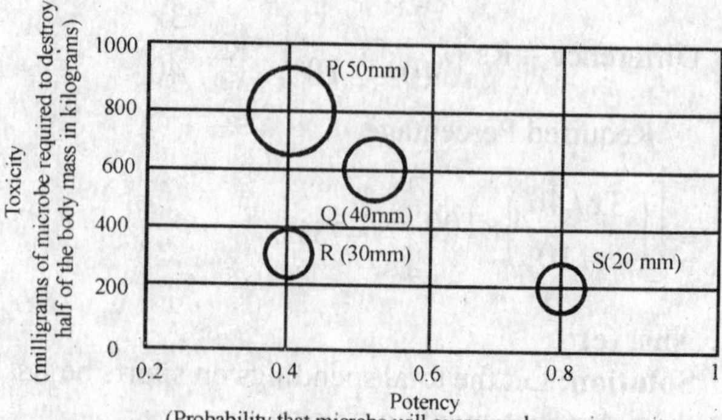

the growth of a single microble surviving human immunity system within 24 hours of entering the body. The danger to human beings varies proportionately with the toxicity, potency and growth attributed to a microbe shown in the figure below:

A pharmaceutical company is contemplating the development of a vaccine against the most dangerous microbe. Which microbe should the company target in its first attempt?

(a) P (b) Q
(c) R (d) S

17. The variable cost (V) of manufacturing a product varies according to the equation V = 4q, where q is the quantity produced. The fixed cost (F) of production of same product reduces with q according to the equation F =100/q. How many units should be produced to minimize the total cost (V+F)?

(a) 5 (b) 4
(c) 7 (d) 6

18. A transporter receives the same number of orders each day. Currently, he has some pending orders (backlog) to be shipped. If he uses 7 trucks, then at the end of the 4th day he can clear all the orders. Alternatively, if he uses only 3 trucks, then all the orders are cleared at the end of the 10th day. What is the minimum number of trucks required so that there will be no pending order at the end of the 5th day?

(a) 4 (b) 5
(c) 6 (d) 7

19. A container originally contains 10 Litres of pure spirit. From this container 1 litre of spirit is replaced with 1 litre of water. Subsequently, I litre of the mixture is again replaced with I litre of water and this process is repeated one more time. How much spirit is now left in the container?

(a) 7.58 litres (b) 7.84 litres
(c) 7 litres (d) 7.29 litres

20. **Few school curricula include a unit on how to deal with bereavement and grief, and yet all students at some point in their lives suffer from losses through death and parting.**

Based on the above passage whcih topic would not be included in a unit on bereavement?

(a) How to write a letter of condolence
(b) What emotional stages are passed thorugh in the healing process
(c) What the leading causes of death are
(d) how to give support to a grieving friend

Answers
Test Paper General Ability

1. **Ans. (c)**
Solution:

2. **Ans. (b)**
Solution:

3. **Ans. (d)**
Solution:

4. **Ans. (d)**
Solution:
If we take its Venn diagram, then we get

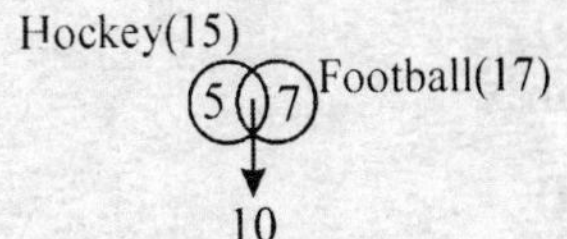

So, persons not playing any game = 25 – 5 – 7 – 10 = 3

5. **Ans. (a)**
Solution:

6. **Ans. (b)**
Solution:

$$\log(P) = \frac{1}{2}\log(Q) = \frac{1}{3}\log(R)$$

$$\Rightarrow P = Q^{\frac{1}{2}} = R^{\frac{1}{3}} = K \text{ (assume)}$$

$$\Rightarrow P = K \ ; \ Q = K^2, \ R = K^3$$

$$Q^2 = \left(K^2\right)^2 = K^4$$

$$= K^3K = RP$$

7. **Ans. (c)**
Solution: Visiting → Contemplate is a transitive verb and hence is followed by a gerund Hence the correct usage of contemplate is verbing form.

8. **Ans. (b)**
Solution: Restrained → Synonym of understand, tentative.The tone of the sentence clearly indicates a word that is similar to understated is needed for the blank. Alternatively, the word should be antonym of strong (fail to make strong impression). Therefore, the best choice is restrained which means controlled/reserved/timid.

9. **Ans: (b)**
Solution: Split → Closest antonym with negative condition separate though same has positive condition. Amalgamate means combine or unite to

form one organization or structure. So the best option here is split. Separate on the other hand, although a close synonym, it is too general to be the best antonym in the given question while Merge is the synonym; Collect is not related.

10. Ans. (a)

Solution: Incomprehensible → Inexplicable means not explicable; that cannot be explained, understood, or accounted for. So the best synonym here is incomprehensible.

11. Ans. (c)

Solution:

12. Ans. (b)

Solution:

13. Ans. (d)

Solution:

14. Ans. (d)

Solution:

5 skilled workes can build a wall in 20 days

1-- 20×5

$= 100$ days

8 semi-skilled workes can build a wall in 25 days

1-- 20×8

$= 200$ days

10 skilled workes can build a wall in 30 days

1-- 30×8

$= 300$ days

A team has 2 skilled, 6 semi-skilled and 5 unskilled works

$$t\left[\frac{2\times1}{100}+\frac{6\times1}{200}+\frac{5\times1}{300}\right]=1$$

$$t\left[\frac{2\times1+6\times1+5\times2}{600}\right]=1$$

$$t=\frac{600}{40}$$

$$=15 \text{ days}$$

15. Ans. (b)

Solution:

16. Ans. (a)

Solution:

As range ∝ toxicity, potency and growth assuming multiplicative mode.

	P	Q	R	S
Toxicity	800	600	300	200
potency	0.4	0.5	0.4	0.8
growth(R^2) (π being common not considered)	2500	1600	900	400
danger	8×10^5	4.8×10^5	1.08×10^5	0.84×10^5

So, most danger: P

17. Ans. (a)

Solution:

Total cost $V + F = 4q + \frac{100}{q} = T$

Put each option get minimum cost else. find minima

$$\frac{dT}{dq} = 4 - \frac{100}{V^2} = 0$$

or, $q = 5$

18. Ans. (c)

Solution:

Let each truck carrier 1 unit

∴ 7 truck 4 day can clear $7 \times 4 = 28$ units and it takes 4 days normal unit (x) and backlog (y)

∴ $28 = 4x + y$

similarly $30 = 10x + y$

⇒ $6x = 2$

or $x = \frac{1}{3}$

$$y = 30 - 10x$$

$$= 30 - \frac{10}{3}$$

$$= \frac{90-10}{3}$$

$$= \frac{80}{3}$$

Now, requirement is no pending order at end of 5th day So, let number of trucks required be z.

∴ $5z = 5x + y$

$$= \frac{5}{3} + \frac{80}{3} = \frac{85}{3}$$

$$z = \frac{85}{3 \times 5}$$

$$= \frac{17}{3} = 5.667$$

So, minimum possible truck = 6

19. Ans. (d)
Solution:

From 1 liter he is replacing $\frac{1}{10}$ each time for there times, so

$$10\left(1 - \frac{1}{10}\right)^3 = 7.29 \text{ liters}$$

20. Ans. (b)
Solution: B option is making visualize all possible emotional states in the healing process i.e. from bereavement and grief.

Reader's Notes

Bibliography

Communication Systems

1. Simon Haykin, Communication System, (New York, John Wiley & Sons, INC, 2001)
2. Simon Haykin, Digital Communications (New York, John Wiley& Sons, INC, 1999)
3. Herbert Taub & Donald L. Schilling, Principles of Communication System (New Delhi Tata McGraw-Hill Publishing Company Limited, 199)
4. James Martin, Telecommunication and the Computer, (New Delhi Prentice-Hall of India Private Limited, 2000)
5. Thiagarajan Viswanathan , Telecommunication Switching Systems and Networks, (New Delhi Prentice-Hall of India Private Limited, 2000)
6. George Kennedy, Electronic Communication Systems,(New Delhi Tata McGraw-Hill Publishing Company Limited, 1993)
7. George Kennedy, Electronic Communication Systems,(New Delhi Tata McGraw-Hill Publishing Company Limited, 1999)
8. Wayne Tomasi, Electronic Communication Systems, Fundamental Through Advanced. (New Delhi Prentice-Hall of India Private Limited, 2005)
9. B.P. Lathi, Communication Systems, (New Delhi Wiley Eastern Limited, 1968)
10. John C. Hancock The Principles of Communication Theory, (New Delhi Tata McGraw-Hill Publishing Company Limited, 1992)
11. R P Singh & S D Sapre, Communication systems (Analog and digital) (New Delhi Tata McGraw-Hill Publishing Company Limited, 1995)
12. Anokh Singh, Principles of communication Engineering, (New Delhi S. Chand & Company LTD, 1984)
13. R.R. Gulati, Monochrome and Colour Television, (New Delhi New age International (P) Limited, Publishers, 2002)
14. Willian C.Y.Lee Mobile Cellular Telecommunications Analog and Digital Systems, (McGraw-Hill,Inc, 1995)
15. Voice and Data Communications Handbook (New Delhi Tata McGraw-Hill Publishing Company Limited, 2000)
16. Sanjeeva Gupta, Dinesh Arora , Satya Bhushan Sarna, Prashant Lonkar, Microwave Engineering, (Delhi, Khanna Publishers, 2003)
17. Dr. D.C. Agarwal Satellite communications, (Delhi Khanna Publishers, 2002)
18. A. Bruce Carlson, Paul-B.Crilly, Ganted C. Ruat Rutledge Communication Systems (MC Graw International Edition 2002)
19. Athanasois Paopulis, S. Unnikrishna Pillai. Probability, Random Variables, and Stochastic Processes (New Delhi, Tata McGraw-Hill Publishing Company Limited, Edition 2006)
20. G.K Mithal , Radio Engineering Khanna Pulisher, new Delhi 1996
21. John G. Proakis, Digital Communications (New Delhi McGraw Hall International Edition, 2001
22. Dr. Dharma Raj Cheruku Satellite Communication (New Delhi Galgotia Publication Private Limited)
23. Bernard Sklar, 'Digital Communications fundamentals and Appliations' (New Delhi Pearson Education in South Asia)
24. Ian A. Glover & Peter M. Grant 'Digital Communications' (New Delhi, Pearson Education in South Asia)
25. William Stallings 'Wireless Communications and Networks' (New Delhi, Pearson Education in South Asia)
26. Simon Haykin & Michael Moher 'Modern Wireless Communications' (New Delhi, Pearson Education in South Asia)
27. K. Sam Shanmugam 'Digital and Analog communication Systems' (New Delhi, John Wiley & Sons (Asia) Private Limited)
28. R. Gopal 'Problems and Solution in Communication Systems' (New Delhi, CBS Publishers & Distributions)
29. Forrest Barker 'Communication Electronics: Systems, Circuits, and Devices' (New Delhi, Prentice-Hall of India Private Limited)
30. Alberto Leon-Garcia & Indra Widjaja 'Communication Networks Fundamental Concepts and Key Architectures' (New Delhi, Tata McGraw-Hill Publishing Company Limited)
31. J.S. Chitode 'Communications Systems-1' (PUNE Technical Publications)
32. Richard A. Williams 'Communication Systems Analysis and Design: A System approach' (New Delhi, Prentice-Hall of India Private Limited)
33. Behroag A. FORUZAN "Data commnication & Networkding" (New Delhi The Mc-Graw Hill Company)
34. J. Das, Sk Mullick and PK chahagee "Digital communication" (New Delhi, New Age international publication)

Electronics Devices and Circuits

1. Mathur Kulshreshtha Chadha Electronic Devices Applications and integrated Circuits (New Delhi Umesh Publications)
2. Allen Mottershead, electronic Devices and circuits (New Delhi Prentice-Hall of Indial Private Limited, 2004)
3. Jacob Millman and Herbert Taub, Pulse, Digital and Switching Waveforms, (New Delhi McGraw-Hall Book Company, 1965)
4. Jimmie J. Cathey, Schaum's out line series Electronic Devices and Circuits (New Delhi Tata McGraw-Hill Publishing Company Limited)
5. Adel S. Sedra and Kenneth C. Smith, Microelectronic Circuits (New York Oxford University Press, 2004)
6. Ben G. Streetman and Sanjay Banerjee, Solid State Electronic Devices(New Delhi Pearson Aducation, PTE Limited Indian Branch, New Delhi)
7. Jacob Millman, and Arvin Grabel, Microelectronics, (New Delhi Tata McGraw-Hill Publishing Company Limited)

8. Jacob Millman, Ph.D., Christos C.HalKias. Ph.D. Integrated Electronics Analog and Digital Circuit and Systems, (New Delhi Tata McGraw-Hill Publishing Company Limited, 1991)
9. S.M. Sze, Physics of Semiconductor Devices, (New YorkA wiley - Interscience Publication John Wiley & Sons, 2004)
10. Robert L. Boylestad Louis Nashelsky, Electronic Devices and Circuit Theory (New Delhi, Published by Pearson Eduaction (Singapore) Private Limited, Indian Branch, 482 F.I.E. Patparganj, 2003)
11. J. B Gupta, Electronic Devices and Circuits (New Delhi S.K. Kataria & Sons, 2004)
12. B.S. Sonde, Introduction to System design using Integrated circuits (New Delhi, 1999)
13. R. Gopal, Problem and Solution in Integrated electronics (New Delhi, CBS Publisher and Distributor)
14. Ramakant A. Gayakwad, OP-AMPS and Linear Integrated Circuits (New Delhi Dhanpat Rai & Sons, 2006)
15. Samuely Y.Liao microwave Devices and Circuits(New Delhi Prentice-Hall, 2000)
16. P. John Paul Micro Electronics (New Delhi Galgotia Publication Private Limited)
17. K. "Gopal" Gopalan 'Introduction To Digital Microelectronic Circuuts' (New Delhi, Pearson Education in South Asia)
18. Theodore F. Bogart Jr., Jeffrey S. Beasley & Guillermo Rico 'Electronic Devices and Circuits' (New Delhi, Pearson Education in South Asia)
19. Pallab Bhattacharya 'Semiconductor Optoelectronic Devices' (New Delhi, Pearson Education in South Asia)
20. Sanjay Sharma 'Electronics Engineering' (New Delhi, S.K. Kataria & Sons)
21. U.A Bakshi and A.P. Godse 'Analog Integrated Circuits' (Techincal Publications)
22. Donald A Neamen 'Electronic Circuits Analysis and Design' Jacob Millman, and Arvin Grabel, Microelectronics, (New Delhi Tata McGraw-Hill Education Private Limited)
23. Jacob Millman, Christos C. Halkias 'Integrated Electronics Analog and Digital Circuits and Systems' (New Delhi Tata McGraw-Hill Publishing Company Limited)
24. Randall L. Geiger Phillip E. Allen and Noel R strader "VLSI Design technique for analog and Digital circuits" (New Delhi, Tata Mc Graw- Hill)
25. Philips E. Allen Douglas R Holbarg "CMOS analog circuit design" (New Delhi, Oxford university press)

Network/Circuit Theory

1. Joseph A. Edminister, M.S.E, Schaum's out line series Theory and Problems of Electric circuit (New Delhi Tata McGraw-Hill Publishing Company Limited, 1991)
2. M.E. Van Valkenburg, Network Analysis (New Delhi Prentice-Hall of India Private Limited, 1989)
3. Franklin F. Kuo, Network Analysis and Synthesis, (John Wiley & Sons (ASIA) Private Limited)
4. T S K V Iyer, Theory and Problems in Circuit Analysis, (New Delhi Tata McGraw-Hill Publishing Company Limited, 2002)
5. M.L. Soni, M.E. and J.C Gupta, M.S., A course in Electrical Circuit Analysis, (New Delhi Dhanpat Rai & Sons, 1990)
6. Willian H. Hayt, Jr. and Jack E. Kemmerly Engineering Circuit Analysis (New Delhi Tata McGraw-Hill Publishing Company Limited, 2001)
7. D. Roy Choudhary, Network and System (New Delhi, Wiley Eastern Ltd.)
8. A Chakrabarti, Circuit theory (Analysis and Synthesis) Dhanpat Rai 2005
9. A sudhakar, Shyammohan S. Palli, Circuit and Networks. (New Delhi Tata McGraw-Hill Publishing Company Limited, 2007)
10. K.S. Suresh Kumar 'Electric Circuits and Networks' (New Delhi, Pearson Education in South Asia)
11. Vincent Del Toro 'Electrical Engineering Fundaments' (New Delhi, Pearson Education in South Asia)
12. D.P Kothari & I J Nagrath 'Basic Electrical Engineering' (New Delhi, Tata McGraw-Hill Publishing Company Limited)
13. Vincent Del Toro 'Electrical Engineering Fundaments' (New Delhi, Pearson Education in South Asia)
14. Ashfaq Husain 'Network and Systems' (New delhi, Khanna Book Publishing Company Private Limited)
15. Charles A. Desoer and Ernest S. Kuh 'Basic Circuit theory'(New Delhi Tata McGraw-Hill Publishing Company Limited, 2001)
16. CL Wadhwa 'Network Analysis' (New Delhi New Age International (P) Limited, Publishers)
17. B.Subramanyam 'Electric Circuit Analysis' (New Delhi I. K. International Publishing House Private Limited)
18. Gopal G Bhise, Prem R. Chandha and Durgesh C. Kulshreshtaha 'Engineering Network Analysis and Filter design' (New Delhi Umesh Publications)
19. Russell M. Mersereau and Joel R. Jackson 'Circuit Analysis A Systems Approach' (New Delhi Pearson Education Private Limited)
20. K. Channa Venkatesh and D. Ganesh Rao "Network Theory" '(New Delhi Pearson Education Pvt. Ltd.)
21. K. Rajewaran "Electrical Circuit Theory" (New Delhi Pearson Education Private Limited)
22. C.P. Kuria Kose "Circuit Theory" (New Delhi Pearson Education Private Limited)
23. Samarajit Ghosh "Network Theory" (New Delhi Pearson Education Private Limited)
24. Charles K Alexander and Mathew Saidikue "Electrical Circuits" (New Delhi Tata McGraw-Hill Publishing Company Limited)
25. James W. Nilsson and Susan A Riedel "Electrical Circuits" (New Delhi Pearson Education Private Limited)

Control System

1. Joseph J. Distefano, III PH.D, Allen R. Stubberud, PH.D and Ivan J. Williams, PH.D. Schaum's Outline of Theory and Problems of Feedback and Control System (New Delhi Tata McGraw-Hill Publishing Company Limited)
2. Benjamin C. Kuo, Automatic control System,(New Delhi Prentice-Hall of India Private Limited, 1990)
3. I.J. Nagrath and G.Gopal, Control System Engineering, (New Delhi New Age International Private Limited, Publishers, 1999)
4. Katsuhiko Ogata, Modern Control Engineering, (New Delhi Prentice-Hall of India, 2000)
5. Dr. S. Das Gupta, Control System Theory (New Delhi Khanna Publishers, 1997
6. M Gopal Digital Control and State Variable Methods (New Delhi Tata McGraw-Hill Publishing Company Limited-2005)
7. Richard C. Dorf & Robert H. Bishop 'Modern Control Systems' (New Delhi, Pearson Education in South Asia)

8. A.K. Jairath 'Problems & Solution of Control Systems' (New Delhi, CBS Publishers & Distributions)
9. Graham C. Goodwin, Stefan F. Graebe & Mario E. Salgado 'Control Systems Design' (New Delhi, Pearson Education in South Asia)
10. Smarajit Ghosh 'Control Systems Theory and Applications' (New Delhi, Pearson Education in South Asia)
11. S.P. Eugene Xavier & Joseph Cyril Babu, J. 'Principles of Control Systems' (New Delhi, S. Chand & Company Limited)
12. Norman S. Nise 'Control Systems Engineering Systems' (New Delhi, John Wiley & Sons (Asia) Private Limited)
13. Dr. Yaduvir Singh & Amit Agrawal 'Control Systems Theory, Problems & Solutions' (New Delhi Galgotia Publication Private Limited)
14. Dr. A. K. Tandan , Dr. A. Subba rao, Parag R. Desai & Dr. S.K. Kulkarni 'Control Engineering' (New Delhi, Dhanpat Rai & Sons)
15. Francis H. Raven 'Automatic Control Engineering' (New Delhi, McGraw-Hill Publishing Company Limited)
16. Dr. H.M. Rai & Dr. S.N. Sivanandam 'Control System Engineering' (New Delhi, Satay prakashan)
17. A.K. Sawhnecy Instrumentation and Process control (New Delhi, Dhanpet Rai and Co.)
18. Santanice Chakraborty "A Text Book of Process control and Contorl Systems" (New Delhi, Dhanpat Rai and Co.)
19. D. Patranbis "Principles of Process Control" (New Delhi, Tata Mc-Graw-Hill)
20. Krishna Kant "Computer Based Industrial control" (New Delhi, PHI Learmy Private Limited)

Digital

1. Albert Paul Malvino, Ph.D. and Jerald A. Brown, Digital Computer Electronics, (New Delhi Tate McGraw-Hill Publishing Company Limited, 1995)
2. Willian I. Fletcher, An Engineering Approach To Digital Design, (New Delhi Prentice-Hall of India Private Liminted, 2003)
3. M. Morris Mano, Digital Design(New Delhi Tate McGraw-Hill Publishing Company Limited, 1995)
4. V. Rajaraman, Ph. D. and T. Radhakrishnan, Ph. D. An introduction to Digital computer Design, (New Delhi Prentice-Hall of India Private Limited, 1988)
5. Albert Paul Malvind, PH.D. and Donald P. Leach, PH.D, Digital Principal and Applications, (New Delhi Tata McGraw-Hill International Editions, 1987)
6. C.K. Nagpal. Elements of Computer Science, (New Delhi Dhanpat Rai & Sons)
7. Herbert Taub, Donald Schilling Digital Integrated Electronics. (New Delhi Tate McGraw-Hill Internation Editions Singapore, 1977)
8. A. Anand Kumar 'Pulse and Digital Circuits' (New Delhi, Prentice-Hall of India Private Limited, 2008)
9. A. Anand Kumar 'Fundaments of Digital Circuits' (New Delhi, Pearson Education in South Asia)
10. M. Morris Mano & Michael D. Ciletti 'Digital Design' (New Delhi, Pearson Education in South Asia)
11. R P Jain 'Modern Digital Electronics' (New Delhi Tata McGraw-Hill Education Private Limited)
12. Albert Paul Malvino and Donald P. Leach 'Digital Principles' (New Delhi Tata McGraw-Hill Editions)
13. Gopalan 'Introduction to Digital Microelectronic Circuits' (New Delhi Tata McGraw-Hill Editions)
14. I.J Nagrath 'Electronics, Analog and Digital' (New Delhi PHI Learning Private Limited)

Microprocessor

1. Ramesh S. Gaonkar, Microprocessor Architecture, Programming, and Applications with the 8085, (Penram International Publishing (India Private Limited)
2. M. Morris Mano, Computer System Architecture, (New Delhi Prentice-Hall of India Private Limited, 2001)
3. B.Ram, Fundaments of Microprocessor and Microcomputers, (New Delhi Dhanpat Rai Publications Private Limited)
4. Douglas V. Hall, Microprocessors and Interfacing Programming and Hardware, (New Delhi Tata McGraw-Hill Publishing Company Limited, 2000)
5. V. Rajaraman, Analog computation and simulation, (New Delhi Prentice-Hall of India Private Limited, 1989)
6. Muhammad Ali Mazidi, Janice Gillispie Mazidi and Rolin D. McKinlay The 8051 Microcontroller and Embedded Systems (New Delhi Prentice-Hall of India Private Limited, 2006)
7. Kenneth J. Ayala The 8051Microcontroller (Thomson Delmar Learning- Penram International Publishing (I) PVT. LTD. Mumbai 1997)
8. K. A Parthasarathy, A. Ramachandran and R. Purushothaman Computer Architecture (Chennai, THOMSON -Vijay Nicole imprints Private Limited, 2004).
9. Andrew S. Tanenbaum 'Computer Networks' (New Delhi, S. Chand & Company Limited)
10. Rafiquzzaman Mohamed & Chandra Rajan 'Modern Computer Architecture' (New Delhi, Galgotia Publications Private Limited)
11. Badri Ram 'Advanced Microprocessors and Interfacing' (New Delhi Tata McGraw-Hill Education Private Limited)
12. Y. Rajasree "Advance Micro Processors" (New Delhi New age International Private Limited)
13. A.K. Mukhopadhyay "Micro processors Micro Computer and their applications" (New Delhi, Narosa publishing House)

Electrical – Electronics Measurement and Instrumentation

1. Umesh Sinha, Electrical & Electronics Measurement and Instrumentation, (New Delhi Satya Prakashan, 1990)
2. A.K. Sawhney, A course In Electrical and Electronic Measurements and Instrumentation, (New Delhi Dhanapat Rai & Sons, 1991)
3. J.B. Gupta, Electronic and Electrical Measurements and Instrumentation (New Delhi S.K. Kataria and Sons, 2007).
4. Albert D. Helfrick & William D. Cooper 'Modern Electronics Instrumentation and Measurement Techniques' (New Delhi, Pearson Education in South Asia)
5. Ernest O. Doebelin 'Measurement Systems Application and Design' (New Delhi, McGraw-Hill Publishing Company Limited)
6. D. Patranabis "Principles of Industrial Instrumentation" (New Delhi, MC Graw Hill Education)

Mathematics

1. Dr. B.S. Grewal Higher Engineering Mathematics (Khanna Publishers)
2. M.L. Aggarwal, Mathematics (Arya Publications, 2006)
3. K.K. Gupta, Ravinder Guru and M.L. Sharma, Prime Mathematics, (Sharma Publications)
4. Manjeet Singh, Pioneer Mathematics (Dhanpat Rai & Company Private Limited, 2002)

5. Parmanand Gupta Comprehensive Mathematics (Laxmi Publications (P) LTD New Delhi, 1999)

Signal System

1. Farooq husain and Mrs. Shahnaz husain, Digital Signal Processing and Applications,(New Delhi Umesh Publications, 2003)
2. Farooq Husain and Mrs. Shahnaz Husain A Text Book of Signals and Systems (New Delhi Umesh Publications, 2002)
3. Smarajit Ghosh 'Signal and Systems'
4. Nagoor Kani 'Signal and System' (New Delhi Tata McGraw-Hill Education Private Limited)
5. S Salivahanan, A Vallaveraj and C Gnanapriya , Digital Signal Processing (New Delhi Tata McGraw-Hill Publishing Company Limited, 2001)
6. John G.Proakis and Dimitris G.Manolakis Digital Signal Processing Principle, Algorithms, and Applications (New Delhi Pearson Education Private Limited, 2003)
7. Alan V. Oppenheim & Alan S. Willsky , Signals and systems, (New Delhi Prentice-Hall of Indial Private Limited, 2000)
8. Alan V. Oppenheim, Ronald W. Schaffer and John R. Buck Discrete-Time Signal Processing (New Delhi PEARSON Prentice-Hall of Indial Private Limited, .2005)
9. Simon Haykin, Barry Van Veen, Signals and Systems (Singapore, John Wiley and Sons (ASIA) Private Limited, 2005)
10. Johnny R. Johnson 'Introduction to Digital Signal Processing' (New Delhi, Prentice-Hall of India Private Limited, 1997)
11. Emmanuel C. Ifeachor & Barrie W. Jervis 'Digital Signal Processing' A Practical Approach (New Delhi, Pearson Education in South Asia)
12. Johnny R. Johnson 'Introduction to Digital Signal Processing' (New Delhi, Prentice-Hall of India Private Limited, 1997)
13. D.P Kothari & I J Nagrath 'Basic Electrical Engineering' (New Delhi, Tata McGraw-Hill Publishing Company Limited)
14. Vincent Del Toro 'Electrical Engineering Fundaments' (New Delhi, Pearson Education in South Asia)
15. Rolf Schaumannm M.S. Ghausi & Kenineth R. Laker 'Design of Analog filters Passive, Active RC, and Switched Capacitor' (New Delhi, Prentice-Hall of India Private Limited)

Analytical, Optical and Biomedical Instrumentation

1. John Gowar, Optical Communication Systems, (New Delhi Prentice-Hall of India Private Limited, 2001)
2. John M Senior, Optical Fiber Communications, Principle and practice(New Delhi Prentice-Hall of India Private Limited, 1992)
3. Govind P. Agrawal "Fiber-optical communication system" (New Delhi, Wiley India)
4. Gerd Keiser "Optical Fiber Communication" (New Delhi Tata McGraw-Hill Publishing Company Limited)
5. Dr. Subir Kumar Sarkar "Optical Fibers and Fiber Optic Communication System" (New Delhi S.Chand and Company)
6. Dr. N. Subrahmavnayam, Brij Lal and DR M.N. Avadhnulu "A textbook of opticics" (New Delhi S. Chand and Company)
7. K.N. Scott and A.K. Mathur "Text book of Biomedical Instrumentation (New Delhi, CBS Publishers & Distributors"
8. Leslie Cromwell, Fred J. Weibell & Erich A. Pfeiffer "Biomedical Instrumentation and Measurements" (New Delhi, Pearson Education in South Asia)
9. Prof. S.K. Venkata Ram "Biomedical Electronics & Instrumentation" (New Delhi Galgotia Publications Private Limited)
10. R.S. Khandpur Analytical Instruments (New Delhi Tate McGraw-Hill Publishing Company Limited, 2004)
11. R.S. Khandpur Biomedical Instrumentation (New Delhi Tate McGraw-Hill Publishing Company Limited, 2002).
12. A.K. Srivastava and P.C. Jain "Instrumental Approach to Chemical Analysis" (New Delhi, S. Chand & Company Limited)
13. Ajay Ghatak and K.Thyagrajan "Introduction to Fiber Optics" (New Delhi, Cambridge university press)
14. J. H. From 2, Vk. Jain "Optical communication Components and System" (New Delhi, Narosa publishing House)
15. John G Webster "Medical Instrumentation Application and Design" (New delhi, Wiley India)

Mechanical Measurement

1. A.K. Tayal & Akash Tayal "Insturmentation, Mechanical Measurmentss & Contorl" (New Delhi Galgotia Publications Private Limited)
2. Dr. S. Kumar "Mechanical Measurements and Contorl" (New Delhi Metropolita Book Company Private Limited)
3. A.K. Sawhney, Puneet Sawhney Mechanical Measurements and Instrumentation (New Delhi Dhanapat Rai & Sons, 2007)
4. D. Patranatirs "Sensors and Transducer" (New Delhi PHI)
5. S.K. Singh "Industrial Instrumentation and Control" (New Delhi, Tata McGraw Hill Education Private Limited)

General Ability

1. Abul Hashem "General English" (New Delhi, Ramesh Publishing House)
2. R.S. Agrawal "Objective Arithmetic" (New Delhi, S. Chand & company Limited)
3. Dr. Ravi Chopra "Reasoning and Reasoning" (New Delhi, Galgotia publication)
4. R.S Agrawal "Verbal and Non Verbal reasonin" (New Delhi, S.Chand & Private Limited)
5. Navratan singh and Ms Manpreet Cheema, "Numerical Ability" (New Delhi, Dhillon Group of Publication)
6. R.S. Agrawal "Quantitative Aptitude" (New Delhi, S.Chand & Company Limited)
7. Sujit Kumar "Reading comprehension for the CAT" (New Delhi Pearson Education Private Limited)
8. Dr. R.S Aggrawal and Vikas Aggrawal "Objective General English" (New Delhi, S.Chand & Company Limited)
9. Dr. R.S Aggrawal, Deepak Aggrawal and Vikas Aggrawal "Quantitative Aptitude" (New Delhi, S. Chand & Company Limited)
10. Bhagwant Swerap sharma, Sumira Aggrawal, Neeta sharma, Monik Ahuja and suchiya Dahall "National Talent Search Examination" (New Delhi, S. Chand & Company Limited)
11. Dr. N.K. Parwal "General Intelligent and Test of Reasoning" (Agra, pritiyogita sahitya)
12. R.D. Sharma "Mathematics for Class IX " (New Delhi, Dhanpat Rai and Company)
13. R.D. Sharma "Mathematics for Class X" (New Delhi, Dhanpat Rai and Company)
14. R.D. Sharma "Mathematics for Class XI" (New Delhi, Dhanpat Rai and Company)
15. R.D. Sharma "Mathematics for Class XII" (New Delhi, Dhanpat Rai and Company).